NEW ZEALAND INVENTORY OF BIODIVERSITY

This volume is dedicated to the memory of
Professor Frank A. Bisby
1945–2011

NEW ZEALAND INVENTORY OF Biodiversity

VOLUME THREE

KINGDOMS

Bacteria, Protozoa, Chromista, Plantae, Fungi

Edited by

DENNIS P. GORDON

CANTERBURY UNIVERSITY PRESS

First published in 2012 by
CANTERBURY UNIVERSITY PRESS
University of Canterbury
Private Bag 4800
Christchurch
NEW ZEALAND

www.cup.canterbury.ac.nz

ISBN 978-1-927145-05-0

A catalogue record for this book is available from
the National Library of New Zealand.

Pre-production by Frith Hughes, Rachel Scott

Printed in China through Bookbuilders

Canterbury University Press gratefully acknowledges grants in aid of publication
from the National Institute of Water & Atmospheric Research (NIWA),
the Ministry of Foreign Affairs & Trade, the Department of
Conservation and the Ministry of Fisheries.

Cover: The silver fern *Cyathea dealbata* (Plantae: Tracheophyta) on a shaded bank in the Wellington Botanic Gardens and (inset) the dog-vomit slime mould *Fuligo septica* (Protozoa: Amoebozoa).
Dennis Gordon and Clive Shirley

Endpapers: Bull kelp *Durvillaea antarctica* on the south coast of Otago Peninsula (Chromista: Ochrophyta).
Dennis P. Gordon, NIWA

Half-title page: Average sea-surface temperature derived from AVHRR satellite. Warmest temperatures (around New Guinea) are ~30° C; coolest temperatures (around Antarctica) are ~-1.5° C.

Title page: Male inflorescences of the kiekie *Freycinetia banksii* (Plantae: Tracheophyta).
Dennis P. Gordon, NIWA

CONTENTS

Relief map of the New Zealand region

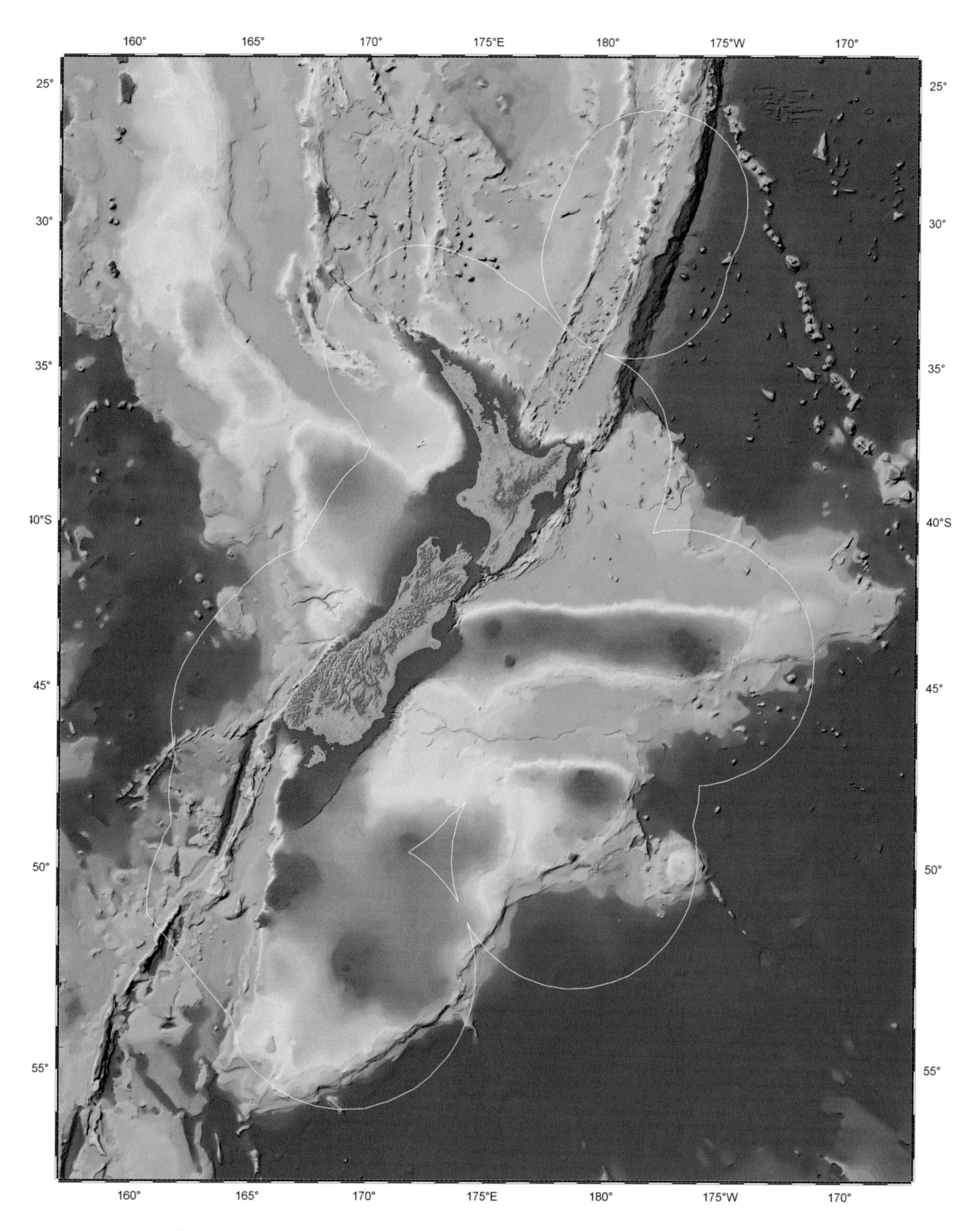

All species inventoried in this volume are found within the marine, terrestrial, freshwater, and fossil habitats bounded by the 200 nautical mile Exclusive Economic Zone (the thin white line). Red indicates the upper kilometre of the ocean, with orange and yellow pertaining to deeper areas of the New Zealand continental mass; blue indicates abyssal depths of 4–6 kilometres, and the purple-magenta of the Kermadec Trench depths of 6–10 kilometres. NIWA

FOREWORD

This third and final volume of the *New Zealand Inventory of Biodiversity* catalogues what we know of New Zealand's biodiversity in some of the most functionally vital groups of all. These include the primary producers that use the sun's energy to form organic nutrients and atmospheric oxygen from inorganic nitrogen, carbon dioxide, and water, and are at the base of all the food webs life depends on – the decomposers that break down and recycle organic waste, and the pathogens that are major drivers of population changes in plant and animal species. Without these organisms, other life-forms, including humans, could not exist. It is to these groups that we are primarily referring when we talk of and attempt to catalogue ecosystem services such as fresh air, clean water and nutrient-rich soils.

This volume also documents our understanding of the evolutionary relationships among species and species groups. Organisms that 30 years ago were all classified as plants because they photosynthesised are distributed among the kingdoms Plantae, Chromista, and Bacteria. For those among us whose knowledge of animal and plant classification is sketchy or outdated, this is a superb update on the high-level classification of life. For example, we are introduced to the kingdom Chromista. This kingdom includes organisms that not too long ago we thought to belong variously among the plants, animals, and fungi.

The Department of Conservation is often thought of as focusing its efforts on the conservation of threatened species, and this is indeed an important part of our work. However, we also have a vital role in conserving healthy, functioning ecosystems both for their intrinsic value and for the environmental services they deliver. This volume, by documenting the diversity of the species that actually provide those services, provides a vital foundation for our future work. New Zealand has a history of disastrous introductions of exotic plants and animals, both deliberately and accidentally. This is probably also true of the mainly microscopic organisms that are listed here. In future, this catalogue and the database that will be developed from it will put us in a much better position to identify potentially problematic newcomers.

As with the animal groups covered in the first two volumes of this series, this volume is only the start – the list includes many known but undescribed species, and there are more as yet undiscovered. We need to support continuing taxonomic research and training to ensure that the cataloguing of our biodiversity is completed over time.

As for the two previous volumes in the series, it is important to acknowledge the massive effort that has gone into the preparation of this volume. Dennis Gordon and the more than 70 chapter authors deserve our sincere thanks and gratitude.

Alastair Morrison
Director-General of Conservation

The New Zealand geological timescale

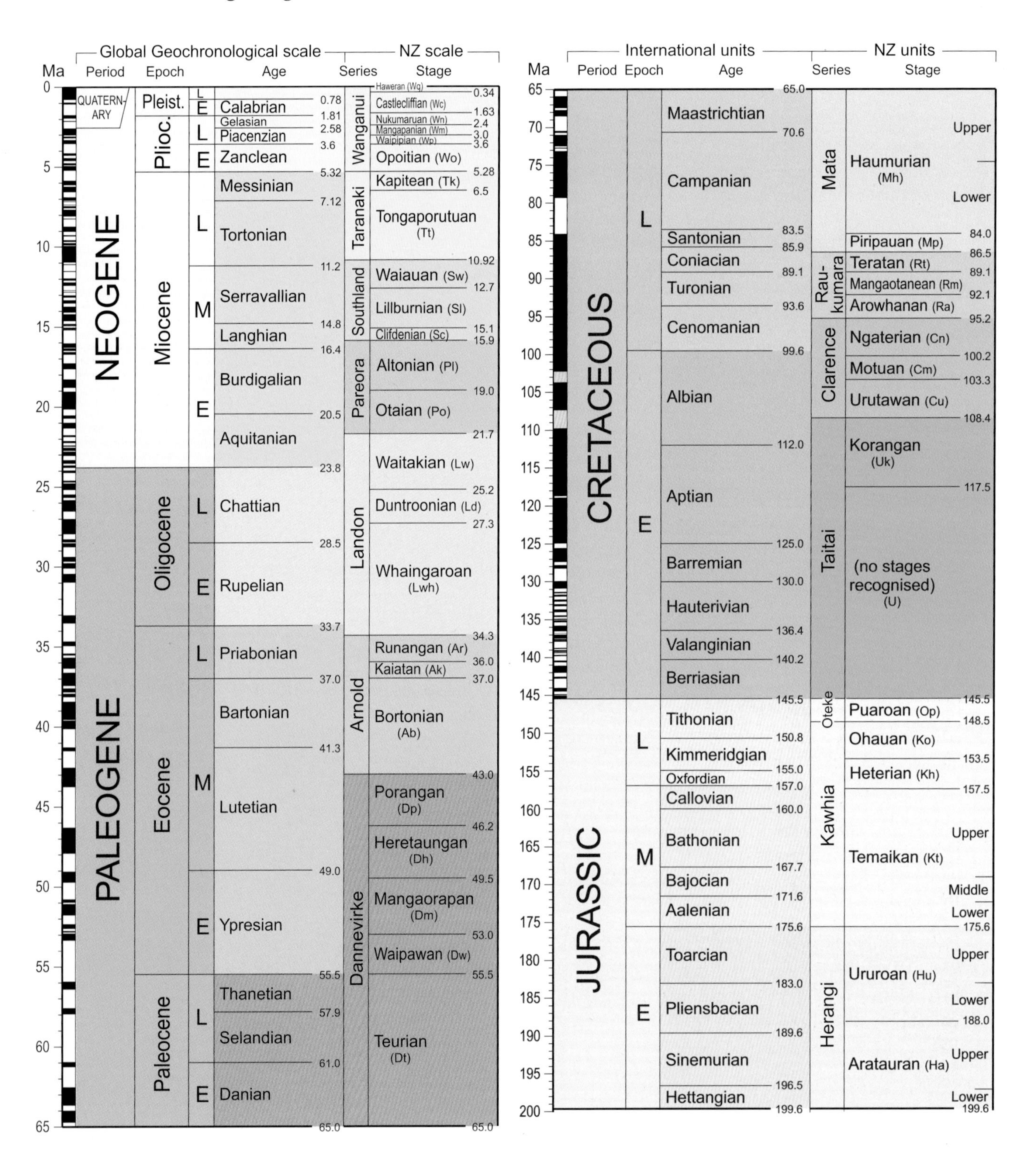

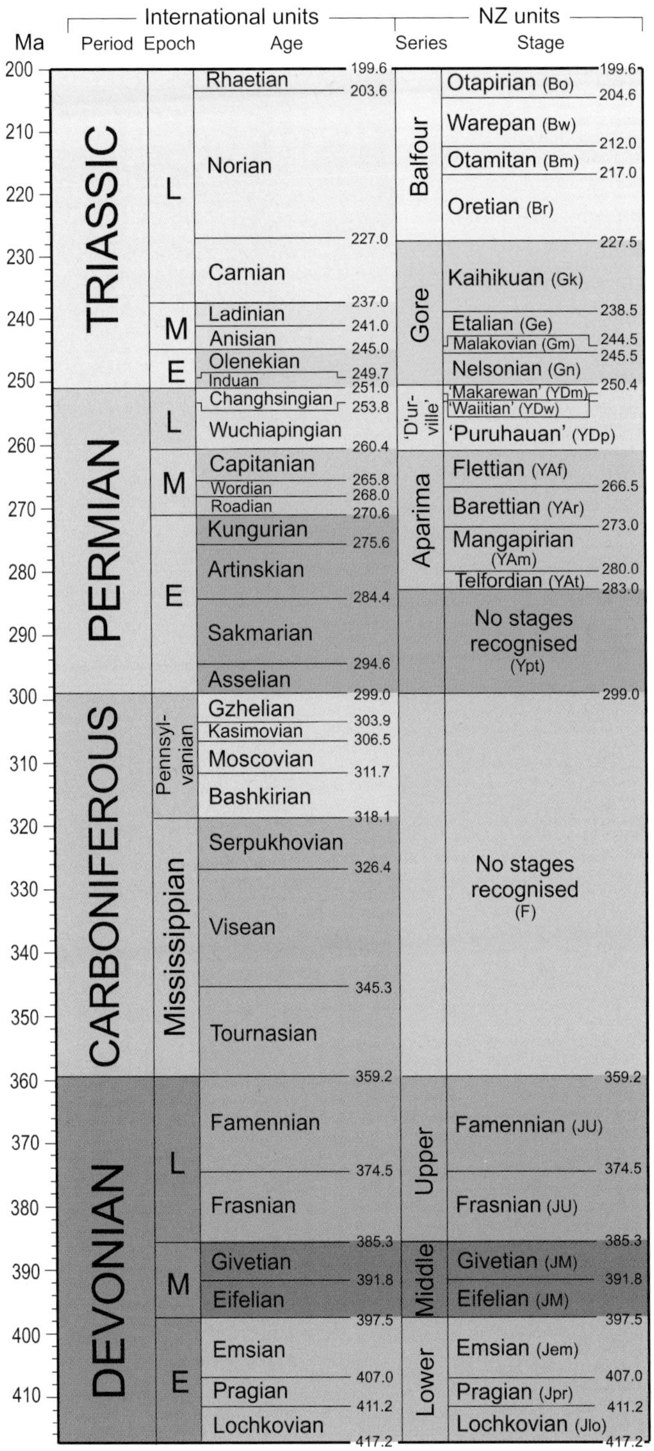

Period	Epoch	Age	Base (Ma)
		(top)	199.6
TRIASSIC	L	Rhaetian	203.6
		Norian	227.0
		Carnian	237.0
	M	Ladinian	241.0
		Anisian	245.0
	E	Olenekian	249.7
		Induan	251.0
PERMIAN	L	Changhsingian	253.8
		Wuchiapingian	260.4
	M	Capitanian	265.8
		Wordian	268.0
		Roadian	270.6
	E	Kungurian	275.6
		Artinskian	284.4
		Sakmarian	294.6
		Asselian	299.0
CARBONIFEROUS	Pennsylvanian	Gzhelian	303.9
		Kasimovian	306.5
		Moscovian	311.7
		Bashkirian	318.1
	Mississippian	Serpukhovian	326.4
		Visean	345.3
		Tournasian	359.2
DEVONIAN	L	Famennian	374.5
		Frasnian	385.3
	M	Givetian	391.8
		Eifelian	397.5
	E	Emsian	407.0
		Pragian	411.2
		Lochkovian	417.2

NZ Series	NZ Stage	Base (Ma)
	(top)	199.6
Balfour	Otapirian (Bo)	204.6
	Warepan (Bw)	212.0
	Otamitan (Bm)	217.0
	Oretian (Br)	227.5
Gore	Kaihikuan (Gk)	238.5
	Etalian (Ge)	244.5
	Malakovian (Gm)	245.5
	Nelsonian (Gn)	250.4
'D'urville'	'Makarewan' (YDm)	
	'Waiitian' (YDw)	
	'Puruhauan' (YDp)	
Aparima	Flettian (YAf)	266.5
	Barettian (YAr)	273.0
	Mangapirian (YAm)	280.0
	Telfordian (YAt)	283.0
	No stages recognised (Ypt)	299.0
	No stages recognised (F)	359.2
Upper	Famennian (JU)	374.5
	Frasnian (JU)	385.3
Middle	Givetian (JM)	391.8
	Eifelian (JM)	397.5
Lower	Emsian (Jem)	407.0
	Pragian (Jpr)	411.2
	Lochkovian (Jlo)	417.2

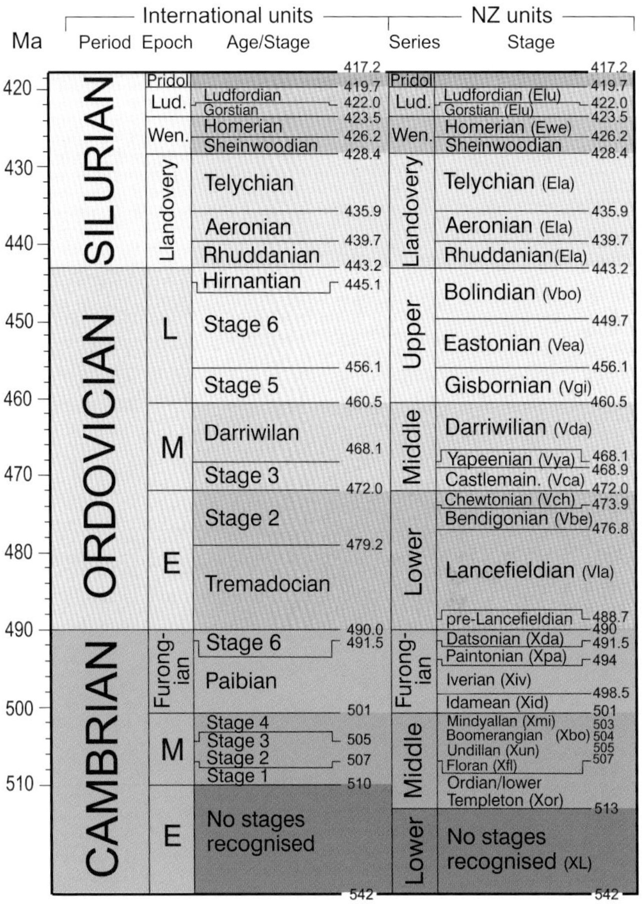

Period	Epoch	Age/Stage	Base (Ma)
		(top)	417.2
SILURIAN	Pridoli		419.7
	Lud.	Ludfordian	422.0
		Gorstian	423.5
	Wen.	Homerian	426.2
		Sheinwoodian	428.4
	Llandovery	Telychian	435.9
		Aeronian	439.7
		Rhuddanian	443.2
ORDOVICIAN	L	Hirnantian	445.1
		Stage 6	456.1
		Stage 5	460.5
	M	Darriwilan	468.1
		Stage 3	472.0
	E	Stage 2	479.2
		Tremadocian	490.0
CAMBRIAN	Furongian	Stage 6	491.5
		Paibian	501
	M	Stage 4	
		Stage 3	505
		Stage 2	507
		Stage 1	510
	E	No stages recognised	542

NZ Series	NZ Stage	Base (Ma)
	(top)	417.2
Pridoli		419.7
Lud.	Ludfordian (Elu)	422.0
	Gorstian (Elu)	423.5
Wen.	Homerian (Ewe)	426.2
	Sheinwoodian	428.4
Llandovery	Telychian (Ela)	435.9
	Aeronian (Ela)	439.7
	Rhuddanian (Ela)	443.2
Upper	Bolindian (Vbo)	449.7
	Eastonian (Vea)	456.1
	Gisbornian (Vgi)	460.5
Middle	Darriwilian (Vda)	468.1
	Yapeenian (Vya)	468.9
	Castlemain. (Vca)	472.0
Lower	Chewtonian (Vch)	473.9
	Bendigonian (Vbe)	476.8
	Lancefieldian (Vla)	488.7
	pre-Lancefieldian	490
Furongian	Datsonian (Xda)	491.5
	Paintonian (Xpa)	494
	Iverian (Xiv)	498.5
	Idamean (Xid)	501
Middle	Mindyallan (Xmi)	503
	Boomerangian (Xbo)	504
	Undillan (Xun)	505
	Floran (Xfl)	507
	Ordian/lower Templeton (Xor)	513
Lower	No stages recognised (XL)	542

International units: Period, Epoch, Age (left chart), Age/Stage (right chart). NZ units: Series, Stage. Ma: 200, 210, 220, 230, 240, 250, 260, 270, 280, 290, 300, 310, 320, 330, 340, 350, 360, 370, 380, 390, 400, 410, 420, 430, 440, 450, 460, 470, 480, 490, 500, 510.

New Zealand divisions, ages of unit boundaries, and error ranges on ages are from Cooper (ed., *The New Zealand Geological Timescale*, Institute of Geological and Nuclear Sciences Monograph 22, 2004), who gives a full description of the stratigraphic basis for the New Zealand scale and its calibration.

INTRODUCTION

A decade is a long time. But it should not be surprising that a 1758-page review and inventory of 'all of life through all of time' in New Zealand, finally involving 237 specialists in 19 countries, has taken nine to eleven years to come to completion. The project was formally launched in February 2000 at the *Species 2000: New Zealand* millennial symposium in Wellington and the reviews that were presented formed the basis of the fuller accounts given in the trilogy of volumes now published. But experts had not yet been found for all groups of organisms in 2000 and the challenge of filling the remaining gaps persisted through the decade. In the end, all the gaps were filled, making this biodiversity inventory the most taxonomically complete for any country.

The first two volumes covered the animal kingdom. The present volume deals with the remaining kingdoms of life and continues the theme of explaining why it is important to know our biodiversity – there are plenty of applications of taxonomic knowledge. Even though there are far fewer named species of microorganisms, algae, plants, and fungi combined compared to the animal kingdom, the challenges encountered in compiling this volume were in many ways greater. Apart from some economically or ecologically important groups, most microscopic organisms in New Zealand have been neglected scientifically. Historical studies were often typified by descriptions, illustrations, and modes of preservation that were inadequate by today's standards and the identities of some species remain uncertain. Classification schemes, too, have changed drastically in the past half century and even the first decade of the 21st century has seen a continuing ferment of ideas (see pages 17–21), as gene sequencing has revealed patterns of relationships among organisms that are often quite counter-intuitive.

What shape does the tree of life take these days? The higher-level classification used in these volumes partitions all of life into six kingdoms, based on a scheme developed by Thomas Cavalier-Smith, Oxford University. Viruses, non-cellular infectious agents that are not living organisms, are excluded from his scheme and this review.[1,2] There is a single prokaryote kingdom (Greek *pro-*, before, and *karyon*, nucleus), in which a cell nucleus is lacking (Bacteria), and five kingdoms of eukaryotic (*eu-*, true) organisms with a cell nucleus – Protozoa, Chromista, Plantae, Fungi, and Animalia. Protozoa is paraphyletic, in that it includes the ancestors of the other four kingdoms, and comprises such forms as euglenoids, trypanosomes, amoebas, collar flagellates, *Giardia*, and some other zooflagellates. Shelly foraminifera and radiolaria, and the ciliates, dinoflagellates, and parasitic sporozoa now all belong, along with water moulds and the golden and brown seaweeds and diatoms, to kingdom Chromista, united by molecular and ultrastructural criteria. The plant kingdom in the 21st century is restricted to glaucophyte algae, green and red algae, and land plants. True fungi include not only chytrids and the familiar pin moulds, yeasts, sac fungi, toadstools, rusts, and their relatives, but also the parasitic microsporidia. Excluded from kingdom Fungi are the water moulds and downy mildews as well as the plasmodiophoras, labyrinthulid slime nets, and thraustochytrids (all Chromista).

How do these components of New Zealand's biodiversity affect our lives and why should they interest governments and various agencies enough to support their study? What can these organisms teach us?

Take bacteria, which have the smallest of cells. You might not think about them much until you get sick or forgotten food in your refrigerator goes off. They are largely invisible yet have an impact beyond their size, not only as

Diversity of New Zealand biota

Kingdom	Total species*
Bacteria	697
Protozoa	539
Chromista	4,303
Plantae	7,548
Fungi	8,395
Animalia	34,636–35,001
Total	**56,118–56,483**

* Approximate

disease organisms but in nature generally. It is a little-known fact, but whether considered in terms of biomass or numbers of individuals, most life on earth consists of bacteria. In the chapter on bacteria in this volume the authors remind us that the most abundant form of biomass today is bacteria in ocean sediments, while the commonest constituents of marine plankton are the photosynthetic cyanobacteria (blue-green 'algae') – the smallest are found in every drop of surface ocean water – that supply a significant fraction of Earth's atmospheric oxygen. It is a common complaint of microbiologists that organisms that have such fundamental and crucial roles in the biosphere are often overlooked.

What we know today about bacteria is mostly based on those that can be grown in pure culture on simple nutrient media and it is not surprising that our knowledge of New Zealand bacterial diversity is based on research that is relevant to human interests. Examples include agents of human, animal, and plant diseases, food-spoilage species, and bacteria used in bioremediation (many species can degrade pollutants). Bacteria are also of interest to biotechnology for their roles in processes used in the manufacturing and services industries (e.g. brewing, baking, cheese production, chemical manufacturing) and as sources of pharmaceuticals (e.g. antibiotics, vaccines, steroids). A number of studies have been carried out in New Zealand on culturable geothermal bacteria (including archaebacteria) as possible sources of heat-tolerant enzymes. Much-photographed tourist spots like the Champagne Pool at Waiotapu, Warbrick Terrace in Waimangu Valley, and the boiling mud pools in Rotorua's Kuirau Park have yielded some novel bacterial diversity including species new to science. From its alps to its deep ocean trenches, New Zealand has such a wealth of varied habitats as to warrant an intensive assessment of microbial diversity in them. Who knows what novel genes and biocatalysts could form the basis of new industries?

It is a matter of great irony that some of the smallest organisms can be seen from outer space – not the cells themselves of course, but their blooms, which in the case of the haptophyte *Emiliania huxleyi* can appear as a milky discolouration of thousands of square kilometres of ocean, meanwhile producing an atmospheric gas that contributes to the formation of acid rain. *Ehux*, as haptophyte specialists familiarly call it, is a tiny single-celled alga. But what is an alga? Traditionally, an alga has been regarded as a lowly kind of plant that has chlorophyll but lacks true stems, roots, leaves, and vascular tissue, thus encompassing both microscopic unicells and giant seaweeds. But even the word 'plant' is a misnomer as algae are distributed across four of the six kingdoms of life.

The 'true-eyed', 'cut-in-half', 'whirling whips', golden algae, yellow algae, dusky algae, hidden plants, 'fasten-to' plants, red plants, green plants – these are the prosaic meanings behind the technical terms that specialists use for the eukaryote algae known respectively as euglenoids, diatoms, dinoflagellates, Chrysophyceae, Xanthophyceae, Phaeophyceae, Cryptophyta, Haptophyta, Rhodophyta, and Chlorophyta. Some are microscopic; others are macroscopic seaweeds. *Macrocystis* kelp can be as long as a blue whale. The true diversity of our microalgae is not known, especially in the sea. In freshwater environments, the best-known microscopic groups are the aesthetically beautiful desmids and diatoms. Their high levels of diversity have much scientific value, which can be applied to studies of biogeography, dispersal, and speciation.

Some of our algae are specialised to live in extreme or unusual habitats, like the several species of snow algae; the resting spores of some can turn snow pinkish-red. *Helicosporidium*, previously thought to be a fungus, is a colourless green-algal parasite of invertebrates. Species of *Cephaleuros* and *Phycopeltis* are wholly terrestrial, living on and within leaves and twigs of plants.

The largest algae are seaweeds, of which the best-known groups are the reds, greens, and browns. There are about 900 species along our coasts, including at least 109 that are new to science and not yet formally described. These are mostly red seaweeds, including a large number of karengo species – related to the algae used for wrapping sushi. There are also many unnamed species of sea lettuce among the green seaweeds. As for the brown seaweeds, of the five known species of the giant bull-kelp genus *Durvillaea*, which is restricted to the Southern Hemisphere, four are present in New Zealand and three of them are endemic.

In coastal waters there are nearly 70 species of microalgae that form coloured blooms, including 'red tides'; 42 of them are toxic or potentially so. All of them are identified in this volume. The toxic species are responsible for several kinds of poisoning, typically through eating shellfish that have been feeding on them. *Karenia brevisulcata*, which caused a massive die-off of marine life in Wellington Harbour in 1998, may be the most toxic dinoflagellate known.

Several other kinds of protists in New Zealand live as parasites in humans and other animals, causing actual disease. These pathogenic species belong to several phyla. Trypanosomes are protozoans (phylum Euglenozoa) that infect native fishes and the introduced rat species. In the Percolozoa is *Naegleria fowleri*, an amoeboflagellate that can cause fatal primary amoebic meningoencephalitis in humans. Introduced *Giardia* (Metamonada) is a well-known cause of infection of the intestine in humans, mammals, and birds, with lots of unpleasant symptoms, but several other species of this phylum also infect people and livestock. The chromist *Cryptosporidium* (Myzozoa) is more difficult to control and can be fatal. *Blastocystis* (Bigyra), another parasite of the intestinal tract, can also infect humans. As problematic as they are, these kinds of organisms are also part of New Zealand's biodiversity.

Protists crop up in the most unlikely places. *Acanthamoeba* and a *Naegleria* have been found in the moisture films of contact-lens storage cases in New Zealand. There have been overseas instances of *Acanthamoeba* causing amoebic keratitis, a painful sight-threatening disease of the eye.

Several kinds of shelly protists and microalgae have skeletons that persist long after death and find their way into the fossil record. The most spectacular examples are foraminifera, radiolaria, diatoms, and haptophytes. Because these groups of fossils are so numerous and widespread and contain particular isotopic signatures, they have numerous scientific and economic applications (particularly to the petroleum industry), including high-resolution stratigraphic dating and paleoenvironmental studies (e.g. interpretation of past ocean circulation and productivity, climates, and sea levels). New Zealand's exposed sequences of latest Cretaceous and Cenozoic marine strata provide one of the most complete records of the past 70 million years, envied in the rest of the world.

On land, the most visible expression of biodiversity is our forests, tussock grasslands, alpine vegetation, and pastures. New Zealand is a global hotspot of endemism among seed plants – of the 2028 indigenous species, 82% are endemic and there are 50 endemic genera. Unusual features of our seed-plant flora include a high proportion of plants that have separate male and female individuals, often small, simple, and drab flowers, and several unusual growth forms. One of our endemic aquatic species, *Trithuria inconspicua* (plate 13a), has the distinction of being our most primitive flowering plant – a veritable dinosaur, albeit tiny. Once thought to be in the same botanical order as the grasses, new research has shown that the family to which it belongs – Hydatellaceae, with a single genus, ten other species in Australia, and one in India – may be considered as 'water lilies with gymnospermous tendencies'![3, 4] Hydatellaceae is now regarded as one of the most ancient living lineages of angiosperms.

High endemism carries over to the bryophytes as well – New Zealand is arguably the most important hotspot of liverwort diversity in the world from the perspectives of species density, degree of endemism (52%), and the presence of a strong archaic element. There are more liverwort species than in the whole of Europe. These, together with mosses and ferns, form the main ground cover and epiphyte vegetation in the wetter rainforests of New Zealand.

New Zealand's macrofossil record of land plants and the microfossil record of pollen and spores are impressive. Compared to most countries, the combined record is potentially rather complete from the Cretaceous onward, with less-continuous but good coverage from the Late Permian through the Mesozoic. What we know about this potential goldmine of information on our paleoflora is just the tip of the iceberg but recent discoveries, which have included fossil flowers and the world's first orchid macrofossils, have been tantalising. The combined plant macro- and microfossil record gives much information on New Zealand's terrestrial paleobiodiversity, vegetation history, and evolution of the flora, with applications to biostratigraphy for industry and studies of climate change.

The final group to be considered in this volume is the fungi, with 8045 named species in New Zealand (of an estimated diversity of 23,300 species, not including Microsporidia). Knowledge of this group is uneven, with work to date focused on ecologically defined groups such as lichens and plant pathogens and on macrofungi. Collectively, New Zealand's fungi show a wonderful diversity of form and function, with a multiplicity of roles and uses – in ecosystems and in industry, in medicines and as food, and we are affected by their causation of plant and animal diseases. Fungi illustrate, par excellence, the link between knowing our biodiversity and applying that knowledge to our economic well-being. For example, BioNET-INTERNATIONAL, the global network for taxonomy, cites a striking example in their Case Study 31 – New Zealand expertise in the taxonomy of the fungal genus *Verticillium* gained overseas market access for New Zealand buttercup squash by providing the evidence needed for quarantine authorities to remove a non-tariff trade barrier.

What next?

New Zealand's biodiversity tally in the three volumes is about 56,200 living species and 14,700 fossil species. Estimates of total undiscovered living biodiversity (bacteria excluded) are in the order of 62,400–65,800 species but there is considerable uncertainty concerning parasitic and commensal protists, parasitic myxozoa, microsporidia, and nematodes, and free-living nematodes and mites. Estimates for these groups inflate the total. It has been calculated that, at the present rate of new species descriptions of New Zealand's marine life (currently averaging fewer than a hundred per year), it will take between170 and 500 years to complete the task[5] and the same timeframe is likely to apply to the biodiversity of the terrestrial environment.

In the meantime, the lists of known species are being incorporated into the New Zealand Organisms Register (NZOR) – a national information infrastructure to integrate and share reliable taxonomic information, not only within New Zealand but also to the global Catalogue of Life and the Global Names Architecture.

Dennis P. Gordon
NIWA, Wellington

1. Cavalier-Smith, T. 2010: Deep phylogeny, ancestral groups and the four ages of life. *Philosophical Transactions of the Royal Society, B, 365*: 111–132.
2. Moreira, D.; López-García, P. 2009: Ten reasons to exclude viruses from the tree of life. *Nature Reviews Microbiology 7*: 306–311.
3. Saarela, J. M.; Rai, H. S.; Doyle, J. A.; Endress, P. K.; Mathews, S.; Marchant, A. D.; Griggs, B. G.; Graham, S. W. 2007: Hydatellaceae identified as a new branch near the base of the angiosperm phylogenetic tree. *Nature 446*: 312–315.
4. Friedmann, W. E. 2008: Hydatellaceae are water lilies with gymnospermous tendencies. *Nature 453*: 94–97.
5. Gordon, D. P. 2011: New Zealand's marine biodiversity. *NZ Science Teacher 128*: 18–21.

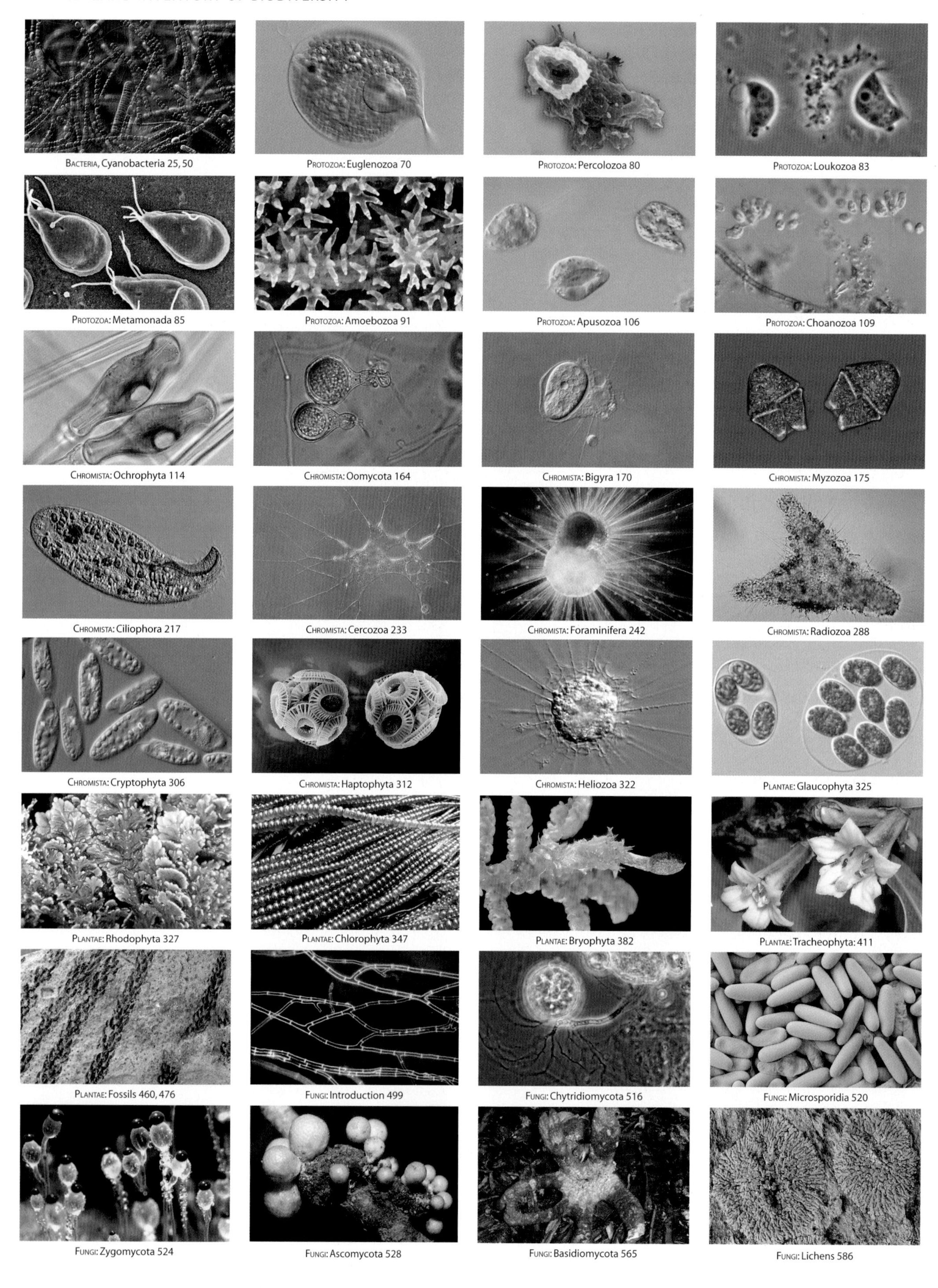

Bacteria, Cyanobacteria 25, 50
Protozoa: Euglenozoa 70
Protozoa: Percolozoa 80
Protozoa: Loukozoa 83
Protozoa: Metamonada 85
Protozoa: Amoebozoa 91
Protozoa: Apusozoa 106
Protozoa: Choanozoa 109
Chromista: Ochrophyta 114
Chromista: Oomycota 164
Chromista: Bigyra 170
Chromista: Myzozoa 175
Chromista: Ciliophora 217
Chromista: Cercozoa 233
Chromista: Foraminifera 242
Chromista: Radiozoa 288
Chromista: Cryptophyta 306
Chromista: Haptophyta 312
Chromista: Heliozoa 322
Plantae: Glaucophyta 325
Plantae: Rhodophyta 327
Plantae: Chlorophyta 347
Plantae: Bryophyta 382
Plantae: Tracheophyta: 411
Plantae: Fossils 460, 476
Fungi: Introduction 499
Fungi: Chytridiomycota 516
Fungi: Microsporidia 520
Fungi: Zygomycota 524
Fungi: Ascomycota 528
Fungi: Basidiomycota 565
Fungi: Lichens 586

SCOPING THE TREE OF LIFE

Humans are born classifiers. We have a natural disposition to want to organise and systematise knowledge, concepts, and things of importance to us, including living organisms. The criteria for classification are many and varied. Every culture and language system has had its own folk taxonomies for naming and classifying animals and plants – which constituted the two major divisions of life from Aristotle until the middle of the 20th century. But nearly a century earlier, Ernst Haeckel, a German contemporary of Charles Darwin and populariser of evolutionary theory, had already proposed a third kingdom, Protista, for organisms that were not plainly either animals or plants or which blurred the distinctions. The boundaries of this third kingdom have fluctuated ever since.

Today we recognise that the fundamental dichotomy between living things is not between animals and plants but between those that lack a cell nucleus (prokaryotes) and those that have one (eukaryotes). In a summary of the evolving views of organism relationships, Cavalier-Smith (2010a) noted that the basic unity of prokaryotes like bacteria and blue-green algae was first clearly recognised by 1875, but it required the advent of the transmission electron microscope, which revealed the internal structure of cells in astonishing detail, to make this distinction unequivocal. By the early 1960s, the prokaryote-eukaryote divide had been clearly articulated. Additional electron-microscope and biochemical procedures led to a spate of work that increasingly illuminated the fine structures and relationships among organisms.

The next major breakthrough was the advent of molecular tools from the 1970s onwards, especially the sequencing of nucleotide bases in ribosomal RNA. Using this molecular information, microbiologists led by Carl Woese of the University of Illinois argued in 1990 for three major groups of life, which they called domains. Two were prokaryotic – Archaea and (Eu-)Bacteria – and the third comprised all eukaryotes (Eukarya). This scheme has been, and remains, very influential, especially among microbiologists, but not all gene trees yield a three-domains view of life (e.g. Rivera & Lake 2004; Cavalier-Smith 2006; Cox et al. 2008). Cavalier-Smith (2002, 2010) also rejects this scheme on ultrastructural, biochemical, and paleontological grounds. It is not used in this volume.

In the meantime, there has been a huge body of work on evolutionary relationships within eukaryotes. After all, all organisms are related genetically and we want to know the details. It's like figuring out your family tree, except that the genealogy we are trying to elucidate began about 800–850 million years ago when the first eukaryote cell(s) evolved from a common ancestor of archaebacteria and eukaryotes (Cavalier-Smith 2009). It's a big and complex family tree, complicated by the 'borrowing' of an α-proteobacterium into symbiotic union with an early eukaryote to become a mitochondrion, and later a cyanobacterium to become a chloroplast in a plant cell. Later such unions, very rare and infrequent in evolution, included the uptake of a red-algal cell into the ancestral chromist, a chloroplast into a euglenozoan, a haptophyte chromist chloroplast into a dinoflagellate, and a chloroplast into a cercozoan, among others, all beautifully summarised and illustrated by Keeling (2004). But this is getting ahead of the story.

In the last decade there has been an emerging view of six major groups – called supergroups – of eukaryotes, not all equal in size but characterised by their major distinctions (e.g. Baldauf 2003; Simpson & Roger 2004; Adl et al. 2005; Keeling et al. 2005). Notwithstanding, there still remained many unanswered questions. A number of taxa were hard to place and there were some overlapping boundaries (Wegener Parfrey et al. 2006; Lane & Archibald 2008). The six supergroups are:

1. Amoebozoa (Greek *amoibe*, change; *zoön*, animal) – composed of lineages of naked and testate amoebae, slime moulds, and relatives, generally well supported in molecular phylogenies; without clear-cut derived morphological characters, although most lineages produce pseudopodia that are broad (lobose).
2. Opisthokonta (Greek *opisthen*, behind; *kontos*, punting pole, oar) – includes animals, fungi, and their unicellular relatives such as collar flagellates, which share (or are derived from organisms with) a single posterior cilium; strongly supported by gene-sequence data.
3. Rhizaria (Greek *rhiza*, root) – a group united only by molecular phylogenies; members include ecologically important and abundant organisms such as foraminifera, with root-like pseudopods, and cercozoans, the latter being significantly understudied.
4. Excavata (Latin *ex-*, out; *cavare*, to make hollow) – unicellular eukaryotes sharing some cytoskeletal features and a distinctive ventral 'excavation' that is a feeding groove, plus forms lacking some of these features that have been identified as relatives by molecular means; it has included such disparate forms as euglenoids, kinetoplastids, diplomonads, parabasalids, and oxymonads and the common ancestry of the group as a whole has not been strongly supported by published gene-sequence data.
5. Chromalveolata (Greek *chroma*, colour; Latin *alveolus*, little cavity) – an assemblage of photosynthetic and non-photosynthetic organisms united by the 'chromalveolate hypothesis', which states that the plastids of chromists (e.g. ochrophyte algae), and alveolates (e.g. ciliates, sporozoan parasites, dinoflagellates, characterised by flattened sacs [alveoli] under the cell membrane) are the product of a single secondary endosymbiosis in the common ancestry of the two groups. Plastids are secondarily lost in several groups or are non-photosynthetic (Keeling 2009).
6. Archaeplastida (Greek *archaios*, old; *plastos*, formed, moulded) – i.e. plants; molecular evidence for a single origin of red-algal, green-algal, and glaucophyte plastids is supported by the structure of plastid genomes and the light-harvesting complex.

Since 2007, additional evidence has been compiled to support the alliance of Rhizaria with Chromalveolata (Burki et al. 2007). A more recent comprehensive phylogenomic analysis was conducted using 143 proteins and 48 taxa including 19 excavates (Hampl et al. 2009). The outcome was a confirmation of an excavate clade and the resolution of relationships among the remaining five supergroups, resulting in two *mega*groupings. Hence Amoebozoa is allied with Opisthokonta as unikonts (ancestrally having a single cilium) and an unnamed grouping (bikonts generally) includes Archaeplastida, Rhizaria, and the chromalveolates (see also Burki et al. 2008). Hence Hampl et al. (2009) identified three groups, i.e. Excavata, unikonts, and Rhizaria/Chromalveolata + Archaeplastida.

How might the resolution of these major branches (clades) of the tree of life help with classification? Cavalier-Smith's (1998) five-kingdom scheme for eukaryotes is retained in this volume but modified (Cavalier-Smith 2010b) as follows:

Supergroups Amoebozoa and Excavata (and the opisthokont phylum Choanozoa, see below) are retained in a basal **kingdom Protozoa**. It is paraphyletic in that it includes the ancestors of the other eukaryote kingdoms. (Abandoning his former opinion that the root of the eukaryote tree was between unikonts and bikonts, Cavalier-Smith (2010b) now argues that it is between what he sees as the most basal group – Euglenozoa – and all other eukaryotes. Accordingly, he separates Euglenozoa from the other excavate phyla.)

Supergroup Opisthokonta includes animals, fungi, and the ancestors of both, found among Choanozoa. These are all unikont/opisthokont, i.e. ancestrally having a single posterior cilium. Because animals and fungi are such different kinds of organisms (!) this major branch of the tree is divided into the **kingdom Animalia** and **kingdom Fungi**, with the basal phylum Choanozoa being included in kingdom Protozoa.

Supergroups Rhizaria and Chromalveolata, having been demonstrated to be related, comprise **kingdom Chromista**.

Supergroup Archaeplastida is generally taken to be equivalent to **kingdom Plantae** in most treatments although Adl et al. (2005) restricted Plantae to land plants (mosses and liverworts, lycopods and ferns, and seed plants).

All this raises the question of what might be construed to constitute a kingdom – a clade and/or a monophyletic grade? Cavalier-Smith (1998), who introduced the name Opisthokonta, remarked that at least one cladist urged him to establish a kingdom-level taxon with this name, but 'A kingdom Opisthokonta would be much less useful than the existing kingdoms Animalia, Fungi and Protozoa as a way of subdividing the living world into manageable groups of similar organisms, i.e. a classification as opposed to a phylogeny' (Cavalier-Smith 1998).

The untidiest part of Cavalier-Smith's five-kingdom schema for eukaryotes is not the splitting of the phylogenetically cohesive Opisthokonta (it makes logical sense to retain the Fungi and Animalia as separate kingdoms), but, arguably, the composition of the paraphyletic Protozoa. Perhaps one should have a separate kingdom for Euglenozoa, another for Amoebozoa, and another for the Excavata. A less satisfactory alternative might be to retain a kingdom Protista (or Protoctista). But even in Haeckel's day Protista did not include ciliates (he had them with the Animalia), and the scheme of Margulis and Schwartz (2001), which has popularised the use of Protista/Protoctista in the USA, includes lower plants and fungi, unhelpfully restricting these two kingdoms solely to land plants and macrofungi respectively. A kingdom Protista thus embraces too much gross disparity in morphology to be taxonomically useful, whereas Protozoa has always been morphologically and taxonomically narrower than Protista as it has excluded the majority of the algae. Therefore, as Cavalier-Smith has pointed out (pers. comm. 2011), it has been a better candidate than Protista for refinement rather than abandonment. Application of the term protist to a multicellular, tissue-forming giant kelp, for example, renders it meaningless.

Adl et al. (2005) presented a nested classification that uses the six supergroup names as the highest ranks of eukaryotes. It was proposed by the International Society of Protistologists but, while ranking all clades, avoided the use of Linnaean rank names (originally kingdom, class, and order for the higher categories of Linnaeus' System of Nature). Insofar as the nested groups comprise a mix of taxon names based on priority (i.e. according to the year of introduction of the name), many individual genera as well as traditional taxon names (family to class) end up having the same rank in Adl et al.'s (2005) hierarchy while having different suffixes or none at all. This is very confusing when these 'group names' (genus to class) are used in isolation without regard to phylogenetic relativity. One of the great benefits of Linnaean ranks and their standardised suffixes is that they instantly relativise taxa that are otherwise unknown to the non-specialist.

The overarching higher-level classification used in these volumes, therefore, is intended to be simultaneously pragmatic and informative of both evolutionary relatedness and morphological dissimilarity. A good part of it, with the exception of Bacteria, is also about to be used in the global Catalogue of Life.

Dennis P. Gordon
NIWA, Wellington

References

ADL, S. M.; SIMPSON, A. G. B.; FARMER, M. A.; ANDERSEN, R. A.; ANDERSON, O. R.; BARTA, J. R.; BOWSER, S. S.; BRUGEROLLE, G.; FENSOME, R. A.; FREDERICQ, S.; JAMES, T. Y.; KARPOV, S.; KUGRENS, P.; KRUG, J.; LANE, C. E.; LEWIS, L. A.; LODGE, J.; LYNN, D. H.; MANN, D. G.; MCCOURT, R. M.; MENDOZA, L.; MOESTRUP, Ø.; MOZLEY-STANDRIDGE, S. E.; NERAD, T. A.; SHEARER, C. A.; SMIRNOV, A. V.; SPIEGEL, F. W.; TAYLOR, M. F. J. R. 2005: The new higher level classification of eukaryotes with emphasis on the taxonomy of protists. *Journal of Eukaryotic Microbiology 52*: 399–451.

BALDAUF, S. L. 2003: The deep roots of eukaryotes. *Science 300*: 1703–1706.

BURKI, F.; SHALCHIAN-TABRIZI, K.; MINGE, M.; SKJAEVELAND, A.; NIKOLAEV, S. I.; JAKOBSEN, K. S.; PAWLOWSKI, J. 2007: Phylogenomics reshuffles the eukaryotic supergroups. *PLoS ONE 2 (e790)*: 1–6.

BURKI, F.; SHALCHIAN-TABRIZI, K.; PAWLOWSKI, J. 2008:. Phylogenomics reveals a new 'megagroup' including most photosynthetic eukaryotes. *Biological Letters 4*: 366–369.

CAVALIER-SMITH, T. 1998: A revised six-kingdom system of life. *Biological Reviews 73*: 203–266.

CAVALIER-SMITH, T. 2002: The neomuran origin of archaebacteria, the negibacterial root of the universal tree and bacterial megaclassification. *International Journal of Systematic and Evolutionary Microbiology 52*: 7–76.

CAVALIER-SMITH, T. 2006: Rooting the tree of life by transition analyses. *Biology Direct 1.19*: 1–83.

CAVALIER-SMITH, T. 2009: Predation and eukaryote cell origins: a coevolutionary perspective. *International Journal of Biochemistry and Cell Biology 41*: 307–322.

CAVALIER-SMITH, T. 2010a: Deep phylogeny, ancestral groups and the four ages of life. *Philosophical Transactions of the Royal Society, B, 365*: 111–132.

CAVALIER-SMITH, T. 2010b: Kingdoms Protozoa and Chromista and the eozoan root of the eukaryotic tree. *Biology Letters 6*: 342–345.

COX, C. J.; FOSTER, P. G.; HIRT, R. P.; HARRIS, S. R.; EMBLEY, T. M. 2008: The archaebacterial origin of eukaryotes. *Proceedings of the National Academy of Sciences of the USA 105*: 20356–20361.

HAMPL, V.; HUG, L.; LEIGH, J. W.; DACKS, J. B.; LANG, B, F.; SIMPSON, A. G. B.; ROGER, A. J. 2009: Phylogenomic analyses support the monophyly of Excavata and resolve relationships among eukaryotic 'supergroups. *Proceedings of the National Academy of Sciences 106*: 3859–3864.

KEELING, P. J. 2004: Diversity and evolutionary history of plastids and their hosts. *American Journal of Botany 91*: 1481–1493.

KEELING, P. J. 2009: Chromalveolates and the evolution of plastids by secondary endosymbiosis. *Journal of Eukaryotic Microbiology 56*: 1–8.

KEELING, P. J.; BURGER, G.; DURNFORD, D. G.; LANG, B. F.; LEE, R. W.; PEARLMAN, R. E.; ROGER, A. J.; GRAY, M. W. 2005: The tree of eukaryotes. *Trends in Ecology and Evolution 20*: 670–676.

LANE, C.E.; ARCHIBALD, J. M. 2008: The eukaryotic tree of life: endosymbiosis takes its TOL. *Trends in Ecology and Evolution 23*: 268–275.

MARGULIS, L.; SCHWARTZ, K. V. 2001: *Five Kingdoms: An Illustrated Guide to the Phyla of Life on Earth*. Third Edn. W. H. Freeman and Company, New York. xxii + 520 p.

RIVERA, M. C.; LAKE, J. A. 2004: The ring of life provides evidence for a genome fusion origin of eukaryotes. *Nature 431*: 152–155.

SIMPSON, A. G. B.; ROGER, A. J. 2005: The real 'kingdoms' of eukaryotes. *Current Biology 14*: R693–R696.

WEGENER PARFREY, L.; BARBERO, E.; LASSER, E.; DUNTHORN, M.; BHATTACHARYA, D.; PETTERSON, D.; KATZ, L. A. 2006: Evaluating support for the current classification of eukaryotic diversity. *PLoS Genetics 2 (e220)*: 2062–2073.

WOESE, C. R.; KANDLER, O.; WHEELIS, M. L. 1990: Towards a natural system of organisms, proposal for the domains Archaea, Bacteria, and Eucarya. *Proceedings of the National academy of Sciences of the USA 87*: 4576–4579.

The diagram on the facing page is a broad-scale depiction of the relationships among the six kingdoms of life and the phyla dealt with in this volume. Each diagram labelled with a three-letter code represents a phylum. We cannot be more precise in depicting relationships within the branches – the tree, or cactus, would need to be shown three-dimensionally and there is in any case some debate concerning the relationships of many of the phyla (explained in the chapters that follow).

Kingdom names are shown beneath each cactus branch to which each pertains. Some of the major groupings of phyla within each kingdom are shown in a smaller font. The two arrowed lines indicate the endosymbiotic origin of a mitochondrion (left arrow) and a chloroplast from bacterial ancestors. The bacterial kingdom includes Cyanobacteria (CYA), Archaebacteria (ARC), and other bacterial phyla (OTH) – not all shown here. The protozoan kingdom includes Euglenozoa (EUG), Percolozoa (PER), Loukozoa (LOU), Metamonada (MET), Amoebozoa (AMO), Apusozoa (APU), and Choanozoa (CHO). The ancestors of the animal and fungal kingdoms are found within Choanozoa. The fungal phyla include Chytridiomycota (CHY), Microsporidia (MIC), Zygomycota (ZYG), Glomeromycota (GLO), Ascomycota (ASC), and Basidiomycota (BAS). The chromist kingdom includes Cercozoa (CER), Foraminifera (FOR), Radiozoa (RAD), Ciliophora (CIL), Myzozoa (MYZ), Ochrophyta (OCH), Bigyra (BIG), Oomycota (OOM), Cryptophyta (CRY), Haptophyta (HAP), and Heliozoa (HEL). The plant kingdom includes Glaucophyta (GLA), Rhodophyta (RHO), Chlorophyta (CHL), Charophyta (CHA), Bryophyta (BRY), and Tracheophyta (TRA).

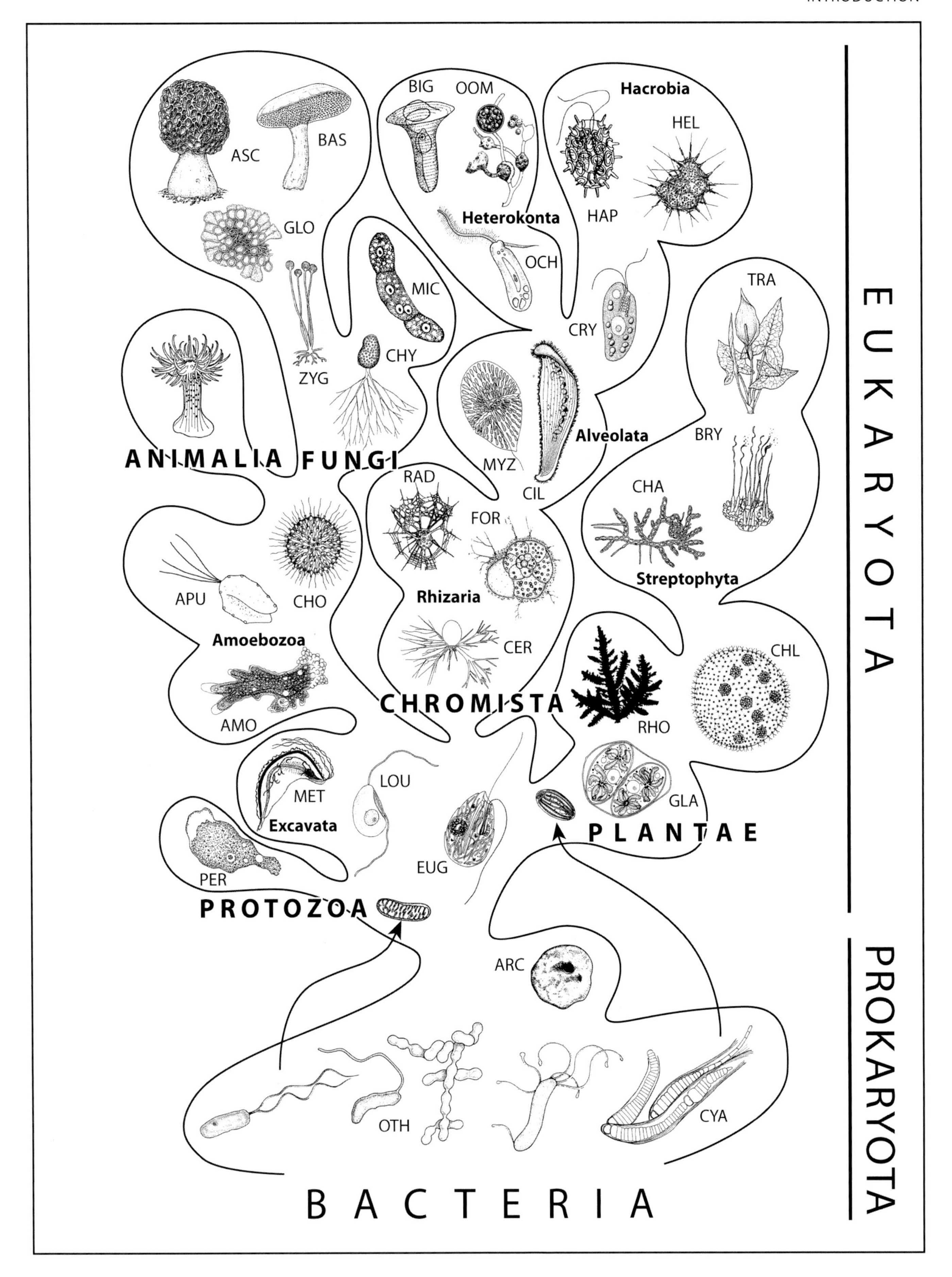

ASC
BAS
GLO
ZYG
BIG
OOM
Heterokonta
OCH
MIC
CHY
Hacrobia
HEL
HAP
CRY
ANIMALIA
FUNGI
MYZ
CIL
Alveolata
TRA
BRY
CHA
Streptophyta
RAD
FOR
Rhizaria
CER
CHROMISTA
APU
CHO
Amoebozoa
AMO
RHO
CHL
GLA
PLANTAE
MET
Excavata
LOU
EUG
PER
PROTOZOA
ARC
OTH
CYA
BACTERIA
EUKARYOTA
PROKARYOTA

Species diversity of New Zealand living Bacteria, Protozoa, Chromista, Plantae, and Fungi

Taxon	Described species	Known undescribed/ undetermined species	Totals	Estimated undiscovered species†
Bacteria	**667**	**30**	**697**	**unknown**
Cyanobacteria	372	30	402	>200
Archaebacteria	1	0*	1	unknown
All other bacteria	294	0*	294	unknown
Protozoa	**507**	**32**	**539**	**570–770**
Euglenozoa	108	7	115	>100
Percolozoa	7	6	13	8
Loukozoa	1	0	1	3
Metamonada	15	3	18	60
Amoebozoa	338	11	349	350–550
Apusozoa	1	0	1	3
Choanozoa	37	5	42	45
Chromista	**4,039**	**264**	**4,303**	**4,595–4,695**
Ochrophyta	1,765	30	1,795	1,260
Oomycota	151	10	161	9
Bigyra	14	1	15	35
Myzozoa	384	57	441	1,370
Ciliophora	340	7	347	1,100
Cercozoa	80	3	83	135
Foraminifera	1,025	11	1,141	205–305
Radiozoa	157	24	181	200
Cryptophyta	17	0	17	30
Haptophyta	82	10	92	210
Heliozoa	24	6	30	40
Plantae	**7,432**	**116**	**7,548**	**1,075–1,175**
Glaucophyta	1	0	1	1
Rhodophyta	472	89	561	35
Chlorophyta	566	19	585	610
Charophyta	527	7	534	65
Bryophyta	1122	0	1,122	62
Tracheophyta	4,744	1	4,745	300–400
Fungi	**8,045**	**350**	**8,395**	**23,525**
Incertae sedis	3	0	3	0
Chytridiomycota s. l.	150	1	151	175
Microsporidia	12	6	18	9,000
Zygomycota	89	0	89	170
Glomeromycota	38	1	39	5
Ascomycota	5,238	79	5,317	9,065
Basidiomycota	2,515	263	2,778	5,110
Totals	**20,690**	**792**	**21,482**	**29,765–30,165§**

† Totals rounded
* None in culture reported by authors
§ Eukaryotes only

Species diversity of New Zealand Bacteria, Protozoa, Chromista, Plantae, and Fungi by major environment as and fossils

Taxon	Marine	Terrestrial	Freshwater	Fossil
Bacteria	**>79**	**306**	**356**	**–**
Cyanobacteria	40	15	352	–†
Archaebacteria	0	0	1*	–
All other bacteria	>39	291	3*	–
Protozoa	**43**	**328**	**194**	**0**
Euglenozoa	17	3	92	0
Percolozoa	0	2	5	0
Loukozoa	0	0	1	0
Metamonada	0	16	2	0
Amoebozoa	3	306	80	0
Apusozoa	1	0	0	0
Choanozoa	22	1	20	0
Chromista	**2,643**	**489**	**1,280**	**4,221**
Ochrophyta	858	16	939	767
Oomycota	4	136	23	0
Bigyra	9	1	6	0
Myzozoa	249	148	49	468
Ciliophora	78	149	175	0
Cercozoa	20	37	53	3
Foraminifera	1141	0	0	2006
Radiozoa	182	0	0	947
Cryptophyta	10	0	7	0
Haptophyta	90	0	2	30
Heliozoa	2	2	26	0
Plantae	**702**	**5,764**	**1,091**	**290**
Glaucophyta	0	0	1	0
Rhodophyta	541	0	20	0
Chlorophyta	156	35	394	18**
Charophyta	0	1	533	7
Bryophyta	0	1,086	35	1§
Tracheophyta	5	4,642	108	264
Fungi	**89**	**8,213**	**191**	**6**
Incertae sedis	0	3	0	5
Chytridiomycota s. l.	6	106	39	0
Microsporidia	3	12	3	0
Zygomycota	0	72	17	0
Glomeromycota	0	39	0	0
Ascomycota	77	5,210	128	0
Basidiomycota	3	2,771	4	1
Totals	**3,556**	**15,100**	**3,118**	**4,517**

† Holocene and older stromatolites from geothermal waters contain species of Phormidium (Cyanobacteria)
* Described from geothermal waters
** Including seven taxa of uncertain affinity
§ Macrofossils only

Species diversity of New Zealand living Bacteria, Protozoa, Chromista, Plantae and Fungi, and percentage endemism

Taxon	Total species	Adventive species	Endemic species	Percentage of all species endemic*	Percentage of indigenous species endemic*
Bacteria	**697**	**unknown**	**unknown**	**unknown**	**unknown**
Cyanobacteria	402	1	19	4.7	4.7
Archaebacteria	1	unknown	unknown	unknown	unknown
All other bacteria	294	unknown	unknown	unknown	unknown
Protozoa	**539**	**23**	**24**	**4.5**	**4.7**
Euglenozoa	115	1	8	7.0	7.0
Percolozoa	13	0	1	7.7	7.7
Loukozoa	1	0	0	0	0
Metamonada	18	9	6	33.3	66.7
Amoebozoa	349	13	6	1.7	1.8
Apusozoa	1	0	1	100	100
Choanozoa	42	0	2	5.6	5.6
Chromista	**4,203**	**287**	**282**	**6.5**	**7.0**
Ochrophyta	1,795	14	55	3.1	3.1
Oomycota	161	106	1	0.6	1.6
Bigyra	15	3	0	0	0
Myzozoa	441	115	32	7.3	9.8
Ciliophora	347	37	50	14.4	16.1
Cercozoa	83	9	10	12.0	13.5
Foraminifera	1,141	3	128	11.2	11.2
Radiozoa	181	0	0	0	0
Cryptophyta	17	0	1	5.9	5.9
Haptophyta	92	0	2	2.2	2.2
Heliozoa	30	0	3	10.0	10.0
Plantae	**7,548**	**2,585**	**2,390**	**31.7**	**48.2**
Glaucophyta	1	0	0	0	0
Rhodophyta	561	15	189	33.7	34.6
Chlorophyta	585	10	29	5.0	5.0
Charophyta	534	1	43	8.1	8.1
Bryophyta	1,122	36	394	35.1	36.3
Tracheophyta	4,745	2,523	1,735	36.6	78.1
Fungi	**8,395**	**1,993**	**1,663**	**19.8**	**26.0**
Incertae sedis	3	0	0	0	0
Chytridiomycota s. l.	151	13	7	4.5	4.9
Microsporidia	18	6	5	27.8	41.7
Zygomycota	89	19	10	11.2	14.3
Glomeromycota	39	2	0	0	0
Ascomycota	5,317	1,485	780	14.7	20.4
Basidiomycota	2,778	481	861	31.0	37.5
Total known	**21,482**	**4,889**	**4,378**	–	–
[Eukaryotes only	20,785	4,888	4,359	21.0	27.4]

* Rounded

ONE

Kingdom BACTERIA

JOHN M. YOUNG, JACKIE AISLABIE, DOROTA M. BRODA, GEOFFREY W. DE LISLE, ELIZABETH W. MAAS, HUGH MORGAN, MAUREEN O'CALLAGHAN, SIMON SWIFT, SUSAN J. TURNER

Bacteria are the smallest and probably the oldest forms of life on Earth. Whether considered in terms of biomass or as numbers of individuals, most life on earth consists of bacteria. As biomass, the most abundant form today is bacteria in ocean sediments, while the commonest constituents of ocean plankton, supplying a significant fraction of the earth's atmospheric oxygen, are cyanobacteria ('blue-green algae'). The diversity of bacteria is indicated by their ability to inhabit the widest range of environments of all organisms – from extremes of heat, cold, acidity, and pressure; growth with and without oxygen; use of the widest range of energy and nutrient sources; and production of the largest and most complex range of metabolites. Bacteria comprise simple cells that lack both a nucleus and internal organelles and are called prokaryotes (Greek *pro-*, before, and *karyon*, nucleus), in contrast to the eukaryote (*eu-*, true) cells of all other life forms (protozoa, chromists, plants, fungi, and animals, which are larger and have complex internal structures.

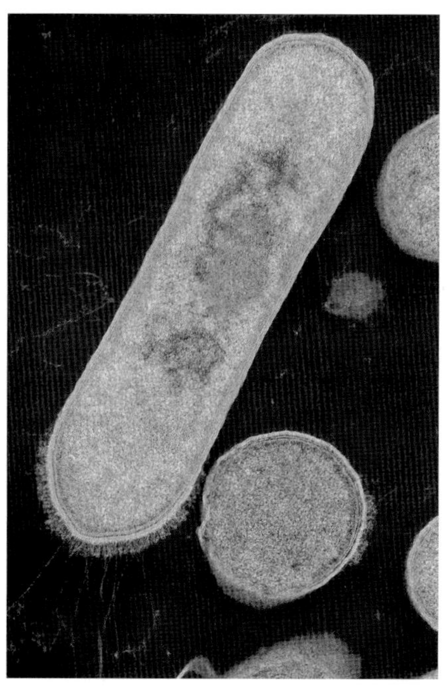

Rod-shaped bacterial cells in longitudinal and cross-section as seen by transmission electron microscopy.
Victoria University of Wellington

Life had its origins on the cooling planet when water first condensed, probably about 3800 Mya (million years ago). The environmental conditions that existed then are a matter of conjecture. It is thought that the earth's atmosphere was a relatively simple gas mixture, probably composed largely of ammonia, carbon dioxide, carbon monoxide, methane, nitrogen, and water vapour. There is an on-going and lively debate as to how life might have evolved from pre-existing chemical structures. Indications of free-living bacteria are found in the oldest fossil record, indicating their presence on Earth very soon after the planet cooled (Knoll 2003). Life as we know it today is based on the interplay between nucleic acids, associated with inheritance mechanisms, and protein enzymes associated with cell metabolism. Bacteria are believed to reflect the ancestry of a very early form of life on Earth, but whether this form arose directly from non-living chemical processes, or whether there were earlier biological processes identifiable as life has not been resolved. Reactions that produced complex biochemicals could have occurred rapidly at temperatures above 100° C in super-heated water. The surface-catalytic properties of clays have been proposed as the basis for a simple replicating biological world prior to organisation based on nucleic acids. The instability of nucleic acid polymers at temperatures near 100° C sets a limit on the earliest starting point for the evolution of life. If bacteria were not the first living organisms, then they can only have been preceded by a group of simpler, short-lived life-forms that they quickly superseded. Microfossils of photosynthetic oscillatorians (Cyanobacteria) have been found in the oldest Australian rock strata at 3600 Mya and thereafter in ever-greater numbers. The similarity of the fossils to bacteria today permits the inference that bacteria then were like bacteria now, with a cell wall that protected protoplasmic processes from the environment, a chromosome comprising a simple circular DNA duplex

Cambrian stromatolites from New York State.
Wikimedia Commons, M. C. Rygel

strand, and ribosomes made up of RNA acid and proteins (perhaps residual structures from an earlier life-form). The similarity between these fossils and modern oscillatorians is so striking that the only doubt concerns the precise date of their earliest appearance. Although the earliest record is of photosynthesising bacteria, it is plausible to suggest that an earlier phase comprised the chemorganotrophic bacteria, which depended on simple chemical or physical reactions as sources of energy.

For almost 2000 million years, bacteria were the only organisms on Earth. This period, before about 1900 Mya, is one in which bacteria adapted and evolved to take advantage of the available environments, and of those generated by their own activity.

The fossil record gives no indication of the extent of on-going evolution until the presence of stromatolites (mounds of alternating layers of sediment and microbial mats) provides evidence of increased oxygen in the atmosphere about 2600 Mya. It seems likely that it was a period of diversification of metabolic processes, with the exploitation of increasingly complex chemical substrates, particularly those generated by biological activity itself. Oxygen is lost to the environment by oxidative reactions, which would eventually result in a completely oxygen-free (anaerobic) environment. Oxygen remains present in the atmosphere only because it is continuously produced in photosynthesis. Specifically, although there were several photosynthetic bacterial groups, it was only the cyanobacteria that produced oxygen as a by-product. The result of oxygen over-production was to render most of the environment toxic to the anaerobic bacteria. These organisms would necessarily have been driven into retreat in anaerobic muds and other microenvironments. The availability of free oxygen at progressively increasing concentrations allowed for the evolution of aerobic respiration. This anabolic process resulted in complete breakdown of organic compounds, with the creation of energy surpluses and greater bacterial activity. It also led to the return of carbon to the atmosphere as CO_2 rather than for it to be unavailable in organic debris.

Bacteria today are marked by a greater level of molecular diversity than is present in all other life-forms – the eukaryotic world – and it is likely that most genetic diversity expressed today has these ancient origins. The diversity expressed by macroorganisms is mainly the consequence of the reorganisation of genes found in the prokaryote world, rather than to evolution of novel eukaryote genes.

The great contribution to biological evolution made by the prokaryote cell was the eukaryote cell. Prokaryote and eukaryote cells represent the greatest division in the biological world. The evolution of the eukaryote cell entailed *the* major step in structural re-engineering at the cellular level (Niklas 1997; Rivera & Lake 2004), about 1900 Mya. No free-living eukaryote cells are known that are capable of sustained anaerobic activity, therefore it is safe to say that this was the latest time that aerobic respiration could have evolved. The eukaryote cell evolved to become gigantic compared with the prokaryote

Estimated number and biomass of bacteria in the world (from Whitman et al. 1998)

Environment	Number of bacterial cells (x 10^{28})	Bacterial carbon (g x 10^{15})
Aquatic habitats	12	2.2
Oceanic subsurface	355	300
Soil	25	25
Terrestrial subsurface	25–250	22–220
Totals	**420–640**	**350–550**

cell, being about 1000 times greater in volume. It is characterised by extensive internal structural specialisation, with a complex membrane system upon which metabolic processes occur. Specialised organelles are associated with energy processes (mitochondria) and with photosynthesis (chloroplasts). Two processes are identified in the evolution of the eukaryote from the prokaryote cell – cell symbiosis and invagination. Detailed modelling suggests that the eukaryote chromosome is the product of a union of two unrelated bacterial ancestors (Rivera & Lake 2004). As well, the fact that chloroplasts and mitochondria have independent genetic mechanisms, their own regulatory DNA, and that this DNA is recognisably bacterial in structure has led to the conclusion that these organelles are the result of symbiotic specialisation of one bacterium associated with capture by another.

Bacterial structure and organisation

With the recognition that most bacteria are uncharacterised, generalities about them can be made only on the basis of the small proportion that have been cultivated and characterised. The key characteristic of the bacteria is their simple cellular organisation. They have no nucleus, no specialised organelles such as mitochondria and chloroplasts, or endoplasmic reticulum. Compared with eukaryotes, the cells of bacteria are small. For example, rod-shaped bacteria are about 0.5–10 micrometres long and about 0.1–2 micrometres in diameter. A typical cell is about 3 x 1 micrometres (= 7×10^{-13} cm^3), corresponding to about 10^{12} cells per millilitre), though some cells may have only 10% of this volume, whereas the largest bacterial cells are visible to the naked eye – *Epulopiscium fishelsoni* and *Thiomargarita namibiensis* are half to three-quarters of a millimetre in diameter (Schulz et al. 1999).

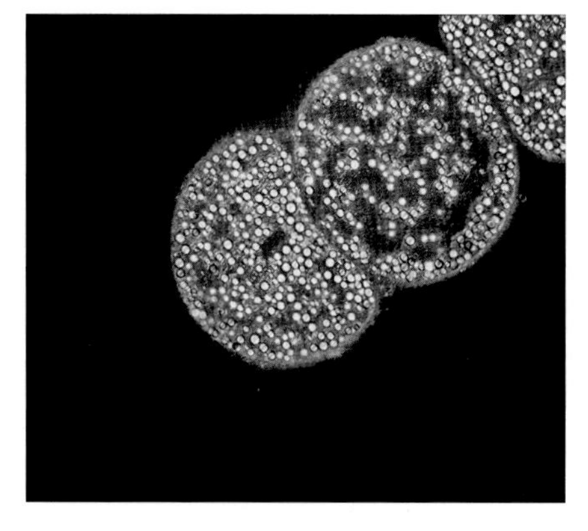

The giant sulphur bacterium *Thiomargarita namibiensis.*

Heide N. Schulz-Vogt, Max Planck Institute for Marine Microbiology, Bremen

Bacteria are simple in structure, being cocci (spheres) (e.g. *Staphylococcus*), rods in straight or spiral forms (e.g. *Pseudomonas*, *Spirochaeta*), or they form extended thread-like mycelial structures with occasional septa (e.g. *Streptomyces*). Some are variable in shape. The Gram staining reaction distinguishes bacteria according to differences in cell-wall structure. Gram-positive bacteria have cell

Summary of New Zealand bacterial diversity

Taxon	Recorded species + subspecies in culture	Estimated total species + subspecies	Estimated adventive species + subspecies	Estimated endemic species + subspecies	Estimated endemic genera
NEGIBACTERIA	178				
EOBACTERIA	1	many	many	few	none?
Eobacteria	1	many	many	few	none?
GLYCOBACTERIA	177				
Cyanobacteria*	–	–	–	–	–
Spirochaetae	0	many	many	few	none?
Sphingobacteria	14	many	many	few	none?
Planctobacteria	0	many	many	few	none?
Proteobacteria	163	many	many	few	none?
UNIBACTERIA	117+1				
Posibacteria	116+1	many	many	few	none?
Archaebacteria	1	many	many	few	none?
Totals	**295+1**	**many**	**many**	**few**	**none?**

* Figures for Cyanobacteria see Chapter 2.

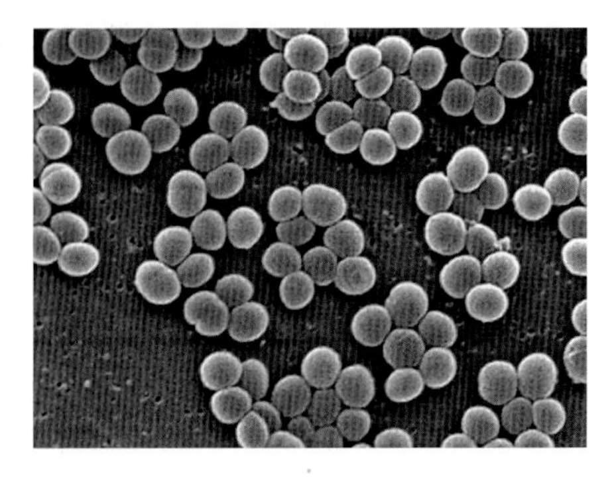

Staphylococcus aureus, the 'golden staph', often part of the normal skin flora but able to cause a range of illnesses from minor skin infections to life-threatening diseases.

Wikimedia Commons, Janice Carr, US Centers for Disease Control and Prevention

walls that contain 30–70% peptidoglycan (amino-acid, amino-sugar) and polysaccharides in an apparently simple structure. Gram-negative bacteria have cell walls containing 10% peptidoglycan, liposaccharides, lipoprotein, and polysaccharides in complex layers. The presence of peptidoglycan in the wall maintains cell shape – bacteria without peptidoglycan in walls are amorphous (e.g. *Chlamydia*). Cell-wall chemistry is associated with important differences in cell function, organisation, and metabolism representative of a major division of the bacteria into the two groups. Some bacteria, the Mollicutes, lack a cell wall. Some of these have an independent existence while others are parasitic in animals (the mycoplasmas) or in plants (the phytoplasmas). Mycoplasmas are culturable with difficulty. So far, phytoplasmas and some related plant parasites have not been cultured.

Bacterial genetic material is highly variable in organisation. It is usually represented by a simple DNA double-stranded circular chromosome of about 3–8 $\times$ 10^6 nucleotide base pairs, 1000–1500 micrometres (1–1.5 millimetres) in length when unfolded. However, some bacteria have two chromosomes of almost equal size and others have one or more satellite chromosomes representing lesser proportions of the total cell DNA. Many, probably most, bacteria contain additional genetic material as plasmids, comprising short lengths of DNA, usually circular in organisation and variable in number up to 40. These may be a few hundred bases or several kilobases in length and represent 2–30% of total genomic material. Megaplasmids, perhaps better considered as satellite chromosomes, more than 1000 kilobases (kb) in length, are present in many bacteria. Some parasitic mycoplasmas, and all phytoplasmas that have become highly adapted in dependent associations with animals and plants, have a reduced (1.3–0.6 $\times$ 10^6) kb genome.

Cell multiplication is by binary fission, not mitotic division; there is no spindle or centriole formation as in eukaryotes. The DNA genome is reproduced as cells divide.

Summary of New Zealand cultured bacterial diversity (bacterial genera) by environment

Taxon	Marine (Chatham Rise)	Terrestrial and freshwater*
NEGIBACTERIA	24	70
EOBACTERIA	0	1
Eobacteria	0	1
GLYCOBACTERIA	24	69
Cyanobacteria†	–	–
Spirochaetae	0	0
Sphingobacteria	7	8
Planctobacteria	0	0
Proteobacteria	17	61
UNIBACTERIA	15	34
Posibacteria	15	33
Archaebacteria‡	–	1
Totals	**39**	**104**
Totals (including estimates of undiscovered species)	**many**	**many**

Isolations from freshwater are invariably of bacteria originally from terrestrial sources.
† Figures for Cyanobacteria (see Chapter 2).
‡ Not recorded.

The capacity to receive functional genes from other bacteria is an important means by which bacterial species can evolve more rapidly than is possible by the mutation of their own genes. It is a mechanism equivalent in function to recombination in eukaryotic cells, but is more adaptive because exchange is promiscuously between unrelated bacteria. DNA sequences can be transferred from one cell to another in processes that are irregular and unpredictable, unlike the structured events associated with meiosis in eukaryotes. Mechanisms of transfer are: conjugation, in which exchange of part of the chromosome occurs between donor and recipient cells and can involve very large chromosomal sections; transduction, in which bacterial viruses mediate the transfer of bacterial DNA between host bacteria; and exchange of plasmid DNA between bacteria. Plasmids are capable of transmission between bacteria and provide a means of rapid horizontal gene transfer, often between unrelated bacteria. Mounting evidence suggests that horizontal gene transfer between unrelated bacteria is widespread. The extent to which transferred DNA is expressed, and hence its significance in the evolution of bacteria and its involvement in creating bacterial diversity, needs to be clarified.

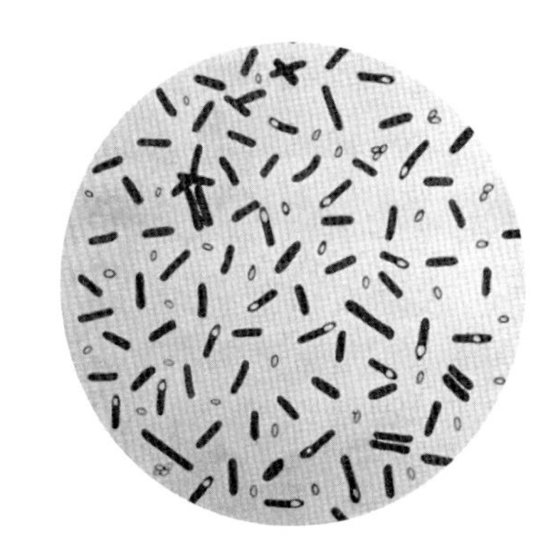

Clostridium botulinum cells with endospores and liberated spores.

Wikimedia Commons, US Centers for Disease Control and Prevention

Ribosomes, central to the mechanism of protein synthesis and universal in living organisms, include two sequences of ribosomal RNA – 5S and 23S in a large subunit, and 16S in a small subunit. DNA ribosomal sequences, especially 16S rDNA, have been widely applied to infer bacterial phylogenetic relationships.

Bacteria are usually motile, swimming by means of one or more long flagella composed of a simple protein structure. Flagella are inserted in rod-shaped bacteria in polar (at the end) or peritrichous (over the surface) orientations. Bacteria can also move across surfaces by 'swarming' or 'twitching' at a rate of up to 1 millimetre per hour. Fimbriae (pili) are macromolecular surface structures on the cell walls of Gram-negative bacteria with diverse roles including attachment to biotic and abiotic surfaces, pathogen specificity, swarming motility, sexual recognition, and DNA transfer and they can be targets for bacteriophage (bacterial virus) attachment.

Some bacteria, predominantly species of *Bacillus* and *Clostridium*, produce endospores as structures for survival, with essential chromosomal DNA being incorporated into a resistant structure. These are not reproductive because there is no multiplication of cell numbers. Endospores are highly resistant to environmental extremes that are lethal to vegetative cells. In the Actinomycetes (which includes *Streptomyces*), many thin-walled, non-resistant spores are produced that grow in filaments, and spread of the spores propagates the bacterium.

Bacterial metabolism is adapted to cope with a wide range of environmental conditions with respect to heat, pH, pressure, and nutrient status. Most groups are restricted in their environmental range. Thus many thermophilic (heat-loving) and halophilic (salt-loving) archaebacteria are incapable of growing in 'ordinary' environments (although they can survive and be cultured from them). Accepting the existence of the proportion of bacteria that have not yet been cultivated and for which no cultural criteria are known, there is now considerable caution in proposing any general conditions for bacterial growth in culture. Typically, culturable soil-inhabiting bacteria have a growth temperature optimum of 27–30° C. Few grow above about 40–45° C. Typical pH optima for growth in characterised bacteria are 6.5–7.5, with a range of pH 4.5–8.5. Water relations set strict limits on bacterial activity. Bacteria grow only in water with a soluble source of nutrients. However, when soluble compounds, including nutrients and necessary salts, are dissolved in concentrations greater then 5%, growth becomes inhibited by the elevated osmotic pressure. When nutrients are present at levels less than 0.05%, they are commonly insufficient to sustain active growth.

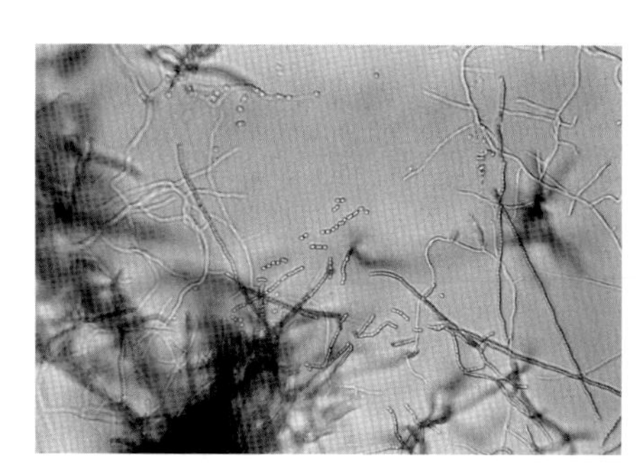

A branching, sporulating *Streptomyces* mycelium.

Wikimedia Commons, David Berd, US Centers for Disease Control and Prevention

All bacteria require energy sources and substrates. Autotrophic bacteria utilise oxidisable inorganic substrates as energy sources. Chemoautotrophic bacteria

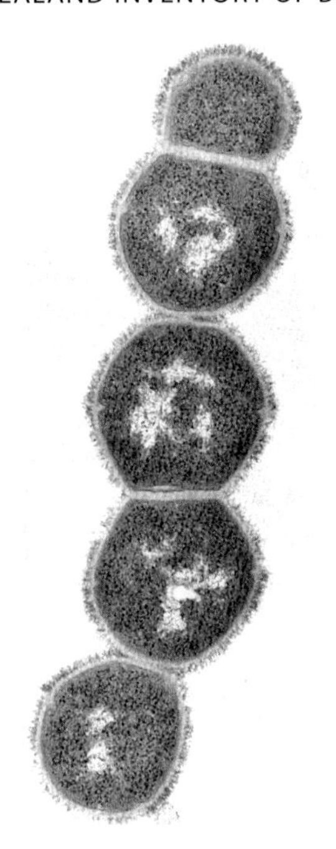

Streptococcus pyogenes, an infectious bacterium with effects ranging from a mild 'strep throat' through skin infections ('impetigo') to life-threatening necrotising fasciitis.
Vincent A. Fischetti, The Rockefeller University, New York

use inorganic substrates such as iron, ammonium, and sulphur as energy sources and use inorganic carbon and nitrogen as sources of structural components. Photoautotrophic bacteria use light as their energy source. However, many bacteria are heterotrophic, requiring organic compounds as energy sources and for cell constituents. They use a wide range of substrates – sugars, alcohols, amino acids, fatty acids, etc. Elemental nutritional requirements have been identified as carbon, nitrogen, phosphorus, potassium, sodium, magnesium, and chlorine (macronutrients), and iron, boron, molybdenum, manganese, zinc, and copper (micronutrients). Some bacteria, the mixotrophs, are photosynthetic in light and heterotrophic in darkness. Some bacteria are fastidious, having a limited nutrient range, and may be exacting, having requirements for specific vitamin compounds. Bacteria may be differentiated according to their electron acceptors and aerobic status – some require oxygen as sole electron acceptor and are obligate aerobes; others may use oxygen but can also use other electron acceptors and are facultative anaerobes, while the remainder do not use oxygen, which may be toxic to them, and are anaerobes. Until recently, oxygen requirement was considered to represent a significant taxonomic distinction. However, molecular studies show that closely related bacteria may be obligate aerobes or facultative anaerobes.

The phytoplasmas are insect-transmitted wall-less bacteria that have not yet been cultured. They are bacteria that have become so highly adapted to a dependent association with animals and plants that they have lost components of their metabolism. These bacteria form a specific association in the intestinal systems of insect species. Sap-feeding species, especially plant-hoppers, can transmit plasmids to plants as alternative hosts, in which they can also multiply. Phytoplasmas appear to do little harm to their insect hosts but can be highly pathogenic to some plants. The plant associations are not specific, so the infected plant host-species depend upon the feeding choice of the insect. However, an insect newly introduced into a region with a different feeding choice may act as a vector between plant species that are already infected and those that have previously been free. Because these bacteria are not culturable, they cannot be described except in terms of sequence characteristics. They are not classified using binominal nomenclature but are named in a special-purpose system as '*Candidatus*', implying that they are candidate species. *Candidatus* nomenclature is applied to all unculturable bacteria.

Bacteria today

What we know about bacteria is mostly based on those that can be grown in pure culture on simple nutrient media (Balows et al. 1992). Bacteria grow in almost all terrestrial and marine environments, and some will grow in environments that are regarded as extreme by comparison with those inhabited by higher organisms. Bacterial diversity exemplifies a wide-ranging and sophisticated biochemistry and an ability to colonise a remarkable range of niches – from the sub-zero deserts of Antarctica to submarine vents at 3000 atmospheres of pressure and 110° C. The extraction and analysis of DNA (particularly ribosomal DNA) from natural environments indicates the presence of a diversity of prokaryotic organisms that far exceeds those already characterised. Estimates are that 1–10% of the major bacterial groups have been cultivated and characterised, and there exist major groups for which no members have been cultured. At present it is not clear why the 'yet to be cultivated' (YBC) bacteria cannot be grown in pure culture. Explanations involve interdependencies of bacteria in communities and the role of organic and inorganic surfaces as necessary for bacterial activity. However, it has not been demonstrated that these bacteria differ in any fundamental way from bacteria already characterised. For example, where significant effort is made to understand the specific requirements of bacteria, as in anaerobic rumen or dental-plaque research, as many as 50% of bacteria have already been cultured. It may be that more sophisticated incubation conditions involving attention to available nutrients, oxygen regulation, acidity, toxicity and temperature will produce conditions suitable for other YBC bacteria.

Bacterial taxonomy

From its origins in the 19th century, bacterial taxonomy has gone through three identifiable historical phases.

Phase 1. Until about 1940, when the principles of bacterial classification were relatively undeveloped, taxonomy was conducted almost entirely by applied bacteriologists who were primarily interested in giving names to organisms associated with some specific human-related activity. Bacterial classification had a strong emphasis on the study of those bacteria that, by their function, were of economic, social, or cultural importance, e.g. human, animal, and plant pathogens, bacteria involved with food products and food spoilage, those associated with nitrogen fixation in agriculture, and those involved with industrial processes, etc. Bacteria were named because they expressed a particular characteristic. Methods of investigation gave rise to classifications that were unsophisticated by modern standards. This process led to a proliferation of genus and species names. In this period were begun the great culture collections, with the accumulation of extensive collections of strains for future study. The value of culture collections as repositories of strains, accumulated from diverse places often for unrelated purposes, was increasingly appreciated for the investigation of biodiversity.

Phase 2. In the period from 1940 to about 1980, studies based on these collections permitted the first organised comparisons of strains using increasing numbers of tests for a multiplicity of characteristics. These studies showed that many names had been applied to the same bacterium and also that many bacteria with a single particular characteristic were otherwise unrelated. This period saw the emergence of taxonomic organisation of bacteria at the species level.

Before 1980, there were about 28,000 previously reported bacterial species names, most of which were uninformative, as names without descriptions or extant representative strains, or were synonyms. A signal step was made in 1980 with the publication of the Approved Lists of Bacterial Names (Skerman et al. 1980, Approved List; Skerman et al. 1989, amended version). The Lists recognised as validly published only those names of bacterial species for which there was a modern description enabling discrimination from other species, and at least one extant strain that could be accepted as the type, or name-bearing, strain. The Approved Lists recorded 1792 validly published species names in 290 genera. Such is the interest in bacterial systematics that, by January 2000, the number had increased to 4303 species in about 933 genera. Euzéby (1997–2006) reported about 6000 distinct species in 1250 genera.

Phase 3. From 1980 onwards, molecular-biological methods have increasingly been used in investigations of molecular diversity and to clarify classification. In this period, applied bacteriologists have continued to contribute bacterial strains representing novel taxa, but bacterial systematists have become involved in the refinement of bacterial classification at generic and higher levels. The systematic study of bacteria from environmental sources has expanded our understanding of the diversity of these organisms.

Bacterial taxonomy comprises the principles and practice of classifying bacteria, in three distinct activities – classification (grouping), nomenclature (naming), and identification (diagnosis).

Classification is the arrangement of strains into natural groups (taxa). These groups may be based either on a consideration of the overall similarities of organisms in phenetic analyses, and therefore are groups about which the greatest number of predictive generalisations can be made, or are based on procedures and systems of analysis considered to show phylogenetic relationships based on their ancestry. Phenotypic information, gathered by the examination of structure and metabolic activity, can be contrasted with genomic information, obtained

The colon bacterium *Escherichia coli*, plated from a sample of pondwater in an Auckland domain duckpond.

Birgit Rhode, Landcare Research

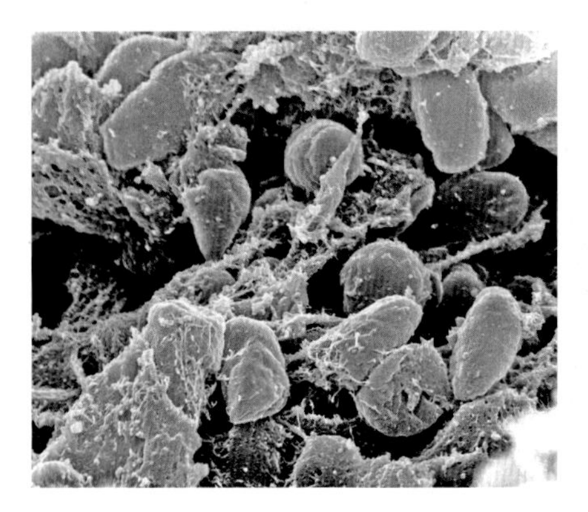

Cells of the bubonic plague bacterium *Yersinia pestis* in the foregut of a flea.
Wikimedia Commons, Rocky Mountain Laboratories, NIH, USA

by the examination of DNA. Both phenotypic and genomic information may be analysed with the intention of generating either phenetic or phylogenetic classifications (Sneath 1989).

Naming involves the allocation of names to circumscribed groups using the International Code of Nomenclature of Prokaryotes (previously the International Code of Nomenclature of Bacteria (Lapage et al. 1992) for taxa at the level of subspecies and above, and by the allocation of names or identifying terms to groups at lower levels.

Diagnosis involves the processes by which unidentified isolates are referred to known taxa.

Classification, naming, and diagnosis are interdependent, for if a classification does not provide descriptions with methods for investigation, it cannot be used for identification in general or to provide identified reference strains for future taxonomic studies. Different approaches lead to different classifications and hence to different nomenclatures (Goodfellow & O'Donnell 1993; Young 2001).

Phenetic classification aims to establish the relationships of organisms based on overall phenotypic characteristics. Early analyses based largely on single-enzyme auxanographic (biochemical) and laboratory reactions that give a very limited representation of bacterial phenotypes have given way to methods that give more general representation of phenotypes, such as comparative sequence analyses of house-keeping and ecological genes and fingerprinting of major components of cell structure.

Phylogenetic classification aims to establish taxa based on the historical and ancestral relationships of organisms. Because bacterial phylogenies can only be inferred using data obtained from organisms existing today, usually by comparative analysis of relatively short nucleic-acid sequences, derived phylogenetic interpretations cannot be confirmed by reference to a fossil record. Notwithstanding the difficulties of authenticating phylogenies, the construction of accurate phylogenies may be possible (Daubin et al. 2003). However, it is increasingly clear that phylogenetic interpretations are influenced by the choice of strains included, the particular molecular sequence investigated, and the particular algorithm used for analysis. Phylogenetic inference based on conserved molecular sequences is not different in principle from the much older and now abandoned concept that 'morphology reflects phylogeny' (Davis & Hayward 1963), except that the 'morphology' is at a molecular level.

Such inferences give an indication of phylogeny in the same way that morphological comparisons do, but the 'true' phylogeny is probably lost in time and is not accessible from any data that can be gathered in the present. As more genes are investigated, even with more sophisticated methods of analysis, it is becoming clear that there may be a limit to analyses, well short of the ancestral bacteria, that comparative sequence analyses, however sophisticated, will not reveal (Martin & Embley 2004).

Polyphasic classification has been proposed as combining the phenetic and phylogenetic approaches (Vandamme et al. 1996) and is generally accepted as the best taxonomic approach. Where several methods are employed, it is assumed that they will support coherent classifications but this is not always the case. Vandamme et al. (1996) proposed a phenetic approach to the discrimination of genera and species, and a phylogenetic approach (based solely on comparative rDNA-sequence analyses) to establish relationships at higher taxonomic levels. Cavalier-Smith (1998, 2002) has argued that bacterial taxonomy must take account of the obvious major phenotypic differences in cell microstructure as part of the higher classification of bacteria. According to this approach, such factors as cell-wall structure, indicated by the Gram reaction, cannot be considered secondary adaptations and are of greater significance than differences in a single sequence such as 16S rDNA. Issues of this kind will be resolved when more information on the genetics of microstructure are better understood. A recent new approach is by Multi Locus Sequence Typing (MLST) (Maiden et al. 1998).

This involves selecting a small number of functional genes that are representative of the chromosome and analysing them collectively either as nucleotide or as protein sequences. This practical method best approaches the goal of natural classification.

In recent times, there has been a tendency to give emphasis to genetic rather than phenotypic information. It is sometimes conjectured that comparative analyses of complete chromosomes may provide definitive answers to taxonomic questions. As yet, the extent to which chromosomal variation occurs between closely related bacteria has been established only for a small number of strains. The substantial differences in chromosome size of as much as 30% reported for closely related strains in *Frankia* indicate high levels of genetic diversity. Furthermore, many reports of individual gene sequences indicate that although there is considerable homology between individual genes, the architecture of the complete gene cassette can be highly variable. Strains may have similar phenotypes with substantially different genetic architecture. These observations, coupled with the fact that the structure of the chromosome is only indirectly informative as to the nature of the expressed phenotype, suggests that the complete chromosome will further only a partial contribution to establishing taxonomic relationships.

A genetic-species concept for bacteria has always proved intractable because, as they are asexually reproducing organisms, definitions will be arbitrary. The problem has been exacerbated in recent times because of the extent to which different methods of examination indicate different relationships and individual strain diversity leads to ill-defined taxa. Current definitions are based on genome relatedness and on a requirement that this be supported by distinctive characters in the phenotype (Stackebrandt et al. 2002).

It should be understood from the above that the advances made in demonstrating the extent of diversity between bacteria has produced improved classifications for many groups, but has also revealed levels of variability that are difficult to codify in a simple binominal naming system of the kind developed for the eukaryotes.

Modern methods are being applied to increasing numbers of bacterial groups from a greater range of environments, but some have yet to be given more than cursory investigation. For example, the Cyanophyceae (blue-green algae) are now formally recognised as bacteria, hence are now called Cyanobacteria or blue-green bacteria. However, their classification and nomenclature has not yet been the subject of the detailed examination needed to incorporate them in Approved Lists. Preliminary studies suggest that they will be the subject of considerable revision when molecular methods are applied. The Cyanobacteria are described in detail in Chapter 2 of this volume.

Representations of evolution and diversity

When phylogenies of organisms were first derived by comparative analyses of ribosomal sequences, there was an expectation that they would give accurate representations of biological ancestry (Fox et al. 1980; Woese 1987). The inferred phylogeny of Woese (1987) captured the imagination of the scientific community. His inferred phylogeny indicated three major evolutionary lines. The eukaryotes comprised all macroorganisms, viz the visible plants, animals, and fungi, and many unicellular organisms. These organisms were distinguished from the two other groups composed of prokaryotic cells. Prokaryotes, collectively referred to here as bacteria, were divided into the Archaebacteria and the Eubacteria. When they were first distinguished, the Archaebacteria were thought to represent an ancestral life form because their 16S rDNA sequence was distinct from the 'eubacteria' and because they were first found associated with the extreme environments of high salt and high temperature, conditions believed to have occurred as life first evolved. However, an alternative, well-argued proposal (Cavalier-Smith 2002) is that they are not so unusual when

Rod-shaped bacterial cells attached to the mouthpart bristles of a New Zealand mayfly.

Andrew Dopheide, Auckland University

more closely examined, and that the Archaebacteria evolved later (about 850 Mya) as adaptations to local conditions.

It is increasingly clear that bacteria are capable of genetic exchange between unrelated taxa at such levels that it seems likely that our species concepts may need reevaluation. An important assumption about bacterial relationships is that present-day taxa share common ancestors in simple lineages (that they are monophyletic) and that we are able to trace their ancestry, in the way supposed to be the case for higher organisms. For this to be possible, evolutionary change is assumed to occur through the natural selection of mutations occurring in a relatively stable genome, and environmental forces function primarily to induce change. According to this view, stability is a function of genome structure, and evolutionary change is the result of environmental pressures. An alternative view is that the bacterial genome is highly mutable, both because mutation rates are high and because horizontal gene transfer is a common activity involving all sites of the genome. According to this view, natural selection plays a primary role both in maintaining the stability of observed taxa, and in producing adaptational changes leading to the evolution of new taxa.

Bacterial endemism

When we consider the populations of bacteria today, the available molecular evidence suggests that free-living species have a global distribution. Although there may be internal diversity within species, strains of *Pseudomonas fluorescens* or *Bacillus thuringiensis* from North America or Australasia do not appear to be readily distinguishable from European ones. Genetic diversity may have arisen within species, but selective effects have maintained essentially the same species phenotype in common microenvironments found around the globe. The likelihood, too, is that the soils of the Carboniferous period were not so different from those of the present that we might expect anything other than the exchange of genes between similar bacterial populations. It is likely that bacteria today are very much like the bacteria of earlier eras.

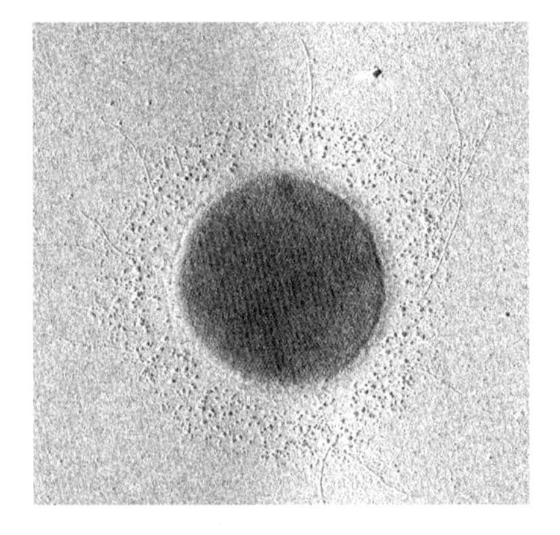

Ignisphaera aggregans, an endemic archaebacterium from a boiling pool in Kuirau Park, Rotorua.
Hugh Morgan

Almost nothing is known of the bacterial taxa present in New Zealand today except for those closely associated with current plant, animal, and human populations. The present record of names for which there are reference strains in culture collections shows 297 species in 103 genera from terrestrial environments (Young 2000). This paucity is certainly the result of the inadequacy of the record rather than that New Zealand's bacterial flora is depauperate.

New Zealand probably became a landmass independent of Gondwana about 80 Mya (Stevens et al. 1988) although there is suggestive evidence from pollen records of some connection, possibly involving island archipelagos as recently as 10 Mya. Whether there exist endemic bacterial taxa unique to New Zealand that evolved in this period of isolation (see Marine Bacteria) is a matter of speculation. Proof of a bacterial flora unique to New Zealand requires not only the demonstration of novel taxa in New Zealand, but also demonstration that such taxa do not exist elsewhere. A moment's reflection will indicate how difficult is such a test of endemism in practice. If endemism has occurred, it is only likely to be expressed as novel species or in lower taxonomic categories (subspecies or infrasubspecies). Ecological niches from which we can hope to find unique endemic bacterial diversity are those associated with isolated extreme environments (for example, the thermal areas) and where co-evolution occurs with elements of the New Zealand endemic flora and fauna. Co-evolution of partners in symbiotic relationship is worthy of study for bacterial endemism. One example for which there is preliminary evidence concerns the *Rhizobium*-legume association. Plants such as pea, bean, and clover form symbiotic relationships with specialised bacteria ('rhizobia') to fix nitrogen from the air to make it available for plant growth in one of the major processes by which inorganic nitrogen is introduced into most ecosystems. Worldwide, there are an estimated 17,000–19,000 legume species whose evolution has centred on the tropics and subtropics. New Zealand has a very small number of native legumes, comprising some 34 endemic woody species in the genera *Carmichaelia* (New Zealand broom), *Clianthus* (kaka beak), *Montigena*, and *Sophora* (kowhai). This group of specialised genera has co-evolved with their nitrogen-fixing bacterial symbionts in isolation from the regions of major legume evolution. A recent study showed that the bacteria are all members of the cosmopolitan rhizobial genus *Mesorhizobium*, closely related to *Rhizobium*. They have co-evolved in a specific symbiosis with the native legumes although they have retained affinities with their overseas relatives. These bacteria have a symbiotic capability only when living in association with legume plants. Introduced weed legumes (acacia, European broom, gorse) are symbiotic with an unrelated genus, *Bradyrhizobium* (Weir et al. 2004). The bacteria appear to be ubiquitous soil inhabitants that retain their symbiotic capability in soil independently of legume plants (J. M. Young unpubl. data). Important additional contributions of nitrogen to the natural environment are those made by the non-legume symbioses of non-rhizobial bacteria of the genus *Frankia* with tutu (*Coriaria*) and matagouri (*Discaria*). Because they are fastidious in their growth requirements, bacteria from these plants have yet to be obtained in pure culture.

Sectioned root nodules of the endemic New Zealand legume kaka beak (*Clianthus puniceus*). The upper nodule, with a *Rhizobium* species, is ineffective; the lower nodule, with a *Mesorhizobium* species, is effective.

Bevan Weir, Landcare Research

Apart from the original Pangean/Gondwanan contribution of bacteria, there is an ongoing addition to New Zealand's bacterial flora from human activity. An estimated 30 and 300 tonnes of soil per year are introduced on aircraft and ships, respectively, via their complements of passengers, animals, and containers. At least 10^{14} bacteria have been delivered annually to New Zealand ecosystems in recent years and there is now no way of knowing with certainty which common free-living bacterial species were present since the Gondwanan separation, and which were recent arrivals.

Interesting New Zealand bacteria include those that have coevolved with native New Zealand macroflora and -fauna. Just as bacterial taxonomy pertains mainly to taxa relevant to human interests, the study of bacterial ecology shows the same bias – research is most strongly supported where there are immediate social outcomes. The sections below describe specific fields of study.

Archaebacteria

The Archaebacteria are a group of bacteria that occupy some the most biologically extreme and hostile environments on Earth. Different species have been described with temperature optima of 110° C, or capacity for growth in acid environments with pH –0.6, or hypersaline conditions equivalent to 3-molar sodium chloride, or in anaerobic (oxygen-free) situations. Until recently, archaebacteria were thought to be obligate to extreme environments. However, investigations of a range of environments show that they are present in varied habitats and some have the capacity to grow in ordinary conditions.

The Archaebacteria are distinguished by characters that differentiate them from all other prokaryotes. Besides unique sequences of nucleotides found in the 16S rRNA molecule, all archaebacteria have unique membrane lipids. These contain ether-linked lipids, compared to the ester-linked lipids universally found in the Eubacteria and eukaryotes. Ether-linked lipids have unusual phytanyl substitutions that are thought to alter the physical properties of the membrane. In some archaebacteria living at high temperatures, the membrane contains a single span of the membrane lipid, a unique structure not found in any other organisms. The cell-wall structures of archaebacteria contain a pseudomurein that is different from the murein or peptidoglycan structures found in almost all of the Eubacteria (Madigan et al. 2000).

The ability to grow under extreme conditions is reflected in the physiology of the different species, which display modes of energy capture not found in any other group of organisms. The ability to oxidise hydrogen anaerobically using carbon dioxide (and occasionally acetate), with the formation of methane as an end product, is found only in methanogenic archaebacteria. The intermediates, enzymes, and co-factors involved in this pathway have no homologues in other kingdoms of life, and are presumed to have evolved independently within the archaebacterial lineage. Methanogens are universally distributed, despite being obligate anaerobes. They are dominant organisms in the intestines of ruminants and caecal fermenting animals, and are the final scavengers of the waste products of anaerobic fermentations of bacteria in sediments and anaerobic sewage treatment systems.

Similarly, some halophilic archaebacteria possess a unique means of capturing light energy, distinct from the photosynthetic systems found in eubacteria and eukaryotes. In these organisms, the membrane-bound pigment bacteriorhodopsin is used to pump protons across the membrane as a means of energy conservation. The sulphur respiration of many thermophilic archaebacteria is also unusual, though this might reflect the growth conditions imposed by culture studies, rather than the normal metabolism of the organism in its natural habitat.

Woese (1987) was the first to identify the archaebacteria as a distinct element of prokaryote diversity and since then more than 93 genera have been described. The 16S rRNA phylogenetic tree demonstrates that there are two major groups of the Archaebacteria – the Euryarchaeota and the Crenarchaeota. The Euryarchaeota contains all methanogens and halophiles and some fermentative hyperthermophiles.

Vent-fluid chimneys on a Kermadec seamount with white bacterial mats.
National Institute of Water & Atmospheric Research

The Crenarchaeota is represented exclusively by hyperthermophiles, most of which either reduce sulphur under anaerobic conditions or oxidise it if they can grow aerobically. When their distinct characteristics were first recognised, and when they were believed to be confined to extreme environments, it was widely believed that they represented an ancestral prokaryote life form. The majority of crenarchaeote isolates have been obtained from submarine hydrothermal vents, frequently 'black smokers', which occur at tectonic plate boundaries. Because these sites may mimic the environment in the primordial world, it was suggested that the Crenarchaeota may most closely represent life on Earth, perhaps 3800 Mya. However, this hypothesis is not universally accepted, and alternative hypotheses have suggested more recent origins (850 Mya) for the Archaebacteria (Cavalier-Smith 2002). Studies in which habitats have been screened using molecular probes for YTB isolates indicate that more than 30% of the total microbial biomass of all marine environments is archaebacterial and from currently uncultivated organisms (DeLong et al. 1999). The same is true in almost all habitats investigated – the Antarctic ocean, temperate and Arctic soils, and the deep subsurface of the Earth's crust all contain numerous genera of uncultivated euryarchaeota and crenarchaeota (Ravenschlag et al. 2001; Schleper et al. 1997).

Molecular probes of some extreme environments suggest the existence of a new major subgroup of the Archaebacteria. Signature sequences of the 16S rRNA of these organisms – provisionally called the Korarchaeota – have been detected in hot pools but also in a wide range of environments elsewhere, and nucleic acid probes derived from these sequences have been used to identify individual cells of the Korarchaeota (Brunk & Eis 1998). The Korarchaeota is probably widespread, not restricted to extreme environments, and constitutes a diverse and as-yet-undescribed prokaryote group.

From a New Zealand perspective, archaebacteria are the cornerstone of the agricultural economy. Methanogens are essential for the well-being and productivity of ruminant herds and flocks and the methane they produce is a major component of the nation's greenhouse-gas emissions.

Bacteria pathogenic to humans

Pathogenic bacteria have caused diseases in humans since prehistoric times. Some are specific to humans while others are shared with other mammals, especially domesticated pets and food animals (Joklik et al. 1992; Quinn et al. 2002). Further novel pathogenic species may be demonstrated if human populations come in closer contact with previously isolated vertebrate species, though this would not be expected to be a common event.

Most bacteria that colonise skin and the gastrointestinal, urinary, and respiratory tracts are usually benign and may be beneficial. Pathogenic bacteria constitute only a small proportion of bacteria and are distributed among less than 1% bacterial genera. Pathogens are commonly associated with skin diseases (e.g. *Streptococcus pyogenes* causing cellulitis), intestinal diseases (e.g. *Campylobacter jejuni* causing diarrhoeal disease), pulmonary diseases (e.g. *Streptococcus pneumoniae* causing pneumonia), and invasive diseases (e.g. *Neisseria meningitidis* causing cerebrospinal meningitis, *Staphylococcus aureus* causing osteomyelitis). Most of these bacteria usually live in harmless associations and when they do cause disease they usually have minor effects. However, in some circumstances that are sometimes not well understood, these bacteria multiply and cause serious diseases. *Staphylococcos aureus* (the 'golden staph') is associated with a variety of diverse skin, urinary, pulmonary, and invasive diseases and can be fatal, although approximately 25% of the healthy human population is colonised with *S. aureus. Neisseria meningitidis* is another human pathogen able to cause significant morbidity and mortality that is also present in up to 30% of healthy humans. Other groups of pathogenic bacteria live in soil and water and usually cause serious diseases only in conditions of poor hygiene or in people weakened from other causes. In New Zealand, bacterial disease outbreaks are usually quickly brought under control. It is likely that most pathogenic species are present in New Zealand in soil, water, or mammalian reservoirs. For example, anthrax was last reported in New Zealand in 1954 (Crump et al. 2001) but because the spores of the pathogen (*Bacillus anthracis*) are very long-lived, it is highly likely that the pathogen is still present. Similarly, although the plague bacillus *Yersinia pestis* has not been reported in New Zealand since 1911 (Anon. 2001), rodent hosts of the pathogen and the fleas necessary for plague transmission within the rodent population exist in New Zealand and may be reservoirs. Records of cases and outbreaks of all notifiable bacterial diseases in New Zealand are publicly reported (www.surv.esr.cri.nz).

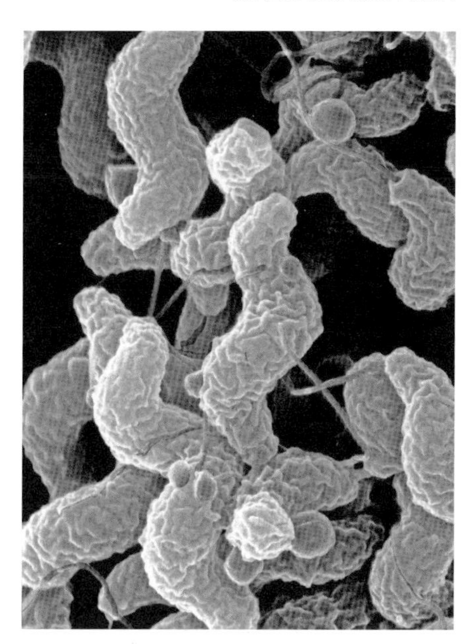

Campylobacter jejuni, a major cause of food-related gastrointestinal illness.
Wikimedia Commons, De Wood & Pooley, USDA

Pathogenic bacteria have been readily controlled using antibiotics. Although resistance to antibiotics was anticipated early in the history of their use, mechanisms of resistance and the relative ease with which bacteria could transmit resistance between strains were not fully understood until recently. Widespread and sometimes ill-considered antibiotic use has resulted in the development of resistance to most antibiotics in most pathogens, producing 'superbugs' (e.g. methicillin- or multidrug-resistant *S. aureus* (MRSA), vancomycin-insensitive *S. aureus* (VISA), and vancomycin-resistant *S. aureus* (VRSA). VRSA has recently been described in the USA (Chang et al. 2003), but unlike the other antibiotic-resistant *S. aureus*, this strain has not yet reached New Zealand. There is a strain of MRSA that is unique to New Zealand, Australia, and the Pacific Islands. *Staphylococcus aureus* strains originally believed to be unique to Samoa are now the predominant MRSA strain causing infections in New Zealand. Antibiotic resistance can arise in New Zealand bacterial populations but statistically it is more likely to occur in the larger bacterial populations overseas. Penicillin-resistant *Streptococcus pneumoniae* in New Zealand has been identified as arising by importation of a pathogenic strain (Bean & Klena 2005).

The epidemiology of human bacterial pathogens has long been the subject of intense study. Recently, molecular methods have provided a range of tools for rapid identification and for detailed analysis of pathogen population structures and distribution. These will be essential if there is to be success in holding pathogens in check in future.

Bacteria pathogenic to animals

Prior to the arrival of Maori in New Zealand, the only mammals present in the country were three species of bats and marine mammals (seals and cetaceans).

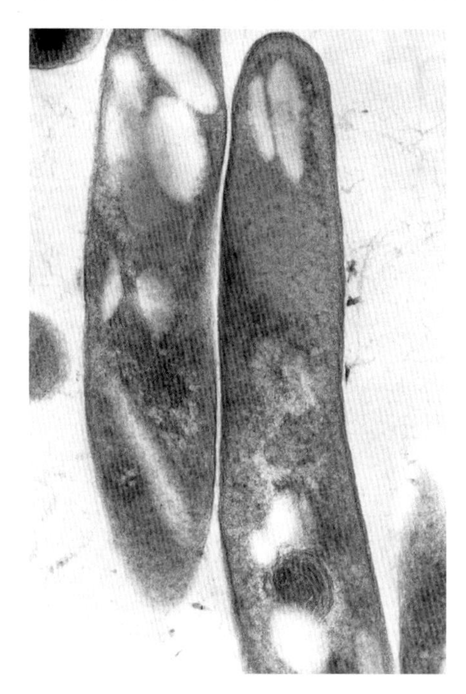

Transmission electron microscope image of *Mycobacterium tuberculosis*, the causative agent of tuberculosis

Wikimedia Commons, Elizabeth White, US Centers for Disease Control and Prevention

Subsequently, a wide range of domestic animals and wildlife has been introduced, and with them have come many associated bacterial pathogens. It is unknown whether any unique bacterial pathogens of vertebrates were present in New Zealand prior to the arrival of humans but, given the country's unique fauna and geographical isolation, it is likely that indigenous bacteria, including pathogens, evolved in association with the native vertebrates. A possible example is *Mycobacterium pinnipedii*, a member of the *Mycobacterium tuberculosis* complex (Cousins et al. 2003). This bacterium appears to be adapted to fur seals (*Arctocephalus forsteri*) in the southern ocean, including New Zealand (Hunter et al. 1998). On rare occasions the pathogen may infect cattle. In contrast, bacterial pathogens that are universally distributed have been cultured from New Zealand's native fauna. An example is the isolation of *Pasteurella multocida* from dying rockhopper penguins (*Eudyptes chrysocome*) from Campbell Island (de Lisle et al. 1990). This bacterium was also isolated from rats at the penguin colony and the serotype of the *P. multocida* isolates from Campbell Island is present throughout the world. Whether the strain was introduced to Campbell Island with the rats or by birds is unknown.

Bacterial pathogens affecting domestic animals in New Zealand have a worldwide distribution. It is likely that most of these pathogens were introduced into the country with the importation of animals. Apart from the introduction of Polynesian rats and dogs with Maori, animal introductions began with European settlement in the 19th century, when biosecurity measures either did not exist or were rudimentary and knowledge of bacterial pathogens was limited. Fortunately, the long sea voyage to New Zealand from Europe was a period of self-imposed quarantine and was probably instrumental in preventing the introduction into the country of some important bacterial pathogens. These include *Pseudomonas mallei*, causing glanders in horses, and *Pasteurella multocida* serotypes B2, E2, causing haemorrhagic septicaemia. Currently, strict biosecurity measures aim to prevent pathogens entering the country but these are not absolute, as evidenced by the recent appearance in New Zealand of *Salmonella* Brandenburg, which causes abortions in sheep. Many important diseases including bovine tuberculosis (*Mycobacterium bovis*) and Johne's disease (*Mycobacterium avium* ssp. *paratuberculosis*) were introduced. Features of these diseases often include chronic infections with long preclinical periods or asymptomatic carrier states.

Mycoplasma mycoides ssp. *mycoides* (small colony), causing bovine contagious pleuropneumonia, was introduced but was quickly eradicated in 1864. Another major bacterial pathogen eradicated from New Zealand livestock was *Brucella abortus*, which was eliminated after a very extensive test and slaughter programme. Whether or not viable spores of *Bacillus anthracis*, causing anthrax mainly in sheep, cattle, and horses, are still present in New Zealand is unknown. As stated above, the last recorded case in New Zealand was in 1954 but this bacterium has the ability to remain viable as spores for decades.

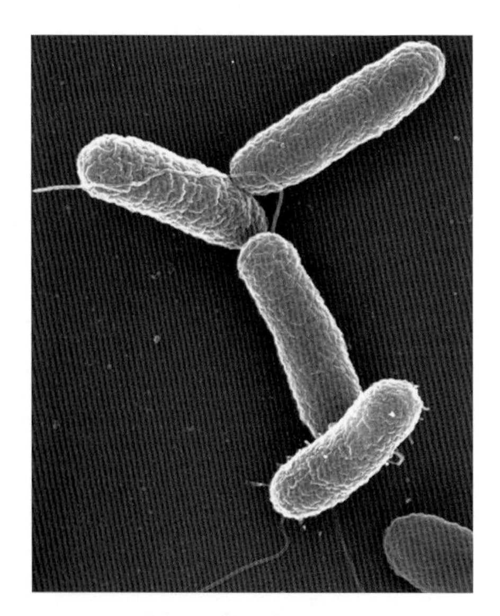

Salmonella typhimurium.

Wikimedia Commons, Volker Brinkmann, Max Planck Institute for Infection Biology, Berlin

The spread of bacterial pathogens between different host species is important both as a source of infection for humans (zoonoses) as well as spread from wildlife to domestic animals. Important bacterial zoonoses in New Zealand include salmonellosis (multiple serotypes), campylobacteriosis (principally *Campylobacter jejuni*), leptospirosis (multiple serovars), verotoxigenic and shiga toxin-producing *Escherichia coli* infections (principally *E. coli* O157), and yersiniosis. With some infections such as toxigenic *E. coli*, contact with animals is an important risk factor. In contrast, consuming food from a retail premise is the major risk factor for campylobacteriosis (Anon. 2003). The failure to eradicate bovine tuberculosis from cattle and farmed deer in New Zealand is the result of the continual spread of infection from wildlife, principally the Australian brushtail possum *Trichosurus vulpecula*, to domestic animals (de Lisle et al. 2001). A marked reduction in the prevalence of bovine tuberculosis in cattle and farmed deer during the last five years is because of a major reduction of possum numbers in selected farming areas by poisoning and trapping coupled with a routine test and slaughter programme for cattle and farmed deer.

A leaf of kiwifruit (*Actinidia deliciosa*) infected with *Pseudomonas syringae* pv *actinidiae* (psa), showing dark angular spotting surrounded by a pale halo.

Bevan Weir, Landcare Research

Bacteria pathogenic to plants

Worldwide, there are about 120 species of plant-pathogenic bacteria in 20 genera (Young et al. 1996) – the Gram-negatives *Acetobacter, Acidovorax, Agrobacterium* (now *Rhizobium*), *Burkholderia, Enterobacter, Erwinia, Herbaspirillum, Pseudomonas, Ralstonia, Rhizobacter, Sphingomonas, Spiroplasma, Xanthomonas*, and *Xylella*, and the Gram-positives *Bacillus, Clavibacter, Curtobacterium, Nocardia, Rhodococcus*, and *Streptomyces*. Some pathogens attack many plant species, so that the total number attacked by bacteria is about 2500 host species (Bradbury 1986). Within some pathogenic species, mainly in *Xanthomonas* and *Pseudomonas*, populations have evolved that are highly specific to small numbers (sometimes one) of host plant species and genera. There are about 350 distinct bacterial pathogens distinguished as pathogenic variants, called pathovars (Young et al. 1992).

In New Zealand, most plant-pathogenic bacteria are associated with introduced horticultural crops. Ten genera – *Agrobacterium* (now *Rhizobium*), *Burkholderia, Erwinia, Herbaspirillum, Pseudomonas, Ralstonia, Xanthomonas, Clavibacter, Rhodococcus*, and *Streptomyces* – representing 19 species, include 63 specific and non-specific pathogens attacking 205 agricultural, horticultural, and ornamental plants (Pennycook *et al.* 1989). Pathogens causing serious diseases are: *Rhizobium radiobacter* (= *Agrobacterium tumefaciens*), causing crown gall on many horticultural plants; *Erwinia amylovora*, which causes fireblight of apple; *Pseudomonas savastanoi* pv. *savastanoi*, causing bacterial knot of olive; *P. syringae* pv. *syringae* and *P. syringae* pv. *persicae*, which cause bacterial blast and bacterial decline of stone-fruit, and *Xanthomonas campestris* pv. *pruni*, causing bacterial spot of stone-fruit. Pathogens not present in New Zealand that are the subject of ongoing quarantine scrutiny are: *Pantoea stewartii*, the pathogen of bacterial wilt of corn; *Curtobacterium flaccumfaciens* pv. *sepedonicus*, causing ring-rot of potato; *Xanthomonas citri*, causing Oriental canker of citrus, and *Xylella fastidiosa*, causing Pierce's disease of grape.

The record of bacterial pathogens of New Zealand's native plants was gathered opportunistically in the past, often from collections made by botanists in the course of recreational tramping. More recently, a survey was made over a three-year period, 1991–93, in which native-plant nurseries were systematically inspected. Some 1,380,000 plants, representing more than 490 endemic species, were examined. From these and the earlier collections, the tally of pathogenic bacteria on native plants comprises only *Erwinia* sp. from *Cordyline* species (ti kouka, cabbage tree), *Pseudomonas syringae* on *Carmichaelia williamsii* (New Zealand broom) and *Dysoxylum spectabile* (kohekohe), *Geranium traversii, Hoheria populnea* (houhere, lacebark), *Laurelia* sp., *Olearia traversii* (akeake), and *Xanthomonas* spp. on *Dysoxylum spectabile* (kohekohe) and on *Metrosideros excelsa* (pohutukawa).

Recently, phytoplasmas have been shown to be closely associated with diseases of native plants and with strawberry. 'Yellow leaf' of *Phormium tenax*, for many years thought to be caused by a virus, a die-back in *Coprosma robusta* (karamu), the widespread death of *Cordyline australis* in the North Island, and an important strawberry disease have all now been linked with '*Candidatus* Phytoplasma australiense', a pathogen also found in Australia (Andersen et al. 2006). Two insect vectors have been identified in New Zealand, both members of the native cixiid genus, *Oliarus*. The pattern of disease spread into the different plant hosts appears to be determined by the feeding choices of the different species (R. E. Beever, pers. comm.)

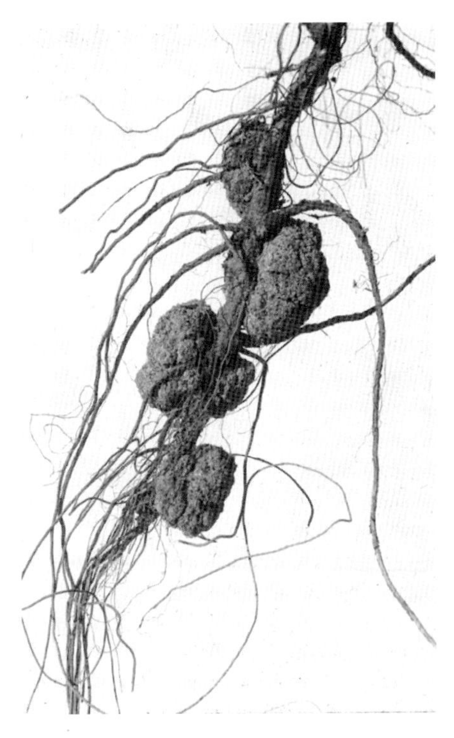

Agrobacterium tumefaciens galls.

Wikimedia Commons, Clemson University

It is difficult to account unequivocally for the paucity of pathogenic bacteria recorded on native plants. After 100 years of observation, mere failure of sighting seems unlikely. The isolation of New Zealand from centres of bacterial interaction and evolution since separation from Gondwana does not explain why omnivorous and ubiquitous pathogens have not been more widely isolated. The

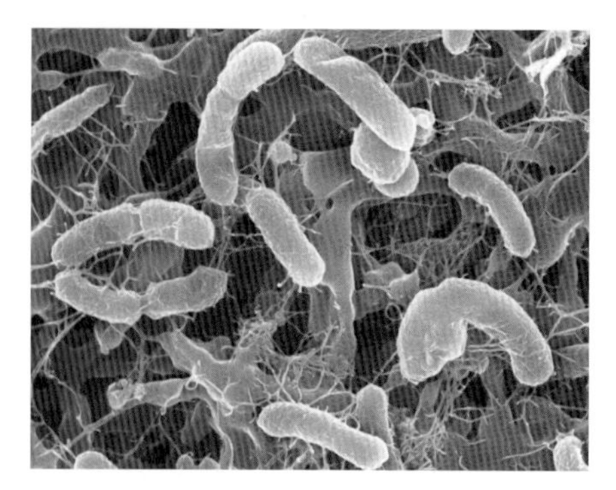

Venenivibrio stagnispumatis from the Champagne Pool, Waiotapu.
Hugh Morgan

difference in susceptibility between wild plants and those artificially selected as cultivars with desirable domestic characters is a plausible explanation that has not yet been subjected to experimental investigation.

Bacteria from geothermal systems

Geothermal systems were the first of New Zealand's natural environments to be studied from a microbiological perspective, the earliest report (Skey 1878) containing a brief description of algal growth in mineral systems of the Taupo volcanic zone. Weed (1889) described the commonest organism present as a 'blue-green alga', *Hypotheothrix laminosa*, i.e. a 'cyanobacterium' that was most probably a member of the genus *Oscillatoria* (Kaplan 1956).

Despite this early interest in New Zealand's thermal systems, further study did not take place until after World War II. Studies then were of the geothermal areas in Rotorua, Taupo, and White Island. These included alkaline-chloride waters with typical pH values of 6.5–8.5 and acidic systems with pH values as low as 1.2. Cyanobacteria predominated in all environments. Specific genera identified were *Fischerella*, *Lyngbya*, *Oscillatoria*, *Phormidium*, and *Synechococcus roseus*, of which *Fischerella*, *Lyngbya*, and *Oscillatoria* were obtained in pure culture (Kaplan 1956). Purple and green sulphur bacteria were also present.

Kaplan (1956) also noted a correlation between the temperature and colour of microbial mats at sites such as Orakei Korako. Mats with colours ranging from white and flesh pink to yellow grew where temperatures were greater than 70° C, while orange, red, and brown mats were generally found at temperatures below 50° C. Green mats were found at all temperatures. These colours were attributed at the time to variants of *Oscillatoria* but recent studies in the USA suggest that different genera predominate in the various colour-temperature zones (Cady & Farmer 1996).

Kaplan's (1956) work was descriptive, mainly using morphological characters to identify specific bacteria, and produced a classification that is now known not to give an accurate expression of cyanobacterial diversity. Nevertheless it laid the foundations for our current understanding of the microbiology of New Zealand's geothermal environments and provided preliminary evidence that the microbial communities in New Zealand geothermal systems are similar to those encountered in other parts of the world.

Recent electron-microscope studies have categorised cyanobacteria in living mats and in silica deposits in active and ancient geothermal systems. Work by Jones et al. (1997, 1998, 1999) supports earlier studies indicating that species of the cyanobacterial genera *Calothrix*, *Phormidium*, and *Synechococcus* dominate in high- and low-temperature systems. However, a major problems with these descriptive methods is that morphological features often provide poor taxonomic resolution and many of the less conspicuous organisms are overlooked.

Recently, molecular methods that do not depend on direct examination of cyanobacterial cultures have been used to characterise the communities in hot-pool fluids, sediments, and microbial mats associated with silica sinter deposits. Although much of this work is still in progress and is yet to be published, the results of preliminary studies (Khan 1998) indicate that the bacterial – including archaebacterial – communities of geothermal systems are much more diverse than originally indicated by microscopy and culture-based studies, and include many yet-to-be cultured organisms.

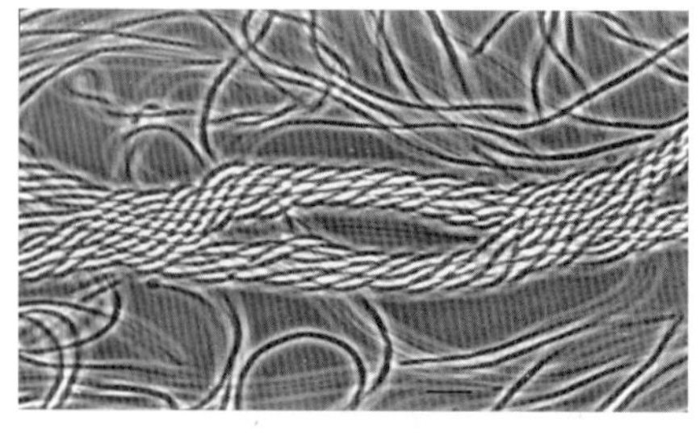

Thermus filiformis from the Clamshell Pool, Waimangu Valley.
Hugh Morgan

Many studies have been made of culturable bacteria (including archaebacteria) from these environments with particular emphasis on thermophilic flora that may provide a source of useful thermotolerant enzymes. The Archaebacteria isolated from geothermal systems are mainly members of the Euryarchaeota. A wide range of mesophilic and thermophilic bacteria, including thermophilic species of *Thermus* and *Caldicelluloseruptor*, have been isolated and the presence of the thermotolerant cyanobacterial species originally isolated by Kaplan (1956) has been confirmed.

Future studies of the microflora of New Zealand's geothermal systems is likely to be driven by two fields of interest – their value as sources of useful thermophilic micro-organisms and metabolic products, and as present-day models for studies of early life on Earth. A common objective of these disparate research areas is that they seek to determine whether these organisms are cosmopolitan in their distribution or are unique to New Zealand.

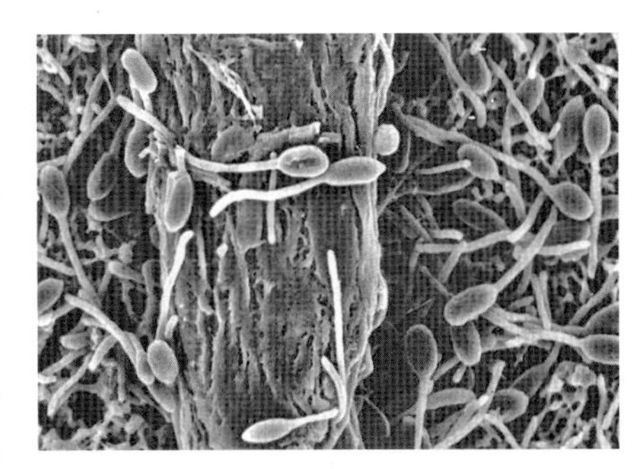

Caldicelluloseruptor saccharolyticus, a cellulose-degrading bacterium from Warbrick terrace, Waimangu Valley.

Hugh Morgan

Marine bacteria

The seas and oceans cover 71% of the earth's surface and contain 97% of the water on Earth. The marine realm is the largest on the planet, but the very low concentrations of dissolved chemicals and the restriction of primary production by photosynthetic bacteria and algae to depths to which sunlight penetrates (less than 30 m), means that its productive capacity is about 25% of the terrestrial realm. Bacterial activity at deep-sea thermal vents makes a minor contribution compared with sunlight. The composition and activity of the large bacterial population forming the deep sea-floor sludge is not understood. Salinity ranges between 3.3 and 3.7% and acidity is between pH 8.2 and 8.5. Most of the ocean's water, 90–95%, is at a temperature of < 5° C, with a warmer surface layer depending on latitude. Surface temperatures range between 25 and 35° C in tropical regions while polar water has an average temperature of 3.7° C with a low of –1.7° C.

Because the seas and oceans of the world are contiguous, marine bacteria are assumed to be cosmopolitan. The same species have been reported from seas in both northern and southern hemispheres, although there may be some locally adapted groups associated with inshore-based macro-organisms. Marine bacteria in the open oceans are present in very low numbers compared with terrestrial environments and are commonly associated with algae or particles of detritus. Bacteria are present in relatively high concentrations in coastal and estuarine waters.

Bacteria isolated from marine environments are halophilic, requiring high salt concentrations for growth, or halotolerant, being capable of growing in a range of salt concentrations. There have been few studies of marine bacteria in the oceans surrounding New Zealand. Isolations from inshore and coastal environments, largely from sediments, are of species belonging to halophilic genera – *Aquimarina, Cellulophaga, Cobetia, Exiguobacterium, Flectobacillus, Glaciecola, Leeuwenhoekiella, Listonella, Marinomonas, Microbacterium, Microbulbifer, Planomicrobium, Pseudoalteromonas, Psychrobacter, Roseobacter, Stappia, Sulfitobacter, Tenacibaculum, Vibrio, Zobellia* – and of halotolerant genera also associated with terrestrial environments, viz *Aneurinibacillus, Arthrobacter, Bacillus, Curtobacterium, Flavobacterium, Halomonas, Isoptericola, Kitasatospora, Micrococcus, Paenibacillus, Planococcus, Pseudomonas, Roseobacter, Rhodobacter, Rhodococcus, Sanguibacter, Stenotrophomonas*, and *Streptomyces*. Other unidentified marine isolates represent hitherto uncharacterised genera. Isolations from oceanic water over the Chatham Rise east of Banks Peninsula to a depth of 3000 metres gave repre-

Estimated number of bacteria in marine habitats

Habitat	Volume, cm^3	Cells/ml x 10^5	Total number of cells x 10^{26}
Continental shelf	2×10^{20}	5	1.0
Open ocean water			
upper 200 m	7×10^{22}	5	350
below 200 m	1×10^{24}	0.5	500
sediment; 0–10 cm	4×10^{19}	4500	180
Total			**1030**

sentatives of halophilic *Alteromonas, Glaciecola, Halomonas, Oceanospirillum,* and *Pseudoalteromonas,* and halotolerant *Pseudomonas* (J. M. Young unpubl.). These data show that the oceans are a strict barrier to intercontinental migration of terrestrial bacteria.

Entomopathogenic bacteria in New Zealand

Relatively few bacterial genera have been shown to play a role in insect diseases in New Zealand. Glare et al. (1993) compiled a checklist of entomopathogenic microbes and nematodes reported in New Zealand to that time, and fewer than ten species of bacteria were listed. The most widely known insect-pathogenic bacteria worldwide belong to the genus *Bacillus,* comprising endospore-forming Gram-positive rods. *Bacillus thuringiensis* is the best-known insect pathogen worldwide and is the basis of a large biopesticide industry. Strains produce a range of intracellular crystal proteins that are toxic to various insect larvae. In New Zealand, *B. thuringiensis* is frequently isolated from environmental samples and is a common soil and plant-surface inhabitant, but it has rarely been recorded as causing disease in nature. However, imported commercial preparations of *B. thuringiensis* have been vital in pest-control efforts where use of chemical pesticides was not an option. In 1998, huge amounts of *B. thuringiensis* serotype *kurstaki* were applied over urban Auckland to eradicate an introduced pest, the spotted tussock moth, in one of the heaviest aerial applications of this bacterium in the history of its use. This subspecies is also a vital component in New Zealand's organic kiwifruit production systems. In 1999, dipteran-active *B. thuringiensis israelensis* was registered for use by the Ministry of Health to treat an outbreak of an introduced mosquito species in the Hawke's Bay region of North Island.

Another *Bacillus* species, *B. popilliae,* has been reported to cause milky disease in several scarab insects in New Zealand. Natural outbreaks of milky disease are found in the New Zealand grass grub *Costelytra zealandica.* Resting spores are ingested by the feeding larvae; once in the gut, the spores germinate and vegetative cells penetrate the body cavity, where extensive multiplication occurs, causing death of the insect. Milky disease has been recorded as infecting more than 50% of larvae in some populations.

Bacillus thuringiensis ssp. *kurstaki* showing, from top to bottom, stages in endospore formation.
Rita B. Moyes, Texas A&M University, and Robert E. Droleskey, USDA

Several species of enterobacteria (facultatively anaerobic Gram-negative rods) have been recorded from diseased insects in New Zealand. *Serratia marcescens* is a red-pigmented bacterium known from ancient times that was recognised as an insect pathogen as early as 1941 (Steinhaus 1941). This species has been described from several scarab species in New Zealand. Two other species, *S. entomophila* and *S. proteamaculans,* are found throughout the pasture environment and cause amber disease of the New Zealand grass grub. Natural epizootics of this disease can regulate pest populations and *S. entomophila* has also been applied on a large scale as a biocontrol agent. While *S. proteamaculans* is found worldwide, *S. entomophila* (Grimont et al. 1988) has been found only in New Zealand.

Most bacterial insect pathogens have been described from pest species that cause significant economic damage to pasture and crops. Many insect species have never been studied and their pathogens remain unknown, suggesting there are potentially many new discoveries yet to be made.

Psychrophilic and psychrotolerant pathogens and food-spoilage bacteria

Traditionally, psychrophilic (cold-loving) and psychrotolerant micro-organisms have been associated with permanently cold, uninhabited natural environments such as deep oceans, land masses of Antarctica and the Arctic, or ice glaciers and lakes in Alpine regions. However, with modern food-processing, refrigeration is increasingly used to extend shelf-life of fresh or minimally processed foods, and, consequently, diverse nutrient-rich habitats for psychrophilic and psychrotolerant micro-organisms are being created in temporarily cold, man-made environments.

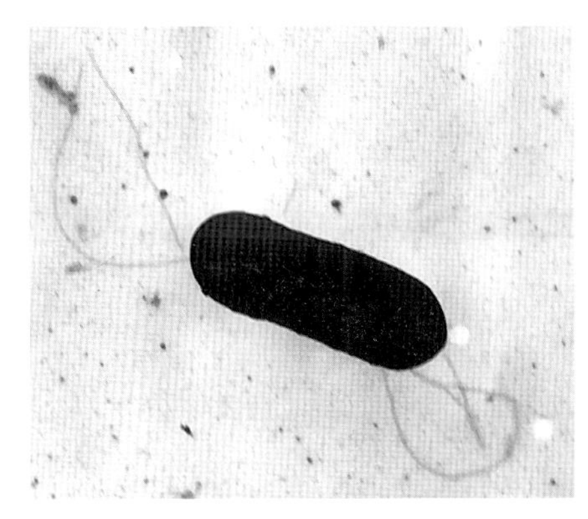

Listeria monocytogenes, the agent responsible for the food-borne illness listeriosis.

Wikimedia Commons, US Centers for Disease Control and Prevention

In the past, storing foods at about 4° C was considered adequate to prevent growth and/or toxin formation by food-borne pathogens, and to prevent premature food spoilage. Since the 1960s, however, it has become clear that many pathogenic bacteria are either capable of growth or are able to survive for prolonged periods, in foods stored at 4° C and below, while other non-pathogenic bacterial species can cause spoilage of foods stored at temperatures near 0° C. Consequently, the majority of food-poisoning and food-spoilage bacteria are now recognised as being either psychrophilic or psychrotolerant.

Species of major psychrotolerant, food-borne pathogens, including *Aeromonas caviae, A. hydrophila, Listeria monocytogenes, Vibrio parahaemolyticus,* and *Yersinia enterocolitica* are known to exist in New Zealand. In addition, some mesophilic species that include psychrotolerant strains (*Salmonella* spp., *Shigella flexneri, Shigella sonnei,* enterotoxigenic *Escherichia coli, Staphylococcus aureus,* and spore-forming *Bacillus cereus*) (Palumbo 1986) are also found in New Zealand, although the degree and extent of psychrotolerance of these strains is generally uncertain. Non-proteolytic strains of *Clostridium botulinum* have been identified in New Zealand using 16S rDNA sequence analysis, but none of these has been shown to carry the botulinal neurotoxin gene (Broda et al. 1998) and, therefore, they are unlikely to cause human food-borne botulism.

It is believed that the majority of species of psychrophilic and psychrotolerant food-spoilage bacteria that occur worldwide are also present in New Zealand. *Brochotrix thermosphacta, Pseudomonas fragi, Shewanella putrefaciens,* and psychrotolerant Enterobacteriaceae (*Enterobacter agglomerans, Hafnia alvei,* and *Serratia liquefaciens*) have been detected in various meat products of New Zealand origin. Many psychrotolerant strains of species of *Acinetobacter, Carnobacterium, Lactobacillus, Lactococcus, Leuconostoc, Micrococcus, Oenococcus, Pediococcus, Pseudomonas, Psychrobacter, Staphylococcus,* and *Weisella* have frequently been found in New Zealand foods. These micro-organisms have rarely been identified to species level. Recently, psychrophilic and psychrotolerant *Clostridium* species (*Clostridium algidicarnis, C. estertheticum, C. gasigenes,* and *C. putrefaciens*) were isolated from New Zealand vacuum-packed chilled meats. These clostridia represent an emerging group of micro-organisms that cause spoilage of these products. Other low-temperature-growing clostridial species such as *C. algidixylanolyticum, C. frigidicarnis,* and *C. vincentii* were also found in New Zealand chill-stored foods. The last-named organism, originally found in deep marine sediments of Antarctica, also occurs in New Zealand farm environments.

A large and increasing proportion of New Zealand food exports depend on refrigeration to reach distant markets. Because the primary-production and health sectors have always been central to New Zealand's economic and social policies, there is a high level of knowledge of risks posed to these sectors by food-borne pathogens and food-spoilage bacteria. However, the on-going viability of the New Zealand food industry hinges on the ability to anticipate the consequences of changes to methods of food processing and preservation. Because these can create new ecological niches that may lead to selection of adapted, novel micro-organisms, there is a need for continuing studies to anticipate threats that psychrophilic and psychrotolerant bacteria present to the health and economic welfare of New Zealanders.

Bacteria for bioremediation

Bacteria from New Zealand sources that are potentially useful for bioremediation, are typically similar to those isolated worldwide. This apparent lack of diversity may be attributed to the use of similar isolation procedures. The biodegradative bacteria most often reported are aerobic heterotrophs. They belong to a number of genera including *Acidovorax, Burkholderia, Mycobacterium, Nocardia, Pseudomonas, Rhodococcus, Sphingomonas,* and *Terrabacter,* and most of have not been identified to species level.

Bacteria degrade a wide range of organic contaminants. Many of the com-

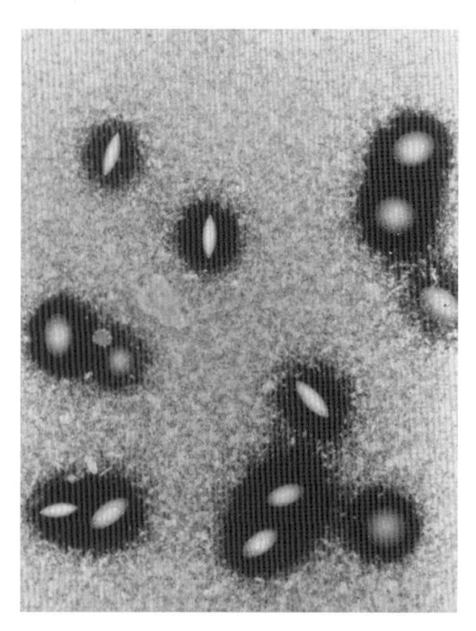

Plate assay developed for the detection of presumptive carbazole-degrading bacteria. Colonies are surrounded by zones of clearing.
Landcare Research

pounds found in petroleum products, such as benzene, toluene, naphthalene, and phenanthrene can serve as the sole source of carbon and energy for bacterial growth, as do pesticides like pentachlorophenol and 2,4-D. Contaminants such as the pesticide atrazine can provide a source of nitrogen, in addition to carbon, for bacterial growth, but other compounds such as DDT or its residues DDD and DDE (DDTr) are not used for bacterial growth. DDTr can, however, be transformed by co-metabolic processes where the bacteria growing at the expense of a growth substrate are able to transform DDT without deriving any energy from it. The bacterium '*Terrabacter* sp. DDE-1', for example, isolated from South Island agricultural soils, extensively degrades DDE when grown on biphenyl (Aislabie et al. 1999). Bacteria vary with respect to their substrate range. While some strains may degrade a wide range of compounds, others do not (Aislabie et al. 1997).

When using low-molecular-weight aromatic hydrocarbons such as naphthalene or toluene as sole carbon source, species of *Pseudomonas* including the very common *P. putida* and *P. fluorescens* are often isolated. Degradation of larger-molecular-weight polyaromatic hydrocarbons like phenanthrene or pyrene, however, is often attributed to strains of *Burkholderia, Mycobacterium,* or *Sphingomonas* (Aislabie et al. 1997). A species of *Sphingomonas* has also been demonstrated to metabolise the nitrogen heterocycle carbazole (Shepherd & Lloyd-Jones 1998), whereas bacteria that degrade the nitrogen heterocycle isoquinoline have been presumptively identified as members of the family Comamonadaceae (Aislabie et al. 1994).

Note

The manuscript of this chapter was completed in 2006. Owing to publication delay, subsequent major advances in bacterial ecology, genetics, and systematics, especially of bacteria from extreme environments, are not reported.

Acknowledgement

Funding from the New Zealand Foundation for Research, Science & Technology supported this contribution.

Authors

Dr Jackie Aislabie Landcare Research, Private Bag 3127, Hamilton, New Zealand [aislabiej@landcareresearch.co.nz] Bacteria for bioremediation

Dr Dorota M. Broda MIRINZ, AgResearch, Private Bag 3123, Hamilton, New Zealand [dorota.broda@agresearch.co.nz] Psychrophilic and psychrotolerant bacteria

Dr Geoffrey W. de Lisle National Centre for Biosecurity and Infectious Disease – Wallaceville, AgResearch, P.O. Box 40-063, Upper Hutt [geoff.delisle@agresearch.co.nz] Animal pathogens

Dr Maureen O'Callaghan NZ Pastoral Agriculture Research Institute, Lincoln, P.O. Box 160, Lincoln, New Zealand [maureen.ocallaghan@agresearch.co.nz] Entomopathogenic bacteria

Dr Elizabeth W. Maas National Institute of Water and Atmosphere, Private Bag 14-901 Kilbirnie, Wellington, New Zealand [e.maas@niwa.co.nz] Marine bacteria

Dr Hugh Morgan Department of Biological Sciences, University of Waikato, Private Bag 3105, Hamilton; New Zealand [h.morgan@waikato.ac.nz] Archaebacteria

Dr Simon Swift University of Auckland, Private Bag 92019, Auckland, New Zealand [s.swift@auckland.ac.nz] Human pathogens

Dr Susan J. Turner University of Auckland, Private Bag 92019, Auckland, New Zealand [s.turner@auckland.ac.nz] Geothermal systems

Dr John M. Young Landcare Research, Private Bag 92170 Auckland, New Zealand [youngj@landcareresearch.co.nz] Introductory sections, Bacterial endemism, Plant pathogens

References

AISLABIE, J.; RICHARDS, N. K.; LYTTLE, T. C. 1994: Description of bacteria able to degrade isoquinoline in pure culture. *Canadian Journal of Microbiology 40*: 555–560.

AISLABIE, J.; HUNTER, D. W. F.; LLOYD-JONES, G. 1997: Polyaromatic compound-degrading bacteria from a contaminated site in New Zealand. *In situ and On-site Bioremediation 4*: 219–224.

AISLABIE, J.; DAVISON, A. D.; BOUL, H. L.; FRANZMANN, P. D.; JARDINE, D. R.; KARUSO, P. 1999: Isolation of *Terrabacter* sp. strain DDE-1 which metabolises DDE when induced with biphenyl. *Applied and Environmental Microbiology 65*: 5607–5611.

ANDERSEN, M. T.; NEWCOMB, M. T.; LEIFTING, L. W.; BEEVER, R. E. 2006: Phylogenetic analysis of "*Candidatus* Phytoplasma australiense" reveals distinct populations in New Zealand. *Phytopathology 96*: 838–845.

ANON. 2001: Surveillance and Control Notes: Intentional release of biologic agents. *New Zealand Public Health Report 8*: 84–85.

ANON. 2003: Notifiable and other diseases in New Zealand. Annual Report 2003. Institute of Environmental Science and Research Limited, Lower Hutt. http://www. surv.esr.cri.nz/PDF_surveillance/AnnSurvRpt/2003AnnualSurvRpt.pdf.

BALOWS, A.; TRÜPER, H. G.; DWORKIN, M.; HARDER, W.; SCHLEIFER, K.-H. (Eds) 1992: The Prokaryotes. A Handbook on the Biology of Bacteria: Ecophysiology, Isolation, Identification, Applications. 2nd edn. Springer-Verlag, New York. xxxv + 4150 p.

BEAN, D. C.; KLENA, J. D. 2005: Characterization of major clones of antibiotic-resistant Streptococcus pneumoniae in New Zealand by multilocus sequence typing. Journal of Antimicrobial Chemotherapy 55: 375–378.

BRADBURY, J. F. 1986: *Guide to Plant Pathogenic Bacteria*. CAB International Mycological Institute, Kew, UK. 332 p.

BRODA, D. M.; BOEREMA, J. A.; BELL, R. G. 1998: A PCR-survey of psychrotrophic *Clostridium botulinum*-like isolates for the presence of BoNT genes. *Letters in Applied Microbiology 27*: 219–223.

BRUNK, C. F.; EIS, N. 1998: Quantitative measure of small-subunit rRNA gene sequences of the kingdom Korarchaeota. *Applied and Environmental Microbiology 64*: 5064–5066.

CADY, S. L.; FARMER, J. D. 1996: Fossilization processes in siliceous thermal springs: trends in preservation along thermal gradients. *CIBA Foundation Symposium 202*: 150–170.

CAVALIER-SMITH, T. 1998: A revised six-kingdom system of life. *Biological Reviews 73*: 203–266.

CAVALIER-SMITH, T. 2002: The neomuran origin of archaebacteria, the negibacterial root of the universal tree and megabacterial classification. *International Journal of Systematic and Evolutionary Microbiology 52*: 7–76.

CHANG, S.; SIEVERT, D. M.; HAGEMAN, J. C.; BOULTON, M. L.; TENOVER, F. C.; DOWNES, F. P.; SHAH, S.; RUDRIK, J. T.; PUPP, G. R.; BROWN, W. J.; CARDO, D.; FRIDKIN, S. K. 2003: Infection with vancomycin-resistant *Staphylococcus aureus* containing the vanA resistance gene. *New England Journal of Medicine 348*: 1342–1347.

COUSINS, D.V.; BASTIDA. R.; CATALDI, A.; QUSE, V.; REDROBE, S.; DOW, S.; DUIGNAN, P.; MURRAY, A.; DUPONT, C.; AHMED, N.; COLLINS, D. M.; BUTLER, W. R.; DAWSON, D.; RODRIGUEZ, D.; LOUREIRO, J.; ROMANO, M. I.; ALITO, A.; ZUMARRAGA, M.; BERNARDELLI, A. 2003: Tuberculosis in seals caused by a novel member of the *Mycobacterium tuberculosis* complex: *Mycobacterium pinnipedii* sp. nov. *International Journal of Systematic and Evolutionary Microbiology 53*: 1305–1314.

CRUMP, J. A.; MURDOCH, D. R.; BAKER, M. G. 2001: Emerging infectious diseases in an island ecosystem: the New Zealand perspective. *Emerging Infectious Diseases 7*: 767–772.

DAUBIN, V.; MORAN, N. A.; OCHMAN, H. 2003: Phylogenetics and the cohesion of bacterial genomes. *Science 301*: 829–832.

DAVIS, P. H.; HAYWARD, V. H. 1963. *Principles of Angiosperm Taxonomy*. Oliver & Boyd, Edinburgh. 558 p.

de LISLE, G. W.; STANISLAWEK, W. L.; MOORS, P. J. 1990: *Pasteurella multocida* infections in rockhopper penguins (*Eudyptes chrysocome*) from Campbell Island, New Zealand. *Journal of Wildlife Diseases 26*: 283–285.

de LISLE, G. W.; MACKINTOSH, C. G.; BENGIS, R. G. 2001: *Mycobacterium bovis* in free-living and captive wildlife, including farmed deer. *Revue scientifique et technique Office International des Epizootes 20*: 86–111.

DeLONG, E. F; TAYLOR, L. T. MARSH, T. L.; PRESTON, C. M. 1999: Visualization and enumeration of marine planktonic Archaea and Bacteria by using polyribonucleotide probes and fluorescent in situ hybridization. *Applied and Environmental Microbiology 65*: 5554–5563.

EUZÉBY, J. 1997–2006: List of bacterial names with standing in nomenclature: a folder available on the Internet (updated: November, 2006; revised URL: http://www. bacterio. cict.fr/). *International Journal of Systematic Bacteriology 47*: 590–592.

FOX, G. E; STACKEBRANDT, E.; HESPELL, R. B.; GIBSON, J.; MANILOFF, J.; DYER, T. A.; WOLFE, R. S.; BALCH, W. E.; TANNER, R. S.; MARGRUM, L. J.; ZABLEN, L. B.; BLAKEMORE, R.; GUPTA, R.; BONEN, L.; LEWIS, B. J.; STAHL, D. A.; LUEHRSEN, K. R.; CHEN, K. N.; WOESE, C. R. 1980: The phylogeny of the prokaryotes. *Science 209*: 457–463.

GLARE, T. R.; O'CALLAGHAN, M.; WIGLEY, P. J. 1993: Checklist of naturally occurring entomopathogenic microbes and nematodes in New Zealand. *New Zealand Journal of Zoology 20*: 95–120.

GOODFELLOW, M.; O'DONNELL, A. G. 1993: The roots of bacterial systematics. Pp. 3–54 *in*: Goodfellow M.; O'Donnell, A. G. (eds), *Handbook of New Bacterial Systematics.* Academic Press, London. 560 p.

GRIMONT, P. A. D.; JACKSON, T. A.; AGERON, E.; NOONAN, M. J. 1988: *Serratia entomophila* sp. nov. associated with amber disease in the New Zealand grass grub, *Costelytra zealandica*. *International Journal of Systematic Bacteriology 38*: 1–6.

HUNTER, J. E.; DUIGNAN, P. J.; DUPONT, C.; FRAY, L.; FENWICK, S. G.; MURRAY, A. 1998: First report of potentially zoonotic tuberculosis in fur seals in New Zealand. *New Zealand Medical Journal 111*: 130–131.

JOKLIK, W. K.; WILLETT, H. P.; AMOS, D. B. (Eds) 1992: *Zinsser's Microbiology*. 20th edn. Appleton & Lange, Norwalk, Connecticut. 1294 p.

JONES, B.; RENAULT, R. W.; ROSEN, M. R. 1997: Vertical zonation of biota in microstromatolites associated with hot springs, North Island, New Zealand. *Palaios 12*: 220–236.

JONES, B.; RENAULT, R. W.; ROSEN, M. R. 1998: Microbial biofacies in hot spring sinters: a model based on Ohaaki Pool, North Island, New Zealand. *Journal of Sedimentary Research 68*: 413–434.

JONES, B.; RENAULT, R. W.; ROSEN, M. R. 1999: Actively growing siliceous oncoids in the Waiotapu geothermal area, North Island, New Zealand. *Journal of the Geological Society 156*: 89–103.

KAPLAN, I. R. 1956: Evidence of microbiological activity in some of the geothermal regions of New Zealand. *New Zealand Journal of Science and Technology 36*: 639–662.

KHAN, L. B. 1998: Microbial Diversity in a Geothermal Hot Pool. MSc Thesis. University of Auckland, Auckland, New Zealand. 138 p.

LAPAGE, S. P.; SNEATH, P. H. A.; LESSEL, E. F.; SKERMAN, V. B. D.; SEELIGER, H. P. R.; CLARK, W. A. 1992: *International Code of Nomenclature of Bacteria (1990 Revision) Bacteriological Code*. [Sneath, P. H. A. (Ed.).] American Society for Microbiology, Washington, D.C..

KNOLL, A. H. 2003: *Life on a Young Planet – The First three Billion Years of Evolution on Earth*. Princeton University Press, Princeton. 287 p.

MADIGAN, M. T.; MARTINKO, J. M.; PARKER, J. (Eds) 2000: Microbial ecology. Pp. 545–572 *in*: *Brock Biology of Microorganisms*. 9th edn. Prentice Hall International. 992 p.

MAIDEN, M. C. J.; BYGRAVES, J. A.; FEIL, E.; MORELLI, G.; RUSSELL, J. E.; URWIN, R.; ZHANG, Q.; ZHOU, J.; ZURTH, K.; CAUGANT, D. A.; FEAVERS, I. M.; ACHTMAN, M.; SPRATT, B. G. 1998: Multilocus sequence typing: a portable approach to the identification of clones within populations of pathogenic microorganisms. *Proceedings of the National Academy of Sciences, USA 95*: 3140–3145.

MARTIN, W.; EMBLEY, T. M. 2004: Early evolution comes full circle. *Nature 431*: 134–135.

NIKLAS, K. J. 1997: *The Evolutionary Biology of Plants*. University of Chicago Press, Chicago. xix + 449 p.

PALUMBO, S. A. 1986: Is refrigeration enough to restrain foodborne pathogens? *Journal of Food Protection 49*: 1003–1009.

PENNYCOOK, S. R.; YOUNG, J. M.; FLETCHER, M. J.; 1989: Bacterial plant pathogens. Pp. 35–83 *in*: Pennycook, S. R. (ed.), *Plant Diseases Recorded In New Zealand, Volume 3*. Plant Diseases Division, DSIR, Auckland. 180 p.

QUINN, P. J.; MARKEY, B. K.; CARTER, M. E.; DONNELLY, W. J.; LEONARD, F. C. 2002: *Veterinary Microbiology and Microbial Disease.* Blackwell Publishing, Oxford. 544 p.

RAVENSCHLAG, K.; SAHM, K.; AMANN, R. 2001: Quantitative molecular analysis of the microbial community in marine Arctic sediments (Svalbard). *Applied and Environmental Microbiology 67*: 387–395.

RIVERA, M. C.; LAKE, J. A. 2004: The ring of life provides evidence for a genome fusion origin of eukaryotes. *Nature 431*: 152–155.

SCHLEPER, C.; HOLBEN, W.; KLENK, H. P. 1997: Recovery of crenarchaeotal ribosomal DNA sequences from freshwater-lake sediments. *Applied and Environmental Microbiology 63*:

321–323.
SCHULZ, H. N.; BRINKHOFF, T.; FERDELMAN, T. G.; HERNÁNDEZ MARINÉ, M.; TESKE, A.; JØRGENSEN, B.B. 1999: Dense populations of a giant sulfur bacterium in Namibian shelf sediments. *Science 284*: 493–495.
SHEPHERD, J. M.; LLOYD-JONES, G. 1998: Novel carbazole degradation genes of *Sphingomonas* CB3: sequence analysis, transcription and molecular ecology. *Biochemical and Biophysical Research Communications. 247*: 129–135.
SKERMAN, V. B. D.; McGOWAN, V.; SNEATH, P. H. A. (Eds) 1980: Approved lists of bacterial names. *International Journal of Systematic Bacteriology 30*: 225–420.
SKERMAN, V. B. D.; McGOWAN, V.; SNEATH, P. H. A. (Eds) 1989: *Approved Lists of Bacterial Names*. American Society for Microbiology, Washington, D.C. 188 p.
SKEY, W. 1878: On the mineral waters of New Zealand. *Transactions of the New Zealand Institute 10*: 423–448.
SNEATH, P. H. A. 1989. Analysis and interpretation of sequence data for bacterial systematics: the view of a numerical taxonomist. *Systematic and Applied Microbiology 12*: 15–31.
STACKEBRANDT, E.; FREDERIKSEN, W.; GARRITY, G. M.; GRIMONT, P. A. D.; KAMPFER, P.; MAIDEN, M. C. J.; NESME, X.; ROSSELLO-MORA, R.; SWINGS, J.; TRÜPER, H. G.; VAUTERIN, L.; WARD, A. C.; WHITMAN, W. B. 2002: Report of the ad hoc committee for the re-evaluation of the species definition in bacteriology. *International Journal of Systematic and Evolutionary Microbiology 52*: 1043–1047.
STEINHAUS, E. A. 1941: A study of bacteria associated with thirty species of insects. *Journal of Bacteriology 42*: 757–789.
STEVENS, G.; McGLONE, M.; McCULLOCH, B. 1988: *Prehistoric New Zealand*. Heinemann Reed, Auckland.
VANDAMME, P.; POT, B.; GILLIS, M.; de VOS, P.; KERSTERS, K.; SWINGS, J. 1996: Polyphasic taxonomy, a consensus approach to bacterial systematics. *Microbiological Reviews 60*: 407–438
WEED, W. H. 1889: The vegetation of hot springs. *American Naturalist 23*: 394–398.
WEIR, B. S.; TURNER, S. J.; SILVESTER, W. B.; YOUNG, J. M. 2004: Unexpectedly diverse *Mesorhizobium* strains and *Rhizobium leguminosarum* nodulate native legume genera of New Zealand, while introduced legume weeds are nodulated by *Bradyrhizobium* species. *Applied and Environmental Microbiology 70*: 5980–5987.
WHITMAN, W. B.; COLEMAN, D. C.; WIEBE, W. J. 1998: Prokaryotes: the unseen majority. *Proceedings of the National Academy of Sciences, USA* 95: 6578–6583.
WOESE, C. R. 1987: Bacterial evolution. *Microbiological Reviews 51*: 221–271.
YOUNG, J. M. 2000: On the need for and utility of depositing reference strains to authenticate microbiological studies (and a list of reference strains for bacteria in New Zealand). *New Zealand Microbiology* 5: 23–28.
YOUNG, J. M. 2001: Implications of alternative classifications and horizontal gene transfer for bacterial taxonomy. *International Journal of Systematic and Evolutionary Microbiology 51*: 945–953.
YOUNG, J. M.; SADDLER, G.; TAKIKAWA, Y.; De BOER, S. H.; VAUTERIN, L.; GARDAN, L.; GVOZDYAK, R. I.; STEAD, D. E. 1996: Names of plant pathogenic bacteria 1864–1995. *Review of Plant Pathology 75*: 721–763.
YOUNG, J. M.; TAKIKAWA, Y.; GARDAN, L.; STEAD, D. E. 1992: Changing concepts in the taxonomy of plant pathogenic bacteria. *Annual Review of Phytopathology 30*: 67–105.

Checklist of Cultured New Zealand Bacteria

Although many bacterial species have been recorded as being present in New Zealand, the accuracy of earlier reports is uncertain. The list below is a record of only those species for which there is at least one reference strain held in a public culture collection. Higher taxonomic levels, at classes and above, are arranged according to the taxonomic scheme of Cavalier-Smith (1998) based on molecular sequence and microstructural considerations. For classification at order and family levels, the classification produced for Bergey's Manual of Determinative Bacteriology (Second edition: http://www.cme.msu.edu/bergeys/), based only on comparative 16S rDNA sequence analyses, is used. The positions of classes recorded in Bergey's scheme are indicated in brackets (=) following Cavalier-Smith's listings. e.g., Class Alphabacteria (= α-Proteobacteria).

SUPERKINGDOM PROKARYOTA
KINGDOM BACTERIA
SUBKINGDOM NEGIBACTERIA
INFRAKINGDOM EOBACTERIA
PHYLUM EOBACTERIA
Class HADOBACTERIA
Order THERMALES
THERMACEAE
Thermus filiformis Hudson et al. 1987

INFRAKINGDOM GLYCOBACTERIA
PHYLUM CYANOBACTERIA
[See separate chapter on Cyanobacteria]

PHYLUM SPHINGOBACTERIA
Class FLAVOBACTERIA
(= Class Flavobacteria; = Class Bacteroidetes; = Class Sphingobacterium)
Order FLAVOBACTERIALES
BACTEROIDACEAE
Bacteroides distasonis Eggerth & Gagnon 1933
Bacteroides fragilis (Veillon & Zuber 1898) Castellani & Chalmers 1919
Bacteroides thetaiotaomicron (Distao 1912) Castellani & Chalmers 1919
Bacteroides vulgatus Eggerth & Gagnon 1933
FLAVOBACTERIACEAE
Bergeyella zoohelcum (Holmes et al. 1987) Vandamme *et al.* 1994
Capnocytophaga canimorsus Brenner et al. 1990
Chryseobacterium indologenes (Yabuuchi et al. 1983) Vandamme et al. 1994
Chryseobacterium meningosepticum (King 1959) Vandamme et al. 1994
Weeksella virosa Holmes et al. 1987
MYROIDACEAE
Myroides odoratus (Stutzer 1929) Vancanneyt et al. 1996
PREVOTELLACEAE
Prevotella intermedia (Holdeman & Moore 1970) Shah & Collins 1990
Prevotella oralis (Loesche et al. 1964) Shah & Collins 1990
SPHINGOBACTERIACEAE
Sphingobacterium multivorum (Holmes et al. 1981) Yabuuchi et al. 1983
Sphingobacterium spiritivorum (Holmes et al. 1982) Yabuuchi et al. 1983

PHYLUM PROTEOBACTERIA
SUBPHYLUM RHODOBACTERIA
Class ALPHABACTERIA
(= α-Proteobacteria)
Order RHIZOBIALES
BRADYRHIZOBIACEAE
Bradyrhizobium japonicum (Kirchner 1896) Jordan 1982
BRUCELLACEAE
Brucella abortus (Schmidt 1901) Meyer & Shaw 1920
Ochrobactrum anthropi Holmes et al. 1988
METHYLOBACTERIACEAE
Methylobacterium mesophilicum (Austin & Goodfellow 1979) Green & Bousfield 1983
RHIZOBACTERIACEAE
Rhizobium leguminosarum (Frank 1879) Frank 1889
Rhizobium loti Jarvis, Pankhurst & Patel 1982
Rhizobium meliloti Dangeard 1926
Rhizobium radiobacter (Beijerinck & van Delden 1902) Young et al. 2001 (=*Agrobacterium tumefaciens* (Smith & Townsend 1907) Conn 1942)
Rhizobium rhizogenes (Riker et al. 1930) Young *et al.* 2001 (= *Agrobacterium rhizogenes* (Riker et al. 1930) Conn 1942)

Order SPHINGOMONADALES
SPHINGOMONADACEAE
Sphingomonas paucimobilis (Holmes et al. 1977) Yabuuchi et al. 1990

Order CAULOBACTERIALES
CAULOBACTERIACEAE
Brevundimonas diminuta (Leifson & Hugh 1954)

Segers et al. 1994
Brevundimonas vesicularis (Büsing et al. 1953) Segers et al. 1994

Class CHROMATIBACTERIA
(includes Class β-Proteobacteria and γ-Proteobacteria)
Class β-Proteobacteria
Order BURKHOLDERIALES
ALCALIGINACEAE
Alcaligenes eutrophus Davis 1969
Alcaligenes denitrificans Rüger & Tan 1983
Alcaligenes faecalis Castenalli & Chalmers 1919
Alcaligenes xylosoxidans (Yabuuchi & Yano 1981) Kiredjian et al. 1986
Bordetella bronchiseptica (Ferry 1912) Moreno-Lopez 1952
Bordetella parapertussis (Eldering & Kendrick 1938) Moreno-Lopez 1952
Bordetella pertussis (Bergey et al. 1923) Moreno-Lopez 1952
Oligella urethralis (Lautrop *et al.* 1970) Rossau et al. 1987
BURKHOLDERIACEAE
Burkholderia andropogonis (Smith 1911) Gillis et al. 1995
Burkholderia cepacia (ex Burkholder 1950) Yabuuchi et al. 1993
Burkholderia gladioli (Severini 1913) Yabuuchi et al. 1993
COMAMONADACEAE
Acidovorax avenae ssp. *cattleyae* (Pavarino 1911) Willems et al. 1992
Acidovorax delafieldii (Davis 1970) Willems et al. 1990
Comamonas acidovorans (den Dooren de Jong 1926) Tamaoka et al. 1987
Comamonas testosteroni (Marcus & Talalay 1956) Tamaoka et al. 1987
OXALOBACTERICEAE
Herbaspirillum rubrisubalbicans (Christopher & Edgerton 1930) Baldani et al. 1996
RALSTONIACEAE
Janthinobacterium lividum (Eisenberg 1891) De Ley et al. 1978
Ralstonia pickettii (Ralston et al. 1973) Yabuuchi et al. 1996
Ralstonia solanacearum (Smith 1896) Yabuuchi et al. 1995

Order NEISSERIALES
NEISSERIACEAE
Eikenella corrodens (Eiken 1958) Jackson & Goodman 1972
Kingella denitrificans Snell & Lapage 1976)
Kingella kingae (Henrikson & Bøvre 1968) Henrikson & Bøvre 1976
Neisseria cinerea (von Lingelsheim 1906) Murray 1939
Neisseria elongata Bøvre & Holten 1970
Neisseria flavescens Branham 1930
Neisseria gonorrhoeae (Zopf 1885) Trevisan 1885)
Neisseria meningitidis (Albrecht & Ghon 1901)
Neisseria mucosa Veron et al. 1959
Neisseria lactamica Hollis et al. 1969
Neisseria polysaccharea Riou & Guibourdenche 1987
Neisseria sicca (von Lingelsheim 1908) Bergey et al. 1923
Neisseria subflava (Flugge 1886) Trevisan 1889
Neisseria weaveri Anderson et al. 1993

Class γ-Proteobacteria
Order CARDIOBACTERIALES
CARDIOBACTERIACEAE
Cardiobacterium hominis Slotnick & Dougherty 1964
Suttonella indologenes (Snell & Lapage 1976) Dewhirst *et al.* 1990

Order LEGIONELLALES
LEGIONELLACEAE
Legionella anisa Gorman et al. 1985
Legionella bozemanii Brenner et al. 1980
Legionella dumoffii Brenner et al. 1980
Legionella feeleii Herwalt et al. 1984
Legionella jordanis Cherry et al. 1982
Legionella longbeachae McKinney et al. 1982
Legionella micdadei Herbert et al. 1980
Legionella pneumophila Brenner et al. 1979

Order PSEUDOMONADALES
MORAXELLACEAE
Acinetobacter calcoaceticus Beijerinck 1911) Bouvet & Grimont 1986
Acinetobacter baumannii Bouvet & Grimont 1986
Acinetobacter lwoffi (Audereau 1940) Bouvet & Grimont 1986
Branhamella catarrhalis (Frosch & Kolle 1896) Catlin 1970
Moraxella (Moraxella) osloensis Bøvre & Henriksen 1967
Moraxella (Moraxella) lacunata (Eyre 1900) Lwoff 1939
Moraxella (Moraxella) nonliquefaciens (Scarlett 1916) Lwoff 1939
Psychrobacter phenylpyruvicus (Bøvre & Henriksen 1967) Bowman et al. 1996
PSEUDOMONADACEAE
Pseudomonas aeruginosa (Schroeter 1872) Migula 1900
Pseudomonas agarici Young 1970
Pseudomonas alcaligenes Monias 1928
Pseudomonas cichorii (Swingle 1925) Stapp 1928
Pseudomonas corrugata (ex Scarlett et al. 1978) Roberts & Scarlett 1981
Pseudomonas fluorescens Migula 1895
Pseudomonas fragi (Eicholz 1902) Huss 1905
Pseudomonas luteola Kodama et al. 1985
Pseudomonas marginalis (Brown 1918) Stevens 1925
Pseudomonas oryzihabitans Kodama et al. 1985
Pseudomonas paucimobilis Holmes et al. 1977
Pseudomonas putida (Trevisan 1889) Migula 1895
Pseudomonas pseudomallei (Whitmore 1913) Haynes 1957
Pseudomonas stutzeri (Lehmann & Newmann 1896) Siderius 1946
Pseudomonas savastanoi (Burkholder 1926) Gardan 1992
Pseudomonas syringae van Hall 1902
Pseudomonas tolaasii Paine 1919
Pseudomonas viridiflava (Burkholder 1930) Dowson 1939
Azotobacter beijerinckii Lipman 1904
Azotobacter chroococcum Beijerinck 1901

Order ALTEROMONADALES
ALTEROMONADACEAE
Shewanella putrefaciens (Lee et al. 1981) MacDonell & Colwell 1986

Order VIBRIONALES
VIBRIONACEAE
Listonella anguillarum (Bergeman 1909) Macdonell & Colwell 1986
Vibrio alginolyticus (Miyamoto et al. 1961) Sakazaki 1968
Vibrio cholerae Pacini 1854
Vibrio furnissii Brenner et al. 1984
Vibrio mimicus Davis et al. 1982
Vibrio parahaemolyticus (Fujino et al. 1951) Sakazaki et al. 1963
Vibrio vulnificus (Reichelt et al. 1979) Farmer 1980

Order AEROMONADALES
AEROMONADACEAE
Aeromonas caviae Popoff 1984
Aeromonas hydrophila (Chester 1901) Stanier 1943
Aeromonas veronii Hickman-Brenner et al.

Order ENTEROBACTERIALES
ENTEROBACTERIACEAE
Citrobacter amalonaticus (Young et al. 1971) Brenner & Farmer 1982
Citrobacter freundii Braak 1928) Werkman & Gillen 1932
Citrobacter koseri Frederikson 1970
Edwardsiella tarda Ewing & McWhorter 1965
Enterobacter aerogenes Hormaeche & Edwards 1960
Enterobacter agglomerans Ewing & Fife 1972
Enterobacter cloacae (Jordan 1890) Hormaeche & Edwards 1960
Enterobacter intermedius Izard et al. 1980
Enterobacter sakazakii Farmer et al. 1980
Erwinia amylovora (Burrill 1882) Winslow et al. 1920
Erwinia carotovora (Jones 1901) Bergey et al. 1923
Erwinia chrysanthemi Burkholder, McFadden & Dimock 1953
Erwinia herbicola (Lohnis 1911) Dye 1964
Erwinia rhapontici (Millard 1924) Burkholder 1948
Escherichia coli (Migula 1895) Castellani & Chalmers 1919
Hafnia alvei Moller 1954
Klebsiella oxytoca (Flugge 1886) Lautrop 1956
Klebsiella pneumoniae (Schroeter 1886) Trevisan 1887
Kluyvera ascorbata Farmer et al. 1981
Leclercia adecarboxylata (Leclerc 1962) Tamura et al. 1987
Morganella morganii (Winslow et al. 1919) Fulton 1943
Pantoea agglomerans (Ewing & Fife 1972) Gavini et al. 1989
Plesiomonas shigelloides (Bader 1954) Habs & Schubert 1962
Proteus mirabilis Hauser 1885
Proteus vulgaris Hauser 1885
Providencia rustigiani Hickman-Brenner et al. 1983
Rahnella aquatilis Izard et al. 1981
Salmonella arizonae (Borman 1957) Kauffman 1964
Salmonella choleraesuis (Smith 1894) Weldin 1927
Salmonella enteritidis (Gaertner 1888) Castellani & Chalmers 1919
Salmonella typhi (Schroeter 1886) Warren & Scott 1930
Salmonella typhimurium (Loefler 1892) Castellani & Chalmers 1919
Serratia entomophila Grimont et al. 1988
Serratia liquefaciens (Grimes & Hennerty 1931) Bascomb et al. 1971
Serratia marcescens Bizio 1923
Serratia proteamaculans (Paine & Stansfield 1919) Grimont et al. 1978
Shigella boydii Ewing 1949
Shigella dysenteriae (Shiga 1898) Castellani & Chalmers 1919
Shigella flexneri Castellani & Chalmers 1919
Shigella sonnei (Levine 1920) Weldin 1927
Yersinia enterocolitica (Schleifstein & Coleman 1939) Frederiksen 1964
Yersinia frederiksenii Ursing et al. 1981
Yersinia kristensenii Bercovier et al. 1981
Yersinia pseudotuberculosis (Pfeiffer 1889) Smith & Thal 1965

Order PASTEURELLALES
PASTEURELLACEAE
Actinobacillus actinomycetemcomitans (Klinger 19120 Topley & Wilson 1929
Actinobacillus equuli (van Straaten 1918) Haupt 1934
Actinobacillus lignieresi Brumpt 1910
Actinobacillus suis van Dorssen & Jaartsveld 1934
Chryseomonas luteola (Kodama et al. 1985) Holmes et al. 1987
Haemophilus aphrophilus Khairat 1940
Haemophilus ducreyi (Neveu-Lemaire 1921) Bergey et al. 1923
Haemophilus haemoglobinophilus (Lehmann & Newmann 1907) Murray 1939
Haemophilus haemolyticus Bergey et al. 1923
Haemophilus influenzae (Lehmann & Newmann 1896) Bergey et al. 1923
Haemophilus parainfluenzae Rivers 1922
Haemophilus paraphrophilus Zinnemann et al. 1968
Mannheimia haemolytica (Newsome & Cross 1932) Angen et al. 1999
Pasteurella aerogenes McAllister & Carter 1974
Pasteurella dagmatis Mutters et al. 1985
Pasteurella multocida (Lehmann & Neumann 1899) Rosenbusch & Merchant 1939
Pasteurella pneumotropica Jawetz 1950

Order XANTHOMONADALES
XANTHOMONADACEAE
Stenotrophomonas maltophilia (Hugh 1981) Palleroni & Bradbury 1993
Xanthomonas campestris (Pammel 1891) Dowson 1939

SUBPHYLUM THIOBACTERIA
Class DELTABACTERIA
(= Class Δ-Proteobacteria)
Order DESULFOVIBRIONALES
DESULFOVIBRIONACEAE
Desulfovibrio desulfuricans (Beijerinck 1895) Kluyver & van Neil 1936

Class EPSILOBACTERIA
(= Class ε-Proteobacteria)
Order CAMPYLOBACTERIALES
CAMPYLOBACTERIACEAE
Campylobacter coli (Doyle 1948) Veron & Chatelain 1973
Campylobacter fetus (Smth & Taylor 1919) Sebald & Veron 1963
Campylobacter jejuni (Jones et al. 1931) Veron & Chatelain 1973
Campylobacter lari corrig. Benjamin *et al.* 1984

Order AQUIFICALES
HYDROGENOTHERMACEAE
Venenivibrio stagnispumantis Hetzer et al. 2008

SUBKINGDOM UNIBACTERIA
PHYLUM POSIBACTERIA
SUBPHYLUM ENDOBACTERIA
Class TEICHOBACTERIA
(= Class Clostridia)
Order CLOSTRIDIALES
ACIDAMINOCOCCACEAE
Veillonella parvula (Veillon & Zuber 1898) Prevot 1933
CLOSTRIDIACEAE
Clostridium algidicarnis Lawson et al.
Clostridium algidixylanolyticum Broda et al.
Clostridium botulinum (van Ermengem 1896) Bergey et al. 1923
Clostridium butyricum Prazmowski 1880
Clostridium difficile (Hall & O'Toole 1935) Prevot 1938
Clostridium frigidicarnis Broda et al.
Clostridium gasigenes Broda et al.
Clostridium limosum Andre 1948
Clostridium novyi (Migula 1894) Bergey et al. 1923
Clostridium perfringens (Veillon & Zuber 1898) Hauduroy et al. 1937
Clostridium sporogenes (Metchnikoff 1908) Bergey et al. 1923
Clostridium tertium (Henry 1917) Bergey et al. 1923
Clostridium tetani (Flugge 1886) Bergey et al. 1923
PEPTOSTREPTOCOCCACEAE
Peptostreptococcus anaerobius (Natvig 1905) Kluyver & van Neil 1936

(= Class Bacilli)
Order BACILLALES
BACILLACEAE
Bacillus cereus Frankland & Frankland 1887
Bacillus circulans Jordan 1890
Bacillus coagulans Hammer 1915
Bacillus licheniformis (Weigemann 1898) Chester 1901
Bacillus megaterium de Bary 1884
Bacillus mycoides Flugge 1886
Bacillus pumilus Meyer & Gottheil 1901
Bacillus stearothermophilus Donk 1920
Bacillus subtilis (Ehrenberg 1835) Cohn 1872
Bacillus thuringiensis Berliner 1915
LISTERIACEAE
Listeria innocua Seeliger 1983
Listeria ivanovii Seeliger et al. 1984
Listeria seeligeri Rocourt & Grimont 1983
Listeria monocytogenes (Murray et al. 1926) Pirie 1940
Listeria welshimeri Rocourt & Grimont 1983
STAPHYLOCOCCACEAE
Gemella haemolysans (Thjotta & Boe 1938) Berger 1960
Staphylococcus aureus Rosenbach 1884
Staphylococcus auricularis Kloos & Schleifer 1983
Staphylococcus capitis Kloos & Schleifer 1975
Staphylococcus cohnii Schleifer & Kloos 1975
Staphylococcus epidermidis (Winslow & Winslow 1908) Evans 1916
Staphylococcus felis Igimi et al. 1989
Staphylococcus haemolyticus Schleifer & Kloos 1975
Staphylococcus hominis Kloos & Schleifer 1975
Staphylococcus intermedius Hajek 1976
Staphylococcus lugdunensis Freney et al. 1988
Staphylococcus saprophyticus (Fairbrother 1940) Shaw et al. 1951
Staphylococcus schleiferi Freney et al. 1988
Staphylococcus sciuri Schleifer & Smith 1976
Staphylococcus simulans Kloos & Schleifer 1975
Staphylococcus warneri Kloos & Schleifer 1975
Staphylococcus xylosus Schleifer & Kloos 1975

Order LACTOBACILLALES
AEROCOCCACEAE
Abiotrophia defectiva (Bouvet et al. 1989) Kawamura et al. 1995
Aerococcus viridans Williams et al. 1953
CARNOBACTERIACEAE
Carnobacterium divergens (Holzapfel & Gerber 1984) Collins et al. 1987
Alloiococcus otitidis Aguirre & Collins 1992
ENTEROCOCCACEAE
Enterococcus avium Collins et al. 1984
Enterococcus casseliflavus Collins et al. 1984
Enterococcus durans Collins et al. 1984
Enterococcus faecalis (Andrewes & Horder 1906) Schleifer & Kilpper-Balz 1984
Enterococcus faecium (Orla-Jensen 1919) Schleifer & Kilpper-Balz 1984
LACTOBACILLACEAE
Lactobacillus brevis (Orla-Jensen 1919) Bergey et al. 1934
Lactobacillus reuteri (Lerche & Reuter 1962) Kandler et al. 1980
Lactobacillus rhamnosus (Hansen 1968) Collins et al. 1989
Lactobacillus sake Katagiri et al. 1934
LEUCONOSTOCACEAE
Leuconostoc mesenteroides (Tsenkovskii 1878) van Teighem 1978
STREPTOCOCCACEAE
Lactococcus lactis (Lohnis 1909) Schleifer et al. 1986
Streptococcus agalactiae Lehmann & Newmann 1896
Streptococcus anginosus (Andrewes & Horder 1906) Smith & Sherman 1938
Streptococcus equi Sand & Jensen 1888
Streptococcus intermedius Prevot 1925
Streptococcus mitis Andrewes & Horder 1906
Streptococcus mutans Clarke 1924
Streptococcus oralis Bridge & Sneath 1982
Streptococcus pneumoniae (Klein 1884) Chester 1901
Streptococcus pyogenes Rosenbach 1884
Streptococcus salivarius Andrewes & Horder 1906
Streptococcus sanguis White & Niven 1946
Streptococcus suis Kilpper-Balz & Schleifer 1987
Streptococcus thermophilus Orla-Jensen 1919
Streptococcus uberis Deirnhoer 1932

Order THERMOANAEROBACTERIALES
INCERTAE SEDIS
Caldicelluloseruptor saccharolyticus Rainey et al. 1995

SUBPHYLUM ACTINOBACTERIA
Class ARTHROBACTERIA
(= Class Actinobacteria)
Order ACTINOMYCETALES
ACTINOMYCETACEAE
Actinomyces neuii Funke et al. 1994
Actinomyces odontolyticus Batty 1958
Actinomyces viscosus (Howell et al. 1965) Georg et al. 1969
Arcanobacterium haemolyticum Collins et al. 1983
Arcanobacterium pyogenes (Glage 1903) Ramos et al. 1997
CELLULOMONADACEAE
Cellulomonas turbata (Erikson 1954) Stackebrandt et al. 1983
CORYNEBACTERIACEAE
Corynebacterium auris Funke et al. 1995
Corynebacterium amycolatum Collins et al. 1988
Corynebacterium aquaticum [Not a valid bacterial name]
Corynebacterium diphtheriae (Kruse 1886) Lehmann & Neumann 1896
Corynebacterium glucuronolyticum Funke et al. 1995
Corynebacterium kutscheri (Migula 1900) Bergey et al. 1923
Corynebacterium minutissimum Collins & Jones 1988
Corynebacterium jeikeium Jackman et al. 1988
Corynebacterium pseudodiphtheriticum Lehmann & Neumann 1896
Corynebacterium pseudotuberculosis (Buchanan 1911) Eberson 1918
Corynebacterium renale (Migula 1900) Ernst 1906
Corynebacterium striatum (Chester 1901) Eberson 1918
Corynebacterium ulcerans (ex Gilbert & Stewart 1927) Riegel et al. 1995
Corynebacterium urealyticum Pitcher et al. 1992
Corynebacterium xerosis Lehmann & Neumann 1896

DERMABACTERIACEAE
Dermabacter hominis Jones & Collins 1989
DERMATOPHILACEAE
Dermatophilus congolensis (Van Saceghem 1915) Gordon 1964
GORDONIACEAE
Gordonia bronchialis (Tsukamura 1971) Stackebrandt et al. 1989
MICROBACTERIACEAE
Clavibacter michiganensis ssp. *michiganensis* (Smith 1910) Davis et al. 1984
— ssp. *insidiosus* (McCulloch 1925) Davis et al. 1984
Clavibacter rathayi (Smith 1913) Davis et al. 1984
MICROCOCCACEAE
Rothia dentocariosa (Onishi 1949) Georg & Brown 1967
Stomatococcus mucilaginosus Bergan & Kocur 1982
NOCARDIACEAE
Rhodococcus equi (Magnusson 1923) Goodfellow & Alderson 1979
Rhodococcus fascian (Tilford 1936) Goodfellow 1984
Nocardia asteroides (Eppinger 1891) Blanchard 1896
PROPIONIBACTERIACEAE
Propionibacterium acnes (Gilchrist 1900) Douglas & Gunter 1946
STREPTOMYCETACEAE
Streptomyces antibioticus (Waksman & Woodruff 1941) Waksman & Henrici 1948
Streptomyces griseoflavus (Krainsky 1914) Waksman & Henrici 1948
Streptomyces lavendulae (Waksman & Curtis 1916) Waksman & Henrici 1948
Streptomyces venezuelae Ehrlich et al. 1948
Streptomyces viridochromogenes (Krainsky 1914) Waksman & Henrici 1948

Order BIFIDOBACTERIALES
BIFIDOBACTERIACEAE
Gardnerella vaginalis Gardner & Dukes 1955

INCERTAE SEDIS
Turicella otitidis Funke et al. 1994

PHYLUM ARCHAEBACTERIA
Class MENDOSICUTES
Order DESULFUROCOCCALES
DESULFUROCOCCACEAE
Ignisphaera aggregans Niederberger et al. 2006

TWO

Phylum
CYANOBACTERIA
blue-green bacteria, blue-green algae

PAUL A. BROADY, FARADINA MERICAN

A species of *Rivularia* (Rivulariaceae).
Paul Broady

Cyanobacteria are commonly called 'blue-green algae' but their prokaryotic cells, including lack of membrane-enclosed chloroplasts, distinguish them from other algae which are all eukaryotic.

Cyanobacteria express a wider range of morphology than any other group of prokaryotes. The smallest and simplest are minute free-living unicells less than one micrometre in diameter. Other more complex forms vary from unicells bound together in colonies by mucilaginous investments, to unbranched chains of cells called trichomes, and trichomes in which cell division in a second plane results in branch formation. Trichomes are also often surrounded by mucilaginous secretions in the form of sheaths, of diverse pigmentation and degree of lamellation. The combination of a trichome and its surrounding sheath is termed a filament. Sheath pigmentation is often yellow-brown, pinkish red, or violet and can mask the characteristic blue-green colour of the enclosed cells.

Although individual cells and trichomes are microscopic, populations of these can be visible as mats, crusts, and gelatinous masses coating rocks, sediments, soils, and vegetation in both aquatic and terrestrial ecosystems. These growths can appear red-brown, purple, blue-green, or almost black, depending on the balance of pigments within the cells and the mucilaginous coatings.

Two types of specialised cells, heterocytes (heterocysts) and akinetes, are formed by species of certain filamentous genera. Akinetes are resting cells – thick-walled spores – that survive in environments unfavourable to vegetative growth. They accumulate nutrients and separate from trichomes on the death of adjacent undifferentiated cells. They germinate to form new trichomes in suitable conditions such as renewed external nutrient supplies and warmer temperatures.

Reproduction is asexual. In most unicellular forms, cells split in two and separate. In some, the parental cell divides into numerous small cells termed baeocytes or endospores. Another variation involves the budding of a cell, termed an exospore, from the apex of the parental cell. Colonies and filaments fragment after size increases through cell division. Short trichomes, capable of gliding motility, are produced by many filamentous forms and these escape from the mucilaginous sheaths of the parent.

In bacteria, it is well known that parasexual processes transfer genetic material between individuals. Transformation, the assimilation of external DNA by a cell, has been shown with cultured strains of cyanobacteria (Saunders 1992) and transduction, the virus-mediated transfer of bacterial genes from one host cell to another, seems likely to be a factor driving their evolution (Fuhrman 2003; Coleman et al. 2006). Conjugation ('mating') systems involving the bacterium *Escherichia coli* have been used to transfer genes to cyanobacteria (Thiel 1994) and there is evidence to suggest conjugation between cultured strains of cyanobacteria (Flores et al. 2008).

In common with most other algae, oxygenic photosynthesis is the dominant way in which cyanobacteria obtain carbon and energy, and the pigment at the centre of the light-capture apparatus is chlorophyll *a*. Cyanobacteria also possess accessory pigments, mostly phycobilins, which increase efficiency of light capture. However, recent discoveries have revealed cyanobacteria, informally called 'prochlorophytes', that contain chlorophyll *b* (Post 1999) and one that contains chlorophyll *d* (Miyashita et al. 2003).

Cyanobacteria contribute significantly to biological nitrogen fixation by using elemental nitrogen to synthesise nitrogen-containing cellular constituents. They are of particular importance in the surface waters of oceans (Stal & Zehr 2008). Their growth is not limited by a low external concentration of chemically combined nitrogen. Cyanobacteria with heterocytes fix nitrogen in aerobic environments. Many which lack these are nitrogen-fixing under anaerobic or microaerobic conditions (Flores & Herrero 1994). Their physiological diversity has been emphasised by a recent discovery in the marine phytoplankton of widespread, nitrogen-fixing unicells that lack oxygenic photosynthesis (Zehr et al. 2008).

Much information is available about cyanobacteria. Broad-ranging accounts of their biology and diversity can be found in textbooks dealing with algae in general, and excellent recent examples are Lee (2008) and Graham et al. (2009). Volumes dedicated to cyanobacteria provide more detailed treatment of, for instance, their general biology (Fay & Van Baalen 1987), molecular biology (Herrero & Flores 2008), and ecology (Whitton & Potts 2000).

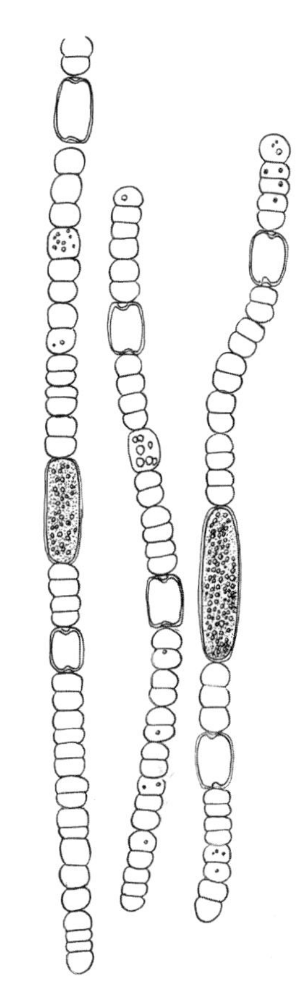

Anabaena inaequalis (Nostocaceae).

Faradina Merican

Evolutionary importance

The fossil record indicates the immense importance of cyanobacteria in the ancient past, certainly from 2000 million years ago (Mya) and with more equivocal evidence from around 3500 Mya (Knoll 2008). They were the dominant photosynthetic organisms in the oceans until about 700 Mya, evident from their extensive fossil remains as stromatolites. These structures form when mats of cyanobacteria trap sediments and deposit calcium carbonate. Motility of the algae enables movement up to the surface through these deposits. When this is repeated many times, a mound is formed which is covered by a thin layer of cyanobacteria. Modern analogues of these ancient structures are found in extreme environments where cyanobacteria are dominant (Stal 2000; Pentecost & Whitton 2000), such as alkaline Lake Van in Turkey, landlocked pools on tropical Indian Ocean islands, and hypersaline Hamelin Pool in Shark Bay, Western Australia.

Oxygenic photosynthesis of cyanobacteria appears to have commenced prior to 2450–2320 Mya, the period when oxygen first accumulated in the atmosphere following oxygenation of the oceans (Knoll 2008). A greatly significant evolutionary consequence was the emergence of eukaryotic life-forms. Photosynthetic eukaryotes then originated from endosymbiotic associations in which cyanobacteria lived within cells of primitive chemoheterotrophic (i.e. non-photosynthesising) eukaryotes. The result was complete integration of a cyanobacterium within a host cell to form the cell organelle we call the chloroplast. Most evidence now suggests that this event occurred just once in evolution (Keeling 2004).

Oxygenation of the atmosphere allowed formation of an ozone layer in the stratosphere. This layer absorbed ultraviolet radiation and enabled colonisation of the land. Although it is commonly considered that green plants were the first colonisers some 450 Mya, fossil and geochemical evidence suggests widespread surface crusts of cyanobacteria as long ago as 2600 Mya (Watanabe et al. 2000). Screening of cells from highly damaging ultraviolet C might have been provided by the yellow-brown pigment scytonemin which is produced in the sheaths of some present-day cyanobacteria (Dillon & Castenholz 1999).

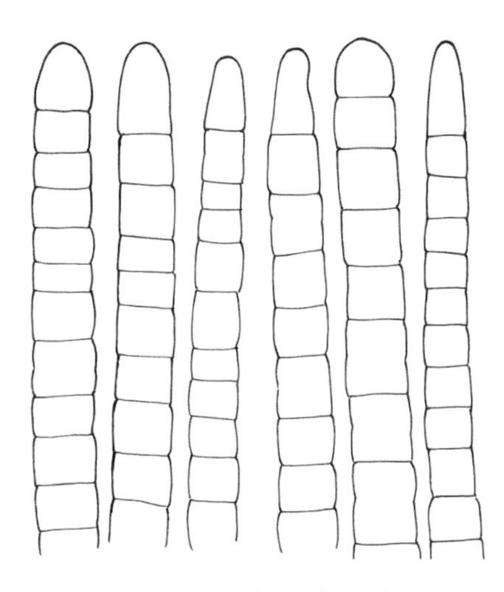

A species of *Phormidium* (Phormidiaceae).

Faradina Merican

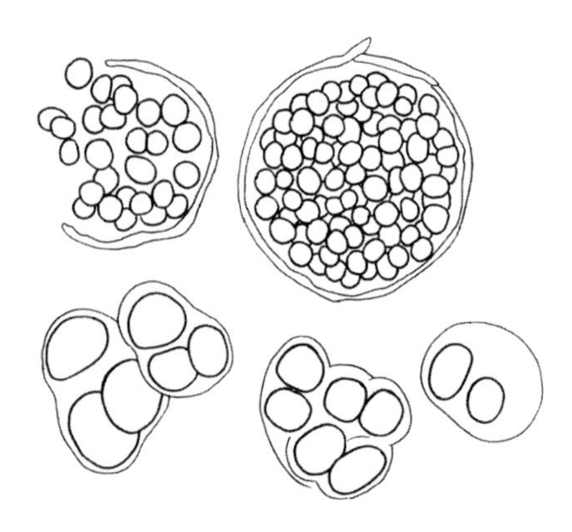

Chroococcidiopsis cf. *kashaii* (Xenococcaceae).
Paul Broady

Present-day distribution

Today, cyanobacteria are usually found with eukaryotic algae in mixed species assemblages. These include phytoplankton suspended in the waters of oceans and lakes, periphyton (associated with surfaces of sediments, rocks, plants, and seaweeds in aquatic environments), and species in terrestrial habitats such as the surface layers of soils. The only restriction to their growth in comparison with eukaryotic algae is a pH limitation of about 4–5, below which their occurrence is rare. Globally, their carbon biomass has been conservatively estimated to be 300 million tonnes, which is about 1/2000 of total global carbon biomass (Garcia-Pichel et al. 2003). About three-quarters of this biomass is in the oceans whilst the remainder is terrestrial. Of that in the oceans, about 50% comprises single cells of *Prochlorococcus* which are less than one micrometre in diameter.

In some extreme habitats, however, cyanobacteria extend their distribution beyond that of eukaryotic algae. In thermal springs worldwide, cyanobacteria are able to grow at temperatures up to about 73° C (Castenholz 1996). Eukaryotic algae are absent above 48° C except for some specialised forms such as *Cyanidium caldarium*, which in acidic conditions occur up to 57° C. In cold temperatures, thick cyanobacterial mats cover sediments in polar-region ponds, lakes, and streams (Taton et al. 2003). They are also often dominant in hypersaline waters in all climatic zones (Javor 1989).

Cyanobacteria are often prominent in extreme terrestrial habitats, surviving repeated desiccation and producing light-protective pigments as shade from often intense solar radiation (Potts 1999). Endolithic communities comprise microorganisms that grow a few millimetres below the surface of suitably translucent rocks. In hot deserts, cyanobacteria are dominant (Bell 1993) and in polar deserts they can be abundant associates in lichen- and green-algal- (chlorophyte-) dominated assemblages (Nienow & Friedmann 1993). Remarkably, unicellular *Chroococcidiopsis* grows within halite (sodium chloride) crystals encrusting the surface of the hyper-arid core of the Atacama Desert, Chile. It is the only organism known to thrive in this extreme environment (Wierzchos et al. 2006). It has been suggested that endolithic species of *Chroococcidiopsis* could be used to colonise Mars and to help transform its environment into one suitable for plants (Friedmann & Ocampo-Friedmann 1995).

Filamentous cyanobacteria often dominate extensive microbiotic crusts, coating soils of hot and cool deserts worldwide (Garcia-Pichel & Belnap 1996; Belnap et al. 2003). In combination with fungal, lichen, and moss populations, they can markedly affect the physicochemical properties of the soil surface, by increasing resistance to erosional forces, for instance. Where translucent stones rest on the soil, cyanobacteria form thick blue-green growths on their undersurfaces (Büdel & Wessels 1991; Nienow & Friedmann 1993). Dark paint-like streaks of algae ('tintenstriche') grow on intermittently irrigated rock walls in mountainous regions of Antarctic (Nienow & Friedmann 1993), temperate (Golubic 1967), and tropical (Büdel 1999) zones.

As well as their widespread occurrence as free-living populations, cyanobacteria are found in mutualistic symbioses with other organisms (Rai et al. 2002). In all these, it is the nitrogen-fixing capacity of usually heterocytous spe-

Diversity by environment

Taxon	Marine + brackish	Fresh-water	Terrestrial
Cyanobacteria	40	352+16	15

Summary of New Zealand cyanobacterial diversity

Taxon	Described species + subspecies/ varieties/ forms	Known undescr./ undetermined species	Estimated unknown species	Adventive species/ subspecies	Endemic species + subspecies/ varieties/ forms	Endemic genera
Cyanobacteria	372+16	30	200+	1	19+1	1

cies that benefits their partner. In lichens, the cyanobacterium can be the only algal partner or can occur in specialised structures called cephalodia when green algae are the phycobiont in the remainder of the lichen. Extracellular associations occur in certain liverworts, hornworts, and the aquatic fern *Azolla*, in which the cyanobacteria live in mucilage-filled cavities, and in about a third of cycad species, where they grow in intercellular spaces within the cortex of specialised coralloid roots. Intracellular associations occur with species of one genus of flowering plants, *Gunnera*; the cyanobacteria grow in cells in the rhizomes at the base of leaf stems. Freshwater species of the diatom genera *Epithemia* and *Rhopalodia* contain unicellular cyanobacterial endosymbionts capable of nitrogen fixation, which aids growth in their typically nitrogen-poor habitats (Prechtl et al. 2004).

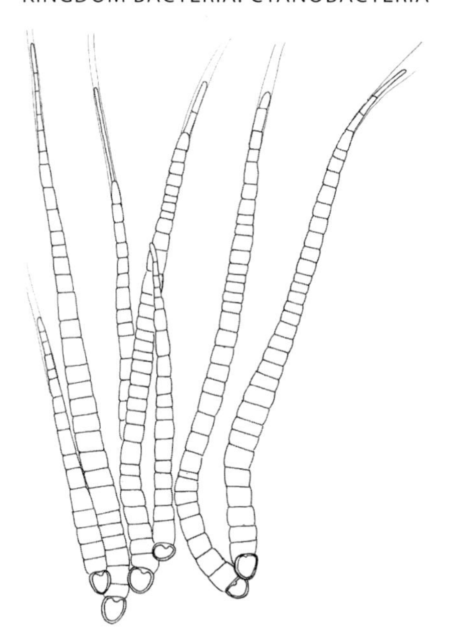

Rivularia cf. *atra* (Rivulariaceae).
Faradina Merican

Classification

There have been several different approaches to the taxonomy of cyanobacteria. The 'traditional approach', exemplified by Geitler (1932) and Desikachary (1959), relies largely on morphology, using light microscopy, of field-collected specimens. Those having the same general form and appearance are placed in the same species. Over 2000 such 'morphospecies' in about 150 genera have been recognised.

The study by Drouet (1981), which reduced these traditional species of cyanobacteria to 62 species in 24 genera, is now generally accepted to have been largely based on the false premise that most 'traditional species' are merely phenotypic variations of the same genotype (Friedmann & Borowitzka 1982; Castenholz 1992). The enthusiasm with which this apparently simple system was received by some researchers has resulted in a loss of valuable information.

Commencing with the study by Stanier et al. (1971), an alternative system has been developed that is based almost entirely on strains in culture. This 'bacteriological approach' uses morphological, physiological, biochemical, and genetic characteristics to redefine the limits of genera using strains, recognised by their culture-collection numbers, mostly without attempting to apply species epithets (Castenholz 2001). Representatives of only about 37 genera are in culture (Staley et al. 1989; Castenholz 1994).

Data from both the traditional and bacteriological approaches are being used in a reorganisation and renaming of families and genera (Anagnostidis & Komárek 1985, 1988; Komárek & Anagnostidis 1989, 1998, 2005). The 'traditional species' have been redistributed amongst approximately 215 genera and, as new data emerge, more changes will follow.

The advent of molecular phylogenetics, in which gene sequences are compared, is challenging taxonomic systems based on previous approaches (Hoffmann et al. 2005). It is now clear that many traditional earlier groupings at the generic and suprageneric levels are unnatural and do not reflect true evolutionary relationships. This is particularly true for simple unicells and unbranched, undifferentiated filamentous forms, which are scattered through phylogenetic trees based on sequences of the small subunit of ribosomal RNA (Turner 1997; Wilmotte & Herdman 2001; Teneva et al. 2005). Heterocyte-forming filamentous species are monophyletic but the traditional orders Nostocales and Stigonematales are polyphyletic (Gugger & Hoffmann 2004). A first attempt has been made to encompass all available genetic and phenotypic data within a new suprageneric classification (Hoffmann et al. 2005).

The morphological species concept has also been challenged by molecular studies (Hayes et al. 2007). There is frequently lack of correlation between genotypic and phenotypic features used to classify populations to species level. Increasing use is being made of a so-called polyphasic approach that utilises both stable phenotypic characters and molecular markers. The species concept in cyanobacteria continues to be debated (Castenholz & Norris 2005; Johansen & Casamatta 2005).

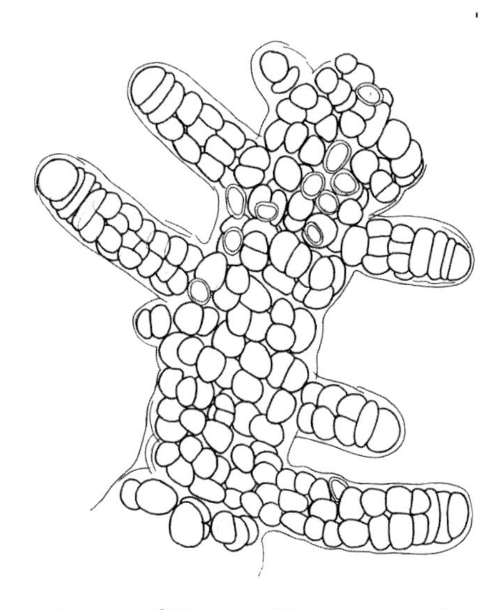

A species of *Stigonema* (Stigonemataceae).
Paul Broady

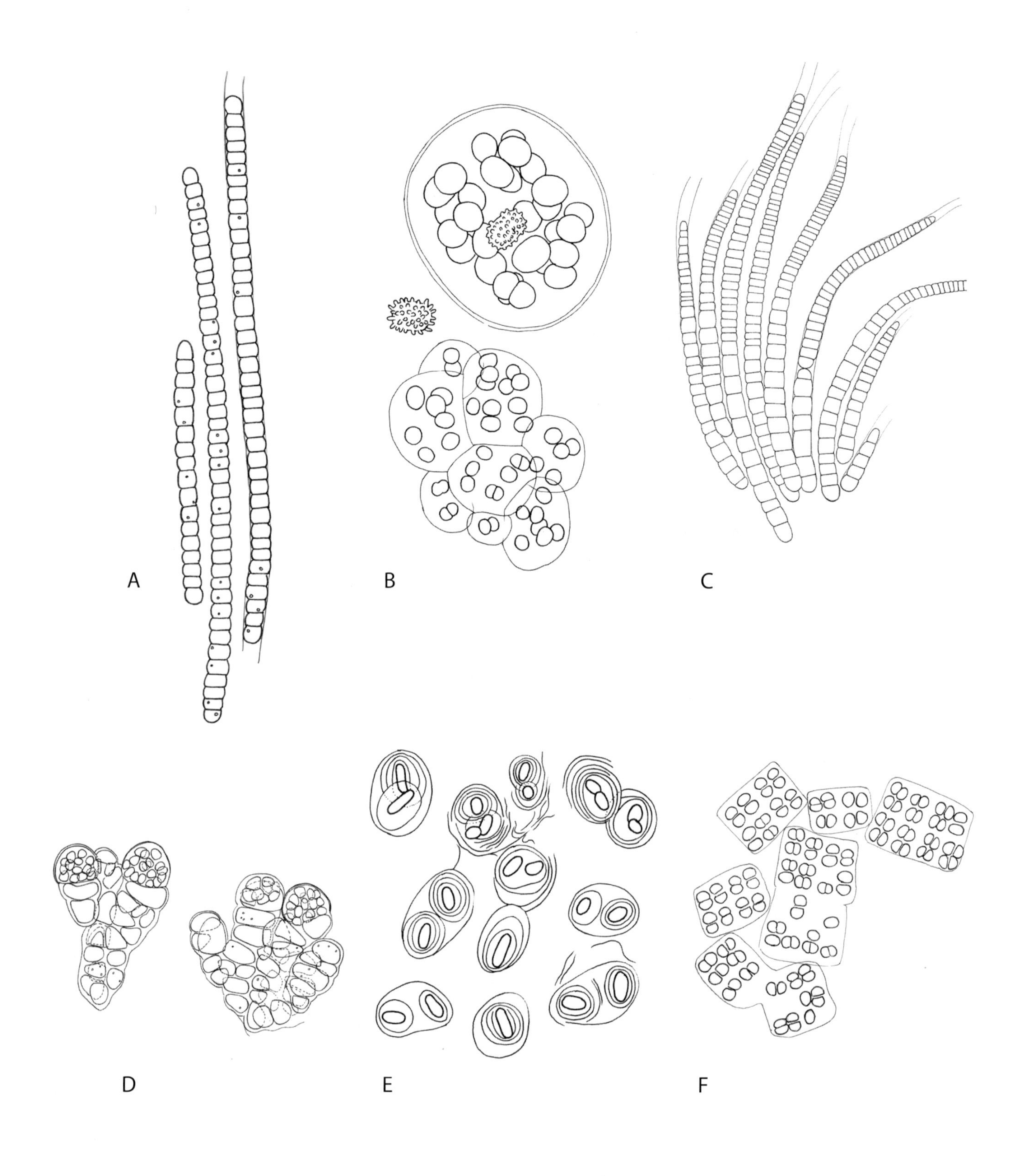

Examples of some New Zealand cyanobacteria.
A: *Leptolyngbya foveolarum* (Phormidiaceae). B: *Asterocapsa* sp. (Chroococcaceae). C: *Phormidiochaete* sp. (Phormidiaceae). D: *Radaisia* sp. (Hyellaceae). E: *Gloeothece* cf. *palea* (Synechococcaceae). F: *Merismopedia punctata* (Merismopediaceae).

A, C, D, F: Faradina Merican; B, E: Paul Broady

Diversity and special features of the New Zealand Cyanobacteria

In compiling the end-chapter checklist, the system of K. Anagnostidis and J. Komárek has been followed (Anagnostidis & Komárek 1985, 1988; Komárek & Anagnostidis 1989, 1998, 2005). This has involved changing the nomenclature of many taxa listed in primary literature of New Zealand records, as their system has established many new combinations. The few identifications that have been made using Drouet's (1981) system have been omitted, as their identity under the new system is unknown.

Most species listed here come from the checklist of New Zealand freshwater algae produced by Cassie (1984) and the Etheredge and Pridmore (1987) taxonomic guide to New Zealand freshwater planktonic cyanobacteria. Records of marine periphyton are from Chapman (1956) and Nelson and Phillips (1996). These should be consulted for literature prior to the mid-1980s. More recent records are from a small but diverse literature.

A total of 413 cyanobacterial species and subspecies, distributed amongst 87 genera, have been recorded from New Zealand freshwater and terrestrial habitats. Records of at least 44 taxa are doubtful. Some 361 species and subspecies are from freshwater and terrestrial habitats, whilst only 41 have been recorded from marine habitats. The total compares with approximately 2000 species and 215 genera known worldwide.

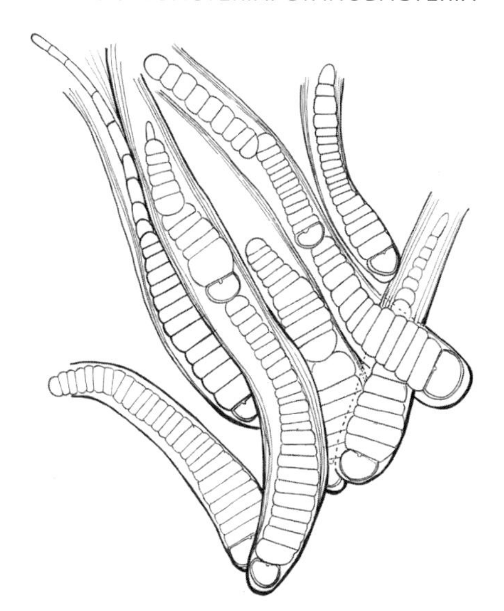

Calothrix parietina (Rivulariaceae).

Faradina Merican

Endemism

Despite a lack of distributional data, there is some evidence that certain cyanobacteria are restricted to tropical or temperate zones and even to more restricted regions (Komárek 1985; Castenholz 1996; Hoffmann 1996), although many are cosmopolitan.

The large majority of taxa recorded from New Zealand are well known elsewhere. However, new species, potentially endemic to New Zealand could have been overlooked for two reasons. First, there are few studies of New Zealand cyanobacteria. Secondly, specimens are often identified using floras and handbooks produced for temperate regions of the Northern Hemisphere. Even when New Zealand specimens differ from species described in those accounts, these differences might have been disregarded in the desire for rapid, albeit erroneous, identification. – a recognised tendency in studies that do not have a specific taxonomic focus. A New Zealand example is the initial identification of *Anabaena oscillarioides* as dominant in the phytoplankton of Lake Rotongaio and its subsequent recognition as a new variety, *Anabaena minutissima* var. *attenuata* (Pridmore & Reynolds 1989).

A small number of species are, so far, known only from New Zealand, for instance *Placoma regularis*, which is widespread in streams (Broady & Ingerfeld 1991) and *Anabaena novizelandica* from peat bog pools (Skuja 1976). The 14 species and three varieties from thermal springs considered by Nash (1938) to be new taxa have never been legitimately published and require reinvestigation.

Alien species

Only one species, *Anabaena planktonica*, is considered to have arrived in New Zealand relatively recently, in the late 1990s (Wood et al. 2005). It has spread rapidly through North Island lakes and has been implicated in toxin production and taste and odour problems in water supplies (Kouzminov et al. 2007). A sensitive molecular assay has been developed for its detection during water-quality monitoring programmes (Rueckert et al. 2007).

Because of inadequate knowledge of both the pre-human and present-day cyanobacterial flora, it is impossible to estimate how many, if any, of the other species recorded from New Zealand have been introduced by humans. It is inevitable that viable propagules of cyanobacteria will have been transported

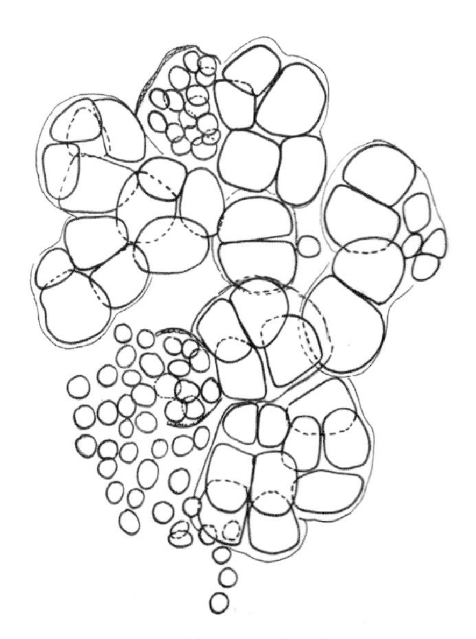

A species possibly related to *Pleurocapsa* (Hyellaceae).

Faradina Merican

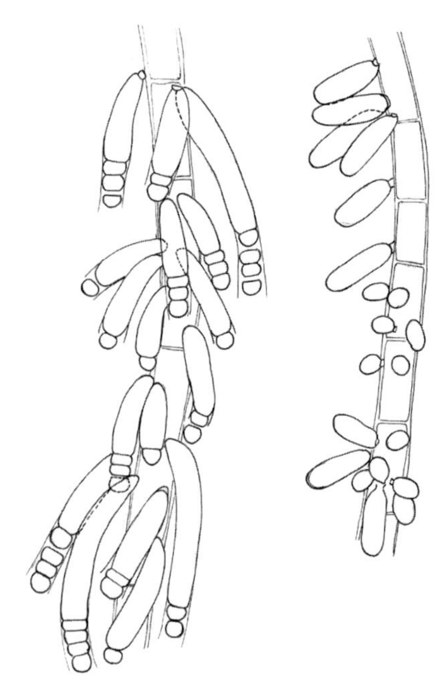

Chamaesiphon confervicolus (Chamaesiphonaceae).
Faradina Merican

to these islands ever since the first human colonisers. It is likely that propagules are now arriving in greater numbers than ever before owing to record numbers of visitors and imports of goods. They will be present on clothes and shoes and on the surfaces of luggage and commercial imports of all kinds. Even if morphologically identical to native species, it is likely that new propagules reaching New Zealand will include a proportion with different genotypes. The possibility that some of these could be a potential nuisance or even toxic cannot be discounted.

Freshwater plankton

New Zealand cyanobacteria have been studied intermittently since planktonic species were first recorded (Nordstedt 1888; Lemmermann 1899). R. Pridmore and M. Etheredge (1987) made a major survey of the taxonomy and distribution of planktonic cyanobacteria in 77 lakes but noted that many lakes have either not been sampled or have been sampled insufficiently to produce comprehensive species lists. There are 776 lakes with at least one axis greater than 0.5 km long (Lowe & Green 1987) as well as numerous smaller lakes and ponds, all of which are likely to support phytoplankton communities.

Pridmore and Etheridge's (1987) cyanobacteria survey and guide to 104 taxa greatly increased knowledge of the distribution of these organisms in New Zealand. Three species – *Microcystis aeruginosa, Anabaena flos-aquae,* and *A. circinalis* – frequently dominate planktonic communities. *Gomphosphaeria lacustris* and *Oscillatoria limnetica* are also often observed, the former being the only common species in the South Island. Twelve species that are common in temperate lakes of the Northern Hemisphere either have not been found or are less common in New Zealand. On average, more species occur in lakes in the North Island than the South Island. This apparent bias is attributable to the higher proportion of usually nutrient-poor glacial lakes in the South Island. When these are omitted, North and South Island lakes contain a similar number of taxa. Most taxa occur across the spectrum from eutrophic to oligotrophic waters, with only five species restricted to nutrient-rich waters. Many commonly reported species, however, have a strong preference for eutrophic conditions, with *G. lacustris* and *Anabaena affinis* being notable exceptions. There is a tendency for heterocytous species to occur more frequently in nitrogen-deficient lakes. Only species of *Microcystis* and *Anabaena* inhabit the humic-stained lakes typical of Westland beech-podocarp forests. However, a richer flora of 32 taxa inhabits four similarly stained Waikato lakes.

Extremely small unicellular phytoplankton 0.2–2 micrometres in diameter, called picophytoplankton, includes typical cyanobacteria and 'prochlorophytes'. They are cosmopolitan in lakes (Reynolds 1997). Although known in New Zealand, their diversity has not been investigated beyond distinguishing *Synechococcus* and *Prochlorococcus* (J. Hall, NIWA, pers. comm.). This lack of detailed investigation probably masks additional diversity, insofar as Komárek and Anagnostidis (1998) assign those taxa formerly known as *Synechococcus* to five species of *Cyanobium* and molecular studies have recognised many clades (e.g. Ernst et al. 2003).

Population dynamics of cyanobacteria in New Zealand lakes are poorly known. Summer maxima are often found during periods of greatest thermal stability of the water column (Viner & White 1987). This is the case in eutrophic Lower Karori Reservoir, Wellington which has a summer bloom of *Anabaena planktonica* (Wood et al. 2010). In some instances cyanobacteria contribute a major proportion of standing crop throughout an entire year. It has been suggested that the unpredictable oceanic climate of New Zealand often prevents similar cyanobacterial occurrence from one year to the next (Pridmore & Etheredge 1987).

These authors reported conspicuous scums and blooms in 33 New Zealand lakes, typically in summer and autumn. The large majority of these lakes are

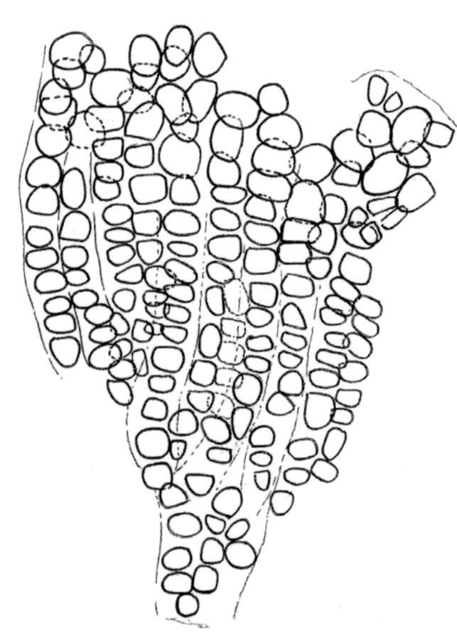

Cyanodermatium sp. 1 (see checklist) (Entophysalidaceae).
Faradina Merican

eutrophic and shallow. At least eight taxa cause blooms, of which several are potentially toxic. Most occurrences are caused by *Anabaena flos-aquae*, *A. spiroides*, and *Microcystis aeruginosa*. Toxic growths of *Nodularia spumigena* are restricted to the brackish South Island lakes Ellesmere (Waihora) and Forsyth (Wairewa). Other recent examples of blooms, usually of unidentified *Anabaena* species, have been noted by various regional councils. Oxidation ponds can also develop massive growths of cyanobacteria, such as a bloom of *Oscillatoria subbrevis* at the Mangere Sewage Treatment Plant, which provoked concern for the wildlife frequenting the ponds (Cassie Cooper 1996).

In the absence of long-term monitoring programmes it is difficult to know how phytoplankton communities might have responded to environmental change. One alternative approach is to study the remains of phytoplankton that have accumulated in sediments. A sediment core from Lake Okaro was used to assess the cyanobacterial species composition of phytoplankton communities back to 1886 (Wood et al. 2008a). A molecular assay and germination of akinetes showed no dramatic change in species composition despite increasing eutrophication and abundance of cyanobacteria since the 1950s. A potentially toxic species of *Anabaena* was detected throughout this period although it has never been identified in Lake Okaro phytoplankton communities.

Toxicity has been reported from a few freshwater ponds (Flint 1966; Cassie 1979) and from Lakes Ellesmere, Forsyth, Johnson, Rotorua, and Roundabout and Butcher's Dam in the South Island (Mulligan 1985; Carmichael et al. 1988; Christoffersen & Burns 2000). In the few New Zealand lakes tested, Christoffersen and Burns (2000) considered toxin levels to be comparable to those elsewhere. They showed the toxins to be capable of killing *Daphnia carinata*, a zooplankton species. Nevertheless, they considered that the relatively low algal biomass in most New Zealand lakes is unlikely to produce major toxic effects in humans and livestock provided that the duration and magnitude of cyanobacterial growth does not increase.

There are increasing reports of death of stock and dogs, however, and in a few specific lakes, such as Lake Forsyth, health warnings owing to cyanobacterial blooms are a regular occurrence during warm, dry summers. Also, increase in abundance and frequency of occurrence of potentially toxic species has been related to eutrophication. For instance, in Lake Rotoiti in central North Island, an absence of cyanobacteria in phytoplankton communities in the mid-1950s changed to late summer dominance of *A. flos-aquae* and *M. aeruginosa* by the mid-1980s (Vincent et al. 1984, 1986). Subsequently, there has been a shift back to diatom dominance, indicating an improvement in water quality (Burns et al. 1997).

The hepatotoxin cylindrospermopsin was first reported from a recreational lake near Wellington in 1999 (Stirling & Quilliam 2001). Identification of the causative organism as *Cylindrospermopsis* sp. was not confirmed. However, *C. raciborskii* and its typical toxins have been positively identified from Lake Waahi (Wood & Stirling 2003) and other Waikato lakes (Ryan et al. 2003). It seems that it may be well-established in New Zealand and a potential threat to water supplies.

The potential danger of misidentification of possibly toxic species is exemplified by a report of a 'porridge-like blue-green soup from a pig farm effluent near the Waikato River'. The person reporting this considered it to be edible *Spirulina* but it was subsequently identified as *Microcystis aeruginosa* by Cassie-Cooper (1996).

A major study of bloom-forming and toxic cyanobacteria in New Zealand has been on-going for the last 10 years. First reports (Wood 2005; Wood et al. 2005, 2006b) noted blooms in 18 more lakes than previously recorded, provided ten new New Zealand records, all from the North Island, of which eight were bloom-forming species, and detected the production of microcystins, nodularins, anatoxin-a and saxitoxins. Anatoxin-a had not been found previously in the Southern Hemisphere (Wood et al. 2007a; Selwood et al. 2007). In Lakes Rotoiti

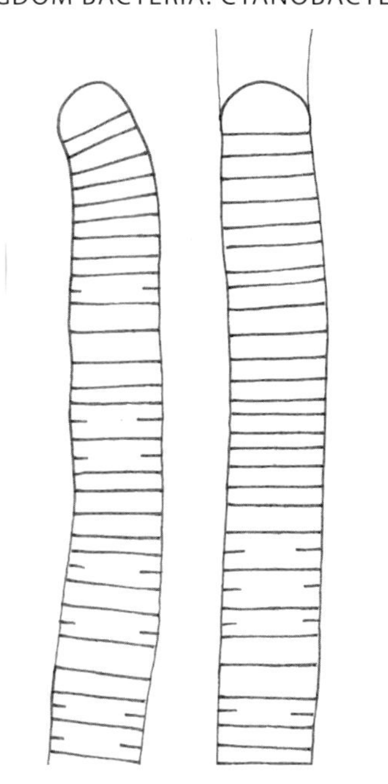

Oscillatoria cf. *simplicissima* (Oscillatoriaceae).

Faradina Merican

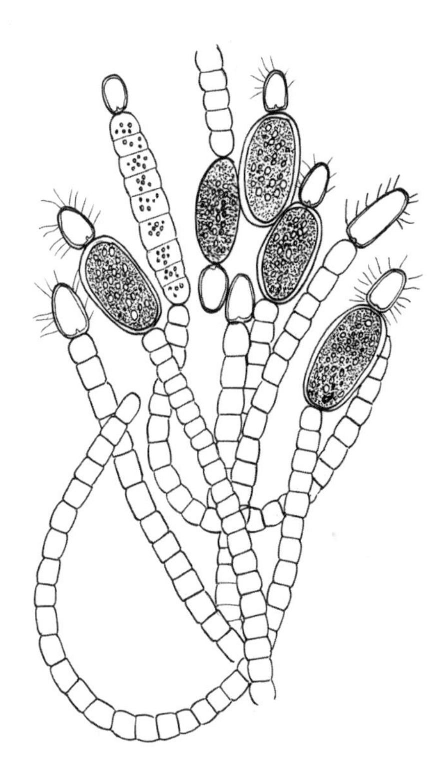

A species of *Cylindrospermum* (Nostocaceae).

Faradina Merican

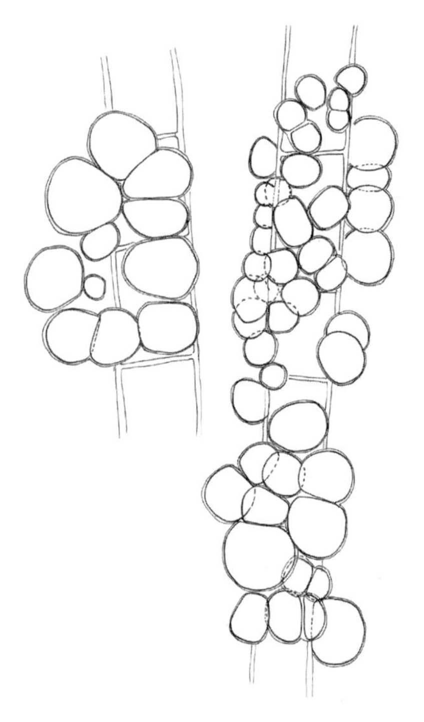

A species of *Xenococcus* (Xenococcaceae).
Faradina Merican

and Rotoehu, microcystins produced by *Microcystis* spp. have been shown to concentrate in liver and muscle tissues of rainbow trout to levels above the daily tolerable intake limit for human consumption recommended by the World Health Organization (Wood et al. 2006a). Species of *Anabaena* that formed blooms in Waikato River were suspected of forming saxitoxins (Kouzminov et al. 2007) but *Scytonema* cf. *crispum* is the first species confirmed to produce these. It occurs as loose aggregates of entangled filaments in two shallow South Island lakes (Smith et al. 2011). This is the first report of saxitoxin production by a species of *Scytonema*.

A range of experimental studies has been undertaken on New Zealand cyanobacteria. Those prior to the mid-1980s were reviewed by Viner (1987) and mostly investigated nitrogen fixation and conditions that lead to cyanobacteria becoming dominant. In 1987, a major international workshop, the 'Forum on Cyanobacterial Dominance', resulted in the publication of several reviews addressing different aspects of this phenomenon (Vincent 1987) as well as a series of detailed accounts of experiments and observations on cyanobacteria in two contrasting eutrophic North Island lakes (Vincent 1989). The latter elucidated reasons for the dominance of *Microcystis aeruginosa* in Lake Okaro and of *Anabaena minutissima* var. *attenuata* in Lake Rotongaio. These were related to contrasting nutrient characteristics in the lakes, the photosynthetic adaptiveness of the cyanobacteria, and their ability to modify their surroundings in ways which enhanced their growth and survival.

Freshwater periphyton

Cyanobacteria are a common component of periphyton (the small organisms and detritus attached to submerged surfaces) in standing and flowing waters of New Zealand. Although the ecology and physiology of periphyton in streams and rivers has been the focus of excellent and detailed studies (e.g. Biggs et al. 1999; Dodds et al. 1999), the taxonomy of groups other than the diatoms is not well known. Identifications of cyanobacteria are usually to genus and include only prominent taxa (e.g. Biggs et al. 1998; Jaarsma et al. 1998). However, undoubtedly overlooked are many other species, including those present in low abundance and with restricted distribution, e.g. epiphytic *Xenococcus* species as well as more widespread taxa that have probably been misidentified, e.g. epilithic *Heteroleibleinia* species. The potential for new discoveries is indicated by an on-going study of cyanobacteria in the Kaituna River and its tributaries on Banks Peninsula (Merican & Broady unpubl.). This study has identified 44 morphospecies as components of visible mats, crusts, and gelatinous colonies; 18 of these are new records for New Zealand.

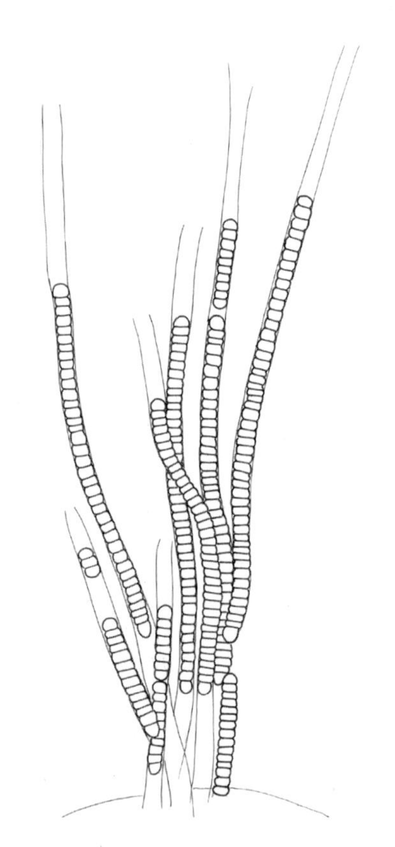

Heteroleibleinia cf. *fontana* (Homoeotrichaceae).
Faradina Merican

Cyanobacteria can produce nuisance proliferations (Biggs & Price 1987; Biggs 1990). For instance, organic pollution can stimulate development of cyanobacterial mats, as seen below the outfall of a Taranaki lactose factory where dark green mats of *Oscillatoria tenuis* developed (Cassie-Cooper 1996). In Southland, cyanobacterial mats of *Oscillatoria*-like species are abundant at 24% of the 68 river sites that are monitored annually (Hamill 1999). During the warm, dry summers of 1998/99 and 1999/2000, when river flows were low, several dogs were killed by toxic mats along the Mataura River, and a similar incident occurred on the Waikanae River, lower North Island (Hamill 2001). Subsequently, toxic mats have caused dog deaths on the Hutt River, lower North Island (Wood et al. 2007b) and on the Waitaki River, South Island (Wood et al. 2010). The former event yielded the first record of homoanatoxin-a in New Zealand and the latter the first record of a microcystin-producing species in mats. A general survey of benthic mats in rivers and lakes throughout New Zealand found anatoxins in seven of 31 environmental samples (Heath et al. 2010). A culture collection of strains of both periphytic and planktonic cyanobacteria including toxic strains is housed at the Cawthron Institute, Nelson (Wood et al. 2008b). This is valuable as a reference collection for identifications and for development of molecular-based tools for toxin detection.

Study of periphyton in New Zealand lakes lags far behind both that of phytoplankton and also periphyton of flowing waters (Hawes & Smith 1994). Although cyanobacteria dominate some communities, knowledge of their diversity is scant and, as in flowing waters, identifications have usually been at generic level (Hawes & Smith 1993, 1994). As aquatic macrophytes are abundant in many New Zealand lakes and occur to a maximum depth, in clear waters, of about 35 metres (Schwarz et al. 2000), periphyton communities of microscopic algae will occur to at least this depth. Macrophytes support epiphytic microalgae including cyanobacteria (Hawes & Schwarz 1996), sometimes in abundance, and below the maximum depth of macrophyte growth periphyton is likely to occur on rock and sediment surfaces. Small ponds are also a potential source of diverse and unusual periphytic cyanobacteria, as indicated by the presence in South Island moraine ponds of unidentified species of two genera, *Stichosiphon* and *Brachytrichia*?, unrecorded elsewhere in New Zealand (Burns et al. 1984).

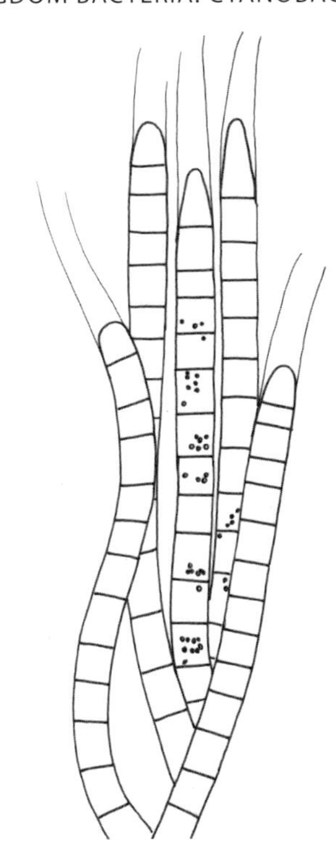

Phormidium inundatum (Phormidiaceae).
Faradina Merican

Cyanobacteria in thermal habitats

New Zealand's numerous and extensive sites of geothermal activity support diverse communities of thermophilic cyanobacteria forming colourful mats both in hot springs and on warm ground around steam vents. These have attracted experts from overseas but there have been few investigations by local researchers other than the extensive taxonomic survey by Cassie (1989) that was inclusive of cooler spring waters and their eukaryotic algal inhabitants. She provided a useful summary of previous work, notable amongst which are the detailed ecological survey by Brock and Brock (1970, 1971) and some physiological studies (Castenholz 1976; Castenholz & Utkilen 1984). Absent from the New Zealand thermophilic flora are high-temperature strains of *Synechococcus lividus* that, in North America, occupy waters up to 73° C (Vincent & Forsyth 1987; Castenholz 1996). In New Zealand, as in most other regions, the maximum temperature of occurrence of cyanobacteria is about 64° C.

Molecular analysis has been used to identify endolithic cyanobacteria in geothermal siliceous rocks (Gaylarde et al. 2006). As well as detecting two species new to the New Zealand flora, there were also many DNA sequences that had no good match with those in public databases.

Laminated sedimentary rock formations known as stromatolites form in hot-spring and geyser systems in the Taupo Volcanic Zone (Jones et al. 2002). Those known as coniform stromatolites resemble similar fossil structures from the Proterozoic Eon. In neutral to alkaline thermal waters the microbiota of these is dominated by species of *Phormidium*, which comprise more than 90% of the biota. The microbiota becomes silicified in opaline silica. Subfossil specimens less than 20,000 years old have also been found.

Godleya alpina (Microchaetaceae), a new genus and species from Arthur's Pass National Park.
Phil Novis

Terrestrial habitats

The diversity of algae, including cyanobacteria, in New Zealand terrestrial habitats is largely unknown. Worldwide, several hundred species of cyanobacteria have been recorded from soils but in New Zealand this habitat has hardly been studied except for the pioneering work of Flint (1968). Observations by the author and research students have revealed a significant flora in soils from low-land to high-alpine regions (Everett 1998; Novis 2001), with five new alpine species and a new genus described from Arthur's Pass National Park alone (Novis & Visnovsky 2011).

Similarly, in other terrestrial habitats, the diversity of algae has barely been investigated. Prominent growths of cyanobacteria include those encrusting rock surfaces in moist mountainous regions, along the supralittoral of rocky shorelines, and in dim-light zones inside the entrances of limestone caverns. Where artificial illumination is provided deep within caves for the comfort of tourists,

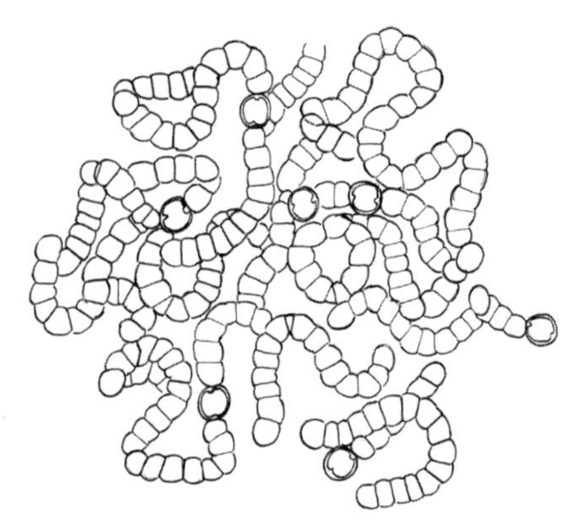

A species of *Nostoc* (Nostocaceae).
Faradina Merican

this can allow growth of algae, which tends to be regarded as unsightly (Cassie-Cooper 1996).

The adaptability of cyanobacteria is emphasised by the occurrence of colonial *Gloeocapsa sanguinea* in an unusual habitat – amongst the wool fibres of dead sheep. Observation of this phenomenon constituted one of the first records of algae in New Zealand (Berggren 1874; Cassie-Cooper 1996).

Cyanobacteria in mutualistic symbioses

Five of the 40 species of the angiosperm genus *Gunnera* are endemic to New Zealand (Webb et al. 1988). These small herbs form a mutualistic symbiosis with *Nostoc* sp. Research on *G. dentata* has examined the process of infection by *Nostoc* (Silvester & McNamara 1976) and nitrogenase and photosynthetic activities of the endosymbiont (Silvester & Smith 1969; Silvester 1976). When associated with *G. dentata*, the cyanobacterium has up to ten times more nitrogenase activity and appears unable to perform photosynthetic carbon fixation. The products of nitrogen-fixation are rapidly translocated to stems and leaves of the host. A species of *Gunnera* introduced from Chile, *G. tinctoria*, is often highly visible owing to its massive leaves on banks of streams and ponds in the city environment. This species is excellent for demonstrating the presence of *Nostoc* in its thick rhizomes.

Two species of the floating aquatic fern *Azolla* are widespread in New Zealand in ponds, ditches, and small lakes. The well-known high nitrogen-fixation capacity of its association with *Trichormus azollae* (syn. *Anabaena azollae*) was confirmed in a comparison with free-living planktonic *Anabaena* in Lake Ngahewa (Kellar & Goldman 1979).

New Zealand has a richly diverse and well-developed lichen flora of more than 1610 species (Galloway, this volume) of which about 8% might be expected to have cyanobacterial photobionts (Büdel 1992). The taxonomy of lichen photobionts is poorly known in general (Tschermak-Woess 1988) and difficult or impossible without study of isolates in culture (Büdel 1992). There is broad scope for their examination in New Zealand lichens.

Amongst New Zealand bryophytes, all four hornwort genera are widespread and contain *Nostoc* in cavities in their thalli (Allison & Child 1975; D. Glenny, Landcare Research, Lincoln pers. comm.). In contrast, thallose liverworts are poorly represented and the occurrence of *Nostoc* in species of *Riccia* and *Riccardia* is equivocal although recorded elsewhere (Grilli Caiola 1992).

Marine habitats

In contrast to freshwater habitats, the open ocean harbours a restricted number of planktonic forms. In New Zealand, eight planktonic species (seven genera) have been recorded in marine habitats, with one species, *Nodularia spumigena*, known to be toxic in brackish waters (Cronberg et al. 2003). The commonest and most abundant species recorded in New Zealand's open seas are the filamentous *Trichodesmium erythraeum* and coccoid picoplanktonic *Synechococcus* sp. and *Prochlorococcus* sp. Elsewhere, numerous ecotypes of the latter two genera have been recorded, each with a particular distribution pattern (e.g. Zwirglmaier et al. 2008). Which ecotypes occur in New Zealand waters has yet to be investigated.

Cyanobacteria attached to surfaces on the seashore and in the sublittoral coastal zone are poorly known and there have been no studies since that of Chapman (1956). Detailed investigations elsewhere, such as that in South Africa (Silva & Pienaar 2000), which yielded 66 taxa from 34 genera, illustrate the potential for finding a much greater range of species in New Zealand, where only 32 species in 21 genera have been recorded.

Geitlerinema ionicum (Pseudanabaenaceae).
Faradina Merican

Fossils

Subfossil stromatolites in thermal waters have been noted above. There is a single published record of more ancient examples in the Miocene (Lindqvist

1994). Stromatolites and oncoids from paleolake Manuherikia in Central Otago represent a short-lived microbially induced carbonate precipitation. This formed under alkaline conditions in an ephemeral sub-basin of the paleolake. Within these structures are well-preserved calcified filament moulds formed by perhaps three species of cyanobacteria, one of which has been tentatively identified as *Scytonema* sp.

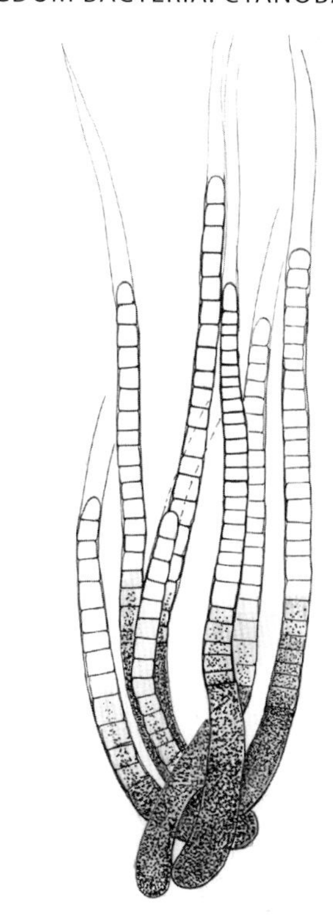

Homoeothrix gracilis (Homoeotrichaceae).
Faradina Merican

Economic importance of cyanobacteria

Negative impacts of cyanobacterial growth

As mentioned above, cyanobacteria are well known as dominant phytoplankton in lakes rich in nutrients (Watson et al. 1997). Nuisance and toxic blooms are increasingly common worldwide, stimulated by eutrophication caused by agricultural activity and urbanisation in catchments. Climate change is a potent catalyst for the further expansion of these blooms (Paerl & Huisman 2008). Periphytic cyanobacteria can also be highly toxic. Monitoring and reducing their occurrence can be costly (Chorus & Bartram 1999).

A high incidence of toxicity of at least 50–75% has been reported for phytoplankton blooms, with frequent livestock intoxications and, less often, human illness and even death (Codd et al. 1995, 1999; Moestrup 1996). There are more than 65 potentially toxigenic cyanobacterial species in 32 genera in marine, brackish, and fresh waters (Cronberg & Annadotter 2006). Most toxicoses involve liver-damaging hepatotoxins. These can also act as tumour promoters in mammals and there is concern that long-term exposure to low concentrations might be damaging. Other toxins are potent neurotoxins.

Cyanobacteria are often major components of biotic communities that can cause deterioration and discolouration of stone surfaces of buildings and monuments (Ortega-Calvo et al. 1995), for instance the Parthenon in Greece (Anagnostidis et al. 1983). However, it can be difficult to distinguish bio-deterioration from the effects of industrial and urban pollution.

Human use of cyanobacteria

Some cyanobacteria have long been used as food or a food supplement (Gantar & Svirčev 2008), for instance colonies of *Nostoc* in China (Potts 2000). More recently, *Arthrospira* (under the name *Spirulina*) has been grown commercially and marketed as a 'health food', rich in vitamins and mineral salts. Of greater importance for alleviation of undernourishment is the development of village-scale production systems in regions where spirulina's high content of readily available protein could be an important supplement to diets (Vonshak 1997). Nutrients for its growth can be supplied from anaerobic waste digesters as part of an integrated system that also produces biogas for cooking and heating.

More use could be made of nitrogen-fixing cyanobacteria in agriculture (Metting et al. 1988; Kerby & Rowell 1992). This would also reduce dependency on fossil fuels used in production and distribution of fertiliser. The natural fertility of tropical rice paddies, at some locations for several thousand years, has often been attributed to nitrogen-fixing cyanobacteria, both free-living and in association with the aquatic fern *Azolla*. Research into improving their use has investigated both stimulation of the growth of naturally occurring species and selection of strains for inoculation into paddies. These strains have good competitive ability and high nitrogen-fixation rates and continuously secrete fixed nitrogen as ammonium (Boussiba 1997). Another approach is genetic modification of strains to resist herbicides and to continue nitrogen fixation even after application of nitrogenous fertiliser (Vaishampayan et al. 1996).

The commercial use of cyanobacteria depends largely on reducing the cost of production by improving efficiency of solar-energy conversion to biomass in outdoor culture systems (Richmond 1992). If this can be achieved, then there is a range of potentially useful products (Apt & Behrens 1999).

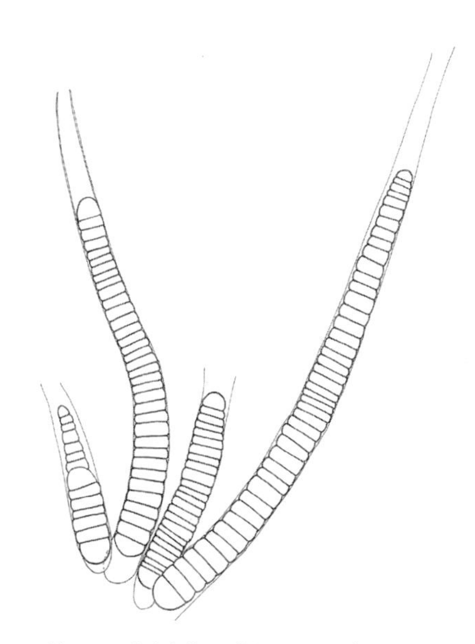

Homoeothrix juliana (Homoeotrichaceae).
Faradina Merican

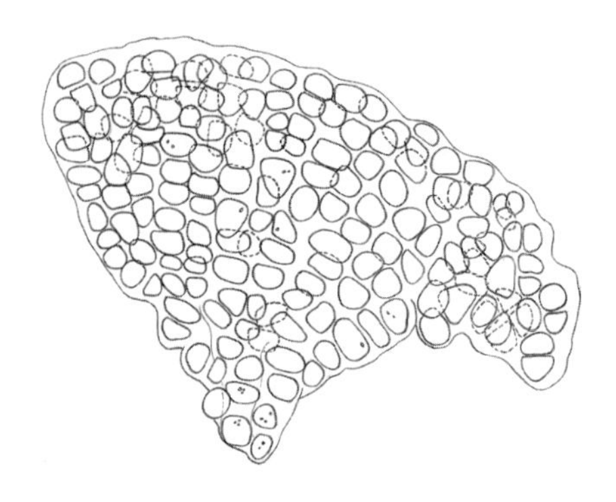

A species of *Hydrococcus* (Hydrococcaceae).
Faradina Merican

Phycobilin pigments can be used as colourants in the food and cosmetic industry and their fluorescent properties make them valuable as labels attached to highly specific molecular probes that can identify particular cell types and proteins. Commercial production of other research tools includes enzymes used in molecular genetics and ^{14}C-labelled amino acids (Kerby & Rowell 1992). Unique chemicals with antibiotic and anticancer activity have been described (Gantar & Svirčev 2008) including one that inactivates HIV. The high viscosity of polysaccharides produced by colonial cyanobacteria suggests that they might be suitable as thickening agents in foods (Huang et al. 1998). Cyanobacteria synthesise and accumulate poly-3-hydroxybutyric acid (PHB) which is already in use for the manufacture of biodegradable, non-toxic plastic (Ardelean 1997). Genetic modification of cyanobacteria is readily performed but has not been used commercially. The potential has been shown by the introduction of a gene coding for production of an insecticide into *Synechococcus*. The growth of mosquito larvae is inhibited when they feed on this transformed cyanobacterium.

Cyanobacteria have a role in the vital search for alternatives to fossil fuels. They can be used for photocatalytic conversion of water to hydrogen and oxygen. However, productivity is presently far too low for this to be economically viable. Research is currently investigating how to improve the efficiency of the process (Sakurai et al. 2008; Dickson et al. 2009). Fuel cells incorporating cyanobacteria have also been investigated (He et al. 2009).

In the field of environmental protection, cyanobacteria are potential biosensors of certain herbicides and could be developed for absorbing mineral nutrients, heavy metals, and organics from polluted waters (Kerby & Rowell 1992; Ardelean 1997).

Gaps in knowledge and scope for future research

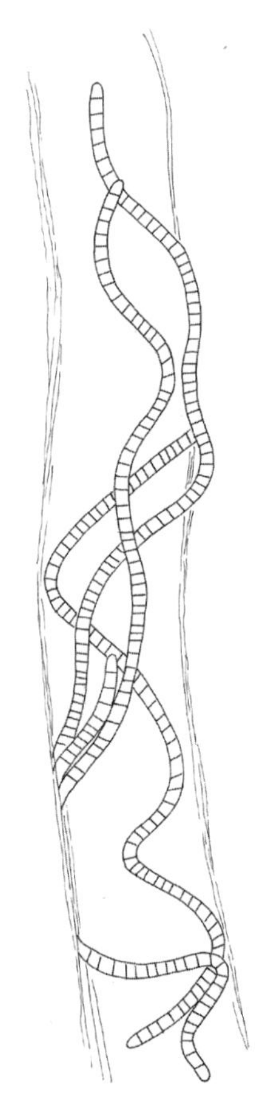

Leibleinia epiphytica (Pseudanabaenaceae).
Faradina Merican

The inventory of New Zealand cyanobacteria is far from complete. Using the current revison of the traditional morphological approach to cyanobacterial systematics (e.g. Komárek & Anagnostidis 1998, 2005), less than 16% of known species from 30% of known genera have been recorded. The similar number of morphospecies, about 360, recorded in the British Isles is considered to be several hundred species short of a full inventory (Whitton 2002) and it is likely that a similar estimate would be true for New Zealand. Except for the few genera that seem to be restricted to tropical regions, it is likely that species of many of the remainder will be present in New Zealand as there is a great diversity of habitats in contrasting climates covering a wide latitudinal and altitudinal spread. Also, a significant proportion of the species determinations made in New Zealand are unreliable or unable to be verified because of inadequate documentation and lack of preserved specimens (Etheredge & Pridmore 1987).

Although freshwater phytoplanktonic cyanobacteria have been the most thoroughly collected, many lakes have not been appropriately sampled. Also, cyanobacteria in periphyton communities of standing and flowing waters and in terrestrial habitats warrant widespread collection.

The identification of morphospecies based on careful and detailed microscopic examination of specimens from the field is essential for completion of a reliable checklist. Publication of thorough descriptions and illustrations of New Zealand specimens, and of new records and poorly known taxa, is the only way confidence in identifications can be established. At present, the literature contains illustrations, from New Zealand specimens, of less than 40% of species. Also, the preservation of well-curated specimens in herbaria would be invaluable for future researchers. Detailed characterisation and description of habitats is equally important, as considerable evidence suggests that distinct morphologically definable populations are restricted within narrow limits to particular ecological conditions (Komárek 1985; Komárek & Anagnostidis 1998). However, it is sobering to consider that the many new genetic entities being discovered merely by 16S rDNA comparisons suggests that estimates of the diversity of cyanobacteria based on morphospecies recognition has been grossly underestimated (Castenholz & Norris 2005).

The shortcomings of the system of classification are admitted (Komárek & Anagnostidis 1998), but its improvement is ongoing, with species concepts being tested using cultures and molecular techniques. However, as only a small proportion of cyanobacteria have been cultured, there is wide scope for attempting to culture others. While the extraction of nucleic acids from cyanobacteria is often difficult, the introduction of new methods (Tillett & Neilan 2000) appropriate for cultures and environmental samples will stimulate research in this area, e.g. Gaylarde et al. (2006). As relatively few 'species' have received critical study in both the field and the laboratory (Whitton 1999), research in New Zealand could contribute to the global integration of classification systems based on field specimens, cultures, and molecular data. It has been emphasised (Hoffmann et al. 2005) that only when this has been achieved will we have a taxonomy of practical use that also reflects the evolutionary relationships of taxa.

A complete inventory of New Zealand species based on a reliable taxonomic system is a vital basis for a wide range of research in ecology, physiology, and biochemistry. If knowledge of the identity of the organisms being studied is inadequate, then the value of many studies is greatly reduced. In the applied sciences, inadequate knowledge of cyanobacterial diversity is compromising 1) benefits arising from recognition of taxa with potential toxicity, and 2) the ability to investigate the full range of New Zealand cyanobacteria for useful bioactive or other chemical products.

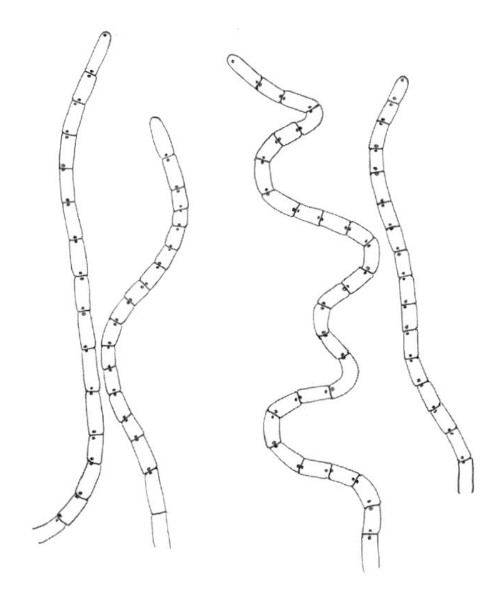

Leptolyngbya cf. *bijugata* (Phormidiaceae).
Faradina Merican

Acknowledgements

The authors thank Shirley Hayward (Lincoln University), Stephen Moore (Landcare Research, Auckland), Keith Hillier (Southland Regional Council), Dr Susanna Wood (Cawthron Institute) and Drs Hoe Chang, Ian Hawes, Julie Hall, and Wendy Nelson (NIWA) for valuable information and helpful discussion. Comments made by Dr E. A. Flint (Landcare Research, Lincoln) greatly improved the manuscript.

Authors

Dr Paul A. Broady School of Biological Sciences, University of Canterbury, Private Bag 4800, Christchurch, New Zealand [paul.broady@canterbury.ac.nz]

Faradina Merican School of Biological Sciences, University of Canterbury, Private Bag 4800, Christchurch, New Zealand [fmm50@uclive.ac.nz]

References

ALLISON, K. W.; CHILD, J. 1975: *The Liverworts of New Zealand*. University of Otago Press, Dunedin. 300 p.

ANAGNOSTIDIS, K.; ECONOMOU-AMILLI, A.; ROUSSOMOUSTAKAKI, M. 1983: Epilithic and chasmolithic microflora (Cyanophyta, Bacillariophyta) from marbles of the Parthenon (Acropolis-Athens, Greece). *Nova Hedwigia 38*: 227–287.

ANAGNOSTIDIS, K.; KOMÁREK, J. 1985: Modern approach to the classification system of cyanophytes. 1 – Introduction. *Algological Studies 38/39*: 291–302.

ANAGNOSTIDIS, K.; KOMÁREK, J. 1988: Modern approach to the classification system of cyanophytes. 5 – Stigonematales. *Algological Studies 59*: 1–73.

APT, K. E.; BEHRENS, P. W. 1999: Commercial developments in microalgal biotechnology. *Journal of Phycology 35*: 215–226.

BELL, R. A. 1993: Cryptoendolithic algae of hot semiarid lands and deserts. *Journal of Phycology 29*: 133–139.

BELNAP, J.; BUDEL, B.; LANGE, O. L. 2003: Biological soil crusts: characteristics and distribution. Pp. 3–30 *in*: Belnap, J.; Lange, O. (eds), *Biological Soil Crusts: Structure, Management and Function*. Springer Verlag, Berlin. xviii + 503 p.

BERGGREN, S. 1874: On the occurrence of *Haematococcus sanguineus* on the wool of a dead sheep. *Transactions and Proceedings of the New Zealand Institute 7*: 369–370.

BIGGS, B. J. F. 1990: Periphyton communities and their environments in New Zealand rivers. *New Zealand Journal of Marine and Freshwater Research 24*: 367–386.

BIGGS, B. J. F.; KILROY, C.; LOWE, R. L. 1998: Periphyton development in three valley segments of a New Zealand grassland river: test of a habitat matrix conceptual model within a catchment. *Archiv für Hydrobiologie 143*: 147–177.

BIGGS, B. J. F.; PRICE, G. M. 1987: A survey of filamentous algal proliferations in New Zealand rivers. *New Zealand Journal of Marine and Freshwater Research 21*: 175–191.

BIGGS, B. J. F.; SMITH, R. A.; DUNCAN, M. J. 1999: Velocity and sediment disturbance of periphyton in headwater streams: biomass and metabolism. *Journal of the North American Benthological Society 18*: 222–241.

BOUSSIBA, S. 1997: Ammonia assimilation and its biotechnological aspects in cyanobacteria. Pp. 35–72 *in*: Rai, A. K. (ed.), *Cyanobacterial Nitrogen Metabolism and Environmental Biotechnology*. Narosa Publishing House, New Delhi. xi + 299 p.

BROADY, P. A.; INGERFELD, M. 1991: *Placoma regulare* sp. nov. (Entophysalidaceae, Cyanobacteria) from New Zealand streams. *Phycologia 30*: 547–555.

BROCK, T. D.; BROCK, M. L. 1970: The algae of Waimangu Cauldron (New Zealand): distribution in relation to pH. *Journal of Phycology 6*: 371–375.

BROCK, T. D.; BROCK, M. L. 1971: Microbiological studies of thermal habitats of the central volcanic region, North Island, New Zealand. *New Zealand Journal of Marine and Freshwater Research 5*: 233–258.

BÜDEL, B. 1992: Taxonomy of lichenized prokaryotic blue-green algae. Pp. 301–324 *in*: Reisser, W. (ed.), *Algae and Symbiosis: Plants, animals, fungi, viruses. Interactions explored.* Biopress Ltd, Bristol. xii + 746 p.

BÜDEL, B. 1999: Ecology and diversity of rock-inhabiting cyanobacteria in tropical regions. *European Journal of Phycology 34*: 361–370.

BÜDEL, B.; WESSELS, D. C. J. 1991: Rock-inhabiting blue-green algae/cyanobacteria from hot arid regions. *Algological Studies 64*: 385–398.

BURNS, N. M.; BUTLER, M. I.; CUTTANCE, P. M. 1984: Invertebrates, macroalgae, and chemical features in morainic ponds near Lakes Tekapo and Ohau, including new distribution records of Crustacea. *New Zealand Journal of Marine and Freshwater Research 18*: 197–210.

BURNS, N. M.; DEELY, J.; HALL, J.; SAFI, K. 1997: Comparing past and present trophic states of seven Central Volcanic Plateau lakes, New Zealand. *New Zealand Journal of Marine and Freshwater Research 31*: 71–87.

CARMICHAEL, W. W.; ESCHEDOR, J. T.; PATTERSON, G. M. L.; MOORE, R. E. 1988: Toxicity and partial structure of a hepatotoxic peptide produced by the cyanobacterium *Nodularia spumigena* Mertens emend. L575 from New Zealand. *Applied and Environmental Microbiology 54*: 2257–2263.

CASSIE, V. 1979: Algae in relation to water quality. *Water and Soil Technical Publication 18*: 21–30

CASSIE, V. 1984: Revised checklist of the freshwater algae of New Zealand (excluding diatoms and charophytes). Part 1 – Cyanophyta, Rhodophyta and Chlorophyta. *Water and Soil Technical Publication 25*: xii + 116 p. + xliv.

CASSIE, V. 1989: A taxonomic guide to thermally associated algae (excluding diatoms) in New Zealand. *Bibliotheca Phycologica 78*: 161–255, 6 pls.

CASSIE, V.; COOPER, R. C. 1989: Algae of New Zealand thermal areas. *Bibliotheca Phycologica 78*: 1–159.

CASSIE-COOPER, V. 1996: *Microalgae. Microscopic marvels.* Riverside Books, Hamilton. 164 p.

CASTENHOLZ, R. W. 1976: The effect of sulphide on the blue-green algae of hot springs. 1. New Zealand and Iceland. *Journal of Phycology 12*: 5–68.

CASTENHOLZ, R. W. 1992: Species usage, concept, and evolution in the Cyanobacteria (blue-green algae). *Journal of Phycology 28*: 737–745.

CASTENHOLZ, R. W. 1994: Group 11. Oxygenic phototrophic bacteria. Pp. 377–425 *in*: Holt, J. G.; Krieg, N. R.; Sneath, P. H. A.; Staley, J. T.; Williams, S. T. (eds), *Bergey's Manual of Determinative Bacteriology*. Williams & Wilkins, Baltimore. xviii + 787 p.

CASTENHOLZ, R. W. 1996: Endemism and biodiversity of thermophilic cyanobacteria. *Beiheft für Nova Hedwigia 112*: 33–47.

CASTENHOLZ, R. W. 2001: General characteristics of the Cyanobacteria. Pp. 474–487 *in*: Boone, D. R.; Castenholz, R. W. (eds), *Bergey's Manual of Bacteriology. Vol. 1. The Archaea and the deeply branching and phototrophic Bacteria*. Springer, New York. xxi +721 p.

CASTENHOLZ, R. W.; NORRIS, T. B. 2005: Revisionary concepts of species in the Cyanobacteria and their applications. *Algological Studies 117*: 53–69.

CASTENHOLZ, R. W.; UTKILEN, H. C. 1984: Physiology of sulfide tolerance in a thermophilic *Oscillatoria*. *Archives of Microbiology 138*: 306–309.

CHAPMAN, V. J. 1956: The marine algae of New Zealand. Part I. Myxophyceae and Chlorophyceae. *Journal of the Linnean Society of London, Botany 55*: 333–501.

CHORUS, I.; BARTRAM, J. (Eds) 1999: *Toxic Cyanobacteria in Water. A guide to their public health consequences, monitoring and management.* E. & F. N. Spon, London. xv +416 p.

CHRISTOFFERSEN, K.; BURNS, C. W. 2000: Toxic cyanobacteria in New Zealand lakes and toxicity to indigenous zooplankton. *Verhandlungen Internationale Vereinigung für Theoretische und Angewandte Limnologie 27*: 3222–3225.

CODD, G. A.; EDWARDS, C.; BEATTIE, K. A.; LAWTON, L. A.; CAMPBELL, D. L.; BELL, S. G. 1995: Toxins from cyanobacteria (blue-green algae). Pp. 1–17 *in*: Wiessner, W.; Schnepf, E.; Starr, R. C. (eds), *Algae, Environment and Human Affairs*. Biopress Ltd., Bristol. xiii + 258 p.

CODD, G. A.; BELL, S. G.; KUNIMITSU, K.; WARD, C. J.; BEATTIE, K. A.; METCALF, J. S. 1999: Cyanobacterial toxins, exposure routes and human health. *European Journal of Phycology 34*: 405–415.

COLEMAN, M. L.; SULLIVAN, M. B.; MARTINY, A. C.; STEGLICH, C.; BARRY, K.; DELONG, E. F.; CHISHOLM, S. W. 2006: Genomic islands and the ecology and evolution of *Prochlorococcus*. *Science 311*: 1768–1770.

CRONBERG, G.; ANNADOTTER, H. 2006: *Manual on Aquatic Cyanobacteria. A photo guide and a synopsis of their toxicology.* International Society for the Study of Harmful Algae, Copenhagen and UNESCO, Paris. 106 p.

CRONBERG, G.; CARPENTER, E. J.; CARMICHAEL, W. W. 2003: Taxonomy of harmful cyanobacteria. Pp. 523–562 *in*: Hallegraeff, G. M.; Anderson, D. M.; Cembella, A. D. (eds), *Manual on Harmful Marine Microalgae*. UNESCO Publishing, Paris. 793 p.

DESIKACHARY, T. V. 1959: *Cyanophyta*. Indian Council of Agricultural Research, New Delhi. x + 686 p.

DICKSON, D. J.; PAGE, C. J.; ELY, R. L. 2009: Photobiological hydrogen production from *Synechocystis* sp. PCC 6803 encapsulated in silica sol-gel. *International Journal of Hydrogen Energy 34*: 204–215.

DILLON, J. G.; CASTENHOLZ, R. W. 1999: Scytonemin, a cyanobacterial sheath pigment, protects against UVC radiation: implications for early photosynthetic life. *Journal of Phycology 35*: 673–681.

DODDS, W. K.; BIGGS, B. J. F.; LOWE, R. L. 1999: Photosynthesis-irradiance patterns in benthic microalgae: variations as a function of assemblage thickness and community structure. *Journal of Phycology 35*: 42–53.

DROUET, F. 1981: Summary of the classification of blue-green algae. *Beiheft für Nova Hedwigia 66*: 133–209.

ERNST, A.; BECKER, S.; WOLLENZIEN, U. I. A.; POSTIUS, C. 2003: Ecosystem-dependent adaptive radiations of picocyanobacteria inferred from 16S rDNA and ITS-1 sequence analysis. *Microbiology 69*: 2430–2443.

ETHEREDGE, M. K.; PRIDMORE, R. D. 1987: The freshwater planktonic blue-greens (Cyanophyta/Cyanobacteria) of New Zealand. A taxonomic guide. *Water and Soil Miscellaneous Publication 111*: 1–122.

EVERETT, A. L. 1998: The ecology and taxonomy of soil algae in Cass Basin, Canterbury, New Zealand. M.Sc. thesis, University of Canterbury, Christchurch. ix + 110 p.

FAY, P.; VAN BAALEN, C. (Eds) 1987: *The Cyanobacteria: A comprehensive review.* Elsevier Science Publishers, Amsterdam. xvi + 453 p.

FLINT, E. A. 1966: Toxic algae in some New Zealand freshwater ponds. *New Zealand Veterinary Journal 14*: 181–185.

FLINT, E. A. 1968: Algae on the surface of some New Zealand soils. *New Zealand Soil Bureau Bulletin 26*: 183–190.

FLORES, E.; HERRERO, A. 1994: Assimilatory nitrogen metabolism and its regulation. Pp. 487–517 *in*: Bryant, D. A. (ed.), *The Molecular Biology of Cyanobacteria*. Kluwer Academic Publishers, Dordrecht. xiv + 881 p., 10 pls.

FLORES, E.; MURO-PASTOR, A. M.; MEEKS, J. C. 2008: Gene transfer to cyanobacteria in the laboratory and in nature. Pp. 45–57 *in*: Herrero, A.; Flores, E. (eds) *The Cyanobacteria: Molecular biology, genomics and evolution*. Caister Academic Press, Norfolk, U.K. xi + 484 p.

FRIEDMANN, E. I.; BOROWITZKA, L. J. 1982: The Symposium on Taxonomic Concepts in Blue-green Algae: towards a compromise with the Bacteriological Code? *Taxon 31*: 673–683.

FRIEDMANN, E. I.; OCAMPO-FRIEDMANN, R. 1995: A primitive cyanobacterium as pioneer microorganism for terraforming Mars. *Advances in Space Research 15*: 243–246.

FUHRMAN, J. 2003: Genome sequences from the sea. *Nature 424*: 1001–1002.

GANTAR, M.; SVIRČEV, Z. 2008: Microalgae and cyanobacteria: food for thought. *Journal of Phycology 44*: 260–268.

GARCIA-PICHEL, F.; BELNAP, J. 1996: Microenvironments and microscale productivity of cyanobacterial desert crusts. *Journal of Phycology 32*: 774–782.

GARCIA-PICHEL, F.; BELNAP, J.; NEUER, S.; SCHANZ, F. 2003: Estimates of global cyanobacterial biomass and its distribution. *Algological Studies 109*: 213–227.

GAYLARDE, P. M.; JUNGBLUT, A.-D.; GAYLARDE, C. C.; NEILAN, B. A 2006: Endolithic phototrophs from an active geothermal region in New Zealand. *Geomicrobiology Journal 23*: 579–587.

GEITLER, L. 1932: Cyanophyceae. Vol. 14 *in*: Rabenhorst, L. (ed.), *Kryptogamen-flora von Deutschland, Österreich und der Schweiz*. Akademisches Verlagsgesellschaft, Leipzig. vi + 1196 p.

GOLUBIC, S. 1967: Algenvegetation der Felsen. Eine ökologische Algenstudie im dinarischen Karstgebiet. Vol. 23 *in*: Elster, H. J.; Ohle, W. (eds) *Die Binnengewässer*. E. Schweizerbart'sche Verlagsbuchhandlung, Stuttgart. 183 p.

GRAHAM, L. E.; GRAHAM, J. M.; WILCOX, L. W. 2009: *Algae*. Benjamin Cummings, San Francisco. xviii + 696 p.

GRILLI CAIOLA, M. 1992: Cyanobacteria in symbioses with bryophytes and tracheophytes. Pp. 231–254 *in*: Reisser, W. (ed.), *Algae and Symbioses*. Biopress Ltd, Bristol. xii + 746 p.

GUGGER, M.; HOFFMANN, L. 2004: Polyphyly of true branching cyanobacteria (Stigonematales).

International Journal of Systematic and Evolutionary Microbiology 52: 1–14.
HAMILL, K. 1999: Dead dogs and toxic algae. Internal report, File 218/2/3. Southland Regional Council, Invercargill.
HAMILL, K. 2001: Toxicity in benthic freshwater cyanobacteria (blue-green algae): first observations in New Zealand. *New Zealand Journal of Marine and Freshwater Research 35*: 1057–1059.
HAWES, I.; SCHWARZ, A.-M. 1996: Epiphytes from a deep-water characean meadow in an oligotrophic New Zealand lake: species composition, biomass and photosynthesis. *New Zealand Journal of Marine and Freshwater Research 36*: 297–313.
HAWES, I.; SMITH, R. 1993: The effect of localised enrichment on the shallow, epilithic periphyton of oligotrophic Lake Taupo, New Zealand. *New Zealand Journal of Marine and Freshwater Research 27*: 365–372.
HAWES, I.; SMITH, R. 1994: Seasonal dynamics of epilithic periphyton in oligotrophic Lake Taupo, New Zealand. *New Zealand Journal of Marine and Freshwater Research 28*: 1–12.
HAYES, P.K.; EL SEMARY, N. A.; SÁNCHEZ-BARACALDO, P. 2007: The taxonomy of cyanobacteria: molecular insights into a difficult problem. Pp. 93-101 *in*: Brodie, J.; Lewis, J. (eds), *Unravelling the Algae: the past, present and future of algal systematics.* CRC, Boca Raton. 376 p.
HE, Z.; JINJUN, K.; MANSFELD, F.; ANGENENT, L. T. 2009: Self-sustained phototrophic microbial fuel cells based on the synergistic cooperation between photosynthetic microorganisms and heterotrophic bacteria. *Environmental Science and Technology 43*: 1648–1654.
HEATH, M. W.; WOOD, S. A.; RYAN, K. G. 2010: Polyphasic assessment of fresh-water benthic mat-forming cyanobacteria isolated from New Zealand. *FEMS Microbiology Ecology 73*: 95–109.
HERRERO, A.; FLORES, E. (Eds) 2008: *The Cyanobacteria: Molecular biology, genomics and evolution.* Caister Academic Press, Norfolk, U.K. xi + 484 p.
HOFFMANN, L. 1996: Geographic distribution of freshwater blue-green algae. *Hydrobiologia 336*: 33–40.
HOFFMANN, L.; KOMÁREK, J.; KAŠTOVSKÝ, J. 2005: System of cyanoprokaryotes (cyanobacteria) – state in 2004. *Algological Studies 117*: 95–115.
HUANG, Z.; YONGDING, L.; PAULSEN, B. S.; KLAVENESS, D. 1998: Studies on polysaccharides from three edible species of *Nostoc* (Cyanobacteria) with different colony morphologies: comparison of monosaccharide compositions and viscosities of polysaccharides from field colonies and suspension cultures. *Journal of Phycology 34*: 962–968.
JAARSMA, N. G.; DE BOER, S. M.; TOWNSEND, C. R.; THOMPSON, R. M.; EDWARDS, E. D. 1998: Characterising food-webs in two New Zealand streams. *New Zealand Journal of Marine and Freshwater Research 32*: 271–286.
JAVOR, B. 1989: *Hypersaline Environments: Microbiology and biogeochemistry.* Springer Verlag, New York. viii + 328 p.
JOHANSEN, J. R.; CASAMATTA, D. A. 2005: Recognizing cyanobacterial diversity through adoption of a new species paradigm. *Algological Studies 117*: 71–93.
JONES, B.; RENAUT, R. W.; ROSEN, M. R.; ANSDELL, K. M. 2002: Coniform stromatolites from geothermal systems, North Island, New Zealand. *Palaios 17*: 84–103.
KEELING, P. J. 2004: Diversity and evolutionary history of plastids and their hosts. *American Journal of Botany 91*: 1481–1493.
KELLAR, P. E.; GOLDMAN, C. R. 1979: A comparative study of nitrogen fixation by the *Anabaena-Azolla* symbiosis and free-living populations of *Anabaena* spp. in Lake Ngahewa, New Zealand. *Oecologia 43*: 269–281.
KERBY, N. W.; ROWELL, P. 1992: Potential and commercial applications for photosynthetic prokaryotes. Pp. 23–265 *in*: Mann, N. H.; Carr, N. G. (eds), *Photosynthetic Prokaryotes.* Plenum Press, New York. xiv + 275 p.
KOMÁREK, J. 1985: Do all cyanophytes have a cosmopolitan distribution? Survey of the freshwater cyanophyte flora of Cuba. *Algological Studies 38/39*: 359–386.
KOMÁREK, J.; ANAGNOSTIDIS, K. 1989: Modern approach to the classification system of cyanophytes. 4 – Nostocales. *Algological Studies 56*: 247–345.
KOMÁREK, J.; ANAGNOSTIDIS, K. 1998: Cyanoprokaryota 1. Teil Chroococcales. Vol. 19/1 *in*: Ettl, H.; Gärtner, G.; Heynig, H.; Mollenhauer, D. (eds), *Süsswasserflora von Mitteleuropa.* Gustav Fischer Verlag, Jena. x + 548 p.
KOMÁREK, J.; ANAGNOSTIDIS, K. 2005: Cyanoprokaryota 2. Teil Oscillatoriales. Vol. 19/2 *in*: Büdel, B.; Krienitz, L.; Gärtner, G.; Schagerl, M. (Eds), *Süsswasserflora von Mitteleuropa.* Elsevier GmbH, München. 759 p.
KNOLL, A. H. 2008: Cyanobacteria and Earth history. Pp. 1-19 *in*: Herrero, A.; Flores, E. (eds), *The Cyanobacteria: Molecular biology, genomics and evolution.* Caister Academic Press, Norfolk, U.K. xi + 484 p.
KOUZMINOV, A.; RUCK, J.; WOOD, S.A. 2007: New Zealand risk management approach for toxic cyanobacteria in drinking water. *Australian and New Zealand Journal of Public Health 31*: 275–281.
LEE, R.E. 2008: *Phycology.* Cambridge University Press, Cambridge. x + 547 p.
LEMMERMANN, E. 1899: Ergebnisse einer Reise nach dem Pacific (H. Schauinsland 1896–97). Plankton Algen. *Abhandlungen herausgegeben vom naturwissenschaftlichen Verein zu Bremen 16*: 313–398.
LINDQVIST, J. 1994: Lacustrine stromatolites and oncoids: Manuherikia Group (Miocene), New Zealand. Pp. 227–254 *in*: Bertrand-Sarfati, J.; Monty, C. (eds), *Phanerozoic Stromatolites II.* Kluwer Academic Publishers, Dordrecht.
LOWE, D. J.; GREEN, J. D. 1987: Origins and development of the lakes. Pp. 1–64 *in*: Viner, A.B. (ed.), *Inland Waters of New Zealand.* Department of Scientific and Industrial Research, Wellington. ix + 494 p.
METTING, B.; RAYBURN, W. R.; REYNAUD, P. A. 1988: Algae and agriculture. Pp. 335–370 *in*: Lembi, C. A.; Waaland, J. R. (eds), *Algae and Human Affairs.* Cambridge University Press, Cambridge. viii + 590 p.
MIYASHITA, H.; IKEMOTO, H.; KURANO, N.; MIYACHI, S. 2003: *Acaryochloris marina* gen. et sp. nov. (Cyanobacteria), an oxygenic photosynthetic prokaryote containing chl *d* as a major pigment. *Journal of Phycology 39*: 1247–1253.
MOESTRUP, Ø. (Ed.) 1996: Proceedings of the First International Congress on Toxic Cyanobacteria (Blue-green Algae) and the First Maj and Tor Nessling Foundation Symposium on 'Recent developments in cyanobacterial research'. *Phycologia 35(6, Suppl.)*: 1353–1496.
MULLIGAN, P. E. 1985: Isolation and characterisation of the *Nodularia spumigena* toxin. MSc thesis. University of Canterbury, Christchurch.
NASH, A. 1938: The cyanophyceae of the thermal regions of Yellowstone National Park, U.S.A., and of Rotorua and Whakarewarewa, New Zealand, with some ecological data. PhD thesis, University of Minnesota. 214 p.
NELSON, W. A.; PHILLIPS, L. 1996: The Lindauer legacy – current names for the Algae Nova-Zelandicae Exsiccatae. *New Zealand Journal of Botany 34*: 553–582.
NIENOW, J. A.; FREIDMANN, E. I. 1993: Terrestrial lithophytic (rock) communities. Pp. 343–412 *in*: Friedmann, E. I. (ed.), *Antarctic Microbiology.* Wiley-Liss, New York. x + 634 p.
NORDSTEDT, O. 1888: Freshwater algae collected by Dr S. Berggren in New Zealand and Australia. *Kungliga Svenska Vetenskapsakademiens Handlingar 22*: 1–98.
NOVIS, P. M. 2001: Ecology and taxonomy of alpine algae, Mt Philistine, Arthur's Pass National Park, New Zealand. PhD thesis, University of Canterbury.
NOVIS, P. M.; VISNOVSKY, G. 2011: Novel alpine algae from New Zealand: Cyanobacteria. *Phytotaxa 22*: 1–24.
ORTEGA-CALVO, J. J.; ARIÑO, X.; HERNANDEZ-MARINE, M.; SAIZ-JIMENEZ, C. 1995: Factors affecting the weathering and colonization of monuments by phototrophic microorganisms. *The Science of the Total Environment 167*: 329–341.
PAERL, H. W.; HUISMAN, J. 2008: Blooms like it hot. *Science 320*: 57–58.
PENTECOST, A.; WHITTON, B. A. 2000: Limestones. Pp. 257–279 *in*: Whitton, B. A.; Potts, M. (eds), *The Ecology of Cyanobacteria—Their diversity in time and space.* Kluwer Academic Publishers, Dordrecht. xviii + 669 p., 32 pls.
PLAZINSKI, J. 1997: Nitrogen metabolism of the symbiotic systems of cyanobacteria. Pp. 95–130 *in*: Rai, A. K. (ed), *Cyanobacterial Nitrogen Metabolism and Environmental Biotechnology.* Narosa Publishing House, New Delhi. xi + 299 p.
POST, A. F. 1999: The prochlorophytes – an algal enigma. Pp. 115–125 *in*: Seckbach, J. (ed.), *Enigmatic Microorganisms and Life in Extreme Environments.* Kluwer Academic Publishers, Dordrecht. xxi + 687 p.
POTTS, M. 1999: Mechanisms of desiccation tolerance in cyanobacteria. *European Journal of Phycology 34*: 319–328.
POTTS, M. 2000: Nostoc. Pp. 465–504 *in*: Whitton, B. A.; Potts, M. (eds), *The Ecology of Cyanobacteria – Their diversity in time and space.* Kluwer Academic Publishers, Dordrecht. xviii, + 669 p., 32 pls.
PRECHTL, J.; KNEIP, C.; LOCKHART, P.; WENDEROTH, K.; MAIER, U. G. 2004: Intracellular spheroid bodies of *Rhopalodia gibba* have nitrogen-fixing apparatus of cyanobacterial origin. *Molecular Biology and Evolution 21*: 1477–1481.
PRIDMORE, R. D.; ETHEREDGE, M. K. 1987: Planktonic cyanobacteria in New Zealand inland waters: distribution and population dynamics. *New Zealand Journal of Marine and Freshwater Research 21*: 491–502.
PRIDMORE, R. D.; REYNOLDS, C. S. 1989: *Anabaena minutissima* var. *attenuata* var. nov. (Cyanophyta/ Cyanobacteria) from New Zealand. *Archiv für Hydrobioogie beiheft für Ergebnisse der Limnologie 32*: 27–33.
RAI, A. M.; BERGMAN, B.; RASMUSSEN, U. (Eds) 2002: *Cyanobacteria in Symbiosis.* Kluwer Academic Publishers, Dordrecht. x + 355 p.
REYNOLDS, C. S. 1997: *Vegetation Processes in the Pelagic: A Model for Ecosystem Theory.* Ecology

Institute, Oldendorf/Luhe. xxvii + 371.

RICHMOND, A. 1992: Mass culture of cyanobacteria. Pp. 181–210 *in*: Mann, N. H.; Carr, N. G. (eds), *Photosynthetic Prokaryotes*. Plenum Press, New York. xiv + 275 p.

RUECKERT, A.; WOOD, S. A.; CARY, S. C. 2007: Development and field assessment of a quantitative PCR for the detection and enumeration of the noxious bloom-former *Anabaena planktonica*. *Limnology and Oceanography: Methods 5*: 474–483.

RYAN, E. F.; HAMILTON, D. P.; BARNES, G. 2003: Recent occurrence of *Cylindrospermopsis raciborskii* in Waikato lakes of New Zealand. *New Zealand Journal of Marine and Freshwater Research 37*: 829–836.

SAKURAI, H.; MASUKAWA, H.; ZHANG, X.; IKEDA, H.; INOUE, K. 2008: Improvement of nitrogenase-based photobiological hydrogen production by cyanobacteria by gene engineering – hydrogenases and homocitrate synthase. Pp. 1277–1280 *in*: Allen, J. F.; Gantt, E.; Golbeck, J. H.; Osmond, B. (eds), *Photosynthesis – Energy from the sun*. Springer, Heidelburg. xxxviii + 1640 p.

SAUNDERS, V. A. 1992: Genetics of the photosynthetic prokaryotes. Pp. 121–152 *in*: Mann, N. H.; Carr, N. G. (eds), *Photosynthetic Prokaryotes*. Plenum Press, New York. xiv + 275 p.

SCHWARZ, A.-M.; HOWARD-WILLIAMS, C.; CLAYTON, J. 2000: Analysis of relationships between maximum depth limits of aquatic plants and underwater light in 63 New Zealand lakes. *New Zealand Journal of Marine and Freshwater Research 34*: 157–174.

SELWOOD, A. I.; HOLLAND, P. T.; WOOD, S. A.; SMITH, K. F.; McNABB, P. S. 2007: Production of anatoxin-a and a novel biosynthetic precursor by the cyanobacterium *Aphanizomenon issatschenkoi*. *Environmental Science and Technology 41*: 506–510.

SILVA, S. M. F.; PIENAAR, R. N. 2000: Benthic marine Cyanophyceae from Kwa-Zulu Natal, South Africa. *Bibliotheca Phycologia 107*: vii + 456 p.

SILVESTER, W. B. 1976: Endophyte adaptation in *Gunnera-Nostoc* symbiosis. Pp. 521–538 *in* Nutman, P. S. (ed.) *Symbiotic nitrogen fixation in plants*. Cambridge University Press, Cambridge. xxviii + 584 p.

SILVESTER, W. B.; McNAMARA, P. J. 1976: The infection process and ultrastructure of the *Gunnera-Nostoc* symbiosis. *New Phytologist 77*: 135–141.

SILVESTER, W. B.; SMITH, D. R. 1969: Nitrogen fixation by *Gunnera-Nostoc* symbiosis. *Nature 224*: 1321.

SKUJA, H. 1976: Zur Kenntnis der Algen Neuseeländischer Torfmoore. *Nova acta Regiae Societatis Scientiarum Upsaliensis, ser. 5, C, 2*: 1–125.

SMITH, F. M. J.; WOOD, S. A.; VAN GINKEL, R.; BROADY, P. A.; GAW, S. 2011: First report of saxitoxin production by a species of the freshwater benthic cyanobacterium, *Scytonema* Agardh. *Toxicon* 57: 566–573.

STAL, L. J. 2000: Cyanobacterial mats and stromatolites. Pp. 61–120 *in*: Whitton, B. A.; Potts, M. (eds), *The Ecology of Cyanobacteria – Their diversity in time and space*. Kluwer Academic Publishers, Dordrecht. xviii + 669 p., 32 pls.

STAL, L. J.; ZEHR, J. P. 2008: Cyanobacterial nitrogen fixation in the ocean: diversity, regulation, and ecology. Pp. 423–446 *in*: Herrero, A.; Flores, E. (eds), *The Cyanobacteria: Molecular biology, genomics and evolution*. Caister Academic Press, Norfolk, U.K. xi + 484 p.

STALEY, J. T.; BRYANT, M. P.; PFENNIG, N.; HOLT, J. G. 1989: *Bergey's Manual of Systematic Bacteriology. Volume 3*. Williams & Wilkins, Baltimore. xxviii + 714 p.

STANIER, R.Y.; KUNISAWA, R.; MANDEL, M.; COHEN-BAZIRE, G. 1971: Purification and properties of unicellular blue-green algae (order Chroococcales). *Bacteriological Review 35*: 171–205.

STIRLING, D. J.; QUILLIAM, M. A. 2001: First report of the cyanobacterial toxin cylindrospermopsin in New Zealand. *Toxicon 39*: 1219–1222.

TATON, A.; GRUBISIC, S.; BRAMBILLA, E.; DE WIT, R.; WILMOTTE, A. 2003: Cyanobacterial diversity in natural and artificial microbial mats of Lake Fryxell (McMurdo Dry Valleys, Antarctica): a morphological and molecular approach. *Applied and Environmental Microbiology 69*: 5157–5169.

TENEVA, I.; DZHAMBAZOV, B.; MLADENOV, R.; SCHIRMER, K. 2005: Molecular and phylogenetic characterization of *Phormidium* species (Cyanoprokaryota) using the cpcB-IGS-cpcA locus. *Journal of Phycology 41*: 188–194.

THIEL, T. 1994: Genetic analysis of cyanobacteria. Pp. 581–611 *in*: Bryant, D.A. (ed.), *The molecular Biology of Cyanobacteria*. Kluwer Academic Publishers, Dordrecht. xiv + 881 p., 10 pls.

TILLETT, D.; NEILAN, B. A. 2000: Xanthogenate nucleic acid isolation from cultured and environmental cyanobacteria. *Journal of Phycology 36*: 251–258.

TSCHERMAK-WOESS, E. 1988: The algal partner. Pp. 39–92 *in*: Galun, M. (ed.) *CRC Handbook of Lichenology. Volume 1*. CRC Press, Boca Raton. 290 p.

TURNER, S. 1997: Molecular systematics of oxygenic photosynthetic bacteria. *Plant Systematics and Evolution (Suppl.) 11*: 13–52.

VAISHAMPAYAN, A.; DEY, T.; AWASTHI, A. K. 1996: Genetic improvement of nitrogen-fixing cyanobacteria in response to modern rice agriculture. Pp. 395–420 *in*: Chaudhary, B. R.; Agrawal, S. B. (eds) *Cytology, Genetics and Molecular Biology of Algae*. SPB Academic Publishing, Amsterdam. viii + 439 p.

VINCENT, W. F. (Ed.) 1987: Dominance of bloom-forming cyanobacteria (blue-green algae). *New Zealand Journal of Marine and Freshwater Research 21*: 361–542.

VINCENT, W. F. (Ed.) 1989: Cyanobacterial growth and dominance in two eutrophic lakes. *Archiv für Hydrobiologie, Ergebnisse der Limnologie 32*: i-v, 1–254.

VINCENT, W. F.; FORSYTH, D. J. 1987: Geothermally influenced waters. Pp. 349–77 *in*: Viner, A. B. (ed.), *Inland Waters of New Zealand*. DSIR, Wellington. ix + 494 p.

VINCENT, W. F.; GIBBS, M. M.; DRYDEN, S. J. 1984: Accelerated eutrophication in a New Zealand lake: Lake Rotoiti, central North Island. *New Zealand Journal of Marine and Freshwater Research 18*: 431–440.

VINCENT, W. F.; SPIGEL, R. H.; GIBBS, M. M.; PAYNE, G. W.; DRYDEN, S. J.; MAY, L. M.; WOODS, P.; PICKMERE, S.; DAVIES, J.; SHAKESPEARE, B. 1986: The impact of the Ohau Channel outflow from Lake Rotorua on Lake Rotoiti. *Taupo Research Laboratory File Report 92*: 1–46.

VINER, A. B. 1987: Cyanobacteria in New Zealand inland waters: experimental studies. *New Zealand Journal of Marine and Freshwater Research 21*: 503–507.

VINER, A. B.; WHITE, E. 1987: Phytoplankton growth. Pp. 191–223 *in*: Viner, A. B. (ed.) *Inland Waters of New Zealand*. DSIR, Wellington. ix + 494 p.

VONSHAK, A. (Ed.) 1997: *Spirulina platensis (Arthrospira): Physiology, cell-biology and biotechnology*. Taylor & Francis, London. xvii + 233 p.

WATANABE, Y.; MARTINI, J. E. J.; OHMOTO, H. 2000: Geochemical evidence for terrestrial ecosystems 2.6 billion years ago. *Nature 408*: 574–578.

WATSON, S. B.; McCAULEY, E.; DOWNING, J. A. 1997: Patterns in phytoplankton taxonomic composition across temperate lakes of differing nutrient status. *Limnology and Oceanography 42*: 487–495.

WEBB, C. J.; SYKES, W. R.; GARNOCK-JONES, P. J. 1988: *Flora of New Zealand, Volume IV. Naturalised pteridophytes, gymnosperms, dicotyledons*. DSIR Botany Division, Christchurch. lxviii + 1365 p.

WHITTON, B. A. 1999: Book review. Cyanoprokaryota 1. Teil Chroococcales. J. Komárek and K. Anagnostidis. *Phycologia 38*: 544.

WHITTON, B. A. 2002: Phylum Cyanophyta (Cyanobacteria). Pp. 25–122 *in*: John, D. M.; Whitton, B. A.; Brook, A. J. (eds), *The Freshwater Algal Flora of the British Isles*. Cambridge University Press, Cambridge. xii + 702 p.

WHITTON, B. A.; POTTS, M. 2000: *The Ecology of Cyanobacteria – Their diversity in time and space*. Kluwer Academic Publishers, Dordrecht. xviii, + 669 p., 32 pls.

WIERZCHOS, J.; ASCASO, C.; McKAY, C. P. 2006: Endolithic cyanobacteria in halite rocks from the hyperarid core of the Atacama Desert. *Astrobiology* 6: 415–422.

WILMOTTE, A.; HERDMAN, M. 2001: Phylogenetic relationships among the Cyanobacteria based on 16S rRNA sequences. Pp. 487–493 *in*: Boone, D. R.; Castenholz, R. W. (eds), *Bergey's Manual of Bacteriology. Vol. 1. The Archaea and the Deeply Branching and Phototrophic Bacteria*. Springer, New York. xxi +721 p.

WOOD, S. A. 2005: Bloom-forming and toxic Cyanobacteria in New Zealand. PhD thesis, Victoria University of Wellington.

WOOD, S. A.; BRIGGS, L. R.; SPROSEN, J.; RUCK, J. G.; WEAR, R. G.; HOLLAND, P. T.; BLOXHAM, M. 2006a: Changes in concentrations of microcystins in rainbow trout, freshwater mussels, and cyanobacteria in Lakes Rotoiti and Rotoehu. *Environmental Toxicology 21*: 205–222.

WOOD, S. A.; CROWE, A. L. M.; RUCK, J. G.; WEAR, R. G. 2005: New records of planktonic cyanobacteria in New Zealand freshwaters. *New Zealand Journal of Botany 43*: 479–492.

WOOD, S. A.; HEATH, M. W.; HOLLAND, P. T.; MUNDAY, R.; McGREGOR, G. B.; RYAN, K. G. 2010: Identification of a benthic microcystin-producing filamentous cyanobacterium (Oscillatoriales) associated with a dog poisoning in New Zealand. *Toxicon 55*: 897–903.

WOOD, S. A.; HOLLAND, P. T.; STIRLING, D. J.; BRIGGS, L. R.; SPROSEN, J.; RUCK, J. G.; WEAR, R. G. 2006b: Survey of cyanotoxins in New Zealand water bodies between 2001 and 2004. *New Zealand Journal of Marine and Freshwater Research 40*: 585–597.

WOOD, S. A.; JENTZSCH, K.; RUECKERT, A.; HAMILTON, D. P.; CARY, S. G. 2008a: Hindcasting cyanobacterial communities in Lake Okaro with germination experiments and genetic analysis. *FEMS Microbiology Ecology 67*: 252–260.

WOOD, S. A.; PRENTICE, M. J.; SMITH, K.; HAMILTON, D. P. 2010: Low dissolved inorganic nitrogen and increased heterocyte frequency: precursors to *Anabaena planktonica* blooms in a temperate, eutrophic reservoir. *Journal of Plankton Research 32*: 1315–1325.
WOOD, S. A.; RASMUSSEN, J. P.; HOLLAND, P. T.; CAMPBELL, R.; CROWE, A. L. M. 2007a: First report of the cyanotoxin anatoxin-a from *Aphanizomenon issatschenkoi* (Cyanobacteria). *Journal of Phycology 43*: 356–365.
WOOD, S. A.; RHODES, L. L.; ADAMS, S. L.; ADAMSON, J. E.; SMITH, K. F.; SMITH, J. F.; TERVIT, H. R.; CARY, S. C. 2008b: Maintenance of cyanotoxin production by cryopreserved cyanobacteria in the New Zealand culture collection. *New Zealand Journal of Marine and Freshwater research 42*: 277–288.
WOOD, S. A.; SELWOOD, A. I.; RUECKERT, A.; HOLLAND, P. T.; MILNE, J. R.; SMITH, K. F.; SMITS, B.; WATTS, L. F.; CARY, C. S. 2007b: First report of homoanatoxin-a and associated dog neurotoxicosis in New Zealand. *Toxicon 50*: 292–301.
WOOD, S. A.; STIRLING, D. J. 2003: First identification of the cylindrospermopsin-producing cyanobacterium *Cylindrospermopsis raciborskii* in New Zealand. *New Zealand Journal of Marine and Freshwater Research 37*: 821–828.
ZEHR, J. P.; BENCH, S. R.; CARTER, B. J.; HEWSON, I.; NIAZI, F.; SHI, T.; TRIPP, H. J.; AFFOURTIT, J. P. 2008: Globally distributed uncultivated oceanic N_2-fixing cyanobacteria lack oxygenic photosystem II. *Science 322*: 1110–1112.
ZWIRGLMAIER, K.; JARDILLIER, L.; OSTROWSKI, M.; MAZARD, S.; GARCZAREK, L.; VAULOT, D.; NOT, F.; MASSANA, R.; ULLOA, O.; SCANLAN, D. J. 2008. Global phylogeography of marine *Synechococcus* and *Prochlorococcus* reveals a distinct partitioning of lineages among ocean biomes. *Environmental Microbiology 10*: 147–161.

Checklist of New Zealand Cyanobacteria

This checklist follows the system of Anagnostidis and Komárek (1985, 1988) and Komárek and Anagnostidis (1989, 1998, 2005). Doubtful records and taxa (?) and new records (*) are indicated thus. Habitat codes: B, brackish-water; M, marine; T, terrestrial. All other taxa are freshwater; taxa found in more than one habitat are indicated, hence T/F signifies a species found in both terrestrial and freshwater habitats. Endemic species are indicated by E, adventive by A. Endemic genera are underlined.

SUPERKINGDOM PROKARYOTA
KINGDOM BACTERIA
SUBKINGDOM NEGIBACTERIA
INFRAKINGDOM GLYCOBACTERIA
PHYLUM CYANOBACTERIA
Class CYANOPHYCEAE
Order CHROOCOCCALES
CHAMAESIPHONACEAE
Chamaesiphon cf. *amethystinus* (Rostaf.) Lemmerm.
Chamaesiphon cf. *britannicus* (Fritsch) Komárek et Anagn.
Chamaesiphon confervicolus A.Braun in Rabenh.
Chamaesiphon hemisphericus Lemmerm. ?
Chamaesiphon incrustans Grunov in Rabenh.
Chamaesiphon cf. *investiens* Skuja
Chamaesiphon cf. *oncobyrsoides* Geitler
Chamaesiphon cf. *subglobosus* (Rostaf.) Lemmerm.
Stichosiphon sp.
CHROOCOCCACEAE
Asterocapsa cf. *divina* Komárek*
Asterocapsa sp.*
Chroococcus aeruginosus Nash nom. nud. ? E
Chroococcus amethystinus Nash nom. nud. ? E
Chroococcus cohaerens (Bréb.) Nägeli
Chroococcus dispersus (Keissl.) Lemmerm.
— var. *minor* G.M.Sm.
Chroococcus distans (G.M.Sm.) Komárk.-Legn. et Cronberg
Chroococcus giganteus W.West
Chroococcus limneticus Lemmerm.
Chroococcus macrococcus (Kütz.) Rabenh.
Chroococcus minimus (Keissl.) Lemmerm.
Chroococcus minor (Kütz.) Nägeli
Chroococcus minutissimus Nash nom. nud. ? E
Chroococcus minutus (Kütz.) Nägeli
— var. *thermalis* J.J.Copel.
Chroococcus novae-zelandiae Nash nom. nud. ? E
Chroococcus sphaericus Nash nom. nud. ? E
Chroococcus turgidus (Kütz.) Nägeli
Chroococcus turicensis (Nägeli) Hansg.
Chroococcus varius A.Braun in Rabenh.
Chroococcus sp.
Coccochloris stagnina Sprengel 1807 ? M
DERMOCARPELLACEAE
Cyanocystis hemisphaerica (Setch. et N.L.Gardner) Kaas
Dermocarpa gracilis Lemmerm. ?
Dermocarpa yellowstonensis (J.J.Copel.) Bourr. ?
Stanieria spherica (Setchell et Gardner) Anagn. et Pantazidou M
ENTOPHYSALIDACEAE
Chlorogloea sp.*
HYDROCOCCACEAE
Cyanodermatium sp.1*
Cyanodermatium sp. 2*
Hydrococcus rivularis Kütz.
Placoma regulare Broady et Ingerfeld
Placoma vesiculosum Schousb. M
HYELLACEAE
Pleurocapsa cf. *minor* Hansg.
Radaisia sp.*
MERISMOPEDIACEAE
Aphanocapsa castagnei (Kütz.) Rabenh. ?
Aphanocapsa conferta (W. et G.S.West) Komárk.-Legn. et Cronberg ?
Aphanocapsa delicatissima W. et G.S.West
Aphanocapsa elachista W. et G.S.West
Aphanocapsa grevillei (Berk.) Rabenh.
Aphanocapsa holsatica (Lemmerm.) Cronberg et Komárek
Aphanocapsa litoralis (Hansg.) Komárek et Anagn.
Aphanocapsa thermalis (Kütz.) Brügger
Coelosphaerium kuetzingianum Nägeli
Merismopedia elegans A.Braun
Merismopedia glauca (Ehrenb.) Nägeli ?
Merismopedia minima Beck
Merismopedia punctata Meyen
Merismopedia smithii DeToni
Merismopedia tenuissima Lemmerm.
Snowella lacustris (Chodat) Komárek et Hindák
Sphaerocavum brasiliense Azevedo et Sant'Anna
Synechocystis aquatilis Sauv.
Synechocystis minuscula Woron.
Synechocystis salina Wislouch M
Synechocystis thermalis J.J.Copel.
Woronichinia compacta (Lemmerm.) Komárek et Hindák
MICROCYSTACEAE
Gloeocapsa atrata Kütz. T/F
Gloeocapsa quaternata (Bréb.) Kütz. ?
Gloeocapsa sanguinea (C.Agardh) Kütz. T/F
Gloeocapsa stegophila (Itzigs.) Rabenh. ?
Gloeocapsa thermalis Lemmerm.
Microcystis aeruginosa (Kütz.) Kütz.
Microcystis botrys Teil.
Microcystis flos-aquae (Wittr.) Kirchn.
Microcystis ichthyoblabe Kütz.
Microcystis montana (Drouet et W.A.Daily) Bourr. ?
Microcystis panniformis Komárek, Komárkova-Legnerová, Sant'Anna, Azevedo et Senna
Microcystis protocystis Crow
Microcystis pulverea (Wood) Forti ?
Microcystis smithii Komárek et Anagn.
Microcystis stagnalis Lemmerm. ?
Microcystis thermalis Nash nom. nud. ? E
Microcystis wesenbergii (Komárek) Komárek in Kondrateva
SYNECHOCOCCACEAE
Aphanothece clathrata W. et G.S.West
Aphanothece conferta Richt. in Hauck et Richt.
Aphanothece elabens (Bréb.) Elenkin
Aphanothece granulosa Nash nom. nud. ? E
Aphanothece microscopica Nägeli
Aphanothece minutissima (W.West) Komárk.-Legn.
Aphanothece nidulans Richt. in Wittr. et Nordst.
— var. *longissima* Nash nom. nud. ? E
— var. *thermalis* Hansg.
Aphanothece saxicola Nägeli
Aphanothece stagnina (Spreng.) A.Braun in Rabenh. M?
Cyanobacterium minervae (J.J.Copel.) Komárek, Kopecky et Cepák
Cyanothece aeruginosa (Nägeli) Komárek T/F
Cyanothece eximia J.J.Copel. ?
Cyanothece rosea (J.J.Copel.) Komárek ?
Dactylococcopsis acicularis Lemmerm. ?
Dactylococcopsis irregularis G.M.Sm. ?
Gloeothece confluens Nägeli
Gloeothece linearis Nägeli M
Gloeothece cf. *palea* (Kütz.) Rabenh.*
Prochlorococcus sp. M
Rhabdoderma compositum (G.M.Sm.) Fedorov
Rhabdoderma lineare Schmidle et Lauterborn
Rhabdogloea smithii (Chodat et F.Chodat) Komárek
Synechococcus lividus J.J.Copel.
Synechococcus sp.
XENOCOCCACEAE
Chroococcidiopsis cf. *kashaii* Friedmann*

Xenococcus sp. 1*
Xenococcus sp. 2*
Xenotholos kerneri (Hansg.) Gold-Morgan, Motejano et Komárek ?

Order OSCILLATORIALES
GOMONOTIELLACEAE
Hormoscilla irregularis Novis & Visnovsky E T
HOMOEOTRICHACEAE
Ammatoidea normannii W. et G.S.West
Heteroleibleinia cf. *fontana* (Hansg.) Anagn. et Komárek
Heteroleiblenia cf. *kossinskajae* (Elenkin) Anagn. et Komárek
Heteroleibleinia cf. *purpurascens* (Hansg.) Anagn. et Komárek
Heteroleibleinia cf. *versicolor* (Wartmann) Gomont
Homoeothrix gracilis (Hansgirg) Komárek et Kovácik
Homoeothrix janthina (Born. et Flah.) Starmach
Homoeothrix juliana (Bornet et Flahault) Kirchner
Homoeothrix varians Geitler
OSCILLATORIACEAE
Blennothrix lyngbyacea (Kütz. ex Gomont) Anagn. et Komárek M
Hormoscilla sp. T
Lyngbya aestuarii Liebm. ex Gomont M
Lyngbya birgei G.M.Sm.
Lyngbya capitata (Desikachary) Anagn.
Lyngbya confervoides C.Agardh ex Gomont M
Lyngbya hieronymusii Lemmerm.
Lyngbya majuscula (Dillw.) Harv. ex Gomont M
Lyngbya cf. *nigra* Agardh ex Gomont
Lyngbya scolecoidea (W. et G.S.West) Bourr.
Lyngbya semiplena J.Agardh ex Gomont M
Oscillatoria anguina (Bory) Gomont
Oscillatoria annae Goor
Oscillatoria biterminalis Nash nom.nud. ? E
Oscillatoria bonnemaisonii Crouan ex Gomont M
Oscillatoria coerulea Kütz.
Oscillatoria contexta Hassall ?
Oscillatoria cruenta Grunov
Oscillatoria curviceps C.Agardh ex Gomont
Oscillatoria decorticans Hassall
Oscillatoria cf. *froelichii* Kütz. ex Gomont
Oscillatoria jenneri (Hassall) Bourr.
Oscillatoria limosa C.Agardh ex Gomont
Oscillatoria margaritifera Kütz. ex Gomont M
Oscillatoria mougeotii Kütz. ex Forti
Oscillatoria platensis var. *non-constricta* (J.C. Banerji) Desikachary
Oscillatoria princeps Vaucher ex Gomont
Oscillatoria proboscidea Gomont ex Gomont
Oscillatoria rotorua Nash nom. nud. ? E
Oscillatoria sancta Kütz. ex Gomont
Oscillatoria cf. *simplicissima* Gomont
Oscillatoria subbrevis Schmidle
Oscillatoria vizagapatensis C.B.Rao
Plectonema tomasinianum Bornet ex Gomont
PHORMIDIACEAE
Arthrospira gigantea (Schmidle) Anagn.
Arthrospira platensis (Nordst.) Gomont
Arthrospira tenuis Brühl et Biswas
Dasygloea amorpha Berkeley ex Gomont
Dasygloea lamyi (Gomont ex Gomont) Senna et Komárek
Jaaginema angustissimum (W. et G.S.West) Anagn. et Komárek
Jaaginema filiforme (J.J.Copel.) Anagn.
Jaaginema geminatum (Menegh. ex Gomont) Anagn. et Komárek
Jaaginema neglectum (Lemmerm.) Anagn. et Komárek M
Jaaginema subtilissimum (Kütz. ex DeToni) Anagn. et Komárek
Jaaginema tenellum (J.J.Copel.) Anagn.
— f. *minor* Nash nom. nud. ? E
Leptolyngbya angustissima (W. et G.S.West) Anagn. et Komárek
Leptolyngbya cf. *bijugata* (Kongisser) Anagn. et Komárek
Leptolyngbya foveolarum (Rabenh. ex Gomont) Anagn. et Komárek
Leptolyngbya fragilis (Gomont) Anagn. et Komárek
Leptolyngbya gelatinosa (Woron.) Anagn. et Komárek
Leptolyngbya laminosa (Gomont) Anagn. et Komárek
Leptolyngbya lurida (Gomont) Anagn. et Komárek
Leptolyngbya margaretheana (Schmidt) Anagn. et Komárek
Leptolyngbya mucicola (Lemmerm.) Anagn. et Komárek
Leptolyngbya nostocorum (Bornet ex Gomont) Anagn. et Komárek
Leptolyngbya purpurascens (Gomont ex Gomont) Anagn. et Komárek
Leptolyngbya ramosa (J.B.Petersen)Anagn. et Komárek
Leptolyngbya rivulariarum (Gomont)Anagn. et Komárek
Leptolyngbya scytonemicola (N.L.Gardner) Anagn. et Komárek
Leptolyngbya tenuis (Gomont) Anagn. et Komárek
Leptolyngbya treleasei (Gomont) Anagn. et Komárek
Leptolyngbya valderiana (Gomont) Anagn. et Komárek
Leptolyngbya weedii (Tilden) Anagn.
Microcoleus chthonoplastes (Mert.) Zanardini ?
Microcoleus gracilis Hassall ?
Microcoleus irriguus (Kütz.) Drouet
Microcoleus paludosus Gomont ex Gomont
Microcoleus steenstrupii J.B.Petersen
Microcoleus vaginatus Gomont ex Gomont T/F
Phormidiochaete sp.*
Phormidium aerugineo-caeruleum (Gomont) Anagn. et Komárek
Phormidium allorgei (Frémy) Anagn. et Komárek
Phormidium ambiguum Gomont ex Gomont
Phormidium amoenum Kütz.
Phormidium arthurensis Novis & Visnovsky E T
Phormidium articulatum (N.L.Gardner) Anagn. et Komárek
Phormidium autumnale (C.Agardh.) Trevisan ex Gomont T/F
Phormidium boryanum (Bory ex Gomont) Anagn. et Komárek
Phormidium breve (Kütz. ex Gomont) Anagn. et Komárek
Phormidium chalybeum (Mert. ex Gomont) Anagn. et Komárek
Phormidium chlorinum (Kütz.) Anagn.
Phormidium corium Gomont M
Phormidium cortianum (Menegh. ex Gomont) Anagn. et Komárek
Phormidium digueti (Gomont) Anagn. et Komárek
Phormidium favosum Gomont ex Gomont
Phormidium formosum (Bory ex Gomont) Anagn. et Komárek
Phormidium hamelii (Frémy) Anagn. et Komárek
Phormidium inundatum Kütz. ex Gomont
Phormidium irregulare Nash nom. nud. ? E
Phormidium laetevirens (Crouan ex Gomont) Anagn. et Komárek M
Phormidium murrayi (W. et G.S.West) Anagn. et Komárek
Phormidium nigrum (Vaucher ex Gomont) Anagn. et Komárek
Phormidium nigro-viride (Thw. ex Gomont) Anagn. et Komárek M
Phormidium novae-zealandiae Nash nom. nud. ? M
Phormidium okenii (C.Agardh ex Gomont) Anagn. et Komárek ?
Phormidium ornatum (Kütz. ex Gomont) Anagn. et Komárek
Phormidium papyraceum Gomont ex Gomont
Phormidium puteale (Mont. ex Gomont) Anagn. et Komárek
Phormidium retzii (C.Agardh) Gomont ex Gomont
Phormidium rotheanum Itzigsohn ?
Phormidium rotoruum Nash nom. nud. ? E
Phormidium submembranaceum Ardiss. et Straff. ex Gomont
Phormidium subuliforme (Kütz. ex Gomont) Anagn. et Komárek
Phormidium tergestinum (Kütz.) Anagn. et Komárek
Phormidium terebriforme (C.Agardh ex Gomont) Anagn. et Komárek
Phormidium uncinatum Gomont ex Gomont
Planktothrix agardhii (Gomont) Anagn. et Komárek
Planktothrix cryptovaginata (Schkorb.) Anagn. et Komárek
Planktothrix rubescens (DC. ex Gomont) Anagn. et Komárek
Porphyrosiphon luteus (Gomont ex Gomont) Anagn. et Komárek M
Porphyrosiphon martensianus (Menegh. ex Gomont) Anagn. et Komárek
Porphyrosiphon notarisii Kütz. ex Gomont
Pseudophormidium golenkinianum (Gomont) Anagn. M
Pseudophormidium purpureum (Gomont) Anagn. et Komárek
Spirulina labyrinthiformis Kütz. ex Gomont
Spirulina laxa G.M.Sm.
Spirulina major Kütz. ex Gomont
Spirulina subtilissima Kütz. ex Gomont M
Symploca laete-viridis Gomont M
Symploca muralis Kütz. ex Gomont
Symploca muscorum Gomont ex Gomont
Symplocastrum friesii (C.Agardh) ex Kirchner
Symplocastrum muelleri (Nägeli ex Gomont) Anagn.
Trichodesmium erythraeum Ehrenb. ex Gomont M
Trichodesmium iwanoffianum Nygaard
Trichodesmium lacustre Kleb.
Tychonema bornetii (Zukal) Anagn. et Komárek
PSEUDANABAENACEAE
Geitlerinema amphibium (C.Agardh ex Gomont) Anagn.
Geitlerinema cf. *ionicum* (Skuja) Anagn.
Geitlerinema jasorvense (Vouk) Anagn.
Geitlerinema splendidum (Grev. ex Gomont) Anagn.
Leibleinia epiphytica (Hieron.) Compère
Limnothrix guttulata (Goor) Umezaki et M. Watanabe
Limnothrix meffertae Anagn.
Limnothrix planctonica (Wolosz.) Meffert
Planktolyngbya bipunctata (Lemmerm.) Anagn. et Komárek
Planktolyngbya lacustris (Lemmerm.) Anagn. et Komárek
Planktolyngbya limnetica (Lemmerm.) Komárková-Legnerová et Cronberg
Pseudanabaena catenata Lauterborn
Pseudanabaena limnetica (Lemmerm.) Komárek
Pseudanabaena mucicola (Naumann et Hub.-Pest.) Schwabe.
Pseudanabaena oblonga Kullberg
SCHIZOTRICHACEAE
Schizothrix arenaria Gomont
Schizothrix calcicola Gomont
Schizothrix hirschii Drouet
Schizothrix lacustris A.Braun ex Gomont M
Schizothrix rubella Gomont
Schizothrix rupicola Tilden
Schizothrix telephoroides Gomont

Schizothrix vaginata (Nägeli) Gomont
Trichocoleus acutissimus (N.L.Gardner) Anagn.
Trichocoleus tenerrimus (Gomont) Anagn. M

Order NOSTOCALES
MICROCHAETACEAE
Coleodesmium wrangelii (C.Agardh) Borzí
Godleya alpina Novis & Visnovsky E T
Microchaete tenera Thur.
Tolypothrix chathamensis Lemmerm.
Tolypothrix distorta Kütz.
— var. *penicillata* (C.Agardh) Lemmerm.
Tolypothrix irregularis Ber. M
Tolypothrix lanata (Desv.) Wartm.
Tolypothrix limbata Thur.
Tolypothrix pseudorexia Novis & Visnovsky E T
Tolypothrix rupestris Wolle
Tolypothrix tenuis Kütz. ex Bornet et Flahault
NOSTOCACEAE
Anabaena aequalis Borge
Anabaena aphanizomenoides Forti
Anabaena catenula var. affinis (Lemmerm.) Geitler
— var. *solitaria* (Kleb.)Geitler
Anabaena circinalis Rabenh. ex Bornet et Flahault
Anabaena flos-aquae (Lyngb.) Bréb. ex Bornet et Flahault
Anabaena inaequalis (Kütz.) Bornet et Flahault
Anabaena lemmermannii P.G.Richt.
Anabaena macrospora Kleb.
Anabaena microspora Kleb.
Anabaena minderi Huber-Pestalozzi
Anabaena miniata Skuja
Anabaena minutissima Lemmerm.
— var. *attenuata* Pridmore et C.S. Reynolds
— var. *australis* Skuja
Anabaena novizelandica Skuja E
Anabaena oscillarioides Bory ex Bornet et Flahault
— var. *angustus* Bharadwaja
— var. *nova-zelandiae* Lemmerm.
Anabaena planktonica Brunnthaler A
Anabaena smithii (Komárek) M.Watanabe
Anabaena sphaerica var. *tenuis* G.S.West
Anabaena spiroides Kleb.
— var. *crassa* Lemmerm.
— var. *tumida* Nygaard
Anabaena tenericaulis Nygaard
Anabaena torulosa (Carm.) Lagerh. M
Anabaena verrucosa J.B.Petersen
Anabaenopsis arnoldii Aptekar
Anabaenopsis circularis (G.S.West) Wolosz. et Miller
Anabaenopsis elenkini Miller
Anabaenopsis flos-aquae (L.) Ralfs
Aphanizomenon flos-aquae (L.) Ralfs ex Bornet et Flahault
Aphanizomenon gracile (Lemmerm.) Lemmerm.
Aphanizomenon issatschenkoi (Usačev) Proshk.-Lavr.
Cylindrospermopsis raciborskii (Woloszynska) Seenayya et Subba Raju
Cylindrospermum comatum Wood
Cylindrospermum licheniforme Kütz.
Cylindrospermum maius Kütz.
Cylindrospermum minutissimum Collins
Cylindrospermum rotoruense Nash nom. nud. ? E
Nodularia harveyana (Thwaites) Nordin et Stein
— var. *sphaerocarpa* (Bornet et Flahault) Elenkin
Nodularia implexa (Bornet et Flahault) Bourr. ?
Nodularia spumigena (Mert.) Nordin et Stein B
— var. *vacuolata* F.E.Fritsch
Nostoc coeruleum Lyngb.
Nostoc carneum C.Agardh
Nostoc commune Vaucher ex Bornet et Flahault T
Nostoc ellipsosporum (Desm.) Rabenh.
Nostoc entophytum Bornet et Flahault M
Nostoc linckia (Roth) Bornet et Thur.
Nostoc microscopicum Carmich.
Nostoc minutissimum Kütz.
Nostoc minutum Desm.
Nostoc muscorum C.Agardh
Nostoc paludosum Kütz.
Nostoc parmelioides Kütz.
Nostoc passerianum Bornet et Thur.
Nostoc pruniforme (C.Agardh) Bornet et Flahault
Nostoc punctiforme (Kütz.) Har.
Nostoc rivulare Kütz.
Nostoc sphaericum Vaucher
Nostoc spongia Lam nom. nud. ?
Nostoc verrucosum (Vaucher) Bornet et Flahault
Nostoc zetterstedtii Aresch.
Raphidiopsis mediterranea Skuja
Trichormus azollae (Strasb.) Komárek et Anagn.
Trichormus thermalis (Vouk) Komárek et Anagn.
Trichormus variabilis (Kütz.) Komárek et Anagn.
RIVULARIACEAE
Calothrix bharadwajae DeToni
Calothrix braunii (A.Braun) Bornet et Flahault
Calothrix calida P.G.Richt. in Kuntze
Calothrix clavata G.S.West
Calothrix confervicola (Roth) C.Agardh M
Calothrix crustacea Thur. M
Calothrix fusca (Kütz.) Bornet et Flahault
Calothrix parasitica (Chauv.) Thur.
Calothrix parietina (Nägeli) Thur. T/F
— var. *thermalis* G.S.West
Calothrix pilosa Harv. M
Calothrix prolifera Flahault M
Calothrix pulvinata C.Agardh
Calothrix scopulorum C.Agardh
Calothrix thermalis (Schwabe) Hansg.
Calothrix vivipara Harv. M
Calothrix weberi Schmidle
Dichothrix baueriana (Grunov) Bornet
Dichothrix gypsophila (Kütz.) Bornet et Flahault
Dichothrix orsiniana (Kütz.) Bornet et Flahault ? M
Gloeotrichia raciborskii Wolosz.
Gloeotrichia ad. *raciborskii* Wolosz.
Gloeotrichia echinulata (Sm.) Richt.
Gloeotrichia natans (Hedw.) Rabenh.
Gloeotrichia pisum Thur.
Isactis plana (Harv.) Thur. M
Rivularia atra Roth
Rivularia australis Harv. M
Rivularia beccariana (DeNot.) Bornet et Flahault
Rivularia haematites (DC.) C.Agardh
Rivularia natans (Hedw.) Welw.
Rivularia polyotis (C.Agardh) Bornet et Flahault M
Rivularia vieillardii (Kütz.) Bornet et Flahault
SCYTONEMATACEAE
Scytonema caldarium Setch.
Scytonema coactile Mont.
Scytonema crispum (C.Agardh) Bornet
Scytonema hofmanni C.Agardh
Scytonema mirabile (Dillwyn) Bornet
Scytonema myochrous (Dillwyn) C.Agardh
Scytonema tolypotrichoides Kütz. M
Scytonema varium Kütz.
Scytonematopsis maxima Novis & Visnovsky E T

Order STIGONEMATALES
FISCHERELLACEAE
Fischerella ambigua (Kütz.) Gomont
Fischerella moniliformis Frémy
MASTIGOCLADACEAE
Brachytrichia quoyi (C.Agardh) Bornet et Flahault M
Brachytrichia sp.?
Hapalosiphon fontinalis (C.Agardh) Bornet
Hapalosiphon hibernicus W. et G.S.West
Hapalosiphon welwitschii W. et G.S.West ? T
Mastigocladus laminosus (C.Agardh) Cohn
Mastigocladus laminosus high-temperature form
NOSTOCHOPSACEAE
Mastigocladopsis repens Mora strain T
Nostochopsis lobatus Wood
STIGONEMATACEAE
Stigonema coactile Mont.
Stigonema figuratum Bornet et Flahault ?
Stigonema hormoides (Kütz.) Bornet et Flahault
Stigonema informe Kütz.
Stigonema mamillosum (Lyngb.) C.Agardh
Stigonema minutum (C.Agardh) Hassall
Stigonema muscicola (Thur.) Borzí
Stigonema ocellatum (Dillwyn) Thur.
Stigonema panniforme (C.Agardh) Kirchn.
Stigonema turfaceum (Berk.) Cooke

THREE

Phylum

EUGLENOZOA

euglenoids, bodonids, trypanosomes

F. R. (BERTHA) ALLISON, PAUL A. BROADY

Scanning electron micrographs of the euglenoids *Phacus brachykentron* (non-New Zealand) and *Euglena gracilis*.
Brian Leander

Euglenozoans include aquatic grass-green euglenids and their colourless euglenoid relatives on the one hand, and the free-living unpigmented bodonids, common in stagnant water, and the famous blood-parasitic trypanosomes, on the other. These two broad groups of euglenozoans used to be separately classified among phyto- ('plant') flagellates and zoo- ('animal') flagellates, but are now known to be closely related. Phytoflagellates were all considered to be part of the plant kingdom owing to the possession of plastids containing green chlorophyll pigments. Ultrastructural and molecular research during the past 30 years, however, has shown that there are at least five categories of plastids, each with different evolutionary origins (but originally involving a cyanobacterial symbiont), so that phytoflagellates are now distributed across three of the six kingdoms recognised by Cavalier-Smith (1998, 1999). Plastids in euglenoids (and most dinoflagellates – phylum Myzozoa) differ from those in some amoebozoans and in kingdoms Plantae and Chromista by being surrounded by three membranes. Molecular-phylogenetic studies provide evidence that green plastids in the photosynthetic euglenoids are derived from an endosymbiotic prasinophyte green alga (Leander 2004; Rogers et al. 2007; Turmel et al. 2009).

Cavalier-Smith (2010) regards Euglenozoa as the most basal eukaryote phylum, differing greatly from all other eukaryotes in at least 11 structural, genomic, and cell-biochemistry attributes. These characters were probably ancestral for Euglenozoa and, to the extent they are found in trypanosomatids, are primitive, not the consequences of a parasitic mode of life. The evidence is that trypanosomatids arose from a bodonid ancestor (Deschamps et al. 2010). Hence the ancestral eukaryote would have been non-amoeboid, having instead a firm but flexible pellicle.

Although many euglenoids have chloroplasts, which contain chlorophylls *a* and *b* and some carotenoid pigments that are shared with green algae and higher plants, they also contain unique carotenoid pigments. Photosynthetic euglenoids also have a pigmented eyespot that allows them to orientate towards light. Euglenoids lack rigid cellulose cell walls, having instead a pellicle made of protein, which lies just below the cell membrane and comprises ribbon-like strips that can be finely grooved (Leander 2011). Pellicles have a corset of microtubules underlying most of the cell surface, and are usually very flexible, so that many euglenoids can change shape with ease. Euglenoids also have at the anterior end a tubular invagination (canal) at the bottom of which is a contractile vacuole (reservoir). Many species can exist in the absence of light, in which case chloroplasts may regress. Feeding is then by ingestion of bacteria or tiny non-living particles into the anterior canal and/or by uptake of dissolved nutrients. Three orders of euglenoids are wholly phagotrophic.

Arising from the base of the canal is a pair of cilia [traditionally called flagella but, as noted in the introduction to this volume, flagella and cilia are the same type of structure and the term flagellum is best restricted to the unrelated structures in bacteria (Cavalier-Smith 2000)]. The cilia bear delicate hairs. Although the basic number of cilia in euglenozoans is two, in some forms only one emerges from the canal; a few parasitic species have more than two. When two emerge, one is often shorter than the other.

Bodonids and trypanosomes are collectively called kinetoplastids, named for the specialised DNA-containing part(s) of the net-like mitochondrion, the kinetoplast, a stainable mass (sometimes more than one) situated near the insertion of the cilia. Kinetoplastids have 1–2 cilia, two in the bodonids and *Ichthyobodo*, and one in trypanosomes. Kinetoplastids also have an anterior canal (cytopharynx) and a pellicular cell wall with a microtubular corset, but it is understandable that, until the advent of the electron microscope, it was not possible to ascertain the several morphological features that euglenoids and kinetoplastids have in common. Biochemical and molecular studies have stunningly confirmed the close relationships between the two groups (Schwartz & Dayhoff 1978; Cavalier-Smith 1981; Von der Heyden et al. 2004). Euglenoids may have arisen from free-living bodonids through the acquisition of plastids by secondary symbiogenesis of a eukaryotic alga, but recent molecular work is equivocal on which is the most basal group of Euglenozoa (Von der Heyden et al. 2004).

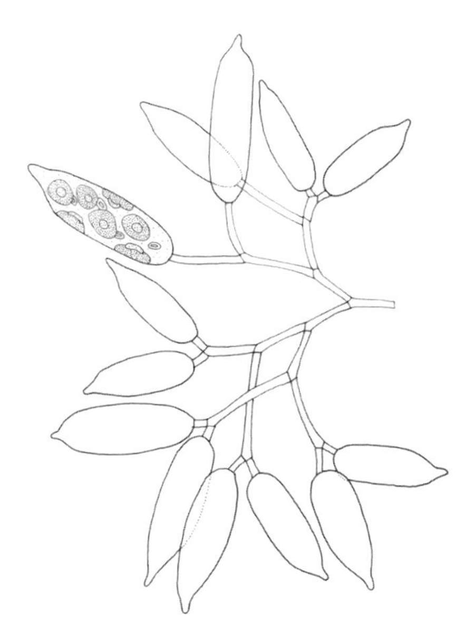

Colacium mucronatum, a colonial euglenoid.
After Bourrelly 1985

Cavalier-Smith (1998) had divided the phylum into two subphyla – Plicostoma and Saccostoma – based on the detailed structure of the feeding apparatus. The former contained the class Euglenoidea (and doubtfully the Diplonemea) and the latter the classes Kinetoplastea and Postgaardea (Cavalier-Smith 2000). The latter group is not yet known in New Zealand. There is now good evidence that Kinetoplastea and Diplonemea are sisters and weaker evidence that Postgaardea are more distant, so the two subphyla are no longer retained (Cavalier-Smith, in litt. 2011). Euglenoid classification below class level (Cavalier-Smith 1993) has been modified in the light of recent molecular work on relationships within the phylum, which supports the recognition of six orders, three wholly phagotrophic (Petalomonadida, Ploeotiida, Peranemida) and three non-phagotrophic, i.e. including mostly photosynthetic forms (Euglenida, Eutreptiida, Rhabdomonadida). Morphological characters used in distinguishing among orders, families, and genera include presence or absence of an eyespot and plastids, the number and length of emergent cilia, the form of the pellicle and the way the cells move, the detailed structure of the canal and feeding apparatus, and the nature and structure of other internal organelles. Most euglenoids exist as unicellular flagellates but a few have stages during which the cells are enclosed within a mucilage capsule (a palmelloid condition). Species of *Colacium*, for example, form bunches, sheets, or branching tree-like colonies of

Summary of New Zealand euglenozoan diversity

Taxon	Described species + subspecies + varieties	Known undescr./ undetermined/ unverified species	Estimated unknown species	Adventive species	Endemic species	Endemic genera
Euglenoidea	98+24	2	>50	0	1	0
Kinetoplastea	10	5	>50	1	7	0
Totals	**108+24**	**7**	**>100**	**1**	**8**	**0**

*Including taxa prefixed with '?'

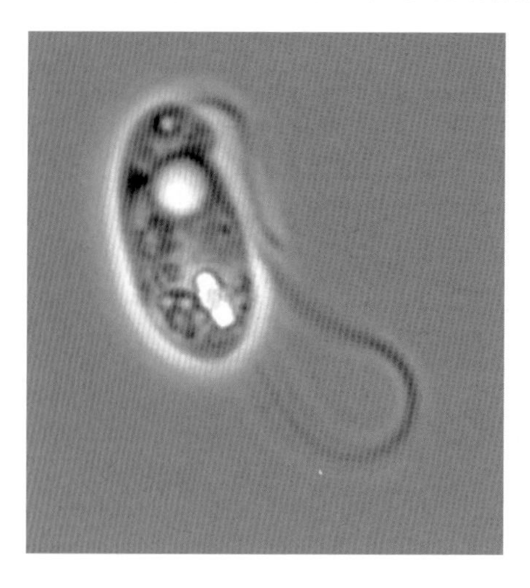

Bodo saltans.
D. J. Patterson, Linda Amaral-Zettler, Mike Peglar, Tom Nerad, per Micro*scope (MBL)

cells. Sexual processes are rare in euglenozoans, having been observed only in a single euglenoid, some African trypanosomes, and one bodonid.

The molecular analysis of Von der Heyden et al. (2004) largely supports the scheme of Moreira et al. (2004) for classifying kinetoplastids, recognising two major branches, one containing three major bodonid groups (Bodonida in the broad sense) and the order Trypanosomatida, and the other comprising a separate order (Protokinetoplastida) for the fish ectoparasite *Ichthyobodo* (not yet reported in New Zealand) and related forms. The free-living bodonids, ectoparasitic ichthyobodos, and blood-parasitic trypanosomes and leishmanias are unique among eukaryotes in having organelles called glycosomes containing the cell's glycolytic (carbohydrate-reducing) enzymes, most of which are absent from the internal fluid of the cell (cytosol), which would be normal for most other organisms.

New Zealand kinetoplastids

Bodonids are ubiquitous creatures of bacteria-rich fresh or brackish waters and soils, coastal sediments, and even dung. In New Zealand, the bodonids are very little known, with a handful of species reported, only two of which are named to species, *Bodo saltans* and *Neobodo designis*. Both are semi-cosmopolitan and may, in fact, be complexes of closely related species. In New Zealand, putative *B. saltans* and *N. designis* have been found in soil and lake sediments in the Waikato district and in rotting seaweeds at Raglan (von der Heyden & Cavalier-Smith 2005). Overseas, *B. saltans* has been reported in the human digestive tract, including in ulcers, after accidental ingestion but it is not pathogenic in humans. Four other bodonids are known from New Zealand, including novel forms from marine sediments at Auckland and a thermal pool near Matamata (von der Heyden et al. 2004). *Bodo saltans* and relatives play an important role in microbial food webs (Dolan & Šimek 1998).

Much better known, at least internationally, for their disease-causing roles, are the trypanosomes and leishmanias. The World Health Organisation ranks these organisms among the top six causes of tropical infectious diseases, e.g. trypanosomiasis (Chagas disease and various forms of sleeping sickness) and leishmaniasis (resulting in skin ulcerations or infected viscera). These diseases are not found in New Zealand but trypanosomes do infect some native fishes and the introduced rat species.

Trypanosomes are blood parasites, and, from the point of view of the parasite, the living conditions are ideal. Parasitism is defined as an intimate and obligatory relationship between two organisms, during which the parasite (usually the smaller) is metabolically dependent on the host. One of the most commonly encountered dependencies is nutrition. Blood is extremely rich in nutrients. In the circulatory system, blood cells and plasma are a favoured habitat for protozoan parasites such as trypanosomes and the haemogregarines and malarias (phylum Myzozoa). The gas content of the blood varies with the host

Diversity by environment

Taxon	Marine	Freshwater	Terrestrial
Euglenoidea	16	92+24	3
Kinetoplastea	11	4	3
Totals	**17**	**96+24**	**6**

* Parasitic species are attributed the major environment of their hosts.
† Some species can occur in more than one environment, reflected in the tabulation. Hence total numbers across all environments will exceed the overall species tally in the previous table.

species and whether it is arterial or venous blood. In humans, for example, there is a 27% difference in oxygen tension between arterial and venous blood, the former being higher. In birds, which have a higher metabolic rate, the difference can be as great as 60%. The carbon dioxide content is remarkably constant in blood, remaining at about 40 millimetres of mercury. The pH is also relatively constant, varying only slightly above or below pH 7.0, the average pH value being 7.4 in most birds and mammals.

The host counteracts parasites by producing antibodies (immunoglobulins). These are complex molecules synthesised in response to antigenic challenge, including invading parasites. However, parasites that live in blood have evolved intricate mechanisms to overcome the deleterious effects of antibodies. The result of the presence of the parasite is that the host develops some degree of acquired immunity. However, because of phases in life-cycles, antibodies produced at one stage may not be effective against another. Untreated African trypanosomiasis usually kills the host. Partial immunity may develop if the host is repeatedly infected, but it wanes without repeated exposure. The antibodies produced against certain trypanosomes kill the parasite by interfering with oxygen consumption.

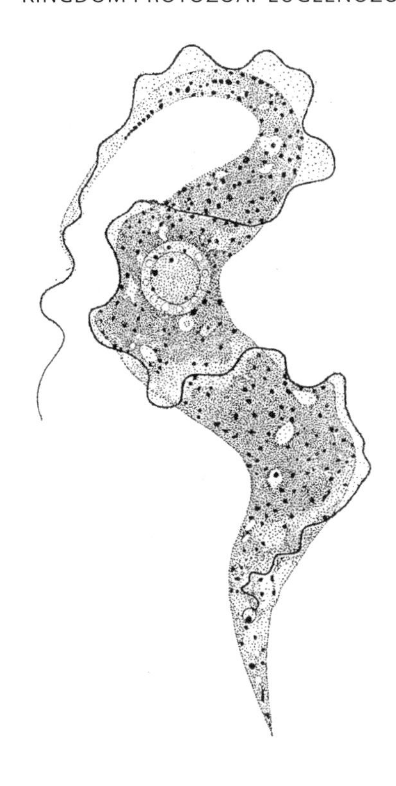

Trypanosoma heptatreti, a parasite of the hagfish *Eptatretus cirrhatus.*
From Laird 1951b

Reproduction in trypanosomes is by simple fission, and, during the life-cycle, individuals can assume more than one form (pleomorphism). A typical life-cycle passes through the following forms, the names of which allude to the position or form of the cilium (*mastigos*, Greek, means whip):

- amastigote – rounded or oval with no trailing cilium, this structure being reduced to a tiny fibril embedded in the cytoplasm;
- promastigote – elongate, with a relatively large nucleus and a trailing cilium originating from a kinetoplast near the anterior end of the cell;
- epimastigote – with the kinetoplast near the nucleus and the cilium arising near it; there is a short undulating membrane;
- trypanomastigote – with a kinetoplast posterior to the nucleus; the cilium arises near it, emerges from the side of the body, runs along an undulating membrane, and is free anteriorly.

Amastigote, promastigote and epimastigote stages occur in the invertebrate vector and the adult trypanomastigote in the primary (vertebrate) host. The mature cell, slightly flattened and torpedo-shaped, can swim in the host's body fluids with little resistance.

The best-known New Zealand example is *Trypanosoma lewisi*, found in rats but able to be cultivated in the laboratory. Doré (1918) was the first to record this species in New Zealand, in 1912, noting that 30% of the rats captured in the vicinity of sewers were infected, compared to 12% from wharves and grain stores. This species may be used to illustrate the trypanosome life-cycle. A rat becomes infected when it licks its fur, thereby ingesting rat-flea (mostly *Nosopsyllus fasciatus*) faeces containing trypanomastigotes. After several hours incubation, trypanosomes appear in the blood as extracellular parasites. These multiply by longitudinal splitting. When the flea feeds on the blood of a rat it takes in trypanomastigotes, which penetrate the epithelial cells lining the flea's stomach. Within the cell the flagellate curls up and becomes rounded. Sporulation by further cell division takes place, giving rise to many small amastigotes. These enlarge, rupturing the cell, and become active in the flea's gut, invading other epithelial cells, thus repeating the sporogonic cycle. Then those that are free in the gut metamorphose into epimastigotes and migrate posteriorly to the rectal region, some simultaneously undergoing longitudinal splitting. In the rectum, the epimastigotes attach to the lining and become pleomorphic while undergoing fission. Finally, some transform into trypanomastigotes and pass out with the host's faeces. In Europe, trypanosomes have at times been widely distributed in livestock. Laird (1951a) reported that, despite blood samples being analysed for such incidence in New Zealand, none had been found. That remains the case.

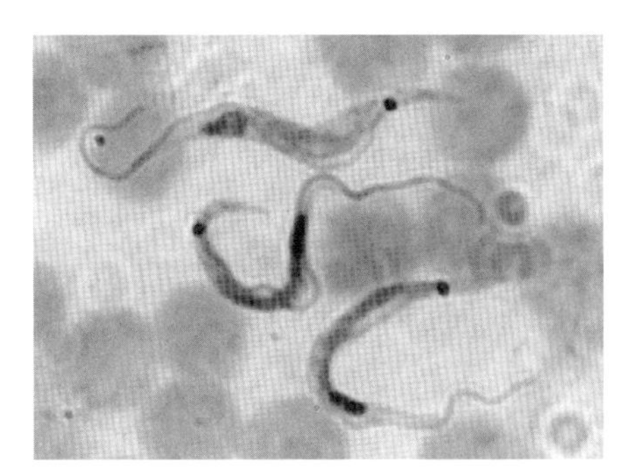

Trypanosoma lewisi in the bloodstream of a rat.
Joe Carney, Lakehead University, Thunder Bay

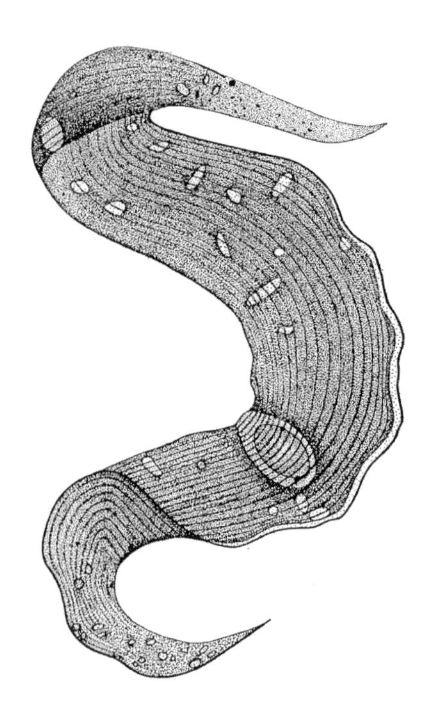

Trypanosoma gargantua, a parasite of the rough skate *Dipturus nasutus*.
From Laird 1951b

Multiplication of trypanomastigotes in the blood of the rat decreases within approximately six days after infection because of a reproduction-inhibiting antibody, ablastin, which interferes with nutritional and respiratory processes, blocking the synthesis of proteins and nucleic acids. The existing flagellates begin to decrease in number because of a trypanocidal antibody that is produced primarily in the rat's spleen and destroys them. A small population persists, comprising slender trypanomastigotes that are infective to fleas but do not undergo further reproduction in the rat host. Survival of antigenic variants cannot repopulate the blood owing to the presence of ablastin. *Trypanosoma lewisi* has no visible deleterious effect on the rat and, owing to the appearance of ablastin and the trypanocidal antibody, infection soon dies out.

Trypanosomes use oxygen in their metabolism but the rate of consumption varies between those found in blood and those in tissue and culture forms, which consume less. Physical and chemical factors influence oxygen consumption. Temperature and availability of sugar have pronounced effects such that temperature is accompanied by greater consumption up to a point. The age of the trypanosomes influences the respiratory rate – young dividing *T. lewisi* show lower consumption than older, non-dividing individuals.

Fishes are the only other known hosts of trypanosomes in New Zealand (having not yet been found in amphibians, reptiles, or birds). Most research on fish parasites was carried out on marine species between 1947 and 1972, particularly shallow-water and inshore species that supported predominantly coastal fishing. From 1973 to 1999, research concentrated on deep-sea and freshwater species. Currently no taxonomic work is being carried out on fish parasites, there being no funded New Zealand specialists.

Seven trypanosome species have been discovered in New Zealand fishes, all originally by Laird (1948, 1951b) (see also Laird 1949), and subsequently by Hewitt and Hine (1972) and Hine et al. (2000), who summarised all records. Known hosts are listed after each parasite species name in the end-chapter checklist. Life-cycles and intermediate hosts of the listed species are not known, but leeches are a likely vector.

New Zealand euglenoids

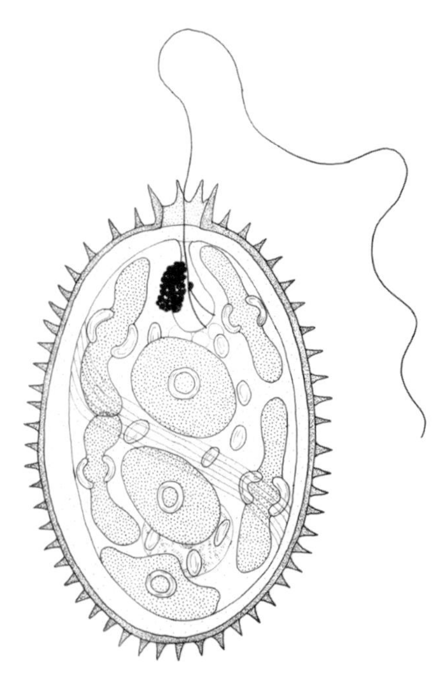

The euglenoid *Trachelomonas hispida* var. *coronata*.
After Bourrelly 1985

Aproximately 124 species and varieties of euglenoids from 20 genera have been recorded from New Zealand. The large majority contain chloroplasts whilst a small number are solely chemoheterotrophic and depend on organic nutrition. Examples of both are amongst the earliest microorganisms to be observed in New Zealand (Maskell 1887). Some 93 species and subspecies are from just three genera – *Euglena, Trachelomonas*, and *Phacus*. Only six species have been found in marine and estuarine habitats. By far the most species are from freshwater habitats. Insofar as there are about 800–1000 species (in 40–50 genera) globally (Bourrelly 1985; Graham et al. 2009), most of which are cosmopolitan, it is expected that considerably more species are in New Zealand and await discovery. Many records listed by Cassie (1984) are unpublished, and these and other published records are unsupported by either descriptions and illustrations or by voucher specimens. This lack reduces confidence in many identifications of these organisms.

It is well known that euglenoids are frequently found in waters that contain high concentrations of dissolved organic matter (Leedale 1967; Graham et al. 2009). This is related to their requirement for vitamins and to their chemoheterotrophic or mixotrophic nutrition. In the former mode, cells feed on a range of microorganisms or absorb organic substances in solution and this is the only mode of nutrition, whilst in the latter heterotrophy is combined with photosynthesis. In New Zealand, the proclivity of euglenoids to occupy environments where there is an abundance of decaying organic matter is exemplified by their being a prominent component of phytoplankton of sewage-treatment ponds. Here their photosynthesis helps oxygenate the water whilst chemoheterotrophic feeding has a role in mineralisation of organic wastes. In the most thoroughly illustrated study of New Zealand euglenoids, Haughey (1968, 1969) found 29 species, including 15 species of *Euglena*, in such ponds in Auckland. Cassie (1983) confirmed this high level of representation by euglenoids and Cassie Cooper (1996) briefly reviewed other studies of oxidation ponds. The importance of *Euglena* species in oxygenating the water in the former Mangere, Auckland, oxidation ponds was demonstrated by Irving and Dromgoole (1986). When the euglenoid populations were impacted by a fungal parasite, effluent quality decreased and pond odours became prevalent (Anderson et al. 1995).

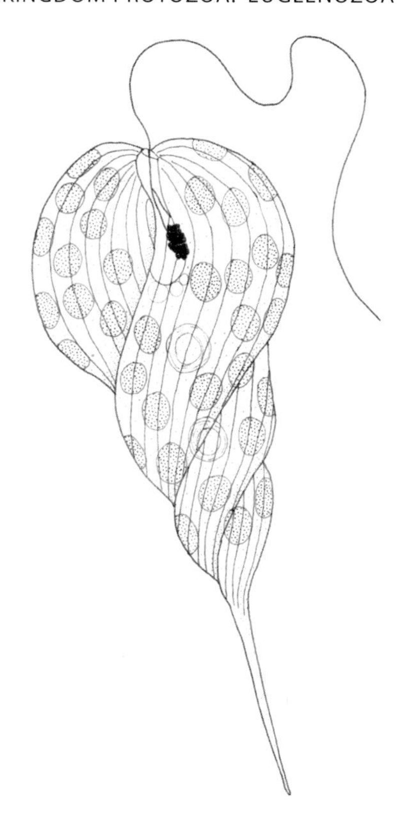

Phacus helikoides.
After Bourrelly 1985

Farm ponds, temporary puddles, and ditches contaminated by livestock are likewise habitats with elevated organic content and are also well-known sources of euglenoids. Deep-green and orange-red blooms can develop in the water and as surface scums. The reddish colour is caused by accumulation in the cells of carotenoid pigments that mask the green chloroplasts, protecting them from high light intensities at the water surface. The development of such blooms in a farm pond in the Waikato district was described and illustrated by Cassie Cooper (1996).

Although the highest populations of euglenoids are found in organically enriched habitats, it would be unusual not to find a few species in most freshwater habitats (Leedale 1967), at least in low numbers. In New Zealand, collections from lakes and ponds have predominantly been of phytoplankton communities. These have often contained euglenoids. Some experts have suggested, however, that there might not be any truly planktonic euglenoids and that they are fundamentally occupants of interfaces (Walne & Kivic 1990). In New Zealand, a greater diversity would likely be recovered from collections of sediments, material associated with submerged plant surfaces, and the specialised neustonic habitat at the water-air interface.

Some feeling for their wide distribution in diverse waterbodies in New Zealand is provided by records from a small pond in Otago (Byars 1960), Lake Rotorua in the North Island (Thomasson 1974), swamp and bog pools (Skuja 1976), dune lakes in Northland (Cassie & Freeman 1980), and eutrophic Lake Pupuke in Auckland (Cassie 1989). The most frequently recorded genus in these studies, and often in others, is *Trachelomonas*, of which 26 species and 16 varieties have been recorded. Traditionally, species of this genus have been recognised by the morphology of the lorica, or case, which surrounds the cell. This rigid container is often elaborately sculptured with spines and warts of various lengths, number, and arrangement. Because the case is easily preserved, in comparison to euglenoids that lack this structure, the apparently low frequency of occurrence of species from other genera could be an artefact of preservation.

One species of *Euglena*, *E. mutabilis*, is characteristic of acidic waters (Leedale 1967). The only published New Zealand record (Forsyth & McColl 1974) is in thermal, eutrophic Lake Rotowhero, where it is a component of dense phytoplankton populations. At the time of the study the water had a pH of 3.1 and a temperature of 30–37° C. In the Canadian Arctic, this species has been found in streams contaminated by acidic mining wastes (Graham et al. 2009). Similarly, in New Zealand, it forms vivid green mats in acidic streams draining coal-mining areas in Westland (Bray et al. 2008). It could be a useful indicator species of such conditions.

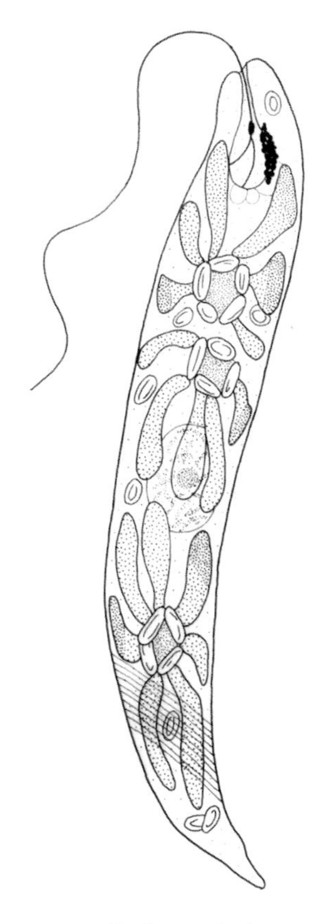

Euglena geniculata.
After Bourrelly 1985

The occurrence of *E. mutabilis* at such a high temperature in Lake Rotowhero seems quite unusual, although no doubt small waterbodies with euglenoids reach such temperatures regularly in tropical regions. However, in two warm springs in the Taupo thermal region, Winterbourn and Brown (1967) considered the absence of euglenoids from the warmer of the two (33–45° C) to be the consequence of it being above the upper limit of tolerance. The cooler spring (22–23° C) contained six species of euglenoids, including both photosynthetic and chemoheterotrophic species.

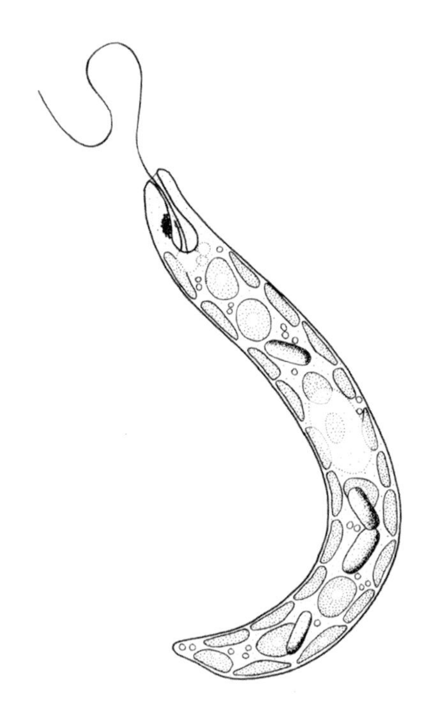

Euglena deses, a euglenoid of marine habitats.
Adapted from various sources

Euglenoids are relatively infrequent components of soil microbial communities. Seven species of *Euglena* and six species from four genera of chemoheterotrophs are described in a flora of terrestrial algae worldwide (Ettl & Gärtner 1995). Undoubtedly species in soils are more widespread in New Zealand than the three present records would suggest. Flint (1968) and Flint and Fineran (1969) both reported an unidentified *Euglena* in mostly acid soils from mainland New Zealand and the Snares Islands, respectively. Novis (2001) noted *E.* cf. *pisciformis* from an alpine organic soil.

In estuarine and marine environments, euglenoids have been reported somewhat incidentally in the course of general studies of plankton. Taylor (1974), for example first recorded *Euglena deses* from New Zealand, in Hauraki Gulf [Cassie (1984) later reported it from freshwater environments] and MacKenzie and Gillespie (1986) and Chang et al. (1990) recorded unidentified euglenoids in plankton from Tasman Bay and Stewart Island, respectively. Other euglenoids form films on the mud- and sandflats of estuaries. Knox and Kilner (1973) and Stefferson (1974) reported species of *Euglena* from the brackish Avon-Heathcote Estuary, Christchurch, where they formed yellowish to light-green surface scums on mud near effluent discharges. Stefferson (1974) summarised all previous observations on euglenoid blooms in the estuary, observed during surveys by the Christchurch Drainage Board. Although photosynthetic, some of the species are also highly heterotrophic, feeding on bacteria and non-living organic particles and nutrients. Some species perform migratory movements within sediments, descending back into the sediment with the incoming tide.

Norris (1961, 1964) described new species of planktonic euglenoids from the New Zealand marine area, including the only known member of the order Ploeotiida in New Zealand, *Entosiphon limenites*, which has also been found in Norwegian waters. *Petalomonas micra* (order Petalomonadida) is nominally endemic, being known only from Wellington Harbour.

Acknowledgement

Dr F. Hoe Chang (National Institute of Water & Atmospheric Research, Wellington) supplied the list of marine euglenoids.

Authors

Dr F. R. (Bertha) Allison Canterbury Museum, Rolleston Avenue, Christchurch, New Zealand. Deceased. Kinetoplastea

Dr Paul A. Broady School of Biological Sciences, University of Canterbury, Private Bag 4800, Christchurch, New Zealand [paul.broady@canterbury.ac.nz] Euglenoidea

References

ANDERSON, S. A.; STEWART, A; TOLICH ALLEN, G. 1995: *Pseudosphaerita euglenae*, a fungal parasite of *Euglena* spp. in the Mangere Oxidation Ponds, Auckland, New Zealand. *New Zealand Journal of Marine and Freshwater Research 29*: 371–379.

BOURRELLY, P. 1985: *Les Algues d'Eau douce. Initiation à la Systématique. Tome III: Les Algues bleues et rouges. Les Eugléniens, Peridiniens et Cryptomonadines.* Société Nouvelle des Éditions Boubée, Paris. 606 p.

BRAY, J. P.; BROADY, P. A.; NIYOGI, D. K.; HARDING, J. S. 2008: Periphyton communities in New Zealand streams impacted by acid mine drainage. *Marine and Freshwater Research 59*: 1084–1091.

BYARS, J. A. 1960: A freshwater pond in New Zealand. *Australian Journal of Marine and Freshwater Research 11*: 222–240.

CASSIE, V. 1983: A guide to algae in oxidation ponds in the Auckland district. *Tane (Journal of Auckland University Field Club) 29*: 119–131.

CASSIE, V. 1984: Revised checklist of the freshwater algae of New Zealand (excluding diatoms and charophytes). Part II. *Water & Soil Technical Publication 26*: xii, 117–250, xiii-lxiv.

CASSIE, V. 1989: Micro-algae of Lake Pupuke, Auckland, New Zealand. *New Zealand Natural Sciences 16*: 39–50.

CASSIE, V.; FREEMAN, P. T. 1980: Observations on some chemical parameters and the phytoplankton of five west coast dune lakes in Northland, New Zealand. *New Zealand Journal of Botany 18*: 299–320.

CASSIE COOPER, V. 1996: *Microalgae – Microscopic Marvels.* Riverside Books, Hamilton. 164 p.

CASSIE COOPER, V. 2001: Recent name changes in eukaryotic freshwater algae of New Zealand. *New Zealand Journal of Botany 39*: 601–616.

CAVALIER-SMITH, T. 1981: Eukaryote kingdoms: seven or nine? *Biosystems 28*: 91–106.

CAVALIER-SMITH, T. 1982: The origins of plastids. *Biological Journal of the Linnean Society 17*: 289–306.

CAVALIER-SMITH, T. 1993: Kingdom Protozoa and its 18 phyla. *Microbiological Reviews 57*: 953–994.

CAVALIER-SMITH, T. 1998: A revised six-kingdom

system of life. Biological Reviews 73: 203–266.
CAVALIER-SMITH, T. 1999: Principles of protein and lipid targeting in secondary symbiogenesis: euglenoid, dinoflagellate, and sporozoan plastid origins and the eukaryote family tree. *Journal of Eukaryotic Microbiology 46*: 347–366.
CAVALIER-SMITH, T. 2000: Flagellate megaevolution. The basis for eukaryote diversification. Pp. 361–390 *in*: Leadbeater, B. S. C.; Green, J. C. (Eds), *The Flagellates — Unity, Diversity and Evolution. [Systematics Association Species Volume Series 59.]* Taylor & Francis for Systematics Association, London.
CHANG, F. H.; ANDERSON, C.; BOUSTEAD, N. C. 1990: First record of a *Heterosigma* (Raphidophyceae) bloom with associated mortality of cage-reared salmon in Big Glory Bay, New Zealand. *New Zealand Journal of Marine and Freshwater Research 24*: 461–469.
DESCHAMPS, P.; LARA, E.; MARANDE, W.; LÓPEZ-GARCÍA, EKELUND, F.; MOREIRA, D. 2010: Phylogenomic analysis of kinetoplastids supports that trypanosomatids arose from within bodonids. *Molecular Biology and Evolution 28*: 53–58.
DOLAN, R. J.; ŠIMEK, K. 1998: Ingestion and digestion of an autotrophic picoplankter, *Synechococcus* by a heterotrophic nanoflagellate, *Bodo saltans*. *Limnology and Oceanography 43*: 1740–1746.
DORÉ, A. B. 1918: Rat trypanosomes in New Zealand. *New Zealand Journal of Science and Technology 1*: 200.
ETTL, H.; GÄRTNER, G. 1995: *Syllabus der Boden-, Luft- und Flechtenalgen.* Gustav Fischer Verlag, Stuttgart. viii + 721 p.
FLINT, E. A. 1968: Algae on the surface of some New Zealand soils. *New Zealand Soil Bureau Bulletin 26*: 183–90.
FLINT, E. A.; FINERAN, B. A. 1969: Observations on the climate, peats and terrestrial algae of the Snares Islands. *New Zealand Journal of Science 12*: 286–301.
FORSYTH, D. J.; McCOLL, R. H. S. 1974: The limnology of a thermal lake: Lake Rotowhero, New Zealand: II. General biology with emphasis on the benthic fauna of chironomids. *Hydrobiologia 44*: 91–104.
GRAHAM, L. E.; GRAHAM, J. M.; WILCOX, L. W. 2009: *Algae*. Benjamin Cummings, San Francisco, xvi + 616 p.
HAUGHEY, A. 1968: The planktonic algae of Auckland sewage treatment ponds. *New Zealand Journal of Marine and Freshwater Research 2*: 721–766.
HAUGHEY, A. 1969: Further algae of Auckland sewage treatment ponds and other waters. *New Zealand Journal of Marine and Freshwater Research 3*: 245–261.
HEWITT, G. C.; HINE, P. M. 1972: Checklist of parasites of New Zealand fishes and of their hosts. *New Zealand Journal of Marine and Freshwater Research 6*: 69–114.
HINE, P. M.; JONES, J. B.; DIGGLES, B. K. 2000: A checklist of the parasites of New Zealand fishes, including previously unpublished records. *NIWA Technical Report 75*: 1–93, + ii.
IRVING, D. E.; DROMGOOLE, F. I. 1986: Algal populations and characteristics of oxygen exchange of effluent samples from a facultative oxidation pond. *New Zealand Journal of Marine and Freshwater Research 20*: 9–16.
KNOX, G. A.; KILNER, A. R. 1973: The ecology of the Avon-Heathcote Estuary. Unpublished Report to the Christchurch Drainage Board by the Estuarine Research Unit, Department of Zoology, University of Canterbury, Christchurch. xxxvi + 358 p.
LAIRD, M. 1948: *Trypanosoma heptatretae* sp.n., a blood parasite of the Hagfish. *Nature 161*: 440–441.
LAIRD, M. 1949: Erratum. *Nature 162*: 279.
LAIRD, M. 1951a: Blood parasites of mammals in New Zealand. *Zoology Publications from Victoria University College 9*: 1–14.
LAIRD, M. 1951b: Studies on the trypanosomes of New Zealand fish. *Proceedings of the Zoological Society of London 121*: 285–309, pls 1–4.
LEANDER, B. S. 2004: Did trypanosomatid parasites have photosynthetic ancestors? *Trends in Microbiology 12*: 251-258.
LEANDER, B. S. (2011) Euglenida – euglenids or euglenoids. http://tolweb.org/Euglenida/97461
LEEDALE, G. F. 1967: *Euglenoid Flagellates*. Prentice-Hall, Englewood Cliffs, N. J. xiii + 242 p.
MacKENZIE, A. L.; GILLESPIE, P.A. 1986: Plankton ecology and productivity, nutrient chemistry, and hydrography of Tasman Bay, New Zealand, 1982–1984. *New Zealand Journal of Marine and Freshwater Research 20*: 365–395.
MASKELL, W. M. 1887: On the fresh-water infusoria of the Wellington District. *Transactions of the New Zealand Institute 20*: 3–19, pls 1–4.
MOREIRA, D.; LÓPEZ-GARCÍA, P.; VICKERMAN, K. 2004: An updated view of kinetoplastid phylogeny using environmental sequences and a closer outgroup: proposal for a new classification of the class Kinetoplastea. *International Journal of Systematic and Evolutionary Microbiology 54*: 1861–1875.
NORRIS, R. E. 1961: Observations on phytoplankton organisms collected on the N.Z.O.I. Pacific Cruise, September 1958. *New Zealand Journal of Science 4*: 162–188.
NORRIS, R. E. 1964: Studies on phytoplankton in Wellington Harbour. *New Zealand Journal of Botany 2*: 258–278.
NOVIS, P. M. 2001: Ecology and taxonomy of alpine algae, Mt Philistine, Arthur's Pass National Park, New Zealand. Unpublished PhD thesis, University of Canterbury, Christchurch. xiii + 273 p.
ROGERS, M. B.; GILSON, P. R.; SU, V.; McFADDEN, G. I.; KEELING, P. J. 2007: The complete chloroplast genome of the chlororachniophyte *Bigelowiella natans*: evidence for independent origins of chlorarachniophyte and euglenid secondary endosymbionts. *Molecular Biology and Evolution 24*: 54-62.
SCHWARTZ, R. M.; DAYHOFF, M. O. 1978: Origins of prokaryotes, eukaryotes, mitochondria and chloroplasts. *Science 199*: 395–403.
SKUJA, H. 1976: Zur Kenntnis der Algen neuseeländischer Torfmoore. *Nova Acta Regiae Societatis Scientarium Upsaliensis, ser. 5, C, 2*: 1–158.
STEFFERSON, D. A. 1974: Distribution of *Euglena obtusa* Schmitz and *E. salina* Liebetanz on the Avon-Heathcote estuary, Christchurch. *Mauri Ora 2*: 85–94.
TAYLOR, F. J. 1974: A preliminary annotated checklist of micro-algae other than diatoms and dinoflagellates from New Zealand coastal waters. *Journal of the Royal Society of New Zealand 4*: 395–400.
TAYLOR, F. J. 1978: Records of marine algae from the Leigh area. Part II: Records of phytoplankton from Goat Island Bay. *Tane (Journal of Auckland University Field Club) 24*: 213–218.
THOMASSON, K. 1974: Rotorua phytoplankton reconsidered (North Island of New Zealand). *Internationale Revue gesamten Hydrobiologie 59*: 703–727.
TURMEL, M.; GAGNON, M. C.; O'KELLY, C. J.; OTIS, C.; LEMIEUX, C. (2009). The chloroplast genomes of the green algae *Pyramimonas*, *Monomastix*, and *Pycnococcus* shed new light on the evolutionary history of prasinophytes and the origin of the secondary chloroplasts of euglenids. *Molecular Biology and Evolution 26*: 631–648.
VON der HEYDEN, S.; CAVALIER-SMITH, T. 2005: Culturing and environmental DNA sequencing uncover hidden kinetoplastid biodiversity and a major marine clade within ancestrally freshwater *Neobodo designis*. *International Journal of Systematic and Evolutionary Microbiology 55*: 2605–2621.
VON der HEYDEN, S.; CHAO, E. E.; VICKERMAN, K.; CAVALIER-SMITH, T. 2004: Ribosomal RNA phylogeny of bodonid and diplonemid flagellates and the evolution of Euglenozoa. *Journal of Eukaryotic Microbiology 51*: 402–416.
WALNE, P. L.; KIVIC, P. A. 1990: Euglenida. Pp. 270–287 *in*: Margulis, L.; Corliss, J. O.; Melkonian, M.; Chapman, D. J. (Eds), *Handbook of Protoctista*. Jones & Bartlett Publishers, Boston. xli + 914 p.
WINTERBOURN, M. J.; BROWN, T. J. 1967: Observations on the faunas of two warm streams in the Taupo thermal region. *New Zealand Journal of Marine and Freshwater Research 1*: 38–50.

Checklist of New Zealand Euglenozoa

Classification follows that of Cavalier-Smith (1993), von der Heyden et al. (2004), and advice from Professor Cavalier-Smith (Oxford University). Abbreviations: A, adventive; E, endemic; F, freshwater; M, marine; T, terrestrial; P, parasitic. Freshwater euglenoid entries are mainly based on a list published by Vivienne Cassie Cooper (Cassie 1984); † = new combination given by Cassie Cooper (2001); a question mark (?) before an entry indicates some uncertainty concerning the identification in the original source literature cited by Cassie. Some unidentified terrestrial euglenoids mentioned in the text but not listed below are included in the tabular summaries. Vertebrate hosts of trypanosome species are listed after the parasite names.

SUPERKINGDOM EUKARYOTA
KINGDOM PROTOZOA
SUBKINGDOM EOZOA
INFRAKINGDOM EUGLENOZOA
PHYLUM EUGLENOZOA
Class KINETOPLASTEA
Subclass METAKINETOPLASTINA
Order BODONIDA
Suborder EUBODONINA
BODONIDAE
Bodo saltans Ehrenberg F/T/M

Suborder PARABODONINA
PARABODONIDAE
Procryptobia sp. indet. von der Heyden et al. 2004 M

Suborder NEOBODONINA
NEOBODONIDAE
Neobodo designis Skuja, 1948 F/T/M
Neobodo sp. von der Heyden & Cavalier-Smith 2005 M
cf. *Rhynchobodo* sp. von der Heyden et al. 2004 F
Gen. indet. 1 et n. sp. von der Heyden et al. 2004 M
Gen. indet. 2 et n. sp. von der Heyden et al. 2004 F

Order TRYPANOSOMATIDA
TRYPANOSOMATIDAE (and hosts)
Trypanosoma caulopsettae Laird, 1951 P M E
  Arnoglossus scapha (Witch, Megrim)
  Rhombosolea plebeia (Sand flounder)
Trypanosoma coelorhynchi Laird, 1951 P M E
  Merluccius australis (Hake)
  Pseudophycis bachus (Red cod)
Trypanosoma congipodi Laird, 1951 P M E
  Congiopodus leucopaecilus (Southern pigfish)
Trypanosoma gargantua Laird, 1951 P M E
  'Raja' sp. (Skate)
Trypanosoma heptatreti Laird, 1948 P M E
  Eptatretus cirrhatus (Common hagfish)
Trypanosoma lewisi (Kent, 1880) P T A
  Rattus norvegicus Norway rat
  Rattus rattus Ship rat
Trypanosoma parapercis Laird, 1951 P M E
  Parapercis colias (Blue cod)
Trypanosoma tripterygium Laird, 1951 P M E
  Ericentrus rubrus (Orange clinid)
  Bellapiscis medius (Twister)
  Forsterygion varium (Variable triplefin)

Class EUGLENOIDEA
Subclass EUGLENIA
Order EUGLENIDA
Suborder EUGLENINA
COLACIIDAE
Colacium calvum F. Stein F
Colacium mucronatum Bourrelly & Chadefaud F
Colacium vesiculosum Ehrenberg F
EUGLENIDAE
Cryptoglena pigra Ehrenberg F
Euglena acus Ehrenberg F
?— var. *angularis* Johnson F
Euglena agilis Carter F
Euglena deses Ehrenberg F/M
Euglena elongata Schewiakof F
Euglena geniculata Dujardin F
?— var. *guttula* Playfair F
?— var. *juvenilis* Playfair F
Euglena gracilis G.A. Klebs F
Euglena haematodes (Ehrenberg) Lemmermann F
Euglena helicoideus (Bernard) Lemmermann F
?*Euglena limosa* N.L. Gardner F
Euglena minuta G.W. Prescott F
Euglena mutabilis F. Schmitz F
Euglena oxyuris Schmarda F
Euglena pisciformis G.A. Klebs F/T?
?— var. *typica* Pringsheim F
Euglena polymorpha P.A. Dangeard F
?*Euglena proxima* P.A. Dangeard F
Euglena salina Liebetanz M
Euglena sanguinea Ehrenberg F
Euglena schmitzii Gojdics F
Euglena spirogyra Ehrenberg F
Euglena torta A. Stokes F
— var. *curta* Playfair F
— var. *obesa* Playfair F
Euglena tripteris (Dujardin) G.A. Klebs F
Euglena truncata L. Walton F
Euglena viridis Ehrenberg F
?*Euglena volvox* Ehrenberg F
Euglena sp. Flint 1968 T
Euglena sp. Flint & Fineran T
Euglenopsis zabra Norris M
Lepocinclis fusiformis (H.J. Carter) Lemmermann F
Lepocinclis marssonii Lemmermann F
Lepocinclis ovum (Ehrenberg) Lemmermann
— var. *butschlii* (Lemmermann) Conrad F
?*Lepocinclis pseudo-ovum* Conrad F
Lepocinclis steinii Lemmermann F
Lepocinclis texta (Dujardin) Lemmermann F
?*Lepocinclis turbinatum* A.M. Cunha
Phacus acuminatus Stokes F
— var. *drezepolskii* Skvortzov F
Phacus agilis Skuja F
Phacus caudatus Huebner F
?*Phacus curvicauda* Svirenko F
Phacus glaber Deflandre F
Phacus helikoides Pochmann F
Phacus hispidula (Eichwald) Lemmermann F
?*Phacus horridus* Pochmann F
?*Phacus longicauda* (Ehrenberg) Dujardin F
Phacus pleuronectes (O.F. Mueller) Dujardin F
?*Phacus platalea* Drezepolski F
Phacus pyrum (Ehrenberg) Stein F
?*Phacus raciborskii* Drezepolski F
?*Phacus rudicola* (Playfair) Pochmann F
Phacus suecicus Lemmermann F
?— var. *cideon* Pochmann F
Phacus tortus (Lemmermann) Skvortzov F
Phacus triqueter (Ehrenberg) Dujardin F
Strombomonas ovalis (Playfair) Deflandre F
?*Strombomonas tetraptera* Balech & Dastugue F
?*Strombomonas urceolata* (Stokes) Deflandre F
Trachelomonas abrupta (Svirenko) Deflandre F
— var. *minor* Deflandre F
Trachelomonas armata (Ehrenberg) Stein F
— var. *duplex* Playfair F
?— var. *longispina* (Playfair) Deflandre F
Trachelomonas australica (Playfair) Deflandre F
Trachelomonas dubia Svirenko var. F
?*Trachelomonas euchlora* (Ehrenberg) Lemmermann † F
?*Trachelomonas fukiensis* Skvortzov F
— var. *australica* Skvortzov † F
Trachelomonas furcata Dolgoff F
Trachelomonas granulosa Playfair F
Trachelomonas hamiltoniana Skvortzov F
?*Trachelomonas hexangulata* (Svirenko) Playfair F
Trachelomonas hispida (Perty) F. Stein † F
— var. *coronata* Lemmermann F
— var. *crenulatocollis* (Maskell) Lemmermann F
— var. *duplex* Deflandre F
?— var. *punctata* Lemmermann F
?— var. *subarmata* Schroeder F
Trachelomonas mirabilis Skvortzov F
?— var. *obesa* (Messikommer) Skvortzov F
— var. *sydneyensis* Skvortzov † F
?*Trachelomonas naviculiformis* Deflandre F
Trachelomonas paludosa Skvortzov F
?— var. *annulata* Skvortzov F
?— var. *sydneyensis* Skvortzov F
Trachelomonas proxima Skvortzov F
— var. *lismorensis* Skvortzov F
Trachelomonas planctonica Svirenko F
?*Trachelomonas pulchella* Drezepolski F
Trachelomonas raciborskii Woloszynska F
?*Trachelomonas recticollis* Deflandre F
?*Trachelomonas rotunda* (Svirenko) Deflandre F
Trachelomonas scabra Playfair F
— var. *pygmeae* Playfair F
?*Trachelomonas schauinslandii* Skvortzov † F
Trachelomonas similis Stokes F
Trachelomonas teres Maskell F
Trachelomonas volvocina Ehrenberg F
— var. *planctonica* Playfair F
Trachelomonas westii Woloszynska F

Order EUTREPTIIDA
EUTREPTIIDAE
Eutreptiella marina Cunha M

Subclass PETALOMONADIA
Order PETALOMONADIDA
PETALOMONADIDAE
Dylakosoma pelophilum H. Skuja F
Petalomonas micra Norris M E
Petalomonas platyrhyncha H. Skuja F

Subclass PERANEMIA
Order PERANEMIDA
PERANEMIDAE
Anisonema acinus Dujardin F
?*Anisonema grande* Ehrenberg F

Dinematomonas griseolum (Perty) P.C. Silva F
?*Peranema cuneatum* Playfair F
Peranema trichophorum (Ehrenberg) Stein F
Urceolus alenitzini Mertens F
Urceolus cyclostomatus (F. Stein) Mertens F
SCYTOMONADACEAE
Scytomonas pusilla F. Stein F

Order PLOEOTIIDA
INCERTAE SEDIS
Entosiphon limenites Norris M

Subclass APHAGIA
Order RHABDOMONADIDA
ASTASIIDAE
Astasia acus (Korshikov) Pringsheim F
RHABDOMONADIDAE
Distigma proteus Ehrenberg F
?*Menoidium gracilis* Playfair F

FOUR

Phylum PERCOLOZOA

Naegleria, acrasid slime moulds, and kin

TONY CHARLESTON, STEVEN L. STEPHENSON

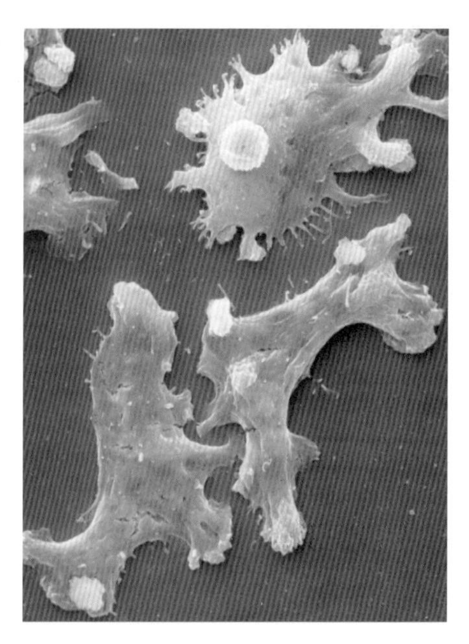

Amoeboid cells of *Naegleria fowleri*, with a circular 'mouth' forming in the cell surface of the upper individual.

Francine Cabral, CSBC, Virginia Commonwealth University

Percolozoans are colourless protozoans that are either strictly amoeboid or flagellate or can transform from amoeboid to flagellate and encysted stages (amoeboflagellates). They have in common a complete lack of Golgi dictyosomes (organelles involved in the secretion and storage of cellular products) and mitochondria with discoid, not tubular, cristae (the internal folds of the mitochondrial membrane that increase the surface area for chemical reactions). The phylum name (Cavalier-Smith 1991) is based on the included genus *Percolomonas* (Latin *percolare*, to percolate), which has not yet been found in New Zealand waters. This filter-feeding flagellate of marine plankton is thought to represent the ancestral type. The phylum has a paucity of species, suggesting to Cavalier-Smith (1993) that the lack of dictyosomes may have been a limiting factor in their ability to diversify.

Cavalier-Smith and Nikolaev (2008) have recognised two subphyla. The Tetramitia includes the ancestral class Heterolobosea, with amoeboflagellates (order Schizopyrenida) and cellular slime moulds (order Acrasida). Some Schizopyrenida, like *Vahlkampfia* and *Pseudovahlkampfia*, and the acrasids lack a flagellated stage. On the other hand, members of order Lyromonadida (*Lyromonas*) and class Percolatea (*Percolomonas*) are not known to have an amoeboid stage. These four-ciliate forms have a ventral feeding groove. The Pharynogomonada, monogeneric for *Macropharyngomonas*, comprises phagotrophic brine-dwelling zooflagellates that also lack an amoeboid phase. Two of their four cilia are forward directed, the others backward directed, the latter being associated with a long ventral groove; a distinct pharynx opens into the groove.

Most percolozoans are bacterial feeders in damp soil and freshwater environments but there are a few marine and parasitic forms, including the most notorious member of the phylum, *Naegleria fowleri*, which can be pathogenic to humans and causes fatal primary amoebic meningoencephalitis (PAM) (Cursons et al. 1978a). While in the amoeboid phase, usually in soil, *N. fowleri* feeds on bacteria, yeasts, and organic debris. Limiting the food supply induces the flagellate phase, which enables the species to disperse to more favourable habitats. Most pathogenic free-living amoebae are associated, in assays, with water with a high coliform content or soil contamination (Brown et al. 1983). *Naegleria fowleri* grows preferentially in warm water. If a swimmer inhales the flagellate phase, it can cross the mucous lining of the nasal cavity and invade the brain, causing haemorrhage, inflammation, and extensive cell and tissue death (necrosis). Death of the individual can occur within 10–14 days after exposure. The disease starts with a headache, sore throat, and mild fever, then progresses over three to four days with increasing headache, stiff neck, and vomiting. By the end of the third day, victims are severely disorientated and comatose. Post-mortem histological sections of brain tissue show large numbers of amoebae, which may also occur in the cerebrospinal fluid (Mandal et al. 1970). In an actual case history in 2000, a 10-year-old girl was diagnosed with PAM after swimming at the Okauia Springs at the bank of the Waihou River. She had severe, persistent headaches, fever, and vomiting and was treated initially for bacterial meningitis but her condition worsened. Cerebrospinal fluid showed the presence of motile amoebae, and culture and molecular analysis identified *N. fowleri*. She died three days after admission (Cursons et al. 2003).

The first recorded case of PAM anywhere in the world was described in 1965 in Australia (John 1982). More than 180 cases have been reported since then and most involve children and young adults in good health (Cabanes et al. 2001). Whereas some overseas cases have contracted the disease from ponds, lakes, and rivers, or even domestic water colonised by the amoebae, most cases in New Zealand (and many others overseas) have involved geothermal waters (Nicoll 1973; Cursons & Brown 1975; Cursons et al. 1976a,b, 1978b) and heated pools (Cursons et al. 1979) – *N. fowleri* lives optimally in water temperatures of 40–45° C. Most New Zealand cases of PAM have been reported from the Hamilton district and the Taupo Volcanic Zone. Maori folklore suggests that 'one should slowly immerse one's body into a thermal pool' (i.e. avoid jumping or diving, which would force water into the nasal sinuses), and it has been speculated that there may be practical reasons, born of experience, for this advice (Brown et al. 1983).

It is uncertain how many species of *Naegleria* are present in New Zealand. Cursons and Brown (1976) definitively isolated both *N. gruberi* (non-pathogenic) and *N. fowleri* from geothermal waters and Gray et al. (1995) isolated four unidentified *Naegleria* species as contaminants of contact-lens storage cases. It is not known if two of these latter species may have been *N. gruberi* and *N. fowleri*. These authors also reported, as co-contaminants of lens storage cases, four unidentified species of *Vahlkampfia*, a genus that is confamilial with *Naegleria*. Based on ribosomal DNA sequences, Cursons et al. (2003) definitively identified four species from geothermal waters in New Zealand – *N. australiensis*, *N. fowleri*, *N. gruberi*, and *N. lovaniensis*. Additionally, De Jonckheere (1994, 2002) described *N. clarki*, isolated from Golden Springs, Rotorua. Altogether, 47 species of *Naegleria* have been distinguished worldwide, mainly on the basis of differences in rDNA (De Jonckheere 2008).

The only other percolozoans known from New Zealand are two acrasid slime moulds (acrasids). Globally, the acrasids are a small group, with only four genera and about a dozen species. The feeding (trophic) phase is amoeboid, producing both lobose and, depending on the species, fine filose extensions (Olive 1975). When conditions are appropriate, amoebae aggregate singly or in small groups to produce an erect, spore-containing 'fruiting body' (sorocarp). This is borne on a thin column of live cells, one of the features that distinguish acrasids from the other cellular slime moulds, the dictyostelids (phylum Amoebozoa), in which the stalk is hollow or filled with dead cells. *Pocheina rosea* (called *Guttulina rosea* in older literature) was first described from dead wood in Russia and later reported from New Zealand. The sorocarp is short-stalked with an apical rose-coloured sphere containing the spores (Hutner 1985). In *Acrasis rosea*, the sorocarp is made up of chains of spores that collectively form an arborescent-like structure.

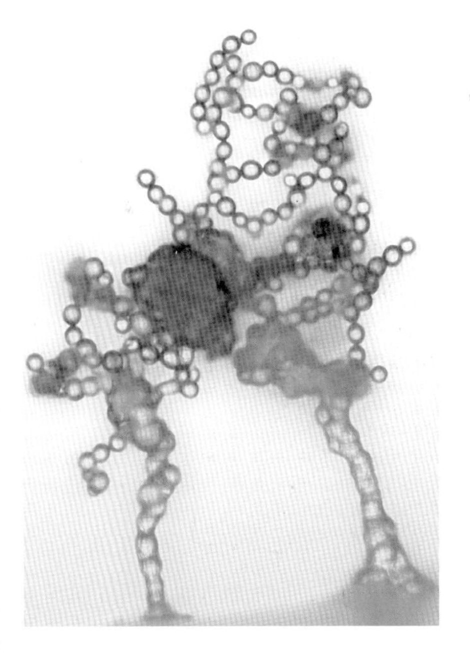

Acrasis rosea, 0.37mm high.
Matt Brown, University of Arkansas, Fayetteville

Summary of New Zealand percolozoan diversity

Taxon	Described species	Known undescr./ undetermined species	Estimated unknown species	Adventive species	Endemic species	Endemic genera
Percolozoa	7	~6	8	0	1?	0

Diversity by environment

Taxon	Marine	Freshwater	Terrestrial
Percolozoa	0	5+6	2

It seems likely other percolozoans may occur in New Zealand but have yet to be reported because the appropriate studies have never been carried out. For example, *Copromyxa* appears to be not uncommon on cow dung (Olive 1975; Stephenson unpubl.).

Authors

Dr W. A. G. (Tony) Charleston Institute of Veterinary, Animal and Biomedical Sciences, Massey University, Palmerston North (Present address: 488 College Street, Palmerston North, New Zealand) [charleston@inspire.net.nz]

Dr Steven L. Stephenson Department of Biological Sciences, University of Arkansas, Fayetteville, Arkansas 72701, USA [slsteph@uark.edu]

References

BROWN, T. J.; CURSONS, R. T. M.; KEYS, E.; MARKS, M.; MILES, M. 1983: The occurrence and distribution of pathogenic free-living amoebae in thermal areas of the North Island of New Zealand. *New Zealand Journal of Marine and Freshwater Research 17*: 59–69.

CABANES, P.-A.; WALLET, F.; PRINGUEZ, E.; PERNIN, P. 2001: Assessing the risk of primary amoebic meningoencephalitis from swimming in the presence of environmental *Naegleria fowleri*. *Applied and Environmental Microbiology 67*: 2927–2931.

CAVALIER-SMITH, T. 1991: Cell diversification in heterotrophic flagellates. Pp. 113–131 *in*: Patterson, D. J.; Larsen, J. (eds), *The Biology of Free-living Heterotrophic Flagellates*. Clarendon Press, Oxford. 518 p.

CAVALIER-SMITH, T. 1993: Kingdom Protozoa and its 18 phyla. *Microbiological Reviews 57*: 953–994.

CAVALIER-SMITH, T.; NIKOLAEV, S. 2008: The zooflagellates *Stephanopogon* and *Percolomonas* are a clade (class Percolatea: phylum Percolozoa). *Journal of Eukaryotic Microbiology 55*: 501–509.

CURSONS, R. T. M.; BROWN, T. J. 1975: The 1968 New Zealand cases of primary amoebic meningoencephalitis –myxomycete or *Naegleria*? *New Zealand Medical Journal 82*: 123–125.

CURSONS, R. T. M.; BROWN, T. J. 1976: Identification and classification of the aetiological agents of primary amebic meningo-encephalitis. *New Zealand Journal of Marine and Freshwater Research 10*: 245–262.

CURSONS, R. T. M.; BROWN, T. J.; BRUNS, B.; TAYLOR, D. E. M. 1976a: Primary amoebic meningo-encephalitis contracted in a thermal tributary of the Waikato River – Taupo: A case report. *New Zealand Medical Journal 84*: 479–481.

CURSONS, R. T. M.; BROWN, T. J.; CULBERTSON, C. G. 1976b: Immunoperoxidase staining of trophozoites in primary amoebic meningo-encephalitis. *Lancet ii*: 479.

CURSONS, R. T. M.; BROWN, T. J.; KEYS, E. A. 1978a: Primary amoebic meningoencephalitis in New Zealand – aetiological agents, distribution, occurrence and control. Pp. 96–109 *in*: Manderson, G. J. (ed.), *Proceedings of the 9th Annual Biotechnology Conference: Aspects of Water Quality and Usage in New Zealand*. Massey University, Palmerston North.

CURSONS, R. T. M.; BROWN, T. J.; KEYS, E. A. 1978b: Diagnosis and identification of the aetiological agents of primary amoebic meningo-encephalitis (PAM). *New Zealand Journal of Medical Laboratory Technology 32*: 11–14.

CURSONS, R. T. M.; BROWN, T. J.; KEYS, E. A. 1979: Primary amoebic meningo-encephalitis in an indoor heat-exchange swimming pool. *New Zealand Medical Journal 90*: 300–301.

CURSONS, R. T. M.; SLEIGH, J.; HOOD, D.; PULLON, D. 2003: A case of primary amoebic meningoencephalitis: North Island, New Zealand. *New Zealand Medical Journal 116(1187)*: 1–5.

DE JONCKHEERE, J. F. 1994: Comparison of partial SSUrDNA sequences suggests revisions of species names in the genus *Naegleria*. *European Journal of Protistology 30*: 333–341.

DE JONCKHEERE, J. F. 2002: A century of research on the amoeboflagellate genus *Naegleria*. *European Journal of Protistology 41*: 309–342.

DE JONCKHEERE, J. F. 2008: *Naegleria*. http://tolweb.org/Naegleria/124653/2008.09.21 *in*: Tree of Life Web Project, Version 21, September 2008 (under construction).

GRAY, T. B.; CURSONS, R. T. M.; SHERWAN, J. F.; ROSE, P. R. 1995: *Acanthamoeba*, bacterial, and fungal contamination of contact lens storage cases. *British Journal of Ophthalmology 79*: 601–605.

HUTNER, S. H. 1985: Class Mycetozoea de Bary, 1859. Pp. 214–227 *in*: Lee, J. J.; Hutner, S. H.; Bovee, E. C. (eds), *An Illustrated Guide to the Protozoa*. Society of Protozoologists, Lawrence. ix + 629 p.

JOHN, D. T. 1982: Primary amebic meningoencephalitis and the biology of *Naegleria fowleri*. *Annual Review of Microbiology 36*: 101–123.

MANDAL, B. N.; GUDEX, D. J.; FITCHETT, M. R.; PULLON, D. H. H.; MALLOCH, J. A.; DAVID, C. M.; APTHORP, J. 1970: Amoebic meningo-encephalitis due to amoebae of the order Myxomycetales (slime mould). *New Zealand Medical Journal 71*: 16–23.

NICOLL, A. M. 1973: Fatal primary amoebic meningoencephalitis. *New Zealand Medical Journal 78(496)*: 108–112.

OLIVE, L. S. 1975: *The Mycetozoans*. Academic Press, New York. 293p.

Checklist of New Zealand Percolozoa

Abbreviations: F, freshwater (including damp soil and films of water); T, terrestrial.

KINGDOM PROTOZOA
SUBKINGDOM EOZOA
INFRAKINGDOM EXCAVATA
PHYLUM PERCOLOZOA
SUBPHYLUM TETRAMITIA
Class HETEROLOBOSEA
Order SCHIZOPYRENIDA
VAHLKAMPFIIDAE
Naegleria australiensis De Jonckheere, 1981 F E
Naegleria clarki De Jonckheere, 1994 F
Naegleria fowleri Carter, 1970 F
Naegleria gruberi Schardinger, 1899 F
Naegleria lovaniensis Stevens, De Jonckheere & Willaert, 1980 F
Naegleria spp. indet. Gray et al. 1995 F
Vahlkampfia spp. indet. (4) Gray et al. 1995 F
Order ACRASIDA
ACRASIDAE
Acrasis rosea L.S. Olive & Stoian. 1960 T
GUTTULINIDAE
Pocheina rosea (Cienkowsky, 1873) T

FIVE

Phylum LOUKOZOA

jakobid and malawimonad excavate flagellates

DENNIS P. GORDON

Loukozoa is a small phylum of two-ciliate, aerobic, free-living excavates that feed on bacteria. The taxon was established by Cavalier-Smith (1999) for jakobids plus Anaeromonadea. Subsequently, Cavalier-Smith (2003) incorporated the latter group into the fundamentally anaerobic and symbiotic Metamonada, making Loukozoa more homogeneous. Thus redefined, Loukozoa comprises only those forms having a shared suite of ultrastructural characteristics, particularly involving the disposition of the Golgi dictyosome and nucleus in relation to the form and orientation of the centrioles and rootlets of the ciliary apparatus.

Two small classes are recognised. The larger is the Jakobea (jakobids) with just two families (O'Kelly 1993); the smaller Malawimonadea has only a single genus. The cilia in Loukozoa are simple, without hairs or scales; one is held anteriorly while the other lies in a ventral feeding groove where it vibrates to draw bacteria in for ingestion. In *Malawimonas*, this posterior cilium has a ventral vane and the mitochondria have discoid cristae (O'Kelly & Nerad 1999); in jakobids the vane is dorsal and the cristae are either tubular or flat, rarely absent. Jakobids have the most bacteria-like mitochondrial genomes known among eukaryotes, a feature that has attracted evolutionary interest in this group.

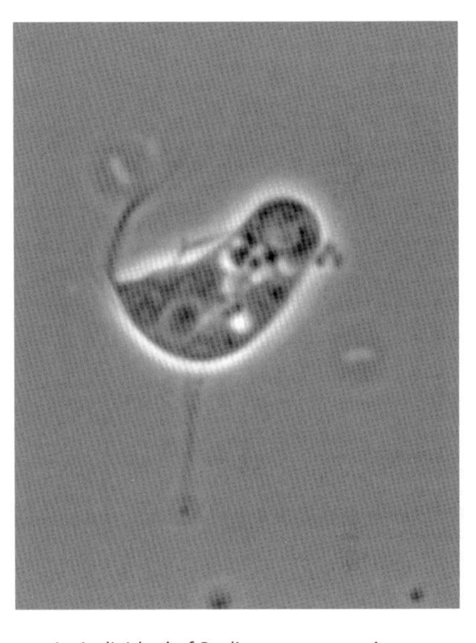

An individual of *Reclinomonas americana*.
David Patterson, per Micro*scope (MBL)

Only a single species of Loukozoa, *Reclinomonas americana*, is so far known from New Zealand (Patterson et al. 2000). It belongs to a loricate family of jakobids, the Histionidae. The lorica is constructed of organic materials and is not easily seen with the light microscope. Electron microscopy reveals that the exterior of the stalked, laterally compressed lorica in *R. americana* is studded with nail-shaped scales. Cells lie in a recumbent position in the lorica with one or both cilia emerging from the cell. The species is found in freshwater habitats in North America as well as New Zealand. Two other genera ascribed to the family have a different-shaped lorica.

The sole species of Jakobidae, *Jakoba libera*, is non-loricate, lacks a stalk, and the anterior flagellum is held in such a way as to resemble a question mark or shepherd's crook. It is widespread in marine habitats and is likely to be found in New Zealand waters.

Malawimonads (two species) have small bean-shaped cells that have been cultured from freshwater sediments and soil. *Malawimonas* lacks most of the unusual 'bacterial-like' features of jakobid mitochondrial genomes and some workers prefer to keep the malawimonads apart from the jakobids. Rodríguez-Ezpeleta et al. (2007) presented evidence suggesting that malawimonads have a sister-group relationship with an assemblage comprising jakobids, Euglenozoa and Heterolobosea, treating 'Jakobozoa' (core jakobids) and 'Malawimonadozoa' as nominal phyla, but these names appear not to have been formally introduced with diagnoses. Cavalier-Smith (2010) highlighted Euglenozoa as having prop-

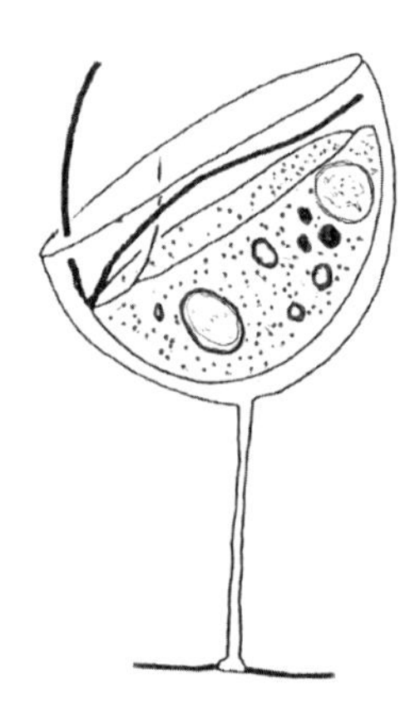

Diagrammatic representation of *Reclinomonas americana* in the stalked lorica (casing) that it secretes around itself.
After Patterson et al. 2000

erties radically different from all other eukaryotes and he includes Heterolobosea in the Percolozoa, which is followed here (see chapter 4). In the supplementary material to his paper, Cavalier-Smith (2010) doubtfully included the 'leaf-like' bilobed Diphyllatea in Loukozoa, but in a private communication (January 2011) he suggested that this group may yet belong to Apusozoa (see chapter 8), in which he previously included them (Cavalier-Smith 2003). Insofar as the ancestor of Percolozoa is likely to have been in Jakobea and that of Metamonada in Malawimonadea (Cavalier-Smith (2003), Loukozoa is paraphyletic.

Cavalier-Smith (2009) introduced new subphyla for Loukozoa – Vanomonada (diagnosed as having at least one vaned cilium in the feeding groove, i.e. Jakobea and Malawimonadea) and Diphyllatia (for its sole class Diphyllatea). If the latter group is determined to belong to the Apusozoa these subphylum names become superfluous.

Additional Loukozoa will certainly be found in New Zealand, both in soils and aquatic habitats. They are most likely to be detected first by environmental gene sequencing.

Author

Dr Dennis P. Gordon National Institute of Water & Atmospheric Research, Private Bag 14901, Kilbirnie, Wellington, New Zealand [d.gordon@niwa.co.nz]

Diversity by environment

Taxon	Marine	Freshwater	Terrestrial
Loukozoa	0	1	0

Summary of New Zealand loukozoan diversity

Taxon	Described species	Known undescribed/ undetermined species	Estimated unknown species	Adventive species	Endemic species	Endemic genera
Loukozoa	1	0	3	0	0	0

References

CAVALIER-SMITH, T. 1999: Principles of protein and lipid targeting in secondary symbiogenesis: euglenoid, dinoflagellate, and sporozoan plastid origins and the eukaryote family tree. *Journal of Eukaryotic Microbiology 46*: 347–366.

CAVALIER-SMITH, T. 2003: The excavate protozoan phyla Metamonada Grassé emend. (Anaeromonadea, Parabasalia, *Carpediemonas*, Eopharyngia) and Loukozoa emend. (Jakobea, *Malawimonas*): their evolutionary affinities and new higher taxa. *International Journal of Systematic and Evolutionary Microbiology 53*: 1741–1758.

CAVALIER-SMITH, T. 2009: Megaphylogeny, cell body plans, adaptive zones; causes and timing of eukaryote basal radiations. *Journal of Eukaryotic Microbiology 56*: 26–33.

CAVALIER-SMITH, T. 2010: Kingdoms Protozoa and Chromista and the eozoan root of the eukaryotic tree. *Biology Letters 6*: 342–345.

PATTERSON, D. J.; VØRS, N.; SIMPSON, A. G. B.; O'KELLY, C. 2000: Residual free-living and predatory heterotrophic flagellates. Pp. 1302–1328 *in*: Lee, J. J.; Leredale, G. F.; Bradbury, P. (eds), *An Illustrated Guide to the Protozoa. Second Edition: Organisms traditionally referred to as Protozoa, or newly discovered groups*. Society of Protozoologists, Lawrence. 2 vols, 1432 p.

O'KELLY, C. J. 1993: The jakobid flagellates: structural features of *Jakoba*, *Reclinomonas*, and *Histiona* and implications for the early diversification of eukaryotes. *Journal of Eukaryotic Microbiology 40*: 627–636.

O'KELLY, C. J.; NERAD, T. A. 1999: *Malawimonas jakobiformis* n. gen., n. sp. (Malawimonadidae) n. fam.), a *Jakoba*-like heterotrophic nanoflagellate with discoidal mitochondrial cristae. *Journal of Eukaryotic Microbiology 46*: 522–531.

RODRÍGUEZ-EZPELETA, N.; BRINKMANN, H.; BURGER, G.; ROGER, A. J.; GRAY, M. W.; PHILIPPE, H.; LANG, B. F. 2007: Toward resolving the eukaryotic tree: the phylogenetic positions of jakobids and cercozoans. *Current Biology 17*: 1420–1425.

Checklist of New Zealand Loukozoa

Abbreviation: F, freshwater.

KINGDOM PROTOZOA
SUBKINGDOM EOZOA
PHYLUM LOUKOZOA
Class JAKOBEA
Order JAKOBIDA
HISTIONIDAE
Reclinomonas americana Flavin & Nerad, 1993 F

SIX

Phylum METAMONADA

oxymonads, parabasalids, *Giardia*, and kin

TONY CHARLESTON, F. R. (BERTHA) ALLISON,
GARETH LLOYD-JONES

Phylum Metamonada embraces a number of zooflagellate groups that were not always recognised as being related. The Greek word *monas* means oneness, in this case alluding to unicells; *meta-*, over, near, implies a grouping. Cavalier-Smith (2003a) has described changing views of the groups now included in the phylum. At one time, actual numbers of cilia (or flagella as they were referred to in flagellate unicells) were important in flagellate classification. Flagellates with six or more were called Polymastiga or Polymastigina (literally 'many whips'), or polymonads, but this group was somewhat heterogeneous. Grassé (1952) referred to most of the zooflagellates with four or more cilia as Metamonadina, but, with the advent of the electron microscope and the discovery of the remarkable diversity of internal structures in protozoans, Honigberg (1973) created a group called Parabasalia for a core cluster of metamonads comprising trichomonads and hypermastigotes. Parabasalids have 'parabasal bodies', comprising a pair of fibres and their associated Golgi dictyosomes (one or a pair), which participate in the synthesis, storage, and transport of proteins, plus other distinctive features. Most have an axial structure, made up of a row of microtubules inrolled to form a simple or spiral tube.

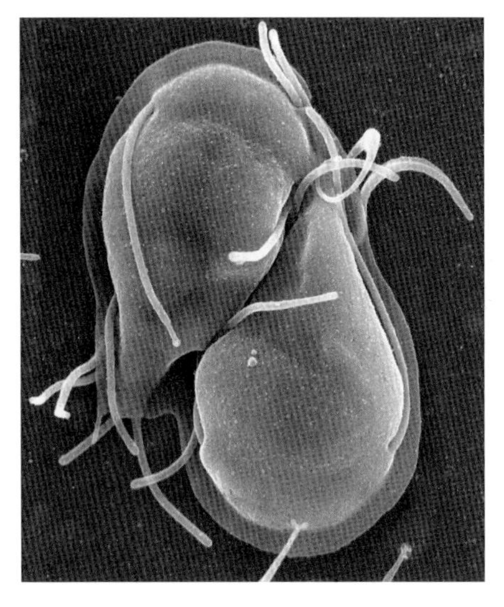

A pair of *Giardia duodenalis* cells that have almost completely separated following cell division.

W. Zacheus Cande, University of California, Berkeley

Cavalier-Smith (1981) raised Parabasalia to phylum rank, retaining a phylum Metamonada for the balance of Grassé's group plus diplomonads like *Giardia*, the most notorious of all metamonads. As constituted, the Metamonada were believed to lack mitochondria, not as an evolutionarily derived feature but as an ancestral feature; thus metamonads were thought to be a primitive group. In fact, metamonads retain mitochondrial relics, including nuclear genes derived from them, and organelles known as hydrogenosomes and mitosomes. Thus, although mitochondria in the strict sense are also lacking from Parabasalia, Cavalier-Smith (1987) regarded hydrogenosomes as having evolved from mitochondria and hence Parabasalia as not closely related to Metamonada. Prior to this, however, he united those groups lacking readily distinguishable mitochondria, including *Entamoeba* and *Pelomyxa*, the spore-forming microsporidians, and the four zooflagellate groups – diplomonads, retortamonads, oxymonads, and parabasalids – as a taxon Archezoa ('ancient animals') (Cavalier-Smith 1983).

Since then, ultrastructural and molecular work has demonstrated that *Entamoeba* has double-membrane-bounded organelles that may have arisen from mitochondria and *Pelomyxa* is only secondarily amitochondriate. Both taxa are now included in phylum Amoebozoa. Microsporidians are related to fungi (Keeling & Doolittle 1996; Germot et al. 1997), and diplomonads also contain relict mitochondria (Tovar et al. 2003). The present classification of phylum Metamonada (Cavalier-Smith 2003a,b) reunites Parabasalia and oxymonads with the balance of Grassé's Metamonadina. Metamonada may have evolved

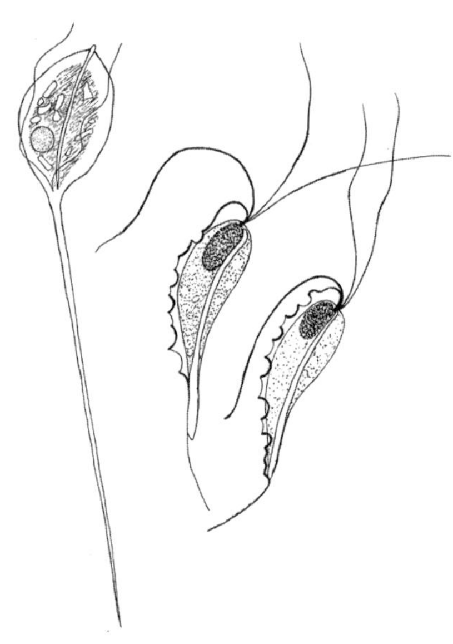

Oxymonas diundulata (left) from the native termite *Kalotermes brouni* and *Trichomonas hoplodactyli* (right) from the native gecko *Woodworthia* sp.

From Nurse 1945 and Percival 1941

from a malawimonad-like ancestor (phylum Loukozoa) by ciliary multiplication and conversion of mitochondria into hydrogenosomes.

There are two subphyla. Subphylum Anaeromadea includes four-ciliate species that lack cristate mitochondria and mostly live in anaerobic conditions. The cilia are usually arranged as two pairs of two, with one or more adhering to the body and trailing posteriorly. *Trimastix* (in its own order), a genus not yet found in New Zealand, is free-living in sediments of lakes and oceans. In contrast, oxymonads (order Oxymonadida) are intestinal parasites of insects (termites and wood-eating cockroaches) and myriapods. One species is known from New Zealand – *Oxymonas diundulata* (Nurse 1945). The host is the endemic *Kalotermes brouni* (Kalotermitidae), the most widespread termite species in New Zealand, extending to the Chatham Islands. It inhabits dry decaying wood of semi-hollow native trees, untreated power poles, fence posts, and buildings. Nurse (1945) obtained specimens from podocarps (*Podocarpus totara* and *Prumnopitys taxifolia*) and broadleaf (*Griselinia littoralis*) on Banks Peninsula, and from pukatea (*Laurelia novae-zelandiae*) and weatherboards of a house near Nelson.

Subphylum Trichozoa, embracing three superclasses, includes several protozoan orders, not all of which have been found in New Zealand. Members of superclass Parabasalia are parasitic or symbiotic in the guts of insects. Bacteria associated with the protozoans digest cellulose ingested when the insect host feeds on wood, liberating sugars that are used by both bacteria, protozoan, and host. Particles of wood are taken up by a posterior ingestive zone of the protozoan's cell membrane and tiny fragments can be seen inside the cell. The parabasal body distinguishes parabasalids from oxymonads, which lack it. Within the Parabasalia, only the order Trichomonadida is known from New Zealand. Trichomonads typically have 4–16 cilia, often associated with supernumerary nuclei (up to 1000 in some members of Calonymphidae). Calonymphids and devescovinids are found only in the hindguts of termites, whereas trichomonadids are widely distributed in invertebrates and vertebrates. Nurse (1945) described *Caduceia calotermidis* (Devescovinidae) and *Trichomonas agilis* (Trichomonadidae) from *Kalotermes brouni*.

Eight other species of Trichomonadida are known from vertebrates in New Zealand, two of them from humans. A number of unidentified *Trichomonas* spp. have also been recorded. Percival (1941) described *Trichomonas hoplodactyli* from a gecko nominally identified as *Hoplodactylus maculatus*. [The particular population, from the shingly margin of the Cass River in Canterbury, is currently recognised to be an undescribed species of *Woodworthia*; it is referred to by Hitchmough et al. (2009, 2010) as '*Hoplodactylus* aff. *maculatus* "Southern Alps"'.] Percival noted that the protozoan was present in large numbers in the hindgut of the reptile. It has also been recorded from geckos on a number of islands in the Wellington region (Ainsworth 1985, cited by McKenna 2003). Motile organisms identified morphologically as *Trichomonas* sp. have also been recorded in fresh faecal

Diversity by environment

Taxon	Marine	Terrestrial	Freshwater
Anaeromonadea	0	1	0
Trichomonadea	0	11	2
Trichonymphea	0	2	0
Trepomonadea	0	1	0
Retortamonadea	0	1	0
Totals	**0**	**16**	**2**

* Parasites are accorded the major habitat of their hosts.

Summary of New Zealand metamonad diversity

Taxon	Described species	Known undescribed/ undetermined species	Estimated unknown species	Adventive species	Endemic species	Endemic genera
Anaeromonadea	1	0	6	0	1	0
Trichomonadea	10	3	45	7	3	0
Trichonymphea	2	0	4	0	2	1
Trepomonadea	1	0	2	1	0	0
Retortamonadea	1	0	3	1	0	0
Totals	**15**	**3**	**60**	**9**	**6**	**1**

smears from Marlborough green geckos (*Naultinus manukanus*) (Gartrell & Hare 2005) and captive-reared tuatara (*Sphenodon punctatus*) (Gartrell et al. 2006). Brace et al. (1953) reported *Monocercomonas batrachorum* (as *Eutrichomastix*) and *Trichomitus batrachorum* (as *Tritrichomonas*) from the rectum of the introduced Green and golden bell frog *Litorea aurea*.

Tritrichomonas foetus has been recorded as causing sporadic breeding problems in beef cattle in New Zealand (e.g. Bruere 1980, as *Trichomonas foetus*). Venereal transmission of the causative organism from infected bulls to susceptible (non-immune) cows can lead to early foetal death and abortion; this can have significant economic impact on cattle enterprises through decreased reproductive efficiency. In the last decade or so *T. foetus* has also been recognised in Europe and the USA as an intestinal parasite of domestic cats that can cause diarrhoea of varying severity (Levy et al. 2003). The organism has recently been identified in cats in New Zealand (Kingsbury et al. 2010).

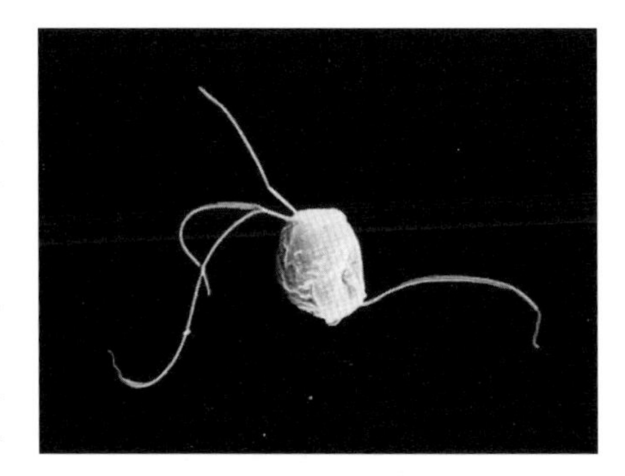

Tritrichomonas foetus.
The Free Dictionary

Trichomonas gallinae is common in the upper digestive tracts of birds, especially columbiformes (doves and pigeons). Although it can be present in nearly 80–90% of pigeons, most show no signs of disease. It is the causative agent of the disease known as 'canker' in pigeons and is usually transmitted to young birds when being fed with regurgitated crop secretions by the parent birds. It can also be transmitted to raptors feeding on infected pigeons, causing the disease traditionally known as 'frounce'. *Trichomonas* sp. has also been reported from the New Zealand native pigeon or kereru (*Hemiphaga novaeseelandiae*); infected birds were in poor condition (Blanks et al. 1995; Gartrell 2002). In the UK and Canada, *T. gallinae* has recently been identified as the cause of an emerging disease of wild finches, with high mortality rates and major effects on bird populations. So far this species has not been recorded in New Zealand.

Pentatrichomonas hominis, which is non-pathogenic, was recorded by Richardson et al. (1943) in stools of 3% of inmates at a Porirua mental hospital. Trichomoniasis, a common sexually transmitted disease caused by *Trichomonas vaginalis,* is of much more concern. This pathogen of the male and female urogenital system infects an estimated 180 million people worldwide each year. Infection ranges from an asymptomatic carrier state to acute inflammatory disease; it generally causes urethritis in men and irritation of the vagina in women, with a yellowy-green frothy discharge and malodour in the latter. Infection in women can also be associated with low birth weight and pre-term delivery of infants. An epidemiological study of the demographic characteristics of women with trichomoniasis reporting to Auckland sexual health clinics in 1998–1999 has been carried out. The incidence was 2.2% of those reporting, the average age of which was 26.5 years. Maori and other Polynesian women were over-represented (Lo et al. 2002).

Histomonas meleagridis, the cause of the disease known as 'black-head', is common in chickens, turkeys, and peafowl; it has also been recorded from red-legged partridge and ring-necked pheasant in New Zealand (McKenna 2010). The organisms primarily affect the caeca and liver. Sporadic disease outbreaks occur in New Zealand, mainly in free-range birds. The mode of transmission is unusual in that it involves carriage of the protozoan in the eggs or larvae of species of *Heterakis,* nematode parasites inhabiting the caeca.

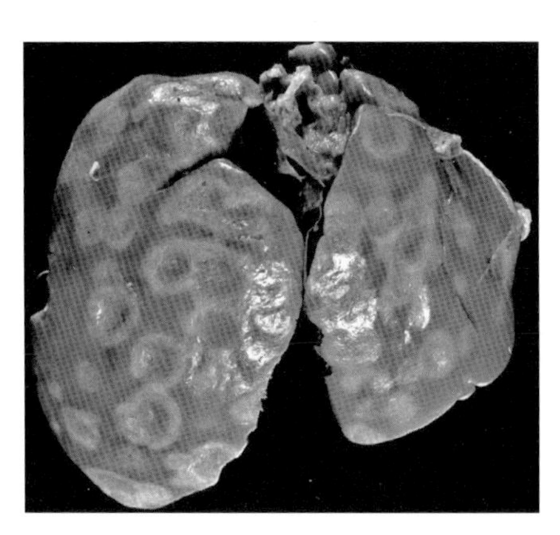

Necrotic lesions in the liver of a turkey infected with blackhead, *Histomonas meleagridis.*
J. E. Ackert, courtesy Kansas State University

Protozoans of the order Trichonymphida (hypermastigotes) have hundreds and even thousands of cilia attached to special bands, generally only one or a few nuclei, and branched or multiple parabasal bodies. These organisms live in the intestines of termites and cockroaches. Two species are known from New Zealand. Helson (1935) described *Spirotrichosoma magna* (the first intestinal protozoan to be described from New Zealand) from the hindgut of the endemic forest termite *Stolotermes ruficeps* and Nurse (1945) described a new genus and species, *Cyclojoenia australis,* from the same termite. *Stolotermes ruficeps* inhabits moist decaying wood of both native and introduced tree species (but not buildings) and, thanks to its commensal protozoans and bacteria, a young

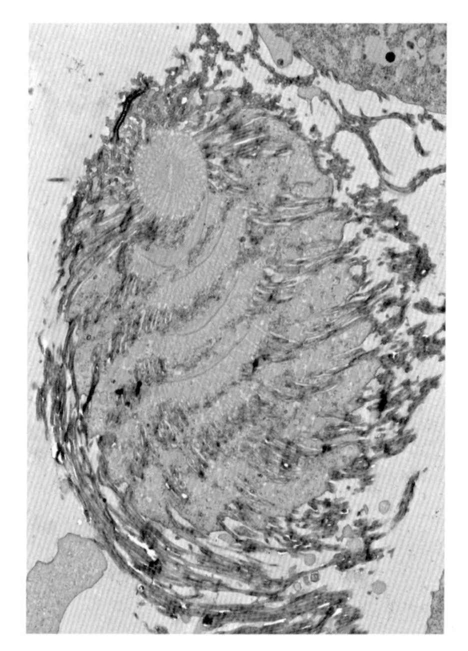

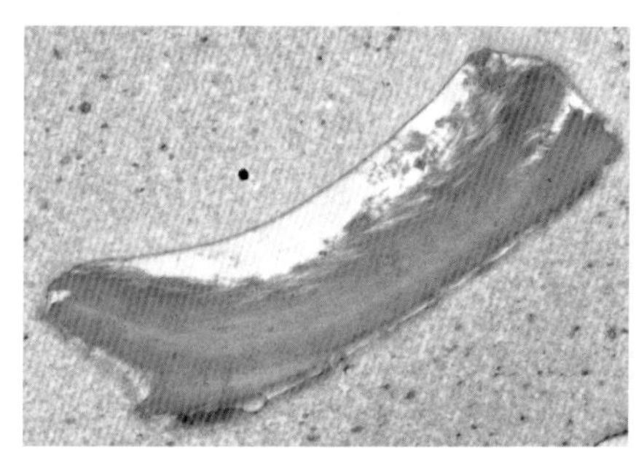

Transmission electron microscope image of a thin section through a parabasalid inhabiting the gut of the native termite *Stolotermes ruficeps* (upper photo). The hair-like structures round the periphery of the cell are spirochaete bacteria. The lower photo shows a fragment of wood cell wall in a food vacuole of the protozoan.

Lloyd Donaldson, Scion

reproductive pair can consume up to 16 cubic centimetres of wood in six months (Morgan 1959). This is more than ten times the amount of wood consumed by a dry-wood-inhabiting species. Typically, *Stolotermes* colonies have several hundred termites and can reach up to 6000 per colony (Thorne & Lenz 2001).

Recently, metagenomic sequencing of DNA extracted from the gut contents of *S. ruficeps* has increased the known diversity of Parabasalia in New Zealand. Sequences were obtained with the following high similarities to published data – 87% for *Monocercomonas colubrorum* (see Hampl et al. 1997), 85% for *Teranympha mirabilis* (see Ohkuma et al. 2005), and 91% and 94% for the genera *Trichomonas* and *Tritrichomonas*, respectively (Philippe & Germot 2000; Torres-Machorro et al. 2005). Together with the more than 1000 bacterial species that inhabit the gut, these parabasalids contribute to the digestion of wood by the termite.

Members of superclass Eopharyngia have 1–6 cilia and typically a cytostome ('cell mouth'), an aperture through which food materials pass into the cell, often via a cytopharynx. If there is more than one cilium, one is trailing and associated with the cytostome. Some species are free-living but most are intestinal parasites. *Giardia duodenalis* (called *G. lamblia* in older literature and also referred to as *G. intestinalis*) is found in the intestine of humans and many other vertebrates. [The name *intestinalis* was introduced first, but as *Cercomonas intestinalis* Lambl, 1859, a junior homonym of an unrelated species, *Cercomonas intestinalis* (Ehrenberg, 1838), hence by the rules of the International Code of Zoological Nomenclature it cannot be used. The next available name is *Giardia duodenalis* (Davaine, 1875), which antedates *G. lamblia* Kunstler, 1882.] Other *Giardia* species, mostly morphologically indistinguishable, have been described from mammals, reptiles, amphibians, and birds. *Giardia* is a highly specialised protozoan, pear-shaped and having a sucker that permits adhesion to the intestinal wall. There is no cytostome, but a pocket between the paired ventral cilia forms a sort of gullet.

The taxonomy of *Giardia*, its host specificity, and the extent to which human infections are zoonotic have been the subject of debate for many years. Numerous species were named after the hosts from which they were isolated but many of these were morphologically indistinguishable and their taxonomic status was uncertain. The development of PCR (polymerase chain reaction)-based techniques has provided a means of unequivocally distinguishing between morphologically similar isolates and clarifying issues of host-specificity (Thompson 2002). It emerged that the *G. duodenalis* morphological group can be subdivided into six genotype assemblages that differ in host-range and host-specificity. Two of these infect humans and a range of other hosts including domestic pets, livestock, and wild animals; the others are more host-specific, with two restricted to dogs, one to cats, one to cattle, sheep, and pigs, and one to domestic and wild rodents. The implications of this research for understanding the epidemiology of *Giardia* infections are discussed in detail by Thompson (2004). He concluded that, while the evidence confirms that animals *can* be a source of human infections, the extent to which this occurs is still uncertain and that an animal origin for water-borne human infections appears to be uncommon. It is clear, however, that the epidemiology of infection and clinical disease is complex and much remains to be discovered. With recent advances in understanding the molecular genetics of *Giardia,* the genetic heterogeneity of isolates from mammals has become apparent and this, together with increasing knowledge of the host specificity of various isolates, has led to the suggestion that the taxonomy of the genus should be revised and that many of the earlier host-related species names should be reinstated (Monis et al. 2008).

While most infections of humans with *Giardia* are asymptomatic, ingestion of cysts, most commonly in drinking water, can result in enteritis, with acute explosive or chronic diarrhoea, abdominal pain, dehydration, and associated symptoms such as nausea, anorexia, and weight loss. The organisms multiply in the small intestine, affecting intestinal permeability and causing inflammation.

Encysted *Giardia* organisms develop, to be passed in the faeces, and these may subsequently contaminate drinking water. As few as 10 cysts can initiate an infection. Today, *Giardia* is the most widespread human intestinal parasite in the world. Approximately 200 million people are estimated to be infected with the parasite globally, with 500,000 new cases reported annually (Hoque et al. 2004). In New Zealand it has been a notifiable disease since 1 July 1996. Before then, surveillance data were collected on an ad hoc basis (van Duivenboden & Walker 1993). In a four-year period from 1996 to 2000, 7818 cases were reported, making it the most commonly notified waterborne disease in New Zealand (Hoque et al. 2002). Most of the cases occurred in the 1–4-year age range, followed by the 25–44 age group and were mostly of European/Pakeha ethnicity. The incidence rate in New Zealand in 2002 was 41.4 per 100,000 population, one of the highest among developed countries (Hoque et al. 2004). Infected humans and animals can contaminate water in streams and on farmland (Chilvers et al. 1998), but, as noted above, the evidence indicates that most water-borne human infections originate from humans (Snel et al. 2009). It has also become clear that the severity of infection in humans is affected by a variety of factors including concurrent infections and differences in virulence between genotype assemblages (Thompson 2004).

Giardia infections have been recorded from a wide range of mammals and birds in New Zealand (McKenna 2009, 2010) but not, so far, from New Zealand reptiles. Infections in young domesticated mammals such as cattle, goats and sheep are often associated with diarrhoea, particularly in intensively reared animals. However, the zoonotic significance of *Giardia* infections in domesticated and wild animals in this country is unclear despite concerns about the potential for transmission from animals to humans.

The Retortamonadida are small (5–20 micrometres) bacteria-feeding intestinal parasites. The best known is *Chilomastix mesnili*, which occurs in humans, monkeys, and pigs. Richardson et al. (1943) reported this species, as cysts, in the stool of an individual in a mental hospital in Porirua.

The potential for research on Metamonada in New Zealand is high, both on taxa that infect humans and economically important vertebrates and the endosymbiotic fauna associated with native arthropods and other animals.

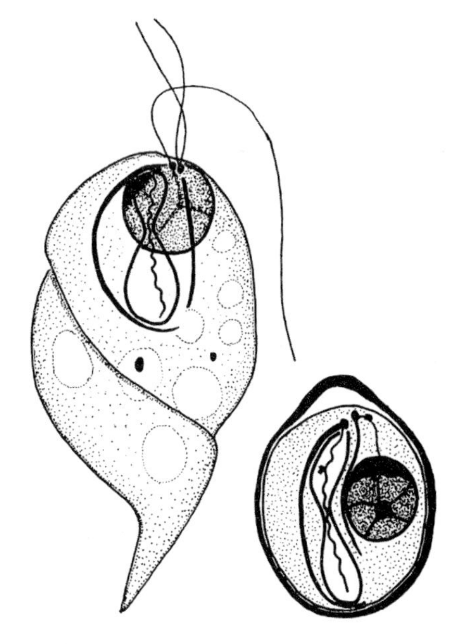

Flagellated and cyst stages of *Chilomastix mesnili* from a human.

After Kofoid & Swezy 1920

Authors

F. R. (Bertha) Allison: late of Canterbury Museum, Rolleston Avenue, Christchurch, New Zealand. Bertha Allison died in May 2010.

Dr W. A. G. (Tony) Charleston: Institute of Veterinary, Animal and Biomedical Sciences, Massey University, Palmerston North (Present address: 488 College Street, Palmerston North, New Zealand) [charleston@inspire.net.nz]

Dr Gareth Lloyd-Jones Scion, Private Bag 3020, Rotorua 3046, New Zealand [gareth.lloyd-jones@scionresearch.com]

References

BRACE, E. C.; CASEY, B. R.; COSSHAM, R. B.; McEWAN, J. M.; McFARLANE, B. G.; MACKEN, J.; MONRO, P. A.; MORELAND, J. M.; NORTHERN, J. B.; STREET, R. J.; YALDWYN, J. C. 1953: The frog, *Hyla aurea* as a source of animal parasites. *Tuatara 5*: 12–21.

BLANKS, R.; BAILEY, K.; GRAHAM, C. 1995: Native wood pigeon – Trichomoniasis. *Kokako 4*: 3.

BRUERE, S. N. 1980: *Trichomonas foetus* infection in a beef herd. *New Zealand Veterinary Journal 30*: 15–16.

CAVALIER-SMITH, T. 1981: Eukaryote kingdoms: seven or nine? *Biosystems 14*: 461–481.

CAVALIER-SMITH, T. 1983: Endosymbiotic origin of the mitochondrial envelope. Pp. 265–279 *in*: Schenk, H. E. A.; Schwemmler, W. (eds), *Endocytobiology II: Intracellular space as oligogenetic ecosystem*. De Gruyter, Berlin. 1071 p.

CAVALIER-SMITH, T. 1987: The simultaneous symbiotic origin of mitochondria, chloroplasts, and microbodies. *Annals of the New York Academy of Sciences 503*: 55–71.

CAVALIER-SMITH, T. 2003a: The excavate protozoan phyla Metamonada Grassé emend. (Anaeromonadea, Parabasalia, *Carpediemonas*, Eopharyngia) and Loukozoa emend. (Jakobea, *Malawimonas*): their evolutionary affinities and new higher taxa. *International Journal of Systematic and Evolutionary Microbiology 53*: 1741–1758.

CAVALIER-SMITH, T. 2003b: Protist phylogeny and the high-level classification of Protozoa. *European Journal of Protistology 39*: 338–348.

CHILVERS, B. L.; COWAN, P. E.; WADDINGTON, D. C.; KELLY, P. J.; BROWN, T. J. 1998: The prevalence of infection of *Giardia* spp. and *Cryptosporidium* spp. in wild animals on farmland, southeastern North Island, New Zealand. *International Journal of Environmental Health Research 8*: 59–64.

GARTRELL, B. 2002: Clinical cases of interest from the wildlife ward at Massey University. *Kokako 9*: 9.

GARTRELL, B. D.; HARE, K. M. 2005: Mycotic

dermatitis with digital gangrene and osteomyelitis, and protozoal intestinal parasitism in Marlborough green geckos (*Naultinus manukanus*). *New Zealand Veterinary Journal 53*: 363–367.
GARTRELL, B. D.; JILLINGS, E.; ADLINGTON, B. A.; MACK, H.; NELSON, N. J. 2006: Health screening for a translocation of captive-reared tuatara (*Sphenodon punctatus*) to an island refuge. *New Zealand Veterinary Journal 54*: 344-349.
GERMOT, A.; PHILIPPE, H.; Le GUYADER, H. 1997: Evidence for loss of mitochondria in Microsporidia from a mitochondrial-type HSP70 in *Nosema locustae*. *Molecular and Biochemical Parasitology 87*: 159–168.
GRASSÉ, P.-P. 1952: Classe des zooflagellés: Zooflagellata ou Zoomastigina (Euflagellata Claus 1887). Généralités. *Traité de Zoologie 1*: 574–578.
HAMPL, V.; CEPICKA, I.; FLEGR, J.; TACHEZY, J.; KULDA, J. 2007: Morphological and molecular diversity of the monocercomonadid genera *Monocercomonas*, *Hexamastix*, and *Honigbergiella* gen. nov. *Protist 158*: 365–383.
HELSON, G. A. H. 1935: *Spirotrichosoma magna* (n. sp.) from a New Zealand termite. *Transactions of the Royal Society of New Zealand 64*: 251–255.
HITCHMOUGH, R. A.; PATTERSON, G. K.; DAUGHERTY, C. H. 2009: New Zealand living and recently extinct reptiles. P. 542 *in*: Gordon, D. P. (ed.), *New Zealand Inventory of Biodiversity Volume One. Kingdom Animalia: Radiata, Lophotrochozoa, Deuterostomia*: in Chapter 24, p. 542. Canterbury University Press, Christchurch. 566 [+ 16] p.
HITCHMOUGH, R. A.; HOARE, J. M.; JAMIESON, H.; NEWMAN, D.; TOCHER, M. D.; ANDERSON, P. J.; LETTINK, M.; WHITAKER, A. H. 2010: Conservation status of New Zealand reptiles, 2009. *New Zealand Journal of Zoology 37*: 203–224.
HONIGBERG, B. M. 1973: Remarks upon trichomonad affinities of certain parasitic protozoa. P. 187 *in*: De Puytorac, P.; Grain, J. (eds), *Progress in Protozoology: Abstracts of Papers Read at the 4th International Congress of Protozoology*. Université de Clermont-Ferrand, Clermont-Ferrand.
HOQUE, M. E.; HOPE, V. T.; SCRAGG, R. 2002: *Giardia* infection in Auckland and New Zealand: trends and international comparison. *New Zealand Medical Journal 115*: 121–123.
HOQUE, M. E.; HOPE, V. T.; SCRAGG, R.; BAKER, M.; SHRESTHA, R. 2004: A descriptive epidemiology of giardiasis in New Zealand and gaps in surveillance data. *The New Zealand Medical Journal 117*: 1–13.
KEELING, P.; DOOLITTLE, W.F. 1996: a-tubulins from early diverging eukaryotic lineages: divergence and evolution of the tubulin family. *Molecular Biology and Evolution 13*: 1297–1305.
KINGSBURY, D. D.; MARKS, S. L.; CAVE, N. J.; GRAHN, R. A. 2010: Identification of *Tritrichomonas foetus* and *Giardia* spp. infection in pedigree show cats in New Zealand. *New Zealand Veterinary Journal 58*: 6–10.
LEVY, M. G.; GOOKIN, J. L.; POORE, M.; BIRKENHEUER, A. J.; DYKSTRA, M. J.; LITAKER, R.W. 2003: *Tritrichomonas foetus* and not *Pentatrichomonas hominis* is the etiological agent of feline trichomonal diarrhea. *Journal of Parasitology 89*: 99–104.
LO, M.; REID, M.; BROKENSHIRE, M. 2002: Epidemiological features of women with trichomoniasis in Auckland sexual health clinics: 1998–99. *The New Zealand Medical Journal 115*: 1–6.
McKENNA, P. B. 2009. An updated checklist of helminth and protozoan parasites of terrestrial mammals in New Zealand. *New Zealand Journal of Zoology*. 36: 89–113.
McKENNA, P. B. 2010. An updated checklist of helminth and protozoan parasites of birds in New Zealand. *WebmedCentral Parasitology. 1(9): WMC00705*: 1–7.
McKENNA, P. B. 2003: An annotated checklist of ecto- and endoparasites of New Zealand reptiles. *Surveillance 30*: 18–25.
MONIS, P. T.; CACCIO, S. M.; THOMPSON, R. C. A. 2008: Variation in *Giardia*: towards a taxonomic revision of the genus. *Trends in Parasitology 25*: 93–-100.
MORGAN, F. D. 1959: The ecology and external morphology of *Stolotermes ruficeps* Brauer (Isoptera: Hoeltermitidae). *Transactions of the Royal Society of New Zealand 86*: 155–-195.
NURSE, F. R. 1945: Protozoa from New Zealand termites. *Transactions of the Royal Society of New Zealand 74*: 305–314, pls 42–44.
OHKUMA, M.; IIDA, T.; OHTOKO, K.; YUZAWA, H.; NODA, S.; VISCOGLIOSI, E.; KUDO, T. 2005: Molecular phylogeny of parabasalids inferred from small subunit rRNA sequences, with emphasis on the Hypermastigea. *Molecular Phylogenetics and Evolution 35*: 646–655.
PERCIVAL, E. 1941: *Trichomonas hoplodactyli* n. sp. from a New Zealand gecko. *Records of the Canterbury Museum 7*: 373–375.
PHILIPPE, H.; GERMOT, A. 2000: Phylogeny of eukaryotes based on ribosomal RNA: long-branch attraction and models of sequence evolution. *Molecular Biology and Evolution 17*: 830–834.
RICHARDSON, L. R.; CLARK, A. E.; RALPH, P. M. 1943: Studies on the entozoa of Man in New Zealand. Part 1.–A preliminary note on the results from the examination of inmates of a mental hospital. *Transactions of the Royal Society of New Zealand 73*: 239–247.
SNEL, S. J.; BAKER, M. G.; KAMALESH VENUGOPAL 2009: The epidemiology of giardiasis in New Zealand, 1997-2006. *New Zealand Medical Journal 122*: 62-75.
THOMPSON, R. C. A. 2002: Towards a better understanding of host-specificity and the transmission of *Giardia*: The impact of molecular epidemiology. Pp. 55–69 *in*: Olson, B. E.; Wallis, P. M. (eds), *Giardia – the Cosmopolitan Parasite*. CAB International, Wallingford, UK. 352 p.
THOMPSON, R. C. A. 2004: The zoonotic significance and molecular epidemiology of *Giardia* and giardiasis. *Veterinary Parasitology 126*: 15–35.
THORNE, B. L.; LENZ, M. 2001: Population and colony structure of *Stolotermes inopus* and *Stolotermes ruficeps* (Isoptera: Stolotermitinae) in New Zealand. *New Zealand Entomologist 24*: 63–70.
TORRES-MACHORRO, A. L.; HERNANDEZ, R.; ALDERETE, J. F.; LOPEZ-VILLASENOR, I. 2009: Comparative analyses among the *Trichomonas vaginalis*, *Trichomonas tenax*, and *Tritrichomonas foetus* 5S ribosomal RNA genes. *Current Genetics 55*: 199–210.
TOVAR, J.; LÉON-AVILA, G.; SÁNCHEZ, L.; SUTAK, R.; TACHEZY, J.; van der GIEZEN, M.; HERNÁNDEZ, M.; MÜLLER, M.; LUCOCQ, J. M. 2003: Mitochondrial remnant organelles of *Giardia* function in iron-sulfur protein maturation. *Nature 426*: 172–176.
van DUIVENBODEN, R.; WALKER, N. 1993: Giardiasis in New Zealand, 1991–93. *Communicable Disease New Zealand 93*: 152–153.

Checklist of New Zealand Metamonada

Abbreviations: A, adventive; E, endemic; T, terrestrial (reflecting the habitat of the host species). Underlined genera are endemic.

KINGDOM PROTOZOA
SUBKINGDOM EOZOA
INFRAKINGDOM EXCAVATA
PHYLUM METAMONADA
SUBPHYLUM ANAEROMONADA
Class ANAEROMONADEA
Order OXYMONADIDA
OXYMONADIDAE
Oxymonas diundulata Nurse, 1945 T E

SUBPHYLUM TRICHOZOA
Superclass PARABASALIA
Class TRICHOMONADEA
Order TRICHOMONADIDA
DEVESCOVINIDAE
Caduceia calotermidis Nurse, 1945 T E
MONOCERCOMONADIDAE
Histomonas meleagridis (Smith, 1895) T A
Monocercomonas batrachorum (Dobell, 1907) F A
TRICHOMONADIDAE
Pentatrichomonas hominis (Davaine, 1860) T A
Trichomitus batrachorum (Perty, 1852) F A
Trichomonas agilis Nurse, 1945 T E
Trichomonas gallinae (Rivolta, 1879) T A
Trichomonas trichodactyli Percival, 1941 T E
Trichomonas vaginalis (Donné, 1836) T A
Trichomonas sp. Blanks et al. 1995 T
Trichomonas sp. Gartrell & Hare 2005 T
Trichomonas sp. Gartrell et al. 2006 T
Tritrichomonas foetus (Riedmuller, 1928) T A

Class TRICHONYMPHEA
Order TRICHONYMPHIDA
JOENIIDAE
Cyclojoenia australis Nurse, 1945 T E
SPIROTRICHOSOMIDAE
Spirotrichosoma magna Helson, 1935 T E

Superclass EOPHARYNGIA
Class TREPOMONADEA
Subclass DIPLOZOA
Order GIARDIIDA
GIARDIIDAE
Giardia duodenalis (Davaine, 1875) T A

Class RETORTAMONADEA
Order RETORTAMONADIDA
RETORTAMONADIDAE
Chilomastix mesnili (Wenyon, 1910) T A

SEVEN

Phylum
AMOEBOZOA

amoebas, pelobionts, slime moulds, and kin

RALF MEISTERFELD, STEVEN L. STEPHENSON,
RUSSELL SHIEL, LORWAI TAN, TONY CHARLESTON

Amoebozoa ('amoeba animals') include the classical amoebas studied by generations of biology students, most slime moulds – formerly classified among the fungi – and a number of less-familiar forms. Most amoebozoans, at least at some stage of their life history, exhibit typical amoeboid motion using lobe-like pseudopodia, in contrast to the long thread-like processes called filopodia that attach to surfaces and draw the cell forward, as in members of the phyla Cercozoa, Foraminifera, and Radiozoa. Formerly, all such protists having some form of pseudopodia or locomotiove protoplasmic flow were included in the class or phylum Sarcodina. Recent discoveries based on electron-microscope and molecular studies have shown that ancestors of amoebozoans were aided in their movement by a cilium. This has been retained in some living forms, and a few species have two cilia or are secondarily multiciliate. [Formerly, these types of cilia were referred to as flagella, a term that is now restricted to the motile filaments of bacteria.]

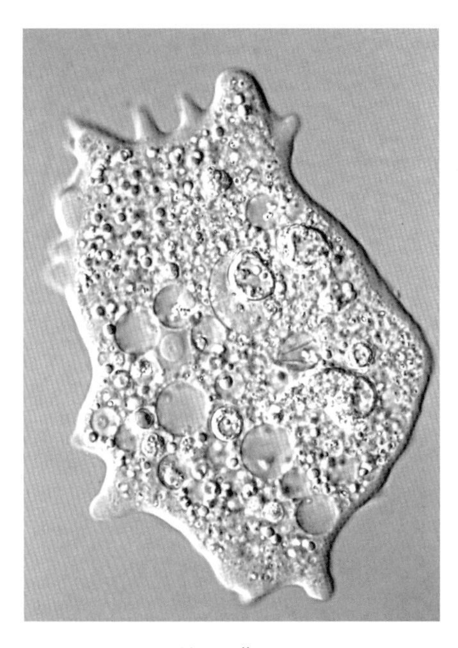

Mayorella sp.
David Patterson, Linda Amaral-Zettler, Mike Peglar, Tom Nerad per MBL (Micro*scope)

Amoebozoans are currently regarded as comprising three subphyla (Cavalier-Smith 2009). Subphylum Conosa comprises the Mycetozoa (slime moulds) and the secondarily mitochondrion-lacking Archamoebae. Subphylum Lobosa includes aerobic amoebas, with or without shells (tests), that typically lack fruiting bodies, have broad pseudopodia (lobes or sheets), and are rarely ciliate. Subphylum Protamoebae comprises only the breviates (Cavalier-Smith et al. 2004; Minge et al. 2009), remarkable amoeboflagellates that have a unique gait among eukaryotes – individuals 'walk' on thin but robust leg-like pseudopodia from the anterior of the cell body that adhere to the substratum while the cell body proceeds forward like a tractor on treads. The filose 'legs' often remain as trailing filaments before retracting into the cell body.

There are eight amoebozoan classes – five in Conosa, two in Lobosa, and a single class of breviates; only the latter is not yet known from New Zealand. The familiar naked amoebas of ponds, aquatic sediments, and soils, as well as a number of testate forms, are included in the Lobosea, the only class with all species devoid of cilia. In polluted waters, including waste-water ponds, free-living amoebas can play a useful role by consuming faecal bacteria. Members of class Flabellinea, among which are the genera *Mayorella* and *Vannella* in New Zealand, include strongly flattened amoebas that have an amorphous surface coating (glycocalyx) that may be finely hairy or scaly. The related marine parasite *Neoparameoba pemaquidensis* causes amoebic gill disease (AGD) in farmed salmon. In New Zealand, it has been responsible for the outbreak of this disease in the introduced Chinook salmon *Oncorhynchus tshawytscha* (Wong et al. 2004) and other species may be affected (Diggles et al. 2002).

Class Variosea includes a varied range of amoeboid and ciliated forms with

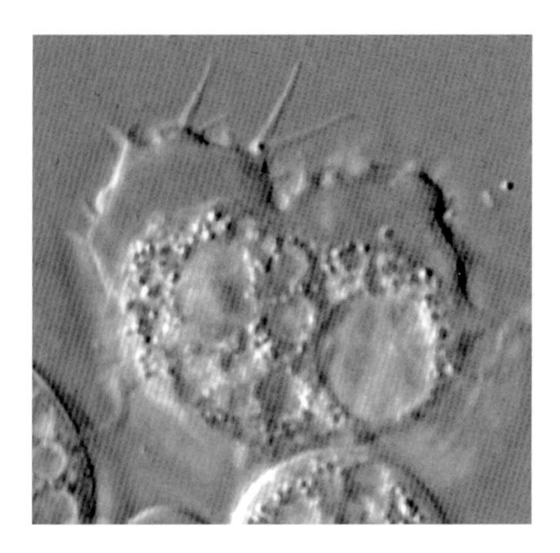

Acanthamoeba sp.
David Patterson, Linda Amaral-Zettler, Mike Peglar, Tom Nerad per MBL (Micro*scope)

resting cysts, among which is the genus *Acanthamoeba*. Overseas, species of *Acanthamoeba* are causative agents of granulomatous amoebic encephalitis, a fatal disease of the nervous sytem, and amoebic keratitis, a painful sight-threatening disease of the eye (Marciano-Cabral & Cabral 2003). In New Zealand, only *Naegleria* (phylum Percolozoa) has been implicated in causing primary amoebic encephalitis (Cursons & Brown 1976; Cursons et al. 1978; Brown et al. 1983), even though *Acanthamoeba* species co-occur in the warm geothermal waters that their virulent strains tolerate or thrive in. Since soil is the preferred habitat of small free-living amoebas, it probably acts as a reservoir, and contamination of geothermal waters may occur via runoff after rain. Pathogenic strains of free-living amoebas can be remarkably ubiquitous, having been reported from tank-fed domestic water, contact-lens cases, air, humidifier systems, swimming pools, throat and nasal cavities, soil, sewage, and almost any aquatic habitat, including seawater. *Acanthamoeba* can be a carrier of *Legionella* bacteria.

Acanthamoeba individuals can also live in most contact-lens storage solutions, which probably explains the occasional association of *Acanthamoeba* with contact lenses and transmission via moist lens cases. A study conducted in Hamilton, New Zealand, showed that contact-lens storage cases can carry biofilms harbouring bacteria, fungi, and a variety of protists including *Naegleria, Hartmannella,* and *Acanthamoeba* (Gray et al. 1995). Infection with the latter can cause inflammation/infection of the cornea, a condition that is not usually fatal but may cause blindness through ulceration (Murdoch et al. 1998).

At least three species of *Acanthamoeba* have been identified from New Zealand (Cursons et al. 1978). Several unidentified species have also been reported (Gray et al. 1995), but whether or not they are conspecific with known species has not been ascertained. More than 20 species names have been ascribed to *Acanthamoeba* (Anon. 2003). A comprehensive review of *Acanthamoeba* species as agents of disease in humans has recently been published that covers all aspects of infection and pathogenicity (Marciano-Cabral & Cabral 2003).

Kirk (1906) described a much-branched plasmodial marine amoeba that he obtained from a gelatinous substance associated with a polychaete egg mass. He named it as a new genus and species, *Myxoplasma rete,* noting similarities with *Labyrinthula*, a chromistan organism, but it may be closer to *Stereomyxa*, a genus of uncertain affinities in the Variosea. Kirk (1906) also obtained a non-

Summary of New Zealand amoebozoan diversity

Taxon	Described species + subspecies	Known undescr./ undetermined species	Estimated unknown species	Adventive species	Endemic species + subspecies	Endemic genera
Conosa	245+1	3	>45	12	?4	?1
Archamoebea	6	1	>20	6	0	0
Variosea	5	2	>10	6	?1	?1
Protostelea	23	0	<5	0	0	0
Dictyostelea	14	0	<5	0	3	0
Myxogastrea	197+1	0	>5	0	0	0
Lobosa	93+8	8	>300–500	1	?2+1	0
Tubulinea	87+8	7	280–480	1	?2+1	0
Flabellinea	6	1	>20	0	0	0
Protamoebae	0	0	5	0	0	0
Breviatea	0	0	5	0	0	0
Totals	338+9	11	>350–550	13	~6+1	?1

plasmodial amoeba from the same sample, naming it *Amoeba agilis*. It may be attributable to the amoebid genus *Stygamoeba*.

The conosan class Archamoebea includes the giant multinucleate amoeba *Pelomyxa*, noted in some New Zealand soil samples (Stout 1961), and some commensals and parasites of vertebrates. Among the latter are species of *Entamoeba*, some of which cause diarrhoea and dysentery in humans while others are non-pathogenic.

While there is very little published information on the occurrence and prevalence of *Entamoeba* and other amoebas in humans in New Zealand, it appears that most of the common species found overseas are present. A 1940s survey of 100 patients in a mental hospital revealed a high prevalence (18%) of what was identified at the time as *E. histolytica* (Richardson et al. 1943a). However, this report preceded by many years the general recognition that the common, non-pathogenic *E. dispar* is a distinct species, morphologically indistinguishable from *E. histolytica* (see Ackers 2002), and it is impossible to say what proportion of the infections were of the former. Interestingly, in a survey of 25 people outside the institution, only one *E. histolytica/dispar* infection was detected (Richardson et al. 1943b).

The considerably higher prevalence of this and some other entozoal infections in the institutionalised patients doubtless reflected the inevitable hygiene problems and increased opportunities for transmission of infections in such circumstances. *Entamoeba histolytica* is potentially highly pathogenic and in some patients leads to the development of hepatic abscesses. A number of such cases have occurred in New Zealand, the patient's histories indicating that the infection was probably acquired overseas (Lane & Nicholson 1984; Ghose 1998); in one reported case, however, this was not so (Brown et al. 1977).

Other, non-pathogenic, amoeba species recorded from humans in New Zealand include *Endolimax nana*, *Entamoeba coli*, and *Iodamoeba buetschlii* (Richardson et al. 1943a,b; Dowling et al. 1999). It is probable that other species, such as *Entamoeba hartmanni*, also occur in New Zealand but there is a lack of published information. *Naegleria* spp. and *Dientamoeba fragilis*, which are certainly present, are reviewed in the present volume in the chapters on Percolozoa and Metamonada (chapters 4 and 6, respectively). One species of *Entamoeba* (*E. morula*) is easily obtainable from the introduced green and golden bell frog

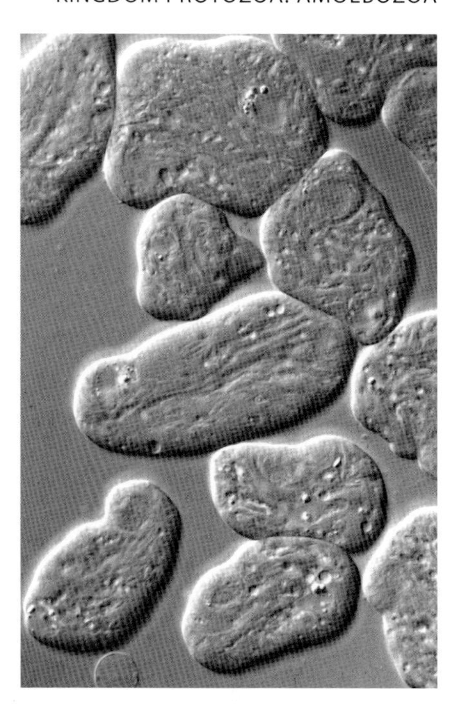

Entamoeba sp.
David Patterson, Linda Amaral-Zettler, Mike Peglar, Tom Nerad per MBL (Micro*scope)

Summary of New Zealand amoebozoan diversity by environment (species + subspecies)†*

Taxon	Marine/ brackish	Terrestrial	Freshwater
Conosa	1	248+1	8
Archamoebea	0	7	1
Variosea	1	7	7
Protostelea	0	23	0
Dictyostelea	0	14	0
Myxogastrea	0	197+1	0
Lobosa	2	58+3	72+7
Tubulinea	1	52+3	67+7
Flabellinea	1	6	5
Protamoebae	0	0	0
Breviatea	0	0	0
Totals	3	306+4	80+7

† Most species can occur in more than one environment, reflected in the tabulation. Hence total numbers across all environments will exceed the overall species tally in the previous table.
* Parasitic species are attributed the major environment of their hosts.

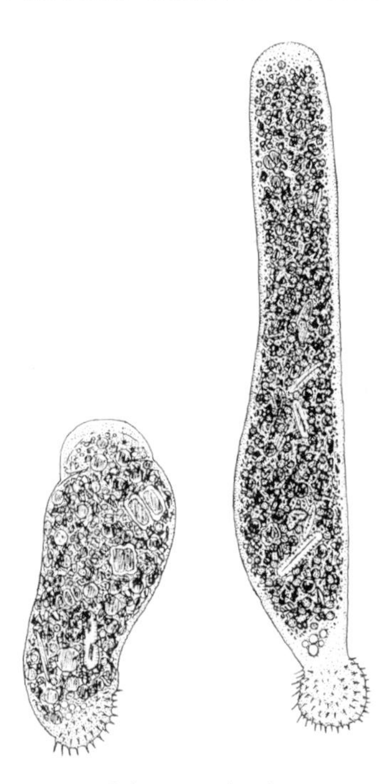

Pelomyxa palustris.
From Bovee 1985

(*Litorea aurea*), in which it is common in the rectum (Brace et al. 1953).

Interestingly, species of *Entamoeba* and *Pelomyxa* lack mitochondria, the 'powerhouses' of the cell that function in cell respiration, and these genera had been regarded as descendants of primitive protozoans (hence the class name Archamoebae). The lack of these important cell organelles is now regarded as a derived – not ancestral – condition, and they are considered to be allied with the mitochondriate amoebas (Cavalier-Smith 1998).

The remaining two conosan classes together comprise the Mycetozoa, made up of the cellular slime moulds (classes Protostelea and Dictyostelea) and plasmodial slime moulds (class Myxogastrea), which are distinctively represented in New Zealand.

The rest of the chapter will go into further detail about the testate amoebas and slime moulds as most work on New Zealand Amoebozoa has been on these groups. The remainder, especially marine forms, have been scarcely studied.

Naked and testate amoebas

These organisms are abundant in New Zealand but very few studies have been published on the naked forms. In the past, some universities imported the large amoeba *Chaos carolinense* (syn. *Chaos chaos*) from biological supply houses in the USA and high-school teachers accessed university cultures for the teaching of biology students. Alternatively, since free-living amoebas are not uncommon in suburban ponds, or even muddy puddles where water has stood for a while, university technicians would capture individuals from local sources and culture them in the laboratory using infusions of split peas, hay, or boiled rice.

Naked and testate amoebas are found in all freshwater habitats, in soils, mosses, and forest litter (Stout 1958, 1961) and on marine beaches. The naked species in the end-chapter checklist are from the published lists in ecological studies carried out by J. D. Stout and colleagues. Stout (1963) noted that grassland topsoil contained 26–124 amoebas in a millilitre of soil water and beech-forest litter 545–9392 individuals. It follows that varied or disturbed soils have modified populations of these and other organisms (Stout 1960). Miller et al. (1955) conducted a before-and-after experiment in which they cleared, burned, and grassed four acres of gorse (*Ulex europaeus*) and manuka (*Leptospermum scoparium*). Following extensive initial changes in numbers and kinds of micro-organisms, there was a gradual return, over 10 months, to conditions not unlike those prevailing before the fire. Protozoan numbers are particularly dependent on an adequate bacterial population.

The diversity of New Zealand testate amoebas is better known. Testates are a polyphyletic group that includes lobose and filose amoebas. All of them have a test or tectum but the filose forms are nowadays regarded as having a different ancestry from the lobose taxa and are included in the phylum Cercozoa. Globally, about 1900 species and subspecies of both types have been described (Decloitre 1986), many of them inadequately, and critical revision will certainly reduce this figure significantly. Many regions of the world like New Zealand, however, are incompletely sampled and numerous new taxa remain to be found.

Apodera vas.
Ralf Meisterfeld

Morphology of testate amoebas

For many species and genera, few if any details are known about the cytoplasm because the thick, opaque shells make study by light or transmission electron microscopy (TEM) difficult. Commonly, there is a zonation of the cytoplasm into an anterior part, close to the aperture, with numerous food and digestive vacuoles, and a denser posterior zone ('chromidium') that stains with basic dyes and contains usually one, rarely two or more, nuclei (e.g. *Arcella*). As viewed using TEM, this is a dense region of rough endoplasmic reticulum with one or more golgi complexes within or more often at the periphery. The latter are involved in the secretion of organic building units and cement and perhaps in the synthesis of siliceous idiosomes – the particles that are extruded to cover the surface of species in the genera *Lesquereusia*, *Quadrulella*, and *Paraquadrula*, for example. Contractile vacuoles responsible for osmoregulation are located lateral to this region. Mitochondria are usually ovoid or spherical and have branched tubular cristae. They are concentrated in the posterior part of the cell. According to the classification of Raikov (1982), two principal types of nuclei are found in testate amoebas – vesicular nuclei, with a generally centrally located nucleolus and sometimes with a few additional very small nucleoli; and ovular nuclei with several to many small nucleoli.

Pseudopodia. Diagnoses of many species (and even genera) are based on empty tests alone. Especially in soil-dwelling species, pseudopodia are difficult to observe. Arcellinida have either fingerlike endolobopodia, which are granular, or completely hyaline ectolobopodia, which can anastomose and often have acute tips (reticulolobopodia). Whether or not taxa can be discriminated by differences in lobose pseudopodia cannot be decided with the data available, because pseudopodial characters can change according to the kind of activity.

Shell. The architecture and composition of the shell have been central in the taxonomy of all shelled amoebas, whether lobose or filose. Four principal types are common – proteinaceous, agglutinate, siliceous, and calcareous. Following the concepts of Ogden (1990) and Meisterfeld (2002a,b), those with a proteinaceous test can be divided into:

- those with a rigid sheet of fibrous material, e.g. Hyalospheniidae;
- those with a test constructed of regularly arranged hollow building units to form an areolate surface, as in Microchlamyiidae and Arcellidae [these shell types have not been described from the filose testate amoebas (Cercozoa)];
- and families that have a more or less flexible membrane enclosing the cytoplasm, e.g. Microcoryciidae.

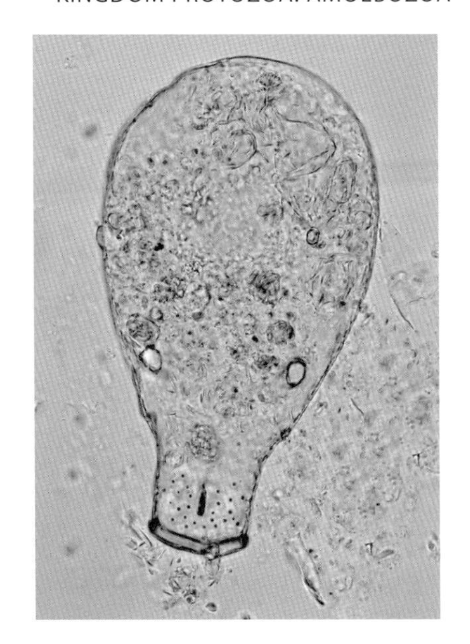

Certesella martiali.
Ralf Meisterfeld

Agglutinated shells of lobose testate amoebas have either a cement matrix of often perforated building units or a sheet-like cement in which foreign materials (xenosomes) like mineral particles (small sand grains) or diatoms are incorporated (suborder Difflugina). The morphology of this cement is of increasing importance for the classification of lobose testate amoebas (Ogden & Ellison 1988; Ogden 1990). In filose amoebas, the amount of organic cement varies and may dominate a shell, or it may be limited to small cement strands to glue the mineral particles together. In all genera studied it is sheet-like.

A few lobose genera (of the family Lesquereusiidae) have siliceous tests, composed of endogenous rods, nails, or rectangular plates that are produced by the organisms themselves (idiosomes). In filose testate amoebas (phylum Cercozoa), a siliceous test composed of secreted plates is characteristic of the order Euglyphida. Here the plates differ in shape, size, and arrangement from genus to genus and also from one species to another and they are often essential for identification.

Calcareous shells are characteristic of the Paraquadrulidae (not yet found in the New Zealand fauna) and the genus *Cryptodifflugia*. The former have rectangular calcite plates bound by an internal sheet of cement, while in the latter a thick layer of calcium phosphate is deposited within an organic template (*C. oviformis*).

Cysts. Under unfavourable environmental conditions, most testaceans produce cysts. A few aquatic genera (e.g. many *Difflugia*) are believed not to form cysts. Resting cysts are formed within the test and have a thick organic wall. Soil and moss species have additionally a special kind of short-term cyst, the precyst. This ability is essential for long-distance transport.

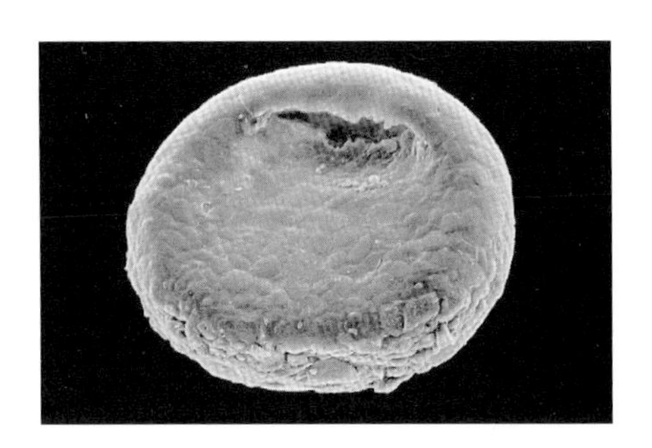

Plagiopyxis declivis, apertural view.
Ralf Meisterfeld

Reproduction

Arcellinida normally reproduce by simple binary fission. The first stage is the construction of an identical daughter shell. Depending on the type of shell, the building material is arranged around a cytoplasmic extension functioning as a template. On completion of the new test, mitosis and fission of the cell take place. Sexual recombination (autogamy) is rare (Mignot & Raikov 1992).

Fossil record

Basic morphotypes of testate amoebas are very old. Tests of Precambrian (742 million years ago) *Paleoarcella athanata* from the Chuar group in the Grand Canyon, Arizona, are strikingly similar to those of present-day *Arcella* species (Porter et al. 2003). These ancient types are distributed worldwide. Younger records are from the Carboniferous (Loeblich & Tappan 1964; Wightman et al. 1994), Triassic (amber) (Poinar et al. 1993,) and Cretaceous (Medioli et al. 1990), but most records are from Quaternary deposits. Examples from New Zealand have been reported by McGlone and Wilmshurst (1999) and Wilmshurst et al. (2002, 2003).

Phylogeny and classification

Thanks to the availability of new molecular methods, our view of the phylogeny and high-level classification of Protozoa is rapidly changing (Cavalier-Smith 2003). This process is ongoing and at present it is not possible to present a classification of testate amoebas that reflects their phylogenetic relationships with certainty (but see Cavalier-Smith & Chao 2003; Cavalier-Smith et al. 2004). As mentioned earlier, testate amoebas do not form a unique taxon but are an artificial construct by taxonomists that combines several evolutionary lineages that are not directly related. It is likely that amoeboid organisms have evolved several times from flagellate ancestors by the loss of the flagellum and that the ability to build a test has developed independently.

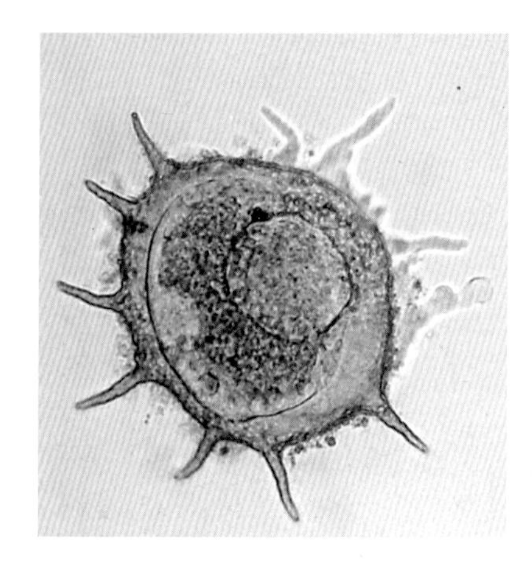

Centropyxis aculeata.
Ralf Meisterfeld

The order Arcellinida is the largest group of testate amoebas, containing about three-quarters of all known species. The first molecular data are now available (Nikolaev et al. 2005), supporting a sister-group relationship of Arcellinida with naked amoebas like *Amoeba, Chaos*, and *Hartmannella*. Classification of the order remains provisional, however. Three suborders are conventionally recognised (Bovee 1985; Meisterfeld 2002a) – Arcellina, Difflugina, and Phryganellina (with reticulolobopodia).

Classification within species differs between different traditions and schools and among 'splitters' and 'lumpers'. One main reason for these different views is the way the authors judge and separate inter- and intraspecific variability, which, especially in agglutinate polymorphic species, can be large. Shell morphology, for example, is affected by environmental factors such as availability of building material, water content of the substratum, and kind and availability of food. Thus, the question arises whether habitat- or region-specific morphs are modifications (ecophenotypes) or genetically distinct (species). The molecular phylogenetic analysis of Wylezich et al. (2002) revealed that some broadly defined morphospecies are not monophyletic in the genus *Euglypha*. A subsequent, more detailed morphological analysis then showed that genetically different isolates often have a distinct morphology, for example in shape and size of siliceous scales in the case of *Euglypha filifera*. Thus these genetic differences may justify giving them species status. With respect to geographical distribution, what looks like support for the 'every species everywhere' concept, might turn out to be a more complex and differentiated geographical patterning. An important and often unresolved practical problem is to find the gaps in a morphological continuum where these new taxa can be delimited.

Ecology

Testate amoebas are common in all freshwater habitats, inclusive of sediments, 'Aufwuchs' (periphyton), plankton, and mosses. Especially speciose collections can be made from wet or moist *Sphagnum* (peatmoss) and soils are also colonised by a characteristic and diverse biota of Arcellinida (Amoebozoa) and Euglyphida (Cercozoa). Availability of water determines the distribution of testate amoebas in mosses (Meisterfeld 1977, 1979; Charman 1997) and soils.

Empty tests remain intact after the death of the amoeba and can be identified in most cases to species. Testate amoebas are an important component in aquatic and terrestrial ecosystems. In many forest soils they can be as important as earthworms. They are consumers of bacteria, fungi, and different algal groups. Testate amoebas may feed selectively. Although some species can perforate fungal mycelia, they usually prefer smaller micro-organisms (Meisterfeld et al. 1992). Larger Heleoperidae and Nebelidae (Arcellinida) are predators of small Euglyphida (e.g. Ogden 1989).

The relatively well-defined niches of several species and the habitat-specific communities are a prerequisite to using testate amoebas in palaeolimnology. Analogous to pollen analysis, subfossil testacean assemblages have been used to investigate the ageing of lakes or to describe changes in the watertable during bog development (see Charman 1997; Charman & Warner 1997; Charman et al. 1999; Charman 2001; and the review by Beyens & Meisterfeld 2001).

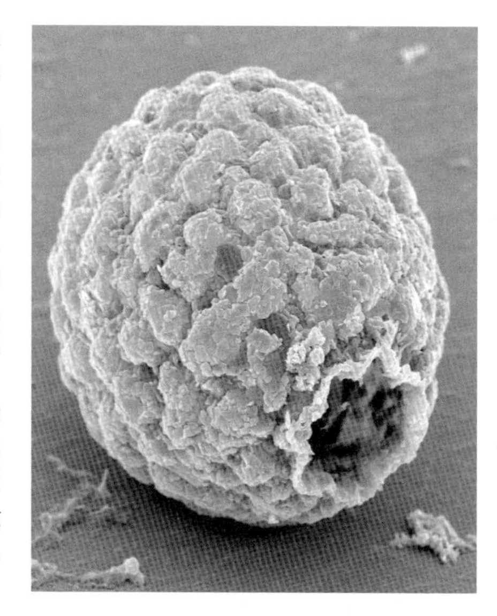

Netzelia tuberculata.
Ralf Meisterfeld

History of research and biogeography of New Zealand testate amoebas

Research on New Zealand testate amoebas started in the early 20th century (Richters 1908; Penard 1911) with the study of a handful of moss samples taken by European expeditions or from bought material (Hoogenraad & De Groot 1948; van Oye 1956). The first protozoologist working in New Zealand on testate amoebas was Stout (1958, 1960, 1961, 1962, 1963, 1978, 1984). As an eminent soil ecologist he was interested more in processes than faunistics and, working mainly with very selective culture methods, his impact on the exploration of species diversity remained low. Additional records were provided by Yeates and Foissner (1995). Hayward et al. (1996), studying a tidal inlet, found some freshwater species washed into the inlet by a small stream. Subsequently, the work of Charman (1997), McGlone and Wilmshurst (1999), and Wilmshurst et al. (2003) has given fresh impetus to the study of bog-dwelling testate amoebas.

Biogeographically, testate amoebas as a group are found worldwide and inhabit almost all substrata from marine sandy beaches and all freshwater biotopes to terrestrial habitats like mosses and soils. There are many ubiquitous species with little habitat specificity, while others live exclusively in one habitat type. Many species seem to have a cosmopolitan distribution but others are more restricted or regional. For instance, several species of Distomatopyxidae, Lamtopyxidae, and Nebelidae occur only in the tropics or in the Southern Hemisphere, where some of them are common, so that any characteristic Gondwanan distribution, for example, is certainly not a result of uneven sampling effort (e.g. Smith & Wilkinson 1986; Meisterfeld & Tan 1998). Although the testate amoeban fauna of New Zealand is largely unknown, most nebelids with a southern distribution (*Alocodera* sp., *Apodera* sp., *Argynnia caudata*, or *Certesella* sp.) are present. The presence of *Feuerbornia lobophora* in New Zealand is remarkable, since it was previously known only from Chile, based on its original description. Van Oye (1956) described five new taxa from New Zealand – the amoebozoans *Difflugia carinata*, *Nebela subsphaerica*, and *N. wailesi magna* and the cercozoans *Amphitrema paparoensis* and *Archerella jollyi*. Of these, only *N. subsphaerica* has since been found elsewhere (Japan). The others seem to be endemic but the justification of *A. paparoensis* and *A. jollyi* has been questioned (Charman 1999). New material from New Zealand should be studied using modern methods to clarify the status of these taxa.

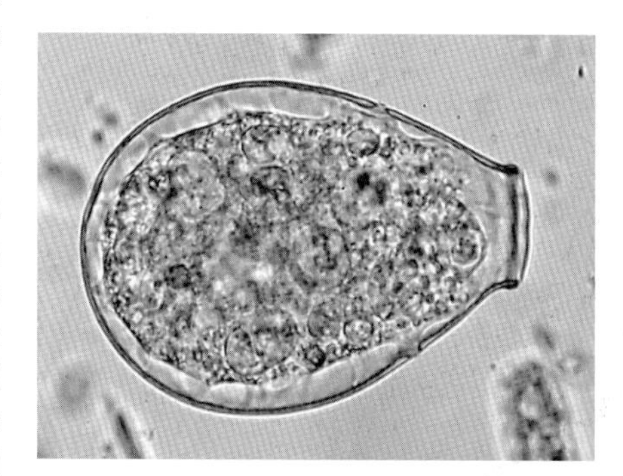

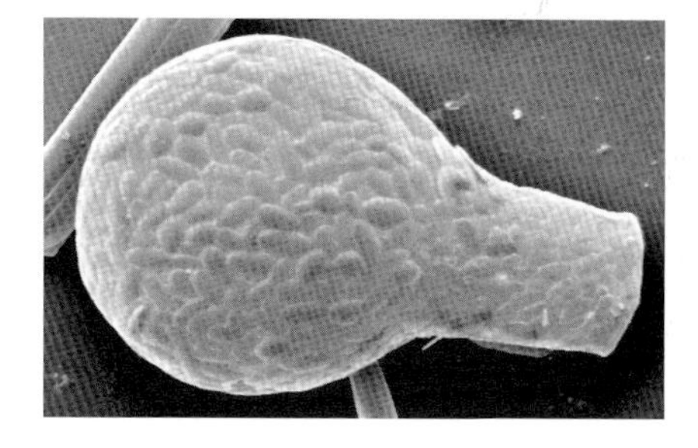

Nebela tincta (upper photo) and *N. lageniformis.*
Ralf Meisterfeld

Gaps in knowledge and scope for future research

Until now, 141 species, subspecies, and forms of all testate amoebas (95 Arcellinida and 46 cercozoan filose testate amoebas) have been reported from New Zealand. This is only 20–30% of the number that one would expect in a

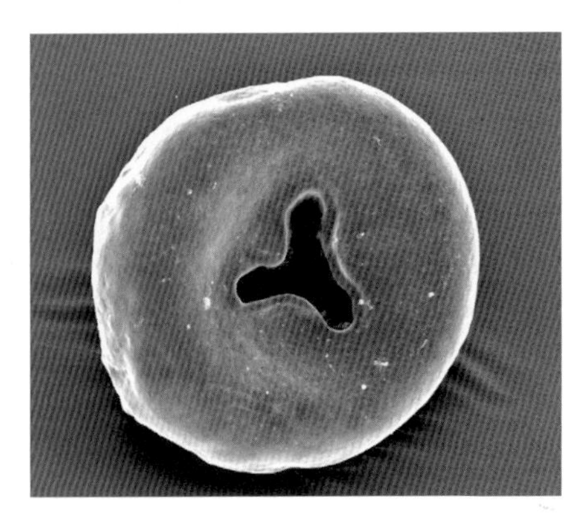

Trigonopyxis arcula, apertural view.
Ralf Meisterfeld

country of comparable area. The most likely reason is undersampling. About 20% of the records are unpublished and result from analysis of a handful of soil samples by R. Meisterfeld.

Obvious gaps in the checklist include marine interstitial species. In European beaches, about 100 species have been found (Meisterfeld 2001). Soil testate amoebas are largely under-represented but also several freshwater genera common in sediments, plankton, and aquatic vegetation are missing (e.g. *Campascus, Cucurbitella, Lesqereusia, Paraquadrula*) or present but have relative few species (e.g. *Arcella, Difflugia*). The testate amoeban fauna of *Sphagnum* bogs has been more completely sampled, but there are peculiarities –for example, the absence of *Hyalosphenia papilio*, the commonest species of the green parts of moist *Sphagnum* lawns in bogs of the Northern Hemisphere.

To fill the gaps in our knowledge of testate diversity most effectively, research should focus on littoral and profundal lake sediments, periphyton communites on aquatic vegetation, marine sandy beaches, and humus of virgin forests on different bedrock.

'Fungal' Protozoa

Amongst the Protozoa there are four groups historically regarded as fungi in the broad sense and typically studied by mycologists. These are traditionally named Acrasiomycota (acrasid slime moulds or acrasids), Dictyosteliomycota (cellular slime moulds or dictyostelids), Myxomycota (plasmodial slime moulds, myxogastrids or myxomycetes, the largest of the four groups), and Plasmodiophoromycota. More than 230 species have been recorded in New Zealand. Following Cavalier-Smith (1998) and Cavalier-Smith et al. (2004), the plasmodiophorids are attributed to phylum Cercozoa and discussed there (chapter 15); likewise the acrasids belong to phylum Percolozoa (Chapter 4); the balance of slime moulds is included in the present chapter among the Amoebozoa in the conosan infraphylum Mycetozoa.

Myxomycetes (class Myxogastrea), on the one hand, and protostelids (class Protostelea) and dictyostelids (class Dictyostelea), on the other, are two phylogenetically distinct groups of phagotrophic bacterivores, usually present and sometimes abundant in terrestrial ecosystems. The two groups have rather similar naked amoeboid stages in their life-cycle but differ in a number of other important respects. The myxomycete life-cycle involves two very different trophic stages – one consisting of uninucleate amoebas, with or without cilia, and the other consisting of a distinctive multinucleate structure, the plasmodium (Martin et al. 1983). Under favourable conditions, the plasmodium gives rise to one or more fruiting bodies containing spores. The fruiting bodies produced by myxomycetes are somewhat suggestive of those produced by higher fungi, although they are considerably smaller (usually no more than 1–2 millimetres tall). The spores of myxomycetes are, in most species, apparently wind-dispersed and complete the life-cycle by germinating to produce the uninucleate myxamoebae (without cilia) or swarm cells (with cilia).

There are about 875 recognised species of myxomycetes (Lado 2001). The majority are probably cosmopolitan, but a few species appear to be confined to the tropics or subtropics and some others have been collected only in temperate regions (Alexopoulos 1963; Farr 1976; Martin et al. 1983). Myxomycetes appear to be particularly abundant in temperate forests, but at least some species apparently occur in any terrestrial ecosystem with plants (and thus plant detritus) present (Stephenson & Stempen 1994). Most of what is known about myxomycete assemblages associated with particular types of terrestrial ecosystems has been derived from studies carried out in temperate regions of the Northern Hemisphere. There have been few papers published on the myxomycetes of New Zealand, and the majority of these (e.g. Cooke 1879; Lister & Lister 1905; Cheesman & Lister 1915; Rawson 1937) appeared more than a half century ago.

Fruiting bodies of the plasmodial slime mould *Diachea leucopodia*.
Steven Stephenson

Mitchell (1992), who compiled a checklist of all myxomycetes known from New Zealand prior to 1989, reported a total of 156 species. A recently published comprehensive taxonomic monograph of the myxomycetes of New Zealand (Stephenson 2003a) listed 185 species (one with two varieties). Stephenson et al. (2009) added eight additional species and described one species new to science that is currently known only from New Zealand. Possibly two or three other as yet undescribed forms are known, each represented by exceedingly limited material (Stephenson, unpubl. data).

In temperate forests of the Northern Hemisphere, myxomycetes are associated with a number of different microhabitats. These include the bark surface of living trees, forest-floor leaf litter, soil, the dung of herbivorous animals, and aerial portions of dead but still standing herbaceous plants. Each of these microhabitats tends to be characterised by a distinct assemblage of myxomycetes (Stephenson 1989). This same situation also seems to be the case in New Zealand. For example, the decaying fronds of nikau palm (*Rhopalostylis sapida*) represent a special microhabitat in which several species (e.g. *Didymium squamulosum*, *Perichaena depressa*, and *Physarum pusillum*), otherwise relatively uncommon in New Zealand forests, often occur with some regularity (Stephenson 2003b).

The assemblages of myxomycetes present in New Zealand are compositionally rather similar to those found in many other regions of the world. Compared to most organisms, myxomycetes seem to show very little evidence of endemism. However, at least a few species appear to be much more abundant in New Zealand than in any other region of the world for which comparative data exist. Prominent examples are *Metatrichia floriformis* and especially *Trichia verrucosa*.

The major difference between the Protostelea and Dictyostelea is that dictyostelids produce multicellular fruiting bodies that contain hundreds of spores while the fruiting bodies of protostelids are very simple, consisting of a stalk and one or at most a few spores. The results obtained from recent molecular studies (e.g. Shadwick et al. 2009) indicate that not all of the taxa previously assigned to the protostelids are necessarily closely related to one another, which means that these organisms do not constitute a monophyletic group. Indeed, some of these taxa were shown to belong to amoebozoan linkages clearly distinct from the 'true' slime moulds. As such, these authors suggested that protosteloid amoeba is a more appropriate term to use for a member of this assemblage of organisms.

For most of their life-cycle, dictyostelids exist as independent amoeboid cells (myxamoebae) that feed upon bacteria, grow, and multiply by binary fission. When the available food supply within a given microsite becomes depleted, numerous myxamoebae aggregate to form a structure called a pseudoplasmodium, within which each cell maintains its integrity. The pseudoplasmodium then produces one or more fruiting bodies (sorocarps) bearing spores. Dictyostelid fruiting bodies are microscopic and rarely observed except in laboratory culture. Under favourable conditions, the spores germinate to release myxamoebae, and the life-cycle begins anew. The spores produced by dictyostelids are embedded in a mucilaginous matrix that dries and hardens. As such, these spores have a rather limited potential for being dispersed by wind (Olive 1975). However, it has been demonstrated that many different animals, ranging from microscopic invertebrates to birds and small mammals (Stephenson & Landolt 1992), can serve as vectors for dictyostelid spores in nature. Approximately 130 species of dictyostelids are known to science. These organisms are most abundant in the surface humus layers of forest soils, but at least some species can be found in most other types of terrestrial habitats.

Until recently, the dictyostelids of New Zealand were rather poorly known, with three species reported by Olive (1975), without specific locality data, as the only records of the group from the entire country. However, during field surveys for myxomycetes carried in 1992 and 1998, soil/litter samples for isolation of dictyostelids were collected throughout New Zealand (Cavender et al. 2002).

Fruiting bodies of the plasmodial slime mould *Didymium squamulosum*.
Steven Stephenson

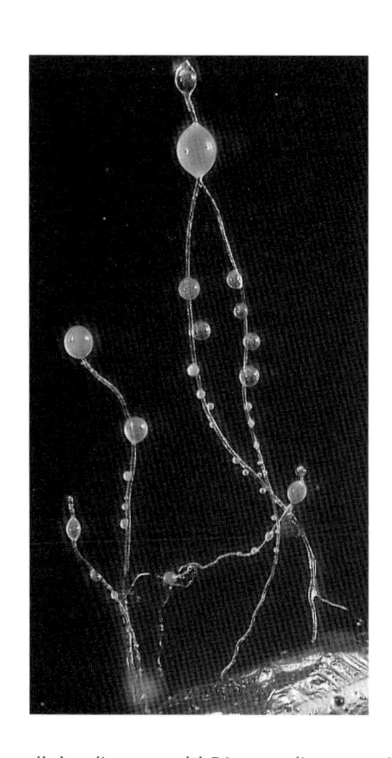

The cellular slime mould *Dictyostelium rosarium*.
Andy Swanson

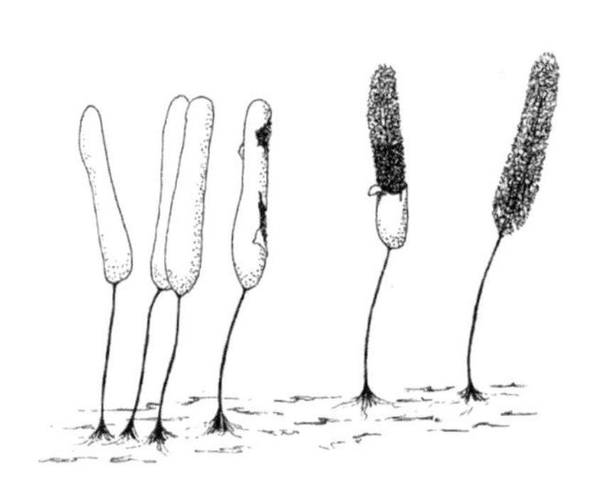

Fruiting bodies of *Stemonitopsis typhina*.
Steven Stephenson

Collecting sites included examples of all major forest types and a range in latitude (35° to 47°S) that encompassed most of the country. Thirteen species of dictyostelids were recovered; all of these occurred at low frequencies and densities. This total included a number of species (e.g. *Dictyostelium mucoroides* and *Polysphondylium violaceum*) that are common and widespread throughout the Northern Hemisphere, as well as several other species that have a more restricted distribution. Among the latter are *Dictyostelium fasciculatum*, not previously known outside of Europe, and *D. rosarium*, previously reported from only a few scattered localities in the Northern Hemisphere. Three of the species were new to science and are currently known to occur only in New Zealand, whereas two others, although also known from elsewhere in the world, were formally described from material collected in New Zealand (Cavender et al. 2002). New Zealand is the most isolated land mass of its size in the world, and the assemblage of dictyostelids present is quite distinctive and seems to reflect this isolation.

Until recently, the information available on the distribution and ecology of protosteloid amoebae in nature was limited to the data obtained as a result of surveys carried out in several small areas of the eastern United States or obtained from a few samples collected from scattered localities throughout the world. However, the Planetary Biodiversity Inventory (PBI) project funded by the National Science Foundation of the United States during the period of 2003 to 2010 provided an opportunity to add a considerable body of new data, including the first large series of records from New Zealand (Spiegel *et al.* 2006; Stephenson, unpubl. data).

There are only about 40 described species of protosteloid amoebae (Shadwick et al. 2009), and all of the those known from an appreciable number of records appear to be cosmopolitan. The primary substrates usually examined for protosteloid amoebae are ground litter (dead plant parts on the group) and aerial litter (dead but still attached plant parts above the ground), and this has been the case for studies carried out in New Zealand. As a general observation, these organisms do not appear to be particularly abundant in the *Nothofagus* forests of the southern portion of the South Island of New Zealand, but samples of litter collected elsewhere throughout the country can be expected to be relatively productive, often with a single sample yielding several different species of protosteloid amoebae.

Authors

Dr W.A. (Tony) Charleston Institute of Veterinary, Animal and Biomedical Sciences, Massey University, Palmerston North. Now 488 College Street, Palmerston North, New Zealand [charleston@inspire.net.nz] Enteric amoebas

Dr Ralf Meisterfeld RWTH Aachen, Institut für Biologie II, Mies-van-der-Rohe-Str. 15, D 52056 Aachen, Germany [meisterfeld@rwth-aachen.de] Testate amoebas

Dr Russell J. Shiel Department of Environmental Biology, University of Adelaide, Adelaide, South Australia 5005, Australia [russell.shiel@adelaide.edu.au] Testate amoebas

Dr Steven L. Stephenson Department of Biological Sciences, SCEN 626, University of Arkansas, Fayetteville, AR 72701, USA [slsteph@uark.edu] Fungal Protozoa

Dr LorWai Tan Department of ENT Surgery, University of Adelaide, Queen Elizabeth Hospital, Woodville South, South Australia 5011, Australia [lorwai.tan@adelaide.edu.au] Testate amoebas

References

ACKERS, J. P. 2002: The diagnostic implications of the separation of *Entamoeba histolytica* and *Entamoeba dispar*. *Journal of Bioscience 27*: 573–578.

ALEXOPOULOS, C. J. 1963: The myxomycetes II. *Botanical Review 29*: 1–78.

ANON. 2003: Free-living Gymnamoebia (Rhizopoda, Lobosea): a checklist of species, with illustrations, advices on identification and guide to literature. [http:amoeba.ifmo.ru/species.htm]

BEYENS, L.; MEISTERFELD, R. 2001: Protozoa: Testate amoebas. Pp. 121–153 *in*: Smol, J. P.; Birks, H. J. B.; Last, W. M. (Eds), *Tracking Environmental Change Using Lake Sediments. Volume 3: Terrestrial, Algal and Siliceous Indicators*. Kluwer, Dordrecht. 371 p.

BOVEE, E. C. 1985: Class Lobosea Carpenter, 1861. Pp. 158–211 *in*: Lee, J. J.; Hutner, S. H.; Bovee, E. C. (Eds), *An Illustrated Guide to the Protozoa*. The Society of Protozoologists, Lawrence. 629 p.

BRACE, E. C.; CASEY, B. R.; COSSHAM, R. B.; McEWAN, J. M.; McFARLANE, B. G.; MACKEN, J.; MONRO, P. A.; MORELAND, J. M.; NORTHERN, J. B.; STREET, R. J.; YALDWYN, J. C. 1953: The frog, *Hyla aurea* as a source of animal parasites. *Tuatara 5*: 12–21.

BROWN, T. J.; CURSONS, R. T. M.; KEYS, E. A.; MARKS, M; MILES, M. 1983: The occurrence and distribution of pathogenic free-living amoebae in thermal areas of the North Island of New Zealand. *New Zealand Journal of Marine and Freshwater Research 17*: 59–69.

BROWN, P.; GASH, D. B.; SHANKS, J.; JENSEN, J. 1977: Endemic amoebic abscess. *New Zealand Medical Journal 85*: 54–55.

CAVALIER-SMITH, T. 1998: A revised six-kingdom system of life. *Biological Reviews 73*: 203–266.

CAVALIER-SMITH, T. 2003: Protist phylogeny and the high-level classification of Protozoa. *European Journal of Protistology 39*: 338–348.

CAVALIER-SMITH, T. 2009: Megaphylogeny, cell body plans, adaptive zones: causes and timing of eukaryote basal radiations. *Journal of Eukaryotic Microbiology 56*: 26–33.

CAVALIER-SMITH, T.; CHAO, E. E.-Y. 2003: Phylogeny and classification of phylum Cercozoa (Protozoa). *Protist 154*: 341–358.

CAVALIER-SMITH, T.; CHAO, E. E.-Y.; OATES, B. 2004: Molecular phylogeny of Amoebozoa and the evolutionary significance of the unikont *Phalansterium*. *European Journal of Protistology 40*: 1–48.

CAVENDER, J. C.; STEPHENSON, S. L.; LANDOLT, J. C.; VADELL, E. M. 2002: Dictyostelid cellular slime moulds in the forests of New Zealand. *New Zealand Journal of Botany 40*: 235–264.

CHARMAN, D. J. 1997: Modelling hydrological relationships of testate amoebae (Protozoa: Rhizopoda) on New Zealand peatlands. *Journal of the Royal Society of New Zealand 27*: 465–483.

CHARMAN, D. J. 1999: Testate amoebae and the fossil record: issues in biodiversity. *Journal of Biogeography 26*: 89–96.

CHARMAN, D. J. 2001: Biostratigraphic and palaeoenvironmental applications of testate amoebae. *Quaternary Science Reviews 20*: 1753–1764.

CHARMAN, D. J.; WARNER, B. G. 1997: The ecology of testate amoebae (Protozoa: Rhizopoda) in oceanic peatlands in Newfoundland, Canada: Modelling hydrological relationships for palaeoenvironmental reconstruction. *Ecoscience 4*: 555–562.

CHARMAN, D. J.; HENDON, D.; PACKMAN, S. 1999: Multiproxy surface wetness records from replicate cores on an ombrotrophic mire: implications for Holocene palaeoclimate records. *Journal of Quaternary Science 14*: 451–463

CHEESMAN, W. N.; LISTER, G. 1915: Mycetozoa of Australia and New Zealand. *Journal of Botany, British and Foreign 53*: 203–212.

COOKE, M. C. 1879: New Zealand fungi. *Grevillea 8*: 54–68.

CURSONS, R. T. M.; BROWN, T. J. 1976: Identification and classification of the aetiological agents of primary amoebic meningo-encephalitis. *New Zealand Journal of Marine and Freshwater Research 10*: 254–262.

CURSONS, R. T. M.; BROWN, T. J.; KEYS, E. A. 1978: Diagnosis and identification of the aetiological agents of primary amoebic meningo-encephalitis (PAM). *New Zealand Journal of Medical Laboratory Technology 32*: 11–14.

DECLOITRE, L. 1986: Statistique mondiale des Thécamoebiens. *Annales de la Société des Sciences Naturelles et d'Archéologie de Toulon et du Var 37*: 171–172.

DIGGLES, B. K.; HINE, P. M.; HANDLEY, S.; BOUSTEAD, N. C. 2002: A handbook of diseases of importance to aquaculture in New Zealand. *NIWA Science and Technology Series No. 49*: 1–200.

DOWLING, J.; RILEY, D.; MORRIS, A. J.; MacCULLOCH, D. 1999: Attempts to detect parasitic causes of diarrhoea. *New Zealand Medical Journal 112*: 104.

FARR, M. L. 1976: Myxomycetes. *Flora Neotropica Monograph No. 16*: 1–305.

GHOSE, R. 1998: Sporadic amoebic liver abscess in the North Island. *New Zealand Medical Journal 111*: 83.

GRAY, T. B.; CURSONS, R. T. M.; SHERWAN, J. F.; ROSE, P. R. 1995: *Acanthamoeba*, bacterial, and fungal contamination of contact lens storage cases. *British Journal of Ophthalmology 79*: 601–605.

HAYWARD, B. W.; GRENFELL, H.; CAIRNS, G.; SMITH, A. 1996: Environmental controls on benthic foraminiferal and thecamoebian associations in a New Zealand tidal inlet. *Journal of Foraminiferal Research 26*: 150–171.

HOOGENRAAD, H. R.; DE GROOT, A. A. 1948: Thecamoebous moss-rhizopods from New Zealand. *Hydrobiologia 1*: 28–44.

KIRK, H. B. 1906: Notes on two marine Gymnomyxa. *Transactions and Proceedings of the New Zealand Institute 39*: 521–523, pls 25, 26.

LADO, C. 2001: Nomenmyx. A Nomenclatural Taxabase of Myxomycetes. *Cuadernos de Trabajo Flora Micológica Ibérica 16*: 1–221.

LANE, M. R.; NICHOLSON, G. J. 1984: Amoebic liver abscess: an Auckland experience. *New Zealand Medical Journal 97*: 187–190.

LISTER, A.; LISTER, G. 1905: Mycetozoa from New Zealand. *Journal of Botany, British and Foreign 43*: 111–114.

LOEBLICH, A. R. Jr, TAPPAN, H. 1964: Protista 2, Sarcodina. *In*: Moore, R.C. (Ed.), *Treatise on Invertebrate Paleontology 1, Part C*. Geological Society of America and University of Kansas Press, Lawrence. Vol. 1, 1–510, Vol. 2, 511–900.

MARCIANO-CABRAL, F.; CABRAL, G. 2003: *Acanthamoeba* spp. as agents of disease in humans. *Clinical Microbiology Reviews 16*: 273–307.

MARTIN, G.W.; ALEXOPOULOS, C.J.; FARR, M.L. 1983: *The Genera of Myxomycetes*. University of Iowa Press, Iowa City. 184 p.

McGLONE, M. S.; WILMSHURST, J. M. 1999: A Holocene record of climate, vegetation change and peat bog development, east Otago, South Island, New Zealand. *Journal of Quaternary Science 14*: 239–254.

MEDIOLI, F. S.; SCOTT, D. B.; COLLINS, E. S.; McCARTHY, F. M. G. 1990: Fossil thecamoebians: present status and prospects for the future. *In*: Hemleben, C.; Kaminski, M. A.; Kuhnt, W.; Scott, D. B. (Eds), *Paleoecology, Biostratigraphy, Paleoceanography and Taxonomy of Agglutinated Foraminifera. NATO Advanced Study Institute Series, ser. C, Mathematical and Physical Sciences 327*: 813–840.

MEISTERFELD, R. 1977: Die horizontale und vertikale Verteilung der Testaceen (Rhizopoda, Testacea) in *Sphagnum*. *Archiv für Hydrobiologie* 79: 319–356.

MEISTERFELD, R. 1979: Clusteranalytische Differenzierung der Testaceenzönosen (Rhizopoda, Testacea) in *Sphagnum*. *Archiv für Protistenkunde 121*: 270–307.

MEISTERFELD, R. 2001: Testate amoebae. Pp. 54–57 *in*: Costello, M.J.; Emblow, C.; White, R. (Eds), *European Register of Marine Species. A check-list of marine species in Europe and a bibliography of guides to their identification. Patrimoines naturels 50*: 1– 463.

MEISTERFELD, R. 2002a: Order Arcellinida. Pp. 827– 860 *in*: Lee, J.J. et al. (eds), *The Illustrated Guide to the Protozoa*. 2nd edn. Allen Press, Lawrence. 1400 p.

MEISTERFELD, R., 2002b: Testate amoebae with filopodia. Pp. 1054–1084 *in*: Lee, J.J. *et al.* (Eds), *The Illustrated Guide to the Protozoa*. 2nd edn. Allen Press, Lawrence. 1400 p.

MEISTERFELD, R.; DOHMEN, C.; MEYER, A.; PANFIL, C.; WIENAND, J. 1992: Influence of food quality and quantity on growth and feeding rates of testate amoebae (Testacealobosia). *European Journal of Protistology 28*: 351.

MEISTERFELD, R.; TAN, L.-W. 1998: First records of testate amoebae (Protozoa: Rhizopoda) from Mount Buffalo National Park, Victoria: preliminary notes. *The Victorian Naturalist 115*: 231–238

MIGNOT, J. P.; RAIKOV, I. B. 1992: Evidence for meiosis in the testate amoeba *Arcella*. *Journal of Protozoology 39*: 287–289.

MILLER, R. B.; STOUT, J. D.; LEE, K. E. 1955: Biological and chemical changes following scrub burning on a New Zealand hill soil. *New Zealand Journal of Science and Technology, ser. B, 37*: 290–313.

MINGE, M. A.; SILBERMAN, J. D.; ORR, R. J. S.; CAVALIER-SMITH, T.; SHALCHIAN-TABRIZI, K.; BURKI, F.; SKJÆVELAND, Å.; JAKOBSEN, K. S. 2009: Evolutionary position of breviate amoebae and the primary eukaryote divergence. *Proceedings of the Royal Society, B, 276*: 597–604.

MITCHELL, D. W. 1992: The Myxomycota of New Zealand and its island territories. *Nova Hedwigia 55*: 231–256.

MURDOCH, D.; GRAY, T. B.; CURSONS, R. T.; PARR, D. 1998: *Acanthamoeba* keratitis in New Zealand, including two cases with in vivo resistance to polyhexamethylene biguanide. *Australian and New Zealand Journal of Ophthalmology 26*: 231–236.

NIKOLAEV, S. I.; MITCHELL, E. A. D.; PETROV, N. B.; BERNEY, C.; FAHRNI, J.; PAWLOWSKI, J. 2005: The testate amoebae (Order Arcellinida Kent, 1880) finally find their home within the Amoebozoa. *Protist 156*: 191–202.

OGDEN, C. G. 1989: The agglutinate shell of *Heleopera petricola* (Protozoa, Rhizopoda), factors affecting its structure and composition. *Archiv für Protistenkunde 137*: 9–24.

OGDEN, C. G. 1990: The structure of the shell wall in testate amoebae and the importance of the organic cement matrix. Pp. 235–237 *in*: Claugher, D. (Ed.), *Scanning Electron Microscopy in Taxonomy and Functional Morphology. [Systematics Association Special Volume 41]* Clarendon Press, Oxford. 315 p.

OGDEN, C. G.; ELLISON, R. L. 1988: The value of the organic cement matrix in the identification of the shells of fossil testate amoebae. *Journal of Micropalaeontology 7*: 233–240.

OLIVE, L. S. 1975: *The Mycetozoans*. Academic Press, New York. 293 p.

PENARD, E. 1911: Sarcodina Rhizopodes d'eau douce. *British Antarctic Expedition 1907–1909, 1(6)*: 203–257, pls 22, 23.

POINAR, G. O.; WAGGONER. B. M.; BAUER, U. 1993: Terrestrial soft-bodied protists and other microorganisms in Triassic amber. *Science 259*: 222–224.

PORTER, S. M.; MEISTERFELD, R.; KNOLL, A. H. 2003: Vase-shaped microfossils from the neoproterozoic Chuar group, Grand Canyon: a classification guided by modern testate amoebae. *Journal of Paleontology 77*: 409–429.

RAIKOV, I. B. 1982: *The Protozoan Nucleus. Morphology and Evolution.* Springer-Verlag, Berlin. 474 p.

RAWSON, S. H. 1937: A list of the Mycetozoa collected chiefly in the vicinity of Dunedin, New Zealand. *Transactions of the Royal Society of New Zealand 66*: 351–353.

RICHARDSON, L. R.; CLARK, A. E.; RALPH, P. M. 1943a: Studies on the Entozoa of Man in New Zealand. Part 1.–A preliminary note on the results from the examination of inmates of a mental hospital. *Transactions of the Royal Society of New Zealand 73*: 239–247.

RICHARDSON, L. R.; CLARK, A. E.; RALPH, P. M. 1943b: Studies on the Entozoa of Man in New Zealand. Part 2.–Results from the examination of a small number of non-clinical individuals. *Transactions of the Royal Society of New Zealand 73*: 248–249.

RICHTERS, F. 1908: Beitrag zur Kenntnis der Moosfauna Australiens und der Inseln des Pazifischen Ozeans. – (Ergebnisse einer Reise nach dem Pacific. Schauinsland 1896–1897). *Zoologisches Jahrbuch Systematik Jena 26*: 196–213.

SHADWICK, L. L.; SPIEGEL, F. W.; SHADWICK, J. D. L.; BROWN, M. W.; SILBERMAN, J. D. 2009: Eumycetozoa = Amoebozoa?: SSUrDNA phylogeny of protosteloid slime molds and its significance for the Amoebozoan supergroup. *PloS ONE 4(8)*: e6754, doi:10.1371/journal.pone.0006754.

SMIRNOV, A. V.; NASSONOVA, E. S.; CHAO, E.; CAVALIER-SMITH, T. 2007: Phylogeny, evolution, and taxonomy of vanellid amoebae. *Protist 158*: 295–324.

SMIRNOV, A. V.; CHAO, E.; NASSONOVA, E. S.; CAVALIER-SMITH, T. 2011: A revised classification of naked lobose amoebae (Amoebozoa: Lobosa). *Protist 162*: 545–570.

SMITH, H. G.; WILKINSON, D. M. 1986: Biogeography of testate rhizopods in the southern temperate and antarctic zones. *Colloque sur les Écosystèmes terrestres subantarctiques 58*: 83–96.

SPIEGEL, F. W.; SHADWICK, J. D.; HEMMES, D. E. 2006: A new ballistosporous species of *Protostelium. Mycologia 98:* 144-148

STEPHENSON, S. L. 1989: Distribution and ecology of myxomycetes in temperate forests. II. Patterns of occurrence on bark surface of living trees, leaf litter, and dung. *Mycologia 81*: 608–621.

STEPHENSON, S. L. 2003a: *Myxomycetes of New Zealand*. Fungal Diversity Press, Hong Kong. 238 p.

STEPHENSON, S. L. 2003b: Myxomycetes associated with decaying fronds of nikau palm (*Rhopalostylis sapida*) in New Zealand. *New Zealand Journal of Botany 41: 311–317.*

STEPHENSON, S. L.; LANDOLT, J. C. 1992: Vertebrates as vectors of cellular slime molds in temperate forests. *Mycological Research 96*: 670–672.

STEPHENSON, S. L.; NOVOZHILOV, Y. K.; SHIRLEY, C.; MITCHELL, D. W. 2009: Additions to the myxomycetes known from New Zealand, including a new species of *Diderma*. *Australian Systematic Botany 23:* 466–472.

STEPHENSON, S. L.; STEMPEN, H. 1994: *Myxomycetes: A Handbook of Slime Molds.* Timber Press, Portland. 183 p.

STOUT, J. D. 1958: Biological studies of some tussock-grassland soils. VII. Protozoa. *New Zealand Journal of Agricultural Research 1*: 974–984.

STOUT, J. D. 1960: Biological studies of some tussock-grassland soils. XVIII. Protozoa of two cultivated soils. *New Zealand Journal of Agricultural Research 3*: 237–244.

STOUT, J. D. 1961: Biological and chemical changes following scrub burning on a New Zealand hill soil. *New Zealand Journal of Science 4*: 739–752.

STOUT, J. D. 1962: An estimation of microfaunal populations in soils and forest litter. *Journal of Soil Science 13*: 314–320.

STOUT, J. D. 1963: The terrestrial plankton. *Tuatara 11*: 57–65.

STOUT, J. D. 1978: Effect of irrigation with municipal water or sewage effluent on the biology of soil cores. Part 2. Protozoan fauna. *New Zealand Journal of Agricultural Research 21*: 11–20.

STOUT, J. D. 1984: The protozoan fauna of a seasonally inundated soil under grassland. *Soil Biology and Biochemistry 16*: 121–125.

VAN OYE, P. 1956: On the thecamoeban fauna of New Zealand with description of four new species and biogeografical [*sic*] discussion. *Hydrobiologia 8*: 16–37.

WIGHTMAN, W. G.; SCOTT, D. B.; MEDIOLI, F. S.; GIBLING, M. R. 1994: Agglutinated foraminifera and thecamoebians from the Late Carboniferous Sydney Coalfield, Nova Scotia: paleoecology, paleoenvironments and paleogeographical implications. *Palaeogeography, Palaeoclimatology, Palaeoecology 106*: 187–202.

WILMSHURST, J. M.; McGLONE, M.S.; CHARMAN, D.J. 2002: Holocene vegetation and climate change in southern New Zealand: linkages between forest composition and quantitative surface moisture reconstructions from an ombrogenous bog. *Journal of Quaternary Science 17*: 653–666.

WILMSHURST, J. M.; WISER, S. K.; CHARMAN, D. J. 2003: Reconstructing Holocene water tables in New Zealand using testate amoebae: differential preservation of tests and implications for the use of transfer functions. *The Holocene 13*: 61–72.

WYLEZICH C.; MEISTERFELD R.; MEISTERFELD S.; SCHLEGEL, M. 2002: Phylogenetic analyses of small subunit ribosomal RNA coding regions reveal a monophyletic lineage of euglyphid testate amoebae (order Euglyphida). *Journal of Eukaryotic Microbiology 49*: 108–118.

YEATES, G. W.; FOISSNER, W. 1995: Testate amoebae as predators of nematodes. *Biology and Fertility of Soils 20*: 1–7.

WONG, F. Y. K.; CARSON, J.; ELLIOTT, N. G. 2004: 18S ribosomal DNA-based PCR identification of *Neoparamoeba pemaquidensis,* the agent of amoebic gill disease in sea-farmed salmonids. *Diseases of Aquatic Organisms 60*: 65–76.

Checklist of New Zealand Amoebozoa

Classification is based on Cavalier-Smith et al. (2004), Smirnov et al. (2007, 2011), and Cavalier-Smith (2009). Abbreviations: A, adventive; E, endemic; S, species with southern distribution in New Zealand; * unpublished records (R. Meisterfeld) from humus collected at Nelson (South Island) and Station Ridge near Orongorongo Valley Field Station (southern North Island) collected by Yeates (see Yeates & Foissner 1995). Habitats of other taxa are indicated as follows: F, fresh waters and wet mosses like *Sphagnum*; M, marine; T, terrestrial (in soil interstitial water, dry mosses, or terrestrial hosts of parasites); P, parasitic or commensal in vertebrates. Endemic genera are underlined (first entry only).

KINGDOM PROTOZOA
SUBKINGDOM SARCOMASTIGOTA
PHYLUM AMOEBOZOA
SUBPHYLUM CONOSA
INFRAPHYLUM ARCHAMOEBAE
Class ARCHAMOEBEA
Order PELOBIONTIDA
ENTAMOEBIDAE
Entamoeba coli (Grassi, 1879) T P A
Entamoeba histolytica Schaudinn, 1903 T P A
Entamoeba dispar Brumpt, 1925 T P A
Entamoeba morula Raff, 1912 T P A
PELOMYXIDAE
Pelomyxa sp. F/T

Order MASTIGAMOEBIDA
ENDOLIMACIDAE
Endolimax nana (Wenyon & O'Connor, 1917) T P A
Iodamoeba buetschlii (Prowazek, 1912) T P A

INFRAPHYLUM UNNAMED
Class VARIOSEA
Order CENTRAMOEBIDA
ACANTHAMOEBIDAE
Acanthamoeba castelanii (Douglas, 1930) F/T A P
Acanthamoeba culbertsoni (Singh & Das, 1970) F/T A P
Acanthamoeba polyphaga (Pushkarew, 1913) F/T A P
Acanthamoeba jacobsi Sawyer, Nerad & Visvesvara, 1992 F/T A P
Acanthamoeba spp. Gray et al. 1995 F/T A P?

VARIOSEA INCERTAE SEDIS
STEREOMYXIDAE?
Myxoplasma rete Kirk, 1906 M E?

INFRAPHYLUM MYCETOZOA
Class PROTOSTELEA
Order PROTOSTELIDA
PROTOSTELIIDAE
Cavostelium apophysatum L.S.Olive T
Clastostelium recurvatum L.S.Olive & Stoian. T
Echinosteliopsis oligospora Reinhardt & L.S.Olive T
Endostelium zonatum (L.S.Olive & Stoian.) W.E. Benn. & L.S.Olive T
Microglomus paxillus L.S.Olive & Stoian. T
Nematostelium gracile (L.S.Olive & Stoian.) L.S.Olive & Stoian. T
Nematostelium ovatum (L.S.Olive & Stoian.) L.S.Olive & Stoian. T
Protosporangium bisporum L.S.Olive & Stoian. T
Protosteliopsis fimicola (L.S.Olive) L.S.Olive & Stoian. T
Protostelium arachisporum L.S.Olive T
Protostelium mycophagum L.S.Olive & Stoian. T
Protostelium nocturnum Spiegel T
Protostelium okumukumu Spiegel, Shadwick & Hemmes T
Protostelium pyriforme L.S.Olive & Stoian. T
Schizoplasmodiopsis amoeboidea L.S.Olive & K.D.Whitney T
Schizoplasmodiopsis micropunctata L.S.Olive & Stoian. T
Schizoplasmodiopsis pseudoendospora L.S.Olive, M.Martin. & Stoian. T
Schizoplasmodiopsis reticulata L.S.Olive & Stoian. T
Schizoplasmodiopsis vulgaris L.S.Olive & Stoian. T
Schizoplasmodium cavostelioides L.S.Olive & Stoian. T
Schizoplasmodium obovatum L.S.Olive & Stoian. T
Soliformovum expulsum (L.S.Olive & Stoian.) Spiegel T
Tychosporium acutostipes Spiegel, D.L.Moore & J.Feldman T

Class DICTYOSTELEA
Order DICTYOSTELIIDA
DICTYOSTELIIDAE
Dictyostelium antarcticum Cavender, S.L.Stephenson, J.C.Landolt & Vadell E?
Dictyostelium aureostipes Cavender, Raper & Norberg T
Dictyostelium australe Cavender, S.L.Stephenson, J.C.Landolt & Vadell T E?
Dictyostelium fasciculatum F.Traub, H.R.Hohl & Cavender T
Dictyostelium giganteum B.N.Singh T
Dictyostelium leptosomum Cavender, S.L.Stephenson, J.C.Landolt & Vadell T
Dictyostelium minutum Raper T
Dictyostelium monochasioides H.Hagiw. T
Dictyostelium mucoroides Bref.
Dictyostelium quercibrachium Cavender, S.L.Stephenson, J.C.Landolt & Vadell T
Dictyostelium rosarium Raper & Cavender T
Polysphondylium anisocaule Cavender, S.L.Stephenson, J.C.Landolt & Vadell T E?
Polysphondylium pallidum Olive T
Polysphondylium violaceum Bref. T

Class MYXOGASTREA
Order PARASTELIDA
Ceratiomyxa fruticulosa (O.F.Müll.) T.Macbr. T
Ceratiomyxa hemisphaerica L.S.Olive & Stoian.

Order ECHINOSTELIIDA
CLASTODERMATACEAE
Clastoderma debaryanum A.Blytt T
ECHINOSTELIACEAE
Echinostelium apitectum K.D.Whitney T
Echinostelium fragile Nann.-Bremek. T
Echinostelium minutum de Bary T

Order LICEIDA
CRIBRARIIDAE
Cribraria argillacea (Pers. ex J.F.Gmel.) Pers. T
Cribraria aurantiaca Schrad. T
Cribraria cancellata (Batsch) Nann.-Bremek. T
Cribraria confusa Nann.-Bremek. & Y.Yamam. T
Cribraria dictydioides Cooke & Balf. f. ex Massee T
Cribraria intricata Schrad. T
Cribraria macrocarpa Schrad. T
Cribraria microcarpa (Schrad.) Pers. T
Cribraria mirabilis (Rostaf.) Massee T
Cribraria persoonii Nann.-Bremek. T
Cribraria splendens (Schrad.) Pers. T
Cribraria violacea Rex T
Cribraria vulgaris Schrad. T
LICEIDAE
Licea biforis Morgan T
Licea capitatoides Nann.-Bremek. & Y.Yamam. T
Licea castanea G.Lister T
Licea eleanorae Ing T
Licea kleistobolus G.W.Martin T
Licea marginata Nann.-Bremek. T
Licea minima Fr. T
Licea operculata (Wingate) G.W.Martin T
Licea parasitica (Zukal) G.W.Martin T
Licea pusilla Schrad. T
Licea pygmaea (Meyl.) Ing T
LISTERELLIDAE
Listerella paradoxa E.Jahn T
LYCOGALIDAE
Dictydiaethalium plumbeum (Schumach.) Rostaf. T
Lycogala epidendrum (L.) Fr. T
Lycogala exiguum Morgan T
Lycogala flavofuscum (Ehrenb.) Rostaf. T
Reticularia liceoides (Lister) Nann.-Bremek. T
Reticularia lycoperdon Bull. T
Tubifera ferruginosa (Batsch) J.F.Gmel. T

Order TRICHIIDA
ARCYRIIDAE
Arcyria affinis Rostaf. T
Arcyria cinerea (Bull.) Pers. T
Arcyria denudata (L.) Wettst. T
Arcyria ferruginea Saut. T
Arcyria incarnata (Pers. ex J.F.Gmel.) Pers. T
Arcyria insignis Kalchbr. & Cooke T
Arcyria major (G.Lister) Ing T
Arcyria minuta Buchet T
Arcyria obvelata (Oeder) Onsberg T
Arcyria oerstedii Rostaf. T
Arcyria pomiformis (Leers) Rostaf. T
Arcyria stipata (Schwein.) Lister T
Arcyria virescens G.Lister T
DIANEMATIDAE
Calomyxa metallica (Berk.) Nieuwl. T
Dianema corticatum Lister T
Dianema depressum (Lister) Lister T
Dianema harveyi Rex T
TRICHIIDAE
Hemitrichia calyculata (Speg.) M.L.Farr T
Hemitrichia clavata (Pers.) Rostaf. T
Hemitrichia leiocarpa (Cooke) Lister T
Hemitrichia pardina (Minakata) Ing T
Hemitrichia serpula (Scop.) Rostaf. ex Lister F
Metatrichia floriformis (Schwein.) Nann.-Bremek. F
Metatrichia vesparia (Batsch) Nann.-Bremek. ex G.W.Martin & Alexop. F
Perichaena chrysosperma (Curr.) Lister T
Perichaena corticalis (Batsch) Rostaf. T
Perichaena depressa Lib. T
Perichaena luteola (Kowalski) Gilbert ex Lado T
Perichaena vermicularis (Schwein.) Rostaf. T
Prototrichia metallica (Berk.) Massee T

Trichia botrytis (J.F.Gmel.) Pers. T
Trichia contorta (Ditmar) Rostaf. T
Trichia crateriformis G.W.Martin T
Trichia decipiens (Pers.) T.Macbr. T
Trichia erecta Rex T
Trichia favoginea (Batsch) Pers. T
Trichia lutescens (Lister) Lister T
Trichia scabra Rostaf. T
Trichia varia (Pers. ex J.F.Gmel.) Pers. T
Trichia verrucosa Berk. T

Order STEMONITIDA
STEMONITIDAE
Collaria arcyrionema (Rostaf.) Nann.-Bremek. ex Lado T
Colloderma oculatum (C.Lippert) G.Lister T
Comatricha alta Preuss T
Comatricha elegans (Racib.) G.Lister T
Comatricha laxa Rostaf. T
Comatricha nigra (Pers. ex J.F.Gmel.) J.Schröt. T
Comatricha pulchella (C.Bab.) Rostaf. T
Comatricha vineatilis Nann.-Bremek. T
Diacheopsis depressa K.S.Thind & T.N.Lakh. T
Diacheopsis metallica Meyl. T
Enerthenema papillatum (Pers.) Rostaf. T
Lamproderma atrosporum Meyl. T
Lamproderma columbinum (Pers.) Rostaf. T
Lamproderma echinulatum (Berk.) Rostaf. T
Lamproderma ovoideum Meyl. T
Lamproderma retirugisporum G.Moreno, H.Singer, Illana & A.Sánchez T
Lamproderma scintillans (Berk. & Broome) Morgan T
Lamproderma splendens Meyl. T
Macbrideola declinata T.E.Brooks & H.W.Keller T
Paradiacheopsis acanthodes (Alexop.) Nann.-Bremek. T
Paradiacheopsis cribrata Nann.-Bremek. T
Paradiacheopsis fimbriata (G.Lister & Cran) Hertel ex Nann.-Bremek. T
Paradiacheopsis solitaria (Nann.-Bremek.) Nann.-Bremek. T
Stemonaria longa (Peck) Nann.-Bremek. T
Stemonitis axifera (Bull.) T.Macbr. T
Stemonitis flavogenita E.Jahn T
Stemonitis fusca Roth var. *fusca* T
Stemonitis fusca var. *nigrescens* (Rex) Torrend T
Stemonitis herbatica Peck T
Stemonitis mussooriensis G.W.Martin, K.S.Thind & Sohi T
Stemonitis pallida Wingate T
Stemonitis smithii T.Macbr. T
Stemonitis splendens Rostaf.
Stemonitis virginiensis Rex T
Stemonitopsis hyperopta (Meyl.) Nann.-Bremek. T
Stemonitopsis typhina (F.H.Wigg.) Nann.-Bremek. T
Symphytocarpus amaurochaetoides Nann.-Bremek. T
Symphytocarpus flaccidus (Lister) Ing & Nann.-Bremek. T

Order PHYSARIDA
DIDYMIIDAE
Diachea leucopodia (Bull.) Rostaf. T
Diachea subsessilis Peck T
Diderma alpinum (Meyl.) Meyl. T
Diderma asteroides (Lister & G.Lister) G.Lister T
Diderma chondrioderma (de Bary & Rostaf.) G.Lister T
Diderma cinereum Morgan T
Diderma deplanatum Fr. T
Diderma donkii Nann.-Bremek. T
Diderma effusum (Schwein.) Morgan T
Diderma globosum Pers. T
Diderma hemisphaericum (Bull.) Hornem. T
Diderma niveum (Rostaf.) T.Macbr. T
Diderma novae-zelandiae S.L.Stephenson et Novozh. T
Diderma ochraceum Hoffm. T
Diderma radiatum (L.) Morgan T
Diderma spumarioides (Fr.) Fr. T
Diderma testaceum (Schrad.) Pers. T
Didymium anellus Morgan T
Didymium bahiense Gottsb. T
Didymium clavus (Alb. & Schwein.) Rabenh. T
Didymium difforme (Pers.) Gray T
Didymium dubium Rostaf. T
Didymium iridis (Ditmar) Fr. T
Didymium melanospermum (Pers.) T.Macbr. T
Didymium nigripes (Link) Fr. T
Didymium serpula Fr. T
Didymium squamulosum (Alb. & Schwein.) Fr. T
Lepidoderma carestianum (Rabenh.) Rostaf. T
Lepidoderma crustaceum Kowalski T
Lepidoderma granuliferum (W.Phillips) R.E.Fr. T
Mucilago crustacea F.H.Wigg. T
PHYSARIDAE
Badhamia apiculospora (Härk.) Eliasson & N.Lundq. T
Badhamia capsulifera (Bull.) Berk. T
Badhamia foliicola Lister T
Badhamia macrocarpa (Ces.) Rostaf.T
Badhamia melanospora Speg. T
Badhamia nitens Berk. T
Badhamia utricularis (Bull.) Berk. T
Craterium aureum (Schumach.) Rostaf. T
Craterium leucocephalum (Pers. ex J.F.Gmel.) Ditmar T
Craterium minutum (Leers) Fr. T
Craterium obovatum Peck T
Fuligo aurea (Penz.) Y.Yamam. T
Fuligo cinerea (Schwein.) Morgan T
Fuligo septica (L.) F.H.Wigg. T
Leocarpus fragilis (Dicks.) Rostaf. T
Physarum albescens Ellis ex T.Macbr. T
Physarum album (Bull.) Chevall. T
Physarum alpestre Mitchel, S.W.Chapm. & M.L.Farr T
Physarum alpinum (Lister & G.Lister) G.Lister T
Physarum bitectum G.Lister T
Physarum bivalve Pers. T
Physarum bogoriense Racib. T
Physarum braunianum de Bary T
Physarum cinereum (Batsch) Pers. T
Physarum citrinum Schumach. T
Physarum compressum Alb. & Schwein. T
Physarum contextum (Pers.) Pers. T
Physarum decipiens M.A.CurtisT
Physarum dictyospermum Lister & G.Lister T
Physarum didermoides (Pers.) Rostaf. T
Physarum flavicomum Berk. T
Physarum globuliferum (Bull.) Pers. T
Physarum gyrosum Rostaf. T
Physarum hongkongense Chao H.Chung T
Physarum lateritium (Berk. & Ravenel) Morgan T
Physarum leucophaeum Fr. T
Physarum leucopus Link T
Physarum licheniforme (Schwein.) Lado T
Physarum limonium Nann.-Bremek. T
Physarum melleum (Berk. & Broome) Massee T
Physarum notabile T.Macbr. T
Physarum nucleatum Rex T
Physarum obscurum (Lister) Ing T
Physarum pezizoideum (Jungh.) Pavill. & Lagarde T
Physarum pusillum (Berk. & M.A.Curtis) G.Lister T
Physarum robustum (Lister) Nann.-Bremek. T
Physarum serpula Morgan T
Physarum straminipes Lister T
Physarum vernum Sommerf. T
Physarum viride (Bull.) Pers. T
Willkommlangea reticulata (Alb. & Schwein.) Kuntze T

SUBPHYLUM LOBOSA
Class TUBULINEA
Order EUAMOEBIDA
AMOEBIDAE
Amoeba agilis Kirk, 1906 E? [*Stygamoeba*?] M
Amoeba proteus Leidy, 1879 F
Chaos carolinense (Wilson, 1902) F A
Metachaos sp. indet. Stout 1961 T
Trichamoeba spp. indet. (2) Stout 1961 T
HARTMANNELLIDAE
Hartmannella spp. indet. (2) Gray et al. 1995 T P

Order ARCELLINIDA
Suborder ARCELLINA
ARCELLIDAE
Arcella arenaria Greeff, 1866 T
Arcella artocrea Leidy, 1876 F
Arcella discoides Ehrenberg, 1871 F
Arcella hemisphaerica Perty, 1852 F
Arcella rotundata var. *aplanata* Deflandre, 1928 F
MICROCHLAMYIDAE
Microchlamys patella (Claparède & Lachmann, 1859) F/T
Microchlamys sylvatica Golemansky, Skarlato & Todorov, 1987* T
MICROCORYCIIDAE
Amphizonella violacea Greeff, 1866 T
Diplochlamys fragilis Penard, 1909 T
Diplochlamys timida Penard, 1909 T
Microcorycia penardi (Penard, 1902) T

Suborder DIFFLUGINA
CENTROPYXIDAE
Centropyxis aculeata (Ehrenberg, 1832) F
Centropyxis aerophila Deflandre, 1929 F/T
Centropyxis cassis (Wallich, 1864) F
Centropyxis constricta (Ehrenberg, 1843) F
Centropyxis discoides (Penard, 1890) F
Centropyxis ecornis (Ehrenberg, 1841) F
Centropyxis elongata (Penard 1890) F/T
Centropyxis horrida Penard, 1911 F
Centropyxis laevigata Penard, 1890 F
Centropyxis minuta Deflandre, 1929 F
Centropyxis orbicularis Deflandre, 1929
Centropyxis platystoma (Penard, 1890) F
Centropyxis sacciformis Hoogenraad & De Groot, 1946* T
Centropyxis sphagnicola Deflandre, 1929* F/T
Centropyxis sylvatica Deflandre, 1929* T
Centropyxis sp.* T
DIFFLUGIIDAE
Difflugia bacillariarum Perty, 1849 F
Difflugia bacillifera Penard, 1890 F
Difflugia bidens Penard, 1902 F
Difflugia bryophila (Penard, 1902* F/T
Difflugia carinata van Oye, 1956 F
Difflugia globulus Jung, 1942 F
Difflugia lacustris (Penard, 1899* F
Difflugia lanceolata Penard, 1890 F
Difflugia lucida Penard, 1890 T
Difflugia manicata Penard, 1902 F
Difflugia oblonga Ehrenberg, 1838 F
Difflugia pristis Penard, 1902 F
Difflugia pulex Penard, 1902 F
Difflugia pyriformis Perty, 1849 F
Zivkovicia bryophila (Penard, 1902) F
Zivkovicia compressa (Carter, 1864) F
HELEOPERIDAE
Heleopera petricola Leidy, 1879 F
Heleopera p. amethystea Penard, 1899 F
Heleopera rosea Penard, 1890* F
Heleopera sordida Penard, 1910 F
Heleopera sphagni (Leidy, 1874) F
Heleopera sylvatica Penard, 1890 F/T
HYALOSPHENIIDAE

Hyalosphenia ovalis Wailes, 1912 F
Hyalosphenia subflava Cash, Wailes & Hopkinson, 1909 F/T
LESQUEREUSIIDAE
Netzelia tuberculata (Wallich, 1864) F
Quadrulella symmetrica (Wallich, 1863)*? F/T
NEBELIDAE
Alocodera cockayni (Penard, 1910) F S
Apodera vas (Certes, 1889) F/T S
A. v. longicollis (Grospietsch, 1969) F S
A. v. obliqua (Grospietsch, 1969) F S
A. v. recticollis (Jung, 1942) F S
Argynnia caudata (Leidy, 1879) F S
Argynnia dentistoma (Penard, 1890) F
Certesella certesi (Penard, 1911) F/T S
Certesella martiali (Certes, 1888) F/T S
Nebela collaris (Ehrenberg, 1848) F/T
Nebela lageniformis Penard, 1890 F/T
Nebela longicollis Penard, 1890 F/T
Nebela militaris Penard, 1890 F/T
Nebela minor Penard, 1893 F/T
Nebela parvula Cash, Wailes & Hopkinson, 1908 F/T
Nebela penardiana Deflandre, 1936 F/T
Nebela subsphaerica van Oye, 1956 F
Nebela tincta (Leidy, 1879) F/T
Nebela tubulata Brown, 1911 F/T
Nebela wailesi (Wailes, 1912) F/T
Nebela w. magna van Oye, 1956 F/T E
Physochila sp.* T

PLAGIOPYXIDAE
Bullinularia indica (Penard, 1907) T
Bullinularia minor Hoogenraad & De Groot, 1948 T
Plagiopyxis callida Penard, 1910 T
Plagiopyxis declivis Bonnet & Thomas, 1955* T
Plagiopyxis intermedia Bonnet, 1959* T
Plagiopyxis minuta Bonnet, 1959* T
TRIGONOPYXIDAE
Cyclopyxis arcelloides Penard, 1902 F
Cyclopyxis dulcis Couteaux & Munsch, 1978* T
Cyclopyxis eurystoma Deflandre, 1929 F/T
Cyclopyxis kahli Deflandre, 1929 F/T
Trigonopyxis arcula (Leidy, 1879) F/T
Trigonopyxis microstoma Decloitre, 1979 F/T

Suborder PHRYGANELLINA
CRYPTODIFFLUGIIDAE
Cryptodifflugia oviformis Penard, 1890 F/T
Cryptodifflugia o. fusca Penard, 1890* F/T
Cryptodifflugia sacculus Penard, 1902* F/T
PHRYGANELLIDAE
Phryganella acropodia (Hertwig & Lesser, 1874) ?F/T
Phryganella nidulus Penard, 1902 F
Phryganella paradoxa Penard, 1902* T
Phryganella p. alta Bonnet & Thomas, 1960* T

Class FLABELLINEA
Order DACTYLOPODIDA
PARAMOEBIDAE
Mayorella vespertilio Penard, 1902 F/T
VEXILLIFERIDAE
Neoparamoeba pemaquidensis (Page, 1970) M

Order VANNELIDA
VANNELLIDAE
Discamoeba guttula (Dujardin, 1835) T
Vannella sp. Stout 1984 F/T

Order DERMAMOEBIDA
STRIAMOEBIDAE
Striamoeba striata (Penard, 1890) F/T

Order THECAMOEBIDA
THECAMOEBIDAE
Thecamoeba verrucosa (Ehrenberg, 1838, emend. Gläser, 1912) F/T

FLABELLINEA INCERTAE SEDIS
HYALODISCIDAE
Hyalodiscus rubicundus Hertwig & Lesser, 1874 F/T

Note
Difflugia capreolata Penard, 1902 *in* Hayward et al. 1996 = *Difflugia oblonga*
Centropyxis excentricus Cushman & Bronnimann, 1948 *in* Hayward et al. 1996 = *C. constricta*
Urnulina compressa in Hayward *et al.* 1996 = *Difflugia bidens* Penard, 1902

EIGHT

Phylum APUSOZOA

thecomonad, ancyromonad, diphylleid zooflagellates

DENNIS P. GORDON

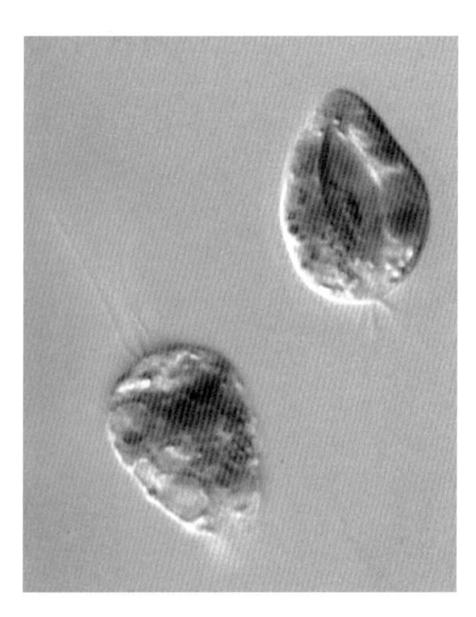

Collodictyon triciliatum (non-New Zealand).
Yuuji Tsukii, Hosei University

Apusozoa is a relatively small phylum of zooflagellates little known to the non-specialist. Thecomonads include two-ciliate, gliding bacterial feeders with a flexible outer sheath ('theca'). Ancyromonads ('planomonads'), also bacterivorous gliders, are strongly flattened with a semi-rigid cell surface supported by an underlying dense layer while micronucleariids lack cilia and have slender, branching pseudopodia. Ancyromonads and micronucleariids together comprise Hilomonadea. Diphylleids, only doubtfully included in the phylum, have two or four cilia and are phagotrophic on protists. Food uptake in Apusozoa may be at the anterior end of the cell or via a groove along one side, but there are no defined oral structures as such. In general, individuals range from 10 to 60 micrometres long. All except Hilomonadea have in common that their mitochondria have tubular or discoidal cristae (the infoldings of the inner mitochondrial membrane); in Hilomonadea the cristae are flat.

The actual composition of the phylum has been modified somewhat since it was first proposed as a subphylum of Neomonada by Cavalier-Smith (1997). At the time, molecular trees suggested that Apusozoa (then somewhat broadly circumscribed) and Choanozoa (which includes collared flagellates) are more closely related to each other than to any other phylum, so Cavalier-Smith united them in the one phylum. Subsequent molecular data moved Cavalier-Smith (2002) to abandon the Neomonada, raising each subphylum to phylum rank, but in different infrakingdoms. Next, Cavalier-Smith and Chao (2003) narrowly defined Apusozoa as comprising a single class (Thecomonadea) with three small orders, collectively characterised by a dense layer in the form of one or a pair of 'thecal plates', underlying much of the cell (plasma) membrane and characteristically curved at the junctions with the soft non-'thecate' parts of the cell membrane. Later the same year, Cavalier-Smith (2003) reincluded in the Apusozoa the order Diphylleida, but in a new class, Diphyllatea, treating the Apusozoa as 'Protozoa incertae sedis' so far as its wider relationships are concerned. Cavalier-Smith et al. (2008) subsequently included a new genus of gliding zooflagellates, *Planomonas* (to replace *Ancyromonas* of authors) grouping it with *Micronuclearia* in a new class, Hilomonadea. Based on the pattern of the ciliary rootlets, Cavalier-Smith and Chao (2010) and Glücksman et al. (2011) perceived the diphylleids were closer to loukozoan excavates like *Malawimonas* but in correspondence (January 2011), Cavalier-Smith wrote of the Diphylleida: 'I would tentatively move this into Apusozoa or place it simply as Protozoa incertae sedis. Unpublished multigene trees show its position is problematic but more likely to be closer to Apusozoa than to malawimonas (the two alternative positions it likes).' Heiss et al. (2010, 2011) reinstated *Ancyromonas* over *Planomonas*.

The phylum name Apusozoa is somewhat meaningless in itself. It is coined after the name of the included genus *Apusomonas*, the species of which are common in soil but none has yet been recorded from New Zealand. The author of *Apusomonas* (Alexeieff 1924) did not explain its etymology.

Class Diphyllatea (and indeed the entire phylum) is solely exemplified in New Zealand by *Collodictyon sphaericum*. The species was first described from Wellington Harbour by Norris (1964). It was later listed by Taylor (1974) and included in Dawson's (1992) index of New Zealand protozoans, but has not been formally collected since, so remains nominally endemic. Cells are spherical, 8.5–11.0 micrometres in diameter, with two unequal pairs of cilia emerging from a cone-shaped depression, the shorter pair about 13 micrometres long with the longer pair double this length. A longitudinal groove via which food items are ingested by phagotrophy (as in other species of *Collodictyon*) has not been observed, but, during cell division, a furrow develops at the apical end of the cell and gradually extends to the posterior end. The cilia are separated, so that two are contributed to each daughter cell, but before the daughter cells fully separate, additional cilia grow alongside. Active cells were seen with ingested *Pyramimonas*, a unicellular green alga.

Collodictyon sphaericum.
From Norris 1964

At the time of its description, *C. sphaericum* was considered to be a chlorophyte belonging to the family Polyblepharidaceae (Norris 1964). Other species of *Collodictyon* live in fresh water (Patterson et al. 2000). In Japanese lakes, *C. triciliatum* is a known grazer on the toxic cyanobacterium *Microcystis aeruginosa* (Nishibe et al. 2002), one of several such species that form nuisance blooms in New Zealand lakes. Though *M. aeruginosa* is an important food for *C. triciliatum*, it seems that the flagellate is not effective as a biological-control organism in this instance.

It is certain that many more members of Apusozoa will be found in New Zealand, both in soils and aquatic habitats. There is evidence for the existence of this group also in deep ocean waters in the Atlantic and Mediterranean (Arndt et al. 2003; López-García et al. 2003), thus Apusozoa should equally be found in New Zealand's extensive deep-water environments.

Author

Dr Dennis P. Gordon National Institute of Water & Atmospheric Research, Private Bag 14901, Kilbirnie, Wellington, New Zealand [d.gordon@niwa.co.nz]

Summary of New Zealand apusozoan diversity

Taxon	Described species	Known undescribed/ undetermined species	Estimated unknown species	Adventive species	Endemic species	Endemic genera
Apusozoa	1	0	3	0	1	0

Diversity by environment

Taxon	Marine	Freshwater/ Terrestrial
Apusozoa	1	0

References

ALEXEIEFF, A. 1924: Notes sur quelques Protistes coprocoles. *Archiv für Protistenkunde 50*: 27–49, pls 1–3.

ARNDT, H.; HAUSMANN, K.; WOLF, M. 2003: Deep-sea heterotrophic nanoflagellates of the Eastern Mediterranean Sea: qualitative and quantitative aspects of their pelagic and benthic occurrence. *Marine Ecology Progress Series 256*: 45–56.

CAVALIER-SMITH, T. 1997: Amoeboflagellates and mitochondrial cristae in eukaryote evolution, megasystematics of the new protozoan subkingdoms Eozoa and Neozoa. *Archiv für Protistenkunde 147*: 237–258.

CAVALIER-SMITH, T. 2002: The phagotrophic origin of eukaryotes and phylogenetic classification of Protozoa. *International Journal of Systematic and Evolutionary Microbiology 52*: 297–354.

CAVALIER-SMITH, T. 2003: The excavate protozoan phyla Metamonada Grassé emend. (Anaeromonadea, Parabasalia, *Carpediemonas*, Eopharyngia) and Loukozoa emend. (Jakobea, *Malawimonas*): their evolutionary affinities and new higher taxa. *International Journal of Systematic and Evolutionary Microbiology 53*: 1741–1758.

CAVALIER-SMITH, T.; CHAO, E. E. 2003: Phylogeny of Choanozoa, Apusozoa, and other Protozoa and early eukaryote megaevolution. *Journal of Molecular Evolution 56*: 540–563.

CAVALIER-SMITH, T.; CHAO, E. E. 2010: Phylogeny and evolution of Apusomonadida (Protozoa: Apusozoa): new genera and species. *Protist 161*: 549–576.

CAVALIER-SMITH, T.; CHAO, E. E.; STECHMANN, A.; OATES, B.; NIKOLAEV, S. 2008: Planomonadida ord. nov. (Apusozoa): ultrastructural affinity with *Micronuclearia podoventralis* and deep divergences within *Planomonas* gen. nov. *Protist 159*: 535–562.

DAWSON, E. W. 1992: The marine fauna of New Zealand: Index to the fauna 1. Protozoa. *New Zealand Oceanographic Institute Memoir 99*: 1–368.

GLÜCKSMAN, E.; SNELL, E. A.; BERNEY, C.; CHAO, E. E.; BASS, D.; CAVALIER-SMITH, T. 2011: The novel marine gliding flagellate genus *Mantamonas* (Mantamonadida) ord. n.: Apusozoa). *Protist 162*: 207–221.

HEISS, A. A.; WALKER, G.; SIMPSON, A. G. B. 2010: Clarifying the taxonomic identity of a phylogenetically important group of eukaryotes: *Planomonas* is a junior synonym of *Ancyromonas*. *Journal of Eukaryotic Microbiology 57*: 285–293.

HEISS, A. A.; WALKER, G.; SIMPSON, A. G. B. 2011: The ultrastructure of *Ancyromonas*, a eukaryote without supergroup affinities. *Protist 162*: 373–393.

LÓPEZ-GARCÍA, P.; PHILIPPE, H.; GAIL, F.; MOREIRA, D. 2003: Autochthonous eukaryotic diversity in hydrothermal sediment and experimental microcolonizers at the Mid-Atlantic Ridge. *Proceedings of the National Academy of Sciences USA 100*: 697–702.

NISHIBE, Y.; KAABATA, Z.; NAKANO, S. 2002: Grazing on *Microcystis aeruginosa* by the heterotrophic flagellate *Collodictyon triciliatum* in a hypertrophic pond. *Aquatic Microbial Ecology 29*: 173–179.

NORRIS, R. E. 1964: Studies on phytoplankton in Wellington Harbour. *New Zealand Journal of Botany 2*: 258–278.

PATTERSON, D. J.; VØRS, N.; SIMPSON, A. G. B.; O'KELLY, C. 2000: Residual free-living and predatory heterotrophic flagellates. Pp. 1302–1328 *in*: Lee, J. J.; Leedale, G. F.; Bradbury, P. (eds), *An Illustrated Guide to the Protozoa. Second Edition. Organisms traditionally referred to as Protozoa or newly discovered groups* Society of Protozoologists, Lawrence. v + 1432 p. [in 2 vols.]

TAYLOR, F. J. 1974: A preliminary annotated checklist of micro-algae other than diatoms and dinoflagellates from New Zealand coastal waters. *Journal of the Royal Society of New Zealand 4*: 395–400.

Checklist of New Zealand Apusozoa

Abbreviations: E, endemic; M, marine.

KINGDOM PROTOZOA
SUBKINGDOM SARCOMASTIGOTA
PHYLUM APUSOZOA
Class DIPHYLLATEA
Order DIPHYLLEIDA
DIPHYLLEIDAE
Collodictyon sphaericum Norris, 1964 M E

NINE

Phylum CHOANOZOA

collar flagellates, nucleariids, and kin

DENNIS P. GORDON

Phylum Choanozoa comprises a morphologically disparate group of non-photosynthetic protozoans that are linked in molecular trees and include organisms said to be at the 'animal-fungal boundary' (Mendoza et al. 2002). At the ultrastructural level, their mitochondria have flattened or discoidal cristae. Cavalier-Smith and Chao (2003) recognised four classes – Cristidiscoidea, Choanoflagellatea, Ichthyosporea, and Corallochytrea – and Shalchian-Tabrizi et al. (2008) added a fifth, Filasterea.

The best-known of these is the choanoflagellates, a highly distinctive group of colourless protozoans that are ubiquitous in aquatic environments. Whether solitary or colonial, free-swimming or attached, all species share the group's most characteristic features – a cell body with a single anterior cilium that is surrounded by a funnel-shaped collar. The collar is made up of microvilli. The form and composition of the cell covering (periplast) vary between species and are important in taxonomy at all levels.

An individual of *Didymoeca costata* showing the central cilium and shorter microvilli inside the lorica.
Barry Leadbeater

The Choanoflagellatea contains two orders (Nitsche et al. 2011). The Craspedida includes species with an organic outer investment that is so thin that cells appear naked and cell division is lateral. Alternatively, the covering may take the form of a firm surrounding theca, in which case cell division is emergent, i.e. the enclosed cell must partly emerge from the theca in order to undergo division outside it. The Acanthoecida includes species with a siliceous basket-like lorica. This comprises a framework of costae, each costa containing a linear series of rod-shaped units known as costal strips. The process of lorica assembly, which is quite astonishing, has been followed in detail using videography and electron microscopy and found to follow certain 'rules' that govern the sequence of events whereby the siliceous strips are positioned within the developing lorica (Frösler & Leadbeater 2009; Leadbeater & Cheng 2010). Siliceous forms are found only in marine and brackish environments (Leadbeater 1985; Leadbeater & Thomsen 2000). The benefit of surrounding a collar cell with a siliceous lorica is to create drag, counteracting the movement caused by the motion of the cilium. A large basket can cancel out movement altogether, helping the cell to remain suspended in the water.

The food of choanoflagellates comprises bacteria (including cyanobacteria) and detritus filtered by the collar microvilli from a current of water created by the motion of the cilium. It beats with a base-to-tip undulation that creates a forwardly directed current of water. Particles are then sifted by the collar and ingested by tongue-like pseudopodia of the cell body just outside the base of the collar. Choanoflagellates are a major component of the so-called microbial loop in food webs, in which their role is that of bacterivory followed by release of inorganic nutrients back into the water.

Species reproduce by splitting of the cell as described above, and sedentary

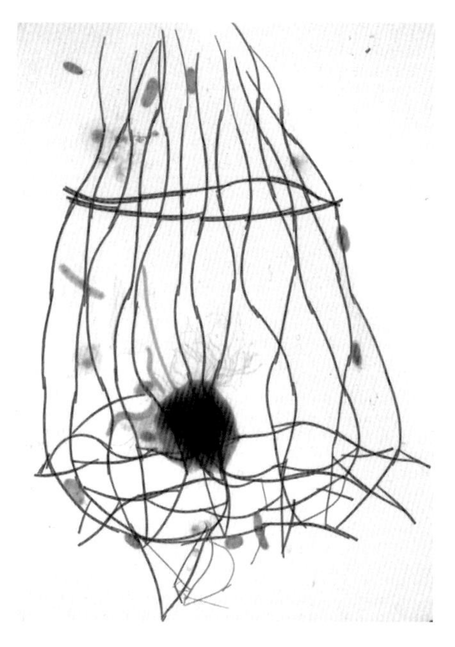

A shadowcast whole mount of an individual of the collar flagellate *Diaphanoeca grandis*, showing the cell body in the lower part of the basket-like lorica that it assembles around itself.
Barry Leadbeater

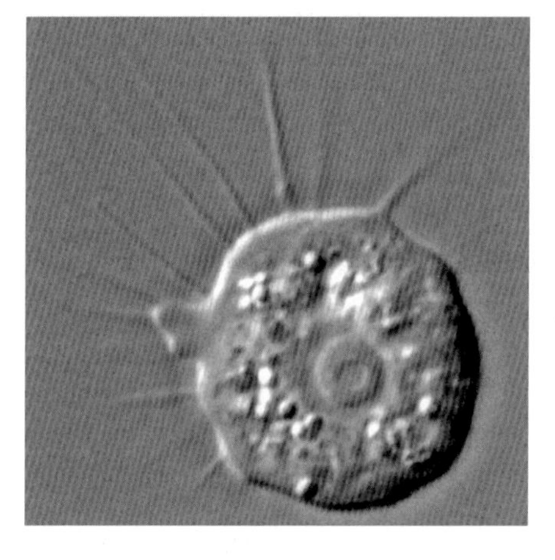

Nuclearia simplex, an amoeboid choanozoan of fresh water and soil water.
Michele Bahr & David J. Patterson per Micro*scope (MBL)

species also produce a naked swarmer that is posteriorly propelled by the cilium. Species with a siliceous lorica may divide longitudinally and produce a naked swarmer, or a juvenile covered with siliceous strips, which is pushed backwards out of the parent's lorica. Sexual reproduction has never been described in choanoflagellates.

In nature, sedentary choanoflagellates can be collected by scraping the sides of aquaria or sampling filamentous algae or other submerged objects. Planktonic species can be collected by the methods described by Moestrup (1979) and Leadbeater (1981). The New Zealand fauna has not been well studied, especially in the freshwater environment, but 31 species have been reported. The first records were those of Maskell (1887, 1888) in the Wellington region. Since then, students of microalgae have reported choanoflagellates, usually among lists of chrysophyte algae, with which these organisms used to be classified (as Craspedophycidae), e.g. Thomasson (1960, 1974), Flint (1966), Hill (1970), Cassie (1974, 1975), and Skuja (1976).

Moestrup (1979) made collections of nanoplankton, including choanoflagellates, from two coastal localities in New Zealand – Leigh and Kaikoura – where the presence of university marine laboratories facilitated sampling and processing. Most species were already known from the North Atlantic Ocean, suggesting a cosmopolitan distribution, perhaps through cooler deeper water at the Equator. Notwithstanding, the existence of a continuous distribution does not preclude the possibility of endemic forms having evolved in certain regions. Bojo (2001) added more species from a study in Blueskin Bay, Otago.

Class Cristidiscoidea is represented in New Zealand by *Nuclearia* and *Rozella*. Unlike the collar flagellates, nucleariids are amoeboid. They are small to medium-sized cyst-forming amoebae, either naked or with a mucus sheath, flattened, or with thin pseudopodia. The latter may be mainly from one margin or the cells may become rounded and the pseudopodia then radiate outwards around the entire margin. Nucleariid amoebae are found in fresh water or soil water where they consume bacteria and detritus. Stout (1958, 1960, 1961) reported *Nuclearia simplex* in hill and tussock soils in New Zealand's North Island. *Rozella* has been included in Oomycota and Chytridiomycota.

Classes Ichthyosporea and Corallochytrea include mostly parasites of vertebrates and, in a few cases, of some aquatic arthropods and molluscs. Life-history stages can include amoeba-like cells, zoospores (flagellate or aflagellate), and even fungus-like hyphae, depending on the taxon (Mendoza et al. 2002). The best-known examples are species of *Dermocystidium* (infecting koi carp) and *Ichthyophonus* (infecting a variety of fishes including salmonids). New Zealand's introduced carassiids and salmonids appear to be free of these organisms and the diseases attributed to them (Diggles et al. 2002).

Ichthyosporea is nevertheless represented in New Zealand by three genera that have typically been included in the fungal class Trichomycetes (Zygomycota) in the orders Amoebidiales (*Paramoebidium*) and Eccrinales (*Enteromyces* and *Taeniella*). *Amoebidium* was the first trichomycete to be cultured (Whisler

Diversity by environment

Taxon	Marine	Freshwater	Terrestrial
Cristidiscoidea*	0	7	0
Choanoflagellatea	20	11	1
Ichthyosporea	2	2	0
Totals	22	20	1

* is found in both fully freshwater and terrestrial (soil water) environments.

Summary of New Zealand choanozoan diversity

Taxon	Described species	Known undescribed/ undetermined species	Estimated unknown species	Adventive species	Endemic species	Endemic genera
Cristidiscoidea	7	0	5	0	0	0
Choanoflagellatea	26	5	20	0	0	1
Ichthyosporea	4	0	20	0	2	0
Totals	37	5	45	0	2	1

1960), which permitted comprehensive studies of its biology and phylogenetic relationships. The fact that it produces amoebae in its life-cycle – not otherwise known in fungi, raised questions about its affinities (Lichtwardt 1986). The absence of chitin in the cell wall (Trotter & Whisler 1965) and the presence of stacked Golgi dictyosomes (Whisler & Fuller 1968) constituted further evidence against inclusion in the fungi. Gene sequencing subsequently demonstrated that Amoebidiales (Benny & O'Donnell 2000; Mendoza et al. 2002) and Eccrinales (Cafaro 2005) belong in the Ichthyosporea. These fungoid forms generally have a tiny thallus-like body about 0.3 to 2.3 millimetres long that is a straight or curved, branched or unbranched, cylindrical filament attached at its base (holdfast) to the host. The thallus is coenocytic, i.e. without cross-walls so that the nuclei are freely distributed. Reproduction takes place when the contents of the thallus round off into spore-like cells or the onset of moulting in the host triggers the production of amoeboid cells. The latter remain motile for a short period before encysting, producing several spores per cyst that are capable of infecting another host.

Enteromyces callianassae, a thalloid choanozoan from the foregut (stomach) of the ghost shrimp *Callianassa filholi*, showing a cluster of thalli producing terminal spores.

Robert W. Lichtwardt, University of Nebraska

Species of *Amoebidium* and *Paramoebidium* are reported to live as commensals of aquatic insects and crustaceans. In New Zealand, endemic *Paramoebidium papillatum* occurs in the hindgut of nesameletid mayfly larvae while *P. bibrachium* is found in leptophlebiid and coloburiscid mayfly larvae (Lichtwardt & Williams 1992). *Enteromyces callianassae* and *Taeniella carcini*, on the other hand, are marine and Pan-Pacific, possibly worldwide. In New Zealand, Williams and Lichtwardt (1990) found *E. callianassae* in the stomach of the ghost shrimp *Callianassa filholi*, and *T. carcini* in both the ghost shrimp and the mud crab *Austrohelice crassa*, at New Brighton, Christchurch.

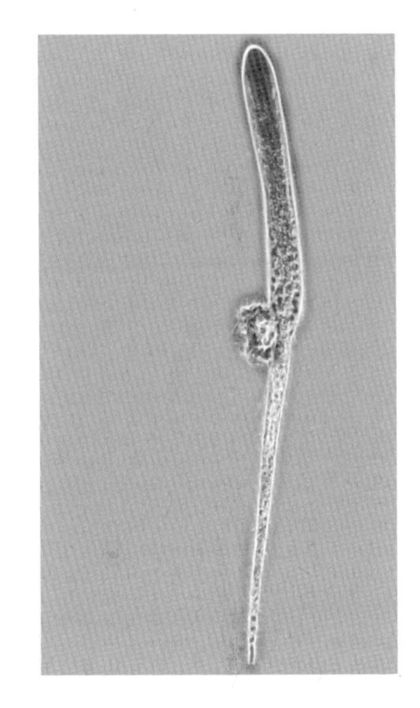

Thallus of *Paramoebidium bibrachium* from the mayfly *Coloburiscus humeralis*; in the middle is a small holdfast by which it was attached to the hindgut.

Robert W. Lichtwardt, University of Nebraska

Evolutionary significance of Choanozoa

The resemblance of choanoflagellates to the collar cells in the ciliated chambers of sponges has long suggested the origin of sponges, and perhaps thereby of all Animalia, from choanoflagellate ancestors (Dujardin 1841; Clark 1866; Afzelius 1961; Leadbeater 1983), although the alternate idea that choanoflagellates may be derived, by simplification, from sponges has also been advanced (Clark 1868; Kent 1878; Maldonado 2004). The demonstration of homology at the ultrastructural level between the choanoflagellate cell and the sponge choanocyte has strengthened the connection between the two groups, as have molecular data, but the precise sequence of events leading from Choanozoa to Animalia has not been clear.

A multi-gene phylogeny has been published by Shalchian-Tabrizi et al. (2008), who sought to clarify the relationships among the various morphologies exhibited among the Choanozoa. They showed that *Ministeria vibrans*, a minute bacteria-eating cell with slender radiating tentacles, is a sister taxon of *Capsaspora*, which also has filose tentacles. Together they comprise the class Filasterea, which in turn is sister to choanoflagellates and animals. These three groups form a branch whose ancestor presumably evolved filose tentacles before these tentacles aggregated as a collar in the choanoflagellate/sponge common ancestor. Ichthyosporea is sister to that common branch. Significantly, the genes responsible for making cadherins – the chemical mediators of cell adhesion in animals – are also found in choanoflagellates, hence the existence of cadherins in choanoflagellates may have contributed to multicellularity and the origin of animals (Abedin & King 2008). At present, the clearest relationship to have been demonstrated by molecular-phylogenetic studies is that choanoflagellates are monophyletic and the sister group of the Animalia, not paraphyletic to them Carr et al. 2008; Nitsche et al. 2011).

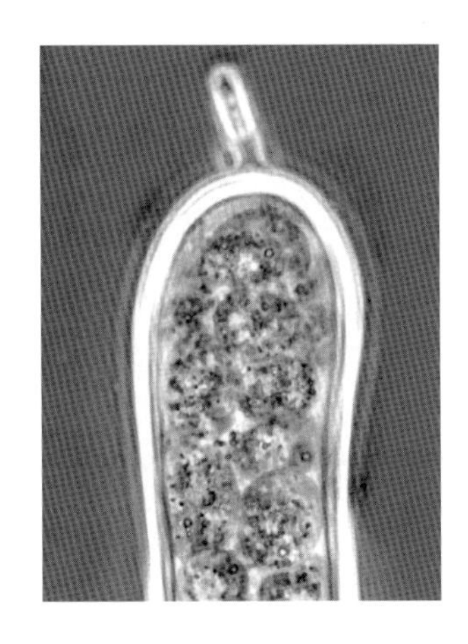

Thallus tip of of *Paramoebidium papillatum* from the mayfly *Nesameletus* sp., containing amoeboid cells just prior to their release.

Robert W. Lichtwardt, University of Nebraska

Cavalier-Smith (1998) coined the name Ichthyosporea for the clade that had earlier been identified as near the animal-fungal divergence (Ragan et al. 1996). Mendoza et al. (2002) sought to substitute the name Mesomycetozoea [based on the clade name mesomycetozoa proposed by Herr et al. (1999)] for Ichythosporea on the grounds that, when Cavalier-Smith (1998) coined it, all its members were fish parasites. With the addition of *Amoebidium* and other taxa not associated with fish, and the discovery of chitin in *Ichthyophonus*, they argued that Mesomycetozoea is a more suitable name. Adl et al. (2005) used Mesomycetozoa for a larger clade nearly synonymous with Choanozoa. In the event, Cavalier-Smith's name has priority and is retained here.

It is now accepted that fungi also have choanozoan ancestry (Cavalier-Smith 1987; 1993; Wainwright et al. 1993); nucleariids appear to be the closest living relatives (Medina et al. 2003; Ruiz-Trillo et al. 2006).

Acknowledgements

The author is grateful to Professor Barry Leadbeater (University of Birmingham), for advice, information, and illustrations, and to Professor Thomas Cavalier-Smith (Oxford University) for advice on classification.

Author

Dr Dennis P. Gordon National Institute of Water and Atmospheric Research, Private Bag 14901, Kilbirnie, Wellington, New Zealand [d.gordon@niwa.co.nz]

References

ABEDIN, M.; KING, N. 2009: The premetazoan ancestry of cadherins. *Science 319*: 946–948.

ADL, S. M.; SIMPSON, A. G. B.; FARMER, M. A.; ANDERSEN, R. A.; ANDERSON, O. R.; BARTA, J. R.; BOWSER, S. S.; BRUGEROLLE, G.; FENSOME, R. A.; FREDERICQ, S.; JAMES, T.Y.; KARPOV, S.; KUGRENS, P.; KRUG, J.; LANE, C. E.; LEWIS, L. A.; LODGE, J.; LYNN, D. H.; MANN, D. G.; McCOURT, R. M.; MENDOZA, L.; MOESTRUP, Ø.; MOZLEY-STANDRIDGE, S. E.; NERAD, T. A.; SHEARER, C. A.; SMIRNOV, A. V.; SPIEGEL, F. W.; TAYLOR, M. F. J. R. 2005: The new higher level classification of eukaryotes with emphasis on the taxonomy of protists. *Journal of Eukaryotic Microbiology 52*: 399–451.

AFZELIUS, B. A. 1961: Flimmer-flagellum of the sponge. *Nature 191*: 1318–1319.

BENNY, G. L.; O'DONNELL, K. 2000: *Amoebidium parasiticum* is a protozoan, not a trichomycete. *Mycologia 92*: 1133–1137.

BOJO, O. 2001: Systematic studies of New Zealand nanoflagellates with special reference to members of the Haptophyta. Unpublished PhD thesis, University of Otago, Dunedin. xi +240 p.

CAFARO, M. J. 2005: Eccrinales (Trichomycetes) are not fungi, but a clade of protists at the early divergence of animals and fungi. *Molecular Phylogenetics and Evolution 35*: 21–34.

CARR, M.; LEADBEATER, B. S. C.; HASSAN, R.; NELSON, M.; BALDAUF, S. L. (2008: Molecular phylogeny of choanoflagellates, the sister group to Metazoa. *Proceedings of the National Academy of Science of the USA 105*: 16641–16646.

CASSIE, V. 1974: Algal flora of some North Island New Zealand lakes including Rotorua and Rotoiti. *Pacific Science 28*: 467–504.

CASSIE, V. 1975: Phytoplankton of Lakes Rotorua and Rotoiti (North Island). Pp. 193–205 *in*: Jolly, V. H.; Brown, J. M. A. (eds), *New Zealand Lakes*. Auckland University Press/Oxford University Press, Auckland. 388 p.

CAVALIER-SMITH, T. 1987: The origin of fungi and pseudofungi. Pp. 339–353 *in*: Rayner, A. D. M.; Brasier, C. M.; Moore, D.M. (eds), *Evolutionary Biology of the Fungi*. [*Symposium of the British Mycological Society 13*.] Cambridge University Press, Cambridge.

CAVALIER-SMITH, T. 1993: Kingdom Protozoa and its 18 phyla. *Microbiological Reviews 57*: 953–994.

CAVALIER-SMITH, T. 1998: Neomonada and the origin of animals and fungi. Pp. 375–407 *in*: Coombs, G. H.; Vickerman, K.; Sleigh, M. A.; Warren, A. (eds), [The Systematics Association Special Volume 56] *Evolutionary Relationships among Protozoa*. Kluwer, Dordrecht. 486 p.

CAVALIER-SMITH, T.; CHAO, E. E.-Y. 2003: Phylogeny and classification of phylum Cercozoa (Protozoa). *Protist 154*: 341–358.

CLARK, H. 1866: Note on the infusoria flagellata and the spongiae ciliatae. *American Journal of Science 1*: 113–114.

CLARK, H. 1868: On the Spongiae ciliatae as *Infusoria flagellata*, or observations on the structure, animality and relationship of *Leucosolenia botryoides* Bowerbank. *Annals and Magazine of Natural History, ser. 5, 4*: 133–142, 188–215, 250–264.

DIGGLES, B. K.; HINE, P. M.; HANDLEY, S.; BOUSTEAD, N. C. 2002: A handbook of disease of importance to aquaculture in New Zealand. *NIWA Science and Technology Series No. 49*: 1–200.

DUJARDIN, F. 1841: *Histoire naturelle du Zoophytes. Infusoires*. Roret, Paris.

FLINT, E. A. 1966: Additions to the checklist of freshwater algae in New Zealand. *Transactions of the Royal Society of New Zealand, Botany 3*: 123–137.

FRÖSLER, J.; LEADBEATER, B. S. C. 2009: Role of the cytoskeleton in choanoflagellate lorica assembly. *Journal of Eukaryotic Microbiology 56*: 167–173.

HILL, C. F. 1970: Phyto- and zooplankton recorded from the Waikato River and hydroelectric lakes between Taupo control gates and the Meremere power station. New Zealand Electricity Department Report, Hamilton. 5 p.

KENT, S. 1878: Notes on the embryology of sponges. *Annals and Magazine of Natural History, ser. 5, 5*: 139–156.

LEADBEATER, B. S. C. 1981: Ultrastructure and deposition of silica in loricate choanoflagellates. Pp. 295–322 *in*: Volcani, B.E.; Simpson, T.L. (eds), *Silicon and Silica Deposition in Biological Systems*. Springer-Verlag, New York.

LEADBEATER, B. S. C. 1983: Life-history and ultrastructure of a new marine species of *Proterospongia* (Choanoflagellida). *Journal of the Marine Biological Association of the United Kingdom 63*: 135–160.

LEADBEATER, B. S. C. 1985: Order 1. Choanoflagellida Kent, 1880. Pp. 106–116 *in*: Lee, J. J.; Hutner, S. H.; Bovee, E. C. (eds), *An Illustrated Guide to the Protozoa*. Society of Protozoologists, Lawrence. ix + 629 p.

LEADBEATER, B. S. C.; CHENG, F. 2010: Costal strip production and lorica assembly in the large tectiform choanoflagellate Diaphanoeca grandis Ellis. *European Journal of Protistology 46*: 96–110.

LEADBEATER, B. S. C.; THOMSEN, H. A. 2000: Order Choanoflagellida. Pp. 14–38 *in*: Lee, J. J.; Leedale, G. F.; Bradbury, P. (eds), *An Illustrated Guide to the Protozoa. Second Edition. Organisms traditionally referred to as Protozoa or newly discovered groups*. Society of Protozoologists, Lawrence. v + 1432 p. [in 2 vols.]

LICHTWARDT, R. W. 1986: *The Trichomycetes: Fungal Associates of Arthropods*. Springer-Verlag, New York. 343 p.

LICHTWARDT, R. W.; WILLIAMS, M. C. 1992: Two new Australasian species of Amoebidiales associated with aquatic insect larvae, and comments on their biogeography. *Mycologia 84*: 376–383.

MALDONADO, M. 2004: Choanoflagellates, choanocytes, and animal multicellularity. *Invertebrate Biology 123*: 1–22.

MASKELL, W. M. 1887: On the fresh-water Infusoria of the Wellington District. *Transactions of the New Zealand Institute 19*: 49–61, pls 3–5.
MASKELL, W. M. 1888: On the fresh-water Infusoria of the Wellington District. *Transactions of the New Zealand Institute 20*: 3–19, pls 1–4.
MEDINA, M.; COLLINS, A. G.; TAYLOR, J. W.; VALENTINE, J. W.; LIPPS, J. H.; AMARAL-ZETTLER, L.; SOGIN, M. L. 2003: Phylogeny of Opisthokonta and the evolution of multicellularity and complexity in Fungi and Metazoa. *International Journal of Astrobiology 2*: 203–211.
MENDOZA, L.; TAYLOR, J. W.; AJELLO, L. 2002: The class Mesomycetozoea: a heterogeneous group of microorganisms at the animal-fungal boundary. *Annual Review of Microbiology 56*: 315–344.
MOESTRUP, Ø. 1979: Identification by electron microscopy of marine nanoplankton from New Zealand, including the description of four new species. *New Zealand Journal of Botany 17*: 61–95.
NITSCHE, F.; CARR, M.; ARNDT, H.; LEADBEATER, B. S. C. 2011: Higher level taxonomy and molecular phylogenetics of the Choanoflagellatea. *Journal of Eukaryotic Microbiology.*
RAGAN, M. A.; GOGGIN, C. L.; CAWTHORN, R. J.; CERENIUS, L.; JAMIESON, A. V. C.; PLOURDE, S. M.; RAND, T. G.; SÖDERHÄLL, K.; GUTELL, R. R. 1996: A novel clade of protistan parasites near the animal-fungal divergence. *Proceedings of the National Academy of Sciences USA 93*: 11907–11912.
RUIZ-TRILLO, I.; ROGER, A. J.; BURGER, G.; GRAY, M. W.; LANG, B. F. 2008: A phylogenomic investigation into the origin of Metazoa. *Molecular Biology and Evolution 25*: 664–672.
SHALCHIAN-TABRIZI, K.; MINGE, M. A.; ESPELUND, M.; ORR, R.; RUDEN, T.; JAKOBSEN, K. S.; CAVALIER-SMITH, T. 2008: Multigene phylogeny of Choanozoa and the origin of animals. *PLoS ONE 3(5) e2098*: 1–7.
SKUJA, H. 1976: Zür Kenntnis der Algen neuseelandischer Torfmoore. *Nova Acta Regiae Societatis Scientiarum Upsaliensis, ser, 5C, 2*: 1–158.
STOUT, J. D. 1958: Biological studies of some tussock-grassland soils. VII. Protozoa. *New Zealand Journal of Agricultural Research 1*: 974–984.
STOUT, J. D. 1960: Biological studies of some tussock-grassland soils. XVIII. Protozoa of two cultivated soils. *New Zealand Journal of Agricultural Research 3*: 237–244.
STOUT, J. D. 1961: Biological and chemical changes following scrub burning on a New Zealand hill soil. *New Zealand Journal of Science 4*: 739–752.
THOMASSON, K. 1960: Some planktic Staurastra from New Zealand. *Botanisker Notiser 113*: 225–245.
THOMASSON, K. 1974: Rotorua phytoplankton reconsidered (North Island of New Zealand). *Internationale Revue der Gesamten Hydrobiologie 59*: 703–727.
TROTTER, M. J.; WHISLER, H. C. 1965: Chemical composition of the cell wall of *Amoebidium parasiticum*. *Canadian Journal of Botany 43*: 869–876.
WAINWRIGHT, P. O.; HINKLE, G.; SOGIN, M. L.; STICKEL, S. K. 1993: Monophyletic origins of the Metazoa, an evolutionary link with the Fungi. *Science 260*: 340–342.
WHISLER, H. C. 1960: Pure culture of the trichomycete *Amoebidium parasiticum*. *Nature 186*: 732–733.
WHISLER, H. C.; FULLER, M. S. 1968: Preliminary observations on the holdfast of *Amoebidium parasiticum*. *Mycologia 60*: 1088–1079.
WILLIAMS, M. C.; LICHTWARDT, R. W. 1990: Trichomycete gut fungi in New Zealand aquatic insect larvae. *Canadian Journal of Botany 68*: 1045–1056.

Checklist of New Zealand Choanozoa

Classification follows Cavalier-Smith and Chao (2003) and Nitsche et al. (2011). Abbreviations: E, endemic; M, marine; T, terrestrial (soil water); P, parasitic.

KINGDOM PROTOZOA
SUBKINGDOM SARCOMASTIGOTA
PHYLUM CHOANOZOA
SUBPHYLUM CRISTIDISCOIDIA
Class CRISTIDISCOIDEA
Order NUCLEARIIDA
NUCLEARIIDAE
Nuclearia simplex Cienowski, 1865 F/T

Order ROZELLIDA
ROZELLIDAE
Rozella chytriomycetis Karling F
Rozella cladochytrii Karling F
Rozella cuculus (E.J.Butler) Sparrow F
Rozella laevis Karling F
Rozella longicollis Karling F
Rozella rhizophlyctidis Karling F

SUBPHYLUM CHOANOFILA
Class CHOANOFLAGELLATEA
Order ACANTHOECIDA
ACANTHOECIDAE
Acanthoeca spectabilis Ellis, 1930 M
STEPHANOECIDAE
Acanthocorbis apoda (Leadbeater, 1972) M
Acanthocorbis haurakiana Thomsen in Thomsen, Buck & Chavez, 1991 M
Apheloecion sp. Bojo 2001 M
Bicosta antennigera Moestrup, 1979 M
Bicosta minor (Reynolds, 1976) M
Calliacantha aff. *natans* (Grøntved, 1956) M
Calotheca alata Thomsen & Moestrup, 1983 M
Cosmoeca norvegica Thomsen in Thomsen & Boonruang, 1984 M
Cosmoeca cf. *subulata* Thomsen in Thomsen & Boonruang, 1984 M
Cosmoeca ventricosa Thomsen in Thomsen & Boonruang, 1984 M
Crinolina isefiordensis Thomsen, 1976 M
Crucispina cruciformis (Leadbeater, 1974) M
Diaphanoeca grandis Ellis, 1930 M
Didymoeca costata (Valkanov, 1970) M
Parvicorbicula cf. *socialis* (Meunier, 1910) M
Parvicorbula n. sp. Moestrup 1979 M
Pleurasiga reynoldsii Throndsen, 1970 M
Polyfibula sphyrelata Thomsen, 1973 M
Syndetophyllum pulchellum (Leadbeater, 1974) M
Order CRASPEDIDA
SALPINGOECIDAE
Codosiga botrytis (Ehrenberg, 1838) F
Diploeca elongata (Fott, 1940) F
Diploeca flava (Korschikov, 1926) F
Monosiga brevipes Kent, 1880 F
Monosiga consociata Kent, 1880 F
Salpingoeca amphoridium Clark, 1867 F
Salpingoeca frequentissima (Zacharias, 1894) F
Salpingoeca fusiformis Kent, 1880 F
Salpingoeca globulosa Zhukov, 1978 F
Salpingoeca inquillata Kent, 1881 F
Salpingoeca marssonii Lemmermann, 1914 F

Class ICHTHYOSPOREA
Order ICHTHYOSPORIDA
AMOEBIDIIDAE
Paramoebidium bibrachium Williams & Lichtwardt, 1990 F P E
Paramoebidium papillatum Lichtwardt & Williams, 1992 F P E
ECCRINIDAE
Enteromyces callianassae Lichtwardt, 1961 M P
Taeniella carcini Léger & Dubosq, 1911 M P

TEN

Phylum

OCHROPHYTA

brown and golden-brown algae, diatoms, silicoflagellates, and kin

MARGARET A. HARPER, VIVIENNE CASSIE COOPER,
F. HOE CHANG, WENDY A. NELSON, PAUL A. BROADY

The 'underwater-forest'-forming kelp *Macrocystis pyrifera* (Phaeophyceae).
Kate Neill, NIWA

Ochrophytes ('pale-yellow plants') are found in almost all environments where there is free water and the range of form exhibited by the group is quite remarkable. The smallest are tiny unicells, including the most abundant phytoplankton organisms in the sea – diatoms; the largest include giant seaweeds, like *Macrocystis* kelp, longer than a blue whale, that forms veritable underwater forests on some Pacific coasts, and the bull kelp *Durvillaea*, which can tolerate the wild ocean swells that buffet southern hemisphere islands and continents. Some ochrophytes live in terrestrial situations where there is adequate dampness for cellular processes and reproduction. Unicellular ochrophytes include free-swimming and non-flagellated forms, which may be naked or covered in protective scales, and some, like the diatoms and silicoflagellates, have exquisite glass-like skeletons of silica. What they generally all have in common, however, is that they are unequally flagellated, i.e. heterokont, at some stage of the life-cycle. Some lineages lack heterokont cells, most notably the diatoms, but centric diatoms produce male gametes with the tinsel (hairy) cilium diagnostic of heterokonts.

Ochrophytes are the most diverse heterokont organisms and mostly have photosynthetic pigments, in contrast to the non-photosynthetic (heterotrophic) chromist phyla Oomycota (water moulds, downy mildews) and Bigyra (opalinids, slime nets, bicosoecids). Chloroplasts are surrounded by four membranes, reflecting their symbiotic origin from a eukaryote (believed to be a red alga). Pigments are characteristically chlorophylls *a* and *c* and the accessory pigment fucoxanthin, also found in haptophytes, which gives ochrophytes a golden-brown or brownish-green colour. Photosynthetic species of the class Chrysophyceae are commonly able to utilise dissolved and/or particulate organic food as well, and are therefore mixotrophic.

Ochrophytes have had a varied history of classification. Formerly, many of the now-included classes were considered phyla (or botanical Divisions) in their own right. The advent of electron microscopy allowed the ultrastructure of plastids, mitochondria, and the tinsel cilium to be revealed in detail. In addition, improved methods of culture of many taxa and studies of pigment biochemistry, began to clarify relationships among the various groups. The discovery of the tinsel cilium in water moulds, labyrinthulas, and some other disparate groups not considered to be 'algal' led to all of them being grouped in one phylum, Heterokontophyta (also known as stramenopiles, or 'straw-bearers' after the tinsel cilium). This grouping is so diverse and disparate, however, that we follow Cavalier-Smith (1998, 2000, 2004), who first proposed the Heterokonta

(Cavalier-Smith 1986), in treating it as an infrakingdom.

In this scheme, we recognise 16 ochrophyte classes, distributed among two subphyla: Phaeista (Cavalier-Smith 1995), with two infraphyla, and Khakista (Cavalier-Smith 2000; Cavalier-Smith & Chao (2006). The phaeistan infraphylum Limnista here comprises four classes, although additional classes are recognised by some authors (usually as further segregates of the Chrysophyceae): Eustigmatophyceae (eustigs), Chrysophyceae (golden-brown algae), Synchromophyceae (*Synchroma*), and Picophagophyceae (*Picophagus* and *Chlamydomyxa*).

In the phaeistan infraphylum Marista 10 classes are provisionally recognised here: Pelagophyceae, Dictyochophyceae, Pinguiophyceae, Raphidophyceae, Chrysomerophyceae, Aurearenophyceae, Xanthophyceae, Phaeothamniophyceae, Schizocladiophyceae, and Phaeophyceae. Not all have been found in New Zealand. It is a measure of the 'hidden' diversity revealed by electron microscopy and molecular studies that many of the included genera in these ochrophyte classes were unrecognised 20 years ago or, like the silicoflagellates, were included among the Chrysophyceae.

Subphylum Khakista comprises just two classes – Bolidophyceae (based on *Bolidomonas*), not yet known in New Zealand, and Bacillariophyceae (Diatomeae or diatoms), which are the most speciose and abundant of ochrophytes.

Mallomonas heterospina (Chrysophyceae).
After Bourrelly 1981

Class Eustigmatophyceae: Eustigs

This class contains small (2–32 micrometres diameter) unicellular coccoid photoautotrophs, some of which produce ciliate zoospores. Formerly classified with the Xanthophyceae, the eustigs were segregated as a class over 40 years ago (Hibberd & Leedale 1970) because of several characters, including chloroplast

Summary of New Zealand ochrophyte diversity

Taxon	Described species + infraspecific taxa	Known undescribed/ undetermined species + infraspecific taxa	Estimated unknown species	Adventive species + infraspecific taxa	Endemic species	Endemic genera
Phaeista	313+9	10	260	12	50	4
Eustigmatophyceae	7	1	5	0	0	0
Chrysophyceae	101+5	4	200	0	7	0
Synchromophyceae	0	0	1	0	0	0
Picophagophyceae	0	0	4	0	0	0
Pelagophyceae	0	0	4	0	0	0
Dictyochophyceae	8	0	5	0	0	0
Pinguiophyceae	0	0	4	0	0	0
Raphidophyceae	4	0	5	0	0	0
Chrysomerophyceae	1	0	0	0	0	0
Aurearenophyceae	0	0	1	0	0	0
Xanthophyceae	45+4	1	10	0	0	0
Phaeothamniophyceae	2	1	0	0	0	0
Schizocladiophyceae	0	0	1	0	0	0
Phaeophyceae	145	3	20	12	50	4
Khakista	1,452+168	20+1	~1,000	2?	5?	1
Bolidophyceae	0	0	1	0	0	0
Bacillariophyceae	1,452+168	*20+1	1,000	2?	5?	1
Totals	**1,765+177**	**30+1**	**1,260**	**~14**	**~55**	**5**

* Not including 100 unpublished taxa of M. Reid.

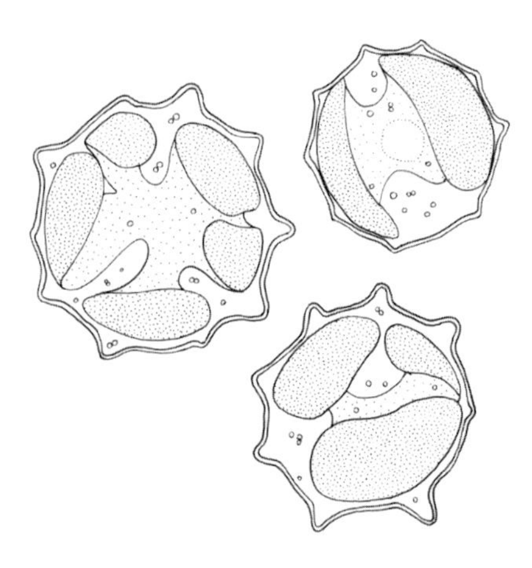

Vischeria stellata (Eustigmatophyceae).
After Ettl 1978

structure, their particular combination of photosynthetic pigments, and especially a unique type of light-sensitive photoreceptor found in their zoospores. This receptor includes a conspicuous orange-red eyespot, composed of a cluster of carotenoid-containing globules, which lies outside the chloroplast (unlike other heterokont algae) close to the swollen base of the long hairy anterior cilium. A smooth posterior cilium is usually either very short or present only as a basal body.

Chlorophyll *a* is present but, unusually for ochrophytes, chlorophyll *c* is lacking. This, combined with possession of the carotenoid violaxanthin, is unique to eustigs. As they can easily be mistaken for coccoid chlorophytes (green algae) or xanthophyceans in light microscopy, pigment analysis, electron microscopy, or molecular genetics can be vital to confirm their identification.

There are about eight genera (Santos 1996) and 18 species, mostly freshwater or terrestrial. Because there is little morphological diversity within the class, it is very difficult to identify species using light microscopy. This has frustrated description of new species and also estimation of biodiversity.

In New Zealand, three species in the genera *Eustigmatos* and *Vischeria* have been found in soils whilst two species of *Chlorobotrys* and one of *Ellipsoidion* are from fresh waters. Amongst marine phytoplankton, an *Ellipsoidion* and an unidentified species of *Nannochloropsis* have been recorded. All are very small and include cells of picoplankton dimensions – that is, tiny planktonic organisms 0.2–2 micrometres in diameter that are the main primary producers in many open-ocean ecosystems. *Nannochloropsis* species are widely used as food in

Summary of New Zealand ochrophyte diversity by environment

Taxon	Marine (marine-brackish) species + infraspecific taxa	Terrestrial species + infraspecific taxa	Freshwater (fresh-brackish) species + infraspecific taxa	Fossil marine species + infraspecific taxa
Phaeista	180	11	145+5	0
Eustigmatophyceae	2	3	6	0
Chrysophyceae	13	1	92+5	0
Synchromophyceae	0	0	0	0
Picophagophyceae	0	0	0	0
Pelagophyceae	0	0	0	0
Dictyochophyceae	8	0	0	0
Pinguiophyceae	0	0	0	0
Raphidophyceae	3	0	1	0
Chrysomerophyceae	1	0	0	0
Aurearenophyceae	0	0	0	0
Xanthophyceae	5	7	44	0
Phaeothamniophyceae	0	0	2	0
Schizocladiophyceae	0	0	0	0
Phaeophyceae	148	0	0	0
Khakista	678+53	5	794+116	767+90
Bolidophyceae	0	0	0	0
Bacillariophyceae†*	678+53 [117+4B]	5	794+116 [80+5B]	767+90 [90+2L]
Totals	**858+53**	**16**	**939+121**	**767+90**

Tallies combine described and known-undescribed/undetermined species.
† There is no overlap in the species lists of living marine-brackish and freshwater-brackish taxa (i.e. names are not duplicated in the lists). Numbers of brackish taxa (B, in parentheses) are included in the totals for Bacillariophyceae.
* Terrestrial diatoms are included in the freshwater checklist. About 5 species are restricted to the terrestrial environment, albeit associated with water films.
L (in parentheses in fossil column) = diatoms having a Tertiary–Recent range with species still living in modern New Zealand seas.

aquaculture and have been proposed for commercial production of eicosapentaenoic acid, an essential dietary fatty acid.

Class Chrysophyceae: Golden-brown algae

Chrysophycean algae owe their name to the golden-yellow to brown colour of their chloroplasts (Greek *chrysos*, gold), the green colour of the chlorophylls being masked by the principal accessory pigment fucoxanthin. As is typical of many heterokont algae, there is a tiny eyespot within the chloroplast and many species produce siliceous body scales. Others are loricate, with the cell body protected by a kind of theca or casing of protein and cellulose. The Chrysophyceae is a highly diverse group, almost exclusively found in fresh water, and includes unicellular and colonial forms. Only a few species are known from snow, soil, and marine habitats. On the whole, freshwater chrysophyceans tend to be dominant in oligotrophic lakes (having low productivity), probably because they can compete against other algal types for phosphorus in conditions where this nutrient is in naturally short supply, although selective grazing by animal plankton like water fleas and copepods can also affect the kinds and numbers of phytoplankton species.

Worldwide, there may be around 200 genera and 1200 species but the taxonomic scope and limits of the Chrysophyceae are being redefined. The group has been split many times in the past 20 years, with the segregation of smaller, discrete classes from it (Preisig 1995; Preisig & Andersen 2000; Kristiansen & Preisig 2001). There are many genera whose affinities are uncertain and, until they are investigated using transmission electron microscopy and gene-sequence analysis, their classification will remain problematic. Notwithstanding, the distinctive 'bottle and cork' resting cysts (statospores) that are produced in vegetative cells of many chrysophyceans is one character that appears to be reliably diagnostic of the class (Cavalier-Smith et al. 1995).

The New Zealand flora is only moderately diverse, with about 105 species, representing most of the various levels of structural organisation. For example, species of *Ochromonas* exhibit the monad (unicellular flagellate) level of organisation, whereas, in species of *Synura*, a number of flagellate cells are linked together to form a colony. Different again is the amoeboid form, in which cells are naked and bear thin pseudopodia called rhizopodia that can take up solid food particles. Species of *Chrysamoeba* have a flagellate stage but this is lacking in *Rhizochrysis* (not known in New Zealand). The palmelloid level of organisation comprises unicells grouped together in a sheath of mucilage (slime), either spherical (up to 4 millimetres diameter) as in *Chrysocapsa* or trailing as in the 30-centimetre-long branching threads of *Hydrurus*, which is found in upland and mountain streams. Coccoid forms are unicells that are non-flagellate, like *Chrysosphaera*, and these may be single or grouped. These levels of organisation have been used as the basis of chrysophycean classification, but, as indicated above, relationships remain obscure among the various orders and families that have been proposed and there is much work yet to be done in clarifying these. The thalloid level of organisation that was formerly attributed to some Chrysophyceae is now recognised as characterising species in the classes Chrysomerophyceae and Phaeothamniophyceae (see below).

One group, comprising *Synura*, *Mallomonas*, and related forms, differs from other Chrysophyceae in having chlorophylls *a* and c_1 (not c_2), no eyespot, a unique flagellar apparatus, and, most obviously, bilaterally symmetrical silica scales. These forms have been segregated as a class Synurophyceae by some authorities. Here we follow Cavalier-Smith (2004) in keeping this group within Chrysophyceae, especially as the distinctive 'bottle and cork' resting cyst is found within both groups.

Overall, the group is under-studied in New Zealand and many more species

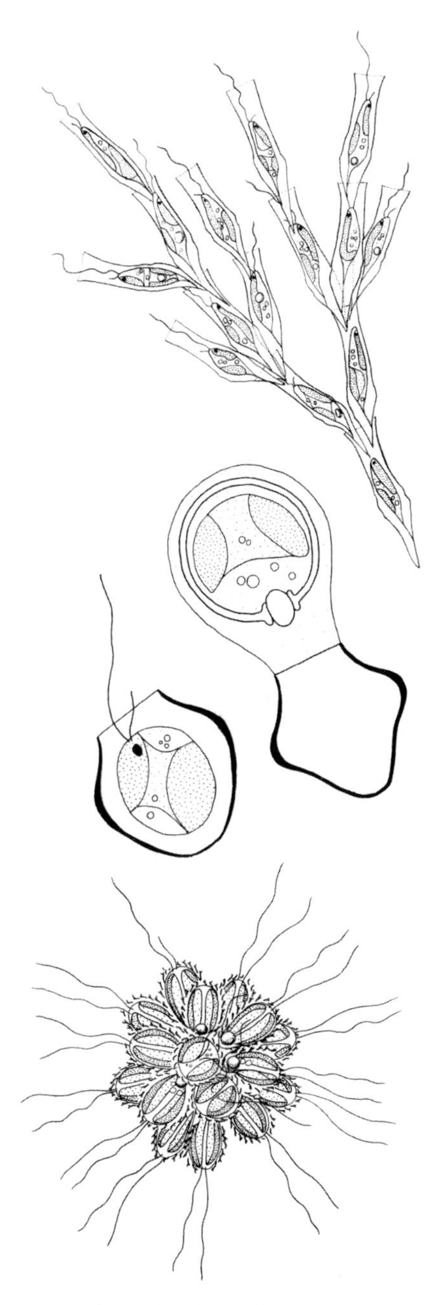

Upper, *Dinobryon divergens*; middle, *Pseudokephyrion conicum*; lower, *Synura uvella* (Chrysophyceae).

Upper, after Kristiansen 2002 and Bourrelly 1981; middle and lower after Bourrelly 1981

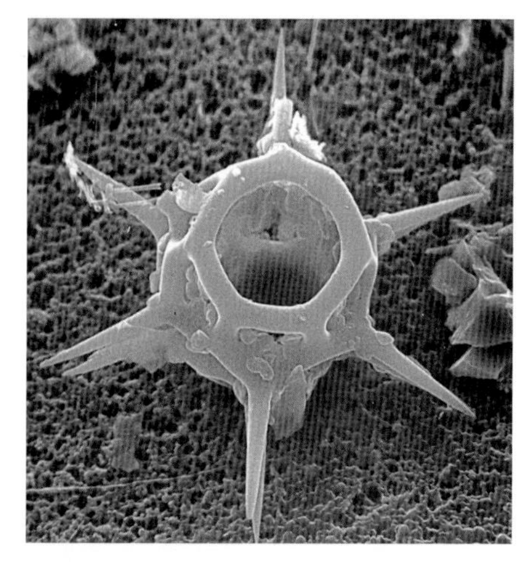

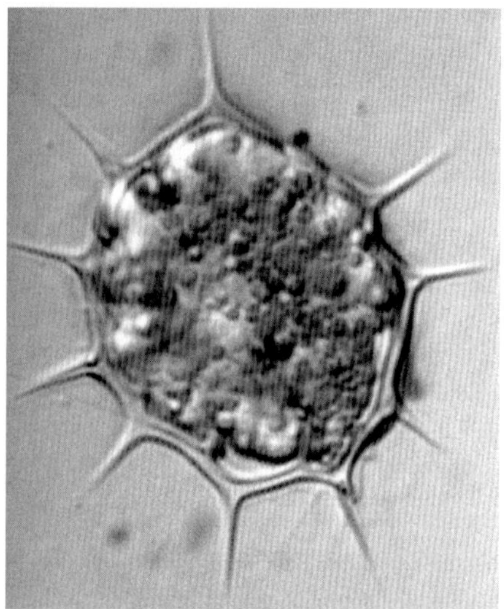

Silicoflagellates *Dictyocha speculum* (upper) and *D. octonaria* (lower) (Dictyochophyceae).
F. Hoe Chang, NIWA

certainly remain to be encountered by those familiar with the class and by applying modern techniques. For example, the attention of experts (Wujek & O'Kelly 1992) resulted in the description of 23 silica-scaled planktonic species from just 30 New Zealand localities using transmission electron microscopy to examine scale structure. This is almost a quarter of the total known flora.

Class Dictyochophyceae: Silicoflagellates and pedinellids

This small group of unicells consists of uniflagellated marine and freshwater forms and extinct species known from the fossil record. Silicoflagellate cells have a basket-like external skeleton of glassy silica. There is only a single cilium, of the tinsel type. In spite of the skeleton, the living species have some amoeboid characteristics, extruding very thin filopodial strands. These may aid movement when the cells, which normally swim, encounter a substratum, or they may be involved in secretion of the skeleton. Chloroplasts are golden-brown or greenish-brown, containing several pigments in addition to chlorophylls *a* and *c* (either c^2 or c^3) and fucoxanthin. During the life-cycle, naked cells are produced that are spherical and lack a skeleton; they appear uniflagellate but there is a vestigial second cilium.

All known silicoflagellates are marine, including the fossil taxa (found only in marine sediments). Because living species are prevalent in cold oceanic waters, or in cooler seasons at low latitudes, fossil silicoflagellates can be used to infer sea temperatures in past geological epochs. Dictyochales is the only order which encompass species that produce a siliceous skeleton. The commonest living species is *Dictyocha speculum*, a somewhat stellate organism that is widespread in the world's oceans. Silicoflagellates (order Dictyochales) generally form a minor component of coastal phytoplankton assemblages around New Zealand (Chang 1983, 1988; Rhodes et al. 1993). There are three living species, each of which has been classified in different genera (including *Distephanus*) in the past. Previously, opinions varied concerning whether two of the species, *Dictyocha fibula* and *D. speculum* and their varieties (cf. Dawson 1992), were intergradational forms. Moestrup and Kelly (2000) considered there to be just one genus. Three species are accepted in this account, each of which (including *D. octonaria*) is found in New Zealand waters. Additionally, on the basis of morphological, life-history, pigment, and sequence data, a new genus, *Vicicitus*, is added to the order. The sole species, *V. globosus*, was formerly included in *Chattonella* (Chang et al. in press).

The order Pedinellales (pedinellids) comprises six genera of flagellates with a single hairy cilium, winged on one side, and tentacles that project from the cell surface. Some species are attached by a stalk; others are free-swimming, with the stalk trailing, and some bear organic scales that may also be spine-like. There are marine and freshwater taxa and nutrition is either heterotrophic, photoautotrophic, or mixotrophic. Three species have been reported in New Zealand coastal waters – *Apedinella radians* (as *A. spinifera*), which is widely distributed in the plankton and is mixotrophic, and *Parapedinella reticulata* and *Pseudopedinella pyriforme* (Bojo 2001).

Edvardsen et al. (2007) erected a new order of Dictyochophyceae, Florenciellales, with two genera. The unicells have one long, forwardly directed cilium and one short cilium. *Pseudochattonella* is characterised by numerous protruding mucocysts that give the cells a warty appearance, while in *Florenciella* the cells lack mucocysts. In New Zealand, only one species, *Pseudochattonella verruculosa* (formerly *Chattonella verruculosa*), is recorded (Chang unpubl.).

In Japan, *Vicicitus globosus* has been associated with respiratory damage to both cultivated and wild fish (Hara et al. 1994) whereas no such problems have been noted in New Zealand. In Europe, *Pseudochattonella verruculosa* is associated with massive fish kills (Edvardsen et al. 2007) and in New Zealand this species was implicated in 2010 in two separate fish-mortality events, in Wellington Harbour (Chang unpubl.) and Marlborough Sounds (L. MacKenzie pers. com.).

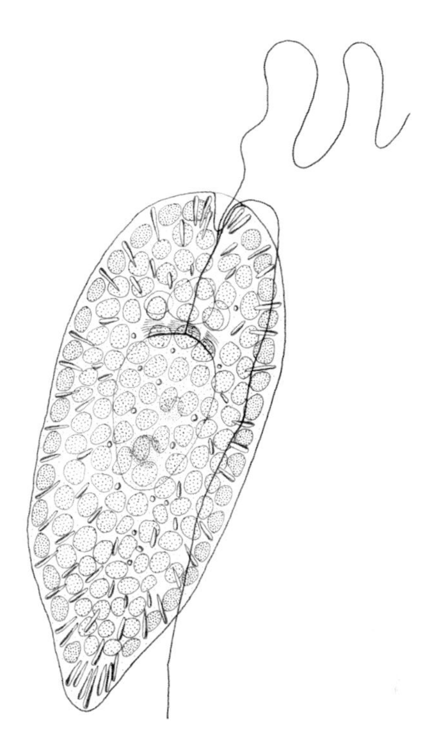

Goniostomum semen (Raphidophyceae).
After Bourrelly 1985

Class Raphidophyceae: Raphidophytes

Raphidophytes are free-swimming flagellates with complex cells that are often relatively large (30–80 micrometres long). The paired cilia emerge at or near the cell apex (instead of laterally as in brown algae) in a shallow groove or 'gullet', with one hair-bearing cilium directed forward and a smooth cilium trailing behind. The cell tends to be dorsoventrally flattened with a curved dorsal side and a flatter ventral side. Chloroplasts are green, yellow-green, or yellow-brown, with chlorophylls *a*, usually both c_1 and c_2, and several accessory pigments but there is no eyespot. Beneath the surface of the cell are numerous ejectosomes (*Heterosigma*), trichocysts, and mucocysts (*Chattonella*, *Fibrocapsa*) that discharge long mucilaginous strands when disturbed. Interestingly, raphidophytes also have flattened sacs under the cell membrane that are virtually indistinguishable from those of alveolate protozoans (Ishida et al. 2000), one of the lines of evidence for the recognition of a major branch of the eukaryote superkingdom, Chromalveolata (Cavalier-Smith 2004).

A small group, the Raphidophyceae used to be classified among the Chrysophyceae. There are only nine genera, four of which are found in New Zealand. Freshwater species, like *Goniostomum semen*, tend to occur in rather acid waters such as small upland pools and bogs. Species of the marine genera *Chattonella* and *Heterosigma* are well-known fish killers in different parts of the world and sometimes give rise to massive blooms. Seven *Chattonella* species are known worldwide, but molecular genetic evidence now suggests that two are conspecific with *C. marina* (Bowers et al. 2006; Demura et al. 2009) and two others are dictyochophytes (Hosoi-Tanabe et al. 2007; Chang et al. in press).

Three genera and four species of raphidophytes are known in New Zealand coastal waters, mostly 6–100 micrometres in cell size. Some have caused major microflagellate blooms (Chang et al. 1990; Rhodes et al 1993). The majority were reported subsequent to a 1989 toxic-bloom episode (Chang et al. 1990). Raphidophytes are often associated with mass mortality of farmed salmon, as in Big Glory Bay, Stewart Island (Boustead et al. 1989). Overseas, losses from episodes of *Heterosigma akashiwo* blooms can exceed millions of dollars. Salmon exposed to this raphidophyte species exhibit impaired respiratory and osmoregulatory gill function, which leads to their death. The ability of this microflagellate to bloom is based on a combination of optimum light and nitrogen availability, especially nitrate (Chang et al. 1993; Chang & Page 1995).

Class Chrysomerophyceae: Chrysomerophytes

This tiny class was first recognised just over a decade ago (Cavalier-Smith et al. 1995). Species exhibit a fairly integrated (compared to typical Chrysophyceae) thalloid organisational structure that is essentially a simple tissue. For example, the golden-brown alga *Giraudyopsis stellifera*, relatively recently reported for New Zealand and the South Pacific for the first time (Broom et al. 1999), produces a flat disc of contiguous cells, one cell layer thick. From the centre of the disc arise short filaments. Filament cells are initially in a single series, but, with increasing length, the filaments become pluriseriate with infrequent branches. Most of these cells in the thickened filaments develop as sporangia, releasing free-swimming biflagellate zoospores at maturity. When these settle, they germinate and divide to produce a series of basal cells that then initiate erect filaments. Chrysomerophytes lack the statospores (resting cysts) that are diagnostic of chrysophytes. They are also characterised by molecular sequence data, eyespots in zoospores, and the pigment violaxanthin (Saunders et al. 1997).

Giraudyopsis stellifera (Chrysomerophyceae).
Tracy Farr, RSNZ

Class Xanthophyceae: Yellow-green algae

The Xanthophyceae are distinguished from the Chrysophyceae, inter alia, in having cell walls, composed of cellulose and pectic compounds, a different suite of pigments (several xanthophylls present but no fucoxanthin), and different resting cysts comprising two equal halves. There are about 100 genera and 600 species worldwide (Ettl 1978). Most live in freshwater habitats such as ponds and lakes but they often comprise a significant part of the soil microflora (Ettle & Gärtner 1995). Some species form thick velvety mats that help to bind sediment in salt marshes and damp soil. Marine species are rare, but some *Vaucheria* species are large enough to rate as seaweeds, albeit small. These are rarely collected (Chapman 1956; Adams 1994).

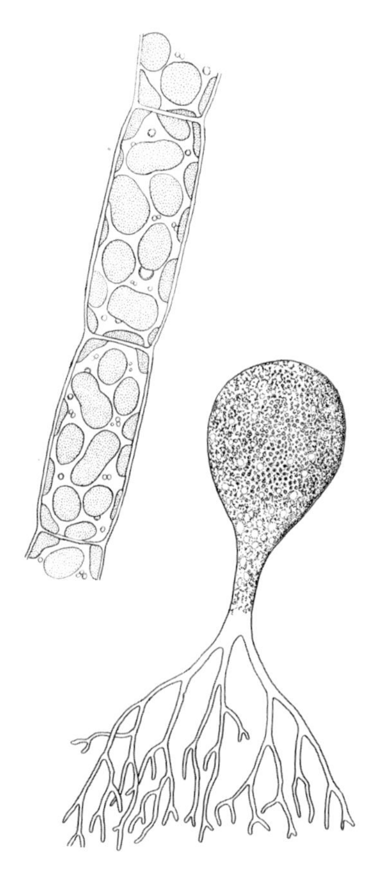

Upper, *Tribonema viride*; lower, *Botrydium granulatum* (Xanthophyceae).
Upper, after Ettl 1978

Like members of the Chrysophyceae, xanthophytes exhibit different levels of organisation. The simplest forms are unicellular flagellates of the order Chloramoebales (not found in New Zealand) but these are few in number of species. Among the 49 known species of New Zealand xanthophyceans, most structural forms are represented. Amoeboid xanthophyceans like *Rhizochloris* have naked cells with pseudopodia and can ingest solid particles such as bacteria and diatoms. *Botrydiopsis* exemplifies a coccoid level of organisation (most xanthophyceans are coccoid), with non-flagellated cells existing singly; some other taxa have coccoid cells linked into colonies. *Tribonema* and relatives have cells arranged in simple or branching filaments. The most striking, and obvious to the naked eye, are the siphonous threads of *Vaucheria*, which consist of multinucleate tubular filaments without cross walls. *Botrydium*, a related form, consists of pinhead-sized vesicles that are anchored in damp soil by root-like rhizoids. The traditional ordinal classification of xanthophyceans, while somewhat unsatisfactory, is based on these levels of structural organisation. Molecular genetic evidence now suggests that the traditional classification is artificial and displays polyphyly at the ranks of order, family, and genus (Negrisolo et al. 2004).

As with the Chrysophyceae, there are certainly many more species to be recorded in New Zealand. Soils are likely to be a rich source of new records (Ettl & Gärtner 1995) but will require application of cultures for their isolation and description. In fresh waters, xanthophyceans are often minor components of periphyton and planktonic communities. The small size and pale colour of many makes them easy to overlook.

Class Phaeothamniophyceae: Phaeothamniophytes

This class was recently established on the basis of gene-sequence, ultrastructural, and pigment data for 16 genera and 25 species. Formerly, these had been included mostly in the Chrysophyceae but one genus was transferred from the Xanthophyceae (Anderson et al. 1998; Bailey et al. 1998; Kristiansen & Preisig 2001). The class is now thought to be more closely related to the Xanthophyceae and Phaeophyceae with which it shares some features of pigmentation and ciliary structure. All are microscopic, with structural organisation ranging from free-living coccoid unicells to unicells dispersed in various arrangements within mucilaginous colonies, or branched filaments of single series of cells.

Just two species have been recorded from New Zealand. The branched filaments of *Phaeothamnion confervicolum* were seen in an undescribed sample taken near Dunedin (Chapman et al. 1957) but details were not published. Mucilaginous colonies of *Stichogloea doederleinii* were recorded in the plankton of two North Island volcanic lakes (Thomasson 1974).

Phaeothamnion confervicolum (Phaeothamniophyceae).
After Fritsch 1965

Class Phaeophyceae: Brown algae

Brown algae are the most conspicuous and probably the best understood group of New Zealand seaweeds. Worldwide, all known species are multicellular and almost all are restricted to the marine environment; none of the five freshwater genera is known to be represented in New Zealand (Bold & Wynne 1985; Wehr 2003; Andersen 2004). The morphology of members of this group varies dramatically, ranging from microscopic uniseriate filaments to bull kelp many metres in length. Brown algae are defined by morphological/ultrastructural and pigment characters. For example, the heterokont cilia of the zoospores are inserted laterally and flagellated cells often possess an eyespot. Chloroplasts are generally discoidal with stalked pyrenoids and contain chlorophylls *a*, *c*1, *c*2, and the accessory pigment fucoxanthin (Hoek et al. 1995).

The traditional systematic framework for the Phaeophyceae was centred on the type of life-cycle – alternation of similar (isomorphic) or dissimilar (heteromorphic) plants or a gametic life cycle, type of gametes (isogamous, anisogamous, oogamous), tissue organisation (filamentous, parenchymatous, pseudoparenchymatous), growth pattern and meristem position (apical, diffuse, intercalary), and the presence or absence of pyrenoids with the chloroplasts (Bold & Wynne 1985; Clayton 1990a,b). In the past two decades, molecular sequence data and phylogenetic analyses have led to a reassessment of evolutionary relationships within the Phaeophyceae and a re-evaluation of character evolution (e.g. Tan & Druehl 1994, 1996; Reviers & Rousseau 1999; Rousseau et al. 2001; Cho et al. 2006; Bittner et al. 2008; Phillips et al. 2008; Silberfeld et al. 2010). Plastid and pyrenoid characters continue to provide phylogenetically informative characters (Peters & Clayton 1998; Rousseau & Reviers 2000; Parente et al. 2000).

The number of orders recognised in the class Phaeophyceae has varied among authors (e.g. Rousseau & Reviers 1999; Cho et al. 2004; Kawai et al. 2007; Bittner et al. 2008; Phillips et al. 2008). Although there is a great range of morphological expression in brown algae, there does not seem to be a simple relationship between phylogeny and complexity of form. The morphologically simple brown algae of the order Ectocarpales in the broad sense have been shown by molecular studies (based on the small subunit of the nuclear ribosomal gene) not to be primitive in a phylogenetic sense (Peters & Clayton 1998). The long-held view of an ancestral simple Ectocarpales and derived complex Fucales has not been supported by molecular phylogenetic studies (Draisma et al. 2001; Rousseau et al. 2001; Cho et al. 2004; Silberfeld et al. 2010). Rousseau et al. (2001) examined representatives from all orders within the class recognised at that time, including taxa from 35 of the 55 families, using nuclear DNA sequence data and phylogenetic analyses. From their study the Dictyotales were inferred to be the first to diverge, followed by the Sphacelariales and the Syringodermatales. The majority of orders of brown algae were clustered in a poorly resolved polytomy that became known as the brown-algal crown radiation (BACR). Draisma et al. (2001) showed the order Discosporangiales to be the earliest diverging order. Cho et al. (2004) established a new order, Ishigeales, for the North Pacific genus *Ishige*, and showed that this order also grouped with the Dictyotales, Syringodermatales, and families Choristocarpaceae and Onslowiaceae. The 'crown' group contained the Ectocarpales, Fucales, Laminariales in addition to other lineages. A series of studies have focused at different taxonomic levels, attempting to resolve phylogenetic relationships within orders and within the BACR (e.g. Yoon et al. 2001; Cho et al. 2006; Lane et al. 2006; Bittner et al. 2008; Phillips et al. 2008). Kawai at al. (2007) reinstated the order Discosporangiales, originally erected by Schmidt (1937), and in Phillips et al. (2008) the orders Onslowiales and Nemodermatales were erected.

Zonaria aureomarginata (Phaeophyceae).
Kate Neill, NIWA

Classification scheme

The following orders are recognised in New Zealand:

Dictyotales. The order is diverse in the New Zealand region, with nine genera listed. The genus *Dictyota* (currently including *Dilophus*) requires taxonomic revision in New Zealand. Members of this order are common in tropical regions

Xiphophora gladiata (upper) and *Hormosira banksii* (Phaeophyceae).
Kate Neill, NIWA

– 14 of the 21 species in the end-chapter checklist are found on the Kermadec Islands.

Sphacelariales. Prud'homme van Reine (1993) reviewed familial and generic circumscription of the Sphacelariales of the world as well as the biogeography of the order. It is very well represented in New Zealand waters with 19 species in five genera.

Syringodermatales. Previously placed in the Cutleriales, the monotypic genus *Microzonia* was recently shown to be a sister taxon of *Syringoderma* (Burrowes et al. 2003).

Fucales. This order is the largest in the Phaeophyceae, comprising about 38 genera worldwide, and is very well represented in New Zealand and Australia. There are nine genera in New Zealand, three of which are endemic, and six of the nine families in the order are present in New Zealand. Clayton (1984) examined the evolutionary relationships of the Fucales and the position of the order within the Phaeophyceae, hypothesing that the order evolved and radiated during the Mesozoic within Australasia. A decade later, Clayton (1994) published a study of the scope and phylogeny of the members of the family Seirococcaceae, based on morphological and ultrastructural characters including aspects of fertilisation. Two recent studies have used molecular-sequence data to examine relationships of the northern hemisphere Fucales (Rousseau et al. 1997) and members of the Fucaceae (Serrao et al. 1999). In both studies, the Fucaceae were strongly supported as a monophyletic group, although the southern hemisphere genus *Xiphophora* was considerably divergent from the northern hemisphere genera. Rousseau et al. (2001) found that *Xiphophora*, usually placed in the Fucaceae, was a sister taxon of *Hormosira*. Cho et al. (2006) established the new monogeneric family Xiphophoraceae. The proposal to merge the Cystoseiraceae within the Sargassaceae proposed by Rousseau and de Reviers (1999) has been supported by the analyses of Cho et al. (2006). The bull kelp genus *Durvillaea*, which is restricted to the southern hemisphere, was at one stage placed in its own order but both morphological and molecular-sequence data place this genus within the Fucales (de Reviers & Rousseau 1999; Rousseau & de Reviers 1999; Cho et al. 2006). Four of the five known species of *Durvillaea* are present in New Zealand and three of them are endemic.

Sporochnales. This small order has its greatest diversity in Australia (Womersley 1987; Clayton 1990b). Four genera are present in New Zealand, of which *Perisporochnus* is endemic to the Three Kings Islands.

Laminariales. Molecular data have changed the understanding of relationships within the Laminariales, particularly in the Alariaceae, Lessoniaceae, and Laminariaceae (Druehl et al. 1997; Yoon et al. 2001; Lane et al. 2006). Of the seven families in the order, only members of the Laminariaceae and Lessoniaceae are native to New Zealand. The introduced kelp *Undaria* is a member of the Alariaceae.

Desmarestiales. Data from all sources (life history, reproduction, anatomy, molecular biology) support the hypothesis that the order Desmarestiales is monophyletic (Rousseau et al. 2001). Peters et al. (1997) examined nuclear DNA-ITS sequences, morphological characters, and ecophysiological traits and concluded that these were all consistent with an austral origin of the family Desmarestiaceae and subsequent radiation into the Northern Hemisphere (see Peters & Breeman 1992).

Scytothamnales. Peters and Clayton (1998) established a new order Scytothamnales for the Scytothamnaceae and Splachnidiaceae. Members of these families possess stellate chloroplasts with a single pyrenoid in the centre of the cells. Molecular sequence data using the nuclear ribosomal cistron provide strong evidence that none of these species belongs to Ectocarpales *sensu lato*. The position of *Bachelotia antillarum* remains unresolved (Müller et al. 1998; Rousseau et al. 2001), although Silberfeld et al. (2010) suggest that it be merged with the Scytothamnales.

Tilopteridales. Phillips et al. (2008) examined relationships among the Cutleriaceae, Tilopteridaceae, and Phyllariaceae and clearly established that the family Cutleriaceae diverges from within the Tilopteridales. In New Zealand there is a single family within this order, Cutleriaceae (previously placed in the Cutleriales), and one genus.

Ralfsiales. There are two families in this order, Ralfsiaceae and Lithodermataceae, both of which are represented in New Zealand. Further research is required on brown crustose algae in New Zealand (Lim et al. 2007).

Ectocarpales sensu lato. The ordinal status of the simple brown algae has been much debated and recent molecular studies have focused attention on some of the long-standing questions about phylogenetic relationships amongst the Ectocarpales, Chordariales, Dictyosiphonales, and Tilopteridales (e.g. Tan & Druehl 1994; Druehl et al. 1997; Peters & Burkhardt 1998; Siemer et al. 1998; Kogame et al. 1999; Rousseau et al. 2001). The consensus now is that there is a close phylogenetic relationship among the Ectocarpales, Chordariales, Dictyosiphonales, and Scytosiphonales and that these orders should be should be merged into Ectocarpales sensu lato. Rousseau and Reviers (1999) have proposed an emended circumscription of the Ectocarpales to include these four orders, based on both cytological and molecular data. The familial placements in this treatment follow Peters and Ramirez (2001) and Raucault et al. (2009.

Carpophyllum plumosum (Phaeophyceae).
Tracy Farr, RSNZ

Historical overview of brown-algal discoveries and research in New Zealand

Adams (1994) presented an overview of the history of marine botany in New Zealand from the time of the first collection of seaweeds on Cook's first voyage to New Zealand in 1769. Esper (1802) described the first brown seaweed from New Zealand waters, *Fucus flexuosum* (now known as *Carpophyllum flexuosum*). Harvey and Hooker (1845) and Harvey (1855) presented the only illustrated treatments of the complete seaweed flora, from mainland New Zealand and the subantarctic islands, prior to that of Adams (1994). French voyages of exploration in the New Zealand region in the early 19th century resulted in many important collections and discoveries of new species (Parsons 1998).

Over a period of more than 60 years from 1841, J. G. Agardh in Lund, Sweden, published many papers in which New Zealand algae were described. Although Agardh never visited New Zealand, he received material that had been initially sent to von Mueller in Melbourne (including Chatham Island collections made by H. H. Travers), as well as receiving collections made by Berggren, a professor from Lund who collected throughout New Zealand during 1874–1875. Towards the end of his career, Agardh was corresponding and exchanging specimens with Christchurch school teacher R. M. Laing. Harald Kylin occupied the same academic position as Agardh, and used the excellent herbarium material of New Zealand seaweeds at Lund to describe several new endemic taxa in the Chordariales (Kylin 1940).

Amateur phycologists played a very important role in the documentation and exploration of the New Zealand seaweed flora in the early 20th century. Laing (1900, 1902, 1926, 1930) prepared a series of reference lists of species and worked on algae of the Subantarctic. The contributions of Victor Lindauer were particularly significant during the mid-20th century (Cassie 1971; Cassie Cooper 1995). The Algae Nova-Zelandicae Exsiccatae distributed by Lindauer between 1939 and 1953 served as a reference set of New Zealand algae in both national and international institutions, with 38 of the 350 sheets having some level of type status (Nelson & Phillips 1996). In addition, Lindauer published a number of papers on algae, particularly the brown algae (Lindauer 1945, 1947, 1949a, 1949b, 1957, 1960). The Phaeophyceae, published as the second part of the *Marine Algae of New Zealand* series by Lindauer et al. (1961), has served as an excellent foundation document and reference text for the brown seaweeds. This volume remains a very important contribution, although the work has been overtaken by an expanded knowledge of the New Zealand flora, largely facilitated by scuba

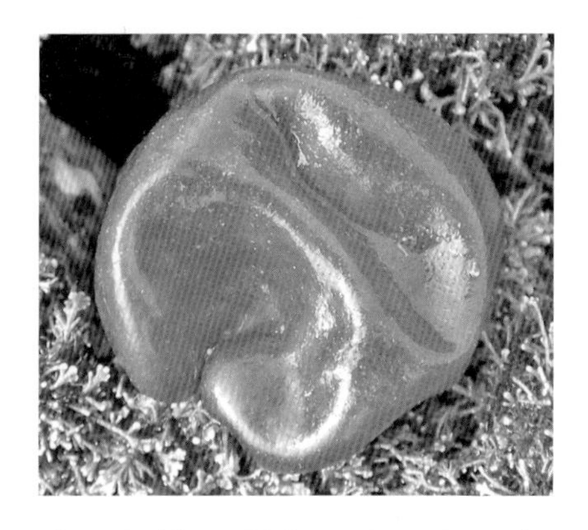

The typical form of *Colpomenia* species on the seashore (Phaeophyceae).
Tracy Farr, RSNZ

diving and by recent research, particularly on the relationships of brown algae.

At the former Botany Division of the DSIR (subsumed by Landcare Research in 1992), Lucy Moore was assigned to work on seaweeds during World War II. She made extensive collections throughout the country and carried out research on particular brown algae, producing a major monograph on the genus *Halopteris*, a study of an unusual growth form of *Macrocystis* in Stewart Island, and a study of a growth form of *Hormosira* (Moore 1942, 1946, 1950, 1951, 1953).

Following the publication of the *Seaweeds of New Zealand* (Adams 1994) there have been three endemic brown algae described from the New Zealand region: *Zonaria aureomarginata* (Phillips & Nelson 1998), *Landsburgia ilicifolia* (Nelson 1999a), and *Marginariella parsonsii* (Nelson 1999b). Asensi et al. (2003) reexamined *Chorda rimosa* collected by d'Urville from the Auckland Islands in 1840, and transferred it to *Adenocystis.* In an examination of *Colpomenia peregrina* from throughout its distributional range, Cho et al. (2005) found two deeply divergent lineages, interpreted as two species, both of which are present in New Zealand, and a further species of *Colpomenia* has been described (Boo et al. 2010).

Major repositories of New Zealand specimens

The major collections of New Zealand Phaeophyceae are housed in the following herbaria: Museum of New Zealand (WELT), Auckland Museum (AK and AKU), and Landcare Research – Manaaki Whenua (CHR) (Holmgren et al. 1990). The collection data from the WELT specimens are databased. There are no recently published manual or computer-based identification keys for the brown algae, although there are keys for six genera (*Cystophora, Durvillaea, Halopteris, Marginariella, Sargassum,* and *Sphacelaria*) presented in Adams (1994). There is also a key to the New Zealand genera of the Chordariaceae and Chordariopsidaceae in Nelson and Adams (1983).

Current known natural diversity

The end-chapter checklist of Phaeophyceae is based on Lindauer et al. (1961), Adams (1994), and a series of regional seaweed floras that provide vouchered lists of species and summarise available information for portions of the New Zealand region (Adams 1972; Adams et al. 1974; South & Adams 1976; Nelson & Adams 1984, 1987; Parsons & Fenwick 1984; Hay et al. 1985; Adams & Nelson 1985; Nelson et al. 1991, 1992, 2002; Neale & Nelson 1998). These have identified taxa requiring study and have increased our understanding of the distribution of species within New Zealand. Authorities follow Brummitt and Powell (1992).

The treatment here lists 148 species of Phaeophyceae in 11 orders. There are four endemic genera, 50 endemic species, and one endemic subspecies. Twelve species are considered to be human-mediated introductions to the New Zealand region.

Special features of the New Zealand Phaeophyceae

Ecology and distribution

Hurd et al. (2004) provided a summary of published data on New Zealand macroalgal ecology (species assemblages and top-down processes controlling macroalgal communities), physiology, and physiological ecology (environmental control or bottom up-processes), and the uses of seaweeds, including aquaculture and seaweed extracts.

A number of large brown algae play very significant roles in nearshore ecosystems. Categorised as 'habitat formers – large and/or aggregated sedentary organisms that characterise a habitat' (Jones & Andrew 1993), kelps can be dominant both in terms of biomass and also in structuring the three-dimensional space of subtidal reefs. In some situations, large forest-forming brown algae may moderate the impact of physical disturbances on biological communities, such as by dampening the impact of storms on subcanopy species and neighbouring intertidal communities (Kennelly 1987). Smaller brown algae can also act as habitat-formers when they grow as dense mats or extensive areas of crusts. The interactions between a mid-intertidal crustose brown alga and invertebrates were studied by Williamson and Creese (1996a,b). Williamson and Rees (1994) presented evidence of complex nutritional interactions between the crust and neighbouring barnacles.

Schiel (1988, 1990) summarised the state of knowledge of macroalgal assemblages in New Zealand. He observed that little work has been carried out on the ecological processes affecting the organisation of algal communities in New Zealand, and what is known 'emphasizes the fact that the effects of predators, grazers, the nature of the substratum, and algal canopies interact in complex ways in the patch dynamics and organization of algal assemblages.' Schiel (1990) concluded that 'these interactions cannot be understood without reference to the specific life history and phenological characteristics of particular species and the demographic consequences of settling and growing under different conditions'.

Macrocystis pyrifera (Phaeophyceae).
Kate Neill, NIWA

Most research associated with marine reserves in New Zealand has focused on the conspicuous fauna (rock lobster, popular fish species, sea urchins) (MacDiarmid & Breen 1993; Cole 1994) and much less attention has been paid to the algae or to the impact of large numbers of visitors on marine communities. On the South Island east coast, Schiel and Taylor (1999) have examined the impact of trampling on intertidal algal assemblages and their recovery after disturbance.

The giant kelp *Macrocystis pyrifera* has a bipolar distribution and in New Zealand waters is found in southern and central areas of the mainland coast, the Chatham Islands, and Auckland and Campbell Islands. Hay (1990a) considered that its distribution was correlated with cool sea-surface temperature. Although temperature may be acting synergistically with other environmental variables, Brown et al. (1997) considered that their observations in Otago provided evidence of nutrient limitation affecting seasonal growth. Other studies on the growth and morphology of *Macrocystis* in New Zealand include the work of Kain (1982) and Nyman et al. (1990, 1993).

Significant features of distribution and biogeography

Laing (1895, 1927) was the first person to attempt an analysis of the biogeography of the New Zealand seaweed flora. Moore (1949, 1961) examined the concept of marine provinces and distributional patterns of New Zealand seaweeds. Parsons (1985a) summarised existing knowledge of the seaweed flora and its relationships, recognising five particular elements within the flora: cosmopolitan, circumpolar, Australasian, South American, and Japanese. In a three-tiered analysis of the macroalgal flora, Nelson (1994) discussed the distribution of 100 large brown-algal species within the New Zealand archipelago and examined the distributional patterns of Southern Hemisphere Laminariales. The most serious limiting factor in a comprehensive analysis of the flora and its relationships is the level of taxonomic knowledge, a point that has been made repeatedly (Parsons 1985a,b; Nelson 1994; Adams 1994).

Distinctive taxa

The most distinctive feature of the seaweeds of New Zealand shores is the predominance of fucalean species, many of which are endemic or restricted to Australasia, and the iconic southern bull kelp species in the genus *Durvillaea*. These algae are large and form a conspicuous presence in the low intertidal and subtidal zones. In contrast with northern Pacific shores, there is a very low diversity of laminarian genera in New Zealand.

Members of the order Dictyotales are a particularly conspicuous part of the flora of the Kermadec Islands. Many of these genera and species are known from other parts of the warm Indian and Pacific Oceans but not from mainland New Zealand. The biomass of brown algae found at the Kermadecs is considerably

Asian kelp *Undaria pinnatifida* (Phaeophyceae).
Kate Neill, NIWA

lower than in more southern parts of the country, in particular on southern shores where the large thalli of species of *Durvillaea*, *Ecklonia*, *Lessonia*, and members of the Fucales predominate. The diversity of algae at the Kermadecs, however, exceeds that of many southern islands (Nelson & Adams 1984; Hay et al. 1985).

Alien species

Twelve of the 34 adventive macroalgae recognised in New Zealand belong to the Phaeophyceae (Adams 1983; Nelson 1999c; Nelson et al. 2004; D'Archino & Nelson 2006). Of these species, eight appear to pose little or no risk to the native flora, apparently not reproducing rapidly or spreading far from their points of introduction. The impact of *Colpomenia durvillaei* is less clear. This is a seasonal species, disappearing for some months. In late winter/early spring it has been observed to form dense populations (e.g. at Mahia, East Cape, and Wellington) and appears to dominate substrata in a broad intertidal band.

The Asian kelp *Undaria pinnatifida* is clearly the most serious marine algal pest species to have entered New Zealand. It combines a very high reproductive output with tolerance of a wide range of growing conditions, enabling it to function as a very successful 'weed'. From its initial introduction and recognition in Wellington Harbour in 1987, it has spread widely, having been recorded from Great Barrier Island, Auckland, Coromandel, Gisborne, Napier, Wellington, Picton/Marlborough Sounds, Golden Bay, Kaikoura, Lyttelton, Timaru, Oamaru, Otago Harbour, Fiordland, Bluff, Stewart Island, and The Snares islands (Hay & Luckens 1987; Hay 1990b; Herbarium – Te Papa). Uwai et al. (2006) reported on genetic diversity in *Undaria* populations in New Zealand and Russell et al. (2008) have investigated the expanding range of the species in southern New Zealand.

Commercial use/commercial potential

Brown algae are used commercially in a range of products – human food (as sea vegetables, dietary supplements, food garnish, salt substitute), stock food for feeding to aquacultured paua (abalone), and seaweed-based fertilisers (domestic and agricultural) – and as a source of the polysaccharide alginate. During the early 1940s, there was interest in the potential for local stocks of *Macrocystis* to sustain commercial harvesting for potash (Rapson et al. 1942).

More recently there has been research into harvesting the brown algae *Durvillaea*, *Ecklonia*, and *Macrocystis* (Schiel & Nelson 1990; Schiel et al. 1997a,b). Zemke-White et al. (1999) presented a summary of the current status of commercial algal use in New Zealand, looking at the legislative environment as well as trends in use, exports, and imports. *Macrocystis pyrifera* and *Durvillaea* species are currently harvested from wild stocks primarily for paua feed. *Ecklonia radiata* is harvested for a variety of health products.

The introduced kelp *Undaria pinnatifida*, which is farmed extensively within Asia and valued highly as a food species, has been investigated as a potential aquacultural crop species in New Zealand. While there is acceptance that this species has naturalised in certain parts of the country, there is concern that farming *Undaria* may increase its spread (Zemke-White et al. 1999). There is interest in the potential for aquaculture of *Lessonia variegata* (Nelson 2005).

Durvillaea species are listed as taonga in the Ngai Tahu Claims Settlement Act 1998 and are to be reserved for traditional uses. This settlement prohibits any commercial use of *Durvillaea* within the Ngai Tahu tribal area. *Durvillaea antarctica* is used to make poha titi, or storage bags for mutton birds. This species of bull kelp has a very tough surface and a honeycomb tissue in the medullary layer that can be split open to create a cavity into which muttonbirds can be placed and preserved, surrounded by their own fat.

Bull kelp *Durvillaea antarctica* (Phaeophyceae).
Kate Neill, NIWA

Gaps in knowledge and scope for future research

Although the brown algae are the best known of all seaweeds in New Zealand, there still remains a considerable amount to discover about them. There is no doubt that there are more species to be discovered. In the past 13 years, new species belonging to some of the largest and most conspicuous genera have been discovered and described, such as two species of *Lessonia* and one each of *Landsburgia, Marginariella,* and *Zonaria,* in addition to a species of bull kelp that remains undescribed, although listed below. An undescribed genus in the Fucales requires further study.

Recent collections and studies of members of the Sphacelariales in New Zealand have shown that the systematics of this group requires further attention (Prud'homme van Reine pers. comm.). Research is currently underway on endophytic brown algae through culture studies, ultrastructural investigations, and molecular sequencing, and a number of new taxa will be recognised as a result of this work (S. Heesch pers. comm.). The crustose brown algae in New Zealand have received very little attention as have the morphologically simple filamentous species. It will be necessary for research on these algae to include a variety of approaches, including studying the growth of species in culture.

Although collections have now been made throughout the New Zealand botanical region, there remain areas of coastline that are largely unexplored because they are difficult of access and the seasonal coverage of collections from many of the offshore islands including the Chatham Islands is very poor.

Class Bacillariophyceae: Diatoms

For sheer beauty, nothing compares with diatoms. Anyone who studies their exquisite glassy skeletons under the microscope is moved to regard human creations as poor in comparison. But in order to view these microalgal marvels, each less than a millimetre long or wide, sophisticated microscopes are needed, and much detailed literature to identify them. The total number of described species of marine and freshwater diatoms worldwide exceeds 24,000 (Fourtanier & Kociolek 2009a,b). Estimates of undiscovered diversity range as high as 200,000 (Mann & Droop 1996), but opinions differ over what constitutes a genus, species, or variety in diatoms. As W. Lauder Lindsay wrote in 1867, 'The species of one botanist is not that of another, and what is a species today may become a variety or perhaps even a genus tomorrow'. Dr Lindsay's comment is as true today as it was in 1867, although molecular-systematic approaches hold promise for clarifying species boundaries in living taxa (Alverson 2008). Diatom taxonomy is in a state of flux. Since the compilation of a revised checklist of New Zealand freshwater taxa in the 1980s (Cassie 1984), there has been a virtual revolution in the approach to diatom taxonomy, resulting in many changes to genera and species (Mann 1999; Cassie Cooper 2001; Fourtanier & Kociolek 2009a,b).

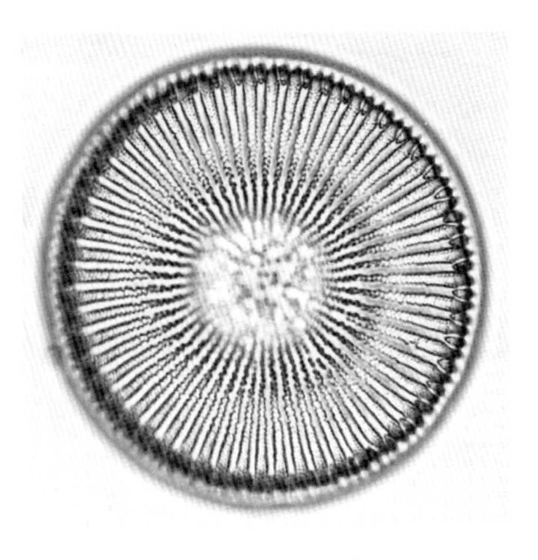

Cyclostephanos novaezeelandiae (Thalassiosirales).
Stuart Stidolph

The study of diatoms has gone hand-in-hand with the development of the microscope. Early diatomists, anxious to impress their acquaintances, vied with each other to produce perfectly arranged slide mounts – with up to 4000 or more acid-cleaned frustules (valves) under one small coverslip (cf. J. D. Möller in Germany). The late Frederick Reed, Christchurch, excelled at this intricate mounting procedure. On one slide he mounted 515 Oamaru diatomite valves under a coverslip 'less than half-an-inch' in diameter (Reed pers. comm.). Whereas diatom taxa were previously diagnosed solely from observations under the light microscope, they now require knowledge from several different disciplines. The advent of electron microscopy, together with advanced techniques in biochemistry and molecular biology, has given rise to a whole new vista of observations on dead and living cells, as well as on patterns of behaviour.

Basically, living diatoms are single cells, each encased by a glass box – shaped like a Petri dish in the case of most centric diatoms or oblong or boat-shaped in the pennates. The box lids are the valves, and the overlapping walls form the girdle or cingulum. Intricate patterns on their siliceous walls provide taxonomists with characters for classification and description. In centrics, the

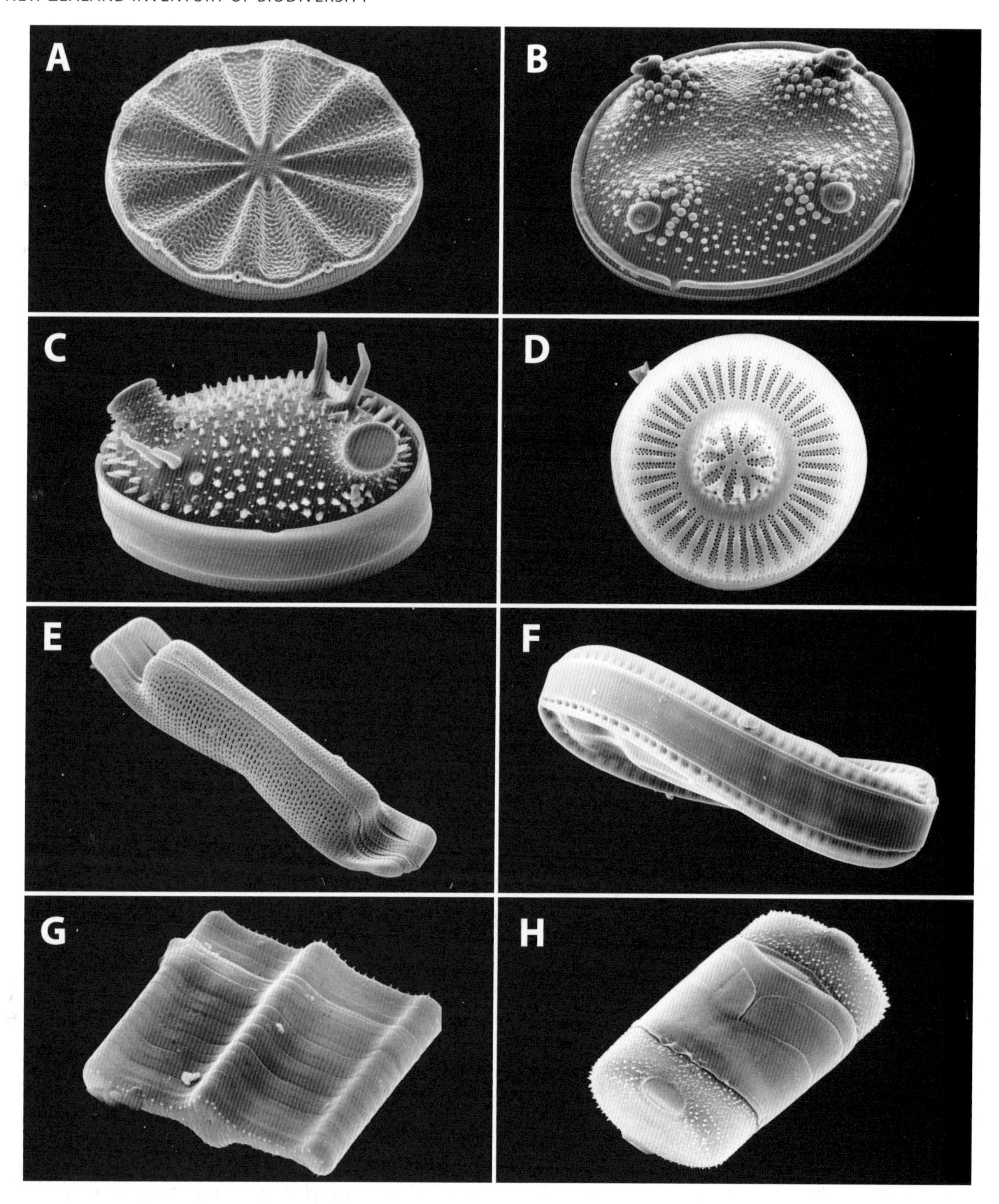

A: *Actinoptychus splendens* (Coscinodiscales). B: *Aulacodiscus petersii* (Coscinodiscales). C: *Cerataulus turgidus* (Triceratiales). D: *Discostella stelligera* (Thalassiosirales). E: *Hantzschia doigiana* (Bacillariales). F: *Surirella contorta* (Surirellales). G: *Tabellaria flocculosa* (Tabellariales). H: *Pleurosira laevis* (Triceratiales).

A–C, E, H, Stuart Stidolph; D, F, G, Vivienne Cassie Cooper

valve originally develops from a point as an annular pattern centre that produces radiating ribs and pores. In pennates the pattern centre is linear and develops ribs in pinnate patterns.

Features of living cells include the golden-brown pigment fucoxanthin, which occurs in chloroplasts with chlorophylls *a*, c_1, and c_2. Cells of raphid pennate diatoms are capable of movement, though they are non-flagellate. The protoplast inside its silica walls pulls on adhesive filaments of slime that are secreted through a slit (raphe), enabling the whole diatom to to slide slowly over surfaces.

Most diatoms grow smaller as they divide. Eventually reaching a stage where they cannot divide any further, they resort to sex – a difficult-to-detect process observed by only a few specialists. Among pennates, cells unite inside gelatinous sheaths; after fusion of haploid amoeboid gametes, the resulting zygotes grow larger again to restart the whole process. The only flagellated cells are male gametes in centrics. Meiosis occurs at gametogenesis. Life-cycles are thus diplontic.

Diatoms are found in almost all aquatic habitats – in the oceans, in lakes, ponds, rivers, streams, ditches, fountains, damp soils, and thermal springs. They occur in open water as components of phytoplankton and on surfaces as periphyton, forming slimy growths. Growth of diatoms in thermal areas is ultimately enhanced by the effects of volcanism – silica released during volcanic eruptions will eventually help to promote their development (Cassie 1989a). Each species has its own requirements for light, temperature, carbon dioxide, and nutrients, and pH is a crucial factor in determining which species will predominate (Van Dam et al. 1994; Smol & Stoemer 2010).

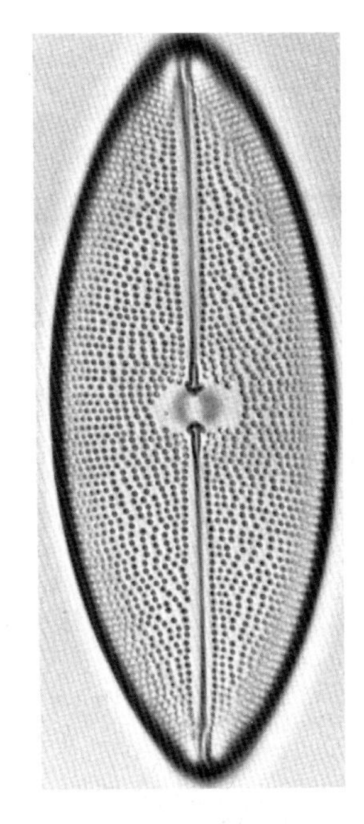

Petroneis marina (Lyrellales).
Stuart Stidolph

The roles and uses of diatoms

Because of their sheer numbers, diatoms play a vital role in nature. They produce oxygen in photosynthesis – vital to the health of the planet – and they can be useful monitors of the degree of water pollution (Cassie 1983a; Kilroy & Bergey 1999). Some tolerate low levels of oxygen whereas others require high levels (Patrick 1977; Van Dam et al. 1994). Additionally, they play a major role in the global cycling of carbon and silicon (Mann 1999); provide food for zooplankton animals like krill and copepods; produce a substance (dimethylsulphoniopropionate or DMSP) that reacts to form dimethyl sulphide, able to seed clouds and form rain; are efficient at recycling silicon and other elements; and are vital partners in some symbiotic relationships – for example, pennate species of *Navicula* and *Nitzschia* within cells of benthic foraminifera (Hoek et al. 1995). In short, as David Attenborough (1995) observed in *The Private Life of Plants*, they are among the most important photosynthetic organisms in the world, particularly as diatoms store lipids rather than starch so they form the original source of much fossil oil (Harwood 2010).

Diatoms are also used directly by humans. Fossil diatoms (for a full account of these see below) forming the mineral diatomite are used as filters for beer, water, and toxic wastes, as well as components of paints, embalming compounds, toothpaste, cement, and glass (Harwood 2010). Diatom remains have a range of scientific applications. They can be used as stratigraphic indicators (Jordan & Stickley 2010) and those in cores from lake and sea floors are used to reconstruct ancient climates and ecosystems (Smol & Stoemer 2010). They also have utility in forensic science (Peabody & Cameron 2010). Up until 1980, 61 New Zealand localities were known to have significant freshwater diatomite deposits (Cassie 1981) and another has been discovered in Pukaki Crater in Auckland City (Hayward 2009).

From the time of their earliest discovery in 1703 to the present day, diatoms have continued to fascinate their viewers, and to cause them to speculate on past geological events. There are a few overseas Jurassic marine records (about 190 million years), with some freshwater genera recorded from the earliest

Cretaceous (140 million years). The earliest detailed records come from the Early Cretaceous about 135 million years ago (Strelnikova 1999; Jordan & Stickley 2010). These genera are virtually all extinct, while Early Tertiary genera are poorly represented today. Although no Eocene freshwater diatoms have been found in New Zealand to date, Eocene marine deposits at Oamaru, consisting mainly of unusual centric taxa, were made famous by early diatomists. A burst of evolution occurred in the pennate diatoms during the Miocene, when they became much more diverse in the fossil record. There is an extensive freshwater deposit at Middlemarch, Central Otago, where a deposit equivalent to almost 50,000 tonnes of diatomite was formed almost entirely by a single species (*Encyonema jordani*, formerly known as *Cymbella jordani*).

Classification of diatoms

For this review, diatoms are named in accord with the internationally recognised Algaebase (Guiry & Guiry 2010), with a few other placements based on recent publications. Higher classification is based on Round et al. (1990), with their classes reduced to subclasses and and subclasses to superorders. Their system (which gave full diagnoses) is preferred to the partial one of Medlin and Kaczmarska based on RNA and cytology (Medlin et al. 2008). The latter system divides diatoms into three classes – Coscinodiscophyceae (most radial centrics), Mediophyceae (bipolar and multipolar centric diatoms plus Thalassiosirales, i.e. radial centrics with central tubular structures), and Bacillariophyceae (the pennate diatoms). They considered three elongate genera, viz *Ardissonea, Climacosphenia,* and *Toxarium,* although of pennate form, also belong to the Mediophyceae. Acceptance of this newer system is considered premature, however, as other evidence indicates that only the raphid Bacillariophyceae are monophyletic (Williams & Kociolek 2007; Theriot et al. 2009), hence we have chosen to group the diatoms into centrics (Coscinophycidae), araphids (Fragilariophycidae), and raphids (Bacillariophycidae) based on their shapes and the presence of raphe slits.

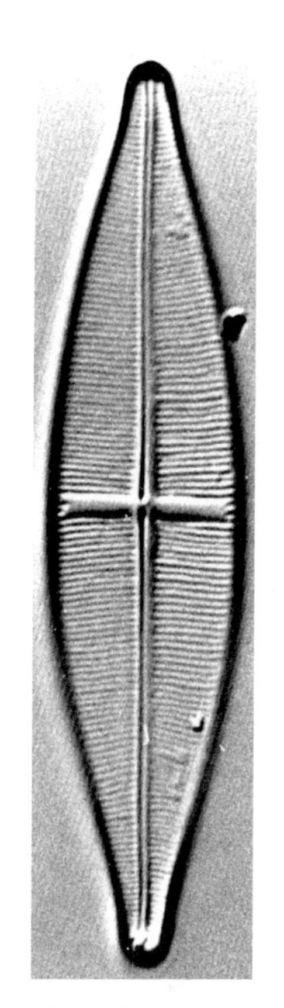

Staurophora salina (Cymbellales).
Stuart Stidolph

Frank Round and colleagues (see Round et al. 1990; Round & Bukhtiyarova 1996; Round & Basson 1997), following electron-microscope studies, divided several genera with many species and a wide variety of structure into 'new' genera. These genera include several older genera that were not widely used, and some new ones that were described with the aid of scanning electron micrographs. Krammer and Lange-Bertalot (1988, 1991a,b) kept to the conservative approach they had earlier outlined in the first volume of their flora (Krammer & Lange-Bertalot 1986) and chose to place most freshwater *Synedra* species in *Fragilaria* because of their similar ultrastructure. Similarly, in their second edition of Naviculaceae in the *Süsswasserflora von Mitteleuropa* (Krammer & Lange-Bertalot 1997), they did not use these 'new' genera. Since then, however, Krammer and Lange-Bertalot have both described some new genera (Kusber & Jahn 2003) and ascribed some diatoms to the new genera of Round et al. (1990). The old genera most affected by division in the checklists are *Navicula* (now split into 29 genera), *Nitzschia* and *Synedra* (10 each), *Achnanthes* (9), *Biddulphia* (8), *Cymbella* and *Fragilaria* (7 each).

Fourtanier and Kociolek's (1999, 2003) *Catalogue of Diatom Genera* was consulted for generic name changes. Species names were compared with Algaebase (Guiry & Guiry 2010) and checked in the the Californian Academy of Sciences online *Catalogue of Diatom Names* (Fourtanier & Kociolek 2009a,b) with its links to *Index Nominorum Algarum* (Silva & Moe 1999; Silva 2009). Older texts also consulted in compiling the species lists are contained in the bibliographies of Cassie (1984, 1989a). Recent research papers and theses with new records for New Zealand are cited below in the references and new taxa from them are included.

In the end-chapter checklists, brackish-water species are segregated by major habitat, with 82 species ascribed to the freshwater list and 118 to the marine list, so as to avoid any duplication of names. This segregation was based on the literature (Foged 1979; Krammer & Lange-Bertalot 1986–1991; Van Dam et al. 1994; Hartley 1996; Cochrane 2002). Those brackish-water taxa (halophil and some mesohalobous taxa) generally found with fully freshwater species are in the freshwater list and those more often found with fully marine species are in the marine list.

Terrestrial diatoms are generally freshwater taxa that can live in water films on mosses or soils. Five taxa in the freshwater list are considered by van Dam et al. (1994) as 'nearly exclusively occurring outside water bodies'. These are *Achnanthes coarctata, Decussata placenta, Encyonema alpinum, Navicula bryophila*, and *Pinnularia lata.* There are 50 other taxa they consider as 'mainly occurring on wet moist or temporarily dry places'. Some of these occur in Cassie's (1989) study of thermal areas which included samples of moist surfaces. Soil diatoms were found in flood deposits in Hawkes Bay (Turner 1997; Cochran *et al*. 2005).

The diatom flora of New Zealand

Four diatom floras exist in the New Zealand region – living marine, fossil marine, living freshwater, and fossil freshwater. The fossil freshwater species are placed in the freshwater checklist, since most freshwater fossil taxa are subfossil taxa from Pleistocene samples and are still living in New Zealand (Cassie 1981). Among these taxa she noted that some occur more frequently in old lake deposits than in living floras, including *Aulacoseira crenulata, Cyclostephanos* (*Stephanodiscus*) *novaezeelandiae, Stoermeria* (*Hydrosera*) *trifoliata* (Kociolek et al. 1996), and *Surirella contorta*. Diatom floras in lake deposits can indicate environmental changes – sea water penetrated to Lake Waikare, Waikato (Harper et al. 1993) and Lake Kohangapiri, Wellington (Cochran et al. 1999); Wainuiomata, Wellington is built on old lake mud (Begg et al. 1993), and diatom blooms occurred in Lake Poukawa following volcanic eruptions (Harper et al. 1986).

There is a much longer time gap (38 million years instead of less than two million years) between Tertiary fossil and living marine taxa than between fossil and living freshwater diatoms. Few species of Oamaru diatoms (< 15%) have been recorded from coastal marine habitats today. Taxa in the Tertiary marine checklist which also occur in one of the recent lists are marked as living (L).

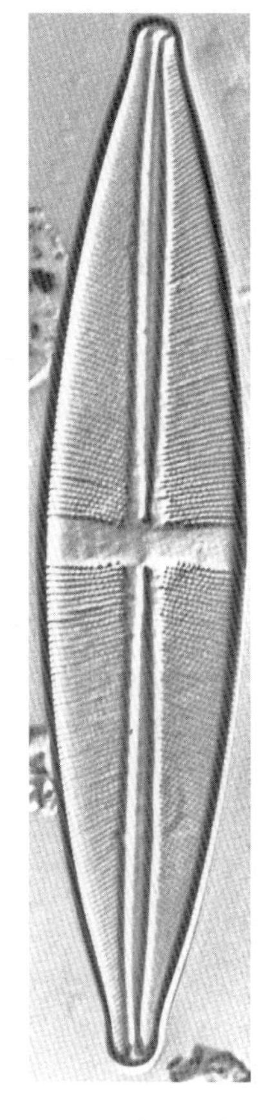

Stauroneis gracilis (Cymbellales).
Stuart Stidolph

Medlin and Kaczmarska (2004) argued that the Mediophycidae (multipolar centrics) are more closely related to pennate diatoms (Bacillariophycidae) than to other centric diatoms (Coscinophycidae) and that the pennates are derived from the Mediophycidae. Theriot et al. (2009) showed a grade of lineages with the most recent group to separate from others being the raphid pennates (excluding *Eunotia*). The earliest known pennate diatoms are a few araphid species from the late Cretaceous (about 94 million years ago). This fits with our finding that only a quarter of the taxa in the New Zealand fossil checklist belong to the Bacillariophycidae, while about 70% of the living marine list and over 90% of the freshwater list do. Marine sediments and diatomites contain some benthic as well as planktonic diatoms depending on the distance from the coast, but workers on living floras are able to sample benthic habitats directly. Although there are more species of pennate diatoms (Bacillariophycidae) in our living floras, centric diatoms are frequently commoner ecologically; for example *Aulacoseira* (Coscinophycidae) is often dominant in the freshwater phytoplankton of the silica-rich Rotorua lakes (Cassie 1978), and *Odontella* (Mediophycidae) in the marine phytoplankton of the Hauraki Gulf (Cassie 1960).

New Zealand collections

Auckland Museum – W. Booth type slides.

NIWA diatom collection, Christchurch – slides from C. Kilroy, L. Bergey, M. Harper, M. Reid, K. Sabbe, S. Stidolph, and W. Vyverman.

V. Cassie Herbarium – preserved samples, slides, light and SEM micrographs in the Allan Herbarium (CHR), Landcare Research, Lincoln.

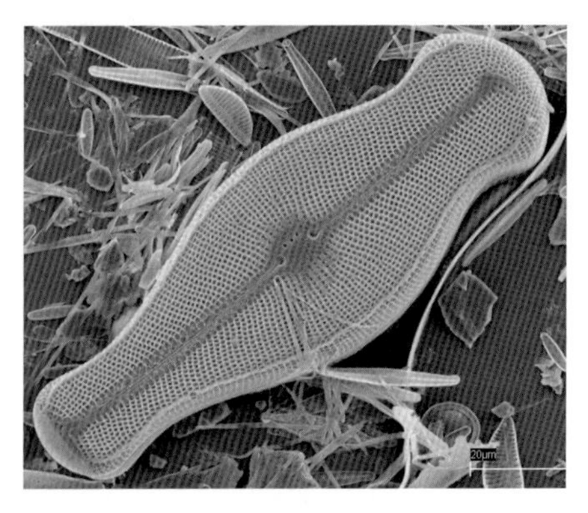

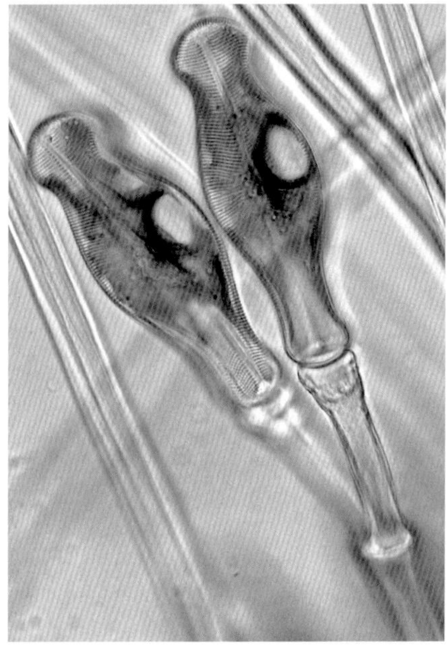

The nuisance alien diatom *Didymosphenia geminata* ('didymo'), showing an isolated cell (upper) and the characteristic twinned arrangement on stalks in life (lower) (Cymbellales).
Cathy Kilroy, NIWA

F. Reed slide collection – 316 named taxa on picked mounts among 1015 slides of Oamaru diatoms, Allan Herbarium, Landcare Research, Lincoln.
Otago Museum, Dunedin – Morris Watt Collection of 680 slides, all from Oamaru Diatomite, and S. H. Rawson Collection of Recent diatoms, comprising 272 slides (both collections catalogued by S. R. Stidolph).
N. Foged Collection – in Botanical Museum of Copenhagen, Denmark (see *Diatom Research* 3: 170–174).

New Zealand freshwater diatoms

As early as 1861, Dr W. Lauder Lindsay gathered diatom-containing samples from South Island localities near Dunedin, in lagoons, marshes, ditches, and ravines, and from moist rock faces and waterfalls. The samples were sent to Britain, where Greville (1866a,b) described 110 species, including three new to science at that time. Later studies followed, with an emphasis on the fossil taxa discovered in diatomite. The findings have been documented in a brief history of freshwater diatom research in New Zealand (Cassie 1983b).

In the first checklist of freshwater diatoms from New Zealand, Skvortzov (1938) listed 260 taxa. Two decades later, Chapman et al. (1957) independently listed 187 taxa and Flint (1966) added another 105. Further lists were compiled by Cassie (1980, 1984), who noted name changes in a later publication (Cassie 2001). The 1984 list has particular utility as it includes references for the records listed under species names widely accepted then and their names as first published. It also gives the localities cited in publications. Since that time there have been several new records, although recent name changes have daunted even the most dedicated diatomists.

The most significant records to date of living diatoms have been made by Skuja (1976; 106 taxa) and Foged (1979; 684 taxa). Skuja's data came from peat-bog samples while Foged's annotated list was assembled from a much wider variety of habitats. It formed the first comprehensive account of New Zealand freshwater taxa, even though it applied only to collections he made from the North Island. One of the diatom world's unsung heroes in New Zealand, Niels Foged braved the cold of winter to enrich our knowledge of the freshwater and brackish-water flora, from diatomite deposits, swamps, bogs, thermal areas, ponds, pools, lakes, springs, rivers, streams, and waterfalls. All his and other known records up to 1983 were included in Cassie's (1984) checklist published in *Bibliotheca Diatomologica.*

Since that time, many studies have appeared, several of which recorded more than 20 taxa that were not identified by earlier workers and included in Cassie's (1984) checklist (see Harvey 1996; Cochran 2002; Reid 2005; Kilroy et al. 2006; Kilroy 2008) including taxa identified by P. Conger (Appendix II in Cassie 1989a). Recent work at NIWA's Christchurch laboratory has resulted in description of some new taxa (Kilroy et al. 2003; Beier & Lange-Bertalot 2007; Kilroy 2008) and fresh approaches to assessing periphyton (Francoer et al. 1998; Biggs & Kilroy 2000).

Much detailed information is contained in unpublished university theses. Unpublished such New Zealand records are annotated in the end-chapter checklist with the word 'thesis' after the entry. The checklist does not include synonyms (see Fourtanier & Kociolek 2009a,b) and unfamiliar names should be checked in the texts quoted in the references.

Endemism and alien species

On the available evidence, the degree of endemism among New Zealand freshwater diatoms identified to date appears to be low – about 2.3% according to Moser et al. (1989). They based their figures on those of Foged (1979), who stated that the vast majority of species are cosmopolitan, ranging far beyond Australasia. This figure contrasts markedly with the estimated endemic element in New Caledonia of 40%. The figure of 2.3% for New Zealand is arbitrary and will undoubtedly change when the local flora is investigated more thoroughly, especially that from offshore islands, mountain lakes, and inaccessible headwaters of streams. The most complete floras are European and North American ones and these are used to identify New Zealand diatoms; so the lists could well include incorrect identifications. Careful morphological, ecological, and genetic studies are revealing more endemism in diatom taxa especially in the Southern Hemisphere (Vanormelingen et al. 2008). Some taxa are clearly distinctive – a new genus, *Eunophora,* has been described from New Zealand and Tasmania (Vyverman et al. 1998, Harper et al. 2009). Four new species of *Actinella* and two of *Fragilariforma* have been described from New Zealand, some of which also occur in Tasmania (Sabbe et al. 2001; Kilroy et al. 2003; Kilroy et al. 2007). Similarly *Discostella tasmanica,* first described from Tasmania, occurs in pre-European sediment from Lake Rotorua, Rotorua. Gene sequencing indicates that '*Eunotia bilunaris*' clones from New Zealand and Tasmania are quite alike but are unlikely to be conspecific with a European clone (Vanormelingen et al. 2007). This trans-Tasman link, first recognised by Foged (1979), led to consideration of Australasian rather than New Zealand endemism by Kilroy (Kilroy et al. 2007). A curious new species is *Frankophila biggsii,* a fragilarioid with a raphe-like slit (Lowe et al. 2006). This genus and *Veigaludwigia* were first recognised in South America, and the latter has since been reported from New Zealand mountain tarns (Van Houtte et al. 2006).

A handful of 'didymo' from the Waiau River.
Cathy Kilroy, NIWA

With increasing means of transporting diatoms round the globe, whether by wind, fish, macrophytes, or in ballast water, it becomes harder to determine if a species is truly indigenous. For instance, Harper (1994) has stated that *Asterionella formosa,* euplanktonic in many North Island lakes, is apparently absent from cores in pre-European sediments of Australia and New Zealand. It is likely to be an introduced species. There is no ambiguity in the case of *Didymosphenia geminata* ('didymo'). First discovered forming nuisance blooms in Southland in 2004, it has since spread to a high proportion of rivers throughout the South Island, though blooms have a more restricted distribution (Kilroy & Unwin 2011). It is classified as an 'unwanted organism'.

Future studies

Although there is now a taxonomic framework as a baseline, it remains for an authoritative flora to be produced. Until we know the complete life-cycles of so-called 'new' taxa, we cannot be sure of exact numbers of genera, species, and varieties in a given diatom flora. The present survey of freshwater diatoms gives a tally of 910 taxa (794 species) in 108 genera (five genera with only brackish species) – considerably higher than Foged's (1979) estimate of 684 taxa.

New Zealand marine diatoms

In the marine environment, diatoms are found on various substrata, like tidal flats, seaweeds, rock faces, and most hard objects, as well as forming a significant, often dominant fraction of the phytoplankton communities of open water, especially in nutrient-rich coastal waters. Phytoplankton comprises a range of microscopic algae in different taxonomic groups. Representatives range in size and can be arbitrarily grouped into three size classes – picoplankton (< 2 micrometres cell diameter), nanoplankton (2–20 micrometres), and micro-plankton (20–200 micrometres) (Sieburth et al. 1978). Worldwide, diatoms are the most species-rich group of microalgae (Margulis et al. 1990; Jeffrey & Vesk 1997). As such, they are very widespread and have global significance in the cycling of many elements, particularly carbon. Recent estimates of global primary production show that 46% is oceanic and 54% terrestrial (Field et al. 1998) in origin. The enormous productivity provided by microalgae in the ocean forms the base of food webs that support directly or indirectly almost the entire animal population of the sea. In the open sea, diatoms are the major producers of 'new' phytoplankton biomass (Falkowski et al. 1998), sustained by fluxes of new

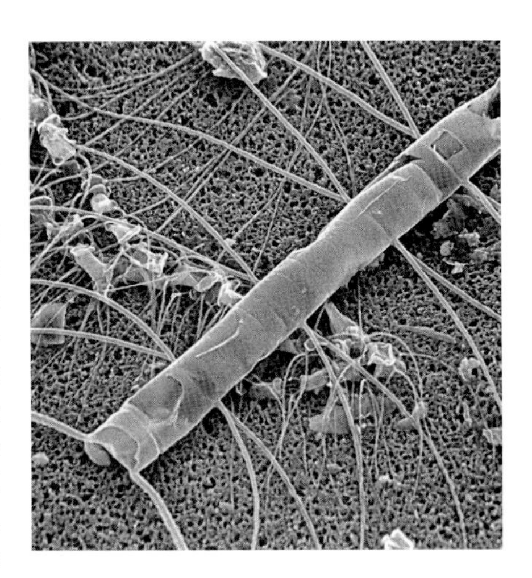

The marine bloom-forming diatom *Chaetoceros affinis* (Chaetocerotales).
F. Hoe Chang, NIWA

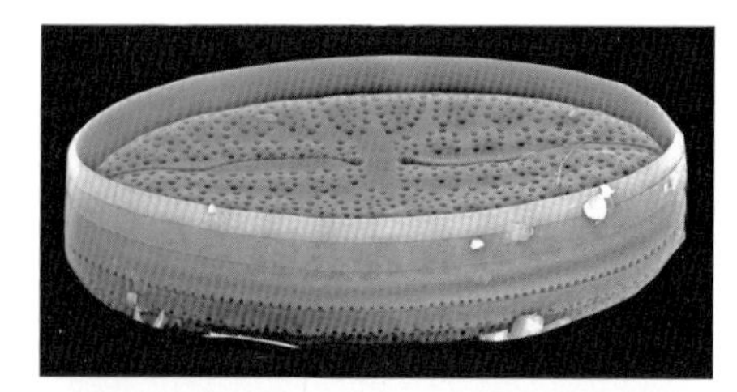

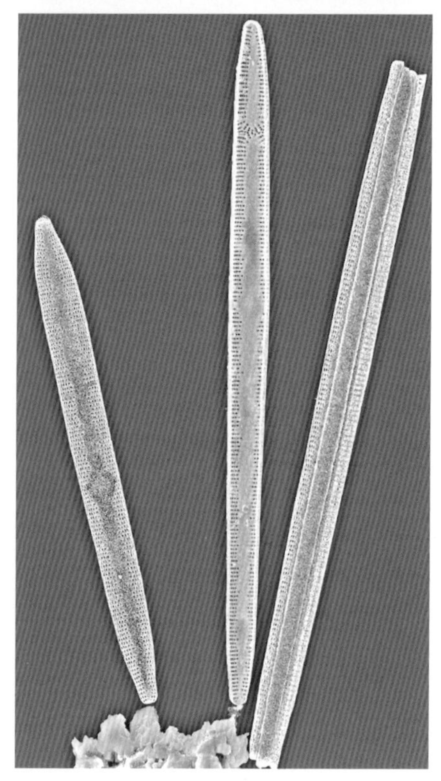

The endemic diatoms *Skeletomastus coelata* (upper) (Mastogloiales) and *Tabularia variostriata* (Fragilariales).

Margaret Harper, Victoria University of Wellington

nitrogen nutrient (NO_3-) from deeper waters through such physical processes as upwelling, and/or tidal/wind mixing, which frequently occurs in either winter or spring (Bradford et al. 1986; Chang et al. 1992; Change et al. 2003). Many species are bloom-forming (e.g. Taylor et al. 1985). In spring, on New Zealand's northeast coast (Chang 2000) and over the Subtropical Convergence near the Chatham Rise (Chang & Gall 1998), diatoms as a group make up the bulk of total carbon biomass (>95%).

Taylor (1970) and Cassie (1980) published lists of New Zealand marine diatoms. Floristic accounts of marine diatoms were made by Stidolph (1980, 1985a) as well as in-depth studies of individual taxa from selected areas (Stidolph 1981, 1985b, 1986, 1988, 1990, 1992, 1993a-d, 1994a,b, 1995, 1998). Enzymes were used by Booth (1981) to study epiphytic diatoms (Booth 1984, 1985a,b, 1986ab). More recent floristic studies include lists of phytoplankton (Safi 2003) and coastal diatoms (Cochran 2002; Harper & Harper 2010). Major revisions of the taxonomy of several groups of diatoms have resulted in a large number of generic and specific name changes. Most notably, the genus *Pseudo-nitzschia*, which has been linked to Amnesic Shellfish Poison (ASP) outbreaks (Holland et al. 2005), was erected for several species of *Nitschia* (Hasle & Syvertsen 1996). Although domoic acid, the causative neurotoxic agent of ASP, has been detected in New Zealand shellfish, no actual ASP outbreak has ever been reported in New Zealand (Chang et al. 1995), even though several ASP species of *Pseudo-nitzschia* are found in coastal waters. Evidence indicates that seven of them are producers of domoic acid and its isomers, e.g. *P. australis* (Chang et al. 1995; Chang 1996), *P. delicatissima*, *P. fraudulenta*, *P. pungens*, *P. multiseries*, *P. turgidula*, and *P. pseudodelicatissima* (Rhodes et al. 1996, 1998). The highest concentrations of domoic acid recorded in New Zealand (600 parts per million) were in the digestive glands of scallops (*Pecten novaezelandiae*) from Northland in 1993 (Rhodes et al. 1998). Toxin production in blooms has been monitored since 1993, preventing unnecessary closure of shellfish beds by the non-toxic *P. heimi* (Rhodes et al. 1998).

A species of *Licmophora* has been reported infesting krill (*Nyctiphanes australis*) in New Zealand's coastal waters, with up to 50–70% of individuals affected at times. The effect on the krill is not known (McClatchie et al. 1990).

The new compilation of New Zealand marine and brackish-water diatoms provided in the end-chapter checklist gives a tally of 730 taxa (677 species) in 154 genera (17 genera with only brackish species). Unpublished New Zealand records from dissertation research are annotated in the checklists with the word 'thesis' after the entry. Many of these species are illustrated with light or scanning electron micrographs in the theses of Levis (1975), Wilkinson (1981), and Booth (1985). Endemism among New Zealand marine diatoms appears very low (<1%), but some taxa are very distinctive: *Skeletomastus coelata* (Harper et al. 2009b), a macroalgal epiphyte, and the two brackish-water species *Hantzschia diogiana* (Stidolph 1993c) and *Tabularia variostriata* (Harper et al. 2009a) have not been seen elsewhere. New Zealand's diatom diversity (199 centric species and 478 pennate species) may be compared with the global figures compiled by Sournia (1995) in his census of the world diatom flora, viz 870–999 species of centric form, and 500–784 species of pennate form, including a large proportion of tychoplanktonic forms (shallow-water plankton originating from the benthos).

Future studies

At present, positive identification of marine diatoms relies on morphological features that can be recognised and characterised using both light and electron microscopy. In the last few years in New Zealand other alternative approaches to identify microalgal taxa have been extended to chemotaxonomy, based on specific biomarker pigments (Chang & Gall 1998), immunofluorescence detection using species-specific immunological techniques (Chang et al. 1999), and nucleotide probe techniques. Although each of these techniques has its limitations, there is great potential combining some of these relatively new approaches with traditional light microscopy to help separation of similar or closely related taxa. For example, isolates of New Zealand *Pseudo-nitzschia* species can be identified using species-specific rRNA nucleotide probes (Rhodes et al. 1998, 2000).

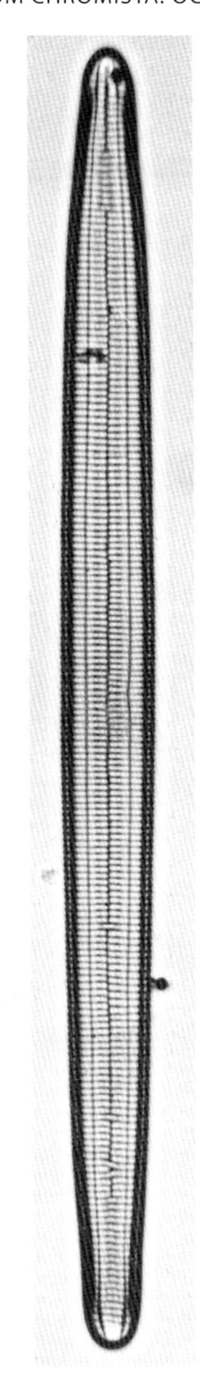

Ardissonea formosa (Ardissoneales).
Stuart Stidolph

Fossil marine diatoms of the New Zealand Tertiary

Fossil diatom floras include more thick-walled diatoms and spores than Recent floras. Oceanic species with thin walls are rarely preserved in the fossil record as salt water slowly dissolves amorphous silica. Diatoms are more likely to be preserved when they are rapidly deposited in sediment following a bloom in upwelling seawater. Diatomites are deposits of fossil diatoms; they often contain evidence of volcanic activity as fine shards of volcanic glass both stimulate diatom growth and protect their walls from dissolution. New Zealand's Oamaru 'diatomite' is renowned among diatomists as it contains many distinctive fossil taxa. It is too impure to be a true diatomite. The Oamaru diatomite was the only source of fossil marine diatoms mentioned in Cassie's (1979) comparison of New Zealand's fossil and recent marine diatom floras. Subsequently, ocean-drilling vessels visited New Zealand in 1983 and 1998 and drilled in offshore sediments, recovering Tertiary and younger sediment cores.

Cores from the 1998 ODP leg 181 sites off the east coast of New Zealand included diatoms from all the major subdivisions of the Tertiary Period (identified by Julianne Fenner) (Carter et al. 1999a-g) except the Eocene – that period is represented by the Oamaru diatomite. Marine diatoms also occur on land in Pliocene mudstones in the Poukawa basin (Harper & Collen 2002). Most of the records in the Checklist are of Oamaru taxa and of taxa from ODP leg 181 and DSDP leg 90; a few marine diatoms found by M. Harper in Tertiary samples (and recorded in student theses at Victoria University of Wellington) are also included.

Paleocene (~ 65 to 54 million years ago)

ODP leg 181 site 1121 off the eastern edge of the Campbell Plateau yielded 19 species of diatoms from Paleocene sediments. These diatoms from the *Hemiaulus incurvus* biostratigraphic zone include several that occur throughout the Tertiary and Quaternary periods.

Eocene (~ 54 to 38 million years ago)

The Oamaru diatomite was formerly considered Oligocene, but Edwards (1991) showed that associated nannofossils and foraminifera indicate that nearly all the diatomaceous material belongs to the latest Eocene (New Zealand's Runangan stage).

The unusual diatoms in the Oamaru diatomite were first noticed in London – a sample of 'kaolin' from Oamaru was sent to the May 1886 Indian and Colonial Exhibition there. Henry Morland examined some with a microscope and saw diatoms. These he gave to Gerald Sturt, who excitedly found several new taxa which he described to the Quekett Microscopical Club two months later (Edwards 1991). With the help of Edmond Grove, he published details of about 100 new taxa in four papers (Grove & Sturt 1886, 1887a-c). Oamaru diatomite was very popular with diatomists in the early 20th century. Most studies concentrated on the many centric diatoms in it. Later, Schrader (1969) specially studied its pennate diatoms and described 45 new taxa. Desikachary and Sreelatha (1989) produced the most complete flora, with descriptions and photographs of 647 taxa. Amateur diatomists in New Zealand have also done significant work. Arthur Doig (in Edwards 1991) studied floral changes in the rock sections, comparing 12 sites. He found *Stephanopyxis* was dominant in older samples and *Coscinodiscus* (which included *Cestodiscus* spp.) in younger samples. Fenner (1986) showed this result to be consistent with a global pattern, as she found similar changes with *Cestodiscus* species increasing from 20% to 50–80% of the flora near the latest Eocene to earliest Oligocene boundary. Similarly, Ross

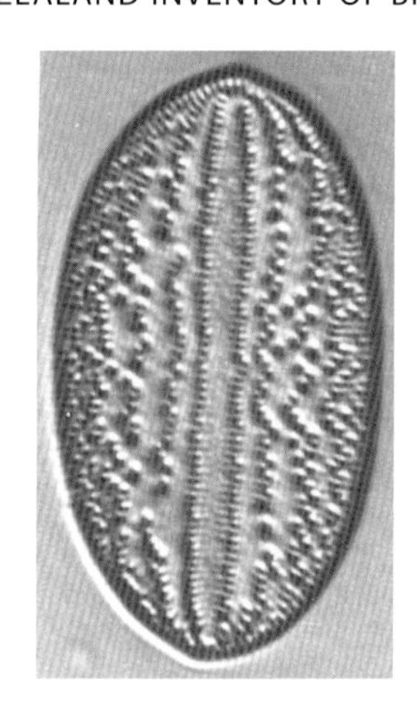

Cocconeis placentula (Achnanthales).
Stuart Stidolph

and Sims (2000) noted that the longer-surviving *Actinodiscus corolla* occurs in both *Stephanopyxis*- and *Coscinodiscus*-dominated samples, but the Paleogene *A. contabulatus* is found (with one possible exception) only in *Stephanopyxis* samples.

Another dedicated amateur, Frederick Reed (Cassie Cooper & Harvey 2003) produced a photographic atlas of 254 taxa (Reed in Edwards 1991) as part of a paleontological bulletin on the Oamaru Diatomite. This bulletin included a checklist of Oamaru taxa published prior to 1965 (Edwards 1991) and also contained detailed information on the geological age, stratigraphic position, and depositional environments of the diatomite. Edwards (1991) concluded that the 'Oamaru diatomite was deposited under quiet, well mixed, marginally tropical to sub-tropical waters, at depths of 75 to 150 m in a basin located some 50 km offshore from a low-lying landmass and adjacent to volcanic shoals that from time to time were surmounted by small short lived, volcanic islands' (Edwards 1991). He did not mention which diatoms recorded from Oamaru are biostratigraphically significant. Four recorded taxa (*Baxterium brunii, Cestodiscus pulchellus* var. *novae zealandiae, Rylandsia inaequiradiata,* and *Pseudotriceratium radiorecticulatum*) first appear in the Late Eocene. Two taxa (*Brightwellia coronata, Pyxilla gracilis*) last appear in the earliest Oligocene. Oamaru taxa in the current checklist are based largely on Desikachary and Sreelatha (1990).

Oligocene (~ 38 to 23 million years ago)
More than 25 diatom species were identified from two ODP sites north and east of the Chatham Rise. Biostratigraphic placement of material as late Oligocene was based partly on the last appearance datum of *Rocella gelida* var. *schraderi* at 322.68 metres depth at site 1124, a datum that occurred about 23.7 million years ago (Carter et al. 1999f). Some phosphatic nodules, considered to be Oligocene on the basis of their ostracod fauna, contained a few benthic diatoms. All of the diatoms were extant species, showing that the nodules included coastal sediment. Also present was the freshwater benthic species *Cocconeis placentula.* Nodules can vary in age and origin, as phosphatisation of the original sediment occurs after deposition (D. McConchie 2000 pers. comm., Centre for Coastal Management, Southern Cross University, NSW).

Miocene (~ 23 to 5 million years ago)
Ciesielski (1986) recorded about 80 taxa as biostratigraphically significant or common in Miocene sediment collected near New Zealand. These came from a core recovered from DSDP leg 90 at Site 594 just south of the Chatham Rise. Ciesielski compared the Miocene floras of this site with those from southwest Atlantic site 513. There, a volcanic ash dated as 8.7 million years old occurs between the first appearance of two *Hemidiscus* species (*H. cuneiformis* and *H. karsteni*) and the last appearance of *Denticulopsis lauta.* This indicates that sediment recovered from around 250 metres at site 594 also dates from the Late Miocene. Other datums indicate that some sediments from below 250 metres were deposited in the Middle Miocene. There is no evidence of early Miocene floras, which are very different as about 80% of diatom taxa change at this time (Baldauf 1992). On ODP Leg 181, diatoms were preserved from the Miocene in cores from five sites and from site 1125. The first appearance of *Denticulopsis dimorpha* indicates that part of the sediment in site 1120 was from the Middle Miocene and the last appearance datum for this and four other species indicate that most of the top 100 metres of the core came from the late Miocene. First and last appearances of diatoms helped distinguish between Middle and Late Miocene sediments from site 1123 and Pliocene sediments from site 1125 (Carter et al. 1999g). According to Cassie Cooper and Harvey (2003), Frederick Reed identified 50 specimens from Waiau Valley diatomite (sent by Roger Griffiths), including *Glyphodiscus strigillatus,* but unfortunately there is no diatom list in Griffiths' (1983) thesis.

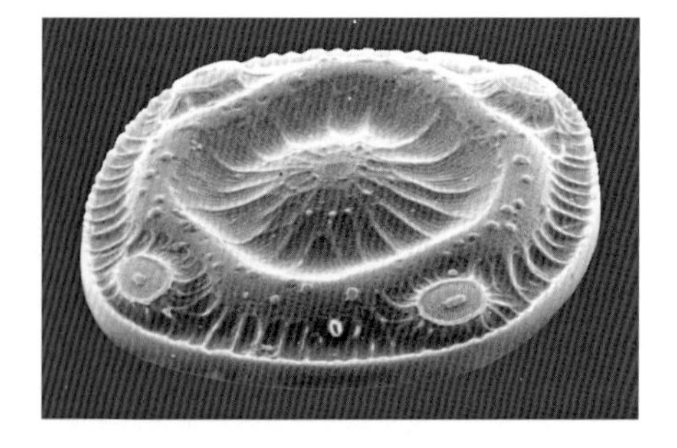

Glyphodiscus strigillatus (Triceratiales).
Stuart Stidolph

Pliocene (~ 5 to 2 million years ago)
Diatom floras from Pliocene sediments have been recorded at eight sites round New Zealand. Oceanic floras occur in much of the sediment recovered from the upper part of DSDP site 594 and from four ODP sites. Coastal floras occur on land in the North Island in the Wanganui and Poukawa basins. Some 54 coastal taxa occurred in a lens of marine diatomite in Pliocene deposits along the Turakina River, Wanganui Basin, (McGuire 1989). A similar but poorly preserved flora (about 35 taxa) is derived from Pliocene mudstone (Raukawa siltstone) in Hawke's Bay. This flora has been redeposited in the freshwater Quaternary sediments of the Poukawa basin (Harper & Collen 2002). Both deposits include a high proportion of coastal species and could well have come from sheltered embayments. The Pliocene is shorter than the preceding periods so it is not surprising that there are fewer appearance datums in it, although diatoms are more likely to be better preserved in these younger sediments.

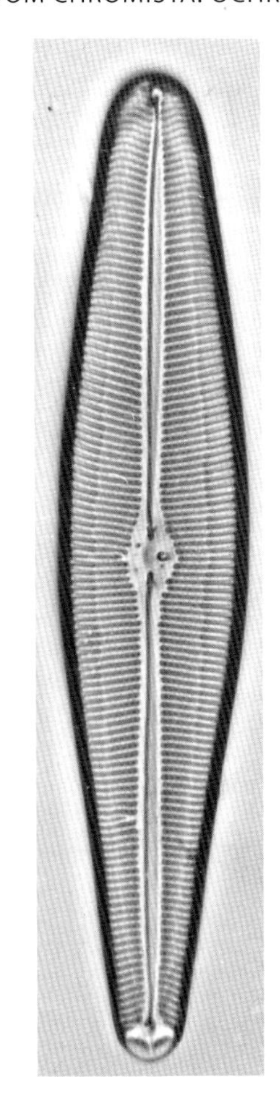
Gomphoneis herculeana (Cymbellales).
Stuart Stidolph

Synthesis
There may well be many other deposits in New Zealand with fossil marine diatoms, such as the diatomaceous ooze associated with the Landren Limestone from which Frederick Reed identified 25 specimens (Cassie Cooper & Harvey 2003). New Zealand has had upwelling currents, siliceous volcanism, and past inundation, all of which promote the formation of marine diatomaceous deposits. Pleistocene sediments have not been included in the fossil checklists as their floras are modern, but useful for establishing past raised sea levels (Shulmeister et al. 1999; Cochran et al. 1999; Ota et al. 1987). Compared with the marine diatoms, very little is known about Tertiary freshwater diatoms in New Zealand, and those identified (see Holden 1983) have all been extant species. There are no records of extinct freshwater genera such as *Mesodictyon*. The Oamaru diatomite, having been deposited close to islands, is likely to include a few freshwater species. Some of the *Aulacodiscus* species in Desikachary and Sreelatha (1989) are quite like the extinct freshwater *Undatodiscus* and *Lobodiscus* species in Khursevich (1994). There is also the possibility of contamination by Recent freshwater diatoms as some samples come from superficial sources and tapwater was used by some of the principal investigators. All the extant freshwater diatom species recorded from Oamaru are likely to be contaminants and are marked with a 'c' in the Checklist. The apparently modern *Stephanodiscus* species would have to be strongly diachronous to be sedimented in the Eocene samples at Oamaru as they have not been recorded elsewhere until the Middle Miocene (Khursevich 1994). Few of New Zealand's fossil marine species are likely to be endemic – most marine taxa are widespread, being dispersed by ocean currents and seabirds. As knowledge of fossil floras depends on the vagaries of preservation it is virtually impossible to establish endemicity.

More new taxa are likely to be found in the Oamaru 'diatomite'. Many of the species in it need careful study with a scanning electron microscope to reveal where they belong in the modern classification of diatoms. Biostratigraphy and paleoecology based on diatoms from ocean cores and other sources in the region will further increase understanding of New Zealand's geological past.

Acknowledgements

Margaret Clayton, Akira Peters, Willem Prud'homme van Reine, Svenja Heesch, and Bruno de Reviers generously responded to enquiries about brown algae and provided assistance to Wendy Nelson and are gratefully acknowledged. Murray Parsons read the text and checklist of macroalgal heterokonts and provided helpful comments.

In compiling the Recent diatom checklists, Pat Kociolek, and Stuart Stidolph supplied relevant references and Ursula Cochran, Mark Harvey, and Sarah Holt gave permission to publish taxonomic names from their unpublished theses. Recent diatom records from the South Island and Stewart Island were kindly supplied by Cathy Kilroy and Liz Bergey,

NIWA, Christchurch. Facilities for preparing the manuscript were made available through courtesy of Landcare Research at Hamilton. The assistance of Carin Burke and Rochelle Holland, Landcare Research, Hamilton, is also gratefully acknowledged, as are the helpful comments of Paul Broady, Bruce Burns, and Elizabeth Flint on aspects of the diatom review. Margaret Harper thanks Leanne Armand, Vivienne Cassie Cooper, Tony Edwards, Juliane Fenner, Elizabeth Fourtanier, Mark Harvey, Tony Jarratt, David McConchie, and Gillian Ruthven for their help with the information on Tertiary marine diatoms. Hoe Chang's review was supported by funding from the New Zealand Foundation for Research, Science and Technology (Contract C01214).

Authors

Dr Paul A. Broady School of Biological Sciences, University of Canterbury, Private Bag 4800, Christchurch, New Zealand [paul.broady@canterbury.ac.nz] Freshwater Eustigmatophyceae, Chrysophyceae, Phaeothamniophyceae, Raphidophyceae, Xanthophyceae

Dr Vivienne Cassie Cooper 1/117 Cambridge Road, Hillcrest, Hamilton. Formerly Landcare Research, Private Bag 3127, Hamilton, New Zealand [viviennecooper@xtra.co.nz] Bacillariophyceae: introductory sections, freshwater diatoms.

Dr F. Hoe Chang National Institute of Water & Atmospheric Research (NIWA), Private Bag 14-901 Kilbirnie, Wellington, New Zealand [h.chang@niwa.co.nz] Marine microalgal Bacillariophyceae, Chrysophyceae, Dictyochophyceae, Raphidophyceae.

Dr Margaret A. Harper School of Geography, Environment and Earth Sciences, Victoria University of Wellington, P.O. Box 600, Wellington, New Zealand [margaret.harper@vuw.ac.nz] Bacillariophyceae: introductory sections, fossil marine diatoms of the New Zealand Tertiary.

Dr Wendy A. Nelson National Institute of Water & Atmospheric Research (NIWA), Private Bag 14-901 Kilbirnie, Wellington, New Zealand [w.nelson@niwa.co.nz] Marine macroalgal Chrysomerophyceae, Phaeophyceae, Xanthophyceae

References

The following list includes works consulted for the preparation of the updated checklist. It should be used in conjunction with the list already published in *Bibliotheca Diatomologica* (Cassie 1984).

ADAMS, N. M. 1972: The marine algae of the Wellington area. *Records of the Dominion Museum 8*: 43–98.

ADAMS, N. M. 1983: Checklist of marine algae possibly naturalised in New Zealand. *New Zealand Journal of Botany 21*: 1–2.

ADAMS, N. M. 1994: *Seaweeds of New Zealand*. Canterbury University Press. 360 p.

ADAMS, N. M.; CONWAY, E.; NORRIS, R. E. 1974: The marine algae of Stewart Island. *Records of the Dominion Museum 8*: 185–245.

ADAMS, N. M.; NELSON, W. A. 1985: The marine algae of the Three Kings Islands. *National Museum of New Zealand Miscellaneous Series 13*: 1–29.

ALVERSON, A. J. 2008: Molecular systematics and the diatom species. *Protist 159*: 339–353.

ANDERSEN, R. A. 2004: Biology and systematics of heterokont and haptophyte algae. *American Journal of Botany 91*: 1508–1522.

ANDERSEN, R. A.; POTTER, D.; BIDIGARE, R. R.; LATASA, M.; ROWAN, K.; O'KELLY, C. J. 1998: Characterization and phylogenetic position of the enigmatic golden alga *Phaeothamnion confervicola*: ultrastructure, pigment composition and partial SSU rDNA sequence. *Journal of Phycology 34*: 286–298.

ASENSI, A.; DELÉPINE, R.; ROUSSEAU, F.; REVIERS, B. de 2003: *Adenocystis rimosa* comb. nov. (Phaeophyceae). *Cryptogamie, Algologique 24*: 83–86.

ATTENBURGH, D. 1995: *The Private Life of Plants*. BBC Books, London. 320 p.

BAILEY, J. C.; BIDIGARE, R. R.; CHRISTENSEN, S. J.; ANDERSEN, R. A. 1998: Phaeothamniophyceae *classis nova*: a new lineage of chromophytes based upon photosynthetic pigments, *rbc*L sequence analysis and ultrastructure. *Protist 149*: 245–263.

BALDAUF, J. G. 1992: Middle Eocene through early Miocene diatom floral turnover. Pp. 310-326 *in*: Prothero, D. R.; Bergen, W. A. (eds), *Eocene-Oligocene Climatic and Biotic Evolution*. Princeton University Press, N.J. 568 p.

BEIER, T.; LANGE-BERTALOT, H. 2007: A synopsis of cosmopolitan, rare and new *Frustulia* species (Bacillariophyceae) from ombrotrophic peat bogs and minerotrophic swamps in New Zealand. *Nova Hedwigia 85*: 73–91.

BEGG, J. G.; MILDENHALL, D. C.; LYON, G. L.; STEPHENSON, W. R.; FUNNELL, R. H.; VAN DISSEN, R. J.; BANNISTER, S.; BROWN, L. J.; PILLANS, B.; HARPER, M. A.; WHITTON, J. 1993: The subsurface Quaternary stratigraphy of Wainuiomata, Lower Hutt City: an integrated study. *New Zealand Journal of Geology and Geophysics 36*: 461–473.

BIGGS, B. J. F.; KILROY, C. 2000: *Stream Periphyton Monitoring Manual*. NIWA, Christchurch. 226 p.

BITTNER, L.; PAYRI, C. E.; COULOUX, A.; CRUAUD, C.; REVIERS, B. de; ROUSSEAU, F. 2008: Molecular phylogeny of the Dictyotales and their position within the Phaeophyceae, based on nuclear, plastid and mitochondrial DNA sequence data. *Molecular Phylogenetics and Evolution 49*: 211–226.

BOJO, O. 2001: Systematic studies of New Zealand nanoflagellates with special reference to members of the Haptophyta. PhD thesis, University of Otago, 240 p.

BOLD, H. C.; WYNNE, M. J. 1985: *Introduction to the Algae: Structure and reproduction*. 2nd edn. Prentice-Hall, Englewood Cliffs, N.J. 720 p.

BOO, S. M.; LEE, K. M.; CHO, G.Y.; NELSON, W. 2011: *Colpomenia claytoniae* sp. nov. (Scytosiphonaceae, Phaeophyceae) based on morphology and mitochondrial cox3 sequences. *Botanica Marina 54*: 159–167.

BOOTH, W. E. 1981: A method for removal of some epiphytic diatoms. *Botanica Marina 24*: 603–609.

BOOTH, W. E. 1984: *Gomphonema novo-zelandicum* sp. nov., an epiphytic diatom with specific host preferences. *New Zealand Journal of Marine and*

Freshwater Research 18: 425–430.
BOOTH, W. E. 1985a: Corrigendum (re. holotype). *New Zealand Journal of Marine and Freshwater Research 19*: 275.
BOOTH, W. E. 1985b: Productivity of diatoms and bacteria epiphytic on marine algae. PhD thesis, University of Auckland, Auckland. 184 p.
BOOTH, W. E. 1986a. Photosynthetic activity of an epiphytic diatom, *Gomphonema novo-zelandicum* and its *Carpophyllum* (Phaeophyta) hosts. *New Zealand Journal of Marine and Freshwater Research 20*: 615–622.
BOOTH, W. E. 1986b: *Navicula climacospheniae* sp. nov., an endophytic diatom inhabitating the stalk of *Climacosphenia monilgera* Ehrenberg. *Nova Hedwigia 42*: 295–300.
BOURRELLY, P. 1981: *Les Algues d'Eau douce. Initiation à la Systématique. Tome II: Les Algues jaunes et brunes. Chrysophycées, Phéophycées, Xanthophycées et Diatomées*. Société Nouvelle des Éditions Boubée, Paris. 517 p.
BOURRELLY, P. 1985: *Les Algues d'Eau douce. Initiation à la Systématique. Tome III: Les Algues bleues et rouges. Les Eugléniens, Peridiniens et Cryptomonadines*. Société Nouvelle des Éditions Boubée, Paris. 606 p.
BOUSTEAD, N.; CHANG, H.; PRIDMORE, R.; TODD, P. 1989: Big Glory Bay algal bloom identified. *Freshwater Catch 39*: 3–4.
BOWERS, H. A.; TOMAS, C.; TENGS, T.; KEPTON, J. W.; LEWITUS, A. J.; OLDACH, D. W. 2006: Raphidophyceae (Chadefaud ex Silva) systematics and rapid identification: sequence analyses and real-time PCR assays. *Journal of Phycology 42*: 1333–1348.
BRADFORD, J. M.; LAPENNAS, P. P.; MURTAGH, R. A.; CHANG, F. H.; WILKINSON, V. 1986: Factors controlling phytoplankton production in greater Cook Strait, New Zealand. *New Zealand Journal of Marine and Freshwater Research 20*: 253–279.
BROOM, J. E.; JONES, W. A.; NELSON, W. A.; FARR, T. J. 1999: A new record of a marine macroalga from New Zealand – *Giraudyopsis stellifera*. *New Zealand Journal of Botany 37*: 751–753.
BROWN, M. T.; NYMAN, M. A.; KEOGH, J. A.; CHIN, N. K. M. 1997: Seasonal growth of the giant kelp *Macrocystis pyrifera* in New Zealand. *Marine Biology 129*: 417–424.
BRUMMITT, R. K.; POWELL, C. E. (Eds) 1992: *Authors of Plant Names*. Royal Botanic Gardens, Kew. 732 p.
BURROWES, R.; ROUSSEAU, F.; MÜLLER, D. G.; REVIERS, B. de 2003: Taxonomic placement of *Microzonia* (Phaeophyceae) in the Syringodermatales based on the *rbc*L and 28S nrDNA sequences. *Cryptogamie, Algologique 24*: 63–73.
CARTER, R. M.; McCAVE, I. N.; RICHTER, C.; CARTER, L. *et al.* 1999a: Site 1119: Drift accretion on Canterbury Slope. *Proceedings of the ODP, Initial Reports 181*: 1-112. [CD-ROM, Ocean Drilling Program, Texas A&M University, College Station, TX 77845-9547, USA; www-odp.tamu.edu/publications/ 181_IR/181ir.htm.]
CARTER, R. M.; McCAVE, I. N.; RICHTER, C.; CARTER, L. *et al.* 1999b: Site 1120: Central Campbell Plateau. *Proceedings of the ODP, Initial Reports 181*: 1-77. [CD-ROM, Ocean Drilling Program, Texas A&M University, College Station, TX 77845-9547, USA; www-odp.tamu.edu/ publications/ 181_IR/181ir.htm.]
CARTER, R. M.; McCAVE, I. N.; RICHTER, C.; CARTER, L. *et al.* 1999c: Site 1121: the Campbell "Drift". *Proceedings of the ODP, Initial Reports 181*: 1-62. [CD-ROM, Ocean Drilling Program, Texas A&M University, College Station, TX 77845-9547, USA; www-odp.tamu.edu/ publications/ 181_IR/181ir.htm).
CARTER, R. M.; McCAVE, I.N.; RICHTER, C.; CARTER, L. *et al.* 1999d: Site 1122: Turbidites with a contourite foundation. *Proceedings of the ODP, Initial Reports 181*: 1-146. [CD-ROM, Ocean Drilling Program, Texas A&M University, College Station, TX 77845-9547, USA; www-odp.tamu.edu/publications /181_IR/181ir.htm).
CARTER, R.M.; McCAVE, I.N.; RICHTER, C.; CARTER, L. *et al.* 1999e: Site 1123: North Chatham Drift – a 20-Ma Record of the Pacific Deep Western Boundary Current. *Proceedings of the ODP, Initial Reports 181*: 1-184. [CD-ROM, Ocean Drilling Program, Texas A&M University, College Station, TX 77845-9547, USA; www-odp.tamu.edu/ publications/181_IR/181ir.htm).
CARTER, R.M.; McCAVE, I.N.; RICHTER, C.; CARTER, L. *et al.* 1999f: Site 1124: Rekohu Drift – from the K/T Boundary to the Deep Western Boundary Current. *Proceedings of the ODP, Initial Reports 181*: 1-137. [CD-ROM, Ocean Drilling Program, Texas A&M University, College Station, TX 77845-9547, USA; www-odp.tamu.edu/publications/ 181_IR/181ir.htm).
CARTER, R.M.; McCAVE, I.N.; RICHTER, C.; CARTER, L. *et al.* 1999g: Site 1125, productivity under the subtropical convergence on North Chatham slope. *Proceedings of the ODP, Initial Reports 181*: 1-137. [CD-ROM, Ocean Drilling Program, Texas A&M University, College Station, TX 77845-9547, USA; www-odp.tamu.edu/publications/ 181_IR/181ir.htm).
CASSIE, V. 1960: Seasonal changes in diatoms and dinoflagellates off the east coast of New Zealand during 1957 and 1958. *New Zealand Journal of Science 3*: 137–172.
CASSIE, V. 1971: Contributions of Victor Lindauer (1888–1964) to New Zealand Phycology. *Journal of the Royal Society of New Zealand 1*: 89–98.
CASSIE, V. 1978: Seasonal changes in phytoplankton densities in four North island lakes, 1973–74. *New Zealand Journal of Marine and Freshwater Research 12*: 153–166.
CASSIE, V. 1979: A comparison between fossil and recent marine diatom floras in New Zealand. Pp. 397-410 *in*: *Proceedings of International Symposium on Marine Biogeography and Evolution in the Southern Hemisphere, Auckland, New Zealand 17–20 July 1978. Department of Scientific and Industrial Research, Information Series 137*: 1-743. [2 vols.]
CASSIE, V. 1980: Additions to the check lists of marine and freshwater diatoms of New Zealand. *Journal of the Royal Society of New Zealand 10*: 215–219.
CASSIE, V. 1981: The fossil and living freshwater diatom flora of New Zealand. Pp. 321-338 *in*: Ross, R. (ed.), *Proceedings of the Sixth Symposium on Recent and Fossil Diatoms, Budapest, 1-5 September 1980*. Koeltz, Koenigstein. 541 p.
CASSIE, V. 1983a: A guide to algae in oxidation ponds in the Auckland district. *Tane 29:* 119-128.
CASSIE, V. 1983b: A history of freshwater diatom research in New Zealand. *Archives of Natural History 11*: 223-231.
CASSIE, V. 1984: Checklist of the freshwater diatoms of New Zealand. *Bibliotheca Diatomologica 4*: 1-129.
CASSIE, V. 1989a: A taxonomic guide to diatoms in thermal habitats in New Zealand. Part 1. Centrales (Diatomophyceae/ Bacillariophyceae. Part 2. Pennales (Diatomophyceae/ Bacillariophyceae). *Bibliotheca Diatomologica 17*: 51-266.
CASSIE, V. 1989b: Microalgae of Lake Pupuke, Auckland, New Zealand. *New Zealand Natural Sciences 16*: 39-50.
CASSIE COOPER, V. 1995: Victor Wilhelm Lindauer (1888–1964): His life and works. *Tuhinga, Records of the Museum of New Zealand Te Papa Tongarewa 1*: 1–14.
CASSIE COOPER, V. 2001: Recent name changes in eukaryotic freshwater algae of New Zealand. *New Zealand Journal of Botany 39*: 601–616.
CASSIE COOPER, V.; HARVEY, M. 2003. Frederick Reed: Pioneer New Zealand diatomist 1909-1995: an account of his life and collections. *Journal of the Royal Society of New Zealand 33*: 571-581.
CAVALIER-SMITH, T. 1986: The kingdom Chromista: Origin and systematics. *Progress in Phycological Research 4*: 309–347.
CAVALIER-SMITH, T. 1995: Membrane heredity, symbiogenesis, and the multiple origins of algae. Pp. 75–114 *in*: Arai, R.; Kato, M.; Doi, Y. (eds), *Biodiversity and Evolution.* National Science Museum Foundation, Tokyo.
CAVALIER-SMITH, T. 1998: A revised six-kingdom system of life. *Biological Reviews 73*: 203–266.
CAVALIER-SMITH, T. 2000: Flagellate megaevolution. The basis for eukaryote diversification. Pp. 361–390 in: Leadbeater, B. S. C.; Green, J.C. (eds), *The Flagellates – Unity, Diversity and Evolution. [The Systematics Association Special Volume, Series 68.]* CRC Press, Boca Raton. 416 p.
CAVALIER-SMITH, T. 2004: Chromalveolate diversity and cell megaevolution: interplay of membranes, genomes and cytoskeleton. Pp. 75–108 *in*: Hirt, R. P.; Horner, D. S. (eds), *Organelles, Genomes and Eukaryote Phylogeny – An Evolutionary Synthesis in the Age of Genomics. [The Systematics Association Special Volume Series 68.]* CRC Press, Boca Raton. [viii], 384 p.
CAVALIER-SMITH, T.; CHAO, E. E.-Y. 2006: Phylogeny and megasystematics of phagotrophic heterokonts (kingdom Chromista). *Journal of Molecular Evolution 62*: 388–420.
CAVALIER-SMITH, T.; CHAO, E. E.; ALLSOPP, M. T. E. P. 1995: Ribosomal RNA evidence for chloroplast loss within heterokonta: pedinellid relationships and a revised classification of ochristan algae. *Archiv für Protistenkunde 145*: 209–220.
CHANG, F. H. 1983: Winter phytoplankton and microzooplankton populations off the coast of Westland, New Zealand, 1979. *New Zealand Journal of Marine and Freshwater Research 17*: 279–304.
CHANG, F. H. 1988: Distribution, abundance, and size composition of phytoplankton off Westland, New Zealand, February 1982. *New Zealand Journal of Marine and Freshwater Research 22*: 345–367.
CHANG, F. H. 1989: *Heterosigma* bloom in Stewart Island, New Zealand. *Red Tide Newsletter 2(2)*: 2.
CHANG, F. H. 1996: Distribution and abundance of *Dinophysis acuminata* (Dinophyceae) and *Pseudonitzschia australis* (Bacillariophyceae) in Kenepuru and Pelorus Sounds, New Zealand. Pp. 93–96 *in*: Yasumoto, T.; Oshima, Y.; Fukuyo, Y. (eds), *Harmful and Toxic Algal Blooms.* Intergovernmental Oceanographic Commission of UNESCO, Paris.
CHANG, F. H. 2000: Seasonal studies of

phytoplankton in the Hauraki Gulf. *Water and Atmosphere 8(2)*: 22–24.

CHANG, F. H.; ANDERSON, C.; BOUSTEAD, N. C. 1990: First record of a *Heterosigma* (Raphidophyceae) bloom with associated mortality of cage-reared salmon in Big Glory Bay, New Zealand. *New Zealand Journal of Marine and Freshwater Research 24*: 461–469.

CHANG, F. H.; GALL, M. 1998: Phytoplankton assemblages and photosynthetic pigments during winter and spring in the Subtropical Convergence region near New Zealand. *New Zealand Journal of Marine and Freshwater Research 32*: 515–530.

CHANG, F.H.; GARTHWAITE, I.; ANDERSON, D.M.; TOWERS, N.; STEWART, R.; MacKENZIE, L. 1999: Immunofluorescent detection of a PSP-producing dinoflagellate, *Alexandrium minutum*, from Bay of Plenty, New Zealand. *New Zealand Journal of Marine and Freshwater Research 33*: 533–543.

CHANG, F. H.; MacKENZIE, L.; TILL, D.; HANNAH, D.; RHODES, L. 1995: The first toxic shellfish outbreaks and the associated phytoplankton blooms in early 1993 in New Zealand. Pp. 145–150 *in*: Lassus, P.; Arzul, G.; Erard-Le Denn, E.; Gentien, P.; Marcaillou-Le-Baut, C. (eds), *Harmful Marine Algal Blooms*. Intercept Ltd, Andover, U.K. xxiv + 878 p.

CHANG, F. H.; McVEAGH, M.; GALL, M.; SMITH, P. In press: *Chattonella globosa* is a member of *Dictyochophyceae*: reassignment to *Vicicitus* gen. nov., based on molecular phylogeny, pigment composition, morphology and life history. *Phycologia.*

CHANG, F. H.; PAGE, M. 1995: Influence of light and three nitrogen sources on growth of *Heterosigma carterae* (Raphidophyceae). *New Zealand Journal of Marine and Freshwater Research 29*: 299–304.

CHANG, F. H.; PRIDMORE, R.; BOUSTEAD, N. 1993: Occurrence and distribution of *Heterosigma* cf. *akashiwo* (Raphidophyceae) in a 1989 bloom in Big Glory Bay, New Zealand. Pp. 675–680 *in*: Smayda, T. J.; Shimizu, Y. (eds), *Toxic Phytoplankton Blooms in the Sea*. Elsevier Science Publishers, Amsterdam. 976 p.

CHANG, F. H.; VINCENT, W. F.; WOODS, P. H. 1992: Nitrogen utilisation by size-fractionated phytoplankton assemblages associated with an upwelling event off Westland, New Zealand. *New Zealand Journal of Marine and Freshwater Research 26*: 287–301.

CHANG, F. H.; ZELDIS, J.; GALL, M.; HALL, J. 2003: Seasonal and spatial variation of phytoplankton assemblages, biomass and cell size from spring to summer across the north-eastern New Zealand continental shelf. *Journal of Plankton Research 35*: 737–758.

CHAPMAN, V. J. 1956: The marine algae of New Zealand. Part I: Myxophyceae and Chlorophyceae. *Journal of the Linnean Society of London 55*: 333–501, pls 24–50.

CHAPMAN, V. J.; THOMPSON, R. H.; SEGAR, E. C. M. 1957: Checklist of the freshwater algae of New Zealand. *Transactions of the Royal Society of New Zealand 84*: 695-747.

CHO, G.Y.; LEE, S. H.; BOO, S. M. 2004: A new brown algal order, Ishigeales (Phaeophyceae), established on the basis of plastid protein-coding *rbc*L, *psa*A and *psb*A region comparisons. *Journal of Phycology 40*: 921–936.

CHO, G.Y.; BOO, S. M.; NELSON, W. A.; CLAYTON, M. N. 2005. Genealogical partitioning and phylogeography of *Colpomenia peregrina* (Scytosiphonaceae, Phaeophyceae) based on plastid *rbcL* and nuclear ribosomal DNA internal transcribed spacer sequences. *Phycologia 44*: 103–111.

CHO, G.Y.; ROUSSEAU, F.; REVIERS, B. de; BOO, S. M. 2006: Phylogenetic relationships within the Fucales (Phaeophyceae) assessed by the photosystem I coding *psa*A sequences. *Phycologia 45*: 512–519.

CIESIELSKI, P. F. 1986: Middle Miocene to Quaternary diatom biostratigraphy of Deep Sea Drilling Project site 594, Chatham Rise, Southwest Pacific. *Initial Reports DSDP 90*: 863-885.

CLAYTON, M. N. 1984: Evolution of the Phaeophyta with particular reference to the Fucales. *Progress in Phycological Research 3*: 11–46.

CLAYTON, M. N. 1990a: Phylum Phaeophyta. Pp. 698–714 *in*: Margulis, L.; Corliss, J. O.; Melkonian, M.; Chapman, D. J. (eds), *Handbook of Protoctista*. Jones and Bartlett, Boston. 914 p.

CLAYTON, M. N. 1990b: Phaeophyta. Pp. 149–182 *in*: Clayton, M. N.; King, R. J. (eds), *Biology of Marine Plants*. Longman Cheshire. 501 p.

CLAYTON, M. N. 1994: Circumscription and phylogenetic relationships of the southern hemisphere family Seirococcaceae (Phaeophyceae). *Botanica Marina 37*: 213–220.

CLEVE, P. T. 1894: Synopsis of the naviculoid diatoms Part 1. *Kongliga Svenska Vetenskaps-Akademiens Handlingar 26*: 1-94.

CLEVE, P. T. 1895: Synopsis of the naviculoid diatoms Part 2. *Kongliga Svenska Vetenskaps-Akademiens Handlingar 27*: 1-219

COCHRAN, U. 2002: Detection of large Holocene earthquakes in the sedimentary record of Wellington, New Zealand, using diatom analysis. PhD thesis, Victoria University of Wellington. 303 p.

COCHRAN, U.; BERRYMAN, K.; MILDENHALL, D.; HAYWARD, B.; SOUTHALL, K.; HOLLIS, C. 2005: Towards a record of Holocene tsunami and storms in northern Hawkes Bay, New Zealand. *New Zealand Journal of Geology and Geophysics 48*: 507–515.

COCHRAN, U.; GOFF, J.; HANNAH, M.; HULL, A. 1999: Relative stability on a tectonically active coast: paleoenvironment during the last 7000 years at lake Kohangapiripiri, Wellington, New Zealand. *Quaternary International 56*: 53-63.

COLE, R. G. 1994: Abundance, size structure, and diver-oriented behaviour of three large benthic carnivorous fishes in a marine reserve in northeastern New Zealand. *Biological Conservation 70*: 93–99.

D'ARCHINO, R.; NELSON, W. A. 2006. Marine brown algae introduced to New Zealand waters: first record of *Asperococcus ensiformis* Chiaje) M.J.Wynne (Phaeophyta, Ectocarpales, Chordariaceae). *New Zealand Journal of marine and freshwater Research 40*: 599–604.

DAWSON, E. W. 1992: The marine fauna of New Zealand: Index to the fauna: 1. Protozoa. *New Zealand Oceanographic Institute Memoir 99*: 1–368.

DEMURA, M.; NOEL, M.-H; KASAO, F.; WATANABE, M. M.; KAWAUCHI, M. 2009: Taxonomic revision of *Chattonella antiqua, C. marina* and *C. ovata* (Raphidophyceae) based on their morphological characteristics and genetic diversity. *Phycologia 48*: 518–535.

DESIKACHARY, T.V.; SREELATHA, P. M. 1989: Oamaru diatoms. *Bibliotheca Diatomologica 19*: 1-330, 145 pls.

DRAISMA, S. G. A.; PRUD'HOMME van REINE, W. F.; STAM, W. F.; OLSEN, J. L. 2001: A reassessment of phylogenetic relationships within the Phaeophyceae based on RUBISCO large subunit and ribosomal DNA sequences. *Journal of Phycology 37*: 586–603.

DRUEHL, L. D.; MAYES, C.; TAN, I. H.; SAUNDERS, G. W. 1997: Molecular and morphological phylogenies of kelp and associated brown algae. *Plant Systematics and Evolution (Suppl.) 11*: 221–235.

EDVARDSEN, B.; EIKREM, W.; SHALCHIAN-TABRIZI, K.; RIISBERG, I.; JOHNSEN, G.; NAUSTVOLL, L.; THRONDSEN, J. 2007: *Verrucophora farcimen* gen. et sp. nov. (Dictyochophyceae, Heterokonta) – a bloom-forming ichthyotoxic flagellate from the Skagerrak, Norway. *Journal of Phycology 43*: 1054–1070.

EDWARDS, A. R. 1991: The Oamaru diatomite. *New Zealand Geological Survey Bulletin 64*: 1-260 + 2 maps.

ESPER, E. J. C. 1802: Icones fucorum cum characteribus systematic, synonimis auctorum et descriptionibus novarum specierum. *Abbildungen der Tange mit beygefugten systematischen Kennzeichen, Anfuhrungen der Schriftsteller und Beschreibungen der neuen Gattungen. Nurnburg (Raspe) 5*: 1–53, pls 112–135.

ETTL, H. 1978: Xanthophyceae 1. *In*: Ettl, H.; Gerloff, J.; Heynig, H. (eds), *Süsswasserflora von Mitteleuropa*, 3. Fischer, Stuttgart. 530 p.

ETTL, H.; GÄRTNER, G. 1995: *Syllabus der Boden-, Luft- und Flechtenalgen.* Gustav Fischer Verlag, Stuttgart. viii + 721 p.

FALKOWSKI, P. G.; BARBER, R. T.; SMETACEK, V. 1998: Biogeochemical controls and feedbacks on ocean primary production. *Science 281*: 200–206.

FENNER, J. 1986: Information from diatom analysis concerning the Eocene–Oligocene boundary. Pp. 283-288 *in*: Pomerol, C. H.; Premoli-Silva, I. (eds), *Terminal Eocene Events*. Elsevier, Amsterdam. 414 p.

FIELD, C. B.; BEHRENFELD, M. J.; RANDERSON, J. T.; FALKOWSKI, P. 1998: Primary production of the biosphere: integrating terrestrial and oceanic components. *Science 281*: 237–240.

FLINT, E. A. 1966: Additions to the freshwater algae of New Zealand. *Transactions of the Royal Society of New Zealand, Botany 17*: 127-134.

FOGED, N. 1979: Diatoms in New Zealand, the North Island. *Bibliotheca Phycologica 47*: 1-225.

FOURTANIER, E.; KOCIOLEK, J. P. 1999: Catalogue of the diatom genera. *Diatom Research 14*: 1-190.

FOURTANIER, E.; KOCIOLEK, J. P. 2003: Addendum to "Catalogue of the diatom genera". *Diatom Research 18*: 245–258.

FOURTANIER, E.; KOCIOLEK, J. P. 2009a: Catalogue of diatom names part I: introduction and bibliography. *Occasional Papers of the Californian Academy of Sciences 156(1)*: x, 1–168.

FOURTANIER, E.; KOCIOLEK, J. P. 2009b: Catalogue of diatom names: Part II: *Abas* through *Bruniopsis. Occasional Papers of the Californian Academy of Sciences 156(2)*: xii, 1–231.

FRANCOER, S. N.; BIGGS, B. J. F.; LOWE, R. L. 1998: Microform bed clusters as refugia for periphyton in a flood-prone headwater stream. *New Zealand Journal of Marine and Freshwater Research 32*: 363-374.

FRITSCH, F.E. 1965: *The Structure and Reproduction of the Algae. Volume I.* University Press, Cambridge. 791 p.

GREVILLE, R. K. 1866a: Descriptions of new and rare diatoms from the tropics and Southern Hemisphere. *Transactions of the Royal Society of Edinburgh 8*: 436-441.

GREVILLE, R. K. 1866b: Descriptions of new and rare diatoms. *Transactions of the Microscopical Society of London, n.s., 14*: 121-134.

GRIFFITHS, R. C. 1983: Tertiary geology of the Wairaki Station area, mid-Waiau Valley, Southland, New Zealand. Dip. Sci. Geology thesis, University of Otago. 100 p.

GROVE, E.; STURT, G. 1886: On a fossil marine diatomaceous deposit from Oamaru, Otago, New Zealand. Part 1. *Journal of the Quekett microscopical Club, ser. 2, 2*: 321-330.

GROVE, E.; STURT, G 1887a: On a fossil marine diatomaceous deposit from Oamaru, Otago, New Zealand. Part 2. *Journal of the Quekett microscopical Club, ser. 2, 3*: 7-12.

GROVE, E.; STURT, G. 1887b: On a fossil marine diatomaceous deposit from Oamaru, Otago, New Zealand. Part 3. *Journal of the Quekett Microscopical Club, ser. 2, 3*: 63-78.

GROVE, E.; STURT, G 1887c: On a fossil marine diatomaceous deposit from Oamaru, Otago, New Zealand. Part 4. *Journal of the Quekett Microscopical Club, ser. 2, 3*: 131-148.

GUIRY, M. D.; GUIRY, G. M. 2010: Algaebase. World-wide electronic publications, National University of Ireland, Galway. [http://www.algaebase.org.]

HARA, Y.; DOI, K.; CHIHARA, M. 1994: Four new species of *Chattonella* (Raphidophyceae, Chromophyta) from Japan. *Japanese Journal of Phycology (Sorui) 42*: 407–420.)

HARPER, M. A. 1994: Did Europeans introduce *Asterionella formosa* Hassall into New Zealand? *Proceedings of the 11th International Diatom Symposium. Memoirs of the Californian Academy of Sciences 17*: 479-484.

HARPER, M. A. 1997: Late Pleistocene diatoms from two holes drilled at Lake Poukawa, Hawkes Bay, North Island, New Zealand. *IPCEE Newsletter 7*: 123-132.

HARPER, M. A.; COLLEN, J. C. 2002: Glaciations, interglaciations and reworked microfossils in Poukawa Basin, New Zealand. *Global and Planetary Change 33*: 243-256.

HARPER, M. A; HARPER, J. F. 2010: Otari and Taputeranga bioblitzes: diatoms – microscopic algae. *Wellington Botanical Society Bulletin 52*: 53–63.

HARPER, M. A.; HOWORTH, R. H.; McLEOD, M. 1986: Late Holocene diatoms in Lake Poukawa: effects of airfall tephra and changes in depth. *New Zealand Journal of Marine and Freshwater Research 20*: 107-118.

HARPER, M. A.; LILLIS, D. A.; McLEA, W. L.; TURNER, G. M. 1993: Diatoms, palynomorphs and the magnetic record of a core from Lake Waikare, North Island, New Zealand. *IPCEE Newsletter 7*: 130–144.

HARPER, M. A.; MANN, D. G.; PATTERSON, J. E. 2009a:Two unusual diatoms from New Zealand: *Tabularia variostriata* a new species and *Eunophora berggrenii. Diatom Research 24*: 291–306.

HARPER, M. A.; PATTERSON, J. E.; HARPER, J. F. 2009b: New diatom taxa from the world's first marine BioBlitz held in New Zealand: *Skeletomastus* a new genus, *Skeletomastus coelatus* nov. comb. and *Pleurosigma inscriptura* a new species, from Cook Strait, New Zealand. *Acta Botanica Croatica 68*: 339–349.

HARTLEY, B.; BARBER, H. G.; CARTER, J. R. 1996: *An Atlas of British Diatoms.* Sims, P. (ed), Biopress, Bristol. 601 p.

HARVEY, M. C. 1996: A paleolimnological study of Lake Ellesmere (Te Waihora), South Island, New Zealand. MSc thesis, University of Canterbury. 277 p.

HARVEY, W. H. 1855: Algae. Pp. 211–266, pls 107–121 *in*: Hooker, J. D., *The Botany of the Antarctic Voyage of H.M. Discovery ships Erebus and Terror in the years 1839–1843. II. Flora Novae-Zelandiae. Part II.* Reeve, London. 574 p.

HARVEY, W. H.; HOOKER, J. D. 1845: Algae. *In*: Hooker, J.D., *The Botany of the Antarctic Voyage of H.M. Discovery ships Erebus and Terror in the years 1839–843. I. Flora Antarctica. Part I. The Botany of Lord Auckland's Group and Campbell's Island.* Reeve, London. 193 p.

HARWOOD, D. M. 2010: Diatomite. Pp. 570–574 *in*: Smol, J. P., Stoemer, E. F. (eds), *The diatoms: applications for the environmental and earth sciences.* 2nd edn. Cambridge University Press, Cambridge U.K. 667 p.

HASLE, G. R.; SYVERTSEN, E. E 1996: Marine diatoms. Pp. 5-385 *in*: Tomas, C. R. (ed.), *Identifying Marine Diatoms and Dinoflagellates.* Academic Press, Harcourt Brace & Co., San Diego. 598 p.

HAY, C. H. 1990a: The distribution of *Macrocystis* (Phaeophyta, Laminariales) as a biological indicator of cool sea surface temperature, with special reference to New Zealand waters. *Journal of the Royal Society of New Zealand 20*: 313–336.

HAY, C. H. 1990b: The dispersal of sporophytes of *Undaria pinnatifida* by coastal shipping in New Zealand and the implication for further dispersal on *Undaria* in France. *British Phycological Journal 25*: 301–313.

HAY, C. H.; ADAMS, N. M.; PARSONS, M. J. 1985: The marine algae of the subantarctic islands of New Zealand. *National Museum of New Zealand, Miscellaneous Series 11*: 1–70.

HAY, C. H.; LUCKENS, P. A. 1987: The Asian kelp *Undaria pinnatifida* (Phaeophyta: Laminariales) found in a New Zealand harbour. *New Zealand Journal of Botany 25*: 329–332.

HAYWARD, B. W 2009. Geology of Pukaki Lagoon. Manukau City Council, Unpublished Report BWH 119/09. 5 p.

HIBBERT, D. J.; LEEDALE, G. F. 1970: Eustigmatophyceae – a new algal class with unique organization of the motile cell. *Nature 225*: 758–780

HILL, N. L. 2003: Paleoseismicity of the eastern section of the Awatere Fault, North-east South Island, New Zealand. MSc thesis, Victoria University of Wellington.

HOEK, C. van den; MANN, D. G.; JAHNS, H. M. 1995: *Algae. An introduction to phycology.* Cambridge University Press, Cambridge. 623 p.

HOLDEN, A. M. 1983: Studies in New Zealand Oligocene and Miocene Plant Macrofossils. PhD thesis, Victoria University of Wellington. 369 p.

HOLLAND, P. T.; SELWOOD, A. I.; MOUNTFORT, D. O.; WILKINS, A. L.; McNABB, P.; RHODES, L. L.; DOUCETTE, G. J.; MIKULSKI, C. M.; KING, K. L. 2005: Isodomoic acid acid C, an unusual Amnesic Shellfish Poisoning toxin from *Pseudo-nitzschia australis. Chemical Research in Toxicology 18*: 814–816.

HOLMGREN, P. K.; HOLMGREN, N. H.; BARNETT, L. C. (Ed.) 1990: Index Herbariorum part I: The Herbaria of the world. *Regnum vegetabile 120*: 1–693.

HOLT, S. 1995: A diatom based paleoenvironmental reconstruction of Late Quaternary sediments from Gebbies Valley, Banks Peninsula. BSc Hons project, School of Earth Sciences, Victoria University of Wellington.

HOSOI-TANABE, S.; HONDA, D.; FUKAYA, S.; OTAKE, I.; INAGAKI, Y.; SAKO, Y. 2007: Proposal of *Pseudochattonella verruculosa* gen. nov., comb. nov. (Dictyochophyceae) for a former raphidophycean alga *Chattonella verruculosa*, based on 18S rDNA phylogeny and ultrastructural characteristics. *Phycological Research 55*: 185–192.

HURD, C. L.; NELSON, W. A.; FALSHAW, R.; NEILL, K. 2004. History, current status and future of marine macroalgae research in New Zealand: taxonomy, ecology, physiology and human uses. *Phycological Research 52*: 80–106.

ISHIDA, K.; CAVALIER-SMITH, T.; GREEN, B. R. 2000: Endomembrane structure and the chloroplast protein targeting pathway in *Heterosigma akashiwo* (Raphidophyceae, Chromista). *Journal of Phycology 36*: 1135–1144.

JEFFREY, S. W.; VESK, M. 1997: Introduction to marine phytoplankton and their pigment signatures. *In:* Jeffrey, S. W.; Mantoura, R. F. C.; Wright, S. W. (eds), *Phytoplankton Pigments in Oceanography: Guidelines to modern methods. [Monographs in Oceanographic Methodology 10.]* UNESCO, Paris. 661 p.

JONES, G. P.; ANDREW, N. L. 1993: Temperate reefs and the scope of seascape ecology. Pp. 63–76 *in*: Battershill, C.N.; Schiel, D.R.; Jones, G.P.; Creese, R.G.; MacDiarmid, A.B. (eds), *Proceedings of the Second International Temperate Reef Symposium.* NIWA Marine, Wellington. 251 p.

JORDAN, R. W.; STICKLEY, C. E. 2010: Diatoms as indicators of paleoceanographic events. Pp. 424–453 *in*: Smol, J. P.; Stoermer, E. F. (eds), *The Diatoms: Applications for the environmental and earth sciences.* 2nd edn. Cambridge University Press, Cambridge U.K. 667 p.

KAIN, J. M. 1982: Morphology and growth of the giant kelp *Macrocystis pyrifera* in New Zealand and California. *Marine Biology 67*: 143–157.

KAWAI, H.; HANYUDA, T.; DRAISMA, S. G. A.; MULLER, D. G. 2007: Molecular phylogeny of *Discosporangium mesarthrocarpum* (Phaeophyceae) with a reinstatement of the order Discosporangiales. *Journal of Phycology 43*: 186–194.

KENNELLY, S. J. 1987: Physical disturbances in an Australian kelp community. II. Effects on understorey species due to differences in kelp cover. *Marine Ecology Progress Series 40*: 155–165.

KHURSEVICH, G. 1994: Evolution of freshwater centric diatoms within the Euroasian continent. Pp. 507-520 *in*: Marino, D.; Montresor, M. (eds), *Proceedings of the 13th International Diatom Symposium.* Biopress, Bristol. 576 p.

KILROY, C. 2008: Diatom communities in New Zealand subalpine mire pools: distribution, ecology and taxonomy of endemic and cosmopolitan taxa. PhD thesis, University of Canterbury. 227 p.

KILROY, C.; BERGEY, L. 1999: Diatoms: biological gems. *Water & Atmosphere 7(3)*: 13-16.

KILROY, C.; BIGGS, B. J. F.; VYVERMAN, W.; BROADY, P. 2006: Benthic diatom communities in subalpine pools in New Zealand: relationships to environmental variables. *Hydrobiologia 561*: 95–110.

KILROY, C.; BIGGS, B. J. F.; VYVERMAN, W. 2007: Rules for macro-organisms applied to micro-organisms: patterns of endemism in benthic freshwater diatoms. *Oikos 116*: 550–564.

KILROY, C.; SABBE, K.; BERGEY, E. A.; VYVERMAN, W.; LOWE, R. 2003: New species of *Fragilariforma* (Bacillariophyceae) from New Zealand and Australia. *New Zealand Journal of Botany 41*: 535–554.

KILROY, C.; UNWIN, M. 2011: The arrival and spread of the bloom-forming, freshwater diatom, *Didymosphenia geminata*, in New Zealand. *Aquatic Invasions 6*: 342–362.

KOCIOLEK, J. P.; ESCOBAR, L.; RICHARDSON, S. 1996: Taxonomy and ultrastructure of *Stoermeria*, a new genus of diatoms (Bacillariophyta). *Phycologia 35*: 70-78.

KOGAME, K.; HORIGUCHI, T.; MASUDA, M.1999: Phylogeny of the order Scytosiphonales (Phaeophyceae) based on DNA sequences of *rbc*L, partial *rbc*S, and partial LSU nrDNA. *Phycologia 38*: 496–502.

KRAMMER, K. 2000: *Diatoms of Europe: Diatoms of the European Inland Waters and Comparable Habitats. Volume 1: The genus* Pinnularia. Koeltz Scientific Books, Koenigstein. 703 p.

KRAMMER, K.; LANGE-BERTALOT, H. 1986: *Die Süßwasserflora von Mitteleuropa Band 2/1. Bacillariophyceae 1 Teil: Naviculaceae*. Gustav Fischer Verlag, Stuttgart, Jena. 876 p.

KRAMMER, K.; LANGE-BERTALOT, H. 1988: *Die Süßwasserflora von Mitteleuropa Band 2/2. Bacillariophyceae 2. Teil: Epithemiaceae, Bacillariaceae, Surirellaceae*. Gustav Fischer Verlag, Stuttgart, Jena. 536 p.

KRAMMER, K.; LANGE-BERTALOT, H. 1991a: *Die Süßwasserflora von Mitteleuropa Band 2/3. Bacillariophyceae 3. Teil: Centrales, Fragilariaceae, Eunotiaceae*. Gustav Fischer Verlag, Stuttgart, Jena. 576 p.

KRAMMER, K.; LANGE-BERTALOT, H. 1991b: *Die Süßwasserflora von Mitteleuropa Gesamtliteraturverzeichnis Band 2/4. Bacillariophyceae 4. Teil: Achnanthaceae, Kritische Ergaenzungen zu Navicula und Gomphonema*. Gustav Fischer Verlag, Stuttgart, Jena. 437 p.

KRAMMER, K.; LANGE-BERTALOT, H. 1997: *Die Süßwasserflora von Mitteleuropa. Band 2/1. Bacillariophyceae 1. Teil: Naviculaceae*. 2nd edn. Gustav Fischer Verlag, Stuttgart, Jena. 876 p.

KRISTIANSEN, J. 2002: Phylum Chrysophyta (Golden Algae). Pp. 214-244 *in*: John, D.M.; Whitton, B. A.; Brook, A. J. (eds), *The Freshwater Algal Flora of the British Isles*. University Press, Cambridge. 714 p.

KRISTIANSEN, J.; PREISIG, H. R. (Eds) 2001: Encyclopedia of chrysophyte genera. *Bibliotheca Phycologica 110*: 1–260.

KUSBER, W. H.; JAHN, R. 2003: Annotated list of diatom names by Horst Lange-Bertalot and co-workers. [http://www.algaterra.org/ Names Version 3.0.pdf]

KYLIN, H. 1940. Die Phaeophyceenordnung Chordariales. *Lunds Universitets Årsskrift 36(9)*: 1–67, pls 1–8.

LAING, R. M. 1895: The algae of New Zealand: their characteristics and distribution. *Transactions and Proceedings of the New Zealand Institute 27*: 297–318.

LAING, R. M. 1900: Revised list of New Zealand seaweeds, Part I. *Transactions and Proceedings of the New Zealand Institute 32*: 57–70.

LAING, R. M. 1902: Revised list of New Zealand seaweeds, Part II. *Transactions and Proceedings of the New Zealand Institute 34*: 384–408.

LAING, R. M. 1926: A reference list of New Zealand marine algae. *Transactions and Proceedings of the New Zealand Institute 57*: 126–185.

LAING, R. M. 1927: The external distribution of the New Zealand marine algae and notes on some algological problems. *Transactions and Proceedings of the New Zealand Institute 58*: 189–201.

LAING, R. M. 1930: A reference list of New Zealand marine algae. Supplement 1. *Transactions and Proceedings of the New Zealand Institute 60*: 575–583.

LANE, C. E.; MAYES, C.; DRUEHL, L. D.; SAUNDERS, G. W. 2006: A multi-gene molecular investigation of the kelp (Laminariales, Phaeophyceae) supports substantial taxonomic re-organization. *Journal of Phycology 42*: 493–512.

LEVIS, L. A. 1975: Marine littoral diatoms in the Leigh area. MSc thesis, University of Auckland. 77 p.

LIM, P.-E.; SAKAGUCHI, M.; HANYUDA, T.; KOGAME, K.; PHANG, S.-M.; KAWAI, H. 2007: Molecular phylogeny of crustose brown algae (Ralfsiales, Phaeophyceae) inferred from *rbc*L sequences resulting in the proposal for Neoralfsiaceae fam. nov. *Phycologia 46*: 456–466.

LINDAUER, V. W. 1945: Note on the brown alga *Ecklonia brevipes* J.Ag. *Transactions and Proceedings of the New Zealand Institute 75*: 394–397.

LINDAUER, V. W. 1947: An annotated list of the brown seaweeds, Phaeophyceae. *Transactions and Proceedings of the New Zealand Institute 76*: 542–566.

LINDAUER, V. W. 1949a: Additions to the marine algae of New Zealand. *Transactions and Proceedings of the New Zealand Institute 77*: 390–393.

LINDAUER, V. W. 1949b: Notes on the marine algae of New Zealand. I. *Pacific Science 3*: 340–352.

LINDAUER, V. W. 1957: A descriptive review of the Phaeophyceae of New Zealand. *Transactions and Proceedings of the New Zealand Institute 85*: 61–74.

LINDAUER, V. W. 1960: New species of Phaeophyceae from New Zealand. *Revue algologique 3*: 161–172.

LINDAUER, V. W.; CHAPMAN, V. J.; AIKEN, M. 1961: The marine algae of New Zealand. II. Phaeophyceae. *Nova Hedwigia 3*: 129–350.

LOWE, R.; MORALES, E.; KILROY, C.; 2006: *Frankophila biggsii* (Bacillariophyceae) a new diatom species from New Zealand. *New Zealand Journal of Botany 44*: 41–46

MacDIARMID, A. B.; BREEN, P. A. 1993: Spiny lobster population change in a marine reserve. Pp. 47–56 *in*: Battershill, C. N.; Schiel, D. R.; Jones, G. P.; Creese, R. G.; MacDiarmid, A. B. (eds), *Proceedings of the Second International Temperate Reef Symposium*. NIWA Marine, Wellington. 251 p.

MANN, D. G. 1999: The species concept in diatoms. *Phycologia 38*: 437-495.

MANN, D. G. DROOP, S. J. M. 1996: Biodiversity, Biogeography and conservation of diatoms. Pp. 19–32 *in*: Kristiansen, J. (ed) *Biogeography of Freshwater algae: Proceedings of the workshop on Biogeography of Freshwater algae*. [*Developments in Hydrobiology 118*.] Kluwer Academic Publishers, Dordrecht. 161 p.

MARGULIS, L.; CORLISS, J. O.; MELKONIAN, M.; CHAPMAN, D. J. 1990: *Handbook of Protoctista*. Jones & Bartlett, Boston. 914 p.

McCLATCHIE, S.; KAWACHI, R.; DALLEY, D. E. 1990: Epizoic diatoms on the euphausiid *Nyctiphanes australis*: consequences for gut-pigment analysis of whole krill. *Marine Biology 104*: 22–232.

McGUIRE, D. M. 1989: Paleomagnetic stratigraphy and magnetic properties of Pliocene strata, Turakina River, New Zealand. PhD thesis, Victoria University of Wellington. 236 p.

MEDLIN, L. K.; KACZMARSKA, I. 2004: Evolution of the diatoms: V. Morphological and cytological support for the major clades and a taxonomic revision. *Phycologia 43*: 245–270.

MEDLIN, L.K.; SATO, S.; MANN, D. G. KOOISTRA, W. H. C. F. 2008: Molecular evidence confirms sister relationship of *Ardissonea*, *Climacosphenia* and *Toxarium* within the biopolar centric diatoms (Bacillariophyta, Mediophyceae), and cladistic analyses confirm that extremely elongated shape has arisen twice in the diatoms. *Journal of Phycology 44*: 1340–1348.

MOESTRUP, Ø.; O'KELLY, C. J. 2000: Class Silicoflagellata Lemmermann, 1901. Pp. 775–782 *in*: Lee, J. J.; Leedale, G. F.; Bradbury, P. (eds), *An Illustrated Guide to the Protozoa. Second Edition. Organisms traditionally referred to as Protozoa, or newly described groups*. Society of Protozoologists, Lawrence. 1432 p. (in 2 vols).

MOORE, L. B. 1942: Observations on the growth of *Macrocystis* in New Zealand with a description of a free-living form. *Transactions of the Royal Society of New Zealand 72*: 333–340.

MOORE, L. B. 1946: Oogamy in the brown alga *Halopteris*. *Nature 157*: 553–554.

MOORE, L. B. 1949: The marine algal provinces of New Zealand. *Transactions of the Royal Society of New Zealand 77*: 187–189.

MOORE, L. B. 1950: A "loose-lying" form of the brown alga *Hormosira*. *Transactions of the Royal Society of New Zealand 78*: 48–53.

MOORE, L. B. 1951: Reproduction in *Halopteris* (Sphacelariales). *Annals of Botany 15*: 265–278.

MOORE, L. B. 1953: Some distribution problems illustrated from brown algae of the genus *Halopteris*. *Seventh Pacific Science Congress 5*: 1–6.

MOORE, L. B. 1961: Distribution patterns in New Zealand seaweeds. *Tuatara 9*: 18–23.

MOSER, G.; LANGE-BERTALOT, H.; METZELTIN, D. 1998: Insel der Endemiten. Geobotanisches Phänomen neukaledonien. *Bibliotheca Diatomologica 38*: 1–464.

MÜLLER, D. G.; PARODI, E. R.; PETERS, A. F. [1998] 1999: *Asterocladon lobatum* gen. et sp. nov., a new brown alga with stellate chloroplast arrangement, and its systematic position judged from nuclear rDNA sequences. *Phycologia 37*: 425–432.

NEALE, D.; NELSON, W. 1998: Marine algae of the west coast, South Island, New Zealand. *Tuhinga 10*: 87–118.

NEGRISOLO, E.; MAISTRO, S.; INCARBONE, M.; MORO, I.; VALLE, L. D.; BROADY, P. A.; ANDREOLI, C. 2004: Morphological convergence characterizes the evolution of Xanthophyceae (Heterokontophyta): evidence from nuclear SSU rDNA and plastidial *rbc*L genes. *Molecular Phylogenetics and Evolution 33*: 156–170.

NELSON, W. A. 1994: Distribution of macroalgae in New Zealand – an archipelago in space and time. *Botanica Marina 37*: 221–233.

NELSON, W. A. 1999a: *Landsburgia ilicifolia* sp. nov. (Cystoseiraceae, Phaeophyta), a new deepwater species endemic to the Three Kings Islands. *New Zealand Journal of Botany 37*: 727–730.

NELSON, W. A. 1999b: *Marginariella parsonsii*

sp. nov. (Seirococcaceae, Phaeophyta), a new species from the Bounty and Antipodes Islands, southern New Zealand. *New Zealand Journal of Botany 37*: 731–735.

NELSON, W. A. 1999c: A revised checklist of marine algae naturalised in New Zealand. *New Zealand Journal of Botany 37*: 355–359.

NELSON, W. A. 2005: Life history and growth in culture of the endemic New Zealand kelp *Lessonia variegata* J.Agardh in response to differing regimes of temperature, photoperiod and light. *Journal of Applied Phycology 17*: 23–28.

NELSON, W. A.; ADAMS, N. M. 1983: A taxonomic revision of the families Chordariaceae and Chordariopsidaceae (Phaeophyta) in New Zealand. *New Zealand Journal of Botany 21*: 77–92.

NELSON, W.A.; ADAMS, N.M. 1984: Marine algae of the Kermadec Islands. *National Museum of New Zealand, Miscellaneous Series 10*: 1–29.

NELSON, W. A.; ADAMS, N. M. 1987: Marine algae of the Bay of Islands area. *National Museum of New Zealand, Miscellaneous Series 16*: 1–47.

NELSON, W. A.; ADAMS, N. M.; FOX, J. M. 1992: Marine algae of the northern South Island. *National Museum of New Zealand, Miscellaneous Series 26*: 1–80.

NELSON, W. A.; ADAMS, N. M.; HAY, C. H. 1991: Marine algae of the Chatham Islands. *National Museum of New Zealand, Miscellaneous Series 23*: 1–58.

NELSON, W. A.; NEILL, K. F.; PHILLIPS, J. A. 2004: First report of *Dictyota furcellata* (C. Agardh) Grev. from New Zealand. *New Zealand Journal of Marine and Freshwater Research* 38: 129–135.

NELSON, W. A.; PHILLIPS, L. 1996: The Lindauer legacy – current names for the Algae Nova-Zelandicae Exsiccatae. *New Zealand Journal of Botany 34*: 553–582.

NELSON, W. A.; VILLOUTA, E.; NEILL, K.; WILLIAMS, G. C.; ADAMS, N. M.; SLIVSGAARD, R. 2002: Marine macroalgae of Fiordland. *Tuhinga 13*: 117–152.

NYMAN, M. A.; BROWN, M. T.; NEUSHUL, M.; HARGER, B. W. W.; KEOGH, J. A. 1993: Mass distribution in the fronds of *Macrocystis pyrifera* from New Zealand and California. *Hydrobiologia 260/261*: 57–65.

NYMAN, M. A.; BROWN, M. T.; NEUSHUL, M.; KEOGH, J. A. 1990: *Macrocystis pyrifera* in New Zealand: testing two mathematical models for whole plant growth. *Journal of Applied Phycology 2*: 249–257.

OTA, Y.; BERRYMAN, K. R.; KASHIMA, K.; ISO, N. 1987: Holocene sediments and vertical tectonic movement near Wairoa, Northern Hawkes Bay, New Zealand. Pp. 38-46 *in*: Ota, Y. (ed.), *Holocene Coastal Tectonics of Eastern North Island New Zealand.* Yokohama National University, Japan. 104 p.

PARENTE, M. I.; FLETCHER, R. L.; NETO, A. I. 2000: New records of brown algae (Phaeophyta) from the Azores. *Hydrobiologia 440*: 153–157.

PARSONS, M. J. 1985a: New Zealand. seaweed flora and its relationships. *New Zealand Journal of Marine and Freshwater Research 19*: 131–138.

PARSONS, M. J. 1985b: Biosystematics of the cryptogamic flora of New Zealand: Algae. *New Zealand Journal of Botany 23*: 663–675.

PARSONS, M. J. 1998: Early collections of seaweeds from New Zealand, particularly Banks Peninsula. Pp. 53–57 *in*: Burrows, C.J. (ed.), *Etienne Raoul and Canterbury Botany 1840–1996.* Canterbury Botanical Society & Manuka Press, Christchurch.

PARSONS, M. J.; FENWICK, G. D. 1984: Marine algae and a marine fungus from Open Bay Islands, Westland. *New Zealand Journal of Botany 22*: 425–432.

PATRICK, R. 1977: Ecology of freshwater diatoms and diatom communities. Pp. 284-382 *in*: Werner, D. (ed.), *The Biology of Diatoms. [Botanical Monographs 13.]* Blackwell, Oxford. 498 p.

PEABODY, A. J.; CAMERON, N. G. 2010: Forensic science and diatoms. Pp. 534–539 *in*: Smol, J. P., & Stoermer, E. F. (eds), *The Diatoms: Applications for the environmental and earth sciences.* 2nd edn. Cambridge University Press, Cambridge U.K. 667 p.

PETERS, A. F.; BREEMAN, A. M. 1992: Temperature responses of disjunct temperate brown algae indicate long-distance dispersal of microthalli across the tropics. *Journal of Phycology 28*: 428–438.

PETERS, A. F.; BURKHARDT, E. 1998: Systematic position of the kelp endophyte *Laminarionema elsbetiae* (Ectocarpales *sensu lato*, Phaeophyceae) inferred from nuclear ribosomal DNA sequences. *Phycologia 37*: 114–120.

PETERS, A. F.; CLAYTON, M. N. 1998: Molecular and morphological investigations of three brown algal genera with stellate plastids: evidence for Scytothamnales *ord. nov.* (Phaeophyceae). *Phycologia 37*: 106–113.

PETERS, A. F.; RAMIREZ, M. E. 2001: Molecular phylogeny of small brown algae, with special reference to the systematic position of *Caepidium antarcticum* (Adenocystaceae, Ectocarpales). *Cryptogamie Algologie 22*: 187–200.

PETERS, A. F.; van OPPEN, M. J. H.; WIENCKE, C.; STAM, W. T.; OLSEN, J. L. 1997: Phylogeny and historical ecology of the Desmarestiaceae (Phaeophyceae) support a southern hemisphere origin. *Journal of Phycology 33*: 294–309.

PHILLIPS, J. A.; NELSON, W. A. 1998: Typification of the Australasian brown alga *Zonaria turneriana* J.Agardh (Firstles) and segregation of *Zonaria aureomarginata* sp. nov., an endemic New Zealand species. *Botanica Marina 41*: 77–86.

PHILLIPS, N.; BURROWES, R.; ROUSSEAU, F.; REVIERS, B. de; SAUNDERS, G. W. 2008: Resolving evolutionary relationships among the brown algae using chloroplast and nuclear genes. *Journal of Phycology 44*: 394–404.

PREISIG, H. R. 1995: A modern concept of chrysophyte classification. Pp. 46–74 *in*: Sandgren, C. D.; Smol, J. P.; Kristiansen, J. (eds), *Chrysophyte Algae. Ecology, phylogeny and development.* Cambridge University Press, Cambridge. 416 p.

PREISIG, H. R.; ANDERSEN, R. A. 2000: Chrysomonada (Class Chrysophyceae Pascher, 1914). Pp. 693–730 *in*: Lee, J.J.; Leedale, G.F.; Bradbury, P. (eds), *An Illustrated Guide to the Protozoa. Second Edition. Organisms traditionally referred to as Protozoa, or newly described groups.* Society of Protozoologists, Lawrence. 1432 p. (in 2 vols).

PRUD'HOMME van REINE, W.F. 1993: Sphacelariales (Phaeophyceae) of the world, a new synthesis. *Korean Journal of Phycology 8*: 145–160.

RACAULT, M.-F. L. P.; FLETCHER, R. L.; REVIERS, B. de; CHO, G.Y.; BOO, S. M.; PARENTE, M. I.; ROUSSEAU, F. 2009: Molecular phylogeny of the brown algal genus *Petrospongium* Nägeli ex Kütz. (Phaeophyceae) with evidence for Petrospongiaceae fam. nov. *Cryptogamie, Algologie 30*: 111–123.

RAPSON, A. M.; MOORE, L. B.; ELLIOTT, I. L. 1942: Seaweed as a source of potash in New Zealand. *New Zealand Journal of Science 23*: 149–170.

REID, M. 2005: Diatom based models for reconstructing past water quality and productivity in New Zealand lakes. *Journal of Paleolimnology 33: 13-35.*

REVIERS, B. de; ROUSSEAU, F. 1999: Towards a new classification of the brown algae. *Progress in Phycological Research 13*: 107–201.

RHODES, L. L.; ADAMSON, J.; SCHOLIN, C. 2000: *Pseudo-nitzschia multistriata* (Bacillariophyceae) in New Zealand. *New Zealand Journal of Marine and Freshwater Research 34*: 463–467.

RHODES, L. L.; HAYWOOD, A. J.; BALLANTINE, W. J.; MacKENZIE, A. L. 1993: Algal blooms and climate anomalies in north-east New Zealand, August–December 1992. *New Zealand Journal of Marine and Freshwater Research 27*: 419–430.

RHODES, L. L.; SCHOLIN, C.; GARTHWAITE, I.; 1998: *Pseudo-nitzschia* in New Zealand and immunoassays in refining marine biotoxin monitoring programmes. *Natural Toxins* 6: 105-111

RHODES, L. L.; WHITE, D.; SYHRE, M.; ATKINSON, M. 1996: *Pseudonitzschia* species isolated from New Zealand coastal waters: domoic acid production in vitro and links with shellfish toxicity. Pp. 155–158 *in*: Yasumoto, T.; Oshima, Y.; Fukuyo, Y. (eds), *Harmful and Toxic Algal Blooms.* Intergovernmental Oceanographic Commission, UNESCO, Paris.

ROUND, F. E.; BASSON, P. W. 1997: A new monoraphid diatom genus (*Pogoneis*) from Bahrain and the transfer of *A. hungarica* and *A. taeniata* to new genera. *Diatom Research 12*: 71–81.

ROUND, F. E.; BUKHTIYAROVA, L. 1996: Four new genera based on *Achnanthes* (*Achnanthidium*) together with a re-definition of *Achnanthidium. Diatom Research 11*: 345–361.

ROUND, F. E.; CRAWFORD, R. M.; MANN, D. G. 1990: *The Diatoms. Biology and Morphology of the Genera.* Cambridge University Press, Cambridge. 747 p.

ROSS, R.; SIMS, P. A. 2000: A revision of *Actinodiscus* Greville, *Craspedoporus* Greville and related genera (Eupodiscaceae). *Diatom Research 15*: 285-347.

ROUSSEAU, F.; BURROWES, R.; PETERS, A.F.; KUHLENKAMP, R.; REVIERS, B. de 2001: A comprehensive phylogeny of the Phaeophyceae based on nrDNA sequences resolves the earliest divergences. *Compte Rendu Hebdomadaire des Séances de l'Académie des Sciences, Paris, Sciences de la Vie 324*: 305–319.

ROUSSEAU, F.; REVIERS, B. de 1999: Circumscription of the order Ectocarpales (Phaeophyceae): bibliographical synthesis and molecular evidence. *Cryptogamie, Algologique 20*: 5–18.

ROUSSEAU, F.; REVIERS, B. de; LECLERC, M-C.; ASENSI, A.; DELÉPINE, R. 2000: Adenocystaceae fam. nov. (Phaeophyceae) based on morphological and molecular evidence. *European Journal of Phycology 35*: 35–43.

ROUSSEAU, F.; LECLERC, M.-C.; REVIERS, B. de 1997: Molecular phylogeny of European Fucales (Phaeophyceae) based on partial large-subunit

rDNA sequence comparisons. *Phycologia 36*: 438–446.
RUSSELL, L. K.; HEPBURN, C. D.; HURD, C. L.; STUART, M. D. 2008: The expanding range of *Undaria pinnatifida* in southern New Zealand: distribution, dispersal mechanisms and the invasion of wave-exposed environments. *Biological Invasions 10*:103–115.
SABBE, K.; VANHOUTTE, K.; HODGSON, D.; BERGEY, L.; VYVERMAN, W. 2001: Six new *Actinella* (Bacillariophyceae) species from Papua New Guinea, Australia and New Zealand: further evidence for widespread endemism in the Australasian region. *European Journal of Phycology 36*: 321–340.
SAFI, F. A. 2003: Microalgal populations of three New Zealand coastal locations: forcing functions and benthic-pelagic links. *Marine Ecology Progress Series 259*: 67-78
SANTOS, L. M. A. 1996: The Eustigmatophyceae: actual knowledge and research perspectives. *Beiheft für Nova Hedwigia 112*: 391–405.
SARMA, P.; CHAPMAN, V. J. 1975: Additions to the checklist of freshwater algae in New Zealand II. *Journal of the Royal Society of New Zealand 5*: 389-412.
SAUNDERS, G. W.; POTTER, D.; ANDERSEN, R. A. 1997: Phylogenetic affinities of the Sarcinochrysidales and the Chrysomeridales (Heterokonta) based on analyses of molecular and combined data. *Journal of Phycology 33*: 310–318.
SCHIEL, D. R. 1988: Algal interaction on subtidal reefs in northern New Zealand: a review. *New Zealand Journal of Marine and Freshwater Research 22*: 481–489.
SCHIEL, D. R. 1990: Macroalgal assemblages in New Zealand: structure, interactions and demography. *Hydrobiologia 192*: 59–76.
SCHIEL, D. R.; NELSON, W. A. 1990: The harvesting of macroalgae in New Zealand. *Hydrobiologia 204/205*: 25–33.
SCHIEL, D. R.; PIRKER, J.; LEES, H. 1997a: Can giant kelp forests be commercially harvested? Part I: Natural production cycle. *Seafood New Zealand 5(10)*: 27–28.
SCHIEL, D. R.; PIRKER, J.; LEES, H. 1997b: Can giant kelp forests be commercially harvested? Part II: Regeneration after harvest. *Seafood New Zealand 5(11)*: 27–28.
SCHIEL, D. R.; TAYLOR, D. I. 1999: Effects of trampling on a rocky intertidal algal assemblage in southern New Zealand. *Journal of Experimental Marine Biology and Ecology 235*: 213–235.
SCHMIDT, O. C. 1937: Choristocarpaceen und Discosporangiaceen. *Hedwigia 77*: 1–4.
SCHRADER, H. J. 1969: Die pennaten Diatomeen aus dem Obereozaen von Oamaru, Neuseeland. *Nova Hedwigia 28*: 1-124, 38 pls.
SERRAO, E. A.; ALICE, L. A.; BRAWLEY, S. H. 1999: Evolution of the Fucaeae (Phaeophyceae) inferred from nrDNA-ITS. *Journal of Phycology 35*: 382–394.
SHULMEISTER, J.; SOONS, J. M.; BERGER, G. W.; HARPER, M. A.; HOLT, S.; MOAR, M.; CARTER, J. A. 1999: Environmental and sea-level changes on Banks Peninsula (Canterbury, New Zealand) through three glaciation-interglaciation cycles. *Palaeogeography, Palaeoclimatology, Palaeoecology 152*: 101-127.
SIEBURTH, J. M.; SMETACEK, V.; LENZ, J. 1978: Pelagic ecosystem structure: heterotrophic compartments of plankton and their relation to plankton size fractions. *Limnology and Oceanography 23*: 1256–1263.
SIEMER, B. L.; STAM, W. T.; OLSEN, J. L.; PEDERSEN, P. M. 1998: Phylogenetic relationships of the brown algal orders Ectocarpales, Chordariales, Dictyosiphonales, and the Tilopteridales (Phaeophyceae) based on RUBISCO large subunit and spacer sequences. *Journal of Phycology 34*: 1038–1048.
SILBERFELD, T.; LEIGH, J. W.; VERBRUGGEN, H.; CRUAUD, C.; REVIERS, B. de; ROSSEAU, F. 2010: A multi-locus time-calibrated phylogeny of the brown algae (Heterokonta, Ochrophyta, Phaeophyceae): Investigating the evolutionary nature of the"brown algal crown radiation". *Molecular Phylogenetics and Evolution 65*: 659–674.
SILVA, P. C.; MOE, R.L. 1999: The Index Nominorum Algarum. *Taxon 48*: 315–353.
SILVA, P. C. 2009. Index Nominum Algarum, University Herbarium, University of California, Berkeley. [http://ucjeps. berkeley.edu/INA.html]
SKUJA, H. 1976: Zur Kenntnis der Algen Neuseelandischer Torfmoore. *Nova Acta Regiae Societatis scientiarun Upsaliensis, ser. VC 2*: 1-158.
SKVORTZOW, B. W. 1938: Notes on the algal flora of New Zealand I – Freshwater diatoms from New Zealand. *Philippine Journal of Science 67*: 167-174.
SMOL, J. P., STOERMER, E. F. 2010: *The Diatoms: Applications for the environmental and earth sciences*. 2nd edn. Cambridge University Press, Cambridge U.K. 667 p.
SOURNIA, A. 1995: Red tide and toxic marine phytoplankton of the world ocean: an inquiry into biodiversity. Pp. 103–112 *in*:Lassus, P.; Arzul, G. Erard, E.; Gentien, P.; Marcaillou, C. (eds), *Harmful Marine Algal Blooms*. Technique et Documentation, Lavoisier, Intercept Ltd.
SOUTH, G. R.; ADAMS, N. M. 1976: Marine algae of the Kaikoura coast. *National Museum of New Zealand, Miscellaneous Series 1*: 1–67.
STERRENBURG, F. A. S. 1991: Studies on the genera *Gyrosigma* and *Pleurosigma* (Bacillariophyceae). The typus generis of *Pleurosigma*, some presumed varieties and imitative species. *Botanica Marina 44*: 561–573.
STERRENBURG, F. A. S. 1992: Studies on the genera *Gyrosigma* and *Pleurosigma* (Bacillariophyceae). The type of the genus *Pleurosigma* and other attenuati sensu Peragallo. *Diatom Research 7*: 137–155.
STIDOLPH, S. R. 1980: A record of some coastal marine diatoms from Porirua Harbour, North Island, New Zealand. *New Zealand Journal of Botany 18*: 379–403.
STIDOLPH, S. R. 1981: *Gyrosigma balticum* var. *turgidum*, a new diatom variety from Porirua Harbour, North Island, New Zealand. *New Zealand Journal of Botany 19*: 405.
STIDOLPH, S. R. 1985a: Some benthic diatoms from Kapiti Island, near Cook Strait, New Zealand. *Nova Hedwigia 41*: 393–417.
STIDOLPH, S. R. 1985b: Occurrence of the diatom *Glyphodiscus stellatus* Greville living in New Zealand coastal waters. *Nova Hedwigia 41*: 495–504.
STIDOLPH, S. R. 1986: *Amphiprora reediana*, a new diatom species from South Island, New Zealand. *Nova Hedwigia 43*: 29–43.
STIDOLPH, S. R. 1988: Observations and remarks on morphology and taxonomy of the diatom genera *Gyrosigma* and *Pleurosigma* W. Smith. *Nova Hedwigia 47*: 377–388.
STIDOLPH, S. R. 1990: *Cavernosa kapitiana*, a new diatom genus and species from Kapiti Island, New Zealand. *Nova Hedwigia 50*: 97–110.
STIDOLPH, S. R. 1992: Observations and remarks on the morphology and taxonomy of the diatom genera *Gyrosigma* Hassall and *Pleurosigma* W. Smith. III. *Gyrosigma sterrenburgii* sp. nov., and *Pleurosigma amara* sp. nov. *Diatom Research 7*: 345–366.
STIDOLPH, S. R. 1993a: Observations and remarks on the morphology and taxonomy of the diatom genera *Gyrosigma* Hassall and *Pleurosigma* W. Smith. II. *Gyrosigma waitangiana* sp. nov. and *Pleurosigma sterrenburgii* sp. nov. *Nova Hedwigia 56*: 139–153.
STIDOLPH, S. R. 1993b: A light and electron microscopical study of *Hyalodiscus pustulatus* A. Schmidt (Bacillariophyceae) from New Zealand marine habitats. *Botanica Marina 36*: 79–86.
STIDOLPH, S. R. 1993c: *Hantzschia doigiana*, a new taxon of brackish-marine diatom from New Zealand coastal waters. *Diatom Research 8*: 465–474.
STIDOLPH, S. R. 1993d: *Hamatusia* nom. nov., a new generic name for the diatom *Amphiprora reediana* Stidolph. *Diatom Research 8*: 481–482.
STIDOLPH, S. R. 1994a: A light and electron microscopical study of the diatom *Aulacodiscus beeveriae* Johnson in Pritchard, from New Zealand coastal marine habitats. *Botanica Marina 37*: 75–82.
STIDOLPH, S. R. 1994b: Observations and remarks on morphology and taxonomy of the diatom genera *Gyrosigma* Hassall and *Pleurosigma* W. Smith. IV. *Gyrosigma fogedii* sp. nov., and some diatoms similar to *G. fasciola* (Ehrenb.) Griffith & Henfrey. *Diatom Research 9*: 213–224.
STIDOLPH, S. R. 1994c: Note: Deposition of diatom holotypes. *Diatom Research 9*: 479.
STIDOLPH, S. R. 1995: A morphological and taxonomic study of the complex of diatoms assigned to *Caloneis brevis* (Gregory) Cleve. *Diatom Research 10*: 165–177.
STIDOLPH, S. R. 1998: A light and electron microscopical study of the diatom *Aulacodiscus petersii* Ehrenberg, from New Zealand coastal marine habitats and identification of the holotype. *Botanica Marina 41*: 399–409.
STRELNIKOVA, N. I.; LASTIVKA, T.V. 1999: The problem of the origin of marine and freshwater diatoms. Pp. 113-123 *in*: Mayama, I.; Mayama, K. (eds), *Proceedings of the 14th Diatom Symposium 1996*. Koeltz, Koenigstein.
SUDA, S.; ATSUMI, M.; HIDEAKI, H. 2002: Taxonomic characterization of a marine *Nannochloropsis* species, *N. oceanica* sp. nov. (Eustigmatophyceae). *Phycologia 41*: 273–279.
TAN, I. H.; DRUEHL, L. D. 1994: A molecular analysis of *Analipus* and *Ralfsia* (Phaeophyceae) suggests the order Ectocarpales is polyphyletic. *Journal of Phycology 30*: 721–729.
TAN, I. H.; DRUEHL, L. D. 1996: A ribosomal DNA phylogeny supports the close evolutionary relationships among the Sporochnales, Desmarestiales, and Laminariales (Phaeophyceae). *Journal of Phycology 32*: 112–118.
TAYLOR, F. J. 1970: A preliminary annotated check list of diatoms from New Zealand coastal waters. *Transactions of the Royal Society of New Zealand, Biological Sciences 12*: 153–174.
TAYLOR, F. J.; TAYLOR, N. J.; WALSBY, J. R. 1985: A bloom of the planktonic diatom, *Cerataulina pelagica*, off the coast of northeastern New Zealand in 1982–3, and its contribution to an associated mortality of fish and benthic fauna.

Internationale Revue der gesamten Hydrobiologie *70*: 773–795.
THERIOT, E.; CANNONE, J. J., GUTELL, R, ALVERSON, A. J. 2009: The limits of nuclear-encoded SSU rDNA for resolving diatom phylogeny. *European Journal of Phycology 44*: 277-290
THOMASSON, K. 1974: Rotorua phytoplankton reconsidered (North Island of New Zealand). *Internationale Revue Gesamten Hydrobiologie*: 703–727.
TURNER, G. M. 1997: Environmental magnetism and magnetic correlation of high resolution lake sediment records from northern Hawkes Bay. *New Zealand Journal of Geology and Geophysics* *40*: 287–298.
UWAI, S.; NELSON, W.; NEILL, K.; WANG, W.D.; AGUILAR-ROSAS, L.E.; BOO, S.M.; KITAYAMA, T.; KAWAI, H. 2006: Genetic diversity in *Undaria pinnatifida* (Laminariales, Phaeophyceae) deduced from mitochondria genes - origins and succession of introduced populations. *Phycologia* 4I5: 687–695.
VAN DAM, H.; MERTENS, A.; SINKELDAM, J. 1994: A coded checklist and ecological indicator values of freshwater diatoms from The Netherlands. *Netherlands Journal of Aquatic Ecology 28*: 117–133.
VAN HOUTTE, K.; VERLEYEN, E.; SABBE, K.; KILROY, C.; STERKEN, M.; VYVERMAN, W. 2006: Congruence and disparity in benthic community structure of small lakes in New Zealand and Tasmania. *Marine and Freshwater Research 57*: 789–801.
VANORMELINGEN, P.; CHEPURNOV, P.; MANN, D. G.; COUSIN, S.; VYVERMAN, W. 2007: Congruence of morphological, reproductive and ITS-rDNA sequence data in some Australasian *Eunotia bilunaris* (Bacillariophyta) *European Journal of Phycology 42*: 61–79.
VANORMELINGEN, P.; VERLEYEN, E.; VYVERMAN, W. 2008: The diversity anad distribution of diatoms: from cosmopolitanism to narrow endemism. *Biodiversity Conservation 17*: 393-405
VYVERMAN, W.; SABBE, K.; MANN, D.; KILROY, C.; VYVERMAN, R.; VANHOUTTE, K.; HODGSON, D. 1998: *Eunophora* gen. nov. (Bacillariophyceae) from New Zealand. Description and comparison with *Eunotia* and amphoroid diatoms. *European Journal of Phycology 33*: 95-111.
WEHR, J. D. 2003. Brown algae. Pp. 757–773 *in*: Wehr, J. D.; Sheath, R. G. (eds), *Freshwater Algae of North America. Ecology and classification.* Academic Press, San Diego. 917 p.
WILKINSON, V. 1981: Production ecology of microphytobenthic populations in the Manukau Harbour. MSc thesis, University of Auckland. 139 p.
WILLIAMS, D. M.; KOCIOLEK, J. P. 2007: Pursuit of a natural classification of diatoms: History, monophyly and rejection of paraphyletic taxa. *European Journal of Phycology 42*: 313-319
WILLIAMSON, J. E.; CREESE, R. G. 1996a: Small invertebrates inhabiting the crustose alga *Pseudolithoderma* sp. (Ralfsiaceae) in northern New Zealand. *New Zealand Journal of Marine and Freshwater Research 30*: 221–232.
WILLIAMSON, J. E.; CREESE, R. G. 1996b: Colonisation and persistence of patches of the crustose brown alga *Pseudolithoderma* sp. (Ralfsiaceae) *Journal of Experimental Marine Biology and Ecology 203*: 191–208.
WILLIAMSON, J. E.; REES, T. A.V. 1994: Nutritional interaction in an alga-barnacle association. *Oecologia 99*: 16–20.
WOMERSLEY, H. B. S. 1987: *The Marine Benthic Flora of Southern Australia. Part II.* Government Printer, Adelaide. 484 p.
WUJEK, D. E.; O'KELLY, C. J. 1992: Silica-scaled Chrysophyceae (Mallomonadaceae and Paraphysomonadaceae) from New Zealand freshwaters. *New Zealand Journal of Botany 30*: 405–414.
YOON, H. S.; BOO, S. M. 1999: Phylogeny of Alariaceae (Phaeophyta) with special reference to *Undaria* based on sequences of the RuBisCo spacer region. *Hydrobiologia 398/399*: 47–55.
YOON, H. S.; LEE, J.Y.; BOO, S. M.; BHATTACHARYA, D. 2001: Phylogeny of Alariaceae, Laminariaceae, and Lessoniaceae (Phaeophyceae) Based on plastid-encoded RuBisCo Spacer and nuclear-encoded ITS sequence comparisons. *Molecular Phylogenetics and Evolution 21*: 231–243.
ZEMKE-WHITE, W. L.; BREMNER, G.; HURD, C. L. 1999: The status of commercial algal utilization in New Zealand. *Hydrobiologia 398/399*: 487–494.

Checklist of New Zealand Ochrophyta

Higher-level classification within the phylum largely follows Cavalier-Smith (1998, 2000, 2004) and Cavalier-Smith and Chao (2006). Abbreviations: A, adventive; E, endemic; F, freshwater; M, marine; T, terrestrial; ? = identification queried; (?) = date queried; * = new record (i.e. not in previous checklists). Endemic genera are underlined (first entry only).

KINGDOM CHROMISTA
SUBKINGDOM HAROSA
INFRAKINGDOM HETEROKONTA
PHYLUM OCHROPHYTA
SUBPHYLUM PHAEISTA
INFRAPHYLUM LIMNISTA
Class EUSTIGMATOPHYCEAE
Order EUSTIGMATALES
CHLOROBOTRYACEAE
Chlorobotrys regularis (West) Bohlin F
Chlorobotrys polychloris Pascher F
EUSTIGMATACEAE
Eustigmatos magnus (J.B.Petersen) D.J.Hibberd T/F
Eustigmatos vischeri D.J.Hibberd T/F
Vischeria stellata (Chodat) Pascher T/F
MONODOPSIDACEAE
Nannochloropsis sp. J. Hall NIWA M
PSEUDOCHARACIOPSIDACEAE
Ellipsoidion salinum Norris 1964 M
Ellipsoidion stichococcoides Pascher F

Class CHRYSOPHYCEAE
Subclass CHRYSOPHYCIDAE
Order CHROMULINALES
CHROMULINACEAE
Anthophysa socialis (From.) W.S.Kent F
Anthophysa vegetans (O.Müll.) F.Stein F
Chromulina flavicans Buetschli F
Dendromonas virgaria (Weisse) F.Stein F
Ochromonas crenata G.A.Klebs F
Ochromonas perlata Doflein F
Oikomonas mutabilis W.S.Kent F
Oikomonas termo W.S.Kent F
Sphaleromantis olivacea Pascher F
Spumella spp. indet. (3) Norris 1964 M
Stipitochrysis monorhiza Korshikov F
Uroglena volvox Ehrenb. F
CHRYSAMOEBACEAE
Chrysamoeba pyrenoidifera Korshikov F
Chrysamoeba radians G.A.Klebs F
Chrysostephanosphaera globulifera Scherff. F
CHRYSOCAPSACEAE
Chrysocapsa flavescens Starmach T/F
Chrysocapsa planctonica Pascher F
Naegeliella flagellifera Correns F
CHRYSOSPHAERACEAE
Chrysosphaera gallica Bourr. F
DINOBRYACEAE
Calycomonas ovalis Wulff 1919 M
Calycomonas vangoorii Conrad 1938 M
Chrysococcus majus Lackey F
Chrysococcus rufescens G.A.Klebs F
Chrysopyxis bipes F.Stein F
Chrysopyxis iwanoffii Lauterborn F
Derepyxis dispar (A.Stokes) Senn var. *gracilipes* Skuja F
Derepyxis stokesii Lemmerm. F
Dinobryon acuminatum Ruttner F
Dinobryon aff. *balticum* Schütt (Moestrup) M
Dinobryon bavaricum O.E.Imhof F
Dinobryon crenulatum West et G.S.West F
Dinobryon cylindricum O.E.Imhof F/M
— var. *alpinum* (O.E.Imhof) H.Bachm. F
— ad. var. *angulatus* (Seligo) Brunnth. F
— var. *palustre* Lemmerm. F
Dinobryon divergens O.E.Imhof F/M
— var. *schauinslandii* (Lemmerm.) Brunnth. F
Dinobryon elegantissimum (Korshikov) Bourr. F
Dinobryon faculiferum (Willén) Willén M
Dinobryon sertularia Ehrenb. F
— var. *protuberans* (Lemmerm.) Willi Krieg. F
Dinobryon sociale Ehrenb. F
Domatomonas cylindrica Lackey F
Domatomonas epiplanctica Thomasson F
Epipyxis kenaiensis D.K.Hilliard et Asmund F
Epipyxis lauterbornii (Lemmerm.) D.K.Hilliard et Asmund F
Epipyxis polymorpha (Lund) D.K.Hilliard et Asmund F
Epipyxis utriculus Ehrenb. var. *pusilla* (Lemmerman) D.K.Hilliard et Asmund F
Kephyrion spirale (Lackey) M.A.Conrad F
Pseudokephyrion conicum (Schiller) Schmid F
Stokesiella epipyxis Pascher F
Stokesiella lepteca (A.Stokes) Lemmerm. F
Stylochrysallis parasitica F.Stein F
PARAPHYSOMONADACEAE

Chrysosphaerella longispina Lauterborn F
Paraphysomonas antarctica E.Takah. M
Paraphysomonas butcheri Pennick et Clarke F/M
Paraphysomonas caelifrica Preisig et D.J.Hibberd F
Paraphysomonas circumvallata Thomsen F
Paraphysomonas foraminifera Lucas M
Paraphysomonas imperforata Lucas F/M
Paraphysomonas vestita (A.Stokes) de Saedeleer F
Spiniferomonas abei E.Takah. F
Spiniferomonas bilacunosa E.Takah. F
Spiniferomonas bourrellyi E.Takah. F
Spiniferomonas coronacircumspina Wujek et Kristiansen F
Spiniferomonas trioralis E.Takah. F
*Tetrasporopsis pelagica** Norris 1961 M
INCERTAE SEDIS
'Ruttnera' pringsheimii Subrahmanyan 1962 M

Order HIBBERDIALES
STYLOCOCCACEAE
*Heliapsis achromatica** Norris 1964 M
Heliapsis mutabilis Pascher F
Heliochrysis radians Pascher F
Lagynion delicatulum Skuja F
Lagynion scherffelii Pascher var. *depressum* Skuja F

Order HYDRURALES
HYDRURACEAE
Hydrurus foetidus (Vill.) Trevis. F

Subclass SYNUROPHYCIDAE
Order SYNURALES
MALLOMONADACEAE
Mallomonas acaroides (Perty) Iwanoff F
Mallomonas akrokomos Ruttner F
Mallomonas alata Asmund F
Mallomonas alpina Ruttner in Pascher F
Mallomonas calceolus Bradley F
Mallomonas conspersa Dürrschm. F E
Mallomonas costata Dürrschm. F
Mallomonas cyathellata Wujek et Asmund F
Mallomonas eoa E.Takah. F
Mallomonas grossa Dürrschm. F E
Mallomonas guttata Dürrschm. F
Mallomonas heterospina (Lund) Asmund F
Mallomonas maculata Bradley F
Mallomonas mangofera Harris et Bradley F
Mallomonas matvienkoae Asmund et Kristiansen F
Mallomonas multisetigera Dürrschm. F
Mallomonas novae-zelandiae Dürrschm. F E
Mallomonas papillosa Harris et Bradley F
Mallomonas perpusilla Dürrschm. F E
Mallomonas plumosa Croome et Tyler F
Mallomonas pumilio Harris et Bradley F
Mallomonas roscida Dürrschm. F E
Mallomonas tongarirensis Dürrschm. F E
Mallomonas tonsurata (Teiling) Willi Krieg. F
Mallomonas villosa Dürrschm. F E
SYNURACEAE
Chrysodidymus synuroideus Prowse F
Synura adamsii G.M.Sm. F
Synura curtispina (Petersen et Hansen) Asmund F
Synura echinulata Korshikov F
Synura mammillosa E.Takah. F
Synura petersenii Korshikov F
Synura sphagnicola Korshikov F
Synura spinosa Korshikov F
Synura uvella Ehrenb. F

INFRAPHYLUM MARISTA
Superclass HYPOGYRISTIA
Class DICTYOCHOPHYCEAE
Order DICTYOCHALES
DICTYOCHACEAE
Dictyocha fibula Ehrenberg 1838 M
Dictyocha octonarium (Ehrenberg 1844) M
Dictyocha speculum (Ehrenberg 1837) M
Vicicitus globosus (Y.Hara et Chihara) Chang M

Order FLORENCIELLALES
INCERTAE SEDIS
Pseudochattonella verruculosa (Y.Hara et Chihara) Hosoi, Tanabe et al. M

Order PEDINELLALES
Suborder ACTINOMONADINEAE
APEDINELLACEAE
Apedinella radians (Lohmann) Campbell M
PEDINELLACEAE
Parapedinella reticulata Pedersen et Thomsen M
Pseudopedinella pyriforme N.Carter M

Superclass RAPHIDOISTIA
Class RAPHIDOPHYCEAE
ORDER RAPHIDOMONADALES
VACUOLARIACEAE
Chattonella marina (Subrahmanyan) Y.Hara et Chihara M
Fibrocapsa japonica Toriumi et Takano M
Goniostomum semen (Ehrenb.) Diesing F
Heterosigma akashiwo (Hada) Hada 1968 M

Superclass FUCISTIA
Class CHRYSOMEROPHYCEAE
Order CHRYSOMERIDALES
INCERTAE SEDIS
Giraudyopsis stellifera P.J.L.Dang. M

Class XANTHOPHYCEAE
Order RHIZOCHLORIDALES
RHIZOCHLORIDACEAE
Stipitococcus urceolatus West et G.S.West F

Order MISCHOCOCCALES
BOTRYDIOPSIDACEAE
Botrydiopsis constricta Broady T/F
Botrydiopsis intercedens Pascher T/F
CHARACIOPSIDIACEAE
Characiopsis longipes (Rabenh.) Borzi F
Characiopsis ovalis Chodat F
Characiopsis subulata (A.Braun) Borzi F
Peroniella planktonica G.M.Sm. F
GLOEOBOTRYDACEAE
Chlorosaccus fluidus Luther F
Gloeobotrys limneticus (G.M.Sm.) Pascher F
MISCHOCOCCACEAE
Mischococcus confervicola Nägeli F
PLEUROCHLORIDACEAE
Chlorocloster dactylococcoides Pascher F
Goniochloris mutica (A.Braun) Fott F
Meringosphaera mediterranea Lohmann 1902 M
Pleurochloris commutata Pascher T/F
Pleurochloris inequalis Pascher T/F
Pleurochloris lobata Pascher T/F
Pseudostaurastrum enorme (Ralfs) Chodat F
— var. *pentaedricum* Prescott F
Pseudostaurastrum lobulatum (Nägeli) Skuja F
SCIADIACEAE
Bumilleriopsis brevis (Gerneck) Printz F
Centritractus africanus Fritsch et Rich F
Centritractus belanophorus Lemmerm. F
Ophiocytium arbuscula (A.Braun) Rabenh. F
Ophiocytium bicuspidatum Lemmerm. F
Ophiocytium capitatum Wolle F
Ophiocytium cochleare Eichw. (A.Braun) F
Ophiocytium majus Nägeli F
Ophiocytium parvulum (Perty) A.Braun F

Order TRIBONEMATALES
TRIBONEMATACEAE
Bumilleria pumila West et G.S.West F
Bumilleria sicula Borzi F
Tribonema affine G.S.West F
Tribonema viride Pascher F
— var. *minor* (Wille) G.S.West F
— var. *sordida* Kütz. F
Xanthonema exile (G.A.Klebs) P.C.Silva T F

Order VAUCHERIALES
BOTRYDIACEAE
Botrydium granulatum (L.) Grev. T
Botrydium vulgare Pascher T
VAUCHERIACEAE
Vaucheria aversa Hassall F
Vaucheria geminata (Vaucher) DC. F
Vaucheria hamata Goetz F
Vaucheria jaoi Ley F
Vaucheria pachyderma J.Walz F
Vaucheria pseudosessilis V.J.Chapm. M
Vaucheria sessilis (Vaucher) DC. F
— var. *hookeri* Kütz. F
Vaucheria syandra Woronin M
Vaucheria terrestris Goetz F
Vaucheria undulata Jao F
Vaucheria velutina C.Agardh M

Class PHAEOTHAMNIOPHYCEAE
Order PHAEOTHAMNIALES
PHAEOTHAMNIACEAE
Phaeothamnion confervicolum Lagerh. F
Stichogloea doederleinii (Schmidle) Wille F

Class PHAEOPHYCEAE
Order DICTYOTALES
DICTYOTACEAE
Dictyopteris kermadecensis (Cotton) Lindauer M E
Dictyopteris repens (Okamura) Boergesen M
Dictyota bartayresiana J.V.Lamour. M
Dictyota dichotoma (Huds.) J.V.Lamour. M
Dictyota divaricata J.V.Lamour. M
Dictyota furcellata (C.Agardh) Grev. M A
Dictyota intermedia Zanardini M
Dictyota kunthii (C.Agardh) Grev. M
Dictyota ocellata J.Agardh M E
Dictyota papenfussii Lindauer M E
Distromium didymothrix Allender & Kraft M
Distromium skottsbergii Levring M
Lobophora variegata (J.V.Lamour.) Womersley M
Padina australis Hauck ? M
Padina fraseri (Grev.) Grev. M
Spatoglossum chapmanii Lindauer M E
Stypopodium australasicum (Zanardini) Allender M
Taonia australasica J.Agardh M
Zonaria aureomarginata J.A.Phillips & W.A.Nelson M E
Zonaria diesingiana J.Agardh* M
Zonaria turneriana J.Agardh M

Order SPHACELARIALES
CLADOSTEPHACEAE
Cladostephus spongiosus (Huds.) C.Agardh f. *verticillatus* M
SPHACELARIACEAE
Herpodiscus bracteatus (Reinke) Draisma, Prud'homme & H.Kawai M
Herpodiscus durvilleae (Lindauer) South M E
Herpodiscus implicatus (Sauv.) Draisma, Prud'homme & H.Kawai M
Herpodiscus pulvinatus (Hook.f. & Harv.) Draisma, Prud'homme & H.Kawai M E
Herpodiscus stewartensis (Lindauer) Draisma, Prud'homme & H.Kawai M E
Sphacelaria brachygonia Mont. M
Sphacelaria limicola Lindauer M E
Sphacelaria rigidula Kütz. M

Sphacelaria solitaria (Pringsheim) Kylin M
Sphacelaria tribuloides Menegh. M
STYPOCAULACEAE
Halopteris brachycarpa Sauv. M
Halopteris congesta (Reinke) Sauv. M
Halopteris funicularis (Mont.) Sauv. M
Halopteris novae-zelandiae Sauv. M E
Halopteris paniculata (Suhr) Prud'homme M
Halopteris platycena Sauv. M
Halopteris virgata (Harv.) N.M.Adams M E
Ptilopogon botryocladus (Hook.f. & Harv.) Reinke M E

Order SYRINGODERMATALES
SYRINGODERMATACEAE
Microzonia velutina (Harv.) J.Agardh M

Order FUCALES
DURVILLAEACEAE
Durvillaea antarctica (Cham.) Har. M
Durvillaea chathamensis C.H.Hay M E
Durvillaea willana Lindauer M E
Durvillaea sp. Antipodes Is. WELT A010343 M E
HORMOSIRACEAE
Hormosira banksii (Turner) Decne. M
NOTHEIACEAE
Notheia anomala Harv. & Bailey M
SARGASSACEAE
Carpophyllum angustifolium J.Agardh M E
Carpophyllum flexuosum (Esper) Grev. M E
Carpophyllum maschalocarpum (Turner) Grev. M E
Carpophyllum plumosum (A.Rich) J.Agardh M E
Cystophora platylobium (Mertens) J.Agardh M
Cystophora retroflexa (Labill.) J.Agardh M
Cystophora scalaris J.Agardh M
Cystophora torulosa (R.Br.) J.Agardh M
Landsburgia ilicifolia W.A.Nelson M E
Landsburgia myricifolia J.Agardh M E
Landsburgia quercifolia (Hook. & Harv.) Harv. M E
Sargassum aquifolium (Turner) C.Agardh M
Sargassum cristaefolium C.Agardh M
Sargassum johnsonii V.J.Chapm. M E
Sargassum scabridum Hook.f. & Harv. M E
Sargassum sinclairii Hook.f. & Harv. M E
Sargassum tahitense (Grunov) Setchell M
Sargassum verruculosum (Mertens) C.Agardh M A
SEIROCOCCACEAE
Marginariella boryana (A.Rich) Tandy M E
Marginariella parsonsii W.A.Nelson M E
Marginariella urvilliana (A.Rich) Tandy M E
XIPHOPHORACEAE
Xiphophora chondrophylla (Turner) Mont. ex Harv. M
Xiphophora gladiata (Labill.) Mont. ssp. *novae-zelandiae* Rice M E

Order SPOROCHNALES
SPOROCHNACEAE
Carpomitra costata (Stackh.) Batters M
Perisporochnus regalis V.J.Chapm. M E
Perithalia capillaris J.Agardh M E
Sporochnus apodus Harv. M
Sporochnus elsieae Lindauer M E
Sporochnus moorei Harv. M
Sporochnus stylosus Harv. M

Order LAMINARIALES
ALARIACEAE
Undaria pinnatifida (Harv.) Suringar M A
LAMINARIACEAE
Macrocystis pyrifera (L.) C.Agardh M
LESSONIACEAE
Ecklonia radiata (C.Agardh) J.Agardh M
Lessonia adamsiae C.H.Hay M E
Lessonia brevifolia J.Agardh M E
Lessonia tholiformis C.H.Hay M E
Lessonia variegata J.Agardh M E

Order DESMARESTIALES
DESMARESTIACEAE
Desmarestia ligulata (Lightf.) J.V.Lamour. M
Desmarestia willii Reinsch M

Order SCYTOTHAMNALES
SCYTOTHAMNACEAE
Scytothamnus australis (J.Agardh) Hook.f. & Harv. M
Scytothamnus fasciculatus (Hook.f. & Harv.) Cotton M
SPLACHNIDIACEAE
Splachnidium rugosum (L.) Grev. M
INCERTAE SEDIS
Bachelotia antillarum (Grunov) Gerlof M

Order TILOPTERIDALES
CUTLERIACEAE
Cutleria mollis Allender & Kraft M
Cutleria multifida (Js.Smith) Grev. M A

Order RALFSIALES
RALFSIACEAE
Hapalospongidion saxigenum Lindauer M E
Pseudolithoderma australis Womersley M
Ralfsia verrucosa (Aresch.) Aresch. M
Ralfsia sp. cf. *confusa* M
Ralfsia sp. cf. *expansa* M

Order ECTOCARPALES
ACINETOSPORACEAE
Feldmannia indica (Sond.) Womersley & Bailey M
Feldmannia irregularis (Kuetz.) Hamel M
Geminocarpus geminatus (Hook.f. & Harv.) Skottsb. M
Hincksia granulosa (Js.Smith) P.C.Silva M
Hincksia michelliae (Harv.) P.C.Silva M
Hincksia sordida (Harv.) P.C.Silva M
Pylaiella littoralis (L.) Kjellm. M
ADENOCYSTACEAE
Adenocystis rimosa (Mont.) Asensi, Delépine, Reviers & Rousseau M
Adenocystis utricularis (Bory) Skottsb. M
Caepidium antarcticum J.Agardh M
CHORDARIACAE
Asperococcus bullosus J.V.Lamour. M A
Asperococcus ensiformis (Delle Chiaje) M.J.Wynne) M A
Chordaria cladosiphon Kuetz. M
Chordariopsis capensis (C.Agardh) Kylin M
Cladothele striarioides (Skottb.) A.D.Zinova M
Corynophlaea cystophorae J.Agardh M
Elachista australis J.Agardh M
Hecatonema stewartensis V.J.Chapm. M E
Herponema hormosirae Lindauer & V.J.Chapm. M E
Herponema maculaeforme (J.Agardh) Laing M E
Laminariocolax macrocystis (A.F.Peters) A.F.Peters M
Leathesia difformis (L.) Aresch. M
Leathesia intermedia V.J.Chapm. M
Leathesia novae-zelandiae Lindauer M E
Microspongium tenuissimum (Hauck) A.F.Peters M
Mikrosyphar pachymeniae Lindauer M E
Myriogloea intestinalis (Harv.) Lindauer, V.J.Chapm. & Aiken M E
Myrionema compactum Lindauer M E
Myrionema strangulans Grev. M
Myriotrichia adriatica Hauck M
Nemacystus novae-zelandiae Kylin M E
Papenfussiella lutea Kylin M
Punctaria latifolia Grev. M A
Striaria attentuata Grev. M A
Tinocladia novae-zelandiae Kylin M E
ECTOCARPACEAE
Ectocarpus bracchiolus Lindauer M E
Ectocarpus chapmanii Lindauer M E
Ectocarpus dellowianus Lindauer M E
Ectocarpus siliculosus (Dillwyn) Lyngb. M
PETROSPONGIACEAE
Petrospongium rugosum (Okamura) Setch. & N.L.Gardner M
SCYTOSIPHONACEAE
Chnoospora minima (K.Hering) Papenf. M A
Colpomenia bullosa (Saunders) Yamada M A
Colpomenia claytoniae S.M.Boo, K.M.Lee, G.Y.Cho & W.A.Nelson M
Colpomenia ecuticulata M.J.Parsons M
Colpomenia peregrina Sauv. M
Colpomenia sinuosa (Roth) Derbes & Solier M
Endarachne binghamiae J.Agardh M
Hydroclathrus clathratus (C.Agardh) M.Howe M A
Petalonia fascia (O.F.Mull.) Kuntze M
Rosenvingea sanctae-crucis Boergesen M A
Scytosiphon lomentaria (Lyngb.) Link M

SUBPHYLUM KHAKISTA
Class BACILLARIOPHYCEAE

FRESHWATER DIATOMS

Vivienne Cassie Cooper and Margaret Harper

Higher classification largely according to Round et al. (1990) and lower according to Algaebase.org. The list includes updates by Biggs and Kilroy (2000), Krammer (2000), Cassie Cooper (2001), Cochran (2002), Harvey 1996, Kilroy et al. (2003) and Reid (2005). Codes additional to those on p. 145: thesis = university thesis as the source of the record; † = new combination, see p. 162; # = taxonomic position uncertain; * = new record (i.e. not in previous main checklists); B = brackish-water species; ‡ = name changed since that published in Cassie (1984).

Subclass COSCINODISCOPHYCIDAE
Superorder BIDDULPHIANAE
Order BIDDULPHIALES
BIDDULPHIACEAE
Stoermeria trifoliata (Cleve) Kociolek, Escobar & Richardson 1996‡ B #

Superorder CHAETOCEROTANAE
Order CHAETOCEROTALES
ACANTHOCEROTACEAE
Acanthoceras zachariasii (Brun) Simonsen 1979‡

Superorder COSCINODISCANAE
Order AULACOSEIRALES
AULACOSEIRACEAE
Aulacoseira alpigena (Grunow) Krammer 1990‡
Aulacoseira ambigua (Grunow) Simonsen 1979‡
Aulacoseira crassipunctata Krammer 1990‡
Aulacoseira crenulata (Ehrenberg) Thwaites 1848
Aulacoseira distans (Ehrenberg) Simonsen 1979‡
Aulacoseira granulata (Ehrenberg) Simonsen 1979‡
— var. *angustissima* (O.Müller) Simonsen 1979‡
Aulacoseira herzogii (Lemmermann) Simonsen 1979‡
Aulacoseira islandica (O.Müller) Simonsen 1979‡
— ssp. *helvetica* (O.Müller) Simonsen 1979‡
Aulacoseira italica (Ehrenberg) Simonsen 1979‡
— var. *tenuissima* (Grunow) Simonsen 1979‡
Aulacoseira lirata (Ehrenberg) Ross 1986‡
Aulacoseira muzzanensis (Meister) Krammer 1991‡
Aulacoseira perglabra (Østrup) Haworth 1988‡
Aulacoseira subarctica (O.Müller) Haworth 1988‡
Aulacoseira subborealis (Nygaard) Denys, Muylaert & Krammer 2003*
Aulacoseira valida (Grunow) Krammer 1991‡

Order COSCINODISCALES
HEMIDISCACEAE
Actinocyclus kuetzingii (Schmidt) Simonsen 1975
Actinocyclus normanii (Gregory) Hustedt 1957* B

Order MELOSIRALES
HYALODISCACEAE
Hyalodiscus lentiginosus John 1982 B
MELOSIRACEAE
Melosira dickiei (Thwaites) Kützing 1849
Melosira lineata (Dillwyn) Agardh 1824 B
Melosira undulata (Ehrenberg) Kützing 1844
Melosira varians Agardh 1827

Order ORTHOSEIRALES
ORTHOSEIRACEAE
Cavernosa kapitiana Stidolph 1990‡ #
Orthoseira dendrophila (Ehrenberg) Crawford‡ 1981
Orthoseira dendroteres (Ehrenberg) Crawford [informal 1990] #
Orthoseira roseana (Rabenhorst) O'Meara 1876‡

Order PARALIALES
RADIALIPLICATACEAE
Ellerbeckia arenaria (Ralfs) Crawford 1988‡
Ellerbeckia baileyii (H.Smith) Crawford & Sims 2007*

Superorder RHIZOSOLENIANAE
Order RHIZOSOLENIALES
RHIZOSOLENIACEAE
Urosolenia eriensis (H.L.Smith) Round & Crawford 1990‡
— var. *morsa* (W. et G.West) Bukhtiyarova 1995‡
Urosolenia longiseta (Zacharias) Edlund & Stoermer 1993‡

Superorder THALASSIOSIRANAE
Order THALASSIOSIRALES
STEPHANODISCACEAE
Cyclostephanos dubius (Fricke) Round 1982‡
Cyclostephanos invisitatus (Hohn & Hellermann) Theriot, Stoemer & Håkansson 1987*
Cyclostephanos novaezeelandiae (Cleve) Round 1982‡
Cyclostephanos tholiformis Stoermer, Håkansson & Theriot 1987*
Cyclotella antiqua W.Smith 1853
Cyclotella atomus Hustedt 1938 B
Cyclotella baikalensis Skvortzow et Meyer 1928
Cyclotella delicatula Hustedt 1952*
Cyclotella distinguenda Hustedt 1927‡
Cyclotella meneghiniana Kützing 1844 B
Cyclotella minima Barber & Carter 1970
Cyclotella operculata (Agardh) Brébisson
Cyclotella striata (Kützing) Grunow 1880 B
Discostella elentari (Alfinito & Tagliaventi) Houk & Klee 2004
Discostella pseudostelligera (Hustedt) Houk & Klee 2004‡
Discostella stelligera (Cleve & Grunow) Houk & Klee 2004
— var. *robusta* (Hustedt) Houk & Klee 2004‡
Discostella tasmanica (Haworth & Tyler) Houk & Klee 2004*
Puncticulata bodanica (Eulenstein) Håkansson 2002*
Puncticulata comta (Ehrenberg) Håkansson 2002‡
Puncticulata radiosa (Grunow) Håkansson 2002
Stephanodiscus alpinus Hustedt 1942
Stephanodiscus atmosphaerica (Ehrenberg) Håkasson & Locker 1981*
Stephanodiscus hantzschii Grunow 1880
Stephanodiscus minutulus (Kützing) Cleve & Möller 1878‡
Stephanodiscus parvus Stoermer & Håkansson 1984*
Stephanodiscus rotula (Kützing) Hendey 1964‡
THALASSIOSIRACEAE
Thalassiosira australiensis (Grunow) Hasle 1989 B?
Thalassiosira hasleae Cassie & Dempsey 1980
Thalassiosira hyperborea (Grunow) Hasle 1989 B
Thalassiosira lacustris (Grunow) Hasle 1977
Thalassiosira weissflogii (Grunow) Fryxell & Hasle 1977 B

Subclass FRAGILARIOPHYCIDAE
Superorder FRAGILARIANAE
Order FRAGILARIALES
FRAGILARIACEAE
Asterionella formosa Hassall 1855 A
— var. *acaroides* Lemmerman 1906‡
Diatoma ehrenbergii Kützing 1844‡
Diatoma hiemale (Lyngbye) Heiberg 1863
Diatoma mesodon Kützing 1844‡
Diatoma moniliforme Kützing 1833* Thesis
Diatoma tenue Agardh 1824‡
Diatoma vulgare Bory 1824
Fragilaria capucina Desmazières 1925
— var. *mesolepta* (Rabenhorst) Rabenhorst 1864
— ssp. *rumpens* (Kützing) Lange-Bertalot 1995‡
— var. *vaucheriae* (Kützing) Lange-Bertalot 1980‡ B
Fragilaria cassubica Witkowski & Lange-Bertalot 1993‡ B
Fragilaria crotonensis Kitton 1869
Fragilaria delicatissima (W.Smith) Lange-Bertalot 1980‡
Fragilaria famelica (Kützing) Lange-Bertalot 1981
Fragilaria fragilarioides (Grunow) Cholnoky 1963‡
Fragilaria goulardii (Brébisson) Lange-Bertalot 1981‡
Fragilaria gracilis Østrup 1910
Fragilaria heidenii Østrup 1910
Fragilaria miniscula (Grunow) Williams & Round 1987‡
Fragilaria montana (Krasske) Lange-Bertalot 1980‡
Fragilaria nanana (Meister) Lange-Bertalot 1980‡
Fragilaria reicheltii (Voigt) Lange-Bertalot 1986‡
Fragilaria tenera (W.Smith) Lange-Bertalot 1980‡
Fragilaria utermoehlii (Hustedt) Lange-Bertalot 1980‡
Fragilariforma cassieae C.Kilroy & E.A.Bergey 2003*
Fragilariaforma constricta (Ehrenberg) Williams & Round 1988 Thesis
Fragilariforma rakiurensis C.Kilroy & E.A.Bergey 2003*
Fragilariforma virescens (Ralfs) Williams & Round 1988‡
— var. *subsalina* (Grunow) Bukhtiyarova 1995‡
Frankophila biggsii Lowe, Morales & Kilroy 2006* E #
Meridion circulare (Greville) Agardh 1831
— var. *constricta* (Ralfs) Brun 1880*
Opephora mutabilis (Grunow) Sabbe & Vyverman 1995* B
Pseudostaurosira brevistriata (Grunow) Williams & Round 1987‡
— var. *inflata* (Pantoscek) Edlund 1994‡
Pseudostaurosira elliptica (Schumann) Edlund, Morales & Spaulding 2006
Pseudostaurosira subsalina (Hustedt) Morales 2000* B
Stauroforma exiguiformis (Lange-Bertalot) Flower 1996*
Staurosira construens (Ehrenberg) Williams & Round 1987‡
— var. *binodis* (Ehrenberg) Hamilton 1992*
— var. *exigua* (W.Smith) Kobayasi 2002‡
— var. *pumila* (Grunow) Kingston 2000
Staurosira elliptica (Schumann) Cleve & Möller 1879
Staurosira venter (Ehrenberg) Cleve & Möller 1879‡
Staurosirella dubia (Grunow) Morales 2006‡
Staurosirella lapponica Williams & Round 1987* Thesis
Staurosirella leptostauron (Ehrenberg) Williams & Round 1987
— var. *rhomboides* (Grunow) Bukhtiyarova 1995* [Andreson et al. 2000]
Staurosirella martyi (Héribaud) Morales 2006‡ [*Martyana martyi*]
Staurosirella pinnata (Ehrenberg) Williams & Round 1987‡
— var. *intercedens* (Grunow) Hamilton 1994‡
— var. *lancettula* (Schumann) Siver & Hamilton 2005‡
Synedrella parasitica (W.Smith) Round & Maidana 2001‡*
Ulnaria acus (Kützing) Aboal 2003‡
Ulnaria biceps (Kützing) Compère 2001‡
Ulnaria capitata (Ehrenberg) Compère 2001‡
Ulnaria oxyrhynchus (Kützing) Aboal 2003‡
Ulnaria ulna (Nitzsch) Compère 2001‡
— var. *aequalis* (Kützing) Aboal 2003‡
— var. *amphirhynchus* (Ehrenberg) Aboal 2003‡
— var. *contracta* (Østrup 1901) Morales 2007‡
— var. *danica* (Kützing 1844)‡ †
— var. *spathulifera* (Grunow) Aboal 2001‡
Ulnaria ungeriana (Grunow) Compère 2001‡

Order TABELLARIALES
TABELLARIACEAE
Tabellaria fenestrata (Lyngbye) Kützing 1844
Tabellaria flocculosa (Ralfs) Kützing 1844
Tabellaria ventricosa Kützing 1844* Thesis
Tetracyclus glans (Ehrenberg) Mills 1935‡

Subclass BACILLARIOPHYCIDAE
Superorder BACILLARIANAE
Order ACHNANTHALES
ACHNANTHACEAE
Achnanthes crosbyana Foged 1979 B
Achnanthes curta (Cleve) Berg 1953*
Achnanthes curvirostrum Brun 1895 B*
Achnanthes densestriata Cleve-Euler 1953*
Achnanthes elata (Leuduger-Fortmorel) Hustedt 1937‡
Achnanthes cf. *holostatica* Hustedt 1936* Thesis
Achnanthes inflata (Kützing) Grunow 1880
Achnanthes quadratea var. *fennica* (Cleve-Euler) Ross 1986*
ACHNANTHIDIACEAE
Achnanthidium affine (Grunow) Czarnecki 1994‡
Achnanthidium biasolettianum (Grunow) Round & Bukhtiyarova 1996*
Achnanthidium coarctata (Brébisson) W.Smith 1885‡ T
Achnanthidium gracillimum (Meister) Lange-Bertalot 2004
Achnanthidium exiguum (Grunow) Czarnecki 1994‡
— var. *elliptica* (Hustedt 1937) † Thesis
— var. *heterovalvum* (Krasske) Czarnecki 1994
Achnanthidium exilis (Kützing) Heiberg 1863
Achnanthidium jackii Rabenhorst 1861*
Achnanthidium kranzii (Lange-Bertalot) Round & Bukhtiyarova 1996*
Achnanthidium minutissimum (Kützing) Czarnecki 1994‡
Achnanthidium pyreniacum (Hustedt) Kobasayi 1997*
Achnanthidium subatomus (Hustedt) Lange-Bertalot 1991*
Achnanthidium trinode Ralfs 1861‡
Eucocconeis alpestris (Brun) Lange-Bertalot 1999*

Thesis
Eucocconeis flexella (Kützing) Cleve 1895‡
Eucocconeis depressa (Østrup) Lange-Bertalot 1999*
Eucocconeis laevis (Østrup) Lange-Bertalot 1999*
Karayevia clevei (Grunow) Round & Bukhtiyarova 1996‡
— var. *rostrata* (Hustedt) Kingston 2000‡
Kolbesia kolbei (Hustedt) Round & Bukhtiyarova 1996‡
Kolbesia ploenensis (Hustedt) Kingston 2000‡
— var. cf. *woldstedtii* (Hustedt) Kelly et al. 2005 Thesis
Kolbesia cf. *suchlandtii* (Hustedt) Kingston 2000* Thesis
Lemnicola hungarica (Grunow) Round & Basson 1997‡
Planothidium conspicuum (Mayer) Aboal 2003‡
Planothidium delicatulum (Grunow) Round & Bukhtiyarova 1996‡ B
Planothidium dubium (Grunow) Round & Bukhtiyarova 1996‡
Planothidium ellipticum (Cleve) Round & Bukhtiyarova 1996* [Edlund et al. 2000]
Planothidium engelbrechti (Cholnoky) Round & Bukhtiyarova 1996* B Thesis
Planothidium frequentissimum (Lange-Bertalot) Round & Bukhtiyarova 1996
Planothidium hauckianum (Grunow) Round & Bukhtiyarova 1996‡ B
— var. *rostratum* (Schulz) Andresen, Stoermer & Kreis 2000‡ B
Planothidium haynaldii (Schaarschmidt) Lange-Bertalot 1999*
Planothidium heteromorphum (Grunow) Lange-Bertalot 1999*
Planothidium hustedtii (Krasske) Lange-Bertalot 2004* Thesis
Planothidium joursacense (Héribaud) Lange-Bertalot 1999*
Planothidium lanceolatum (Brébisson) Lange-Bertalot 1999‡
Planothidium lemmermannii (Hustedt) Morales 2006‡ B
Planothidium peragallii (Brun & Héribaud) Round & Bukhtiyarova 1996* Thesis
Planothidium pericavum (Carter) Lange-Bertalot 1999*
Planothidium robustius Lange-Bertalot 1999*
Planothidium rostratum (Østrup) Lange-Bertalot 1999*
Psammothidium curtissimum (Carter) Aboal 2003* Thesis
Psammothidium grischunum (Wuthrich) Bukhtiyarova & Round 1996* Thesis
Psammothidium levanderi (Hustedt) Czarnecki & Edlund 1995‡
Psammothidium marginulatum (Grunow) Bukhtiyarova & Round 1996* Thesis
Psammothidium montanum (Krasske) Mayuma 2002‡
Psammothidium oblongellum (Østrup) Van de Vijer 2002‡
Psammothidium sacculum (Carter) Bukhtiyarova 1996*
Psammothidium subatomoides (Hustedt) Bukhtiyarova & Round 1996* #
Rossithidium linearis (W. Smith) Bukhtiyarova & Round 1996‡
Rossithidium petersenii (Hustedt) Bukhtiyarova & Round 1996‡
Rossithidium pusillum (Grunow) Bukhtiyarova & Round 1996‡
COCCONEIDACEAE
Cocconeis disculus (Schumann) Cleve 1882* B
Cocconeis molesta var. *crucifera* Grunow 1880*
Cocconeis neodiminuta Krammer 1991‡
Cocconeis neothumensis Krammer 1991‡
Cocconeis pediculus Ehrenberg 1838
Cocconeis placentula Ehrenberg 1838 B
— var. *euglypta* (Ehrenberg) Grunow 1884 B
— var. *klinoraphis* Geitler 1927
— var. *lineata* (Ehrenberg) Van Heurck 1880 B
— var. *pseudolineata* Geitler 1927
— var. *rouxii* (Heribaud) Cleve 1895* Thesis
— var. *tenuistriata* Geitler 1932*

Order BACILLARIALES
BACILLARIACEAE
Denticula elegans Kützing 1844*
Denticula kuetzingii Grunow 1862
Denticula subtilis Grunow 1862
Denticula tenuis Kützing 1844
Grunowia sinuata (Thwaites) Rabenhorst 1864‡
Grunowia solgensis (Cleve-Euler) Aboal 2003‡
Grunowia tabellaria (Grunow) Rabenhorst 1864‡
Hantzschia amphilepta (Grunow) Lange-Bertalot 1993
Hantzschia amphioxys (Ehrenberg) Grunow 1880
— var. *major* Grunow 1880
— var. *robusta* Østrup 1910
Hantzschia elongata (Hantzsch) Grunow 1877
Hantzschia spectabilis (Ehrenberg) Hustedt 1959 B
Hantzschia vivacior Lange-Bertalot 1993
Nitzschia acicularis (Kützing) W.Smith 1853
— var. *closterioides* Grunow 1862*
Nitzschia acidoclinata Lange-Bertalot 1976*
Nitzschia aequorea Hustedt 1939*
Nitzschia agnita Hustedt 1957* B Thesis
Nitzschia alpina Hustedt 1943*
Nitzschia amphibia Grunow 1862
— f. *frauenfeldii* (Grunow) Lange-Bertalot 1987*
Nitzschia apiceconica Kufferath 1956*
Nitzschia brevissima Grunow 1881‡ B
Nitzschia capitellata Hustedt 1922* B Thesis
Nitzschia clausii Hantzsch 1860* B
Nitzschia closterium (Ehrenberg) W.Smith 1853‡ B
Nitzschia communis Rabenhorst 1860
Nitzschia commutata Grunow 1880 B
Nitzschia dissipata (Kützing) Grunow 1862
— var. *media* (Hantzsch) Grunow 1881*
Nitzschia dubia W.Smith 1853 B
Nitzschia fasciculata (Grunow) Grunow 1881 B
Nitzschia flexa Schumann 1862
Nitzschia fonticola Grunow 1879‡
Nitzschia fossilis (Grunow) Grunow 1881*
Nitzschia frustulum (Kützing) Grunow 1880 B [*N. austriaca* Hustedt 1959]
Nitzschia gracilis Hantzsch sensu Lange-Bertalot 1976
Nitzschia harderii Hustedt 1949
Nitzschia hantzschiana Rabenhorst 1860*
Nitzschia hierosolymitana D.Mann 1980*
Nitzschia homburgiensis Lange-Bertalot 1978*
Nitzschia inconspicua Grunow 1862 B
Nitzschia intermedia Hantzsch 1880
Nitzschia irresoluta Hustedt 1942
Nitzschia lacunarum Hustedt 1922* B
Nitzschia lacuum Lange-Bertalot 1980*
Nitzschia lamprocampa Hantzsch 1880*
Nitzschia linearis (Agardh) W.Smith 1853
— var. *subtilis* (Kützing) Hustedt 1923
— var. *suecica* Grunow 1880‡*
Nitzschia lorenziana Grunow 1879‡ B
Nitzschia microcephala Grunow 1878*
Nitzschia nana Grunow 1881‡
Nitzschia nova-zealandia Inglis 1881*
Nitzschia palea (Kützing) W.Smith 1856
— var. *debilis* (Kützing) Grunow 1880*
— var. *tenuirostris* Grunow 1881*
Nitzschia paleacea Grunow 1881
Nitzschia paleaeformis Hustedt 1950 Thesis
Nitzschia parvula W.Smith 1853 B
Nitzschia perminuta (Grunow) Peragallo 1903*
Nitzschia persuadens Cholnoky 1961 B*
Nitzschia pseudosigma Hustedt 1937
Nitzschia pusilla (Kützing) Grunow emend Lange-Bertalot 1976‡
Nitzschia recta Hantzsch 1861–1879 B
Nitzschia rosentockii Lange-Bertalot 1980* Thesis
Nitzschia rostellata Hustedt 1922*
Nitzschia ruttneri Hustedt 1938
Nitzschia scalaris (Ehrenberg) W.Smith 1853 B
Nitzschia semirobusta Lange-Bertalot 1993‡
Nitzschia sigmoidea (Nitzsch) W.Smith 1853
Nitzschia sociablilis Hustedt 1957
Nitzschia spathulata Brébisson 1853
Nitzschia subacicularis Hustedt 1922‡
Nitzschia sublinearis Hustedt 1921*
Nitzschia terrestris (Peterson) Hustedt 1934*
Nitzschia tropica Hustedt 1949*
Nitzschia umbonata (Ehrenberg) Lange-Bertalot 1978 B
Nitzschia vermicularis (Kützing) Hantzsch 1860
Tryblionella acuminata W.Smith 1853 B Thesis
Tryblionella acuta (Cleve) D.Mann 1990‡
Tryblionella aerophila (Hustedt) D.Mann 1990*
Tryblionella angustata W.Smith 1853‡
Tryblionella apiculata Gregory 1857‡ B
Tryblionella debilis Arnott 1873‡
Tryblionella gracilis W.Smith 1853‡ B
— var. *yarrensis* (Grunow)
Tryblionella hungarica (Grunow) Frenguelli 1942‡ B
Tryblionella victoriae Grunow 1862‡ B

Order CYMBELLALES
ANOMOENEIDACEAE
Anomoeoneis sphaerophora (Ehrenberg) Pfitzer 1871 B
— var. *sculpta* (Ehrenberg) Müller 1900 B
CYMBELLACEAE
Afrocymbella beccari (Grunow) Krammer 2003
Cymbella affinis Kützing 1844
Cymbella aspera (Ehrenberg) Cleve 1894
Cymbella australica (Schmidt) Cleve 1894‡
Cymbella borealis Cleve 1891
Cymbella cistula (Ehrenberg) Kirchner 1878
Cymbella cursiformis Hufford & Collins 1972‡
Cymbella cymbiformis Agardh 1830
Cymbella ehrenbergii Kützing 1844
Cymbella heteropleura (Ehrenberg) Kützing 1844*
Cymbella hungarica (Grunow) Pantocsek 1902‡
Cymbella hustedtii Krasske 1923
Cymbella kolbei Hustedt 1949
Cymbella lanceolata (Ehrenberg) Kirchner 1878
Cymbella leptoceroides Hustedt 1942
Cymbella leptoceros (Ehrenberg) Kützing 1844
Cymbella lindsayana Greville 1865
Cymbella novazeelandiana Krammer 2002‡
Cymbella pseudodelicatula Hustedt 1942
Cymbella pseudolapponica Cleve-Euler 1955*
Cymbella rainierensis Sovereign 1963* Thesis
Cymbella reinhardtii Grunow 1875
Cymbella similis Krasske 1932
Cymbella sphaerophora Cleve-Euler 1934*
Cymbella subovalis Barber 1962 #
Cymbella tumescens Cleve-Euler 1939*
Cymbella tumida (Brébisson) Van Heurck 1880
Cymbella turgidula Grunow 1875
Cymbella yarrensis (Schmidt) Cleve 1894 B
Cymbopleura amphicephala (Naegeli) Krammer 2003‡
Cymbopleura cuspidata (Kützing) Krammer 2003‡
Cymbopleura hybrida (Grunow) Krammer 2003‡
Cymbopleura naviculiformis (Auerswald) Krammer 2003‡

Cymbopleura rupicola (Grunow) Krammer 2005* Thesis
Cymbopleura subcuspidata (Krammer) Krammer 2005
Delicata delicatula (Kützing) Krammer 2003‡ #
Encyonema alpinum (Grunow) D.Mann 1990‡ T
Encyonema auerwaldsii Rabenhorst 1853‡
Encyonema blanchensis Krammer 1997* Thesis
Encyonema elginense (Krammer) D.Mann 1990‡
Encyonema gaeumannii (Meister) Krammer 1997* Thesis
Encyonema hebridicum Grunow 1891‡
Encyonema javanicum (Hustedt) D.Mann 1990‡
Encyonema jordanii (Grunow) Mills 1934‡
Encyonema lunatum (W. Smith) van Heurck 1896*
Encyonema mesianum (Cholnoky) D.Mann 1990*
Encyonema minutum (Hilse) D.Mann 1990‡
Encyonema muelleri (Hustedt) D.Mann 1990‡
Encyonema neogracile Krammer 1997‡
Encyonema rugosum (Hustedt) D.Mann 1990‡
Encyonema silesiacum (Bleisch) D.Mann 1990
Encyonema tasmaniense Krammer 1997* Thesis
Encyonema triangulum (Ehrenberg) Kützing 1849‡
Encyonema turgidum (Gregory) Grunow 1875*
Encyonopsis aequalis (W.Smith) Krammer 1997*
Encyonopsis blanchensis Krammer 1997*
Encyonopsis cesatii (Rabenhorst) Krammer 1997‡
Encyonopsis delicatissima (Hustedt) Krammer 1997*
Encyonopsis falaisiensis (Grunow) Krammer 1997‡
Encyonopsis kriegeri (Krasske) Krammer 1997‡
Encyonopsis microcephala (Grunow) Krammer 1997‡
Encyonopsis palustris (Hustedt) Krammer 1997‡
Navicella pusilla (Grunow) Krammer 1997‡
Placoneis elginensis (Gregory) Cox 1987‡
— var. *cuneata* (Møller ex Foged) Haworth & Kelly informal 2003 #
Placoneis exigua (Gregory) Mereschkowsky 1903‡
Placoneis gastrum (Ehrenberg) Mereschkowsky 1903‡
Placoneis neglecta Lowe 2004
Placoneis paraundulata T.Ohsuka 2002
Placoneis placentula (Ehrenberg) Mereschkowsky 1903‡
— var. *jenisseyensis* (Grunow) Bukhtiyarova 1995‡
Placoneis undulata (Østrup) Lange-Bertalot 2000‡
GOMPHONEMATACEAE
Didymosphenia geminata (Lyngbye) M.Schmidt 1899 A
Gomphoneis herculeana (Ehrenberg) Cleve 1894
Gomphoneis minuta Kociolek & Stoermer 1988
— var. *cassieae* Kociolek & Stoermer 1988 E?
Gomphonema acuminatum Ehrenberg 1832
— var. *trigonocephala* (Ehrenberg) Grunow 1880
Gomphonema affine Kützing 1844
Gomphonema angustatum (Kützing) Rabenhorst 1864
— var. *aequalis* (Gregory) Cleve 1894
— var. *linearis* Hustedt 1930
Gomphonema angustum Agardh 1831‡
Gomphonema augur Ehrenberg 1840
— var. *sphaerophorum* (Ehrenberg) Grunow 1878
— var. *turris* (Ehrenberg) Lange-Bertalot 1985
Gomphonema berggrenii (Cleve) Cleve 1894
— var. *capitata* Frenguelli 1926
Gomphonema calcifugum Lange-Bertalot & Reichardt 1999‡
Gomphonema clavatum Ehrenberg 1832‡
— var. *acuminatum* (Peragallo) †
Gomphonema clevei Fricke 1902
Gomphonema contraturris Lange-Bertalot & Reichardt 1993
Gomphonema dichotum Kützing 1833‡
Gomphonema gracile Ehrenberg 1838
Gomphonema helveticum Brun 1895
Gomphonema insigne Gregory 1856
Gomphonema kobayasii Kociolek & Kingston 1999* Thesis
Gomphonema lagenula Kützing 1844*
Gomphonema minutum (Agardh) Agardh 1831‡
Gomphonema olivaceum (Hornemann) Brébisson 1838
Gomphonema pala Reichardt 2001‡
Gomphonema parvulum (Kützing) Kützing 1849
— var. *micropus* (Kützing) Cleve 1894
Gomphonema procerum Reichardt & Lange-Bertalot 1991
Gomphonema productum (Grunow) Lange-Bertalot & Reichardt 1993‡
Gomphonema pumilum (Grunow) Reichardt & Lange-Bertalot 1991‡
Gomphonema subtile Ehrenberg 1843
Gomphonema truncatum Ehrenberg 1832‡
Gomphonema undulatum Hustedt 1930
Gomphonema ventricosum Gregory 1856
Gomphonema vibrio Ehrenberg 1841‡
— var. *pulvinatum* (Braun) Ross 1986*
Gomphosphenia tackei (Hustedt) Lange-Bertalot 1995
Gomphosphenia tenerrima (Hustedt) Reichardt 1999‡
— var. *nunguaensis* (Foged 1966)‡ †
Reimeria sinuata (Gregory) Kociolek & Stoermer 1987‡
RHOICOSPHENIACEAE
Rhoicosphenia abbreviata (Agardh) Lange-Bertalot 1985‡
— var. *elongata* (Cleve-Euler) †*
Rhoicosphenia marina (W.Smith) Schmidt 1889* B

Order LYRELLALES
LYRELLACEAE
Petroneis japonica (Heiden) D.Mann 1990‡

Order MASTOGLOIALES
MASTOGLOIACEAE
Aneumastus strosei (Oestrup) D.Mann 1990
Mastogloia braunii Grunow 1863 B
Mastogloia elliptica (Agardh) Cleve 1895 B
— var. *dansei* (Thwaites) Cleve 1895
Mastogloia exigua Lewis 1861 B
Mastogloia grevillei W.Smith 1856
Mastogloia lacustris (Grunow) Van Heurck 1880‡
Mastogloia lanceolata Thwaites 1856 B
Mastogloia pseudoexigua Cholnoky 1956*
Mastogloia recta Hustedt 1942 B
Mastogloia smithii Thwaites ex W.Smith 1856 B
Mastogloia subrobusta Hustedt 1942

Order NAVICULALES
AMPHIPLEURACEAE
Frustulia amphipleuroides (Grunow) Cleve-Euler 1934‡
Frustulia aotearoa Lange-Bertalot & Beier 2007* E?
Frustulia apicola Amosse 1932
Frustulia cassieae Lange-Bertalot & Beier 2007*
Frustulia crassinervia (Brébisson) Lange-Bertalot & Krammer 1996*
Frustulia erifuga Lange-Bertalot & Krammer 1996*
Frustulia gondwana Lange-Bertalot & Beier 2007*
Frustulia mageliesmontana Cholnoky 1957
Frustulia maoriana Lange-Bertalot & Beier 2007* E?
Frustulia nana Moser 1998*
Frustulia pangaeopsis Lange-Bertalot 2001
Frustulia rhomboides (Ehrenberg) De Toni 1891
— var. *lineolata* (Ehrenberg) Cleve 1894
Frustulia saxonica Rabenhorst 1851‡
Frustulia spicula Amosse 1932
Frustulia vulgaris (Thwaites) De Toni 1891
Frustulia weinholdii Hustedt 1937
BERKELEYACEAE
Berkeleya pellucida (Kützing) Giffen 1970
Parlibellus protracta (Grunow) Witkowski 2000‡ B
BRACHYSIRACEAE
Brachysira brebissonii Ross 1966‡
Brachysira lehmanniae Lange-Bertalot & Moser 1994*
Brachysira microcephala (Grunow) Compère 1986‡
Brachysira neoexilis Lange-Bertalot 1994
Brachysira procera Lange-Bertalot 1994* Thesis
Brachysira serians (Brébisson) Round & Mann 1981‡
Brachysira styriaca (Grunow) Ross 1986
Brachysira vitrea (Grunow) Ross 1966‡
Brachysira wygaschii Lange-Bertalot 1994*
Brachysira zellensis (Grunow) Round & Mann 1981‡
CALVINULACEAE
Cavinula cocconeiformis (Gregory) Mann & Stickle 1990‡
Cavinula jaernfeltii (Hustedt) Mann & Stickle 1990‡
Cavinula lacustris (Gregory) Mann & Stickle 1990‡
Cavinula lapidosa (Krasske) Lange-Bertalot 1996‡
Cavinula pseudoscutiformis (Hustedt) Mann & Stickle 1990‡
Cavinula pusio (Cleve) Lange-Bertalot 2004‡
Cavinula scutelloides (W.Smith) Lange-Bertalot 1996
Cavinula variostriata (Krasske) Mann & Stickle 1990‡
COSMIONEIDACEAE
Cosmioneis incognita (Krasske) Lange-Bertalot 2004‡
Cosmioneis pusilla (W.Smith) Mann & Stickle 1990‡ B
DIADESMIDACEAE
Diadesmis confervacea Kützing 1844‡
Diadesmis contenta (Grunow) D.Mann 1990‡
Diadesmis gallica W.Smith 1857*
Diadesmis peregrina W.Smith 1857
Luticola cohnii (Hilse) D.Mann 1990‡ B
Luticola goeppertiana (Bleisch) D.Mann 1990‡
Luticola mutica (Kützing) D.Mann 1990‡
Luticola muticoides (Hustedt) D.Mann 1990‡
Luticola muticopsis (Van Heurck) D.Mann 1990‡
Luticola saxophila (Bock) D.Mann 1990‡
Luticola ventricosa (Kützing) D.Mann 1990‡
DIPLONEIDACEAE
Diploneis boldtiana Cleve 1891*
Diploneis elliptica (Kützing) Cleve 1891
Diploneis marginistriata Hustedt 1922
Diploneis munda (Janisch) Cleve 1894
Diploneis oblongella (Naegeli) Cleve-Euler 1922
Diploneis oculata (Brébisson) Cleve 1894
Diploneis ovalis (Hilse) Cleve 1891
Diploneis parma Cleve 1891
Diploneis puella (Schumann) Cleve 1894
Diploneis suborbicularis (Gregory) Cleve 1894*
Diploneis subovalis Cleve 1894
NAVICULACEAE
Adlafia bryophila (Petersen) Moser, Lange-Bertelot & Metzeltin 1998‡ T
Adlafia minuscula (Grunow) Lange-Bertalot 1999* Thesis
Caloneis aequatorialis Hustedt 1931
Caloneis amphisbaena (Bory) Cleve 1894
Caloneis bacillaris (Gregory) Cleve 1894
Caloneis bacillum (Grunow) Cleve 1894
Caloneis budensis (Grunow) Krammer 1985‡
Caloneis clevei (Lagerstedt) Cleve 1894‡
— var. *carteriana* Foged 1979
Caloneis dubia Krammer 1987‡
Caloneis incognita Hustedt 1937
Caloneis lauta Carter & Bailey-Watts 1980*
Caloneis leptosoma (Grunow) Krammer 1985‡

Caloneis malayensis Hustedt 1942
Caloneis molaris (Grunow) Krammer 1985*
Caloneis patagonica Cleve 1894
Caloneis schumanniana (Grunow) Cleve 1894‡
Caloneis silicula (Ehrenberg) Cleve 1984‡*
— var. *inflata* (Grunow) Cleve 1894
— var. *subundulata* (Grunow) Mayer 1917* Thesis
— var. *truncatula* (Grunow) Cleve 1894*
Caloneis tenuis (Gregory) Krammer 1985‡
Caloneis thermalis (Grunow) Krammer 1985*
Caloneis undosa Krammer 1987‡
Capartogramma crucicula (Grunow) Ross 1963
Chamaepinnularia begeri (Krasske) Lange-Bertalot 1996
Chamaepinnularia bremensis (Hustedt) Lange-Bertalot 1996* Thesis
Chamaepinnularia soehrensis (Krasske) Lange-Bertalot 1996
Eolimna minima (Grunow) Lange-Bertalot 1998‡
Eolimna muraloides (Hustedt) Lange-Bertalot & Kulikovskiy 2010*
Eolimna subminuscula (Manguin) Lange-Bertalot & Metzeltin 1998
Eolimna submuralis (Hustedt) Lange-Bertalot & Kulikovskiy 2010*
Geissleria decussis (Østrup) Lange-Bertalot & Metzelten 1996‡ B
Geissleria paludosa (Hustedt) Lange-Bertalot & Metzelten 1996‡
Geissleria similis (Krasske) Lange-Bertalot & Metzelten 1996‡
Hippodonta capitata (Ehrenberg) Lange-Bertalot, Metzeltin & Witkowski 1996‡
Hippodonta hungarica (Grunow) Lange-Bertalot, Metzeltin & Witkowski 1996‡
Hippodonta lueneburgensis (Grunow) Lange-Bertalot, Metzeltin & Witkowski 1996* Thesis
Kobayasiella jaagii (Meister) Lange-Bertalot 1999‡
Kobayasiella madumensis (Jorgensen) Lange-Bertalot 1999
Kobayasiella parasubtilissima (Kobayasi & Nagumo) Lange Bertalot 1999* Thesis
Kobayasiella subtilissima (Cleve) Lange Bertalot 1999‡
Mayamaea agrestis (Hustedt) Lange-Bertalot 2001* Thesis
Mayamaea atomus (Kützing) Lange-Bertalot 1997* Thesis
Mayamaea recondita (Hustedt) Lange-Bertalot 1997* Thesis
Microcostatus kuelbsii (Lange-Bertalot) Lange-Bertalot 1999*
Navicula angusta Grunow 1860*
Navicula barberiana Foged 1979
Navicula bicephala Hustedt 1952
Navicula capitoradiata Germain 1981‡ B
Navicula cari Ehrenberg 1836
Navicula cassieana Foged 1979
Navicula cholnokyana Foged 1975
Navicula cincta (Ehrenberg) Ralfs 1861
Navicula cinctaeformis Hustedt 1937 B
Navicula cryptocephala Kützing 1844
Navicula cryptotenella Lange-Bertalot 1985
Navicula dahurica Skvortzow 1937
Navicula difficillima Hustedt 1950*
Navicula erifuga Lange-Bertalot 1985 B*
Navicula exilis Kützing 1844‡
Navicula feuerbornii Hustedt 1936
Navicula flanatica var. *ammophila* (Grunow) Ross 1986*
Navicula gottlandica Grunow 1857
Navicula gregaria Donkin 1861 B
Navicula harderii Hustedt 1949‡
Navicula heimansioides Lange-Bertalot 1993*
Navicula lanceolata (Agardh) Ehrenberg 1838 B
Navicula libonensis Schoeman 1970 B
Navicula magnifica Hustedt 1934
Navicula margalithii Lange-Bertalot 1985* Thesis
Navicula medioconvexa Hustedt 1961
Navicula menisculus Schumann 1867
Navicula microcari Lange-Bertalot 1993 Thesis
Navicula microcephala Grunow 1880
Navicula notanda Østrup 1913‡
Navicula notha Wallace 1960*
Navicula oblonga (Kützing) Kützing 1844
Navicula obsoleta Hustedt 1942*
Navicula perminuta Grunow 1880
Navicula platystoma Ehrenberg 1838
Navicula cf. *praeterita* Hustedt 1945* Thesis
Navicula pseudostundii Barber & Carter 1971
Navicula radiosa Kützing 1844
Navicula recens (Lange-Bertalot) Lange-Bertalot 1985* B Thesis
Navicula rhynchocephala Kützing 1844
— var. *elongata* Grunow 1860
Navicula rostellata Kützing 1844‡
Navicula roteana (Rabenhorst) Grunow 1880*
Navicula schroeteri Meister 1932
Navicula seminuloides Hustedt 1937*
— var. *sumatrensis* Hustedt 1937*
Navicula sequens Barber & Carter 1971
Navicula slesvicensis Grunow 1880‡ B
Navicula splendicula van Landingham 1975*
Navicula subrhyncocephala Hustedt 1935*
Navicula subrotundata Hustedt 1945*
Navicula tenelloides Hustedt 1937
Navicula tripunctata (Müller) Bory 1822‡
Navicula upsaliensis (Grunow) Peragallo 1903*
Navicula veneta Kützing 1844‡
Navicula viridula (Kützing) Ehrenberg 1838
— var. *linearis* Hustedt 1937
Navicula virihensis Cleve-Euler 1953
Navicula vulpina Kützing 1844
Navicula woltereckii Hustedt 1934
Naviculadicta cosmopolitiana Lange-Bertalot 2000*
Naviculadicta hambergii (Hustedt) Lange-Bertalot & Metzeltin 1995‡
Naviculadicta laterostrata (Hustedt) Lange-Bertalot & Metzeltin 1994
Naviculadicta obdurata Hohn & Hellerman) Lange-Bertalot & Metzeltin 1996‡
Naviculadicta vaucheriae (Petersen) Lange-Bertalot 1994
Nupela chilensis (Krasske) Lange-Bertalot 1994* Thesis
Nupela paludigena (Scherer) Lange-Bertalot 1994*
NEIDIACEAE
Neidium affine (Ehrenberg) Pfitzer 1871
Neidium amphigomphus (Ehrenberg) Pfitzer 1871‡
Neidium ampliatum (Ehrenberg) Krammer 1985‡
Neidium bisulcatum (Lagerstedt) Cleve 1894
Neidium dubium (Ehrenberg) Cleve 1894
Neidium gracile Hustedt 1938
Neidium hitchcockii (Ehrenberg) Cleve 1894
Neidium iridis (Ehrenberg) Cleve 1894
Neidium longiceps (Gregory) Ross 1947‡
Neidium productum (W.Smith) Cleve 1894
Neidium sacoense Reimer 1966*
PINNULARIACEAE
Diatomella balfouriana Greville 1855
Diatomella parva Manguin in Kociolek & Reviers 1996
Pinnularia acoricola Hustedt 1934
Pinnularia acrosphaeria W.Smith 1853
— var. *laevis* (M. Peragallo & Héribaud) Cleve 1895
Pinnularia acuminata W.Smith 1853‡
Pinnularia anglica Krammer 1992
Pinnularia appendiculata (Agardh) Cleve 1895
Pinnularia barberiana Foged 1979
— var. *opouraensis* Foged 1979
Pinnularia biceps Gregory 1856*
Pinnularia bilobata Cleve-Euler 1955
Pinnularia bogotensis (Grunow) Cleve 1895
Pinnularia borealis Ehrenberg 1843
— var. *rectangularis* Carlson 1913
Pinnularia brauniana (Grunow) Studnicka 1888‡
Pinnularia brebissonii var. *acuta* Cleve-Euler 1955‡
— var. *bicuneata* Grunow 1880
Pinnularia brevicostata Cleve 1891
— var. *sumatrana* Hustedt 1934
Pinnularia brevistriata Cleve 1891
Pinnularia cardinalis (Ehrenberg) W.Smith 1853
Pinnularia chapmaniana Foged 1979
Pinnularia cuneata (Østrup) Cleve-Euler 1915
Pinnularia curtistriata Hustedt 1942
Pinnularia dactylus Ehrenberg 1843
Pinnularia decrescens (Grunow) Krammer 2000‡
Pinnularia distinguenda (Cleve) Cleve 1891
Pinnularia divergens W.Smith 1853
Pinnularia divergentissima (Grunow) Cleve 1895
Pinnularia dubitalis Hustedt 1949
Pinnularia episcopalis Cleve 1891
Pinnularia gentilis (Donkin) Cleve 1891
Pinnularia gibba Ehrenberg 1841
— var. *sancta* (Grunow) Meister 1932
Pinnularia globiceps Gregory 1856
Pinnularia graciloides Hustedt 1942
— var. *triundulata* (Fontell) Krammer 2000
Pinnularia hartleyana Greville 1865
Pinnularia hemiptera (Kützing) Rabenhorst 1853
Pinnularia hudsonensis Ross 1947
Pinnularia hustedtii Meister 1934
Pinnularia lagerstedtii (Cleve) Cleve-Euler 1934
Pinnularia lata (Brébisson) Rabenhorst 1853 T
Pinnularia legumen (Ehrenberg) Ehrenberg 1843
— var. *florentina* (Grunow) Cleve 1895
Pinnularia macilenta Ehrenberg 1843* Thesis
Pinnularia maior (Kützing) Rabenhorst 1853
— var. *kuetzingii* A.Cleve 1891
Pinnularia mesolepta (Ehrenberg) W.Smith 1853
Pinnularia microstauron (Ehrenberg) Cleve 1891
— var. *ambigua* Meister 1912
— var. *australis* Manguin 1854
— var. *diminuta* (Grunow) Mayer 1913‡
Pinnularia neomajor var. *frequentis* Krammer 2000
Pinnularia nodosa (Ehrenberg) W.Smith 1856
Pinnularia nobilis (Ehrenberg) Ehrenberg 1843
Pinnularia notata Heiden & Kolbe 1928
Pinnularia obscura Krasske 1932
Pinnularia opouraensis Foged 1979
Pinnularia oriunda Krammer 1992*
Pinnularia ovata Krammer 2000*
Pinnularia polyonca (Brébisson) W.Smith 1856
— var. *stidolphii* Krammer 2000
Pinnularia rangoonensis Grunow 1895
Pinnularia rupestris Hantzsch 1861‡
Pinnularia schroederi (Hustedt) Krammer 1985‡
Pinnularia schroeterae Krammer 2000
Pinnularia segariana Foged 1979 E?
Pinnularia stidolphii Krammer 2000
Pinnularia stomatophora (Grunow) Cleve 1891
Pinnularia streptoraphe Cleve 1891
Pinnularia subbrevistriata Krammer 2000‡
Pinnularia subcapitata Gregory 1856‡
— var. *elongata* Krammer 1992* Thesis
— var. *paucistriata* (Grunow) Cleve 1895
Pinnularia subgibba Krammer 1992‡
Pinnularia sundaensis Hustedt 1938
Pinnularia transversiformis Krammer 2000
Pinnularia triumvirorum var. *linearis* Hustedt 1934
— var. *ventricosa* Hustedt 1934
Pinnularia tropica Hustedt 1949
Pinnularia viridis (Nitzsch) Ehrenberg 1843
— var. *semicruciata* (Ehrenberg) Cleve 1895
Pinnularia woodiana Foged 1979

PLEUROSIGMATACEAE
Gyrosigma acuminatum (Kützing) Rabenhorst 1853 B
— var. *brebissonii* (Grunow) Cleve 1894 B
Gyrosigma attenuatum (Kützing) Rabenhorst 1853
Gyrosigma exilis (Grunow) Reimer 1966* B Thesis
Gyrosigma nodiferum (Grunow) Reimer 1966
Gyrosigma parkerii (Harrison) Elmore 1924*
Gyrosigma peisonis (Grunow) Hustedt 1930* B
Gyrosigma obtusatum (Sullivan & Wormley) Boyer 1922
Gyrosigma strigilis (W.Smith) Griffith & Henfrey 1855* B
Gyrosigma turgidum Stidolph 1988
Pleurosigma crookii Inglis 1881
Pleurosigma salinarum Grunow 1880* B
SELLAPHORACEAE
Fallacia helensis (Schulz) D.Mann 1990* Thesis
Fallacia naumannii (Hustedt) D.Mann 1990*
Fallacia pygmaea (Kützing) Stickle & D.Mann 1990 B
Fallacia vittata var. *rotoehuensis* Foged 1979‡
Sellaphora americana (Ehrenberg) Mann 1989‡
Sellaphora bacillum (Ehrenberg) Mann 1989‡
Sellaphora gregoryana (Cleve & Grunow) Metzeltin & Lange-Bertalot 1998‡
Sellaphora laevissima (Kützing) Mann 1989‡
— var. *semiaperta* (Barber & Carter 1971)‡ †
Sellaphora parapupula Lange-Bertalot 1996‡
Sellaphora pupula (Kützing) Mereschkowsky 1902‡
— var. *elliptica* (Hustedt) Bukhtiyarova 1995
Sellaphora rectangularis (Grunow) Lange-Bertalot & Metzeltin 1996
Sellaphora seminulum (Grunow) Mann 1989‡
Sellaphora stroemii (Hustedt) Kobayasi 2002* Thesis
Sellaphora ventroconfusa (Lange-Bertalot) Metzeltin & Lange-Bertalot 1998‡
Sellaphora verecundiae Lange-Bertalot 1994 Thesis
STAURONEIDACEAE
Craticula cuspidata (Kützing) D.Mann 1990‡
Craticula halophila (Grunow) D.Mann 1990‡ B
Craticula molestiformis (Hustedt) Mayama 1999*
Craticula perrotettii Grunow 1868‡
Craticula riparia (Hustedt) Lange-Bertalot 1993‡* Thesis
Craticula submolesta (Hustedt) Lange-Bertalot 1996*
Stauroneis acuta W.Smith 1853
— var. *inflata* (Heiden) Frenguelli 1926
— var. *terryana* Tempère 1894*
Stauroneis agrestis Petersen 1915
Stauroneis anceps Ehrenberg 1843
Stauroneis borrichii (Petersen) Lund 1946
Stauroneis constricta Ehrenberg 1843
Stauroneis frauenfeldiana (Grunow) Heiden 1903
Stauroneis fulmen Brightwell 1859
— var. *capitata* Heiden 1903
Stauroneis gracilis Ehrenberg 1841
Stauroneis javanica (Grunow) Cleve 1894
Stauroneis kriegerii Patrick 1966
Stauroneis obtusa Lagerstedt 1873
Stauroneis pachycephala Cleve 1881
Stauroneis phoenicenteron (Nitzsch) Ehrenberg 1843
Stauroneis producta Grunow 1880
Stauroneis reicheltii Heiden 1903*
Stauroneis rotundata Greville 1866
Stauroneis scaphulaeformis Greville 1866
Stauroneis siberica (Grunow) Lange-Bertalot & Krammer 1996
Stauroneis smithii Grunow 1860‡
Stauroneis thermicola (Petersen) Lund 1946* Thesis

Order RHOPALODIALES
RHOPALODIACEAE
Epithemia adnata (Kützing) Brébisson 1838‡
— var. *porcellus* (Kützing) Patrick 1975‡
— var. *proboscidea* (Kützing) Hendey 1954‡
— var. *saxonica* (Kützing) Patrick 1975‡
Epithemia argus (Ehrenberg) Kützing 1844
Epithemia cistula (Ehrenberg) Ralfs 1861
Epithemia frickei Krammer 1987‡
Epithemia goeppertiana Hilse 1860‡
Epithemia hyndmanii W.M.Smith
Epithemia smithii Carruthers 1864
Epithemia sorex Kützing 1844
Epithemia turgida (Ehrenberg) Kützing 1844
— var. *capitata* Fricke 1904
— var. *granulata* (Ehrenberg) Brun 1880
— var. *westermannii* (Ehrenberg) Grunow 1862
Rhopalodia brebissonii Krammer 1987*
Rhopalodia gibba (Ehrenberg) O. Müller 1895
Rhopalodia gibberula (Ehrenberg) O. Müller 1895 B
— var. *producta* (Grunow) O. Müller 1900
— var. *vanheurckii* O. Müller 1900
Rhopalodia novaezelandiae Hustedt 1913
— var. *ventricosa* Skvortzow 1938
Rhopalodia operculata (Agardh) Håkansson 1979
Rhopalodia rupestris (W.Smith) Krammer 1987‡ B

Order SURIRELLALES
ENTOMONEIDACEAE
Entomoneis wilkinsonia (Stidolph 1985) Cassie 1989*
SURIRELLACEAE
Campylodiscus hibernicus Ehrenberg 1845‡
Cymatopleura solea (Brébisson) W.Smith 1851
— var. *apiculata* (W.Smith) Ralfs 1861
Stenopterobia anceps (Lewis) Brébisson 1896
Stenopterobia curvula (W.Smith) Krammer 1987‡
Stenopterobia delicatissima (Lewis) Brébisson 1896‡
Stenopterobia densestriata (Hustedt) Krammer 1987*
Stenopterobia sigmatella (Gregory) Ross 1996
Surirella amphioxys W.Smith 1856‡ B
Surirella angusta Kützing 1844
Surirella bifrons Ehrenberg 1843‡
Surirella biseriata Brébisson 1836
— var. *bicuspidata* Cleve-Euler 1922
Surirella bohemica Maly 1895* Thesis
Surirella brebissonii Krammer & Lange-Bertalot 1987 # B
— var. *kuetzingii* Krammer & Lange-Bertalot 1987*
Surirella celebesiana Hustedt 1922
Surirella coei Cholnoky 1960
Surirella contorta Kitton 1875 B
Surirella elegans Ehrenberg 1843
Surirella engleri var. *constricta* O. Müller 1904*
Surirella fimbriata Hustedt 1942
Surirella fluviicygnorum John 1983‡
Surirella gracilis (W.Smith) Grunow 1862
Surirella guatimalensis Ehrenberg 1854
Surirella inducta Schmidt 1882
Surirella linearis W.Smith 1853
— var. *constricta* Grunow 1862
— var. *helvetica* (Brun) Meister 1912
Surirella minuta Brébisson 1849
Surirella ovalis Brébisson 1838 B
— var. *excelsa* Müller 1899
Surirella patella Kützing 1844* Thesis
Surirella rattrayi Schmidt 1885
Surirella robusta Ehrenberg 1841
Surirella rorata Frenguelli 1935
Surirella rudis Hustedt 1922
Surirella splendida (Ehrenberg) Kützing 1844
Surirella tenera Gregory 1856
— var. *nervosa* Schmidt 1875‡
— var. *splendidula* Schmidt 1875*
Surirella tenuissima Hustedt 1913
Surirella turgida W.Smith 1853

Order THALASSIOPHYSSALES
CATENULACEAE
Amphora commutata Grunow 1880 B
Amphora delicatissima Krasske 1930
Amphora fogediana Krammer 1985
Amphora libyca Ehrenberg 1840‡
Amphora montana Krasske 1932*
Amphora normanii Rabenhorst 1864
Amphora oligotraphenta Lange-Bertalot 1996‡
Amphora ovalis (Kützing) Kützing 1844
Amphora pediculus (Kützing) Grunow 1880
Amphora schroederi Hustedt 1921
Amphora subturgida Hustedt 1938
Amphora veneta Kützing 1844 B
Hamatusia reediana (Stidolph) Stidolph 1993‡ B #

INCERTAE SEDIS (Naviculoid)
Decussata placenta (Ehrenberg) Lange-Bertalot & Metzeltin 2000‡ T #

Superorder EUNOTIANAE
Order EUNOTIALES
EUNOTIACEAE
Actinella aotearoaia Lowe, Biggs & Francoeur 2001* E?
Actinella brasiliensis Grunow 1881
Actinella indistincta Vyverman & Bergey 2001*
Actinella parva Vanhoutte & Sabbe 2001*
Actinella pulchella Sabbe & Hodgson 2001*
Amphicampa mirabilis Ehrenberg 1861‡
— var. *transsylvanica* (Pantocsek 1892)‡ †
Eunophora berggrenii (Cleve) D.Mann, Harper & Levkov 2009‡
Eunophora oberonica Vyverman & Hodgson 1998*
Eunotia arcus Ehrenberg 1837
Eunotia bidens Ehrenberg 1843‡
Eunotia bidentula W.Smith 1856
Eunotia bigibba Kützing 1849
Eunotia bilunaris (Ehrenberg) Schaarschmidt 1881‡
Eunotia camelus Ehrenberg 1841
Eunotia cancellata Berg 1953*
Eunotia cassiae Barber 1962
Eunotia cristagalli Cleve 1891
Eunotia didyma Grunow in Hustedt 1914
Eunotia diodon Ehrenberg 1837
Eunotia eruca Ehrenberg 1844
Eunotia exigua (Brébisson) Rabenhorst 1864
— var. *tenella* (Grunow) Nörpel & Alles 1991‡
Eunotia faba (Ehrenberg) Grunow 1881‡
Eunotia fallax Cleve 1895
Eunotia flexuosa (Brébisson) Kützing 1849
Eunotia formica Ehrenberg 1843
Eunotia garusica Cholnoky 1952
Eunotia glacialis Meister 1912
Eunotia gracillima (Krasske) Norpel 1996‡
Eunotia guyanensis (Ehrenberg) De Toni 1992‡
Eunotia hexaglyphis Ehrenberg 1854
Eunotia implicata Nörpel et al. 1991
Eunotia inaequalis Peragallo 1914*
Eunotia incisa Gregory 1854*
Eunotia indica Grunow 1865
Eunotia inflata (Grunow) Nörpel, Lange-Bertalot & Alles 1991‡
Eunotia meisteri Hustedt 1930
Eunotia microcephala Krasske 1932‡
Eunotia minor (Kützing) Grunow 1881‡
Eunotia monodon Ehrenberg 1843
— var. *tropica* Hustedt 1927
Eunotia mucophila (Lange-Bertalot & Nörpel-Schempp) Lange-Bertalot 2005
Eunotia muscicola var. *perminuta* (Grunow) Nörpel & Lange-Bertalot 1991
— var. *tridentula* Nörpel & Lange-Bertalot 1991
Eunotia naegelii Migula 1907
Eunotia ophidacampa Cleve 1882

Eunotia papilio (Ehrenberg) Hustedt 1911
Eunotia parallela Ehrenberg 1843
Eunotia pectinalis (Kützing) Rabenhorst 1864
— var. *undulata* (Ralfs) Rabenhorst 1864
— var. *ventralis* (Ehrenberg) Hustedt 1911
Eunotia polydentula (Brun) Hustedt 1932*
Eunotia praerupta Ehrenberg 1843
Eunotia quaternaria Ehrenberg 1841
Eunotia rhomboidea Hustedt 1950
Eunotia similis Hustedt 1933
Eunotia soleirolii (Kützing) Rabenhorst 1864*
Eunotia subaequalis Hustedt 1933
Eunotia sudetica Müller 1898
— var. *hamuraensis* Foged 1979
Eunotia tecta Krasske 1939*
Eunotia trigibba Hustedt 1913
Eunotia triodon Ehrenberg 1837
Eunotia veneris (Kützing) De Toni 1892

MARINE DIATOMS
F. Hoe Chang and Margaret Harper

Classification and notations as for freshwater diatoms. Some records are sourced from Stidolph (1988), Sterrenburg (1991, 1992), Cochran (2002), and Harper & Harper (2010).

Subclass COSCINODISCOPHYCIDAE
Superorder BIDDULPHIANAE
Order ANAULALES
ANAULACEAE
Anaulus balticus Simonsen 1959* Thesis
Anaulus birostratus Grunow 1887
Anaulus mediterraneus Grunow 1882
Eunotogramma laevis Grunow 1883* B
Eunotogramma marinum (W.Smith) Peragallo 1908*
Eunotogramma rectum Salah 1955* Thesis

Order BIDDULPHIALES
BIDDULPHIACEAE
Biddulphia alternans (Bailey) van Heurck 1885
Biddulphia antediluviana (Ehrenberg) van Heurck 1883*
Biddulphia biddulphiana (W.Smith) Boyer 1900
Biddulphia brachiolatum (Brightwell) Taylor 1970
Biddulphia picturatum (Greville) Taylor 1970
Biddulphia spinosum (Bailey) Boyer 1900
Biddulphia tridens (Ehrenberg) Ehrenberg 1840*
Euodiella semicircularis (Brightwell) Sims 2000*
Leudugeria janischii (Grunow) Van Heurck 1896
Neohuttonia reichardtii (Grunow) Kuntze 1898*
Terpsinoë americana (Bailey) Ralfs 1861 B
Trigonium reticulum (Ehrenberg) Simonsen 1974

Order HEMIAULALES
BELLEROCHEACEAE
Bellerochea malleus (Brightwell) van Heurck 1895
HEMIAULACEAE
Cerataulina pelagica (Cleve) Hendey 1937*
Eucampia antarctica (Castracane) Manguin 1915*
Eucampia zodiacus Ehrenberg 1839
Hemiaulus hauckii Grunow 1882
Hemiaulus sinensis Greville 1865

Order TRICERATIALES
PLAGIOGRAMMACEAE
Dimeregramma maculatum (Cleve) Frenguelli f. *mercuryensis* Foged 1979
Dimeregramma minor (Gregory) Ralfs 1861
— var. *nana* (Gregory) Van Heurck 1885*
Plagiogramma appendiculatum Giffen 1975*
Plagiogramma interruptum (Gregory) Ralfs 1861
Plagiogramma pulchellum Greville 1859
— var. *pygmaeum* (Greville) Peragallo 1901*
Plagiogramma robertsianum Greville 1863
Plagiogramma staurophorum (Gregory) Heiberg 1863
Plagiogramma validum Greville 1859
TRICERATIACEAE
Auliscus pruinosus Bailey 1853
Auliscus sculptus (W.Smith) Ralfs 1861
Auliscus splendidus Rattray 1888 Thesis
Cerataulus turgidus (Ehrenberg) Ehrenberg 1843
Glyphodiscus stellatus Greville 1862*
Odontella aurita (Lyngbye) C. Agardh 1832
— var. *obtusa* (Kützing) Denys 1982
Odontella mobiliensis (Bailey) Grunow 1884
Odontella sinensis (Greville) Grunow 1884
Pleurosira laevis (Ehrenberg) Compère 1982* B Thesis
Triceratium balearicum f. *biquadrata* (Janisch) Hustedt 1930
Triceratium broeckii Leuduger-Fortmorel 1879*
Triceratium dictyotum Ross & Sims 1990
— var. *rawsonii* (Wood 1961) †
Triceratium distinctum Janisch 1855*
Triceratium dubium Brightwell 1859
Triceratium favus Ehrenberg 1839
Triceratium obtusum Ehrenberg 1844
Triceratium pentacrinus (Ehrenberg) Wallich 1858
Triceratium receptum Schmidt 1885*
Triceratium robertsianum Greville 1863

Superorder CHAETOCEROTANAE
Order CHAETOCEROTALES
ATTHEYACEAE
Attheya armata (T. West) Crawford 1994
CHAETOCEROTACEAE
Bacteriastrum delicatulum Cleve 1897
Bacteriastrum furcatum Shadbolt 1854
Bacteriastrum hyalinum Lauder 1864
Chaetoceros affinis Lauder 1864 B
Chaetoceros atlanticus Cleve 1873*
— f. *audax* (Schütt) Gran 1904*
Chaetoceros borealis Bailey 1854
Chaetoceros brevis Schütt 1895
Chaetoceros concavicornis Manguin 1917
Chaetoceros constrictus Gran 1897*
Chaetoceros contorta Schütt 1895
Chaetoceros convolutus Castracane 1886
Chaetoceros costatus Pavillard 1911
Chaetoceros criophilus Castracane 1886
Chaetoceros curvisetus Cleve 1889*
Chaetoceros danicus Cleve 1889
Chaetoceros debilis Cleve 1894
Chaetoceros decipiens Cleve 1873
Chaetoceros diadema (Ehrenberg) Gran 1897
Chaetoceros dichaeta Ehrenberg 1844
Chaetoceros didymus Ehrenberg 1845
Chaetoceros difficilis Cleve 1900
Chaetoceros eibenii (Grunow) Meunier ex van Heurck 1882
Chaetoceros holsaticus Schütt 1895*
Chaetoceros laciniosus Schütt 1895
Chaetoceros lorenzianus Grunow 1863
Chaetoceros neglectus Karsten 1906*
Chaetoceros peruvianus Brightwell 1856*
Chaetoceros pseudocurvisetus Manguin 1910
Chaetoceros protuberans Lauder 1864
Chaetoceros secundus Cleve 1873
Chaetoceros seychellarum var. *australis* Manguin 1968
Chaetoceros similis Cleve 1896
Chaetoceros simplex Ostenfeld 1901
Chaetoceros socialis Lauder 1864
Chaetoceros teres Cleve 1896
Chaetoceros vanheurcki Gran 1900
Chaetoceros vistulae Apstein 1909 B

Order LEPTOCYLINDRALES
LEPTOCYLINDRACEAE
Leptocylindrus danicus Cleve 1889
Leptocylindrus mediterraneus (Peragallo) Hasle 1975
Leptocylindrus minimus Gran 1915*

Superorder CORETHRANAE
Order CORETHRALES
CORETHRACEAE
Corethron criophilum Castracane 1886*

Superorder COSCINODISCANAE
Order ASTEROLAMPRALES
ASTEROLAMPRACEAE
Asteromphalus challengerensis Castracane 1886*
Asteromphalus flabellatus (Brébisson) Greville 1859
Asteromphalus heptactis (Brébisson) Ralfs 1861*
Asteromphalus hookeri Ehrenberg 1844

Order COSCINODISCALES
AULACODISCACEAE
Aulacodiscus argus (Ehrenberg) Schmidt 1886
Aulacodiscus aucklandicus Grunow 1876
Aulacodiscus beeveriae Johnson ex Pritchard 1861
Aulacodiscus brownei Norman ex Pritchard 1861
Aulacodiscus kittoni Arnott 1861
— var. *johnsonii* (Arnott) Rattray 1888
Aulacodiscus petersii Ehrenberg 1845*
COSCINODISCACEAE
Coscinodiscus argus Ehrenberg 1838
Coscinodiscus asteromphalus Ehrenberg 1844
Coscinodiscus centralis Ehrenberg 1844
Coscinodiscus concinnus W.Smith 1856
Coscinodiscus curvatulus Grunow in Schmidt 1878
Coscinodiscus denarius var. *variolatus* (Castracane) Rattray 1890*
Coscinodiscus fimbriatus Ehrenberg 1854
Coscinodiscus gigas Ehrenberg 1841
Coscinodiscus granii Gough 1905
Coscinodiscus janischii Schmidt 1878
Coscinodiscus marginatus Ehrenberg 1843
Coscinodiscus oculus-iridis Ehrenberg 1854
Coscinodiscus pacificus (Grunow) Rattray 1890*
Coscinodiscus radiatus Ehrenberg 1841
Coscinodiscus wailesii Gran & Angst 1931
Cosmiodiscus elegans Greville 1866
GOSSLERIELLACEAE
Gossleriella punctata Wood, Crosby & Cassie 1959
HELIOPELTACEAE
Actinoptychus adriaticus Grunow 1863
Actinoptychus punctatus Ehrenberg 1843
Actinoptychus senarius (Ehrenberg) Ehrenberg 1843
Actinoptychus splendens (Shadbolt) Ralfs 1861
Actinoptychus vulgaris Schumann 1867
HEMIDISCACEAE
Actinocyclus barklyi var. *aggregatus* Rattray 1890
Actinocyclus curvatulus Janisch 1878*
Actinocyclus octonarius Ehrenberg 1838
— var. *ralfsii* (W.Smith) Hendey 1954
— var. *sparsus* (Gregory) Hendey 1954
— var. *tenellus* (Brébisson) Hendey 1954
Actinocyclus pyrotechnicus Deby 1890*
Actinocyclus subtilis (Gregory) Ralfs in Pritchard 1861*
Hemidiscus cuneiformis Wallich 1860
Roperia tesselata (Roper) Grunow 1889*

Order MELOSIRALES
HYALODISCACEAE
Hyalodiscus pustulatus Schmidt 1889
Hyalodiscus radiatus var. *maximus* (Petit) Cleve 1880
Hyalodiscus scoticus (Kützing) Grunow 1879
Hyalodiscus subtilis Bailey 1854
Podosira maxima (Kützing) Grunow 1879*
Podosira montagnei Kützing 1844*
Podosira stelliger (Bailey) A.Mann 1907*

MELOSIRACEAE
Melosira moniliformis (O.Müller) Agardh 1824 B
Melosira nummuloides (Dilwyn) Agardh 1824* B
PSEUDOPODOSIRACEAE #
Pseudopodosira westii (W.Smith) Scheschukova-Poretzkeya 1964 #
STEPHANOPYXIDACEAE
Stephanopyxis cruciata (Ehrenberg) Tempère & Peragallo 1889
Stephanopyxis orbicularis Wood, Crosby & Cassie 1959
Stephanopyxis turris (Greville & Arnott) Ralfs 1861

Order PARALIALES
PARALIACEAE
Paralia sulcata (Ehrenberg) Cleve 1873

Superorder CYMATOSIRANAE
Order CYMATOSIRALES
CYMATOSIRACEAE
Cymatosira belgica Grunow 1881*
Cymatosira elliptica Salah 1955* B Thesis
Plagiogrammopsis crawfordii Witkowski 2001* Thesis
Plagiogrammopsis mediaequatus Gardner & Crawford 1994* B
Plagiogrammopsis vanheurckii (Grunow) Hasle, von Stosch & Syvertsen 1983* B

Superorder LITHODESMIANAE
Order LITHODESMIALES
LITHODESMIACEAE
Ditylum brightwelli (T. West) Grunow 1883
Helicotheca tamesis (Shrubsole) Ricard 1987
Lithodesmium undulatum Ehrenberg 1839

Superorder RHIZOSOLENIANAE
Order RHIZOSOLENIALES
RHIZOSOLENIACEAE
Dactyliosolen antarcticus Castracane 1886*
Dactyliosolen blavyanus (Peragallo) Hasle 1975*
Guinardia delicatula (Cleve) Hasle 1996*
Guinardia flaccida (Castracane) Peragallo 1892
Guinardia striata (Stolterfoth) Hasle 1996*
Neocalyptrella robusta (Norman) Hernández-Becerril & Meave 1997
Proboscia alata (Brightwell) Sundström 1986
Proboscia indica (Peragallo) Herandez-Becerril 1995
Proboscia truncata (Karsten) Nöthig & Ligowski 1991*
Pseudosolenia calcaravis (Schulze) Sundström 1986
Rhizosolenia chunii Karsten 1905*
Rhizosolenia curvata Zacharias 1905*
Rhizosolenia hebetata Bailey 1856
— f. *hiemalis* Gran 1904
— f. *semispina* (Hensen) Gran 1908
Rhizosolenia imbricata Brightwell 1858
Rhizosolenia setigera Brightwell 1858
Rhizosolenia simplex Karsten 1905*
Rhizosolenia styliformis Brightwell 1858
INCERTAE SEDIS
Haynaldella antiqua (Pantocsek) Pantocsek 1892 #
Haynaldella strigillata (Witt) †* #

Superorder THALASSIOSIRANAE
Order THALASSIOSIRALES #
LAUDERIACEAE
Lauderia annulata Cleve 1873
Lauderia borealis Gran 1900
SKELETONEMATACEAE
Detonula confervacea (Cleve) Gran 1900*
Detonula pumila (Castracane) Schütt 1896
Skeletonema costatum (Greville) Cleve 1873
STEPHANODISCACEAE
Cyclotella caspia Grunow 1878*
Cyclotella stylorum Brightwell 1860 B
THALASSIOSIRACEAE
Ehrenbergiulva granulosa (Grunow) Witkowski, Lange-Bertalot & Metzeltin 2004 B #
Planktoniella florea Wood 1959
Planktoniella sol (Wallich) Schutt 1893
Porosira glacialis (Grunow) Jørgensen 1905*
Shionodiscus oestrupii (Ostenfeld) Alverson, Kang & Theriot 2006*
Thalassiosira angulata (Gregory) Hasle 1978 B
Thalassiosira augustelineata (Schmidt) Fryxell & Hasle 1977*
Thalassiosira baltica (Grunow) Ostenfeld 1901*B
Thalassiosira eccentrica (Ehrenberg) Cleve emend Fryxell & Hasle 1972
Thalassiosira gracilis (Karsten) Hustedt 1958
Thalassiosira hyalina (Grunow) Gran 1897
Thalassiosira lentiginosa (Janisch) Fryxell 1977*
Thalassiosira leptopus (Grunow) Hasle & Fryxell 1977
Thalassiosira lineata Jousé 1968
Thalassiosira pacifica Gran & Angst 1930*
Thalassiosira oliverana (O'Meara) Makarova & Nikolaev 1984*
Thalassiosira rotula Meunier 1915
Thalassiosira subtilis (Ostenfeld) Gran 1900
Thalassiosira tumida (Janisch) Hasle 1971*

Subclass FRAGILARIOPHYCIDAE
Superorder FRAGILARIANAE
Order ARDISSONEALES
ARDISSONEACEAE
Ardissonea crystallina (Agardh) Grunow 1880*
Ardissonea formosa (Hantzsch) Grunow 1880*
Ardissonea fulgens (Greville) Grunow 1880

Order CLIMACOSPHENIACEAE
CLIMACOSPHENIACEAE
Climacosphenia moniligera Ehrenberg 1841

Order CYCLOPHORALES
CYCLOPHORACEAE
Cyclophora tenuis Castracane 1878* Thesis
ENTOPYLACEAE
Entopyla australis Ehrenberg 1848
Entopyla ocellata (Arnott) Grunow 1862
— var. *calaritana* Fricke 1902
— var. *pulchella* (Arnott) Fricke 1902

Order FRAGILARIALES
FRAGILARIACEAE
Asterionellopsis glacialis (Castracane) Round 1990*
Asteroplanus kariana (Grunow) Gardner & Crawford 1997
Bleakeleya notata (Grunow) Round 1990
Catacombas gaillonii (Bory) Williams & Round 1986
Ctenophora pulchella (Ralfs) Williams & Round 1986 B
Fragilaria striatula Lyngbye 1819 B
Hyalosynedra laevigata (Grunow) Williams & Round 1986* Thesis
Martyana schulzii (Brockman) Snoejis 1991* B Thesis
Opephora cf. *burchardtiae* Witkowski et al. 1998* Thesis
Opephora guenter-grassi (Witkowski & Lange-Bertalot) Sabbe & Vyverman 1995* B Thesis
Opephora marina (Gregory) Petit 1888*
Opephora pacifica (Grunow) Petit 1888*
Podocystis adriatica (Kützing) Ralfs 1861
Podocystis spathulatum (Shadbolt) Van Heurck 1896
Synedra frauenfeldii Grunow 1862
Synedra superba Kützing 1844
Tabularia fasciculata (Agardh) Williams & Round 1986 B
Tabularia investiens (W.Smith) Williams & Round 1986*
Tabularia parva (Kützing) Williams & Round 1986 B
Tabularia variostriata M.Harper 2009* B E?
Tabularia waernii Snoeijs 1991* B
Trachysphenia australis Petit 1877

Order LICMOPHORALES
LICMOPHORACEAE
Licmophora communis (Heiberg) Grunow 1881* Thesis
Licmophora ehrenbergii (Kützing) Grunow 1867* Thesis
Licmophora flabellata (Carmichael) Agardh 1831
Licmophora gracilis (Ehrenberg) Grunow 1867
— var. *anglica* (Kützing) Peragallo 1901*
Licmophora juergensii Agardh 1832* Thesis
Licmophora luxuriosa Heiden & Kolbe 1928
Licmophora lyngbyei (Kützing) Grunow 1867
Licmophora paradoxa (Lyngbye) Agardh 1828-1836
— var. *tincta* (Agardh) Hustedt 1931
Licmophora pfannkucheae Giffen 1970*

Order PROTORAPHIDALES
PROTORAPHIDACEAE
Pseudohimantidium pacificum Hustedt 1941*

Order RHABONEMATALES
RHABDONEMATACEAE
Rhabdonema adriaticum Kützing 1844
Rhabdonema arcuatum (Lyngbye) Kützing 1844*
Rhabdonema crassum Hendey 1964*
Rhabdonema hamuliferum Kitton 1877
Rhabdonema minutum Kützing 1844 B
Rhabdonema torelli Cleve 1873*

Order RHAPHONEIDALES
PSAMMODISCACEAE
Psammodiscus nitidus (Gregory) Round & Mann 1980
RHAPHONEIDACEAE
Delphineis karstenii (Boden) Fryxell 1978*
Delphineis minutissima (Hustedt) Simonsen 1987* B
Delphineis surirella (Ehrenberg) Andrews 1980*
— var. *australis* (Grunow) Andrews 1980
Delphineis surirelloides (Simonsen) Andrews 1980*
Diplomenora cocconeiformis (Schmidt) Blazé 1984*
Rhaphoneis amphiceros (Ehrenberg) Ehrenberg 1844
Rhaphoneis belgica (Grunow) Grunow 1885
Rhaphoneis nitida (Gregory) Grunow 1867*
Rhaphoneis superba (Janisch) Grunow 1862* Thesis

Order STRIATELLALES
STRIATELLACEAE
Grammatophora angulosa Ehrenberg 1839
— var. *islandica* (Ehrenberg) Grunow 1881*
Grammatophora arcuata Ehrenberg 1853
Grammatophora arnottii Grunow 1881
Grammatophora gibberula Kützing 1844
Grammatophora hamulifera Kützing 1844
Grammatophora longissima Petit 1877
Grammatophora macilenta W.Smith 1862
Grammatophora marina (Lyngbye) Kützing 1844
Grammatophora oceanica Ehrenberg 1840
— var. *subtilissima* (Bailey) De Toni 1894.
Grammatophora serpentina (Ralfs) Ehrenberg 1844
Grammatophora undulata Ehrenberg 1840
Hyalosira interrupta (Ehrenberg) Navarro 1991* #
Microtabella delicatula (Kützing) Round 1990*
Striatella delicatula var. *gibbosa* Østrup 1904* Thesis
Striatella unipunctata (Lyngbye) Agardh 1832

Order THALASSIONEMATALES

THALASSIONEMATACEAE
Thalassionema frauenfeldii (Grunow) Tempère & Peragallo 1910
Thalassionema nitzschioides (Grunow) Mereschowsky 1902
— var. *lanceolatum* (Grunow) Heiden 1928*
Thalassiothrix antarctica Schimper 1905*
Thalassiothrix longissima Cleve & Grunow 1880
Trichotoxon reinboldii (van Heurck) Williams & Round 1987

Order TOXARIALES
TOXARIACEAE
Toxarium undulatum Bailey 1854

INCERTAE SEDIS
Catillus cf. *subimpletus* (Peragallo) Hendey 1977* # Thesis
Hyalinella lateripunctata Witkowski 2000* #

Subclass BACILLARIOPHYCIDAE
Superorder BACILLARIANAE
Order ACHNANTHALES
ACHNANTHACEAE
Achnanthes angustata Greville 1859* B
Achnanthes brevipes Agardh 1824 B
— var. *intermedia* (Kützing) Cleve 1895* B
Achnanthes entrancensis Foged 1978* B
Achnanthes cf. *fimbriata* (Grunow) Ross 1963* Thesis
Achnanthes glabrata Grunow 1883
— var. *auklandica* Grunow 1880
Achnanthes hyperboreoides Witkowski, Metzeltin & Lange-Bertalot 2000* B Thesis
Achnanthes longipes Agardh 1824
Achnanthes parvula Kützing 1844* B
ACHNANTHIDIACEAE
Astartiella cf. *punctifera* (Hustedt) Witkowski et al. 1998* # Thesis
Astartiella wellsiae (Reimer) Witkowski & Lange-Bertalot 1998* B #
*Planothidium diplopunctatum** (Simonsen) Witkowski 2000 Thesis
Planothidium cf. *lilljeborgei* (Grunow) Witkowski 2000* B Thesis
Planothidium quarnerensis (Grunow) Witkowski, Lange-Bertalot & Metzeltin 2000
Psammothidium reversum (Lange-Bertalot & Krammer) Bukhitiyarova & Round 1996* B Thesis
COCCONEIDACEAE
Campyloneis grevillei (W.Smith) Grunow 1867
Cocconeis australis Petit 1877
Cocconeis britannica Naegeli 1849*
Cocconeis californica (Grunow) Grunow 1880*
Cocconeis capensis (Cholnoky) Witkowski 2000*
Cocconeis costata Gregory 1855
— var. *hexagona* Grunow 1880*
Cocconeis convexa Giffen 1987*
Cocconeis dirupta Gregory 1857
— var. *antarctica* Grunow 1880
— var. *flexella* (Janisch & Rabenhorst) Grunow 1880* Thesis
Cocconeis discrepans Schmidt 1894*
Cocconeis distans Gregory 1855
Cocconeis fasciolata (Ehrenberg) Brown 1920
Cocconeis finnmarchica Grunow1880*
Cocconeis heteroidea Hantzsch 1863
— var. *curvirotunda* (Tempère) Cleve 1895
Cocconeis interrupta Grunow 1863
Cocconeis notata Petit 1877
Cocconeis orbicularis Frenguelli & Orlando 1958*
Cocconeis pellucida Grunow 1863
Cocconeis peltoides Hustedt 1939* B
Cocconeis pseudodiruptoides Foged 1975*
Cocconeis pseudomarginata Gregory 1857
— var. *intermedia* Grunow 1868
Cocconeis sancti-paulii Heiden & Kolbe 1928
Cocconeis scutellum Ehrenberg 1838 B
— var. *parva* (Grunow) Cleve 1896*
Cocconeis speciosa Gregory 1855*
Cocconeis stauroneiformis (W.Smith) Okuno 1957
Cocconeis sublittoralis Hendey 1951*
Psammococconeis disculoides (Hustedt) Garcia 2001* #

Order BACILLARIALES
BACILLARIACEAE
Bacillaria paradoxa Gmelin 1788 B
Cymbellonitzschia cf. *szulczewskii* Witkowski et al. 2000* Thesis
Denticula neritica Holmes & Croll 1982*
Fragilariopsis curta (Van Heurck) Hustedt 1957*
Fragilariopsis kerguelensis (O'Meara) Hustedt 1952*
Fragilariopsis oceanica (Cleve) Hasle 1965
Giffenia cocconeiformis (Grunow) Round & Basson 1998 B
Hantzschia doigiana Stidolph 1993* B E?
Hantzschia virgata (Roper) Grunow 1880 B
— var. *leptocephala* Østrup 1910* B
Nitzschia amabilis Suzuki 2010*
Nitzschia angularis W.Smith 1853*
Nitzschia australis (Peragallo) Mann 1937
Nitzschia distans Gregory 1857*
Nitzschia filiformis (W.Smith) van Heurck 1896* B
Nitzschia frigida Grunow 1880*
Nitzschia fusiformis Grunow 1880*
Nitzschia irregularis Ross & Abdin 1949* B
Nitzschia liebethruthii Rabenhorst 1864* B
Nitzschia longissima (Brébisson) Ralfs 1861
Nitzschia marina Grunow 1880
Nitzschia martiana (Agardh) van Heurck 1896* Thesis
Nitzschia obtusa W.Smith 1853 B
Nitzschia ovalis Arnott 1880* B
Nitzschia pellucida Grunow 1880*
Nitzschia scalpelliformis (Grunow) Grunow 1880 B
Nitzschia spathulata W.Smith 1853*
Nitzschia semirobusta Lange-Bertalot 1993*
Nitzschia sigma (Kützing) W.Smith 1853 B
— var. *rigida* Grunow 1878
Nitzschia socialis Gregory 1857*
Nitzschia subacuta Hustedt 1922
Nitzschia tubicola Grunow 1880* B
Nitzschia vitrea Norman 1861* B
Psammodictyon panduriforme (Gregory) D.Mann 1990
— var. *continua* (Grunow) Snoeijis 1998* Thesis
— var. *lata* (Witt.) †*
— var. *minor* (Gregory) Haworth & Kelly informal 2003 # Thesis
Pseudo-nitzschia americana (Hasle) Fryxell in Hasle 1993*
Pseudo-nitzschia australis Frenguelli 1939*
Pseudo-nitzschia delicatissima (Cleve) Heiden 1928*
Pseudo-nitzschia fraudulenta (Cleve) Hasle 1993*
Pseudo-nitzschia heimii Manguin 1957*
Pseudo-nitzschia lineola (Cleve) Hasle 1993*
Pseudo-nitzschia multiseries (Hasle) Hasle 1995*
Pseudo-nitzschia multistriata (Takano) Takano 1993*
Pseudo-nitzschia pseudodelicatissima (Hasle) Hasle 1996*
Pseudo-nitzschia pungens (Grunow ex Cleve) Hasle 1993
Pseudo-nitzschia seriata (Cleve) Peragallo 1900
Pseudo-nitzschia turgidula (Hustedt) Hasle 1993*
Tryblionella apiculata Gregory 1857* B
Tryblionella balatonis (Grunow) D.Mann 1990* B Thesis
Tryblionella circumsuta (Bailey) Ralfs 1861 B
Tryblionella coarctata (Grunow) D.Mann 1990* B
Tryblionella granulata (Grunow) D.Mann 1990* B
Tryblionella levidensis W.Smith 1856
Tryblionella marginulata (Grunow) D.Mann 1990*
Tryblionella navicularis (Brébisson) Ralfs 1861* B
Tryblionella perversa (Grunow) D.Mann 1990* B
Tryblionella punctata Wm Smith 1853

Order CYMBELLALES
ANOMOENEIDACEAE
Staurophora salina (W.Smith) Mereschkowsky 1903* B
RHOICOSPHENIACEAE
Gomphonemopsis exigua (Kützing) Medlin 1986 B
Gomphonemopsis novo-zelandicum (Booth 1985)*
Gomphoseptatum aestuarii (Cleve) Medlin 1986*
Rhoicosphenia marina (W.Smith) Schmidt 1889* B
Rhoicosphenia adolphii Schmidt 1899*
Rhoicosphenia genuflexa (Kützing) Medlin 1982

Order LYRELLALES
LYRELLACEAE
Lyrella approximata (Greville) D.Mann 1990*
— f. *manca* (Schmidt 1874)
Lyrella atlantica (Schmidt) D.Mann 1990
Lyrella circumsecta (Grunow) D.Mann 1990
Lyrella clavata (Gregory) D.Mann 1990*
Lyrella hennedyi (W.Smith) Stickle & D.Mann 1990
Lyrella inhalata (Schmidt) D.Mann 1990
Lyrella lyra (Ehrenberg) Karajeva 1978
Lyrella nebulosa (Gregory) D.Mann 1990*
Lyrella novaeseelandiae (Hustedt) D.Mann 1990
Lyrella praetexta (Ehrenberg) D.Mann 1990
Lyrella sandriana (Grunow) D.Mann 1990*
Lyrella subirroratoides (Hustedt) D.Mann 1990
Lyrella sulcifera (Hustedt) Witkowski 1994
Petroneis granulata (Bailey) D.Mann 1990*
Petroneis humerosa (Brébisson) Stickle & D.Mann 1990
Petroneis latissima (Gregory) Stickle & D.Mann 1990 B
Petroneis marina (Ralfs) D.Mann 1990 B
— var. *novae-zealandiae* (Grunow)

Order MASTOGLOIALES
MASTOGLOIACEAE
Mastogloia angulata Lewis 1861
Mastogloia baldjikiana Grunow 1893* Thesis
Mastogloia barbadensis (Greville) Cleve 1895*
Mastogloia binotata (Grunow) Cleve 1895
Mastogloia cribrosa Grunow 1860*
Mastogloia fimbriata (Brightwell) Grunow 1863* Thesis
Mastogloia gieskesii Cholnoky 1963*
Mastogloia marginulata Grunow 1867
Mastogloia mediterranea Hustedt 1933* Thesis
Mastogloia paradoxa Grunow 1878
Mastogloia pumila (Grunow) Cleve 1895 B
Mastogloia quinquecostata Grunow 1860* B
Mastogloia rhombus (Petit) Cleve & Grove 1891
Skeletomastus coelatus (Grunow) M. Harper 2009* E?

Order NAVICULALES
AMPHIPLEURACEAE
Amphiprora angustata Hendey 1964*
Amphiprora sulcata O'Meara 1871* Thesis
BERKELEYACEAE
Berkeleya antarctica Grunow 1880*
Berkeleya fragilis Greville 1827
Berkeleya micans (Lynbye) Grunow 1868
Berkeleya parasitica var. *novaezelandiae* (Grun) De Toni 1891
Berkeleya rutilans (Trentepohl) Grunow 1880
Berkeleya scopulorum (Brébisson) Cox 1979*

— var. *belgica* (van Heurck) R.Ross 1984*
Climaconeis inflexa (Brébisson) Cox 1982* Thesis
Parlibellus bennikei Witkowski 2000* Thesis
Parlibellus berkeleyi (Kützing) Cox 1988*
Parlibellus delognei (vanHeurck) Cox 1988*
Parlibellus hamulifera (Grunow) Cox 1988* B
Parlibellus perytii Witkowski et al. 2000*
Parlibellus plicatus (Donkin) Cox 1988*
Parlibellus rhombicula (Hustedt) Witkowski 2000
DIPLONEIDACEAE
Diploneis aestuari Hustedt 1939*
Diploneis boldtiana Cleve 1891*
Diploneis bomboides (Schmidt) Cleve 1894*
— var. *madagascarensis* Cleve 1894* Thesis
Diploneis bombus (Ehrenberg) Ehrenberg 1854 B
Diploneis campylodiscus Grunow 1875*
Diploneis coffaeformis (Schmidt) Cleve 1894* Thesis
Diploneis constricta (Grunow) Cleve 1894
Diploneis crabro (Ehrenberg) Ehrenberg 1854
— var. *pandura* (Brébisson) Cleve 1894
Diploneis didyma (Ehrenberg) Ehrenberg 1854* B
Diploneis entomon (Ehrenberg) Cleve 1894
Diploneis eudoxia (Schmidt) Jorgensen 1905* Thesis
Diploneis fusca (Gregory) Cleve 1894
Diploneis gemmata (Greville) Cleve 1894
Diploneis graeffii (Grunow) Cleve 1894
Diploneis incurvata (Gregory) Cleve 1894
Diploneis interrupta (Kützing) Cleve 1894 B
Diploneis litoralis (Donkin) Cleve 1894
Diploneis mediterranea (Grunow) Cleve 1894
Diploneis nitescens (Gregory) Cleve 1894
Diploneis notabilis (Greville) Cleve 1894
Diploneis novaeseelandiae (Schmidt) Hustedt 1937*
Diploneis papula (Schmidt) Cleve 1894* Thesis
Diploneis schmidtii Cleve 1894*
Diploneis smithii (Brébisson) Cleve 1894 B
— var. *pumila* (Grunow) Hustedt 1937 B
— var. *rhombica* Mereschkowsky 1902
Diploneis vacillans (Schmidt) Cleve 1894*
— *var. renitens* (Schmidt) Cleve 1894*
Diploneis weissflogii (Schmidt) Cleve 1894
NAVICULACEAE
Caloneis africana (Griffen) Stidolph 1994*
Caloneis biseriata (Petit) Cleve 1894
Caloneis fusioides (Grunow) Heiden & Kolbe 1928* Thesis
Caloneis liber (W.Smith) Cleve 1894
Caloneis linearis (Grunow) Boyer 1927
Caloneis subsalina (Donkin) Hendey 1951* B
Caloneis westii (W.Smith) Hendey 1964 B
Cymatoneis circumvallata Cleve 1894*
Diademoides luxuriosa (Greville) Kemp & Paddock 1990
Haslea britannica (Hustedt) Witkowski 2000*
Haslea crucigera (W.Smith) Simonsen 1974
Haslea pallidum (Riznyk) Poulin & Massé 2004*
Mastoneis biformis (Grunow) Cleve 1894*
Meuniera membranacea (Cleve) Silva 1996*
Navicula abscondita Hustedt 1939*
Navicula aucklandica Grunow 1894
Navicula britannica Hustedt & Aleem 1951*
Navicula cancellata Donkin 1872
Navicula centraster Cleve 1895
Navicula climacopheniae Booth 1986*
Navicula decussata Ehrenberg 1843
Navicula digitoradiata (Gregory) Ralfs 1861 B
Navicula directa (W.Smith) Ralfs 1861*
— var. *remota* Grunow 1879
Navicula distans (W.Smith) Ralfs 1861
Navicula duerrenbergiana Hustedt 1934*
Navicula fortis (Gregory) Ralfs 1861*
Navicula fromenterae Cleve 1881*
Navicula grevilleana Hendey 1964* Thesis
Navicula hamiltonii Witkowski 2000*
Navicula hochstetteri Grunow 1863
Navicula inflexa (Gregory) Ralfs 1861
Navicula longa (Gregory) Ralfs 1861*
Navicula lusoria Giffen 1975*
Navicula luzonensis Hustedt 1942*
Navicula meniscoides Hustedt 1955*
Navicula microdigitoradiata Lange-Bertalot 1993* Thesis
Navicula cf. *normaloides* Cholnoky 1968* Thesis
Navicula opima (Grunow) Grunow 1876*
Navicula cf. *palpebralis* Brébisson 1853* Thesis
Navicula pavillardii Hustedt 1939*
Navicula pennata Schmidt 1876* Thesis
Navicula peregrina (Ehrenberg) Kützing 1844 B
Navicula perminuta Grunow 1880* B
Navicula phyllepta Kützing 1844* B Thesis
Navicula pinnata Pantocsek 1889
Navicula ramosissima (Agardh) Cleve 1895*
Navicula rudiformis Hustedt 1964*
Navicula rusticensis Lobban 1984*
Navicula salinarum Grunow 1880 B
— var. *minima* Kolbe 1927* B Thesis
Navicula salinicola Hustedt 1939*
Navicula cf. *stachurae* Witkowski 2000* Thesis
Navicula subcarinata (Grunow) Hendey 1951
Navicula aff. *taedens* Cholnoky 1968*
Navicula tairuaensis Foged 1979*
Navicula uniseriata Østrup 1913*
Navicula utlandshoerniensis van Landingham 1975* B Thesis
Navicula cf. *wasmundii* Witkowski et al. 1996*
Navicula wunsamiae Witkowski 2000
Navicula zostereti Grunow 1860*
Pinnunavis (Pinnuavis) elegans (W.Smith) Okuno 1975* B
Pseudogomphonema kamschaticum (Grunow) Medlin 1986*
Rhoikoneis [Rhoiconeis] sponsalia (Giffen) Medlin 1985*
Seminavis delicatula Wachnicka & Gaiser 2007*
Seminavis eulensteinii (Grunow) Danielidis & Economou-Amilli 2003
Seminavis macilenta (Gregory) Danielidis & D.Mann 2002* Thesis
Seminavis strigosa (Hustedt) Danielidis & Economou-Amilli 2008 Thesis
Seminavis ventricosa (Gregory) Garcia-Baptista 1993*
Seminavis witkowskii Wachnicka & Gaiser 2007*
Trachyneis aspera (Ehrenberg) Cleve 1894
— var. *contermina* (Schmidt) Cleve 1894
— var. *oblonga* (Bailey) Cleve 1894
— var. *robusta* (Petit) Cleve 1894
Trachyneis johnsoniana (Greville) Cleve 1894
Trachyneis velata (Schmidt) Cleve 1894*
Veigaludwigia sp. VanHoutte et al. 2006 #
PINNULARIACEAE
Pinnularia claviculus (Gregory) Rabenhorst 1864*
Pinnularia cf. *quadratarea* (Schmidt) Cleve 1895* B Thesis
— var. *fluminensis* (Grunow) Cleve 1895* Thesis
Pinnularia trevelyana (Donkin) Rabenhorst 1864 B
Pinnularia yarrensis (Grunow) Jurilj 1957 B
PHAEODACTYLACEAE
Phaeodactylum tricornutum Bohlin 1897*B
PLAGIOTROPIDACEAE
Plagiotropis antarctica (Grunow) Kuntze 1898*
Plagiotropis lata (Cleve) Kuntze 1898*
Plagiotropis lepidoptera (Gregory) Kuntze 1898
Plagiotropis pusilla (Gregory) Kuntze 1898*
PLEUROSIGMATACEAE
Gyrosigma balticum (Ehrenberg) Rabenhorst 1853 B
Gyrosigma beaufortianum Hustedt 1955* B Thesis
Gyrosigma diminutum var. *constricta* (Grunow) Cleve 1894* Thesis
Gyrosigma exmium (Thwaites) Boyer 1927* B
Gyrosigma fasciola (Ehrenberg) Griffith et Henfrey 1856* B
Gyrosigma fogedii Stidolph 1994* B
Gyrosigma foxtonia Stidolph 1988* B
Gyrosigma hippocampus (Ehrenberg) Hassall 1845 B
Gyrosigma lineare var. *longissima* (Cleve) Cleve 1894* Thesis
Gyrosigma macrum (W.Smith) Griffth & Henfrey 1856 B
Gyrosigma mediterraneum (Cleve) Cleve 1894*
Gyrosigma nodiferum (Grunow) Reimer 1966* B
Gyrosigma prolongatum (W.Smith) Griffth & Henfrey 1856*
Gyrosigma rectum (Donkin) Cleve 1894* Thesis
Gyrosigma scalproides (Rabenhorst) Cleve 1894
Gyrosigma spencerii (Quekett) Griffth & Henfrey 1856 B
Gyrosigma sterrenburgii Stidolph 1992*
Gyrosigma stidolphii Sterrenburg 1992*
Gyrosigma tenuissimum (W.Smith) Griffith & Henfrey 1856* B
Gyrosigma turgidum (Stidolph) Stidolph 1988*
Gyrosigma waitangiana Stidolph 1993*
Gyrosigma wansbeckii (Donkin) Cleve 1894* B
Gyrosigma wormleyi (Sullivant) Boyer 1922
Pleurosigma acus Mann 1925*
Pleurosigma aestuarii (Brébisson) W.Smith 1853* B
Pleurosigma amara Stidolph 1992*
Pleurosigma angulatum (Quekett) W.Smith 1853 B
Pleurosigma australe Grunow 1867
Pleurosigma decorum W.Smith 1853*
Pleurosigma directum Grunow 1880
Pleurosigma elongatum W.Smith 1852*
Pleurosigma falcatum (Donkin) Rabenhorst 1864
Pleurosigma formosum W.Smith 1852
Pleurosigma intermedium W.Smith 1853
Pleurosigma inscriptura M. Harper 2009*
Pleurosigma marinum Donkin 1858
Pleurosigma naviculaceum Brébisson 1854
Pleurosigma normanii Ralfs 1861
Pleurosigma rhombeum Grunow 1880
Pleurosigma rigidum W.Smith 1853
— var. *giganteum* (Grunow) Cleve 1894
Pleurosigma sterrenburgii Stidolph 1993*
Pleurosigma stidolphii Sterrenburg 1991*
Pleurosigma strigosum W.Smith 1852* B
PROSCHKINIACEAE
Proschkinia complanata (Grunow) D.Mann 1990* B
SELLAPHORACEAE
Fallacia clepsidroides Witkowski 1994* Thesis
Fallacia cryptolyra (Brockmann) Stickle & D.Mann 1990
Fallacia florinae (Møller) Witkowski 1993* Thesis
Fallacia forcipata (Greville) Stickle & D.Mann 1990* B
Fallacia inscriptura (Hendey) Witkowski 2000*
Fallacia litoricola (Hustedt) D.Mann 1990
Fallacia melanocephala (Gifffen) Witkowski 2000* B Thesis
Fallacia nummularia (Greville) D.Mann 1990* Thesis
Fallacia ny (Cleve) D.Mann 1990* B Thesis
Fallacia nyella (Hustedt) D.Mann 1990*
Fallacia pseudoforcipata (Hustedt) D.Mann 1990* Thesis
Fallacia oculiformis (Hustedt) D.Mann 1990* B
Fallacia scaldensis Sabbe & Muylaert 1999*
Fallacia subforcipata (Greville) Stickle & D.Mann 1990* Thesis
Fallacia tenera (Hustedt) D.Mann 1990 B
SCOLIONEIDACEAE
Scolioneis tumida (Brébisson) D.Mann 1990 B
SCOLIOTROPIDACEAE
Biremis ambigua (Cleve) D.Mann 1990* B
Biremis lucens (Hustedt) Sabbe, Witkowski &

Vyverman 1995* B Thesis
STAURONEIDACEAE
Stauroneis koniamboensis (Manguin) Moser 1998*
Stauroneis marina Hustedt 1955*
INCERTAE SEDIS
Cocconeiopsis fraudulenta (Schmidt) Witkowski 2000* #
Cocconeiopsis lubetii (König 1959) Witkowski informal 2000* #
Cocconeiopsis regularis (Hustedt) Witkowski 2000* #
Cocconeiopsis wrightii (O'Meara) Witkowski 2000* #

Order RHOPALODIALES
RHOPALODIACEAE
Rhopalodia acuminata var. *protracta* (Grunow) Krammer 1987 B
Rhopalodia constricta (Brébisson) Krammer 1987 B
Rhopalodia musculus (Kützing) Müller 1899 B
— var. *mirabilis* Fricke 1905

Order SURIRELLALES
AURICULACEAE
Auricula complexa (Gregory) Cleve 1894* Thesis
ENTOMONEIDACEAE
Entomoneis alata (Ehrenberg) Ehrenberg 1845 B
Entomoneis costata (Hustedt) Reimer 1975* B Thesis
Entomoneis kjellmanii var. *kariana* (Grunow) Poulin & Cardinal 1983
Entomoneis ornata (Bailey) Reimer 1975* B
Entomoneis cf. *paludosa* (W.Smith) Reimer 1975* B
Entomoneis pulchra (Bailey) Reimer 1975 B
Entomoneis surirelloides (Hendey) Poulin *et. al.* 1987* Thesis
SURIRELLACEAE
Campylodiscus birostratus Deby 1891*
Campylodiscus brightwellii Grunow 1862
Campylodiscus decorus Brébisson 1854*
Campylodiscus echeneis Ehrenberg 1840 B
Campylodiscus fastuosus Ehrenberg 1845
Campylodiscus incertus Schmidt 1875
Campylodiscus ralfsii W.Smith 1853
Campylodiscus samoensis Grunow 1875
Campylodiscus taeniatus Schmidt 1877
Petrodictyon gemma (Ehrenberg) D.Mann 1990`
Plagiodiscus nervatus Grunow 1867* Thesis
Surirella comis Schmidt 1876*
Surirella eximia Greville 1857
Surirella fastuosa (Ehrenberg) Kützing 1840
— var. *recedens* (Schmidt) Cleve 1878
Surirella filholii Petit 1877
Surirella fluviicygynorum John 1983*
Surirella imperfecta Hustedt 1925*
Surirella neumeyeri Janisch 1877
Surirella smithii Ralfs 1861
Surirella striatula Turpin 1828 B

Order THALASSIOPHYSALES
CATENULACEAE
Amphora acuta Gregory 1857
Amphora acutiuscula Kützing 1844
Amphora arcta Schmidt 1886
Amphora aspera Petit 1895
Amphora australiensis John 1981* B Thesis
Amphora bigibba Grunow 1875
— var. *interrupta* Grunow 1875*
Amphora coffeaeformis (Agardh) Kützing 1844 B
— var. *borealis* (Kützing) Cleve 1895*
— var. *perpusilla* (Grunow) Cleve 1895* Thesis
Amphora commutata Grunow 1880*
Amphora crassa Gregory 1857
Amphora cristata Petit 1877
Amphora cymbaphora Cholnoky 1960*
Amphora cymbifera Gregory 1857
Amphora decussata Grunow 1867* Thesis
Amphora exigua Gregory 1857* B
Amphora eunotia Cleve 1873*
Amphora foveauxiana (Petit) Taylor 1970
Amphora graeffeana Hendey 1972
Amphora granulata Gregory 1875*
Amphora helensis Giffen 1973*
Amphora hyalina Kützing 1844
Amphora immarginata T. Nagumo 2003*
Amphora janischii Schmidt 1875
Amphora aff. *kolbei* Aleem 1950*
Amphora laevis var. *minuta* Cleve 1895* Thesis
Amphora laevissima Gregory 1857*
Amphora lineolata Ehrenberg 1843*
— var. *sinensis* (Schmidt) Cleve 1895
Amphora marina W.Smith emend Van Heurck 1880
Amphora mexicana Schmidt 1875
Amphora obtusa Gregory 1857
Amphora ocellata Donkin 1861*
Amphora pannucea Griffen 1984*
Amphora proboscidea Gregory 1857
Amphora aff. *profusa* Giffen 1971* Thesis
Amphora pseudohyalina Simonsen 1960*
Amphora pseudoproteus Wachnicka & Gaiser 2007*
Amphora rhombica Kitton 1876*
Amphora robusta Gregory 1857
Amphora sarniensis var. *sinuata* (Greville) Cleve 1895
Amphora sinuata Greville 1863
Amphora spectabilis Gregory 1857
Amphora terroris Ehrenberg 1853 B
Amphora turgida Gregory 1857
Amphora weissflogii Schmidt 1875
Catenula adhaerens Mereschkowsky 1902* B
Lunella aff. *bisecta* Snoejis 1996* #

Order EUNOTIALES
EUNOTIACEAE
Colliculoamphora reichardtiana (Grunow) Williams & Reed 2006

FOSSIL MARINE DIATOMS

M. Harper

Classification and notations as for freshwater taxa. Unless otherwise indicated, diatoms come from the Oamaru diatomite. Abbreviations: #, family uncertain; C, common to Oamaru and other Tertiary samples; D, ocean drillholes – ODP 181 (94 taxa), DSDP 90 (20 taxa); V, Victoria University samples, F, freshwater; L, 'living' and in above checklists, r, reworked or contaminant.

Subclass COSCINODISCOPHYCIDAE
Superorder BIDDULPHIANAE
Order ANAULALES
ANAULACEAE
Anaulus biostratus (Grunow) Grunow 1887 L
Anaulus fossus (Grove & Sturt) Grunow 1888
Anaulus obtusus Bone & Gosden 1968
Anaulus subconstrictus Grove & Sturt 1887
Eunotogramma weissei Ehrenberg 1855
— var. *producta* Grove & Sturt 1887
Porpeia guadriceps Bailey 1861

Order BIDDULPHIALES
BIDDULPHIACEAE
Biddulphia antediluviana (Ehrenberg) Van Heurck 1885 L
Biddulphia birostrum Brun 1891
Biddulphia capuzina Schmidt 1888
Biddulphia cornuta Brun 1894
Biddulphia dissipata Grove & Sturt 1887
Biddulphia elegantula Greville 1865
— var. *polygibba* Pantocsek 1886
Biddulphia fragilis Bone & Gosden 1967
Biddulphia grovei Pantocsek 1889
Biddulphia lata Grove & Sturt 1887
Biddulphia miraculosa Brun 1892
Biddulphia novaezealandiae Wise 1952
Biddulphia oamaruensis Grove & Sturt 1886
Biddulphia obtusa (Kützing) Ralfs 1861 L
Biddulphia punctata Greville 1864
Biddulphia rhombus var. *fossilis* Brun 1915
Biddulphia rigida Schmidt 1888
Biddulphia ruthenica Witt 1885
Biddulphia scutum Bone & Gosden 1967
Biddulphia spinosum (Bailey) Boyer 1900 L
Biddulphia sturtii Wise 1952
Biddulphia tenera Grove & Sturt 1887
Biddulphia tridens (Ehrenberg) Ehrenberg 1840 C L
Biddulphia vittata Grove & Sturt 1887
Euodiella bicornigerum (Hanna) Sims 2000 #
Isthmia enervis Ehrenberg 1838
Isthmia nervosa Kützing 1844
Leudugeria janischii (Grunow) Tempère 1896 L #
— var. *crenulata* Desikachary & Sreelatha 1989
Lisitzinia ornata Jousé 1978 D #
Medlinia weissei (Grunow) Sims 1998
Neograya argonauta (Grove & Brun) Kuntze 1898 #
Neohuttonia alternans (Grove & Sturt) Kuntze 1898
Neohuttonia reedii (Bone & Gosden) Desikachary & Sreelatha 1989
Neohuttonia reversa (Grove & Sturt) Desikachary & Sreelatha 1989
Neohuttonia tripartita (Barker & Meakin) Desikachary & Sreelatha 1989
Neohuttonia virgata (Grove & Sturt) Kuntze 1898
Pseudotriceratium radiosoreticulatum (Grunow) Fenner 1985
Terpsinoe americana (Bailey) Ralfs 1861 L
— var. *trigona* (Pantocsek) Brun 1890
Trigonium americanum (Ralfs) Hustedt 1959
Trigonium arcticum (Brightwell) Cleve 1868
— var. *japonica* (Grunow) Desikachary & Sreelatha 1989
— var. *kerguelense* (Grunow) Desikachary & Sreelatha 1986
— var. *pentagonalis* (Tempère & Peragallo) Desikachary & Sreelatha 1989
— var. *quadrata* (Grunow) Desikachary & Sreelatha 1989
Trigonium crenulatum (Grove & Sturt) Hustedt 1959
Trigonium dissimile (Grunow) Mann 1925
Trigonium dobreeanum (Norman) Hustedt 1959
— var. *novazealandiae* (Grove & Sturt) Hustedt 1959
Trigonium formosum (Greville) Hustedt 1959
Trigonium gallapagense (Cleve) Desikachary & Sreelatha 1989
Trigonium glandarium (Schmidt) Hustedt 1959
Trigonium inelegans (Greville) Mann 1925
— var. *macropora* (Grunow) Desikachary & Sreelatha 1989
— var. *tubulosa* (Brun) Desikachary & Sreelatha 1989
Trigonium lautourianum (Grove) Hustedt 1959
Trigonium lineatum (Greville) Desikachary & Sreelatha 1989
Trigonium luculentum Hustedt 1959
Trigonium pileolus var. *jutlandica* Grunow 1884
Trigonium pseudonervatum (Grove & Sturt) Hustedt 1959
Trigonium reticulum (Ehrenberg) Simonsen 1974 L
Trigonium stokesianum (Greville) Hustedt 1959
Trigonium tabellarium (Brightwell) Mann 1907
Trigonium taeniatum Hustedt 1959
Trigonium undatum (Grunow) Hustedt 1959
Trigonium verecundum Hustedt 1959

Order HEMIAULALES
HEMIAULACEAE

Baxteriopsis brunii (Van Heurck) Karsten 1928
Briggera capitata (Greville) Ross & Sims 1985
Briggera ornithocephala (Greville) Ross & Sims 1985
Eucampia antarctica (Castracane) Manguin 1914 D L
Fontigonium rectangulare (Grove & Sturt) Sims & Hendey 1996 #
Hemiaulus altar Brun 1896
Hemiaulus angustus Greville 1865
Hemiaulus barbadensis Grunow 1884
Hemiaulus cf. *beatus* Fenner 1991 D
Hemiaulus caracteristicus Hajos 1976 D
Hemiaulus dissimilis Grove & Sturt 1887
Hemiaulus incisus Hajos 1976 D
Hemiaulus incurvus Schibkova 1959 D
Hemiaulus kristofferseni Fenner 1991 D
Hemiaulus podagrosus (Greville) Grunow 1884
Hemiaulus polycystinorum Ehrenberg 1854 C
— var. *mesolepta* Grunow 1884 D
Hemiaulus polymorphus Grunow 1884 C
Hemiaulus rossicus Pantocsek 1889 D
Hemiaulus subacutus Grunow 1884 C
Hemiaulus tenuicornis var. *novaezealandiae* Grunow 1886
Hemiaulus unicornutus Brun 1891
Hemiaulus velatus sensu Fenner 1991 D
Hemiaulus sp. 2 Fenner 1991 D
Kittonia elaborata (Grove & Sturt) Grove & Sturt 1887
Kittonia gigantea (Greville) De Toni 1894
Monobrachia unicornutus (Brun) Schrader & Fenner 1976 #
Pseudorutilaria monile Grove & Sturt 1894
Riedelia altar (Brun) Schrader & Fenner 1976
Riedelia clavigera (Schmidt) Schrader & Fenner 1976 B
Riedelia lyriformis (Greville) Schrader & Fenner 1976 C
Trinacria aries Schmidt 1886 D
Trinacria fragilis Grunow 1888 C
Trinacria insipiens Witt 1886
Trinacria lingulata (Greville) Grove & Sturt 1887
Trinacria senta (Witt) Sims & Ross 1988 D
Trinacria simulacrum Grove & Sturt 1887 C
Trinacria solenoceros (Ehrenberg) van Landingham 1971
Trinacria ventricosa Grove & Sturt 1887

Order TRICERATIALES
ISODISCACEAE #
Isodiscus coronalis Brun 1895 #
Isodiscus debyi (Grove & Sturt) Rattray 1888 #
Isodiscus mirificus Rattray 1888 #
PLAGIOGRAMMACEAE
Dimerogramma fulvum (Gregory) Ralfs 1861
Glyphodesmis interspiralis Brun 1891
Plagiogramma atomus Greville 1863
Plagiogramma interruptum (Gregory) Ralfs 1897 L
Plagiogramma neogradense Pantocsek 1886
Plagiogramma schraderii Desikachary & Sreelatha 1989
Plagiogramma tesselatum Greville 1897
TRICERATIACEAE
Auliscus accedens Rattray 1888
Auliscus barbadensis Greville 1887
Auliscus coincidens Schmidt 1892
Auliscus confluens Grunow 1875
Auliscus convolutus Rattray 1888
Auliscus dubius Tempère 1913
Auliscus elegans Greville 1863
— var. *subpunctata* Rattray 1888
Auliscus ellipticus Schmidt 1890
Auliscus fenestratus Grove & Sturt 1887
Auliscus grevillei Janisch 1887
Auliscus hardmanianus Greville 1886
— var. *bifurcata* Rattray 1888
Auliscus incertus Schmidt 1886
Auliscus inflatus Grove & Sturt 1887
Auliscus intermedius Grove & Sturt 1888
— var. *simplex* Rattray 1888
Auliscus interruptus var. *sparsa* Rattray 1888
Auliscus intestinalis Schmidt 1886
Auliscus lacunosus Grove & Sturt 1887
Auliscus lineatus Grove & Sturt 1887
Auliscus lunatus Grove & Sturt 1888
Auliscus macraeanus Greville 1863
Auliscus moronensis Greville 1864
Auliscus normanianus Greville 1864
Auliscus oamaruensis Grove & Sturt 1887
Auliscus ovalis Arnott 1861
Auliscus pectinatus Rattray 1888
Auliscus polyphemus Schmidt 1890
Auliscus propinquus Grove & Sturt 1887
Auliscus pruinosus Bailey 1853
— var. *robustus* Grunow 1890
Auliscus punctatus Bailey 1853
— var. *circumducta* Rattray 1888
Auliscus raeanus Rattray 1888
Auliscus sculptus (W.Smith) Ralfs 1861 L
— var. *mergens* (Rattray) Desikachary & Sreelatha 1989
Auliscus sigillum Brun 1892
Auliscus stoeckhardtii Janisch 1861
— var. *aspera* Grove 1896
— var. *inconspicua* Rattray 1888
— var. *subpunctata* Rattray 1888
Auliscus subcaelatus Rattray 1888
Auliscus subspeciosus Rattray 1888
Auliscus tenuistriatus Hustedt 1944
Auliscus transpennatus Brun 1891
Auliscus tripinnatus Bone & Gosden 1968
Cerataulus californicus Schmidt 1888
Cerataulus johnsonianus (Greville) Cleve 1885
Cerataulus marginatus Grove & Sturt 1887
Cerataulus rotundus Tempère & Brun 1890
Cerataulus subangulatus Grove & Sturt 1886
Cerataulus turgidus (Ehrenberg) Ehrenberg 1843 L
Glyphodiscus bipunctatus Schmidt 1890 #
Glyphodiscus grunowii var. *lacunosus* Tempère 1891 #
Glyphodiscus nancoorense (Grunow) Kolbe 1957 #
Glyphodiscus stellatus Greville 1862 L #
— var. *grunowii* (Schmidt) De Toni 1894 1882 #
Glyphodiscus strigillatus Schmidt 1890 L #
Grovea pedalis (Grove & Sturt) Schmidt 1890 #
Lampriscus shadboltianus (Greville) Peragallo 1897
Meretrosulus grayii Bone & Gosden 1967 #
Monopsia mammosa Grove & Sturt 1887 #
Odontella aurita (Lyngbye) C. Agardh 1832 L
Odontella sinensis (Greville) Grunow 1884 L
Pleurosira laevis (Ehrenberg) Compère 1982 L
Pseudauliscus anceps Rattray 1888
Pseudauliscus diffusus Rattray 1888
Pseudauliscus johnsonianus (Greville) Rattray 1888
Pseudauliscus notatus (Greville) Rattray 1888
Pseudauliscus pulvinatus (Cleve) Rattray 1888
Pseudauliscus spinosus (Christian) Rattray 1888
Rattrayella oamaruensis (Grunow) De Toni 1889 #
Rattrayella robusta Bone & Gosden 1967 #
Rattrayella simbirskianus (Grunow) Desikachary & Sreelatha 1989 #
Sheshukovia excavata (Heiberg) Nikolaev & Harwood 2001 C
Triceratium amplexum Schmidt 1882
Triceratium anostomosans Grove 1890
Triceratium antipodum Pantocsek 1892
Triceratium areolatum Greville 1861
Triceratium auliscoides Grove & Sturt 1887
Triceratium barbadense Greville 1861 C
Triceratium barkerii Desikachary & Sreelatha 1989
Triceratium bergonii Tempère & Brun 1889
Triceratium bimarginatum Grove & Sturt 1887
Triceratium broeckii Leuduger-Fortmorel 1879 L
Triceratium bullatum Grove 1890
Triceratium capitatum Greville 1861
Triceratium castellatum West 1860
— var. *fracta* Grunow 1891
Triceratium castelliferum Grunow 1888
Triceratium centralis (Boyer) van Landringham 1968
Triceratium coelatum (Janisch) Proshkina-Lavrenko 1949
Triceratium columbi Witt 1886
Triceratium concinnum Greville 1864
Triceratium condecorum Ehrenberg 1844
Triceratium cordiferum Grove & Sturt 1887
Triceratium crebrestriatum Grove 1890
Triceratium cuneatum Schmidt 1874
Triceratium denticulatum Schmidt 1882
Triceratium dictyotum Ross & Sims 1990 L
Triceratium distictum Barker & Meakin 1949
Triceratium divisum Grunow 1883
Triceratium eccentricum Grove & Sturt 1887
Triceratium exornatum Greville 1865
Triceratium fallaciosum Grunow 1889
Triceratium favus Ehrenberg 1839 C V L
— var. *maxima* Grunow 1883
Triceratium forresteri Tempère 1890
Triceratium grande Brightwell 1853
Triceratium grayii Grove & Sturt 1889
Triceratium grunowii Janisch 1885
Triceratium inconspicuum var. *trilobatum* Fenner 1984 D
Triceratium intermedium Grove & Sturt 1887
Triceratium irregulare Greville 1864
Triceratium labyrinthaeum Greville 1861
Triceratium majus Grove & Sturt 1888
Triceratium mirabile Jousé 1949 D
Triceratium montereyi Brightwell 1853
— var. *primordialis* Brun 1891
Triceratium morlandii Grove & Sturt 1887
— var. *aperta* Grunow 1889
Triceratium moronense Greville 1865
Triceratium neglectum Greville 1865
Triceratium nitescens Greville 1865
Triceratium oamaruense Grove & Sturt 1887
— var. *sparsipunctata* Grove 1890
Triceratium obesum Greville 1864
Triceratium pantocsekii Schmidt 1886
Triceratium papillatum Grove & Sturt 1887
Triceratium pavimentosum Castracane 1886
Triceratium perryanum Luard & Witt 1888
Triceratium plenum Grove & Sturt 1889
Triceratium plumosum Greville 1864
Triceratium productissimum Bergon 1890
Triceratium pulvinar Schmidt 1888
Triceratium reedii Desikachary & Sreelatha 1989
Triceratium repletum Greville 1866
Triceratium robertsianum Greville 1863 L
Triceratium rotundatum Greville 1861
Triceratium rugosum Grove & Sturt 1887
Triceratium schmidtii Janisch 1885
Triceratium secedens Schmidt 1888
Triceratium simplex Brun 1889
Triceratium triorbicum Schmidt 1888
Triceratium trisulcum Bailey 1861
— var. *cuneata* Fricke 1902
— var. *producta* Luard & Witt 1888
Triceratium undulans Barker & Meakin 1845
Triceratium unguiculatum Greville 1864
Triceratium venosum Brightwell 1856
Triceratium ventriculosum Schmidt 1886
Triceratium websterii Tempère 1890

Superorder CHAETOCEROTANAE
Order CHAETOCEROTALES

CHAETOCEROTACEAE
Chaetoceros dicladia Castracane 1886
Chaetoceros didymus Ehrenberg 1845 L
Chaetoceros leudugerii Brun 1894
Spore genera cf. *Chaetocerotales*
Chasea bicornis Hanna 1934
Dossetia lacerta (Forti) Hanna 1932 #
Epithelion hungaricum Pantocsek 1892 #
Goniothecium decoratum Brun 1891 #
Goniothecium odontella Ehrenberg 1844 C #
Hercotheca mamillaris Ehrenberg 1844 #
Hercotheca oamaruensis Desikachary & Sreelatha 1989 #
Kentrodiscus fortii deflandre & Rampi 1937 #
Liradiscus ellipticus Greville 1865 #
Liradiscus furcatus Grove 1890 #
Liradiscus marginatus Grove 1890 #
Liradiscus ovalis Greville 1865 #
Muelleriopsis limbiata (Ehrenberg) Hendey 1972 #
Periptera petiolata Andrews 1979 #
Poretzkia umbonata Glesner 1962 #
Xanthiopyxis acrolopha Forti 1912 C #
Xanthiopyxis hystrix Forti 1913 #
Xanthiopyxis ovalis Lohmann 1938 D #
Xanthiopyxis panduraeformis Pantocsek 1886 #
Xanthiopyxis specticularis Hanna 1927 #

Superorder COSCINODISCANAE
Order ARACHNOIDISCALES
ARACHNOIDISCEAE
Arachnoidiscus clarus Brown 1933
Arachnoidiscus deficiens (Grove) Brown 1933
Arachnoidiscus ehrenbergii Bailey 1849
Arachnoidiscus indicus Ehrenberg 1854
Arachnoidiscus japonicus (Shadbolt) Pritchard 1852
Arachnoidiscus lepidus Brown 1933
Arachnoidiscus oamaruensis (Schmidt) Brown 1933

Order ASTEROLAMPRALES
ASTEROLAMPRACEAE
Asterolampra affinis Greville 1862
Asterolampra decora Greville 1862
— var. *concentrica* Rattray 1890
Asterolampra grevillei (Wallich) Greville 1860
Asterolampra insignis Schmidt 1888 C
Asterolampra marylandica Ehrenberg 1844 C
Asterolampra punctifera (Grove) Hanna 1927
Asterolampra uraster Grove & Sturt 1887
Asterolampra vulgaris Greville 1862
Asterolampra wisei Desikachary & Sreelatha 1989
Rylandsia inaequiradiata Barker & Meakin 1945

Order COSCINODISCALES
AULACODISCACEAE
Aulacodiscus amoenus Greville 1864
— var. *minor* Grove & Sturt 1888
— var. *subdecora* Rattray 1888
— var. *sparsipunctata* Desikachary & Sreelatha 1989
Aulacodiscus angulatus Greville 1863
— var. *hungarica* (Pantocsek) Rattray 1888
— var. *plana* Rattray 1888
Aulacodiscus argus (Ehrenberg) Schmidt 1886 L
Aulacodiscus barbadensis Ralfs 1861
Aulacodiscus bullatus Wise 1956
Aulacodiscus cellulosus Grove & Sturt 1887
— var. *plana* Grove & Sturt 1887
Aulacodiscus comberi Arnott 1861
Aulacodiscus coronatus Grove 1888
Aulacodiscus crux Ehrenberg 1844
— var. *subsquamosa* Grunow 1888
Aulacodiscus decorus Greville 1864
— var. *stoschii* (Janisch) Rattray 1888
Aulacodiscus dispersus Rattray 1888
Aulacodiscus elegans Grove & Sturt 1887
Aulacodiscus exiguus var. *undulata* Rattray 1888
Aulacodiscus granuloprocessus Bone & Gosden 1968
Aulacodiscus hustedtii Ross & Sims 1970
Aulacodiscus huttonii Grove & Sturt 1887
Aulacodiscus intumescens Rattray 1888
Aulacodiscus janischii Grove & Sturt 1887
— var. *areolata* Rattray 1888
Aulacodiscus kilkellyanus Greville 1863
— var. *sparsa* (Greville) Rattray 1888
Aulacodiscus kittoni var. *africanus* (Cottam) Rattray 1888
Aulacodiscus margaritaceus Ralfs 1861
— var. *debyi* (Pantocsek) Rattray 1888
— var. *undosa* Grove & Sturt 1888
Aulacodiscus meakinii Woodward & Flint 1969
Aulacodiscus nigricans Tempère & Brun 1889
Aulacodiscus oamaruensis Grunow 1888
Aulacodiscus patens Rattray 1888
Aulacodiscus patulus Grunow 1888
Aulacodiscus polygonus Grunow 1886
Aulacodiscus probabilis Schmidt 1876
Aulacodiscus radiosus Grove & Sturt 1887
Aulacodiscus rattrayii Grove & Sturt 1887
— var. *convexa* (Grove & Sturt) Rattray 1888
Aulacodiscus rawsonii Barker & Meakin 1949
Aulacodiscus reticulatus Pantocsek 1886
Aulacodiscus scaber Ralfs 1861
Aulacodiscus schmidtii Witt 1885
Aulacodiscus solittianus var. *nova-zealandica* Grove & Sturt 1887
Aulacodiscus spectabilis Greville 1863
Aulacodiscus subrimosus Grunow 1888
Aulacodiscus tabernaculum Brun 1895
Aulacodiscus temperi Schmidt 1890
Aulacodiscus tener (Witt) Hustedt 1958
Aulacodiscus tumulosus Bone & Gosden 1968
Aulacodiscus umbonatus var. *dirupta* Grove & Sturt 1888
COSCINODISCAEAE
Brightwellia coronata (Brightwell) Ralfs 1861
Cestodiscus convexus Castracane 1886 D #
Cestodiscus johnsonianus Greville 1865 #
Cestodiscus novaezealandicus (Grove) Fenner 2001 C #
Cestodiscus peplum Brun 1891 D #
Cestodiscus pulchellus Rattray 1890 #
Cestodiscus subtilis Ehrenberg 1841 #
Coscinodiscus angulatus Greville 1864
Coscinodiscus argus Ehrenberg 1838 L
— var. *subtraducens* Rattray 1890
Coscinodiscus asteroides Truan & Witt 1888
Coscinodiscus asteromphalus Ehrenberg 1844 L
Coscinodiscus bulliens Schmidt 1878 C
Coscinodiscus centralis Ehrenberg 1838 L
Coscinodiscus concavus var. *punctatus* Grove 1890
Coscinodiscus conclusus Ehrenberg 1844
Coscinodiscus confusus Rattray 1890
Coscinodiscus curvatulus Grunow 1878 L
— var. *minor* (Ehrenberg) Grunow 1885
Coscinodiscus debilis Rattray 1890
Coscinodiscus decrescens Grunow 1878
Coscinodiscus decussatus Grove & Sturt 1890
Coscinodiscus denarius Schmidt 1878
Coscinodiscus densus Grove & Sturt 1890
Coscinodiscus diversus Grunow 1884
Coscinodiscus galapagensis (Grunow) Rattray 1890
Coscinodiscus herculus Brun 1891
Coscinodiscus heteroporus Ehrenberg 1844
Coscinodiscus inaequisculptus Rattray 1890
Coscinodiscus intumescens Pantocsek 1896
Coscinodiscus jacksonii Desikachary & Sreelatha 1989
Coscinodiscus lactucae-patella Bone & Gosden 1967
Coscinodiscus lacunosus Grove 1890
Coscinodiscus luctuosus Grove 1890
Coscinodiscus marginatus Ehrenberg 1841 C L
Coscinodiscus megacentrum Grove 1890
Coscinodiscus micans Schmidt 1889
Coscinodiscus oamaruensis Grove & Sturt 1887
Coscinodiscus oculus-iridis Ehrenberg 1839 D L
— var. *loculifera* Rattray 1890
Coscinodiscus ovicentrum Grove 1893
Coscinodiscus partitus Grove & Sturt 1890
— var. *nebulosa* Grove 1891
Coscinodiscus patellaeformis Greville 1861
Coscinodiscus perforatus Ehrenberg 1844
— var. *cellulosa Grunow* 1884
Coscinodiscus radiatus Ehrenberg 1839 L
Coscinodiscus radiosus Grunow 1883
Coscinodiscus rhombicus Castracane 1886 D
Coscinodiscus splendidulus Rattray 1890
Coscinodiscus subconcavus Grunow 1878
Coscinodiscus subdivicus Luard & Witt 1888
Coscinodiscus sublineatus (Grunow) Rattray 1890
Coscinodiscus superbus Hardman in Rattray 1890
— var. *novazealandica* Grove 1890 C
Coscinodiscus symmetricus Greville 1861
Coscinodiscus traducens Rattray 1890
Coscinodiscus uralensis Jousé 1948
Coscinodiscus vigilans Schmidt 1888
Coscinodiscus weissflogii Schmidt 1890
Cosmiodiscus insignis Jousé 1961 D #
Cosmiodiscus intersectus (Brun) Jousé 1961 D #
Cosmiodiscus normanianus Greville 1866 #
Craspedodiscus coscinodiscus Ehrenberg 1844
Craspedodiscus ellipticus (Grunow) Gombos 1982
Fenestrella convexa Brun 1891
Fenestrella gloriosa Brun 1891
Neobrunia mirabilis (Brun) Kuntze D #
Porodiscus hirsutus Grove & Sturt 1887 #
Porodiscus interruptus Grove & Sturt 1888 #
Porodiscus spinoradiata Brun 1891 #
Stellarima microstrias (Ehrenberg) Hasle & Sims 1986
Stellarima primalabiata (Gombos) Hasle & Sims 1986 D
Stellarima steinyi (Hanna) Hasle & Sims 1986 C
Symbolophora amblyoceros (Ehrenberg) Hasle & Sims 1986 #
HELIOPELTACEAE
Actinodiscus contabulatus (Schmidt) Ross & Sims 2000 #
Actinodiscus corolla (Brun) Ross & Sims 2000 #
Actinoptychus biformis Brun 1890
Actinoptychus clavatus Brun 1890
Actinoptychus constrictus Grove & Sturt 1887
Actinoptychus decorans Schmidt 1890
Actinoptychus fuscus Grove & Sturt 1888
Actinoptychus grovei Thomas 1890
Actinoptychus heliopelta Grunow 1883
Actinoptychus interpositus Brun 1893
Actinoptychus maculatus (Grove & Sturt) Schmidt 1888
Actinoptychus maculosus Pantocsek 1892
Actinoptychus marmoreus Brun 1890
Actinoptychus moelleri Grunow 1888
Actinoptychus nitidus (Greville) Schmidt 1874
Actinoptychus oamaruensis Grunow 1883
Actinoptychus papilio Brun 1889
Actinoptychus pericavatus var. *oamaruensis* Schmidt 1892
Actinoptychus pulchellus (Grunow) Wolle 1890
— var. *tenera* Grove & Sturt 1887
Actinoptychus racemosus Schmidt 1890
Actinoptychus senarius (Ehrenberg) Ehrenberg 1843 L
— var. *barbadensis* (Schmidt) Desikachary & Sreelatha 1989
Actinoptychus simbirskianus Schmidt 1875
Actinoptychus splendens (Shadbolt) Ralfs 1861 L

— var. *glabrata* (Grunow) Pantocsek 1886
Actinoptychus thumii (Schmidt) Hanna 1932
Actinoptychus trilunatus Brun 1893
Actinoptychus vulgaris Schumann 1867 L
Actinoptychus wittianus Janisch 1888
— var. *hexagona* Truan & Witt 1888
Anthodiscina floreatus (Grove & Sturt) Silva 1970
Lepidodiscus imperialis Witt 1886
Sturtiella elegans (Grove & Sturt) Simonsen & Schrader 1974
Trossulus elegantulus (Grove & Sturt) Ross & Sims 2002
HEMIDISCACEAE
Actinocyclus actinochilus (Ehrenberg) Simonsen 1982 C
Actinocyclus curvatulus Janisch 1878 D L
Actinocyclus ellipticus Grunow 1883 D
Actinocyclus ingens Rattray 1890 C
— var. *nodus* Baldauf 1980 D
— var. *ovalis* Gersonde 1990 D
— f. *planus* Whiting & Schrader 1985 D
Actinocyclus karstenii Van Heurck 1909 D
Actinocyclus kuetzingii (Schmidt) Simonsen 1975 C L
Actinocyclus labyrinthicus Pantocsek 1886
Actinocyclus normanii (Gregory) Hustedt 1957 r L
Actinocyclus octonarius Ehrenberg 1838 C V L
— var. *tenellus* (Brébisson) Hendey 1954 L
Actinocyclus rotula Brun 1891
Azpeita biradiata (Greville) Sims 1986
Azpeita endoi (Kanaya) Sims & Fryxell 1991 D
Azpeita grovei Sims 1986
Azpeita nodulifer (Schmidt) Fryxell & Sims 1986 C
Azpeita tabularis (Grunow) Fryxell & Sims 1986 D
Azpeita tuberculata (Greville) Sims 1986
Azpeita vetustissimus (Pantocsek) Sims 1986 C
Hemidiscus cuneiformis Wallich 1860 D L
Hemidiscus karstenii Jousé 1962 D
Hemidiscus triangularis (Jousé) Harwood & Maruyama 1992 D
ROCELLACEAE
Rocella gelida (Mann) Bukry 1978
— var. *schraderi* (Bukry) Barron 1983
Rocella praenitida (Fenner) Fenner 1986
Rocella vigilans (Kolbe) Fenner 1984

Order ETHMODISCALES
ETHMODISCACEAE
Ethmodiscus rex (Wallich) Hendey 1953

Order MELOSIRALES
ENDICTYACEAE
Endictya robustus (Greville) Hanna & Grant 1926
HYALODISCACEAE
Hyalodiscus dubiosus var. *curvans* (Rattray) Desikachary & Sreelatha 1989
Hyalodiscus franklinii (Ehrenberg) Ross 1986
— var. *robusta* (Grove & Sturt) Desikachary & Sreelatha 1989
Hyalodiscus laevis Ehrenberg 1845
Hyalodiscus punctatus Schmidt 1889
Hyalodiscus radiatus (O'Meara) Grunow 1880
— var. *arctica* Grunow 1884
— var. *maximus* (Petit) Cleve & Grunow 1880 L
— var. *striata* Desikachary & Sreelatha 1989
Hyalodiscus rossii Desikachary & Sreelatha 1989
Hyalodiscus cf. *scoticus* (Kützing) Grunow 1879 D L
Hyalodiscus valens Schmidt 1889
Podosira argus Grunow 1878
Podosira corolla Schmidt 1889
Podosira maxima (Kützing) Grunow 1879 L
Podosira ovoidea Hanna & Grant 1926
Podosira spinoradiata Brun 1889
Podosira stelliger (Bailey) Mann 1907 L
MELOSIRACEAE
Distephanosira architecturalis (Brun) Glesner 1992
Melosira cristata Pantocsek 1889
Melosira deblockii Van Heurck 1909
Melosira exspectata Schmidt 1892
Melosira major Grove 1892
Melosira moniliformis (Müller) Agardh 1824 L
Melosira orbifera Brun 1892
Melosira cf. *polaris* Grunow 1884 V
Melosira pontificalis Brun 1893
Melosira praeclara Schmidt 1892
Melosira saturnalis Brun 1892
Melosira truncata Grove 1893
PSEUDOPODOSIRACEAE #
Pseudopodosira bella Posnova 1964 #
Pseudopodosira hyalina (Jousé) Vechina 1961 #
Pseudopodosira westii (W.Smith) Scheschukova-Poretzkeya 1964 C V L #
STEPHANOPYXIDACEAE
Pyxidicula cruciata Ehrenberg 1834 #
Pyxidicula oblonga (Ehrenberg) Kützing 1849 #
Stephanonycites corona (Ehrenberg) Komura 1999 #
Stephanopyxis barbadensis (Greville) Grunow 1884
Stephanopyxis ferox (Greville) Ralfs 1861
Stephanopyxis grossecellulata Pantocsek 1886 D
Stephanopyxis grunowii Grove & Sturt 1888
Stephanopyxis hannai Hajos 1975
Stephanopyxis hyalinomarginata Hajos 1976 D
Stephanopyxis monstrosa (Forti) Desikachary & Sreelatha 1989
Stephanopyxis permarginata Grove & Tempère 1890
Stephanopyxis spinossima Grunow 1884 D
Stephanopyxis turris (Greville & Arnott) Ralfs 1861 C L
— var. *brevispina* Grunow 1884
— var. *intermedia* Grunow 1884 C
— f. *nuda* Pantocsek 1886
— var. *parvispina* Grunow 1884
Stephanopyxis valida (Grunow) Peragallo 1903

Order PARALIALES
PARALIACEAE
Paralia crenulata (Grunow) Gleser 1992
Paralia fausta (Schmidt) Sims & Crawford 2002
Paralia cf. *siberica* (Schmidt) Crawford & Sims 1990
Paralia cf. *sulcata* (Ehrenberg) Cleve 1873 C V L
— var. *biseriata* Grunow 1884
RADIALIPLICATACEAE
Anuloplicata ornata (Grunow) Glesner 1992
Bipalla oamaruensis (Grove & Sturt) Glesner 1992 [*Ellerbeckia*]
Ellerbeckia clavigera (Grunow) Crawford & Sims 2006
Ellerbeckia sol (Ehrenberg) Crawford & Sims 2006

Order STICTODISCALES
STICTODISCACEAE
Stictodiscus arcuatus Barker & Meakin 1945
Stictodiscus buryanus Greville 1862
Stictodiscus californicus Greville 1861
— var. *areolata* Grunow 1882
— var. *nitida* Grove & Sturt 1887
Stictodiscus conspersus Barker & Meakin 1945
Stictodiscus gibbosus (Grove & Sturt) Hustedt 1940
Stictodiscus grunowii Luard & Witt 1888
Stictodiscus haitianus var. *major* Desikachary & Sreelatha 1989
Stictodiscus hardmanianus Greville 1865
— var. *megapora* Grove & Sturt 1887
Stictodiscus harrisonianus (Norman & Greville) Castracane 1886
Stictodiscus inelegans Hustedt 1944
Stictodiscus kamischevensis Chenevière 1934
Stictodiscus kittonianus Greville 1861
Stictodiscus novaezealandiae Grunow 1888
Stictodiscus parallelus (Ehrenberg) Grove & Sturt 1887
Stictodiscus pulcherrimus Hustedt 1944
Stictodiscus splendidus Hustedt 1944

Superorder CYMATOSIRANAE
Order CYMATOSIRALES
CYMATOSIRACEAE
Cymatosira acremonica Schrader 1969
Rossiella symmetrica Fenner 1984 D
RUTILARIACEAE
Rutilaria lanceolata Grove & Sturt 1886
Rutilaria philippinarum Cleve & Grove 1891
Rutilaria radiata Grove & Sturt 1886
Rutilaria tenuicornis ssp. *tenuis* (Grove & Sturt) Ross 1977
Syndetoneis amplectans (Grove & Sturt) Grunow 1889 #
— var. *major* (Grove & Sturt) Mills 1935

Order LITHODESMIALES
LITHODESMIACEAE
Ditylum grovei Brun 1893

Superorder RHIZOSOLENIANAE
Order RHIZOSOLENIALES
PYXILLIDAE
Gyrodiscus radiosus Barker & Meakin 1945
Pyxilla dubia Grunow 1882
Pyxilla johnsoniana Greville 1865
Pyxilla prolongata Brun 1893
Pyxilla reticulata Grove & Sturt 1887 C
Pyxilla russica Pantocsek 1892
RHIZOSOLENIACEAE
Proboscia cretacea (Hajos & Stradner) Jordan & Priddle 1991 D
Pterotheca aculeifera (Grunow) Van Heurck 1896 C #
Pterotheca carinifera (Grunow) Forti 1909 #
Pterotheca danica (Grunow) Forti 1909 #
Rhizosolenia bergonii Peragallo 1892 D
Rhizosolenia hebetata f. *hiemalis* Gran 1904 D L
Rhizosolenia styliformis Brightwell 1858 D L
Stephanogonia aculeata Pantocsek 1889 #
Stephanogonia novazealandica Grunow 1888 #

Superorder THALASSIOSIRANAE
Order THALASSIOSIRALES
STEPHANODISCACEAE
Cyclotella castracanii Brun 1891
Cyclotella stylorum Brightwell 1856 r L
Stephanodiscus cf. *niagarae* Ehrenberg 1845 r L F
THALASSIOSIRACEAE
Porosira glacialis (Grunow) Jørgensen 1905 L
Shionodiscus oestrupii (Ostenfeld) Alverson, Kang & Theriot 2006 L D
Thalassiosira baltica (Grunow) Ostenfeld 1901 L
Thalassiosira convexa var. *aspinosa* Schrader 1974 D
Thalassiosira eccentrica (Ehrenberg) Cleve 1903 C V L
Thalassiosira insigna (Jousé) Harwood & Maruyama 1992 D
Thalassiosira inura Gersonde 1991 D
Thalassiosira kolbei (Jousé) Gersonde 1990 D
Thalassiosira lentiginosa (Janisch) Fryxell 1977 D L
Thalassiosira leptopus (Grunow) Hasle & Fryxell 1977 C V L
Thalassiosira lineata Jousé 1968 D
Thalassiosira miocenica Schrader 1974 D
Thalassiosira oliverana var. *sparsa* Harwood & Maruyama 1992 D

Subclass FRAGILARIOPHYCIDAE
Superorder FRAGILARIANAE
Order CYCLOPHORALES
ENTOPYLACEAE
Entopyla australis (Ehrenberg) Ehrenberg 1848 L
— var. *incurvata* (Arnott) Fricke 1902

Entopyla oamaruensis Schrader 1969

Order FRAGILARIALES
FRAGILARIACEAE
Fragilaria oamaruensis Schrader 1969
Opephora gemmata (Grunow) Hustedt 1931
Synedra baculus Gregory 1857
Synedra grovei Grunow 1889
Synedra jouseana Scheschukova-Poretzkaya 1962 D
Synedra oamaruensis Schrader 1969

Order RHABDONEMATALES
RHABDONEMATACEAE
Rhabdonema adriaticum Kützing 1844 L
Rhabdonema arcuatum (Agardh) Kützing 1844 L

Order RHAPHONEIDALES
PSAMMODISCACEAE
Psammodiscus nitidus (Gregory) Round & Mann 1980 L
RHAPHONEIDACEAE
Lancineis lancettula (Grunow) Andrews 1990
Rhaphoneis amphiceros Ehrenberg 1884 C L
Rhaphoneis elongata (Schrader) Andrews 1975
Rhaphoneis lepta Schrader 1969
Drewsandria nodulifer (Grove & Sturt) Sims & Ross 1996 #

Order STRIATELLALES
STRIATELLACEAE
Grammatophora marina (Lyngbye) Kützing 1844 L
Grammatophora oamaruensis Grunow 1889
Grammatophora totorae Schrader 1969

Order THALASSIONEMATALES
THALASSIONEMATACEAE
Thalassionema bacillaris (Heiden) Kolbe 1955 D
Thalassionema nitzschioides (Grunow) Van Heurck 1896 D L
— var. *parva* Heiden & Kolbe 1928 D
Thalassiothrix longissima Cleve & Grunow 1880 D L

Order TOXARIALES
TOXARIACEAE
Toxarium undulatum Bailey 1854 L

INCERTAE SEDIS
Cavitatus jouseanus (Scheschukova-Poretskaya) Williams 1989 D
Clavicula polymorpha Grunow & Pantocsek 1886
— var. *delicatula* Pantocsek 1886
Drepanotheca bivittata (Grunow & Pantocsek) Schrader 1969
— var. *elongata* (Brun) Desikachary & Sreelatha 1989
Drepanotheca macra Schrader 1969 #
Helminthopsidella drepanodes (Schrader) Silva 1970
Helminthopsidella elachista (Schrader) Silva 1970
Helminthopsidella ortha (Schrader) Silva 1970
Helminthopsidella weissflogii (Van Heurck) Silva 1970
Lamella oculata Brun 1894
Leptoscaphos punctatus (Grove & Sturt) Schrader 1969
Tubulariella pistillaris (Brun) Silva 1970
Tubulariella tabellarioides (Schrader) Silva 1970
Tubulariella totarae (Schrader) Silva 1970

Subclass BACILLARIOPHYCIDAE
Order ACHNANTHALES
COCCONEIDACEAE
Campyloneis grevillei (W.Smith) Grunow 1867 L
Campyloneis totarae (Brun) Schrader 1969
Cocconeis antiqua Tempère & Brun 1889
— var. *fossilis* Cleve 1894
Cocconeis costata Gregory 1855 L
Cocconeis cyclophora Grunow 1880
— var. *schraderi* Desikachary & Sreelatha 1989
Cocconeis cymatoda Schrader 1969
Cocconeis eocaenica Schrader 1969
Cocconeis gymna Schrader 1969
Cocconeis jacksonii Schrader 1969
Cocconeis oamaruensis Schrader 1969
Cocconeis pellucida Grunow 1863 L
— var. *nankoorensis* Grunow 1867
Cocconeis pseudomarginata Gregory 1865 L
— var. *intermedia* Grunow 1868 L
Cocconeis reedii Desikachary 1989
Cocconeis scutellum Ehrenberg 1838 L
Cocconeis vitrea Brun 1891
Cocconeis voluta Brun 1894

Order BACILLARIALES
BACILLARIACEAE
Alveus marina (Grunow) Kaczmarska & Fryxell 1996 D
Denticulopsis dimorpha (Schrader) Simonsen 1979 D
Denticulopsis hustedtii (Simonsen & Kanaya) Simonsen 1979 D
Denticulopsis praedimorpha Barron 1982 D
Fragilariopsis barroni (Gersonde) Gersonde & Bárcena 1998 D
Fragilariopsis doliolus (Wallich) Medlin & Sims 1993 D
Fragilariopsis aff. *donahuensis* (Schrader) Censarek & Gersonde 2002 D
Fragilariopsis fossilis (Frenguelli) Medlin & Sims 1993 D
Fragilariopsis interfrigidaria (McCollum) Gersonde & Bárcena 1998 D
Fragilariopsis januaria (Schrader) Bohaty 2003 D
Fragilariopsis kerguelensis (O'Meara) Hustedt 1952 D L
Fragilariopsis miocenica (Burckle) Censarek & Gersonde 2002 D
Fragilariopsis praeinterfrigidaria (McCollum) Gersonde & Bárcena 1998 D
Fragilariopsis reinholdii (Kanaya) Zielinski & Gersonde 2002 D
Fragilariopsis rhombica (O'Meara) Hustedt 1952 D
Fragilariopsis weaveri (Ciesielski) Gersonde & Bárcena 1998 D
Mediaria splendida Scheschukova-Poretzkaya 1962 D #
Nitzschia bilobata var. *oamaruensis* Desikachary & Sreelatha 1989
Nitzschia challengerii Schrader 1973 D
Nitzschia denticuloides Schrader 1976 D
Nitzschia groveana (Grove & Sturt) De Toni 1892
— var. *cavernosa* (Brun) Peragallo 1903
Nitzschia grovei Grunow 1888
Nitzschia gruendleri Grunow 1878
Nitzschia marina Grunow 1880 D L
Nitzschia oamaruensis Schrader 1969
Nitzschia palustris Hustedt 1934 F r
Nitzschia porteri Frenguelli 1949 D
Nitzschia praefossilis Schrader 1973 D
Nitzschia praereinholdii Schrader 1973 D
Nitzschia pseudokerguelensis Schrader 1976 D
Nitzschia vermiculata Castracane 1886
Nitzschia weissflogii Grunow 1880
Nitzschia sp. 17 Schrader 1976

Order DICTYONEIDALES
DICTYONEIDACEAE
Dictyoneis jamaicensis (Greville) Cleve 1890
Dictyoneis marginata (Lewis) Cleve 1890
— var. *janischii* (Castracane) Cleve 1894

Order LYRELLALES
LYRELLACEAE
Lyrella hennedyi f. *granulata* (Grunow) † 1964
Lyrella illustrioides (Hustedt) D.Mann 1990
Lyrella mediopartita (Grove) D.Mann 1990
Lyrella novaeseelandiae (Hustedt) D.Mann 1990 L
Lyrella oamaruensis (Grunow) D.Mann 1990
Lyrella praetexta (Ehrenberg) D.Mann 1990 L
Lyrella sandriana (Grunow) D.Mann 1990 L
Lyrella spectabilis (Gregory) D.Mann 1990
— var. *oamaruensis* (Grove) †
Lyrella subirroratoides var. *fossilis* (Schrader) †
Lyrella variolata (Cleve) D.Mann 1990
Petroneis japonica (Heiden) D.Mann 1990 L F
Petroneis transfuga (Grunow) D.Mann 1990

Order MASTOGLOIALES
MASTOGLOIACEAE
Mastogloia archaia Schrader 1969
Mastogloia barbadensis (Greville) Cleve 1895 L
Mastogloia binotata (Grunow) Cleve 1895 L
Mastogloia lamprosticta (Gregory) Desikachary & Sreelatha 1989
Mastogloia oamaruensis Cleve 1895
Mastogloia schraderii Desikachary & Sreelatha 1989
Mastogloia splendida (Gregory) Cleve & Möller 1879

Order NAVICULALES
AMPHIPLEURACEAE
Amphiprora cornuta Chase 1887 #
DIPLONEIDACEAE
Diploneis adiaphana Schrader 1969
Diploneis aestiva var. *oamaruensis* (Cleve) Desikachary & Sreelatha 1989
Diploneis chersonensis (Grunow) Cleve 1892
Diploneis crabro (Ehrenberg) Ehrenberg 1844 L
— var. *subelliptica* Cleve 1894
Diploneis debyi (Pantocsek) Cleve 1894
Diploneis eudoxia (Schmidt) Jørgensen 1905
Diploneis exemta (Schmidt) Cleve 1890
Diploneis gemmata (Greville) Cleve 1894 L
Diploneis haploa Schrader 1969
Diploneis interrupta (Kützing) Cleve 1894 V L
Diploneis nitescens (Gregory) Cleve 1894 L
Diploneis oamaruensis (Cleve) Mills 1934
Diploneis placida (Schmidt) Hustedt 1938
Diploneis praestes (Schmidt) Cleve 1894
Diploneis rouxioides Schrader 1969
Diploneis smithii (Brébisson) Cleve 1894 L
Diploneis totarae Schrader 1969
NAVICULACEAE
Caloneis biconstricta (Grove & Sturt) Cleve 1894
Caloneis dispersa (Grove & Sturt) Cleve 1894
Navicula braeumleri Pantocsek 1886
Navicula brasiliensis Grunow 1863 C
— var. *bicuneata* Cleve 1895
Navicula carinifera Grunow 1874
Navicula conjuncta Hustedt 1966
Navicula decora Grove & Sturt 1887
Navicula definita Grove & Sturt 1887
— var. *intermedia* Cleve 1894
Navicula destituta Hustedt 1966
Navicula eocaenica Schrader 1969
Navicula grovei Cleve 1893
Navicula halionata Pantocsek 1886
Navicula hochstetteri Grunow 1863 L
Navicula inelegans Grove & Sturt 1887
Navicula inexacta Mann 1925
— var. *symmetrica* (Cleve) Desikachary & Sreelatha 1989
Navicula interlineata Grove & Sturt 1886
Navicula jacksoni Schrader 1969
Navicula kappa Cleve 1894
Navicula labrobacata Bone & Gosden 1968

Navicula lampra Schrader 1969
Navicula lepta Schrader 1969
Navicula lorenzii (Grunow) Hustedt 1961
Navicula meridiepacifica Hustedt 1966
Navicula opaia Schrader 1969
Navicula pacha Schrader 1969
Navicula pararabda Schrader 1969
Navicula pseudodispersa Schrader 1969
Navicula pseudomediopartita Schrader 1969
Navicula pulchella Heiden 1903
Navicula sparsipunctata Grove & Sturt 1886
Navicula stercusmuscarum Cleve 1895
Navicula suavis Cleve & Grove in Cleve 1894
Navicula totarae Schrader 1969
Navicula trilineata Grove & Sturt 1886
Trachyneis aspera (Ehrenberg) Cleve 1894 L
PINNULARIACEAE
Oestrupia interrupta Schrader 1969
Oestrupia lineata Schrader 1969
Oestrupia powellii (Lewis) Heiden 1906
Pinnularia excellens Cleve 1891
— var. *interrupta* Cleve 1895
Pinnularia lobata (Grove & Sturt) Cleve 1895
Pinnularia yarrensis (Grunow) Jurilj 1957 L
PLAGIOTROPIDACEAE
Plagiotropis maxima (Gregory) Kuntze 1898
PLEUROSIGMATACEAE
Gyrosigma compactum (Greville) Cleve 1894
SCOLIOTROPIDACEAE
Progonoia didomatia Schrader 1969
Progonoia margina-lineata (Grove & Sturt) Schrader 1969
Progonoia margina-punctata (Grove & Sturt) Schrader 1969
Progonoia thaumasia Schrader 1969

Order SURIRELLALES
SURIRELLACEAE
Psammodictyon panduriforme (Gregory) D.Mann 1990 D L
Surirella aff. *comis* Schmidt 1874 D L

Order THALASSIOPHYSALES
CATENULACEAE
Amphora acuta Gregory 1857 L
Amphora antiqua Cleve & Grove 1895
Amphora cingulata Cleve 1878
Amphora crassa Gregory 1857 L
— var. *interlineata* (Grove & Sturt) Cleve 1895
Amphora dura Mann 1925
Amphora egregia Ehrenberg 1895
Amphora exornata Janisch 1876
Amphora fimbriata Cleve & Grove 1895
Amphora gruendleri var. *robusta* Cleve 1895
Amphora jacksonii Schrader 1969
Amphora labuensis Cleve 1895
Amphora lepta Schrader 1969
Amphora oamaruensis Schrader 1969
Amphora obtusa Gregory 1857 L
— var. *oamaruensis* Schrader 1969
— var. *oceanica* (Castracane) Cleve 1895
Amphora ocellata var. *oamaruensis* Cleve 1895
Amphora ovalis (Kützing) Kützing 1844 Fr L
Amphora pecten Brun 1891
Amphora prisca Cleve & Grove 1895
Amphora prismatica Cleve 1895
Amphora punctata Schrader 1969
Amphora rectilineata Cleve & Grove 1895
Amphora spectabilis Gregory 1857 L
Amphora sturtii Grunow 1888
Amphora subpunctata Grove & Sturt 1887
Amphora tesselata Grove & Sturt 1888
Amphora vetusta Cleve & Grove 1895
Amphora zebrata Tempère & Brun 1889

INCERTAE SEDIS
Lophotheca megala Schrader 1969 #
Lophotheca micra Schrader 1969 #
Rouxia naviculoides Schrader 1973 D #

Superorder EUNOTIANAE
Order EUNOTIALES
EUNOTIACEAE
Colliculoamphora reedii (Schrader) Williams & Reed 2006
Eunotia glacialis Meister 1912
Eunotia grovei Desikachary & Sreelatha 1989
Eunotia marina Schrader 1969

New combinations: Class Bacillariophyceae (diatoms)

Note. All but two of these new combinations link subspecific taxa with parent taxa that have already been transferred to new genera.

Achnanthidium exiguum var. *elliptica* (Hustedt) M.A.Harper *comb. nov.* Basionym: *Achnanthes exigua* var. *elliptica* F. Hustedt 1937, Arch. Hydrobiol., Suppl.15: 197, pl. 9, figs 8, 9.

Amphicampa mirabilis var. *transsylvanica* (Pantocsek) M.A.Harper *comb. nov.* Basionym: *Eunotia transylvanica (transsilvanica)* J. Pantocsek 1892, Beitr. Kenntn. Foss. Bacill. Ungarns. 3: 3, fig. 36. Synonym: *Eunotia serpentina* var. *transsilvanica (transylvanica)* (Pantocsek) Hustedt in Schmidt et al. 1913.

Fallacia vittata var. *rotoehuensis* (Foged) M.A.Harper *comb.nov.* Basionym: *Navicula vittata* var. *rotoehuensis* N. Foged 1979, Diatoms of New Zealand, the North Island. Bibliotheca Phycologica 47: 128 p., 48 pls; p. 82, pl. 25, fig. 18, pl. 28, fig. 6

Gomphonema clavatum var. *acuminatum* (Peragallo & Héribaud) M.A.Harper *comb. nov.* Basionym: *Gomphonema subclavatum* var. *acuminata* M. Peragallo & Héribaud 1893, Les Diatomées d'Auvergne. p. 55, pl. 3, fig. 8. Synonym: *Gomphonema montanum* var. *acuminata (acuminatum)* (M. Peragallo & Héribaud) Mayer 1928

Gomphosphenia tenerrima var. *nunguaensis* (Foged) M.A.Harper *comb. nov.* Basionym: *Gomphonema wulasiense* var. *nunguaensis* N. Foged 1966, Det Kong. Dansk.Vidensk., Biol. Shrifter 15: p. 109, 146, pl. 21, fig. 9

Gomphonemopsis novozelandicum (Booth) M.A.Harper *comb. nov.* Basionym: *Gomphonema novo-zelandicum* W.E.Booth 1985, N. Z. Jl Mar. Freshw. Res. 19: 275. 1985, Corrigendum to Booth 1984, N. Z. Jl Mar. Freshw. Res. 18: 426, fig. 1a-4d.

Haynaldella strigillata (O.Witt in Schmidt et al.) M.A.Harper *comb. nov.*.Basionym: *Coscinodiscus strigillatus* O.N.Witt in Schmidt et al. 1888, Atlas der Diat. Taf. 138, fig. 20. Synonym: *Haynaldia strigillata* (Witt in Schmidt *et al.*) Hanna 1934. Note: *Haynaldia* illegitimate as prior use for genus in family Poaceae.

Lyrella approximata f. *manca* (Schmidt) M.A.Harper *comb. nov.* Basionym: *Navicula hennedyi (hennedii)* var. *manca* Schmidt in Schmidt *et al.* 1874, Atlas Diat. Taf. 3, fig. 17. Synonym: *Navicula approximata* f. *manca* (Schmidt) Hustedt 1964.

Lyrella hennedyi f. *granulata* (Grunow in Schmidt et al.) M.A.Harper *comb. nov.* Basionym: *Navicula hennedyi* var. *granulata* Grunow in Schmidt et al. 1874, Atlas Diat. Taf. 3, fig. 3. Synonym: *Navicula hennedyi (hennedii)* f. *granulata* (Grunow in Schmidt et al.) Hustedt 1964.

Lyrella spectabilis var. *oamaruensis* (Grove) M.A.Harper *comb. nov.* Basionym: *Navicula spectabilis* var. *oamaruensis* Grove in A. Schmidt 1896, Atlas Diat.Taf. 204, fig. 15. Synonym: *Navicula spectabilis* var. *oamaruensis* Grove in Schmidt et al. 1905,

Lyrella subirroratoides var. *fossilis* (Schrader) M.A.Harper *comb. nov.* Basionym: *Navicula subirroratoides* var. *fossilis* Schrader 1969, Beih. Nova Hedw. 28: 52, pl. 20, figs 1, 2

Petroneis marina var. *novae-zealandiae* (Grunow) M.A.Harper *comb. nov.* Basionym: *Navicula cluthensis* var. *novaezealandiae* Grunow 1884, Denkschr. Kais. Akad. Wiss. Math. Naturwiss. Classe, Wien. 48: p. 52 (104). Synonyms: *Navicula marina* var. *novaezealandiae* (Grunow) F.W. Mills 1934, *Navicula punctulata* var. *novaezealandiae* (Grunow) Cleve 1895,

Psammodictyon panduriforme var. *lata* (O.N.Witt.) M.A.Harper *comb. nov.* Basionym: *Tryblionella lata* O.N.Witt 1873, J. Mus. Godeffroy 1: 66; pl. 8, fig. 6. Synonyms: *Nitzschia lata* (Witt) Lagerstedt 1876 *Nitzschia panduriformis* var. *lata* (Witt) Cleve & Möller 1878

Rhoicosphenia abbreviata var. *elongata* (A. Cleve-Euler) M.A.Harper *comb. nov.* Basionym: *Rhoicosphenia curvata* var. *elongata* A. Cleve-Euler 1953, K. Svenska Vetensk. Akad. Handl., Fjärde. Ser. 4: 52, fig. 601d,e.

Sellaphora laevissima var. *semiaperta* (H.G. Barber and J.R. Carter 1971) M.A.Harper *comb. nov.* Basionym: *Navicula laevissima* var. *semi-aperta* Barber & Carter 1971, J. Quek. Microsc. Club 32: 82, fig 97.

Triceratium dictyotum var. *rawsonii* (E.J.F.Wood 1961) M.A.Harper *comb. nov.* Basionym: *Biddulphia reticulata* var. *rawsonii* E.J.F.Wood 1961, Trans. Roy. Soc. N. Z. 88: 702, pl. 57, fig. 9

Tryblionella gracilis var. *yarrensis* (Grunow in Cleve and Grunow 1880) M.A Harper comb. nov. Basionym: *Nitzschia tryblionella* var. *yarrensis* A. Grunow in Cleve & Grunow 1880, K. Svenska-Vetensk. Akad. Hand. 17: 69. Comment: Species *Nitzschia tryblionella* Hantzsch in Rabenhorst 1848–1860, synonymous with *Tryblionella gracilis* W. Smith 1853.

Ulnaria ulna var. *danica* (F.T. Kützing 1844) M.A.Harper *comb. nov.* Basionym: *Synedra danica* F. T. Kützing 1844, Kies. Bacill. 66: pl. 14, fig. 13. Synonyms: *Synedra ulna* var. *danica* (Kützing) Grunow in Van Heurck 1885. *Synedra ulna* f. *danica* (Kützing) Hustedt 1957. *Synedra ulna* ssp. *danica* (Kützing) Skabichevskii 1959. *Fragilaria ulna* var. *danica* (Kützing) Kalinsky 1982. *Fragilaria danica* (Kützing) Lange-Bertalot in Lange-Bertalot & Metzeltin 1996.

ELEVEN

Phylum
OOMYCOTA
water moulds, downy mildews

ROSS E. BEEVER, ERIC H. C. McKENZIE,
SHAUN R. PENNYCOOK, STANLEY E. BELLGARD,
MARGARET A. DICK, PETER K. BUCHANAN

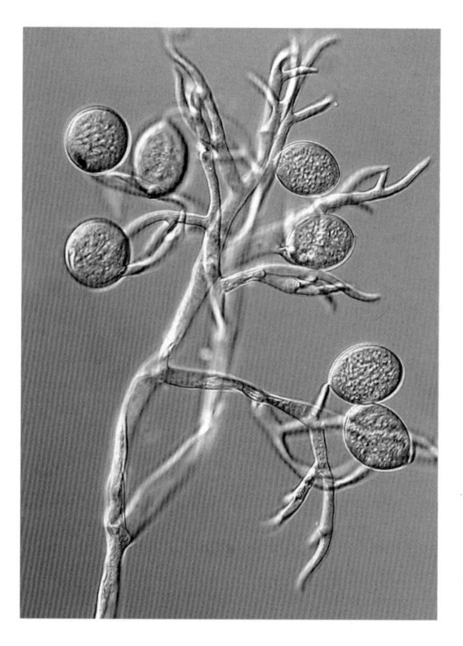

Peronospora conglomerata, with sporangia, from a leaf of *Geranium molle*.
Jerry A. Cooper, Landcare Research

Oomycota are mostly fungoid chromists, some of which have major ecological and economic significance. The majority are multicellular and used to be allied with fungi. Like fungi, they extend fungus-like threads (hyphae) into decaying matter or the cells and tissues of living hosts, including humans (Bulaji et al. 1999). Currently one of two non-photosynthetic heterokont phyla, the name Oomycota is used here following the Dictionary of Fungi. An alternative name is Pseudofungi, coined by Cavalier-Smith (1986). Three classes are recognised (Cavalier-Smith 2004) – Bigyromonadea (with the sole genus *Developayella*), the fungus-mimicking Hyphochytrea (*Pirsonia* and hyphochytrids), and Peronosporea (Oomycetes).

Chloroplasts appear to have been lost secondarily in the Oomycota. Inasmuch as the group is heterokont and lacks chitin and β-glucan in cell walls, it is well placed among the chromists. The most important exemplars in New Zealand are plant pathogens (oomycetes such as the downy mildews, collar rots, and damping-off species) and a few others, altogether about 150 presently recorded species but total diversity is likely to be much higher.

Members of class Oomycetes are characterised by biflagellate zoospores and walls composed of a cellulose-like material. They vary from unicellular to species with extensive, mostly nonseptate mycelia and are widespread in aquatic and terrestrial environments. Ecologically, they may be saprobes (feeding on decaying or non-living matter) or parasites, attacking fungi, other oomycetes, plants, fish and occasionally other animals.

Because they cause many economically important diseases, some pathogenic members are relatively well known. The genus *Phytophthora* (literally 'plant destroyer') contains many pathogens of importance to native and cultivated plants and to human history. The famine resulting from failure of the potato crop in Ireland in 1845–1847, and the subsequent death or migration of a million people, was caused by *Phytophthora infestans*, late blight of potato and a member of the Pythiales. In New Zealand, this species affects potato and tomato, and is one of several species of the cosmopolitan genus *Phytophthora* to occur in this country, many parasitic on aerial parts of plants and forming wind-dispersed sporangia while others are soil-borne.

Phytophthora cinnamomi is widespread and common in New Zealand soils under exotic and native forests and has been responsible for the death of a range of trees including shelterbelt species, avocado, *Pinus*, *Camellia*, and *Agathis* (Newhook 1959; Newhook 1970; Pennycook 1989). Damage by *P. cinnamomi* is typically localised when plants become highly stressed during summer droughts preceded by a warm wet winter. The species is thought to have been introduced to New Zealand by Maori as it is known in the Pacific on *Colocasia esculenta* (taro), a crop plant brought by Maori for cultivation in this country (Johnston et al. 2003). Assessments of impacts on native forests suggest that this oomycete has had long-term effects on establishment, regeneration, and spread of dominant forest species such as kauri and beech. Establishment of kauri seedlings was found to improve by use of the fungicide Ridomil at *Phytophthora*-infested sites (Horner 1984). While mycorrhizal beech seedlings are less susceptible to this pathogen, the natural pattern of regeneration of beech may have been disrupted by mortality of non-mycorrhizal beech seedlings due to *P. cinnamomi* (Johnston et al. 2003). In avocado orchards, the serious disease caused by *P. cinnamomi* has been overcome by routine trunk injections and sprays of phosphorous acid (Giblin et al. 2005).

Molecular detection, assisted by morphological study, has led recently to increased recognition of new species of *Phytophthora* in New Zealand, even on introduced plant hosts. Dick et al. (2006) described *P. captiosa* and *P. fallax*, which cause a crown dieback of *Eucalyptus* in New Zealand, suspecting that these species originated from Australia where *Eucalyptus* is native; the first record of *P. fallax* in Australia was reported by Cunnington et al. (2010). *Phytophthora kernoviae*, on the other hand, was described from Cornwall, England, in 2005, causing serious disease of *Fagus sylvatica* (European beech) and the introduced *Rhododendron ponticum*. While suspected to have originated in Asia or South America, it was recorded in New Zealand soon after, first on *Annona cherimola* (cherimoya) and then in soils under native and exotic forests in several parts of the North Island including earlier records (as '*Phytophthora* sp.') from the 1950s (Ramsfield et al. 2009). In New Zealand, pathogenicity of this species is limited to the single record on cherimoya. To address quarantine concerns, exports of New Zealand logs to Australia must comply with Australian AQIS import requirements by providing a '*Phytophthora kernoviae* ... Pest Area Freedom Declaration'.

Recent major epidemics by invasive pathogenic *Phytophthora* species have heightened global awareness of biosecurity threats arising from trade in contaminated plants and plant products. In most cases, invasive *Phytophthora* species cannot be eradicated owing to the time interval between initial invasion and eventual detection via host disease symptoms, as well as the difficulty to disrupt dispersal of propagules. *Phytophthora ramorum*, likely spread through the nursery trade (Brasier & Jung 2006), is responsible for the serious and widespread 'sudden oak death' in the western USA and Europe. In California, the aerially dispersed *P. ramorum* is pathogenic to more than 90 forest species, with high levels of mortality in susceptible oak species (Hüberli et al. 2008). These authors have also evaluated susceptibility of components of the New Zealand flora should *P. ramorum* arrive here, identifying five species, including *Fuschia excorticata* (tree fuchsia), *Leptospermum scoparium* (Manuka), and *Nothofagus fusca* (red beech), that are particularly susceptible.

Oogonia (female sex organs) of *Phytophthora fallax*, which causes crown die-back of *Eucalyptus* in Australasia.

Margaret A. Dick, Scion

Summary of New Zealand Oomycota diversity

Taxon	Described species + infraspecific taxa	Known undescribed undetermined species	Estimated undiscovered species	Adventive species	Endemic species	Endemic genera
Bigyromonadea	0	0	0	0	0	0
Hyphochytrea	7	0	20	0	0	0
Peronosporea	144+1	10	50	106+1	1	0
Totals	**151+1**	**10**	**70**	**106+1**	**1**	**0**

Diversity by environment

Taxon	Marine/ brackish	Fresh-water	Terrestrial
Bigyromonadea	0	0	0
Hyphochytrea	0	0	7
Peronosporea	4	23	129+1
Totals	**4**	**23**	**136+1**

* Three species are found in wet terrestrial settings are are list in both the freshwater and terrestrial columns

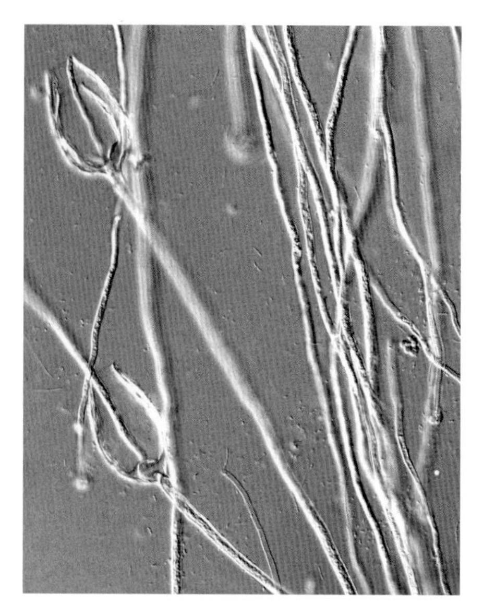

Phytophthora inundata, with nested empty sporangia and proliferation of the hypha beyond the sporangium.

Margaret A. Dick, Scion

Kauri dieback disease has been recognised as a significant threat to *Agathis australis* (kauri). The causal agent *Phytophthora* taxon 'Agathis' (PTA) has been recorded in several parts of Northland and Auckland including in forests containing the iconic giant trees. Previously described as *P. heveae* (Gadgil 1974), the pathogen causes a collar rot characterised by bleeding trunk lesions near the ground, yellowing foliage, and tree death. Trees of a broad age range are susceptible (Beever et al. 2009). As part of the current response, extensive public education led by Auckland Council and MAF seeks to minimise transfer of the soil-borne pathogen on boots and equipment, with boot-wash stations established at the entrance to the more popular tracks in kauri forests. Soil sampling in forests for molecular detection of PTA is in progress to determine distribution of the pathogen.

Other damaging pathogenic species of *Phytophthora* in New Zealand include *P. cactorum* on apples and *Pinus radiata* seedlings, *P. fragariae* on strawberries, and *P. citrophthora* on citrus (Dingley 1969; Beever et al. 2006; Reglinski et al. 2009).

Pythium, another cosmopolitan oomycete genus in soils and fresh water, causes damping-off diseases of zoospore-infected seedlings and sometimes older plants. Some aquatic species parasitise algae; others are saprobes growing on insect cadavers or decaying vegetation. The genus is relatively well known in New Zealand, with a guide to methods of isolation and preservation, a taxonomic key, and illustrated microscopic features of 27 species (Robertson 1970). *Pythium* is typically more frequent in cultivated than in forest soils (Domsch et al. 1980), and populations of pathogenic species on pasture have been reported to increase following spray irrigation of pasture with dairy-shed effluent (Waipara & Hawkins 2000). *Pythium oligandrum* is both plant pathogen and mycoparasite, causing a 'black compost' disease and significantly reduced yields of *Agaricus bisporus*, the commercially cultivated button mushroom (Godfrey et al. 2003). It is also parasitic on other *Pythium* species and on several mould fungi including *Trichoderma* species.

The causal agent of 'swamp disease' of horses was first recognised as *Pythium* by Austwick and Copland (1974) and later described by others as *Pythium insidiosum*. Known mainly from the tropics and subtropics, this species causes pythiosis, a skin disease of dogs, horses, cattle, and occasionally humans (Fraco & Parr 1997). These authors reported the first New Zealand record of *P. insidiosum*, causing keratitis, a severe inflammation of the eye's cornea. The affected patient, a fit 28 year-old male, sustained the infection following a day 'playing ball' in a hot pool. The pathogen was totally resistant to all antifungal agents and treatment involved surgical excision. Subsequently, *P. insidiosum* has been recorded in New Zealand from a horse (White 2006).

Peronospora and *Plasmopara* species (causing downy mildews) and *Albugo* species (the white rusts) include many other obligate plant pathogens. Downy mildews of economic importance in New Zealand include *Peronospora viciae* on peas, *P. antirrhini* on snapdragons, *P. sparsa* on boysenberries, *P. destructor* on onions, and *Plasmopara viticola* on grape. These oomycetes form wind-dispersed mitosporangia that germinate to produce hyphae or zoospores. *Albugo candida*, the white-rust pathogen of crucifers, also attacks the native *Lepidium ruderale* (Cook's scurvy grass) and is likely to be one of the factors leading to the decline of this species in the wild and its present threatened status (Baker 1955; Armstrong 2007).

Other oomycetes are parasites of filamentous fungi, algae, amoebae, and pollen grains. Several species of *Olpidiopsis* and *Rozella*, parasites of *Pythium*, *Achlya*, and other oomycetes, were isolated by Karling (1966, 1968b) from New Zealand soils. Water moulds (e.g. *Achlya* and *Saprolegnia*) include pathogens of algae, fish and their eggs, and invertebrates, as well as saprobes. Several saprobic species have been isolated from New Zealand pond water and soil samples (e.g. Karling 1968c; Elliott 1968).

Sporangiate *Hyaloperonospora parasitica* from a seedling leaf of *Brassica oleracea* var. *botrytis*.

Jerry A. Cooper, Landcare Research

Haptoglossa elegans, described from New Zealand (Barron 1990), is an especially curious aquatic oomycete that parasitises rotifers. Sunken spores of this fungus germinate to produce a so-called 'gun cell', which, when triggered by a passing rotifer, explodes and launches a harpoon-like missile, with protoplasm attached, into the animal.

Species belonging to class Hyphochytrea (hyphochytrids or Hyphochytridiomycetes) resemble the fungal chytrids (Chytridiomycetes) in having uniflagellate zoospores and living in fresh water and soil. However, they differ in the nature of the locomotory cilium (flagellum) and associated ultrastructure, being of the tinsel type in the former and whiplash in chytrids. They are common in New Zealand soils where a small number of saprobes and two parasites of oomycete fungi have been reported (Karling 1968a).

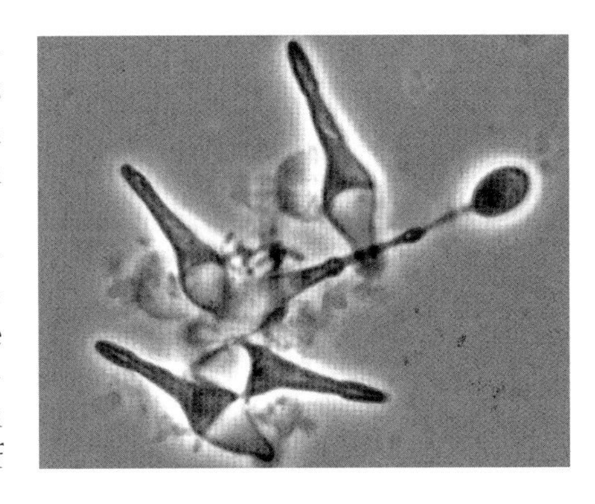

Gun cells of *Haptoglossa mirabilis*.
George L. Barron, University of Guelph

Authors

Dr Ross E. Beever Landcare Research, Private Bag 92170, Auckland, New Zealand. Deceased.

Dr Stanley E. Bellgard Landcare Research, Private Bag 92170, Auckland, New Zealand [bellgards@LandcareResearch.co.nz]

Dr Peter Buchanan Landcare Research, Private Bag 92170, Auckland, New Zealand [buchananp@LandcareResearch.co.nz]

Dr Margaret A. Dick Scion, NZ Forest Research Institute Ltd, 49 Sala Street, Rotorua, New Zealand [margaret.dick@scionresearch.com]

Dr Eric H. C. McKenzie Landcare Research, Private Bag 92170, Auckland, New Zealand [mckenziee@LandcareResearch.co.nz]

Dr Shaun R. Pennycook Landcare Research, Private Bag 92170, Auckland, New Zealand [pennycooks@LandcareResearch.co.nz]

References

ARMSTRONG, T. 2007. Molecular detection and pathology of the oomycete *Albugo candida* (white rust) in threatened coastal cresses. *Department of Conservation Research & Development Series 274*:1–18.

AUSTWICK, P. K. C.; COPLAND. J. W. 1974. Swamp cancer. *Nature 250*: 84.

BAKER, S. D. 1955: The genus *Albugo* in New Zealand. *Transactions of the Royal Society of New Zealand 82*: 987–993.

BARRON, G. L. 1990. A new and unusual species of *Haptoglossa*. *Canadian Journal of Botany 68*: 435–438.

BEEVER, R. E.; RAMSFIELD, T. D.; DICK, M. A.; PARK, D.; FLETCHER, M. J.; HORNER, I. J. 2006: Molecular Characterisation of New Zealand isolates of the fungus *Phytophthora* [MBS305]. *Landcare Research Contract Report LC05-6/155*: 1–43.

BEEVER, R. E.; WAIPARA, N. W.; RAMSFIELD, T. D., DICK, M. A.; HORNER I. J. 2009. Kauri (*Agathis australis*) under threat from *Phytophthora*? Pp. 74–85 *in*: Goheen, E. M., Frankel, S. J. (tech. coords.), *Phytophthoras in Forests and Natural Ecosystems.*[*Proceedings of the Fourth Meeting of the IUFRO Working Party S07.02.09. General Technical Report PSW-GTR-221.*] USDA Forest Service, Albany, California. 334 p.

BRASIER, C. M.; JUNG, T. 2006: Recent developments in *Phytophthora* diseases of trees in natural ecosystems in Europe. Pp. 5–16 *in*: Brasier, C. M.; Jung, T. (eds), *Progress in Research on* Phytophthora *Diseases in Forest Trees*. Forest Research, Farnham. 188 p.

BULAJI, N.; VELIMIROVI, S.; VUKOJEVI, J.; NONKOVI, Z.; JOVANOVI, D.; KUCERA, I.; ILI, S.; BRAJUSKOVI, G.; BOKUN, R.; PAVLIEVI, G.; TRNJAK, Z. 1999: Fungus-like hyphochytrids associated with human disease. *Acta Pathologica, Microbiologica et Immunologica Scandinavica 107*: 833–836.

CAVALIER-SMITH, T. 1986: The kingdom Chromista, origin and systematics. *Progress in Phycological research 4*: 309–347.

CAVALIER-SMITH, T. 2004: Chromalveolate diversity and cell megaevolution: interplay of membranes, genomes and cytoskeleton. Pp. 75–108 *in*: Hirt, R.P.; Horner, D.S. (eds), *Organelles, Genomes and Eukaryote Phylogeny – An Evolutionary Synthesis in the Age of Genomics*. [*The Systematics Association Special Volume Series 68.*] CRC Press, Boca Raton. [viii], 384 p.

CUNNINGTON, J. H.; SMITH, I. W.; DE ALWIS, S.; JONES, R. H.; IRVINE, G. 2010. First record of *Phytophthora fallax* in Australia. *Australasian Plant Disease Notes 5*: 96–97.

DICK, M. A.; DOBBIE, K.; COOKE, D. E. L.; BRASIER, C. M. 2006. *Phytophthora captiosa* sp. nov. and *P. fallax* sp. nov. causing crown dieback of Eucalyptus in New Zealand. *Mycological Research 110*: 393–404. doi:10.1016/j.mycres.2006.01.008

DINGLEY, J.M. 1969: Records of plant diseases in New Zealand. *New Zealand Department of Scientific and Industrial Research Bulletin 192*: 1–298.

DOMSCH, K. H.; GAMS, W.; ANDERSON, T.-H. 1980: *Compendium of Soil Fungi*. Academic Press, London. Vol. 1, viii + 860 p.; Vol. 2, vi + 406 p.

ELLIOTT, R. F. 1968: Morphological variation in New Zealand Saprolegniaceae 2. *Saprolegnia terrestris* Cookson and *S. australis* sp. nov. *New Zealand Journal of Botany 6*: 94–105.

FRACO, D. M.; PARR, D. 1997: Case Report. *Pythium insidiosum* keratitis. *Australian and New Zealand Journal of Ophthalmology 25*: 177–179.

GADGIL, P. D. 1974. *Phytophthora heveae*, a pathogen of kauri. *New Journal of Forestry Science 4*: 59–63.

GIBLIN, F.; PEGG, K.; WILLINGHAM, S.; ANDERSON, J.; COATES, L.; COOKE, T.; DEAN, J.; SMITH, L. 2005: *Phytophthora* revisited. New Zealand and Australia Avocado Grower's Conference '05. 20–22 September 2005. Tauranga, New Zealand. 7 p.

GODFREY, S. A. C.; MONDS, R. D.; LASH, D. T.; MARSHALL, J. W. 2003. Identification of *Pythium oligandrum* using species-specific ITS rDNA PCR oligonucleotides. *Mycological Research 107*: 790–796.

HORNER, I. J. 1984. The role of *Phytophthora cinnamomi* and other fungal pathogens in the establishment of kauri and kahikatea. MSc Thesis, University of Auckland.

HÜBERLI, D.; LUTZY, B.; VOSS, B.; CALVER, M. C.; ORMSBY, M.; GARBELOTTO, M. 2008: Susceptibility of New Zealand flora to *Phytophthora ramorum* and pathogen sporulation potential: an approach based on the precautionary principle. *Australasian Plant Pathology 37*: 615–625.

JOHNSTON, P. R.; HORNER, I. J.; BEEVER, R. E. 2003: *Phytophthora cinnamomi* in New Zealand native forests. Pp. 41–48 *in*: McComb, J. A.; Hardy, G. E. St J.; Tommerup, I. C. (eds), Phytophthora *in Forests and Natural Ecosystems.*

Murdoch University Print, Perth. vii + 292 p.
KARLING, J. S. 1966 [1965]: Some zoosporic fungi of New Zealand I. *Sydowia 19*: 213226.
KARLING, J. S. 1968a [1966]: Some zoosporic fungi of New Zealand. IX. Hyphochytridiales or Anisochytridiales. *Sydowia 20*: 137–143.
KARLING, J. S. 1968b [1966]: Some zoosporic fungi of New Zealand. XII. Olpidiopsidaceae, Sirolpidiaceae and Lagenidiaceae. *Sydowia 20*: 190–199.
KARLING, J. S. 1968c [1966]: Some zoosporic fungi of New Zealand. XIII. Thraustochytriaceae, Saprolegniaceae and Pythiaceae. *Sydowia 20*: 226–234.
NEWHOOK, F. J. 1959: The association of *Phytophthora* spp. with mortality of *Pinus radiata* and other conifers. I. Symptoms and epidemiology in shelter belts. *New Zealand Journal of Agricultural Research 2*: 808–843.
NEWHOOK, F. J. 1970: *Phytophthora cinnamomi* in New Zealand. Pp. 173–176 *in*: Tousson, T. A.; Bega, B.V.; Nelson, P. E. (eds), *Root Disease and Soil-borne Pathogens*. University of California Press, Berkeley. 252 p.
PENNYCOOK, S. R. 1989: *Plant Diseases Recorded in New Zealand*. Plant Diseases Division, DSIR. Auckland. Vol. 1, 276 p.; Vol. 2, 502 p.; Vol. 3, 180 p.
RAMSFIELD, T. D.; DICK, M. A.; BEEVER, R. E.; HORNER I. J.; McALONAN, M. J.; HILL, C.F. 2009: *Phytophthora kernoviae* in New Zealand. Pp. 47–53 *in*: Goheen, E. M., Frankel, S. J. (tech. coords.), *Phytophthoras in Forests and Natural Ecosystems. [Proceedings of the Fourth Meeting of the IUFRO Working Party S07.02.09. General Technical Report PSW-GTR-221.]* USDA Forest Service, Albany, California. 334 p.
REGLINSKI, T.; SPIERS, T. M.; DICK, M. A.; TAYLOR, J. T.; GARDNER, J. 2009: Management of phytophthora root rot in radiata pine seedlings. *Plant Pathology 58*: 723–730.
ROBERTSON, G. I. 1970: The genus *Pythium* in New Zealand. *New Zealand Journal of Botany 18*: 73–102.
WAIPARA, N. W.; HAWKINS, S. K. 2000: The effect of dairy-shed effluent irrigation on the ccurrence of plant pathogenic *Pythium* species in pasture. *In*: Zydenbos, S. M. (ed.), *New Zealand Plant Protection 53*: 436–440.
WHITE, P. 2006: Report: 20th New Zealand Fungal Foray, Westport. [http://www.funnz.org.nz/forays/20/report.htm]

Checklist of New Zealand Oomycota

Major environments are denoted by: F, freshwater; M, marine; T, terrestrial; F/T denotes a species that lives in a moist or wet terrestrial setting. Life habits and hosts are denoted by P, parasite; Pl, plant; V, vertebrate. E endemic species; A, adventive (naturalized alien); species not indicated by either notation are native (naturally indigenous but not endemic).

KINGDOM CHROMISTA
PHYLUM OOMYCOTA
Class HYPHOCHYTREA
Order HYPHOCHYTRIALES
HYPHOCHYTRIACEAE
Hyphochytrium catenoides Karling T
Hyphochytrium oceanicum Karling T
RHIZIDIOMYCETACEAE
Rhizidiomyces apophysatus Zopf T
Rhizidiomyces bivellatus Nabel T
Rhizidiomyces hansonii Karling T
Rhizidiomyces hirsutus Karling T
Rhizidiomyes saprophyticus (Karling) Karling T

Class PERONOSPOREA
Order LEPTOMITALES
LEPTOLEGNIELLACEAE
Aphanomycopsis punctata Karling T
Leptolegniella exoospora W.D.Kane T
Leptolegniella keratinophila Huneycutt T
INCERTAE SEDIS
Cornumyces pygmaeus (Zopf) M.W.Dick T

Order MYZOCYTIOPSIDALES
ECTROGELLACEAE
Haptoglossa elegans G.L.Barron T E
Haptoglossa heterospora Drechsler T
MYZOCYTIOPSIDACEAE
Myzocytiopsis microspora (Karling) M.W.Dick T
Myzocytium proliferum Schenk M
Myzocytium rabenhorstii (Zopf) M.W.Dick F
Syzygangia sp. sensu M.W.Dick 2001 F/T
SIROLPIDIACEAE
Sirolpidium bryopsidis (de Bruyne) H.E.Petersen M

Order OLPIDIOPSIDALES
OLPIDIOPSIDACEAE
Olpidiopsis achlyae McLarty F
Olpidiopsis aphanomycis Cornu F
Olpidiopsis brevispinosa Whiffen F
Olpidiopsis pythii (E.J.Butler) Karling F
Olpidiopsis saprolegniae (A.Braun) Cornu var. *saprolegniae* F
INCERTAE SEDIS
Gracea gracilis (E.J.Butler) M.W.Dick T

Order PERONOSPORALES
ALBUGINACEAE
Albugo bliti (Biv.) Kuntze T A
Albugo candida (J.F.Gmel.:Pers.) Kuntze T A
Albugo centaurii (Hansf.) Cif. & Biga T A
Albugo portulacae (DC.) Kuntze T A
Albugo tragopogonis (Pers.:Pers.) Gray T A
Albugo trianthemae G.W.Wilson T A
PERONOSPORACEAE
Basidiophora entospora Roze & Cornu T A
Bremia lactucae Regel T A
Peronospora alsinearum Casp. T A
Peronospora alta Fuckel T A
Peronospora anemones Tramier T A
Peronospora antirrhini J.Schröt. T A
Peronospora calotheca de Bary T A
Peronospora candida Fuckel T A
Peronospora conglomerata Fuckel T A
Peronospora destructor (Berk.) Fr. T A
Peronospora dianthi de Bary T A
Peronospora digitalis Gäum. T A
Peronospora farinosa (Fr.:Fr.) Fr. T A
— f.sp. *betae* Byford T A
Peronospora ficariae Tul. ex de Bary T A
Peronospora grisea (Unger) Unger T
Peronospora jaapiana Magnus T A
Peronospora knautiae Fuckel ex J.Schröt. T A
Peronospora lamii A. Braun T A
Peronospora lepidii (McAlpine) G.W.Wilson T A
Peronospora manshurica (Naumov) Syd. ex Gäum. T A
Peronospora mesembryanthemi Verwoerd T A
Peronospora myosotidis de Bary T
Peronospora obovata Bonord. T A
Peronospora ornithopi Gäum. T A
Peronospora parasitica (Pers.:Fr.) Fr. T A
Peronospora rumicis Corda T A
Peronospora sparsa Berk. T A
Peronospora trifoliorum de Bary T A
Peronospora viciae (Berk.) Casp. T A
Plasmopara geranii (Peck) Berl. & De Toni T A
Plasmopara halstedii (Farl.) Berl. & De Toni T A
Plasmopara viticola (Berk. & M.A.Curtis) Berl. & De Toni T A
Pseudoperonospora cubensis (Berk. & M.A.Curtis) Rostovzev T A

Order PYTHIALES
PYTHIACEAE
Halophytophthora sp. M
Phytophthora cactorum (Lebert & Cohn) J.Schröt. T A
Phytophthora captiosa M.A. Dick & Dobbie T A
Phytophthora cinnamomi Rands T A
Phytophthora citricola Sawada T A
Phytophthora citrophthora (R.E.Sm. & E.H.Sm.) Leonian T A
Phytophthora cryptogea Pethybr. & Laff. T A
Phytophthora europaea E.M. Hansen & T. Jung T A
Phytophthora erythroseptica Pethybr. T A
Phytophthora fallax Dobbie & M.A. Dick T A
Phytophthora fragariae Hickman T A
Phytophthora gonapodyides (H.E.Petersen) Buisman T A
Phytophthora hibernalis Carne T A
Phytophthora infestans (Mont.) de Bary T A
Phytophthora inflata Caroselli & Tucker T A
Phytophthora inundata Brasier, Sánch. Hern. & S.A. Kirk T A
Phytophthora kernoviae Brasier Beales and S.M. Kirk T
Phytophthora medicaginis E.M. Hansen & D.P. Maxwell T A
Phytophthora meadii McRae T A
Phytophthora megasperma Drechsler T A
Phytophthora multivesiculata Ilieva, Man in 't Veld, W. Veenb.-Rijks & R. Pieters T A
Phytophthora multivora P.M. Scott & T. Jung T A
Phytophthora nicotianae Breda de Haan T A
Phytophthora palmivora E.J. Butler T A
Phytophthora primulae J.A.Toml. T A
Phytophthora plurivora T. Jung & T.I. Burgess T A
Phytophthora syringae (Kleb.) Kleb. T A
Phytophthora sp. Apple-Cherry T A
Phytophthora sp. Asparagus T A
Phytophthora taxon Agathis T A
Phytophthora taxon PgChlamydo T A
Phytophthora taxon Raspberry T A
Phytophthora taxon Salixsoil T A
Phytophthora taxon Walnut T A
Pythium acanthicum Drechsler T A
Pythium afertile Kanouse & T.Humphrey T A

Pythium anandrum Drechsler T A
Pyhtium aphanodermatum (Edson) Fitz. T A
Pythium aquatile Höhnk T A
Pythium arrhenomanes Drechsler T A
Pythium butleri Subram. T A
Pythium chamaityphon Sideris T A
Pythium coloratum Vaartaja T A
Pythium debaryanum auct. non R.Hesse T A
Pythium echinulatum V.D.Matthews T A
Pythium erinaceum G.I.Robertson T A
Pythium gracile Schenk T A
Pythium graminicola Subraman. T A
Pythium inflatum V.D.Matthews T A
Pythium insidiosum De Cock, L.Mend., A.A.Padhye, Ajello & Kaufman T A
Pythium intermedium de Bary T A
Pythium irregulare Buisman T A
Pythium mastophorum Drechsler T A
Pythium megalacanthum de Bary T A
Pythium middletonii Sparrow T A
Pythium monospermum Pringsh. T A
Pythium myriotylum Drechsler T A
Pythium oligandrum Drechsler T A
Pythium paroecandrum Drechsler T A
Pythium rostratum E.J.Butler T A
Pythium spinosum Sawada T A
Pythium splendens Hans Braun T A
Pythium tenue Gobi T A
Pythium torulosum Coker & P.Patt. T A
Pythium ultimum Trow T A
Pythium undulatum H.E.Petersen T A
Pythium vanterpoolii V.Kouyeas & H.Kouyeas T A
Pythium vexans de Bary T A

Order SALILAGENIDIALES
HALIPHTHORACEAE
Haliphthoros sp. M

Order SAPROLEGNIALES
LEPTOLEGNIACEAE
Aphanomyces cochlioides Drechsler T A
Aphanomyces euteiches Drechsler T
Aphanomyces laevis de Bary F/T
Aphanomyces ovidestruens Gicklh. T
Aphanomyces phycophilus de Bary F
Aphanomyces raphani J.B.Kendr. T A
Aphanomyces stellatus de Bary F
Leptolegnia caudata de Bary T
SAPROLEGNIACEAE
Achlya caroliniana Coker F
Achlya flagellata Coker F
Achlya hypogyna Coker & Pemberton F
Achlya klebsiana Pieters T
Achlya prolifera Nees T
Achlya treleaseana (Humphrey) Kauffman F
Brevilegnia longicaulis T.W.Johnson T
Dictyuchus monosporus Leitg. T
Isoachlya unispora Coker & Couch F
Pythiopsis cymosa de Bary F/T
Saprolegnia australis R.F.Elliott F
Saprolegnia diclina Humphrey F
Saprolegnia ferax (Gruith.) Kütz. T
Saprolegnia litoralis Coker F
Saprolegnia terrestris Cookson ex R.L.Seym. T
Sommerstorffia spinosa Arnautov F
Thraustotheca clavata (de Bary) Humphrey F

Order SCLEROSPORALES
VERRUCALVACEAE
Sclerophthora macrospora (Sacc.) Thirum., C.G.Shaw & Naras. T A

Subclass INCERTAE SEDIS
Order ROZELLOPSIDALES
PSEUDOSPHAERITACEAE
Sphaerita dangeardii Chatton & Brodsky F
Sphaerita endogena P.A.Dang. F
ROZELLOPSIDACEAE
Rozellopsis inflata (E.J.Butler) Karling ex Cejp T

TWELVE

Phylum BIGYRA

opalinids, *Blastocystis*, slime nets, thraustochytrids, bicosoecids

DENNIS P. GORDON, TONY CHARLESTON,
SERENA L. WILKENS, ERIC H. C. McKENZIE

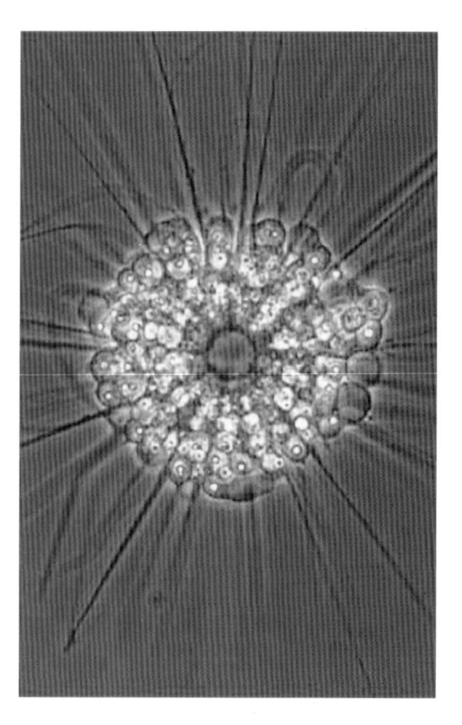

Actinophrys sol.
David J. Patterson, per Micro*scope (MBL).

Phylum Bigyra comprises microscopic microbial feeders or heterotrophs that absorb nutrients from their surroundings. The phylum name means 'twice-ringed' (Latin, *bi-*, twice, and *gyrus*, circle or ring), and alludes to a feature seen only by viewing ultrathin sections of cells below the level where the paired ciliary hairs (not all species have them) are inserted. Here, the base of the cilia pair has a distinctive profile in cross-section that is a signature of the phylum and of some other genera. There are other congruent structural features (like the complete absence of plastids) and molecular analyses support a relationship among these groups. The phylum name was introduced by Cavalier-Smith (1998), who currently recognises three subphyla – Opalozoa, Bicosoecia, and Sagenista, collectively containing six classes (Cavalier-Smith 2010).

Subphylum Opalozoa contains four of these classes, only one of which, Proteromonadea, is not yet known from New Zealand. Of the others, class Nucleohelea contains taxa that used to be classified among the Heliozoa ('sun animals'). One of the best known of these is *Actinophrys sol*. Actinophryids are especially common in lakes and rivers and some are found in marine and soil habitats as well. Their unicells are essentially spherical and they lack a shell or test. There is no cilium, but numerous stiff pseudopodia (axopodia) radiate outwards from the nucleus, supported by stiff tapering axonemes that are formed by double spirals of microtubules. The axopodia contain extrusible bodies (extrusomes) that adhere to passing prey, including small flagellates, diminutive ciliates, and microalgae. The outer portion of the cell is filled with many tiny vacuoles, which assist in flotation. Populations of *Actinophrys sol*, its cells 40–60 micrometres in diameter, have been reported from varied New Zealand localities such as the crater lakes of Mayor Island (Bayly et al. 1956) and seasonally inundated grassland soils in the Wairarapa (Stout 1984). Under unfavourable conditions, the organism will form a cyst, which is multi-walled and covered in siliceous scales (Mikrjukov & Patterson 2001).

Class Opalinea in New Zealand is represented by only two species. Indeed, there are only five genera worldwide, containing about 400 species, mostly living in frog guts, but also in reptiles, fishes, and a few invertebrates. The surface of opalines, which are relatively large unicells usually exceeding 100 micrometres diameter, is covered by rows of cilia that arise anteriorly at the cell's morphological centre. For this reason, they were formerly classified among the ciliates (phylum Ciliophora), as protociliates, but they lack several distinctive features of ciliates such as dimorphic macro- and micronuclei (though they can be bi- or multinucleated). Opalines are thought to have evolved in frogs in Gondwana prior to the break-up of that supercontinent in the Late Jurassic. They now have a worldwide distribution (Delvinquier & Patterson 2000). The three species of Australian hylid frogs that were introduced to New Zealand brought with them their unicellular and metazoan parasites (Brace et al. 1953; Hicks 1974), including two species of *Protoopalina*. Interestingly, native New Zealand frogs of the genus *Leiopelma* lack opalinids, even though *Leiopelma* ancestors may have lived in Asia where *Protoopalina* is common. This is probably because, lacking aquatic larvae, *Leiopelma* species have no pathway for opalinid reinfection (Hicks 1974).

Class Blastocystea is solely represented in New Zealand by *Blastocystis hominis*, a parasite of the intestinal tracts of humans and many animal species (Tan 2008). Its significance as a pathogen has been the subject of debate for many years as has its biology and taxonomic status. Its presence in a person may be completely asymptomatic but, on the other hand, its abundance in stool samples of some patients with diarrhoea has led to it being implicated as the causal agent (Al-kaissi & Al-Magdi 2009). Although originally thought to be a yeast – cilia are completely lacking – it has the characteristics of an anaerobic protozoan and molecular evidence supports a relationship with other Chromista (Silberman et al. 1996; Tan 2008). *Blastocystis hominis*, *Pythium insidiosum* (Fraco & Parr 1997), and, overseas, hyphochytrids (Bulajić et al. 1999) are the only Heterokonta known to infect humans.

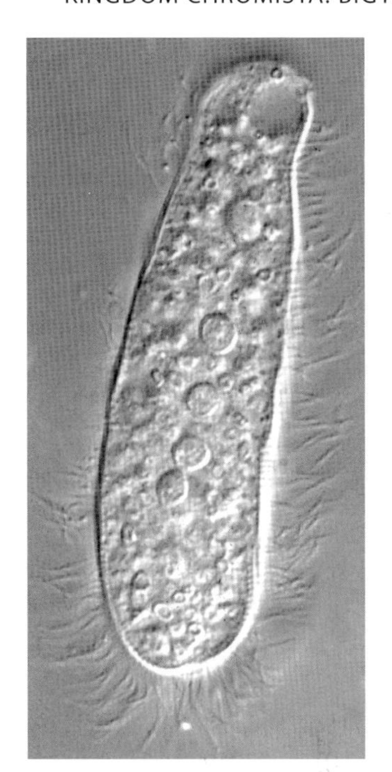

A typical opalinid (non-New Zealand).
J. Craig Bailey, UNC Wilmington

The organism is extraordinarily pleomorphic and variable in size. In faecal (stool) samples, cells are spherical, averaging 4–15 micrometres in diameter but sometimes much larger, with a large central body filled with amorphous material, surrounded by a peripheral band of cytoplasm. These and several other forms (sometimes found in faeces) are seen in laboratory cultures. A cystic stage has also been described and it is likely that this is the stage involved in the transmission of infection through faecal contamination of food or water. A possible life-cycle and patterns of transmission between animals and humans have been proposed by Tan (2008). There is still much that is not known about the basic biology of *B. hominis* but there is now considerable evidence to indicate that it is, in fact, a complex of species varying in host specificity and pathogenicity (Scicluna et al. 2006; Tan 2008). For example, a recent phylogenetic study of *Blastocystis* isolates from humans and animals, cited by Tan (2008), indicated that they could be potentially divided into 12 or more species. A comparative study of isolates from humans with and without diarrhoea revealed both phenotypic and genotypic differences between them (Tan et al. 2008).

It has been suggested that immigrants, refugees, and adopted children from developing countries appear to have higher incidences of *B. hominis* infection than do adults and children raised from birth in their new community (Stenzel & Boreham 1996). However, the incidence of 'pathological levels' of *B. hominis*

Summary of New Zealand bigyran diversity

Taxon	Described species	Known undescribed/ undetermined species	Estimated unknown species	Adventive species	Endemic species	Endemic genera
Nucleohelea	1	0	3	0	0	0
Proteromonadea	0	0	2	0	0	0
Opalinea	2	0	2	2	0	0
Blastocystea	1	0	2	1	0	0
Bicoecia	2	1	20	0	0	0
Labyrinthulea	8	0	6	0	0	0
Totals	**14**	**1**	**35**	**3**	**0**	**0**

Diversity by major environment*

Taxon	Marine	Freshwater	Terrestrial
Nucleohelea	0	1	0
Proteromonadea	0	0	0
Opalinea	0	2	0
Blastocystea	0	0	1
Bicoecia	0	3	0
Labyrinthulea	9	0	0
Totals	**9**	**6**	**1**

* Parasites are attributed the major environment of their host

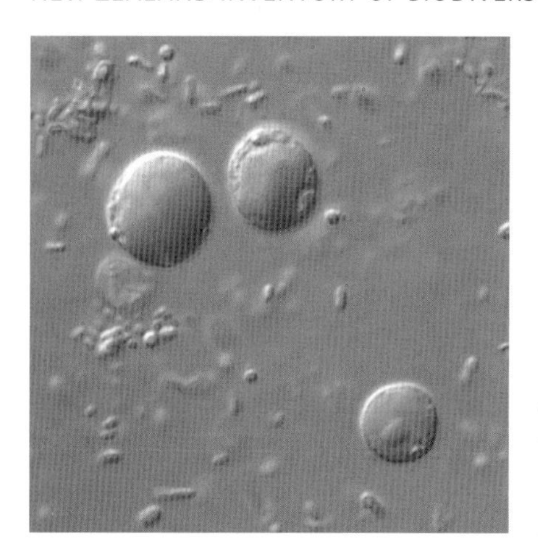

Blastocystis hominis cells, variable in size, with a large central vacuole surrounded by a thin layer of cytoplasm containing up to six nuclei.
CDC-DPDx, USA

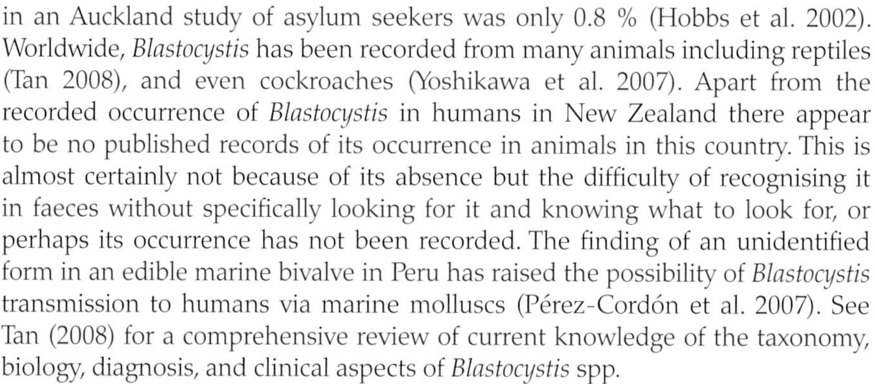

in an Auckland study of asylum seekers was only 0.8 % (Hobbs et al. 2002). Worldwide, *Blastocystis* has been recorded from many animals including reptiles (Tan 2008), and even cockroaches (Yoshikawa et al. 2007). Apart from the recorded occurrence of *Blastocystis* in humans in New Zealand there appear to be no published records of its occurrence in animals in this country. This is almost certainly not because of its absence but the difficulty of recognising it in faeces without specifically looking for it and knowing what to look for, or perhaps its occurrence has not been recorded. The finding of an unidentified form in an edible marine bivalve in Peru has raised the possibility of *Blastocystis* transmission to humans via marine molluscs (Pérez-Cordón et al. 2007). See Tan (2008) for a comprehensive review of current knowledge of the taxonomy, biology, diagnosis, and clinical aspects of *Blastocystis* spp.

Subphylum Bicosoecia and its sole class Bicoecea comprise bicosoecids and related forms. Only three bicosoecids have been recorded to date in New Zealand, all inhabitants of fresh water. They exist as solitary cells, attached by their smooth cilium in an organic 'shell' (lorica) shaped like a drinking cup (Greek, *bicos*). They are capable of ingesting bacteria and other microorganisms through phagocytosis.

Subphylum Sagenista, with its sole class Labyrinthulea, comprises two subclasses of marine heterotrophic organisms, Labyrinthulidae and Thraustochytridae (Cavalier-Smith 2001). Some authors (e.g. Leander et al. 2004) recognise a third group, the aplanochytrids. The trophic stage of labyrinthulids comprises spindle-shaped cells that glide through channels of a proteinaceous polysacharide network (or slime net) that does not completely surround the cells of aplanochytrids (Leander & Porter 2001). The slime matrix is produced from organelles called sagenosomes.

Labyrinthulids are parasites, commensals, or mutualists of marine algae and sea grasses. In contrast, thraustochytrids are rarely found on these living plants, but are common in the neritic and oceanic water column and sediments, including the deep sea (Raghukumar 2002). Members of both groups may play an important role as saprobes and are commonly present on dead plant material, for example macroalgae and submerged mangrove leaves. Labyrinthulids have been also reported to cause diseases in animals (Raghukumar 2002). Although isolated from a variety of habitats all over the world, only a few species have been reported from New Zealand including one species in Labyrinthulidae and eight in Thraustochytridae (Pennycook & Galloway 2004). The sole New Zealand species of *Labyrinthula* has been implicated in the wasting of the eelgrass *Zostera* in the Auckland area (Armiger 1964).

Thraustochytrids are microscopic microbes that superficially resemble the fungal chytrids. They are essentially immobile except for the biflagellate zoospore stage (Kirk et al. 2001; Leander & Porter 2001). They can play dual roles in nature as bacterial feeders when in an amoeboid form, and organic material degraders in their thallus form (Raghukumar 2002). In coastal environments they play a role in the breakdown of plant leaves such as those of mangroves, by colonising and penetrating them and degrading them with their enzymes. Whereas the labyrinthulids have a slime matrix around their cells, in thraustochytrids this matrix is produced only from sagenosomes at the base of the structure containing the zoospores.

A *Bicosoeca petiolata* colony with loricate cells.
William Bourland, per Micro*scope (MBL)

Economic importance of Bigyra

Owing to uncertainty concerning the relative importance of Blastocystis in relation to other actual or potential causes of gastroenteritis, it is difficult to quantify the negative importance of this organism in human health. On the other hand, thraustochytrids have demonstrated economic value as sources of biochemicals important to human health and aquaculture (Lewis et al. 1999; Raghukumar 2008). Some strains produce high biomass of omega-3 polyunsaturated fatty acids (PUFAs) and docosahexaenoic acid and are emerging as sources of other PUFAs such as arachidonic acid and oils that can be targeted for specific uses. Fish oils are one of the main sources of PUFAs and, with the decline in commercial fish stocks, thraustochytrids are a possible alternative source. Indeed, several thraustochytrid-based products are already on the market. Other potential areas for biotechnology application include enzymes, polysaccharides, antioxidants and secondary metabolites.

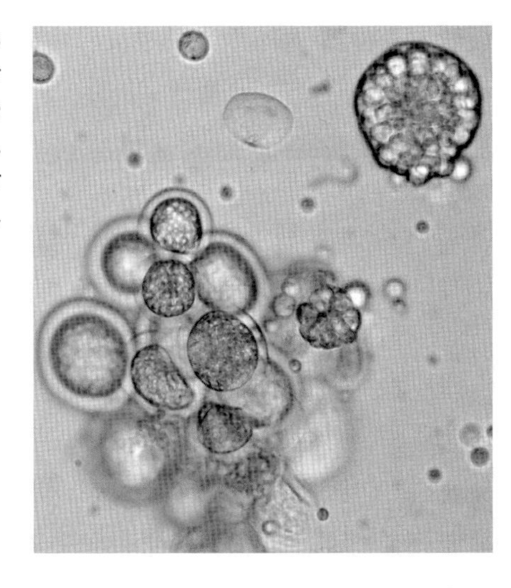

Thraustochytrium sp. in various stages of maturation, with sporangium at upper right releasing zoospores.

Serena Wilkens

Authors

Dr W. A. G. (Tony) Charleston Institute of Veterinary, Animal and Biomedical Sciences, Massey University, Palmerston North. Present address: 488 College Street, Palmerston North, New Zealand [charleston@inspire.net.nz] *Blastocystis*

Dr Dennis P. Gordon National Institute of Water & Atmospheric Research (NIWA), Private Bag 14901, Kilbirnie, Wellington, New Zealand [d.gordon@niwa.co.nz] introduction, opalinids, bicosoecids

Dr Eric H. C. McKenzie Landcare Research, Private Bag 92170, Auckland, New Zealand [mckenziee@landcareresearch.co.nz] Labyrinthulidae

Dr Serena L. Wilkens National Institute of Water and Atmospheric Research (NIWA), Private Bag 14901 Kilbirnie, Wellington, New Zealand [s. wilkens@niwa.co.nz] Thraustochytridae

References

AL-KAISSI, E.; AL-MAGDI, K. J. 2009: Pathogenicity of *Blastocystis hominis* in relation to entropathogens in gastroenteritis cases in Baghdad. *European Journal of Scientific Research 25*: 606–613.

ARMIGER, L. 1964: An occurrence of *Labyrinthula* in New Zealand *Zostera*. *New Zealand Journal of Botany 2*: 3-9.

BAYLY, I. A. E.; EDWARDS, J. S.; CHAMBERS, T. C. 1956: The crater lakes of Mayor Island. *Tane (Journal of the Auckland University Field Club) 7*: 36–46.

BRACE, E. C.; CASEY, B. R.; COSSHAM, R. B.; McEWAN, J. M.; McFARLANE, B. G.; MACKEN, J.; MONRO, P. A.; MORELAND, J. M.; NORTHERN, J. B.; STREET, R. J.; YALDWYN, J. C. 1953: The frog, *Hyla aurea* as a source of animal parasites. *Tuatara 5*: 12–21.

BULAJI, N.; VELIMIROVI, S.; VUKOJEVI, J.; NONKOVI, Z.; JOVANOVI, D.; KUCERA, I.; ILI, S.; BRAJUSKOVI, G.; BOKUN, R.; PAVLIEVI, G.; TRNJAK, Z. 1999: Fungus-like hyphochytrids associated with human disease. *Acta Pathologica, Microbiologica et Immunologica Scandinavica 107*: 833–836.

CAVALIER-SMITH, T. 1998: A revised six-kingdom system of life. *Biological Reviews 73*: 203-266.

CAVALIER-SMITH, T. 2001: What are fungi? Pp. 3–37 in: McLaughlin, D. J.; McLaughlin, E. G.; Lemche, P. A. (eds), *The Mycota Part VII. Systematics and Evolution Part A.* Springer-Verlag, Berlin. 386 p.

CAVALIER-SMITH, T. 2010: Kingdoms Protozoa and Chomista and the eozoan root of the eukaryotic tree. *Biology Letters 6*: 342–345.

DELVINQUIER, B. L. J.; PATTERSON, D. J. 2000: Order Slopalinida. Pp. 754-759 *in*: Lee, J.J.; Leedale, G.F.; Bradbury, P. (eds), *An Illustrated Guide to the Protozoa. Second Edition. Organisms traditionally referred to as Protozoa, or newly discovered groups.* Society of Protozoologists, Lawrence. v +1432 p. (2 vols).

FRACO, D. M.; PARR, D. 1997: Case report. *Pythium insidiosum* keratitis. *Australian and New Zealand Journal of Ophthalmology 25*: 177–179.

HICKS, B. R. 1974: Relationships of the opalinida with particular reference to *Protoopalina* (Protozoa). *Mauri Ora 2*: 3–6.

HOBBS, M.; MOOR, C.; WANSBROUGH, T.; CALDER, L. 2002. The health status of asylum seekers screened by Auckland Public Health in 1999 and 2000. *Journal of the New Zealand Medical Association 115(1160)*: 1–7.

KIRK, P. M.; CANNON, P. F.; DAVID, J. C.; STALPERS, J. A. 2001: *Ainsworth & Bisby's Dictionary of the Fungi.* Ninth edn. CAB International, Wallingford.

LEANDER, C. A.; PORTER, D.; LEANDER, B. S. 2004: Comparative morphology and molecular phylogeny of aplanochytrids (Labyrinthulomycota). *European Journal of Protistology 40*: 317–328.

LEANDER, C.; PORTER, D. 2001: The Labyrinthulomycota is comprised of three distinct lineages. *Mycologia 93*: 459–464.

LEWIS, T. E.; NICHOLS, P. D.; McMEEKIN, T. A. 1999: The biotechnological potential of thraustochytrids. *Marine Biotechnology 1*: 580–587.

MIKRJUKOV, K. A.; PATTERSON, D. J. 2001: Taxonomy and phylogeny of Heliozoa. III. Actinophryids. *Acta Protozoologica 40*: 3–25.

PENNYCOOK, S. R.; GALLOWAY, D. J. 2004: Checklist of New Zealand 'Fungi'. Pp. 401–488 *in*: McKenzie, E.H.C. (ed.), *Introduction to the Fungi of New Zealand.* Fungal Diversity Press, Hong Kong.

PÉREZ-CORDÓN, G.; ROSALES, M. J.; GAVIRA, M. d. M.; VALDEZ, R. A.; VARGAS, F.; CÓRDOVA, O. 2007: Finding of *Blastocystis* sp. in bivalves of the genus *Donax*. *Revista Peruana de Biología 14*: 301–302.

RAGHUKUMAR, S. 2002: Ecology of the marine protists, the Labyrinthulomycetes (Thraustochytrids and Labyrinthulids). *European Journal of Protistology 38*: 127–145.

RAGHUKUMAR, S. 2008: Thraustochytrid marine protists: production of PUFAs and other emerging technologies. *Marine Biotechnology 10*: 631–640.

SCICLUNA, S.; TAWARI, B.; CLARK, C. G. 2006: DNA barcoding of *Blastocystis*. *Protist 157*: 77–85.

SILBERMAN, J. D.; SOGIN, M. L.; LEIPE, D. D.; CLARK, C. G. 1996: Human parasite finds taxonomic home. *Nature 380*: 398.

STENZEL, D. J. L.; BOREHAM, P. F. L. 1996: *Blastocystis hominis* revisited. *Clinical Microbiology Review 9*: 61–79.

STOUT, J. D. 1984: The protozoan fauna of a seasonally inundated soil under grassland. *Soil Biology and Biochemistry 16*: 121–125.

TAN, K. S. W. 2008: New insights on classification, identification, and clinical relevance of *Blastocystis* spp. *Clinical Microbiology Reviews 21008*: 639–665.

TAN, T. C.; SURESH, K. G.; SMITH, H.V. 2008: Phenotypic and genotypic characterization of *Blastocystis hominis* isolates implicates subtype 3 as a subtype with pathogenic potential. *Parasitology Research 104*: 85--93.

YOSHIKAWA, H.; WU, Z.; HOWE, J.; HASHIMOTO, T.; NG, G.-C.; TAN, K. S. W. 2007: Ultrastructural and phylogenetic studies

Checklist of New Zealand Bigyra

Species habitats are indicated by F, freshwater; M, marine; T, terrestrial; and life habits and hosts by P, parasite; V, vertebrate. E endemic species, A, adventive.

KINGDOM CHROMISTA
SUBKINGDOM HAROSA
INFRAKINGDOM HETEROKONTA
PHYLUM BIGYRA
SUBPHYLUM OPALOZOA
Superclass NUCLEOHELEA
Class NUCLEOHELEA
Order ACTINOPHRYALES
ACTINOPHRYACEAE
Actinophrys sol Ehrenberg, 1840 F

Superclass OPALINATA
Class OPALINEA
Order OPALINALES
OPALINACEAE
Protoopalina australis Metcalf, 1923 PV F A
Protoopalina hylarum (Raff, 1911) PV F A

Class BLASTOCYSTEA
Order BLASTOCYSTALES
BLASTOCYSTACEAE
Blastocystis hominis Brumpt, 1912 PV T A

SUBPHYLUM BICOSOECIA
Class BICOECEA
Order BICOECALES
BICOSOECACEAE
Bicosoeca lacustris J.Clark F
Bicosoeca petiolata (F.Stein) Pringsh. F
Bicosoeca sp. Bojo 2001 F

SUBPHYLUM SAGENISTA
Class LABYRINTHULEA

Subclass LABYRINTHULIDAE
Order LABYRINTHULALES
LABYRINTHULACEAE
Labyrinthula macrocystis Cienk. M

Subclass THRAUSTOCHYTRIDAE
Order THRAUSTOCHYTHRIALES
THRAUSTOCHYTRIACEAE
Schizochytrium aggregatum S.Goldst. & Belsky M
Thraustochytrium cf. *aureum* S.Goldst. M
Thraustochytrium kinnei A.Gaertn. M
Thraustochytrium multirudimentale S.Goldst. M
Thraustochytrium proliferum Sparrow M
Thraustochytrium roseum S.Goldst. M
Thraustochytrium striatum Joach.Schneid. M
Ulkenia visurgensis (Ulken) A.Gaertn. M

THIRTEEN

Phylum
MYZOZOA

dinoflagellates, perkinsids, ellobiopsids, sporozoans

F. HOE CHANG, TONY CHARLESTON, PHILIP B. McKENNA,
CHRISTOPHER D. CLOWES, GRAEME J. WILSON, PAUL A. BROADY

Phylum Myzozoa ('sucking life') unites two very distinctive groups of organisms, the best-known of which are the dinoflagellates (mostly marine, free-living, and photosynthetic) and the apicomplexans (including sporozoans), which are wholly parasitic in invertebrates and vertebrates. Members of both groups are commonly or ancestrally suctorial feeders. This mode of ingestions is achieved by an apparatus that is best developed in the Apicomplexa, named for an apical complex of fibrils, microtubules, vacuoles, and other organelles at one end of the cell. Myzozoans are ancestrally biciliate but the economically important Sporozoa (which includes malarial parasites, piroplasms, and coccidia) is largely non-ciliate.

Cavalier-Smith (1987) united the two groups as Miozoa, based on the idea, then current, that dinoflagellates and coccidian Sporozoa share an unusual form of meiotic cell division (single-step instead of two-step), but this has since been disproved for both groups. Given the misleading name, Cavalier-Smith and Chao (2004) changed Miozoa to Myzozoa. Myzocytosis is the name originally given to a method of feeding in certain dinoflagellates, whereby a peduncular structure is inserted into the surface of a prey organism to suck out its cell contents into food vacuoles (in contrast to the more widespread engulfing of whole cells by phagocytosis). The last common ancestor of dinoflagellates and apicomplexans was almost certainly a myzocytotic predatory flagellate.

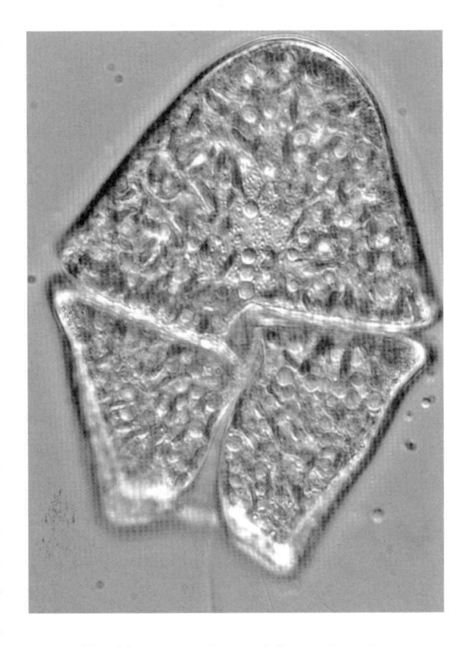

Akashiwo sanguinea, a bloom-forming dinoflagellate.
Hwan Su Yoon, per Micro*scope (MBL)

Myzozoans are evolutionarily related to ciliates (phylum Ciliophora) by having membranous alveoli in their cell cortex (but some groups lack them) (Saldarriaga et al. 2004). The cortex is essentially the outer body layer, which is extraordinarily complex in ciliates. This will be explained in the following chapter on Ciliophora; suffice it to say here that in ciliates and myzozoans the cell membrane is internally supported by membranous sacs (alveoli) (except in several specialised regions of the surface), hence these protozoans are collectively called alveolates. Alveolates also have tubular mitochondrial cristae (infoldings of the inner mitochondrial membrane). Collectively, alveolates and heterokonts (often called chromalveolates) can be said to have cells that are marvellously more complex, and immensely more disparate, than those of animals and plants. Chromalveolates are thought to have evolved when a phagotrophic biciliate protozoan enslaved and merged with a unicellular red alga more than 530 million years ago (Cavalier-Smith 2004). Both groups lost photosynthesis and plastids (usually containing chlorophyll and other pigments) several times and therefore exhibit phototrophic (algae), phagotrophic (e.g. ciliates), parasitic (e.g. *Plasmodium*), and saprotrophic nutritional strategies. The plastid of sporozoans is non-pigmented and therefore non-photosynthetic (Cavalier-Smith 1999), but

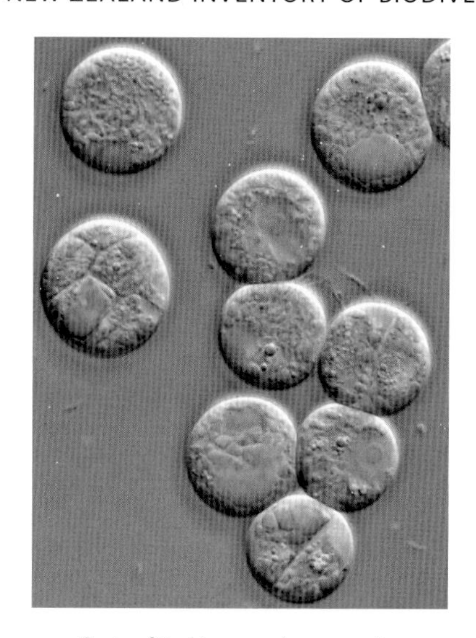

Cysts of *Perkinsus*, a clam parasite.
David J. Patterson, Linda Amaral-Zettler, Mike Peglar, Tom Nerad per Micro*scope (MBL)

the recent discovery of *Chromera* gives evidence that the earliest apicomplexan was photosynthetic (Moore et al. 2008). This facultative symbiont has typical features of alveolates, being phylogenetically related to apicomplexa, but uses a plastid, containing chlorophyll *a*, as its primary energy source.

Apicomplexa is one of two subphyla of Myzozoa, the other being Dinozoa (dinoflagellates and relatives) (Cavalier-Smith 2004; Cavalier-Smith & Chao 2004). An alternative classification, with details of generic attributions to the level of family, is given in Lee et al. (2000).

Subphylum Dinozoa: Protalveolates (*Perkinsus*, ellobiopsids)

The Dinozoa (Greek *dinos*, whirling, rotation, eddy) includes two infraphyla, the Protalveolata, probably paraphyletic (Cavalier-Smith & Chao 2004), and the Dinoflagellata. The former includes free-living predatory flagellates and intracellular-parasitic forms in which the alveoli tend to be discrete and inflated (in contradistinction to the Apicomplexa in which they tend to be highly flattened and cover most of the cell surface). Of the four included classes – Colponemea (*Algovora, Colponema*), Myzomonadea (*Alphamonas, Chilovora, Voromonas*), Perkinsea (*Parvilucifera, Perkinsus, Phagodinium, Rastrimonas*), and Ellobiopsea (*Ellobiopsis, Thalassomyces*), only the Myzomonadea is not yet known in New Zealand.

Perkinsus olseni (Perkinsea), a parasite of bivalve molluscs (clams), has a wide distribution around the world, including in New Zealand, where it affects cockles (*Austrovenus stutchburyi*), wedge shells (*Macomona liliana*), pipi (*Paphies australis*), and the ark clam *Barbatia novaezealandiae* (Diggles et al. 2002; Murrell et al. 2002). The parasite causes a disease known as perkinsosis, which is internationally notifiable. Whereas the infection can be fatal to some overseas shellfish, native species seem not to suffer mortalities, although soft yellow abscesses may sometimes be evident in the flesh. A recent New Zealand-wide survey found that *P. olseni* occurs mainly in cockles around the Waitemata Harbour, along the east coast north of Auckland, and in Kaipara Harbour. Dungan et al. (2007) noted infections of 10–84% of sampled clams in affected populations in the eastern Auckland area, with extensive lesions seen histologically in gill, mantle, foot, and visceral connective tissues. The parasite is able to spread via flagellate sporozoites

Summary of New Zealand myzozoan diversity

Taxon	Described species + infraspecific taxa	Known undescribed/ undetermined species	Estimated unknown species	Adventive species	Endemic species	Endemic genera
Dinozoa	271+1	11	255	0	0	0
Colponemea	1	0	2	0	0	0
Myzomonadea	0	0	3	0	0	0
Perkinsea	1	0	4	0	0	0
Ellobiopsida	2	0	10	0	0	0
Syndinea	0	0	15	0	0	0
Noctilucea	2	0	1	0	0	0
Peridinea	265+1	11	220	0	0	0
Apicomplexa	113	46	1,115	~118	~32	0
Apicomonadea	0	0	5	0	0	0
Coccidea	100	28	>100	~111	~12	0
Gregarinea	4	13	>1,000	>1	>16	0
Haematozoa	9	5	10	~6	~4	0
Totals	**384+1**	**57**	**>1,370**	**~115**	**~32**	**0**

('zoospores'), there being no sexual phase. It has a suctorial apical complex with which it feeds on host cells.

Species of *Thalassomyces* (Ellobiopsea) are remarkable parasites of oceanic euphausiid and amphipod crustaceans. At the time of their discovery, there was some debate as to whether ellobiopsids were fungi or protozoans. *Thalassomyces marsupii*, which infects the amphipod *Parathemisto gaudichaudii*, can exceed two millimetres in size, having a branching structure and a system of root-like cell processes that penetrate the host and anchor it. The parasite root is embedded in the host's nerve cord, causing localised swelling. It also affects the exoskeleton, causing scarring. The most striking effect, however, is when the parasite's branches produce spherical clusters of cells, presumably the dispersal stage, in the brood-pouch of the amphipod. Curiously, these spheres are about the same size as the amphipod's eggs, but the host is not otherwise harmed and is still able to reproduce. *Thalassomyces marsupii* was first found in amphipods captured some distance northwest of Campbell Island, but it is widespread in most oceans like its crustacean host (Kane 1964). Several species of euphausiid are similarly affected by *T. fagei* (Webber et al. 2010). Zoospores of *Thalassomyces* are biflagellated, with one circumferential cilium and one trailing, reminiscent of dinoflagellates, but the typical dinoflagellate nucleus has not been observed. Gene sequencing clearly allies *Thalassomyces* with alveolate protozoans (Silberman et al. 2004).

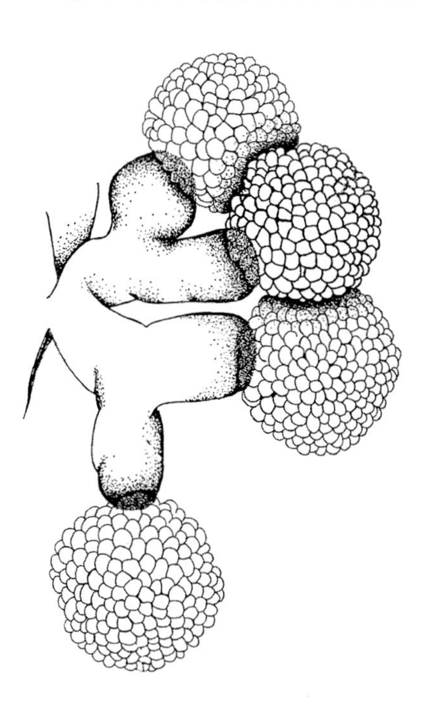

Thalassomyces marsupii, a fungoid ellobiopsid.
From Kane 1964

Subphylum Dinozoa: Dinoflagellates

Dinoflagellates are a remarkably diverse and complex group of flagellates. In cell size they span almost precisely the same range as diatoms, 2–2000 micrometres, but most fall within the range of 5–200 micrometres. Most are unicellular; some form tiny chains by repeated division of cells. Photosynthetic forms typically contain chlorophylls *a* and c_2, but some have other pigment combinations (e.g. chlorophylls *a*, c_2, and c_3, chlorophylls *a* and *b*, or chlorophyll *a* and phycobilin), owing to chloroplast capture from some other alga. Peridinin appears to be the only indicator accessory pigment, except for a few *Karenia* species that lack peridinin and have fucoxanthin, 19`-hexanoyl-oxyfucoxanthin and/or

Summary of New Zealand myzozoan [and acritarch] diversity by environment*

Taxon	Marine/brackish	Terrestrial	Freshwater	Fossil
Achritarcha	0	0	0	41
Dinozoa	241	0	41+1	427
Colponemea	0	0	1	0
Myzomonadea	0	0	0	0
Perkinsea	1	0	0	0
Ellobiopsida	2	0	0	0
Syndinea	0	0	0	0
Noctilucea	2	0	0	0
Peridinea	236	0	40+1	420
Apicomplexa	8	148	3	0
Apicomonadea	0	0	0	0
Coccidea	6	121	1	0
Gregarinea	1	15	1	0
Haematozoa	1	12	1	0
Totals	**249**	**148**	**49+1**	**468**

*Species + infraspecific taxa

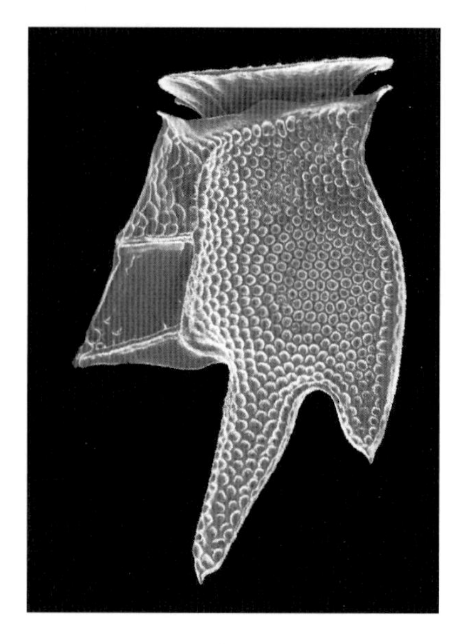

Dinophysis tripos.
Stephanie Valentin, David J. Patterson, per Micro*scope (MBL)

19`-butanoyl-oxyfucoxanthin as the main accessory pigment (Daugbjerg et al. 2000). Non-photosynthetic forms often have a pale pinkish colouration, such as that observed in *Noctiluca*, resulting from the presence of minute carotenoid globules at the cell periphery. Dinoflagellates exist either as armoured forms, with thick cellulose thecal plates, or as naked forms without any visible plates under the light microscope. The nucleus is unique. It has been termed 'mesokaryotic', as it has characteristics intermediate between the coiled DNA areas of prokaryotic bacteria and the well-defined eukaryotic nucleus. Permanently condensed chromosomes throughout the cell cycle represent a striking difference from typical eukaryotic chromosomes. These chromosomes, however, are devoid of histones. The biflagellate motile phase of dinoflagellates is distinctive, owing to the dissimilarity of the cilia [in the terminology of Cavalier-Smith (2000), who restricts the term flagellum to bacterial flagella]. In some species both cilia appear from the cell anterior, in others from the posterior, but in most from the ventral surface. In most cases one cilium encircles the cell transversely, generally in a groove (cingulum). Ribbon-like, it provides the propulsive and spinning force for the cell; the other smooth, longitudinal, cilium trails within a furrow called a sulcus and presumably acts as a rudder for steering.

Dinoflagellates are found in a wide variety of environments, reflected in their diversity of form and an extensive fossil record dating back several hundred million years (Hackett et al. 2004). As swimming cells, they can flourish under conditions that are unsuitable for many non-motile phytoplankton species, a success that owes in part to unique behaviour patterns, like daily vertical migration through the water column. Some are toxic, others are predatory or parasitical, and some are bioluminescent and emit light. Symbiotic species of *Symbiodinium* in coral and anemone polyps are known as zooxanthellae. They transfer a significant amount of fixed carbon to the polyp host, which exposes its tentacles and surrounding tissues to the light needed by the dinoflagellate. As with other photosynthetic eukaryotes, dinoflagellates acquired their plastids through endosymbiosis, but only the dinoflagellates have undergone tertiary endosymbiosis, which is the engulfment of an alga containing a secondary plastid (Cavalier-Smith 1999; Yoon et al. 2005) – the commonest plastid type, containing the pigment peridinin, is bounded by three membranes.

The alveoli of dinoflagellates (generally called amphiesmal vesicles in dinoflagellate literature) are the sites where the cellulose thecal plates are contained. In thecate orders (Gonyaulacales, Peridiniales, Dinophysiales, Prorocentrales) the theca is contained in relatively few alveoli, with a pattern of plates that can be determined relatively easily. Plate tabulation is thus the most important diagnostic character for the armoured dinoflagellates. Athecate taxa, however, notably the order Gymnodiniales, but also Syndiniales, Noctilucales, etc.) often contain hundreds of alveoli. As a consequence, thecate taxa are easier to classify than athecate ones (Saldarriaga et al. 2004).

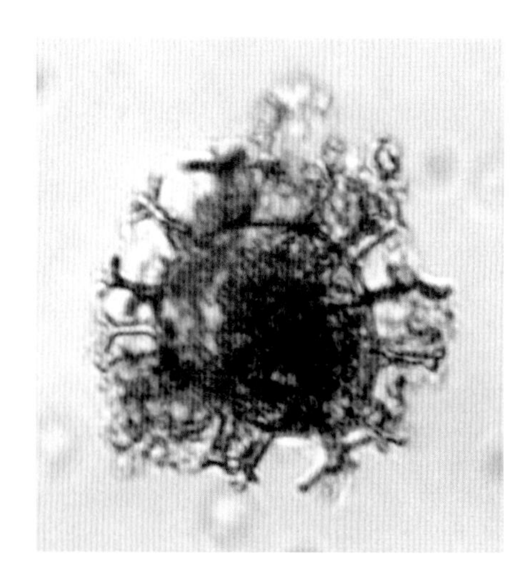

A cyst of the marine dinoflagellate *Gonyaulax scrippsae.*
F. Hoe Chang

Some dinoflagellates form resting cysts as part of their sexual life cycle, most notably towards the end of a bloom (usually when nutrients become depleted and/or temperature drops below optimum). In the sexual phase, two reproductive cells (gametes) fuse to form a zygote (planozygote) (e.g. Silva & Faust 1995; Chang et al. 2010). In the case of thecate species, the outer wall of the zygote further thickens and the cell becomes either warty-looking or spiny (hypnozygote). In most species this wall is made up of organic matter that chemically appears similar to the wall material in spores and pollen grains (sporopollenin) and is resistant to natural decay (Goodman 1987). Some have siliceous or calcareous walls. In most species, the hypnozygote (resting cyst) remains dormant for a period of time before it germinates to establish a new motile population under favourable conditions. In unfavourable conditions, the resting cysts can remain viable in the sediment for a long time (Dale 1983). Cysts found in seabed sediments can provide a record of species in the area. Thus, dinoflagellate cysts hidden in the seabed, particularly those of toxic origin, can be a source of harmful algal outbreaks. Fossilised dinoflagellate cysts from as far back as the Triassic period have been identified in sediments (see the section 'Fossil dinoflagellates and acritarchs').

Marine dinoflagellates of New Zealand

Some 240 taxa (63 genera) have been recorded from New Zealand (end-chapter checklist). Globally, there are about 4000 described species in 550 genera (Graham et al. 2009). Knowledge of New Zealand dinoflagellate diversity is continually increasing. Since the publication of checklists by Taylor (1974) and Dawson (1992), there have been major revisions of the taxonomy of several groups of dinoflagellates, resulting in name changes. For example, the unarmoured genus _Gymnodinium_ was split into four, with the establishment of three new genera, namely _Akashiwo, Karenia,_ and _Karlodinium_ (Daugbjerg et al. 2000). Subsequently, using morphological and molecular sequencing data, another new genus, _Takayama,_ was split from _Gymnodinium_ (de Salas et al. 2003). The checklist also reflects the numbers of new distributional records that have added species to the EEZ in the past decade.

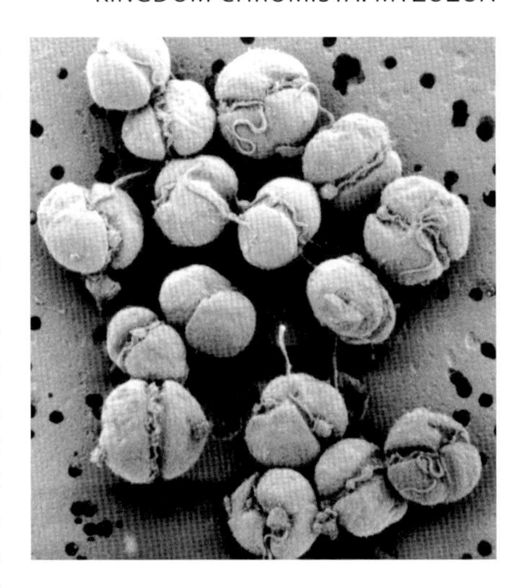

The bloom-forming dinoflagellate _Karenia brevisulcata._

F. Hoe Chang

In Marlborough Sounds on the northern South Island, New Zealand, seventeen dinoflagellate cyst types capable of seeding plankton dinoflagellate populations were listed by Baldwin (1987). Between 2003 and 2005, surface sediment samples collected from fourteen ports and harbours of New Zealand showed 22 different cyst types (Chang et al. 2008). The port and harbour surveys were part of large-scale baseline surveys for the Ministry of Agriculture and Forestry (MAF) and Biosecurity New Zealand (BNZ). Sun and McMinn (1994) and McMinn and Sun (1994) identified 21 cyst types on the Chatham Rise, east of New Zealand. In their studies, the distribution of dinoflagellate cysts in the Subtropical Front (STF) showed a marked response to this major oceanographic boundary. The total number of cyst types recorded in New Zealand is approximately 46 (see table below). Except for a few species (e.g. _Dalella chathamense, Impagidinium_ spp., _Nematosphaeropsis lemniscata, Operculodinium centrocarpum, Pyxidinopsis reticulata_) most cyst types found in New Zealand have an identifiable living planktonic form. In comparison, 60 cyst types have been recorded in Australia (McMinn et al. 2010). Worldwide, about 150 extant species of marine dinoflagellates are known to form resting cysts, but fewer than 80 of these have the potential to become fossilised (Head 1996).

In the Marlborough Sounds _Lingulodinium polyedrum_ (= _Gonyaulax polyedra_) was the only potentially harmful species identified (Baldwin 1987), while in the port and harbour surveys as many as six harmful or potentially harmful species were detected – four paralytic shellfish poisoning (PSP-producing) (_Gymnodinium_

Checklist of New Zealand dinoflagellate cyst types in Recent sediments

PERIDINIIDA
_Peridinium stellatum_*
Peridinium sp.*
Protoperidinium cf. _americanum_+
Protoperinium cf. _avellana_+
Protoperidinium claudicans+
Protoperidinium cf. _conicoides_*
_Protoperinium conicum_*+**
_Protoperidinium leonis_*+
_Protoperidinium oblongum_*+**
_Protoperidinium pentagonum_*+**
Protoperidinium cf. _punctulatum_*+
_Protoperidinium subinerme_*+**
Protoperidinium spp.+**
_Scrippsiella trochoidea_*+
Scrippsiella regalis†
GONYAULACIDA
Alexandrium cf. _affine_+
Alexandrium catenella+
Alexandrium minutum+
Alexandrium ostenfieldii
Alexandrium tamarense+
Bitectatodinium tepikiense**
Dalella chathamense**
Gonyaulax cf. _digitalis_*
_Gonyaulax grindleyi_*
Gonyaulax scrippsae+
Spiniferites mirabilis** (= _G. spinifera_ complex+)
Spiniferites ramosus**)
Gonyaulax sp.*
Impagidinium aculeatum**
Impagidinium paradoxum**
Impagidinium patulum**
Impagidinium sphaericum**
Impagidinium strialatum**
Impagidinium velorum**
Impagidinium sp.
_Lingulodinium polyedrum_*+
Nematosphaeropsis lemniscata**
Operculodinium centrocarpum**
Protoceratium reticulatum+**
GYMNODINIIDA
Gymnodinium catenatum+
_Polykrikos schwartzii_*+
DIPLOPSIDA
Diplopsalis sp.*
INCERTAE SEDIS
Pyxidinopsis reticulata**
Pyxidinopsis sp.*
Two unknown spp.+

*Baldwin 1987; +Chang et al. 2010; **McMinn & Sun 1994; Sun & McMinn 1994; † Edwards pers. comm.

Coloured-bloom-forming/'red-tide' and toxic/potentially harmful marine and brackish-water dinoflagellate and other microalgal species in New Zealand

Microalgal group (Phylum/Class)	No. of bloom-forming species (listed below table)	No. of potentially harmful/ toxic species (listed below table)
CYANOBACTERIA	2	1
EUGLENOZOA		
Euglenoidea	1	0
OCHROPHYTA		
Dictyochophyceae	2	2
Raphidophyceae	2	3
Bacillariophyceae	21	6
MYZOZOA		
Noctilucea	1	0
Peridinea	30	29
CRYPTOPHYTA		
Cryptophyceae	1	0
HAPTOPHYTA		
Coccolithophyceae	6	1
CHLOROPHYTA		
Pyramimonadophyceae	2	0

Cyanobacteria
Coloured bloom/red tide: *Synechococcus* sp., *Trichodesmium erythraeum*
Toxic: *Nodularia spumigena*

Euglenozoa
Euglenoidea
Coloured bloom: *Eutreptiella marina*

Ochrophyta
Dictyochophyceae (silicoflagellates)
Toxic: *Pseudochattonella verruculosa* [*Chattonella verruculosa*], *Vicicitus globosus* [*Chattonella globosa*]

Raphidophyceae (raphidophytes)
Coloured bloom: *Fibrocapsa japonica, Heterosigma akashiwo*
Toxic: *Chattonella marina, Fibrocapsa japonica, Heterosigma akashiwo*

Bacillariophyceae (diatoms)
Coloured bloom: *Asterionella glacialis, Aureodiscus kittoni, Cerataulina pelagica, Chaetoceros armatum, C. peruvianus, C. pseudocurvisetus, C. socialis, Coscinodiscus wailesii, Ditylum brightwellii, Fragilariopsis kerguelensis, Guinardia striata, Lauderia annulata, Leptocylindricus danicus, Nitzschia longissima, Odontella mobiliensis, Pseudo-nitzschia australis, P. fraudulenta, P. pungens* f. *pungens, Rhizosolenia setigera, Thalassionema nitzschioides, Thalassiosira decipiens*
Toxic: *Amphora coffaeiformis, Pseudo-nitzschia australis, P. delicatissima, P. fraudulenta, P. pseudodelicatissima, P. turgidula*

Myzozoa
Noctilucea (sea sparkles)
Coloured bloom/red tide: *Noctiluca scintillans*

Peridinea (other dinoflagellates)
Coloured bloom/red tide: *Akashiwo sanguinea, Alexandrium angustitabulatum, A. camurascutulum, A. catenella, A. minutum, Amphidinium carterae, A. trulla, Ceratium furca, C. fusus, Dinophysis acuminata, Gonyaulax hyalina, G. polygramma, G. spinifera, Gymnodinium aureolum, G. catenatum, Gymnodinium impudicum, Heterocapsa triquetra, Karenia bidigitata, K. brevisulcata, K. concordia, K. papilionacea, K. selliformis, Lingulodinium polyedrum, Ostreopsis siamensis, Protoceratium reticulatum, Prorocentrum gracile, P. micans, P. triestinum, Scrippsiella trochoidea, Takayama pulchella*
Toxic: *Alexandrium angustitabulatum, A. camurascutulum, A. catenella, Alexandrium minutum, A. ostenfeldii, A. tamarense, Amphidinium carterae, A. trulla, Coolia monotis, Dinophysis acuminata, D. acuta, Dinophysis fortii, D. tripos, Gambierdiscus toxicus, Gymnodinium catenatum, G. aureolum, Karenia bidigitata, K. brevisulcata, K. concordia, K. papilionacea, K. selliformis, Karlodinium veneficum, Ostreopsis lenticularis, O. ovata, O. siamensis, Pfiesteria piscicida, Pseudopfiesteria shumwayae, Protoceratium reticulatum, Prorocentrum lima*

Cryptophyta
Cryptophyceae (cryptophytes)
Coloured bloom: *Chilomonas paramoecium*

Haptophyta
Coccolithophyceae
Coloured bloom: *Emiliania huxleyi, Gephyrocapsa oceanica, Phaeocystis globosa, P. pouchetii, Prymnesium calathiferum, Umbellosphaera tenuis*
Toxic: *Prymnesium calathiferum*

Chlorophyta
Pyramimonadophyceae
Coloured bloom: *Pyramimonas orientalis, Pyramimonas* sp.

catenatum, *Alexandrium tamarense*, *A. catenella*, *A. minutum*) and two yessotoxin- and yessotoxin-like-producing species (*Protoceratium reticulatum*, *Lingulodinium polyedrum*) (Chang et al. 2008). Cysts of another PSP-producing species, *A. ostenfeldii*, have also been found by MacKenzie et al. (1996) on both the east and west coasts of New Zealand.

Most dinoflagellates are components of marine phytoplankton, with a much smaller number of freshwater species. Some, in both environments, are wholly parasitic. Globally, while diatoms are the most species-rich group (Margulis et al. 1990; Jeffrey & Vesk 1997), dinoflagellates rank second in numbers of species, half of which are heterotrophic. Microflagellates (comprising several taxonomic groups) collectively have far fewer species. Marine microalgae are found in surface waters of all oceans and seas and have global significance in the cycling of many elements, particularly carbon. Estimates of global primary production show 46% is oceanic, and 54% terrestrial (e.g. Field et al. 1998).

Dinoflagellates generally thrive in warm and relatively stable environments. Some are capable of daily vertical migration and are able to exploit nutrient-rich deep water below the sunlit euphotic zone (Eppley & Harrison 1974; Chang 1988, Chang et al. 2003). As a consequence, this group usually dominates in summer, especially when nutrients in the upper water column become depleted (Dodge 1982; Bradford & Chang 1987; Chang et al. 2003). In New Zealand in summer, dinoflagellates can provide more than 90% of the total carbon biomass in inshore blooms (Chang 1988). The dominance of dinoflagellates in summer has important implications for secondary producers – dinoflagellates are known to have greater nutritional value than diatoms of equal volume and chlorophyll *a*. More recent studies have confirmed the higher carbon content of dinoflagellates than diatoms (Chan 1980; Hitchcock 1982) and have revealed higher protein, carbohydrate, and sometimes lipid content in dinoflagellates.

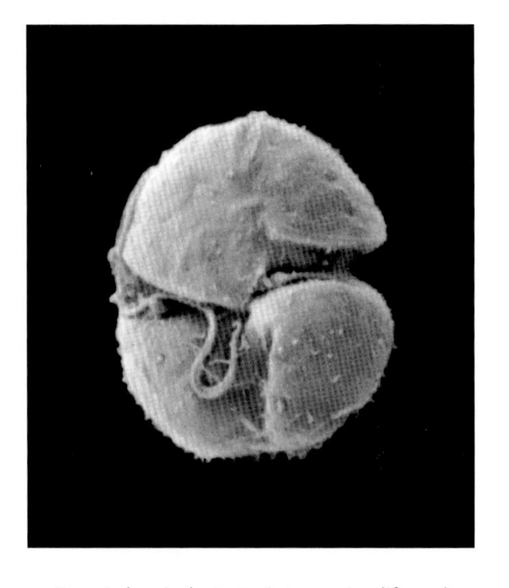

Karenia brevisulcata, toxic to marine life and causing human respiratory distress.
F. Hoe Chang

The majority of blooms, 'red ' or 'brown tides' recorded in New Zealand are harmless but on several occasions potent toxins produced by a small number of dinoflagellates (the harmful-algal-bloom or HAB species) have been reported to cause either mass mortalities of marine life (Chang & Ryan 1985, 2004; Chang et al. 1990; Chang 1999a,b; MacKenzie et al. 1996), or to find their way from shellfish to humans, causing such illnesses as neurotoxic shellfish poisoning (NSP), paralytic shellfish poisoning (PSP), diarrhetic shellfish poisoning (DSP), and ciguatera seafood poisoning (CSP) (Chang et al. 1995; Chang 2004). In New Zealand waters, only NSP and PSP have been implicated in causing human illnesses (Bates et al. 1993; Chang et al. 1995, 1997). A toxic dinoflagellate from Wellington Harbour, *Karenia brevisulcata* (first named as *Gymnodinium*), was found to be associated with marine-life mortalities and also outbreaks of human respiratory illness (Chang 1999a,b; Chang et al. 2001).

Bloom-forming species in New Zealand waters involve several algal groups (68 species in total), with five out of nine classes represented by harmful species (see table on next page). Dinoflagellates as a group attract attention as they have the greatest number of harmful taxa (29 species frequently associated with coastal blooms), followed by diatoms (six species) and three microflagellate classes (six species) (Chang 2004).

Harmful algal blooms (HAB)

Since the 1980s, a range of phytoplankton taxa have been demonstrated to make toxins in New Zealand coastal waters. The total number of toxic or potentially harmful species recorded from New Zealand is 42 (about 3% of the phytoplankton taxa of the EEZ, somewhat higher than the 2% or so harmful species estimated globally (Sournia 1995)). Because of the health risks associated with the toxic algal episodes, shellfish and water samples collected from discrete sites in New Zealand have been regularly monitored by the industries and the government regulatory authorities after the 1993 toxic events. The following is a range of negative effects commonly associated with harmful outbreaks in New Zealand.

Neurotoxic shellfish poisoning (NSP). The first reported episode of NSP in New Zealand, in early 1993, resulted in neurological symptoms in more than 180 people, with numerous cases of human respiratory illnesses in eastern North Island (Bates et al. 1993; Chang et al. 1995). These outbreaks led to a complete ban on shellfish harvesting throughout New Zealand. During the NSP episode, a mixed assemblage of species resembling *Karenia brevis* (a known NSP organism in the USA) and *K. mikimotoi* were recorded (Chang 1995; MacKenzie et al. 1995). In 2002, *Karenia* sp., subsequently described as *K. concordia* (Chang & Ryan 2004), was found in areas where NSP was reported (Chang 1995; Chang et al. 2008). In Hawkes Bay during the same period, another new *K. brevis* look-alike (*K. papilionacea*) was discovered and later described by Haywood et al. (2004). Since the 1993 NSP event, there has been no recurrence of brevetoxin contamination associated with *Karenia* species in the same region.

Paralytic shellfish poisoning (PSP). The 1993 outbreak was the first reported in New Zealand, based on detection of PSP toxin in Bay of Plenty shellfish. Apparently *Alexandrium minutum* was the causative organism (Chang et al. 1995, 1997). The taxonomy of this group of PSP-producing dinoflagellates is the subject of much wrangling. [Related *Alexandrium tamarense* happens to be the type species of the genera *Gonyaulax* and *Protogonyaulax*.] Most PSP-producing species so far recorded from New Zealand belong to *Alexandrium*, of which five – *A. angustitabulatum* (Cembella et al. 1988), *A. minutum* (Chang et al. 1997), *A. catenella*, *A. ostenfeldii*, and *A. tamarense* (MacKenzie et al. 2004) are known to be toxic. For the first time since mid-winter 2000, another taxon, *Gymnodinium catenatum*, has been associated with a series of PSP outbreaks on both the North Island east and west coasts (MacKenzie & Adamson 2000; Chang 2001), and is the only gymnodinioid species known to cause PSP.

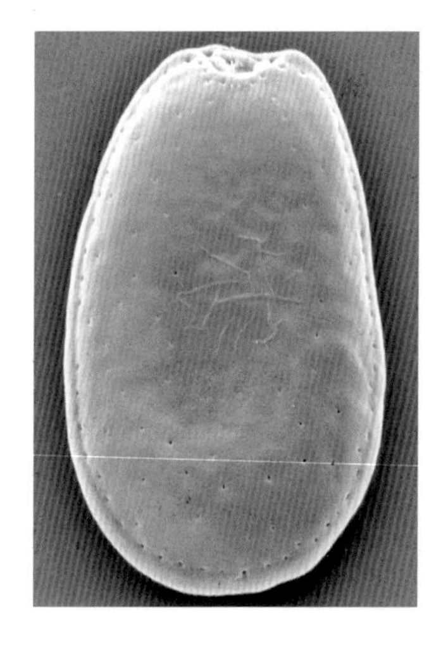

Prorocentrum lima, a form with two large plates and no flagellar grooves.
Shauna Murray, per Micro*scope (MBL)

Diarrhetic shellfish poisoning (DSP). No direct DSP episode has ever been reported from New Zealand but the causative toxin, okadaic acid, has been detected in shellfish extracts gathered from around New Zealand, in particular the Marlborough Sounds (Baldwin 1992; Chang et al. 1995; Chang 1996a; MacKenzie et al. 1998) and Akaroa, Canterbury (McNabb & Holland 2003). *Dinophysis acuminata*, *D. acuta*, *D. fortii*, and *Prorocentrum lima* are the few common species found in New Zealand that produce okadaic acid and also pectenotoxin (Rhodes & Syhre 1995; MacKenzie et al. 2005). *Protoceratium reticulatum* and *Gonyaulax spinifera* both common New Zealand species, are known to produce yessotoxins (MacKenzie et al. 1998; Rhodes et al. 2006). Yessotoxins had previously been included among toxins related to DSP, but there is now some doubt that it should be in this category (Yasumoto & Satake 1998). To date, no studies have highlighted any toxic effects of yessotoxins when they are present in humans but negative effects on mice when injected intraperitoneally, and the fact that yessotoxins are not destroyed by heating and freezing, mean that several countries including New Zealand, Japan, and those in Europe regulate the levels of these toxins in shellfish.

Ciguatera seafood poisoning (CSP). Causative organisms of CSP are epiphytic/benthic dinoflagellates, and *Gambierdiscus toxicus* is the key species associated with outbreaks overseas (Yasumoto et al. 1971). The commonest epiphytic dinoflagellate recorded on the northeast coast of New Zealand is *Ostreopsis siamensis* (Chang et al. 2000). It produces a palytoxin derivative, which is as toxic as palytoxin itself (a named biological warfare agent) in intraperitoneal mouse bioassays (Rhodes et al. 2002) and which has been associated with deaths of kina (*Evechinus chloroticus*) in Northland during the summers of 2004 and 2005 (Shears & Ross 2009). Other closely related and potentially harmful dinoflagellates recorded from New Zealand include *Ostreopsis lenticularis*, *O. ovata*, *Coolia monotis*, and *Gambierdiscus toxicus* (Chang et al. 2000). *Coolia monotis* in New Zealand has tested as toxic to larvae of both the brine shrimp *Artemia salina* and virgin paua *Haliotis virginea* (Rhodes & Thomas 1997). Although shellfish have consistently tested positive for low levels of unidentified lipid-soluble toxins in *Ostreopsis*-infected areas (Chang et al. 2000), no ciguatera outbreak has ever been reported in New Zealand.

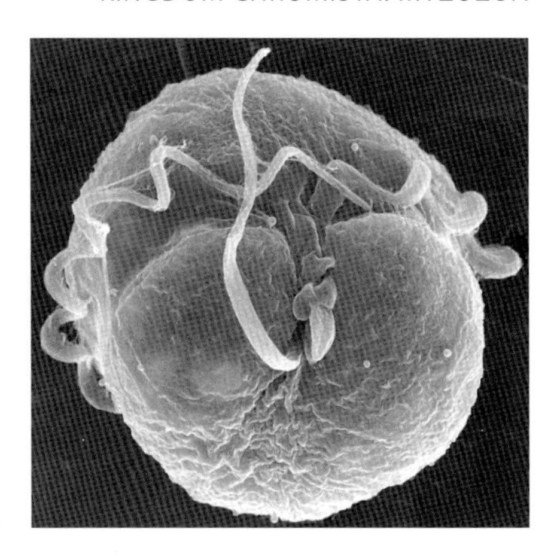

Antapical view of *Pseudopfiesteria shumwayae*.
Patrice Mason, Virginia Institute of Marine Science

Ichthyotoxicity (fish killing). Fish-killing species include dinoflagellates and other microflagellates. The commonest dinoflagellate taxa associated with fish kills in New Zealand include *Gymnodinium aureolum* (previously reported as *Gyrodinium aureolum*) from Banks Peninsula (Chang 1996b), *Karenia brevisulcata* from Wellington Harbour (Chang 1999a,b), *K. selliformis* from Foveaux Strait (Haywood et al. 2004), and *K. concordia* from Hauraki Gulf (Chang & Ryan 2004; Chang et al. 2008). Toxins produced by *Karenia brevisulcata* exhibit high cytotoxicity in a wide range of marine organisms (Chang 1999; Chang et al. 2001). A bloom in Wellington Harbour in 1998 caused massive kills of fish, invertebrates, and seaweeds and caused human respiratory distress. It appears to be the most toxic dinoflagellate known.

Pfiesteriids have been identified in New Zealand by the use of PCR (Polymerase Chain Reaction)-based DNA probes that have detected both *Pfiesteria piscicida* and *Pseudopfiesteria shumwayae* (Rhodes et al. 2002b, 2005). Each is present throughout New Zealand and both are residents of Tasman Bay's well-flushed estuaries and Canterbury's brackish lakes. Both occur in a wide range of salinities and temperatures. DNA sequence data indicate that *P. shumwayae* may actually comprise a suite of genetically closely related species, whereas *P. piscicida* is genetically identical with isolates from the USA (Rhodes et al. 2005). The presence of *Pfiesteria piscicida* ('the cell from hell' in some emotive overseas media reports) in New Zealand is not considered an immediate risk to fish or human health given the current low to moderate nutrient concentrations in New Zealand's estuaries and brackish lakes. Increases in nutrient loadings could pose a risk, however, as it has in eastern US estuaries (Rhodes & Adamson 2005).

Harmful algal blooms and unusual weather patterns

During the period from 1982 to 2002, eight major or 'exceptional' blooms were reported in New Zealand inshore waters (Chang & Ryan 1985, 2004; Taylor et al. 1985; Chang et al. 1990, 1995, 2001, 2008; Rhodes et al. 1993; MacKenzie et al. 1996; Chang 1999a,b; MacKenzie & Adamson 2000). These blooms all shared something in common – they were noticeable to the general public, caused fish or invertebrate mortality, toxicity to humans, and spread across large coastal areas of New Zealand (Chang & Uddstrom 2000; Chang et al. 2001, 2008). Six out of eight of these blooms were dominated by dinoflagellates and virtually all coincided with major El Niño-Southern Oscillation (ENSO) events in the two decades from1 1982 to 2002. During this period, the Southern Oscillation Index (SOI) strengthened (with more negative values) as the Interdecadal Pacific Oscillation (IPO) index shifted from negative to positive (Chang & Mullan 2003) (see figure below). Apart from several nuisance 'slime' events reported during the La Niña events (Hurley 1982), no major harmful algal blooms occurred in the three decades prior to the 1980s (with weakened SOI in El Niños) when the IPO index turned negative (see figure below). Likewise no major HAB was observed during the last several El Niños when the IPO index reversed to negative phase since 2002. It is likely that in the next few years the conditions will remain the same as IPO continues to be negative. It is clear that strengthening of local winds (northwesterlies on the North Island northeast coast) during El Niño (Zeldis et al. 2000), in particular when the IPO is positive, leads to an increase in upwelling intensity and surface-nutrient enrichment during a season (summer) that is normally nutrient-poor and thus encourage the development of harmful algal blooms in these regions (Chang & Mullan 2003; Chang et al. 2008).

The most notable blooms resulted in the recognition of a substantial number of newly reported HAB organisms, either undescribed species or new records to

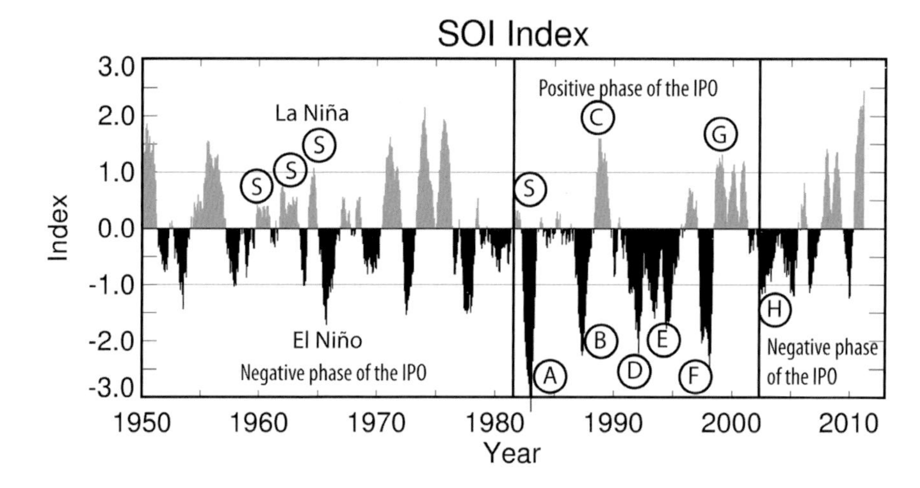

Major harmful algal blooms (HAB) 1950–2010: S, nuisance 'slime' events; A–H, other major HAB events reported in 1982–2002 (according to the temporal sequence in the table on the next page). ENSO events and IPO phases are marked.

New Zealand. Altogether, eight dinoflagellate species have been described as new to science in New Zealand, with a disproportionately large number of *Karenia* species – *Alexandrium angustitabulatum* (Balech 1995), *Karenia brevisulcata* (Chang 1999b), *Alexandrium camurascutulum* (MacKenzie & Todd 2002), *Amphidinium trulla* (Murray et al. 2003), *K. concordia* (Chang & Ryan 2004), *K. bidigitata, K. selliformis,* and *K. papilionacea* (Haywood et al. 2004). The presence of these new dinoflagellate species accounted for > 45% of the major phytoplankton recorded and most show negative impacts on either marine life or humans.

Future studies on toxic and non-toxic dinoflagellates

Positive identification relies on definitive morphological features that can be recognised and characterised using both light and electron microscopy, and on DNA sequence data. In the last few years in New Zealand, alternative approaches to identifying microalgal taxa have included chemotaxonomy, based on specific biomarker pigments (Chang & Gall 1998; Chang et al. 2003), immunofluorescence detection using species-specific immunological techniques (Chang et al. 1999), and oligonucleotide-probe techniques (Rhodes et al. 1998). Although each of these has its limitations, there is great potential in combining the new approaches with traditional techniques to discriminate closely related taxa, in particular morphologically similar taxa. For example, the New Zealand isolates of *Alexandrium minutum* can be identified from other morphologically similar species such as *A. lusitanicum* using an immunofluorescent technique, and isolates of *Gymnodinium, Karenia,* and *Takayama* species can be differentiated using species-specific rRNA probes (Rhodes et al. 2004; de Salas et al. 2005).

Likewise, the use of 18S ribosomal RNA has determined that *Symbiodinium* in the New Zealand sea anemone *Anthopleura aureoradiata* consists of at least five genotypically distinct sequences representing a new subclade in the phylogeny of the genus (Birkenstock 2001). This needs further investigation.

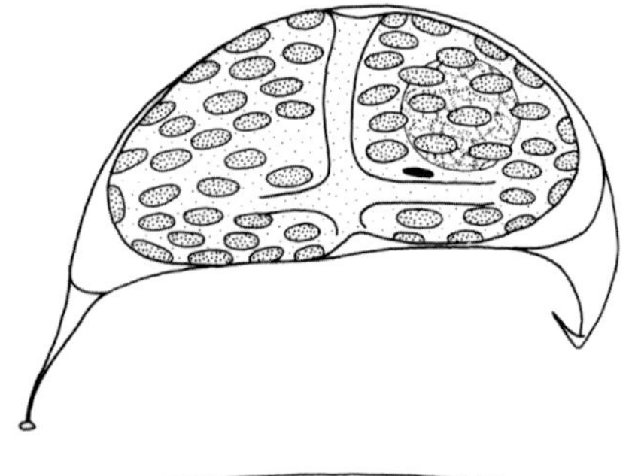

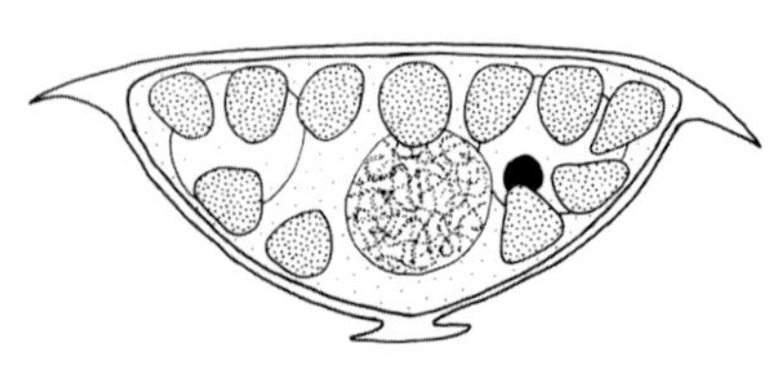

Freshwater dinoflagellates *Cystodinium cornifax* (upper) and *Dinococcus oedogonii* (lower).
After Bourrelly 1985

Freshwater dinoflagellates

The total worldwide diversity of freshwater dinoflagellates comprises about 240 species in about 32 genera (Bourrelly 1985). In New Zealand, 40 species and one subspecies have been recorded. This number is close to the 54 species in 13 genera that Lewis and Dodge (2002) described from the British Isles, a region of similar area and climate. They noted that it is likely that many more species remain to be discovered as studies are extended to additional waterbodies. Carty (2003) made similar comments with regard to the dinoflagellate diversity of North America, where about 150 species have been found to date. Undoubtedly this is also the case in New Zealand.

Although most freshwater dinoflagellates possess chloroplasts and are capable of photosynthesis, an increasing number are known to be mixotrophic –

that is, they have the additional capacity to feed on organic matter. This is usually achieved by ingesting prey cells whole or by sucking out their contents. Other dinoflagellates can feed only in this way. Among the New Zealand species, *Ceratium hirundinella* is a well-known mixotroph whilst *Gymnodinium helveticum* lacks chloroplasts and feeds by ingestion.

A fascinating variation on the theme of mixotrophy is so-called kleptoplastidy, as seen in *Gymnodinium aeruginosum* (Schnepf et al. 1989). It obtains temporary functional chloroplasts by ingesting cryptomonad cells, then digesting the nucleus and nucleomorph and keeping other cell components. Thus, chloroplasts of this species are unlike those of most other photosynthetic dinoflagellates in being blue-green, reflecting the pigment content of the prey, namely chlorophyll *a* and phycobilins. In New Zealand, *G. aeruginosum* has been found in swamp pools similar to its recorded preferences overseas.

The large majority of records are from the North Island, in particular from the Rotorua district (Cassie 1984), where Thomasson (1974) listed 14 species from nine Rotorua lakes. Although they can form visible populations, often they are low in numbers in water samples and this would lead to their being overlooked.

The majority of records of dinoflagellates in New Zealand, as elsewhere, are of motile species from phytoplankton communities of lakes. In addition, a significant number are from small pools in bogs and swamps, where they swim freely in the water or occur amongst aquatic vegetation and organic detritus of the sediments.

Worldwide, the commonest bloom-forming genera are *Ceratium* and *Peridinium*. Four species noted as very common by Pollingher (1987) – *Ceratium hirundinella, Peridinium bipes, P. cinctum,* and *P. willei* – all occur in New Zealand. *Ceratium hirundinella* is the most frequently recorded species, with more than 30 records in lake phytoplankton. Visible blooms have been seen in Lake Pupuke,

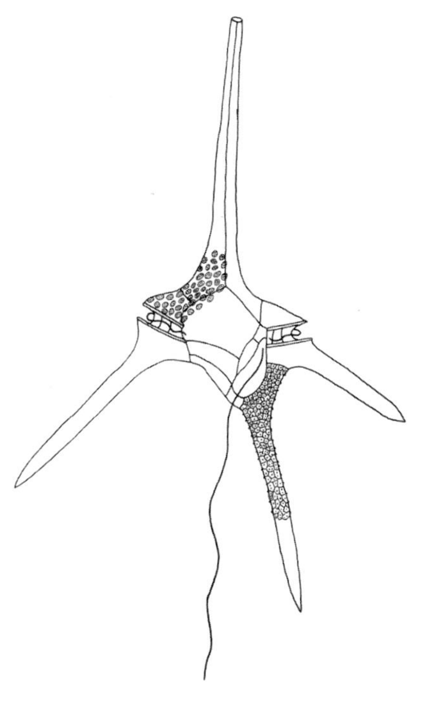

Ceratium hirundinella.
After Bourrelly 1985

Association of exceptional blooms/toxic episodes and unusual weather patterns between 1982 and 2002

Year / type of weather pattern	Bloom organisms (taxa) all microalgal groups	Toxic events	Geographical distribution
1982–83 El Niño	*Cerataulina pelagica*[1] *Prymnesium calathiferum*[2] *Alexandrium angustitabulatum*[3]	Fish and marine fauna affected	NE North Island
1986–87 El Niño	*Amphidinium* sp.[4] *Cryptomonas* sp.[4]	— —	NE North Island
1988–89 La Niña	*Heterosigma akashiwo*[5]	Salmon kills	Big Glory Bay, Stewart Island
1992–93 El Niño	*Fibrocapsa japonica*[6] *Karenia concordia*[7] *Karenia papilionacea*[8] *Alexandrium minutum*[9] *Pseudo-nitzschia australis*[9]	— NSP, HRD NSP PSP ASP*	NE North Island
1994–95 El Niño	*Karenia selliformis*[10]	Fish and marine fauna killed	South Island southeast coast and Foveaux Strait
1998–99 El Niño	*Karenia brevisulcata*[11,12]	Marine life killed, HRD	Central east coast, North Island
2000 La Niña	*Gymnodinium catenatum*[13]	PSP	West coast and lower east coast of North Island
2002 El Niño	*Karenia concordia*[7]	Fish and marine fauna killed	NE North Island

[1]Taylor et al. 1985; [2]Chang & Ryan 1985; [3]Balech 1995; [4]Chang unpubl.; [5]Chang et al. 1990; [6]Rhodes et al. 1993; [7]Chang et al. 2004; [8]Haywood et al. 2004 ; Chang 2010 ; [9]Chang et al. 2005 ; [10]MacKenzie & Adamson 2000; [11]Chang 1999; [12]Chang et al. 2001; [13]MacKenzie pers. comm. [*No ASP recorded during 1992–93 outbreaks.] HRD = human respiratory distress.

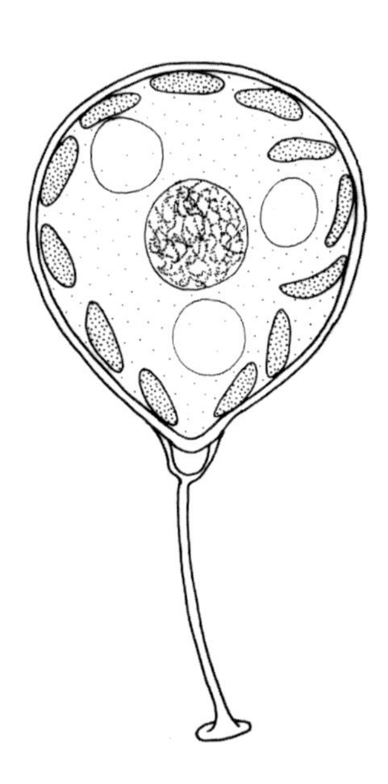

Stalked non-motile form of *Stylodinium globosum*.
After Bourrelly 1985

Auckland, where they can appear brown and impart a bad taste to the water (Cassie Cooper 1996). In nutrient-enriched Lake Hayes, Otago, there has been a recent shift in species composition of the phytoplankton community from cyanobacteria and desmids to *C. hirundinella* (Baye et al. 2008). This species was absent in 1969–72 and 1994–95 but in July 2006 comprised 91% of the biovolume of a bloom. Another bloom in summer 2007 coincided with observations of poorly conditioned brown trout which is of concern in a lake with high recreational value. The species could have a competitive advantage – its mixotrophic ability allows it to feed on particulate organic material in order to supplement inorganic nutrient supply. Also, it is strongly motile and able to migrate to deeper nutrient-enriched water at night, returning to illuminated surface water during the day. It might become commoner and more prominent in additional lakes if their nutrient loadings increase because of human activities. Freshwater dinoflagellates have not been associated with toxin production (Carty 2003).

Whereas there are many marine dinoflagellates that parasitise a wide range of organisms including fish, in fresh waters just a single species, *Oodinium limneticum*, is known to parasitise fish but the species has been found only in North America (Bourrelly 1985; Carty 2003). However, New Zealand has parasitic dinoflagellates that feed on filamentous chlorophyte algae such as *Oedogonium* and *Spirogyra*. Species of *Cystodinedria*, *Cystodinium*, *Dinococcus*, and *Stylodinium* produce amoeboid stages that consume the cell contents of their hosts. They have complex life-cycles that are incompletely understood. In fact, some 'species' may be stages in the life-cycles of others, having been incorrectly erected as species in their own right (Cachon & Cachon 1987). They all have a dominant non-motile stage, which, in the case of *Dinococcus* and *Stylodinium*, bears a stalk for attachment.

Worldwide, only four species of dinoflagellates, all coccoid forms, have been found on moist rocks and soils (Ettl & Gärtner 1995). There is an uncertain identification (Skuja 1976) of just one of these, *Gloeodinium montanum*, in various swamps. It is possible that this is just a non-motile stage of some other typically flagellate species. For instance, *Hemidinium nasutum* has been noted to divide whilst in a non-motile '*Gloeodinium*' phase (Lewis & Dodge 2002).

It can be concluded that much more knowledge of New Zealand dinoflagellates could be usefully gained. Their diversity is incompletely known, their ecology has been barely addressed, and studies on particular species could provide valuable clarification of life-cycles.

Fossil dinoflagellates and acritarchs

Dinoflagellates have a long fossil record and there are approximately 2000 fossil species worldwide. They are rare in the Paleozoic but are common from the Triassic onwards. Dinoflagellate fossils comprise the cysts that enable dinoflagellates to survive unfavourable environmental conditions. One class of cysts is the hystrichospheres, which look like spiny balls. Also included in the following account are acritarchs, which first occur much earlier in the fossil record (Precambrian). Some may represent dinoflagellate hystrichospheres but acritarchs are said to lack some characteristic dinoflagellate features, including the archaeopyle or excystment pore through which the dinoflagellate exits the cyst, hence exactly what acritarchs were is not known with certainty. Ellegaard et al. (2002), however, described an acritarchous resting cyst in an athecate dinoflagellate; it had a germination opening identical to that found in acritarchs, supporting the idea that athecate dinoflagellates are represented by acritarchs in the older geological record, predating the first morphologically identifiable dinoflagellate cysts.

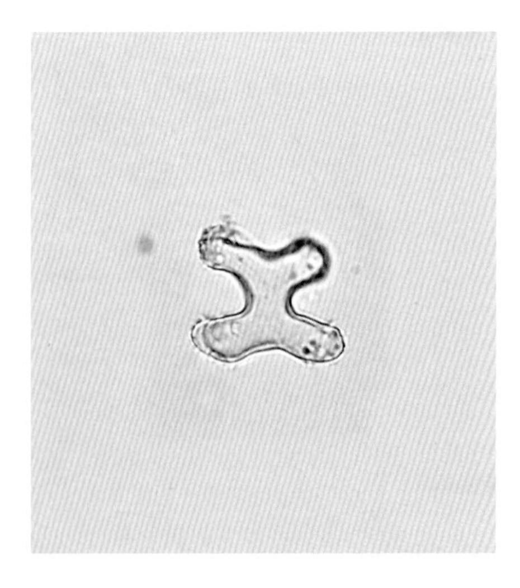

Paucilobimorpha spinosa (Acritarcha). Latest Eocene, Kakanui River, Otago.
Christopher Clowes

Early reports of New Zealand fossil algae, predominantly dinoflagellate cysts, include those of Deflandre (1933), Couper (1960) (a passing mention only), and a string of publications by Graeme Wilson, beginning in 1967 and spanning a forty-year career. Additional papers followed in the 1980s (Clowes & Morgans 1984; Clowes 1985) and, since about 1997, contributions have come from a small number of additional authors, including Schiøler and Wilson (1998), Roncaglia et al. (1999), Crouch (2001), and Willumsen (2003). Of the several papers dealing with detailed taxonomy, the two most significant contributions are those of Wilson (1967, 1988), which together include full systematic descriptions and stratigraphic ranges for more than 35 dinoflagellate species. Numerous other papers describe smaller numbers of new taxa.

The earliest practical dinoflagellate zonal scheme for New Zealand sediments is that of Wilson (1967), an informal eight- or nine-fold division based upon species of Wetzelielloideae. Extending this early work, Wilson (1984) published a formal biozonation that proposed a sequence of 13 zones extending from the Cretaceous–Tertiary boundary to the top of the Eocene. Subsequently, more detailed revisions were undertaken of the Late Cretaceous–Tertiary (Wilson 1987; Roncaglia et al. 1999, Schiøler & Wilson 1998; Crampton et al. 2000; Willumsen 2003) and of the mid-Paleocene to Middle Eocene interval (Wilson 1988; Crouch 2001). Much of this work is summarised in Cooper (2004).

A number of publications and various unpublished reports list assemblages from various localities, and further information is available in the Fossil Record File (FRF) that is maintained jointly by the Geological Society of New Zealand and GNS Science. The short review given here is the first attempt to provide a taxonomic synthesis of the taxonomic breadth covered by these sources; accordingly, a comprehensive bibliography of New Zealand fossil dinoflagellates and acritarchs is included in the references.

The FRF is a registration scheme for recording fossil localities in New Zealand and nearby regions, including South West Pacific Islands and the Ross Sea region of Antarctica. The paper-based records stored in regional 'master files' at Auckland, Victoria, Canterbury, and Otago universities and at GNS Science record locality and stratigraphic information for almost 90,000 fossil collection sites. A project is currently underway to digitise the Fossil Record File and enter it into a computer database, known as the Fossil Record Electronic Database (FRED), operated by GNS Science with the assistance of staff from Auckland, Victoria, Canterbury and Otago universities.

For the present review, data were extracted from FRED in June 2009 and manipulated electronically to yield both a list of taxa and the known age constraints on the rock units from which they are recorded. After the conversion of many (though probably not all) old names to currently recognised taxonomy, and elimination of duplicates, 536 species representing nearly 300 genera are recognised from more than 70,000 individual records throughout mainland

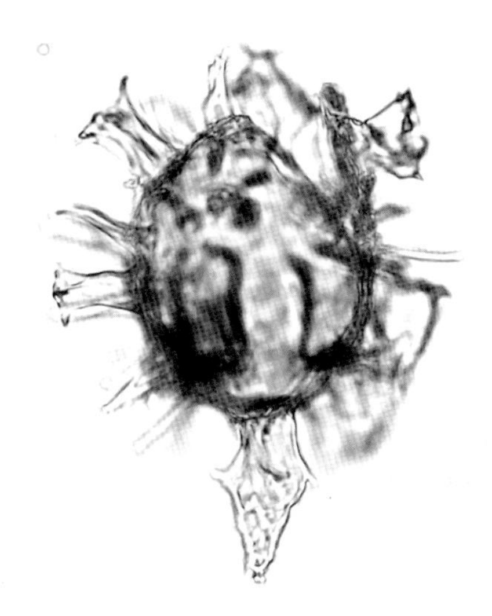

Achilleodinium biformoides, a hystrichosphere. Middle Eocene, Hampden Beach, Otago.
Christopher Clowes

Taxon diversity of acritarchs and fossil dinoflagellates in New Zealand

Genera appearing by end of:	Acritarcha	Gonyaulacida	Peridiniida	Other dinoflagellates
Permian	2	0	0	0
Triassic	9	2	0	2
Jurassic	10	48	2	3
Early Cretaceous	12	60	8	3
Late Cretaceous	23	112	30	4
Palocene	23	120	33	4
Eocene	24	131	37	4
Oligocene	24	131	37	4
All records in the FRF	**41**	**184**	**50**	**5**

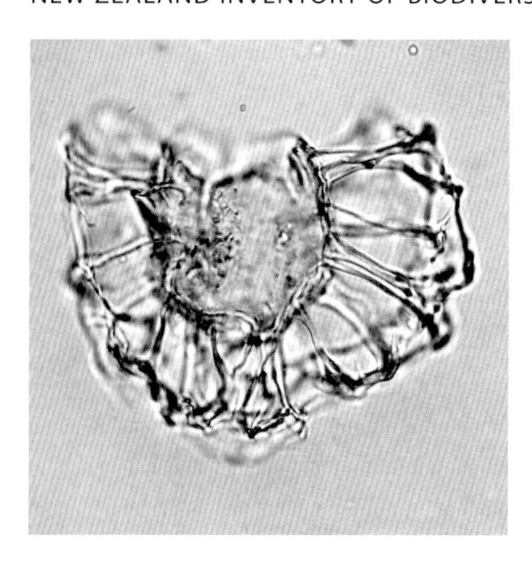

Cyst of the dinoflagellate *Glaphyrocysta* cf. *retiintexta*. Middle Eocene, Hampden Beach, Otago.
Christopher Clowes

and near-shore New Zealand. A translation table was built from the unique binominals contained in the raw data and used to map the FRF names onto current names. Although such mapping is technically straightforward, taxonomic ideas and nomenclature have changed considerably over the intervening years and it was not always possible to be certain which taxon was recorded. Many records had to be mapped to 'unknown dinoflagellate' or similar. A particular difficulty arose where an 'old' taxon has been split into two or more, and it is not known which of the FRF records refer to which of the new taxa. In most cases the 'unknown sp.' records have been omitted from the results presented here. However, they are retained where they provide the only record of a genus or higher-level taxon.

In the FRF, age data are often derived from known constraints on the containing rock units and may not be updated from subsequent paleontological or other analyses. Thus the ranges are to be regarded as intervals *within which* the actual stratigraphic ranges of the taxa occur, and are not to be interpreted literally. An attempt was made to avoid this problem by excluding all range data where the lower and upper age bounds did not fall within a single stage but, even so, reworking, cavings, etc., produced some results that are manifestly incorrect. For these reasons, ranges are not provided for all taxa in the end-chapter fossil checklist.

Diversity of the paleoflora

The earliest palynomorphs recorded in the FRF are rare indeterminate acritarchs from the Precambrian, notably the Balloon Formation of the upper Waingaro Valley. *Micrhystridium* is recorded from the Permian (Makarewan Stage) but more frequent reports do not appear until the Triassic when a number of acritarchs, identified as *Baltisphaeridium, Leiofusa, Leiosphaeridia,* and *Veryhachium,* are recorded. Although none of these generic assignments should be taken too literally, they provide an approximate guide to the kinds of taxa represented in the oldest New Zealand records of this group.

The Middle Triassic Etalian Stage also provides the oldest evidence of the Dinophyceae, with the first appearance of *Nannoceratopsis*. No species assignments are recorded, but the genus is distinctive and has been stably defined for a long time, so the identification is likely to be sound. Possible *Suessia* and *Sverdrupiella* occurrences are recorded from the Late Triassic Otamitan Stage (middle Norian), with ?*Heibergella* and more reliable records of *Sverdrupiella* in the following Warepan Stage (late Norian).

Gonyaulacalean dinoflagellates diversify thereafter, with *Belodinium, Carnarvonodinium, Chlamydophorella, 'Cleistosphaeridium'* (possibly *Systematophora*), *Endoscrinium, Pareodinia, Rigaudella, Scriniodinium, Stiphrosphaeridium,* and *Tubotuberella* all making an appearance by the end of the Middle Jurassic. About 160 species, representing more than eighty genera, are known by the end of the Early Cretaceous.

The earliest potential records of Peridiniales are compromised by poor age control; well-dated records do not appear until near the end of the Late Jurassic (*Diconodinium* and ?*Spinidinium*). Diversity remains relatively low through the Early Cretaceous, with only about twenty species representing eight or so genera being reliably known by the end of that epoch. The group rapidly diversifies thereafter.

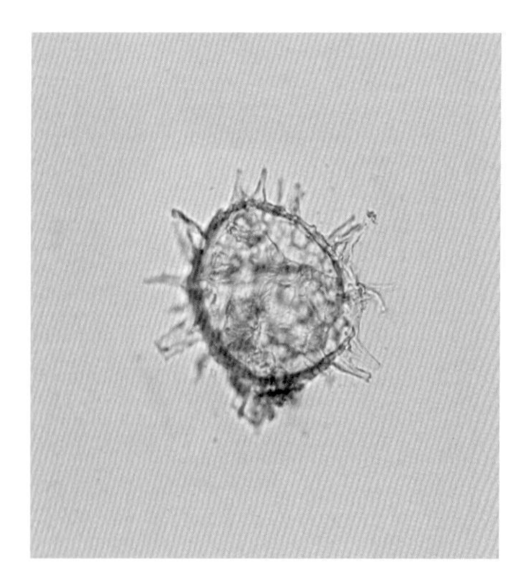

Cyst of the dinoflagellate *Cordosphaeridium* aff. *funiculatum*. Latest Eocene to Oligocene, Cape Foulwind, Westland.
Christopher Clowes

The numbers of recorded genera for each of the major groups discussed herein, is shown in the preceding table. Note that, while the table gives a clue to generic diversity through time, it should not be interpreted too literally. Some intervals have certainly been far better sampled than others, perhaps having been better represented in outcrop to begin with. In particular, the Neogene has been less widely searched for marine palynomorphs than the Late Cretaceous and Paleogene in the past, although current research is giving the younger New Zealand rocks more scrutiny.

Subphylum Apicomplexa: apicomonads and sporozoans

In apicomplexans, the alveoli tend to be highly flattened and cover most of the cell surface. There are four included classes. Apicomonadea is the sole class of infraphylum Apicomonada, consisting of parasites or myzocytotic (suctorial) predators; there are no known New Zealand representatives. The remaining classes (Conoidasina and Aconoidasina) belong to infraphylum Sporozoa, comprising the most widespread and important parasitic Protozoa. All are obligate parasites of vertebrates or invertebrates and are non-ciliate when feeding. All form infective bodies called spores.

Sporozoan parasites of vertebrates in New Zealand

The apical complex at the anterior ends of these organisms is not fully understood in all aspects of its functioning. Insofar as vertebrate sporozoans generally develop within host cells, often particular types of cell, it is likely that components of the apical complex are involved in cell-recognition and/or cell-entry. However, many of the groups infecting invertebrates are not intracellular parasites, so, at least in these, the apical complex must have other functions.

Sporozoan morphology varies between groups and changes during life cycles, but there is usually at least one point in their development where the elongate, banana-shaped organisms that are typical of the group occur. These lack cilia, but the cells move very effectively by flexing and gliding movements. (Cilia are found only in the flagellated male gametes in some groups.) Given the vast range of hosts and habitats in which these parasites live, it is not surprising that life-cycles vary considerably, although they also have many features in common. Some are direct and relatively simple, involving only one host (homoxenous), while others are indirect, requiring development in two different host species in succession for their completion (heteroxenous). Some of these involve two vertebrate hosts, others a vertebrate and an invertebrate. The life-cycles of a few species, such as *Toxoplasma gondii*, are even more complicated, with several different modes of transmission. While many apicomplexans are highly host-specific and confined to a single host species or a small number of closely related ones, others are able to infect a wide range of hosts.

The account that follows is concerned primarily with the sporozoans infecting New Zealand's terrestrial mammals and birds and, to a lesser extent, reptiles. Invertebrate infections will be mentioned mainly in connection with life-cycles of species infecting vertebrates. An important point to note is that almost all the terrestrial mammals found in New Zealand were introduced, so the sporozoan parasites found in them were imported also. Many of these animals are domesticated and commercially important, or are widespread pests, and their protozoan parasites are comparatively well known. On the other hand, relatively little is known about protozoans infecting other introduced mammals and birds, or native birds and reptiles. Much remains to be discovered.

Brief reviews of the groups and major genera or species of sporozoan parasites are given, together with comments on topics of particular interest in a New Zealand context and those that warrant further investigation. The vast majority of sporozoans dealt with here belong to the order Eucoccidiorida (Levine 1988), and these will be dealt with first.

Order Eucoccidiorida (Eimeriida): families with direct life-cycles

Eimeriidae

Two of the genera in this family, *Eimeria* and *Isospora*, are very common and widespread in New Zealand, as they are elsewhere in the world. Each has many species, mostly highly host-specific, infecting only one host or a few closely related ones. Eimeriids are commonly known as coccidia, and some of these, in certain circumstances, can cause serious disease (coccidiosis). Most develop in the cells lining the intestines of their hosts, although some infect other organs

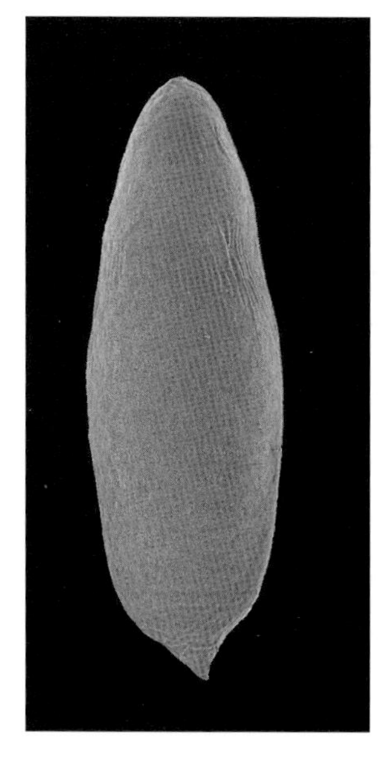

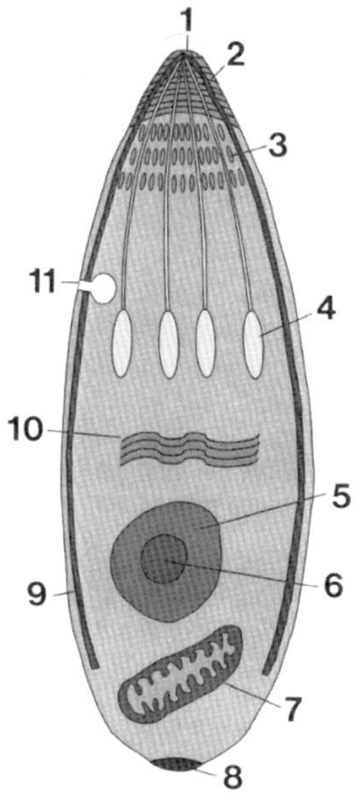

Upper: Surface view of a gregarine apicomplexan. Lower: General structure of an apicomplexan – 1 polar ring of microtubules, 2 conoid (funnel of rods in class Conoidasida), 3 micronemes, 4 rhoptries, 5 nucleus, 6 nucleolus, 7 mitochondrion, 8 posterior ring, 9 alveoli (supporting the cell membrane), 10 Golgi dictyosomes, 11 micropore (diminished mouth). The vesicles of the apical complex (rhoptries and micronemes) secrete enzymes that allow the parasite to enter a host cell.

Adapted from Wikimedia Commons

such as the bile ducts of the liver (e.g. *E. stiedai* in rabbits) or the kidney (e.g. *E. truncata* in geese). Life-cycles follow the same general pattern; this is outlined below and also introduces some of the terminology used in connection with sporozoan life cycles. The terminology is complicated by the fact that there are two parallel sets of terms for some developmental stages used by different groups of scientists working on sporozoans. In the summary of the basic coccidian life-cycle that follows, both are given to try to prevent confusion arising from reading publications from the two groups.

In the basic coccidian life-cycle, infection is initiated by ingestion of a sporulated oocyst containing sporozoites (explained below). Sporozoites are released in the gut and immediately invade host cells. Interestingly, they are generally highly selective about the type of cell they invade, indicating that they have sophisticated cell-recognition capabilities. One or more (usually two) cycles of asexual multiplication follow, involving a type of multiple division termed schizogony (or merogony). This leads to the invasion and destruction of increasing numbers of host cells. During the actual multiplication process, the parasite is known as a schizont (or meront) with the individual organisms that develop and burst out of the host cell being termed schizozoites (or merozoites). Parasite numbers can be greatly increased, in some species by hundreds or thousands of times. Schizogony is followed by sexual reproduction, termed gametogony (or gamogony). This involves the invasion of yet more host cells by the last generation of schizozoites, leading to the development and fusion of gametes (or gamonts) to produce a zygote which then secretes a resistant layer around itself to become an oocyst.

The oocyst is released and passed out of the host into the environment where a further phase of asexual multiplication must take place for it to become infective. This is termed sporulation (or sporogony); it requires oxygen and occurs most readily in warm, moist conditions and may take a few days or even weeks, depending on temperature and the species concerned. Sporogony results in the development within the oocyst of a variable number of smaller cyst-like structures called sporocysts. In each of these, infective sporozoites develop. The numbers of sporocysts that develop, and of sporozoites in each sporocyst, differ between genera. The sporozoites can initiate a new infection if the oocyst is swallowed by a suitable susceptible host, so completing the cycle.

Oocysts of different species differ in size, shape, and some other characteristics. They are all microscopic and mostly range from about 10 to 35 micrometres in their largest diameter. However, although some can be identified to species level on the basis of size, shape, and other characteristics before sporulation, for many this is not possible and it is necessary to examine sporulated oocysts for more subtle differences between them, a task requiring some expertise. The oocyst wall is very highly protective and, particularly after sporulation, oocysts can survive for many months as long as they are kept moist and shielded from ultraviolet radiation.

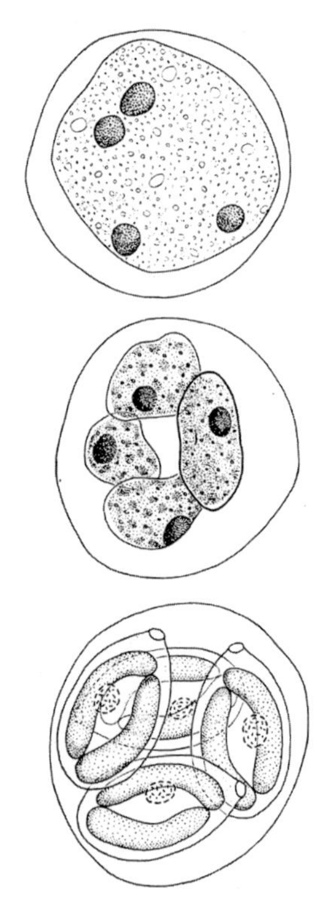

Stages in the development of *Epieimeria anguillae* in a New Zealand eel (*Anguilla australis*).
From Hine 1975

Eimeria. Worldwide, well over a thousand species of *Eimeria* have been described. Their sporulated oocysts are characterised by having four sporocysts, each containing two sporozoites. They are extremely common in mammals, particularly ruminants, lagomorphs (rabbits and hares), and rodents, and in some groups of birds, especially galliforms (fowl-like birds) and anseriforms (ducks, geese, and swans). Other species occur in cold-blooded vertebrates and a few are found in invertebrates. Hine (1975) reported *Epieimeria anguillae* from New Zealand eels (*Anguilla* species) and Hine et al. (2000) recorded an *Eimeria* from southern blue whiting (*Micromesistius australis*). Parasitised taxa are usually host to several species of *Eimeria*, which are not only generally very highly host-specific but also site-specific, developing in particular locations and cell types. Outbreaks of coccidiosis are relatively common in some domesticated species, particularly young cattle, goats, and poultry. Preventing coccidiosis is a major concern and cost to the poultry industry. However, infection also results in the development of protective immunity, so that the disease is rarely seen in older animals.

Eimeria is well-represented in New Zealand, particularly in domesticated ruminants and poultry. Some 69 species have been identified and there are additional records in which the species was not identified (see chapter checklist). Eleven species have been recorded from cattle, 11 from sheep, 14 from goats (McKenna 2009) (including three that were first described from New Zealand – Soe & Pomroy 1992), six from pigs (McKenna 1975, 2009), six from rabbits (Bull 1953; McKenna 2009), and five from chickens (McKenna 2010). *Eimeria* infections are widespread in the brushtail possum (*Trichosurus vulpecula*) in New Zealand (Stankiewicz et al. 1996, 1997a,b). Whether or not these are all *E. trichosuri*, recently described from New Zealand possums (O'Callaghan & O'Donoghue 2001), is unclear. A few other species have been recorded in New Zealand but, it appears, none from wild ruminants or ducks and geese, either indigenous or introduced. This probably reflects more a lack of examination of these and other hosts for coccidia than total absence.

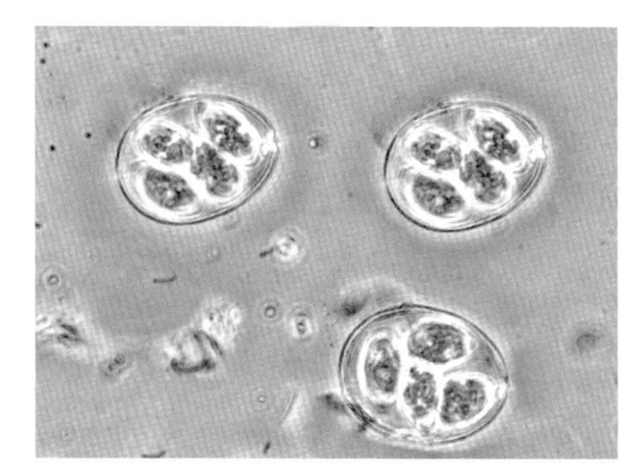

Oocysts of *Eimeria maxima*, which infects poultry.
Agricultural research Service, USDA, per Wikimedia Commons

Of particular interest in a New Zealand context is the discovery of at least two species of *Eimeria* in northern brown kiwi (*Apteryx mantelli*) (Charleston unpubl.; Jakob-Hoff et al. 1999). These have been associated with severe and sometimes fatal intestinal disease in captive-reared kiwi chicks. Coccidial oocysts, generally in low numbers, have also been found to be widespread in healthy wild kiwi in the northern half of the North Island (Jakob-Hoff et al. 1999; Jakob-Hoff 2001). Whether these coccidia can be of significance to kiwi health in the wild is, at this stage, unclear. Coccidia have also been observed in histopathology sections of the kidney of a kiwi. These were recorded as *Eimeria truncata* (Thompson & Orr 1991) but it is highly likely that they were a different, undescribed species, as *E. truncata* is known only from geese. The significance of this infection to kiwi is also not known. Research to characterize the coccidian species involved and their development in the tissues of kiwi is currently in progress (K. J. Morgan pers. comm. 2011).

A species of *Eimeria* has been recorded in the gallbladder of the common gecko *Hoplodactylus maculatus* (Ainsworth 1985, cited by McKenna 2003).

Isospora. Worldwide, about 250 species of this genus have been described but, to date, only eight have been recorded from New Zealand, together with a few records of unidentified species. While the life-cycles of *Isospora* species are broadly similar to those of *Eimeria* (although some have an optional indirect cycle), the sporulated oocysts are distinguishable by the presence of paired sporocysts, each containing four sporozoites. It should be noted, however, that many other genera in the order Eucoccidiorida have oocysts with the same configuration and are said to have 'isosporan-type' oocysts.

Species of *Isospora* are commonly found in cats and dogs in New Zealand (two species in each) (McKenna & Charleston 1980a,b) and a single species has also been recorded from pigs and one from ferrets (McKenna 2009). Overseas, species of *Isospora* are very common in passerine and other birds, but so far in New Zealand there are only four records – from a greenfinch (*Carduelis chloris*), blackbird (*Turdus merula*), kokako (*Callaeas cinereus*) and an ostrich (*Struthio camelus*) (McKenna 1999, 2001, 2010). Again, this paucity of records almost certainly reflects a lack of observations rather than the absence of these parasites.

There are undoubtedly more species of *Eimeria* and *Isospora* to be found in both native and introduced species in New Zealand, particularly in birds. Coccidial oocysts are sometimes found during routine examination of faecal samples from native birds, more frequently in samples from young birds (B. A. Adlington et al. unpubl.), but little is known about them or their significance to the birds. They are of some concern to those working with endangered species although they do not necessarily pose a significant health risk in the wild.

Although the coccidian *Cyclospora cayatenensis* has been recorded in humans in New Zealand on four occasions, all of these reports relate to people returning from overseas (Dougherty 1996; Ockleford et al. 1997).

Atoxoplasmatidae

Species of *Atoxoplasma* infect birds. They are similar in many ways to *Isospora* and have similar oocysts, but asexual multiplication (schizogony) occurs in cells of the lymphoid-macrophage system in the bloodstream and other tissues, as well as in intestinal cells; the sexual stages develop in the intestinal epithelium. There have been a number of reports of *Atoxoplasma* or *Atoxoplasma*-like infections in birds in New Zealand (see checklist) although it is not always certain which species they represent or, in some cases, whether they are actually *Atoxoplasma* and not some other sporozoan. They are not easy to identify.

There is some disagreement among experts whether or not *Atoxoplasma* should be given separate generic status or included in *Isospora*. Levine (1982) resurrected the genus for isosporan coccidia of birds in which schizogony occurs in cells in both the bloodstream and intestine, and placed it in its own family. Levine (1988) listed 19 species but noted that they may not all be valid. Overseas, *Atoxoplasma* infections have been associated with clinical disease in a variety of birds. Coccidial infections with isosporan-type oocysts have caused deaths of substantial numbers of captive-reared chicks of the endangered hihi (stitchbird, *Notiomystis cincta*) (Twentyman 2001). Examination of tissues from some of the early cases suggested that a species of *Atoxoplasma* might be involved but, despite further work, it is still not clear if this is so or whether the parasites concerned are species of *Isospora* (Twentyman 2001). More research is needed but it is extremely difficult when one is dealing with endangered species and what are probably host-specific parasites, as this precludes the possibility of studying experimental infections using commoner alternative hosts.

Cryptosporidiidae

The principal genus in this family is *Cryptosporidium*, an extremely widespread and common parasite, particularly of mammals but also of birds, reptiles, and fish. Whereas only a small number of species of low host-specificity were recognised previously, molecular characterisation of different isolates in the last few years has led to a radical revision of the taxonomy of this genus. As a result, twenty species are now accepted as valid, with two named species being found in fish, one in amphibians, two in reptiles, three in birds, and 12 in mammals (Plutzer & Karanis 2009). A major difficulty with this genus is that, with few exceptions, the parasites found in various hosts are morphologically indistinguishable. For many years it was considered that *Cryptosporidium parvum* and *C. muris* were the common parasites of mammals, and *C.baileyi* and *C. meleagridis* the species that parasitise birds. What has emerged from molecular characterisation of isolates is that what was considered *C. parvum* actually includes a number of species of differing host-specificity and that *C. meleagridis* may also infect some mammals as well as birds (Plutzer & Karanis 2009).

All *Cryptosporidium* species are extremely small parasites, most of which live in the brush border, the layer of microvilli on the surface of some types of epithelial cells, mainly in the intestine and respiratory tracts of their hosts. As with *Eimeria*, development includes schizogony followed by gametogony, but in this case sporulation occurs within the host so that the oocysts are already infective when they are passed out. These oocysts are generally only about 4–7 micrometres in their longest diameter, depending on the species, and contain four sporozoites that are not enclosed in a sporocyst. The oocysts can be difficult to detect and special techniques are used for diagnosis (see Fayer et al. 2000). They are extremely resistant to chemical agents, including most disinfectants and sterilising agents, and can also survive for long periods in cool, moist conditions and in water. In *C. parvum*, about 20% of the oocysts produced are thin-walled and these excyst spontaneously within the host, resulting in autoinfection.

In New Zealand, *Cryptosporidium* infections have been recorded from a variety of domesticated animals, particularly ruminants, although in most cases the species concerned have not been identified (McKenna 2009); it is likely that most of them are *C. parvum*, as has been found in infections of young foals (Grinberg et al. 2003, 2009). However, almost half of 423 human *Cryptosporidium*-infected faecal specimens were found to contain *C. hominis* oocysts (Learmonth et al. 2004). Infections with *Cryptosporidium* species have also been found in a range of wild animals including possums, ship rats, and house mice (Chilvers et al. 1998). Little is known of the prevalence or significance of infection in domestic or wild birds in New Zealand. An outbreak of clinical disease has been recorded in aviary birds (Belton & Powell 1987) and oocysts have been detected in the faeces of chickens, ostriches, and a few wild species (Chilvers et al. 1998; McKenna 2010). There are no records of *Cryptosporidium* from New Zealand reptiles or amphibians.

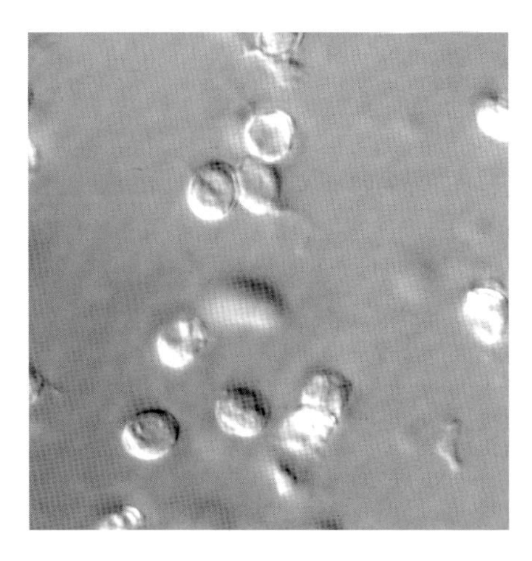

Oocysts of *Cryptosporidium parvum* from mouse faeces.

H. D. A. Lindquist, US Environmental Protection Agency, per Wikimedia Commons

Cryptosporidium infections are often the cause of diarrhoea in very young domesticated animals, commonly in association with other intestinal infections that may include other protozoa, bacteria, or viruses. Older animals rarely show clinical signs of infection. Human infections associated with diarrhoea are also common but they are not confined to the young (Carter 1984, 1986; Ungar 1990). In animals and people with a normally functioning immune system, spontaneous recovery is usual, although oocyst shedding may continue for several weeks. On the other hand, people with Acquired Immune Deficiency Syndrome (AIDS), or on immunosuppressive therapy, or affected by some types of lymphoid cancer, can develop uncontrollable infections that may prove fatal (Stark et al. 2009). Most infections, in both animals and humans, are undiagnosed and it is probable that, as has been found in serological studies overseas, most New Zealanders are infected at some time in their lives.

The incidence of cryptosporidiosis in humans in New Zealand is relatively high compared to other developed countries (Snel et al. 2009). As elsewhere, cases have been traced to contact with farm animals (Stefanogiannis et al. 2001) or contaminated drinking water (Duncanson et al. 2000) and for many years it was considered that most human infections were caused by *C. parvum* and directly or indirectly of animal origin. There was evidence to suggest, however, that there might be distinct human and animal populations of *Cryptosporidium*. Molecular genetics has confirmed this to be the case and with *C. hominis* established as a distinct species (Morgan-Ryan et al. 2002), the epidemiology of human infections has become much clearer. Studies of human cases in New Zealand have shown that infection rates vary through the year, with the *C. parvum* bovine genotype commoner in rural areas and predominating in spring at the time when newborn farm animals are most numerous; in contrast *C. hominis* is commoner in urban areas and more prevalent in late summer (Learmonth et al. 2003, 2004; Snel et al. 2009). It appears, however, that *C. hominis* is not only infectious for humans as it has been recorded overseas from ruminants, pigs and other species (Plutzer & Karanis 2009).

Molecular studies will undoubtedly lead to the identification of further species of *Cryptosporidium* in New Zealand and shed further light on the epidemiology of both animal and human infections in all its complexity.

Klossiellidae

Klossiella equi, an unusual and relatively harmless coccidial parasite of the kidneys of horses, has been recorded in New Zealand but it is not known how widespread or common it is (Orr & Black 1996).

Families with indirect (heteroxenous) life-cycles

Sarcocystidae

This family includes several genera, some of which are widespread and common in New Zealand. All have indirect life-cycles characterised by the development of cysts containing infective organisms in the tissues of intermediate hosts. Both the intermediate and definitive hosts are vertebrates linked by predator-prey or scavenger relationships.

Toxoplasma gondii. Of all Sporozoa, this is the most extraordinary and interesting. It is capable of infecting virtually all mammal species and some birds. It is common throughout the world (Dubey 2008). The vast majority of infections, however, cause little or no observable disease although sometimes the effects can be serious. In New Zealand, infection has so far been recorded in humans and in eleven domesticated and wild mammalian and five avian species (McKenna 2009, 2010) but it undoubtedly occurs in others. The life-cycle is relatively complicated as there are three common ways in which the parasite can be transmitted.

1. The definitive hosts are cats (*Felis* species), both domestic and wild, a fact that was discovered only about 40 years ago. In these, *Toxoplasma* behaves as a typical coccidian, multiplying in cells of the intestine and, particularly in young non-immune cats, then producing large numbers of oocysts that are passed unsporulated in the faeces. Previously infected, immune cats produce far fewer oocysts, if any, when reinfected. Once sporulated, oocysts are infective to other animals (including cats) and can survive for months in moist conditions. Oocysts are of the 'isosporan' type (i.e. with two sporocysts each enclosing four sporozoites).

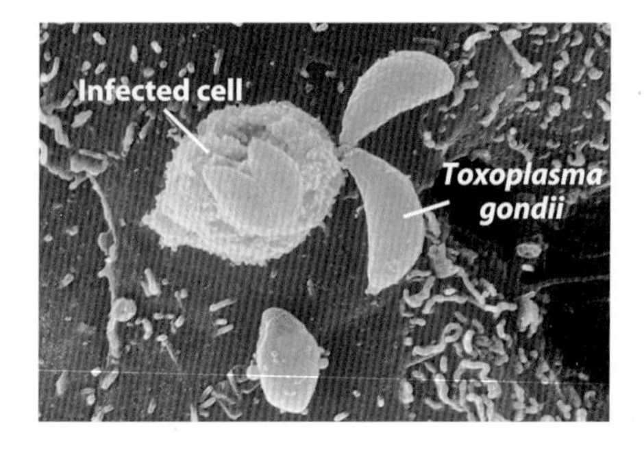

A host cell infected by *Toxoplasma gondii*.
Pathogen Profile Dictionary

2. In previously uninfected (non-immune) hosts, following ingestion of oocysts (or tissue cysts, see below), the organism multiplies rapidly in a succession of host cells, spreading via the blood stream into many organs and tissues. During this rapid multiplication phase, the organisms are referred to as *tachyzoites*. This is followed by a phase of slower multiplication in which the parasite accumulates in large numbers in cells that become tissue cysts. At this stage, the organisms are termed *bradyzoites* and unlike the tachyzoites they are relatively resistant to digestive enzymes. Tissue cysts are found in various organs and can persist for long periods, sometimes years. They do not normally cause any disease at this stage but are infective to any animal that eats them through predation or scavenging (or, in the case of humans, by deliberate eating of uncooked infected meat or accidental contamination of other food with bradyzoites from meat). If this animal (or person) is not already immune, the same cycle of multiplication occurs, ending with the production of tissue cysts containing bradyzoites. Only in cats does the parasite invade intestinal cells and give rise to oocysts.

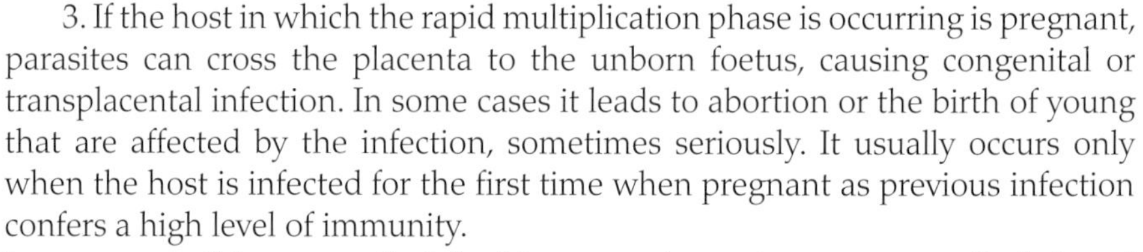

3. If the host in which the rapid multiplication phase is occurring is pregnant, parasites can cross the placenta to the unborn foetus, causing congenital or transplacental infection. In some cases it leads to abortion or the birth of young that are affected by the infection, sometimes seriously. It usually occurs only when the host is infected for the first time when pregnant as previous infection confers a high level of immunity.

In summary, this means that herbivores such as sheep are usually infected by ingesting oocysts, but carnivores and omnivores, including humans (and cats), can be infected by either ingesting oocysts *or* bradyzoites in tissue cysts. Congenital infection can occur in a wide range of species, including humans.

The significance of *Toxoplasma* infections in domesticated animals and humans in New Zealand has been reviewed elsewhere (Charleston 1994). In all species, the great majority of infections are not associated with clinical disease and pass unnoticed. However, *Toxoplasma* has long been recognised as an important parasite of sheep, commonly causing abortion (Hartley & Marshall 1957), stillbirth, and perinatal death of lambs. Goats and other species can be similarly affected. A live vaccine was developed in New Zealand from an attenuated strain of the parasite for use in sheep (Wilkins et al. 1988) and vaccination of ewes is now widely practised. Sporadic disease also occurs in other domesticated animals. Very little is known about the prevalence of *Toxoplasma* infection in wild animals, including rodents, in New Zealand although it is probably very common given the widespread occurrence of feral cats (Collins & Charleston 1979).

Surveys in New Zealand and many other countries have shown that a high proportion of adult humans have antibodies to *Toxoplasma*, indicating past infection. Prevalence generally increases with age. Clinical disease occurs sporadically but there is little information on how often this actually occurs, either from primary or congenital infection in New Zealand. Congenital infection can have serious consequences. An estimate of the risk of congenital toxoplasmosis based on estimated rates at which New Zealand women develop antibodies to *Toxoplasma* for the first time while pregnant, indicating newly acquired infection, has been published (Moor et al. 2000) but the actual incidence of clinical problems following congenital infection is unknown. Pregnant women can be tested for antibodies to determine whether or not they are immune or susceptible but this is not routine. It is often assumed that cats are the main source of human infections but infection from handling or eating raw or inadequately cooked meat containing bradyzoites is also possible and it is not known which is the more important. There are relatively simple precautions that can be taken to minimise any risk. There is, as yet, no vaccine for use in humans (Dubey 2008). For an excellent review of toxoplasmosis in all species, the monograph by Dubey and Beattie (1988) is recommended.

Neospora caninum. This parasite was first described just over 20 years ago, as causing severe disease in dogs overseas (Dubey et al. 1988). It had been seen previously but mistaken for *Toxoplasma* or other related sporozoans. Subsequently, it was recorded in dogs in many countries, and a wide range of domesticated and laboratory animals were shown to be susceptible to infection, either naturally or experimentally. Clinical disease in dogs has also been recorded in New Zealand (Patitucci et al. 1997). At about the time that it was recognised as causing disease in dogs, *Neospora* was also found, both in New Zealand and overseas, to be an important cause of abortion and neonatal disease in cattle. It has also been recorded from a range of other domesticated and wild animal species (Dubey 1999).

The parasite is similar in many ways to *Toxoplasma*. The occurrence of congenital transmission, often leading to serious disease in dogs and cattle, was recognised early on. Further, it has since become clear that at least in cattle, congenital transmission is by far the commonest route, often (but not invariably) resulting in abortion or disease in the newborn calf. However, for several years after its discovery, other ways in which the parasite could be transmitted was a puzzle. Then it was found that, after experimental infection, dogs passed out unsporulated oocysts similar to those of *Toxoplasma*, showing that dogs are the definitive (final) hosts of the parasite (McAllister et al. 1998). It is now also known that, as with *Toxoplasma*, infections begin with rapid multiplication in tissues, followed by the development of tissue cysts containing bradyzoites that are infective to other animals. Initially these were found to develop in nervous tissues (Dubey 1992), but later research indicated that, at least in cattle and dogs, they can also occur in other tissues including muscle (Peters et al. 2001) and placenta (Dijkstra et al. 2001). It appears that many dogs in New Zealand have antibodies to *Neospora*, indicating that infection is probably widespread (Antony & Williamson 2001). The isolation and molecular characterisation of *Neospora* from New Zealand cattle has recently been reported (Okeoma et al. 2004).

There is still much that is not known about this parasite and the epidemiology of the infection, including the question as to whether *Neospora* is infectious to humans. For reviews of the infection in cattle and dogs in New Zealand, see Reichel and Pomroy (1996) and Antony and Williamson (2001).

Hammondia. Isosporan-type oocysts identified as those of *Hammondia heydorni* have been recorded from the faeces of dogs in New Zealand (McKenna & Charleston 1980b) but a similar feline species, *Hammondia hammondi*, has not. Species of *Hammondia* form cysts in the muscles of their intermediate hosts, which, in the case of *H. heydorni*, are probably ruminants (Dubey 1993).

Oocysts of *Hammondia*, *Toxoplasma*, and *Neospora* are morphologically indistinguishable and there are a number of similarities in the life-cycles of these genera. There has therefore been some debate about their validity and relationships (Mehlhorn & Heydorn 2000; Heydorn & Mehlhorn 2002). This is a complex issue, not only because the organisms are so similar, but because not every detail of the life-cycles and host ranges of *Hammondia* and *Neospora* are known. A further complication is that, when it comes to studies carried out some years ago (before the recognition of *Neospora* as a separate organism, for example), it is not always possible to be certain now which parasite was actually present, or whether more than one was present at the same time. In a recent review of the biology of these parasites, however, including their antigenic and molecular (DNA) characteristics, it was concluded that the three genera are indeed distinct (Dubey et al. 2002a,b). There is, nevertheless, still much to be discovered about them.

In this connection, it is interesting to note that the original identification of *H. heydorni* oocysts in the faeces of dogs in New Zealand (McKenna & Charleston 1980b) preceded the identification of *N. caninum* that is now acknowledged to be present. However, a recent study of DNA extracted from oocysts from the faeces of a New Zealand dog showed these to be of *H. heydorni*, confirming that both it and *N. caninum* occur in dogs in this country (Ellis & Pomroy 2003).

Besnoitia. Species of *Besnoitia* form cysts containing bradyzoites in the tissues of mammalian (or reptile) intermediate hosts which, when eaten by the definitive host, give rise to unsporulated isosporan-type oocysts. The definitive hosts are species of *Felis*, including domestic and wild species in different life cycles. One *Besnoitia* species has been recorded in New Zealand infecting domestic cats and infective to Norway rats (intermediate hosts) (McKenna & Charleston 1980c). In the rat, the cysts develop on the peritoneal surface of the intestine and appear as small white spots. The identity of the causative *Besnoitia* species has not been identified with certainty but is likely to be *B. wallacei*, for which rodents are known to be intermediate hosts.

Sarcocystis. Globally, more than 200 species of *Sarcocystis* have been described. The generic name derives from the fact that they form cysts in muscles of their intermediate hosts. It was not until 1972, however, that the first life-cycle was published. Before that it was not even universally agreed what type of protist the spores represented. After electron-microscope studies showed they were morphologically similar to sporozoans, and it was discovered that *Toxoplasma* behaved as a coccidian in cats, transmission experiments were carried out that finally led to the first life-cycle being described. It is now known that all species of *Sarcocystis* have obligatory two-host life-cycles involving carnivorous definitive hosts and herbivorous or omnivorous intermediate hosts. Definitive hosts are infected by eating muscle containing sarcocysts in which there are bradyzoites that develop directly into sexual stages in the intestinal epithelium of the definitive host. Oocysts develop that sporulate within the host and then break down to release the enclosed sporocysts, each containing four sporozoites; these pass out in the faeces. If they are ingested by the appropriate intermediate host, schizogony occurs in certain cells, often those lining blood vessels, after which muscle cells are invaded and the sarcocysts develop.

Since the discovery of the first life-cycle, many more have been elucidated and it is now known that there are species of *Sarcocystis* involved in many predator-prey relationships, with mammals and birds (e.g. hawks and owls) as definitive hosts. *Sarcocystis* species are highly host-specific with respect to intermediate hosts and generally infect only one definitive host or a few closely related species.

The prevalence and significance of *Sarcocystis* infections in New Zealand mammals has been reviewed in some detail previously (Charleston 1994). Eight identified species of *Sarcocystis* have been recorded in New Zealand, all but one of them in cattle, sheep, and goats (intermediate hosts) with either dogs or cats as definitive hosts (McKenna 2009). The other species is *S. muris* in mice, which is transmitted by cats (McKenna & Charleston 1980d). Sarcocysts of unknown or unspecified species have also been recorded in several other introduced mammalian species including horses, pigs, alpacas, red deer, Norway rats, and rabbits (McKenna 2009). There is also a record of sarcocysts in the muscles of a New Zealand pigeon (*Hemiphaga novaeseelandiae*) (Johnstone & Cork 1993) and in the native short-tailed bat (*Mystacina tuberculata*) (Duignan et al. 2003), which raises intriguing questions as to whether or not these are endemic or introduced species and, in either case, what the definitive host(s) might be.

The sarcocysts of some species will, in time, grow large enough to be visible to the naked eye. These are termed *macroscopic* sarcocysts; others remain forever *microscopic*, i.e. visible only under a microscope. Macroscopic cysts of species infecting domesticated ruminants are of some concern to meat producers as they are unsightly in meat and can result in it being downgraded so that it cannot be sold for human consumption. However, these sarcocysts are not infectious to humans so there is no risk to human health – cats are the definitive hosts. There are two species of *Sarcocystis* for which humans are the definitive hosts (cattle and pigs being their intermediate hosts) but these have not been recorded in New Zealand. Some species of *Sarcocystis*, notably those that cycle between dogs and ruminants, can cause an acute clinical disease in the ruminant intermediate hosts associated with multiplication of the parasite in the cells lining blood vessels. Although the parasites concerned are very common in New Zealand, actual cases of disease have not been observed under farming conditions (Charleston 1994).

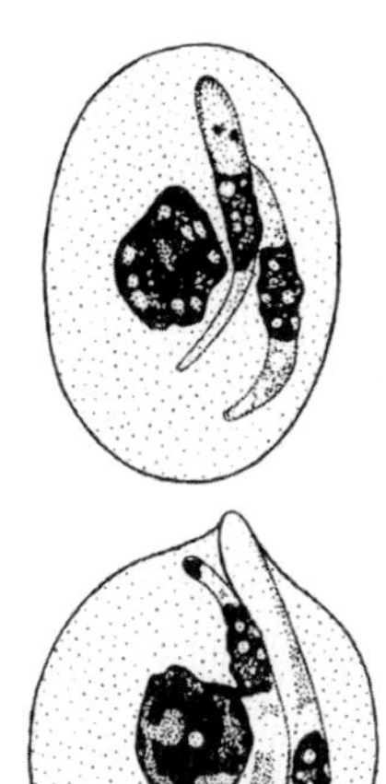

Haemogregarina bigemina individuals within red blood cells of the variable triplefin (*Forsterygion varium*).
From Laird 1953

Haemogregarinidae

Haemogregarines are found in blood cells of a wide range of vertebrates including fish, reptiles, amphibians, birds, and mammals. More than 300 species have been described from around the world. They all have indirect life-cycles and are transmitted by blood-sucking invertebrates of various kinds including mosquitoes, mites, and leeches. In general terms, schizogony and gamont formation occur in the vertebrate host, and oocyst and sporozoite formation in the invertebrate host. In the life-cycles of species of *Haemogregarina* and *Hepatozoon*, definitive hosts usually become infected by being bitten by an infected invertebrate host, but in some species of *Hepatozoon* infection can also be acquired by ingesting the intermediate host.

Hepatozoon lygosomarum, a parasite of the common New Zealand skink *Oligosoma nigriplantare*, is of particular interest – its life-cycle involves parasitic mites (Allison & Desser 1981). *Haemogregarina* species have also been recorded from a number of other New Zealand reptiles, including tuatara, and from several fish species (Laird 1952, 1953; Dawson 1992; Hine et al. 2000). Species of *Hepatozoon* have been recorded in New Zealand from rabbits (*H. cuniculi*) and mice (*H. musculi*) (Laird 1951a) but nothing is known about how common or significant they are. A species of *Hepatozoon* in northern brown kiwi (*Apteryx mantelli*) was reported by Jakob-Hoff et al. (2000) and Jakob-Hoff (2001) that was subsequently described as *H. kiwii* (Peirce et al. 2003). This, along with *H. albatrossi*, which was recorded in the blood films of albatrosses from Campbell Island (McKenna 2005), are the only known avian species of *Hepatozoon* in birds in New Zealand.

Order Haemospororida (Haemosporidida)

This order includes the sporozoan parasites that are found in the blood of vertebrate hosts and are transmitted by blood-sucking insects such as mosquitoes, midges, and sandflies.

Plasmodiidae

Plasmodium. The best-known members of the family are species of *Plasmodium*, the malaria parasites, of which over 170 species have been described from mammals, birds, and reptiles. Although occasional human cases of malaria are seen in New Zealand, these are always acquired overseas, most commonly from Papua New Guinea, the Solomon Islands, and Southeast Asia (Ikram et al. 1984; Ingram et al. 1988; Shew et al. 1995; Kriechbaum & baker 1996; Bradley 1999). As there are no suitable mosquito vectors (species of *Anopheles*) in New Zealand, human malaria cannot be transmitted here. The potential for introduction of a suitable cold-tolerant vector from Australia has been discussed (Boyd & Weinstein 1996). A number of species of *Plasmodium* have been recorded from birds, both native and introduced (Laird 1950; Reed 1997; Anon. 2000; McKenna 2010). In addition, *P. lygosomae* has been described from the native skink *Oligosoma moco* (Laird 1951b). It is not known which of the native or introduced mosquito species present in New Zealand are the vectors of these parasites.

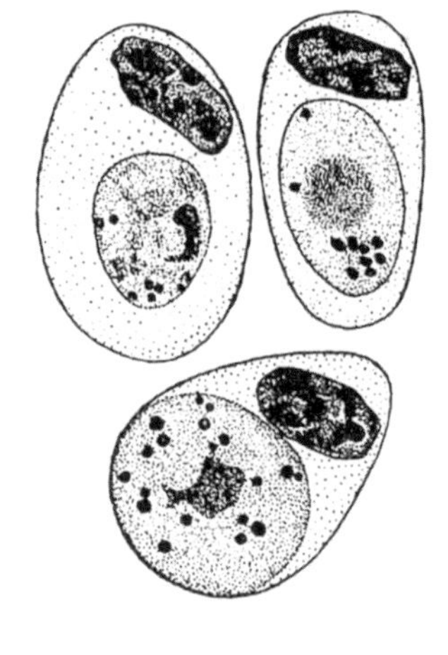

Developmental stages of *Plasmodium relictum* individuals within red blood cells of the sparrow (*Passer domesticus*).
From Laird 1950

Surveys of the prevalence of antibodies to *Plasmodium* indicate that infections are particularly common in penguins in the New Zealand subregion and in other parts of the world where disease outbreaks and mortalities have been recorded on a number of occasions (Graczyk et al. 1995). It has been suggested that infection with *P. relictum* may have been the cause of the large numbers of deaths of adult yellow-eyed penguins (*Megadyptes antipodeus*) on Otago Peninsula in 1989 (Graczyk et al. 1995). *Plasmodium relictum* has also been recorded from a small number of introduced passerine species (Laird 1950; McKenna 2010). Two other *Plasmodium* species, *P. cathemerium* and *P. elongatum*, have been recorded from aviary and wild birds and there are additional records of unidentified species (see McKenna 2010 and the chapter checklist).

There is considerable scope for research on the malarial parasites of birds in New Zealand as very little is known about the prevalence of infection in birds other than penguins. Derraik et al. (2008) summarised occurrences of avian malaria in native birds and reported *Plasmodium* infection in mohua (yellowhead, *Mohoua ochrocephala*) translocated from the wild. In general, the significance of infections to bird populations, including penguins, is not known and the mosquito vectors not identified although it has been speculated that *Culex quinquefasciatus* may be responsible for transmitting the infection to yellow-eyed penguins on Otago Peninsula (Graczyk et al. 1995) and indigenous *C. pervigilans* is strongly implicated as the vector in the case of mohua. An interesting question that has been raised at various times but probably impossible to answer, is whether the introduction of birds from the Northern Hemisphere in the 19th century also introduced avian malaria or exotic species of *Plasmodium*, leading to the dramatic reductions in native bird numbers that were observed at the time. However, the fact that *P. relictum* has been reported throughout the world from 270 species of birds in 51 families (Bennett et al. 1982, cited by Levine 1988) suggests that at least this species could perhaps be native to New Zealand, provided that one or more of the native mosquito species is a suitable vector. It is also possible that there are distinct species native to New Zealand yet to be discovered.

There appear to be no reports of *Plasmodium* in New Zealand reptiles since that of Laird (1951b) and clearly there is a need for further research on this subject.

Haemoproteus. Life-cycles are indirect, with vertebrate hosts being birds, reptiles, and amphibia, and invertebrate hosts comprising blood-sucking Diptera of various kinds. More than 165 species have been described worldwide, the majority from birds, but it is not clear how many of these are valid species, and the life-cycles of only a few are known. In New Zealand, only one species, *H. danilewsky* (*H. danilewskii* in Levine 1988), has been identified and only from three introduced bird species – blackbird (*Turdus merula*), song thrush (*T. philomelos*), and possibly the skylark (*Alauda arvensis*) (Doré 1920, 1921; cited by Laird 1950). However, unidentified species of *Haemoproteus* have also been recorded in the New Zealand robin (*Petroica australis*) and the Australasian gannet (*Morus serrator*) (McKenna 2010).

Laird (1950) commented that the specific identity of the species Doré (1921) described from the skylark required confirmation. Apart from those mentioned above, there do not appear to be any more recent records of *Haemoproteus* in either introduced or native birds, probably reflecting a lack of observations rather than absence of the parasite. Where known, the invertebrate hosts of *Haemoproteus* species are blood-sucking Diptera of the families Hippoboscidae or Ceratopogonidae (e.g. *Culicoides*). Since the latter genus is not present in New Zealand, the likely vectors of the parasite in this country would be flies of the hippoboscid genera *Ornithomya* and *Ornithoica*, reported from introduced and native passerine birds in New Zealand (Bishop & Heath 1998).

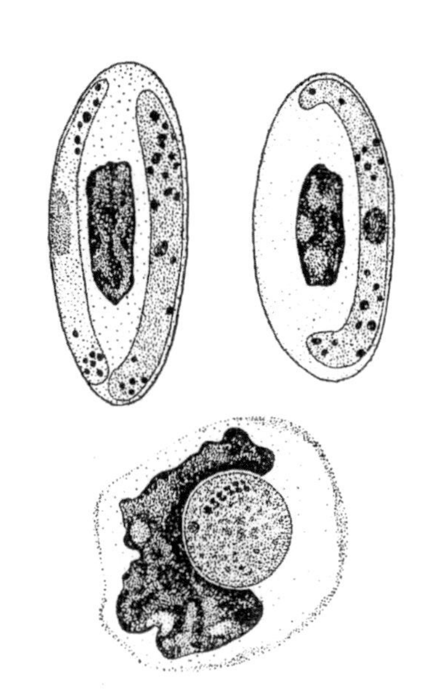

Individuals of *Haemoproteus danilewsky* in red blood cells of the song thrush (*Turdus philomelos*, upper) and chaffinch (*Fringilla coelebs*, lower).

From Laird 1950

Leucocytozoon. Species of *Leucocytozoon* also infect birds; those in the type subgenus *Leucocytozoon* are transmitted by blood-sucking flies in the family Simuliidae (Desser & Bennett 1993), generally known as black flies, although in New Zealand known colloquially as sand flies. In the host bird, schizogony takes place in various organs followed by gamont formation in blood cells. These are taken up by the simuliid vector in which oocysts and sporozoites develop. Once these have formed, they are injected into the bird host when the fly feeds. More than 65 species of *Leucocytozoon* have been described globally. Two of these have been recorded in New Zealand with an unidentified species being found in the yellow-eyed penguin (*Megadyptes antipodes*) (Alley et al. 2005). Laird (1950) recorded *L. fringillinarum* from a chaffinch, *Fringilla coelebs*. The other, *L. tawaki*, was originally described from the Fiordland crested penguin (*Eudyptes pachyrhynchus*) (Fallis et al. 1976). Although the latter parasite was found to develop in several species of *Austrosimulium* (Fallis et al. 1976), later work indicated that the principal vector is *Austrosimulium ungulatum* (Allison et al. 1978). Details of the life-cycle have been worked out. In the host bird, schizogony takes place in liver and kidney cells and gametogony in blood cells; the development of sporozoites in the vector takes about 10–12 days (Allison et al. 1978; Desser & Allison 1979). A very high prevalence of infection in Fiordland crested penguin chicks was found and infection was also transmitted to a juvenile blue penguin (*Eudyptula minor*) (Allison et al. 1978). The chicks are probably not infected until 2–3 weeks after hatching, when they move away from the nest and form small groups or creches; before this they are covered during the day by the brood pouch of the male parent (Warham 1974). The heavy parasitaemias in chicks and the light 'chronic' infections in mature birds are typical of other species of *Leucocytozoon*.

There appear to have been no further records of *L. fringillinarum* in New Zealand although it has been recorded from 146 species overseas (Bennett et al. 1982, cited by Levine 1988). In the study that led to the discovery of *L. tawaki*, blood smears from 43 native and introduced birds from a wide range of families were also examined, but no other blood parasites were found. It is interesting to note that *L. tawaki* has since been recorded from the jackass penguin (*Spheniscus demersus*) in South Africa (Earle et al. 1992). In some classifications, *Leucocytozoon* is placed in a separate family, Leucocytozoidae.

Order Piroplasmorida (Piroplasmida)

The organisms in this order develop in red blood cells (erythrocytes) and, in some cases, other cells of vertebrates. These sporozoans are commonly known as piroplasms and are transmitted by ticks.

Theileriidae

Species of *Theileria*, of which about 40 have been described, are parasites of mammals. They multiply in lymphocytes and other cells such as tissue macrophages before entering erythrocytes. When these are taken up by a suitable ixodid tick, further multiplication occurs and organisms invade the salivary glands. The parasite cells are then injected into the vertebrate host when the next stage in the tick life-cycle feeds. Some *Theileria* species are highly pathogenic and cause serious diseases in domestic livestock in some parts of the world, particularly Asia and Africa. Most are transmitted by particular species of tick so that their distribution is limited by the distribution of suitable vectors. Only *Theileria orientalis* has been recorded in New Zealand, infecting cattle and red deer (Anon. 1999; McKenna 2009). The species was first reported in 1984 (James et al. 1984); the tick vector is *Haemaphysalis longicornis*, an exotic ixodid introduced into Northland about 1900 or perhaps earlier and now found mainly in warmer parts of the North Island. It is the only livestock tick so far established in New Zealand. *Theileria orientalis* has been found on a number of occasions in cattle from the north of the North Island, almost invariably as an incidental finding during routine diagnostic haematology. Although it is sometimes associated with mild anaemia, the infection is essentially non-pathogenic (Thompson 1991).

Babesiidae

More than 100 species of *Babesia* have been described, the great majority from mammals, with a small number from birds and reptiles. They multiply in the red blood cells of their vertebrate hosts and are generally highly host-specific. Some species are well known parasites of livestock, and domestic animals such as dogs. They can cause serious disease, variously known as babesiosis, tick fever, or red-water fever. None of the species infecting mammals is present in New Zealand and the only potential tick vector, *Haemaphysalis longicornis*, is not capable of transmitting them. Indeed, until relatively recently, it would have been correct to say that New Zealand is free from species of *Babesia* of any kind. In December 2000, however, it was reported during haematological examination of northern brown kiwi chicks, some of which were anaemic, that a blood parasite was found that was identified as a species of *Babesia*. Subsequently, low-level infections have also been detected in blood smears from other kiwi from Northland (Jakob-Hoff et al. 2000). The parasite was formally described as *B. kiwiensis* by Peirce et al. (2003).

Kiwi are commonly infested with ticks in the North Island, usually *Ixodes anatis*, and it is likely, although yet to be proved, that this is the vector of this infection. Given the precarious state of kiwi populations, further research into this parasite, its life-cycle and significance, and the epidemiology of infections would seem warranted. The current intensive management of kiwi in various regions may provide an opportunity for such research. There would also be worldwide scientific interest, as very little is known about the biology of any of the species of *Babesia* described from birds.

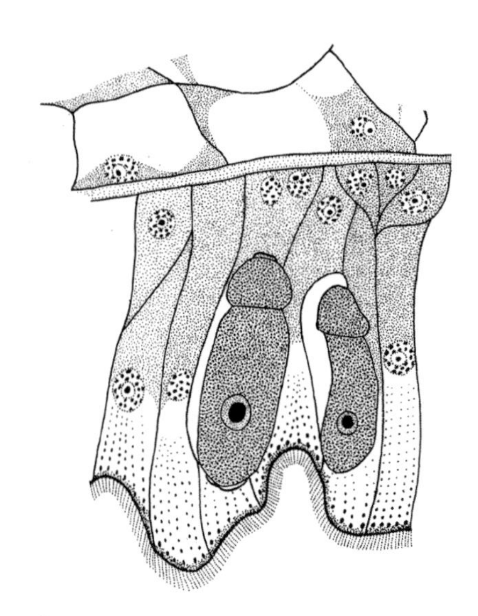

An undescribed species of *Stenophora* in gut epithelial cells of a millipede.
From Fitzgerald 1952

Gregarine sporozoans of invertebrates

A wide range of invertebrates play host to sporozoans (Levine 1985). Marine worms, especially polychaetes, are common hosts of gregarines (found only in invertebrates) and primitive coccidia. Earthworms are very often infected with gregarines of the genus *Monocystis* and its relatives. Monocystids have been noted as very common in both native and introduced earthworms in New Zealand but none has been formally described and published and there appear to be no unpublished reports dealing with any species (K. Lee, pers. comm.).

Jones (1976) reported a species of *Nematopsis* (Porosporidae) from the green-lip mussel *Perna canaliculus*. Marine crustaceans have their own gregarines, as do myriapods (centipedes and millipedes) and opilionids (harvestmen and daddy-longlegs). Insects, especially those associated with water, commonly harbour gregarines. Very few records of invertebrate gregarines have been made in a New Zealand context, however. Other records are few and scattered and restricted to arthropods, mostly insects.

Dumbleton (1949) described *Diplocystis oxycani* in larvae of the hepalid moth *Wiseana cervinata*. Trophozoites are found attached to the wall of the intestine or are free in the body cavity. Allison (1969) described three new species of eugregarines from the grass-grub (the larva of the beetle *Costelytra zealandica*) – *Stictospora costelytra, Euspora zelandica,* and *Euspora* sp., of which *E. zelandica* and *S. costelytra* occur near the midgut caeca and *E.* sp. near the Malpighian tubules. The three species are abundant in the third instar stage of the insect; they complete their life-cycles before the instars moult and are not present in the prepupae, pupae, or adults. The life-history of these gregarines is closely correlated with the life-history of the beetle. Ultrastructural details of the spore of one of these gregarines were elaborated by Archibald (1985).

Crumpton (1974) described *Hirmocystis pterygospora* and *Stylocephalus* sp. from the larva of the sand scarab *Pericoptus truncatus* (Coleoptera), where they live in the midgut caeca. Jensen (1979) discovered gregarines (reportedly *Schistocephalus*, but this is a tapeworm genus and the name may be a *lapsus* for *Stylocephalus*) in several species of Odonata (dragonflies and damselflies) – *Austrolestes colensonis, Procordulia grayi, P. smithii,* and *Xanthocnemis zealandica*. By the end of summer, almost all adults were parasitised by the large sporozoans (1.5 by 1.0 millimetre), which occupied mostly the midgut region of the intestine. The effects of these parasites on host populations is not known.

Zervos (1989) compared the incidence of two species of gregarines (possibly *Gregarina* species) in two endemic cockroaches, *Celatoblatta peninsularis* and *Parellipsidion pachycercum*. Whereas incidence, intensity, and cyst production peaked only in early winter in *C. peninsularis*, there were three peaks in *P. pachycercum* (early autumn, midwinter, midsummer). Prevalence of infection was low – 11% in the former, 4% in the latter.

Fitzgerald (1952) described four new species of gregarine from arthropods in the Dunedin area. *Enterocystis* (Enterocystidae) occupied the midgut of an unidentified cockroach found beneath the bark of the tree fuchsia (*Fuchsia excorticata*). Under-stone lagriine tenebrionid beetles (*Pheloneis aeratus, P. curtulus,* and *P. turgidulus*) hosted a new species of *Gregarina*. The same three beetle species and *Mimeopus otagoensis* (as *Cilibe*) (Tenebrionidae) also hosted an undescribed species of *Ophryocystis* (Ophryocystidae), but these were found in the Malpighian tubules. *Stenophora* n. sp. was found in 75% of 59 individuals of an unidentified millipede, together with an additional gregarine of uncertain systematic position (possibly another stenophorid). Fitzgerald (1952) also gave details of development of the parasites and their effects on their hosts. The fact that all four gregarine species were collected from arthropods at one locality (Pine Hill, Dunedin, with additional specimens from south Dunedin), hints at a likely high gregarine diversity overall, should further arthropod species be investigated from a wide range of localities.

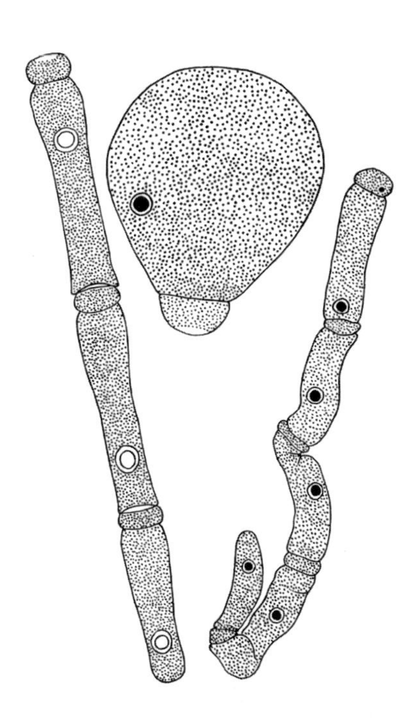

Individuals, some fused, of an undescribed species of *Gregarina* in tenebrionid beetles of the genus *Pheloneis*.

From Fitzgerald 1952

Concluding remarks on Apicomplexa

The vast majority of sporozoans recorded to date in New Zealand are those found in introduced mammals, particularly domesticated or pest species. However, even among mammals, especially feral, there are many species for which there are few or no records. Similarly, many of the species recorded overseas from introduced birds have yet to be found in New Zealand or have been reported only once. A number of sporozoans have been found in native birds and reptiles but, with few exceptions, very little is known about their life-cycles or significance and much remains to be discovered. There is therefore considerable scope for, and a need for, further research into these organisms in both native and introduced species, not only because of their intrinsic biological interest, but also because of their possible impact on New Zealand's native vertebrate fauna, much of which is scarce and under pressure of one kind or another.

Acknowledgements

Thanks are due to the late F. R. (Bertha) Allison for reading drafts of the review of Sporozoa and for providing additional information on several taxa, with valuable comments and suggestions. Some taxonomic details were also checked in The Coccidia of the World, compiled by Donald W. Duszynski, Steve J. Upton, and Lee Couch, supported by NSF Grant PEET DEB 9521687; available at: http://biology.unm.edu/biology/coccidia/table.html. Lesley Rhodes (Cawthron Institute), improved the dinoflagellate review and Brett Mullan updated the Southern Oscillation Index figure.

Authors

Dr Paul A. Broady School of Biological Sciences, University of Canterbury, Private Bag 4800, Christchurch, New Zealand [paul.broady@canterbury.ac.nz] Freshwater dinoflagellates

Dr F. Hoe Chang National Institute of Water & Atmospheric Research (NIWA), Private Bag 14-901, Kilbirnie, Wellington, New Zealand [h.chang@niwa.co.nz] Marine dinoflagellates

Dr W. A. G. (Tony) Charleston Institute of Veterinary, Animal and Biomedical Sciences, Massey University, Palmerston North (Present address: 488 College Street, Palmerston North, New Zealand) [charleston@inspire.net.nz] Apicomplexa

Dr Christopher D. Clowes School of Earth Sciences, Victoria University of Wellington, PO Box 600, Wellington 6140, New Zealand [chris.clowes@vuw.ac.nz] Fossil dinoflagellates and acritarchs text and checklist

Dr Philip B. McKenna Gribbles Veterinary Pathology, PO Box 536, Palmerston North, New Zealand [phil.mckenna@gribbles.co.nz] Apicomplexa

Dr Graeme J. Wilson GNS Science, PO Box 30368, Lower Hutt, New Zealand [g.wilson@gns.cri.nz] Fossil dinoflagellates and acritarchs checklist

References

The following bibliography includes, inter alia, a comprehensive list of all substantive publications pertaining to New Zealand acritarchs and fossil dinoflagellates.

ALLEY, M. R.; GARTRELL, B. D.; MORGAN, K. J. 2005: Leucocytozoonosis in yellow-eyed penguins, *Megadyptes antipodes*. *Kokako 26*: 30–31.

ALLISON, F. R. 1969: A study of the eugregarines of the grass-grub (larva of *Costelytra zealandica* (White), Melolonthinae), with a descripton of three new species. *Parasitology 59*: 663–682.

ALLISON, F. R.; DESSER, S. S. 1981: Developmental stages of *Hepatozoon lygosomarum* (Doré 1919) comb. n. (Protozoa: Haemogregarinidae) a parasite of the New Zealand skink *Leiolopisma nigriplantare*. *Journal of Parasitology 67*: 852–858.

ALLISON, F. R.; DESSER, S. S.; WHITTEN, L. K. 1978: Further observations on the life cycle and vectors of the haemosporidian *Leucocytozoon tawaki* and its transmission to the Fiordland crested penguin. *New Zealand Journal of Zoology 5*: 371–374.

ANON 1999: Parasites in deer erythrocytes. Suspected exotic disease investigations. *Surveillance 26(3)*: 21.

ANON 2000: First record of avian malaria in kiwi. *Kokako 7*: 11–12.

ANTONY, A.; WILLIAMSON, N. B. 2001: Recent advances in understanding the epidemiology of *Neospora caninum* in cattle. *New Zealand Veterinary Journal 49*: 42–47.

ARCHIBALD, R. D. 1985: [Abstract.] A eugregarine spore with a similar ultrastructure to some coccidian spores. *New Zealand Journal of Zoology 12*: 451.

BALDWIN, R. P. 1987: Dinoflagellate resting cysts isolated from sediments in Marborough Sounds, New Zealand. *New Zealand Journal of Marine and Freshwater Research 21*: 543–553.

BALDWIN, R. P. 1992: Cargo vessel water as a vector for the spread of toxic phytoplankton species to New Zealand. *Journal of the Royal Society of New Zealand 22*: 229–242.

BALECH, E. 1995: *The Genus Alexandrium Halim (Dinoflagellata)*. Sherkin Island Marine Station, Sherkin Island, Co. Cork. 151 p.

BATES, M.; BAKER, M.; WILSON, N.; LANE, L.; HANDFORD, S. 1993: Epidemiologic overview of the New Zealand shellfish toxicity outbreak. *In*: Jasperse, J. A. (ed.), *Marine Toxins and New Zealand Shellfish. Royal Society of New Zealand Miscellaneous Series 24*: 35–40.

BAYE, T. K.; SCHALLENBERG, M.; MARTIN, C. E. 2008: Investigation of nutrient limitation status and nutrient pathways in Lake Hayes, Otago, New Zealand: a case study for integrated lake assessment. *New Zealand Journal of Marine and Freshwater Research 42*: 285–295.

BELTON, D. J.; POWELL, I. B. 1987: Cryptosporidiosis in lovebirds (*Agapornis* sp.). *New Zealand Veterinary Journal 35*: 15.

BENNETT, G. F.; WHITEWAY, M.; WOODWORTH-LYNAS, C. 1982: A host-parasite catalogue of the avian haematozoa. *Memorial University, Newfoundland Occasional Papers in Biology 5*: ix, 1–243.

BIRKENSTOCK, T. A. 2001: Genetic diversity of symbiotic algae harboured by the sea-anemone *Anthopleura aureoradiata*. Unpublished MSc thesis, University of Auckland, Auckland. 69 p.

BISHOP, D. M.; HEATH, A. C. G. 1998: Checklist of ectoparasites of birds in New Zealand. *Surveillance 25*: 13–31.

BOURRELLY, P. 1985: *Les Algues d'Eau Douce. Initiation à la Systématique. Tome III: Les Algues Bleues et Rouges. Les Eugléniens, Peridiniens et Cryptomonadines. Réimpression revue et augmentée.* Société Nouvelle des Éditions Boubé, Paris. 606 p.

BOYD, A. M.; WEINSTEIN, P. 1996: *Anopheles annulipes* Walker s.l. (Diptera: Culicidae), an under-rated temperate climate malaria vector? *New Zealand Entomologist 19*: 35–41.

BOLCH, C. J.; HALLEGRAEFF, G. M. 1990: Dinoflagellate cysts in recent marine sediments from Tasmania, Australia. *Botanica Marina 33*: 173–192.

BRADFORD, J. M.; CHANG, F. H. 1987: Standing stocks and productivity of phytoplankton off Westland, New Zealand, February 1982. *New Zealand Journal of Marine and Freshwater Research 21*: 71–90.

BRADLEY, J. 1999: A case of malaria. *New Zealand Medical Journal 112*: 105–106.

BULL, P. C. 1953: Parasites of the wild rabbit *Oryctolagus cuniculus* (L.) in New Zealand. *New Zealand Journal of Science and Technology 34B*: 341–372.

CACHON, J.; CACHON, M. 1987: Parasitic dinoflagellates. Pp. 571–610 *in*: Taylor, F. J. R. (ed.), *The Biology of Dinoflagellates*. Blackwell Scientific Publications, Oxford.

CARTER, M. J. 1984: *Cryptosporidium* infestation. *New Zealand Medical Journal 97*: 743.

CARTER, M. J. 1986: *Cryptosporidium*: an important cause of gastrointestinal disease in immunocompetent patients. *New Zealand Medical Journal 99*: 101–103.

CARTY, S. 2003: Dinoflagellates. Pp. 685–714 *in*: Wehr, J. D.; Sheath, R. G. (eds), *Freshwater algae of North America. Ecology and Classification.* Academic Press, San Diego.

CASSIE, V. 1961: Marine phytoplankton in New Zealand waters. *Botanica Marina 2 (Suppl.)*: 1–54.

CASSIE, V. 1984: Revised checklist of the freshwater algae of New Zealand (excluding diatoms and charophytes). Part II – Chlorophyta, Chromophyta and Pyrrhophyta, Rhaphidophyta and Euglenophyta. *Water and Soil Technical Publication 26*: lxiv, 1–134.

CASSIE COOPER, V. 1996: *Microalgae –Microscopic Marvels*. Riverside Books, Hamilton. 164 p.

CAVALIER-SMITH, T. 1987: The origin of eukaryote and archaebacterial cells. *Annals of the New York Academy of Sciences 503*: 17–54.

CAVALIER-SMITH, T. 1999: Principles of protein and lipid targeting in secondary symbiogenesis: euglenoid, dinoflagellate, and sporozoan plastid origins and the eukaryote family tree. *Journal of Eukaryotic Microbiology 46*: 347–366.

CAVALIER-SMITH, T. 2000: Flagellate megaevolution. The basis for eukaryote diversification. Pp. 361–390 *in*: Leadbeater, B. S. C.; Green, J. C. (eds), *The Flagellates – Unity, Diversity and Evolution. [Systematics Association Species Volume Series 59.]* Taylor & Francis for Systematics Association, London.

CAVALIER-SMITH, T. 2004: Chromalveolate diversity and cell megaevolution: interplay of membranes, genomes and cytoskeleton. Pp. 75–108 *in*: Hirt, R. P.; Horner, D. S. (eds), *Organelles, Genomes and Eukaryote Phylogeny – An Evolutionary Synthesis in the Age of Genomics. [The Systematics Association Special Volume Series 68.]* CRC Press, Boca Raton. [viii], 384 p.

CAVALIER-SMITH, T.; CHAO, E. E. 2004: Protalveolate phylogeny and systematics and the origins of Sporozoa and dinoflagellates (phylum Myzozoa nom. nov.). *European Journal of Protistology 40*: 185–212.

CEMBELLA, A. D.; THERRIAULT, J.-C.; BELAND, P. 1988: Toxicity of cultured isolates and natural populations of *Protogonyaulax tamarensis* from the St. Lawrence estuary. *Toxic Shellfish Research 7*: 611–621.

CHAN, A. T. 1980: Comparative physiological study of marine diatoms and dinoflagellates in relation to irradiance and cell size. II. Relationship between photosynthesis, growth and carbon/chlorophyll *a* ratio. *Journal of Phycology 16*: 428–432.

CHANG, F. H. 1988: Distribution, abundance, and size composition of phytoplankton off Westland, New Zealand, February 1982. *New Zealand Journal of Marine and Freshwater Research 22*: 345–367.

CHANG, F. H. 1995: The first records of *Gymnodinium* sp. nov. (cf. *breve*) (Dinophyceae) and other harmful phytoplankton species in the early 1993 blooms in New Zealand. Pp. 27–32 *in*: Lassus, P.; Arzul, G.; Erard-Le Denn, E.; Gentien, P.; Marcaillou-Le Baut, C. (eds), *Harmful Algal Blooms*. Intercept Ltd, Andover, U.K. xxiv + 878 p.

CHANG, F. H. 1996a: Distribution and abundance of *Dinophysis acuminata* (Dinophyceae) and *Pseudonitzschia australis* (Bacillariophyceae) in Kenepuru and Pelorus Sounds, New Zealand. Pp. 93–96 *in*: Yasumoto, T.; Oshima, Y.; Fukuyo, Y. (eds), *Harmful and Toxic Algal Blooms.* Intergovernmental Oceanographic Commission, UNESCO.

CHANG, F. H. 1996b: A review of knowledge of a group of closely related, economically important toxic *Gymnodinium/Gyrodinium* (Dinophyceae) species in New Zealand. *Journal of the Royal Society of New Zealand 26*: 382-394.

CHANG, F. H. 1999a: Phytoplankton blooms around Wellington. *Aquaculture Update 23*: 11–12.

CHANG, F. H. 1999b: *Gymnodinium brevisulcatum* sp. nov. (Gymnodiniales, Dinophyceae), a new species isolated from the 1998 summer toxic bloom in Wellington Harbour, NewZealand. *Phycologia 38*: 377–384.

CHANG, F. H. 2001: Harmful algal blooms contribute to paralytic shellfish poisoning. *Biodiversity Update 2*: 7.

CHANG, F. H. 2004: Marine harmful microalgae of the South Pacific with special emphasis on bloom-forming species in Australasia. *Japanese Journal of Phycology 52 (Suppl.)*: 49–56.

CHANG, F. H. 2011: Toxic effects of three closely-related dinoflagellates, *Karenia concordia, K. brevisulcata* and *K. mikimotoi* (Gymnodiniales, Dinophyceae) on other microalgal species. *Harmful Algae 10*: 181–187.

CHANG, F. H.; ANDERSON, C.; BOUSTEAD, N. C. 1990: First record of a *Heterosigma* (Raphidophyceae) bloom with associated mortality of cage-reared salmon in Big Glory Bay, New Zealand. *New Zealand Journal of*

Marine and Freshwater Research 24: 461–469.
CHANG, F. H.; ANDERSON, D. M.; KULIS, D. M.; TILL, D. G. 1997: Toxin production of *Alexandrium minutum* (Dinophyceae) from the Bay of Plenty, New Zealand. *Toxicon 35*: 393–409.
CHANG, F. H.; CHISWELL, S. M.; UDDSTROM, M. J. 2001: Occurrence and distribution of *Karenia brevisulcata* (Dinophyceae) during the 1998 summer toxic outbreaks on the central east coast of New Zealand. *Phycologia 40*: 215–222.
CHANG, F. H.; GALL, M. 1998: Phytoplankton assemblages and photosynthetic pigments during winter and spring in the Subtropical Convergence region near New Zealand. *New Zealand Journal of Marine and Freshwater Research 32*: 515–530.
CHANG, F. H.; GARTHWAITE, I.; ANDERSON, D. M.; TOWERS, N.; STEWART, R.; MacKENZIE, L. 1999: Immunofluorescent detection of a PSP-producing dinoflagellate, *Alexandrium minutum*, from Bay of Plenty, New Zealand. *New Zealand Journal of Marine and Freshwater Research 33*: 533–543.
CHANG, F. H.; HALL, J.; CUMMING, A. 2010: The life history of a harmful, epiphytic dinoflagellate, *Ostreopsis siamensis* (Dinophyceae), isolated from the north-eastern coast of New Zealand. Pp. 31–33 *in*: Ho, K.-C.; Zhou, M. J.; Qi, Y. Z. (eds), *Harmful Algae 2008: Proceedings of 13th International Conference on Harmful Algae*. ISSHA and Environmental Publication House, Hong Kong. [ix] + 278 p.
CHANG, F. H.; MacKENZIE, L.; TILL, D.; HANNAH, D.; RHODES, L. 1995: The first toxic shellfish outbreaks and the associated phytoplankton blooms in early 1993 in New Zealand. Pp. 145–150 *in*: Lassus, P.; Arzul, G.; Erard-Le Denn, E.; Gentien, P.; Marcaillou-Le-Baut, C. (eds), *Harmful Marine Algal Blooms*. Intercept Ltd, Andover, U.K. xxiv + 878 p.
CHANG, F. H.; MULLAN, B. 2003: Occurrence of major harmful algal blooms in New Zealand: is there a link with climate variations? *The Climate Update 53(11)*: 4.
CHANG, F. H.; MULLAN, B. CASSIE-COOPER, V. 2002: Downstream effects of climatic forcing such as ENSO and other longer term climatic shifts on algal blooms in New Zealand. P. 48 *in*: Steidinger, K. A. (ed.), *Abstracts of 10th International Conference on Harmful Algae, October 21–25, 2002*. Florida Fish and Wildlife Conservation Commission, St Petersburg. 352 [+ 24] p.
CHANG, F. H.; RYAN, K. G. 1985: *Prymnesium calathiferum* sp. nov. (Prymnesiophyceae), a new species isolated from Northland, New Zealand. *Phycologia 24*: 191–198.
CHANG, F. H.; RYAN, K. 2004: *Karenia concordia* sp. nov. (Gymnodiniales, Dinophyceae), a new nonthecate dinoflagellate isolated from the New Zealand northeast coast during the 2002 harmful algal bloom events. *Phycologia 43*: 552–562.
CHANG, F. H.; SHIMIZU, Y.; HAY, B.; STEWART, R.; MACKAY, G.; TASKER, R. 2000: Three recently recorded *Ostreopsis* spp. (Dinophyceae) in New Zealand: temporal and regional distribution in the upper North Island from 1995 to 1997. *New Zealand Journal of Marine and Freshwater Research 34*: 29–39.
CHANG, F. H.; STEWART, R.; INGLIS, G.; FITRIDGE, I. 2008: Dinoflagellate cysts from New Zealand ports and harbours withemphasis on the distribution of harmful species. Pp. 168–170 *in*: Moestrup, Ø. et al. (eds), Proceedings, 12th International Conference on Harmful Algae. International Society for the Study of Harmful Algae and Intergovernmental Oceanographic Commission of UNESCO, Copenhagen. 397 p.
CHANG, F. H.; UDDSTROM, M. 2002: Unusual weather patterns and exceptional harmful blooms in New Zealand. Pp. 31–36 *in*: Ho, K. C.; Lam, I. H.Y.; Yu, T. S.; Wang, Z. D. (eds), *Prevention and Management of Harmful Algal Blooms in the South China Sea*. The Association on Harmful Algal Blooms in the South China Sea, Hong Kong. 423 p.
CHANG, F. H.; UDDSTROM, M. J.; PINKERTON, M. H.; RICHARDSON, K. M. 2008: Characterising the 2002 toxic *Karenia concordia* (Dinophyceae) outbreak and its development using satellite imagery on the north-eastern coast of New Zealand. *Harmful Algae 7*: 532–544.
CHANG, F. H.; ZELDIS, J.; GALL, M.; HALL, J. 2003: Seasonal and spatial variation of phytoplankton assemblages, biomass and cell size from spring to summer across the north-eastern New Zealand continental shelf. *Journal of Plankton Research 25*: 737–758.
CHARLESTON, W. A. G. 1994: *Toxoplasma* and other protozoan infections of economic importance in New Zealand. *New Zealand Journal of Zoology 21*: 67–81.
CHILVERS, B. L.; COWAN, P. E.; WADDINGTON, D. C.; KELLY, P. J.; BROWN, T. J. 1998: The prevalence of infection of *Giardia* spp. and *Cryptosporidium* spp. in wild animals on farmland, southeastern North Island, New Zealand. *International Journal of Environmental Health Research 8*: 59–64.
CLOWES, C. D. 1985: *Stoveracysta*, a new gonyaulacacean dinoflagellate genus from the Upper Eocene and Lower Oligocene of New Zealand. *Palynology 9*: 27–35, pls 1–2.
CLOWES, C. D. 2009: Dinoflagellate taxonomy and biostratigraphy of the Mid- to Late Eocene and Early Oligocene of New Zealand. PhD thesis in geology, Victoria University of Wellington. 210 p., 94 pls.
CLOWES, C. D.; MORGANS, H. E. G. 1984: Micropaleontology of the Runangan-Whaingaroan (Eocene–Oligocene) Totara Limestone, Kakanui River, New Zealand. *New Zealand Geological Survey Record 3*: 30–40, 2 pls.
CLOWES, C. D.; WILSON, G. J. 2006: Some new species of *Corrudinium* Stover & Evitt 1978 (Dinophyceae) from the Eocene of New Zealand. *New Zealand Journal of Geology and Geophysics 49*: 399–408.
COLLINS, G. H.; CHARLESTON, W. A. G. 1979: Studies on *Sarcocystis* species: 1. Feral cats as definitive hosts for Sporozoa. *New Zealand Veterinary Journal 27*: 80–84.
COOK, R. A.; CROUCH, E. M.; RAINE, J. I.; STRONG, C. P.; URUSKI, C. I.; WILSON, G. J. 2006: Initial review of the biostratigraphy and petroleum system around the Tasman Sea hydrocarbon-producing basins. *APPEA Journal 2006*: 201–213.
COOPER, R. A. (Ed.) 2004: The New Zealand geologic timescale. *Institute of Geological and Nuclear Sciences Monograph 22*: 1–284.
COUPER, R. A. 1960: New Zealand Mesozoic and Cainozoic plant microfossils. *New Zealand Geological Survey Paleontological Bulletin 32*: 1–87.
COX, F. E. G. 1991: Systematics of parasitic protozoa. Pp. 55–80 *in*: Kreier, J. P. (ed.), *Parasitic Protozoa, Volume 1*. 2nd edn. Academic Press, San Diego & London.
CRAMPTON, J. S.; MUMME, T.; RAINE, J. I.; RONCAGLIA, L.; SCHIØLER, P.; STRONG, C. P.; TURNER, G.; WILSON, G. J. 2000: Revision of the Piripauan and Haumurian local stages and correlation of the Santonian–Maastrichtian (Late Cretaceous) in New Zealand. *New Zealand Journal of Geology and Geophysics 43*: 309–333.
CRAMPTON, J. S.; SCHIØLER, P.; RONCAGLIA, L. 2006: Detection of Late Cretaceous eustatic signatures using quantitative biostratigraphy. *Geological Society of America Bulletin 118*: 975–990.
CROUCH, E. M. 2001: Environmental change at the time of the Paleocene–Eocene biotic turnover. *LPP Contributions Series 14*: 1–216.
CROUCH, E. M.; DICKENS, G. R.; BRINKHUIS, H.; AUBRY, M.-P.; HOLLIS, C. J.; ROGERS, K. M.; VISSCHER, H. 2003: The *Apectodinium* acme and terrestrial discharge during the Paleocene-Eocene thermal maximum: new palynological, geochemical and calcareous nannoplankton observations at Tawanui, New Zealand. *Palaeogeography, Palaeoclimatology, Palaeoecology 194*: 387–403.
CROUCH, E. M.; HOLLIS, C. J. 1996: Paleogene palynomorph and radiolarian biostratigraphy of DSDP leg 29, sites 280 and 281 South Tasman Rise. *Institute of Geological & Nuclear Sciences Science Report 96/19*: 1–46, 7 pls.
CRUMPTON, W. J. 1974: Eugregarines from the larva of the sand scarab (*Pericoptus truncatus* Fabricius; Scarabaeidae). *Journal of the Royal Society of New Zealand 4*: 319–326.
DAUGBJERG, N.; HANSEN, G.; LARSEN, J.; MOESTRUP, Ø. 2000: Phylogeny of some of the major genera of dinoflagellates based on ultrastructure and partial LSU rDNA sequence data, including the erection of three new genera of unarmoured dinoflagellates. *Phycologia 39*: 302–317.
DAWSON, E. W. 1992: The marine fauna of New Zealand: Index to the fauna: 1. Protozoa. *New Zealand Oceanographic Institute Memoir 99*: 1–368.
DEFLANDRE, G. 1933: Note préliminaire sur un péridinien fossile *Lithoperidinium oamaruense* n.g., n.sp. *Bulletin de la Société zoologique de France 58*: 265–273.
de JERSEY, N. J.; RAINE, J. I. 1990: Triassic and earliest Jurassic miospores from the Murihiku Supergroup, New Zealand. *New Zealand Geological Survey Paleontological Bulletin 62*: 1–164.
DERRAIK, J. G. B.; TOMPKINS, D. M.; ALLEY, M. R.; HOLDER, P.; ATKINSON, T. 2008: Epidemiology of an avian malaria outbreak in a native bird species (*Mohoua ochrocephala*) in New Zealand. *Journal of the Royal Society of New Zealand 38*: 237–242.
De SALAS, M. F.; RHODES, L. L.; MacKENZIE, L. A.; ADAMSON, J. E.; PONIKLA, K. 2005: The gymnodinioid genera *Karenia* and *Takayama* (Dinophyceae) in New Zealand coastal waters. *New Zealand Journal of Marine and Freshwater Research 39*: 135–139.
DESSER, S. S.; ALLISON, F. R. 1979: Aspects of the sporogonic development of *Leucocytozoon tawaki* of the Fiordland crested penguin in its primary vector, *Austrosimulium ungulatum*: an ultrastructural study. *Journal of Parasitology 65*: 737–744.
DESSER, S. S.; BENNETT, G. F. 1993: The genera *Leucocytozoon*, *Haemoproteus*, and *Hepatocystis*. Pp. 273–307 *in*: Kreier, J. P. (ed.), *Parasitic Protozoa Volume 4*. 2nd edn. Academic Press, San

Diego & London.
DIGGLES, B. K.; HINE, P.M.; HANDLEY, S.; BOUSTEAD, N. C. 2002: A handbook of diseases of importance to aquaculture in New Zealand. *NIWA Science and Technology Series 49*: 1–200.
DIJKSTRA, T.; EYSKER, M.; SCHARES, G.; CONRATHS, F. J.; WOUDA, W.; BARKEMA, H. W. 2001: Dogs shed *Neospora caninum* oocysts after ingestion of naturally infected bovine placenta but not after ingestion of colostrum spiked with *Neospora caninum* tachyzoites. *International Journal for Parasitology 31*: 747–752.
DODGE, J. D. 1982: *Marine Dinoflagellates of the British Isles*. HMSO, London. 303 p.
DORÉ, A. B. 1920: Notes on some avian haematozoa observed in New Zealand. *New Zealand Journal of Science and Technology 3*: 10–12.
DORÉ, A. B. 1921: Notes on malarial infection in the imported skylark. *New Zealand Journal of Science and Technology 4*: 126–129.
DOUGHERTY, M. 1996: *Cyclospora cayatenensis*: an emerging intestinal pathogen in New Zealand. *New Zealand Journal of Medical Laboratory Science 50*: 12–13.
DUBEY, J. P. 1992: Neosporosis – a newly recognised protozoan infection. *Comparative Pathology Bulletin 24*: 4–6.
DUBEY, J. P. 1993: *Toxoplasma, Neospora, Sarcocystis*, and other tissue cyst-forming coccidia of humans and animals. Pp. 1–158 *in*: Kreier, J. P. (ed.), *Parasitic Protozoa Volume 6*. 2nd edn. Academic Press, San Diego & London.
DUBEY, J. P. 1999: Recent advances in *Neospora* and neosporosis. *Veterinary Parasitology 84*: 349–367.
DUBEY, J. P. 2008: The history of *Toxoplasma gondii* – the first hundred years. *Journal of Eukaryotic Microbiology 55*: 467–475.
DUBEY, J. P.; BARR, B. C.; BARTA, J. R.; BJERKÅS, I.; BJÖRKMAN, C.; BLAGBURN, B. L.; BOWMAN, D. D.; BUXTON, D.; ELLIS, J. T.; GOTTSTEIN, B.; HEMPHILL, A.; HILL, D. E.; HOWE, D. K.; JENKINS, M. C.; KOBAYASHI, Y.; KOUDELA, B.; MARSH, A. E.; MATTSON, J. G.; McALLISTER, M. M.; MODRÝ, D.; OMATA, Y.; SIBLEY, L. D.; SPEER, C. A.; TREES, A. J.; UGGLA, A.; UPTON, S. J.; WILLIAMS, D. J. L.; LINDSAY, D. S. 2002b: Redescription of *Neospora caninum* and its differentiation from related coccidia. *International Journal for Parasitology 32*: 929–946.
DUBEY, J. P.; BEATTIE, C. P. 1988: *Toxoplasmosis of Animals and Man*. CRC Press, Boca Raton. 220 p.
DUBEY, J. P.; CARPENTER, J. L.; SPEER, C. A.; TOPPER, M. J.; UGGLA, A. 1988: Newly recognised fatal protozoan disease of dogs. *Journal of the American Veterinary Medical Association 192*: 1269–1285.
DUBEY, J. P.; HILL, D. E.; LINDSAY, D. S.; JENKINS, M. C.; UGGLA, A.; SPEER, C. A. 2002a: *Neospora caninum* and *Hammondia heydorni* are separate species. *Trends in Parasitology 18*: 66–69.
DUIGNAN, P.; HORNER, G.; O'KEEFE, J. 2003: Infectious and emerging diseases of bats, and health status of bats in New Zealand. *Surveillance 30(1)*: 15–18.
DUMBLETON, L. J. 1949: *Diplocystis oxycani* n.sp. – a gregarine parasite of *Oxycanus cervinatus* Walk. *Transactions of the Royal Society of New Zealand 77*: 530–532.
DUNCANSON, M.; RUSSELL, N.; WEINSTEIN, P.; BAKER, M.; SKELLY, C.; HEARNDEN, M.; WOODWARD, A. 2000: Rates of notified cryptosporidiosis and quality of drinking water supplies in Aotearoa, New Zealand. *Water Research (Oxford) 34*: 3804–3812.
DUNGAN, C. F.; REECE, K. S.; MOSS, J. A.; HAMILTON, R. M.; DIGGLES, B. K. 2007: *Perkinsus olseni* in vitro isolates from the New Zealand clam *Austrovenus stutchburyi*. *Journal of Eukaryotic Microbiology 54*: 263–270.
EARLE, R. A.; BENNETT, G. F.; BROSSY, J. J. 1992: First African record of *Leucocytozoon tawaki* (Apicomplexa: Leucocytozoidae) from the jackass penguin *Spheniscus demersus*. *South African Journal of Zoology 27*: 89–90.
ELLEGAARD, M.; DALE, B.; AMORIM, A. 2002: The acritarchous resting cyst of the athecate dinoflagellate *Warnowia* cf. *rosea* (Dinophyceae). *Phycoologia 40*: 542–546.
ELLIS, J. T.; POMROY, W. E. 2003: *Hammondia heydorni* oocysts in the faeces of a greyhound. *New Zealand Veterinary Journal 51*: 38–39.
EPPLEY, R. W.; HARRISON, W. G. 1974: Physiological ecology of *Gonyaulax polyedra*: a red water dinoflagellate of Southern California. Pp. 11–12 *in*: Le Cicero, U. R. (ed.), *Proceedings of the First International Conference on Toxic Dinoflagellate Blooms*. MSTF, Wakefield, Mass.
ETTL, H.; GÄRTNER, G. 1995: *Syllabus der Boden-, Luft- und Flechtenalgen*. Gustav Fischer Verlag, Stuttgart. viii + 721 p.
FALLIS, A. M.; BISSET, S. A.; ALLISON, F. R. 1976: *Leucocytozoon tawaki* n. sp. (Eucoccidea: Leucocytozoidae) from the penguin *Eudyptes pachyrhynchus*, and preliminary observations on its development in *Austrosimulium* spp. (Diptera: Simuliidae). *New Zealand Journal of Zoology 3*: 11–16.
FAYER, R.; MORGAN, U.; UPTON, S. J. 2000: Epidemiology of *Cryptosporidium*: transmission, detection and identification. *International Journal for Parasitology 30*: 1305–1322.
FENSOME, R. A.; MacRAE, R. A.; WILLIAMS, G. L. 2008: DINOFLAJ2, Version 1. *American Association of Stratigraphic Palynologists, Data Series no. 1*.
FENSOME, R. A.; TAYLOR, F. J. R.; NORRIS, G.; SARJEANT, W. A. S.; WHARTON, D. I.; WILLIAMS, G. L. 1993: A classification of fossil and living dinoflagellates. *Micropaleontology Press Special Paper 7*: 1–351.
FIELD, C. B.; BEHRENFELD, M. J.; RANDERSON, J. T.; FALKOWSKI, P. 1998: Primary production of the biosphere: integrating terrestrial and oceanic components. *Science 281*: 237–240.
FITZGERALD, N. W. 1952: Some new gregarines. Unpublished MSc thesis, University of Otago, Dunedin. 75 p., 46 pls.
GOODMAN, D. K. 1987: Dinoflagellate cysts in ancient and modern sediments. Pp. 649–722 *in*: Taylor, F. J. R. (ed.), *The Biology of Dinoflagellates*. Blackwell Scientific Publications, London. xii + 785 p.
GRACZYK, T. K.; COCKREM, J. F.; CRANFIELD, M. R.; DARBY, J. T.; MOORE, P. 1995: Avian malaria seroprevalence in wild New Zealand penguins. *Parasite 2*: 401–405.
GRINBERG, A.; OLIVER, L.; LEARMONTH, J. J.; LEYLAND, M.; ROE, W.; POMROY, W. E. 2003: Identification of *Cryptosporidium parvum* 'cattle' genotype from a severe outbreak of neonatal foal diarrhoea. *Veterinary Record 153*: 628–631.
GRINBERG, A.; POMROY, W. E.; CARSLAKE, H. B.; SHI, Y.; GIBSON, I. R.; DRAYTON, B. M. 2009: A study of neonatal cryptosporidiosis of foals in New Zealand. *New Zealand Veterinary Journal 57*: 284–289.
HACKETT, J. D.; ANDERSON, D. M.; ERDNER, D. L.; BHATTACHARYA, D. 2004: Dinoflagellates: a remarkable evolutionary experiment. *American Journal of Botany 91*: 1523–1534.
HARTLEY, W. J.; MARSHALL, S. C. 1957: Toxoplasmosis as a cause of ovine perinatal mortality. *New Zealand Veterinary Journal 5*: 119–124.
HASKELL, T. R.; WILSON, G. J. 1975: Palynology of Sites 280–284, DSDP Leg 29, off southeastern Australia and western New Zealand. *In*: Kennett, J. P., Houtz, R. E. et al. (eds). *Initial Reports of the Deep Sea Drilling Project 29*: 723–741.
HAYWOOD, A. J.; STEIDINGER, K. A.; TRUBY, E. W.; BERGQUIST, P. R.; BERGQUIST, P. L.; ADAMSON, J.; MacKENZIE, L. 2004: Comparative morphology and molecular phylogenetic analysis of three new species of the genus *Karenia* (Dinophyceae) from New Zealand. *Journal of phycology 40*: 165–179.
HEAD, M. J. 1996; Modern dinoflagellate cysts and their biological affinities. Pp. 1197–1248, *in*: Jansonius, J.; McGregor, D. C. (eds), *Palynology:Principles and applications*. American Association of Stratigraphic Palynologists Foundation, Dallas. 1330 p. [3 vols.]
HELBY, R.; WIGGINS, V. D.; WILSON, G. J. 1987: The circum-Pacific occurrence of the Late Triassic dinoflagellate *Sverdrupiella*. *Australian Journal of Earth Sciences 34*: 151–152.
HELBY, R.; WILSON, G. J. 1988: A new species of *Sverdrupiella* Bujak & Fisher (Dinophyceae) from the Late Triassic of New Zealand. *New Zealand Journal of Botany 26*: 117–122.
HELBY, R.; WILSON, G. J.; GRANT-MACKIE, J. A. 1988: A preliminary biostratigraphic study of Middle to Late Jurassic dinoflagellate assemblages from Kawhia, New Zealand. *Memoirs of the Association of Australasian Paleontologists 5*: 125–166.
HERZER, R. H.; SYKES, R.; KILLOPS, S. D.; FUNNELL, R. H.; BURGRAFF, D. R.; TOWNEND, J.; RAINE, J. I.; WILSON, G. J. 1999: Cretaceous carbonaceous rocks from the Norfolk Ridge system, Southwest Pacific: implications for regional petroleum potential. *New Zealand Journal of Geology and Geophysics 42*: 57–73.
HEYDORN, A. O.; MEHLHORN, H. 2002: *Neospora caninum* is an invalid species name: an evaluation of facts and statements. *Parasitology Research 88*: 175–184.
HINE, P. M. 1975: *Eimeria anguillae* Léger & Hollande, 1922 parasitic in New Zealand eels. *New Zealand Journal of Marine and Freshwater Research 9*: 239–243.
HINE, P. M.; JONES, J. B.; DIGGLES, B. K. 2000: A checklist of the parasites of New Zealand fishes, including previously unpublished records. *NIWA Technical Report 75*: 1–93 [+ 2].
HITCHCOCK, G. L. 1982: A comparative study of the size-dependent organic composition of marine diatoms and dinoflagellates. *Journal of Plankton Research 4*: 363–377.
HOLLIS, C. J.; BEU, A. G.; RAINE, J. I.; STRONG, C. P.; TURNBULL, I. M.; WAGHORN, D. B.; WILSON, G. J. 1997: Integrated biostratigraphy of Cretaceous–Paleogene strata on Campbell Island, southwest Pacific. *Institute of Geological & Nuclear Sciences Science Report 97/25*: 1–56.
HOLLIS, C. J.; FIELD, B. D.; JONES, C. M.; STRONG, C. P.; WILSON, G. J.; DICKENS, G. R. 2005: Biostratigraphy and carbon isotope

stratigraphy of uppermost Cretaceous–lower Cenozoic Muzzle Group in middle Clarence Valley, New Zealand. *Journal of the Royal Society of New Zealand 35*: 345–383.

HURLEY, D. E. 1982: The Nelson slime: observations on past occurrences. *NZOI Oceanographic Summary 20*: 1–11.

IKRAM, R. B.; FAOAGALI, J. L.; PALTRIDGE, G. P. 1984: Malaria in Christchurch. *Pathology 16*: 431–433.

INGRAM, R. J. H.; ELLIS-PEGLER, R. B.; GALLER, L. H. 1988: A New Zealand death from malaria. *New Zealand Medical Journal 101*: 548.

JAKOB-HOFF, R. 2001: Establishing a health profile for the North Island brown kiwi (*Apteryx australis mantelli*). *Kokako 8(2)*: 6–9.

JAKOB-HOFF, R.; BUCHAN, B.; BOYLAND, M. 1999: Kiwi coccidia – North Island survey results. *Kokako 6(1)*: 3–5.

JAKOB-HOFF, R.; TWENTYMAN, C. M.; BUCHAN, B. 2000: Clinical features associated with a haemoparasite of North Island brown kiwi. *Kokako 7(2)*: 11–12.

JAMES, M. P.; SAUNDERS, B. W.; GUY, L. A.; BROOKBANKS, E. O.; CHARLESTON, W. A. G.; UILENBERG, G. 1984: *Theileria orientalis*, a blood parasite of cattle. First report in New Zealand. *New Zealand Veterinary Journal 32*: 154–156.

JEFFREY, S. W.; VESK, M. 1997: Introduction to marine phytoplankton and their pigment signatures. Pp. 37–84 *in*: Jeffrey, S. W.; Mantoura, R. F. C.; Wright, S. W. (Eds), *Phytoplankton Pigments in Oceanography: Guidelines to Modern Methods*. UNESCO Publishing, Paris. 661 p.

JENSEN, A. L. 1979: Some protozoan and acarine parasites from New Zealand Odonata. *Mauri Ora 7*: 147–149.

JOHNSTONE, A. C.; CORK, S. C. 1993: Diseases of aviary and native birds in New Zealand. *Surveillance 20(3)*: 35–36.

JONES, J. B. 1976: *Nematopsis* n.sp. (Sporozoa: Gregarinia) in *Perna canaliculus*. *New Zealand Journal of Marine and Freshwater Research 9*: 567–568.

KANE, J. E. 1964: *Thalassomyces marsupii*, a new species of ellobiopsid parasite on the hyperiid amphipod *Parathemisto gaudichaudii* (Guér.). *New Zealand Journal of Science 7*: 289–303.

KRIECHBAUM, A. J.; BAKER, M. G.; 1996: The epidemiology of imported malaria in New Zealand 1980–92. *New Zealand Medical Journal 109*: 405–407.

LAIRD, M. 1950: Some blood parasites of New Zealand birds. *Zoology Publications from Victoria University College 5*: 1–20.

LAIRD, M. 1951a: Blood parasites of mammals in New Zealand. *Zoology Publications from Victoria University College 9*: 1–14.

LAIRD, M. 1951b: *Plasmodium lygosomae* n. sp., a parasite of a New Zealand skink, *Lygosoma moco* (Gray 1839). *Journal of Parasitology 37*: 183–189.

LAIRD, M. 1952: New haemogregarines from New Zealand marine fishes. *Transactions of the Royal Society of New Zealand 79*: 589–600.

LAIRD, M. 1953: The Protozoa of New Zealand intertidal zone fishes. *Transactions of the Royal Society of New Zealand 81*: 79–143.

LAIRD, M. G.; BASSETT, K. N.; SCHIØLER, P.; MORGANS, H. E. G.; BRADSHAW, J. D.; WEAVER, S. D. 2003: Paleoenvironmental and tectonic changes across the Cretaceous/Tertiary boundary at Tora, southeast Wairarapa, New Zealand: a link between Marlborough and Hawke's Bay. *New Zealand Journal of Geology and Geophysics 46*: 275–293.

LEARMONTH, J. J.; IONAS, G.; PITA, A. B.; COWIE, R. S. 2003: Identification and genetic characterization of *Giardia* and *Cryptosporidium* strains in humans and dairy cattle in the Waikato region of New Zealand. *Water Science and Technology 47*: 21–26.

LEARMONTH, J. J.; IONAS, G.; EBBETT, K. A.; KWAN, E. S. 2004: Genetic characterization and transmission cycles of *Cryptosporidium* species isolated from humans in New Zealand. *Applied and Environmental Microbiology 70*: 3973–3978.

LEE, J. J.; LEEDALE, G. F.; BRADBURY, P. 2000: *An Illustrated Guide to the Protozoa. Second edition. Organisms traditionally referred to as Protozoa, or newly discovered groups*. Society of Protozoologists, Lawrence. 1432 p. (2 vols).

LEVINE, N. D. 1982: The genus *Atoxoplasma* (Protozoa: Apicomplexa). *Journal of Parasitology 68*: 719–723.

LEVINE, N. D. 1985: Phylum II. Apicomplexa Levine, 1970. Pp. 322–374 *in*: Lee, J. J.; Hutner, S. H.; Bovee, E. C. (eds), *An Illustrated Guide to the Protozoa*. Society of Protozoologists, Lawrence. ix +629 p.

LEVINE, N. D. 1988: *The Protozoan Phylum Apicomplexa*. CRC Press Inc., Boca Raton, Florida, USA. 2 vols.

LEWIS, J. M.; DODGE, J. D. 2002: Phylum Pyrrhophyta (Dinoflagellates). Pp. 186-207 *in*: John, D. M.; Whitton, B. A.; Brook, A. J. (eds), *The Freshwater Algal Flora of the British Isles*. Cambridge University Press, Cambridge, U.K.

MacKENZIE, L; ADAMSON, J. 2000: The biology and toxicity of *Gymnodinium catenatum*: an overview with observations from the current bloom. *Proceedings of the Marine Biotoxin Workshop No. 14*: 13–22. [Marine Biotoxin Technical Committee, Wellington.]

MacKENZIE, L.; BEUZENBERG, V.; HOLLAND, P.; McNABB, P.; SUZUKI, T.; SELWOOD, A. 2005: Pectenotoxin and okadaic acid-based toxin profiles in *Dinophysis acuta* and *Dinophysis acuminata* from New Zealand. *Harmful Algae 4*: 75–86.

MacKENZIE, L.; de SALAS, M.; ADAMSON, J.; BEUZENGERG, V. 2004: The dinoflagellate genus *Alexandrium* (Halim) in New Zealand coastal waters: comparative morphology, toxicity and molecular genetics. *Harmful Algae 3*: 71–92.

MacKENZIE, L.; HAYWOOD, A.; ADAMSON, J.; TRUMAN, P.; TILL, D.; SEKI, T.; SATAKE, M.; YASUMOTO, T. 1996: Gymnodimine contamination of shellfish in New Zealand. Pp. 97–100 *in*: Yasumoto, T.; Oshima, Y.; Fukuyo, Y. (eds), *Harmful and Toxic Algal Blooms*. Intergovernmental Oceanographic Commission, UNESCO, Paris.

MacKENZIE, L.; RHODES, L.; TILL, D.; CHANG, F. H.; KASPAR, H.; HAYWOOD, A.; KAPA, J.; WALKER, B. 1995: A *Gymnodinium* sp. bloom and the contamination of shellfish with lipid soluble toxins in New Zealand, Jan.–April 1993. Pp. 795–800 *in*: Lassus, P.; Arzul, G.; Erard-Le Denn, E.; Gentien, P.; Marcaillou-Le Baut, C. (eds), *Harmful Algal Blooms*. Intercept Ltd, Andover, U.K. xxiv + 878 p.

MacKENZIE, L., SIMS, I., BEUZENBERG, V., GILLESPIE, P. 2002: Mass accumulation of mucilage caused by dinoflagellate polysaccharide exudates in Tasman Bay, New Zealand. *Harmful Algae 1*: 69–83.

MacKENZIE, L.; TODD, K. 2002: *Alexandrium camurascutulum* sp. nov. (Dinophyceae): a new dinoflagellate species from New Zealand. *Harmful Algae 1*: 295–300.

MacKENZIE, L.; TRUMAN, P.; SATAKE, M.; YASUMOTO, T.; ADAMSON, J.; MOUNTFORT, D.; WHITE, D. 1998: Dinoflagellate blooms and associated DSP-toxicity in shellfish in New Zealand. Pp. 74–77 *in*: Reguera, B.; Blanco, J.; Fernandez, M.; Wyatt, T. (eds), *Harmful algae*. Xunta de Galicia and Intergovernmental Oceanographic Commission, UNESCO, Paris. xv + 635 p.

MacKENZIE, L.; WHITE, D.; OSHIMA, Y.; KAPA, J. 1996: The resting cyst and toxicity of *Alexandrium ostenfeldii* (Dinophyceae) in New Zealand. *Phycologia 35*: 148–155.

MANDRA, Y. T.; BRIGGER, A. L.; MANDRA, H. 1973: Preliminary report on a study of fossil silicoflagellates from Oamaru, New Zealand. *Occasional Papers of the Californian Academy of Sciences 107*: 1–11.

MARGULIS, L.; CORLISS, J. O.; MELKONIAN, M.; CHAPMAN, D. J. 1990: *Handbook of Protoctista*. Jones & Bartlett, Boston, 914 p.

MAZENGARB, C.; WILSON, G. J.; SCOTT, G. H. 1991: A Miocene debris flow deposit, Puketoro Station, Raukumara Peninsula. *New Zealand Geological Survey Record 43*: 107–111.

McALLISTER, M. M.; DUBEY, J. P.; LINDSAY, D. S.; JOLLEY, W. R.; WILLS, R. A.; McGUIRE, A. M. 1998: Dogs are definitive hosts of *Neospora caninum*. *International Journal for Parasitology 28*: 1473–1478.

McINTYRE, D. J.; WILSON, G. J. 1966: Preliminary palynology of some Antarctic tertiary erratics. *New Zealand Journal of Botany 4*: 315–321.

McKENNA, P. B. 1975: The identity and occurrence of coccidia species in some New Zealand pigs. *New Zealand Veterinary Journal 23*: 99–101.

McKENNA, P. B. 1999: Some new host-parasite records. *Surveillance 26(4)*: 5–6.

McKENNA, P. B. 2001: Register of new host parasite records. *Surveillance 28(4)*: 15–16.

McKENNA, P. B. 2003: An annotated checklist of ecto- and endoparasites of New Zealand reptiles. *Surveillance 30(3)*: 18–25.

McKENNA, P. B. 2005: Register of new host parasite records. *Surveillance 32(4)*: 7–8.

McKENNA, P. B. 2009. An updated checklist of helminth and protozoan parasites of terrestrial mammals in New Zealand. *New Zealand Journal of Zoology*. 36: 89–113.

McKENNA, P. B. 2010. An updated checklist of helminth and protozoan parasites of birds in New Zealand. *WebmedCentral Parasitology 1(9)* WMC00705: 1–29.

McKENNA, P. B.; CHARLESTON, W. A. G. 1980a: Coccidia (Protozoa: Sporozoasida) of cats and dogs. Part 1. Identity and prevalence in cats. *New Zealand Veterinary Journal 28*: 86–88.

McKENNA, P. B.; CHARLESTON, W. A. G. 1980b: Coccidia (Protozoa: Sporozoasida) of cats and dogs IV. Identity and prevalence in dogs. *New Zealand Veterinary Journal 28*: 128–130.

McKENNA, P. B.; CHARLESTON, W. A. G. 1980c: Coccidia (Protozoa: Sporozoasida) of cats and dogs III. The occurrence of a species of *Besnoitia* in cats. *New Zealand Veterinary Journal 28*: 120–122.

McKENNA. P. B.; CHARLESTON, W. A. G. 1980d: Coccidia (Protozoa: Sporozoasida) of cats and dogs. Part II. Experimental induction of *Sarcocystis* infections in mice. *New Zealand Veterinary Journal 28*: 117–119.

McMINN, A.; SUN, X. 1994: Recent dinoflagellate cysts from the Chatham Rise, Southern Ocean,

east of New Zealand. *Palynology 18*: 41–53.
McMINN, A.; BOLCH, C. J. S.; de SALAS, M. F.; HALLEGRAEFF, G. M. 2010: Recent dinoflagellate cysts. Pp. 260–292 *in*: Hallegraeff, G. M.; Bolch, C. J. S.; Hill, D. R. A.; Jameson, I.; LeRoi, J.-M.; McMinn, A.; Murray, S.; de Salas, M. F.; Saunders, K. (eds), *Algae of Australia: Phytoplankton of temperate coastal waters*. CSIRO Publishing, Melbourne. 432 p.
McNABB, P.; HOLLAND, P.T. 2003: Using liquid chromatography mass spectrometry to manage shellfish harvesting and protect public health. Pp. 179–186 *in*: Villaba, A.; Reguera, B.; Romalde, J.; Beiras, L. R. (eds), *Proceedings of the 4th International Conference on Molluscan Shellfish Safety*. Xunta de Galicia & Intergovernmental Oceanographic Commission of UNESCO, Vigo & Paris. 620 p.
MEHLHORN, H.; HEYDORN, A. O. 2000: *Neospora caninum*: Is it really different from *Hammondia heydorni* or is it a strain of *Toxoplasma gondii*? An opinion. *Parasitology Research 86*: 169–178.
MOOR, C.; STONE, P.; PURDIE, G.; WEINSTEIN, P. 2000: An investigation into the incidence of toxoplasmosis in pregnancy. *New Zealand Medical Journal 113*: 29–32.
MOORE, P. R.; ISAAC, M. J.; MAZENGARB, C.; WILSON, G. J. 1989: Stratigraphy and structure of Cretaceous (Neocomian-Maastrichtian) sedimentary rocks in the Anini-Okaura Stream area, Urewera National Park, New Zealand. *New Zealand Journal of Geology and Geophysics 32*: 515–526.
MOORE, R. B.; OBORNIK, M.; JANOUŠKOVEC, J.; CHRUDIMSKÝ, T.; VANCOVÁ, M.; GREEN, D. H.; WRIGHT, S. W.; DAVIES, N. W.; BOLCH, C. J. S.; HEIMANN, K.; ŠLAPETA, J.; HOEGH-GULDBERG, O.; LOGSDON, J. M. Jr; CARTER, D. A. 2008: A photosynthetic alveolate closely related to apicomplexan parasites. *Nature 451*: 959–963.
MORGAN, U. M; CONSTANTINE, C. C; FORBES, D. A.; THOMPSON, R. C. A. 1997: Differentiation between human and animal isolates of *Cryptosporidium parvum* using rDNA sequencing and direct PCR analysis. *Journal of Parasitology 83*: 825–830.
MORGAN-RYAN, U. M.; FALL, A.; WARD, L. A.; HIJJAWI, N.; SULAIMAN, L. S.; FAYER, R.; THOMPSON, R. C.; OLSON, M.; LAL, A.; XIAO, L. 2002: *Cryptosporidium hominis* n. sp. (Apicomplexa: Cryptosporidiidae) from *Homo sapiens*. *Journal of Eukaryotic Microbiology 49*: 433–440.
MORGANS, H. E. G.; EDWARDS, A. R.; SCOTT, G. H.; GRAHAM, I. J.; KAMP, P. J. J.; MUMME, T. C.; WILSON, G. J.; WILSON, G. S. 1999: Integrated stratigraphy of the Waitakian-Otaian Stage boundary stratotype, Early Miocene, New Zealand. *New Zealand Journal of Geology and Geophysics 42*: 581–614.
MURRAY, S.; DAUGBJERG, N.; FLØ JØRGENSEN, M.; RHODES, L. 2003: *Amphidinium* revisited: II. Resolving species boundaries in the *Amphidinium operculatum* species complex (Dinophyceae), including the descriptions of *Amphidinium trulla* sp. nov. and *Amphidinium gibbosum* comb. nov. *Journal of Phycology 40*: 366–382.
O'CALLAGHAN, M. G.; O'DONOGHUE, P. J. 2001: A new species of *Eimeria* (Apicomplexa: Eimeriidae) from the brushtail possum, *Trichosurus vulpecula* (Diprotodontia: Phalangeridae). *Transactions of the Royal Society of South Australia 125*: 129–132.
OCKLEFORD, A; DOWLING, J; MORRIS, A. J. 1997: *Cyclospora cayetanensis* diarrhoea in a traveller. *New Zealand Medical Journal 110*: 404.
OKEOMA, C. M.; WILLIAMSON, N. B.; POMROY, W. E.; STOWELL, K. M.; GILLESPIE, L. M. 2004: Isolation and molecular characterisation of *Neospora caninum* in New Zealand. *New Zealand Veterinary Journal 52*: 364–370.
ORR, M.; BLACK, A. 1996: Animal Health Laboratory Network review of diagnostic cases – October to December 1995. *Surveillance 23(1)*: 3–5.
PATITUCCI, A. N.; ALLEY, M. R.; JONES, B. R.; CHARLESTON, W. A. G. 1997: Protozoal encephalomyelitis of dogs involving *Neospora caninum* and *Toxoplasma gondii* in New Zealand. *New Zealand Veterinary Journal 45*: 231–235.
PEIRCE, M. A.; JAKOB-HOFF, R. M.; TWENTYMAN, C. 2003: New species of haematozoa from Apterygidae in New Zealand. *Journal of Natural History 37*: 1797–1804.
PERKINS, F. O.; BARTA, J. R.; CLOPTON, R. E.; PEIRCE, M. A. UPTON, S. J. 2000: Phylum Apicomplexa Levine, 1970. Pp. 190–369 *in*: Lee, J. J.; Leedale, G. F.; Bradbury, P. (eds), *An Illustrated Guide to the Protozoa. Second Edition. Organisms traditionally referred to as Protozoa, or newly discovered groups*. Society of Protozoologists, Lawrence. Vol. 1, 689 p.
PETERS, M.; LÜTKEFELS, E.; HECKEROTH, A. R.; SCHARES, G. 2001: Immunohisto-chemical and ultrastructural evidence for *Neospora caninum* tissue cysts in skeletal muscles of naturally infected dogs and cattle. *International Journal for Parasitology 31*: 1144–1148.
PLUTZER, J.; KARANIS, P. 2009: Genetic polymorphism in *Cryptosporidium* species: An update. *Veterinary Parasitology 165*: 187–199.
POLLINGHER, U. 1987: Freshwater ecosystems. Pp. 502–529 *in*: Taylor, F. J. R. (ed.), *The Biology of Dinoflagellates*. Blackwell Scientific Publications, Oxford.
RAINE, J. I.; WILSON, G. J. 1988: Palynology of the Mt Somers (South Island, New Zealand) early Cenozoic sequence (Note). *New Zealand Journal of Geology and Geophysics 31*: 385–390.
REED, C. 1997: Avian malaria in New Zealand dotterel. *Kokako 4*: 3.
REICHEL, M. P.; POMROY, W. E. 1996: *Neospora* spp. and bovine abortions. *Second Pan Pacific Veterinary Conference: Proceedings of cattle sessions*: 21–27.
RHODES, L. L.; ADAMSON, J. 2001: *Pfiesteria* – a dinoflagellate to respect. *Seafood New Zealand 10*: 34–35.
RHODES, L. L.; ADAMSON, J. E.; RUBLEE, P.; SCHAEFFER, E. 2005: Geographic distribution of *Pfiesteria piscicida* and *P. shumwayae* (Pfiesteriaceae) in Tasman Bay and Canterbury, New Zealand (2002–2003). *New Zealand Journal of Marine and Freshwater Research*.
RHODES, L. L.; BURKHOLDER, J. M.; GLASGOW, H. B.; RUBLEE, P. A.; ALLEN, C.; ADAMSON, J. E. 2002b: *Pfiesteria shumwayae* (Pfiesteriaceae) in New Zealand. *New Zealand Journal of Marine and Freshwater Research 36*: 621–630.
RHODES, L.; HAYWOOD, A.; ADAMSON, J.; PONIKLA, K.; SCHOLIN, C. 2004: DNA probes for the detection of *Karenia* species in New Zealand's coastal waters. Pp. 273–275 *in*: Steidinger, K. A.; Landsberg, J. H.; Tomas, C. R.; Vargo, G. A. (eds), *Harmful Algae 2002*. Florida Institute of Oceanography & Intergovernmental Oceanographic Commission of UNESCO, St Petersburg, Florida. xx + 588 p.
RHODES, L. L.; HAYWOOD, A. J.; BALLANTINE, W. J.; MacKENZIE, A. L. 1993: Algal blooms and climate anomalies in north-east New Zealand, August–December 1992). *New Zealand Journal of Marine and Freshwater Research 27*: 419–430.
RHODES, L. L.; McNABB, P.; de SALAS, M.; BRIGGS, L.; BEUZENBERG, V.; GLADSTONE, M. 2006: Yessotoxin production by *Gonyaulax spinifera*. *Harmful Algae 5*: 148–155.
RHODES, L. L.; SCHOLIN, C.; GARTHWAITE, I.; HAYWOOD, A.; THOMAS, A. 1998: Domoic acid producing *Pseudo-nitzschia* species educed by whole cell DNA probe-based and immunochemical assays. Pp. 274–277 *in*: Reguera, B.; Blanco, J.; Fernandez, M. L.; Wyatt, T. (eds), *Harmful Algae*. Xunta de Galicia and Intergovernmental Oceanographic Commission, UNESCO, Paris. xv + 635 p.
RHODES, L. L.; SYHRE, M. 1995: Okadaic acid production by a New Zealand *Prorocentrum lima* isolate. *New Zealand Journal of Marine & Freshwater Research 29*: 367–370.
RHODES, L. L.; THOMAS, A. E. 1997: *Coolia monotis* (Dinophyceae): a toxic epiphytic microalgal species found in New Zealand. *New Zealand Journal of Marine and Freshwater Research 31*: 139–141.
RHODES, L. L.; TOWERS, N.; BRIGGS, L.; MUNDAY, R.; ADAMSON, J. 2002a: Uptake of palytoxin-like compounds by feeding shellfish with *Osteopsis siamensis* (Dinophyceae). *New Zealand Journal of Marine and Freshwater Research 36*: 631–636.
RONCAGLIA, L. 2000: A new dinoflagellate species from the Upper Cretaceous of New Zealand – a morphological intermediate between three genera. *Alcheringa 24*: 135–146.
RONCAGLIA, L.; FIELD, B. D.; RAINE, J. I.; SCHIØLER, P; WILSON, G. J. 1998: A dinoflagellate approach to correlation of Piripauan–Haumurian (Upper Cretaceous) sections from the South Island of New Zealand. *In*: Smelror, M. (ed.), 1998: Abstracts from the Sixth International Conference on Modern and Fossil Dinoflagellates, Dino 6. *Rapport botanisk serie/Norges teknisk-naturvitenskapelige universitet Vitenskapsmuseet 1998-1*: 129–130.
RONCAGLIA, L.; FIELD, B. D.; RAINE, J. I.; SCHIØLER, P; WILSON, G. J. 1999: Dinoflagellate biostratigraphy of Piripauan-Haumurian (Upper Cretaceous) sections from the northeast South Island, New Zealand. *Cretaceous Research 20*: 271–314.
RONCAGLIA, L.; SCHIØLER, P. 1997: Dinoflagellate biostratigraphy of Piripauan–Haumurian sections in southern Marlborough and northern Canterbury, New Zealand. *Institute of Geological & Nuclear Sciences Science Report 97/09*: 1–50.
RONCAGLIA, L.; SCHIØLER, P. 1999: *Alterbidinium austrinum* Roncaglia et Schiøler, sp. nov., a new dinoflagellate from the Conway Siltstone (Upper Cretaceous), southern Marlborough, New Zealand. *Review of Palaeobotany and Palynology 106*: 121–129.
SALDARRIAGA, J. F.; TAYLOR, F. J. R.; CAVALIER-SMITH, T.; MENDEN-DEUER, S.; KEELING, P.J. 2004: Molecular data and the evolutionary history of dinoflagellates. *European Journal of Protistology 40*: 85–111.
SCHIØLER, P.; CRAMPTON, J.; KING, P. 2000: Palynostratigraphic analysis of a measured

section through the Karekare, Owhena and Whangai formations (Upper Cretaceous) at Waitahaia River –"Owhena Stream" (Raukumara Peninsula). *Institute of Geological & Nuclear Sciences Science Report 2000/31*: 1–24.

SCHIØLER, P.; CRAMPTON, J.; KING, P. 2001: Palynostratigraphic analysis of a measured section through the Karekare, Tahora and Whangai formations (Upper Cretaceous) at Koranga Stream (Raukumara Peninsula). *Institute of Geological & Nuclear Sciences Science Report 2001/27*: 1–20.

SCHIØLER, P.; CRAMPTON, J. S.; LAIRD, M. G. 2002: Palynofacies and sea-level changes in the Middle Coniacian–Late Campanian (Late Cretaceous) of the East Coast Basin, New Zealand. *Palaeogeography, Palaeoclimatology, Palaeoecology 188*: 101–125.

SCHIØLER, P.; ROGERS, K. M.; SYKES, R.; HOLLIS, C. J.; ILG, B. R.; MEADOWS, D.; RONCAGLIA, L.; URUSKI, C. I. 2010: Palynofacies, organic geochemistry and depositional environment of the Tartan Formation (Late Paleocene), a potential source rock in the Great South Basin, New Zealand. *Marine and Petroleum Geology 27*: 351–369.

SCHIØLER, P.; RONCAGLIA, L.; WILSON, G. J. 2001: *Alterbidinium? novozealandicum*, a new dinoflagellate from the Herring Formation (Upper Cretaceous), southern Marlborough, New Zealand. *Neues Jahrbuch für Geologie und Paläontologie, Abhandlungen 219*: 139–152.

SCHIØLER, P.; WILSON, G. J. 1998: Dinoflagellate biostratigraphy of the middle Coniacian-lower Campanian (Upper Cretaceous) in south Marlborough, New Zealand. *Micropaleontology 44*: 313–349.

SCHNEPF, E.; WINTER, S.; MOLLENHAUER, D. 1989: *Gymnodinium aeruginosum* (Dinophyta): a blue-green dinoflagellate with a vestigial, anucleate, cryptophycean endosymbiont. *Plant Systematics and Evolution 164*: 75–92.

SHEARS, N. T.; ROSS, P. M. 2009: Blooms of benthic dinoflagellates of the genus *Ostreopsis*; an increasing and ecologically important phenomenon on temperate reefs in New Zealand and worldwide. *Harmful Algae 8*: 916–925.

SHEW, R.; WONG, C.; THOMAS, M. G.; ELLIS-PEGLER, R. B.; 1995: Imported malaria in Auckland in 1993. *New Zealand Medical Journal 108*: 380–382.

SILBERMAN, J. D.; COLLINS, A. G.; GERSHWIN, L.-A.; JOHNSON, P. J.; ROGER, A. J. 2004: Ellobiopsids of the genus *Thalassomyces* are alveolates. *Journal of Eukaryotic Microbiology 51*: 246–252.

SILVA, E. S.; FAUST, M. A. 1995: Small cells in the life history of dinoflagellates (Dinophyceae): a review. *Phycologia 34*: 396–408.

SKUJA, H. 1976: Zur Kenntnis der Algen Neuseeländischer Torfmoore. *Nova acta Regiae Societatis Scientiarum Upsaliensis, ser. 5, C(2)*: 1–125.

SLUIJS, A.; BRINKHUIS, H.; WILLIAMS, G. L.; FENSOME, R. A. 2009: Taxonomic revision of some Cretaceous–Cenozoic spiny organic-walled peridiniacean dinoflagellate cysts. *Review of Palaeobotany and Palynology 154*: 34 –53.

SNEL, S. J.; BAKER, M. G.; KAMALESH V. 2009: The epidemiology of cryptosporidiosis in New Zealand, 1997-2006. *New Zealand Medical Journal 122*: 47–61.

SOE, A. K.; POMROY, W. E. 1992: New species of *Eimeria* (Apicomplexa: Eimeriidae) from the domesticated goat *Capra hircus* in New Zealand. *Systematic Parasitology 23*: 195–202.

SOURNIA, A. 1995: Red tide and toxic marine phytoplankton of the world ocean: an inquiry into biodiversity. Pp. 103–112 *in*: Lassus, P.; Arzul, G.; Erard-Le Denn, E.; Gentien, P.; Marcaillou-Le-Baut, C. (eds), *Harmful Marine Algal Blooms*. Intercept Ltd, Andover, U.K. xxiv + 878 p.

STANKIEWICZ, M.; COWAN, P. E; HEATH, D. D. 1997b: Endoparasites of brushtail possums (*Trichosurus vulpecula*) from the South Island, New Zealand. *New Zealand Veterinary Journal 45*: 257–260.

STANKIEWICZ, M; HEATH, D. D; COWAN, P. E. 1997a: Internal parasites of possums (*Trichosurus vulpecula*) from Kawau Island, Chatham Island and Stewart Island. *New Zealand Veterinary Journal 45*: 247–50.

STANKIEWICZ, M.; JOWETT, G. H.; ROBERTS, M. G.; HEATH, D. D.; COWAN, P.; CLARK, J. M.; JOWETT, J.; CHARLESTON, W. A. G. 1996: Internal and external parasites of possums (*Trichosurus vulpecula*) from forest and farmland, Wanganui, New Zealand. *New Zealand Journal of Zoology 23*: 345–353.

STARK, D.; BARRATT, J. L. N.; van HAL, S.; MARRIOTT, D.; HARKNESS, J.; ELLIS, J. T. 2009. Clinical significance of enteric protozoa in the immunosuppressed human population. *Clinical Microbiology Reviews 22*: 634–650.

STEFANOGIANNIS, N.; McLEAN, M.; MIL, H. van 2001: Outbreak of cryptosporidiosis linked with a farm event. *New Zealand Medical Journal 114*: 519–521.

STRONG, C. P.; HOLLIS, C. J.; WILSON, G. J. 1995: Foraminiferal, radiolarian, and dinoflagellate biostratigraphy of Late Cretaceous to Middle Eocene pelagic sediments (Muzzle Group), Mead Stream, Marlborough, New Zealand. *New Zealand Journal of Geology and Geophysics 38*: 171–212.

SUN, X.; McMINN, A. 1994: Recent dinoflagellate cyst distribution associated with the Subtropical Convergence on the Chatham Rise, east of New Zealand. *Marine Micropaleontology 23*: 345–356.

SUTHERLAND, R.; HOLLIS, C. J.; NATHAN, S.; STRONG, C. P.; WILSON, G. J. 1996: Age of Jackson Formation proves late Cenozoic allochthony in South Westland, New Zealand. *New Zealand Journal of Geology and Geophysics 39*: 559–563.

TAYLOR, F. J. 1974: A preliminary annotated check list of dinoflagellates from New Zealand coastal waters. *Journal of the Royal Society of New Zealand 4*: 193–202.

TAYLOR, F. J.; TAYLOR, N. J.; WALSBY, J. R. 1985: A bloom of the planktonic diatom, *Cerataulina pelagica*, off the coast of north-eastern New Zealand in 1982–3, and its contribution to an associated mortality of fish and benthic fauna. *International Revue der gesamtem Hydrobiologie und Hydrographie 70*: 773–795.

TAYLOR, F. J. R. 1987: Taxonomy and classification. Pp. 723–731 *in*: Taylor, F. J. R. (ed.), *The Biology of Dinoflagellates*. Blackwell Scientific Publications, Oxford. xii + 785 p.

THOMASSON, K. 1974: Rotorua phytoplankton reconsidered (North Island of New Zealand). *Internationale Revue Gesamten Hydrobiologie 59*: 703–727.

THOMPSON, J. 1991: Theileriasis in New Zealand. *Surveillance 18(5)*: 21–23.

THOMPSON, J.; ORR, M. 1991: Animal Health Laboratory Network: Review of diagnostic cases – January to March 1991. *Surveillance 18(2)*: 3–5.

TWENTYMAN, C. M. 2001: A study of coccidial parasites in the hihi (*Notiomystis cincta*). MVSc thesis, Massey University, Palmerston North.

UNGAR, B. L. P. 1990: Cryptosporidiosis in humans (*Homo sapiens*). Pp. 59–82 *in*: Dubey, J. P.; Speer, C. A.; Fayer, R. (eds), *Cryptosporidiosis of Man and Animals*. CRC Press, Boca Raton. vii + 199 p.

VIDAL, G. 1982: Microfossil studies. *In*: Cooper, R. A.; Grindley, G. W. (eds), Late Proterozoic to Devonian sequences of southeastern Australia, Antarctica and New Zealand and their correlation. *Geological Society of Australia Special Publication 9*: 55–56.

WARHAM, J. 1974: The Fiordland crested penguin *Eudyptes pachyrhynchus*. *Ibis 116*: 1–27.

WEBBER, W. R. 2010: Order Euphausiacea: Krill. Pp. 170–179, 224 *in*: Gordon, D. P. (ed.), *New Zealand Inventory of Biodiversity Volume Two. Kingdom Animalia: Chaetognatha, Ecdysozoa, Ichnofossils.* Canterbury University Press, Christchurch. 528 [+ 16] p.

WILKINS, M. F.; O'CONNELL, E. O.; TE PUNGA, W. A. 1988: Toxoplasmosis in sheep. III. Further evaluation of a live *Toxoplasma gondii* vaccine to prevent lamb losses and reduce congenital infection following experimental oral challenge. *New Zealand Veterinary Journal 36*: 86–89.

WILSON, G. J. 1988: Early Cretaceous dinoflagellate assemblages from Torlesse rocks near Ethelton, North Canterbury. *New Zealand Geological Survey Record 35*: 38–43, 2 pls.

WILLUMSEN, P. S. 2003: Marine palynology across the Cretaceous-Tertiary boundary in New Zealand. PhD thesis in geology, Victoria University of Wellington. 387 p.

WILLUMSEN, P. S. 2004: Two new species of the dinoflagellate cyst genus *Carpatella* Grigorovich, 1969 from the Cretaceous-Tertiary transition in New Zealand. *Journal of Micropalaeontology 23*: 119–125.

WILLUMSEN, P. S.; WILSON, G. J.; HOLLIS, C. J.; SCHIØLER, P.; HANNAH, M. J.; FIELD, B. D.; STRONG, C. P. 2004: Palynofacies and paleoenvironmental changes across the Cretaceous–Tertiary boundary in New Zealand. *In*: Abstracts XI IPC 2004, Universidad de Cordoba. *Polen 14*: 197–198.

WILSON, G. J. 1967a: Some new species of Lower Tertiary dinoflagellates from McMurdo Sound, Antarctica. *New Zealand Journal of Botany 5*: 57–83.

WILSON, G. J. 1967b: Microplankton from the Garden Cove Formation, Campbell Island. *New Zealand Journal of Botany 5*: 223–240.

WILSON, G. J. 1967c: Some species of *Wetzeliella* Eisenack (Dinophyceae) from New Zealand Eocene and Paleocene strata. *New Zealand Journal of Botany 5*: 469–497.

WILSON, G. J. 1968a: Palynology of some Lower Tertiary coal measures in the Waihao district, South Canterbury, New Zealand. *New Zealand Journal of Botany 6*: 56–62.

WILSON, G. J. 1968b: On the occurrence of fossil microspores, pollen grains, and microplankton in bottom sediments of the Ross Sea, Antarctica. *New Zealand Journal of Marine and Freshwater Research 2*: 381–389.

WILSON, G. J. 1972: Age of the Garden Cove Formation, Campbell Island. *New Zealand Journal of Geology and Geophysics 15*: 184–185.

WILSON, G. J. 1973: Palynology of the Middle Pleistocene Te Piki bed, Cape Runaway, New

Zealand. *New Zealand Journal of Geology and Geophysics 16*: 345–354.

WILSON, G. J. 1975: Palynology of deep-sea cores from DSDP Site 275, southeast Campbell Plateau. *In*: Kennett, J.P.; Houtz, R.E. et al. (eds). *Initial reports of the Deep Sea Drilling Project 29*: 1031–1035.

WILSON, G. J. 1976: Permian palynomorphs from the Mangarewa Formation, Productus Creek, Southland, New Zealand. *New Zealand Journal of Geology and Geophysics 19*: 136–140.

WILSON, G. J. 1977: A new species of *Svalbardella* Manum (Dinophyceae) from the Eocene of New Zealand. *New Zealand Journal of Geology and Geophysics 20*: 563–566.

WILSON, G. J. 1978a: The dinoflagellate species *Isabelia druggii* (Stover) and *I. seelandica* (Lange): their association in the Teurian of Woodside Creek, Marlborough, New Zealand. *New Zealand Journal of Geology and Geophysics 21*: 75–80.

WILSON, G. J. 1978b: *Kaiwaradinium*, a new dinoflagellate genus from the Late Jurassic of North Canterbury, New Zealand. *New Zealand Journal of Geology and Geophysics 21*: 81–84, fig. 1-8.

WILSON, G. J. 1982a: Abstracts of unpublished N.Z. Geological survey reports on fossil dinoflagellates 1981–1982. *New Zealand Geological Survey Report PAL 52*: 1–20.

WILSON, G. J. 1982b: Early Tertiary dinoflagellates from Chalky Island, SW Fiordland. *New Zealand Geological Survey Report PAL 56*: 1–12.

WILSON, G. J. 1982c: Dinoflagellate assemblages from the Puaroan, Ohauan and Heterian stages (Late Jurassic) Kaiwara Valley, North Canterbury, N.Z. *New Zealand Geological Survey Report PAL 59*: 1–20.

WILSON, G. J. 1982d: Eocene and Oligocene dinoflagellate assemblages from the Oamaru area, North Otago, N.Z. *New Zealand Geological Survey Report PAL 60*: 1–20.

WILSON, G. J. 1982e: Early Tertiary dinoflagellate assemblages from the Oxford area, Central Canterbury and their bearing on the age of associated basaltic volcanics. *New Zealand Geological Survey Report PAL 61*: 1–10, 2 pls.

WILSON, G. J. 1984a: Two new dinoflagellates from the Late Jurassic of North Canterbury, New Zealand. *Journal of the Royal Society of New Zealand 14*: 215–221.

WILSON, G. J. 1984b: A new Paleocene dinoflagellate cyst from the Chatham Islands, New Zealand. *New Zealand Journal of Botany 22*: 545–547.

WILSON, G. J. 1984c: Some new dinoflagellate species from the New Zealand Haumurian and Piripauan stages (Santonian-Maastrichtian, Late Cretaceous). *New Zealand Journal of Botany 22*: 549–556.

WILSON, G. J. 1984d: New Zealand Late Jurassic to Eocene dinoflagellate biostratigraphy – a summary. *Newsletters on Stratigraphy 13*: 104–117.

WILSON, G. J. 1985: Dinoflagellate biostratigraphy of the Eocene Hampden Section, North Otago, New Zealand. *New Zealand Geological Survey Record 8*: 93–101, pl. 1, fig. 1-19.

WILSON, G. J. 1986: New Zealand Triassic dinoflagellates – a preliminary survey. *Geological Society of New Zealand Newsletter 71*: 48–49.

WILSON, G. J. 1987: Dinoflagellate biostratigraphy of the Cretaceous-Tertiary boundary, mid-Waipara River section, North Canterbury, New Zealand. *New Zealand Geological Survey Record 20*: 8–15.

WILSON, G. J. 1988a: Paleocene and Eocene dinoflagellate cysts from Waipawa, Hawkes Bay, New Zealand. *New Zealand Geological Survey Paleontological Bulletin 57*: 1–96, 26 pls.

WILSON, G. J. 1988b: Early Cretaceous dinoflagellate assemblages from Torlesse rocks near Ethelton, North Canterbury. *New Zealand Geological Survey Record 35*: 38–43, 2 pls.

WILSON, G. J. 1989a: Dinoflagellate assemblages from the Urewera Group. *New Zealand Journal of Geology and Geophysics 32*: 525–526.

WILSON, G. J. 1989b: Marine palynology. *In*: Barrett, P. J. (ed.), 1989: Antarctic Cenozoic history from the CIROS-1 drillhole, McMurdo Sound. *Department of Scientific and Industrial Research Bulletin 245*: 129–133, pl. 1.

WILSON, G. J.; HELBY, R. 1987: A probable Oxfordian dinoflagellate assemblage from North Canterbury, New Zealand. *New Zealand Geological Survey Record 20*: 119–125.

WILSON, G. J.; McMILLAN, S. G. 1996: Late Cretaceous-Tertiary stratigraphic sections of coastal Otago, South Island, New Zealand: a summary of biostratigraphic and lithostratigraphic data. *Institute of Geological and Nuclear Sciences Science Report 96/39*: 1–182.

WILSON, G. J.; MOORE, P. R.; 1988: Cretaceous–Tertiary boundary in the Te Hoe River area, western Hawkes Bay. *New Zealand Geological Survey Record 35*: 34–37.

WILSON, G. J.; MOORE, P. R.; ISAAC, M. J. 1988: Age of greywacke basement in the Urewera Ranges, eastern North Island. *New Zealand Geological Survey Record 35*: 29–33.

WILSON, G. J.; MORGANS, H. E. G.; MOORE, P. R. 1989: Cretaceous-Tertiary boundary at Tawanui, southern Hawkes Bay, New Zealand. *New Zealand Geological Survey Record 40*: 29–40.

WILSON, G. J.; SCHIØLER, P.; HILLER, N.; JONES, C. M. 2005: Age and provenance of Cretaceous marine reptiles from the South Island and Chatham Islands, New Zealand. *New Zealand Journal of Geology and Geophysics 48*: 377–387.

WINTERBOURN, M. J.; BROWN, T. J. 1967: Observations on the faunas of two warm streams in the Taupo thermal region. *New Zealand Journal of Marine and Freshwater Research 1*: 38–50.

WIZEVICH, M. C.; THRASHER, G. P.; BUSSELL, M. R.; WILSON, G. J.; COLLEN, J. D. 1992: Evidence for marine deposition in the Late Cretaceous Pakawau Group, northwest Nelson. *New Zealand Journal of Geology and Geophysics 35*: 363–369.

YASUMOTO, T.; HASHIMOTO, Y.; BAGNIS, R.; RANDALL, J. E.; BANNER, A. H. 1971: Toxicity of surgeon fishes. *Bulletin of the Japanese Society of Scientific Fisheries 43*: 1021–1026.

YASUMOTO, T.; SATAKE, M. 1998: New toxins and their toxicological evaluations. Pp. 461–464 *in*: Reguera, B.; Blanco, J.; Fernandez, M. L.; Wyatt, T. (eds), *Harmful Algae*. Xunta de Galicia and Intergovernmental Oceanographic Commission, UNESCO, Paris. xv + 635 p.

YOON, H. S.; HACKETT, J. D.; VAN DOLAH, F. M.; NOSENKO, T.; LIDIE, K. L.; BHATTACHARYAH, D. 2005: Tertiary endosymbiosis driven genome evolution in dinoflagellate algae. *Molecular Biology and Evolution 22*: 1299–1308.

ZELDIS, J.; GALL, M.; UDDSTROM M.; GREIG, M. 2000: *La Niña* shuts down upwelling in northeastern New Zealand. *Water and Atmosphere 8(2)*: 15–18.

ZERVOS, S. 1989: Stadial and seasonal occurrence of gregarines and nematomorphs in two New Zealand cockroaches. *New Zealand Journal of Zoology 16*: 143–146.

Checklist of New Zealand living Myzozoa

The single-letter codes indicate the following: A, adventive (introduced and naturalised) species; E, endemic species. Habitat codes (for hosts in the case of sporozoan parasites) are: F, freshwater; M, marine; T, terrestrial. In the case of parasitic species, the hosts are indicated as follows: B, bird; Fi, fish; I, invertebrate; Ma, mammal; R, reptile. A question mark (?) indicates uncertainty concerning the whole binominal (genus and species). Where the occurrence of an organism has been recorded only at generic level, the host concerned is indicated in full. The overall higher-level classification for the phylum and dinoflagellates is that of Cavalier-Smith (2004) and Cavalier-Smith and Chao (2004). The list of marine dinoflagellate species updates that given in Dawson (1992), which should be consulted for synonyms. Dinoflagellate families and genera mostly follow Algaebase but family endings are based on the International Code on Zoological Nomenclature, rather than the Botanical Code. Classification for Apicomplexa is that presented by Perkins et al. (2000), which updates that in Levine (1985, 1988), with alternative names in parentheses (Cox 1991; Cavalier-Smith & Chao 2004).

KINGDOM CHROMISTA
SUBKINGDOM HAROSA
INFRAKINGDOM ALVEOLATA
PHYLUM MYZOZOA
SUBPHYLUM DINOZOA
INFRAPHYLUM PROTALVEOLATA
Class COLPONEMEA
Order COLPONEMIDA
Colponema loxodes Stein, 1878 F

Class PERKINSEA
Order PERKINSIDA
PERKINSIDAE
Perkinsus olseni Lester & Davis, 1981 M

Class ELLOBIOPSEA
Order ELLOBIOPSIDA
ELLOBIOPSIDAE
Thalassomyces fagei (Boschma, 1949) M
Thalassomyces marsupii Kane, 1964 M

INFRAPHYLUM DINOFLAGELLATA
Superclass DINOKARYOTA
Class NOCTILUCEA
Order NOCTILUCIDA
NOCTILUCIDAE
Noctiluca scintillans (Macartney, 1810) M
PRONOCTILUCIDAE
Pronoctiluca acuta (Lohmann, 1913) M

Class PERIDINEA
Subclass PERIDINOIDIA
Order PROROCENTRIDA
PROROCENTRIDAE
Mesoporus adriaticus (Schiller, 1928) M
Prorocentrum balticum (Lohmann, 1908) M
Prorocentrum compressum Bailey, 1850 M
Prorocentrum dentatum Stein, 1883 M
Prorocentrum gracile Schütt, 1895 M
Prorocentrum lima (Ehrenberg, 1859) M
Prorocentrum micans Ehrenberg, 1833 M
Prorocentrum marina (Cienkowski, 1881) M
Prorocentrum rostratum Stein, 1883 M
Prorocentrum rotundatum Schiller, 1918 M
Prorocentrum triestinum Schiller, 1918 M

Order PERIDINIIDA
Suborder PERIDINIINA
CALCIODINELLIDAE
Scrippsiella trochoidea Stein, 1883 M
Thoracosphaera heimii (Lohmann, 1920) M
CENTRODINIIDAE
Centrodinium pacificum Rampi, 1950 M
Centrodinium splendidum (Rampi, 1941) M
GLENODINIIDAE
Glenodinium sp. indet. Taylor 1973, 1978 M
GLENODINIOPSIDIDAE
Glenodiniopsis steinii (Lemmermann, 1900) F
Sphaerodinium cinctum (Ehrenberg, 1838) F
HETEROCAPSIDAE
Heterocapsa illdefina (Herman & Sweeney, 1976) M
Heterocapsa niei (Loeblich, 1968) M
Heterocapsa triquetra (Ehrenberg, 1840) M
OXYTOXIDAE
Corythodinium compressum Kofoid, 1907 M
Corythodinium elegans Pavillard, 1916 M
Oxytoxum gracile Schiller, 1937 M
Oxytoxum laticeps Schiller, 1937 M
Oxytoxum longiceps Schiller, 1937 M
Oxytoxum pachyderme Schiller, 1937 M
Oxytoxum scolopax Stein, 1883 M
Oxytoxum sphaeroideum Stein, 1883 M
Oxytoxum turbo Kofoid, 1907 M
PERIDINIIDAE
Peridiniopsis borgei Lemmermann, 1910 F
Peridiniopsis dinobryonis (Woloszynska, 1916) F
Peridiniopsis elpatiewskyi (Ostenfeld, 1907) F
Peridiniopsis penardii (Lemmermann, 1900) F
Peridiniopsis quadridens (Stein, 1883) F
Peridinium aciculiferum Lemmermann, 1908 F
Peridinium africanum Lemmermann, 1907 F
Peridinium ampulliforme (Wood, 1954) M
Peridinium bipes Stein, 1883 F
Peridinium cinctum (Müller, 1773) F
Peridinium lomnickii Woloszynska, 1916 F
— var. *wierjeskii* Woloszynska, 1916 F
Peridinium raciborskii Woloszynska, 1912 F
Peridinium striolatum Playfair, 1919 F
Peridinium sydneyense Thomasson, 1974 F
Peridinium umbonatum Stein, 1883 F
Peridinium wierjeskii Woloszynska, 1916 F
Peridinium willei Huitfeldt-Kaas, 1900 F
PFIESTERIIDAE
Pfiesteria piscicida Steidinger & Burkholder *in* Steidinger, Burkholder, Glasgow, Hobbs, Garrett, Truby, Noga & Smith, 1996 M
Pseudopfiesteria shumwayae (Glasgow & Burkholder in Glasgow, Burkholder, Morton & Springer, 2001) M
Pfiesteria n. sp. Rhodes & Rublee pers. comm. M
PODOLAMPADIDAE
Podolampas curvatus Schiller, 1937 M
Podolampas elegans Schütt, 1895 M
Podolampas palmipes Stein, 1883 M
Podolampas spinifera Okamura, 1912 M
PROTOPERIDINIIDAE
Diplopsalis lenticula Bergh, 1882 M
Oblea rotunda (Lebour, 1922) M
Protoperidinium affine Balech, 1958 M
Protoperidinium applanatum (Mangin, 1915) M
Protoperidinium breve (Paulsen, 1905) M
Protoperidinium brevipes (Paulsen, 1908) M
Protoperidinium brochii (Kofoid & Swezy, 1921) M
Protoperidinium claudicans (Paulsen, 1907) M
Protoperidinium conicoides (Paulsen, 1905) M
Protoperidinium conicum (Gran, 1900) M
Protoperidinium crassipes (Kofoid, 1907) M
Protoperidinium curtum (Balech, 1958) M
Protoperidinium curvipes (Ostenfeld, 1903) M
Protoperidinium decipiens (Jörgensen, 1899) M
Protoperidinium depressum (Bailey, 1855) M
Protoperidinium diabolum (Cleve, 1900) M
Protoperidinium divergens (Ehrenberg, 1840) M
Protoperidinium granii (Ostenfeld, 1906) M
Protoperidinium humile (Schiller, 1937) M
Protoperidinium latissimum Kofoid, 1907 M
Protoperidinium leonis (Pavillard, 1916) M
Protoperidinium marielebourae (Paulsen, 1930) M
Protoperidinium oblongum (Aurivillius, 1898) M
Protoperidinium obtusum (Karsten, 1906) M
Protoperidinium oceanicum (Vanhöffen, 1897) M
Protoperidinium ovatum Pouchet, 1883 M
Protoperidinium pallidum (Ostenfeld, 1899) M
Protoperidinium pedunculatum (Schütt, 1895) M
Protoperidinium pellucidum Bergh, 1881 M
Protoperidinium pentagonum (Gran, 1902) M
Protoperidinium cf. *punctulatum* (Paulsen, 1907) M
Protoperidinium pyriforme (Paulsen, 1904) M
Protoperidinium quarnerense (Schröder, 1900) M
Protoperidinium roseum (Paulsen, 1904) M
Protoperidinium steinii (Jörgensen, 1899) M
Protoperidinium stellatum Wall, 1968 M
Protoperidinium subinerme (Paulsen, 1904) M
Protoperidinium turbinatum Mangin, 1974 M
Protoperidinium variegatum (Peters, 1928) M
THECADINIIDAE
Thecadinium petasatum (Herdman, 1922) M

Subclass DINOPHYSOIDIA
Order DINOPHYSIDA
DINOPHYSIDIDAE
Dinophysis acuminata Claparède & Lachmann, 1859 M
Dinophysis acuta Ehrenberg, 1839 M
Dinophysis amandula Sournia, 1973 M
Dinophysis caudata Kent, 1881 M
Dinophysis fortii Pavillard, 1923 M
Dinophysis ovum Schütt, 1895 M
Dinophysis recurva Kofoid & Skogsberg, 1928 M
Dinophysis sacculus Stein, 1883 M
Dinophysis spinosa Rampi, 1950 M
Dinophysis tripos Gourret, 1883 M

Dinophysis truncata Cleve, 1901 M
Histioneis hyalina Kofoid & Michener, 1911 M
Histioneis paulsenii Kofoid, 1907 M
Histioneis variabilis Schiller, 1933 M
Ornithocercus sp. indet. Stein 1883 M
Phalacroma apicatum Kofoid & Skogsberg, 1928 M
Phalacroma pulchellum Lebour, 1906 M
Phalacroma rotundatum (Claparède & Lachmann, 1859) M

Subclass GONYAULACOIDIA
Order GONYAULACIDA
CERATIIDAE
Ceratium arietinum Cleve, 1900 M
Ceratium axiale Kofoid, 1907 M
Ceratium azoricum Cleve, 1900 M
Ceratium bigelowi Kofoid, 1907 M
Ceratium breve (Ostenfeld & Schmidt, 1902) M
Ceratium buceros (Zacharias, 1906) M
Ceratium candelabrum (Ehrenberg, 1859) M
Ceratium claviger Kofoid, 1907 M
Ceratium concilians Jörgensen, 1920 M
Ceratium contrarium (Gourret, 1883) M
Ceratium declinatum Karsten, 1907 M
Ceratium extensum (Gourret, 1883) M
Ceratium falcatiforme Jörgensen, 1920 M
Ceratium falcatum (Kofoid, 1907) M
Ceratium furca (Ehrenberg, 1833) M
Ceratium furcoides (Levander, 1900) F
Ceratium fusus (Ehrenberg, 1833) M
Ceratium gibberum Gourret, 1883 M
Ceratium gravidum Gourrett, 1883 M
Ceratium hexacanthum Gourret, 1883 M
Ceratium hirundinella (Müller, 1773) F
Ceratium horridum Gran, 1902 M
Ceratium inflatum (Kofoid, 1907) M
Ceratium karsteni Pavillard, 1907 M
Ceratium lineatum (Ehrenberg, 1853) M
Ceratium longirostrum Gourret, 1883 M
Ceratium macroceros (Ehrenberg, 1840) M
Ceratium massiliense Gourret, 1883 M
Ceratium minutum Jörgensen, 1920 M
Ceratium pentagonum Gourret, 1883 M
Ceratium petersii Nielsen, 1934 M
Ceratium platycorne von Daday, 1888 M
Ceratium porrectum Karsten, 1907 M
Ceratium pulchellum Schröder, 1906 M
Ceratium ranipes Cleve, 1900 M
Ceratium semipulchellum (Jörgensen, 1920) M
Ceratium setaceum Jörgensen, 1911 M
Ceratium symmetricum Pavillard, 1905 M
Ceratium teres Kofoid, 1907 M
Ceratium tripos (Müller, 1776) M
CERATOCORYIDAE
Ceratocorys gourreti Paulsen, 1930 M
CLADOPYXIDIDAE
Acanthodinium spinosum (Kofoid, 1907) M
Palaeophalacroma unicinctum Schiller, 1928 M
GONIODOMATIDAE
Gambierdiscus toxicus Adachi & Fukuyo, 1979 M
Goniodoma polyedricum (Pouchet, 1883) M
Goniodoma sphaericum (Murray & Whitting, 1899) M
GONYAULACIDAE
Alexandrium affine (Inoue & Fukuyo *in* Fukuyo, Yoshida & Inoue, 1985) M
Alexandrium angustitabulatum Taylor *in* Balech, 1995 M
Alexandrium catenella (Whedon & Kofoid, 1936) M
Alexandrium camurascutulum MacKenzie & Todd, 2002 M
Alexandrium cf. *cohorticula* Balech, 1985/ *tariyavanichi* Balech, 1994 M
Alexandrium concavum (Gaarder, 1954) M
Alexandrium margalefii Balech, 1992 M
Alexandrium minutum Halim, 1960 M
Alexandrium ostenfeldii (Paulsen, 1904) M
Alexandrium pseudogonyaulax (Biecheler, 1952) M
Alexandrium tamarense (Lebour, 1925) M
Gonyaulax alaskensis Kofoid, 1911 M
Gonyaulax diegensis Kofoid, 1911 M
Gonyaulax digitale (Pouchet, 1883) M
Gonyaulax grindleyi Reinecke, 1967 M
Gonyaulax hyalina Ostenfeld & Schmidt, 1902 M
Gonyaulax inflata (Kofoid, 1907) M
Gonyaulax minima Matzenauer, 1933 M
Gonyaulax monacantha Pavillard, 1916 M
Gonyaulax polygramma Stein, 1883 M
Gonyaulax cf. *sphaeroidea* Kofoid, 1911 M
Gonyaulax spinifera (Claparède & Lachmann, 1859) M
Gonyaulax turbynei Murray & Whitting, 1899 M
Lingulodinium polyedrum (Stein, 1883) M
Protoceratium reticulatum (Claparède & Lachmann, 1859) M
HETERODINIIDAE
Heterodinium dubium Rampi, 1941 M
Heterodinium inaequale (Schiller, 1937) M
Heterodinium minutum Kofoid & Michener, 1911 M
KARENIIDAE
Karenia bidigitata Haywood & Steidinger *in* Haywood, Steininger, Truby, Bergquist, Bergquist, Adamson & McKenzie, 2004 M
Karenia brevisulcata (Chang, 1999) M
Karenia concordia Chang & Ryan, 2004 M
Karenia mikimotoi (Miyake & Kominami *in* Oda, 1935) M
Karenia papilonacea Haywood & Steidinger *in* Haywood, Steininger, Truby, Bergquist, Bergquist, Adamson & McKenzie, 2004 M
Karenia selliformis Haywood, Steidinger & MacKenzie *in* Haywood, Steininger, Truby, Bergquist, Bergquist, Adamson & McKenzie, 2004 M
Karenia umbella de Salas, Bolch & Hallegraeff, 2004 M
Karlodinium veneficum (Ballantine, 1956) M
Takayama helix de Salas, Bolch, Botes & Hallegraeff *in* de Salas, Bolch, Botes, Nash, Wright & Halegraeff, 2003 M
Takayama pulchella (Larsen, 1994) M
Takayama tasmanica de Salas, Bolch & Hallegraeff *in* de Salas, Bolch, Botes, Nash, Wright & Halegraeff, 2003 M
OSTREOPSIDIDAE
Coolia monotis Meunier, 1919 M
Ostreopsis siamensis Schmidt, 1901 M
Ostreopsis lenticularis Fukuyo, 1981 M
Ostreopsis ovata Fukuyo, 1981 M
PYROPHACIDAE
Fragilidium subglobosum von Stosch, 1969 M
Pyrophacus horologium Stein, 1883 M
Pyrophacus steinii Schiller, 1935 M

Order GYMNODINIIDA
Suborder GYMNODINIINA
GYMNODINIIDAE
Akashiwo sanguinea (Hirasaka, 1922) M
Amphidinium acutissimum Schiller, 1933 M
Amphidinium acutum Lohmann, 1920 M
Amphidinium aloxalocium Norris, 1961 M
Amphidinium amphidinioides (Geitler, 1924) M
Amphidinium carteri Hulburt, 1957 M
Amphidinium emarginatum Diesing, 1866 M
Amphidinium cf. *extensum* Wulff, 1916 M
Amphidinium flagellans Schiller, 1928 M
Amphidinium cf. *massartii* Rhodes & Murray pers. comm. M
Amphidinium operculatum Claparède & Lachmann, 1859 M
Amphidinium sphenoides Wulf, 1916 M
Amphidinium trulla Murray, Jorgensen, Daubjerg & Rhodes, 2004 M
Cochlodinium brandtii Wulff, 1916 M
Cochlodinium polykrikoides Margelef, 1961 M
Gymnodinium aeruginosum F. Stein F
Gymnodinium aureolum (Hulburt, 1957) M
Gymnodinium catenatum Graham, 1943 M
Gymnodinium cinctum Kofoid & Swezy, 1921 M
Gymnodinium diamphidium Norris, 1961 M
Gymnodinium flavum Kofoid & Swezy, 1921 M
Gymnodinium fuscum (Ehrenberg, 1835) F
Gymnodinium galeaeforme Matzenauer, 1933 M
Gymnodinium grammaticum Pouchet, 1887 M
Gymnodinium hamulus Kofoid & Swezy, 1921 M
Gymnodinium helveticum Penard, 1891 F
Gymnodinium impudicum (Fraga & Bravo, 1995) M
Gymnodinium leptum Norris, 1961 M
Gymnodinium cf. *marinum* Kent, 1881 M
Gymnodinium minor Lebour, 1917 M
Gymnodinium nanum Schiller, 1928 M
Gymnodinium pulchellum Larsen, 1994 M
Gymnodinium punctatum Pouchet, 1887 M
Gymnodinium pygmaeum Lebour, 1925 M
Gymnodinium simplex Lohmann, 1908 M
Gymnodinium uberrimum (Allman, 1854) F
Gymnodinium varians Maskell, 1887 M
Gyrodinium apidiomorphum Norris, 1961 M
Gyrodinium biconicum Kofoid & Swezy, 1921 M
Gyrodinium hyalinum (Schilling, 1891) M
Gyrodinium kofoidii Norris, 1961 M
Gyrodinium lachryma (Meunier, 1910) M
Gyrodinium ovum (Schütt, 1895) M
Gyrodinium phorkorium Norris, 1961 M
Gyrodinium spirale (Bergh, 1881) M
Katodinium bohemicum (Fott, 1938) F
Katodinium glaucum (Lebour, 1917) M
Katodinium hyperxanthum (Harris, 1940) F
Katodinium vorticellum (Stein, 1883) F
Togula jolla Jørgensen, Murray & Daubjerg, 2004 M
Torodinium teredo (Pouchet, 1885) M
HEMIDINIIDAE
Hemidinium nasutum Stein, 1878 F
POLYKRIKIDAE
Polykrikos kofoidii Chatton, 1914 M
Polykrikos schwartzii Bütschli, 1873 M
PYROCYSTIDIDAE
Pyrocystis lunula (Schütt, 1895) M

Suborder DINOCOCCINA
DINOCOCCIDAE
Cystodinedria inermis (Geitler, 1943) F
Cystodinium cornifax (Schilling, 1891) F
Cystodinium iners Geitler, 1928 F
Cystodinium sp. indet. Taylor 1978 M
Dinococcus bicornis (Woloszynska, 1919) F
Dinococcus oedogonii (Richter, 1897) F
Gloeodinium montanum Klebs, 1912 F
Phytodinium simplex Klebs, 1912 F
Stylodinium globosum Klebs, 1912 F
Tetradinium javanicum Klebs, 1912 F

Suborder PTYCHODISCINA
PTYCHODISCIDAE
Ptychodiscus noctiluca Stein, 1883 M

Suborder CHYTRIODINIINA
CHYTRIODINIIDAE
Paulsenella chaetoceratis (Paulsen, 1911) M

Subclass SUESSIOIDIA
Order SUESSIIDA
SYMBIODINIIDAE
Symbiodinium sp. Davy pers. comm. 2005 M
WOLOSZYNSKIIDAE

Woloszynskia neglecta (Schilling, 1891) F
Woloszynskia pascheri (Suchlandt, 1916) F
Woloszynskia ordinata (Skuja, 1939) F

Subclass OXYRRHIA
Order OXYRRHIDA
OXYRRHIDIDAE
Oxyrrhis marina Dujardin, 1841 M

SUBPHYLUM APICOMPLEXA
INFRAPHYLUM SPOROZOA
Class CONOIDASIDA (COCCIDEA)
Subclass COCCIDIASINA
Order EUCOCCIDIORIDA (EIMERIIDA)
ATOXOPLASMATIDAE
Atoxoplasma adiei (Aragão, 1911) TB A?
Atoxoplasma paddae (Aragão, 1911) TB A?
Atoxoplasma spp. indet. (~2) TB E/A
Kaka (*Nestor meridionalis*)
?Stitchbird (hihi) (*Notiomystis cincta*)
Greenfinch (*Carduelis chloris*)
House sparrow (*Passer domesticus*)
CRYPTOSPORIDIIDAE
Cryptosporidium parvum Tyzzer, 1912 TB/Ma A
Probably = *Cryptosporidium* sp. indet.
Brushtail possum (*Trichosurus vulpecula*)
House mouse (*Mus musculus*)
Ship rat) (*Rattus rattus*)
Dog (*Canis familiaris*)
Horse (*Equus caballus*)
House cat (*Felis catus*)
Cattle (*Bos taurus*)
Goat (*Capra hircus*)
Sheep (*Ovis aries*)
Red deer (*Cervus elaphus*)
Pig (*Sus scrofa*)
Man (*Homo sapiens*)
House sparrow (*Passer domesticus*)
Lovebird (*Agapornis* sp.)
Song thrush (*Turdus philomelos*)
Starling (*Sturnus vulgaris*)
Domestic fowl (*Gallus domesticus*)
Ostrich (*Struthio camelus*)]
EIMERIIDAE
Cyclospora cayatenensis Ortega, Gilman & Sterling, 1994 TMa A
Eimeria acervulina Tyzzer, 1929 TB A
Eimeria ahsata Honess, 1942 TMa A
Eimeria alabamensis Christensen, 1941 TM A
Eimeria alijevi Musaev, 1970 TMa A
Eimeria alpacae Guerrero, 1967 TMa A
Eimeria apsheronica Musaev, 1970 TMa A
Eimeria arloingi (Marotel, 1905) Martin 1909 TMa A
Eimeria auburnensis Christensen & Porter, 1939 TMa A
Eimeria bovis (Züblin, 1908) TMa A
Eimeria brasiliensis Torres & Ramos, 1939 TMa A
Eimeria brunetti Levine, 1942 TB A
Eimeria bukidnonensis Tubangui, 1931 TMa A
Eimeria canadensis Bruce, 1921 TMa A
Eimeria capralis Soe & Pomroy, 1992 TMa A
Eimeria caprina Lima, 1979 TMa A
Eimeria caprovina Lima, 1980 TMa A
Eimeria cerdonis Vetterling, 1965 TMa A
Eimeria charlestoni Soe & Pomroy, 1992 TMa A
Eimeria christenseni Levine, Ivens & Fritz, 1962 TMa A
Eimeria crandallis Honess, 1942 TMa A
Eimeria cylindrica Wilson, 1931 TMa A
Eimeria debliecki Douwes, 1921 TMa A
Eimeria duodenalis Norton, 1967 TB A
Eimeria ellipsoidalis Becker & Frye, 1929 TMa A
Eimeria faurei (Moussu & Marotel, 1902) TMa A
Eimeria flavescens Marotel & Guilhon, 1941 TMa A
Eimeria furonis Hoare, 1927 TMa A
Eimeria granulosa Christensen, 1938 TMa A
Eimeria hirci Chevalier, 1966 TMa A
Eimeria intricata Spiegl, 1925 TMa A
Eimeria irresidua Kessel & Jankiewicz, 1931 TMa A
Eimeria jolchijevi Musaev, 1970 TMa A
Eimeria kofoidi Yakimoff & Matikaschwili, 1936 TB A
Eimeria labbeana Pinto, 1928 TB A
Eimeria lamae Guerrero, 1967 TMa A
Eimeria leuckarti (Flesch, 1883) Reichenow, 1940 TMa A
Eimeria macropodis Wenyon & Scott, 1925 TMa A
Eimeria masseyensis Soe & Pomroy, 1992 TMa A
Eimeria maxima Tyzzer, 1929 TB A
Eimeria media Kessel, 1929 TMa A
Eimeria meleagridis Tyzzer, 1927 TB A
Eimeria meleagrimitis Tyzzer, 1927 TB A
Eimeria necatrix Johnson, 1930 TB A
Eimeria neodebliecki Vetterling, 1965 TMa A
Eimeria ninakohlyakimovae Yakimoff & Rastegaieff, 1930 TMa A
Eimeria ovina Krylov, 1961 TMa A
Eimeria ovinoidalis McDougald, 1979 TMa A
Eimeria pacifica Ormsbee, 1939 TB A
Eimeria pallida Christensen, 1938 TMa A
Eimeria parva Kotlán, Mócsy & Vajda, 1929 TMa A
Eimeria perforans (Leuckart, 1879) TMa A
Eimeria phasiani Tyzzer, 1929 TB A
Eimeria piriformis Kotlan & Pospesch, 1934 TMa A
Eimeria porci Vetterling, 1965 TMa A
Eimeria punctata Landers, 1955 TMa A
Eimeria punoensis Guerrero, 1967 TMa A
Eimeria scabra Henry, 1931 TMa A
Eimeria stiedai (Lindemann, 1895) TMa A
Eimeria subspherica Christensen, 1941 TMa A
Eimeria suis Nöller, 1921 TMa A
Eimeria tenella (Railliet & Lucet, 1891) TB A
Eimeria trichosuri O'Callaghan & O'Donohue, 2001 TMa A
Eimeria tunisiensis Musaev & Mamedova, 1981 TMa A
Eimeria weybridgensis Norton, Joyner & Catchpole, 1974 TMa A
Eimeria wyomingensis Huizinga & Winger, 1942 TMa A
Eimeria zuernii (Rivolta, 1878) TMa A
Eimeria sp. indet. MFi
Southern blue whiting (*Micromesistius australis*) Hine et al. 2000
Eimeria spp. indet. (~8) M/TB/Ma/R A/E
Brushtail opossum (*Trichosurus vulpecula*)
Fallow deer (*Dama dama*)
Red deer (*Cervus elaphus*)
Northern brown kiwi (*Apteryx mantelli*)
Yellow-eyed penguin (*Megadyptes antipodes*)
Black stilt (*Himantopus novaezelandiae*)
New Zealand pigeon (*Hemiphaga novaeseelandiae*) Peafowl (*Pavo cristatus*)
Common gecko (*Hoplodactylus maculatus*)
Epieimeria anguillae Léger & Hollande, 1922 FFi
Isospora canis Nemeséri, 1959 TMa A
Isospora felis Wenyon, 1923 TMa A
Isospora laidlawi Hoare, 1927 TMa A
Isospora ohioensis Dubey, 1975 TMa A
Isospora rivolta (Grassi, 1879) TMa A
Isospora suis Biester, 1934 TMa A
Isospora turdi Schwalbach, 1959 TB A
Isospora spp. indet. (~3) TB/Ma A
Cattle (*Bos taurus*)
Greenfinch (*Carduelis chloris*)
Ostrich (*Struthio camelus*)
HAEMOGREGARINIDAE
Haemogregarina bigemina Laveran & Mesnil, 1901 MFi
Haemogregarina coelorhynchi Laird, 1952 MFi
Haemogregarina hoplichthys Laird, 1952MFi
Haemogregarina leptoscopi Laird, 1952 MFi E
Haemogregarina tuatarae Laird, 1950 TR E
Haemogregarina spp. indet. (~2) TR E
Common gecko (*Hoplodactylus maculatus*)
Duvaucel's gecko (*Hoplodactylus duvaucelii*)
Pacific gecko (*Hoplodactylus pacificus*)
Common skink (*Oligosoma nigriplantare*)
Speckled skink (*Oligosoma infrapunctatum*)
Spotted skink (*Oligosoma lineoocellatum*)
Hepatozoon? acanthoclini Laird, 1953 MFi E
Hepatozoon cuniculi (Sangiorgi, 1914) Wenyon, 1926 TMa A
Hepatozoon lygosomarum (Doré, 1919) TR E
Hepatozoon musculi (Porter, 1908) Wenyon, 1926 TMa A
Hepatozoon kiwii Peirce, Jakob-Hoff & Twentyman, 2003 TB E
KLOSSIELLIDAE
Klossiella equi Baumann, 1946 TMa A
SARCOCYSTIDAE
Besnoitia cf. wallacei (Tadros & Laarman, 1976) TMa A
Hammondia heydorni (Tadros & Laarman, 1976) TMa A
Neospora caninum Dubey, Carpenter, Speer, Topper & Uggla, 1988 TMa A
Sarcocystis arieticanis Heydorn, 1985 TMa A
Sarcocystis capracanis Fischer, 1979 TMa A
Sarcocystis cruzi (Hasselman, 1923) TMa A
Sarcocystis gigantea (Railliet, 1886) TMa A
Sarcocystis hirsuta Moulé, 1888 TMa A
Sarcocystis medusiformis Collins, Atkinson & Charleston, 1979 TMa A
Sarcocystis muris (Railliet, 1886) TMa A
Sarcocystis tenella (Railliet, 1886) TMa A
Sarcocystis spp. indet. (~7) TB/Ma A
Sporocysts: Dog (*Canis familiaris*)
House cat (*Felis catus*)
Sarcocysts: Alpaca (*Lamos pacos*)
Ship rat (*Rattus rattus*)
Horse (*Equus caballus*)
Norway rat (*Rattus norvegicus*)
Pig (*Sus scrofa*)
Rabbit (*Oryctolagus cuniculus*)
Red deer (*Cervus elaphus*)
New Zealand pigeon (*Hemiphaga novaeseelandiae*)
Toxoplasma gondii (Nicolle & Manceaux, 1908) TB/Ma A
Toxoplasma spp. indet. (~4) TB/Ma A/E
Kaka (*Nestor meridionalis*)
Little spotted kiwi (*Apteryx owenii*)
Squirrel monkey (*Saimiri* sp.)
Dama wallaby (*Macropus eugenii*)

[Class GREGARINEA]
Subclass GREGARINASINA
Order EUGREGARINORIDA (EUGREGARINIDA)
Suborder ASEPTATORINA
DIPLOCYSTIDAE
Diplocystis oxycani Dumbleton, 1949 TI E
ENTEROCYSTIDAE
Enterocystis n. sp. Fitzgerald 1952 T E
MONOCYSTIDAE
Gen. et spp. indet. (>1) K. Lee pers. comm. TI A?/E?

Suborder SEPTATORINA
ACTINOCEPHALIDAE
Stictospora costelytrae Allison, 1969 TI E
DIDYMOPHYIDAE
Euspora zealandica Allison, 1969 TI E
Euspora sp. TI E
GREGARINIDAE

Gregarina n. sp. Fitzgerald 1952 T E
Pheloneis spp.
Gregarina? spp. indet. (2) Zervos 1989 TI 2E
HIRMOCYSTIDAE
Hirmocystis pteryospora Crumpton, 1974 TI E
POROSPORIDAE
Nematopsis n. sp. Jones 1976 MI E
STENOPHORIDAE
Stenophora n. sp. Fitzgerald 1952 T E
Diplopoda
Gen. et. sp. indet. Fitzgerald 1952 T E
Diplopoda
STYLOCEPHALIDAE
Stylocephalus sp. indet. Crumpton 1974 TI E
?*Stylocephalus* sp. indet. Jensen 1979 FI E

Order NEOGREGARINORIDA
OPHRYOCYSTIDAE
Ophryocystis n. sp. Fitzgerald 1952 T E
Tenebrionidae

Class ACONOIDASIDA (HAEMATOZOA)
Order HAEMOSPORORIDA
(HAEMOSPORIDIDA)
PLASMODIIDAE
Haemoproteus danilewsky Kruse, 1890 TB A?
Leucocytozoon fringillinarum Woodcock, 1910 TB A
Leucocytozoon tawaki Fallis, Bisset & Allison, 1976 MB
Plasmodium cathemerium Hartman, 1927 TB A?
Plasmodium elongatum Huff, 1930 TB E?
Plasmodium lygosomae Laird, 1951 TR E
Plasmodium relictum (Grassi & Feletti, 1891) TB A?
Plasmodium spp. indet. (~5) F/TB/Ma A/E
Man (*Homo sapiens*)
Grey duck (*Anas superciliosa superciliosa*)
Mohua (yellowhead) (*Mohoua ochrocephala*)
New Zealand pipit (*Anthus novaeseelandiae*)
Skylark (*Alauda arvensis*)
Nothern brown kiwi (*Apteryx mantelli*)

Order PIROPLASMORIDA (PIROPLASMIDA)
BABESIIDAE
Babesia kiwiensis Peirce, Jakob-Hoff & Twentyman, 2003 TB E
THEILERIIDAE
Theileria orientalis (Yakimov & Sudachenkov, 1931) TMa A

Checklist of New Zealand fossil Myzozoa and Acritarcha

Compiled by C. Clowes and G. J. Wilson. The records below are taken from the published literature and the Fossil Record File of GNS Science and the Geological Society of New Zealand. Stratigraphic ranges are given where known. Standard three-letter abbreviations are given for epochs, e.g. Pal = Paleoecene, Eoc = Eocene, Oli = Oligocene, etc., with E, M, and L for Early, Middle, and Late. Three-letter abbreviations are also given for periods, e.g. Tri = Triassic, Jur = Jurassic, Cre = Cretaceous.

PHYLUM INCERTAE SEDIS
ACRITARCHA
Anthosphaeridium convolvuloides Cookson & Eisenack, 1968
Baltisphaeridium sp.
Bavlinella faveolata Shepeleva, 1962
Brazilea parva (Cookson & Dettmann, 1959)
Chomotriletes circulus (Wolff, 1934)
Chuaria circularis Walcott, 1899
Circulisporites monilis (Balme & Segroves, 1966)
Circulisporites parvus de Jersey, 1962
Crassosphaera sp.
Cyclodictyon paradoxum (Cookson & Eisenack, 1958)
Cyclopsiella sp.
Disphaeria macropyla Cookson & Eisenack, 1960a LCre
Fromea amphora Cookson & Eisenack, 1958
Fromea apiculata (Cookson & Eisenack, 1960)
Fromea chytra (Drugg, 1967)
Gorgonisphaeridium sp.
Lecaniella dictyota Cookson & Eisenack, 1962
Leiofusa sp.
Leiosphaeridia sp.
Maculatasporites sp.
Membranosphaera sp.
Micrhystridium sp.
Nummus sp.
Orygmatosphaeridium holtedahlii (Timofeev, 1966)
Palaeostomocystis ovata (Wilson, 1967)
Paralecaniella indentata (Deflandre & Cookson, 1955)
Paucilobimorpha apiculata (Cookson & Eisenack, 1962)
Paucilobimorpha inaequalis (Marshall & Partridge, 1988)
Paucilobimorpha incurvata (Cookson & Eisenack, 1962)
Paucilobimorpha spinosa (Cookson, 1965)
Paucilobimorpha tripa de Coninck, 1986
Psiloschizosporis scissus (Balme & Hennelly, 1956)
Rhombodella paucispina (Alberti, 1961)
Schizocystia sp.
Schizophacus rugulatus (Cookson & Dettmann, 1959)
Schizosporis reticulatus Cookson & Dettmann, 1959
Sigmopollis sp.
Tetraporina sp.
'*Valensiella*' *clathrodermum* (Deflandre & Cookson, 1955)
Vendotaenia sp.
Veryhachium reductum (Deunff, 1959)

PHYLUM MYZOZOA
SUBPHYLUM DINOZOA
Superclass DINOKARYOTA
Class PERIDINEA
Subclass PERIDINOIDIA
Order PERIDINIDA
PERIDINIIDAE
DEFLANDREINAE
Abratopdinium kerguelense Mao & Mohr, 1992 LCre
?*Alterbidinium distinctum* (Wilson, 1967a) Eoc
?*Alterbidinium novozealandicum* Schiøler, Roncaglia & Wilson, 2001
?*Alterbidinium pentaradiata* (Cookson & Eisenack, 1965) Pal
Alterbidinium acutulum (Wilson, 1967b) LCre
Alterbidinium austrinum Roncaglia & Schiøler, 1999
Alterbidinium longicornutum Roncaglia, Field, Raine, Schiøler & Wilson, 1999 LCre
Alterbidinium minus (Alberti, 1959) LCre
Alterbidinium varium Kirsch, 1991 LCre
Amphidiadema denticulata Cookson & Eisenack, 1960 LCre
Amphidiadema nucula (Cookson & Eisenack, 1962) LCre
Amphidiadema rectangularis (Cookson & Eisenack, 1962) LCre
Cerodinium dartmoorium (Cookson & Eisenack, 1965) Pal
Cerodinium diebelii (Alberti, 1959) LCre
Cerodinium nielsii Willumsen, 2011 EPal
Cerodinium medcalfii (Stover, 1974) Pal
Cerodinium obliquipes (Deflandre & Cookson, 1955)
Cerodinium speciosum (Alberti, 1959) Pal
Cerodinium striatum (Drugg, 1967) Pal
Chatangiella madura Lentin & Williams, 1976
Chatangiella packhamii Marshall, 1990a LCre
Chatangiella serratula (Cookson & Eisenack, 1958) LCre
Chatangiella tripartita (Cookson & Eisenack, 1960)
Chatangiella verrucosa (Manum, 1963)
Chatangiella victoriensis (Cookson & Manum, 1964) LCre
Deflandrea antarctica Wilson, 1967 Eoc
Deflandrea convexa Wilson, 1988 Eoc
Deflandrea cygniformis Pöthe de Baldis, 1966 Eoc
Deflandrea delineata Cookson & Eisenack, 1965 Pal–Eoc
Deflandrea denticulata Alberti, 1959 Pal
Deflandrea flounderensis Stover, 1974
Deflandrea foveolata Wilson, 1984 Pal
Deflandrea galeata (Lejeune-Carpentier, 1942) LCre
Deflandrea heterophlycta Deflandre & Cookson, 1955 Eoc
Deflandrea leptodermata Cookson & Eisenack, 1965 Eoc
Deflandrea phosphoritica Eisenack, 1938 Eoc–Oli
Deflandrea robusta Deflandre & Cookson, 1955 Eoc
Deflandrea scabrata Wilson, 1988 Eoc
Deflandrea truncata Stover, 1974 Eoc
Eucladinium kaikourense Schiøler & Wilson, 1998 LCre
Eucladinium madurense (Cookson & Eisenack, 1970)
Gippslandia extensa (Stover, 1974)
Hexagonifera glabra Cookson & Eisenack, 1961 LCre
?*Isabelidinium rhombovale* (Cookson & Eisenack, 1970) LCre
Isabelidinium bakeri (Deflandre & Cookson, 1955) Pal
Isabelidinium belfastense (Cookson & Eisenack, 1961) LCre
Isabelidinium campbellensis (Wilson, 1967) LCre
Isabelidinium cingulatum Wilson, 1988 Pal
Isabelidinium cooksoniae (Alberti, 1959b)
Isabelidinium glabrum (Cookson & Eisenack, 1969) LCre
Isabelidinium greenense Marshall, 1990 LCre
Isabelidinium korojonense (Cookson & Eisenack, 1958) LCre
Isabelidinium marshallii Roncaglia, 2000
Isabelidinium microarmum (McIntyre, 1975) LCre

Isabelidinium papillum Sumner, 1992 LCre
Isabelidinium pellucidum (Deflandre & Cookson, 1955) LCre
Isabelidinium thomasii (Cookson & Eisenack, 1961)
Isabelidinium variabile Marshall, 1988 LCre
Lentinia sp.
Magallanesium asymmetricum (Wilson, 1967) Eoc–Oli
Magallanesium balmei (Cookson & Eisenack, 1962)
Magallanesium densispinatum (Stanley, 1965) Pal
Magallanesium macmurdoense (Wilson, 1967) Eoc–Oli
?*Manumiella cretacea* (Cookson, 1956) LCre
Manumiella conorata (Stover, 1974) LCre
Manumiella druggii (Stover, 1974) LCre
Manumiella lata (Cookson & Eisenack, 1968) LCre
Manumiella rotunda Wilson, 1988 Pal
Manumiella seelandicum (Lange, 1969) LCre
Nelsoniella aceras Cookson & Eisenack, 1960 LCre
Nelsoniella semireticulata Cookson & Eisenack, 1960 LCre
Nelsoniella tuberculata Cookson & Eisenack, 1960 LCre
?*Palaeocystodinium rhomboides* (Wetzel, 1933) LCre
Palaeocystodinium australinum (Cookson, 1965) Pal
Palaeocystodinium bulliforme Ioannides, 1986
Palaeocystodinium golzowense Alberti, 1961 Pal
Palaeocystodinium granulatum (Wilson, 1967) LCre
Palaeocystodinium lidiae (Górka, 1963)
Pierceites sp.
Satyrodinium bengalense Lentin & Manum, 1986 L Cretaceous]
Satyrodinium haumuriense (Wilson, 1984) LCre
?*Senegalinium dilwynense* (Cookson & Eisenack, 1965) Pal
Spinidinium echinoideum (Cookson & Eisenack, 1960) LCre
Spinidinium styloniferum Cookson & Eisenack, 1962
Trithyrodinium evittii Drugg, 1967 Pal
Trithyrodinium striatum Benson, 1976 LCre
Trithyrodinium suspectum (Manum & Cookson, 1964) LCre
Trithyrodinium vermiculatum (Cookson & Eisenack, 1961) LCre
Volkheimeridium lanterna (Cookson & Eisenack, 1970)
Vozzhennikovia angulata Wilson, 1988 Pal
Vozzhennikovia apertura (Wilson, 1967) Eoc–Oli
Vozzhennikovia rotunda (Wilson, 1967) Eoc–Oli
Vozzhennikovia spinulosa Wilson, 1984 LCre
Xenikoon australis Cookson & Eisenack, 1960 LCre
Ovoidineinae
Ascodinium acrophorum Cookson & Eisenack, 1960
Ascodinium parvum (Cookson & Eisenack, 1958) LCre
Epelidosphaeridia pentagona Morgan, 1980
Leberidocysta chlamydata (Cookson & Eisenack, 1962) LCre
Ovoidinium ovale (Cookson & Eisenack, 1970a)
Palaeoperidiniinae
Diconodinium cristatum Cookson & Eisenack, 1974
Diconodinium glabrum Eisenack & Cookson, 1960 LCre
Diconodinium multispinum (Deflandre & Cookson, 1955) LCre
Diconodinium psilatum Morgan, 1977 LCre
Diconodinium vitricorne Roncaglia, Field, Raine, Schiøler & Wilson, 1999 LCre
Dioxya armata Cookson & Eisenack, 1958
Laciniadinium inflatum (Eisenack & Cookson, 1960)
Laciniadinium tenuistriatum (Eisenack & Cookson, 1960) LCre
Luxadinium sp.
Palaeohystrichophora infusorioides Deflandre, 1935 LCre
Palaeoperidinium pyrophorum (Ehrenberg, 1838) Pal
Phthanoperidinium echinatum Eaton, 1976
Phthanoperidinium eocenicum (Cookson & Eisenack, 1965) Eoc
Phthanoperidinium geminatum Bujak, 1980
Subtilisphaera sp.
Incertae sedis
Maduradinium pentagonum Cookson & Eisenack, 1970a
Wetzeliellinae
Apectodinium homomorphum (Deflandre & Cookson, 1955) Pal–Eoc
Apectodinium hyperacanthum (Cookson & Eisenack, 1965)
Charlesdowniea coleothrypta (Williams & Downie, 1966b) Eoc
Charlesdowniea edwardsii (Wilson, 1967) Eoc
Dracodinium granulatum (Wilson, 1967) Eoc
Dracodinium waipawaense (Wilson, 1967) Eoc
Rhombodinium glabrum (Cookson, 1956) Eoc
Rhombodinium subtile Wilson, 1988 Eoc
Wetzeliella articulata Wetzel, 1938b Eoc
Wetzeliella hampdenensis Wilson, 1967c Eoc
Wetzeliella spinulosa Wilson, 1988 Eoc
Wilsonidium echinosuturatum (Wilson, 1967) Eoc
Wilsonidium lineidentatum (Deflandre & Cookson, 1955) Eoc
Wilsonidium ornatum (Wilson, 1967) Eoc
Wilsonidium tabulatum (Wilson, 1967) Eoc
PROTOPERIDINIIDAE
Protoperidiniinae
Lejeunecysta hyalina (Gerlach, 1961) Pal–Eoc
Lejeunecysta kammae Willumsen, 2011 Pal
Phelodinium magnificum (Stanley, 1965) LCre.

Subclass DINOPHYSOIDIA
Order NANNOCERATOPSIIDA
NANNOCERATOPSIIDAE
Nannoceratopsis pellucida Deflandre, 1939 LJur

Subclass GONYAULACOIDIA
Order GONYAULACIDA
AREOLIGERIDAE
Adnatosphaeridium filiferum (Cookson & Eisenack, 1958)
Areoligera senonensis Lejeune-Carpentier, 1938 Paleocene
?*Canningia rotundata* Cookson & Eisenack, 1961 Late Cre
Canninginopsis bretonica Marshall, 1990 LCre
Canninginopsis denticulata Cookson & Eisenack, 1962
Canninginopsis intermedia Morgan, 1980
Cassidium filosum Wilson, 1988 Pal
Cassidium fragile (Harris, 1965) Pal
Chiropteridium lobospinosum Gocht, 1960 Oli–Mio
Circulodinium colliveri (Cookson & Eisenack, 1960)
Circulodinium densebarbatum (Cookson & Eisenack, 1960) LJur
Circulodinium distinctum (Deflandre & Cookson, 1955) LCre
Circulodinium paucispinum (Davey, 1969)
Cleistosphaeridium ancyreum (Cookson & Eisenack, 1965) Eoc
Cleistosphaeridium diversispinosum Davey, Downie, Sarjeant & Williams, 1966
Cleistosphaeridium placacanthum (Deflandre & Cookson, 1955) Eoc
Cyclonephelium clathromarginatum Cookson & Eisenack, 1962
Cyclonephelium compactum Deflandre & Cookson, 1955 LCre
Cyclonephelium crassimarginatum Cookson & Eisenack, 1974
Cyclonephelium vannophorum Davey, 1969a
Glaphyrocysta marlboroughensis Schiøler & Wilson, 1998 LCre
Glaphyrocysta pastielsii (Deflandre & Cookson, 1955)
Glaphyrocysta retiintexta (Cookson, 1965a) Pal–Eoc
Glaphyrocysta semitecta (Bujak, 1980)
Glaphyrocysta texta (Bujak, 1976) Eoc
Membranophoridium aspinatum Gerlach, 1961
Membranophoridium perforatum Wilson, 1988 Eoc
Riculacysta sp.
Schematophora obscura Wilson, 1988 Eoc
Schematophora speciosa Deflandre & Cookson, 1955 Eoc
Senoniasphaera edenensis Marshall, 1990 LCre
Senoniasphaera inornata (Drugg, 1970)
Senoniasphaera rotundata Clarke & Verdier, 1967
Tenua sp.
CERATIIDAE
Endoceratium dettmannae (Cookson & Hughes, 1964)
Endoceratium ludbrookiae (Cookson & Eisenack, 1958) Cre
Endoceratium turneri (Cookson & Eisenack, 1958) ECre
Muderongia mcwhaei Cookson & Eisenack, 1958
Muderongia tetracantha (Gocht, 1967) ECre
Odontochitina costata Alberti, 1961 Cre
Odontochitina cribropoda Deflandre & Cookson, 1955 LCre
Odontochitina operculata (Wetzel, 1933) Cre
Odontochitina porifera Cookson, 1956 LCre
Odontochitina spinosa Wilson, 1984 LCre
Xenascus asperatus Stover & Helby, 1987
Xenascus ceratioides (Deflandre, 1937) LCre
CLADOPYXIIDAE
Cladopyxidium saeptum (Morgenroth, 1968)
Druggidium sp.
Fibradinium sp.
Gillinia hymenophora Cookson & Eisenack, 1960 LCre
Glyphanodinium sp.
Microdinium cassiculus Wilson, 1984 LCre
Microdinium dentatum Vozzhennikova, 1967
Microdinium ornatum Cookson & Eisenack, 1960
Microdinium reticulatum Vozzhennikova, 1967
Microdinium setosum Sarjeant, 1966
Microdinium sp. 1 Châteauneuf 1980/Clowes, MS
GONIODOMIDAE
Goniodominae
Heteraulacacysta sp.
Helgolandiniinae
Tuberculodinium vancampoae (Rossignol, 1962) [Pleistocene]
Pyrodiniinae
Biconidinium sp.
Dinopterygium cladoides Deflandre, 1935
Dinopterygium tuberculatum (Eisenack & Cookson, 1960)
Eisenackia circumtabulata Drugg, 1967 Paleocene
Eisenackia crassitabulata Deflandre & Cookson, 1955 Eocene
Eisenackia margarita (Harland, 1979)
Eisenackia reticulata (Damassa, 1979b) LCre
Homotryblium aculeatum Williams, 1978
Homotryblium tasmaniense Cookson & Eisenack, 1967 Eoc
Homotryblium tenuispinosum Davey & Williams, 1966
Hystrichosphaeridium arborispinum Davey & Williams, 1966
Hystrichosphaeridium patulum Davey & Williams, 1966
Hystrichosphaeridium recurvatum (White, 1842)
Hystrichosphaeridium salpingophorum Deflandre, 1935

Hystrichosphaeridium tubiferum (Ehrenberg, 1838) Pal–Eoc
Polysphaeridium zoharyi (Rossignol, 1962)
Taleisphaera sp.
GONYAULACIDAE
CRIBROPERIDINIINAE
Achilleodinium biformoides (Eisenack, 1954) Eoc
Achilleodinium cf. *biformoides* Clowes pers. obs.
'*Apteodinium*' *australiense* (Deflandre & Cookson, 1955) Eoc–Oli
Carpatella cornuta Grigorovich, 1969 Pal
Cordosphaeridium fibrospinosum Davey & Williams, 1966 Pal–Eoc
Cordosphaeridium gracile (Eisenack, 1954) Eoc
Cordosphaeridium inodes (Klumpp, 1953) Eoc
?*Cribroperidinium edwardsii* (Cookson & Eisenack, 1958) LCre
?*Cribroperidinium muderongense* (Cookson & Eisenack, 1958) LCre
Cribroperidinium apione (Cookson & Eisenack, 1958)
Cribroperidinium wetzeli (Lejeune-Carpentier, 1939)
Cribroperidinium wilsonii (Yun, 1981)
Damassadinium californicum (Drugg, 1967)
Damassadinium crassimuratum (Wilson, 1988) Eoc
Diphyes colligerum (Deflandre & Cookson, 1955) Eoc
Diphyes ficusoides Islam, 1983b
Disphaerogena irregularis (Wilson, 1988) Eoc
Florentinia ferox (Deflandre, 1937) LCre
Hapsocysta peridictya (Eisenack & Cookson, 1960)
Hystrichokolpoma bullatum Wilson, 1988 Eoc
Hystrichokolpoma cinctum Klumpp, 1953 Eoc
Hystrichokolpoma rigaudiae Deflandre & Cookson, 1955 Eoc
Hystrichokolpoma spinosum Wilson, 1988 Eoc
Hystrichokolpoma wilsonii Biffi & Manum, 1988 Eoc
Kallosphaeridium sp.
Kenleyia sp.
Lingulodinium machaerophorum (Deflandre & Cookson, 1955) Oli–Mio
Operculodinium centrocarpum (Deflandre & Cookson, 1955) Eoc–Oli
Operculodinium tiara (Klumpp, 1953) Eoc-Oli
Samlandia chlamydophora Eisenack, 1954 Eoc
Samlandia delicata Wilson, 1988 Eoc
Samlandia reticulifera Cookson & Eisenack, 1965 Eoc
Spongodinium sp. Clowes pers. obs.
Stoveracysta kakanuiensis Clowes, 1985 Oli
Stoveracysta ornata (Cookson & Eisenack, 1965) Eoc
Thalassiphora pelagica (Eisenack, 1954) Eoc
Turbiosphaera filosa (Wilson, 1967) Pal–Eoc
Turbiosphaera galatea Eaton, 1976 Eoc
GONYAULACINAE
Achomosphaera antleriformis Schiøler, 1993
Achomosphaera crassipellis (Deflandre & Cookson, 1955) Eoc
Achomosphaera ramulifera (Deflandre, 1937)
Achomosphaera regiensis Corradini, 1973 LCre
Achomosphaera sagena Davey & Williams, 1966
Bitectatodinium tepikiense Wilson, 1973 Ple
Cannosphaeropsis sp.
?*Corrudinium vermiculatum* (Wilson, 1988) Eoc
Corrudinium obscurum Wilson, 1988 Eoc
Corrudinium regulare Clowes & Wilson, 2006 Eoc
Hafniasphaera septata (Cookson & Eisenack, 1967) Pal–Eoc
Hystrichosphaeropsis sp.
Hystrichostrogylon sp.
Impagidinium agremon Willumsen, 2011 LCre
Impagidinium cassiculus Wilson, 1988 Eoc
Impagidinium cavea Willumsen, 2011 LCre
Impagidinium crassimuratum Wilson, 1988 Eoc
Impagidinium crouchiae Willumsen, 2011 LCre
Impagidinium dispertitum (Cookson & Eisenack, 1965) Eoc
Impagidinium elegans (Cookson & Eisenack, 1965) Eoc
Impagidinium hannahii Willumsen, 2011 LCre
Impagidinium maculatum (Cookson & Eisenack, 1961) Eoc
Impagidinium parvireticulatum Wilson, 1988 Eoc
Impagidinium patulum (Wall, 1967)
Impagidinium victorianum (Cookson & Eisenack, 1965) Eoc–Oli
Nematosphaeropsis balcombiana Deflandre & Cookson, 1955 Eoc–Oli
Pentadinium sp.
Psaligonyaulax sp.
Pterodinium cingulatum (Wetzel, 1933) LCre
Rottnestia borussica (Eisenack, 1954) Eoc
Spiniferites bentorii (Rossignol, 1964) Ple
Spiniferites bulloideus (Deflandre & Cookson, 1955)
Spiniferites granulatus (Davey, 1969)
Spiniferites membranaceus (Rossignol, 1964)
Spiniferites mirabilis (Rossignol, 1964)
Spiniferites pseudofurcatus (Klumpp, 1953)
Spiniferites ramosus (Ehrenberg, 1838)
Spiniferites ?ramous (Ehrenberg, 1838)
Tectatodinium pellitum Wall, 1967 Eoc–Oli
Tubotuberella apatela (Cookson & Eisenack, 1960) LJur
Ynezidinium waipawaense Wilson, 1988 Eoc
LEPTODINIINAE
Areosphaeridium dictyostilum (Menéndez, 1965) Eoc
Areosphaeridium partridgei (Stover & Williams, 1995) Eoc
Areosphaeridium pectiniforme (Gerlach, 1961)
Conosphaeridium abbreviatum Wilson, 1984 LCre
Conosphaeridium striatoconum (Deflandre & Cookson, 1955) LCre
Conosphaeridium tubulosum Cookson & Eisenack, 1969
Cooksonidium capricornum (Cookson & Eisenack, 1965) Eoc
Ctenidodinium sp.
Cymososphaeridium benmorense Schiøler & Wilson, 1998 LCre
Emmetrocysta sp.
Endoscrinium attadalense (Cookson & Eisenack, 1958)
Gonyaulacysta sp.
Herendeenia postprojecta Stover & Helby, 1987
Kleithriasphaeridium secatum Schiøler & Wilson, 1998 LCre
?*Leptodinium tenuicornutum* Cookson & Eisenack, 1962 LCre
Leptodinium ambiguum (Deflandre, 1939)
Leptodinium clathratum (Cookson & Eisenack, 1960)
Lithodinia sp.
Litosphaeridium siphoniphorum (Cookson & Eisenack, 1958)
Lophocysta sp.
Meiourogonyaulax bulloidea (Cookson & Eisenack, 1960)
Occisucysta sp.
Oligosphaeridium complex (White, 1842) Cre
Oligosphaeridium pulcherrimum (Deflandre & Cookson, 1955) Cre
Perisseiasphaeridium sp.
Rhynchodiniopsis serrata (Cookson & Eisenack, 1958)
Rigaudella aemula (Deflandre, 1939) LJur
Rigaudella filamentosum (Cookson & Eisenack, 1958)
Sirmiodinium grossii Alberti, 1961 ECre
Spiniferella cornuta (Gerlach, 1961) LCre
Stiphrosphaeridium sp.
?*Systematophora septata* Wilson, 1988 Eoc
Systematophora areolata Klement, 1960 ECre
INCERTAE SEDIS
Actinotheca aphroditae Cookson & Eisenack, 1960 LCre
Aireiana sp.
Arachnodinium antarcticum Wilson & Clowes, 1982 Eoc–Oli
Belodinium dysculum Cookson & Eisenack, 1960 LJur
Callaiosphaeridium asymmetricum Deflandre & Courteville, 1939 LCre
Chytroeisphaeridia chytreoides (Sarjeant, 1962a) ECre
Cometodinium whitei (Deflandre & Courteville, 1939)
Coronifera oceanica Cookson & Eisenack, 1958 LCre
Coronifera striolata (Deflandre, 1937)
Fibrocysta axialis (Eisenack, 1965)
Fibrocysta bipolaris (Cookson & Eisenack, 1965) Pal–Eoc
Hemiplacophora semilunifera Cookson & Eisenack, 1965
Hurunuia maxwellii Wilson, 1984 ECre
Hystrichodinium pulchrum Deflandre, 1935 Cre
Hystrichodinium ramoides Alberti, 1961 LCre
Kaiwaradinium buccinatum Wilson, 1978 ECre
Kaiwaradinium ramosum Wilson, 1984 ECre
Kiokansium unituberculatum (Tasch, 1964)
Komewuia sp.
Melitasphaeridium choanophorum (Deflandre & Cookson, 1955)
Pervosphaeridium pseudhystrichodinium (Deflandre, 1937) LCre
Pyxidinopsis crassimurata Wilson, 1988 Eoc
Pyxidinopsis delicata Wilson, 1988 Eoc
Pyxidinopsis epakros Willumsen, 2011 Cre
Pyxidinopsis everriculum Willumsen, 2011 Cre
Pyxidinopsis meadensis Willumsen, 2011 Cre
Pyxidinopsis waipawaensis Wilson, 1988 Eoc
Saturnodinium sp.
Scriniodinium ceratophorum Cookson & Eisenack, 1960 ECre
Scriniodinium crystallinum (Deflandre, 1939) LJur
Scriniodinium parvimarginatum (Cookson & Eisenack, 1958)
Scriniodinium playfordii Cookson & Eisenack, 1960 LJur
Sentusidinium sp.
?*Sepispinula ambigua* (Deflandre, 1937)
Sepispinula ancorifera (Cookson & Eisenack, 1960) LJur
Sepispinula huguoniotii (Valensi, 1955)
Sirmiodiniopsis sp.
Stephodinium sp.
Surculosphaeridium cribrotubiferum (Sarjeant, 1960)
Trichodinium castanea Deflandre, 1935 LCre
Trichodinium pellitum Eisenack & Cookson, 1960
Xenicodinium sp.
Yalkalpodinium scutum Morgan, 1980 ECre
GYMNODINIIDAE
Gymnodinium sp.
PAREODINIIDAE
BROOMEINAE
Batioladinium sp.
Broomea sp.
Kalyptea sp.
PAREODINIINAE
Pareodinia ceratophora Deflandre, 1947d
SCRINIOCASSIIDAE
Scriniocassis sp.
INCERTAE SEDIS
Atopodinium sp.
Balcattia cirrifera Cookson & Eisenack, 1974

Balteocysta perforata (Davey, 1978) LCre
?*Batiacasphaera reticulata* (Davey, 1969) LCre
Batiacasphaera cassiculus Wilson, 1988 Eocene
Batiacasphaera grandis Roncaglia, Field, Raine, Schiøler & Wilson, 1999 LCre
Batiacasphaera kekerengensis Schiøler & Wilson, 1998 LCre
Batiacasphaera rugulata Schiøler & Wilson, 1998 LCre
Batiacasphaera subtilis Stover & Helby, 1987 LCre
Caligodinium aceras (Manum & Cookson, 1964)
Carnarvonodinium sp.
?*Cassiculosphaeridia intermedia* Slimani, 1994
Cassiculosphaeridia magna Davey, 1974 LCre
Cassiculosphaeridia reticulata Davey, 1969
Cerebrocysta sp.
Chlamydophorella delicata Hultberg, 1985 LCre
Chlamydophorella discreta Clarke & Verdier, 1967 LCre
Chlamydophorella nyei Cookson & Eisenack, 1958 LCre
Chlamydophorella wallala Cookson & Eisenack, 1960 LJur
Cobricosphaeridium sp.
Dapsilidinium sp.
Dingodinium cerviculum Cookson & Eisenack, 1958 ECre
Dingodinium jurassicum Cookson & Eisenack, 1958 ECre
Distatodinium sp.
Downiesphaeridium sp.
Duosphaeridium nudum (Cookson, 1965)
Ellipsoidictyum sagena (Duxbury, 1980) LCre
Elytrocysta druggii Stover & Evitt, 1978 LCre
Exochosphaeridium arnace Davey & Verdier, 1973 LCre
Exochosphaeridium bifidum (Clarke & Verdier, 1967) LCre
Exochosphaeridium phragmites Davey, Downie, Sarjeant & Williams, 1966 LCre
Flamingoia sp.
Heibergella sp.
Heslertonia striata (Eisenack & Cookson, 1960)
Heterosphaeridium heteracanthum (Deflandre & Cookson, 1955)
Impletosphaeridium sp.
Membranilarnacia polycladiata Cookson & Eisenack, 1963 LCre
Octodinium askiniae Wrenn & Hart, 1988 Eoc
Papuadinium apiculatum (Cookson & Eisenack, 1960b)
Prolixosphaeridium capitatum (Cookson & Eisenack, 1960) LJur
Prolixosphaeridium conulum Davey, 1969
Prolixosphaeridium granulosum (Deflandre, 1937)
Prolixosphaeridium parvispinum (Deflandre, 1937) ECre
Rhiptocorys veligera (Deflandre, 1937)
Sverdrupiella sp.
Tanyosphaeridium xanthiopyxides (Wetzel, 1933) LCre
Trigonopyxidia ginella (Cookson & Eisenack, 1960) LCre
Valensiella sp.
Wallodinium glaessneri (Cookson & Eisenack, 1960)
Xiphophoridium alatum (Cookson & Eisenack, 1962) LCre
Xiphophoridium asteriforme Yun, 1981
INCERTAE SEDIS
Mendicodinium sp.
COMPARODINIIDAE
Valvaeodinium sp.
DOLLIDINIIDAE
Horologinella lineata (Cookson & Eisenack, 1962)

Order PTYCHODISCIDA
PTYCHODISCIDAE
DINOGYMNIINAE
Dinogymnium acuminatum Evitt, Clarke & Verdier, 1967
Dinogymnium nelsonense (Cookson, 1956) LCre
Dinogymnium westralium (Cookson & Eisenack, 1958) LCre

Subclass SUESSIOIDIA
Order SUESSIIDA
SUESSIIDAE
Suessia sp.

*Following Sluijs et al. 2009, *Magallanesium* is treated as a junior synonym of *Spinidinium*

Holocene dinoflagellate cysts from New Zealand surficial sediments

The following checklist contains fossil and subfossil taxa not in the above list. Species records are based on the following sources (listed in references): McMinn & Sun 1994; Sun & McMinn 1994; A. R. Edwards pers. comm.

PHYLUM MYZOZOA
SUBPHYLUM DINOZOA
Superclass DINOKARYOTA
Class PERIDINEA
Subclass PERIDINOIDIA
Order PERIDINIIDA
Suborder PERIDINIINA
CALCIODINELLIDAE
Scrippsiella regalis (Gaarder, 1954)

Subclass GONYAULACOIDIA
Order GONYAULACIDA
INCERTAE SEDIS
Dalella chathamense McMinn & Sun, 1994
Impagidinium aculeatum (Wall, 1967)
Impagidinium paradoxum (Wall, 1967)
Nematosphaeropsis lemniscata Bujak, 1984

INCERTAE SEDIS
Pyxidinopsis reticulata McMinn & Sun, 1994

FOURTEEN

Phylum
CILIOPHORA

ciliates

WILHELM FOISSNER, TONY CHARLESTON,
MARTIN KREUTZ, DENNIS P. GORDON

Ciliates are among the most complex unicellular organisms. Far from being simple, the bodies of ciliates are analogous to those of metazoans, with highly specialised organelles performing functions of movement, ingestion, digestion, water balance, excretion, reproduction, sensing the environment, and defence. The diversity of size and morphology among ciliates is remarkable. The largest species (funnel-like *Stentor* can reach two millimetres high and worm-like *Spirostomum ambiguum* may be four millimetres long) can exceed the smallest (*Nivaliella plana*, a 20-micrometre-long soil inhabitant) by several orders of magnitude of cell volume. Although most of the approximately 10,000 described species are free-living, some are parasitic, such as *Balantidium coli*, which can cause human disease. Other ciliates induce diseases in fish and are a problem for aquaculturists. Ciliates of cloven-hoofed animals stabilise the huge numbers of bacteria in the rumen and gut. Ciliate feeding strategies include herbivory, fungivory, and predation. Some ciliates, like *Paramecium bursaria*, have symbiotic bacteria and algae ('zoochlorellae'). The products of photosynthesis are beneficial to the protozoan host, which can bring the algae to optimal light conditions and supply carbon dioxide and shelter. A number of ciliates are sessile. They may be stalked or unstalked and some build a protective case (lorica) around the cell.

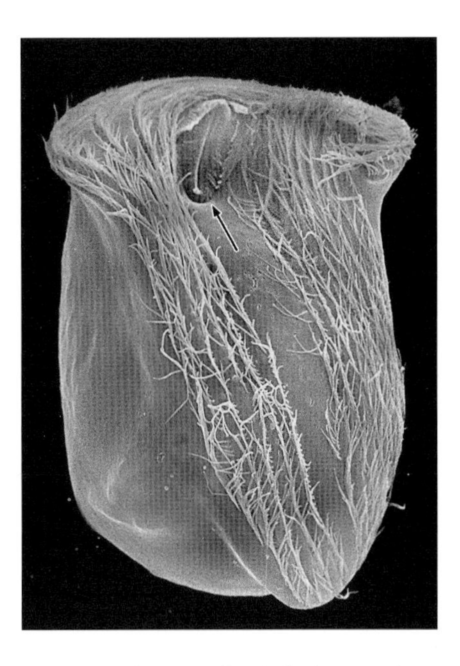

Phascolodon vorticella (Phyllopharyngea). It has only two small ciliary fields leading to the anterior mouth (arrow), and feeds on planktonic lake microalgae (70 µm).
Wilhelm Foissner

Among eukaryotes, ciliates are distinctive in having a macronucleus and a micronucleus (sometimes more than one of each kind) of very different size and function within the same cytoplasm – a condition termed heterokaryotic and elsewhere known only among a few foraminifera at certain stages of their life-history. The macronucleus, which is usually highly polyploid (having several or multiple sets of chromosomes), divides during asexual reproduction and controls mainly somatic functions (e.g. RNA synthesis). The diploid micronucleus is active mainly during sexual reproduction, when mating pairs exchange genetic material during conjugation.

Ciliates are evolutionarily related to dinoflagellates and apicomplexans (previous chapter) in having membranous sacs (alveoli) in their cell cortex (lacking or reduced in size, however, in a number of groups). The cortex is essentially the outer body layer, which, in ciliates, is extraordinarily complex. The cell membrane is internally supported by the alveoli, except in several specialised regions of the surface. The basic component of the cortex is the kinetid, a complex composed of the basal body (kinetosome) of a cilium and its associated fibrils. The fibrils link kinetids of the same and adjacent kineties (rows of cilia) as an organised system. The arrangement of cilia (which can be fused into larger organelles called cirri and adoral membranelles) on the body and in the oral region is significant in taxonomy. The reader is urged to consult general works like those of Corliss (1979) and Lynn and Small (2000) for details of cellular construction and function in these remarkable organisms.

Coleps hirtus (Prostomatea), an armoured ciliate with calcified plates that have elliptical openings through which the cilia (not shown) emerge; benthoplanktonic, feeding on other ciliates (60 µm.
Wilhelm Foissner

Ciliates have long been recognised to be a monophyletic group, but a definitive classification has not yet been achieved. In the 1960s, identification of major lineages was based on cell morphology and details of the cortex (especially arrangement of ciliary rows) and oral structures as revealed by silver staining. In the 1970s, electron-microscope studies revealed details of cell ultrastructure, especially of kinetids and their fibrillar associates. With the advent of gene-sequencing techniques, ribosomal RNA genes have been used to test the reliability of classifications arrived at using previous evidence. The two ciliophoran subphyla, Postciliodesmatophora and Intramacronucleata, and six out of eleven classes (Karyorelictea, Heterotrichea, Litostomatea, Phyllopharyngea, Nassophorea, and Colpodea) are well supported by both molecules and morphology; the remaining five classes (Spirotrichea, Armophorea, Plagiopylea, Prostomatea, Oligohymenophorea) are not (Lynn 2003a).

Ciliates may be found almost anywhere there is liquid water, but different forms predominate in various habitats. On the other hand, some species are such environmental generalists as to be found in marine, freshwater, and terrestrial habitats. Generally, ciliates in soils tend to be small forms that can generate resistant cysts in order to survive periods of dryness. Ciliates abound in freshwater environments, especially those that have been organically enriched, like sewage and oxidation ponds. 'Infusoria' is a term that was used historically for ciliates (and some other protozoans) of nutrient-rich organic (e.g. hay) infusions. Marine plankton, also, can have a diverse ciliate fauna, especially the loricate forms known as tintinnids.

The New Zealand ciliate fauna

Ciliates of New Zealand are relatively little studied and therefore poorly known. The few illustrated accounts focused on freshwater forms in the 1890s (Kirk 1886, 1887; Maskell 1886, 1887; Schewiakoff 1892, 1893) and the 1950s (Bary 1950a,b; Barwick et al. 1955) and unillustrated work was published on soil ciliates during the 1950s through 1980s (e.g. Stout 1952, 1955a,b, 1958, 1960a, 1961, 1962, 1978, 1984). A study of parasites in the gut of the common green and golden bell frog (*Litorea aurea*) yielded the ciliate *Nyctotherus cordiformis* (Brace et al. 1953) and rumen and gut ciliates of New Zealand introduced mammals were reported by Clarke (1964, 1968), Clarke et al. (1982), and Fairley (1996). A study was made on the marine planktonic tintinnids in the 1980s (Burns 1983) and the ciliate parasites of fishes have attracted some attention (Laird 1953; Diggles et al. 2002).

Taken together, these data and some soil samples recently studied by W. Foissner (see below), reveal about 347 ciliate species, of which 50 may be endemic to New Zealand. Unfortunately, few ciliate catalogues are available worldwide to show faunal comparisons. European catalogues, for example, show 696 ciliate species in Austria (Foissner & Foissner 1988), 585 species in Slovakia (Matiset al. 1996), 188 species in Italy (Diniet al. 1995), and about 500 species in Bulgaria (Detcheva 1992). Even these compilations do not show the real number of species likely to be present, demonstrating our ignorance about this exceptional

Summary of New Zealand ciliate diversity

Taxon	Described species + subspecies*	Known undet. species	Estimated unknown species	Adventive species	Endemic species	Endemic genera
Postciliodesmatophora	14+1	2	>100	0	4?	0
Intramacronucleata	326+3	5	>1,000	37	46?+1?	0
Totals	**340+4**	**7**	**>1,100**	**37**	**50?+1?**	**0**

* Not including forms (f).

group of minute but important organisms (see Foissner 1997a, 2000, for more detailed discussion). The known New Zealand fauna is thus only a small fraction of that likely to exist.

Freshwater ciliates

The first accounts of freshwater ciliates in New Zealand were published in the late 19th century when Kirk (1886), Maskell (1886, 1887), and Schewiakoff (1892, 1893) described and illustrated known and new taxa. However, most samples were taken from a small area of New Zealand, viz the Wellington District. More than half a century elapsed before there were any further studies. Bary (1950a,b), working in the Wellington region, added many new records to the fauna and also summarised the earlier accounts. Kirk had briefly reported 13 species of the stalked ciliate *Vorticella*, describing and illustrating two as new – *V. zealandica* (freshwater) and *V. oblonga* (marine). Maskell's lists and descriptions covered about 70 species, 25 of which were regarded as new, together with five new varieties and one new genus, *Thurophora* (= *Lembadion*). Schewiakoff described 11 new species and five new genera. Thus, among them, these three authors reported almost 100 species, of which 38 were described for the first time.

Underscoring the scientific inadequacy of the earlier works, Bary remarkably found none of Maskell's purported new species or any of those described by Kirk and Schewiakoff, even though he made collections from several of the streams sampled by Maskell (1886, 1887). Bary (1950b) himself reported 28 species, none of which was new, and he regarded the fauna as being essentially cosmopolitan. However, 28 species is a very low number, which suggests that his sampling and/or identifications were insufficient. Reporting on a graduate class project carried out at Victoria University (College), Barwick et al. (1953) illustrated 21 taxa, 11 of which were new records for New Zealand, giving locality details and methods of infusion-culture.

It is clear that, in spite of the paucity of studies in New Zealand, freshwater ciliates are common, diverse, and grossly understudied, especially outside the Wellington area.

Marine ciliates

In an annotated index of the known marine protozoa of New Zealand, Dawson (1992) listed 66 named species and five of uncertain identity, 42 of which were tintinnids. Tintinnids are common loricate microplankton organisms of coastal and oceanic waters. The lorica (made of protein, polysaccharide, or both) cements silica grains, diatom frustules, or coccoliths to its matrix, making it hard and resistant. As the living tintinnid cell is frequently destroyed during collection, it is the lorica that has historically formed the basis of classification.

Pelagic tintinnids are primary feeders of nanoplankton, including bacteria, small flagellates, coccolithophorids (Haptophyta), and dinoflagellates (Myzozoa). Cassie (1961) reported several species from New Zealand waters, but the most detailed study was that of Burns (1983), who illustrated loricas by scanning

Summary of New Zealand ciliate diversity by environment

Taxon	Marine/ brackish	Terrestrial	Freshwater
Postciliodesmatophora	3	5	12+1
Intramacronucleata	75	144+1	163+3
Totals	**78**	**149+1**	**175+4**

* Several species occur in more than one environment hence the total across all environments will exceed total diversity of 346 species and four subspecies.

Parasitic ciliates are accorded the major habitat of their hosts.

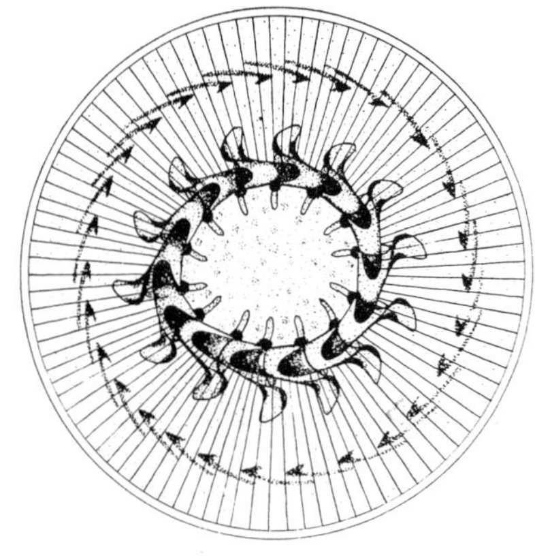

Trichodina parabranchicola (Oligohymenophorea). Upper, lateral view of whole organism from a gill of the olive rockfish (*Acanthoclinus fuscus*, drawn from stained specimen). Lower, aboral view of the internal skeletal complex.
From Laird 1953

electron microscopy and gave distributional data on each species in relation to hydrological features of the marine environment in which they were found. Other common planktonic ciliates are *Cyclotrichium meunieri* and *Myrionecta rubra* (*Mesodinium rubrum* in older literature), which cause seasonal red-water blooms in harbours and coastal waters (Bary & Stuckey 1950; Bary 1951, 1953a,b; Taylor 1973; MacKenzie & Gillespie 1986; Chang 1994).

Marine benthic ciliates have scarcely been noted in New Zealand, especially free-living forms, about which nothing is known. On the other hand, some commensal and parasitic forms have been reported, giving evidence of the fascinating range of ciliate microhabitats and hosts, e.g. *Haematophagus megapterae* on baleen plates of the humpback whale (Woodcock & Lodge 1921); *Endosphaera engelmanni*, hyperparasitic in the ciliate *Trichodina multidentis* from gills of the twister *Bellapiscis medius* (Laird 1953); and *Paranophrys elongata*, an endocommensal in the gut of the sea urchin *Evechinus chloroticus* (McRae 1959).

The most problematic forms are those that cause actual diseases in fishes. In reviewing all known parasites of New Zealand marine fishes, and latterly of economic marine invertebrates, Boustead (1982), Hine et al. (2000), and Diggles et al. (2002) reported several species of parasitic ciliates. *Ichthyophthirius multifiliis*, which occurs globally in wild and ornamental freshwater fishes, causes white spot disease. This is the only known parasite to cause significant losses in eel culture in New Zealand (Boustead 1982), despite being easily controlled. Another type of white spot disease affects wild and cultured marine fishes. This is caused by *Cryptocaryon irritans*. In New Zealand it has been observed in captive snapper (*Pagrus auratus*) and probably occurs in wild fish along the northeast coast of the North Island. It may also be present in imported marine ornamental fish (Hine 1982). Ciliates of the genus *Trichodina*, easily recognised by their distinctive ring of 'denticles' and their 'flying-saucer' shape, cause trichodiniasis, which can affect most species of wild and cultured marine fish including snapper (*Pagrus auratus*) and turbot (*Colistium nudipinnis*). Although wild fish may harbour heavy infestations of trichodinids, these usually do not cause disease (Laird 1953). However, ciliate numbers can build up quickly in confined fish, especially at higher temperatures, and skin and gill lesions can develop that may become infected by bacteria and fungi (Diggles 2000).

Soil ciliates

Diversity of ciliates is high in terrestrial habitats, with about 1000 species known worldwide (Foissner 1997a, 1998; Foissner et al. 2002). Ecologically, most soil ciliates live in the fresh and slightly decomposed litter layer, where abundances of up to 10,000 individuals per gram of dry mass of litter are reached. Ciliates are thus important primary decomposers and humus producers. In the humus horizon and in mineral soil where testate amoebae dominate, active ciliates are rare, although many cysts (dormant stages) are present. Most soil ciliates feed on bacteria (39%) or are predatory (34%) or omnivorous (20%). Some, however, are strictly mycophagous (fungal-feeding) and highly characteristic for terrestrial habitats; a few are anaerobic, providing a simple tool to assess the soil-oxygen regime (Foissner 1998, 1999). About 70% of soil ciliates are cosmopolitan, while others have a more or less restricted distribution (Foissner 2000).

In New Zealand, soil protozoa were studied mainly by John D. Stout, who worked through the 1950s–1980s in the New Zealand Soil Bureau of the former DSIR and was one of the leading soil protozoologists of this period, publishing important still-cited reviews (Stout & Heal 1967; Stout 1980; Stout et al. 1982). He was interested mainly in ecology, and thus his species lists are not as complete as taxonomists might wish. However, he was a careful worker and most of his identifications appear sound. Stout's species lists contain many unidentified taxa because he recognised – outstanding for that time – that soil and freshwater habitats have few species in common (Stout 1952) and many of the species he

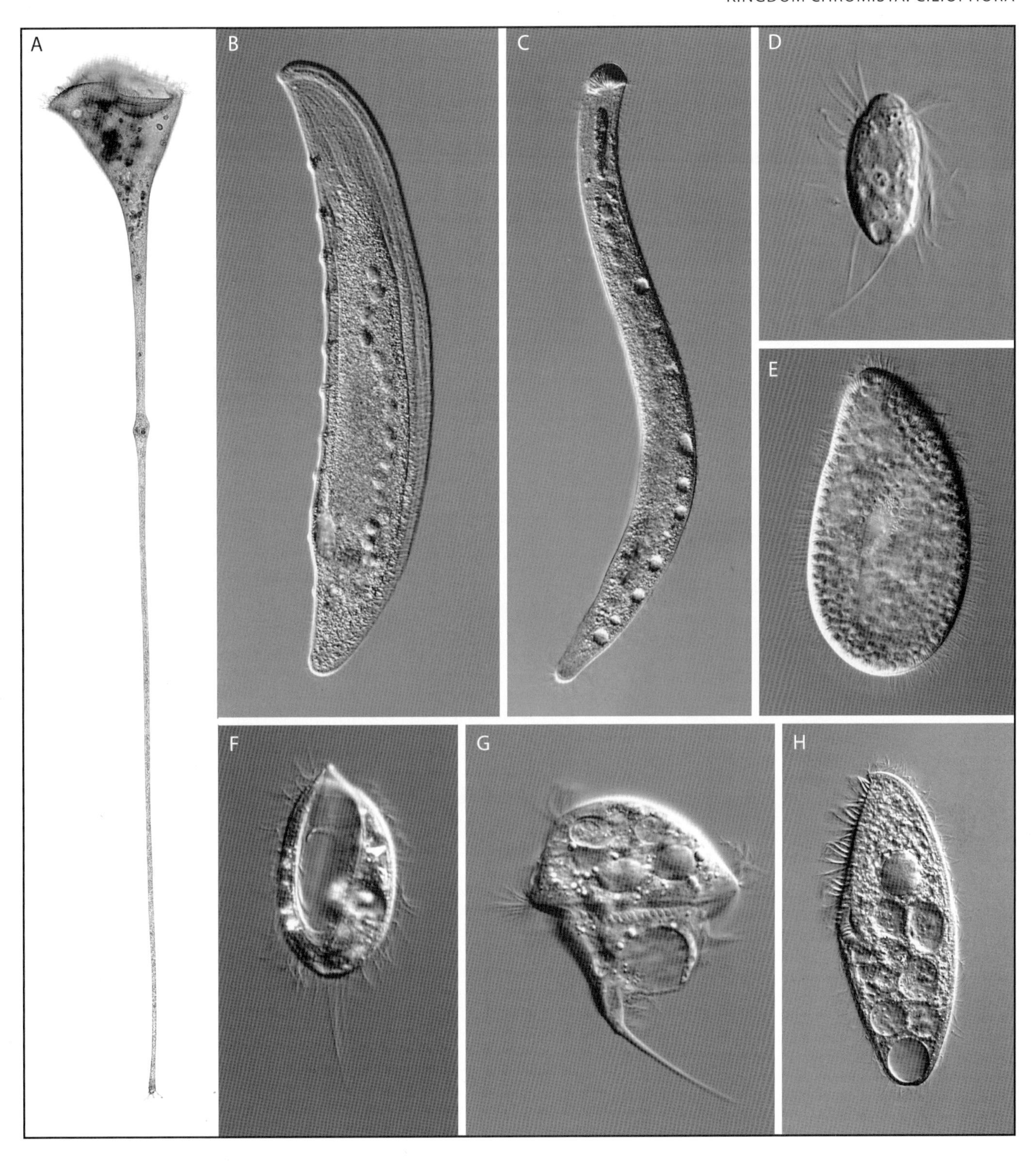

Some conspicuous freshwater ciliates occurring both in New Zealand and in *Sphagnum* ponds in Germany. A: *Stentor coeruleus*. Bluish in life owing to minute cortical granules (see Checklist footnote [h]) (4 mm fully extended). B: *Loxophyllum meleagris*. Flattened and leaflike, it glides on the surface of organic debris and feeds on other protists. The macronucleus comprises a chain of nodules in the body midline (ca. 400 µm long). C: *Homalozoon vermiculare*, a vermiform, strongly contractile ciliate with the mouth at the convex anterior end, opening widely to ingest other ciliates. The macronucleus is composed of a chain of nodules in the body midline, with many contractile vacuoles along the body margin (1 mm when fully extended). D: *Cyclidium glaucoma*, a minute ciliate with comparatively long cilia. It feeds on bacteria digested in globular food vacuoles (30 µm). E: *Paramecium bursaria*, green in life owing to symbiotic green algae ('zoochlorellae') (100 µm). F: *Lembadion lucens*, with a large mouth occupying most of the ventral side (60 µm). G: *Caenomorpha medusula*, anaerobic, with a well-developed caudal spine (150 µm). H: *Blepharisma steinii*, red in life owing to cortical granules (150 µm).

Martin Kreutz

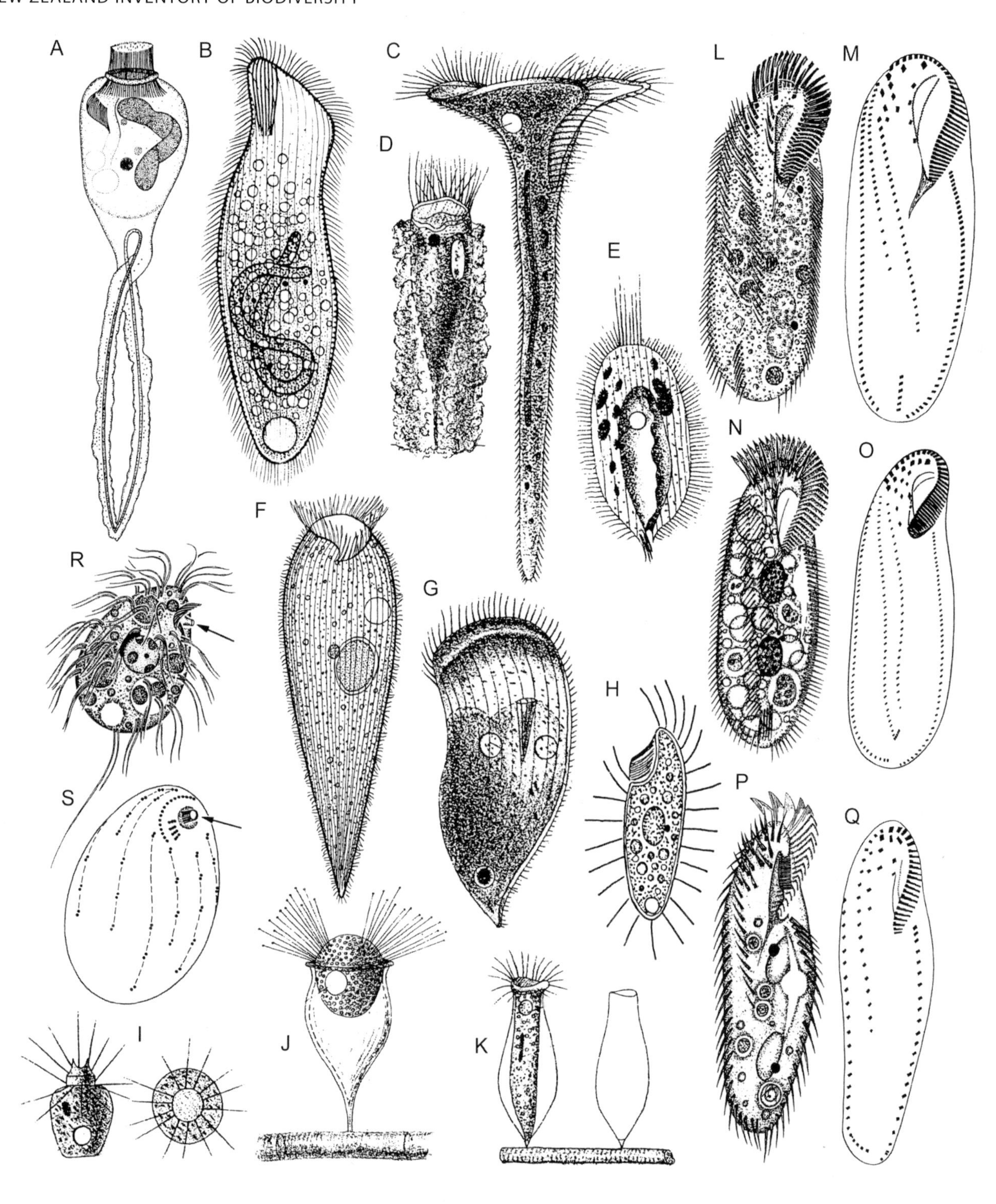

Some remarkable marine (A), freshwater (B–K), and soil (L–S) ciliates originally described from New Zealand (lengths in micrometres). The pairs L–M, N–O, P–Q, R–S are shown before (left) and after silver impregnation, in ventral view except S (lateral). A: *Calipera longipes*, from gills of the striped clingfish *Tracheloschismus melobesia* (140 µm). B: *Cranotheridium taeniatum* (170 µm). C: *Stentor gracilis* (830 µm). D: *Tintinnidium emarginatum* (100 µm). E: *Lembadion lucens* (62 µm). F: *Meseres stentor* (130 µm). G: *Phascolodon elongatus* (130 µm). H: *Cyrtolophosis elongata* (30 µm). I: *Mesodinium phialinum*, lateral and apical views (18 µm). J: *Acineta elegans* on alga (100 µm). K: *Cothurina amphorella* on alga (70 µm). L, M: *Pseudouroleptus buitkampi*, a conspicuous hypotrichous ciliate with a unique cirral pattern, previously known only from an alpine pasture soil in Austria, later found at New Zealand site 8 (150 µm). N, O: *Orthoamphisiella grelli*, an inconspicuous hypotrichous ciliate previously known only from a soil in Antarctica, later found at New Zealand site 10 (80 µm). P, Q: *Keronopsis tasmaniensis*, a conspicuous hypotrichous ciliate with a unique cirral pattern, previously known only from soil in a Tasmanian hop field, later found at New Zealand site 9 (180 µm). R, S: *Pseudoplatyophrya saltans*, an inconspicuous cosmopolitan colpodide ciliate that feeds exclusively on fungal hyphae and spores, which are penetrated by a minute, highly complex feeding tube (arrows), as described by Foissner (1993) (15 µm).

From Laird 1953 (A); Schewiakoff 1892, 1893 (B, F, H); Maskell 1887, 1888 (C–E, G, I–K); Foissner 1982 (L, M); Eigner & Foissner 1993 (N, O); Blatterer & Foissner 1988 (P, Q); Foissner 1988 (R, S)

observed were probably undescribed. It was only 30 years later that this was fully acknowledged (Foissner 1987).

More recently, only Yeates et al. (1991), Yeates and Foissner (1995), and Foissner (1987, 1994) have worked on New Zealand soil protozoans.

Diversity and ecology of New Zealand soil ciliates

A total of 106 soil ciliate species, including one possibly undescribed species and five suctorians identified to genus level only, are presently recognised to occur in New Zealand. Most of them were recorded by J. D. Stout; 34 were new records from three samples W. Foissner investigated in 1987 and 1994 (site descriptions, see below). A diversity of 112 species is pretty low compared with the total number of described soil ciliates globally (almost 1000; Foissner 1998; Foissner et al. 2002) and their estimated actual diversity (up to 2000 species – Foissner 1997b). It is also low when compared with the rather high number of samples investigated by Stout (about 70 as calculated from Stout's papers; see References) because, for instance, two samples from the Murray River floodplain in Australia contained 110 species (Foissner 2000). Stout, being mainly an ecologist, left many species unidentified and used a rather ineffective culture method. Hence, the finding of 34 new records in only three samples indicates that further investigations will undoubtedly reveal many more species, described and undescribed.

None of the known New Zealand soil ciliates is endemic or was originally described from the region. However, some probably have a restricted Gondwanan distribution, namely *Keronopsis tasmaniensis*, discovered by Blatterer and Foissner (1988) in Tasmania, and *Orthoamphisiella grelli*, discovered by Eigner and Foissner (1993) on Gough Island, Antarctica. On the other hand, some species previously known only from the Holarctic (Foissner 1998) were rediscovered in the few samples studied from New Zealand, viz *Amphisiella quadrinucleata* and *Pseudouroleptus buitkampi*.

Knowledge about the ecology of New Zealand's soil ciliates is rather limited, although Stout's papers provide valuable insights. However, all data were obtained with highly selective culture methods. Thus, for instance, Stout never recognised the abundance of mycophagous ciliates in New Zealand soils (*Grossglockneria acuta*, *Mykophagophrys terricola*, *Pseudoplatyophrya nana*, and *P. saltans*). Generally, the community structure is very similar to that from other regions of the world (Foissner 1987, 1998), i.e. hypotrichous and colpodid species dominate. Species number per sample is also in the usual range (Foissner 1997a). However, a very humous sample from site 10 (see below), which contained many and rare testate amoebae (Meisterfeld pers. comm.), harboured only 32 ciliate species. Possibly many did not survive prolonged air-drying, as is the case with rain-forest species (Foissner 1997b).

Description of historical sampling sites and methods

The numbers accorded to sites sampled by J. D. Stout and W. Foissner refer to those given in the end-chapter checklist. Descriptions are brief, providing only the most important features. Unfortunately, Stout gave few details in some of his papers.

Site 1 (Stout 1958) comprised three tussock-grassland soils – Omarama soil near Alexandra, a fine sandy loam high in exchangeable bases and with a very high base saturation with reduced and senile tussock cover; Tekoa soil near Bealey, a silt loam of medium base status formerly covered by beech forest but now under tussock and introduced grasses; and Taupo soil near Waiouru, a sand silt of medium base status and with a vigorous plant cover of tussock and other grasses.

Site 2 (Stout 1960a) comprised litter and topsoil samples from the Wairouru and Broken River area – tussock grassland and an adjoining 18-year-old pasture; tussock grassland and an adjoining field which had been under crop cultivation

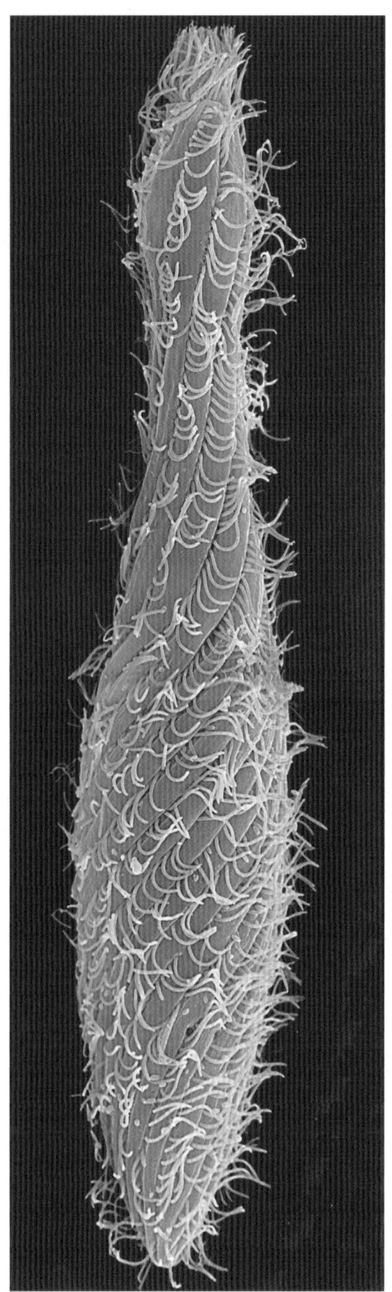

Lacrymaria olor (Litostomatea) is highly contractile, up to 1.2 millimetres long when fully extended, with the ciliary rows running meridionally. It lives in the organic mud of stagnant and running waters, feeding on flagellates and small ciliates.
Wilhelm Foissner

for two years; undisturbed native tussock grassland; an adjoining year-old pasture; and an adjoining crop field in its first year of cultivation.

Site 3 (Stout 1961) comprised litter and topsoil samples from lightly and heavily burnt areas of gorse (*Ulex europaeus*) and manuka (*Leptospermum scoparium*) on the Taita Experimental Station of the former New Zealand Soil Bureau that was at Lower Hutt, North Island.

Site 4 (Stout 1984), 4.5 kilometres southwest of Castlepoint, southeast North Island, was in hill country and received an annual rainfall of 1200 millimetres. The grassland area sampled was at the foot of a slope formed on colluvium dissected by a tributary of Ngakauau Stream. For most of the spring, summer and autumn the site is well drained and may become very dry during the summer and early autumn, but during winter the run-off from the adjacent slope is ponded, and water to a depth of several centimetres tends to lie over much of the site. Nine topsoil (up to 25 millimetres) samples from four adjacent pasture areas (*Lolium perenne, Trifolium repens*) were investigated.

Site 5 (Stout 1956a) protozoa from a 'flourishing beech litter culture' were cultivated in various media. Stout's paper also described experiments with freshwater protozoan communities.

Site 6 (Stout 1978; Cairns et al. 1978) comprised 0–10-centimetre topsoil samples from experimental soil cores of 10 very different (pH 4.7–6.6, percentage carbon 2.7–34.5) pastures in North Island. The test plots were irrigated with either tapwater or effluent from a biological sewage-treatment plant.

Site 7 pertains to the species mentioned in the papers by Stout (1955a) – an experimental study with three *Colpoda* species from soil of unknown origin); Stout (1955b) – greenhouse soil at Cawthron Institute; Stout (1962) – grassland soils and forest litter; and Yeates et al. (1991) – pasture at Silverstream (Lower Hutt), forest at Taita (Lower Hutt), and pasture and forest at Kaitoke (Upper Hutt). None of these papers contains full species lists.

Site 8 was at Mt Herbert, Banks Peninsula, about 900 metres above sea-level, comprising litter and topsoil (0–5 centimetres) from a tussock grassland, pH 6.0. A sample was collected on 23 December 1986 by Dr Wolfgang Petz (Salzburg University) and investigated by W. Foissner on 19 August 1987.

Site 9 was at Birdlings Flat, Gemstone Beach, Banks Peninsula, about 100 metres inland from the shore. The sample, which comprised roots and soil from grass tufts, pH 5.3, was collected on 28 December 1986 by Dr Wolfgang Petz (Salzburg University) and investigated by W. Foissner on 19 August 1987.

Site 10 was at the Orongorongo Vally research site (south Wairarapa, North Island). The three samples comprised humus litter (fresh and partially decayed with many fine roots) and topsoil (0–7 centimetres, pH 5.5) from a forest dominated by silver beech (*Nothofagus menziensii*) at about 800 metres above sea-level. They were collected in August 1994 by Peter Berben (Landcare Research) and investigated by W. Foissner on 13 October 1994.

Most such protozoa are not easily recognised among the innumerable soil particles and many are probably in a dormant (cyst) stage most of their lives. Thus, culture methods are required to make them visible. Stout used a simple technique for most of his investigations (Stout et al. 1982): 'The most convenient medium is a firm agar (2% of a good commercial agar) in a 10-cm petri dish. Add 10 g (wet wt) of soil to the surface of the agar on one side of the dish. Add 20 ml of sterile distilled water carefully to the other side. By keeping the soil to one side, the protozoa are able to migrate to the clear translucent part of the dish and can be observed directly under the microscope with a dry or water immersion lens. Distilled water enhances excystment [hatching], and the soil supplies the nutrients necessary for the bacteria, which grow as a film over the agar surface. Incubation temperature can be varied within limits. The most suitable temperature is from 15 to 20°C.'

Obviously, this is some kind of soil infusion and might explain why Stout, and others who used similar methods, never recognised the abundance of mycophagous ciliates, because most soil fungi do not readily grow under submerged conditions.

Another technique was thus developed – the so-called 'non-flooded petri-dish method' (Foissner 1987). Briefly, this simple method involves placing 10–50 grams of terrestrial material in a petri dish (10–15 centimetres in diameter) and saturating but not flooding it with distilled water. Such cultures are analysed for ciliates by inspecting about two millilitres of the run-off on days 2, 7, 14, 21, and 28. The non-flooded petri-dish method is selective, i.e. probably only a small proportion of the resting cysts present in a sample are reactivated and undescribed species or species with specialised demands are very likely undersampled (Foissner 1997a). Thus, the real number of species, described and undescribed, in the samples investigated is probably much higher. Unfortunately, a better method for broad analysis of soil ciliates is not known. However, about 800 new ciliate species have been discovered during the past 20 years using this simple technique (Foissner 1998).

Rumen and gut ciliates

An astonishing diversity of ciliate species inhabit the guts of herbivores. They live in a mutualistic relationship with their hosts and play an important role in the digestion of plant matter. Best known and most studied are those inhabiting the rumen and reticulum of ruminants. Ciliates are also found in the alimentary tract of a wide range of non-ruminant herbivores and omnivores including, for example, the stomach of camelids (pseudo-ruminants), hippopotamuses, capybara, and macropodid marsupials, and the colon of hindgut fermenters such as horses, tapirs, rhinoceroses, chimpanzees, gorillas, and elephants.

Reviews of the biology of rumen ciliates and their role in ruminant digestion have been provided by Clarke (1977), Williams and Coleman (1992), and Dehority (1993). Taxonomically, rumen ciliates fall into two groups, viz the orders Entodiniomorphida and Vestibuliferida (Lynn & Small 2000). Fourteen to eighteen genera and more than 250 entodiniomorph species have been described on morphological grounds, most of them from the rumen of various animals. Similarly, more than 100 species of vestibuliferids in about 18 genera have been recorded. Molecular techniques will undoubtedly affect the status and numbers of species. Recently, many endemic genera and species have been described from the alimentary tract of Australian marsupial herbivores (Cameron & O'Donoghue 2004).

Rumen ciliates live in highly complex relationships with each other, rumen bacteria, and their hosts. Although some are predatory on other ciliates, most are directly involved in the digestion of the same plant materials as the rumen bacteria with which they compete and which they also ingest. The ciliates can make up a large proportion of the microbial biomass in the rumen and contribute 50% or more of total microbial fermentation products. Nevertheless, defaunation experiments involving chemical removal of the ciliate fauna have shown that they are not essential for the survival and normal functioning of the host. Nor does their removal have a marked effect on animal growth and production because of compensatory increases in bacterial populations and activity in the rumen and increased hindgut digestion.

A substantial amount of research on the role of rumen ciliates in digestion, factors affecting their population dynamics, and their relationship to bloat in cattle has been carried out in New Zealand, notably between the mid-1960s and early 1980s. In the process, ciliates encountered were identified to genus or species level (see checklist). Thirty-five species belonging to eight entodiniomorph and four vestibuliferid genera were described from cattle (Clarke 1964), nine entodiniomorph species (but no vestibuliferids) were recorded from wild red deer (*Cervus elaphus*) (Clarke 1968), and four entodiniomorph genera were recorded from sheep (Clarkeet al. 1982). Three of the species found in deer had also been found in cattle in New Zealand. Since that time little has been published on this subject and much remains to be discovered about the ciliate fauna of both farmed and feral ruminants. However, with current concerns about the potential for rumen gases, particularly methane, to contribute to global warming, there is renewed interest in the role of rumen bacteria and ciliates in their production. The complexity and intimacy of the relationship between methanogenic bacteria (particularly archaebacteria) and rumen ciliates has become increasingly evident in recent years. That a substantial proportion of the methane generated in the rumen is attributable to methanogenic bacteria living on the surface of or within rumen ciliates has been known for some time (Newbold et al. 1995; Tokura et al. 1999) and information is emerging on which particular bacterial-ciliate associations are most involved (e.g. Chagan et al. 2004; Ranilla et al. 2007). To add further to the complexity of these associations, there is also evidence for the horizontal transfer of genes, particularly those involved in metabolism, from rumen eubacteria and archaebacteria to rumen ciliates (Ricard et al. 2006).

Much less has been published worldwide about the ciliates inhabiting the large intestine of herbivores or their significance to digestion. Some species found in the large intestine of ruminants have also been found in the rumen, while others have not. Those present in the large intestine of ruminants in New Zealand do not appear to have been identified. The ciliates of hindgut fermenters such as equids and elephants are generally distinct from those found in ruminants. Most of those described from horses, for example, belong to genera different from those found in ruminants. Nothing is known of those present in equids in New Zealand. There have been very few studies of the role of ciliates in digestion in equids or their interactions with hindgut bacterial and fungal populations. In one it was found that defaunation of ponies resulted in a small decrease in dry matter (but not cellulose) digestibility and an increase in fungal populations in the colon (but not the caecum); there was no effect on bacterial populations (Moore & Dehority 1993).

Balantidium coli is a common inhabitant of the large intestine of pigs worldwide. *Balantidium* organisms have also been recorded from a wide variety of other hosts including humans, other primates, horses, rodents, reptiles, birds, and some invertebrates. While some of these have been given species names, their relationships are not fully understood and many may prove to be synonyms (Levine 1985; Zaman 1993). It is widely considered that human infections are caused by *B. coli* and, in most cases, can be ascribed to direct or indirect contact with pigs. However, comparisons of the antigenicity and *in vitro* cultural characteristics of isolates from humans and other sources, and some epidemiological evidence, suggest this is an oversimplification (see review by Zaman 1993). The organisms are transmitted by means of the relatively resistant cyst stage which is passed in the faeces. Most infections are asymptomatic and the organisms remain within the gut lumen although they may be deep in the crypts of the mucosa. In some circumstances, most often in association with the high carbohydrate diet that encourages the build-up of *Balantidium* populations, organisms penetrate the mucosa and invade the submucosal tissues, causing damage and inflammation. Ulceration of the gut and clinical enteritis may result (Zaman 1993).

Balantidium coli is common in pigs in New Zealand, and sporadic disease cases occur (Anon 1980; Fairley 1996). Infections of calves have also been reported (Dewes 1959), although Levine (1985) suggests that *Balantidium* organisms recorded in ruminants may actually be *Buxtonella sulcata*. It appears that no *Balantidium* infections of humans have been recorded in New Zealand (Fairley 1996). Suspected *Balantidium* organisms have been observed histologically in ostriches (Hooper et al. 1999).

Acknowledgements

This review was partly supported by the Austrian Science Foundation (Project P-12367-BIO). The technical assistance of Dr Eva Herzog, Mag. Maria Pichler and Mag. Birgit Peukert is gratefully acknowledged.

Authors

Dr W. A. G. (Tony) Charleston Institute of Veterinary, Animal and Biomedical Sciences, Massey University, Palmerston North (Present address: 488 College Street, Palmerston North, New Zealand) [charleston@inspire.net.nz] Rumen and gut ciliates

Dr Wilhelm Foissner FB Organismische Biologie, Universität Salzburg, Hellbrunnerstrasse 34, A-5020 Salzburg, Austria [wilhelm.foissner@sbg.ac.at] Introduction, soil ciliates

Dr Dennis P. Gordon National Institute of Water & Atmospheric Research (NIWA), Private Bag 14-901 Kilbirnie, Wellington, New Zealand [d.gordon@niwa.co.nz] Aquatic ciliates

Dr Martin Kreutz Private Laboratory, Magdeburger Strasse 2, D-78467 Konstanz, Germany [makreu@gmx.de] Soil ciliates

References

ANON 1980: Invermay Animal Health Laboratory Report. *Surveillance 7(1)*: 24–26.

BARWICK, R. E.; BEVERIDGE, P. J.; BRAZIER, R. G.; CLOSE, R. I.; HIRSCHFELD, N.; PILLAI, S.; RAMSAY, G. W.; ROBINSON, E. S.; STEVENS, G. R.; TODD, I. M. 1955: Some freshwater ciliates from the Wellington area including eleven species recorded from N.Z. for the first time. *Tuatara 5*: 87–99.

BARY, B. M. 1950a: Studies on the freshwater ciliates of New Zealand. Part I. A general morphology of *Bursaria truncatella* Müller. *Transactions of the Royal Society of New Zealand 78*: 301–310, pl. 36.

BARY, B. M. 1950b: Studies on the freshwater ciliates of New Zealand. Part II. An annotated list of species from the neighbourhood of Wellington. *Transactions of the Royal Society of New Zealand 78*: 311–323, pls 37–39.

BARY, B. M. 1951: Sea-water discolouration. *Tuatara 4*: 41–46.

BARY, B. M. 1953a: Records from New Zealand of seawater discolouration, and a discussion of its significance in the study of water movements. *In*: *Report of the Seventh Science Congress, Royal Society of New Zealand, Christchurch, May 15–21 (inclusive), 1951*. RSNZ, Wellington. vi + 256 p.

BARY, B. M. [1952] 1953b: Sea-water discolouration by living organisms. *New Zealand Journal of Science and Technology B, 34*: 393–407.

BARY, B. M.; STUCKEY, R. G. 1950: An occurrence in Wellington Harbour of *Cyclotrichium meunieri* Powers, a ciliate causing red water, with some additions to its morphology. *Transactions of the Royal Society of New Zealand 78*: 86–92, pl. 13.

BERGER, H. 1999: Monograph of the Oxytrichidae (Ciliophora, Hypotrichia). *Monographiae Biologicae 78*: 1–1080.

BLATTERER, H.; FOISSNER, W. 1988: Beitrag zur terricolen Ciliatenfauna (Protozoa, Ciliophora) Australiens. *Stapfia (Linz) 17*: 1–84.

BOUSTEAD, N. C. 1982: Fish diseases recorded in New Zealand, with a discussion on potential sources and certification procedures. *Fisheries Research Division Occasional Publication 34*: 1–19.

BRACE, E .C.; CASEY, B. R.; COSSHAM, R. B.; McEWAN, J. M.; McFARLANE, B. G.; MACKEN, J.; MONRO, P. A.; MORELAND, J. M.; NORTHERN, J. B.; STREET, R. J.; YALDWYN, J. C. 1953: The frog, *Hyla aurea* as a source of animal parasites. *Tuatara 5*: 12–21.

BURNS, D. A. 1983: The distribution and morphology of tintinnids (ciliate protozoans) from the coastal waters around New Zealand. *New Zealand Journal of Marine and Freshwater Research 17*: 387–406.

CAIRNS, A.; DUTCH, M. E.; GUY, E. M.; STOUT, J. D. 1978: Effect of irrigation with municipal water or sewage effluent on the biology of soil cores. *New Zealand Journal of Agricultural Research 21*: 1–9.

CAMERON, S. L.; O'DONOGHUE, P. J. 2004: Phylogeny and biogeography of the 'Australian' trichostomes (Ciliophora: Litostomatea). *Protist 155*: 215–235.

CASSIE, V. 1961: Marine phytoplankton in New Zealand waters. *Botanica Marina 2 (Suppl.)*: 1–54.

CHAGAN, I.; USHIDA, K. 2004: Detection of methanogens and proteobacteria from a single cell of rumen ciliate protozoa. *Journal of General and Applied Microbiology 50*: 203-212.

CHANG, F. H. 1994: New Zealand: Major shellfish poisoning (NSP) in early '93. *Harmful Algae News 1994(8)*: 1–2.

CLARKE, R. T. J. 1964: Ciliates of the rumen of domestic cattle (*Bos taurus* L.). *New Zealand Journal of Agricultural Research 7*: 248–257.

CLARKE, R. T. J. 1968: The ophryoscolecid ciliates of red deer (*Cervus elaphus* L.) in New Zealand. *New Zealand Journal of Science 11*: 686–692.

CLARKE, R. T. J. 1977: Protozoa in the rumen ecosystem. Pp. 251–275 *in*: Clarke, R. T. J.; Bauchop, T. (eds), *Microbial Ecology of the Gut*. Academic Press, London. xvii + 410 p.

CLARKE, R. T. J.; ULLYATT, M. J.; JOHN, A. 1982: Variation in numbers and mass of ciliate protozoa in the rumens of sheep fed chaffed alfalfa (*Medicago sativa*). *Applied and Environmental Microbiology 43*: 1201–1204.

CORLISS, J. O. 1979: *The Ciliated Protozoa. Characterization, classification, and guide to the literature*. Pergamon Press, Oxford. 455 p.

DAWSON, E. W. 1992: The marine fauna of New Zealand: Index to the fauna 1. Protozoa. *New Zealand Oceanographic Institute Memoir 99*: 1–368.

DEHORITY, B. A. 1993: The rumen protozoa. Pp. 1–42 *in*: Kreier, J. P.; Baker, J. R. (eds), *Parasitic Protozoa. Volume 3*. 2nd edn. Academic Press, San Diego. 333 p.

DETCHEVA, R. B. 1992: Protozoa, Ciliophora. *Catalogi Faunae Bulgaricae 1*: 1–134.

DEWES, H. F. 1959: An occurrence of *Balantidium coli* in calves. *New Zealand Veterinary Journal 7*: 42.

DIGGLES, B. K. 2000: Chemotherapy of *Trichodina* sp. infections in cultured turbot *Colistium nudipinnis* with notes on the susceptibility to *Trichodina* sp. of fish with abnormal pigmentation. *New Zealand Journal of Marine and Freshwater Research 34*: 645–652.

DIGGLES, B. K.; HINE, P. M.; HANDLEY, S.; BOUSTEAD, N. C. 2002: A handbook of diseases of importance to aquaculture in New Zealand. *NIWA Science and Technology Series 49*: 1–200.

DINI, F.; LUCCHESI, P.; MACCHIONI, G. 1995: 'Protozoa'. *In*: Minelli, A.; Ruffo, S.; La Posta, S. (eds), *Checklist delle Specie della Fauna Italiana 1*: 1–92.

EIGNER, P.; FOISSNER, W. 1993: Divisional morphogenesis in *Orthoamphisiella stramenticola* Eigner & Foissner, 1991 and *O. grelli* nov. spec. (Ciliophora, Hypotrichida). *Archiv für Protistenkunde 143*: 337–345.

FAIRLEY, R. A. 1996: Infectious agents and parasites of New Zealand pigs transmissible to humans. *Surveillance 23*: 17–18.

FOISSNER, W. 1982: Ökologie und Taxonomie der Hypotrichida (Protozoa: Ciliophora) einiger österreichischer Böden. *Archiv für Protistenkunde 126*: 19–143.

FOISSNER, W. 1987: Soil protozoa: fundamental problems, ecological significance, adaptations in ciliates and testaceans, bioindicators, and guide to the literature. *Progress in Protistology 2*: 69–212.

FOISSNER, W. 1988: Gemeinsame Arten in der terricolen Ciliatenfauna (Protozoa: Ciliophora) von Australien und Afrika. *Stapfia (Linz) 17*: 85–133.

FOISSNER, W. 1993: Colpodea (Ciliophora). *Protozoenfauna 4*: x, 1–798.

FOISSNER, W. 1994: Soil protozoa as bioindicators in ecosystems under human influence. Pp. 147–193 *in*: Darbyshire, J.F. (ed.), *Soil Protozoa*. CAB International, Oxford. 224 p.

FOISSNER, W. 1997a: Global soil ciliate (Protozoa, Ciliophora) diversity: a probability-based approach using large sample collections from Africa, Australia and Antarctica. *Biodiversity and Conservation 6*: 1627–1638.

FOISSNER, W. 1997b: Soil ciliates (Protozoa: Ciliophora) from evergreen rain forests of Australia, South America and Costa Rica:

diversity and description of new species. *Biology and Fertility of Soils 25*: 317–339.
FOISSNER, W. 1998: An updated compilation of world soil ciliates (Protozoa, Ciliophora), with ecological notes, new records, and descriptions of new species. *European Journal of Protistology 34*: 195–235.
FOISSNER, W. 1999: Soil protozoa as bioindicators: pros and cons, methods, diversity, representative examples. *Agriculture, Ecosystems and Environment 74*: 95–112.
FOISSNER, W. 2000: Protist diversity: estimates of the near-imponderable. *Protist 150*: 363–368.
FOISSNER, W.; BERGER, H. 1999: Identification and ontogenesis of the *nomen nudum* hypotrichs (Protozoa: Ciliophora) *Oxytricha nova* (= *Sterkiella nova* sp. n.) and *O. trifallax* (= *S. histriomuscorum*). *Acta Protozoologica 38*: 215–248.
FOISSNER, W.; FOISSNER, I. 1988: Stamm: Ciliophora. *Catalogus Faunae Austriae 1c*: 1–147.
FOISSNER, W.; LEIPE, D. 1995: Morphology and ecology of *Siroloxophyllum utriculariae* (Penard, 1922) n. g., n. comb. (Ciliophora, Pleurostomatida) and an improved classification of pleurostomatid ciliates. *Journal of Eukaryotic Microbiology 42*: 476–490.
FOISSNER, W.; AGATHA, S.; BERGER, H. 2002: Soil ciliates (Protozoa, Ciliophora) from Namibia (Southwest Africa), with emphasis on two contrasting environments, the Etosha Pan and the Namib Desert. Part I: Text and line drawings. Part II: Photographs. *Denisia 5*: 1–1459.
FOISSNER, W.; BERGER, H.; KOHMANN, F. 1992: Taxonomische und ökologische Revision der Ciliaten des Saprobiensystems – Band II: Peritrichia, Heterotrichida, Odontostomatida. *Informationsberichte des Bayerischen Landesamtes für Wasserwirtschaft 5/92*: 1–502.
FOISSNER, W.; BERGER, H.; KOHMANN, F. 1994: Taxonomische und ökologische Revision der Ciliaten des Saprobiensystems – Band III: Hymenostomata, Prostomatida, Nassulida. *Informationsberichte des Bayerischen Landesamtes für Wasserwirtschaft 1/94*: 1–548.
FOISSNER, W.; BLATTERER, H.; BERGER, H; KOHMANN, F. 1991: Taxonomische und ökologische Revision der Ciliaten des Saprobiensystems – Band I: Cyrtophorida, Oligotrichida, Hypotrichia, Colpodea. *Informationsberichte des Bayerischen Landesamtes für Wasserwirtschaft 1/91*: 1–478.
FOISSNER, W.; BLATTERER, H.; BERGER, H; KOHMANN, F. 1995: Taxonomische und ökologische Revision der Ciliaten des Saprobiensystems – Band IV: Gymnostomatea, *Loxodes*, Suctoria. *Informationsberichte des Bayerischen Landesamtes für Wasserwirtschaft 1/95*: 1–540.
GORDON, D. P. 1972: Biological relationships of an intertidal bryozoan population. *Journal of Natural History 6*: 503–514.
HINE, P. M. 1982: Potential disease in snapper farming. *In*: Smith, P. J.; Taylor, J. L. (eds), *Prospects for Snapper Farming and Reseeding in New Zealand. Fisheries Research Division Occasional Publication 37*: 21–23.
HINE, P. M.; JONES, J. B.; DIGGLES, B. K. 2000: A checklist of the parasites of New Zealand fishes, including previously unpublished records. *NIWA Technical Report 75*: 1–93 [+ 2].
HOOPER, C.; WILLIAMSON, C.; FRASER, T; JULIAN, A.; HILL, F.; THOMPSON, J.; COOKE, M.; McKENNA, P. 1999. Quarterly review of diagnostic cases. *Surveillance 26*: 16–17.
HUTTON, F. W. 1878: On a new infusorian parasitic on *Patella argentea*. *Transactions of the New Zealand Institute 11*: 330.
JOHNSTON, T. H. 1938: Parasitic infusoria from Macquarie Island. *Scientific Reports of the Australasian Antarctic Expedition 1911-14. Series C. 1(3)*: 1–12.
KAHL, A. 1930: Urtiere oder Protozoa. I. Wimpertiere oder Ciliata (Infusoria). 1. Allgemeiner Teil und Prostomata. *Die Tierwelt Deutschlands 18*: 1–180.
KAHL, A. 1931: Urtiere oder Protozoa. I. Wimpertiere oder Ciliata (Infusoria). 2. Holotricha außer den im ersten Teil behandelten Prostomata. *Die Tierwelt Deutschlands 21*: 181–398.
KAHL, A. 1932: Urtiere oder Protozoa. I. Wimpertiere oder Ciliata (Infusoria). 3. Spirotricha. *Die Tierwelt Deutschlands 25*: 399–650.
KAHL, A. 1935: Urtiere oder Protozoa. I. Wimpertiere oder Ciliata (Infusoria). 4. Peritricha und Chonotricha. *Die Tierwelt Deutschlands 30*: 651–886.
KIRK, T. W. 1886: On some new Vorticellae collected in the neighbourhood of Wellington. *Transactions of the New Zealand Institute 18*: 215–217.
KIRK, T.W. 1887: New infusoria from New Zealand. *Annals and Magazine of Natural History 19*: 439–441.
KOFOID, C. A.; CAMPBELL, A. S. 1929: A conspectus of the marine and freshwater ciliata belonging to the suborder Tintinnoinea, with descriptions of new species principally from the AGASSIZ expedition to the eastern tropical Pacific, 1904–1905. *University of California Publications in Zoology 34*: 1–403.
KREUTZ, M.; FOISSNER, W. 2006: The *Sphagnum* ponds of Simmelreid in Germany: a biodiversity hot-spot for microscopic organisms. *Protozoological Monographs 3*: 1–267.
LAIRD, M. 1953: The protozoa of New Zealand intertidal zone fishes. *Transactions of the Royal Society of New Zealand 81*: 79–143.
LEVINE, N. D. 1985: *Veterinary Protozoology*. Iowa State University Press, Ames. ix + 414 p.
LYNN, D. H. 2003a: Morphology or molecules: How do we identify the major lineages of ciliates (phylum Ciliophora)? *European Journal of Protistology 39*: 356–364.
LYNN, D. H. 2003b: The ciliate resource archive. [http://www.uoguelph.ca/~ciliates]
LYNN, D. H.; SMALL, E. B. 2000: Phylum Ciliophora Doflein, 1901. Pp. 371–656 *in*: Lee, J. J.; Leedale, G. F.; Bradbury, P. (eds), *An Illustrated Guide to the Protozoa, Second Edition. Organisms traditionally referred to as Protozoa, or newly discovered groups*. Society of Protozoologists, Lawrence. v + 1432 p. (2 vols).
MacKENZIE, A. L.; GILLESPIE, P. A. 1986: Plankton ecology and productivity, nutrient chemistry, and hydrology of Tasman Bay, New Zealand, 1982–1984. *New Zealand Journal of Marine and Freshwater Research 20*: 365–395.
MASKELL, W. M. [1885] 1886: On the fresh-water infusoria of the Wellington District. *Transactions of the New Zealand Institute 19*: 49–61, pls 3–5.
MASKELL, W.M. [1886] 1886: On the fresh-water infusoria of the Wellington District. *Transactions of the New Zealand Institute 20*: 3–19, pls 1–4.
MATIS, D.; TIRJAKOVÁ, E.; STLOUKAL, E. 1996: Ciliophora in the database of Slovak fauna [Nálevníky (Ciliophora) v databanke fauny slovenska.] *Folia Faunistica Slovaca 1*: 3–37.
McRAE, A. 1959: *Evechinus chloroticus* (Val.) an endemic New Zealand echinoid. *Transactions of the Royal Society of New Zealand 86*: 205–267, pls 12, 13.
MOORE, B. E.; DEHORITY, B. A. 1993: Effects of diet and hindgut defaunation on diet digestibility and microbial concentrations in the cecum and colon of the horse. *Journal of Animal Science 71*: 3350–3358.
NEWBOLD, C. J.; LASSALAS, B.; JOUANY, J. P. 1995: The importance of methanogens associated with ciliate protozoa in ruminal methane production in vitro. *Letters in Applied Microbiology 21*: 230-234.
RANILLA, M. J.; JOUANY, J.-P; MORGAVI, D. P. 2007: Methane production and substrate degradation by rumen microbial communities containing single protozoal species in vitro. *Letters in Applied Microbiology 45*: 675-680.
RICARD, G.; McEWEN, N. R.; DUTILH, B. E.; JOUANY, J. P.; MACHEBOEUF, D.; MITSUMORI, M.; McINTOSH, F. M.; MICHALOWSKI, T.; NAGAMINE, T.; NELSON, N.; NEWBOLD, C. J.; NSABINAMA, E.; TAKENAKA, A.; THOMAS, N. A.; USHIDA, K.; HACKSTEIN, J. H. P.; HUYNEN, M. A. 2006: Horizontal gene transfer from bacteria to rumen ciliates indicates adaptation to their anaerobic, carbohydrates-rich environment. *BMC GENOMICS 7:22*: 1–13.
SCHEWIAKOFF, W. 1892: Ueber die geographische Verbreitung der Süsswasser-Protozoën. *Verhandlungen des naturhistorisch-medizinischen Vereins zu Heidelberg, n.s,. 4*: 544–567.
SCHEWIAKOFF, W. 1893: Über die geographische Verbreitung der Süsswasser-Protozoën. *Zapiski Imperatorskoi Akademii Nauk, ser. 7* [*Mémoires de l'Académie Impériale des Sciences de St-Pétersbourg*] *41*: 1–201.
STOUT, J. D. 1952: Protozoa and the soil. *Tuatara 4*: 103–107.
STOUT, J. D. 1954a: Some observations on the ciliate fauna of an experimental meat digestion plant. *Transactions of the Royal Society of New Zealand 82*: 199–211.
STOUT, J. D. 1954b: The ecology, life history and parasitism of *Tetrahymena* [*Paraglaucoma*] *rostrata* (Kahl) Corliss. *Journal of Protozoology 1*: 211–215.
STOUT, J. D. 1955a: Environmental factors affecting the life history of three soil species of *Colpoda* (Ciliata). *Transactions of the Royal Society of New Zealand 82*: 1165–1188.
STOUT, J. D. 1955b: The effect of partial steam sterilization on the protozoan fauna of a greenhouse soil. *Journal of General Microbiology 12*: 237–240.
STOUT, J. D. 1956a: *Saprophilus muscorum* Kahl, a tetrahymenal ciliate. *Journal of Protozoology 3*: 28–30.
STOUT, J. D. 1956b: Excystment of *Frontonia depressa* (Stokes) Penard. *Journal of Protozoology 3*: 31–32.
STOUT, J. D. 1956c: Reaction of ciliates to environmental factors. *Ecology 37*: 178–191.
STOUT, J. D. 1958: Biological studies of some tussock-grassland soils VII. Protozoa. *New Zealand Journal of Agricultural Research 1*: 974–984.
STOUT, J. D. 1960a: Morphogenesis in the ciliate *Bresslaua vorax* Kahl and the phylogeny of the Colpodidae. *Journal of Protozoology 7*: 26–35.
STOUT, J. D. 1960b: Biological studies of some

tussock-grassland soils XVIII. Protozoa of two cultivated soils. *New Zealand Journal of Agricultural Research 3*: 237–244.
STOUT, J. D. 1961: Biological and chemical changes following scrub burning on a New Zealand hill soil. *New Zealand Journal of Agricultural Research 4*: 740–752.
STOUT, J. D. 1962: An estimation of microfaunal populations in soils and forest litter. *Journal of Soil Science 13*: 314–320.
STOUT, J. D. 1978: Effect of irrigation with municipal water or sewage effluent on the biology of soil cores. *New Zealand Journal of Agricultural Research 21*: 11–20.
STOUT, J. D. 1980: The role of Protozoa in nutrient cycling and energy flow. *Advances in Microbial Ecology 4*: 1–50.
STOUT, J. D. 1984: The protozoan fauna of a seasonally inundated soil under grassland. *Soil Biology and Biochemistry 16*: 121–125.
STOUT, J. D.; HEAL, O. W. 1967: Protozoa. Pp. 149–195 *in*: Burges, A.; Raw, F. (eds), *Soil Biology*. Academic Press, New York. 532 p.
STOUT, J. D.; BAMFORTH, S. S.; LOUSIER, J. D. 1982: Protozoa. Pp. 1103–1120 *in*: Page, A. L. (ed.), *Methods of Soil Analysis, Part 2. Chemical and Microbiological Properties*. [Agronomy Monograph No. 9 (2nd edn).] ASA-SSSA, Madison. 1159 p.
TAYLOR, F. J. 1973: Phytoplankton and nutrients in the Hauraki Gulf approaches. Pp. 485–482 *in*: Fraser, R. (comp.), *Oceanography of the South Pacific 1972*. N.Z. National Commission for UNESCO, Wellington. 524 p.
TOKURA, M.; CHAGAN, I.; USHIDA, K.; KOJIMA, Y. 1999: Phylogenetic study of methanogens associated with rumen ciliates. *Current Microbiology 39*: 123-128.
WILLIAMS, A. G.; COLEMAN, G. S. 1992: *The Rumen Protozoa*. Springer-Verlag, New York. xii + 441 p.
WOODCOCK, H. M.; LODGE, O. 1921: Protozoa. Part 1.– Parasitic Protozoa. *Natural History Reports. British Antarctic (Terra Nova) Expedition, 1910, Zoology 6(1)*: 1–124., pls 1–3.
YEATES, G. W.; FOISSNER, W. 1995: Testate amoebae as predators of nematodes. *Biology and Fertility of Soils 20*: 1–7.
YEATES, G. W.; BAMFORTH, S. S.; ROSS, D. J.; TATE, K. R., SPARLING, G. P. 1991: Recolonization of methyl bromide sterilized soils under four different field conditions. *Biology and Fertility of Soils 11*: 181–189.
ZAMAN, V. 1993: *Balantidium coli*. Pp. 43–63 *in*: Kreier, J. P.; Baker, J. R. (eds), *Parasitic Protozoa. Volume 3*. 2nd edn. Academic Press, San Diego. 333 p.

Checklist of New Zealand Ciliophora

The list contains 308 ciliate species from the literature plus 34 new records of soil ciliates (marked by asterisks) from three samples (sites 8–10) investigated by W. Foissner in 1987 and 1994. Classification is based on Lynn and Small (2000) and Lynn (2003a,b). Abbreviations are as follows: E, endemic; F, freshwater; M, marine; T, terrestrial. Marine species (including parasitic species) are mainly those listed by Dawson (1992) for the Exclusive Economic Zone (EEZ).

For soil ciliates (T), the site number from where samples were collected [see Description of historical sampling sites and methods (in text)] is given after each entry; † signifies a doubtful record, i.e. a typical freshwater species not yet found by W. Foissner in terrestrial habitats;.* signifies a new record for New Zealand from sites 8, 9, 10; • signifies a species described in detail by Stout (1954a,b, 1956a,b, 1960a).

Basically the checklist is non-annotated, that is, species are listed as given in the original publications and synonymies have been only partially removed, mainly in those groups for which detailed revisions are available. See Berger (1999), Foissner and Foissner (1988), Foissner (1993, 1998), Foissner and Leipe (1995), Foissner et al. (1991, 1992, 1994, 1995, 2002), Kahl (1930, 1931, 1932, 1935), and Kofoid and Campbell (1929) for names and dates of combining authors and literature. Nomenclature has also been adapted to these reviews. Usually, only taxa identified to species level have been included. Note also:

(a) A junior synonym of *S. minus* Roux (1901) according to Foissner et al. (1992) Found by Stout (1978) in sewage-irrigated soil.

(b) Very likely a senior synonym of *Acineria uncinata* Tucolesco according to Foissner et al. (1995).

(c) The author of this species is not readily discoverable.

(d) Likely misidentified *V. astyliformis* [see Foissner et al. (1992) for detailed revision of this type of peritrich].

(e) A a misidentified species of the *Vorticella infusionum* complex (see Foissner et al. 1992), as evident from the figures in Stout (1954a), which show a horseshoe shaped macronucleus in the transverse axis of the cell.

(f) Status uncertain.

(g) Likely a misidentification as this is a freshwater species.

(h) Kahl (1932) suggested synonymy with *S. coeruleus* (Pallas), but the large size (up to 3.125 mm) and the filiform posterior body half indicates that it could be a distinct, probably endemic, species because many observations have shown that *S. coeruleus* is smaller (up to 2.0 mm) and stouter (Foissner et al. 1992). However, Kreutz and Foissner (2006) have shown that the European *S. coeruleus* is indeed very similar to Maskell's species when observed under optimal conditions).

KINGDOM CHROMISTA
SUBKINGDOM HAROSA
INFRAKINGDOM ALVEOLATA
PHYLUM CILIOPHORA
SUBPHYLUM POSTCILIODESMATOPHORA
Class KARYORELICTEA
Order LOXODIDA
LOXODIDAE
Loxodes rostrum (Müller, 1773) F

Class HETEROTRICHEA
Order HETEROTRICHIDA
BLEPHARISMIDAE
Blepharisma hyalinum Perty, 1849 F/T 1,3,9
Blepharisma lateritium (Ehrenberg, 1831) F/T 4
Blepharisma steini Kahl, 1932 F/T 1,3,5,6
FOLLICULINIDAE
?*Ascobius* sp. Gordon 1972 M
Echinofolliculina mortenseni Dons, 1935 M E?
cf. *Lagotia expansa* (Levinsen, 1893) M
SPIROSTOMIDAE
Spirostomum ambiguum ambiguum (Müller, 1786) F
Spirostomum a. major Roux, 1909 F (f)
Spirostomum intermedium Kahl, 1932 F/T 6 (a)
STENTORIDAE
Stentor attenuatus Maskell, 1887 F (h)
Stentor gracilis Maskell, 1886 (? *S. roeseli*) F E?
Stentor multiformis (Müller, 1786)† F/T 3
Stentor polymorphus (Müller, 1773) F
Stentor roeselii Ehrenberg, 1835 F
Stentor striatus Maskell, 1886 (? *S. coeruleus*) F E?

Order LICNOPHORIDA
LICNOPHORIDAE
Licnophora setifera Maskell, 1886 F E? (very curious species)

SUBPHYLUM INTRAMACRONUCLEATA
Class SPIROTRICHEA
Subclass PHACODINIIDIA
Order PHACODINIIDA
PHACODINIIDAE
Phacodinium metchnikoffi (Certes, 1891) T 3,5

Subclass HYPOTRICHIA
Order EUPLOTIDA
ASPIDISCIDAE
Aspidisca cicada (Müller, 1786) F (formerly *A. costata*)
Aspidisca lynceus (Müller, 1773) F
Aspidisca turrita (Ehrenberg, 1831) F
EUPLOTIDAE
Euplotes aediculatus (Pierson, 1943) F
Euplotes charon (Müller, 1786) F/M
Euplotes muscicola Kahl, 1932 F/T 3
Euplotes patella (Müller, 1773) F

Subclass CHOREOTRICHIA
Order TINTINNIDA
ASCAMPBELLIELLIDAE
Acanthostomella gracilis (Brandt, 1896) M
Acanthostomella minutissima Kofoid & Campbell, 1929 M
CODONELLIDAE
Codonella elongata Kofoid & Campbell, 1929 M
Codonella robusta Kofoid & Campbell, 1929 M
Tintinnopsis cylindrica Daday, 1887 M
Tintinnopsis laevigata Kofoid & Campbell, 1929 M
Tintinnopsis minuta Wailes, 1925 M
Tintinnopsis parvula Jörgensen, 1912 M
Tintinnopsis radix (Imhof, 1886) M
Tintinnopsis rapa Meunier, 1910 M
Tintinnopsis rotundata Jörgensen, 1899 M
Tintinnopsis sacculus Brandt, 1896 M
CODONELLOPSIDAE
Codonellopsis morchella Cleve, 1899 M
Luminella pacifica (Kofoid & Campbell, 1929) M
Stenosemella nivalis (Meunier, 1910) M
CYTTAROCYLIDIDAE
Cyttarocylis eucecryphalus (Haeckel, 1887) M
Cyttarocylis magna (Brandt, 1906) M
DICTYOCYSTIDAE
Dictyocysta dilatata (Brandt, 1906) M
Dictyocysta fenestrata Kofoid & Campbell, 1929 M
Dictyocysta lata Kofoid & Campbell, 1929 M
Dictyocysta reticulata Kofoid & Campbell, 1929 M
Dictyocysta tiara Haeckel, 1873 M
EPIPLOCYLIDIDAE
Epiplocylis acuminata (Daday, 1887) M
Epiplocylis blanda Jörgensen, 1924 M
Epiplocylis inflata Kofoid & Campbell, 1929 M
Epiplocylis lata Kofoid & Campbell, 1929 M
METACYLIDIDAE
Climacocylis scalaria (Brandt, 1906) M
Coxliella fasciata (Kofoid, 1905) M
Helicostomella kiliensis (Laackmann, 1906) M
PETALOTRICHIDAE
Petalotricha ampulla (Fol, 1881) M
Petalotricha serrata Kofoid & Campbell, 1929 M
RHABDONELLIDAE
Protorhabdonella curta (Cleve, 1901) M
Rhabdonella amor (Cleve, 1899) M
Rhabdonella torta Kofoid & Campbell, 1929) M
TINTINNIDAE
Amphorides brandtii (Jörgensen, 1924) M
Dadayiella ganymedes (Entz, 1884) M
Eutintinnus macilentus (Jörgensen, 1924) M
Eutintinnus rugosus (Kofoid & Campbell, 1929) M
Tintinnidium fluviatile (Stein, 1863) F
Tintinnidium emarginatum Maskell, 1888 F (? *T. semiciliatum* Sterki, 1879)
UNDELLIDAE
Proplectella fastigata (Jörgensen, 1924) M
XYSTONELLIDAE
Favella ehrenbergii (Claparède & Lachmann, 1858) M
Xystonella clavata Jörgensen, 1924 M
Xystonella treforti (Daday, 1887) M

Subclass STICHOTRICHIA
Order SPORADOTRICHIDA
AMPHISIELLIDAE
Amphisiella quadrinucleata Berger & Foissner, 1989* T 8
Orthoamphisiella grelli Eigner & Foissner, 1993* T 10
Pseudouroleptus buitkampi (Foissner, 1982)* T 8
Tetrastyla oblonga Schewiakoff, 1892 F E?
SPIROFILIIDAE
Stichotricha remex (Hudson, 1875) F
Stichotricha secunda Perty, 1849 F
KAHLIELLIDAE
Engelmanniella mobilis (Engelmann, 1862) T 1,2,6
Kahliella simplex (Horvath, 1934) F/T
Psilotricha acuminata Stein, 1859 F
KERONIDAE
Kerona pediculus (Müller, 1773) F
Keronopsis tasmaniensis Blatterer & Foissner, 1988* T 9
Keronopsis sp. (n. sp.?)* T 10
Paraholosticha lichenicola Gellért, 1955 T 2

Order SPORADOTRICHIDA
OXYTRICHIDAE
Cyrtohymena quadrinucleata (Dragesco & Njiné, 1971)* F/T 9
Gastrostyla steinii Engelmann, 1862 F/T
Histrio acuminatus Maskell, 1886 F E? (doubtful)
Onychodromus grandis Stein, 1859† F/T 5
Oxytricha fallax Stein, 1859† F/T 3,4,6,7
Oxytricha granulifera Foissner & Adam, 1983* T 9,10
Oxytricha minor (Maskell, 1887) F (as *Opisthotricha parallela* var. *minor*)
Oxytricha parallela Engelmann, 1862 F
Oxytricha setigera Stokes, 1891 F/T 1,3–6,9
Steinia platystoma (Ehrenberg, 1831) F
Sterkiella histriomuscorum-complex Foissner & Berger, 1999* F/T 9,10
Stylonychia mytilus (Müller, 1773) F/T 4
Stylonychia notophora Stokes, 1888 F
Stylonychia putrina Stokes, 1885 F
Tachysoma pellionellum (Müller, 1773)† F/T 1–5,7
Urosoma acuminata (Stokes, 1887)* F/T 8
Urosomoida agiliformis Foissner, 1982* F/T 9
TRACHELOSTYLIDAE
Gonostomum affine (Stein, 1859) F/T 1–6,8–10
Hemisincirra inquieta Hemberger, 1985* T 8

Order UROSTYLIDA
UROSTYLIDAE
Holosticha muscorum (Kahl, 1932) T 1–4,7
Uroleptus musculus (Kahl, 1932) F/T 2,3,6
Uroleptus piscis (Müller, 1773)† F/T 4,6

Subclass OLIGOTRICHIA
Order HALTERIIDA
HALTERIIDAE
Halteria chlorelligera Kahl, 1932 F
Halteria grandinella (Müller, 1773) F/T 1,3–6
Meseres cordiformis Schewiakoff, 1892 F E?
Meseres stentor Schewiakoff, 1892 F E?
STROBILIDIIDAE
Strobilidium adhaerens Schewiakoff, 1892 F (? *S. gyrans*)

Order STROMBIDIIDA
STROMBIDIIDAE
Strombidium claparedi Kent, 1882 F (? *Strobilidium gyrans*)
Strombidium intermedium Maskell, 1887 F E? (? *Strobilidium gyrans*)
Strombidium sulcatum Claparède & Lachmann, 1859 M

Order ODONTOSTOMATIDA
EPALXELLIDAE
Epalxella mirabilis (Roux, 1901) F

Class ARMOPHOREA
Order ARMOPHORIDA
METOPIDAE
Metopus es (Müller, 1776) F
Brachonella spiralis (Smith, 1897) F
CAENOMORPHIDAE
Caenomorpha medusula Perty, 1852 F

Order CLEVELANDELLIDA
NYCTOTHERIDAE
Nyctotherus cordiformis (Ehrenberg, 1838) F

Class LITOSTOMATEA
Subclass HAPTORIA
Order CYCLOTRICHIDA
MESODINIIDAE
Mesodinium phialinum Maskell, 1887 F E?
Myrionecta rubra (Lohmann, 1908) M

Order HAPTORIDA
DIDINIIDAE
Cyclotrichium meunieri Powers, 1932 M
ENCHELYIDAE
Haematophagus megapterae Woodcock & Lodge, 1920 M
HOMALOZOONIDAE
Homalozoon vermiculare (Stokes, 1887) F
LACRYMARIIDAE

Chaenea crassa Maskell, 1887 F E?
Lacrymaria olor (Müller, 1786)† F/T 1,3
Lacrymaria filiformis Maskell, 1886 F E? (as *Trachelocerca*)
PSEUDOHOLOPHRYIDAE
Pseudoholophrya terricola Berger, Foissner & Adam, 1984* T 9
SPATHIDIIDAE
Arcuospathidium atypicum (Wenzel, 1953)* T 10
Arcuospathidium muscorum (Dragesco & Dragesco-Kernéis, 1979)* T 8,9
Bryophyllum loxophylliforme Kahl, 1931 T 3
Cranotheridium taeniatum Schewiakoff, 1892 F E?
Epispathidium amphoriforme (Greeff, 1888) T 3
Epispathidium ascendens (Wenzel, 1955)* T 9
Epispathidium papilliferum (Kahl, 1930) T 3
Spathidium spathula (Müller, 1773) T 4
TRACHELIIDAE
Dileptus americanus Kahl, 1931 T 1
Dimacrocaryon amphileptoides (Kahl, 1931) T 3
Dileptus anguillula Kahl, 1931 T 1–3
Dileptus anser (Müller, 1773) F
Dileptus binucleatus Kahl, 1931 T 3
Dileptus conspicuus Kahl, 1931 T 3
Dileptus irregularis (Maskell, 1887) F E? (as *Amphileptus*)
Dileptus margaritifer (Ehrenberg, 1838) F/T 4
Trachelius tracheloides (Maskell, 1887) F E? (as *Amphileptus*)
Trachelius ovum (Ehrenberg, 1831) F (probably incl. *Amphileptus rotundus* Maskell, 1887)
TRACHELOPHYLLIDAE
Trachelophyllum pusillum (Perty, 1852) F/T 4 (b)

Order PLEUROSTOMATIDA
LITONOTIDAE
Acineria incurvata Dujardin, 1841 F
Amphileptus irregularis (Maskell, 1888) F E? (see *Trachelius ovum*)
Litonotus fasciola (Müller, 1773) F
Litonotus muscorum (Kahl, 1931) T 3
Loxophyllum meleagris (Müller, 1773) F
Siroloxophyllum utriculariae (Penard, 1922)† F/T 1,2

Subclass TRICHOSTOMATIA
Order VESTIBULIFERIDA
BALANTIDIIDAE
Balantidium coli (Malmsten, 1857) A T P
ISOTRICHIDAE
Dasytricha ruminantium Schuberg, 1888 A T P
Dasytricha sp. indet. Clarke et al. 1982 A T P
Isotricha intestinalis Stein, 1859 A T P
Isotricha prostoma Stein, 1859 A T P

Order ENTODINIOMORPHIDA
Suborder ARCHISTOMATINA
BUETSCHLIIDAE
Buetschlia parva Schuberg, 1888 A T P

Suborder BLEPHAROCORYTHINA
BLEPHAROCORYTHIDAE
Charonina equi (Hsiung, 1930) A T P

Suborder ENTODINIOMORPHINA
OPHRYOSCOLECIDAE
Diplodinium anacanthum f. *anisacanthum* (Dogiel, 1925) A T P
Diplodinium a. f. *diacanthum* (Dogiel, 1925) A T P
Diplodinium a. f. *monacanthum* (Dogiel, 1925) A T P
Diplodinium a. f. *pentacanthum* (Dogiel, 1925) A T P
Diplodinium a. f. *tetracanthum* (Dogiel, 1925) A T P
Diplodinium a. f. *triacanthum* (Dogiel, 1925) A T P
Diplodinium costatum Dogiel, 1925 A T P
Diplodinium dogieli Kofoid & MacLennan, 1932 A T P
Entodinium bicarinatum da Cunha, 1914 A T P
Entodinium biconcavum Kofoid & MacLennan, 1930 A T P
Entodinium dilobum Dogiel, 1927 A T P
Entodinium dubardi Buisson, 1923 A T P
Entodinium indicum Kofoid & MacLennan, 1930 A T P
Entodinium lobosospinosum Dogiel, 1927 A T P
Entodinium longinucleatum Dogiel, 1925 A T P
Entodinium nanellum Dogiel, 1922 A T P
Entodinium ovinum Dogiel, 1927 A T P
Entodinium rostratum Fiorentini, 1889 A T P
Entodinium sp. indet. Clarke et al. 1982 A T P
Eodinium bilobosum (Dogiel, 1927) A T P
Eodinium lobatum Kofoid & MacLennan, 1932 A T P
Eodinium posterovesiculatum (Dogiel, 1927) A T P
Epidinium caudatum (Fiorentini, 1889) A T P
Epidinium ecaudatum (Fiorentini, 1889) A T P
Epidinium e. f. *bicaudatum* Dogiel 1927 A T P
Epidinium e. f. *parvicaudatum* Dogiel, 1927 A T P
Epidinium e. f. *quadricaudatum* Dogiel, 1927 A T P
Epidinium e. f. *tricaudatum* Dogiel, 1927 A T P
Epidinium sp. indet. Clarke et al. 1982 A T P
Eremoplastron bovis (Dogiel, 1927) A T P
Eremoplastron brevispinum Kofoid & MacLennan, 1932 A T P
Eremoplastron monolobum (Dogiel, 1927) A T P
Eremoplastron rostratum (Fiorentini, 1889) A T P
Eudiplodinium maggii (Fiorentini, 1889) A T P
Eudiplodinium sp. indet. Clarke et al. 1982 A T P
Metadinium medium Awerinzew & Mutafowa, 1914 A T P
Metadinium tauricum (Dogiel & Federowa, 1925) A T P
Ostracodinium dilobum Dogiel, 1927 A T P
Ostracodinium mammosum (Railliet, 1890) A T P
Ostracodinium rugoloricatum Kofoid & MacLennan, 1932 A T P

Class PHYLLOPHARYNGEA
Subclass PHYLLOPHARYNGIA
Order CHLAMYDODONTIDA
CHILODONELLIDAE
Chilodonella uncinata (Ehrenberg, 1838) F/T 4
Odontochlamys gouraudi Certes, 1891 F/T 1–4,6
Odontochlamys wisconsinensis (Kahl, 1931) T 3
Phascolodon elongatus Maskell, 1887 F E?
Trithigmostoma bavariensis (Kahl, 1931) T 1
Trithigmostoma cucullulus (Müller, 1786)† F/T 3,4
GASTRONAUTIDAE
Gastronauta membranaceus Bütschli, 1889 F/T 1,3,6
Dysteria astyla (Maskell, 1887) F E? (as *Aegyria*)
Dysteria distyla (Maskell, 1887) F E? (as *Aegyria*)

Subclass SUCTORIDA
Order EXOGENIDA
EPHELOTIDAE
Ephelota gemmipara (Hertwig, 1876) M
OPHRYODENDRIDAE
Ophryodendron macquariae Johnston, 1938 M
PARACINETIDAE
Paracineta crenata f. *pachytheca* Collin, 1912 M
Paracineta limbata f. *convexa* Dons, 1921 M
PODOPHRYIDAE
Podophrya fixa (Müller, 1786) F
Sphaerophrya magna Maupas, 1881† F/T 4
Sphaerophrya terricola Foissner, 1986* T 8
METACINETIDAE
Metacineta angularis (Maskell, 1888) F E? (as *Acineta*)
Metacineta mystacina (Ehrenberg, 1831) F

Order ENDOGENIDA
ENDOSPHAERIDAE
Endosphaera engelmanni Entz, 1896 F/M
ACINETIDAE
Acineta elegans Maskell, 1886 F E?
Acineta flos Maskell, 1887 F E?
Acineta lasanicola Maskell, 1887 F E?
Acineta simplex Maskell, 1887 F E?
Acineta speciosa Maskell, 1887 F E?
Acineta tulipa Maskell, 1887 F E?
Trematosoma complanata (Gruber, 1884) F
TRICHOPHRYIDAE
Trichophrya epistylidis Claparède & Lachmann, 1859 F

Class NASSOPHOREA
Order SYNHYMENIIDA
SCAPHIDIODONTIDAE
Chilodontopsis muscorum Kahl, 1931 T 3

Order NASSULIDA
NASSULIDAE
Nassula ambigua tumida Maskell, 1887 F E? † T 3

Order MICROTHORACIDA
MICROTHORACIDAE
Drepanomonas pauciciliata Foissner, 1987* T 8
Drepanomonas revoluta Penard, 1922 F/T 3,4,6
Drepanomonas sphagni Kahl, 1931 F/T 3,6
Leptopharynx costatus Mermod, 1914 F/T 1–10
Microthorax simulans (Kahl, 1926) T 3,4,10

Class COLPODEA
Order BRYOMETOPIDA
BRYOMETOPIDAE
Bryometopus pseudochilodon Kahl, 1932 T 1–3,5
KREYELLIDAE
Kreyella muscicola Kahl, 1931 T 1–3,6
TECTOHYMENIDAE
Pseudokreyella terricola Foissner, 1985* T 10

Order BRYOPHRYIDA
BRYOPHRYIDAE
Bryophrya bavariensis (Kahl, 1931) T 4
Parabryophrya penardi (Kahl, 1931)* T 9

Order BURSARIOMORPHIDA
BURSARIIDAE
Bursaria truncatella Müller, 1773 F/T

Order COLPODIDA
COLPODIDAE
Bresslaua vorax Kahl, 1931 • T 3,4,6
Colpoda aspera Kahl, 1926* T 8
Colpoda cucullus (Müller, 1773) F/T 1–7,8
Colpoda henneguyi Fabre-Domergue, 1889* T 10
Colpoda inflata (Stokes, 1884) T 1–10
Colpoda lucida Greeff, 1888* T 8,10
Colpoda magna (Gruber, 1879) F/T 3
Colpoda maupasi Enriques, 1908 F/T 3,8–10
Colpoda steinii Maupas, 1883 F/T 1–10
Tillina enormis Maskell, 1886 F E? (?*Ophryoglena*)
Tillina inequalis Maskell, 1886 F E?
MARYNIDAE
Mycterothrix tuamotuensis (Balbiani, 1887) T 3
GROSSGLOCKNERIIDAE
Grossglockneria acuta Foissner, 1980* T 8,10
Mykophagophrys terricola (Foissner, 1985)* T 8-10
Nivaliella plana Foissner, 1980* T 8–10
Pseudoplatyophrya nana (Kahl, 1926)* T 8–10
Pseudoplatyophrya saltans Foissner, 1988* T 8–10

Order CYRTOLOPHOSIDIDA
CYRTOLOPHOSIDIDAE
Cyrtolophosis elongata (Schewiakoff, 1892)* F/T 8,9
Cyrtolophosis mucicola Stokes, 1885 F/T 1,3–7,10
Pseudocyrtolophosis alpestris Foissner, 1980* T 9,10
PLATYOPHRYIDAE

Platyophrya macrostoma Foissner, 1980* T 8
Platyophrya spumacola Kahl, 1927* T 10
Platyophrya vorax Kahl, 1926 F/T 7,10

COLPODEA INCERTAE SEDIS
Pseudoglaucoma muscorum (Kahl, 1931) T 1–3,6

Class PROSTOMATEA
Order PRORODONTIDA
COLEPIDAE
Coleps elongatus Ehrenberg, 1830 F
Coleps hirtus (Müller, 1786) F
PLAGIOCAMPIDAE
Urotricha globosa Schewiakoff, 1892 F
HOLOPHRYIDAE
Holophrya discolor Ehrenberg, 1833 F
Holophrya teres (Ehrenberg, 1833) F
PRORODONTIDAE
Prorodon microstoma Stout, 1954 F E?
Prorodon sulcatus Maskell, 1886 F E?

PROSTOMATEA INCERTAE SEDIS
Cryptocaryon irritans Brown, 1951 M

Class OLIGOHYMENOPHOREA
Subclass PENICULIA
Order PENICULIDA
Suborder FRONTONIINA
FRONTONIIDAE
Frontonia depressa (Stokes, 1886) • T 1,3–5,10
Frontonia fusca (Quennerstedt, 1869) M

Suborder PARAMECIINA
PARAMECIIDAE
Paramecium aurelia-complex (Müller, 1773) F
Paramecium bursaria (Ehrenberg, 1831) F
Paramecium caudatum Ehrenberg, 1833 F
Paramecium trichium Stokes, 1885 F
LEMBADIONIDAE
Lembadion bullinum (Müller, 1786) F
Lembadion lucens (Maskell, 1887) F (as *Thurophora*)
UROCENTRIDAE
Urocentrum turbo (Müller, 1786) F

Subclass SCUTICOCILIATIA
Order PHILASTERIDA
CINETOCHILIDAE
Cinetochilum margaritaceum (Ehrenberg, 1830) F/T 1,3,4,6,7
Sathrophilus muscorum (Kahl, 1931) • T 1–4,6,9
ORCHITOPHRYIDAE
Paranophrys elongata (Biggar & Wenrich, 1932) M
PSEUDOCOHNILEMBIDAE
Pseudocohnilembus pusillus (Quennerstedt, 1869) F/T 4
URONEMATIDAE
Homalogastra setosa Kahl, 1926 T 1–4,6
Uronema marinum Dujardin, 1841 M

INCERTAE SEDIS
Kahlilembus fusiformis (Kahl, 1926) F/T 1–3,7

Order PLEURONEMATIDA
CTEDOCTEMATIDAE
Ctedoctema acanthocryptum Stokes, 1884 F
CYCLIDIIDAE
Cyclidium brandoni Kahl, 1931 F E? (*Pleuronema cyclidium* Maskell, 1886)
Cyclidium glaucoma Müller, 1773 F/T 5–7
Cyclidium muscicola Kahl, 1931 T 1–3,7,10
Pleuronema coronatum Kent, 1881 F/M
Pleuronema crassum Dujardin 1841 F

Subclass HYMENOSTOMATIA
Order HYMENOSTOMATIDA
Suborder TETRAHYMENINA
TETRAHYMENIDAE
Tetrahymena pyriformis-complex Ehrenberg, 1830† F/T 3,6
Tetrahymena rostrata (Kahl, 1926) • F/T 1–4
SPIROZONIDAE
Stegochilum fusiforme Schewiakoff, 1892 F
TURANIELLIDAE
Colpidium colpoda (Losana, 1829) F
Colpidium striatum Stokes, 1886 F
Dexiostoma campylum (Stokes, 1886) F
GLAUCOMIDAE
Glaucoma colpidium Schewiakoff, 1892 F
Glaucoma scintillans Ehrenberg, 1830 F
OPHRYOGLENIDAE
Ophryoglena flava (Ehrenberg, 1833) F

Suborder OPHRYOGLENINA
OPHRYOGLENIDAE
Ophryoglena flava (Ehrenberg, 1833) F
ICHTHYOPHTHIRIIDAE
Ichthyophthirius multifiliis Fouquet, 1876 F/M

Subclass PERITRICHIA
Order SESSILIDA
ELLOBIOPHRYIDAE
Caliperia longipes Laird, 1953 M E?
SCYPHIDIIDAE
Scyphidia (Gerda) acanthoclini Laird, 1953 M E?
EPISTYLIDIDAE
Campanella flavicans (Ehrenberg, 1831) F
Epistylis anastatica (Linnaeus, 1767) F
Epistylis leucoa Ehrenberg, 1838 F
Pyxicola cothurnioides Kent, 1880 F
Pyxicola pyxidiformis (D'Udekem, 1862) F
LAGENOPHRYIDAE
Lagenophrys cochinensis Santhakumari & Gopalan, 1980 M A
OPERCULARIIDAE
Opercularia cylindrata Wrzesniowski, 1870 F
Opercularia frondicola Precht, 1936 F/T 3 (c)
Opercularia nutans (Ehrenberg, 1831) F
Opercularia parallela Maskell, 1886 F E?
VAGINICOLIDAE
Cothurnia compressa f. *ovata* Dons, 1921 M
Cothurnia curvula Entz, 1884 M
Cothurnia grandis Perty, 1852 M (g)
Cothurnia maritima nodosa (Claparède & Lachmann, 1858) M
Cothurnia patellae Hutton, 1878 M E?
Cothurnia valvata (Stokes, 1893) F
Platycola decumbens (Ehrenberg, 1830) F
Platycola d. intermedia Maskell, 1887 F E?
Platycola donsi Kahl, 1933 M E?
Platycola longicollis Kent, 1881 F
Thuricola valvata (Wright, 1858) F
Vaginicola amphorella (Maskell, 1887) F E?
Vaginicola crystallina Ehrenberg, 1830 F
Vaginicola elongata (Fromentel, 1874) F
Vaginicola parallela (Maskell, 1887) F E?
VORTICELLIDAE
Carchesium polypinum (Linnaeus, 1758) F
Vorticella annularis Müller, 1773 F
Vorticella aperta Fromentel, 1876 F
Vorticella astyliformis Foissner, 1981* T 9,10
Vorticella brevistyla D'Udekem, 1864 F
Vorticella campanula Ehrenberg, 1831 F
Vorticella citrina Müller, 1773 F
Vorticella convallaria (Linnaeus, 1758) F/T 6
Vorticella cratera Kent, 1881 F
Vorticella elongata Fromentel, 1876 F
Vorticella longifilum Kent, 1881 F
Vorticella marina Greeff, 1870 M
Vorticella microstoma Ehrenberg, 1830 • F/M/T 6 (e)
Vorticella mortenseni Dons, 1921 M
Vorticella nebulifera Müller, 1773 M
Vorticella n. similis Noland & Finley, 1931 F
Vorticella oblonga Kirk, 1886 M E?
Vorticella patellina Müller, 1776 F
Vorticella striata Dujardin, 1841 F/T 2,3–7 (d)
Vorticella s. octava Stokes, 1885 F/T 6 (d)
Vorticella zealandica Kirk, 1887 F E?
ZOOTHAMNIIDAE
Zoothamnium affine granulatum Maskell, 1886 F E?
Zoothamnium cienkowskii Wrzesniowski, 1877 M
Zoothamnium limpidum Maskell, 1887 F E?
ASTYLOZOONIDAE
Astylozoon pyriformis Schewiakoff, 1892 F E?

Order MOBILIDA
TRICHODINIDAE
Trichodina multidentis Laird, 1953 M E?
Trichodina parabranchiola Laird, 1953 M E?
URCEOLARIIDAE
Urceolaria gaimardiae Johnston, 1938 M E?

Class PLAGIOPYLEA
Order PLAGIOPYLIDA
PLAGIOPYLIDAE
Plagiopyla varians Maskell, 1887 F E?

FIFTEEN

Phylum CERCOZOA

cercomonads, filose testate amoebae, Phaeodaria, plasmodiophoras, *Gromia*, haplosporidians, and kin

DENNIS P. GORDON, BENJAMIN K. DIGGLES,
RALF MEISTERFELD, CHRISTOPHER J. HOLLIS,
PETER K. BUCHANAN

Cercozoa means 'tailed animals' and alludes to the locomotory cilium in many of the species. The phylum is extremely morphologically diverse, encompassing ancestrally biciliate 'zooflagellates', euglyphid and other filose testate amoebae, shelled phaeodarians (which used to be classified as radiolaria), phytomyxean plant parasites like *Plasmodiophora*, the invertebrate-parasitic Acetosporea, and *Gromia*. In contrast with the parasitic forms, the free-living cercozoan zooflagellates have been very little studied, even though some of them, like the cercomonads, are the most ubiquitous flagellates in soil and important predators of bacteria in virtually all aquatic habitats (Cavalier-Smith & Chao 2003; Bass & Cavalier-Smith 2004; Karpov et al. 2006). The groups that currently make up the Cercozoa used to be distributed among several protozoan phyla, and even in the fungi in the case of the plasmodiophoras.

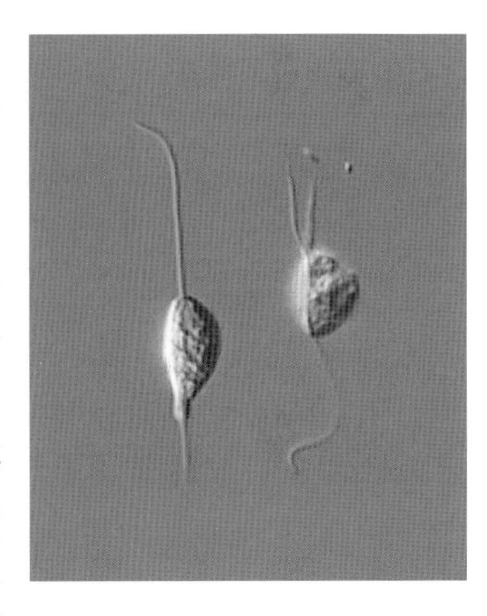

Two individuals of *Cercomonas paravarians* (Cercomonadida).
David Bass, University of Oxford

Partly synonymous with the former Rhizopoda, the phylum Cercozoa was the first major eukaryote group to be recognised primarily as a result of molecular-phylogenetic analyses. Until recently, the phylum suffered from a corresponding lack of shared ultrastructural characters (Cavalier-Smith & Chao 2003; Hoppenrath & Leander 2006a; Adl et al. 2005; Simpson & Patterson 2006), but Cavalier-Smith et al. (2008) discovered two unusual structures in the transitional region (the most proximal part) of the cilium, comprising shared derived features (synapomorphies) for the phylum. Leander (2008) has remarked on the remarkable degree of convergent evolution that exists among microbial eukaryotes. Whether they live in planktonic environments, interstitial environments, or the intestines of animal hosts, protozoans and microbial chromists from several different phyla can superficially resemble each other in quite striking ways. As indicated above, such is the case among the many cercozoan taxa that morphologically resemble species of Euglenozoa, Amoebozoa, Ochrophyta, Bigyra, Foraminifera, Radiozoa, and Heliozoa, among others. At its core, Cercozoa comprises a group of soft-bodied free-living phagotrophs with two pervasive but not universal properties – the propensity to glide on surfaces by a posterior cilium and/or to protrude slender pseudopodia (filopodia), often branching (rhizopodia) and sometimes anastomosing as a net (undulipodia).

The classification of the phylum used here is a combination of that proposed by Cavalier-Smith and Chao (2003), Bass et al. (2009a), and Howe et al. (2011), comprising two subphyla and eleven classes, not all of which have been found in New Zealand. Subphylum Filosa, in part representing the old taxon Zooflagellata, includes naked and shelled, ciliated and filopodial, free-living forms; subphylum

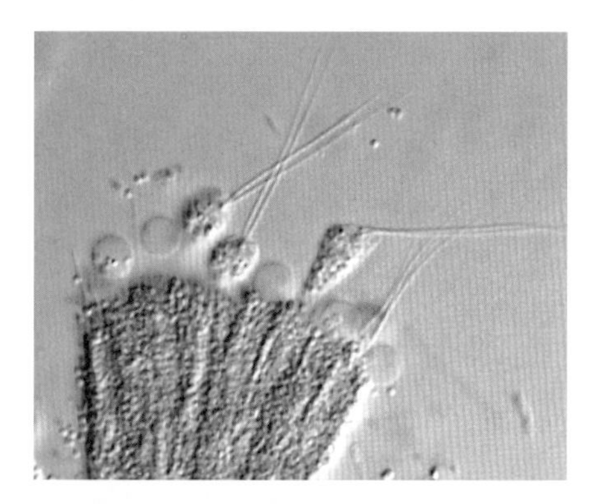

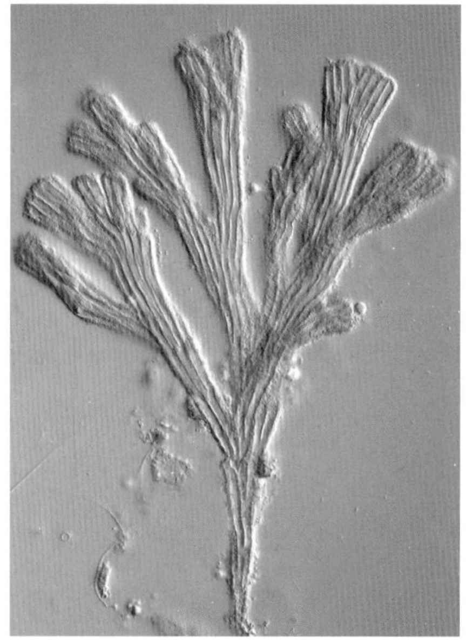

A whole colony and close-up of a branch tip of the freshwater cercozoan *Rhidipodendron splendidum* (Spongomonadida).
William Bourland, per Micro*scope (MBL)

Endomyxa includes mostly parasitic forms. Filosa was introduced by Cavalier-Smith and Chao (2003) with varied diagnostic characters depending largely on the form and locomotion of the cell. Endomyxa (from the Greek *endo-*, within, and *myx-*, slime), was coined by Cavalier-Smith (2002) 'because they are typically plasmodial endoparasites of other eukaryotes.'

The New Zealand cercozoan fauna comprises only 83 species and is patchily known, with some groups better studied than others.

Subphylum Filosa

Class Granofilosea is possibly represented by the semi-cosmopolitan marine epizoite *Wagnerella borealis*, although its precise affinities are still unknown. It has been traditionally classified in the Heliozoa. Bass et al. (2009a) suggested that the genus 'may, like desmothoracids, turn out to granofilosean'. It is a tiny stalked protist with a spherical 'head' from which radiating spicules project. It was first recorded in New Zealand waters on the tubes of a spirorbine polychaete (Dons 1921; Dawson 1992).

Records of class Sarcomonadea in New Zealand are few, even though its species include the dominant flagellates of soils, including members of the orders Cercomonadida and the recently recognised Glissomonadida (Howe et al. 2009). Maskell (1887) described *Cercomonas grandis* from unnamed freshwater localities in the Wellington region. The species seems not to have been reported in New Zealand since, but Myl'nikov and Karpov (2004) included it in a global list of *Cercomonas* species. Subsequently, Cavalier-Smith and Karpov (in press) have changed it to *Neocercomonas*. Karpov et al. (2006) sequenced and provisionally illustrated two unnamed species of Cercomonadidae (New Zealand localities not noted), including one in the newly circumscribed genus *Paracercomonas*. These, and two additional species, were formally described by Bass et al. (2009b). Maskell (1887) also recorded the enigmatic flagellate then known as *Heteromita lens*, subsequently designated by Patterson and Zölffel (1992) as the type species of their new genus *Kamera*. The affinities of *Kamera lens* are entirely unknown but this widespread species is listed at the end of the Cercozoa checklist so that it is not overlooked.

Summary of New Zealand living cercozoan diversity

Taxon	Described species + subspecies	Known undescribed/ undetermined species	Estimated unknown species	Adventive species	Endemic species	Endemic genera
Filosa	66	1+1	120	0	6?	0
Chlorarachnea	0	0	2	0	0	0
Granofilosea	1	0	20	0	0	0
Sarcomonadea	5	0	50	0	2?	0
Metromonadea	0	0	4	0	0	0
Thecofilosea	18	1	14	0	3	0
Imbricatea	38+2	0+1	22	0	0	0
Incertae sedis	4	0	8	0	1	0
Endomyxa	13	2	15	9?	4	0
Proteomyxidea	1	0	7	0	0	0
Phytomyxea	9	0	2	9?	0	0
Gromiidea	1	0	1	0	0	0
Ascetosporea	2	2	5	0	4	0
Incertae sedis	1	0	0	0	0	0
Totals	80+2	3+1	135	9?	10?	0

Members of class Thecofilosea comprise cells that are surrounded by a flexible secreted covering (tectum) or a rigid test with one or two apertures through which hyaline filopodia emerge. These pseudopodia fold like a jackknife when retracted. The best-known examples, in the order Tectofilosida, are such genera as *Amphitrema* and *Pseudodifflugia*, and the family Chlamydophryidae with a flexible membrane (not yet found in New Zealand). In contrast, members of the order Euglyphida (class Imbricatea) have a rigid test of overlapping silica scales (for classification, see Meisterfeld 2000). Members of these two testate orders resemble the testate lobose amoebae of the protozoan phylum Amoebozoa, and the general comments on the ecology and history of discovery of testate amoebae given in the Amoebozoa chapter (Chapter 7) apply here. Testate amoebae with filopodia have been studied using molecular methods (Bhattacharya et al. 1995; Burki et al. 2002; Wylezich et al. 2002). Available evidence suggests that the ability to build a test emerged at least three times during evolution among the Cercozoa (*Pseudodifflugia*, Euglyphida, and *Gromia*). The sequence data of Wylezich et al. (2002), based on several genera, show that Euglyphida is monophyletic and the euglyphid testate amoebae form a sister group to flagellates like *Cercomonas*. The siliceous body-plates are considered to be homologous in all Euglyphida.

Phaeodarians also feature silica in the test. The group is solely marine and used to be classified among the radiolarians (phylum Radiozoa) (Bass et al. 2005) but a recent molecular study of protein genes squarely places phaeodarians in the Cercozoa (Polet et al. 2004). Along with silica, the phaeodarian test also incorporates organic matter and there can be traces of other minerals. There is a central capsule housing the nucleus as in radiolarians, but this is thicker and has a single major opening (the astropylum, a complicated type of cytopharynx), usually with two smaller openings (parapylae) at the opposite pole. The astropylum tends to be cone-like and connects the central capsule with the outer, extracapsular, test. Prey organisms are captured by spine-supported axopodia and the filopodial extensions of the cytoplasm, ingested through the astropylum, and digested outside or inside the central capsule. Another characteristic feature of phaeodarians is the presence of a phaeodium, which consists of darkly pigmented waste products, usually in the region of the astropylum.

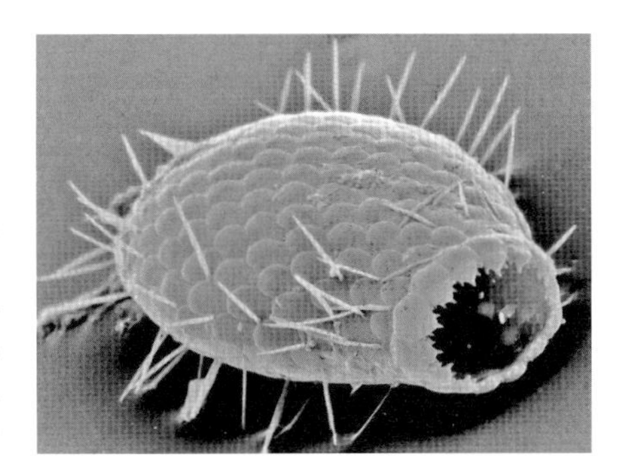

Euglypha strigosa (Euglyphida), a testate cercozoan of moss and soil.
Ralf Meisterfeld

Summary of New Zealand cercozoan diversity by environment

Taxon	Marine	Freshwater†	Terrestrial†	Fossil
Filosa	14	49+2	30+3	3
Chlorarachnea	0	0	0	0
Granofilosea	1	0	0	0
Sarcomonadea	0	1	4	0
Metromonadea	0	0	0	0
Thecofilosea	11	8	0	3
Imbricatea	3	35+2	26+3	0
Incertae sedis	0	5	0	0
Endomyxia	6	3	7	0
Proteomyxidea	0	1	1	0
Phytomyxea	1	2	6	0
Gromiidea	1	0	0	0
Ascetosporea	4	0	0	0
Incertae sedis	0	1	0	0
Totals	**20**	**53+2**	**37+3**	**3**

† Some species can occur in more than one environment, reflected in the tabulation. Hence total numbers across all environments will exceed the overall species tally in the previous table.
* This is taken to be the environment of the major host in the case of parasitic species.

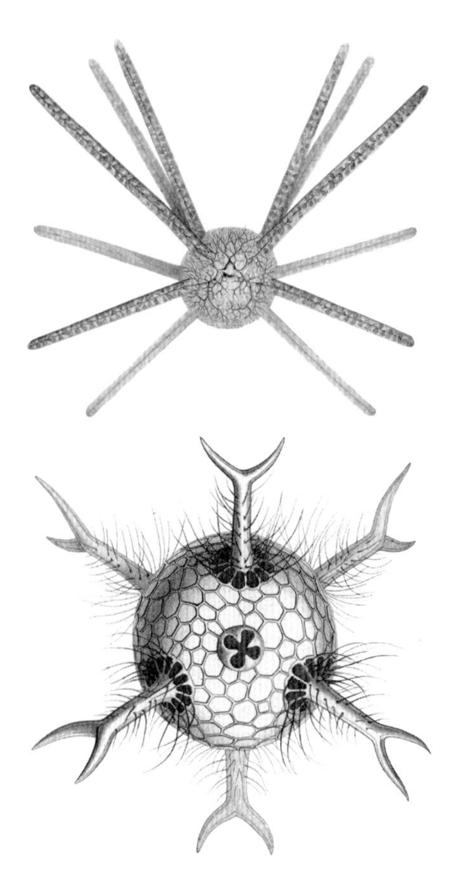

Two phaeodarians from marine plankton – *Coelothauma duodenum* (Phaeodendrida) (upper) and *Circospathis sexfurca* (Phaeocalpida).
From Haeckel 1887

Despite the siliceous test in many species, there are only sparse records of phaeodarians in Recent sediments. A number have no skeleton at all, or have one represented by loose spines. When a complete skeleton is developed, its bars are hollow (rather than solid as in polycystine radiolarians), and its silica is supported by an organic matrix, for which reason phaeodarians are rarely found in sediments. Even so, the Southwest Pacific is one of very few regions where fossil phaeodarians have been recovered – initially in Neogene sediments in the north Tasman Sea (Dumitrica 1973) and, more recently, in Late Cretaceous sediments from DSDP site 275, with three species found (Dumitrica & Hollis 2004). This discovery was highly significant as these species comprise some of the earliest-known phaeodarians. Nine living species have been reported in the seas around New Zealand (Haeckel 1887; Tibbs & Tibbs 1986; Dawson 1992).

A single species of cryomonad is known in New Zealand waters – *Protaspa* (formerly *Protaspis*) *tanyopsis*, found only once, off Curtis Island on the Kermadec Ridge (Norris 1961).

Ebriids (ebridians) have an internal skeleton of silica and, not surprisingly, have a good fossil record, but there are only two monotypic genera. The group has long defied precise classification and suggested relationships have linked them to silicoflagellates, dinoflagellates, and radiolarians among others, or simply as incertae sedis. Thus ebriid affinities seemed to be with any of three major chromistan groups – rhizarians, heterokonts, or alveolates. Phylogenetic analysis of ribosomal DNA sequences have recently demonstrated unequivocally that ebriids belong to the Cercozoa, close to the cryomonads (Hoppenrath & Leander 2006b).

The better-known of the two living species is *Ebria tripartita*, in which the skeleton has three branches. Cells are phagotrophic and range from 25 to 40 micrometres in length. The cilia, which are inserted in the cell near the apex, are hard to see. The name *Ebria*, from the Latin *ebrius*, inebriated, alludes to the distinctive swimming mode. *Ebria tripartita* is widespread in coastal waters of the world and is of ecological interest because cell concentrations can occasionally reach high numbers, making it a significant grazer of phytoplankton, especially nannoplanktonic diatoms and also dinoflagellates. It was reported in New Zealand waters by Cassie (1961) in the Hauraki Gulf, but is also known from the Eocene Oamaru diatomite (Mandra et al. 1973; Dawson 1992).

More problematic but probably also belonging to the Thecofilosea is the Pompholyxophryidae. Traditionally classified among the Heliozoa, members of this family have spherical cells with not only finely radiating axopodal spines but a variety of siliceous elements that cover the cell surface. In *Pompholyxophrys* they are spherical. These structures are diagnostic of species but are so tiny they must be viewed using electron microscopy. Nichols and Dürrschmidt (1985) reported three species in New Zealand lakes and ponds where they feed on algae and detritus.

Class Imbricatea is represented in New Zealand not only by the Euglyphida (above) but also order Spongomonadida, comprising two genera of colonial freshwater flagellates whose cells are embedded in granular mucus (Maskell 1886, 1887). Colonies are fan-shaped in species of *Rhipidodendron* and globular or cylindrical in *Spongomonas*. These genera have defied precise classification in the past.

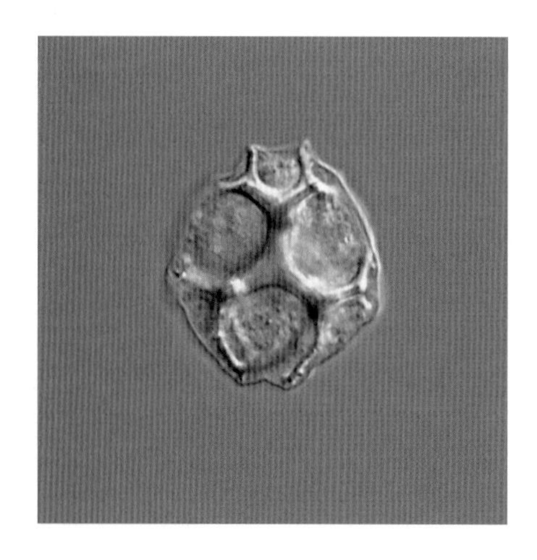

Ebria tripartita (Ebriida).
David Patterson, per Micro*scope (MBL)

Subphylum Endomyxa

Endomyxa includes two classes of free-living, often predatory, amoeboid forms and two classes of obligate parasites. Class Proteomyxidea in New Zealand is exemplified by *Biomyxa vagans*, found in mosses or wet grassland soils (Stout 1962, 1984). Superficially, this species resembles a naked foraminiferan, and putative relatives resemble slime moulds. Highly mobile and changeable, its filopodia branch and anastomose.

Species of class Phytomyxea live as endobiotic parasites in a wide range of organisms including flowering plants, brown algae, diatoms and oomycetes (Neuhauser et al. 2010). In New Zealand, there are nine species in seven genera, of which the best-known is *Plasmodiophora brassicae*, which causes club root of crucifers. The disease manifests itself when roots of cabbages and other brassicas become grossly swollen. The infectious stage of the organism is biciliate zoospores that penetrate host root hairs and develop into microscopic multinucleate plasmodia. Each plasmodium forms sporangia that produce either swimming zoospores or aggregations of thick-walled resting spores. These can persist in the soil and, when conditions are right, germinate into a zoospore. Several races of this pathogen have been reported from New Zealand (Lammerink 1965). Similarly adventive *Plasmodiophora elaeagni* infects exotic *Elaeagnus*.

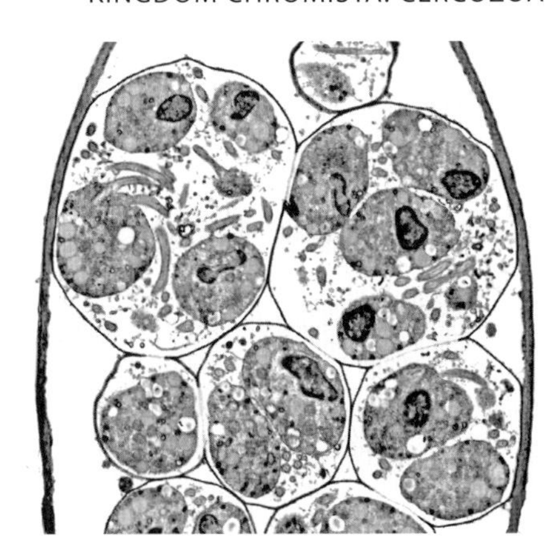

Transmission electron micrograph of secondary zoospores of *Plasmiodophora brassicae* (Plasmodiophorida) in a root hair of Chinese cabbage.

Plasmodiophorid Home Page [www.ohio.edu/people/braselto/plasmos/], courtesy of James P. Braselton, University of Ohio

Other plasmodiophorids in New Zealand infect a range of hosts. *Ligniera pilorum* attacks perennial ryegrass *Lolium perenne*, a significant component of winter pasture that is also useful in erosion control. *Polymyxa graminis* infects wheat and *Spongospora subterranea* causes powdery scab of potatoes. *Polymyxa* and *Spongospora* species are also economically significant as vectors of plant viruses (Bulman et al. 2011) and much research has gone into investigating methods of control (e.g. Falloon et al. 2008). *Sorosphaera veronicae* attacks introduced *Veronica* and species of *Woronina*, unusually, infect chromistans. *Woronina polycystis* attacks mycelia of the water mould *Saprolegnia* and *W. pythii* infects other oomycetes of the genus *Pythium*. Part of the life-cycle of plasmodiophorids is obligate intracellular, preventing growth in pure culture. Current research at Lincoln University, New Zealand, is focusing on in vitro culturing of the pathogen with host callus tissue to facilitate future molecular analysis of these important plant pathogens (Bulman et al. 2011). *Tetramyxa parasitica* is a widespread plasmodiophorid of marine and brackish-water angiosperms in different parts of the world. In New Zealand it has been found infecting *Ruppia megacarpa* in South Canterbury (Karling 1968). It is likely that other marine phytomyxids occur in New Zealand. For example, Maier et al. (2000) discovered that their new phytomyxid *Maullinia ectocarpii*, found to infect the filamentous brown alga *Ectocarpus siliculosus* in Chile, was capable of infecting a laboratory culture of the same species from New Zealand as well as New Zealand *Acinetospora crinita*. While *M. ectocarpii* has not yet been found in New Zealand, it has been found in southeastern Australia (Neuhauser et al. 2011). These authors suspect that a phytomyxid may be capable of transmitting viruses in brown algae, such as the virus that causes epidemic die-back in the kelp *Ecklonia radiata*.

Class Gromiidea comprises free-living, mostly marine forms. The best-known example is cosmopolitan *Gromia oviformis*. It is relatively large, achieving five millimetres in diameter (two millimetres is common), and thus quite visible to the naked eye. It has an organic test, but, unlike other testate cercozoans, and the testate lobose amoebozoans, the cell body is multinucleate and hence plasmodial, as in other Endomyxa. It is common in New Zealand in the intertidal zone, under rocks, in *Corallina* turf, or associated with kelp holdfasts, but it can be found at shelf depths and also in muddy and sandy sediments (Hedley 1962; Hedley & Bertaud 1962; Hedley et al. 1967). Stout (1984) even found individuals in seasonally flooded grassland five metres above sea level at Castlepoint on the southeast coast of North Island. Salinity was nil and the overall protozoan fauna was a mixture of soil and freshwater forms. Hedley (1962) wondered if more than one species is represented in the marine environment, and the form encountered by Stout (1984), tolerating fresh water, highlights the extraordinary range of environments in which *Gromia* is found.

In marine habitats, small individuals tend to be spherical, whereas large individuals are kidney-shaped or bilobed, or even quite irregular. The test is muddy-brown and shiny. Branching filose pseudopodia radiate across the substratum either side of the aperture for a distance equivalent to test diameter. When *Gromia* is about to reproduce, numerous small nuclei, each with associated

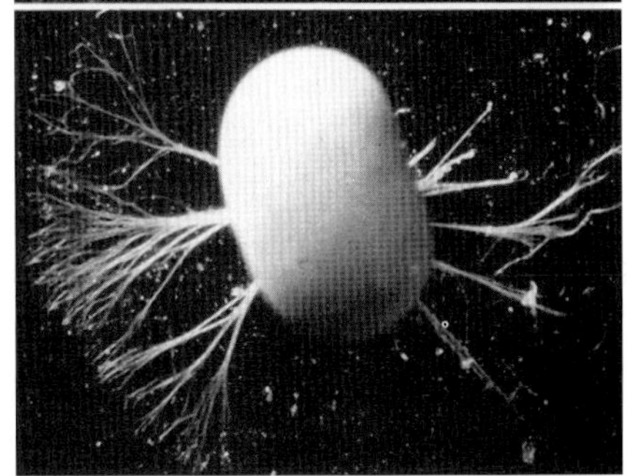

Gromia oviformis (Gromiida).

From Hedley et al. 1967

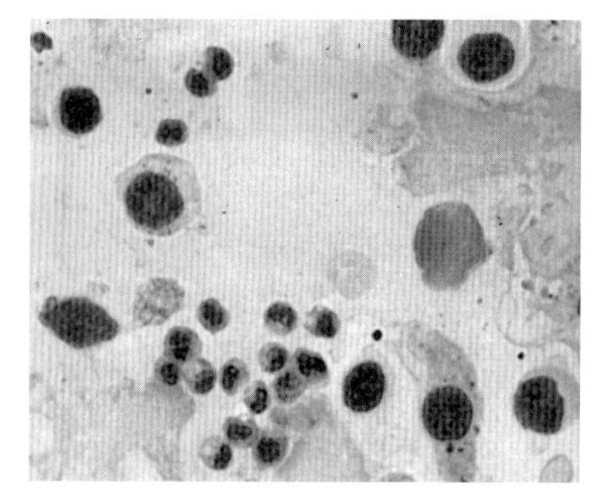

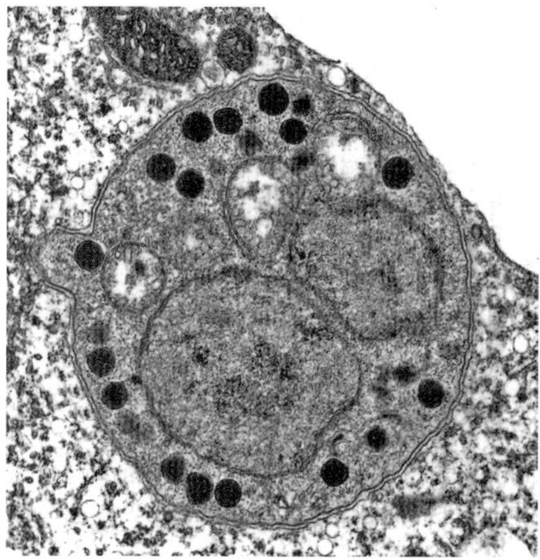

Bonamia exitiosa (Haplosporida) in the tissues of a Bluff oyster (*Ostrea chilensis*). Upper, a stained preparation showing a cluster of cells at bottom with uninfected blood cells elsewhere in the field. Lower, a binucleate cell of *Bonamia* with numerous haplosporosomes of unknown function.

Benjamin K. Diggles and P. Michael Hine

cytoplasm, differentiate in the upper part of the test into a creamy layer of uniciliate zoospores that are subsequently released.

Class Ascetoporea comprises mostly molluscan parasites. There are four orders, only one of which, Haplosporida (phylum Haplosporidia in some treatments), has so far been recorded in New Zealand. Haplosporidans are histozoic (inhabiting tissues) or coelozoic (inhabiting the digestive canal). Their cells contain organelles called haplosporosomes – spherical or vermiform, electron-dense structures with an outer bounding membrane and a separate interior membrane. The function of the haplosporosomes is unknown. The best-known haplosporidan in New Zealand is *Bonamia exitiosa*, the cause of bonamiosis in the Bluff oyster *Ostrea chilensis* (see Cranfield et al. 2005) Although first recognised in New Zealand in 1986, the causative organism was not formally named until 15 years later (Hine et al. 2001; Berthe & Hine 2004). It is a tiny (2–3 micrometres diameter) intracellular parasite of oyster haemocytes (blood cells) and is transmitted directly, from oyster to oyster. When *Bonamia* enters an uninfected oyster, the haemocytes recognise it as foreign and engulf it. Once inside the haemocyte, *Bonamia* feeds on the cytoplasm of the cell and grows, dividing many times until up to 24 parasites may be found in one haemocyte. When the cell bursts, the released parasites are engulfed by other haemocytes, carrying on the cycle. The oyster stops producing eggs or sperm and instead puts energy into producing more haemocytes, further favouring the parasite. Eventually the oyster dies of exhaustion. *Bonamia exitiosa* appears to have an annual pattern of infection. In Foveaux Strait it is usually hard to detect in late winter/early spring, and numbers usually increase with the spawning of predominantly male oysters in November–January (Hine 1991). However, *B. exitiosa* appears to rely heavily on the lipid reserves in oyster eggs, and it is during absorption of unspawned eggs, in the 25% of female oysters that do not spawn, that it uses host lipid and reaches its highest levels (Diggles et al. 2002a). A second *Bonamia*-like organism has since been found in Foveaux Strait oysters (P. M. Hine & B. K. Diggles pers. obs.). The phylogenetic relationships of *Bonamia* have exercised researchers for some time. The work of Carnegie et al. (2000) and Hine et al. (2009) supports inclusion in the Haplosporida.

Howell (1967) described a haplosporidan – *Urosporidium constantae* – that is associated with a trematode (fluke) parasite of *Ostrea chilensis* in Tasman Bay. The protozoan was noted as causing total mortality of those embryonic cercaria larvae of the fluke that it parasitised. Owing to the difficulty of obtaining the haplosporidan in the wild, it was not deemed to be a prospective biological-control agent for use against trematode infection of the oyster in Foveaux Strait.

The third, and most enigmatic, New Zealand haplosporidan remains undescribed, but represents a new genus and species that taxonomically falls at the very base of the Haplosporida (Reece et al. 2004; Hine et al. 2009). It affects juvenile and adult paua (*Haliotis iris*), causing haplosporidosis (Diggles et al. 2002b; Hine et al. 2002; Reece & Stokes 2003). The symptoms are lethargy, loss of the righting reflex, wasting of the foot, and chronic mortalities (80–90%) of juveniles during the summer months. This disease agent has been observed only once, in aquacultured individuals in 2000, and has not been recorded since. This suggests that paua may be an accidental host and that its usual hosts may be other molluscs, perhaps one of the many species of intertidal gastropods. There is no known effective method of treatment. Since wild-caught adult paua may be reservoirs of infection, broodstock paua ought to be kept separate from cultured juveniles (Diggles et al. 2002b).

In summary, the Cercozoa is an ecologically and economically important phylum of protists in New Zealand that is significantly understudied. This review touches on the known diversity, information about which is scattered in the literature. At least 135 more species can be expected in New Zealand's aquatic and terrestrial environments, but this figure should be taken as conservative.

Authors

Dr Peter K. Buchanan Landcare Research, Private Bag 92-170 Auckland, New Zealand [buchananp@landcareresearch.co.nz] Plasmodiophorida

Dr Benjamin K. Diggles DigsFish Services Pty Ltd, 32 Bowsprit Crescent, Banksia Beach, Queensland 4507, Australia [ben@digsfish.com] Haplosporida

Dr Dennis P. Gordon National Institute of Water and Atmospheric Research, Private Bag 14901 Kilbirnie, Wellington, New Zealand [d.gordon@niwa.co.nz] Introduction, miscellaneous groups

Dr Christopher J. Hollis Institute of Geological and Nuclear Sciences, P.O. Box 30-368, Lower Hutt, New Zealand [c.hollis@gns.cri.nz] Phaeodaria

Dr Ralf Meisterfeld RWTH Aachen, Institut für Biologie II, Mies-van-der-Rohe-Str. 15, D 52056 Aachen, Germany [meisterfeld@rwth-aachen.de] Tectofilosida, Euglyphida

References

ADL, S. M.; SIMPSON, A. G. B.; FARMER, M. A.; ANDERSEN, R. A.; ANDERSON, O. R.; BARTA, J. R.; BOWSER, S. S.; BRUGEROLLE, G.; FENSOME, R. A.; FREDERICQ, S.; JAMES, T.Y.; KARPOV, S.; KUGRENS, P.; KRUG, J.; LANE, C. E.; LEWIS, L. A.; LODGE, J.; LYNN, D. H.; MANN, D. G.; McCOURT, R. M.; MENDOZA, L.; MOESTRUP, Ø.; MOZLEY-STANDRIDGE, S. E.; NERAD, T. A.; SHEARER, C. A.; SMIRNOV, A.V.; SPIEGEL, F. W.; TAYLOR, M. F. J. R. 2005: The new higher level classification of eukaryotes with emphasis on the taxonomy of protists. *Journal of Eukaryotic Microbiology 52*: 399–451.

BASS, D.; CAVALIER-SMITH, T. 2004: Phylum-specific environmental DNA analysis reveals remarkably high global biodiversity of Cercozoa (Protozoa). *International Journal of Systematic and Evolutionary Microbiology 54*: 2393–2404.

BASS, D.; CHAO, E. E.-Y.; NIKOLAEV, S.; YABUKI, A.; ISHIDA, K.; BERNEY, C.; PAKZAD, U.; WYLEZICH, C.; CAVALIER-SMITH, T. 2009a: Phylogeny of novel naked filose and reticulose Cercozoa: Granofilosea cl. n. and Proteomyxidea revised. *Protist 160*: 75–109.

BASS, D.; HOWE, A. T.; MYLNIKOV, A. P.; VICKERMAN, K.; CHAO, E. E.; SMALLBONE, J. E.; SNELL, J.; CABRAL, C. Jr; CAVALIER-SMITH, T. 2009b: Phylogeny and classification of Cercomonadida (Protozoa, Cercozoa): *Cercomonas, Eocercomonas, Paracercomonas*, and *Cavernomonas* gen. nov. *Protist 160*: 483–521.

BASS, D.; MOREIRA, D.; LÓPEZ-GARCÍA, P.; POLET, S.; CHAO, E. E.; von der HEYDEN, S.; PAWLOVSKI, J.; CAVALIER-SMITH, T. 2005: Polyubiquitin insertions and the phylogeny of Cercozoa and Rhizaria. *Protist 156*: 149–161.

BERTHE, F. C. J.; HINE, P. M. 2004: *Bonamia exitiosa* Hine et al., 2001 is proposed instead of *B. exitiosus* as the valid name of *Bonamia* sp. infecting flat oysters *Ostrea chilensis* in New Zealand. *Diseases of Aquatic Organisms 57*: 18.

BHATTACHARYA, D.; HELMCHEN, T.; MELKONIAN, M. 1995: Molecular evolutionary analyses of nuclear-encoded small subunit ribosomal RNA identify an independent rhizopod lineage containing the Euglyphina and the Chlorarachniophyta. *Journal of Eukaryotic Microbiology 42*: 65–69.

BULMAN, S.; CANDY, J. M.; FIERS, M.; LISTER, R.; CONNER, A. J.; EADY, C. C. 2011: Genomics of biotrophic, plant-infecting plasmodiophorids using in vitro dual cultures. *Protist 162*: 449–461.

BURKI, F.; BERNEY, C.; PAWLOWSKI, J. 2002: Phylogenetic position of *Gromia oviformis* Dujardin inferred from nuclear encoded small subunit ribosomal DNA. *Protist 153*: 251–260.

CARNEGIE, R. B.; BARBER, B. J.; CULLOTY, S. C.; FIGUERAS, A. J.; DISTEL, D. L. 2000: Development of a PCR assay for detection of the oyster pathogen *Bonamia ostreae* and support for its inclusion in the Haplosporidia. *Diseases of Aquatic Organisms 42*: 199–206.

CASSIE, V. 1961: Marine phytoplankton in New Zealand waters. *Botanica Marina 2 (Suppl.)*: 1–54.

CAVALIER-SMITH, T. 1998: A revised six-kingdom system of life. *Biological Reviews 73*: 203–266.

CAVALIER-SMITH, T. 2002: The phagotrophic origin of eukaryotes and phylogenetic classification of Protozoa. *International Journal of Systematic and Evolutionary Microbiology 52*: 297–354.

CAVALIER-SMITH, T.; CHAO, E. E.-Y. 2003: Phylogeny and classification of phylum Cercozoa (Protozoa). *Protist 154*: 341–358.

CAVALIER-SMITH, T.; LEWIS, R.; CHAO, E. E.; OATES, B.; BASS, D. 2008: Morphology and phylogeny of *Sainouron acronematica* sp. n. and the ultrastructural unity of Cercozoa. *Protist 159*: 591–620.

COOPER, R. A. (Comp.) 2004: New Zealand Geological Timescale 2004/2 wallchart. *Institute of Geological and Nuclear Sciences Information Series 64*.

CRANFIELD, H. J.; DUNN, A.; DOONAN, I. J.; MICHAEL, K. P. 2005: *Bonamia exitiosa* epizootic in *Ostrea chilensis* from Foveaux Strait, southern New Zealand between 1986 and 1992. *ICES Journal of Marine Science 62*: 3–13.

DAWSON, E. W. 1992: The marine fauna of New Zealand: Index to the fauna 1. Protozoa. *New Zealand Oceanographic Institute Memoir 99*: 1–368.

DIGGLES, B. K.; HINE, P. M.; HANDLEY, S.; BOUSTEAD, N. C. 2002a: A handbook of diseases of importance to aquaculture in New Zealand. *NIWA Science and Technology Series 49*: 1–200.

DIGGLES, B. K.; NICHOL, J.; HINE, P. M.; WAKEFIELD, S.; COCHENNEC-LAUREAU, N.; ROBERTS, R. D.; FREIDMAN, C. S. 2002b: Pathology of cultured paua *Haliotis iris* infected with a novel haplosporidian parasite, with some observations on the course of the disease. *Diseases of Aquatic Organisms 50*: 219–231.

DONS, C. 1921: Papers from Dr. Th. Mortensen's Pacific Expedition 1914–1916. V. Notes sur quelques Protozoaires marins. *Videnskabelige Meddelelser fra Dansk Naturhistorisk Forening i Kjøbenhavn 73*: 49–84.

DUMITRICA, P. 1973: Phaeodarian Radiolaria in Southwest Pacific sediments cored during Leg 21 of the Deep Sea Drilling Project. *In*: Burns, R.E.; Andrews, J.E. et al. (eds). *Initial Reports of the Deep Sea Drilling Project 21*: 751–785.

DUMITRICA, P.; HOLLIS, C. J. 2004: Maastrichtian Challengeridae (phaeodarian radiolaria) from deep sea sediments of SW Pacific. *Revue de Micropaléontologie 47*: 127–134.

FALLOON, R. E. 2008: Control of powdery scab ot potato: towards integrated disease management. *American Journal of Potato Research 85*: 253–260.

HAECKEL, E. 1887: Report on the Radiolaria collected by HMS Challenger during the years 1873–76. First part. – Porulosa (Spumellaria and Acantharia). Second part. Osculosa (Nassellaria and Phaeodaria). *Reports of the Scientific Results of the Voyage of H.M.S. Challenger 1873–76, 40, Zoology 18*(1): viii + i-viii + i-clxxxviii, 1–188; (2): 889–1803; (3): pls 1–140, 1 map.

HEDLEY, R. H. 1962: *Gromia oviformis* (Rhizopodea) from New Zealand with comments on the fossil Chitinozoa. *New Zealand Journal of Science 5*: 121–136.

HEDLEY, R. H.; BERTAUD, W. S. 1962: Electron-microscopic observations of *Gromia oviformis* (Sarcodina). *Journal of Protozoology 9*: 79–87.

HEDLEY, R. H.; HURDLE, C. M.; BURDETT, I. D. J. 1967: The marine fauna of New Zealand: intertidal Foraminifera of the *Corallina officinalis* zone. *New Zealand Department of Scientific and Industrial Research Bulletin 180* [*New Zealand Oceanographic Institute Memoir 38*]: 1–86.

HINE, P. M. 1991: Annual pattern of infection by *Bonamia* sp. in New Zealand flat oysters, *Tiostrea chilensis*. *Aquaculture 93*: 241–251.

HINE, P. M.; CARNEGIE, R. B.; BURRESON, E. M.; ENGELSMA, M.Y. 2009: Inter-relationships of haplosporidians deduced from ultrastructural studies. *Diseases of Aquatic Organisms 83*: 247–256.

HINE, P. M.; COCHENNEC-LAUREAU, N.; BERTHE, F. C. J. 2001: *Bonamia exitiosus* n. sp. (Haplosporidia) infecting flat oysters *Ostrea chilensis* (Philippi) in New Zealand. *Diseases of Aquatic Organisms 47*: 63–72.

HINE, P. M.; WAKEFIELD, S.; DIGGLES, B. K.; WEBB, V.; MAAS, E. W. 2002: The ultrastructure of a haplosporidian containing Rickettsiae, associated with mortalities among cultured paua *Haliotis iris*. *Diseases of Aquatic Organisms 49*: 207–219.

HOPPENRATH, M.; LEANDER, B. S. 2006a: Dinoflagellate, euglenid, or cercomonad? The ultrastructure and molecular phylogenetic position of *Protaspis grandis* n. sp. *Journal of eukaryotic Microbiology 53*: 327–342.

HOPPENRATH, M.; LEANDER, B. S. 2006b: Ebriid phylogeny and the expansion of the Cercozoa. *Protist 157*: 279–290.

HOWE, A. T.; BASS, D.; SCOBLE, J. M. ; LEWIS, R.; VICKERMAN, K.; ARNDT, H.; CAVALIER-SMITH, T. 2011: Novel cultured protists identify deep-branching environmental DNA clades of Cercozoa: new genera *Tremula, Micrometopion, Minimassisteria, Nudifila, Peregrinia*. *Protist 162*: 332–372.

HOWE, A. T.; BASS, D.; VICKERMAN, K.; CHAO, E. E.; CAVALIER-SMITH, T. 2009: Phylogeny, taxonomy, and astounding genetic diversity of Glissomonadida ord. nov., the dominant gliding zooflagellates in soil (Protozoa: Cercozoa). *Protist 160*: 159–189.

HOWELL, M. 1967: The trematode, *Bucephalus longicornutus* (Manter, 1954) in the New Zealand mud oyster, *Ostrea lutaria*. *Transactions of the Royal Society of New Zealand, Zoology 8*: 221–237, pls 1–5.

KARLING, J. S. [1966] 1968: Some zoosporic fungi of New Zealand. XI. Plasmodiophorales. *Sydowia 20*: 151–156.

KARPOV, S. A.; BASS, D.; MYLNIKOV, A. P.; CAVALIER-SMITH, T. 2006: Molecular phylogeny of Cercomonadidae and kinetid patterns of *Cercomonas* and *Eocercomonas* gen. nov. (Cercomonadida, Cercozoa). *Protist 157*: 125–158.

LAMMERINK, J. 1965: Six pathogenic races of *Plasmodiophora brassicae* Wor. in New Zealand. *New Zealand Journal of Agricultural Science 8*: 156–164.

LEANDER, B. S. 2008: A hierarchical view of convergent evolution in microbial eukaryotes. *Journal of Eukaryotic Microbiology 55*: 59–68.

MAIER, I.; PARODI, E.; WESTERMEIER, R.; MÜLLER, D. G. 2000: *Maullinia ectocarpii* gen. et sp. nov. (Plasmodiophorea), an intracellular parasite in *Ectocarpus siliculosus* (Ectocarpales, Phaeophyceae) and other filamentous brown algae. *Protist 151*: 225–238.

MANDRA, Y. T.; BRIGGER, A. L.; MANDRA, H. 1973. Preliminary report on a study of fossil silicoflagellates from Oamaru, New Zealand. *Occasional Papers of the California Academy of Sciences No. 107*: 1–11.

MASKELL, W. M. 1886: On the fresh-water infusoria of the Wellington District. *Transactions of the New Zealand Institute 19*: 49–61, pls 3–5.

MASKELL, W. M. 1887: On the fresh-water infusoria of the Wellington District. *Transactions of the New Zealand Institute 20*: 3–19, pls 1–4.

MEISTERFELD, R. 2000: Testate amoebae with filopodia. Pp. 1054–1084 *in*: Lee, J. J.; Leedale, G. F.; Bradbury, P. (Eds), *An Illustrated Guide to the Protozoa. Second Edition. Organisms traditionally referred to as Protozoa, or newly described groups*. Society of Protozoologists, Lawrence. 1432 p. (in 2 vols).

MYL'NIKOV, A. P.; KARPOV, S. A. 2004: Review of diversity and taxonomy of cercomonads. *Protistology 3*: 201–217.

NEUHAUSER, S.; BULMAN, S.; KIRCHMAIR, M. 2010: Plasmodiophorids: the challenge to understand soil-borne, obligate biotrophs with a multi-phasic life cycle. Pp 51–78 *in*: Gherbawy, Y.; Voigt, K. (eds), *Molecular Identification of Fungi*. Springer, Berlin. 522 p.

NEUHAUSER, S.; KIRCHMAIR, M.; GLEASON, F. H. 2011: Ecological roles of the parasitic phytomyxids (plasmodiophorids) in marine ecosystems – a review. *Marine and Freshwater Research 62*: 365–371.

NICHOLLS, K. H.; DÜRRSCHMIDT, M. 1985: Scale structure and taxonomy of some species of *Raphidocystis, Raphidiophrys*, and *Pompholyxophrys* (Heliozoea) including descriptions of six new taxa. *Canadian Journal of Zoology 63*: 1944–1961.

NORRIS, R. E. 1961: Observations on plankton organisms collected on the N.Z.O.I. Pacific cruise, September 1958. *New Zealand Journal of Science 4*: 162–188.

PATTERSON, D. J.; ZÖLFFEL, M. 1992: Heterotrophic flagellates of uncertain taxonomic position. Pp. 427–475 in: Patterson, D. J.; Larsen, J. (eds), *The Biology of Free-Living Heterotrophic Flagellates*. Clarendon Press, Oxford. 518 p.

POLET, S.; BERNEY, C.; FAHRNI, J.; PAWLOWSKI, J. 2004: Small-subunit ribosomal RNA gene sequences of Phaeodaria challenge the monophyly of Haeckel's Radiolaria. *Protist 155*: 53–63.

REECE, K. S.; SIDDALL, M. E.; STOKES, N. A.; BURRESON, E. M. 2004: Molecular phylogeny of the Haplosporidia based on two independent gene sequences. *Journal of Parasitology 90*: 1111–1122.

REECE, K. S.; STOKES, N. A. 2003: Molecular analysis of a haplosporidian parasite from cultured New Zealand abalone *Haliotis iris*. *Diseases of Aquatic Organisms 53*: 61–66.

SIMPSON, A. G. B.; PATTERSON, D. J. 2006: Diversity of microbial eukaryotes Pp. 7–30 *in*: Katz, L. A.; Bhattacharya, D. (eds), *Genomics and Evolution of Microbial Eukaryotes*. Oxford University Press, Oxford. 256 p.

STOUT, J. D. 1962: An estimation of microfaunal populations in soils and forest litter. *Journal of Soil Science 13*: 314–320.

STOUT, J. D. 1984: The protozoan fauna of a seasonally inundated soil under grassland. *Soil Biology and Biochemistry 16*: 121–125.

TIBBS, J. F.; TIBBS, S. D. 1986: Biology of the Antarctic Seas 16. Further studies of the Phaeodaria (Protozoa: Radiolaria) of the Antarctic seas. *Antarctic Research Series 41*: 167–202.

WYLEZICH C.; MEISTERFELD R.; MEISTERFELD, S.; SCHLEGEL, M. 2002: Phylogenetic analyses of small subunit ribosomal RNA coding regions reveal a monophyletic lineage of euglyphid testate amoebae (order Euglyphida). *Journal of Eukaryotic Microbiology 49*: 108–118.

YEATES, G. W.; FOISSNER, W. 1995: Testate amoebae as predators of nematodes. *Biology and Fertility of Soils 20*: 1–7.

Checklist of New Zealand Cercozoa

Classification follows Cavalier-Smith and Chao (2003). Abbreviations: E, endemic; S, species with southern distribution in New Zealand; * unpublished records (R. Meisterfeld) from humus collected at Nelson (South Island) and Station Ridge near Orongorongo Valley Field Station (southern North Island) collected by Yeates (see Yeates & Foissner 1995). Habitats of taxa are indicated as follows: F, fresh waters and/or wet habitats like Sphagnum moss; M, marine and brackish; T, terrestrial (in soil water or in terrestrial hosts); P, parasitic or commensal in invertebrates (in) or plants (Pl).

KINGDOM CHROMISTA
SUBKINGDOM HAROSA
INFRAKINGDOM RHIZARIA
PHYLUM CERCOZOA
SUBPHYLUM FILOSA
INFRAPHYLUM MONADOFILOSA
Class GRANOFILOSEA
Order INCERTAE SEDIS
GYMNOSPHAERIDAE?
Wagnerella borealis Merechowsky, 1878 M

Class SARCOMONADEA
Order CERCOMONADIDA
CERCOMONADIDAE
Cercomonas media Bass & Cavalier-Smith *in* Bass et al., 2009 T
Cercomonas paravarians Bass & Cavalier-Smith *in* Bass et al., 2009 T
Neocercomonas grandis (Maskell, 1887) F E
Paracercomonas minima Mylnikov, 1985 T
Paracercomonas vonderheydeni Bass & Cavalier-Smith *in* Bass et al., 2009 T

Class THECOFILOSEA
Order TECTOFILOSIDA
AMPHITREMIDAE
Amphitrema paparoensis (van Oye, 1956) F E
Amphitrema wrightianum Archer, 1869 F
Archerella flavum (Archer, 1877) F
Archerella jollyi (van Oye, 1956) F E
CHLAMYDOPHRYIDAE
Chlamydophrys stercorea Cienkowski, 1876 F
Lecythium sp. F
PSEUDODIFFLUGIIDAE
Pseudodifflugia fulva Penard, 1902 F
Pseudodifflugia gracilis Schlumberger, 1845 F

Subclass PHAEODARIA
Order PHAEOCYSTIDA
AULACANTHIDAE
Aulodendron australe Haeckel, 1887 E

Order PHAEOSPHAERIDA
SAGOSPHAERIDAE
Sagoscena lampadophora Haecker, 1905
Sagoscena ornata Haeckel, 1887

Order PHAEOCALPIDA
CASTANELLIDAE
Castanarium huxleyi Haeckel, 1887
Castanidium bromleyi Haeckel, 1887
CIRCOPORIDAE
Circospathis sexfurca (Haeckel, 1887)

Order PHAEODENDRIDA
COELODENDRIDAE
Coelodrymus lappulatus Haeckel, 1887
COELOGRAPHIDIDAE
Coelothauma duodenum Haeckel, 1887

Order PHAEOGROMIDA
MEDUSETTIDAE
Gazelletta orthonema Haeckel, 1887

Subclass INCERTAE SEDIS
Order CRYOMONADIDA
PROTASPIDAE
Protaspa tanyopsis (Norris, 1961) M E

Order EBRIIDA
EBRIIDAE
Ebria tripartita (Schumann, 1867) M

Class IMBRICATEA
Order SPONGOMONADIDA
SPONGOMONADIDAE
Rhipidodendron huxleyi Saville-Kent, 1882 F
Rhipidodendron splendidum Stein, 1956 F
Spongomonas discus Stein, 1878 F
Spongomonas sacculus Saville-Kent, 1880 F

Order THAUMATOMONADIDA
THAUMATOMONADIDAE
Thaumatomastix bipartita Beech & Moestrup, 1986 M
Thaumatomastix salina (Birch-Anderson, 1973) M
Thaumatomastix tripus (Takahashi et Hara) Beech & Moestrup M

Order EUGLYPHIDA
CYPHODERIIDAE
Cyphoderia ampulla (Ehrenberg, 1840) F
EUGLYPHIDAE
Assulina muscorum Greeff, 1888 FT
Assulina seminulum (Ehrenberg, 1848) F
Euglypha acanthophora (Ehrenberg, 1843) FT
Euglypha bryophila Brown, 1911* FT
Euglypha ciliata (Ehrenberg, 1848) FT
Euglypha ciliata f. *glabra* Cash, Wailes & Hopkinson, 1915 FT
Euglypha compressa Carter, 1864 FT
Euglypha compressa f. *glabra* Cash, Wailes & Hopkinson, 1915 FT
Euglypha cristata Leidy, 1874 F
Euglypha denticulata Brown, 1912 FT
Euglypha filifera Penard, 1890* FT
Euglypha laevis Perty, 1849 FT
Euglypha polylepis Bonnet, 1959* T
Euglypha rotunda Wailes & Penard, 1911 FT
Euglypha rotunda small form Wailes, 1915* FT
Euglypha scutigera Wailes & Penard, 1911 FT
Euglypha simplex Decloitre, 1965* T
Euglypha strigosa (Ehrenberg, 1871) FT
Euglypha strigosa f. *glabra* Wailes & Penard, 1911 FT
Euglypha tuberculata Dujardin, 1841 F
Placocista spinosa (Carter, 1865) Leidy, 1879 F
Sphenoderia fissirostris Penard, 1890 F
Sphenoderia minuta Deflandre, 1931* FT
Sphenoderia rhombophora Bonnet, 1966* T
Tracheleuglypha dentata (Vejdovsky, 1882) F
Trachelocorythion pulchellum (Penard, 1890* FT
TRINEMATIDAE
Corythion delamarei Bonnet & Thomas, 1960* T
Corythion dubium Taranek, 1882 FT
Corythion d. gigas Thomas, 1954* FT
Playfairina valkanovi Golemansky, 1966* FT
Trinema complanatum Penard, 1890* FT
Trinema c. platystoma Schönborn, 1964* T
Trinema enchelys (Ehrenberg, 1838)? FT
Trinema galeata (Penard, 1902) FT
Trinema grandis Chardez, 1960* FT
Trinema lineare Penard, 1890 FT

MONADOFILOSA INCERTAE SEDIS
Feuerbornia lobophora Jung, 1942* FT S

INFRAPHYLUM INCERTAE SEDIS
POMPHOLYXOPHRYIDAE
Pompholyxophrys exigua Hertwig & Lesser, 1874 F
Pompholyxophrys ossea Dürrschmidt in Nicholls & Dürrschmidt, 1985 F E
Pompholyxophrys ovuligera Penard, 1904 F

SUBPHYLUM ENDOMYXA
Class PROTEOMYXIDEA
Order ACONCHULINIDA
BIOMYXIDAE
Biomyxa vagans Leidy, 1879 FT

Class PHYTOMYXEA
Order PLASMODIOPHORIDA
PLASMODIOPHORIDAE
Ligniera pilorum Fron & Gaillat, 1925 T PPl
Plasmodiophora brassicae Woronin, 1877 T PPl
Plasmodiophora elaeagni Schröter, 1886 [1889]. T PPl
Polymyxa graminis Ledingham, 1939 T PPl
Sorosphaera veronicae (Schröter, 1886) [1889] T PPl
Spongospora subterranea (Wallroth, 1842) T PPl
Tetramyxa parasitica Goebel, 1884 M PPl
Woronina polycystis Cornu, 1872 T PPl
Woronina pythii Goldie-Smith, 1956 T PPl

Class GROMIIDEA
Order GROMIIDA
GROMIIDAE
Gromia oviformis Dujardin, 1835 M

Class ASCETOSPOREA
Order HAPLOSPORIDA
HAPLOSPORIDIIDAE
Bonamia exitiosa Hine, Cochennec-Laureau & Berthe, 2001 M Pin E
Urosporidium constantae Howell, 1967 M PIn E
Gen. nov. et n. sp. Diggles et al. 2002 M PIn E
Gen. et sp. indet. Hine & Diggles pers. obs. M PIn E

PHYLUM INCERTAE SEDIS
Kamera lens (Müller, 1773) F

Checklist of New Zealand fossil Cercozoa

Abbreviations for Stage names are based on Cooper (2004).

PHYLUM CERCOZOA
SUBPHYLUM FILOSA
Superclass MONADOFILOSA
Class THECOFILOSEA
Subclass PHAEODARIA
Order PHAEOGROMIDA
CHALLENGERIDAE
Challengeron takhashii Dumitrica & Hollis, 2004 Mh
Challengeron sp. Dumitrica & Hollis 2004 Mh
Protocystis pacifica Dumitrica & Hollis, 2004 Mh

SIXTEEN

Phylum FORAMINIFERA

foraminifera, xenophyophores

BRUCE W. HAYWARD, OLE S. TENDAL, ROWAN CARTER, HUGH R. GRENFELL, HUGH E. G. MORGANS, GEORGE H. SCOTT, C. PERCY STRONG, JESSICA J. HAYWARD

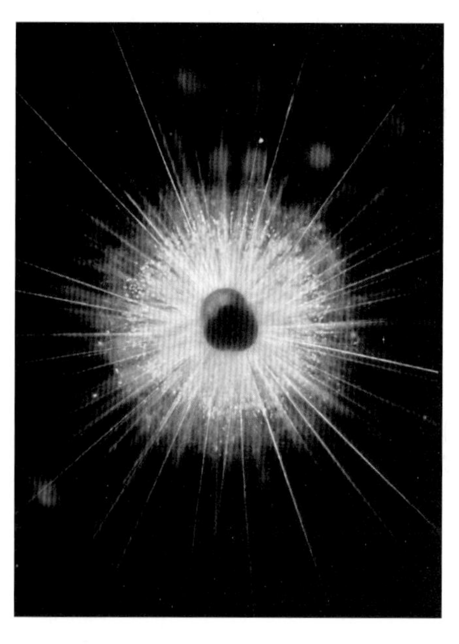

Live planktic *Globigerinoides ruber*; chambered test at centre with radiating spines and surrounded by a mesh of fine cytoplasmic reticulopodia that entrap food.

David Lea, UCSB

Foraminifera ('hole-bearers') are among the most abundant and scientifically important groups of organisms on earth, yet the general public is unaware of their existence, relatively few biologists know much about them, and the biology of living species is poorly known. The shells of recently dead foraminifera are so abundant that they form a thick blanket of sediment over one third of the surface of the planet (as *Globigerina* ooze on the ocean floor). Although foraminifera are single-celled and mostly microscopic, the shells of some of the larger species form the rock used to build the pyramids of Egypt and others are used to make necklaces on tropical islands. The importance of fossil foraminifera in biostratigraphy, paleoenvironmental studies, and isotope geochemistry derives from their ubiquity in most marine rocks, often as large, well-preserved, diverse assemblages, and has resulted in their being the most studied group of fossils worldwide and in New Zealand. Because modern foraminifera have attracted little interest from biologists, paleontologists have been forced to undertake most studies, including genetic research, on the living fauna.

Foraminifera are marine and estuarine-dwelling protozoans that mostly produce chambered shells (of calcite, aragonite, or sediment grains), commonly preserved in sedimentary rocks as fossils. [Fig. 1] Their evolutionary relationships have been clarified in the last decade thanks to gene sequencing. Foraminifera belong to the eukaryote supergroup Rhizaria (Archibald & Keeling 2004; Bass et al. 2005), in turn recently amalgamated with chromalveolate protists (Hackett et al. 2007; Burki et al. 2007, 2008; Reeb et al. 2009) into an expanded kingdom Chromista (Cavalier-Smith 2010). Within the Rhizaria, Foraminifera frequently appear more closely related to the Cercozoa in gene trees than to the Radiozoa with which they have been united as phylum Retaria (Cavalier-Smith 2002, 2003). Within the Foraminifera, Cavalier-Smith (2003) recognises three classes – the naked Athalamea (e.g. *Reticulomyxa*), not yet known in New Zealand, Polythalamea (chambered foraminifera), and Xenophyophorea. The latter are true giants of the protozoan world, forming clumps up to 25 cm high, made up mostly of mineral grains and other foreign particles.

The classification of chambered foraminifera has been developed by paleontologists and is based solely on shell characters since the biology of most species is not known (see Loeblich & Tappan 1987). Within the Polythalamea, 14 orders are currently recognised (Loeblich & Tappan 1992); members of 13 orders are present in New Zealand waters today and the one extinct order (Fusulinida) occurs in New Zealand's Paleozoic fossil record.
The orders are:

Allogromiida – proteinaceous test or outer membrane, Cambrian–Recent.

Astrorhizida – agglutinated test, branched tubular forms, Cambrian–Recent.
Lituolida – multilocular, agglutinated test, Cambrian–Recent.
Trochamminida – multilocular, agglutinated, non-canaliculate test, low trochospiral forms, Carboniferous–Recent.
Textulariida – canaliculate, agglutinated test with magnesium calcite cement, Jurassic–Recent.
Fusulinida – test of microgranular calcite, Silurian–Permian.
Spirillinida – test of an optically single crystal of calcite or aragonite, test growth by marginal accretion, Permian–Recent.
Carterinida – test of rod-like secreted calcite crystals, Eocene–Recent.
Miliolida – test of largely imperforate, porcellaneous, high-magnesium calcite, Carboniferous–Recent.
Lagenida – test of optically radiate calcite, Carboniferous–Recent.
Robertinida – test of hyaline, perforate aragonite, Triassic–Recent.
Globigerinida – test of perforate, optically radiate, hyaline, lamellar calcite, planktic (planktonic) habit, Jurassic–Recent.
Buliminida – test of hyaline, perforate, lamellar calcite, high trochospiral, triserial, biserial or uniserial forms, Triassic–Recent.
Rotaliida – test of hyaline, perforate, lamellar calcite, mostly low trochospiral or planispiral forms, Triassic–Recent.

Worldwide, an estimated 100,000 foraminiferal species, both living and fossil, have been described (*see* Ellis and Messina, Catalogue of Foraminifera – since 1942, now online). Studies suggest that many species, especially deep-water ones, are more widely spread around the world than many early workers conceived and thus many thousands of species have been described and named more than once from different regions, including New Zealand. As a result, there is enormous taxonomic confusion worldwide, except in the most-studied groups of planktics (planktonic species) and deep-sea benthics (seafloor species). It is therefore difficult to assess world foraminiferal diversity, but there are probably 5000–10,000 living species, and perhaps another 20,000–30,000 extinct fossil species.

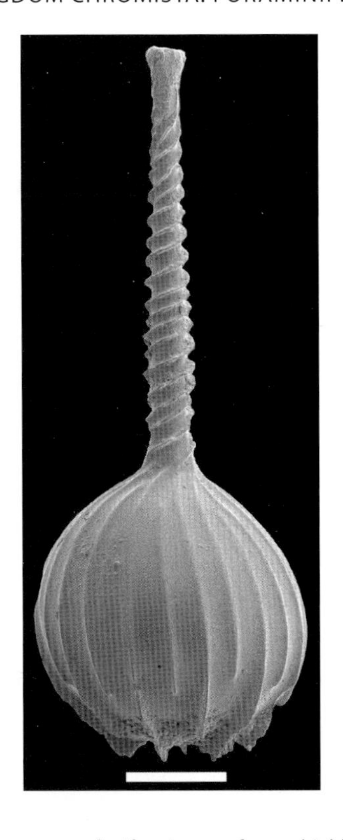

Lagena peculiariformis, one of many highly ornamented, one-chambered species of Lagenida in the EEZ. BWH166/15, Pukaki Saddle, Campbell Plateau, Recent. Scale bar 0.1 millimetre.

150 years of New Zealand foraminiferal research

Study of New Zealand's living and fossil foraminifera has evolved through a number of phases, generally overlapping, since its simple beginnings 150 years ago (Mantell 1850):

1850s–1910s Global scientific expeditions and microscopists' curiosity
1930s–1960s Early petroleum exploration
1940s Establishment of local biostratigraphic stages
1950s–1970s Used in geological mapping by New Zealand Geological Survey
1960s–1970s Blossoming of planktic foraminiferal biostratigraphy
1960s–1990s Foraminiferal paleoenvironment assessments

Summary of New Zealand foraminiferan diversity

Taxon	Described species/ subspecies*	Known undescribed/ undetermined species/ subspecies	Estimated unknown species/ subspecies	Adventive species/ subspecies	Endemic species/ subspecies	Endemic genera
Polythalamea	1,018	116	200–300	3	125	2
Xenophyophorea	7	0	>5	0	3	0
Totals	**1,025**	**116**	**~205–305**	**3**	**128**	**2**

* Not including forms (f.).

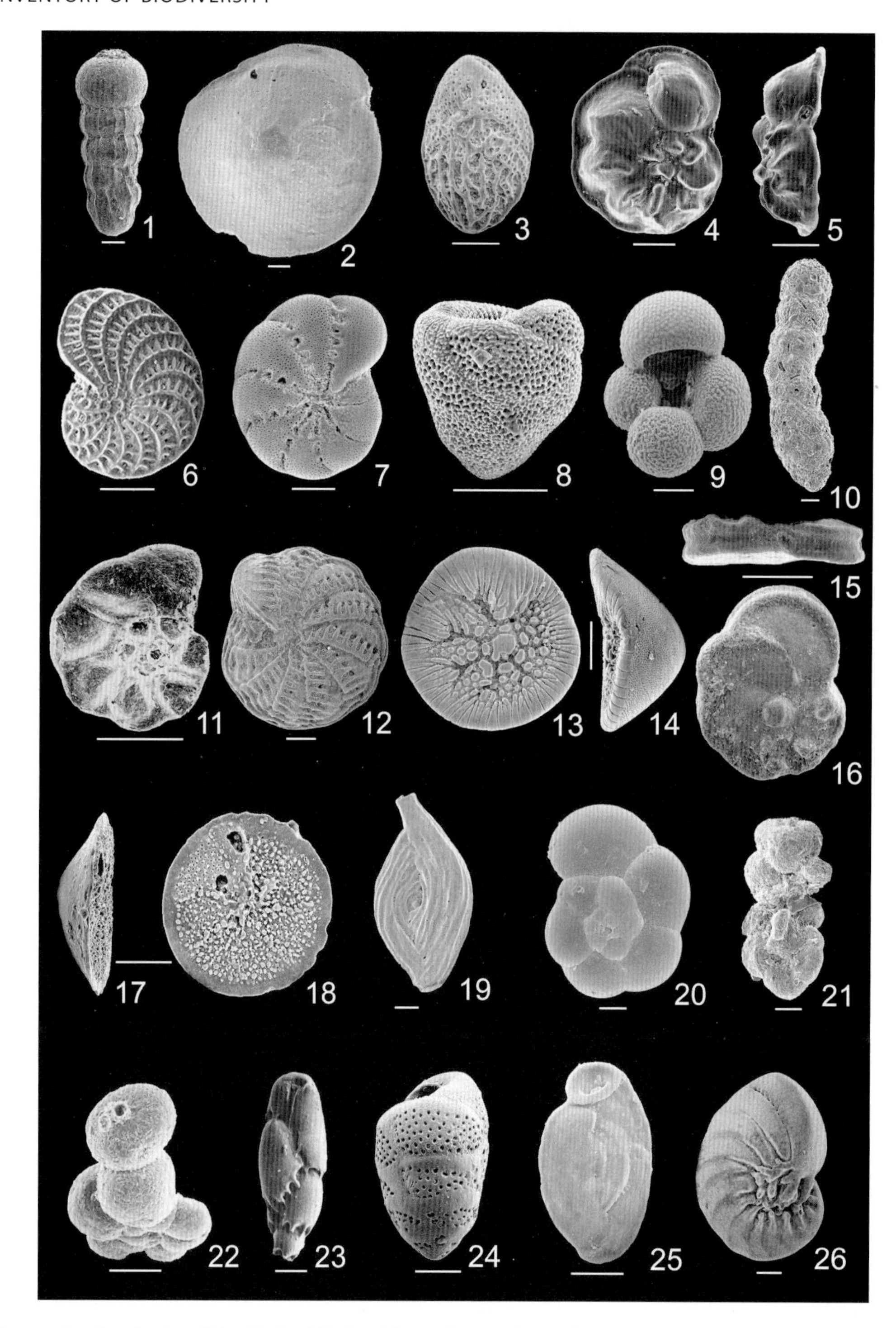

Scanning electron micrographs of a selection of New Zealand fossil and Recent foraminifera. Unless otherwise specified, all scale bars = 0.1 millimetre. New Zealand Fossil Record File numbers (Xnn/fnnn) are given for fossil specimens. Figured specimen catalogue numbers are from Auckland War Memorial Museum (AK) or GNS Science (all others).

1. *Amphimorphinella butonensis*. BWH120/3, R09/f107, Matheson Bay, Auckland, Early Miocene. 2. *Amphistegina* sp. FP4836, R10/f9001A, Waiheke Island, Auckland, Early Miocene. 3. *Bolivina reticulata*. FP4038, R11/f107, Hillsborough Bay, Auckland, Early Miocene. 4, 5. *Discorbinella deflata*. FP4776, Chatham Islands, Recent. 6. *Elphidium novozealandicum*. FP4655, Sound, Fiordland, Recent. 7. *Elphidium vellai*. TF1658/1, Queen Charlotte Sound, Marlborough, Recent. 8. *Glabratellina kermadecensis*. TF1660/1, Kermadec Islands, Recent. 9. *Globigerina bulloides*. FP2846, Kawerua, Northland, Recent. 10. *Haeuslerella pliocenica*. BWH134/6, ODP1120B-2H-5, 69–71 centimetres, Chatham Rise, Pliocene. 11. *Jadammina macrescens*. AK130794, Upper Waitemata Harbour, Auckland, Recent. 12. *Notorotalia finlayi*. BWH133/5, Wangaehu Valley, Wairarapa, Late Pliocene. 13, 14. *Pileolina zealandica*. BWH125/18, Kawau Island, Auckland, Recent. 15, 16. *Planulinoides norcotti*. BWH129/14, Cavalli Islands, Northland, Recent. 17, 18. *Rosalina vitrizea*. BWH129/12, Whangateau Harbour, Auckland, Recent. 19. *Spiroloculina antillarum*. BWH125/2, Kermadec Islands, Recent. 20. *Trochammina inflata*. AK92416, Purakanui Inlet, Otago, Recent. 21, 22. *Trochamminita salsa*. 21, AK76407, Oparara Estuary, Westland, Recent; 22, AK130800, Upper Waitemata Harbour, Auckland, Recent. 23. *Virgulinella fragilis*. From Grindell and Collen (1976), VF1119, Wellington Harbour, Recent. 24. *Virgulopsis turris*. FP4633, Pauatahanui Inlet, Wellington, Recent. 25. *Wiesnerella auriculata*. FP2981, Cavalli Islands, Northland, Recent. 26. *Zeaflorilus parri*. AK131714, Wanganui Bight, Recent.

1970s–1980s Deep Sea Drilling Program
1970s–1990s The oil crises and Cretaceous–Cenozoic Basins Programme
1970s–2000s Ecological distribution of Recent foraminifera
1980s–2000s Impact of catastrophic events on foraminiferal faunas; paleoceanography and isotope studies
2000s Foraminiferal documentation of Holocene earthquakes and human impacts.

These phases closely mirror worldwide trends. The first major contributions to the documentation of New Zealand's foraminiferal faunas came from the results of a number of global scientific expeditions. The Austrian *Novara* Expedition (1858–1859) led to the first descriptions of New Zealand fossil foraminifera (Karrer 1864; Stache 1864). The British *Challenger* (1873–1876) and *Terra Nova* (1910) Expeditions produced the first extensive documentation and descriptions of New Zealand's modern foraminifera (Brady 1884; Heron-Allen & Earland 1922). Much of the research on foraminifera in the 19th and early 20th centuries in New Zealand and worldwide was actually undertaken by amateur microscopists, who were fascinated by the beautifully ornamented shells and their enormous diversity.

The study of foraminifera took off in the early 20th century with the advent of the petrol-driven motor-car and the exploration for subsurface petroleum resources. Work in North America showed that the evolutionary succession of fossil foraminifera could be used to tell the age of strata. Because of their small size, fossil foraminifera could be recovered from mere chips of rock brought up from the drilling face in exploration wells and thereby greatly assist in the search for oil. In New Zealand, this new economic value of foraminiferal studies led to the employment, in the 1930s, of the country's first professional foraminiferal micropaleontologist, Harold Finlay. Ever since, foraminifera have played a major role in servicing petroleum-exploration endeavours in New Zealand. This has been the main economic reason for the training and employment of foraminiferal micropaleontologists and for research on New Zealand's foraminiferal biostratigraphy.

Initially (1930s–1950s), New Zealand benthic foraminifera were used mostly for dating Cretaceous and Cenozoic strata. Each region of the world, including New Zealand, has its own somewhat distinctive benthic faunas with local time ranges that needed to be determined. Thus overseas work could not be applied to New Zealand benthics, and Finlay set about documenting the local time ranges of key taxa (e.g. Finlay 1939a). He used these, together with molluscs, to revise and firmly establish the New Zealand Cenozoic time scale of local stages (Finlay & Marwick 1947) that had been proposed several decades earlier (Thomson 1916). Since their establishment, the local stages and time scale have been the subject of considerable foraminiferal biostratigraphic research (e.g. Hornibrook & Harrington 1957; Scott 1971; Crundwell 2004; Morgans 2004) to improve dating precision and correlation with the International Cenozoic time scale, while maintaining a relatively stable system for application by New Zealand geologists.

Summary of New Zealand foraminiferan diversity by environment

Taxon	Marine	Freshwater	Terrestrial	Fossil*
Polythalamea	1,134	0	0	2,006
Xenophyophorea	7	0	0	0
Totals	**1,141**	**0**	**0**	**2,006**

* Not including forms (f.).

The sound foraminiferal biostratigraphic basis for the New Zealand Cenozoic stages was responsible for major advances in our understanding of New Zealand geology in the 1950s–1970s. The most significant of these advances was the geological mapping coverage, strongly underpinned by foraminiferal age determinations, of the whole country at a scale of 1:250,000.

In the 1960s–1970s, planktic foraminiferal studies blossomed worldwide with the recognition by hydrocarbon-industry micropaleontologists of their cosmopolitan distributional patterns and their potential for global biostratigraphic correlation of the Cretaceous and Cenozoic. Because of excellent local Cenozoic marine sequences, New Zealand work on planktic foraminiferal biostratigraphy was among the leading world research (e.g. Scott 1969; Jenkins 1971; Hornibrook & Edwards 1971). Planktics rapidly replaced the benthics as the major group of foraminifera used for dating and in recognising and refining ages in terms of New Zealand stages.

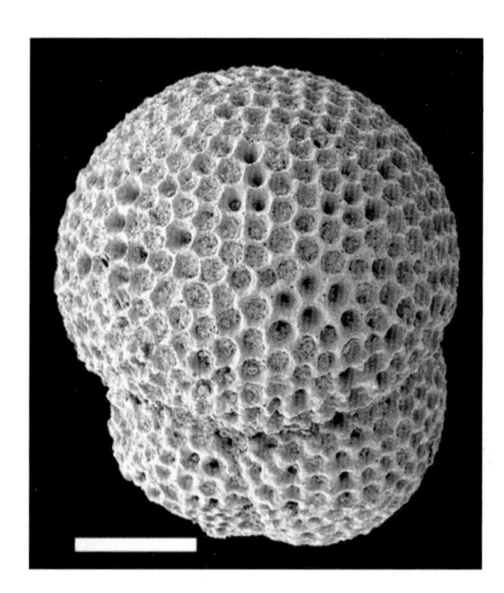

Globigerinoides trilobus, one of many species of planktic foraminifera useful in dating New Zealand's sedimentary rocks. ODP Site 1125, Chatham Rise, Pliocene. Scale bar 0.1 millimetre.

The international Deep Sea Drilling Project (DSDP) of the late 1960s–1980s provided a major boost to planktic foraminiferal studies. Three legs cored the seafloor of the New Zealand region (Leg 21, 1970–71; Leg 29, 1973; Leg 90, 1982–83) and provided excellent material for high-resolution planktic biostratigraphic studies (e.g. Kennett 1978; Srinivasan & Kennett 1981; Hornibrook 1982; Martini & Jenkins 1986; Scott 1992). These have greatly improved the precision and international correlation of Neogene planktic foraminiferal dating in the less well-developed faunas preserved in on-land sequences.

The use of fossil foraminiferal faunas to provide paleoenvironmental data, and particularly the depth of accumulation of Cenozoic sediments, began in the 1960s (e.g. Vella 1962a) and has become progressively more quantitative and refined (e.g. Scott 1970; Hayward & Buzas 1979; Hayward & Brook 1994; Abbott 1997; Hayward 2004). In an effort to improve the resolution of paleodepth and paleo-sea-level assessments of local fossil faunas, there have been an increasing number of studies, particularly in the 1990s–2000s, on the ecological distribution of Recent foraminifera around New Zealand (e.g. Lewis 1979; Hayward 1982; Hayward et al. 1999a; 2010a).

In the mid- to late 1970s, the New Zealand Government reacted to international oil crises by setting up a state-run petroleum exploration agency (Petrocorp) and encouraging overseas companies to explore for hydrocarbons here. The increased demand for foraminiferal servicing of exploration wells was to last for over a decade and led to the employment of two of the present authors. Another by-product of the crises was the initiation of the New Zealand Geological Survey's Cretaceous–Cenozoic Basins Programme (1980s–1990s), which dominated foraminiferal research at that agency for fifteen years and led to major reviews of the biostratigraphic dating and paleodepth assessments of most of the country's late Cretaceous and Cenozoic sedimentary basins (e.g. Hayward 1990a; Raine et al. 1993).

New Zealand's excellent exposures of late Cretaceous and Cenozoic marine sequences have attracted a number of studies in recent decades (1980s–2000s) on the impact of global astronomical or paleoclimatic events on microfossils, including foraminifera (Strong 1977, 2000; Brooks et al. 1986; Kaiho et al. 1996; Hollis & Strong 2003). During the same time period, international and New Zealand foraminiferal studies have moved into paleoceanographic and paleoclimatic research (1980s–2000s). These studies have largely arisen because of the high-quality, high-resolution sequences of deep-sea planktic foraminifera obtained from hydraulic piston coring by the Ocean Drilling Program, ODP, and the later legs of its ancestor, DSDP. These include DSDP Leg 90 (two holes in the New Zealand region) and ODP Leg 181 (seven holes off the east coast of New Zealand, 1998). These studies have mostly focused on oxygen and carbon isotopes in the carbonate of the foraminiferal shells and paleoclimate-related changes in the planktic assemblages (e.g. Nelson et al. 1993; Scott et al. 1995; Weaver et al. 1998; Sabaa et al. 2004; Schaefer et al. 2005; Crundwell et al. 2008;

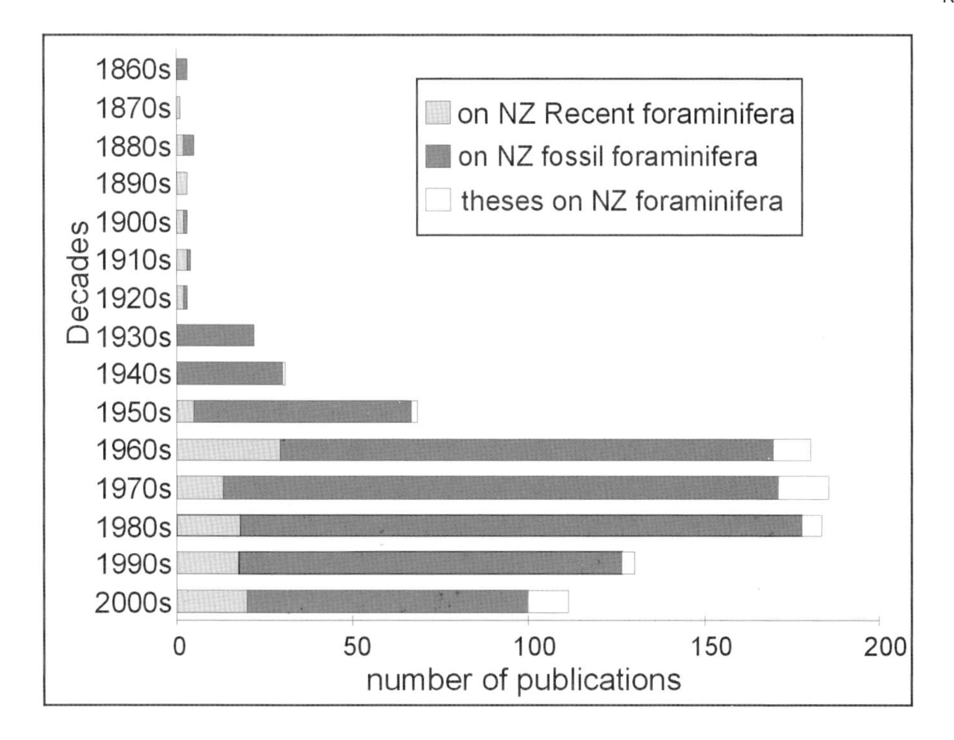

Graph showing the number of publications (papers and monographs, not abstracts) and theses (at least one chapter) on New Zealand foraminifera per decade, reflecting trends in the number of researchers and recent changes in the balance between basic and applied foraminiferal research.

Hayward et al. 2008a).

In the last decade (2000s) there has been a worldwide swing into the field of applied Holocene studies of nearshore foraminiferal records. This shift has also been seen in New Zealand with application of these studies to understanding: a) the recurrence times of large earthquake events and tsunamis (e.g. Goff et al. 2000; Hayward et al. 2006c, 2007b, 2010a,b); b) the history and causes of late Holocene sea-level rise (e.g. Gehrels et al. 2008); and c) human impacts on the the health of the coastal marine environment (e.g. Hayward et al. 2004c, 2006b, 2008b; Cochran et al. 2005; Matthews et al. 2006; Grenfell et al. 2007).

The number of publications on New Zealand foraminifera per decade [Fig. 2] is a crude indicator of the history of study and the number of active researchers through time. Not surprisingly there was a steady increase in studies and workers from the 1930s through the 1960s, but there has been a major decline in publication numbers in the 1990s–2000s. The decline can be correlated in part with a decrease in New Zealand foraminiferal researchers owing to retirement or redundancy, but is also attributable to increased commercial servicing work and a trend towards larger, more detailed and more time-consuming studies, often involving census work. The histograms clearly show the emergence and blossoming of fossil studies from the 1930s, and the resurgence of studies on Recent foraminifera from the 1950s.

New Zealand Recent foraminifera

Historical overview

The earliest report of Recent foraminifera (Brady 1884) necessarily deals with material that would have been not only living when collected but also 'recently dead', i.e. approximating the latter part of the Holocene (past ~3000 years). This is because it is not easy or routinely possible to determine from bottom sediments what was actually alive or dead. Many of Brady's figured specimens were obtained from the five sediment samples collected from New Zealand waters (one from Port Nicholson, two west of Cook Strait, and two east of North Island) by the *Challenger* expedition. Soon afterwards, an enthusiastic microscopist, Haeusler (1887), listed foraminifera from beach samples from around the Hauraki Gulf. Hutton (1904) used these two works to compile the first checklist of living New

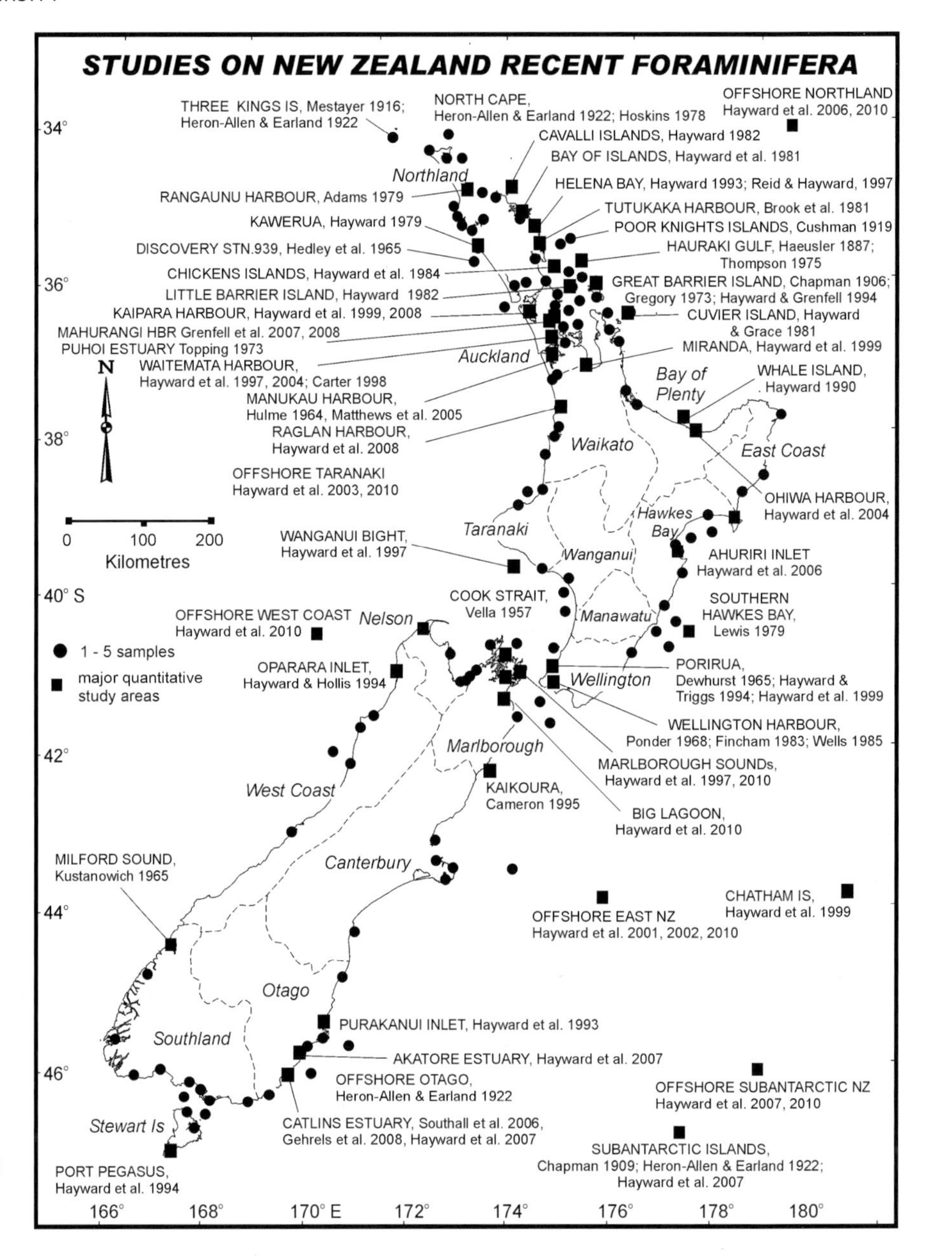

Map of New Zealand showing the location of samples or marine areas from which Recent foraminiferal faunas have been documented. All references are listed in Hayward et al. (1999a, 2010a).

Zealand foraminifera in his *Faunae Novae Zealandiae* (190 species and 'varieties').

During the following two decades, miscellaneous collections from off Northland and the subantarctic islands were identified by a local enthusiast (Mestayer 1916) or sent overseas for description by foraminiferal specialists (e.g. Chapman 1909; Cushman 1919). The most significant early work came when the English foraminiferal duo of Heron-Allen and Earland (1922) described the faunas (including 24 new species) from 12 samples taken from New Zealand waters during the British Antarctic (*Terra Nova*) Expedition of 1910–13. Again the samples were from northern New Zealand, East Cape, and the subantarctic region.

From 1922 to the early 1950s, there was a hiatus in studies on living New Zealand foraminifera, with just occasional references in overseas monographs. Vella's (1957) pivotal work in the central New Zealand region (yielding 54 new species) heralded renewed vigour in studies on the local fauna. It stimulated Hulme's promising work (e.g. Hulme 1964), which was tragically cut short by his death while testing his own design of underwater breathing apparatus in Wellington Harbour. Meanwhile, a group of visiting and resident scientists at the then New Zealand Oceanographic Institute (part of the later National Institute of Water & Atmospheric Research) were particularly active (e.g. Kustanowich 1963, 1965; Hedley et al. 1965, 1967; Eade 1967a; Lewis 1970). From the 1960s on, a number of graduate student theses, supervised by Paul Vella at Victoria University of Wellington and Murray Gregory at Auckland University, were completed. Most document the present-day ecological distribution of New Zealand's living foraminifera to provide information useful in interpreting the depositional environments of fossil faunas.

The foraminifera of New Zealand salt marshes were first documented by American micropaleontologist F. B. Phleger (1970). Numerous subsequent salt-marsh studies have been summarised by Hayward and Hollis (1994) and Hayward et al. (1999a). During the 1980s–1990s, Hayward and colleagues at Auckland Museum and subsequently Geomarine Research in Auckland undertook a series of studies on the biodiversity, ecological distribution, and biogeography of New Zealand's inner- and mid-shelf foraminifera, [Fig. 3] summarised in Hayward et al. 1999a). Knowledge of New Zealand's deep-water foraminifera is largely derived from the taxonomic studies of Saidova (1975), who described 56 new species from the New Zealand region, and from the depth-distribution studies of Lewis (1979) and Hayward et al. (2001, 2003, 2006, 2007), summarised in Hayward et al. (2010a). The planktic foraminifera have been documented in several studies, with the most recent summary by Hayward (1983).

Major checklists of the Recent foraminiferal fauna of New Zealand were published by Eade (1967b) and Dawson (1992), recording 957 and 1294 species, respectively. Neither of these checklists was corrected for synonyms, and many species, especially early records, have been recorded under more than one name. In compiling the checklist presented here, we have endeavoured to check every recorded name in Dawson (1992) and to eliminate synonyms. This reduced the number to 746 species. We also took the opportunity to add subsequent new records of 300 species from the recent reviews of New Zealand shallow- and deep-water foraminifera by Hayward et al. (1999a, 2010a).

Current recent diversity

The current total (herein) now stands at 1134 species and subspecies (including 116 undescribed or unidentified taxa) from the Exclusive Economic Zone (EEZ) (see end-chapter checklist). Of these, only 23 species are restricted to brackish or slightly brackish environments (Hayward & Hollis 1994); an estimated 600 species live at shelf depths (shallower than 200 metres), 400 species at bathyal depths (200–2000 metres), and approximately 400 species at abyssal depths (deeper than 2000 metres). [There is some overlap in species ranges.]

Agglutinated species constitute approximately 80% of the small brackish fauna, with the remainder being members of the order Rotaliida. In contrast, calcareous buliminids, rotaliids, lagenids, and miliolids (about 20% each) dominate the shelf fauna, with only about 15% of the fauna being agglutinated. With increasing depths through the bathyal and upper abyssal, miliolids decrease in diversity and abundance, whereas lagenids and agglutinated species increase. Below the Carbonate Compensation Depth (about 4500–5000 metres), all calcareous shells dissolve and the fauna is entirely composed of agglutinated species (orders Astrorhizida, Lituolida, Textulariida, and Trochamminida).

The ecological distribution of brackish-water foraminifera is most strongly influenced by salinity and tidal height, whereas those in shallow, normal marine environments are influenced by a variety of factors including wave and current energy, oxygen concentrations, substratum, and depth-related variables (Hayward et al. 1999a). The distribution of bathyal and abyssal benthic foraminifera is crudely zoned by depth and strongly influenced by surface-water phytoplankton productivity (carbon flux), oxygen concentration, salinity and temperature characteristics of bottom-water masses, and calcium carbonate saturation (Hayward et al. 2010a).

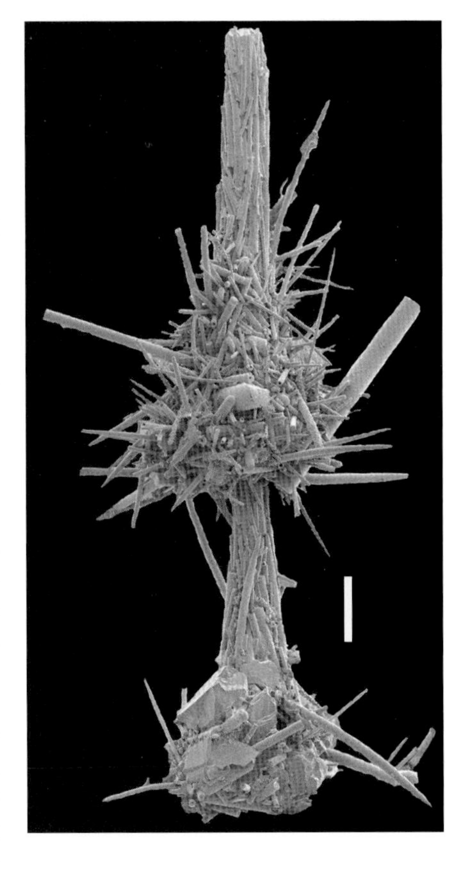

Reophax hispidulus, a deep-water species that makes its shell from silt grains or, in this case, sponge spicules that are stuck together. BWH181/2, South Canterbury Bight, Recent. Scale bar 0.1 millimetre.

Some 29 species of planktic foraminifera live in the top few hundred metres of oceanic waters in New Zealand's EEZ, with their geographic distribution strongly influenced by water temperature (Weaver et al. 1998) and their abundance by phytoplankton productivity. Species diversity is highest in the subtropical region around the Kermadec Islands, decreasing progressively southwards to just two or three species living in the subantarctic region (Hayward 1983).

Brackish-water and shallow marine benthic foraminifera also show a progressive decrease in diversity from north to south through the country. Some 22 brackish species are present in Northland, reducing to 11 in Southland and Stewart Island and five on the Chatham Islands; and 304 shallow-water marine species live off the east coast of Northland under the influence of the warm East Auckland current, reducing to 193 around the southern end of the South Island (Hayward et al. 1999a). Lower diversities of shallow-water species are recorded around New Zealand's outlying islands (Kermadec Islands 47 species; Chatham Islands 125 species; Auckland and Campbell Islands 78 species).

Pileolina calcarata, one of a number of endemic species described from the British (Terra Nova) Expedition of 1910–13. BWH127/16, Cavalli Islands, Northland, Recent. Scale bar 0.1 millimetre.

Collection repositories

The major repositories of Recent New Zealand Foraminifera are (in decreasing order of collection size): GNS Science (incorporating the former New Zealand Geological Survey), Lower Hutt (Vella, Hulme, Hoskins, Hayward collections); Auckland War Memorial Museum (Mestayer, Hayward et al. collections); Natural History Museum, London (Brady, Heron-Allen & Earland collections); Geology Department, University of Auckland (Gregory, Topping, Adams, Thompson, Carter, Hayward et al. collections); National Institute of Water and Atmospheric Research, Wellington (Hedley, Eade, Lewis collections); Geology Department, Victoria University of Wellington (Vella, Collen, Ponder, Fincham, Wells collections); Geology Department, University of Canterbury (Cameron collection).

New Zealand fossil foraminifera

Historical overview

The first published record of foraminifera from New Zealand was of several species collected from the Oligocene Totara Limestone near Oamaru by Walter Mantell and identified for him by Rupert Jones in London (Mantell 1850). However, the first major studies and description of foraminifera from New Zealand were of two rich mid-Cenozoic fossil faunas from samples (Early Miocene, Orakei Greensand, Auckland; Early Oligocene, Whaingaroa Siltstone, Raglan) collected by Hochstetter during the *Novara* Expedition in 1856 and studied by Felix Karrer and Guido Stache in Vienna (Karrer 1864; Stache 1864). Of the 145 species described, 65 are considered valid (Hornibrook 1971), with many recognised in the fossil record in other parts of the world.

There was a long gap before further work was undertaken on New Zealand's fossil foraminifera, with the commissioning of Frederick Chapman, National Museum of Victoria, to produce a New Zealand Geological Survey Paleontology Bulletin on the Cenozoic foraminifera and ostracods (Chapman 1926). This was followed in the 1930s by several short papers by another Victorian, Walter Parr, describing further elements of the Cenozoic foraminifera.

In 1932, the brilliant New Zealand molluscan systematist Harold Finlay (recognised as the 'father of New Zealand micropaleontology'), unemployed and destitute in the midst of the Great Depression, switched his studies to fossil foraminifera. He began as an employee of Vacuum Oil Coy, which was exploring for petroleum around Gisborne, and later (1937 onwards) joined the New Zealand Geological Survey (NZGS), where he remained until his untimely death in 1951. While at NZGS, Finlay described in highly abbreviated fashion 241 species of the Cretaceous and Cenozoic foraminifera (Finlay 1939a,b,c, 1940, 1947) that he considered to be key taxa for biostratigraphic subdivision. They formed a pivotal part of his work with molluscan colleague Jack Marwick, in which they firmly established the modern New Zealand Late Cretaceous and Cenozoic biostratigraphic stages (e.g. Finlay & Marwick 1947). Prior to about 1970, however, most workers including Finlay considered that the majority of foraminifera (like the molluscs he had previously specialised in) were endemic to each region of the world. Thus many foraminiferal species described and named from New Zealand are now being recognised as having much wider distributions and are frequently synonyms of species named earlier from other regions.

Following Finlay's death, his place at NZGS as New Zealand's only foraminiferal micropaleontologist was taken by Norcott Hornibrook. Over the next 40 years (1950s–1980s), Hornibrook undertook a number of major taxonomic and biostratigraphic studies on the Cenozoic (e.g. Hornibrook 1961, 1968, 1971, 1996, Hornibrook et al. 1989), describing 136 new foraminiferal species from New Zealand. At NZGS, Hornibrook gradually built up around him a highly productive team of foraminiferal researchers, who serviced most of the enormous dating requirements of the Survey's four-mile and one-mile mapping projects in the 1960s–1970s and of New Zealand's petroleum exploration boom in the 1970s–1980s while undertaking their own foraminiferal research projects. Perhaps Hornibrook's most substantial contribution to New Zealand's foraminiferal studies was his establishment and curation of a national repository, reference collection, and biostratigraphic data centre for New Zealand foraminifera. Together with the New Zealand Fossil Record File, it is now the envy of every other micropaleontological research centre in the world.

Since the 1950s, the NZGS micropaleontology group (now GNS Science) has mostly focused on improving the biostratigraphic resolution of New Zealand foraminifera, with most Cenozoic work on planktic faunas (e.g. Scott 1971; Hornibrook 1981; Cifelli & Scott 1986; Scott et al. 1990; Morgans et al. 1999). Biometric studies have been used on both planktic and benthic lineages in an attempt to improve the rigour of their biostratigraphic application (e.g. Scott 1965, 1974). Studies on Cretaceous foraminifera have recognised mainly cosmopolitan species (benthic and planktic) in New Zealand rocks (e.g. Webb 1971; Hornibrook et al. 1989). In more recent decades there has been increased use of benthic foraminifera in quantitative paleoecological and paleogeographical studies (e.g. Scott 1970; Hayward & Buzas 1979; Hayward 1986).

Chrysalogonium conica, a cosmopolitan deep-water species that became extinct during the mid-Pleistocene Climate Transition about 1 million years ago. BWH143/19, ODP Site 1125, Chatham Rise, Pliocene. Scale bar 0.1 millimetre.

Two internationally prominent New Zealand foraminiferal workers, who had short stints early in their careers in Hornibrook's laboratory as well as in petroleum exploration companies before moving into University positions, are Paul Vella (Victoria University of Wellington) and the late Graham Jenkins (University of Canterbury). Jenkins made a major contribution to understanding Cenozoic planktic foraminiferal diversity and biostratigraphy of New Zealand and the Southwest Pacific (e.g. Jenkins 1971). Vella made significant contributions to knowledge of New Zealand Neogene foraminifera and their paleoecology (e.g. Vella 1961, 1962b) and helped train a number of prominent foraminiferal researchers whose graduate and subsequent research made major contributions concerning New Zealand's foraminifera of Eocene, Oligocene (e.g. Srinivasan 1966), late Miocene (e.g. Kennett 1966; Gibson 1967), and Pliocene (e.g. Collen 1972) ages.

Since the birth of indigenous New Zealand foraminiferal micropaleontology in the 1930s, most contributions to our knowledge of taxon diversity has come from research at NZGS or the universities (particularly Victoria University of Wellington). At times, overseas petroleum exploration companies have brought in their own micropaleontologists to provide biostratigraphic services and two of these expanded their work to describe some fossil foraminifera – Dorreen (1948) described 19 new species from the Late Eocene and Stoneley (1962) described 19 new species from the Early Cretaceous of New Zealand.

The first record of foraminifera in New Zealand Triassic rocks was by Benson and Chapman (1938). Apart from several other minor records, the only major descriptive works on the Triassic foraminifera have been by Strong (1984a,b) on faunas from North and South Islands, respectively. The only published records

Number of species in each order recorded from different New Zealand time periods

Order	Recent	Cenozoic	Late Cretaceous	Early Cretaceous	Jurassic	Triassic–Permian	Total
Allogromiida	2	0	1	0	0	0	3
Astrorhizida	75	7	18	0	3	0	95
Lituolida	108	74	47	12	11	1	231
Trochamminida	24	3	5	2	1	0	34
Textulariida	34	97	10	2	1	0	133
Fusulinida	0	0	0	0	?1	44	45
Spirillinida	13	5	3	0	0	3	22
Carterinida	1	0	0	0	0	0	1
Miliolida	110	71	0	0	0	4	154
Lagenida	433	353	69	3	61	9	805
Robertinida	8	19	1	1	1	0	29
Globigerinida	28	205	39	7	0	0	259
Buliminida	96	332	18	2	1	0	388
Rotaliida	176	442	39	18	0	0	564
Total species	**1108**	**1608**	**250**	**47**	**80**	**61**	**2761**
NZ type localities	**144**	**604**	**25**	**21**	**4**	**2**	**805**

Time ranges of some species extend across several columns, hence rows do not total to the sum given at right.

of Jurassic foraminifera from New Zealand are of small faunas from the Late Jurassic – one from Uruti-1 well, north Taranaki (Hornibrook 1953) and the other from Port Waikato (Hornibrook et al. 1989).

The only Paleozoic foraminifera so far recognised in New Zealand are from several isolated Permian localities in Northland, north Otago, and Canterbury. They were first reported by Hornibrook (1951) from Marble Bay, near Whangaroa in Northland. The fauna from this locality has been extensively examined in thin sections in the 1990s with around 50 species now recorded (Vachard & Ferrière 1991; Leven & Grant-Mackie 1997). South Island Permian faunas are less rich and have been documented by Hornibrook and Shu (1965), Hada and Landis (1995), and Leven and Campbell (1998).

The end-chapter checklist is the first attempted compilation of New Zealand's fossil foraminifera, resulting in a tally of 2006 species and subspecies.

Cenozoic

Some 1661 species are recorded from New Zealand's Cenozoic strata. [Table] The best-studied group, the planktics (Globigerinida), have 7.5 times as many species in the Cenozoic record (206) than at the present day (29). This is a result of both evolution and a warmer climate throughout most of the Cenozoic, with consequently greater planktic diversity, comprising 38–41 species present throughout most of the Miocene and Early Pliocene.

In the benthics there are approximately three times as many species recorded from the Cenozoic than the Recent in the orders Textulariida, Buliminida, and Rotaliida. These are the three best-studied groups of benthics in the New Zealand Cenozoic record and these figures provide a good indication of the probable total diversity of the Cenozoic benthic fauna – approximately three times the Recent (3000–4000 species). This suggests a mean species duration for the Cenozoic benthic foraminifera in New Zealand of about 20 million years.

As with the planktics, a greater diversity of shallow-water benthics was

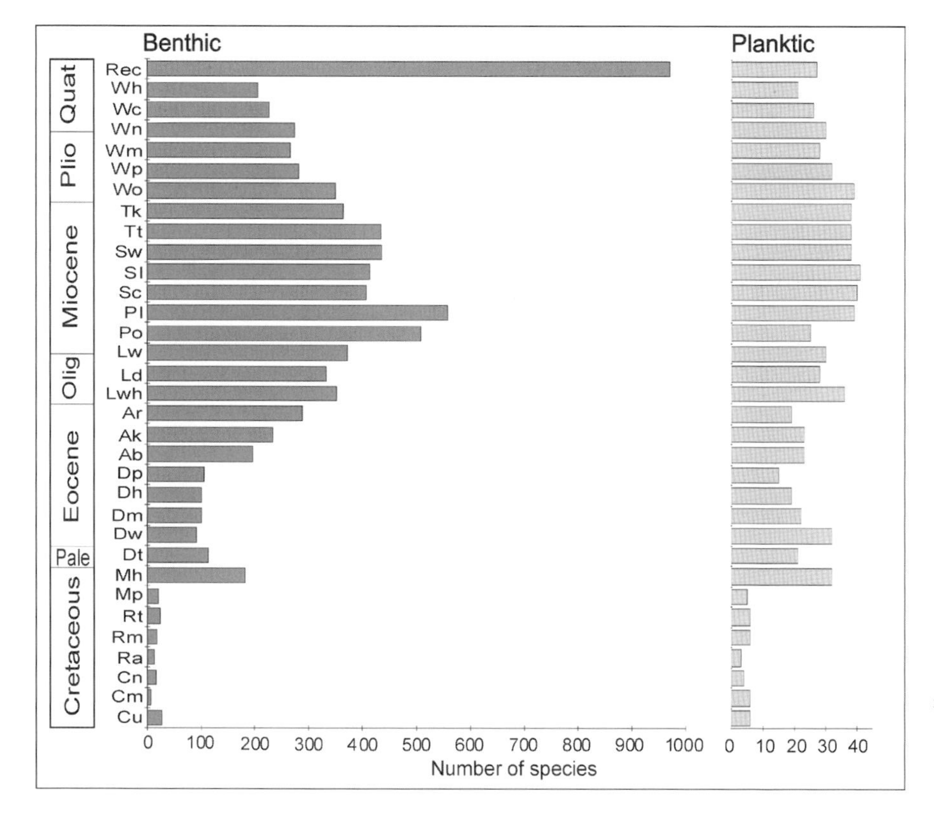

Number of benthic and planktic foraminiferal species in each local stage of the Cretaceous and Cenozoic of New Zealand, based on the total recorded time range of each species (see checklist).

present throughout much of the Cenozoic than at present, because of the warmer sea-surface temperatures.

The full original fossil benthic foraminiferal diversity will never be known because of taphonomic loss. This is the natural disintegration of the foraminiferal shell following the organism's death, usually because of the bacterial decay of the organic cement or lining that holds the shell together while the organism is alive. Taphonomic loss of shells is well documented on the seafloor in many modern agglutinated species (e.g. Hayward et al. 2001) and its effect can be seen in the number of recorded fossil species compared with Recent ones in the orders Allogromiida, Astrorhizida, Trochamminida, Carterinida, and possibly to a lesser extent in Lituolida (see Table). The porcellaneous and aragonitic shells of the order Miliolida and Robertinida are also susceptible to taphonomic loss because of preferential solution or weathering. The lower proportion of Spirillinida, Miliolida, and Lagenida in the recorded Cenozoic fauna compared to the Recent is probably partly because of a lack of detailed documentation, as few are common enough to be useful in biostratigraphic dating or paleoecological assessments.

The pattern of recorded species diversity in each New Zealand stage through the Cenozoic [Fig. 4] can be interpreted as resulting from several factors:

1) taphonomic loss and less intense faunal documentation – best illustrated in the difference between Pleistocene and Recent species diversity;
2) warmer sea-surface temperatures in the early and middle Cenozoic – best illustrated by the declining diversity from late Miocene to Pleistocene;
3) lower latitudinal temperature gradient and less-stratified ocean waters in the Paleogene – illustrated by the lower diversities in the Paleocene and Eocene;
4) varying intensities of detailed faunal documentation in different parts of the stratigraphic column – illustrated by lower diversities in less-documented Paleocene, Early Eocene, and Middle Miocene [Clifdenian (Sc) and Lillburnian (Sl) Stages], and higher diversities in the most intensely documented Early and Late Miocene [Otaian (Po), Altonian (Pl), and Tongaporutuan (Tt) Stages] (e.g. Kennett 1966; Hayward & Buzas 1979).

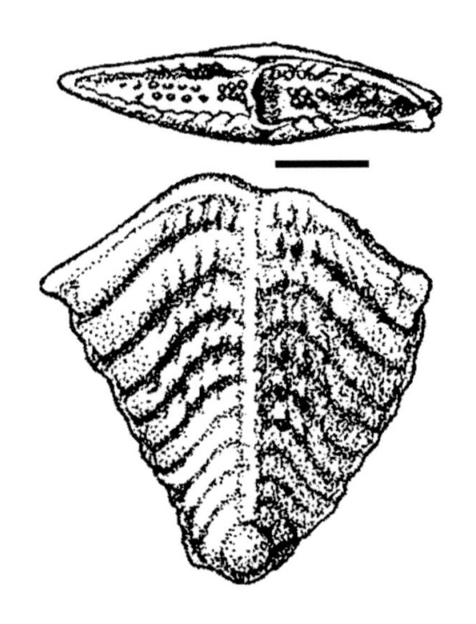

Quasibolivinella finlayi, an endemic Oligocene–Miocene species. TF1593/1, holotype, Tengawai River, Canterbury, Miocene. Scale bar 0.1 millimetre.

Most marine environments and their foraminiferal faunas are well represented in the New Zealand Cenozoic fossil record, except for brackish-water and Late Neogene abyssal faunas. Shallow-water facies are not particularly common in the Paleocene and Eocene, however.

As in the Recent, there are recognisable biogeographical differences in the faunas around New Zealand in the Neogene with decreasing diversity from north to south, but these have not been documented in detail. Distinct biogeographical differences around New Zealand are minimal in the Paleogene, when there was a lower temperature gradient between equatorial and polar waters.

Cretaceous

Some 297 species of foraminifera are recorded from the New Zealand Cretaceous. Apart from the 15 million-year-long latest Cretaceous Stage (Haumurian, Mh), recorded diversity is particularly low (see Table). In contrast to the Cenozoic, New Zealand Cretaceous strata lack well-exposed, structurally simple, continuously fossiliferous sequences. Diverse, well-preserved foraminiferal assemblages are patchy in occurrence and mostly confined to the latest Cretaceous (Haumurian). It is suggested that our relatively poor Cretaceous foraminiferal record is because of induration, structural complexity, unfavorable paleoenvironments, and selective preservation (Hornibrook et al. 1989). A detailed account of Cretaceous foraminiferal faunas, stage by stage, is given in Hornibrook et al. (1989, p. 45).

Planktic foraminifera began to diversify in the Cretaceous with 46 species recorded from New Zealand. The benthic fauna is dominated by members of the orders Lagenida, Lituolida, and Rotaliida, with subordinate numbers of Buliminida, Astrorhizida, and Textulariida.

The Cretaceous ended with a major extinction event in many phyla (e.g. dinosaurs, ammonites, belemnites) thought to have been caused by a bolide impact with Earth. The foraminiferal record across the Cretaceous–Tertiary (K–T) boundary has been studied in many northern South Island sequences (e.g. Strong 1977, 2000). It is marked by the extinction of most Cretaceous planktic foraminifera at a thin iridium-rich clay layer (K–T boundary clay) and the demise of a few possible 'survivor' planktic taxa a short distance above. In contrast, one third of New Zealand's Late Cretaceous benthic foraminifera (65 species) survived this holocaust. These survivors are a mixture from all orders, with a predominance of deep-water species.

Triassic and Jurassic

Only two small faunas, totalling seven species, have been recorded from the Jurassic in New Zealand. Both are from the latest Jurassic, Puaroan Stage, in western North Island, where preservation is moderately good and shows promise that further collecting is likely to yield more faunas of greater diversity and interest.

Despite the small amount of study so far devoted to New Zealand's Triassic foraminifera (Strong 1984a,b), the fauna comprises by far the most diverse (72 species) and well-preserved yet recorded from the Southern Hemisphere and is comparable with the best faunas known from the Northern Hemisphere. Foraminifera are present in all New Zealand Triassic stages and are known from both the Southland and Kawhia Synclines in South and North Islands, respectively. A similar number of additional species have been observed from Torlesse samples in the North Island, but are so far undocumented. The wholly benthic fauna is dominated by lagenids and comprises a mix of cosmopolitan and probable endemic species (Strong 1984a).

Paleozoic

Although fossil foraminifera occur throughout the Paleozoic in other parts of the world, so far they have been reported only from Permian rocks in New Zealand, probably because of the highly indurated or metamorphosed state of most of the older sedimentary rocks (Hornibrook et al. 1989). There are 44 fusulines and 17 smaller benthic species. Faunas from within isolated blocks in the South Island Torlesse are of Middle Permian age (Leven & Campbell 1998), whereas those in Northland are Late Permian (Leven & Grant-Mackie 1997).

Members of the Fusulinida (an extinct Paleozoic order) are thus the most significant component of New Zealand's Permian fauna. Fusulinids are largely confined to Northern Hemisphere localities, but occur in exotic blocks of limestone within the Waipapa and Torlesse Terranes of Northland and southern South Island. Most of New Zealand's Permian strata (e.g. Brook Street and Maitai Terranes) lack them. Fusuline foraminifera are regarded as being of warm-water, Tethyan affinity, whereas most of the New Zealand region was at cool, southern high latitudes at this time. New Zealand's fusulines provide the best available evidence for a far-northern (East Asian) origin for the Torlesse and Waipapa Terranes, which subsequently were moved southwards by plate tectonics before colliding with and suturing to ancestral New Zealand in the early Cretaceous.

Yabeina (Lepidolina) multiseptata, one of the few Paleozoic foraminifera (order Fusulinida) recorded from New Zealand. From Hornibrook 1989. Marble Bay, Northland, Permian. Scale bar 1 millimetre.

Collection repositories

The major repository of fossil New Zealand foraminifera is GNS Science (incorporating the former New Zealand Geological Survey), Lower Hutt (Finlay, Hornibrook, Jenkins, Scott, Strong, Stoneley, Webb, Hayward specimens, types of most fossil species described from New Zealand, and 30,000 fossil foraminiferal faunas). Other major repositories are (in decreasing order of collection size): Geology Department, Victoria University of Wellington (Vella, Collen, Srinivasan, Kennett, Gibson collections); Geology Department, University of Auckland (graduate student research collections); Natural History Museum, Vienna (Karrer and Stache types); Geology Department, University of Canterbury (graduate student research collections); Smithsonian Institution, Washington, DC (Dorreen types, Hayward & Buzas specimens). The main database of foraminiferal identifications is the computerised record of the New Zealand Fossil Record File managed by the Institute of Geological and Nuclear Sciences, holding records from some 50,000 samples.

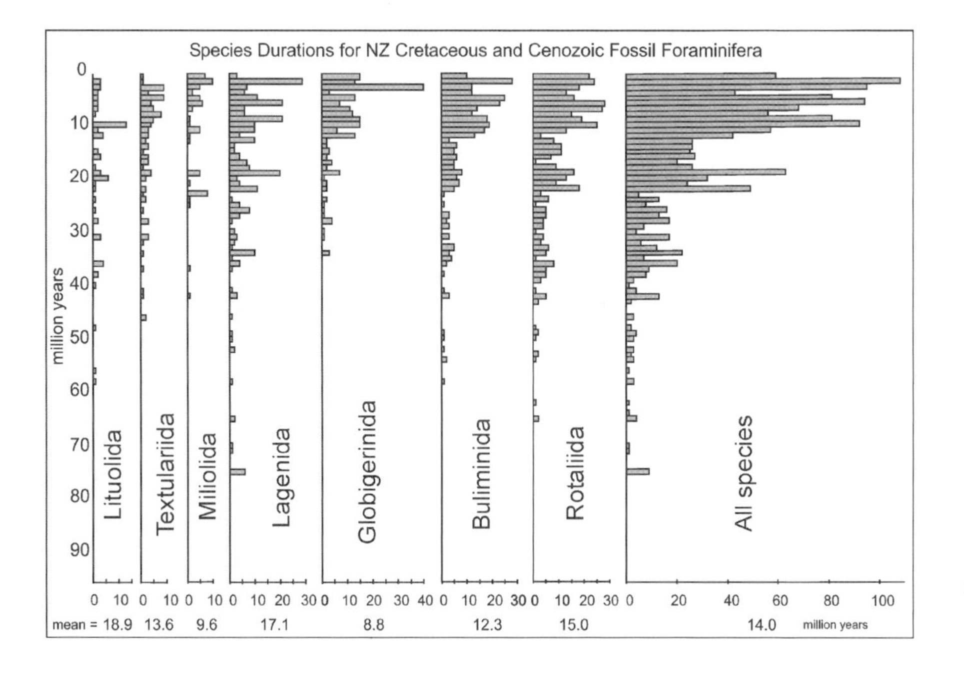

Species durations in New Zealand of Cretaceous and Cenozoic foraminifera grouped by ordinal rank, based on the total recorded time range of each species (see checklist).

Features of the New Zealand foraminiferan fauna

Endemism

Because of the confusion in foraminiferal taxonomy worldwide, it is difficult to achieve an accurate estimate of the endemic component of New Zealand's foraminiferal faunas. The most rigorously reviewed part of the living fauna is that which is shallower than 100 metres (Hayward et al. 1999a). Here, 14% of the fauna (50 species) is considered to be endemic. No brackish-water species is endemic, probably a result of worldwide dispersal on the feet of migratory sea birds. In the total Recent end-chapter species list, 13% (125 species) are indicated as endemic. This is likely to be a considerable overestimate, with many of the deeper-water species that have been described from the region and not yet recorded from outside it – for example, 56 species described by Saidova (1975) – being probable synonyms of more widespread species. Experience suggests that in reality there are fewer endemics in deep water (> 500 metres) than in shallow.

No estimate is available for the endemic component in the fossil Cretaceous and Cenozoic faunas other than a suggestion that it was probably similar to the present day at about 10–15%, and comprising mostly shallow-water species.

In the Recent species list, the majority of endemics are from the orders Miliolida and Rotaliida, with 25% (28 species) and 20% (35 species), respectively, currently considered endemic. Not surprisingly, no planktic species (order Globigerinida), either Recent or fossil, is endemic.

There are no subfamilies or higher taxonomic groups endemic to New Zealand and just two endemic genera: *Zeaflorilus* (two species, Oligocene–Recent) and the extinct *Haeuslerella* (seven species, Oligocene–Early Pleistocene). However, New Zealand does seem to be the centre of diversity and origin of the Southern Hemisphere family Notorotaliidae (3 genera, 55 species, about 95% endemic, Mid-Eocene–Recent). The eight living species of *Notorotalia* (all endemic) are a major component of New Zealand's shallow-water fauna. New Zealand also seems to be the centre of diversity of the glabratellid genus *Pileolina* (8 species, 75% endemic, Oligocene–Recent).

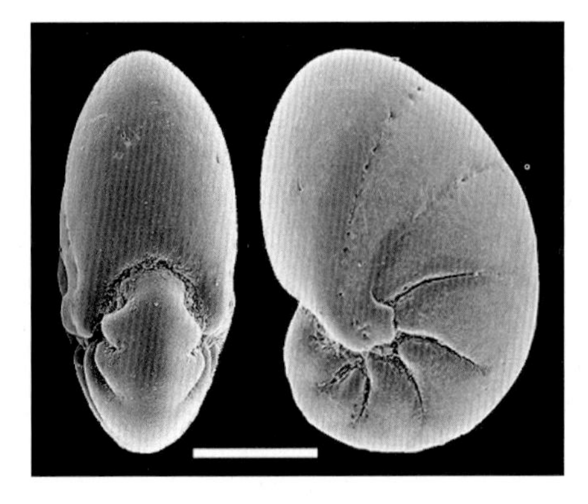

Nonionellina flemingi, a widespread endemic species. BWH63/21, off Kapiti Island, Wellington, Recent. Scale bar 0.1 millimetre.

Contrary to intuition, many of the most abundant and widespread shallow-water species around New Zealand today (Hayward et al. 1999a) are endemics (e.g. *Elphidium novozealandicum, Nonionellina flemingi, Notorotalia finlayi, Pileolina zealandica, Spiroplectinella proxispira, Virgulopsis turris*, and *Zeaflorilus parri*. Only a small number of living endemic shallow-water species are rare with restricted geographic distributions (e.g. *Discorbinella deflata, Glabratellina kermadecensis, Pileolina gracei, Planulinoides norcotti, Rosalina vitrizea*). One living endemic species is known from each of the Kermadec, Chatham, and Auckland Islands (Hayward et al. 1999a), as well as an endemic *Elphidium* species from the Chatham Islands in the Early Eocene (Hayward et al. 1997).

Species longevity

The excellent New Zealand Late Cretaceous and Cenozoic marine sedimentary record has the potential to provide some of the highest-quality data anywhere in the world on benthic species duration (longevity) and the speed of evolutionary changes. These data are only now beginning to be tapped. Studies of the living shallow-water species that have a fossil record in New Zealand show that they have a mean partial species duration of 21 million years (My). Species from shallower habitats tend to have shorter species durations than those from deeper waters (Hayward et al. 1999a).

The well-studied New Zealand species of the common shallow-water genus *Elphidium* have species durations of 1–39 My, with a mean of 14.5 My (Hayward et al. 1997). These studies and those on the shallow-water family Bolivinellidae (Hayward 1990b) suggest that common cosmopolitan species generally have much longer species durations (10–40 My) than rarer, geographically confined species (1–10 My).

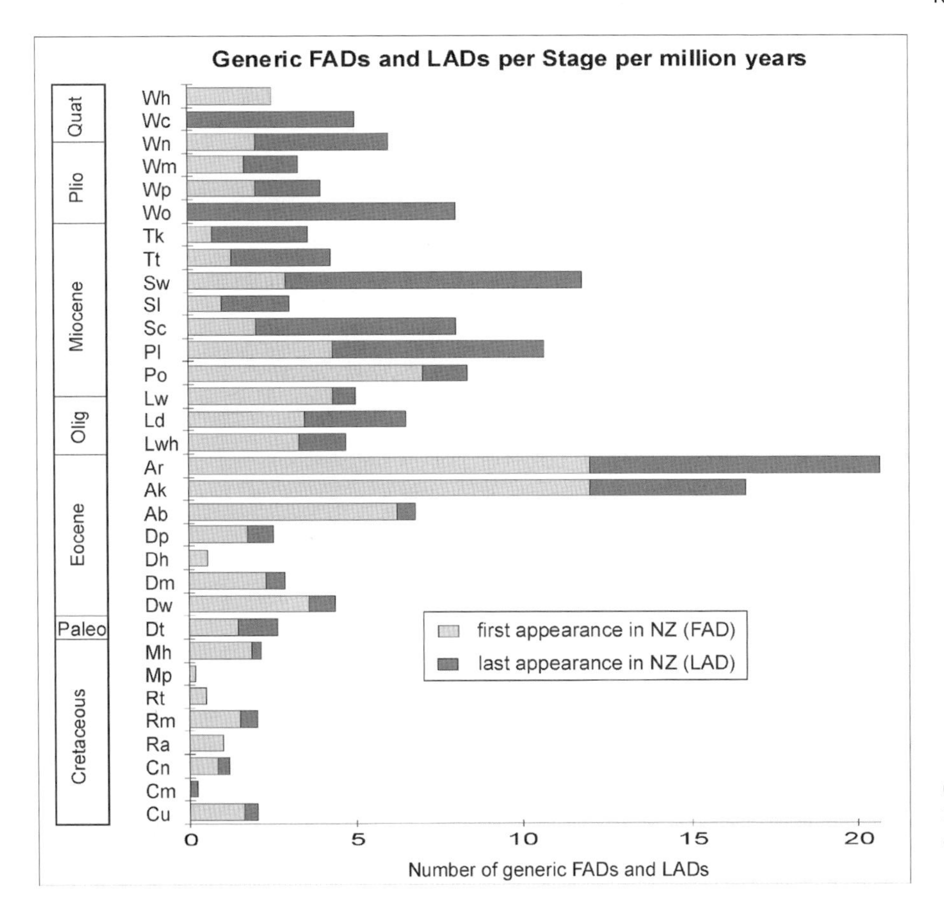

Number of Last and First Appearance Datums (LADs and FADs) of foraminiferal genera in New Zealand per local stage per million years, based on the total recorded time range of each genus (see checklist).

Based on the time ranges compiled for the fossil species (see checklist), the mean species duration for all Cretaceous and Cenozoic foraminifera in New Zealand is 14 My, with a range of 0.5 to 95 My. [Fig. 5] Members of the planktic order Globigerinida have the shortest species durations (mean 8.8 My) and those of the benthic order Lituolida have the longest (mean 19 My).

Unusual habitat associations

Foraminifera live in all marine and brackish environments, from the extremes of salt-water incursion up estuaries (e.g. *Trochamminita salsa*) to the floor of the deepest parts of the ocean. One modern species (*Jadammina macrescens*) is largely confined to living in salt marshes above mean high-tide level, with its greatest abundances close to extreme high-water spring, which is inundated by the tide only several times a year (Hayward et al. 1999b). Another marsh-restricted species (*Trochammina inflata*) has been found living in a saline lake (salinity about 17%) well above tide level on Kapiti Island. Perhaps the most unusual occurrence is that of the fossil species *Amphimorphinella butonensis*, which has been recorded in New Zealand only in early Miocene sediment around an inferred ancient bathyal methane seep (Eagle et al. 2000).

Adventives

The paucity of pre-1950 studies and/or samples of New Zealand Recent foraminifera makes it difficult to identify species that have been introduced by human-related mechanisms, such as hull-fouling or ship's ballast. On the basis of their unusual distributional pattern around New Zealand, three species (*Elphidium vellai*, *Siphogenerina raphana*, and *Virgulinella fragilis*) have been identified as recent marine invaders (Hayward et al. 1999a). All three lack a fossil record and live in sheltered harbours in very slightly brackish conditions typical of many overseas ports, where ballast water together with some bottom sediment would have been taken on by ships.

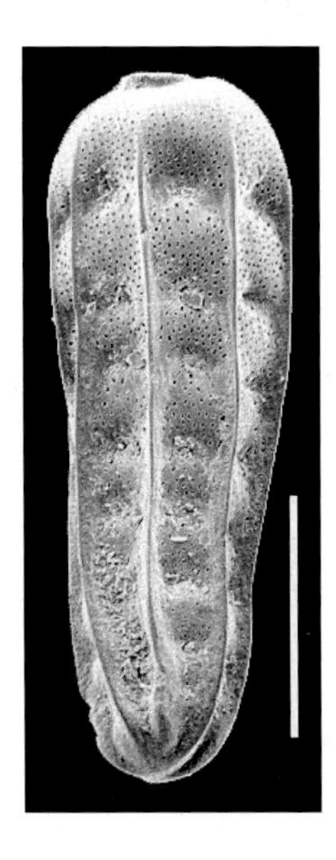

Siphogenerina raphana, sporadic around the North Island and common only in Waitemata Harbour. Probably adventive following European colonisation. FP4634, Kapiti Island, Wellington, Recent. Scale bar 0.1 millimetre.

Applications of New Zealand foraminiferal research

Biostratigraphy

In New Zealand and around the world, foraminifera have been studied in considerable detail because of their value in biostratigraphic dating of marine sedimentary strata. Although macrofossils are useful for rapid dating in the field, they are generally far less abundant or widespread than microfossils, and only a few macrofossil groups (e.g. graptolites, ammonites, and belemnites) have the worldwide species distributions of pelagic (open-ocean) microfossils.

Despite the presence of other useful microfossil groups, foraminifera are still the most valuable group overall for dating New Zealand's Cenozoic marine strata. This is because:

1. New Zealand foraminifera have received more detailed biostratigraphic study (e.g. Hornibrook et al. 1989).
2. Foraminifera are larger than most other biostratigraphically useful microfossil groups (e.g. nannofossils, diatoms, radiolaria, dinoflagellates, pollens, spores), making them less susceptible to reworking and current winnowing.
3. Foraminifera live in more diverse habitats (both planktic and benthic) than other microfossils (mostly planktic) and are therefore present in most marine strata.
4. Foraminifera have a wider diversity of shell composition (e.g. calcareous and agglutinated) than other microfossil groups and are therefore present both above and below the Carbonate Compensation Depth (CCD), and in strata where there has been differential solution.

Not surprisingly, each microfossil group, including foraminifera, provides higher-resolution dating in some parts of the Cenozoic than others. Around New Zealand, planktic foraminifera provide higher-resolution dating than other groups through most of the Miocene and Pliocene, whereas they offer less biostratigraphic refinement than calcareous nannofossils in the Paleocene, Oligocene–earliest Miocene, and Pleistocene.

Planktics are the most used group of foraminifera for dating New Zealand's Cenozoic strata. This is because their pelagic lifestyle results in a wider distribution in seafloor sediments, and because their time ranges are more accurately documented, largely from complete oceanic sequences in deep-sea drillholes. Planktic foraminifera are particularly valued for inter-regional and worldwide correlation of strata, because of their cosmopolitan distribution in latitude-parallel climate zones.

Chiloguembelina cubensis, a common fossil planktic species. BWH141/62, Chatham Rise, Oligocene. Scale bar 0.1 millimetre.

Planktic foraminifera mostly live in oceanic water and thus increase in abundance from shallow nearshore localities out to deeper water. Many shallow marine rocks lack age-diagnostic planktic foraminifera and can be dated only by the benthics. Sedimentary rocks deposited at depths below the CCD (about 4500–5000 metres) also lack planktic foraminifera, because their calcareous shells dissolve on the seafloor, and must be dated using non-calcareous microfossils such as radiolaria, dinoflagellates, or agglutinated benthic foraminifera. Thus benthic foraminifera complement the planktics in biostratigraphic dating of New Zealand's late Cretaceous and Cenozoic strata. Unlike planktic foraminiferal species, which have similar time ranges world-wide, the time ranges of benthic species used for dating in New Zealand (e.g. Hornibrook et al. 1989) are local. Only the time ranges of many of the long-lived, deep-sea benthics are similar worldwide (e.g. van Moorkhoven et al. 1986). Benthic foraminifera suffer because they have more ecologically or biogeographically restricted distributions than planktics. Thus the absence of a benthic species is a more unreliable biostratigraphic indicator than the absence of a planktic species in an oceanic assemblage.

Most foraminiferal dating of New Zealand's Cretaceous and Cenozoic strata utilises the first- and last-appearance datums (FADs and LADs) of individual species. [Fig. 6] These datums are a combination of evolutionary, immigration, and extinction events, which are strongly influenced by paleoclimate and paleoceanography. Accurate numeric ages are unknown for most of these events in New Zealand. The number of generic LADs and FADs per New Zealand stage per million years indicates that the late Eocene is the period of greatest faunal turnover, possibly resulting from paleoceanographic changes initiated by the opening of a deep-water oceanic gap between Australia and Antarctica at this time.

Major worldwide foraminiferal faunal turnover events that are valuable in biostratigraphic dating of New Zealand rocks are:

1. The near instantaneous extinction at the Cretaceous–Tertiary boundary (65 Mya) of virtually all planktic and a large proportion of the shallow-water benthic foraminifera – thought to be caused by bolide impact (e.g. Hornibrook et al. 1989).
2. The rapid extinction near the end of the Paleocene (56 Mya) of numerous deep-water benthics – attributed to abrupt warming of deep-ocean water-masses (e.g. Kaiho et al. 1996; Hancock et al. 2003).
3. The gradual extinction through the Late Eocene and Early Oligocene of numerous planktic foraminifera – the result of worldwide climatic cooling (e.g. Hornibrook et al. 1989).
4. The rapid extinction in the mid-Pleistocene Climatic Transition (1.1–0.6 Mya) of a group of anoxic deep-water benthics – coinciding with an abrupt increase in the severity of the ice ages (e.g. Hayward 2002; Hayward et al. 2007c).

Foraminifera have proved to be of little use in dating New Zealand's Triassic, Jurassic, and early Cretaceous rocks. On the other hand, the small number of Permian foraminiferal faunas in New Zealand have been particularly valuable in dating these rocks and in understanding their complicated history.

Phylogeny

Because of their small size and abundance in almost all marine sedimentary rocks, fossil foraminifera are ideal for evolutionary studies, as they are less subject to the 'incompleteness of the fossil record', which is often cited as an excuse by neontologists to ignore the 'ground-truth' fossils. The excellent Late Cretaceous and Cenozoic marine-sedimentary layers that have been uplifted and exposed by erosion around New Zealand provide ideal foraminiferal sequences for study from a wide range of water depths and environments.

The bathyal and upper-abyssal foraminiferal-ooze sequences on the ocean floor in the EEZ provide high-resolution, near-continuous Neogene sequences that are particularly suitable for evolutionary studies of planktic foraminifera in the Southwest Pacific (e.g. Malmgren & Kennett 1981; Wei & Kennett 1988; Tabachnick & Bookstein 1990; Wei 1994a,b; Schneider & Kennett 1995). These data are relevant to as-yet-unresolved questions about isolating mechanisms and speciation models for this important group of oceanic zooplankters (Cifelli & Scott 1986) and their evolutionary patterns.

The golden opportunity to study the phylogeny of New Zealand's benthic foraminifera is largely unrealised. The gradual evolutionary changes in the shape of the shells of *Haeuslerella* and *Bolivinita* have been documented in biometric studies and are used in dating (e.g. Scott 1965, 1978). Study of the Cenozoic fossil record of *Elphidium* in New Zealand shows a complex phylogenetic and paleobiogeographical history, which is far beyond any that could have been predicted from knowledge of the modern fauna of 11 species or subspecies. There are 25 *Elphidium* taxa in New Zealand's Cenozoic record, documenting a mix of successful immigration events, local evolution, ecological shifts, biogeographic changes, and extinctions (Hayward et al. 1997).

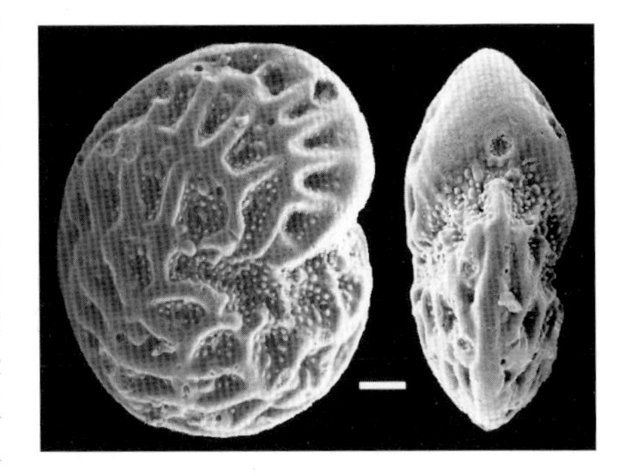

Elphidium hampdenense, an Eocene deep-water oceanic species. All modern species of this genus are shallower than 100 m. FP4366, Hampden Beach, Otago, Middle Eocene. Scale bar = 0.1 millimetre.

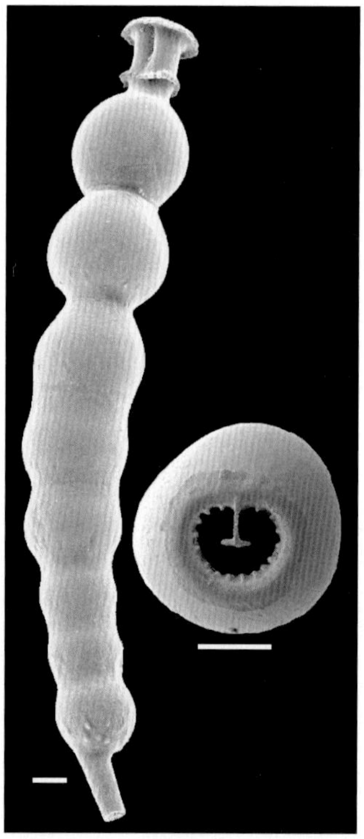

Paleoenvironmental analysis

All fossil groups can provide information about the paleoenvironment in which they lived and in which they were buried and fossilised. Foraminiferal shells are more valuable than most because of their presence and often abundance in most late Cretaceous and Cenozoic marine sedimentary rocks in New Zealand; they are larger than most other common microfossils and therefore less susceptible (but not immune) to post-mortem transport. Further, most are infaunal benthic species that are less likely than pelagics or epifauna to have suffered post-mortem transport, and thus they can provide direct information about the depositional environment of the sedimentary rock in which they occur.

This paleoenvironmental information is particularly valuable for: understanding the evolving paleogeography and tectonic history of marine sedimentary basins and their resource potential; the study of sequence stratigraphy and eustatic sea-level cycles; quantifying late Quaternary sea-level changes and earthquake-induced vertical displacements; and studying the changing paleoceanography of the world's oceans.

Paleo-water-depth assessments based on fossil foraminifera also provide important information for geologists. Water depth in itself has no influence on the distributional patterns of benthic foraminifera, but many of the factors that do (e.g. wave energy, substratum, food supply) are indirectly correlated with depth. Thus modern benthic foraminiferal faunas exhibit a rough depth-related zonation from the seashore to the abyss. This zonation is documented around New Zealand in waters shallower than 100 metres (e.g. Hayward et al. 1999a) and at outer-shelf to abyssal depths (e.g. Lewis 1979; Hayward et al. 2010a). The documented modern depth ranges of foraminiferal assemblages and individual taxa are utilised to estimate paleodepth of burial of their fossil counterparts (e.g. Hayward 1986, 2004, 2010a) on the assumption that the depth habitats of taxa are stable through time. Vella (1962a) circumvented the paucity of data then available on depth distributions of modern taxa by inferring the depth order of assemblages from their stratigraphic position. In applying this approach, he used benthic-assemblage biofacies as the principal evidence for recognition of cyclic excursions in relative sea level during Pliocene–Pleistocene time. His pioneering use of foraminiferal depth biofacies to detect cyclothems has been extended and refined with census data to provide detailed mappings of changes in relative sea level (e.g. Haywick & Henderson 1991; Naish & Kamp 1997; Kamp et al. 1998).

The value of regional analyses of fossil benthic foraminiferal faunas for identifying eustatic movements of sea level was shown with a study of Late Miocene–Early Pliocene sequences by Kennett (1965, 1967, 1968) in which a major low-sea-level event was detected in the latest Miocene. Subsequent research (e.g. Hsu et al. 1973) showed that the foraminiferal signature was a local response to global events associated with the Mediterranean Messinian 'salinity crisis'. Hodell and Kennett (1986) gave a southern-hemisphere paleoclimatic scenario for the event Kennett had noted two decades earlier from benthic foraminiferal studies.

Paleoenvironment assessments based on the documented ecological distribution of modern foraminifera become less reliable with increasing age of the fossil fauna. Assessments on Neogene faunas (younger than 25 Mya) are generally accepted as moderately reliable, as many of the species and most of the genera are still living. Pre-Oligocene assessments of paleoenvironments are less precise and rely less on the documented distribution of modern foraminifera. This is because of major global oceanic changes that occurred in the Paleogene, which resulted in a number of foraminiferal genera changing their ecologic niches in the late Eocene and Oligocene. In the New Zealand fossil record are examples of some Mid-Eocene deep-water taxa becoming shallow-water restricted, e.g. *Elphidium* (Hayward et al. 1997), and of some taxa that lived at shelf depths becoming restricted to mid-bathyal and greater depths (> 600 metres), e.g. *Pleurostomella*, *Stilostomella*, and diverse Lagenida.

Siphonodosaria pomuligera, an extinct cosmopolitan species named in 1864 from New Zealand specimens collected by Ferdinand von Hochstetter during the *Novara* expedition. BWH134/24, ODP Site 1125, Chatham Rise, Early Pleistocene. Scale bar 0.1 millimetre.

Detailed studies of modern salt-marsh foraminifera in New Zealand indicate that their distribution is most strongly controlled by tidal exposure and salinity (e.g. Hayward et al. 1999a,b, 2004e). Salt-marsh foraminiferal faunas are preserved in Holocene coastal sediments. They are now in common use to document accurately former sea levels over the last 8000 years. These are being applied to better understand the rapid acceleration of sea level rise around New Zealand during the 20th century (e.g. Gehrels et al. 2008) and to document the recurrence times of major earthquake displacements around various parts of the New Zealand tectonically-active coastline (e.g. Goff et al. 2000; Hayward et al. 2004a,c, 2006b,c, 2007b, 2008b, 2010b,c).

An additional value of the modern estuarine and harbour foraminiferal studies that are now being utilised worldwide is to document and monitor the ecological impacts of human activities. This is because foraminiferal shells are preserved in the young sedimentary record and provide an insight into prehuman conditions as well as the changes that have occurred during various human activities on the adjacent land or in the marine environment. In New Zealand, foraminiferal studies have been used to look at the impacts of effluent disposal from freezing works and sewage treatment plants into Manukau Harbour (Matthews et al. 2005), of forest clearance and urban sprawl around Auckland city (Hayward et al. 2004c, 2006b), of the establishment of oyster farms in Mahurangi Harbour (Grenfell et al. 2007), and of the establishment of invasive Asian date mussels and chord grass communities (Hayward et al. 2008b).

A further area where foraminifera have considerable potential in paleoenvironmental assessment is in paleoceanography. Deep-sea benthic foraminiferal distributions are strongly influenced by changes in food flux and bottom oxygen conditions that record changes in nutrient supply from land runoff, upwelling, and surface front movements, and changes in the distribution and character of deep-ocean water-masses. Deciphering these paleoceanographic histories captured by benthic foraminiferal faunas in deep-sea cores from around New Zealand is in its infancy (e.g. Hayward et al. 2004b,d, 2005)

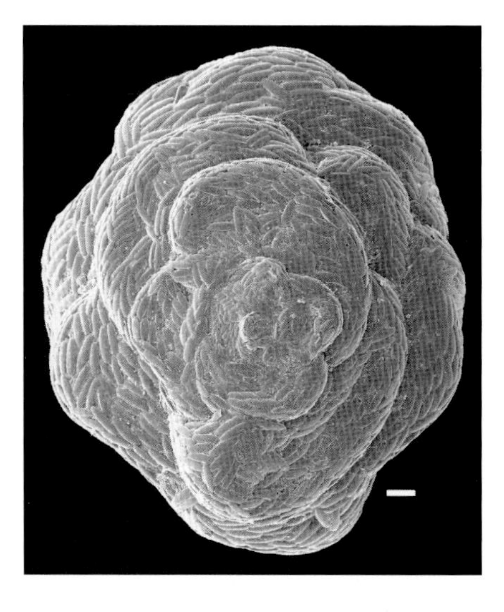

Carterina spiculotesta, the only member of the order Carterinida found in New Zealand. FP2986, Cavalli Islands, Northland, Recent. Scale bar 0.1 millimetre.

Paleoclimate

Various lines of study on fossil foraminifera provide information on paleoclimate, particularly on the temperature of the surface waters of the oceans. The presence or absence of various groups of tropical to subtropical, shallow-water benthic foraminifera (e.g. larger foraminifera like *Amphistegina*, Bolivinellidae) in the New Zealand fossil record has been used to help interpret and graph the mean annual sea-surface temperatures around New Zealand through the Cenozoic (e.g. Hornibrook 1992). Paleoclimate greatly influenced the taxonomic diversity of the planktic and shallow-water benthic foraminifera around New Zealand during the Cenozoic, with greater diversity in warmer seas than cooler. New Zealand's Neogene record of the number of first- and last-appearance datums (FADs & LADs) per stage per million years clearly shows a predominance of FADs (local arrivals) during the Oligocene–Early Miocene (Lwh–Po) overall warming trend and a predominance of LADs (local extinctions) during Mid-Miocene–Pleistocene (Sc–Wc) overall cooling.

Studies east of New Zealand and elsewhere (e.g. Weaver et al. 1997; King & Howard 2001, 2003) have shown that there is a close correlation between the composition of planktic foraminiferal assemblages that are present in modern seafloor sediments and the seasonal sea-surface temperatures above. This correlation allows transfer-function or modern analog methods to be used to compute former sea-surface temperatures based on Quaternary oceanic planktic foraminiferal assemblages and to use these to study the cyclically changing paleoceanography (e.g. Weaver et al. 1998; Schaefer et al. 2005; Wilson et al. 2005; Crundwell et al. 2008; Hayward et al. 2008a).

The oxygen-isotope geochemistry of foraminiferal shells, both planktic and deep-sea benthic, provide the resource for the most widespread oceanic

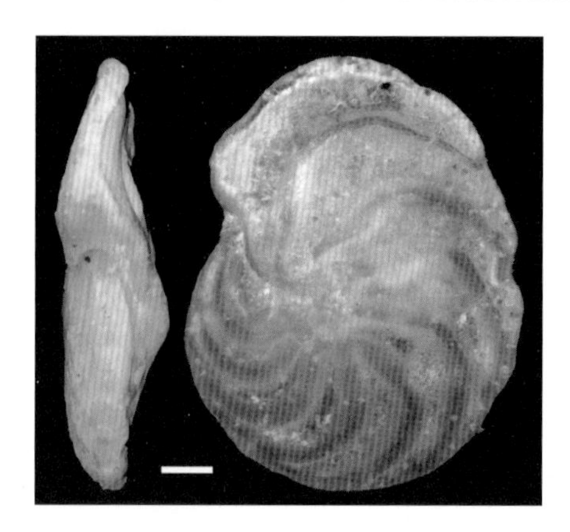

Cibicides wuellerstorfi, cosmopolitan, deep-water, widely used in oxygen- and carbon-isotope studies of past ocean-floor environmental conditions. BWH194/18, Great Barrier Island, Recent. Scale bar 0.2 millimetre.

paleoclimate studies so far undertaken. Hundreds of thousands of oxygen-isotope analyses on foraminiferal shells have been undertaken worldwide, including around New Zealand (e.g. Nelson et al. 1993, 2000; Neil et al. 2004), to reveal the cyclical temperature history of both sea-surface water (planktics) and seafloor water (benthics) through the Cretaceous and Cenozoic, and particularly in the Quaternary. These studies provide the most compelling evidence for the history of the astronomically controlled global climate cycles extending back well before the onset of the more severe climate cycles of the Ice Ages.

Gaps in knowledge and scope for future research

Because of their high economic and geological value in dating, the diversity of New Zealand's Late Cretaceous and Cenozoic fossil foraminifera is probably better documented than any other fossil group, other than molluscs. Because of their geological value in paleoenvironmental assessments, the diversity of New Zealand's Recent foraminifera is now reasonably well known also. After accounting for numerous synonyms, about 1108 Recent species are currently recognised in New Zealand waters. The Recent shallow-water and bathyal faunas have been well studied, except for those in Fiordland and around Stewart, Antipodes, Bounty, and Kermadec Islands. The abyssal fauna, especially that below the CCD, is still not well documented and is likely to yield another 200–300 species.

The fossil census undertaken for this chapter recognises approximately 1650 species from New Zealand's Late Cretaceous and Cenozoic. This is probably no more than 60% of the total number of fossil species likely to be present, as many species-rich but biostratigraphically valueless groups have been largely ignored (e.g. Miliolida, unilocular Lagenida), and some parts of the Cenozoic record have received no comprehensive faunal study.

Only 126 species have been recorded from the New Zealand Triassic, Jurassic, and Early Cretaceous and 44 species from the Paleozoic. This diversity is clearly a small proportion of that likely during these time periods. More searching and processing, possibly by new techniques, of the less-indurated and less-metamorphosed lithologies will undoubtedly greatly increase knowledge of New Zealand foraminiferal diversity in older rocks.

The Late Cretaceous and Cenozoic planktic foraminiferal diversity (order Globigerinida) of New Zealand and the world is undoubtedly the most studied and best documented for any group of fossils. There are 259 planktic species recorded from New Zealand, with only a small number of rare taxa unidentified.

Whilst there are several hundred benthic species awaiting specific identification or description (predominantly Cenozoic fossils), most are likely to have already been described elsewhere in the world and need tracking down. The major problem facing benthic foraminiferal taxonomy is the enormous over-description of the world's Cenozoic and Recent species. It is a major hindrance to unravelling patterns of dispersal, biogeography, evolution, biostratigraphic correlation, and world paleoceanography. Several attempts are under way to unify the taxonomy of the commoner, and largely cosmopolitan, deep-sea Cenozoic benthic species, but the taxonomy of the enormously diverse shallow-water benthic foraminifera of the world is in chaos. Another factor contributing to the mayhem is the current morphotypic, non-phyletic nature of most benthic foraminiferal classification. Issues of parallel and convergent evolution have hardly been addressed and will be difficult but revealing to unravel with detailed monographic world studies.

A great need is for numerous world monographs on different families or genera to unify the divergent taxonomies of various regions of the world (e.g. Hayward 1990b). New Zealand needs to share the load by contributing with monographs on taxonomic groups with rich indigenous histories (e.g. Notorotaliidae, Glabratellidae, *Rectuvigerina*, *Bolivinita*, *Elphidium*) and by upgrading the highly cryptic descriptions of many of the species described from

New Zealand by Finlay. One of the greatest hurdles to these studies is in assembling type material for study and comparison from repositories in dozens of countries worldwide.

Another useful, but interim, approach would be to unify the benthic foraminiferal taxonomies for various time periods with New Zealand's nearer neighbours in the Southern Hemisphere, viz Australia and South America. Immediate useful spin-offs would be better understanding of circum-Antarctic paleoceanography and its role in the dispersal and evolution of benthic foraminifera.

Whilst the recent publication of several monographs (e.g. Hornibrook et al. 1989; Scott et al. 1990; Hayward et al. 1999a, 2010a) greatly assists with the identification of large sections of the New Zealand foraminiferal fauna, there is a real need for an easily searchable computer database with full illustrations, time ranges, geographic and ecologic ranges, and documented key features of New Zealand's fossil and Recent foraminifera. This would greatly assist the advancement of all other areas of foraminiferal research, which require the firm base of accurate species identification. A start to this task is the CD-ROM database on Recent shallow-water foraminifera of the Southwest Pacific (Albani et al. 2001).

Among the areas requiring specific research attention in the near future are biostratigraphic, paleoecologic, taphonomic, and biological studies, in addition to taxonomy. Detailed studies are required to improve the precision of biostratigraphic age determinations and to calibrate foraminiferal events in the Southwest Pacific through the global paleomagnetic time-scale and climate cycles to numeric ages, similar to that recently completed by Crundwell (2004) and Crundwell and Nelson (2007).

In paleoecology, there is considerable need to improve precision and understanding of the limiting factors of paleodepth assessments of bathyal and abyssal faunas. One fundamental approach to this is to improve knowledge of the depth distribution of modern deep-sea benthic foraminifera around New Zealand. More innovative approaches are required to establish better the paleoenvironmental, and particularly paleodepth, information contained within Cretaceous and Paleogene fossil faunas. Improved paleoenvironmental interpretations also require further detailed study on taphonomic processes of post-mortem transport and mixing, and particularly the post-mortem or early-burial disintegration or dissolution of foraminiferal shells in salt-marsh and deep-sea environments.

Research on the biology, population dynamics, and trophic status of modern foraminifera is almost non-existent in New Zealand. It is a necessary prerequisite if the enormous potential of foraminifera is to be realised in environmental monitoring and in documenting very recent environmental change, particularly in coastal and estuarine ecosystems. Because of their widespread distribution, enormous numbers, and preservable shells, foraminifera have a major advantage over most other groups of organisms for shallow marine-environmental monitoring.

Other more esoteric areas of great potential would be research that combines phylogenetic models based on the rich fossil record with genetic studies on living species, paleobiogeographic research including dispersal mechanisms and paleoceanographic migration routes related to continental drift, and the largely unsolved problem of planktic foraminiferal speciation in pelagic environments where allopatric models do not work.

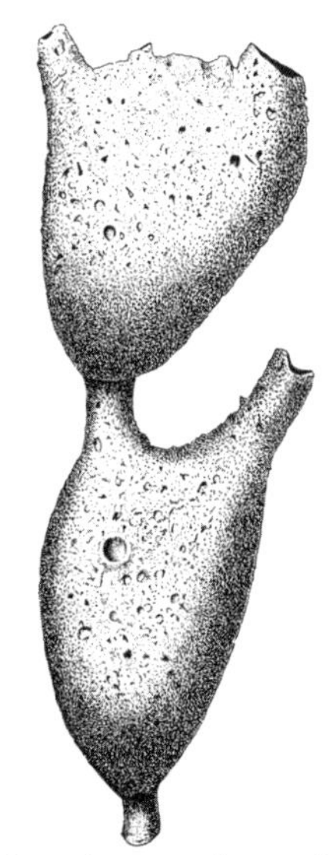

Aschemonella scabra, a small xenophyophore with a test of a only a few millimetres in size, consisting of several inflated chambers joined by short stolons.

From Brady 1879

Class Xenophyophorea

Xenophyophores are the giants in the world of shelled protists. They comprise part of the macrofauna and megafauna at bathyal and abyssal depths. Representatives of the group were first encountered by the British *Challenger* Expedition (1873–76), especially in the Central Pacific, and described by Haeckel (1889) as a

Syringammina fragilissima,with a test of numerous radiating tubes (up to 38 millimetres diameter).

Ole Tendal

new group of keratose sponges living in symbiosis with certain hydroids. A few years later, more material was provided by the German Valdivia Expedition (1898–99) from the western Indian Ocean, the Dutch Siboga Expedition (1899–1900) from Indonesian waters, and the US Albatross Expeditions (1891, 1899–1900, 1904–05) from the Central Pacific. The material from these expeditions allowed Schulze (1906, 1907) to redescribe most of Haeckel's species, describe one new species, define the group (giving it the name Xenophyophora), and reconsider its taxonomic position, placing it within the large complex of rhizopodean protozoans. More details of the early history of the group and the features of our modern understanding were given by Tendal (1972, 1996, 1997).

Xenophyophore organisation is basically plasmodial. The protoplasm is richly dichotomously branched and in some species anastomosing. Branches are 30–200 micrometres (μm) in diameter and each branch is enveloped in a transparent organic sheath. The cytoplasm contains numerous nuclei, 2–10 μm in diameter, and various inclusions, the most numerous and obvious of these being crystals of barite (barium sulphate) 2–5 μm in size. The crystals are called granellae and the term for the whole complex of protoplasm and organic sheath is granellare. More or less intertwined with granellare are dark strings or masses composed of numerous fecal pellets (stercomata) 10–20 μm in size. The strings or masses are called stercomare and are 35–540 μm in diameter. Granellare and stercomare are surrounded by an agglutinated test made of foreign particles, such as mineral grains, sponge-spicule fragments, and radiolarian and foraminiferan tests. It is the wide-reaching branching of the granellare and the comprehensive test construction that permit xenophyophores a size range spanning a few millimetres to about 25 centimetres maximum dimension despite the very low volume of plasma (< 5% of the specimen volume).

Diagnostic characters used in identifying and describing xenophyophores include the manner in which the test is constructed and the nature of the agglutinated particles. Apart from the terms defined above, the terminology is the same as for foraminifera. Further details on the morphology of xenophyophores can be found in Tendal (1972, 1989) and Gooday and Tendal (2000).

Knowledge of the biology of xenophyophores is somewhat fragmentary. The pseudopodia should be reticulopodia and the few, insufficient observations are consistent with this view. Food is fine particles taken up in the seafloor boundary layer, and a species' position on the seafloor is in accordance with uptake either from the sediment surface or from the water directly above, depending on test morphology. Growth seems to be fast. Reproductive mode is poorly known. Xenophyophores are particularly numerous in regard to numbers of species and their abundance in regions with high sea-surface productivity. There are only a few observations that indicate predation on members of the group. Species with the test constructed as a spatial latticework house a fauna of foraminifers and small metazoans and, probably because they add a three-dimensional structure to the otherwise even seafloor, they also influence positively the diversity of smaller organisms at the centimetre-scale just around and under the test. Information on xenophyophore biology is given by Tendal (1972, 1985), Tendal and Gooday (1981), Levin and Thomas (1988), Levin (1991), Gooday et al. (1993), and Hughes and Gooday (2004).

The group is widely distributed in the world's oceans, roughly at latitudes between 50° N and 50° S. Comparatively few records are from the Indian Ocean, but the fauna there is as rich as in the Pacific and Atlantic Oceans. Few records and species are known from Arctic and Antarctic seas. Xenophyophores have been reported from sublittoral (five metres) to hadal (7700 metres) depths. The very shallow records need confirmation and the very deep ones are few, the main bathymetrical distribution being from about 700 to 5000 metres. Further details are given by Tendal (1972, 1996).

Over time, xenophyophores have been placed by various authors at all taxonomic levels from family to phylum, in some cases even in an 'Incertae sedis' position. The view that their closest relatives are to be found among agglutinating foraminifers has been advocated directly or indirectly by a number of authors (Goës, Pearcy, Brady, Schulze, Rhumbler, Cushman, Loeblich & Tappan, and Tendal, see Tendal 1972) and has recently been strongly emphasised through molecular investigations by Pawlowski et al. (2003). In view of a number of unresolved phylogenetic questions, the position as a traditional class seems at present to be the best pragmatic solution (Tendal 1996; Gooday & Tendal 2000, with a key to families and genera; Cavalier-Smith 2003).

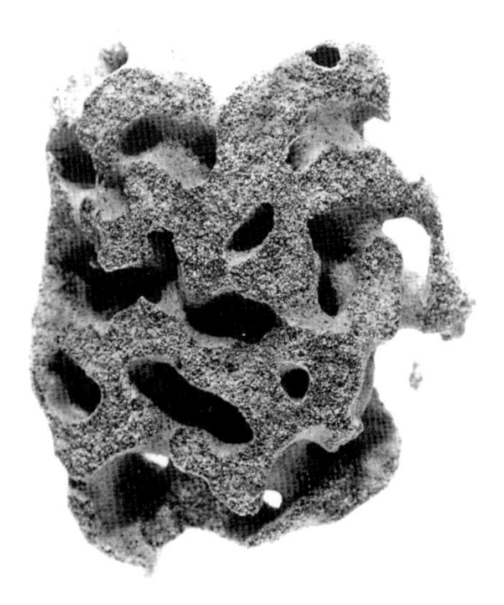

Reticulammina novazealandica, which can achieve a test size of six centimetres diameter.
Ole Tendal

New Zealand xenophyophores

Seven xenophyophore species have been recorded from New Zealand waters (Tendal 1975; Tendal & Lewis 1978, with a key to species of the genera *Reticulammina* and *Syringammina*; Tendal 1981, 1996). Three of these are endemic to the region, viz *Reticulammina lamellata, R. maini,* and *R. novazealandica*. *Syringammina tasmanensis*, formerly thought to be endemic (Hedley 1966; Lewis 1966), also occurs off Tasmania (O. Kamenskaya pers. comm.). *Reticulammina labyrinthica* is recorded also in the North Atlantic and Indian Oceans, while *Syringammina fragilissima* seems widely distributed in the North Atlantic and possibly also occurs in the Pacific Ocean. For these latter two species, slight differences have been noticed between the New Zealand and North Atlantic populations and a forthcoming revision may lead to the erection of separate species. *Aschemonella scabra* has been found in very deep water, but as dead tests only, and its identity and occurrence await further confirmation.

Apart from those of *Aschemonella scabra*, the sampling localities are on the eastern, northern, and western continental slopes of New Zealand and on the Lord Howe Rise and Chatham Rise (Tendal 1981). All the species hitherto found are very fragile, and a lack of records south of New Zealand is probably only because of the paucity of investigations with relevant gear or the difficulty in identifying these animals, especially if fragmented. The bathymetric distribution of the New Zealand fauna is roughly between 500 and 1300 metres, with most records within the 1000 metre depth contour.

Judging from the relatively low number of records, the actual diversity and abundance of xenophyophores on the slopes and rises around New Zealand is probably higher than known today and perhaps another half dozen species might be expected. On a global basis, this is the region with by far the most species living in easily accessible outer-shelf and upper-bathyal depths, and it therefore represents a unique area for the study of xenophyophores. In situ photographs show some species to be very common in some places, and detailed studies in these areas would be rewarding, probably revealing an ecological importance of xenophyophores much like that known for deep seamounts (Levin *et al.* 1986; Levin and Thomas 1998). Almost 60 species of xenophyophores have been described worldwide.

Acknowledgements

We acknowledge the contributions made by all the many researchers who over the years have documented the diversity and myriad applications of New Zealand's foraminifera, particularly Harold Finlay, Norcott Hornibrook, Paul Vella, and Graham Jenkins. This review and compilation would not have been possible without the foresight many decades ago of N. Hornibrook in establishing the national collections and catalogues of foraminifera at the New Zealand Geological Survey (now GNS Science).

We thank Ian Raine for generating the initial list of fossil foraminifera from the computerised New Zealand Fossil Record File, which has evolved over many months into the checklist presented here.

Authors

Rowan Carter Geomarine Research, 49 Swainston Rd, St Johns, Auckland, New Zealand

Dr Hugh R. Grenfell Geomarine Research, 49 Swainston Rd, St Johns, Auckland, New Zealand [h.grenfell@geomarine.org.nz]

Dr Bruce W. Hayward Geomarine Research, 49 Swainston Rd, St Johns, Auckland, New Zealand [b.hayward@geomarine.org.nz]

Dr Jessica J. Hayward Geomarine Research, 49 Swainston Rd, St Johns, Auckland, New Zealand [jjh276@cornell.edu]

Hugh E. G. Morgans Institute of Geological and Nuclear Sciences, PO Box 30-368, Lower Hutt, New Zealand [h.morgans@gns.cri.nz]

George H. Scott Institute of Geological and Nuclear Sciences, PO Box 30-368, Lower Hutt, New Zealand [g.scott@gns.cri.nz]

Dr C. Percy Strong Institute of Geological and Nuclear Sciences, PO Box 30-368, Lower Hutt, New Zealand [c.strong@gns.cri.nz]

Dr Ole Secher Tendal Zoological Museum, Danish State Museum of Natural History, University of Copenhagen, Universitetsparken 15, DK-2100 Copenhagen O, Denmark

References

ABBOTT, S. T. 1997: Foraminiferal paleobathymetry and mid-cycle architecture of mid-Pleistocene depositional sequences, Wanganui Basin, New Zealand. *Palaios 12*: 267–281.

ALBANI, A.; HAYWARD, B. W.; GRENFELL, H. R. 2001: *Taxomorph: Foraminifera from the South-west Pacific. An inter-active catalogue.* University of New South Wales and Australian Biological Resources Study *CD-ROM.*

ARCHIBALD, J. M.; KEELING, P. J. 2004: Actin and ubiquitin protein sequences support a cercozoan/ foraminiferan ancestry for the plasmodiophorid plant pathogens. *Journal of Eukaryotic Microbiology 51*: 113–118.

BASS, D.; MOREIRA, D.; LÓPEZ-GARCÍA, P.; POLET, S.; CHAO, E. E.; von der HEYDEN, S.; PAWLOVSKI, J.; CAVALIER-SMITH, T. 2005: Polyubiquitin insertions and the phylogeny of Cercozoa and Rhizaria. *Protist 156*: 149–161.

BENSON, W. N.; CHAPMAN, F. 1938: Note of the occurrence of radiolarian limestone among the older rocks of south-western Otago. *Transactions of the Royal Society of New Zealand 38*: 373–374.

BRADY, H. B. 1884: Report on the foraminifera dredged by HMS Challenger during the years 1873–1876. *Report on the Scientific Results of the Voyage of H.M.S. Challenger, Zoology 9*: 1–814 p., 115 pls.

BURKI, F.; SHALCHIAN-TABRIZI, K.; PAWLOWSKI, J. 2008: Phylogenomics reveals a new 'megagroup' including most photosynthetic eukaryotes. *Biology Letters 4*: 366–369.

BURKI, F.; SHALCHIAN-TABRIZI, K.; MINGE, M.; SKJÆVELAND, Å.; NIKOLAEV, S. I.; JAKOBSEN, K. S.; PAWLOWSKI, J. 2007: Phylogenomics reshuffles the eukaryotic supergroups. *PLoS ONE 8, e790*: 1–6.

CAVALIER-SMITH, T. 2002: The phagotrophic origin of eukaryotes and phylogenetic classification of Protozoa. *International Journal of Systematic and Evolutionary Microbiology 52*: 297–354.

CAVALIER-SMITH, T. 2003: Protist phylogeny and the high-level classification of Protozoa. *European Journal of Protistology 39*: 338–348.

CAVALIER-SMITH, T. 2010: Kingdoms Protozoa and Chromista and the eozoan root of the eukaryotic tree. *Biology Letters 6*: 342–345.

CHAPMAN, F. 1909: Report on the foraminifera from the Subantarctic Islands of New Zealand. Pp. 312–371 *in*: Chilton, C. (Ed.), *The Subantarctic Islands of New Zealand.* Volume 1. Philosophical Institute of Canterbury and Government Printer, Wellington.

CHAPMAN, F. 1926: The Cretaceous and Tertiary foraminifera of New Zealand, with an appendix on the Ostracoda. *New Zealand Geological Survey Paleontological Bulletin 11*: 1–119, 22 pls.

CIFELLI, R. I.; SCOTT, G. H. 1986: Stratigraphic record of the Neogene globorotalid radiation (planktonic Foraminiferida). *Smithsonian Contribution to Paleobiology 58*: 1–101.

COCHRAN, U.; BERRYMAN, K.; MILDENHALL, D. C.; HAYWARD, B. W.; SOUTHALL, K.; HOLLIS, C. 2005: Towards a record of Holocene tsunami and storms for northern Hawke's Bay, New Zealand. *New Zealand Journal of Geology and Geophysics 48*: 507–515.

COLLEN, J. D. 1972: New foraminifera from the Pliocene and Pleistocene of the Wanganui Basin, New Zealand. *Journal of the Royal Society of New Zealand 2*: 373–382.

CRUNDWELL, M. P. 2004: Miocene (Pareore, Southland and Taranaki Series). *In*: Cooper, R. A. (ed.), *The New Zealand Geological Timescale. Institute of Geological and Nuclear Sciences Monograph 22*: 165–194.

CRUNDWELL, M. P.; NELSON, C. S. 2007: A magnetostratigraphically-constrained biochronology for late Miocene bolboforms and planktic foraminifers in the temperate Southwest Pacific. *Stratigraphy 4*: 1–34.

CRUNDWELL, M.; SCOTT, G.; NAISH, T.; CARTER, L. 2008: Glacial–interglacial ocean climate variability from planktonic foraminifera during the Mid-Pleistocene transition in the temperate Southwest Pacific, ODP Site 1123. *Palaeogeography, Palaeoclimatology, Palaeoecology 260*: 202–229.

CUSHMAN, J. A. 1919: Recent foraminifera from off New Zealand. *Proceedings of the United States National Museum 56*: 593–640.

DORREEN, J. M. 1948: A foraminiferal fauna from the Kaiatan Stage (Upper Eocene) of New Zealand. *Journal of Paleontology 22*: 281–300.

DAWSON, E. W. 1992: The marine fauna of New Zealand: Index to the Fauna: 1. Protozoa. *New Zealand Oceanographic Institute Memoir 99*: 1–368.

EADE, J. V. 1967a: A checklist of Recent New Zealand foraminifera. *New Zealand Oceanographic Institute Memoir 44*: 1–72.

EADE, J. V. 1967b: New Zealand Recent foraminifera of the families Islandiellidae and Cassidulinidae. *New Zealand Journal of Marine and Freshwater Research 1*: 421–454.

EAGLE, M. K.; HAYWARD, B. W.; GRANT-MACKIE, J. A.; GREGORY, M. 2000: Fossil communities in an early Miocene transgressive sequence, Mathesons Bay, Leigh, Auckland. *Tane (Journal of the Auckland University Field Club) 37*: 43–67.

FINLAY, H. J. 1939a: New Zealand foraminifera: Key species in stratigraphy – No. 1. *Transactions of the Royal Society of New Zealand 68*: 504–533.

FINLAY, H. J. 1939b: New Zealand foraminifera: Key species in stratigraphy – No. 2. *Transactions of the Royal Society of New Zealand 69*: 89–128.

FINLAY, H. J. 1939c: New Zealand foraminifera: Key species in stratigraphy – No. 3. *Transactions of the Royal Society of New Zealand 69*: 309–329.

FINLAY, H. J. 1940: New Zealand foraminifera: Key species in stratigraphy – No. 4. *Transactions of the Royal Society of New Zealand 69*: 448–472.

FINLAY, H. J. 1947: New Zealand foraminifera: Key species in stratigraphy – No. 5. *New Zealand Journal of Science and Technology B28*: 259–292.

FINLAY, H. J.; MARWICK, J. 1947: New divisions of the New Zealand Upper Cretaceous and Tertiary. *New Zealand Journal of Science and Technology B28*: 228–236.

GEHRELS, W. R.; HAYWARD, B. W.; NEWNHAM, R. M.; SOUTHALL, K. E. 2008: A 20th century acceleration in sea-level rise in New Zealand. *Geophysical Research Letters 35, L02717*: 1–5.

GIBSON, G. W. 1967: Foraminifera and

stratigraphy of the Tongaporutuan stage in the Taranaki coastal and six other sections. Part 1– Systematics and distribution. *Transactions of the Royal Society of New Zealand (Geology) 5*: 1–70.
GOFF, J. R.; ROUSE, H. L.; JONES, S. L.; HAYWARD, B. W.; COCHRAN, U.; McLEA, W. W.; DICKINSON, W. W.; MORLEY, M. S. 2000: Evidence for an earthquake and tsunami about 3100–3400 years ago, and other catastrophic saltwater inundations recorded in a coastal lagoon, New Zealand. *Marine Geology 170*: 231–249.
GOODAY, A. J.; TENDAL, O. S. 2000: Class Xenophyophorea. Pp. 1086–1097 *in*: Lee, J. J.; Leedale, G. F.; Bradbury, P. (eds), *An Illustrated Guide to the Protozoa. Second edition. Organisms traditionally referred to as Protozoa, or newly discovered groups*. Society of Protozoologists, Lawrence. 1432 p. (2 vols).
GOODAY, A. J.; BETT, B. J.; PRATT, D. N. 1993: Direct observation of episodic growth in an abyssal xenophyophore (Protista). *Deep-Sea Research I, 40*: 2131–2143.
GRENFELL, H. R.; HAYWARD, B. W.; HORROCKS, M. 2007: Foraminiferal record of ecological impact of deforestation and oyster farms, Mahurangi Harbour, New Zealand. *Marine and Freshwater Research 58*: 475–491.
HACKETT, J. D.; YOON, H. S.; LI, S.; REYES-PRIETO, A.; RÜMMELE, S. E.; BHATTACHARYA, D. 2007: Phylogenomic analysis supports the monophyly of cryptophytes and haptophytes and the association with chromalveolates. *Molecular Biology and Evolution 24*: 1702–1713.
HADA, S.; LANDIS, C. A. 1995: Te Akatarawa Formation – an exotic oceanic-continental margin terrane within the Torlesse-Haast Schist transition zone. *New Zealand Journal of Geology and Geophysics 38*: 349–359.
HAECKEL, E. 1889: Report on the deep-sea Keratosa. *Report on the Scientific Results of the Voyage of H.M.S. Challenger during the years 1873–1876, Zoology 32*: 1–92.
HAEUSLER, R. 1887: Notes on some foraminifera from the Hauraki Gulf. *Transactions and Proceedings of the New Zealand Institute 19*: 196–200.
HANCOCK, H. J. L.; DICKENS, G. R.; STRONG, C. P.; HOLLIS, C. J.; FIELD, B. D. 2003: Foraminiferal and carbon isotope stratigraphy through the Paleocene-Eocene transition at Dee Stream, Marlborough, New Zealand. *New Zealand Journal of Geology and Geophysics 46*: 1–19.
HAYWARD, B. W. 1982: Associations of benthic foraminifera (Protozoa: Sarcodina) of inner shelf sediments around the Cavalli Islands, north-east New Zealand. *New Zealand Journal of Marine and Freshwater Research 16*: 27–56.
HAYWARD, B. W. 1983: Planktic foraminifera (Protozoa) in New Zealand waters: a taxonomic review. *New Zealand Journal of Zoology 10*: 63–74.
HAYWARD, B. W. 1986: A guide to paleoenvironmental assessment using New Zealand Cenozoic foraminiferal faunas. *New Zealand Geological Survey Report Pal 109*: 1–73.
HAYWARD, B. W. 1990a: Use of foraminiferal data in analysis of Taranaki Basin, New Zealand. *Journal of Foraminiferal Research 20*: 71–83.
HAYWARD, B. W. 1990b: Taxonomy, paleogeography and evolutionary history of the Bolivinellidae (Foraminiferida). *New Zealand Geological Survey Paleontological Bulletin 63*: 1–132, 19 pls.
HAYWARD, B. W. 2002: Late Pliocene to middle Pleistocene extinctions of deep-sea benthic foraminifera ("*Stilostomella* extinction") in the South-west Pacific. *Journal of Foraminiferal Research 32*: 274–306.
HAYWARD, B. W. 2004 Foraminifera-based estimates of paleobathymetry using modern analogue technique, and the subsidence history of the early Miocene Waitemata Basin. *New Zealand Journal of Geology and Geophysics 47*: 749–768.
HAYWARD, B. W.; BROOK, F. J. 1994: Foraminiferal paleoecology and initial subsidence of the early Miocene Waitemata Basin, Waiheke Island, Auckland. *New Zealand Journal of Geology and Geophysics 37*: 11–24.
HAYWARD, B. W.; BUZAS, M. A. 1979: Taxonomy and paleoecology of early Miocene benthic foraminifera of northern New Zealand and the north Tasman Sea. *Smithsonian Contributions to Paleobiology 36*: 1–154, 28 pls.
HAYWARD, B. W.; HOLLIS, C. J. 1994: Brackish foraminifera in New Zealand: a taxonomic and ecologic review. *Micropaleontology 40*: 185–222.
HAYWARD, B. W.; CARTER, R.; GRENFELL, H. R.; HAYWARD, J. J. 2001: Depth distribution of Recent deep-sea benthic foraminifera east of New Zealand, and their potential for improving paleobathymetric assessments of Neogene microfaunas. *New Zealand Journal of Geology and Geophysics 44*: 555–587.
HAYWARD, B. W.; COCHRAN, U.; SOUTHALL, K.; WIGGINS, E.; GRENFELL, H. R.; SABAA, A. T.; SHANE, P. A. R.; GEHRELS, R. 2004a: Micropalaeontological evidence for the Holocene earthquake history of the eastern Bay of Plenty, New Zealand. *Journal of Quaternary Science Reviews 23*: 1651–1667.
HAYWARD, B. W.; GRENFELL, H. R.; CARTER, R.; HAYWARD, J. J. 2004b: Benthic foraminiferal proxy evidence for the Neogene palaeoceanographic history of the Southwest Pacific, east of New Zealand. *Marine Geology 205*: 147–184.
HAYWARD, B. W.; GRENFELL, H. R.; NICHOLSON, K.; PARKER, R.; WILMHURST, J.; HORROCKS, M.; SWALES, A.; SABAA, A. T. 2004c: Foraminiferal record of human impact on intertidal estuarine environments in New Zealand's largest city. *Marine Micropaleontology 53*: 37–66.
HAYWARD, B. W.; GRENFELL, H. R.; REID, C. M.; HAYWARD, K. A. 1999a: Recent New Zealand shallow-water benthic foraminifera: Taxonomy, ecologic distribution, biogeography, and use in paleoenvironmental assessment. *Institute of Geological and Nuclear Sciences Monograph 21*: 1–258, 17 pls.
HAYWARD, B. W.; GRENFELL, H. R.; SABAA, A. T. 2003: Recent benthic foraminifera from offshore Taranaki, New Zealand. *New Zealand Journal of Geology and Geophysics 46*: 489–518.
HAYWARD, B. W.; GRENFELL, H. R.; SABAA, A. T.; CARTER, R.; COCHRAN, U.; LIPPS, J. H.; SHANE, P. A. R.; MORLEY, M. S. 2006c: Micropaleontological evidence of large earthquakes in the past 7200 years in southern Hawke's Bay, New Zealand. *Quaternary Science Reviews 25*: 1186–1207.
HAYWARD, B. W.; GRENFELL, H. R.; SABAA, A. T.; HAYWARD, C. M.; NEIL, H. L. 2006a: Ecologic distribution of benthic foraminifera, offshore north-east New Zealand. *New Zealand Journal of Marine and Freshwater Research 36*: 332–354.
HAYWARD, B. W.; GRENFELL, H. R.; SABAA, A. T.; KAY, J.; DAYMOND-KING, R.; COCHRAN, U. 2010b: Holocene subsidence at the transition between strike-slip and subduction on the Pacific-Australian plate boundary, Marlborough Sounds. *New Zealand: Quaternary Science Reviews 29*: 648–661.
HAYWARD, B. W.; GRENFELL, H. R.; SABAA, A. T.; MORLEY, M. S. 2008b: Ecological impact of the introduction to New Zealand of Asian date mussels and cordgrass – the foraminiferal, ostracod and molluscan record. *Estuaries and Coasts 31*: 941–959.
HAYWARD, B. W.; GRENFELL, H. R.; SABAA, A.; MORLEY, M. S.; HORROCKS, M. 2006b: Impact and timing of increased freshwater runoff into sheltered harbour environments around Auckland City, New Zealand. *Estuaries and Coasts 29*: 165–182.
HAYWARD, B. W.; GRENFELL, H. R.; SABAA, A. T.; NEIL, H. L. 2007a: Factors influencing the distribution of Subantarctic deep-sea benthic foraminifera, Campbell and Bounty Plateaux, New Zealand. *Marine Micropaleontology 62*: 141–166.
HAYWARD, B. W.; GRENFELL, H. R.; SABAA, A. T.; NEIL, H.; BUZAS, M. A., 2010a: Recent New Zealand deep-water benthic foraminifera: taxonomy, ecologic distribution, biogeography, and use in paleoenvironmental assessment. *Institute of Geological & Nuclear Sciences Monograph 26*: 1–363.
HAYWARD, B. W.; GRENFELL, H. R.; SABAA, A. T.; SIKES, E. L. 2005: Deep-sea benthic foraminiferal record of the mid-Pleistocene transition in the South-west Pacific. *In*: Head, M. J.; Gibbard, P. L. (eds), *Early–Middle Pleistocene Transitions: The Land-Ocean Evidence. Geological Society (London), Special Publication 247*: 85–115.
HAYWARD, B. W.; GRENFELL, H. R.; SABAA, A. T.; SOUTHALL, K. E.; GEHRELS, W. R. 2007b: Foraminiferal evidence of Holocene fault displacements in coastal South Otago, New Zealand. *Journal of Foraminiferal Research 37*: 344–359.
HAYWARD, B. W.; GRENFELL, H. R.; SCOTT, D. B. 1999b: Tidal range of marsh foraminifera for determining former sea-level heights in New Zealand. *New Zealand Journal of Geology and Geophysics 42*: 395–413.
HAYWARD, B. W.; HOLLIS, C. J.; GRENFELL, H. R. 1997: Recent Elphidiidae (Foraminiferida) of the South-west Pacific and fossil Elphidiidae of New Zealand. *New Zealand Geological Survey Paleontological Bulletin 72*: 1–166.
HAYWARD, B. W.; KAWAGATA, S.; GRENFELL, H. R.; SABAA, A. T.; O'NEILL, T. 2007c: The last global extinction in the deep sea during the mid-Pleistocene climate transition. *Paleoceanography* 22, *PA3103*: 1–14.
HAYWARD, B. W.; SABAA, A. T.; GRENFELL, H. R. 2004d: Benthic foraminifera and the late Quaternary (last 150 ka) paleoceanographic and sedimentary history of the Bounty Trough, east of New Zealand. *Palaeogeography, Palaeoclimatology, Palaeoecology 211*: 59–93.
HAYWARD, B. W.; SCOTT, G. H.; CRUNDWELL, M. P.; KENNETT, J. P.; CARTER, L.; NEIL, H.; SABAA, A. T.; WILSON, K.; RODGER, J. S.; SCHAEFER, G.; GRENFELL, H. R.; LI, Q. 2008a: The effect of submerged plateaux on Pleistocene gyral circulation and sea-surface temperatures in the Southwest Pacific. *Planetary and Global Change 63*: 309–316.
HAYWARD, B. W.; SCOTT, G. H.; GRENFELL,

H. R.; CARTER, R.; LIPPS, J. H. 2004e: Techniques for estimation of tidal elevation and confinement (~salinity) histories of sheltered harbours and estuaries using benthic foraminifera: Examples from New Zealand. *The Holocene 14*: 218–232.

HAYWARD, B. W.; WILSON, K.; MORLEY, M. S.; COCHRAN, U.; GRENFELL, H. R.; SABAA, A. T.; DAYMOND-KING, R. 2010c: Microfossil record of the Holocene evolution of coastal wetlands in a tectonically-active region of New Zealand. *The Holocene 20*: 405–421.

HAYWICK, D. W.; HENDERSON, R. A. 1991: Foraminiferal paleobathymetry of Plio-Pleistocene cyclothemic sequences, Petane Group, New Zealand. *Palaios 6*: 586–599.

HEDLEY, R. H. 1966: Appendix. Cytological notes on *Syringaminna tasmanensis*. Pp. 123–124 *in*: Lewis, K. B. A giant foraminifer: a new species of *Syringammina* from the New Zealand region. *New Zealand Journal of Science 9*: 114–124.

HEDLEY, R. H.; HURDLE, C. M.; BURDETT, I. D. J. 1965: A foraminiferal fauna from the western continental shelf, North Island, New Zealand. *New Zealand Oceanographic Institute Memoir 25*: 1–46.

HEDLEY, R. H.; HURDLE, C. M.; BURDETT, I. D. J. 1967: The marine fauna of New Zealand. Intertidal foraminifera of the *Corallina officinalis* zone. *New Zealand Oceanographic Institute Memoir 38*: 1–88.

HERON-ALLEN, E.; EARLAND, A. 1922: Protozoa, Part 2. Foraminifera. *Natural History Reports of the British Antarctic ("Terra Nova") Expedition, 1910, 6(2)*: 25–268, 8 pls.

HODELL, D. A.; KENNETT, J. P. 1986: Late Miocene – Early Pliocene stratigraphy and paleoceanography of the South Atlantic and Southwest Pacific Oceans: a synthesis. *Paleoceanography 1*: 285–311.

HOLLIS, C. J.; STRONG, C. P. 2003: Biostratigraphic review of the Cretaceous/Tertiary boundary transition, mid Waipara River section, North Canterbury, New Zealand. *New Zealand Journal of Geology and Geophysics 46*: 243–253.

HORNIBROOK, N. de B. 1951: Permian fusulinid foraminifera from the North Auckland Peninsula, New Zealand. *Transactions of the Royal Society of New Zealand 79*: 319–321.

HORNIBROOK, N. de B. 1953: Jurassic foraminifera from New Zealand. *Transactions of the Royal Society of New Zealand 81*: 375–378.

HORNIBROOK, N. de B. 1961: Tertiary foraminifera from Oamaru District (N.Z). Part 1 – Systematics and distibution. *New Zealand Geological Survey Paleontological Bulletin 34*: 1–194, 28 pls.

HORNIBROOK, N. de B. 1968: A Handbook of New Zealand Microfossils. (Foraminifera and Ostracoda). *DSIR Information Series 62*: 1–136.

HORNIBROOK, N. de B. 1971: Revision of the Oligocene and Miocene foraminifera from New Zealand, described by Karrer and Stache in the reports of the"Novara"Expedition (1864). *New Zealand Geological Survey Paleontological Bulletin 43*: 1–85, 13 pls.

HORNIBROOK, N. de B. 1981: *Globorotalia* (planktic Foraminiferida) in the Late Pliocene and Early Pleistocene of New Zealand. *New Zealand Journal of Geology and Geophysics 24*: 263–292.

HORNIBROOK, N. de B. 1982: Late Pliocene and Pleistocene *Globorotalia* (Foraminiferida) from DSDP Leg 29, Site 284, Southwest Pacific. *New Zealand Journal of Geology and Geophysics 25*: 83–99.

HORNIBROOK, N. de B. 1992: New Zealand Cenozoic marine paleoclimates: A review based on the distribution of some shallow water and terrestrial biota. Pp. 83–106 *in*: Tsuchi, R.; Ingle, J. C. (eds), *Pacific Neogene: Environment, Evolution and Events*. University of Tokyo Press. 257 p.

HORNIBROOK, N. de B 1996: New Zealand Eocene and Oligocene benthic foraminifera of the family Notorotaliidae. *Institute of Geological and Nuclear Sciences Monograph 12*: 1–52.

HORNIBROOK, N. de B.; EDWARDS, A. R. 1971: Integrated planktonic foraminiferal and calcareous nannoplankton datum levels in the New Zealand Cenozoic. *Proceedings of the 2nd International Planktonic Conference, Vol. 1*: 649–657.

HORNIBROOK, N. de B.; HARRINGTON, H. J. 1957: The status of the Wangaloan Stage. *New Zealand Journal of Science and Technology B38*: 655–670.

HORNIBROOK, N. de B.; SHU, Y. K. 1965: Fusuline limestone in the Torlesse Group near Benmore Dam, Waitaki. *Transactions of the Royal Society of New Zealand, Geology 3*: 99–137.

HORNIBROOK, N. de B.; BRAZIER, R. C.; STRONG, C .P. 1989: Manual of New Zealand Permian to Pleistocene foraminiferal biostratigraphy. *New Zealand Geological Survey Paleontological Bulletin 56*: 1–175.

HSU, K. J.; CITA, M. B.; RYAN, W. B. F. 1973: The origin of the Mediterranean evaporites. *Initial Reports of the Deep Sea Drilling Project 13*: 1203–1231.

HUGHES, J. A.; GOODAY, A. J. 2004: Associations between living benthic foraminifera and dead tests of *Syringammina fragilissima* (Xenophyophorea) in the Darwin Mounds region (NE Atlantic). *Deep-Sea Research I, 51*: 1741–1758.

HULME, S. G. 1964: Recent foraminifera from Manukau Harbour, Auckland, New Zealand. *New Zealand Journal of Science 7*: 305–340.

HUTTON, F. W. 1904: *Index Faunae Novae Zelandiae*. Dulau & Co., London. viii + 372 p.

JENKINS, D. G. 1971: New Zealand Cenozoic planktonic foraminifera. *New Zealand Geological Survey Paleontological Bulletin 42*: 1–278.

KAIHO, K.; ARINOBU, T.; ISHIWATARI, R.; MORGANS, H. E. G.; OKADA, H.; TAKEDA, N.; TAZAKI, K.; ZHOU, G.; KAJIWARA, Y.; MATSUMOTO, R.; HIRAI, A.; NIITSUMA, N.; WADA, H. 1996: Latest Paleocene benthic foraminiferal extinction and environmental changes at Tawanui, New Zealand. *Paleoceanography 11*: 447–465.

KAMP, P. J. J.; JOURNEAUX, T. D.; MORGANS, H. E. G. 1998: Cyclostratigraphy of middle Pliocene mid shelf to upper slope strata, eastern Wanganui Basin (New Zealand); correlations to the deep sea isotope record. *Sedimentary Geology 117*: 165–192.

KARRER, F. 1864: Die foraminiferen-fauna des Tertiaren grunsandsteines der Orakei-bei, Auckland. *Novara Expedition 1857–59 Report, Wien, Geologie Theil 1(2)*: 69–86.

KENNETT, J. P. 1965: The Kapitean Stage (Upper Miocene) of New Zealand. Unpublished Ph.D. thesis, Victoria University of Wellington.

KENNETT, J. P. 1966: Stratigraphy and fauna of the type section of the Kapitean Stage, Greymouth, New Zealand. *Transactions of the Royal Society of New Zealand (Geology) 4*: 1–77.

KENNETT, J. P. 1967: Recognition and correlation of the Kapitean Stage (upper Miocene, New Zealand). *New Zealand Journal of Geology and Geophysics 10*: 1051–1063.

KENNETT, J. P. 1968: Paleo-oceanographic aspects of the foraminiferal zonation in the upper Miocene-lower Pliocene of New Zealand. *Giornale di Geologia 35*: 143–156.

KENNETT, J. P. 1978: The development of planktonic biogeography in the Southern Ocean during the Cenozoic. *Marine Micropaleontology 3*: 301–345.

KING, A. L.; HOWARD, W. R. 2001: Seasonality of foraminiferal flux in sediment traps at Chatham Rise, SW Pacific: implications for paleotemperature estimates. *Deep-Sea Research I 48*: 1687–1708.

KING, A. L.; HOWARD, W. R. 2003: Planktonic foraminiferal flux seasonality in Subantarctic sediment traps: A test for paleoclimate reconstructions. *Paleoceanography* 18,1019: 1–17.

KUSTANOWICH, S. 1963: Distibution of planktonic foraminifera and surface sediments of the south-west Pacific Ocean. *New Zealand Journal of Geology and Geophysics 6*: 534–565.

KUSTANOWICH, S. 1965: Foraminifera of Milford Sound. *In*: Skerman, T. M. (Ed.), *Studies of a Southern Fiord. New Zealand Oceanographic Institute Memoir 17*: 49–63.

LEVEN, E. J.; CAMPBELL, H. J. 1998: Middle Permian (Murgabian) fusuline faunas, Torlesse Terrane, New Zealand. *New Zealand Journal of Geology and Geophysics 41*: 149–156.

LEVEN, E. J.; GRANT-MACKIE, J. A. 1997: Permian fusulinid foraminifera from Wherowhero Point, Northland, New Zealand. *New Zealand Journal of Geology and Geophysics 40*: 473–486.

LEVIN, L. A. 1991: Interaction between metazoans and large, agglutinating protozoans: Implications for the community structure of deep-sea benthos. *American Zoologist 31*: 886–900.

LEVIN, L. A.; DEMASTER, D. J.; McCANN, L. D.; THOMAS, C. L. 1986: Effects of giant protozoans (class Xenophyophorea) on deep-seamount benthos. *Marine Ecology Progress Series 29*: 99–104.

LEVIN, L. A.; THOMAS, C. L. 1988: The ecology of xenophyophores (Protista) on eastern Pacific seamounts. *Deep-Sea Research 35*: 2003–2027.

LEWIS, K. B. 1966. A giant foraminifer: a new species of *Syringammina* from the New Zealand region. *New Zealand Journal of Science 9*: 114–124. (With an appendix by R. H. HEDLEY).

LEWIS, K. B. 1970: A key to the Recent genera of Foraminiferida. *New Zealand Oceanographic Institute Memoir 45*: 1–88.

LEWIS, K. B. 1979: Foraminifera on the continental shelf and slope off southern Hawke's Bay, New Zealand. *New Zealand Oceanographic Institute Memoir 84*: 1–45.

LOEBLICH, A. R.; TAPPAN, H. 1987: *Foraminiferal Genera and their Classification*. Van Norstrand Reinhold Co., New York. 1182 p. (2 vols).

LOEBLICH, A. R.; TAPPAN, H. 1992: Present status of foraminiferal classification. Pp. 93–102 *in*: Takayanagi, Y.; Saito, T. (eds), *Studies in Benthic Foraminifera. Proceedings of the Fourth Symposium on Benthic Foraminifera, Sendai, 1990*. Tokai University Press, Kanagawa.

MALMGREN, B. A.; KENNETT, J. P. 1981: Phyletic gradualism in a Late Cenozoic planktonic foraminiferal lineage: DSDP Site 284: Southwest Pacific. *Paleobiology 7*: 230–240.

MANTELL, G. A. 1850: Notice of the remains of the *Dinornis* and other birds and of fossils and

rock-specimens, recently collected by Mr Walter Mantell in the middle island of New Zealand; with additional notes on the northern island. *Quaterly Journal of the Geological Society of London* 6: 319–342.
MARTINI, E.; JENKINS, D. G. 1986: Biostratigraphic syntheis of Deep Sea Drilling Project Leg 90 in the South-west Pacific. *Initial Reports of the Deep Sea Drilling Project 90*: 1459–1470.
MATTHEWS, A.; GRENFELL, H. R.; HAYWARD, B. W.; HORROCKS, M. 2005: Foraminiferal record of sewage outfall impacts on the inner Manukau Harbour, Auckland, New Zealand. *New Zealand Journal of Marine and Freshwater Research 39*: 193–215.
MESTAYER, R. L. 1916: List of foraminifera dredged from 15 feet south of Big King at 98 fathom depth. *Transactions of the New Zealand Institute 48*: 128–130.
MORGANS, H. E. G. 2004: Paleogene (Dannevirke, Arnold, Landon Series). *In*: R. A. Cooper (ed.), *The New Zealand Geological Timescale. Institute of Geological and Nuclear Sciences Monograph 22*: 125–161.
MORGANS, H. E. G.; EDWARDS, A. R.; SCOTT, G. H.; GRAHAM, I. J.; KAMP, P. J. J.; MUMME, T. C.; WILSON, G. J.; WILSON, G. S. 1999: Integrated stratigraphy of the Waitakian-Otaian stage boundary stratotype, early Miocene, New Zealand. *New Zealand Journal of Geology and Geophysics 42*: 581–614.
NAISH, T.; KAMP, P. J. J. 1997: Foraminiferal depth paleoecology of Late Pliocene shelf sequences and system tracts, Wanganui Basin, New Zealand. *Sedimentary Geology 110*: 237–255.
NEIL, H.; CARTER, L.; MORRIS, M. 2004: Thermal isolation of Campbell Plateau, New Zealand, by the Antarctic Circumpolar Current over the past 130 kyr. *Paleoceanography* 19, PA4008: 1–17.
NELSON, C. S.; COOKE, P. J.; HENDY, C. H.; CUTHBERTSON, A. M. 1993: Oceanographic and climatic changes over the past 160,000 years at the Deep Sea Drilling Project Site 594 off southeastern New Zealand, South-west Pacific Ocean. *Paleoceanography 8*: 435–458.
NELSON, C. S.; HENDY, I. L.; NEIL, H. L.; HENDY, C. H.; WEAVER, P. P. E. 2000: Last glacial jetting of cold waters through the Subtropical Convergence zone in the Southwest Pacific off eastern New Zealand and some geological implications. *Palaeogeography, palaeoclimatology, Palaeoecology 156*: 103–121.
PAWLOWSKI, J.; HOLZMANN, M.; FAHRNI, J.; RICHARDSON, S. L. 2003: Small subunit ribosomal DNA suggests that the Xenophyophorean *Syringammina corbicula* is a foraminiferan. *The Journal of Eukaryotic Microbiology 50*: 483–487.
PHLEGER, F. B. 1970: Foraminiferal populations and marine marsh processes. *Limnology and Oceanography 15*: 522–534.
RAINE, J. I.; STRONG, C. P.; WILSON, G. J. 1993: Biostratigraphic revision of petroleum exploration wells, Great South Basin, New Zealand. *Institute of Geological and Nuclear Sciences Science Report 93/32*: 1–146.
REEB, V. C.; PEGLAR, M. T.; YOON, H. S.; BAI, J. R.; WU, M.; SHIU, P.; GRAFENBERG, J. L.; REYES-PRIETO, A.; RÜMMELE, S. E.; GROSS, J.; BHATTACHARYA, D. 2009: Interrelationships of chromalveolates within a broadly sample tree of photosynthetic protists. *Molecular Phylogenetics and Evolution 53*: 202–211.
SABAA, A. T.; SIKES, E. L.; HAYWARD, B. W.; HOWARD, W. R. 2004: Pliocene sea surface temperature changes in ODP Site 1125, Chatham Rise, east of New Zealand. *Marine Geology 205*: 113–125.
SAIDOVA, K. M. 1975: *Bentosniye foraminifery Tikhogo Okeana*. P. P. Shirshov Institute of Oceanology, Academy of Sciences of the USSR, Moscow. 875 p. (in 3 vols), 116 pls.
SCHAEFER, G.; RODGER, J. S.; HAYWARD, B. W.; KENNETT, J. P.; SABAA, A. T.; SCOTT, G. H. 2005: Planktic foraminiferal and sea surface temperature record during the last 1 myr across the Subtropical Front, Southwest Pacific. *Marine Micropaleontology 54*: 191–212.
SCHNEIDER, C. E.; KENNETT, J. P. 1996: Isotopic evidence for interspecific habitat differences during evolution of the Neogene planktonic foraminiferal clade *Globoconella*. *Paleobiology 22*: 282–303.
SCHULZE, F. E.1906: Die Xenophyophoren der "Siboga"-Expedition. *Siboga Expeditie 4b*: 1–18.
SCHULZE, F. E. 1907: Die Xenophyophoren, eine besondere Gruppe der Rhizopoden. *Wissenschaftliche Ergebnisse der deutschen Tiefsee-Expedition 1898–1899 11*: 1–55.
SCOTT, G. H. 1965: Utility of *Haeuslerella* Parr (Foraminifera) in New Zealand middle Tertiary stratigraphy. *New Zealand Geological Survey Paleontological Bulletin 38*: 1–48.
SCOTT, G. H. 1969: Biometric study of the lower Miocene *Globigerinoides* from New Zealand, Trinidad and Europe: A review. Pp. 603–610 *in*: *Proceedings of the 1st International Conference on Planktonic Microfossils, Geneva, 1967*. E. J. Brill, Leiden. 422 + 745 p.
SCOTT, G. H. 1970: Miocene foraminiferal biotopes in New Zealand: Waitemata Group, Kaipara, Northland. *New Zealand Journal of Geology and Geophysics 13*: 316–342.
SCOTT, G. H. 1971: Revision of Hutchinsonian, Awamoan and Altonian Stages (lower Miocene, New Zealand) – 1. *New Zealand Journal of Geology and Geophysics 14*: 705–726.
SCOTT, G. H. 1974: Biometry of the foraminiferal shell. Pp. 55–151 *in*: Hedley, R. H.; Adams, C. L. (eds), *Foraminifera*. Vol. 1. Academic Press, London.
SCOTT, G. H. 1978: Shell design in *Bolivinita quadrilatera*, *B. pohana* and *B. compressa* (Foraminiferida). *New Zealand Journal of Geology and Geophysics 21*: 617–634.
SCOTT, G. H. 1992: Planktonic foraminiferal biostratigraphy (Altonian-Tongaporutuan Stages, Miocene) at DSDP Site 593, Challenger Plateau, Tasman Sea. *New Zealand Journal of Geology and Geophysics 35*: 501–513.
SCOTT G. H.; BISHOP S.; BURT B. J. 1990: Guide to some Neogene globorotalids (Foraminiferida) from New Zealand. *New Zealand Geological Survey Paleontological Bulletin 61*: 1–135.
SCOTT, G. H.; NELSON, C. S.; STONE, H. H. 1995: Planktic foraminiferal events in early Miocene zones N6 and N7 at southwest Pacific DSDP site 593: relation with climatic changes in oxygen isotope zone Mi1b. *Marine Micropaleontology 25*: 29–45.
SOUTHALL, K. E.; GEHRELS, W. R.; HAYWARD, B. W., 2006: Foraminifera in a New Zealand salt marsh and their suitability as sea-level indicators. *Marine Micropaleontology* 60: 167–179.
SRINIVASAN, M. S. 1966: Descriptions of new species and notes on taxonomy of foraminifera from the Upper Eocene and Lower Oligocene of New Zealand. *Transactions of the Royal Society of New Zealand* (*Geology*) 3: 231–256.
SRINIVASAN, M. S.; KENNETT, J. P. 1981: Neogene planktonic foraminiferal biostratigraphy and evolution: equatorial to subantarctic, South Pacific. *Marine Micropaleontology 6*: 499–533.
STACHE, G. 1864: Die Foraminiferen der Tertiaran mergel des Whaingaroa-Hafens (Prov. Auckland). *Novara Expedition 1857–59 Report, Wien, Geologie Theil 1*(2): 159–304.
STONELEY, H. M. M. 1962: New foraminifera from the Clarence Series (lower Cretaceous) of New Zealand. *New Zealand Journal of Geology and Geophysics 5*: 592–616.
STRONG, C. P. 1977: Cretaceous Tertiary Boundary at Woodside Creek, North-eastern Marlborough. *New Zealand Journal of Geology and Geophysics* 20: 687–696.
STRONG, C. P. 1984a: Triassic foraminifera from Southland Syncline, New Zealand. *New Zealand Geological Survey Paleontological Bulletin 92*: 1–60, 12 pls.
STRONG, C. P. 1984b: Triassic foraminifera in the North Island. *Geological Society of New Zealand Newsletter 65*: 19–21.
STRONG, C. P. 2000: Cretaceous–Tertiary foraminiferal succession at Flaxbourne River, Marlborough, New Zealand. *New Zealand Journal of Geology and Geophysics 43*: 1–20.
TABACHNICK, R. E.; BOOKSTEIN, F. L. 1990: The structure of individual variation in Miocene *Globorotalia*. *Evolution 44*: 416–434.
TENDAL, O. S. 1972: A monograph of the Xenophyophoria (Rhizopodea, Protozoa). *Galathea Report 12*: 7–99.
TENDAL, O. S. 1975: The xenophyophores of New Zealand (Rhizopodea, Protozoa). *Tuatara 21*: 92–97.
TENDAL, O. S. 1981: New records of xenophyophores from the upper slope around New Zealand. *New Zealand Journal of Marine and Freshwater Research 15*: 285–287.
TENDAL, O. S. 1985: Xenophyophores (Protozoa, Sarcodina) in the diet of *Neopilina galatheae*. *Galathea Report 16*: 95–98.
TENDAL, O. S. 1989: Phylum Xenophyophora. Pp. 135–138 *in*: Margulis, L.; Corliss, J. O.; Melkonian, M.; Chapman, D. J. (eds), *Handbook of the Protoctista*. Jones & Bartlett Publishers, Boston. 914 p.
TENDAL, O. S. 1996: Synoptic checklist and bibliography of the Xenophyophorea (Protista), with a zoogeographical survey of the group. *Galathea Report 17*: 79–101.
TENDAL, O. S. 1997: The xenophyophores – from enigma to dominating fauna element. *Deep-Sea Newsletter 26*: 14–15.
TENDAL, O. S.; LEWIS, K. B. 1978: New Zealand xenophyophores: upper bathyal distribution, photographs of growth position, and a new species. *New Zealand Journal of Marine and Freshwater Research 12*: 197–203.
TENDAL, O. S.; GOODAY, A. J. 1981: Xenophyophoria (Rhizopoda, Protozoa) in bottom photographs from the bathyal and abyssal NE Atlantic. *Oceanologica Acta 4*: 415–422.
THOMSON, J. A. 1916: On stage names applicable to the divisions of the Tertiary in New Zealand. *Transactions of the New Zealand Institute 48*: 28–40.
VACHARD, D.; FERRIÈRE, J. 1991: An assemblage with *Yabeina* (fusulinid foraminifera) from the Midian (upper Permian) of Whangaroa area (Orua Bay), New Zealand. *Revue de Micropaléontologie 34*: 201–230.

VAN MORKHOVEN, F. P. C. M.; BERGGREN, W. A.; EDWARDS, A. S. 1986: Cenozoic cosmopolitan deep-water benthic foraminifera. *Bulletin du Centre de la Recherche pour l'Exploration et Production, Elf Aquitaine, Memoire 11*: 1–650.
VELLA, P. 1957: Studies in New Zealand foraminifera; Part I– Foraminifera from Cook Strait. Part II– Upper Miocene to Recent species of the genus *Notorotalia. New Zealand Geological Survey Paleontological Bulletin 28*: 1–64, 9 pls.
VELLA, P. 1961: Upper Oligocene and Miocene uvigerinid foraminifera from Raukumara Peninsula, New Zealand. *Micropaleontology 7*: 467–483.
VELLA, P. 1962a: Determining depths of New Zealand Tertiary seas. *Tuatara 10*: 19–40.
VELLA, P. 1962b: Biostratigraphy and paleoecology of Mauriceville District, New Zealand. *Transactions of the Royal Society of New Zealand (Geology) 1*: 183–199.
WEAVER, P. E.; CARTER, L.; NEIL, H. L. 1998: Response of surface water masses and circulation to late Quaternary climate change east of New Zealand. *Paleoceanography 13*: 70–83.
WEAVER, P. P. E.; NEIL, H.; CARTER, L. 1997: Sea surface temperature estimates from the Southwest Pacific based on planktonic foraminifera and oxygen isotopes. *Paleogeography, Paleoclimatology, Paleoecology 131*: 241–256.
WEBB, P. N. 1971: New Zealand late Cretaceous (Haumurian) foraminifera and stratigraphy: a summary. *New Zealand Journal of Geology and Geophysics 14*: 795–828.
WEI, K.-Y. 1994a: Stratophenetic tracing of phylogeny using SIMCA pattern recognition technique: a case study of the Late Neogene planktic foraminifera *Globoconella* clade. *Paleobiology 20*: 52–65.
WEI, K.-Y. 1994b: Allometric heterochrony in the Pliocene-Pleistocene planktic foraminiferal clade *Globoconella. Paleobiology 20*: 66–84.
WEI, K-Y.; KENNETT, J. P. 1988: Phyletic gradualism and punctuated equilibrium in the Late Neogene planktonic foraminiferal clade *Globoconella. Paleobiology 14*: 345–363.
WILSON, K.; HAYWARD, B. W.; SABAA, A. T.; SCOTT, G. H.; KENNETT, J. P. 2005: A one-million year history of a north-south segment of the Subtropical Front, east of New Zealand. *Palaeoceanography 20, PA2004*: 1–10.

Checklist of Recent New Zealand Foraminifera

Foraminifera in the following checklists pertain only to those in the Exclusive Economic Zone. For the Recent fauna, single-letter codes indicate endemic species (E), apparently adventive species (A), species present only around the Kermadec Islands (K), new records for New Zealand (*) and species records unable to be confirmed in this review (#). Endemic genera are underlined (first entry only). For the fossil fauna, species are followed by time ranges using the standard abbreviations for New Zealand stages. Because of the historic plethora of taxonomic names introduced globally it is not possible at this time to provide any accurate estimate of endemism in fossil species. Phylum- and class-level classification follow Cavalier-Smith (2003). Genus, family, and superfamily classification largely follows Loeblich and Tappan (1987), with the ordinal classification after Loeblich and Tappan (1992). The classification of Xenophyophorea follows Tendal (1996) and Gooday and Tendal (2000).

KINGDOM CHROMISTA
SUBKINGDOM HAROSA
INFRAKINGDOM RHIZARIA
PHYLUM FORAMINIFERA
Class POLYTHALAMEA
Order ALLOGROMIIDA
ALLOGROMIIDAE
Placopsilinella aurantiaca Earland, 1934
Shepheardella taeniformis Siddall, 1880 A

Order ASTRORHIZIDA
ASTRORHIZOIDEA
ASTRORHIZIDAE
Astrorhiza arenaria Carpenter, 1877#
Pelosina cylindrica Brady, 1884#
Pelosina rotundata Brady, 1879#
Pelosina variabilis Brady, 1879#
BATHYSIPHONIDAE
Bathysiphon alba (Heron-Allen & Earland, 1932)
Bathysiphon argenteus Heron-Allen & Earland, 1913
Bathysiphon filiformis M. Sars, 1872
Bathysiphon saeva (Saidova, 1975)
HEMISPHAERAMMINIDAE
Crithionina granum Goës, 1894#
Crithionina hispida Flint, 1899#
Crithionina mamilla Goës, 1894#
Crithionina pisum Goës, 1896#
Crithionina rugosa Goës, 1896#
Hemisphaerammina bradyi Loeblich & Tappan, 1957#
Hemisphaerammina depressa Heron-Allen & Earland, 1932
Hemisphaerammina lens Goës, 1896#
Iridia diaphana Heron-Allen & Earland, 1914
Tholosina bulla (Brady, 1881)#
Tholosina protea Heron-Allen & Earland, 1932#
Tholosina vesicularis (Brady, 1879)
HIPPOCREPINELLIDAE
Hippocrepinella alba Heron-Allen & Earland, 1932
POLYSACCAMMINIDAE
Polysaccammina ipohalina Scott, 1976
PSAMMOSPHAERIDAE
Psammophax consociata Rhumbler, 1931
Psammosphaera bowmani Heron-Allen & Earland, 1912
Psammosphaera fusca Schulze, 1875
Psammosphaera parva Flint, 1899
Psammosphaera rustica Heron-Allen & Earland, 1912
Psammosphaera testacea Flint, 1899
Storthosphaera albida Schulze, 1875
RHABDAMMINIDAE
Dendronina arborescens Heron-Allen & Earland, 1922 E
Dendronina limosa Heron-Allen & Earland, 1922 E
Dendrophrya kermadecensis Saidova, 1975 E K
Marsipella chapmani Heron-Allen & Earland, 1922
Marsipella cylindrica Brady, 1882
Marsipella elongata Norman, 1878
Rhabdammina abyssorum M. Sars, 1869
Rhabdammina antarctica Saidova, 1975*
Rhabdammina cornuta (Brady, 1879)#
Rhabdammina discreta Brady, 1881#
Rhabdammina linearis Brady, 1879#
Rhabdammina major de Folin, 1887
Rhabdammina scabra Höglund, 1947
Rhizammina algaeformis Brady, 1879
Rhizammina spp. indet. (2)
SACCAMMINIDAE
Astrammina limnicola Hofker, 1972#
Brachysiphon corbuliniformis Chapman, 1906
Lagenammina arenulata (Skinner, 1961)
Lagenammina bulbosa (Chapman & Parr, 1937)
Lagenammina difflugiformis (Brady, 1879)
Lagenammina spiculata (Skinner, 1961)
Pseudothurammina limnetis (Scott & Medioli, 1980)
Saccammina alba Hedley, 1962
Saccammina sphaerica M. Sars, 1872
Technitella harrisii (Heron-Allen & Earland, 1914)
Technitella legumen Norman, 1878*
Thurammina albicans Brady, 1879
Thurammina aff. favosa Flint, 1899*
Thurammina papillata Brady, 1879#
Thurammina p. castanea Heron-Allen & Earland, 1917#
SCHIZAMMINIDAE
Jullienella zealandica Hayward & Gordon, 1984
HIPPOCREPINOIDEA
HIPPOCREPINIDAE
Botellina labyrinthica Brady, 1881#
Hyperammina cylindrica Parr, 1950
Hyperammina elongata alba Saidova, 1975# E
Hyperammina e. elongata Brady, 1878
Hyperammina friabilis Brady, 1884#
Hyperammina kermadecensis Saidova, 1975# E K
Hyperammina laevigata Wright, 1891
Hyperammina mestayeri Cushman, 1919 E
Hyperammina novaezealandiae Heron-Allen & Earland, 1922 E
Jaculella acuta Brady, 1879#
Jaculella obtusa Brady, 1882#
Saccorhiza echinata Saidova, 1975#
Saccorhiza ramosa (Brady, 1879)
Saccorhiza sp. indet.
KOMOKIOIDEA
KOMOKIIDAE
Normanina elgata Saidova, 1975#
Normanina ultrabyssalica Saidova, 1975# E K

Order LITUOLIDA
AMMODISCOIDEA
AMMODISCIDAE
Ammodiscus exsertus Cushman, 1910
Ammodiscus gullmarensis Höglund, 1948
Ammodiscus incertus (d'Orbigny, 1839)#
Ammodiscus mestayeri Cushman, 1919
Ammodiscus pacificus Cushman & Valentine, 1930
Ammodiscus planorbis Höglund, 1947
Ammodiscus profundissimus Saidova, 1970*
Ammodiscus tenuis Brady, 1881#
Ammolagena clavata (Jones & Parker, 1860)
Ammovertellina prima Suleymanov, 1959*
Glomospira elongata Collins, 1958
Glomospira cf. *fijiensis* Brönnimann, Whittaker & Zaninetti, 1992
Glomospira gordialis (Jones & Parker, 1860)
Tolypammina horrida Cushman, 1919 E
Tolypammina vagans (Brady, 1879)#
Tolypammina sp. Hedley, Hurdle & Burdett 1967
Turritellella shoneana (Siddall, 1878)#
Usbekistania charoides (Jones & Parker, 1860)
ATAXOPHRAGMIOIDEA
GLOBOTEXTULARIIDAE
Globotextularia anceps Brady, 1884
Liebusella goesi Hoglund, 1947
Liebusella improcera Loeblich & Tappan, 1994*
Liebusella soldanii (Jones & Parker, 1860)
Ruakituria magdaliformis (Schwager, 1866)*
Ruakituria pseudorobusta Kennett, 1967
TEXTULARIELLIDAE
Textulariella barretti (Jones & Parker, 1863)
HAPLOPHRAGMIOIDEA
AMMOSPHAEROIDINIDAE
Adercotryma glomeratum (Brady, 1878)
Ammosphaeroidina sphaeroidiniformis (Brady, 1884)
Cystammina pauciloculata (Brady, 1879)
Recurvoidatus parcus Saidova, 1970
Recurvoides contortus Earland, 1934
Recurvoides crassus Zheng, 1988*
Recurvoides rotundus Saidova, 1975#
DISCAMMINIDAE
Ammoscalaria compressa (Cushman & McCulloch, 1939)*
Ammoscalaria georgescotti Hayward, 2010
Ammoscalaria pseudospiralis (Williamson, 1858)#
Ammoscalaria tenuimargo (Brady, 1882)
Discammina compressa (Goës, 1882)
Glaphyrammina americanus Cushman, 1910#
HAPLOPHRAGMOIDIDAE
Buzasina galeata (Brady, 1881)
Buzasina ringens (Brady, 1879)
Cribrostomoides bradyi Cushman, 1910*
Cribrostomoides crassimargo Norman, 1892
Cribrostomoides jeffreysii (Williamson, 1858)
Cribrostomoides spiculolega (Parr, 1950)*
Cribrostomoides subglobosus (Cushman, 1910)
Cribrostomoides subtrullissatus (Parr, 1950)
Cribrostomoides wiesneri (Parr, 1950)
Evolutinella rotulatum (Brady, 1881)#
Haplophragmoides manilaensis Andersen, 1953*
Haplophragmoides neobradyi Uchio, 1960*
Haplophragmoides pusillus Collins, 1974
Haplophragmoides sphaeriloculum Cushman, 1910
Haplophragmoides wilberti Andersen, 1953
Labrospira spiculotesta (Zheng, 1979)
Trochamminita salsa (Cushman & Brönnimann, 1948)
Veleroninoides scitulus (Brady, 1881)#
HORMOSINOIDEA
HORMOSINIDAE
Archimerismus subnodosa Brady, 1884#
Cuneata arctica (Brady, 1881)
Hormosina globulifera Brady, 1879
Hormosina pilulifera (Brady, 1884)
Hormosinella distans Brady, 1881
Hormosinella guttifera (Brady, 1881)
Reophax aduncus Brady, 1882#
Reophax bacillaris Brady, 1881#
Reophax catenatus Höglund, 1947
Reophax dentaliniformis Brady, 1881
Reophax eunetus Jensen, 1905
Reophax fusiformis (Williamson, 1858)
Reophax hispidulus Cushman, 1930*
Reophax micaceus Earland, 1934#
Reophax nodulosus Brady, 1879
Reophax pseudodistans Cushman, 1919 E
Reophax scorpiurus Montfort, 1808#
Reophax scotti Chaster, 1892
Reophax spiculifer Brady, 1879
Reophax subfusiformis Earland, 1933
Reophax sp. indet.
Scherochorella moniliforme (Siddall, 1886)
LITUOLOIDEA
LITUOLIDAE
Ammobaculites agglutinans (d'Orbigny, 1846)
Ammobaculites calcareus (Brady, 1884)#
Ammobaculites cf. *catenulatus* Cushman & McCulloch, 1939
Ammobaculites crassiformis Zheng, 1988
Ammobaculites exiguus Cushman & Brönnimann, 1948
Ammobaculites filiformis Earland, 1934
Ammobaculites microformis Saidova, 1970
Ammobaculites paradoxus Clark 1994
Ammobaculites villosus Saidova, 1975 E
Ammomarginulina ensis Wiesner, 1931
Ammotium directum (Cushman & Brönnimann, 1948)*
Ammotium fragile Warren, 1957
Eratidus foliaceus (Brady, 1881)
PLACOPSILINIDAE
Ammocibicoides notalnus Saidova, 1975 E
Placopsilina cenomana d'Orbigny, 1850#
LOFTUSIOIDEA
CYCLAMMINIDAE
Alveolophragmium zealandicum Vella, 1957 E
Cyclammina cancellata Brady, 1879#
Cyclammina rotundidorsata (Hantken, 1876)#
Cyclammina pusilla Brady, 1884
Cyclammina trullissata Brady, 1879
RZEHAKINOIDEA
RZEHAKINIDAE
Miliammina fusca (Brady,1870)
Miliammina obliqua Heron-Allen & Earland, 1930
SPIROPLECTAMMINOIDEA
NOURIIDAE
Nouria harrisii Heron-Allen & Earland, 1914#
Nouria polymorphinoides Heron-Allen & Earland, 1914
PSEUDOBOLIVINIDAE
Parvigenerina arenacea Heron-Allen & Earland, 1922*
Parvigenerina heronalleni Seiglie, 1964#
Parvigenerina inflata arenacea (Heron-Allen & Earland, 1922)# E
Pseudobolivina antarctica Wiesner, 1931
SPIROPLECTAMMINIDAE
Spiroplectammina biformis (Parker & Jones, 1865)
Spiroplectammina carteri Hayward, 2010
Spiroplectella earlandi Barker, 1960#
Spirotextularia fistulosa (Brady, 1884)
Vulvulinoides benignus Saidova, 1975# E
VERNEUILINOIDEA
VERNEUILINIDAE
Gaudryina anaticula Saidova, 1975# E
Gaudryina convexa (Karrer, 1865)
Gaudryina ferruginea Heron-Allen & Earland, 1922# E
Gaudryina quadrangularis Bagg, 1908#

Order TROCHAMMINIDA
TROCHAMMINOIDEA
TROCHAMMINIDAE
Alterammina alternans (Earland, 1934)
Jadammina macrescens (Brady, 1870)
Paratrochammina bartrami (Hedley, Hurdle & Burdett, 1967) E
Paratrochammina challengeri (Brönnimann & Whittaker, 1988)
Paratrochammina simplissima (Cushman & McCulloch, 1939)
Polystomammina nitida (Brady, 1881)
Portatrochammina bipolaris Brönnimann & Whittaker, 1980
Portatrochammina sorosa (Parr, 1950)
Pseudotrochammina dehiscens (Frerichs, 1969)
Rotaliammina adaperta Rhumbler, 1938
Rotaliammina cf. *mayori* Cushman, 1924#
Rotaliammina ochracea (Williamson, 1858)
Rotaliammina sigmoidea Wells, 1985 E
Tritaxis challengeri (Hedley, Hurdle & Burdett, 1964)
Tritaxis fusca (Williamson, 1858)#
Trochammina cf. *curvativa* Saidova, 1975
Trochammina inflata (Montagu, 1808)
Trochammina? moniliformis Heron-Allen & Earland, 1922# E
Trochammina multiloculata Höglund, 1947
Trochammina nana (Brady, 1881)#
Trochammina pusilla Höglund, 1947
Trochammina tasmanica Parr, 1950
Trochammina? uviformis Grzybowski, 1901#
Trochamminella conica (Parker & Jones, 1865)
Trochamminopsis xishaensis (Zheng, 1988)

Order TEXTULARIIDA
TEXTULARIOIDEA
EGGERELLIDAE
Dorothia scabra (Brady, 1884)
Eggerella advena (Cushman, 1922)#
Eggerella bradyi (Cushman, 1911)
Eggerelloides scaber (Williamson, 1858)
Karreriella albida Saidova, 1975# E
Karreriella bradyi (Cushman, 1911)
Karreriella novangliae (Cushman, 1922)
Karrerulina conversa (Grzybowski, 1901)
Martinottiella communis (d'Orbigny, 1826)
Martinottiella omnia Saidova, 1975
Multifidella nodulosa (Cushman 1927)*
Verneuilinulla propinqua (Brady, 1884)#
TEXTULARIIDAE
Bigenerina nodosaria d'Orbigny, 1826#
Planctostoma luculenta (Brady, 1884)
Siphotextularia aperturalis (Cushman 1911)
Siphotextularia blacki Vella, 1957 E
Siphotextularia flintii (Cushman, 1911)
Siphotextularia foliosa Zheng, 1988
Siphotextularia fretensis Vella 1957 E
Siphotextularia mestayerae Vella 1957 E
Siphotextularia rolshauseni Phleger & Parker, 1951
Siphotextularia sp. indet.
Spiroplectinella proxispira (Vella,1957) E
Textularia agglutinans d'Orbigny, 1839
Textularia candeiana d'Orbigny, 1839
Textularia earlandi Parker, 1952
Textularia lythostrota (Schwager, 1866)
Textularia monstrata Saidova, 1975# E
Textularia pseudogramen Chapman & Parr, 1937
Textularia stricta Cushman, 1911
Textularia subantarctica Vella, 1957 E
Textularia torquata Parker, 1952
VALVULINIDAE
Cylindroclavulina bradyi (Cushman, 1911)
Pseudoclavulina obscura (Chaster, 1892)#
Pseudoclavulina serventyi (Chapman & Parr, 1935)*

Order SPIRILLINIDA
Suborder SPIRILLININA
PATELLINIDAE
Patellina corrugata Williamson, 1858
Patellinoides conica Heron-Allen & Earland, 1932*
PLANISPIRILLINIDAE
Planispirillina denticulata (Brady, 1884)
SPIRILLINIDAE
Mychostomina peripora Zheng, 1988
Mychostomina revertens (Rhumbler, 1906)
Spirillina decorata Brady, 1884#
Spirillina denticulogranulata Chapman, 1907
Spirillina inaequalis Brady, 1879#
Spirillina margaritifera Williamson, 1858#
Spirillina novaezealandiae Chapman, 1909 E
Spirillina obconica Brady, 1879#
Spirillina aff. *selseyensis* Heron-Allen & Earland, 1909
Spirillina aff. *tuberculata* Brady, 1879#
Spirillina vivipara Ehrenberg, 1843

Order CARTERINIDA
CARTERINIDAE
Carterina spiculotesta (Carter, 1877)

Order MILIOLIDA
Suborder MILIOLINA
CORNUSPIROIDEA
CORNUSPIRIDAE
Cornuspira carinata (Costa, 1856)
Cornuspira expansa Chapman, 1915
Cornuspira foliacea (Philippi, 1844)
Cornuspira lacunosa (Brady, 1884)
Cornuspira planorbis Schultze, 1853
Cornuspirella diffusa (Heron-Allen & Earland, 1913)
MILIOLOIDEA
HAUERINIDAE
Cribromiliolinella subvalvularis (Parr, 1950)#
Flintoides labiosa (d'Orbigny, 1839)
Hauerina fragilissima (Brady, 1884)#
Hauerina pacifica Cushman, 1917
Massilina brodiei Hedley, Hurdle & Burdett, 1967 E
Massilina granulocostata (Germeraad, 1946)
Massilina secans (d'Orbigny, 1826)#
Miliolinella schauinslandi (Rhumbler, 1906)
Miliolinella subrotunda (Montagu, 1803)
Miliolinella vigilax Vella, 1957 E
Nevillina coronata (Millett, 1898)
Nummoloculina contraria (d'Orbigny, 1846)
Parrina bradyi (Millett, 1898)#
Pseudoflintina triquetra (Brady, 1879)
Pyrgo clypeata (d'Orbigny, 1846)
Pyrgo comata (Brady, 1881)
Pyrgo denticulata (Brady, 1884)
Pyrgo depressa (d'Orbigny, 1826)
Pyrgo elongata (d'Orbigny, 1826)
Pyrgo inornata (d'Orbigny, 1846)
Pyrgo laevis Defrance, 1824#
Pyrgo lucernula (Schwager, 1866)#
Pyrgo murrhina (Schwager, 1866)
Pyrgo notalna Saidova, 1975 E#
Pyrgo oligocenica Cushman, 1935
Pyrgo ringens (Lamarck, 1804)
Pyrgo serrata (Bailey, 1862)
Pyrgo tasmanensis Vella, 1957 E
Pyrgo williamsoni (Silvestri, 1923)
Pyrgoella irregularis (d'Orbigny, 1839)#
Pyrgoella sphaera (d'Orbigny, 1839)
Quinqueloculina agglutinans d'Orbigny, 1839
Quinqueloculina auberiana d'Orbigny, 1839
Quinqueloculina bicornis (Walker & Jacob, 1798)
Quinqueloculina bicostatensis Saidova, 1975 E
Quinqueloculina bicostoides Vella, 1957 E
Quinqueloculina boueana d'Orbigny, 1846
Quinqueloculina carinatastriata (Wiesner, 1923)
Quinqueloculina cooki Vella, 1957 E
Quinqueloculina delicatula Vella, 1957 E
Quinqueloculina incisa Vella, 1957 E
Quinqueloculina n.sp. aff. *intricata* Terquem, 1878* E
Quinqueloculina kapitiensis Vella, 1957 E
Quinqueloculina neosigmoilinoides Kennett, 1966 E
Quinqueloculina notalnella Saidova, 1975# E
Quinqueloculina oblonga (Montagu, 1803)
Quinqueloculina parkeri (Brady, 1881)* K
Quinqueloculina parvaggluta Vella, 1957 E
Quinqueloculina rebeccae Vella, 1957 E
Quinqueloculina seminula (Linnaeus, 1758)
Quinqueloculina suborbicularis d'Orbigny, 1826
Quinqueloculina subpolygona Parr, 1945
Quinqueloculina vellai Saidova, 1975 E
Quinqueloculina venusta Karrer, 1868
Sigmoilina edwardsi (Schlumberger, 1887)#
Sigmoilina laevigata Saidova, 1975 E
Sigmoilina ovata Sidebottom, 1904#
Sigmoilina sigmoidea (Brady, 1884)#
Sigmoilopsis celata (Costa, 1855)#
Sigmoilopsis elliptica (Galloway & Wissler, 1927)
Sigmoilopsis finlayi Vella, 1957 E
Sigmoilopsis schlumbergeri (Silvestri, 1904
Sigmoilopsis wanganuiensis Vella, 1957 E
Siphonaperta crassa Vella, 1957 E
Siphonaperta macbeathi Vella, 1957 E
Spirosigmoilina pusilla (Earland, 1934)
Spirosigmoilina tenuis (Czjzek, 1848)
Triloculina brongniartii d'Orbigny, 1826#
Triloculina chrysostoma (Chapman, 1909) E
Triloculina idae Vella, 1957 E
Triloculina insignis (Brady, 1881)
Triloculina striatotrigonula Parr, 1941
Triloculina tricarinata d'Orbigny, 1826
Triloculina trigonula (Lamarck, 1804)
Triloculinella hornibrooki (Vella, 1957) E
Triloculinella cf. *obliquinodus* Riccio, 1950*
Triloculinella politus Saidova, 1975 E#
Tubinella funalis (Brady, 1884)
Tubinella inornata (Brady, 1884)#
SORITIDAE
Sorites marginalis (Lamarck, 1816) K
SPIROLOCULINIDAE
Inaequalina disparilis (Terquem, 1878)
Nummulopyrgo globulus (Hofker, 1976)
Planispirinoides bucculentus (Brady, 1884)#
Planispirinoides b. placentiformis (Brady, 1884)#
Spiroloculina antillarum d'Orbigny, 1839 K
Spiroloculina carinata Fornasini, 1903
Spiroloculina communis Cushman & Todd, 1944
Spiroloculina novozealandica Cushman & Todd, 1944 E*
Spiroloculina subaequa McCulloch, 1977
NUBECULARIOIDEA
DISCOSPIRINIDAE
Discospirina italica (Costa, 1856)
FISCHERINIDAE
Fischerina antarctica (Chapman, 1909) E
Fischerina pellucida Millett, 1898
Fischerinella diversa McCulloch, 1977*
Planispirinella exigua (Brady, 1879)#
Wiesnerella auriculata (Egger, 1893)
OPHTHALMIDIIDAE
Cornuloculina inconstans (Brady, 1879)
Cornuloculina margaritifera (Heron-Allen & Earland, 1922)
Edentostomina cultrata (Brady, 1881)
NUBECULARIIDAE
Calcituba polymorpha Roboz, 1884
Cornuspiramia cf. *antillarum* (Cushman, 1928)#
Nodophthalmidium simplex (Cushman & Todd, 1947)
Nubecularia decorata Heron-Allen & Earland, 1915#
Nubecularia lucifuga Defrance, 1825#
Nubecularia tubulosa Heron-Allen & Earland, 1915#
Webbina sp. indet.*

Order LAGENIDA
NODOSARIOIDEA
NODOSARIIDAE
Botuloides pauciloculus Zheng, 1979*
Dentalina acuta (d'Orbigny, 1846)
Dentalina catenulata (Brady, 1884)#
Dentalina cuvieri (d'Orbigny, 1826)
Dentalina decepta Bagg, 1912
Dentalina mutabilis (Costa, 1855)#
Dentalina mutsui Hada, 1931
Enantiodentalina muraii Uchio, 1953*
Frondicularia annularis d'Orbigny, 1846#
Frondicularia cf. *californica* Cushman & McCulloch, 1950 #
Frondicularia compta Brady, 1879
Frondicularia scottii Heron-Allen & Earland, 1922 E#
Glandulonodosaria ittai Loeblich & Tappan, 1953
Grigelis neopyrula McCulloch, 1981*
Grigelis orectus Loeblich & Tappan, 1994
Grigelis semirugosa (d'Orbigny, 1846)*
Laevidentalina advena (Cushman, 1923)
Laevidentalina aphelis Loeblich & Tappan, 1987
Laevidentalina ariena (Patterson & Pettis, 1986)*
Laevidentalina badenensis (d'Orbigny, 1846)*
Laevidentalina communiensis (Saidova, 1975) E#
Laevidentalina elegans (d'Orbigny, 1846)*
Laevidentalina guttifera (d'Orbigny, 1846)*
Laevidentalina haueri Neugeboren, 1856*
Laevidentalina inornata (d'Orbigny, 1846)
Laevidentalina mucronata (Neugeboren, 1856)#
Laevidentalina notalnella (Saidova, 1975) E
Laevidentalina obliquensis (Saidova, 1975) E#
Laevidentalina sidebottomi (Cushman, 1933)*
Lingulina grandis Cushman, 1917
Mucronina n.sp. aff. *advena* (Cushman, 1923)*
Neolingulina veijoensis McCulloch, 1977*
Nodosaria inflexa Reuss, 1866
Nodosaria nebulosa (Ishizaki, 1943)
Nodosaria pellita Heron-Allen & Earland, 1922 E
Nodosaria simplex Silvestri, 1872
Nodosaria subsoluta (Cushman, 1923)
Pseudolingulina bassensis (Parr, 1950)*
Pseudolingulina bradii (Silvestri, 1903)#
Pseudonodosaria brevis (d'Orbigny, 1846)*
Pseudonodosaria comatula (Cushman, 1923)
Pyramidulina luzonensis (Cushman, 1921)
Pyramidulina pauciloculata (Cushman & Grey, 1946)
Pyramidulina perversa (Schwager, 1866)
Pyramidulina raphanistrum (Linnaeus, 1758)#
VAGINULINIDAE
Amphicoryna georgechapronierei Yassini & Jones, 1995
Amphicoryna hirsuta (d'Orbigny, 1826)
Amphicoryna h. f. *sublineata* (Brady, 1884)
Amphicoryna proxima (Silvestri, 1872)#
Amphicoryna scalaris (Batsch, 1791)
Amphicoryna separans (Brady, 1884)
Astacolus crepidulus (Fichtel & Moll, 1798)
Astacolus insolitus (Schwager, 1866)
Astacolus neolatus Vella, 1957 E
Astacolus vellai Saidova, 1975 E
Hemirobulina angistoma (Stache, 1864)
Hemirobulina hydropica (Hornibrook, 1961) E
Hyalinonetrion gracillima (Seguenza, 1862)
Lagena acuticosta Reuss, 1862#
Lagena alticostata Cushman, 1913 #
Lagena authentica McCulloch, 1977
Lagena blomaeformis Yassini & Jones, 1995
Lagena chasteri Millet, 1901

Lagena clavata (d'Orbigny, 1846) #
Lagena costata (Williamson, 1858)
Lagena crenata Parker & Jones, 1865
Lagena cf. *doveyensis* Haynes, 1973
Lagena eccentrica Sidebottom, 1912
Lagena flatulenta Loeblich & Tappan, 1953
Lagena foveolata Reuss, 1863#
Lagena gibbera Buchner, 1940
Lagena hertwigiana undulata Sidebottom, 1912#
Lagena hispida Reuss, 1858
Lagena hispidula Cushman, 1913
Lagena interrupta Williamson, 1848
Lagena cf. *koreana* McCulloch, 1977
Lagena laevicostata Cushman & Gray, 1946
Lagena laevicostatiformis McCulloch, 1981
Lagena laevis (Montagu, 1803)
Lagena l. baggi Cushman & Gray, 1946
Lagena l. distoma Silvestra, 1900#
Lagena lyellii (Seguenza, 1862)
Lagena nebulosa (Cushman, 1923)
Lagena oceanica Albani, 1974
Lagena orbignyana clathrata Brady, 1884
Lagena paradoxa Sidebottom, 1912
Lagena peculiariformis Albani & Yassini, 1995
Lagena cf. *peculiaris* (Cushman & McCulloch, 1950)
Lagena peterroyi Yassini & Jones, 1995
Lagena plumigera Brady, 1881
Lagena protea Chaster, 1892#
Lagena semistriata (Williamson,1848)#
Lagena sphaerula Silvestri, 1902#
Lagena spicata (Cushman & McCulloch, 1950)
Lagena spiratiformis McCulloch, 1981
Lagena spumosa Millett, 1902#
Lagena striata (d'Orbigny, 1839) #
Lagena s. strumosa Reuss, 1858
Lagena substriata Williamson, 1848#
Lagena sulcata (Walker & Jacob, 1798)
Lagena tokiokai Uchio, 1962
Lagena tubulata Sidebottom, 1912
Lagena spp. indet. (2)
Lenticulina anaglypta (Loeblich & Tappan, 1987)
Lenticulina antarctica Parr, 1950
Lenticulina asymetrica (Saidova, 1975) E#
Lenticulina australis (Parr, 1950)
Lenticulina calcar (Linnaeus, 1767)
Lenticulina cultrata (Montfort, 1808)
Lenticulina denticulifera (Cushman, 1913)
Lenticulina erratica Hornibrook, 1961 E
Lenticulina foliata (Stache, 1864)*
Lenticulina formosa (Cushman, 1923)*
Lenticulina gibba (d'Orbigny, 1826)
Lenticulina limbosa (Reuss, 1863)
Lenticulina orbicularis (d'Orbigny, 1826)
Lenticulina subgibba Parr, 1950
Lenticulina submamilligera (Cushman, 1917)
Lenticulina suborbicularis Parr, 1950
Lenticulina tasmanica Parr, 1950
Lenticulina thalmanni (Hessland, 1943)#
Lenticulina sp. indet.*
Marginulina augensiensis Saidova, 1975#
Marginulina costata (Batsch, 1791)#
Marginulina gummi Saidova, 1975# E
Marginulina obesa (Cushman, 1923)
Marginulina striata d'Orbigny, 1852
Marginulinopsis bradyi (Goës, 1894)
Marginulinopsis tenuis (Bornemann, 1855)
Neolenticulina variabilis (Reuss, 1850)
Planularia spinipes (Cushman, 1913)
Planularia magnifica falciformis Thalmann, 1937
Procerolagena attonita (McCulloch, 1977)*
Procerolagena distomargaritifera (Rymer Jones 1872)
Procerolagena elongata (Ehrenberg, 1844)
Procerolagena gracilis (Williamson 1848)#
Procerolagena meridionalis (Wiesner, 1931)
Procerolagena multilatera (McCulloch, 1977)
Saracenaria altifrons (Parr, 1950)
Saracenaria italica Defrance, 1824
Saracenaria latifrons (Brady, 1884)
Saracenaria spinosa Eichenberg, 1935
Vaginulina inflata Parr, 1950
Vaginulina spinigera Brady, 1881
Vaginulina vertebralis Parr, 1932
Vaginulinopsis gnamptina Loeblich & Tappan, 1994
Vaginulinopsis reniformis (d'Orbigny, 1846)
Vaginulinopsis sublegumen Parr, 1950*
Vaginulinopsis tasmanica Parr, 1950
POLYMORPHINOIDEA
ELLIPSOLAGENIDAE
Bifarilaminella advena (Cushman, 1923)
Cushmanina desmophora (Jones, 1872)#
Cushmanina striatopunctata (Parker & Jones, 1865)*
Exsculptina eccentrica (Sidebottom, 1912)
Exsculptina exsculpta (Brady, 1881)
Favulina favosopunctata (Brady, 1881)
Favulina hexagona (Williamson, 1848)
Favulina hexagoniformis (McCulloch, 1997)*
Favulina melo (d'Orbigny, 1839)
Favulina melosquamosa (McCulloch, 1977)*
Favulina squamosa (Montagu, 1803)
Favulina vadosa (McCulloch, 1977)*
Fissurina aligeria caudimarginata McCulloch, 1977
Fissurina alveolata alveolata (Brady, 1884)#
Fissurina a. separans (Sidebottom, 1912)#
Fissurina annectens (Burrows & Holland, 1895)
Fissurina antiqua Yassini & Jones, 1995
Fissurina apiculata punctulata (Sidebottom, 1912)
Fissurina armatum (Thompson, 1956)#
Fissurina auriculata (Brady, 1881)#
Fissurina auriculata ssp. indet. (Sidebottom 1913)
Fissurina auriglobosa (McCulloch, 1977)
Fissurina balteata McCulloch, 1977
Fissurina biancae Seguenza, 1862#
Fissurina bicarinata villosa (Heron-Allen & Earland, 1922)# E
Fissurina bispinosa (Heron-Allen & Earland, 1932)*
Fissurina biumbonata McCulloch, 1977
Fissurina circularis Todd, 1954
Fissurina clathrata (Brady, 1884)
Fissurina clypeatomarginata crassa (Sidebottom, 1912)
Fissurina compressiformis McCulloch 1977
Fissurina contusa colomboensis McCulloch, 1977
Fissurina crassiannulata Collins, 1973
Fissurina cf. *crassiformis* McCulloch, 1981
Fissurina crassiporosa McCulloch, 1977
Fissurina crucifera McCulloch, 1977
Fissurina cf. *curvitubulosa* McCulloch, 1977
Fissurina danica (Madsen, 1895)#
Fissurina d. pendulum (Heron-Allen & Earland, 1922) E
Fissurina earlandi Parr, 1950
Fissurina enderbiensis (Chapman, 1909) E
Fissurina cf. *eumarginata limpida* McCulloch, 1977
Fissurina evoluta McCulloch, 1977
Fissurina e. laticlava McCulloch, 1977
Fissurina evolutiformis McCulloch, 1977
Fissurina evolutiquetra McCulloch, 1977
Fissurina fasciata (Egger, 1857)#
Fissurina f. carinata (Sidebottom, 1906)
Fissurina f. faba (Balkwill & Millet, 1884)#
Fissurina fimbriata (Brady, 1881)#
Fissurina f. occulosa (Sidebottom, 1912)#
Fissurina aff. *fissicarinata* Parr, 1950
Fissurina formosa (Schwager, 1866)#
Fissurina aff. *formosa* (Schwager, 1866) #
Fissurina globosocaudata Albani & Yassini, 1995
Fissurina cf. *infera lipposa* McCulloch, 1977
Fissurina cf. *infimabrocha* Loeblich & Tappan, 1994
Fissurina labiosa McCulloch, 1977
Fissurina lacunata (Burrows & Holland, 1895)#
Fissurina laevigata Reuss, 1850
Fissurina lagenoides (Williamson, 1858)#
Fissurina l. nuda Chapman, 1909 E
Fissurina laureata (Heron-Allen & Earland, 1932)
Fissurina lauretiformis McCulloch, 1977
Fissurina longispina (Brady, 1881)
Fissurina longula McCulloch, 1977*
Fissurina lucida (Williamson, 1848)
Fissurina malcolmsonii (Wright, 1911)
Fissurina margaritiana McCulloch, 1981
Fissurina marginata (Montagu, 1803)
Fissurina m. elegans (Sidebottom, 1912)#
Fissurina metaporosa McCulloch, 1977
Fissurina novaezelandiae (Thompson, 1956)#
Fissurina nudiformis McCulloch, 1977
Fissurina omniperforata McCulloch, 1977
Fissurina opaca McCulloch, 1977
Fissurina orbignyana Seguenza, 1862
Fissurina o. selseyensis (Heron-Allen & Earland, 1909)#
Fissurina o. unicostata (Sidebottom, 1912)
Fissurina o. variabilis (Wright, 1891)
Fissurina o. walleriana (Wright, 1891)#
Fissurina ornata (Williamson, 1858)#
Fissurina pulchella (Brady, 1867) #
Fissurina p. hexagona (Heron-Allen & Earland, 1916)#
Fissurina prolata (McCulloch, 1977)
Fissurina pyrum (Thompson, 1956)#
Fissurina quadrata (Williamson, 1858)
Fissurina q. carinata (Chapman, 1909) E
Fissurina quadriporosa McCulloch, 1977*
Fissurina revertens (Heron-Allen & Earland, 1932)
Fissurina rizzae Seguenza, 1862#
Fissurina sacculiformis Jones, 1984
Fissurina semialata (Balkwill & Millett, 1884)
Fissurina semimarginata (Reuss, 1870)
Fissurina seminiformis (Schwager, 1866)#
Fissurina solidata McCulloch, 1977
Fissurina cf. *southbayensis* McCulloch, 1977
Fissurina spectabile (Thompson, 1956)#
Fissurina spinosa (Sidebottom, 1912)
Fissurina spinulata McCulloch, 1977
Fissurina squamosomarginata (Parker & Jones, 1865)
Fissurina staphyllearia Schwager, 1866
Fissurina cf. *striolata* (Sidebottom, 1912)
Fissurina submarginata (Boomgart, 1949)#
Fissurina takapuniense (Thompson, 1956)#
Fissurina tobagoensis McCulloch, 1981
Fissurina varum (Thompson, 1956)#
Fissurina wiesneri Barker, 1960
Fissurina yokoyamae (Millett, 1895)#
Fissurina spp. indet. (20)
Galwayella trigonomarginata (Parker & Jones, 1865)*
Galwayella trigonoorbignyana (Balkwill & Millett, 1884)
Galwayella sp. indet.*
Homalohedra liratiformis (McCulloch, 1977)
Laculatina quadrilatera (Earland, 1934)*
Lagenosolenia anteroalatiformis Albani & Yassini, 1995
Lagenosolenia bermaguiensis Yassini & Jones, 1995
Lagenosolenia cf. *bicariniformis* McCulloch, 1977*
Lagenosolenia bilagenoides McCulloch, 1977
Lagenosolenia confossa McCulloch, 1977
Lagenosolenia cf. *confossa* McCulloch, 1977
Lagenosolenia cf. *dubiosa* McCulloch, 1977
Lagenosolenia elliptica Yassini & Jones, 1995
Lagenosolenia eucerviculata McCulloch, 1977
Lagenosolenia exquisita McCulloch, 1977
Lagenosolenia falcata (Chaster, 1892)*
Lagenosolenia habrotes McCulloch, 1977
Lagenosolenia incomposita Patterson & Pettis, 1986
Lagenosolenia ineffecta McCulloch, 1977

Lagenosolenia marginatoperforata (Seguenza, 1880)
Lagenosolenia pacifica (Parr, 1950)
Lagenosolenia cf. *pressa* McCulloch, 1977
Lagenosolenia radiata striatula (Cushman, 1913)#
Lagenosolenia sanpedroensis (McCulloch, 1977)
Lagenosolenia scintillans (McCulloch, 1977)
Lagenosolenia sigmoidella timmsensis (Cushman & Gray, 1946)
Lagenosolenia squamosoalata (Brady, 1881)*
Lagenosolenia strigimarginata Loeblich & Tappan, 1994
Lagenosolenia tenuistriatiformis McCulloch, 1977*
Lagenosolenia aff. *wollongongensis* Yassini & Jones, 1994*
Lagenosolenia 13 spp.
Lagnea honshuensis (McCulloch, 1977)*
Lagnea neosigmoidella (McCulloch, 1977)
Oolina ampulladistoma (Jones, 1872)#
Oolina ampulliformis McCulloch, 1981
Oolina ampullineata McCulloch, 1977
Oolina apiopleura (Loeblich & Tappan, 1953)
Oolina borealis Loeblich & Tappan, 1954
Oolina botelliformis (Brady, 1881)#
Oolina caudigera (Wiesner, 1931)
Oolina cf. *cervicata* McCulloch, 1977
Oolina collaripolygonata Albani & Yassini, 1989*
Oolina aff. *confluenta* McCulloch, 1977*
Oolina costata (Williamson, 1958)
Oolina emaciata (Reuss, 1863)
Oolina felsinea (Fornasini, 1894)
Oolina globosa (Montagu, 1803)
Oolina g. lineatopunctata (Heron-Allen & Earland, 1922)#
Oolina globosa ssp. (Montagu 1803)
Oolina heteromorpha Parr, 1950
Oolina cf. *intercalata* Jones, 1984
Oolina lineata (Williamson, 1848)
Oolina ovum (Ehrenberg, 1843)#
Oolina piriformis Yassini & Jones, 1995
Oolina scalariformis McCulloch, 1977
Oolina setosa (Earland, 1934)
Oolina stelligera (Brady, 1881)
Oolina striatopunctata (Parker & Jones, 1865)#
Oolina tasmanica Parr, 1950
Oolina variata (Brady, 1881)#
Oolina williamsoni (Alcock, 1865)#
Oolina spp. indet. (7)
Palliolatella aradisiformis Albani & Yassini, 1989
Palliolatella bradyformis McCulloch, 1977*
Palliolatella lacunata paucialveolata Albani & Yassini, 1989
Palliolatella quadrirevertens (McCulloch, 1977)
Palliolatella sp. indet.
Parafissurina acarinata McCulloch, 1977
Parafissurina biconicoformis McCulloch, 1977
Parafissurina caledoniana McCulloch, 1981
Parafissurina curta Parr, 1950
Parafissurina cf. *deuterolegna* Loeblich & Tappan, 1994
Parafissurina discalis McCulloch, 1977
Parafissurina faceta McCulloch, 1977
Parafissurina inaequilateralis (Wright, 1886)#
Parafissurina lata (Wiesner, 1931)
Parafissurina lateralis (Cushman, 1913)
Parafissurina laticaudata McCulloch, 1977
Parafissurina limpidiformis McCulloch, 1977
Parafissurina marginoradiata McCulloch, 1977*
Parafissurina neocurta McCulloch, 1977*
Parafissurina paulula McCulloch, 1977
Parafissurina procidua McCulloch, 1977
Parafissurina reniformis (Sidebottom, 1913)*
Parafissurina schlichti (Silvestri, 1902)#
Parafissurina cf. *semidevestiva* McCulloch, 1977
Parafissurina sublata Parr, 1950
Parafissurina subovata (Parr, 1950)*
Parafissurina ventricosa (Silvestri, 1904)
Parafissurina wiesneri Parr, 1950
Parafissurina spp. indet. (8)
Pseudofissurina mccullochae Jones, 1994
Pseudofissurina metaconica (McCulloch, 1977)
Pseudofissurina unicostata (Sidebottom, 1912)
Reussoolina strangeri Loeblich & Tappan, 1994
Vasicostella rara (McCulloch, 1977)
Ventrostoma scaphaeformis (Parr, 1950)
Ventrostoma unguis (Heron-Allen & Earland, 1913)
Wiesnerina baccata (Heron-Allen & Earland, 1922) E
Wiesnerina scarabaeus (Heron-Allen & Earland, 1922) E
GLANDULINIDAE
Entolingulina sp. indet.#
Entomorphinoides cf. *karenae* McCulloch, 1977
Entomorphinoides opposita McCulloch, 1977
Glandulina ovula d'Orbigny, 1846
Glandulina sp. indet.
Laryngosigma cf. *compacta* McCulloch, 1977
Laryngosigma hyalascidia Loeblich & Tappan, 1953#
Laryngosigma williamsoni (Terquem, 1878)#
Seabrookia pellucida Brady, 1890#
POLYMORPHINIDAE
Francuscia extensa (Cushman, 1923)*
Globulina gibba (d'Orbigny, 1826)#
Globulina inaequalis Reuss, 1850
Globulina minuta (Roemer, 1838)
Globulina rotundata (Bornemann, 1855)#
Guttulina austriaca d'Orbigny, 1846#
Guttulina bartschi Cushman & Ozawa, 1930
Guttulina communis (d'Orbigny, 1826)*
Guttulina irregularis (d'Orbigny, 1846)#
Guttulina ovata (d'Orbigny, 1826)
Guttulina regina (Brady, Parker & Jones, 1870)
Guttulina silvestrii (Cushman & Ozawa, 1930)
Guttulina yabei Cushman & Ozawa, 1929
Lingulosigmomorphina sanata Saidova, 1975# E
Polymorphinella executa Saidova, 1975# E
Pseudopolymorphina cf. *australis* (d'Orbigny, 1839)
Pseudopolymorphina tortuosa Vella, 1957 E
Pyrulina angusta (Egger, 1857)#
Pyrulina cylindroides (Roemer, 1838)#
Pyrulina fusiformis (Roemer, 1838)*
Pyrulina gutta (d'Orbigny, 1839)
Ramulina globulifera Brady, 1879
Ramulina laevis Jones, 1875#
Sigmoidella elegantissima (Parker & Jones, 1865)
Sigmoidella cf. *kagaensis* Cushman & Ozawa, 1928
Sigmoidella pacifica Cushman & Ozawa, 1928
Sigmoidella silvestrii Cushman & Ozawa, 1930#
Sigmomorphina lacrimosa Vella, 1957 E
Sigmomorphina rhomboidalis Vella, 1957 E#

Order ROBERTINIDA
CERATOBULIMINOIDEA
CERATOBULIMINIDAE
Ceratobulimina jonesiana (Brady, 1881)*#
Lamarckina haliotidea (Heron-Allen & Earland, 1911)#
Lamarckina cf. *tuberculata* (Balkwill & Wright, 1885)#
EPISTOMINIDAE
Hoeglundina elegans (d'Orbigny, 1826)
ROBERTINOIDEA
ROBERTINIDAE
Cerobertina tenuis (Chapman & Parr, 1937)#
Robertina subcylindrica (Brady, 1881)#
Robertinoides bradyi (Cushman & Parker, 1936)#
Robertinoides oceanicus (Cushman & Parker, 1947)

Order GLOBIGERINIDA
GLOBIGERINOIDEA
GLOBIGERINIDAE
Beella digitata Brady, 1879
Globigerina bulloides d'Orbigny, 1826
Globigerina calida Parker, 1962
Globigerina falconensis Blow, 1959
Globigerina humilis (Brady, 1884)
Globigerina quinqueloba Natland, 1938
Globigerinella aequilateralis (Brady, 1879)
Globigerinoides conglobatus (Brady, 1879)
Globigerinoides ruber (d'Orbigny, 1839)
Globigerinoides sacculifer (Brady, 1877)
Orbulina universa d'Orbigny, 1839
Sphaeroidinella dehiscens (Parker & Jones, 1865)
HASTERIGINIDAE
Hastigerina pelagica (d'Orbigny, 1839)
GLOBOROTALIOIDEA
CANDEINIDAE
Candeina nitida d'Orbigny, 1839
Globigerinita glutinata (Egger, 1893)
Globigerinita iota Parker, 1962
Globigerinita uvula (Ehrenberg, 1861)
GLOBOROTALIIDAE
Globorotalia bermudezi Rogl & Bolli, 1973
Globorotalia cavernula Bé, 1967
Globorotalia crassula Cushman & Stewart, 1930
Globorotalia hirsuta (d'Orbigny, 1839)
Globorotalia inflata (d'Orbigny, 1839)
Globorotalia scitula (Brady, 1882)
Globorotalia truncatulinoides (d'Orbigny, 1839)
Globorotalia tumida (Brady, 1877)
Neogloboquadrina incompta (Cifelli, 1961)
Neogloboquadrina dutertrei (d'Orbigny, 1839)
Neogloboquadrina pachyderma (Ehrenberg, 1861)
Pulleniatina obliquiloculata (Parker & Jones, 1862)

Order BULIMINIDA
BOLIVINITOIDEA
BOLIVINITIDAE
Abditodentrix pseudothalmanni (Boltovskoy & Guissani de Kahn, 1981)
Bolivinita quadrilatera (Schwager, 1866)
BOLIVINOIDEA
BOLIVINIDAE
Bolivina alata (Seguenza, 1862)
Bolivina arta MacFadyen, 1931 E
Bolivina cacozela Vella, 1957 E
Bolivina compacta Sidebottom, 1905
Bolivina earlandi Parr, 1950
Bolivina glutinata Egger, 1893
Bolivina neocompacta McCulloch, 1981
Bolivina pseudolobata Yassini & Jones, 1995*
Bolivina pseudoplicata Heron-Allen & Earland, 1930
Bolivina pusilla Schwager, 1866
Bolivina pygmaea (Brady, 1881)
Bolivina robusta Brady, 1881
Bolivina seminuda (Cushman, 1911)
Bolivina spathulata (Williamson, 1858)
Bolivina striatula Cushman, 1922
Bolivina variabilis (Williamson, 1958)
BULIMINOIDEA
BULIMINELLIDAE
Buliminella elegantissima (d'Orbigny, 1839)
BULIMINIDAE
Bulimina elongata d'Orbigny, 1826
Bulimina exilis Brady, 1884
Bulimina gibba Fornasini, 1902
Bulimina marginata f. *acaenapeza* Loeblich & Tappan, 1994
Bulimina m. f. *acanthia* Costa, 1856
Bulimina m. f. *aculeata* d'Orbigny, 1826
Bulimina m. f. *marginata* d'Orbigny, 1826
Bulimina striata d'Orbigny, 1826
Bulimina subornata Brady, 1884*
Bulimina truncana Gumbel, 1868
Globobulimina pacifica Cushman, 1927
Globobulimina turgida (Bailey, 1851)

Praeglobobulimina pupoides (d'Orbigny, 1846)
Praeglobobulimina spinescens (Brady, 1884)
REUSSELLIDAE
Reussella spinulosa (Reuss, 1850)
SIPHOGENERINOIDIDAE
Euloxostoma bradyi (Asano, 1938)
Loxostomina costulata (Cushman, 1922)
Pseudobrizalina lobata (Brady, 1881)
Rectobolivina bifrons (Brady, 1881)
Saidovina karrierana (Brady, 1881)
Siphogenerina columellaris (Brady, 1881)
Siphogenerina dimorpha (Parker & Jones, 1865)
Siphogenerina raphana (Parker & Jones, 1865) A
Siphogenerina semistriata (Schubert, 1911)
Spiroloxostoma glabra Millett, 1903
UVIGERINIDAE
Neouvigerina ampullacea Brady, 1884
Neouvigerina hispida (Schwager, 1866)
Neouvigerina interrupta (Brady, 1879)
Neouvigerina proboscidea (Schwager, 1866)
Trifarina angulosa (Williamson, 1858)
Trifarina bradyi Cushman, 1923
Trifarina gracilis (Vella, 1957) E
Trifarina occidentalis (Cushman, 1923)
Trifarina pacifica (Albani, 1974)
Uvigerina bifurcata d'Orbigny, 1839
Uvigerina hollocki Thalmann, 1950#
Uvigerina hornibrooki Boersma, 1984
Uvigerina mediterranea Hofker, 1932*
Uvigerina peregrina Cushman, 1923
Uvigerina pygmaea d'Orbigny, 1826#
CASSIDULINOIDEA
CASSIDULINIDAE
Cassidulina angulosa surtida Saidova, 1975
Cassidulina carinata Silvestri, 1896
Cassidulina reniforme Norvang, 1945
Cassidulina planata Saidova, 1975 E
Cassidulina spiniferiformis McCulloch, 1977
Ehrenbergina aspinosa Parr, 1950
Ehrenbergina carinata Eade, 1967 E
Ehrenbergina aff. *crassitrigona* Nomura, 1999*
Ehrenbergina glabra Heron-Allen & Earland, 1922
Ehrenbergina hystrix Brady, 1881
Ehrenbergina mestayeri Cushman, 1922 E
Ehrenbergina trigona Goës, 1896
Evolvocassidulina belfordi Nomura, 1983
Evolvocassidulina bradyi (Norman, 1881)
Evolvocassidulina orientalis (Cushman, 1922) E
Evolvocassidulina tenuis (Phleger & Parker, 1951)
Favocassidulina australis Eade, 1967 E
Globocassidulina canalisuturata Eade, 1967 E
Globocassidulina crassa (d'Orbigny, 1839)
Globocassidulina decorata (Sidebottom, 1910)
Globocassidulina elegans (Sidebottom, 1910)
Globocassidulina gemma (Todd, 1954)
Globocassidulina minuta (Cushman, 1933)
Globocassidulina notalnella (Saidova, 1975) E
Globocassidulina producta (Chapman & Parr, 1937)
Globocassidulina spherica Eade, 1967 E
Globocassidulina subglobosa (Brady, 1881)
Globocassidulina tumida Heron-Allen & Earland, 1922 E
Islandiella cf. *smechovi* (Voloshinova, 1952)
Lernella inflata (Le Roy, 1944)
Paracassidulina sagamiensis Asano & Nakamura, 1937
DELOSINOIDEA
CAUCASINIDAE
Francesita advena (Cushman, 1922)
FURSENKOINOIDEA
FURSENKOINIDAE
Cassidella bradyi (Cushman, 1922)
Fursenkoina acuta (d'Orbigny, 1846)
Fursenkoina complanata (Egger, 1893)
Fursenkoina schreibersiana (Czjzek, 1848)
Rutherfordoides rotundata (Parr, 1950)
Sigmavirgulina tortuosa (Brady, 1881)
VIRGULINELLIDAE
Virgulinella fragilis Grindell & Collen, 1976 A
TURRILINOIDEA
STAINFORTHIIDAE
Virgulopsis turris (Heron-Allen & Earland, 1922) E

Order ROTALIIDA
ACERVULINOIDEA
ACERVULINIDAE
Acervulina inhaerens Schulze, 1854
Gypsina vesicularis (Parker & Jones, 1860)
Sphaerogypsina globulus (Reuss, 1848)
HOMOTREMATIDAE
Miniacina alba (Carter, 1877)
Miniacina miniacea (Pallas, 1766)
ASTERIGINOIDEA
AMPHISTEGINIDAE
Amphistegina papillosa Said, 1949
ASTERIGERINATIDAE
Eoeponidella pulchella (Parker, 1952)
CHILOSTOMELLOIDEA
ALABAMINIDAE
Svratkina australiensis (Chapman, Parr & Collins, 1934)
CHILOSTOMELLIDAE
Chilostomella oolina Schwager, 1878
Chilostomella ovoidea Reuss, 1850
GAVELLINELLIDAE
Discanomalina coronata (Parker & Jones, 1857)
Discanomalina semipunctata (Bailey, 1851)
Gyroidella planata Saidova, 1975#
Gyroidina danvillensis Howe & Wallace, 1932
Gyroidina kawagatai (Ujiié, 1995)
Gyroidina orbicularis (d'Orbigny, 1826)
Gyroidina soldanii d'Orbigny, 1826
Hanzawaia aff. *grossepunctata* (Earland, 1934)
Hanzawaia cf. *wilcoxensis* (Cushman & Pontin, 1932)
HETEROLEPIDAE
Anomalinoides colligerus (Chapman & Parr, 1937)
Anomalinoides glabratus (Cushman, 1924)
Anomalinoides cf. *nonionoides* (Parr, 1932)
Anomalinoides semicribratus (Beckmann, 1954)
Anomalinoides sphericus (Finlay, 1940) E
Anomalinoides tasmanicus (Parr, 1950)
Anomalinoides spp. indet. (3)
KARRERIIDAE
Karreria maoria (Finlay, 1939)
ORIDORSALIDAE
Oridorsalis umbonatus (Reuss, 1851)
OSANGULARIIDAE
Osangularia bengalensis (Schwager, 1866)
Osangularia culter (Parker & Jones, 1865)*
QUADRIMORPHINIDAE
Quadrimorphina laevigata (Phleger & Parker, 1951)
TRICHOHYALIDAE
Aubignyna perlucida (Heron-Allen & Earland, 1913)
DISCORBINELLOIDEA
DISCORBINELLIDAE
Colonimilesia coronata (Heron-Allen & Earland 1932)
Discorbinella bertheloti (d'Orbigny, 1839)
Discorbinella complanata (Sidebottom, 1918)
Discorbinella deflata (Finlay, 1940) E
Discorbinella subcomplanata (Parr, 1950)
Discorbinella timida Hornibrook, 1961
Discorbinella vitrevoluta Hornibrook, 1961 E
Discorbitina pustulata (Heron-Allen & Earland 1913)
Laticarinina altocamerata (Heron-Allen & Earland, 1922) E
Laticarinina pauperata (Parker & Jones, 1865)
PARRELLOIDIDAE
Parrelloides umbonatus Saidova, 1975 E#
PSEUDOPARRELLIDAE
Alexanderina viejoensis McCulloch, 1977
Eilohedra vitrea (Parker, 1953)
Epistominella exigua (Brady, 1884)
Nuttallides bradyi (Earland, 1934)
Nuttallides umbonifera (Cushman, 1933)
Planulinoides biconcavus (Parker & Jones, 1862)
Planulinoides norcotti Hedley, Hurdle & Burdett, 1967 E
DISCORBOIDEA
BAGGINIDAE
Baggina cf. *philippinensis* Loeblich & Tappan, 1994
Cancris oblongus (Williamson, 1858)
Rugidia simplex Collis, 1974
Valvulineria minuta (Schubert, 1904)
BRONIMANNIIDAE
Bronnimannia disparilis (Heron-Allen & Earland, 1922) E
DISCORBIDAE
Neoeponides schreibersii (d'Orbigny, 1846)
Trochulina dimidiatus (Jones & Parker, 1862)
EPONIDIDAE
Alabaminella weddellensis (Earland, 1936)
Eponides repandus (Fichtel & Moll, 1798)
Ioanella tumidula (Brady, 1884)
Poroeponides lateralis (Terquem, 1878)* K
Porogavelinella ujiiei Kawagata, 1999
HELENINIDAE
Helenina anderseni (Warren, 1957)
MISSISSIPPINIDAE
Stomatorbina concentrica (Parker & Jones, 1864)
Mississippina omuraensis Shuto, 1953
PLACENTULINIDAE
Patellinella inconspicua (Brady, 1884)
ROSALINIDAE
Gavelinopsis hamatus Vella, 1957 E
Gavelinopsis praegeri (Heron-Allen & Earland, 1913)
"Gavelinopsis" umbonifer (Parr, 1950)
Neoconorbina cavalliensis Hayward, Grenfell, Reid & Hayward, 1999
Neoconorbina concinna (Brady, 1884)
Neoconorbina pacifica Hofker, 1951
Neoconorbina terquemi (Rzehak, 1888)
Planodiscorbis rarescens (Brady, 1884)
Rosalina araucana d'Orbigny, 1839#
Rosalina bradyi (Cushman, 1915)
Rosalina irregularis (Rhumbler, 1906)
Rosalina paupereques Vella, 1957 E
Rosalina tofuana Saidova, 1975#
Rosalina vilardeboana d'Orbigny, 1839#
Rosalina vitrizea Hornibrook, 1961 E
Tretomphalus planus Cushman, 1924
SPHAEROIDINIDAE
Sphaeroidina bulloides d'Orbigny, 1826
GLABRATELLOIDEA
GLABRATELLIDAE
Conorbella clarionensis (McCulloch, 1977)
Conorbella pulvinata (Brady, 1884) K
Glabratella margaritaceus (Earland, 1933)
Glabratellina kermadecensis Hayward, Grenfell, Reid & Hayward, 1999 E K
Pileolina calcarata (Heron-Allen & Earland, 1922) E
Pileolina gracei Hayward, Grenfell, Reid & Hayward, 1999 E
Pileolina harmeri (Heron-Allen & Earland, 1922) E
Pileolina patelliformis (Brady, 1884)
Pileolina radiata Vella, 1957 E
Pileolina zealandica Vella, 1957 E
Planoglabratella opercularis (d'Orbigny, 1826)
HERONALLENIIDAE
Heronallenia gemmata Earland, 1934
Heronallenia laevis Parr, 1950
Heronallenia lingulata (Burrows & Holland, 1895)

Heronallenia nodulosa (McCulloch, 1977)
Heronallenia parva Parr, 1950*
Heronallenia polita Parr, 1950*
Heronallenia pulvinulinoides (Cushman, 1915)
Heronallenia translucens Parr, 1945
Heronallenia unguiculata Loeblich & Tappan, 1994
NONIONOIDEA
NONIONIDAE
Astrononion novozealandicum Cushman & Edward, 1937
Astrononion stelligerum (d'Orbigny, 1839)
Bermudezinella profunda (Saidova, 1975)*
Haynesina depressula (Walker & Jacob, 1798)
Laminonion tumidum (Cushman & Edwards, 1937)
Melonis affinis (Reuss, 1851)
Melonis pompilioides (Fichtel & Moll, 1798)
Melonis sphaeroides Voloshinova, 1958
Nonion pacificum (Cushman, 1924)
Nonionella auris (d'Orbigny, 1839)
Nonionella magnalingua Finlay, 1940 E
Nonionellina flemingi (Vella, 1957) E
Nonionoides grateloupi (d'Orbigny, 1826)
Nonionoides turgida (Williamson, 1858)
Pseudononion granuloumbilicatum Zheng, 1979
Pullenia bulloides (d'Orbigny, 1826)
Pullenia quinqueloba (Reuss, 1851)
Pullenia salisburyi Stewart & Stewart, 1930
Pulleniella asymmetrica Ujiié, 1990*
Subanomalina guadalupensis McCulloch, 1977
Zeaflorilus parri (Cushman, 1936) E
SPIROTECTINIDAE
Spirotectina crassa Saidova, 1975*
PLANORBULINOIDEA
CIBICIDAE
Cibicicoides fumeus Saidova, 1975# E
Cibicides bradyi (Trauth, 1918)
Cibicides cicatricosus (Schwager, 1866)
Cibicides corticatus Earland, 1934
Cibicides deliquatus Finlay, 1940
Cibicides dispars (d'Orbigny, 1839)
Cibicides fumeus (Saidova, 1975)
Cibicides grosseperforatus (Van Morkhoven & Berggren, 1986)
Cibicides lobatulus (Walker & Jacob, 1798)
Cibicides mundulus (Brady, Parker & Jones, 1888)
Cibicides neoperforatus Hornibrook, 1989
Cibicides pachyderma (Rzehak, 1886)
Cibicides planus Saidova, 1975*
Cibicides refulgens Montfort, 1808
Cibicides robertsonianus (Brady, 1881)
Cibicides subhaidingerii (Parr, 1950)
Cibicides temperata Vella, 1957 E
Cibicides tesnersianus (Saidova, 1975) E
Cibicides variabilis (d'Orbigny, 1826)
Cibicides vehemenus (Saidova, 1975)
Cibicides wuellerstorfi (Schwager, 1866)
Dyocibicides biserialis Cushman & Valentine, 1930#
Dyocibicides primitiva Vella, 1957# E
Dyocibicides uniserialis Thalmann, 1933#
CYMBALOPORIDAE
Cymbaloporetta bradyi (Cushman, 1924)
Cymbaloporetta squammosa (d'Orbigny, 1826)#
PLANORBULINIDAE
Planorbulina acervalis Brady, 1884
Planorbulina mediterranensis d'Orbigny, 1826#
PLANULINIDAE
Hyalinea asiana Huang, 1972
Planulina ariminensis d'Orbigny, 1826
Planulina renzi Cushman & Stainforth, 1945
Planulina sinuosa (Sidebottom, 1918)
Planulina aff. *subinflata* Bandy, 1949
VICTORIELLIDAE
Biarritzina proteiforma (Goës, 1882)
Carpentaria monticularis Carter, 1877*
Rupertina pustulosa Hatta, 1992
ROTALIOIDEA
ELPHIDIIDAE
Elphidium advenum f. *limbatum* (Chapman, 1907)
Elphidium a. f. *maorium* Hayward, 1997
Elphidium charlottense (Vella, 1957)
Elphidium crispum (Linnaeus, 1758)
Elphidium excavatum f. *clavatum* Cushman, 1930
Elphidium e. f. *excavatum* (Terquem, 1875)
Elphidium e. f. *oirgi* Hayward, 1997 E
Elphidium e. f. *williamsoni* Haynes, 1873
Elphidium gunteri Cole, 1931
Elphidium novozealandicum Cushman, 1936 E
Elphidium reticulosum Cushman, 1933 K
Elphidium vellai Hayward, 1997 E
Notorotalia aucklandica Vella, 1957 E
Notorotalia depressa Vella, 1957 E
Notorotalia finlayi Vella, 1957 E
Notorotalia hornibrooki Hayward, Grenfell, Reid & Hayward, 1999 E
Notorotalia inornata Vella, 1957 E
Notorotalia olsoni Vella, 1957 E
Notorotalia profunda Vella, 1957 E
Notorotalia zelandica Finlay, 1949 E
ROTALIIDAE
Ammonia aoteana (Finlay, 1940)
Ammonia pustulosa (Albani & Barbero, 1982)
Ammonia n.sp.*
SIPHONINOIDEA
SIPHONINIDAE
Siphonina bradyana Cushman, 1927*
Siphonina tubulosa Cushman, 1924

Class XENOPHYOPHOREA
Order PSAMMINIDA
PSAMMINIDAE
Reticulammina labyrinthica Tendal, 1972
Reticulammina lamellata Tendal, 1972 E
Reticulammina maini Tendal & Lewis, 1978 E
Reticulammina novazealandica Tendal, 1972 E
SYRINGAMMINIDAE Tendal, 1972
Aschemonella scabra Brady, 1879
Syringammina fragilissima Brady, 1883
Syringammina tasmanensis Lewis, 1966

Checklist of New Zealand fossil Foraminifera

Species entries are followed by time ranges using the standard abbreviations for New Zealand stages. Because of the historic plethora of taxonomic names introduced globally it is not possible at this time to provide any accurate estimate of endemism in fossil species. Endemic species are indicated by E; endemic genera are underlined (first entry only). Genus, family, and superfamily classification largely follows Loeblich and Tappan (1987), with the ordinal classification after Loeblich and Tappan (1992).

PHYLUM FORAMINIFERA
Class POLYTHALAMEA
Order ALLOGROMIIDA
ALLOGROMIIDAE
Nodellum velascoensis (Cushman, 1926) Mh

Order ASTRORHIZIDA
ASTRORHIZOIDEA
BATHYSIPHONIDAE
Bathysiphon cylindrica (Glaessner, 1937) Ra-Dt
Bathysiphon filiformis M. Sars, 1872 Mh-Rec
Bathysiphon robusta (Grzybowski, 1897) Mh
Bathysiphon sp. indet. Rt-Rm
Nothia grilli (Noth, 1951) Mh
RHABDAMMINIDAE
Dendrophyra excelsa Grzybowski, 1898 Mh
Rhabdammina abyssorum Sars, 1869 Po-Rec
Rhabdammina annulata Gryzbowski, 1896 Mh
Rhabdammina linearis Brady, 1879 Mh
Rhizammina algaeformis Brady, 1879 Mh-Rec
Rhizammina excelsa (Grzybowski, 1897) Mh
SACCAMMINIDAE
Placentammina placenta (Grzybowski, 1898) Mh
Saccammina elongata (Wiesner, 1931) Mh
Saccammina sp. indet. Bm
Saccammina sphaerica Brady, 1871 Mh
Technitella archaeonitida Stainforth & Stevenson, 1946 Mh
Technitella sp. indet. Po
Thurammina papillata Brady, 1879 Mh
SILICOTUBIDAE
Silicotuba grzybowskii (Dylazanka, 1923) Ra-Lwh
HIPPOCREPINOIDEA
HIPPOCREPINIDAE
Hippocrepina barksdalei (Tappan, 1957) Mh
Hippocrepina sp. indet. Bm
Hyperammina elongata Brady, 1878 Mh-Rec
Hyperammina latissima (Grzybowski, 1898) Mh
Hyperammina sp. indet. Bm

ORDER LITUOLIDA
AMMODISCOIDEA
AMMODISCIDAE
Ammodiscus archimedis (Stache, 1864) E Ab-Tt
Ammodiscus cretaceus (Reuss, 1845) Ra-Dt
Ammodiscus finlayi Parr, 1935 E Dt-Tt
Ammodiscus gaultinus Berthelin, 1880 Cu-Cn
Ammodiscus pennyi Cushman & Jarvis, 1928 Ra-Dt
Ammodiscus sp. indet. Bm
Ammovertella sp. indet. Bm
Glomospira gordialis (Jones & Parker, 1860) Ra-Mh
Glomospira sp. indet. Perm
Tolypammina vagans (Brady, 1879) Mh
Tolypammina sp. indet. Bw

Usbekistania charoides (Jones & Parker, 1860) Cn-Rec
ATAXOPHRAGMIOIDEA
ATAXOPHRAGMIIDAE
Arenobulimina sp. indet. Lw-Pl
GLOBOTEXTULARIIDAE
Matanzia varians (Glaessner, 1937) Mh-Dt
Ruakituria mahoenuica (Finlay, 1939) E Ab-Wo
HAPLOPHRAGMIOIDEA
AMMOSPHAEROIDINIDAE
Budashevaella multicamerata (Voloshinova & Budasheva, 1961) Mh-Dt
Thalmannammina sp. indet. Cn
Thalmannammina subturbinata (Grzybowski, 1897) Ra-Dt
HAPLOPHRAGMIIDAE
Haplophragmium lueckei (Cushman & Hedberg) Mh
Haplophragmium spp. indet. (2)
HORMOSINOIDEA
HORMOSINIDAE
Hormosina imitator (Finlay, 1947) E Dw
Hormosina ovula (Grzybowski, 1896) Mh-Ab
Hormosina trinitatensis Cushman & Renz, 1946 Mh
Pseudonodosinella nodulosa (Brady, 1879) Mp-Mh
Reophax constrictus (Reuss, 1874) Mh
Reophax eominutus Kristan-Tollmann, 1964 Bw
Reophax horridus (Schwager, 1865) Bw
Reophax regularis Hugland, 1947 Dt
Reophax sp. indet. Mh
LITUOLOIDEA
HAPLOPHRAGMOIDIDAE
Cribrostomoides trinitatensis Cushman & Jarvis, 1928 Dt
Haplophragmoides arenatus Crespin, 1963 Bm
Haplophragmoides eggeri Cushman, 1926 Dt
Haplophragmoides kirki Wickenden, 1932 Mh
Haplophragmoides medwayensis (Parr, 1935) E Tt
Haplophragmoides suborbicularis (Grzybowski, 1896) Ra-Rt
Haplophragmoides topagorukensis Tappan, 1962 Cn
Haplophragmoides spp. indet. (5)
LITUOLIDAE
Ammobaculites agglutinans (d'Orbigny, 1846) Mh
Ammobaculites clifdenensis Hornibrook, 1961 E Lw-Sl
Ammobaculites coprolithiformis (Schwager, 1868) Mh
Ammobaculites duncani Schroeder, 1968 Bm
Ammobaculites fragmentarius Cushman, 1927 Mh
Ammobaculites korangaensis Stoneley, 1962 E Cm
Ammobaculites taitaiica Stoneley, 1962 E Uk-Cu
Ammobaculites spp. indet. (6)
Ammomarginulina beggi Strong, 1984 E Bm
Ammomarginulina stephensoni (Cushman, 1933) Mh
Ammomarginulina sp. indet. Ab
Flabellammina compressa (Beissel, 1891) Mh
Flabellammina jacksoni Hornibrook, 1961 E Ak-Lwh
Flabellammina spp. indet. (2)
Triplasia marwicki Loeblich & Tappan, 1952 Ak-Pl
LITUOTUBIDAE
Lituotuba lata (Grzybowski, 1897) Mh
Trochamminoides acervulatus (Grzybowski, 1896) Mh
LOFTUSIOIDEA
CYCLAMMINIDAE
Cyclammina amplectens Grzybowski, 1898 Dt-Ar
Cyclammina elegans Cushman & Jarvis, 1932 Mh-Dt
Cyclammina incisa (Stache, 1864) E Ab-Sw
Cyclammina rotundata Chapman & Crespin, 1930 Po
Cyclammina rotundidorsata (Hantken, 1876) Dp-Ar
Reticulophragmium paupera (Chapman, 1904) Dt
SPIROPLECTAMMINOIDEA
SPIROPLECTAMMINIDAE
Bolivinopsis compta Finlay, 1947 E Dt-Ab
Bolivinopsis cubensis (Cushman & Bermudez, 1937) Dw-Rec
Bolivinopsis spectabilis (Grzybowski, 1898) Mh-Dt
Bolivinopsis trinitatensis (Cushman & Renz, 1948) Dw-Ab
Bolivinopsis spp. indet.
Spiroplectammina dentata Alth, 1850 Mh
Spiroplectammina laevis cretosa Cushman, 1932 Mh
Spiroplectammina piripaua Finlay, 1939 E Mh-Dt
Spiroplectammina semicomplanata (Carsey, 1964) Mh
Spiroplectammina steinekei Finlay, 1939 E Mh-Dt
Spiroplectammina sp. indet. Rt-Mp
Vulvulina bortonica Finlay, 1947 E Ab-Ar
Vulvulina bueningi Finlay, 1939 E Dt-Dt
Vulvulina espinosa Finlay, 1947 E Dm
Vulvulina granulosa Finlay, 1947 E Ar-Ld
Vulvulina haeringensis (Guembel, 1868) Dw-Dp
Vulvulina jablonskii Finlay, 1939 E Ld-Sw
Vulvulina pennatula (Batsch, 1791) Lw-Tt
Vulvulina spinosa Cushman, 1927 Ab-Lwh
RZEHAKINOIDEA
RZEHAKINIDAE
Rzehakina epigona (Rzehak, 1985) Ra-Dt
VERNEUILINOIDEA
CONOTROCHAMMINIDAE
Conotrochammina depressa Finlay, 1947 E Dw-Dm
Conotrochammina whangaia Finlay, 1940 E Mh-Dt
PROLIXOPLECTIDAE
Karrerulina aegra Finlay, 1940 E Mh-Ak
Karrerulina bortonica Finlay, 1940 E Dt-Ar
Karrerulina clarentia Finlay, 1940 E Cu-Cn
Karrerulina obscura Srinivasan, 1966 E Ab-Tk
Karrerulina sp. indet. Mp-Dw
Karrerulina urutawica Stoneley, 1962 E Cu
Orientalia sp. indet. Bw
Plectina agrestior Finlay, 1939 E Ab-Ak
Plectina quennelli Finlay, 1939 E Ab
TRITAXIIDAE
Tritaxia alpina (Cushman, 1936) Dm-Tk
Tritaxia olssoni (Finlay, 1939) E Po-Pl
VERNEUILINIDAE
Bermudezina sp. indet. Dt
Gaudryina convexa (Karrer, 1864) E Po-Rec
Gaudryina fenestrata Finlay, 1939 E Po-Tt
Gaudryina healyi Finlay, 1939 E Mh
Gaudryina kingi Parr, 1936 Tt-Wn
Gaudryina miniscula Finlay, 1940 E Ar-Ld
Gaudryina proreussi Finlay, 1939 E Ab-Ak
Gaudryina quadrazea Hornibrook, 1961 E Ld-Sw
Gaudryina reliqua Finlay, 1939 E Dw-Dm
Gaudryina reussi Stache, 1864 E Ab-Po
Gaudryina whangaia Finlay, 1939 E Dt
Gaudryina spp. indet. (4)
Gaudryinella campbelli Strong, 1984 E Bm
Gaudryinella pseudoserrata Cushman, 1932 Ra
Gaudryinella p. extensa Cushman, 1949 Ra-Rm
Spiroplectinata annectans (Parker & Jones, 1863) Cn
Spiroplectinata capitosa (Cushman, 1933) Rt-Mp
Spiroplectinata williamsi Stoneley, 1962 E Cu
Spiroplectinata sp. indet. Cu
Uvigerinammina jankoi Majzon, 1943 Mh
Verneuilina browni Finlay, 1939 E Lwh-Sc
Verneuilina novozealandica Cushman, 1936 E Lwh-Sc

Order TROCHAMMINIDA
TROCHAMMINOIDEA
TROCHAMMINIDAE
Trochammina albertensis Wickenden, 1932 Mh
Trochammina enouraensis Asano, 1958 Po
Trochammina globigeriniformis (Parker & Jones, 1865) Mh
Trochammina maoria Stoneley, 1962 E Cn-Cm
Trochammina novozealandica Stoneley, 1962 E Cu
Trochammina proteus Karrer, 1866 Mh
Trochammina spp. indet. (2)

Order TEXTULARIIDA
TEXTULARIOIDEA
EGGERELLIDAE
Arenodosaria antipoda (Stache, 1864) E Ab-Sw
Arenodosaria kaiataensis Dorreen, 1948 E Ar-Ld
Arenodosaria turris Kennett, 1967 E Tt
Arenodosaria sp. indet. Ar-Lwh
Dorothia agrestis Finlay, 1939 E Dh-Ab
Dorothia biformis Finlay, 1939 E Mh
Dorothia bulletta Carsey, 1926 Dt-Dh
Dorothia elongata Finlay, 1940 E Mh
Dorothia indentata (Cushman and Jarvis, 1928) Ak-Po
Dorothia minima (Karrer, 1864) E Ak-Sc
Dorothia scabra (Brady, 1884) Po-Rec
Dorothia stephensoni Cushman, 1946 Mh
Dorothia trochoides (d'Orbigny, 1852) Mh
Dorothia spp. indet. (5)
Eggerella bradyi (Cushman, 1911) Tt-Rec
Eggerella columna Finlay, 1940 E Dt
Eggerella decepta Finlay, 1939 E Dw-Dm
Eggerella ihungia Finlay, 1940 E Lwh-Sc
Karreriella alticamera Cushman & Stainforth, 1945 Ar-Sl
Karreriella bradyi (Cushman, 1911) Ab-Rec
Karreriella chilostoma (Reuss, 1852) Ab-Sw
Karreriella cushmani Finlay, 1940 E Po-Wo
Karreriella cylindrica Finlay, 1940 E Sw-Wn
Karreriella fastigata Kennett, 1967 E Tt \
Karreriella kennetti Srinivasan, 1966 E Ar-Lwh
Karreriella novangliae (Cushman, 1922) Wo-Rec
Karreriella novozealandica Cushman, 1936 Dp-Po
Karreriella ututaica Stoneley, 1962 E Cu
Karreriella sp. indet. Cm
Marssonella oxycona (Reuss, 1954) Mh-Dt
Martinottiella clarae Gibson, 1967 E Sw-Tk
Martinottiella communis (d'Orbigny, 1846) Dp-Rec
Martinottiella levis (Finlay, 1939) E Dt-Ab
Martinottiella massami Gibson, 1967 E Sw-Tk
Martinottiella variabilis (Schwager, 1866) Wm-Wc*
Martinottiella weymouthi Finlay, 1939 E Ab-Tk
PSEUDOGAUDRYINIDAE
Migros medwayensis (Parr, 1935) E Po-Tk
Migros sp. indet. Mh
TEXTULARIIDAE
Bigenerina burri Finlay, 1947 E Dw
Bigenerina sp. indet. Dm-Dp
Haeuslerella decepta Hornibrook, 1961 E Lw-Pl
Haeuslerella finlayi Vella, 1963 E Tt-Wo
Haeuslerella hectori Finlay, 1939 E Lw-Po
Haeuslerella morgani (Chapman, 1926) E Sl-Wo
Haeuslerella parri Finlay, 1939 E Tk-Wn
Haeuslerella pliocenica (Finlay, 1939) E Tt-Wn
Haeuslerella pukeuriensis Parr, 1935 E Pl-Sl
Haeuslerella textilariformis (Stache, 1864) E Lwh-Ld
Semivulvulina capitata (Stache, 1864) E Lwh-Ld
Semivulvulina ihungia Finlay, 1947 E Pl-Sc
Semivulvulina prisca Finlay, 1947 E Ab
Semivulvulina waitakia Finlay, 1939 E Ld-Lw
Siphotextularia acutangula Finlay, 1939 E Dp-Ab
Siphotextularia awamoana Finlay, 1939 E Lw-Sw
Siphotextularia bolivina Hornibrook, 1961 E Lw-Sw
Siphotextularia cordis Hornibrook, 1961 E Ar-Lwh
Siphotextularia dawesi Kennett, 1967 E Tt-Wo
Siphotextularia emaciata Hornibrook, 1961 E Ld-Sw
Siphotextularia finlayi Hornibrook, 1961 E Ab-Ld
Siphotextularia ihungia Finlay, 1940 E Po-Tt
Siphotextularia kreuzbergi Finlay, 1940 E Lw-Tt
Siphotextularia lajollaensis (Lalicker, 1935) Mh
Siphotextularia lornensis Hornibrook, 1961 E Ak-Ar
Siphotextularia rolshauseni Phleger & Parker, 1951 Wm-Rec

Siphotextularia subcylindrica Finlay, 1940 E Sc-Wo
Siphotextularia wairoana Finlay, 1939 E Wo-Wc
Siphotextularia wanganuia Finlay, 1939 E Wp-Wc
Siphotextularia sp. indet. Dw
Textularia aorangi Dorreen, 1948 E Ar
Textularia awamoana Hornibrook, 1961 E Pl-Sw
Textularia awazea Finlay, 1939 E Pl
Textularia barnwelli Kennett, 1966 E Tk-Wo
Textularia crater Kennett, 1967 E Tt
Textularia cuneazea Hornibrook, 1961 E Po-Pl
Textularia cuspis Finlay, 1939 E Ab-Lwh
Textularia eyrei Finlay, 1947 E Dm
Textularia foeda Reuss, 1846 Bm
Textularia gladizea Finlay, 1947 E Sc-Tt
Textularia hayi Karrer, 1864 E Ar-Sw
Textularia kapitea Finlay, 1947 E Tt-Wo
Textularia leuzingeri Cushman & Renz, 1941 Lwh-Wo
Textularia lythostrota (Schwager, 1866) Wc-Rec
Textularia marsdeni Finlay, 1939 E Ar-Sw
Textularia miozea Finlay, 1939 E Pl-Tt
Textularia ototara Hornibrook, 1961 E Ak-Sw
Textularia plummerae Lalicker, 1935 Mh-Ar
Textularia pozonensis Cushman & Renz, 1941 Po
Textularia pseudomiozea Finlay, 1947 E Lwh-Po
Textularia saggitula Defrance, 1824 Po
Textularia semicarinata Hornibrook, 1961 E Po-Pl
Textularia subantarctica Vella, 1957 E Wn-Rec
Textularia subrhombica Stache, 1864 E Ak-Tk
Textularia walcotti Srinivasan, 1966 E Lwh
Textularia zeaggluta Finlay, 1939 E Dm-Ld
Textularia spp. indet.
VALVULINIDAE
Clavulina anglica Cushman, 1936 Dt
Clavulina capitosa (Cushman, 1933) Rm-Mp
Clavulina maqfiensis (LeRoy, 1953) Dw-Dp
Cylindroclavulina bradyi (Cushman, 1911) Lwh-Rec
Tritaxilina zealandica Finlay, 1939 E Dm-Tt

Order FUSULINIDA
ARCHAESPHAEROIDEA
ARCHAESPHAERIDAE
Diplosphaerina inaequalis (Derville, 1931) Perm
ENDOTHYROIDEA
BRADYINIDAE
Bradyina? sp. indet. Perm
FUSULINOIDEA
NEOSCHWAGERINIDAE
Colania douvillei (Ozawa, 1925) Perm
Neoschwagerina cf. *craticulifera* (Schwager, 1883) Perm
Neoschwagerina margaritae Deprat, 1913 Perm
Yabeina ampla Skinner & Wilde, 1966 Perm
Yabeina archaica Dutkevich, 1967 Perm
Yabeina globosa (Yabe, 1906) Perm
Yabeina multiseptata (Deprat, 1913) Perm
Yabeina parvula Skinner & Wilde, 1966 Perm
Yabeina shiraiwensis (Ozawa, 1925) Perm
OZAWAINELLIDAE
Pseudokahlerina compressa Sosnina, 1968 Perm
Rauserella cf. *breviscula* Sosnina, 1968 Perm
Reichelina cf. *lamarensis* Skinner & Wilde, 1955 Perm
Reichelina cf. *turgida* Sheng, 1963 Perm
Reichelina media Miklukho-Maclay, 1954 Perm
Sichotenella cf. *sutschanica* Toumanskaya, 1953 Perm
SCHUBERTELLIDAE
Codonofusiella cf. *nana* Erk, 1941 Perm
Codonofusiella cf. *schubertelloides* Sheng, 1956 Perm
Dunbarula nana Kochansky-Devide & Ramovs, 1955 Perm
Minojapanella elongata Fujimoto & Kanuma, 1953 Perm
Schubertella? sp. indet. Perm
SCHWAGERINIDAE
Chusenella cf. *cheni* Skinner & Wilde, 1966 Perm
Chusenella urulungensis Wang, Sheng & Zhang, 1981 Perm
Parafusulina cuniculata (Igo, 1967) Perm
Parafusulina japonica (Gumbel, 1874) Perm
Parafusulina j. curta Leven & Campbell 1998 Perm
Parafusulina j. deprati Leven & Campbell 1998 Perm
STAFFELLIDAE
Nankinella cf. *orbicularia* Lee, 1934 Perm
VERBEEKINIDAE
Kahlerina cf. *globiformis* Sosnina, 1968 Perm
Kahlerina pachytheca Kochansky-Devide & Ramovs, 1955 Perm
GEINITZINOIDEA
GEINITZINIDAE
Lunucammina uralica (Souleimanov, 1949) Perm
PACHYPHLOIIDAE
Pachyphloia cf. *iranica* Bozorgnia, 1973 Perm
Pachyphloia ovata Lange, 1925 Perm
PALEOTEXTULARIOIDEA
PALAEOTEXTULARIIDAE
Climacammina cf. *valvulinoides* Lange, 1925 Perm
Cribrogenerina obesa Lange, 1925 Perm
Dagmarita sp. indet. Perm
Deckerella sp. indet. Perm
Globivalvulina cf. *vonderschmitti* Reichel, 1945 Perm
Globivalvulina graeca Reichel, 1945 Perm
Palaeotextularia sp. indet. Perm
Paraglobivalvulina? sp. Perm
TETRATAXOIDEA
ABADEHELLIDAE
Abadehella cf. *coniformis* Okimura & Ishii, 1975 Perm
TETRATAXIDAE
Tetrataxis cf. *maxima* Schellwien, 1898 Perm
Tetrataxis? *inflata* Kristan, 1957 Bm

Order SPIRILLINIDA
Suborder INVOLUTININA
INVOLUTINIDAE
Neochemigordius grandis (Ozawa, 1925) Perm
Neochemigordius japonica (Ozawa, 1925) Perm
Neohemigordius cf. *zaninettiae* (Altiner, 1978) Perm

Suborder SPIRILLININA
PATELLINIDAE
Patellina corrugata Williamson, 1858 Ar-Rec
Patellina piripaua Finlay, 1939 E Mh
Patellina spp. indet. (3)
SPIRILLINIDAE
Spirillina plana Möller, 1879 Po-Pl
Spirillina vivipara Ehrenberg, 1843 Po-Rec
Spirillina sp. indet. Po-Pl

Order MILIOLIDA
Suborder MILIOLINA
CORNUSPIROIDEA
CORNUSPIRIDAE
Cornuspira archimedis Stache, 1864 E Lw-Pl
Cornuspira foliacea (Philippi, 1844) Pl-Rec
Cornuspiroides striolata (Brady, 1882) Po-Rec
HEMIGORDIOPSOIDEA
BAISALINIDAE
Baisalina pulchra Reitlinger, 1965 Perm
HEMIGORDIOPSIDAE
Agathammina sp. indet. Perm
Gordiospira fragilis Heron-Allen & Earland, 1932 Pl
Hemigordiopsis renzi Reichel, 1945 Perm
MILIOLOIDEA
HAUERINIDAE
Articulina parri Cushman, 1944 Pl
Cruciloculina ericsoni Loeblich & Tappan, 1957 Tt
Hauerina notoensis Dorreen, 1948 E Ar
Massilina granulocostata (Germeraad, 1946) Po-Rec
Massilina speciosa (Karrer, 1868) Pl
Massilina subaequalis (Parr, 1936) Tt
Miliolinella subrotunda (Montagu, 1803) Wc-Rec
Miliolinella vellai Srinivasan, 1966 E Ak-Ar
Pyrgo bulloides (d'Orbigny, 1826) Ak-Pl
Pyrgo depressa (d'Orbigny, 1826) Pl-Rec
Pyrgo elongata (d'Orbigny, 1826) Tk
Pyrgo inornata (d'Orbigny, 1846) Po-Rec
Pyrgo laevis Defrance, 1824 Pl
Pyrgo lucernula (Schwager, 1866) Pl
Pyrgo murrhina (Schwager, 1866) Tt-Rec
Pyrgo tasmanensis Vella, 1957 E Tt-Rec
Quinqueloculina agglutinans d'Orbigny, 1839 Wc-Rec
Quinqueloculina angulostriata Cushman & Valentine, 1930 Po-Pl
Quinqueloculina buchiana d'Orbigny, 1846 Po-Pl
Quinqueloculina contorta d'Orbigny, 1846 Tk
Quinqueloculina incisa Vella, 1957 E Tk-Rec
Quinqueloculina kapitiensis Vella, 1957 E Sw-Rec
Quinqueloculina oblonga (Montagu, 1803) Tt-Rec
Quinqueloculina plana d'Orbigny, 1850 Po-Pl
Quinqueloculina pygmaea Reuss, 1850 Tk
Quinqueloculina schroekingeri Karrer, 1868 Po-Pl
Quinqueloculina seminula (Linnaeus, 1758) Po-Rec
Quinqueloculina singletoni Crespin, 1950 Ld-Pl
Quinqueloculina suborbicularis d'Orbigny, 1826 Lw-Rec
Quinqueloculina triangularis d'Orbigny, 1846 Tt-Rec
Quinqueloculina venusta Karrer, 1868 Tt-Rec
Quinqueloculina vulgaris d'Orbigny, 1826 Po-Pl
Quinqueloculina waimea Kennett, 1966 E Tk
Quinqueloculina zealandica Srinivasan, 1966 E Ak-Lwh
Sigmoilina victoriensis Cushman, 1946 Ak-Sl
Sigmoilopsis compressa Hornibrook, 1958 E Pl-Sl
Sigmoilopsis finlayi Vella, 1957 E Wn-Rec
Sigmoilopsis gavini Gibson, 1967 E Tt
Sigmoilopsis neocelata Hornibrook, 1958 E Ld-Pl
Sigmoilopsis schlumbergeri (Silvestri, 1904) Po-Rec
Sigmoilopsis wanganuiensis Vella, 1957 E Wp-Rec
Sigmoilopsis zeaserus Vella, 1963 E Tk-Wp
Siphonaperta macbeathi Vella, 1957 E Wo-Rec
Siphonaperta vellai Hornibrook, 1961 E Pl-Sw
Spirosigmoilina tenuis (Czjzek, 1848) Ab-Rec
Triloculina chrysostoma (Chapman, 1909) E Wn-Rec
Triloculina insignis (Brady, 1881) Wm-Rec
Triloculina oculina d'Orbigny, 1846 Pl
Triloculina tricarinata d'Orbigny, 1826 Po-Rec
Triloculina trigonula (Lamarck, 1804) Po-Rec
Triloculinella elizabethae (Srinivasan, 1966) E Ak-Ar
Triloculinella hornibrooki (Vella, 1957) E Pl-Rec
Tubinella funalis (Brady, 1884) Pl-Rec
Wellmanella kaiata Finlay, 1947 E Ak
SPIROLOCULINIDAE
Inaequalina disparilis (Terquem, 1878) Lw-Rec
Spiroloculina canaliculata d'Orbigny, 1846 Ab-Tk
Spiroloculina kennetti Gibson, 1967 E Tt
Spiroloculina novozealandica Cushman & Todd, 1944 Po-Rec
NEBECULARIOIDEA
FISCHERINIDAE
Wiesnerella auriculata (Egger, 1893) Pl-Rec
Wiesnerella mackayi Hornibrook, 1961 E Ar
OPHTHALMIDIIDAE
Ophthalmidium sp. indet. Zwingli & Kübler, 1870 Pl-Tt
SORITOIDEA
MILIOLIPORIDAE
Kamurana? sp. indet. Perm
PENEROPLIDAE
Peneroplis mauii Dorreen, 1948 E Ar
Peneroplis sp. indet. Sw
SORITIDAE
Marginopora sp. indet. Sw

Order LAGENIDA
NODOSARIOIDEA
LAGENIDAE
Lagena acuticosta Reuss, 1862 Lwh-Rec
Lagena costata (Williamson, 1858) Lwh-Rec
Lagena distoma Parker & Jones, 1864 Ar-Rec
Lagena ellipsoidalis Schwager, 1878 Mh
Lagena elongata (Ehrenberg, 1854) Ar-Wo
Lagena fornasinii Buchner, 1940 Sw
Lagena foveolata Reuss, 1863 Mh
Lagena geometrica Reuss, 1863 Mh
Lagena gracilis Williamson, 1848 Pl-Tk
Lagena gracillima (Seguenza, 1862) Mh-Tk
Lagena heronalleni Earland, 1934 Tt
Lagena hispida Reuss, 1863 Mh-Rec
Lagena hispidula Cushman, 1913 Po-Rec
Lagena laevis (Montagu, 1803) Pl-Rec
Lagena malcomsonii (Wright, 1911) Pl
Lagena marginatoperforata Seguenza, 1880 Pl
Lagena meridonalis (Wiesner, 1931) Mh
Lagena nebulosa (Cushman, 1923) Tk-Rec
Lagena paradoxa Sidebottom, 1912 Lw-Rec
Lagena perlucida (Montagu, 1803) Tt-Wo
Lagena pliocenica Cushman and Gray, 1946 Tt
Lagena plumigera (Brady, 1881) Ld-Rec
Lagena spicata Cushman & McCulloch, 1950 Tt-Rec
Lagena spiralis Brady, 1884 Lw-Pl
Lagena staphyllearia Schwager, 1866 Sw
Lagena striata (d'Orbigny, 1839) Ld-Pl
Lagena strumosa Reuss, 1858 Tk-Rec
Lagena substriata Williamson, 1848 Po-Rec
Lagena sulcata Walker & Jacob, 1798 Mh-Rec
Lagena tenuistriata Stache, 1864 E Lwh-Ld
Lagena trigonomarginata Parker and Jones, 1865 Pl
Lagena vulgaris Williamson, 1946 Mh
Lagena williamsoni Harvey & Bailey, 1854 Tt
Procerolagena amphora (Willliamson, 1848) Wo
Procerolagena clavata (d'Orbigny, 1846) Pl-Rec
NODOSARIIDAE
Amphimorphinella amchitkaensis Todd, 1953 Dw-Pl*
Amphimorphinella butonensis Keyzer, 1953 Po
Awhea tetragona (Costa, 1855) Lw-Wn
Awhea tosta (Schwager, 1866) Pl-Wc
Chrysalogonium asperelum (Neugeboren, 1852) Dm-Wp
Chrysalogonium bortonicum (Hornibrook, 1961) E Dw-Ar
Chrysalogonium calomorphum (Reuss, 1866) Ab-Wc
Chrysalogonium ciperense Cushman & Stainforth, 1945 Ar*
Chrysalogonium conicum (Neugeboren, 1852) Ab-Wn
Chrysalogonium crassitestum (Schwager, 1866) Tt-Wn
Chrysalogonium deceptorium (Schwager, 1866) Lwh-Wc
Chrysalogonium equisetiformis (Schwager, 1866) Po-Wn
Chrysalogonium gomphiformis (Schwager, 1866) Wo-Wn
Chrysalogonium karreri (Hantken, 1868) Wo*
Chrysalogonium lamellatum (Cushman & Stainforth, 1945) Po-Wn*
Chrysalogonium laeve Cushman & Bermudez, 1936 Ar-Pl*
Chrysalogonium nuttalli (Hedberg, 1937) Tt-Wn*
Chrysalogonium polystomum (Schwager, 1866) Wo-Wn*
Chrysalogonium rudis (d'Orbigny, 1846) Po-Wn*
Chrysalogonium stimuleum (Schwager, 1866) Dw-Wc
Chrysalogonium texanum Cushman, 1936 Dm-Pl*
Chrysalogonium n.sp. aff. *calamorphum* (Reuss, 1866) Wn*
Chrysalogonium n.sp. aff. *crassitestum* (Schwager, 1866) Ab*
Chrysalogonium n.sp. Wo*
Citharina strigillata (Reuss, 1846) Mh
Cribronodosaria sp. indet. Wm
Dentalina aculeata d'Orbigny, 1840 Mh
Dentalina advena (Cushman, 1923) Wo
Dentalina albatrossi Cushman, 1923 Po-Rec
Dentalina armata (Neugeboren, 1852) Mh
Dentalina basiplanata Cushman, 1938 Mh-Dt
Dentalina catenulata Reuss (Brady, 1884) Mh
Dentalina cf. *clavulaeformis* Schwager, 1878 Wn
Dentalina confluens Reuss, 1862 Mh
Dentalina crinata Plummer, 1931 Mh-Dt
Dentalina cuvieri (d'Orbigny, 1826) Tt-Rec
Dentalina decepta (Bagg, 1912) Tt-Tk
Dentalina frobisherensis Loeblich and Tappan, 1953 Wn
Dentalina gracilescens Schwager, 1866 Pl
Dentalina legumen Reuss, 1846 Mh
Dentalina lorneiana d'Orbigny, 1840 Mh
Dentalina mucronata (Neugeboren, 1856) Wn
Dentalina mutabilis (Costa, 1855) Tk-Rec*
Dentalina mutsui Hada, 1931 Wn-Rec
Dentalina obliqua (Linneaus, 1797) Sw-Tk
Dentalina obliquecostata (Stache, 1864) E Ab-Pl
Dentalina solvata Cushman, 1938 Mh
Dentalina spirans Cushman, 1940 Mh
Dentalina spirostriolata (Cushman, 1917) Po-Pl
Dentalina subcostata (Chapman, 1926) Dp-Pl
Dentalina tauricornis (Schwager, 1866) Wn
Dentalina torta Terquem, 1858 Bm
Dentalina translucens Parr, 1950 Wn
Dentalina sp. indet. Bm
Dentalinoides sp. indet. Dh-Pl
Frondicularia cf. *bassensis* Parr, 1950 Tt*
Frondicularia brizoides Gerke, 1961 Bm
Frondicularia bulla Belford, 1960 Mh
Frondicularia goldfussi Reuss, 1860 Rm-Rt
Frondicularia compta Brady Wn
Frondicularia mucronata Karrer, 1867 Mh-Dt
Frondicularia otamitaensis Strong, 1984 E Bm
Frondicularia rakauroana (Finlay, 1939) E Mh
Frondicularia rhaetica Kristan-Tollmann, 1964 Bm
Frondicularia rhombiformis Mamontova, 1957 Bm
Frondicularia scotti Heron-Allen & Earland, 1922 E Pl-Rec
Frondicularia steinekei Finlay, 1939 E Mh
Frondicularia tenuissima Hantken, 1875 Ab-Pl
Frondicularia teuria Finlay, 1939 E Dt
Frondicularia triangularis (Styx, 1975) Bm
Frondicularia spp. indet. (3)
Glandulonodosaria ambigua (Neugeboren, 1856) Ab-Wc
Glandulonodosaria glandigena (Schwager, 1866) Ab-Wn*
Glandulonodosaria trincherasensis (Bermudez, 1949) Lwh-Wn
Glandulonodosaria n.sp. Wm-Wn*
Grigelis neopyrula (McCulloch, 1981) Wn*
Grigelis orectus Loeblich & Tappan, 1994 Sw-Rec
Grigelis semirugosa (d'Orbigny, 1846) Lw-Rec
Grigelis n.sp. Sw-Wc*
Laevidentalina aphelis Loeblich & Tappan, 1987 Wn-Rec
Laevidentalina ariena (Patterson and Pettis, 1986) Wn-Rec
Laevidentalina badenensis (d'Orbigny, 1846) Mh-Rec
Laevidentalina communiensis Saidova, 1975 Wn-Rec
Laevidentalina communis (d'Orbigny, 1826) Sw-Wo
Laevidentalina elegans (d'Orbigny, 1846) Po-Rec*
Laevidentalina haueri Neugeboren, 1856 Lw-Rec*
Laevidentalina notalnella Saidova, 1975 Wn-Rec
Laevidentalina obliquensis Saidova, 1975 Wn-Rec
Laevidentalina plebeia (Reuss, 1855) Wn
Laevidentalina sidebottomi (Cushman, 1933) Wn-Rec
Laevidentalina guttifera (d'Orbigny, 1846) Dt-Rec*
Lagenoglandulina annuilata (Stache, 1864) E Dp-Pl
Lingulina aghdarbandi Oberhauser, 1960 Bm
Lingulina ampliata Loeblich & Tappan, 1950 Br
Lingulina avellanoides (Kreuzberg, 1930) Pl-Tt
Lingulina bartrumi Chapman, 1926 E Lwh-Tt
Lingulina decipiens (Stache, 1864) E Ar-Pl
Lingulina esseyana Deecke, 1956 Bm
Lingulina evansi Hornibrook, 1953 E B
Lingulina tenera Bornemann, 1854 Bm
Lingulina vitrea Heron-Allen & Earland, 1932 Pl
Lingulina spp. indet. (2)
Mucronina compressa (Costa, 1855) Tt-Wc*
Mucronina dumontana (Reuss, 1861) Tt-Wc*
Mucronina hasta (Parker, Jones & Brady, 1865) Wo-Wn
Mucronina miocenica (Cushman, 1926) Po-Wn*
Mucronina monacantha (Reuss, 1850) Ab-Sw*
Mucronina silvestriana (Thalmann, 1952) Wn*
Mucronina spatulata (Costa, 1855) Lwh-Wc*
Mucronina n.sp. aff. *advena* (Cushman, 1923) Wc-Rec
Mucronina n.sp. Dw
Neolingulina veijoensis McCulloch, 1977 Wn-Rec
Neugeborena longiscata (d'Orbigny, 1846) Mh-Wc
Neugeborena ovicula (d'Orbigny, 1826) Ar-Wc*
Nodosaria affinis Reuss, 1846 Mh-Wo
Nodosaria apheiloculata Tappan, 1955 Bm
Nodosaria aspera Reuss, 1846 Mh-Dt
Nodosaria candela Franke, 1936 Bm
Nodosaria distans Reuss, 1855 Mh
Nodosaria doliolaris Parr, 1950 Po-Wn*
Nodosaria holocostata (Kristan-Tollmann, 1964) Bm
Nodosaria lamnulifera Thalmann, 1950 Tt-Wo
Nodosaria larina Tappan, 1951 Bm
Nodosaria limbata d'Orbigny, 1840 Mh
Nodosaria nebulosa (Ishizaki, 1943) Pl-Rec
Nodosaria nitidana Brand, 1964 Bm
Nodosaria obscura Reuss, 1845 Mh
Nodosaria oculina (Terquem & Berthelin, 1875) Br
Nodosaria pellita Heron-Allen and Earland, 1922 Wn-Rec
Nodosaria ploechingeri (Oberauser, 1960) Bm
Nodosaria pyrula d'Orbigny, 1826 Ab-Wn
Nodosaria sceptrum Reuss, 1863 Cn
Nodosaria simplex Silvestri, 1872 Wn-Rec
Nodosaria velascoensis Cushman, 1928 Mh
Nodosaria weaveri Finger & Lipps, 1990 Tt-Wn*
Orthomorphina jedlitschkai (Thalmann, 1937) Wo-Wn
Orthomorphina laevis (Cushman & Bermudez, 1937) Wo-Wn*
Orthomorphina multicosta (Neugeboren, 1856) Wo-Wn*
Orthomorphina perversa (Schwager, 1866) Ab-Wc
Parafrondicularia antonina (Karrer, 1878) Sw-Wc*
Plectofrondicularia concava (:Liebus, 1902) Po-Sw*
Plectofrondicularia digitalis (Neugeboren, 1850) Pl-Sc*
Plectofrondicularia irregularis (Neugeboren, 1850) Pl*
Plectofrondicularia pohana Finlay, 1947 E Sl--Wo
Plectofrondicularia proparri Finlay, 1947 E Lwh-Po
Pseudoglandulina symmetrica (Stache, 1864) E Dt-Wc
Pseudolingulina bassensis (Parr, 1950) Lw-Rec
Pseudolingulina bradii (Silvestri, 1903) Wn*
Pseudonodosaria brevis (d'Orbigny, 1846) Dm-Rec*
Pseudonodosaria comatula (Cushman, 1923) Wn-Rec
Pseudonodosaria cylindracea (Reuss, 1845) Mh
Pseudonodosaria lata (Tappan, 1951) Bm
Pseudonodosaria manifesta (Reuss, 1851) Mh

Pseudonodosaria tenuis (Bornemann, 1964) Bm
Pseudonodosaria vulgata multicamerata (Kristan-Tollman, 1964) Bm
Pyramidulina acuminata Hantken, 1875 Po-Wn
Pyramidulina callosa (Stache, 1864) E Dm-Wm
Pyramidulina hochstetteri (Schwager, 1866) Dw-Wn
Pyramidulina luzonensis (Cushman, 1921) Wn-Rec
Pyramidulina pauciloculata (Cushman, 1917) Sw-Rec
Pyramidulina raphanistriformis (Gümbel, 1862) Bm
Pyramidulina raphanus (Linnaeus, 1758) Dw-Wo
Pyramidulina substrigata (Stache, 1864) E Ab-Pl
Tristix permiana Cole, 1961 Bm
Tristix sp. indet. Lw-Wo
VAGINULINIDAE
Amphicoryna bradii (Silvestri, 1902) Sw-Wc
Amphicoryna georgechapronierei Yassini & Jones, 1995 Wn-Rec
Amphicoryna hispida (d'Orbigny, 1846) Lwh-Rec
Amphicoryna leurodeira Loeblich and Tappan, 1994 Wn*
Amphicoryna prora Finlay, 1940 E Lwh-Pl
Amphicoryna proxima (Silvestri, 1872) Wn-Rec
Amphicoryna scalaris (Batsch, 1791) Lwh-Rec
Amphicoryna separans (Brady, 1884) Wn-Rec*
Amphicoryna sublineata (Brady, 1884) Wn-Rec
Astacolus bradyi Cushman, 1917 Ar-Rec
Astacolus compressus (Stache, 1864) E Ak-Pl
Astacolus crepidulus (Fichtel & Moll, 1798) Sw-Rec
Astacolus cretaceus (Cushman, 1937) Mh
Astacolus dorbignyi (Roemer, 1839) Bm
Astacolus haasti (Stache, 1864) E Lwh-Pl
Astacolus hemiselena Kristan-Tollmann, 1964 Bm
Astacolus insolitus (Schwager, 1866) Po-Rec
Astacolus inspissatus (Loeblich & Tappan, 1950) Op
Astacolus judyae Hornibrook, 1961 E Lwh-Ld
Astacolus neolatus Vella, 1957 Wn-Rec
Astacolus pediacus Tappan, 1955 Bo
Astacolus vellai Saidova, 1975 Wn-Rec
Astacolus vetustus (d'Orbigny, 1850) Bm
Astacolus spp. indet. (4)
Citharina inaequistriata (Terquem, 1864) Bm
Citharina plumoides (Plummer, 1927) Mh-Dt
Citharina suturalis (Cushman, 1937) Mh
Darbyella tosaensis Takayanagi, 1953 Pl
Hemirobulina hydropica (Hornibrook, 1961) E Ar-Rec
Lenticulina acutiangulata (Terquem, 1964) Bm
Lenticulina australis (Parr, 1950) Lw-Rec
Lenticulina barretti (Srinivasan, 1966) E Ak-Lwh
Lenticulina calcar (Defrance, 1818) Pl-Rec
Lenticulina callifera (Stache, 1864) E Ar-Pl
Lenticulina colorata (Stache, 1864) E Ar-Pl
Lenticulina costata (Fichtell & Moll, 1798) Lw-Wp
Lenticulina cultrata (Montfort, 1808) Po-Rec
Lenticulina dicampyla (Franz, 1894) Lwh-Wp
Lenticulina discus (Brotzen, 1965) Dt
Lenticulina dorothiae (Finlay, 1939) E Ak-Pl
Lenticulina echinata (d'Orbigny, 1846) Po-Wo
Lenticulina erratica Hornibrook, 1961 E Lwh-Rec
Lenticulina foliata (Stache, 1864) E Ak-Rec
Lenticulina gibba (d'Orbigny, 1839) Ar-Rec
Lenticulina gottingensis (Bornemann, 1964) Bm
Lenticulina gyroscalpra (Stache, 1864) E Ab-Pl
Lenticulina hampdenense (Hornibrook, 1961) E Dh-Ak
Lenticulina insulsa (Cushman, 1947) Mh-Dt
Lenticulina iota (Cushman, 1923) Lwh-Wc
Lenticulina lenticula (Stache, 1864) E Lwh-Po
Lenticulina limbosa (Reuss, 1863) Sw-Rec
Lenticulina loculosa (Stache, 1864) E Ak-Wp
Lenticulina lucida (Cushman, 1923) Lwh-Pl
Lenticulina macrodisca (Reuss, 1863) Mh-Dt
Lenticulina magna (Chapman, 1926) E Po
Lenticulina mamilligera (Karrer, 1864) E Lw-Wp
Lenticulina mironovi (Dain 1948) Bm
Lenticulina muensteri (Roemer, 1960) Bm
Lenticulina navarroensis (Plummer, 1927) Mh
Lenticulina nitida (Reuss, 1863) Po-Pl
Lenticulina oligostegia (Reuss, 1860) Mh
Lenticulina orbicularis (d'Orbigny, 1826) Tt-Rec
Lenticulina paparoaensis (Srinivasan, 1966) E Ak-Lwh
Lenticulina planula (Galloway & Heminway, 1941) Lwh-Wo
Lenticulina pseudocalcarata (Stache, 1864) E Ab-Pl
Lenticulina pseudocrassa Hornibrook, 1961 E Po-Pl
Lenticulina pseudomamilligera (Plummer, 1927) Dt
Lenticulina pusilla (Stache, 1864) E Lwh-Po
Lenticulina regina (Karrer, 1864) E Po-Pl
Lenticulina rotulata (Lamarck, 1804) Mh
Lenticulina spissocostata (Cushman, 1938) Mh
Lenticulina subalata (Reuss, 1854) Tt
Lenticulina subgibba Parr, 1950 Tt-Rec
Lenticulina suborbicularis Parr, 1950 Sw-Rec
Lenticulina taettowata (Stache, 1864) E Ak-Lw
Lenticulina vortex (Fitchel & Moll, 1803) Lwh-Pl
Lenticulina wisselmanni Bettenstaedt, 1952 Cn
Lenticulina spp. indet. (8)
Marginulina augensiensis Saidova, 1975 Wn-Rec
Marginulina costata (Batsch, 1791) Wn-Rec
Marginulina gummi Saidova, 1975 Wn-Rec
Marginulina glabra d'Orbigny, 1826 Ak-Tt
Marginulina navarroana Cushman, 1937 Mh
Marginulina obesa (Cushman, 1923) Sw-Wo
Marginulina obliquesuturata (Stache, 1864) E Ab-Pl
Marginulina striata d'Orbigny Wn-Rec
Marginulina subbullata Hantken, 1875 Ab-Sc
Marginulina texasensis Cushman, 1937 Mh-Dt
Marginulina sp. indet. Bm
Marginulinopsis allani (Finlay, 1939) E Ar-Pl
Marginulinopsis bradyi (Goess, 1894) Wn-Rec
Marginulinopsis curvatura (Cushman, 1938) Mh
Marginulinopsis curvisepta (Cushman & Goudkoff, 1944) Rt-Mp
Marginulinopsis phragmites Loeblich & Tappan, 1950 Op
Marginulinopsis spinobesa Finlay, 1947 E Dp-Ab
Marginulinopsis spinulosus (Stache, 1864) E Ab-Ld
Marginulinopsis tenuis (Bornemann, 1855) Wn-Rec
Marginulinopsis trinitatensis (Cushman, 1937) Mh
Neoflabellina jarvisi (Cushman, 1935) Mh
Neoflabellina praereticulata Hiltermann, 1952 Mh
Neoflabellina reticulata (Reuss, 1851) Mh
Neoflabellina rugosa (d'Orbigny, 1840) Mh
Neoflabellina semireticulata (Cushman & Jarvis, 1928) Dt
Neolenticulina variabilis (Reuss, 1850) 1866 Tt-Rec
Palmula bensoni Finlay, 1939 E Ak
Palmula bivium Finlay, 1939 E Ab-Ak
Palmula deslongchampsi (Terquem, 1864) Br-Bm
Palmula marshalli (Chapman, 1926) E Ab
Palmula taranakia Finlay, 1939 E Ar-Wo
Planularia bzurae Pozaryska, 1957 Dt
Planularia formosa Keijzer, 1945 Lw-Tk
Planularia halophora (Stache, 1864) E Ar-Lw
Planularia pauperata Jones & Parker, 1860 Bm
Planularia tricarinella (Reuss, 1863) Wn-Rec
Planularia sp. indet. Wo
Saracenaria arcuatula (Stache, 1864) E Ab-Sw
Saracenaria colei Srinivasan, 1966 E Ak-Ar
Saracenaria italica Defrance, 1824 Pl-Rec
Saracenaria kellumi Dorreen, 1948 E Ar-Pl
Saracenaria latifrons (Brady, 1884) Pl-Rec
Saracenaria obesa Cushman & Todd, 1945 Pl
Saracenaria oxfordiana Tappan, 1955 Op
Saracenaria triangularis (d'Orbigny, 1840) Mh
Saracenaria sp. indet. Bm
Vaginulina alazanensis Nuttall Wn*
Vaginulina awamoana Hornibrook, 1961 E Pl
Vaginulina elegans d'Orbigny, 1826 Dt-Tt
Vaginulina inflata Parr, 1950 Wn-Rec
Vaginulina jurassica (Gumbel, 1862) Op
Vaginulina legumen Linnaeus, 1758 Po-Wo
Vaginulina neglecta (Karrer, 1864) E Po-Wn
Vaginulina spinigera Brady, 1881 Wn-Rec
Vaginulina subelegans Parr, 1950 Sw-Wc
Vaginulina vertebralis Parr, 1932 Wn-Rec
Vaginulina tenuissima Heron-Allen & Earland, 1932 Lw-Rec
Vaginulina vagina (Stache, 1864) E Ab-Pl
Vaginulina sp. indet. Br
Vaginulinopsis albatrossi (Cushman, 1923) Wn*
Vaginulinopsis asperuliformis (Nuttall, 1930) Ab-Ar
Vaginulinopsis carinata Kennett, 1967 E Pl-Wn
Vaginulinopsis clifdenensis Hornibrook, 1961 E Po-Wo
Vaginulinopsis cristellata (Stache, 1864) E Lwh-Lw
Vaginulinopsis ensis (Reuss, 1845) Mh
Vaginulinopsis gnamptina Loeblich & Tappan, 1994 Wn-Rec
Vaginulinopsis hochstetteri (Stache, 1864) E Ab-Ld
Vaginulinopsis hutchinsoni Hornibrook, 1961 E Po-Pl
Vaginulinopsis interrupta (Stache, 1864) E Ak-Ld
Vaginulinopsis marshalli (Finlay, 1939) E Dm-Dp
Vaginulinopsis marwicki Finlay, 1947 E Ap-Ak
Vaginulinopsis mokauensis Hornibrook, 1961 E Lw-Sl
Vaginulinopsis recta (Karrer, 1864) E Lw-Pl
Vaginulinopsis reniformis (d'Orbigny, 1846) Wn-Rec
Vaginulinopsis spinulosus (Stache, 1864) E Ab-Ld
Vaginulinopsis sublegumen Parr, 1950 Wn-Rec*
Vaginulinopsis tasmanica Parr, 1950 Wn-Rec
Vaginulinopsis waiparaensis (Finlay, 1939) E Dw-Dp
Vaginulinopsis zeacarinata Kennett, 1967 E Sc-Wo
Vaginulinopsis spp. indet. (4)
POLYMORPHINOIDEA
ELLIPSOLAGENIDAE
Bifarilaminella advena (Cushman, 1923) Wo-Rec
Cushmanina desmophora (Rymer Jones, 1872) Ar-Pl
Cushmanina striatopunctata (Parker and Jones, 1865) Lw-Pl
Fissurina alveolata (Brady, 1884) Po-Pl
Fissurina annectens (Burrows and Holland, 1895) Tk
Fissurina aperta Seguenza, 1862 Po-Pl
Fissurina aureoligera (Buchner, 1940) Po-Pl
Fissurina biancae Seguenza, 1862 Po-Pl
Fissurina bicarinata Terquem, 1882 Sw-Tt
Fissurina bradii Silvestri, 1902 Po-Pl
Fissurina carinata Reuss, 1863 Mh
Fissurina clathrata (Brady, 1884) Lw-Rec
Fissurina clypeatomarginata (Jones, 1872) Po-Pl
Fissurina crenulata (Cushman, 1913) Po-Pl
Fissurina fimbriata (Brady, 1881) Pl
Fissurina formosa (Schwager, 1866) Ar-Pl
Fissurina kerguelenensis Parr, 1950 Tt
Fissurina laevigata Reuss, 1850 Po-Rec
Fissurina lagenoides (Williamson, 1858) Lw-Pl
Fissurina lucida (Williamson, 1848) Ld-Rec
Fissurina marginata (Montagu, 1803) Mh-Rec
Fissurina marylandica Nogan, 1964 Mh
Fissurina orbignyana Seguenza, 1862 Mh-Rec
Fissurina quadrata (Williamson, 1858) Lw-Rec
Fissurina staphyllearia Schwager, 1866 Sw
Fissurina submarginata Boomgaart, 1949 Sw-Wo
Fissurina wrightiana (Brady, 1881) Tk
Fissurina yokoyamae (Millett, 1895) Ld-Wm
Oolina botelliformis (Brady, 1881) Po-Pl
Oolina collaris (Cushman, 1913) Po-Pl
Oolina felsinea (Fornasini, 1901) Po-Pl
Oolina globosa (Montagu, 1803) Ar-Pl
Oolina hexagona (Williamson, 1848) Ar-Rec
Oolina hispida (Reuss, 1863) Ar-Pl
Oolina squamosa (Montagu, 1803) Lw-Pl

Parafissurina botelliformis (Brady, 1881) Pl-Pl
Parafissurina lateralis (Cushman, 1913) Wo
Ventrostoma scaphaeformis (Parr, 1950) Pl-Rec
GLANDULINIDAE
Glandulina ovula d'Orbigny, 1846 Wn-Rec*
Seabrookia pellucida Brady, 1890 Sw-Rec
POLYMORPHINIDAE
Bullopora collarata Kristan-Tollmann, 1964 Bm
Bullopora globulata Barnard, 1949 Bm
Bullopora rugosa Paalzow, 1932 Bm
Eoguttulina ovigera (Terquem, 1864) Br
Globulina gibba (d'Orbigny, 1826) Ab-Lw
Globulina lacrima Reuss, 1946 Mh
Globulina minima Bornemann, 1855 Tt
Globulina minuta (Roemer, 1838) Wo
Globulina polita (Terquem, 1882) Ab-Lw
Globulina prisca Reuss, 1863 Mh
Globulina subsphaerica (Berthelin, 1880) Mh
Guttulina adhaerens (Olszewski, 1875) Mh
Guttulina austriaca d'Orbigny, 1846 Ld-Pl
Guttulina bartschi Cushman & Ozawa, 1930 Tt-Rec
Guttulina caudita d'Orbigny, 1900 Mh
Guttulina clifdenensis Parr, 1937 Sc
Guttulina fissurata Stache, 1864 E Ab-Pl
Guttulina hantkeni Cushman & Ozawa, 1930 Mh
Guttulina otiakensis Hornibrook, 1961 E Ld
Guttulina problema d'Orbigny, 1826 Ab-Rec
Guttulina regina (Brady, Parker & Jones, 1870) Lwh-Rec
Guttulina seguenzana (Brady, 1884) Ab-Rec
Guttulina trigonula (Reuss, 1846) Mh-Dt
Guttulina yabei Cushman & Ozawa, 1929 Ab-Rec
Polymorphina complanata d'Orbigny, 1846 Ar-Lwh
Polymorphina incavata Stache, 1864 E Lwh-Po
Polymorphina lingulata Stache, 1864 E Ar-Lw
Polymorphina marshalli Hornibrook, 1961 E Ld-Lw
Polymorphina marwicki Hornibrook, 1961 E Ld-Po
Polymorphina waitakiensis Hornibrook, 1961 E Ld-Po
Polymorphina sp. indet. Ak-Ar
Pseudopolymorphina allani Hornibrook, 1961 E Ld
Pseudopolymorphina cuyleri Plummer, 1931 Mh
Pseudopolymorphina digitata (d'Orbigny, 1843) Mh
Pseudopolymorphina parri Hornibrook, 1961 E Lwh-Ld
Pyrulina cylindroides (Roemer, 1838) Mh-Lw
Pyrulina fusiformis (Roemer, 1838) Ld-Rec
Pyrulina labiata (Schwager, 1866) Po-Wm*
Pyrulinoides metensis (Terquem, 1864) Bw
Ramulina aculeata Wright, 1886 Cu-Cn
Ramulina globulifera Brady, 1879 Po-Rec
Ramulina navarroana Cushman, 1938 Mh
Sigmoidella bortonica Finlay, 1939 E Dh
Sigmoidella elegantissima (Parker & Jones, 1870) Lwh-Rec
Sigmoidella kagaensis Cushman & Ozawa, 1928 Ld-Rec
Sigmoidella pacifica Cushman & Ozawa, 1928 Po-Rec
Sigmomorphina haeusleri Parr & Collins, 1937 Ld
Sigmomorphina lacrimosa Vella, 1957 E Sw-Rec
Sigmomorphina lornensis Hornibrook, 1961 E Ak-Ar
Sigmomorphina obesa Hornibrook, 1961 E Ar-Ld
Sigmomorphina pernaeformis (Stache, 1864) E Dp-Lwh
Sigmomorphina williamsoni (Terquem, 1878) Ar-Wn
Sigmomorphina sp. indet. Bm
ROBULOIDOIDEA
ICHTHYOLARIIDAE
Cryptoseptida fragilis (Sellier de Civrieux & Dessauvagie, 1965) Perm
Frondina appressaria Sosnina, 1978 Perm
Frondina permica de Civrieux & Dessauvagie, 1965 Perm
Protonodosaria? sp. indet. Perm
Grillina grilli Kristan-Tollmann, 1964 Bm
Ichthyolaria squamosa (Terquem & Berthelin, 1875) Bm
Ichthyolaria sulcata (Bornemann, 1854) Bm
Ichthyolaria terquemi (d'Orbigny, 1850) Bm
PARTISANIIDAE
Partisania cf. *flangensis* Sosnina, 1978 Perm
ROBULOIDIDAE
Robuloides cf. *acutus* Reichel, 1945 Perm
Robuloides cf. *lens* Reichel, 1945 Perm
Robuloides cf. *reicheli* (Reitlinger, 1965) Perm
SYZRANIIDAE
Rectostipulina quadrata Jenny-Deshusses, 1985 Perm

Order ROBERTINIDA
CERATOBULIMINOIDEA
CERATOBULIMINIDAE
Ceratobulimina awamoana Hornibrook, 1961 E Po-Sl
Ceratobulimina kellumi Finlay, 1939 E Lwh-Tt
Ceratobulimina lornensis Finlay, 1939 E Ak-Ar
Ceratobulimina spp. indet. (2)
Ceratocancris clifdenensis Finlay, 1939 E Po-Tt
Ceratocancris hortalveus Finlay, 1947 E Ab-Ak
Ceratolamarckina clarki Srinivasan, 1966 E Ak-Lwh
Lamarckina novozealandica Dorreen, 1948 E Ak-Lwh
Lamarckina turgida Dorreen, 1948 E Ak-Lwh
Praelamarckina sp. indet. Op
Vellaena zealandica Srinivasan, 1966 E Ar-Lwh
Zelamarkina excavata Collen, 1972 E Wm
EPISTOMINIDAE
Hoeglundina bensoni (Stoneley, 1962) E Cu
Hoeglundina elegans (d'Orbigny, 1826) Ab-Rec
CONORBOIDOIDEA
CONORBOIDIDAE
Colomia austrotrochus Taylor, 1964 Mh
ROBERTINOIDEA
ROBERTINIDAE
Cerobertina bartrumi Finlay, 1939 E Po-Sw
Cerobertina crepidula Finlay, 1939 E Lw-Sw
Cerobertina kakahoica Finlay, 1939 E Ab-Lwh
Cerobertina mahoenuica Finlay, 1939 E Lw-Sl
Robertina lornensis Finlay, 1939 E Ab-Ar
Robertina pukeuriensis Hornibrook, 1961 E Po-Tt

Order GLOBIGERINIDA
GLOBIGERINOIDEA
GLOBIGERINIDAE
Beella digitata (Brady, 1879) Sc
Globigerina ampliapertura Bolli, 1957 Ar-Lwh
Globigerina angiporoides Hornibrook, 1965 Lwh
Globigerina a. minima Jenkins, 1966 Dp-Ak
Globigerina apertura Cushman, 1918 Wo-Wp
Globigerina boweri Bolli, 1957 Dm-Dp
Globigerina brazieri Jenkins, 1966 Lw
Globigerina brevis Jenkins, 1966 Ar-Lwh
Globigerina bulloides d'Orbigny, 1826 Lwh-Rec
Globigerina ciperoensis angulisuturalis Bolli, 1957 Lwh-Lw
Globigerina c. angustiumbilicata Bolli, 1957 Lwh-Tt
Globigerina c. ciperoensis Bolli, 1954 Lwh-Pl
Globigerina druryi Akers, 1955 Sw-Tt
Globigerina eamesi Blow, 1959 Lwh-Wc
Globigerina euapertura Jenkins, 1960 Lwh-Lw
Globigerina eugubina Luterbacher & Premoli Silva, 1964 Dt
Globigerina falconensis Blow, 1959 Pl-Rec
Globigerina foliata Bolli, 1957 Pl-Wn
Globigerina fringa Subbotina, 1950 Dt
Globigerina higginsi (Bolli, 1957) Dm-Ab
Globigerina juvenilis Bolli, 1957 Ld-Wq
Globigerina labiacrassata Jenkins, 1966 Lwh-Lw
Globigerina linaperta Finlay, 1939 Dp-Ar
Globigerina nepenthes Todd, 1957 Sw-Wo
Globigerina ouachitaensis Howe & Wallace, 1932 Ab-Pl
Globigerina parabulloides Blow, 1959 Tt-Wo
Globigerina pauciloculata Jenkins, 1966 Dt
Globigerina praebulloides Blow, 1959 Sw-Wo
Globigerina praeturritilina Blow & Banner, 1962 Ar-Lwh
Globigerina quinqueloba Natland, 1938 Po-Rec
Globigerina simplissima (Blow, 1979) Dt
Globigerina spiralis Bolli, 1957 Dt-Dw
Globigerina trilocularis d'Orbigny, 1826 Tk
Globigerina triloculinoides Plummer, 1926 Dt-Dm
Globigerina venezuelana Hedberg, 1937 Pl
Globigerina woodi Jenkins, 1960 Lw-Wn
Globigerina w. connecta Jenkins, 1964 Lw-Pl
Globigerina w. decoraperta Takayanagi & Saito, 1962 Sl-Wq
Globigerinatella insueta Cushman & Stainforth, 1945 Sc
Globigerinella aequilateralis (Brady, 1879) Tt-Rec
Globigerinella obesa (Bolli, 1957) Lwh-Lw
Globigerinoides altiapertur us Bolli, 1957 Po-Pl
Globigerinoides apertasuturalis Jenkins, 1960 Lwh-Tt
Globigerinoides bisphericus Todd, 1954 Pl-Sc
Globigerinoides conglobatus (Brady, 1879) Wm-Rec
Globigerinoides inusitatus Jenkins, 1966 Ld-Lw
Globigerinoides obliquus Bolli, 1957 Sc-Wm
Globigerinoides o. extremus Bolli & Bermudez, 1965 Wn
Globigerinoides ruber (d'Orbigny 1839) Pl-Wq
Globigerinoides sacculifer (Brady, 1877) Tt-Wp
Globigerinoides sicanus de Stefani, 1950 Sc
Globigerinoides trilobus (Reuss, 1850) Pl-Wn
Neoacarinina blowi Thompson, 1973 Wc
Orbulina suturalis (Brönnimann, 1951) Sl-Sw
Orbulina universa d'Orbigny, 1839 Sl-Rec
Praeorbulina circularis (Blow, 1956) Sc-Sl
Praeorbulina curva (Blow, 1956) Sc
Praeorbulina glomerosa (Blow, 1956) Sc-Sl
Sphaeroidinella seminulina (Schwager, 1866) Tt-Wo
Sphaeroidinellopsis disjuncta (Finlay, 1940) Pl-Sw
Sphaeroidinellopsis grimsdalei (Keijzer, 1957) Sl-Sw
Sphaeroidinellopsis kochi (Caudry, 1934) Sl
Sphaeroidinellopsis paenedehiscens Blow, 1969 Tt
HASTIGERINIDAE
Hastigerina pelagica (d'Orbigny, 1839) Wo-Rec
GLOBOROTALIOIDEA
CANDEINIDAE
Candeina zeocenica Hornibrook & Jenkins, 1965 Dp-Lwh
Globigerinita glutinata (Egger, 1893) Sw-Rec
Globigerinita iota Parker, 1962 Lw-Rec
Globigerinita uvula (Ehrenberg, 1861) Lwh-Rec
Tenuitella gemma Jenkins, 1966 Ar-Lwh
Tenuitella insolita Jenkins, 1966 Ak-Lwh
CATAPSYDRACIDAE
Catapsydrax dissimilis (Cushman & Bermudez, 1937) Lw-Pl
Catapsydrax echinatus Bolli, 1957 Ab-Ar
Catapsydrax martini (Blow & Banner, 1962) Ar-Lwh
Catapsydrax parvulus Bolli, Loeblich & Tappan, 1957 Sl-Tt
Catapsydrax stainforthi Bolli, Loeblich & Tappan, 1957 Pl-Sc
Catapsydrax unicavus Bolli, Loeblich & Tappan, 1957 Po-Pl
Globoquadrina altispira (Cushman & Jarvis, 1936) Sl-Sw
Globoquadrina dehiscens (Chapman, Parr & Collins, 1934) Lw-Tt
Globoquadrina langhiana Cita & Gelati, 1960 Sc-Sl
Globoquadrina larmeui Akers, 1955 Pl
Globoquadrina tripartita (Koch, 1926) Ld-Pl
Globorotaloides cancellata (Pessagno, 1963) Lwh-Lw
Globorotaloides suteri Bolli, 1957 Ak-Sl

Globorotaloides testarugosus (Jenkins, 1960) Lwh-Ld
Globorotaloides turgidus (Finlay, 1939) Dt-Ak
EOGLOBIGERINIDAE
Globoconusa daubjergensis (Brönnimann, 1953) Dt
GLOBOROTALIIDAE
Fohsella kugleri (Bolli, 1957) Ld-Po
Fohsella peripheroacuta (Blow & Banner, 1966) Sw
Fohsella peripheroronda (Blow & Banner, 1966) Pl-Sw
Globorotalia aculeata Jenkins, 1966 Ab-Ak
Globorotalia amuria Scott, Bishop & Burt, 1990 Sc-Sw
Globorotalia cavernula Bé, 1967 Wh-Rec
Globorotalia centralis Cushman & Bermudez, 1937Ab-Ar
Globorotalia challengeri Srinivasan & Kennett, 1981 Sw
Globorotalia conica Jenkins, 1960 Sc-Sw
Globorotalia crassacarina Scott, Bishop & Burt, 1990 Wn-Wc
Globorotalia crassaconica Hornibrook, 1981 Wo-Wp
Globorotalia crassaformis (Galloway & Wissler, 1927) Tk-Wn
Globorotalia crassula Cushman & Stuart, 1930 Wn-Rec
Globorotalia crozetensis Thompson, 1973 Wo-Wc
Globorotalia explicationis Jenkins, 1967 Sl-Tk
Globorotalia hirsuta (d'Orbigny, 1839) Wc-Rec
Globorotalia incognita Walters, 1965 Po-Pl
Globorotalia increbescens (Bandy, 1949) Ab-Ak
Globorotalia inflata (d'Orbigny 1839) Wo-Rec
Globorotalia i. triangula Theyer, 1973 Wp-Wn
Globorotalia juanai Bermudez & Bolli, 1969 Tt-Tk
Globorotalia kingmai Scott, Bishop & Burt, 1990 Pl
Globorotalia margaritae Bolli & Bermudez, 1965 Tk-Wo
Globorotalia menardii (Parker, Jones & Brady, 1865) Sw-Wn
Globorotalia minutissima Bolli, 1957 Po-Sc
Globorotalia miotumida Jenkins, 1960 Sl-Tk
Globorotalia miozea Finlay, 1939 Pl-Sc
Globorotalia m. wabagensis Bellford, 1984 Pl-Sl
Globorotalia mons Hornibrook, 1982 Tk-Wo
Globorotalia munda Jenkins, 1966 Lwh-Ld
Globorotalia opima Bolli, 1957 Lwh-Ld
Globorotalia panda Jenkins, 1960 Sl-Tt
Globorotalia paniae Scott, Bishop & Burt, 1990 Sc-Sw
Globorotalia pliozea Hornibrook, 1982 Tk-Wo
Globorotalia praehirsuta Blow, 1969 Wp
Globorotalia praemenardii Cushman & Stainforth, 1945 Sc-Sl
Globorotalia praescitula Blow, 1959 Pl-Pl
Globorotalia pseudobulloides (Plummer, 1926) Dt
Globorotalia puncticulata (Deshayes, 1832) Wo-Wp
Globorotalia puncticuloides Hornibrook, 1981 Wp-Wc
Globorotalia scitula (Brady, 1882) Sw-Rec
Globorotalia sphericomiozea Walters, 1965 Tk
Globorotalia subconomiozea Bandy, 1975 Wo-Wp
Globorotalia tosaensis Takayanagi & Saito, 1962 Wp-Wc
Globorotalia truncatulinoides (d'Orbigny, 1839) Wc-Rec
Globorotalia tumida (Brady, 1877) Wo-Rec
Globorotalia zealandica (Hornibrook, 1961) Pl-Sc
Neogloboquadrina acostaensis (Blow, 1959) Tk-Wo
Neogloboquadrina dutertrei (d'Orbigny, 1839) Tt-Rec
Neogloboquadrina humerosa (Takayanagi & Saito, 1962) Tk
Neogloboquadrina incompta (Cifelli, 1961) Tt-Rec
Neogloboquadrina pachyderma (Ehrenberg, 1861) Tt-Rec
Paragloborotalia barisanensis (Le Roy, 1939) Sc-Sl
Paragloborotalia bella (Jenkins, 1967) Pl-Sc
Paragloborotalia continuosa (Blow, 1959) Sl-Tt
Paragloborotalia mayeri (Cushman & Ellisor, 1939) Sc-Tt
Paragloborotalia nana (Bolli, 1957) Ak-Sc
Paragloborotalia nympha (Jenkins, 1967) Sw-Tt
Paragloborotalia pseudocontinuosa (Jenkins, 1967) Lwh-Sl
Paragloborotalia semivera (Hornibrook, 1961) Lwh-Sc
Planorotalites australiformis Jenkins, 1966 Dw-Dp
Planorotalites laevigata Bolli, 1957 Dt-Dh
Planorotalites pseudomenardii Bolli, 1957 Dt-Dw
Planorotalites pseudoscitula Glaessner, 1937 Dw
Planorotalites renzi Bolli, 1957 Ab-Ak
Turborotalia griffinae (Blow, 1979) Dm
Turborotalia praecentralis (Haque, 1966) Dm
Turborotalia reissi Loeblich & Tappan, 1957 Dw
PULLENIATINIDAE
Pulleniatina primalis Banner & Blow, 1967 Tk-Wo
TRUNCOROTALOIDIDAE
Acarinina acarinata (Subbotina, 1953) Dt
Acarinina aquiensis (Loeblich & Tappan, 1957) Dw
Acarinina convexa Subbotina, 1953 Dt
Acarinina cuneicamerata (Blow, 1979) Dm
Acarinina esnaensis (Le Roy, 1953) Dw
Acarinina mckannai (White, 1928) Dt-Ab
Acarinina soldadoensis (Brönnimann, 1952) Dw-Dh
Acarinina spinoinflata (Bandy, 1949) Ab
Acarinina strabocella Loeblich & Tappan, 1957 Dw
Globigerapsis index (Finlay, 1939) Ab-Ar
Globigerapsis i. barri (Bronnimann, 1952) Ab-Ar
Globigerapsis semiinvoluta (Keijzer, 1945) Ar
Morozovella acuta (Toulmin, 1941) Dw
Morozovella aequa Cushman & Renz, 1942 Dw
Morozovella a. bullata Jenkins, 1966 Dw
Morozovella a. marginodentata Subbotina, 1953 Dw
Morozovella aequa rex Martin, 1943 Dw-Dh
Morozovella angulata (White, 1928) Wp
Morozovella apanthesma Loeblich & Tappan, 1957 Dt-Dw
Morozovella crater Finlay, 1939 E Dm-Dh
Morozovella c. caucasica Glaessner, 1937 Dm-Dh
Morozovella dolabrata (Jenkins, 1966) Dw-Dh
Morozovella gracilis Bolli, 1957 Dw
Morozovella velascoensis (Cushman, 1925) Dw
Morozovella v. parva Rey, 1955 Dw
Pseudogloboquadrina primitiva (Finlay, 1947) Dw-Ak
Testacarinata inconspicua Howe, 1939 Ab-Ak
Truncorotaloides collactea (Finlay, 1939) Dm-Ak
Truncorotaloides mayoensis Brönnimann & Bermudez, 1953 Dh
Truncorotaloides pseudotopilensis (Subbotina, 1953) Dw-Ab
Truncorotaloides topilensis (Cushman, 1925) Ab
GLOBOTRUNCANOIDEA
GLOBOTRUNCANIDAE
Abathomphalus intermedius (Bolli, 1951) Mh
Abathomphalus mayaroensis (Bolli, 1951) Mh
Dicarinella canaliculata (Reuss, 1854) Rm-Rt
Globotruncana cretacea (d'Orbigny, 1840) Rm-Mp
Globotruncana inornata Bolli, 1957 Rm-Mp
Globotruncana rosetta (Carsey, 1926) Mh
Globotruncanella havanensis (Voorwijk, 1937) Mh
Globotruncanella petaloides (Gandolfi, 1955) Mh
Globotruncanita stuarti (de Lapparent, 1918) Mh
Rugotruncana circumnodifer (Finlay, 1940) Mh
RUGOGLOBIGERINIDAE
Rugoglobigerina macrocephala Brönnimann, 1952 Mh
Rugoglobigerina milamensis Smith & Pessagno, 1973 Mh
Rugoglobigerina pennyi Brönnimann, 1952 Mh
Rugoglobigerina pilula Belford, 1960 Rm-Mp
Rugoglobigerina pustulata Brönnimann, 1952 Mh
Rugoglobigerina rotundata Brönnimann, 1952 Mh
Rugoglobigerina rugosa (Plummer, 1927) Mh
Rugoglobigerina sp. indet. Mh
HANTKENINOIDEA
CASSIGERINELLIDAE
Cassigerinella chipolensis (Cushman & Ponton, 1932) Lwh-Sc
GLOBANOMALINIDAE
Globanomalina compressa (Plummer, 1926) Dt
Globanomalina wilcoxensis (Cushman & Ponton, 1932) Dw
Pseudohastigerina micra (Cole, 1927) Dw-Lwh
HANTKENINIDAE
Hantkenina alabamensis Cushman, 1925 Ak-Ar
Hantkenina australis Finlay, 1939 E Ab-Ak
HETEROHELICOIDEA
CHILOGUEMBELINIDAE
Chiloguembelina crinita (Glaessner, 1937) Dt-Dw
Chiloguembelina cubensis (Palmer, 1934) Lwh-Ld
Chiloguembelina ototara (Finlay, 1940) Ab-Lwh
Chiloguembelina subtriangularis Beckmann, 1957 Dt
Chiloguembelina trinitatensis (Cushman & Renz, 1942) Dw
Chiloguembelina waiparaensis Jenkins, 1966 Dt-Dw
Streptochilus pristinus Brönnimann & Resig, 1971 Ld-Lw
Zeauvigerina parri Finlay, 1939 Dm-Ak
Zeauvigerina teuria Finlay, 1947 Dt
Zeauvigerina zelandica Finlay, 1939 Dw-Ar
Zeauvigerina spp. indet. (2) Lw
GUEMBELITRIIDAE
Guembelitria cretacea Cushman, 1933 Mh-Dt
Guembelitria harrisi Tappan, 1940 Cn
Guembelitria triseriata (Terquem, 1882) Dw-Ld
HETEROHELICIDAE
Gublerina glaessneri Brönnimann & Browne, 1953 Mh
Gublerina sp. indet. Mh
Heterohelix glabrans (Cushman, 1938) Mh
Heterohelix globulosa (Ehrenberg, 1840) Ra-Mh
Heterohelix navarroensis Loeblich, 1951 Mh
Heterohelix pulchra (Brotzen, 1936) Mh
Heterohelix punctulata (Cushman, 1938) Mh
Heterohelix striata (Ehrenberg, 1840) Mh
Pseudotextularia deformis (Kikoine, 1948) Mh
Pseudotextularia elegans (Rzehak, 1891) Mh
PLANOMALINOIDEA
GLOBIGERINELLOIDIDAE
Globigerinelloides messinae (Brönnimann, 1952) Mh
Globigerinelloides multispina (Lalicker, 1948) Mh
Globigerinelloides subcarinata Brönnimann, 1952 Mh
Globigerinelloides volutus (White, 1928) Mh
ROTALIPOROIDEA
HEDBERGELLIDAE
Hedbergella cf. *gorbachikae* (Longoria, 1974) Cu-Cm
Hedbergella delrioensis (Carsey, 1926) Cu-Ra
Hedbergella holmdelensis Olsson, 1964 Rm-Mp
Hedbergella hoterivica (Subbotina, 1953) Cu-Cm
Hedbergella monmouthensis (Olsson, 1960) Mh
Hedbergella planispira (Tappan, 1940) Cu-Cn
Hedbergella sliteri Huber, 1990 Mh
Hedbergella sp. indet.
Praeglobotruncana novozealandica Stoneley, 1962 Cu
Whiteinella brittonensis (Loeblich & Tappan, 1961) Cn-Ra

Order BULIMINIDA
BOLIVINOIDEA
BOLIVINIDAE
Bolivina acerosa Cushman, 1936 Lw-Wo
Bolivina affiliata Finlay, 1939 E Pl-Wn
Bolivina alata (Seguenza, 1862) Lw-Rec
Bolivina albatrossi Cushman, 1922 Sl-Wn
Bolivina barnwelli Finlay, 1947 E Sl-Wo
Bolivina beyrichi Reuss, 1851 Lwh-Wn
Bolivina cacozela Vella, 1957 E Wo-Rec

Bolivina finlayi Hornibrook, 1961 E Lwh-Sl
Bolivina fyfei Hornibrook, 1958 E Tk-Wn
Bolivina hornibrooki Collen, 1972 E Pl-Wc
Bolivina incrassata Reuss, 1851 Mh
Bolivina lapsus Finlay, 1939 E Lwh-Wn
Bolivina lutosa Finlay, 1947 E Sl-Sw
Bolivina mahoenuica Hornibrook, 1961 E Lw-Sl
Bolivina mantaensis Cushman, 1929 Po-Pl
Bolivina mitcheli Gibson, 1967 E Po-Tk
Bolivina moodyensis Cushman & Todd, 1945 Ak-Lwh
Bolivina neocompacta McCulloch, 1981 Wc-Rec
Bolivina parri Cushman, 1936 Wo-Wc
Bolivina petiae Gibson, 1967 E Sw-Wo
Bolivina plicatella Cushman, 1930 Ab-Wn
Bolivina p. mera Cushman & Ponton, 1932 Po-Pl
Bolivina pontis Finlay, 1939 E Ar-Lwh
Bolivina pseudolobata Yassini & Jones, 1995 Pl-Rec
Bolivina pseudoplicata Heron-Allen & Earland, 1930 Sl-Rec
Bolivina pukeuriensis Hornibrook, 1961 E Po-Pl
Bolivina punctatostriata Kreuzberg, 1930 Pl
Bolivina pusilla Schwager, 1866 Po-Rec*
Bolivina pygmaea (Brady, 1881) Tt-Rec*
Bolivina reticulata Hantken, 1875 Lwh-Sw
Bolivina robusta Brady, 1881 Wm-Rec
Bolivina seminuda (Cushman, 1911) Tt-Rec
Bolivina semitruncata Hornibrook, 1961 E Ld-Sw
Bolivina silvestrina Cushman, 1936 Po-Pl
Bolivina spathulata (Williamson, 1858) Sl-Rec
Bolivina srinivasani (Kennett, 1967) E Wo
Bolivina subcompacta Finlay, 1947 E Lw-Sw
Bolivina targetensis Hornibrook, 1961 E Po-Pl
Bolivina tectiformis Cushman, 1926 Lwh-Lw
Bolivina turbiditorum Vella, 1963 E Tk
Bolivina variabilis (Williamson, 1858) Ld-Rec
Bolivina vellai Collen, 1972 E Wo-Wc
Bolivina wanganuiensis Collen, 1972 E Wo-Wc
Bolivina watti Gibson, 1967 E Sw-Wo
Bolivina zedirecta Finlay, 1947 E Pl-Tk
Bolivina spp. indet. (10)
Loxostomoides bortonica (Finlay, 1939) E Dw-Ar
Tappanina glaessneri (Finlay, 1947) E Dw
Tappanina olsoni (Hornibrook, 1961) E Ak-Lwh
BOLIVINOIDIDAE
Bolivinoides delicatulus Cushman, 1927 Dt
Bolivinoides dorreeni Finlay, 1940 E Mh
Bolivinoides draco (Marsson, 1878) Mh
Bolivinoides petersoni Brotzen, 1945Mh
BOLIVINITOIDEA
BOLIVINITIDAE
Abditodentrix pseudothalmanni (Boltovskoy & Guissani de Kahn, 1981) Po-Rec
Bolivinita compressa Finlay, 1939 E Tt-Wn
Bolivinita elegantissima Boomgart, 1949 Sw-Wo
Bolivinita finlayi Kennett, 1967 E Tk-Wo
Bolivinita pliobliqua Vella, 1963 E Tt
Bolivinita pliozea Finlay, 1939 E Tt-Wc
Bolivinita pohana Finlay, 1939 E Tt-Wo
Bolivinita quadrilatera (Schwager, 1866) Tt-Rec
Bolivinita spp. indet. (3)
BULIMINOIDEA
BULIMINELLIDAE
Buliminella browni Finlay, 1939 E Dw-Dh
Buliminella carseyae Plummer, 1931 Mh
Buliminella creta Finlay, 1939 E Dt
Buliminella elegantissima d'Orbigny, 1839 Pl-Rec
Buliminella grata Parker & Bermudez, 1937 Lw-Po
Buliminella missilis Vella, 1963 E Lwh-Tk
Buliminella spicata Cushman & Parker, 1942 Lw-Po
Buliminella sp. indet. Sl-Sw
BULIMINIDAE
Bulimina bortonica Finlay, 1939 E Ab
Bulimina bremneri Finlay, 1940 E Po-Wo
Bulimina elongata d'Orbigny, 1826 Wn-Rec
Bulimina exilis Brady, 1884 Tt-Rec
Bulimina forticosta Finlay, 1940 E Ar-Lwh
Bulimina mapiria Finlay, 1940 E Tt
Bulimina marginata f. *acanthia* Costa, 1856 Wc-Rec
Bulimina m. f. *aculeata* d'Orbigny, 1826 Tk-Rec
Bulimina m. f. *marginata* d'Orbigny, 1826 Wo-Rec
Bulimina midwayensis Cushman & Parker, 1936 Mh-Dt
Bulimina miolaevis Finlay, 1940 E Lwh-Tk
Bulimina navarroensis Cushman & Parker, 1946 Dt
Bulimina pahiensis Finlay, 1940 E Dp-Ab
Bulimina prolixa Cushman & Parker, 1935 Mh-Dw
Bulimina quadrata Plummer, 1927 Dt
Bulimina rakauroana Finlay, 1940 E Mh
Bulimina scobinata Finlay, 1940 E Lwh-Lw
Bulimina senta Finlay, 1940 E Lw-Tt
Bulimina serratospina Finlay, 1947 E Dw-Dh
Bulimina striata d'Orbigny, 1826 Ab-Rec
Bulimina truncana Gumbel, 1868 Lw-Rec
Bulimina truncanella Finlay, 1940 E Dw-Tt
Bulimina tuxpamensis Cole, 1928 Dt-Dh
Bulimina vellai Gibson, 1967 E Tt
Bulimina spp. indet. (4)
Globobulimina kickapooensis Cole, 1938 Mh-Dt
Globobulimina montereyana (Kleinpell, 1938) Po-Tt
Globobulimina pacifica Cushman, 1927 Ld-Rec
Globobulimina perversa (Cushman, 1921) Tt
Globobulimina pupula (Stache, 1864) Ab-Tt
Globobulimina turgida (Bailey, 1851) Ld-Rec
Praeglobobulimina pupoides (d'Orbigny, 1846) Sw-Tt
Praeglobobulimina spinescens (Brady, 1884) Tt-Rec
PAVONINIDAE
Finlayina hornibrooki Hayward & Morgans, 1981 E Lwh
Pavonina triformis Parr, 1933 Pl-Sw
REUSSELLIDAE
Chrysalidinella costata (Heron-Allen & Earland, 1924) Pl
Reusella szajnochae (Grzybowski, 1896) Mh
Reussella attenuata Hornibrook, 1961 E Ar-Ld
Reussella finlayi Dorreen, 1948 E Ak-Ld
Reussella spinulosa (Reuss, 1850) Po-Rec
Reussella sp. indet. Ld
SIPHOGENERINOIDIDAE
Rectobolivina hangaroana Finlay, 1940 E Tt
Rectobolivina maoria Finlay, 1939 E Pl-Wn
Rectobolivina maoriella Finlay, 1939 E Lw-Tt
Rectobolivina parvula Finlay, 1939 E Sc-Tt
Rectobolivina striatula (Cushman, 1917) Tt-Rec
Rectuvigerina clifdenensis Hornibrook, 1989 E Ar-Lwh
Rectuvigerina ongleyi (Finlay, 1939) E Lw-Tk
Rectuvigerina pohana (Finlay, 1939) E Tt-Wo
Rectuvigerina postprandia (Finlay, 1939) E Ar-Lwh
Rectuvigerina prisca (Finlay, 1939) E Ab-Ak
Rectuvigerina rerensis (Finlay, 1939) E Lw-Sc
Rectuvigerina ruatoria (Vella, 1961) E Pl-Sw
Rectuvigerina striata (Schwager, 1866) Wo-Wn*
Rectuvigerina striatissima (Stache, 1864) E Lwh-Ld
Rectuvigerina vesca (Finlay, 1939) E Sl-Sw
Rectuvigerina spp. indet. (3) Dw
Saidovina karreriana (Brady, 1881) Wq-Rec
Saidovina subangularis (Brady, 1881) Sc
Tritubulogenerina waiparaensis (Hornibrook, 1958) E Mh-Dt
Tubulogenerina haywardi Gibson, 1989 E Pl
UVIGERINIDAE
Kolesnikovella australis (Heron-Allen & Earland, 1924) Ab-Tt
Kolesnikovella zealandica Kennett, 1967 E Tt-Tk
Neouvigerina ampullacea (Brady, 1884) Tt-Rec
Neouvigerina aotea Vella, 1961 E Pl
Neouvigerina auberiana d'Orbigny, 1839 Tt-Rec
Neouvigerina bellula Vella, 1963 E Wo
Neouvigerina canariensis (d'Orbigny, 1839) Lwh-Wn
Neouvigerina eketahuna Vella, 1963 E Tt-Tk
Neouvigerina hispida (Schwager, 1866) Sl-Rec
Neouvigerina interrupta (Brady, 1879) Pl-Rec
Neouvigerina moorei Vella, 1961 E Sl-Sw
Neouvigerina plebeja Vella, 1961 E Po-Sl
Neouvigerina p. waiapuensis Vella, 1961 E Po
Neouvigerina proboscidea (Schwager, 1866) Po-Rec
Neouvigerina toddi Vella, 1961 E Sc
Neouvigerina vadescens (Cushman, 1933) Wo-Rec
Trifarina angulosa (Williamson, 1858) Tt-Rec
Trifarina bradyi Cushman, 1923 Ab-Rec
Trifarina continua Hornibrook, 1961 E Lw-Pl
Trifarina costornata (Hornibrook, 1961) E Po-Sw
Trifarina edwardsi Srinivasan, 1966 E Ak-Ar
Trifarina elliptica (Dorreen, 1948) E Ak-Lwh
Trifarina esuriens Hornibrook, 1961 E Lwh-Wn
Trifarina juniornata Hornibrook, 1961 E Po-Sw
Trifarina kaiata Srinivasan, 1966 E Ak-Lwh
Trifarina occidentalis (Cushman, 1923) Tt-Rec
Trifarina oligocaenica (Andreae, 1894) Ld-Po
Trifarina ototara (Hornibrook, 1961) E Ak-Ld
Trifarina parva Hornibrook, 1961 E Ab-Sl
Trifarina tortuosa (Hornibrook, 1961) E Lw-Sw
Trifarina spp. indet. (7)
Uvigerina bortotara (Finlay, 1939) E Ab-Lwh
Uvigerina b. costata Dorreen, 1948 E Ab-Ar
Uvigerina bradleyi (Srinivasan, 1966) E Ak-Ar
Uvigerina bradyana Fornasini, 1900 Tt-Rec
Uvigerina delicatula (Vella, 1961) E Tt-Wo
Uvigerina dorreeni Finlay, 1939 E Ld-Po
Uvigerina euteres (Vella, 1961) E Tt-Tk
Uvigerina gargantua (Vella, 1961) E Sl
Uvigerina lutorum (Vella, 1963) E Sw-Wo
Uvigerina mata (Vella, 1961) E Pl-Sc
Uvigerina m. zealta (Vella, 1961) E Sl-Sw
Uvigerina maynei Chapman, 1926 E Lwh-Lw
Uvigerina minima (Vella, 1961) E Pl-Sl
Uvigerina mioschwageri Finlay, 1939 E Sl-Sw
Uvigerina miozea Finlay, 1939 E Lw-Sc
Uvigerina paeniteres Finlay, 1939 E Pl-Tt
Uvigerina peregrina Cushman, 1923 Tt-Rec
Uvigerina picki (Vella, 1961) E Lw-Po
Uvigerina pliozea (Vella, 1963) E Wo-Wn
Uvigerina primigena (Vella, 1961) E Sl
Uvigerina pseudojavana (Vella, 1961) E Tt
Uvigerina pygmea d'Orbigny, 1826 Tt-Wn
Uvigerina rodleyi (Vella, 1963) E Sw-Wc
Uvigerina r. tutamoides (Vella, 1963) E Wo-Wp
Uvigerina semiteres (Vella, 1961) E Sl
Uvigerina taranakia (Vella, 1961) E Tt-Tk
Uvigerina tutamoea (Vella, 1961) E Sc-Tt
Uvigerina t. altonica (Vella, 1961) E Pl
Uvigerina wanzea (Finlay, 1939) E Dp-Ab
Uvigerina whakatua (Vella, 1961) E Pl
Uvigerina zeacuminata (Vella, 1961) E Pl-Tk
Uvigerina zelamina (Vella, 1961) E Sw-Tk
Uvigerina zelandica (Vella, 1963) E Wo
Uvigerina spp. indet. (12)
CASSIDULINOIDEA
CASSIDULINIDAE
Burseolina pacifica (Cushman, 1925) Sl-Wc
Cassidulina carapitana Hedberg, 1937 Pl
Cassidulina carinata Silvestri, 1896 Ld-Rec
Cassidulina laevigata d'Orbigny, 1826 Ar-Rec
Cassidulina margareta Karrer, 1877 Po-Pl
Cassidulina monstruosa Voloshinova, 1952 Po
Cassidulina norvangi Thalmann, 1952 Po-Rec
Ehrenbergina aspinosa Parr, 1950 Tk-Rec
Ehrenbergina bicornis Brady, 1888 Lw-Po
Ehrenbergina carinata Eade, 1967 E Tk-Rec
Ehrenbergina fyfei Finlay, 1939 E Tt-Tk
Ehrenbergina marwicki Finlay, 1939 E Po-Pl
Ehrenbergina mestayeri Cushman, 1922 E Wo-Rec
Ehrenbergina osbornei Finlay, 1939 E Sl-Sw

Ehrenbergina trigona Goës, 1896 Wc-Rec
Ehrenbergina sp. indet. Pl-Sl
Evolvocassidulina belfordi Nomura, 1983 Po-Rec
Evolvocassidulina bradyi (Norman, 1881) Lwh-Rec
Evolvocassidulina chapmani Parr, 1931 Po-Pl
Evolvocassidulina orientalis (Cushman, 1922) Lwh-Rec
Globocassidulina arata (Finlay, 1939) E Po-Tt
Globocassidulina canalisuturata Eade, 1967 Tk-Rec
Globocassidulina cuneata (Finlay, 1940) E Lwh-Wo
Globocassidulina minuta (Cushman, 1933) Sw-Rec
Globocassidulina producta (Chapman & Parr, 1937) Ld
Globocassidulina pseudocrassa (Hornibrook, 1961) E Lwh-Lw
Globocassidulina subglobosa (Brady, 1881) Dh-Rec
DELOSINOIDEA
CAUCASINIDAE
Baggatella inconspicua Howe, 1939 Dt-Lwh
Francesita advena (Cushman, 1922) Wm-Rec
EOUVIGERINOIDEA
LACOSTEINIDAE
Lacosteina gouskovi Marie, 1945 Mh
FURSENKOINOIDEA
FURSENKOINIDAE
Cassidella bradyi (Cushman, 1922) Tt-Rec
Fursenkoina bramletti (Galloway & Morrey, 1929) Lw-Tt
Fursenkoina complanata (Egger, 1893) Tt-Rec
Fursenkoina schreibersiana (Czjzek, 1848) Ar-Rec
Fursenkoina texturata (Brady, 1884) Po-Wc*
Fursenkoina vellai Kennett, 1966 E Tt-Wo
Rutherfordoides rotundata (Parr, 1950) Tt-Wn
Sigmavirgulina tortuosa (Brady, 1881) Ld-Pl
Virguloides wellmani Srinivasan, 1966 E Ab-Lwh
LOXOSTOMATOIDEA
BOLIVINELLIDAE
Bolivinella australis Cushman, 1929 Pl
Bolivinella profolium Hayward, 1982 E Pl-Sw
Inflatobolivinella lilliei (Hayward, 1982) E Sw-Tt
Inflatobolivinella subrugosa zealandica Hayward, 1990 E Lwh-Lw
Nodobolivinella nodosa Hayward, 1990 Lwh-Ld
Quasibolivinella finlayi Hayward, 1982 E Ld-Sw
Quasibolivinella taylori Quilty, 1981 Lwh
Rugobolivinella bensoni (Hayward, 1982) E Lw-Pl
Rugobolivinella rugosa (Howe, 1930) Lwh-Tt
Rugobolivinella scotti Hayward, 1990 E Pl-Tt
LOXOSTOMATIDAE
Aragonia ouezzanensis (Rey, 1955) Dt
Aragonia tenera Finlay, 1947 E Dw
Aragonia zelandica Finlay, 1939 E Dm-Ab
Loxostomum goodwoodensis Hornibrook, 1961 E Pl
Loxostomum limonense (Cushman, 1926) Dt
Loxostomum pakaurangiense Hornibrook, 1958 E Po-Sw
Loxostomum truncatum Finlay, 1947 E Sw-Tt
Loxostomum spp. indet. (3)
PLEUROSTOMELLOIDEA
PLEUROSTOMELLIDAE
Amplectoductina multicostata (Galloway & Morrey, 1929) Dw-Wn*
Ellipsodimorphina reussi (Berthelin, 1880) Cu
Ellipsoglandulina compacta Hillebrandt, 1962, Ar*
Ellipsoglandulina labiata (Schwager, 1866) Mh-Wc
Ellipsoglandulina velascoensis Cushman, 1936 Mh*
Ellipsoidella dacica Neagu, 1968 Ab*
Ellipsoidella heronalleni Storm, 1929 Dt-Dw*
Ellipsoidella pleurostomelloides Heron-Allen & Earland, 1910 Ar-Wn*
Ellipsoidella salmojraghii (Martinotti, 1924) Dt-Ar*
Ellipsoidella subtuberosa Liebus, 1928 Dt-Wn*
Ellipsoidina abbreviata Seguenza, 1859 Ab-Wn*
Ellipsoidina ellipsoides Seguenza, 1859 Ab-Wn*
Ellipsopolymorphina rimosa (Cushman & Bermudez, 1937) Wn
Ellipsopolymorphina rostrata (Silvestri, 1904) Tt-Wo
Ellipsopolymorphina russitanoi Silvestri, 1904 Wo-Wc*
Ellipsopolymorphina zuberi (Grzybowski, 1896) Wn-Wc*
Ellipsopolymorphina n.sp. Tt-Wn*
Neopleurostomella polymorpha (Popescu & Crihan, 2005) Wc
Nodosarella frequens (Storm, 1929) Dt-Pl*
Nodosarella inaequalis (Silvestri, 1901) Dw-Wn*
Nodosarella lorifera (Halkyard, 1918) Dh-Pl*
Nodosarella rotundata (d'Orbigny, 1846) Dt-Wc*
Nodosarella subnodosa (Guppy, 1894) Ab
Pleurostomella acuminata Cushman, 1922 Pl-Wc
Pleurostomella acuta Hantken, 1875 Dw-Wn*
Pleurostomella alazanensis Cushman, 1925 Dw-Wc*
Pleurostomella alternans Schwager, 1866 Dt-Wc
Pleurostomella bierigi Palmer & Bermudez, 1934 Ar-Ld
Pleurostomella bolivinoides Schubert, 1911 Sl-Sw
Pleurostomella brevis Schwager, 1866 Dw-Wn
Pleurostomella concava Hermelin, 1991 Wc*
Pleurostomella frons Todd, 1957 Tt-Wc*
Pleurostomella gracillima (Cushman, 1933) Mh-Ar
Pleurostomella incrassata Hantken, 1883 Dt-Ab*
Pleurostomella nuttalli Cushman & Seigfus, 1939 Ar*
Pleurostomella subnodosa (Reuss, 1851) Ab-Wc
Pleurostomella pleurostomella (Silvestri, 1904) Lw–Wc*
Pleurostomella sapperi Schubert, 1911 Wo-Wc
Pleurostomella tenuis Hantken, 1883 Ar-Wn
Pleurostomella wadowicensis Grzybowskii, 1896 Dw-Wc*
Pleurostomella n.sp. aff. *alazanensis* (Cushman, 1925) Wn*
STILOSTOMELLOIDEA
STILOSTOMELLIDAE
Siphonodosaria alexanderi (Cushman, 1936) Mh-Dt
Siphonodosaria basicarinata (Hornibrook, 1961) E Dm-Ld
Siphonodosaria bradyi (Cushman, 1927) Wo*
Siphonodosaria consobrina (d'Orbigny, 1846) Po-Tt*
Siphonodosaria cooperensis (Cushman, 1933) Dw-Ab*
Siphonodosaria curvatura (Cushman, 1939) Po-Tt*
Siphonodosaria dentaliniformis (Cushman & Jarvis, 1934) Tt-Wc
Siphonodosaria gracillima (Cushman & Jarvis, 1934) Ld-Wc
Siphonodosaria hochstetteri (Schwager, 1866) Dw-Sc
Siphonodosaria insecta (Schwager, 1866) Ld-Po
Siphonodosaria jacksonensis (Cushman & Applin, 1926) Mh-Wc
Siphonodosaria lepidula (Schwager, 1866) Ld-Wc
Siphonodosaria lohmanni (Kleinpell, 1938) Po-Wc
Siphonodosaria longispina (Egger, 1900) Ab-Wo
Siphonodosaria minuta (Cushman, 1938) Lw-Sw
Siphonodosaria paucistriata (Galloway & Morrey, 1931) Ab-Wn
Siphonodosaria pomuligera (Stache, 1864) E Dt-Wc
Siphonodosaria stephensoni (Cushman, 1936) Ab*
Siphonodosaria subspinosa (Cushman, 1943) Ab-Lwh*
Siphonodosaria tauricornis (Schwager, 1866) Wn-Wc*
Siphonodosaria typa (Seguenza, 1880) Dt*
Siphonodosaria n.sp. aff. *adolphina* (d'Orbigny, 1846) Wn
Stilostomella decurta (Bermudez, 1937) Tt*
Stilostomella fistuca (Schwager, 1866) Tt-Wc
Stilostomella parexilis (Cushman & Stewart, 1930) Wo-Wc*
Stilostomella rugosa Guppy, 1894 Dp*
Stilostomella n.sp. Wo*
Strictocostella advena (Cushman & Laiming, 1931) Lwh-Wn
Strictocostella hyugaensis (Ishizaki, 1943) Wn-Wc
Strictocostella japonica (Ishizaki, 1943) Wn*
Strictocostella joculator (Finlay, 1947) E Po-Tt
Strictocostella matanzana (Palmer & Bermudez, 1936) Dh-Wc*
Strictocostella modesta (Bermudez, 1937) Wo-Wn
Strictocostella scharbergana (Neugeboren, 1856) Ar-Wc
Strictocostella spinata (Cushman, 1934) Wo-Wc
Strictocostella n.sp. Wo-Wn*
Strictocostella n.sp. Wn
Strictocostella sp. indet. Wn*
TURRILINOIDEA
STAINFORTHIIDAE
Hopkinsina mioindex Finlay, 1947 E Sl-Wn
Virgulopsis costata Hornibrook, 1961 E Lw-Sl
Virgulopsis junior Hornibrook, 1961 E Po-Sc
Virgulopsis pustulata Finlay, 1939 E Lw-Sw
Virgulopsis reticulata Hornibrook, 1961 E Lw-Pl
Virgulopsis turris (Heron-Allen & Earland, 1922) Po-Rec
TURRILINIDAE
Neobulimina canadaensis Cushman & Wickenden, 1928 Cn
Neobulimina sp. indet. Mh

Order ROTALIIDA
ACERVULINOIDEA
ACERVULINIDAE
Sphaerogypsina globulus (Reuss, 1843) Po-Rec
Sphaerogypsina sp. indet. Ar-Sw
HOMOTREMATIDAE
Sporadotrema sp. indet. Sc
ASTERIGERINOIDEA
ALFREDINIDAE
Anomalina aotea Finlay, 1940 E Dw-Tt
Anomalina clarentica Stoneley, 1962 E Cu
Anomalina hectori Stoneley, 1962 E Cu
Anomalina macphersoni Stoneley, 1962 E Cu
AMPHISTEGINIDAE
Amphistegina aucklandica Karrer, 1864 E Lw-Sw
Amphistegina eyrensis Larsen, 1978 E Dm-Dh
Amphistegina hauerina d'Orbigny, 1846 Sc-Sw
Amphistegina mamilla (Fichtel & Moll, 1798) Sc
ASTERIGERINATIDAE
Asterigerinella gallowayi Bandy, 1944 Ar
Eoeponidella pulchella (Parker, 1952) Tt-Rec
Eoeponidella scotti Hayward, 1979 E Po-Pl
Eoeponidella zealandica (Hornibrook, 1961) E Ar-Ld
ASTERIGERINIDAE
Asterigerina cyclops Dorreen, 1948 E Ak-Lwh
Asterigerina lornensis Finlay, 1939 E Ak-Ar
Asterigerina waiareka Finlay, 1939 E Ak-Lwh
Asterigerina sp. indet. Dm
EPISTOMARIIDAE
Nuttallides bradyi (Earland, 1934) Lw-Rec*
Nuttallides carinotruempyi Finlay, 1947 E Dt-Dh
Nuttallides cretatruempyi Finlay, 1947 E Dt-Dh
Nuttallides florealis (White, 1928) Dt
Nuttallides tholus Finlay, 1940 E Mh
Nuttallides truempyi (Nuttall, 1930) Dp-Ab
Nuttallides umbonifera (Cushman, 1933) Sl-Rec
LEPIDOCYCLINIDAE
Lepidocyclina dilatata dilatata (Michelotti, 1851) Po
Lepidocyclina ephippoides Jones & Chapman, 1900 Po
Nephrolepidina hornibrooki Matsumaru, 1971 Sl
Nephrolepidina howchini (Chapman & Crespin, 1932) Sc
Nephrolepidina orakiensis (Karrer, 1864) E Lw-Sw
Nephrolepidina waikukuensis Chapronière, 1984 E Sl-Sw

CHILOSTOMELLOIDEA
ALABAMINIDAE
Alabamina australis Belford, 1960 Mh
Alabamina creta (Finlay, 1940) E Mh-Dt
Alabamina tenuimarginata (Chapman, Parr & Collins, 1934) Dm-Wo
CHILOSTOMELLIDAE
Abyssamina poagi Schniker & Tjalsma, 1980 Dm-Ab
Abyssamina quadrata Schniker & Tjalsma, 1980 Ab
Allomorphina spp. indet. (2)
Allomorphina conica Cushman & Todd, 1947 Dt-Dh
Allomorphina cretacea Reuss, 1851 Mh
Allomorphina cubensis Palmer & Bermudez, 1936 Lw-Sw
Allomorphina pacifica Cushman & Todd, 1949 Po-Pl
Allomorphina trigona Reuss, 1850 Dh-Sw
Allomorphina trochoides (Reuss, 1845) Mh
Allomorphina whangaia Finlay, 1939 E Mh-Dt
Chilostomella czizeki Reuss, 1850 Po-Pl
Chilostomella globata Galloway & Heminway, 1941 Po-Pl
Chilostomella oolina Schwager, 1878 Po-Rec
Chilostomella ovoidea Reuss, 1850 Po-Rec
Chilostomelloides oviformis (Sherborn & Chapman, 1886) Po-Pl
COLEITIDAE
Coleites gagei Hornibrook, 1961 E Ar
Coleites sp. indet. Dp-Ab
GAVELINELLIDAE
Gavelinella ammonoides (Reuss, 1845) Cn
Gavelinella beccariiformis (White, 1928) Mh-Dt
Gavelinella complanata (Reuss, 1851) Rm-Mp
Gavelinella emaciata Stoneley, 1962 E Cu
Gavelinella flemingi Stoneley, 1962 E Cu
Gavelinella intermedia (Berthelin, 1880) Cn
Gavelinella marwicki Stoneley, 1962 E Cu
Gavelinella moorei Stoneley, 1962 E Cu
Gavelinella sp. indet. Cu
Gyroidina bradyi (Trauth, 1918) Sl-Rec
Gyroidina danvillensis Howe & Wallace, 1932 Dw-Wn
Gyroidina jenkinsi Srinivasan, 1966 E Ak-Lwh
Gyroidina wellmani Stoneley, 1962 E Cu
Gyroidina sp. indet. Lwh-Wo
Gyroidinoides allani (Finlay, 1939) E Lwh-Lw
Gyroidinoides girardanus (Reuss, 1851) Dm-Lw
Gyroidinoides globosus (von Hagenow, 1842) Rm-Dt
Gyroidinoides hornibrooki Stoneley, 1962 E Cu
Gyroidinoides kawagatai Ujiié, 1995 Lw-Rec
Gyroidinoides naranjoensis (White, 1928) Mh
Gyroidinoides neosoldanii (Brotzen, 1936) Dp-Wo
Gyroidinoides perampla (Cushman & Stainforth, 1945) Dp-Ab
Gyroidinoides primitivus Hofker, 1957 Cu-Ra
Gyroidinoides prominulus (Stache, 1864) E Ab-Wn
Gyroidinoides pulisukensis (Saidova, 1975) Tt-Rec
Gyroidinoides scrobiculatus (Finlay, 1939) E Ab-Ar
Gyroidinoides soldanii (d'Orbigny, 1826) Lw-Rec
Gyroidinoides stineari (Finlay, 1939) E Tt-Wn
Gyroidinoides subangulata (Plummer, 1927) Dt
Gyroidinoides subzelandicus Hornibrook, 1961 E Lw-Sw
Gyroidinoides zelandicus (Finlay, 1939) E Lwh-Wn
Hanzawaia grossepunctata (Earland, 1934) Wc-Rec
Hanzawaia laurisae (Mallory, 1959) Po-Pl
Lingulogavelinella asterigerinoides (Plummer, 1931) Cn
Notoplanulina rakauroana (Finlay, 1939) E Mh
Nummodiscorbis novozealandicus Hornibrook, 1961 E Ak-Sw
Orithostella indica Scheibnerova, 1974 Cu-Cm
Rotaliatina sulcigera (Stache, 1864) E Ab-Ld
Rotaliatina sp. indet. Ab-Ak
GLOBOROTALITIDAE
Globorotalites micheliniana (d'Orbigny, 1840) Mh
HETEROLEPIDAE
Anomalinoides alazanensis Nuttall, 1932 Pl-Wo
Anomalinoides awamoana Hornibrook, 1961 E Lw-Pl
Anomalinoides capitata (Guembel, 1868) Dw-Ak
Anomalinoides chathamensis Strong, 1979 E Mh
Anomalinoides eoglabra Finlay, 1940 E Mh-Lwh
Anomalinoides eosuturalis Finlay, 1940 E Dw-Lw
Anomalinoides fasciatus (Stache, 1864) E Ab-Po
Anomalinoides finlayi Stoneley, 1962 E Cu
Anomalinoides globulosus (Chapman & Parr, 1937) Po-Pl
Anomalinoides macraglabra (Finlay, 1940) E Ar-Sw
Anomalinoides miosuturalis Finlay, 1940 E Ld-Sl
Anomalinoides nobilis Brotzen, 1948 Dt
Anomalinoides orbiculus (Stache, 1864) E Dw-Po
Anomalinoides parvumbilius (Finlay, 1940) E Lwh-Wn
Anomalinoides pinguiglabra (Finlay, 1940) E Lwh-Tt
Anomalinoides piripaua (Finlay, 1939) E Mh-Dt
Anomalinoides pseudogrosserugosa Colon, 1945 Lwh-Sc
Anomalinoides rubiginosus (Cushman, 1926) Mh-Dt
Anomalinoides semicribrata (Beckmann, 1953) Po-Pl
Anomalinoides semiteres (Finlay, 1940) E Ab-Ar
Anomalinoides sphericus (Finlay, 1940) E Sl-Rec
Anomalinoides subnonionoides (Finlay, 1940) E Lwh-Tk
Anomalinoides vitrinodus (Finlay, 1940) E Po-Wo
Anomalinoides spp. indet. (2) Mh
KARRERIIDAE
Karreria maoria (Finlay, 1939) E Po-Rec
ORIDORSALIDAE
Oridorsalis umbonatus (Reuss, 1851) Ab-Rec
OSANGULARIIDAE
Charltonina acutimarginata (Finlay, 1940) E Mh-Dt
Osangularia bengalensis (Schwager, 1866) Sw-Rec
Osangularia culter (Parker & Jones, 1865) Ab-Rec
Osangularia lens Brotzen, 1940 Dw-Dh
Osangularia navarroana (Cushman, 1938) Mh-Dt
Osangularia plummerae Brotzen, 1940 Dw-Pl
QUADRIMORPHINIDAE
Quadrimorphina allomorphinoides (Reuss, 1860) Mh-Dp
Quadrimorphina pescicula Saidova, 1975 Wm-Rec
Quadrimorphina profunda Schnitker & Tjalsma, 1980 Dw-Dp
Quadrimorphina sp. indet. Dw-Dp
TRICHOHYALIDAE
Buccella lotella Hornibrook, 1961 E Ak-Sw
Buccella sp. indet. Po-Pl
DISCORBOIDEA
BAGGINIDAE
Baggina ampla (Finlay, 1940) E Lw-Wp
Baggina brevior (Finlay, 1940) E Pl
Baggina spp. indet. (2) Lwh-Ld
Cancris compressus Finlay, 1940 E Ab-Lwh
Cancris laevinflatus Hornibrook, 1961 E Lwh-Sw
Cancris lateralis Finlay, 1940 E Lwh-Tk
Cancris l. minima Srinivasan, 1966 E Ar-Lwh
Cancris oblongus (Williamson, 1858) Tk-Rec
Valvulineria saulcii (d'Orbigny, 1839) Po-Wc
Valvulineria stachei Hornibrook, 1961 E Ar
Valvulineria teuriensis Loeblich & Tappan, 1939 Mh-Dt
BUENINGIIDAE
Bueningia creeki Finlay, 1939 E Ld-Tt
CONORBINIDAE
Conorbina sp. indet. Cu
DISCORBIDAE
Discorbis balcombensis Chapman, Parr & Collins, 1934 Ar-Sw
Discorbis perditus Strong, 1979 E Rt-Mh
Discorbis rajnathi Srinivasan, 1966 E Lwh
Discorbis sp. indet. Mh
Trochulina dimidiatus (Jones & Parker, 1862) Tt-Rec
EPONIDIDAE
Alabaminella weddellensis (Earland, 1936) Po-Rec
Eponides bollii Cushman & Renz, 1946 Mh
Eponides broeckhianus (Karrer, 1878) Ab-Wp
Eponides concinna Brotzen, 1960 Mh
Eponides cribrorepandus (Asano & Uchio, 1951) Ak-Rec
Eponides dorsopustulatus Dorreen, 1948 E Ab-Ar
Eponides lornensis Finlay, 1939 E Ab-Lw
Eponides motuensis Stoneley, 1962 E Cu
Eponides repandus (Fichtel & Moll, 1803) Ak-Rec
Eponides tethycus Dorreen, 1948 E Ab-Ar
Eponides sp. indet. Ak
Hofkerina semiornata (Howchin, 1889) Lwh-Pl
Ioanella tumidula (Brady, 1884) Sl-Rec
Porogavelinella ujiiei Kawagata, 1999 Pl-Rec
MISSISSIPPINIDAE
Stomatorbina concentrica (Parker & Jones, 1864) Dm-Rec
PLACENTULINIDAE
Patellinella inconspicua (Brady, 1884) Tk-Rec
ROSALINIDAE
Gavelinopsis praegeri (Heron-Allen & Earland, 1913) Tt-Rec
Gavelinopsis pukeuriensis Hornibrook, 1961 E Po-Wc
Gavelinopsis zealandica (Hornibrook, 1961) E Ld-Sw
Gavelinopsis sp. indet. Lwh-Sc
Neoconorbina terquemi (Rzehak, 1888) Po-Rec
Neoconorbina spp. indet. (2)
Planodiscorbis rarescens (Brady, 1884) Lwh-Rec
Pseudopatellinoides primus Krasheninnikov, 1958 Po
Rosalina augur Hornibrook, 1961 E Po-Sw
Rosalina bradyi (Cushman, 1915) Sw-Rec
Rosalina concinna (Brady, 1884) Po-Pl
Rosalina irregularis (Rhumbler, 1906) Tt-Rec
Rosalina paupereques Vella, 1957 E Wc-Rec
Rosalina vitrizea Hornibrook, 1961 E Lw-Rec
Rosalina sp. indet. Dt
Semirosalina deflata Hornibrook, 1961 E Pl
Semirosalina inflata Hornibrook, 1961 E Pl
SPHAEROIDINIDAE
Sphaeroidina bulloides d'Orbigny, 1826 Ar-Rec
Sphaeroidina variabilis Reuss, 1851 Ab-Ar
Sphaeroidina sp. indet. Dm
DISCORBINELLOIDEA
DISCORBINELLIDAE
Colonimilesia coronata (Heron-Allen & Earland, 1932) Ab-Rec
Discorbinella apposita Finlay, 1940 E Dm-Lwh
Discorbinella bertheloti (d'Orbigny 1839) Ld-Rec
Discorbinella complanata (Sidebottom, 1918) Ar-Rec
Discorbinella deflata (Finlay, 1940) E Sc-Rec
Discorbinella galera (Finlay, 1940) E Po-Pl
Discorbinella jugosa Finlay, 1940 E Dt-Ar
Discorbinella lepida Hornibrook, 1961 E Po-Wc
Discorbinella scopos (Finlay, 1940) E Lwh-Pl
Discorbinella stachi (Asano, 1951) Po-Pl
Discorbinella subcomplanata (Parr, 1950) Tk-Rec
Discorbinella timida Hornibrook, 1961 E Po-Wc
Discorbinella turgida (Finlay, 1940) E Lwh-Pl
Discorbinella vitrevoluta (Hornibrook, 1961) E Pl-Rec
Discorbinella sp. indet. Mh
Laticarinina altocamerata (Heron-Allen & Earland, 1922) E Ak-Rec
Laticarinina nitida (Hornibrook, 1961) E Pl
Laticarinina pauperata (Parker & Jones, 1865) Ab-Rec
Laticarinina sp. indet. Pl-Wc
PLANULINOIDIDAE
Planulinoides sp. indet. Po-Pl

PSEUDOPARRELLIDAE
Eilohedra levicula (Resig, 1958) Wc-Rec
Epistominella cassidulinoides Hornibrook, 1961 E Lwh-Pl
Epistominella exigua (Brady, 1884) Po-Rec
Epistominella iota Hornibrook, 1961 E Lw-Wm
Epistominella minima (Cushman, 1933) Pl-Wc*
GLABRATELLOIDEA
BULIMINOIDIDAE
Buliminoides bantamensis Yabe & Asano, 1937 Tk
Buliminoides williamsonianus (Brady, 1881) Pl-Rec
Elongobula chattonensis Finlay, 1939 E Ld
Elongobula creta Finlay, 1939 E Mh
Elongobula lawsi Finlay, 1939 E Lwh
GLABRATELLIDAE
Glabratella crassa Dorreen, 1948 E Dm-Sw
Glabratella finlayi Srinivasan, 1966 E Ar
Glabratella quadrata Hornibrook, 1961 E Lwh-Sw
Glabratella sp. indet. Sw
Glabratellina sigali Seiglie & Bermudez, 1965 Po-Pl
Pijpersia gracilis Srinivasan, 1966 E Ar
Pijpersia spp. indet. (2)
Pileolina calcarata (Heron-Allen & Earland, 1922) E Ld-Rec
Pileolina patelliformis (Brady, 1884) Sl-Rec
Pileolina radiata Vella, 1957 E Pl-Rec
Pileolina turbinata (Terquem, 1882) Lwh
Pileolina zealandica Vella, 1957 E Po-Rec
Pileolina sp. indet. E Sw
Planoglabratella semiopercularis (Hornibrook, 1961) E Po-Pl
HERONALLENIIDAE
Heronallenia pulvinulinoides (Cushman, 1915) Sc-Rec
Heronallenia wilsoni (Heron-Allen & Earland, 1922) Ak-Wn
NONIONOIDEA
ALMAENIDAE
Almaena sp. indet.
NONIONIDAE
Astrononion fijiensis Cushman & Edwards, 1937 Po-Pl
Astrononion impressum Hornibrook, 1961 E Ab-Pl
Astrononion neefi (Vella, 1963) E Tt-Wo
Astrononion novozealandicum Cushman & Edwards, 1937 E Wo-Rec
Astrononion parki Hornibrook, 1961 E Ar-Wo
Astrononion poriensis (Vella, 1963) E Wo-Wn
Astrononion pusillum Hornibrook, 1961 E Ar-Rec
Astrononion stelligerum (d'Orbigny, 1839) Po-Wn
Astrononion tumidum Cushman & Edwards, 1937 Po-Rec
Astrononion vadorum Vella, 1962 E Wp-Wm
Astrononion sp. indet. Wn
Florilus olsoni Kennett, 1966 E Tt
Haynesina depressula (Walker & Jacob, 1798) Lwh-Rec
Melonis affinis (Reuss, 1851) Lw-Rec
Melonis dorreeni (Hornibrook, 1961) E Ab-Tt
Melonis halkyardi (Cushman, 1936) Dt-Wp
Melonis lutorum Vella, 1962 E Wo
Melonis maorica (Stache, 1864) E Ab-Tk
Melonis pacimaoricum Srinivasan, 1966 E Dw-Lw
Melonis pompilioides (Fichtel & Moll, 1798) Lw-Rec
Melonis simplex (Karrer, 1864) E Lw-Tk
Melonis zeobesus Vella, 1962 E Tt-Wm
Melonis spp. indet. (2) Mh
Nonion cassidulinoides Hornibrook, 1961 E Lwh-Wm
Nonion deceptrix Hornibrook, 1961 E Lwh-Wm
Nonion iota Finlay, 1940 E Ab-Ak
Nonion mexicanum Cushman, 1933 Mh
Nonion tanumia (Finlay, 1940) E Mh-Dt
Nonionella auris (d'Orbigny, 1839) Po-Rec
Nonionella gemina Hornibrook, 1961 E Lw-Wn
Nonionella magnalingua Finlay, 1940 E Ld-Rec
Nonionella novozealandica Cushman, 1936 E Lwh-Tt
Nonionella robusta Plummer, 1931 Mh
Nonionella satiata Finlay, 1940 E Pl
Nonionella soldadoensis Cushman & Renz, 1942 Dt
Nonionella zenitens Finlay, 1940 E Dw-Tk
Nonionellina flemingi (Vella 1957) E Sw-Rec
Nonionoides grateloupi (d'Orbigny, 1826) Tt-Rec
Nonionoides turgida (Williamson, 1858) Wp-Rec
Pullenia bulloides (d'Orbigny, 1846) Dt-Rec
Pullenia coryelli White, 1929 Mh-Dw
Pullenia cretacea Cushman, 1936 Mh-Ab
Pullenia quinqueloba (Reuss, 1851) Dt-Rec
Pullenia salisburyi Stewart & Stewart, 1930 Wo-Rec
Spirotecta pellicula Belford, 1961 Mh
Zeaflorilus parri (Cushman, 1936) Sl-Rec
Zeaflorilus stachei (Cushman, 1936) Ld-Sl
NUMMULITOIDEA
ASTEROCYCLINIDAE
Asterocyclina hornibrooki Cole, 1967 Ar
Asterocyclina speighti (Chapman, 1932) Dw-Ab
DISCOCYCLINIDAE
Discocyclina sp. indet. Ar
NUMMULITIDAE
Cycloclypeus carpenteri Brady, 1881 Po-Sw
Heterostegina borneensis van der Vlerk, 1930 Lw-Sw
Heterostegina sp. indet. Ab
Nummulitella polystylata Dorreen, 1948 E Ab-Ar
Operculina kawakawaensis Chapman & Parr, 1938 E Lwh-Po
Operculina mbalavuensis Cole, 1945 Sw
PLANORBULINOIDEA
CIBICIDIDAE
Cibicides amoenus Finlay, 1940 E Pl-Wo
Cibicides araroa Kennett, 1967 E Tt
Cibicides beaumontianus (d'Orbigny, 1840) Mh
Cibicides bradyi (Trauth, 1918) Ld-Rec
Cibicides catillus Finlay, 1940 E Ld-Tt
Cibicides cenop (Woodruff, 1985) Sl-Wm
Cibicides cicatricosus (Schwager, 1866) Sc-Wp
Cibicides collinsi Finlay, 1940 E Dm-Lwh
Cibicides cookei Cushman & Garrett, 1938 Dt-Lw
Cibicides corticatus Earland, 1934 Wc-Rec
Cibicides deliquatus Finlay, 1940 E Sl-Rec
Cibicides dispars (d'Orbigny, 1839) Sw-Rec
Cibicides finlayi Gibson, 1967 E Sw-Wp
Cibicides gibsoni Kennett, 1966 E Sw-Wo
Cibicides hampdenensis Hornibrook, 1961 E Dw-Ak
Cibicides havanensis Cushman & Bermudez, 1937 Lwh
Cibicides ihungia Finlay, 1940 E Ab-Wo
Cibicides karreriformis Hornibrook, 1961 E Lwh
Cibicides kullenbergi Parker, 1953 Po-Wp
Cibicides lobatulus (Walker & Jacob, 1798) Po-Rec
Cibicides maculatus (Stache, 1864) E Ak-Ld
Cibicides mediocris Finlay, 1940 E Ld-Tt
Cibicides molestus Hornibrook, 1961 E Ab-Wp
Cibicides mundulus (Brady, Parker & Jones, 1888) Pl-Rec
Cibicides neoperforatus Hornibrook, 1976 E Tt-Wn
Cibicides notocenicus Dorreen, 1948 E Ab-Sc
Cibicides novozelandicus (Karrer, 1864) E Lwh-Tt
Cibicides pachyderma (Rzehak, 1886) Ld-Rec
Cibicides perforatus (Karrer, 1864) E Lw-Wo
Cibicides porrodeliquatus Kennett, 1967 E Tt-Wo
Cibicides praecursorius (Schwager, 1883) Dt
Cibicides praemundulus (Berggren & Miller, 1986) Dt
Cibicides pronovozelandicus Srinivasan, 1966 E Ak-Lwh
Cibicides protemperatus Hornibrook, 1961 E Ar-Tt
Cibicides pseudoconvexus Parr, 1938 Dt-Lwh
Cibicides pseudoungerianus (Cushman, 1922) Sw-Wo
Cibicides refulgens de Montfort, 1808 Po-Rec
Cibicides robertsonianus (Brady, 1881) Dm-Rec
Cibicides rugosa Phleger & Parker, 1951 Sc-Rec
Cibicides semiperforatus Hornibrook, 1961 E Ab
Cibicides subhaidingerii (Parr, 1950) Ar-Rec
Cibicides temperata Vella, 1957 E Wo-Rec
Cibicides thiara (Stache, 1864) E Ar-Lw
Cibicides thiaracuta Hornibrook, 1958 E Pl-Tt
Cibicides tholus Finlay, 1939 E Ab
Cibicides truncanus (Guembel, 1868) Dp-Lwh
Cibicides variabilis (d'Orbigny, 1826) Sl-Rec
Cibicides verrucosus Finlay, 1940 E Ar-Pl
Cibicides victoriensis Chapman, Parr & Collins, 1934 Pl-Sc
Cibicides vortex Dorreen, 1948 E Ab-Sw
Cibicides whitei Martin, 1943 Dt
Cibicides wuellerstorfi (Schwager, 1866) Pl-Rec
Cibicides spp. indet. (21)
Dyocibicides biserialis Cushman & Valentine, 1930 Lwh-Rec
Dyocibicides primitiva Vella, 1957 E Ld-Rec
CYMBALOPORIDAE
Fabiania sp. indet. Dt
Halkyardia bartrumi Parr, 1934 Ak-Lwh
PLANORBULINIDAE
Planorbulina macphersoni Finlay, 1947 E Dm-Ar
Planorbulinella plana (Heron-Allen & Earland, 1924) Po-Sw
Planorbulinella zelandica Finlay, 1947 E Po-Tt
Planorbulinella sp. indet. Ld
PLANULINIDAE
Hyalinea balthica (Schroeter, 1873) Wp
Planulina costata (Hantken, 1875) Lw
Planulina crassa Galloway & Heminway, 1941 Po-Pl
Planulina renzi Cushman & Stainforth, 1945 Lw-Rec
Planulina sinuosa (Sidebottom, 1918) Tt-Rec
VICTORIELLIDAE
Biarritzina carpenteriaeformis (Halkyard, 1919) Po
Carpenteria balaniformis Gray, 1858 Pl
Carpenteria monticularis Carter, 1877 Ak-Lwh
Victoriella conoidea (Rutten, 1914) Lwh-Pl
Wadella globiformis (Chapman, 1926) E Ak-Pl
Wadella hamiltonensis (Glaessner & Wade, 1959) Ar
Wadella sp. indet. Ab
ROTALIOIDEA
CHAPMANINIDAE
Crespinina kingscotensis Wade, 1955 Ak-Lw
Sherbornina atkinsoni Chapman, 1922 Ld-Po
Sherbornina cuneimarginata zealandica Hayward, 1981 E Pl
Sherbornina sp. indet. Ak
ELPHIDIIDAE
Elphidium aculeatum norcotti Hayward, 1997 E Lw-Po
Elphidium a. subrotatum Hornibrook, 1961 E Pl-Sw
Elphidium advenum advenum (Cushman, 1922) Ab-Sw
Elphidium a. limbatum (Chapman, 1907) Ar-Rec
Elphidium a. macelliforme McCulloch, 1981 Tt-Wo
Elphidium a. maorium Hayward, 1997 E Ar-Rec
Elphidium carteri Hayward, 1997 Lw-Sw
Elphidium charlottense (Vella, 1957) E Lwh-Rec
Elphidium crispum (Linneus, 1758) Ar-Rec
Elphidium c. waiwiriense Hayward, 1997 E Lw-Po
Elphidium excavatum clavatum Cushman, 1930 Po-Rec
Elphidium e. excavatum (Terquem, 1875) Wo-Rec
Elphidium e. williamsoni Haynes, 1973 Wn-Rec
Elphidium hampdenensis Finlay, 1940 E Dh
Elphidium ingressans Dorreen, 1948 E Ar-Lwh
Elphidium kanoum Hayward, 1979 E Ld-Sl
Elphidium matanginuiense Hayward, 1997 E Dw-Dm

Elphidium matauraense Hayward, 1997 E Lw-Pl
Elphidium aff. *novozealandicum* Cushman, 1936 Sw-Wo
Elphidium novozealandicum Cushman, 1936 E Tk-Rec
Elphidium oceanicum Cushman, 1933 Wc
Elphidium pseudoinflatum Cushman, 1936 Ar-Sw
Elphidium saginatum Finlay, 1940 E Dp
Elphidium schencki Cushman & Dusenbury, 1934 Ab-Ar
Elphidium wadeae (Hornibrook, 1961) E Lwh-Po
NOTOROTALIIDAE
Cribrorotalia chattonensis Hornibrook, 1996 E Ld
Cribrorotalia dorreeni Hornibrook, 1961 E Lwh-Tt
Cribrorotalia keari Hornibrook, 1962 E Ar-Lwh
Cribrorotalia longwoodensis Hornibrook, 1962 E Lwh-Ld
Cribrorotalia lornensis Hornibrook, 1961 E Ak-Ar
Cribrorotalia miocenica (Cushman, 1936) Po-Pl
Cribrorotalia obesa Hornibrook, 1961 E Po-Pl
Cribrorotalia okokoensis Hornibrook, 1962 E Ar-Lwh
Cribrorotalia ornatissima (Karrer, 1864) E Lw-Tt
Cribrorotalia tainuia (Dorreen, 1948) E Ak-Lwh
Discorotalia aranea (Hornibrook, 1958) E Po-Sw
Discorotalia tenuissima (Karrer, 1864) E Lwh-Sw
Notorotalia aparimae Hornibrook, 1996 E Lwh
Notorotalia biconvexa Hornibrook, 1961 E Po-Sw
Notorotalia bortonica Hornibrook, 1996 E Dp-Ak
Notorotalia braxtonensis Hornibrook, 1996 E Lwh
Notorotalia briggsi Collen, 1972 E Wp-Wn
Notorotalia depressa Vella, 1957 E Wp-Rec
Notorotalia finlayi Vella, 1957 E Wp-Rec
Notorotalia hornibrooki Hayward, Grenfell, Reid & Hayward, 1999 E Tk-Rec
Notorotalia hurupiensis Vella, 1957 E Sw-Wp
Notorotalia inornata Vella, 1957 E Wn-Rec
Notorotalia kaiata Hornibrook, 1996 E Ak-Ar
Notorotalia kingmai Vella, 1957 E Wo-Wm
Notorotalia kiripakae Hornibrook, 1996 E Ar
Notorotalia kohaihae Hornibrook, 1996 E Lwh-Ld
Notorotalia macinnesi Kennett, 1966 E Tt
Notorotalia mammiligera Kennett, 1967 E Tt
Notorotalia mangaoparia Vella, 1957 E Wm-Wn
Notorotalia olsoni Vella, 1957 E Wo-Rec
Notorotalia pliozea Vella, 1957 E Wo-Wm
Notorotalia powelli Finlay, 1939 E Po-Pl
Notorotalia praeserrata Hornibrook, 1996 E Lwh-Ld
Notorotalia pristina Vella, 1957 E Tt-Tk
Notorotalia profunda Vella, 1957 E Wm-Rec
Notorotalia rotunda Vella, 1961 E Wn
Notorotalia ruatangatae Hornibrook, 1996 E Lwh
Notorotalia serrata Finlay, 1939 E Lw-Pl
Notorotalia spinosa (Chapman, 1926) E Ld-Pl
Notorotalia stachei Finlay, 1939 E Lwh
Notorotalia s. macrostachei Hornibrook, 1996 E Lwh-Ld
Notorotalia taranakia Vella, 1957 E Tt-Wn
Notorotalia targetensis Hornibrook, 1961 E Po-Pl
Notorotalia uttleyi Hornibrook, 1961 E Ar-Lwh
Notorotalia waitakiensis Hornibrook, 1992 E Lw
Notorotalia whaingaroica Hornibrook, 1996 E Lwh
Notorotalia wilsoni Hornibrook, 1958 E Pl-Sw
Notorotalia zelandica Finlay, 1939 E Wm-Rec
Notorotalia z. mangaoparia Vella, 1957 E Wm
Notorotalia spp. indet. (6)
MIOGYPSINIDAE
Miogypsina cushmani Vaughan, 1924 Po-Pl
Miogypsina globulina (Mitchelotti, 1841) Po
Miogypsina intermedia Drooger, 1952 Po-Pl
ROTALIIDAE
Ammonia nanus (Hornibrook, 1961) E Lw-Wm
Ammonia aoteana (Finlay, 1940) Tk-Rec
Pararotalia mackayi (Karrer, 1864) E Lwh-Sc
Rotalia fastigata Collen, 1972 E Wq
Rotalia wanganuiensis Hornibrook, 1989 E Wn-Wc
SIPHONINOIDEA
SIPHONINIDAE
Siphonina australis Cushman, 1927 Ar-Wn

SEVENTEEN

Phylum RADIOZOA

radiolaria

CHRISTOPHER J. HOLLIS, VANESSA LÜER, YOSHIAKI AITA, DEMETRIO BOLTOVSKOY, RIE S. HORI, BARRY M. O'CONNOR, ATSUSHI TAKEMURA

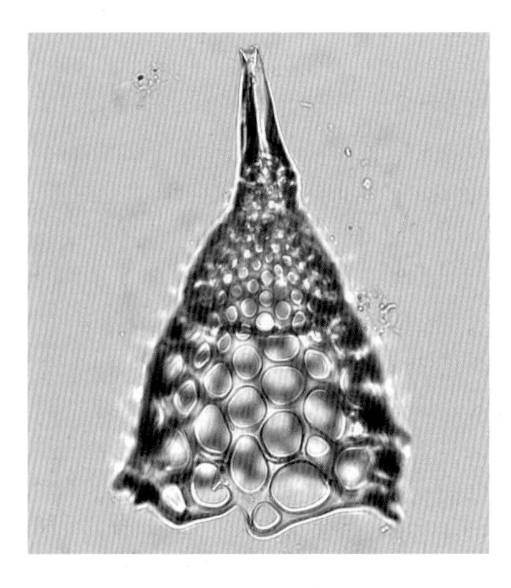

A Quaternary nassellarian, *Lamprocyclas hannai* (Pterocorythidae), ODP Site 1123.

Vanessa Lüer

Radiozoa ('ray animals'), commonly referred to as radiolaria, are exquisite planktonic organisms with ornate skeletons of opaline silica or the mineral celestite (strontium sulphate). Most have radial symmetry – with spectacular variations of form based on spherical, ellipsoidal, discoidal, and conical skeletal frameworks. Individuals vary in size from hundredths of millimetres to several millimetres. Colonial forms can be very much larger, with tests organised simply as an association of loose rods or there may be no hard parts at all. The cell is organised into a porous central capsule that contains the nucleus, mitochondria, and other important organelles that carry out core cell functions like respiration and reproduction. Radiating from this capsule are ray-like spokes. Pseudopodia (actinopodia or axopodia) surround these spokes, capturing prey, disposing of wastes, and reacting to external stimuli, aided by irregular rhizopodia. Outside the central capsule, surrounding the spokes, the cytoplasm is frothy, with bubble-like alveoli, presumably a flotation device, disappearing when agitated and reforming when the individual has sunk to some depth. This part of the cytoplasm includes fewer mitochondria, digestive vacuoles, and, depending on the species, algal symbionts.

Radiolaria are wholly marine and holoplanktonic (i.e. having their whole life-cycle in the plankton), with a fossil record extending back to the Cambrian. In this chapter we follow the scheme of Cavalier-Smith (2003) (see also Bass et al. 2005) for higher-level classification and that of Riedel (1967), as modified by Boltovskoy (1998), and De Wever et al. (2001) for ordinal and family-level classification.

The radiolaria collected by the 1872–76 *Challenger* expedition were classified by German biologist Ernst Haeckel (1887) into three groups – Acantharia, Phaeodarea, and Polycystinea. Although they all share a perforated central capsule and axopodia, there are some significant differences. The central capsule contains numerous pores in the siliceous polycystines and celestite acantharians but has generally only three pores in phaeodarians (which have a mixed siliceous-organic skeleton). Individual celestite crystals can also be found in some polycystines and their swarmers. Phaeodarians have a complicated cytopharynx (lacking in the other groups) and a structure called the phaeodium that consists of balls of darkly pigmented waste products. They also lack endosymbionts whereas polycystines and acantharians can host them. Recent molecular data confirm that phaeodarians are most closely related to testate Cercozoa (Polet 2004).

Despite their small size, radiolaria have great utility to geologists. As with the remains of other marine plankton, they have distributions on the sea floor that broadly represent a long-term average of living plankton in overlying water masses, and the distributions of living plankton species broadly correspond to

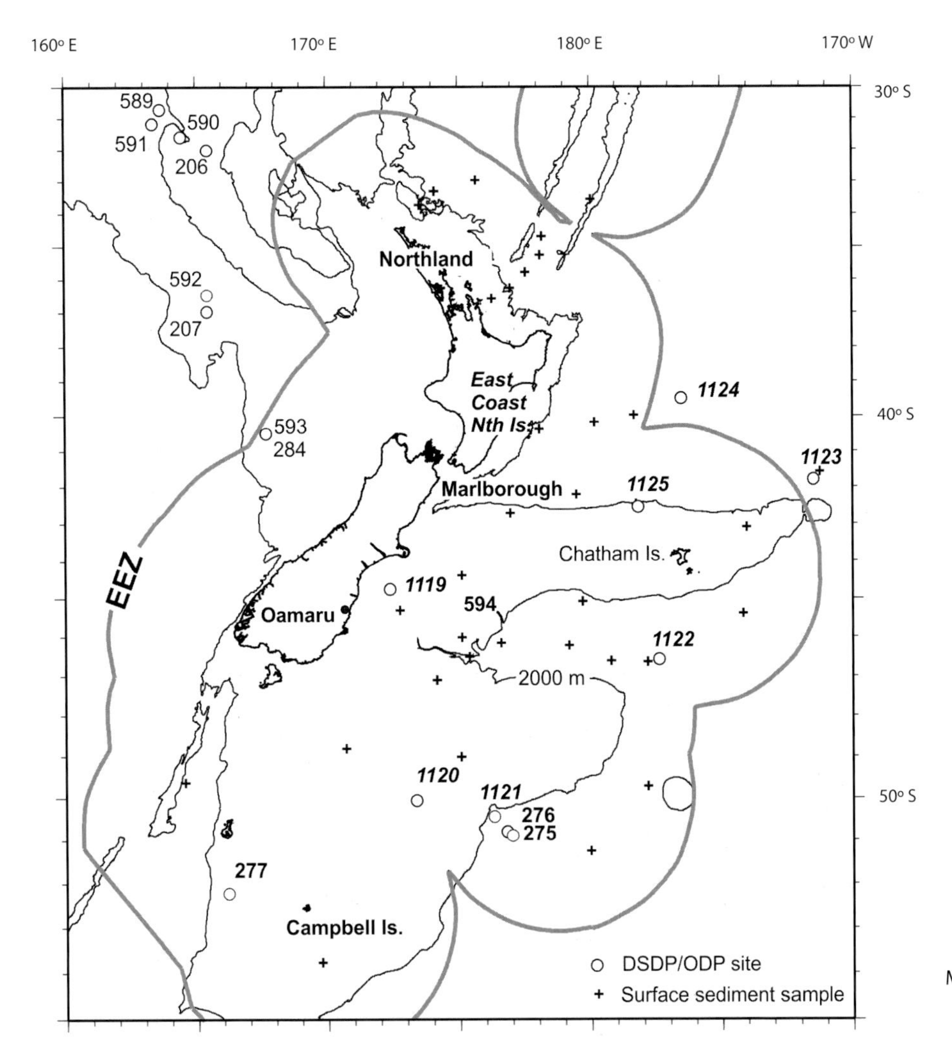

Map of New Zealand region showing drill sites and localities mentioned in the text.

patterns of oceanic water-masses. Thus, we can learn much about prehistoric oceans by studying changes in planktonic microfossils in layers of seafloor sediment, either those still preserved in ocean basins or those now uplifted and accessible as surface outcrops of sedimentary rock.

Of the two groups of zooplankton widely used in paleoceanography, the siliceous radiolaria have two advantages over their calcareous-shelled cousins, the planktonic foraminifera. First, the resistance of their opaline tests to dissolution makes them particularly useful in productive coastal areas, in polar waters, and in the deep ocean where calcium carbonate dissolution rates are high. Secondly, inventories of the modern plankton indicate there are almost 190 radiolarian

Summary of New Zealand Recent radiozoan diversity

Taxon	Described species + subspecies	Known undescribed/ undetermined species	Estimated unknown species	Adventive species	Endemic species	Endemic genera
Spumellaria	60	11	75	0	0	0
Collodaria	16	0	20	0	0	0
Entactinaria	1	0	5	0	0	0
Nassellaria	80+4	13	100	0	0	0
Totals	**157+4**	**24**	**200**	**0**	**0**	**0**

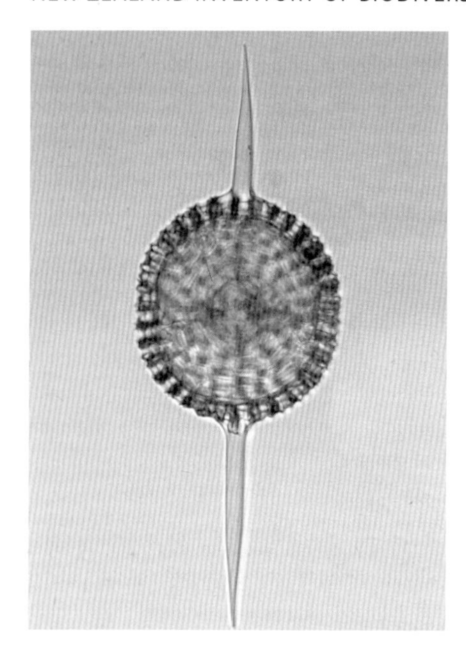

A Late Eocene spumellarian with polar spines, *Axoprunum pierinae* (Saturnalidae), DSDP Site 277.
Chris Hollis

species in New Zealand waters (Boltovskoy 1987; Hollis & Neil 2005) but only 29 species of planktonic foraminifera (Hayward 1983, and this volume). The contrast is more striking south of the Subtropical Convergence, with about 100 radiolarian species versus 2–3 planktonic foraminiferal species. These numbers agree well with the known or assumed worldwide diversity of these two groups – more than 1000 extant radiolarian species (Polycystina + Phaeodaria) versus about 50 planktonic foraminifera. With variations in the relative abundance of species being widely used as paleoceanographic indicators, this greater diversity provides a more finely divided scale by which to measure environmental and oceanographic change.

Despite this potential, studies of fossil and modern planktonic foraminifera far outnumber studies of radiolaria in the New Zealand region. Until fairly recently (Hollis & Neil 2005), the only inventories of the modern fauna have been as part of broader studies (Moore 1978; Boltovskoy 1987), while there have been only localised studies of Neogene radiolaria on land (Ashby 1986; O'Connor 1997a,b) and offshore (Caulet 1986; Carter et al. 1999; Lüer et al. 2008) as part of the Deep Sea Drilling Project (DSDP) and Ocean Drilling Program (ODP). Most detailed study has focused on Late Cretaceous and Paleogene radiolaria. The potential for radiolarian studies in these strata was first demonstrated by DSDP studies of Late Cretaceous–Paleocene assemblages from the north Tasman Sea (Dumitrica 1973a,b) and Late Cretaceous assemblages from the eastern Campbell Plateau (Pessagno 1975). Subsequently, the rich Late Cretaceous to Middle Eocene radiolarian succession in the pelagic limestones of eastern Marlborough has been the subject of numerous detailed studies (e.g. Hollis 1993, 1997; Strong et al. 1995; Hollis et al. 2003a,b, 2005a). Related taxonomic, biostratigraphic, and paleoceanographic studies have been undertaken of Paleocene to Oligocene radiolarian assemblages from deep-sea cores on the western and eastern margins of the Campbell Plateau and Lord Howe Rise (Hollis et al. 1997a; O'Connor 2001; Hollis 2002) and outcrops on Campbell Island (Hollis et al. 1997b). Detailed taxonomic studies have also been undertaken on Eocene to Early Miocene radiolarian assemblages in Northland (O'Connor 1994, 1997a,b, 1999a) and the rich Late Eocene assemblages in the Oamaru Diatomite (O'Connor 1999b).

Earlier Mesozoic and Paleozoic radiolarian assemblages have been studied mainly with the aim of dating basement rocks that are often not amenable to age determination by other methods (e.g. Feary & Hill 1978; Aita & Spörli 1994). It is increasingly evident, however, that paleobiogeographic interpretation of these assemblages provides important insights into the origin and migration of New Zealand's basement terranes (e.g. Aita & Spörli 1992; Sutherland & Hollis 2001). The paleobiogeography of Cenozoic radiolarian assemblages in the Southwest Pacific has helped to constrain models for the evolution of the Southern Ocean (Lazarus & Caulet 1993).

Summary of Recent radiozoan diversity by environment (species + subspecies)

Taxon	Marine	Freshwater	Terrestrial	Fossil
Albaillellaria	0	0	0	22
Latentifistularia	0	0	0	5
Spumellaria	72	0	0	259
Collodaria	16	0	0	3
Entactinaria	1	0	0	97
Nassellaria	93+4	0	0	561+10
Totals	**182+4**	**0**	**0**	**947+10**

New Zealand radiolarian diversity

Here we review records of modern and fossil radiolaria within the limits of New Zealand's 200-nautical-mile Exclusive Economic Zone (EEZ). Southwest Pacific DSDP sites investigated by Dumitrica (1973a) and Petrushevskaya (1975) are not included because they are outside the EEZ.

Under the current classification, the Radiozoa comprises three classes – Acantharia, Sticholonchea, and Polycystinea. The Sticholonchea has not reported from New Zealand. The genus *Sticholonche* used to be classified in the Heliozoa, a group formerly allied with radiolaria in Actinopoda. It is a curious form with a heart-shaped central capsule, tangential and radial siliceous spicules, and long oar-like pseudopodia. Our inventory is dominated by polycystine radiolaria of the orders Spumellaria and Nassellaria. Both orders have a rich fossil record in the New Zealand region. The fossil record of the polycystine order Collodaria is restricted to isolated siliceous spicules (Boltovskoy 1998). The Paleozoic order Albaillellaria is known from basement rocks in the Whangaroa area, Northland (e.g. Takemura et al. 1999), and Red Rocks in Wellington (Grapes et al. 1990; Begg & Mazengarb 1996).

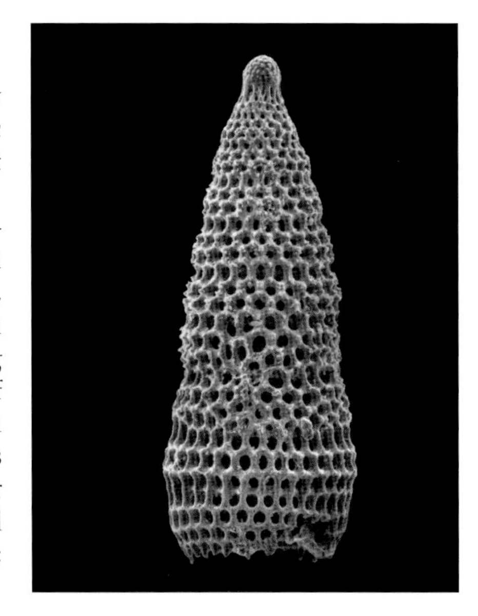

Amphipyndax stocki (Amphipyndacidae), a nassellarian showing the typical multisegmented Cretaceous form. This species has one of the longest geological ranges in the fossil record, appearing in the Middle Cretaceous (c. 100 Ma), surviving the Cretaceous–Paleogene mass extinction, and finally going extinct in the Paleocene (c. 60 Ma). Late Cretaceous, from a dolomite concretion, Whangai Formation, Mara quarry, eastern Wairarapa (U26/836555).

Chris Hollis

Neogene and Recent radiolaria

Approximately 50 radiolarian species were recorded from plankton or bottom-sediment samples from the New Zealand region during several Pacific and Antarctic oceanographic surveys of the late 19th and mid-20th centuries (Haeckel 1887; Hays 1965; Petrushevskaya 1967). Dawson (1992) compiled a useful synonymic list of these species. Moore (1978) added several new records, including the first for the Pleistocene, as part of a biogeographic study of radiolarian assemblages in Recent and late Pleistocene sediments. Boltovskoy (1987) carried out a more comprehensive regional survey of modern radiolaria as part of a biogeographic study of radiolarian assemblages in 18 sediment cores from western Pacific sites between the Equator and 65° S. Recently, Hollis and Neil (2005) reported on the distribution of radiolaria in 33 surface-sediment samples from offshore eastern New Zealand between latitudes 33° S and 54° S. Currently, 186 radiolarian species are recorded from Recent or Holocene sediment or plankton surveys.

Radiolaria are relatively sparse in New Zealand's onshore Neogene sedimentary sequences but some local formations have produced assemblages of biostratigraphic importance (Ashby 1985, 1986; O'Connor 1997a,b, 2000). Richer Neogene assemblages have been reported from offshore eastern New Zealand at DSDP Site 594 (Caulet 1986; Nelson et al. 1993), ODP Sites 1123 and 1224 (Carter et al. 1999; Lüer et al. 2008), and Kasten core Y8 (Lüer et al. 2009). Overall, 267 radiolarian species have been recorded from the Quaternary, 87 species from the Pliocene, and 202 species from the Miocene.

Late Cretaceous and Paleogene radiolaria

Marshall (1916) first hinted at the richness of New Zealand's Late Cretaceous–Paleogene radiolarian fauna when he noted abundant siliceous microfossils in thin sections of the Amuri and Mahurangi Limestones, but it was not until the latter half of the 20th century that the potential of the group for biostratigraphic and paleoenvironmental studies in this time interval was realised.

Paleogene assemblages were described from DSDP sites drilled west of New Zealand's EEZ by Dumitrica (1973a), Petrushevskaya (1975), and Caulet (1986). Pessagno (1975) described the first Late Cretaceous (Campanian–Maastrichtian) assemblages from DSDP Site 275, within the EEZ on the eastern flank of the Campbell Plateau. No studies of onshore sequences were undertaken over this period, although radiolaria were noted in pelagic sediments from Campbell Island (Beggs 1978) and in the Late Cretaceous Whangai Formation – a siliceous mudstone distributed throughout eastern New Zealand (Moore 1988). In the

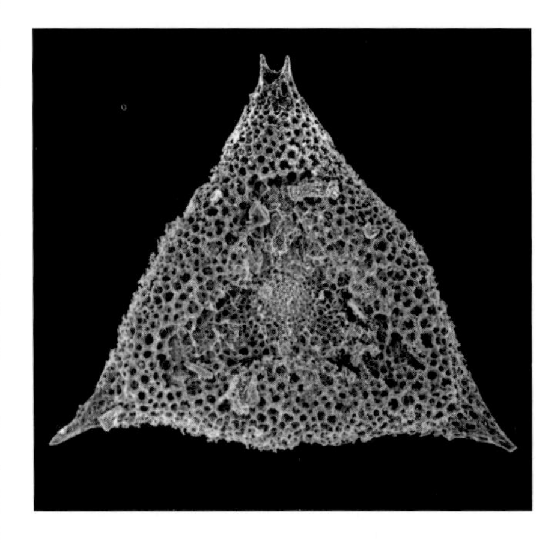

A Late Cretaceous spumellarian, *Spongotripus* sp. (Spongodiscidae). From a dolomite concretion, Whangai Formation, Mara quarry, eastern Wairarapa (U26/836555).

Chris Hollis

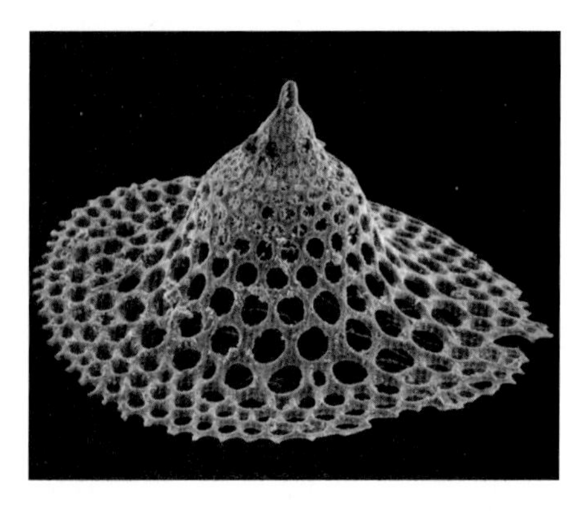

A Late Cretaceous nassellarian, *Neosciadiocapsa jenkinsi* (Neosciadiocapsidae). From a dolomite concretion, Whangai Formation, Mara quarry, eastern Wairarapa (U26/836555).
Chris Hollis

1990s, students from the University of Auckland began reporting the results of detailed doctoral studies of Late Cretaceous and early Paleogene assemblages from eastern Marlborough (Hollis 1991, 1993, 1997; Strong et al. 1995) and late Paleogene assemblages from Northland (O'Connor 1993, 1994, 1997a,b, 1999b).

The eastern Marlborough sequence provides the best-known global record of radiolarian evolution across the Cretaceous–Tertiary boundary (Hollis et al. 1995; Hollis 1996; Hollis et al. 2003a,b). Biostratigraphic studies have been undertaken on subantarctic Paleogene assemblages from DSDP Site 277 and Campbell Island (Hollis et al. 1997a,b) while O'Connor (1999a) has made a thorough taxonomic study of the rich Late Eocene radiolarian assemblages in the Oamaru Diatomite. Subsequent studies have focused on the rich Paleocene assemblages from ODP Site 1121 (O'Connor 2001; Hollis 2002) and the Paleocene–Eocene succession outcropping in eastern Marlborough (Hollis et al. 2005a,b).

Late Cretaceous and Paleogene radiolaria have been reported from pelagic sediments within the Tangihua and Matakaoa volcanic complexes of Northland and East Cape (Hollis & Hanson 1991; Spörli & Aita 1994), from dredged sediments offshore Northland (Herzer et al. 1997), and from limestones of southern Westland (Sutherland et al. 1996).

In summary, 211 radiolarian species are recorded from the Oligocene, 211 from the Eocene, 177 from the Paleocene, and 86 from the Late Cretaceous.

Mesozoic–Paleozoic radiolaria from basement terranes

Reconnaissance studies of Paleozoic–Mesozoic radiolaria in New Zealand began soon after the discovery that radiolaria could be extracted from hard siliceous mudstone and chert using dilute hydrofluoric acid (Pessagno & Newport 1972). This technique, together with the development of the scanning electron microscope as a routine research tool, opened the door to a new era of radiolarian research worldwide. Radiolarian biostratigraphy and paleobiogeography have since played a pivotal role in determining the age and history of accreted terranes, particularly in the circum-Pacific region.

In New Zealand, however, the focus has usually been on selective identification of key species of biostratigraphic or paleobiogeographic utility. Therefore, the inventory of the pre-Late Cretaceous radiolarian species is considered to underestimate greatly the actual diversity. By way of comparison, O'Dogherty (1994) listed 303 species from the Middle Cretaceous of Italy and Spain while Meyerhoff Hull (1997) listed 476 species from the Late Jurassic of western North America. Moreover, very few studies have included a systematic review of local records, so it is difficult to determine whether the morphotypes recorded in the same or similar open nomenclature in different studies are distinct species. Therefore, while it is certain that the 250–300 species recorded from Late Permian to Early Cretaceous rocks is an underestimate of radiolarian diversity, it is also likely that some records in the checklist are duplications.

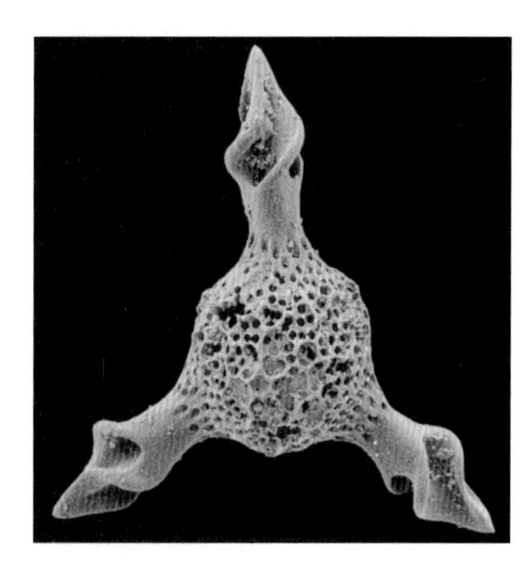

A Late Triassic entactinarian, *Capnuchosphaera* cf. *deweveri* (Capnuchosphaeridae). From a phosphorite nodule, Houghton Bay, Wellington (R27/f232).
John Simes, GNS Science

Aita and Spörli (1992) reviewed previous studies of Paleozoic and Mesozoic radiolaria from New Zealand, with an emphasis on their significance for tectonics and paleobiogeography. Subsequent studies of Torlesse Terrane rocks have yielded Triassic radiolaria from the Rimutaka Ranges (Aita & Spörli 1994) and Jurassic radiolaria from the Ekatahuna area (Aita 1995). Hori and colleagues have followed up early discoveries (Aita & Grant-Mackie 1992) with new studies of Jurassic radiolaria from Murihiku Terrane in North and South Islands (Hori et al. 1996, 1997). Similarly, pioneering studies of radiolaria in the Waipapa Terrane (Prin 1984; Adachi 1988; Spörli et al. 1989; Caridroit & Ferrière 1988) have been followed by detailed studies in the Whangaroa area, Northland (Sakai et al. 1998; Takemura et al. 1998, 1999). These latter studies focused on a bed-by-bed investigation of radiolarian and conodont assemblages and geochemistry through a potential Permian–Triassic boundary sequence at Arrow Rocks. They included the first systematic treatment of Permian radiolaria for New Zealand (Takemura et al. 1999). Aita and Bragin (1999) discussed the paleobiogeographic significance of new species recognised in Triassic assemblages from the Northland and Southland (see also Ito et al. 2000).

In summary, 284 species have been reported from the Paleozoic and earlier Mesozoic (i.e. pre-Late Cretaceous) of New Zealand. More than 80% of these species have been reported from Jurassic or Triassic rocks (120 and 123 species, respectively), with the remainder coming from Middle to Late Permian rocks in Northland and Wellington (33 species) and Early Cretaceous rocks from eastern North Island (17 species).

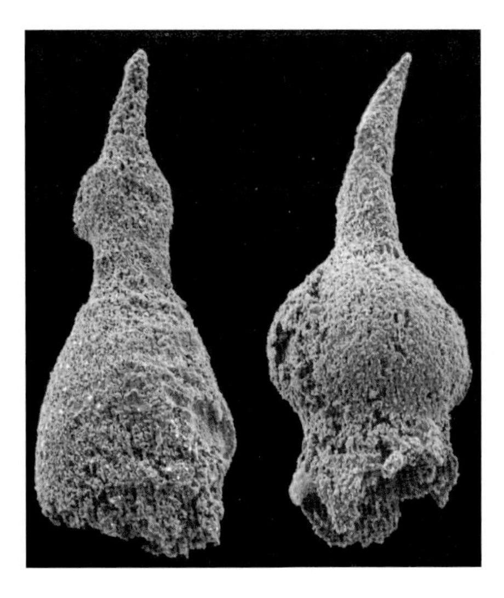

Late Permian albaillellarians, *Follicuculus monacanthus* (left) and *F. ventricosus* (Follicucullidae), from radiolarian chert, Red Rocks, Wellington (R27/f144).
J. Simes

Gaps in knowledge and scope for future research

There are undoubtedly large gaps in the knowledge of radiolarian biodiversity and paleobiodiversity in the New Zealand region. There have been no local studies of the distribution of modern radiolaria in New Zealand water masses. There is huge scope for linking plankton studies with bottom-sediment surveys to aid in understanding the oceanic influences on marine productivity and biodiversity. Studies of the modern local fauna are also essential for improving paleoenvironmental interpretations based on fossil assemblages.

Knowledge of the Cenozoic fauna has advanced greatly with the publication of recent studies of Quaternary radiolaria from offshore eastern New Zealand (Lüer et al. 2008, 2009). Tied with previous DSDP and ODP legs in the South Pacific and Subantarctic, there is great potential for radiolarian research to contribute to better understanding of the Cenozoic evolution of the Southern Ocean.

Knowledge of Mesozoic and Paleozoic radiolaria may well remain patchy as rich assemblages are seldom recovered regularly through a sequence – rock sequences with good stratigraphy contain sparse assemblages (i.e. Murihiku Terrane) while the most diverse assemblages tend to occur in complexly deformed oceanic sequences (Torlesse and Waipapa Terranes). However, as shown in the study by Sutherland and Hollis (2001), new discoveries have the potential to have major impact on models of regional tectonic evolution. Because of this potential, we recommend more thorough taxonomic study of older radiolarian assemblages better to constrain paleobiogeographic inferences.

Acknowledgements

We thank GNS Science colleagues Giuseppe Cortese, Hamish Campbell, and John Simes for suggesting improvements to the manuscript and checking the species lists, and Marianna Terezow for editing the SEM images.

Authors

Dr Yoshiaki Aita Department of Geology, Faculty of Agriculture, Utsunomiya University, Utsunomiya 321, Japan [aida@cc.utsunomiya-u.ac.jp]

Dr Demetrio Boltovskoy Departamento de Ciencias Biológicas, Facultad de Ciencas Exactas y Naturales, Universidad de Buenos Aires, 1428 Buenos Aires, and CONICET, MACN, Argentina [demetrio@ege.fcen.uba.ar]

Dr Christopher J. Hollis GNS Science, P.O. Box 30-368, Lower Hutt, New Zealand [c.hollis@gns.cri.nz]

Dr Rie S. Hori Department of Earth Sciences, Faculty of Science, Ehime University, Matsuyama 790-77, Japan [shori@sci.ehime-u.ac.jp]

Dr Vanessa Lüer Geowissenschaften, Universität Bremen, Klagenfurter Strasse, D-28359 Bremen, Germany [vlueer@uni-bremen.de]

Dr Barry M. O'Connor Department of Geology, University of Auckland, Private Bag 92-018, Auckland, New Zealand [b.oconnor@auckland.ac.nz]

Dr Atsushi Takemura Geoscience Institute, Hyogo University of Teacher Education, Yashiro-cho, Kato-gun, Hyogo 673-1494, Japan [takemura@sci.hyogo-u.ac.jp]

References (and bibliography of studies of New Zealand radiolarians)

ADACHI, M. 1988: Permian radiolarian-bearing cherts from the Marble Bay area, North Island, New Zealand. *Chigaku Zasshi (Journal of Geography) 97*: 90–92.

AITA, Y. 1995: Middle Jurassic Radiolaria from Ekatahuna, North Island, New Zealand. *In*: Moore, P. R. (ed.), A Jurassic chert-limestone-spilite association near Ekatahuna, New Zealand. *Journal of the Royal Society of New Zealand 25*: 113–114.

AITA, Y.; BRAGIN, N.Yu. 1999: Non-Tethyan Triassic radiolaria from New Zealand and northeastern Siberia. *Geodiversitas 21*: 503–526.

AITA, Y.; GRANT-MACKIE, J. A. 1992: Late Jurassic Radiolaria from the Kowhai Point Siltstone, Murihiku terrane, North Island, New Zealand. Pp. 375–382 *in*: Ishizaki, K.; Saito, T. (eds), *Centenary of Japanese Micropaleontology*. Terra Scientific Publishing Company, Tokyo. ix + 480 p.

AITA, Y.; SPÖRLI, K. B. 1992: Tectonic and paleobiogeographic significance of radiolarian microfaunas in the Permian to Mesozoic basement rocks of the North Island, New Zealand. [*In*: Aitchison, J.; Murchey, B. L. (eds), Special Issue: Significance and application of Radiolaria to terrane analysis.] *Palaeogeography, Palaeoclimatology, Palaeoecology 96*: 103–125.

AITA, Y.; SPÖRLI, K. B. 1994: Late Triassic Radiolaria from the Torlesse Terrane, Rimutaka Range, North Island, New Zealand. *New Zealand Journal of Geology and Geophysics 37*: 155–162.

ASHBY, J. N. 1985: Upper Neogene radiolarian zonation for the East Coast Deformed Belt, New Zealand. *In*: Cooper, R. A. (ed.), Hornibrook Symposium, 1985: Extended abstracts. *New Zealand Geological Survey Record 9*: 5–7.

ASHBY, J. N. 1986: Late Neogene Radiolaria from the East Coast Deformed Belt, New Zealand. Unpublished PhD thesis, Victoria University of Wellington. 222 p.

BASS, D.; MOREIRA, D.; LÓPEZ-GARCÍA, P.; POLET, S.; CHAO, E. E.; von der HEYDEN, S.; PAWLOVSKI, J.; CAVALIER-SMITH, T. 2005: Polyubiquitin insertions and the phylogeny of Cercozoa and Rhizaria. *Protist 156*: 149–161.

BEGG, J. G.; MAZENGARB, C. 1996: Geology of the Wellington area: sheets R27, R28, and part Q27, scale 1:50,000. *Institute of Geological & Nuclear Sciences Geological Map 22*: 1 sheet + 128 p.

BEGGS, J. M. 1978: Geology of the metamorphic basement and Late Cretaceous to Oligocene sedimentary sequence of Campbell Island, southwest Pacific Ocean. *Journal of the Royal Society of New Zealand 8*: 161–177.

BLOME, C.; MOORE, P.; SIMES, J.; WATTERS, W. A. 1987: Late Triassic Radiolaria from phosphatic concretions in the Torlesse Terrane, Kapiti Island, Wellington. *New Zealand Geological Survey Record 18*: 103–109.

BOLTOVSKOY, D. 1987: Sedimentary record of radiolarian biogeography in the equatorial to Antarctic western Pacific Ocean. *Micropaleontology 33*: 267–281.

BOLTOVSKOY, D. 1998: Classification and distribution of South Atlantic Recent polycystine Radiolaria. *Palaeontologica Electronica 1*: 1–116. [http://palaeoelectronica. org/1998_2/boltovskoy/issue2.htm]

CAMPBELL, H. J.; KORSCH, R. J.; FOLEY, L. A.; ORR, T. O. H. 1988: Radiolaria of Middle Jurassic to Early Cretaceous ages from the Torlesse Complex, eastern Tararua Range, New Zealand. *New Zealand Journal of Geology and Geophysics 31*: 121–123.

CARIDROIT, M.; FERRIÈRE, J. 1988: Premières datations précises du Paléozoique par radiolaires en Nouvelle-Zélande; intérêts géologique et paléontologique. *Comptes rendus des Séances de l'Académie des Sciences, sér. 2, 306* : 321–326.

CARTER, R. M.; McCAVE, I. N.; RICHTER, C.; CARTER, L. et al. 1999: *Proceedings of the Ocean Drilling Program, Initial Reports 181* [CD-ROM].

CAULET, J. P. 1986: Radiolarians from the Southwest Pacific. *In*: Kennett, J. P. et al. (eds). *Initial Reports of the Deep Sea Drilling Project 90*: 835–861.

CAVALIER-SMITH, T. 2003: Protist phylogeny and the high-level classification of Protozoa. *European Journal of Protistology 39*: 338–348.

CROUCH, E. M.; HOLLIS, C. J. 1996: Paleogene palynomorph and radiolarian biostratigraphy of DSDP Leg 29, sites 280 and 281, South Tasman Rise. *Institute of Geological & Nuclear Sciences Science Report 96/19*: 1–46.

DAWSON, E. W. 1992: The marine fauna of New Zealand: Index to the fauna 1. Protozoa. *New Zealand Oceanographic Institute Memoir 99*: 1–368.

DE WEVER, P.; DUMITRICA, P.; CAULET, J.-P.; NIGRINI, C.; CARIDROIT, M. 2001: *Radiolarians in the Sedimentary Record*. Gordon & Breach Science Publishers, Amsterdam. 533 p.

DUMITRICA, P. 1973a: Paleocene Radiolaria, DSDP Leg 21. *In*: Burns, R. E. *et al.* (eds). *Initial Reports of the Deep Sea Drilling Project 21*: 787–817.

DUMITRICA, P. 1973b: Phaeodarian Radiolaria in Southwest Pacific sediments cored during Leg 21 of the Deep Sea Drilling Project. *In*: Burns, R. E. *et al.* (Eds). *Initial Reports of the Deep Sea Drilling Project 21*: 751–785.

DUMITRICA, P.; HOLLIS, C. J. 2004: Maastrichtian Challengeridae (phaeodarian radiolaria) from deep sea sediments of SW Pacific. *Revue de Micropaléontologie 47*: 127–134.

FEARY, D. A.; HILL, P. H. 1978: Mesozoic Radiolaria from cherts in the Raukumara Peninsula, New Zealand. *New Zealand Journal of Geology and Geophysics 21*: 363–373.

FEARY, D. A.; PESSAGNO, E. A. 1980: An Early Jurassic age for chert within the Early Cretaceous Oponae Melange (Torlesse Supergroup), Raukumara Peninsula, New Zealand. *New Zealand Journal of Geology and Geophysics 23*: 623–628.

FOLEY, L. A.; KORSCH, R. J.; ORR, T. O. H. 1986: Radiolaria of Middle Jurassic to Early Cretaceous ages from the Torlesse Complex, eastern Tararua Range, New Zealand. *New Zealand Journal of Geology and Geophysics 29*: 481–490.

GEORGE, A. D. 1993: Radiolarians in offscraped seamount fragments, Aorangi Range, New Zealand. *New Zealand Journal of Geology and Geophysics 36*: 185–199.

GORDON, D. P.; BALLANTINE, W. J. 1977: Cape Rodney to Okakari Point Marine Reserve. Review of knowledge and bibliography to December 1976. *Tane (Journal of the Auckland University Field Club) 22 (Suppl.)*: 1–146.

GRAPES, R. H.; LAMB, S. M.; CAMPBELL, H. J.; SPÖRLI, K. B.; SIMES, J. E. 1990: Geology of the red rocks-turbidite association, Wellington Peninsula, New Zealand. *New Zealand Journal of Geology and Geophysics 33*: 377–391.

HAECKEL, E. 1887: Report on the Radiolaria collected by H.M.S. Challenger during the years 1873–1876. *Report on the Scientific Results of the Voyage of the H.M.S. Challenger, Zoology 18*: 1–1803, 140 pls. [In 2 parts + Atlas.]

HAYWARD, B. W. 1983: Planktic foraminifera (Protozoa) in New Zealand waters: a taxonomic review. *New Zealand Journal of Zoology 10*: 63–74.

HAYS, J. D. 1965: Radiolaria and late Tertiary and Quaternary history of Antarctic Seas. *In*: Llano, G. A. (ed.). *Biology of Antarctic Seas II. American Geophysical Union, Antarctic Research Series 5*: 125–184.

HERZER, R. H.; CHAPRONIERE, G. C. H.; EDWARDS, A. R.; HOLLIS, C. J.; PELLETIER, B.; RAINE, J. I.; SCOTT, G. H.; STAGPOOLE, V.; STRONG, C. P.; SYMONDS, P.; WILSON, G. J.; ZHU, H. 1997: Seismic stratigraphy and structural history of the Reinga Basin and its margins, southern Norfolk Ridge system. *New Zealand Journal of Geology and Geophysics 40*: 425–451.

HOLLIS, C. J. 1991: Latest Cretaceous to Late Paleocene Radiolaria from Marlborough (New Zealand) and DSDP Site 208. PhD thesis, University of Auckland, Auckland. 308 p.

HOLLIS, C. J. 1993: Latest Cretaceous to Late Paleocene radiolarian biostratigraphy: A new zonation from the New Zealand region. *Marine Micropaleontology 21*: 295–327.

HOLLIS, C. J. 1996: Radiolarian faunal change through the Cretaceous–Tertiary transition of eastern Marlborough, New Zealand. Pp. 173–204 *in*: MacLeod, N.; Keller, G. (eds), *Cretaceous-Tertiary Mass Extinctions: Biotic and Environmental Changes*. Norton Press, New York. 575 p.

HOLLIS, C. J. 1997: Cretaceous–Paleocene Radiolaria from eastern Marlborough, New Zealand. *Institute of Geological and Nuclear Sciences Monograph 17*: 1–152.

HOLLIS, C. J. 2002: Biostratigraphy and paleoceanographic significance of Paleocene radiolarians from offshore eastern New Zealand. *Marine Micropaleontology 46*: 265–316.

HOLLIS, C. J.; HANSON, J. A. 1991: Well-preserved late Paleocene Radiolaria from Tangihua Complex, Camp Bay, eastern Northland. *Tane (Journal of the Auckland University Field Club) 33*: 65–76.

HOLLIS, C. J.; NEIL, H. 2005: Sedimentary record of radiolarian biogeography, offshore eastern New Zealand. *New Zealand Journal of Marine and Freshwater Research 39*: 165–192.

HOLLIS, C. J.; BEGGS, J. M.; CROUCH, E. M.; RAINE, J. I.; STRONG, C. P.; TURNBULL, I. M.; WILSON, G. J. 1997b: Late Cretaceous–Paleogene biostratigraphy of Campbell Island, southern Campbell Plateau, southwest Pacific. *Institute of Geological & Nuclear Sciences Science Report 97/25*: 1–56.

HOLLIS, C. J.; DICKENS, G. R.; FIELD, B. D.; JONES, C. J.; STRONG, C. P. 2005a: The Paleocene–Eocene transition at Mead Stream, New Zealand: a southern Pacific record of early Cenozoic global change. *Palaeogeography, Palaeoclimatology, Palaeoecology 215*: 313–343.

HOLLIS, C. J.; FIELD, B. D.; JONES, C. M.; STRONG, C. P.; WILSON, G. J. 2005b: Biostratigraphy and carbon isotope stratigraphy of Paleocene-lower Eocene Muzzle Group in Muzzle and Bluff Streams, middle Clarence valley, New Zealand. *Journal of the Royal Society of New Zealand 35*: 345–383.

HOLLIS, C. J.; RODGERS, K. A.; PARKER, R. J. 1995: Siliceous plankton bloom in the earliest Tertiary of Marlborough, New Zealand. *Geology 23*: 835–838.

HOLLIS, C. J.; RODGERS, K. A.; STRONG, C. P.; FIELD, B. D.; ROGERS, K. M. 2003a: Paleoenvironmental changes across the Cretaceous/Tertiary boundary in the northern Clarence Valley, southeastern Marlborough, New Zealand. *New Zealand Journal of Geology and Geophysics 46*: 209–234.

HOLLIS, C. J.; STRONG, C. P.; RODGERS, K. A.; ROGERS, K. M. 2003b: Paleoenvironmental changes across the Cretaceous/Tertiary boundary at Flaxbourne River and Woodside Creek, eastern Marlborough, New Zealand. *New Zealand Journal of Geology and Geophysics 46*: 177–197.

HOLLIS, C. J.; WAGHORN, D. B.; STRONG, C. P.; CROUCH, E. M. 1997a: Integrated Paleogene biostratigraphy of DSDP Site 277 (Leg 29): Foraminifera, calcareous nannofossils, Radiolaria, and palynomorphs. *Institute of Geological and Nuclear Sciences Science Report 97/07*: 1–87.

HORI, R. S.; AITA, Y.; GRANT-MACKIE, J. A. 1996: Preliminary report on Lower Jurassic radiolaria of Gondwana origin from the Kawhia coast, New Zealand. *The Island Arc 5*: 104–113.

HORI, R. S.; CAMPBELL, J. D.; GRANT-MACKIE, J. A. 1997: A new Early Jurassic radiolarian fauna from the Murihiku Supergroup of the Otago coast, New Zealand. *New Zealand Journal of Geology and Geophysics 40*: 397–399.

ITO, M.; AITA, Y.; HADA, S. 2000: New radiolarian age information for the Chrystalls Beach Complex, southwest of Dunedin, New Zealand. *New Zealand Journal of Geology and Geophysics 43*: 349–354.

LAZARUS, D.; CAULET, J. P. 1993: Cenozoic Southern Ocean reconstructions from sedimentologic, radiolarian, and other microfossil data. *In*: Kennett, J. P.; Warnke, D. A. (eds), *The Antarctic Paleonvironment: A Perspective on Global Change. Part 2. Antarctic Research Series 60*: 145–174.

LÜER, V.; HOLLIS, C. J.; WILLEMS, H. 2008: Late Quaternary radiolarian assemblages as indicators of paleoceanographic changes north of the Subtropical Front, offshore eastern New Zealand, southwest Pacific. *Micropaleontology 54*: 49–69.

LÜER, V.; CORTESE, G.; NEIL, H. L.; HOLLIS, C.J.; WILLEMS, H. 2009: Radiolarian-based sea surface temperatures and paleoceanographic changes during the Late Pleistocene-Holocene in the subantarctic southwest Pacific. *Marine Micropaleontology 70*: 151–165.

MARSHALL, P. 1916: The younger limestones of New Zealand. *Transactions of the New Zealand Institute 48*: 87–99.

MEYERHOFF HULL, D. 1997: Upper Jurassic Tethyan and southern Boreal radiolarians from western North America. *Micropaleontology 43 (Suppl. 2)*: 1–202.

MOORE, P. R. 1983: Chert-bearing formations from New Zealand. *In*: Iijima, A.; Hein, J. R.; Siever, R. (eds), *Siliceous Deposits of the Pacific Region. Developments in Sedimentology 6*: 93–108.

MOORE, P. R. 1988: Stratigraphy, composition, and environment of deposition of the Whangai Formation and associated Late Cretaceous–Paleocene rocks, eastern North Island, New Zealand. *New Zealand Geological Survey Bulletin 100*: 1–82.

MOORE, T. C. 1978: The distribution of radiolarian assemblages in the modern and ice-age Pacific. *Marine Micropaleontology 3*: 229–266.

NELSON, C. S.; COOKE, P. J.; HENDY, C. H.; CUTHBERTSON, A. M. 1993: Oceanographic and climatic changes over the past 160,000 years at Deep Sea Drilling Site 594 off southeastern New Zealand, southwest Pacific Ocean. *Paleoceanography 8*: 435–458.

O'CONNOR, B. M. 1993: Radiolaria from the Mahurangi Limestone, Northland, New Zealand. MSc Thesis, University of Auckland. 136 p.

O'CONNOR, B. M. 1994: Seven new radiolarian species from the Oligocene of New Zealand. *Micropaleontology 40*: 337–350.

O'CONNOR, B. M. 1997a: Lower Miocene Radiolaria from Te Kopua Point, Kaipara Harbour, New Zealand. *Micropaleontology 43*: 63-100.

O'CONNOR, B. M. 1997b: New Radiolaria from the Oligocene and Early Miocene of New Zealand. *Micropaleontology 43*: 101–128.

O'CONNOR, B. M. 1999a: Radiolaria from the Late Eocene Oamaru Diatomite, South Island, New Zealand. *Micropaleontology 45*: 1–55.

O'CONNOR, B. M. 1999b: Distribution and biostratigraphy of latest Eocene to latest Oligocene Radiolaria from Mahurangi Limestone, Northland, New Zealand. *New Zealand Journal of Geology and Geophysics 42*: 489–511.

O'CONNOR, B. M. 2000: Stratigraphic and geographic distribution of recently described Eocene–Miocene Radiolaria from the southwest Pacific. *Micropaleontology 46*: 189–228.

O'CONNOR, B. M. 2001: *Buryella* (Radiolaria, Artostrobiidae) from DSDP Site 208 and ODP Site 1121, and an emendation to the South Pacific radiolarian zonation. *Micropaleontology 47*: 1–22.

O'DOGHERTY, L. 1994: Biochronology and paleontology of Mid-Cretaceous radiolarians from Northern Apennines (Italy) and Betic Cordillera (Spain). *Mémoires de Géologie (Lausanne) 21*: 1–415.

PESSAGNO, E. A. 1975: Upper Cretaceous Radiolaria from DSDP Site 275. *In*: Kennett, J. *et al.* (eds). *Initial Reports of the Deep Sea Drilling Project 29*: 1011–1029.

PESSAGNO, E. A.; NEWPORT, R. L. 1972: A technique for extracting Radiolaria from radiolarian cherts. *Micropaleontology 18*: 231–234.

PETRUSHEVSKAYA, M. G. 1967: Radiolyarii otryadov Spumellaria i Nassellaria antarkticheskoi oblasti. *Issledovaniya Fauny Morei 4(12): Resultaty Biologicheskikh Issledovanii Sovetskoi Antarkticheskoi Ekspeditsii 1955–1958, 3*: 1–186.

PETRUSHEVSKAYA, M. G. 1975: Cenozoic radiolarians of the Antarctic, Leg 29, DSDP. *In*: Kennett, J. P. *et al.* (eds). *Initial Reports of the Deep Sea Drilling Project 29*: 541–675.

POLET, S.; BERNEY, C.; FAHRNI, J.; PAWLOWSKI, J. 2004: Small-subunit ribosomal RNA gene sequences of Phaeodaria challenge the monophyly of Haeckel's Radiolaria. *Protist 155*: 53–63.

PRIN, W. F. 1984: Radiolaria from cherts of Auckland area.Unpublished MSc thesis, University of Auckland. 85 p.

RIEDEL, W. R. 1967: Subclass Radiolaria. Pp. 291–298 in: Harland, W. B.; Smith, A. G.; Wilcock, B. (eds.), *The Fossil Record*. Geological Society of London, London. 827 p.

SAKAI, T.; AITA, Y.; HIGUCHI, Y.; HORI, R. S.; KODAMA, K.; TAKEMURA, A.; CAMPBELL, H. J.; GRANT-MACKIE, J. A.; HOLLIS, C. J.; SPÖRLI, K. B. 1998: Mesozoic phosphatic and calcareous nodules containing well-preserved radiolarian fauna from the North Island, New Zealand. *Journal of the Geological Society of Japan 104*: v–vi.

SPÖRLI, K. B.; AITA, Y. 1994: Tectonic significance of Late Cretaceous Radiolaria from the obducted Matakaoa volcanics, East Cape, North Island, New Zealand. *Geoscience Reports of Shizuoka University 20*: 115–134.

SPÖRLI, K. B.; AITA, Y.; GIBSON, G. W. 1989: Juxtaposition of Tethyan and non-Tethyan Mesozoic radiolarian faunas in melanges, Waipapa Terrane, North Island, New Zealand. *Geology 17*: 753–756.

STRONG, C. P.; HOLLIS, C. J.; WILSON, G. J. 1995: Foraminiferal, radiolarian and dinoflagellate biostratigraphy of Late Cretaceous to Middle Eocene pelagic sediments (Muzzle Group), Mead Stream, Marlborough, New Zealand. *New Zealand Journal of Geology and Geophysics 38*: 171–212.

SUTHERLAND, R.; HOLLIS, C. J. 2001: Cretaceous demise of the Moa Plate and strike-slip motion at the Gondwana margin. *Geology 29*: 279–282.

SUTHERLAND, R.; HOLLIS, C. J.; NATHAN, S.; STRONG, C. P.; WILSON, G. J. 1996: Age of Jackson Formation proves late Cenozoic allochthony in South Westland, New Zealand. *New Zealand Journal of Geology and Geophysics 39*: 559–563.

TAKEMURA, A.; AITA, Y.; HORI, R. S.; HIGUCHI, Y.; SPÖRLI, K. B.; CAMPBELL, H. J.; KODAMA, K.; SAKAI, T. 1998. Preliminary report on the lithostratigraphy of Arrow Rocks and geologic age of the northern part of the Waipapa Terrane, New Zealand. *News of Osaka Micropaleontologists, Special Volume 11*: 47–57.

TAKEMURA, A.; MORIMOTO, T.; AITA, Y.; HORI, R. S.; HIGUCHI, Y.; SPÖRLI, K. B.; CAMPBELL, H. J.; KODAMA, K.; SAKAI, T. 1999: Permian Albaillellaria (Radiolaria) from a limestone lens at the Arrow Rocks in the Waipapa Terrane (Northland, New Zealand). *Geodiversitas 21*: 751–765.

TIBBS, J. F.; TIBBS, S. D. 1986: Biology of the Antarctic Seas 16. Further studies on the Phaeodaria (Protozoa: Radiolaria) of the Antarctic Seas. *Antarctic Research Series 41*: 167–202.

Checklist of New Zealand Recent Radiozoa

Radiolaria in the following checklist are only those found in New Zealand's EEZ. Many of these species are also listed in the Cretaceous–Cenozoic list that follows owing to their occurrence in mainly Neogene sediments. Species names are followed by author, abbreviated reference to New Zealand records, and reference content (List, Figure, Descriptive comments). Reference abbreviations: B87, Boltovskoy 1987; C86, Caulet 1986; CM99, Carter et al. 1999; GB77, Gordon & Ballantine 1977; HN05, Hollis & Neil 2005; Pet67, Petrushevskaya 1967.

KINGDOM CHROMISTA
SUBKINGDOM HAROSA
INFRAKINGDOM RHIZARIA
PHYLUM RADIOZOA
Class POLYCYSTINEA
Order SPUMELLARIA
ACTINOMMIDAE
Actinomma antarcticum (Haeckel, 1887) C86 B87 CM99 HN05
Actinomma delicatulum (Dogiel, 1952) Pet67 C86 B87 HN05
Actinomma leptodernum (Jörgensen, 1900) B87 CM99 HN05
Actinomma medianum Nigrini, 1967 B87 HN05
Actinomma sol Cleve, 1900 B87 HN05
Anomalocantha dentata (Mast, 1910) B87 HN05
Carposphaera acanthophora (Popofsky, 1912) B87
Cenosphaera compacta Haeckel, 1887 Pet67
Cenosphaera cristata Haeckel, 1887 B87 HN05
Cenosphaera elysia Haeckel, 1887 B87
Cenosphaera? vesparia Haeckel, 1887 B87
Cromyechinus antarctica (Dreyer, 1889) B87 HN05 CM99
Druppatractus irregularis Popofsky, 1912 B87 C86 CM99 HN05
Haliometta miocenica (Campbell & Clark, 1944) C86 HN05
Heliaster hexagonium Hollande & Enjumet, 1960 B87
Hexacontarium sp. Boltovskoy 1987 B87
Hexacontium arachnoidale Hollande & Enjumet, 1960 HN05
Hexacontium armatum/hostile gr. Cleve, 1900 B87 HN05
Hexacontium entacanthum Jörgensen, 1899 B87
Hexacontium laevigatum Haeckel, 1887 B87 HN05
Hexadoras octahedrum Haeckel, 1887 B87
Plegmosphaera lepticali Renz, 1976 B87
Prunopyle titan Campbell & Clark, 1944a C86 HN05
Prunulum? coccymelium Haeckel, 1887 B87
Spongosphaera streptacantha Haeckel, 1862 B87
Thecosphaera inermis (Haeckel, 1860) B87 HN05
Xyphostylus trogon (Haeckel, 1887) B87
Actinommid sp. A Boltovskoy & Riedel 1981 B87
ASTROSPHAERIDAE
Acanthosphaera dodecastyla Mast, 1910 B87
Acanthosphaera pinchuda Boltovskoy & Riedel, 1981 B87
Acanthosphaera? gibbosa Haeckel, 1887 B87
Cladococcus cervicornis Haeckel, 1887 B87
Cladococcus stalactites Haeckel, 1887 B87
STYLOSPHAERIDAE
Stylatractus disetanius Haeckel, 1887 Hk87
Stylosphaera melpomene Haeckel, 1887 B87
COCCODISCIDAE
Didymocyrtis tetrathalamus (Haeckel, 1887) B87 CM99 HN05
Ommatodiscus pantanelli Carnevale 1908 B87
HELIODISCIDAE
Heliodiscus asteriscus Haeckel, 1862 B87 CM99 HN05
LARNACILLIDAE
Circodiscus microporus (Stöhr, 1880) Oc97b HN05
Phorticium clevei (Jörgensen, 1900) Pet67 B87 Oc97b CM99
Phorticium pylonium gr. Haeckel, 1887 Oc97b HN05
LITHELIIDAE
Lithelius nautiloides Popofsky, 1908 Oc99b B87 CM99 HN05
Lithelius? riedeli Petrushevskaya, 1967 Pet67
Pylospira? octopyle Haeckel, 1887 B87
Spirema melonia Haeckel, 1887 HN05
Tholospira cervicornis gr. Haeckel, 1887 Pet67 HN05
PYLONIDAE
Dipylissa bensoni Dumitrica, 1988 HN05
Larcopyle buetschlii Dreyer, 1889 Oc97b B87 HN05
Larcospira quadrangula Haeckel, 1887 B87 HN05
Octopyle stenozona Haeckel, 1887 B87 CM99 HN05
Pylolena armata gr. Haeckel, 1887 B87 HN05
SPONGODISCIDAE
Amphirhopalum ypsilon Haeckel, 1887 B87 CM99 HN05
Dictyocoryne profunda Ehrenberg, 1860 B87 HN05
Euchitonia elegans/furcata (Ehrenberg, 1872) B87 HN05
Euchitonia? sp. B87
Porodiscus sp. A sens. Nigrini & Moore 1979 Pet67 HN05
Schizodiscus spatangus Dogiel, 1950 HN05
Spongaster aff. *pentas* Riedel & Sanfilippo, 1970 B87
Spongodiscus resurgens Ehrenberg, 1854 B87 HN05
Spongopyle osculosa (= *S. setosa*) Dreyer, 1889 Pet67 B87 CM99 HN05
Spongotrochus glacialis Popofsky, 1908 B87 C86 CM99 HN05
Stylochlamydium/Porodiscus sp. gr. sens. Boltovskoy & Riedel 1980 Pet67 B87 HN05
Stylodictya aculeata Jörgensen, 1905 HN05
Stylodictya multispina (= *S. validispina*) Haeckel, 1860 Pet67 B87 HN05
Styptosphaera? spumacea Haeckel, 1887 B87 HN05
Tessarastrum straussii Haeckel, 1887 B87
SPONGURIDAE
Spongocore cylindrica (= *S. puella*) Haeckel, 1860 B87 HN05
Spongurus aff. *elliptica* (Ehrenberg, 1872) B87
Spongurus pylomaticus Riedel, 1957 B87 HN05
Spongurus? sp. Petrushevskaya, 1967 B87 HN05
THOLONIDAE
Cubotholus spp. B87 HN05

Order COLLODARIA
COLLOSPHAERIDAE
Acrosphaera lappacea (Haeckel, 1887) B87, HN05
Acrosphaera murrayana (Haeckel, 1887) B87
Acrosphaera spinosa (Haeckel, 1860) B87, HN05
Collosphaera huxleyi Mueller, 1855 B87, HN05
Collosphaera macropora Popofsky, 1917 B87
Collosphaera polygona Haeckel, 1887 B87 C86
Collosphaera tuberosa Haeckel, 1887 B87
Collozoum inerme Haeckel, 1862 Hk87
Siphonosphaera fragilis Haeckel, 1887 Hk87
Siphonosphaera martensi Brandt, 1905 B87 HN05
Siphonosphaera polysiphonia Haeckel, 1887 B87 HN05
Solenosphaera polymorpha (Haeckel, 1887) B87
Solenosphaera quadrata (Ehrenberg, 1872) B87
Solenosphaera zanguebarica (Ehrenberg, 1872) B87
SPHAEROZOIDAE
Raphidozoum australe Haeckel, 1887 Hk87
THALLASSICOLLIDAE
Thallassicolla australis Haeckel, 1887 Hk87

Order ENTACTINARIA
SATURNALIDAE
Axoprunum stauraxonium Haeckel, 1887 B87 HN05

Order NASSELLARIA
ACANTHODESMIIDAE
Acanthodesmia vinculata (Müller, 1857) B87
Semantis micropora Popofsky, 1908 Pet67
Zygocircus productus (Hertwig, 1879) B87 HN05
ACROPYRAMIDIDAE
Artostrobus annulatus (Bailey, 1856) HW97, B87
Cornutella profunda Ehrenberg, 1854 Pet67 B87 Oc97b CM99 HN05
Peripyramis circumtexta Haeckel, 1887 B87 HN05
ARTOSTROBIIDAE
Botryostrobus aquilonaris (Bailey, 1856) B87 C86 HN05
Botryostrobus auritus/australis gr. (Ehrenberg, 1844) B87 C86 CM99 HN05
Phormostichoartus corbula (Harting, 1863) C86 B87 CM99
Phormostichoartus platycephala (Ehrenberg, 1873) Pet67 C86
Siphocampe arachnaea (Ehrenberg, 1861) Pet67 B87
Siphocampe lineata (Ehrenberg, 1838) B87
Spirocyrtis scalaris Haeckel, 1887 C86 B87 HN05
Tricolocampe cylindrica Haeckel, 1887 HN05
CANNOBOTRYIDAE
Botryopyle dictyocephalus Haeckel, 1887 B87
Saccospyris antarctica (Haecker, 1907) B87 C86 HN05
Saccospyris conithorax Petrushevskaya, 1965 B87
CARPOCANIIDAE
Carpocanarium papillosum (Ehrenberg, 1872) B87 HN05
Carpocanistrum spp. B87 CM99 HN05
EUCYRTIDIIDAE
Cyrtopera laguncula Haeckel, 1887 B87
Dictyocephalus? papillosus? (Ehrenberg, 1872) Pet67
Dictyophimus hirundo (Haeckel, 1887) B87 CM99 HN05
Dictyophimus infabricatus Nigrini, 1968 B87 HN05
Dictyophimus aff. *tripus* Haeckel, 1862 B87
Eucyrtidium acuminatum acuminatum (Ehrenberg, 1844) B87 CM99 HN05
Eucyrtidium a. octocolum (Haeckel, 1887) B87 HN05
Eucyrtidium anomalum (Haeckel, 1860) B87
Eucyrtidium calvertense Martin, 1904 C86, CM99, HN05
Eucyrtidium hexastichum (Haeckel, 1887) B87
Eucyrtidium teuscheri Haeckel, 1887 Pet67 C86 CM99 HN05

Gondwanaria dogieli (Petrushevskaya & Kozlova, 1972) CM99 HN05
Lipmanella dictyoceras (= *L. virchowii*) (Haeckel, 1860) B87 HN05
Litharachnium tentorium Haeckel, 1860 B87 HN05
Lithocampe? eupora (Ehrenberg, 1872) Pet67
Lithocampe? furcaspiculata (Popofsky, 1908) Pet67
Lithopera bacca Ehrenberg, 1872 B87, CM99
Lithostrobus? botryocyrtis Haeckel, 1887 Pet67
Sethoconus tabulatus (Ehrenberg, 1873) B87
Stichopilium annulatum Popofsky, 1913 Hays65
Stichopilium bicorne Haeckel, 1887 B87
Stichopilium variabile Popofsky, 1908 HN05
Theocorys veneris Haeckel, 1887 B87
Theopilium tricostatum (Haeckel, 1887) B87
LOPHOPHAENIDAE
Antarctissa cylindrica Petrushevskaya, 1967 HN05
Antarctissa denticulata (incl. *A strelkovi*) (Ehrenberg, 1844) C86 B87 HN05
Antarctissa longa (Popofsky, 1908) Pet67
Arachnocorallium calvata gr. Petrushevksaya, 1971 B87
Cladoscenium? ancoratum Haeckel, 1887 B87
Clathrocorys teuscheri Haeckel, 1887 B87
Deflandrella sp. GB77
Helotholus histricosa Jörgensen, 1905 B87 HN05
Hexaplagia collaris Haeckel, 1887 Hk87
Lampromitra huxleyi (Haeckel, 1879) Hk87
Lampromitra quadricuspis Haeckel, 1887 B87
Lithomelissa? borealis (Ehrenberg, 1872) B87
Lophophaena aff. *apiculata* Ehrenberg, 1973 B87
Lophophaena aff. *capito* Ehrenberg, 1873 B87
Lophophaena hispida (Ehrenberg, 1872 B87
Lophophaena aff. *nadezdae* Petrushevskaya, 1971 B87
Peromelissa phalacra (Haeckel, 1887) B87
Phormacantha hystrix (Jörgensen, 1905) B87
Pseudocubus obeliscus Haeckel, 1887 B87
Pseudodictyophimus gracilipes (Bailey, 1856) B87 HW97 HN05
Pteroscenium pinnatum Haeckel, 1887 B87
Trisulcus aff. *testudus* Petrushevskaya, 1971 B87
PTEROCORYTHIDAE
Androcyclas gamphonycha (= *L. hannai*) (Jörgensen, 1899) B87 HN05
Anthocyrtidium ophirense (Ehrenberg, 1872) B87
Anthocyrtidium zanguebaricum (Ehrenberg, 1872) B87
Lamprocyclas maritalis maritalis Haeckel, 1887 B87 CM99 HN05
Lamprocyclas m. polypora Nigrini, 1967 HN05
Lamprocyrtis nigriniae (Caulet, 1971) CM99 HN05
Pterocorys clausus (Popofsky, 1913) B87
Pterocorys hertwigii (Haeckel, 1887) B87
Pterocorys zancleus (Müller, 1858) B87, CM99, HN05
Pterocorys sp. HN05
Theocorythium trachelium diannae (Haeckel, 1887) B87 HN05
Theocorythium t. trachelium (Ehrenberg, 1872) B87 CM99
SETHOPHORMIDIDAE
Sethophormis rotula Haeckel, 1887 B87
THEOPILIIDAE
Corocalyptra cervus (Ehrenberg, 1872) HN05
Corocalyptra columba (Haeckel, 1887) B87
Corocalyptra aff. *danaes* Haeckel, 1887 B87
Cycladophora bicornis (Popoksky, 1908) B87 CM99 HN05
Cycladophora davisiana (Ehrenberg, 1861) B87 CM99 HN05
Theocalyptra cornuta Bailey, 1856 B87
TRIOSPYRIDIDAE
Amphispyris reticulata (Ehrenberg, 1872) B87
Ceratospyris angulata (Popofsky, 1913) B87
Lophospyris pentagona pentagona (Ehrenberg, 1872) B87
Phormospyris stabilis antarctica (Haeckel, 1907) C86 B87 HN05
Phormospyris s. scaphipes (Haeckel, 1887) HN05
Phormospyris s. stabilis (Goll, 1976) HN05
Tholospyris baconiana Goll, 1972 B87
Tholospyris fornicata Popofsky, 1913 B87
Tholospyris tripodiscus Haeckel, 1887 B87
Spyrid sp. 1 B87
ULTRANAPORIDAE
Pterocanium praetextum gr. (Ehrenberg, 1872) B87 CM99 HN05
Pterocanium trilobum (Haeckel, 1860) B87 CM99 HN05

Checklist of New Zealand Late Cretaceous and Cenozoic Radiozoa

Radiolaria in the following checklist are only those found in New Zealand's EEZ. Solely Recent species are excluded. Species names are followed by author, abbreviated reference to New Zealand records, reference content (List, Figure, Descriptive comments), age range using standard abbreviations for New Zealand stages (except for Q, Quaternary). Family-level classification follows that of De Wever et al. (2001). Reference abbreviations: B87, Boltovskoy 1987; C86, Caulet 1986; CM99, Carter et al. 1999; DH04, Dumitrica & Hollis 2004; H65, Hays 1965; HD05, Hollis et al. 2005; HN05, Hollis & Neil 2005; Ho97, 02, Hollis 1997, 2002; HR03, Hollis et al. 2003; HW97, Hollis, Waghorn et al. 1997; LH08, Lüer et al. 2008; LC09, Lüer et al. 2009; Oc97a,b, 99a,b, 00, 01, O'Connor 1997a,b, 1999a,b, 2000, 2001; Pes75, Pessagno 1975; Pet67, 75, Petrushevskaya 1967, 1975; SH95, Strong et al. 1995; TT86, Tibbs & Tibbs 1986.

Phylum RADIOZOA
Class POLYCYSTINEA
Order SPUMELLARIA
ACTINOMMIDAE
Druppatractus hastatus Blueford, 1982 LH08 LC09 Q
Druppatractus variabilis Dumitrica, 1973 LH08 LC09 Q
Heliosoma echinaster Haeckel, 1887 LH08 Q
Hexacontium aristarchi (Haeckel, 1887) LH08 LC09 Q
Hexastylus triaxonius Haeckel, 1887 LH08 LC09 Q
Rhizoplegma boreale (Cleve, 1899) LC09 Q
Stylatractus sp. D, sens. Nakaseko & Nishimura, 1982 LH08 Q
Actinomma antarcticum (Haeckel, 1887) C86 B87 CM99 HN05 LH08 LC09 Tt-k–Q
Actinomma arcadophorum Haeckel, 1887 LH08 LC09 Q
Actinomma delicatulum (Dogiel, 1952) Pet67 C86 B87 HN05 LH08 LC09 Q
Actinomma leptodernum (Jörgensen, 1900) Hays65 B87 CM99 HN05 LH08 LC09 Q
Actinomma leptoderma longispina Cortese & Bjørklund, 1998 LH08 LC09 Q
Actinomma medianum Nigrini, 1967 B87 HN05 LH08 LC09 Q
Actinomma sol Cleve, 1900 B87 HN05 LH08 LC09 Q
Actinomma tetrapyla (Hays, 1965) C86 Sc-w–Tt-k
Actinomma trinacria (Haeckel, 1860) LH08 LC09 Q
Actinosphaera acantophora (Popofsky, 1912) LH08 LC09 Q
Anomalacantha dentata (Mast, 1910) B87 HN05 LH08 LC09 Q
Axoprunum cf. *liostylum* (Ehrenberg, 1873) LH08 Q
Carposphaera cf. *capillacea* (Haeckel, 1887) LH08 LC09 Q
Cenosphaera cristata Haeckel, 1887 B87 HN05 LH08 LC09 Q
Cenosphaera aff. perforata Haeckel, 1887 LH08 LC09 Q
Cenosphaera spp. group (incl. C. elysia) Boltovskoy, 1998 B87 LH08 LC09 Q
Cromyechinus antarctica (Dreyer, 1889) B87 HN05 CM99 LH08 LC09 Sl–Q
Diploplegma aff. *somphum* Sanfilippo & Riedel, 1973 Ho02 Dt
Druppatractus hastatus Blueford, 1982 LH08 LC09 Q
Druppatractus irregularis Popofsky, 1912 B87 C86 CM99 HN05 LH08 LC09 Tt-k–Q
Druppatractus variabilis Dumitrica, 1973 LH08 LC09 Q
Ellipsoxiphus? cf. *attractus* Haeckel, 1887 SH95 Dm–Ab
Haliometta miocenica (Campbell & Clark, 1944) C86 HN05 LH08 LC09 Tt-k–Q
Haliomma gr. b Hollis 1997 Ho97 Dt–Ab
Haliomma teuria Hollis, 1997 Ho97 Dt
Heliosoma echinaster Haeckel, 1887 LH08 Q
Hexadorium cf. *magnificum* Campbell & Clark, 1944b Ho97 Dt
Hexacontium aristarchi (Haeckel, 1887) LH08 LC09 Q
Hexacontium armatum/hostile gr. Cleve, 1900 B87

HN05 LH08 LC09 Q
Hexacontium enthacanthum Jørgensen, 1899 B87 LH08 LC09 Q
Hexacontium laevigatum Haeckel, 1887 B87 HN05 LH08 LC09 Q
Hexastylus triaxonius Haeckel, 1887 LH08 LC09 Q
Lonchosphaera spicata gr. Popofsky, 1908 C86 Tt-k–Q
Prunocarpus sp. A Hollis 1997 Ho97 Mh–Dt
Prunopyle adelstoma Kozlova & Gorbovetz, 1966 HW97 Ho02 Dt–Dw
Prunopyle cf. *decipiens* (Stöhr, 1880) HW97 Ak–Lwh
Prunopyle fragilis (Stöhr, 1880) HW97 Oc99b Ak–Lw
Prunopyle hayesi Chen, 1975 HW97 CM99 Ak–Q
Prunopyle occidentalis Clark & Campbell, 1942 Ho02 Dt–Dt
Prunopyle polyacantha (Campbell & Clark, 1944) HW97 Oc99b Ak–Po
Prunopyle aff. *polyacantha* Campbell & Clark, 1944a Oc99b Ar
Prunopyle tetrapila Hays, 1965 HW97 CM99 Oc99b Ar–Po
Prunopyle titan Campbell & Clark, 1944a C86 HN05 LH08 LC09 Sc-w–Q
Prunopyle cf. *titan* Campbell & Clark, 1944a HW97 Oc99b Ak–Ld
Prunopyle? sp. O'Connor 1997 Oc97b Po
Rhizoplegma boreale (Cleve, 1899) LC09 Q
Sphaeropyle langii Dreyer, 1889 CM99 Tk–Q
Sphaeropyle robusta Kling, 1973 CM99 Sl–Q
Spongoplegma churchii (Campbell & Clark 1944) CM99 Tt
Stauroxiphos communis Carnevale, 1908 C86 Sc-w–Tt-k
Stylatractus sp. D, sens. Nakaseko & Nishimura, 1982 LH08 Q
Stylosphaera melpomene Haeckel, 1887 B87 LH08 LC09 Q
Thecosphaera inermis (Haeckel, 1860) B87 HN05 LH08 LC09 Q
Thecosphaera larnacium Sanfilippo & Riedel, 1973 Ho02 Dt
Thecosphaera miocenica Nakaseko, 1955 CM99 Sw
Thecosphaera ptomatus Sanfilippo & Riedel, 1973 Ho02 Dt
Tricorporisphaera bibula O'Connor, 1999a Oc00 Ar-Lwh
STYLOSPHAERIDAE
Amphisphaera aotea Hollis, 1993 Ho97 Dt
Amphisphaera coronata s.l. (Ehrenberg, 1873) Ho97 Dt–Lwh
Amphisphaera goruna (Sanfilippo & Riedel, 1973) Ho97 Dt–Dw
Amphisphaera kina Hollis, 1993 Ho97 Dt
Amphisphaera macrosphaera (Nishimura, 1992) Ho97 Dt–Ak-r
Amphisphaera aff. *magnaporulosa* (Clark & Campbell, 1942) Ho05 Dt
Amphisphaera priva (Foreman, 1978) Ho97 Mh–Dt
Amphisphaera radiosa (Ehrenberg, 1854) HW97 Ak–Po
Amphisphaera spinulosa (Ehrenberg, 1873) SH95 Dm-p–Ab
Amphisphaera aff. *spinulosa* (Ehrenberg, 1873) Oc99a HW97 Ak–Lwh
Lithomespilus coronatus Squinabol, 1904 Ho97 Mh–Dw
Protoxiphotractus perplexus Pessagno, 1973 Ho97 Mh–Dt
Protoxiphotractus wilsoni Hollis, 1997 Ho97 Mh–Dt
Stylatractus neptunus gr. Haeckel, 1887 B87 CM99 HN05 LH08 LC09 Sw–Q
Stylatractus universus (= *Stylosphaera angelina*) Hays, 1965 Oc97b C86 CM99 LH08 Po–Q
Stylatractus aff. *universus* Hays, 1965 C86 Sc-w –Q
Stylosphaera aff. *hastata* (Campbell & Clark, 1944b) Ho97 Mh–Dt
Stylosphaera hexaxyphophora (Clark & Campbell, 1942) Oc99a Ar
Stylosphaera minor Clark & Campbell, 1942 Ho97 Dt–Lwh
Stylosphaera pusilla Campbell & Clark, 1944b Ho97 Mh–Dt
Stylosphaera timmsi (Campbell & Clark, 1944a) CM99 Tk–Q
Stylosphaera sp. A O'Connor 1999b Oc99b Lwh–Lw
Stylosphaera sp. B O'Connor 1999b Oc99b eLwh–Lw
Styptosphaera? *spumacea* Haeckel, 1887 B87 HN05 LH08 LC09 Q
COCCODISCIDAE
Didymocyrtis antepenultima (Riedel & Sanfilippo, 1970) C86 CM99 Tt-k
Didymocyrtis bassanii (Carnevale, 1908) Oc97b Po
Didymocyrtis laticonus (Riedel, 1959) C86 CM99 Sc-w–Tt-k
Didymocyrtis penultima (Riedel, 1957) CM99 Tt
Didymocyrtis prismatica (Haeckel, 1887) Oc99b Lwh–Po
Didymocyrtis tetrathalamus (Haeckel, 1887) B87 CM99 HN05 LH08 LC09 Tk–Q
Heliosestrum? sp. Ho97 Mh–Dt
Heliostylus sp. (2) SH95 Oc99b Dm–Lw
Lithocyclia angusta (Riedel, 1959) Oc99b Lwh
Periphaena alveolata (Lipman, 1949) HD05 Dt–Dw
Periphaena decora Ehrenberg, 1873 HW97, CM99, Oc99b Dm-p–Tt
Periphaena? *duplus* (Kozlova & Gorbovetz, 1966) HW97 Dt–Dw
Periphaena heliastericus (Clark & Campbell, 1942) HW97 CM99 Oc99a Dw–Po
Periphaena cf. *saturnalis* (Clark & Campbell, 1942) Oc99b Lwh
Phacostaurus? *quadratus* Nishimura, 1992 Ho02 Dt
CONOCARYOMMIDAE
Conocaryomma aff. *universa* (Pessagno, 1976) Ho97 Dt
Conocaryomma stilloformis (Lipman, 1960) HD05 Dw
ENTAPIIDAE
Entapium regulare Sanfilippo & Riedel, 1973 Ho02 Dt
Zealithapium anoectum (Riedel & Sanfilippo, 1970) Oc99a SH95 Dm-p–Ab
Zealithapium mitra (Ehrenberg, 1873) Oc99a HW97 Ab–Ar
Zealithapium oamaru O'Connor, 1999 Oc00 Ar–Lwh
HAGIASTRIIDAE
Orbiculiforma australis Pessagno, 1975 Pes75 Mh–Dt
Orbiculiforma campbellensis Pessagno, 1975 Pes75 Mh–Dt
Orbiculiforma renillaeformis (Campbell & Clark, 1944) Ho97 Ho02 Mh–Dt
HELIODISCIDAE
Astrophacus linckiaformis (Clark & Campbell, 1942) SH95 Dm-p–Ab
Heliodiscus asteriscus Haeckel, 1862 B87 CM99 HN05 LH08 LC09 Tt–Q
Heliodiscus cf. *heliastericus* Clark & Campbell, 1942 Oc99b Lwh–Ld
Heliodiscus inca Clark & Campbell, 1942 HW97 Oc99a Ak–Lwh
Heliodiscus tunicatus O'Connor, 1997 Oc00 Lwh
Heliodiscus sp. Hollis 1997 Ho97 Mh–Dt
LARNACILLIDAE
Circodiscus ellipticus (Stöhr, 1880) C86 Sc-w– Tt-k
Circodiscus microporus (Stöhr, 1880) Oc97b HN05 LH08 LC09 Po–Q
Phorticium clevei (Jörgensen, 1900) Pet67 B87 Oc97b CM99 LH08 LC09 Po–Q
Phorticium pylonium gr. Haeckel, 1887 Oc97b HN05 LH08 LC09 Po–Q
Plectodiscus circularis (Clark & Campbell, 1942) Ho02 Dt
Plectodiscus runanganus O'Connor, 1999 Oc00 Ab–Ar
LITHELIIDAE
Lithelius aff. *foremanae* Sanfilippo & Riedel, 1973 Ho97 Dt
Lithelius cf. *hexaxyphophorus* (Clark & Campbell, 1942) Ho97 Dt
Lithelius marshalli Hollis, 1997 Ho97 Dt
Lithelius minor gr. Jörgensen, 1889 Ho97 Oc99b HN05 LH08 LC09 Mh–Q
Lithelius? aff. *minor* Jörgensen, 1889 Ho97 Dt
Lithelius nautiloides Popofsky, 1908 Oc99b B87 CM99 HN05 LH08 LC09 Ar–Q
Lithelius sp. CM99 Tt–Q
Lithocarpium sp A. LH08 LC09 Q
Ommatodiscus aff. *haeckeli* Stöhr, 1880 HW97 Oc99b LH08 LC09 Ak–Q
Pylospira sp. CM99 Q
Streblacantha circumtexta Jørgensen, 1900 LH08 LC09 Q
PATULIBRACHIIDAE
Amphibrachium aff. *sansalvadorensis* Pessagno, 1971 Ho97 Mh–Dt
Paronaella sp. 1 Pessagno 1975 Pes75 Mh
Paronaella? sp. 2 Pessagno 1975 Pes75 Mh
Paronaella? sp. 3 Pessagno 1975 Pes75 Mh
Patulibracchium sp. Ho97 Mh–Dt
Patulibracchium sp. 1 Pessagno 1975 Pes75 Mh
PHASELIFORMIDAE
Phaseliforma laxa Pessagno, 1972 Ho97 Mh–Dt
Phaseliforma subcarinata Pessagno, 1975 Ho97 Mh–Dt
PYLONIDAE
Dipylissa bensoni Dumitrica, 1988 HN05 LH08 LC09 Q
Larcopyle buetschlii Dreyer, 1889 Oc97b B87 HN05 LH08 LC09 Po–Q
Larcopyle sp. CM99 Lwh–Q
Larcopyle? sp. Oc97b Po
Larcospira quadrangula Haeckel, 1887 B87 HN05 LH08 LC09 Q
Larcospira sp. CM99 Lwh–Tk
Octopyle stenozona Haeckel, 1887 B87 CM99 HN05 LH08 LC09 Tt–Q
Palaeotetrapyle muelleri Dumitrica, 1988 Ho97 Dt
Pylolena armata gr. Haeckel, 1887 B87 HN05 LH08 LC09 Q
Pylospira? *octopyle* Haeckel, 1887 B87 LH08 LC09 Q
Tetrapyle octacantha Mueller, 1858 B87 CM99 HN05 LH08 LC09 Tt–Q
SPONGODISCIDAE
Amphirhopalum ypsilon Haeckel, 1887 B87 CM99 HN05 LH08 LC09 Q
Dictyocoryne profunda Ehrenberg, 1860 B87 Oc99b HN05 LH08 LC09 Lwh–Q
Dictyocoryne truncatum (Ehrenberg, 1861) Oc97b LH08 Po–Q
Euchitonia elegans/furcata gr. (Ehrenberg, 1872) B87 HN05 LH08 LC09 Q
Euchitonia sp. CM99 Sl–Q
Flustrella charlestonensis (Clark & Campbell, 1945) Oc99a Ar–Lwh
Flustrella cretacea (Campbell & Clark, 1944) Ho97 Mh–Dt
Flustrella ruesti (Campbell & Clark, 1944) Ho97 Mh–Dt

Flustrella sp. Oc99b Lwh–Lw
Porodiscus sp. A sens. Nigrini & Moore 1979 Pet67 HN05 LH08 LC09 Q
Porodiscus (?) sp. B, sensu Nigrini & Moore, 1979 LH08 LC09 Q
Hymeniastrum sp. Oc97b Po
Rhopalastrum tritelum O'Connor, 1997 Oc00 Lwh
Rhopalastrum? sp. Ho02 Dt
Schizodiscus spatangus Dogiel, 1950 HN05 LH08 LC09 Q
Sethodiscus macrococcus Haeckel, 1887 LH08 LC09 Q
Spongaster tetras tetras Ehrenberg, 1860 CM99 Tt
Spongocore cylindrica (= *S. puella*) Haeckel, 1860 B87 HN05 LH08 LC09 Q
Spongodiscus alveatus (Sanfilippo & Riedel, 1973) Ho97 Ho02 Dt
Spongodiscus americanus Kozlova & Gorbovetz, 1966 Oc99b Lwh
Spongodiscus cruciferus Clark & Campbell, 1942 Oc99a HW97 Dm–Ar
Spongodiscus cf. *cruciferus* Clark & Campbell, 1942 Oc99a Ar
Spongodiscus maculatus Clark & Campbell, 1945 Oc99b Ar–Ld
Spongodiscus nitidus (Sanfilippo & Riedel, 1973) Oc99b Ar–Lw
Spongodiscus pulcher Clark & Campbell, 1945 Oc99b Ar–Ld
Spongodiscus resurgens Ehrenberg, 1854 B87 HN05 LH08 LC09 Q
Spongodiscus rhabdostylus (Ehrenberg, 1873) Ho97 Oc99b Mh–Lwh
Spongodiscus aff. *rhabdostylus* (Ehrenberg, 1973) Oc99a,b Ar–Lwh
Spongopyle osculosa (= *S. setosa*) Dreyer, 1889 Pet67 B87 CM99 Oc99b HN05 LH08 LC09 Ak–Q
Spongopyle? *sanfilippoae* O'Connor, 2001 O01 Ho02 Dt
Spongopyle spiralis Bjørklund & Kellogg, 1972 Oc99b Ar–Lwh
Spongopyle sp. Oc99b Ar–Lwh
Spongotripus sp. Ho97 Mh–Dt
Spongotrochus antiquus (Campbell & Clark, 1944b) Ho97 Mh–Dt
Spongotrochus antoniae O'Connor, 1997 Oc00 Po
Spongotrochus cf. *antoniae* O'Connor, 1997 Oc99b Lwh
Spongotrochus glacialis Popofsky, 1908 B87 C86 Oc97b CM99 HN05 LH08 LC09 Po–Q
Spongotrochus cf. *glacialis* Popofsky, 1908 Ho97 Dt
Spongotrochus cf. *polygonatus* (Campbell & Clark, 1944) Ho97 Mh–Dt
Spongotrochus (?) *venustum* (Bailey, 1856) LH08 LC09 Q
Stauralastrum spp. Oc99b Lwh
Stylochlamydium asteriscus Haeckel, 1887 LH08 LC09 Q
Stylodictya aculeata Jørgensen, 1905 HN05 LH08 LC09 Q
Stylodictya multispina (= *S. validispina*) Haeckel, 1860 Pet67 B87 Oc97b HN05 LH08 LC09 Po–Q
Stylodictya rosella Petrushevskaya & Kozlova, 1972 Ho02 Dt
Stylodictya cf. *sexispinata* Clark & Campbell, 1942 Ho97 Dt
Stylodictya targaeformis (Clark & Campbell, 1942) Oc99b Ar–Lwh
Stylodictya cf. *variabilis* Bjørklund & Kellogg, 1972 Oc99a Ar–Lwh
Stylodictya (?) sp. sens. Nakaseko & Nishimura 1982 LH08 Q
Tholodiscus densus (Kozlova & Gorbovetz, 1966) Ho97 Mh–Dt
Tholodiscus cf. *ocellatus* (Ehrenberg, 1875) Ho97 Ho02 Mh–Dt
Tholodiscus splendens (Ehrenberg, 1875) Ho02 Dt
Tholodiscus cf. *targaeformis* (Clark & Campbell, 1942) Ho97 Dt

SPONGURIDAE
Amphicraspedum murrayanum Haeckel, 1887 Oc99b Ar–Lwh
Amphicraspedum prolixum gr. Sanfilippo & Riedel, 1973 Oc99a Dt–Ar
Amphicraspedum prolixum s.s. Sanfilippo & Riedel, 1973 HW97 SH95 Dw–Dm-p
Amphymenium amphistylium Haeckel, 1887 CM99 Tk
Amphymenium concentricum (Lipman, 1960) Ho97 Mh
Amphymenium splendiarmatum Clark & Campbell, 1942 Oc99b Ak–Lw
Amphymenium cf. *splendiarmatum* (Clark & Campbell, 1942) Ho97 Ho02 Dt
Diartus hughesi (Campbell & Clark, 1944a) C86 CM99 Tt-k
Middourium regulare (Borissenko, 1958) HD05 Dt–Dw
Monobrachium irregulare (Nishimura, 1992) HD05 Dt–Dw
Ommatogramma sp. Oc99b Lwh–Lw
Ommatogramma? sp. Oc97b Po
Prunobrachium kennetti Pessagno, 1975 Ho97 Mh–Dt
Prunobrachium longum Pessagno, 1975 Pes75 Mh
Prunobrachium sibericum (Lipman, 1960) Pes75 Mh
Prunobrachium? *aucklandensis* Pessagno, 1975 Pes75 Mh
Spongoprunum cf. *markleyense* Clark & Campbell, 1942 Ho97 Dt
Spongurus bilobatus gr. Clark & Campbell, 1942 Ho02 HD05 Dt–Dw
Spongurus aff. elliptica (Ehrenberg, 1872) B87 LH08 Q
Spongurus pylomaticus Riedel, 1957 B87 HN05 LH08 LC09 Q
Spongurus aff. *prolixum* (Sanfilippo & Riedel, 1973) Ho97 Dt
Spongurus spongiosus (Lipman, 1960) Ho97 Mh–Dt
Spongurus? sp. Petrushevskaya, 1967 B87 HN05 LH08 LC09 Q
Spongurus? sp. B Nigrini & Lombari 1984 Oc97b Po

THOLONIDAE
Cubotholus spp. B87 HN05 LH08 LC09 Q

INCERTAE SEDIS
Peritiviator? *dumitricai* Nishimura, 1992 Ho02 Dt
Peritiviator cf. *labyrinthi* Pessagno, 1976 Ho97 Mh–Dt

Order COLLODARIA

COLLOSPHAERIDAE
Acanthosphaera actinota (Haeckel, 1862) LH08 LC09
Acanthosphaera dodecastyla Mast, 1910 B87 LH08 LC09 Q
Acanthosphaera pinchuda Boltovskoy & Riedel, 1981 B87 LH08 LC09 Q
Acrosphaera arktios (Nigrini, 1970) LH08 LC09 Q
Acrosphaera cyrtodon (Haeckel, 1887) LH08 LC09 Q
Acrosphaera inflata Haeckel, 1887 LH08 LC09 Q
Acrosphaera lappacea (Haeckel, 1887) B87 HN05 LH08 LC09 Q
Acrosphaera murrayana (Haeckel, 1887) B87 LH08 LC09 Q
Acrosphaera aff. *murrayana* (Haeckel, 1887) C86 Sc-w
Acrosphaera? sp. Oc97b Po
Acrosphaera? *mercurius* Lazarus, 1992 LH08 LC09
Acrosphaera spinosa (Haeckel, 1860) B87 HN05 LH08 LC09
Collosphaera confossa Takahashi, 1991 LH08 LC09
Collosphaera huxleyi Müller, 1855 B87 HN05 LH08 LC09 Q
Collosphaera macropora Popofsky, 1917 B87 LH08 LC09 Q
Collosphaera polygona Haeckel, 1887 B87 C86 Sc-w–Q
Siphonosphaera martensi Brandt, 1905 B87 HN05 LH08 LC09 Q
Siphonosphaera socialis (= *S. polysiphonia*) Haeckel, 1887 B87 LH08 LC09 Q
Solenosphaera (=*Disolenia*) *zanguebarica* (Ehrenberg, 1872) B87 LH08 LC09 Q

Order ENTACTINARIA

OROSPHAERIDAE
Orodapis sp. CM99 Q
Orosphaera huxleyii Haeckel, 1887 LH08 Q

SATURNALIDAE
Acanthocircus campbelli (Foreman, 1968) Ho97 Mh
Acanthocircus ellipticus (Campbell & Clark, 1944b) Ho97 Mh–Dt
Axoprunum? aff. *bispiculum* (Popofsky, 1912) Ho02 Dt
Axoprunum? *irregularis* Takemura, 1992 HW97 Lwh
Axoprunum pierinae (Clark & Campbell, 1942) Oc99a CM99 Ho02 Dt–Lwh
Axoprunum stauraxonium Haeckel, 1887 B87 HN05 LH08 LC09 Q
Saturnalis circularis Haeckel, 1887 Oc99b CM99 LH08 LC09 Lwh–Q
Saturnalis kennetti Dumitrica, 1973 Ho97 Dt
Spongosaturninus sp. 1 Pessagno 1975 Pes75 Mh
Spongosaturninus sp. 2 Pessagno 1975 Pes75 Mh
Stylacontarium aquilonium (Hays, 1965) CM99 Tk–Q
Stylacontarium bispiculum Popofsky, 1912 HW97 CM99 LH08 LC09 Ar–Q

Order NASSELLARIA

ACANTHODESMIIDAE
Acanthodesmia vinculata (Müller, 1857) B87 LH08 Q
Tympanidium binoctonum Haeckel, 1887 Oc97b Po
Zygocircus buetschlii Haeckel, 1887 HW97 Oc99b Ar–Ld
Zygocircus cf. *piscicaudatus* Popofsky, 1913 LH08 Q
Zygocircus productus (Hertwig, 1879) Bo87 HN05 LH08 LC09 Q
Zygocircus p. tricarinatus Goll, 1979 Oc97b Po

ACROPYRAMIDIDAE
Archipilium macroporus (Haeckel, 1887) LH08 LC09 Q
Archipilium? spp. Oc97b, Oc99b Lwh–Po
Artostrobus annulatus (Bailey, 1856) HW97, B87 LH08 LC09 Ak-r–Q
Artostrobus jörgenseni Petrushevskaya, 1967 LH08 LC09 Q
Artostrobus cf. *pretabulatus* Petrushevskaya, 1975 HW97 LH08 Ak-r–Q
Artostrobus pusillus (Ehrenberg, 1873) Ho97 Mh–Ld
Bathropyramis magnifica (Clark & Campbell, 1942) Oc99a Ho02 Dt–Lwh
Bathropyramis sanjoaquinensis Campbell & Clark, 1944b Ho97 Mh–Dt
Bathropyramis scalaris (Ehrenberg, 1873) SH95 Dw
Bathropyramis woodringi Campbell & Clark, 1944 LH08 LC09 Q
Bathropyramis sp. Oc99b Lwh–?Lw
Cinclopyramis sp. Pes75 Mh
Conarachnium polyacanthum (Popofsky, 1913) LH08 Q
Cornutella bimarginata Haeckel, 1887 LH08 LC09 Q
Cornutella californica Campbell & Clark, 1944b Ho97 Oc99b Mh–Lw
Cornutella profunda Ehrenberg, 1854 Pet67 B87

Oc97b CM99 HN05 LH08 LC09 Ar–Q
Peripyramis circumtexta Haeckel, 1887 B87 HN05 LH08 LC09 Q
Plectopyramis dodecomma Haeckel, 1887 LH08 LC09 Q
Plectopyramis sp. Oc99b Ar–Ld
AMPHIPYNDACIDAE
Amphipternis alamedaensis (Campbell & Clark, 1944) Ho97 Mh–Ab
Amphipternis clava (Ehrenberg, 1873) Dum73 Dt
Amphipyndax aff. *conicus* Nakaseko & Nishimura, 1981 Ho97 Mh–Dt
Amphipyndax stocki gr. (Campbell & Clark, 1944b) Ho97 Ho02 Mh–Dt
Cyrtolagena aglaolampa (Takahashi, 1991) Oc97b Po
Cyrtolagena cf. *aglaolampa* (Takahashi, 1991) Oc99b Lwh
Sticholagena sp. A Ho02 Dt
ARCHAEODICTYOMITRIDAE
Archaeodictyomitra cf. *lamellicostata* (Foreman, 1968) Ho97 Mh–Dt
Archaeodictyomitra? sp. Oc99b Lwh–Po
Dictyomitra andersoni (Campbell & Clark, 1944b) Ho97 Mh–Dt
Dictyomitra multicostata Zittel, 1876 Ho97 Mh–Dt
Dictyomitra aff. *rhadina* Foreman, 1968 Ho97 Mh–Dt
Mita regina (Campbell & Clark, 1944b) Ho97 Mh–Dt
Mita cf. *regina* (Campbell & Clark, 1944b) Ho97 Mh–Dt
Mita sp. A HD05 Dw
ARTOSTROBIIDAE
Botryostrobus aquilonaris (Bailey, 1856) Hays65 B87 C86 HN05 LH08 LC09 Tt-k–Q
Botryostrobus auritus/australis gr. (Ehrenberg, 1844) B87 C86 CM99 HN05 LH08 LC09 Sc-w–Q
Botryostrobus bramlettei bramlettei (Campbell & Clark, 1944a) C86 Sc-w–Tt-k
Botryostrobus b. pretumidulus Caulet, 1979 C86 Sc-w–Q
Botryostrobus b. seriatus (Jörgensen, 1905) C86 Tt-k–Q
Botryostrobus b. tumidulus (Bailey, 1856) C86 Q
Botryostrobus hollisi O'Connor, 1997b Oc00 Po
Botryostrobus miralestensis (Campbell & Clark, 1944a) Oc97b C86 CM99 Po–Sl
Botryostrobus? *parsonsae* Hollis, 1997 Ho97 Mh–Dt
Botryostrobus spp. C86 CM99 Sc-w–Tt-k
Dictyoprora amphora gr. (Haeckel, 1887) Oc99b SH95 Dm-p–?Lwh
Dictyoprora gibsoni O'Connor, 1994 Oc00 Ar–Po
Dictyoprora mongolfieri (Ehrenberg, 1854) SH95 Oc99b CM99 Dm–Po
Dictyoprora nigriniae O'Connor, 2000 Oc00 ?Ar–Lwh
Dictyoprora urceolus (Haeckel, 1887) Oc99b Dm–Lwh
Eribotrys? *johnsoni* Petrushevskaya, 1977 Pet77 Ho02 Mh–Dt
Phormostichoartus corbula (Harting, 1863) C86 B87 CM99 Sc-w–Q
Phormostichoartus doliolum (Riedel & Sanfilippo, 1971) C86 Tt-k
Phormostichoartus fistula Nigrini, 1977 HW97 C86 Lwh–Sc-w
Phormostichoartus marylandicus (Martin, 1904) Oc99b Lwh–Lw
Phormostichoartus platycephala (Ehrenberg, 1873) Pet67 C86 LH08 LC09 Sc-w–Q
Phormostichoartus? *strongi* Hollis, 1997 Ho97 Mh–Dt
Phormostichoartus sp. Oc99b Lwh–?Ld
Plannapus? *aitai* O'Connor, 2000 Oc00 Ar–Lwh
Plannapus hornibrooki O'Connor, 1999 Oc00 Ab–Lwh
Plannapus mauricei O'Connor, 1999 Oc00 Ab–Lwh
Plannapus microcephalus (Haeckel, 1887) Oc99b Lwh–Lw
Plannapus? sp. Oc99b Lwh
Siphocampe acephala gr. (Ehrenberg, 1875) HW97 Ab–Lwh
Siphocampe cf. *acephala* (Ehrenberg, 1875) Oc99a Ar
Siphocampe (= Lithomitra) *arachnea* (Ehrenberg, 1861) Pet67 B87 LH08 LC09 Q
Siphocampe altamontensis (Campbell & Clark, 1944b) Ho97 Mh–Dt
Siphocampe cf. *altamontensis* (Campbell & Clark, 1944b) Ho97 Mh–Dt
Siphocampe "elizabethae" sens. Nigrini 1977) HW97 Dw–Lwh
Siphocampe bassilis (Foreman, 1968) Ho02 Dt
Siphocampe grantmackiei O'Connor, 1997b Oc00 Po
Siphocampe missilis O'Connor, 1994 Oc00 ?Ab–Lwh
Siphocampe modeloensis (Campbell & Clark, 1944a) C86 Sc-w–Tt-k
Siphocampe nodosaria (Haeckel, 1887) Oc99b Ho02 LH08 LC09 Dt–Q
Siphocampe aff. *nodosaria* (Haeckel, 1887) Oc99a Ar
Siphocampe quadrata (Petrushevskaya & Kozlova, 1972) Oc99b Ho02 Dt–Lwh
Siphostichartus praecorona Nigrini, 1977 Oc97b Po
Siphostichartus aff. *praecorona* Nigrini, 1977 Oc99b Lwh–?Ld
Spirocyrtis greeni O'Connor, 1999 Oc99 Ar–Lwh
Spirocyrtis aff. *greeni* O'Connor, 1999 Ho02 Dt
Spirocyrtis gyroscalaris Nigrini, 1977 LH08 Q
Spirocyrtis aff. *gyroscalaris* Nigrini, 1977 Ho02 Dt
Spirocyrtis proboscis O'Connor, 1994 Oc94 Lwh
Spirocyrtis scalaris gr. Haeckel, 1887 C86 B87 HN05 LH08 LC09 Q
Spirocyrtis subscalaris Nigrini, 1977 C86 LH08 LC09 Sc-w–Q
Spirocyrtis subtilis Petrushevskaya & Kozlova, 1972 Oc97b C86 Po–Sc-w
Spirocyrtis aff. *subtilis* Petrushevskaya & Kozlova, 1972 Oc99b Lwh
Spirocyrtis sp. A HW97 Oc97b Ak–Lwh
Theocampe cf. *vanderhoofi* Campbell & Clark, 1944b Ho97 Mh–Dt
Tricolocampe cylindrica (=*Siphocampe lineata* gr., *Lithomitra lineata*) Haeckel, 1887 Pet67 B87 HN05 LH08 LC09 Q
BEKOMIDAE
Bekoma bidartensis Riedel & Sanfilippo, 1971 SH95 HD05 Dw
Bekoma campechensis Foreman, 1973 SH95 HD05 Dt
Bekoma divaricata Foreman, 1973 SH95 HD05 Dw
CANNOBOTRYIDAE
Acrobotrys sp. Oc99b Lwh–Po
Bisphaerocephalina sp. sens. Petrushevskaya, 1965 LH08 Q
Botryocampe cf. *inflata* (Bailey, 1856) LH08 LC09 Q
Botryocella cribrosa gr. (Ehrenberg, 1873) HW97 C86 Ak-r–Tt-k
Botryocella pauciperforata O'Connor, 1999 Oc00 Ar–Lwh
Botryocephalina armata Petrushevskaya, 1965 LH08 Q
Botryocyrtis scutum (Harting, 1863) LH08 Q
Botryopyle dictyocephalus Haeckel, 1887 B87 Oc99b LH08 LC09 Lwh–Q
Botryopyle dionisi Petrushevskaya, 1975 C86 Sc-w
Centrobotrys petrushevskayae Sanfilippo & Riedel, 1971 Oc99b Lwh
Saccospyris antarctica (Haecker, 1907) B87 C86 HN05 LH08 LC09 Q
Saccospyris (= *Botryocampe*) *conithorax* Petrushevskaya, 1965 B87 LC09 Q
Saccospyris preantarctica Petrushevskaya, 1975 C86 Sc-w–Q
CARPOCANIIDAE
Carpocanium rubyae O'Connor, 1997b Oc00 Po–uTt-k
Carpocanistrum acutidentatum Takahashi, 1991 LH08 LC09 Q
Carpocanarium papillosum gr. (Ehrenberg, 1872) B87 HN05 LH08 LC09 Q
Carpocanistrum sp. B87 CM99 HN05 LH08 LC09 Po-l–Q
Carpocanium sp. Oc97b Oc99b Oc00 Lwh–Po
Carpocanopsis ballisticum O'Connor, 1999 Oc99a Ar–?Ld
Carpocanopsis bramlettei Riedel & Sanfilippo, 1971 CM99 Sc
Carpocanopsis cingulata Riedel & Sanfilippo, 1971 Oc97b CM99 Po–Tt
Carpocanopsis cristatum? (Carnevale, 1908) CM99 Tk–Wo
Carpocanopsis favosum (Haeckel, 1887) Oc97b Po
CUNICULIFORMIDAE
Cassideus aff. *mariae* Nishimura, 1992 Ho02 HD05 Dt–Dw
EUCYRTIDIIDAE
Anthocyrtis aff. *mespilus* Nishimura, 1992 Ho02 Dt
Anthocyrtoma sp. Oc99b Lwh
Aphetocyrtis gnomabox Sanfilippo & Caulet, 1998 CM99 Lwh–Po
Artophormis fluminafauces O'Connor, 1999 Oc00 Ar–Lwh
Artophormis gracilis Riedel, 1959 Oc99b Lwh–Ld
Aspis murus Nishimura, 1992 Ho02 Dt
Aspis velutochlamydosuarus Nishimura, 1992 Ho02 Dt
Aspis sp. A Ho02 Dt
Buryella dumitricai Petrushevskaya, 1977 Ho97 Ho02 Dt
Buryella foremanae Petrushevskaya, 1977 Ho97 Ho02 Dt
Buryella aff. *foremanae* Petrushevskaya, 1977 Ho02 Dt
Buryella granulata (Petrushevskaya, 1977) Ho97 Ho02 Dt–Dm
Buryella cf. *granulata* Petrushevskaya, 1977 HR03 Dt
Buryella helenae O'Connor, 2001 O01 Ho02 Dt
Buryella? *insensis* (Kozlova, 1983) Ho02 Dt
Buryella kaikoura Hollis, 1997 Ho97 O01 Dt
Buryella pentadica Foreman, 1993 O01 Ho02 Dt
Buryella petrushevskayae O'Connor, 2001 O01 Ho02 Dt
Buryella tetradica tetradica Foreman, 1973 Ho97 Oc01 Ho02 Dt–Dh
Buryella t. tridica O'Connor, 2001 O01 Ho02 Dt
Calocyclas? *nakasekoi* Takemura & Ling, 1997 CM99 Ar
Calocycloma ampulla (Ehrenberg, 1854) SH95 HW97 Dw–Dm-h
Calocycloma castum (Haeckel, 1887) SH95 Dm–Dm
Cyrtocapsa campi Campbell & Clark, 1944b Ho97 Mh–Dt
Cyrtocapsa cornuta Haeckel, 1887 C86 CM99 Oc97b Po–Sc-w
Cyrtocapsa cylindroides (Principi, 1909) Oc97b Po
Cyrtocapsa livermorensis (Campbell & Clark, 1944b) Ho97 Mh–Dt
Cyrtocapsa osculum O'Connor, 1997 Oc00 Ar–Ld
Cyrtocapsa tetrapera Haeckel, 1887 Oc97b CM99 Po
Cyrtocapsella elongata (Nakaseko, 1963) C86 Sc-w–Tt-k
Cyrtocapsella japonica (Nakaseko, 1963) C86 CM99 Po-Pl–Wo
Cyrtocapsella robusta Abelmann, 1990 HW97 Lwh

Cyrtocapsella tetrapera (Haeckel, 1887) C86 Sc-w–Tt-k
Cyrtopera laguncula Haeckel, 1887 Oc99b B87 LH08 LC09 Lwh–Q
Dicolocapsa microcephala Haeckel, 1887 CM99 Lwh
Dictyophimus cf. *archipilium* Petrushevskaya, 1975 HW97 Lwh
Dictyophimus bicornis (Ehrenberg, 1847) B87 LH08 LC09 Q
Dictyophimus cf. *borisenkoi* (Nishimura, 1992) Ho02 Dt
Dictyophimus cf. *constrictus* Nishimura, 1992 Ho02 Dt
Dictyophimus crisiae Ehrenberg, 1854 C86 LH08 Tt-k–Q
Dictyophimus hirundo (Haeckel, 1887) B87 CM99 HN05 LH08 LC09 Sw–Q
Dictyophimus infabricatus Nigrini, 1968 B87 HN05 LH08 LC09 Q
Dictyophimus killmari (Renz, 1974) LH08 Q
Dictyophimus pocillum Ehrenberg, 1873 Oc99b Lwh–?Lw
Dictyophimus? aff. *campana* (Ehrenberg, 1873) Ho02 Dt
Dictyophimus? aff. *constrictus* Nishimura, 1992 Oc99b Lwh
Dictyophimus? *okadai* Nishimura, 1992 Ho02 Dt
Dictyophimus aff. *pocillum* Ehrenberg, 1873 Ho97 Dt
Dictyophimus splendens (Campbell & Clark, 1944a) C86 CM99 Sc-w–Wo
Dictyophimus sp. CM99 Tt
Eucyrtidium acuminatum acuminatum (Ehrenberg, 1844) B87 CM99 HN05 LH08 LC09 Q
Eucyrtidium acuminatum octocolum (Haeckel, 1887) B87 HN05 LH08 LC09 Q
Eucyrtidium anomalum (Haeckel, 1860) B87 LH08 LC09 Q
Eucyrtidium hexastichum (Haeckel, 1887) B87 LH08 LC09 Q
Eucyrtidium cf. *acuminatum* (Ehrenberg, 1844) Oc97b Po
Eucyrtidium antiquum Caulet, 1991 HW97, CM99 Lwh–Po
Eucyrtidium calvertense Martin, 1904 C86, CM99, HN05 LH08 Sw–Q
Eucyrtidium cf. *cheni* Takemura, 1992 SH95 Dm-p–Ab
Eucyrtidium cienkowskii gr. Haeckel, 1887 Oc97b C86 CM99 Po–Tt-k
Eucyrtidium hexagonatum Haeckel, 1887 LH08 LC09 Q
Eucyrtidium mariae Caulet, 1991 Oc99a HW97 Ar
Eucyrtidium matuyamai Hays, 1970 CM99 LH08 LC09 Q
Eucyrtidium cf. matuyamai Hays, 1970 LH08 Q
Eucyrtidium montiparum Ehrenberg, 1873 Oc99a Ar
Eucyrtidium punctatum (Ehrenberg, 1844) C86 Oc97b CM99 Po–Tt-k
Eucyrtidium spinosum Takemura, 1992 CM99 Oc99a HW97 Ak to Lwh
Eucyrtidium teuscheri Haeckel, 1887 Pet 67 C86 CM99 HN05 LH08 LC09 Sc-w–Q
Eucyrtidium ventriosum O'Connor, 1999 Oc00 Ab–Lwh
Eucyrtidium yatsuoense Nakaseko, 1963 CM99 Sl–Sw
Eucyrtidium spp. (= *Eucyrtidium* spp. 1 gr.) B87 LH08 LC09 Q
Gondwanaria dogieli (Petrushevskaya & Kozlova, 1972) CM99 HN05 LH08 LC09 Tk–Q
Gondwanaria sp. CM99 Q
Lipmanella bombus (Haeckel1887) LH08 LC09 Q
Lipmanella dictyoceras (= *L. virchowii*) (Haeckel, 1860) B87 HN05 LH08 LC09 Q
Lipmanella xiphephorum (Jørgensen, 1900) LH08 Q
Litharachnium tentorium Haeckel, 1860 B87 HN05 LH08 LC09 Q
Lithocampe aff. *subligata* Stöhr, 1880 Ho97 Mh–Dt
Lithocampe wharanui Hollis, 1997 Ho97 Mh–Dt
Lithocampe? *furcaspiculata* (Popofsky, 1908) Pet67 LH08 LC09 Q
Lithopera bacca Ehrenberg, 1872 B87 CM99 LH08 Pl–Q
Lithopera neotera Sanfilippo & Riedel, 1970 C86 CM99 Sc-w–Tk
Lithopera renzae Sanfilippo & Riedel, 1970 CM99 Sl
Lithostrobus hexagonalis Haeckel, 1887 LH08 LC09 Q
Lithostrobus longus (= *L. wero* Hollis) Grigorjeva, 1975 Ho97 Ho02 Dt
Lithostrobus cf. *longus* Grigorjeva, 1975 Ho02 Dt
Lophoconus biaurita (Ehrenberg, 1873) CM99 Po-Pl–Wp
Phormocyrtis alexandrae O'Connor, 1997 CM99 Oc00 Po–Sl
Phormocyrtis cf. *alexandrae* O'Connor, 1997 Oc99b HW97 Lwh–Lw
Phormocyrtis cubensis (Riedel & Sanfilippo, 1971) HD05 Dw
Phormocyrtis ligulata Clark & Campbell, 1942 SH95 Oc99b Dm-p–Ld
Phormocyrtis striata exquisita (Kozlova & Gorbovetz, 1966) SH95 HD05 Dt–Dw
Phormocyrtis s. praeexquisita Nishimura, 1992 Ho02 Dt
Phormocyrtis s. striata (Brandt, 1935) SH95 Oc99b HD05 Dw–Dh
Phormocyrtis turgida (Krasheninnikov, 1960) HD05 Dw
Phormocyrtis vasculum O'Connor, 1994 Oc00 Lwh–?Lw
Phormocyrtis? sp. Oc97b Po
Pterocodon? *lex* Sanfilippo & Riedel, 1979 Ho02 Dt
Pterocodon poculum Nishimura, 1992 Ho02 HD05 Dt
Pterocodon? aff. *tenellus* Foreman, 1973 Ho02 Dt
Pterocyrtidium parebarbadense Kozlova, 1983 Ho02 Dt
Pteropilium sp. Oc99a Ar
Pteropilium? sp. Oc99b Ho02 Dt–Lwh
Pterosyringium hamata O'Connor, 1999 Oc00 ?Ab–?Lw
Rhopalosyringium? *magnificum* Campbell & Clark, 1944b HR03 Dt
Sethoconus tabulatus (Ehrenberg, 1873) B87 LH08 LC09 Q
Sethocyrtis n. sp. A SH95 HW97 Dm–Ab
Siphostichartus corona (Haeckel, 1887) C86 Sc-w
Stichocorys armata (Haeckel, 1887) CM99 Sc
Stichocorys coronata (Carnevale, 1908) Oc97b Po
Stichocorys delmontensis (Campbell & Clark, 1944a) Oc97b C86 CM99 Po–Pl–Wo
Stichocorys johnsoni Caulet, 1986 CM99 Sw
Stichocorys negripontensis O'Connor, 1997 Oc00 Lwh–Lw
Stichocorys aff. *negripontensis* O'Connor, 1997 Oc97a Lwh
Stichocorys peregrina (Riedel, 1953) C86 CM99 Sw–Wo-m
Stichocorys wolffii Haeckel, 1887 CM99 Sc
Stichocorys sp. CM99 Sl–Wp
Stichomitra bertrandi Cayeux, 1897 Ho97 Mh–Dt
Stichomitra carnegiensis (Campbell & Clark, 1944b) Ho97 Mh–Dt
Stichomitra cf. *carnegiensis* (Campbell & Clark, 1944) Ho97 Dt
Stichomitra cathara Foreman, 1968 HR03 Dt
Stichomitra grandis (Campbell & Clark, 1944) Ho97 Mh–Dt
Stichopilium (=*Eucyrtidium*) *annulatum* Popofsky, 1913 H65 LH08 LC09 Q
Stichopodium biconicum (Vinassa, 1900) C86 Sc-w–Q
Stichopodium inflatum (Kling, 1973) C86 Tt-k–Wo-m
Theocorys bianulus O'Connor, 1997 Oc00 Lwh–?Lw
Theocorys longithorax Petrushevskaya, 1975 Oc99b Lwh–Po
Theocorys perforalvus O'Connor, 1997 Oc00 Ab–Lw
Theocorys? *phyzella* Foreman, 1973 HD05 Dw
Theocorys? aff. *phyzella* Foreman, 1973 Ho02 HD05 Dt–Dw
Theocorys? cf. *phyzella* Foreman, 1973 HD05 Dt–Dw
Theocorys puriri O'Connor, 1997 Oc00 Lwh–Po
Theocorys redondoensis (Campbell & Clark, 1944) CM99 Sl–Wo
Theocorys aff. *redondoensis* (Campbell & Clark, 1944) Oc99b Lwh
Theocorys spongoconum Kling, 1971 CM99 Po-Pl–Sl
Theocorys veneris Haeckel, 1887 B87 LH08 LC09 Q
Theocorys sp. A CM99 Tk–Wo
Theopilium tricostatum (Haeckel, 1887) B87 LH08 LC09 Q
Triacartus? sp. A Ho02 Dt
LOPHOCYRTIIDAE
Lophocyrtis (*Apoplanius*) *aspera* (Ehrenberg, 1873) HW97 Oc99a CM99 Dt–Ld
Lophocyrtis (*Cyclampterium*) *hadra* Riedel & Sanfilippo, 1986 HW97 Ab–Lwh
Lophocyrtis (*C.*) *millowi* Riedel & Sanfilippo, 1971 HW97 Lwh
Lophocyrtis (*Lophocyrtis*) aff. *exitelus* Sanfilippo, 1990 Oc99b Lwh
Lophocyrtis (*L.*) *haywardi* O'Connor, 1999 Oc00 Ab–Lwh
Lophocyrtis (*L.*) *jacchia hapsis* Sanfilippo & Caulet, 1998 Oc99a Ar
Lophocyrtis (*L.*) cf. *jacchia* (Ehrenberg, 1873) SH95 Dm-p–Ab
Lophocyrtis (*L.*?) *semipolita* (Clark & Campbell, 1942) Oc99a HW97 Dm–Lwh
Lophocyrtis (*Paralampterium*?) *galenum* Sanfilippo, 1990 Oc97b Po
Lophocyrtis (*P.*?) *inaequalis* O'Connor, 1997a Oc00 Lwh–Ld
Lophocyrtis (*P.*?) *longiventer* (Chen, 1975) HW97 Oc99b CM99 Ab–Lwh
Lophocyrtis (*P.*) *dumitricai* Sanfilippo, 1990 HW97, Oc99b, CM99 Dm-p to Lwh
Lophocyrtis (*P.*?) cf. *longiventer* (Chen, 1975) Oc99b Lwh
Lophocyrtis? *auriculaleporis* (Clark & Campbell, 1942) SH95 Dw–Ab
LOPHOPHAENIDAE
Acanthocorys? sp. A Ho02 Dt
Acidnomelos? *laevis* Petrushevskaya, 1977 Pet77 Mh
Amphiplecta acrostoma Haeckel, 1887 LH08 LC09 Q
Antarctissa cylindrica Petrushevskaya, 1967 HN05 LH08 LC09 Q
Antarctissa denticulata (incl. *A strelkovi*) (Ehrenberg, 1844) C86 B87 HN05 LH08 LC09 Q
Antarctissa cf. *denticulata* (Ehrenberg, 1844) LC09 Q
Antarctissa ewingi Chen, 1975 C86 Q
Antarctissa longa (Popofsky, 1908) Pet67 LC09 Q
Antarctissa sp. Oc97b Po
Arachnocorallium calvata gr. Petrushevskaya, 1971 B87 LC09 Q
Arachnocorys umbellifera Haeckel, 1862 LH08 LC09 Q
Callimitra atavia Goll, 1979 HW97 Oc99b Ar–Ld
Callimitra solocicribrata Takahashi, 1991 LH08 Q
Ceratocyrtis galeus (Cleve, 1899) LH08 LC09 Q
Ceratocyrtis mashae? Bjørklund, 1976 Oc99a Ar

Ceratocyrtis robustus? Bjørklund, 1976 Oc99a Ar
Ceratocyrtis? volubilis Petrushevskaya, 1977 Pet77 Mh
Ceratocyrtis sp. Oc99a Ar–?Ld
Ceratocyrtis? sp. B Oc99a Lwh
Cladoscenium aff. *fulcratum* Haeckel, 1887 Ho02 Dt
Cladoscenium limbatum Jørgensen, 1905 LH08 Q
Cladoscenium cf. *tricolpium* (Haeckel, 1887) LH08 LC09 Q
Cladoscenium? ancoratum Haeckel, 1887 B87 LH08 LC09 Q
Clathrocanium coarctatum Ehrenberg, 1860 LH08 LC09 Q
Clathrocanium sphaerocephalum Haeckel, 1887 Oc97b Po
Clathrocanium sp. Oc99b Lwh–Ld
Clathrocorys teuscheri (= *C. murrayi*) Haeckel, 1887 B87 LH08 LC09 Q
Clathrocorys? sp. Oc97b Po
Clathrocorys? sp. LH08 Q
Clathromitra pterophormis Haeckel, 1887 LH08 Q
Clathromitra sp. sensu Nakaseko & Nishimura, 1982 LH08 Q
Corythomelissa adunca (Sanfilippo & Riedel, 1973) SH95 Ho02 Dt–Dw
Corythomelissa horrida Petrushevskaya, 1975 CM99 Lwh
Helotholus histricosa Jörgensen, 1905 Oc97b B87 HN05 LH08 LC09 Po–Q
Lampromitra cracenta Takahashi, 1991 LH08 LC09 Q
Lampromitra quadricuspis Haeckel, 1887 B87 LH08 LC09 Q
Lampromitra schultzei (Haeckel, 1887) LH08 LC09 Q
Lithomelissa challengerae Chen, 1975 HW97 Ar
Lithomelissa ehrenbergi Bütschli, 1882 HW97 C86 Ar–Sc-w
Lithomelissa aff. *ehrenbergi* Bütschli, 1882 Oc99b Ar–?Ld
Lithomelissa gelasinus O'Connor, 1997 HW97 Oc00 Ab–Lw
Lithomelissa cf. *gelasinus* O'Connor, 1997 Oc99a Ar–Lwh
Lithomelissa aff. *haeckeli* Bütschli, 1882 Oc99b Lwh
Lithomelissa cf. *haeckeli* Bütschli, 1882 Oc99a HW97 Ar
Lithomelissa cf. *heros* Campbell & Clark, 1944b Ho97 Mh–Dt
Lithomelissa cf. *hertwigi* Bütschli, 1882 Oc99a Ar–Lwh
Lithomelissa? aitai Hollis, 1997 Ho97 Mh–Dt
Lithomelissa? borealis (Ehrenberg, 1872) B87 LH08 LC09 Q
Lithomelissa? hoplites Foreman, 1968 Ho97 Mh–Dt
Lithomelissa hystrix Jørgensen, 1900 LH08 LC09 Q
Lithomelissa laticeps Jørgensen, 1905 LH08 LC09 Q
Lithomelissa lautouri O'Connor, 1999 Oc00 ?Ab–Lwh
Lithomelissa maureenae O'Connor, 1997 Oc00 Ar–Lwh
Lithomelissa cf. *mitra* Bütschli, 1882 Oc99 Ar–Lwh
Lithomelissa? sakaii O'Connor, 2000 Oc00 Ak–Lwh
Lithomelissa sphaerocephalis Chen, 1975 HW97 Ar–Lwh
Lithomelissa stigi Bjørklund, 1976 LH08 LC09 Q
Lithomelissa thoracites Haeckel, 1862 LH08 LC09 Q
Lithomelissa (?) sp. A. Petrushevskaya, 1967 LC09 Q
Lophophaena butschlii (Haeckel, 1887) LH08 LC09 Q
Lophophaena sp. Ho02 Dt–Lwh
Lophophaena? capito Ehrenberg, 1873 HW97 Ar–Lwh
Lophophaena hispida (Ehrenberg, 1872) B87 LH08 LC09 Q
Lophophaena mugaica (Grigorjeva, 1975) Ho02 Dt–Dw
Lophophaena? polycyrtis (Campbell & Clark, 1944b) Pes75 Mh–Dt
Lophophaena tekopua O'Connor, 1997 Oc00 Lwh–Po
Lophophaena? sp. Oc97b Po
Mitrocalpis araneafera Popofsky, 1908 LC09 Q
Neosemantis distephanus (Haeckel, 1887) LH08 LC09 Q
Peridium longispinum Jørgensen, 1900 LH08 LC09 Q
Peridium cf. *longispinum* Benson, 1966 LH08 Q
Peridium sp. Oc99a Ar
Peromelissa phalacra (Haeckel, 1887) B87 LH08 LC09 Q
Phormacantha hystrix (Jørgensen, 1905) B87 LH08 LC09 Q
Plectacantha sp. Benson, 1966 LH08 LC09 Q
Pseudocubus obeliscus Haeckel, 1887 B87 LH08 LC09 Q
Pseudocubus vema (Hays, 1965) C86 Q
Pseudodictyophimus (= *Dictyophimus*) *platycephalus* (Haeckel, 1887) LH08 LC09 Q
Pseudodictyophimus gracilipes (Bailey, 1856) B87 HW97 Oc99b HN05 LH08 LC09 Ak–Q
Psilomelissa? sp. A Ho02 Dt
Trisulcus triacanthus Popofsky, 1913 LH08 LC09 Q
Tripodiscinus clavipes (Clark & Campbell, 1942) HW97 Ak-r
Velicucullus fragilis O'Connor, 1999 Oc00 Ak–Ar
NEOSCIADIOCAPSIDAE
Ewingella? sp. Oc99a Ar
Microsciadiocapsa? sp. Pes75 Mh
Neosciadiacapsa? gyrus Petrushevskaya, 1977 Pet77 Ho02 Mh–Dt
Neosciadiacapsa jenkinsi Pessagno, 1975 Ho97 Mh–Dt
Verutotholus doigi O'Connor, 1999 Oc00 Ab–Ld
Verutotholus cf. *doigi* O'Connor, 1999 Oc99a Ar
Verutotholus edwardsi O'Connor, 1999 Oc00 Ar–Ld
Verutotholus mackayi O'Connor, 1999 Oc00 Ar–Lwh
Gen. et sp. indet. Oc99b Lwh
PTEROCORYTHIDAE
Albatrossidium sp. CM99 Ar–Lwh
Anthocyrtidium adiaphorum Sanfilippo & Riedel, 1992 Oc99b Lwh
Anthocyrtidium angulare Nigrini, 1971 CM99 Tt–Q
Anthocyrtidium ehrenbergi (Stöhr, 1880) C86 Oc97b CM99 Po–Q
Anthocyrtidium marieae O'Connor, 1997 CM99 Oc00 Po
Anthocyrtidium aff. *marieae* O'Connor, 1997 Oc00 ?Ld–?Lw
Anthocyrtidium odontatum O'Connor, 1994 Oc00 Lwh–Lw
Anthocyrtidium ophirense (Ehrenberg, 1872) B87 LH08 LC09 Q
Anthocyrtidium zanguebaricum (Ehrenberg, 1872) B87 LH08 LC09 Q
Anthocyrtidium aff. *pliocenica* (Seguenza, 1880) Oc97b Po
Anthocyrtidium stenum Sanfilippo & Riedel, 1992 Oc99b Ar–Lwh
Anthocyrtidium sp. Oc99b Lwh–Ld
Calocycletta (*Calocycletta*) *parva* Moore, 1972 Oc99b Lwh–?Lw
Calocycletta (*C.*) *robusta* Moore, 1971 Oc99b Lwh–Ld
Calocycletta (*C.*) cf. *virginis* Haeckel, 1887 Oc97b Po
Calocycletta (*Calocyclopsis*) *serrata* Moore, 1972 Oc97b Po
Cryptocarpium bussonii gr. (Carnevale, 1908) HW97 Ak–Lwh
Cryptocarpium? cf. *ornatum* (Ehrenberg, 1873) Ho97 HW97 Ho02 Mh–Ab
Lamprocyclas hannai (Campbell & Clark, 1944a) C86 CM99 LH08 LC09 Sc-w–Q
Lamprocyclas margatensis (Campbell & Clark, 1944) Oc97b C86 Po–Tt-k
Lamprocyclas maritalis maritalis Haeckel, 1887 B87 CM99 HN05 LH08 LC09 Po-Pl–Q
Lamprocyclas m. polypora Nigrini, 1967 B87 HN05 LH08 LC09 Q
Lamprocyclas aff. *maritalis* gr. Haeckel, 1887 C86 Tt-k
Lamprocyclas matakohe O'Connor, 1994 Oc00 HW97 Ak–Po
Lamprocyclas particollis O'Connor, 1999 Oc00 HW97 Ak–Lwh
Lamprocyclas sp. indet. Oc97b CM99 Ar–Q
Lamprocyrtis heteroporos (Hays, 1965) C86 CM99 Tt-k–Q
Lamprocyrtis nigriniae (Caulet, 1971) CM99 HN05 LH08 LC09 Q
Lamprocyrtis sp. Oc97b Po
Myllocercion acineton Foreman, 1968 Ho97 Mh–Dt
Myllocercion aff. *echtus* (Empson-Morin, 1981) Ho97 Mh–Dt
Podocyrtis (*Podocyrtis*) *acelles* Sanfilippo & Riedel, 1992 SH95 HD05 Dw–Dm
Podocyrtis (*P.*) *papalis* Ehrenberg, 1847 SH95 Oc99b HD05 Dw–Lwh
Pterocorys clausus (Popofsky, 1913) B87 LH08 LC09 Q
Pterocorys hertwigii (Haeckel, 1887) B87 LH08 Q
Pterocorys macroceras (Popofsky, 1913) LH08 LC09 Q
Pterocorys minythorax (Nigrini, 1968) LH08 LC09 Q
Pterocorys sabae (Ehrenberg, 1872) LH08 Q
Pterocorys sp. sens. Nakaseko & Nishimura1982 LC09 Q
Pterocorys zancleus (Müller, 1858) B87, CM99, HN05 LH08 LC09 Q
Stichopilium bicorne Haeckel, 1887 B87 LH08 Q
Stichopilium variabile var. *davisianoides* Petrushevskaya, 1967 LH08 LC09 Q
Theocapsomma amphora (Campbell & Clark, 1944) Ho97 Mh–Dt
Theocapsomma erdnussa (Empson-Morin, 1981) Ho97 Mh–Dt
Theocorythium trachelium trachelium (Ehrenberg, 1872) B87 CM99 LH08 LC09 Q
Theocorythium t. dianae (Haeckel, 1887) B87 HN05 LH08 LC09 Q
Theocorythium vetulum Nigrini, 1971 CM99 Tt–Q
Theocyrtis aff. *tuberosa* Riedel, 1959 SH95 HW97 Dp–Lwh
Theocyrtis tuberosa Riedel, 1959 Oc99b Lwh
Theocyrtis sp. CM99 Ar–Po
SETHOPHORMIDIDAE
Sethophormis rotula Haeckel, 1887 B87 LH08 LC09 Q
Valkyria pukapuka O'Connor, 1997 Oc00 Lwh–Po
THEOCOTYLIDAE
Lamptonium? aff. *colymbus* Foreman, 1973 HD05 Dw
Lamptonium fabaeforme fabaeforme (Krasheninnikov, 1960) HD05 Dw
Lamptonium pennatum Foreman, 1973 SH95 HD05 Dw
Lamptonium? aff. *pennatum* Foreman, 1973 Oc99a Ar
Thyrsocyrtis (*Thyrsocyrtis?*) *pinguisicoides* O'Connor, 1999 Oc00 HW97 Ar–Lwh
Thyrsocyrtis (*Thyrsocyrtis?*) sp. Oc00 Ar
THEOPERIDAE
Eusyringium fistuligerum (Ehrenberg, 1873) SH95 Oc99b Ab–Lwh
Eusyringium lagena (Ehrenberg, 1873) HW97 SH95

CM99 Ab–Lwh
Eusyringium? woodsidensis Hollis, 1997 Ho97 Mh–Dt
Lychnocanium alma O'Connor, 1999 Oc00 Ar–Lwh
Lychnocanium amphitrite (Foreman, 1973) HW97 Oc99b CM99 Dw–Lwh
Lychnocanium auxilla (Foreman, 1973) HD05 Dw
Lychnocanium bellum (Clark & Campbell, 1942) SH95 HW97 HD05 Dw–Ar
Lychnocanium aff. *carinatum* Ehrenberg, 1875 HD05 Dw
Lychnocanium conicum Clark & Campbell, 1942 Oc97b Po
Lychnocanium cf. *conicum* Clark & Campbell, 1942 HW97 Lwh
Lychnocanium elongatum (Vinassa de Regny, 1900) Oc99b CM99 Lw–Po
Lychnocanium grande Campbell & Clark, 1944a C86 Sc-w–Tt-k
Lychnocanium neptunei O'Connor, 1997 Oc00, HW97 Dm-p–Ld
Lychnocanium satelles (Kozlova & Gorbovetz, 1966) Ho02 HD05 Dt–Dw
Lychnocanium trifolium Riedel & Sanfilippo, 1971 Oc99b Lwh–Lwh
Lychnocanium tripodium Ehrenberg, 1875 HW97 Ak-r
Lychnocanium aff. *tripodium* Ehrenberg, 1875 HD05 Dt–Dw
Lychnocanium waiareka O'Connor, 1999 Oc00 Ab–Lwh
Lychnocanium waitaki O'Connor, 1999 Oc00 Ab–Lwh
Lychnocanium zhamoidai Kozlova, 1999 SH95 HD05 Dt–Dm
Lychnocanium sp. B SH95 HD05 Dt–Dw
Lychnocanoma nipponica nipponica Nakaseko, 1963 CM99 Sl
Lychnocanoma n. sakai Morley & Nigrini, 1995 CM99 Tt–Wo
Lychnocanoma parallelipes Motoyama, 1996 CM99 Tt–Tk
Lychnocanoma sp. CM99 Ar–Wo
Sethochytris babylonis gr. (Clark & Campbell, 1942) Oc99b CM99 Ho02 HD05 Dt–Ld
Sethochytris cavipodis O'Connor, 1999 Oc00 ?Ab–?Lwh

THEOPILIIDAE
Clathrocyclas aff. *aurelia* Clark & Campbell, 1945 Oc97b Po
Clathrocyclas australis Hollis, 1997 Ho97 Ho02 Dt
Clathrocyclas bicornis Hays, 1965 C86 Tt-k–Q
Clathrocyclas cabrilloensis Campbell & Clark, 1944a C86 Sc-w–Tt-k
Clathrocyclas? dolioliformis Petrushevskaya, 1977 Pet77 Mh
Clathrocyclas? elongatus Petrushevskaya, 1977 Pet77 Mh
Clathrocyclas humerus Petrushevskaya, 1975 SH95 HW97 C86 Dw–Tt-k
Clathrocyclas universa Clark & Campbell, 1942 SH95 HW97 Dt–Lwh
Clathrocyclas? sp. A Ho02 Dt
Clathrocyclas spp. Oc99a,b Ar–Lwh
Corocalyptra alcmenae Haeckel, 1887 LC09 Q
Corocalyptra kruegeri Popofsky, 1908 LH08 Q
Corocalyptra cervus (Ehrenberg, 1872) HN05 LH08 LC09 Q
Corocalyptra columba (Haeckel, 1887) B87 LH08 LC09 Q
Cycladophora? auriculaleporis (Clark & Campbell, 1942) HW97 Oc99b Dm–?Ld
Cycladophora bicornis (Popoksky, 1908) B87 CM99 HN05 LH08 LC09 Tt–Q
Cycladophora cf. *bicornis* (Popofsky, 1908) Ho02 Dt
Cycladophora aff. *cosma* Lombari & Lazarus, 1988 Ho02 Dt
Cycladophora davisiana (Ehrenberg, 1861) B87 CM99 HN05 LH08 LC09 Q
Cycladophora pliocenica (= about *bicornis* Hays) Lombari & Lazarus, 1988 CM99 Tk–Q
Cycladophora aff. *pseudoadvena* (Kozlova, 1999) Ho02 Dt
Cycladophora sp. Oc99b, CM99 LH08 Lwh–Q
Eurystomoskevos cauleti O'Connor, 1999 Oc00 Ho97 Ab–Lwh
Eurystomoskevos petrushevskaae Caulet, 1991 Oc99b HW97 Ho02 Dt–Lwh
Eurystomoskevos sp. A Ho02 Dt

TRIOSPYRIDIDAE
Amphispyris reticulata (Ehrenberg, 1872) B87 LH08 Q
Dendrospyris aff. *anthocyrtoides* (Bütschli, 1882) Oc99 Ar–Lwh
Dendrospyris aff. *binapertonis* Goll, 1968 Oc99 Ar–Lw
Dendrospyris bursa Sanfilippo & Riedel, 1973 Oc97b Po
Dendrospyris inferispina Goll, 1968 Oc99 Ar–Lwh
Dendrospyris pododendros (Carnevale, 1908) Oc97b Po
Dendrospyris tumidula (Kozlova & Gorbovetz, 1966) Oc99b Ar–Lwh
Dendrospyris spp. indet. (4) Oc99a Ar–Lwh
Dendrospyris "De1" gr. Goll 1968 Oc99b Lwh–Po
Dorcadospyris alata (Riedei, 1959) CM99 Pl–Sc
Dorcadospyris argisca (Ehrenberg, 1873) Oc99b Ar–Ld
Dorcadospyris ateuchus (Ehrenberg, 1873) Oc99b Lwh–Po
Dorcadospyris circulus (Haeckel, 1887) Oc99b Lwh
Dorcadospyris aff. *confluens* (Ehrenberg, 1873) Ho97 Dt
Dorcadospyris costatescens Goll, 1969 Oc99a Ar
Dorcadospyris mahurangi O'Connor, 1994 CM99 Oc00 Lwh–Po
Dorcadospyris praeforcipata Moore, 1971 Oc97b Po
Dorcadospyris sp. Oc99a Ar
Gorgospyris? spp. Oc97b Po
Hexaspyris sp. Oc99b Lwh
Liriospyris clathrata (Ehrenberg, 1973) Oc99a Ar
Liriospyris? spp. indet. (2) Oc97b Oc99a CM99 Ar–Sc
Lophospyris pentagona pentagona (Ehrenberg, 1872) B87 LH08 LC09 Q
Lophospyris pentagona aff. *pentagona* (Ehrenberg, 1872) Oc97b Po
Lophospyris sp. Oc99b Lwh–Ld
Patagospyris confluens (Ehrenberg, 1873) HD05 Dt–Dw
Petalospyris fovealata Ehrenberg, 1854 HD05 Dt–Dw
Petalospyris senta (Kozlova & Gorbovetz, 1966) HD05 Dt–Dw
Phormospyris stabilis antarctica (Haecker, 1907) C86 B87 HN05 LH08 LC09 Q
Phormospyris s. capoi Goll, 1976 LH08 Q
Phormospyris s. aff. *capoi* Goll, 1976 Oc97b Po
Phormospyris s. scaphipes (Haeckel, 1887) Oc97b HN05 LH08 LC09 Po–Q
Phormospyris s. stabilis (Goll, 1976) Oc97b HN05 LH08 LC09 Po–Q
Phormospyris? spp. indet. (4) Oc99b Lwh–Po
Rhodospyris aff. *anthocyrtis* (Haeckel, 1887) Oc99b Lwh–Ld
Rhodospyris? sp. Oc99b Lwh–Ld
Tholospyris newtoniana (Haeckel, 1887) Oc97b Po
Tholospyris rhombus Goll, 1972 LH08 LC09 Q
Tholospyris sp. gr. sens. Takahashi, 1991 LH08 Q
Tholospyris? sp. Oc99a Ar–Lwh
Triceraspyris cf. antarctica (Haecker, 1907) LH08 LC09 Q
Tristylospyris triceros (Ehrenberg, 1873) HW97, Oc99b, CM99 Ar–Po

ULTRANAPORIDAE
Pterocanium audax (Riedel, 1953) Oc99b CM99 Lwh–Wo
Pterocanium praetextum gr. (Ehrenberg, 1872) B87 CM99 HN05 LH08 LC09 Q
Pterocanium prismatium Riedel, 1957 CM99 Q
Pterocanium trilobum (Haeckel, 1860) B87 CM99 HN05 LH08 LC09 Tk–Q
Pterocanium sp. A LH08 LC09 Q
Pterocanium sp. CM99 Tt–Q
Pterocanium? sp. Oc99b Lwh–Ld

INCERTAE SEDIS
Tepka perforata Sanfilippo & Riedel, 1973 Oc97b Po

Checklist of New Zealand Permian to Early Cretaceous Radiozoa

Radiolaria in the following checklist are only those recorded in published studies from New Zealand's EEZ. To account for the many species recorded in open nomenclature we have: 1) combined informally differentiated species within a single assemblage or associated assemblages (e.g. *Stichomitra* sp. A, B, C are listed as *Stichomitra* spp.); 2) listed separately morphotypes from different assemblages with the same name in open nomenclature (e.g. *Droltus* sp.). Species names are followed by: author, abbreviated reference to New Zealand records, reference content (List, Figure, Descriptive comments), age range (P, Permian; T, Triassic; J, Jurassic; K, Cretaceous; E, M, L = Early, Middle, Late). Reference abbreviations: A88, Adachi 1988; AB99, Aita & Bragin 1999; AG92, Aita & Grant-Mackie 1992; AS92, Aita & Spörli 1992; AS94, Aita & Spörli 1994; BM96, Begg & Mazengarb 1996; BMS87, Blome et al. 1987; C&F88, Caridroit & Ferrière 1988; FH78, Feary & Hill 1978; FKO86, Foley et al. 1986; FP80, Feary & Pessagno 1980; G90, Grapes et al. 1990; G93, George 1993; HAG96, Hori et al. 1996; HCG97, Hori et al. 1997; ITO00, Ito et al. 2000; SAG89, Spörli et al. 1989; TAK98, Takemura et al. 1998; TAK99, Takemura et al. 1999.

PHYLUM RADIOZOA
Class POLYCYSTINEA
Order ALBAILLELLARIA
ALBAILLELLIDAE
Albaillella excelsa Ishiga, Kito & Imoto, 1982 C&F88 LP
Albaillella levis Ishiga, Kito & Imoto, 1982 C&F88 LP
Albaillella triangularis Ishiga, Kito & Imoto, 1982 TAK98 LP
Albaillella sp. A88 P
Neoalbaillella ornithoformis Takemura & Nakaseko, 1981 C&F88 LP
FOLLICUCULLIDAE
Follicucullus aff. *scholasticus* Ormiston & Babcock, 1979 C&F88 MP–LP
Follicucullus bipartitus Caridroit & De Wever, 1984 TAK98 LP
Follicucullus charveti Caridroit & De Wever, 1984 TAK98 LP
Follicucullus dorsoconvexus (Kozur, 1993) TAK99 MP–LP
Follicucullus monacanthus Ishiga & Imoto, 1980 G90 MP
Follicucullus orthogonus Caridroit & De Wever, 1984 C&F88 LP
Follicucullus porrectus Rudenko, 1984 TAK99 MP–LP
Follicucullus scholasticus Ormiston & Babcock, 1979 TAK99 MP–LP
Follicucullus sphaericus Takemura, 1999 TAK99 MP–LP
Follicucullus ventricosus Ormiston & Babcock, 1979 TAK99 MP–LP
cf. *Follicucullus ventricosus* Ormiston & Babcock, 1979 G90 MP–LP
Follicucullus whangaroaensis Takemura, 1999 TAK99 MP–LP
Pseudoalbaillella fusiformis (Holdsworth & Jones, 1980) TAK99 MP
Pseudoalbaillella aff. *longicornis* Ishiga & Imoto, 1980 TAK99 MP
Pseudoalbaillella? spp. indet. (2) TAK98 MP
cf. *Pseudoalbaillella* sp. G90 MP

Order LATENTIFISTULARIA
LATENTIFISTULIDAE
Latentifistula kamigoriensis (DeWever & Caridroit, 1984) C&F88 LP–LP
Latentifistula? sp. G90 LP–LP
Latentifistulidae indet. BM96 LP–LP
ORMISTONELLIDAE
Nazarovella gracilis De Wever & Caridroit, 1984 C&F88 LP
PSEUDOLITHELIIDAE
Hegleria sp. TAK98 MP–LP

Order SPUMELLARIA
ANGULOBRACCHIIDAE
Paronaella spp. indet. (4) BM96 LT–LJ
Paronaella? sp. HAG96 EJ
EMILUVIIDAE
Emiluvia sp. AG92 LJ
Emiluvia? sp. A HAG96 EJ
Plafkerium? sp. AB99 MT
HAGIASTRIDAE
Archaeotriastrum? aff. *hirsutum* De Wever, 1982 HAG96 EJ
Archaeotriastrum? sp. HAG96 EJ
Crucella aff. *prisca* Kozur & Mostler, 1990 HAG96 EJ
Crucella spp. HAG96 EJ
Hagiastrinae gen. et spp. indet. FH78 EJ
Pseudocrucella hettangica (Kozur & Mostler, 1990) HAG96 EJ
Pseudocrucella cf. *sanfilippoae* (Pessagno, 1977) AG92 LJ
Pseudocrucella spp. indet. (2) HAG96 EJ
Sophia? sp. B HAG96 EJ
ORBICULIFORMIDAE
Orbiculiforma? *multifora* De Wever, 1982 HAG96 EJ
Orbiculiforma trispinula Carter, 1988 HAG96 EJ
Orbiculiforma sp. A Carter et. al. 1988 HCG97 EJ
Orbiculiformidae gen. et spp. indet. FH78 EJ
PANTANELLIIDAE
Betraccium deweveri Pessagno & Blome, 1980 BMS87 LT
Betraccium aff. *deweveri* Pessagno & Blome, 1980 AS94 LT
Betraccium cf. *deweveri* Pessagno & Blome, 1980 AS92 LT
Betraccium maclearni Pessagno & Blome, 1980 BMS87 LT
Betraccium smithi Pessagno, 1979 G90 LT
Betraccium spp. indet. (2) BM96 LT
Betraccium? spp. indet. (2) AS94 LT
Cantalum cf. *alium* Blome, 1984 AS92 LT
Cantalum globosum Blome, 1984 AS92 LT
Cantalum aff. *globosum* Blome, 1984 AS92 LT
Cantalum holdsworthi Pessagno, 1979 G90 LT
Cantalum spp. indet. (2) AS92 LT–EJ
Cantalum? sp. BM96 LT
Capnodoce sp. AS92 LT
Capnodoce? sp. G90 LT
Cecrops septemporatus (Parona, 1890) AS92 EK
Gorgansium morganense Pessagno & Blome, 1980 FP80 EJ
Gorgansium sp. AS92 LT
Pantanellium cf. *haidaense* Pessagno & Blome, 1980 SAG89 EJ
Pantanellium riedeli Pessagno, 1977 AG92 LJ–EK
Pantanellium suessi (Dunikowski, 1882) HAG96 EJ
Pantanellium aff. *tanuense* Pessagno & Blome, 1980 SAG89 EJ
Pantanellium sp. HAG96 EJ–LJ
Pantanellium? spp. HAG96 EJ
Zartus jurassicus Pesagno & Blome, 1980 AM95 MJ
PARVIVACCIDAE
Acaeniotyle diaphorogona Foreman, 1973 AG92 LJ–EK
PATULIBRACCHIIDAE
Homeoparonaella aff. *elegans* (De Wever, 1982) HAG96 EJ
Patulibracchiinae gen. et spp. indet. Feary & Hill FH78 EJ
PHASELIFORMIDAE
?*Phaseliforma carinata* Pessagno, 1976 G93 EK
PRAECONOCARYOMMIDAE
Praeconocaryomma aff. *magnimamma* (Rüst, 1898) FP80 EJ
RELINDELLIDAE
cf. *Tetraspongodiscus nazarovi* (Kozur & Mostler, 1979) G90 LT
SPONGODISCIDAE
Spongotrochus sp. A HAG96 EJ
Spongotrochus sp. B HAG96 EJ
VEGHICYCLIIDAE
Veghicyclia sp. G90 LT
XIPHOSTYLIDAE
Tripocyclia sp. AG92 LJ
cf. *Tripocyclia*? *japonica* Nakaseko & Nishimura, 1979 G90 LT

Order ENTACTINARIA
CAPNUCHOSPHAERIDAE
Capnuchosphaera colemani Blome, 1983 G90 LT
Capnuchosphaera cf. *colemani* Blome, 1983 AS94 LT
Capnuchosphaera deweveri Kozur & Mostler, 1979 AS94 LT
Capnuchosphaera cf. *deweveri* Kozur & Mostler, 1979 AS94 LT
Capnuchosphaera cf. *mexicana* Pessagno, 1979 G90 LT
Capnuchosphaera cf. *soldierensis* Blome, 1983 G90 LT
Capnuchosphaera triassica De Wever, 1979 BM96 LT
Capnuchosphaera cf. *triassica* De Wever, 1979 G90 LT
Capnuchosphaera cf. *tricornis* De Wever, 1979 AS94 LT
Sarla cf. *natividadensis* Pessagno, Finch & Abbott, 1979 AS94 LT
Sarla sp. AS92 LT
Sarla? spp. indet. (2) AS92 LT
Weverella sp. G90 LT
CENTROCUBIDAE

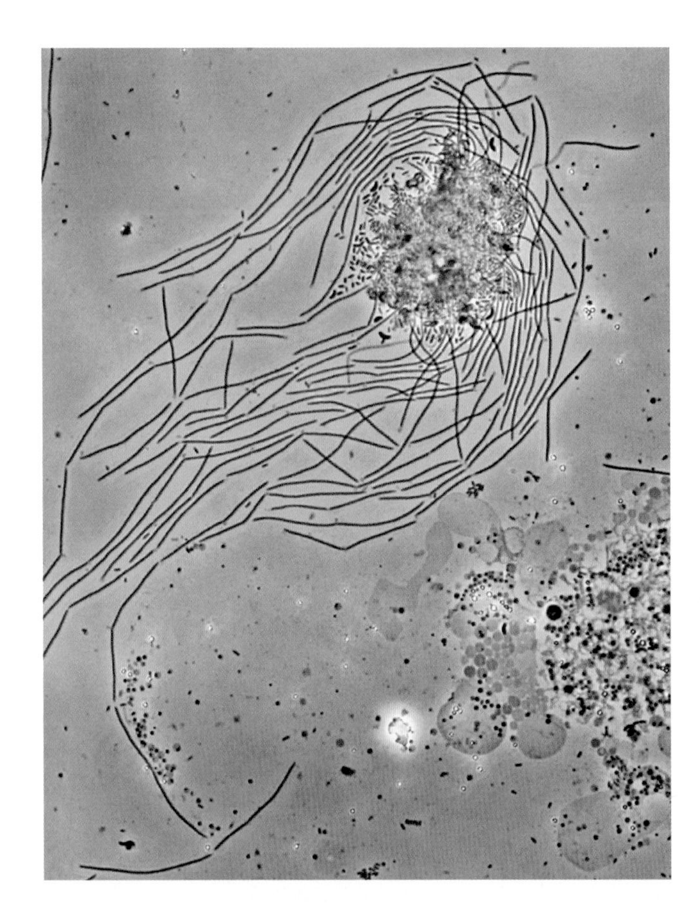

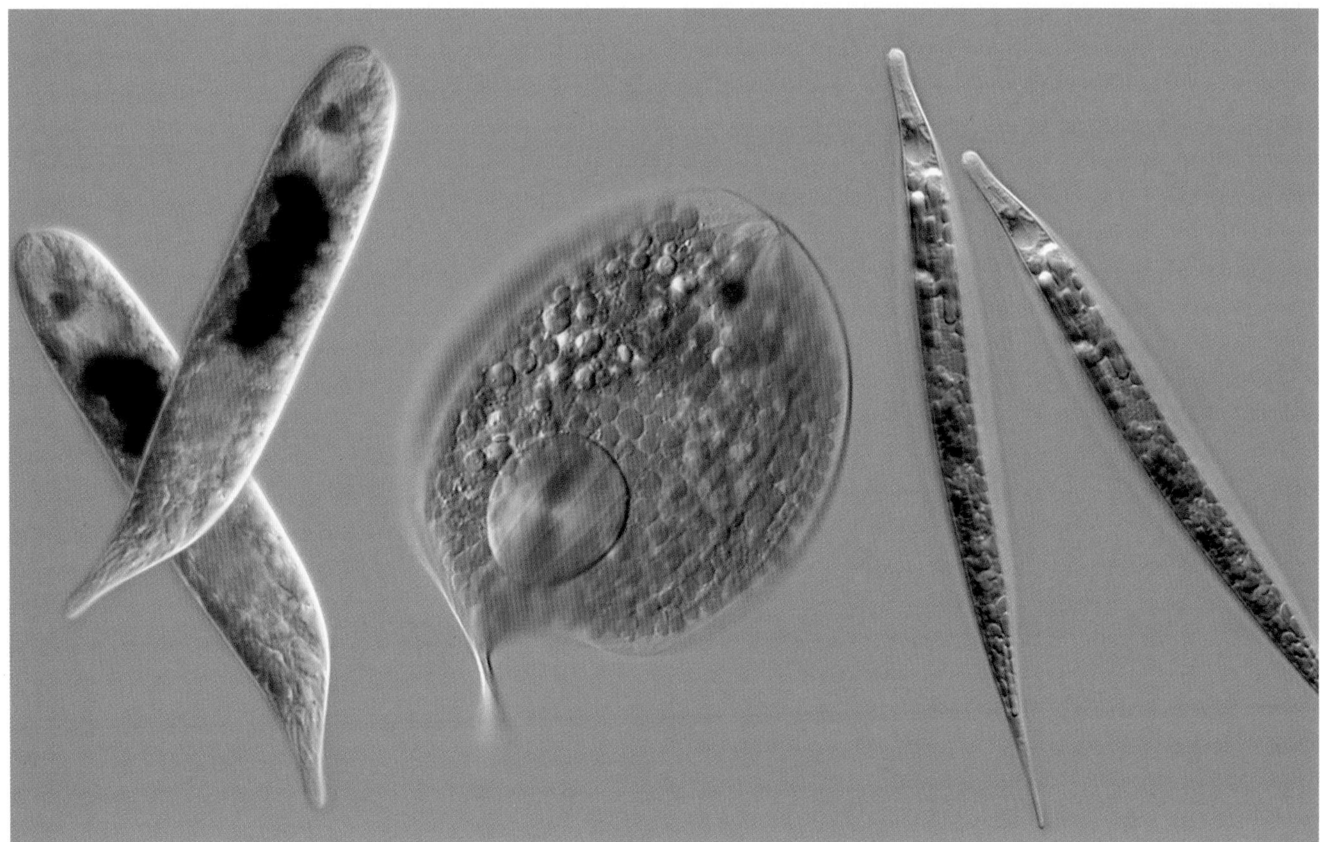

Plate 1a (top left) BACTERIA Sphingobacteria: filaments of *Flectobacillus* sp.; (top right) Cyanobacteria: abundant *Anabaena inaequalis* and an *Oscillatoria* (right of centre). Andrew Dopheide, University of Auckland; Faradina Merican, University of Canterbury

Plate 1b PROTOZOA Euglenozoa: from left, *Euglena* sp., *Phacus gigas*, and *Lepocinclis acus* (all the same scale). Richard E. Triemer, Michigan State University

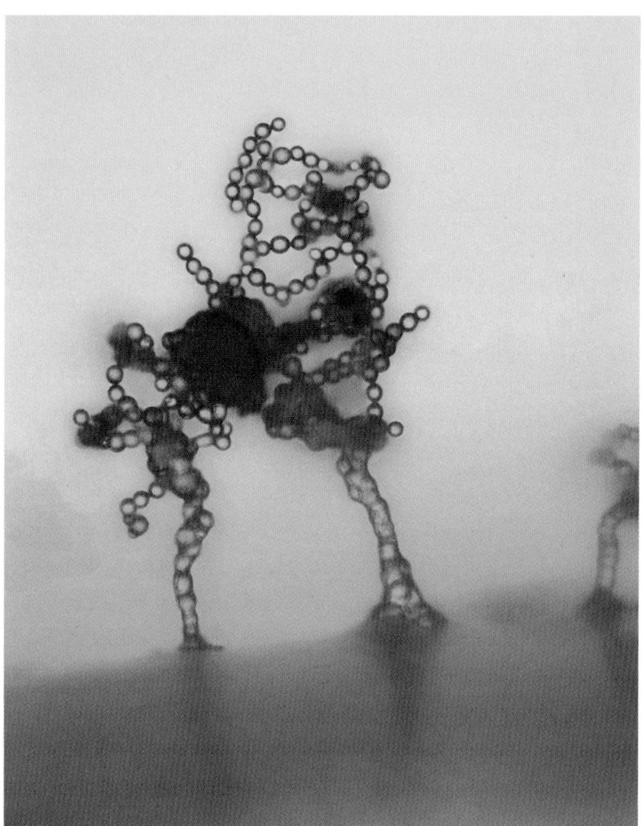

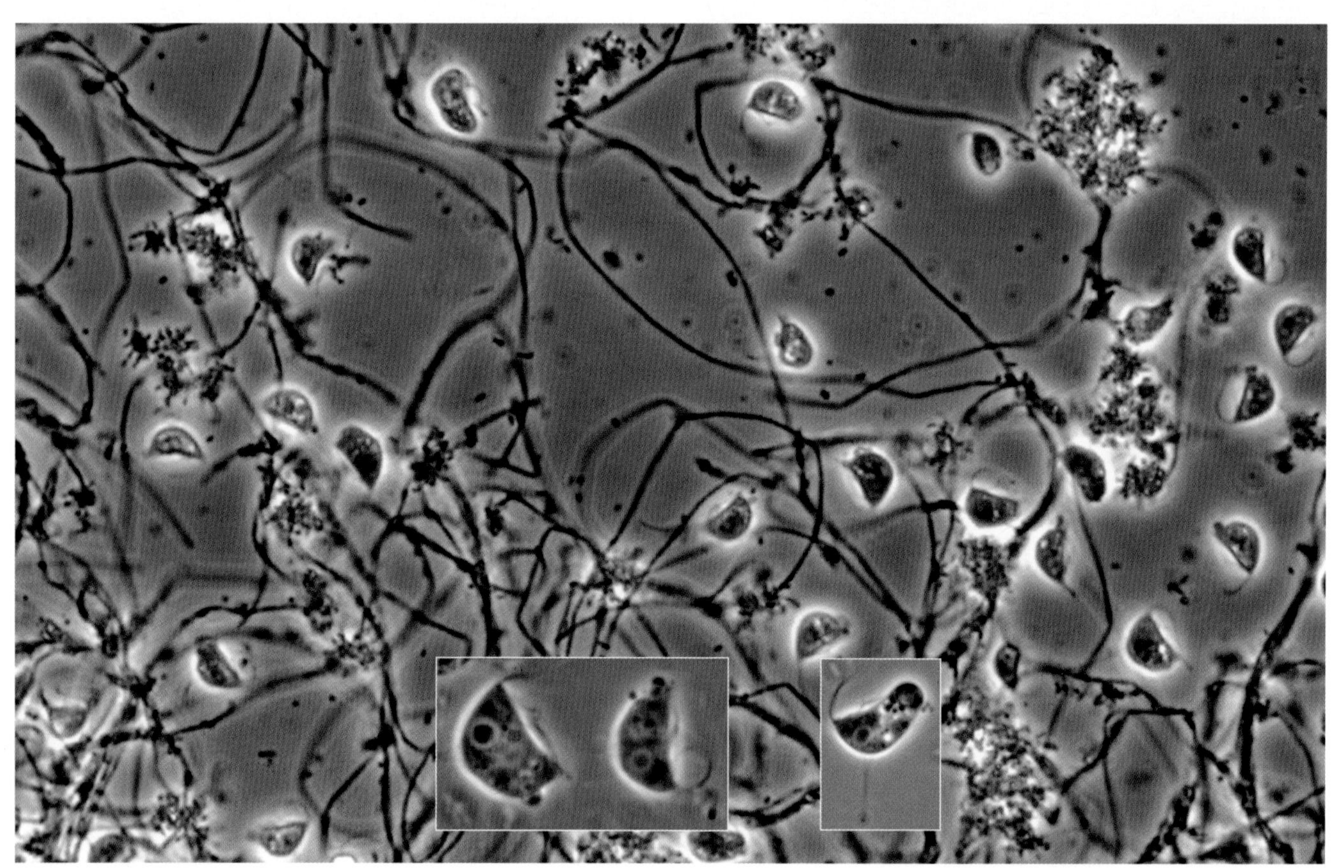

Plate 2a (top) PROTOZOA Percolozoa: colourised SEM of *Naegleria fowleri* amoebae (left) and false-colour image of the acrasid slime mould *Acrasis rosea* (right). Francine Cabral, Virginia Commonwealth University; Matthew Brown, University of Arkansas, Fayetteville

Plate 2b PROTOZOA Loukozoa: jakobid *Reclinomonas americana*. David J. Patterson, per Micro*scope, MBL, Woods Hole

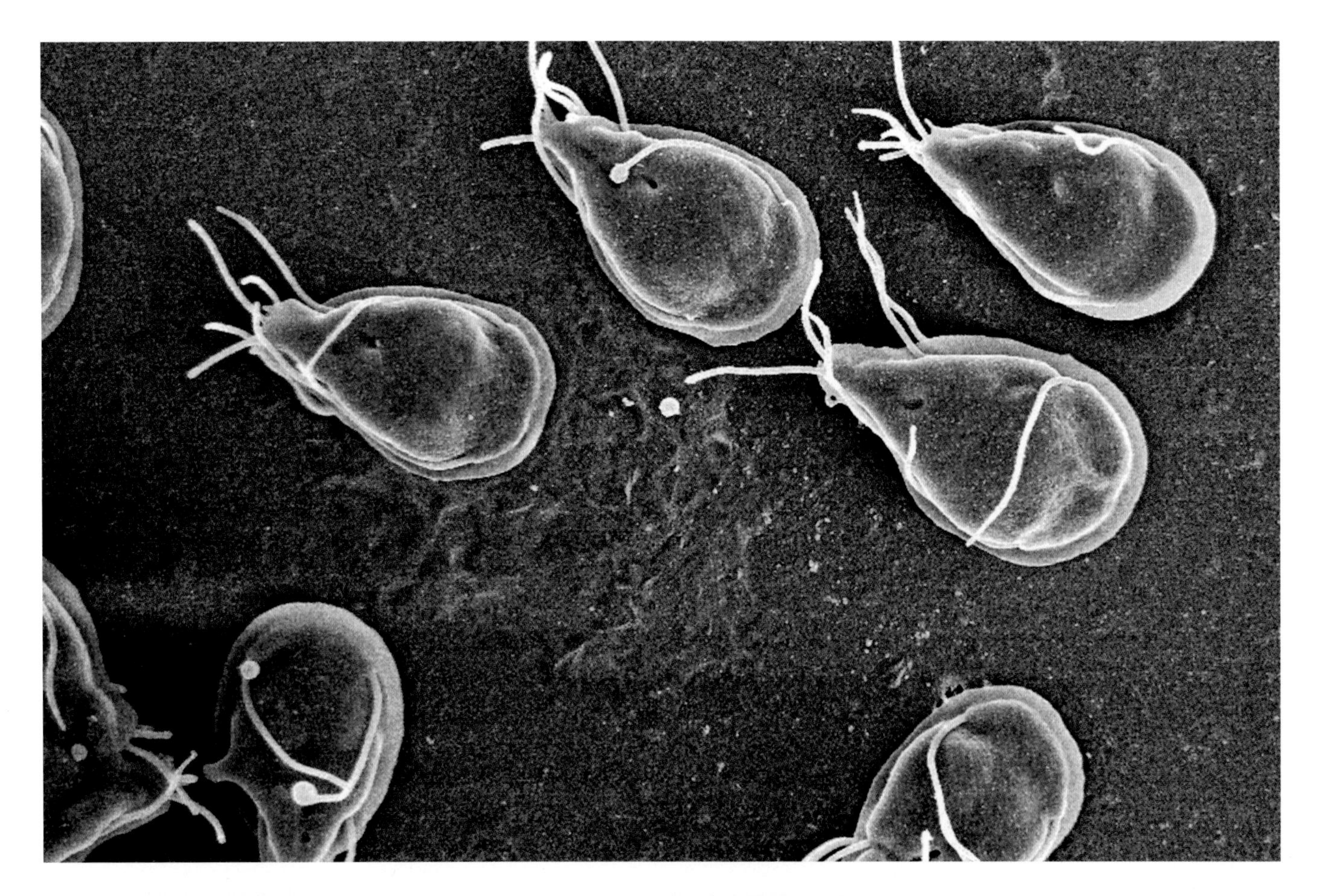

Plate 3a (top) PROTOZOA Metamonada: colourised SEM of *Giardia duodenalis*. W. Zacheus Cande, University of California, Berkeley
Plate 3b PROTOZOA Amoebozoa: developing fruiting bodies of slime mould *Ceratiomyxa fruticulosa*. Clive Shirley

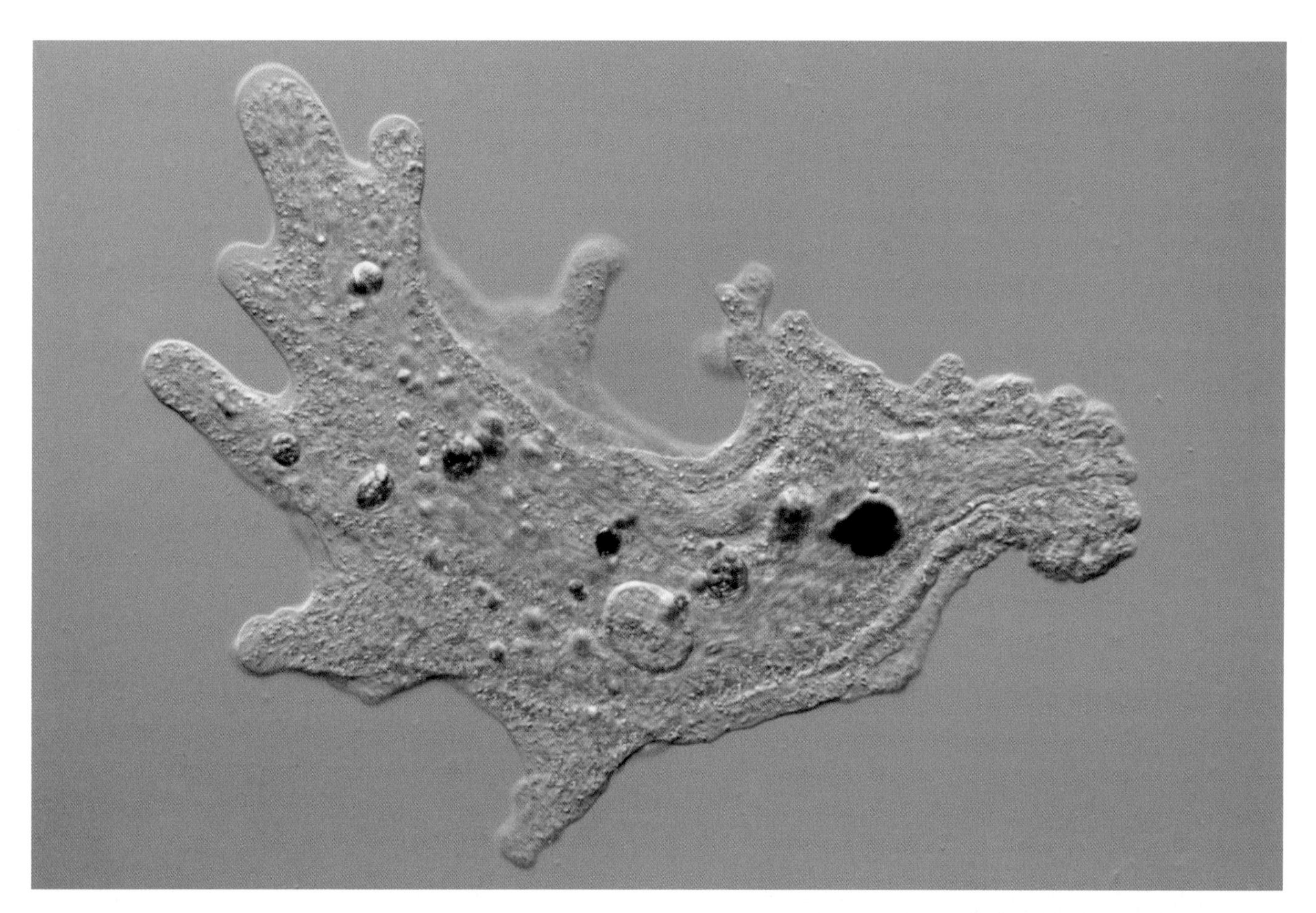

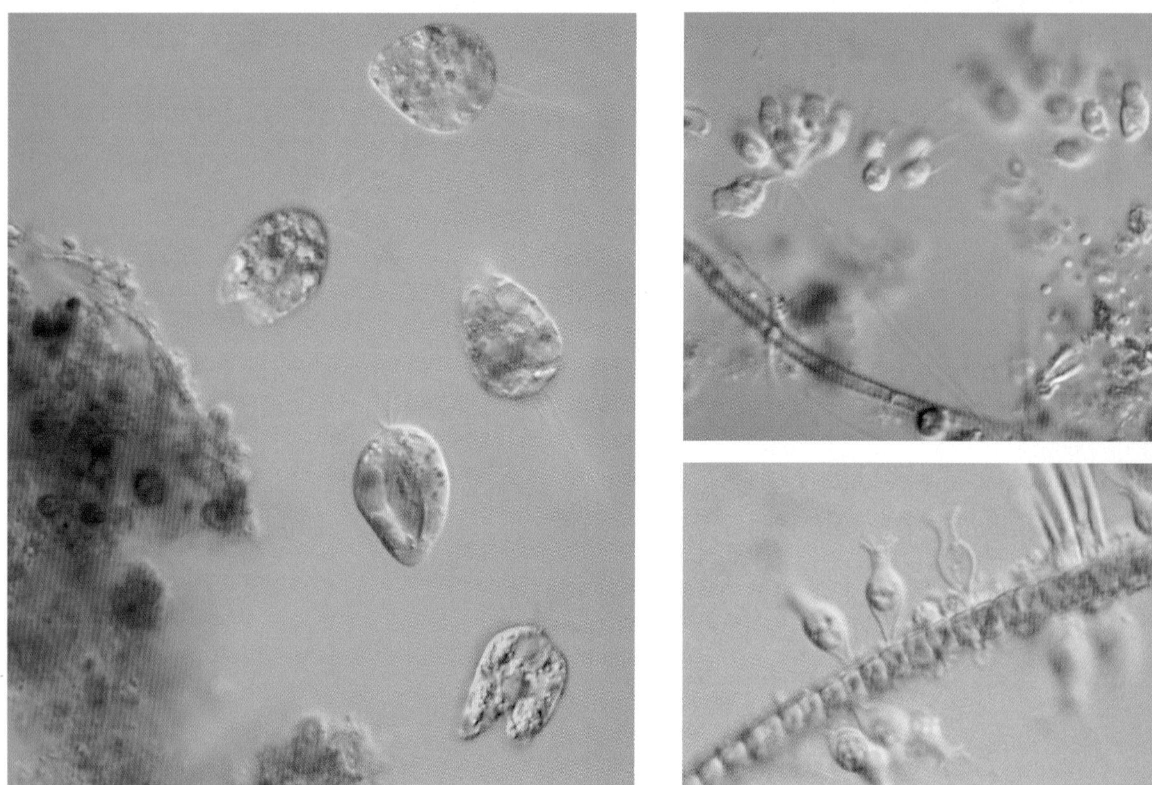

Plate 4a (top) PROTOZOA Amoebozoa: *Amoeba proteus*. Wim van Egmond/Visuals Unlimited Inc.
Plate 4b PROTOZOA (left) Apusozoa: non-New Zealand diphylleid *Collodictyon triciliatum;* (right) Choanozoa: choanoflagellates *Codosiga botrytis* (upper) and *Salpingoeca* sp. (lower). Protist Information Server, Hosei University

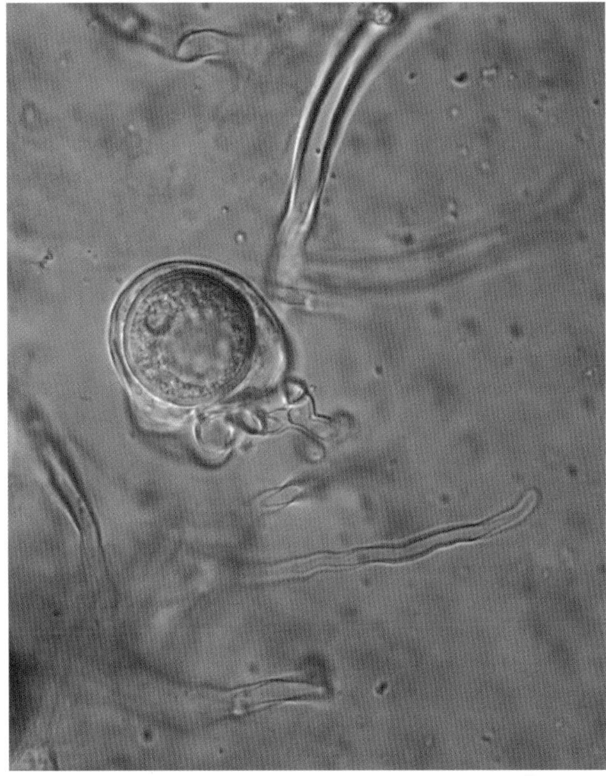

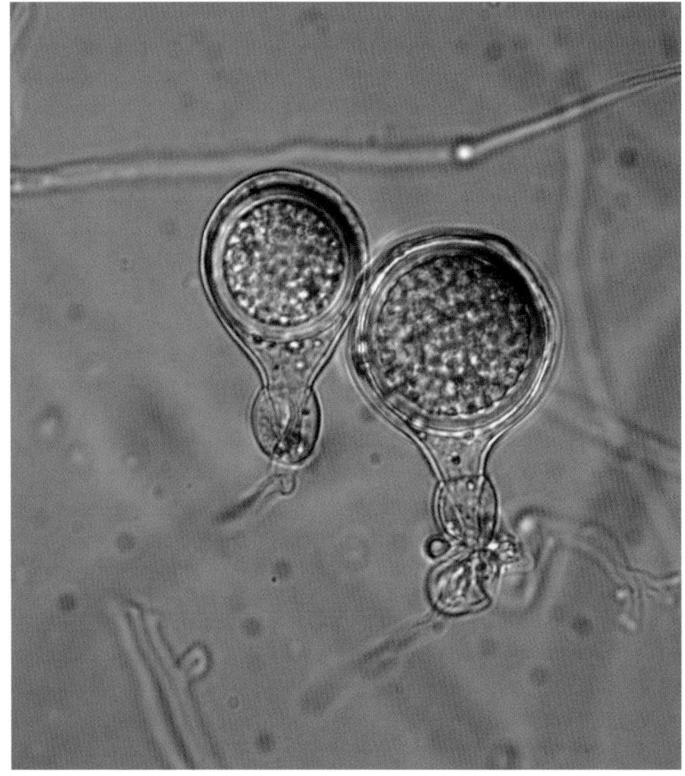

Plate 5a (top) CHROMISTA Ochrophyta: marine diatom *Licmophora flabellata*. Wim van Egmond/Visuals Unlimited Inc.
Plate 5b CHROMISTA Oomycota: oospores of *Phytophthora* taxon 'Agathis' (left) and *P. fallax* (right). Margaret A. Dick, Scion

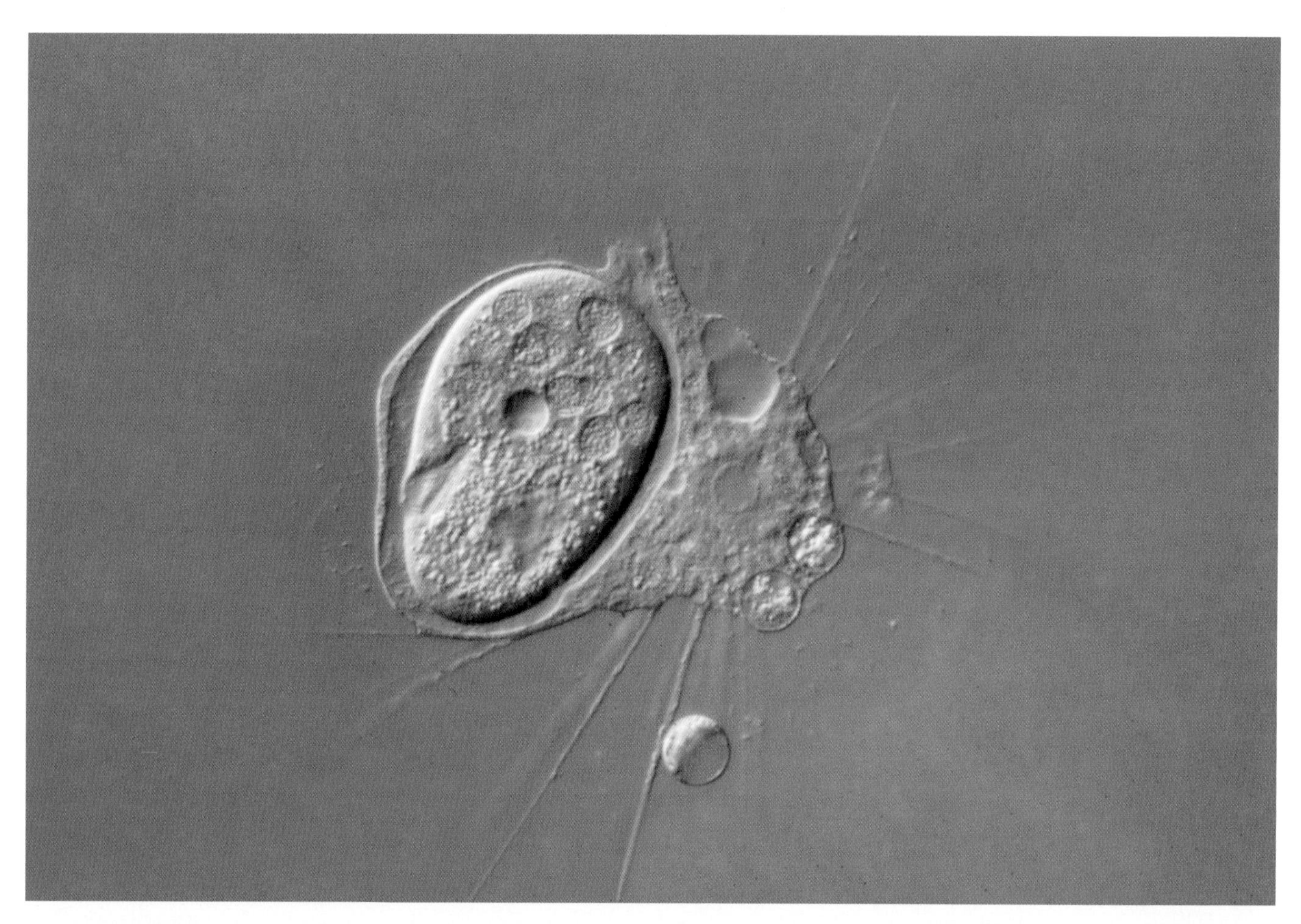

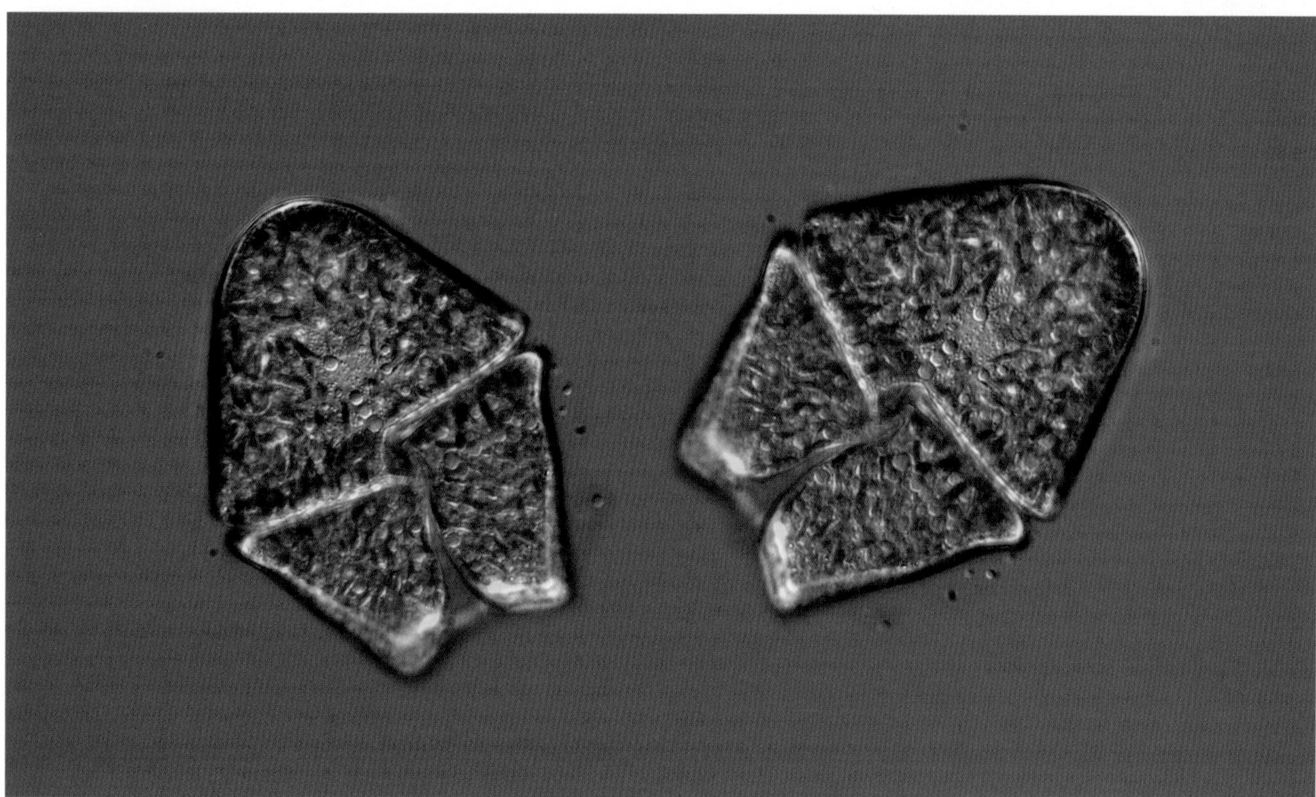

Plate 6a (top) CHROMISTA Bigyra: *Actinophrys sol* with captured ciliate. David J. Patterson, per Microscope, MBL, Woods Hole
Plate 6b CHROMISTA Myzozoa: bloom-forming dinoflagellate, *Akashiwo sanguinea*. Hwan Su Yoon, per Micro*scope, MBL

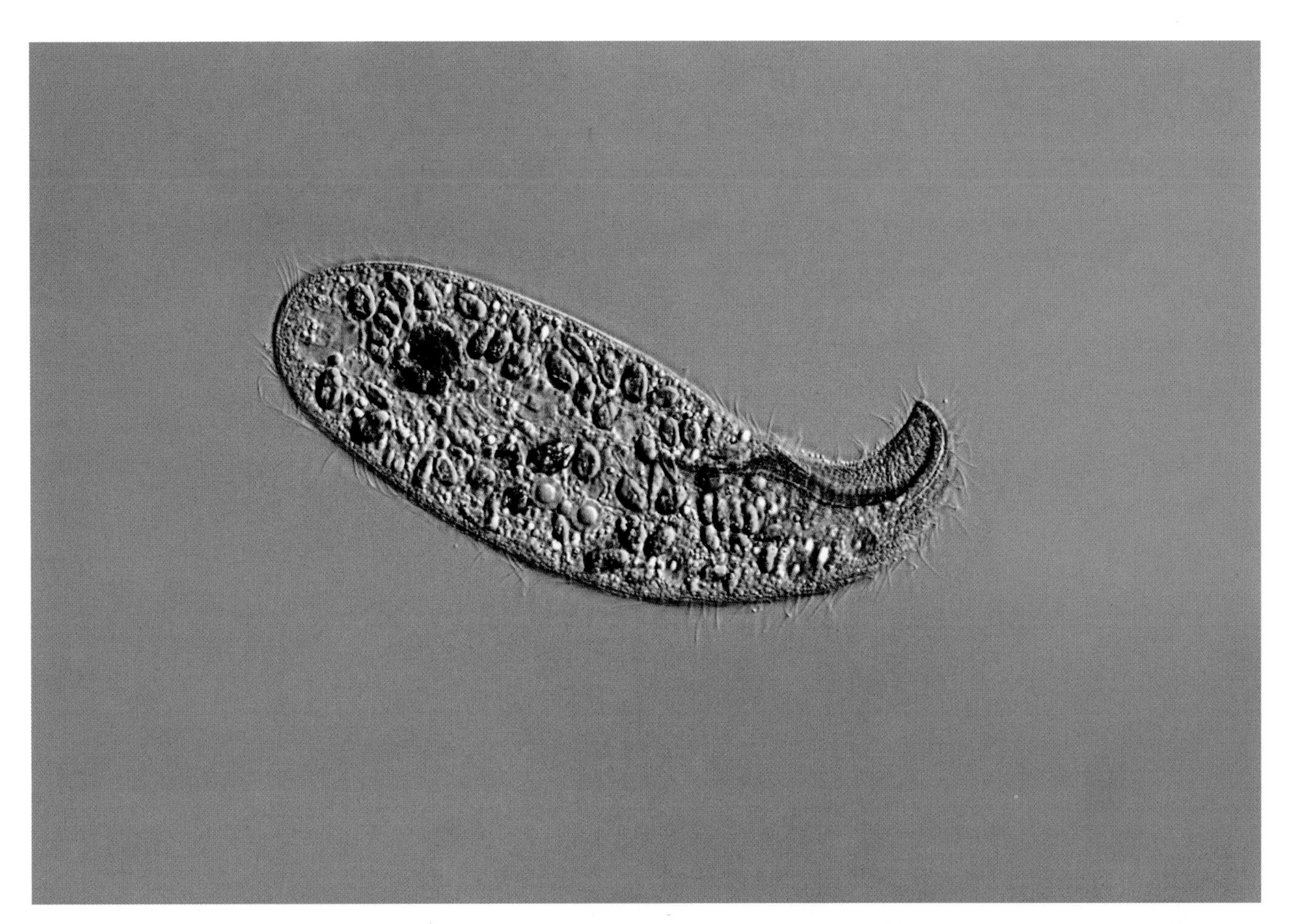

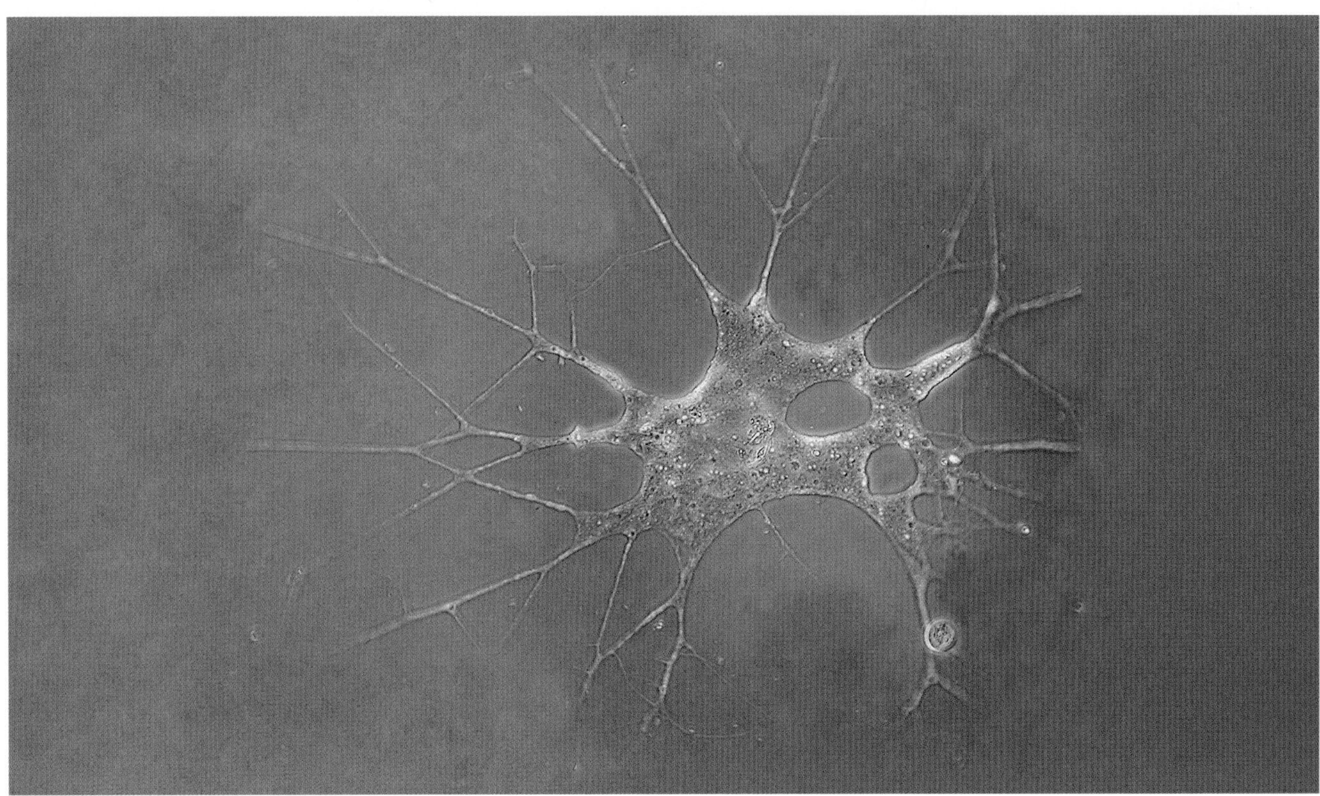

Plate 7a (top) CHROMISTA Ciliophora: *Loxodes rostrum*, ~200 micrometres long, with symbiotic algae. Martin Kreutz, Germany
Plate 7b CHROMISTA Cercozoa: *Biomyxa vagans*. Josef Brief, per Micro*scope, MBL, Woods Hole

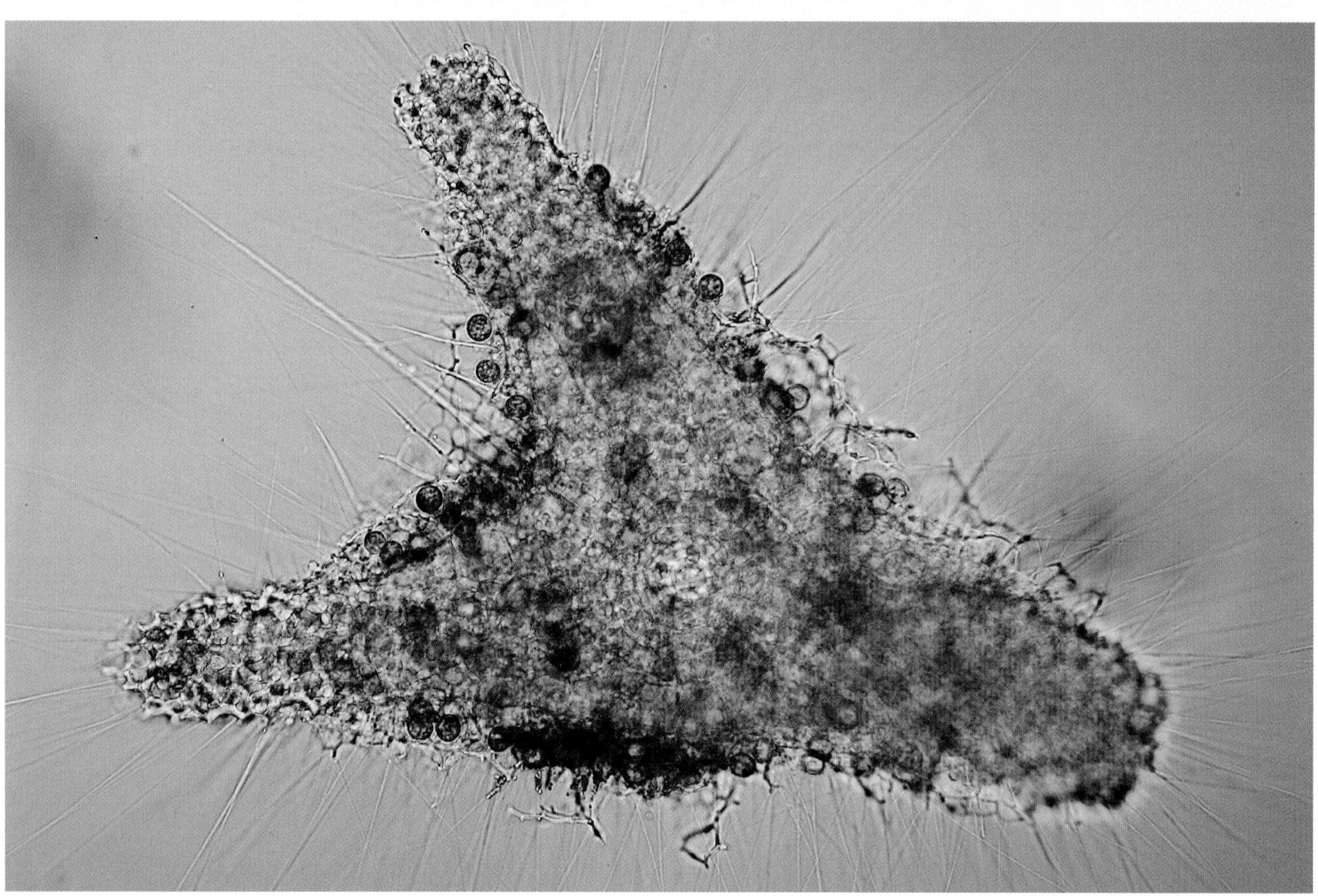

Plate 8a (top) CHROMISTA Foraminifera: a miscellany of New Zealand foraminifera (for identifications, see page 601). John Whalan
Plate 8b CHROMISTA Radiozoa: non-New Zealand *Euchitonia elegans*, with symbiotic algae. Rie S. Hori, Ehime University, and Akihiro Tuji, National Museum of Nature and Science, Tsukuba

Plate 9a (top) CHROMISTA Cryptophyta: *Cryptomonas paramoecium.* David J. Patterson, per Micro*scope, MBL, Woods Hole
Plate 9b CHROMISTA Haptophyta: coccolithophore *Emiliania huxleyi.* F. Hoe Chang, NIWA

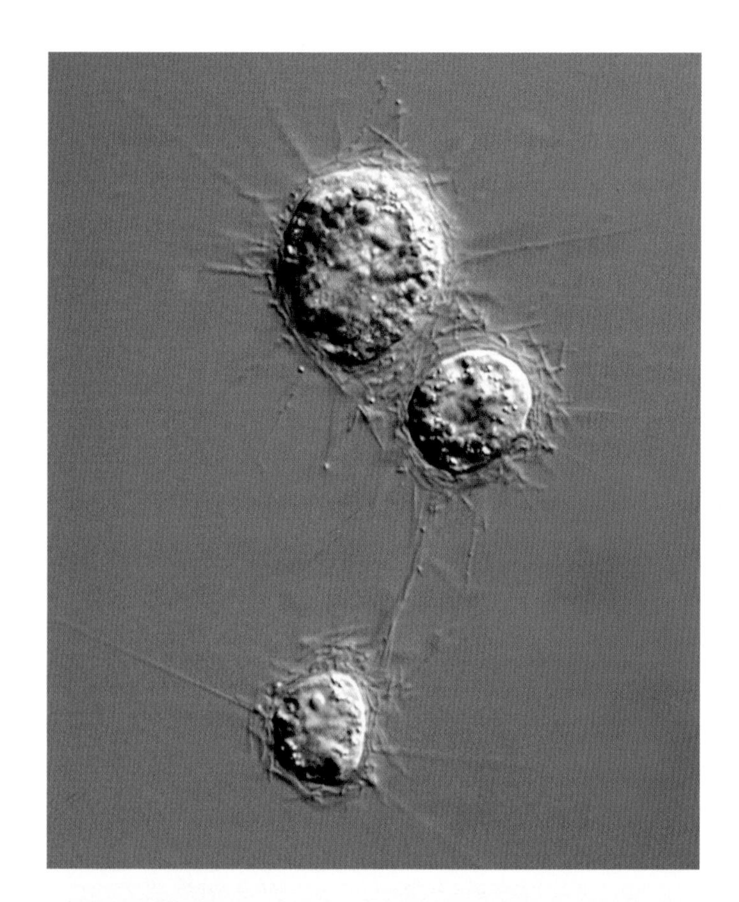

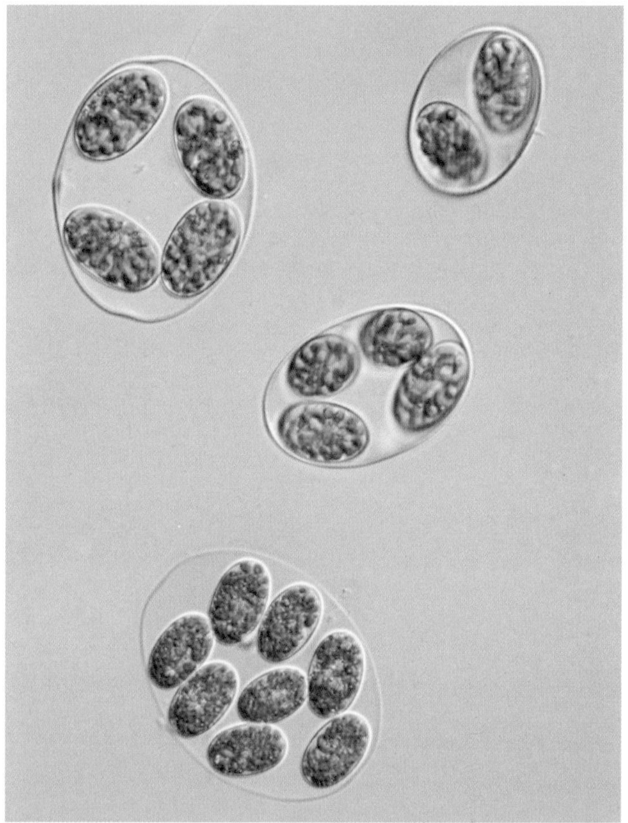

Plate 10a (top) CHROMISTA Heliozoa: sun protist *Raphidiophrys ambigua* (left); PLANTAE Glaucophyta: *Glaucocystis nostochinearum* (right) Martin Kreutz, per Micro*scope (MBL); Protist Information Server, Hosei University

Plate 10b PLANTAE Rhodophyta: red seaweed *Plocamium cirrhosum*. Roger Grace

Plate 11a (top) PLANTAE Chlorophyta: green seaweed *Chaetomorpha coliformis*. Wendy Nelson, NIWA
Plate 11b PLANTAE Charophyta: freshwater desmid alga *Micrasterias rotata*. Wim van Egmond/Visuals Unlimited Inc.

Plate 12a (top) PLANTAE Bryophyta: liverwort *Lepidolaena clavigera* (left) and moss *Cryptopodium bartramioides* (right). Bill Malcolm, Micro-Optics Ltd, Nelson
Plate 12b PLANTAE Tracheophyta: adder's-tongue fern *Ophioglossum petiolatum* (left) and tanekaha (celery pine) *Phyllocladus trichomanoides*.
Jeremy Rolfe; Dennis Gordon

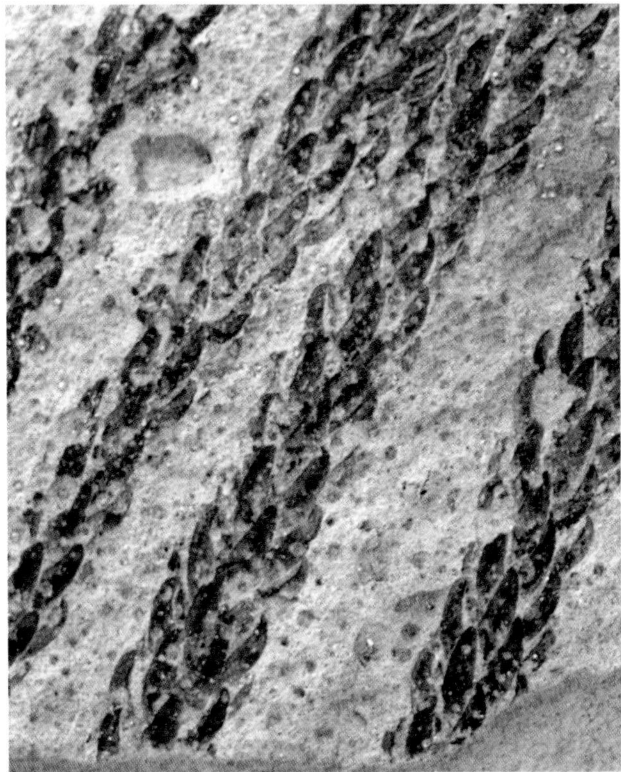

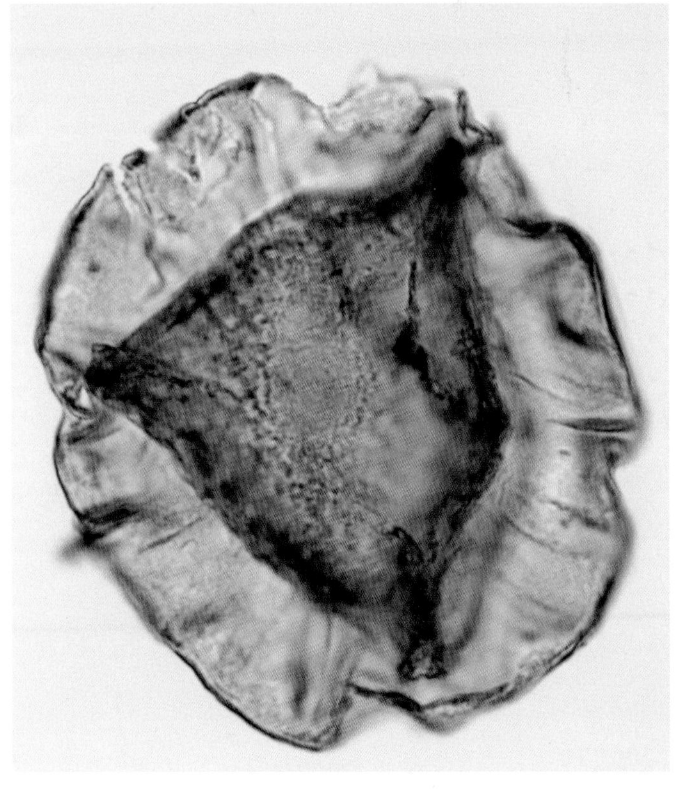

Plate 13a (top) PLANTAE Tracheophyta: New Zealand's 'dinosaur flower' *Hydatella inconspicua*, with female and male inflorescences. Dennis P. Gordon, NIWA
Plate 13b PLANTAE Tracheophyta: (left) early Miocene *Araucaria* sp.; (right) a Jurassic pollen. Michael S. Pole, Brisbane Botanic Gardens; Ian Raine, GNS Science

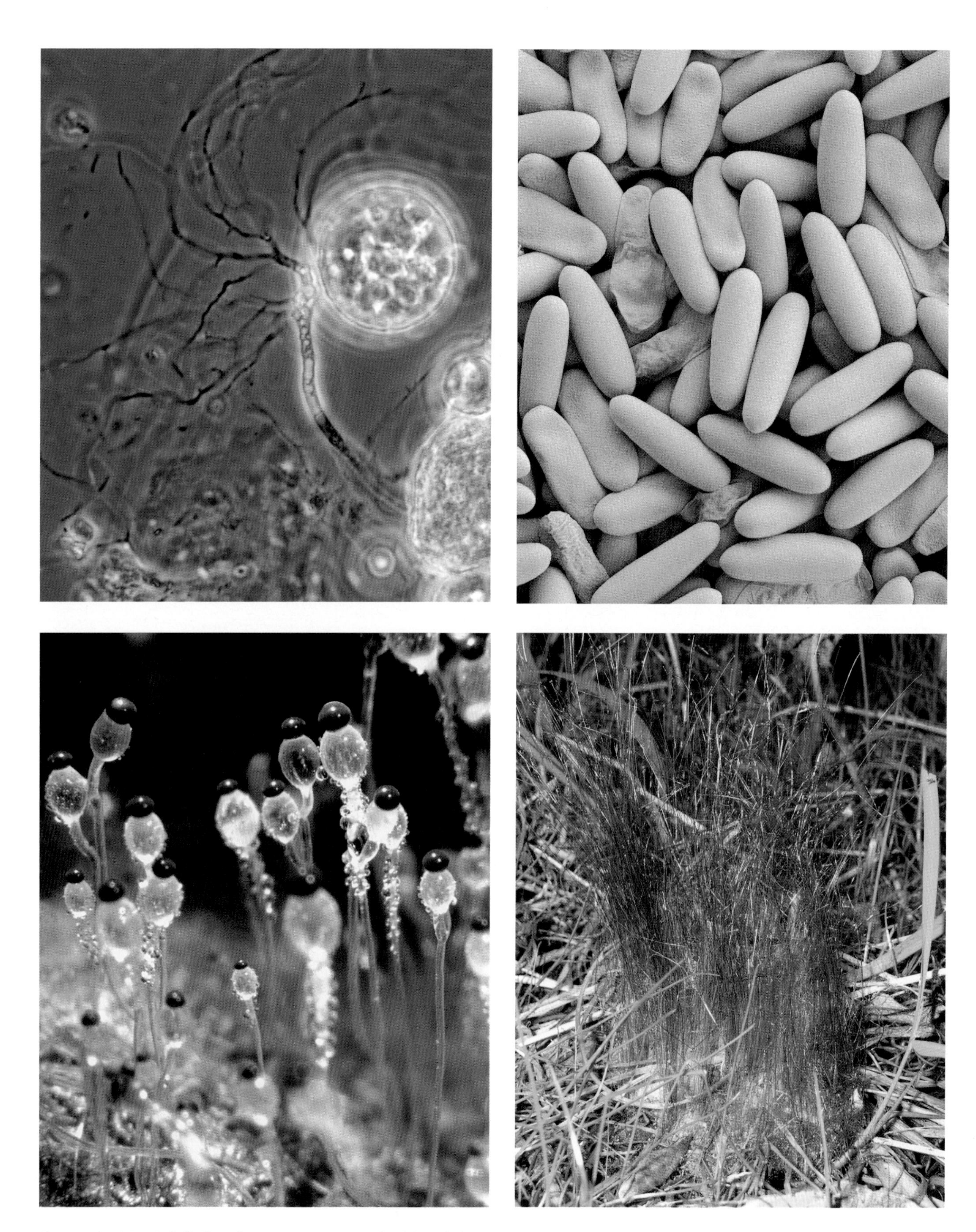

Plate 14a (top) FUNGI (left) Chytridiomycota: non-New Zealand chytrid showing typical morphology of a thallus/sporangium with rhizoids; (right) Microsporidia: colourised field-emission scanning micrograph of an undescribed non-New Zealand microsporidian. D. J. Patterson & Aimee Ladermann, per Micro*scope (MBL); Miroslav Hylis, Charles University, Prague

Plate 14b FUNGI Zygomycota: *Pilobolus kleini* (left) and *Phycomyces nitens*. George L. Barron, University of Guelph and Jerry A. Cooper, Landcare Research, Lincoln

Plate 15a FUNGI Ascomycota: beech strawberry *Cyttaria gunnii*. Peter K. Buchanan, Landcare Research, Auckland
Plate 15b FUNGI Ascomycota: lichen *Pseudocyphellaria colensoi*. Janet M. Ledingham, University of Otago

Plate 16a FUNGI Basidiomycota: *Entoloma hochstetteri*. Peter K. Buchanan, Landcare Research, Auckland
Plate 16b FUNGI Basidiomycota: white rust *Puccina caricina* on tree nettle (left) and the smut *Ustilago agropyri* on a species of the grass genus *Rytidosperma*. Jerry A. Cooper, Landcare Research, Lincoln

Welirella weveri Dumitrica, Kozur & Mostler, 1980 G90 LT
ENTACTINIIDAE
Entactinasphaera aubouini Caridroit & De Wever, 1986 C&F88 LP
Entactinasphaera cimelia Nazarov & Ormiston, 1985 C&F88 MP
EPTINGIIDAE
Cryptostephanidium longispinosum (Sashida, 1991) AB99 ET to MT
Eptingium? aff. *onesimos* Carter, 1993 HCG97 EJ
Eptingium spp. indet. (2) AB99 MT–LT
Eptingium? spp. indet. (2) ITO00 MT–LT
Ferresium aff. *contortum* Blome, 1984 BMS87 LT
Ferresium cf. *contortum* Blome, 1984 AS92 LT
Ferresium loganense Blome, 1984 BMS87 LT
Ferresium aff. *loganense* Blome, 1984 AS92 LT
Ferresium spp. BMS87 LT
Perispyridium sp. AG92 LJ
Perispyridium? sp. HAG96 EJ
Spongostephanidium? sp. AS92 LT
cf. *Xenorum* sp. G90 LT
ETHMOSPHAERIDAE
Ethmosphaeridae? indet. cf. *Cenosphaera* G90 LT
HINDEOSPHAERIDAE
Hindeosphaera sp. AB99 MT
Pseudostylosphaera sp. AB99 MT
Pseudostylosphaera? sp. ITO00 MT
MULTIARCUSELLIDAE
Austrisaturnalidae indet. G90 LT
PALAEOSCENIDIIDAE
Parentactinia inerme Dumitrica, 1978 AB99 MT
Parentactinia pugnax Dumitrica, 1978 AB99 MT
PENTACTINOCARPIIDAE
Pentactinorbis mostleri Dumitrica, 1978 AB99 MT
Pentactinorbis sp. AB99 MT
cf. *Pentactinocarpus* sp. Grapes *et al.* G90 LT
SATURNALIDAE
Acanthocircus cf. *dicranacanthas* (Squinabol, 1914) SAG89 LJ–EK
Acanthocircus rotundus Blome, 1984 AS94 LT
Acanthocircus trizonalis (Rüst, 1898) SAG89 LJ–EK
Acanthocircus sp. AS94 LT
Pseudoheliodiscus sp. A HAG96 EJ
Pseudoheliodiscus spp. indet. (2) AS94 LT–EJ
SPONGOSATURNALOIDIDAE
Parasaturnalis sp. FP80 EJ
INCERTAE SEDIS
Cariver dorsoconvexus Kozur, 1991 TAK98 LP
Dumitricasphaera sp. BM96 LT
Glomeropyle aurora Aita *in* Aita & Bragin, 1999 AB99 ET–MT
Glomeropyle boreale Bragin *in* Aita & Bragin, 1999 AB99 MT
Glomeropyle? *galagala* Aita *in* Aita & Bragin, 1999 AB99 MT
Glomeropyle grantmackiei Aita *in* Aita & Bragin, 1999 AB99 MT
Glomeropyle mahinepuaensis Aita *in* Aita & Bragin, 1999 AB99 MT
Glomeropyle poinui Aita *in* Aita & Bragin, 1999 AB99 MT
Glomeropyle waipapaensis Aita *in* Aita & Bragin, 1999 AB99 MT
cf. *Heliosoma*? *riedeli* Kozur & Mostler, 1981 G90 LT
Kahlerosphaera spp. indet. (2) AB99 MT–LT
Kahlerosphaera? spp. G90 LT
Norispongus? sp. ITO00 MT–MT
Renila? sp. G90 LT
Spongopallium contortum Dumitrica, Kozur & Mostler, 1980 AB99 MT
Spongopallium sp. AB99 MT
Spongopallium? cf. *contortum* Dumitrica, Kozur & Mostler, 1980 ITO00 MT
Spongopallium? sp. ITO00 MT
Spongostaurus aff. *cruciformis* Carter, 1988 HAG96 EJ
Spongostaurus sp. C HAG96 EJ
Spongostaurus? sp. B HCG97 EJ
Spongostylus carnicus Kozur & Mostler, 1979 G90 LT
Spongostylus cf. *carnicus* Kozur & Mostler, 1979 AS94 LT
Stauracontium? *trispinosum* (Kozur & Mostler, 1979) G90 LT
Spumellaria gen. et spp. indet. (5) ITO00 MT–LJ

Order NASSELLARIA
ACROPYRAMIDIDAE
Cornutella? sp. AS94 LT
ARCHAEODICTYOMITRIDAE
Archaeodictyomitra spp. indet. (4) AG92 LJ–EK
Dictyomitra spp. FH78 EJ
Mita? sp. G93 EK
Xipha cf. *striata* Blome, 1984 AS94 LT
Xipha sp. AS94 LT
ARCHAEOSEMANTIDAE
Archaeosemantis sp. AB99 ET–MT
BAGOTIDAE
Bagotum spp. indet. (2) HAG96 EJ–LJ
Bagotum? spp. AS92 LJ
Droltus? *probosus* Pessagno & Whalen, 1982 AS92 MJ
Droltus cf. *probosus* Pessagno & Whalen, 1982 AM95 MJ
Droltus robustospinosus Kozur & Mostler, 1990 HAG96 EJ
Droltus spp. indet. (3) HAG96 EJ–MJ
CANOPTIDAE
Canoptum sp. FP80 LT–EJ
Laxtorum atliense Blome, 1984 BMS87 LT
Laxtorum cf. *atliense* Blome, 1984 BMS87 LT
Laxtorum hindei Blome, 1984 BMS87 LT
Laxtorum spp. indet. (2) SAG89 LT–EJ
Laxtorum? sp. AB99 MT
CARPOCANIIDAE
Diacanthocapsa sp. G93 EK
Tricolocapsa sp. FKO86 LJ
CUNICULIFORMIDAE
Cuniculiformis cf. *aristotelis* De Wever, 1982 SAG89 EJ
Cuniculiformis plinius De Wever, 1982 SAG89 EJ
Cuniculiformis cf. *plinius* De Wever, 1982 AS92 EJ
Goestlingella sp. AB99 MT
EUCYRTIDIELLIDAE
?*Eucyrtidiellum ptyctum* (Riedel & Sanfilippo, 1974) FKO86, AS92 LJ
Eucyrtidiellum cf. *quinatum* Takemura, 1986 AS92, AM95 MJ
EUCYRTIDIIDAE
Eucyrtis? sp. SAG89 EJ
Stichocapsa robusta Matsuoka, 1984 SAG89 MJ–LJ
Stichocapsa spp. indet. (2) AS92 LJ–EK
Stichomitra spp. indet. (3) AG92 LJ
Thetis aff. *undulata* De Wever, 1982 HAG96 EJ
Thetis spp. HAG96 EJ
Thetis? spp. indet. (2) HAG96 EJ
FOREMANELLINIDAE
Foremanellina aranea Dumitrica, 1982 AB99 MT
Foremanellina expansolabrum Dumitrica, 1982 AB99 MT
Foremanellina sp. AB99 MT
Recoaroella sp. AB99 MT
Riedelius sp. HAG96 EJ
HSUIDAE
Hsuum maxwelli Pessagno, 1977 AG92 LJ
Hsuum sp. HCG97 EJ
Parahsuum aff. *nitidum* (Pessagno & Whalen, 1982) AS92, AM95 MJ
Parahsuum cf. *officerense* (Pessagno & Whalen, 1982) AS92, AM95 MJ
Parahsuum spp. AS92, AM95 MJ
Parahsuum ? spp. AM95 MJ
LIVARELLIDAE
Livarella sp. AS92 LT
NEOSCIADIOCAPSIDAE
Squinabolella sp. BM96 LT
PARVICINGULIDAE
Parvicingula aff. *burnsensis* Pessagno & Whalen, 1982 AS92, AM95 MJ
Parvicingula aff. *rothwelli* Pessagno, 1977 AG92 LJ
Parvicingula spp. indet. (4) AG92 MJ–LJ
Ristola altissima (Rüst, 1885) SAG89 LJ–LJ
Ristola procera Pessagno, 1977 SAG89 LJ–LJ
PLAGIACANTHIDAE
Deflandrella manica De Wever & Caridroit, 1984 C&F88 LP
Deflandrella obesum (De Wever & Caridroit, 1984) C&F88 LP
Deflandrella trifustis (De Wever & Caridroit, 1984) C&F88 LP
Tripilidium? sp. FH78 EJ
POULPIDAE
Gigi cf. *fustis* De Wever, 1982 HCG97 EJ
Hozmadia cf. *reticulata* Dumitrica, Kozur & Mostler, 1980 AB99 MT
Hozmadia sp. AB99 ET–MT
cf. *Hozmadia* sp. G90 LT
cf. *Neopylentonema* sp. G90 LT
Poulpus sp. AB99 MT
Poulpus? sp. A HCG97 EJ
Saitoum sp. HAG96 EJ
PSEUDODICTYOMITRIDAE
Corum regium Blome, 1984 AS94 LT
Corum sp. AS94 LT
Latium sp. AS94 LT
cf. *Latium* sp. G90 LT
Pseudodictyomitra depressa Baumgartner, 1984 AS92 EK
Pseudodictyomitra primitiva Matsuoka & Yao, 1985 FKO86, AS92 LJ
Pseudodictyomitra? sp. G93 EK
SETHOCAPSIDAE
Sethocapsa? *subcrassitestata* Aita & Okada, 1986 AS92 EK
SPONGOCAPSULIDAE
Obesacapsula sp. AG92 LJ
Spongocapsula spp. AS92 LJ–EK
SPONGOLOPHOPHENIDAE
Triassospongocyrtis sp. AB99 MT
SYRINGOCAPSIDAE
Canesium cf. *lentum* Blome, 1984 AS94 LT–LT
Podocapsa amphitreptera Foreman, 1973 AS92 LJ–EK
TRIPEDURNULIDAE
Pseudopoulpus sp. AS94 EJ
ULTRANAPORIDAE
Bipedis sp. B HAG96 EJ
Bipedis? sp. A HAG96 EJ
Jacus? sp. HAG96 EJ–MJ
Napora sp. AG92 LJ
Silicarmiger costatus costatus Dumitrica, Kozur & Mostler, 1980 AB99 MT
XITIDAE
Xitus? sp. AG92 LJ
WILLIRIEDELLIDAE
Cryptamphorella spp. indet. (2) AG92 LJ
Zhamoidellum sp. FKO86 LJ
INCERTAE SEDIS
Raoultius calcar De Wever, 1982 HAG96 EJ
Raoultius sp. HAG96 EJ
Raoultius? sp. HAG96 EJ
cf. *Siphocampium* sp. G90 LT
Nassellaria gen. et spp. indet. (5) ITO00 MT–EK

EIGHTEEN

Phylum CRYPTOPHYTA

cryptomonads, katablepharids

F. HOE CHANG, PAUL A. BROADY

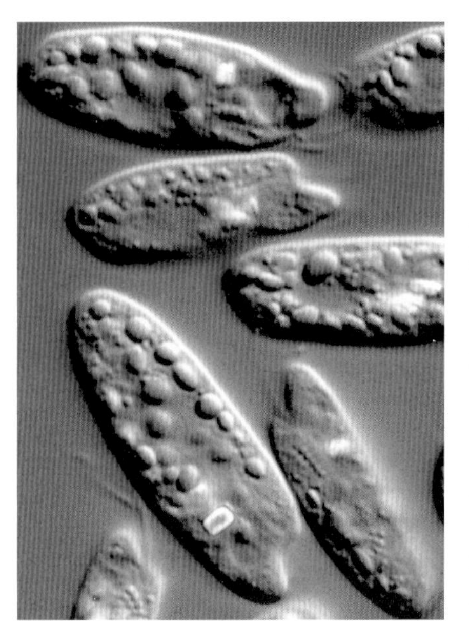

The freshwater cryptomonad *Cryptomonas paramoecium.*
David J. Patterson, per Micro*scope (MBL)

Cryptophytes constitute a highly distinctive group of flagellates that, as the name suggests, can be overlooked and under-appreciated, and this is certainly true of their diversity, which is likely to be significantly more than what is known. They are important in that they are generally regarded as high-quality food for aquatic herbivores (Dunstan et al. 2005). As photosynthetic organisms they contribute to aquatic primary productivity and as non-toxic planktonic flagellates they constitute important prey in the food chain. Although often tiny (many species are less than 20 micrometres in size), cryptophytes are extraordinarily complex.

At a general level, the cryptophyte cell is ovoid and asymmetrical but flattened on one side (ventral) where there is an apical or subapical furrow or gullet, or a combination of both, with two unequal cilia emerging from a depression at its apical end. Both cilia are hairy, having bristle-like mastigonemes 1–2 micrometres long (Jeffrey & Vesk 1990), the longer cilium having two rows and the shorter one row, and they may also bear extremely small organic scales (Kugrens et al. 2000). A feature that sets apart the photosynthetic cryptophytes from other microalgae is their pigments. They are the only group apart from blue-green algae (Cyanobacteria) and the red algae (Rhodophyta) and glaucophytes (Glaucophyta) to have phycobilins (either reddish phycoerythrin or bluish phycocyanin) as accessory photosynthetic pigments, located in the spaces enclosed by the thylakoid-sac membranes within the chloroplast (Gantt et al. 1971). The chloroplast of cryptophytes (generally one, sometimes two per cell) was ancestrally derived from the symbiotic uptake of a red-algal cell. Evidence for this capture is based not only on the unusual accessory pigments but the presence of a nucleus-like structure – the nucleomorph – located between the membranes that surround the chloroplast. The nucleomorph is surrounded by a double membrane and contains its own DNA and a nucleolus-like structure. There is in addition, in photosynthetic cryptophytes, chloroplast DNA, found in numerous small bodies (nucleoids) scattered throughout the chloroplast, mitochondrial DNA, and the host cell's own nuclear DNA, making the typical cryptophyte cell four-genomic (Cavalier-Smith 2004).

Most cryptophytes are photosynthetic, but some are colourless heterotrophs like the cryptomonads *Cyathomonas* and some *Cryptomonas*, and also the goniomonads and katablepharids. The non-photosynthetic genus *Chilomonas* has been shown to comprise species of *Cryptomonas* that lack chloroplasts (Hoef-Emden 2005), *Cyathomonas* may be a modern analogue of the ancestral cryptophyte. Phagotrophy supplements nutrition in some photosynthetic species and these can therefore be termed mixotrophs. Because of the mixture of photosynthetic and heterotrophic taxa, cryptophytes are also referred to as cryptomonads.

It is the plastids and other internal structures that make the cryptophyte cell (like that of some other chromists) marvellously more complex than cells in animals and plants. For example, cryptophytes have intricate structures called ejectosomes lining the gullet. These are constructed differently from the otherwise similar trichocysts of dinoflagellates and raphidophytes, having a tightly coiled cylinder of material contained inside. When triggered, ejectosomes explosively discharge their scrolls, which instantaneously unroll, causing the cryptophyte cell to lurch suddenly backwards. This movement may be an avoidance reaction.

Among other characters, the cell in most species is enclosed by a stiff proteinaceous covering (periplast) made up of a ridged layer superimposed on an inner periplast layer of thin proteinaceous plates, with the plasma membrane sandwiched in between (Wetherbee et al. 1987; Brett et al. 1994). The periplast maintains the shape of the cell in the absence of a microtubular cytoskeleton. In addition to these, or instead of them, scales and fibrillar material may coat the cell surface. The main photosynthetic pigments are chlorophylls *a* and c_2 (chlorophyll *b* is never present), phycobiliproteins, and carotenoids; the presence of a specific cryptomonad carotenoid, alloxanthin (Gieskes & Kraay 1983), makes it possible to detect these cells by use of HPLC (high-performance liquid chromatography) pigment analysis. An eyespot may be present in the chloroplast or lacking and the accompanying pyrenoid may protrude from the chloroplast. There is a contractile vacuole, the ciliary rootlet apparatus is unique, and, uniquely, starch reserves are accumulated neither in the chloroplast, nor in the cytoplasm of the cell, but in the compartment between chloroplast endoplasmic reticulum and the chloroplast envelope (Graham et al. 2009). Near the gullet, are two structures called Maupas bodies, which may be involved in the removal and digestion of superfluous membrane.

Cavalier-Smith (2004) divided the phylum into two subphyla – the Cryptomonada with classes Cryptophyceae and Goniomonadea, and the Leucocrypta with classes Leucocryptea and Telonemea (Cavalier-Smith 2007). Only the latter is not yet known in New Zealand. Subsequently, Cavalier-Smith (2010, supplementary material) dropped the subphylum names and substituted Katablepharidea [Katablepharidophyceae herein] for Leucocryptea.

This latter group, widely referred to as katablepharids, is characterised by genera such as *Katablepharis* and *Leucocryptos*. They lack the proteinaceous periplast, having instead an extracellular sheath that also lines the cilia. As a consequence, these appear thick and lack mastigonemes. The ejectosomes are also structured somewhat differently internally. Katablepharids had long been classified as a subgroup of cryptophytes based on similarities observed by light microscopy then later reclassified as incertae sedis based on ultrastructural studies. Recent molecular-phylogenetic analyses (Okamoto & Inouye 2005; Okamoto et al. 2009; Reeb et al. 2009) support a sister-group relationship with the balance of cryptophytes.

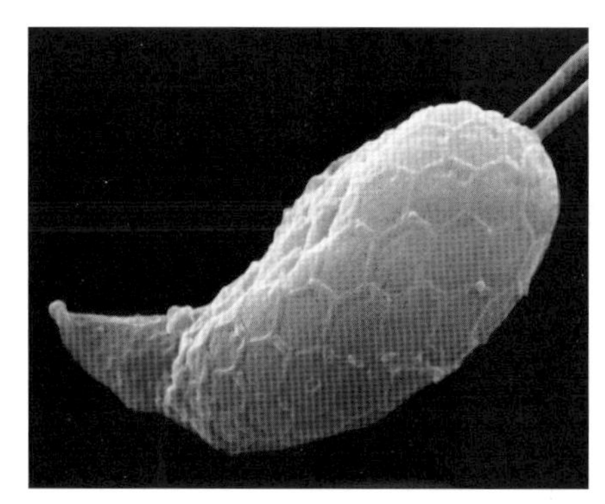

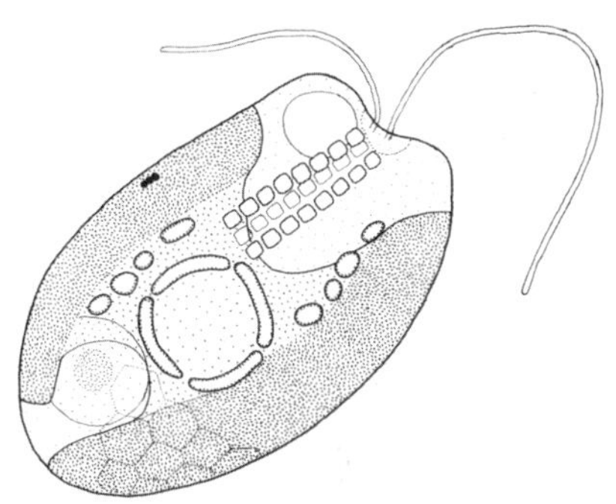

Marine *Plagioselmis prolonga* (upper) and freshwater *Chroomonas nordstedtii* (Cryptomonadales) showing hexagonal periplast plates.

Reconstructed SEM: F. Hoe Chang, NIWA
Drawing after Kugrens & Clay 2003

Summary of New Zealand cryptophyte diversity

Taxon	Described species	Known undetermined/ undescribed species	Estimated unknown species	Adventive species	Endemic species	Endemic genera
Cryptophyceae	14	0	20	0	1	0
Goniomonadophyceae	1	0	1	0	0	0
Katablepharidophyceae	2	0	5	0	0	0
Telonemea	0	0	4	0	0	0
Totals	**17**	**0**	**30**	**0**	**1**	**0**

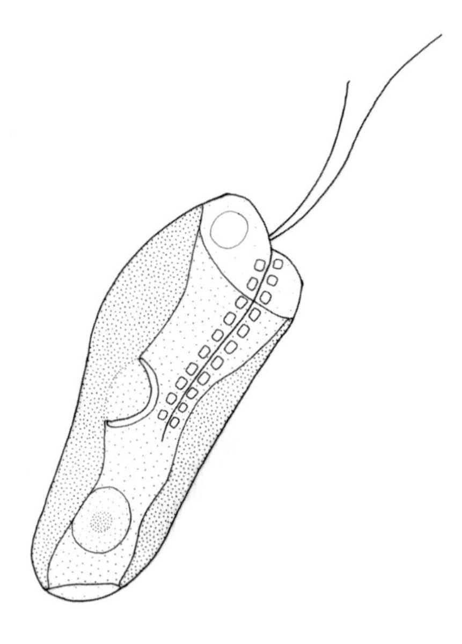

The freshwater cryptomonad *Cryptochrysis commutata*.
Adapted from various sources

There has been an issue about the spelling of the genus name *Katablepharis*. In the first description, Skuja (1939) used *Kathablepharis*. According to the International Code of Botanical Nomenclature (ICBN), however, this must be corrected to *Katablepharis* because of the name's etymology (Greek *kata*, downwards, *blepharis*, eyelash, alluding to the trailing cilium). On the other hand, by the rules of the equivalent Zoological Code (ICZN), the original spelling must be retained regardless of etymology. Okamoto et al. (2009) informally proposed acceptance of the botanical spelling under the ICZN and *Katablepharis* seems to have acceptance in current usage. Cavalier-Smith (2010, supplementary material) insisted that the four cryptophyte classes should be treated under the ICBN.

Global estimates of the number of cryptophyte genera range from 12–23, with about 200 species divided equally between marine and freshwater habitats. A molecular study of cryptic diversity of protists in two freshwater ponds, one oxic, the other suboxic, failed to detect cryptophytes in the latter, suggesting that they may be sensitive to oxygen-depleted environments (Šlapeta et al. 2005). A single species has been recorded from soil (Ettl & Gärtner 1995). Most cryptophytes are unicells, but some can form sessile encapsulated (palmelloid or tetrasporal) stages. One genus has a simple filamentous thallus. Globular resting cysts with thick cell walls formed under unfavourable conditions have been suggested to facilitate survival (Lichtlé 1979).

Seventeen cryptophyte species have been recorded in New Zealand, ten of them marine (Norris 1961, 1964; Taylor 1978; Chang 1983, 1988; Bradford et al. 1987) and five freshwater (Maskell 1887, 1888; Schewiakoff 1893; Chapman et al. 1957; Etheridge 1983; Flint 1966; Green 1968, 1974; Haughey 1968; Johnstone 1972; Forsyth & McColl 1975; Sarma & Chapman 1975; Lam 1977; Cassie 1974, 1975, 1978; Hutchinson & Oliver 1978; Waikato Valley Authority 1979; Hughes & McColl 1980). Cassie (1984) listed all known New Zealand freshwater species and summarised all occurrences. Dawson (1992) listed all known marine species in an annotated bibliography. Unidentified taxa have also been reported in blooms in the Marlborough Sounds and Tasman Bay (MacKenzie et al. 1986a), at a salmon farm in Big Glory Bay, Stewart Island (Chang et al. 1990), in the Subtropical Convergence region near New Zealand (Chang & Gall 1998), and in Hauraki Gulf on the northeast coast (Chang et al. 2003, 2008).

Freshwater forms are found in lakes, often in poorly illuminated deep water, and nutrient-rich temporary pools. Marine species are found in tidal pools, puddles of brackish water, the interstitial water of sandy beaches, and in the open ocean where they can dominate spring plankton blooms. In June 1979, a research voyage of the New Zealand Oceanographic Institute (now part of NIWA) off Westland recorded *Hillea marina* at densities of 1.8 million cells per litre of seawater (Chang 1983). At one deepwater location, *H. marina* cells were twice as abundant as all other phytoplankton combined (Batson 2003). In general, cryptophytes are by far the most widely distributed flagellates recorded in New Zealand waters and can contribute to more than 50% of the total cell carbon biomass in coastal waters, in particular in summer (Chang 1983, 1988; Chang et al. 2003). Unlike toxin-producing cyanobacteria, dinoflagellates, raphidophytes,

Summary of New Zealand cryptophyte diversity by environment

Taxon	Marine	Terrestrial	Freshwater
Cryptophyceae	9	0	5
Goniomonadophyceae	0	0	1
Katablepharidophyceae	1	0	1
Telonemea	0	0	0
Totals	**10**	**0**	**7**

and dictyochophytes, cryptophyte blooms are not known to be harmful to freshwater or marine life.

The katablepharids are cosmopolitan phagotrophs, feeding on both bacteria and microalgae, and play an important role in the aquatic microbial ecology of both marine and freshwater environments. There are two species in New Zealand, one of them originally described as *Goniomonas elongata* (Maskell 1888), but Cavalier-Smith (pers. comm.) advises that *G. elongata* is 'the wrong shape for a *Goniomonas*; it is likely to be a *Katablepharis*' (see end-chapter checklist). The sole known New Zealand marine katablepharid, *Leucocryptos marina*, is widespread in the world's oceans.

The katablepharid *Leucocryptos marina* (reconstructed SEM image).
Hoe Chang

Cryptophytes are important mediators of blooms of other plankton organisms. It has been known since 1969, for example, that the bloom-forming ciliate protozoan *Myrionecta rubra* (often referred to as *Mesodinium rubrum*) contains cryptophyte chloroplasts. These had been interpreted to represent a highly reduced form of cryptophyte living in a symbiotic relationship with its ciliate host (Hoek et al. 1995) but it is now known that *M. rubra* ingests cryptophytes and 'steals' their chloroplasts (Gustafson et al. 2000). Experiments have shown that an individual ciliate can obtain 20 new chloroplasts in about 30 minutes. Interestingly, cryptophytes are not necessarily the only or major source of prey in feeding experiments in terms of the numbers of prey cells ingested. On the other hand, photosynthesis appears to be the main source of nutrition for *M. rubra*, hence a regular supply of cryptophyte chloroplasts is necessary. Although newly captured chloroplasts are retained in an active state, every time a *M. rubra* cell divides, it loses half of its complement of chloroplasts, hence uptake of new ones is always necessary (Yih et al. 2004). Recent work has shown that *M. rubra* also maintains active cryptophyte nuclei for up to 30 days (Johnson et al. 2007). Chloroplast function and replication depend on gene expression by these stolen nuclei. As many as eight chloroplasts, originating from several cryptophyte cells, can be serviced by a single cryptophyte nucleus.

Myrionecta rubra blooms are common around the world, including in New Zealand waters (MacKenzie & Gillespie 1986b; Chang 1994, 1999). Although appearing as a 'red tide', the blooms are not toxic and the redness results from the captured chloroplasts. Cryptophytes may also be found to form an 'endosymbiotic association' with dinoflagellates. *Gymnodinium acidotum*, *Karlodinium veneficum*, and *Dinophysis* spp. have been reported to capture cryptophyte chloroplasts through mixotrophy, with phagotrophic ingestion of prey (Wilcox & Wedemayer 1984; Lucas & Vesk 1990; Li et al. 1996). A recent culture study showed that the source of cryptophyte chloroplasts in *Dinophysis acuminata* was *Myrionecta rubra* (Park et al. 2006). Prey abundance, particularly the abundance of cryptophytes, has been suggested to drive the formation of toxic *K. veneficum* blooms in eutrophic environments (Jason et al. 2008).

Globally, it is anticipated that, because of the small size of many cryptophytes, especially as nanoplankton (2–20 micrometres maximum size), many more species are likely to be discovered. Cryptophytes are very difficult to identify using light microscopy alone and in recent years their taxonomy and systematics have been revolutionised by the use of ultrastructural, molecular-sequence, and pigment data. In the British Isles, where 15 species have been recognised using light microscopy, the freshwater cryptophyte flora is regarded as poorly known (Novarino 2002). Environmental gene sampling suggests hidden diversity among the cryptophyte classes globally (e.g. Bråte et al. 2010; Pawlowski et al. 2011), including in the deep sea or the water column above it. A thorough study of New Zealand aquatic environments, therefore, using a combination of modern and traditional techniques, is likely to triple, at least, the known diversity of cryptophytes.

Authors

Dr Paul A. Broady School of Biological Sciences, University of Canterbury, Private Bag 4800, Christchurch, New Zealand [paul.broady@canterbury.ac.nz] Freshwater cryptophytes

Dr F. Hoe Chang National Institute of Water & Atmospheric Research (NIWA), Private Bag 14-901 Kilbirnie, Wellington, New Zealand [h.chang@niwa.co.nz] Marine cryptophytes

References

ADOLF, J. E.; BACHVAROFF, T.; PLACE, A. R. 2008: Can cryptophyte abundance trigger toxic *Karlodinium veneficum* blooms in eutrophic estuaries? *Harmful Algae 8*: 119–128.

BATSON, P. 2003: *Deep New Zealand. Blue Water, Black Abyss*. Canterbury University Press, Christchurch. 240 p.

BRADFORD, J. M.; CHANG, F. H.; BALDWIN, R.; CHAPMAN, B.; DOWNES, M.; WOODS, P. 1987: Hydrology, plankton, CHANG, F. H.; GALL, M. 1998: Phytoplankton assemblages and photosynthetic pigments during winter and spring in the Subtropical Convergence region near New Zealand. *New Zealand Journal of Marine and Freshwater Researh 32*: 515–530. and nutrients in Pelorus Sound, New Zealand, New Zealand, July 1981 and May 1982. *New Zealand Journal of Marine and Freshwater Research 21*: 223–233.

BRÅTE, J.; KLAVENESS, D.; RYGH, T.; JAKOBSEN, K. S.; SHALCHIAN-TABRIZI, S. 2010: Telonemia-specific environmental 18S rDNA PCR reveals unknown diversity and multiple marine-freshwater colonizations. *BMC Microbiology 10, 168*: 1–9.

BRETT, S. J.; PERASSO, L.; WETHERBEE, R. 1994: Structure and development of the cryptomonad periplast: a review. *Protoplasma 181*: 106–122.

CASSIE, V. 1974: Algal flora of some North Island New Zealand lakes including Rotorua and Rotoiti. *Pacific Science 28*: 467–504.

CASSIE, V. 1975: Phytoplankton of Lakes Rotorua and Rotoiti (North Island). Pp. 193–205 *in*: Jolly, V. H.; Brown, J. M. A. (eds), *New Zealand Lakes*. Auckland University Press/Oxford University Press, Auckland. 388 p.

CASSIE, V. 1978: Seasonal changes in phytoplankton densities in four North Island lakes, 1973–1974. *New Zealand Journal of Marine and Freshwater Research 12*: 153–166.

CASSIE, V. 1984: Revised checklist of the freshwater algae of New Zealand (excluding diatoms and charophytes). Part II – Chlorophyta, Chromophyta and Pyrrophyta, Rhaphidophyta and Euglenophyta. *Water and Soil Technical Publication 26*: i–xii, 117–250, xiii–lxiv.

CAVALIER-SMITH, T. 2004: Chromalveolate diversity and cell megaevolution: interplay of membranes, genomes and cytoskeleton. Pp. 75–108 *in*: Hirt, R. P.; Horner, D. S. (eds), *Organelles, Genomes and Eukaryote Phylogeny – An evolutionary synthesis in the age of genomics*. [*The Systematics Association Special Volume Series 68*.] CRC Press, Boca Raton. [viii] + 384 p.

CAVALIER-SMITh, T. 2007: Evolution and relationships of algae – major branches of the tree of life. Pp. 21–55 *in*: Brodie, J.; Lewis, J. (eds), *Unravelling the Algae: The past, present, and future of algal systematics*. [*The Systematics Association Special Volume Series 75*.] CRC Press, Boca Raton. [xix] + 376 p.

CAVALIER-SMITH, T. 2010: Kingdoms Protozoa and Chromista and the eozoan root of the eukaryotic tree. *Biology Letters 6*: 342–345.

CHANG, F. H. 1983: Winter phytoplankton and microzooplankton populations off the west coast of Westland, New Zealand, 1979. *New Zealand Journal of Marine and Freshwater Research 17*: 279–304.

CHANG, F. H. 1988: Distribution, abundance, and size composition of phytoplankton off Westland, New Zealand, Febraury 1982. *New Zealand Journal of Marine and Freshwater Research 22*: 345–367.

CHANG, F. H. 1994: New Zealand: Major shellfish poisoning (NSP) in early '93. *Harmful Algae News 1994(8)*: 1–2.

CHANG, F. H. 1999: Phytoplankton blooms around Wellington. *[NIWA] Aquaculture Update 23*: 1–12.

CHANG, F. H.; ANDERSON, C.; BOUSTEAD, N. C. 1990: First record of a *Heterosigma* (Raphidophyceae) bloom with associated mortality of cage-reared salmon in Big Glory Bay, New Zealand. *New Zealand Journal of Marine and Freshwater Research 24*: 461–469.

CHANG, F. H.; GALL, M. 1998: Phytoplankton assemblages and photosynthetic pigments during winter and spring in the Subtropical Convergence region near New Zealand. *New Zealand Journal of Marine and Freshwater Research 32*: 515–530.

CHANG, F. H.; UDDSTROM, M. J.; PINKERTON, M. H.; RICHARDSON, K. M. 2003: Characterizing the 2002 toxic *Karenia concordia* (Dinophyceae) outbreak and its development using satellite imagery on the north-eastern coast of New Zealand. *Harmful Algae 7*: 532–544.

CHANG, F. H.; ZELDIS, J.; GALL, M.; HALL, J. 2003: Seasonal and spatial variation of phytoplankton assemblages, biomass and cell size from spring to summer across the north-eastern New Zealand continental shelf. *Journal of Plankton Research 25*: 737–758.

CHAPMAN, V. J.; THOMPSON, R. H.; SEGAR, E. C. M. 1957: Checklist of the fresh-water algae of New Zealand. *Transactions of the Royal Society of New Zealand 84*: 695–747.

DAWSON, E. W. 1992: The marine fauna of New Zealand: Index to the fauna: 1. Protozoa. *New Zealand Oceanographic Institute Memoir 99*: 1–368 p.

DUNSTAN, G. A.; BROWN, M. R.; VOLKMAN, J. K. 2005: Cryptophyceae and Rhodophyceae; chemotaxonomy, phylogeny and application. *Phytochemistry 21*: 2557–2570.

ETHERIDGE, M. K. 1983: The seasonal biology of the phytoplankton in Lake Maratoto and Lake Rotomahana. Unpublished MSc thesis, University of Waikato, Hamilton. 264 p.

ETTL, H.; GÄRTNER, G. 1995: *Syllabus der Boden-, Luft- und Flectenalgen*. Gustav Fischer Verlag, Stuttgart. viii + 721 p.

FLINT, E. A. 1966: Additions to the checklist of freshwater algae in New Zealand. *Transactions of the Royal Society of New Zealand, Botany 3*: 123–137.

FORSYTH, D. J.; McCOLL, R. H. S. 1975: Limnology of Lake Ngahewa, North Island, New Zealand. *New Zealand Journal of Marine and Freshwater Research 9*: 311–332.

GANTT, E.; EDWARDS, M. R.; PROVASOLI, L. 1971: Chloroplast structure of the Cryptophyceae. Evidence of phycobiliproteins within intrathylakoidal spaces. *Journal of cell Biology 48*: 280–290.

GIESKES, W. W. C.; KRAAY, G. W. 1983: Dominance of Cryptophyceae during the phytoplankton spring bloom in the central North Sea detected by HPLC analysis of pigments. *Marine Biology 75*: 179–185.

GRAHAM, L. E.; GRAHAM, J. M.; WILCOX, L. W. 2009: *Algae*. Benjamin Cummings, San Francisco. xvi + 616 p.

GREEN, J. D. 1968: Limnological studies on a Waitakere Reservoir. Unpublished MSc thesis, University of Auckland. 195 p.

GREEN, J. D. 1974: The limnology of a New Zealand reservoir with particular reference to the life histories of the copepods *Boeckella propinqua* Sars and *Mesocyclops leuckarti* Claus. *Internationale Revue der gesamten Hydrobiologie 59*: 441–487.

GUSTAFSON, D. E. Jr; STOEKER, D. K.; JOHNSON, M. D.; VAN HEUKELEM, W. F.; SNEIDER, K. 2000: Cryptophyte algae are robbed of their organelles by the marine ciliate *Mesodinium rubrum*. *Nature 405*: 1049–1052.

HAUGHEY, A. 1968: The planktonic algae of Auckland sewage treatment ponds. *New Zealand Journal of Marine and Freshwater Research 2*: 721–766.

HOEF-EMDEN, K. 2005: Multiple independent losses of photosynthesis and differing evolutionary rates in the genus *Cryptomonas* (Cryptophyceae): combined phylogenetic analyses of DNA sequences of the nuclear and nucleomorph ribosomal operons. *Journal of molecular Evolution 60*: 183–195.

HOEK, C. van den; MANN, D. G.; JAHNS, H. M. 1995: *Algae: An introduction to Phycology*. Cambridge University Press, Cambridge. xiv + 623 p.

HUGHES, H. R.; McCOLL, R. H. S. 1980: Aquatic weed control in Lake Wanaka. Report by the Lake Wanaka Working Party to the Officials Committee on Eutrophication. *DSIR Information Series 143*: 1–50.

HUTCHINSON, E. G.; OLIVER, D. A. 1987: A survey of oxidation ponds in the Auckland region. *Auckland Regional Authority Works Division Report –*: 1–114.

JEFFREY, S. W.; VESK, M. 1990: Cryptophyta. Pp. 86–95 *in*: Clayton, M. N.; King, R. J. (eds), *Biology of Marine Plants*. Longman Cheshire,

Melbourne.
JOHNSON, M. D.; OLDACH, D.; DELWICHE, C. F.; STOEKER, D. K. 2007: Retention of transcriptionally active cryptophyte nuclei by the ciliate *Myrionecta rubra*. *Nature 445*: 426–428.
JOHNSTONE, I. M. 1972: Limnology of Western Springs, Auckland, New Zealand. *New Zealand Journal of Marine and Freshwater Research 6*: 298–328.
KRISTIANSEN, J. 2002: Phylum Chrysophyta (Golden Algae). Pp. 214–244 *in*: John, D.M.; Whitton, B. A.; Brook, A. J. (eds), *The Freshwater Algal Flora of the British Isles.* University Press, Cambridge. 714 p.
KUGRENS, P.; CLAY, B. L. 2003: Cryptomonads. Pp. 716–755 *in*: Wehr, J. D.; Sheath, R. G. (eds), *Freshwater Algae of North America*. Academic Press, San Diego. 917 p.
KUGRENS, P.; LEE, R. E.; HILL, D. R. A. 2000: Order Cryptomonadida Senn, 1900. Pp. 1111–1125 *in*: Lee, J. L.; Leedale, G. F.; Bradbury, P. (eds), *An Illustrated Guide to the Protozoa. Second Edition. Organisms traditionally referred to as Protozoa, or newly discovered groups*. Society of Protozoologists, Lawrence. 1432 p. (2 vols).
LAM, C. W. Y. 1977: Blue-green algae in the Waikato River. Unpublished PhD thesis, University of Auckland, Auckland. 238 p.
LI, A.; STOECKER, D. K.; COATS, D. W.; ADAM, E. J. 1996: Ingestion of fluorescently labelled and phycoerythrin-containing prey by mixotrophic dinoflagellates. *Aquatic Microbial Ecology 10*: 139–147.
LICHTLÉ, C. 1979: Effects of nitrogen deficiency and light of high intensity on *Cryptomonas rufescens* (Cryptophyceae). I. Cell and photosynthetic transformations and encystment. *Protoplamma 101*: 283–299.
LUCAS, I.; VESK, M. 1990: The fine structure of two photosynthetic species of *Dinophysis* (Dinophysiales, Dinophyceae). *Journal of Phycology 26*: 345–357.
MacKENZIE, A. L.; GILLESPIE, P. A. 1986a: Plankton ecology and productivity, nutrient chemistry, and hydrography of Tasman Bay, New Zealand, 1982–1984. *New Zealand Journal of Marine and Freshwater Research 20*: 365–395.
MacKENZIE, A. L.; KASPAR, H. F.; GILLESPIE, P. A. 1986b: Some observations on phytoplankton species composition, biomass, and productivity in Kenepuru Sound, New Zealand. *New Zealand Journal of Marine and Freshwater Research 20*: 397–405.
MASKELL, W. M. 1887: On the freshwater Infusoria of the Wellington District. *Transactions and Proceedings of the New Zealand Institute 19*: 49–61, pls 3–5.
MASKELL, W. M. 1888: On the freshwater Infusoria of the Wellington District. *Transactions of the New Zealand Institute 20*: 3–19, pls 1–4.
NORRIS, R. E. 1961: Observations on plankton organisms collected on the N.Z.O.I. Pacific Cruise, September 1958. *New Zealand Journal of Science 4*: 162–188.
NORRIS, R. E. 1964: Studies on phytoplankton in Wellington Harbour. *New Zealand Journal of Botany 2*: 258–278.
NOVARINO, G. 2002: Phylum Cryptophyta (cryptomonads). Pp. 180–185 *in*: John, D. M.; Whitton, B. A.; Brook, A. J. (eds), *The Freshwater Algal Flora of the British Isles*. Cambridge University Press, Cambridge. vii + 702 p.
OKAMOTO, N.; CHANTANGSI, C.; HORÁK, A.; LEANDER, B. S.; KEELING, P. J. 2009: Molecular phylogeny and description of the novel katablepharid *Roombia truncata* gen. et sp. nov., and establishment of the Hacrobia taxon nov. *PLoS ONE 4(9), e7080*: 1–11.
OKAMOTO, N.; INOUYE, I. 2005: The katablepharids are a distant sister group of the cryptophyta: a proposal for Katablepharidophyta divisio nova/ Kathablepharida phylum novum based on SSU rDNA and beta-tubulin phylogeny. *Protist 156*: 163–179.
PARK, M. G.; KIM, S.; KIM, H. S.; MYUNG, G.; KANG, Y. G.; YIH, W. 2006: First successful culture of the marine dinoflagellate *Dinophysis acuminata*. *Aquatic Microbial Ecology 45*: 101–106.
PAWLOWSKI, J.; CHRISTEN, R.; LECROQ, B.; BACHAR, D.; SHAHBAZKIAZKIA, H. R.; AMARAL-ZETTLER, L.; GUILLOU, L. 2011: Eukaryotic richness in the abyss: insights from Pyrotag sequencing. *PLoS ONE 6(4), e18169*: 1–10.
REEB, V. C.; PEGLAR, M. T.; YOON, H. S.; BAI, J. R.; WU, M.; SIU, P.; GRAFENBERG, J. L.; REYES-PRIETO, A.; RÜMMELE, S. E.; GROSS, J.; BHATTACHARYA, D. 2009: Interrelationships of chromalveolates within a broadly sampled tree of photosynthetic protists. *Molecular Phylogenetics and Evolution 53*: 202–211.
SARMA, P.; CHAPMAN, V. J. 1975: Additions to the checklist of freshwater algae of New Zealand II. *Journal of the Royal Society of New Zealand 5*: 289–312.
SCHEWIAKOFF, W. 1893: Uber die geographische Verbreitung der Süsswasser-Protozoen. *Verhandlungen des Naturhistorisch-medizinischen Vereins zu Heidelberg, n.f. 4*: 544–567.
SKUJA, H. 1939: Beitrag zur Algenflora Lettlands. II. *Acta Horta Botanici Universitatis Latviensis 11/12* : 41–169.
ŠLAPETA, J.; MOREIRA, D.; LÓPEZ-GARCIA, P. 2005: The extent of protist diversity: insights from molecular ecology of freshwater eukaryotes. *Proceedings of the Royal Society, B, 272*: 2073–2081.
TAYLOR, F. J. 1978: Records of marine algae from the Leigh area. Part II: Records of phytoplankton from Goat Island Bay. *Tane (Journal of the Auckland University Club) 24*: 213–218.
WAIKATO VALLEY AUTHORITY 1979: The Waikato River: a water resource study. *Water and Soil Technical Publication 11*: 1–225.
WETHERBEE, R.; HILL, D. R. A.; BRETT, S. J. 1987: The structure of the periplast components and their association with the plasma membrane in a cryptomonad flagellate. *Canadian Journal of Botany 65*: 1019–1026.
WILCOX, L. W.; WEDEMEYER, G. J. 1984: *Gymnodinium acidotum* Nygaard (Pyrrophyta), a dinoflagellate with an endosymbiotic cryptomonad. *Journal of Phycology 20*: 236–242.
YIH, W.; KIM, H. S.; JEONG, H. J.; MYUNG, G.; KIM, Y. G. 2004: Ingestion of cryptophyte cells by the marine photosynthetic ciliate *Mesodinium rubrum*. *Aquatic Microbial Ecology 36*: 165–170

Checklist of New Zealand Cryptophyta

Abbreviations: F, freshwater; M, marine.

KINGDOM CHROMISTA
SUBKINGDOM HACROBIA
PHYLUM CRYPTOPHYTA
Class CRYPTOPHYCEAE
Order CRYPTOMONADALES
CRYPTOMONADACEAE
Chroomonas minutissima (Norris) Taylor M E
Chroomonas nordstedtii Hansg. F
Chroomonas pulex Pascher F
Cryptochrysis commutata Pascher F
Cryptomonas mikrokuamosa Norris M
Cryptomonas paramoecium (Ehrenb.) F
Cryptomonas profunda Butcher M
Isoselmis obconica Butcher M
Plagioselmis prolonga Butcher M
Plagioselmis punctata Butcher M
Rhodomonas baltica Hill & Wetherbee M
Teteaulax acuta (Butcher) D.R.A.Hill M
HILLEACEAE
Hillea marina Butcher M

Order TETRAGONIDIALES
TETRAGONIDIACEAE
Tetragonidium verrucatum Pascher F

Class GONIOMONADOPHYCEAE
Order GONIOMONADALES
GONIOMONADACEAE
Goniomonas truncata (Fresenius) Stein F

Class KATABLEPHARIDOPHYCEAE
Order KATABLEPHARIDALES
KATABLEPHARIDACEAE
'Goniomonas' elongata (Maskell) F [= *Katablepharis*?]
Leucocryptos marina Braarud M

NINETEEN

Phylum
HAPTOPHYTA

haptophytes

LESLEY L. RHODES, ANTHONY R. EDWARDS,
OTHMAN BOJO, F. HOE CHANG

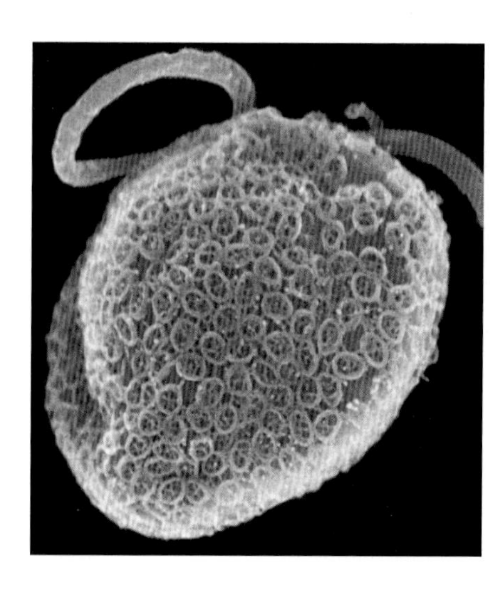

Prymnesium calathiferum, a bloom-forming species implicated in fish kills. The cell shows two cilia, a short median haptonema, and a surface covered by tiny basket-like scales.
F. Hoe Chang, NIWA

Among the notable components of New Zealand's marine phytoplankton are the golden-brown, scale-bearing microalgae known as haptophytes. They are ubiquitous in nearshore and oceanic waters and two species are also known from New Zealand's freshwater environments. Haptophyte cells are typically less than 20 micrometres in length, although there are exceptions. There are two cilia (used editorially in this volume in preference to flagella) (*Chrysochromulina quadrikonta* is an exception with four), a haptonema of varying length, and an exoskeleton usually formed of finely sculptured organic scales. In many species the scales form a template for the deposition of biologically controlled calcium carbonate. The green photosynthetic pigments chlorophyll *a* and different combinations of the forms of chlorophyll *c*, such as c_1, c_2 or c_2, c_3, are present but are masked by large amounts of the photosynthetically active carotenoid fucoxanthin, 19′-butanoyloxyfucoxanthin and 19′-hexanoyloxyfucoxanthin (Jeffrey & Wright 1994). About seven species in the haptophyte genera *Chrysochromulina* and *Prymnesium* produce substances that are toxic to fish, especially when the cells are experiencing phosphorus deficiency.

The phylum name derives from the characteristic haptonema, a short, thread-like organelle that protrudes anteriorly from the cell and which was originally thought to be a third cilium. Although, like cilia, it has axial microtubules, these differ in number and placement, there is an electron-dense central core, and ribosomes may be present. In species of the genus *Chrysochromulina* the haptonema is commonly relatively long (up to 180 micrometres in *C. strobilus*), whereas it is shorter in species of *Prymnesium*. In *Phaeocystis* and the calcareous scale-bearing coccolithophores, the haptonema is short or even rudimentary and in *Imantonia* it is reduced to an internal remnant.

Videography of some longer haptonemas have shown them acting as feeding tools, collecting food particles (e.g. bacteria) from spine scales and rolling them up to the tip, then inserting the bolus through the cell membrane (Kawachi et al. 1991). Many haptophytes are mixotrophic, carrying out photosynthesis during daylight hours and exhibiting heterotrophy (absorbing dissolved organic molecules) during dark periods, as well as ingesting and digesting food particles (Jones et al. 1993). The haptonema can also anchor cells to a substratum in the flagellated stage.

Many haptophytes exhibit non-flagellated life stages, including amoeboid forms, in culture. The amoeboid cells of *Chrysochromulina ericina* (Parke et al. 1956) and *C. quadrikonta* (Rhodes et al. 1994b) produce three or four spherical daughter cells that appear to constitute a resting stage. *Phaeocystis*, a common genus in New Zealand coastal and oceanic waters, is easily identified in the palmelloid stage of the life-cycle when the cells are embedded in a gelatinous capsule. Motile cells produce distinctive filamentous appendages, mostly five-rayed (nine-rayed in non-colonial *P. scrobiculata*) and composed of a-chitin crystal (Chrétiennot-Dinet et al. 1997), that are important in diagnostics. Common in New Zealand, *P. scrobiculata* also differs from the other species in having a much longer haptonema and cilia and a different scale structure.

The unmineralised scales of most haptophytes allow ready identification to species level using transmission electron microscopy (TEM). The mineralised (calcareous) scales of the coccolithophore group can be readily identified by scanning electron microscopy (SEM) and, on a less secure but more straightforward basis, by light microscopy. Coccolithophores, especially members of the genera *Emiliania* and *Gephyrocapsa* (Isochrysidales) are common in New Zealand's coastal and oceanic waters. Rare taxa with naked cells are known but coccolithophore taxonomy is otherwise based mostly on the structural and morphological character of their coccoliths. Taxonomic identification at the species level is usually straightforward but can be complicated by cells with dimorphic, etched, malformed, or phenotypic coccoliths. In addition, many cells produce different kinds of coccoliths during different parts of their life-cycle (as occasionally revealed by 'combination' coccospheres). Many coccolithophore cells can also be tagged as being either of the HET (heterococcolithophore) or HOL (holococcolithophore) type. In general, HET coccoliths are built from robust interlocked calcitic laths during the diploid phase whereas HOL coccoliths are formed from minute stacked rhombohedral crystallites during the haploid phase. Because of their differing potentials for disaggregation and dissolution, HOL coccoliths are normally far less abundant than HET coccoliths in samples from the water column, seafloor sediments, and marine rocks.

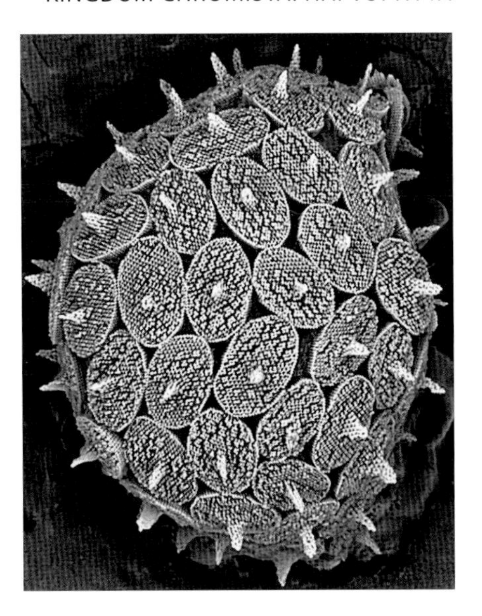

A cell of *Helicosphaera carteri* in the haploid phase, with HOL-type coccoliths.

F. Hoe Chang, NIWA

Current classification of haptophytes is still a work in progress, and until 2007 recognised the class names Pavlovophyceae and Prymnesiophyceae (Cavalier-Smith 1993, 2000, 2004; Edvardsen et al. 2000; Andersen 2004). A recent review of the literature (Silva et al. 2007) retains two classes while changing the nomenclature of Prymnesiophyceae to Coccolithophyceae. Classification at ordinal level and below is controversial owing to the need to reconcile taxonomic treatment of fossil and living species (cf. Bown & Young 1997; Young & Bown 1997; Anon. 2001). Haptophytes are also refered to as prymnesiophytes, a valid name under the International Code of Botanical Nomenclature based on the genus *Prymnesium*.

History of studies of the New Zealand region

Living Haptophyta have largely been ignored until fairly recently in New Zealand, mainly because of their small size and the need for EM to identify them to species level with confidence. The genus *Chrysochromulina* was first recorded by light microscopy in New Zealand waters by Norris (1961), with further observations of haptophytes being made by Cassie (1961) and Norris (1964). Taylor (1974) noted that three strains of *Chrysochromulina* had been cultured at

Summary of New Zealand Recent haptophyte diversity

Taxon	Described species	Known undescribed/ undetermined species	Estimated unknown species	Adventive species	Endemic species	Endemic genera
Pavlovophyceae	1	1	100	0	0	0
Coccolithophyceae	81	9	200	0	2	0
Totals	**82**	**10**	**210**	**0**	**2**	**0**

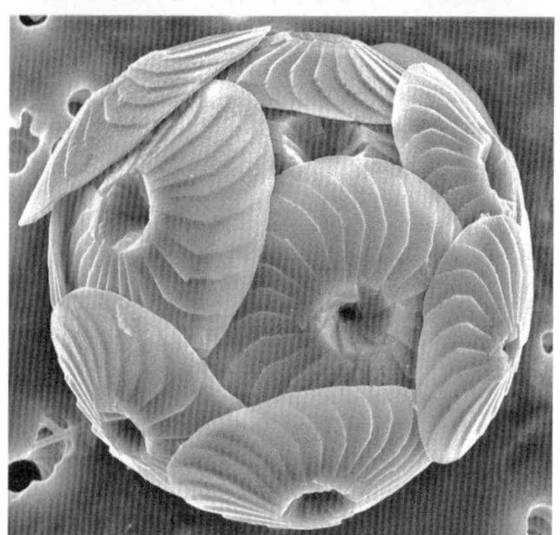

Acanthoica quattrospina (upper) and *Calcidiscus leptoporus*.
F. Hoe Chang, NIWA

the University of Auckland's Leigh research station in eastern Northland and that the coccolithophore *Emiliania huxleyi* was sometimes abundant in nearby coastal waters (Taylor 1978).

In 1974, Danish biologist Øjvind Moestrup collected plankton samples from Leigh (North Island) and Kaikoura (South Island) and characterised thirteen haptophyte species using EM, one of which was new to science (Moestrup 1979). His unusual new monotypic genus *Petasaria*, with siliceous scales, can now also be included among the haptophytes – Yoshida et al. (2006) have shown that a close relative is a member of the Prymnesiales. Based on a literature survey, Dawson (1992) compiled a list of all species recorded in the New Zealand region. About the same time Burns (1977) investigated the phenotypes and dissolution morphologies of the coccoliths of living cells of *Emiliania huxleyi* and *Gephyrocapsa* species obtained from different parts of the southwestern Pacific. An often-overlooked EM-based study by Nishida (1979) of the coccolithophores from plankton traverses in the Pacific Ocean revealed the presence of several other species not previously known from the New Zealand region.

The first record of coccolithophores in the New Zealand region was provided by Murray and Renard (1891) in their report on the voyage of HMS *Challenger*. They recorded them from both deep-sea sediments and the overlying surface waters of the Tasman Sea. Their sketch of the microfossils found in a pelagic calcareous ooze from a central New Zealand area now known as the Challenger Plateau included unnamed specimens of several taxa attributed to the living species *Calcidiscus leptoporus, Coccolithus pelagicus braarudi, Rhabdosphaera claviger*, and *Scyphosphaera apsteinii*. Some 75 years later, Edwards (1968a,b) undertook a mostly unpublished reconnaissance study of the coccolith content of New Zealand's oceanic sediments as an essential preliminary to assessing Cenozoic marine climate. An ocean-wide study of subfossil Pacific Ocean coccoliths (McIntyre et al. 1970) indicated the occurrence of several previously unrecorded coccolithophores in the New Zealand region. Of particular local interest at that time was their postulate that *Coccolithus pelagicus*, a distinctive cool-water species, had recently become extinct in the Southern Hemisphere. This possibility had not seemed likely in the New Zealand region because *C. pelagicus* was known to be common in some of New Zealand's very young oceanic sediments (Edwards 1968a). It was eventually disproved by Nishida (1979). During the 1970s, several papers were produced on haptophytes in New Zealand marine sediments – the taxonomy of *Pontosphaera* species in deep-sea sediments north of New Zealand (Burns 1973a), the latitudinal distribution and paleoenvironmental significance of coccoliths in deep-sea sediments around New Zealand (Burns 1973b), and changes in the coccolith content and composition of surficial Tasman Sea sediments west of Raglan (Burns 1974, 1975).

Only two freshwater haptophytes – *Pavlova gyrans* and *Prymnesium saltans* – have been reported in New Zealand, both by Lineham (1983) during a study of eutrophication in Lake Ellesmere (Cassie 1984).

The Haptophyta gained notoriety in New Zealand in 1981 following the suffocation of benthic marine life by a mucilaginous bloom in Tasman Bay, Nelson, that was attributed to *Phaeocystis pouchetii* (Chang 1983). A species of

Haptophyte diversity by environment

Taxon	Marine	Freshwater	Terrestrial	Fossil/subfossil
Pavlovophyceae	1	1	0	0
Coccolithophyceae	89	1	0	30
Totals	90	2	0	30

Phaeocystis was certainly present although recent work has also implicated the mucilage-forming dinoflagellate *Protoceratium reticulatum* (Mackenzie et. al. 2002). Fish kills attributed to *Prymnesium calathiferum* followed in Northland (Chang 1985; Chang & Ryan 1985). Unexplained blooms that have caused vast stretches of foam along the Canterbury coastline at different times through the 1990s almost certainly involved *Phaeocystis*.

In 1993, New Zealand responded to a toxic gymnodinioid dinoflagellate bloom by instituting marine-biotoxin monitoring programmes to underpin the safety of seafood products and thus protect both the shellfish industry and public health (Rhodes et al. 1998). This initiative was accompanied by research programmes that focussed on the causative agents of biotoxins, funded at the time by the Foundation for Research, Science and Technology and the Ministry of Agriculture and Forestry. The resulting studies (Rhodes et al. 1993; Rhodes et al. 1994a,b; Rhodes & Burke 1996), which included two PhD theses (Rhodes 1994; Bojo 2001) and the analysis of monitoring data from the biotoxin programmes, showed the Haptophyta to be a persistent though relatively small component of the phytoplankton in New Zealand's coastal waters.

New Zealand has been spared the enormous economic losses that other countries have experienced in recent years, for example, the massive fish kills in Scandinavia in the 1990s (Dahl et al. 1997). Most locally recorded blooms have involved the nontoxic coccolithophores, but known toxin producers inhabiting the South Pacific region include *Chrysochromulina leadbeateri, C. polylepis, C. quadrikonta* (low toxicity), *Prymnesium calathiferum*, and *P. patelliferum* (Rhodes 1994; LeRoi 2000; Bojo 2001).

One of the earliest attempts to estimate past marine climate – now the discipline of paleoclimatology – used the known distribution of modern coccolithophores, such as *Coccolithus pelagicus*, in New Zealand waters (Edwards 1968a,b). Fossil coccoliths of *Emiliania huxleyi* have been recorded from New Zealand's shallow marine and deep-sea sediments (Edwards 1995) and records date its first regional appearance at 270,000 years ago (Nelson 1988). Coccoliths of the *Gephyrocapsa oceanica* species complex appeared in New Zealand some 850,000 years ago (Edwards 1995) and climatic changes can be inferred from the species present. For example, *Gephyrocapsa* distributions in a sediment core from a Wellington drill hole suggest that temperatures warmer than the present day prevailed in central New Zealand during early postglacial time (Edwards 1992).

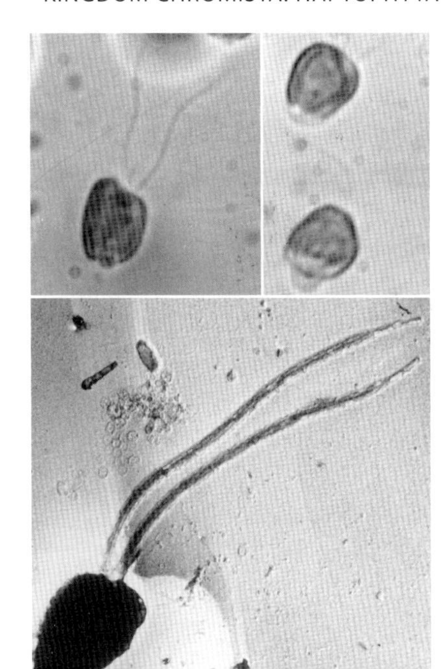

Individual cells of *Prymnesium calathiferum* as visualised by light microscopy (upper pair of photos) and shadow-cast transmission electron microscopy. The very short, thin haptonema is visible in two of the images.

F. Hoe Chang, NIWA

Diversity of New Zealand Haptophyta

Globally, numerous genera and species of cosmopolitan haptophytes have been described. Thomsen et al. (1994) suggested 81 genera – 11 unmineralised with about 80 species, and 70 mineralised comprising more than 200 coccolithophorid species. Some 152 coccolithophores are listed by Jordan et al. (2000). We recognise 92 haptophyte species in New Zealand waters.

Of the 55 species described in the genus *Chrysochromulina* to date, 21 have been identified in New Zealand coastal waters (Bojo 2001) and several more have been observed in the East Australian current (Hallegraeff 1983). The ability of many of the haptophytes to develop resting forms could assist in their dispersal via ships' ballast water (Hallegraeff & Bolch 1992) and we can expect to detect all species in New Zealand eventually. Transport of haptophytes over long distances within New Zealand has already been demonstrated on seaweed-bearing mussel spat (Rhodes & Burke 1996).

An interesting species occurring in New Zealand waters is quadriflagellate *Chrysochromulina quadrikonta*. This species was first observed in 1988 in Tokyo Bay, Japan (Kawachi & Inouye 1993) and has since been recorded in Australian and New Zealand waters (Rhodes 1994). Ultrastructural observations of scale morphology and cell shape suggested that *C. quadrikonta* was related to *C. ericina*,

Sketches from life of an undivided cell (centre) and fission stages of an undescribed oyster-larva parasite informally referred to as 'Pavlomulina'.
Othman Bojo

and it was hypothesised that the large quadriflagellate species was a polyploid form of *C. ericina*. Subsequently, DNA sequencing data has shown them to be genetically distinct species (Bojo 2001).

Chrysochromulina palpebralis (Seoane et al. 2009) was for many years simply referred to as the 'eye-lash' *Chrysochromulina*. It was first recorded in New Zealand by Moestrup (1979) on the basis of unmineralised scales. Another novel species, still unnamed and referred to as 'Pavlomulina', was first identified in Japanese waters (M. Kawachi pers. comm.) and has since been found off northeastern North Island. The microalga attaches to oyster larvae, inserts a feeding tube (possibly the haptonema) into the larva, and detaches only when the cell has become round and bloated, presumably from the uptake of nutrients (Rhodes pers. obs.).

One of the most widely studied members of the Haptophyta is the cosmopolitan bloom-forming genus *Phaeocystis*. Species have been recognised both as nuisance taxa and as ecologically important members of the phytoplankton and the genus has been found almost everywhere from polar to tropical regions (Guillard & Hellebust 1971). Despite being widely studied, there are many ambiguities regarding the taxonomic status of *Phaeocystis* that remain unresolved. Evidence from molecular analysis, however, suggests that at least six distinct species exist (Medlin et al. 1994; Zingone et al. 1999; Bojo 2001). Based on cladistic support for a monophyletic group containing only *Phaeocystis* species, and the genetic divergence between clades within the Haptophyta, sufficient grounds were established to erect a new order Phaeocystales to accommodate *Phaeocystis* (Edvardsen et al. 2000). Currently, six species are acknowledged – *P. pouchetii*, *P. antarctica*, *P. globosa*, *P. scrobiculata*, *P. jahnii*, and *P. cordata*. *Phaeocystis globosa* (sensu lato), *P. scrobiculata*, and *P. antarctica* (in offshore Subantarctic waters of New Zealand) are common in New Zealand waters. On the basis of genetic evidence, Medlin et al. (1994) were the first to propose that *P. globosa* and *P. pouchetii* were separate species. They also showed that colonial *Phaeocystis* from the Antarctic was genetically distinct from the other two species (Medlin & Zingone 2007).

The biologically, environmentally, and geologically important coccolithophores are well known because of their distinctive mineralised scales, called coccoliths (Perch-Nielsen 1985a,b; Winter & Siesser 1994; Bown 1998; Young et al. 2003). They were first recognised in chalk in 1836 by the venerable German microscopist Christian Gottfried Ehrenberg and observed in the plankton in 1857 by biologist G. C. Wallich who in 1877 gave the first species descriptions (Siesser in Winter & Siesser 1994). They differ from other haptophytes in having cells (coccospheres) with one or more exoskeletal layers of calcareous scales (coccoliths) formed on an organic-scale template. During major bloom events, they cause the seawater to turn a characteristic milky colour, visible from space. Common New Zealand bloom-formers are *Emiliania huxleyi* and *Gephyrocapsa oceanica*; blooms of *Syracosphaera* and *Pleurochrysis* have also been recorded (Rhodes 1994).

About 40 coccolithophore species had been recorded living in New Zealand waters by the middle of the last decade. Since 2009, eight additional species have been found in samples collected from both Subtropical and Subantarctic water on the east coast and from a transect in the Tasman Sea on the west coast (Chang & Law unpubl.). These species form part of an ongoing ocean-acidification study funded by the Ministry of Fisheries.

Individual coccolithophore orders are still primarily based on morphologically obvious structural differences in scale mineralisation (Young et al. 2003). As might be expected, many of the coccolithophores recognised to date from the presence of their coccoliths in New Zealand's subfossil sediments have already been recorded as living in the region; it seems likely that the remainder will eventually also join the list. As far as is known, no haptophytes other than coccolithophores and their coccoliths have been preserved in New Zealand's subfossil or fossil sediments.

Gephyrocapsa oceanica, a common bloom-forming species.
F. Hoe Chang, NIWA

Environmental and economic significance of haptophytes

Haptophytes produce large quantitities of the atmospheric gas dimethyl sulphide (DMS), which constitutes a small but significant amount of the total sulphur flux of the planet (Gibson et al. 1990). DMS also makes an important contribution to the formation of acidic rain (Keller et al. 1989; Davidson & Marchant 1992).

Ecologically, species of *Phaeocystis* are particularly significant (Estep et al. 1990). As well as producing large amounts of DMS, and with it acrylic acid (a bactericide and algicide – Sieburth 1960), *Phaeocystis* species produce an ultraviolet-B-absorbing compound that may counteract increases in physiological stress on cells because of UV, including decrease in photosynthetic efficiency, cell reproduction, and cell survival (Marchant et al. 1991; Davidson & Marchant 1992; Karentz & Spero 1995).

Coccolithophore blooms of *Emiliania huxleyi* can constitute a nuisance, occasionally making swimming unattractive in eutrophic waters overseas. Its blooms may also result in sedimentation of their calcareous scales or coccoliths. In living coccolithophore populations, enough coccoliths and coccosphere-rich faecal pellets can fall to the seabed to form a fine-grained pelagic calcareous ooze, especially below oligotrophic oceanic waters. Eventually, given sufficient time and sediment loading, such coccolith-rich deposits can become a chalk or limey marl; thus the informal term 'chalk-formers' for these organisms. Analysis of the coccolith or nannofossil content of a wide range of different kinds of marine sediment and rock usually provides valuable data regarding their age, depositional environment, provenance, and subsequent (post-depositional) history; making coccoliths by analogy 'geological microchips'.

The most famous example of chalk strata formed by coccoliths is the White Cliffs of Dover, but coccoliths have also contributed to the fine-grained Amberley Limestone and Oxford Chalk of Canterbury in the south Island and the 'argillaceous limestone' of Northland. New Zealand's Oamaru Diatomite, a tuffaceous Late Eocene open-water deposit famous for its fossil diatom content, is also rich in diverse coccolith remains (Stradner & Edwards 1968; Edwards 1991).

Coccoliths and their related or associated nannofossils, such as the diverse extinct genera *Chiasmolithus* and *Discoaster*, are very useful for determining the age, source, depositional environment, and post-depositional history of New Zealand's Cenozoic and Late Cretaceous sediments and rocks. They can also provide valuable information on the source of certain cultural or historical materials such as fireplace (hangi) stones, poorly documented moa or penguin bones, and fossil shells.

Commercially, certain haptophyte species are used in aquaculture industries where mass production of cultures provides food for bivalve and fish larvae (Laing & Ayala 1990). Cryopreservation methods to ensure a constant supply of selected microalgal strains are available to underpin the hatchery production of shellfish dependent on this feed (Rhodes et al. 2006).

Haptophytes are also increasingly used as sources of pigments, polysaccharides, vitamins, and fatty acids (Borowitzka 1992; Radmer & Parker 1994) and their potential in pharmaceutical industries is promising (Wyman & Goodman 1993; Gerwick et al. 1994).

Novel species of haptophytes continue to be discovered in New Zealand waters, and more taxonomic studies, based on molecular-genetic as well as morphological criteria, are needed. Other important diagnostic information such as ecological, physiological, and biochemical data carry weight in solving the many problems of systematics, including clarification of species boundaries, and deserve further investigation. Studies on the distribution, ecology, taxonomy, and variability of New Zealand coccolithophores would greatly assist in the climatic

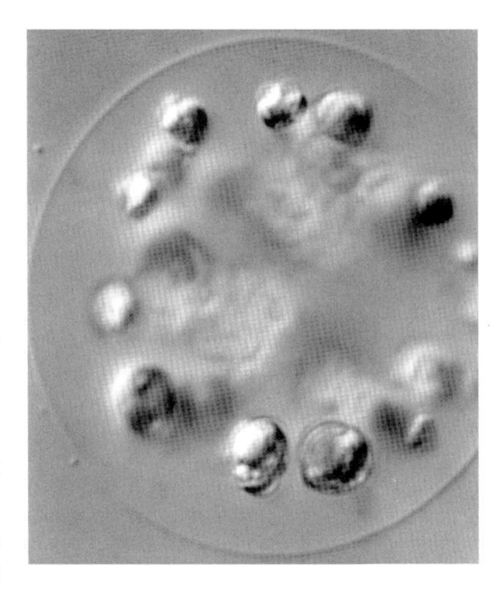

Phaeocystis globosa, a colonial species whose cells are embedded in mucilage. Blooms can produce unpleasant foam that may accumulate on beaches.
F. Hoe Chang, NIWA

A cell of *Emiliania huxleyi*. Blooms of this species may occupy thousands of square kilometres of ocean and can be seen from space.
F. Hoe Chang, NIWA

and environmental interpretation of fossil coccolith assemblages.

The modes of feeding shown by haptophytes are not yet fully understood. For example, 'Pavlomulina' cells engorge themselves after attaching to oyster larvae, presumably by feeding on the host, but the details are not known. *Phaeocystis* is another group that is still under scrutiny, but knowledge of life-cycles and molecular data of some New Zealand strains are still lacking because of the difficulty in establishing live cultures. *Phaeocystis* may prove useful as an indicator of coastal eutrophication.

Valuable areas for future research include studies of the impacts of land run-off on coastal phytoplankton production. For example, in New Zealand the lack of cobalt and selenium in some regions has led to the addition of these chemicals to fertilisers, which are then spread over farmland by aerial topdressing. Selenium is a requirement for some haptophytes (Harrison et al. 1988) and cobalt additions have been shown to cause selective increases in *Chrysochromulina* species, in turn linked to *Chrysochromulina* blooms causing massive fish kills in Scandinavia (Sangfors 1998). The potential effect of these additions on coastal phytoplanktonic biomass and toxicity needs assessing.

The use of chemical analyses to determine the presence of ichthyotoxins would help fin-fish farmers in their risk assessments. New Zealand's marine-biotoxin management programmes have moved away from mouse bioassays towards the use of instrumentation, for example, LC-MS (liquid chromatography mass spectrometry). A method for detecting ichthytoxins is still needed.

Acknowledgements

The Ministry of Fisheries is gratefully acknowledged for permission to reproduce SEM images of coccoliths obtained during the funded project ZBD200811.

Authors

Dr Othman Bojo Faculty of Resource Science and Technology, University Malaysia Sarawak, 94300 Kota Samarahan, Sarawak, Malaysia [bothman@ppp.unimas.my]

Dr F. Hoe Chang National Institute for Water & Atmospheric Research, Private Bag 14901, Kilbirnie, Wellington, New Zealand [h.chang@niwa.co.nz]

Dr Anthony R. Edwards Stratigraphic Solutions, Waikanae, New Zealand (Deceased.)

Dr Lesley L. Rhodes Cawthron Institute, Private Bag 2, Nelson, New Zealand [lesley.

References

ANDERSEN, R. A. 2004: Biology and systematics of heterokont and haptophyte algae. *American Journal of Botany 91*: 1508–1522.

ANON. 2001: *JODC Taxonomic Code of Marine Organisms*. http://jdoss1.jodc.go. jp/cgi-bin/2001/organism_taxonomic /index.html.

BOJO, O. 2001: Systematic studies of New Zealand nanoflagellates with special reference to members of the Haptophyta. Unpublished PhD thesis, University of Otago, Dunedin. 240 p.

BOROWITZKA, M. A. 1992: Algal biotechnology products and processes – matching science with economics. *Journal of Applied Phycology 4*: 267– 279.

BOWN, P. R. (Ed.) 1998: *Calcareous Nannofossil Biostratigraphy*. British Micropalaeontological Society Publications Series, Chapman & Hall (Kluwer Academic Publishers), Dordrecht. 328 p.

BOWN, P. R.; YOUNG, J. R. 1997: Proposals for a revised classification system for calcareous nannoplankton. *Journal of Nannoplankton Research 19*: 15–47.

BURNS, D. A. 1973a: Structural analysis of flanged coccoliths in sediments from the south west Pacific Ocean. *Revista Española de Micropaleontologia 5*: 147–160.

BURNS, D. A. 1973b: The latitudinal distribution and significance of calcareous nannofossils in the bottom sediments of the South-west Pacific Ocean (Lat. 15–55°S) around New Zealand. Pp. 221–228 *in*: Fraser, R. (ed), *Oceanography of the South Pacific 1972*. New Zealand National Commission for UNESCO, Wellington. 524 p.

BURNS, D. A. 1974: Changes in the carbonate component of Recent sediments with depth: a guide to paleoenvironmental interpretation. *Marine Geology 16*: M13–M19.

BURNS, D. A. 1975: The abundance and species composition of nannofossil assemblages in sediments from continental shelf to offshore basin, western Tasman Sea. *Deep-Sea Research 22*: 425–431.

BURNS, D. A. 1977: Phenotypes and dissolution morphotypes of the genus *Gephyrocapsa* Kamptner and *Emiliania huxleyi* (Lohmann). *New Zealand Journal of Geology and Geophysics 20*: 143–155.

BURNS, R. E.; ANDREWS, J. E.; et al. 1973: Site 206, Site 207. *Initial Reports of the Deep Sea Drilling Project 21*: 103–270.

CASSIE, V. 1961: Marine phytoplankton in New Zealand waters. *Botanica Marina 2 (Suppl.)*: 1–55.

CASSIE, V. 1984: Revised checklist of the freshwater algae of New Zealand (excluding diatoms and charophytes). Part II. *Water & Soil Technical Publication 26*: i-xii, 117–250, xiii-lxiv.

CAVALIER-SMITH, T. 1993: Kingdom Protozoa and its 18 phyla. *Microbiological Reviews 57*: 953–994.

CAVALIER-SMITH, T. 2000: Flagellate megaevolution. The basis for eukaryote diversification. Pp. 361–390 *in*: Leadbeater, B.

S. C.; Green, J. C. (eds), *The Flagellates – Unity, Diversity and Evolution*. [*Systematics Association Species Volume Series 59*.] CRC Press, Boca Raton. 416 p.

CAVALIER-SMITH, T. 2004: Chromalveolate diversity and cell megaevolution: interplay of membranes, genomes and cytoskeleton. Pp. 75–108 *in*: Hirt, R. P.; Horner, D. S. (eds), *Organelles, Genomes and Eukaryote Phylogeny – An Evolutionary Synthesis in the Age of Genomics*. [*The Systematics Association Special Volume Series 68*.] CRC Press, Boca Raton. [viii], 384 p.

CAVALIER-SMITH, T. 2010: Kingdoms Protozoa and Chromista and the eozoan root of the eukaryotic tree. *Biology Letters 6*: 342–345.

CHANG, F. H. 1983: The mucilage-producing *Phaeocystis pouchetii* (Prymnesiophyceae), cultured from the 1981 'Tasman Bay slime'. *New Zealand Journal of Marine and Freshwater Research 17*: 165–168.

CHANG, F. H. 1985: Preliminary toxicity test of *Prymnesium calathiferum* n. sp. isolated from New Zealand. Pp. 109–112 *in*: Anderson, D. M.; White, A. W.; Baden, D. G. (eds), *Proceedings of the Third International Conference on Toxic Dinoflagellates*. Elsevier, New York. 561 p.

CHANG, F. H.; RYAN, K. G. 1985: *Prymnesium calathiferum* (Prymnesiophyceae), a new species isolated from Northland, New Zealand. *Phycologia 24*: 191–198.

CHRETIENNOT-DINET, M.-J.; GIRAUD-GUILLE, M.-M.; VAULOT, D.; PUTEAUX, J.-L.; SAITO, Y.; CHANZY, H. 1997: The chitinous nature of filaments ejected by *Phaeocystis* (Prymnesiophyceae). *Journal of Phycology 33*: 666–672.

DAHL, E.; EDVARDSEN, B.; EIKREM, W. 1997: *Chrysochromulina* blooms in the Skagerrak after 1988. Pp. 104–105 *in*: Reguera, B.; Blanco, J.; Fernandez, M. L.; Wyatt, T. (eds),*Harmful Microalgae*. Xunta de Galicia and Intergovernmental Oceanographic Commission of UNESCO, Santiago de Compostela. 635 p.

DAVIDSON, A. T.; MARCHANT, H. J. 1992: The biology and ecology of *Phaeocystis* (Prymnesiophyceae). Pp. 1–45 *in*: Round, F. E.; Chapman, D. J. (eds), *Progress in Phycological Research, Volume 8*. Biopress Ltd, Bristol. 278 p.

DAWSON, E. W. 1992: The marine fauna of New Zealand: Index to the fauna: 1. Protozoa. *New Zealand Oceanographic Institute Institute Memoir 99*: 1–368.

EDVARDSEN, B.; EIKREM, W.; GREEN, J. C.; ANDERSON, R. A.; MOON-van der STAAY, S.Y.; MEDLIN, L. K. 2000: Phylogenetic reconstructions of the Haptophyta inferred from 18S ribosomal DNA sequences and available morphological data. *Phycologia 39*: 19–35.

EDWARDS, A. R. 1968a: The calcareous nannoplankton – evidence for New Zealand Tertiary marine climate. *Tuatara 16*: 26–31

EDWARDS, A. R. 1968b: Marine climates in the Oamaru District during late Kaiatan to early Whaingaroan time. *Tuatara 16*: 75–79.

EDWARDS, A. R. 1982: Calcareous nannofossils. *In*: Hoskins, R. H. (ed.). Stages of the New Zealand marine Cenozoic: a synopsis. *New Zealand Geological Survey Report 107*: 23–27.

EDWARDS, A. R. 1991: The Oamaru Diatomite. *New Zealand Geological Survey Paleontological Bulletin 64*: 1–240.

EDWARDS, A. R. 1992: The calcareous nannofossils of Miramar stratigraphic drillhole. *Institute of Geological and Nuclear Sciences Report G166*: 39–41.

EDWARDS, A. R. 1995: Appendix. Fossil record of the coccolithophores *Gephyrocapsa oceanica* and *Emiliania huxleyi* (Prymnesiophyceae = Haptophyceae) in the New Zealand region. *New Zealand Journal of Marine and Freshwater Research 29*: 357.

EDWARDS, A. R.; PERCH-NEILSEN, K. 1975: Calcareous nannofossils from the southern Southwest Pacific, Deep Sea Drilling Project, Leg 29. *Initial reports of the Deep Sea Drilling Project 29*: 469–539.

ESTEP, K. W.; NEJSTGAARD, J. C.; SKJOLDAL, H. R.; REY, F. 1990: Predation of copepods upon natural populations of *Phaeocyctis pouchetii* as a function of the physiological state of the prey. *Marine Ecology Progress Series 67*: 235–249.

GERWICK, W. H.; ROBERTS, M. A.; PROTEAU, P. J.; CHEN, J.-L. 1994: Screening cultured marine microalgae for anticancer-type activity. *Journal of Applied Phycology 6*: 143–149.

GIBSON, J. A. E.; GARRICK, R. C.; BURTON, H. R.; McTAGGART, A. R. 1990: Dimethylsulfide and the alga *Phaeocystis pouchetii* in antarctic coastal waters. *Marine Biology 104*: 339–346.

GUILLARD, R. R. K.; HELLEBUST, J. A. 1971: Growth and the production of extracellular substances by two strains of *Phaeocystis poucheti*. *Journal of Phycology 7*: 330–338.

HALLEGRAEFF, G. M. 1983: Scale-bearing and loricate nanoplankton from the east Australian current. *Botanica Marina 26*: 493–515.

HALLEGRAEFF, G. M. 1984: Coccolithophorids (calcareous nanoplankton) from Australian waters. *Botanica marina 27*: 229–247.

HALLEGRAEFF, G. M.; BOLCH, C. J. 1992: Transport of diatom and dinoflagellate resting spores in ship's ballast water: implications for plankton biogeography and aquaculture. *Journal of Plankton Research 14*: 1067–1084.

HARRISON, P. J.; YU, P. W.; THOMPSON, P. A.; PRICE, N. M.; PHILLIPS, D. J. 1988: Survey of selenium requirements in marine phytoplankton. *Marine Ecology Progress Series 47*: 89–96.

JEFFREY, S. W.; WRIGHT, S. W. 1994: Photosynthetic pigments in the Prymnesiophyceae. *in*: Green, J. C.; Leadbeater, B. S. C. (eds), *The Haptophyte Algae*. Clarendon Press, Oxford, p. 111–132.

JONES, H. L. J.; LEADBEATER, B. S. C.; GREEN, J. C. 1993: Mixotrophy in marine species of *Chrysochromulina* (Prymnesiophyceae): ingestion and digestion of a small green flagellate. *Journal of the Marine Biological Association of the United Kingdom 73*: 283–296.

JORDAN, R. W.; BROERSE, A. T. C.; HAGINO, K.; KINKEL, H.; SPRENGEL, C.; TAKAHASHI, K.; YOUNG, J. R. 2000: Taxon lists for studies of modern nannoplankton. *Marine Micropaleontology 39*: 309–314.

KARENTZ, D.; SPERO, H. J. 1995: Response of a natural *Phaeocystis* population to ambient fluctuation of UVB radiation caused by Antarctic ozone depletion. *Journal of Plankton Research 17*: 1771–1789.

KAWACHI, M.; INOUYE, I. 1993: *Chrysochromulina quadrikonta* sp. nov., a quadriflagellate member of the genus *Chrysochromulina* (Prymnesiophyceae = Haptophyceae). *Japanese Journal of Phycology 41*: 221–230.

KAWACHI, M.; INOUYE, I.; MAEDA, O.; CHIHARA, M. 1991: The haptonema as a food-capturing device: observations on *Chrysochromulina hirta* (Pymnesiophyceae). *Phycologia 30*: 563–573.

KELLER, M. D.; BELLOWS, W. K.; GUILLARD, R. L. 1989: Dimethyl sulfide production and marine phytoplankton. Pp. 167–182 *in*: Salzman, E.; Cooper, W. J. (eds), *Biogenic Sulfur in the Environment*. [Volume 393.] American Chemical Society, Washington, D.C. 572 p.

LAING, I.; AYALA, F. 1990: Commercial mass culture techniques for producing microalgae. Pp. 447–477 *in*: Akatsuka, I. (ed.), *Introduction to Applied Phycology*. SPB Academic Publishing, The Hague. 683 p.

LeROI, J.-M. 2000: Nanoflagellates of southern Tasmanian waters: taxonomy, toxicology and distribution. Unpublished MSc thesis, University of Tasmania, Hobart.

LINEHAM, I. W. 1983: Eutrophication of Lake Ellesmere: a study of phytoplankton. Unpublished PhD thesis, University of Canterbury, Christchurch.

McINTYRE, A.; BÉ, A. W. H.; ROCHE, M. B. 1970: Modern Pacific Coccolithophorida: a paleontological thermometer. *Transactions of the New York Academy of Sciences, series 2, 32*: 720–731.

MacKENZIE, L.; SIMS, I.; BEUZENBERG, V.; GILLESPIE, P. 2002: Mass accumulation of mucilage caused by dinoflagellate polysaccharide exudates in Tasman Bay, New Zealand. *Harmful Algae 1*: 69–84.

MARCHANT, H. J.; DAVIDSON, A. T.; KELLY, G. 1991: UV-B protecting pigments in the alga *Phaeocystis pouchetii* from Antarctica. *Marine Biology 109*: 391–395.

MEDLIN, L. K.; LANGE, M.; BAUMANN, M. E. M. 1994: Genetic differentiation among three colony-forming species of *Phaeocystis*: further evidence for the phylogeny of the Prymnesiophyta. *Phycologia 33*: 199–212.

MEDLIN, L.; ZINGONE, A. 2007: A taxonomic review of the genus *Phaeocystis*. *Biogeochemistry 83*: 3–18.

MOESTRUP, Ø. 1979: Identification by electron microscopy of marine nanoplankton from New Zealand, including the description of four new species. *New Zealand Journal of Botany 17*: 61–95.

MURRAY, J.; RENARD, A. F. 1891: *Report on the Scientific Results of the Voyage of H.M.S. Challenger during the years 1873–76. Deep-Sea Deposits*. HMSO, London. xxiv + 525 p., 29 pls, 43 charts, 22 diagrams.

NELSON, C. S. 1988: Revised age of a Quaternary tephra at DSDP site 594 off eastern South Island and some implications for correlation. *Geological Society of New Zealand Newsletter 82*: 35–40.

NISHIDA, S. 1979: Atlas of Pacific nannoplanktons. News of Osaka micro-paleontologists. *Micropaleontological Society of Osaka, Special Paper 3*: 1–31, 23 pls.

NORRIS, R. E. 1961: Observations on phytoplankton organisms collected on the N.Z.O.I. Pacific cruise, September 1958. *New Zealand Journal of Science 4*: 162–188.

NORRIS, R. E. 1964: Studies on phytoplankton in Wellington Harbour. *New Zealand Journal of Botany 2*: 258–278.

PARKE, M.; MANTON, I.; CLARK, B. 1956: Studies on marine flagellates. 3. Three further species of *Chrysochromulina*. *Journal of the Marine Biological Association of the United Kingdom 35*: 387–414.

PERCH-NIELSEN, K 1985a: Mesozoic Calcareous nannofossils. Pp. 329–426 *in*: Bolli, H. M.; Saunders, J. B.; Perch-Nielsen, K. (eds), *Plankton Stratigraphy*. Cambridge Earth Science Series, Cambridge University Press, Cambridge. viii +

1032 p.
PERCH-NIELSEN, K. 1985b: Cenozoic Calcareous nannofossils. Pp. 427–554 *in*: Bolli, H. M.; Saunders, J. B.; Perch-Nielsen, K. (eds), *Plankton Stratigraphy*. Cambridge Earth Science Series, Cambridge University Press, Cambridge. viii + 1032 p.
RADMER, R. J.; PARKER, B. C. 1994: Commercial applications of algae; opportunities and constraints. *Journal of Applied Phycology 6*: 93–98.
RHODES, L. 1994: Prymnesiophytes of New Zealand's coastal waters: taxonomy, physiology and ecology. Unpublished PhD thesis, Massey University, Palmerston North, New Zealand. 153 [+ xv] p.
RHODES, L. L.; BURKE, B. 1996: Morphology and growth characteristics of *Chrysochromulina* species (Haptophyceae = Prymnesiophyceae) isolated from New Zealand waters. *New Zealand Journal of Marine and Freshwater Research 30*: 91–103.
RHODES, L. L.; EDWARDS, A. R.; PEAKE, B. M.; MARWICK, S.; MacKENZIE, A. L. 1994a: The coccolithophores *Gephyrocapsa oceanica* and *Emiliania huxleyi* (Prymnesiophyceae) in New Zealand's coastal waters. *New Zealand Journal of Marine and Freshwater Research 29*: 345–357.
RHODES, L. L.; HAYWOOD, A. J.; BALLANTINE, W. J.; MacKENZIE, A. L. 1993: Algal blooms and climate anomalies in north-east New Zealand, August to December, 1992. *New Zealand Journal of Marine and Freshwater Research 27*: 419–430.
RHODES, L. L.; O'KELLY, C. J.; HALL, J. A. 1994b: Comparison of growth characteristics of New Zealand isolates of the prymnesiophytes *Chrysochromulina quadrikonta* and *C. camella* with those of the ichthyotoxic species *C. polylepis*. *Journal of Plankton Research 16*: 69–82.
RHODES, L. L.; SCHOLIN, C.; GARTHWAITE, I. 1998: *Pseudo-nitzschia* in New Zealand and the role of DNA probes and immunoassays in refining marine biotoxin monitoring programmes. *Natural Toxins 6*: 105–111.
RHODES, L. L.; SMITH, J.; TERVIT, R.; ROBERTS, R.; ADAMSON, J.; ADAMS, S.; DECKER, M. 2006: Cryopreservation of economically valuable marine micro-algae in the classes Bacillariophyceae, Chlorophyceae, Cyanophyceae, Dinophyceae, Haptophyceae, Prasinophyceae and Rhodophyceae. *Cryobiology 52*: 152–156.
SANGFORS, M. 1998: Are synergistic effects of acidification and eutrophication causing excessive algal growth in Scandinavian waters? *Ambio 17*: 296.
SEOANE, S., EIKEM, W., PIENAAR, R., EDVARDSEN, B. 2009: *Chrysochromulina palpebralis* sp. nov. (Prymnesiophyceae): a haptophyte possessing two alternative morphologies. *Phycologia 48*: 165–176.
SIEBURTH, J. M. 1960: Acrylic acid, an 'antibiotic' principle in *Phaeocystis* blooms in Antarctic waters. *Science 132*: 676–677.
SILVA, P. C.; THRONDSEN, J.; EIKREM, W. 2007: Revisiting the nomenclature of haptophytes. *Phycologia 46*: 471–475.
STRADNER, H.; EDWARDS, A. R. 1968: Electron microscopic studies on upper Eocene coccoliths from the Oamaru Diatomite, New Zealand. *Jahrbuch der Geologischen Bundesanstalt Sonderband 13*: 1–66, 48 pls.
TAYLOR, F. J. 1974: A preliminary annotated check list of micro-algae other than diatoms and dinoflagellates from New Zealand coastal waters. *Journal of the Royal Society of New Zealand 4*: 395–400.
TAYLOR, F. J. 1978: Records of marine algae from the Leigh area. *Tane (Journal of the Auckland University Field Club) 24*: 211–222.
THOMSEN, H. A.; BUCK, K. R.; CHAVEZ, F. P. 1994: Haptophytes as components of marine phytoplankton. Pp. 187–208 *in*: Green, J. C.; Leadbeater, B. S. C. (eds), *The Haptophyte Algae*. [*The Systematics Association Special Volume Series 51*.] Oxford Science Publications, Clarendon Press, Oxford. 446 p.
WINTER, A.; SIESSER, W. G. (Eds) 1994: *Coccolithophores*. Cambridge University Press, Cambridge. 242 p.
WYMAN, C. E.; GOODMAN, B. J. 1993: Biotechnology for production of fuels, chemicals and materials from biomass. *Applied Biochemistry and Biotechnology 39/40*: 41–59.
YOSHIDA, M.; NOËL, M.-H.; NAKAYAMA, T. ; NAGANUMA, T.; INOUYE, I. 2006: A haptophytes bearing siliceous scales: ultrastructure and phylogenetic position of *Hyalolithus neolepis* gen. et sp. nov. (Prymnesiophyceae, Haptophyta). *Protist 157*: 213–234.
YOUNG, J. R.; BOWN, P. R. 1997: Higher classification of calcareous nannofossils. http://www.nhm.ac.uk/hosted_sites/ina/taxcatalog/families.htm
YOUNG, J.; GEISEN, M.; CROS, L.; KLEIJNE, A.; SPRENGEL, C.; PROBERT. I.; OSTERGAARD, J. 2003: A guide to extant coccolithophore taxonomy. *Journal of Nannoplankton Research, Special Issue 1*: 1–125.
ZINGONE, A.; CHRÉTIENNOT-DINET, M.; LANGE, M.; MEDLIN, L. 1999: Morphological and genetic characterization of *Phaeocystis cordata* and *P. jahnii* (Prymnesiophyceae), two new species from the Mediterranean Sea. *Journal of Phycology 35*: 1322–1337.

Checklist of New Zealand living Haptophyta

Higher-level classification and terminology follow Cavalier-Smith (2004, 2010) and Silva et al. (2007). Families and orders follow Algaebase (algaebase.org). The checklist is made up of species listed by Norris (1961, 1964), Moestrup (1979), Dawson (1992), Rhodes (1994), Bojo (2001), and Chang and Law (unpubl.). Species indicated by an asterisk are those that have been isolated from plankton samples and confirmed by electron microscopy. All species are marine unless otherwise indicated by F, freshwater.

KINGDOM CHROMISTA
SUBKINGDOM HACROBIA
PHYLUM HAPTOPHYTA
Class PAVLOVOPHYCEAE
Order PAVLOVALES
PAVLOMULINACEAE
Gen. nov. et n. sp. 'Pavlomulina' L. L. Rhodes
PAVLOVACEAE
Pavlova gyrans Butcher F

Class COCCOLITHOPHYCEAE
Order ISOCHRYSIDALES
GEPHYROCAPSACEAE
Emiliania huxleyi (Lohmann) Hay & Mohler*
Gephyrocapsa cf. *caribbeanica* Boudreaux & Hay*
Gephyrocapsa ericsonii McIntyre & Bé*
Gephyrocapsa oceanica Kamptner*
Imantonia rotunda Reynolds*
NOELAERHABDACEAE
Reticulofenestra parvula (Okada & McIntyre) Biekart*

Order PHAEOCYSTALES
PHAEOCYSTACEAE
Phaeocystis antarctica Karst
Phaeocystis globosa Scherffel*
Phaeocystis pouchetii (Hariot) Lagerheim*
Phaeocystis scrobiculata Moestrup* E
Phaeocystis cf. *jahnii* Zingone*

Order PRYMNESIALES
PRYMNESIACEAE
Chrysochromulina acantha Leadbeater & Manton*
Chrysochromulina alifera Parke & Manton*
Chrysochromulina apheles Moestrup & Thomsen*
Chrysochromulina brevifilum Parke, Manton & Clarke*
Chrysochromulina camella Leadbeater & Manton*
Chrysochromulina campanulifera Manton & Leadbeater*
Chrysochromulina chiton Parke, Manton & Clarke*
Chrysochromulina cymbium Leadbeater & Manton*
Chrysochromulina ephippium Parke, Manton & Clarke*
Chrysochromulina ericina Parke & Manton*
Chrysochromulina aff. *fragilis* Manton*
Chrysochromulina hirta Manton*
Chrysochromulina latilepis Manton*
Chrysochromulina leadbeateri Estep, Davis, Hargraves & Sieburth*
Chrysochromulina mactra Manton*
Chrysochromulina mantoniae Leadbeater*
Chrysochromulina minor Parke, Manton & Clarke*
Chrysochromulina novae-zelandiae Moestrup* E
Chrysochromulina pachycylindra Manton, Oates & Course*
Chrysochromulina palpebralis Seoane, Eikrem, Pienaar & Edvardsen*
Chrysochromulina parkeae Green & Leadbeater*
Chrysochromulina pringshemii Parke & Manton*
Chrysochromulina pyramidosa Thomsen*
Chrysochromulina quadrikonta Kawachi & Inouye*
Chrysochromulina simplex Estep, Davis, Hargraves & Sieburth*
Chrysochromulina spinifera Pienaar & Norris*
Chrysochromulina cf. *kappa* Parke & Manton

Chrysochromulina cf. *throndsenii* Eikrem*
Chrysochromulina cf. *vexillifera* Manton & Oates*
Corymbellus aureus Green*
Petasaria heterolepis Moestrup
Prymnesium calathiferum Chang & Ryan*
Prymnesium parvum Carter
Prymnesium saltans Massart F

Order COCCOLITHALES
CALCIDISCACEAE
Calcidiscus leptoporus (Murray & Blackman) Kamptner
Oolithus antillarum (Cohen) Reinhardt
Umbilicosphaera foliosa (Kamptner ex Kleijne) Geisen*
Umbilicosphaera hulburtiana Gaarder*
Umbilicosphaera sibogae (Weber-van Bosse) Gaarder
COCCOLITHACEAE
Coccolithus pelagicus (Wallich) Schiller ssp. *braarudii* (Gaarder) Geisen, Billard, Broerse, L. Cros, Probert & J.R.Young
HALOPAPPACEAE
Halopappus adriaticus Schiller
HYMENOMONADACEAE
Hymenomonas globosa (Magne) Gyral & Fresnel*
PLEUROCHRYSIDACEAE
Pleurochrysis sp. Rhodes 1994*

Order SYRACOSPHAERALES
CALCIOSOLENIACEAE
Calciosolenia brasiliensis (Lohmann) Young
Calciosolenia murrayi Gran
RHABDOSPHAERACEAE
Acanthoica acanthifera Lohmann ex Lohmann*
Acanthoica quattrospina Lohmann
Algirosphaera robusta (Lohmann) Norris
Aspidorhabdus stylifer (Lohmann) Hay & Towe
Discosphaera tubifera (Murray & Blackman) Ostenfeld
Rhabdosphaera claviger Murray & Blackman
SYRACOSPHAERACEAE
Coronosphaera binodata (Kamptner) Gaarder
Coronosphaera mediterranea (Lohmann) Gaarder
Lohmannosphaera paucoscyphos Schiller
Michaelsarsia elegans Gran*
Michaelsarsia splendens Lohmann
Ophiaster hydroideus (Lohmann) Lohmann
Syriacosphaera corolla Lecal
Syracosphaera coronata Schiller
Syracosphaera corrugis Okada & McIntyre*
Syracosphaera dalmatica Kamptner
Syracosphaera histrica Kamptner
Syracosphaera cf. *lamina*
Syracosphaera molischii Schiller
Syracosphaera nodosa Kamptmer
Syracosphaera cf. *pirus* Halldal & Markali*
Syracosphaera prolongata Gran ex Lohmann*
Syracosphaera pulchra Lohmann .

Order ZYGODISCALES
HELICOSPHAERACEAE
Helicopontosphaera kamptneri Hay & Mohler
Helicosphaera carterii (Wallich) Kamptner
PONTOSPHAERACEAE
Pontosphaera alboranensis Bartolini
Pontosphaera caelamenisa Lecal-Schlauder
Pontosphaera grani Gaarder
Pontosphaera japonica (Takayama) Nishida
Pontosphaera messinae Bartolini
Scyphosphaera apsteinii Lohmann
Poricalyptra aurisinae (Kamptner) Kleijne

Order INCERTAE SEDIS
ALISPHAERACEAE
Alisphaera unicornis Okada & McIntyre*
Polycrater galapagensis Manton & Oates
BRAARUDOSPHAERACEAE
Braarudosphaera bigelowii (Gran & Braarud) Deflandre
CALYPTROSPHAERACEAE
Calyptrolithina wettsteinii (Kamptner) Norris*
Calyptrosphaera oblonga Lohmann*
Helladosphaera vavilovii (Borsetti & Cati) Young & Kleijne
Sphaerocalyptra quadridentata (Schiller) Deflandre*
Turrisphaera sp. Moestrup 1979*
PAPPOSPHAERACEAE
Papposphaera lepida Tangan
Papposphaera sp.
UMBELLOSPHAERACEAE
Umbellosphaera corolla (Lecal) Gaarder*
Umbellosphaera irregularis Paasche
Umbellosphaera tenuis (Kamptner) Paasche*

Checklist of coccolithophores from New Zealand surficial marine sediments

The following checklist contains fossil and subfossil taxa and possibly some living species not in the above list. Higher-level classification and terminology follow Cavalier-Smith (2004, 2010) and Silva et al. (2007). Families and orders follow Algaebase (algaebase.org). Species records are based on the following sources (listed in references): Edwards 1968a, 1982, unpubl.; Edwards & Perch-Neilsen 1975; McIntyre et al. 1970; Burns 1973a,b, 1975; Burns et al. 1973.

KINGDOM CHROMISTA
SUBKINGDOM HACROBIA
PHYLUM HAPTOPHYTA
Class COCCOLITHOPHYCEAE
Order COCCOLITHALES
CALCIDISCACEAE
Calcidiscus leptoporus (Murray & Blackman) Loeblich & Tappan
Hayaster perplexus (Bramlette & Riedel) Bukry
Oolithotus antillarum (Cohen) Reinhardt
Umbilicosphaera sibogae (Weber-van Bosse) Gaarder
COCCOLITHACEAE
Coccolithus pelagicus (Wallich) Schiller f. *braarudii* (Gaarder) Geisen, Billard, Broerse, L.Cros, Probert & J.R.Young
Cruciplacolithus cf. *neohelis* (McIntyre & Bé) Reinhardt

Order ISOCHRYSIDALES
GEPHYROCAPSACEAE
Emiliania huxleyi (Lohmann) Hay & Mohler
Gephyrocapsa ericsonii McIntyre & Bé
Gephyrocapsa oceanica Kamptner

Order SYRACOSPHAERALES
CALCIOSOLENIACEAE
Calciosolenia sp. (*Scapholithus* sp. auctt.)
RHABDOSPHAERACEAE
Discosphaera tubifer (Murray & Blackman) Ostenfeld
Rhabdosphaera clavigera Murray & Blackman
SYRACOSPHAERACEAE
Syracosphaera histrica Kamptner
Syracosphaera pulchra Lohmann

Order ZYGODISCALES
HELICOSPHAERACEAE
Helicosphaera carterii (Wallich) Kamptner
Helicosphaera wallichii (Lohmann) Okada & McIntyre
PONTOSPHAERACEAE
Pontosphaera alboranensis Bartolini
Pontosphaera discopora Schiller
Pontosphaera distincta (Bramlette & Sullivan) Burns
Pontosphaera japonica (Takayama) Nishida
Pontosphaera messinae Bartolini
Pontosphaera multipora (Kamptner) Roth
Pontosphaera pacifica Burns
Pontosphaera syracusana Lohmann
Scyphosphaera apsteinii Lohmann

INCERTAE SEDIS
BRAARUDOSPHAERACEAE
Braarudosphaera bigelowii (Gran & Braarud) Deflandre
CERATOLITHACEAE
Ceratolithus cristatus Kamptner
CALYPTROSPHAERACEAE
Syracolithus schilleri (Kamptner) Norris
UMBELLOSPHAERACEAE
Umbellosphaera irregularis Paasche
Umbellosphaera tenuis (Kamptner) Paasche

TWENTY

Phylum
HELIOZOA
sun protists

DENNIS P. GORDON

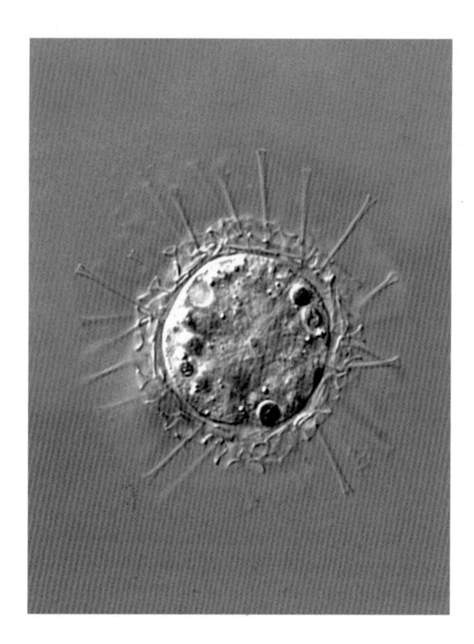

The sun protist *Raphidiocystis tubifera*, with its cell body protected by a layer of scales and radiating spicules.
William Bourland, per Micro*scope (MBL)

Sun protists are ubiquitous, tiny, unicellular organisms imaginatively named for their resemblance to the sun (Greek, *helios*). Mostly spherical in shape, the cells of a majority of species are surrounded by fine rays that project from the surface. As formerly constituted, Heliozoa united a group of protists that superficially resembled each other by having the same overall radiating form but which differed ultrastructurally and phylogenetically. They had in common the rays (axopodia) that terminate in the cell interior, either on the nucleus or on some other structure, and the lack of an internal skeleton or capsule. Heliozoa thus comprised four well-circumscribed groups – centrohelids (the majority), actinophryids, desmothoracids, and gymnosphaerids. The latter three groups have been demonstrated to belong elsewhere in the Chromista and phylum Heliozoa now comprises only the class Centrohelea, with upwards of 15 genera and 95 described species worldwide (Cavalier-Smith & von der Heyden 2007). Once classified among sarcodine protozoans, phylogenomic studies link Heliozoa with cryptophytes and haptophytes (Burki et al. 2009).

Centroheleans have in the middle of the cell body a centroplast or central granule that anchors the rays, with the nucleus placed off-centre. The centroplast is made up of a tripartite disc sandwiched between two hemispheres of dense material. Each axopodial ray, a type of pseudopodium, is straight and slender and supported internally by axonemes – geometric arrays of microtubules that confer some stiffness and which, in cross section, form complex hexagonal and triangular patterns. The thin layer of cytoplasm that surrounds the axoneme contains extrusible bodies (extrusomes) that are involved in the capture of food. Axopodia can extent or retract, sometimes quickly, by assembly (polymerisation) and disassembly of the microtubules (Mikrjukov et al. 2000), facilitating their additional roles in sensation, movement, and attachment.

Cell size is small, typically less than 100 micrometres diameter. Some have mucous stalks. The majority of species have a coat, often double, of tangential or radial siliceous rods, plate-like scales, or tube-like, trumpet-like, or spine-like radial spicules, and the axopodia must project through this coat. On the other hand, one species has a naked cell surface and members of one genus have a mucous cell envelope. Most species are found in freshwater habitats but centrohelids can also be found in the sea and in terrestrial habitats. They are passive predators of other protists (flagellates, ciliates, microalgae), chytrid zoospores, and very small animals. Reproduction is by simple binary fission; budding and biciliated cells have also been reported. Many species produce cysts.

New Zealand heliozoan diversity is poorly known. Although heliozoans are often routinely noted to be present in aquatic samples (e.g. Stout 1973; James 1991; Levine et al. 2005), they are rarely identified to family, let alone species, which is not surprisingly given the microscopical and taxonomic challenges.

The late John D. Stout of New Zealand's former Soil Bureau was one exception, reporting on the presence of many named protists, including the centrohelid *Raphidiophrys elegans* or *Raphidiophrys* sp. in the soils of South Island tussock grassland (Stout 1958) and seasonally inundated coastal grassland (Stout 1984), and even in recovering topsoil a month after the burning of gorse (*Ulex europaeus*) and manuka (*Leptospermum scoparium*) scrub (Stout 1961). The most definitive studies to date on New Zealand heliozoans are those of Nicholls and Dürrschmidt (1985) and Dürrschmidt (1987a,b) using electron microscopy. They recorded 23 taxa, including two subspecies and one undescribed species, mainly in the genera *Acanthocystis* and *Raphidiophrys*. Nicholls and Dürrschmidt (1985) did not note the occurrence of *R. elegans* in New Zealand, but their samples were from lakes and swamps whereas Stout's records of this species were from soils. His identifications were solely based on light microscopy, however.

In their study of molecular phylogeny, scale evolution, and taxonomy of centrohelids, Cavalier-Smith and von der Heyden (2007) analysed genes from cultured heliozoans and environmental gene libraries. The latter yielded evidence of at least five additional taxa from New Zealand marine, freshwater, and terrestrial environments, indicating significant hidden diversity of this group in New Zealand.

Some New Zealand taxa ascribed to Heliozoa in the past belong to other phyla. The semi-cosmopolitan marine epizoite *Wagnerella borealis* is common in coastal waters (Dons 1921; Dawson 1992). Bass et al. (2009) suggested that the genus 'may, like desmothoracids, turn out to granofilosean' and it is here included in the Cercozoa along with the freshwater species of *Pompholyxophrys* reported by Nicholls and Dürrschmidt (1985). *Actinophrys sol*, noted by Stout (1958, 1961, 1984) as a common inhabitant of the same soils in which he recorded *Raphidiophrys*, belongs to phylum Bigyra.

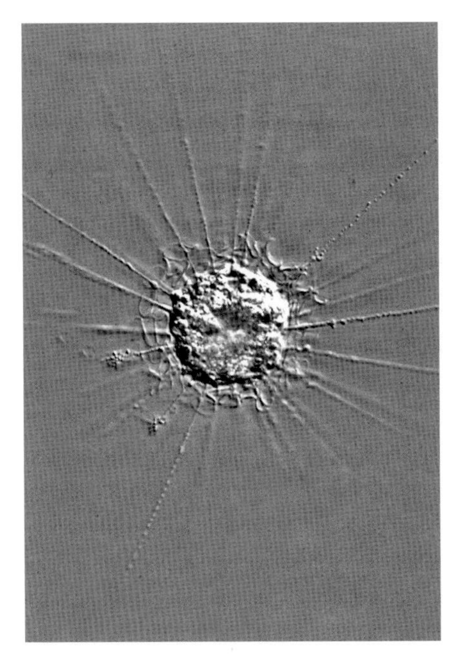

Raphidiophrys elegans.
Martin Kreutz, Private Laboratory, Konstanz

Summary of New Zealand heliozoan diversity

Taxon	Described species + subspecies	Known undetermined/ undescribed species	Estimated unknown species	Adventive species	Endemic species + subspecies	Endemic genera
Heliozoa	24+1	6	40	0	3+1	0

Diversity by environment*

Taxon	Marine	Freshwater	Terrestrial
Heliozoa	2	26+1	2

* Some species can occur in more than one environment, but the ones listed here are scored only for the environment in which they were found in New Zealand.

Acknowledgement

I am grateful to Tom Cavalier-Smith for advice on the taxonomic placement of some taxa previously attributed to Heliozoa.

Author

Dr Dennis P. Gordon National Institute of water & Atmospheric Research, Private Bag 14901, Kilbirnie, Wellington, New Zealand [d.gordon@niwa.co.nz]

References

BASS, D.; CHAO, E. E.-Y.; NIKOLAEV, S.; YABUKI, A.; ISHIDA, K.; BERNEY, C.; PAKZAD, U.; WYLEZICH, C.; CAVALIER-SMITH, T. 2009: Phylogeny of novel naked filose and reticulose Cercozoa: Granofilosea cl. n. and Proteomyxidea revised. *Protist 160*: 75–109.

BURKI, F.; INAGAKI, Y.; BRÅTE, J.; ARCHIBALD, J. M.; KEELING, P. J.; CAVALIER-SMITH, T.; SAKAGUCHI, M.; HASIMOTO, T.; HORAK, A.; KUMAR, S.; KLAVENESS, D.; JAKOBSEN, K. S.; PAWLOWSKI, J.; SHALCHIAN-TABRIZI, K. 2009: Large-scale phylogenomic analyses reveal that two enigmatic protist lineages, Telonemia and Centroheliozoa, are related to photosynthetic chromalveolates. *Genome Biology and Evolution 1*: 231–238.

CAVALIER-SMITH, T.; von der HEYDEN, S. 2007: Molecular phylogeny, scale evolution and taxonomy of centrohelid heliozoa. *Molecular Phylogenetics and Evolution* 44: 1186–1203.

DAWSON, E. W. 1992: The marine fauna of New Zealand: Index to the fauna 1. Protozoa. *New Zealand Oceanographic Institute Memoir 99*: 1–368.

DONS, C. 1921: Papers from Dr. Th. Mortensen's Pacific Expedition 1914–1916. V. Notes sur quelques Protozoaires marins. *Videnskabelige Meddelelser fra Dansk Naturhistorisk Forening i Kjøbenhavn 73*: 49–84.

JAMES, M. R. 1991: Sampling and preservation methods for the quantitative enumeration of microplankton. *New Zealand Journal of Marine and Freshwater Research 25*: 305–310.

LEVINE, S. N.; ZEHRER, R. F.; BURNS, C. W. 2005: Impact of resuspended sediment on zooplankton feeding in Lake Waihola, New Zealand. *Freshwater Biology 50*: 1515–1536.

NICHOLLS, K. H.; DÜRRSCHMIDT, M. 1985: Scale structure and taxonomy of some species of *Raphidocystis*, *Raphidiophrys*, and *Pompholyxophrys* (Heliozoea) including descriptions of six new taxa. *Canadian Journal of Zoology 63*: 1944–1961.

DÜRRSCHMIDT, M. 1987a: An electron microscopical study of freshwater Heliozoa (genus *Acanthocystis*, Centrohelidia) from Chile, New Zealand, Malaysia and Sri Lanka. II. *Archiv für Protistenkunde 133*: 21–48.

DÜRRSCHMIDT, M. 1987b: An electron microscopical study of freshwater Heliozoa (genus *Acanthocystis*, Centrohelidia) from Chile, New Zealand, Malaysia and Sri Lanka. III. *Archiv für Protistenkunde 133*: 49–80.

MIKRJUKOV, K. A. 1995: Revision of species composition of the genus *Choanocystis* (Centroheliozoa), Sarcodina) and its members in Eastern Europe. *Zoologicheskii Zhurnal 74*: 3–17.

MIKRJUKOV, K. A. 1996: Revision of genera and species composition in lower Centroheliozoa. II. Family Raphidiophryidae n. fam. *Archiv für Protistenkunde 147*: 205–212.

MIKRJUKOV, K. A. 1997: Revision of genera and species composition of the family Acanthocystidae (Centroheliozoa; Sarcodina). *Zoologicheskii Zhurnal 76*: 1–13.

MIKRJUKOV, K. A.; SIEMENSMA, F. J.; PATTERSON, D. J. 2000: Phylum Heliozoa. Pp. 860–871 *in*: Lee, J. J.; Leedale, G. F.; Bradbury, P. (eds), *An Illustrated Guide to the Protozoa. Second Edition. Organisms traditionally referred to as Protozoa, or newly discovered groups*. Society of Protozoologists, Lawrence. Vol. 2, v + 690–1432 p.

SIEMENSMA, F. J. 1991: Heliozoea. Pp. 171–290 *in*: Page, F. C.; Siemensma, F. J. (eds), *Protozoenfauna. Bd 2. Nachte Rhizopoda und Heliozoea*. Fisher Verlag, Stuttgart. 309 p.

STOUT, J. D. 1958: Biological studies of some tussock-grassland soils. VII. Protozoa. *New Zealand Journal of Agricultural Research 1*: 974–984.

STOUT, J. D. 1961: Biological and chemical changes following scrub burning on a New Zealand hill soil. *New Zealand Journal of Science 4*: 740–752.

STOUT, J. D. 1984: The protozoan fauna of a seasonally inundated soil under grassland. *Soil Biology and Biochemistry 16*: 121–125.

STOUT, V. M. 1973. The freshwater environment. Pp. 229–250 *in*: Williams, G. R. (ed.), *The Natural History of New Zealand – An ecological survey*. A. H. & A. W. Reed, Wellington. viii + 434 p., 40 pls.

Checklist of New Zealand Heliozoa

Classification follows that of Cavalier-Smith and von der Heyden (2007). Some of the binominals published in Nicholls and Dürrschmidt (1985) and Dürrschmidt (1987a,b) are modified according to the treatments of Siemensma (1991) and Mikrjukov (1995, 1996, 1997).

Kingdom CHROMISTA
SUBKINGDOM HACROBIA
PHYLUM HELIOZOA
Class CENTROHELEA
Order CENTROHELIDA
Suborder ACANTHOCYSTINA
ACANTHOPHRYIDAE
Acanthocystis bicornis Dürrschmidt, 1987 F
Acanthocystis cornuta Dürrschmidt, 1987 F
Acanthocystis paliformis Dürrschmidt, 1987 F
Acanthocystis pectinata Penard, 1889 F
Acanthocystis penardii Wailes, 1925 F
Acanthocystis pulchra Dürrschmidt, 1985 F
Acanthocystis rasilis Dürrschmidt, 1987 F E
Acanthocystis striata Nicholls, 1983 F
Acanthocystis tubata Dürrschmidt, 1987 F
Acanthocystis sp. Dürrschmidt 1987 F E
Raphidocystis flabellata flabellata (Dürrschmidt, 1987) F
Raphidocystis f. novaezelandiae Dürrschmidt, 1987 F E
Raphidocystis tubifera Penard, 1904 F
'Unidentified helio 16' Cavalier-Smith & von der Heyden 2007 F
'Unidentified helio 17' Cavalier-Smith & von der Heyden 2007 T
RAPHIDIOPHRYIDAE
Polyplacocystis brunii (Penard, 1903) M
Raphidiophrys ambigua Penard, 1904 F
Raphidiophrys elegans Hertwig & Lesser, 1874 T
Raphidiophrys intermedia Penard, 1904 F
Raphidiophrys marginata Siemensma, 1981 F
Raphidiophrys orbicularis ovalis Nicholls & Dürrschmidt, 1985 F
'Unidentified helio 10' Cavalier-Smith & von der Heyden 2007 M

Suborder PTEROCYSTINA
CHOANOCYSTIDAE
Choanocystis cordiformis (Dürrschmidt, 1987) F
Choanocystis rhytidos (Dürrschmidt, 1987) F
Choanocystis rotoairense (Dürrschmidt, 1987) F E
Choanocystis rotundata (Nicholls, 1983) F
'Unidentified helio 2' Cavalier-Smith & von der Heyden 2007 F
PTEROCYSTIDAE
Pterocystis foliacea (Dürrschmidt, 1985) F
Pterocystis pinnata (Nicholls, 1983) F
Pterocystis pyriformis (Dürrschmidt, 1987) F
'Unidentified helio 5' Cavalier-Smith & von der Heyden 2007 F

TWENTY-ONE

Phylum
GLAUCOPHYTA
glaucophytes

PAUL A. BROADY

This very small phylum of the plant kingdom contains about eight genera of freshwater algae that exist as flagellate unicells or small colonies of up to sixteen cells. Representatives of only three genera (*Cyanophora*, *Glaucocystis*, and *Gloeochaete*) are reasonably common in soft waters of acid bogs (Kies 1992; Graham et al. 2009).

Glaucophytes are of particular interest because, as the phylum name implies, the cells contain blue-green plastids, often referred to as cyanelles, that resemble unicellular cyanobacteria both in pigment content and structure. In the latter respect, the cyanelles are remarkable in having retained vestiges of a peptidoglycan wall between the two membranes that surrounded the original endosymbiont from which they are presumably derived (Kugrens et al. 1999). However, the cyanelle genome is less than four per cent of the size of that found in *Synechocystis*, a free-living, unicellular cyanobacterium, and has very limited genetic autonomy. About 90 per cent of the soluble proteins of cyanelles are encoded by the host nucleus, having been transferred there during the evolutionary transition to becoming a plastid. The storage of photosynthetic product as starch is within the host cytoplasm. Cyanelles cannot be cultured independent of the host cell and are not simply cyanobacterial endosymbionts as was once thought.

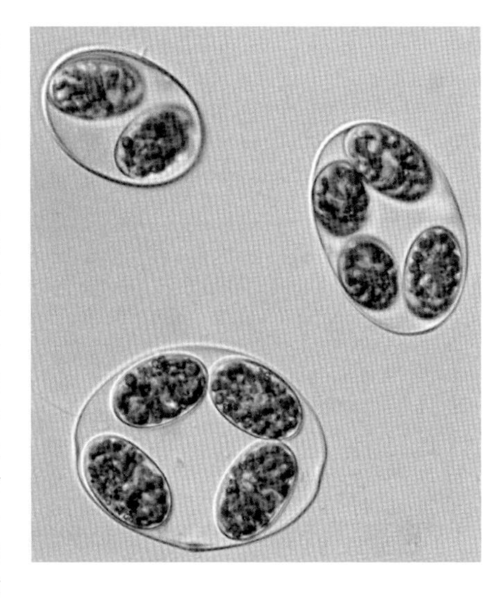

Colonies of *Glaucocystis nostochinearum*.
Protist Information Server, Hosei University

DNA sequencing suggests that the phylum is of ancient origin. Despite being of no current economic interest, glaucophytes are extremely important for understanding the origin of plastids of photosynthetic eukaryotes and this aspect has received considerable scientific attention (Graham et al. 2009). The three lineages of glaucophyte, red, and green algae could all be derived from an ancestral eukaryote that captured a plastid by the process of endosymbiosis of a cyanobacterium (Keeling 2004). This monophyletic relationship is suggested by the double membrane surrounding their plastids, by the flat cristae of their mitochondria, and by several multigene trees (Inouye & Okamota 2005). However, this hypothesis remains contentious and chloroplasts could have been gained on more than one occasion. For instance, other multi-gene trees do not strongly support monophyly (e.g. Nozaki et al. 2007). Also, phylogenetic analyses using nuclear genes could make glaucophyte, red, and green algae appear more closely related than they really are as there is evidence that a significant number of these genes is derived from cyanobacteria (Stiller 2007).

Only the colony-forming *Glaucocystis nostochinearum* has been recorded in New Zealand (Cassie 1984a). It is cosmopolitan in temperate regions. Colonies consist of four to sixteen cells enclosed by a persistent, dilated mother-cell wall that eventually disintegrates to release the daughter cells. Each cell typically contains about sixteen bright blue-green cyanelles. The species was found in a boggy pool of unknown location in the South Island (Skuja 1976). It is the only glaucophyte to have been identified in Australia (Day et al. 1995) and is one of just two glaucophytes known from the British Isles (Whitton 2002).

Glaucocystis nostochinearum and several other glaucophytes occupy similar ecological niches (Kies 1992). It seems likely that more species will be discovered in New Zealand, especially since their preferred, somewhat acidic, boggy habitats are widespread.

Diversity by environment

Taxon	Marine	Freshwater	Terrestrial
Glaucophyta	0	1	0

Summary of New Zealand glaucophyte diversity

Taxon	Described species/ subspecies	Known undetermined/ undescribed/ unverified species	Estimated unknown species	Adventive species	Endemic species	Endemic genera
Glaucophyta	1	0	1	0	0	0

Author

Dr Paul A. Broady: School of Biological Sciences, University of Canterbury, Private Bag 4800, Christchurch 8140, New Zealand [paul.broady@canterbury.ac.nz]

References

CASSIE, V. 1984a: Revised checklist of the freshwater algae of New Zealand (excluding diatoms and charophytes). Part II — Chlorophyta, Chromophyta and Pyrrhophyta, Rhaphidophyta and Euglenophyta. *Water and Soil Technical Publication No. 26*: i-xii, 1–134, xiii-lxiv.

DAY, S. A.; WICKHAM, R. P.; ENTWISLE, T. J.; TYLER, P. A. 1995: Bibliographic checklist of non-marine algae in Australia. *Flora of Australia Supplementary Series, No. 4*: i-vii, 1–276.

GRAHAM, L. E.; GRAHAM, J. M.; WILCOX, L. W. 2009: *Algae*. Benjamin Cummings, San Francisco, xvi + 616 p.

INOUYE, I.; OKAMOTO, N. 2005: Changing concepts of a plant: current knowledge on plant diversity and evolution. *Plant Biotechnology 22*: 505–514.

KEELING, P. J. 2004: Diversity and evolutionary history of plastids and their hosts. *American Journal of Botany 91*: 1481–1493.

KIES, L. 1992: Glaucocystophyceae and other protists harbouring procaryotic endocytobionts. Pp. 353–377 *in*: Reisser, W. (ed.), *Algae and Symbioses.* Biopress Ltd, Bristol. xii + 746 p.

KUGRENS, P.; CLAY, B. L.; MEYER, C. J.; LEE, R. E. 1999: Ultrastructure and description of *Cyanophora biloba*, sp. nov., with additional observations on (Glaucophyta). *Journal of Phycology 35*: 844–854.

NOZAKI, H.; ISEKI, M.; HASEGAWA, M.; MISAWA, K.; NAKADA, T.; SASAKI, N.; WATANABE, M. 2007: Phylogeny of primary photosynthetic eukaryotes as deduced from slowly evolving nuclear genes. *Molecular Biology and Evolution 24*: 1592–1595.

SKUJA, H. 1976: Zur Kenntnis der Algen Neuseeländischer Torfmoore. *Nova acta Regiae Societatis Scientiarum Upsaliensis, ser. 5, C(2)*: 1–125.

STILLER, J. W. 2007: Plastid endosymbiosis, genome evolution and the origin of green plants. *Trends in Plant Science 12*: 391–396.

WHITTON, B. A. 2002: Phylum Glaucophyta. P. 613 *in*: John, D. M.; Whitton, B. A.; Brook, A. J. (eds), *The Freshwater Algal Flora of the British Isles.* Cambridge University Press, Cambridge, U.K. xii + 702 p.

Checklist of New Zealand Glaucophyta

KINGDOM PLANTAE
SUBKINGDOM RHODOPLANTAE
Phylum GLAUCOPHYTA
Class GLAUCOPHYCEAE
Order GLAUCOCYSTALES
Glaucocystis nostochinearum Itzigsohn **F**

TWENTY-TWO

Phylum RHODOPHYTA

red algae

WENDY A. NELSON

Red algae are a very diverse and ancient plant lineage. The earliest fossils of putative red algae date from about two billion years ago and are superficially similar to living taxa in the Porphyridiales (Tappan 1976; Saunders & Hommersand 2004). Fossils resembling extant red algae also have been reported from Proterozoic and Neoproterozoic strata (Butterfield et al. 1990; Xiao et al. 1998). The fossil bangialean red alga *Bangiomorpha pubescens*, dated to 1200 million years ago (Ma), is virtually indistinguishable from modern material of *Bangia* (Butterfield et al. 1990; Butterfield 2000). The presence of at least two distinct spore-producing phases in *Bangiomorpha* and the similarity of these to sexual phases in modern *Bangia* provide evidence for the existence of eukaryotic sex in *Bangia*-like organisms by at least this date (Butterfield 2000). Additional fossil evidence of early Bangiales comes from deposits dated circa 500 Ma (Xiao et al. 1998) and 425 Ma (Campbell 1980).

For the majority of fossil red algae, the diagnostic features required for unequivocal placement in families or genera have not been preserved and hence evolutionary interpretation of the material is difficult. There is evidence, however, that the Bangiales and four major lineages of florideophyte red algae were established by 600–550 Ma (Zhang et al. 1998). Unfortunately, few taxa are represented in the fossil record, the commonest of which are members of the Corallinaceae, dating from the Cretaceous (130 Ma) onwards. Fossils that can be identified as belonging to the Peyssonneliaceae appear in the Late Jurassic (160 Ma) (Wray 1977 in Saunders & Hommersand 2004), providing evidence of the early divergence among orders within the florideophyte lineages.

Pterosiphonia pennata (Florideophyceae, Ceramiales).
Roberta D'Archino, NIWA

The antiquity of the red algae has also been inferred from molecular studies (Lim et al. 1986; Ragan et al. 1994; Yoon et al. 2004). Rhodophyta are reported to be more divergent among themselves, when assessed by the divergence of small-subunit ribosomal DNA sequences within the most conservative regions of the molecule, than are fungi, or green algae and green plants together (Ragan et al. 1994). Although comparisons of levels of divergence can be misleading, given that homologous genes in different lineages can evolve at unequal rates, the red algae show high diversity in a range of genes (Ragan et al. 1994; Medlin et al. 1997). Divergences in small-subunit ribosomal gene sequences among strains of cyanidiophytes were found by Gross et al. (2001) to be equivalent to divergences in florideophyes at the ordinal and lineage level. Yoon et al. (2004) estimated the following divergences – red and green lines c. 1500 Ma, Cyanidiales (class Cyanidiophyceae) c. 1370 Ma, bangiophycean lines largely divergent by c. 1200 Ma (by the time Chlorophyta had diverged from charophytes/land plants), florideophytes and bangiophytes divergent prior to 800 Ma (time of the spilt between charophytes and land plants), and major florideophyte groups divergent prior to the first appearance of land plants (c. 460 Ma).

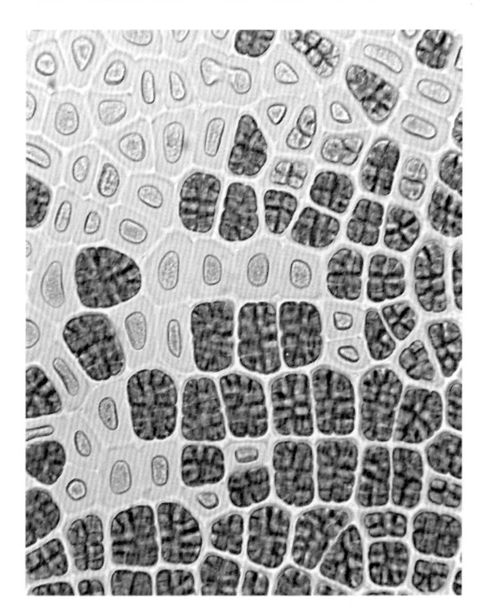

Pyropia cinnamomea, showing darker vegetative cells and lighter reproductive cells (Bangiophyceae, Bangiales).
Wendy Nelson, NIWA

Red algae are highly diverse in their morphological and ultrastructural characters – ranging from single cells to highly complex parenchymatous thalli. Members of the phylum Rhodophyta are characterised by plastids containing the pigments phycoerythrin, phycocyanin, and allophycocyanins, and by the absence of centrioles and cilia (flagella) (Kraft & Woelkerling 1990). Cell structure and the diversity of developmental processes and reproductive strategies in the red algae have been reviewed in Cole and Sheath (1990). The red algae are largely marine in their distribution, and widespread from cold-water and temperate regions through to the tropics.

Red-algal systematics has been undergoing a period of change and instability. Within the florideophyte lines there have been very significant discoveries within the past 20 or so years that have substantially altered the number of orders recognised and the understanding of how these orders relate to one another.

The taxonomic treatment of the red algae by Kylin (1956) was enormously significant and remained a key factor shaping hypotheses about red algae and their phylogeny until the 1980s. Kylin's work was based on female reproductive anatomy and post-fertilisation events, and recognised six orders within the Florideophyceae (as subclass Florideophycidae). This work built on the earlier work of both Schmitz (1892) and Oltmanns (1904–05).

Work on ultrastructure and cytochemistry of pit plugs, as well as the morphology and number of pit-plug cap layers, provided an additional line of evidence with which to examine various phylogenetic hypotheses, and, in particular, the monophyly of certain groups of red algae (Pueschel & Cole 1982; Pueschel 1987, 1989, 1994). Cladistic methods of character analysis were employed by Gabrielson et al. (1985) and Gabrielson and Garbary (1987) to consider the relationships amongst recognised red-algal orders. Molecular sequence data are an increasingly important source of systematic information at all taxonomic levels.

Over the past two decades, many new red-algal orders have been proposed, including Palmariales (Guiry 1978), Batrachospermales and Hildenbrandiales (Pueschel & Cole 1982), Corallinales (Silva & Johansen 1986), Ahnfeltiales (Maggs & Pueschel 1989), Gracilariales (Fredericq & Hommersand 1989), Plocamiales (Saunders & Kraft 1994), Rhodogorgonales (Fredericq & Norris 1995), Halymeniales (Saunders & Kraft 1996), Balbianiales (Sheath & Müller 1999), Balliales (Choi et al. 2000), Colaconematales (Harper & Saunders 2002), Thoreales (Müller et al. 2002), a resurrected and emended Nemastomales (Saunders & Kraft 2002), Pihiellales (Huisman et al. 2003), Acrosymphytales and Sebdeniales (Withall & Saunders 2007), Rhodachylales (West et al. 2008), Rufusiales Zuccarello et al. (2008), Peyssonneliales (Krayesky et al. 2009). Cyanidioschyzonales F. D. Ott (2009), and Sporolithales (Le Gall et al. 2009).

Supraordinal diversity and taxonomy in the red algae have been reviewed recently by Saunders and Hommersand (2004) and Yoon et al. (2006), and

Summary of New Zealand red-algal diversity

Taxon	Described species	Known undescribed species	Estimated unknown species	Adventive species	Endemic species	Endemic genera
Cyanidiophyceae	1	0	0	0	0	0
Compsopogonophyceae	10	0	0	0	5	2
Porphyridiophyceae	1	0	0	0	0	0
Stylonematophyceae	2	0	0	0	0	0
Bangiophyceae	11	36	5	0	7	3
Florideophyceae	447	53	30	15	178	16
Totals	**472**	**89**	**35**	**15**	**190**	**21**

two synoptic reviews of red-algal genera have been published (Schneider & Wynne 2007; Wynne & Schneider 2010). The treatment of Yoon et al. (2006), followed here, identified seven red-algal lineages at the class level, with the class Cyanidiophyceae as the earliest-diverging clade of red algae.

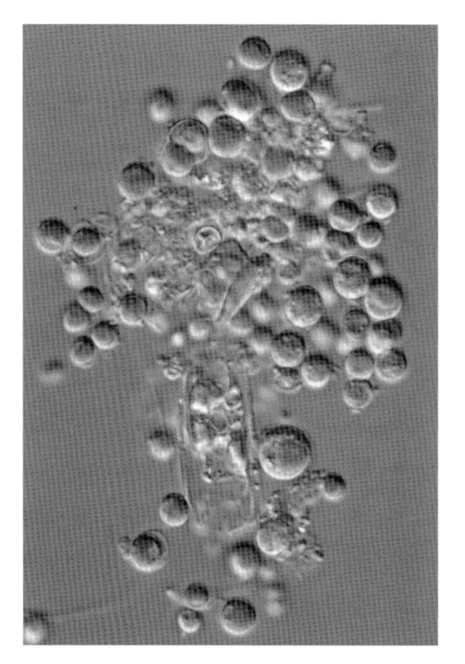

Unicells of *Cyanidium caldarium*, a bluish-green red alga found in some North Island geothermal areas (Cyanidiophyceae, Cyanidiales).

David J. Patterson, per Micro*scope (MBL)

The red-algal classes and orders

Subphylum Cyanidiophytina
Class Cyanidiophyceae

The Cyanidiophyceae are unicellular red algae that inhabit extreme environments. These are the only eukaryotes known to survive in extreme acidophilic/thermophilic environments (pH 0.5–3, with temperatures up to 56° C) and have been reported from volcanoes, hot springs, and acidic soils around the world, including in Java (Indonesia), the Kamchatka Peninsula (Russia), Yellowstone National Park (USA), and Naples (Italy). In New Zealand, *Cyanidium caldarium* was first reported by Kaplan (1956) and subsequent studies have recorded it from a number of North Island thermal sites (Brock & Brock 1970, 1971; Cassie & Cooper 1989).

Morphologically, *Cyanidium* and its allies have thick proteinaceous cell walls, endospores, heterotrophic capacity, and a Golgi-endoplasmic reticulum association (although common amongst other eukaryotes this is rare in the red algae). Cyanidiophytes also have the smallest known genomes of any phototrophic eukaryotes (Muravenko et al. 2001). A number of recent studies have examined cytomorphology, biochemistry, and molecular-sequence data in members of this group (e.g. Albertano et al. 2000; Cozzolino et al. 2000; Gross et al. 2001; Pinto et al. 2003).

Although the order Cyanidiales was first proposed by Christensen (1962), based on their biliprotein composition, the relationships of these algae within the infrakingdom Rhodoplantae have been unclear and engendered considerable speculation (e.g. Seckbach 1987). Yoon et al. (2002a,b, 2004) estimated the time at which this lineage diverged to be about 1370 Ma.

At present two orders are recognised (Cyanidiales and Cyanidioschyzonales). It is likely, however, that there will be further taxonomic revision in the future given the level of genetic diversity amongst isolates revealed by studies to date (Gross et al. 2001; Pinto et al. 2003; Nozaki et al. 2005; Lehr et al. 2007). Although the phylum Cyanidiophyta was established by Doweld (2001) and supported by analyses of Saunders and Hommersand (2004), Yoon et al. (2006) proposed the recognition of two red-algal subphyla – Cyanidiophytina and Rhodophytina – the latter encompassing the rest of the rhodophyte classes.

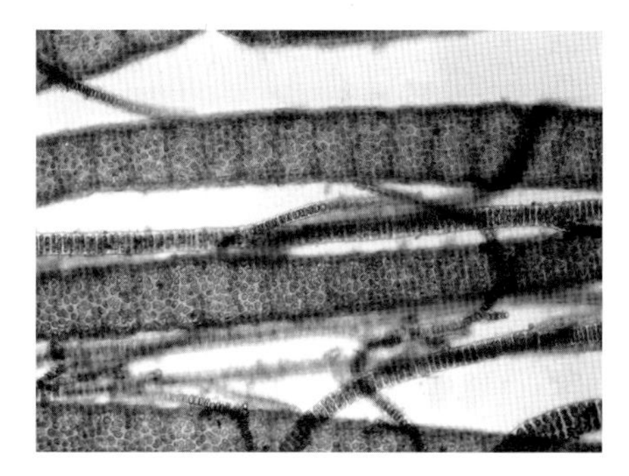

Compsopogon caeruleus, a freshwater species (Compsopogonophyceae, Compsopogonales).

Donna Ainsworth, Taranaki Regional Council

Subphylum Rhodophytina
Class Compsopogonophyceae

Two of the three orders placed in this class are recorded from New Zealand.

Summary of New Zealand red-algal diversity by environment

Taxon	Marine	Terrestrial	Freshwater	Fossil
Cyanidiophyceae	0	0	1	0
Compsopogonophyceae	9	0	1	0
Porphyridiophyceae	1	0	0	0
Stylonematophyceae	2	0	0	0
Bangiophyceae	47	0	0	0
Florideophyceae	482	0	18	0
Totals	**541**	**0**	**20**	**0**

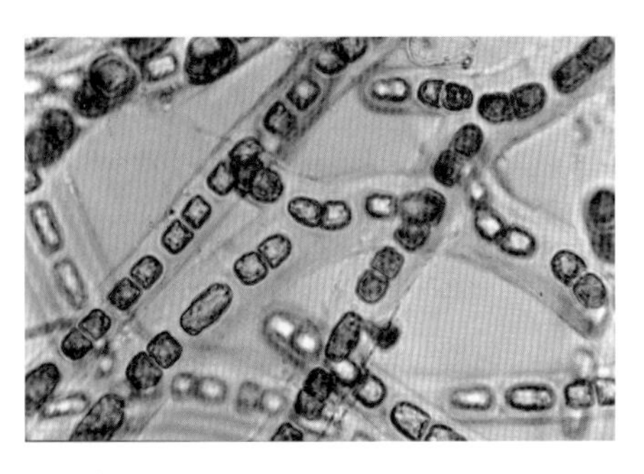

Common in brackish parts of the lower Clutha River, *Chroodactylon ornatum* has blue-green cells embedded in transparent sheaths (Stylonematophyceae, Stylonematales).

Stephen Moore, Landcare Research

Compsopogonales. In New Zealand this order is represented by a single freshwater genus, *Compsopogon*, which is multicellular with complex morphology.

Erythropeltidales. This marine order comprises multicellular taxa forming discs, blades, or filaments. Research in New Zealand has revealed taxa with morphology convergent with members of the Bangiales, and two new genera in the Erythropeltidales (*Pyrophyllon, Chlidophyllon*) have been established for species previously placed in *Porphyra* (Nelson et al. 2003)

Class Porphyridiophyceae

Porphyridiales. The order includes unicellular or 'pseudofilamentous' forms from marine, freshwater, and moist terrestrial habitats. Very little attention has been paid to marine red-algal unicells in New Zealand, although *Porphyridium purpureum* has been recorded (Nelson & Ryan 1988).

Class Rhodellophyceae

No members of this class have been reported to date from New Zealand.

Class Stylonematophyceae

Stylonematales. Two genera are reported from New Zealand (*Chroodactylon* and *Stylonema*), but they are infrequently collected and poorly known locally.

Class Bangiophyceae

Bangiales. The single order recognised in this class is particularly well represented in New Zealand. Members possess multicellular thalli that are either filaments or blades and have a heteromorphic life history, alternating with a filamentous phase (conchocelis). The order is monophyletic and remarkably divergent, with only a single family, Bangiaceae. Internationally it is likely that the number of taxa in this order has been significantly underestimated; the true diversity in this group was obscured by the recognition of only two genera (*Bangia* and *Porphyra*) until relatively recently (Broom et al. 2004; Nelson et al. 2006a). Research in New Zealand has resulted in two endemic genera, *Dione* and *Minerva*, described for filamentous taxa (Nelson et al. 2005). A major revision of the Bangiales (Sutherland et al. in press) has resulted in fifteen genera being recognised, seven filamentous and eight foliose, in including five newly described and two resurrected genera.

Arthrocardia corymbosa, a geniculate (jointed) calcified red alga (Florideophyceae, Corallinales).

Kate Neill, NIWA

Class Florideophyceae

Saunders and Hommersand (2004) recognised four subclasses, three of which are present in New Zealand – Hildenbrandiophycidae, Nemaliophycidae, and Rhodymeniophycidae. There are no New Zealand members of the subclass Ahnfeltiophycidae (orders Ahnfeltiales, Pihiellales). Le Gall and Saunders (2007) established an additional subclass, the Corallinophycidae.

Subclass Hildenbrandiophycidae contains a single order:

Hildenbrandiales. This order was established on the basis of the ontogeny of tetrasporangial conceptacles and pit-plug characters. The placement of the endemic New Zealand genus *Apophlaea* in this order was confirmed by molecular sequence analyses (Saunders & Bailey 1999; Sherwood & Sheath 2003).

Subclass Corallinophycidae:

Corallinales. Members of this order have calcium carbonate incorporated in their cell walls. There are geniculate (jointed) and non-geniculate (non-jointed) members, the latter often referred to as crustose corallines. Genicula are non-homologous structures that evolved independently within at least three tribes and thus corallines should not be separated on the basis of the presence or absence of genicula (Bailey & Chapman 1998). Recent research on this order in New Zealand has focused primarily on non-geniculate taxa (Woelkerling & Nelson 2004; Harvey et al. 2005; Broom et al. 2008; Farr et al. 2009).

Sporolithales. This order was established by Le Gall and Saunders in Le Gall et al. (2009). It is characterised by possession of tetrasporangia in calcified sporangial compartments and by cruciately divided tetrasporangia.

Subclass Nemaliophycidae:

Acrochaetiales. This order is poorly known in New Zealand. Once regarded as 'primitive', the position of this order has been contentious. Recent analyses of the nuclear small-subunit ribosomal RNA (nSSU) gene suggest that it is a highly derived order (Saunders & Kraft 1997; Harper & Saunders 1998).

Balbianiales. This order was proposed by Sheath and Müller (1999) for the freshwater genera *Balbiania* and *Rhododraparnaldia*, which share a range of morphological and ultrastructural features. Molecular data placed the Balbianiales in a clade containing the Acrochaetiales, Nemaliales, and the Palmariales.

Balliales. Choi et al. (2000) based this order on *Ballia callitricha*, a species previously placed in the Ceramiales. Anatomical and molecular characters placed this species closer to members of the Acrochaetiales, Nemaliales, and Palmariales. There is a single monogeneric family, with three species present in New Zealand.

Batrachospermales. Established by Pueschel and Cole (1982), this order is found only in fresh water. Members have a heterotrichous life-history phase, lack tetraspore production, and have a two-layered pit plug (Entwisle & Necchi 1992; Vis et al. 1998). Two of the four families are represented in New Zealand including the monotypic Psilosiphonaceae, previously thought to be endemic to Australia.

Colaconematales. Established by Harper and Saunders (2002), this order of marine species currently contains a single monogeneric family. The order was segregated from the Achrochaetiales as members were shown to have closer affinities to Nemaliales than to the Achrochaetiales or Palmariales.

Nemaliales. At one stage this was a very large order, but most families have been moved to other orders leaving only the Galaxauraceae, Liagoraceae, and the recently segregated Scinaiaceae (Huisman et al. 2004). The order is well represented in the Kermadec Islands flora.

Palmariales. Established by Guiry (1978), the Palmariales is distinguished by both unique tetrasporogenesis and life-history characters. The relationships between this order and members of the Acrochaetiales require further attention (Harper & Saunders 2002).

Subclass Rhodymeniophycidae currently contains the following twelve orders, all of which are represented in New Zealand:

Acrosymphytales. Established by Withall & Saunders (2007) this order has two New Zealand representatives, *Acrosymphyton* and *Schimmelmannia*.

Bonnemaisoniales. Life-history studies led Feldmann and Feldmann (1942) to propose this order, which has also been supported by ultrastructural studies (Pueschel & Cole 1982).

Ceramiales. This order contains almost half the genera and over a third of all red-algal species (Kraft & Woelkerling 1990; Saunders & Kraft 1997). The long-held notion that the Ceramiales is the most advanced of all the red-algal orders is no longer supported. Recent work places the Ceramiales within 'lineage 4' of the red algae (Saunders & Hommersand 2004). Choi et al. (2002), combining anatomical and molecular sequence data, provided evidence that the Dasyaceae and Delesseriaceae are polyphyletic. Recently nine families have been recognised in the order, seven of which are present in New Zealand (Choi et al. 2008).

Gelidiales. This order was initially proposed by Kylin (1923) although later workers returned it to the Nemaliales (e.g. Dixon 1961). The ultrastructural work of Pueschel and Cole (1982), anatomical studies (Hommersand & Fredericq 1988), and molecular studies (Ragan et al. 1994; Freshwater et al. 1994) all support recognition of this order. Historically, circumscription of genera within this group

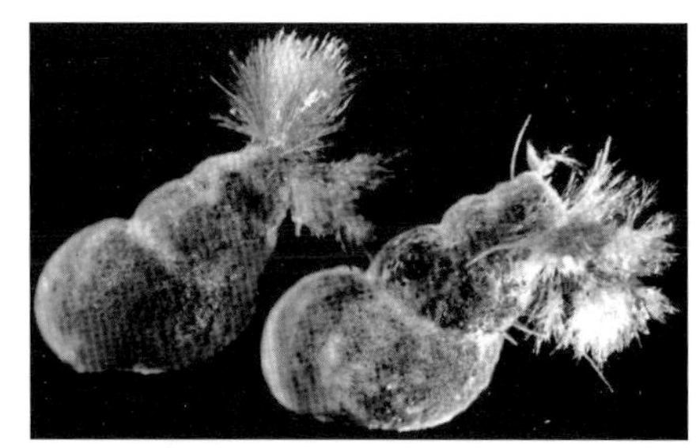

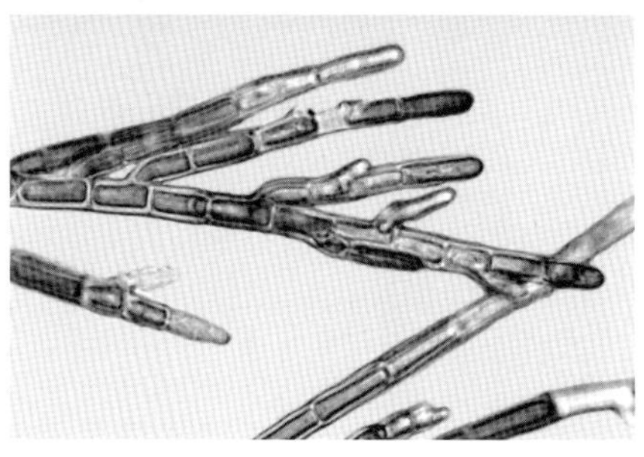

Fresh/brackish-water snails *Potamopyrgus antipodum* bearing filamentous tufts of *Audouinella hermannii*, enlarged in the lower photo (Florideophyceae, Acrochaetiales).

Stephen Moore, Landcare Research

Caloglossa vieillardii, with one blade showing reproductive structures (Florideophyceae, Ceramiales).

Roberta D'Archino, NIWA

Grateloupia stipitata (Florideophyceae, Halymeniales).
Roberta D'Archino, NIWA

has been controversial and although the need to revise generic concepts has been reinforced by sequencing studies (e.g. Freshwater et al. 1995; Millar & Freshwater 2005; Nelson et al. 2006b), there is also clear support for the separation of *Gelidium* and *Pterocladia* and the distinction between *Pterocladia* and *Pterocladiella* (Santelices & Hommersand 1997; Patwary et al. 1998). The diversity and phylogenetic relationships of New Zealand members of the Gelidiales have recently been examined (Nelson et al. 2006b).

Gigartinales. This became a 'mega-order' of more than 40 families when Kraft and Robbins (1985) merged the Cryptonemiales into the Gigartinales as the first step in evaluating the complex of families. Several new orders have since been established, removing some of the families, and the process of examination of familial relationships will result in further changes (Saunders & Kraft 1997; Saunders & Hommersand 2004).

Gracilariales. Cytological, anatomical, and molecular data all support the recognition of this order, established by Fredericq and Hommersand (1989) (Ragan et al. 1994; Freshwater et al. 1994; Gurgel & Fredericq 2004). The Gracilariales is well represented in New Zealand with two families, four genera, and 11 species.

Halymeniales. Saunders and Kraft (1996) placed two families in this order, the Halymeniaceae and the Sebdeniaceae, previously attributed to the Cryptonemiales and Gigartinales, respectively. Saunders and Kraft (2002) moved *Tsengia* from the Nemastomataceae and established the family Tsengiaceae in the Halymeniales. Most recently the Sebdeniaceae has been moved to its own order (Withall & Saunders 2007). The New Zealand members of this order require further investigation. Russell et al. (2009) examined the genera *Pachymenia* and *Aeodes*.

Nemastomatales. Saunders et al. (2004) considered this to be a strongly supported group consisting of two families, Nemastomataceae and Schizymeniaceae, both of which are present in New Zealand but remain poorly known.

Peyssonneliales. Only recently established (Krayesky et al. 2009), this order has a single family. The species present in New Zealand are poorly known.

Plocamiales. Established by Saunders and Kraft (1994), this order contains three families, Plocamiaceae, Sarcodiaceae, and the monotypic Pseudoanemoniaceae. The New Zealand species of *Sarcodia* have recently been investigated (Rodríguez-Prieto et al. 2011), and further work is required to clarify the relationships among these families.

Rhodymeniales. Although Kraft and Robbins (1985) questioned whether this order should be maintained, molecular data provide support for its continued recognition, with its affinities lying most closely with the Halymeniales and Sebdeniales (Saunders & Kraft 1996; Withall & Saunders 2007; Le Gall et al. 2008). Saunders et al. (1999) examined families within the order using both molecular data and reproductive features. They established the family Faucheaceae, emended the defining characters of the other families in the order, and reassigned genera. Le Gall et al. (2008) further revised the order and etsbalished an additional two families, the Fryeellaceae and Hymenocladiaceae. This order is very well represented in New Zealand with 18 genera and 34 species, 15 of which are endemic. Dalen et al. (2009) reported *Cephalocystis* in New Zealand, a genus previously considered endemic to Australia.

Sebdeniales. Recently erected (Withall & Saunders 2007), this order is represented in New Zealand by a single, northern deep-water species that has been rarely collected, *Sebdenia lindaueri*.

Plocamium cartilagineum (Florideophyceae, Plocamiales).
Kate Neill, NIWA

Diversity of New Zealand Rhodophyta

Historical overview of discoveries/research

Prior to Adams' (1994) illustrated treatment of the New Zealand seaweed flora, only Harvey and Hooker (1845) and Harvey (1855) had produced complete, illustrated accounts of New Zealand seaweeds (subantarctic islands and mainland New Zealand). Seaweeds were collected for scientific study on Cook's first voyage to New Zealand in 1769. Thirteen species of seaweed from New Zealand were described by Dawson Turner between 1808 and 1819 (Turner 1808–1819). Early voyages by French explorers resulted in a number of collections of seaweeds described by Bory (1826–29) and Montagne (1842, 1845). During the middle to latter half of the 19th century, Harvey at Trinity College in Dublin, J. D. Hooker in London, and J. G. Agardh in Lund, Sweden, played very significant roles in describing the New Zealand seaweed flora. In addition, research in other parts of Europe contributed to the knowledge of the New Zealand flora and endemic red algae (e.g. Heydrich 1893a,b).

'Gigartina' atropurpurea (Florideophyceae, Gigartinales).
Roberta D'Archino, NIWA

Amateur phycologists played a very important role in the documentation and exploration of the New Zealand seaweed flora in the early 20th century. Laing (1900, 1902, 1926, 1930) prepared a series of reference lists of species and worked on algae of the New Zealand subantarctic islands (Laing 1909). In addition, he published work on *Delesseria*, various Ceramiaceae, a revision of the order Bangiales, and a revision of the genus *Gigartina* with Gourlay (Laing 1897, 1905, 1928; Laing & Gourlay 1929, 1931). Laing sent specimens to, and sought advice from, both J. G. Agardh in Lund and W. A. Setchell at the University of California.

The extensive holdings of New Zealand algae in the Agardh herbarium at Lund were used by Kylin in his research on red algae, and he produced several works specifically on New Zealand taxa (e.g. Kylin 1929, 1933) as well as including New Zealand species in some of his broader accounts of particular orders (e.g., Kylin 1931, 1932).

The contributions of Victor Lindauer were particularly significant during the mid-20th century (Cassie 1971; Cassie Cooper 1995). The visit to the Bay of Islands by Professor Josephine Tilden in 1935 initiated Lindauer's long interest in seaweeds. Encouraged by Professor W. A. Setchell at the University of California at Berkeley, Lindauer began making up sets of exsiccatae. The Algae Nova-Zelandicae Exsiccatae distributed by Lindauer between 1939 and 1953 served as a reference set of New Zealand algae in both national and international institutions, with 38 of the 350 sheets having some level of type status (Nelson & Phillips 1996). In addition, Lindauer published papers on the red algae *Champia* and *Lenormandia* (Lindauer 1938; Lindauer & Setchell 1946). Although Setchell did not publish extensively on New Zealand red algae, he visited New Zealand on more than one occasion and corresponded with Laing and Lindauer as well as with Lucy Moore and Lucy Cranwell, and his advice and assistance were highly appreciated (and acknowledged by the naming of *Gloiodermatopsis setchellii* Lindauer).

The outbreak of World War II and the requirement to find supplies of agarophytes fueled the work of Lucy Moore working at Botany Division, DSIR. She made extensive collections around the country and, with the assistance of Nancy Adams and Ruth Mason, built up a collection of marine algae at the Lincoln herbarium (CHR) (Thiers 2011). She focused particularly on the agar-bearing genera *Pterocladia* and *Gelidium* (Moore 1944–1946) while being involved in a broader examination of seaweeds and their potential economic uses (Moore 1941, 1966). After attending the Pacific Science Congress in New Zealand in 1949, Professor G. F. Papenfuss, from the University of California at Berkeley, made an extended visit. During the summer he collected in the company of many local seaweed enthusiasts and these collections yielded a rich harvest of taxonomic studies by his doctoral students at Berkeley (e.g. Scagel 1953; Wagner 1954; Norris 1957; Fan 1961; Hommersand 1963; Sparling 1957; Searles 1968; Chiang 1970). The Swedish phycologist Tore Levring visited New Zealand in 1948 for two months (March–April) and later published two papers in which he described 16 new species of red algae from the New Zealand region (Levring 1949, 1955). The locality of the type specimens of many of these taxa remained unknown until recently (Nelson & Phillips 2001).

At the University of Auckland, Professor V. J. Chapman and coworkers published sections of the red-algal flora in four parts of the Marine Algae of New Zealand series (Chapman 1969; Chapman & Dromgoole 1970; Chapman & Parkinson 1974; Chapman 1979). A fifth volume was intended to cover the Ceramiales but the series was not completed. As noted by Parsons (1985a,b), many of the genera covered in Chapman's treatments require extensive revision as the volumes were essentially compilations from widely scattered literature and contained little original research.

There are very few monographic treatments of New Zealand red algae, although work on the following genera has been published – *Medeiothamnion, Ptilothamnion* (Gordon-Mills 1977), *Plocamium* (South & Adams 1979; Wynne 2002), *Synarthrophyton* (Townsend 1979, Woelkerling & Foster 1989), *Brongniartella* (Parsons 1980), *Schmitzia* (Hawkes 1982a), *Acrosymphyton* (Hawkes 1982b), *Hummbrella* (Hawkes 1983a), *Apophlaea* (Hawkes 1983b), Bonnemaisoniales (Bonin & Hawkes 1987, 1988a,b), *Gracilaria* (Nelson 1987; Bird et al. 1990), *Polysiphonia* (Adams 1991), *Nesophila* (Nelson & Adams 1993), Bangiales (Nelson 1993; Nelson et al. 2001; Broom et al. 2002; Nelson et al. 2006a; Nelson & Broom 2010; Sutherland et al. in press), *Curdiea* (Nelson & Knight 1997; Nelson et al. 1999b), *Pleurostichidium* (Phillips 2000); *Lenormandia/Adamsiella* (Phillips 2002a,b), genera of Delesseriaceae (Millar & Nelson 2002; Lin et al. 2007, 2010, in press); *Pyrophyllon* and *Chlidophyllon* (Nelson et al. 2003), *Euptilota/Aristoptilon* (Hommersand et al. 2004), Gelidiales (Nelson et al. 2006b), *Pachymenia* and *Aeodes* (Russell et al. 2009), *Glaphyrosiphon* (Hommersand et al. 2010), members of the Kallymeniaceae including *Psaromenia* and *Ectophora* (D'Archino et al. 2010, 2011), *Sarcodia* (Rodriguez-Prieto et al. 2011), and Gigartinaceae (Nelson & Broom 2008; Nelson et al. 2011). Molecular data are contributing significantly to understanding the relationships of the New Zealand flora and the diversity that remains to be described. Systematic studies carried out in Australia have also contributed to knowledge about New Zealand red algae (e.g. Huisman & Kraft 1982; King & Puttock 1989; Guiry & Womersley 1993; Johansen & Womersley 1994; Womersley 1994, 1996, 1998, 2003; Entwisle & Foard 1997a,b; 1999a,b; West et al. 2001; Zuccarello & West 2006).

Major repositories of New Zealand specimens

The major collections of New Zealand Rhodophyta are housed in the following herbaria: Te Papa (WELT), Auckland Museum (AK and AKU), Manaaki Whenua Landcare (CHR) (Thiers 2011). The collection data from the specimens housed in the herbarium at Te Papa are being entered onto an electronic database. There are no recently published keys to aid identification of the red algae although there are keys for a few genera presented in Adams (1991, 1994).

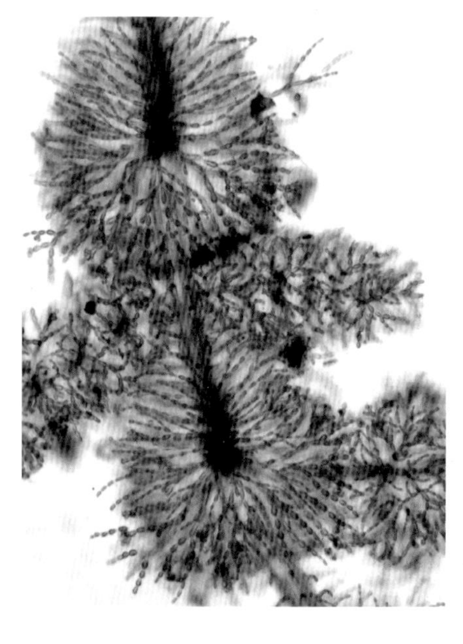

Batrachospermum sp., a native freshwater species (Florideophyceae, Batrachospermales).
Stephen Moore, Landcare Research

Current known natural diversity

The end-chapter checklist is based on Adams (1994), specimens housed in WELT, and a series of regional seaweed floras that provide vouchered lists of species and summarise available information for selected parts of the New Zealand region (Adams 1972; Adams et al. 1974; South & Adams 1976; Nelson & Adams 1984; Parsons & Fenwick 1984; Hay et al. 1985; Adams & Nelson 1985; Nelson & Adams 1987; Nelson et al. 1991; Nelson et al. 1992; Neale & Nelson 1998; Nelson et al. 2002). These have identified taxa requiring study and have increased understanding of the distribution of species within New Zealand. The records of the freshwater species in the genus *Batrachospermum* are from Entwisle and Foard (1997a,b; 1999a,b). Abbreviations for authorities follow the International Plant Names Index list of Authors (www.ipni.org).

This treatment lists 561 species of marine and freshwater red algae, placed in six classes, four subclasses, 28 orders, 56 families, and 223 genera. There are 21 endemic genera and 190 endemic species. At least 36 undescribed taxa are signalled in the Bangiales, distinguished on the basis of unique 18S rDNA sequences (e.g. Nelson 1993; Nelson & Broom 1997; Broom et al. 1999, 2002, 2004; Nelson et al. 2001, 2006a; Sutherland et al. in press).

Studies on the Gigartinaceae over recent years have resulted in emended generic concepts, with many species previously placed in *Gigartina* being moved to the genera *Chondracanthus, Sarcothalia, Mazzaella,* and *Chondrus* (Hommersand et al. 1993, 1994, 1999; Nelson et al. 2011). These studies have included some representatives of New Zealand taxa. However, the results from this work do not yet allow the placement of all New Zealand members of the Gigartinaceae into clearly defined genera. There is a need to resolve both generic and species concepts for New Zealand members of this family given the ubiquity of the family throughout New Zealand, as well as the commercial interest in these carrageenophytes.

Mid-tidal *Apophlaea sinclairii*, an endemic genus and species (Florideophyceae, Hildenbrandiales).
Tracy Farr, RSNZ

Special features of the New Zealand Rhodophyta

Ecology and distribution

Red algae are found from the high to middle intertidal zone (e.g. species of *Porphyra, Bangia, Apophlaea*) to deep coastal waters where there is sufficient penetration of sunlight to allow photosynthesis to occur. The deepest seaweeds known from the New Zealand region are non-geniculate coralline crusts dredged from 200 m in the clear warm waters around the Kermadec Islands.

Although red algae do not grow as large as brown seaweeds, they display a very wide range of growth forms from single-celled *Porphyridium,* through delicate filaments (members of the Acrochaetiales, Ceramiales), to very thick leathery thalli (e.g. *Pachymenia crassa*), or wiry tough clumps (e.g. *Melanthalia abscissa, Apophlaea lyallii*). Members of the family Delesseriaceae include some of the most delicate thalli found in the red algae. In contrast, some non-geniculate corallines form heavily calcified unattached growths resembling small corals. Known as rhodoliths ('red stones') or maerl, they can form beds providing habitat for other marine life. Rhodolith beds are found throughout the world's oceans and are well developed at several locations around the New Zealand coast.

Significant features of distribution/biogeography

The antiquity of the red algae and their diversity of evolutionary histories present a very complex setting in which to develop hypotheses about biogeographic relationships. Parsons (1985a) summarised work on the biogeography of the New Zealand macroalgal flora, presenting five major groupings – cosmopolitan, circumpolar, Australasian, South American, and Japanese. Hommersand (1986, 2007) hypothesised about the timing of distributional events that link the floras of the Western Cape region of South Africa with Tasmania, southern Australia, and New Zealand. Using evidence from *rbc*L (ribulose-1,5 biphosphate carboxylase) sequence analyses and morphological studies, Hommersand et al. (1999) considered that there is support for the origin of the Gigartinaceae in the Pacific Ocean along the eastern and southern edge of Gondwana, with a later distribution along the Pacific coast of South and North America.

The poor state of knowledge and the critical lack of monographic studies of most red-algal taxa in New Zealand place a severe constraint on biogeographic analyses of the flora. Entwisle and Huisman (1998) observed that phylogenetic hypotheses are an integral part of good systematics and cautioned that phycologists should be extremely circumspect in proposing biogeographic hypotheses. Referring to the Australian flora they stated 'it would be premature to encourage biogeographic studies except for those few groups where we can infer phylogenies.'

Sporolithon durum, a crustose coralline in the form of a rhodolith (Florideophyceae, Sporolithales).
Tracy Farr, RSNZ

Special features

The diversity of life histories and variations in reproductive biology are greater in the Rhodophyta than in any other algal phylum (Hawkes 1990). There can

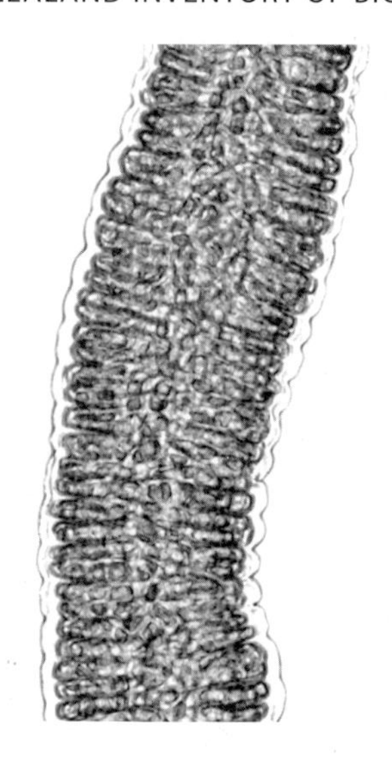

Part of a filament of the endemic genus and species *Dione arcuata* (Bangiophyceae, Bangiales).
Tracy Farr, RSNZ

be both morphological and karyological alternation of phases, and species may be monoecious or dioecious. There are a number of asexual or accessory pathways of development (e.g. asexual spores or propagules, vegetative propagation, apomeiosis, apogamy, parthenogenesis), and these may be facultative or obligate in some or all individuals in a population or in some or all populations of a species (Hawkes 1990). The number of asexual and sexual modes available to 'simple' red algae such as members of the Bangiales (Nelson et al. 1999a) is remarkable. In some Bangiales, the blade phase is known to reproduce via archeospores, endospores, zygotospores, agamospores, and neutral spores, and the conchocelis phase has been reported to reproduce via archeospores, conchospores, neutral conchospores, and protothalli, although not all reproductive modes are reported for all species.

In the majority of members of the Florideophyceae, zygote amplification occurs after fertilisation, achieved by the development of the carposporophyte, a diploid mitosporangial generation. Searles (1980) hypothesised that the lack of motile sperm and the inherent inefficiency of the fertilisation system in red algae favoured the evolution of a triphasic life history. Hommersand and Fredericq (1990) considered the evolutionary trends in the nutrition of the carposporophyte and concluded that 'similar structures and nutritional strategies have evolved at different times in different groups of red algae'. As Hawkes (1990) noted, most life-history theory is based on animals and seed plants and there is a need to examine the complexities of red-algal reproductive systems in relation to the body of theory.

Alien species

Of the 37 non-indigenous species of seaweed currently recognised in New Zealand, 15 are red algae (Nelson 1999; McIvor et al. 2001; D'Archino et al. 2007), including *Grateloupia turuturu*, a species originating in Asia that has been highly invasive in other countries. Ten of the species belong to the order Ceramiales, including seven species of *Neosiphonia* and *Polysiphonia*. In some cases they are locally abundant but restricted in their distribution (e.g. *Chondria harveyana*). One species appears to form a substantial biomass in the Auckland region (Orakei Basin and Manukau Harbour), but it is not clear how long the populations have been present in these habitats or how widespread the species has become. Investigations of polysaccharide chemistry and *rbc*L (ribulose-1,5-bisphosphate carboxylase oxygenase) gene sequences place this species within the Solieriaceae. In addition, there is a species of *Gracilaria* in Manukau, Waitemata, and Kaipara harbours that is very similar in appearance to *G. chilensis* but which differs from this species in its chemistry and RFLP (restriction fragment-length polymorphism) patterns (Candia et al. 1999; Wilcox et al. 2001).

Commercial potential and use

Hurd et al. (2004) reviewed the human use of algae in New Zealand. Red algae are the only source of the complex classes of polysaccharides known as carrageenans and agars, which have wide use both in food products and in various industrial and pharmaceutical applications (Falshaw et al. 1997). These molecules are large and complex and cannot be synthesised artificially, and there is worldwide interest in both harvesting of wild stocks and aquaculture of algae bearing these compounds. In New Zealand, the harvest of wild stocks of *Pterocladia* (primarily *P. lucida* and also *Pterocladiella capillacea*) has occurred for more than 60 years, particularly near Ahipara, in the Bay of Islands, Bay of Plenty/East Cape, and the Wairarapa. McCormack (1990) examined methods for stock assessment of wild populations.

Pterocladiella capillacea (Florideophyceae, Gelidiales).
Roberta D'Archino, NIWA

New Zealand has one of the most diverse floras of carrageenan-containing algae in the world. Research on the polysaccharide chemistry of these species has been under way since the work of Pickmere et al. (1973, 1975) and Parsons et al. (1977), in which differences in the isomers of carrageenan present in the gametophyte and sporophyte phases as well as interspecific differences were examined. New methods have been developed to study red-algal polysaccharides (Stevenson & Furneaux 1991) and a number of genera have been examined, including *Gracilaria* (Miller & Furneaux 1987a,b), *Gelidium* (Nelson et al. 1994), *Champia* (Miller et al. 1996), *Pachymenia* (Miller et al. 1997), and *Curdiea* (Falshaw et al. 1998). The biosynthesis of agar in *Gracilaria* has been examined (Hemmingson et al. 1996a, b; Hemmingson & Furneaux 1997).

The agarophyte genus *Gracilaria* has received more research attention than any other red-algal genus in the New Zealand flora, with studies on its ecology and potential for aquaculture as well as systematics and reproduction (Nelson 1987, 1989; Nelson & Ryan 1991; Ryan & Nelson 1991; Bird et al. 1990; Pickering et al. 1990, 1993; Intasuwan et al. 1993; Candia et al. 1999). Knowledge about most other potential agarophytes or carrageenophytes is very scant, with little information available about phenology, reproduction, or ecology (Nelson & Knight 1997; Nelson et al. 1999b).

Maori place high cultural value on karengo (*Porphyra* species). Karengo is listed as a taonga (cultural treasure) in the Ngai Tahu Deed of Settlement. There is strong interest in understanding the biology of New Zealand species both to determine sustainable harvest of wild stocks and also with regard to the potential for enhancing wild populations or aquaculture. Although some field work in the Kaikoura region has examined the impact of harvest methods on yield and on regeneration after harvest (Nelson & Conroy 1989), it is clear that the diversity of species present in New Zealand means that a great deal more work on the field ecology of these species is required in order to understand this resource.

Porphyra sp., a species of edible karengo (Bangiophyceae, Bangiales).
Wendy Nelson, NIWA

Gaps in knowledge and scope for future research

There is a critical need for monographic studies of genera and families of New Zealand red algae. It is difficult to select the most problematic groups – there are serious shortfalls in virtually all families. The non-geniculate coralline algae have been acknowledged as poorly known in New Zealand and recent attention has been directed both to coralline taxonomy (Woelkerling & Nelson 2004) and to the identification of non-geniculate species from northern and central New Zealand (Harvey et al. 2005; Farr et al. 2009). Non-geniculate corallines are present throughout low intertidal and subtidal shores and have been clearly implicated in the settlement and development of invertebrates such as paua (*Haliotis*) (Nelson 2009). Given their apparent ecological significance and ubiquity it is important that we have a better understanding of their distribution and of regional and ecological variability in the species diversity of this group.

The New Zealand members of the Gigartinaceae are very diverse and their systematics is very poorly understood at present. This is at a time when there are world shortages of particular isomers of carrageenan and strong interest in the potential for aquaculture of these seaweeds. At present the lack of systematic, biological, and ecological information is a serious bottleneck in the development of the economic potential of these algae. Research on the Bangiales in New Zealand has uncovered unexpected diversity and further work is required to describe the many taxa that have been revealed through biodiversity screening using molecular sequencing (e.g., Broom et al. 2004; Nelson et al. 2006a; Sutherland et al. in press). Given the cultural significance of karengo and its use internationally as a food crop, there is a need to gain a fuller understanding of these species.

As so few taxa have received detailed attention, it is not clear how the current estimates of species numbers reflect the actual biodiversity of New Zealand red algae.

Acknowledgements

I would like to express my gratitude to Jenn Dalen, Roberta D'Archino, Tim Entwistle, John Huisman, Max Hommersand, Louise Phillips, and Joe Zuccarello for providing information; to Jenn Dalen, Roberta D'Archino, and the late Nancy Adams and Murray Parsons, for reading the text and lists and providing helpful comments; and to Kate Neill, Roberta D'Archino, Tracy Farr, Stephen Moore, and Donna Ainsworth for providing images.

Author

Dr Wendy A. Nelson National Institute for Water & Atmospheric Research, Private Bag 14-901, Kilbirnie, Wellington, New Zealand [w.nelson@niwa.co.nz]

References

ADAMS, N. M. 1972: The marine algae of the Wellington area. *Records of the Dominion Museum 8*: 43–98.

ADAMS, N. M. 1991: The New Zealand species of *Polysiphonia* Greville (Rhodophyta). *New Zealand Journal of Botany 29*: 411–427.

ADAMS, N. M. 1994: *Seaweeds of New Zealand.* Canterbury University Press, Christchurch. 360 p.

ADAMS, N. M.; CONWAY, E.; NORRIS, R. E. 1974: The marine algae of Stewart Island. *Records of the Dominion Museum 8*: 185–245.

ADAMS, N. M.; NELSON, W. A. 1985: The marine algae of the Three Kings Islands. *National Museum of New Zealand Miscellaneous Series 13*: 1–29.

ALBERTANO, P.; CINIGLIA, C.; PINTO, G.; POLLIO, A. 2000: The taxonomic position of *Cyanidium*, *Cyanidioschyzon* and *Galdieria*: an update. *Hydrobiologia 433*: 137–143.

BAILEY, J. C.; CHAPMAN, R. L. 1998: A phylogenetic study of the Corallinales (Rhodophyta) based on nuclear small-subunit rRNA gene sequences. *Journal of Phycology 34*: 692–705.

BIRD, C. J.; NELSON, W. A.; RICE, E. L.; RYAN, K. G.; VILLEMUR, R. 1990: A critical comparison of *Gracilaria chilensis* and *G. sordida* (Rhodophyta, Gracilariales). *Journal of Applied Phycology 2*: 375–382.

BONIN, D. R.; HAWKES, M. W. 1987: Systematics and life histories of New Zealand Bonnemaisoniaceae (Bonnemaisoniales, Rhodophyta): 1. The genus *Asparagopsis*. *New Zealand Journal of Botany 25*: 577–590.

BONIN, D. R.; HAWKES, M. W. 1988a: Systematics and life histories of New Zealand Bonnemaisoniaceae (Bonnemaisoniales, Rhodophyta): 2. The genus *Delisea*. *New Zealand Journal of Botany 26*: 619–632.

BONIN, D. R.; HAWKES, M. W. 1988b: Systematics and life histories of New Zealand Bonnemaisoniaceae (Bonnemaisoniales, Rhodophyta): 3. The genus *Ptilonia*. *New Zealand Journal of Botany 26*: 633–634.

BORY de SAINT-VINCENT, J. B. 1826–1829: Botanique, Cryptogamie. *In*: Duperrey, L. I., *Voyage autour de Monde, exécuté par ordre du Roi, sur la corvette de sa Majesté, la Coquille pendant les années 1822, 1823, 1824, et 1825.* Paris. 300 p.

BROCK, T. D.; BROCK, M. L. 1970: The algae of Waimangu cauldron (New Zealand): distribution in relation to pH. *Journal of Phycology 6*: 371–375.

BROCK, T. D.; BROCK, M. L. 1971: Microbiological studies of thermal habitats of the central volcanic region, North Island, New Zealand. *New Zealand Journal of Marine and Freshwater Research 5*: 233–258.

BROOM, J. E.; JONES, W. A.; HILL, D. F.; KNIGHT, G. A.; NELSON, W. A. 1999: Species recognition in New Zealand *Porphyra* using 18s rDNA sequencing. *Journal of Applied Phycology 11*: 421–428.

BROOM, J. E.; NELSON, W. A.; YARISH, C.; JONES, W. A.; AGUILAR ROSAS, R.; AGUILAR ROSAS, L. E. 2002: A reassessment of the taxonomic status of *Porphyra suborbiculata* Kjellm., *Porphyra carolinensis* Coll et J.Cox and *Porphyra lilliputiana* W. A.Nelson, G. A.Knight et M. W.Hawkes (Bangiales, Rhodophyta) based on molecular and morphological data. *European Journal of Phycology 37*: 227–235.

BROOM, J. E. S.; FARR, T. J.; NELSON, W. A. 2004: Phylogeny of the *Bangia* flora of New Zealand suggests a southern origin for *Porphyra* and *Bangia* (Bangiales, Rhodophyta). *Molecular Phylogenetics and Evolution 31*: 1197–1207.

BROOM, J. E. S.; HART, D. R.; FARR, T. J.; NELSON, W. A.; NEILL, K. F.; HARVEY, A. H.; WOELKERLING, W. J. 2008. Utility of *psb*A and nSSU for phylogenetic reconstruction in the Corallinales based on New Zealand taxa. *Molecular Phylogenetics & Evolution 46*: 958–973.

BUTTERFIELD, N. I. 2000: *Bangiomorpha pubescens* n.gen., n.sp.: implications for the evolution of sex, multicellularity, and the Mesoproterozoic/Neoproterozoic radiation of eukaryotes. *Paleobiology 26*: 386–404.

BUTTERFIELD, N. J.; KNOLL, A. H.; SWETT, K. 1990: A bangiophyte red alga from the Proterozoic of Arctic Canada. *Science 250*: 104–107.

CAMPBELL, S. E. 1980: *Palaeoconchocelis starmachii*, a carbonate boring microfossil from the Upper Silurian of Poland (425 million years old): implications for the evolution of the Bangiaceae (Rhodophyta). *Phycologia 19*: 25–36.

CANDIA, A.; GONZALEZ, M. A.; MONTOYA, R.; GOMEZ, P.; NELSON, W. 1999: A comparison of ITS RFLP patterns for *Gracilaria* (Rhodophyceae, Gracilariales) populations from Chile and New Zealand and an examination of interfertility of Chilean morphotypes. *Journal of Applied Phycology 11*: 185–193.

CASSIE, V. 1971: Contributions of Victor Lindauer (1888–1964) to New Zealand Phycology. *Journal of the Royal Society of New Zealand 1*: 89–98.

CASSIE COOPER, V. 1995: Victor Wilhelm Lindauer (1888–1964): His life and works. *Tuhinga, Records of the Museum of New Zealand Te Papa Tongarewa 1*: 1–14.

CASSIE, V.; COOPER, R. C. 1989: Algae of New Zealand thermal areas. *Bibliotheca Phycologica 18*: 1–159.

CHAPMAN, V. J. 1969: *The Marine Algae of New Zealand. Part III: Rhodophyceae. Issue 1: Bangiophycidae and Florideophycidae (Nemalionales, Bonnemaisoniales, Gelidiales).* Cramer, Lehre. Pp. 1–113, pls 1–38.

CHAPMAN, V. J. 1979: *The Marine Algae of New Zealand. Part III: Rhodophyceae. Issue 4: Gigartinales.* Cramer, Vaduz. Pp. 279–510, pls 95–183.

CHAPMAN, V. J.; DROMGOOLE, F. I. 1970: *The Marine Algae of New Zealand. Part III: Rhodophyceae. Issue 2: Florideophycidae: Rhodymeniales.* Cramer, Lehre. Pp. 114–154, pls 39–50.

CHAPMAN, V. J.; PARKINSON, P. G. 1974: *The Marine Algae of New Zealand. Part III: Rhodophyceae. Issue 3: Cryptonemiales.* Cramer, Lehre. Pp. 155–278, pls 51–94.

CHIANG, Y.-M. 1970: Morphological studies of red algae of the family Cryptonemiaceae. *University of California Publications in Botany 58*: 1–95.

CHOI, H.-G.; KRAFT, G. T.; SAUNDERS, G. W. 2000: Nuclear small subunit rDNA sequences from *Ballia* spp. (Rhodophyta): proposal of the Balliales ord. nov., Balliaceae fam. nov., *Ballia nana* sp. nov. and *Inkyuleea* gen. nov. (Ceramiales). *Phycologia 39*: 272–287.

CHOI, H.-G.; KRAFT, G. T.; LEE, I. K.; SAUNDERS, G. W. 2002: Phylogenetic analyses of anatomical and nuclear SSU rDNA sequence data indicate that the Dasyaceae and Delesseriaceae (Ceramiales, Rhodophyta) are polyphyletic. *European Journal of Phycology 37*: 551–569.

CHOI, H.-G.; KRAFT, G. T.; KIM, H.-S.; GUIRY, M. D.; SAUNDERS, G. W. 2008: Phylogenetic relationships among lineages of the Ceramiaceae (Ceramiales, Rhodophyta) based on nuclear small subunit rDNA sequence data. *Journal of Phycology 44*: 1033–1048.

CHRISTIANSEN, T. 1962: Alger. *Botanik. Bind II. Systematisk Botanik 2*: 1–178.

COLE, K. M.; SHEATH, R. G. (Eds) 1990: *Biology of the Red Algae*. Cambridge University Press, Cambridge. 525 p.

COZZOLINO, S.; CAPUTO, P.; DE CASTRO, O.; MORETTI, A.; PINTO, G. 2000: Molecular variation in *Galdieria sulphararia* (Galdieri) Merola and its bearing on taxonomy. *Hydrobiologia 433*: 145–151.

DALEN, J. L.; NELSON, W. A.; GUIRY, M. D. 2009; New macroalgal record for the New Zealand region: *Cephalocystis furcellata* (Rhodymeniales, Rhodophyta). *Phycologia 48*: 66–69.

D'ARCHINO, R.; NELSON, W. A.; ZUCCARELLO, G. C. 2007: Invasive marine red alga introduced to New Zealand waters: first record of *Grateloupia turuturu* (Halymeniaceae, Rhodophyta). *New Zealand Journal of Marine and Freshwater Research 41*: 35–42.

D'ARCHINO, R.; NELSON, W. A.; ZUCCARELLO, G. C. 2010: *Psaromenia* gen. nov. (Kallymeniaceae, Rhodophyta): a new genus for *Kallymenia berggrenii*. *Phycologia 49*: 73–85.

D'ARCHINO, R.; NELSON, W. A.; ZUCCARELLO, G. C. 2011: Diversity and complexity in New Zealand Kallymeniaceae (Rhodophyta): recognition of the genus *Ectophora* and description of *E. marginata* sp. nov. *Phycologia 50*: 241–255.

DIXON, P. S. 1961: On the classification of the Florideae with particular reference to the position of the Gelidiaceae. *Botanica Marina 3*: 1–6.

DOWELD, A. 2001: *Prosyllabus Tracheophytorum. Tentamen systematis plantarum vascularium (Tracheophyta)*. GEOS, Moscow. [32], lxxx, 33–110 p.

ENTWISLE, T. J.; FOARD, H. J. 1997a: *Batrachospermum* (Batrachospermales, Rhodophyta) in Australia and New Zealand: new taxa and emended circumscriptions in section *Aristata, Batrachospermum, Turfosa* and *Virescentia*. *Australian Systematic Botany 10*: 331–380.

ENTWISLE, T. J.; FOARD, H. J. 1997b: Corrigendum to: *Batrachospermum* (Batrachospermales, Rhodophyta) in Australia and New Zealand: new taxa and emended circumscriptions in section *Aristata, Batrachospermum, Turfosa* and *Virescentia*. *Australian Systematic Botany 10*: 331–380.

ENTWISLE, T. J.; FOARD, H. J. 1999a: *Sirodotia* (Batrachospermales, Rhodophyta) in Australia and New Zealand. *Australian Systematic Botany 12*: 605–613.

ENTWISLE, T. J.; FOARD, H. J. 1999b: *Batrachospermum* (Batrachospermales, Rhodophyta) in Australia and New Zealand: New taxa and records in sections *Contorta* and *Hybrida*. *Australian Systematic Botany 12*: 615–633.

ENTWISLE, T. J.; HUISMAN, J. 1998: Algal systematics in Australia. *Australian Systematic Botany 11*: 203–214.

ENTWISLE, T. J.; NECCHI, O. Jr 1992: Phylogenetic systematics of the freshwater red algal order Batrachospermales. *Japanese Journal of Phycology 40*: 1–13.

FALSHAW, R.; FURNEAUX, R. H.; STEVENSON, D. E. 1998: Agars from nine species of red seaweed in the genus *Curdiea* (Gracilariaceae, Rhodophyta). *Carbohydrate Research 308*: 107–115.

FALSHAW, R.; SLIM, G. C.; NELSON, W. A. 1997: Marine hydrocolloids: commercial slimes, gums and jellies. *Seafood New Zealand 5(3)*: 77–80.

FAN, K. C. 1961: Morphological studies of the Gelidiales. *University of California Publications in Botany 32*: 315–368.

FARR, T.; BROOM, J.; HART, D.; NEILL, K.; NELSON, W. 2009: Common coralline algae of northern New Zealand: an identification guide. *NIWA Information Series 70*: 1–248.

FELDMANN, J.; FELDMANN, G. 1942: Recherches sur les Bonnemaisoniacees et leur alternance de générations. *Annales des Sciences naturelles, sér. Botanique, 11*: 75–175.

FREDERICQ, S.; HOMMERSAND, M. H. 1989: Proposal of the Gracilariales *ord. nov.* (Rhodophyta) based on an analysis of the reproductive development of *Gracilaria verrucosa*. *Journal of Phycology 25*: 213–227.

FREDERICQ, S.; NORRIS, J. N. 1995: A new order (Rhodogorgonales) and a new family (Rhodogorgonaceae of red algae composed of two tropical calciferous genera, *Renouxia* gen. nov. and *Rhodogorgon*. *Cryptogamie Botanique 5*: 316–331.

FRESHWATER, D. W.; FREDERICQ, S.; BUTLER, B. S.; HOMMERSAND, M. H.; CHASE, M. W. 1994: A gene phylogeny of the red algae (Rhodophyta) based on plastid *rbc*L. *Proceedings of the National Academy of Sciences USA 91*: 7281–7285.

FRESHWATER, D. W.; FREDERICQ, S.; HOMMERSAND, M. H. 1995. A molecular phylogeny of the Gelidiales (Rhodophyta) based on analysis of plastid *rbc*L nucleotide sequences. *Journal of Phycology 31*: 616–632.

GABRIELSON, P. W.; GARBARY, D. J. 1987: A cladistic analysis of Rhodophyta: Florideophycidean orders. *British Phycological Journal 22*: 125–138.

GABRIELSON, P. W.; GARBARY, D. J.; SCAGEL, R. F. 1985: The nature of the ancestral red alga: inferences from a cladistic analysis. *BioSystems 18*: 335–346.

GARBARY, D. J.; GABRIELSON, P. W. 1990: Taxonomy and evolution. Pp. 47–498 *in*: Cole, K. M.; Sheath, R. G. (eds), *Biology of the Red Algae*. Cambridge University Press, Cambridge. 525 p.

GORDON-MILLS, E. 1977: Two new species of marine algae from Stewart Island, New Zealand, *Medeiothamnion norrisii* and *Ptilothamnion rupicolum* (Ceramiaceae, Rhodophyta). *Phycologia 16*: 79–85.

GROSS, W.; HEILMANN, I.; LENZE, D.; SCHNARRENBERGER, C. 2001: Biogeography of the Cyanidiaceae (Rhodophyta) based on 18S ribosomal RNA sequence data. *European Journal of Phycology 36*: 275–280.

GUIRY, M. D. 1978: The importance of sporangia in the classification of the Florideophycidae. Pp. 111–144 *in*: Irvine, D. E. G.; Price, J. H. (eds), *Modern Approaches to the Taxonomy of Red and Brown Algae*. [*Systematics Association Special Volume* 10.] Academic Press, London. 484 p.

GUIRY, M. D.; WOMERSLEY, H. B. S. 1993: *Capreolia implexa* gen. et sp. nov. (Gelidiales, Rhodophyta) in Australia and New Zealand; an intertidal mat-forming alga with an unusual life history. *Phycologia 32*: 266–277.

GURGEL, C. F. D.; FREDERICQ, S. 2004: Systematics of the Gracilariaceae (Gracilariales, Rhodophyta): a critical assessment based on *rbc*L sequence analyses. *Journal of Phycology 40*: 138–159.

HARPER, J. T.; SAUNDERS, G. W. 1998: A molecular systematic investigation of the Acrochaetiales (Florideophycidae, Rhodophyta) and related taxa based on nuclear small-subunit ribosomal DNA sequence data. *European Journal of Phycology 33*: 221–229.

HARPER, J. T; SAUNDERS, G. W. 2002: A re-classification of the Acrochaetiales based on molecular and morphological data, and establishment of the Colaconematales ord. nov. (Florideophyceae, Rhodophyta). *European Journal of Phycology 37*: 463–476.

HARVEY, A.; WOELKERLING, W.; FARR, T.; NEILL, K.; NELSON, W. 2005: Coralline algae of central New Zealand: an identification guide to common 'crustose' species. *NIWA Information Series 57*: 1–145.

HARVEY, W. H. 1855: Algae. Pp. 211–266, pls 107–121 *in*: Hooker, J. D. *The Botany of the Antarctic Voyage of H.M. Discovery ships Erebus and Terror in the years 1839–1843. II. Flora Novae-Zelandiae. Part II.* London. 378 p, pls 71–130.

HARVEY, W. H.; HOOKER, J. D. 1845: Algae. Pp. 1–193 *in*: Hooker, J. D. *The Botany of the Antarctic Voyage of H.M. Discovery ships Erebus and Terror in the years 1839–1843. I. Flora Antarctica. Part I. The Botany of Lord Auckland's Group and Campbell's Island.* London. 208 p.

HAWKES, M. W. 1982a: *Schmitzia evanescens* sp. nov. (Rhodophyta, Gigartinales) a new species of Calosiphoniaceae from the northeastern New Zealand coast. *Journal of Phycology 18*: 368–378.

HAWKES, M. W. 1982b: *Acrosymphyton firmum* sp. nov. (Rhodophyta, Cryptonemiales) a new subtidal red alga from New Zealand: Developmental morphology and distribution of the gametophyte. *Journal of Phycology 18*: 447–454.

HAWKES, M. W. 1983a: *Hummbrella hydra* Earle (Rhodophyta, Gigartinales): seasonality, distribution, and development in laboratory culture. *Phycologia 22*: 403–13.

HAWKES, M. W. 1983b: Anatomy of *Apophlaea sinclairii* — an enigmatic red alga endemic to New Zealand. *Japanese Journal of Phycology 31*: 55–64.

HAWKES, M. W. 1990: Reproductive strategies. Pp. 455–476 *in*: Cole, K. M.; Sheath, R. G. (eds), *Biology of the Red Algae*. Cambridge University Press, Cambridge. 525 p.

HAY, C. H.; ADAMS, N. M.; PARSONS, M. J. 1985: The marine algae of the subantarctic islands of New Zealand. *National Museum of New Zealand, Miscellaneous Series 11*: 1–70.

HEMMINGSON, J. A.; FURNEAUX, R. H.; MURRAY-BROWN, V. L. 1996a: Biosynthesis of agar polysaccharides in *Gracilaria chilensis* Bird, McLachlan et Oliveira. *Carbohydrate Research 287*: 101–115.

HEMMINGSON, J. A.; FURNEAUX, R. H.; WONG, H. 1996b: In vivo conversion of 6-*O*-sulfo-L-galactopyranosyl residues to 3,6-anhydro-L-galactopyranosyl residues in *Gracilaria chilensis* Bird, McLachlan et Oliveira. *Carbohydrate Research 296*: 285–292.

HEMMINGSON, J. A.; FURNEAUX, R. H. 1997: Biosynthetic activity and galactan composition in different regions of the thallus of *Gracilaria chilensis* Bird, McLachlan et Oliveira. *Botanica Marina 40*: 351–357.

HEYDRICH, F. 1893a: Vier neue Florideen von Neu-Seeland. *Berichte der Deutsche Botanische gesellschaff 11*: 75–79.

HEYDRICH, F. 1893b: *Pleurostichidium*, ein neues Genus der Rhodomeleen. *Berichte der Deutsche Botanische gesellschaff 11*: 344–348.

HOMMERSAND, M. H. 1963: The morphology and classification of some Ceramiaceae and Rhodomelaceae. *University of California Publications in Botany 32*: 165–366.

HOMMERSAND, M. H. 1986: The biogeography of the South African marine red algae: a model. *Botanica Marina 29*: 257–270.

HOMMERSAND, M. H. 2007. Global biogeography and relationships of the Australian marine macroalgae. Pp. 511–542 *in*: McCarthy, P. M.; Orchard, A. E. (eds), *Algae of Australia – Introduction*. ABRS/CSIRO Publishing, Melbourne. xvi + 727 p.

HOMMERSAND, M. H.; DE CLERCK, O.; COPPEJANS, E. 2004: A morphological study and taxonomic revision of *Euptilota* (Ceramiaceae, Rhodophyta). *European Journal of Phycology 39*: 369–393.

HOMMERSAND, M. H.; FREDERICQ, S. 1988: An investigation of cystocarp development in *Gelidium pteridifolium* with a revised description

of the Gelidiales (Rhodophyta). *Phycologia 27*: 254–272.
HOMMERSAND, M. H.; FREDERICQ, S. 1990: Sexual reproduction and cystocarp development. Pp. 305– 346 *in*: Cole, K. M.; Sheath, R. G. (eds), *Biology of the Red Algae*. Cambridge University Press, Cambridge. 525 p.
HOMMERSAND, M. H.; FREDERICQ, S.; FRESHWATER, D. W. 1994: Phylogenetic systematics and biogeography of the Gigartinaceae (Gigatinales, Rhodophyta) based on sequence analysis of *rbc*L. *Botanica Marina 37*: 193–203.
HOMMERSAND, M. H.; FREDERICQ, S.; FRESHWATER, D. W.; HUGHEY, J. 1999: Recent developments in the systematics of the Gigartinaceae (Gigartinales, Rhodophyta) based on *rbc*L sequence analysis and morphological evidence. *Phycological Research 47*: 139–151.
HOMMERSAND, M. H.; GUIRY, M. D.; FREDERICQ, S.; LEISTER, G. L. 1993: New perspectives in the taxonomy of the Gigartinaceae (Gigartinales, Rhodophyta). *Hydrobiologia 260/261*: 105–120.
HOMMERSAND, M. H.; LEISTER, G. L.; RAMIREZ, M. E.; GABRIELSON, P. W.; NELSON, W. A. 2010: A morphological and phylogenetic study of *Glaphyrosiphon* gen. nov. Halymeniaceae, Rhodophyta) based on *Grateloupia intestinalis* with descriptions of two new species: *Glaphyrosiphon lindaueri* sp. nov. from New Zealand and *Glaphyrosiphon chilensis* sp. nov. from Chile. *Phycologia* 49: 554–573.
HUISMAN, J. M.; KRAFT, G. T. 1982: *Deucalion* gen. nov. and *Anisoschizus* gen. nov. (Ceramiaceae, Ceramiales), two new propagule-forming red algae from southern Australia. *Journal of Phycology 18*: 177–192.
HUISMAN, J. M.; SHERWOOD, A. R.; ABBOTT, I. A. 2003: Morphology, reproduction, and the 18S rRNA gene sequence of *Pihiella liagoraciphila* gen. et sp. nov. (Rhodophyta), the so-called 'monosporangial discs' associated with members of the Liagoraceae (Rhodophyta) and proposal of the Pihiellales ord. nov. *Journal of Phycology 39*: 978–987.
HUISMAN, J. M.; ABBOTT, I. A.; SHERWOOD, A. R. 2004: Large subunit rDNA gene sequences and reproductive morphology reveal *Stenopeltis* to be a member of the Liagoraceae (Nemaliales, Rhodophyta), with a description of *Akalaphycus* gen. nov. *European Journal of Phycology 39*: 257–272.
HURD, C. L.; NELSON, W. A.; FALSHAW, R.; NEILL, K. 2004: History, current status and future of marine macroalgae research in New Zealand: taxonomy, ecology, physiology and human uses. *Phycological Research 52*: 80–106.
INTASUWAN, S.; GORDON, M. E.; DAUGHERTY, C. H.; LINDSAY, G. C. 1993: Assessment of allozyme variation among New Zealand populations of *Gracilaria chilensis* (Gracilariales, Rhodophyta) using starch-gel electrophoresis. *Hydrobiologia 260/261*: 159–165.
JOHANSEN, H. W.; WOMERSLEY, H. B. S. 1994: *Jania* (Corallinales, Rhodophyta) in southern Australia. *Australian Systematic Botany 7*: 605–625.
KAPLAN, I. R. 1956: Evidence of microbial activity in some of the geothermal regions of New Zealand. *New Zealand Journal of Science and Technology 37B*: 639–662.
KING, R. J.; PUTTOCK, C. F. 1989: Morphology and taxonomy of *Bostrychia* and *Stictosiphonia* (Rhodomelaceae, Rhodophyta). *Australian Systematic Botany 2*: 1–73.
KRAFT, G. T.; ROBBINS, P. A. 1985: Is the order Cryptonemiales (Rhodophyta) defensible? *Phycologia 24*: 67–77.
KRAFT, G. T.; WOELKERLING, W. J. 1990: Rhodophyta. Pp. 41–85 *in*: Clayton, M. N.; King, R. J. (eds), *Biology of Marine Plants*. Longman Cheshire, Melbourne. vii + 501 p.
KRAYESKY, D. M.; NORRIS, J. N., GABRIELSON, P. W., GABRIELA, D., FREDERICQ, S. 2009. A new order of red algae based on the Peyssonneliaceae, with an evaluation of the ordinal classification of the Florideophyceae (Rhodophyta). *Proceedings of the Biological Society of Washington 122*: 364–391.
KYLIN, H. 1923: Studien über die Entwicklungsgeschichte der Florideen. *Kongliga Svenska Vetenskapsakademiens Handlingar 63*: 1–139.
KYLIN, H. 1929: Die Delesseriaceen Neu-Seelands. *Lunds Universitets Årsskrift 25(2)*: 1–15, pls 1–12.
KYLIN, H. 1931: Die Florideenordnung Rhodymeniales. *Lunds Universitets Årsskrift 27(11)*: 1–48, pls 1–20.
KYLIN, H. 1932: Die Florideenordnung Gigartinales. *Lunds Universitets Årsskrift 28(8)*: 1–88, pls 1–28.
KYLIN, H. 1933: On three species of Delesseriaceae from New Zealand. *Transactions of the New Zealand Institute 63*: 109–111.
KYLIN, H. 1956: *Die Gattungen der Rhodophyceen.* CWK Gleerups Forlag, Lund. 673 p.
LAING, R. M. 1897: Notes on several species of *Delesseria*, one being new. *Transactions and Proceedings of the New Zealand Institute 29*: 446–450.
LAING, R. M. 1900: Revised list of New Zealand seaweeds. Part 1. *Transactions and Proceedings of the New Zealand Institute 29*: 446–450.
LAING, R. M. 1902: Revised list of New Zealand seaweeds, Part II. *Transactions and Proceedings of the New Zealand Institute 34*: 384–408.
LAING, R. M. 1905: On the New Zealand species of Ceramiaceae. *Transactions and Proceedings of the New Zealand Institute 37*: 384–408.
LAING, R. M. 1909: The marine algae of the subantarctic islands of New Zealand. Pp. 493–527 *in*: Chilton, C. (ed.), *The Subantarctic Islands of New Zealand. Reports on the geo-physics, geology, zoology and botany of the islands lying to the south of New Zealand, based mainly on observations and collections made during an expedition in the government steamer "Hinemoa" (Captain J. Bollons) in November, 1907.* The Philosophical Institute of Canterbury, Christchurch. 2 vols, 848 p.
LAING, R. M. 1926: A reference list of New Zealand marine algae. *Transactions and Proceedings of the New Zealand Institute 57*: 126–185.
LAING, R. M. 1928: New Zealand Bangiales (*Bangia, Porphyra, Erythrotrichia* and (?) *Erythrocladia*). *Transactions and Proceedings of the New Zealand Institute 59*: 33–59.
LAING, R. M. 1930: A reference list of New Zealand marine algae. Supplement 1. *Transactions and Proceedings of the New Zealand Institute 60*: 575–583.
LAING, R. M.; GOURLAY, H. W. 1929: The New Zealand species of *Gigartina.* Part 1. *Transactions and Proceedings of the New Zealand Institute 60*: 102–135.
LAING, R. M.; GOURLAY, H. W. 1931: The New Zealand species of *Gigartina.* Part II (foliose forms). *Transactions and Proceedings of the New Zealand Institute 62*: 1–22, pls 16–19.
LE GALL, L.; SAUNDERS, G. W. 2007: A nuclear phylogeny of the Florideophyceae (Rhodophyta) inferred from combined EF2, small subunit and large subunit ribosomal DNA: Establishing the new red algal subclass Corallinophycidae. *Molecular Phylogenetics and Evolution 43*: 1118–1130.
LE GALL, L.; DALEN, J. L.; SAUNDERS, G. W. 2008: Phylogenetic analyses of the red algal order Rhodymeniales supports recognition of the Hymenocladiaceae fam. nov., Fryeellaceae fam. nov., and *Neogastroclonium* gen. nov. *Journal of Phycology* 44: 1556–1571.
LE GALL, L.; PAYRI, C. E.; BITTNER, L.; SAUNDERS, G. W. 2009: Multigene phylogenetic analyses support recognition of the Sporolithales ord. nov. *Molecular Phylogenetics and Evolution 54*: 302–305.
LEHR, C. R.; FRANK, S. D.; NORRIS, T. B.; D'IMPERIO, S.; KALININ, A.V.; TOPLIN, J. A.; CASTENHOLZ, R. W.; MCDERMOTT, T. R. 2007: Cyanidia (Cyanidiales) population diversity and dynamics in an acid-sulfate-chloride spring in Yellowstone National Park. *Journal of Phycology* 43: 3–14.
LEVRING, T. 1949: Six new marine algae from New Zealand. *Transactions and Proceedings of the Royal Society of New Zealand 77*: 394–397.
LEVRING, T. 1955: Contributions to the marine algae of New Zealand. I: Rhodophyta: Goniotrichales, Bangiales, Nemalionales and Bonnemaisoniales. *Arkiv för Botanik 3*: 407–432.
LIM, B.-L.; KAWAI, H.; HORI, H.; OSAWA, S. 1986: Molecular evolution of 5S ribosomal RNA from red and brown algae. Japanese *Journal of Genetics 61*: 169–176.
LIN, S.-M.; HOMMERSAND, M. H.; NELSON, W. A. 2007: An assessment of *Haraldiophyllum* (Delesseriaceae, Rhodophyta) based on *rbc*L and LSU sequence analysis and morphological evidence, including *H. crispatum* J. D.Hooker et Harvey) comb. nov. from New Zealand. *European Journal of Phycology* 42: 391–408
LIN, S.-M.; NELSON, W. A. 2010: Systematic revision of the genus *Phycodrys* Delesseriaceae, Rhodophyta) from New Zealand with the description of *P. novae-zelandiae* sp. nov., *P. franiae* sp. nov. and *P. adamsiae* sp. nov. *European Journal of Phycology 45*: 200–214.
LIN, S.-M.; NELSON, W. A.; HOMMERSAND, M. H. (In press) *Hymenenopsis heterophylla* gen. et sp. nov. (Delesseriaceae, Rhodophyta) from New Zealand, based on a red alga previously known as *Hymenena palmata* f. *marginata sensu* Kylin, with emphasis on its cystocarp development. *Phycologia*.
LINDAUER, V. W. 1938: Notes on a new species of New Zealand *Champia*. *Transactions of the Royal Society of New Zealand 67*: 411–413.
LINDAUER, V. W.; SETCHELL, W. A. 1946: Note on a new species of red alga, *Lenormandia coronata*. *Transactions of the Royal Society of New Zealand 76*: 66–67.
MAGGS, C. A.; PUESCHEL, C. M. 1989: Morphology and development of *Ahnfeltia plicata* (Rhodophyta): proposal of Ahnfeltiales ord. nov. *Journal of Phycology 25*: 333–351.
McCORMACK, M. I. 1990: Handbook for stock assessment of agar seaweed *Pterocladia lucida*; with a comparison of survey techniques. *New Zealand Fisheries Technical Report 24*: 1–36.
McIVOR, L.; MAGGS, C. A.; PROVAN, J.; STANHOPE, M. J. 2001: *rbc*L sequences reveal multiple cryptic introductions of the Japanese red alga *Polysiphonia harveyi*. *Molecular Ecology 10*: 911–919.
MEDLIN, L. K., KOOISTRA, W. H. C. F.; POTTER, D.; SAUNDERS, G. W.; ANDERSEN, R. A. 1997: Phylogenetic relationships of the 'golden algae' (haptophytes, heterokont chromophytes) and

their plastids. *Plant Systematics and Evolution (Suppl.) 11*: 187–219.

MILLAR, A. J. K.; NELSON, W. A. 2002: The genera *Nancythalia humilis* gen. et sp. nov. and *Abroteia suborbiculare* (Harv.) Kylin (Delesseriaceae, Rhodophyta) from New Zealand. *Phycologia 41*: 245–253.

MILLAR, A. J. K.; FRESHWATER, D. W. 2005: Morphology and molecular phylogeny of the marine algal order Gelidiales (Rhodophyta) from New South Wales, including Lord Howe and Norfolk Islands. *Australian Systematic Botany 18*: 215–263

MILLER, I. J.; FALSHAW, R.; FURNEAUX, R. H. 1996: A polysaccharide from the red seaweed *Champia novae-zelandiae* Rhodymeniales, Rhodophyta. *Hydrobiologia 326/327*: 505–509.

MILLER, I. J.; FALSHAW, R.; FURNEAUX, R. H.; HEMMINGSON, J. A. 1997: Variations in the consitutent sugars of the polysaccharides from New Zealand species of *Pachymenia* (Halymeniaceae). *Botanica Marina 40*: 119–127.

MILLER, I. J.; FURNEAUX, R. H. 1987a: The chemical substitution of the agar-type polysaccharide from *Gracilaria secundata* f. *pseudoflagellifera* (Rhodophyta). *Hydrobiologia 151/152*: 523–529.

MILLER, I. J.; FURNEAUX, R. H. 1987b: Chemical characteristics of the galactans from the forms of *Gracilaria secundata* from New Zealand. *Botanica Marina 30*: 427–435.

MONTAGNE, C. 1842: *Prodromus Generum Specierumque Phycearum Novarum, in itinere ad polum antarcticum ... collectarum.* Paris. 16 p.

MONTAGNE, C. 1845: Plantes cellulaires. *In*: Hombron, J. B.; Jacquinot, H. *Voyage au Pole Sud et dans l'Océanie sur les corvettes l'Astrolabe et la Zéleé ... pendant les annees 1837–1838–1839–1840, sous le commandement de M. J. Dumont-D'Urville: Botanique.* Paris. Vol. 1, 349 p.; Atlas, Botanique: Cryptogamie, 20 pls.

MOORE, L. B. 1941: The economic importance of seaweeds. *New Zealand Department of Scientific and Industrial Research Bulletin 85*: 1–40.

MOORE, L. B. 1944: New Zealand seaweed for agar-manufacture. *New Zealand Journal of Science and Technology 27, B*: 183–209.

MOORE, L. B. 1945: The genus *Pterocladia* in New Zealand. *Transactions of the Royal Society of New Zealand 74*: 332–342.

MOORE, L. B. 1946: New Zealand seaweed for agar-manufacture. *New Zealand Journal of Science and Technology 27, B*: 311–317.

MOORE, L. B. 1966: The economic importance of seaweeds. *New Zealand Department of Scientific and Industrial Research 85*: 1–40. [Reprint of 1941 bulletin, with updated bibliography.]

MÜLLER, K. M.; SHERWOOD, A. R.; PUESCHEL, C. M.; GUTELL, R. R.; SHEATH, R. G. 2002: A proposal for a new red algal order, the Thoreales. *Journal of Phycology 38*: 807–820.

MURAVENKO, O. V.; SELYAKH, I. O.; KONONENKO, N. V; STADNICHUK, I. N. 2001: Chromosome numbers and nuclear DNA contents in the red microalgae *Cyanidium caldarium* and three *Galdieria* species. *European Journal of Phycology 36*: 227–232.

NEALE, D., NELSON, W. 1998: Marine algae of the west coast, South Island, New Zealand. *Tuhinga 10*: 87–118.

NELSON, W. A. 1987: The New Zealand species of *Gracilaria* Greville (Rhodophyta, Gigartinales). *New Zealand Journal of Botany 25*: 87–98.

NELSON, W. A. 1989: Phenology of *Gracilaria sordida* W. Nelson populations. Reproductive status, plant and population size. *Botanica Marina 32*: 41–51.

NELSON, W. A. 1993: Epiphytic species of *Porphyra* (Bangiales, Rhodophyta) from New Zealand. *Botanica Marina 36*: 525–534.

NELSON, W. A. 1999: A revised checklist of marine algae naturalised in New Zealand. *New Zealand Journal of Botany 37*: 355–359.

NELSON, W. A. 2009. Calcified macroalgae – critical to coastal ecosystems and vulnerable to change: a review. *Marine and Freshwater Research 60*:787–801.

NELSON, W. A.; ADAMS, N. M. 1984: Marine algae of the Kermadec Islands. *National Museum of New Zealand, Miscellaneous Series 10*: 1–29.

NELSON, W. A.; ADAMS, N. M. 1987: Marine algae of the Bay of Islands area. *National Museum of New Zealand, Miscellaneous Series 16*: 1–47.

NELSON, W. A.; ADAMS, N. M. 1993: *Nesophila hoggardii* gen. et sp. nov. (Rhizophyllidaceae, Rhodophyta) from offshore islands of northern New Zealand. *Museum of New Zealand Records 1*: 1–7.

NELSON, W. A.; ADAMS, N. M.; FOX, J. M. 1992: Marine algae of the northern South Island. *National Museum of New Zealand, Miscellaneous Series 26*: 1–80.

NELSON, W. A.; ADAMS, N. M.; HAY, C. H. 1991: Marine algae of the Chatham Islands. *National Museum of New Zealand, Miscellaneous Series 23*: 1–58.

NELSON, W. A.; BRODIE, J.; GUIRY, M. D. 1999a: Terminology used to describe reproduction and life history stages in the genus *Porphyra* (Bangiales, Rhodophyta). *Journal of Applied Phycology 11*: 407–410.

NELSON, W. A.; BROOM, J. E. 1997: New Zealand *Porphyra* species: systematics, species diversity, endemism. *Phycologia 36*: 77–78.

NELSON, W. A.; BROOM, J. E. 2008; New Zealand (Gigartinaceae Rhodophyta): resurrecting *Gigartina grandifida* J.Agardh, endemic to the Chatham Islands. *New Zealand Journal of Botany 45*: 177–187.

NELSON, W. A.; BROOM, J. E. S. 2010: The identity of *Porphyra columbina* Bangiales, Rhodophyta) originally described from the New Zealand subantarctic islands. *Australian Systematic Botany 23*: 16–26.

NELSON, W. A.; BROOM, J. E.; FARR, T. J. 2001: Four new species of *Porphyra* (Bangiales, Rhodophyta) from the New Zealand region. *Cryptogamie Algologie 22*: 263–284.

NELSON, W. A.; BROOM, J. E.; FARR, T. J. 2003: *Pyrophyllon* and *Chlidophyllon* (Erythropeltidales, Rhodophyta), two new genera for obligate epiphytic species previously placed in *Porphyra*, and a discussion of the orders Erythropeltidales and Bangiales. *Phycologia 42*: 308–315.

NELSON, W. A.; CONROY, A. M. 1989: Effect of harvest method and timing on yield and regeneration of karengo (*Porphyra* spp.) (Bangiales, Rhodophyta) in New Zealand *Journal of Applied Phycology 1*: 277–283.

NELSON, W. A.; FARR, T. J.; BROOM, J. E. 2005: *Dione* and *Minerva*, two new genera from New Zealand circumscribed for basal taxa in the Bangiales (Rhodophyta). *Phycologia 44*: 139–145.

NELSON, W. A., FARR, T. J., BROOM, J. E. S. 2006a: Phylogenetic relationships and generic concepts in the red order Bangiales: challenges ahead. *Phycologia 45*: 249–259.

NELSON, W. A.; FARR, T. J.; BROOM, J. E. S. 2006b: Phylogenetic diversity of New Zealand Gelidiales as revealed by *rbc*L sequence data. *Journal of Applied Phycology 18*: 653–661.

NELSON, W. A.; KNIGHT, G. A. 1997: Reproductive structures in *Curdiea coriacea* (Hook. f. et Harv.) V. J.Chapm. (Gracilariales, Rhodophyta) including the first report of spermatangia for the genus. *New Zealand Journal of Botany 35*: 195–202.

NELSON, W. A.; KNIGHT, G. A.; FALSHAW, R.; FURNEAUX, R. H.; FALSHAW, A.; LYNDS, S. M. 1994: Characterisation of the enigmatic, endemic red alga *Gelidium allanii* (Gelidiales) from northern New Zealand – morphology, distribution, agar chemistry. *Journal of Applied Phycology 6*: 497–507.

NELSON, W. A.; KNIGHT, G. A.; FALSHAW, R. 1999b: A new agarophyte, *Curdiea balthazar* sp. nov. (Gracilariales, Rhodophyta), from the Three Kings Islands, northern New Zealand. *Hydrobiologia 398/399*: 57–63.

NELSON, W. A.; LEISTER, G. L.; HOMMERSAND, M. H. 2011: *Psilophycus alveatus* gen. et comb. nov., a basal taxon in the Gigartinaceae Rhodophyta) from New Zealand. *Phycologia 50*: 219–231.

NELSON, W. A.; PHILLIPS, L. 1996: The Lindauer legacy – current names for the Algae Nova-Zelandicae Exsiccatae. *New Zealand Journal of Botany 34*: 553–582.

NELSON, W. A.; PHILLIPS, L. 2001: Locating the type specimens of New Zealand marine algae described by Levring. *New Zealand Journal of Botany 39*: 349–353.

NELSON, W. A.; RYAN, K. G. 1988: *Porphyridium purpureum* (Bory) Drew et Ross (Porphyridiales, Rhodophyceae) – first record of a marine unicellular red alga in New Zealand. *Journal of the Royal Society of New Zealand 18*: 127–128.

NELSON, W. A.; RYAN, K. G. 1991: Comparative study of reproductive development in two species of *Gracilaria* (Gracilariales, Rhodophyta) – II. Carposporogenesis. *Cryptogamic Botany 2*: 234–241.

NELSON, W. A.; VILLOUTA, E.; NEILL, K.; WILLIAMS, G. C.; ADAMS, N. M.; SLIVSGAARD, R. 2002: Marine macroalgae of Fiordland. *Tuhinga 13*: 117–152.

NORRIS, R. E. 1957: Morphological studies on the Kallymeniaceae. *University of California Publications in Botany 28*: 251–334.

NOZAKI, H.; MATSUZAKI, M.; MISUMI, O.; KUROIWA, H.; HIGASHIYAMA, T.; KUROIWA, T. 2005: Phylogenetic implications of the CAD complex from the primitive red alga *Cyanidioschyzon merolae* (Cyanidiales, Rhodophyta). *Journal of Phycology 41*: 652–657.

OLTMANNS, F. 1904–05: *Morphologie und Biologie der Algen*. Gustav Fischer, Jena. Vol. 1, vi + 733 p.; Vol. 2, vi + 396 p.

OTT, F. D. 2009: *Handbook of the Taxonomic Names Associated with the Non-marine Rhodophycophyta*. J. Cramer in der Gebrüder Borntraeger Verlagsbuchhandlung, Stuttgart. Pp. [i-xiii] xiv-xxiv, [1]–969, [2].

PARSONS, M. J. 1980: The morphology and taxonomy of *Brongniartella* Bory *sensu* Kylin (Rhodomelaceae, Rhodophyta). *Phycologia 19*: 273–295.

PARSONS, M. J. 1985a: New Zealand seaweed flora and its relationships. *New Zealand Journal of Marine and Freshwater Research 19*: 131–138.

PARSONS, M. J. 1985b: Biosystematics of the cryptogamic flora of New Zealand: Algae. *New Zealand Journal of Botany 23*: 663–675.

PARSONS, M. J.; FENWICK, G. D. 1984: Marine algae and a marine fungus from Open Bay Islands, Westland. *New Zealand Journal of Botany 22*: 425–432.

PARSONS, M. J.; PICKMERE, S. E.; BAILEY, R. W. 1977: Carrageenan composition in New Zealand species of *Gigartina* (Rhodophyta): Geographic variation and interspecific differences. *New Zealand Journal of Botany 15*: 589–595.
PATWARY, M. U.; SENSEN, C. W.; MACKAY, R. M.; VAN DER MEER, J. P. 1998: Nucleotide sequences of small-subunit and internal transcribed spacer regions of nuclear rRNA genes support the autonomy of some genera of the Gelidiales (Rhodophyta). *Journal of Phycology 34*: 299–305.
PHILLIPS, L. E. 2000: Taxonomy of the New Zealand endemic genus *Pleurostichidium* (Rhodomelaceae, Rhodophyta) *Journal of Phycology 36*: 773–786.
PHILLIPS, L. E. 2002a: Taxonomy and molecular phylogeny of the red algal genus *Lenormandia* (Rhodomelaceae, Ceramiales). *Journal of Phycology 38*: 184–208.
PHILLIPS, L. E. 2002b: Taxonomy of *Adamsiella* L. E.Phillips et W. A.Nelson, gen. nov. and *Epiglossum* Kützing (Rhodomelaceae, Ceramiales). *Journal of Phycology 38*: 209–229.
PICKERING, T. D.; GORDON, M. E.; TONG, L. J. 1990: Seasonal growth, density, reproductive phenology and agar quality of *Gracilaria sordida* (Gracilariales, Rhodophyta) at Mokomoko Inlet, New Zealand. *Hydrobiologia 204/205*: 253–262.
PICKERING, T. D.; GORDON, M. E.; TONG, L. J. 1993: Effect of nutrient pulse concentration and frequency on growth of *Gracilaria chilensis* plants and levels of epiphytic algae. *Journal of Applied Phycology 5*: 525–533.
PICKMERE, S. E.; PARSONS, M. J.; BAILEY, R. W. 1973: Composition of *Gigartina* carrageenan in relation to sporophyte and gametophyte stages of the life cycle. *Phytochemistry 12*: 2441–2444.
PICKMERE, S. E.; PARSONS, M. J.; BAILEY, R. W. 1975: Variations in carrageenan levels and composition in three New Zealand species of *Gigartina*. *New Zealand Journal of Science 18*: 585–590.
PINTO, G.; ALBERTANO, P.; CINIGLIA, C.; COZZOLINO, S.; POLLIO, A.; YOON, H. S.; BHATTACHARYA, D. 2003: Comparative approaches to the taxonomy of the genus *Galdieria* Merola (Cyanidales, Rhodophyta). *Cryptogamie Algologie 24*: 13–32.
PUESCHEL, C. M. 1987: Absence of cap membranes as a characteristic of pitplugs of some red algal orders. *Journal of Phycology 23*: 150–156.
PUESCHEL, C. M. 1989: An expanded survey of the ultrastructure of red algal pit plugs. *Journal of Phycology 25*: 625–636.
PUESCHEL, C. M. 1994: Systematic significance of the absence of pit plug cap membranes in the Batrachospermales (Rhodophyta). *Journal of Phycology* 30: 310–315.
PUESCHEL, C. M.; COLE, K. M. 1982: Rhodophycean pit plugs: an ultrastructural survey with taxonomic implications. *American Journal of Botany 69*: 703–720.
PUESCHEL, C. M.; TRICK, H. N.; NORRIS, J. N. 1992: Fine structure of the phylogenetically important marine alga *Rhodogorgon carriebowensis* (Rhodophyta, Batrachospermales?). *Protoplasma 166*: 78–88.
RAGAN, M. A.; BIRD, C. J.; RICE, E. L.; GUTELL, R. R.; MURPHY, C. A.; SINGH, R. K. 1994: A molecular phylogeny of the marine red algae (Rhodophyta) based on the nuclear small-subunit rRNA gene. *Proceedings of the National Academy of Sciences 91*: 7276–7280.
RODRIGUEZ-PRIETO, C.; LIN, S.-M.; NELSON, W. A.; HOMMERSAND, M. H. 2011: Developmental morphology of *Sarcodia montagneana* and *S. grandifolia* from New Zealand and a phylogeny of *Sarcodia* (Sarcodiaceae, Rhodophyta) based on *rbc*L sequence analysis. *European Journal of Phycology* 46: 153–170.
RUSSELL, L. K.; HURD, C. L.; NELSON, W. A.; BROOM, J. E. 2009: An examination of *Pachymenia* and *Aeodes* Halymeniaceae, Rhodophyta) in New Zealand: the transfer of two species of *Aeodes* in South Africa to *Pachymenia*. *Journal of Phycology* 45: 1389–1399.
RYAN, K. G.; NELSON, W. A. 1991: Comparative study of reproductive development in two species of *Gracilaria* (Gracilariales, Rhodophyta) — I. Spermatiogenesis. *Cryptogamic Botany 2*: 229–233.
SANTELICES, B.; HOMMERSAND, M. H. 1997: *Pterocladiella*, a new genus in the Gelidiaceae (Gelidiales, Rhodophyta). *Phycologia 36*: 114–119.
SAUNDERS, G. W.; BAILEY, J. C. 1999: Molecular systematic analyses indicate that the enigmatic *Apophlaea* is a member of the Hildenbrandiales (Rhodophyta, Florideophycidae). *Journal of Phycology 35*: 171–175.
SAUNDERS, G. W.; HOMMERSAND, M. H. 2004: Assessing red algal supraordinal diversity and taxonomy in the context of contemporary systematic data. *American Journal of Botany 91*: 1494–1507.
SAUNDERS, G. W.; KRAFT, G. T. 1994: Small-subunit of rRNA gene sequences from representatives of selected families of the Gigartinales and Rhodymeniales (Rhodophyta). 1. Evidence for the Plocamiales *ord. nov. Canadian Journal of Botany 72*: 1250–1263.
SAUNDERS, G. W.; KRAFT, G. T. 1996: Small-subunit of rRNA gene sequences from representatives of selected families of the Gigartinales and Rhodymeniales (Rhodophyta). 2. Recognition of the Halymeniales *ord.nov. Canadian Journal of Botany 74*: 694–707.
SAUNDERS, G. W.; KRAFT, G. T. 1997: A molecular perspective on red algal evolution: focus on the Florideophycidae. Pp. 115–138 *in*: Bhattacharya, D. (ed.), *Origins of Algae and their Plastids*. Springer Biology, Vienna. x + 287 p.
SAUNDERS, G. W.; KRAFT, G. T. 2002: Two new Australian species of *Predaea* (Nemastomataceae, Rhodophyta), with taxonomic recommendations for an emended Nemastomatales and expanded Halymeniales. *Journal of Phycology 38*: 1245–1260.
SAUNDERS, G. W.; CHIOVITTI, A.; KRAFT, G. T. 2004: Small-subunit rDNA sequences from representatives of selected families of the Gigartinales and Rhodymeniales (Rhodophyta). 3. Delineating the Gigartinales sensu stricto. *Canadian Journal of Botany 82*: 43–74.
SAUNDERS, G. W.; STRACHAN, I. M.; KRAFT, G. T. 1999: The families of the order Rhodymeniales (Rhodophyta): a molecular-systematic investigation with a description of Faucheaceae *fam.nov*. *Phycologia 38*: 23–40.
SCAGEL, R. F. 1953: A morphological study of some dorsiventral Rhodomelaceae. *University of California Publications in Botany 27*: 1-108.
SCHMITZ, F. 1892: [6. Klasse Rhodophyceae] 2. Unterklasse Florideae. Pp. 16–23 *in*: Engler, A. (ed.), *Syllabus der Vorlesungen über specielle und medicinsch-pharmaceutische Botanik. Grosse Ausgabe*. Gebrüder Borntraeger, Berlin. 184 p.
SCHNEIDER, C. W.; WYNNE, M. J. 2007: A synoptic review of the classification of red algal genera a half century after Kylin's "Die Gattungen der Rhodophyceesn". *Botanica Marina 50*: 197–249.
SEARLES, R. B. 1968: Morphological studies of red algae of the order Gigartinales. *University of California Publications in Botany 43*: 1-86.
SEARLES, R. B. 1980: The strategy of the red algal life history. *American Naturalist 115*: 113–120.
SECKBACH, J. 1987: Evolution of eukaryotic cells via bridge algae. *Annals of the New York Academy of Sciences 503*: 424–437.
SHEATH, R. G.; MÜLLER, K. M. 1999: Systematic status and phylogenetic relationships of the freshwater genus *Balbiania* (Rhodophyta). *Journal of Phycology 35*: 855–864.
SHERWOOD, A. R.; SHEATH, R. G. 2003: Systematics of the Hildenbrandiales (Rhodophyta): gene sequence and morphometric analyses of global collections. *Journal of Phycology 39*: 409–422.
SILVA, P. C.; JOHANSEN, H. W. 1986: A reappraisal of the order Corallinales (Rhodophyceae). *British Phycological Journal 21*: 245–254.
SILVA, P. C.; BASSON, P. W.; MOE, R. L. 1996: *Catalogue of the Benthic Marine Algae of the Indian Ocean*. University of California Press, Berkeley. 1259 p.
SOUTH, G. R.; ADAMS, N. M. 1976: Marine algae of the Kaikoura coast. *National Museum of New Zealand, Miscellaneous Series 1*: 1–67.
SOUTH, G. R.; ADAMS, N. M. 1979: A revision of the genus *Plocamium* Lamouroux (Rhodophyta, Gigartinales) in New Zealand. *Phycologia* 18: 120–132.
SPARLING, S. R. 1957: The structure and reproduction of some members of the Rhodymeniaceae. *University of California Publications in Botany 29*: 319–396.
STEVENSON, T. T.; FURNEAUX, R. H. 1991: Chemical methods for the analysis of sulphated galactans from red algae. *Carbohydrate Research 210*: 277–298.
SUTHERLAND, J.; LINDSTROM, S.; NELSON, W.; BRODIE, J.; LYNCH, M.; HWANG, M. S.; CHOI, H. G.; MIYATA, M.; KIKUCHI, N.; OLIVEIRA, M.; FARR, T.; NEEFUS, C.; MOLS-MORTENSEN, A.; MILSTEIN, D.; MÜLLER, K. (In press) A new look at an ancient order: generic revision of the Bangiales. *Journal of Phycology*.
TAPPAN, H. 1976: Possible eucaryotic algae (Bangiophycidae) among early Proterozoic microfossils. *Geological Society of America Bulletin 87*: 633–639.
THIERS, B. 2011 [continuously updated]: *Index Herbariorum: A global directory of public herbaria and associated staff.* New York Botanical Garden's Virtual Herbarium. [http://sweetgum.nybg.org/ih/.]
TOWNSEND, R. A. 1979: *Synarthrophyton*, a new genus of Corallinaceae (Cryptonemiales, Rhodophyta) from the southern hemisphere. *Journal of Phycology 15*: 251–259.
TURNER, D. 1808–1819: *Fuci, sive Plantarum Fucorum Generi a Botanicis Ascriptarium Icones Descriptiones et Historia*. J. & A. Arch, London. 258 pls, in 4 vols.
VIS, M. L.; SAUNDERS, G. W.; SHEATH, R. G.; DUNSE, K.; ENTWISLE, T. J. 1998: Phylogeny of the Batrachospermales (Rhodophyta) inferred from *rbc*L and 18S ribosomal DNA gene sequences. *Journal of Phycology 34*: 341–350.
WAGNER, F. S. 1954: Contributions to the morphology of the Delesseriaceae. *University of California Publications in Botany 27*: 279-340.
WEST, J. A.; ZUCCARELLO, G. C.; KAMIYA, M. 2001. Reproductive patterns of *Caloglossa* species

(Delesseriaceae, Rhodophyta) from Australia and New Zealand: multiple origins of asexuality in *C. leprieurii*. Literature reivew on apomixis, mixed-phase, bisexuality and sexual compatibility. *Phycological Research 49*: 183–200.
WEST, J. A.; SCOTT, J. L.; WEST, K. A.; KARSTEN, U.; CLAYDEN, S. L.; SAUNDERS, G. L. 2008: *Rhodachlya madagascarensis* gen. et sp. nov.: a distinct acrochaetioid represents a new order and family (Rhodachlyales ord. nov., Rhodachlyaceae fam. nov.) of the Florideophyceae (Rhodophyta). *Phycologia 47*: 203–212.
WILCOX, S.; BLOOR, S.; HEMMINGSON, J. A.; FURNEAUX, R. H.; NELSON, W. A. 2001: The presence of gigartinine in New Zealand *Gracilaria*. *Journal of Applied Phycology 13*: 409–413.
WITHALL, R. D.; SAUNDERS, G. W. 2007: Combining small and large subunit ribosomal DNA genes to resolve relationships among orders of the Rhodymeniophycidae (Rhodophyta): recognition of the Acrosymphytales ord. nov. and Sebdeniales ord. nov. *European Journal of Phycology 41*: 379–394.
WOELKERLING, W. J.; FOSTER, M. S. 1989: A systematic and ecographic account of *Synarthrophyton schielianum* sp. nov. (Corallinaceae, Rhodophyta) from the Chatham Islands. *Phycologia 28*: 39-60.
WOELKERLING, W. J.; NELSON, W. A. 2004: A baseline summary and analysis of the taxonomic biodiversity of coralline red algae (Corallinales, Rhodopyta) recorded from the New Zealand region. *Cryptogamie Algologie 25*: 39–106.
WOMERSLEY, H. B. S. 1994: *The Marine Benthic Flora of Southern Australia, Part IIIA*. ABRS, Canberra. 508 p.
WOMERSLEY, H. B. S. 1996: *The Marine Benthic Flora of Southern Australia, Part IIIB*. ABRS, Canberra. 392 p.
WOMERSLEY, H. B. S. 1998: *The Marine Benthic Flora of Southern Australia, Part IIIC*. ABRS, Canberra. 535 p.
WOMERSLEY, H. B. S. 2003: *The Marine Benthic Flora of Southern Australia, Part IIID*. ABRS, State Herbarium of South Australia, Adelaide. 533 p.
WRAY, J. L. 1977: *Calcareous Algae*. Elsevier, Amsterdam. 185 p.
WYNNE, M. J. 2002: *Plocamium cirrhosum* comb. nov. (Plocamiales, Rhodophyta) to replace *P. costatum*. *New Zealand Journal of Botany 40*: 137–142.
WYNNE, M. J., SCHNEIDER, C. W. 2010. Addendum to the synoptic review of red algal genera. *Botanica Marina 53*: 291–299.
XIAO, S.; ZHANG, Y.; KNOLL, A. H. 1998: Three-dimensional preservation of algae and animal embryos in a Neoproterozoic phosphorite. *Nature 391*: 553–558.
YOON, H. S.; HACKETT, J. D.; BHATTACHARYA, D. 2002a: A single origin of the peridinin- and fucoxanthin-containing plastids in dinoflagellates through tertiary endosymbiosis. *Proceedings of the National Academy of Sciences, USA 99*: 11724–11729.
YOON, H. S.; HACKETT, J. D.; PINTO, G.; BHATTACHARYA, D. 2002b: The single, ancient origin of chromist plastids. *Proceedings of the National Academy of Sciences, USA 99*: 15507–15512.
YOON, H. S.; HACKETT, J. D.; CINIGLIA, C.; PINTO, G.; BHATTACHARYA, D. 2004: A molecular timeline for the origin of photosynthetic eukaryotes. *Molecular Biology and Evolution 21*: 809–818.
YOON, H. S.; MULLER, K. M.; SHEATH, R. G.; OTT, F. D.; BHATTACHARYA, D. 2006. Defining the major lineages of red algae (Rhodophyta). *Journal of Phycology 42*: 482–492.
ZHANG, Y.; YIN, L.; XIAO, S.; KNOLL, A. H. 1998: Permineralized fossils from the terminal Proterozoic Doushantuo Formation, South China. *Journal of Paleontology 72 (Suppl. 4)*: 1–52.
ZUCCARELLO, G. C; WEST, J. A. 2006: Molecular phylogeny of the subfamily Bostrychioideae (Ceramiales, Rhodophyta): subsuming *Stictosiphonia* and highlighting polyphyly in species of *Bostrychia*. *Phycologia 45*: 24–36.
ZUCCARELLO, G. C.; WEST, J. A.; KIKUCHI, N. 2008: Phylogenetic relationships within the Stylonematales (Stylonematophyceae, Rhodohyta): biogeographic patterns do not apply to *Stylonema alsidii*. *Journal of Phycology 44*: 384–393.

Checklist of New Zealand Rhodophyta

Endemic species are indicated by E, adventive species by A. Endemic genera are underlined (first entry only, if more than one species). Freshwater species are indicated by F. All other species are marine.

KINGDOM PLANTAE
SUBKINGDOM BILIPHYTA
INFRAKINGDOM RHODOPLANTAE
PHYLUM RHODOPHYTA
SUBPHYLUM CYANIDIOPHYTINA
Class CYANIDIOPHYCEAE
Order CYANIDIALES
CYANIDIACEAE
Cyanidium caldarium (Tilden) Geitler F

SUBPHYLUM RHODOPHYTINA
Class COMPSOPOGONOPHYCEAE
Order COMPSOPOGONALES
COMPSOPOGONACEAE
Compsopogon caeruleus (Balb. ex C.Agardh) Mont. F

Order ERYTHROPELTIDIALES
ERYTHROTRICHIACEAE
<u>*Chlidophyllon*</u> *kaspar* (W.A.Nelson & N.M.Adams) W.A.Nelson E
Erythrocladia irregularis Rosenv.
Erythrotrichia bangioides Levring E
Erythrotrichia carnea (Dillwyn) J.Agardh
Erythrotrichia foliiformis South & N.M.Adams
Erythrotrichia hunterae N.L.Gardner E
<u>*Pyrophyllon*</u> *cameronii* (W.A.Nelson) W.A.Nelson E
Pyrophyllon subtumens (J.Agardh ex Laing) W.A.Nelson E
Sahlingia subintegra (Rosenv.) Kornmann

Class PORPHYRIDIOPHYCEAE
Order PORPHYRIDIALES
PORPHYRIDIACEAE
Porphyridium purpureum (Bory) K.M.Drew & R.Ross

Class STYLONEMATOPHYCEAE
Order STYLONEMATALES
STYLONEMATACEAE
Chroodactylon ornatum (C.Agardh) Basson
Stylonema alsidii (Zanardini) K.M.Drew

Class BANGIOPHYCEAE
Order BANGIALES
BANGIACEAE
'*Bangia*' spp. indet. (>9) See note 1.
Clymene coleana (W.A.Nelson) W.A.Nelson E
Clymene OTA sensu Sutherland et al. in press E
<u>*Dione*</u> *arcuata* W.A.Nelson E
<u>*Lysithea*</u> *adamsiae* (W.A.Nelson) W.A.Nelson E
<u>*Minerva*</u> *aenigmata* W.A.Nelson E
Pyropia cinnamomea (W.A.Nelson) W.A.Nelson E
Pyropia columbina (Mont.) W.A.Nelson
Pyropia pulchella (Ackland, J.A.West, J.L.Scott & Zuccarello) T.J.Farr & J.E.Sutherland
Pyropia rakiura (W.A.Nelson) W.A.Nelson
Pyropia suborbiculata (Kjellm.) J.E.Broom, H.G. Choi, M.S. Hwang & W.A. Nelson
Pyropia virididentata (W.A.Nelson) W.A.Nelson E
'*Porphyra*' *woolhousiae* Harv.
'*Porphyra*' spp. undescr. (>25) See note 2.

Class FLORIDEOPHYCEAE
Subclass HILDENBRANDIOPHYCIDAE
Order HILDENBRANDIALES
HILDENBRANDIACEAE
<u>*Apophlaea*</u> *lyallii* Hook.f. & Harv. E
Apophlaea sinclairii Hook.f. & Harv. E
Hildenbrandia dawsonii (Ardre) Hollenb.
Hildenbrandia kerguelensis (Askenasy) Y.M.Chamb.
Hildenbrandia lecannellieri Har.

Subclass CORALLINOPHYCIDAE
Order CORALLINALES
CORALLINACEAE
Amphiroa anceps (Lam.) Decne.
Arthrocardia corymbosa (Lam.) Decne.
Arthrocardia wardii (Harv.) Aresch.
Corallina armata Hook.f. & Harv. E
Corallina hombronii (Mont.) Mont ex Kütz.
Corallina officinalis L.
Corallina spp. undescr. (Farr et al. 2009)
Hydrolithon farinosum (J.V.Lamour.) Penrose & Y.M.Chamb.
Hydrolithon improcerum (Foslie & Howe) Foslie
Hydrolithon onkodes (Heydr.) Penrose & Woelk.
Hydrolithon rupestris (Foslie) Penrose
Hydrolithon samoense (Foslie) Keats & Y.M.Chamb.
Jania affinis Harv.
Jania micarthrodia J.V.Lamour.
Jania novae-zelandiae Harv. E
Jania pistillaris Mont. E
Jania rosea (Lam.) Decne.
Jania sagittata (J.V.Lamour.) Blainv.
Jania verrucosa J.V.Lamour.
Jania sp. aff. *ungulata*
Lithophyllum carpophylli (Heydr.) Heydr. E
Lithophyllum corallinae (P.Crouan & H.Crouan) Heydr.
Lithophyllum detrusum Foslie E
Lithophyllum johansenii Woelk. & S.J.Campb.
Lithophyllum jugatum (Foslie) W.H.Adey E
Lithophyllum pustulatum (J.V.Lamour.) Foslie

Lithophyllum riosmenae A.Harv. & Woelk.
Lithophyllum stictaeforme (Aresch.) Hauck
Lithophyllum tuberculatum (Foslie) W.H.Adey E
Lithoporella melobesioides (Foslie) Foslie
Mastophora pacifica (Heydr.) Foslie
Neogoniolithon brassica-florida (Harv.) Setch. & L.R.Mason
Pneophyllum coronatum (Rosanoff) Penrose
Pneophyllum fragile Kütz.
Spongites tunicatus Penrose
Spongites yendoi (Foslie) Y.M.Chamb.
HAPALIDIACEAE
Choreonema thuretii (Bornet) F.Schmitz
Lithothamnion asperulum (Foslie) Foslie E
Lithothamnion crispatum Hauck
Liththamnion muelleri Lenorm. ex Rozanov
Lithothamnion proliferum Foslie
Melobesia explanata (Foslie) W.H.Adey & Lebednik E
Melobesia leptura Foslie E
Melobesia membranacea (Esper) J.V.Lamour.
Melobesia rosanoffii (Foslie) M.Lemoine
Mesophyllum engelhartii (Foslie) W.H.Adey
Mesophyllum erubescens (Foslie) M.Lemoine
Mesophyllum incisum (Foslie) W.H.Adey
Mesophyllum insigne (Foslie) W.H.Adey E
Mesophyllum macroblastum (Foslie) W.H.Adey
Mesophyllum printzianum Woelk. & A.Harv.
Phymatolithon repandum (Foslie) Wilks & Woelk.
Synarthrophyton patena (Hook.f. & Harv.) R.A.Towns.
Synarthrophyton schielianum Woelk. & M.S.Foster E
Synarthrophyton/Mesophyllum patena/englehartii spp. complex

Order SPOROLITHALES
SPOROLITHACEAE
Heydrichia homalopasta R.A.Towns. & Borow.
Heydrichia woelkerlingii R.A.Towns., Y.M.Chamb. & Keats
Sporolithon durum (Foslie) R.A.Towns. & Woelk.

Subclass NEMALIOPHYCIDAE
Order ACROCHAETIALES
ACROCHAETIACEAE
Acrochaetium leptonemoides Levring E
Acrochaetium neozeelandicum Levring E
Acrochaetium porphyrae (K.M.Drew) G.M.Sm.
Audouinella dictyotae Collins (Woelk.)
Audouinella hermannii (Roth) Duby F

Order BALBIANIALES
BALBIANIACEAE
Balbiania meiospora Skuja F

Order BALLIALES
BALLIACEAE
Ballia callitricha (C.Agardh) Kütz.
Ballia pennoides E.M.Woll.
Ballia sertularioides (Suhr) Papenf.

Order BATRACHOSPERMALES
BATRACHOSPERMATACEAE
Batrachospermum anatinum Sirodot F
Batrachospermum antipodites Entwisle F
Batrachospermum arcuatum Kylin F
Batrachospermum atrum (Huds.) Harv. F
Batrachospermum campyloclonum Skuja ex Entwisle & Foard F
Batrachospermum confusum (Bory) Hassall F
Batrachospermum discorum Entwisle & Foard F
Batrachospermum kraftii Entwisle & Foard F
Batrachospermum pseudogelatinosum Entwisle & Vis F
Batrachospermum terawhiticum Entwisle & Foard F
Batrachospermum theaguum Entwisle & Foard F
Batrachospermum virgatodecaisneanum Sirodot F
Nothocladus lindaueri Skuja F
Sirodotia delicatula Skuja F
Sirodotia suecica Kylin F
PSILOSIPHONACEAE
Psilosiphon scoparium Entwisle F

Order COLACONEMATALES
COLACONEMATACEAE
Colaconema caespitosum (J.Agardh) Jackelman, Stegenga & J.J.Bolton

Order NEMALIALES
GALAXAURACEAE
Dichotomaria marginata (J.Ellis & Sol.) Lam.
Galaxaura cohaerens Kjellm.
Galaxaura filamentosa R.C.Y.Chou
Galaxaura rugosa (J.Ellis & Sol.) J.V.Lamour.
Tricleocarpa cylindrica (J.Ellis & Sol.) Huisman & Borow.
LIAGORACEAE
Ganonema farinosum (J.V.Lamour.) K.C.Fan & Yung C.Wang
Helminthora australis J.Agardh ex Levring
Helminthora lindaueri Desikachary
Helminthocladia australis Harv.
Helminthocladia densa (Harv.) F.Schmitz & Hauptfl.
Helminthocladia dotyi Womersley
Helminthocladia sp. sensu Adams (1994)
Liagora harveyana Zeh
Liagora sp. sensu Adams (1994)
Nemalion helminthoides (Velley) Batters
SCINAIACEAE
Nothogenia fastigiata (Bory) P.G.Parkinson
Nothogenia pseudosaccata (Levring) P.G.Parkinson E
Nothogenia pulvinata (Levring) P.G.Parkinson E
Scinaia australis (Setch.) Huisman
Scinaia berggrenii (Levring) Huisman E
Scinaia firma Levring E

Order PALMARIALES
PALMARIACEAE
Palmaria decipiens (Reinsch) R.W.Ricker
RHODOTHAMNIELLACEAE
Camontagnea hirsuta (E.M.Woll.) Woelk. & Womersley
Camontagnea oxyclada (Mont.) Pujals

Subclass RHODYMENIOPHYCIDAE
Order ACROSYMPHYTALES
ACROSYMPHYTACEAE
Acrosymphyton firmum M.W.Hawkes E
Schimmelmannia sp. sensu Adams (1994)

Order BONNEMAISONIALES
BONNEMAISONIACEAE
Asparagopsis armata Harv.
Asparagopsis taxiformis (Delile) Trevis.
Delisea compressa Levring E
Delisea elegans J.V.Lamour.
Delisea plumosa Levring
Delisea pulchra (Grev.) Mont.
Ptilonia mooreana Levring E
Ptilonia willana Lindauer

Order CERAMIALES
CALLITHAMNIACEAE
Aristoptilon mooreanum (Lindauer) Hommers. & W.A.Nelson E
Callithamnion brachygonum Hook.f & Harv.
Callithamnion colensoi Hook.f & Harv.
Callithamnion consanguineum Hook.f & Harv.
Callithamnion crispulum Harv.
Callithamnion cryptopterum Kütz.
Callithamnion gracile Hook.f & Harv.
Callithamnion puniceum Hook.f & Harv.
Crouania willae R.E.Norris
Crouania sp. cf. *capricornica* Saenger & E.M.Woll.
Euptilota formosissima (Mont.) Kütz. E
Euptilota sp. nov. (northern) E
CERAMIACEAE
Acrothamnion sp. sensu Adams (1994)
Amoenothamnion planktonicum E.M.Woll.
Antithamnion decipiens (J.Agardh) Athanas.
Antithamnion hubbsii E.Y.Dawson
Antithamnion pectinatum (Mont.) Brauner
Antithamnionella adnata (J.Agardh) N.M.Adams
Antithamnionella breviramosa (E.Y.Dawson) E.M.Woll.
Antithamnionella flagellata (Boergesen) I.A.Abbott
Antithamnionella ternifolia (Hook.f. & Harv.) Lyle A
Balliella pseudocorticata (E.Y.Dawson) D.N.Young
Centroceras clavulatum (C.Agardh) Mont.
Ceramium apiculatum J.Agardh E
Ceramium aucklandicum Kütz. E
Ceramium chathamense Feldm.-Maz. E
Ceramium clarionense Setch. & N.L.Gardner
Ceramium cliftonianum J.Agardh
Ceramium codii (H.Richards) Feldm.-Maz.
Ceramium comptum Børgesen
Ceramium diaphanum (Lightf.) Roth
Ceramium discorticatum Heydr. E
Ceramium divergens J.Agardh E
Ceramium filiculum Harv. ex Womersley
Ceramium laingii Reinbold E
Ceramium lenticulare Womersley
Ceramium nanum Kuehne
Ceramium rubrum (Huds.) C.Agardh
Ceramium spyridioides Feldm.-Maz.
Ceramium stichidiosum J.Agardh
Ceramium subdichotomum Weber Bosse
Ceramium subverticillatum (Grunow) Weber Bosse
Ceramium tasmanicum (Kütz.) Womersley
Ceramium uncinatum Harv. E
Ceramium vestitum Harv. E
Gayliella flaccida (Harv. ex Kütz.) T.O.Cho & L.J.McIvor
Microcladia novae-zelandiae J.Agardh E
Microcladia pinnata J.Agardh
Microcladia sp. sensu Adams (1994)
Perithamnion ceramoides J.Agardh
Platythamnion sp. sensu Adams (1994)
Pterothamnion antarcticum (Kylin) Moe & Silva
Pterothamnion confusum (J.Agardh) Athanas.
Pterothamnion simile (Hook.f. & Harv.) Nageli
Pterothamnion squarrulosum (Harv.) Athanas. & Kraft
Skeletonella nelsoniae A.Millar & De Clerck E
Trithamnion vulgare E.M.Woll.
DASYACEAE
Colacodasya inconspicua (Reinsch) F.Schmitz
Dasya baillouviana (Gmelin) Mont.
Dasya collabens Hook.f. & Harv. E
Dasya subtilis Lindauer E
Dasya sp. sensu Adams (1994)
Heterosiphonia concinna (Hook.f. & Harv.) Reinbold E
Heterosiphonia punicea (Mont.) Kylin E
Heterosiphonia squarrosa (Hook.f. & Harv.) Falkenb. E
Heterosiphonia tessellata (Harv.) Reinbold E
DELESSERIACEAE
Abroteia suborbiculare (Harv.) Kylin E
Acrosorium ciliolatum (Harv.) Kylin
Acrosorium decumbens (J.Agardh) Kylin
Apoglossum montagneanum (J.Agardh) J.Agardh E
Apoglossum oppositifolium (Harv.) J.Agardh E
Caloglossa ogasawaraensis Okamura
Caloglossa vieillardii (Kütz.) Setch.
Delesseria crassinervia Mont. E

Delesseria nereifolia Harv. E
Erythroglossum undulatissimum (J.Agardh) Kylin E
Erythroglossum sp. sensu Adams (1994)
Gonimophyllum insulare Wagner E
Haraldiophyllum crispatum (Hook.f & Harv.) S.-M. Lin Lin, Hommers. & W.A.Nelson E
Hymenena affinis (Harv.) Kylin
Hymenena curdieana (Harv.) Kylin
Hymenena durvillaei (Bory) Kylin
Hymenena harveyana (J.Agardh) Kylin
Hymenena multipartita (Hook.f & Harv.) Kylin
Hymenena palmata (Harv.) Kylin
Hymenena variolosa(Harv.) Kylin
Hymenenopsis heterophylla S.-M.Lin, W.A.Nelson & Hommers. E
Hypoglossum sp. sensu Adams (1994)
Laingia hookeri (Lyall) Kylin E
Marionella prolifera (Kylin) Wagner E
Martensia fragilis Harv.
Myriogramme sp. sensu Adams (1994)
Nancythalia humilis A.Millar & W.A.Nelson E
Nitophyllum sp. sensu Adams (1994)
Phitymophora linearis (Laing) Kylin
Phycodrys adamsiae S.-M.Lin & W.A.Nelson E
Phycodrys franiae S.-M.Lin & W.A.Nelson E
Phycodrys novae-zelandiae S.-M.Lin & W.A.Nelson E
Platyclinia 'purpurea' sensu Adams (1994)
Platyclinia sp. sensu Adams (1994)
Schizoseris dichotoma (Hook.f & Harv.) Kylin
Schizoseris griffithsia (Suhr) M.J.Wynne
Schizoseris sp. aff. *Myriogramme gattyana* (J.Agardh) Kylin
Schizoseris sp. Bounty Islands sensu Adams (1994) E
Schizoseris sp. sensu Adams (1994)
Taenioma nanum (Kütz.) Papenf.
RHODOMELACEAE
Adamsiella angustifolia (Harv.) L.E.Phillips & W.A.Nelson E
Adamsiella chauvinii (Harv.) L.E.Phillips & W.A.Nelson E
Adamsiella lorata L.E.Phillips & W.A.Nelson E
Adamsiella melchiori L.E.Phillips & W.A.Nelson E
Amplisiphonia pacifica Hollenb.
Aphanocladia delicatula (Hook.f. & Harv.) Falkenb.
Bostrychia arbuscula Hook.f. & Harv. E
Bostrychia flagellifera E.Post
Bostrychia gracilis (R.J.King & Puttock) Zuccarello & J.A.West E
Bostrychia harveyi Mont.
Bostrychia hookeri (Harv.) Hook.f. & Harv.
Bostrychia intricata (Bory) Mont.
Bostrychia moritziana (Sond. ex Kütz.) J.Agardh
Bostrychia simpliciuscula Harv. ex J.Agardh
Bostrychia vaga Hook.f. & Harv.
Brongniartella australis (C.Agardh) Schmitz
Chondria arcuata Hollenb.
Chondria harveyana (J.Agardh) De Toni A
Chondria lanceolata Harv.
Chondria macrocarpa Harv. E
Cladhymenia coronata (Lindauer & Setch.) Saenger E
Cladhymenia lyallii Harv. E
Cladhymenia oblongifolia Harv. E
Colacopsis lophurellae Kylin
Dasyclonium adiantiformis (Decne.) Scagel E
Dasyclonium bifurcatum Scagel E
Dasyclonium bipartitum (Hook.f. & Harv.) Scagel E
Dasyclonium flaccidum (Harv.) Falkenb.
Dasyclonium harveyanum (Decne. ex Harv.) Scagel E
Dasyclonium incisum (J.Agardh) Kylin
Dasyclonium ovalifolium (Hook.f. & Harv.) Scagel E
Dipterosiphonia heteroclada (J.Agardh) Falkenb. E
Echinothamnion hystrix (Hook.f. & Harv.) Kylin
Echinothamnion lyallii (Hook.f. & Harv.) Kylin E
Echinothamnion sp. sensu Adams (1994)
Herposiphonia ceratoclada (Mont.) Reinbold
Herposiphonia clavata M.J.Wynne
Herposiphonia sp. sensu Adams (1994)
Janczewskia sp.
Laurencia brongniartii J.Agardh
Laurencia distichophylla J.Agardh E
Laurencia elata (C.Agardh) Hook.f. & Harv.
Laurencia gracilis Hook.f. & Harv.
Laurencia thyrsifera J.Agardh
Lembergia allanii (Lindauer) Saenger E
Lophosiphonia prostrata (Harv.) Falkenb.
Lophosiphonia sp. sensu Adams (1994)
Lophurella caespitosa (Hook.f. & Harv.) Falkenb. E
Lophurella hookeriana (J.Agardh) Falkenb.
Lophurella periclados (Sond.) Schmitz
Metamorphe colensoi (Hook.f. & Harv.) Falkenb. E
Microcolax botryocarpa (Hook.f. & Harv.) F.Schmitz
Neosiphonia apiculata (Hollenb.) Masuda & Kogame
Neosiphonia harveyi (J.W.Bailey) M.-S.Kim, H.-G. Choi, Guiry & G.W.Saunders A
Perrinia ericoides (Harv.) Womersley
Placophora binderi (J.Agardh) J.Agardh
Pleurostichidium falkenbergii Heydr. E
Polysiphonia abscissoides Womersley
Polysiphonia adamsiae Womersley
Polysiphonia aterrima Hook.f. & Harv. E
Polysiphonia brodiei (Dillwyn) Spreng. A
Polysiphonia constricta Womersley A
Polysiphonia decipiens Mont.
Polysiphonia hancockii E.Y.Dawson
Polysiphonia isogona Harv.
Polysiphonia morrowii Harv. A
Polysiphonia muelleriana J.Agardh E
Polysiphonia pernacola N.M.Adams E
Polysiphonia rhododactyla Harv. E
Polysiphonia rudis Hook.f. & Harv.
Polysiphonia scopulorum Harv.
Polysiphonia senticulosa Harv. A
Polysiphonia sertularioides (Grateloup) J.Agardh A
Polysiphonia strictissima Hook.f. & Harv. E
Polysiphonia subtilissima Mont. A
Polysiphonia sp. sensu Adams 1994
Pterosiphonia pennata (C.Agardh) Sauv.
Sporoglossum lophurellae Kylin
Streblocladia glomerulata (Mont.) Papenf. E
Symphyocladia marchantioides (Harv.) Falkenb.
Vidalia colensoi (Hook.f. & Harv.) J.Agardh E
SPYRIDIACEAE
Spyridia dasyoides Sond.
Spyridia filamentosa (Wulfen) Harv.
WRANGELIACEAE
Anotrichium crinitum (Kütz.) Baldock
Dasyptilon pellucidum (Harv.) Feldmann E
Deucalion levringii (Lindauer) Huisman & Kraft
Griffithsia antarctica Hook.f. & Harv.
Griffithsia crassiuscula C.Agardh A
Griffithsia monilis Harv.
Griffithsia teges Harv.
Griffithsia traversii (J.Agardh) Baldock E
Gymnothamnion elegans (Schousb.) J.Agardh
Lophothamnion hirtum (Hook.f. & Harv.) Womersley
Medeiothamnion lyallii (Harv.) Gordon E
Medeiothamnion norrissii Gordon-Mills E
Ochmapexus minimus (Harv.) Womersley
Ptilothamnion rupicolum Gordon-Mills E
Ptilothamnion schmitzii Heydr. E
Spermothamnion sp.
Spongoclonium pastorale Laing E
Spongoclonium sp. sensu Adams (1994)
Wrangelia penicillata (C.Agardh) C.Agardh
Wrangelia sp. sensu Adams (1994)

Order GELIDIALES
GELIDIACEAE
Capreolia implexa Guiry & Womersley
Gelidium allanii V.J.Chapm.
Gelidium caulacantheum J.Agardh
Gelidium hommersandii A.Millar & Freshwater
Gelidium longipes J.Agardh E
Gelidium microphyllum (Crosby-Sm.) Kylin E
Gelidium sp. sensu Nelson et al. (2006) E
Pterocladia lucida (Turner) J.Agardh
Pterocladiella capillacea (S.G.Gmel.) Santel. & Hommers.

Order GIGARTINALES
CALOSIPHONIACEAE
Schmitzia evanescens M.W.Hawkes E
CAULACANTHACEAE
Catenella fusiformis (J.Agardh) Skottsb.
Catenella nipae Zanardini
Caulacanthus ustulatus (Turner) Kütz.
Taylorophycus filiformis Searles E
CYSTOCLONIACEAE
Craspedocarpus erosus (Hook.f. & Harv.) Schmitz E
Rhodophyllis acanthocarpa (Harv.) J.Agardh
Rhodophyllis lacerata Hook.f. & Harv.
Rhodophyllis membranacea (Harv.) Hook.f. & Harv.
GIGARTINACEAE
Chondracanthus chapmanii (Hook.f. & Harv.) Fredericq E
Gigartina clavifera J.Agardh E
Gigartina dilatata (Hook.f. & Harv.) N.M.Adams E
Gigartina divaricata Hook.f. & Harv. E
Gigartina grandifida J.Agardh E
Gigartina laingii Lindauer E
Gigartina macrocarpa J.Agardh E
Gigartina minima V.J.Chapm. E
Gigartina pachymenioides Lindauer E
Gigartina sp.'Lindauer Exsicc. No. 164' E
Gigartina sp. subantarctic sensu Adams (1994) E
Gigartina sp. Bounty Is. sensu Adams (1994) E
Gigartina sp. Three Kings Is. sensu Adams (1994) E
'Gigartina' ancistroclada Mont. E
'Gigartina' atropurpurea (J.Agardh) J.Agardh
Iridaea lanceolata Harv.
Iridaea tuberculosa (Hook.f. & Harv.) Leister
Iridaea sp. sensu Adams (1994) E
Psilophycus alveata (Turner) W.A.Nelson, Leister & Hommers. E
Rhodoglossum latissimum (Hook.f. & Harv.) J.Agardh
Rhodoglossum sp.
Sarcothalia circumcincta J.Agardh E
Sarcothalia lanceata (J.Agardh) Hommersand E
Sarcothalia livida (Turner) Grev. E
Sarcothalia marginifera (J.Agardh) Hommersand E
'Sarcothalia' decipiens (Hook.f & Harv.) Hook.f. & Harv. E
GLOIOSIPHONIACEAE
Hypnea charoides J.V.Lamour.
Hypnea cornuta (Kütz.) J.Agardh A
Hypnea esperi Bory
Hypnea nidifica J.Agardh
KALLYMENIACEAE
Callocolax neglectus Schmitz ex Batters
Callophyllis angustifrons (Hook.f. & Harv.) South & N.M.Adams E
Callophyllis atrosanguinea (Hook.f & Harv.) Har.
Callophyllis calliblepharoides J.Agardh E
Callophyllis decumbens J.Agardh E
Callophyllis hombroniana (Mont.) Kütz. E
Callophyllis laingiana A.Millar E
Callophyllis ornata (Mont.) Kütz. E
Callophyllis variegata (Bory) Kütz.

Ectophora depressa J.Agardh E
Ectophora marginata D'Archino & W.A.Nelson E
Glaphrymenia pustulosa J.Agardh
Psaromenia berggrenii (J.Agardh) D'Archino, W.A.Nelson & Zuccarello E
Pugetia delicatissima R.E.Norris
Rhizopogonia asperata (Harv.) Kylin E
Thamnophyllis laingii (J.Agardh) R.E.Norris E
PHACELOCARPACEAE
Phacelocarpus labillardieri (Turner) J.Agardh
Phacelocarpus sessilis Harv. ex J.Agardh
PHYLLOPHORACEAE
Gymnogongrus furcatus (Hook.f. & Harv.) Kütz. E
Gymnogongrus humilis Lindauer
Gymnogongrus torulosus (Hook.f. & Harv.) F.Schmitz E
Stenogramma interruptum (C.Agardh) Mont. ex Harv.
RHIZOPHYLLIDACEAE
Nesophila hoggardii W.A.Nelson & N.M.Adams E
SOLIERIACEAE
Placentophora colensoi (Hook.f. & Harv.) Kraft E
Solieria robusta (Grev.) Kylin
'Solieria' sp. WELT A020843 A

Order PEYSSONNELIALES
PEYSSONNELIACEAE
Peyssonnelia capensis Mont.
Peyssonnelia novae-hollandiae Kütz.
Peyssonnelia rugosa Harv. E
Peyssonnelia sp. sensu Adams (1994)
Sonderopelta coriacea Womersley & Sinkora

Order GRACILARIALES
GRACILARIACEAE
Curdiea balthazar W.A.Nelson, G.A.Knight & R.Falshaw E
Curdiea codioides V.J.Chapm. E
Curdiea coriacea (Hook.f. & Harv.) V.J.Chapm. E
Curdiea flabellata V.J.Chapm. E
Gracilaria chilensis Bird, McLachlan & Oliveira
Gracilaria pulvinata Skottsb.
Gracilaria secundata Harv.
Gracilaria truncata Kraft E
Gracilaria sp. Manukau Harbour A
Melanthalia abscissa (Turn.) Hook.f. & Harv. E
PTEROCLADIOPHYLLACEAE
Pterocladiophila hemisphaerica K.C.Fan & Papenf.

Order HALYMENIALES
HALYMENIACEAE
Aeodes nitidissima J.Agardh
Cryptonemia latissima J.Agardh
Cryptonemia sp. sensu Adams (1994)
Cryptonemia cf. *umbraticola* E.Y.Dawson
Glaphyrosiphon intestinalis (Harv.) Leister & W.A.Nelson E
Glaphyrosiphon lindaueri W.A.Nelson & P.W.Gabrielson E
Grateloupia aucklandica Mont. E
Grateloupia longifolia Kylin E
Grateloupia prolifera J.Agardh E
Grateloupia stipitata J.Agardh E
Grateloupia turuturu Yamada A
Grateloupia urvilleana (Mont.) Parkinson E
Grateloupia sp.'unknown' sensu Adams (1994)
Halymenia latifolia Kütz.
Pachymenia crassa Lindauer E
Pachymenia dichotoma J.Agardh E
Pachymenia laciniata J.Agardh E
Pachymenia lusoria (Grev.) J.Agardh E
Polyopes? sensu Lindauer ANZE 345
Prionitis decipiens (Mont.) J.Agardh
TSENGIACEAE
Tsengia feredayae (Harv.) Womersley & Kraft
Tsengia laingii (Kylin) Womersley & Kraft

Order NEMASTOMALES
NEMASTOMACEAE
Catenellopsis oligarthra (J.Agardh) V.J.Chapm. E
Nemastoma laciniata J.Agardh E
Predaea sp. sensu Adams (1994)
SCHIZYMENIACEAE
Schizymenia novae-zelandiae J.Agardh E
Schizymenia? sp. Kermadecs sensu Adams (1994)
Schizymenia? sp. Bounty Is. sensu Adams (1994)

Order PLOCAMIALES
PLOCAMIACEAE
Plocamium angustum (J.Agardh) Hook.f. & Harv.
Plocamium cartilagineum (L.) P.S.Dixon
Plocamium cirrhosum (Turner) M.J.Wynne E
Plocamium hamatum J.Agardh
Plocamium leptophyllum Kütz.
Plocamium microcladioides South & N.M.Adams E
Plocamium sp. sensu Adams (1994)
Plocamiocolax sp.
PSEUDOANEMONIACEAE
Hummbrella hydra Earle
SARCODIACEAE
Sarcodia grandifolia Levring E
Sarcodia montagneana (Hook.f. & Harv.) J.Agardh
Trematocarpus acicularis (J.Agardh) Kylin E

Order RHODYMENIALES
CHAMPIACEAE
Champia affinis (Hook.f. & Harv.) J.Agardh A
Champia chathamensis V.J.Chapm. & Dromgoole E
Champia laingii Lindauer E
Champia novae-zelandiae (Hook.f. & Harv.) J.Agardh E
Champia parvula (C.Agardh) Harv.
Champiocolax sp. nov. E
FAUCHEACEAE
Cenacrum subsutum R.W.Ricker & Kraft
Fauchea sp. sensu Adams (1994)
Gloioderma saccatum (J.Agardh) Kylin E
Gloiocolax novae-zelandiae Sparling E
Gloiodermatopsis setchellii Lindauer E
Webervanbossea tasmanensis Womersley
HYMENOCLADIACEAE
Hymenocladia sanguinea (Harv.) Sparling E
LOMENTARIACEAE
Gelidiopsis intricata (C.Agardh) Vickers
Lomentaria caespitosa (Harv.) V.J.Chapm. E
Lomentaria saxigena V.J.Chapm. E
Lomentaria secunda (Hook.f. & Harv.) V.J.Chapm. E
Lomentaria umbellata (Hook.f. & Harv.) Yendo E
RHODYMENIACEAE
Botryocladia skottsbergii (Boergesen) Levring
Cephalocystis furcellata (J.Agardh) A.Millar, G.W.Saunders, I.M.Strachan & Kraft
Chrysymenia ornata (J.Agardh) Kylin
Chrysymenia?polydactyla Hook.f. & Harv. E
Coelarthrum decumbens Huisman
Gloiosaccion brownii Harv.
Rhodymenia dichotoma Hook.f. & Harv. E
Rhodymenia foliifera Harv.
Rhodymenia hancockii E.Y.Dawson
Rhodymenia leptophylla J.Agardh
Rhodymenia linearis J.Agardh E
Rhodymenia novazelandica E.Y.Dawson E
Rhodymenia obtusa (Grev.) Womersley
Rhodymenia sonderi P.C.Silva
Rhodymenia sp. sensu Adams (1994)
Rhodymeniocolax sp.

Order SEBDENIALES
SEBDENIACEAE
Sebdenia lindaueri Setchell ex V.J.Chapm. E

INCERTAE SEDIS
Carpococcus linearis J.Agardh E
'Gelidium' ceramoides Levring E

Note 1. On the basis of 18S ribosomal DNA sequences, more than nine species of unnamed filamentous Bangiales can be recognised in New Zealand collections. These are referred to here as 'Bangia'.

Note 2. More than 25 distinct species of undescribed thallose Bangiales can be recognised on the basis of both 18S ribosomal DNA sequences and various morphological, anatomical, and growth characteristics. These are referred to here as 'Porphyra' although a number of these are known to belong to other segregate genera (Sutherland et al. in press).

TWENTY-THREE

Phyla CHLOROPHYTA and CHAROPHYTA

green algae

PAUL A. BROADY, ELIZABETH A. FLINT,
WENDY A. NELSON, VIVIENNE CASSIE COOPER,
MARY D. DE WINTON, PHILIP M. NOVIS

Green algae include unicellular, filamentous, and frondose plants of relatively simple construction; that is, the plant body (thallus) is not differentiated into tissues and lacks multicellular embryos. Neverthless, green algae have long been recognised to be evolutionarily related to land plants (embryophytes) because of the common possession of chlorophylls *a* and *b* and starch in the chloroplast. There are two major lineages of green algae. One, the chlorophyte clade (branch), includes most traditional green algae; the other, the 'charophyte' clade, contains a smaller number of green-algal taxa that includes both structurally simple forms and some with a more complex plant body that has been called parenchymatous (Lewis & McCourt 2004). The chlorophyte clade is also known as the UTC clade, referring to the major classes Ulvophyceae, Trebouxiophyceae, and Chlorophyceae. The charophyte clade, often called Charophyceae, itself contains five or six distinct groups, each of which has also been ranked as a class. One of them – and there is still some debate as to which – constitutes a sister group to land plants. The various viewpoints are discussed, among others, by Lemieux et al. (2000, 2007), Karol et al. (2001), Turmel et al. (2003, 2007), Lewis and McCourt (2004), McCourt et al. (2004), Hall and Delwiche (2007), and Finet et al. (2010).

Streptophyta is the name given to the clade that includes land plants and their green-algal ancestor (Bremer 1985). It has been used as a taxon name at the level of phylum or subphylum (Streptophytina). The treatment used here defers to the editorial scope of the present volume wherein land plants are classified in two phyla, a monophyletic Bryophyta (see Graham & Gray 2001; Nishiyama et al. 2004) and a phylum of vascular plants, the Tracheophyta. In this schema Streptophyta may be conceived of as an infrakingdom or superphylum in which a paraphyletic Charophyta includes the ancestor of land plants.

The number of classes of green algae that are recognised, and their delineation, are in a state of flux, with recent treatments attempting to incorporate data from molecular and ultrastructural studies. In traditional systems of classification for the green algae, thallus organisation was used as the principal basis for distinguishing orders (e.g. Bold & Wynne 1985). This meant, for example, that unbranched filamentous species (Ulotrichales) were separated from those with branched filaments (Chaetophorales), frondose taxa (Ulvales), or those with siphonous construction (Caulerpales, Dasycladales). Recent research has shown, however, that cell shape, thallus form, uni- or multinucleate

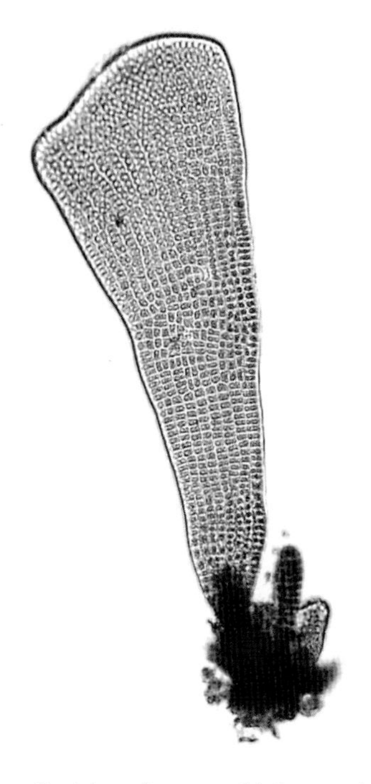

Prasiola sp. from an oxidation pond at Christchurch Wastewater Treatment Plant (Chlorophyta, Trebouxiophyceae, Prasiolales).
Paul Broady

Filaments of *Klebsormidium crenulatum* (Charophyta, Klebsormidiophyceae, Klebsormidiales).
Phil Novis

cells, plastid morphology, and other features, are evolutionarily convergent at higher levels. For example, unbranched filaments are found in genera such as *Ulothrix, Uronema,* and *Klebsormidium,* genera now classified in three different classes. Parallelism is being repeatedly demonstrated, especially amongst taxa of simple morphology, and once well-accepted genera are being found to contain species that lie within distinctly different evolutionary lineages. *Chlorella* is an excellent example (Huss et al. 1999). Molecular data suggest that 19 taxa traditionally assigned to *Chlorella* are distributed between two classes and that only four species should be retained within *Chlorella*. The smallest spherical and ellipsoidal coccoid green algae found in fresh water (less than three micrometres diameter) are frequently identified as '*Chlorella*-like' during ecological studies. These belong to at least six genera distributed between two classes (Krienitz et al. 1999; Neustupa et al. 2009). Other examples are found in the genera *Carteria* (Nozaki et al. 2003) and *Radiofilum* (Novis et al. 2010). The classification scheme used in this chapter is more or less based on that in Lewis and McCourt (2004).

Estimates of the total number of recognised species of green algae in all habitats range from about 15,000–17,000 species whilst the actual total is conjectured to be in the range 34,000–124,000 (Norton et al. 1996). In this chapter, the highly diverse freshwater and terrestrial green algae will be described first, followed by an account of marine species – unicells and seaweeds.

Freshwater and terrestrial green algae

Global diversity of known freshwater and terrestrial green algae comprises about 8000 species in 520 genera (Bourrelly 1990), including about 3000 species in 43 genera of desmids. Of these, there are more than 600 species in about 180 genera in terrestrial habitats (Ettl & Gärtner 1995).

The morphology of freshwater and terrestrial species varies widely (Hoek et al. 1995; Graham et al. 2009). There is a bewildering range of unicells (see below for an account of the desmids), ranging in size from those less than two

Summary of New Zealand green-algal diversity

Taxon	Described species + infraspecific taxa	Known undescr./ undet. species	Estimated unknown species	Adventive species + infra-specific taxa	Endemic species + infraspecific taxa	Endemic genera
CHLOROPHYTA	566+42	19	610	10+1	29+8	2
Mamiellophyceae	3	0	30	0	0	0
Nephroselmidophyceae	2	0	15	0	0	0
Pyramimonadophyceae	11	2	30	0	0	0
Chlorodendrophyceae	1	0	5	0	0	0
Chlorophyceae	287+29	4	350	1	4+8	0
Ulvophyceae	181+11	11	60	9+1	24	2
Trebouxiophyceae	81+2	2	120	0	1	0
CHAROPHYTA	527+213	7	65	1?	43+41	0
Mesostigmatophyceae	0	0	3	0	0	0
Chlorokybophyceae	0	0	2	0	0	0
Klebsormidiophyceae	9+2	0	3	0	1	0
Zygnemophyceae	496+207	5	50	0	39+41	0
Coleochaetophyceae	6	0	2	0	0	0
Charophyceae	16+4	2	5	1?	3	0
Totals	1,093+255	26	675	11?+1	72+49	2

micrometres in diameter to single multinucleate cells up to 1.5 millimetres long, from non-motile, walled 'coccoid' forms to scale-coated flagellates, from simple spheres, needles, and crescents to intricate symmetrical shapes, and from free-floating cells to those attached to substrata by stalks. Colonies, too, can be non-motile or propelled by long cilia, with cell numbers from two to thousands whilst these can be closely joined or dispersed through gelatinous mucilage. Cells attached end-to-end form filaments that can be unbranched or branched in a variety of patterns. This growth-form reaches its most complex in the structurally unique stoneworts (Characeae) (see below). This name derives from the deposition of limey calcite on the cell walls of many species. A few green algae develop a sheet-like morphology one or two cells thick.

Coleochaete orbicularis, which has a sheet-like morphology (Charophyta, Coleochaetophyceae, Coleochaetales).

Phil Novis

Any sample from an aquatic or terrestrial habitat is likely to contain between one and several tens of species of green algae. Even where algal growths are not immediately visible, microscopic examination or culture techniques will usually reveal a rich diversity. They are prominent when seen as massive growths of filaments in streams and ponds. Divers can encounter underwater meadows of Characeae covering lake sediments, especially in hard, alkaline waters, whilst mountaineers often traverse snow coloured by green algae in alpine and polar regions of the world. Although not usually visible to the unaided eye, they are ubiquitous as members of the phytoplankton community, suspended in the water of lakes (Happey-Wood 1988).

Terrestrial species are obvious as green crusts on walls and fences in the city environment. Even at the furthest-south locations of algae in Antarctica (86°30′ S), a microscopic filamentous species occurs as small vivid-green patches at the ice surface below a thin veneer of moraine (Broady & Weinstein 1998). However, green algae do not always appear green, as accumulations of photoprotective carotenoid pigments can mask the chlorophyll, as in orange crusts of filamentous *Trentepohlia* and red snow algae.

New Zealand green-algal diversity by environment

Taxon	*Marine brackish species + infraspecific taxa	Terrestrial species + infraspecific taxa	*Freshwater/ brackish species + infraspecific taxa	Fossil species
CHLOROPHYTA	156+2	35+5	394+34	11
Mamiellophyceae	3	0	0	0
Nephroselmidophyceae	2	0	0	0
Pyramimonadophyceae	13	0	0	7
Chlorodendrophyceae	1	0	0	0
Chlorophyceae	3	10	278+28	3
Ulvophyceae	126+2	20+5	46+4	1
Trebouxiophyceae	8	5	70+2	0
CHAROPHYTA	0	1+5	533+208	7
Mesostigmatophyceae	0	0	0	0
Chlorokybophyceae	0	0	0	0
Klebsormidiophyceae	0	1	8+2	0
Zygnemophyceae	0	0+5	501+202	6
Coleochaetophyceae	0	0	6	0
Charophyceae	0	0	18+4	1
INCERTAE SEDIS	0	0	0	7
Totals	156+2	36+10	927+242	25

* There is no overlap in the species lists of living marine-brackish and freshwater-brackish taxa (i.e. names are not duplicated in the lists).

In soils, unicellular, simple-filamentous, and colonial forms are widespread and often abundant contributors to surface mats and films. A conservative estimate of their productivity, based on sparse data, is at least 80 kilograms of carbon per hectare per year (Andersen 1998). Green algae live in soils from the cold deserts of Antarctica (Broady 1996) to hot arid regions (Johansen 1993).

As the photosynthetic partner in mutualisms they are common in lichens (Tschermak-Woess 1988). Species of coccoid *Trebouxia* and *Pseudotrebouxia* are found in 50–70% of all lichens, and about 24 additional chlorophyte genera are amongst about 40 genera of algae in the remainder. In fresh waters, endosymbionts are also almost exclusively coccoid chlorophytes, these being unicellular *Chlorella* and *Chlorella*-like species found within ciliate protists, the cnidarian *Hydra,* and sponges (Reisser & Widowski 1992). This contrasts with marine mutualisms that involve a greater range of algal taxa.

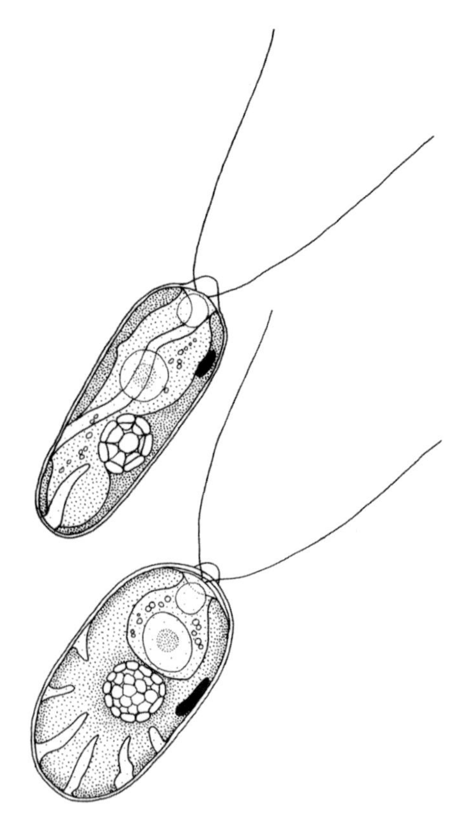

Chlamydomonas rima (left) and *C. chlorostellata* (Chlorophyta, Chlorophyceae, Chlamydomonadales).
After Flint & Ettl 1966

Importance to humans – benefits

As oft-abundant members of algal communities in natural freshwater and terrestrial habitats, green algae play a vital role by influencing the physicochemical and biotic features of their environment. Their importance for humans has ranged from contributions via food webs and the production of harvested wild freshwater fish to the development of high-technology products in the modern world (Pulz & Gross 2004; Raja et al. 2008).

Green microalgae have entered the realm of commerce (Apt & Behrens 1999). Dried biomass or cell extracts from *Chlorella* represent one of the dominant commercial opportunities in microalgal biotechnology, with the products supplying the health-food market. The cyanobacterium *Spirulina* is generally well-known as a source of dietary protein but *Chlorella* and the colonial chlorophyte *Scenedesmus* can also provide protein of 80% the quality of casein if treated correctly (Becker 1994). In animal tests, there is evidence that dietary supplements of these algae can help maintain low blood cholesterol. Although photoautotrophic growth would seem intuitively to be the obvious way to produce biomass, *Chlorella* and other microalgae can also be grown heterotrophically in the dark, for instance by the fermentation of glucose. In Japan, over 500 tonnes per year are produced in this way.

The high-value pigment astaxanthin is added to fish food in order to provide an attractive colour to the flesh of farm-grown salmon and trout. *Haematococcus,* a flagellate unicell, is an abundant producer of this pigment (Hagen et al. 2000; Domínguez-Bocanegra et al. 2004) and forms the basis of successful enterprises. Pigments from other freshwater chlorophytes have been incorporated into poultry feed and cosmetics (Del Campo et al. 2007).

Microalgae, including the chlorophyte unicells *Chlamydomonas* and *Neochloris,* are ideal sources of stable isotopically labelled compounds for medical diagnostic tests (Behrens et al. 1994). These commercial applications could be enhanced and expanded in the future as a result of intensive research on genetic-transformation techniques using *Chlamydomonas* and *Chlorella*.

Coccoid and colonial green algae are often dominant members of phytoplankton communities in sewage oxidation ponds where they play an integral role in providing oxygen for the aerobic breakdown of organic matter to inorganic consitituents (Oswald 1988). Experimental high-rate ponds have been developed in which the shallow water is vigorously circulated by paddles. An application of these is the removal of nutrients from animal wastes with around 70% of nitrogen and 50% of phosphorus being absorbed by the algae (Fallowfield et al. 1999). The algal biomass can be harvested, dried, and used in animal feed. It has been successfully fed to poultry and pigs at up to 75% of the feed preparation (Becker 1994). Most interest, however, is now focused on energy production from these algae, which are a source of a wide range of biofuels (Demirbas 2010; Pittman et al. 2011). A recent development that greatly increases productivity and processing rates is the biocoil in which wastewater and algae, often *Chlorella*, are circulated around a coil of transparent plastic tubing. In this case, the algal product has been tested as fuel for an engine that, when coupled to a generator, can produce electric power. Another approach is to use shallow streams with the algae growing on submerged surfaces from which they can be more easily harvested (Hoffmann 1998). The branched, filamentous chlorophyte *Stigeoclonium* has been tested in such systems.

Colonial *Botryococcus braunii* has been the subject of considerable interest as a potential fuel source that could be a partial replacement for fossil fuel (Tenaud et al. 1989; Li & Qin 2005; Weiss et al. 2010). One physiological form can accumulate lipids up to 75% of its dry weight. Mass growths of related species in the past have contributed to high-grade oil shales and coals. *Botryococcus braunii* can form surface blooms in nutrient-rich lakes and its ease of harvest could be coupled to both fuel production and nutrient-reduction in the water body. Another source of energy could be hydrogen obtained by the biophotolysis of water (Greenbaum 1988; Melis & Happe 2004). The unicellular chlorophytes *Scenedesmus* and *Chlamydomonas* have been shown to be capable of using light energy to split water into hydrogen and oxygen. Research is underway to genetically engineer *Chlamydomonas reinhardtii* so that it achieves this trick under a broader range of conditions (e.g. Kruse & Hankamer 2010).

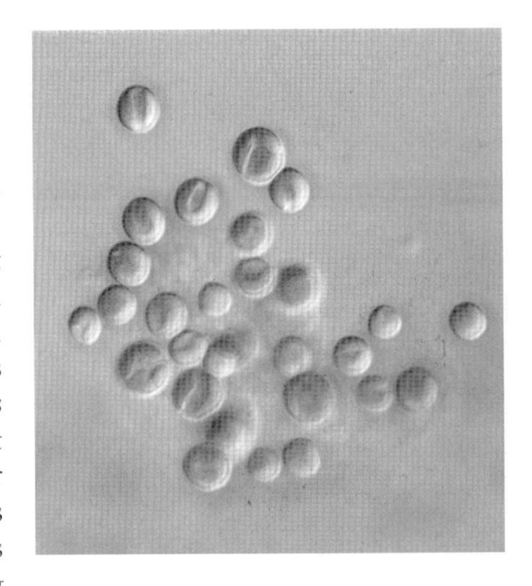

Unicells of *Chlorella* sp. (Chlorophyta, Trebouxiophyceae, Chlorellales).
William Bourland, per Micro*scope (MBL)

Those with the urge to colonise space are excited by the prospect of using *Chlorella* aboard space-craft as a component of life-support sytems that would process wastes whilst providing oxygen and high-protein nourishment (Wharton et al. 1988).

Algae are used increasingly for testing water quality (Skulberg 1995; Haglund 1997). They are well suited to the need for relatively inexpensive short-term assessment procedures that are defensible during proceedings associated with application of regulatory laws. Green algae act as environmental indicators, both by taxonomic analysis of algal communities collected from the field and in laboratory-based tests. The latter include specific strains in bioassays of nutrient availability or toxin levels. The algal clone NIVA-CHL 1 '*Selenastrum capricornutum*' is probably the most commonly used test alga worldwide. The importance of a well-founded stable taxonomy is emphasised by the subsequent realisation that this important alga should really be assigned to *Raphidocelis subcapitata*. It and the colonial *Scenedesmus subspicatus* are used in ISO standardized algal assays for determination of the toxic effects of chemicals on the growth of planktonic algae (Chen et al. 2005).

Studies of algal floras in mine drainage areas show that green algae are more tolerant of metal pollution than are cyanobacteria and diatoms. Novis and Harding (2007) listed 14 chlorophyte and seven charophyte taxa from habitats influenced by acid-mine drainage. Green algae have been investigated for use in binding and removal of metal ions from polluted waters (Greene & Darnall 1990; Gupta & Rastogi 2008). Research, often utilising coccoid chlorophytes, has shown surfaces of algae to be a mosaic of metal-ion binding sites with affinities for metals including aluminium, copper, chromium, and even gold. Copper binding to algae has been successfully applied to treatment of electroplating wastewaters. Despite such apparent successes, there is still doubt whether the technique would be useful for anything other than small volumes of contaminated water (Becker 1994).

Green algae in soils have been investigated as bioremediation agents (Lukešová 2001) for the modification of soil structure to optimise aggregate stability and reduce wind and water erosion (Metting et al. 1988). A potential has been recognised for the mass culture of those species that produce copious polysaccharide extracellular mucilages and the application of these to surfaces of agricultural soils.

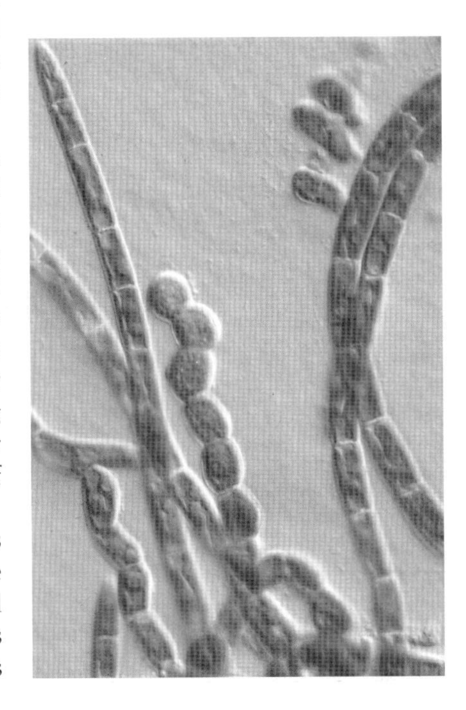

Stigeoclonium cf. *helveticum* (Chlorophyta, Chlorophyceae, Chaetophorales).
Phil Novis

Freshwater green algae include excellent examples of experimental organisms used to increase understanding of fundamental plant characteristics, e.g. species of *Chlamydomonas* (Trainor & Cain 1986). Classic experiments investigat-

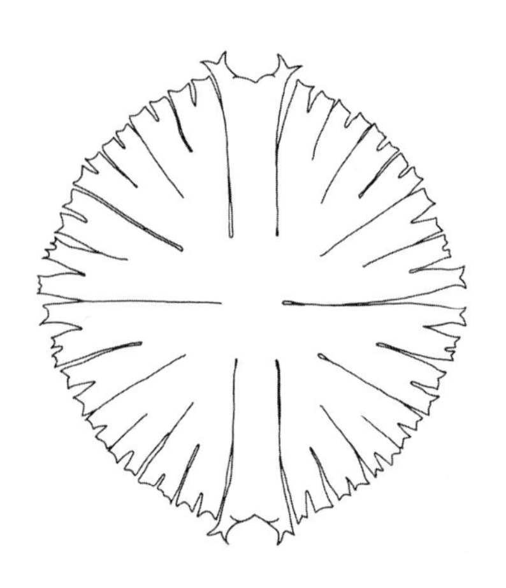

Micrasterias rotata (Charophyta, Zygnemophyceae, Desmidiales).
From Croasdale & Flint 1986

ing the biochemistry of photosynthesis have used cultures of *Chlorella*, whilst *Chlamydomonas* now provides a model system for studies of its genomic control (Davies & Grossman 1998). Additionally, *Chlamydomonas* and *Volvox* aid investigation of the genetic basis of many other cellular attributes (Merchant et al. 2007). The intricately shaped unicellular desmid *Micrasterias* has contributed to understanding of cellular differentiation, including production and organisation of cellulose microfibrils during cell-wall development. The giant cells of *Chara* and related Characeae are ideal for insertion of microelectrodes in electrophysiological studies of the control of ion flow across membranes. Major advances have been made in revealing the origin of sexual reproduction in plants by study of other charophyte algae, in particular the minute disc-like epiphyte *Coleochaete* (Graham et al. 2000).

Importance to humans – negative impacts

Although toxic species of cyanobacteria and dinoflagellates are the most well-known examples of negative effects of algae for humans there can be adverse impacts of growths of freshwater green algae.

Filamentous species are amongst the major 'weed' algae of fresh waters when they form free-floating surface mats, like the water net *Hydrodictyon*, or extensive thick growths of *Cladophora* (sometimes known as blanket weed) attached to rock or concrete surfaces (Lembi et al. 1988; Biggs 1996). In water bodies receiving excessive nutrients because of human activities, algal growths can have adverse effects on recreational swimming, fishing, and boating and can impede water flow in irrigation channels and drainage ditches. Regulated water-flows downstream from hydropower dams often develop proliferations of filamentous green algae (Blinn et al. 1998). Following their death and decay, they can impart unpleasant tastes and odours to drinking water and the oxygen deficit in streamwater can stress fish. Control of algal growth can be costly, as through the application of herbicides and mechanical harvesting, but alternatively a profitable harvest of fish (e.g. *Tilapia* and grass carp) can be made when these are used in biological control.

Chlorophytes have been implicated in human illness (Stein & Borden 1984; Thiele & Bergmann 2002). Unicellular *Prototheca* is an unusual parasite, which molecular genetics has confirmed to be a non-photosynthetic chlorophyte (Huss et al. 1999). It is found free-living in aquatic habitats but can cause lesions in humans, dogs, cats, cattle, and fruit bats and, in Australia, ulcerative dermatitis of platypus. Morphologically similar cells containing green chloroplasts have also been seen in infected animal tissues. The trebouxiophyte *Helicosporidium*, previously thought to be a fungus, is a colourless green-algal parasite of invertebrates (Tartar et al. 2002). Other, commoner, chlorococcalean soil algae, which can be found in samples of air and house dust, have stimulated clinical responses in allergenic individuals.

Species of three genera – *Cephaleuros*, *Stomatochroon*, and *Phyllosiphon* – parasitise leaves in moist regions of the tropics and subtropics (Joubert & Rijkenberg 1971; Thompson & Wujek 1997). *Cephaleuros virescens* infects 105 species from 40 families of vascular plants and has a reputation for causing serious economic damage to *Citrus*, tea, coffee, and cocoa crops. Filaments of *Stomatochroon* ramify through leaf tissue and can block stomates. *Phyllosiphon* is unusual in that only its spores contain chloroplasts. The filaments found in leaves of Araceae are nutritionally dependent on the host plant.

New Zealand freshwater and terrestrial green-algal diversity

Some 1108 species and 254 separate (additional) infraspecific taxa of freshwater aquatic and terrestrial chlorophytes and charophytes from 194 genera have been recorded in New Zealand. Desmids (see below) comprise 444 (46%) of the total species and 197 (77%) of the total varieties but these are distributed amongst just 26 (13%) of the total genera.

Overall, the New Zealand green-algal flora contains about 38% of the genera and about 11% of the species known worldwide. Total freshwater and terrestrial green-algal species diversity in New Zealand seems rather less than in Australia (Day et al. 1995), with about 1470 species, and the British Isles (Whitton et al. 1998), with about 1430 species. In both the latter, as in New Zealand, desmids predominate and comprise about 55% of the total flora. There are substantial uncertainties in all these estimates and the floras of Australia and the British Isles are acknowledged as being poorly known (John et al. 2002; Huisman & Saunders 2007).

In New Zealand, only three orders have received extensive collection and intensive taxonomic treatment, these being the Desmidiales or desmids (Croasdale & Flint 1986, 1988; Croasdale et al. 1994), the Chaetophorales (mostly filaments with branches) (Sarma 1986), and the Charales (Wood & Mason 1977; de Winton et al. 2007; Casanova et al. 2007). Detailed work has also been carried out on *Microspora*, *Oedogonium*, Zygnemataceae, and *Klebsormidium* in New Zealand (Novis 2003, 2004a,b, 2006). Studies on the latter, as well as the snow alga *Chlainomonas* and species of *Chlamydomonas*, *Stichococcus*, and *Pseudococcomyxa*, have included molecular data, and these tend to indicate recent dispersal events to New Zealand (Novis et al. 2008).

Despite the huge contribution to knowledge of the desmids, an expert in this group has noted that this group is still incomplete (Brook 1995). Also, Sarma (1986) outlined shortcomings of his study and considered that there are likely to be more chaetophoralean taxa in New Zealand. These comments, together with the lack of detailed treatment of all other orders, and the almost complete neglect of soils and other terrestrial habitats, suggests that the New Zealand flora is at least as poorly known as those of Australia and the British Isles.

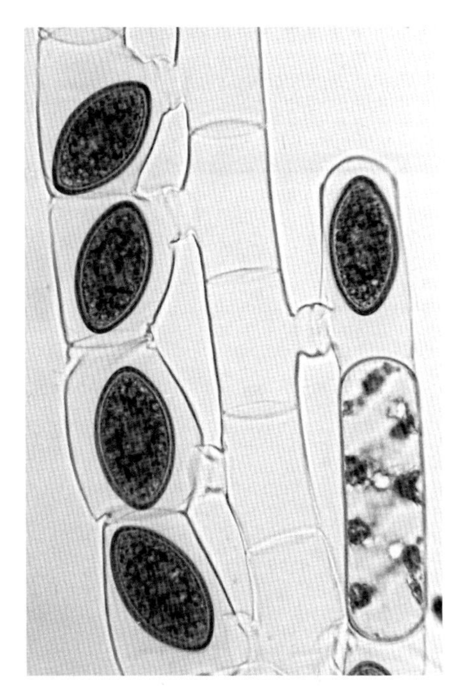

Conjugation and zygospores in *Spirogyra circumlineata* (Charophyta, Zygnemophyceae, Zygnematales).
Phil Novis

Desmids

Desmids (Charophyta) are almost wholly restricted to freshwater habitats. They are distinguished by the absence of cilia (flagella) and by conjugation – a form of sexual reproduction wherein the contents of two cells (usually vegetative cells) fuse, forming a thick-walled zygospore which, after a resting period, germinates to produce two (rarely four) individuals. Most genera are unicellular; a few are unbranched filaments in which cells remain attached after cell division. The name desmid, derived from the Greek word for chain, is a misnomer, inasmuch as unicellular forms far outnumber the chain-forming species that gave the group their name.

The earliest microscopists were confused about the identity of desmids, classifying them first among protozoans and then with diatoms. Ralfs (1848) showed conclusively that they are green plants and his monograph *The British Desmidieae* became the starting point internationally for the naming of desmids. There are two groups. Saccoderm desmids have rod or oblong shapes, their structure is relatively simple, their cell wall is smooth and composed of one piece, and they lack a median constriction and pores. The saccoderms are now known to be polyphyletic, being phylogenetically interspersed with zygnematalean filamentous taxa. Chloroplast shape is a better indicator of relatedness than habit, such as whether they are filamentous or unicellular (McCourt 1995).

Placoderm desmids display a bewildering variety of shapes and wonderful symmetry, and include some of the most beautiful microscopic objects. Most of them have a median constriction, dividing the cell into two symmetrical halves (semicells), one being the mirror-image of the other and, as a result of cell division, one semicell is a generation older than the other. The cell wall is of two halves, is perforated by pores, and is often ornamented by granules, verrucae, or spines, all having a definite pattern. Cells have three planes of symmetry at right angles to one another, and should therefore be examined in front, side, and end views.

As with other freshwater algae, the study of desmids in New Zealand has

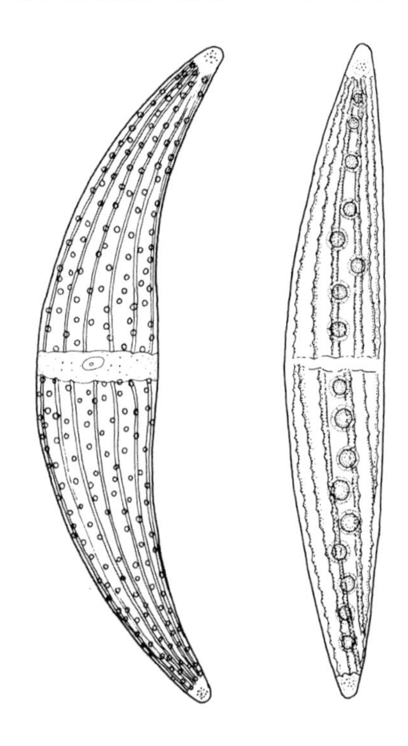

Closterium ehrenbergii (left) and *C. lanceolatum* (Charophyta, Zygnemophyceae, Desmidiales).
From Croasdale & Flint 1986

been sporadic, lagging behind much other botanical research. Desmids lack obvious economic value, and their significance as indicators of water quality has been overlooked. Specialists of New Zealand desmids have been phycologists living abroad as well as resident phycologists.

Grunow (1870) seems to have been the first to identify any New Zealand desmids when he named three species growing among *Nitella hyalina* collected by Ferdinand von Hochstetter from Lake Pupuke, Auckland. Eminent phycologists who followed later were Nordstedt (1887, 1888), who studied 300 samples collected by Sven Berggren in 1874–75 between Bluff in the south and Ohaeawai in the north. E. Lemmermann worked on samples from Wakatipu, Glenorchy, and a lake on Chatham Island, all containing desmids. In the 1930s, H. Skuja identified algae sent by Betty Flint from Canterbury (Flint 1938; Burrows 1977) and Westland (Flint 1979). Hugo Osvald collected from mires in North and South islands during 1950–51. Thomasson (1960) described desmids in samples from five lakes near Rotorua, then, while visiting New Zealand in 1968, he collected from ten lakes in the North Island (Thomasson 1972, 1973, 1974a,b) and nine lakes in the South Island (Thomasson 1980). Skuja (1976) described algae including desmids in samples from bogs. Rolf Grönblad worked on New Zealand material but died before publishing his results. He left five manuscripts that the late Professor Hannah Croasdale (Dartmouth College, U.S.) published between 1964 and 1971. Afterwards, she added to observations on the New Zealand material, integrating it with the results earlier phycologists had published. In due course, this labour became *Flora of New Zealand – Desmids* (Croasdale & Flint 1986, 1988; Croasdale et al. 1994). In 1991, Rupert Lenzenweger collected from South Island tarns and identified the desmids (Croasdale et al. 1994; Flint 1996). Fumanti and Alfinito (2004) described a new species of *Brachytheca* from Lake Monowai; the genus was previously monotypic for a New Guinean species.

William Miles Maskell was the first resident naturalist to study New Zealand desmids, at first in Canterbury. He continued for a while after moving to Wellington (Maskell 1881, 1883, 1886, 1888, 1889) but then switched to working on scale insects, becoming a world authority on them. W. I. Spencer (1881) worked on desmids in Napier and corresponded with Maskell to whom he gave his material when, as a physician, he could not spare the time for desmids. After a lapse of nearly 40 years, another resident, W. M. Mather, included desmids in her study of algae in the Hutt Valley for her masters degree (Mather 1928). Forty years later, Vivienne Cassie began a series of papers on phytoplankton of lakes near Rotorua (Cassie 1969a, 1974, 1978, 1980). Baars-Kloos (1976), and Etheredge (1983, 1986) included in their theses desmids in North Island lakes. Desmids were also included in the papers by Burns and Mitchell (1974), Flint (1977), Jolly (1977), Paerl et al. (1979), Vincent (1983), Donald (1993), and Flint and Williamson (1998, 1999, 2005).

About 43 genera and 3000 species of desmids are known worldwide, the exact number varying according to desmidiologists' personal opinions (Prescott et al. 1975; Gerrath 1993). In New Zealand, 26 genera, 444 species, and 197 varieties have been recorded. *Ancylonema*, growing on ice and snow, and *Oocardium* in lime-rich streams, will probably be found here in due course, but the ten genera peculiar to the tropics are unlikely to survive so far south of the equator. As in some other countries, *Cosmarium* and *Staurastrum* are represented by the most species – 131 and 97, respectively. There are 49 and 34 species, respectively, belonging to *Closterium* and *Euastrum*.

While writing the account of the desmids for the *Flora of New Zealand*, Professor Croasdale remarked on the number of species of which the nominate or typical variety had not yet been recorded here. For *Cosmarium* and *Staurastrum* the numbers are 19% and 16%, of the species, respectively. Comparable numbers reported for the Australian desmid flora are 12% for both genera (Day et al. 1995). Professor Croasdale did not comment on the possible significance of her observations.

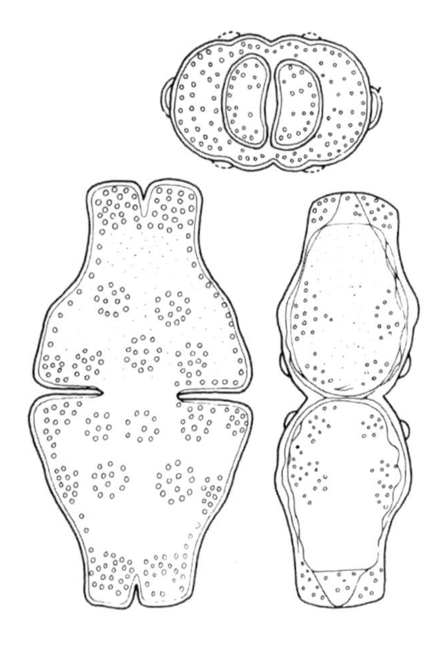

Euastrum sinuosum var. *gemmulosum* (Charophyta, Zygnemophyceae, Desmidiales).
From Croasdale & Flint 1986

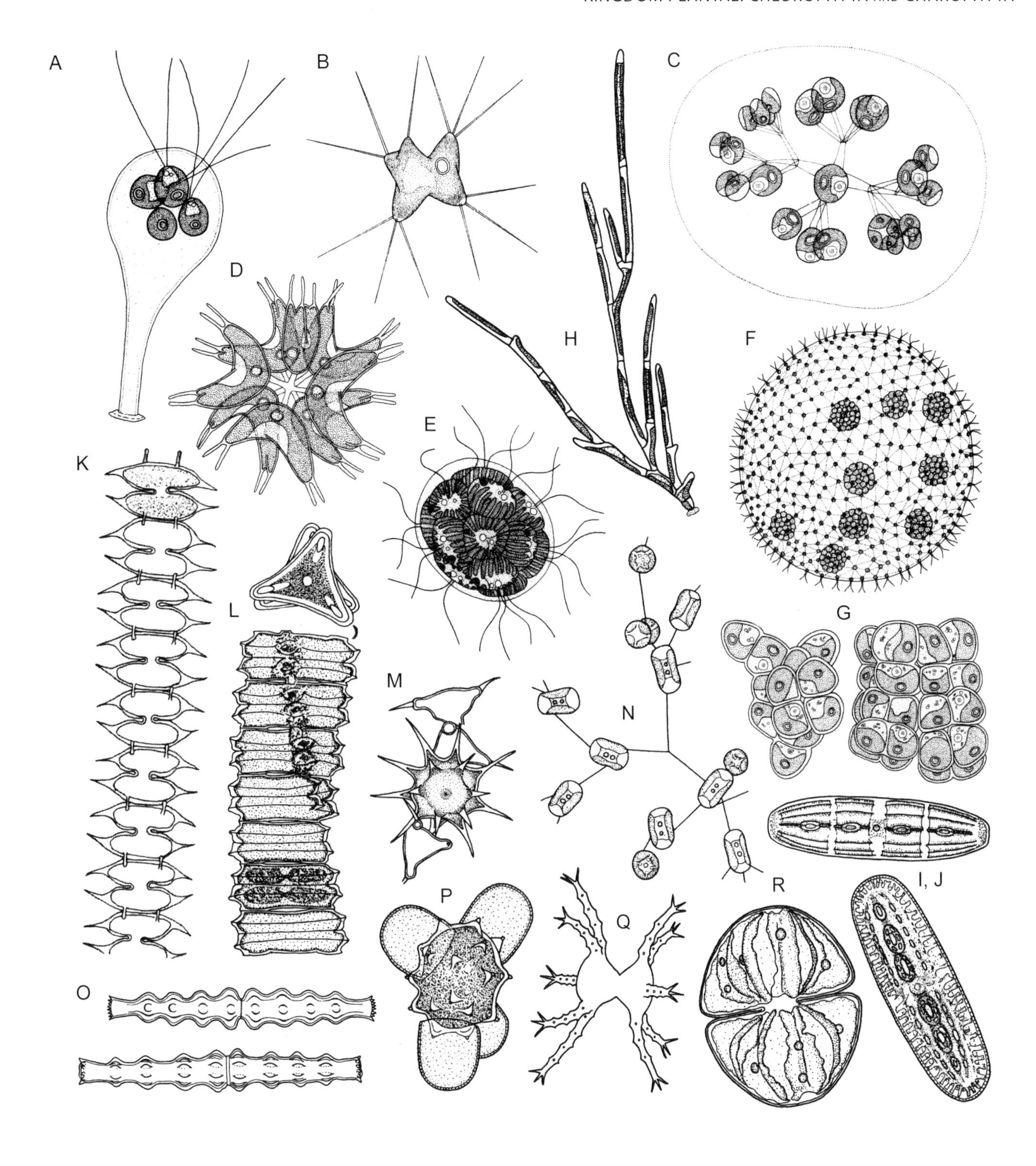

Chlorophyta, Chlorophyceae, Chlamydomonadales: A, *Apiocystis brauniana*. B, *Polyedriopsis spinulosa*. C, *Dictyosphaerium pulchellum*. D, *Sorastrum spinulosum*. E, *Pandorina morum*. F, *Volvox aureus*. Chlorophyceae, Chaetophorales: G, *Desmococcus vulgaris*. Trebouxiophyceae, Microthamniales: H, *Microthamnion strictissimum*. Charophyta, Zygnemophyceae, Zygnematales: I, J, *Netrium interruptum*, *N. oblongum*. Zygnemophyceae, Desmidiales: K, *Onychonema laeve* var. *laeve*. L, *Desmidium swartzii*. M, *Staurodesmus unicornis* var. *unicornis*. N, *Cosmocladium constrictum*. O, *Pleurotaenium nodosum*. P, *Actinotaenium cucurbita*. Q, *Staurastrum limneticum* var. *burmense*. R, *Cosmarium ralfsii*, var. *ralfsii*.

A, E–H, after Bourrelly 1990; B, D, after Hindák 1980; C, after Hindák 1980; I, J, O, after Croasdale & Flint 1986; K– M, Q, after Croasdale et al. 1994; N, P, R, after Croasdale & Flint 1988

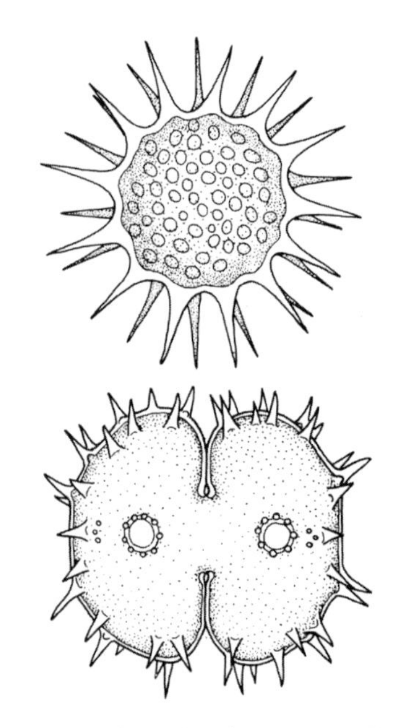

Zygospore (upper) and vegetative cell of *Xanthidium aculeatum* (Charophyta, Zygnemophyceae), Desmidiales).
From Croasdale & Flint 1988

There are 75 endemic taxa – 34 species and 41 varieties. *Cosmarium, Staurastrum, Euastrum*, and *Xanthidium* have 22, 18, 11, and 6 endemic taxa, respectively. No endemic desmid genera have been reported yet.

Krieger (1932) perceived that different desmid floras typified ten geographic regions worldwide. Coesel (1996) reviewed this observation in the light of research since 1933. He noted that the most distinctive desmid regions are Indo-Malaysia/northern Australia, tropical America, and equatorial Africa. The desmid floras of East Asia, New Zealand/southern Australia, and North America are less pronounced.

Taxa in common between New Zealand and southern Australia include the following [asterisked taxa are recorded more often in the North Island than the South Island]: *Euastrum bullatum, E. longicolle* var. *australicum, E. sphyroides**, *Xanthidium octonarium**, *Staurastrum tangaroae**, and *S. victoriense* var. *tasmanicum*. Other taxa occur in New Zealand, Australia, and Indonesia (e.g. *Cosmarium amplum**, *C. capitulum* var. *australe, Staurastrum sagittarium**, and *Desmidium baileyi* var. *undulatus**). This latter taxon and *Closterium compactum* are recorded from New Caledonia (Carter 1922). Other taxa characteristic of the tropics include *Pleurotaenium ovatum**, species and varieties of *Triploceras**, and *Staurastrum leptocanthum**. Common to New Zealand and South America are *Cosmarium tetraophthalmum* var. *patagonicum* and *Staurastrum trifidum* var. *porrectum*.

Desmids survive neither desiccation nor seawater and rarely, if ever, form resistant zygospores, features that make it difficult to explain their presence in New Zealand consequent upon a turbulent geological history after breaking away from Gondwana in the Late Cretaceous–Early Paleocene. New Zealand's isolation from other land masses, its significant submergence in the Oligocene, extensive North Island volcanism in the Miocene and Pliocene, and glaciation during the ice ages would have affected survivability. An alternative to survival as botanical relics is Trans-Tasman dispersal from Australia in wind-borne aerosols or carriage by birds, but these vectors remain to be proven.

There are few records of fossil desmids in New Zealand. Cranwell and von Post (1936) mentioned desmids, diatoms, testate amoebae, and crustaceans in peat and McGlone and Wilmshurst (1999) found *Cosmarium* sp. and zygospores of *Micrasterias* spp. in Holocene peat (3659 years BP 1960).

Some fungi parasitise desmids but few have been described from New Zealand. Maskell (1881) illustrated *Staurastrum sexangulare* var. *productum* (as *Didymocladon stella*) with what may have been chytrids. The same desmid was recorded infected by chytrids in a tarn (Flint & Williamson 1998). *Cosmarium undulatum* in a pool at Taita, Lower Hutt, was infected by the cytridiomycete fungus *Olpidium saccatum* (Karling 1965).

Most desmids prefer oligotrophic or mesotrophic habitats in which the calcium content is low and the pH is neutral. Coesel (1998) analysed the desmid flora of the Netherlands according to the preferred trophic status of the taxa (i.e. oligotrophic, mesotrophic, eutrophic), acidity (i.e. below pH 6.5, above pH 7.5, or neutral pH 6.5–7.5), life-form (i.e. aerophytic on moist terrestrial surfaces, benthic on soil or aquatic plants, or tychoplanktonic or genuinely planktonic and thus free-floating in large bodies of water), value of taxa as indicators of ecosystem maturity, and status in the Red List of endangered species. At the present rate of progress, it will be a long time before a similar list can be compiled for the desmid flora of New Zealand.

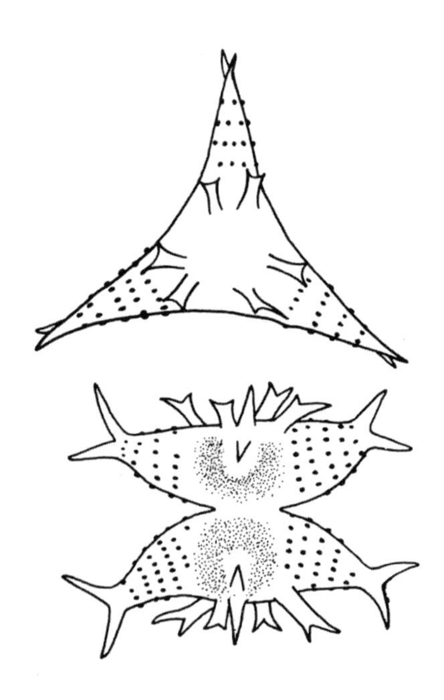

Different profiles of *Staurastrum arcuatum* (Charophyta, Zygnemophyceae, Desmidiales).
From Croasdale et al. 1994

Characeae

Characeae is a distinctive family of charophytes that superficially resemble higher plants. They have erect stems supporting regular whorls of branchlets and are anchored in the substratum by rhizoids. They have a fossil record, giving evidence of their origination more than 420 million years ago. Today, they are found in a wide range of continental waters, temporary and permanent, flowing and still, fresh to brackish, and even hypersaline waters. The Greek name *chara* is taken to mean 'pleasure of the water'. The common name stoneworts alludes to the tendency for some species to accumulate a conspicuous coating of calcite in habitats rich in carbonate ions; New Zealand waters are generally too low in alkalinity, however, for any such conspicuous encrusting of the plants.

The general structure of Characeae is that of a long stem of large internodal cells (some of the largest cells in the plant kingdom), alternating with a group of nodal cells from which whorls of branchlet cells originate. In some *Chara* species, the internodal cell becomes reinforced with a secondary layer of cells (cortication) that also originate from nodal cells. Characeae exhibit both monoecious and dioecious breeding systems, with antheridia and (oogonia on the same plant in the former. In both cases, hardened diploid propagules, oospores, develop from fertilised oogonia and are analogous to seeds. Oospores can be distributed by water currents and waterfowl, with the propagules able to survive passage through a bird's digestive tract. Other methods of reproduction and dispersal in Characeae include specialised and non-specialised vegetative propagules and starch bodies (bulbils), which form on the rhizoids.

Chara fibrosa with oospores before release.

Mary de Winton

Traditionally, Characeae have been classified according to cell-development patterns, branching architecture, and breeding system, culminating in the publication of a world monograph (Wood & Imahori 1965). This work recognised 81 characean species in six genera. There are likely to be more species in that account than are presently recognised, however, as some workers have established that reproductive isolation within the genus *Chara* is not always distinguished by significant morphological characteristics using Wood and Imahori's criteria (McCracken et al. 1966; Proctor 1971a,b). In addition, Wood and Imahori's practice of lumping similar monoecious and dioecious entities into one species has been questioned (Meiers et al. 1999). Genetic data, such as *rbc*L (ribulose-1,5 biphosphate carboxylase) nucleotide sequences, have also been used to elucidate taxonomy (Casanova et al. 2007) and phylogeny. More recently, workers have recognised that oospore size and ornamentation of the outer wall are taxonomically useful, especially in the genus *Nitella* (De Winton et al. 2007); Casanova et al. 2007).

In New Zealand, 18 species are recognised, in the genera *Chara, Nitella, Lamprothamnium,* and *Tolypella*. Most of the species belong to *Chara* and *Nitella*, which exhibit their maximum development in clear, low-productivity water bodies. *Lamprothamnium* and *Tolypella* are each represented by a single species and are restricted to coastal water bodies. *Lamprothamnium* is the only New Zealand member of the family that requires some degree of salinity. The earliest scientific collections of New Zealand Characeae were made in 1841 by botanist J. D. Hooker. Numerous other early collectors included Characeae among their herbarium specimens (e.g. W. Colenso, T. R. Ralph, T. Kirk, and T. F. Cheeseman) and many specimens were sent to northern-hemisphere workers, notably Sir W. Hooker and A. Braun. During the early 1900s, many botanists made only incidental collections. Interest in Characeae was revived in the latter half of the century, with extensive collections made by R. Mason concurrent with R. D. Wood's classification of Characeae. The most significant advance in understanding of New Zealand's species came in the 1960s and 1970s when the two workers collaborated and Wood visited New Zealand, resulting in publication of the 'Characeae of New Zealand' (Wood & Mason 1977). This work provided an impetus into investigations of their distribution and ecology by J. S. Clayton and colleagues, and a simplified key to the common Characeae was published (Clayton & Wells 1980). This work confirmed the existence of two previously uncertain taxa and posited the questionable status of a further three. Nomenclatural changes for several New Zealand species followed a reassessment of the related Tasmanian Characeae (Van Raam 1995). Taxonomic uncertainties in the New Zealand flora were further explored during a study of ooospore characteristics (de Winton et al. 2007) and a revision was made of the *Nitella*

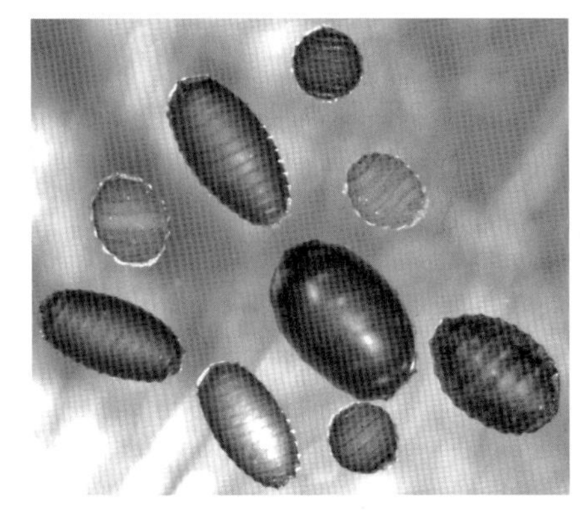

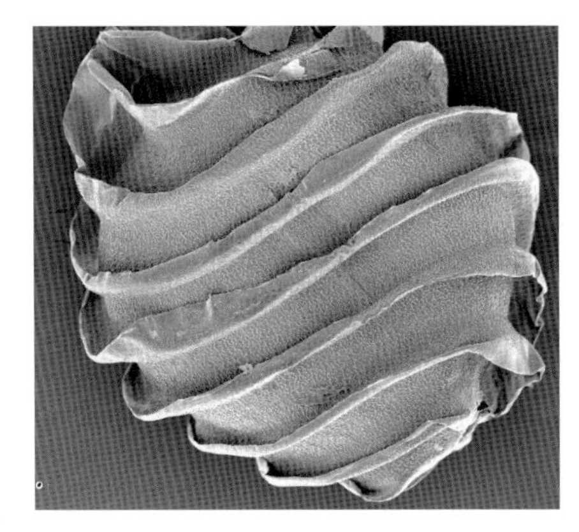

Upper, oospores of multiple species of Characeae. Lower, oospore of *Nitella tricellularis*.

Chrystal Kelly, Mary de Winton

Nitella opaca, fertile and sterile branches.
John Clayton, NIWA

hookeri complex that contained morphologically diverse members (Casanova et al. 2007). The major repository of of New Zealand Characeae specimens is Landcare Research Herbarium (CHR), Lincoln, with extensive collections also housed at Waikato University (WAIK).

Wood and Mason (1977) concluded that there were no endemic species of New Zealand Characeae, which they regarded as essentially a subset derived from the Australian flora. Since then, the circumscription of five New Zealand species previously included in the *Nitella hookeri* species complex (Casanova et al. 2007) has delineated three endemic taxa (*N. claytonii, N. masonae,* and *N. tricellularis*), along with a virtual endemic (*N. hookeri*) shared only with the remote Indian Ocean Island of Kerguelen. An ecorticate *Chara* species related to the Australian *C. muelleri* but destinguished by *rbc*L nucleotide sequences (K. Karol pers comm. 2008) is another possible endemic. Nevertheless, the low overall level of endemism of New Zealand Characeae contrasts with the high proportion of endemics amongst higher plants and probably stems from a relatively recent arrival, geologically speaking, of Characeae to this country, with continued introduction from other land masses (especially Australia) through dispersal by migratory waterfowl (de Lange 1997). Certainly, endemism within the *N. hookeri* complex is considered to represent recent and rapid evolution (Casanova et al. 2007).

Only four of New Zealand's species are dioecious – *Chara australis, Nitella opaca, N. subtilissima,* and *N.* sp. aff. *cristata*, probably reflecting the reduced vagility of dioecious species, which require the proximity, in water bodies, of plants of both sexes to be sexually reproductive. Indeed, only female plants of *N. subtilissima* and *N. opaca* have been reported in New Zealand and these species have constrained distributions. Interestingly, *N. opaca* is unrecorded in Australia, Malaysia, or Oceania, but is recorded from South America.

Species of *Chara* and *Nitella* are the most ubiquitous indigenous plants of the submerged flora in New Zealand water bodies. They contribute to diverse species-rich floral assemblages at the margins of lakes and commonly accompany taller native plants in the mid-depth range, but Characeae may also form meadows beyond the depth limit (ten metres) of vascular plants. Characean meadows frequently comprise the deepest vegetation in deep clear lakes, where they can extend to depths of 34 metres or more, with light availability being a key determinant of maximum depth (Schwarz et al. 2000). Zonation of Characeae with lake depth has been related to the amount of subsurface light received (Schwarz et al. 2002) and specific differences in photosynthetic metabolism (Sorrell et al. 2001). *Chara australis* and *Nitella claytonii* were measured as having the lowest light requirements, with a median lower depth limit equivalent to less than 5% of irradiance at the lake subsurface (Schwarz et al. 2002). The greatest threat to this community comes from the dual impacts of reduced water quality and invasive alien species. Increased turbidity of lakes from catchment modification has resulted in a decrease in the available littoral area for Characeae. Concurrently, the introduction of competitive adventive weeds excludes Characeae from lake mid-depth regions.

The most widespread species in New Zealand are *Chara australis, Nitella pseudoflabellata,* and *N.* sp. aff. *cristata*. The latter, a sister taxon to *N. hookeri,* could not be accurately distinguished in the absence of fertile material until the recent treatment of this complex (Casanova et al. 2007), so that historical distribution is often erroneously attributed to '*N. hookeri*'. Seven species have restricted distributions. *Chara* sp. aff. *muelleri, Nitella claytonii,* and *N. subtilissima* are found only in South Island inland water bodies. *Nitella opaca* is restricted to central North Island lakes. *Lamprothamnium macropogon* occurs in coastal water bodies with significant though variable levels of salinity. *Tolypella nidifica* has been collected from two sites – Lake Forsyth, Canterbury, in the 1960s and gravel ponds at Oreti Beach, Invercargill, in 2009. Only two uncertain records exist for *Chara vulgaris*, both from Hawke's Bay in the late 1800s. In recent years, the same species has been found growing in a Hawke's Bay culture facility for waterlilies although it is possible this occurrence was separately introduced with imported material. Cultures of the latter two species are being maintained at NIWA.

Determination of several New Zealand Characeae is still required. As mentioned above, evidence from *rbc*L nucleotide sequences distinguishes an ecorticate species related to *C. muelleri* in keeping with doubts over the identity of the taxa (as *Chara braunii*) expressed by Wood and Mason (1977). Plants similar to *N.* sp. aff. *cristata* are also found in southern Australia and revision of the *N. cristata* complex is required to clarify nomenclature (M. Casanova pers. comm. 2007). Lastly, variation in oospore morphology within New Zealand specimens of *N. pseudoflabellata* indicate that additional taxa are present (de Winton et al. 2007).

Insofar as Characeae are an important component of the submerged flora of New Zealand's water bodies, they have lately generated scientific interest owing to their possible role as bioindicators for long-term trends in water quality. They have an international reputation for ameliorating negative impacts on water quality and they play a role in the remediation of degraded lakes. Understanding the Characeae in New Zealand is important for ascertaining their actual and potential roles in freshwater ecology.

Endemism

Most freshwater and terrestrial green algae species in New Zealand have been recorded from elsewhere. Nevertheless, the true level of endemism is impossible to assess as many identifications will have been made using keys and floras based on specimens from Europe and North America. There is a recognised tendency for researchers to assign their specimens to the closest species found in available floras, even if differences are present. In the absence of floras based on New Zealand specimens this is a likely outcome.

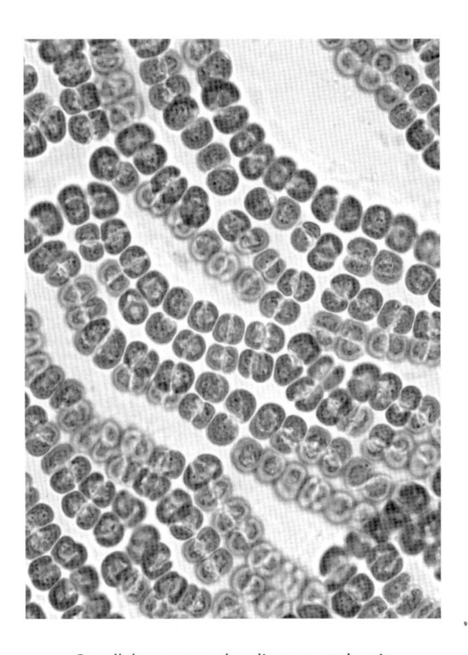

Parallela novae-zelandiae, an endemic freshwater alga (Chlorophyta, Chlorophyceae, Chlamydomonadales).

Paul Broady

Conversely, there is also a tendency to create possibly unjustified new species and varieties for specimens that may vary only slightly from those already known. For instance, this could apply to some of the five new species and eight new varieties erected for New Zealand chaetophoralean algae by Sarma (1986).

Even when a new genus is erected for a distinctively different alga, its apparent absence elsewhere can be the result of inadequate investigation. An example is *Parallela novae-zelandiae* Flint, which was known only from New Zealand (Flint 1974) until found in Brazil (Leite Sant'Anna & Bicudo 1979) and North Korea (Mrozińska 1990). Molecular data have also been used to show that *P. novae-zelandiae* is related to another species from overseas that has been transferred to *Parallela transversale* (Novis et al. 2010). The first discovery of this alga in a New Zealand farm pond suggests that other novel algae could readily be discovered by those with curiosity and the necessary knowledge.

Any accurate assessment of levels of endemism will have to await far more thorough knowledge of the New Zealand flora at least, and additionally that of many other world regions at best, based on the full range of techniques now available, from detailed light microscopy to molecular genetics.

Alien freshwater species

Considering the wide range of other organisms that have been introduced into New Zealand by human activity, including chlorophyte seaweeds (Dromgoole 1975), it would be remarkable if freshwater and terrestrial green algae were not amongst them. Most are surely destined to go undetected owing to lack of knowledge of the composition of both the prehuman and the present flora.

The increased eutrophication of many lakes since European colonisation and the modification of wide areas of soil by clearance of native vegetation and fertilisation could have resulted in rich development of species for which no or very limited areas of habitat were previously available. Although natural dispersal of propagules by wind or birds might bring such algae to New Zealand, soil traces on the feet of millions of visitors and on the packaging and surfaces of

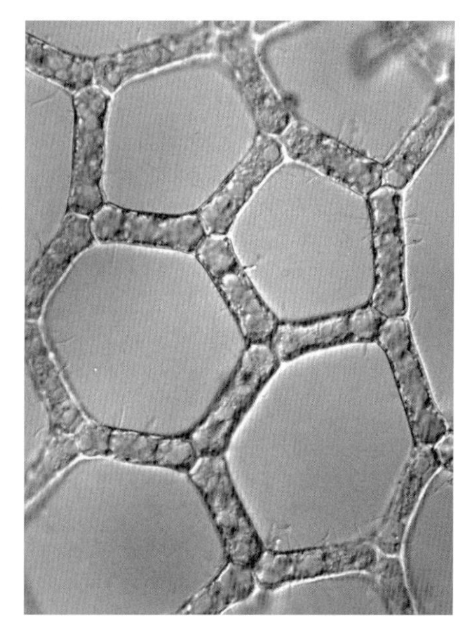

Water net *Hydrodictyon reticulatum* (Chlorophyta, Chlorophyceae, Chlamydomonadales).
Protist Information Server, Hosei University

huge volumes of imported materials would also act as a ready source.

Detection of new introductions requires a species to be distinctive and to become readily apparent through vigorous growth. It probably also needs to become a sufficient nuisance to invoke calls for expert identification. The net-like species *Hydrodictyon reticulatum* provides a prime illustration of this; it is the only certain example of a human-mediated introduction of a freshwater green alga to New Zealand. It was first identified in 1986 in an ornamental pond near Tauranga, Bay of Plenty, where it formed large weedy growths (Coffey & Miller 1988). However, it had been noted since about 1980 in a downstream commercial fish hatchery used for quarantine and culture of imported fish and water plants. The source of the alga was possibly Singapore, in a consignment of ornamental fish. It spread rapidly through lakes and waterways in the central North Island and Bay of Plenty, causing proliferations that had an economic impact (Hawes et al. 1991). Unusually, in New Zealand it has not been seen to undergo sexual reproduction, a process that elsewhere results in the formation of distinctive resistant zygotes. Its nuisance level led to experimental studies of growth requirements, which indicated an ability to tolerate quite low inorganic nitrogen concentrations (Hawes & Smith 1993a; Hall & Cox 1995). However, the weed problem appears not to have developed further and water net is no longer seen as a threat (I. Hawes pers. comm.). Local blooms of this species have been noted in Hawke's Bay (e.g. Karamu Stream) and Wairarapa (e.g. Bartons Lagoon) in the last decade (P. Champion pers. comm.). Possibly other introductions have gone unnoticed because of merely mild growth.

Phytoplankton

A wide range of unicellular, colonial, and occasionally filamentous green-algal species have been reported as dominants in the phytoplankton communities of a variety of lakes (Flint 1975; Viner & White 1987; Cassie Cooper 1996). They reach maximum numbers usually in summer and infrequently in winter. Summer communities can differ markedly, even in lakes of similar trophic status, as exemplified by oligotrophic lakes Taupo and Waikaremoana (Vincent 1983). In the former, unicellular chlorococcaleans are important contributors to algal biomass whereas in the latter mucilaginous colonies of *Sphaerocystis schroeteri* dominate.

They have value as indicators of lake trophic status as their abundance and diversity can increase with nutrient enrichment. In Lake Rotoiti, between 1955 and 1984, they became more prominent in summer phytoplankton communities, in parallel with other indicators of nutrient enrichment (Vincent et al. 1984). These algae were amongst the indicators used as evidence of a return to somewhat less enriched conditions in lakes Rotoiti and Rotorua, and of little change in five other Central Volcanic Plateau lakes up to 1995 (Burns et al. 1997). Hypereutrophic sewage oxidation ponds are particularly rich in coccoid and colonial green algae other than desmids. Of a total of 69 algal taxa recorded from 10 Auckland oxidation ponds, 35 were these types (Cassie 1983).

Detailed floristic studies of phytoplankton of individual New Zealand lakes by expert taxonomists are few. Often only dominant taxa are listed in reports of studies that are primarily process-orientated. Cassie Cooper (1994) cited several, for instance those pertaining to Lakes Ohakuri (Coulter et al. 1983), Okaro (Dryden & Vincent 1986), and Waikaremoana (Howard-Williams et al. 1986), in which total phytoplankton taxa numbered 10, 30, and 20 respectively, which included three, eight, and nine green algae.

A much wider range of species has been noted in the few relatively recent studies where the primary focus is taxonomic and floristic. A rare example is the study of ten lakes in the Rotorua district by Thomasson (1974b). A total of 628 phytoplankton taxa were identified, with between 45 and 155 in individual lakes. Of the overall total, 107 were green algae other than desmids and 170 were desmids. The non-desmids comprised 9–34 taxa in individual lakes. It is noteworthy that 64 green-algal taxa occurred in just a single lake and only 12 taxa in four or more lakes. There can be significant differences in species occurrences even within a single lake district. Other taxa would certainly have been overlooked, as Thomasson sampled each lake once in late summer, used nets with a minimum mesh size of 10 micrometres, and examined only preserved specimens. Additional taxa would include those requiring observation of living specimens for identification, those occurring at other times of the year, and those small enough to pass easily through a ten-micrometre mesh.

These latter forms are taxonomically difficult. Such small phytoplankters include the picoplankton (0.2–2 micrometres diameter) and the smaller members of the nanoplankton (2–20 micrometres diameter). They encompass a wide taxonomic diversity of eukaryotic and prokaryotic algae, including coccoid and flagellate green algae. Although known to be important in terms of biomass and productivity in New Zealand lakes (Paerl 1977, 1978), their taxonomy has barely been addressed. There appears to have been just one detailed descriptive account (Dempsey et al. 1980) of a single two-micrometre-diameter coccoid chlorophyte, *Kermatia pupukensis* (= *Neodesmus pupukensis*), known only from Lake Pupuke in Auckland (Cassie 1989a). Other studies group these small algae according to broad size and shape categories without attempting identification even to the level of division (e.g. Dryden & Vincent 1986).

Periphyton

Assemblages of algae associated with diverse substrata in lakes, ponds, and flowing waters are termed periphyton. Their habitats include the surfaces of sediments, rocks, and plants and even other algae and animals such as snails. Green algae can dominate these assemblages and it would be unusual to find none.

In general, the species-level taxonomy and distribution of green algae other than desmids in New Zealand periphyton are very poorly known. An exception is the monographic treatment of the branched filamentous forms by Sarma (1986). He described and illustrated 52 species from 20 genera found attached to other algae, aquatic angiosperms, and stones in a variety of flowing and standing waters. The most species-rich genera were *Stigeoclonium* (17 species), which forms bright green tufts of short filaments, and *Coleochaete* (six species). More-recent studies by Novis (2003, 2004a,b, 2006) provided detailed illustrated descriptions of new records of *Microspora*, *Oedogonium*, Zygnemataceae, and *Klebsormidium*. Otherwise, species identifications of unicellular periphyton (e.g *Characium*), colonies (e.g. *Pediastrum*), and unbranched filaments (e.g. *Ulothrix*), are scattered in the literature (Cassie 1984a,b) and usually unaccompanied by descriptions and illustrations. The reliability of many of these must be doubted. In studies with an ecological emphasis, for example those of epilithic periphyton of Lake Taupo (Hawes & Smith 1993b) and epiphytes on deep-water Characeae (Hawes & Schwarz 1996), identifications are usually made to genus level and then only for dominant taxa and morphologically distinctive specimens. This is also true of the many recent investigations of river and stream periphyton.

Rivers and streams can develop proliferations of filamentous algae following prolonged periods of low flows. This can lead to an unattractive appearance, reduced oxygen levels, and clogging of water-extraction pipes (Biggs 1985). During the summer drought of February 1983, green algae comprised seven of the ten taxa found to dominate at 167 sites with proliferations in a New Zealand-wide survey (Biggs & Price 1987). In a later survey (Biggs 1990), the composition of the community and its biomass were related to water conductivity, temperature, and nutrients. High-biomass sites had more than 40 grams per square metre dry weight of algae, with the coarse filaments of *Cladophora* and *Rhizoclonium* dominant, in waters of relatively high conductivity. A focus on distributional patterns in an individual catchment (Biggs et al. 1998) showed filamentous green algae to dominate riffles at mid-catchment and lowland sites, possibly as a result

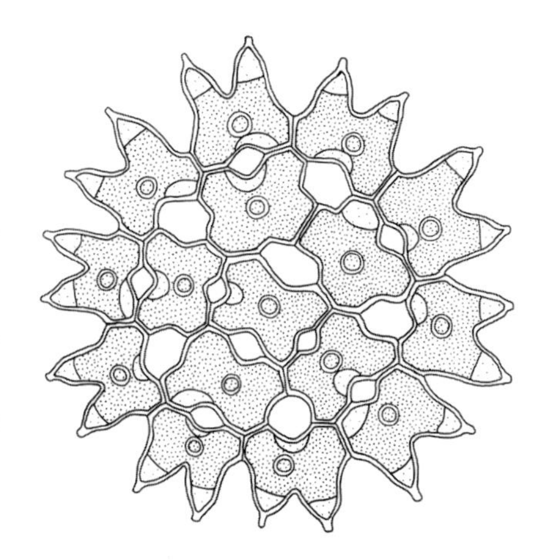

Pediastrum duplex (Chlorophyta, Chlorophyceae, Chlamydomonadales).
After Bourrelly 1990

Chara globularis meadow in Lake Waikaremoana.
John Clayton, NIWA

of increased nitrate concentrations with progression downstream.

These and other studies of periphyton in New Zealand flowing waters have contributed greatly to general global understanding of their dynamics (Biggs 1996). Application of this knowledge to management of flowing waters in New Zealand has been stimulated by the production of a water-quality monitoring kit and associated manual (Biggs et al. 2002) that can be used by farmers and other concerned people. Identification of common taxa by technical staff of monitoring agencies is being eased by guides to the genera of algae (e.g. Biggs & Kilroy 2000; Moore 2000).

Among the most important primary producers of lakebeds are Characeae. Their meadows can range to 35 metres depth (Schwarz et al. 2000). The bottom limit for *Chara corallina*, for example, is controlled mainly by light intensity and the need for photosynthetic gains to exceed, or at least balance, respiratory losses over an annual cycle (Schwarz et al. 1996); in this species, net photosynthetic gain is achieved at a light intensity of less than 0.5 % of full summer sunlight. In Lake Waikaremoana, species of *Chara* and *Nitella* occur in 80% of the macrophyte zone around the lake margin, forming a continuous meadow at 9–16.5 metres deep, and contributing a large proportion of the 578 tonnes of carbon fixed each year by the macrophytes. Similar meadows occur in high-altitude South Island lakes (De Winton et al. 1991) and in major lakes of Fiordland (Wells et al. 1998). They provide extensive surfaces for the growth of microscopic epiphytic algae that are grazed by small invertebrates, while the macrophytes themselves enter the food web only following decomposition. In Lake Coleridge, the epiphytes, mostly diatoms, are vigorously grazed by snails that in turn are significant in the diet of predatory fish (Hawes & Schwarz 1996).

Thermal waters

Although no green-algal species is strictly thermophilic, a few can be found at unusually high temperatures associated with North Island thermal springs, pools, and lakes (Cassie 1989b; Cassie & Cooper 1989) and several prefer acidic habitats. Filaments of *Schizomeris leibleinii* occurred at 37° C at the fringes of an even hotter water flow, and orange crusts of *Trentepohlia* are common on dead stems and branches of steam-enshrouded vegetation (Sarma 1986). Thick, green, bottom-dwelling wefts of filamentous *Zygogonium* enhance the beauty of Emerald Lakes on Mt Tongariro (Vincent & Forsyth 1987). Lake Rotowhero is particularly unusual – it has acid (pH 3.1) nutrient-enriched water at a temperature of 30–35° C that, throughout the year, is a uniform bright green. This dense phytoplankton community is dominated by two species of coccoid *Chlorella* (Forsyth & McColl 1974).

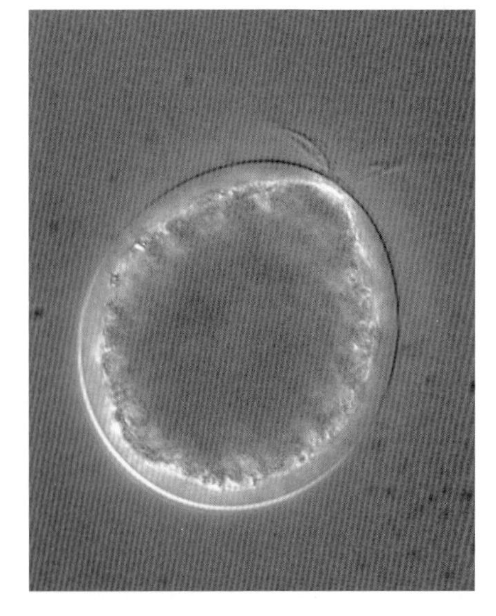

Chlainomonas kolii.
Phil Novis

Snow algae

Snow coloured by abundant populations of algae is well known, particularly in alpine regions of the South Island during summer (Thomas & Broady 1997). About ten taxa, some of dubious identity, were recorded up to 1998 (Novis 2002a). The major taxa are usually green algae but the true identities of many remain elusive. Often the snow becomes a visible pinkish red when resting spores dominate the community. It is only relatively recently that studies elsewhere have used cultures to link vegetative cells to spores (e.g. Hoham et al. 1979; Ling & Seppelt 1998). Only by doing so is it possible to make confident identifications.

Despite the widespread occurrence of snow algae in New Zealand, detailed investigation of their ecology and taxonomy has begun only relatively recently, resulting in the observation of previously unrecorded taxa, including a species of flagellate unicell, *Chlainomonas kolii*, previously known only from North America (Novis 2002a,b). Its identity and its affinity with North American specimens have been confirmed using gene sequencing (Novis et al. 2008). A survey of North American sites in 2005 failed to reveal the presence of *C. kolii*, although *C. rubra* was still common, and the former species may no longer exist there owing to a higher spring snowline. The ecology of this species seems unusual among snow algae and its apparent decline is a conservation issue. Other New Zealand green algae species cultured from snow are *Chlamydomonas rubroleosa* and *Raphidonema nivale* (Novis 2002a).

Terrestrial algae

In general, the diversity of terrestrial algae in New Zealand is very poorly known. Detailed taxonomic treatment has been given only to members of the Trentepohliaceae (Sarma 1986). Of these, *Trentepohlia*, with 11 species, is the most diverse and apparent. Its bright orange-red crusts on boulders in stream gullies are familiar to travellers through mountain regions of both main islands and similar growths cover wooden power poles and fence posts in many parts of the country. *Cephaleuros* is less well known but is widespread in the north of the North Island, on and within leaves and twigs of many exotic and native host plants. The leaf spots formed by *C. parasiticus* on the leaves of the endemic protead *Knightia excelsa* are shown in botanical paintings (Harvey 1969). The orange discs of the epiphyte *Phycopeltis* usually require a hand-lens for observation but have been recorded on 13 fern, four conifer, and 19 flowering plant species.

Trentepohlia aurea filaments.
Protist Information Server, Hosei University

It would not be a leap of the imagination to hypothesise that the great majority of soil surfaces in New Zealand support whole communities of algae, each of which contains about 5–20 species of greens. Also, the species composition of the communities is likely to differ widely between soils of different physicochemical characteristics (Hoffmann 1989). An indication of the abundance of green algae in New Zealand soils is given by one quantitative investigation in which they comprised more than 87% in all samples (Ramsay & Ball 1983).

Worldwide, there have been numerous floristic studies of soil algae. One review (Metting 1981) listed 195 of these and no doubt there have been many since that review. Soil algae comprise most of the 600 species of green algae described in the first comprehensive flora of terrestrial algae worldwide (Ettl & Gärtner 1995).

In New Zealand, however, knowledge of soil-algal communities is scant, despite their being implicated in damaging the cricket square in an international match (Longley 1999)! There have been five published accounts of preliminary surveys that list or describe soil algae (Flint 1958; Flint & Ettl 1966; Flint 1968; Flint & Fineran 1969; MacEntee et al. 1977). In these, most identifications are taken only to genus level and only 19 green-algal taxa are recorded. However, the presence of a far richer flora is suggested in unpublished reports in which 24 and 26 species were recovered from subalpine (Everett 1998) and alpine (Novis 2001) soils, respectively, each sampled at a single location in Canterbury. Further studies might find that the large majority of taxa are morphologically indistinguishable from those known elsewhere. In fact, a strain of soil alga isolated from Lincoln was genetically indistinguishable from *Chlorella vulgaris* from overseas (Novis et al. 2009) and the same was true for *Chlamydomonas pseudogloeogama* (also found in the Czech Republic) (Novis et al. 2008). It is equally likely, however, that amongst some of these widely distributed species there will be genetically distinct strains, some of which might have valuable applications in biotechnology.

Gaps in knowledge and scope for further research

Despite the large number of non-desmid green algae recorded from New Zealand, the reliability of a likely substantial proportion of them is open to some doubt and urgently needs reassessment. Many records are unaccompanied by written descriptions and illustrations and are not associated with voucher specimens. Most identifiers will have used manuals and floras based on North American or European specimens. The degree to which New Zealand material might differ from related taxa in those regions requires detailed investigation using morphological and molecular approaches.

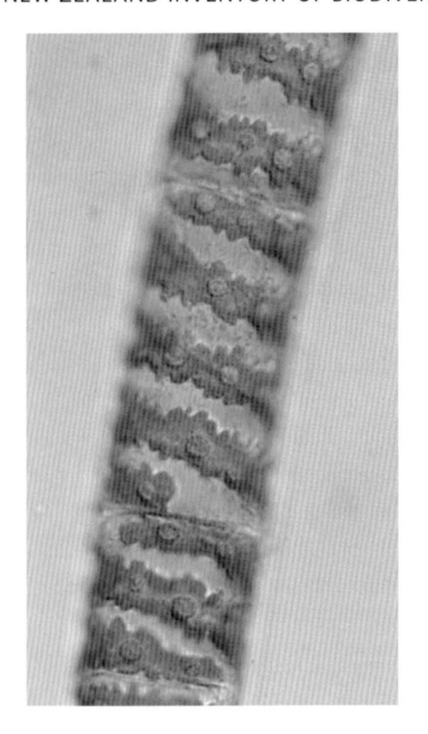

Spirogyra cf. *decimina* (Charophyta, Zygnemophyceae, Zygnematales).
Phil Novis

Although non-desmid green algae in all habitats need further collection and critical taxonomic assessment, priorities will have to be established. Because of their importance in nuisance growths and their potential as indicators of water quality, filamentous forms in flowing waters would be a useful focus for species-level taxonomy. Many have received attention in recent years (Novis 2003, 2004a,b, 2006) but some taxa remain in need of study.

In lakes, the ubiquitous, morphologically simple unicells in the 1–5 micrometres size range can be major contributors to biomass, and undoubtedly will often include green algae. The diversity of this morphological category requires elucidation in New Zealand waters, a task that will require skills in electron microscopy, molecular genetics, and pigment analysis. Also, many additional species would probably be found in lake periphyton by thorough floristic studies.

Although it is easy to assume that New Zealand soil algae are identical to those elsewhere, such an assumption is founded on sparse knowledge. New Zealand's distinctive combination of climate and soils might select for unusual soil communities. Even if species are morphologically similar to those in other regions, their adaptation to the indigenous environment might be reflected by unique genotypes. Such soil algae are most likely to be found in the extensive and diverse areas that have remained largely unmodified by human activities.

Amongst the wide diversity of New Zealand green algae are likely to be strains with potential for biotechnology. Modern applications like those outlined above await investigation in New Zealand. One could envisage the use of green algae to remove pollutants such as heavy metals from tannery effluent. Also, there could be much greater development of the use of algal biomass harvested from waste-water treatment plants. This biomass, if of a suitable strain of alga, could contain a high-value product such as the pigment astaxanthin or could be converted to liquid biofuel. A pilot scheme at Christchurch Wastewater Treatment Plant is currently investigating the latter possibility [see http://www.niwa.co.nz/__data/assets/pdf_file/0005/105593/Water-and-Atmosphere-July2010.pdf].

Screening programmes for strains brought into culture need to be instigated in order to recognise those of maximum potential. Support is needed for the establishment of a significant culture collection of both green and other freshwater and terrestrial algae, if those from overseas are not to be solely relied on. The considerable expense in isolation of strains, in terms of time and resources, is lost when small, personal research collections are discarded at the end of a research programme. Many strains might have application very different from the reason they were originally isolated.

Whether any New Zealand green-algal species are under threat of regional extinction is debatable. In the current situation of poor knowledge of diversity and distributional patterns it is impossible to provide a realistic estimate. However, it is considered that widespread loss of habitats from environmental degradation, nutrient enrichment, changes in the water table, drainage schemes, and urbanisation are a threat to algae in Europe (Norton et al. 1996). These pressures have been and continue to be prevalent in New Zealand also.

Marine Chlorophyta

There are far fewer green seaweeds and marine-planktonic chlorophytes than there are terrestrial and freshwater algae. Nevertheless, they are important and distinctive, especially the macroalgae. The account that follows deals with them on a taxonomic basis by major class.

Chlorophyceae: The vast majority of members of this class are found in fresh water although there are some marine and terrestrial species. Among the plankton, most naked green flagellates range from about one to 40 micrometres in length. Only one genus (*Dunaliella*) and two species have been recorded from New Zealand waters.

Ulvophyceae: This class contains unicellular, multicellular, and siphonocladous non-flagellate green algae, the majority of which are marine or brackish in distribution. The order Ulvales is best known for the genus *Ulva*, with both tubular and blade growth forms, previously known as *Enteromorpha* and *Ulva* respectively (Hayden et al. 2003). All members of the Cladophorales have a siphonocladous level of organisation, i.e. the filaments are uniseriate, either branched or unbranched, and are composed of multinucleate cells. The order Bryopsidales is characterised by a siphonous organisation. Each thallus is essentially a single multinucleate cell containing a central vacuole with a thin layer of cytoplasm. These cells are called coenocytes. Parts of the coenocyte are sometimes sealed off by plugs of cell-wall material, for example when gametangia are formed in *Codium*. One major order (Bryopsidales in the broad sense) was the subject of a cladistic analysis based on morphological, anatomical, reproductive, and cellular characters (Vroom et al. (1998). Almost all members of the Dasycladales are marine and tropical or subtropical in distribution and encrusted with lime (calcium carbonate). Because of this there are a number of fossil genera. The only New Zealand member of the group is a species of *Parvocaulis* from the Kermadec Islands. The order Ulotrichales is poorly known in New Zealand although represented by six genera.

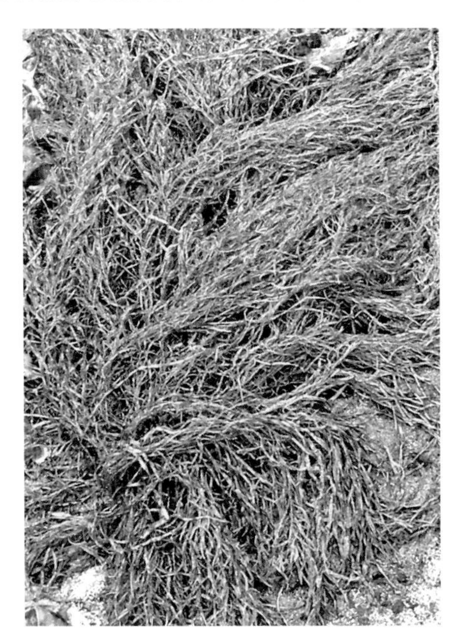

Ulva compressa (Chlorophyta, Ulvophyceae, Ulvales).

Wendy Nelson

Prasinophytes: One of the key features of this group of mostly marine-planktonic flagellates is that, in many species, both the cilia (flagella) and the cell body are covered by minute organic scales that are often highly ornamented. The broad variety of prasinophyte algae have generally been grouped in a single class, Prasinophyceae, but recent work has clarified the status of the component clades, requiring some taxon rearrangements and revised status. For example, *Pedinomonas* nests within the trebouxiophyte order Chlorellales (Turmel et al. 2009) and Marin and Melkonian (2010) gave grounds for segregating the Mamiellales – one of the ecologically most important groups of marine picophytoplankton – and two new orders as class Mamiellophyceae. Additionally, according to Nakayama et al. (2007), the class name Prasinophyceae, which was typified by *Prasinocladus marina*, cannot stand because it is synonymous with *Tetraselmis* (Norris et al. 1980), which is now placed in the Chlorodendrophyceae (Massjuk 2006). The genus *Nephroselmis*, also previously placed in the Prasinophyceae, typifies class Nephroselmidophyceae. The balance of former prasinophytes not included in the above classes can be attributed to Pyramimonadophyceae.

In New Zealand, six genera (11 species) of prasinophytes have been recorded from coastal embayments and the open ocean. These include two 'palmelloid' forms (Pyramimonadophyceae). *Halosphaera viridis* has a large non-motile (up to one millimeter in diameter) globular palmelloid stage in the life-cycle (Parke & Hartog-Adam 1965). This palmelloid form has often been captured among animal plankton in plankton net-tows off the South Island west coast (R. Murdoch pers. comm.). It is so abundant in the open ocean that they are an important source of food for planktonic and nektonic organisms like larvae of the commercial fish *Macruronus novaezelandiae* (hoki) (Chang 2001). *Palmophyllum umbracola*, a crustose palmelloid alga found in northern waters (Kermadecs and Poor Knight Islands), has recently been recognised as being most closely related to Prasinococcales and belonging to a distinct and early diverging lineage of green algae, comprising a new order and family erected (Zechman et al. 2010).

Trebouxiophyceae: Marine species have been the subject of a recent study (Heesch et al. in press). Species of *Prasiola* and *Rosenvingiella* (order Prasiolales), have been placed provisionally in this class until more data become available.

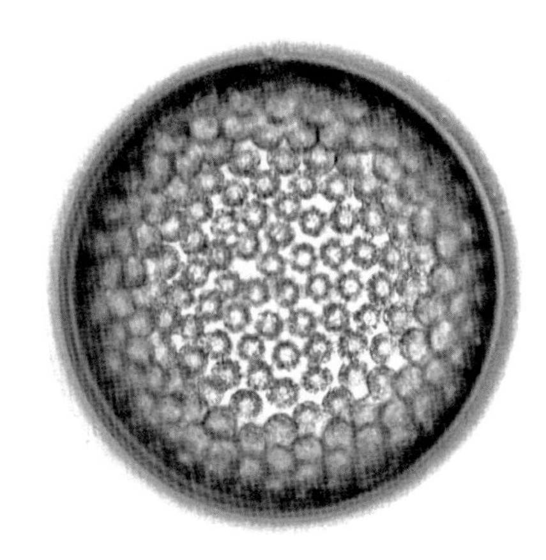

Multicellular palmelloid stage of *Halosphaera viridis* (Chlorophyta, Pyramimonadophyceae, Pyramimonadales).

F. Hoe Chang, NIWA

Historical overview of discoveries and research on New Zealand marine Chlorophyta

In the first of a series of floras of the macroalgae of New Zealand, Chapman (1956) treated the Myxophyceae (so-called blue-green algae, now known as Cyanobacteria) and the Chlorophyceae. He described many new species, subspecies, varieties, forms, and ecads of green algae, placing a number into

Caulerpa flexilis (Chlorophyta, Ulvophyceae, Bryopsidales).
Kate Neill, NIWA

genera that he had established (*Gemina, Lobata, Rama*). He recognised 92 endemic species in a flora of 161 species (excluding the members of *Vaucheria* which are now classified in the chromistan class Xanthophyceae). For a number of genera, Chapman described many new entities that subsequent authors have had difficulty recognising.

Since Chapman's (1956) treatment there have been no monographic studies of any New Zealand green macroalgae. The underlying problems in the systematics of New Zealand green algae were highlighted by Adams (1994) in her account of the seaweed flora. The earlier description of New Zealand members of *Codium* by Dellow (1952, 1953) remains the only thorough monographic treatment of a green seaweed genus in New Zealand. Later, Cassie (1969b) (née Dellow) reported the occurrence of *Pseudobryopsis* in New Zealand. Nielsen (1987) reported on green algae growing within calcareous shells and described an endemic monotypic genus, *Ovillaria catenata*. Nelson and Ryan (1986) reported the genus *Palmophyllum* in northern New Zealand, describing a new endemic species, *P. umbracola*. A major study of the genetic diversity of Ulvaceae in New Zealand was undertaken to establish baseline information, particularly about the distribution of members of the genus *Ulva* (Heesch et al. 2007, 2009). Twenty-four distinct taxa were recognised from this study including 19 species of *Ulva*, four species of *Umbraulva*, and one species of *Gemina*. This research has provided a platform for the future work on the taxonomy of New Zealand species.

Major repositories of New Zealand specimens

The major collections of New Zealand green seaweeds are housed at the herbaria at Te Papa (WELT), Auckland Museum (AK), and Landcare Research (CHR) (Thiers 2011).

Current known natural diversity

The end-chapter checklist of species is based on Chapman (1956), Adams (1994), and published regional seaweed flora lists (Adams 1972; Adams et al. 1974; South & Adams 1976; Nelson & Adams 1984, 1987; Parsons & Fenwick 1984; Hay et al. 1985; Adams & Nelson 1985; Nelson et al. 1991, 1992; Neale & Nelson 1998), as well as recent collections housed at WELT. Authorities follow Brummitt and Powell (1992). In this chapter 135 species and two additional subspecies of marine green seaweeds are listed, of which two genera and 30 species are endemic.

The taxonomic knowledge of the New Zealand green macroalgae on which this list is based is seriously deficient. The most speciose genera are also the ones with the most serious taxonomic problems (e.g. *Ulva, Monostroma, Cladophora, Bryopsis,* and *Chaetomorpha*). In addition, type specimens of some taxa described by Chapman (1956) have not been located. For example, *Bryopsis kermadecensis*, described as endemic to the Kermadec Islands and known only from the type collection, cannot be verified because the specimen is missing. Some genera and/or species are well represented in collections because they are readily distinguished in the field and thus the distributional information about them may be considered to be sound (e.g. some species of *Codium, Caulerpa,* and *Chaetomorpha*). In some other genera (e.g. *Monostroma, Cladophora*) the treatment given here is based on earlier work that cannot be regarded as reliable.

Chaetomorpha coliformis (Chlorophyta, Ulvophyceae, Cladophorales).
Wendy Nelson

Ecology and distribution

The highest-occurring macroalgae on the seashore are members of the genera *Prasiola* and *Rosenvingiella,* which grow in the spray zone of the upper intertidal, often amongst the guano of seabird rookeries. Elsewhere, green macroalgae are found throughout the intertidal and subtidal zones. The prostrate alga *Palmophyllum umbracola* is the deepest-recorded green alga in New Zealand, found at 70 metres depth in the clear waters of the Kermadec Islands (Nelson & Adams 1984).

In addition to growing epilithically, green algae can be found growing within rock and shell (Nielsen 1987) and also as epiphytes or endophytes. Green-algal endophytes (growing within other algae and plants) are associated with a number of host species, particularly certain red algae, but to date there has been no study of these in the New Zealand region. Green endophytes are particularly conspicuous in collections made from Auckland and Campbell Islands. In Canada, recent studies using molecular sequence data conclusively identified the endophytes *Chlorochytrium* and *Codiolum* as being the alternate life-history phases of species of *Acrosiphonia* (Sussmann et al. 1999), and such may be the case in New Zealand.

Species of *Ulva* are tolerant of the variations in salinity typical of estuarine habitats, and are also found in upper intertidal pools that experience the extremes of freshwater input (through runoff or rain water) and evaporation and high-salinity stress. Populations of *Ulva* species have been recorded growing very abundantly in some habitats, apparently in response to human modifications of the environment (Tauranga Harbour, Avon Heathcote Estuary) in a phenomenon that is now internationally known as a 'green tide' (Taylor 1999). Identifying the species involved in green tides can be complex, as not only are the systematics of the genera involved poorly understood, but it appears that species have highly variable morphology (Steffensen 1976).

Significant features of distribution and biogeography

It would be most unwise to use this list of species as a basis for detailed analysis of biogeographic trends, making biodiversity comparisons between regions, or for identifying endemism 'hotspots' or biodiversity priorities; the parlous state of the underlying systematics would compromise the value of any such analysis.

Despite the serious systematic problems with these green algae, there are nevertheless some interesting regional features in the green-algal flora. At the Kermadec Islands, warm-water affinities of the flora are reflected in the presence of *Caulerpa racemosa*, *C. webbiana*, *Codium arabicum*, *C. geppiorum*, *Parvocaulis parvulus*, and *Boodlea composita*, all species that are widespread in tropical and subtropical waters and not found on mainland New Zealand. Species of *Codium* around the mainland reflect regional distributional patterns, with the prostrate species *C. cranwelliae* restricted to northern North Island, *C. convolutum* widespread around North, South, Stewart, and Chatham Islands, and *C. dimorphum* occurring from Banks Peninsula south to the subantarctic islands.

Alien marine species

Ten green macroalgae are considered to be introduced to New Zealand – eight species of *Ulva* and one species of *Umbraulva* (Heesch et al. 2009) and a subspecies of *Codium fragile* (Trowbridge 1995). The impact and risk presented by these species is far from clear.

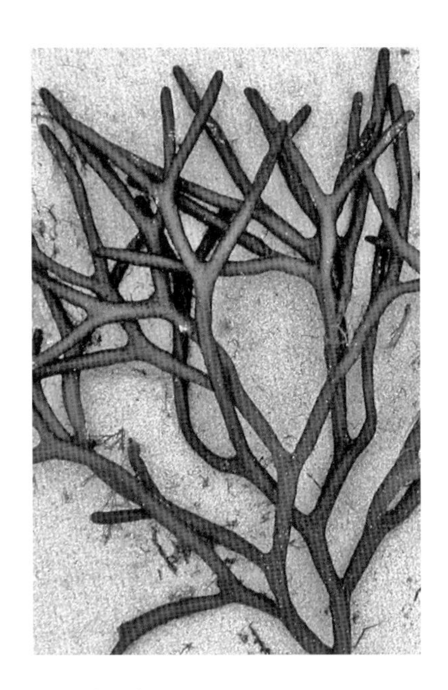

Codium fragile.
Wendy Nelson

Commercial use/potential

There is no commercial use of green seaweeds in New Zealand at present, although some marine-farming sites (from Coromandel through to Stewart Island) have been licensed for farming the sea lettuce *Ulva* (Zemke-White et al. 1999).

Gaps in knowledge and scope for future research on green seaweeds

In the absence of modern systematic studies it is very hard to assess the total number of green seaweeds in New Zealand waters. Although there have been collections made throughout the region, without a sound systematic framework it is difficult to know how well the existing collections encompass the actual diversity. The more remote parts of the New Zealand region (Kermadec Islands, Chatham Islands, subantarctic islands) have been relatively infrequently visited, and seasonally representative collections have not been made. Parsons (1985)

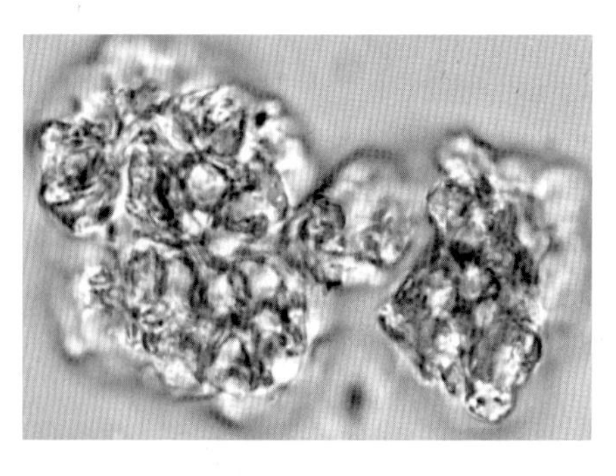

Botryococcus braunii, cluster of preserved cell walls (width of field ca. 60 micrometres). Newvale Mine, Waimumu Coalfield, Southland (F45/f394, Early Miocene).
Ian Raine, GNS Science

considered that once the green algae are revised fewer species will be recognised in New Zealand, although there are also unrecorded species and new species that will require description.

Fossil green algae

Up to 25 different kinds of fossils, not all named, have been listed as definite or possible green algae (Raine & Mildenhall 2008; Chris Clowes pers. comm.). A list of these is given following the end-chapter checklist of living green algae. A marine macrofossil, *Shonabellia verrucosa*, attributed to the ulvophyte family Codiaceae, was described from the New Zealand Triassic near Benmore Dam (Retallack 1983). A number of the microfossils are only doubtfully included in the Chlorophyta; several of the genera listed as incertae sedis appear in lists of microfossil fungi.

Acknowledgements

Thanks are due to Jenn Dalen for assistance in the preparation of the species checklist, to the late Nancy Adams and Murray Parsons for reading the text and checklist and for helpful comments, and to Christiaan van den Hoek, John Huisman, Ruth Nielsen, and Roberta Cowan, who generously responded to enquiries and provided valuable assistance. The editor of the *New Zealand Journal of Botany* is thanked for permission to use Plates 2 and 4 (Volume 37, 1999). Helpful comments on the desmid review were provided by Matt S. McGlone and Janet M. Wilmshurst, and D. B. Williamson made drawings of desmids. We are also grateful for permission from the publishers to use published images of desmids. Hoe Chang (NIWA) gave information on marine planktonic microalgae. Chris Clowes, Victoria University of Wellington, included green-algal taxa in a list of microfossils provided to the editor of this volume.

Authors

Dr Paul A. Broady School of Biological Sciences, University of Canterbury, Private Bag 4800, Christchurch, New Zealand [paul.broady@canterbury.ac.nz] Freshwater and terrestrial non-desmid green algae

Dr Vivienne Cassie Cooper Formerly Landcare Research, Private Bag 3127, Hamilton; now 1/117 Cambridge Road, Hillcrest, Hamilton, New Zealand [viviennecooper@xtra.co.nz] Freshwater non-desmid green algae

Dr Mary D. de Winton National Institute of Water & Atmospheric Research (NIWA), P.O. Box 11115, Hamilton, New Zealand [m.dewinton@niwa.co.nz] Characeae

Dr Elizabeth A. Flint Formerly Landcare Research, P.O. Box 69, Lincoln; now Fitzroy of Merivale Resthome, 4 McDougall Avenue, Merivale, Christchurch, New Zealand. Desmids

Dr Wendy A. Nelson National Institute of Water & Atmospheric Research (NIWA), Private Bag 14-901 Kilbirnie, Wellington, New Zealand [w.nelson@niwa.co.nz] Marine Chlorophyta

Dr Philip M. Novis Landcare Research, P.O. Box 40, Lincoln, New Zealand [novisp@landcareresearch.co.nz] Freshwater green algae and phylogeny

References

ADAMS, N. M. 1972: The marine algae of the Wellington area. *Records of the Dominion Museum 8*: 43–98.

ADAMS, N. M. 1994: *Seaweeds of New Zealand*. Canterbury University Press, Christchurch. 360 p.

ADAMS, N. M.; CONWAY, E.; NORRIS, R. E. 1974: The marine algae of Stewart Island. *Records of the Dominion Museum 8*: 185–245.

ADAMS, N. M.; NELSON, W. A. 1985: The marine algae of the Three Kings Islands. *National Museum of New Zealand Miscellaneous Series 13*: 1–29.

ANDERSEN, R. A. 1998: What to do with Protists? *Australian Systematic Botany 11*: 185–201.

APT, K. E.; BEHRENS, P. W. 1999: Commercial developments in microalgal biotechnology. *Journal of Phycology 35*: 215–226.

BAARS-KLOOS, J. 1976: Phytoplankton in Lake Rotorua and Lake Okareka and its interaction with macrophytes. Unpublished Ph.D. thesis, University of Waikato, Hamilton. 152 p.

BECKER, E. W. 1994: *Microalgae: Biotechnology and Microbiology*. Cambridge University Press, Cambridge. vii + 293 p.

BEHRENS, P. W.; SICOTTE, V. J.; DELENTE, J. 1994: Microalgae as a source of stable isotopically labelled compounds. *Journal of Applied Phycology 6*: 113–121.

BIGGS, B. J. F. 1985: Algae: a blooming nuisance in rivers. *Soil and Water 21(2)*: 27–31.

BIGGS, B. J. F. 1990: Periphyton communities and their environments in New Zealand rivers. *New Zealand Journal of Marine and Freshwater Research 24*: 367–386.

BIGGS, B. J. F. 1996: Patterns in benthic algae of streams. Pp. 31–56 *in*: Stevenson, R. J.; Bothwell, M. L.; Lowe, R. L. (eds), *Algal Ecology – Freshwater Benthic Ecosystems* Academic Press, San Diego. xxvi + 753 p.

BIGGS, B. J. F.; KILROY, C. 2000: Stream periphyton monitoring manual. NIWA (for New Zealand Ministry for the Environment), Christchurch. x + 226 [+ 8] p.

BIGGS, B. J. F.; KILROY, C.; LOWE, R. L. 1998: Periphyton development in three valley segments of a New Zealand grassland river: test of a habitat matrix conceptual model within a catchment. *Archiv für Hydrobiologie 143*: 147–177.

BIGGS, B. J. F.; KILROY, C.; MULCOCK, C.; SCARSBROOK, M. 2002: New Zealand stream health monitoring and assessment kit. Stream monitoring manual. Version 2. *NIWA Technical Report 111*: i-iv, 1–190.

BIGGS, B. J. F.; PRICE, G. M. 1987: A survey of filamentous algal proliferations in New Zealand rivers. *New Zealand Journal of Marine and Freshwater Research 21*: 175–191.

BLINN, D. W.; SHANNON, J. P.; BENENATI, P. L.; WILSON, K. P. 1998: Algal ecology in tailwater stream communities: the Colorado River below Glen Canyon Dam, Arizona. *Journal of Phycology 34*: 734–740.

BOLD, H. C.; WYNNE, M. J. 1985: *Introduction to the Algae: structure and reproduction*. 2nd edn. Prentice-Hall, Englewood Cliffs, N. J. 720 p.

BOURRELLY, P. 1990: *Les Algues d'Eau Douce. Initiation á la systématique. Tome 1: Les algues vertes*. Société Nouvelle des Éditions Boubée, Paris. 572 p.

BREMER, K. 1985: Summary of green plant phylogeny and classification. *Cladistics 1*: 369–385.

BROADY, P. A. 1996: Diversity, distribution and dispersal of Antarctic terrestrial algae. *Biodiversity and Conservation 5*: 1307–1335.

BROADY, P. A.; WEINSTEIN, R. N. 1998: Algae, lichens and fungi in La Gorce Mountains, Antarctica. *Antarctic Science 10*: 376–385.

BROOK, A. J. 1995: Book review. Flora of New Zealand, Freshwater Algae, Chlorophyta, Desmids Volume III. *New Zealand Journal of Marine and Freshwater Research 29*: 132–133.

BRUMMITT, R. K.; POWELL, C. E. (Eds) 1992: *Authors of Plant Names*. Royal Botanic Gardens, Kew. 732 p.

BURNS, C. W.; MITCHELL, S. F. 1974: Seasonal succession and vertical distribution of phytoplankton in Lake Hayes and Lake Johnson, South Island, New Zealand. *New Zealand Journal of Marine and Freshwater Research 8*: 167–209.

BURNS, N. M.; DEELY, J.; HALL, J.; SAFI, K. 1997: Comparing past and present trophic states of seven Central Volcanic Plateau lakes, New Zealand. *New Zealand Journal of Marine and Freshwater Research 31*: 71–87.

BURROWS, C. J. (Ed.) 1977: *Cass: History and Science in the Cass District, Canterbury, New Zealand*. Department of Botany, University of Canterbury. 418 p.

CARTER, N. 1922: Freshwater algae. *In*: A systematic account of plants collected in New Caledonia and the Isle of Pines by Mr. R. H. Compton M. A. in 1914. Part 3. *Journal of the Linnean Society, London, Botany 46*: 47–68.

CASANOVA, M. T.; de WINTON, M. D.; KAROL, K. G.; CLAYTON, J. S. 2007: *Nitella hookeri* A. Braun (Characeae, Charophyceae) in New Zealand and Australia: implications for endemism, speciation and biogeography. *Charophytes 1*: 2–18.

CASSIE, V. 1969a: Seasonal variation in phytoplankton from Lake Rotorua and other inland waters, New Zealand, 1966–67. *New Zealand Journal of Marine and Freshwater Research 3*: 98–123.

CASSIE, V. 1969b: A free-floating *Pseudobryopsis* (Chlorophyceae) from New Zealand. *Phycologia 8*: 71–76.

CASSIE, V. 1974: Algal flora of some North Island New Zealand lakes including Rotorua and Rotoiti. *Pacific Science 28*: 467–504.

CASSIE, V. 1978: Seasonal changes in phytoplankton densities in four North Island lakes, 1973–1974. *New Zealand Journal of Marine and Freshwater Research 12*: 153–166.

CASSIE, V. 1980: Bibliography of the freshwater algae of New Zealand. *New Zealand Journal of Botany 18*: 433–447.

CASSIE, V. 1983: A guide to algae in oxidation ponds in the Auckland district. *Tane (Journal of the Auckland University Field Club) 29*: 119–132.

CASSIE, V. 1984a: Revised checklist of the freshwater algae of New Zealand (excluding diatoms and charophytes). Part 1 – Cyanophyta, Rhodophyta and Chlorophyta. *Water and Soil Technical Publication No. 25*: lxiv, 1–116.

CASSIE, V. 1984b: Revised checklist of the freshwater algae of New Zealand (excluding diatoms and charophytes). Part II — Chlorophyta, Chromophyta and Pyrrhophyta, Rhaphidophyta and Euglenophyta. *Water and Soil Technical Publication No. 26*: lxiv, 1–134.

CASSIE, V. 1989a: Micro-algae of Lake Pupuke, Auckland, New Zealand. *New Zealand Natural Sciences 16*: 39–50.

CASSIE, V. 1989b: A taxonomic guide to thermally associated algae (excluding diatoms) in New Zealand. *Bibliotheca Phycologica 78*: 161–255, 6 pls.

CASSIE COOPER, V. 1994: Freshwater algal research in New Zealand: an update. *Archives of Natural History 21*: 113–130.

CASSIE COOPER, V. 1996: *Microalgae – Microscopic Marvels*. Riverside Books, Hamilton. 164 p.

CASSIE, V.; COOPER, R. C. 1989: Algae of New Zealand thermal areas. *Bibliotheca Phycologica 78*: 1–159.

CHANG, F. H. 2001: A mystery of the sea finally solved. *Biodiversity Update 4*: 7.

CHAPMAN, V. J. 1956: The marine algae of New Zealand. Part I: Myxophyceae and Chlorophyceae. *Journal of the Linnean Society of London 55*: 333–501, pls 24–50.

CHEN, C.-Y.; CHEN, S/-L.; CHRISTENSEN, R. 2005: Individual and combined toxicity of nitriles and aldehydes to *Raphidocelis subcapitata*. *Environmental Toxicology and Chemistry 24*: 1067–1073.

CLAYTON, J. S.; WELLS, R. 1980: Key to the common Characeae of New Zealand. *New Zealand Journal of Botany 18*: 569–570.

COESEL, P. F. M. 1996: Biogeography of desmids. *Hydrobiologia 336*: 41–63.

COESEL, P. F. M. 1998: Sieralgen en natuurwaarden. *Wetenschappelijke Mededeling KNNV 224*: 1–56.

COFFEY, B. T.; MILLER, S. T. 1988: *Hydrodictyon reticulatum* (L.) Lagerheim (Chlorophyta): a new genus record from New Zealand. *New Zealand Journal of Botany 26*: 317–320.

COULTER, G. W.; DAVIES, J.; PICKMERE, S. 1983: Seasonal limnological change and phytoplankton production in Ohakuri, a hydro-electric lake on the Waikato River. *New Zealand Journal of Marine and Freshwater Research 17*: 169–83.

CRANWELL, L. M.; POST, L. von 1936: Post-Pleistocene pollen diagrams from the Southern Hemisphere, I: New Zealand. *Geografiska Annaler 18*: 308–347.

CROASDALE, H.; FLINT, E. A. 1986: *Flora of New Zealand. Freshwater Algae, Chlorophyta, Desmids. Volume I.* Government Printer, Wellington. xii + 132 p., 27 pls.

CROASDALE, H.; FLINT, E. A. 1988: *Flora of New Zealand. Freshwater Algae, Chlorophyta, Desmids. Volume II.* Botany Division, DSIR, Christchurch. x + 147 p., 61 pls.

CROASDALE, H.; FLINT, E. A.; RACINE, M. M. 1994: *Flora of New Zealand. Freshwater Algae, Chlorophyta, Desmids. Volume III.* Manaaki Whenua Press, Lincoln. x + 218 p., 146 pls.

DAVIES, J. P.; GROSSMAN, A. R. 1998: The use of *Chlamydomonas* (Chlorophyta: Volvocales) as a model algal system for genome studies and the elucidation of photosynthetic processes. *Journal of Phycology 34*: 907–917.

DAY, S. A.; WICKHAM, R. P.; ENTWISLE, T. J.; TYLER, P. A. 1995: Bibliographic checklist of non-marine algae in Australia. *Flora of Australia, Supplementary Series 4*: 1–276.

de LANGE, P. J. 1997: *Gratiola pedunculata* (Scrophulariacae): a new addition to the New Zealand flora. *New Zealand Journal of Botany 35*: 317–322.

DEL CAMPO, J. A.; GARCÍA-GONZÁLEZ, M.;

GUERRERO, M. G. 2007: Outdoor cultivation of microalgae for carotenoid production: current state and perspectives. *Applied Microbiology and Biotechnology 74*: 1163–1174.
DELLOW, U.V. 1952: The genus *Codium* in New Zealand. Part I. Systematics. *Transactions of the Royal Society of New Zealand 80*: 119–141.
DELLOW, U.V. 1953: The genus *Codium* in New Zealand. Part II. Ecology, geographic distribution. *Transactions of the Royal Society of New Zealand 80*: 237–243.
DEMIRBAS, A. 2010: Use of algae as biofuel sources. *Energy Conversion and Management 51*: 2738–2749.
DEMPSEY, G. P.; LAWRENCE, D.; CASSIE, V. 1980: The ultrastructure of *Chlorella minutissima* Fott et Nováková. *Phycologia 19*: 13–19.
de WINTON, M. D.; CLAYTON, J. S.; WELLS, R. D. S.; TANNER, C. C.; MILLER, S. T. 1991: Submerged vegetation of Lakes Sumner, Marion, Katrine, Taylor, and Sheppard in Canterbury, New Zealand. *New Zealand Journal of Marine and Freshwater Research 25*: 145–151.
de WINTON, M. D.; DUGDALE, T. M.; CLAYTON, J. S. 2007: An identification key for oospores of the extant chrophytes. *New Zealand Journal of Botany 45*: 463–476.
DOMÍNGUEZ-BOCANEGRA, A. R.; LEGARRETA, I. G.; JERONIMO, F. M.; CAMPOCOSIO, A. T. 2004: Influence of environmental and nutritional factors in the production of astaxanthin from *Haematococcus pluvialis*. *Bioresource Technology 92*: 209–214.
DONALD, R. 1993: Bay of Plenty Regional Council Natural Environment Monitoring Network Freshwater Ecology Programme. Lakes Component 1991/92. Whakatane. *Bay of Plenty Regional Council Technical Report 40*: 1–112.
DROMGOOLE, F. I. 1975: Occurrence of *Codium fragile* subspecies *tomentosoides* in New Zealand waters. *New Zealand Journal of Marine and Freshwater Research 9*: 257–264.
DRYDEN, S. J.; VINCENT, W. F. 1986: Phytoplankton species of Lake Okaro, Central North Island. *New Zealand Journal of Marine and Freshwater Research 20*: 191–198.
ETHEREDGE, M. K. 1983: The seasonal biology of the phytoplankton in Lake Maratoto and Lake Rotomahana. Unpublished M.Sc. thesis, University of Waikato, Hamilton. 264 p.
ETHEREDGE, M. K. 1986: The phytoplankton communities of nine lakes, Waikato, New Zealand. Unpublished Ph.D. thesis, University of Waikato, Hamilton. 3 vols.
ETTL, H.; GÄRTNER, G. 1995: *Syllabus der Boden-, Luft- und Flechtenalgen*. Gustav Fischer Verlag, Stuttgart. viii + 721 p.
EVERETT, A. L. 1998: The ecology and taxonomy of soil algae in Cass Basin, Canterbury, New Zealand. Unpublished MSc thesis, University of Canterbury, Christchurch. ix + 110 p.
FALLOWFIELD, H. J.; MARTIN, N. J.; CROMAR, N. J. 1999: Performance of a batch-fed high rate algal pond for animal waste treatment. *European Journal of Phycology 34*: 231–137.
FINET, C.; TIMME, R. E.; DELWICHE, C. F.; MARLÉTAZ, F. 2010: Multigene phylogeny of the green lineage reveals the origin and diversification of land plants. *Current Biology 20*: 2217–2222.
FLINT, E. A. 1938: A preliminary study of the phytoplankton in Lake Sarah (New Zealand). *Journal of Ecology 26*: 353–358.
FLINT, E. A. 1958: Biological studies of some tussock-grassland soils IX. Algae: preliminary observations. *New Zealand Journal of Agricultural Research 1*: 991–997.
FLINT, E. A. 1968: Algae on the surface of some New Zealand soils. *New Zealand Soil Bureau Bulletin 26*: 183–190.
FLINT, E. A. 1974: *Parallela*, a new genus of freshwater Chlorophyta in New Zealand. *New Zealand Journal of Botany 12*: 357–364.
FLINT, E. A. 1975: Phytoplankton in some New Zealand lakes. Pp. 163–192 *in*: Jolly, V. H.; Brown, J. M. A. (eds), *New Zealand Lakes*. Auckland University Press, Auckland. 388 p.
FLINT, E. A. 1977: Phytoplankton in seven monomictic lakes in Rotorua, New Zealand. *New Zealand Journal of Botany 15*: 197–208.
FLINT, E. A. 1979: Comments on the phytoplankton and chemistry of three monomictic lakes in Westland National park, New Zealand. *New Zealand Journal of Botany 17*: 127–134.
FLINT, E. A. 1996: Some additions to the desmid flora of New Zealand. *New Zealand Journal of Botany 34*: 547–551.
FLINT, E. A.; ETTL, H. 1966: Some new and uncommon *Chlamydomonas* species from New Zealand. *New Zealand Journal of Botany 4*: 418–433.
FLINT, E. A.; FINERAN, B. A. 1969: Observations on the climate, peats and terrestrial algae of The Snares Islands. *New Zealand Journal of Science 12*: 286–301.
FLINT, E. A.; WILLIAMSON, D. B. 1998: Desmids (Chlorophyta) in a small tarn in Central Canterbury, New Zealand. *Algological Studies 91*: 71–100.
FLINT, E. A.; WILLIAMSON, D. B. 1999: Desmids (Chlorophyta) in a small tarn in Central Canterbury, New Zealand. *New Zealand Journal of Botany 37*: 541–551.
FLINT, E. A.; WILLIAMSON, D. B. 2005: Desmids (Chlorophyta), including two new species and three new varieties, in two swamps, a lake, and a tarn in the South Island of New Zealand. *New Zealand Journal of Botany 43*: 285–300.
FORSYTH, D. J.; McCOLL, R. H. S. 1974: The limnology of a thermal lake: Lake Rotowhero, New Zealand: II. General biology with emphasis on the benthic fauna of chironomids. *Hydrobiologia 44*: 91–104.
FRIEDL, T. 1997: The evolution of the green algae. Pp. 87–101 *in*: Bhattacharya, D. (ed.), *Origins of Algae and their Plastids*. Springer-Verlag, Wien. 287 p.
FUMANTI, B.; ALFINITO, S. 2004: *Brachytheca inopinata* (Desmidiaceae, Zygnematophyceae), a new desmid from New Zealand. *Phycologia 43*: 455–458.
GERRATH, J. F. 1993: The biology of desmids: a decade of progress. Progress in Phycological Research 9: 78–192.
GRAHAM, L. E.; COOK, M. E.; BUSSE, J. S. 2000: The origin of land plants: body plan changes contributing to a major evolutionary radiation. *Proceedings of the National Academy of Sciences* 97: 4535–4540.
GRAHAM, L. E.; GRAHAM, J. M.; WILCOX, L. W. 2009: *Algae*. Benjamin Cummings, San Francisco, xvi + 616 p.
GRAHAM, L. E.; GRAY, J. 2001: The origin, morphology, and ecophysiology of early embryophytes: neontological and paleontological perspectives. Pp. 140–158 *in*: Gensel, P. G.; Edwards, D. (eds), *Plants Invade the Land: Evolutionary and environmental perspectives*. Columbia University Press, New York. 512 p.
GREENBAUM, E. 1988: Energetic efficiency of hydrogen photoevolution by algal water splitting. *Biophysical Journal 54*: 365–368.
GREENE, B.; DARNALL, D. W. 1990: Microbial oxygenic photoautotrophs (cyanobacteria and algae) for metal-ion binding. Pp. 277–302 *in*: Ehrlich, H. L.; Brierley, C. L. (eds), *Microbial Mineral Recovery*. McGraw-Hill, New York. x + 454 p.
GRUNOW, A. 1870: Algae. *Reise der Novara, Botanischer Theil 1 (1)*: 1–104.
GUPTA, V. K.; RASTOGI, A. 2008: Biosorption of lead from aqueous solutions by green algae *Spirogyra* species: kinetics and equilibrium studies. *Journal of Hazardous Materials 152*: 407–414.
HAGEN, C.; GRÜNEWALD, K.; SCHMIDT, S.; MÜLLER, J. 2000: Accumulation of secondary carotenoids in flagellates of *Haematococcus pluvialis* (Chlorophyta) is accompanied by an increase in per unit chlorophyll productivity of photosynthesis. *European Journal of Phycology 35*: 75–82.
HAGLUND, K. 1997: The use of algae in aquatic toxicity assessment. *Progress in Phycological Research 12*: 181–212.
HALL, J. A.; COX, N. 1995: Nutrient concentrations as predictors of nuisance *Hydrodictyon reticulatum* populations in New Zealand. *Journal of Aquatic Plant Management 33*: 68–74.
HALL, J. D.; DELWICHE, C. F. 2007: In the shadow of giants; systematics of the charophyte green algae. Pp. 155–169 *in*: Brodie, J.; Lewis, J. (eds), *Unravelling the Algae: the past, present, and future of algal systematics*. [*Systematics Association Special Volume Series 75*.] CRC Press, Boca Raton. xvii, 376 p.
HAPPEY-WOOD, C. M. 1988: Ecology of freshwater planktonic green algae. Pp. 175–226 *in*: Sandgren, C. D. (ed.), *Growth and Reproductive Strategies of Freshwater Phytoplankton*. Cambridge University Press, Cambridge. vi + 442 p.
HARVEY, N. B. 1969: *New Zealand Botanical Paintings*. Whitcombe & Tombs Ltd, Christchurch. 86 p., 40 pls.
HAWES, I.; HOWARD-WILLIAMS, C.; WELLS, R.; CLAYTON, J. 1991: Invasion of water net, *Hydrodictyon reticulatum*: the surprising success of an aquatic plant new to our flora. *New Zealand Journal of Marine and Freshwater Research 25*: 227–229.
HAWES, I.; SCHWARZ, A.-M. 1996: Epiphytes from a deep-water characean meadow in an oligotrophic New Zealand lake: species composition, biomass and photosynthesis. *Freshwater Biology 36*: 297–313.
HAWES, I.; SMITH, R. 1993a: Influence of environmental factors on the growth in culture of a New Zealand strain of the fast-spreading alga *Hydrodictyon reticulatum* (water-net). *Journal of Applied Phycology 5*: 437–445.
HAWES, I.; SMITH, R. 1993b: Seasonal dynamics of epilithic periphyton in oligotrophic Lake Taupo, New Zealand. *New Zealand Journal of Marine and Freshwater Research 28*: 1–12.
HAY, C. H.; ADAMS, N. M.; PARSONS, M. J. 1985: The marine algae of the subantarctic islands of New Zealand. *National Museum of New Zealand, Miscellaneous Series 11*: 1–70.
HAYDEN, H. S.; BLOMSTER, J.; MAGGS, C. A.; SILVA, P. C.; STANHOPE, M. J.; WAALAND, J. R. 2003. Linnaeus was right all along: *Ulva* and

Enteromorpha are not distinct genera. *European Journal of Phycology* 38: 277–294.
HEESCH, S.; BROOM, J.; NEILL, K.; FARR, T.; DALEN, J.; NELSON, W. 2007: Genetic diversity and possible origins of New Zealand populations of *Ulva*. *Final Research Report for Ministry of Fisheries Research Project ZBS2004-08*. NIWA, Wellington. 203 p.
HEESCH, S.; BROOM, J. E.; NEILL, K.; FARR, T. J.; DALEN, J.; NELSON, W. A. 2009: *Ulva*, *Umbraulva* and *Gemina*: genetic survey of New Zealand Ulvaceae reveals diversity and introduced species. *European Journal of Phycology 44*: 143–154.
HEESCH, S.; SUTHERLAND, J. E.; NELSON, W. A. (In press) Marine Prasiolales (Trebouxiophyceae, Chlorophyta) from New Zealand and the Balleny Islands, with descriptions of *Prasiola novaezelandiae* sp. nov. and *Rosenvingiella australis* sp. nov. *Phycologia*.
HINDÁK, F. 1980: Studies on the chlorococcal algae (Chlorophyceae). II. *Biologické Práce 26*: 1–195.
HINDÁK, F. 1988: Studies on the chlorococcal algae (Chlorophyceae). IV. *Biologické Práce 34*: 1–263.
HOEK, C. van den; MANN, D. G.; JAHNS, H. M. 1995: *Algae: an introduction to phycology*. Cambridge University Press, Cambridge. xiv + 623 p.
HOFFMANN, L. 1989: Algae of terrestrial habitats. *The Botanical Review 55*: 77–105.
HOFFMANN, J. P. 1998: Wastewater treatment with suspended and nonsuspended algae. *Journal of Phycology 34*: 757–763.
HOHAM, R. W.; ROEMER, S. C.; MULLET, J. E. 1979: The life history and ecology of the snow alga *Chloromonas brevispina* comb. nov. (Chlorophyta, Volvocales). *Phycologia 18*: 55–70.
HOWARD-WILLIAMS, C.; LAW, K.; VINCENT, C. L.; DAVIES, J.; VINCENT, W. F. 1986: Limnology of Lake Waikaremoana with special reference to littoral and pelagic primary producers. *New Zealand Journal of Marine and Freshwater Research 20*: 583–597.
HUISMAN, J. M.; SAUNDERS, G. W. 2007: Phylogeny and classification of the Algae. Pp. 66–103 *in*: McCarthy, P. M.; Orchard, A. E. (eds), *Algae of Australia – Introduction*. ABRS/CSIRO Publishing, Melbourne. xvi + 727 p.
HUSS, V. A. R.; FRANK, C.; HARTMANN, E. C.; HIRMER, M.; KLOBOUCEK, A.; SEIDEL, B. M.; WENZELER, P.; KESSLER, E. 1999: Biochemical taxonomy and molecular phylogeny of the genus *Chlorella* sensu lato (Chlorophyta). *Journal of Phycology 35*: 587–598.
JOHANSEN, J. R. 1993: Cryptogamic crusts of semiarid and arid lands of North America. *Journal of Phycology 29*: 140–147.
JOHN, D. M.; WHITTON, B. A.; BROOK, A. J. (Eds) 2002: *The Freshwater Algal Flora of the British Isles. An identification guide to freshwater and terrestrial algae.* Cambridge University Press, Cambridge. xii + 702 p.
JOLLY, V. H. 1977: The comparative limnology of some New Zealand lakes. 2. Plankton. *New Zealand Journal of Marine and Freshwater Research 11*: 307–340.
JOUBERT, J. J.; RIJKENBERG, F. H. J. 1971: Parasitic green algae. *Annual Review of Phytopathology 9*: 45–64.
KARLING, J. S. 1965: Some zoosporic fungi of New Zealand. 1. *Sydowia. Annales Mycologici, ser. 2, 19*: 213–226.
KAROL, K. G.; McCOURT, R. M.; CIMINO, M. T.; DELWICHE, C. F. 2001: The closest living relatives of land plants. *Science 294*: 2351–2353.
KRIEGER, W. 1932: Die Desmidiaceen der Deutschen Limnologischen Sunda-Expedition. *Archiv für Hydrobiologie 11 (Suppl.)*: 129–230.
KRIENITZ, L.; TAKEDA, H.; HEPPERLE, D. 1999: Ultrastructure, cell wall composition, and phylogenetic position of *Pseudodictyosphaerium jurisii* (Chlorococcales, Chlorophyta) including a comparison with other picoplanktonic green algae. *Phycologia 38*: 100–107.
KRUSE, O.; HANKAMER, B. 2010: Microalgal hydrogen production. *Current Opinion in Biotechnology 21*: 238–243.
LEE, R. E. 1999: *Phycology*. Cambridge University Press, Cambridge. x + 614 p.
LEITE SANT'ANNA, C.; BICUDO, R. M. T. 1979: Record of *Parallela* (Chlorococcales, Chlorophyceae) in Brazil. *Rickia 8*: 101–104.
LEMBI, C. A.; O'NEAL, S. W.; SPENCER, D. F. 1988: Algae as weeds: economic impact, ecology, and management alternatives. Pp. 455–481 *in*: Lembi, C. A.; Waaland, J. R. (eds), *Algae and Human Affairs*. Cambridge University Press, Cambridge. viii + 590 p.
LEMIEUX, C.; OTIS, C.; TURMEL, M. 2000: Ancestral chloroplast genome in *Mesostigma viride* reveals an early branch of green plant evolution. *Nature 403*: 649–652.
LEMIEUX, C.; OTIS, C.; TURMEL, M. 2007: A clade uniting the green algae *Mesostigma viride* and *Chlorokybus atmophyticus* represents the deepest branch of the Streptophyta in chloroplast genome-based phylogenies. *BMC Biology 5,2*: 1–17.
LEWIS, L. A.; McCOURT, R. M. 2004: Green algae and the origin of land plants. *American Journal of Botany 91*: 1535–1556.
LI, Y.; QIN, J. G. 2005: Comparison of growth and lipid content in three *Botryococcus braunii* strains. *Journal of Applied Phycology 17*: 551–556
LING, H. U.; SEPPELT, R. D. 1998: Snow algae of the Windmill Islands, continental Antarctica 3. *Chloromonas polyptera* (Volvocales, Chlorophyta). *Polar Biology 20*: 320–324.
LONGLEY, G. 1999: Culliman profits from NZ bungle. *The Press* (Christchurch), Monday, 1 March, p. 19.
LUKEŠOVÁ, A. 2001: Soil algae in brown coal and lignite post-mining areas in central Europe (Czech Republic and Germany). *Restoration Ecology 9*: 341–350.
MacENTEE, F. J.; BOLD, H. C.; ARCHIBALD, P. A. 1977: Notes on some edaphic algae of the South Pacific and Malaysian areas, with special reference to *Pseudotetraedron polymorphum* gen. et sp. nov. *Soil Science 124*: 161–166.
MARIN, B.; MELKONIAN, M. 2010: Molecular phylogeny and classification of the Mamiellophyceae class. nov. (Chlorophyta) based on the sequence comparisons of the nuclear- and plastid-encoded rRNA operons. *Protist 161*: 304–336.
MASKELL, W. M. 1881: Contributions towards a list of the New Zealand Desmidieae. *Transactions and Proceedings of the New Zealand Institute 13*: 297–317, pls 11, 12.
MASKELL, W. M. 1883: On the New Zealand Desmidieae. Additions to catalogue and notes on various species. *Transactions and Proceedings of the New Zealand Institute 15*: 237–258, pls 24, 25.
MASKELL, W. M. 1886: On a new variety of desmid. *Transactions and Proceedings of the New Zealand Institute 18*: 325.
MASKELL, W. M. 1887: On the fresh-water infusoria of the Wellington District. *Transactions of the New Zealand Institute 19*: 49–61, pls 3–5.
MASKELL, W. M. 1888: On the fresh-water infusoria of the Wellington District. *Transactions of the New Zealand Institute 20*: 3–19, pls 1–4.
MASKELL, W. M. 1889: Further notes on the Desmidieae of New Zealand, with descriptions of new species. *Transactions and Proceedings of the New Zealand Institute 21*: 3–32, pls 1–6.
MASSJUK, N. P. 2006: Chlorodendrophyceae class nov. (Chlorophyta, Viridiplantae) in the Ukrainian flora: 1. The volume, phylogenetic relations and taxonomical status. *Ukrainian Journal of Botany 63*: 601–614. [In Ukrainian.]
MATHER, W. M. 1928: Freshwater algae of the Hutt Valley. Unpublished M.Sc. thesis, Victoria University of Wellington. 118 p.
McCOURT, R. M. 1995: Green algal phylogeny. *Trends in Ecology and Evolution 10*: 159–163.
McCOURT, R. M.; DELWICHE, C. F.; KAROL, K. G. 2004: Charophyte algae and land plant origins. *Trends in Ecology and Evolution 19*: 661–666.
McCRACKEN, M. D.; PROCTOR, V. W.; HOTCHKISS, A. T. 1966: Attempted hybridization between monoecious and dioecious clones of *Chara*. *American Journal of Botany 53*: 937–940.
McGLONE, M. S.; WILMSHURST, J. M. 1999: A Holocene record of climate, vegetation change and peat bog development, east Otago, New Zealand. *Journal of Quaternary Science 14*: 239–254.
MEIERS, S. T.; PROCTOR, V. W.; CHAPMAN, R. L. 1999: Phylogeny and biogeography of *Chara* (Charophyta) inferred from 18S rDNA sequences. *Australian Journal of Botany 47*: 347–360.
MELIS, A.; HAPPE, T. 2004: Trails of green alga hydrogen research - from Hans Gaffron to new frontiers. *Photosynthesis Research 80*: 401–409.
MERCHANT, S. S.; PROCHNIK, S. E.; VALLON, O.; HARRIS, E. H.; KARPOWICZ, S. J. et al. 2007: The *Chlamydomonas* genome reveals the evolution of key animal and plant functions. *Science 318*: 245–250.
METTING, B. 1981: The systematics and ecology of soil algae. *The Botanical Review 47*: 195–312.
METTING, B.; RAYBURN, W. R.; REYNAUD, P. A. 1988: Algae and agriculture. Pp. 335–370 *in*: Lembi, C. A.; Waaland, J. R. (eds), *Algae and Human Affairs*. Cambridge University Press, Cambridge. viii + 590 p.
MOORE, S. C. 2000: *Photographic Guide to the Freshwater Algae of New Zealand*. Otago Regional Council, Dunedin. 77 p.
MROZIŃSKA, T. 1990: Aerophytic algae from North Korea. *Algological Studies 58*: 29–47.
NAKAYAMA, T.; SUDA, S.; KAWACHI, M.; INOUYE, I. 2007: Phylogeny and ultrastructure of *Nephroselmis* and *Pseudoscourfieldia* (Chlorophyta), including the description of *Nephroselmis anterostigmatica* sp. nov. and a proposal for the Nephroselmidales ord. nov. *Phycologia 46*: 680–697.
NEALE, D.; NELSON, W. 1998: Marine algae of the west coast, South Island, New Zealand. *Tuhinga 10*: 87–118.
NELSON, W. A. 1999: A revised checklist of marine algae naturalised in New Zealand. *New Zealand Journal of Botany 37*: 355–359.
NELSON, W. A.; ADAMS, N. M. 1984: Marine algae of the Kermadec Islands. *National Museum of New Zealand, Miscellaneous Series 10*: 1–29.

NELSON, W. A.; ADAMS, N. M. 1987: Marine algae of the Bay of Islands area. *National Museum of New Zealand, Miscellaneous Series 16*: 1–47.
NELSON, W. A.; ADAMS, N. M.; FOX, J. M. 1992: Marine algae of the northern South Island. *National Museum of New Zealand, Miscellaneous Series 26*: 1–80.
NELSON, W. A.; ADAMS, N. M.; HAY, C. H. 1991: Marine algae of the Chatham Islands. *National Museum of New Zealand, Miscellaneous Series 23*: 1–58.
NELSON, W. A.; RYAN, K. G. 1986: *Palmophyllum umbracola* sp. nov. (Chlorophyta) from offshore islands of northern New Zealand. *Phycologia 25*: 168–177.
NIELSEN, R. 1987: Marine algae within calcareous shells from New Zealand. *New Zealand Journal of Botany 25*: 425–348.
NEUSTUPA, J.; NĚMCOVÁ, Y.; ELIÁŠ, M.; ŠKALOUD, P. 2009: *Kalinella bambusicola* gen. et sp. nov. (Trebouxiophyceae, Chlorophyta), a novel coccoid *Chlorella*-like subaerial alga from Southeast Asia. *Phycological research 57*: 159–169.
NISHIYAMA, T.; WOLF, P. G.; KUGITA, M.; SINCLAIR, R. B.; SUGITA, M.; SUGIURA, C.; WAKASUGI, T.; YAMADA, K.; YOSHINAGA, K.; YAMAGUCHI, K.; UEDA, K.; HASEBE, M. 2004: Chloroplast phylogeny indicates that bryophytes are monophyletic. *Molecular Biology and Evolution 21*: 1813–1819.
NORDSTEDT, O. 1887: Über die von Prof. S. Berggren auf Neu-Zeeland gesammelten süsswasser Algen. *Botanisches Zentralblatt 31*: 321–322.
NORDSTEDT, O. 1888: Freshwater algae collected by Dr S. Berggren in New Zealand and Australia. *Kungliga svenska Vetenskapsakademiens Handlingar 22*: 1–98.
NORRIS, R. E.; HORI, T.; CHIHARA, M. 1980: Revision of the genus *Tetraselmis* (class Prasinophyceae). *Botanical Magazine, Tokyo 93*: 317–339.
NORTON, T. A.; MELKONIAN, M.; ANDERSEN, R. A. 1996: Algal biodiversity. *Phycologia 35*: 308–326.
NOVIS, P. M. 2001: Ecology and taxonomy of alpine algae, Mt Philistine, Arthur's Pass National park, New Zealand. Unpublished Ph.D. thesis, University of Canterbury, Christchurch. xiii + 273 p.
NOVIS, P. M. 2002a: New records of snow algae for New Zealand, from Mt Philistine, Arthur's Pass National Park. *New Zealand Journal of Botany 40*: 297–312.
NOVIS, P. M. 2002b: Ecology of the snow alga *Chlainomonas kolii* (Chlamydomonadales, Chlorophyta) in New Zealand. *Phycologia 41*: 280–292.
NOVIS, P. M. 2003: A taxonomic survey of *Oedogonium* (Oedogoniales, Chlorophyta) in the South Island and Chatham Islands, New Zealand. *New Zealand Journal of Botany 41*: 335–358.
NOVIS, P. M. 2004a: New records of *Spirogyra* and *Zygnema* in New Zealand. *New Zealand Journal of Botany 42*: 139–152.
NOVIS, P. M. 2004b: A taxonomic survey of *Microspora* (Chlorophyceae, Chlorophyta) in New Zealand, using field material. *New Zealand Journal of Botany 42*: 153–165.
NOVIS, P. M. 2006: Taxonomy of *Klebsormidium* (Klebsormidiales, Charophyceae) in New Zealand streams and the significance of low-pH habitats. *Phycologia 45*: 293–301.
NOVIS, P. M.; BEER, T.; VALLANCE, J. 2008: New records of microalgae from the New Zealand alpine zone, and their distribution and dispersal. *New Zealand Journal of Botany 46*: 347–366.
NOVIS, P. M.; HALLE, C.; WILSON, B.; TREMBLAY, L. A. 2009: Identification and characterization of freshwater algae from a pollution gradient using rbcL sequencing and toxicity testing. *Archives of Environmental Contamination and Toxicology 57*: 504–514.
NOVIS, P. M.; HARDING, J. S. 2007: Extreme acidophiles: freshwater algae associated with acid mine drainage. Pp. 443–463 *in*: Seckbach, J. (ed.), *Algae and Cyanobacteria in Extreme Environments. Cellular origins and life in extreme environments. Volume 14*. Springer Verlag, Dordrecht. xxxii + 805 p.
NOVIS, P. M.; LORENZ, M.; BROADY, P. A.; FLINT, E. A. (2010). *Parallela* Flint: its phylogenetic position in the Chlorophyceae and the polyphyly of *Radiofilum* Schmidle. *Phycologia 49*: 373–383.
NOZAKI, H.; MISUMI, O.; TSUNEYOSHI, K. 2003: Phylogeny of the quadriflagellate Volvocales (Chlorophyceae). *Molecular Phylogenetics and Evolution 29*: 58–66.
OSWALD, W. J. 1988: The role of microalgae in liquid waste treatment and reclamation. Pp. 255–281 *in*: Lembi, C. A.; Waaland, J. R. (eds), *Algae and Human Affairs*. Cambridge University Press, Cambridge. viii + 590 p.
PAERL, H. W. 1977: Ultraphytoplankton biomass and production in some New Zealand lakes. *New Zealand Journal of Marine and Freshwater Research 11*: 297–305.
PAERL, H. W. 1978: Effectiveness of various counting methods in detecting viable phytoplankton. *New Zealand Journal of Marine and Freshwater Research 12*: 67–72.
PAERL, H. W.; PAYNE, G. W.; McKENZIE, A. L.; KELLAR, P. E.; DOWNES, M. T. 1979: Limnology of nine Westland beech forest lakes. *New Zealand Journal of Marine and Freshwater Research 13*: 47–57.
PARKE, M.; HARTOG-ADAM, I. den 1965: Three species of *Halosphaera*. *Journal of the Marine Biological Association of the United Kingdom 56*: 537–594.
PARSONS, M. J. 1985: New Zealand seaweed flora and its relationships. *New Zealand Journal of Marine and Freshwater Research 19*: 131–138.
PARSONS, M. J.; FENWICK, G. D. 1984: Marine algae and a marine fungus from Open Bay Islands, Westland. *New Zealand Journal of Botany 22*: 425–432.
PITTMAN, J. K.; DEAN, A. P.; OSUNDEKO, O. 2011: The potential of sustainable algal biofuel production using wastewater resources. *Bioresource Technology 102*: 17–25.
PRESCOTT, G. W.; CROASDALE, H. T.; VINYARD, W. C. 1975–1983: *A Synopsis of North American Desmids. Pt. 2, Desmidiaceae: Placodermae: Section 1*. University of Nebraska Press, Lincoln. 275 p., 57 pls.
PROCTOR, V. W. 1971a: Taxonomic significance of monoecism and dioecism in the genus *Chara*. *Phycologia 10*: 299–307.
PROCTOR, V. W. 1971b: *Chara globularis* Thuiller (=*C. fragilis* Devaux.): breeding patterns within a cosmopolitan complex. *Limnology and Oceanography 16*: 422–436.
PULZ, O.; GROSS, W. 2004: Valuable products from biotechnology of microalgae. *Applied Microbiology and Biotechnology 65*: 635–648.
RAINE, D. C.; MILDENHALL, D. C.; KENNEDY, E. M. 2008: New Zealand fossil spores and pollen: an illustrated catalogue. 3rd edn. *GNS Science Miscellaneous Series No. 4*. [http://www.gns.cri.nz/what/earthhist/fossils/spore_pollen/catalog/index.htm.]
RAJA, R.; HEMAISWARYA, S.; KUMAR, N. A.; SRIDHAR, S.; RENGASAMY, R. 2008: A perspective on the biotechnological potential of microalgae. *Critical Reviews in Microbiology 34*: 77–88.
RALFS, J. 1848: *The British Desmidieae*. Reeve, Benham & Reeve, London. xxii + 226 p., 35 pls.
RAMSAY, A. J.; BALL, K. T. 1983: Estimation of algae in New Zealand pasture soil and litter by culturing and by chlorophyll *a* extraction. *New Zealand Journal of Science 26*: 493–503.
REISSER, W.; WIDOWSKI, M. 1992: Taxonomy of eukaryotic algae endosymbiotic in freshwater associations. Pp. 21–40 *in*: Reisser, W. (ed.), *Algae and Symbioses*. Biopress Ltd, Bristol. xii + 746 p.
RETALLACK, G. J. 1983: Middle Triassic megafossil marine algae and land plants from near Benmore Dam, southern Canterbury, New Zealand. *Journal of the Royal Society of New Zealand 13*: 129–154.
SARMA, P. 1986: The freshwater Chaetophorales of New Zealand. *Beihefte zur Nova Hedwigia 58*: xii, 1–169, 143 pls.
SCHWARZ, A.-M.; HAWES, I.; de WINTON, M. 2002: Species-specific depth zonation in New Zealand charophytes as a function of light availability. *Aquatic Botany 72*: 209–217.
SCHWARZ, A.-M.; HAWES, I.; HOWARD-WILLIAMS, C. 1996: The role of photosynthesis/light relationships in determining lower depth limits of Characeae in South Island, New Zealand lakes. *Freshwater Biology 35*: 69–80.
SCHWARZ, A.-M.; HOWARD-WILLIAMS, C.; CLAYTON, J. 2000: Analysis of relationships between maximum depth limits of aquatic plants and underwater light in 63 New Zealand lakes. *New Zealand Journal of Marine and Freshwater Research 34*: 157–174.
SKUJA, H. 1976: Zür Kenntnis der Algen neuseeländischer Torfmoore. *Nova Acta Regiae Societatis, Scientiarum Upsaliensis, ser. 5, C, 2*: 1–158.
SKULBERG, O. M. 1995: Use of algae for testing water quality. Pp. 181–199 *in*: Wiessner, W.; Schnepf, E.; Starr, R. C. (eds), *Algae, Environment and Human Affairs*. Biopress Ltd, Bristol. xiii + 258 p.
SORRELL, B.; HAWES, I.; SCHWARZ, A.; SUTHERLAND, D. 2001: Inter-specific differences in photosynthetic carbon uptake, photosynthate partitioning and extracellular organic carbon release by deep-water characean algae. *Freshwater Biology 46*: 453–464.
SOUTH, G. R.; ADAMS, N. M. 1976: Marine algae of the Kaikoura coast. *National Museum of New Zealand, Miscellaneous Series 1*: 1–67.
SPENCER, W. I. 1882: On the fresh-water algae of New Zealand. *Transactions and Proceedings of the New Zealand Institute 14*: 287–289, pl. 23.
STEFFENSEN, D. A. 1976: Morphological variation in *Ulva* in the Avon-Heathcote Estuary, Christchurch. *New Zealand Journal of Marine and Freshwater Research 10*: 329–341.
STEIN, J. R.; BORDEN, C. A. 1984: Causative and beneficial algae in human disease conditions: a review. *Phycologia 23*: 485–501.
SUSSMANN, A. V.; MABLE, B. K.; DeWREEDE, R. E.; BERBEE, M. L. 1999: Identification of

green algal endophytes as the alternate phase of *Acrosiphonia* (Codiolales, Chlorophyta) using ITS1 and ITS2 ribosomal DNA sequence data. *Journal of Phycology 35*: 607–614.
TARTAR, A.; BOUCIAS, D. G.; ADAMS, B. J.; BECNEL, J. J. 2002: Phylogenetic analysis identifies the invertebrate pathogen *Helicosporidium* sp. as a green alga (Chlorophyta). *International Journal of Systematic and Evolutionary Microbiology 52*: 273–279.
TAYLOR, R. 1999: The green tide threat in the UK – a brief overview with particular reference to Langstone Harbour, south coast of England and the Ythan estuary, east coast of Scotland. *Botanical Journal of Scotland 51*: 195–203.
TENAUD, M.; OHMORI, M.; MIYACHI, S. 1989: Inorganic carbon and acetate assimilation in *Botryococcus braunii* (Chlorophyta). *Journal of Phycology 25*: 662–667.
THIELE, D.; BERGMANN, A. 2002: Protothecosis in human medicine. *International Journal of Hygiene and Environmental Health 204*: 297–302.
THIERS, B. 2011 [continuously updated]: *Index Herbariorum: A global directory of public herbaria and associated staff.* New York Botanical Garden's Virtual Herbarium. [http://sweetgum.nybg.org/ih/.]
THOMAS, W. H.; BROADY, P. A. 1997: Distribution of coloured snow and associated algal genera in New Zealand. *New Zealand Journal of Botany 35*: 113–117.
THOMASSON, K. 1960: Some planktic *Staurastra* from New Zealand. *Botaniska Notiser 113*: 225–245.
THOMASSON, K. 1972: Some planktic *Staurastra* from New Zealand. 2. *Svensk Botanisk Tidskrift 66*: 257–274.
THOMASSON, K. 1973: *Actinotaenium, Cosmarium* and *Staurodesmus* in the plankton of Rotorua lakes. *Svensk Botanisk Tidskrift 67*: 127–141.
THOMASSON, K. 1974a: Some planktic *Staurastra* from New Zealand. 3. *Svensk Botanisk Tidskrift 68*: 33–50.
THOMASSON, K. 1974b: Rotorua phytoplankton reconsidered (North Island of New Zealand). *Internationale Revue Gesamten Hydrobiologie 59*: 703–727.
THOMASSON, K. 1980: Antipodal algal annotations. *Nova Hedwigia 33*: 919–931.
THOMPSON, R. H.; WUJEK, D. E. 1997: *Trentepohliales:* Cephaleuros, Phycopeltis, *and* Stomatochroon. Science Publishers, Enfield, New Hampshire. x + 149 p.
TRAINOR, F. R.; CAIN, J. R. 1986: Famous algal genera. I. *Chlamydomonas*. *Progress in Phycological Research 4*: 81–127.
TROWBRIDGE, C. D. 1995: Establishment of the green alga *Codium fragile* ssp. *tomentosoides* on New Zealand rocky shores: current distribution and invertebrate grazers. *Journal of Ecology 83*: 949–965.
TSCHERMAK-WOESS, E. 1988: The algal partner. Pp. 39–92 *in*: Galun, M. (ed.), *CRC Handbook of Lichenology. Volume 1.* CRC Press, Boca Raton. 290 p.
TURMEL, M.; OTIS, C.; LEMIEUX, C. 2003: The mitochondrial genome of *Chara vulgaris*: insights into the mitochondrial DNA architecture of the last common ancestor of green algae and land plants. *The Plant Cell 15*: 1888–1903.
TURMEL, M.; POMBERT, J. F.; CHARLEBOIS, P.; OTIS, C.; LEMIEUX, C. 2007: The green algal ancestry of land plants as revealed by the chloroplast genome. *International Journal of Plant Sciences 168*: 679–689.
TURMEL, M.; OTIS, C.; LEMIEUX, C. 2009 : The chloroplast genomes of the green algae *Pedinomonas minor, Parachlorella kessleri*, and *Oocystis solitaria* reveal a shared ancestry between Pedinomonadales and Chlorellales. *Molecular Biology and Evolution 26*: 2317–2331.
VAN RAAM, J. C. 1995: *The Characeae of Tasmania*. J. Cramer, Berlin. 81 p.
VINCENT, W. F. 1983: Phytoplankton production and winter mixing: contrasting effects in two oligotrophic lakes. *Journal of Ecology 71*: 1–20.
VINCENT, W. F.; FORSYTH, D. J. 1987: Geothermally influenced waters. Pp. 349–377 *in*: Viner, A. B. (ed.), *Inland Waters of New Zealand*. Department of Scientific and Industrial Research, Wellington. ix + 494 p.
VINCENT, W. F.; GIBBS, M. M.; DRYDEN, S. J. 1984: Accelerated eutrophication in a New Zealand lake: Lake Rotoiti, central North Island. *New Zealand Journal of Marine and Freshwater Research 18*: 431–440.
VINER, A. B.; WHITE, E. 1987: Phytoplankton growth. Pp. 191–223 *in*: Viner, A. B. (ed.) *Inland Waters of New Zealand.* Department of Scientific and Industrial Research, Wellington. ix + 494 p.
VROOM, P. S.; SMITH, C. M.; KEELEY, S. C. 1998: Cladistics of the Bryopsidales: a preliminary analysis. *Journal of Phycology 34*: 351–360.
WEISS, T. L.; JOHNSTON, J. S.; FUJISAWA, K.; SUMIMOTO, K.; OKADA, S.; CHAPPELL, J.; DEVARENNE, T. P. 2010: Phylogenetic placement, genome size, and GC content of the liquid-hydrocarbon-producing green microalga *Botryococcus braunii* strain Berkeley (Showa) (Chlorophyta). *Journal of Phycology 46*: 534–540.
WELLS, R. D. S.; CLAYTON, J. S.; de WINTON, M. D. 1998: Submerged vegetation of Lakes Te Anau, Manapouri, Monowai, Hauroko, and Poteriteri, Fiordland, New Zealand. *New Zealand Journal of Marine and Freshwater Research 32*: 621–638.
WHARTON, R. A. Jr; SMERNOFF, D. T.; AVERNER, M. M. 1988: Algae in Space. Pp. 485–509 *in*: Lembi, C. A.; Waaland, J. R. (eds), *Algae and Human Affairs*. Cambridge University Press, Cambridge. viii + 590 p.
WHITTON, B. A.; JOHN, D. M.; JOHNSON, L. R.; BOULTON, P. N. G.; KELLY, M. G.; HAWORTH, E. Y. 1998: A coded list of freshwater algae of the British Isles. *Land Ocean Interaction Study Publication No. 222*: 1–274.
WOOD, R. D.; IMAHORI, K. 1965: *Monograph of the Characeae. Vol. 1, A. Revision of the Characeae.* J. Cramer Verlag, Weinheim. 904 p.
WOOD, R. D.; MASON, R. 1977: Characeae of New Zealand. *New Zealand Journal of Botany 15*: 87–180.
ZECHMAN, F. W.; VERBRUGGEN, H.; LELIAERT, F.; ASHWORTH, M.; BUCHHEIM, M. A.; FAWLEY, M. W.; SPALDING, H.; PUESCHEL, C. M.; BUCHHEIM, J. A.; VERGHESE, B.; HANISAK, M. D. 2010: An unrecognized ancient lineage of green plants persists in deep marine waters. *Journal of Phycology 46*: 1288–1295.
ZEMKE-WHITE, W. L.; BREMNER, G.; HURD, C. L. 1999: The status of commercial algal utilization in New Zealand. *Hydrobiologia 398/399*: 487–494.

Checklist of New Zealand Chlorophyta and Charophyta

The checklist combines elements of the systems of classification of Foerster (1982) (for desmids), Bourrelly (1990) and Lewis and McCourt (2004) and is an update of previously published checklists (Cassie 1984a, 1984b). Endemic species are indicated by E, adventive species by A; ? indicates an unconfirmed identification; * indicates a new record (i.e. not in previous checklists); † new combination (not in previous checklists). Endemic genera are underlined (first entry only, if more than one species). Marine/brackish species are indicated by M and terrestrial species by T. All other species are freshwater.

KINGDOM PLANTAE
SUBKINGDOM VIRIDIPLANTAE
INFRAKINGDOM CHLOROPHYTA
PHYLUM CHLOROPHYTA
Class PYRAMIMONADOPHYCEAE
Order PTEROSPERMALES
PTEROSPERMACEAE
Pterosperma sp. Bojo 2001 M

Order PYRAMIMONADALES
HALOSPHAERACEAE
Halosphaera viridis Schmitz M
PYRAMIMONADACEAE
Cymbomonas tetramitiformis Schiller M
Pyramimonas amylifera Conrad M
Pyramimonas cirolanae Pennick M
Pyramimonas disomata Butcher ex McFadden, Hill & Wetherbee M
Pyramimonas grossii Parke M
Pyramimonas janetae R.E.Norris M
Pyramimonas longicauda Van Meel M
Pyramimonas cf. *moestrupi* McFadden M
Pyramimonas orientalis Butcher ex McFadden, Hill & Wetherbee M
Pyramimonas virginica Pennick M

Order PALMOPHYLLALES

PALMOPHYLLACEAE
Palmophyllum umbracola W.A.Nelson & K.G.Ryan M

Class MAMIELLOPHYCEAE
Order MAMIELLALES
MAMIELLACEAE
Mantoniella squamata (Manton & Parke) Desikachary M
Mamiella gilva (Parke & Rayns) Moestrup M
Micromonas pusilla (Butcher) Manton & Parke M

Class NEPHROSELMIDOPHYCEAE
Order NEPHROSELMIDALES
NEPHROSELMIDACEAE
Nephroselmis pyriformis (N.Carter) Ettl M
Nephroselmis rotundata (N.Carter) Fott M

Class CHLORODENDROPHYCEAE
Order CHLORODENDRALES
CHLORODENDRACEAE
Tetraselmis marinus (Cienkowski) R.E.Norris, Hori & Chihara M

Class CHLOROPHYCEAE
Order CHLAMYDOMONADALES
ASTEROCOCCACEAE
Asterococcus limneticus G.M.Smith
?*Asterococcus superbus* (Cienkowski) Scherffel
Chlamydocapsa ampla (Kützing) Fott†
Chlamydocapsa bacillus (Teiling) Fott
?*Chlamydocapsa grevillei* (Berkeley) Fott
?*Chlamydocapsa planctonica* (W.& G.S.West) Fott †
Pseudosphaerocystis lacustris (Lemmermann) Novakova
CHLAMYDOMONADACEAE
Brachiomonas submarina Bohlin
Carteria globosa Korshikov
Chlainomonas kolii (Hardy & Curl) Hoham
Chlainomonas rubra (J.R.Stein & R.C.Brooke) Hoham
Chlamydomonas aculeata Korshikov†
Chlamydomonas antarctica Wille
Chlamydomonas bacillus Pascher & Jahoda
Chlamydomonas chlorostellata E.A.Flint & H.Ettl T
?*Chlamydomonas debaryana* Goroszhankin
Chlamydomonas globosa J.W.Snow
?*Chlamydomonas kleinii* Schmidle
Chlamydomonas leiostraca (Strehlow) H.Ettl
Chlamydomonas moewusii Gerloff var. *microstigmata* (J.W.G.Lund) H.Ettl T
Chlamydomonas monadina (Ehrenberg) F.Stein
Chlamydomonas nivalis (F.A.Bauer) Wille
Chlamydomonas novae-zelandiae Playfair
Chlamydomonas obscura Playfair
— var. *ovata* Playfair
?*Chlamydomonas obtusa* A.Braun
Chlamydomonas penium Pascher
Chlamydomonas pseudogloeogama Gerloff
Chlamydomonas reinhardtii P.A.Dangeard T
Chlamydomonas rima E.A.Flint & H. Ettl T
Chlamydomonas sanguinea Lagerheim
Chlamydomonas snowiae Printz
?*Chlamydomonas steinii* Goroschankin
Chlamydomonas subangulosa F.E.Fritsch & R.P.John T
Chlamydomonas taurangensis Playfair
Chlorogonium elongatum P.A.Dangeard
Chlorogonium euchlorum Ehrenberg
Chlorogonium minimum Playfair var. *obesum* Playfair
Chloromonas cryophila Hoham & Mullet
Chloromonas rubreolosa Ling & Seppelt
?*Haematococcus capensis* Pocock var. *novae-zelandiae* Pocock*
?*Haematococcus pluvialis* (Flotow) Wille
?*Haematococcus zimbabwiensis* Pocock ssp. *novae-zelandiae* Pocock*
Sphaerellopsis fluviatilis (F.Stein) Pascher
CHLORANGIELLACEAE
Chlorangiella pygmaea (Ehrenberg) P.C.Silva
Stylosphaeridium stipitatum (H.Bachmann) L.Geitler & N.I.Gimesi
CHLOROCOCCACEAE
Ankyra ancora (G.M.Smith) Fott
?*Bracteacoccus anomalus* (E.J.James) Starr
?*Bracteacoccus irregularis* (J.B.Petersen) Starr†
?*Bracteacoccus minor* (Chodat) Petrova T
Characium acuminatum (A. Braun) A.Braun
Characium curvatum G.M.Smith
Characium ensiforme Herman
Characiopodium pseudopolymorphum (Philipose) G.L.Floyd & S.Watanabe†
Chlorochytrium lemnae Cohn
Chlorococcum tatrense Archibald T
Desmatractum bipyramidatum (Chodat) Pascher
Korshikoviella gracilipes (F.D.Lambert) P.C.Silva†
Planktosphaeria gelatinosa G.M.Smith
Polyedriopsis spinulosa Schmidle
?*Spongiochloris spongiosa* Starr
Tetraedron gigas (Wittrock) Hansgirg
Tetraedron gracile (Reinsch) Hansgirg
Tetraedron limneticum O.Borge
Tetraedron lobulatum (Nägeli) Hansgirg
Tetraedron minimum (A. Braun) Hansgirg
— var. *apiculato-scrobiculatum* (Reinsch) Skuja
Tetraedron muticum (A.Braun) Hansgirg
Tetraedron planctonicum G.M.Smith
Tetraedron regulare Kützing
— var. *granulare* Prescott
Tetraedron trigonum (Nägeli) Hansgirg
— var. *gracile* Reinsch
— var. *inerme* Hansgirg
Tetraedron trilobatum (Reinsch) Hansgirg
Tetracystis tetrasporum (Archibald & Bold) R.M.Brown & H.C.Bold
CHLOROSARCINACEAE
Chlorosarcinopsis minor (Gerneck) Herndon
COCCOMYXACEAE
?*Coccomyxa confluens* (Kütz.) Fott
Coccomyxa lacustris Chodat
Coccomyxa subglobosa Pascher
Dispora crucigenioides Printz
DICTYOSPHAERACEAE
Botryococcus braunii Kütz.
Botryosphaerella sudetica (Lemmermann) P.C.Silva†
Dictyochlorella reniformis (Korshikov) P.C.Silva
Dictyochloropsis splendida var. *gelatinosa* Tschermak-Woess
Dictyochloropsis symbiontica Tschermak-Woess
Dictyosphaerium ehrenbergianum Nägeli
Dictyosphaerium planctonicum Tiffany & Ahlstrom
Dictyosphaerium pulchellum H.C.Wood
Dictyosphaerium subsolitarium Van Goor†
Dimorphococcus lunatus A.Braun†
?*Lobocystis planctonica* (Tiffany & Ahlstrom) Fott
Westella botryoides (W.West) DeWildeman
Westellopsis linearis (G.M.Smith) C.C.Jao
DUNALIELLACEAE
Dunaliella euchlora Lerche M
Dunaliella salina (Dunal) Teodoresco M
HORMOTILACEAE
Tetracladus mirabilis Svirenko
HYDRODICTYACEAE
Hydrodictyon reticulatum (L.) Lagerheim A
Pediastrum angulosum (Ehrenberg) Meneghini
Pediastrum biradiatum Meyen
Pediastrum boryanum (Turpin) Meneghini
— var. *brevicorne* A.Braun
Pediastrum duplex Meyen
— var. *gracillimum* W. & G.S.West
Pediastrum integrum Nägeli
Pediastrum orbitale Komarek
Pediastrum simplex (Meyen) Lemmermann
Pediastrum tetras (Ehrenberg) Ralfs
Sorastrum spinulosum Nägeli
MICRACTINIACEAE
Golenkinia radiata (Chodat) Wille
Micractinium bornhemiense (Conrad) Korshilov†
Micractinium pusillum Fresenius
Micractinium quadrisetum (Lemmermann) G.M.Smith
PALMELLACEAE
?*Palmella crouenta* C.Agardh
?*Palmella miniata* Leiblein
Parallela novae-zelandiae E.A.Flint
Sphaerocystis planctonica (Korshikov) Bourrelly
Sphaerocystis schroeteri Chodat
PHACOTACEAE
Dysmorphococcus variabilis Takeda
Phacotus lenticularis (Ehrenberg) F.Stein
Pteromonas aculeata Lemmermann
Pteromonas aequiciliata (Gickelhorn) Bourrelly
Pteromonas angulosa (H.J.Carter) Lemmermann†
Pteromonas cordiformis Lemmermann
?*Wislouchiella planktonica* Skvortzov
POLYBLEPHARIDACEAE
?*Polytomella citri* Kater
?*Spermatozopsis exultans* Korshikov
PROTOSIPHONACEAE
Protosiphon botryoides Klebs
RADIOCOCCACEAE
Actinastrum gracillimum G.M.Smith
Actinastrum hantzschii Lagerheim
Actinastrum fluviatile (Schroeder) Fott
— var. *subtile* Woloszynska
Coccomyxa gloeobotrydiformis Reisigl T
Coelastrum cambricum W.Archer
— ?var. *intermedium* (Bohlin) G.S.West
Coelastrum microporum Nägeli
Coelastrum pulchrum Schmidle var. *cruciatum* (Kammerer) Komarek
Coelastrum reticulatum (P.A.Dangeard) Senn
Coelastrum sphaericum Nägeli
Coenocystis quadriguloides Fott
Coenocystis reniformis Korshikov
Coenocystis subcylindrica Korshikov
Crucigenia fenestrata (Schmidle) Schmidle
?*Crucigenia lauterbornii* (Schmidle) Schmidle
Crucigenia quadrata C.Morren
Crucigenia tetrapedia (Kirchner) W.& G.S.West
Crucigeniella rectangularis (Nägeli) Komarek†
Crucigeniella truncata (G.M.Smith) Komarek
Desmodesmus abundans (Kirchner) Hegewald†
Desmodesmus armatus (Chodat) Hegewald†
— var. *longispina* (Chodat) Hegewald†
— var. *subalternans* (G.M.Smith) Hegewald†
Desmodesmus brasiliensis (Bohlin) Hegewald†
?*Desmodesmus costato-granulatus* (Skuja) Hegewald†
Desmodesmus denticulatus (Lagerheim) An, Friedl & Hegewald†
— var. *linearis* (Hansgirg) Hegewald†
Desmodesmus intermedius (Roll) Hegewald†
Desmodesmus maximus (W. & G.S.West) Hegewald †
Desmodesmus opoliensis (P. Richter) Hegewald†
Desmodesmus pannonicus (Hortobagyi) Hegewald†
Desmodesmus serratus (Corda) An. Friedl & Hegewald†
Desmodesmus spinosus (Chodat) Hegewald†
Desmodesmus subsapicatus (Chodat) Hegewald & A.Schmidt†
Didymogenes palatina Schmidle†
Elakatothrix biplex (Nygaard) Hindak
Elakatothrix gelatinosa Wille

Elakatothrix gloeocystiformis Korshikov
Elakatothrix parvula (W.Archer) Hindak
Elakatothrix spirochroma (Reverdin) Hindak
Enallax acutiformis (Schroeder) Hindak
Enallax costatus Schmidle
Gloeocystis papuana (Watanabe) Ettl & Gärtner T
Gloeocystis polydermatica (Kütz.) Hindak†
Keratococcus bicaudatus (A. Braun) J.B.Petersen†
Neodesmus pupukensis (Kalina & Puncochárová) Hegewald & Hanagata
Radiococcus nimbatus (DeWildeman) Schmidle
Radiococcus planctonicus (W. & G.S.West) J.W.G.Lund
Scenedesmus acuminatus (Lagerheim) Chodat
Scenedesmus acutiformis Schroeder
Scenedesmus arcuatus (Lemmermann) Lemmermann
Scenedesmus bijuga (Turpin) Lagerheim
— ?var. *alternans* (Reinsch) Borge
?*Scenedesmus dimorphus* (Turpin) Kütz.
Scenedesmus ecornis (Ehrenberg) Chodat
Scenedesmus obliquus (Turpin) Kütz.
Scenedesmus obtusus Meyen
?*Scenedesmus quadricauda* (Turpin) Brébisson
Tetrallantos lagerheimii Teiling
Tetrachlorella alternans (G.M.Smith) Korshikov
?*Tetrastrum divergens* G.M.Smith
Tetrastrum elegans Playfair
Willea irregularis (Wille) Schmidle†
SPONDYLOMORACEAE
Pyrobotrys casinoensis (Playfair) P.C.Silva†
TETRASPORACEAE
Apiocystis brauniana Nägeli
Chaetopeltis barbata (Bohlin) Wille
Chaetopeltis orbicularis Berthold
Paulschulzia pseudovolvox (Schulz) Skuja
Paulschulzia tenera (Korshikov) J.W.G.Lund
Schizochlamys gelatinosa A. Braun
Tetraspora gelatinosa (Vaucher) Desvaux†
Tetraspora lemmermannii Fott
Tetraspora lubrica (Roth) C.Agardh
VOLVOCACEAE
Eudorina cylindrica Korshikov
Eudorina elegans Ehrenberg
Eudorina unicocca G.M.Smith
?*Gonium multicoccum* Pocock*
Gonium pectorale O.F.Müller
Gonium sociale (Dujardin) Warming
?*Mastigosphaera gobii* Schewiakoff
Pandorina charkowiensis Korshikov
Pandorina morum (O.F.Müll.) Bory
?*Stephanosphaera pluvialis* Cohn
Volvox aureus Ehrenberg
Volvox barberi W.Shaw
Volvox globator (Linnaeus) Ehrenberg
Volvox perglobator Powers
Volvox tertius Meyer
Volvulina steinii Playfair

Order SPHAEROPLEALES
SPHAEROPLEACEAE
Sphaeroplea chapmanii Sarma

Order OEDOGONIALES
OEDOGONIACEAE
Bulbochaete angulosa Wittrock & P.Lundell
Bulbochaete anomala Pringsheim
Bulbochaete gigantea Pringsheim
Bulbochaete intermedia deBary
?*Bulbochaete minor* A.Braun
Bulbochaete mirabilis Wittrock
Bulbochaete punctulata (Nordstedt) Hirn
?*Bulbochaete repanda* Wittrock
?*Bulbochaete setigera* (Roth) C.Agardh
?*Oedogonium acrosporum* de Bary
?*Oedogonium borisianum* (Le Clerc) Wittrock
?*Oedogonium boscii* (Le Clerc) Wittrock
Oedogonium braunii Kütz. var. *grande* Novis E
Oedogonium capillare (L.) Kütz.
Oedogonium cf. *capillare* (L.) Kütz.
?*Oedogonium capilliforme* Kütz.
?*Oedogonium cardiacum* (Hassall) Witrock
Oedogonium chapmanii Tiffany
?*Oedogonium ciliatum* (Hassall) Pringsheim
?*Oedogonium crenulatum* Wittrock
Oedogonium crispum (Hassall) Wittrock
— var. *crispum* f. *obesum* (Wittrock) Mrozińska
— var. *gracilescens* Wittrock
Oedogonium cyathigerum Wittrock var. *laevis* Novis E
Oedogonium cymatosporum Wittrock & Nordst. var. *areoliferum* C.C.Jao
Oedogonium didymum Novis E
Oedogonium gracilius (Wittrock) Tiffany
Oedogonium macrandrium Wittrock var. *hohennackerii* (Wittrock) Tiffany
?*Oedogonium oelandicum* Wittrock
Oedogonium platygynum Wittrock
— var. *novae-zelandiae* Hirn E
?*Oedogonium princeps* (Hassall) Wittrock
Oedogonium pringsheimii C.E.Cramer
— cf. var. *goczalkowicensis* Mrozińska
— cf. var. *nordstedtii* Wittrock
— var. *nordstedtii* f. *pachydermatosporum* (Nordst.) Hirn
Oedogonium cf. *pseudomitratum* Prescott
?*Oedogonium pusillum* Kirchner
Oedogonium sp. aff. *pyrulum* Wittrock
Oedogonium sexangulare Cleve
— var. *maius* Wille
Oedogonium southlandiae Novis E
?*Oedogonium stellatum* Wittrock
Oedogonium subdissimile C.C.Jao
— var. *submasculum* Novis E
?*Oedogonium suecicum* Wittrock
?*Oedogonium undulatum* (Brébisson) A.Braun
Oedogonium aff. *wissmanii* Skinner

Order CHAETOPELTIDALES
CHAETOPELTIDACEAE
Chaetopeltis barbata (Bohlin) Wille
Chaetopeltis orbicularis Berthold

Order CHAETOPHORALES
APHANOCHAETACEAE
Aphanochaete polychaete (Hansgirg) F.E.Fritsch
Aphanochaete repens A. Braun
Aphanochaete vermiculoides Wolle
CHAETOPHORACEAE
Chaetophora attenuata Hazen var. *claytonii* P.Sarma E
Chaetophora elegans (Roth) C.Agardh
Chaetophora incrassata (Hudson) Hazen†
Chaetophora tuberculosa (Roth) C.Agardh
Cloniophora paihiae (A.K.M.N.Islam) P.Sarma E
Desmococcus vulgaris (Nägeli) Brandt† E
Diaphragma radiosum Geitler
Draparnaldia mutabilis (Roth) Bory†
Draparnaldiopsis taylori P.Sarma T
Epibolium dermaticola Printz
Gloeoplax weberi Schmidle
?*Leptosira chapmanii* (P.Sarma) H.Ettl & G.Gärtner
?*Leptosira jollyi* (P.Sarma) H.Ettl & G. Gärtner
?*Leptosira terricola* (Bristol) Borzi†
Protoderma beesleyi (Fritsch) Printz
Protoderma viride Kütz.
Stigeoclonium amoenum Kütz.
— var. *aucklandicum* A.K.M.N.Islam
— var. *novizelandicum* Nordstedt
Stigeoclonium biosolettianum (Kütz.) Kütz.
Stigeoclonium elongatum (Hassall) Kütz.
Stigeoclonium farctum Berthold
Stigeoclonium fasciculare Kütz.
— var. *amrutii* P.Sarma E
Stigeoclonium islamii P.Sarma
?*Stigeoclonium lubricum* (Dillwyn) Kütz.
— ?var. *nairnii* P.Sarma E
— ?var. *nathanii* P.Sarma
Stigeoclonium nudiusculum (Kütz.) Kütz.
— var. *lamii* P.Sarma E
Stigeoclonium pachydermum Prescott
Stigeoclonium pusillum (Lyngbye) Kütz.
Stigeoclonium segarae A.K.M.N.Islam
Stigeoclonium setigerum Kütz.
Stigeoclonium stagnatile (Hazen) Collins
Stigeoclonium subsecundum (Kützing) Kütz.
— var. *tenuis* Nordstedt
Stigeoclonium tenue (C.Agardh) Kütz.
— var. *uniforme* (C.Agardh) Kütz.
Stigeoclonium thermale A.Braun
Stigeoclonium variabile Nägeli
Stromatella monostromatica (P.J.L.Dang.) Kornmann & Sahling M
CHAETOSPHAERIDIACEAE
Chaetosphaeridium globosum (Nordstedt) Klebahn†
Chaetotheke reptans Dueringer
Conochaete comosa Klebahn
?*Conochaete klebahnii* Schmidle
Conochaete polytricha (Nordstedt) Klebahn
?*Dicoleon nordstedtii* Klebahn

Class ULVOPHYCEAE
Order BRYOPSIDALES
BRYOPSIDACEAE
*Bryopsis derbesioides*V.J.Chapm. M E
Bryopsis gemellipara J.Agardh M
Bryopsis kermadecensis V.J.Chapm. M E
*Bryopsis lindaueri*V.J.Chapm. M E
Bryopsis myosuroides Kütz. M
Bryopsis plumosa (Huds.) C.Agardh M
Bryopsis rhizoidea V.J.Chapm. M E
Bryopsis vestita J.Agardh M
Pseudobryopsis planktonica Cassie M E
CAULERPACEAE
Caulerpa articulata Harv. M
Caulerpa brownii (C.Agardh) Endlicher M
Caulerpa fastigiata Mont. M
Caulerpa flexilis J.V.Lamour. M
Caulerpa geminata (Harv.) Weber Bosse M
Caulerpa longifolia C.Agardh M
Caulerpa racemosa (Forssk.) J.Agardh M
Caulerpa sertularioides (S.Gmelin) Howe M
Caulerpa webbiana Mont. M
CODIACEAE
Codium bursa (L.) C.Agardh M
Codium cranwelliae Setch. M E
Codium dichotomum (Huds.) S.F.Gray f. *novozelandicum* M
Codium dimorphum Sved. M
Codium fragile (Suringar) Har. ssp. *fragile* (Goor) P.C.Silva M A
— ssp. *maclovianae* Maggs M
— ssp. *novae-zelandiae* (J.Agardh) P.C.Silva M
Codium geppiorum O.C.Schmidt M
Codium gracile (O.C.Schmidt) Dellow M
Codium platyclados R.Jones & Kraft M
Codium spongiosum Harv. M
DERBESIACEAE
Derbesia novae-zelandiae V.J.Chapm. M E
Pedobesia clavaeformis (J.Agardh) MacRaild & Womersley M
OSTREOBIACEAE
Ostreobium quekettii Bornet & Flahault M

Order CLADOPHORALES

ANADYOMENACEAE
Microdictyon mutabile Dellow M E
Microdictyon umbilicatum (Velley) Zanardini M
CLADOPHORACEAE
Chaetomorpha aerea (Dillwyn) Kütz. M
Chaetomorpha capillaris sensu N.M.Adams M
Chaetomorpha coliformis (Mont.) Kütz. M
Chaetomorpha elongata V.J.Chapm. M E
Chaetomorpha linum (O.F.Mueller) Kütz. M
Chaetomorpha pallida V.J.Chapm. M E
Chaetomorpha valida (Hook.f. & Harv.) Kütz. M
Cladophora callicoma Kütz.
Cladophora colensoi Harv. M E
Cladophora crinalis Harv. M
Cladophora crispata Kütz.
Cladophora daviesii Harv. M E
Cladophora feredayi Harv. M
Cladophora flavida Kütz.
Cladophora glomerata (L.) Kütz.
— var. *ornata* Lemmermann
Cladophora herpestica (Mont.) Kütz. M
Cladophora hochstetteri Grunow
Cladophora incompta (Hook.f. & Harv.) Hook.f. & Harv. M
?*Cladophora lyallii* Hooker
Cladophora montagneana Kütz. M NR - WELT A026624
Cladophora prolifera (Roth) Kütz. M
Cladophora sericea (Huds.) Kütz. M
Cladophora socialis Kütz. M
?*Cladophora stewartensis* F. Brand
Cladophora subsimplex Kütz. M
Cladophora vagabunda s.l. (L.) C.Hoek M
Cladophora valonioides (Sond.) Kütz. M
Cladophora verticillata (Hook.f. & Harv.) J.Agardh M
Rhizoclonium ambiguum (Hook.f. & Harv.) Kütz. M
Rhizoclonium berggrenianum Hauck
Rhizoclonium crispum Kütz.
Rhizoclonium curvatum V.J.Chapm. M
?*Rhizoclonium fissum* F. Brand
?*Rhizoclonium fontanum* Kütz.
?*Rhizoclonium hieroglyphicum* (C.Agardh) Kütz. var. *waikitense* Hauck
Rhizoclonium kerneri Stockm. M
Rhizoclonium implexum (Dillwyn) Kütz. M
Rhizoclonium riparium (Roth) Harv. M
Rhizoclonium tortuosum (Dillwyn) Kütz. M
Wittrockiella lyallii (Harv.) C.Hoek, Ducker & Womersley M
Wittrockiella salina V.J.Chapm. M E
SIPHONOCLADACEAE
Boodlea composita (Harv.) F.Brand M

Order DASYCLADALES
POLYPHYSACEAE
Parvocaulis parvulus (Solms) S.Berger et al. M

Order TRENTEPOHLIALES
TRENTEPOHLIACEAE
Cephaleuros lagerheimii Schmidle T
Cephaleuros minimus Karsten T
Cephaleuros parasiticus Karsten T
Cephaleuros virescens Kunze T
Phycopeltis expansa (A.V.Jennings) P.Sarma T
Phycopeltis irregularis (Schmidle) Wille T
Phycopeltis prostrata Schmidle emend. P.Sarma T
Trentepohlia abietina (Flotow) Hansgirg T
Trentepohlia arborum (C.Agardh) Hariot T
Trentepohlia aurea (Linnaeus) Martius T
Trentepohlia bossei De Wildeman T
— var. *brevicellulis* Cribb T
— var. *samoensis* Wille T
Trentepohlia effusa (Krempelhuber) Hariot T
Trentepohlia flintii P.Sarma T
Trentepohlia jolithus (L.) Martius T
— var. *anthonyi* P.Sarma T
— var. *crassior* Nordstedt T
Trentepohlia monilia De Wildeman T
Trentepohlia odorata Wittrock var. *compacta* Cribb T
Trentepohlia peruana (Kütz.) P.Sarma T
Trentepohlia rigidula (Müller) Hariot T
— var. *lynchii* P.Sarma T
Trentepohlia umbrina (Kütz.) Bornet T

Order ULOTRICHALES
ACROSIPHONACEAE
Spongomorpha pacifica (Mont.) Kütz. M
Urospora penicilliformis (Roth) Areschoug M
CYLINDROCAPSACEAE
Cylindrocapsa geminella Wolle
DICRANOCHAETACEAE
Dicranochaetae reniformis Hieronymus
ENDOSPHAERACEAE
Eugomontia sacculata Kornmann M
Eugomontia stelligera R.Nielsen M E
MICROSPORACEAE
?*Microspora abbreviata* (Rabenhorst) Lagerheim
Microspora amoena (Kütz.) Rabenhorst
— var. *novae-zelandiae* Novis E
?*Microspora floccosa* (Vaucher) Thuret
Microspora manifesta Novis E
?*Microspora pachyderma* (Wille) Lagerheim
Microspora quadrata Hazen
Microspora rotundata Novis E
Microspora stagnorum (Kütz.) Lagerheim
Microspora wittrocki (Wille) Lagerheim
Microspora sp. A Novis 2004
MONOSTROMATACEAE
Blidingia minima (Nägeli) Kylin M
Monostroma antarcticum V.J.Chapm. M E
Monostroma applanatum Gain M
Monostroma crepidinum Farlow M
Monostroma latissimum Wittr. M
Monostroma lindaueri V.J.Chapm. M E
Monostroma membranacea W. & G.S.West
Monostroma nitidum Wittr. M
Monostroma pacificum V.J.Chapm. M E
Monostroma parvum V.J.Chapm. M E
ULOTRICHACEAE
Binuclearia tatrana Wittrock
Binuclearia tectorum (Kütz.) Beger
Geminella amphigranulata (Skuja) Ramanathan †
Geminella interrupta (Turpin) Lagerheim
Geminella minor (Nägeli) Heering
Geminella mutabilis (Brébisson) Wille
?*Gloeotila capensis* Grunow
?*Gloeotila pelagica* (Nygaard) Skuja
— ?var. *novae-zelandiae* Skuja
Gloeotila pulchra Skuja
Interfilum paradoxum Chodat & Topali
Planctonema lauterbornii Schmidle
Radiofilum conjunctivum Schmidle
Radiofilum transversalis (Brébisson) Ramanathan†
Stichococcus bacillaris Nägeli T
Stichococcus cylindricus Butcher M
Ulothrix aequalis Kütz.
— var. *cylindricum* (Prescott) H.S.Forest
Ulothrix limnetica Lemmermann
Ulothrix moniliformis Kütz.
Ulothrix novae-zelandiae V.J.Chapm. M E
Ulothrix radicans Kütz.
Ulothrix subflaccida Wille M
Ulothrix tenerrima (Kütz.) Kütz.
Ulothrix zonata (Weber & Mohr) Kütz.
Uronema elongatum Hodgetts

Order ULVALES
CAPSOSIPHONACEAE
Capsosiphon aurea V.J.Chapm. M E
GOMONTIACEAE
Gomontia polyrhiza (Lagerh.) Bornet & Flahault M
PHAEOPHILACEAE
Phaeophila dendroides (P.Crouan & H.Crouan) Batters M
ULVACEAE
Gayralia oxysperma (Kuetz.) K.L.Vinog. ex Scagel et al. M
Gemina clavata V.J.Chapm. M
Gemina enteromorphoidea V.J.Chapm. M
Gemina letterstedtoidea V.J.Chapm. M
Gemina ulvoidea V.J.Chapm. M
Letterstedtia insignis Aresch. M
Lobata foliosa V.J.Chapm. M
Percursaria percursa (C.Agardh) Rosenv. M
Schizomeris leibleinii Kütz.
Ulva armoricana P.Dion, B.de Reviers & G.Coat M A
Ulva californica Wille M A
Ulva compressa L. M A
Ulva fasciata Delile M
Ulva flexuosa Wulfen M A
Ulva intestinalis L. M A
Ulva lactuca L. M A
Ulva linza L. M
Ulva pertusa Kjellm. M A
Ulva procera (K.Ahlner) H.S.Hayden et al. M
Ulva prolifera O.F.Mull. M
Ulva ralfsii Harv. M
Ulva rigida C.Agardh M
Ulva sp. 1 sensu Heesch et al. 2009 M A
Ulva sp. 2 sensu Heesch et al. 2009 M
Ulva sp. 4 sensu Heesch et al. 2009 M
Ulva sp. 6 sensu Heesch et al. 2009 M
Ulva sp. 9 sensu Heesch et al. 2009 M
Ulva sp. 10 sensu Heesch et al. 2009 M
Umbraulva olivascens (P.J.L.Dang.) E.H.Bae & I.K.Lee M A
Umbraulva 'Kermadecs' sensu Heesch et al. 2009 M E
Umbraulva 'Northland' sensu Heesch et al. 2009 M E
Umbraulva 'Auckland Islands' sensu Heesch et al. 2009 M E
ULVELLACEAE
Acrochaete endostraca R.Nielsen M E
Entocladia cingens Setch. & N.L.Gardner M
Entocladia perforans (Huber) Levring M
Entocladia polymorpha (G.S.West) G.M.Smith
Entocladia rivulariae V.J.Chapm. M E
Entocladia russelliae V.J.Chapm. M E
Entocladia viridis Reinke M
Epicladia testarum (Kylin) R.Nielsen M
Ochlochaete hystrix Thwaites ex Harv. M
Ovillaria catenata R.Nielsen M E
Syncoryne reinkei R.Nielsen & P.M.Pedersen M
Ulvaella sp. M

Order INCERTAE SEDIS
'*Sporocladopsis novae-zelandiae* V.J.Chapm.' M

Class TREBOUXIOPHYCEAE
Order MICROTHAMNIALES
MICROTHAMNIACEAE
Microthamnion kuetzingianum Nägeli
Microthamnion strictissimum Rabenhorst T

Order PRASIOLALES
PRASIOLACEAE
Prasiola crispa (Lightf.) Kütz. M
Prasiola delicatula V.J.Chapm. E
Prasiola skottsbergii sensu V.J.Chapm.
Prasiola snareana V.J.Chapm. M E
Prasiola 'stipitata' Suhr M
Prasiola novaezelandiae S.Heesch, J.E.Sutherland & W.A.Nelson M E

Prasiola sp.'Antipodes Island' Heesch et al. 2011 M
Prasiolopsis ramosa Vischer
Rosenvingiella polyrhiza (Rosenv.) P.C.Silva M
Rosenvingiella australis S.Heesch, J.E.Sutherland & W.A.Nelson M
Rosenvingiella constricta (Setch. & N.L.Gardner) P.C.Silva M

Order CHLORELLALES
CHLORELLACEAE
Chlorella ellipsoidea Gerneck
?*Chlorella parasitica* (Brandt) Beijerinck
?*Chlorella protothecoides* W.Krueger
Chlorella sphaerica Tschermak-Woess
Chlorella vulgaris Beijerinck
Choricystis minor (Skuja) Fott
Elliptochloris reniformis (Watanabe) Ettl & Gärtner T
OOCYSTACEAE
Ankistrodesmus bibraianus (Reinsch) Korshikov†
Ankistrodesmus falcatus (Corda) Ralfs
Ankistrodesmus fusiformis Corda
Ankistrodesmus gracilis (Reinsch) Korshikov†
?*Ankistrodesmus irregularis* G.M.Smith
Ankistrodesmus spiralis (Turner) Lemmermann
Chlorolobion braunii (Nägeli) Komarek†
Closteriopsis acicularis (G.M.Smith) J.H.Belcher & E.M.F.Swale
Closteriopsis longissima (Lemmermann) Lemmermann
Cryocystis brevispina (F.E.Fritsch) Kol†
Dactylococus bicaudatus A.Braun T
Dicanthos belenophorus Korshikov
Echinosphaerella limnetica G.M.Smith
Eremosphaera gigas (W.Archer) Fott†
Eremosphaera viridis deBary
Fusola viridis (Snow) Hindak†
Franceia ovalis (France) Lemmermann
Granulocystis verrucosa (Roll) Hindak†
Kirchneriella arcuata G.M.Smith
Kirchneriella contorta (Schmidle) Bohlin
— var. *elongata* (G.M.Smith) Komarek
Kirchneriella lagerheimii Teiling
Kirchneriella obesa (W.West) Schmidle
Kirchneriella obtusa (Korshikov) Komarek†
Lagerheimia chodatii Bernard†
Lagerheimia citriformis (Snow) Collins†
Lagerheimia genevensis (Chodat) Chodat†
Lagerheimia subsalsa Lemmermann†
Monoraphidium contortum (Thuret) Legnerova†
Monoraphidium convolutum (Corda) Legnerova†
Monoraphidium dybowskii (Wolozynska) Hindak & Legnerova
Monoraphidium griffithsii (Berkeley) Legnerova
Monoraphidium irregulare (G.M.Smith) Legnerova
Monoraphidium minutum (Nägeli) Legnerova
Monoraphidium mirabile (W. & G.S.West) Pankow †
Monoraphidium setiforme (Nygaard) Legnerova
Monoraphidium tortile (W. & G.S.West) Legnerova
Muriella terrestris J.B.Petersen
Nephrocytium agardhianum Nägeli
Nephrocytium limneticum (G.M.Smith) G.M.Smith
Nephrocytium lunatum W.West
Oocystis borgei J.Snow
Oocystis elliptica W.West
Oocystis lacustris Chodat
Oocystis marssonii Lemmermann
Ooystis naegeli A.Braun
Oocystis parva (H.B.Ward) W. & G.S.West
Oocystis pusilla Hansgirg
Oocystis rhomboidea Fott
Oocystis solitaria Wittrock
— ?var. *apiculata* Printz
Oocystis submarina Lagerheim
Oonephris obesa (W. & G.S.West) Fott†
Prototheca wickerhamii Tubaki & Soneda
Pseudococcomyxa simplex (Mainx) Fott T
Quadrigula lacustris (Chodat) G.M.Smith
Saturnella saturna (Steinecke) Fott
Scotiella spinosa Geitler
Scotiellopsis terrestris (Reisigl) Punčochářová & Kalina T
Siderocelis ornata (Fott) Fott
Treubaria crassispina G.M.Smith
?*Treubaria schmidlei* (Schroeder) Fott & Komarek
?*Trochiscia reticularis* (Reinsch) Hansgirg
PEDINOMONADACEAE
?*Pedinomonas minor* Korshikov
INCERTAE SEDIS
Helicosporidium parasiticum Keilin

INFRAKINGDOM STREPTOPHYTA
PHYLUM CHAROPHYTA
Class ZYGNEMOPHYCEAE
Order ZYGNEMATALES
ZYGNEMATACEAE
Debarya glyptosperma (DeBary) Wittrock
Debarya polyedrica Skuja
Mougeotia capucina (deBary) C.Agardh
?*Mougeotia laetivirens* (A.Braun) Wittrock
?*Mougeotia laevis* (Kütz.) W.Archer
?*Mougeotia recurva* (Hassall) DeToni
?*Mougeotia viridis* (Kütz.) Wittrock
Mougeotiopsis calospora Palla
?*Spirogyra aphanosculpta* Skuja
Spirogyra canaliculata Segar E
Spirogyra chuniae C.C.Jao
Spirogyra circumlineata Transeau
Spirogyra clavata Segar E
?*Spirogyra communis* (Hassall) Kütz.
?*Spirogyra crassa* (Kütz.) Czurda
?*Spirogyra daedalea* Lagerheim
?*Spirogyra ellipsospora* Transeau
Spirogyra fluviatilis Hilse
Spirogyra cf. *gautier-lievrae* Kadlub.
Spirogyra cf. *jaoii* S.H.Li
?*Spirogyra longata* (Vaucher) Kütz.
Spirogyra cf. *lutetiana* P.Petit
Spirogyra majuscula Kütz.
?*Spirogyra nitida* (Dillwyn) Link
?*Spirogyra parvula* (Transeau) Czurda
?*Spirogyra pellucida* (Hassall) Kütz.
Spirogyra polymorpha Kirchn.
?*Spirogyra porticalis* (O.F.Müll.) Cleve
?*Spirogyra quadrata* (Hassall) P.Petit
Spirogyra rugulosa Iwanoff
?*Spirogyra sahnii* Randhawa
?*Spirogyra semiornata* C.C.Jao
Spirogyra singularis Nordstedt
Spirogyra tenuissima (Hassall) Kütz.
Spirogyra varians (Hassall) Kütz.
?*Spirogyra verruculosa* C.C.Jao
Spirogyra cf. *willei* Skuja
Spirogyra cf. *wuchanensis*
?*Zygnema cruciatum* (Vaucher) C.Agardh
Zygnema eumetabletos Skuja E
?*Zygnema insigne* (Hassall) Kütz.
Zygnema pawneanum Taft
?*Zygnema pectinatum* (Vaucher) Czurda
?*Zygnema stellinum* (Vaucher) Czurda
Zygnemopsis pachyderma Skuja E
Zygogonium ericetorum Kütz.
Zygogonium kumaoense Randhawa
Zygogonium stictosporum (Skuja) Kadlubowska E
MESOTAENIACEAE
Cylindrocystis brébissonii Menegh. ex DeBary var. *brébissonii*
— var. *minor* West & West T
Cylindrocystis crassa DeBary var. *crassa*
— var. *elliptica* West & West T
— var. *skujae* Croasdale E
Mesotaenium chlamydosporum DeBary var. *chlamydosporum* T
— var. *violescens* (DeBary) Willi Krieg.
Mesotaenium macrococcum (Kütz.) Roy et Bisset var. *macrococcum* T
— var. *minus* (DeBary) Compère
Netrium digitus (Ehrenb.) Itzigs. & Rothe var. *digitus*
— var. *lamellosum* (Brébisson) Grönblad
— var. *latum* Hustedt
— var. *naegelii* (Brébisson) Willi Krieg.
— var. *parvum* (Borge) Willi Krieg.
Netrium interruptum (Brébisson) Lütkem.
Roya obtusa (Brébisson) West & West var. *obtusa*
— var. *montana* West & West
Spirotaenia condensata Brébisson in Ralfs
Spirotaenia obscura Ralfs

Order DESMIDIALES
Suborder CLOSTERIINEAE
GONATOZYGACEAE
Gonatozygon aculeatum Hastings
Gonatozygon brébissonii DeBary var. *brébissonii*
— var. *minutum* (W.West) West & West
Gonatozygon kinahanii (W.Archer) Rabenh.
Gonatozygon monotaenium DeBary
Gonatozygon pilosum Wolle
Genicularia spirotaenia (DeBary) DeBary
PENIACEAE
Penium cylindrus (Ehrenb.) ex Brébisson in Ralfs var. *cylindrus*
— var. *attenuatum* Racib.*
Penium margaritaceum (Ehrenb.) ex Brébisson in Ralfs
Penium polymorphum (Perty) Perty
Penium spirostriolatum J.Barker
CLOSTERIACEAE
Closterium abruptum W.West var. *abruptum*
— var. *africanum* (Fritsch & Rich) Willi Krieg.
Closterium acerosum (Schrank) Ehrenb. ex Ralfs var. *acerosum*
— var. *angolense* West et West
— var. *elongatum* Breb.
Closterium aciculare T.West
Closterium acutum Brébisson in Ralfs var. *acutum*
— var. *linea* (Perty) West et West
— var. *variabile* (Lemmerm.) Willi Krieg.
Closterium archerianum Cleve in P.M.Lundell
Closterium attenuatum Ralfs
Closterium braunii Reinsch
Closterium closterioides (Ralfs) Louis et Peeters var. *closterioides*
— var. *intermedium* (Roy et Bisset) Ruzicka
Closterium compactum Nordst.
Closterium cornu Ehrenb. ex Ralfs
Closterium costatum Corda ex Ralfs var. *costatum*
— var. *borgei* (Willi Krieg.) Ruzicka
Closterium cynthia DeNot.
Closterium decorum Brébisson
Closterium dianae Ehrenb. ex Ralfs var. *dianae*
— var. *arcuatum* (Brébisson) Rabenh.
— var. *compressum* Klebs
— var. *pseudodianae* (Roy) Willi Krieg.
Closterium didymotocum Ralfs
Closterium directum W.Archer
Closterium ehrenbergii Menegh. ex Ralfs var. *ehrenbergii*
— var. *immane* Wolle
— var. *malinvernianum* (De Not.) Rabenh.*
Closterium exiguum West et West
Closterium forte E.A.Flint et D.B.Will.* E
Closterium gracile Brébisson ex Ralfs var. *gracile*
— var. *elongatum* West et West
Closterium idiosporum West et West
Closterium incurvum Brébisson

Closterium intermedium Ralfs
Closterium jenneri Ralfs
Closterium juncidum Ralfs
Closterium kuetzingii Brébisson var. *kuetzingii*
— var. *procerum* Skuja E
— var. *vittatum* Nordst.
Closterium lanceolatum Kütz. ex Ralfs
Closterium leibleinii Kütz. ex Ralfs
Closterium lineatum Ehrenb. ex Ralfs
Closterium lunula (O.F.Müll.) Nitzsch ex Ralfs var. *lunula*
— var. *massartii* (DeWild.) Willi Krieg.
Closterium macilentum Brébisson
Closterium moniliferum (Bory) Ehrenb. ex Ralfs
Closterium navicula (Brébisson) Lütkem.
Closterium nematodes Joshua
Closterium nordstedtii Chodat var. *polystichum* (Nygaard) Ruzicka
Closterium okaritoense E.A.Flint & D.B.Will. E
Closterium parvulum Nägeli var. *parvulum*
— var. *angustum* West et West
Closterium praelongum Brébisson var. *praelongum*
— var. *brevius* (Nordst.) Willi Krieg.
Closterium pronum Brébisson
Closterium ralfsii Brébisson in Ralfs var. *gracilius* (Mask.) Willi Krieg.
— var. *hybridum* Rabenh.
Closterium rostratum Ehrenb. ex Ralfs
Closterium selenaeum Mask. E
Closterium selenastroides Roll
Closterium setaceum Ehrenb. ex Ralfs
Closterium strigosum Brébisson
Closterium striolatum Ehrenb. ex Ralfs
Closterium tumidum Joshua var. *nylandicum* Grönblad*
Closterium venus Kütz. ex Ralfs var. *venus*
— var. *westii* Willi Krieg.
Closterium wallicherii W.B.Turner

Suborder DESMIDIINEAE
DESMIDIACEAE
Actinotaenium adelochondrum (Elfing) Teiling var. *parvum* E.A.Flint & D.B.Will. E
Actinotaenium cordanum (Brébisson) Ruzicka & Pouzer
Actinotaenium cruciferum (DeBary) Teiling
Actinotaenium cucurbita (Brébisson) Teiling
Actinotaenium cucurbitinum (Bisset) Teiling
Actinotaenium didymocarpum (P.M.Lundell) Coesel et Delfos
Actinotaenium diplosporum (P.M.Lundell) Teiling
Actinotaenium globosum (Bulnh.) Kurt Först. ex Compère
Actinotaenium inconspicuum (West & West) Teiling
Actinotaenium incurvum (Grönblad) Rino*
Actinotaenium pyramidatum (West & West) Teiling
Actinotaenium rufescens (Cleve) Teiling
Actinotaenium spinospermum (Joshua) Kouwets et Coesel
Actinotaenium subglobosum (Nordst.) Teiling
Actinotaenium subtile (West & West) Teiling
Actinotaenium turgidum (Brébisson) ex Ruzicka et Pouzar
Actinotaenium wollei (W. & G.S.West) Teiling
Brachytheca inopinata Fumanti et Alfinito E
Cosmarium amoenum Brébisson in Ralfs var. *amoenum*
— var. *intumescens* Nordst.
— var. *mediolaeve* Nordst.
Cosmarium amplum Nordst.
Cosmarium angulosum Brébisson var. *angulosum*
— var. *concinnum* (Rabenh.) West & West
Cosmarium annulatum (Nägeli) DeBary var. *elegans* Nordst.
Cosmarium arctoum Nordst.
Cosmarium askenasyi Schmidle
Cosmarium asphaerosporum Nordst. var. *productum* Nordst. E
Cosmarium binum Nordst.
Cosmarium bioculatum Brébisson in Ralfs var. *bioculatum*
— var. *depressum* (Schaarschm.) Schmidle
Cosmarium bireme Nordst.
Cosmarium blyttii Wille var. *blyttii*
— var. *novae-sylvae* West et West
Cosmarium boeckii Wille var. *isthmolaeve* Skuja ex Kouwets
Cosmarium botrytis Menegh. ex Ralfs
Cosmarium brasiliense (Wille) Nordst. var. *taphrosporum* Nordst. E
Cosmarium broomei Thwaites in Ralfs
Cosmarium caelatum Ralfs var. *caelatum*
— var. *spectabile* (DeNot.) Nordst.
Cosmarium calcareum Wittr.
Cosmarium candianum Delponte
Cosmarium capitulum Roy & Bisset var. *australe* G.S.West*
Cosmarium clepsydra Nordst.
Cosmarium conspersum Ralfs
Cosmarium contractum Kirchn. var. *contractum*
— var. *ellipsoideum* (Elfving) West & West
— var. *minutum* (Delponte) West & West
— var. *retusum* (West et West) Willi Krieg. & Gerloff
Cosmarium crassipelle Boldt var. *ornatum* E.A.Flint & D.B. ill. E
— var. *noduliferum* E.A.Flint & D.B.Will.* E
Cosmarium crenatum Ralfs
Cosmarium cucumis Corda ex Ralfs
Cosmarium cymatium (West & West) Willi Krieg. et Gerloff
Cosmarium decedens (Reinsch) Racib. var. *decedens* T
— var. *sinuosum* (P.M.Lundell) Racib.
Cosmarium depressum (Nägeli) P.M.Lundell var. *depressum*
— var. *achondrum* (Boldt) West & West
— var. *planctonicum* Reverdin
Cosmarium dickii Coesel*
Cosmarium difficile Lütkem.
Cosmarium distichum Nordst.
Cosmarium dorsitruncatum (Nordst.) G.S.West
Cosmarium euryophrys Skuja var. *euryophrys* E
— var. *hemisphaericum* Skuja E
Cosmarium eutelium Skuja E
Cosmarium exiguum W. Archer var. *exiguum*
— var. *maius* (Nordst.) Willi Krieg. & Gerloff
Cosmarium formosulum (W.E.Hoffm.) in Nordst.
Cosmarium foveatum Schmidle
Cosmarium geminatum P.M.Lundell
Cosmarium granatum Brébisson ex Ralfs var. *granatum*
— var. *rotundatum* Willi Krieg.
Cosmarium hammeri Reinsch var. *hammeri*
— var. *protuberans* West & West
— var. *subbinale* Nordst. E
Cosmarium holmiense P.M.Lundell
Cosmarium hornavanense Gutw. var. *dubovianum* (Lütkem.) Ruzicka
Cosmarium humile (Gay) Nordst. in DeToni
Cosmarium impressulum Elfving
Cosmarium javanicum Nordst.
Cosmarium laeve Rabenh. var. *laeve*
— var. *westii* Willi Krieg. & Gerloff
Cosmarium lapponicum Borge
Cosmarium lundellii Delponte var. *lundellii*
— var. *corruptum* (W.B.Turner) West & West
— var. *ellipticum* West & West
Cosmarium magnificum Nordst.
Cosmarium margaritatum (P.M.Lundell) Roy & Bisset
Cosmarium margaritiferum Menegh. ex Ralfs
Cosmarium maskellii Croasdale E
Cosmarium meneghinii Brébisson in Ralfs
Cosmarium microsphinctum Nordst.
Cosmarium minimum West & West
Cosmarium moniliferum Reinsch
Cosmarium moniliforme (Turpin) Ralfs
Cosmarium monochondrum Nordst. var. *fallax* Ruzicka
Cosmarium multiordinatum West & West var. *rotundatum* Willi Krieg.
Cosmarium nasutum Nordst. var. *subcirculare* Skuja E
— var. *subnasutum* (Racib.) Nordst.
Cosmarium nitidulum DeNot.
Cosmarium norimbergense Reinsch
Cosmarium novae-semliae Wille var. *sibiricum* Boldt
Cosmarium obliquum Nordst.
Cosmarium obsoletum (Hantzsch) Reinsch var. *punctatum* Mask.
— var. *sitvense* Gutw.
Cosmarium obtusatum Schmidle
Cosmarium ochthodes Nordst. var. *ochthodes*
—var. *amoebum* W.West
Cosmarium pachydermum P.M.Lundell var. *pachydermum*
— var. *indicum* M.O.P.Iyengar et Vimala Bai
Cosmarium paludicola E.A.Flint & D.B.Will. E
Cosmarium parvulum Brébisson
Cosmarium perfissum G.S.West
Cosmarium phaseolus Brébisson in Ralfs var. *phaseolus*
— var. *omphalum* (Schaarschm.) Racib.
Cosmarium plicatum Reinsch.
Cosmarium polygonum (Nägeli) W.Archer in Pritchard*
Cosmarium portianum W.Archer
Cosmarium prometopidion Skuja E
Cosmarium pseudamoenum Wille var. *basilare* Nordst.
Cosmarium pseudarctoum Nordst.
Cosmarium pseudoexiguum Racib. var. *pseudoexiguum*
— var. *clausum* Skuja E
Cosmarium pseudonitidulum Nordst.
Cosmarium pseudopachydermum Nordst.
Cosmarium pseudophaseolus Brühl & Biswas
Cosmarium pseudoprotuberans Kirchn. var. *angustius* Nordst.
— var. *notatum* Skuja E
Cosmarium pseudopyramidatum P.M.Lundell var. *pseudopyramidatum*
— var. *excavatum* (Nordst.) Willi Krieg. & Gerloff
— var. *extensum* (Nordst.) Willi Krieg. & Gerloff
— var. *umbonulatum* (Nordst.) Willi Krieg. & Gerloff
Cosmarium pseudoretusum Ducellier var. *inaequalipellicum* (West & West) Willi Krieg. & Gerloff
Cosmarium punctulatum Brébisson var. *punctulatum*
— var. *granulusculum* (Roy & Bisset) West & West
— var. *subpunctulatum* (Nordst.) Børgesen
Cosmarium pusillum (Brébisson) W. Archer
Cosmarium pyramidatum Brébisson in Ralfs var. *pyramidatum*
— var. *stenonotum* (Nordst.) Klebs
Cosmarium quadratulum (Gay) DeToni
— var. *quadratum* Ralfs ex Ralfs
— var. willei (Schmidle) Willi Krieg. & Gerloff
Cosmarium quadrifarium P.M.Lundell var. *quadrifarium*
— var. *hexastichum* (P.M.Lundell) Kurt Först.
— var. *octastichum* (Nordst.) Kurt Först.
Cosmarium quadriverrucosum West & West var. *quadriverrucosum*
— var. *supraornatum* Skuja
Cosmarium quadrum P.M.Lundell var. *quadrum*

— var. *sublatum* (Nordst.) West & West
Cosmarium ralfsii Brébisson in Ralfs var. *ralfsii*
— var. *montanum* Racib.
Cosmarium rectangulare Grun.
Cosmarium regnelli Wille var. *regnelli*
— var. *minimum* B. Eichler & Gutw.
Cosmarium regnesi Reinsch var. *regnesi*
— var. *montanum* Schmidle
Cosmarium reniforme (Ralfs) W.Archer var. *reniforme*
— var. *compressum* Nordst.
Cosmarium repandum Nordst.
Cosmarium retusiforme (Wille) Gutwe var. *incrassatum* Gutwe
Cosmarium sexangulare P.M.Lundell var. *sexangulare*
— var. *minus* Roy & Bisset
Cosmarium speciosum P.M.Lundell var. *speciosum*
— var. *simplex* Nordst.
Cosmarium sphagnicolum West & West
Cosmarium sphalerostichum Nordst.
Cosmarium sportella Brébisson in Kütz. var. *sportella*
— var. *subnudum* West & West
Cosmarium stigmosum (Nordst.) Willi Krieg.
Cosmarium subarctoum (Lagerh.) Racib.
Cosmarium subcontractum West & West var. *detritum* (Playfair) Willi Krieg. & Gerloff
Cosmarium subcrenatum Hantzsch in Rabenh.
Cosmarium subcucumis Schmidle
Cosmarium subcyclicum Mask. E
Cosmarium subgranatum (Nordst.) Lütkem.
Cosmarium subprotumidum Nordst.*
Cosmarium subquadratum Nordst. var. *subquadratum*
— var. *genuosum* (Nordst.) Willi Krieg. & Gerloff E
Cosmarium subspeciosum Nordst. var. *subspeciosum*
— var. *validius* Nordst.
Cosmarium subtumidum Nordst.
Cosmarium sumatranum Willi Krieg.
Cosmarium tatricum Racib. var. *novizelandicum* Nordst.
Cosmarium tenue W. Archer
Cosmarium tetracentrotum Skuja E
Cosmarium tetraophthalmum Brébisson in Ralfs var. *tetraophthalmum*
— var. *patagonicum* Borge
Cosmarium tinctum Ralfs var. *tinctum*
— var. *intermedium* Nordst.
Cosmarium trilobulatum Reinsch var. *trilobulatum*
Cosmarium tuddalense Strøm var. *tuddalense**
— var. *australe* Skuja E
Cosmarium turnerianum Mask. E
Cosmarium umbilicatum Lütkem.*
Cosmarium undulatum Corda ex Ralfs
Cosmarium variabile Mask. E
Cosmarium variolatum P.M.Lundell var. *tortum* E.A.Flint & D.B.Will.* E
— var. *skujae* Croasdale E
— var. *umbonatum* Skuja E
Cosmarium venustum (Brébisson) W.Archer var. *venustum*
— var. *basichondrum* (Nordst.) Willi Krieg. & Gerloff
— var. *excavatum* (B.Eichler) West & West*
— var. *induratum* Nordst.
Cosmarium vogesiacum Lemmerm.
Cosmarium westii Bernard
Cosmocladium constrictum W. Archer ex Joshua
Cosmocladium saxonicum DeBary
Euastrum ansatum Ehrenb. var. *ansatum*
— var. *supraposituum* Nordst.
Euastrum bidentatum Nägeli
Euastrum binale (Turpin) Ehrenb. ex Ralfs var. *binale*
— var. *gutwinski* (Schmidle) Homfeld
Euastrum brevisinuosum (Nordst.) Kouwets var. *brevisinusosum*
— var. *dissimile* (Nordst.) Kouwets
Euastrum bullatum Playfair
Euastrum cornubiense West & West var. *medianum* (Nordst.) Willi Krieg.
Euastrum crassicolle P.M.Lundell var. *dentiferum* Nordst.
Euastrum cuneatum Jenner in Ralfs var. *cuneatum*
— var. *solum* Nordst.
Euastrum denticulatum (Kirchn.) Gay var. *denticulatum*
— var. *quadrifarium* Willi Krieg.
— var. *bengalicum* Lagerh.
Euastrum dubium Nägeli var. *dubium*
— var. *ornatum* Wolosz.
Euastrum elegans (Brébisson) Kütz. ex Ralfs
Euastrum elongatum (Nordst.) Willi Krieg.
Euastrum erosum P.M.Lundell var. *evolution* Cedergr.
Euastrum euteles Skuja E
Euastrum haplos Skuja E
Euastrum holocystoides Nordst. E
Euastrum incrassatum Nordst.
Euastrum insulare (Wittr.) Roy var. *insulare*
— var. *lacustre* (Messik.) Willi Krieg.
Euastrum irregulare Mask. E
Euastrum lagynion Skuja E
Euastrum laticolle G.S.West
Euastrum longicolle Nordst. var. *longicolle*
— var. *australicum* Playfair
Euastrum mammatum Mask. var. *mammatum* E
— var. *ellipticum* Mask. E
Euastrum obesum Joshua var. *obesum*
— var. *trapezicum* (Börges.) Willi Krieg.
Euastrum oblongum (Grev.) Ralfs
Euastrum praemorsum (Nordst.) Schmidle
Euastrum quadrioculatum West et West
Euastrum rotundum Mask. E
Euastrum sinuosum (Lenorm.) W.Archer var. *sinuosum*
— var. *gangense* (W.B.Turner) Willi Krieg.
— var. *gemmulosum* Mask. E
— var. *germanicum* (Racib.) Willi Krieg.
— var. *subjenneri* West & West
Euastrum sphyroides Nordst.
Euastrum subalpinum Messik.
— var. *hybridum* Skuja E
Euastrum subangulosum Skuja E
Euastrum turneri W.West
Euastrum verrucosum Ehrenb. var. *alatum* Wolle*
— var. *alpinum* (Hub.-Pest.) Willi Krieg.
Haplotaenium minutum (Ralfs) Bando var. *elongatum* (West & West) Bando
Haplotaenium rectum (Delponte) Bando
Micrasterias alata G.C.Wall.
Micrasterias decemdentata (Nägeli) W.Archer
Micrasterias denticulata Brébisson ex Ralfs var. *denticulata*
— var. *angulosa* (Hantzsch) West & West
Micrasterias mahabuleshwarensis J.Hobson var. *mahabuleshwarensis*
— var. *ampullacea* (Mask.) Nordst.
Micrasterias papillifera Brébisson in Ralfs
Micrasterias pinnatifida (Kütz.) ex Ralfs
Micrasterias radiosa Ralfs var. *radiosa*
— var. *evoluta* (Nordst.) Grönblad
Micrasterias rotata (Grev.) Ralfs ex Ralfs
Micrasterias schweinfurthii Cohn
Micrasterias subdenticulata (Nordst.) Willi Krieg. var. *subdenticulata*
— var. *nobilis* Skuja E
Micrasterias suboblonga Nordst. var. *tecta* Willi Krieg
Micrasterias thomasiana W.Archer var. *thomasiana*
— var. *notata* (Nordst.) Grönblad
Micrasterias tropica Nordst. var. *indivisa* (Nordst.) B. Eichler & Racib.
Micrasterias zeylanica Fritsch
Pleurotaenium ehrenbergii (Brébisson) DeBary var. *ehrenbergii*
— var. *contractum* E.A.Flint & D.B.Will. E
— var. *curtum* Willi Krieg.*
— var. *elongatum* (W.West) W.West
— var. *undulatum* Schaarschm.
Pleurotaenium nodosum (Bailey) P.M.Lundell
Pleurotaenium ovatum (Nordst.) Nordst. var. *ovatum*
— var. *tumidum* (Mask.) G.S.West
Pleurotaenium quantillum (W.B.Turner) West & West
Pleurotaenium trabecula (Ehrenb.) ex Nägeli var. *trabecula*
— var. *elongatum* Cedergr.
Pleurotaenium tridentulum (Wolle) W.West var. *subturgidum* Skuja E
Pleurotaenium truncatum (Brébisson) Nägeli
Staurastrum alternans Brébisson in Ralfs var. *alternans*
— var. *subalternans* Mask. E
Staurastrum amplum Skuja E
Staurastrum anatinum Cooke & Wills var. *anatinum*
— var. *lagerheimii* (Schmidle) West & West
Staurastrum arctiscon (Ehrenb.) P.M.Lundell var. *glabrum* West & West
Staurastrum arcuatum Nordst.
Staurastrum armigerum Brébisson var. *armigerum*
— var. *furcigerum* (Brébisson) Teiling
Staurastrum asperatum Grönblad var. *minus* Behre
Staurastrum assurgens Nordst.
Staurastrum aureolatum Playfair
Staurastrum avicula Brébisson in Ralfs var. *avicula*
— var. *subarcuatum* (Wolle) West & West
Staurastrum bacillare Brébisson in Ralfs var. *obesum* P.M.Lundell
Staurastrum bibrachiatum (Reinsch) Grönblad et Scott
Staurastrum bicorne Hauptfl.
Staurastrum brachiatum Ralfs var. *brachiatum*
— var. *gracilius* Mask.
— var. *tenerrimum* Skuja E
Staurastrum capitulum Brébisson in Ralfs var. *australe* Skuja E
— var. *tumidiusculum* (Nordst.) West & West
Staurastrum cingulum (West & West) G.M.Smith var. *cingulum*
— var. *obesum* G.M.Smith
Staurastrum coarctatum Brébisson var. *coarctatum*
— var. *solitarium* E.A.Flint E
Staurastrum controversum Brébisson in Ralfs
Staurastrum cyclacanthum West & West var. *submanfeldtioides* Scott & Prescott
Staurastrum denticulatum (Nägeli) W.Archer
Staurastrum dilatatum (Ehrenb.) Ralfs
Staurastrum dispar Brébisson
Staurastrum disputatum West & West var. *sinense* (Lütkem.) West & West
Staurastrum dorsuosum Nordst. E
Staurastrum ellipticum W.West var. *ellipticum*
— var. *minus* Skuja
Staurastrum excavatum West & West
Staurastrum floriferum W. & G.S.West
Staurastrum furcatum (Ehrenb.) Brébisson var. *furcatum*
— var. *renardi* (Reinsch) Nordst.
— var. *asymmetricum* Grönblad & Scott*
Staurastrum fusoideum Croasdale & E.A.Flint E
Staurastrum grande Bulnh. var. *parvum* W.West
Staurastrum hexacerum (Ehrenb.) Wittr.
— var. *semicirculare* Wittr.
Staurastrum hirsutum (Ehrenb.) Brébisson in Ralfs
Staurastrum inconspicuum Nordst.
Staurastrum johnsonii West et West var. *altius* Fritsch et Rich
Staurastrum kosmos Skuja E
Staurastrum laeve Ralfs

Staurastrum leptacanthum Nordst.
Staurastrum leptocladum Nordst. var. *elegans* G.S.West
— var. *insigne* West & West
— var. *smithii* Grönblad
Staurastrum limneticum Schmidle var. *limneticum*
— var. *aculeatum* Lemmerm.
— var. *burmense* West & West
— var. *rectum* Lemmerm. E
Staurastrum longibrachiatum (Borge) Gutw.
Staurastrum longipes (Nordst.) Teiling var. *longipes*
— var. *contractum* Teiling
— var. *evolutum* (West & West) Thomasson
Staurastrum longiradiatum West & West
Staurastrum lunatum Ralfs var. *lunatum*
— var. *planctonicum* West & West
Staurastrum magnifurcatum Scott & Grönblad
Staurastrum manfeldtii Delponte var. *manfeldtii*
— var. *annulatum* West & West
Staurastrum margaritaceum (Menegh.) Ralfs
Staurastrum messikommeri Lundb.
Staurastrum militare Behre
Staurastrum muricatum Brébisson in Ralfs
Staurastrum muticum (Brébisson) Ralfs var. *muticum*
— var. *convexum* (Scott & Prescott) Croasdale
— var. *subcurtum* (Nordst.) Croasdale
— var. *victoriense* G.S.West
Staurastrum natator W.West
Staurastrum neglectum G.S.West
Staurastrum obversum Prescott
Staurastrum octoverrucosum Scott et Grönblad var. *octoverrucosum*
— var. *simplicius* Scott et Grönblad
Staurastrum orbiculare Ralfs var. *orbiculare*
— var. *depressum* Roy et Bisset
— var. *hibernicum* West et West
— var. *protractum* Playfair*
Staurastrum osvaldii Skuja var. *osvaldii* E
— var. *brevibrachiatum* E.A.Flint* E
Staurastrum paradoxum Meyen in Ralfs
Staurastrum pelagicum West et West
Staurastrum pileatum Delponte var. *inflatum* Mask.
Staurastrum pingue Teiling
Staurastrum planctonicum Teiling var. *planctonicum*
— var. *bulbosum* (W.West) Thomasson
— var. *ornatum* (Grönblad) Teiling
Staurastrum polymorphum Brébisson in Ralfs emend. Nordst.
Staurastrum proboscideum (Brébisson) W.Archer var. *proboscideum*
— var. *altum* Boldt
Staurastrum pseudassurgens Mask. *pseudassurgens*
— var. *robustum* E.A.Flint* E
Staurastrum pseudoligacanthum Mask. E
Staurastrum pseudosebaldi Wille var. *pseudosebaldi*
— var. *tonsum* (Nordst.) Croasdale
Staurastrum pseudotetracerum (Nordst.) West & West
Staurastrum punctulatum Brébisson in Ralfs var. *punctulatum*
— var. *pygmaeum* (Brébisson) West & West
— var. *subproductum* West & West
Staurastrum retusum W.B.Turner*
Staurastrum rosei Playfair var. *novizelandicum* Thomasson E
— var. *stemmatum* Scott & Prescott
Staurastrum rotula Nordst.
Staurastrum sagittarium Nordst.
Staurastrum sebaldi Reinsch var. *sebaldi*
— var. *ornatum* Nordst.
— var. *productum* West & West
Staurastrum sexangulare (Bulnh.) P.M.Lundell var. *sexangulare*
— var. *attenuatum* W.B.Turner
— var. *productum* Nordst.
Staurastrum smithii Teiling
Staurastrum sonthalianum W.B.Turner
Staurastrum spinuliferum Mask. E
Staurastrum splendidum Mask. E
Staurastrum spongiosum Brébisson ex Ralfs*
Staurastrum striolatum (Nägeli) W. Archer var. *striolatum*
— var. *acutius* Mask. E
Staurastrum subamoenum Mask. E
Staurastrum subarmigerum Roy & Bisset
Staurastrum subavicula West & West
Staurastrum subbrébissonii Schmidle
Staurastrum subcruciatum Cooke & Wills
Staurastrum subdenticulatum Nordst.
Staurastrum subgracillimum West & West
Staurastrum sublaevispinum West & West
Staurastrum subnudibrachiatum West & West
Staurastrum subpolymorphum Borge
Staurastrum subradians Rich
Staurastrum tangaroae Thomasson
Staurastrum tetracerum Ralfs
Staurastrum tohopekaligense Wolle var. *tohopekaligense*
— var. *brevispinum* G.M.Smith
— var. *nonanum* (W.B.Turner) Schmidle
— var. *robustum* Scott & Prescott
Staurastrum trifidum Nordst. var. *porrectum* Croasdale & Scott*
Staurastrum trihedrale Wolle
Staurastrum victoriense G.S.West var. *victoriense*
— var. *tasmanicum* Croasdale & E.A.Flint
Staurodesmus angulatus (W.West) Teiling
Staurodesmus brevispina (Brébisson) Croasdale
Staurodesmus clepsydra (Nordst.) Teiling
Staurodesmus connatus (P.M.Lundell) Thomasson
Staurodesmus convergens (Ehrenb.) Teiling var. *convergens*
— var. *laportei* Teiling
— var. *ralfsii* Teiling
Staurodesmus crispus (W.B.Turner) Compère var. *americanus* (Scott & Grönblad) Croasdale
— var. *obesus* (West & West) Croasdale
Staurodesmus cuspidatus (Brébisson) Teiling var. *cuspidatus*
— var. *curvatus* (W.West) Teiling
Staurodesmus dejectus (Brébisson ex Ralfs) Teiling var. *dejectus*
— var. *apiculatus* (Brébisson) Teiling
Staurodesmus dickiei (Ralfs) var. *dickiei*
— var. *circularis* (W.B.Turner) Croasdale
— var. *rhomboideus* (West & West) S. Lill.
Staurodesmus extensus (Borge) Teiling var. *extensus*
— var. *joshuae* (Gutw.) Teiling
Staurodesmus glaber (Ehrenb.) Teiling var. *glaber*
— var. *debaryanus* (Nordst.) Teiling
— var. *limnophilus* Teiling
Staurodesmus grandis (Bulnh.) Teiling var. *grandis*
— var. *parvus* (West & West) Teiling
Staurodesmus incus (Brébisson) Teiling var. *incus*
— var. *ralfsii* (West &West) Teiling
Staurodesmus indentatus (West &West) Teiling
Staurodesmus isthmosus (Heimerl) Croasdale
Staurodesmus lanceolatus (W.Archer) Croasdale in Teiling
Staurodesmus leptodermus (P.M.Lundell) Teiling var. *leptodermus*
— var. *corniculatus* (P.M.Lundell) Thomasson
— var. *subcorniculatus* (Rich) Teiling
Staurodesmus mamillatus (Nordst.) Teiling var. *mamillatus*
— var. *maximus* (W.West) Teiling
Staurodesmus megacanthus (P.M.Lundell) Thunmark var. *megacanthus*
—var. *orientalis* (Scott & Prescott) Teiling*
Staurodesmus mucronatus (Ralfs) Croasdale var. *mucronatus*
— var. *delicatulus* (G.S.West) Teiling
— var. *parallelus* (Nordst.) Teiling
— var. *subtriangularis* (West & West) Croasdale
Staurodesmus omearae (W.Archer) Teiling var. *omearae*
— var. *minutus* (West & West) Teiling
Staurodesmus osvaldii (Skuja) Compère & Bicudo E
Staurodesmus pachyrhynchus (Nordst.) Teiling var. *pachyrhynchus*
— var. *nothogeae* Skuja E
— var. *pseudopachyrhynchus* (Wolle) Teiling
— var. *tenerus* (Grönblad) Teiling
Staurodesmus patens (Nordst.) Croasdale
Staurodesmus phimus (W.B.Turner) Thomasson var. *phimus*
— var. *hebridensis* (West & West) Teiling
Staurodesmus pterosporum (P.M.Lundell) Bourrelly
Staurodesmus quiriferus (West & West) Teiling var. *quiriferus*
— var. *minor* (Iréné-Marie) Croasdale
Staurodesmus skujae Croasdale & E.A.Flint E
Staurodesmus spencerianus (Mask.) Teiling var. *spencerianus*
— var. *triangulatus* (Willi Krieg.) Teiling
Staurodesmus spetsbergensis (Nordst.) Teiling var. *spetsbergensis*
— var. *floriniae* Teiling
— var. *limneticus* Teiling
Staurodesmus triangularis (Lagerh.) Teiling var. *triangularis*
— var. *limneticus* Teiling
Staurodesmus tumidus (Brébisson ex Ralfs) Teiling
Staurodesmus unicornis (W.B.Turner) Thomasson var. *gracilis* (M.O.P.Iyengar et Vimala Bai) Teiling
Tetmemorus brébissonii (Menegh.) Ralfs
Tetmemorus granulatus (Brébisson) Ralfs
Tetmemorus laevis (Kütz.) ex Ralfs var. *laevis*
Triploceras gracile Bailey var. *gracile*
— var. *aculeatum* (Nordst.) Mask.
— var. *bidentatum* Nordst.
— var. *laticeps* (Nordst.) Willi Krieg. E
— var. *tridentatum* (Mask.) Willi Krieg. E
Triploceras verticillatum Bailey var. *superbum* (Mask.) DeToni E
Xanthidium aculeatum Ehrenb. in Ralfs
Xanthidium armatum Brébisson in Ralfs var. *armatum*
— var. *basidentatum* Nordst.
Xanthidium bifidum (Brébisson) Deflandre var. *latodivergens* (W.West) Croasdale
Xanthidium cristatum (Brébisson) in Ralfs
Xanthidium dilatatum Nordst. E
Xanthidium fasciculatum Ehrenb. ex Ralfs var. *perornatum* Nordst. E
Xanthidium hastiferum W.B.Turner var. *hastiferum*
— var. *inevolutum* Nordst.
— var. *javanicum* (Nordst.) W.B.Turner
Xanthidium inchoatum Nordst.
Xanthidium intermedium Mask. E
Xanthidium multigibberum (Nordst.) Skuja E
Xanthidium octonarium Nordst.
Xanthidium simplicius Nordst.
Xanthidium smithii W.Archer var. *smithii*
— var. *maius* (Ralfs) West & G.S.West
Xanthidium submicracanthum E.A.Flint & D.B.Will.* E
Xanthidium variabile (Nordst.) West & West var. *variabile*
— var. *scabrum* Skuja E

Desmidiaceae – Filamentous forms

Bambusina brébissonii Kütz.
Desmidium aptogonum Brébisson ex Kütz.
Desmidium asymmetricum Grönblad
Desmidium baileyi (Ralfs) Nordst. var. *baileyi*

— var. *coelatum* (Kirchn.) Nordst.
— var. *undulatum* (Mask.) Nordst.
Desmidium coarctatum Nordst.
Desmidium grevillii (Kütz.) DeBary
Desmidium occidentale West & West var. *concatenatum* Skuja E
Desmidium swartzii Agardh ex Ralfs
Groenbladia neglecta (Racib.) Teiling
Hyalotheca dissiliens (J.E.Smith) Brébisson ex Ralfs var. *dissiliens*
— var. *hians* Wolle
— var. *tatrica* Racib.
Hyalotheca dubia (Kütz.) Ralfs
Hyalotheca hians Nordst.
Hyalotheca mucosa (Mert.) Ehrenb. ex Ralfs
Onychonema filiforme (Ehrenb. ex Ralfs) Roy et Bisset
Onychonema laeve Nordst. var. *micracanthum* Nordst.
Phymatodocis nordstedtiana Wolle var. *nordstedtiana*
— var. *novizelandica* Nordst. E
Sphaerozosma aubertianum W.West var. *aubertianum*
— var. *compressum* Rich.
Sphaerozosma vertebratum (Brébisson) Ralfs
Spondylosium compressum (Mask.) Croasdale & E.A.Flint E
Spondylosium ligatum (West & West) Skuja var. *pleurotum* Skuja E
Spondylosium panduriforme (Heimerl) Teiling
Spondylosium papillosum West & West
Spondylosium planum (Wolle) West & West
Spondylosium pulchellum W.Archer var. *pulchellum*
— var. *bambusinoides* (Wittr.) P.M.Lundell
Spondylosium pulchrum (Bailey) W.Archer
Teilingia excavata (Ralfs ex Ralfs) Bourrelly ex Compère
Teilingia granulata (Roy et Bisset) Bourrelly ex Compère

Class KLEBSORMIDIOPHYCEAE
Order KLEBSORMIDIALES
KLEBSORMIDIACEAE
Klebsormidium acidophilum Novis E
Klebsormidium crenulatum (Kütz.) H.Ettl & G.Gärtner†
— ?var. *corticulatum* (Rabenhorst) H.Ettl & G.Gärtner†
Kebsormidium dissectum (Gay) Lokhorst
Klebsormidium elegans Lokhorst T
Klebsormidium flaccidum (Kütz.) Silva, Mattox & Blackwell†
— var. *varia* (Kütz.) Silva, Mattox & Blackwell †
Klebsormidium mucosum (J.B.Petersen) Silva, Mattox & Blackwell†
Klebsormidium oedogonioides (Skuja) H.Ettl & G. Gärtner
Klebsormidium subtile (Kütz.) Tracanna†
Raphidonema nivale Lagerheim

Class COLEOCHAETOPHYCEAE
Order COLEOCHAETALES
COLEOCHAETACEAE
Coleochaete irregularis Pringsheim
Coleochaete leve (Bailey) P.Sarma & V.J.Chapman
Coleochaete orbicularis Pringsheim
Coleochaete pulvinata A.Braun
Coleochaete scutata Brébisson
Coleochaete soluta (Brébisson) Pringsheim

Class CHAROPHYCEAE
Order CHARALES
CHARACEAE
Chara australis Brown
— var. *nobilis* (A.Braun) R.D.Wood
Chara fibrosa (Agardh ex Bruzelius) R.D.Wood
— var. *acanthopitys* (A.Braun) Zaneveld
Chara globularis Thuillier
Chara vulgaris (Linnaeus) R.D.Wood A?
Chara sp. aff. *muelleri* (A.Braun) F.Mueller
Lamprothamnium macropogon (A. Braun) J.L.Ophel
Nitella claytonii M.T.Casanova in Casanova et al. E
Nitella hookeri A.Braun
Nitella hyalina (DeCandolle) Agardh
Nitella leonhardii R.D.Wood
Nitella masonae (R.D.Wood) M.T.Casanova in Casanova et al. E
Nitella opaca (Bruzelius) Agardh
Nitella pseudoflabellata pseudoflabellata (A.Braun) R.D.Wood
—var. *conformis* (Nordstedt) R.D.Wood
—var. *mucosa* (Nordstedt) Bailey
Nitella stuartii A.Braun
Nitella sp. aff. *cristata*
Nitella subtilissima A.Braun
Nitella tricellularis C.F.O.Nordstedt in T.F.Allen E
Tolypella nidifica (O.Müller) R.D.Wood

Checklist of New Zealand fossil green algae

The entries below are derived from Raine et al. (2008) and records compiled by Chris Clowes, Victoria University of Wellington. Record File of GNS Science and the Geological Society of New Zealand.

INFRAKINGDOM CHLOROPHYTA
PHYLUM CHLOROPHYTA
Class PYRAMIMONADOPHYCEAE
Order PTEROSPERMALES
CYMATIOSPHAERACEAE
Cymatiosphaera sp.
PTEROSPERMELLACEAE
Pterospermella sp.

Order PYRAMIMONADALES
HALOSPHAERACEAE
Pachysphaera sp.
TASMANACEAE
Pleurozonaria sp.
Tasmanites sp.

INCERTAE SEDIS
Tytthodiscus sp.
Gen. et sp. indet.

Class CHLOROPHYCEAE
Order CHLAMYDOMONADALES
DICTYOSPHAERACEAE
Botryococcus braunii Kütz. Cretaceous–Recent
HYDRODICTYACEAE
Pediastrum boryanum (Turpin) Meneghini Neogene–Recent
RADIOCOCCACEAE
Scenedesmus sp.

Class ULVOPHYCEAE
Order BRYOPSIDALES
CODIACEAE
Shonabellia verrucosa Retallack

PHYLUM INCERTAE SEDIS
Chordecystia chalasta Foster 1979 Permian
Quadrisporites horridus Hennelly 1958 emend. Potonie & Lele 1961 Permian
Schizophacus cooksoniae (Pocock 1962) Pierce 1976 Cretaceous
Schizophacus parvus (Cookson & Dettmann 1959) Pierce 1976 Cretaceous
Schizophacus rugulatus (Cookson & Dettmann 1959) Pierce 1976 Cretaceous
Schizosporis sp. Mildenhall 1994 Cretaceous
Schizosporis reticulatus Cookson & Dettmann 1959 emend. Pierce 1976 Cretaceous

INFRAKINGDOM STREPTOPHYTA
PHYLUM CHAROPHYTA
Class ZYGNEMOPHYCEAE
Order ZYNEMATALES
ZYGNEMATACEAE
Mougeotia sp.
Ovoidites ligneolus Potonié
Spirogyra sp.
Zygnemaceae sp. a Raine 1987
Zygnemaceae sp. b Raine 1987
Zygnemaceae sp. c Raine 1987

Class CHAROPHYCEAE
CHARACEAE
Gen. et sp. indet.

TWENTY-FOUR

Phylum BRYOPHYTA

hornworts, liverworts, mosses

DAVID GLENNY, ALLAN J. FIFE

Hornwort *Anthoceros laminiferus* (Anthocerotales: Anthocerotaceae).
Bill Malcolm, Micro-Optics Ltd

Bryophytes may be thought of as the amphibians of the plant world, being dependent on water for successful living. For example, water is necessary for reproduction (sperm transfer). Additionally, bryophytes lack true roots for conducting moisture from soil and they lack a waxy cuticle to reduce water loss from stem and leaf surfaces. Most bryophytes are thus considerably dependent on larger vascular plants to create an evenly cool and moist environment, and today they may have achieved their greatest diversity since the evolution of the angiosperms (Schuster 1981). They have exploited the forest environment very effectively in New Zealand, making up the main forest-floor cover in high-rainfall areas instead of the herbaceous vascular plants that cover forest floors in many other countries.

The immediate ancestor of the bryophytes was probably a charophyte green alga whose closest living relative is *Coleochaete* (Mischler & Churchill 1984; Wheeler 2000). The time of origin of the bryophytes is believed to have been during the Silurian period about 400 million years ago. The earliest liverwort fossils are a tiny fragment of a *Riccia*-like plant from the early Devonian (Edwards et al. 1995) and *Pallavicinites devonicus*, a *Metzgeria*-like fossil from the late Devonian (Stewart 1983). All bryophytes are haploid-dominant plants; that is, the conspicuous green phase of the bryophyte life-cycle has a single set of chromosomes, while the smaller, generally inconspicuous non-photosynthetic sporophyte phase (seta and spore-bearing capsule) that is attached to it is diploid.

The liverworts, mosses, and hornworts are three equally distinct bryophyte groups (Schofield 1985). Of these, the hornworts are the smallest group with only about 200 species world-wide. *Takakia*, a bryophyte of Asia and North America described as a liverwort genus in 1958 and only later found fruiting (Smith & Davison 1993), has characteristics of both liverworts (particularly *Haplomitrium*) and mosses (particularly *Andreaea*) and it has been suggested that it should be recognised as a fourth group (Schuster 1997). Liverworts share with mosses the following features – many chloroplasts per cell, typically a capsule on a seta (but the seta may be short), leaves often present, and, if so, then they are in three rows (in mosses this is true but is not usually obvious), and the haploid gametophyte phase is dominant. Inasmuch as liverworts comprise both leafy and thallose forms, they are morphologically a more diverse group than the mosses, which are all leafy.

Liverworts differ from mosses in a number of important respects. When leafy, they never have a midvein in the leaf. The seta in liverworts is short-lived, collapsing after the capsule opens, whereas in mosses it persists after the capsule opens. The capsule in most liverworts opens by four slits running down its length, whereas in mosses the capsule opens by an apical pore (except in the moss genus *Andreaea* where the capsule opens by four slits as in liverworts). Liverwort capsules, with few exceptions, have elaters mixed with the spores, whereas mosses (including *Andreaea*) never have elaters. The calyptra of a liverwort is soft and fleshy and does not persist, whereas in mosses it is membranous and persistent. The rhizoids of liverworts are unicellular, whereas in mosses they are multicellular. Liverworts contain oil bodies in most cells, whereas these are absent from mosses.

Hornworts have in common with liverworts that they are thallose, they have elaters and spores inside the capsule, and the rhizoids are unicellular. Hornworts differ from liverworts in having only one or two chloroplasts per cell rather than many (liverworts resemble mosses in this respect). Hornworts have no seta distinct from the capsule (liverworts and mosses share the feature of a distinct seta), they lack oil bodies, and have a capsule that is a tapering needle-like structure that opens by one or two vertical slits. Hornworts have in common with mosses stomata in the capsule wall.

DNA-sequencing studies involving bryophytes have been attempted, with several aims – 1) to ascertain relationships among the mosses, liverworts, and hornworts, in particular which group, if any, diverged first from the green algae; 2) to find whether the vascular plants evolved from within one of the bryophyte groups or diverged separately from the green algae; and 3) to establish the characteristics of the earliest liverworts. These questions have not been answered with any certainty.

Small sample size appeared initially to be limiting the resolution of molecular data; for example, Hedderson et al. (1996) used only nine moss, eight liverwort, and two hornwort species and a later study concluded that the result is very dependent on which species are used to represent each group (Duff & Nickrent 1999). It seems likely that mutations occurring since the time of divergence of the three bryophyte groups has overwritten much of the information that was originally in the DNA (Lewis et al. 1997). Two methods that may overcome this 'noise' problem are sequencing of whole-chloroplast genomes and the use of repeats and inversions in the DNA that may have changed less often than single base mutations.

Haplomitrium gibbsiae with a sporophyte (left) and small pearl-like antheridia (which contain sperm) near the tip of the branch at right (Calobryales: Haplomitriaceae).

Bill Malcolm, Micro-Optics Ltd

Liverworts and hornworts

Their early fossil record is poor. The Early Devonian *Riccia*-like fossil of Edwards et al. (1995) may belong to the Marchantiales, otherwise the most recent confirmed fossil of that order is from the Triassic (Stewart 1983). No leafy liverworts of the order Jungermanniales are known before the Tertiary (Stewart 1983). The lack of earlier fossils in these groups may well indicate their late divergence. In New Zealand, a Late Triassic liverwort, placed in the form-genus *Marchantites*, was a common ground cover at a Southland locality (Pole & Raine 1994) and similar forms were reported from the Middle Cretaceous of the Clarence River (Parrish et al. 1998).

Summary of New Zealand living bryophyte diversity

Taxon	Described species and sub-specific taxa	Known undescr./undet. species and sub-specific taxa	Estimated unknown species	Adventive species and subspecific taxa	Endemic species and sub-specific taxa	Endemic genera
Anthocerotopsida	13+0	0	2	0	8	0
Marchantiopsida	593+39	0	50	6	276+53	14
Bryopsida	516+26	0	10	30	110+12	9
Totals	1122+65	0	62	36	394+65	23

Triangular thalli of the floating liverwort *Ricciocarpos natans* (Marchantiales: Ricciaceae) with the water fern *Azolla filiculoides* and duckweed *Lemna minor*.
Bill Malcolm, Micro-Optics Ltd

Both leafy and thallose forms exist even within the order Metzgeriales. There are two competing theories on whether the ancestor of the liverworts was leafy or thallose. One theory is that it was a leafy liverwort close to *Haplomitrium*, and that this leafy type gave rise to thallose liverworts by loss of the leaves. This theory orginated with Wettstein (1903–1908) and was accepted by Verdoorn (1932), Evans (1939), and Schuster (1981) and Grolle (1983), and is the basis for the classification scheme currently accepted that places *Haplomitrium* at the beginning of the classification and the Marchantiales at the end (Grolle 1983). The other main view is called the antithetic theory, and claims that the ancestral liverwort had a very simple sporophyte. *Riccia* capsules lack a seta and elaters and are embedded in the gametophytic thallus, and *Riccia* was seen by proponents of this theory (e.g. Bower 1908) as closest to the ancestral type.

Two cladistic analyses have given support to this theory that puts the thallose liverworts at the beginning of the classification. The first is an analysis of all the major bryophyte groups using 51 morphological characters that found the root of the liverworts to be in the Marchantiidae (Mischler & Churchill 1985). The second is a cladistic analysis of the order Marchantiales based on 43 morphological characters which found *Riccia* to be basal in the order Marchantiales (Bischler 1998). However, a DNA-sequencing study by Boissier-Dubayle et al. (2002) found a result consistent with multiple cases of morphological specialisation in the Marchantiidae and they concluded that the simple sporophyte structure in *Riccia* resulted from secondary loss.

The liverworts consist of seven orders – Calobryales, Jungermanniales, Treubiales, Metzgeriales, Monocleales, Marchantiales, and Sphaerocarpales. All of these with the exception of the Sphaerocarpales are represented in New Zealand. The higher-level classification most used is that of Grolle (1983). Schuster (1984) provided a discussion of classifications, along with an alternative classification and detailed descriptions of the liverwort morphology.

There are about 6000 species of liverworts and hornworts worldwide (von Konrat 2000) in 360 genera and 72 families (Grolle 1983). Some 4000 of these species are leafy liverworts of the order Jungermanniales. They are most diverse in the temperate zone of both hemispheres, although epiphytic species in the family Lejeunaceae are most richly represented in the wet tropics. Hornworts comprise about 200 species worldwide in two families and six genera (Grolle 1983).

Diversity of New Zealand liverworts and hornworts

A total of 606 species in 155 genera and 49 families is currently accepted as occurring in the New Zealand region (numbers current at April 2006). The New Zealand region is defined to include the Kermadec and subantarctic islands but not Macquarie Island (Australian territory). The largest genera in New Zealand are *Chiloscyphus* (38 species), *Heteroscyphus* (31 species), *Riccardia* (28 species), *Telaranea* (26 species), and *Frullania* (25 species). New Zealand liverworts range in size from *Schistochila appendiculata*, with exceptional plants 1.2 m long

Summary of New Zealand bryophyte diversity by environment

Taxon	Marine	Freshwater	Terrestrial	Fossil (macrofossil)	Fossil (spores)*
Anthocerotopsida	0	1	12	0	-
Marchantiopsida	0	4	590+37	1	-
Bryopsida	0	30+3	484+26	0	-
Totals	**0**	**35+3**	**1,086+63**	**1**	**65**

* Fossil spores attributed to bryophytes (see Raine & Mildenhall, this volume) have not been further categorised with certainty.

(Schuster & Engel 1985) and leaves 9 mm long, to the tiny creeping epiphyte *Diplasiolejeunea pusilla* with leaves only 0.2 mm long.

New Zealand is arguably the most important 'hotspot' of liverwort diversity in the world when considered from the perspectives of species density, degree of endemism, and the presence of a strong archaic element in the flora. New Zealand has more species than the whole of Europe, and has the highest density of species for any country for which a recent checklist is available (table below).

New Zealand's liverwort and hornwort flora is also remarkable for its size in relation to the vascular-plant flora (see following table). The reasons for this are likely to be several – New Zealand's geographical isolation makes it more difficult for the seeds of vascular plants to disperse from abroad to New Zealand than the spores of liverworts; New Zealand's climate is very favourable to liverworts; and New Zealand's vascular-plant flora is poor in small herbaceous species that would compete with liverworts in the forest environment.

Schistochila appendiculata, a 'giant' among liverworts (Jungermanniales: Schistochilaceae).
John Braggins

Historical overview of taxonomic studies

The history of work on the New Zealand liverwort and hornwort flora was reviewed by Fife (1985). Archibald Menzies, on the Vancouver expedition in 1791, collected the first New Zealand liverworts at Dusky Sound and 24 species were described by William Hooker. Joseph Hooker collected on Campbell and Auckland Islands on James Ross's expedition to the subantarctic in the ships *Erebus* and *Terror* in 1841, with the result that 41 New Zealand liverwort species were described from specimens collected on these islands. The most recent complete account of the New Zealand liverworts was Hooker's *Handbook of the New Zealand Flora* (Hooker 1864–1867) which contains 212 species. Berggren and Goebel visited New Zealand in the same century and described species. William Colenso collected for Franz Stephani of Geneva and later described many new species himself.

Amy Hodgson published 19 papers covering most of the liverwort flora over a 31-year period from 1941–1972 (for a listing of these see Hamlin 1972). While she named many new species, her papers are not revisions of the genera she covered. Hamlin published a checklist with full synonymy in 1972 and amendments in 1973 and 1975. K. W. Allison did much collecting during his

Density of species for some geographical areas, ranked by species density

Geographical area	No. of liverwort and hornwort species	Log area $km^2 \times 10^6$	Ratio species/log area
New Zealand	606	5.4	112
Taiwan[1]	484	4.5	108
Borneo[2]	617	5.9	105
China[3]	667	6.2	96
Korea[4]	222	2.3	95
Philippines[5]	505	5.5	92
Japan[6]	530	5.6	92
Australia incl. Tasmania[7]	610	6.9	84
Madagascar[8]	435	5.8	75
North America[9]	c. 500	7.3	69
Europe[10]	453	7	62
Bhutan[11]	284	4.7	61
Tasmania[12]	282	4.8	58
Brazil[13]	352	6.9	51

Sources: 1, Piippo 1992; 2, Menzel 1998; 3, Piippo 1990; 4, Yamada & Choe 1997; 5, Tan & Engel 1986; 6, Mizutani 1984; 7, Scott & Bradshaw 1986; 8, Grolle 1995; 9, Fife 1985; 10, Grolle & Long 2000; 11, Long & Grolle 1990; 12, Jarman & Fuhrer 1995; 13, Yano 1995.

Ratio of liverwort species to vascular-plant species for some geographical areas

Geographical area	No. of liverwort and hornwort species	Indigenous vascular-plant species	Ratio of liverwort and hornwort species to vascular-plant species
New Zealand[1]	606	2,221	27%
Tasmania[2]	282	1,200	24%
Japan[3]	530	3,900	14%
Taiwan[4]	484	3,577	14%
Bhutan[5]	284	5,000	6%
Korea[6]	222	4,500	5%
Madagascar[7]	435	10,800	4%
Australia incl. Tasmania[8]	580	20,000	3%

Sources: 1, Breitwieser et al. this volume; 2, Curtis & Morris 1993; 3, Numata 1974; 4, Li et al. 1979; 5, Anon. 2000a; 6, Anon. 2000b; 7, Richard-Vindard & Battistina 1972; 8, Crisp et al. 1999. Liverwort and hornwort species totals as in the preceding table.

career as a forester between 1927 and 1976 (Macmillan 1978) but published little, although his herbarium collection makes it clear he was aware of a number of unrecognised species. His main published contribution is in the form of a popular guide (Allison & Child 1975). Although compiled by John Child, this book drew heavily on Allison's knowledge. Ella Campbell revised the hornwort flora in a series of papers (Campbell 1981, 1982a,b, 1984, 1993, 1995a,b, 1997). While at Otago University, George Scott coauthored a revision of *Treubia* (Schuster & Scott 1969) and revised *Zoopsis* (Scott 1969). In 1970 he moved to Australia where he continued work on the Australasian flora, his largest published contributions being a guide to the South Australian liverworts and hornworts (Scott 1985) and an Australian checklist (Scott & Bradshaw 1986). Rudy Schuster (University of Massachusetts, Amherst) New Zealand on a Fulbright Scholarship in the early 1960s, subsequently publishing revisions of New Zealand liverwort families and genera, and the New Zealand liverwort flora has figured prominently in his biogeographical writings (Schuster 1979, 1981, 1982). Schuster is presently working on a four-volume account of the austral liverworts (Schuster 2000, 2002). John Engel (Field Museum, Chicago) began publishing on the New Zealand liverwort flora in the 1960s (Engel 1968) and has also continued to publish revisions of New Zealand liverwort families and genera since, including a monograph of *Telaranea* (Engel & Merrill 2004). Currently, John Engel and David Glenny are coauthoring a three-volume New Zealand liverwort flora that will be the first complete flora since Hooker's of 1867; the first volume has been published (Engel & Glenny 2008).

John Braggins at Auckland University has had three students who revised New Zealand liverwort genera – Elizabeth Brown who worked on *Riccardia* (Brown & Braggins 1989), Matt Renner who partly revised *Radula* (Renner 2005), and Matt von Konrat who has revised the genus *Frullania* in a series of papers. David Glenny and Bill Malcolm published an interactive key to the liverwort and hornwort genera (Glenny & Malcolm 2005).

Major collections in New Zealand are held at Lincoln (CHR, c. 20,000 specimens), Wellington (WELT), Dunedin (OTA), Massey University (MPN), and Auckland (AK).

The number of funded people working on the New Zealand flora currently probably amounts to about two 2.5 full-time researchers. Amateur interest has been a feature of New Zealand bryology. Interest in the liverworts and hornworts is not as strong as for the mosses, being held back by the inaccessibility of the taxonomic literature that has been published in overseas journals. Publications are needed that bring together and interpret the revisionary work done on the New Zealand flora in the last 30 years.

Asterella australis with three archegoniophores on the underside of which are four sporophytes, each of which is enclosed in a cage-like sheath (pseudoperianth) (Marchantiales: Aytoniaceae).
Clive Shirley

Endemism and biogeography

New Zealand's liverwort and hornwort flora holds strong biogeographic interest. The two groups constitute a flora dating from the Devonian, 400 million years ago (Ma), older than the flowering-plant flora dating from the mid-Cretaceous (110 Ma), and the possibility exists that some distributions reflect Paleozoic and Mesozoic landmass patterns. The archaic element of New Zealand's bryophyte flora is proving to be important in DNA studies that attempt to trace the origin of land plants.

One family (2% of the New Zealand families) is endemic (Jubulopsidaceae). It is monotypic and the species is uncommon in New Zealand although widespread.

Of the New Zealand genera, 14 (9%) are endemic – *Allisonia, Amphilophocolea, Bragginsella, Cryptostipula, Echinolejeunea, Herzogianthus, Jubulopsis, Kymatolejeunea, Lamellocolea, Lembidium, Megalembidium, Pseudolophocolea, Seppeltia, Stolonivector,* and *Xenothallus*. All of these are monotypic with the exception of *Herzogianthus* and *Lembidium*, which have two species each. Only one endemic genus (*Jubulopsis*) is epiphytic and it is most commonly found on the papery bark of *Metrosideros* and *Leptospermum*. *Echinolejeunea* is epiphyllic, usually on fern fronds. The remaining 11 genera are terrestrial. Ten of the genera belong to the leafy order Jungermanniales and three are thallose genera of the order Metzgeriales.

Metzgeria hamata (Metzgeriales: Metzgeriaceae).
Clive Shirley

Some 284 (48%) indigenous New Zealand species are endemic. This proportion exceeds that for ferns and fern allies (45%) but is much less than that for seed plants (81%) (see next chapter), and higher than the 23% endemism of the mosses (see this chapter). Few estimates of the endemicity of other hepatic floras are available, but 20% of the Australian hepatic flora appears to be endemic (Glenny & Malcolm 2005).

Liverwort spores are probably easily wind-dispersed, but survival of spores in the atmosphere and successful establishment in already-occupied habitats are probably the two limiting factors to dispersal between southern-hemisphere landmasses (Schuster 1979). Van Zanten and Gradstein (1988) tested the resistance of spores of some Neotropical liverworts to ultraviolet radiation, freezing, and desiccation. Only four of 86 species survived the UV test and they concluded that 'long-range dispersal via dry air streams at high altitude is impossible for most hepatics.' However, they did not rule out lower-altitude (< 3000 m) wet-airstream dispersal. The correlation they found between monoecy and trans-oceanic distribution suggests that such dispersal has taken place.

About 90% of New Zealand liverwort and hornwort species are dioecious and so depend on fertilisation between plants to produce spores, using water films for transfer of sperm from male to female sites of reproduction. Liverworts are able to co-exist with other liverwort species, with the result that in New Zealand rainforests, a terrestrial bryophyte mat may contain a dozen species, and constant associations of species can be seen.

There is a strong representation in New Zealand of a southern-hemisphere terrestrial liverwort flora. An example of this distinctive southern flora is the family Schistochilaceae. The family contains the large genus *Schistochila*, the smaller genera *Pachyschistochila* and *Paraschistochila*, and the monotypic genera *Perssoniella* of New Caledonia and *Pleurocladopsis* of southern South America. The family has 30 species in New Zealand, 12 species in Australia, and 21 species in South America (Schuster & Engel 1977), thinning out in Southeast Asia and the Pacific, reaching Taiwan, Samoa, and Fiji, and westward as far as Africa. In the Pacific tropics, New Guinea has the greatest number of species (18) (Piippo 1984). Of the 30 New Zealand species, 21 are endemic, six species are shared with Australia, and one with South America. New Zealand has the largest number of species in *Pachyschistochila*, a largely alpine genus.

A second example of such a southern-hemisphere family is the Lepidolaenaceae. This family is composed of three genera. *Gackstroemia* has five endemic

Treubia lacunosa (Treubiales: Treubiaceae).
Clive Shirley

species in South America, two species in New Zealand, and one in Australia (*G. weindorferi*, shared with New Zealand). *Lepidogyna* has one species in South America and one in New Zealand, each regionally endemic. *Lepidolaena* has five species in New Zealand, six in Australia, and one on Norfolk Island (Grolle 1967). Four of the six species of *Lepidolaena* are shared by Australia and New Zealand and one New Zealand species is endemic. No phylogenetic analysis has been undertaken, but Schuster (1981) believed that the origin of the family was in West Antarctica or the Campbell Plateau. A further pattern is seen in genera such as *Zoopsis* and *Treubia* sensu lato, with distributions centred on the Pacific.

As the four examples above illustrate, the affinities of the New Zealand liverwort and hornwort flora lie mainly with temperate Australia (including Tasmania). Some 247 species (41% of New Zealand's flora) and 130 (83%) genera occur also in Australia. Eight genera are shared solely with Tasmania – *Brevianthus, Dendromastigophora, Eoisotachis, Eotrichocolea, Isolembidium, Isophyllaria, Neogrollea,* and *Trichotemnoma*. All of these genera are monotypic.

Five genera are shared only by New Zealand and South America (*Archeophylla, Austrolophozia, Cephalolobus, Lepidogyna,* and *Phyllothallia*), and in those five genera, no species are shared. Only seven species species are shared exclusively by New Zealand and South America. Some 94 genera are shared non-exclusively by New Zealand and South America, but just six of these are shared exclusively by New Zealand, Australia, and South America (*Gackstroemia, Hepatostolonophora, Leptophyllopsis, Nephelolejeunea, Pachyglossa* and *Triandrophyllum*). Only 23 species are shared by New Zealand and the Falkland Islands (Engel 1990).

Some 72 species (12% of New Zealand's flora) are found also in Polynesia. Two small genera, *Goebeliella* and *Chloranthelia,* are shared solely with New Caledonia. Links to South East Asia at species level are very slight, with only about 5% of New Zealand species occurring in Asia (Piippo 1992), and most of these are widespread. However, at genus level, the links are stronger, as illustrated by *Zoopsis* and *Treubia*.

A few species are circum-subantarctic in distribution e.g. *Austrofossombronia australis* of Australia, New Zealand, Kerguelen, Gough, and Crozet Islands (Schuster 1994). *Neohodgsonia mirabilis* is found only in New Zealand, Gough Island, and Tristan da Cunha (Bischler 1998), a striking disjunction. Seven species are of northern-hemisphere origin – *Anthelia juratzkana, Ptilidium ciliare, Marsupella sparsifolia, Scapania nemorosa, S. undulata, Lophozia excisa,* and *Calypogeia sphagnicola* and all but *Lophozia excisa* have their sole southern-hemisphere representation in New Zealand. They are all mountain species with restricted New Zealand distributions and are probably geologically recent arrivals that have crossed the equator.

Archaism

New Zealand has a large number of families and genera in the two largest orders of the liverworts that are believed to be archaic. New Zealand has three of the 12 known species of *Haplomitrium* (order Calobryales), indicated by some phylogenetic analyses as the most primitive living liverworts, or at least of the superorder Jungermannidae. The diversity of species in *Haplomitrium* in New Zealand is equalled only in the Himalayas, another hotspot of liverwort diversity (Engel 1982). In the Jungermanniales, *Archeophylla, Eotrichocolea, Isophyllaria, Lepicolea, Temnoma, Triandrophyllum,* and *Trichotemnoma* are in what are commonly regarded as the most primitive families of the order. They are mostly small genera of one or two species and are found on infertile soils in cool, wet places such as occur on the heathlands of southern Stewart Island. They are endemic or shared with either Tasmania and/or southern South America. In the Metzgeriales, *Allisonia, Phyllothallia,* and *Verdoornia* are similarly regarded as primitive members of the order. They are taxonomically isolated and endemic or shared with Australia and South America. Also in this order are *Xenothallus* and *Seppeltia* – both endemic, monotypic, and rare.

These primitive elements may have evolved in the Gondwana supercontinent during the Paleozoic and Mesozoic and been continuously present in New Zealand since the Mesozoic. Alternatively, they may have dispersed to New Zealand from West Antarctica during the Tertiary. Hill and Scriven (1995) have described the forests of West Antarctica during the Mesozoic, when West Antarctica was at c. 80° S latitude and experiencing a cool-temperate climate and continuous dark in winter. The forest was sparse and composed of conifers, with cycads and ferns forming an understorey. By the Late Cretaceous, the vegetation in the disturbed rift zone of West Antartica (adjacent to New Zealand) was a sparse mixture of *Nothofagus* and spire-like conifers such as *Libocedrus*. Such forests in Westland now comprise a habitat for a rich terrestrial liverwort flora, with a strong representation of the archaic elements mentioned above. What needs to be considered is how the archaic liverwort element of New Zealand could have survived the 'Oligocene bottleneck' when New Zealand was either very small or completely submerged (Pole 1994; Cooper & Cooper 1995) and the Miocene period when New Zealand was at a higher latitude and experienced a subtropical climate (Fleming 1979).

The most taxonomically isolated element in the flora is *Monoclea forsteri*, which belongs to an order of two species, the other comprising *M. gottscheana* of South America, Central America, and the Caribbean. Whereas it is usual to find taxonomically isolated taxa to be rare within their range, *Monoclea forsteri* is a very common species of stream banks throughout both main islands of New Zealand. The species occurs also in Norfolk Island and doubt exists over whether it has been found once in Tasmania (Arnell 1962; Campbell 1987). Meissner et al. (1998) compared DNA sequences from the *trn*T-*trn*L spacer regions and the *trn*L intron and found 9% and 6% differences respectively between New Zealand and South American *Monoclea* and concluded that this supported differentiation with the genus within the last 80–100 My. Only when sequences for more taxa become available will it be possible to put these differences into a context to determine more precisely the time of separation of the two species. Newer DNA sequences (Wheeler 2000) have indicated that *Monoclea* may lie within the Marchantiales, but this evidence has to be weighed against its morphology, which shows many distinctive features.

The thallose Marchantiales are thought to show a Pangean distribution (Schuster 1981; Wheeler 2000), with the greatest diversity and archaism being in continental areas with a Mediterranean climate (Bischler 1998). The New Zealand marchantialean flora has only one genus that is not widespread – *Neohodgsonia* of New Zealand, Gough, and Tristan da Cunha islands.

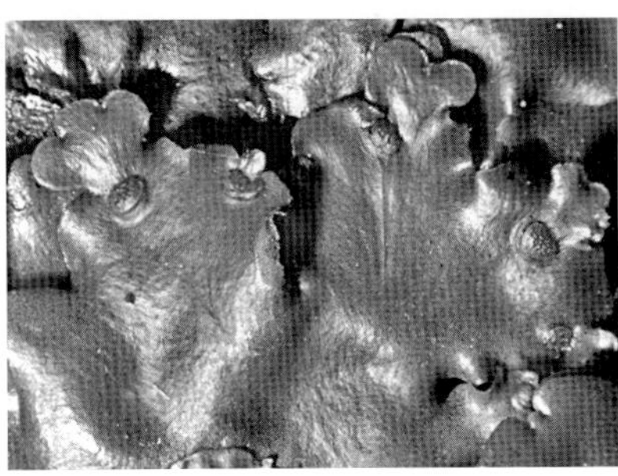

The large thallose liverwort *Monoclea forsteri* (Monocleales: Monocleaceae). At top is a dehiscing sporophyte; the lower photo shows thallus lobes with the button-like male receptacles that contain sperm-bearing antheridia.

Bill Malcolm, Micro-Optics Ltd

Distribution in New Zealand

Most non-alpine liverwort and hornwort species have wide distributions within New Zealand, to about the same degree that is seen in the non-alpine vascular-plant flora. Southern limits of northern New Zealand species appear to be climatically controlled (e.g. *Dendroceros granulatus* with a southern limit in Taranaki). Northern limits may result from the lack of suitable mountain habitat in the North Island (e.g. *Schistochila monticola* of the South Island and subantarctic islands). The Kermadec Islands have a subtropical element not otherwise present in New Zealand, but no endemic species. The subantarctic islands have a large flora for their size (a combined flora of about 175 species), including species that are uncommon further north but only one endemic species (*Riccardia umida*). The Chatham Islands have a relatively small liverwort and hornwort flora (about 45–50 species) with no endemic species. There are two species endemic to the central North Island associated with volcanic soil and rock habitats (*Schistochila pellucida* and *Herzogianthus vaginatus*). The South Island and Stewart Island mountains have many alpine species that are restricted in their distributions. For instance, *Austrolophozia paradoxa* is an alpine species common in fault trenches in Otago and Southland but not known north

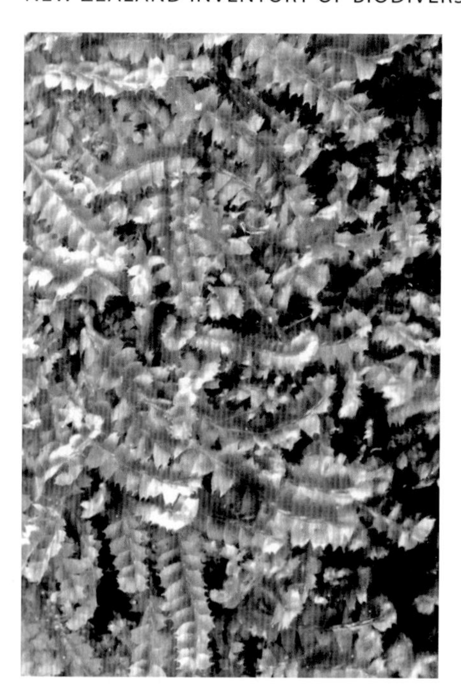

Chiloscyphus coalitus (Jungermanniales: Geocalycaceae).
Clive Shirley

of Arthur's Pass. The alpine liverwort flora differs from the alpine vascular-plant flora in that it has not undergone rapid and recent speciation. Whereas many of the largest vascular-plant genera (e.g. *Aciphylla, Celmisia, Epilobium, Gentianella, Hebe,* and *Ranunculus*) have their greatest diversity in the alpine zone, this is only the case in *Pachyschistochila*, a radiation of eight alpine endemic species. Ranked in order according to numbers of species, the five largest alpine liverwort genera are *Chiloscyphus, Heteroscyphus, Telaranea, Riccardia,* and *Frullania*, but collectively they comprise only a minor alpine component. There is an archaic element on wet peat soils over granite in southern Stewart Island, the subantarctic islands, Fiordland, and the granite ranges of Nelson (e.g. *Eoisotachis, Isophyllaria,* and *Pachyglossa*).

Ecology

Liverworts and hornworts are a significant part of the diversity and biomass in wetter New Zealand forests. The wet forests of New Zealand have the greatest indigenous diversity of liverwort and hornwort species because of the evenness of rainfall punctuated by only short droughts. They support a rich ground flora on tree bases, soil banks, overhangs, and rotten logs, and a strong epiphyllic flora. The epiphytic flora is not as well developed as in the wet tropics.

Ecological studies of New Zealand bryophytes are few, and have most commonly been community analyses (e.g. Tangney 1988). Frey and Beever (1995) analysed the ecology of the dendroid life-form that is common in both New Zealand mosses and liverworts.

Threat status

Liverworts and hornworts are included in a threatened-plant list that has been maintained by the Department of Conservation since 2002 (Hitchmough 2002) and was updated in 2005. There is just one hornwort (*Phaeoceros hirticalyx*) on the list and 140 described liverwort species – 23% of the flora. The list is divided into six categories and the largest of these is the Nationally Critical category. Species in this category are known from very few sites (three or fewer), but are not known to be in decline. Twelve species are classified in the Nationally Endangered or Nationally Vulnerable categories, which imply that a species is considered to have declined, in most cases owing to habitat loss in lowland areas. The size of the Data Deficient category is a reflection of the poor state of knowledge of the true distribution of many liverwort species, and some of the species considered Nationally Critical are likely to move to the Sparse or Range Restricted categories with continuing search effort. Some 35 names on the list are of undescribed taxa or taxa whose identity is uncertain. Most of these are in the Data Deficient category. This data-deficient group also reflects a lack of knowledge. The 2005 unpublished list contains 140 described species or varieties, while the 2002 list contained 90. This increase mainly results from a concerted effort by New Zealand bryologists to identify those species that may be threatened. It is likely that it will take several decades, however, before the threatened-species list stabilises to reflect accurately the true threat status of the liverwort species.

Unnamed element

In all, 123 liverwort and hornwort species, subspecies, and varieties have been described in the 33 years since the appearance of Hamlin's (1972) checklist, and 46 described species have been newly recorded for New Zealand in this time. This rate amounts to about five species per year for the period. It will probably continue over the next 10–15 years and the size of the liverwort and hornwort flora, thought to be about equal to that of the mosses in 1985 (Fife 1985) will be significantly larger than the better-known moss flora (see the account of mosses, below).

Adventive (naturalised-alien) species

New Zealand has only five species previously thought to be introduced –

Lunularia cruciata (Campbell 1966) and *Riccia bifurca*, *R. ciliata*, *R. crystallina*, and *R. glauca* (Campbell 1975, 1977). *Marchantia polymorpha* is probably also introduced. Of these, only *Lunularia* is abundant. This situation is in marked opposition to that for the flowering plants in New Zealand where the established adventive flora is about equal in size to the native flora. Part of the reason for this may be the exceptionally high density of species already present, nevertheless it is puzzling that species of open, alternately wet and dry habitats (e.g. *Riccia*) have not colonised the new pastoral landscapes of New Zealand to a greater degree. New Zealand has 'exported' to Britain the endemic species *Lophozia herzogiana* (Crundwell & Smith 1990) and *Telaranea tetradactyla* (Engel & Merrill 2004). *Chiloscyphus semiteres* (Finch et al. 2000) and *Heteroscyphus fissitipus* (Blackstock & Long 2002), both native to Australia and New Zealand, were first recorded in Europe in 1965 and 2002, respectively.

Gemma cups of the introduced thallose liverwort *Lunularia cruciata* (Marchantiales: Lunulariaceae).

Clive Shirley

Gaps in knowledge and scope for future research on hornworts and liverworts

Revisions have yet to be carried out on about 20% of the genera. Some are in progress (on the large genera *Chiloscyphus* and *Frullania*). The family in which the largest amount of work remains to be done is the Lejeuneaceae.

Over the last 30 years, Schuster and Engel have extensively remodelled the genera and families that are in the New Zealand flora. The largest such genus remodelling was by Engel and Schuster (1985), in which *Lophocolea* was made a subgenus of *Chiloscyphus*, involving about 50 new combinations for Australasian species, a change not accepted by northern-hemisphere authors (e.g. Grolle 1995; Grolle & Long 2000) but supported since by DNA sequence data (He-Nygren & Piippo 2003). In a series of papers, Rudolf Schuster has segregated *Dendromastigophora* from *Mastigophora* (Schuster 1987), *Austrofossombronia* from *Fossombronia* (Schuster 1994), *Nothogymnomitrion* from *Herzogobryum* (Schuster 1996), and *Solenostoma* from *Jungermannia* and *Calyptrocolea* from *Adelanthus* (Schuster 2002). Hässel de Menendez (1996) placed *Clasmatocolea*, a southern-hemisphere temperate genus of about 20 species, into *Chiloscyphus*. Schuster and Grolle have been the two hepaticologists with the greatest knowledge of families and genera worldwide. Schuster has used the intuitive phylogenetic method of the pre-cladistic European school to explain his remodelling of genera and families. The first cladistic analysis of a southern hemisphere family, the Balantiopsidaceae, was published using morphological data (Engel & Merrill 1997). The southern-hemisphere liverwort flora should prove a rich ground for cladistic analyses using DNA, and will no doubt lead to some interesting biogeographic conclusions.

Mosses

The diversity of mosses is often little appreciated by lay people, probably because of the relatively small size of moss plants. Like other cryptogams, mosses are generally too diminutive as individual plants to capture the attention of any but the most observant non-specialist. However, in their aggregate, and in combinations with hepatics, lichens, and other cryptogams, mosses are a significant and diverse component of New Zealand's vegetation, comprising a sizeable fraction of the indigenous flora. One cannot fail to notice mosses hanging from the trunks and branches of forest trees, forming lush and extensive turves on streamside rocks, dominating the ground vegetation of bogs and pakihi, and enveloping decaying logs, stumps, and rocks.

Mosses have an undeserved reputation for being taxonomically difficult and hard to identify. For the most part, the delimitation of moss taxa is no harder than for flowering plants. Families, genera, and indeed the great majority of species, can be recognised in the field. Field identifications are facilitated by the routine use of a hand lens, with those in the x10–14 range being most useful.

The giant moss *Dawsonia superba* (Polytrichales: Polytrichaceae). The image below is an apical view of two stems with splash cups (perigonia).
Allan Fife

The difficulties perceived by non-specialists are largely a matter of scale. While in some instances, critical morphological characters are observable only under the compound microscope or even a scanning electron microscope, features that cannot be detected to some degree under a hand lens or dissecting scope are the exception, rather than the rule.

Their conspicuousness in aquatic and near-aquatic habitats provides a clue as to why mosses are relatively small. As with hepatics and hornworts, the conspicuous green moss plant is the haploid gametophyte, and fertilisation (and consequent production of the spore-producing capsule) is dependent upon the presence of free water for the dispersal of the mitotically produced, motile male gametes. The capsules (sporophytes), resulting from the fusion of the sperm and egg cells, are epiphytic on the female plants and in most species persist for many months. Secondly, because mosses lack vascular tissues, they are dependent upon the availability of free water for metabolic processes and nutrient absorption. These two factors place severe limitations both upon the size of moss plants and upon the habitats they are able to occupy.

At the mouth of the mature capsule in the majority of mosses is a morphologically complex and hygroscopic structure, the peristome, which regulates the release of the meiotically produced haploid spores from the capsule. Details of peristome structure provide much of the basis for the higher-level classification of the mosses. The presence of the peristome, readily observed under a handlens, is the most obvious of several morphological features that distinguish mosses from hepatics and hornworts.

Diversity of New Zealand mosses and comparisons to other regions

The known New Zealand moss flora comprises 516 species in 206 genera and 61 families. A checklist of accepted moss species is appended at the end of this chapter. The species nomenclature derives from the current names in the Allan Herbarium's Plant Names Database (current at April 2006) and reflect the taxonomic opinions of Allan Fife. The assignment of genera to higher taxa follows Vitt (1984), but the order Bryales is not broken down into suborders. Neither taxonomic nor nomenclatural synonyms are included. The known diversity of New Zealand moss taxa compares to a worldwide estimate of 12,760 species (Crosby et al. 2000) distributed among approximately 700 genera (H. Crum pers. comm. 1977) and approximately 90 families (Vitt 1984).

Inventories of species and higher-level taxa provide a simple measure of biodiversity. Given the current level of knowledge of the New Zealand moss flora, the above figures are unlikely to change dramatically in the future. It can, however, be safely predicted that a modest number of indigenous species await discovery and that the number of adventive taxa (currently estimated at 30 species) will continue to grow.

An inherent difficulty in comparing the richness of the New Zealand bryoflora to that of other geographic regions is the paucity of islands of similar size located in temperate latitudes and with comparably well-documented bryofloras. The three main islands of New Zealand span roughly 12 degrees and 30 minutes of latitude (34°30′ to 47° S) and have a combined area of about 265,000 square kilometres. For comparative purposes, two regions that are closest in equivalence are Tasmania and the British Isles.

Tasmania, while lying at a similar latitudinal range as the northern half of the South Island, is a much smaller landmass, with an area about 45% that of the South Island and only about 26% that of the main three New Zealand islands. The moss flora of Tasmania has been studied to a comparable degree to that of New Zealand and an annotated checklist has been published (Dalton et al. 1991). These authors critically reviewed all the published accounts of Tasmanian mosses and verified records, mostly using specimens lodged in the Tasmanian Herbarium (HO); they recognised 361 species (including adventives) in 142 genera.

The British Isles lie at a higher latitude but are roughly comparable to New

Comparison of moss specific and generic diversity for New Zealand and selected geographic regions

Region	No. of species	No. of genera	Area km^2 x 10^3	Log area	No. spp./ log area	No. spp./ 1000 km^2	No. genera/ log area	No. genera/ 1000 km^2
New Zealand	514	206	265	5.42	95.2	1.95	38.2	0.78
Tasmania[1]	361	143	68	4.83	74.7	5.31	29.6	2.10
NSW[2]	527	171	801	5.90	89.3	0.66	20.0	0.21
Chile[3]	778	203	752	5.88	132.3	1.03	34.5	0.27
UK/Ireland[4]	692	175	314	5.50	125.8	2.20	31.8	0.56

1, Dalton et al. 1991; 2, Ramsay 1984; 3, He 1998; 4, Smith 1978.

Zealand in climate. The main islands (including Ireland) span only 9° of latitude (c. 50° to 59°) but have a landmass (about 313,000 square kilometres) about 118% that of New Zealand. The flora of the British Isles is extremely well known and the account by Smith (1978) provides an excellent basis for comparison. Smith (1978) treated 692 species distributed in 115 genera.

A third regional, but non-insular, flora that is instructive to compare with New Zealand's is that of Chile. Chile has an area of c. 752,000 square kilometres, about 2.8 times that of New Zealand. A checklist of Chilean mosses, based primarily on specimen-supported literature reports, was provided by He (1998). Apart from a modest number of new combinations (five), He made no independent taxonomic judgement of reported taxa. He's list includes 778 species distributed in 203 genera and 63 families. He (1998, p. 104) stated that 'Chile holds great promise of new bryological discoveries in South America because of its extensive south temperate regions' and suggested that taxonomic information on Chilean bryophytes is poor relative to other South American countries. However, it is quite unclear whether further exploration and taxonomic study of Chilean mosses will result in a net increase or a decrease in the size of this flora.

Figures of 527 species and 171 genera of mosses in the Australian state of New South Wales were provided by Ramsay (1984), and Catcheside (1980) estimated the South Australia flora to be between 179 and 200 species. While figures for New South Wales are included in the table below, the differences in topography, climate, and area make comparison with the flora of New Zealand of limited value. No attempt has been made to extract data from other Australian regions from sources such as Scott and Stone (1976) and Streimann and Curnow (1989).

If numbers of taxa are compared to raw area figures, there is a strong tendency for regions with smaller areas to appear disproportionately taxon-rich. Dividing taxon numbers by the logarithm of the landmass area (expressed as square kilometres) compensates for this bias. To facilitate comparison of the figures presented for mosses to those of Glenny (this chapter) for hepatics, this method is employed here, giving coefficients of generic and species richness for New Zealand, Tasmania, New South Wales, the British Isles, and Chile.

The number of genera (206) present in New Zealand exceeds the raw figure for all regions selected for comparison; 206 divided by the log of 265,000 (5.42) yields a generic diversity coefficient of 38.2, a figure markedly greater than that for the four regions used for comparison, which range from 34.5 (Chile) to 29.0 (New South Wales). The reasons for this high level of generic diversity are not readily apparent.

At the species level, 514 divided by 5.42 yields a coefficient of 95.2, a figure greater than that calculated for Tasmania (74.7) and New South Wales (89.3), but less than for the British Isles (125.8) or Chile (132.3). Because an estimate of the number of adventive species was not available for all regions, these were not eliminated from the calculations. Although infraspecific taxa were not included in the calculations, their inclusion might go some way to reduce a taxonomic bias towards excessive 'lumping' or 'splitting'.

Cryptopodium bartramioides, a monotypic endemic genus (Bryales: Rhizogoniaceae).
Bill Malcolm, Micro-Optics Ltd

Endemism and phytogeographic elements in the New Zealand moss flora
The degree of endemism, determined at a variety of taxonomic levels, is often used by biogeographers to provide a measure of the relative isolation of biotas. The geographic distributions of mosses are generally broader than those of flowering plants (Crum 1972; Schofield, cited by Schuster 1983, p. 500) and consequently the rate of endemism at both the genus and species levels for a specific geographic region is lower among mosses than among flowering plants.

There are no endemic moss families in New Zealand. Ten genera (*Beeveria, Bryobeckettia, Canalohypopterygium, Cladomnion, Crosbya, Cryptopodium, Dichelodontium, Hypnobartlettia, Mesotus,* and *Tetracoscinodon*) are considered endemic to New Zealand. With the exception of ditypic *Crosbya*, all of these genera are monotypic.

Crum (1991) proposed the segregation of *Camptochaete aciphylla* into a monotypic genus *Fifea*, and this genus was accepted (and considered a New Zealand endemic) by both Fife (1995) and Tangney (1997). However, an unpublished morphological analysis by Tangney (1994, p. 216) shows the only unambiguous character defining *Fifea* is the acuminate leaf apices, a character too weak to justify generic separation.

Hypnobartlettia was proposed by Ochyra (1985), and is accepted here as a monotypic genus in the Amblystegiaceae (it was placed by Ochyra in a segregate family). Its relationships to other genera in the Amblystegiaceae require clarification. Jessica Beever (pers. comm.) believes that *Hypnobartlettia fontana* is merely an extreme environmental modification of *Cratoneuropsis relaxa*, but her interpretation is not accepted here.

At the species level, 110 species (Fife unpubl.) of the 486 species of indigenous mosses are presently considered to be endemic. This figure is approximately 22.6% of the indigenous moss flora and is considerably lower than the 47% cited for hepatic species (Glenny, this chapter) and merely one-fourth of the level of endemism recorded for indigenous angiosperm species (next chapter). The level of species endemism is dramatically lower than that for hepatics; this difference is possibly related to the fact that fewer mosses than hepatics are restricted to hyper-moist habitats.

One of the more distinctive endemic species to be discovered in recent years is *Epipterygium opararense*, described from the Oparara Valley of Nelson (Fife & Shaw 1990). It belongs to a genus of about 10 species with a centre of diversity in or near Central America; no other member of *Epipterygium* is known from the Southern Hemisphere. The species is known from a single population and provides an example of a taxonomically well-documented moss species that is both exceedingly rare and phytogeographically anomalous in the New Zealand flora. Such discoveries suggest that the moss flora of New Zealand, while well known compared to groups such as the hepatics or lichens, is likely to continue to yield modest numbers of surprises. While endemic taxa have intrinsic interest, their phytogeographic significance can be appreciated only when the phylogeny of the taxonomic group to which they belong has been elucidated. The recent description of *Calomnion brownseyi* from the west coast of the South Island as part of a revision and phylogenetic analysis of the genus *Calomnion* (Vitt 1995) demonstrates this point.

The number of species shared with mainland Australia and/or Tasmania (sometimes extending northwards to New Caledonia and/or New Guinea) is estimated to be 127, or about 25% of the total flora. This figure excludes 10 infraspecific taxa (unpublished data) that exhibit an Australasian distribution and is a substantially lower estimate than the 148 species calculated from figures presented by Zanten (1983).

The large number of species exhibiting an austral or circum-subantarctic distribution is a conspicuous feature of the New Zealand moss flora. Such taxa are defined here as occurring in New Zealand, the New Zealand subantarctic islands, and southern South America (including Juan Fernandez) and may also

Epipterygium opararense (Bryales: Bryaceae).
Allan Fife

be present in Australia/Tasmania, South Africa, and the subantarctic islands of the Indian Ocean. Some 99 species (19% of the total flora) are estimated to belong to this element; a substantial number of these 99 species extend northwards in South America along the Andean Cordillera, with many reaching alpine regions of tropical South America (Griffin et al. 1982). Four species of austral distribution occur on subantarctic islands such as Marion Island or Kerguelen but appear to be unreported from South America.

Another feature of the New Zealand moss flora is the large number of taxa with bipolar distributions. For the most part, these are species widespread in temperate or high-latitude regions in the northern hemisphere and present in one or more temperate or cool-temperate southern-hemisphere regions. Some 69 species in the New Zealand flora exhibit this pattern of distribution, with a disproportionately large number belonging to hypnobryalean families. Future studies, especially using molecular techniques, may show that some species included in the above figure are, in fact, introductions. Weedy species such as *Brachythecium salebrosum* and *B. rutabulum* merit closer study. However, the majority of the 69 species interpreted here as bipolar are associated with montane to alpine habitats. They usually occur in relatively intact native vegetation and give every indication of being indigenous species.

Modern biogeographic studies of the New Zealand moss flora

The cladistic biogeography of southern-hemisphere mosses has attracted little attention, although Vitt and coworkers have presented two pioneering cladistic studies of groups strongly represented in New Zealand.

Vitt (1983) and Vitt and Ramsay (1985a,b) published a seminal series of publications dealing with some 23 species of *Macromitrium* of Australasian distribution. Their attempts to relate the phylogeny of a species-rich group of Australasian mosses to regional Tertiary and Quaternary tectonic and climatic events were elegant and could be further refined using molecular-systematic methods. At the conclusion of their studies, Vitt and Ramsay (1985b) presented a series of corollaries by which their hypotheses might be further tested.

Vitt's (1995) revision of the genus *Calomnion*, a genus of nine species distributed in Australasia and the Pacific (including Malesia), used similar methods to attempt to correlate speciation and dispersal events with tectonic and volcanic events in the South Pacific. *Calomnion* includes two New Zealand species (one endemic) and is placed in a monotypic family whose relationships are difficult to ascertain owing largely to the lack of a peristome. Vitt's attempts to trace dispersal used cladograms based on morphological features. We here consider that Vitt's revision provides a model of the type of research needed for a better understanding of the relationships of the New Zealand moss flora and its endemics.

Adventive (naturalised-alien) moss species in New Zealand

The number of unquestionably adventive moss taxa seems surprisingly low, but this figure is likely to increase moderately with further study. Thirty-one species and five genera are currently considered to be entirely adventive in New Zealand, reducing to 483 and 201 the number of indigenous species and genera, respectively. One variety (*Pleuridium subulatum* var. *subulatum*) is also considered to be adventive. The criteria for determining the adventive status of a taxon are always somewhat subjective, involving a number of factors including an assessment of the plant's habitat. Given the minimal use of mosses for horticultural purposes (excluding *Sphagnum*) in New Zealand, all introductions have presumably been accidental rather than deliberate. Fourteen of the 31 documented adventive moss species inhabit soil in disturbed areas, while a smaller number (*Brachythecium albicans*, *B. campestre*, *Calliergonella cuspidata*, *Kindbergia praelonga*, *Pseudoscleropodium purum*, and *Rhytidiadelphus squarrosus*) occur as mats among grasses and herbs in waste areas such as rough lawns and road verges.

Macromitrium ramsayae, with pleated skirt-like calyptrae covering some of the sporophytes (Bryales: Orthotrichaceae).
Jeremy Rolfe

A shoot of the alien moss *Rhytidiadelphus triquetrus* (Bryales: Hylocomiaceae).
Bill Malcolm, Micro-Optics Ltd

Four adventive species are worthy of additional comment. *Sphagnum subnitens* was first reported from New Zealand by Dobson (1975) and it is perhaps the most invasive adventive moss in New Zealand. *Sphagnum subnitens* is documented from the West Coast of the South Island between Westport and about 30 km southwest of Haast and extends from the coast inland for about 50 km. *Sphagnum subnitens* occurs in pakihi, in pine plantations, and on alluvial terraces disturbed by logging and repeated burning. It is sometimes known as 'weasel moss' and usually occurs in association with *S. cristatum*, the *Sphagnum* species harvested for horticultural use. *Sphagnum subnitens* appears to be rapidly extending its range in the South Island and its abundance frequently increases following the commercial harvesting of *S. cristatum* (R. Buxton, pers. comm.). *Sphagnum subnitens* is widespread in the oceanic regions of the northern hemisphere and isozyme or nucleotide sequence data could be used to investigate the geographic origins of the New Zealand population.

Rhytidiadelphus squarrosus is a northern-hemisphere species, first recorded in New Zealand by Child and Allison (1975), that is now well documented throughout western South Island from Nelson Lakes south to Fiordland, in the Dunedin area, and on Stewart Island. It is restricted mainly to roughly mown grassy areas and poses no particular threat to undisturbed native vegetation. The fact that this widespread, successful, and very distinctive weed has not yet been recorded from the North Island suggests that it is probably a recent introduction that has not yet crossed Cook Strait. A second species of *Rhytidiadelphus*, *R. triquetrus*, was discovered at St Arnaud in Nelson Lakes National Park in 1997 by Jean Espie. Subsequent visits to the collection site by A. Fife have shown it to form nearly pure wefts up to a metre square and to be invading mats of *Acrocladium chlamydophyllum* and *Ptychomnion aciculare*. Although it seems confined to a very small area (apparently less than 100 m in diameter, as of October 1998) at St Arnaud, it gives the impression of being potentially more competitive with indigenous mosses than *R. squarrosus*.

In the northern hemisphere it is an abundant ground-cover species in boreal forests. This fact, plus the high level of tourist activity at St Arnaud strongly suggests that *R. triquetrus* arrived there on camping equipment. Given its ecology in the northern hemisphere, and Fife's very preliminary observations, this species may have the potential to alter the species composition in ground-level vegetation in New Zealand *Nothofagus* forest and its occurrence at St Arnaud should be closely monitored.

Pseudoscleropodium purum is perhaps unique in New Zealand's adventive flora for having been first reported (from Mt Eden, Auckland, in 1930) as a new species, *Brachythecium cymbifolium*, by Dixon and Sainsbury (1933). *Pseudoscleropodium purum* is now widespread on both main islands. It occupies disturbed habitats such as roughly mown road and track verges but, like *R. squarrosus*, shows little ability to invade intact native vegetation. It is of modest economic value, as it is frequently used as a mulch on potplants by florists.

At least two moss species indigenous to New Zealand (*Orthodontium lineare* and *Campylopus introflexus*) are adventive in Europe and actively expanding their ranges there.

Long-distance dispersal in the development of the New Zealand moss flora

Moss taxa widely distributed in the temperate and subantarctic portions of the Southern Hemisphere have been objects of interest by botanists since the publication of the bryological results of the *Erebus* and *Terror* Expedition by Wilson (1845, 1847, 1854, 1859). Disjunctions at various taxonomic levels, involving taxa shared between New Zealand and its outlying islands and southern South America, have been the object of particular interest.

As noted above, 99 (19%) of the 514 New Zealand moss species appear to have an austral distribution, as defined above. The great majority of these occur in temperate South America. Attempts to explain the New Zealand–South American disjunctions can roughly be characterised by two types. The first argument suggests that such disjunctions are the result of Late Tertiary or Pleistocene long-distance dispersal events. It is here considered that, for mosses, the weight of evidence favours the long-distance dispersal explanation for the bulk of New Zealand–South American species-level disjunctions.

The second argument interprets the isolated populations as relicts of pre-Tertiary distributions. The leading proponent of this explanation for New Zealand–South America bryophyte disjunctions, R. Schuster (1979, 1981, 1983), has argued at length that the overall patterns of many southern-hemisphere hepatic distributions are relicts of a Gondwanan flora previously extensively distributed in Antarctica and forced to extinction there by a southward drift of the Antarctic plate. These arguments suggest that taxa have become stranded on rafting portions of Gondwana that have drifted apart since Late Cretaceous–Early Tertiary times. Schuster (1979, 1981, 1983) argued that Antarctic populations were eliminated by Middle to Late Tertiary glaciation, further isolating Australasian and South American populations; he applied these arguments primarily at supraspecific levels. Many of the hepatic genera and families exhibiting austral distributions, according to Schuster (1979, 1981, 1983), possess primitive morphological features and he proposed pre-Tertiary origins for these taxa. In regard to the hepatics, and particularly when applied to disjunctions at the family and generic levels, Schuster's arguments are compelling and are largely endorsed by Glenny (this chapter).

Gigaspermum repens, with large, sessile sporophyte capsules (Bryales: Gigaspermaceae).
David Glenny

The next several paragraphs consider evidence that supports long-distance dispersal as an explanation for species-level disjunctions involving New Zealand mosses. The existence of strong correlations between spore resistance to desiccation and wet freeze-thaw cycles and the breadth of distribution within the temperate regions in the Southern Hemisphere has been convincingly demonstrated by Zanten (1983 and other publications) for many southern-hemisphere moss species. Zanten has shown that the spores of New Zealand endemic moss species have a lower resistance to both desiccation and desiccation/freeze-thaw cycles than do New Zealand species that also occur in Australia and/or South America. He argued that long-distance dispersal is far more likely to occur in species with spores smaller than 25 micrometres with good resistance to desiccation and freezing, in species with an autoicous sexuality, and in species capable of surviving in exposed habitats such as bare soil or tree trunks. He also argued convincingly that long-distance dispersal is much more likely when the source and the target landmasses are in equivalent latitudinal bands, as are New Zealand and southern South America (Patagonia).

The presence of strong and persistent westerly winds in the 'roaring forties' is too well known to require detailed documentation here. A weather balloon released from Christchurch was shown to travel between New Zealand and South America in less than a week (Mason 1971, as quoted by Raven & Raven 1976, p. 73) at altitudes of around 12,000 metres. The same balloon made eight circuits of the globe in 102 days between c. 30° and 60° S latitude. Raven and Raven (1976) have argued in favour of eastwards long-distance dispersal of flowering plants (Onagraceae) on west-to-east air flows and specifically suggested that the predominantly Australasian species *Epilobium hirtigerum* may have become established in South America by such means. Zanten (1983) noted that the Orchidaceae of New Zealand have a substantially lower rate of endemism than other flowering-plant groups and attributed this to their minute seeds being more susceptible to long-distance dispersal by wind.

It is considered here that the richness of the moss floras of several islands of young geological age in the southern Indian and Pacific oceans provides irrefutable evidence that recent long-range dispersal events involving mosses has been a commonplace event in the 'roaring forties' in Late Tertiary or Pleistocene time.

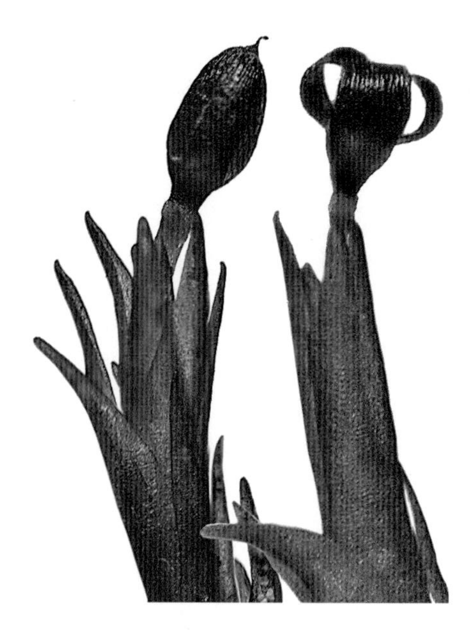

Undehisced and dehisced sporophytes of the granite moss *Andreaea mutabilis* (Andreaeales: Andreaeaceae).
Bill Malcolm, Micro-Optics Ltd

Of the 80 species of mosses recorded by Zanten (1971) from Marion and Prince Edward islands (about 47° S, 38° E), at least 41 species are shared with New Zealand. These islands also have a hepatic flora of some 36 species (Grolle 1971). According to McDougall (1971), Marion and Prince Edward islands are the summits of exposed basaltic volcanoes located on the mid-Indian Ocean Ridge, of which the oldest basalts are only 276,000 ± 30,000 years old.

Macquarie Island (54°30′ S, 159° E) is an island 'totally oceanic in origin' (Selkirk et al. 1990, p. 40). The earliest subaerial deposits are of Middle to Late Pleistocene age (ibid., p. 49) and most probably about 700,000 years old (Selkirk, pers. comm. 1996). Macquarie Island has a moss flora of 77 species (Seppelt 1990). All but three species (*Andreaea gainii*, *Ulota phyllantha*, and *Verrucidens tortifolius*) found there occur on the main islands of New Zealand. Of the 51 hepatic species reported by Seppelt, 45 are named to species level. All but two (*Lepidolaena magellanica* and *Plagiochila ratkowskiana*) are shared with mainland New Zealand.

Geological evidence concerning the period of continuously exposed land surfaces at other subantarctic islands (e.g. Kerguelen, Campbell, and Auckland islands) is less clear. Evidence from Kerguelen (about 47° S, 38° E) presented by Chastain (1958) suggests that a large volcanic eruption at the end of the Eocene (about 40 mya), followed by extensive glaciation, probably produced the present extent and form of that island. Kerguelen has a moss flora of at least 64 species (Brotherus 1906).

The present form of Campbell Island (c. 52°30′ S, 169° E) is a consequence of Pliocene (c. 7 Ma) volcanic activity, producing both marine tuff and terrestrial lava flows, followed by extensive Pleistocene glaciation. The volcanic deposits are underlain by marine limestone of probable Cretaceous age and by 'schist' of Mesozoic age (Oliver et al. 1950, p. 14); the limestone was 'gently folded and probably elevated slightly above sea level' in the mid-Oligocene, but before then it is doubtful if any of the limestone was emergent. Campbell Island has a reported moss flora of 119 species (Vitt 1974). The Auckland Islands (c. 50° 45′ S, 166° E) have a geological history similar to that of Campbell Island. According to Clark & Dingwall (1985), their present form is the result of Oligocene–Miocene volcanic eruptions, with the present islands being the 'eroded remains of cones of two basaltic volcanoes'. The volcanics are atop basement granites of Cretaceous (95 mya) age. The Auckland Islands have a moss flora of about 115 species (Vitt 1979).

Entosthodon laxus (Funariaceae) is a species of essentially austral (circum-subantarctic) distribution. It is an occupier of wet humic soil and most often occurs at the margins of alpine streams or seepages. It is autoicous (having male and female organs on the same plant but on separate branches), frequently fruiting, and has spores of 25–35 micrometres diameter (Fife 1987). Its distribution is well documented; it is known from both main islands of New Zealand, the Auckland Islands, Macquarie Island, southeastern mainland Australia (Australian Alps), Tasmania, Kerguelen, Marion Island, and Crozet Islands, and Chile, Bolivia, Peru, Ecuador, and Venezuela. In tropical South America (Bolivia northward) *E. laxus* occurs only at elevations above 3000 metres, with most records from páramo and puna vegetation at or above 4000 metres. The widespread occurrence of *Entosthodon laxus* on oceanic islands provides a convincing example of a moss dispersing between southern oceanic landmasses via prevailing westerly winds. Its presence in the páramos of the northern Andes, a vegetation unit of Pleistocene age (Van der Hammen 1974), further attests to the mobility of this species.

Frey et al. (1999) compared nucleotide sequences in disjunctive (Australasian–South American) populations of the epiphytic, autoicous, and small-spored moss *Lopidium concinum*. They noted virtually no differences in three non-coding sequences from chloroplast DNA. They interpreted this lack of sequence difference as evidence for no evolutionary change in 60–80 million years of disrupted gene flow, postulating such genetic stability for a single species in an arguably highly derived family (Hypopterygiaceae). *Lopidium* is a genus of moderate size [with six '4-star' species cited by Crosby et al. (2000) and 16 species cited by Brotherus (1925)], widespread in both New and Old World tropics and southern temperate regions. Frey et al.'s (1999) arguments are here considered to lack plausibility; their experimental results could easily be interpreted as evidence in support of gene flow, via long-distance dispersal events, between Australasian and South American populations. Hence, long-distance dispersal events during Late and post-Tertiary periods have played a major, and perhaps a predominant role in the evolution of New Zealand's moss flora. It is probable that recent long-distance dispersal events obscure underlying and older (cf. Schuster 1983, p. 555) relationships between southern-hemisphere moss floras. Research approaches that might serve to differentiate recent and ancient disjunction patterns are suggested below.

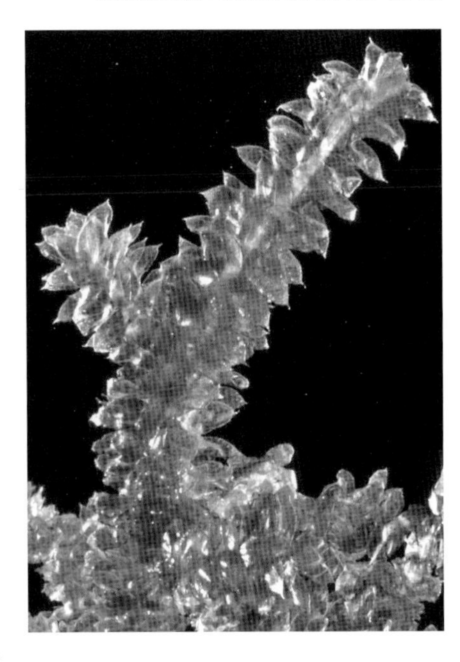

Cyclodictyon blumeanum (Bryales: Hookeriaceae).
Jeremy Rolfe

Future directions for study of the New Zealand moss flora

Despite the extensive application of molecular phylogenetic methods for the study of angiosperms, the application of such methods to southern-hemisphere moss taxa has been exceedingly limited. The application of such methods, together with cladistic biogeography, holds the greatest promise for increasing understanding of the history of New Zealand moss flora and that of other related regions. Such methods also hold promise for differentiating disjunctions resulting from recent dispersal events and those indicative of more ancient separations. The following are some thoughts on groups worthy of such study.

The Ptychomniaceae is a family of seven genera (Vitt 1984) and probably only 15 species (Crosby et al. 2000). The family has maximum generic diversity in New Zealand where six of the seven genera occur. Three of the New Zealand genera (*Cladomnion, Dichelodontium, Tetraphidopsis*) are monotypic and the first two of these are endemic. The family is most species-rich in New Zealand, Patagonia, and Juan Fernandez. It also occurs in Tasmania, mainland Australia, southern Brazil, Marion and Crozet islands, New Caledonia, Malesia, and probably Sri Lanka and Mindanao. A single widespread species extends into Polynesia, reportedly as far east as Rapa (Whittier 1976). A cladistic study of this family using molecular-systematic methods seems a particularly worthwhile project in the context of southern-hemisphere biogeography.

The Rhizogoniaceae is a moderate-sized family of 10 genera (Vitt 1984), six of which occur in New Zealand. Its distribution is predominantly in the Southern Hemisphere. A detailed systematic study of this family on a global scale could significantly increase understanding of regional phytogeographic relations. It is likely that *Pyrrhobryum mnioides*, a species reported from New Zealand, south-eastern Australia and Tasmania, and southern South America and Venezuela (cf. Scott & Stone 1976; Brotherus 1924, p. 428), consists of two subspecies, one South American and the other Australasian, with the main differences being leaf size and degree of leaf contortion in dry material (Fife 1995).

New Zealand is a favourable base for the phylogenetic study of other predominantly austral moss taxa, including the Lembophyllaceae (partially completed by Tangney (1994, 1997 and other publications), and the Seligeriaceae, especially *Blindia* (see Bartlett & Vitt 1986). The Cryphaeaceae and Hypopterygiaceae are relatively genus-rich in New Zealand, while the large and predominantly tropical Hookeriaceae are well represented by six genera (including one endemic) and 15 species. Other moss families/genera with particular potential for shedding light on southern-hemisphere phytogeographic relations include the Dicnemonaceae, Lepyrodontaceae, and Hypnodendraceae. The former two have attracted recent taxonomic attention from Allen (1987a,b, 1989, 1999) and the last was superbly monographed by Touw (1971). The Hypopterygiaceae, while not represented by many species in New Zealand, contains three of the four genera recognised by Vitt (1984) as well as the recently erected *Canalohypopterygium*. Three (*Canalohypopterygium, Catharomnion,*

Cyathophorum) of the five genera occurring in New Zealand are monotypic and appear to represent highly derived taxa within the family. Each of these groups could profitably be investigated using molecular characters.

Macrofossil mosses have not been recorded from New Zealand but there are some 65 groups of fossil spores that have been attributed to bryophytes. These are listed by J. I. Raine and D. C. Mildenhall in their account of New Zealand fossil pollen and spores in this volume.

Overall, then, the general outline and taxonomy of the New Zealand moss flora is well known and a modern and fully illustrated moss Flora is currently in preparation. Published phylogenetic and geographic studies of the moss flora of New Zealand and biogeographically related regions in the Southern Hemisphere using modern methods are few. A greater understanding of this rich flora is to be sought through the application of molecular, cladistic, and monographic studies. There is large and increasing international interest in the relationships of New Zealand mosses. Given the significance of the New Zealand flora in the context of southern hemisphere biogeography, it is hoped that more of these studies can be done within New Zealand, perhaps in cooperation with overseas workers.

Economic importance of bryophytes

In general terms, bryophytes play a significant 'ecosystem-service' role in forest catchments, regarded as 'water-purification plants' for the removal of atmospheric particulates.

Pharmacological studies have been conducted on some New Zealand liverworts to find bactericides and fungicides (R. Tangney pers. comm.).

Note

The manuscript of this chapter was completed in 2006. Subsequent changes in classification resulting mainly from phylogenetic work using DNA sequencing are not evaluated here. One subsequent publication that directly contradicts statements in this chapter concerns the status of *Hypnobartlettia* (Beever, J. E.; Fife, A. J. 2008: *Hypnobartlettia fontana* is an environmental form of *Cratoneuropsis relaxa* (Bryophyta: Amblystegiaceae). *New Zealand Journal of Botany: 46*: 341–345), wherein *Hypnobartlettia* was placed in the synonymy of *Cratoneuropsis*.

Acknowledgements

Ilse Breitwieser is thanked for helpful comments on drafts of this chapter. Aaron Wilton and Sue Gibb help in the production of the moss checklist and reference citations and Christine Bezar made editorial improvements. The preparation of this chapter was supported by the New Zealand Foundation for Research, Science, and Technology.

Authors

Dr Allan J. Fife Landcare Research, P.O. Box 69, Lincoln, New Zealand
[fifea@landcareresearch.co.nz] Mosses

Dr David Glenny Landcare Research, P.O. Box 69, Lincoln, New Zealand
[fifea@landcareresearch.co.nz] Liverworts and hornworts

References

ALLEN, B. H. 1987a: A revision of the Dicnemonaceae (Musci). *Journal of the Hattori Botanical Laboratory 62*: 1–100.

ALLEN, B. H. 1987b: A revision of the genus *Mesotus* (Musci: Dicranaceae). *Journal of Bryology 14*: 441–452.

ALLEN, B. H. 1989: The genus *Dicnemon* (Musci: Dicnemonaceae) in New Caledonia. *Hikobia 10*: 245–267.

ALLEN, B. H. 1999: A revision of the moss genus *Lepyrodon* (Leucondontales, Lepyrodontaceae). *Bryobrothera 5*: 23–48.

ALLISON, K. W.; CHILD, J. 1975: *The Liverworts of New Zealand*. Otago University Press, Dunedin. 300 p.

ANON. 2000a: *The Flora of Bhutan*. http://www.rbge.org.uk/research/bhutan.htm. [Accessed 8 September 2000.]

ANON. 2000b: *The Flora of Korea*. http://vod.shinhyon-e.ed.seoul.kr/WINDOW/window/win00018.htm. [Accessed 8 September 2000.]

ARNELL, S. 1962: Hepaticae collected by Dr. O. H. Selling in Central Australia, Tasmania and New Zealand in 1949. *Botaniska Notiska 115*: 311–317.

BARTLETT, J. K.; VITT, D. H. 1986: A survey of the species in the genus *Blindia* (Bryopsida, Seligeriaceae). *New Zealand Journal of Botany 24*: 203–246.

BISCHLER, H. 1998: Systematics and evolution of the genera of Marchantiales. *Bryophytorum Bibliotheca 51*: 1–201.

BLACKSTOCK, T. H.; LONG, D. G. 2002: *Heteroscyphus fissistipus* (Hook.f. & Taylor) Schiffn. Established in south-west Ireland, new to the Northern Hemisphere. *Journal of Bryology 24*: 147–150.

BOISSELIER-DUBAYLE, M.-C.; LAMBOURDIÈRE, J.; BISCHLER, H. 2002: Molecular phylogenies support multiple morphological reductions in the liverwort subclass Marchantiidae (Bryophyta). *Molecular Phylogenetics and Evolution 24*: 66–77.

BOWER, F. O. 1908: *The Origin of a Land Flora. A theory based upon the facts of alternation*. MacMillan & Co. Ltd, London. 727 p.

BROTHERUS, V. F. 1906: Die Laubmoose. *Deutsche Südpolar-Expedition 8(1)*: 82–96, pls 7, 8.

BROTHERUS, V. F. 1924–1925: Musci. *In*: Engler, A.; Prantl, K. (eds), *Die Natürlichen Pflanzenfamilien*. 2nd edn. Engelmann, Leipzig. Vol. 10, 1924, pp. 143–478; Vol. 11, 1925, 542 p.

BROWN, E. A.; BRAGGINS, J. E. 1989: A revision of the genus *Riccardia* S.F.Gray in New Zealand with notes on the genus *Aneura* Dum. *Journal of the Hattori Botanical Laboratory 66*: 1–132.

CAMPBELL, E. O. 1966: *Lunularia* in New Zealand. *Tuatara 13*: 31–42.

CAMPBELL, E. O. 1975: Notes on the liverwort family Ricciaceae in New Zealand. *Tuatara 21*: 121–129.

CAMPBELL, E. O. 1977: Further notes on the liverwort family Ricciaceae in New Zealand. *Tuatara 22*: 222–232.

CAMPBELL, E. O. 1981: Notes on some Anthocerotae of New Zealand. *Tuatara 25*: 7–13.

CAMPBELL, E. O. 1982a: Notes on some Anthocerotae of New Zealand (2). *Tuatara 25*: 65–70.

CAMPBELL, E. O. 1982b: Notes on some Anthocerotae of New Zealand (3). *Tuatara 26*: 20–26.

CAMPBELL, E.O. 1984: Notes on some Anthocerotae of New Zealand (4). *Tuatara 27*: 105–120.

CAMPBELL, E. O. 1986: Notes on some Anthocerotae of New Zealand (5). *Tuatara 28*: 83–94.

CAMPBELL, E. O. 1987: *Monoclea* (Hepaticae); distribution and number of species. *The Bryologist 90*: 371–373.

CAMPBELL, E. O. 1993: Some name changes in New Zealand Hepaticae and Anthocerotae. *New Zealand Journal of Botany 31*: 341–346.

CAMPBELL, E. O. 1995: Name changes in Australasian *Megaceros* (Anthocerotae). *New Zealand Journal of Botany 33*: 279–283.

CAMPBELL, E. O. 1995: *Phaeoceros delicatus*, a new species of Anthocerotae from New Zealand. *New Zealand Journal of Botany 33*: 285–290.

CATCHESIDE, D. G. 1980: *Mosses of South Australia*. Government Printer, Adelaide. 364 p.

CHASTAIN, A. 1958: La flore et la végétation des Iles de Kerguelen. *Mémoires du Muséum national d'Histoire naturelle, sér. B, 11*: 1–136.

CHILD, J.; ALLISON, K. W. 1975: *Rhytidiadelphus squarrosus* (Hedw.) Warnst.: an addition to the New Zealand moss flora. *New Zealand Journal of Botany 13*: 321.

CLARK, M. R.; DINGWALL, P. R. 1985: *Conservation of Islands in the Southern Ocean: A Review of the Protected Areas of Insulantarctica*. IUCN, Gland & Cambridge. 193 p.

COOPER, A.; COOPER, R. A. 1995: The Oligocene bottleneck and New Zealand biota: genetic record of a past environmental crisis. *Proceedings of the Royal Society of London, ser. B, 261*: 293–302.

CRISP, M. D.; WEST, J. D.; LINDER, H. P. 1999: Biogeography of the terrestrial flora. Pp. 321–367 *in*: Orchard, A. E. (ed.), *Flora of Australia. Volume 1*. 2nd edn. ABRS/CSIRO, Canberra. 669 p.

CROSBY, M. R.; MAGILL, R. E.; ALLEN, B.; HE, S. 2000: *A Checklist of the Mosses*. Missouri Botanical Garden, St Louis. http://www.mobot.org/mobot/tropicos/most/checklists.html. [Accessed 26 July 2000]

CRUM, H. A. 1972: The geographic origins of the mosses of North America's eastern deciduous forest. *Journal of the Hattori Botanical Laboratory 35*: 269–298.

CRUM, H. A. 1991: A partial clarification of the Lembophyllaceae. *Journal of the Hattori Botanical Laboratory 69*: 313–322.

CRUNDWELL, A. C.; SMITH, A. J. E. 1990: *Lophozia herzogiana* Hodgson & Grolle in southern England, a liverwort new to Europe. *Journal of Bryology 15*: 653–658.

CURTIS, W. M.; MORRIS, D. I. 1993: *The Student's Flora of Tasmania. Part 1. Gymnospermae. Angiospermae: Ranunculaceae to Myrtaceae.* St Davids Park Publishing, Hobart. 240 p.

DALTON, P. J.; SEPPELT, R. D.; BUCHANAN, A. M. 1991: An annotated checklist of Tasmanian mosses. Pp. 15–32 *in*: Banks, M. R.; Curtis, W. M. (eds.), *Aspects of Tasmanian Botany – A tribute to Winifred Curtis.* Royal Society of Tasmania, Hobart. 247 p.

DIXON, H. N.; SAINSBURY, G. O. K. 1933: New and rare species of New Zealand mosses. *Journal of Botany 71*: 213–220, 244–251.

DOBSON, A. T. 1975: *Sphagnum subnitens, S. squarrosum*, and *Drepanocladus revolvens* in New Zealand mires. *New Zealand Journal of Botany 13:* 169–171.

DUFF, R. J.; NICKRENT, D. L. 1999: Phylogenetic relationships of land plants using mitochondrial small-subunit rDNA sequences. *American Journal of Botany 86*: 372–386.

EDWARDS, D.; DUCKETT, J.G.; RICHARDSON, J. B. 1995: Hepatic characters in the earliest land plants. *Nature 374*: 635–636.

ENGEL, J. J. 1968: A taxonomic monograph of the genus *Balantiopsis* (Hepaticae). *Nova Hedwigia 16*: 83–130.

ENGEL, J. J. 1982: Hepaticopsida. Pp. 271–305 *in*: Parker, S.P. (ed.), *Synopsis of Living Organisms.* McGraw-Hill, New York. 2 vols, 1232 p.

ENGEL, J.J. 1990: Falkland Islands (Islas Malvinas) Hepaticae and Anthocerotophyta: a taxonomic and phytogeographic study. *Fieldiana, n.s., Botany 25*: 1–209.

ENGEL, J. J.; GLENNY, D. 2008: A liverwort and hornwort flora of New Zealand. Volume 1. *Systematic Monographs in Botany 110*: 1–882.

ENGEL, J. J.; MERRILL, G. L. S. 1997: Austral Hepaticae 22. The genus *Balantiopsis* in New Zealand, with observations on extraterritorial taxa and a phylogeny of *Balantiopsis* and the family Balantiopsaceae (Jungermanniales). *Fieldiana, n.s., Botany 37*: 1–62.

ENGEL, J. J.; MERRILL, G. L. S. 2004: Austral Hepaticae 35. A taxonomic and phylogenetic study of *Telaranea* (Lepidoziaceae), with a monograph of the genus in temperate Australasia and commentary on extra-Australasian taxa. *Fieldiana, n.s., Botany 44*: 1–265.

ENGEL, J. J.; SCHUSTER, R. M. 1984 [1985]: An overview and evaluation of the genera of Geocalycaceae subfamily Lophocoleoideae (Hepaticae). *Nova Hedwigia 39*: 385-445.

EVANS, A. W. 1939: The classification of the Hepaticae. *Botanical Review 5*: 49–96.

FIFE, A. J. 1985: Biosystematics of the cryptogamic flora of New Zealand: bryophytes. *New Zealand Journal of Botany 23*: 645–662.

FIFE, A. J. 1987: Taxonomic and nomenclatural observations on the Funariaceae. 4. A review of *Entosthodon laxus* with incidental notes on *E. obtusifolius*. *Bryologist 89*: 302–390.

FIFE, A. J. 1995: Checklist of the mosses of New Zealand. *Bryologist 98*: 313–337.

FIFE, A. J.; SHAW, A. J. 1990: *Epipterygium* (Musci: Bryaceae) new to Australasia, with the description of *E. opararense*, sp. nov. *New Zealand Journal of Botany 28*: 375–379.

FINCH, R. A.; FISK, R. J.; STRAUSS, D. F.; STEVENSON, C. R. 2000: *Lophocolea semiteres* new to East Anglia. *Journal of Bryology 22*: 146–148.

FLEMING, C. A. 1979: *The Geological History of New Zealand and its Life*. Auckland University Press. 141 p.

FREY, W.; BEEVER, J. E. 1995: Dendroid bryophyte communities of New Zealand. *Nova Hedwigia 61*: 323–354.

FREY, W.; STECH, M.; MEISSNER, K. 1999: Chloroplast DNA-relationship in palaeoaustral *Lopidium concinnum* (Hook.) Wils. (Hypopterygiaceae, Musci). An example of stenoevolution in mosses. Studies in austral temperate rain forest bryophytes 1. *Plant Systematics and Evolution 218*: 67–75.

GLENNY, D.; MALCOLM, W. 2005: *Key to Australasian Liverwort and Hornwort Genera*. ABRS Identification Series, Canberra. CD-ROM.

GRIFFIN, D. III; GRADSTEIN, S. R.; J. AGUIRRE, C. 1982: Studies on Colombian cryptogams XVII. On a new antipodal element in the neotropical páramos – *Dendocryphaea latifolia* sp. nov. (Musci). *Acta Botanica Neerlandica 31*: 175–184.

GROLLE, R. 1967: Monographie der Lepidolaenaceae. *Journal of the Hattori Botanical Laboratory 30*: 1–53.

GROLLE, R. 1971: Hepaticopsida. Pp. 228–236 *in*:

van Zinderen Bakker, E. M.; Winterbottom, J. M.; Dyers, R. A. (eds), *Marion and Prince Edward Islands*. Balkema, Cape Town. 427 p.

GROLLE, R. 1983: Nomina generica Hepaticarum: references, types and synonymies. *Acta Botanica Fennica 121*: 1–62.

GROLLE, R. 1995: The Hepaticae and Anthocerotae of the East African Islands. An annotated catalogue. *Bryophytorum Bibliotheca 48*: 1–178.

GROLLE, R.; LONG, D. G. 2000: An annotated checklist of the Hepaticae and Anthocerotae of Europe and Macaronesia. *Journal of Bryology 22*: 103–140.

HAMLIN, B. G. 1972: Hepaticae of New Zealand, Parts I and II. Index of binomials and preliminary checklist. *Records of the Dominion Museum 7*: 243–366.

HAMLIN, B. G. 1973: Hepaticae of New Zealand, Part III. Additions and corrections to the Index of Binomials. *Records of the Dominion Museum 8*: 139–152.

HAMLIN, B. G. 1975: Hepaticae of New Zealand, Part IV. Further additions and corrections to the Index of Binomials. *Records of the National Museum of New Zealand 1*: 87–89.

HÄSSEL de MENENDEZ, G. G. 1996: Reduction of *Clasmatocolea* Spruce and *Xenocephalozia* Schust. to the synonymy of *Chiloscyphus* Corda. *Nova Hedwigia 63*: 493–516.

HE, S. 1998: A checklist of the mosses of Chile. *Journal of the Hattori Botanical Laboratory 85*: 103–189.

HEDDERSON, T. A.; CHAPMAN, R. L.; ROOTES, W. L. 1996: Phylogenetic relationships of bryophytes inferred from nuclear encoded rRNA gene sequences. *Plant Systematics and Evolution 200*: 213–224.

HE-NYGRÉN, X.; PIIPPO, S. 2003: Phylogenetic relationships of the generic complex *Chiloscyphus-Lophocolea-Heteroscyphus* (Geocalycaceae, Hepaticae): insights from three chloroplast genes and morphology. *Annales Botanica Fennici 40*: 317–329.

HILL, R. S.; SCRIVEN, L. J. 1995: The angiosperm-dominated woody vegetation of Antarctica – a review. *Review of Palaeobotany and Palynology 86*: 175–198.

HITCHMOUGH, R. A. (Ed.) 2002: *New Zealand Threat Classification System Lists*. Biodiversity Recovery Unit, Department of Conservation, Wellington. 210 p.

HOOKER, J. D. 1864–1867: *Handbook of the New Zealand flora*. Reeve, London. 798 p.

JARMAN, S.; FUHRER, B.A. 1995: *Mosses and Liverworts of Rainforest in Tasmania*. CSIRO & Forestry Tasmania, Melbourne. 134 p.

KONRAT, M. von 2000: Liverworts. http://www.sbs.auckland.ac.nz/ biology_web_pages/ nzplants/ liverwort_intro.htm. [Accessed 14 September 2000.]

LEWIS, L. A.; MISCHLER, B. D.; VILGALYS, R. 1997: Phylogenetic relationships of the liverworts (Hepaticae), a basal embryophyte lineage, inferred from nucleotide sequence data of the chloroplast gene rbcL. *Molecular Phylogenetics and Evolution 7*: 377–393.

LI, H.-L.; LIU, T.-S.; HUANG, T.-C.; KOYAMA, T.; DeVOL, C. E. 1979: *Flora of Taiwan. Volume* 6. Epoch Publishing, Taipei. 665 p.

LONG, D. G.; GROLLE, R. 1990: Hepaticae of Bhutan. II. *Journal of the Hattori Botanical Laboratory 68*: 381–440.

MACMILLAN, B. H. 1978: Kenneth Wilway Allison 1894–1976. *New Zealand Journal of Botany 16*: 169–172.

McDOUGALL, I. 1971: Geochronology. Pp. 72–77 *in*: van Zinderen Bakker, E. M.; Winterbottom, J. M.; Dyers, R. A. (eds), *Marion and Prince Edward Islands*. Balkema, Cape Town. 427 p.

MEISSNER, K.; FRAHM, J.-P.; STECH, M.; FREY, W. 1998: Molecular divergence patterns and infrageneric relationship of *Monoclea* (Monocleales, Hepaticae). *Nova Hedwigia 67*: 289–302.

MENZEL, M. 1998: Annotated catalogue of the Hepaticae and Anthocerotae of Borneo. *Journal of the Hattori Botanical Laboratory 65*: 145–206.

MISCHLER, B. D.; CHURCHILL, S. P. 1984: A cladistic approach to the phylogeny of the 'bryophytes'. *Brittonia 36*: 406–424.

MIZUTANI, M. 1984: Check list of Japanese Hepaticae and Anthocerotae, 1983. *Proceedings of the Bryological Society of Japan 3*: 155–163.

NUMATA, M. (Ed.) 1974: *The Flora and Vegetation of Japan*. Kodansha, Tokyo. 294 p.

OCHYRA, R. 1985: *Hypnobartlettia fontana* gen. et sp. nov. (Musci: Hypnobartlettiaceae fam. nov.), a unique moss from New Zealand. *Lindbergia 11*: 2–8.

OLIVER, R. L.; FINLAY, H. J.; FLEMING, C. A. 1950: The geology of Campbell Island. *Cape Expedition Series, Bulletin 3*: 1–62.

PARRISH, J. T.; DANIEL, I. L.; KENNEDY, E. M.; SPICER, R. A. 1988: Paleoclimatic significance of mid-Cretaceous floras from the Middle Clarence Valley, New Zealand. *Palaios 13*: 149–159.

PIIPPO, S. 1984: Bryophyte flora of the Huon Peninsula, Papua New Guinea. III. *Annales Botanica Fennica 21*: 21–48.

PIIPPO, S. 1990: Annotated catalogue of Chinese Hepaticae and Anthocerotae. *Journal of the Hattori Botanical Laboratory 68*: 1–192.

PIIPPO, S. 1992: On the phyto-geographical affinities of temperate and tropical Asiatic and Australasian hepatics. *Journal of the Hattori Botanical Laboratory 71*: 1–35.

POLE, M. 1994: The New Zealand flora — entirely long-distance dispersal? *Journal of Biogeography 21*: 625–635.

POLE, M. S.; RAINE, J. I. 1994: Triassic plant fossils from Pollock Road, Southland, New Zealand. *Alcheringa 18*: 147–159.

RAMSAY, H. P. 1984: Census of New South Wales mosses. *Telopea 2*: 455–533.

RAVEN, P. H.; RAVEN, T. E. 1976: The genus *Epilobium* (Onagraceae) in Australasia: a systematic and evolutionary study. *Bulletin New Zealand Department of Scientific and Industrial Research 216*: 1–321.

RENNER, M. A. M. 2005 Additions to the *Radula* (Radulaceae: Hepaticae) floras of New Zealand and Tasmania. *Journal of the Hattori Botanical Laboratory 97*: 39–79.

RICHARD-VINDARD, G.; BATTISTINA, R. 1972: *Biogeography and Ecology of Madagascar*. Junk, The Hague. 765 p.

SCHOFIELD, W. F. 1985: *Introduction to Bryology*. MacMillan Publishing, New York. 447 p.

SCHUSTER, R. M. 1979: On the persistence and dispersal of transantarctic Hepaticae. *Canadian Journal of Botany 57*: 2179–2225.

SCHUSTER, R. M. 1981: Paleoecology, origin, distribution through time, and evolution of Hepaticae and Anthocerotae. Pp. 129–189 *in*: Niklas, K. J. (ed.), *Paleobotany, Paleoecology, and Evolution*. Praeger, New York. Vol. 2, 279 p.

SCHUSTER, R. M. 1982: Generic and family endemism in the hepatic flora of Gondwanaland: origins and causes. *Journal of the Hattori Botanical Laboratory 52*: 3–35.

SCHUSTER, R. M. 1983: Phytogeography of the Bryophyta. Pp. 463–626 *in*: Schuster, R. M. (ed.), *New Manual of Bryology*. Hattori Botanical Laboratory, Nichinan. Vol. 1, pp. 1–626.

SCHUSTER, R. M. (Ed.) 1984: *New Manual of Bryology. Volume 2*. Hattori Botanical Laboratory, Japan. Pp. 627–1295.

SCHUSTER, R. M. 1987: Phylogenetic studies on Jungermanniidae, II. Mastigophoraceae and Chaetophyllopsidaceae. *Memoirs of the New York Botanical Gardens 45*: 733–748.

SCHUSTER, R. M. 1994: Studies on Metzgeriales. III. The classification of the Fossombroniaceae and on *Austrofossombronia* Schust. gen. n. *Hikobia 11*: 439–449.

SCHUSTER, R. M. 1996: Studies on antipodal Hepaticae XII. Gymnomitriaceae. *Journal of the Hattori Botanical Laboratory 80*: 1–147.

SCHUSTER, R. M. 1997: On *Takakia* and the phylogenetic relationships of the Takakiales. *Nova Hedwigia 64*: 281–310.

SCHUSTER, R. M. 2000: Austral Hepaticae I. *Nova Hedwigia Beiheft 118*: 1–524.

SCHUSTER, R. M. 2002: Austral Hepaticae II. *Nova Hedwigia Beiheft 119*: 1–606.

SCHUSTER, R. M.; ENGEL, J. J. 1977: Austral Hepaticae V. The Schistochilaceae of South America. *Journal of the Hattori Botanical Laboratory 42*: 273-423.

SCHUSTER, R. M.; ENGEL, J. J. 1985: Austral Hepaticae V(2). Temperate and subantarctic Schistochilaceae of Australasia. *Journal of the Hattori Botanical Laboratory 58*: 255–539.

SCHUSTER, R. M.; SCOTT, G. A. M. 1969: A study of the family Treubiaceae (Hepaticae: Metzgeriales). *Journal of the Hattori Botanical Laboratory 32*: 219–268.

SCOTT, G. A. M. 1969: The New Zealand species of *Zoopsis* (Hepaticae). *Records of the Dominion Museum 6*: 159–174.

SCOTT, G. A. M. 1985: Southern Australian Liverworts. *Australian Flora and Fauna Series 2*: 1–216.

SCOTT, G. A. M.; BRADSHAW, J. A. 1986: Australian liverworts (Hepaticae): annotated list of binomials and check-list of published species with bibliography. *Brunonia 8*: 1–171.

SCOTT, G. A. M.; STONE, I. G. 1976: *The Mosses of Southern Australia*. Academic Press, London. 510 p.

SELKIRK, P. M.; SEPPELT, R. D.; SELKIRK, D. R. 1990: *Subantarctic Macquarie Island: Environment and Biology*. Cambridge University Press, Cambridge. xii + 285 p.

SEPPELT, R. D. 1990: Bryophytes recorded from Macquarie Island. Pp. 253–255 *in*: Selkirk, P.M.; Seppelt, R. D.; Selkirk, D. R. (eds), *Subantarctic Macquarie Island: Environment and Biology*. Cambridge University Press, Cambridge. 285 p.

SMITH, A. J. E. 1978: *The Moss Flora of Britain and Ireland*. Cambridge University Press, Cambridge. 706 p.

SMITH, D. K.; DAVISON, P. G. 1993: Antherida and sporophytes in *Takakia ceratophylla* (Mitt.) Grolle: evidence for reclassification among the mosses. *Journal of the Hattori Botanical Laboratory 23*: 263–271.

STEWART, W. N. 1983: *Paleobotany and the Evolution of Plants*. Cambridge University Press, Cambridge. 405 p.

STREIMANN, H.; CURNOW, J. 1989: Catalogue of mosses of Australia and its external territories. *Australian Flora and Fauna Series 10*: viii, 1–479.

TAN, B. C.; ENGEL, J. J. 1986: An annotated checklist of Philippine Hepaticae. *Journal of the Hattori Botanical Laboratory 60*: 283–355.

TANGNEY, R. S. 1988: Ecological studies of a

marine terrace sequence in the Waitutu Ecological District of southern New Zealand. Part 2: The bryophyte communities. *Journal of the Royal Society of New Zealand 18*: 59-78.
TANGNEY, R. S. 1994: A revision of *Camptochaete* Reichardt Lembophyllaceae (Musci). Unpublished PhD thesis, University of Otago, Dunedin. 282 p.
TANGNEY, R. S. 1997: A generic revision of the Lembophyllaceae. *Journal of the Hattori Botanical Laboratory 81*: 123–153.
TOUW, A. 1971: A taxonomic revision of the Hypnodendraceae (Musci). *Blumea 19*: 211–354.
Van der HAMMEN, T. 1974: The Pleistocene changes of vegetation and climate in tropical South America. *Journal of Biogeography 1*: 3–26.
VERDOORN, F. 1932: Classification of Hepatics. Pp. 413–422 in: *Manual of Bryology*. M. Nijhoff, The Hague. viii + 486 p.
VITT, D. H. 1974: A key and synopsis of the mosses of Campbell Island, New Zealand. *New Zealand Journal of Botany 12*: 185–210.
VITT, D. H. 1979: The moss flora of the Auckland Islands, New Zealand, with a consideration of habitats, origins, and adaptations. *Canadian Journal of Botany 57*: 2226–2263.
VITT, D. H. 1983: The New Zealand species of the pantropical genus *Macromitrium* (Orthotrichaceae: Musci): taxonomy, phylogeny and phytogeography. *Journal of the Hattori Botanical Laboratory 54*: 1–94.
VITT, D. H. 1984: Classification of the Bryopsida. Pp. 696–759 *in*: Schuster, R. M. (ed.), *New Manual of Bryology*. Hattori Botanical Laboratory, Nichinan. Vol. 2, pp. 627–1295
VITT, D. H., 1995: The genus *Calomnion* (Bryopsida): taxonomy, phylogeny, and biogeography. *Bryologist 98*: 338–358.
VITT, D. H.; RAMSAY, H. P. 1985a: The *Macromitrium* complex in Australasia (Orthotrichaceae: Bryopsida). Part I. Taxonomy and phylogenetic relationships. *Journal of the Hattori Botanical Laboratory 59*: 325–451.
VITT, D. H.; RAMSAY, H. P. 1985b: The *Macromitrium* complex in Australasia (Orthotrichaceae: Bryopsida). Part II. Distribution, ecology and paleogeography. *Journal of the Hattori Botanical Laboratory 59*: 453–468.
von KONRAT, M. 2000: Review of Rudolph M. Schuster. Austral Hepaticae Part I. Nova Hedwigia Beiheft 118:vii + 1-524, with 211 figures. J. Cramer in der Gebrüder Borntraeger Verlagsbuchbehandlung, Berlin and Stuttgart. *The Bryologist 105*: 505–506.
WETTSTEIN, R. von. 1903–1908: *Handbuch der Systematischen Botanik. II. Band*. F. Deuticke, Leipzig & Wien. 577 p.
WHEELER, J. A. 2000: Molecular phylogenetic reconstructions of the Marchantioid liverwort radiation. *The Bryologist 103*: 314–333.
WHITTIER, H. O. 1976: *Mosses of the Society Islands*. University Presses of Florida, Gainesville. 420 p.
WILSON, W. 1845: Musci. Pp. 117–143 *in*: Hooker, J. D. *The Botany of the Antarctic Voyage of H.M. Discovery Ships* Erebus *and* Terror, *in the Years 1839–1843, under the command of Captain Sir James Clark Ross. Part I. Flora Antarctica. Part I. Botany of Lord Auckland's Group and Campbell's Island*. Reeve, London. 574 p.
WILSON, W. 1847: Musci. Pp. 395–423 *in*: Hooker, J. D. *The Botany of the Antarctic Voyage of H.M. Discovery Ships* Erebus *and* Terror, *in the Years 1839–1843, under the command of Captain Sir James Clark Ross. Part I. Flora Antarctica. Part II. Botany of Fuegia, the Falklands, Kerguelen's Land etc.* Reeve, London. 574 p.
WILSON, W. 1854: Musci. Pp. 57–125 *in*: Hooker, J. D. *The Botany of the Antarctic Voyage of H.M. Discovery Ships* Erebus *and* Terror, *in the Years 1839–1843, under the command of Captain Sir James Clark Ross. Part II. Flora Novae-Zelandiae*. Reeve, London. Vol. 2, 378 p.
WILSON, W. 1859: Musci. Pp. 160–221 *in*: Hooker, J.D. *The Botany of the Antarctic Voyage of H.M. Discovery Ships* Erebus *and* Terror, *in the Years 1839–1843, under the command of Captain Sir James Clark Ross. Part III. Flora Tasmaniae*. Reeve, London. Vol. 2, 422 p.
YAMADA, K.; CHOE, D.-M. 1997: A checklist of Hepaticae and Anthocerotae in the Korean Peninsula. *Journal of the Hattori Botanical Laboratory 81*: 281–306.
YANO, O. 1995: A new additional annotated checklist of Brazilian bryophytes. *Journal of the Hattori Botanical Laboratory 78*: 137–182.
ZANTEN, B. O. van 1971: Musci. Pp. 173–227 *in*: van Zinderen Bakker, E. M.; Winterbottom, J. M.; Dyers, R. A. (eds), *Marion and Prince Edward Islands*. Balkema, Cape Town. 427 p.
ZANTEN, B. O. van 1983: Possibilities of long-range dispersal in bryophytes with special reference to the southern hemisphere. *Sonderbände des Naturwissenschaftlichen Vereins in Hamburg 7*: 49–64.
ZANTEN, B. O. van; GRADSTEIN, S.R. 1988: Experimental dispersal geography of Neotropical liverworts. *Beiheft zur Nova Hedwigia 90*: 41–94.

Checklist of New Zealand Bryophyta

Endemic species and subspecific taxa are signified by E, naturalised ones by A. Endemic genera are underlined (first entry only). Wholly aquatic species are indicated by F.

KINGDOM PLANTAE
SUBKINGDOM VIRIDIPLANTAE
INFRAKINGDOM EMBRYOPHYTA
PHYLUM BRYOPHYTA
Class ANTHOCEROTOPSIDA
Order ANTHOCEROTALES
ANTHOCEROTACEAE
Anthoceros laminiferus Steph. E
Anthoceros muscoides Colenso E
Dendroceros granulatus Mitt.
Dendroceros validus Steph. E
Megaceros denticulatus (Lehm.) Steph.
Megaceros flagellaris (Mitt.) Steph. F E
Megaceros giganteus (Lehm. & Lindenb.) Steph. E
Megaceros leptohymenius (Hook.f. & Taylor) Steph. E
Megaceros pellucidus (Colenso) E.A.Hodgs.
Phaeoceros carolinianus (Michx.) Prosk.
Phaeoceros coriaceus (Steph.) E.O.Campb. E
Phaeoceros delicatus E.O.Campb & Outred E
Phaeoceros hirticalyx (Steph.) Haseg.

Class MARCHANTIOPSIDA
Order CALOBRYALES
HAPLOMITRIACEAE
Haplomitrium gibbsiae (Steph.) R.M.Schust.
Haplomitrium minutum (E.O.Campb.) J.J.Engel & R.M.Schust. E
Haplomitrium ovalifolium R.M.Schust. E

Order JUNGERMANNIALES
ACROBOLBACEAE
Acrobolbus cinerascens (Lehm. & Lindenb.) Bastow
Acrobolbus concinnus (Mitt.) Steph.
Acrobolbus lophocoleoides (Mitt.) Schiffn. E
Acrobolbus ochrophyllus (Hook.f. & Taylor) R.M.Schust.
Acrobolbus spinifolius R.M.Schust. E
Austrolophozia paradoxa R.M.Schust. E
Goebelobryum unguiculatum (Hook.f. & Taylor) Grolle
Lethocolea pansa (Taylor) G.A.M.Scott & K.Beckmann
Marsupidium epiphytum Colenso E
Marsupidium knightii Mitt.
Marsupidium perpusillum (Colenso) E.A.Hodgs. E
Marsupidium setulosum (Mitt.) Watts
Marsupidium surculosum (Nees) Schiffn
Tylimanthus diversifolius E.A.Hodgs.
Tylimanthus saccatus (Hook.) Mitt. E
Tylimanthus tenellus (Taylor) Mitt.
ADELANTHACEAE
Adelanthus bisetulus (Steph.) Grolle
Adelanthus falcatus (Hook.) Mitt.
Adelanthus gemmiparus (R.M.Schust.) E.A.Hodgs.
Adelanthus occlusus (Hook.f. & Tayloar) Carrington
Wettsteinia schusteriana Grolle E
ANTHELIACEAE
Anthelia juratzkana (Limpr.) Trevis.
BALANTIOPSACEAE
Austroscyphus nitidissimus (R.M.Schust.) R.M.Schust. E
Austroscyphus phoenicorhizus (Grolle) R.M.Schust.
Balantiopsis convexiuscula Berggr.
Balantiopsis diplophylla (Hook.f. & Taylor) Mitt. var. *diplophylla*
— var. *hockenii* (Berggr.) J.J.Engel & G.L.Sm. E
Balantiopsis lingulata R.M.Schust. E
Balantiopsis montana (Colenso) J.J.Engel & G.L.Sm. E
Balantiopsis rosea Berggr. E
Balantiopsis tumida Berggr.
Balantiopsis verrucosa J.J.Engel & G.L.Sm. E
Eoisotachis stephanii (Salm.) R.M.Schust. E
Isotachis intortifolia (Hook.f. & Taylor) Gottsche E
Isotachis lyallii Mitt. E
Isotachis minima Pearson E
Isotachis montana Colenso
Isotachis olivacea R.M.Schust. E
Isotachis plicata J.J.Engel E
Isotachis westlandica (E.A.Hodgs.) R.M.Schust. E
BREVIANTHACEAE
Brevianthus flavus (Grolle) R.M.Schust. & J.J.Engel
CALYPOGEIACEAE
Calypogeia sphagnicola (S.W.Arnell & J.Perss.) Warnst. & Loeske
Mnioloma novae-zelandiae J.J.Engel E

CEPHALOZIACEAE
Cephalozia austrigena (R.M.Schust.) R.M.Schust. E
Cephalozia badia (Gottsche) Steph. E
Cephalozia ciliolata R.M.Schust. E
Cephalozia pachygyna R.M.Schust. E
Metahygrobiella drucei R.M.Schust. E
CEPHALOZIELLACEAE
Allisoniella nigra subsp. *novaezelandiae* R.M.Schust. E
Allisoniella recurva R.M.Schust. E
Allisoniella scottii (R.M.Schust.) R.M.Schust. E
Cephalomitrion aterrimum (Steph.) R.M.Schust. var. *aterrimum*
Cephaloziella aenigmatica R.M.Schust. E
Cephaloziella byssacea (Roth) Warnst. subsp. *byssacea*
Cephaloziella crassigyna (R.M.Schust.) R.M.Schust. E
Cephaloziella densifolia R.M.Schust. var. *densifolia* E
— var. *dubia* R.M.Schust. E
Cephaloziella exigua R.M.Schust.
Cephaloziella exiliflora (Taylor) R.M.Schust.
Cephaloziella grandiretis (R.M.Schust.) R.M.Schust. E
Cephaloziella hispidissima R.M.Schust.
Cephaloziella invisa R.M.Schust. E
Cephaloziella muelleriana R.M.Schust. E
Cephaloziella nothogena R.M.Schust. E
Cephaloziella pellucida R.M.Schust. E
Cephaloziella pseudocrassigyna R.M.Schust. E
Cephaloziella pulcherrima R.M.Schust. subsp. *pulcherrima* E
—subsp. *sphagnicola* R.M.Schust.
Cephaloziella subspinosa R.M.Schust. E
Cephaloziella varians (Gottsche) Steph.
CHAETOPHYLLOPSACEAE
Chaetophyllopsis whiteleggei (Carrington & Pearson) R.M.Schust.
Herzogianthus sanguineus R.M.Schust. E
Herzogianthus vaginatus (Herzog) R.M.Schust. E
FRULLANIACEAE
Frullania allanii E.A.Hodgs.
Frullania anomala E.A.Hodgs. E
Frullania aterrima (Hook.f. & Taylor) Hook.f. & Taylor var. *aterrima*
— var. *lepida* E.A.Hodgs. E
— var. *rostrata* (R.M.Schust.) S.Hatt. E
Frullania chevalieri (R.M.Schust.) R.M.Schust.
Frullania deplanata Mitt.
Frullania engelii S.Hatt.
Frullania falciloba Lehm.
Frullania fugax (Hook.f. & Taylor) Taylor
Frullania incumbens Mitt.
Frullania junghuhniana Gottsche
Frullania media (E.A.Hodgs.) S.Hatt. E
Frullania monocera (Hook.f. & Taylor) Taylor
Frullania nicholsonii E.A.Hodgs. E
Frullania patula Mitt. E
Frullania ptychantha Mont. E
Frullania pycnantha (Hook.f. & Taylor) Taylor E
Frullania reptans Mitt. E
Frullania rostellata Mitt. E
Frullania rostrata (Hook.f. & Taylor) Hook.f. & Taylor
Frullania scandens Mont. E
Frullania setchellii Pearson E
Frullania solanderiana Colenso E
Frullania spinifera Taylor
Frullania squarrosula (Hook.f. & Taylor) Taylor
Frullania subhampeana E.A.Hodgs.
Frullania svihlana S.Hatt. E
GEOCALYCACEAE
Amphilophocolea sciaphila R.M.Schust. E
Bragginsella anomala R.M.Schust. E
Chiloscyphus aculeatus Mitt. E
Chiloscyphus aperticaulis J.J.Engel F E
Chiloscyphus australis Hook.f. & Taylor
Chiloscyphus austrigenus subsp. *okaritanus* (Steph.) J.J.Engel F E
Chiloscyphus bispinosus (Hook.f. & Taylor) J.J.Engel & R.M.Schust.
Chiloscyphus calcareus (Steph.) J.J.Engel & R.M.Schust. E
Chiloscyphus cheesemanii (Steph.) J.J.Engel & R.M.Schust.
Chiloscyphus chlorophyllus (Hook.f. & Taylor) Mitt.
Chiloscyphus dallianus (Steph.) J.J.Engel & R.M.Schust. E
Chiloscyphus echinellus (Lindenb. & Gottsche) Mitt.
Chiloscyphus erectifolius (Steph.) Steph. E
Chiloscyphus erosus J.J.Engel E
Chiloscyphus fissistipulus (Steph.) J.J.Engel & R.M.Schust.
Chiloscyphus fulvus (Steph.) J.J.Engel & R.M.Schust. E
Chiloscyphus hattorii J.J.Engel E
Chiloscyphus helmsianus (Steph.) J.J.Engel & R.M.Schust. E
Chiloscyphus herzogii J.J.Engel & R.M.Schust. E
Chiloscyphus inflexifolius (Steph.) J.J.Engel & R.M.Schust.
Chiloscyphus insularis (Steph.) J.J.Engel & R.M.Schust. E
Chiloscyphus lentus (Hook.f. & Taylor) J.J.Engel & R.M.Schust.
Chiloscyphus leucophyllus (Hook.f. & Taylor) Gottsche, Lindenb. & Nees E
Chiloscyphus minor (Nees) J.J.Engel & R.M.Schust.
Chiloscyphus mittenianus (Colenso) J.J.Engel var. *mittenianus* E
— var. *obtusus* J.J.Engel E
— var. *symmetricus* J.J.Engel E
Chiloscyphus multipennus (Hook.f. & Taylor) J.J.Engel & R.M.Schust. E
Chiloscyphus muricatus (Lehm.) J.J.Engel & R.M.Schust.
Chiloscyphus novae-zeelandiae (Lehm. & Lindenb.) J.J.Engel & R.M.Schust. var. *novae-zeelandiae*
— var. *grandistipulus* (Schiffn.) J.J.Engel E?
— var. *meridionalis* (Steph.) J.J.Engel E?
Chiloscyphus pallidus (Mitt.) J.J.Engel & R.M.Schust.
Chiloscyphus parvispinus (Hook.f. & Taylor) J.J.Engel E
Chiloscyphus perpusillus (Hook.f. & Taylor) J.J.Engel
Chiloscyphus semiteres (Lehm.) Lehm. var. *semiteres*
— var. *canaliculatus* (Gottsche, Lindenb. & Nees) J.J.Engel
Chiloscyphus spiniferus (Hook.f. & Taylor) J.J.Engel & R.M.Schust. E
Chiloscyphus subporosus (Mitt.) J.J.Engel & R.M.Schust. var. *subporosus*
— var. *inflexifolius* (Steph.) J.J.Engel E
Chiloscyphus tuberculatus J.J.Engel E
Chiloscyphus variabilis (Steph.) J.J.Engel & R.M.Schust.
Chiloscyphus villosus (Steph.) J.J.Engel & R.M.Schust.
Clasmatocolea crassiretis (Herzog) Grolle E
Clasmatocolea humilis (Hook.f. & Taylor) Grolle var. *humilis*
Clasmatocolea inflexispina (Hook.f. & Taylor) J.J.Engel E
Clasmatocolea notophylla (Hook.f. & Taylor) Grolle
Clasmatocolea strongylophylla (Hook.f. & Taylor) Grolle
Clasmatocolea vermicularis (Lehm.) Grolle
Geocalyx caledonicus Steph.
Hepatostolonophora paucistipula (Rodway) J.J.Engel
Hepatostolonophora rotata (Hook.f. & Taylor) J.J.Engel var. *rotata*
— var. *perssonii* (R.M.Schust.) J.J.Engel E
Heteroscyphus allodontus (Hook.f. & Taylor) J.J.Engel & R.M.Schust.
Heteroscyphus ammophilus (Colenso) R.M.Schust. E
Heteroscyphus argutus (Reinw., Blume & Nees) Schiffn.
Heteroscyphus biciliatus (Hook.f. & Taylor) J.J.Engel
Heteroscyphus billardierei (Schwaegr.) Schiffn.
Heteroscyphus circumdentatus (W.Martin & E.A.Hodgs.) J.J.Engel & R.M.Schust. E
Heteroscyphus coalitis (Hook.) Schiffn.
Heteroscyphus colensoi (Mitt.) Schiffn.
Heteroscyphus compactus (Colenso) R.M.Schust. E
Heteroscyphus cuneistipulus (Steph.) Schiffn. E
Heteroscyphus cymbaliferus (Hook.f. & Taylor) J.J.Engel & R.M.Schust.
Heteroscyphus decipiens (Gottsche) J.J.Engel & R.M.Schust.
Heteroscyphus erraticus (W.Martin & E.A.Hodgs.) J.J.Engel & R.M.Schust. E
Heteroscyphus furcistipulus (E.A.Hodgs.) J.J.Engel & R.M.Schust. E
Heteroscyphus hastatus (E.A.Hodgs.) J.J.Engel & R.M.Schust. E
Heteroscyphus knightii (Steph.) J.J.Engel & R.M.Schust.
Heteroscyphus lingulatus (Colenso) J.J.Engel & R.M.Schust. E
Heteroscyphus lyallii (Mitt.) R.M.Schust.
Heteroscyphus mononuculus J.J.Engel E
Heteroscyphus multispinus (E.A.Hodgs. & Allison) J.J.Engel & R.M.Schust.
Heteroscyphus normalis (Steph.) R.M.Schust. E
Heteroscyphus physanthus (Hook.f. & Taylor) Schiffn.
Heteroscyphus planiusculus (Hook.f. & Taylor) J.J.Engel
Heteroscyphus polycladus (Hook.f. & Taylor) R.M.Schust. E
Heteroscyphus renistipulus (Steph.) Schiffn. E
Heteroscyphus sinuosus (Hook.) Schiffn.
Heteroscyphus splendidus (E.A.Hodgs.) J.J.Engel & R.M.Schust. E
Heteroscyphus supinus (Hook.f. & Taylor) R.M.Schust.
Heteroscyphus triacanthus (Hook.f. & Taylor) Schiffn.
Lamellocolea granditexta (Steph.) J.J.Engel E
Leptophyllopsis laxus (Mitt.) R.M.Schust.
Leptoscyphus innovatus (E.A.Hodgs.) J.J.Engel
Pachyglossa tenacifolia (Hook.f. & Taylor) Herzog & Grolle F
Pseudolophocolea denticulata R.M.Schust. & J.J.Engel E
Saccogynidium australe (Mitt.) Grolle
Saccogynidium decurvum (Mitt.) Grolle
Stolonivector fiordlandiae (E.A.Hodgs.) J.J.Engel E
Stolonivector waipouensis J.J.Engel E
GOEBELIELLACEAE
Goebeliella cornigera (Mitt.) Steph.
GYMNOMITRIACEAE
Acrolophozia pectinata R.M.Schust. E
Gymnomitrion cuspidatum (Berggr.) R.M.Schust. E
Gymnomitrion strictum (Berggr.) R.M.Schust. var. *strictum* E
— var. *inaequalis* R.M.Schust. E
Herzogobryum atrocapillum (Hook.f. & Taylor) Grolle
Herzogobryum filiforme R.M.Schust. E
Herzogobryum molle Grolle
Herzogobryum teres (Carrington & Pearson) Grolle
Herzogobryum vermiculare (Schiffn.) Grolle
Marsupella sparsifolia subsp. *childii* R.M.Schust. E
Marsupella sprucei (Limpr.) Bernet

Nothogymnomitrion erosum (Carrington & Pearson) R.M.Schust.
HERBERTACEAE
Herbertus oldfieldianus (Steph.) Rodway
Triandrophyllum subtrifidum (Hook.f. & Taylor) Fulford & Hatcher
Triandrophyllum symmetricum J.J.Engel E
JACKIELLACEAE
Jackiella curvata Allison & E.A.Hodgs.
JUBULOPSIDACEAE E
Jubulopsis novae-zelandiae (E.A.Hodgs. & S.W.Arnell) R.M.Schust. E
JUNGERMANNIACEAE
Anastrophyllum novae-zelandiae R.M.Schust. E
Anastrophyllum papillosum J.J.Engel & Braggins E
Anastrophyllum schismoides (Mont.) Steph.
Andrewsianthus confusus (R.M.Schust.) R.M.Schust. E
Andrewsianthus hodgsoniae (R.M.Schust.) R.M.Schust. E
Andrewsianthus perigonialis (Hook.f. & Taylor) R.M.Schust. E
Cephalolobus squarrosus R.M.Schust.
Chandonanthus squarrosus (Hook.) Schiffn.
Cryptochila acinacifolia (Hook.f. & Taylor) Grolle E
Cryptochila grandiflora (Lindenb. & Gottsche) Grolle
Cryptochila nigrescens (Steph.) Grolle E
Cryptochila pseudocclusa (E.A.Hodgs.) R.M.Schust. E
Cryptostipula inundata R.M.Schust. E
Jamesoniella colorata (Lehm.) Schiffn.
Jamesoniella kirkii Steph. E
Jamesoniella monodon (Lehm.) N. Kitag.
Jamesoniella tasmanica (Hook.f. & Taylor) Steph.
Lophozia autoica R.M.Schust. var. *autoica*
Lophozia subalpina (R.M.Schust.) R.M.Schust. E
Lophozia bicrenata (Schmid.) Dumort.
Lophozia druceae Grolle & E.A.Hodgs. E
Lophozia excisa (Dicks.) Dumort.
Lophozia herzogiana E.A.Hodgs. & Grolle E
Lophozia multicuspidata (Hook.f. & Taylor) Grolle E
Lophozia nivicola R.M.Schust. E
Lophozia pumicicola Berggr. E
Solenostoma inundatum (Hook.f. & Taylor) Mitt.
Solenostoma orbiculatum (Colenso) R.M.Schust.
Solenostoma novazelandiae R.M.Schust. E
Solenostoma totopapillosum (E.A.Hodgs.) R.M.Schust. E
LEJEUNEACEAE
Acrolejeunea allisonii Gradst. E
Acrolejeunea mollis (Hook.f. & Taylor) Schiffn. E
Acrolejeunea securifolia (Endl.) Steph. subsp. *securifolia*
Archilejeunea olivacea (Hook.f. & Taylor) Steph. E
Archilejeunea planiuscula (Mitt.) Steph.
Austrolejeunea fragilis (R.M.Schust.) R.M.Schust. E
Austrolejeunea hispida R.M.Schust. E
Austrolejeunea olgae (R.M.Schust.) R.M.Schust.
Cheilolejeunea albovirens (Hook.f. & Taylor) E.A.Hodgs.
Cheilolejeunea campbelliensis (Steph.) R.M.Schust.
Cheilolejeunea comitans (Hook.f. & Taylor) R.M.Schust.
Cheilolejeunea hamlinii Grolle E
Cheilolejeunea implexicaulis (Hook.f. & Taylor) R.M.Schust. E?
Cheilolejeunea intertexta (Lindenb.) Steph.
Cheilolejeunea mimosa (Hook.f. & Taylor) R.M.Schust.
Cheilolejeunea novaezelandiae R.M.Schust. E
Cheilolejeunea tenella (Taylor) J.J.Engel & Tan
Cololejeunea cardiocarpa (Mont.) R.M.Schust.
Cololejeunea cucullifolia (Herzog) E.A.Hodgs. E
Cololejeunea ellipsoidea R.M.Schust. E
Cololejeunea falcidentata R.M.Schust. E
Cololejeunea fragilis R.M.Schust. E
Cololejeunea hodgsoniae (Herzog) E.A.Hodgs. E
Cololejeunea inflexifolia R.M.Schust. E
Cololejeunea laevigata (Mitt.) R.M.Schust.
Cololejeunea mamillata (Angstr.) R.M.Schust.
Cololejeunea minutissima (Sm.) Schiffn.
Cololejeunea pulchella (Mitt.) R.M.Schust. var. *pulchella* E
— var. *stylifera* R.M.Schust. E
Colura pulcherrima var. *bartlettii* Ast E
Colura saccophylla E.A.Hodgs. & Herzog E
Diplasiolejeunea plicatiloba (Hook.f. & Taylor) Grolle
Diplasiolejeunea pusilla Grolle
Drepanolejeunea aucklandica Steph.
Drepanolejeunea vesiculosa subsp. *euvesiculosa* Herzog
Echinolejeunea papillata (Mitt.) R.M.Schust. E
Harpalejeunea filicuspis (Steph.) Mizut.
Harpalejeunea latitans (Hook.f. & Taylor) Grolle
Kymatolejeunea bartlettii Grolle E
Lejeunea albiflora Colenso E
Lejeunea anisophylla Mont.
Lejeunea cyanophora R.M.Schust. E
Lejeunea epiphylla Colenso
Lejeunea flava (Sw.) Nees
Lejeunea gracilipes (Taylor) Steph.
Lejeunea helmsiana Steph. E
Lejeunea primordialis (Hook.f. & Taylor) Taylor
Lejeunea tumida Mitt.
Leptolejeunea elliptica subsp. *subacuta* (A.Evans) R.M.Schust.
Lopholejeunea colensoi Steph. E
Lopholejeunea plicatiscypha (Hook.f. & Taylor) Steph.
Mastigolejeunea anguiformis (Hook.f. & Taylor) B.Thiers & Gradst. E
Metalejeunea cucullata (Reinw., Blume & Nees) Grolle
Nephelolejeunea conchophylla Grolle
Nephelolejeunea hamata Grolle
Nephelolejeunea papillosa Glenny E
Ptychanthus stephensonianus (Mitt.) Steph.
Rectolejeunea denudata R.M.Schust. E
Rectolejeunea ocellata Herzog E
Siphonolejeunea nudipes (Hook.f. & Taylor) Herzog
Stenolejeunea acuminata R.M.Schust. E
LEPICOLEACEAE
Lepicolea attenuata (Mitt.) Steph.
Lepicolea scolopendra (Hook.) Trevis.
LEPIDOLAENACEAE
Gackstroemia alpina R.M.Schust. E
Gackstroemia weindorferi (Herzog) Grolle
Lepidogyna hodgsoniae (Grolle) R.M.Schust. E
Lepidolaena berggrenii E.A.Hodgs. E
Lepidolaena clavigera (Hook.) Trevis. E
Lepidolaena palpebrifolia (Hook.f.) Trevis. E
Lepidolaena reticulata (Hook.f. & Taylor) Trevis.
Lepidolaena taylorii (Gottsche) Trevis.
LEPIDOZIACEAE
Acromastigum anisostomum (Lehm. & Lindenb.) A.Evans
Acromastigum brachyphyllum A.Evans E
Acromastigum cavifolium R.M.Schust.
Acromastigum colensoanum (Mitt.) A.Evans
Acromastigum cunninghamii (Steph.) A.Evans
Acromastigum marginatum E.A.Hodgs. E
Acromastigum mooreanum (Steph.) E.A.Hodgs.
Acromastigum verticale (Steph.) E.A.Hodgs.
Bazzania adnexa (Lehm. & Lindenb.) Trevis. var. *adnexa*
— var. *aucklandica* (Lindenb. & Gottsche) J.J.Engel & G.L.Sm. E
Bazzania hochstetteri (Reicht.) E.A.Hodgs. E
Bazzania involuta (Mont.) Trevis. var. *involuta*
— var. *submutica* (Lindenb. & Gottsche) J.J.Engel & G.L.Sm. E
Bazzania monilinervis (Lehm. & Lindenb.) Trevis.
Bazzania nitida (Web.) Grolle
Bazzania nova J.J.Engel & G.L.Sm.
Bazzania novae-zelandiae (Mitt.) Besch. & C.Massal.
Bazzania tayloriana (Mitt.) Kuntze
Chloranthelia berggrenii (Herzog) R.M.Schust. E
Drucella integristipula (Steph.) E.A.Hodgs.
Hygrolembidium acrocladum (Berggr.) R.M.Schust.
Hygrolembidium australe (Steph.) Grolle E
Hygrolembidium rigidum R.M.Schust. & J.J.Engel E
Hygrolembidium triquetrum J.J.Engel & R.M.Schust. E
Isolembidium anomalum var. *cucullatum* (E.A.Hodgs.) J.J.Engel & R.M.Schust.
Kurzia calcarata (Steph.) Grolle E
Kurzia compacta (Steph.) Grolle
Kurzia helophila R.M.Schust. E
Kurzia hippuroides (Hook.f. & Taylor) Grolle var. *hippuroides*
— var. *ornata* J.J.Engel & G.L.Sm. E
Kurzia tenax (Grev.) Grolle
Lembidium longifolium R.M.Schust. E
Lembidium nutans (Hook.f. & Taylor) Mitt. E
Lepidozia acantha J.J.Engel E
Lepidozia bidens J.J.Engel E
Lepidozia bisbifida Steph. E
Lepidozia concinna Colenso E
Lepidozia digitata Herzog E
Lepidozia elobata R.M.Schust. E
Lepidozia fugax J.J.Engel
Lepidozia glaucescens J.J.Engel E
Lepidozia glaucophylla (Hook.f. & Taylor) Gottsche, Lindenb. & Nees
Lepidozia hirta Steph.
Lepidozia kirkii Steph. E
Lepidozia laevifolia (Hook.f. & Taylor) Gottsche, Lindenb. & Nees var. *laevifolia*
— var. *acutiloba* J.J.Engel E
— var. *alpina* J.J.Engel & R.M.Schust. E
Lepidozia microphylla (Hook.) Lindenb. E
Lepidozia novae-zelandiae var. *heterostipa* R.M.Schust. E
— var. *minima* R.M.Schust. E
— var. *novae-zelandiae* E
Lepidozia obtusiloba Steph. var. *obtusiloba*
— var. *parvula* J.J.Engel E
Lepidozia ornata J.J.Engel E
Lepidozia pendulina (Hook.) Lindenb.
Lepidozia procera Mitt.
Lepidozia serrulata J.J.Engel E
Lepidozia setigera Steph. E
Lepidozia spinosissima (Hook.f. & Taylor) Mitt.
Lepidozia ulothrix (Schwaegr.) Lindenb.
Megalembidium insulanum (W.Martin & E.A.Hodgs.) R.M.Schust. E
Neogrollea notabilis E.A.Hodgs.
Paracromastigum drucei (R.M.Schust.) R.M.Schust. E
Paracromastigum fiordlandiae R.M.Schust. & J.J.Engel E
Paracromastigum fissifolium (Steph.) R.M.Schust. E
Paracromastigum furcifolium (Steph.) R.M.Schust.
Paracromastigum kirkii (Steph.) R.M.Schust. E
Paracromastigum macrostipum (Steph.) R.M.Schust. E
Pseudocephalozia lepidozioides R.M.Schust.
Pseudocephalozia paludicola R.M.Schust.
Psiloclada clandestina Mitt.
Psiloclada major (R.M.Schust.) J.J.Engel E
Telaranea elegans (Colenso) J.J.Engel & G.L.Sm.
Telaranea fragilifolia (R.M.Schust.) J.J. Engel & G.L.Sm. E
Telaranea gibbsiana (Steph.) E.A.Hodgs. E

Telaranea granulata J.J.Engel & G.L.Sm. E
Telaranea herzogii (E.A.Hodgs.) E.A.Hodgs. E
Telaranea hodgsoniae J.J.Engel & G.L.Sm.
Telaranea inequalis J.J.Engel & G.L.Sm. E
Telaranea lindbergii (Gottsche) J.J.Engel & G.L.Sm. var. *lindbergii* E
— var. *complanata* J.J.Engel & G.L.Sm. E
— var. *mellea* J.J.Engel & G.L.Sm. E
Telaranea martinii (E.A.Hodgs.) R.M.Schust. E
Telaranea meridiana (E.A.Hodgs.) E.A.Hodgs. E
Telaranea nivicola R.M.Schust. E
Telaranea pallescens (Grolle) J.J.Engel & G.L.Sm.
Telaranea paludicola (E.A.Hodgs.) J.J.Engel & G.L.Sm. E
Telaranea patentissima (Hook.f. & Taylor) E.A.Hodgs. var. *patentissima*
— var. *ampliata* J.J.Engel & G.L.Sm. E
— var. *zebrina* J.J.Engel & G.L.Sm. E
Telaranea pennata J.J.Engel & G.L.Sm. E
Telaranea praenitens (Lehm. & Lindenb.) E.A.Hodgs. var. *praenitens* E
— var. *dentifolia* J.J.Engel & G.L.Sm. E
Telaranea pulcherrima (Steph.) R.M.Schust. subsp. *pulcherrima* E
Telaranea quadriseta (Steph.) J.J.Engel & G.L.Sm.
Telaranea quinquespina (J.J.Engel & G.L.Sm.) J.J.Engel & G.L.Sm.
Telaranea remotifolia (E.A.Hodgs.) E.A.Hodgs.
Telaranea tetradactyla (Hook.f. & Taylor) E.A.Hodgs. E
Telaranea tetrapila (Taylor) J.J.Engel & G.L.Sm. var. *tetrapila* E
— var. *cancellata* (Colenso) J.J.Engel & G.L.Sm. E
Telaranea trilobata (R.M.Schust.) J.J.Engel & G.L.Sm. E
Telaranea tuberifera J.J.Engel & R.M.Schust. E
Zoopsidella caledonica (Steph.) R.M.Schust.
Zoopsis argentea (Hook.f. & Taylor) Hook.f. var. *argentea*
— var. *flagelliforme* (Colenso) R.M.Schust. E
Zoopsis ceratophylla (Spruce) Hamlin E
Zoopsis leitgebiana (Carrington & Pearson) Bastow
Zoopsis macrophylla R.M.Schust. E
Zoopsis nitida Glenny, Braggins & R.M.Schust. E
Zoopsis setulosa Leitg. E
MASTIGOPHORACEAE
Dendromastigophora flagellifera (Hook.f.) R.M.Schust.
PLAGIOCHILACEAE
Acrochila biserialis (Lehm. & Lindenb.) Grolle
Pedinophyllum monoicum (Steph.) Grolle
Plagiochila annotina Lindenb.
Plagiochila baileyana Steph.
Plagiochila banksiana Gottsche var. *banksiana* E
— var. *echinophora* Inoue & R.M.Schust. E
Plagiochila baylisii Inoue & R.M.Schust. E
Plagiochila bazzanioides J.J.Engel & G.L.Sm. E
Plagiochila caducifolia Inoue & R.M.Schust. E
Plagiochila circinalis (Lehm. & Lindenb.) Lehm. & Lindenb.
Plagiochila circumdentata Steph. E
Plagiochila deltoidea Lindenb. E
Plagiochila fasciculata Lindenb.
Plagiochila fragmentissima Inoue & R.M.Schust. E
Plagiochila fruticella (Hook.f. & Taylor) Hook.f. & Taylor E
Plagiochila fuscella (Hook.f. & Taylor) Taylor & Hook.f. E
Plagiochila gigantea (Hook.) Dumort. E
Plagiochila gregaria (Hook.f. & Taylor) Hook.f. & Taylor E
Plagiochila incurvicolla (Hook.f. & Taylor) Hook.f. & Taylor E
Plagiochila intertexta Hook.f. & Taylor
Plagiochila kermadecensis J.J.Engel & G.L.Sm. E
Plagiochila lyallii Mitt. var. *lyallii*
— var. *quinquespina* (Steph.) Inoue & R.M.Schust. E
Plagiochila obscura Colenso E
Plagiochila pacifica Mitt. E
Plagiochila pleurata (Hook.f. & Taylor) Hook.f. & Taylor var. *pleurata*
— var. *arguta* Steph. E
Plagiochila radiculosa Mitt. E
Plagiochila ramosissima (Hook.) Lindenb. E
Plagiochila retrospectans (Nees) Lindenb.
Plagiochila rutlandii Steph. E
Plagiochila stephensoniana Mitt.
Plagiochila strombifolia Taylor
Plagiochilion conjugatus (Hook.) R.M.Schust.
Plagiochilion prolifer (Mitt.) R.M.Schust. E
PORELLACEAE
Porella elegantula (Mont.) E.A.Hodgs.
Porella pulcherrima S.Hatt. E
PTILIDIACEAE
Ptilidium ciliare (L.) Hampe
RADULACEAE
Radula acutiloba Steph.
Radula aneurysmalis (Hook.f. & Taylor) Gottsche, Lindenb. & Nees
Radula australiana K.Yamada
Radula buccinifera (Hook.f. & Taylor) Gottsche, Lindenb. & Nees
Radula compacta Castle
Radula cordiloba Taylor subsp. *erigens* M.Renner & Braggins E
Radula dentifolia Grolle
Radula grandis Steph.
Radula javanica Gottsche, Lindenb. & Nees
Radula marginata Taylor
Radula physoloba Mont.
Radula plicata Mitt.
Radula ratkowskiana K.Yamada
Radula retroflexa Taylor
Radula sainsburiana E.A.Hodgs. & Allison
Radula multiamentula E.A.Hodgs.
Radula tasmanica Steph.
Radula uvifera (Hook.f. & Taylor) Gottsche, Lindenb. & Nees
SCAPANIACEAE
Blepharidophyllum vertebrale (Gottsche, Lindenb. & Nees) C.Massal.
Clandarium xiphophyllum (Grolle) R.M.Schust.
Diplophyllum dioicum R.M.Schust. E
Diplophyllum domesticum (Gottsche) Steph. var. *domesticum*
— var. *icari* J.J.Engel & G.L.Sm. E
Diplophyllum gemmiparum J.J.Engel & G.L.Sm. E
Diplophyllum novum J.J.Engel & G.L.Sm. E
Diplophyllum verrucosum R.M.Schust.
SCHISTOCHILACEAE
Pachyschistochila altissima (E.A.Hodgs.) R.M.Schust. subsp. *altissima* E
Pachyschistochila berggrenii J.J.Engel & R.M.Schust. E
Pachyschistochila childii R.M.Schust. & J.J.Engel E
Pachyschistochila colensoana (Steph.) R.M.Schust. & J.J.Engel E
Pachyschistochila latiloba R.M.Schust. & J.J.Engel E
Pachyschistochila nivicola R.M.Schust. & J.J.Engel E
Pachyschistochila papillifera (R.M.Schust.) R.M.Schust. & J.J.Engel E
Pachyschistochila parvistipula (Rodway) R.M.Schust. & J.J.Engel
Pachyschistochila subhyalina (R.M.Schust.) R.M.Schust. & J.J.Engel var. *subhyalina* E
— var. *grandidentata* J.J.Engel & R.M.Schust. E
Pachyschistochila succulenta J.J.Engel & R.M.Schust.
Pachyschistochila trispiralis (R.M.Schust.) R.M.Schust. & J.J.Engel
Pachyschistochila virescens (R.M.Schust.) R.M.Schust. & J.J.Engel E
Paraschistochila conchophylla (E.A.Hodgs. & Allison) R.M.Schust. E
Paraschistochila pinnatifolia (Hook.) R.M.Schust.
Paraschistochila tuloides (Hook.f. & Taylor) R.M.Schust.
Schistochila appendiculata (Hook.) Trevis. E
Schistochila balfouriana (Hook.f. & Taylor) Steph.
Schistochila chlorophylla (Colenso) J.J.Engel & R.M.Schust. E
Schistochila ciliata (Mitt.) Steph. E
Schistochila glaucescens (Hook.) A.Evans E
Schistochila kirkiana Steph. E
Schistochila lehmanniana (Lindenb.) Carrington & Pearson
Schistochila monticola R.M.Schust. E
Schistochila muricata E.A.Hodgs. & Allison E
Schistochila nitidissima R.M.Schust. E
Schistochila nobilis (Hook.) Trevis. E
Schistochila pellucida R.M.Schust. & J.J.Engel E
Schistochila pluriciliata R.M.Schust. & J.J.Engel E
Schistochila pseudociliata R.M.Schust.
Schistochila repleta (Hook.f. & Taylor) Steph. E
TRICHOCOLEACEAE
Archeophylla schusteri (E.A.Hodgs. & Allison) R.M.Schust. E
Eotrichocolea polyacantha (Hook.f. & Taylor) R.M.Schust.
Isophyllaria attenuata (Rodway) E.A.Hodgs.
Leiomitra julacea J.J.Engel E
Leiomitra lanata (Hook.) R.M.Schust. E
Temnoma angustifolium R.M.Schust. E
Temnoma palmatum (Lindb.) R.M.Schust. var. *palmatum*
— var. *cuneatum* R.M.Schust. E
— var. *laxifolium* R.M.Schust. E
— var. *pseudospiniferum* R.M.Schust. E
Temnoma paucisetigerum R.M.Schust. E
Temnoma pulchellum (Hook.) Mitt. E
Temnoma quadrifidum (Mitt.) E.A.Hodgs. & Allison E
Temnoma quadripartitum (Hook) Mitt. var. *quadripartitum*
— var. *pseudopungens* R.M.Schust. E
— var. *randii* (S.Arnell) R.M.Schust.
Trichocolea hatcheri E.A.Hodgs.
Trichocolea mollissima (Hook.f. & Taylor) Gottsche
Trichocolea rigida R.M.Schust.
TRICHOTEMNOMACEAE
Trichotemnoma corrugatum (Steph.) R.M.Schust.

Order MARCHANTIALES
AYTONIACEAE
Asterella australis (Hook.f. & Taylor) Verd. E
Asterella tenera (Mitt.) R.M.Schust.
Plagiochasma rupestre (J.R. & G.Forst.) Steph.
Reboulia hemisphaerica subsp. *australis* R.M.Schust. E
LUNULARIACEAE A
Lunularia cruciata (L.) Dumort. A
MARCHANTIACEAE
Dumortiera hirsutula (Sw.) Nees
Marchantia berteroana Lehm. & Lindenb.
Marchantia foliacea Mitt.
Marchantia macropora Mitt. E
Marchantia pileata Mitt.
Marchantia polymorpha var. *aquatica* Nees A
Neohodgsonia mirabilis (Perss.) Perss.
RICCIACEAE
Riccia bifurca Hoffm. A
Riccia bullosa Link
Riccia ciliata Hoffm. A
Riccia crozalsii Levier
Riccia crystallina L. A

Riccia fluitans L. F
Riccia glauca L. A
Riccia sorocarpa Bisch.
Ricciocarpos natans (L.) Corda F
TARGIONIACEAE
Targionia hypophylla L.

Order METZGERIALES
ALLISONIACEAE
<u>*Allisonia*</u> *cockaynei* (Steph.) R.M.Schust. E
ANEURACEAE
Aneura alterniloba (Hook.f. & Taylor) Gottsche, Lindenb. & Nees
Aneura lobata subsp. *australis* R.M.Schust. E
Aneura novaeguineensis Hewson
Aneura orbiculata Colenso E
Aneura pinguis (L.) Dumort.
Aneura subaquatica R.M.Schust. E
Riccardia aequicellularis (Steph.) Hewson
Riccardia aequitexta (Steph.) E.A.Brown E
Riccardia alba (Colenso) E.A.Brown E
Riccardia alcicornis (Hook.f. & Taylor) Trevis.
Riccardia asperulata R.M.Schust. E
Riccardia australis (Hook.f. & Leveille) E.A.Brown E
Riccardia bipinnatifida (Colenso) Hewson E
Riccardia breviala E.A. Brown E
Riccardia cochleata (Hook.f. & Taylor) Kuntze
Riccardia colensoi (Steph.) W.Martin
Riccardia crassa (Schwaegr.) C.Massal.
Riccardia eriocaula (Hook.) C. Massal.
Riccardia exilis E.A.Brown
Riccardia furtiva E.A.Brown & Braggins E
Riccardia intercellula E.A.Brown E
Riccardia lobulata (Colenso) E.A.Hodgs.
Riccardia marginata (Colenso) Pearson E
Riccardia multicorpora E.A.Brown E
Riccardia nitida (Colenso) E.A.Hodgs.
Riccardia papulosa (Steph.) E.A.Brown E
Riccardia pennata E.A.Brown E
Riccardia perspicua E.A.Brown E
Riccardia pseudodendroceros R.M.Schust. E
Riccardia pusilla (Steph.) E.A.Brown
Riccardia umida E.A.Brown E
FOSSOMBRONIACEAE
Austrofossombronia australis (Mitt.) R.M.Schust.
Fossombronia pusilla (L.) Nees
Fossombronia reticulata Steph. E
Fossombronia wondraczekii (Corda) Dumort.
Petalophyllum australe Colenso E
Petalophyllum hodgsoniae Crandall-Stotler & C.H.Ford E
HYMENOPHYTACEAE
Hymenophyton flabellatum (Labill.) Trevis.
Hymenophyton leptopodum (Hook.f. & Taylor) A.Evans
METZGERIACEAE
Austrometzgeria saccata (Mitt.) Kuwah.
Metzgeria alpina R.M.Schust. & J.J.Engel E
Metzgeria bartlettii Kuwah. E
Metzgeria consanguinea Schiffn.
Metzgeria crassipilus (Lindb.) A.Evans
Metzgeria flavovirens Colenso E
Metzgeria furcata (L.) Dumort.
Metzgeria leptoneura Spruce
Metzgeria rigida Lindb.
Metzgeria scobina Mitt.
Metzgeria submarginata M.L.So
PALLAVICINIACEAE
Jensenia connivens (Colenso) Grolle
Pallavicinia innovans Steph. E
Pallavicinia lyellii (Hook.) Gray
Pallavicinia tenuinervis (Hook.f. & Taylor) Trevis. E
Pallavicinia xiphoides (Hook.f. & Taylor) Trevis.
Podomitrium phyllanthus (Hook.) Mitt.
Seppeltia succuba Grolle
Symphyogyna hymenophyllum (Hook.) Mont. & Nees
Symphyogyna subsimplex Mitt.
Symphyogyna tenuinervis (Hook.f. & Taylor) Grolle E
Symphyogyna undulata Colenso E
<u>*Xenothallus*</u> *vulcanicolus* R.M.Schust. E
PHYLLOTHALLIACEAE
Phyllothallia nivicola E.A.Hodgs. E
VERDOORNIACEAE
Verdoornia succulenta R.M.Schust. E

Order MONOCLEALES
MONOCLEACEAE
Monoclea forsteri Hook. E

Order TREUBIALES
TREUBIACEAE
Treubia lacunosa (Colenso) Prosk.
Treubia lacunosoides T.Pfeiffer, W.Frey & M.Stech E
Treubia pygmaea R.M.Schust. E

Class BRYOPSIDA
Subclass SPHAGNIDAE
Order SPHAGNALES
SPHAGNACEAE
Sphagnum australe Mitt.
Sphagnum compactum DC.
Sphagnum cristatum Hampe
Sphagnum falcatulum Besch.
Sphagnum novo-zelandicum Mitt.
Sphagnum perichaetiale Hampe
Sphagnum simplex Fife E
Sphagnum squarrosum Crome
Sphagnum subnitens Russ. & Warnst. A

Subclass ANDREAEIDAE
Order ANDREAEALES
ANDREAEACEAE
Andreaea acutifolia Hook.f. & Wilson subsp. *acutifolia*
— subsp. *acuminata* (Mitt.) Vitt
Andreaea australis F.Muell.
Andreaea mutabilis Hook.f. & Wilson
Andreaea nitida Hook.f. & Wilson
Andreaea subulata Harv.

Subclass BRYIDAE
Order POLYTRICHALES
POLYTRICHACEAE
Atrichum androgynum (Müll.Hal.) A.Jaeger
Dawsonia superba Grev. var. *superba*
Dendroligotrichum dendroides (Hedw.) Broth.
Notoligotrichum australe (Hook.f. & Wilson) G.L.Sm.
Notoligotrichum bellii (Broth.) G.L.Sm. E
Notoligotrichum crispulum (Hook.f. & Wilson) G.L.Sm.
Oligotrichum tenuirostre (Hook.) A.Jaeger E
Pogonatum subulatum (Brid.) Brid.
Polytrichadelphus magellanicus (Hedw.) Mitt.
Polytrichastrum alpinum (Hedw.) G.L.Sm.
Polytrichastrum formosum (Hedw.) G.L.Sm.
Polytrichastrum longisetum (Brid.) G.L.Sm.
Polytrichum commune Hedw.
Polytrichum juniperinum Hedw.

Order TETRAPHIDALES
TETRAPHIDACEAE
Tetrodontium brownianum (Dicks.) Schwägr.
CALOMNIACEAE
Calomnion brownseyi Vitt & H.A.Miller E
Calomnion complanatum (Hook.f. & Wilson) Lindb.

Order BRYALES
AMBLYSTEGIACEAE
Acrocladium chlamydophyllum (Hook.f. & Wilson) Müll.Hal. & Broth.
Amblystegium varium (Hedw.) Lindb.
Amblystegium serpens (Hedw.) Bruch & Schimp.
Calliergidium austro-stramineum (Müll.Hal.) E.B.Bartram
Calliergon richardsonii (Mitt.) Warnst.
Calliergonella cuspidata (Hedw.) Loeske A
Campyliadelphus polygamus (Bruch & Schimp.) Kanda
Campyliadelphus stellatus (Hedw.) Kanda (1976)
Cratoneuron filicinum (Hedw.) Spruce
Cratoneuropsis relaxa (Hook.f. & Wilson) Broth.
Drepanocladus aduncus (Hedw.) Warnst.
Drepanocladus brachiatus (Mitt.) Dixon
<u>*Hypnobartlettia*</u> *fontana* Ochyra E
Leptodictyum riparium (Hedw.) Warnst.
Sanionia uncinata (Hedw.) Warnst.
Scorpidium revolvens (Sw.) Rubers
Straminergon stramineum (Brid.) Hedenäs
Warnstorfia fluitans (Hedw.) Loeske
Warnstorfia sarmentosa (Wahlenb.) Hedenäs
ARCHIDIACEAE
Archidium elatum Dixon & Sainsbury E
AULACOMNIACEAE
Aulacomnium palustre (Hedw.) Schwägr.
BARTRAMIACEAE
Bartramia alaris Dixon & Sainsbury
Bartramia crassinervia Mitt. E
Bartramia mossmaniana Müll.Hal.
Bartramia papillata Hook.f. & Wilson
Bartramia robusta Hook.f. & Wilson E
Breutelia affinis (Hook.) Mitt.
Breutelia elongata (Hook.f. & Wilson) Mitt.
Breutelia pendula (Sm.) Mitt.
Conostomum curvirostre (Mitt.) Mitt.
Conostomum pentastichum (Brid.) Lindb.
Conostomum pusillum Hook.f. & Wilson var. *pusillum*
— var. *otagoensis* Fife E
Philonotis pyriformis (R.Br.bis) Wijk & Marg.
Philonotis scabrifolia (Hook.f. & Wilson) Braithw.
Philonotis tenuis (Taylor) Reichardt
Plagiopus oederiana (Sw.) H.A.Crum & Anders.
BRACHYTHECIACEAE
Brachythecium albicans (Hedw.) Bruch & Schimp. A
Brachythecium allisonii Fife E
Brachythecium campestre (Müll.Hal.) Bruch & Schimp. A
Brachythecium fontanum Fife E
Brachythecium paradoxum (Hook.f. & Wilson) A.Jaeger
Brachythecium plumosum (Hedw.) Bruch & Schimp.
Brachythecium rutabulum (Hedw.) Bruch & Schimp.
Brachythecium salebrosum (F.Weber & D.Mohr) Bruch & Schimp.
Brachythecium subpilosum (Hook.f. & Wilson) A.Jaeger
Brachythecium velutinum (Hedw.) Bruch & Schimp. A
Eriodon cylindritheca (Dixon) Sainsbury E
Eurhynchium asperipes (Mitt.) Dixon
Eurhynchium praelongum (Hedw.) Hook. A
Eurhynchium speciosum (Brid.) Jur. A
Palamocladium sericeum (A.Jaeger) Müll.Hal.
Platyhypnidium austrinum (Hook.f. & Wilson) M.Fleisch.
Pseudoscleropodium purum (Hedw.) M.Fleisch. A
Rhynchostegium laxatum (Mitt.) Paris
Rhynchostegium muriculatum (Hook.f. & Wilson) Reichardt
Rhynchostegium tenuifolium (Hedw.) Reichardt
Scorpiurium cucullatum (Mitt.) Hedenäs
BRYACEAE

Brachymenium preissianum (Hampe) A.Jaeger
Bryum algovicum var. *rutheanum* (Warnst.) Crundw.
Bryum amblyodon Müll.Hal.
Bryum appressifolium Broth.
Bryum argenteum Hedw.
Bryum billardierei Schwägr. var. *billardierei*
— var. *platyloma* Mohamed
Bryum blandum Hook.f. & Wilson subsp. *blandum*
Bryum caespiticium Hedw.
Bryum campylothecium Taylor
Bryum capillare Hedw.
Bryum clavatum (Schimp.) Müll.Hal.
Bryum coronatum Schwägr.
Bryum crassum Hook.f. & Wilson
Bryum creberrimum Taylor
Bryum dichotomum Hedw.
Bryum duriusculum Hook.f. & Wilson
Bryum funkii Schwägr.
Bryum harriottii R.Br.bis E
Bryum laevigatum Hook.f. & Wilson
Bryum mucronatum Mitt.
Bryum pallescens Schwägr.
Bryum perlimbatum Cardot
Bryum pseudotriquetrum (Hedw.) P.Gaertn., E.Meyer & Scherb.
Bryum radiculosum Brid. A
Bryum rubens Mitt. A
Bryum ruderale Crundw. & Nyholm A
Bryum sauteri Bruch & Schimp.
Bryum tenuidens Dixon & Sainsbury E
Bryum tenuisetum Limpr. A
Epipterygium opararense Fife & A.J.Shaw E
Leptobryum pyriforme (Hedw.) Wilson
Leptostomum inclinans R.Br.
Leptostomum macrocarpum (Hedw.) Bach.Pyl.
Orthodontium lineare Schwägr.
Plagiobryum novae-seelandiae Broth. E
Pohlia australis A.J.Shaw & Fife E
Pohlia camptotrachela (Renauld & Cardot) Broth. A
Pohlia cruda (Hedw.) Lindb.
Pohlia elongata Hedw.
Pohlia nutans (Hedw.) Lindb.
Pohlia ochii Vitt
Pohlia tenuifolia (A.Jaeger) Broth.
Pohlia wahlenbergii (F.Weber & D.Mohr) Andr.
Schizymenium bryoides Hook.
BUXBAUMIACEAE
Buxbaumia aphylla Hedw.
Buxbaumia novae-zelandiae Dixon E
CALYMPERACEAE
Syrrhopodon armatus Mitt.
CLIMACIACEAE
Climacium dendroides (Hedw.) F.Weber & D.Mohr
CRYPHAEACEAE
Cryphaea acuminata Hook.f. & Wilson E
Cryphaea chlorophyllosa Müll.Hal. E
Cryphaea parvula Mitt.
Cryphaea tenella (Schwägr) Müll.Hal.
Cyptodon dilatatus (Hook.f. & Wilson) Paris & Schimp. E
Dendrocryphaea tasmanica (Mitt.) Broth.
CYRTOPODIACEAE
Cyrtopus setosus (Hedw.) Hook.f.
DALTONIACEAE
<u>*Crosbya*</u> *nervosa* (Hook.f. Wilson) Vitt E
Crosbya straminea (Beckett) Vitt E
Daltonia splachnoides (Sm.) Hook. & Taylor
DICNEMONACEAE
Dicnemon calycinum (Hook.) Schwägr.
Dicnemon dixonianum B.H.Allen E
Dicnemon semicryptum Müll.Hal. E
<u>*Mesotus*</u> *celatus* Mitt. E
DICRANACEAE
Campylopodium lineare (Hook.f.) Dixon E
Campylopodium medium (Duby) Giese & J.-P.Frahm
Campylopus acuminatus var. *kirkii* (Mitt.) Frahm
Campylopus bicolor (Müll.Hal.) Hook.f. & Wilson var. *bicolor*
Campylopus clavatus (R.Br.) Hook.f. & Wilson
Campylopus introflexus (Hedw.) Brid.
Campylopus purpureocaulis Dusén
Campylopus pyriformis (Schultz) Brid.
Chorisodontium aciphyllum (Hook.f. & Wilson) Broth.
Dicranella cardotii (R.Br.bis) Dixon
Dicranella dietrichiae (Müll.Hal.) A.Jaeger
Dicranella gracillima (Beckett) Paris E
Dicranella heteromalla (Hedw.) Schimp. A
Dicranella jamesonii (Mitt.) Broth.
Dicranella vaginata var. *clathrata* (Hook.f. & Wilson) Sainsbury
Dicranoweisia antarctica (Müll.Hal.) Paris
Dicranoweisia spenceri Dixon & Sainsbury E
Dicranoloma billardierei (Brid.) Paris
Dicranoloma dicarpum (Nees) Paris
Dicranoloma fasciatum (Hedw.) Paris E
Dicranoloma menziesii (Hook.f. & Wilson) Paris
Dicranoloma obesifolium (R.Br.bis) Broth. E
Dicranoloma platycaulon Dixon
Dicranoloma plurisetum Dixon E
Dicranoloma robustum (Hook.f. & Wilson) Paris
Dicranum trichopodum Mitt.
Holodontium strictum (Hook.f. & Wilson) Ochyra
Holomitrium perichaetiale (Hook.) Brid.
Kiaeria pumila (Mitt.) Ochyra
Pseudephemerum nitidum (Hedw.) Reimers E
Sclerodontium pallidum (Hook.) Schwägr.
Trematodon flexipes Mitt.
Trematodon mackayi (R.Br.bis) Broth.
Trematodon suberectus Hook.f.
DITRICHACEAE
Ceratodon purpureus (Hedw.) Brid.
Chrysoblastella chilensis (Mont.) Reimers
Distichium capillaceum (Hedw.) Bruch & Schimp.
Ditrichum brachycarpum Hampe
Ditrichum brevirostre (R.Br.bis) Broth.
Ditrichum brotherusii (R.Br.bis) Seppelt
Ditrichum buchananii (R.Br.bis) Broth. E
Ditrichum cylindricarpum (Müll.Hal.) F.Muell.
Ditrichum difficile (Duby) M.Fleisch.
Ditrichum flexicaule (Schwägr.) Hampe
Ditrichum punctulatum Mitt.
Ditrichum rufo-aureum (Hampe) J.H.Willis
Ditrichum strictum (Hook.f. & Wilson) Hampe
Eccremidium minutum (Mitt.) I.G.Stone & G.A.M.Scott
Eccremidium pulchellum (Hook.f. & Wilson) Müll. Hal.
Pleuridium arnoldii (R.Br.bis) Paris
Pleuridium nervosum (Hook.) Mitt.
Pleuridium subulatum (Hedw.) Rabenh.
Saelania glaucescens (Hedw.) Bom. & Broth.
Trichodon cylindricus (Hedw.) Schimp. A
ECHINODIACEAE
Echinodium hispidum (Hook.f. & Wilson) Reichardt
Echinodium umbrosum (Mitt.) A.Jaeger var. *umbrosum*
— var. *glauco-viride* (Mitt.) S.P.Churchill
ENCALYPTACEAE
Encalypta rhaptocarpa Schwägr.
Encalypta vulgaris Hedw.
ENTODONTACEAE
Entodon truncorum Mitt. E
EPHEMERACEAE A
Ephemerum serratum (Hedw.) Hampe A
Ephemerum sessile (Bruch) Müll.Hal. A
Micromitrium tenerum (Bruch & Schimp.) Crosby A
EPHEMEROPSACEAE
Ephemeropsis trentepohlioides (Renner) Sainsbury
ERPODIACEAE
Erpodium glaucum (Wilson) I.G.Stone
FABRONIACEAE
Fabronia australis Hook.
Ischyrodon lepturus (Taylor) Schelpe
FISSIDENTACEAE
Fissidens adianthoides Hedw.
Fissidens anisophyllus Dixon E
Fissidens asplenioides Hedw.
Fissidens berteroi (Mont.) Müll.Hal.
Fissidens blechnoides J.E.Beever E
Fissidens bryoides Hedw. A
Fissidens curvatus Hornsch. var. *curvatus*
— var. *inclinabilis* (Dixon) Beever
Fissidens dealbatus Hook.f. & Wilson
Fissidens dubius P.Beauv. A
Fissidens exilis Hedw. A
Fissidens hylogenes Dixon E
Fissidens integerrimus Mitt.
Fissidens leptocladus Rodway
Fissidens linearis Brid. var. *linearis*
— var. *angustifolius* (Dixon) I.G.Stone E
Fissidens megalotis Müll.Hal.
Fissidens oblongifolius Hook.f. & Wilson var. *oblongifolius*
— var. *capitatus* (Hook.f. & Wilson) Hook.f. & Wilson E
— var. *hyophilus* (Mitt.) J.E.Beever & I.G.Stone
Fissidens pallidus Hook.f. & Wilson
Fissidens perangustus Broth.
Fissidens rigidulus Hook.f. & Wilson var. *rigidulus*
— var. *pseudostrictus* J.E.Beever E
Fissidens strictus Hook.f. & Wilson
Fissidens taxifolius Hedw. A
Fissidens taylorii Müll.Hal. var. *taylorii*
— var. *epiphytus* (Allison) I.G.Stone & J.E.Beever
— var. *sainsburianus* (Dixon) Allison
Fissidens tenellus Hook.f. & Wilson var. *tenellus*
— var. *australiensis* (A.Jaeger) J.E.Beever & I.G.Stone
Fissidens waiensis J.E.Beever E
FUNARIACEAE
<u>*Bryobeckettia*</u> *bartlettii* (Fife) Fife E
Entosthodon apophysatus (Taylor) Mitt.
Entosthodon laxus (Hook.f. & Wilson) Mitt.
Entosthodon muehlenbergii (Turn.) Fife
Entosthodon productus Mitt.
Entosthodon radians (Hedw.) Müll.Hal.
Entosthodon subnudus (Taylor) Fife var. *subnudus*
— var. *gracilis* (Hook.f. & Wilson) Fife
— var. *subcuspidatus* (Broth.) Fife E
Funaria hygrometrica Hedw.
Goniomitrium acuminatum Hook. & Wilson
Physcomitrella patens subsp. *readeri* (Müll.Hal.) B.C.Tan
Physcomitrium pusillum Hook.f. & Wilson E
Physcomitrium pyriforme (Hedw.) Hampe
GIGASPERMACEAE
Gigaspermum repens (Hook.) Lindb.
GRIMMIACEAE
Coscinodon calyptratus (Drumm.) Kindb.
Grimmia anodon Bruch & Schimp.
Grimmia australis (Dixon & Sainsbury) J.M.Muñoz & Ochyra E
Grimmia incrassicapsulis B.G. Bell
Grimmia laevigata (Brid.) Brid.
Grimmia longirostris Hook.
Grimmia plagiopoda Hedw.
Grimmia pulvinata var. *africana* (Hedw.) Hook.f. & Wilson
Grimmia reflexidens Müll.Hal.
Grimmia trichophylla Grev.
Grimmia wilsonii H.Greven E
Racomitrium crispulum (Hook.f. & Wilson) Hook.f. & Wilson
Racomitrium crumianum Fife E

Racomitrium curiosissimum Bednarek-Ochyra & Ochyra E
Racomitrium lanuginosum (Hedw.) Brid.
Racomitrium pruinosum (Hook.f. & Wilson) Müll. Hal.
Racomitrium ptychophyllum (Mitt.) Hook.f.
Racomitrium striatipilum Cardot
Schistidium apocarpum (Hedw.) Bruch & Schimp.
Schistidium rivulare (Brid.) Podp. var. *rivulare*
— var. *subflexifolium* (Müll.Hal.) Fife E
HEDWIGIACEAE
Hedwigia ciliata (Hedw.) P. Beauv.
Hedwigia integrifolia P. Beauv.
Rhacocarpus purpurascens (Brid.) Paris
HOOKERIACEAE
Achrophyllum dentatum (Hook.f. & Wilson) Vitt & Crosby
Achrophyllum quadrifarium (J. E. Sm.) Vitt & Crosby E
Beeveria distichophylloides (Broth. & Dixon) Fife E
Calyptrochaeta apiculata (Hook.f. & Wilson) Vitt var. *apiculata*
— var. *tasmanica* (Broth.) Fife
Calyptrochaeta brownii (Dixon) J.K.Bartlett
Calyptrochaeta cristata (Hedw.) Desv. E
Calyptrochaeta flexicollis (Mitt.) Vitt
Cyclodictyon blumeanum (Müll.Hal.) Kuntze
Distichophyllum adnatum (Hook.f. & Wilson) Broth.
Distichophyllum crispulum (Hook.f. & Wilson) Mitt.
Distichophyllum kraussei (Lorentz) Mitt.
Distichophyllum microcarpum (Hedw.) Mitt.
Distichophyllum pulchellum (Hampe) Mitt. var. *pulchellum*
— var. *ellipticifolium* Sainsbury E
Distichophyllum rotundifolium (Hook.f. & Wilson) Müll.Hal. & Broth.
Sauloma tenella (Hook.f. & Wilson) Mitt.
HYLOCOMIACEAE
Hylocomium splendens (Hedw.) Bruch & Schimp.
Rhytidiadelphus squarrosus (Hedw.) Warnst. A
Rhytidiadelphus triquetrus (Hedw.) Warnst. A
HYPNACEAE
Ctenidium pubescens (Hook.f. & Wilson) Broth.
Ectropothecium sandwichense (Hook. & Arn.) Mitt.
Fallaciella gracilis (Hook.f. & Wilson) H.A.Crum
Fallaciella robusta Tangney & Fife E
Hypnum chrysogaster Müll.Hal.
Hypnum cupressiforme Hedw. var. *cupressiforme*
— var. *filiforme* Brid.
— var. *lacunosum* Brid.
Orthothecium strictum Lorentz
HYPNODENDRACEAE
Braithwaitea sulcata (Hook.) A.Jaeger
Hypnodendron arcuatum (Hedw.) Mitt. E
Hypnodendron colensoi (Hook.f. & Wilson) Mitt. E
Hypnodendron comatum (Müll.Hal.) Touw E
Hypnodendron comosum (Labill.) Mitt. var. *comosum*
— var. *sieberi* (Müll.Hal.) Touw
Hypnodendron kerrii (Mitt.) Paris E
Hypnodendron marginatum (Hook.f. & Wilson) A.Jaeger E
Hypnodendron menziesii (Hook.) Paris subsp. *menziesii*
Hypnodendron spininervium (Hook.) A.Jaeger subsp. *spininervium* E
HYPOPTERYGIACEAE
Canalohypopterygium tamariscinum (Hedw.) Kruijer E
Catharomnion ciliatum (Hedw.) Hook.f. & Wilson
Cyathophorum bulbosum (Hedw.) Müll.Hal.
Dendrohypopterygium filiculaeforme (Hedw.) Kruijer E
Hypopterygium didictyon Müll.Hal. (1859)
Hypopterygium discolor Mitt. (1867)
Hypopterygium tamarisci (Sw.) Müll.Hal.
Lopidium concinnum (Hook.) Hook.f. & Wilson
LEMBOPHYLLACEAE
Camptochaete aciphylla Dixon & Sainsbury E
Camptochaete angustata (Mitt.) Reichardt E
Camptochaete arbuscula (Sm.) Reichardt var. *arbuscula*
— var. *tumida* Tangney E
Camptochaete deflexa (Wilson) A.Jaeger
Camptochaete pulvinata (Hook.f. & Wilson) A.Jaeger E
Lembophyllum divulsum (Hook.f. & Wilson) Lindb.
Lembophyllum clandestinum (Hook.f. & Wilson) Lindb.
LEPYRODONTACEAE
Lepyrodon australis Broth.
Lepyrodon lagurus (Hook.) Mitt.
LESKEACEAE
Haplohymenium pseudo-triste (Müll.Hal.) Broth.
Lindbergia maritima Lewinsky E
LEUCOBRYACEAE
Leucobryum candidum (P.-Beauv.) Wilson
MEESIACEAE
Meesia uliginosa Hedw.
METEORIACEAE
Papillaria crocea (Hampe) A.Jaeger
Papillaria flavo-limbata (Müll.Hal. & Hampe) A.Jaeger
Papillaria flexicaulis (Wilson) A.Jaeger
Papillaria leuconeura (Müll.Hal.) A.Jaeger
Papillaria nitens (Hook.f. & Wilson) Sainsbury
Weymouthia cochlearifolia (Schwägr.) Dixon
Weymouthia mollis (Hedw.) Broth.
MITTENIACEAE
Mittenia plumula (Mitt.) Lindb.
MNIACEAE
Plagiomnium novae-zealandiae (Colenso) T.J.Kop. E
NECKERACEAE
Leptodon smithii (Hedw.) F.Weber & D.Mohr
Neckera laevigata Hook.f. & Wilson E
Neckera pennata Hedw.
Pendulothecium auriculatum (Hook.f. & Wilson) Enroth & S.He E
Pendulothecium oblongifolium (Hook.f. & Wilson) Enroth & S.He E
Pendulothecium punctatum (Hook.f. & Wilson) Enroth & S.He E
Thamnobryum pandum (Hook.f. & Wilson) I.G.Stone & G.A.M.Scott
Thamnobryum pumilum (Hook.f. & Wilson) B.C.Tan
ORTHOTRICHACEAE
Macrocoma tenue (Hook. & Grev.) Vitt subsp. *tenue*
Macromitrium angulatum Mitt.
Macromitrium brevicaule (Besch.) Broth.
Macromitrium gracile (Hook.) Schwägr. E
Macromitrium grossirete Müll.Hal. E
Macromitrium helmsii Paris E
Macromitrium ligulaefolium Broth.
Macromitrium ligulare Mitt.
Macromitrium longipes (Hook.) Schwägr. E
Macromitrium longirostre (Hook.) Schwägr.
Macromitrium microstomum (Hook. & Grev.) Schwägr.
Macromitrium orthophyllum Mitt. E
Macromitrium prorepens (Hook.) Schwägr. E
Macromitrium ramsayae Vitt E
Macromitrium retusum Hook.f. & Wilson E
Macromitrium submucronifolium Müll.Hal. & Hampe E
Muelleriella angustifolia (Hook.f. & Wilson) Dusén E
Muelleriella aucklandica Vitt E
Muelleriella crassifolia (Hook.f. & Wilson) Dusén subsp. *crassifolia*
Orthotrichum assimile Müll.Hal.
Orthotrichum calvum Hook.f. & Wilson E
Orthotrichum cupulatum Brid.
Orthotrichum cyathiforme R.Br.bis E
Orthotrichum graphiomitrium Beckett E
Orthotrichum hortense Bosw.
Orthotrichum rupestre Schwägr. var. *rupestre*
— var. *papillosum* Lewinsky E
Orthotrichum sainsburyi Allison E
Orthotrichum tasmanicum Hook.f. & Wilson var. *tasmanicum*
— var. *parvithecum* (R.Br.bis) Lewinsky E
Schlotheimia campbelliana Müll.Hal. E
Schlotheimia knightii Müll.Hal. E
Ulota lutea (Hook.f. & Wilson) Mitt.
Ulota membranata Malta
Ulota perichaetialis (Sainsbury) Goffinet E
Ulota viridis Venturi
Zygodon gracillimus M.Fleisch.
Zygodon hookeri Hampe
Zygodon intermedius Bruch & Schimp.
Zygodon menziesii (Schwägr.) Arn. var. *menziesii*
— var. *angustifolius* Malta E
Zygodon minutus Müll.Hal. & Hampe
Zygodon obtusifolius Hook.
Zygodon rufescens (Hampe) Broth.
PHYLLOGONIACEAE
Catagonium nitens (Brid.) Cardot subsp. *nitens*
Orthorrhynchium elegans (Hook.f. & Wilson) Reichardt
PLAGIOTHECIACEAE
Isopterygiopsis pulchella (Hedw.) Z.Iwats.
Isopterygium limatum (Hook.f. & Wilson) Broth.
Isopterygium minutirameum (Müll.Hal.) A.Jaeger
Plagiothecium lucidum (Hook.f. & Wilson) Paris
Plagiothecium lamprostachys (Hampe) A.Jaeger
Pseudotaxiphyllum distichaceum (Mitt.) Z.Iwats.
Pseudotaxiphyllum falcifolium (Hook.f. & Wils.) He
PLEUROPHASCACEAE
Pleurophascum ovalifolium Fife & P.J.Dalton E
POTTIACEAE
Acaulon integrifolium Müll.Hal.
Aloina bifrons (De Not.) Delgadillo
Anoectangium bellii Dixon E
Barbula calycina Schwägr.
Barbula convoluta Hedw. A
Barbula unguiculata Hedw. A
Bryoerythrophyllum jamesonii (Taylor) H.A.Crum
Bryoerythrophyllum recurvirostrum (Hedw.) P.C.Chen
Calyptopogon mnioides (Schwägr.) Broth.
Chenia leptophylla (Müll.Hal.) R.H.Zander
Crossidium davidai Catches.
Crossidium geheebii (Broth.) Broth.
Didymodon australasiae (Hook. & Grev.) R.H.Zander
Didymodon calycinus Dixon E
Didymodon lingulatus (Hook.f. & Wilson) Broth. E
Didymodon tophaceus (Brid.) Lisa
Didymodon torquata (Taylor) Catches.
Didymodon weymouthii (R.Br.bis) R.H.Zander E
Gymnostomum calcareum Nees & Hornsch.
Hennediella arenae subsp. *petriei* (Broth.) R.H.Zander E
Hennediella heimii (Hedw.) R.H.Zander
Hennediella macrophylla (R.Br.bis) Paris E
Hymenostylium recurvirostrum (Hedw.) Dixon
Hyophila novae-seelandiae Dixon & Sainsbury E
Leptodontium interruptum (Mitt.) Broth. E
Microbryum davallianum (Drake) R.H.Zander
Microbryum starckeanum (Hedw.) R.H.Zander
Pseudocrossidium crinitum (Schultz) R.H.Zander var. *crinitum*
— var. *obscurum* (Dixon) B.H.Macmill. & Fife E
Pterygoneurum ovatum (Hedw.) Dixon
Syntrichia anderssonii (Ångstr.) R.H.Zander
Syntrichia antarctica (Hampe) R.H.Zander

Syntrichia brevisetacea (F.Muell.) R.H.Zander
Syntrichia laevipila Brid.
Syntrichia pagorum (Milde) J.J.Amann
Syntrichia papillosa (Wilson) Jur.
Syntrichia phaea (Hook.f. & Wilson) R.H.Zander E
Syntrichia pygmaea (Dusén) R.H.Zander
Syntrichia robusta (Hook. & Grev.) R.H.Zander
Syntrichia rubella (Hook.f. & Wilson) R.H.Zander
Syntrichia rubra (Mitt.) R.H.Zander var. *rubra*
— var. *subantarctica* (Sainsbury) R.H.Zander E
Syntrichia ruralis (Hedw.) F.Weber & D.Mohr
Syntrichia serrata (Dixon) R.H.Zander
Tetracoscinodon irroratum (Hook.f.) R.H.Zander E
Tortella flavovirens (F. Muell.) Broth.
Tortella fragilis (Drumm.) Limpr.
Tortella knightii (Mitt.) Broth.
Tortella mooreae Sainsbury E
Tortula acaulon (With.) R.H.Zander A
Tortula areolata (Knight) Fife E
Tortula atrovirens (Sm.) Lindb.
Tortula marginata (Bruch & Schimp.) Spruce A
Tortula maritima (R.Br.bis) R.H.Zander
Tortula mucronifolia Schwägr. A
Tortula muralis Hedw.
Tortula truncata (Hedw.) Mitt. A
Tortula viridipila Dixon & Sainsbury E
Trichostomum brachydontium Bruch
Trichostomum imshaugii (Vitt) R.H.Zander E
Trichostomum tenuirostre (Hook. & Taylor) Lindb.
Tridontium tasmanicum Hook.f.
Triquetrella papillata (Hook.f. & Wilson) Broth.
Triquetrella tasmanica (Broth.) Granzow
Weissia austro-crispa (Beckett) I.G.Stone
Weissia controversa Hedw.
Weissia patula (Knight) Fife E
Willia calobolax (Müll.Hal.) Lightowlers
PTEROBRYACEAE
Pulchrinodus inflatus (Hook.f. & Wilson) B.H.Allen
Trachyloma diversinerve Hampe
Trachyloma planifolium (Hedw.) Brid.
PTYCHOMITRIACEAE
Ptychomitrium australe (Hampe) A.Jaeger
PTYCHOMNIACEAE
Cladomnion ericoides (Hook.) Hook.f. & Wilson E
Dichelodontium nitidum (Hook.f. & Wilson) Broth. E
Glyphothecium sciuroides (Hook.) Hampe
Hampeella alaris (Dixon & Sainsbury) Sainsbury
Hampeella pallens (Sande Lac.) M.Fleisch. var. *pallens*
— var. *symmetrica* Sainsbury
Ptychomnion aciculare (Brid.) Mitt.
Ptychomnion densifolium (Brid.) A.Jaeger
Tetraphidopsis pusilla (Hook.f. & Wilson) Dixon
RACOPILACEAE
Racopilum cuspidigerum var. *convolutaceum* (Müll. Hal.) Zanten & Dijkstra
Racopilum robustum Hook.f. & Wilson E
Racopilum strumiferum (Müll.Hal.) Mitt. E
RHABDOWEISIACEAE
Amphidium cyathicarpum (Mont.) Broth.
RHIZOGONIACEAE
Cryptopodium bartramioides (Hook.) Brid. E
Goniobryum subbasilare (Hook.) Lindb.
Hymenodon pilifer Hook.f. & Wilson
Leptotheca gaudichaudii Schwägr.
Pyrrhobryum bifarium (Hook.) Manuel
Pyrrhobryum mnioides subsp. *contortum* (Hook.f. & Wilson) Fife
Pyrrhobryum paramattense (Müll.Hal.) Manuel
Rhizogonium distichum (Sw.) Brid.
Rhizogonium novae-hollandiae (Brid.) Brid.
Rhizogonium pennatum Hook.f. & Wilson
SELIGERIACEAE
Blindia contecta (Hook.f. & Wilson) Müll.Hal.
Blindia immersa (E.B.Bartram & Dixon) Sainsbury E
Blindia lewinskyae J.K.Bartlett & Vitt E
Blindia magellanica Schimp.
Blindia martinii Sainsbury E
Blindia robusta Hampe
Blindia seppeltii J.K.Bartlett & Vitt E
Seligeria cardotii R.Br.bis
Seligeria diminuta (R.Br.bis) Dixon E
SEMATOPHYLLACEAE
Rhaphidorrhynchium amoenum (Hedw.) M.Fleisch.
Rhaphidorrhynchium leucocytus (Müll.Hal.) Broth.
Rhaphidorrhynchium tenuirostre (Hook.) Broth.
Sematophyllum contiguum (Mitt.) Mitt.
Sematophyllum homomallum (Hampe) Broth.
Sematophyllum jolliffii (Hook.f. & Wilson) Dixon
Wijkia extenuata (Brid.) H.A.Crum
SPLACHNACEAE
Tayloria callophylla (Müll.Hal.) Mitt.
Tayloria octoblepharum (Hook.) Mitt.
Tayloria purpurascens (Hook.f. & Wilson) Broth. E
THUIDIACEAE
Pseudoleskea imbricata (Hook.f. & Wilson) Broth.
Thuidium cymbifolium (Dozy & Molk.) Dozy & Molk.
Thuidium furfurosum (Hook.f. & Wilson) Reichardt
Thuidium laeviusculum (Mitt.) A.Jaeger
Thuidium sparsum (Hook.f. & Wilson) Reichardt var. *sparsum*
TIMMIACEAE
Timmia norvegica J.E.Zetterst.

TWENTY-FIVE

Phylum TRACHEOPHYTA

vascular plants

ILSE BREITWIESER, PATRICK J. BROWNSEY,
PHILIP J. GARNOCK-JONES, LEON R. PERRIE, AARON D. WILTON

Vascular plants (lycophytes, ferns, and seed plants) are the most species-rich of all green-plant lineages. Vascular plants are characterised by the presence of specialised tissue – xylem and phloem – for structural support and long-distance movement of water and nutrients throughout the plant body, which is typically a highly structured, dominant spore-bearing (sporophyte) phase. In addition their living members all differ from bryophytes in having true roots (sometimes lost secondarily) and branching sporophytes. Roots, vascular tissue, and branching allow the attainment of much greater size and the production of multiple sporangia. In many (some lycophytes, a few ferns, and all seed plants), the sporangia have become specialised as male (producing microspores only) and female (producing megaspores only). The sex of the spores is governed by the type of sporangium they are produced in, rather than by the presence of a Y (male) or X (female) chromosome segregating at meiosis, as appears to be universally the case in those bryophytes that have unisexual gametophytes.

Manuka, *Leptospermum scoparium* (Magnoliopsida, Myrtaceae).
Dennis Gordon, NIWA

The discovery of remarkable similarities shared by bryophytes, ferns, and seed plants, in particular the interpolation of mitotic cell divisions between fertilisation and meiosis to create a multicellular, sporophyte, life-cycle stage, suggests that only a single lineage of green plants successfully colonised land. The transition from an aquatic to a terrestrial environment was a period of remarkable morphological innovation. Based upon similarities in their photosynthetic pigments, cell structure, and spermatozoid morphology, it is widely accepted that land plants evolved from a green-algal ancestor, and the major lineages of land plants had diversified by the end of the Devonian period some 360 million years ago. The bryophytes, pteridophytes, and gymnosperms dominated the land for most of this time, yielding to the angiosperms only in the last 100 million years or so.

Our understanding of the evolutionary history of land plants has been revised by a multitude of phylogenetic studies in recent years. These studies refute the prevailing view that the horsetails and ferns form an evolutionary transition between the bryophytes and seed plants. Instead, recent studies support three monophyletic lineages of vascular plants corresponding to the lycophytes, seed plants, and a clade that includes the horsetails, psilophytes (whisk ferns), and all eusporangiate and leptosporangiate ferns. This latter lineage (ferns including horsetails) emerges as the closest relative of the seed-plant clade. Within the seed plants, many relationships are now well established, while clarity on the others is still emerging, including the relationships of gnetophytes among living gymnosperms, and the order of branching of the earliest lineages of the flowering plants.

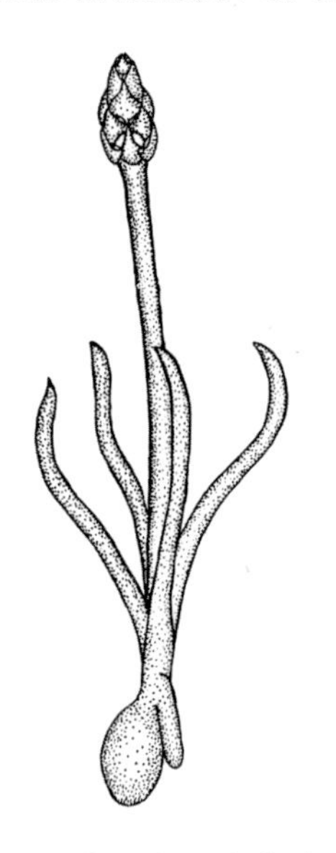

The diminutive lycophyte *Phylloglossum drummondii*, which dies back to a tuber in hot summer months (Lycopodiopsida, Lycopodiaceae).
From Dawson 1953

The various members of this lineage, such as lycophytes, ferns, gymnosperms, and flowering plants, are easy to recognise as monophyletic because of their shared characters. Judging from spore microfossils, vascular plants probably diverged from a bryophyte ancestor (or bryophyte-like, insofar as vascular elements in living bryophytes are not homologous with the tracheids and sieve cells of vascular plants) in the late Early Silurian (c. 430 million years ago). A little later, by the early Late Silurian, the earliest unequivocal vascular-plant macrofossils occur. These plants are known as rhyniophytoids. Since the Silurian is scarcely represented in New Zealand and is marine, fossils of such plants are unlikely ever to be found here.

Lycophytes and ferns

Until the 1990s, the collective terms 'pteridophytes' or 'ferns and fern allies' were widely used for a disparate group of plants, with about 12,000 living species worldwide (Roos 1996), that were thought to fall into four main groups – the Lycopsida, Psilotopsida, Equisetopsida, and Filicopsida. These plants are distinguished from all other vascular land plants by having a regular alternation between free-living gametophytic (gamete-producing) and sporophytic (spore-producing) stages, the first being very small and often cryptic, the latter being the more typical 'fern' plants that are familiar to botanists and lay people alike. The sporophyte plants are generally differentiated into stem, leaves, and roots, have well-developed vascular tissue, and produce single-celled spores. The gametophytes are small, generally undifferentiated, structures that are either green and surface-living or colourless and partially subterranean, with endosymbiotic fungi (mycotrophic). They produce motile sperm cells from antheridia, and egg cells in flask-shaped archegonia.

Since the 1990s, however, there has been a revolution in our understanding of the evolution of the different lineages that make up the vascular plants (see Pryer et al. 2004 for references to the earlier literature). It is now clear that a deep phylogenetic dichotomy occurred about 400 million years ago separating a group known as the lycophytes (clubmosses, spikemosses, and quillworts) from all other vascular-plant lineages.

Summary of New Zealand living tracheophyte diversity

The tallies below are based on full species only, not subspecies, and do not include hybrids.

Taxon	Described species	Known undescr./ undet. species	Estimated unknown species	Adventive species (fully naturalised and casual)	Endemic species	Endemic genera
LYCOPHYTINA	16	0	0	3	2	0
Lycopodiopsida	16	0	0	3	2	0
EUPHYLLOPHYTINA	4728	1	300–400	2520	1733	53
Monilophytae	231	1	7	49	85	3
Psilotopsida	11	0	0	0	3	0
Equisetopsida	3	0	0	3	0	0
Marattiopsida	1	0	0	0	0	0
Polypodiopsida	216	1	7	46	82	3
Spermatophytae	4497	0	300–400	2471	1648	50
Ginkgoopsida	1	0	0	1	0	0
Pinopsida	60	0	0	40	20	2
Magnoliopsida	4436	0	300–400	2430	1628	48
Totals	**4744**	**1**	**300–400**	**2523**	**1735**	**53**

Lycophytes have a long fossil record extending back to at least the Early Devonian. They are monophyletic and a sister-group to the rest of the vascular plants (Bateman 1996). Worldwide, about 1200 species, mostly relatively small herbaceous plants, are still extant, representing the evolutionary remnant of a group that was much more widespread in the Carboniferous and then included large forest trees. They are characterised by having spirally arranged leaves with single unbranched veins (microphylls) and thick-walled sporangia borne on the upper surface. Some families are also heterosporous and others have a distinctive ligule on the leaves. Gametophytes are generally subterranean, mycotrophic, and non-green.

The quillwort *Isoetes alpina* (Lycopodiopsida, Isoetaceae).
Dennis Gordon, NIWA

A second dichotomy in the vascular-plant lineage resulted in the separation of the seed plants (spermatophytes) from the ferns (monilophytes). Within the ferns, four monophyletic lineages are now recognised – whisk ferns and ophioglossoid ferns diverging from the marattioid ferns, horsetails, and leptosporangiate ferns. Unexpectedly, therefore, whisk ferns and horsetails, which were previously known as 'fern allies' (Psilotopsida and Equisetopsida, respectively) are now regarded as part of the fern lineage.

Whisk ferns comprise a single family of about 16 extant species that show highly reduced characters. The aerial stems and rhizomes branch dichotomously, but they lack roots and the leaves are reduced, even scale-like. The sporangia are thick walled and are fused into two or three-chambered synangia. Gametophytes are generally subterranean, mycotrophic, and non-green. Although showing some superficial similarities to the earliest vascular plants, the fossil history of the group is virtually unknown.

Ophioglossoid ferns also comprise a single family with up to 100 extant species. They show considerable morphological simplification, with the leaf being divided into a blade-like photosynthetic portion and a highly reduced spike-like fertile portion. They have unbranched roots and lack root hairs, a trait shared with the whisk ferns that lack roots altogether. Gametophytes are subterranean, mycotrophic and non-green. Like whisk ferns, they have a poor fossil record.

Marattioid ferns include six genera and 100 species (Murdock 2008). They have distinctive polycyclic dictyosteles, and sori comprising a double row of free sporangia, or sporangia fused into a synangium. They have a fossil record dating back to the upper Carboniferous.

New Zealand tracheophyte diversity by major environment*

Taxon	Marine	Freshwater	Terrestrial
LYCOPHYTINA	0	2	14
Lycopodiopsida	0	2	14
EUPHYLLOPHYTINA	5	106	4628
Monilophytae	0	5	226
Psilotopsida	0	0	11
Equisetopsida	0	0	3
Marattiopsida	0	0	1
Polypodiopsida	0	5	211
Spermatophytae	5	101	4402
Ginkgoopsida	0	0	1
Pinopsida	0	0	60
Magnoliopsida	5	101	4341
Totals	**5**	**108**	**4642**

*There is some overlap between fully terrestrial and fully freshwater, with some species occurring in both environments.

Horsetails today comprise a single genus of about 15 species, forming the evolutionary remnant of a group that extends back to the Devonian and was much more widespread and diverse in earlier geological periods. Plants are characterised by jointed stems, whorled leaves, and thin-walled sporangia arranged in specialised cones, bearing spores and elaters. They no longer occur naturally in the Australasian region.

The vast bulk of monilophytes are leptosporangiate ferns. This is a group that has leaves with branching veins (megaphylls) bearing the sporangia on the margins or undersides, and is characterised by sporangia that develop from a single cell, have sporangial walls one cell thick, open by means of a distinct annulus, and usually produce 64 spores. The other four lineages of ferns are eusporangiate – bearing sporangia that develop from several cells, have thick sporangial walls, lack an annulus, and produce large numbers of spores. However, eusporangiate ferns as a whole are paraphyletic.

Most families of leptosporangiate ferns produce homosporous spores (all alike), but a few are heterosporous (microspores and megaspores). There is an extensive fossil record and the group has a long fossil history back to the Carboniferous. However, some families are undoubtedly of much more recent origin; those with a horizontal or oblique annulus are older lineages, whilst the Polypodiales (leptosporangiate ferns with a vertical annulus) may not have arisen until the lower Cretaceous, based on fossil evidence (Pryer et al. 2004), or the Jurassic, based on molecular age estimates (Schneider et al. 2004). There is, in fact, increasingly good correlation between the evolutionary history of the Filicopsida determined from the fossil record (Collinson 1996) and phylogenies determined from molecular and morphological data (Hasebe et al. 1994, 1995; Pryer et al. 1995).

The black tree fern or mamaku, *Cyathea medullaris* (Polypodiopsida, Cyatheaceae).
Dennis Gordon, NIWA

Ferns vary enormously in size from tiny filmy ferns only one or two centimetres in length to tall tree ferns with trunks of 20 metres or more in height and fronds of six metres or more in length. Ferns occur in terrestrial and freshwater habitats throughout the world except for the very driest deserts and polar and high-alpine regions. They reach their greatest diversity in the wet tropics. In addition to spores, some sporophyte plants can reproduce asexually by bulbils or proliferous buds on the fronds or roots, by tubers, or from subterranean gemmae on the rhizomes.

Gametophyte plants also show a high degree of diversity (Atkinson 1973), although they are less familiar to the majority of botanists because of their small size and cryptic nature. Some eusporangiate ferns have subterranean, mycotrophic, non-green gametophytes, but most leptosporangiate ferns have green, surface-living prothalli. These are very frequently heart-shaped, short-lived, and only one or two centimetres long, but some are linear or strap-shaped and continue growing for more than one season; others, especially in Hymenophyllaceae, Vittariaceae, and Grammitidaceae, are filamentous and frequently bear gemmae. Many bear antheridia and archegonia on the same prothallus, but others produce male and female gametes on separate plants. There are mechanisms for promoting out-breeding, and mating systems in ferns are generally more complex than was previously realised.

The higher-level classification of pteridophytes is still the subject of much debate. Through the 1970s and 1980s, taxonomic concepts were guided by the classifications proposed by Nayar (1970), Holttum (1973), Crabbe et al. (1975), Lovis (1977), Pichi Sermolli (1977), and Tryon & Tryon (1982). The work of Kramer and Green (1990) gained general acceptance for a while, but a new classification of extant ferns by Smith et al. (2006), based on phylogenetic developments since that time, has become widely adopted. It is followed here with some minor exceptions. Classification of the lycophytes, which was not dealt with by Smith et al. (2006), follows Christenhusz et al. (2011).

A wealth of information on the taxonomy, ecology, and biology of pteridophytes can be found in the published accounts of major symposia held over the last three decades (Jermy et al. 1973; Dyer 1979; Dyer & Page 1985; Camus et al. 1996).

Diversity of New Zealand lycophytes and ferns

A total of 248 species of extant lycophytes and ferns is currently accepted as occurring in the New Zealand Botanical Region (excluding Macquarie Island) – an area extending from the Kermadec Islands in the north, through North, South, Stewart, and Chatham Islands, to the Auckland Islands and Campbell Island. This total includes 196 native species (of which one is still undescribed) and 52 that are believed to have been introduced since European colonisation.

In the Checklist, the species are grouped into 31 families, of which Selaginellaceae, Equisetaceae, and Onocleaceae are represented only by introduced species, and 76 genera, of which 16 are represented only by introduced species. In addition, 18 subspecific taxa are recognised, of which one is introduced and one undescribed.

On the basis of native species alone, the largest families are Hymenophyllaceae (27 species), Blechnaceae (23 species), Aspleniaceae (20 species), and Pteridaceae (17 species). The most diverse genera are *Hymenophyllum* (21 species), *Asplenium* (19 species and six subspecies), *Blechnum* (18 species), and *Grammitis* (10 species and two subspecies).

Historical overview of New Zealand lycophyte and fern taxonomy

The first European collections of New Zealand ferns were made by Banks and Solander on Cook's first voyage in 1769. Solander wrote descriptions of 57 species, and 22 of them were engraved by Banks's artists. However, the proposed *Primitiae Florae Novae Zelandiae* was never completed, and the first published descriptions had to await the return of the Forsters from Cook's second voyage and the appearance of two small books by Georg Forster (1786a,b). Unfortunately Forster's descriptions were inadequate in many ways, and New Zealand pteridology got off to a very poor start. The Forsters' collections did, however, provide the basis for what was probably the first published illustration of a New Zealand fern – the plate of *Tmesipteris tannensis* in Schrader's *Journal für die Botanik* (Bernhardi 1801).

Subsequent expeditions continued to collect ferns from New Zealand and slowly built up a picture of the rich diversity to be found. In 1791, Archibald Menzies visited Dusky Sound during Vancouver's voyage, and the ferns he collected became the basis for the lavishly produced *Icones Filicum* (Hooker & Greville 1831). The collection and study of Australian ferns by Labillardière (1806) and Robert Brown (1810) also contributed significantly to later work in New Zealand.

The first attempt to produce a systematic catalogue of New Zealand ferns was made by a Frenchman, Achille Richard (1832). He never visited New Zealand himself but listed and described some 57 species of ferns collected by d'Urville and Lesson during the voyages of Duperrey in 1824 and d'Urville himself in 1827. From about 1840 onwards there was a steady increase in knowledge and understanding of New Zealand ferns as overseas collectors made longer visits (e.g. Cunningham 1837) and other botanists settled in the country (e.g. Colenso 1842, 1846).

The greatest boost to New Zealand pteridology resulted from the visit of Joseph Hooker during the Antarctic voyage of James Clark Ross. Three major publications resulted from Hooker's own collections on this expedition and from the material that was sent to him at Kew by local botanists (Hooker 1844, 1855, 1860). Eventually, he compiled all his knowledge of the New Zealand flora into a single volume, *Handbook of the New Zealand Flora*, a seminal work that described 135 species of ferns (Hooker 1867).

From then on, resident New Zealand botanists sustained interest in New Zealand pteridology with a series of popular and scientific publications (Thom-

Tmesipteris sigmatifolia (Psilotopsida, Psilotaceae).
Dennis Gordon, NIWA

King fern *Ptisana salicina* (Marattiopsida, Marattiaceae).
Leon Perrie

son 1882; Field 1890; Cheeseman 1906, 1925; Dobbie 1921). However, by 1906, when Cheeseman's first Flora was published, 156 species were recognised and the rapid increase in new knowledge had come to an end. In the next 50 years, only a handful of new species were added until a new Flora series was initiated in the 1950s.

The first volume of *Flora of New Zealand* (Allan 1961) described 164 lycophyte and fern species and stimulated the preparation of a series of taxonomic revisions, initially by local botanists, and then increasingly by overseas workers, especially those working on Malesian or Pacific floras. Groups that have been critically investigated in the last 40 years include *Cyathea* (Holttum 1964; Brownsey 1979), *Lastreopsis* (Tindale 1965), *Doodia* (Parris 1972, 1980a), *Adiantum* (Parris & Croxall 1974; Parris 1980b), *Tmesipteris* (Chinnock 1975; Perrie et al. 2010), *Grammitis* (Parris & Given 1976; Parris 1980a), *Lindsaea* (Kramer & Tindale 1976), Thelypteridaceae (Holttum 1977), *Asplenium* (Brownsey 1977a, 1983; 1985a; Brownsey & Jackson 1984; Brownsey & de Lange 1997), *Botrychium* (Braggins 1980), *Hypolepis* (Brownsey & Chinnock 1984), *Deparia* (Kato 1984), *Pyrrosia* (Hovenkamp 1986), Lycopodiaceae (Øllgaard 1987), *Pilularia* (Large & Braggins 1989; Nagalingum et al. 2008), *Cheilanthes* (Chambers & Farrant 1991), *Microsorum* (Large et al. 1992; Nooteboom 1997), *Davallia* (Nooteboom 1994; von Konrat et al. 1999), *Blechnum* (Chambers & Farrant 1996, 1998), *Polystichum* (Perrie et al. 2003a,b), *Nephrolepis* (de Lange et al. 2005; Hovenkamp & Miyamoto 2005), and *Ptisana* (Murdock 2008). Some taxonomic work has been the subject of university theses but not yet published – in particular, investigation of *Pteris* (Braggins 1975), *Isoetes* (Marsden 1979), *Christella* (Davison 1995) and *Dicksonia* (Lewis 2001). Several other papers have described new species or documented new records for the flora. This massive resurgence of interest in fern taxonomy since 1961 has seen the known flora increase from 164 species in 1961 to the present 196 native species. Brownsey et al. (1985a) published a revised classification of the New Zealand lycophytes and ferns, and a full synonymic checklist of species. A key to the identification of genera appeared two years later (Brownsey & Galloway 1987), and a comprehensive, illustrated guide to the whole fern and lycophyte flora by Brownsey and Smith-Dodsworth (1989, 2000) replaced the then standard work on the New Zealand fern flora (Crookes 1963).

Major collections of New Zealand lycophytes and ferns are held in AK, BM, CHR, K, and WELT, with significant numbers of type specimens also in B, FI, GOET, P and W (see Holmgren et al. 1990 for full details of herbaria listed here by acronyms). Images of type specimens of ferns and lycophytes held in New Zealand institutions are available online:

AK: http://www.aucklandmuseum.com/databases/general/basicsearch.aspx?D atasetID=107
CHR: http://bcd.landcareresearch.co.nz
WELT: http://collections.tepapa.govt.nz

Hybridism

Hybrids are a characteristic feature of many plant groups, and ferns are no exception. A total of 56 hybrid combinations, for which there are supporting herbarium specimens, are now recorded, although this number will undoubtedly increase with astute observation and further collecting. Cockayne and Allan (1934) listed 41 hybrid fern combinations amongst 491 wild hybrids in the New Zealand flora as a whole. Several of their reported combinations, especially those in Hymenophyllaceae, now seem unlikely, but, on the other hand, Brownsey (1985b) showed that, in genera that had been critically revised, the number of combinations was equal to or greater than that suggested by Cockayne and Allan.

The overwhelming number of hybrid ferns are found in *Asplenium* where 30 combinations have now been identified with some confidence (Brownsey 1977b, 1985a,b; Brownsey & de Lange 1997; Perrie et al. 2005), suggesting that these are very closely related species that have diverged only relatively recently. Eight combinations are recorded in *Polystichum* (Perrie et al. 2003a, b), six in *Hypolepis* (Brownsey & Chinnock 1984), four in *Blechnum* (Tindale 1972; Chambers & Farrant 1998), two each in *Doodia* (Parris 1972; de Lange et al. 2004) and *Grammitis* sensu lato (Parris 1977; de Lange & Rolfe 2011), and one each in *Leptopteris* (Brownsey 1981), *Hymenophyllum* (Daellenbach 1982), *Pteris* (Braggins unpubl.), and *Dryopteris* (Lovis unpubl.). The combinations in *Grammitis* were originally described as intergeneric hybrids between *Ctenopteris heterophylla* and *Grammitis billardierei* and *G. ciliata* respectively, but recent molecular evidence suggests these species are congeneric, even though the taxonomic changes have yet to be made (Ranker et al. 2004). The combination in *Dryopteris* is unusual in being the only example of a fern hybrid that has apparently originated in New Zealand from two introduced species (*D. affinis* and *D. filix-mas*). The recently described *Asplenium* ×*lucrosum* is truly remarkable in being a hybrid between the New Zealand endemic *A. bulbiferum* and the Norfolk Island endemic *A. dimorphum*. It may have originated in England in the 19th century, becoming widely cultivated in much of the world and a casual adventive in New Zealand (Perrie et al. 2005).

Hybrids are unknown amongst any of the New Zealand lycophytes or eusporangiate lineages of ferns, and are confined to the leptosporangiate ferns. However, amongst modern leptosporangiate families (i.e. those with a vertical annulus) comprising genera of least two sympatric native species, only Polypodiaceae apparently lacks hybrids.

Many fern hybrids are thought to be effectively sterile, unlike those in flowering plants where hybrids often show some degree of fertility. All the known combinations in *Asplenium* and *Polystichum*, and all but one in *Hypolepis*, have misshapen spores that are assumed to be inviable, although no detailed germination experiments have been carried out. In *Doodia* there are reports of both normal and abnormal spores in hybrids (Parris 1972; de Lange et al. 2004), whereas in *Leptopteris*, *Pteris*, and *Blechnum* the spores appear normal. In *Doodia* and *Leptopteris* there is evidence of inviability but in *Pteris* and *Blechnum* the extent of the wild populations of hybrids suggest that they may be at least partially fertile (Brownsey 1985b).

Asplenium x*lucrosum*, a hybrid between New Zealand *A. bulbiferum* and Norfolk Island *A. dimorphum* (Polypodiopsida, Aspleniaceae).

Leon Perrie

Spore morphology

A comprehensive atlas of New Zealand pteridophyte spores has been prepared by Large and Braggins (1991). This very thorough work provides descriptions and light-microscope and SEM photographs of the spores of nearly all of the species found in New Zealand.

Cytology

Chromosome counts have been obtained for about 83% of the native fern taxa. An annotated and fully referenced list of all chromosome counts made on New Zealand ferns was published by Dawson et al. (2000) and supplemented by de Lange et al. (2004). About 45% of those investigated are diploid and 55% polyploid, which is comparable to other temperate pteridophyte floras (Walker 1979). Only about 2% are apogamous (reproducing asexually as a result of an incomplete meiosis), which is rather more surprising. The major groups for which counts are still required are Lycopodiaceae, *Grammitis*, *Blechnum*, and taxa that are confined to offshore islands.

Gametophyte morphology

Almost no work has been done on the gametophytes of New Zealand pteridophytes in recent years. However, the pioneer studies of Holloway (1918, 1920, 1921, 1935, 1939, 1944) on the gametophytes of Lycopodiaceae, Psilotaceae, and Hymenophyllaceae remain classics of their time.

Phylogeny and genetic diversity

Analyses of genetic diversity, and phylogenies of New Zealand ferns based on molecular data, were very much in their infancy when the millennial symposium that launched this inventory project was held in Wellington in 2000. Since then, results from DNA sequencing and AFLP DNA-fingerprinting in several genera and families have begun to challenge many of our previous ideas on evolution and relationships in New Zealand ferns. In *Polystichum*, the polymorphic species previously known as *P. richardii* has been shown to be an allopolyploid complex of four separate evolutionary lineages (Perrie et al. 2003a), whereas plants from the Chatham Islands, once thought to represent a distinct species, appear not to constitute an evolutionary lineage separate from *P. vestitum* (Perrie et al. 2003b). In *Microsorum*, it is apparent that the three New Zealand species currently assigned to this genus are not closely related to *Microsorum* sensu stricto, but no generic name is currently available for them (Schneider et al. 2006). The monotypic genus *Anarthropteris* has been shown to nest within *Loxogramme* and has been reinstated as *L. dictyopteris* (Kreier & Schneider 2006). In Gleicheniaceae, preliminary investigation suggests that the taxonomy of New Zealand *Gleichenia* may be more complex than presently recognised (Perrie et al. 2007). In Blechnaceae, the New Zealand species of *Doodia* are nested within a paraphyletic *Blechnum*. Nevertheless several groupings of *Blechnum* species have been recovered, consistent with previous evidence from morphology, hybridization, and base chromosome numbers (Shepherd et al. 2007a). Amongst New Zealand Pteridaceae, a study by Bouma et al. (2010) found that the indigenous species of *Adiantum* were not closely related to the type species, reinforcing the need for a global investigation of generic boundaries in this group.

Most work in New Zealand has centred on the complex evolutionary history of the genus *Asplenium* (Perrie & Brownsey 2005a–c; Shepherd et al. 2008a). Phylogenetic analyses of chloroplast sequence data indicate that the New Zealand species of *Asplenium* are not monophyletic. Three well-supported subgroups, represented by *A. bulbiferum*, *A. flaccidum*, and *A. obtusatum*, comprise an austral group within which hybridisation is common. Species such as *A. trichomanes*, *A. flabellifolium*, *A. pauperequitum*, and *A. polyodon*, which apparently do not hybridise in New Zealand, have closer affinities to non-New Zealand species. Molecular techniques have also shown that genetic variation within the morphologically variable species *A. hookerianum* is not concordant with differences in pinnule morphology and that only one species should be recognised. The auto- and allopolyploid origins of several octoploid species from a range of tetraploid species have also been more clearly defined (Shepherd et al. 2008b). In the first case of its kind, Perrie et al. (2010a) showed that the octoploids *A. cimmeriorum* and *A. gracillimum* had each originated multiple times by allopolyploidy from *A. bulbiferum* and *A. hookerianum*, and that these genetically distinguishable allo-octoploids were broadly sympatric over large areas.

The filmy fern *Trichomanes endlicherianum* (Polypodiopsida, Hymenophyllaceae).
Leon Perrie

Numerous investigations have been conducted around the world on the large filmy fern family Hymenophyllaceae. These confirm the monophyly of *Trichomanes* and *Hymenophyllum*. It seems that rbcL data are still inadequate for resolving relationships within *Hymenophyllum* (Hennequin et al. 2003) but they are informative within *Trichomanes* (Pryer et al. 2001; Dubuisson et al. 2003). Using additional loci, Ebihara et al. (2004) confirmed the existence of of these two clades, but concluded that they are very different to each other. The *Hymenophyllum* clade shows little genetic divergence and has probably diversified more recently than the *Trichomanes* clade. However, with specific reference to New Zealand, it is clear that the monotypic genus *Cardiomanes* groups with *Hymenophyllum*, rather than *Trichomanes* as previously assumed. Based on molecular and morphological evidence, Ebihara et al. (2006) propose recognising a single genus in *Hymenophyllum* (including *Cardiomanes*), but eight separate genera in *Trichomanes*, of which only *Polyphlebium* and *Abrodictyum* are represented in New Zealand. This work requires further evaluation.

The classification and phylogeny of tree ferns has long been controversial, but a combination of morphological and molecular evidence now suggests that

Cyatheaceae comprises at least three main clades – *Alsophila, Cyathea,* and *Sphaeropteris* (Korall et al. 2006). If accepted, this would leave New Zealand with a single species of *Sphaeropteris* (currently *Cyathea medullaris*) and the rest in *Alsophila*.

Kidney fern *Cardiomanes reniforme* (Polypodiopsida, Hymenophyllaceae).
Leon Perrie

Endemism and biogeography

Of the 196 native species of lycophytes and ferns in New Zealand, 87 (44%) are endemic. This compares with the much higher figure of 82% for the angiosperms and gymnosperms. Seven of the 18 subspecific taxa are also endemic. At higher taxonomic levels, endemism is very low. There are no endemic families and only three endemic genera – *Cardiomanes, Leptolepia,* and *Loxsoma* (all monotypic). Indeed, the first two may, in future research, be found to nest within *Hymenophyllum* and *Dennstaedtia* respectively.

The affinities of 105 widespread species then recognised were reviewed by Brownsey (2001a). Some 94 of the 105 widespread species (90%) occur elsewhere in temperate Australasia, including Norfolk Island and Lord Howe Island. Some 56 species (53%) occur in tropical regions of Australia, SE Asia, and the Pacific, 15 species (14%) are shared with southern Africa, and 14 species (13%) with the circumantarctic islands and South America.

From an analysis of this evidence, the known fossil history of lycophytes and ferns in New Zealand and worldwide, and the present-day distributions within New Zealand (see below), Brownsey (2001a) argued that most of the fern flora had arrived in New Zealand relatively recently by long-distance dispersal. Some New Zealand ferns have also dispersed across the Tasman Sea to Australia, and possibly elsewhere. While this perspective conflicts with earlier ideas of New Zealand's ferns belonging to lineages that were resident before the break-up of Gondwana (e.g. Lovis 1959), recent fossil and phylogenetic data suggest that many modern fern families had not evolved by this time. This recent-dispersal hypothesis has been supported by subsequent phylogenetic and molecular-dating studies in *Polystichum* (Perrie et al. 2003c), *Asplenium* (Perrie & Brownsey 2005a), *Dicranopteris, Gleichenia,* and *Sticherus* (Perrie et al. 2007), and in pairs of closely related New Zealand and overseas species representative of the New Zealand fern flora as a whole (Perrie & Brownsey 2007). In *Asplenium hookerianum,* dispersal from mainland New Zealand has occurred on multiple occasions to Australia and the Chatham Islands (Shepherd et al. 2009; Perrie et al. 2010b).

Ecology and distribution within New Zealand

Lycophytes and ferns occur throughout New Zealand and can be found in almost all terrestrial and freshwater habitats except the very highest alpine regions. Brownsey's (2001a) analysis of the distributions of 194 native species showed that they fall into several distinctive, repeating patterns.

In general, lycophyte and fern distributions are very widespread – more than half the species extend across more than half of both islands. Distributions are not significantly correlated with geographic barriers such as Cook Strait or the main mountain ranges. They are, however, strongly correlated with temperature, rainfall, and geothermal activity. The greatest proportion of endemic species show a predominantly southern distributional pattern. However, few lycophytes and ferns extend into the alpine zone – only about 10% of the total flora compared with more than 30% of the flowering plants. The main endemic element comprises species that live in cool lowland-to-montane forest. By contrast, lycophytes and ferns with a northern distributional pattern are predominantly tropical and Australasian species that have established themselves to varying degrees in northern New Zealand but appear to be restricted by temperature from spreading southwards. Geographically restricted endemic ferns are very rare compared to other vascular plants. Wardle (1991) listed 578 endemic vascular plants (27% of the native flora) restricted to one of nine phytogeographical regions within New Zealand, but only 11 endemic lycophytes and ferns (6% of the native flora) are confined to one of these same regions. The only areas with geographically restricted endemic ferns are the Kermadec Islands (4 species), the Chatham Islands (1 species), the far north of the North Island (5 species), and the central South Island (1 species).

In the most recent review of the conservation status of the New Zealand vascular-plant flora (de Lange et al. 2009), four ferns and lycophytes were listed as Nationally Critical, two as Nationally Endangered, and two as Nationally Vulnerable. These are described and illustrated by de Lange et al. (2010). A further 33 species are categorised as At Risk.

There have been rather few studies of the ecology of lycophytes and ferns in New Zealand, either individually or on a more general scale. Parris (1976) identified the component species that were associated with different habitats throughout the country, and discussed the contribution of different biogeographic elements to each habitat. Brownsey (1996) looked at the ferns of the Wellington region in a similar way and identified the different species that could be found in coastal to subalpine habitats. Bannister (1984) examined frost resistance in New Zealand ferns and concluded that native species may be unable to develop as great a degree of frost hardiness as their northern-hemisphere counterparts. Norton (1994) investigated the relationship between pteridophytes and topography in lowland podocarp forest in South Westland, whilst Lehmann et al. (2002) used climate and landform variables to model and predict the distribution of 43 species of ferns throughout New Zealand. The latter case study provides an approach for identification of biodiversity hotspots, and setting targets for biodiversity assessment and restoration programmes.

Blechnum vulcanicum
(Polypodiopsida, Blechnaceae).
Leon Perrie

Investigation of the ecology of individual species has tended to concentrate on culturally important, or rare and threatened species. Those which have been examined in any detail include *Botrychium* (Braggins 1980; Kelly 1994), *Asplenium chathamense* (Brownsey 1985a), *A. cimmeriorum* (Brownsey & de Lange 1997), *A. pauperequitum* (Brownsey & Jackson 1984; de Lange & Cameron 1999; Cameron et al. 2006), *Blechnum blechnoides* (Chambers & Farrant 1996), *Davallia tasmanii* (von Konrat et al. 1999), *Macrothelypteris torresiana* (de Lange & Crowcroft 1997), *Ophioglossum petiolatum* (de Lange 1988), *Pellaea calidirupium* (Brownsey & Lovis 1990), *Pleurosorus rutifolius* (Given 1972), *Sticherus flabellatus* (Given 1982), *Pteridium esculentum* (McGlone et al. 2005), and *Tmesipteris horomaka* (Perrie et al. 2010).

The importance of tree ferns as a host for epiphytic vascular plants has long been recognised. An early pioneer study by Pope (1924, 1926) demonstrated how some of New Zealand's largest forest trees establish themselves first on tree-fern trunks, which provide an ideal substratum for their juvenile growth. The question of whether these plants (sometimes called 'strangling' epiphytes) then smother the host or merely outlive it has remained an unresolved issue. Page and Brownsey (1986) suggested that the skirts of dead fronds on many tree ferns provide a defence against epiphytes and climbing plants. Beever (1984) investigated the frequency and cover of moss species on tree-fern trunks and showed that there was significant variation in the communities on different species of tree fern. More recently, Gillman and Ogden (2005) have shown that shedding whole fronds from tree ferns, or producing a skirt of fronds, are alternative strategies that can reduce competition from terrestrial and epiphytic seedlings, respectively.

Shepherd et al. (2007b) found that the genetic diversity in *Asplenium hookerianum* was distributed throughout the range of this species, rather than being concentrated in particular regions (cf. glacial refugia). They inferred that populations of *A. hookerianum* survived the last glacial throughout the country and, because this species does not occur in grassland, survival of woody vegetation must have been similarly widespread.

Introduced species

Brownsey (1988a) described and documented the first records of 23 introduced pteridophytes in New Zealand. Additional records of newly discovered adventive species (Webb et al. 1995; Heenan et al. 1998, 1999, 2002b, 2004a, 2008) have since increased this number to 52 species, together with one hybrid that has arisen in cultivation (Perrie et al. 2005).

Most introduced fern species are known from isolated or occasional records, often as a result of escape from cultivation. About half are recorded only as casuals. However, some species are potentially invasive and a few have become serious weeds. The most significant of these are *Selaginella kraussiana,* which is widespread along stream banks and damp forests where it replaces native ferns and bryophytes, and *Equisetum arvense,* an extremely invasive weed in high-rainfall and riverbank sites (Brownsey et al. 1985b). *Dryopteris affinis* and *D. filix-mas* are very widespread weeds, *Osmunda regalis* is well established in many swampy areas of the northern North Island, and *Azolla pinnata* is common on ponds and lakes in northern regions, but these ferns pose less of a threat than *Selaginella* and *Equisetum.* Potentially much more serious is the water-fern *Salvinia molesta,* an aggressive tropical weed that has been largely eradicated in New Zealand by careful management. A rather different threat is posed by *Asplenium ×lucrosum* which has been confused with *A. bulbiferum* and used extensively in revegetation projects. It has the capacity to become self-sustaining because of its vegetative bulbil production (Perrie et al. 2005).

The introduced spikemoss *Selaginella kraussiana* (Lycopodiopsida, Selaginellaceae).
Dennis Gordon, NIWA

Identification and cultivation, and economic, medicinal, and recreational uses

There has been huge popular interest in New Zealand ferns ever since the Victorian fern craze of the late 19th century – an era of Wardian cases, nature-printing, the production of elaborate albums of pressed ferns, and the use of fern designs in inlaid marquetry (Goulding 1977; Brownsey 1990, 2001b; Ide 1999). In response to this interest, the last five decades have seen regular publication of identification guides and illustrated books to the commoner species (Stevenson 1959; Hamlin 1963; Chinnock & Heath 1981; Firth et al. 1986; Crowe 1994).

Interest in nomenclature has resulted in compilations of common names (Nicol 1997) and Maori names (Beever 1991), and the derivation of Latin names of ferns (Brownsey 1988b). Books on the cultivation of New Zealand ferns have always been popular (Fisher 1984; Goudey 1988; Van der Mast & Hobbs 1998), and increasingly there is a revival of interest in the use of New Zealand ferns as food (Crowe 1990) and in Maori healing (Riley 1994). However, for general information about pteridophytes, the *Encyclopaedia of Ferns* (Jones 1987) is a wonderful compendium of information about diverse aspects of ferns from all over the world.

New Zealand seed plants

Seed plants arose 370 million years ago in the Late Devonian and they are the most numerous and ecologically successful of all plants (Niklas 1997). A seed is a remarkable structure. In the angiosperms (flowering plants), several genetically different individuals contribute tissue or genetic material to the mature seed. The seed coat or testa is contributed by the maternal sporophyte (the visible plant), and is diploid. The endosperm of the seed is a unique triploid tissue formed following the union of one paternal and two maternal genomes. It is a nutritive tissue present in most seeds but sometimes very reduced. The embryo itself is diploid and forms the next sporophyte individual. It is the product of the union of an egg and a sperm, each from a unisexual multicellular haploid gametophyte. Gametophytes are highly reduced in flowering plants – the female megagametophyte or embryo sac usually comprises seven cells (eight nuclei) and the male microgametophyte or pollen grain comprises two cells, one of

Kauri, *Agathis australis* (Pinopsida, Araucariaceae).
From Poole & Adams 1964

Hutu, *Ascarina lucida* (Magnoliopsida, Chloranthaceae).
From Poole & Adams 1964

which divides to form two sperm cells. Seeds of gymnosperms (conifers and close allies) are different, most notably in that their endosperm is not triploid but the haploid remnant of the larger megagametophyte. Seeds are the sites of internal fertilisation, and provide nutritive tissue to support and nourish the new seedling as it establishes. They are capable of remarkable feats of dispersal and dormancy (Raven et al. 1998).

Seeds, and the fruits that contain them, are also important items in human life. All the staple grains, e.g. rice, wheat, and barley, are one-seeded fruits. But seed plants are essential to human life and activity in many other ways. The seed plants are the familiar green components of wild, modified, and managed landscapes. They provide many of the things that are important to humans and other animal species – food and shelter, timber and paper, drugs, clothing, and ornamental plants. Also, along with other plants, algae, and photosynthetic bacteria, they sustain other life by converting sunlight into chemical energy and providing atmospheric oxygen. It is surprising then that the diversity of New Zealand's seed plants is still imperfectly catalogued and that seed plants are not given more emphasis in education, research, and conservation.

Within the seed plants, two groups are traditionally recognised – gymnosperms (Greek *gymnos*, naked; *spermos*, seed) and angiosperms (Greek *angeion*, container), with the seeds of the latter enclosed in an ovary. Traditionally, and based on the fossil record, botanists have believed that the angiosperms arose from within the gymnosperms. Some recent molecular studies (e.g. Denton et al. 1998; Raubeson 1998; Soltis et al. 2002) provide evidence for an alternative view – that the angiosperms and at least the living gymnosperms are sister groups, each having evolved independently from a common ancestor. This is interesting because, if living gymnosperms are a sister-group to the angiosperms, they must both be of the same age. But the earliest fossil flowering plant is *Archaefructus*, an aquatic plant of Early Cretaceous age (Sun et al. 2002), whereas fossils belonging to gymnosperm groups that are extant today extend back to late Paleozoic (Axsmith et al. 2003), so there is probably a long gap in the angiosperm fossil record. The 733 living species of gymnosperms are classified into four main orders (Judd et al. 1999), the largest of which is the Coniferales. All of New Zealand's 20 native gymnosperms are conifers, classified in three families (Hart 1987; Kelch 1997, 1998).

Angiosperms are also characterised by living companion cells in their inner bark that help organise the flow of nutrients, by the special triploid nutritive tissue (endosperm) mentioned above, and by their most obvious feature, the flower, which is variously modified and provides the important characters for classification of the group. Angiosperms comprise about 271,500 living species (Mabberley 2008).

Traditionally, the angiosperms have been divided into dicots (with two cotyledons in the seedling, broad, net-veined leaves, and flower parts mostly in fours or fives) and monocots (with one cotyledon, leaves that are mostly narrow and parallel-veined, and flower parts in threes). In recent years, this picture has changed. Angiosperm relationships have been the focus of a coordinated research effort that has used both molecular and morphological tools (Chase et al. 1993; Soltis et al. 1997; Davies et al. 2004) to understand their pattern of evolution. Research results have shown that, while the monocots are all related to each other, the dicots are not such a natural grouping. Some dicots (Eudicotyledons) are more closely related to the monocots than they are to other dicots (so-called basal angiosperms), because dicots as a whole do not have a unique common ancestor. The familiar division of flowering plants into dicots and monocots has therefore to be abandoned in favour of a more informative arrangement.

Molecular cladistic studies have also shown that some remarkably dissimilar plants are related. Recent publications by the Angiosperm Phylogeny Group (1998, 2003, 2009) provide a new phylogenetic classification system.

New Zealand molecular systematists have begun asking where our plants fit in these revised angiosperm classifications (e.g. Mitchell & Heenan 2000; Wagstaff & Dawson 2000; Wagstaff & Breitwieser 2002; Meudt & Bayly 2008; Wagstaff 2010a). In some cases, alternative generic placements (e.g. Albach & Chase 2001; Heenan et al. 2002a) or tribal placements (Wagstaff & Breitwieser 2002) have been proposed. In others, molecular data have not supported proposed monotypic segregate genera (Wagstaff et al. 1999; Mitchell & Heenan 2000; Wagstaff & Wege 2002).

Phylogenetic studies have also assessed critically the assumption that much of the New Zealand flora is an 80-million-year-old relic of Gondwana, and in many cases we have been forced to conclude that it is not so old (e.g. Winkworth et al. 2002; Smissen et al. 2003a; Mummenhoff et al. 2004; Winkworth et al. 2005). Molecular evidence indicates there is merit in Pole's (1994) suggestion that most of the New Zealand flora arrived here by long-distance dispersal since the late Cenozoic, although the notion of a complete Oligocene drowning remains controversial.

Alseuosmia quercifolia (Magnoliopsida, Alseuosmiaceae).
Dennis Gordon

The phylogenetic effort has been conducted alongside an enthusiastic search for new species. Field botanists, ecologists, and horticulturists have played an increasing role in the recognition of new taxa. This has been a stimulus to the rate of discovery, as evidenced in the many informal nicknames for supposed unnamed species that have become widely accepted and applied to New Zealand plants. Although these nicknames continue to cause nomenclatural difficulties for those who would uncritically accept them in publications, the existence of a list with nicknamed new entities (Druce 1993; Courtney 1999) is a valuable indicator of those groups in which more work is needed. It appears that somewhere between 15 and 20% of the known seed plants in New Zealand have not been named.

Classifications should be open to objective and critical testing. Speciation is the point at which netlike patterns of relationships among individuals in a population become branching patterns of relationships between populations. The process of speciation is considered to leave evidence of its occurrence in various forms, for example, unique new morphological characters, fixed allozyme differences, or DNA-sequence changes that are characteristic of new lineages. The different species concepts that are available all provide means to seek patterns that are evidence for or against the occurrence of the process of speciation. Considerable progress has been made in recent years in testing nicknamed entities and in publishing new species (e.g., Heenan 2009a, b; Heenan & de Lange 2011a, b; Heenan et al. 2008).

Diversity of New Zealand seed plants

About 4500 seed-plant species are indigenous and exotic in New Zealand. The end-chapter checklist records the names of 1 species of Ginkgoopsida, 60 species of Pinopsida, and 4436 species of Magnoliopsida. New Zealand has 2026 indigenous species of seed plants; 1648 (82%) species are endemic; five of these species are presumed to be extinct. About half of New Zealand's seed-plant flora is exotic, totalling 2471 species; 1753 (71%) exotic species are fully naturalised. More than 25,000 additional species are estimated to be in cultivation.

All 20 species of indigenous gymnosperms are endemic. The family Araucariaceae is represented by *Agathis*, the Cupressaceae by *Libocedrus*, the Podocarpaceae by *Dacrycarpus*, *Dacrydium*, *Halocarpus*, *Lepidothamnus*, *Manoao*, *Podocarpus*, *Prumnopitys*, and *Phyllocladus*. The genera *Manoao* and *Halocarpus* are endemic to New Zealand. The 40 exotic species belong to the families Araucariaceae, Cupressaceae, Ephedraceae, Pinaceae, and Taxaceae. Nearly a third (32%) of the exotic gymnosperms belong to the genus *Pinus*.

Halocarpus kirkii (Pinopsida, Podocarpaceae).
From Poole & Adams 1964

New Zealand has representatives of 204 families and 1249 genera of angiosperms. New Zealand has no endemic angiosperm families, but there are 50 endemic genera although the monophyly, relationships, and taxonomic status of many of them needs review. Almost half of the families (93 families) are represented by one genus only. The largest angiosperm family is the Asteraceae (Compositae) with 544 species, followed by the Poaceae (Gramineae) with 464 species, the Cyperaceae with 226 species, and the Plantaginaceae with 201 species. The largest families according to number of indigenous species are the Compositae with 292 species, followed by the Gramineae with 188 species, the Cyperaceae with 178 species, and the Plantaginaceae with 153 species. The largest families according to number of exotic species are the Gramineae with 276 species, followed by the Compositae with 252 species, the Fabaceae (Leguminosae) with 155 species, and the Rosaceae with 117 species.

The largest angiosperm genus ranked by number of indigenous and exotic species is *Veronica*, with at least 121 indigenous and 17 naturalised. Next is *Carex* with 107 species (81 indigenous; 26 naturalised). The largest genus ranked by exotic species is *Juncus* with 34 species. The largest endemic genus is *Raoulia* with 23 species.

Indigenous species

New Zealand lies on two continental plates that rafted from the Gondwana supercontinent some 80 million years (Ma) ago. Therefore it represents a land-mass of great age. New Zealand is renowned for its exceptional ecological richness and disharmonic biota, which may have been brought about by its long isolation, the great range in altitude and rainfall, and its diverse landforms, rocks, and soils. Following separation from Australia, New Zealand drifted northwards into warmer climates but remained close enough to other landmasses such as New Caledonia to gain new plants by long-distance dispersal. A distinctive tropical, woody element of New Zealand's flora can be traced to this period. Many of its members, including palms, pittosporums, coprosmas, and araliads, adapted to cooler climates or survived the following cool periods only in the far north.

Beginning 10 Ma ago, movement of the two continental plates thrust up tall mountain chains, thus creating new alpine habitats. There is much debate over the origin of New Zealand's alpine flora. Undoubtedly, some plants already in New Zealand adapted to the new alpine areas whereas others arrived from or via Australia or Antarctica. The Apiaceae, Asteraceae, Plantaginaceae, and other families rapidly diversified in the alpine zone. Dispersal of plants to New Zealand by strong westerly winds, by floating in seawater, or on the feet of waterbirds still provides a continual supply of new species.

Male flower of mat-forming *Coprosma petriei* (Magnoliopsida, Rubiaceae).
Dennis Gordon, NIWA

New Zealand's flora shows many unusual features. For example, New Zealand has a high proportion of plants that have separate male and female individuals (Webb et al. 1999) and New Zealand has become a centre for research in plant sexuality. Unisexuality has been linked with the unspecialised pollinator fauna, as a mechanism that prevents within-plant pollinations (Heine 1937; Lloyd 1985). Godley (1975, 1979) showed that a high proportion of this gender dimorphism had probably not evolved in New Zealand, leading Lloyd (1985) to propose that immigrant selection was involved in about 85% of the examples of the evolution of New Zealand's unisexual plant groups.

Compared with their overseas relatives, the flowers of native plants are often small, white, and regular, with short tubes and spreading petals. Such simple flowers are hypothesised to suit New Zealand's pollinating insects (Lloyd 1985). In some seed-plant families the ancestors of today's natives probably had simple flowers when they arrived in New Zealand, enabling them to diversify rapidly here. In other groups, like *Melicytus* (Powlesland 1984), ancestors of New Zealand's simple-flowered species lost their complex pollination mechanisms during later evolution in New Zealand (Newstrom & Robertson 2005). In contrast, flowers of some plants on New Zealand's subantarctic islands are larger and more brighly coloured than their mainland relatives (Wagstaff et al. 2011), in spite of absence or paucity of specialized pollinators there. This feature has not yet been adequately explained.

Some families have shown remarkable adaptability by evolving special growth-forms to cope with new habitats. This is particularly true of New Zealand's alpine cushion plants, mountain scree plants, and lowland divaricating shrubs. New Zealand's flora has few annuals, deciduous plants, or geophytes, and this is thought to result from an equable and unpredictable climate (Godley 1975). The absence of the scleromorphic element so characteristic of Australia is thought to be a result of higher soil fertilities in New Zealand's rapidly eroding landscapes (Flannery 1994; Groves 1994), which place slow-growing scleromorphic immigrants at a disadvantage in competition with mesomorphic species (Lloyd 1985).

The rare chlorophyll-lacking saprophyte *Thismia rodwayi* (Magnoliopsida, Burmanniaceae).

Ken Grange, NIWA

Naturalised species

Polynesian settlers, the ancestors of today's Maori people, brought plants with them intentionally or unintentionally, some of which, like *Sigesbeckia orientalis*, became established as wild plants. Some weedy plants characteristic of the Northern Hemisphere, such as *Rorippa palustris*, were already established here when Banks and Solander collected them in 1769, and perhaps had been able to establish following human disturbance. Large-scale introduction of European crops and garden plants after 1840 brought many attendant weeds and the number has steadily grown year by year (e.g. Heenan et al. 1998, 1999, 2002b, 2004a), with only a few losses. Origins and biology of the naturalised flora were reviewed by Healy (1969, 1973), Esler (1988), and Webb et al. (1988).

The Plant Names Database, housed at Landcare Research and made accessible through the Ngā Tipu Aotearoa/New Zealand Plants website, records about as many naturalised species as indigenous species. This supports Darwin's (1859) view that European wildlife 'would become thoroughly naturalized ... and would exterminate many of the natives'. The success of naturalised plants in New Zealand does not imply superiority, but rather reflects adaptations to the prevailing European land-uses in New Zealand and the paucity of native ruderal and annual species. Where human disturbance is absent, naturalised plants generally fail to establish. With generally improved border controls and a public awareness of the dangers of introducing foreign species, it is likely that future additions to the naturalised flora will come mostly from horticultural plants that are already in New Zealand (Webb et al. 1988).

Historical overview of studies on New Zealand seed plants

Maori plant-taxonomic knowledge is significant (Beever 1991). It relates particularly to naming of plants used for food, medicine, and timber, and having artistic or cultural significance. Maori have detailed knowledge of variability in traditional economic species, for example, harakeke (species of *Phormium*) where they have selected and named many cultivars (Scheele & Walls 1994).

Banks and Solander were the first scientific botanists to visit New Zealand, but their descriptions of the plants they collected were never published. New Zealand's first Floras were published by Johannes and Georg Forster, who were naturalists on Cook's second voyage. The *Characteres Generum Plantarum* (Forster & Forster 1775) described 37 genera and 134 species from New Zealand. The *Florulae Insularum Australium Prodromus*, published 11 years later, described 141 species of New Zealand plants and listed others from Solander's manuscript. The botanists on Cook's and other early voyages of exploration catalogued and described over 10% of the vascular-plant flora (Allan 1961).

Parapara, *Pisonia brunoniana* (Magnoliopsida, Nyctaginaceae).

Dennis Gordon, NIWA

After colonisation, local botanists sent plant specimens to Kew for naming by J. D. Hooker, to other European centres, and to F. J. H. von Mueller in Melbourne. Hooker visited New Zealand in 1841 and published two New Zealand Floras in the following decades (Hooker 1852–1855, 1864–1867).

From 1869 to 1928, the establishment of the New Zealand Institute, development of museums, and growing confidence and independence of local scientists led to publication of botanical research in New Zealand and the development of a resident capability in botany. Kirk (1899) and Cheeseman (1906, 1925) published the first Floras by resident botanists.

Soon after its inception in 1926, the Department of Scientific and Industrial Research (DSIR) took responsibility for Floras. The Flora series started by the Botany Division in 1949 is now complete (Edgar & Connor 2000). The plant taxonomic community is now planning for the next generation of taxonomic treatments of the flora – an electronic Flora will allow a novel, more reproduceable approach to monographic revisions, Flora writing, and other taxonomic work.

Major collections of New Zealand seed plants are held at AK, BM, CHR, K, OTA and WELT herbaria. About half the types determining the application of names to New Zealand plants are held in overseas herbaria.

Current seed plant taxonomic research in New Zealand

Until recently the taxonomic research of seed plants at the Allan Herbarium (Landcare Research) and Museum of New Zealand Te Papa Tongarewa was funded through an Outcome-Based-Investment (OBI), in part, by the New Zealand Ministry of Science and Innovation and formerly by the Foundation for Research, Science and Technology. In 2010, the 'Defining New Zealand's Land Biota' OBI was transferred to a Backbone contract that covered the preservation, maintenance, and development of Nationally Significant Database assets and associated research activities. This contract is now included in Landcare Research's Core Funding. Some research on seed plants is funded by New Zealand's universities and the Auckland Museum, and the Department of Conservation funds research on rare and threatened seed plants.

Celmisia spectabilis (Magnoliopsida, Asteraceae).
Dennis Gordon

Plant systematics research associated with the Allan Herbarium and Te Papa seeks to determine authoritative names, provide descriptions, and document the distribution of key indigenous and naturalised plant groups selected with the guidance of end-users. These resulting taxonomic studies allow end-users to identify plant groups and access other known biological information such as geographic distributions, reproductive biology, ecological links and taxonomic significance of characters. The taxonomic treatments underpin a range of other biological sciences, principally in the biosecurity and biodiversity sectors. Research also encompasses phylogenetic studies that reveal the evolutionary history, relationships, and genetic diversity of plants, and assess the conservation status of key taxa. Both the taxonomic and phylogenetic studies fit into global initiatives to document the biodiversity of widespread taxonomic groups and to reconstruct the Tree of Life.

Plant systematics staff at the Allan Herbarium are responsible for writing New Zealand Floras. The latest completed Flora volume is the second edition of *Flora of New Zealand Grasses* (Edgar & Connor 2010). A new seed-plant flora project is now underway (see below; http://www.nzflora.info). The first treatment published in this new electronic Flora of New Zealand is a new treatment of *Hypericum* (Heenan 2010). Following the publication of *Seeds of New Zealand Gymnosperms and Dicotyledons* (Webb & Simpson 2001), work on seeds focuses now on a seed atlas of New Zealand monocots. Floras and other guides are essential to many economic pursuits, and are of major scientific, cultural, and social value.

Biosystematics research aims at understanding evolutionary processes and relationships. Genetic-diversity research has focused mainly on the Asteraceae (Smissen et al. 2003b, 2004, 2006, 2007; Smissen & Breitwieser 2008; Breitwieser et al. 2010), but also on a range of other plant groups such as *Phormium* (Smissen et al. 2008; Smissen & Heenan 2010) and *Simplicia* (Smissen et al. 2011). Phylogenetic analyses, especially of molecular data, provide insight into the origin and diversity of New Zealand's plants, leading eventually to a more stable and predictive classification of the flora. Recent research has unravelled phylogenetic relationships in numerous families, including Asteraceae (Wagstaff & Breitwieser 2004; Wagstaff et al. 2006; Wagstaff et al. 2011), Ericaceae (Wagstaff

2010a), Malvaceae (Wagstaff 2010b), Podocarpaceae (Wagstaff 2004), and Stylidiaceae (Wagstaff & Wege 2002).

Several major revisions were recently completed: of *Gentianella* (Glenny 2004), *Ourisia* (Meudt 2006), and *Plantago* (Meudt, submitted). Revision of New Zealand *Veronica* at species level is also complete (Garnock-Jones 1993, as *Heliohebe*; Garnock-Jones & Lloyd 2004, as *Parahebe*; Bayly & Kellow 2006, as *Hebe*; Meudt 2008). Three speciose genera lacking modern taxonomic treatments – *Cardamine* (P. Heenan pers. comm.), *Craspedia* (I. Breitwieser), and *Myosotis* (H. Meudt and C. Lehnebach pers. comm.) – are currently being revised.

Long-term collaborative research between Landcare Research, University of Canterbury, and researchers in Spain, the US, and Australia focuses on systematics and evolution of Gnaphalieae (Asteraceae). Contributions to the increased understanding of relationships in the New Zealand Gnaphalieae have been provided in morphology, anatomy, taxonomy, flavonoid profiles, cytology, isozymes, intergeneric hybridisation, flowering phenology, and molecular systematics (Haase et al. 1993; McKenzie 2001; Breitwieser & Ward 2003 and references therein; Smissen et al. 2003b, 2004; McKenzie et al. 2004, 2008; Bayer et al. 2007; Ward et al. 2009; Smissen et al. 2011), and in the genera *Leucogenes* (Molloy 1995; Smissen & Breitwieser 2008), *Anaphalioides* (Glenny 1997), *Rachelia* (Ward et al. 1997), *Helichrysum* (Smissen et al. 2006, 2007), *Ozothamnus* (Breitwieser & Ward 1997; Schoenberger 2002), *Euchiton* (Ward & Breitwieser 1998; Flann 2005; Flann et al. 2008), *Argyrotegium* (Ward et al. 2003), and *Craspedia* (Ford 2004; Ford et al. 2007; Breitwieser et al. 2010).

Veronica (Plantaginaceae) is New Zealand's largest seed-plant genus. It has 121 indigenous species in a single clade nested among the widespread northern-hemisphere species of *Veronica*. The New Zealand species were until recently classified in several genera – *Hebe, Parahebe, Heliohebe, Chionohebe,* and *Leonohebe* – but the segregate genera are not all monophyletic and their recognition along with northern segregates *Pseudolysimachion* and *Synthyris* (Albach 2008; Albach et al. 2004; Garnock-Jones et al. 2007) would restrict *Veronica* to a circumscription that is clearly paraphyletic. The aims of recent research on the *Veronica* complex, *Ourisia* (Meudt 2006), and *Plantago* (Meudt in press), carried out by a multidisciplinary team centered at the Museum of New Zealand Te Papa Tongarewa, have been to determine the morphological and chemical characteristics, distribution, and habitat preferences of all the species in Plantaginaceae (Garnock-Jones & Lloyd 2004; Bayly & Kellow 2006 and references therein; Meudt 2008; Meudt in prep.; Mitchell et al. 2007; Taskova et al. 2010) and to test hypotheses about their variation and evolution (Wagstaff et al. 2002; Meudt & Bayly 2008).

New species, mainly rare and threatened, have been described and taxonomic information provided for numerous genera, e.g. *Sophora* (Heenan et al. 2004b), *Cyperus* (Heenan & de Lange 2005), *Hoheria* (Heenan et al. 2005), *Olearia* (Heenan et al. 2008), *Leptinella* (Heenan 2009a), *Pachycladon* (Heenan 2009b), *Myoporum* (Heenan & de Lange 2011a), and *Lepidium* (Heenan & de Lange 2011b). A significant contribution in cytology is Dawson's (2000) chromosome index of New Zealand's indigenous seed plants.

Exotic species are a growing threat to New Zealand's indigenous flora. Therefore naturalised species, important to biosecurity and primary industries, have to be well documented. New records of naturalised or casual dicotyledons and pteridophytes in New Zealand have been documented in checklists (Webb et al. 1995; Heenan et al. 1998, 1999, 2002b, 2004a, 2008).

Other research on the systematics of seed plants is undertaken at universities and the Department of Conservation. Most universities have ongoing programmes of research in molecular systematics (Lockhart et al. 2001; Smissen et al. 2003a; Gemmill et al. 2002; Gardner et al. 2004; Tay et al., 2009 a, b; Woo et al. 2011; Pelser et al. 2010; Prebble et al. 2011) and some are active also in taxonomy (e.g. Ward et al. 2003; Garnock-Jones & Lloyd 2004; Garnock-Jones et al., 2007).

Gentianella corymbosa subsp. *corymbosa* (Magnoliopsida, Gentianaceae).
Dennis Gordon, NIWA

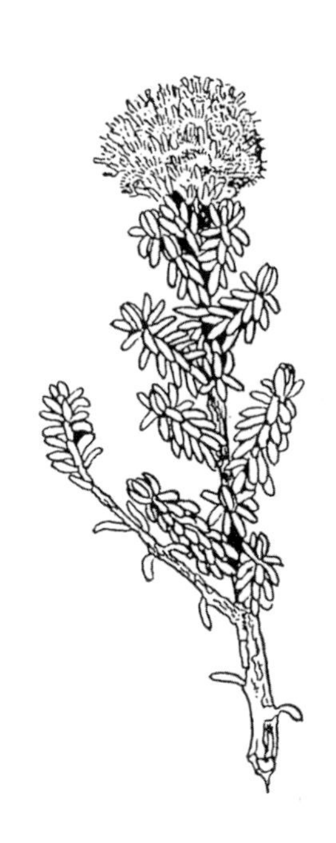

Ozothamnus leptophyllus (Magnoliopsida, Asteraceae).
From Poole & Adams 1964

Veronica elliptica (Magnoliopsida, Plantaginaceae).
Philip Garnock-Jones

Although the Department of Conservation is not specifically charged with taxonomic research, it has maintained a role of supporting staff time and contracting specialist help to resolve: unnamed entities believed to be seriously under threat and unlikely to be resolved in the immediate future; those taxa that are part of existing work programmes, such as Recovery Plans; and those for which there has been shown to be a requirement for further taxonomic research. Examples include *Carex* (de Lange & Heenan 1997; Heenan et al. 1997; Heenan & de Lange 1997), *Crassula* (de Lange et al. 2008), *Hibiscus* (Craven et al. 2011), *Ranunculus* (de Lange & Murray 2008), *Pittosporum serpentinum* (de Lange 1998b, 2003), *Veronica* (as *Hebe*: de Lange 1996, 1997, 1998a; de Lange & Rolfe 2008), and *Simplicia* (Smissen et al. 2011). Stemming from urgently identified research issues, a major revision of New Zealand *Lepidium* is well underway (P. J. de Lange & P. B. Heenan pers. comm.), while a full revision of the *Kunzea ericoides* complex is now nearing completion (P. J. de Lange pers. comm.). Further, through the auspices of the New Zealand Plant Conservation Network, Department of Conservation staff, in collaboration with other agencies, have produced a Threatened Plant Book (de Lange et al. 2010) that by its very nature contains a wealth of taxonomic information.

Gaps in knowledge and scope for future research

There are only about five 'full-time equivalents' (FTEs) of actual research time on seed-plant taxonomy for employed scientists in New Zealand. Additionally, four permanent full-time academics in New Zealand include seed-plant systematics in their research programmes. Given that New Zealand has a highly endemic flora and is considered an important hot-spot of endemism (Mittermeier et al. 1999), and given the wide range of applications for seed-plant studies (in ecology, forestry, agriculture, conservation, ecotourism, pharmacy, toxicology, natural-products chemistry, weed control, biological control, biotechnology, plant breeding, biosecurity, forensic botany, and ethnobotany), the seed plants are deserving of greater study. Some 15–20% of the known seed plants are still undescribed and need to be critically examined, preferably in the context of a monograph. A new seed-plant Flora needs to be written (Allan's Flora is 50 years old) as well as other identification aids. Seed-plant introductions need to be continuously recorded and correctly identified and various high-quality information systems need to be developed (e.g. databases, GIS, interactive keys, web-based information, and an electronic Flora). The good news is that the development of an electronic Flora for New Zealand is now underway (http://www.nzflora.info). This project will provide a dynamic, continually updated, electronically-based Flora of New Zealand. It will be based on new systematic research and will bring together information from our network of databases and online resources. Users will have easy access to the most authoritative, accurate, and up-to-date information on New Zealand flowering plants, gymnosperms, ferns, bryophytes, marine and freshwater algae, and lichens. Participating institutions are Landcare Research, Te Papa, and NIWA. Other institutions may join later.

Acknowledgements

We acknowledge support from the former New Zealand Foundation for Research, Science and Technology.

PB and LP thank Barbara Parris and John Lovis for many helpful comments and additions to the account of New Zealand pteridophyte species.

Authors*

Dr Ilse Breitwieser Allan Herbarium, Landcare Research, P.O. Box 40, Lincoln, New Zealand [breitwieseri@landcareresearch.co.nz]

Dr Patrick J. Brownsey Museum of New Zealand Te Papa Tongarewa, P.O. Box 467, Wellington, New Zealand [patb@tepapa.govt.nz]

Prof. Philip J. Garnock-Jones School of Biological Sciences, Victoria University of Wellington, P.O. Box 600, Wellington, New Zealand [phil.garnock-jones@vuw.ac.nz]

Dr Leon R. Perrie Museum of New Zealand Te Papa Tongarewa, P.O. Box 467, Wellington, New Zealand [leonp@tepapa.govt.nz]

Dr Aaron Wilton Allan Herbarium, Landcare Research, P.O. Box 40, Lincoln, New Zealand [wiltona@landcareresearch.co.nz]

* Authors of this chapter are listed alphabetically on the first page; in all other multi-authored chapters in this volume authors are listed according to the number of column centimetres each contributed to the chapter (including the checklist).

References

ALLAN, H. H. 1961: *Flora of New Zealand, Volume 1, Indigenous Tracheophyta.* Government Printer, Wellington. 1085 p.

ALBACH, D. C.; CHASE, M. W. 2001: Paraphyly of *Veronica* (Veronicaceae: Scrophulariaceae): evidence from internal transcribed spacer (ITS) sequences of nuclear ribosomal DNA. *Journal of Plant Research 114*: 9–18.

ANGIOSPERM PHYLOGENY GROUP 1998: An ordinal classification for the families of flowering plants. *Annals of the Missouri Botanical Garden 85*: 531–553.

ANGIOSPERM PHYLOGENY GROUP 2003: An update of the Angiosperm Phylogeny Group classification for the orders and families of flowering plants: APG II. *Botanical Journal of the Linnean Society 141*: 399–436.

ANGIOSPERM PHYLOGENY GROUP 2009: An update of the Angiosperm Phylogeny Group classification for the orders and families of flowering plants: APG III. *Botanical Journal of the Linnean Society 161*: 105–121.

ATKINSON, L. R. 1973: The gametophyte and family relationships. Pp. 73–90 *in*: Jermy, A. C.; Crabbe, J. A.; Thomas, B. A. (eds), *The Phylogeny and Classification of the Ferns.* The Linnean Society of London, London. 284 p.

AXSMITH, B. J.; SERBET, R.; KRINGS, M.; TAYLOR, T. N.; TAYLOR, E. L.; MAMAY, S. H. 2003: The enigmatic Paleozoic plants *Spermopteris* and *Phasmatocycas* reconsidered. *American Journal of Botany 90*: 1585–1595.

BANNISTER, P. 1984: The seasonal course of frost resistance in some New Zealand pteridophytes. *New Zealand Journal of Botany 22*: 557–563.

BATEMAN, R. M. 1996: An overview of Lycophyte phylogeny. Pp. 405–415 *in*: Camus, J. M.; Gibby, M.; Johns, R. J. (eds), *Pteridology in Perspective.* Royal Botanic Gardens, Kew. 700 p.

BAYER, R.; BREITWIESER, I.; WARD J. M.; PUTTOCK, C. 2007: Gnaphalieae. Pp. 246–284 *in*: Kubitzki, K. (ed.), *The Families and Genera of Vascular Plants. Flowering Plants – Dicotyledons: Compositae.* Springer-Verlag, Berlin. 635 p.

BAYLY, M. J.; KELLOW, A. V. 2006: *An Illustrated Guide to New Zealand Hebes.* Te Papa Press, Wellington. 350 p.

BAYLY, M. J.; KELLOW, A. V.; MITCHELL, K. A.; MARKHAM, K. R.; de LANGE, P. J.; HARPER, G. E.; GARNOCK-JONES, P. J.; BROWNSEY, P. J. 2002: Descriptions and flavonoid chemistry of new taxa in *Hebe* sect. *Subdistichae* (Scrophulariaceae). *New Zealand Journal of Botany 40*: 571–602.

BEEVER, J. 1991: A dictionary of Maori plant names. 2nd edn. *Auckland Botanical Society Bulletin 20*: 1–75.

BEEVER, J. E. 1984: Moss epiphytes of tree ferns in a warm temperate forest, New Zealand. *Journal of the Hattori Botanical Laboratory 56*: 89–95.

BERNHARDI, J. J. 1801: Tentamen alterum filices in genera redigendi. *Journal für die Botanik (Schrader) 1800*: 121–136.

BOUMA, W. L. M.; RITCHIE, P.; PERRIE, L. R. 2010: Phylogeny and generic taxonomy of the New Zealand Pteridaceae ferns from chloroplast *rbc*L DNA sequences. *Australian Systematic Botany 23*: 143–151.

BRAGGINS, J. E. 1975: Studies on the New Zealand, and some related, species of *Pteris* L. Unpublished PhD thesis, University of Auckland, Auckland.

BRAGGINS, J. E. 1980: Some studies on the New Zealand species of *Botrychium* Sw. (Ophioglossaceae). *New Zealand Journal of Botany 18*: 353–366.

BREITWIESER, I.; WARD, J. M. 1997: Transfer of *Cassinia leptophylla* to *Ozothamnus. New Zealand Journal of Botany 35*: 125–128.

BREITWIESER, I.; WARD, J. M. 2003: Phylogenetic relationships and character evolution in New Zealand and selected Australian Gnaphalieae (Compositae) inferred from morphological and anatomical data. *Botanical Journal of the Linnean Society 141*: 183–203.

BREITWIESER, I.; SMISSEN, R. D.; FORD, K. A. 2010: A test of reproductive isolation among three sympatric species of *Craspedia* (Asteraceae: Gnaphalieae) at Mt Arthur in New Zealand. *New Zealand Journal of Botany 48*: 1–7

BREITWIESER, I.; BROWNSEY, P.; FORD, K.; GLENNY, D.; NELSON, W.; HEENAN, P.; WILTON, A. (Eds) 2010: *Flora of New Zealand.* Online Edition. Accessed at www.nzflora.info.

BROWN, R. 1810: *Prodromus florae Novae Hollandiae et Insulae van Diemen.* Johnson, London. 590 p.

BROWNSEY, P. J. 1977a: A taxonomic revision of the New Zealand species of *Asplenium. New Zealand Journal of Botany 15*: 39–86.

BROWNSEY, P. J. 1977b: *Asplenium* hybrids in the New Zealand flora. *New Zealand Journal of Botany 15*: 601–637.

BROWNSEY, P. J. 1979: *Cyathea cunninghamii* in New Zealand. *New Zealand Journal of Botany 17*: 97–107.

BROWNSEY, P. J. 1981: A biosystematic study of a wild population of *Leptopteris* hybrids in New Zealand. *New Zealand Journal of Botany 19*: 343–352.

BROWNSEY, P. J. 1983: *Asplenium terrestre* and two *Asplenium* hybrids: new fern records for Australia. *Muelleria 5*: 219–221.

BROWNSEY, P. J. 1985a: *Asplenium chathamense* – a new fern species from the Chatham Islands, New Zealand. *New Zealand Journal of Botany 23*: 135–140.

BROWNSEY, P. J. 1985b: Biosystematics of the cryptogamic flora: pteridophytes. *New Zealand Journal of Botany 23*: 681–685.

BROWNSEY, P. J. 1988a: Pteridophyta. Pp. 2–37 *in*: Webb, C. J.; Sykes, W. R.; Garnock-Jones, P. J. *Flora of New Zealand, Vol. IV, Naturalised Pteridophytes, Gymnosperms and Dicotyledons.* DSIR, Christchurch. 1365 p.

BROWNSEY, P. J. 1988b: Fern names and their origins. *Wellington Botanical Society Bulletin 44*: 19–25.

BROWNSEY, P. J. 1990: The Lucy Cranwell Lecture 1990: New Zealand ferns and allied plants – being a further contribution to making known the botany of New Zealand. *Auckland Botanical Society Journal 46*: 38–60.

BROWNSEY, P. J. 1996: Ferns of the Wellington region. *Wellington Botanical Society Bulletin 47*: 2–11.

BROWNSEY, P. J. 2001a: New Zealand's pteridophyte flora – plants of ancient lineage but recent arrival? *Brittonia 53*: 284–303.

BROWNSEY, P. J. 2001b: Ferns – the glory of the forest. *New Zealand Geographic 49*: 64–82.

BROWNSEY, P. J.; CHINNOCK, R. J. 1984: A taxonomic revision of the New Zealand species of *Hypolepis. New Zealand Journal of Botany 22*: 43–80.

BROWNSEY, P. J.; de LANGE, P. J. 1997: *Asplenium cimmeriorum* – a new fern species from New Zealand. *New Zealand Journal of Botany 35*: 283–292.

BROWNSEY, P. J.; GALLOWAY, T. N. H. 1987: A key to the genera of New Zealand ferns and allied plants. *National Museum of New Zealand Miscellaneous Series 15*: 1–31.

BROWNSEY, P. J.; GIVEN, D. R.; LOVIS, J. D. 1985a: A revised classification of New Zealand pteridophytes with a synonymic checklist of species. *New Zealand Journal of Botany 23*: 431–489.

BROWNSEY, P. J.; JACKSON, P. J. 1984: *Asplenium pauperequitum* – a new species from the Poor Knights Islands, New Zealand. *New Zealand Journal of Botany 22*: 315–321.

BROWNSEY, P. J.; LOVIS, J. D. 1990: *Pellaea calidirupium* – a new fern species from New Zealand. *New Zealand Journal of Botany 28*: 197–205.

BROWNSEY, P. J.; MOSS, T. C.; SNEDDON, B.V. 1985b: Cone production in *Equisetum arvense*. *Wellington Botanical Society Bulletin 42*: 16–21.

BROWNSEY, P. J.; SMITH-DODSWORTH, J. C. 1989: *New Zealand Ferns and Allied Plants*. David Bateman Ltd, Auckland. 168 p.

BROWNSEY, P. J.; SMITH-DODSWORTH, J. C. 2000: *New Zealand Ferns and Allied Plants*, 2nd edn. David Bateman Ltd, Auckland. 168 p.

CAMERON, E. K.; de LANGE, P. J.; PERRIE, L. R.; BROWNSEY, P. J.; CAMPBELL, H. J.; TAYLOR, G. A.; GIVEN, D. R.; BELLINGHAM, R. M. 2006: A new location for the Poor Knights spleenwort (*Asplenium pauperequitum*, Aspleniaceae) on The Forty Fours, Chatham Islands, New Zealand. *New Zealand Journal of Botany 44*: 199–209.

CAMUS, J. M.; GIBBY, M.; JOHNS, R. J. (Eds) 1996: *Pteridology in Perspective*. Royal Botanic Gardens, Kew. 700 p.

CAVALIER-SMITH, T. 1998: A revised six-kingdom system of life. *Biological Reviews 73*: 203–266.

CHAMBERS, T. C.; FARRANT, P. A. 1991: A re-examination of the genus *Cheilanthes* (Adiantaceae) in Australia. *Telopea 4*: 509–557.

CHAMBERS, T. C.; FARRANT, P. A. 1996: *Blechnum blechnoides* (Bory) Keys. (Blechnaceae), formerly *B. banksii* (Hook.f.) Mett. ex Diels, a fern from salt-spray habitats of New Zealand and Chile. *New Zealand Journal of Botany 34*: 441–445.

CHAMBERS, T. C.; FARRANT, P. A. 1998: The *Blechnum procerum* ('*capense*') (Blechnaceae) complex in New Zealand. *New Zealand Journal of Botany 36*: 1–19.

CHASE, M. W.; REVEAL, J. L. 2009: A phylogenetic classification of the land plants to accompany APG III. *Botanical Journal of the Linnean Society 161*: 122–127.

CHASE, M. W.; SOLTIS, D. E.; OLMSTEAD, R. G.; MORGAN, D.; LES, D. H. et al. 1993: Phylogenetics of seed plants: An analysis of nucleotide sequences from the plastid gene rbcL. *Annals of the Missouri Botanical Garden 80*: 528–580.

CHEESEMAN, T. F. 1906: *Manual of the New Zealand Flora*. Government Printer, Wellington. 1199 p.

CHEESEMAN, T. F. 1925: *Manual of the New Zealand Flora*, 2nd edn. Government Printer, Wellington. 1163 p.

CHINNOCK, R. J. 1975: The New Zealand species of *Tmesipteris*. *New Zealand Journal of Botany 13*: 743–768.

CHINNOCK, R. J.; HEATH, E. 1981: *Mobil New Zealand Nature Series. Common Ferns and Fern Allies*. Reed, Wellington. 48 p.

CHRISTENHUSZ. M. J.; XIAN-CHUN, Z.; SCHNEIDER, H. 2011: A linear sequence of extant families and genera of lycophytes and ferns. *Phytotaxa 19*: 7–54.

COCKAYNE, L.; ALLAN, H. H. 1934: An annotated list of groups of wild hybrids in the New Zealand flora. *Annals of Botany 48*: 1–55.

COLENSO, W. 1842: Description of some new ferns lately discovered in New Zealand. *Tasmanian Journal of Natural Science 1*: 375–379.

COLENSO, W. 1846: A classification and description of some newly discovered ferns, collected in the Northern Island of New Zealand, in the summer of 1841–2. *Tasmanian Journal of Natural Science 2*: 161–189.

COLLINSON, M. E. 1996:*'What use are fossil ferns?'* – 20 years on: with a review of the fossil history of extant pteridophyte families and genera. Pp. 349–394 *in*: Camus, J. M.; Gibby, M.; Johns, R. J. (eds), *Pteridology in Perspective*. Royal Botanic Gardens, Kew. 700 p.

COURTNEY, S. 1999: *A Checklist of Indigenous Vascular Plants of New Zealand*. Updated and expanded from A.P. Druce's 9th revision, September 1993. Unpublished checklist, Department of Conservation, Nelson / Marlborough Conservancy.

CRABBE, J. A.; JERMY, A. C.; MICKEL, J. T. 1975: A new generic sequence for the pteridophyte herbarium. *Fern Gazette 11*: 141–162.

CRAVEN, L. A.; de LANGE, P. J.; LALLY,T. R.; MURRAY, B. G.; JOHNSON, S. B. 2011: A taxonomic re-evaluation of *Hibiscus trionum* (Malvaceae) in Australasia. *New Zealand Journal of Botany 49*: 27–40.

CROOKES, M. E. 1963: *New Zealand Ferns, 6th edn, incorporating illustrations and original work by H. B. Dobbie*. Whitcombe & Tombs, Christchurch. 408 p.

CROWE, A. 1990: *Native Edible Plants of New Zealand*. Hodder & Stoughton, Auckland. 194 p.

CROWE, A. 1994: *Which Native Fern?*Viking, Auckland. 64 p.

CUNNINGHAM, A. 1837: Florae insularum Novae Zelandiae precursor, or a specimen of the botany of the islands of New Zealand. *Companion to the Botanical Magazine 2*: 358–378.

DAELLENBACH, M. 1982: A biosystematic study in the *Mecodium sanguinolentum* (Forst.f.) Presl species group. Unpublished BSc Hons project, University of Canterbury, Christchurch.

DARWIN, C. 1859: *On the Origin of Species by Means of Natural Selection, or the preservation of favoured races in the struggle for life*. John Murray, London. 502 p.

DAVIES, T. J.; BARRACLOUGH, T. G.; CHASE, M. W.; SOLTIS, P. S.; SOLTIS, D. E.; SAVOLAINEN, V. 2004: Darwin's abominable mystery: Insights from a supertree of the angiosperms. *Proceedings of the National Academy of Sciences 101*: 1904–1909.

DAVISON, A. C. 1995: Studies on the genus *Christella* Léveillé in New Zealand. Unpublished MSc thesis, University of Auckland, Auckland.

DAWSON, J. W. 1953: A key to the New Zealand lycopods. *Tuatara 5*: 6–11.

DAWSON, M. I. 2000: Index of chromosome numbers of indigenous New Zealand spermatophytes. *New Zealand Journal of Botany 38*: 47–150.

DAWSON, M. I.; BROWNSEY, P. J.; LOVIS, J. D. 2000: Index of chromosome numbers of indigenous New Zealand pteridophytes. *New Zealand Journal of Botany 38*: 25–46.

de LANGE, P. J. 1988: *Ophioglossum petiolatum* Hook. in a Reserve near Kawhia. *Wellington Botanical Society Bulletin 44*: 4–7.

de LANGE, P. J. 1996: *Hebe bishopiana* (Scrophulariaceae) – an endemic species of the Waitakere Ranges, west Auckland, New Zealand. *New Zealand Journal of Botany 34*: 187–194.

de LANGE, P. J. 1997: *Hebe brevifolia* (Scrophulariaceae) – an ultramafic endemic of the Surville Cliffs, North Cape, New Zealand. *New Zealand Journal of Botany 35*: 1–8.

de LANGE, P. J. 1998a: *Hebe perbella* (Scrophulariaceae) – a new and threatened species from western Northland, North Island, New Zealand. *New Zealand Journal of Botany 36*: 399–406.

de LANGE, P. J. 1998b: *Pittosporum ellipticum* subsp. *serpentinum* (Pittosporaceae) – a new ultramafic endemic from the Surville Cliffs, North Cape, New Zealand. *New Zealand Journal of Botany 36*: 389–397.

de LANGE, P. J. 2003: *Pittosporum serpentinum* (Pittosporaceae) – a new combination for an ultramafic endemic of the Surville Cliffs, North Cape, New Zealand. *New Zealand Journal of Botany 41*: 725–726.

de LANGE, P. J.; CAMERON, E. K. 1999: The vascular flora of Aorangi Island, Poor Knights Islands, northern New Zealand. *New Zealand Journal of Botany 37*: 433–468.

de LANGE, P. J.; CROWCROFT, G. M. 1997: *Macrothelypteris torresiana* (Thelypteridaceae) at North Cape, North Island, New Zealand – a new southern limit for a tropical fern. *New Zealand Journal of Botany 35*: 555–558.

de LANGE, P. J.; GARDNER, R. O; SYKES, W. R.; CROWCROFT, G. M.; CAMERON, E. K.; STALKER, F.; CHRISTIAN, M. L.; BRAGGINS, J. E. 2005: Vascular flora of Norfolk Island: some additions and taxonomic notes. *New Zealand Journal of Botany 43*: 563–596.

de LANGE, P. J.; HEENAN, P. B. 1997: *Carex ophiolithica* (Cyperaceae): a new ultramafic endemic from the Surville Cliffs, North Cape, New Zealand. *New Zealand Journal of Botany 35*: 429–436.

de LANGE, P. J.; HEENAN, P. B.; NORTON, D. A.; ROLFE, J.; SAWYER, J. 2010: *Threatened plants of New Zealand*. Canterbury University Press, Christchurch. 471 p.

de LANGE, P. J.; MURRAY, B. G.; DATSON, P. M. 2004: Contributions to a chromosome atlas of the New Zealand flora – 38. Counts for 50 families. *New Zealand Journal of Botany 42*: 873–904.

de LANGE, P. J.; MURRAY, B. G. 2008: *Ranunculus ranceorum*, a new name and rank for *Ranunculus recens* var. *lacustris* G.Simpson, an elusive, rarely seen buttercup of the Fiordland lakes, South Island, New Zealand. *New Zealand Journal of Botany 46*: 1–11.

de LANGE, P. J.; HEENAN, P. B.; KEELING, D. J.; MURRAY, B. G.; SMISSEN, R.; SYKES, W. R. 2008: Biosystematics and conservation: a case study with two enigmatic and uncommon species of *Crassula* from New Zealand. *Annals of Botany 101*: 881–899.

de LANGE, P. J.; NORTON, D. A.; COURTNEY, S. P.; HEENAN, P. B.; BARKLA, J. W.; CAMERON, E. K. 2009: Threatened and uncommon plants of New Zealand (2008 revision). *New Zealand Journal of Botany 47*: 61–96.

de LANGE, P. J.; ROLFE, J. R. 2008: *Hebe saxicola* (Plantaginaceae) – a new threatened species from western Northland, North Island, New Zealand. *New Zealand Journal of Botany 46*: 531–545.

de LANGE, P. J.; ROLFE, J. R. 2011: Vascular flora of Maungaraho Rock Scenic Reserve. *Wellington Botanical Society Bulletin 53*: 11–22.

DENTON, A. L.; McCONAUGHY, B. L.; HALL, B. D. 1998: Land plant phylogeny estimation using

RNA polymerase II sequences. *American Journal of Botany 85 (Suppl.)*: 123.
DOBBIE, H. B. 1921: *New Zealand Ferns*, 2nd edn. Whitcombe & Tombs, Auckland. 394 p.
DRUCE, A. P. 1993: Indigenous Vascular Plants of New Zealand. Unpublished checklist held at Landcare Research, Lincoln.
DUBUISSON, J.-Y.; HENNEQUIN, S.; DOUZERY, E. J. P.; CRANFILL, R. B.; SMITH, A. R.; PRYER, K. M. 2003: rbcL phylogeny of the fern genus *Trichomanes* (Hymenophyllaceae), with special reference to neotropical taxa. *International Journal of Plant Science 164*: 753–761.
DYER, A. F. (Ed.) 1979: *The Experimental Biology of Ferns*. Academic Press, London. 657 p.
DYER, A. F.; PAGE, C. N. (Eds) 1985: *Biology of Pteridophytes*. Royal Society of Edinburgh, Edinburgh. 474 p.
EBIHARA, A.; DUBUISSON, J.-Y.; IWATSUKI, K.; HENNEQUIN, S.; ITO, M. 2006: A taxonomic revision of Hymenophyllaceae. *Blumea 51*: 221–280.
EBIHARA, A.; HENNEQUIN, S.; IWATSUKI, K.; BOSTOCK, P. D.; MATSUMOTO, S.; JAMAN, R.; DUBUISSON, J.-Y.; ITO, M. 2004: Polyphyletic origin of *Microtrichomanes* (Prantl) Copel. (Hymenophyllaceae), with a revision of the species. *Taxon 53*: 935–948.
EDGAR, E.; CONNOR, H. E. 2000: *Flora of New Zealand. Volume 5. Grasses.* Manaaki Whenua Press, Lincoln. 650 p.
ESLER, A. E. 1988: The naturalisation of plants in urban Auckland 5. Success of the alien species. *New Zealand Journal of Botany 26*: 565–584.
FIELD, H. C. 1890: *The Ferns of New Zealand*. Willis, Wanganui. 164 p.
FIRTH, S.; FIRTH, M.; FIRTH, E. 1986: *Ferns of New Zealand*. Hodder & Stoughton, Auckland. 80 p.
FISHER, M. E. 1984: *Gardening with New Zealand Ferns*. Collins, Auckland. 119 p.
FLANN, C. 2005: Systematics of *Euchiton* (Gnaphalieae: Asteraceae) with a focus on alpine taxa in Australia and New Zealand. Unpublished PhD thesis, The University of Melbourne, Melbourne.
FLANN, C.; BREITWIESER, I.; WARD, J. M.; WALSH, N. G.; LADIGES, P.Y. 2008: Morphometric study of *Euchiton traversii* complex (Gnaphalieae: Asteraceae). *Australian Systematic Botany 21*: 178–191
FLANNERY, T. F. 1994: *The Future Eaters: an ecological history of the Australasian lands and people*. Reed, Melbourne. 423 p.
FORD, K. A. 2004: Origin and biogeography of New Zealand *Craspedia* (Compositae: Gnaphalieae). Unpublished MSc thesis, University of Canterbury, Christchurch.
FORD, K. A.; WARD, J. M.; SMISSEN, R. D.; WAGSTAFF, S. J.; BREITWIESER,I. 2007: Phylogeny and biogeography of *Craspedia* (Asteraceae : Gnaphalieae) based on ITS, ETS and psbA-trnH sequence data. *Taxon 56*: 783–794
FORSTER, J. G. A. 1786a: *Florulae Insularum Australium Prodromus*. Dietrich, Göttingen. 103 p.
FORSTER, J. G. A. 1786b: *De Plantis Esculentis Insularum Oceani Australis*. Haude & Spener, Berlin. 80 p.
FORSTER, J. R.; FORSTER, G. 1775: *Characteres Generum Plantarum quas in itinere ad Insulas Maris Australis Collegerunt, Descripserunt, Delinearunt, Annis MDCCLXXII–MDCCLXXV*. B. White; T. Cadell & P. Elmsely, London.
GARDNER, R. C.; de LANGE, P. J.; KEELING, D. J.; BOWALA, T.; BROWN H. A.; WRIGHT, S. D. 2004: A late Quaternary phylogeography for *Metrosideros* (Myrtaceae) in New Zealand inferred from chloroplast DNA haplotypes. *Biological Journal of the Linnean Society 83*: 399–412.
GARNOCK-JONES, P. J.; LLOYD, D. G. 2004: A taxonomic revision of *Parahebe* (Plantaginaceae) in New Zealand. *New Zealand Journal of Botany 42*: 181–232.
GARNOCK-JONES, P. J.; ALBACH, D.; BIGGS, B. G. 2007: Botanical names in Southern Hemisphere *Veronica* (Plantaginaceae): sect. *Detzneria*, sect. *Hebe*, and sect. *Labiatoides*. *Taxon 56*: 571–582.
GEMMILL, C. E. C.; ALLAN, G.; WAGNER, W. L; ZIMMER, E. A. 2002: Evolution of insular pacific *Pittosporum* (Pittosporaceae): Origin of the Hawaiian radiation. *Molecular Phylogenetics and Evolution 22*: 31–42.
GILLMAN, L. N.; OGDEN, J. 2005: Microsite heterogeneity in litterfall risk to seedlings. *Austral Ecology 30*: 497–504.
GIVEN, D. R. 1972: *Pleurosorus rutifolius* (R.Br.) Fée (Aspleniaceae) in New Zealand. *New Zealand Journal of Botany 10*: 495–506.
GIVEN, D. R. 1982: Records of *Sticherus flabellatus* (R.Br.) H. St John (Pteridophyta – Gleicheniaceae) from South Island, New Zealand. *New Zealand Journal of Botany 20*: 381–385.
GLENNY, D. 1997: A revision of the genus *Anaphalioides* (Asteraceae: Gnaphalieae). *New Zealand Journal of Botany 35*: 451–477.
GLENNY, D. 2004: A revision of the genus *Gentianella* in New Zealand. *New Zealand Journal of Botany 42*: 361–530.
GODLEY, E. J. 1975: Flora and vegetation. Pp. 177–229 *in*: Kuschel, G. (ed.), *Biogeography and Ecology in New Zealand.* W. Junk, The Hague. xvi + 689 p., + folding charts.
GODLEY, E. J. 1979: Flower biology in New Zealand. *New Zealand Journal of Botany 17*: 441–466.
GOUDEY, C. J. 1988: *A Handbook of Ferns for Australia and New Zealand*. Lothian Publishing, Melbourne. 212 p.
GOULDING, J. H. 1977: Early publications and exhibits of New Zealand ferns and the work of Eric Craig. *Records of the Auckland Institute and Museum 14*: 63–79.
GROVES, R. H. 1994: *Australian Vegetation*, 2nd edn. Cambridge University Press, Cambridge. x + 562 p.
HAASE, P.; BREITWIESER, I.; WARD J. M. 1993: Genetic relationships of *Helichrysum dimorphum* (*Inuleae –Compositae*) with *H. filicaule*, *H. depressum* and *Raoulia glabra* as resolved by isozyme analysis. *New Zealand Journal of Botany 31*: 59–64.
HAMLIN, B. 1963: *Nature in New Zealand. Native Ferns*. Reed, Wellington. 64 p.
HART, J. A. 1987: A cladistic analysis of conifers: Preliminary results. *Journal of the Arnold Arboretum 68*: 269–307.
HASEBE, M.; OMORI, T.; NAKAZAWA, M.; SANO, T.; KATO, M.; IWATSUKI, K. 1994: *rbcL* gene sequences provide evidence for the evolutionary lineages of leptosporangiate ferns. *Proceedings of the National Academy of Sciences USA 91*: 5730–5734.
HASEBE, M.;WOLF, P. G.; PRYER, K. M.; UEDA, K.;ITO, M.; SANO, R.; GASTONY, G.;YOKOYAMA, J.; MANHART, J. R.; MURAKAMI, N.; CRANE, E. H.; HAUFLER, C. H.; HAUK, W. 1995: Fern phylogeny based on *rbcL* nucleotide sequences. *American Fern Journal 85*: 134–181.
HEALY, A. J. 1969: The adventive flora in Canterbury. Pp. 261–333 *in*: Knox, G. A. (ed.), *Natural History of Canterbury.* Reed, Wellington. 620 p.
HEALY, A. J. 1973: Weeds in New Zealand. *Proceedings of the 4th Asian-Pacific Weed Science Society Conference 1973*: 105–114.
HEENAN, P. B. 2009a: *Leptinella conjuncta* (Asteraceae), a diminutive new species from arid habitats in the South Island, New Zealand. *New Zealand Journal of Botany 47*: 127–132.
HEENAN, P. B. 2009b: A new species of *Pachycladon* (Brassicaceae) from limestone in eastern Marlborough, New Zealand. *New Zealand Journal of Botany 47*: 155–161.
HEENAN, P. B. 2010: *Hypericum*. In BREITWIESER, I; BROWNSEY, P.; FORD, K.; GLENNY, D.; HEENAN, P.; NELSON, W.; WILTON, A. (eds.) *Flora of New Zealand*. Online Edition at www.nzflora.info.
HEENAN, P. B.; BREITWIESER, I.; GLENNY, D. S.; de LANGE, P. J.; BROWNSEY, P. J. 1998: Checklist of dicotyledons and pteridophytes naturalised or casual in New Zealand: additional records 1994–96. *New Zealand Journal of Botany 36:* 155–162.
HEENAN, P. B.; DAWSON, M. I.; REDMOND, D. N.; WAGSTAFF, S. J. 2005: Relationships of the New Zealand mountain ribbonwoods (*Hoheria glabrata* and *H. lyallii*: Malvaceae), based on molecular and morphological data. *New Zealand Journal of Botany 43:* 527–549.
HEENAN, P. B.; DAWSON, M. I.; WAGSTAFF, S. J. 2004b: The relationship of *Sophora* sect. *Edwardsia* (Fabaceae) to *Sophora tomentosa*, the type species of the genus *Sophora*, observed from DNA sequence data and morphological characters. *Botanical Journal of the Linnean Society 146:* 439–446.
HEENAN, P. B.; de LANGE, P. J. 2005: *Cyperus insularis* (Cyperaceae), a new species of sedge from northern New Zealand. *New Zealand Journal of Botany 43:* 351–359.
HEENAN, P. B.; de LANGE, P. J. 1997: *Carex dolomitica* (Cyperaceae), a new rare species from New Zealand. *New Zealand Journal of Botany 35*: 423–428.
HEENAN, P. B.; de LANGE, P. J. 2011a: *Myoporum semotum* (Scrophulariaceae), a new tree species from the Chatham Islands, *New Zealand. New Zealand Journal of Botany 49*: 17–26.
HEENAN, P. B.; de LANGE, P. J. 2011b: *Lepidium peregrinum* (Brassicaceae) – a new addition to the New Zealand Flora *New Zealand Journal of Botany*. In press
HEENAN, P. B.; de LANGE, P. J.; CAMERON, E. K.; CHAMPION, P. D. 2002b: Checklist of dicotyledons, gymnosperms, and pteridophytes naturalised or casual in New Zealand: additional records 1999–2000. *New Zealand Journal of Botany 40*: 155–174.
HEENAN, P. B.; de LANGE, P. J.; CAMERON, E. K.; OGLE, C. C.; CHAMPION, P. D. 2004a: Checklist of dicotyledons, gymnosperms, and pteridophytes naturalised or casual in New Zealand: additional records 2001–2003. *New Zealand Journal of Botany 42*: 797–814.
HEENAN, P. B.; de LANGE, P. J.; CAMERON, E. K.; PARRIS, B. S. 2008: Checklist of dicotyledons, gymnosperms, and pteridophytes naturalised or casual in New Zealand: additional records 2004–06. *New Zealand Journal of Botany*

45: 257–283.

HEENAN, P. B.; de LANGE, P. J.; GLENNY, D. S.; BREITWIESER, I.; BROWNSEY, P. J.; OGLE, C. C. 1999: Checklist of dicotyledons, gymnosperms, and pteridophytes naturalised or casual in New Zealand: additional records 1997–1998. *New Zealand Journal of Botany 37*: 629–642.

HEENAN, P. B.; de LANGE, P. J.; MURRAY, B. G. 1997: *Carex tenuiculmis* comb. et stat. nov. (Cyperaceae), a threatened red-leaved sedge from New Zealand. *New Zealand Journal of Botany 35*: 159–165.

HEENAN, P. B.; MITCHELL, A. D.; KOCH, M. 2002a: Molecular systematics of the New Zealand *Pachycladon* (Brassicaceae) complex: generic circumscription and relationships to *Arabidopsis* sens. lat. and *Arabis* sens. lat. *New Zealand Journal of Botany 40*: 543–562.

HEENAN, P. B.; de LANGE, P. J.; HOULISTON, G. C.; BARNAUD, A.; MURRAY, B. G. 2008: *Olearia telmatica* (Asteraceae), a new and previously overlooked tree species endemic to the Chatham Islands. *New Zealand Journal of Botany 46*: 567–583.

HEINE, E. M. 1937: Observations on the pollination of New Zealand flowering plants. *Transactions and proceedings of the Royal Society of New Zealand 67*: 133–148.

HENNEQUIN, S.; EBIHARA, A.; ITO, M.; IWATSUKI, K.; DUBUISSON, J.-Y. 2003: Molecular systematics of the fern genus *Hymenophyllum s.l.* (Hymenophyllaceae) based on chloroplastic coding and non-coding regions. *Molecular Phylogenetics and Evolution 27*: 283–301.

HOLLOWAY, J. E. 1918: The prothallus and young plant of *Tmesipteris*. *Transactions and Proceedings of the New Zealand Institute 50*: 1–44.

HOLLOWAY, J. E. 1920: Studies in the New Zealand species of the genus *Lycopodium*. Part IV – the structure of the prothallus of the five species. *Transactions and Proceedings of the New Zealand Institute 52*: 193–239.

HOLLOWAY, J. E. 1921: Further studies on the prothallus, embryo and young sporophyte of *Tmesipteris*. *Transactions and Proceedings of the New Zealand Institute 53*: 386–422.

HOLLOWAY, J. E. 1935: The gametophyte of *Phylloglossum drummondii*. *Annals of Botany 49*: 513–520.

HOLLOWAY, J. E. 1939: The gametophyte, embryo, and young rhizome of *Psilotum triquetrum* Swartz. *Annals of Botany, n.s., 3*: 313–336.

HOLLOWAY, J. E. 1944: The gametophyte, embryo and developing sporophyte of *Cardiomanes reniforme* (Forst.) Presl. *Transactions and Proceedings of the New Zealand Institute 74*: 196–206.

HOLMGREN, P. K.; HOLMGREN, N. H.; BARNETT, L. C. 1990: Index Herbariorum. Part I: The herbaria of the world, 8th ed. *Regnum Vegetabile 120*: 1–693.

HOLTTUM, R. E. 1964: The tree ferns of the genus *Cyathea* in Australasia and the Pacific. *Blumea 12*: 241–274.

HOLTTUM, R. E. 1973: Posing the problems. Pp. 1–10 *in*: Jermy, A. C.; Crabbe, J. A.; Thomas, B. A. (eds), *The Phylogeny and Classification of the Ferns*. The Linnean Society of London, London. 284 p.

HOLTTUM, R. E. 1977: The family Thelypteridaceae in the Pacific and Australasia. *Allertonia 1*: 169–234.

HOOKER, J. D. 1844: *Flora Antarctica*. Reeve, London. 574 p.

HOOKER, J. D. 1852–1855: *Flora Novae-Zelandiae*. Reeve, London. 312 and 378 p.

HOOKER, J. D. 1860: *Flora Tasmaniae*. Reeve, London. 359 and 422 p.

HOOKER, J. D. 1864–1867: *Handbook of the New Zealand Flora*. Reeve, London. 798 p.

HOOKER, W. J.; GREVILLE, R. K. 1831: *Icones filicum*. Treuttel & Wurtz, London.

HOVENKAMP, P. 1986: A monograph of the fern genus *Pyrrosia*. *Leiden Botanical Series 9*: 1–310.

HOVENKAMP, P.; MIYAMOTO, F. 2005: A conspectus of the native and naturalized species of *Nephrolepis* (Nephrolepidaceae) in the world. *Blumea 50*: 279–322.

IDE, J. M. 1999: The ferns of the Baden-Powell desk. *Pteridologist 3*: 79–86.

JERMY, A. C.; CRABBE, J. A.; THOMAS, B. A. (Eds) 1973: *The Phylogeny and Classification of the Ferns*. The Linnean Society of London, London. 284 p.

JONES, D. L. 1987: *Encyclopaedia of Ferns*. Lothian Publishing, Melbourne. 433 p.

JUDD, W. S.; CAMPBELL, C. S.; KELLOG, E. A.; STEVENS, P. F. 1999: *Plant Systematics, a Phylogenetic Approach*. Sinauer, Sunderland. 464 p.

KATO, M. 1984: A taxonomic study of the athyrioid fern genus *Deparia* with main reference to the Pacific species. *Journal of the Faculty of Science, University of Tokyo, Section 3, 13*: 375–429.

KELCH, D. G. 1997: The phylogeny of the Podocarpaceae based on morphological evidence. *Systematic Botany 22*: 113–131.

KELCH, D. G. 1998: Phylogeny of Podocarpaceae: Comparison of evidence from morphology and 18S rDNA. *American Journal of Botany 85*: 975–985.

KELLY, D. 1994: Demography and conservation of *Botrychium australe*, a peculiar, sparse, mycorrhizal fern. *New Zealand Journal of Botany 32*: 393–400.

KIRK, T. 1899: *The Students' Flora of New Zealand*. Government Printer, Wellington. 408 p.

KORALL, P.; PRYER, K. M.; METZGAR, J. S.; SCHNEIDER, H.; CONANT, D. S. 2006: Tree ferns: monophyletic groups and their relationships as revealed by four protein-coding plastid loci. *Molecular Phylogenetics and Evolution 39*: 830–845.

KRAMER, K. U.; GREEN, P. S. (Eds) 1990: *The Families and Genera of Vascular Plants. Vol. 1. Pteridophytes and Gymnosperms*. Springer-Verlag, Berlin. 404 p.

KRAMER, K. U.; TINDALE, M. D. 1976: The lindsaeoid ferns of the Old World VII. Australia and New Zealand. *Telopea 1*: 91–128.

KREIER, H.-P.; SCHNEIDER, H. 2006: Reinstatement of *Loxogramme dictyopteris*, based on phylogenetic evidence, for the New Zealand endemic fern, *Anarthropteris lanceolata* (Polypodiaceae: Polypodiidae). *Australian Systematic Botany 19*: 309–314.

LABILLARDIÈRE, J. J. H. de 1806: *Novae Hollandiae Plantarum Specimen*. Huzard, Paris. 130 p.

LARGE, M. F.; BRAGGINS, J. E. 1989: An assessment of characters of taxonomic significance in the genus *Pilularia* (Marsileaceae): with particular reference to *P. americana*, *P. novae-hollandiae*, and *P. novae-zelandiae*. *New Zealand Journal of Botany 27*: 481–486.

LARGE, M. F.; BRAGGINS, J. E. 1991: *Spore Atlas of New Zealand Ferns and Fern Allies*. SIR Publishing, Wellington. 167 p.

LARGE, M. F.; BRAGGINS, J. E.; GREEN, P. S. 1992: A new combination for *Polypodium pustulatum* Forster f. (Polypodiaceae). *New Zealand Journal of Botany 30*: 207–208, 372 (erratum).

LEHMANN, A.; LEATHWICK, J. R.; OVERTON, J. M. 2002: Assessing New Zealand fern diversity from spatial predictions of species assemblages. *Biodiversity and Conservation 11*: 2217–2238.

LEWIS, R. 2001: Molecular studies on the New Zealand tree ferns. Unpublished PhD thesis, Massey University, Palmerston North.

LLOYD, D. G. 1985: Progress in understanding the natural history of New Zealand plants. *New Zealand Journal of Botany 23*: 707–722.

LOCKHART, P. J.; McLENACHAN, P. A.; HAVELL, D.; GLENNY, D.; HUSON, D.; JENSEN, U. 2001: Phylogeny, radiation, and transoceanic dispersal of New Zealand alpine buttercups: molecular evidence under split decomposition. *Annals of the Missouri Botanical Garden 88*: 458–477.

LOVIS, J. D. 1959: The geographical affinities of the New Zealand pteridophyte flora. *British Fern Gazette 10*: 1–7.

LOVIS, J. D. 1977: Evolutionary patterns and processes in ferns. *Advances in Botanical Research 4*: 230–424.

MABBERLEY, D. J. 2008: *Mabberley's Plant-book. A portable dictionary of plants, their classifications, and uses*. 3rd edition. Cambridge Unviersity Press. 1040 p.

MARSDEN, C. R. 1979: Morphology and taxonomy of *Isoetes* in Australasia, India, north-east and south-east Asia, China and Japan. Unpublished PhD thesis, University of Adelaide, Adelaide.

McGLONE, M. S.; WILMSHURST, J. M.; LEACH, H. M. 2005: An ecological and historical review of bracken (*Pteridium esculentum*) in New Zealand, and its cultural significance. *New Zealand Journal of Ecology 291*: 165–184.

McKENZIE, R. J. 2001: *Intergeneric hybridisation in New Zealand Gnaphalieae (Compositae)*. Unpublished PhD thesis, University of Canterbury, Christchurch.

McKENZIE, R. J.; WARD, J. M.; LOVIS, J. D.; BREITWIESER, I. 2004: Morphological evidence for natural intergeneric hybridisation in the New Zealand Gnaphalieae (Compositae): *Anaphalioides bellidioides* x *Ewartia sinclairii*. *Botanical Journal of the Linnean Society 145*: 59–75.

McKENZIE, R. J.; WARD, J. M.; BREITWIESER, I. 2008: Hybridization beyond the F1 generation between the New Zealand endemic everlastings *Anaphalioides bellidioides* and *Ewartia sinclairii* (Asteraceae, Gnaphalieae). *Plant Systematics and Evolution 273*: 13–24

MEUDT, H. M. 2006: A revision of the genus *Ourisia* (Plantaginaceae). *Systematic Botany Monographs 77*: 1–188.

MEUDT, H. M. 2008: Taxonomic revision of the snow hebes (*Veronica s.l.*, Plantaginaceae). *Australian Systematic Botany 21*: 387–421.

MEUDT, H. M. 2011: AFLP data reveal a history of auto- and allopolyploidy in New Zealand endemic species of *Plantago* (Plantaginaceae): New perspectives on a taxonomically-challenging group. *International Journal of Plant Sciences 172*: 220–237.

MEUDT, H. M.; BAYLY, M. J. 2008: Phylogeographic patterns in the Australasian genus *Chionohebe* (*Veronica s.l.*, Plantaginaceae) based on AFLP and chloroplast DNA sequences. *Molecular*

Phylogenetics and Evolution 47: 319–338.
MITCHELL, A. D.; HEENAN, P. B. 2000: Systematic relationships of New Zealand endemic Brassicaceae inferred from rDNA sequence data. *Systematic Botany25*: 98–105.
MITCHELL, K. A.; KELLOW, A.V.; BAYLY, M. J.; MARKHAM, K. R.; BROWNSEY, P. J.; GARNOCK-JONES, P. J. 2007: Composition and distribution of leaf flavonoids in *Hebe* and *Leonohebe* (Plantaginaceae) in New Zealand – 2. 'Apertae', 'Occlusae', and 'Grandiflorae'. *New Zealand Journal of Botany 45*: 329–392.
MITTERMEIER, R. A.; MYERS, N.; GIL, P. R.; MITTERMEIER, C. G. 1999: *Hotspots: Earth's biologically richest and most endangered terrestrial ecoregions.* CEMEX, Conservation International & Agrupacion Sierra Madre, Monterrey. 430 p.
MOLLOY, B. P. J. 1995: Two new species of *Leucogenes* (Inuleae: Asteraceae) from New Zealand, and typification of *L. grandiceps. New Zealand Journal of Botany 33*: 53–63.
MUMMENHOFF, K.; LINDER, P.; FRIESEN, N.; BOWMAN, J. L.; LEE, J.-Y.; FRANZKE, A. 2004: Molecular evidence for bicontinental hybridogenous genomic constitution in *Lepidium* sensu stricto (Brassicaceae) species from Australia and New Zealand. *American Journal of Botany 91*: 254–261.
MURDOCK, A. G. 2008: A taxonomic revision of the eusporangiate fern family Marattiaceae, with a description of a new genus *Ptisana. Taxon 57*: 737–755.
NAGALINGUM, S.; NOWAK, M. D.; PRYER , K. M. 2008: Assessing phylogenetic relationships in extant homosporous ferns (Salviniales), with a focus on *Pilularia* and *Salvinia*. *Botanical Journal of the Linnean Society 157*: 673–685.
NAYAR, B. K. 1970: A phylogenetic classification of the homosporous ferns. *Taxon 19*: 229–236.
NEWSTROM, L.; ROBERTSON, A. 2004: Progress in understanding pollination systems in New Zealand. *New Zealand Journal of Botany 43*: 1–59.
NICOL, E. R. 1997: *Common Names of Plants in New Zealand*. Manaaki Whenua Press, Lincoln. 115 p.
NIKLAS, K. 1997: *Evolutionary Biology of Plants.* Universityof Chicago Press, Chicago. 470 p.
NOOTEBOOM, H. P. 1994: Notes on Davalliaceae II. A revision of the genus *Davallia*. *Blumea 39*: 151–214.
NOOTEBOOM, H. P. 1997: The microsoroid ferns (Polypodiaceae). *Blumea 42*: 261–395.
NORTON, D. A. 1994: Relationships between pteridophytes and topography in a lowland South Westland podocarp forest. *New Zealand Journal of Botany 32*: 401–408.
ØLLGAARD, B. 1987: A revised classification of the Lycopodiaceae s. lat. *Opera Botanica 92*: 153–178.
PAGE, C. N.; BROWNSEY, P. J. 1986: Tree fern skirts: a defence against climbers and large epiphytes. *Journal of Ecology 74*: 787–796.
PARRIS, B. S. 1972: The genus *Doodia* R.Br. (Blechnaceae: Filicales) in New Zealand. *New Zealand Journal of Botany 10*: 585–604.
PARRIS, B. S. 1976: Ecology and biogeography of New Zealand pteridophytes. *Fern Gazette 11*: 231–245.
PARRIS, B. S. 1977: A naturally occurring intergeneric hybrid in Grammitidaceae (Filicales): *Ctenopteris heterophylla* x *Grammitis billardieri*. *New Zealand Journal of Botany 15*: 597–599.
PARRIS, B. S. 1980a: Further notes on *Doodia, Grammitis* and *Blechnum* (Filicales). *New Zealand Journal of Botany 18*: 145–147.
PARRIS, B. S. 1980b: *Adiantum hispidulum* Swartz and *A. pubescens* Schkuhr (Adiantaceae: Filicales) in New Zealand. *New Zealand Journal of Botany 18*: 505–506.
PARRIS, B. S.; CROXALL, J. P. 1974: *Adiantum viridescens* Colenso in New Zealand. *New Zealand Journal of Botany 12*: 227–233.
PARRIS, B. S.; GIVEN, D. R. 1976: A taxonomic revision of the genus *Grammitis* Sw. (Grammitidaceae: Filicales) in New Zealand. *New Zealand Journal of Botany 14*: 85–111.
PELSER, P. B.; KENNEDY, A. H.; TEPE, E. J.; SHIDLER, J. B.; NORDENSTAM, B.; KADEREIT, J. W.; WATSON, L. E. 2010: Patterns and causes of incongruence between plastid and nuclear Senecioneae (Asteraceae) phylogenies. *American Journal of Botany 97*: 856–873.
PERRIE, L. R.; BAYLY, M. J.; LEHNEBACH, C. A.; BROWNSEY, P. J. 2007: Molecular phylogenetics and molecular dating of the New Zealand Gleicheniaceae. *Brittonia 59*: 129–141.
PERRIE, L. R.; BROWNSEY, P. J. 2005a: Insights into the biogeography and polyploid evolution of New Zealand *Asplenium* from chloroplast DNA sequence data. *American Fern Journal 95*: 1–21.
PERRIE, L. R.; BROWNSEY, P. J. 2005b: Genetic variation is not concordant with morphological variation in the fern *Asplenium hookerianum* sensu lato (Aspleniaceae). *American Journal of Botany 92*: 1559–1564.
PERRIE, L. R.; BROWNSEY, P. J. 2005c: New Zealand *Asplenium* (Aspleniaceae: Pteridophyta) revisited – DNA sequencing and AFLP fingerprinting. *Fern Gazette 17*: 235–242.
PERRIE, L. R.; BROWNSEY, P. J. 2007: Molecular evidence for long-distance dispersal in the New Zealand pteridophyte flora. *Journal of Biogeography 34*: 2028–2038.
PERRIE, L. R.; BROWNSEY, P. J.; LOCKHART, P. J.; LARGE, M. F. 2003a: Evidence for an allopolyploid complex in New Zealand *Polystichum* (Dryopteridaceae). *New Zealand Journal of Botany 41*: 189–215.
PERRIE, L. R.; BROWNSEY, P. J.; LOCKHART, P. J.; LARGE, M. F. 2003b: Morphological and genetic diversity in the New Zealand fern *Polystichum vestitum* (Dryopteridaceae), with special reference to the Chatham Islands. *New Zealand Journal of Botany 41*: 581–602.
PERRIE, L. R.; BROWNSEY, P. J.; LOCKHART, P. J.; BROWN, E. A.; LARGE, M. F. 2003c: Biogeography of temperate Australasian *Polystichum* ferns as inferred from chloroplast sequence and AFLP. *Journal of Biogeography 30*: 1729–1736.
PERRIE, L. R.; BROWNSEY, P. J.; LOVIS, J. D. 2010: *Tmesipteris horomaka,* a new octoploid species from Banks Peninsula. *New Zealand Journal of Botany 48*: 15–29.
PERRIE, L. R.; SHEPHERD, L. D.; BROWNSEY, P. J. 2005: *Asplenium* x*lucrosum* nothosp. nov. *Plant Systematics and Evolution 250*: 243–257.
PERRIE, L. R.; SHEPHERD, L. D.; de LANGE, P. J.; BROWNSEY, P. J. 2010a: Parallel polyploid speciation: distinct sympatric gene-pools of recurrently derived allo-octoploid *Asplenium* ferns. *Molecular Ecology 19*: 2916–2932.
PERRIE, L. R.; OHLSEN, D. J.; SHEPHERD, L. D.; GARRETT, M.; BROWNSEY, P. J.; BAYLY, M. J. 2010b: Tasmanian and Victorian populations of the fern *Asplenium hookerianum* result from independent dispersals from New Zealand. *Australian Systematic Botany 23*: 387–392.
PICHI SERMOLLI, R. E. G. 1977: Tentamen pteridophytorum genera in taxonomicum ordinem redigendi. *Webbia 31*: 313–512.
POLE, M. 1994: The New Zealand flora – entirely long-distance dispersal? *Journal of Biogeography 21*: 625–635.
POOLE, A. L.; ADAMS, N. M. 1964: *Trees and Shrubs of New Zealand*. Government Printer, Wellington. 250 p.
POPE, A. 1924: The role of the tree fern in the New Zealand bush. Part I. *New Zealand Journal of Science and Technology 7*: 52–61.
POPE, A. 1926: The role of the tree fern in the New Zealand bush. Part II. *New Zealand Journal of Science and Technology 8*: 85–98.
POWLESLAND, M. H. 1984: Reproductive biology of three species of *Melicytus* (Violaceae) in New Zealand. *New Zealand Journal of Botany 22*: 81–94.
PREBBLE, J., CUPIDO, C.; MEUDT. H. M.; GARNOCK-JONES, P. J. 2011: Biogeography and a first phylogeny of *Wahlenbergia* (Campanulaceae). *Molecular Phylogenetics and Evolution 59*: 636–648.
PRYER, K. M.; SCHUETTPELZ, E.; WOLF, P. G.; SCHNEIDER, H.; SMITH, A. R.; CRANFILL, R. 2004: Phylogeny and evolution of ferns (Monilophytes) with a focus on the early leptosporangiate divergences. *American Journal of Botany 91*: 1582–1598.
PRYER, K. M.; SMITH, A. R.; HUNT, J. S.; DUBUISSON, J.-Y. 2001: rbcL data reveal two monophyletic groups of filmy ferns (Filicopsida: Hymenophyllaceae). *American Journal of Botany 88*: 1118–1130.
PRYER, K. M.; SMITH, A. R.; SKOG, J. E. 1995: Phylogenetic relationships of extant ferns based on morphology and *rbcL* sequences. *American Fern Journal 85*: 205–282.
RANKER, T. A.; SMITH, A. R.; PARRIS, B. S.; GEIGER, J. M. O.; HAUFLER, C. H.; STRAUB, S. C. K.; SCHNEIDER, H. 2004: Phylogeny and evolution of grammitid ferns (Grammitidaceae): a case of rampant morphological homoplasy. *Taxon 53*: 415–428.
RAUBESON, L. D. 1998: Chloroplast DNA structural similarities shared by conifers and Gnetales: coincidence or common ancestry? *American Journal of Botany 85 (Suppl.)*: 153.
RAVEN, P. H.; EVERT, R. F.; EICHORN, S. E. 1998: *Biology of Plants*. 6th edn. Freeman, New York. 944 p.
RICHARD, A. 1832: Essai d'une flore de la Nouvelle-Zélande. Pp. 1–376 *in*: Lesson, A.; Richard, A. *Voyage de Découvertes de l'Astrolabe, Botanique*. Tastu, Paris.
RILEY, M. 1994: *Maori Healing and Herbal.* Viking Sevenseas, Paraparaumu. 528 p.
ROOS, M. 1996: Mapping the world's pteridophyte diversity – systematics and floras. Pp. 29–42 *in*: Camus, J. M.; Gibby, M.; Johns, R. J. (eds), *Pteridology in Perspective*. Royal Botanic Gardens, Kew. 700 p.
SCHEELE, S.; WALLS, G. 1994: *Harakeke – The Rene Orchiston Collection. Revised Edition, based on information provided by Rene Orchiston.* Manaaki Whenua Press, Lincoln. 24 p.
SCHNEIDER, H.; KREIER, H-P.; PERRIE, L. R.; BROWNSEY, P. J. 2006: The relationships of *Microsorum* (Polypodiaceae) species occurring in New Zealand. *New Zealand Journal of Botany 44*: 121–127.
SCHNEIDER, H.; SCHUETTPELZ, E.; PRYER, K. M.; CRANFILL, R.; MAGALLION, S.; LUPIA, R. 2004: Ferns diversified in the shadow of angiosperms. *Nature 428*: 553–557.

SHEPHERD, L. D.; PERRIE, L. R.; PARRIS, B. S.; BROWNSEY, P. J. 2007a: A molecular phylogeny for the New Zealand Blechnaceae ferns from analyses of chloroplast trnL-trnF DNA sequences. *New Zealand Journal of Botany 45*: 67–80.

SHEPHERD, L. D.; PERRIE, L. R.; BROWNSEY, P. J. 2007b: Fire and ice: volcanic impacts on the phylogeography of the New Zealand forest fern *Asplenium hookerianum*. *Molecular Ecology 16*: 4536–4549.

SHEPHERD, L. D.; HOLLAND, B. R.; PERRIE, L. R. 2008a: Conflict amongst chloroplast DNA sequences obscures the phylogeny of a group of *Asplenium* ferns. *Molecular Phylogenetics and Evolution 48*: 176–187.

SHEPHERD, L. D.; PERRIE, L. R.; BROWNSEY, P. J. 2008b: Low-copy nuclear DNA sequences reveal a predominance of allopolypoids in a New Zealand *Asplenium* fern complex. *Molecular Phylogenetics and Evolution 49*: 240–248.

SHEPHERD, L. D.; de LANGE, P. J.; PERRIE, L. R. 2009: Multiple colonizations of a remote oceanic archipelago by one species: how common is long-distance dispersal? *Journal of Biogeography 36*: 1972–1977.

SMISSEN, R. D.; BREITWIESER, I. 2008: Species relationships and genetic variations in the New Zealand endemic *Leucogenes* (Asteraceae: Gnaphalieae). *New Zealand Journal of Botany 46*: 65–76.

SMISSEN, R D.; HEENAN, P. B. 2010: A taxonomic appraisal of the Chatham Island flax (*Phormium tenax*) using morphological and DNA fingerprint data. *Australian Systematic Botany 23*: 371–380.

SMISSEN, R. D.; BREITWIESER, I.; WARD, J. M. 2004: Phylogenetic implications of trans-specific chloroplast DNA sequence polymorphism in New Zealand Gnaphalieae. *Plant Systematics and Evolution 249*: 37–53.

SMISSEN, R. D.; BREITWIESER, I.; WARD, J. M. 2007: Genetic characterisation of hybridisation between the New Zealand everlastings *Helichrysum lanceolatum* and *Anaphalioides bellidioides* (Asteraceae: Gnaphalieae). *Botanical Journal of the Linnean Society 154*: 89–97

SMISSEN, R. D.; BREITWIESER, I.; WARD, J. M. 2006: Genetic diversity in the New Zealand endemic species *Helichrysum lanceolatum* (Asteraceae: Gnaphalieae). *New Zealand Jounal of Botany 44*: 237–247.

SMISSEN, R. D.; GALBANY-CASALS, M.; BREITWIESER, I. 2011: Ancient allopolyploidy in the everlasting daisies (Asteraceae: Gnaphalieae) – complex relationships among extant clades. *Taxon 60*: 649–622.

SMISSEN, R. D.; GARNOCK-JONES, P. J.; CHAMBERS, G. K. 2003a: Phylogenetic analysis of ITS sequences suggests a Pliocene origin for the bipolar distribution of *Scleranthus* (Caryophyllaceae). *Australian Systematic Botany 16*: 301–315.

SMISSEN, R. D.; HEENAN, P. B.; HOULISTON, G. 2008: Genetic and morphological evidence for localised interspecific gene flow in *Phormium* (Hemerocallidaceae). *New Zealand Journal of Botany 46*: 287–297.

SMISSEN, R. D.; BREITWIESER, I.; WARD, J. M.; McLENACHAN, P. A.; LOCKHART, P. J. 2003b: Use of ISSR profiles and ITS sequences to study biogeography of alpine cushion plants in the genus *Raoulia* (Asteraceae). *Plant Systematics and Evolution 239*: 79–94.

SMISSEN, R. D.; de LANGE P. J.; THORSEN, M. J.; OGLE C. C. 2011: Species delimitation and genetic variation in the rare New Zealand endemic grass genus *Simplicia*. *New Zeaalnd Journal of Botany 49*: 187–199.

SMITH, A. R.; PRYER, K. M.; SCHUETTPELTZ, E.; KORALL, P.; SCHNEIDER, H.; WOLF, P. G. 2006: A classification for extant ferns. *Taxon 55*: 705–731.

SOLTIS, D. E.; SOLTIS, P. S.; NICKRENT, D. L.; JOHNSON, L. A.; HAHN, W. J. et al. 1997: Phylogenetic relationships among angiosperms inferred from 18S rDNA sequences. *Annals of the Missouri Botanical Garden 84*: 1–49.

SOLTIS, D. E.; SOLTIS, P. S.; ZANIS, M. J. 2002: Phylogeny of seed plants based on evidence from eight genes. *American Journal of Botany 89*: 1670–1681.

STEVENSON, G. 1959: *A Book of Ferns*, 2nd edn. Paul's Book Arcade, Hamilton. 168 p.

SUN, G.; JI, Q.; DILCHER, D. L.; ZHENG, S.; NIXON, K. C.; WANG, X. 2002: Archaefructaceae, a new basal angiosperm family. *Science 296*: 899–904.

TASKOVA, R. M.; KOKUBUN, T.; RYAN, K. G.; GARNOCK-JONES, P. J.; JENSEN, S. R. 2010: Phenylethanoid and iridoid glycosides in the New Zealand snow hebes (*Veronica*, Plantaginaceae). *Chemical and Pharmaceutcial Bulletin 58*: 703–711.

TAY, M. L., MEUDT, H. M., GARNOCK-JONES, P. J., RITCHIE, P. A. 2010a: DNA sequences from three genomes reveal multiple long-distance dispersals and non-monophyly of sections in Australasian *Plantago* (Plantaginaceae). *Australian Systematic Botany 23*: 47–68.

TAY, M. L., MEUDT, H. M., GARNOCK-JONES, P. J., RITCHIE, P. A. 2010b: Testing species limits of New Zealand *Plantago* (Plantaginaceae) using internal transcribed spacer (ITS) DNA sequences. *New Zealand Journal of Botany* 48(3): 205–224.

THOMSON, G. M. 1882: *The Ferns and Fern Allies of New Zealand*. Wise, Dunedin. 132 p.

TINDALE, M. D. 1965: A monograph of the genus *Lastreopsis* Ching. *Contributions from the New South Wales National Herbarium 3*: 249–339.

TINDALE, M. D. 1972: Pteridophyta. Pp. 39–94 *in*: Beadle, N. C. W.; Evans, O. D.; Carolin, R. C. (eds), *Flora of the Sydney Region*. Reed, Sydney. 724 p.

TRYON, R. M.; TRYON, A. F. 1982: *Ferns and Allied Plants*. Springer-Verlag, New York. 857 p.

Van der MAST, S.; HOBBS, J. 1998: *Ferns for New Zealand Gardens*. Godwit, Auckland. 95 p.

von KONRAT, M. J.; BRAGGINS, J. E.; de LANGE, P. J. 1999: *Davallia* (Pteridophyta) in New Zealand, including description of a new subspecies of *D. tasmanii*. *New Zealand Journal of Botany 37*: 579–593.

VENTER, S. J. 2002: *Dracophyllum marmoricola* and *Dracophyllum ophioliticum* (Ericaceae), two new species from north-west Nelson, New Zealand. *New Zealand Journal of Botany 40*: 39–47.

WAGSTAFF, S. J. 2004: Evolution and biogeography of the austral genus *Phyllocladus* (Podocarpaceae). *Journal of Biogeography 31*: 1569–1577.

WAGSTAFF, S. J.; BAYLY, M. J.; GARNOCK-JONES, P. J.; ALBACH, D. C. 2002: Classification, origin, and diversification of the New Zealand hebes (Scrophulariaceae). *Annals of the Missouri Botanical Garden 98*: 38–63.

WAGSTAFF, S. J.; BREITWIESER, I. 2002: Phylogenetic relationships of New Zealand Asteraceae inferred from ITS sequences. *Plant Systematics and Evolution 231*: 203–224.

WAGSTAFF, S. J.; BREITWIESER, I. 2004: Phylogeny and classification of *Brachyglottis* (Senecioneae: Asteraceae): an example of a rapid species radiation in New Zealand. *Systematic Botany 29*: 1003–1010.

WAGSTAFF, S. J.; DAWSON, M. I. 2000: Classification, origin, and patterns of diversification of *Corynocarpus* (Corynocarpaceae) inferred from DNA sequences. *Systematic Botany 25*: 134–149.

WAGSTAFF, S. J.; HEENAN, P. B.; SANDERSON, M. J. 1999: Classification, origins, and patterns of diversification in New Zealand Carmichelinae (Fabaceae). *American Journal of Botany 86*: 1346–1356.

WAGSTAFF, S. J.; WEGE, J. 2002: Patterns of diversification in New Zealand Stylidiaceae. *American Journal of Botany 89*: 865–874.

WAGSTAFF, S. J.; BREITWIESER, I.; ITO M. 2011: Evolution and biogeography of *Pleurophyllum* (Astereae, Asteraceae), a small genus of endemic megaherbs from the subantarctic islands. *American Journal of Botany. 98*: 62–75.

WAGSTAFF, S. J.; BREITWIESER I.; SWENSON, U. 2006: Origin and relationships of the circumpacific genus *Abrotanella* (Asteraceae) inferred from DNA sequences. *Taxon 55*: 95–106

WAGSTAFF, S. J.; DAWSON, M. I.; VENTER, S.; MUNZINGER, J.; CRAYN, D. M.; STEANE, D. A.; LEMSON, K. L. 2010a: Origin, diversification, and classification of the Australasian genus *Dracophyllum* (Richeeae, Ericaceae). *Annals of the Missouri Botanical Garden. 97*: 235–258

WAGSTAFF, S. J.; MOLLOY, B. P. J.; TATE, J. A. 2010b: Evolutionary significance of long-distance dispersal and hybridisation in the New Zealand endemic genus *Hoheria* (Malvaceae). *Australian Systematic Botany 23*: 112–130.

WALKER, T. G. 1979: The cytogenetics of ferns. Pp. 87–132 *in*: Dyer, A. F. (ed.), *The Experimental Biology of Ferns*. Academic Press, London. 657 p.

WARD, J. M.; BREITWIESER, I. 1998: New combinations in *Euchiton* Cass. (Compositae – Gnaphalieae). *New Zealand Journal of Botany 36*: 303–304.

WARD, J. M.; BREITWIESER, I.; LOVIS, J. D. 1997: *Rachelia glaria* (Compositae), a new genus and species from the South Island of New Zealand. *New Zealand Journal of Botany 35*: 145–154.

WARD, J. M.; BREITWIESER, I.; FLANN, C. 2003: *Argyrotegium*, a new genus of Gnaphalieae (Compositae). *New Zealand Journal of Botany 41*: 603–611.

WARD, J.; BAYER, R.; BREITWIESER, I.; SMISSEN, R.; GALBANY-CASALS, M.; UNWIN, M. 2009: Gnaphalieae. In: Funk, V.; Susanna, A.; Stuessy, T.; Bayer, R. (eds), *Systematics, Evolution, and Biogeography of Compositae*. IAPT, Vienna, 965 p.

WARDLE, P. 1991: *Vegetation of New Zealand*. Cambridge University Press, Cambridge. 672 p.

WEBB, C. J.; LLOYD, D. G.; DELPH, L. F. 1999: Gender dimorphism in indigenous New Zealand seed plants. *New Zealand Journal of Botany 37*: 119–130.

WEBB, C. J.; SIMPSON, M. J. A. 2001: *Seeds of New Zealand Gymnosperms and Dicotyledons*. Manuka Press: Christchurch. 428 p.

WEBB, C. J.; SYKES, W. R.; GARNOCK-JONES, P. J. 1988: *Flora of New Zealand. Volume 4. Naturalised Pteridophytes, Gymnosperms, Dicotyledons*. Botany Division, DSIR, Christchurch. 1365 p.

WEBB, C. J.; SYKES, W. R.; GARNOCK-JONES,

P. J.; BROWNSEY, P. J. 1995: Checklist of dicotyledons, gymnosperms, and pteridophytes naturalised or casual in New Zealand: additional records 1988–1993. *New Zealand Journal of Botany 33*: 151–182.
WINKWORTH, R. C.; WAGSTAFF, S.J .; GLENNY, D.; LOCKHART, P. J. 2002: Plant dispersal N.E.W.S. from New Zealand. *Trends in Ecology & Evolution 17*: 514–520.
WINKWORTH, R. C.; WAGSTAFF, S. J.; GLENNY, D.; LOCKHART, P. 2005: Evolution of the New Zealand mountain flora: Origins, diversification and dispersal. *Organisms, Diversity & Evolution 5*: 237–247.
WOO, V. L.; FUNKE, M. M.; SMITH, J. F.; LOCKHART, P. J.; GARNOCK-JONES, P. J. 2011: New World origins of Southwest Pacific Gesneriaceae: multiple movements across and within the South Pacific. *International Journal of Plant Sciences 172*: 434–457.

Checklist of Recent New Zealand Tracheophyta

Recognition of phylum Tracheophyta follows Cavalier-Smith (1998). Accordingly, subclasses in Chase and Reveal (2009) are herein recognised as classes. Classification of the ferns mostly follows Smith et al. (2006), and the lycophytes (orders and below) Christenhusz et al. (2011). Classification of the seed plants follows that of Mabberley (2008), with the ordinal sequence for flowering plants based on Chase and Reveal (2009). The Checklist of seed plants was extracted from the Plant Names Database of Landcare Research, Lincoln, New Zealand on 15 September 2011. The data in the Plant Names Database are being continually updated and curated, and are being made available as a searchable database on the Ngā Tipu o Aotearoa–New Zealand plants web site (http://nzflora.landcareresearch.co.nz). Abbreviations: A, adventive (fully naturalised); Cas, casual [i.e. including species that are: passively regenerating only in the immediate vicinity of a cultivated parent plant; more widespread but known only as isolated or a few individuals; garden escapes persisting only 2–3 years; garden discards persisting vegetatively but not spreading sexually or asexually (Heenan et al. 1998, 1999)]; C, Chatham Islands only; E, endemic; K, Kermadec Islands only; F, freshwater (i.e. mostly permanently dependent on free-standing/free-flowing water, whether submerged, floating, or emergent, at least for typical populations); M, marine (including estuaries and saline lagoons); *, unpublished record. Endemic genera are underlined (first entry only). References or voucher specimens are given for recognised, but as yet undescribed, taxa and for previously unpublished records.

KINGDOM PLANTAE
SUBKINGDOM VIRIDIPLANTAE
INFRAKINGDOM CORMOPHYTA
PHYLUM TRACHEOPHYTA
SUBPHYLUM LYCOPHYTINA
Class LYCOPODIOPSIDA
Order ISOETALES
ISOETACEAE
Isoetes alpina Kirk F E
Isoetes kirkii A.Braun F E

Order LYCOPODIALES
LYCOPODIACEAE
Huperzia australiana (Herter) Holub
Huperzia varia (R.Br.) Trevis.
Lycopodiella cernua (L.) Pic.Serm.
Lycopodiella diffusa (R.Br.) B.Øllg.
Lycopodiella lateralis (R.Br.) B.Øllg.
Lycopodiella serpentina (Kunze) B.Øllg.
Lycopodium deuterodensum Herter
Lycopodium fastigiatum R.Br.
Lycopodium scariosum G.Forst.
Lycopodium volubile G.Forst.
Phylloglossum drummondii Kunze

Order SELAGINELLALES A
SELAGINELLACEAE A
Selaginella kraussiana (Kunze) A.Braun A
Selaginella martensii Spring A
Selaginella moellendorffii Hieron. A

SUBPHYLUM EUPHYLLOPHYTINA
INFRAPHYLUM MONILOPHYTAE
Class PSILOTOPSIDA
Order PSILOTALES
PSILOTACEAE
Psilotum nudum (L.) P.Beauv.
Tmesipteris elongata P.A.Dang.
Tmesipteris horomaka Perrie, Brownsey & Lovis E
Tmesipteris lanceolata P.A.Dang.
Tmesipteris sigmatifolia Chinnock
Tmesipteris tannensis (Spreng.) Bernh. E

Order OPHIOGLOSSALES
OPHIOGLOSSACEAE
Botrychium australe R.Br.
Botrychium biforme Colenso E
Botrychium lunaria (L.) Sw.
Ophioglossum coriaceum A.Cunn.
Ophioglossum petiolatum Hook.

Class EQUISETOPSIDA A
Order EQUISETALES A
EQUISETACEAE A
Equisetum arvense L. A
Equisetum fluviatile L. Cas
Equisetum hyemale L. A

Class MARATTIOPSIDA
Order MARATTIALES
MARATTIACEAE
Ptisana salicina (Sm.) Murdock

Class POLYPODIOPSIDA
Order OSMUNDALES
OSMUNDACEAE
Leptopteris hymenophylloides (A.Rich.) C.Presl E
Leptopteris superba (Colenso) C.Presl E
Osmunda regalis L. A
Todea barbara (L.) T.Moore

Order HYMENOPHYLLALES
HYMENOPHYLLACEAE
<u>*Cardiomanes*</u> *reniforme* (G.Forst.) C.Presl E
Hymenophyllum armstrongii (Baker) Kirk E
Hymenophyllum atrovirens Colenso E
Hymenophyllum bivalve (G.Forst.) Sw.
Hymenophyllum cupressiforme Labill.
Hymenophyllum demissum (G.Forst.) Sw. E
Hymenophyllum dilatatum (G.Forst.) Sw. E
Hymenophyllum flabellatum Labill.
Hymenophyllum flexuosum A.Cunn. E
Hymenophyllum frankliniae Colenso E
Hymenophyllum lyallii Hook.f.
Hymenophyllum malingii (Hook.f.) Mett. E
Hymenophyllum minimum A.Rich.
Hymenophyllum multifidum (G.Forst.) Sw.
Hymenophyllum peltatum (Poir.) Desv.
Hymenophyllum pulcherrimum Colenso E
Hymenophyllum rarum R.Br.
Hymenophyllum revolutum Colenso E
Hymenophyllum rufescens Kirk E
Hymenophyllum sanguinolentum (G.Forst.) Sw. E
Hymenophyllum scabrum A.Rich. E
Hymenophyllum villosum Colenso E
Trichomanes colensoi Hook.f. E
Trichomanes elongatum A.Cunn. E
Trichomanes endlicherianum C.Presl
Trichomanes strictum Hook. et Grev. E
Trichomanes venosum R.Br.

Order GLEICHENIALES
GLEICHENIACEAE
Dicranopteris linearis (Burm.f.) Underw.
Gleichenia alpina R.Br.
Gleichenia dicarpa R.Br.
Gleichenia microphylla R.Br.

Sticherus cunninghamii (Hook.) Ching E
Sticherus flabellatus (R.Br.) H. St. John
Sticherus tener (R.Br.) Ching

Order SCHIZAEALES
LYGODIACEAE
Lygodium articulatum A.Rich. E
SCHIZAEACEAE
Schizaea australis Gaudich.
Schizaea bifida Willd.
Schizaea dichotoma (L.) Sm.
Schizaea fistulosa Labill.

Order SALVINIALES
MARSILEACEAE
Marsilea mutica Mett. F A
Pilularia novae-hollandiae A.Braun F
SALVINIACEAE
Azolla filiculoides Lam. F
Azolla pinnata R.Br. F A
Salvinia molesta D.S.Mitch. F A

Order CYATHEALES
CYATHEACEAE
Cyathea colensoi (Hook.f.) Domin E
Cyathea cooperi (F.Muell.) Domin Cas
Cyathea cunninghamii Hook.f.
Cyathea dealbata (G.Forst.) Sw. E
Cyathea kermadecensis W.R.B.Oliv. E K
Cyathea medullaris (G.Forst.) Sw.
Cyathea milnei Hook.f. E K
Cyathea smithii Hook.f. E
DICKSONIACEAE
Dicksonia fibrosa Colenso E
Dicksonia lanata Colenso E
Dicksonia squarrosa (G.Forst.) Sw. E
LOXSOMATACEAE
Loxsoma cunninghamii A.Cunn. E

Order POLYPODIALES
ASPLENIACEAE
Asplenium aethiopicum (Burm.f.) Bech. Cas
Asplenium appendiculatum (Labill.) C.Presl subsp. *appendiculatum*
— subsp. *maritimum* (Brownsey) Brownsey E
Asplenium bulbiferum G.Forst. E
Asplenium chathamense Brownsey EC
Asplenium cimmeriorum Brownsey et de Lange E
Asplenium flabellifolium Cav.
Asplenium flaccidum G.Forst. subsp. *flaccidum*
— subsp. *haurakiense* Brownsey E
Asplenium gracillimum Colenso
Asplenium hookerianum Colenso
Asplenium lamprophyllum Carse E
Asplenium lyallii (Hook.f.) T.Moore E
Asplenium northlandicum (Brownsey) Ogle
Asplenium oblongifolium Colenso E
Asplenium obtusatum G.Forst.
Asplenium pauperequitum Brownsey et P.J. Jacks. E
Asplenium polyodon G.Forst.
Asplenium richardii (Hook.f.) Hook.f. E
Asplenium scleroprium Hombr. E
Asplenium shuttleworthianum Kunze K
Asplenium trichomanes L. subsp. *quadrivalens* D.E.Mey. emend. Lovis
— subsp. nov. WELT P000354–000359
Asplenium xlucrosum Perrie et Brownsey (= *Asplenium bulbiferum* x *A.dimorphum*) Cas
Phyllitis scolopendrium (L.) Newman A
Pleurosorus rutifolius (R.Br.) Fée
BLECHNACEAE
Blechnum blechnoides (Bory) Keyserl.
Blechnum chambersii Tindale
Blechnum colensoi (Hook.f.) N.A.Wakef. E
Blechnum discolor (G.Forst.) Keyserl. E
Blechnum durum (T.Moore) C.Chr. E
Blechnum filiforme (A.Cunn.) Ettingsh. E
Blechnum fluviatile (R.Br.) Salomon
Blechnum fraseri (A.Cunn.) Luerss.
Blechnum membranaceum (Hook.) Diels E
Blechnum minus (R.Br.) Ettingsh.
Blechnum montanum T.C.Chambers et P.A.Farrant E
Blechnum nigrum (Colenso) Mett. E
Blechnum norfolkianum (Heward) C.Chr.
Blechnum novae-zelandiae T.C.Chambers et P.A.Farrant E
Blechnum patersonii (R.Br.) Mett. Cas
Blechnum penna-marina (Poir.) Kuhn subsp. *alpina* T.C.Chambers et P.A.Farrant
Blechnum procerum (G.Forst.) Sw. E
Blechnum punctulatum Sw. Cas
Blechnum triangularifolium T.C.Chambers et P.A.Farrant E
Blechnum vulcanicum (Blume) Kuhn
Doodia aspera R.Br.
Doodia australis (Parris) Parris
Doodia milnei Carruth. E K
Doodia mollis Parris E
Doodia squarrosa Colenso E
DAVALLIACEAE
Davallia griffithiana Hook. A
Davallia mariesii Baker Cas
Davallia tasmanii Field subsp. *tasmanii* E K
— subsp. *cristata* von Konrat, Braggins & de Lange E
DENNSTAEDTIACEAE
Dennstaedtia davallioides (R.Br.) T.Moore Cas
Histiopteris incisa (Thunb.) J.Sm.
Hypolepis amaurorachis (Kunze) Hook.
Hypolepis ambigua (A.Rich.) Brownsey et Chinnock E
Hypolepis dicksonioides (Endl.) Hook.
Hypolepis distans Hook.
Hypolepis lactea Brownsey et Chinnock E
Hypolepis millefolium Hook. E
Hypolepis rufobarbata (Colenso) N.A.Wakef. E
Leptolepia novae-zelandiae (Colenso) Diels E
Microlepia strigosa (Thunb.) C.Presl Cas
Paesia scaberula (A.Rich.) Kuhn E
Pteridium esculentum (G. Forst.) Cockayne
DRYOPTERIDACEAE
Arachniodes aristata (G.Forst.) Tindale K
Cyrtomium falcatum (L.f.) C.Presl A
Dryopteris affinis (Lowe) Fraser-Jenk. A
Dryopteris carthusiana (Vill.) H.P.Fuchs Cas
Dryopteris cycadina (Franch. et Sav.) C.Chr. Cas
Dryopteris dilatata (Hoffm.) A.Gray A
Dryopteris erythrosora (D.C.Eaton) Kuntze Cas
Dryopteris filix-mas (L.) Schott A
Dryopteris inaequalis (Schltdl.) Kuntze Cas
Dryopteris kinkiensis Koidz. ex Tagawa Cas
Dryopteris sieboldii (Van Houtte ex Mett.) Kuntze Cas
Dryopteris stewartii Fraser-Jenk. Cas
Lastreopsis glabella (A.Cunn.) Tindale E
Lastreopsis hispida (Sw.) Tindale
Lastreopsis microsora (Endl.) Tindale subsp. *pentangularis* (Colenso) Tindale E
Lastreopsis velutina (A.Rich.) Tindale E
Lastreopsis n. sp. E K (Tindale 1965) WELT P001716, P008537-008538
Polystichum cystostegia (Hook.) J.B.Armstr. E
Polystichum lentum (D.Don) T.Moore Cas
Polystichum neozelandicum Fée subsp. *neozelandicum* E
— subsp. *zerophyllum* (Colenso) Perrie E
Polystichum oculatum (Hook.) J.B.Armstr. E
Polystichum polyblepharum (Kunze) C.Presl Cas
Polystichum proliferum (R.Br.) C.Presl A
Polystichum setiferum (Forssk.) Woyn. A
Polystichum silvaticum (Colenso) Diels E
Polystichum vestitum (G.Forst.) C.Presl E
Polystichum wawranum (Szyszyl.) Perrie E
Rumohra adiantiformis (G.Forst.) Ching
GRAMMITIDACEAE
Ctenopteris heterophylla (Labill.) Tindale
Grammitis billardierei Willd.
Grammitis ciliata Colenso E
Grammitis givenii Parris E
Grammitis gunnii Parris
Grammitis magellanica Desv. subsp. *magellanica*
— subsp. *nothofageti* Parris
Grammitis patagonica (C.Chr.) Parris
Grammitis poeppigiana (Mett.) Pic.Serm.
Grammitis pseudociliata Parris
Grammitis rawlingsii Parris E
Grammitis rigida Hombr. E
LINDSAEACEAE
Lindsaea linearis Sw.
Lindsaea trichomanoides Dryand.
Lindsaea viridis Colenso E
Odontosoria chinensis (L.) J.Sm. Cas
LOMARIOPSIDACEAE
Nephrolepis cordifolia (L.) C.Presl A
Nephrolepis flexuosa Colenso
Nephrolepis brownii (Desv.) Hovenkamp & Miyam. K
ONOCLEACEAE A
Onoclea sensibilis L. Cas
POLYPODIACEAE
Loxogramme dictyopteris (Mett.) Copel. E
Microsorum novae-zealandiae (Baker) Copel. E
Microsorum pustulatum (G.Forst.) Copel. subsp. *pustulatum*
Microsorum scandens (G.Forst.) Tindale
Niphidium crassifolium(L.) Lellinger Cas
Platycerium bifurcatum (Cav.) C.Chr. Cas
Polypodium vulgare L. A
Pyrrosia eleagnifolia (Bory) Hovenkamp E
PTERIDACEAE
Adiantum aethiopicum L.
Adiantum capillus-veneris L. A
Adiantum cunninghamii Hook. E
Adiantum diaphanum Blume
Adiantum formosum R.Br.
Adiantum fulvum Raoul E
Adiantum hispidulum Sw.
Adiantum raddianum C.Presl A
Adiantum viridescens Colenso E
Anogramma leptophylla (L.) Link
Cheilanthes distans (R.Br.) Mett.
Cheilanthes lendigera (Cav.) Sw. Cas
Cheilanthes sieberi Kunze subsp. *sieberi*
Pellaea calidirupium Brownsey et Lovis
Pellaea falcata (R.Br.) Fée
Pellaea rotundifolia (G.Forst.) Hook. E
Pellaea viridis (Forssk.) Prantl Cas
Pteris argyraea T.Moore Cas
Pteris comans G.Forst.
Pteris cretica L. A
Pteris dentata Forssk. subsp. *flabellata* (Thunb.) Runemark Cas
Pteris macilenta A.Rich. E
Pteris pacifica Hieron. Cas
Pteris saxatilis (Carse) Carse E
Pteris tremula R.Br.
Pteris vittata L. A
TECTARIACEAE
Arthropteris tenella (G.Forst.) Hook.f.
THELYPTERIDACEAE
Christella dentata (Forssk.) Brownsey et Jermy
Cyclosorus interruptus (Willd.) H.Ito
Macrothelypteris torresiana (Gaudich.) Ching
Pneumatopteris pennigera (G.Forst.) Holttum
Thelypteris confluens (Thunb.) C.V.Morton

WOODSIACEAE
Athyrium filix-femina (L.) Roth A
Athyrium otophorum (Miq.) Koidz. Cas
Cystopteris fragilis (L.) Bernh. A
Cystopteris tasmanica Hook.
Deparia petersenii (Kunze) M.Kato subsp. *congrua* (Brack.) M.Kato
Deparia tenuifolia (Kirk) M.Kato E
Diplazium australe (R.Br.) N.A.Wakef.

INFRAPHYLUM SPERMATOPHYTAE
Class GINKGOOPSIDA Cas
Order GINKGOALES Cas
GINKGOACEAE Cas
Ginkgo biloba L. Cas

Class PINOPSIDA
Order PINALES
ARAUCARIACEAE
Agathis australis (D.Don) Lindl. ex Loudon E
Araucaria bidwillii Hook. Cas
Araucaria heterophylla (Salisb.) Franco A
CUPRESSACEAE
Callitris oblonga A.Rich. & Rich. Cas
Callitris rhomboidea R.Br. ex Rich. & A.Rich. A
Chamaecyparis lawsoniana (Murray) Parl. A
Cryptomeria japonica (Thunb. ex L.f.) D.Don A
Cunninghamia lanceolata (Lamb.) Hook. Cas
Cupressus lusitanica Mill. A
Cupressus macrocarpa Hartw. ex Gordon A
Cupressus sempervirens L. A
Libocedrus bidwillii Hook.f. E
Libocedrus plumosa (D.Don) Sarg. E
Sequoia sempervirens (D.Don) Endl. A
Thuja plicata Donn ex D.Don Cas
EPHEDRACEAE Cas
Ephedra campylopoda C.A.Mey. Cas
PINACEAE
Abies grandis (Douglas ex D.Don) Lindl. A
Abies nordmanniana (Steven) Spach A
Cedrus deodara (Roxb.) G.Don Cas
Larix decidua Mill. A
Larix kaempferi (Lamb.) Carrière Cas
Picea abies (L.) H.Karst. A
Picea sitchensis (Bong.) Carrière A
Pinus banksiana Lamb. A
Pinus canariensis C.Sm. Cas
Pinus contorta Loudon A
Pinus elliottii Engelm. Cas
Pinus halepensis Mill. A
Pinus mugo Turra A
Pinus muricata D.Don A
Pinus nigra J.F.Arnold A
Pinus patula Schltdl. & Cham. A
Pinus pinaster Aiton A
Pinus pinea L. Cas
Pinus ponderosa Douglas ex C.Lawson A
Pinus radiata D.Don A
Pinus strobus L. A
Pinus sylvestris L. A
Pinus taeda L. A
Pseudotsuga menziesii (Mirb.) Franco A
Tsuga heterophylla (Raf.) Sarg. Cas
PODOCARPACEAE
Dacrycarpus dacrydioides (A.Rich.) de Laub. E
Dacrydium cupressinum Lamb. E
Halocarpus bidwillii (Kirk) Quinn E
Halocarpus biformis (Hook.) Quinn E
Halocarpus kirkii (Parl.) Quinn E
Lepidothamnus intermedius (Kirk) Quinn E
Lepidothamnus laxifolius (Hook.f.) Quinn E
Manoao colensoi (Hook.) Molloy E
Phyllocladus alpinus Hook.f. E
Phyllocladus toatoa Molloy E
Phyllocladus trichomanoides D.Don E
Podocarpus acutifolius Kirk E
Podocarpus elatus R.Br. ex Endl. Cas
Podocarpus hallii Kirk E
Podocarpus nivalis Hook. E
Podocarpus totara G.Benn. ex D.Don E
Prumnopitys ferruginea (D.Don) de Laub. E
Prumnopitys taxifolia (D.Don) de Laub. E
TAXACEAE A
Taxus baccata L. A

Class MAGNOLIOPSIDA
Order NYMPHAEALES
HYDATELLACEAE
Trithuria inconspicua Cheeseman F E
NYMPHAEACEAE A
Nuphar lutea (L.) Sm. F A
Nymphaea alba L. F A
Nymphaea capensis Thunb. F Cas
Nymphaea mexicana Zucc. F Cas

Superorder AUSTROBAILEYANAE Cas
Order AUSTROBAILEYALES Cas
SCHISANDRACEAE Cas
Schisandra sphenanthera Rehder & E.H.Wilson Cas

Superorder CHLORANTHANAE
Order CHLORANTHALES
CHLORANTHACEAE
Ascarina lucida Hook.f. E

Superorder MAGNOLIANAE
Order CANELLALES
WINTERACEAE
Pseudowintera axillaris (J.R.Forst. & G.Forst.) Dandy E
Pseudowintera colorata (Raoul) Dandy E
Pseudowintera insperata Heenan & de Lange E
Pseudowintera traversii (Buchanan) Dandy E

Order LAURALES
ATHEROSPERMATACEAE
Laurelia novae-zelandiae A.Cunn. E
CALYCANTHACEAE Cas
Chimonanthus praecox (L.) Link Cas
LAURACEAE
Beilschmiedia tarairi (A.Cunn.) Benth. & Hook.f. ex Kirk E
Beilschmiedia tawa (A.Cunn.) Benth. & Hook.f. ex Kirk E
Beilschmiedia tawaroa A.E.Wright E
Cassytha paniculata R.Br.
Cassytha pubescens R.Br. A
Cinnamomum camphora (L.) J.S.Pres. Cas
Cinnamomum loureiroi Nees Cas
Laurus nobilis L. A
Litsea calicaris (Sol. ex A.Cunn.) Benth. & Hook.f. ex Kirk E
Persea americana Mill. A
MONIMIACEAE
Hedycarya arborea J.R.Forst. & G.Forst. E

Order MAGNOLIALES
ANNONACEAE A
Annona cherimola Mill. A
MAGNOLIACEAE
Liriodendron tulipifera L. Cas
Magnolia grandiflora L. Cas
Magnolia sieboldii K.Koch Cas
Magnolia stellata (Siebold & Zucc.) Maxim. Cas
Michelia doltsopa Buch.-Ham. ex DC. Cas

Order PIPERALES
ARISTOLOCHIACEAE Cas
Aristolochia elegans Mast. Cas
Aristolochia sempervirens Forssk. Cas
PIPERACEAE
Macropiper excelsum (G.Forst.) Miq. E
Macropiper melchior Sykes E
Peperomia tetraphylla (G.Forst.) Hook. & Arn.
Peperomia urvilleana A.Rich. E
SAURURACEAE Cas
Houttuynia cordata Thunb. Cas
Saururus cernuus L. Cas

Superorder LILIANAE [Monocots]
Order ALISMATALES
ALISMATACEAE A
Alisma lanceolatum With. F A
Alisma plantago-aquatica L. F A
Hydrocleys nymphoides (Humb. & Bonpl.) Buchenau F A
Sagittaria montevidensis Cham. & Schltdl. F A
Sagittaria platyphylla (Engelm.) J.G.Sm. F Cas
Sagittaria sagittifolia L. F Cas
APONOGETONACEAE A
Aponogeton distachyos L.f. F A
ARACEAE
Alocasia brisbanensis (F.M.Bailey) Domin A
Arisarum vulgare Targ.-Tozz. Cas
Arum italicum Mill. A
Colocasia esculenta (L.) Schott A
Dracunculus vulgaris Schott A
Landoltia punctata (G.Mey.) Les & D.J.Crawford F A
Lemna minor L. F A
Pistia stratiotes L. F Cas
Wolffia australiana (Benth.) Hartog & Plas F
Xanthosoma sagittifolium (L.) Schott A
Zantedeschia aethiopica (L.) Spreng. A
Zantedeschia albomaculata (Hook.f.) Baill. A
HYDROCHARITACEAE A
Egeria densa Planch. F A
Elodea canadensis Michx. F A
Hydrilla verticillata (L.f.) Royle F A
Lagarosiphon major (Ridl.) Moss ex Wager F A
Ottelia ovalifolia (R.Br.) Rich. F A
Vallisneria australis S.W.L.Jacobs & Les F Cas
Vallisneria spiralis L. F A
JUNCAGINACEAE
Triglochin palustris L.
Triglochin striata Ruiz & Pav.
POTAMOGETONACEAE
Lepilaena bilocularis Kirk F
Potamogeton cheesemanii A.Benn. F
Potamogeton crispus L. F A
Potamogeton ochreatus Raoul F
Potamogeton suboblongus Hagstr. F E
Stuckenia pectinata (L.) Börner F
Zannichellia palustris L. F
RUPPIACEAE
Ruppia megacarpa R.Mason M
Ruppia polycarpa R.Mason F
ZOSTERACEAE
Zostera muelleri Asch. M

Order ASPARAGALES A
AGAPANTHACEAE
Agapanthus praecox Willd. A
ALLIACEAE A
Allium ampeloprasum L. A
Allium cepa L. A
Allium neapolitanum Cirillo A
Allium porrum L. A
Allium roseum L. A
Allium triquetrum L. A
Allium vineale L. A
Ipheion uniflorum (Lindl.) Raf. A
Nothoscordum gracile (Aiton) Stearn A
Tulbaghia violacea Harv. A
AMARYLLIDACEAE A
Amaryllis belladonna L. A

Clivia miniata (Lindl.) J.F.W.Bosse Cas
Cyrtanthus elatus (Jacq.) Traub A
Leucojum aestivum L. A
Narcissus ×*incomparabilis* Mill. A
Narcissus ×*odorus* L. A
Narcissus bulbocodium L. A
Narcissus papyraceus Ker Gawl. Cas
Narcissus poeticus L. A
Narcissus pseudonarcissus L. A
Narcissus tazetta L. A
Nerine filifolia Baker A
Nerine sarniensis (L.) Herb. Cas
Zephyranthes candida Herb. A
ASPARAGACEAE
Agave americana L. A
Arthropodium bifurcatum Heenan, A.D.Mitch. & de Lange E
Arthropodium candidum Raoul E
Arthropodium cirratum (G.Forst.) R.Br. E
Asparagus aethiopicus L. A
Asparagus asparagoides (L.) Druce A
Asparagus officinalis L. A
Asparagus plumosus Baker A
Asparagus scandens Thunb. A
Cordyline australis (G.Forst.) Endl. E
Cordyline banksii Hook.f. E
Cordyline fruticosa (L.) A.Chev. A
Cordyline indivisa (G.Forst.) Endl. E
Cordyline obtecta (Graham) Baker
Cordyline pumilio Hook.f. E
Cordyline terminalis A
Danae racemosa (L.) Moench Cas
Dracaena draco (L.) L. Cas
Eucomis comosa (Houtt.) Wehrh. Cas
Furcraea foetida (L.) Haw. A
Furcraea parmentieri (Roezl ex Ortgies) García-Mendoza A
Herpolirion novae-zelandiae Hook.f.
Hyacinthoides non-scripta (L.) Chouard ex Rothm. A
Hyacinthus orientalis L. A
Lachenalia aloides (L.f.) Engl. Cas
Lachenalia bulbifera (Cirillo) Asch. & Graebn. Cas
Lomandra longifolia Labill. Cas
Muscari armeniacum Baker A
Nolina longifolia (Karw. ex Schult.f.) Hemsl. Cas
Ornithogalum umbellatum L. A
Polygonatum multiflorum (L.) All. Cas
Ruscus aculeatus L. Cas
Scilla peruviana L. A
Veltheimia bracteata Harv. ex Baker Cas
Yucca gloriosa L. A
ASPHODELACEAE
Aloe arborescens Mill. A
Aloe ciliaris Haw. A
Aloe maculata All. A
Asphodelus fistulosus L. A
Bulbine semibarbata (R.Br.) Haw. A
Bulbinella angustifolia (Cockayne & Laing) L.B.Moore E
Bulbinella gibbsii Cockayne E
Bulbinella hookeri (Hook.) Cheeseman E
Bulbinella modesta L.B.Moore E
Bulbinella rossii (Hook.f.) Cheeseman E
Bulbinella talbotii L.B.Moore E
Kniphofia linearifolia Baker Cas
Kniphofia rooperi (T.Moore) Lem. Cas
Kniphofia rufa Baker Cas
ASTELIACEAE
Astelia banksii A.Cunn. E
Astelia chathamica (Skottsb.) L.B.Moore E
Astelia fragrans Colenso E
Astelia graminea L.B.Moore E
Astelia grandis Hook.f. ex Kirk E
Astelia linearis Hook.f. E
Astelia nervosa Hook.f. E
Astelia nivicola Cockayne ex Cheeseman E
Astelia petriei Cockayne E
Astelia skottsbergii L.B.Moore E
Astelia solandri A.Cunn. E
Astelia subulata (Hook.f.) Cheeseman E
Astelia trinervia Kirk E
Collospermum hastatum (Colenso) Skottsb. E
Collospermum microspermum (Colenso) Skottsb. E
Collospermum spicatum (Colenso) Skottsb. E
HEMEROCALLIDACEAE
Dianella haematica Heenan & de Lange E
Dianella latissima Heenan & de Lange E
Dianella nigra Colenso E
Hemerocallis fulva (L.) L. A
Hemerocallis lilioasphodelus L. Cas
Phormium cookianum Le Jol. E
Phormium tenax J.R.Forst. & G.Forst. E
HYPOXIDACEAE A
Hypoxis capensis (L.) Vines & Druce Cas
Hypoxis glabella R.Br. A
IRIDACEAE
Aristea ecklonii Baker A
Babiana stricta Ker Gawl. A
Chasmanthe bicolor (Gasp.) N.E.Br. A
Chasmanthe floribunda (Salisb.) N.E.Br. A
Crocosmia ×*crocosmiiflora* (G.Nicholson) N.E.Br. A
Crocosmia paniculata (Klatt) Goldblatt A
Crocus flavus Weston A
Dierama pendulum (L.f.) Baker A
Dietes bicolor (Lindl.) Sweet ex G.Don Cas
Dietes grandiflora N.E.Br. Cas
Dietes iridioides (L.) Sweet ex Klatt Cas
Freesia laxa (Thunb.) Goldblatt & J.C.Manning A
Freesia refracta (Jacq.) Klatt A
Gladiolus dalenii Van Geel A
Gladiolus undulatus L. A
Gynandriris setifolia (L.f.) R.C.Foster A
Iris foetidissima L. A
Iris germanica L. A
Iris japonica Thunb. Cas
Iris laevigata Fisch. & C.A.Mey. A
Iris orientalis Mill. A
Iris pseudacorus L. A
Iris spuria L. A
Ixia maculata L. A
Ixia paniculata D.Delaroche A
Ixia polystachya L. A
Libertia cranwelliae Blanchon, B.G.Murray & Braggins E
Libertia edgariae Blanchon, B.G.Murray & Braggins E
Libertia flaccidifolia Blanchon & J.S.Weaver E
Libertia grandiflora (R.Br.) Sweet E
Libertia ixioides (G.Forst.) Spreng. E
Libertia micrantha A.Cunn. E
Libertia mooreae Blanchon, B.G.Murray & Braggins E
Libertia peregrinans Cockayne & Allan E
Melasphaerula ramosa (L.) N.E.Br. A
Moraea flaccida (Sweet) Steud. A
Romulea minutiflora Klatt Cas
Romulea rosea (L.) Eckl. A
Schizostylis coccinea Backh. & Harv. A
Sisyrinchium californicum (Ker Gawl.) W.T.Aiton A
Sisyrinchium iridifolium Kunth A
Sisyrinchium striatum Sm. A
Sparaxis bulbifera (L.) Ker Gawl. A
Sparaxis grandiflora (D.Delaroche) Ker Gawl. A
Sparaxis tricolor (Schneev.) Ker Gawl. A
Tritonia crocata (L.) Ker Gawl. A
Tritonia lineata (Salisb.) Ker Gawl. A
Watsonia ardernei Sander A
Watsonia borbonica (Pourr.) Goldblatt Cas
Watsonia marginata Ker Gawl. A
Watsonia meriana (L.) Mill. A
Watsonia versefeldii J.W.Mathews & L.Bolus Cas
Watsonia zeyheri L.Bolus Cas
ORCHIDACEAE
Acianthus sinclairii Hook.f. E
Adelopetalum tuberculatum (Colenso) D.L.Jones, M.A.Clem. & Molloy
Adenochilus gracilis Hook.f. E
Anzybas carsei (Cheeseman) D.L.Jones & M.A.Clem.
Anzybas rotundifolius (Hook.f.) D.L.Jones & M.A.Clem. E
<u>Aporostylis</u> bifolia (Hook.f.) Rupp & Hatch E
Bletilla striata (Thunb.) Rchb.f. A
Calochilus herbaceus Lindl.
Calochilus paludosus R.Br.
Calochilus robertsonii Benth.
Corunastylis nuda (Hook.f.) D.L.Jones & M.A.Clem.
Corunastylis pumila (Hook.f.) D.L.Jones & M.A.Clem. E
Corybas cheesemanii (Hook.f. ex Kirk) Kuntze E
Cryptostylis subulata (Labill.) Rchb.f.
Cyrtostylis oblonga Hook.f. E
Cyrtostylis rotundifolia Hook.f. E
<u>Danhatchia</u> australis (Hatch) Garay & Christenson E
Diplodium alobulum (Hatch) D.L.Jones, Molloy & M.A.Clem. E
Diplodium alveatum (Garnet) D.L.Jones & M.A.Clem.
Diplodium brumale (L.B.Moore) D.L.Jones, Molloy & M.A.Clem. E
Diplodium trullifolium (Hook.f.) D.L.Jones, Molloy & M.A.Clem. E
Drymoanthus adversus (Hook.f.) Dockrill E
Drymoanthus flavus St George & Molloy E
Earina aestivalis Cheeseman E
Earina autumnalis (G.Forst.) Hook.f. E
Earina mucronata Lindl. E
Epidendrum ibaguense Kunth. Cas
Gastrodia cunninghamii Hook.f. E
Gastrodia minor Petrie E
Gastrodia sesamoides R.Br.
Hymenochilus tanypodus (D.L.Jones, Molloy & M.A.Clem) D.L.Jones, M.A.Clem. & Molloy E
Hymenochilus tristis (Colenso) D.L.Jones, M.A.Clem. & Molloy E
Ichthyostomum pygmaeum (Sm.) D.L.Jones, M.A.Clem. & Molloy
Linguella puberula (Hook.f.) D.L.Jones, M.A.Clem. & Molloy E
Microtis arenaria Lindl.
Microtis oligantha L.B.Moore E
Microtis parviflora R.Br.
Microtis unifolia (G.Forst.) Rchb.f. E
<u>Molloybas</u> cryptanthus (Hatch) D.L.Jones & M.A.Clem. E
Myrmechila formicifera (Fitzg.) D.L.Jones & M.A.Clem.
Myrmechila trapeziformis (Fitzg.) D.L.Jones & M.A.Clem.
Nematoceras acuminatum (M.A.Clem. & Hatch) Molloy, D.L.Jones & M.A.Clem. E
Nematoceras dienemum (D.L.Jones) D.L.Jones, M.A.Clem. & Molloy
Nematoceras hypogaeum (Colenso) Molloy, D.L.Jones & M.A.Clem. E
Nematoceras iridescens (Irwin & Molloy) Molloy, D.L.Jones & M.A.Clem. E
Nematoceras longipetalum (Hatch) Molloy, D.L.Jones & M.A.Clem. E
Nematoceras macranthum Hook.f. E
Nematoceras orbiculatum (Colenso) Molloy, D.L.Jones & M.A.Clem. E
Nematoceras panduratum (Cheeseman) Molloy, D.L.Jones & M.A.Clem. E

Nematoceras papa (Molloy & Irwin) Molloy, D.L.Jones & M.A.Clem. E
Nematoceras papillosum (Colenso) Molloy, D.L.Jones & M.A.Clem. E
Nematoceras rivulare (A.Cunn.) Hook.f. E
Nematoceras trilobum Hook.f. E
Orthoceras novae-zeelandiae (A.Rich.) M.A.Clem., D.L.Jones & Molloy E
Petalochilus alatus (R.Br.) D.L.Jones & M.A.Clem.
Petalochilus bartlettii (Hatch) D.L.Jones & M.A.Clem. E
Petalochilus calyciformis R.S.Rogers E
Petalochilus carneus (R.Br.) D.L.Jones & M.A.Clem.
Petalochilus chlorostylus (D.L.Jones, Molloy & M.A.Clem.) D.L.Jones & M.A.Clem. E
Petalochilus minor (Hook.f.) D.L.Jones & M.A.Clem.
Petalochilus nothofageti (D.L.Jones, Molloy & M.A.Clem.) D.L.Jones & M.A.Clem. E
Petalochilus saccatus R.S.Rogers
Petalochilus variegatus (Colenso) D.L.Jones & M.A.Clem.
Plumatichilos tasmanicum (D.L.Jones) Szlach.
Prasophyllum colensoi Hook.f. E
Prasophyllum hectorii (Buchanan) Molloy, D.L.Jones & M.A.Clem. E
Pterostylis agathicola D.L.Jones, Molloy & M.A.Clem. E
Pterostylis areolata Petrie E
Pterostylis auriculata Colenso E
Pterostylis australis Hook.f. E
Pterostylis banksii A.Cunn. E
Pterostylis cardiostigma D.Cooper E
Pterostylis cernua D.L.Jones, Molloy & M.A.Clem. E
Pterostylis cycnocephala Fitzg.
Pterostylis foliata Hook.f.
Pterostylis graminea Hook.f. E
Pterostylis humilis R.S.Rogers E
Pterostylis irsoniana Hatch E
Pterostylis irwinii D.L.Jones, Molloy & M.A.Clem. E
Pterostylis micromega Hook.f. E
Pterostylis montana Hatch E
Pterostylis nutans R.Br.
Pterostylis oliveri Petrie E
Pterostylis paludosa D.L.Jones, Molloy & M.A.Clem. E
Pterostylis patens Colenso E
Pterostylis porrecta D.L.Jones, Molloy & M.A.Clem. E
Pterostylis silvicultrix (F.Muell.) Molloy, D.L.Jones & M.A.Clem. E
Pterostylis venosa Colenso E
Simpliglottis cornuta (Hook.f.) Szlach. E
Simpliglottis valida (D.L.Jones) Szlach.
Singularybas oblongus (Hook.f.) Molloy, D.L.Jones & M.A.Clem. E
Spiranthes novae-zelandiae Hook.f. E
Spiranthes sinensis (Pers.) Ames
Stegostyla atradenia (D.L.Jones, Molloy & M.A.Clem.) D.L.Jones & M.A.Clem.
Stegostyla iridescens (R.S.Rogers) D.L.Jones
Stegostyla lyallii (Hook.f.) D.L.Jones & M.A.Clem.
Sullivania minor (R.Br.) D.L.Jones & M.A.Clem.
Taeniophyllum norfolkianum D.L.Jones, B.Gray et M.A.Clem.
Thelymitra aemula Cheeseman E
Thelymitra brevifolia Jeanes
Thelymitra carnea R.Br.
Thelymitra colensoi Hook.f. E
Thelymitra cyanea (Lindl.) Benth.
Thelymitra formosa Colenso
Thelymitra hatchii L.B.Moore E
Thelymitra ixioides Sw.
Thelymitra longifolia J.R.Forst. & G.Forst.
Thelymitra malvina M.A.Clem., D.L.Jones & Molloy
Thelymitra matthewsii Cheeseman
Thelymitra nervosa Colenso E
Thelymitra pauciflora R.Br.
Thelymitra pulchella Hook.f. E
Thelymitra purpureo-fusca Colenso E
Thelymitra sanscilia Irwin ex Hatch
Thelymitra tholiformis Molloy & Hatch E
Thelymitra venosa R.Br.
Townsonia deflexa Cheeseman E
Waireia stenopetala (Hook.f.) D.L.Jones, M.A.Clem. & Molloy E
Winika cunninghamii (Lindl.) M.A.Clem., D.L.Jones & Molloy E
XERONEMATACEAE
Xeronema callistemon W.R.B.Oliv. E

Order DIOSCOREALES
BURMANNIACEAE
Thismia rodwayi F.Muell.
DIOSCOREACEAE A
Dioscorea communis (L.) Caddick & Wilkin A

Order LILIALES
ALSTROEMERIACEAE A
Alstroemeria aurea Graham A
Alstroemeria ligtu hybrid Cas
Alstroemeria pulchella L.f. A
Bomarea multiflora (L.f.) Mirb. A
COLCHICACEAE
Colchicum autumnale L. A
Iphigenia novae-zelandiae (Hook.f.) Baker E
LILIACEAE A
Cardiocrinum giganteum (Wall.) Makino A
Lilium formosanum A.Wallace A
Lilium lancifolium Thunb. A
Lilium longiflorum Thunb. Cas
LUZURIAGACEAE
Luzuriaga parviflora (Hook.f.) Kunth E
RIPOGONACEAE
Ripogonum scandens J.R.Forst. & G.Forst. E

Order PANDANALES
PANDANACEAE
Freycinetia banksii A.Cunn. E

Order ARECALES
ARECACEAE (PALMAE)
Archontophoenix cunninghamiana H. Wendl. & Drude Cas
Cocos nucifera L. Cas
Howea forsteriana (F.Muell. & H.Wendl.) Becc. Cas
Livistona australis (R.Br.) Mart. A
Phoenix canariensis Chabaud A
Rhopalostylis baueri (Seem.) H.Wendl. & Drude
Rhopalostylis sapida H.Wendl. & Drude E
Trachycarpus fortunei (Hook.) H.Wendl. A

Order COMMELINALES A
COMMELINACEAE A
Gibasis pellucida (M.Martens & Galeotti) D.R.Hunt Cas
Gibasis schiediana (Kunth) D.R.Hunt A
Tradescantia albiflora Kunth A
Tradescantia cerinthoides Kunth A
Tradescantia fluminensis Vell. A
Tradescantia virginiana L. A
HAEMODORACEAE A
Anigozanthos flavidus DC. Cas
Wachendorfia thyrsiflora L. F A
PONTEDERIACEAE A
Eichhornia crassipes (Mart.) Solms F A

Order POALES
BROMELIACEAE Cas
Dyckia frigida (Linden) Hook.f. Cas
Ochagavia carnea (Beer) L.B.Sm. & Looser Cas
CENTROLEPIDACEAE
Centrolepis ciliata (Hook.f.) Druce E
Centrolepis minima Kirk E
Centrolepis pallida (Hook.f.) Cheeseman E
Centrolepis strigosa (R.Br.) Roem. & Schult. A
Gaimardia setacea Hook.f.
CYPERACEAE
Baumea arthrophylla (Nees) Boeck.
Baumea articulata (R.Br.) S.T.Blake F
Baumea complanata (Berggr.) S.T.Blake E
Baumea juncea (R.Br.) Palla F
Baumea rubiginosa (Spreng.) Boeck. F
Baumea tenax (Hook.f.) S.T.Blake E
Baumea teretifolia (R.Br.) Palla F
Bolboschoenus caldwellii (V.J.Cook) Soják
Bolboschoenus fluviatilis (Torr.) Soják
Bolboschoenus medianus (V.J.Cook) Soják
Carex acicularis Boott E
Carex albula Allan E
Carex allanii Hamlin E
Carex appressa R.Br.
Carex astonii Hamlin E
Carex berggrenii Petrie E
Carex bichenoviana Boott A
Carex breviculmis R.Br.
Carex brownii Tuck. A
Carex buchananii Berggr. E
Carex calcis K.A.Ford E
Carex capillacea Boott
Carex carsei Petrie E
Carex chathamica Petrie E
Carex cirrhosa Berggr. F E
Carex cockayneana Kük. E
Carex colensoi Boott E
Carex comans Berggr. E
Carex coriacea Hamlin E
Carex cremnicola K.A.Ford E
Carex dallii Kirk F E
Carex decurtata Cheeseman E
Carex demissa Hornem. A
Carex devia Cheeseman E
Carex diandra Schrank
Carex dipsacea Berggr. E
Carex dissita Sol. ex Boott E
Carex divisa Huds. A
Carex divulsa Stokes A
Carex dolomitica Heenan & de Lange E
Carex druceana Hamlin E
Carex echinata Murray
Carex edgariae Hamlin E
Carex elingamita Hamlin E
Carex enysii Petrie E
Carex fascicularis Boott E
Carex filamentosa Petrie E
Carex flacca Schreb. A
Carex flagellifera Colenso E
Carex flaviformis Nelmes F E
Carex forsteri Wahlenb. E
Carex fretalis Hamlin E
Carex gaudichaudiana Kunth
Carex geminata Schkuhr E
Carex goyenii Petrie E
Carex hectorii Petrie E
Carex hirsutella Mack. A
Carex hirta L. A
Carex impexa K.A.Ford E
Carex inopinata V.J.Cook E
Carex inversa R.Br.
Carex iynx Nelmes A
Carex kaloides Petrie E
Carex kermadecensis Petrie E
Carex kirkii Petrie E
Carex lachenalii Schkuhr
Carex lambertiana Boott E
Carex lessoniana Steud. E

Carex libera (Kük.) Hamlin E
Carex litorosa L.H.Bailey E
Carex longebrachiata Boeck. A
Carex longiculmis Petrie E
Carex longii Mack. A
Carex lurida Wahlenb. A
Carex maculata Boott Cas
Carex maorica Hamlin F E
Carex muelleri Petrie E
Carex muricata L. A
Carex ochrosaccus (Cheeseman) Hamlin E
Carex ophiolithica de Lange & Heenan E
Carex otrubae Podp. A
Carex ovalis Gooden. A
Carex pallescens L. A
Carex paniculata L. A
Carex pendula Huds. A
Carex petriei Cheeseman E
Carex pleiostachys C.B.Clarke E
Carex pseudocyperus L.
Carex pterocarpa Petrie E
Carex pumila Thunb.
Carex punctata Gaudin A
Carex pyrenaica Wahlenb.
Carex raoulii Boott E
Carex resectans Cheeseman E
Carex riparia Curtis A
Carex rubicunda Petrie E
Carex scoparia Schkuhr ex Willd. A
Carex secta Boott F E
Carex sectoides (Kük.) Edgar E
Carex sinclairii Boott E
Carex solandri Boott E
Carex spicata Huds. A
Carex spinirostris Colenso E
Carex subdola Boott E
Carex sylvatica Huds. A
Carex tenuiculmis (Petrie) Heenan & de Lange F E
Carex ternaria Boott E
Carex testacea Sol. ex Boott E
Carex trachycarpa Cheeseman E
Carex traversii Kirk T E
Carex trifida Cav.
Carex uncifolia Cheeseman E
Carex ventosa C.B.Clarke E
Carex virescens Muhl. ex Willd. Cas
Carex virgata Sol. ex Boott E
Carex vulpinoidea Michx. A
Carex wakatipu Petrie E
Carpha alpina R.Br.
Cyperus albostriatus Schrad. A
Cyperus congestus Vahl A
Cyperus eragrostis Lam. A
Cyperus esculentus L. A
Cyperus gunnii Hook.f. A
Cyperus insularis Heenan & de Lange E
Cyperus involucratus Rottb. A
Cyperus kyllingia Endl. A
Cyperus longus L. A
Cyperus polystachyos Rottb. A
Cyperus prolifer Lam. Cas
Cyperus rotundus L. A
Cyperus sanguinolentus Vahl A
Cyperus ustulatus A.Rich. E
Eleocharis acuta R.Br.
Eleocharis gracilis R.Br.
Eleocharis neozelandica C.B.Clarke ex Kirk E
Eleocharis pusilla R.Br. E
Eleocharis sphacelata R.Br. E
Ficinia nodosa (Rottb.) Goetgh., Muasya & D.A.Simpson
Ficinia spiralis (A.Rich.) Muasya & de Lange E
Fimbristylis squarrosa Vahl
Gahnia lacera (A.Rich.) Steud. E
Gahnia pauciflora Kirk E
Gahnia procera J.R.Forst. & G.Forst. E
Gahnia rigida Kirk E
Gahnia setifolia (A.Rich.) Hook.f. E
Gahnia xanthocarpa (Hook.f.) Hook.f. E
Isolepis aucklandica Hook.f.
Isolepis australiensis (Maiden & Betche) K.L.Wilson A
Isolepis basilaris Hook.f. E
Isolepis caligenis (V.J.Cook) Soják E
Isolepis cernua (Vahl) Roem. & Schult.
Isolepis crassiuscula Hook.f.
Isolepis distigmatosa (C.B.Clarke) Edgar E
Isolepis fluitans (L.) R.Br.
Isolepis habra (Edgar) Soják
Isolepis inundata R.Br.
Isolepis levynsiana Muasya & D.A.Simpson A
Isolepis marginata (Thunb.) A.Dietr. A
Isolepis pottsii (V.J.Cook) Soják E
Isolepis praetextata (Edgar) Soják E
Isolepis prolifera (Rottb.) R.Br.
Isolepis reticularis Colenso E
Isolepis sepulcralis Steud. A
Isolepis setacea (L.) R.Br. A
Isolepis subtilissima Boeck.
Kyllinga brevifolia Rottb. A
Lepidosperma australe (A.Rich.) Hook.f. E
Lepidosperma filiforme Labill.
Lepidosperma laterale R.Br.
Machaerina sinclairii (Hook.f.) T.Koyama
Morelotia affinis (Brongn.) S.T.Blake E
Oreobolus impar Edgar E
Oreobolus pectinatus Hook.f. E
Oreobolus strictus Berggr. E
Rhynchospora capitellata (Michx.) Vahl Cas
Rhynchospora globularis (Chapm.) Small Cas
Schoenoplectus californicus (C.A.Mey.) Palla A
Schoenoplectus pungens (Vahl) Palla
Schoenoplectus tabernaemontani (C.C.Gmel.) Palla
Schoenus apogon Roem. & Schult.
Schoenus brevifolius R.Br.
Schoenus carsei Cheeseman
Schoenus concinnus (Hook.f.) Hook.f. E
Schoenus fluitans Hook.f.
Schoenus maschalinus Roem. & Schult.
Schoenus nitens (R.Br.) Roem. & Schult.
Schoenus pauciflorus (Hook.f.) Hook.f. E
Schoenus tendo (Banks & Sol. ex Hook.f.) Hook.f. E
Scirpus georgianus R.M.Harper A
Scirpus polystachyus F.Muell.
Tetraria capillaris (F.Muell.) J.M.Black
Uncinia affinis (C.B.Clarke) Hamlin E
Uncinia angustifolia Hamlin E
Uncinia astonii Hamlin E
Uncinia aucklandica Hamlin E
Uncinia banksii Boott E
Uncinia caespitosa Boott E
Uncinia clavata (Kük.) Hamlin E
Uncinia distans Colenso ex Boott E
Uncinia divaricata Boott E
Uncinia drucei Hamlin E
Uncinia egmontiana Hamlin E
Uncinia elegans (Kük.) Hamlin E
Uncinia ferruginea Boott E
Uncinia filiformis Boott
Uncinia fuscovaginata Kük. E
Uncinia gracilenta Hamlin E
Uncinia hookeri Boott
Uncinia involuta Hamlin E
Uncinia laxiflora Petrie E
Uncinia leptostachya Raoul E
Uncinia longifructus (Kük.) Petrie E
Uncinia nervosa Boott
Uncinia obtusata Colenso E
Uncinia obtusifolia Heenan E
Uncinia perplexa Heenan & de Lange E
Uncinia purpurata Petrie E
Uncinia rubra Boott E
Uncinia rupestris Raoul E
Uncinia scabra Boott E
Uncinia silvestris Hamlin E
Uncinia sinclairii Boott E
Uncinia strictissima (Kük.) Petrie E
Uncinia uncinata (L.f.) Kük.
Uncinia viridis (C.B.Clarke) Edgar E
Uncinia zotovii Hamlin E
JUNCACEAE
Juncus acuminatus Michx. A
Juncus acutiflorus Ehrh. ex Hoffm. A
Juncus acutus L. A
Juncus amabilis Edgar A
Juncus ambiguus Guss. A
Juncus antarcticus Hook.f.
Juncus anthelatus (Wiegand) R.E.Brooks A
Juncus articulatus L. A
Juncus australis Hook.f.
Juncus brachycarpus Engelm. A
Juncus bufonius L. A
Juncus bulbosus L. A
Juncus caespiticius E.Mey.
Juncus canadensis J.Gay ex Laharpe A
Juncus capitatus Weigel A
Juncus conglomeratus L. A
Juncus continuus L.A.S.Johnson A
Juncus dichotomus Elliott A
Juncus distegus Edgar E
Juncus edgariae L.A.S.Johnson & K.L.Wilson E
Juncus effusus L. A
Juncus ensifolius Wikstr. A
Juncus filicaulis Buchenau A
Juncus flavidus L.A.S.Johnson A
Juncus fockei Buchenau A
Juncus gerardii Loisel. A
Juncus holoschoenus R.Br.
Juncus homalocaulis F.Muell. A
Juncus imbricatus Laharpe A
Juncus inflexus L. A
Juncus kraussii Hochst.
Juncus lomatophyllus Spreng. A
Juncus microcephalus Kunth A
Juncus novae-zelandiae Hook.f. E
Juncus pallidus R.Br.
Juncus pauciflorus R.Br.
Juncus planifolius R.Br.
Juncus polyanthemus Buchenau A
Juncus prismatocarpus R.Br.
Juncus procerus E.Mey. A
Juncus pusillus Buchenau
Juncus sarophorus L.A.S.Johnson
Juncus scheuchzerioides Gaudich.
Juncus sonderianus Buchenau A
Juncus squarrosus L. A
Juncus subnodulosus Schrank A
Juncus subsecundus N.A.Wakef. A
Juncus tenuis Willd. A
Juncus usitatus L.A.S.Johnson
Juncus vaginatus R.Br. Cas
Luzula banksiana E.Mey. E
Luzula campestris (L.) DC. A
Luzula celata Edgar E
Luzula colensoi Hook.f. E
Luzula congesta (Thuill.) Lej. A
Luzula crenulata Buchenau E
Luzula crinita Hook.f.
Luzula decipiens Edgar E
Luzula flaccida (Buchenau) Edgar A
Luzula leptophylla Buchenau & Petrie E
Luzula multiflora (Retz.) Lej. A
Luzula picta A.Rich. E
Luzula pumila Hook.f. E
Luzula rufa Edgar E

Luzula sylvatica (Huds.) Gaudin Cas
Luzula traversii (Buchenau) Cheeseman E
Luzula ulophylla (Buchenau) Cockayne & Laing E
Marsippospermum gracile (Hook.f.) Buchenau E
Rostkovia magellanica (Lam.) Hook.f.
POACEAE (GRAMINEAE)
Achnatherum caudatum (Trin.) S.W.L.Jacobs & J.Everett A
Achnatherum petriei (Buchanan) S.W.L.Jacobs & J.Everett E
Agrostis capillaris L. A
Agrostis castellana Boiss. & Reut. A
Agrostis dyeri Petrie E
Agrostis gigantea Roth A
Agrostis imbecilla Zotov E
Agrostis magellanica Lam.
Agrostis muelleriana Vickery
Agrostis muscosa Kirk E
Agrostis oresbia Edgar E
Agrostis pallescens Cheeseman E
Agrostis personata Edgar E
Agrostis petriei Hack. E
Agrostis stolonifera L. A
Agrostis subulata Hook.f. E
Aira caryophyllea L. A
Aira cupaniana Guss. A
Aira elegans Kunth A
Aira praecox L. A
Alopecurus aequalis Sobol. A
Alopecurus geniculatus L. A
Alopecurus myosuroides Huds. Cas
Alopecurus pratensis L. A
Ammophila arenaria (L.) Link A
Amphibromus fluitans Kirk
Andropogon virginicus L. A
Anemanthele lessoniana (Steud.) Veldkamp E
Anthosachne aprica (Á.Löve & Connor) C.Yen & J.L.Yang E
Anthosachne falcis (Connor) Barkworth & S.W.L.Jacobs E
Anthosachne multiflora (Banks & Sol. ex Hook.f.) C.Yen & J.L.Yang
Anthosachne sacandros (Connor) Barkworth & S.W.L.Jacobs E
Anthosachne scabra (R.Br.) Nevski A
Anthosachne solandri (Steud.) Barkworth & S.W.L.Jacobs E
Anthoxanthum aristatum Boiss. A
Anthoxanthum odoratum L. A
Aristida longespica Poir. Cas
Aristida ramosa R.Br. A
Aristida vagans Cav. A
Arrhenatherum elatius (L.) J.Presl & C.Presl A
Arundo donax L. A
Australopyrum calcis Connor & Molloy E
Australopyrum pectinatum (Labill.) Á.Löve A
Australopyrum retrofractum (Vickery) Á.Löve A
Austroderia fulvida (Buchanan) N.P.Bakrer & H.P.Linder E
Austroderia richardii (Endl.) N.P.Barker & H.P.Linder E
Austroderia splendens (Connor) N.P.Barker & H.P.Linder E
Austroderia toetoe (Zotov) N.P.Barker & H.P.Linder E
Austroderia turbaria (Connor) N.P.Barker & H.P.Linder E
Austrostipa bigeniculata (Hughes) S.W.L.Jacobs & J.Everett A
Austrostipa blackii (C.E.Hubb.) S.W.L.Jacobs & J.Everett A
Austrostipa flavescens (Labill.) S.W.L.Jacobs & J.Everett A
Austrostipa nitida (Summerh. & C.E.Hubb.) S.W.L.Jacobs & J.Everett A
Austrostipa nodosa (S.T.Blake) S.W.L.Jacobs & J.Everett A
Austrostipa ramosissima (Trin.) S.W.L. Jacobs & J. Everett A
Austrostipa rudis (Spreng.) S.W.L.Jacobs & J.Everett A
Austrostipa scabra (Lindl.) S.W.L.Jacobs & J.Everett A
Austrostipa stipoides (Hook.f.) S.W.L.Jacobs & J.Everett
Austrostipa stuposa (Hughes) S.W.L.Jacobs & J.Everett A
Austrostipa verticillata (Nees ex Spreng.) S.W.L.Jacobs & J.Everett A
Avena barbata Link A
Avena byzantina K.Koch Cas
Avena fatua L. A
Avena sativa L. A
Avena sterilis L. A
Avena strigosa Schreb. A
Axonopus fissifolius (Raddi) Kuhlm. A
Bambusa multiplex (Lour.) Schult. & Schult.f. A
Bambusa oldhamii Munro A
Bothriochloa bladhii (Retz.) S.T.Blake A
Bothriochloa macra (Steud.) S.T.Blake A
Brachypodium distachyon (L.) P.Beauv. Cas
Brachypodium pinnatum (L.) P.Beauv. A
Brachypodium sylvaticum (Huds.) P.Beauv. A
Briza maxima L. A
Briza media L. A
Briza minor L. A
Briza rufa (J.Presl) Steud. A
Briza uniolae (Nees) Nees ex Steud. Cas
Bromus arenarius Labill. A
Bromus brevis Nees A
Bromus catharticus Vahl Cas
Bromus commutatus Schrad. A
Bromus diandrus Roth A
Bromus erectus Huds. A
Bromus hordeaceus L. A
Bromus inermis Leyss. A
Bromus japonicus Thunb. A
Bromus lithobius Trin. A
Bromus madritensis L. A
Bromus racemosus L. A
Bromus rubens L. Cas
Bromus secalinus L. Cas
Bromus sitchensis Trin. A
Bromus stamineus E.Desv. A
Bromus sterilis L. A
Bromus tectorum L. A
Bromus valdivianus Phil. A
Bromus willdenowii Kunth A
Buchloe dactyloides (Nutt.) Engelm. A
Calamagrostis epigejos (L.) Roth Cas
Catapodium rigidum (L.) C.E.Hubb. A
Cenchrus caliculatus Cav.
Cenchrus clandestinus (Hochst. ex Chiov.) Morrone A
Cenchrus latifolius (Spreng.) Morrone A
Cenchrus longisetus M.C.Johnst. A
Cenchrus macrourus (Trin.) Morrone A
Cenchrus purpureus (Schumach.) Morrone A
Cenchrus setaceus (Forssk.) Morrone A
Chimonobambusa quadrangularis (Franceschi) Makino A
Chionochloa acicularis Zotov E
Chionochloa antarctica (Hook.f.) Zotov E
Chionochloa australis (Buchanan) Zotov E
Chionochloa beddiei Zotov E
Chionochloa bromoides (Hook.f.) Zotov E
Chionochloa cheesemanii (Hack.) Zotov E
Chionochloa conspicua (G.Forst.) Zotov E
Chionochloa crassiuscula (Kirk) Zotov E
Chionochloa defracta Connor E
Chionochloa flavescens Zotov E
Chionochloa flavicans Zotov E
Chionochloa juncea Zotov E
Chionochloa lanea Connor E
Chionochloa macra Zotov E
Chionochloa nivifera Connor & K.M.Lloyd E
Chionochloa oreophila (Petrie) Zotov E
Chionochloa ovata (Buchanan) Zotov E
Chionochloa pallens Zotov E
Chionochloa rigida (Raoul) Zotov E
Chionochloa rubra Zotov E
Chionochloa spiralis Zotov E
Chionochloa teretifolia (Petrie) Zotov E
Chionochloa vireta Connor E
Chloris gayana Kunth A
Chloris truncata R.Br. A
Coix lacryma-jobi L. Cas
Cortaderia jubata (Lemoine) Stapf A
Cortaderia selloana (Schult. & Schult.f.) Asch. & Graebn. A
Critesion glaucum (Steud.) Á.Löve A
Critesion hystrix (Roth) Á.Löve A
Critesion jubatum (L.) Nevski A
Critesion marinum (Huds) Á.Löve A
Critesion murinum (L.) Á.Löve A
Critesion secalinum (Schreb.) Á.Löve A
Cynodon dactylon (L.) Pers. A
Cynodon transvaalensis Burtt Davy Cas
Cynosurus cristatus L. A
Cynosurus echinatus L. A
Dactylis glomerata L. A
Deschampsia cespitosa (L.) P.Beauv.
Deschampsia chapmanii Petrie E
Deschampsia flexuosa (L.) Trin. A
Deschampsia gracillima Kirk E
Deschampsia pusilla Petrie E
Deschampsia tenella Petrie E
Deyeuxia aucklandica (Hook.f.) Zotov E
Deyeuxia avenoides (Hook.f.) Buchanan E
Deyeuxia lacustris Edgar & Connor E
Deyeuxia quadriseta (Labill.) Benth.
Deyeuxia youngii (Hook.f.) Buchanan E
Dichelachne crinita (L.f.) Hook.f.
Dichelachne inaequiglumis (Hack.) Edgar & Connor
Dichelachne lautumia Edgar & Connor E
Dichelachne micrantha (Cav.) Domin
Dichelachne rara (R.Br.) Vickery A
Dichelachne sieberiana Trin. & Rupr. A
Digitaria aequiglumis (Hack. & Arechav.) Parodi A
Digitaria ciliaris (Retz.) Koeler A
Digitaria ischaemum (Schreb.) Muhl. A
Digitaria sanguinalis (L.) Scop. A
Digitaria setigera Roem. & Schult. A
Digitaria violascens Link A
Diplachne fusca (L.) P.Beauv. ex Roem. & Schult. Cas
Distichlis spicata (L.) Greene A
Echinochloa crus-galli (L.) P.Beauv. A
Echinochloa crus-pavonis (Kunth) Schult. A
Echinochloa esculenta (A.Braun) H.Scholz A
Echinochloa microstachya (Wiegand) Rydb. A
Echinochloa oryzoides (Ard.) Fritsch Cas
Echinochloa telmatophila P.W.Michael & Vickery A
Echinopogon ovatus (G.Forst.) P.Beauv.
Ehrharta calycina Sm. A
Ehrharta erecta Lam. A
Ehrharta longiflora Sm. A
Ehrharta villosa (L.f.) Schult. & Schult.f. A
Eleusine indica (L.) Gaertn. A
Eleusine tristachya (Lam.) Lam. A
Elytrigia pycnantha (Godr.) Á.Löve A
Elytrigia repens (L.) Nevski A
Entolasia marginata (R.Br.) Hughes A
Eragrostis amabilis (L.) Wight & Arn. ex Nees A
Eragrostis brownii (Kunth) Nees ex Hook. & Arn. A

Eragrostis cilianensis (All.) Janch. A
Eragrostis curvula (Schrad.) Nees A
Eragrostis diffusa Buckley A
Eragrostis leptostachya (R.Br.) Steud. A
Eragrostis mexicana (Hornem.) Link A
Eragrostis minor Host Cas
Eragrostis pilosa (L.) P.Beauv. A
Eragrostis plana Nees A
Festuca actae Connor E
Festuca contracta Kirk
Festuca coxii (Petrie) Hack. E
Festuca deflexa Connor E
Festuca filiformis Pourr. A
Festuca luciarum Connor E
Festuca madida Connor E
Festuca matthewsii (Hack.) Cheeseman E
Festuca multinodis Petrie & Hack. E
Festuca novae-zelandiae (Hack.) Cockayne E
Festuca ovina L. A
Festuca rubra L. A
Festuca ultramafica Connor E
Gastridium ventricosum (Gouan) Schinz & Thell. A
Glyceria declinata Bréb. F A
Glyceria fluitans (L.) R.Br. F A
Glyceria maxima (Hartm.) Holmb. F A
Glyceria plicata (Fr.) Fr. F A
Glyceria striata (Lam.) Hitchc. A
Hainardia cylindrica (Willd.) Greuter A
Helictotrichon pubescens (Huds.) Pilg. Cas
Hemarthria uncinata R.Br. Cas
Hierochloe brunonis Hook.f. E
Hierochloe cuprea Zotov E
Hierochloe equiseta Zotov E
Hierochloe fusca Zotov E
Hierochloe novae-zelandiae Gand. E
Hierochloe recurvata (Hack.) Zotov E
Hierochloe redolens (Vahl) Roem. & Schult.
Himalayacalamus falconeri (Munro) Keng f. A
Holcus lanatus L. A
Holcus mollis L. A
Hordeum vulgare L. A
Imperata cheesemanii Hack. E
Imperata cylindrica (L.) Räusch. A
Isachne globosa (Thunb.) Kuntze
Koeleria cheesemanii (Hack.) Petrie E
Koeleria novozelandica Domin E
Koeleria riguorum Edgar & Gibb E
Lachnagrostis ammobia Edgar E
Lachnagrostis billardierei (R.Br.) Trin.
Lachnagrostis elata Edgar E
Lachnagrostis filiformis (G.Forst.) Trin.
Lachnagrostis glabra (Petrie) Edgar E
Lachnagrostis leptostachys (Hook.f.) Zotov E
Lachnagrostis littoralis (Hack.) Edgar E
Lachnagrostis lyallii (Hook.f.) Zotov E
Lachnagrostis pilosa (Buchanan) Edgar E
Lachnagrostis striata (Colenso) Zotov E
Lachnagrostis tenuis (Cheeseman) Edgar E
Lachnagrostis uda Edgar E
Lagurus ovatus L. A
Leersia oryzoides (L.) Sw. A
Lepturus repens (G.Forst.) R.Br.
Leymus arenarius (L.) Hochst. A
Leymus racemosus (Lam.) Tzvelev A
Lolium multiflorum Lam. A
Lolium perenne L. A
Lolium remotum Schrank A
Lolium rigidum Gaudin A
Lolium temulentum L. A
Megathyrsus maximus (Jacq.) B.K.Simon & S.W.L.Jacobs A
Melica minuta L. A
Melinis repens (Willd.) Zizka Cas
Microlaena avenacea (Raoul) Hook.f.
Microlaena carsei Cheeseman E
Microlaena polynoda (Hook.f.) Hook.f. E
Microlaena stipoides (Labill.) R.Br.
Milium effusum L. Cas
Miscanthus nepalensis (Trin.) Hack. A
Miscanthus sinensis Andersson A
Nardus stricta L. A
Nassella neesiana (Trin. & Rupr.) Barkworth A
Nassella tenuissima (Trin.) Barkworth A
Nassella trichotoma (Nees) Hack. ex Arechav. A
Oplismenus hirtellus (L.) P.Beauv.
Panicum capillare L. A
Panicum dichotomiflorum Michx. A
Panicum huachucae Ashe A
Panicum lindheimeri Nash A
Panicum miliaceum L. A
Panicum schinzii Hack. A
Panicum sphaerocarpon Elliott A
Parapholis incurva (L.) C.E.Hubb. A
Parapholis strigosa (Dumort.) C.E.Hubb. A
Paspalum conjugatum P.J.Bergius A
Paspalum dilatatum Poir. A
Paspalum distichum L. A
Paspalum orbiculare G.Forst. A
Paspalum paniculatum L. A
Paspalum pubiflorum E.Fourn. A
Paspalum scrobiculatum L.
Paspalum urvillei Steud. A
Paspalum vaginatum Sw. A
Pentapogon quadrifidus (Labill.) Baill. A
Phalaris angusta Trin. A
Phalaris aquatica L. A
Phalaris arundinacea L. A
Phalaris canariensis L. A
Phalaris minor Retz. A
Phalaris paradoxa L. A
Phleum pratense L. A
Phragmites australis (Cav.) Trin. ex Steud. A
Phyllostachys aurea Rivière & C.Rivière A
Phyllostachys bambusoides Siebold & Zucc. A
Phyllostachys edulis (Carrière) J.Houz. Cas
Phyllostachys heterocycla (Carrière) S.Matsum. Cas
Phyllostachys nigra (Lodd.) Munro A
Piptatherum miliaceum (L.) Coss. A
Pleioblastus auricomus (Mitford) D.C.McClint. A
Pleioblastus chino (Franch. & Sav.) Makino A
Pleioblastus gramineus (Bean) Nakai A
Pleioblastus hindsii (Munro) Nakai A
Pleioblastus variegatus (Miq.) Makino A
Poa acicularifolia Buchanan E
Poa alpina L. Cas
Poa anceps G.Forst. E
Poa annua L. A
Poa antipoda Petrie E
Poa astonii Petrie E
Poa aucklandica Petrie E
Poa billardierei (Spreng.) St.-Yves
Poa breviglumis Hook.f. E
Poa buchananii Zotov E
Poa bulbosa L. A
Poa celsa Edgar E
Poa chathamica Petrie E
Poa cita Edgar E
Poa cockayneana Petrie E
Poa colensoi Hook.f. E
Poa compressa L. A
Poa dipsacea Petrie E
Poa foliosa (Hook.f.) Hook.f.
Poa hesperia Edgar E
Poa imbecilla Spreng. E
Poa incrassata Petrie E
Poa infirma Kunth A
Poa intrusa Edgar E
Poa kirkii Buchanan E
Poa labillardierei Steud. A
Poa lindsayi Hook.f. E
Poa litorosa Cheeseman
Poa maia Edgar E
Poa maniototo Petrie E
Poa matthewsii Petrie E
Poa nemoralis L. A
Poa novae-zelandiae Hack. E
Poa palustris L. A
Poa pratensis L. A
Poa pusilla Berggr. E
Poa pygmaea Buchanan E
Poa ramosissima Hook.f. E
Poa remota Forselles Cas
Poa schistacea Edgar & Connor E
Poa senex Edgar E
Poa sieberiana Spreng. A
Poa spania Edgar & Molloy E
Poa sublimis Edgar E
Poa subvestita (Hack.) Edgar E
Poa sudicola Edgar E
Poa tennantiana Petrie E
Poa tonsa Edgar E
Poa trivialis L. A
Poa xenica Edgar & Connor E
Polypogon fugax Nees ex Steud. A
Polypogon monspeliensis (L.) Desf. A
Polypogon viridis (Gouan) Breistr. A
Pseudosasa japonica (Siebold & Zucc. ex Steud.) Makino ex Nakai A
Puccinellia distans (Jacq.) Parl. A
Puccinellia fasciculata (Torr.) E.P.Bicknell A
Puccinellia raroflorens Edgar E
Puccinellia rupestris (With.) Fernald & Weath. A
Puccinellia stricta (Hook.f.) C.H.Blom
Puccinellia walkeri (Kirk) Allan E
Pyrrhanthera exigua (Kirk) Zotov E
Rostraria cristata (L.) Tzvelev Cas
Rytidosperma auriculatum (J.M.Black) Connor & Edgar A
Rytidosperma australe (Petrie) Connor & Edgar
Rytidosperma biannulare (Zotov) Connor & Edgar E
Rytidosperma buchananii (Hook.f.) Connor & Edgar E
Rytidosperma caespitosum (Gaudich.) Connor & Edgar A
Rytidosperma clavatum (Zotov) Connor & Edgar E
Rytidosperma corinum Connor & Edgar E
Rytidosperma erianthum (Lindl.) Connor & Edgar A
Rytidosperma geniculatum (J.M.Black) Connor & Edgar A
Rytidosperma gracile (Hook.f.) Connor & Edgar
Rytidosperma horrens Connor & Molloy E
Rytidosperma laeve (Vickery) Connor & Edgar A
Rytidosperma maculatum (Zotov) Connor & Edgar E
Rytidosperma merum Connor & Edgar E
Rytidosperma nigricans (Petrie) Connor & Edgar E
Rytidosperma nudum (Hook.f.) Connor & Edgar E
Rytidosperma penicillatum (Labill.) Connor & Edgar A
Rytidosperma petrosum Connor & Edgar E
Rytidosperma pilosum (R.Br.) Connor & Edgar A
Rytidosperma pulchrum (Zotov) Connor & Edgar E
Rytidosperma pumilum (Kirk) Connor & Edgar
Rytidosperma racemosum (R.Br.) Connor & Edgar A
Rytidosperma setifolium (Hook.f.) Connor & Edgar E
Rytidosperma telmaticum Connor & Molloy E
Rytidosperma tenue (Petrie) Connor & Edgar E
Rytidosperma tenuius (Steud.) A.Hansen & Sunding A
Rytidosperma thomsonii (Buchanan) Connor & Edgar E
Rytidosperma unarede (Raoul) Connor & Edgar E
Rytidosperma viride (Zotov) Connor & Edgar E
Saccharum officinarum L. A
Sacciolepis indica (L.) Chase A

Sasa palmata (Burb.) E.G.Camus A
Sasaella ramosa (Makino) Makino A
Schedonorus arundinaceus (Schreb.) Dumort. A
Secale cereale L. A
Semiarundinaria fastuosa (Mitford) Makino A
Sesleria albicans Schult. Cas
Sesleria autumnalis (Scop.) F.W.Schultz Cas
Setaria italica (L.) P.Beauv. A
Setaria palmifolia (J.König) Stapf A
Setaria parviflora (Poir.) Kerguélen A
Setaria pumila (Poir.) Roem. & Schult. A
Setaria sphacelata (Schumach.) Stapf & C.E.Hubb Cas
Setaria verticillata (L.) P.Beauv. A
Setaria viridis (L.) P.Beauv. A
Sieglingia decumbens (L.) Bernh. A
Simplicia buchananii (Zotov) Zotov E
Simplicia laxa Kirk E
Sorghum bicolor (L.) Moench A
Sorghum halepense (L.) Pers. A
Spartina alterniflora Loisel. M A
Spartina anglica C.E.Hubb. M A
Spinifex sericeus R.Br.
Sporobolus africanus (Poir.) A.Robyns & Tournay A
Sporobolus cryptandrus (Torr.) A.Gray A
Sporobolus elongatus R.Br. A
Stenostachys deceptorix Connor E
Stenostachys enysii (Kirk) Barkworth & S.W.L.Jacobs E
Stenostachys gracilis (Hook.f.) Connor E
Stenostachys laevis (Petrie) Connor E
Stenotaphrum secundatum (Walter) Kuntze A
Themeda triandra Forssk. A
Thinopyrum intermedium (Host) Barkworth & D.R.Dewey A
Thinopyrum junceiforme (Á.Löve & D.Löve) Á.Löve A
Trisetum antarcticum (G.Forst.) Trin. E
Trisetum arduanum Edgar & A.P.Druce
Trisetum drucei Edgar E
Trisetum flavescens (L.) P.Beauv. A
Trisetum lasiorhachis (Hack.) Edgar E
Trisetum lepidum Edgar & A.P.Druce E
Trisetum serpentinum Edgar & A.P.Druce E
Trisetum spicatum (L.) K.Richt.
Trisetum tenellum (Petrie) A.W.Hill E
Trisetum youngii Hook.f. E
Triticum aestivum L. A
Triticum compactum Host Cas
Urochloa mutica (Forssk.) T.Q.Nguyen A
Urochloa panicoides P.Beauv. A
Vulpia bromoides (L.) Gray A
Vulpia fasciculata (Forssk.) Fritsch Cas
Vulpia myuros (L.) C.C.Gmel. A
Zea mays L. Cas
Zizania latifolia (Griseb.) Stapf A
Zotovia acicularis Edgar & Connor E
Zotovia colensoi (Hook.f.) Edgar & Connor E
Zotovia thomsonii (Petrie) Edgar & Connor E
Zoysia minima (Colenso) Zotov E
Zoysia pauciflora Mez E
RESTIONACEAE
Apodasmia similis (Edgar) B.G.Briggs & L.A.S.Johnson E
Empodisma minus (Hook.f.) L.A.S.Johnson & D.F.Cutler
Sporadanthus ferrugineus de Lange, Heenan & B.D.Clarkson E
Sporadanthus traversii (F.Muell.) F.Muell. ex Kirk E
SPARGANIACEAE
Sparganium subglobosum Morong F
TYPHACEAE
Typha orientalis C.Presl F

Order ZINGIBERALES A
CANNACEAE A
Canna indica L. A
MUSACEAE A
Ensete ventricosum (Welw.) Cheesman A
ZINGIBERACEAE A
Hedychium flavescens Roscoe A
Hedychium gardnerianum Ker Gawl. A

Superorder CERATOPHYLLANAE A
Order CERATOPHYLLALES A
CERATOPHYLLACEAE A
Ceratophyllum demersum L. F A

Superorder BUXANAE A
Order BUXALES A
BUXACEAE A
Buxus microphylla Siebold & Zucc. Cas
Buxus sempervirens L. A

Superorder PROTEANAE
Order PROTEALES
PLATANACEAE A
Platanus ×hispanica Mill. ex Münchh. A
PROTEACEAE
Banksia aemula R.Br. Cas
Banksia ericifolia L.f. Cas
Banksia integrifolia L.f. A
Banksia serrata L.f. Cas
Embothrium coccineum J.R.Forst. & G.Forst. A
Grevillea aspleniifolia R.Br. ex Salisb. A
Grevillea robusta R.Br. A
Hakea drupacea (C.F.Gaertn.) Roem. & Schult. A
Hakea eriantha R.Br. Cas
Hakea gibbosa Cav. A
Hakea salicifolia (Vent.) B.L.Burtt A
Hakea sericea Schrad. & J.C.Wendl. A
Knightia excelsa R.Br. E
Leucadendron argenteum (L.) R.Br. A
Lomatia ilicifolia R.Br. Cas
Macadamia integrifolia Maiden & Betche Cas
Macadamia tetraphylla L.A.S.Johnson A
Protea subvestita N.E.Br. Cas
Stenocarpus salignus R.Br. Cas
Stenocarpus sinuatus (Loudon) Endl. Cas
Telopea oreades F.Muell. A
Telopea speciosissima (Sm.) R.Br. Cas
Toronia toru (A.Cunn.) L.A.S.Johnson & B.G.Briggs E

Superorder RANUNCULANAE
Order RANUNCULALES
BERBERIDACEAE A
Berberis bealei Fortune Cas
Berberis aquifolium Pursh A
Berberis aristata DC. Cas
Berberis darwinii Hook. A
Berberis glaucocarpa Stapf A
Berberis japonica (Thunb.) R.Br. A
Berberis lomariifolia (Takeda) Laferr. Cas
Berberis soulieana C.K.Schneid. A
Berberis vulgaris L. A
Berberis wilsonae Hemsl. A
Nandina domestica Thunb. Cas
Vancouveria hexandra (Hook.) Morr. & Decne Cas
LARDIZABALACEAE A
Akebia quinata (Houtt.) Decne. A
PAPAVERACEAE A
Argemone ochroleuca Sweet A
Chelidonium majus L. A
Corydalis cheilanthifolia Hemsl. Cas
Corydalis lutea (L.) DC. A
Corydalis ochroleuca Koch Cas
Dicentra scandens (D.Don) Walp. Cas
Eomecon chionantha Hance Cas
Eschscholzia californica Cham. A
Fumaria bastardii Boreau A
Fumaria capreolata L. A
Fumaria densiflora DC. A
Fumaria muralis W.D.J.Koch A
Fumaria officinalis L. A
Glaucium corniculatum (L.) Rudolph A
Glaucium flavum Crantz A
Macleaya cordata (Willd.) R.Br. Cas
Meconopsis cambrica (L.) Vig. Cas
Papaver aculeatum Thunb. Cas
Papaver argemone L. A
Papaver atlanticum (Ball) Coss. A
Papaver dubium L. A
Papaver hybridum L. A
Papaver nudicaule L. Cas
Papaver pseudorientale (Fedde) Medw. Cas
Papaver rhoeas L. A
Papaver somniferum L. A
Romneya coulteri Harv. Cas
RANUNCULACEAE
Aconitum napellus L. Cas
Anemone ×hybrida Paxton A
Anemone coronaria L. Cas
Anemone nemorosa L. A
Anemone tenuicaulis (Cheeseman) Parkin & Sledge E
Aquilegia vulgaris L. A
Caltha novae-zelandiae (Hook.f.) W.A.Weber E
Caltha obtusa (Cheeseman) W.A.Weber E
Caltha palustris L. Cas
Ceratocephala pungens Garn.-Jones
Clematis afoliata Buchanan E
Clematis cunninghamii Turcz. E
Clematis flammula L. A
Clematis foetida Raoul E
Clematis forsteri J.F.Gmel. E
Clematis grewiiflora DC. Cas
Clematis marata J.B.Armstr. E
Clematis marmoraria Sneddon E
Clematis montana DC. A
Clematis paniculata J.F.Gmel. E
Clematis petriei Allan E
Clematis quadribracteolata Colenso E
Clematis tangutica (Maxim.) Korsh. Cas
Clematis terniflora DC. A
Clematis tibetana Kuntze A
Clematis vitalba L. A
Consolida ajacis (L.) Schur A
Eranthis hyemalis (L.) Salisb. Cas
Helleborus foetidus L. Cas
Helleborus orientalis Lam. Cas
Myosurus minimus L.
Nigella damascena L. A
Ranunculus acaulis Banks & Sol. ex DC.
Ranunculus acraeus Heenan & P.J.Lockhart E
Ranunculus acris L. A
Ranunculus altus Garn.-Jones E
Ranunculus amphitrichus Colenso
Ranunculus arvensis L. A
Ranunculus biternatus Sm.
Ranunculus brevis Garn.-Jones E
Ranunculus buchananii Hook.f. E
Ranunculus bulbosus L. A
Ranunculus carsei Petrie E
Ranunculus cheesemanii Kirk E
Ranunculus crithmifolius Hook.f. E
Ranunculus enysii Kirk E
Ranunculus ficaria L. A
Ranunculus flammula L. A
Ranunculus foliosus Kirk E
Ranunculus glabrifolius Hook.
Ranunculus godleyanus Hook.f. E
Ranunculus gracilipes Hook.f. E
Ranunculus grahamii Petrie E
Ranunculus haastii Hook.f. E

Ranunculus insignis Hook.f. E
Ranunculus kirkii Petrie E
Ranunculus limosella F.Muell. ex Kirk F E
Ranunculus lyallii Hook.f. E
Ranunculus macropus Hook.f. E
Ranunculus maculatus Cockayne & Allan E
Ranunculus membranifolius (Kirk) Garn.-Jones E
Ranunculus mirus Garn.-Jones E
Ranunculus multiscapus Hook.f. E
Ranunculus muricatus L. A
Ranunculus nivicola Hook.f. E
Ranunculus ophioglossifolius Vill. A
Ranunculus pachyrrhizus Hook.f. E
Ranunculus parviflorus L. A
Ranunculus pilifera (F.J.F.Fisher) Heenan & P.J.Lockhart E
Ranunculus pinguis Hook.f. E
Ranunculus ranceorum de Lange E
Ranunculus recens Kirk E
Ranunculus reflexus Garn.-Jones E
Ranunculus repens L. A
Ranunculus royi G.Simpson E
Ranunculus sardous Crantz A
Ranunculus sceleratus L. A
Ranunculus scrithalis Garn.-Jones E
Ranunculus sericophyllus Hook.f. E
Ranunculus sessiliflorus DC. A
Ranunculus simulans Garn.-Jones E
Ranunculus stylosus H.D.Wilson & Garn.-Jones E
Ranunculus subscaposus Hook.f. E
Ranunculus ternatifolius Kirk E
Ranunculus trichophyllus Chaix F A
Ranunculus urvilleanus Cheeseman E
Ranunculus verticillatus Kirk E
Ranunculus viridis H.D.Wilson & Garn.-Jones E
Thalictrum aquilegiifolium L. Cas
Thalictrum minus L. Cas

Superorder MYROTHAMNANAE
Order GUNNERALES
GUNNERACEAE
Gunnera densiflora Hook.f. E
Gunnera dentata Kirk E
Gunnera hamiltonii Kirk E
Gunnera monoica Raoul E
Gunnera prorepens Hook.f. E
Gunnera tinctoria (Molina) Mirb. A

Superorder DILLENIANAE Cas
Order DILLENIALES Cas
DILLENIACEAE Cas
Hibbertia scandens (Willd.) Gilg. Cas

Superorder SAXIFRAGANAE
Order SAXIFRAGALES
APHANOPETALACEAE Cas
Aphanopetalum resinosum Endl. Cas
CRASSULACEAE
×*Sedadia amecamecana* (Praeger) Moran Cas
Aeonium ×*floribundum* A.Berger Cas
Aeonium ×*velutinum* Praeger A
Aeonium arboreum (L.) Webb & Berthel. A
Aeonium canariense Webb & Berthel. Cas
Aeonium haworthii (Salm-Dyck) Webb & Berthel. A
Aeonium haworthii Hybrids A
Aeonium undulatum Webb & Berthel. A
Aeonium urbicum (C.A.Sm.) Webb & Berthel. A
Bryophyllum ×*houghtonii* (D.B.Ward) P.I.Forst. A
Bryophyllum daigremontianum Raym.-Hamet & Perrier Cas
Bryophyllum delagoense (Eckl. & Zeyh.) Schinz A
Bryophyllum fedtschenkoi (Raym.-Hamet & H.Perrier) Lauz.-March. Cas
Bryophyllum pinnatum (Lam.) Oken A
Cotyledon orbiculata L. A
Crassula alata (Viv.) A.Berger Cas
Crassula arborescens (Mill.) Willd. Cas
Crassula biplanata Haw. Cas
Crassula coccinea L. A
Crassula colligata Toelken
Crassula colorata (Nees) Ostenf. Cas
Crassula cotyledonis Thunb. Cas
Crassula decumbens Thunb. A
Crassula dejecta Jacq. Cas
Crassula fallax Friedrich Cas
Crassula helmsii (Kirk) Cockayne E
Crassula kirkii (Allan) A.P.Druce & Given E
Crassula manaia A.P.Druce & Sykes E
Crassula mataikona A.P.Druce E
Crassula moschata G.Forst.
Crassula multicaulis (Petrie) A.P.Druce & Given E
Crassula multicava Lem. A
Crassula muscosa L. Cas
Crassula orbicularis L. Cas
Crassula ovata (Mill.) Druce Cas
Crassula peduncularis (Sm.) F.Meigen
Crassula pellucida L. Cas
Crassula pubescens Thunb. Cas
Crassula pupurata (Hook.f.) Domin
Crassula ruamahanga A.P.Druce E
Crassula sarmentosa Harv. Cas
Crassula schmidtii Regel Cas
Crassula sieberiana (Schult. & Schult.f.) Druce
Crassula sinclairii (Hook.f.) A.P.Druce & Given E
Crassula spathulata Thunb. A
Crassula tetragona L. A
Echeveria ×*imbricata* Deleuil ex E. Morr. Cas
Echeveria amoena de Smet ex E.Morr. Cas
Echeveria multicaulis Rose Cas
Echeveria secunda Booth A
Echeveria setosa Rose & Purpus A
Graptopetalum paraguayense (N.E.Br.) Walth. Cas
Greenovia aurea (C.A.Sm.) Webb & Berthel. Cas
Kalanchoe blossfeldiana Poelln. Cas
Kalanchoe grandiflora Wight & Arn. A
Sedum ×*rubrotinctum* R.T.Clausen Cas
Sedum acre L. A
Sedum album L. A
Sedum commixtum Moran & Hutchison Cas
Sedum dasyphyllum L. A
Sedum kimnachii V.V.Byalt A
Sedum mexicanum Britton A
Sedum moranense Kunth A
Sedum oreganum Nutt. Cas
Sedum pachyphyllum Rose Cas
Sedum praealtum A.DC. A
Sedum rupestre L. A
Sedum sediforme (Jacq.) Pau Cas
Sedum spectabile Boreau A
Sedum spurium M.Bieb. A
Sempervivum tectorum L. Cas
Umbilicus rupestris (Salisb.) Dandy Cas
GROSSULARIACEAE A
Ribes nigrum L. A
Ribes odoratum H.L.Wendl. A
Ribes rubrum L. A
Ribes sanguineum Pursh A
Ribes uva-crispa L. A
HALORAGACEAE
Gonocarpus aggregatus (Buchanan) Orchard E
Gonocarpus incanus (A.Cunn.) Orchard E
Gonocarpus micranthus Thunb.
Gonocarpus montanus (Hook.f.) Orchard E
Haloragis aspera Lindl. A
Haloragis erecta (Banks ex Murray) Oken E
Myriophyllum aquaticum (Vell.) Verdc. F A
Myriophyllum pedunculatum Hook.f. F
Myriophyllum propinquum A.Cunn. F
Myriophyllum robustum Hook.f. F E
Myriophyllum simulans Orchard F A
Myriophyllum triphyllum Orchard F E
Myriophyllum votschii Schindl. F E
HAMAMELIDACEAE Cas
Liquidambar styraciflua L. Cas
SAXIFRAGACEAE A
Heuchera sanguinea Engelm. Cas
Saxifraga ×*urbium* D.A.Webb Cas
Saxifraga exarata Vill. Cas
Saxifraga stolonifera Meerb. A
Tellima grandiflora (Pursh) Douglas ex Lindl. Cas

Superorder ROSANAE
Order VITALES A
VITACEAE A
Cissus striata Ruiz & Pav. Cas
Parthenocissus inserta (J.Kern.) Fritsch A
Parthenocissus tricuspidata (Siebold & Zucc.) Planch. Cas
Vitis vinifera L. A

Order CELASTRALES
CELASTRACEAE
Celastrus orbiculatus Thunb. A
Euonymus europaeus L. A
Euonymus fortunei (Turcz.) Hand.-Mazz. Cas
Euonymus japonicus Thunb. A
Euonymus pendulus Wall. Cas
Euonymus phellomanus Loes. Cas
Maytenus boaria Molina A
Stackhousia minima Hook.f. E

Order CUCURBITALES
BEGONIACEAE A
Begonia corallina Carr. Cas
Begonia foliosa Kunth Cas
Begonia grandis Dryand. Cas
Begonia ×*Semperflorens-Cultorum* hybrids A
CORIARIACEAE
Coriaria angustissima Hook.f. E
Coriaria arborea Linds. E
Coriaria kingiana Colenso E
Coriaria lurida Kirk
Coriaria plumosa W.R.B.Oliv. E
Coriaria pottsiana W.R.B.Oliv. E
Coriaria pteridoides W.R.B.Oliv. E
Coriaria sarmentosa G.Forst. E
CORYNOCARPACEAE
Corynocarpus laevigatus J.R.Forst. & G.Forst. E
CUCURBITACEAE
Bryonia cretica L. A
Citrullus lanatus (Thunb.) Matsum. & Nakai A
Cucumis myriocarpus Naudin A
Cucurbita ficifolia Bouche A
Cucurbita maxima Duchesne A
Cucurbita pepo L. A
Ecballium elaterium (L.) A.Rich. A
Sechium edule (Jacq.) Sweet. A
Sicyos australis Endl.

Order FABALES
FABACEAE (LEGUMINOSAE)
Acacia baileyana F.Muell. A
Acacia dealbata Link A
Acacia decurrens Willd. A
Acacia elata Benth. A
Acacia fimbriata A.Cunn. ex G.Don. Cas
Acacia floribunda (Vent.) Willd. A
Acacia longifolia (Andrews) Willd. A
Acacia maidenii F.Muell. Ca
Acacia mearnsii De Wild. A
Acacia melanoxylon R.Br. A
Acacia paradoxa DC. A
Acacia parramattensis Tindale A
Acacia podalyriifolia G.Don Cas
Acacia pravissima F.Muell. Cas

Acacia prominens A.Cunn ex G.Don Cas
Acacia riceana Hensl. Cas
Acacia saligna (Labill.) H.L.Wendl. Cas
Acacia schinoides Benth. Cas
Acacia stricta (Andrews) Willd. Cas
Acacia ulicifolia (Salisb.) Court Cas
Acacia verticillata (L'Hér.) Willd. A
Albizia julibrissin Durazz. A
Anthyllis vulneraria L. Cas
Caesalpinia decapetala (Roth) Alston A
Caesalpinia spinosa (Molina) Kuntze Cas
Calicotome spinosa (L.) Link A
Calliandra brevipes Benth. A
Calliandra tweedii Benth. Cas
Callistachys lanceolata Vent. A
Canavalia rosea (Sw.) DC.
Carmichaelia appressa G.Simpson E
Carmichaelia arborea (G.Forst.) Druce E
Carmichaelia astonii G.Simpson E
Carmichaelia australis R.Br. E
Carmichaelia carmichaeliae (Hook.f.) Heenan E
Carmichaelia compacta Petrie E
Carmichaelia corrugata Colenso E
Carmichaelia crassicaulis Hook.f. E
Carmichaelia curta Petrie E
Carmichaelia glabrescens (Petrie) Heenan E
Carmichaelia hollowayi G.Simpson E
Carmichaelia juncea Hook.f. E
Carmichaelia kirkii Hook.f. E
Carmichaelia monroi Hook.f. E
Carmichaelia muelleriana Regel
Carmichaelia muritai (A.W.Purdie) Heenan E
Carmichaelia nana (Hook.f.) Hook.f. E
Carmichaelia odorata Benth. E
Carmichaelia petriei Kirk E
Carmichaelia stevensonii (Cheeseman) Heenan E
Carmichaelia torulosa (Kirk) Heenan E
Carmichaelia uniflora Kirk E
Carmichaelia vexillata Heenan E
Carmichaelia williamsii Kirk E
Cassia leptophylla Vog. Cas
Castanospermum australe A.Cunn. ex Mudie Cas
Cercis siliquastrum L. Cas
Chamaecytisus palmensis (H.Christ) F.A.Bisby & K.W.Nicholls A
Clianthus maximus Colenso E
Clianthus puniceus (G.Don) Sol. ex Lindl. E
Coronilla varia L. A
Crotalaria agatiflora Schweinf. A
Crotalaria lunata Bedd. ex Polhill Cas
Cytisus multiflorus (L'Hér.) Sweet A
Cytisus scoparius (L.) Link A
Dipogon lignosus (L.) Verdc. A
Dorycnium hirsutum Ser. Cas
Erythrina ×sykesii Barneby & Krukoff A
Erythrina caffra Thunb. Cas
Erythrina crista-galli L. A
Galega officinalis L. A
Genista linifolia L. A
Genista monspessulana (L.) L.A.S.Johnson A
Genista stenopetala Webb & Berthel. A
Genista tinctoria L. A
Gleditsia triacanthos L. Cas
Goodia lotifolia Salisb. A
Hedysarum coronarium L. A
Hesperis matronalis L. A
Indigofera decora Lindl. A
Indigofera heterantha Brandis Cas
Kennedia rubicunda (Schneev.) Vent. A
Laburnum anagyroides Medik. A
Lathyrus aphaca L. Cas
Lathyrus grandiflorus Sibth. & Sm. A
Lathyrus japonicus Willd. Cas
Lathyrus latifolius L. A
Lathyrus nissolia L. A
Lathyrus odoratus L. A
Lathyrus pratensis L. A
Lathyrus sphaericus Retz. A
Lathyrus sylvestris L. A
Lathyrus tingitanus L. A
Lens culinaris Medik. Cas
Lotus angustissimus L. A
Lotus corniculatus L. A
Lotus pedunculatus Cav. A
Lotus suaveolens Pers. A
Lotus tenuis Willd. A
Lotus tetragonolobus L. Cas
Lupinus albus L. Cas
Lupinus angustifolius L. A
Lupinus arboreus Sims A
Lupinus luteus L. A
Lupinus polyphyllus Lindl. A
Medicago arabica (L.) Huds. A
Medicago arborea L. A
Medicago glomerata Balb. A
Medicago lupulina L. A
Medicago minima (L.) Bartal. A
Medicago nigra (L.) Krock. A
Medicago sativa L. A
Medicago scutellata (L.) Mill. Cas
Melilotus albus Medik. A
Melilotus indicus (L.) All. A
Melilotus officinalis (L.) Lam. A
Montigena novae-zelandiae (Hook.f.) Heenan E
Onobrychis viciifolia Scop. Cas
Ononis campestris W.D.J.Koch A
Ononis repens L. A
Ornithopus perpusillus L. A
Ornithopus pinnatus (Mill.) Druce A
Ornithopus sativus Brot. A
Paraserianthes lophantha (Willd.) I.C.Nielsen A
Parochetus communis D.Don A
Phaseolus coccineus L. A
Phaseolus lunatus L. A
Piptanthus laburnifolius (D.Don) Stapf Cas
Pisum sativum L. A
Podalyria sericea R.Br. A
Psoralea pinnata L. A
Pueraria lobata (Willd.) Ohwi Cas
Pultenaea daphnoides J.C.Wendl. A
Robinia pseudoacacia L. A
Scorpiurus muricatus L. A
Senna didymobotrya (Fresn.) H.S.Irwin & Barneby Cas
Senna multiglandulosa (Jacq.) H.S.Irwin & Barneby A
Senna pendula (Willd.) H.S.Irwin & Barneby Cas
Senna septemtrionalis (Viv.) H.S.Irwin & Barneby A
Senna tora (L.) Roxb. Cas
Sophora cassioides (Phil.) Sparre Cas
Sophora chathamica Cockayne E
Sophora fulvida (Allan) Heenan & de Lange E
Sophora godleyi Heenan & de Lange E
Sophora howinsula (W.R.B.Oliv.) P.S.Green Cas
Sophora longicarinata G.Simpson & J.S.Thomson E
Sophora microphylla Aiton E
Sophora molloyi Heenan & de Lange E
Sophora prostrata Buchanan E
Sophora tetraptera J.S.Mill. E
Spartium junceum L. A
Trifolium ambiguum M.Bieb. Cas
Trifolium angustifolium L. A
Trifolium arvense L. A
Trifolium aureum Pollich A
Trifolium campestre Schreb. A
Trifolium cernuum Brot. A
Trifolium dubium Sibth. A
Trifolium fragiferum L. A
Trifolium glomeratum L. A
Trifolium hirtum All. A
Trifolium hybridum L. A
Trifolium incarnatum L. A
Trifolium medium L. A
Trifolium micranthum Viv. A
Trifolium ochroleucon Huds. A
Trifolium ornithopodioides L. A
Trifolium pratense L. A
Trifolium repens L. A
Trifolium resupinatum L. A
Trifolium retusum L. A
Trifolium scabrum L. A
Trifolium squamosum L. Cas
Trifolium striatum L. A
Trifolium subterraneum L. A
Trifolium suffocatum L. A
Trifolium tomentosum L. A
Ulex europaeus L. A
Ulex minor Roth A
Vicia cracca L. A
Vicia disperma DC. A
Vicia faba L. A
Vicia hirsuta (L.) Gray A
Vicia lathyroides L. A
Vicia lutea L. A
Vicia narbonensis L. Cas
Vicia sativa L. A
Vicia tetrasperma (L.) Schreb. A
Vicia villosa Roth A
Virgilia oroboides (P.J.Bergius) Salter Cas
Wisteria sinensis (Sims) Sweet A
POLYGALACEAE A
Polygala myrtifolia L. A
Polygala serpyllifolia Hosé A
Polygala verticillata L. A
Polygala virgata Thunb. A
Polygala vulgaris L. A
QUILLAJACEAE cas
Quillaja saponaria Molina Cas

Order FAGALES
BETULACEAE A
Alnus glutinosa (L.) Gaertn. A
Alnus viridis (Chaix) DC. A
Betula pendula Roth A
Corylus avellana L. Cas
CASUARINACEAE A
Allocasuarina littoralis (Salisb.) L.A.S.Johnson Cas
Casuarina cunninghamiana Miq. A
Casuarina glauca Spreng. A
FAGACEAE A
Castanea sativa Mill. Cas
Fagus sylvatica L. Cas
Quercus acutissima Carruth. Cas
Quercus cerris L. A
Quercus ilex L. A
Quercus palustris Muenchh. Cas
Quercus robur L. A
Quercus rubra L. A
JUGLANDACEAE A
Juglans ailantifolia Carrière A
Juglans regia L. A
Pterocarya ×rehderiana C.K.Schneider Cas
NOTHOFAGACEAE
Nothofagus antarctica (G.Forst.) Oerst. Cas
Nothofagus fusca (Hook.f.) Oerst. E
Nothofagus menziesii (Hook.f.) Oerst. E
Nothofagus solandri (Hook.f.) Oerst. E
Nothofagus truncata (Colenso) Cockayne E

Order MALPIGHIALES
ELATINACEAE
Elatine gratioloides A.Cunn. E
EUPHORBIACEAE
Acalypha wilkesiana Müll.Arg. Cas
Aleurites fordii Hemsl. Cas

Aleurites moluccana (L.) Willd. A
Baloghia inophylla (G.Forst.) P.S.Green A
Chamaesyce hirta (L.) Millsp. A
Chamaesyce maculata (L.) Small A
Chamaesyce nutans (Lag.) Small A
Euphorbia amygdaloides L. A
Euphorbia characias L. A
Euphorbia cyparissias L. A
Euphorbia depauperata A.Rich. A
Euphorbia dulcis L. Cas
Euphorbia exigua L. A
Euphorbia glauca G.Forst. E
Euphorbia helioscopia L. A
Euphorbia lambii Svent. Cas
Euphorbia lathyris L. A
Euphorbia mauritanica L. Cas
Euphorbia mellifera Aiton Cas
Euphorbia milii Des Moul. Cas
Euphorbia myrsinites L. Cas
Euphorbia peplus L. A
Euphorbia pithyusa L. Cas
Euphorbia platyphyllos L. A
Euphorbia pulcherrima Willd. ex Klotzsch Cas
Euphorbia schillingii Radcl.-Sm. Cas
Euphorbia segetalis L. A
Euphorbia stricta L. A
Homalanthus polyandrus (Müll.Arg.) Cheeseman E
Homalanthus populifolius Graham A
Mercurialis annua L. A
Ricinus communis L. A
HYPERICACEAE
Hypericum androsaemum L. A
Hypericum calycinum L. A
Hypericum canariense L. Cas
Hypericum gramineum G.Forst.
Hypericum henryi H.Lév. & Vaniot A
Hypericum humifusum L. A
Hypericum involutum (Labill.) Choisy
Hypericum kouytchense H.Lév. A
Hypericum linariifolium Vahl A
Hypericum minutiflorum Heenan E
Hypericum montanum L. A
Hypericum mutilum L. A
Hypericum olympicum L. Cas
Hypericum perforatum L. A
Hypericum pulchrum L. A
Hypericum pusillum Choisy
Hypericum rubicundulum Heenan E
Hypericum tetrapterum Fr. A
LINACEAE
Linum bienne Mill. A
Linum catharticum L. A
Linum monogynum G.Forst. E
Linum trigynum L. A
Linum usitatissimum L. A
Reinwardtia indica Dumort. Cas
OCHNACEAE A
Ochna serrulata (Hochst.) Walp. A
PASSIFLORACEAE
Passiflora antioquiensis H.Karst. Cas
Passiflora apetala Killip Cas
Passiflora caerulea L. A
Passiflora edulis Sims A
Passiflora jorullensis Kunth Cas
Passiflora mixta L.f. A
Passiflora pinnatistipula Cav. A
Passiflora tarminiana Coppens & V.E.Barney Cas
Passiflora tetrandra Banks ex DC. E
Passiflora tripartita (Juss.) Poir. A
PHYLLANTHACEAE
Phyllanthus amarus Schumach. & Thonn. Cas
Poranthera alpina Cheeseman ex Hook.f. E
Poranthera microphylla Brongn.
PUTRANJIVACEAE Cas
Drypetes deplanchei (Brongn. & Griseb.) Merr. Cas
SALICACEAE A
Azara microphylla Hook.f. A
Dovyalis hebecarpa (Gardner) Warb. Cas
Idesia polycarpa Maxim. A
Oncoba spinosa Forsk. Cas
Populus alba L. A
Populus deltoides Marshall A
Populus nigra L. A
Populus tremula L. A
Populus trichocarpa Hook. A
Populus yunnanensis Dode A
Salix ×pontederana Willd. Cas
Salix alba L. A
Salix babylonica L. A
Salix caprea L. Cas
Salix cinerea L. A
Salix daphnoides Vill. A
Salix elaeagnos Scop. A
Salix fragilis L. A
Salix glaucophylloides Fernald A
Salix gracilistyla Miq. A
Salix matsudana Koidz. A
Salix purpurea L. A
Salix viminalis L. A
VIOLACEAE
Hybanthus monopetalus (Schult.) Domin Cas
Melicytus alpinus (Kirk) Garn.-Jones E
Melicytus chathamicus (F.Muell.) Garn.-Jones E
Melicytus crassifolius (Hook.f.) Garn.-Jones E
Melicytus dentatus (R.Br. ex DC.) Molloy & Mabb. Cas
Melicytus drucei Molloy & B.D.Clarkson E
Melicytus flexuosus Molloy & A.P.Druce E
Melicytus lanceolatus Hook.f. E
Melicytus macrophyllus A.Cunn. E
Melicytus micranthus (Hook.f.) Hook.f. E
Melicytus novae-zelandiae (A.Cunn.) P.S.Green
Melicytus obovatus (Kirk) Garn.-Jones E
Melicytus ramiflorus J.R.Forst. & G.Forst.
Viola arvensis Murray A
Viola banksii K.R.Thiele & Prober A
Viola cunninghamii Hook.f.
Viola filicaulis Hook.f. E
Viola hederacea Labill. A
Viola lyallii Hook.f. E
Viola odorata L. A
Viola riviniana Rchb. A
Viola sieboldii Maxim. Cas
Viola tricolor L. A

Order OXALIDALES
CUNONIACEAE
Ackama nubicola de Lange E
Ackama rosifolia A.Cunn. E
Ceratopetalum gummiferum Sm. Cas
Cunonia capensis L. Cas
Weinmannia racemosa L.f. E
Weinmannia silvicola Sol. ex A.Cunn. E
ELAEOCARPACEAE
Aristotelia fruticosa Hook.f. E
Aristotelia serrata (J.R.Forst. & G.Forst.) W.R.B.Oliv. E
Elaeocarpus dentatus (J.R.Forst. & G.Forst.) Vahl E
Elaeocarpus hookerianus Raoul E
Elaeocarpus reticulatus Sm. Cas
OXALIDACEAE
Oxalis articulata Savigny A
Oxalis bowiei Lindl. Cas
Oxalis brasiliensis Lodd. Cas
Oxalis chnoodes Lourteig A
Oxalis corniculata L. A
Oxalis debilis Kunth A
Oxalis dillenii Jacq. Cas
Oxalis exilis A.Cunn.
Oxalis fontana Bunge Cas
Oxalis hirta L. A
Oxalis incarnata L. A
Oxalis latifolia Kunth A
Oxalis magellanica G.Forst.
Oxalis megalorrhiza Jacq. Cas
Oxalis perennans Haw. A
Oxalis pes-caprae L. A
Oxalis polyphylla Jacq. Cas
Oxalis purpurea L. A
Oxalis rubens Haw.
Oxalis tetraphylla Cav. Cas
Oxalis thompsoniae B.J.Conn & P.G.Richards A
Oxalis tuberosa Molina A
Oxalis vallicola (Rose) R.Knuth A
Oxalis versicolor L. A

Order ROSALES
CANNABACEAE
Cannabis sativa L. A
Celtis australis L. A
Celtis sinensis Pers. Cas
Humulus lupulus L. A
ELAEAGNACEAE A
Elaeagnus ×reflexa C.Morren & Decne. A
Elaeagnus macrophylla Thunb. Cas
MORACEAE
Broussonetia papyrifera (L.) L'Hér. ex Vent. Cas
Fatoua pilosa Gaudich. Cas
Ficus carica L. A
Ficus macrophylla Desf. ex Pers. Cas
Ficus pumila L. A
Ficus rubiginosa Vent. A
Maclura pomifera (Raf.) C.K.Schneid. A
Streblus banksii (Cheeseman) C.J.Webb E
Streblus heterophyllus (Blume) Corner E
Streblus smithii (Cheeseman) Corner E
RHAMNACEAE
Cryptandra amara Sm. Cas
Discaria toumatou Raoul E
Frangula purshiana (DC.) E.Cooper A
Phylica gnidioides Eckl. & Zeyh. Cas
Pomaderris amoena Colenso E
Pomaderris apetala Labill.
Pomaderris aspera DC. A
Pomaderris hamiltonii L.B.Moore E
Pomaderris kumeraho A.Cunn. E
Pomaderris paniculosa F.Muell. ex Reissek
Pomaderris phylicifolia Lodd. ex Link
Pomaderris prunifolia A.Cunn. ex Fenzl
Pomaderris rugosa Cheeseman E
Rhamnus alaternus L. A
ROSACEAE
Acaena agnipila Gand. A
Acaena anserinifolia (J.R.Forst. & G.Forst.) J.B.Armstr. E
Acaena buchananii Hook.f. E
Acaena caesiglauca (Bitter) Bergmans E
Acaena dumicola B.H.Macmill. E
Acaena echinata Nees Cas
Acaena emittens B.H.Macmill. E
Acaena fissistipula Bitter E
Acaena glabra Buchanan E
Acaena hirsutula Bitter E
Acaena inermis Hook.f. E
Acaena juvenca B.H.Macmill. E
Acaena magellanica (Lam.) M.Vahl
Acaena microphylla Hook.f.
Acaena minor (Hook.f.) Allan E
Acaena novae-zelandiae Kirk
Acaena pallida (Kirk) Allan
Acaena profundeincisa (Bitter) B.H.Macmill. E
Acaena rorida B.H.Macmill. E
Acaena saccaticupula Bitter E
Acaena tesca B.H.Macmill. E
Alchemilla gracilis Opiz Cas

Alchemilla mollis (Buser) Rothm. Cas
Amelanchier lamarckii F.G.Schroed. Cas
Aphanes arvensis L. A
Aphanes australiana (Rothm.) Rothm. A
Aphanes inexspectata W.Lippert A
Chaenomeles speciosa (Sweet) Nakai A
Cotoneaster adpressus Bois. Cas
Cotoneaster bullatus Bois Cas
Cotoneaster conspicuus Marquand A
Cotoneaster dammeri C.K.Schneid. Cas
Cotoneaster divaricatus Rehder & E.H.Wilson Cas
Cotoneaster franchetii Bois A
Cotoneaster frigidus Lindl. A
Cotoneaster glaucophyllus Franch. A
Cotoneaster horizontalis Decne. A
Cotoneaster lacteus W.W.Sm. A
Cotoneaster microphyllus Lindl. A
Cotoneaster pannosus Franch. A
Cotoneaster simonsii Baker A
Crataegus monogyna Jacq. A
Cydonia oblonga Mill. A
Cydonia sinensis Thouin Cas
Eriobotrya japonica (Thunb.) Lindl. A
Exochorda racemosa (Lindl.) Rehder Cas
Filipendula rubra (Hill) B.L.Rob. Cas
Filipendula ulmaria (L.) Maxim. Cas
Filipendula vulgaris Moench A
Geum albiflorum (Hook.f.) Scheutz E
Geum aleppicum Jacq. A
Geum cockaynei (F.Bolle) Molloy & C.J.Webb E
Geum divergens Cheeseman E
Geum leiospermum Petrie E
Geum pusillum Petrie E
Geum uniflorum Buchanan E
Geum urbanum L. A
Hagenia abyssinica J.F.Gmel. Cas
Kerria japonica (L.) DC. A
Malus ×atrosanguinea (hort. ex Späth) C.K.Schneid. Cas
Malus ×domestica Borkh. A
Photinia davidsoniae Rehder & E.H.Wilson A
Physocarpus opulifolius (L.) Maxim. ex Koehne Cas
Potentilla ×ananassa (Duchesne ex Rozier) Mabb. Cas
Potentilla anglica Laichard. A
Potentilla anserinoides Raoul E
Potentilla argentea L. A
Potentilla crantzii (Crantz) Fritsch Cas
Potentilla indica (Andrews) Th.Wolf A
Potentilla nepalensis Raf. Cas
Potentilla norvegicae L. Cas
Potentilla recta L. A
Potentilla reptans L. A
Prunus ×domestica L. A
Prunus armeniaca L. Cas
Prunus avium L. A
Prunus campanulata Maxim. A
Prunus cerasifera Ehrh. A
Prunus cerasus L. A
Prunus dulcis (Mill.) D.A.Webb Cas
Prunus laurocerasus L. A
Prunus lusitanica L. A
Prunus mahaleb L. A
Prunus persica (L.) Batsch A
Prunus serotina Ehrh. Cas
Prunus serrulata Lindl. A
Prunus spinosa L. Cas
Pyracantha angustifolia (Franch.) C.K.Schneid. A
Pyracantha crenatoserrata (Hance) Rehder A
Pyracantha crenulata (D.Don.) M.Roem. A
Pyrus communis L. A
Rhaphiolepis ×delacourii André Cas
Rhaphiolepis indica (L.) Lindl. Cas
Rhaphiolepis umbellata (Thunb.) Makino A
Rosa ×centifolia L. Cas
Rosa ×wichurana Crép. A
Rosa canina L. A
Rosa chinensis Jacq. Hybrids Cas
Rosa gallica L. A
Rosa micrantha Sm. A
Rosa moschata hybrids A
Rosa multiflora Thunb. A
Rosa pimpinellifolia L. A
Rosa roxburghii Tratt. Cas
Rosa rubiginosa L. A
Rosa rugosa Thunb. A
Rosa sempervirens hybrids A
Rosa tomentosa Sm. Cas
Rubus amplificatus E.Lees Cas
Rubus argutus Link A
Rubus australis G.Forst. E
Rubus caesius L. Cas
Rubus cardiophyllus Lefèvre & P.J.Müll. A
Rubus cissburiensis W.C.Barton & Ridd. A
Rubus cissoides A.Cunn. E
Rubus echinatus Lindl. A
Rubus errabundus W.C.R.Watson Cas
Rubus erythrops Edees & A.Newton A
Rubus flagellaris Willd. A
Rubus fruticosus L. A
Rubus idaeus L. A
Rubus laciniatus Willd. A
Rubus leptothyrsos G.Braun A
Rubus mollior L.H.Bailey A
Rubus mucronulatus Boreau A
Rubus nemoralis P.J.Müll. A
Rubus ostryifolius Rydb. A
Rubus parvus Buchanan E
Rubus phoenicolasius Maxim. A
Rubus polyanthemus Lindeb. A
Rubus procerus P.J.Müll. ex Boulay A
Rubus rosifolius Sm. A
Rubus rugosus Sm. A
Rubus schmidelioides A.Cunn. E
Rubus squarrosus Fritsch E
Rubus tuberculatus Bab. A
Rubus ulmifolius Schott A
Rubus vestitus Weihe & Nees A
Sanguisorba minor Scop. A
Sorbaria tomentosa (Lindl.) Rehder A
Sorbus ×latifolia (Lam.) Pers. A
Sorbus aucuparia L. A
Spiraea cantoniensis Lour. A
Spiraea douglasii Hook. Cas
Spiraea japonica L.f. A
Stephanandra tanakae (Franch. & Sav.) Franch. & Sav. Cas
Stranvaesia davidiana Decne. Cas
ULMACEAE Cas
Ulmus glabra Mill. Cas
URTICACEAE
Australina pusilla Gaudich.
Elatostema rugosum A.Cunn. E
Parietaria debilis G.Forst.
Parietaria judaica L. A
Parietaria officinalis L. A
Pilea nummulariifolia (Sw.) Wedd. Cas
Pouzolzia australis (Endl.) Friis & Wilmot-Dear
Soleirolia soleirolii (Req.) Dandy A
Urtica aspera Petrie E
Urtica australis Hook.f. E
Urtica dioica L. A
Urtica ferox G.Forst. E
Urtica incisa Poir.
Urtica linearifolia (Hook.f.) Cockayne E
Urtica membranacea Poir. ex Savigny Cas
Urtica urens L. A

Order ZYGOPHYLLALES Cas
ZYGOPHYLLACEAE Cas
Tribulus terrestris L. Cas

Order BRASSICALES
BRASSICACEAE
Alliaria petiolata (M.Bieb.) Cavara & Grande Cas
Alyssum alyssoides (L.) L. A
Alyssum saxatile L. A
Arabidopsis thaliana (L.) Heynh. A
Armoracia rusticana (Lam.) Gaertn., B.Mey. & Scherb. A
Barbarea intermedia Boreau A
Barbarea stricta Andrz. A
Barbarea verna (Mill.) Asch. A
Barbarea vulgaris R.Br. A
Brassica fruticulosa Cirillo A
Brassica juncea (L.) Czern. A
Brassica napus L. A
Brassica nigra (L.) W.D.J.Koch A
Brassica oleracea L. A
Brassica oxyrrhina (Coss.) Coss. A
Brassica rapa L. A
Brassica tournefortii Gouan A
Cakile edentula (Bigelow) Hook. A
Cakile maritima Scop. A
Camelina alyssum (Mill.) Thell. A
Camelina sativa (L.) Crantz Cas
Capsella bursa-pastoris (L.) Medik. A
Cardamine bilobata Kirk E
Cardamine corymbosa Hook.f. E
Cardamine debilis Banks ex DC. E
Cardamine depressa Hook.f. E
Cardamine flexuosa With. A
Cardamine hirsuta L.
Cardamine lacustris (Garn.-Jones & P.N.Johnson) Heenan E
Cardamine latior Heenan E
Cardamine pratensis L. A
Cardamine subcarnosa (Hook.f.) Allan E
Carrichtera annua (L.) DC. A
Cheiranthus cheiri L. A
Descurainia sophia (L.) Prantl A
Diplotaxis muralis (L.) DC. A
Diplotaxis tenuifolia (L.) DC. A
Erophila verna (L.) Chevall. A
Eruca vesicaria (L.) Cav. Cas
Erysimum cheiranthoides L. A
Heliophila coronopifolia L. Cas
Hirschfeldia incana (L.) Lagr.-Foss. A
Hymenolobus procumbens (L.) Schinz & Thell. A
Iberis amara L. A
Iberis umbellata L. A
Lepidium africanum (Burm.f.) DC. A
Lepidium banksii Kirk E
Lepidium bonariense L. A
Lepidium campestre (L.) R.Br. A
Lepidium densiflorum Schrad. A
Lepidium desvauxii Thell. A
Lepidium didymum L. A
Lepidium divaricatum W.T.Aiton Cas
Lepidium draba L. A
Lepidium flexicaule Kirk
Lepidium heterophyllum Benth. A
Lepidium hyssopifolium Desv. A
Lepidium kirkii Petrie E
Lepidium naufragorum Garn.-Jones & D.A.Norton E
Lepidium obtusatum Kirk E
Lepidium oleraceum G.Forst. ex Sparrm.
Lepidium peregrinum Thell. A
Lepidium pseudohyssopifolium Hewson Cas
Lepidium pseudotasmanicum Thell. A
Lepidium ruderale L. A
Lepidium sativum L. A
Lepidium sisymbrioides Hook.f. E
Lepidium solandri Kirk E

Lepidium squamatum Forssk. A
Lepidium tenuicaule Kirk E
Lepidium virginicum L. A
Lobularia maritima (L.) Desv. A
Lunaria annua L. A
Malcolmia maritima (L.) R.Br. Cas
Matthiola incana (L.) R.Br. A
Matthiola longipetala (Vent.) DC. Cas
Nasturtium microphyllum Boenn. ex Rchb. A
Nasturtium officinale R.Br. A
Neslia paniculata (L.) Desv. Cas
Notothlaspi australe Hook.f. E
Notothlaspi rosulatum Hook.f. E
Pachycladon cheesemanii Heenan & A.D.Mitch. E
Pachycladon crenatus Philipson E
Pachycladon enysii (Cheeseman) Heenan & A.D.Mitch. E
Pachycladon exile (Heenan) Heenan & A.D.Mitch. E
Pachycladon fasciarium Heenan E
Pachycladon fastigiatum (Hook.f.) Heenan & A.D.Mitch. E
Pachycladon latisiliquum (Cheeseman) Heenan & A.D.Mitch. E
Pachycladon novae-zelandiae (Hook.f.) Hook.f. E
Pachycladon stellatum (Allan) Heenan & A.D.Mitch. E
Pachycladon wallii (Carse) Heenan & A.D.Mitch. E
Raphanus raphanistrum L. A
Raphanus sativus L. A
Rapistrum rugosum (L.) All. A
Rorippa amphibia (L.) Besser A
Rorippa divaricata (Hook.f.) Garn.-Jones & Jonsell
Rorippa laciniata (F.Muell.) L.A.S.Johnson
Rorippa palustris (L.) Besser F
Rorippa sylvestris (L.) Besser A
Sinapis alba L. A
Sinapis arvensis L. A
Sisymbrium altissimum L. A
Sisymbrium erysimoides Desf. Cas
Sisymbrium irio L. Cas
Sisymbrium officinale (L.) Scop. A
Sisymbrium orientale L. A
Sisymbrium polyceratium L. A
Thlaspi arvense L. A
CARICACEAE A
Carica pubescens Lenné & K.Koch A
CLEOMACEAE A
Cleome hassleriana Chodat A
LIMNANTHACEAE A
Limnanthes douglasii R.Br. Cas
RESEDACEAE A
Reseda alba L. A
Reseda lutea L. A
Reseda luteola L. A
Reseda odorata L. Cas
TROPAEOLACEAE A
Tropaeolum majus L. A
Tropaeolum pentaphyllum Lam. A
Tropaeolum speciosum Poepp. & Endl. A
Tropaeolum tuberosum Ruiz & Pav. Cas

Order CROSSOSOMATALES
IXERBACEAE
Ixerba brexioides A.Cunn. E
STAPHYLEACEAE Cas
Staphylea emodi Wall. Cas

Order GERANIALES
FRANCOACEAE Cas
Francoa sonchifolia Cav. Cas
GERANIACEAE
Erodium botrys (Cav.) Bertol. A
Erodium cicutarium (L.) L'Hér. A
Erodium malacoides (L.) L'Hér. A
Erodium moschatum (L.) L'Hér. A
Erodium trifolium (Cav.) Guitt. Cas
Geranium aequlae (Bab.) Aedo Cas
Geranium australe Nees
Geranium brevicaule Hook.f.
Geranium dissectum L. A
Geranium endressii J.Gay Cas
Geranium gardneri de Lange A
Geranium homeanum Turcz.
Geranium lucidum L. Cas
Geranium maderense Yeo A
Geranium microphyllum Hook.f. E
Geranium molle L. A
Geranium phaeum L. Cas
Geranium potentilloides L'Hér. ex DC.
Geranium pratense L. A
Geranium purpureum Vill. A
Geranium pusillum L. A
Geranium retrorsum L'Hér. ex DC.
Geranium robertianum L. A
Geranium sessiliflorum Cav.
Geranium solanderi Carolin
Geranium traversii Hook.f. E
Geranium wallichianum D.Don. ex Sweet Cas
Geranium yeoi Aedo & Muñoz Garm. A
Pelargonium ×asperum Willd. A
Pelargonium ×domesticum L.H.Bailey A
Pelargonium ×fragrans Willd. A
Pelargonium ×hortorum L.H.Bailey A
Pelargonium capitatum (L.) Aiton Cas
Pelargonium crispum (L.) L'Hér. Cas
Pelargonium inodorum Willd.
Pelargonium panduriforme Eckl. & Zeyh. Cas
Pelargonium peltatum (L.) L'Hér. A
Pelargonium radens H.E.Moore Cas
Pelargonium tomentosum Jacq. A
Pelargonium vitifolium (L.) L'Hér. A
MELIANTHACEAE A
Melianthus major L. A

Order MALVALES
CISTACEAE A
Cistus creticus L. A
Cistus ladanifer L. A
Cistus laurifolius L. A
Cistus psilosepalus Sweet A
Cistus salvifolius L. Cas
MALVACEAE
Abutilon grandifolium (Willd.) Sweet A
Abutilon megapotamicum A.St.-Hil. & Naudin A
Abutilon theophrasti Medik. A
Alcea ficifolia L. Cas
Alcea rosea L. A
Anisodontea capensis (L.) D.M.Bates Cas
Dombeya torrida (J.F.Gmel.) Bamps Cas
Entelea arborescens R.Br. E
Hibiscus diversifolius Jacq.
Hibiscus mutabilis L. Cas
Hibiscus richardsonii Sweet ex Lindl.
Hibiscus syriacus L. Cas
Hibiscus tiliaceus L. A
Hibiscus trionum L. A
Hibiscus trionum 'diploid New Zealand naturalised race'
Hoheria angustifolia Raoul E
Hoheria equitum Heads E
Hoheria glabrata Sprague & Summerh. E
Hoheria lyallii Hook.f. E
Hoheria populnea A.Cunn. E
Hoheria sexstylosa Colenso E
Lagunaria patersonia (Andrews) G.Don A
Lavatera olbia L. A
Lavatera trimestris L. A
Malope trifida Cav. Cas
Malva arborea L. A
Malva assurgentiflora (Kellogg) M.F. Ray A
Malva linnaei M.F.Ray A
Malva moschata L. A
Malva neglecta Wallr. A
Malva nicaeensis All. A
Malva parviflora L. A
Malva sylvestris L. A
Malva verticillata L. Cas
Malvaviscus arboreus Cav. A
Modiola caroliniana (L.) G.Don A
Pavonia hastata Cav. A
Plagianthus divaricatus J.R.Forst. & G.Forst. E
Plagianthus regius (Poit.) Hochr. E
Sida rhombifolia L. A
Sidalcea malviflora (DC.) A.Gray ex Benth. Cas
Sparmannia africana L.f. A
Sphaeralcea philippiana Krapov. Cas
THYMELAEACEAE
Daphne bholua Buch.-Ham. ex D.Don Cas
Daphne laureola L. A
Daphne oleoides Schreb. Cas
Kelleria childii Heads E
Kelleria croizatii Heads E
Kelleria dieffenbachii (Hook.) Endl. E
Kelleria laxa (Cheeseman) Heads E
Kelleria lyallii (Hook.f.) Berggr. E
Kelleria multiflora (Cheeseman) Heads E
Kelleria paludosa Heads E
Kelleria tessellata Heads E
Kelleria villosa Berggr. E
Pimelea acra C.J.Burrows & de Lange E
Pimelea actea C.J.Burrows E
Pimelea aridula Cockayne E
Pimelea bicolor Colenso E
Pimelea buxifolia Hook.f. E
Pimelea carnosa C.J.Burrows E
Pimelea concinna Allan E
Pimelea cryptica C.J.Burrows & Enright E
Pimelea declivis C.J.Burrows E
Pimelea dura C.J.Burrows E
Pimelea eremitica C.J.Burrows E
Pimelea gnidia (J.R.Forst. & G.Forst.) Willd. E
Pimelea ignota C.J.Burrows & Courtney E
Pimelea longifolia Sol. ex Wikstr. E
Pimelea lyallii Hook.f. E
Pimelea mesoa C.J.Burrows E
Pimelea microphylla Colenso E
Pimelea notia C.J.Burrows & Thorsen E
Pimelea oreophila C.J.Burrows E
Pimelea orthia C.J.Burrows & Thorsen E
Pimelea poppelwellii Petrie E
Pimelea prostrata (J.R.Forst. & G.Forst.) Willd. E
Pimelea pseudolyallii Allan E
Pimelea pulvinaris C.J.Burrows E
Pimelea sericeovillosa Hook.f. E
Pimelea sporadica C.J.Burrows E
Pimelea hirta C.J.Burrows A
Pimelea suteri Kirk E
Pimelea telura C.J.Burrows E
Pimelea tomentosa (J.R.Forst. & G.Forst.) Druce E
Pimelea traversii Hook.f. E
Pimelea urvilleana A.Rich. E
Pimelea villosa Sol. ex Sm. E
Pimelea xenica C.J.Burrows E

Order MYRTALES
LYTHRACEAE A
Cuphea hyssopifolia Kunth A
Cuphea ignea A.DC. Cas
Cuphea lanceolata W.T.Aiton
Heimia salicifolia Link Cas
Lythrum hyssopifolia L. A
Lythrum junceum Banks & Sol. A
Lythrum portula (L.) D.A.Webb A
Lythrum salicaria L. A
Punica granatum L. Cas

MELASTOMATACEAE A
Heterocentron elegans (Schltdl.) Kuntze Cas
Heterocentron subtriplinervium (Link & Otto) A.Braun & C.D.Bouché Cas
Tibouchina paratropica Cogn. Cas
Tibouchina urvilleana (DC.) Cogn. A
MYRTACEAE
Agonis flexuosa (Willd.) Sweet Cas
Agonis juniperina Schauer Cas
Angophora costata (Gaertn.) Britten Cas
Astartea fascicularis (Labill.) DC. Cas
Callistemon linearis (Schrad. & J.C.Wendl.) DC. A
Chamelaucium uncinatum Schauer A
Corymbia ficifolia (F.Muell.) K.D.Hill & L.A.S.Johnson Cas
Eucalyptus botryoides Sm. A
Eucalyptus cinerea Benth. A
Eucalyptus cypellocarpa L.A.S.Johnson Cas
Eucalyptus delegatensis R.T.Baker A
Eucalyptus dendromorpha (Blakely) L.A.S.Johnson & Blaxell Cas
Eucalyptus elata Dehnh. Cas
Eucalyptus eugenioides Spreng. A
Eucalyptus fastigata H.Deane & Maiden A
Eucalyptus globulus Labill. A
Eucalyptus grandis W.Hill A
Eucalyptus gunnii Hook.f. A
Eucalyptus leucoxylon F.Muell. Cas
Eucalyptus macarthurii H.Deane & Maiden Cas
Eucalyptus muelleriana A.W.Howitt Cas
Eucalyptus nicholii Maiden & Blakely Cas
Eucalyptus nitens (H.Deane & Maiden) Maiden A
Eucalyptus obliqua L'Hér. A
Eucalyptus ovata Labill. A
Eucalyptus pilularis Sm. A
Eucalyptus piperita Sm. Cas
Eucalyptus pulchella Desf. A
Eucalyptus punctata DC. Cas
Eucalyptus regnans F.Muell. A
Eucalyptus resinifera Sm. Cas
Eucalyptus robusta Sm. A
Eucalyptus saligna Sm. A
Eucalyptus sideroxylon Woolls. Cas
Eucalyptus sieberi L.A.S.Johnson Cas
Eucalyptus tenuiramis Miq. A
Eucalyptus tereticornis Sm. A
Eucalyptus viminalis Labill. A
Kunzea ericoides (A.Rich.) Joy Thomps.
Kunzea sinclairii (Kirk) W.Harris E
Leptospermum laevigatum (Gaertn.) F.Muell. A
Leptospermum minutifolium C.T.White Cas
Leptospermum petersonii F.M.Bailey Cas
Leptospermum polygalifolium Salisb. Cas
Leptospermum scoparium J.R.Forst. & G.Forst.
Leptospermum spectabile Joy Thomps. Cas
Leptospermum variabile Joy Thomps. Cas
Lophomyrtus bullata (Sol. ex A.Cunn.) Burret E
Lophomyrtus obcordata (Raoul) Burret E
Lophostemon confertus (R.Br.) Peter G.Wilson & J.T.Waterh. Cas
Luma apiculata (DC.) Burret Cas
Melaleuca armillaris Sm. Cas
Melaleuca ericifolia Sm. Cas
Melaleuca hypericifolia Sm. Cas
Melaleuca leucadendra (L.) L. Cas
Melaleuca procera Craven
Melaleuca styphelioides Sm. Cas
Metrosideros albiflora Sol. ex Gaertn. E
Metrosideros bartlettii J.W.Dawson E
Metrosideros carminea W.R.B.Oliv. E
Metrosideros colensoi Hook.f. E
Metrosideros collina A.Gray Cas
Metrosideros diffusa (G.Forst.) Sm. E
Metrosideros excelsa Sol. ex Gaertn. E
Metrosideros fulgens Sol. ex Gaertn. E
Metrosideros kermadecensis W.R.B.Oliv. E
Metrosideros parkinsonii Buchanan E
Metrosideros perforata (J.R.Forst. & G.Forst.) A.Rich. E
Metrosideros robusta A.Cunn. E
Metrosideros umbellata Cav. E
Myricaria germanica (L.) Desv. A
Neomyrtus pedunculata (Hook.f.) Allan E
Psidium cattleianum Sabine A
Psidium guajava L. A
Syncarpia glomulifera Nied. Cas
Syzygium australe (Link) B.Hyland A
Syzygium floribundum F.Muell. Cas
Syzygium maire (A.Cunn.) Sykes & Garn.-Jones E
Syzygium paniculatum Gaertn. Cas
Syzygium smithii (Poir.) Nied. A
Tristaniopsis laurina (Sm.) Peter G.Wilson & J.T.Waterh. Cas
Ugni molinae Turcz. A
ONAGRACEAE
Clarkia amoena (Lehm.) A.Nelson & J.F.Macbr. A
Clarkia unguiculata Lindl. Csa
Epilobium alsinoides A.Cunn. E
Epilobium angustum (Cheeseman) P.H.Raven & Engelhorn E
Epilobium astonii (Allan) P.H.Raven & Engelhorn E
Epilobium billardiereanum DC.
Epilobium brevipes Hook.f. E
Epilobium brunnescens (Cockayne) P.H.Raven & Engelhorn E
Epilobium chionanthum Hausskn. E
Epilobium chlorifolium Hausskn. E
Epilobium ciliatum Raf. A
Epilobium confertifolium Hook.f. E
Epilobium crassum Hook.f.
Epilobium forbesii Allan E
Epilobium glabellum G.Forst. E
Epilobium gracilipes Kirk E
Epilobium gunnianum Hausskn.
Epilobium hectorii Hausskn. E
Epilobium hirtigerum A.Cunn.
Epilobium insulare Hausskn. E
Epilobium komarovianum H.Lév.
Epilobium macropus Hook. E
Epilobium margaretiae Brockie E
Epilobium matthewsii Petrie E
Epilobium melanocaulon Hook. E
Epilobium microphyllum A.Rich. E
Epilobium montanum L. A
Epilobium nerteroides A.Cunn. E
Epilobium nummulariifolium A.Cunn. E
Epilobium obscurum Schreb. A
Epilobium pallidiflorum A.Cunn.
Epilobium parviflorum Schreb. A
Epilobium pedunculare A.Cunn. E
Epilobium pernitens Cockayne & Allan E
Epilobium petraeum Heenan E
Epilobium pictum Petrie E
Epilobium porphyrium G.Simpson E
Epilobium pubens A.Rich. E
Epilobium purpuratum Hook.f. E
Epilobium pycnostachyum Hausskn. E
Epilobium rostratum Cheeseman E
Epilobium rotundifolium G.Forst. E
Epilobium tasmanicum Hausskn.
Epilobium tetragonum L. A
Epilobium wilsonii Cheeseman
Fuchsia boliviana Carrière A
Fuchsia denticulata Ruiz & Pav. Cas
Fuchsia excorticata (J.R.Forst. & G.Forst.) L.f. E
Fuchsia magellanica Lam. A
Fuchsia paniculata Lindl. Cas
Fuchsia perscandens Cockayne & Allan E
Fuchsia procumbens A.Cunn. E
Gaura lindheimeri Engelm. & A.Gray Cas
Ludwigia palustris (L.) Elliott F A
Ludwigia peploides (Kunth) P.H.Raven F A
Oenothera affinis Cambess. A
Oenothera biennis L. A
Oenothera drummondii Hook. Cas
Oenothera fruticosa L. Cas
Oenothera glazioviana Micheli A
Oenothera indecora Cambess. A
Oenothera parviflora L. A
Oenothera rosea Aiton A
Oenothera speciosa Nutt. A
Oenothera stricta Ledeb. ex Link A
Oenothera tetraptera Cav. Cas

Order SAPINDALES
ANACARDIACEAE
Rhus typhina L. A
Schinus molle L. A
Schinus terebinthifolius Raddi Cas
Toxicodendron succedaneum (L.) Kuntze A
MELIACEAE
Dysoxylum spectabile (G.Forst.) Hook.f. E
Melia azedarach L. Cas
Toona sinensis (A.Juss.) M.Roem. Cas
RUTACEAE
Citrus limon (L.) Burm.f. Cas
Citrus reticulata Blanco Cas
Citrus sinensis (L.) Osbeck A
Coleonema pulchellum I.Williams Cas
Correa alba Andrews Cas
Leionema nudum (Hook.) Paul G.Wilson E
Melicope simplex A.Cunn. E
Melicope ternata J.R.Forst. & G.Forst. E
Ptelea trifoliata L. Cas
Zanthoxylum simulans Hance Cas
SAPINDACEAE
Acer buergerianum Miq. Cas
Acer cappadocicum Gled. Cas
Acer negundo L. A
Acer palmatum (Thunb.) Murray Cas
Acer pseudoplatanus L. A
Aesculus hippocastanum L. A
Aesculus indica Colebr. ex Wall. Cas
Alectryon excelsus Gaertn. E
Cardiospermum grandiflorum Sw. Cas
Dodonaea viscosa Jacq.
Litchi chinensis Sonn. Cas
SIMAROUBACEAE A
Ailanthus altissima (Mill.) Swingle A

Superorder BERBERIDOPSIDANAE Cas
Order BERBERIDOPSIDALES Cas
BERBERIDOPSIDACEAE Cas
Berberidopsis corallina Hook.f. Cas

Superorder CARYOPHYLLANAE
Order CARYOPHYLLALES
AIZOACEAE
Carpobrotus chilensis (Molina) N.E.Br. A
Carpobrotus edulis (L.) N.E.Br. A
Carpobrotus glaucescens (Haw.) Schwantes
Delosperma cooperi (Hook.f.) L.Bolus Cas
Disphyma australe (W.T.Aiton) N.E.Br. E
Disphyma clavellatum (Haw.) Chinnock
Disphyma papillatum Chinnock E
Dorotheanthus bellidiformis (Burm.f.) N.E.Br. Cas
Drosanthemum floribundum (Haw.) Schwantes A
Erepsia heteropetala (Haw.) Schwantes Cas
Lampranthus glaucus (L.) N.E.Br. A
Lampranthus multiradiatus (Jacq.) N.E.Br. Cas
Lampranthus spectabilis (Haw.) N.E.Br. A
Mesembryanthemum crystallinum L. Cas
Oscularia caulescens (Mill.) Schwantes Cas
Ruschia geminiflora (Haw.) Schwantes A
Tetragonia implexicoma (Miq.) Hook.f.

Tetragonia tetragonioides (Pall.) Kuntze
AMARANTHACEAE
Achyranthes aspera L. A
Alternanthera denticulata R.Br.
Alternanthera nahui Heenan & de Lange
Alternanthera philoxeroides (Mart.) Griseb. A
Alternanthera pungens Kunth Cas
Amaranthus albus L. A
Amaranthus caudatus L. A
Amaranthus cruentus L. A
Amaranthus deflexus L. A
Amaranthus graecizans L. A
Amaranthus hybridus L. A
Amaranthus lividus L. A
Amaranthus powellii S.Watson A
Amaranthus retroflexus L. A
Amaranthus spinosus L. A
Amaranthus viridis L. A
Atriplex australasica Moq.
Atriplex billardierei (Moq.) Hook.f.
Atriplex buchananii (Kirk) Cheeseman E
Atriplex cinerea Poir.
Atriplex halimus L. A
Atriplex hollowayi de Lange & D.A.Norton E
Atriplex hortensis L. A
Atriplex patula L. A
Atriplex prostrata Boucher ex DC. A
Atriplex rosea L. A
Beta vulgaris L. A
Chenopodium album L. A
Chenopodium bonus-henricus L. A
Chenopodium capitatum (L.) Asch. A
Chenopodium detestans Kirk E
Chenopodium erosum R.Br. A
Chenopodium ficifolium Sm. Cas
Chenopodium foliosum (Moench) Asch. Cas
Chenopodium giganteum D.Don A
Chenopodium glaucum L.
Chenopodium murale L. A
Chenopodium opulifolium Schrad. ex W.D.J.Koch & Ziz A
Chenopodium pumilio R.Br. A
Chenopodium vulvaria L. A
Dysphania ambrosioides (L.) Mosyakin & Clemants A
Dysphania pusilla (Hook.f.) Mosyakin & Clemants E
Einadia allanii (Aellen) Paul G.Wilson E
Einadia nutans (R.Br.) A.J.Scott A
Einadia triandra (G.Forst.) A.J.Scott E
Einadia trigonos (Schult.) Paul G.Wilson
Gomphrena celosioides Mart. Cas
Kochia scoparia (L.) Schrad. A
Rhagodia triandra (Forst.f.) Aellen
Salsola kali L. A
Salsola ruthenica Iljin Cas
Sarcocornia quinqueflora (Bunge ex Ung.-Sternb.) A.J.Scott
Suaeda novae-zelandiae Allan E
BASELLACEAE A
Anredera cordifolia (Ten.) Steenis A
CACTACEAE A
Austrocylindropuntia subulata (Muehlenpf.) Backeb. Cas
Cylindropuntia fulgida (Engelm.) F.M.Knuth Cas
Opuntia cylindrica (Lam.) DC. Cas
Opuntia ficus-indica (L.) Mill. A
Opuntia monacantha Haw. A
Opuntia robusta Pfeiff. Cas
CARYOPHYLLACEAE
Agrostemma githago L. Cas
Arenaria serpyllifolia L. A
Cerastium arvense L. A
Cerastium fontanum Baumg. A
Cerastium glomeratum Thuill. A
Cerastium semidecandrum L. A
Cerastium tomentosum L. A
Colobanthus acicularis Hook.f. E
Colobanthus affinis (Hook.) Hook.f.
Colobanthus apetalus (Labill.) Druce
Colobanthus brevisepalus Kirk E
Colobanthus buchananii Kirk E
Colobanthus canaliculatus Kirk E
Colobanthus hookeri Cheeseman E
Colobanthus masonae L.B.Moore E
Colobanthus monticola Petrie E
Colobanthus muelleri Kirk E
Colobanthus muscoides Hook.f.
Colobanthus squarrosus Cheeseman
Colobanthus strictus Cheeseman E
Colobanthus wallii Petrie
Dianthus armeria L. A
Dianthus barbatus L. A
Dianthus deltoides L. A
Dianthus plumarius L. A
Gypsophila australis (Schltdl.) A.Gray A
Gypsophila muralis L. Cas
Gypsophila paniculata L. A
Herniaria glabra L. Cas
Herniaria hirsuta L. A
Holosteum umbellatum L. A
Illecebrum verticillatum L. Cas
Minuartia hybrida (Vill.) Schischk. A
Moehringia trinervia (L.) Clairv. Cas
Moenchia erecta (L.) Gaertn. A
Paronychia brasiliana DC. A
Petrorhagia prolifera (L.) P.W.Ball & Heywood A
Petrorhagia velutina (Guss.) P.W.Ball & Heywood A
Polycarpon tetraphyllum (L.) L. A
Sagina apetala Ard. A
Sagina procumbens L. A
Sagina subulata (Sw.) Presl A
Saponaria officinalis L. A
Scleranthus annuus L. A
Scleranthus biflorus (J.R.Forst. & G.Forst.) Hook.f.
Scleranthus brockiei P.A.Williamson
Scleranthus fasciculatus (R.Br.) Hook.f. A
Scleranthus uniflorus P.A.Williamson E
Silene armeria L. A
Silene chalcedonica (L.) E.H.L.Krause Cas
Silene coeli-rosa (L.) Godr. A
Silene conica L. A
Silene coronaria (L.) Clairv. A
Silene dioica (L.) Clairv. A
Silene disticha Willd. A
Silene flos-cuculi (L.) Clairv. A
Silene gallica L. A
Silene italica (L.) Pers. Cas
Silene latifolia Poir. A
Silene noctiflora L. A
Silene nutans L. A
Silene pendula L. A
Silene viscaria (L.) Jess. Cas
Silene vulgaris (Moench) Garcke A
Spergula arvensis L. A
Spergularia bocconei (Scheele) Graebn. A
Spergularia marina (L.) Griseb.
Spergularia media (L.) C.Presl
Spergularia rubra (L.) J.Presl & C.Presl A
Spergularia tasmanica (Kindb.) L.G.Adams
Stellaria alsine Grimm A
Stellaria decipiens Hook.f. E
Stellaria elatinoides Hook.f. E
Stellaria gracilenta Hook.f. E
Stellaria graminea L. A
Stellaria holostea L.
Stellaria media (L.) Vill. A
Stellaria neglecta Weihe Cas
Stellaria parviflora Hook.f.
Stellaria roughii Hook.f. E
Vaccaria hispanica (Mill.) Rauschert Cas
DROSERACEAE
Dionaea muscipula Ellis Cas
Drosera arcturi Hook.
Drosera auriculata Backh. ex Planch.
Drosera binata Labill.
Drosera burmanni Vahl Cas
Drosera capensis L. A
Drosera filiformis Raf. Cas
Drosera peltata Thunb.
Drosera pygmaea DC.
Drosera spatulata Labill.
Drosera stenopetala Hook.f. E
NYCTAGINACEAE
Bougainvillea glabra Choisy A
Mirabilis jalapa L. A
Pisonia brunoniana Endl.
PHYTOLACCACEAE A
Phytolacca americana L. A
Phytolacca clavigera W.W.Sm. A
Phytolacca octandra L. A
PLUMBAGINACEAE A
Armeria alliacea (Cav.) Hoffmanns. & Link Cas
Armeria maritima (Mill.) Willd. Cas
Limonium bonduellii (F.Lestib.) Kuntze A
Limonium companyonis Kuntze Cas
Limonium sinuatum (L.) Mill. Cas
Plumbago auriculata Lam. Cas
POLYGONACEAE
Aconogonon campanulatum (Hook.f.) H.Hara Cas
Aconogonon molle (D.Don) Hara Cas
Antenoron filiforme (Thunb.) Roberty & Vautier A
Bistorta vaccinifolia (Wall.) Greene Cas
Emex australis Steinh. A
Fagopyrum dibotrys (D.Don) Hara Cas
Fagopyrum esculentum Moench A
Fallopia aubertii (L.Henry) Holub A
Fallopia convolvulus (L.) Á.Löve A
Fallopia japonica (Houtt.) Ronse Decraene A
Fallopia sachalinensis (F.Schmidt) Ronse Decr. A
Muehlenbeckia astonii Petrie E
Muehlenbeckia australis (G.Forst.) Meisn.
Muehlenbeckia axillaris (Hook.f.) Endl.
Muehlenbeckia complexa (A.Cunn.) Meisn.
Muehlenbeckia ephedroides Hook.f. E
Oxyria digyna (L.) Hill Cas
Persicaria capitata (Buch.-Ham. ex D.Don) H.Gross A
Persicaria chinensis (L.) H.Gross Cas
Persicaria decipiens (R.Br.) K.L.Wilson F
Persicaria hydropiper (L.) Spach A F
Persicaria lapathifolia (L.) Gray A
Persicaria maculosa Gray A
Persicaria odorata (Lour.) Soják Cas
Persicaria orientalis (L.) Spach A
Persicaria perfoliata (L.) H.Gross Cas
Persicaria prostrata (R.Br.) Sojak F A
Persicaria punctata (Elliott) Small F A
Persicaria strigosa (R.Br.) Gross F A
Persicaria wallichii Greuter & Burdet A
Polygonum arenastrum Boreau A
Polygonum aviculare L. A
Polygonum plebeium R.Br. A
Rheum rhabarbarum L. A
Rumex acetosa L. A
Rumex acetosella L. A
Rumex brownii Campd. A
Rumex conglomeratus Murray A
Rumex crispus L. A
Rumex flexuosus Sol. ex G.Forst.
Rumex frutescens Thouars A
Rumex neglectus Kirk E
Rumex obtusifolius L. A
Rumex pulcher L. A
Rumex sagittatus Thunb. A

Rumex tenuifolius (Wallr.) Á.Löve A
Rumex vesicarius L. Cas
PORTULACACEAE
Calandrinia compressa Schrad. ex DC. A
Calandrinia menziesii (Hook.) Torr. & A.Gray A
Claytonia perfoliata Donn ex Willd. A
Claytonia sibirica L. Cas
Hectorella caespitosa Hook.f. E
Montia angustifolia Heenan E
Montia calycina (Colenso) Pax & K.Hoffm. E
Montia campylostigma (Heenan) Heenan E
Montia drucei (Heenan) Heenan E
Montia erythrophylla Heenan (Heenan) E
Montia fontana L. F
Montia racemosa (Buchanan) Heenan E
Montia sessiliflora (G.Simpson) Heenan E
Portulaca grandiflora Hook. Cas
Portulaca oleracea L. A
Portulaca pilosa L. Cas
Talinum paniculatum (Jacq.) Gaertn. Cas
TAMARICACEAE A
Tamarix chinensis Lour. A
Tamarix parviflora DC. Cas

Superorder SANTALANAE
Order SANTALALES
BALANOPHORACEAE
Dactylanthus taylorii Hook.f. E
LORANTHACEAE
Alepis flavida (Hook.f.) Tiegh. E
Ileostylus micranthus (Hook.f.) Tiegh. E
Muellerina celastroides (Sieber ex Schult. & Schult.f.) Tiegh.
Peraxilla colensoi (Hook.f.) Tiegh. E
Peraxilla tetrapetala (L.f.) Tiegh. E
Trilepidea adamsii (Cheeseman) Tiegh. E
Tupeia antarctica (G.Forst.) Cham. & Schltdl. E
SANTALACEAE
Exocarpos bidwillii Hook.f. E
Korthalsella clavata (Kirk) Cheeseman E
Korthalsella lindsayi (Oliv.) Engl. E
Korthalsella salicornioides (A.Cunn.) Tiegh. E
Mida salicifolia A.Cunn. E
Viscum album L. Cas

Superorder ASTERANAE
Order CORNALES
CORNACEAE A
Cornus capitata Wall. A
HYDRANGEACEAE A
Deutzia crenata Siebold & Zucc. Cas
Hydrangea macrophylla (Thunb.) Ser. A
Philadelphus ×cymosus Rehder Cas
Philadelphus mexicanus Schltdl. A

Order ERICALES
ACTINIDIACEAE A
Actinidia chinensis Planch. Cas
Actinidia deliciosa (A.Chev.) C.F.Liang & A.R.Ferguson A
Actinidia eriantha Benth. Cas
BALSAMINACEAE A
Impatiens balsamina L. Cas
Impatiens glandulifera Royle A
Impatiens niamniamensis Gilg Cas
Impatiens sodenii Engl. A
Impatiens textorii Miq. Cas
Impatiens walleriana Hook.f. A
CLETHRACEAE Cas
Clethra arborea W.T.Aiton Cas
EBENACEAE Cas
Diospyros lotus L. Cas
Diospyros virginiana L. Cas
ERICACEAE
Acrothamnus colensoi (Hook.f.) Quinn E
Androstoma empetrifolia Hook.f. E
Arbutus unedo L. A
Archeria racemosa Hook.f. E
Archeria traversii Hook.f. E
Calluna vulgaris (L.) Hull A
Cyathodes pumila Hook.f. E
Daboecia cantabrica (Huds.) K.Koch. A
Dracophyllum acerosum Berggr. E
Dracophyllum adamsii Petrie E
Dracophyllum arboreum Cockayne E
Dracophyllum densum W.R.B.Oliv. E
Dracophyllum elegantissimum S.Venter E
Dracophyllum filifolium Hook.f. E
Dracophyllum fiordense W.R.B.Oliv. E
Dracophyllum kirkii Berggr. E
Dracophyllum latifolium A.Cunn. E
Dracophyllum lessonianum A.Rich. E
Dracophyllum longifolium (J.R.Forst. & G.Forst.) R.Br. E
Dracophyllum marmoricola S.Venter E
Dracophyllum menziesii Hook.f. E
Dracophyllum muscoides Hook.f. E
Dracophyllum oliveri Du Rietz E
Dracophyllum ophioliticum S.Venter E
Dracophyllum palustre Cockayne ex W.R.B.Oliv. E
Dracophyllum patens W.R.B.Oliv. E
Dracophyllum pearsonii Kirk E
Dracophyllum politum (Cheeseman) Cockayne E
Dracophyllum pronum W.R.B.Oliv. E
Dracophyllum prostratum Kirk E
Dracophyllum pubescens Cheeseman E
Dracophyllum recurvum Hook.f. E
Dracophyllum rosmarinifolium (G.Forst.) R.Br. E
Dracophyllum scoparium Hook.f. E
Dracophyllum sinclairii Cheeseman E
Dracophyllum strictum Hook.f. E
Dracophyllum subulatum Hook.f. E
Dracophyllum townsonii Cheeseman E
Dracophyllum traversii Hook.f. E
Dracophyllum trimorphum W.R.B.Oliv. E
Dracophyllum uniflorum Hook.f. E
Dracophyllum urvilleanum A.Rich. E
Dracophyllum verticillatum Labill. E
Dracophyllum viride W.R.B.Oliv. E
Epacris alpina Hook.f. E
Epacris pauciflora A.Rich. E
Epacris purpurascens R.Br. A
Erica arborea L. A
Erica baccans L. A
Erica caffra L. A
Erica carnea L. Cas
Erica cinerea L. A
Erica lusitanica Rudolphi A
Erica melanthera L. Cas
Erica tetralix L. Cas
Erica vagans L. A
Gaultheria antipoda G.Forst. E
Gaultheria colensoi Hook.f. E
Gaultheria crassa Allan E
Gaultheria depressa Hook.f.
Gaultheria macrostigma (Colenso) D.J.Middleton E
Gaultheria mucronata (L.f.) Hook. & Arn. A
Gaultheria nubicola D.J.Middleton E
Gaultheria oppositifolia Hook.f. E
Gaultheria paniculata B.L.Burtt & A.W.Hill E
Gaultheria parvula D.J.Middleton E
Gaultheria rupestris (L.f.) D.Don E
Gaultheria shallon Pursh A
Leptecophylla juniperina (J.R.Forst. & G.Forst.) C.M.Weiller
Leucopogon fasciculatus (G.Forst.) A.Rich. E
Leucopogon fraseri A.Cunn.
Leucopogon nanum M.I.Dawson & Heenan E
Leucopogon parviflorus (Andrews) Lindl.
Leucopogon xerampelinus de Lange, Heenan & M.I.Dawson E
Leucothoe axillaris (Lam.) D.Don Cas
Pentachondra pumila (J.R.Forst. & G.Forst.) R.Br.
Pieris japonica (Thunb.) D.Don ex G.Don Cas
Rhododendron decorum Franch. Cas
Rhododendron maddenii Hook.f. Cas
Rhododendron ponticum L. A
Sprengelia incarnata Sm.
Vaccinium corymbosum L. Cas
POLEMONIACEAE A
Cobaea scandens Cav. A
Collomia cavanillesii Hook. & Arn. A
Gilia capitata Sims. Cas
Navarretia squarrosa (Eschsch.) Hook. & Arn. A
Phlox drummondii Hook. A
Phlox paniculata L. A
Phlox subulata L. Cas
Polemonium caeruleum L. A
PRIMULACEAE
Anagallis arvensis L. A
Anagallis minima (L.) E.H.L.Krause A
Ardisia crenata Sims Cas
Cyclamen persicum Mill. Cas
Elingamita johnsonii G.T.S.Baylis
Lysimachia mauritiana Lam. Cas
Lysimachia nummularia L. A
Lysimachia vulgaris L. A
Myrsine aquilonia de Lange & Heenan E
Myrsine argentea Heenan & de Lange E
Myrsine australis (A.Rich.) Allan E
Myrsine chathamica F.Muell. E
Myrsine coxii Cockayne E
Myrsine divaricata A.Cunn. E
Myrsine kermadecensis Cheeseman E
Myrsine nummularia Hook.f. E
Myrsine oliveri Allan E
Myrsine salicina Heward ex Hook.f. E
Myrsine umbricola Heenan & de Lange E
Primula malacoides Franch. A
Primula prolifera Wall. Cas
Primula vulgaris Huds. A
Samolus repens (J.R.Forst. & G.Forst.) Pers.
Samolus valerandii L. Cas
SAPOTACEAE
Planchonella costata (Endl.) Pierre
SARRACENIACEAE Cas
Sarracenia alata (Alph. Wood) Alph. Wood Cas
Sarracenia flava L. Cas
Sarracenia leucophylla Raf. Cas
Sarracenia purpurea Cas
THEACEAE Cas
Camellia forrestii (Diels) Cohen-Stuart Cas
Camellia grijsii Hance Cas
Camellia japonica L. Cas

Order BORAGINALES
BORAGINACEAE
Amsinckia calycina (Moris) Chater A
Borago officinalis L. A
Brunnera macrophylla (Adams) I.M.Johnst. A
Cerinthe major L. A
Cynoglossum amabile Stapf & Drumm. A
Cynoglottis barrelieri (All.) Vural & Kit Tan A
Echium candicans L.f. A
Echium pininana Webb & Berthel. A
Echium plantagineum L. A
Echium vulgare L. A
Echium wildpretii Hook.f. Cas
Heliotropium arborescens L. Cas
Lithospermum arvense L. A
Lithospermum purpurocaeruleum L. A
Myosotidium hortensium (Decne.) Baill. E
Myosotis albosericea Hook.f. E
Myosotis angustata Cheeseman E
Myosotis antarctica Hook.f.

Myosotis arnoldii L.B.Moore E
Myosotis arvensis (L.) Hill A
Myosotis australis R.Br.
Myosotis brevis de Lange & Barkla E
Myosotis brockiei L.B.Moore & M.J.A.Simpson E
Myosotis capitata Hook.f. E
Myosotis cheesemanii Petrie E
Myosotis colensoi (Kirk) J.F.Macbr. E
Myosotis concinna Cheeseman E
Myosotis discolor Pers. A
Myosotis drucei (L.B.Moore) de Lange & Barkla E
Myosotis elderi L.B.Moore E
Myosotis eximia Petrie E
Myosotis explanata Cheeseman E
Myosotis forsteri Lehm. E
Myosotis glabrescens L.B.Moore E
Myosotis glauca (G.Simpson & J.S.Thomson) de Lange & Barkla E
Myosotis goyenii Petrie E
Myosotis laeta Cheeseman E
Myosotis laingii Cheeseman E
Myosotis laxa Lehm. F A
Myosotis lyallii Hook.f. E
Myosotis lytteltonensis (Laing & A.Wall) de Lange E
Myosotis macrantha (Hook.f.) Benth. & Hook.f. E
Myosotis matthewsii L.B.Moore E
Myosotis monroi Cheeseman E
Myosotis oreophila Petrie E
Myosotis petiolata Hook.f. E
Myosotis pulvinaris Hook.f. E
Myosotis pygmaea Colenso E
Myosotis rakiura L.B.Moore E
Myosotis saxatilis Petrie E
Myosotis saxosa Hook.f. E
Myosotis scorpioides L. A
Myosotis spathulata G.Forst. E
Myosotis stricta Roem. & Schult. A
Myosotis suavis Petrie E
Myosotis sylvatica Hoffm. A
Myosotis tenericaulis Petrie E
Myosotis traversii Hook.f. E
Myosotis uniflora Hook.f. E
Myosotis venosa Colenso E
Omphalodes nitida Hoffmanns. & Link Cas
Omphalodes verna Moench Cas
Pentaglottis sempervirens (L.) L.H.Bailey A
Phacelia campanularia A.Gray Cas
Phacelia tanacetifolia Benth. A
Symphytum asperum Lepech. A
Symphytum officinale L. A
Trachystemon orientalis (L.) G.Don Cas
Wigandia caracasana Kunth Cas

Order GARRYALES Cas
GARRYACEAE Cas
Aucuba japonica Thunb. Cas

Order GENTIANALES
APOCYNACEAE
Araujia hortorum E.Fourn. A
Asclepias curassavica L. Cas
Catharanthus roseus (L.) G.Don A
Gomphocarpus fruticosus (L.) W.T.Aiton A
Gomphocarpus physocarpus E.Mey. Cas
Mandevilla laxa (Ruiz & Pav.) Woodson Cas
Nerium oleander L. A
Oxypetalum caeruleum (D.Don) Decne. Cas
Parsonsia capsularis (G.Forst.) R.Br. E
Parsonsia heterophylla A.Cunn. E
Parsonsia praeruptis Heads & de Lange E
Vinca major L. A
Vinca minor L. A
GENTIANACEAE
Blackstonia perfoliata (L.) Huds. A
Centaurium erythraea Rafn. A
Centaurium tenuiflorum (Hoffm. & Link) Fritsch ex Janch. A
Gentianella amabilis (Petrie) Glenny E
Gentianella angustifolia Glenny E
Gentianella antarctica (Kirk) T.N.Ho & S.W.Liu E
Gentianella antipoda (Kirk) T.N.Ho & S.W.Liu E
Gentianella astonii (Petrie) T.N.Ho & S.W.Liu E
Gentianella bellidifolia (Hook.f.) Holub E
Gentianella calcis Glenny & Molloy E
Gentianella cerina (Hook.f.) T.N.Ho & S.W.Liu E
Gentianella chathamica (Cheeseman) T.N.Ho & S.W.Liu E
Gentianella concinna (Hook.f.) T.N.Ho & S.W.Liu E
Gentianella corymbifera (Kirk) Holub E
Gentianella decumbens Glenny E
Gentianella divisa (Kirk) Glenny E
Gentianella filipes (Cheeseman) T.N.Ho & S.W.Liu E
Gentianella gibbsii (Petrie) T.N.Ho & S.W.Liu E
Gentianella grisebachii (Hook.f.) T.N.Ho E
Gentianella impressinervia Glenny E
Gentianella lilliputiana (C.J.Webb) Glenny E
Gentianella lineata (Kirk) T.N.Ho & S.W.Liu E
Gentianella luteoalba Glenny E
Gentianella magnifica (Kirk) Glenny E
Gentianella montana (G.Forst.) Holub E
Gentianella patula (Kirk) Holub E
Gentianella saxosa (G.Forst.) Holub E
Gentianella scopulorum Glenny E
Gentianella serotina (Cockayne) T.N.Ho & S.W.Liu E
Gentianella spenceri (Kirk) T.N.Ho & S.W.Liu E
Gentianella stellata Glenny E
Gentianella tenuifolia (Petrie) T.N.Ho & S.W.Liu E
Gentianella vernicosa (Cheeseman) T.N.Ho & S.W.Liu E
Sebaea ovata (Labill.) R.Br.
LOGANIACEAE
Geniostoma rupestre J.R.Forst. & G.Forst.
Logania depressa Hook.f. E
Mitrasacme montana Hook.f.
Mitrasacme novae-zelandiae Hook.f. E
RUBIACEAE
Asperula orientalis Boiss. & Hohen. Cas
Coffea arabica L. Cas
Coprosma acerosa A.Cunn. E
Coprosma acutifolia Hook.f. E
Coprosma arborea Kirk E
Coprosma arcuata Colenso E
Coprosma areolata Cheeseman E
Coprosma atropurpurea (Cockayne & Allan) L.B.Moore E
Coprosma chathamica Cockayne E
Coprosma cheesemanii W.R.B.Oliv. E
Coprosma ciliata Hook.f. E
Coprosma colensoi Hook.f. E
Coprosma crassifolia Colenso E
Coprosma crenulata W.R.B.Oliv. E
Coprosma cuneata Hook.f. E
Coprosma decurva Heads E
Coprosma depressa Colenso ex Hook.f. E
Coprosma distantia (de Lange & R.O.Gardner) de Lange E
Coprosma divaricata A.Cunn. E
Coprosma dodonaeifolia W.R.B.Oliv. E
Coprosma dumosa (Cheeseman) G.T.Jane E
Coprosma elatirioides de Lange & A.S.Markey E
Coprosma foetidissima J.R.Forst. & G.Forst. E
Coprosma fowerakeri D.A.Norton & de Lange E
Coprosma grandifolia Hook.f. E
Coprosma intertexta G.Simpson E
Coprosma linariifolia Hook.f. E
Coprosma lucida J.R.Forst. & G.Forst. E
Coprosma macrocarpa Cheeseman E
Coprosma microcarpa Hook.f. E
Coprosma neglecta Cheeseman E
Coprosma niphophila Orchard
Coprosma obconica Kirk E
Coprosma parviflora Hook.f. E
Coprosma pedicellata Molloy, de Lange & B.D.Clarkson E
Coprosma perpusilla Colenso
Coprosma petiolata Hook.f. E
Coprosma petriei Cheeseman E
Coprosma polymorpha W.R.B.Oliv. E
Coprosma propinqua A.Cunn. E
Coprosma pseudociliata G.T.Jane E
Coprosma pseudocuneata W.R.B.Oliv. ex Garn.-Jones & Elder E
Coprosma repens A.Rich.
Coprosma rhamnoides A.Cunn. E
Coprosma rigida Cheeseman E
Coprosma robusta Raoul E
Coprosma rotundifolia A.Cunn. E
Coprosma rubra Petrie E
Coprosma rugosa Cheeseman E
Coprosma serrulata Hook.f. ex Buchanan E
Coprosma solandri Kirk E
Coprosma spathulata A.Cunn. E
Coprosma talbrockiei L.B.Moore & R.Mason E
Coprosma tayloriae A.P.Druce ex G.T.Jane E
Coprosma tenuicaulis Hook.f. E
Coprosma tenuifolia Cheeseman E
Coprosma virescens Petrie E
Coprosma waima A.P.Druce E
Coprosma wallii Petrie in Cheeseman E
Galium aparine L. A
Galium debile Desv. A
Galium divaricatum Lam. A
Galium humifusum M.Bieb. A
Galium mollugo L. A
Galium murale (L.) All. A
Galium palustre L. A
Galium perpusillum (Hook.f.) Allan E
Galium propinquum A.Cunn.
Galium trilobum Colenso E
Galium uliginosum L. A
Galium verum L. A
Leptostigma setulosa (Hook.f.) Fosberg E
Nertera balfouriana Cockayne E
Nertera ciliata Kirk E
Nertera cunninghamii Hook.f. E
Nertera depressa Banks & Sol. ex Gaertn.
Nertera dichondrifolia (A.Cunn.) Hook.f. E
Nertera scapanioides Lange E
Nertera villosa B.H.Macmill. & R.Mason E
Sherardia arvensis L. A

Order LAMIALES
ACANTHACEAE
Acanthus mollis L. A
Avicennia marina (Forssk.) Vierh. M
Hygrophila angustifolia R.Br. Cas
Justicia carnea Lindl. Cas
Mackaya bella Harv. C as
Strobilanthes anisophyllus (Lodd.) T.Anderson A
Thunbergia alata Bojer A
Thunbergia coccinea Wall. Cas
Thunbergia grandiflora (Roxb. ex Rottler) Roxb. Cas
BIGNONIACEAE
Campsis ×tagliabuana (Vis.) Rehder A
Catalpa bignonioides Walter A
Distictis buccinatoria (DC.) A.H.Gentry A
Eccremocarpus scaber Ruiz & Pav. A
Jacaranda mimosifolia D.Don Cas
Macfadyena unguis-cati (L.) A.H.Gentry Cas
Pandorea jasminoides (Lindl.) Schum. Cas
Pandorea pandorana (Andrews) Steenis A
Pithecoctenium crucigerum (L.) A.H.Gentry Cas
Podranea ricasoliana (Tanfani) Sprague A

Pyrostegia venusta (Ker Gawl.) Miers Cas
Radermachera pentandra Hemsl. Cas
Tecoma stans (L.) Juss. ex Knuth Cas
Tecomanthe speciosa W.R.B.Oliv. E
Tecomaria capensis (Thunb.) Spach A
CALCEOLARIACEAE
Calceolaria tripartita Ruiz & Pav. A
Jovellana repens (Hook.f.) Kraenzl. E
Jovellana sinclairii (Hook.) Kraenzl. E
GESNERIACEAE
Rhabdothamnus solandri A.Cunn. E
LAMIACEAE (LABIATAE)
Acinos arvensis (Lam.) Dandy A
Agastache foeniculum (Pursh) Kuntze Cas
Ajuga reptans L. A
Ballota nigra L. A
Calamintha nepeta (L.) Savi A
Callicarpa rubella Lindl. Cas
Cedronella canariensis (L.) Webb & Berthel. A
Clerodendrum myricoides (Hochst.) Vatke Cas
Clerodendrum trichotomum Thunb. A
Clinopodium vulgare L. A
Colquhounia coccinea Wall. A
Elsholtzia flava (Benth.) Benth. Cas
Galeopsis ladanum L. Cas
Galeopsis tetrahit L. A
Glechoma hederacea L. A
Lamium album L. A
Lamium amplexicaule L. A
Lamium confertum Fr. A
Lamium galeobdolon (L.) L. A
Lamium hybridum Vill. A
Lamium maculatum L. A
Lamium purpureum L. A
Lavandula angustifolia Mill. Cas
Lavandula dentata L. A
Lavandula stoechas L. A
Leonotis nepetifolia (L.) R.Br. Cas
Leonotis ocymifolia (Burm.f.) Iwarsson A
Leonurus cardiaca L. A
Lycopus europaeus L. F A
Marrubium vulgare L. A
Melissa officinalis L. A
Mentha arvensis L. A
Mentha australis R.Br. A
Mentha cunninghamii Benth. E
Mentha pulegium L. A
Mentha spicata L. A
Mentha suaveolens Ehrh. A
Nepeta cataria L. A
Nepeta mussinii Henckel A
Ocimum basilicum L. A
Origanum vulgare L. A
Phlomis fruticosa L. A
Phlomis russeliana (Sims) Benth. A
Plectranthus argentatus S.T.Blake Cas
Plectranthus behrii Compton Cas
Plectranthus ciliatus E.Mey. A
Plectranthus ecklonii Benth. A
Plectranthus grandis (Cramer) Willems A
Plectranthus ornatus Codd Cas
Plectranthus parviflorus Willd.
Plectranthus saccatus Benth. Cas
Prunella laciniata (L.) L. Cas
Prunella vulgaris L. A
Rosmarinus officinalis L. Cas
Salvia aurea L. A
Salvia azurea Lam. Cas
Salvia farinacea Benth. A
Salvia glutinosa L. Cas
Salvia guaranitica A.St.-Hil. ex Benth. Cas
Salvia involucrata Cav. Cas
Salvia microphylla Kunth A
Salvia nemorosa L. A
Salvia officinalis L. A
Salvia pseudococcinea Jacq. Cas
Salvia reflexa Hornem. A
Salvia repens Benth. A
Salvia rutilans Carr. Cas
Salvia sclarea L. A
Salvia splendens Sellow ex Wied-Neuw. Cas
Salvia uliginosa Benth. A
Salvia verbenaca L. A
Salvia verticillata L. Cas
Satureja hortensis L. Cas
Scutellaria indica L. Cas
Scutellaria minor Huds. A
Scutellaria novae-zelandiae Hook.f. E
Stachys annua (L.) L. Cas
Stachys arvensis (L.) L. A
Stachys byzantina K.Koch A
Stachys germanica L. Cas
Stachys palustris L. A
Stachys sylvatica L. A
Tetradenia riparia (Hochst.) Codd Cas
Teucridium parvifolium Hook.f. E
Teucrium hircanicum L. A
Teucrium scorodonia L. A
Thymus pulegioides L. A
Thymus vulgaris L. A
Vitex lucens Kirk E
Westringia fruticosa (Willd.) Druce Cas
LENTIBULARIACEAE
Pinguicula grandiflora Lam. Cas
Utricularia arenaria A.DC. F Cas
Utricularia australis R.Br.
Utricularia delicatula Cheeseman E
Utricularia dichotoma Labill.
Utricularia geminiscapa Benj. F A
Utricularia gibba L. F A
Utricularia livida E.Mey. F A
Utricularia sandersonii Oliv. Cas
OLEACEAE
Forsythia ×intermedia Zabel Cas
Forsythia suspensa (Thunb.) Vahl Cas
Fraxinus excelsior L. A
Fraxinus ornus L. A
Jasminum beesianum Forrest & Diels A
Jasminum humile L. A
Jasminum mesnyi Hance A
Jasminum officinale L. A
Jasminum polyanthum Franch. A
Ligustrum lucidum W.T.Aiton A
Ligustrum ovalifolium Hassk. A
Ligustrum sinense Lour. A
Ligustrum vulgare L. A
Nestegis apetala (Vahl) L.A.S.Johnson
Nestegis cunninghamii (Hook.f.) L.A.S.Johnson E
Nestegis lanceolata (Hook.f.) L.A.S.Johnson E
Nestegis montana (Hook.f.) L.A.S.Johnson E
Olea europaea L. A
Osmanthus heterophyllus (G.Don) P.S.Green Cas
Syringa vulgaris L. A
OROBANCHACEAE
Euphrasia australis Petrie E
Euphrasia cheesemanii Wettst. E
Euphrasia cockayneana Petrie E
Euphrasia cuneata G.Forst. E
Euphrasia disperma Hook.f. E
Euphrasia drucei Ashwin E
Euphrasia dyeri Wettst. E
Euphrasia integrifolia Petrie E
Euphrasia laingii Petrie E
Euphrasia monroi Hook.f. E
Euphrasia nemorosa (Pers.) Wallr. A
Euphrasia petriei Ashwin E
Euphrasia repens Hook.f. E
Euphrasia revoluta Hook.f. E
Euphrasia townsonii Petrie E
Euphrasia wettsteiniana Du Rietz E
Euphrasia zelandica Wettst. E
Lathraea clandestina L. Cas
Micrargeria filiformis (Schumach. & Thonn.) Hutch. & Dalziel Cas
Orobanche minor Sm. A
Parentucellia latifolia (L.) Caruel A
Parentucellia viscosa (L.) Caruel A
PAULOWNIACEAE A
Paulownia tomentosa (Thunb.) Steud. A
PHRYMACEAE
Glossostigma cleistanthum W.R.Barker F
Glossostigma diandrum (L.) Kuntze F E
Glossostigma elatinoides Benth. ex Hook.f. F
Mazus arenarius Heenan, P.N.Johnson & C.J.Webb E
Mazus novaezeelandiae W.R.Barker E
Mazus pumilio R.Br.
Mazus radicans (Hook.f.) Cheeseman E
Mazus reptans N.E.Br. Cas
Mimulus guttatus DC. F A
Mimulus luteus L. Cas
Mimulus moschatus Lindl. F A
Mimulus repens R.Br.
PLANTAGINACEAE
Antirrhinum majus L. A
Antirrhinum orontium L. A
Callitriche antarctica Engelm. ex Hegelm. F
Callitriche aucklandica R.Mason E
Callitriche hamulata W.D.J.Koch F A
Callitriche heterophylla Pursh F A
Callitriche muelleri Sond.
Callitriche petriei R.Mason F E
Callitriche stagnalis Scop. F A
Chaenorhinum minus (L.) Lange Cas
Chaenorhinum origanifolium (L.) Fourr. Cas
Cymbalaria muralis P.Gaertn.; B.Mey. & Scherb. A
Digitalis purpurea L. A
Gratiola nana Benth.
Gratiola pedunculata R.Br.
Gratiola pubescens R.Br.
Gratiola sexdentata R.Cunn. ex A.Cunn.
Kickxia elatine (L.) Dumort. A
Kickxia spuria (L.) Dumort. Cas
Linaria alpina (L.) Mill. Cas
Linaria arvensis (L.) Desf. A
Linaria genistifolia (L.) Mill. A
Linaria maroccana Hook.f. A
Linaria pelisseriana (L.) Mill. Cas
Linaria platycalyx Boiss. Cas
Linaria purpurea (L.) Mill. A
Linaria repens (L.) Mill. A
Linaria triornithophora (L.) Willd. Cas
Linaria vulgaris Mill. A
Lophospermum erubescens D.Don A
Maurandya barclaiana Lindl. Cas
Ourisia caespitosa Hook.f. E
Ourisia confertifolia Arroyo E
Ourisia crosbyi Cockayne E
Ourisia glandulosa Hook.f. E
Ourisia macrocarpa Hook.f. E
Ourisia macrophylla Hook. E
Ourisia modesta Diels E
Ourisia montana Buchanan E
Ourisia remotifolia Arroyo E
Ourisia sessilifolia Hook.f. E
Ourisia simpsonii (L.B.Moore) Arroyo E
Ourisia spathulata Arroyo E
Ourisia vulcanica L.B.Moore E
Plantago afra L. Cas
Plantago aristata Michx. A
Plantago aucklandica Hook.f. E
Plantago australis Lam. A
Plantago coronopus L. A
Plantago debilis R.Br. A
Plantago lanceolata L. A

Plantago lanigera Hook.f.
Plantago major L. A
Plantago obconica Sykes E
Plantago raoulii Decne. E
Plantago scabra Moench A
Plantago spathulata Hook.f. E
Plantago triandra Berggr. E
Plantago triantha Spreng.
Plantago unibracteata Rahn E
Veronica adamsii Cheeseman E
Veronica agrestis L. A
Veronica albicans Petrie E
Veronica americana Benth. A
Veronica amplexicaulis J.B.Armstr. E
Veronica anagallis-aquatica L. A
Veronica angustissima (Cockayne) Garn.-Jones E
Veronica annulata (Petrie) Cockayne ex Cheeseman E
Veronica arganthera (Garn.-Jones, Bayly, W.G.Lee & Rance) Garn.-Jones E
Veronica armstrongii Johnson ex J.B.Armstr. E
Veronica arvensis L. A
Veronica barkeri Cockayne E
Veronica baylyi Garn.-Jones E
Veronica benthamii Hook.f. E
Veronica biggarii Cockayne E
Veronica birleyi N.E.Br. E
Veronica bishopiana Petrie E
Veronica bollonsii Cockayne E
Veronica brachysiphon (Summerh.) Bean E
Veronica breviracemosa W.R.B.Oliv. E
Veronica buchananii Hook.f. E
Veronica calcicola (Bayly & Garn.-Jones) Garn.-Jones E
Veronica canterburiensis J.B.Armstr. E
Veronica catarractae G.Forst. E
Veronica catenata Pennell A
Veronica chamaedrys L. A
Veronica chathamica Buchanan E
Veronica cheesemanii Benth. E
Veronica chionohebe Garn.-Jones E
Veronica ciliolata (Hook.f.) Cheeseman
Veronica cockayneana Cheeseman E
Veronica colensoi Hook.f. E
Veronica colostylis Garn.-Jones E
Veronica corriganii (Carse) Garn.-Jones E
Veronica cryptomorpha (Bayly, Kellow, G.Harper & Garn.-Jones) Garn.-Jones E
Veronica cupressoides Hook.f. E
Veronica decora (Ashwin) Garn.-Jones E
Veronica decumbens J.B.Armstr. E
Veronica densifolia (F.Muell.) F.Muell.
Veronica dieffenbachii Benth. E
Veronica dilatata (G.Simpson & J.S.Thomson) Garn.-Jones E
Veronica diosmifolia A.Cunn. E
Veronica elliptica G.Forst.
Veronica epacridea Hook.f. E
Veronica evenosa Petrie E
Veronica filiformis Sm. A
Veronica flavida (Bayly, Kellow & de Lange) Garn.-Jones E
Veronica gibbsii Kirk E
Veronica glaucophylla Cockayne E
Veronica haastii Hook.f. E
Veronica hectorii Hook.f. E
Veronica hederifolia L. A
Veronica hookeri (Buchanan) Garn.-Jones E
Veronica hookeriana Walp. E
Veronica hulkeana F.Muell. E
Veronica insularis Cheeseman E
Veronica javanica Blume Cas
Veronica jovellanoides Garn.-Jones & de Lange E
Veronica kellowiae Garn.-Jones E
Veronica lanceolata Benth. E
Veronica lavaudiana Raoul E
Veronica leiophylla Cheeseman E
Veronica ligustrifolia R.Cunn ex A.Cunn. E
Veronica lilliputiana Stearn E
Veronica linifolia Hook.f. E
Veronica lyallii Hook.f. E
Veronica lycopodioides Hook.f. E
Veronica macrantha Hook.f. E
Veronica macrocalyx J.B.Armstr. E
Veronica macrocarpa Vahl E
Veronica masoniae (L.B.Moore) Garn.-Jones E
Veronica melanocaulon Garn.-Jones E
Veronica mooreae (Heads) Garn.-Jones E
Veronica murrellii (G.Simpson & J.S.Thomson) Garn.-Jones E
Veronica notialis Garn.-Jones E
Veronica obtusata Cheeseman E
Veronica ochracea (Ashwin) Garn.-Jones E
Veronica odora Hook.f. E
Veronica officinalis L. Cas
Veronica pareora (Garn.-Jones & Molloy) Garn.-Jones E
Veronica parviflora Vahl E
Veronica pauciramosa (Cockayne & Allan) Garn.-Jones E
Veronica pentasepala (L.B.Moore) Garn.-Jones E
Veronica perbella (de Lange) Garn.-Jones E
Veronica peregrina L. Cas
Veronica persica Poir. A
Veronica petriei (Buchanan) Kirk E
Veronica phormiiphila Garn.-Jones E
Veronica pimeleoides Hook.f. E
Veronica pinguifolia Hook.f. E
Veronica planopetiolata G.Simpson & J.S.Thomson E
Veronica plebeia R.Br.
Veronica polita Fr. A
Veronica poppelwellii Cockayne E
Veronica propinqua Cheeseman E
Veronica pubescens Benth. E
Veronica pulvinaris (Hook.f.) Cheeseman E
Veronica punicea Garn.-Jones E
Veronica quadrifaria Kirk E
Veronica rakaiensis J.B.Armstr. E
Veronica raoulii Hook.f. E
Veronica rigidula Cheeseman E
Veronica rivalis Garn.-Jones E
Veronica rupicola Cheeseman E
Veronica salicifolia G.Forst.
Veronica salicornioides Hook.f. E
Veronica scopulorum (Bayly. de Lange & Garn.-Jones) Garn.-Jones E
Veronica scrupea Garn.-Jones E
Veronica scutellata L. A
Veronica senex (Garn.-Jones) Garn.-Jones E
Veronica serpyllifolia L. A
Veronica simulans Garn.-Jones E
Veronica societatis (Bayly & Kellow) Garn.-Jones E
Veronica spathulata Benth. E
Veronica speciosa R.Cunn. ex A.Cunn.
Veronica spectabilis (Garn.-Jones) Garn.-Jones E
Veronica stenophylla Steudel E
Veronica stricta Banks & Sol. ex Benth. E
Veronica strictissima (Kirk) Garn.-Jones E
Veronica subalpina Cockayne E
Veronica subfulvida (G.Simpson & J.S.Thomson) Garn.-Jones E
Veronica tairawhiti (B.D.Clarkson & Garn.-Jones) Garn.-Jones E
Veronica tetragona Hook. E
Veronica tetrasticha Hook.f. E
Veronica thomsonii (Buchanan) Cheeseman E
Veronica topiaria (L.B.Moore) Garn.-Jones E
Veronica townsonii Cheeseman E
Veronica traversii Hook.f. E
Veronica treadwellii (Cockayne & Allan) Garn.-Jones E
Veronica trifida Petrie E
Veronica triphyllos L. A
Veronica truncatula Colenso E
Veronica tumida Kirk E
Veronica urvilleana (W.R.B.Oliv.) Garn.-Jones E
Veronica venustula Colenso E
Veronica verna L. A
Veronica vernicosa Hook.f. E
Veronica zygantha Garn.-Jones E
SCROPHULARIACEAE
Alonsoa meridionalis (L.f.) Kuntze Cas
Buddleja davidii Franch. A
Buddleja dysophylla (Benth.) Radlk. A
Buddleja globosa Hope A
Buddleja loricata Leeuwenberg Cas
Buddleja madagascariensis Lam. A
Buddleja salvifolia (L.) Lam. A
Eremophila debilis (Andrews) Chinnock A
Limosella curdieana F.Muell. Cas
Limosella lineata Glück
Myoporum insulare R.Br. A
Myoporum kermadecense Sykes E
Myoporum laetum G.Forst. E
Myoporum semotum Heenan & de Lange E
Nemesia floribunda Lehm. A
Nemesia strumosa Benth. A
Phygelius capensis E.Mey. ex Benth. A
Scrophularia auriculata L. A
Scrophularia grandiflora DC. Cas
Scrophularia nodosa L. A
Verbascum blattaria L. A
Verbascum creticum (L.) Cav. A
Verbascum phlomoides L. Cas
Verbascum pulverulentum Vill. Cas
Verbascum thapsus L. A
Verbascum virgatum Stokes A
STILBACEAE Cas
Halleria lucida L. Cas
TETRACHONDRACEAE
Tetrachondra hamiltonii Petrie ex Oliv. E
VERBENACEAE
Lantana camara L. A
Lantana montevidensis (Spreng.) Briq. A
Phyla nodiflora (L.) Greene A
Stachytarpheta ×adulterina Urb. & Ekman Cas
Verbena africana (R.Fern. & Verdc.) P.W.Michael A
Verbena bonariensis L. A
Verbena brasiliensis P.Vell. A
Verbena litoralis Kunth A
Verbena officinalis L. A
Verbena rigida Spreng. A
Verbena tenuisecta Briq. Cas

Order SOLANALES
CONVOLVULACEAE
Calystegia marginata R.Br.
Calystegia sepium (L.) R.Br. A
Calystegia silvatica (Kit.) Griseb. A
Calystegia soldanella (L.) R.Br.
Calystegia tuguriorum (G.Forst.) R.Br. ex Hook.f.
Convolvulus arvensis L. A
Convolvulus cneorum L. Cas
Convolvulus fractosaxosa Petrie E
Convolvulus graminetinus R.W.Johnson Cas
Convolvulus sabatius Viv. A
Convolvulus verecundus Allan E
Convolvulus waitaha (Sykes) Heenan, Molloy & de Lange E
Cuscuta campestris Yunck. A
Cuscuta epithymum (L.) L. A
Cuscuta europaea L. Cas
Cuscuta planiflora Ten. Cas
Cuscuta suaveolens Ser. Cas

Dichondra brevifolia Buchanan
Dichondra micrantha Urb. A
Dichondra repens J.R.Forst. & G.Forst.
Ipomoea alba L. A
Ipomoea batatas (L.) Lam. A
Ipomoea cairica (L.) Sweet
Ipomoea indica (Burman) Merr. A
Ipomoea pes-caprae (L.) Sweet
Ipomoea purpurea (L.) Roth A
Wilsonia backhousei Hook.f.
SOLANACEAE
Atropa bella-donna L. A
Brugmansia ×candida Pers. A
Brugmansia sanguinea (Ruiz & Pav.) D.Don A
Brugmansia suaveolens (Humb. & Bonpl. ex Willd.) Bercht. & J.Presl A
Brunfelsia pauciflora (Cham. & Schltdl.) Benth. Cas
Capsicum annuum L. Cas
Cestrum aurantiacum Lindl. A
Cestrum elegans (Brongn.) Schltdl. A
Cestrum fasciculatum (Schltdl.) Miers A
Cestrum nocturnum L. A
Cestrum parqui L'Hér. A
Datura ferox L. A
Datura stramonium L. A
Hyoscyamus niger L. A
Iochroma australe Griesb. Cas
Iochroma gesnerioides (Kunth) Miers Cas
Iochroma grandiflorum Benth. A
Lycium barbarum L. A
Lycium ferocissimum Miers A
Nicandra physalodes (L.) Gaertn. A
Nicotiana ×sanderae W.Watson Cas
Nicotiana alata Link & Otto A
Nicotiana glauca Graham A
Nicotiana langsdorffii Weinm. Cas
Nicotiana longiflora Cav. Cas
Nicotiana rustica L. A
Nicotiana sylvestris Speg. & Comes A
Nicotiana tabacum L. A
Nierembergia repens Ruiz & Pav. Cas
Petunia ×hybrida A.Vilm. A
Petunia parviflora Juss. Cas
Physalis peruviana L. A
Physalis philadelphica Lam. A
Physalis pubescens L. A
Salpichroa origanifolia (Lam.) Thell. A
Salpiglossis sinuata Ruiz & Pav. A
Schizanthus pinnatus Ruiz & Pav. A
Solandra maxima (Sesse & Moc.) P.S.Green Cas
Solanum ×procurrens A.C.Leslie Cas
Solanum betaceum Cav. A
Solanum carolinense L. A
Solanum chacoense Bitter Cas
Solanum chenopodioides Lam. A
Solanum crispum Ruiz & Pav. A
Solanum diflorum Vell. A
Solanum dulcamara L. A
Solanum furcatum Dunal A
Solanum laciniatum Aiton
Solanum laxum Spreng. A
Solanum linnaeanum Hepper & P.-M.L.Jaeger A
Solanum lycopersicum L. A
Solanum marginatum L.f. A
Solanum mauritianum Scop. A
Solanum melongena L. Cas
Solanum muricatum W.T.Aiton A
Solanum nigrum L. A
Solanum nodiflorum Jacq.
Solanum physalifolium Rusby A
Solanum pseudocapsicum L. A
Solanum rantonnei Carrière A
Solanum robustum J.C.Wendl. Cas
Solanum rostratum Dunal A
Solanum sisymbrifolium Lam. A
Solanum torvum Sweet Cas
Solanum tuberosum L. A
Solanum villosum Mill. A
Solanum wendlandii Hook.f. Cas
Vestia foetida (Ruiz & Pav.) Hoffmanns. Cas

Order APIALES
APIACEAE (UMBELLIFERAE)
Aciphylla anomala Allan E
Aciphylla aurea W.R.B.Oliv. E
Aciphylla cartilaginea Petrie E
Aciphylla colensoi Hook.f. E
Aciphylla congesta Cheeseman E
Aciphylla crenulata J.B.Armstr. E
Aciphylla crosby-smithii Petrie E
Aciphylla dieffenbachii (F.Muell.) Kirk E
Aciphylla dissecta (Kirk) W.R.B.Oliv. E
Aciphylla divisa (Cheeseman) Cheeseman E
Aciphylla dobsonii Hook.f. E
Aciphylla ferox W.R.B.Oliv. E
Aciphylla glaucescens W.R.B.Oliv. E
Aciphylla hectorii Buchanan E
Aciphylla hookeri Kirk E
Aciphylla horrida W.R.B.Oliv. E
Aciphylla indurata Cheeseman E
Aciphylla inermis W.R.B.Oliv. E
Aciphylla intermedia Petrie E
Aciphylla kirkii Buchanan E
Aciphylla lecomtei J.W.Dawson E
Aciphylla leighii Allan E
Aciphylla lyallii Hook.f. E
Aciphylla monroi Hook.f. E
Aciphylla montana Armstr. E
Aciphylla multisecta Cheeseman E
Aciphylla pinnatifida Petrie E
Aciphylla polita (Kirk) Cheeseman E
Aciphylla scott-thomsonii Cockayne & Allan E
Aciphylla similis Cheeseman E
Aciphylla simplex Petrie E
Aciphylla spedenii Cheeseman E
Aciphylla squarrosa J.R.Forst. & G.Forst. E
Aciphylla stannensis J.W.Dawson E
Aciphylla subflabellata W.R.B.Oliv. E
Aciphylla takahea W.R.B.Oliv. E
Aciphylla traillii Kirk E
Aciphylla traversii (F.Muell.) Hook.f. E
Aciphylla trifoliolata Petrie E
Aciphylla verticillata W.R.B.Oliv. E
Actinotus novae-zelandiae Petrie E
Aegopodium podagraria L. A
Ammi majus L. A
Anethum graveolens L. Cas
Angelica pachycarpa Lange A
Anisotome acutifolia (Kirk) Cockayne E
Anisotome antipoda Hook.f. E
Anisotome aromatica Hook.f. E
Anisotome brevistylis (Hook.f.) Poppelw. E
Anisotome capillifolia (Cheeseman) Cockayne E
Anisotome cauticola J.W.Dawson E
Anisotome deltoidea (Cheeseman) Cheeseman E
Anisotome filifolia (Hook.f.) Cockayne & Laing E
Anisotome flexuosa J.W.Dawson E
Anisotome haastii (F.Muell. ex Hook.f.) Cockayne & Laing E
Anisotome imbricata (Hook.f.) Cockayne E
Anisotome lanuginosa (Kirk) J.W.Dawson E
Anisotome latifolia Hook.f. E
Anisotome lyallii Hook.f. E
Anisotome pilifera (Hook.f.) Cockayne & Laing E
Anthriscus caucalis M.Bieb. A
Apium graveolens L. A
Apium nodiflorum (L.) Lag. A
Apium prostratum Labill. ex Vent.
Bowlesia tropaeolifolia Gillies & Hook. A
Bupleurum subovatum Spreng. A
Bupleurum tenuissimum L. A
Carum carvi L. Cas
Centella asiatica (L.) Urb. Cas
Centella uniflora (Colenso) Nannf.
Chaerophyllum basicola (Heenan & Molloy) K.F.Chung E
Chaerophyllum colensoi (Hook.f.) K.F.Chung
Chaerophyllum novae-zelandiae K.F.Chung
Chaerophyllum ramosum (Hook.f.) K.F.Chung
Chaerophyllum temulum L. A
Ciclospermum leptophyllum (Pers.) Sprague A
Conium maculatum L. A
Coriandrum sativum L. A
Cryptocarya obovata R.Br. Cas
Cryptotaenia canadensis (L.) DC. Cas
Daucus carota L. A
Daucus glochidiatus (Labill.) Fisch., C.A.Mey. & Avé-Lall.
Eryngium amethystinum L. Cas
Eryngium campestre L. A
Eryngium pandanifolium Cham. & Schltdl. A
Eryngium vesiculosum Labill.
Foeniculum vulgare Mill. A
Gingidia baxterae (J.W.Dawson) C.J.Webb E
Gingidia decipiens (Hook.f.) J.W.Dawson E
Gingidia enysii (Kirk) J.W.Dawson E
Gingidia flabellata (Kirk) J.W.Dawson E
Gingidia grisea Heenan E
Gingidia montana (J.R.Forst. & G.Forst.) J.W.Dawson
Gingidia trifoliolata (Hook.f.) J.W.Dawson E
Heracleum mantegazzianum Sommier & Levier A
Heracleum sphondylium L. A
Lignocarpa carnosula (Hook.f.) J.W.Dawson E
Lignocarpa diversifolia (Cheeseman) J.W.Dawson E
Lilaeopsis novae-zelandiae (Gand.) A.W.Hill
Lilaeopsis ruthiana Affolter
Melanoselinum decipiens (Schrad. & J.C.Wendl.) Hoffm. A
Oenanthe aquatica (L.) Poir. A
Oenanthe pimpinelloides L. A
Oenanthe sarmentosa DC. Cas
Orlaya grandiflora (L.) Hoffm. Cas
Pastinaca sativa L. A
Petroselinum crispum (Mill.) A.W.Hill A
Scandia geniculata (G.Forst.) J.W.Dawson E
Scandia rosifolia (Hook.f.) J.W.Dawson E
Scandix pecten-veneris L. Cas
Schizeilema allanii Cheeseman E
Schizeilema cockaynei (Diels) Cheeseman E
Schizeilema colensoi Domin E
Schizeilema exiguum (Hook.f.) Domin E
Schizeilema haastii (Hook.f.) Domin E
Schizeilema hydrocotyloides (Hook.f.) Domin E
Schizeilema nitens (Petrie) Domin F E
Schizeilema pallidum (Kirk) Domin E
Schizeilema reniforme (Hook.f.) Domin E
Schizeilema roughii (Hook.f.) Domin E
Schizeilema trifoliolatum (Hook.f.) Domin E
Sison amomum L. A
Stilbocarpa lyallii J.B.Armstr. E
Stilbocarpa polaris (Hombr. & Jacquinot) A.Gray
Stilbocarpa robusta (Kirk) Cockayne E
Torilis arvensis (Huds.) Link A
Torilis japonica (Houtt.) DC. A
Torilis nodosa (L.) Gaertn. A
ARALIACEAE
Aralia californica S.Watson Cas
Fatsia japonica (Thunb.) Decne. & Planch. A
Hedera colchica (K.Koch) K.Koch Cas
Hedera helix L. A
Hydrocotyle bowlesioides Mathias & Constance A
Hydrocotyle dissecta Hook.f. E
Hydrocotyle elongata A.Cunn. E
Hydrocotyle heteromeria A.Rich. E

Hydrocotyle hydrophila Petrie F E
Hydrocotyle leucocephala Cham. & Schltdl. Cas
Hydrocotyle microphylla A.Cunn. E
Hydrocotyle moschata G.Forst. E
Hydrocotyle novae-zeelandiae DC. E
Hydrocotyle pterocarpa F.Muell.
Hydrocotyle robusta Kirk E
Hydrocotyle sulcata C.J.Webb & P.N.Johnson F E
Hydrocotyle tripartita R.Br. ex A.Rich. A
Hydrocotyle umbellata L. Cas
Hydrocotyle verticillata Thunb. A
Meryta sinclairii (Hook.f.) Seem. E
Pseudopanax arboreus (Murray) Philipson E
Pseudopanax chathamicus Kirk E
Pseudopanax colensoi (Hook.f.) Philipson E
Pseudopanax crassifolius (Sol. ex A.Cunn.) K.Koch E
Pseudopanax discolor (Kirk) Harms E
Pseudopanax ferox Kirk E
Pseudopanax gilliesii Kirk E
Pseudopanax kermadecensis (W.R.B.Oliv.) Philipson E
Pseudopanax laetus (Kirk) Philipson E
Pseudopanax lessonii (DC.) K.Koch E
Pseudopanax linearis (Hook.f.) K.Koch E
Pseudopanax macintyrei (Cheeseman) Wardle E
Raukaua anomalus (Hook.) A.D.Mitch., Frodin & Heads E
Raukaua edgerleyi (Hook.f.) Seem. E
Raukaua simplex (G.Forst.) A.D.Mitch., Frodin & Heads E
Schefflera actinophylla (Endl.) Harms Cas
Schefflera arboricola (Hayata) Merr. Cas
Schefflera digitata J.R.Forst. & G.Forst. E
Tetrapanax papyrifer (Hook.) K.Koch A
Tieghemopanax sambucifolius R.Vig. Cas
GRISELINIACEAE
Griselinia littoralis Raoul E
Griselinia lucida G.Forst. E
PENNANTIACEAE
Pennantia baylisiana (W.R.B.Oliv.) G.T.S.Baylis E
Pennantia corymbosa J.R.Forst. & G.Forst. E
PITTOSPORACEAE
Hymenosporum flavum (Hook.) F.Muell. Cas
Pittosporum anomalum Laing & Gourlay E
Pittosporum colensoi Hook.f. E
Pittosporum cornifolium A.Cunn. E
Pittosporum crassicaule Laing & Gourlay E
Pittosporum crassifolium Banks & Sol. ex A.Cunn. E
Pittosporum dallii Cheeseman E
Pittosporum divaricatum Cockayne E
Pittosporum ellipticum Kirk E
Pittosporum eugenioides A.Cunn. E
Pittosporum fairchildii Cheeseman E
Pittosporum huttonianum Kirk E
Pittosporum kirkii Hook.f. ex Kirk E
Pittosporum lineare Laing & Gourlay E
Pittosporum obcordatum Raoul E
Pittosporum patulum Hook.f. E
Pittosporum pimeleoides A.Cunn. E
Pittosporum ralphii Kirk E
Pittosporum rigidum Hook.f. E
Pittosporum serpentinum (de Lange) de Lange E
Pittosporum tenuifolium Sol. ex Gaertn. E
Pittosporum turneri Petrie E
Pittosporum umbellatum Banks & Sol. ex Gaertn. E
Pittosporum undulatum Vent. A
Pittosporum virgatum Kirk E

Order AQUIFOLIALES A
AQUIFOLIACEAE A
Ilex aquifolium L. A

Order ASTERALES
ALSEUOSMIACEAE
Alseuosmia banksii A.Cunn. E
Alseuosmia macrophylla A.Cunn. E
Alseuosmia pusilla Colenso E
Alseuosmia quercifolia A.Cunn. E
Alseuosmia turneri R.O.Gardner E
ARGOPHYLLACEAE
Corokia buddleioides A.Cunn. E
Corokia cotoneaster Raoul E
Corokia macrocarpa Kirk E
ASTERACEAE (COMPOSITAE)
Abrotanella caespitosa Petrie ex Kirk E
Abrotanella fertilis Swenson E
Abrotanella inconspicua Hook.f. E
Abrotanella linearis Berggr. E
Abrotanella muscosa Kirk E
Abrotanella patearoa Heads E
Abrotanella pusilla (Hook.f.) Hook.f. E
Abrotanella rostrata Swenson E
Abrotanella rosulata (Hook.f.) Hook.f. E
Abrotanella spathulata (Hook.f.) Hook.f. E
Achillea filipendulina Lam. Cas
Achillea millefolium L. A
Achillea nobilis L. Cas
Achillea ptarmica L. A
Acroptilon repens (L.) DC. Cas
Aetheorhiza bulbosa (L.) Cass Cas
Ageratina adenophora (Spreng.) R.M.King & H.Rob. A
Ageratina riparia (Regel) R.M.King & H.Rob. A
Ageratum houstonianum Mill. A
Ambrosia artemisiifolia L. A
Ambrosia tenuifolia Spreng. Cas
Ammobium alatum R.Br. Cas
Anaphalioides alpina (Cockayne) Glenny E
Anaphalioides bellidioides (G.Forst.) Glenny E
Anaphalioides hookeri (Allan) Anderb. E
Anaphalioides subrigida (Colenso) Anderb. E
Anaphalioides trinervis (G.Forst.) Anderb. E
Anthemis arvensis L. A
Anthemis cotula L. A
Anthemis punctata Vahl Cas
Anthemis tinctoria L. Cas
Arctium lappa L. A
Arctium minus (Hill) Bernh. A
Arctotheca calendula (L.) Levyns A
Arctotis ×hybrida hort. Cas
Arctotis stoechadifolia P.J.Bergius A
Argyranthemum frutescens (L.) Sch.Bip. A
Argyrotegium mackayi (Buchanan) J.M.Ward & Breitw.
Argyrotegium nitidulum (Hook.f.) J.M.Ward & Breitw.
Artemisia absinthium L. A
Artemisia annua L. Cas
Artemisia arborescens L. A
Artemisia dracunculus L. Cas
Artemisia verlotiorum Lamotte A
Aster amellus L. Cas
Aster lanceolatus Willd. A
Aster novae-angliae L. A
Aster novi-belgii hybrids A
Aster subulatus Michx. A
Baccharis halimifolia L. A
Bartlettina sordida (Less.) R.M.King & H.Rob. A
Bellis perennis L. A
Bidens frondosa L. A
Bidens pilosa L. A
Bidens tripartita L. A
Brachyglottis adamsii (Cheeseman) B.Nord. E
Brachyglottis arborescens W.R.B.Oliv. E
Brachyglottis bellidioides (Hook.f.) B.Nord. E
Brachyglottis bidwillii (Hook.f.) B.Nord. E
Brachyglottis bifistulosa (Hook.f.) B.Nord. E
Brachyglottis buchananii (J.B.Armstr.) B.Nord. E
Brachyglottis cassinioides (Hook.f.) B.Nord. E
Brachyglottis cockaynei (G.Simpson & J.S.Thomson) B.Nord. E
Brachyglottis compacta (Kirk) B.Nord. E
Brachyglottis elaeagnifolia (Hook.f.) B.Nord. E
Brachyglottis greyi (Hook.f.) B.Nord. E
Brachyglottis haastii (Hook.f.) B.Nord. E
Brachyglottis hectorii (Buchanan) B.Nord. E
Brachyglottis huntii (F.Muell.) B.Nord. E
Brachyglottis kirkii (Kirk) C.J.Webb E
Brachyglottis lagopus (Raoul) B.Nord. E
Brachyglottis laxifolia (Buchanan) B.Nord. E
Brachyglottis monroi (Hook.f.) B.Nord. E
Brachyglottis myrianthos (Cheeseman) D.G.Drury E
Brachyglottis pentacopa (D.G.Drury) B.Nord. E
Brachyglottis perdicioides (Hook.f.) B.Nord. E
Brachyglottis repanda J.R.Forst. & G.Forst. E
Brachyglottis revoluta (Kirk) B.Nord. E
Brachyglottis rotundifolia J.R.Forst. & G.Forst. E
Brachyglottis saxifragoides (Hook.f.) B.Nord. E
Brachyglottis sciadophila (Raoul) B.Nord. E
Brachyglottis southlandica (Cockayne) B.Nord. E
Brachyglottis stewartiae (J.B.Armstr.) B.Nord. E
Brachyglottis traversii (F.Muell.) B.Nord. E
Brachyglottis turneri (Cheeseman) C.J.Webb E
Brachyscome humilis G.Simpson & J.S.Thomson E
Brachyscome iberidifolia Benth. Cas
Brachyscome linearis (Petrie) Druce E
Brachyscome longiscapa G.Simpson & J.S.Thomson E
Brachyscome montana G.Simpson E
Brachyscome perpusilla (Steetz) J.M.Black Cas
Brachyscome pinnata Hook.f. E
Brachyscome radicata Hook.f. E
Brachyscome sinclairii Hook.f. E
Calendula arvensis L. A
Calendula officinalis L. A
Callistephus chinensis (L.) Nees A
Calotis lappulacea Benth. A
Carduus acanthoides L. A
Carduus nutans L. A
Carduus pycnocephalus L. A
Carduus tenuiflorus Curtis A
Carthamus lanatus L. A
Carthamus tinctorius L. Cas
Cassinia aculeata (Labill.) R.Br. A
Celmisia adamsii Kirk E
Celmisia allanii W.Martin E
Celmisia alpina (Kirk) Cheeseman E
Celmisia angustifolia Cockayne E
Celmisia argentea Kirk E
Celmisia armstrongii Petrie E
Celmisia bellidioides Hook.f. E
Celmisia bonplandii (Buchanan) Allan E
Celmisia brevifolia Cockayne E
Celmisia cockayneana Petrie E
Celmisia cordatifolia Buchanan E
Celmisia coriacea (G.Forst.) Hook.f. E
Celmisia dallii Buchanan E
Celmisia densiflora Hook.f. E
Celmisia discolor Hook.f. E
Celmisia dubia Cheeseman E
Celmisia durietzii Cockayne & Allan E
Celmisia gibbsii Cheeseman E
Celmisia glabrescens Petrie E
Celmisia glandulosa Hook.f. E
Celmisia gracilenta Hook.f. E
Celmisia graminifolia Hook.f. E
Celmisia haastii Hook.f. E
Celmisia hectorii Hook.f. E
Celmisia hieraciifolia Hook.f. E
Celmisia holosericea (G.Forst.) Hook.f. E
Celmisia hookeri Cockayne E
Celmisia inaccessa Given E
Celmisia incana Hook.f. E
Celmisia insignis W.Martin E
Celmisia laricifolia Hook.f. E

Celmisia lateralis Buchanan E
Celmisia lindsayi Hook.f. E
Celmisia lyallii Hook.f. E
Celmisia mackaui Raoul E
Celmisia macmahonii Kirk E
Celmisia major Cheeseman E
Celmisia markii W.G.Lee & Given E
Celmisia monroi Hook.f. E
Celmisia morganii Cheeseman E
Celmisia parva Kirk E
Celmisia petriei Cheeseman E
Celmisia philocremna Given E
Celmisia polyvena G.Simpson & J.S.Thomson E
Celmisia prorepens Petrie E
Celmisia ramulosa Hook.f. E
Celmisia rupestris Cheeseman E
Celmisia rutlandii Kirk E
Celmisia semicordata Petrie
Celmisia sessiliflora Hook.f. E
Celmisia similis Given E
Celmisia sinclairii Hook.f. E
Celmisia spectabilis Hook.f. E
Celmisia spedenii G.Simpson E
Celmisia thomsonii Cheeseman E
Celmisia traversii Hook.f. E
Celmisia verbascifolia Hook.f. E
Celmisia vespertina Given E
Celmisia viscosa Hook.f. E
Celmisia walkeri Kirk E
Centaurea calcitrapa L. A
Centaurea cineraria L. A
Centaurea cyanus L. A
Centaurea jacea L. Cas
Centaurea maculosa Lam. A
Centaurea melitensis L. A
Centaurea montana L. Cas
Centaurea nigra L. A
Centaurea scabiosa L. A
Centaurea solstitialis L. A
Centipeda aotearoana N.G.Walsh E
Centipeda cunninghamii (DC.) A.Braun & Asch.
Centipeda elatinoides (Less.) Benth. & Hook. ex O.Hoffm.
Centipeda minima (L.) A.Braun & Asch.
Chamaemelum nobile (L.) All. A
Chondrilla juncea L. A
Chrysanthemoides monilifera (L.) Norl. A
Chrysanthemum ×grandiflorum Ramat. Cas
Chrysanthemum carinatum Schousb. Cas
Chrysanthemum coronarium L. A
Chrysanthemum segetum L. A
Cichorium intybus L. A
Cirsium arvense (L.) Scop. A
Cirsium brevistylum Cronquist A
Cirsium palustre (L.) Scop. A
Cirsium vulgare (Savi) Ten. A
Conyza bilbaoana J.Rémy A
Conyza bonariensis (L.) Cronquist A
Conyza canadensis (L.) Cronquist A
Conyza parva Cronquist A
Conyza sumatrensis (Retz.) E.H.Walker A
Coreopsis lanceolata L. A
Coreopsis tinctoria Nutt. A
Coreopsis verticillata L. Cas
Cosmos bipinnatus Cav. A
Cosmos sulphureus Cav. Cas
Cotula australis (Spreng.) Hook.f.
Cotula coronopifolia L.
Cotula turbinata L. Cas
Cotula vulgaris Levyns Cas
Craspedia incana Allan E
Craspedia lanata (Hook.f.) Allan E
Craspedia major (Hook.f.) Allan E
Craspedia minor (Hook.f.) Allan E
Craspedia robusta (Hook.f.) Cockayne E
Craspedia uniflora G.Forst. E
Crepis capillaris (L.) Wallr. A
Crepis foetida L. A
Crepis setosa Haller f. A
Crepis vesicaria L. A
Cyanthillium cinereum (L.) H.Rob. Cas
Cynara cardunculus L. A
Cynara scolymus L. A
Dahlia coccinea Cav. Cas
Dahlia excelsa Benth. A
Damnamenia vernicosa (Hook.f.) Given E
Delairea odorata Lem. A
Dittrichia graveolens (L.) Greuter A
Dolichoglottis lyallii (Hook.f.) B.Nord. E
Dolichoglottis scorzoneroides (Hook.f.) B.Nord. E
Doronicum plantagineum L. Cas
Echinops ritro L. Cas
Embergeria grandifolia (Kirk) Boulos E
Erechtites hieraciifolia (L.) Raf. ex DC. A
Erechtites valerianifolia (Link ex Spreng.) DC. A
Erigeron annuus (L.) Pers. A
Erigeron karvinskianus DC. A
Erigeron philadelphicus L. Cas
Euchiton audax (D.G.Drury) Holub E
Euchiton delicatus (D.G.Drury) Holub
Euchiton ensifer (D.G.Drury) Holub E
Euchiton involucratus (G.Forst.) Holub
Euchiton japonicus (Thunb.) Holub
Euchiton lateralis (C.J.Webb) Breitw. & J.M.Ward
Euchiton limosus (D.G.Drury) Holub E
Euchiton paludosus (Petrie) Holub E
Euchiton polylepis (D.G.Drury) Breitw. & J.M.Ward E
Euchiton ruahinicus (D.G.Drury) Breitw. & J.M.Ward E
Euchiton sphaericus (Willd.) Holub
Euchiton traversii (Hook.f.) Holub
Eupatorium cannabinum L. A
Euryops abrotanifolius (L.) DC. Cas
Euryops chrysanthemoides (DC.) B.Nord. Cas
Euryops pectinatus (L.) Cass. Cas
Euryops virgineus Less. Cas
Eutrochium purpureum (L.) E.E.Lamont A
Ewartia sinclairii (Hook.f.) Cheeseman E
Facelis retusa (Lam.) Sch.Bip. A
Felicia amelloides (L.) Voss Cas
Felicia fruticosa (L.) H.Nicholson A
Felicia petiolata (Harv.) N.E.Br. Cas
Filago pyramidata L. Cas
Filago vulgaris Lam. Cas
Gaillardia ×grandiflora Van Houtte A
Galinsoga parviflora Cav. A
Galinsoga quadriradiata Ruiz & Pav. A
Gamochaeta americana (Mill.) Wedd. A
Gamochaeta calviceps (Fernald) Cabrera A
Gamochaeta coarctata (Willd.) Kerg. A
Gamochaeta pensylvanica (Willd.) Cabrera A
Gamochaeta purpurea (L.) Cabrera A
Gamochaeta simplicicaulis (Willd. ex Spreng.) Cabrera A
Gamochaeta subfalcata (Cabrera) Cabrera A
Gazania linearis (Thunb.) Druce A
Gazania rigens (L.) Gaertn. A
Gerbera jamesonii Adlam Cas
Gnaphalium uliginosum L. Cas
Guizotia abyssinica (L.f.) Cass. Cas
Gymnocoronis spilanthoides (Don) DC. F A
Haastia pulvinaris Hook.f. E
Haastia recurva Hook.f. E
Haastia sinclairii Hook.f. E
Helenium autumnale L. Cas
Helenium puberulum DC. A
Helianthus annuus L. A
Helianthus salicifolius A.Dietr. A
Helianthus tuberosus L. A
Helichrysum argyrophyllum (DC.) N.A.Wakef. Cas
Helichrysum coralloides (Hook.f.) Benth. & Hook.f. E
Helichrysum cymosum (L.) D.Don Cas
Helichrysum depressum (Hook.f.) Benth. & Hook.f. E
Helichrysum dimorphum Cockayne E
Helichrysum filicaule Hook.f. E
Helichrysum intermedium G.Simpson E
Helichrysum lanceolatum (Buchanan) Kirk E
Helichrysum parvifolium Yeo E
Helichrysum petiolare Hilliard & B.L.Burtt A
Helichrysum plumeum Allan E
Helminthotheca echioides (L.) Holub A
Hieracium argillaceum group A
Hieracium lepidulum (Stenstr.) Omang A
Hieracium murorum L. A
Hieracium pollichiae Sch.Bip. A
Hieracium sabaudum L. A
Hypochaeris glabra L. A
Hypochaeris radicata L. A
Hypochaeris tweediei (Hook. & Arn.) Cabrera Cas
Inula conyzae (Griess.) Meikle A
Inula helenium L. A
Inula orientalis Lam. Cas
Iva xanthiifolia Nutt. Cas
Jacobaea aquatica (Hill) P.Gaertn., B.Mey. & Scherb. A
Jacobaea maritima (L.) Pelser & Meijden A
Jacobaea vulgaris Gaertn. A
Kirkianella novae-zelandiae (Hook.f.) Allan E
Lactuca saligna L. A
Lactuca sativa L. A
Lactuca serriola L. A
Lactuca virosa L. A
Lagenifera barkeri Kirk
Lagenifera cuneata Petrie E
Lagenifera lanata A.Cunn. E
Lagenifera montana Hook.f.
Lagenifera petiolata Hook.f. E
Lagenifera pinnatifida Hook.f. E
Lagenifera pumila (G.Forst.) Cheeseman E
Lagenifera stipitata (Labill.) Druce
Lagenifera strangulata Colenso E
Lapsana communis L. A
Leontodon autumnalis L. A
Leontodon taraxacoides (Vill.) Mérat A
Leptinella albida (D.G.Lloyd) D.G.Lloyd & C.J.Webb E
Leptinella atrata (Hook.f.) D.G.Lloyd & C.J.Webb E
Leptinella calcarea (D.G.Lloyd) D.G.Lloyd & C.J.Webb E
Leptinella conjuncta Heenan E
Leptinella dendyi (Cockayne) D.G.Lloyd & C.J.Webb E
Leptinella dioica Hook.f. E
Leptinella dispersa (D.G.Lloyd) D.G.Lloyd & C.J.Webb E
Leptinella featherstonii F.Muell. E
Leptinella filiformis (Hook.f.) D.G.Lloyd & C.J.Webb E
Leptinella goyenii (Petrie) D.G.Lloyd & C.J.Webb E
Leptinella intermedia (D.G.Lloyd) D.G.Lloyd & C.J.Webb E
Leptinella lanata Hook.f. E
Leptinella maniototo (Petrie) D.G.Lloyd & C.J.Webb E
Leptinella minor Hook.f. E
Leptinella nana (D.G.Lloyd) D.G.Lloyd & C.J.Webb E
Leptinella pectinata (Hook.f.) D.G.Lloyd & C.J.Webb E
Leptinella plumosa Hook.f. E
Leptinella potentillina F.Muell. E
Leptinella pusilla Hook.f. E

Leptinella pyrethrifolia (Hook.f.) D.G.Lloyd & C.J.Webb E
Leptinella rotundata (Cheeseman) D.G.Lloyd & C.J.Webb E
Leptinella serrulata (D.G.Lloyd) D.G.Lloyd & C.J.Webb E
Leptinella squalida Hook.f. E
Leptinella tenella (A.Cunn.) D.G.Lloyd & C.J.Webb E
Leptinella traillii (Kirk) D.G.Lloyd & C.J.Webb E
Leucanthemum ircutianum DC. Cas
Leucanthemum maximum (Ramond) DC. A
Leucanthemum vulgare Lam. A
Leucogenes grandiceps (Hook.f.) Beauverd E
Leucogenes leontopodium (Hook.f.) Beauverd E
Leucogenes neglecta Molloy E
Leucogenes tarahaoa Molloy E
Liatris spicata (L.) Willd. Cas
Ligularia clivorum Maxim. Cas
Logfia gallica (L.) Coss. & Germ. A
Logfia minima (Sm.) Dumort. A
Madia capitata Nutt. Cas
Madia sativa Molina A
Matricaria discoidea DC. A
Matricaria recutita L. Cas
Mauranthemum paludosum (Poir.) Vogt & Oberpr. A
Microseris scapigera (Sol. ex A.Cunn.) Sch.Bip.
Mycelis muralis (L.) Dumort. A
Olearia adenocarpa Molloy & Heenan E
Olearia albida (Hook.f.) Hook.f. E
Olearia allomii Kirk E
Olearia angustifolia Hook.f. E
Olearia arborescens (G.Forst.) Cockayne & Laing E
Olearia argophylla (Labill.) Benth. Cas
Olearia avicenniifolia (Raoul) Hook.f. E
Olearia bullata H.D.Wilson & Garn.-Jones E
Olearia chathamica Kirk E
Olearia cheesemanii Cockayne & Allan E
Olearia colensoi Hook.f. E
Olearia coriacea Kirk E
Olearia crebra E.K.Cameron & Heenan E
Olearia crosby-smithiana Petrie E
Olearia cymbifolia (Hook.f.) Cheeseman E
Olearia fimbriata Heads E
Olearia fragrantissima Petrie E
Olearia furfuracea (A.Rich.) Hook.f. E
Olearia gardneri Heads E
Olearia hectorii Hook.f. E
Olearia ilicifolia Hook.f. E
Olearia lacunosa Hook.f. E
Olearia laxiflora Kirk E
Olearia lineata (Kirk) Cockayne E
Olearia lirata (Sims) Hutch. Cas
Olearia lyallii Hook.f. E
Olearia moschata Hook.f. E
Olearia nummulariifolia (Hook.f.) Hook.f. E
Olearia odorata Petrie E
Olearia oporina (G.Forst.) Hook.f. E
Olearia pachyphylla Cheeseman E
Olearia paniculata (J.R.Forst. & G.Forst.) Druce E
Olearia phlogopappa (Labill.) DC. Cas
Olearia polita H.D.Wilson & Garn.-Jones E
Olearia quinquevulnera Heenan E
Olearia rani (A.Cunn.) Druce E
Olearia semidentata Decne. E
Olearia solandri (Hook.f.) Hook.f. E
Olearia telmatica Heenan & de Lange E
Olearia thomsonii Cheeseman E
Olearia townsonii Cheeseman E
Olearia traversiorum (F.Muell.) Hook.f. E
Olearia virgata (Hook.f.) Hook.f. E
Onopordum acanthium L. A
Onopordum tauricum Willd. Cas
Osteospermum fruticosum (L.) Norl. A
Osteospermum jucundum (E.Phillips) Norl. A
Othonna capensis L.H.Bailey A
Ozothamnus leptophyllus (G.Forst.) Breitw. & J.M.Ward E
Pachystegia insignis (Hook.f.) Cheeseman E
Pachystegia minor (Cheeseman) Molloy E
Pachystegia rufa Molloy E
Pericallis ×hybrida B.Nord. A
Petasites fragrans (Vill.) C.Presl A
Picris angustifolia DC.
Picris burbidgeae S.Holzapfel
Picris hieracioides L. Cas
Pilosella aurantiaca (L.) F.W.Schultz & Sch.Bip. A
Pilosella caespitosa (Dumort.) P.D.Sell & C.West A
Pilosella officinarum Vaill. A
Pilosella piloselloides (Vill.) Sojak A
Pleurophyllum criniferum Hook.f. E
Pleurophyllum hookeri Buchanan E
Pleurophyllum speciosum Hook.f. E
Pseudognaphalium luteoalbum (L.) Hilliard & B.L.Burtt
Ptilostemon afer (Jacq.) Greuter A
Pulicaria dysenterica (L.) Bernh. A
Rachelia glaria J.M.Ward & Breitw. E
Raoulia albosericea Colenso E
Raoulia apicinigra Kirk E
Raoulia australis Hook.f. ex Raoul E
Raoulia beauverdii Cockayne E
Raoulia bryoides Hook.f. E
Raoulia buchananii Kirk E
Raoulia cinerea Petrie E
Raoulia eximia Hook.f. E
Raoulia glabra Hook.f. E
Raoulia goyenii Kirk E
Raoulia grandiflora Hook.f. E
Raoulia haastii Hook.f. E
Raoulia hectorii Hook.f. E
Raoulia hookeri Allan E
Raoulia mammillaris Hook.f. E
Raoulia monroi Hook.f. E
Raoulia parkii Buchanan E
Raoulia petriensis Kirk E
Raoulia rubra Buchanan E
Raoulia subsericea Hook.f. E
Raoulia subulata Hook.f. E
Raoulia tenuicaulis Hook.f. E
Raoulia youngii (Hook.f.) Beauverd E
Roldana petasitis (Sims) H.Robinson & Brettell A
Rudbeckia fulgida Aiton Cas
Rudbeckia hirta L. Cas
Rudbeckia laciniata L. A
Santolina chamaecyparissus L. Cas
Senecio angulatus L.f. A
Senecio australis Willd.
Senecio banksii Hook.f. E
Senecio bipinnatisectus Belcher A
Senecio biserratus Belcher
Senecio carnosulus (Kirk) C.J.Webb E
Senecio crassiflorus (Poir.) DC. Cas
Senecio diaschides D.G.Drury A
Senecio dunedinensis Belcher E
Senecio elegans L. A
Senecio esleri C.J.Webb A
Senecio glastifolius L.f. A
Senecio glaucophyllus Cheeseman E
Senecio glomeratus Poir.
Senecio hauwai Sykes E
Senecio hispidulus A.Rich.
Senecio hypoleucus Benth. Cas
Senecio kermadecensis Belcher E
Senecio lautus G.Forst. ex Willd.
Senecio linearifolius A.Rich. A
Senecio macroglossus DC. Cas
Senecio marotiri C.J.Webb E
Senecio minimus Poir.
Senecio quadridentatus Labill.
Senecio radiolatus F.Muell. E
Senecio repangae de Lange & B.G.Murray E
Senecio rufiglandulosus Colenso E
Senecio scaberulus (Hook.f.) D.G.Drury E
Senecio serpens G.D.Rowley Cas
Senecio skirrhodon DC. A
Senecio sterquilinus Ornduff E
Senecio sylvaticus L. A
Senecio vulgaris L. A
Senecio wairauensis Belcher E
Sigesbeckia orientalis L. A
Silybum marianum (L.) Gaertn. A
Solenogyne dominii L.G.Adams A
Solenogyne gunnii (Hook.f.) Cabrera A
Solenogyne mikadoi (Koidz.) Koidz. Cas
Solidago canadensis L. A
Solidago gigantea Aiton Cas
Solidago virgaurea L. Cas
Soliva anthemifolia (Juss.) Sweet A
Soliva sessilis Ruiz & Pav. A
Sonchus arvensis L. A
Sonchus asper (L.) Hill A
Sonchus kirkii Hamlin E
Sonchus oleraceus L. A
Stuartina muelleri Sond. A
Tagetes erecta L. A
Tagetes minuta L. A
Tagetes patula L. A
Tagetes tenuifolia Cav. Cas
Tanacetum parthenium (L.) Sch.Bip. A
Tanacetum vulgare L. A
Taraxacum hamatum Raunk. A
Taraxacum insigne Ekman ex Wiinst. & K.Jess. Cas
Taraxacum lambinonii Soest A
Taraxacum magellanicum Sch.Bip.
Taraxacum officinale F.H.Wigg. A
Tithonia rotundifolia (Mill.) S.F.Blake Cas
Tolpis barbata (L.) Gaertn. A
Tragopogon dubius Scop. A
Tragopogon porrifolius L. A
Tragopogon pratensis L. A
Traversia baccharoides Hook.f. E
Tripleurospermum inodorum Sch.Bip. A
Tussilago farfara L. A
Ursinia anthemoides Gaertn. Cas
Vellereophyton dealbatum (Thunb.) Hilliard & B.L.Burtt A
Vittadinia australis A.Rich. E
Vittadinia cuneata DC. A
Vittadinia dissecta (Benth.) N.T.Burb. Cas
Vittadinia gracilis (Hook.f.) N.T.Burb. A
Vittadinia muelleri N.T.Burb. Cas
Xanthium spinosum L. A
Xanthium strumarium L. A
Xerochrysum bracteatum (Vent.) Tzvelev A
CALYCERACEAE Cas
Acicarpha tribuloides Juss. Cas
CAMPANULACEAE
Adenophora triphylla (Thunb.) A.DC. Cas
Campanula medium L. A
Campanula portenschlagiana Roem. & Schult. Cas
Campanula poscharskyana Degen A
Campanula rapunculoides L. A
Campanula rapunculus L. A
Campanula rotundifolia L. A Cas
Campanula versicolor Andrews
Colensoa physaloides (A.Cunn.) Hook.f. E
Isotoma fluviatilis (R.Br.) F.Muell. ex Benth.
Jasione montana L. Cas
Legousia speculum-veneris (L.) Chaix. Cas
Lobelia ×gerardii Sauv. Cas
Lobelia anceps L.f.
Lobelia angulata G.Forst. E
Lobelia arenaria (Hook.f.) Heenan & de Lange E
Lobelia benthamii F.Muell. Cas

Lobelia carens Heenan E
Lobelia erinus L. A
Lobelia fatiscens Heenan E
Lobelia fugax Heenan, S.P.Courtney & P.N.Johnson E
Lobelia glaberrima Heenan E
Lobelia ionantha Heenan E
Lobelia linnaeoides (Hook.f.) Petrie E
Lobelia macrodon (Hook.f.) Lammers E
Lobelia pedunculata R.Br. A
Lobelia perpusilla Hook.f. E
Lobelia roughii Hook.f. E
Lobelia siphilitica L. Cas
Pratia purpurascens (R.Br.) E.Wimm. A
Solenopsis laurentia (L.) C.Presl A
Trachelium caeruleum L. Cas
Wahlenbergia akaroa J.A.Petterson E
Wahlenbergia albomarginata Hook. E
Wahlenbergia cartilaginea Hook.f. E
Wahlenbergia congesta (Cheeseman) N.E.Br. E
Wahlenbergia littoricola P.J.Sm.
Wahlenbergia matthewsii Cockayne E
Wahlenbergia planiflora P.J.Smith A
Wahlenbergia pygmaea Colenso E
Wahlenbergia ramosa G.Simpson E
Wahlenbergia rupestris G.Simpson E
Wahlenbergia stricta (R.Br.) Sweet Cas
Wahlenbergia violacea J.A.Petterson E
GOODENIACEAE
Scaevola gracilis Hook.f.
Selliera microphylla Colenso E
Selliera radicans Cav. E
Selliera rotundifolia Heenan E
MENYANTHACEAE
Liparophyllum gunnii Hook.f. F
Menyanthes trifoliata L. F Cas
Nymphoides geminata (R.Br.) Kuntze F A
Nymphoides peltata (S.G.Gmel.) Kuntze F Cas
ROUSSEACEAE
Carpodetus serratus J.R.Forst. & G.Forst. E
STYLIDIACEAE
Donatia novae-zelandiae Hook.f.
Forstera cristis Glenny & Courtney E
Forstera mackayi Allan E
Forstera purpurata Glenny E
Forstera sedifolia G.Forst. E
Forstera tenella Hook.f. E
Oreostylidium subulatum (Hook.f.) Berggr. E
Phyllachne clavigera (Hook.f.) F.Muell. E
Phyllachne colensoi (Hook.f.) Berggr.
Phyllachne rubra (Hook.f.) Cheeseman E

Order DIPSACALES A
ADOXACEAE A
Sambucus nigra L. A
Sambucus pubens Michx. A
Viburnum japonicum Spreng. Cas
Viburnum opulus L. Cas
Viburnum plicatum Thunb. A
Viburnum tinus L. A
CAPRIFOLIACEAE A
Abelia ×grandiflora (André) Rehder Cas
Centranthus macrosiphon Boiss. Cas
Centranthus ruber (L.) DC. A
Dipsacus fullonum L. A
Dipsacus sativus (L.) Honck. Cas
Kolkwitzia amabilis Graebn. A
Leycesteria formosa Wall. A
Lonicera ×americana (Mill.) K.Koch A
Lonicera japonica Thunb. A
Lonicera nitida E.H.Wilson A
Lonicera periclymenum L. A
Pterocephalus lasiospermus Link Cas
Scabiosa anthemifolia Eckl. & Zeyh. Cas
Scabiosa atropurpurea L. A
Scabiosa caucasica M.Bieb. Cas
Symphoricarpos albus (L.) S.F.Blake A
Symphoricarpos orbiculatus Moench A
Valerianella carinata Loisel. A
Valerianella locusta (L.) Laterr. A
Weigela florida (Bunge) A.DC. Cas

Order ESCALLONIALES A
ESCALLONIACEAE A
Escallonia bifida Link & Otto A
Escallonia rubra (Ruiz & Pav.) Pers. A

Order PARACRYPHIALES
PARACRYPHIACEAE
Quintinia acutifolia Kirk E
Quintinia elliptica Hook.f. E
Quintinia serrata A.Cunn. E

TWENTY-SIX

Plant MACROFOSSILS

MICHAEL S. POLE

Araucaria sp. (Araucariaceae) (Bannockburn, Early Miocene). The species shows shoot and seed similarities with *A. cunninghamii* and *A. heterophylla* (Norfolk pine) although the form of the epidermal cells suggests some difference from both those and related living species.

Michael Pole

With the first decade of the new millennium and the International Year of Biodiversity (2010) behind us, and 2011–2010 established as the Decade of Biodiversity, this is a good time to take stock of Earth's biodiversity. But with so many living species being lost on a daily basis, why study fossil plants? It is often said 'the present is the key to the past', but paleontologists will point out that the past is also the key to the future. For instance, using fossils to understand how vegetation reacted to changes in boundary conditions in the past (e.g. changes in carbon dioxide levels) is the only way we have to check the accuracy of models for predicting change into the future. As a basis for prediction, a thorough understanding of fossil floras and vegetation is particularly valuable.

Compared to most countries, New Zealand has a potentially rather complete record of its terrestrial vegetation over the last 80 million years or so. This is a result of New Zealand's being a geologically young and active landscape with an unbroken sedimentary record during this period. The potential goldmine of floristic data, however, has scarcely been touched. To make the most of what we have will require focusing on at least two aspects of the science, one being techniques of recovering the fossils, and the other on protocols for identification.

The commonest plant macrofossils in collections throughout New Zealand consist of leaf impressions. In this mode, the outline of the leaf may be visible, along with varying details of leaf venation. But many leaves simply look like countless others. Identifying this kind of material is equivalent to visiting a living rainforest, scooping up a bag of leaf litter, and identifying the leaves within it. In reasonably low-diversity vegetation like that of New Zealand today, this is not a great problem. But a closer analogue for much of New Zealand's Cenozoic era would be southern-hemisphere mesothermal forests, and in this case, even when you know what trees are growing in the region, identifying isolated leaves (or leaflets) is difficult. For New Zealand's Cenozoic fossil record, there is no known list of plants against which to compare the fossils, but it is probably reasonable to consider plants now growing in New Zealand, Australia, New Caledonia, New Guinea, and Patagonia as candidates. Of course, many of the plants preserved as fossils may represent extinct endemic genera that have no close living relatives. The point here is that identification of fossil leaves requires much ground-work to be done on the morphology of living plants. The data retrieved from such a study must be ordered in some taxonomically meaningful way. The identification of fossil plant fragments (like leaves) with extant taxa must be considered as a hypothesis, and the characters used clearly stated and referenced back to the database of living plants. Only in this way can one be confident of identification.

In the past, plant macrofossil collection in New Zealand has involved the hand collection of leaves or fruits by eye. However, with appropriate processing, carbonaceous mud, or even coal, which may show little to the eye in the field, will provide a wealth of organically preserved plant fragments. These have well-preserved cellular detail, typically the morphology of the epidermis as reflected in the cuticle (Holden 1987; Kovar et al. 1987; Pole 1996). This level of preservation provides more characters than a simple surface impression. Once again, a database and retrieval system is required that emphasises these characters in extant taxa.

To classify plant fossils, at least two systems are in use that, unfortunately, look very similar – the Linnaean system and morphotaxa (McNeill et al. 2007). The latter essentially comprise what used to be called organ- or form-genera). An example of a morphotaxon is *Laurophyllum*. This name may be used if the identity of a fossil is known to be Lauraceae, but the actual genus is unknown. An extreme morphotaxon is *Phyllites*, a rubbish-bin taxon for almost any angiosperm leaf fragment. Because both systems involve Latin binominals, they can be confused, and in texts and lists may be treated as equals, which they are not. In the form-generic system, the taxon is only differentially diagnosed from other taxa in that system. For instance, the conifer form-genus *Podozamites* is used for multi-veined conifer leaves. The genus is not comparable with the Linnaean genus *Agathis*, because *Podozamites* cannot be diagnosed from *Agathis*. One cannot talk about *Agathis* and *Podozamites* in the same locality in the biodiversity sense of two genera. One cannot talk about *Podozamites* becoming extinct in a global sense because *Agathis* in the fossil record could be assigned to *Podozamites*.

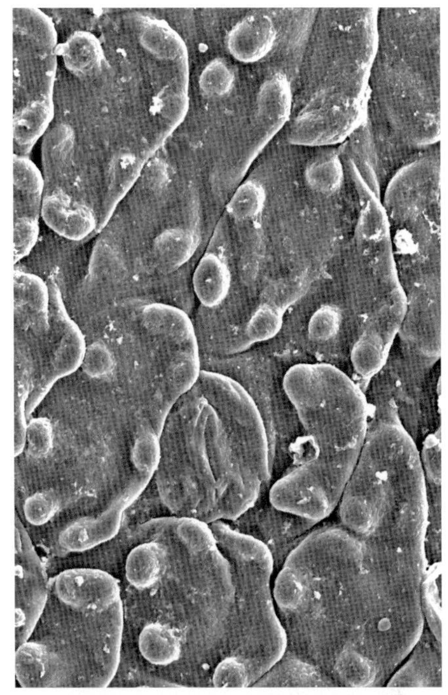

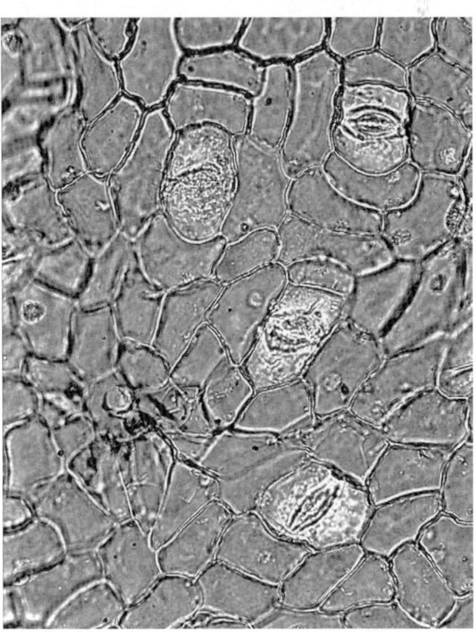

Preserved leaf cuticles, showing epidermal cells and one or more stomata of two species of Lauraceae as visualised by scanning electron microscopy and light microscopy. Upper, an undescribed species (Mangonui, Early–Middle Miocene); lower, *Endiandra* sp. (Blue Lake, Early Miocene).

Michael Pole

New Zealand plant-macrofossil diversity

New Zealand's plant-macrofossil record was listed thoroughly by Mildenhall (1970), who separated valid names from the many nomina nuda. The update presented here builds on Mildenhall's work. Many of the valid species names, while correct in the sense of nomenclature, are far from reliable in any taxonomic sense. Most descriptions of plant fossils before Mildenhall's review do not include any reasonable discussion supporting the identification – that is, diagnostic characters are not stated. Indeed, many of the fossils are so poorly preserved that there are probably not enough characters ever to identify them. Many of Ettingshausen's (1887, 1891), Penseler's (1930), and Unger's (1864) records fall in this category. Most of these fossils are of mere historical interest and even the whereabouts of some of the early type material is unknown. Gregg (1975) listed the Ettingshausen specimens held at the Canterbury Museum and located some specimens in the Otago Museum, but also noted some illustrated specimens that could not be found. He also reviewed Ettingshausen's localities and cited the latest opinions on their ages. Raine (1982) reported on the present whereabouts of some of Arber's (1917) material and cleared up some problems concerning their provenance. A review of plant fossils in New Zealand was published by Raine and Pocknall (1983) and Cieraad and Lee (2006) collated a database on New Zealand fossil fern records.

The summary below and the end-chapter checklist (listing specimens identified to genus level or lower) cover the published record of New Zealand plant macrofossils, a term that here includes fragments of plants other than their pollen and spores (see the next chapter). This category includes a size fraction that some workers call 'mesofossils'. In conventional terminology, 'plant' macrofossils may include algae and fungi. The few that are known from New Zealand are mentioned in other chapters in this volume, but, in passing, these include the following – a Triassic marine codiacean alga, *Shonabellia verrucosa* (Retallack 1983b); a range of stromatolite morphologies from Miocene lacustrine sediments (Douglas 1987; Lindqvist 1994); several kinds of fungal remains (see page 508); a Late Triassic liverwort (form-genus *Marchantites*), which was a common ground cover at a Southland locality (Pole & Raine 1994) and similar forms were reported from the Early Cretaceous of the Clarence River (Parrish et al. 1998).

The present review does not extend to the late Pleistocene or Holocene where all records are thought to be of extant taxa. Suffice it to say that macrofossils have been documented from these periods by authors including Burrows (1995, 1997), Burrows and Russell (1990), Burrows et al. (1993), Newnham and Lusk (1990), Lintott and Burrows (1973), and Simpson and Burrows (1978), and Kennedy et al. (2008). The macrofossil remains of forest destroyed by the Taupo eruption (AD 232, Sparks et al. 1995) were documented by Clarkson et al. (1988). There are at least six unpublished or partly published important theses dealing with plant macrofossils (Mildenhall 1968; Holden 1983; Daniel 1989; Broekhuisen 1984, Kennedy 1993, 1998). This review does not include species that are described only in theses. The end-chapter bibliography attempts to cover all publications relevant to a summary of New Zealand's plant macrofossil record, in which identification has been made to family level or below, and includes references to works not cited in the text.

Horsetails

The oldest sphenopsids (Apocalamitaceae) reported in New Zealand are Permian *Equisetites* sp. (McQueen 1954a,b) and Triassic *Neocalamites carrerei* (Retallack 1980). The group was also present in the Jurassic (Arber 1917) and Cretaceous (Parrish et al. 1998).

Ferns

Fossil Osmundaceae were first identified in New Zealand on the basis of the silicified 'trunks' *Osmundites dunlopi* and *O. gibbiana* (Kidston & Gwynne-Vaughan 1907, 1908, 1909, 1910) with a third species, *O. aucklandica*, being added by Marshall (1926). Miller (1967) considered the genus invalid and transferred the species to a new genus, *Osmundacaulis*, and at the same time synonymised *O. aucklandica* with *O. dunlopi*. Tidwell (1994) placed the two remaining species into separate genera, hence *Millerocaulis dunlopii* and *Ashicaulis gibbiana*. Most of the New Zealand species of the sterile-frond form-genus *Cladophlebis* probably belong in the Osmundaceae. Retallack (1983b) pointed out that *Cladophlebis australis*, a name commonly applied to New Zealand Jurassic specimens, should be restricted to the Triassic, and the Jurassic specimens were therefore of an unnamed species. Fertile fronds of *Cladophlebis* form are placed in *Todites* and clearly belong in the Osmundaceae. *Todites* has been found in New Zealand's Triassic, but not the Jurassic.

Dipteridaceae is known in the Jurassic from the frond-genera *Dictyophyllum* (Arber 1917) and *Goepertella* (Rees 1993). Rozefelds et al. (1992) confirmed the presence of *Lygodium* sp. (Schizaeaceae) from New Zealand based on both fronds and sporangia. Pole (1997a) noted that their age was probably Paleocene rather than the Eocene recorded.

Pole (1992b) described *Pneumatopteris* sp. (Thelypteridaceae) and *Blechnum* cf. *procerum* (Blechnaceae) from the Early Miocene of Central Otago. [There are several errors in his paper – figures 3 and 4 are misplaced with respect to their captions; the caption of fig. 3 should refer to a, b, and c, not c, d, and e, while the formal description of *Pneumatopteris* should read *Pneumatopteris* sp. cf. *P. pennigera*. See also Collinson (2001).] As stated in the text, there is overlap in morphology with the living species, but the shape of the pinnae apices differs.

Fertile specimens of *Davallia walkeri* are known from the Early Miocene (Conran et al. 2010).

Cycads

Small frond fragments from the Late Cretaceous have been described as *Macrozamia*, a genus extant in Australia, and the extinct *Pterostoma*. *Pterostoma* is the only cycad known in New Zealand's Cenozoic (Hill & Pole 1994). It arguably belongs in the Zamiaceae. The 'possible cycad' described by Pole (1992d), is almost certainly a specimen of this genus.

Macrofossil taxon diversity

Taxon/ Group	Described macrofossil taxa	Unnamed macrofossil taxa (species + subspecies)
Chlorophyta	1	0
Bryophyta	1	0
Tracheophyta	264	93
Lycopsida	1	2
Psilotopsida	0	0
Equisetopsida	6	3
'Filicopsida'	47	17
Gymnosperms	101+5	33
Angiosperms	109	38
Totals	**266**	**93**

Ginkgos

Extinct Ginkgoales (*Sphenobaiera robusta*) are known from the Triassic (Retallack 1980). The leaves described by Pole and Raine (1994) as *Desmiophyllum* cf. *indicum* might be ginkgoid. Cuticle of *Ginkgo* sp. has been described from the Late Cretaceous (Pole & Douglas 1999).

Extinct coniferous groups

Glossopteris (Glossopteridales) is known from several leaf fragments in the Permian (Mildenhall 1970a, 1975a). Cordaitales is represented by a small fragment from the Permian identified as *Noeggerathiopsis* by McQueen (1954a); Meyen (1987) regarded southern-hemisphere identifications of this genus as more likely to be *Cordaites*. McLoughlin and Drinnan (1996) maintained *Noeggerathiopsis* but regarded it a member of the Cordaitales.

Peltaspermales was one of the major plant groups in New Zealand during the Triassic and included *Antevsia, Lepidopteris, Peltaspermum, Townrovia,* and *Pachydermophyllum* (Retallack 1977, 1979, 1981, 1983a,b, 1985; Pole & Raine 1994).

Corystospermales was another such group in the Triassic and included *Dicroidium, Johnstonia, Pilophorosperma,* and *Pteruchus* (Retallack 1983b; Pole & Raine 1994). Further lists of Triassic plants recovered in New Zealand are found in Retallack (1977, 1979, 1981, 1983a,b, 1985).

Pentoxylales are represented by two taxa. The leaf form-genus *Taeniopteris* is common in Triassic–Jurassic assemblages in New Zealand and comprises a simple morphology that is known to have evolved in several plant groups. It is not known to which of these the Triassic species belonged, while Cretaceous *T. batleyensis* had sori and is clearly a fern. At least one of the Jurassic species, *T. daintreei*, is likely to belong to the family Pentoxylaceae (Drinnan & Chambers 1985), although Blaschke and Grant-Mackie (1976) gave reasons to be cautious about this placement. Pentoxylacean plants produced cones called *Carnoconites*, also found in New Zealand (Harris 1962), and formed stems known as *Pentoxylon*. New Zealand records have not been published but the group appears to be present in a private collection (Pole pers. observ. 2000).

Bennettitales, common in the Jurassic, has been implicated in angiosperm origins. The group was represented in New Zealand by the form-genera *Otozamites, Ptilophyllum,* and *Zamites* (McQueen 1956). The placement of these form-genera in the Bennettitales follows Watson and Sincock (1992).

The Palyssiaceae, an enigmatic group based on long conical structures, includes one species from New Zealand, *Palyssia bartrumi*. It had been assumed that these structures were produced by conifers, but a recent review of the species has suggested otherwise (Parris et al. 1995). On the basis of this work, Miller (1999) has ruled out a coniferous origin.

Coniferales

New Zealand's araucarian record has been reviewed by Pole (2008). The oldest example of Araucariaceae in New Zealand may be Jurassic foliage placed in the form genus *Podozamites* (Arber 1917). Probable araucarian wood (Stopes 1914), cones (Mildenhall & Johnston 1971), and foliage have been reported from the Cretaceous. The foliage includes *Araucaria* (e.g. Bose 1973; Edwards 1926) as well as the extinct genus *Araucarioides* (Pole 1995). Two species of *Agathis* have been reported from the Albian (Cretaceous) of the Clarence River (Parrish et al. 1998), but their description has not been published (see Daniel 1989). To be convincing they need to be distinguished from *Araucarioides* and *Wollemia*. There was a major extinction of Araucariaceae at the Cretaceous–Paleocene boundary and *Agathis*, or an ancestral form, may immediately follow this event (Pole & Vajda 2009). *Agathis* is present in the Early Miocene of the Gore Lignite (Lee et al. 2007). The lower Pleistocene of Northland (the Bayly's Beach locality of Mildenhall et al. 1992) has the oldest New Zealand microfossils that are with certainty *Agathis* (Pole unpubl.) and appear to be conspecific with the endemic modern kauri *A. australis*.

The family Pinaceae is nominally represented by a specimen of Cretaceous wood named as *Planoxylon hectori* (Stopes 1916), suggested to belong to the Abietineae (firs). Seward (1919) highlighted its combination of characters as being found in both the Araucariaceae and Abietineae, hence the familial relationships are best regarded as uncertain.

Fossils attributed to Podocarpaceae are rather diverse. One of the most important early conifer fossils, Jurassic *Mataia podocarpoides*, includes epidermal detail and reproductive structures that had suggested placement in the Podocarpaceae or in a closely related but extinct group (Townrow 1967). Gregg (1975) has pointed out the rather confused status of the species; here it is treated as valid. *Mataia podocarpoides* has foliage that is covered by the form '*Elatocladus conferta*', but *E. conferta* is not likely to be a synonym of *M. podocarpoides* as *E. conferta* describes a very general type of conifer foliage that could have occurred in several taxa.

Retallack (1981) recorded the leaves *Heidiphyllum elongatum* and the cones *Telemachus lignosus* from the Triassic of New Zealand. The conclusion that these organs represented the leaves and female strobilus of the same plant is now strongly supported by the evidence of co-occurrence in many localities (e.g. Anderson 1978; Axsmith et al. 1998). Retallack (1981) regarded both cones and leaves as belonging to the extinct family Volziaceae. *Heidiphyllum* is a form genus with cuticle that is too poorly known to be of much taxonomic use (Anderson & Anderson 1989). Axsmith et al. (1998) regarded *Heidiphyllum* and another form, *Notophytum* (from Antarctica), as being the same or closely related biological species but in different preservational modes. The stem anatomy of *Notophytum* has suggested affinities with the Podocarpaceae (Meyer-Berthaud & Taylor 1991). *Telemachus* is certainly very different from any extant podocarps as is the case for the epidermal characters of *Mataia podocarpoides* leaves. Both may be 'stem taxa' of the Podocarpaceae in the sense of Doyle and Donoghue (1993), but one must be cautious of extending the definition of Podocarpaceae by their inclusion.

Mildenhall (1976) described a podocarp cone from the Cretaceous of the Wairarapa and organically preserved foliage is common thereafter (e.g. Pole 1995, 1998b). Cretaceous and Paleocene podocarps include extinct genera. The oldest records of extant New Zealand conifer species are from the Early Miocene of Central Otago, including *Dacrycarpus dacrydioides*, *Lepidothamnus intermedia*, and *Prumnopitys taxifolia* (Pole 1992c, 1997b). *Libocedrus* and *Phyllocladus* are present, and may include extant species. Three species of *Podocarpus* are known, but all are distinct from extant New Zealand species. In the Early Miocene there are also genera now extinct in New Zealand but still surviving elsewhere, including *Acmopyle*, *Araucaria*, *Papuacedrus*, and *Retrophyllum*. Unlike in the late Cretaceous–Paleogene, there are no entirely extinct genera known in the Miocene. The total generic diversity of conifer macrofossils from the Central Otago and Southland Miocene is now 12 (Pole 2007a). Ettingshausen's (1887, 1891) record of *Ginkgocladus* from the middle Cenozoic is most likely a *Phyllocladus*, as he himself hinted at. Younger deposits contain representatives of locally extinct conifer genera, as well as some first-occurrences of extant species. *Dacrydium* is present from the late Miocene of Tadmor (south of Nelson) and the Pliocene of Karamea. Both of these deposits also contain *Acmopyle* (Pole 2007b). Fossil Podocarpaceae wood is common in some localities (e.g. Lindqvist & Isaac 1991; Moore & Wallace 2000; Sutherland 2003).

The genus *Paahake*, attributed to the Taxaceae or Taxodiaceae by Pole (1998b), was assigned to the Cupressaceae (in the broader sense, including Taxodiaceae) by Hill and Brodribb (1999). Since this transfer was made without justification or comment as to why Taxaceae could not be the family, their reassignment is rejected here. The family Taxaceae is represented in the Miocene (Pole 1997b, 2007a) and by the very unusual *Mataoria* (Pole & Moore 2011) that may have had phyllodinous foliage.

Sequoiadendron novae-zeelandiae was described as *Sequoia* by Ettingshausen

(1887) based on gross morphology but was regarded as *Arthrotaxis* by Florin (1940). Pole (1995) recovered epidermal details and pointed out that, owing to a later segregation of the living *Sequoia* species to include *Sequoiadendron*, the fossils needed to be renamed. Hill and Brodribb (1999) regarded the fossil as *Arthrotaxis*, but this is clearly incorrect insofar as epidermal details are concerned; *Arthrotaxis* has monocyclic to partially dicyclic stomates, while *Sequoiadendron* and the fossil have dicyclic stomates. The fossil differs from extant *Sequoiadendron* in not having sloping walls separating the inner and outer subsidiary cells, and perhaps the fossil might be an extinct genus. However, *Austrosequoia* from the Cretaceous (Cenomanian or ?Turonian) of Queensland (Peters & Christophel 1978) and Oligocene of Tasmania (Hill et al. 1993) is almost identical to *Sequoiadendron*, and may in fact be the same. It is clear that the genus, or something very like it, was in the region during the Cretaceous. Little may be gained by placing the New Zealand species in a new genus at this stage.

Gnetales is probably represented by leaf cuticle in the Early Miocene and Paleocene (Pole 2008, 2010; Pole & Vajda 2009). It is comparable to but distinct from *Gnetum*.

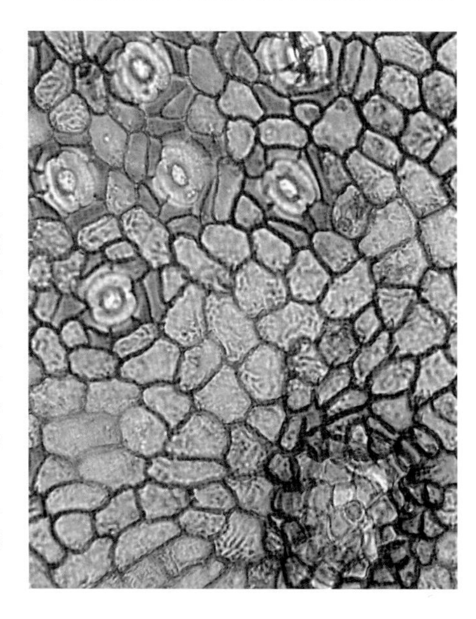

Leaf cuticle of probable *Elaeocarpus* (Elaeocarpaceae) (Blue Lake, Early Miocene).
Michael Pole

Angiosperms

Arecaceae (palms) has been reported several times. Mildenhall (1968) noted a possible palm fruit from the late Cretaceous. Undoubted palm fruits were described from the Miocene of Northland as *Cocos zeylandica* (Berry 1926). However, these 'coconuts' are very small (about 30–40 mm in diameter) and there is little to support their placement in the currently monospecific tropical genus *Cocos*. A later suggestion (Endt & Hayward 1997) is that they may belong in the montane South American genus *Parajubaea*. Ettingshausen (1887, 1891) described fronds of *Seaforthia* from the Miocene of Central Otago. A reinvestigation of what was possibly the original locality yielded fronds, fruits, and flower-heads (Pole 1993c) that compare well with the extant nikau *Rhopalostylis sapida*, but there are enough differences to suggest that they are not the same species. They might well belong in another genus; Pole (1993c) placed them in the organ genus *Phoenicites*, a choice not now considered appropriate. Cuticle is present on palm fronds from another Miocene locality in Central Otago (St Bathans) and based on this evidence they certainly seem to belong in another genus (Pole unpubl.). Dispersed Arecaceae cuticle is present in the Early Miocene of Southland (Pole 2007d) and calamoid palm leaves and fruits have been described from the late Eocene as *Lepidocaryopsis zeylanicus* (Hartwich et al. 2010).

Asteliaceae (*Astelia* family) is present in the Early Miocene as cuticle (Pole 2007d).

Musaceae (banana family) is nominally represented by *Haastia speciosa* (Ettingshausen 1887), constituting the earliest record of a monocot from New Zealand. Mildenhall (1972a) pointed out that this name was already in use and diagnosed a new genus for the fossil, *Pakawaua*. He discussed its probable placement in Musaceae.

Ripogonaceae (supplejack) is known from Early Miocene leaves of *Ripogonum scandens* at Bannockburn (Pole 1993d), making this one of the oldest extant New Zealand angiosperm species to be found as a fossil. Subsequently, further leaves were discovered with cuticle, confirming the identification. The species also occurs (with epidermal detail) at Foulden Hills (Pole 1996). [Note that the caption to fig. 11 in this publication was incomplete. Figs 11c and d should be labelled as comparable views of extant *R. scandens* cuticle.] A restudy of Holden's (1982a) specimen of *Cinnamomum miocenicum* suggests that it is also a *Ripogonum* and that the records of *Coriaria arborea* and *C. latepetiolata* might also be attributable to *Ripogonum* (Pole unpubl.).

Poaceae (grasses) is likely represented in the Early Miocene by apparent seeds with awns (Pole (1993h).

Orchidaceae is now known from the Early Miocene (Foulden Maar) based

Group of female 'cones' of '*Casuarina*' *avenacea* (Casuarinaceae) (Bannockburn, Early Miocene).
Michael Pole

on two types of orchid leaf – *Dendrobium* and *Earina* – the first macrofossils of the family (Conran et al. 2009).

Dispersed cuticle of possible Pandanaceae is present in the Early Miocene of Southland (Pole 2007d).

Typhaceae (raupo) leaf cuticle and six types of *Typha* seed are known from the Early Miocene (Pole 2007d).

Argophyllaceae (*Corokia* and relatives) leaf cuticle is known from the Early Miocene (Pole 2008).

Atherospermataceae (pukatea and relatives) is represented by several types of leaf cuticle from the Early Miocene (Pole 2008).

Avicenniaceae (Acanthaceae in APG III) is represented by mangrove fruits (Campbell 2002) and wood (Sutherland 1994, 2003) have been reported.

Betulaceae (birch family) is nominally represented by some leaves from the Late Cretaceous (Pole 1992a). Based on their gross morphology, there is little doubt that they would be attributed to this family if found in the Northern Hemisphere. Betulaceae is not present in the extant indigenous flora of New Zealand, though a Plio-Pleistocene leaf, *Betulites couperi*, has been attributed to this family (McQueen 1954c). Hill and Dettmann (1996) suggested that the Late Cretaceous leaves belonged to a 'fagalean complex' that gave rise to *Nothofagus* in the south and Fagaceae-Betulaceae in the Asian region. This is certainly possible, but this concept needs to be defined, especially as the Maastrichtian age of the ?Betulaceae fossils would imply that the complex was contemporaneous with *Nothofagus*.

Casuarinaceae (sheoakes) is more certain. 'Cones' and foliage were described from the Miocene of Central Otago and Southland by Campbell and Holden (1984) as *Casuarina avenacea* (in division Cryptostomae) and *C. stellata* (in division Gymnostomae), respectively. At around the same time, Johnson (1980, 1982, 1988) split extant *Casuarina* into three additional genera, with division Cryptostomae becoming *Casuarina* sensu stricto, *Allocasuarina*, and *Ceuthostoma*, and division Gymnostomae forming *Gymnostoma*. The formal recombination '*Gymnostoma stellata*' has not been made. Wilson and Johnson (1989) concluded that *C. avenacea* fell into *Allocasuarina*, but later Johnson (1991) reassessed this fossil and suggested it was more likely *Casuarina* s. str. The distinctive 'articles' and cuticle of *Gymnostoma* are present in the Early Miocene of the Southland coal fields (Pole 2008).

Pod of a legume resembling *Serianthes* (Fabaceae) (Bannockburn, Early Miocene).
Michael Pole

The only certain macrofossil of Cunoniaceae is an Early Miocene inflorescence identified as ?*Weinmannia racemosa* (Pole 1993h). Many of the mummified leaves from the Manuherikia Group are likely to be Cunoniaceae but have not yet been clearly distinguished from the Elaeocarpaceae. Flowers from the Early Miocene have been provisionally identified as *Acsmithia* sp. (Pole et al. 2003).

It is now certain that the Early Miocene leaves identified by Pole (1993d) as *Elaeocarpus/Sloanea* are *Elaeocarpus* (Elaeocarpaceae). They are distinct from the extant New Zealand species but similar to *Elaeocarpus costatus* of Lord Howe Island.

Ericaceae (heath family) is known from New Zealand based latest Oligocene–Early Miocene leaves of *Cyathodophyllum novae-zelandiae* (tribe Styphelieae) and *Richeaphyllum waimumuensis* (tribe Richeeae) (Jordan et al. 2010). Other leaf cuticle was illustrated by Pole (2008).

Euphorbiacae (spurge family) may be present in the Manuherikia Group. Impressions of Early Miocene leaves were reported by Pole (1993h) and suggested to belong to this family but they could not be placed in an extant genus so were left unnamed. A similar leaf (although differing in detail) from Foulden Diatomite was described as being either *Mallotus* or *Macaranga* (Pole 1996). Lee et al. (2010) found further specimens from the same locality and isolated cuticle that supported the identification. In the absence of characters allowing the fossils to be placed in either *Mallotus* or *Macaranga* they described a new genus and species, *Malloranga fouldenensis*, to cover both options. They also described a fruits and an inflorescence with in situ *Nyssapollenites endobalteus* pollen as further proof of the family.

Fabaceae is likely represented by large legumes and bipinnate foliage from the Early Miocene. They are clearly not related to any extant New Zealand genus and, although considered to belong in the Mimosoideae, the fruits have been placed in the organ genus *Parvileguminophyllum* (Pole et al. 1989; Pole 1992e).

Griseliniaceae is represented by leaf cuticle of *Griselinia* from the Early Miocene (Pole 2008).

Lauraceae is the oldest extant New Zealand angiosperm family known from macrofossils, dating from at least the Maastrichtian as leaf cuticle (Pole & Douglas 1999) and flowers (Cantrill et al. 2011). Lauraceae includes some of the most abundant macrofossils throughout the Cenozoic, occurring as leaves and dispersed cuticle (Pole 2007c). Their placement into genera is difficult but has been aided by surveys of extant Lauraceae cuticle (Christophel & Rowett 1996; Christophel et al. 1996). Holden (1982a) identified some Middle Miocene impressions as *Cryptocarya* based on similar forms with cuticular preservation it is likely that his identification will turn out to be correct. The three extant species of *Beilschmiedia* and *Litsea calicaris* have not yet been identified as fossils but a species of *Beilschmiedia* is present in the Pliocene (Pole 2007b).

Meliaceae (mahogany family) is represented by several types of leaf cuticle from the Early Miocene (Pole 2008).

Menispermaceae, the moonseed family, is currently extinct in New Zealand, but leaf cuticle is known from the Early Miocene (Pole 2008).

Monimiaceae is also represented by several types of leaf cuticle from the Early Miocene (Pole 2008).

Myricaceae (sweet gale) is not found in New Zealand today. The leaf *Dryandra comptoniaefolia* was synonymised, with some reservations, with *Comptonia diforme* by Berry (1906)

Myrtaceae (myrtles) is present in the New Zealand Eocene (Pole 1994) and common in the Miocene. One of the surprises is *Eucalyptus*, which has been found as leaves and a 'gumnut' (Holden 1983b; Pole 1993f) as well as possible wood (Sutherland 2003). *Metrosideros* is present as leaves and also as fruits (Pole 1993f, Pole et al. 2008). Several other species are present as mummified leaves, fragments of cuticle, or impressions. Identification to genera is very difficult but common rainforest genera like *Syzygium* or *Eugenia* are probably represented (Pole *et al.* 2008).

Myrsinaceae (Primulaceae in APG III) is represented by a leaf from the Miocene of Foulden Hills attributed to *Ardisia* (Pole 1996). Since then, several types of dispersed leaf cuticle have been documented from the Early Miocene (Pole 2008). The small-leaved *Myrsine waihiensis* was described from late Miocene of Mataora (Pole & Moore 2011).

Nothofagaceae (southern beeches), as might be expected, has a rich record in New Zealand. Unger's (1864) record of *Fagus* was confirmed as *Nothofagus* based on the preparation of cuticle from the leaf surface (Kovar *et al.* 1987). Ettingshausen (1887, 1891) described some species as *Fagus*, and these were transferred to *Nothofagus* by Oliver (1950). Oliver (1936) described seven species in *Fagus, Nothofagus,* and *Parafagus* from Dunedin but these almost certainly comprise two species at most, both in *Nothofagus*. Campbell (1985) illustrated *Nothofagus* leaves and seeds from the same locality. Holden (1982b) described four species of *Nothofagus* from Murchison, including *N. novae-zealandiae*. Again, these are probably just one, representing a new species that was much larger than *N. novae-zealandiae*. *Nothofagus novae-zealandiae* occurred in Central Otago, but there are varietal differences between some localities (Pole 1993g). Extinct species of *Nothofagus* existed in New Zealand as recently as the Early Pleistocene (McQueen 1954). Pole (1992a) described *Nothofagus* from the Late Cretaceous, based on the presence of leaves with regular, straight lateral veins and a general association with *Nothofagus* pollen. If cuticle is ever isolated, they may prove to be of an extinct genus.

Leaf of *Ardisia* sp. (Myrsinaceae) (Foulden Hills, Early Miocene).

Michael Pole

Paracryphiaceae, formerly regarded as monotypic for *Paracryphia* and endemic to New Caledonia (but now including *Quintinia*), is present in the Early

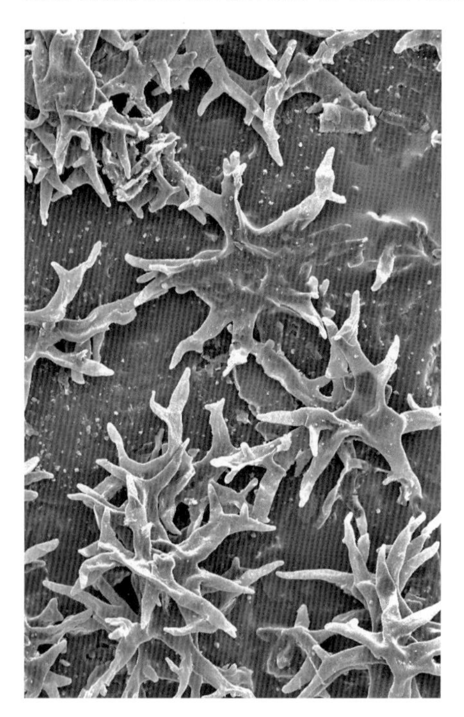

Scanning electron micrograph of the leaf cuticle of *Notothixos* sp. (Santalaceae) with branched hairs called trichomes (Grey Lake, Early Miocene).
Michael Pole

Miocene of Southland (Pole 2009) based on *Paracryphia*.

Phellinaceae, with sole living genus *Phelline* endemic to New Caledonia, is also present in the Early Miocene of Southland (Pole 2009).

Platanaceae (plane trees) is almost certainly represented by leaves of Cenomanian age (Pole 1992a).

Impressions of *Muehlenbeckia* (Polygonacae) were recorded from the Early Miocene by Pole (1993d).

Fossil Proteaceae in New Zealand were summarised by Pole (1998a) and more taxa have been discovered since. The oldest examples are from the Paleocene, and include *Lomatia novae-zelandiae*. The Early Miocene contains taxa such as *Banksia, Helicia, Lomatia, Macadamia, Musgravea, Persoonieaephyllum,* and *Placospermum* (Carpenter 1994; Pole 2008; Carpenter et al. 2010a,b), which are now restricted to the Australian rainforest. The oldest record of an extant species, *Knightia excelsa*, comes form the Pliocene of the Karamea district (Pole 2007b) a record slightly southwest of its current southern limit in Marlborough (Wardle 1991). The same locality also contains *Beauprea*.

Santalaceae is based on leaf cuticle of *Notothixos* and *Amphorogyne* from the Early Miocene (Pole 2008, 2009).

Sapindaceae is known from several types of leaf cuticle from the Early Miocene (Pole 2008).

Strasburgeriaceae, formerly regarded as monotypic for *Strasburgeria* and endemic to New Caledonia (but now including *Ixerba*), is known from *Strasburgeria*-like Early Miocene leaf cuticle (Pole 2008).

Winteraceae is represented by several types of leaf cuticle from the Early Miocene (Pole 2008). All are quite distinct from extant New Zealand *Pseudowintera* and appear closer to genera found today on New Caledonia. The oldest macrofossil record of *Pseudowintera* is from the Pliocene (Pole 2007b).

Conclusions

New Zealand now has a rather long list of described plant macrofossils, although many of these are probably worthless as reliable additions to biogeography. The last few years, however, have shown that fossil deposits occur in New Zealand that are as good as the best in the world. Focusing on these deposits will add a significant amount to our knowledge of vegetation and climate processes of the times they represent. Notwithstanding, these fossils will generate more questions than answers in the short term. For instance, despite the wide range of well-preserved leaves in the Early Miocene, very few of them seem to represent extant species. All the major forest trees and shrubs in New Zealand are represented in the M. Pole cuticle reference collection, and these would be recognisable if they were found as fossils. Were New Zealand's living species only rare components of Early Miocene vegetation, not yet evolved, or were they (or their ancestors) not even in the country? There are still enormous time gaps in our knowledge of macrofossils; the Eocene–Oligocene, for example, is barely represented. It is very likely that, with the right techniques, these gaps will start to fill.

Throughout the Cenozoic, New Zealand was a low-lying small island (or group of smaller islands) of extremely oceanic character. There would have been no topographically induced extreme highs or lows of rainfall, and none of the extremes of temperature that are found in the interiors of large landmasses. The very low topography implies that altitudinal zonation may have been insignificant. It also means that New Zealand's vegetation throughout the Cenozoic was tracking, rather directly, global climate changes. Added to a potentially very complete plant fossil record, this likelihood means that New Zealand has global significance in understanding the evolution of atmospheric circulation. New Zealand is, in effect, one of the world's best natural laboratories. Documenting species, fossil and living, is the first step in understanding global systems.

Author

Dr Michael S. Pole Queensland Herbarium, Brisbane Botanic Gardens Mt Coot-tha, Mt Coot-tha Rd, Toowong QLD 4066, Australia [murikihu@yahoo.com]

References and bibliography

ANGIOSPERM PHYLOGENY GROUP 2009: An update of the Angiosperm Phylogeny Group classification for the orders and families of flowering plants: APG III. *Botanical Journal of the Linnean Society 161*: 105–121.

ANDERSON, H. M. 1978: *Podozamites* and associated cones and scales from the Upper Triassic Molteno Formation, Karoo Basin, South Africa. *Palaeontologica Africana 21*: 57–77.

ANDERSON, J. M.; ANDERSON, H. M. 1989: *Palaeoflora of Southern Africa, Molteno Formation (Triassic) Vol. 2, Gymnosperms (excluding Dicroidium)*. A. A. Balkema, Rotterdam. 567 p.

ARBER, E. A. N. 1917: The earlier Mesozoic floras of New Zealand. *New Zealand Geological Survey Palaeontological Bulletin 6*: 1–80.

AXSMITH, B. J.; TAYLOR, T. N.; TAYLOR, E. L. 1998: Anatomically preserved leaves of the conifer *Notophytum krauselii* (Podocarpaceae) from the Triassic of Antarctica. *American Journal of Botany 85*: 704–713.

BELL, S.; HARRINGTON, H. J.; MCKELLAR, I. C. 1956: Lower Mesozoic plant fossils from Black Jacks, Waitaki River, South Canterbury. *Transactions of the Royal Society of New Zealand 83*: 663–72.

BERRY, E. W. 1906: Living and fossil species of *Comptonia*. *The American Naturalist 40*: 485–524.

BERRY, E. W. 1926: *Cocos* and *Phymatocaryon* in the Pliocene of New Zealand. *American Journal of Science 12*: 181–184.

BLASCHKE, P. M.; GRANT-MACKIE, J. A. 1976: Mesozoic leaf genus *Taeniopteris* at Port Waikato and Clent Hills, New Zealand. *New Zealand Journal of Geology and Geophysics 19*: 933–941.

BOSE, M. N. 1973: *Araucaria haastii* Ettingshausen from Shag Point, New Zealand. *The Palaeobotanist 22*: 76–80.

BROEKHUIZEN, P. 1984: Studies on the Huriwai Group flora, Port Waikato, *Taeniopteris* Brongniart and *Cladophlebis* Brongniart. MSc Thesis, Department of Botany, University of Auckland.

BURROWS, C. J. 1995: A macrofossil flora from sediments in a lagoon marginal to Lake Coleridge, Canterbury, New Zealand. *New Zealand Journal of Botany 33*: 519–522.

BURROWS, C. J. 1997: A macrofossil flora from early Aranuian lake-bed deposits, Doubtful River, Waiau-uha catchment, North Canterbury, New Zealand. *New Zealand Journal of Botany 35*: 545–553.

BURROWS, C. J.; McCULLOCH, B.; TROTTER, M. M. 1981: The diet of moas based on gizzard contents samples from Pyramid Valley, North Canterbury, and Scaifes Lagoon, Lake Wanaka, Otago. *Records of the Canterbury Museum 9*: 309–336.

BURROWS, C. J.; RANDALL, P.; MOAR, N. T.; BUTTERFIELD, B. G. 1993: Aranuian vegetation history of the Arrowsmith Range, Canterbury 111. Vegetation changes in the Cameron, upper South Ashburton, and Paddle Hill Creek catchments. *New Zealand Journal of Botany 31*: 147–174.

BURROWS, C. J; RUSSELL, J. B. 1990: Aranuian vegetation history of the Arrowsmith Range, Canterbury 1. Pollen diagrams, plant macrofossils, and buried soils from Prospect Hill. *New Zealand Journal of Botany 28*: 323–345.

CAMPBELL, J. D. 1985: Casuarinaceae, Fagaceae, and other plant macrofossils from Kaikorai leaf beds (Miocene) Kaikorai Valley, Dunedin, New Zealand. *New Zealand Journal of Botany 23*: 311–20.

CAMPBELL, J. D. 2002: Angiosperm fruit and leaf fossils from Miocene silcrete, Landslip Hill, northern Southland, New Zealand. *Journal of the Royal Society of New Zealand 32*: 149–154.

CAMPBELL, J. D.; FORDYCE, R. E.; GREBNEFF, A.; MAXWELL, P. 1991: Coconuts, coconuts, coconuts. *Geological Society of New Zealand Newsletter 92*: 37.

CAMPBELL, J. D.; HOLDEN, A. M. 1984: Miocene casuarinacean fossils from Southland and Central Otago, New Zealand. *New Zealand Journal of Botany 22*: 159–67.

CANTRILL, D. J.; RAINE, J. I. 2006: *Wairarapaia mildenhallii* gen. et sp. nov., a new araucarian cone related to *Wollemia* from the Cretaceous (Albian–Cenomanian) of New Zealand. *International Journal of Plant Science 167*: 1259–1269.

CANTRILL, D. J.; WANNTORP, L.; DRINNAN, A. N. 2011: Mesofossil flora from the Late Cretaceous of New Zealand. *Cretaceous Research 32*: 164–173.

CARPENTER, R. J. 1994: Cuticular morphology and aspects of the ecology and fossil history of North Queensland rainforest Proteaceae. *Botanical Journal of Linnean Society 116*: 249–303.

CARPENTER, R. J., JORDAN, G. J., LEE, D. E., HILL, R. S. 2010a: Leaf fossils of *Banksia* (Proteaceae) from New Zealand: An Australian abroad. *American Journal of Botany 97*: 288–297.

CARPENTER, R. J.; BANNISTER, J. M.; JORDAN, G. J.; LEE, D. E. 2010b: Leaf fossils of Proteaceae tribe Persoonieae from the Late Oligocene–Early Miocene of New Zealand. *Australian Systematic Botany 23*: 1–15.

CHRISTOPHEL, D. C.; ROWETT, A. I. 1996: *Leaf and Cuticle Atlas of Australian Leafy Lauraceae*. Australian Biological Resources Study, Canberra. 217 p.

CHRISTOPHEL, D. C.; KERRIGAN, R.; ROWETT, A. I. 1996: The use of cuticular features in the taxonomy of the Lauraceae. *Annals of the Missouri Botanical Garden 83*: 419–432.

CIERAAD, E.; LEE, D. E. 2006: The New Zealand fossil record of ferns for the past 85 million years. *New Zealand Journal of Botany 44*: 143–170.

CLARKSON, B. R.; PATEL, R. N.; CLARKSON, B. D. 1988: Composition and structure of forest overwhelmed at Pureora, central North Island, New Zealand, during the Taupo eruption (c. AD 130). *Journal of the Royal Society of New Zealand 18*: 417–436.

COLLINSON, M. E. 1996: *'What use are fossil ferns?'* – 20 years on: with a review of the fossil history of extant pteridophyte families and genera. Pp. 349–394 *in*: Camus, J. M.; Gibby, M.; Johns, R. J. (eds), *Pteridology in Perspective*. Royal Botanic Gardens, Kew. 700 p.

COLLINSON, M. E. 2001: Cainozoic ferns and their distribution. *Brittonia 53*: 173–235.

CONRAN, J. G.; BANNISTER, J. M.; LEE, D. E. 2009: Earliest orchid macrofossils: Early Miocene *Dendrobium* and *Earina* (Orchidaceae: Epidendroideae) from New Zealand. *American Journal of Botany 96*: 466–474.

CONRAN, J. G.; KAULFUSS, U.; BANNISTER, J. M., MILDENHALL, D. C.; LEE, D. E. 2010: *Davallia* (Polypodiales: Davalliaceae) macrofossils from Early Miocene Otago (New Zealand) with in situ spores. *Review of Palaeobotany and Palynology 162*: 84–94.

CRIE, L. 1888: Sur les affinités des flores jurassiques et triassiques de l'Australie et de la Nouvelle-Zélande. *Compte Rendu de l'Académie des Sciences, Paris 107*: 1014–1015.

CRIE, L. 1889: Beitrage zur Kenntniss der fossilen Flora einiger Inseln des Südpacifischen und Indischen Ocean. *Palaeontologia, Abh. 5, NF 1(2)*: 77–91.

CRIE, L. 1889: *Exposition paléophytique — paléontologie des colonies Françaises*. Université de Paris, Paris. 32 p.

DANIEL, I. L. 1989: Taxonomic investigation of elements from the Early Cretaceous megaflora from the Middle Clarence Valley, New Zealand. Unpublished PhD thesis, University of Canterbury, Christchurch.

DETTMANN, M. E.; POCKNALL, D. T.; ROMERO, E. J.; ZAMALOA, M.delC. 1990: *Nothofagidites* Erdtman ex Potonié, 1960; a catalogue of species with notes on the paleogeographic distribution of *Nothofagus* Bl. (southern beech). *New Zealand Geological Survey Paleontological Bulletin 60*: 1–79.*

DOUGLAS, B. J. 1986: Lignite resources of Central Otago. *New Zealand Energy Research and Development Committee Publication P104*: 1–367.

DOYLE, J. A.; DONOGHUE, M. J. 1993: Phylogenies and angiosperm diversification. *Paleobiology 19*: 141–167.

DRINNAN, A. N.; CHAMBERS, T. C. 1985: A reassessment of *Taeniopteris daintreei* from the Victorian Early Cretaceous: a member of the Pentoxylales and a significant Gondwanaland plant. *Australian Journal of Botany 33*: 89–100.

EDWARDS, A. R.; HORNIBROOK, N. deB.; RAINE, J. I.; SCOTT, G. H.; STEVENS, G. R.; STRONG, C. P.; WILSON, G J. 1988: A New Zealand Cretaceous–Cenozoic geological time scale. *New Zealand Geological Survey Record 35*: 135–149.

EDWARDS, W. N. 1934: Jurassic plants from New Zealand. *Annals and Magazine of Natural History, ser. 10, 13*: 81–109.

EDWARDS, W. N. 1926: Cretaceous plants from Kaipara, New Zealand. *Transactions of the New Zealand Institute 56*: 121–128.

ENDT, D.; HAYWARD, B. 1997: Modern relatives of New Zealand's fossil coconuts from high altitude South America. *New Zealand Geological Society Newsletter 113*: 67–70.

ETTINGSHAUSEN, C. von 1887: Beitrage zur

Kenntniss der Fossilen Flora Neuseelands. *Denkschriften der Akademie der Wissenschaften, Wien 53*: 143–194.
ETTINGSHAUSEN, C. von 1891: Contributions to the knowledge of the fossil flora of New Zealand. *Transactions of the New Zealand Institute 23*: 237–310.
EVANS, W. P. 1931: A fossil *Nothofagus* (*Nothofagoxylon*?) from the Central Otago coal-measures. *Transactions of the New Zealand Institute 62*: 98.
EVANS, W. P. 1936: Microstructure of New Zealand lignites. Part 3: Lignites apparently not altered by igneous action. A: Coal Creek Flat, Roxburgh, Central Otago. *New Zealand Journal of Science and Technology 17*: 649–658.
EVANS, W. P. 1937: Note on the flora which yielded the Tertiary lignites of Canterbury, Otago and Southland. *New Zealand Journal of Science and Technology 19*: 188–193.
FLORIN, R. 1940: The Tertiary fossil conifers of south Chile and their phytogeographical significance. *Kungliga Svenska Vetenskapakademiens Handlingar 19*: 1–107.
GREGG, D. R. 1975: Type and figured specimens of fossil plants in the Canterbury Museum. *Records of the Canterbury Museum 9*: 259–276.
HARRIS, T. M. 1962: The occurrence of the fructification *Carnoconites* in New Zealand. *Transactions of the Royal Society of New Zealand Geology 1*: 17–27.
HARTWICH, S. J.; CONRAN, J. G.; BANNISTER, J. M.; LINDQVIST, J. K.; LEE, D. E. 2010: Calamoid fossil palm leaves and fruits (Arecaceae: Calamoideae) from Late Eocene Southland, New Zealand. *Australian Systematic Botany 23*:131–140.
HILL, R. S. 1993: Taxodiaceous macrofossils from Tertiary and Quaternary sediments in Tasmania. *Australian Systematic Botany 6*: 237–249.
HILL, R. S.; BRODRIBB, T. J. 1999: Southern conifers in time and space. *Australian Journal of Botany 47*: 639–696.
HILL, R. S.; DETTMANN, M. E. 1996: Origin and diversification of the genus *Nothofagus*. Pp. 11–23 *in*: Veblen, T. T.; Hill, R. S.; Read, J. (eds), *The Ecology and Biogeography of Nothofagus Forest.* Yale University Press, New Haven. 403 p.
HILL, R. S.; POLE, M. S. 1994: Two new species of *Pterostoma* R. S. Hill from Cenozoic sediments in Australasia. *Review of Palaeobotany and Palynology 80*: 123–130.
HOLDEN, A. M. 1982a: Fossil Lauraceae and Proteaceae from the Longford Formation, Murchison, New Zealand. *Journal of the Royal Society of New Zealand 12*: 79–80.
HOLDEN, A. M. 1982b: Fossil *Nothofagus* from the Longford Formation, Murchison, New Zealand. *Journal of the Royal Society of New Zealand 12*: 65–77.
HOLDEN, A. M. 1983a: Studies in New Zealand Oligocene and Miocene plant macrofossils. Unpublished PhD thesis, Victoria University of Wellington, Wellington.
HOLDEN, A. M. 1983b: *Eucalyptus*-like leaves from Miocene rocks in Central Otago. *Abstracts. Pacific Science Association 15th Congress*: 103.
HOLDEN, A. M. 1987: *Vegetation-lithotype Correlations in New Zealand Lignites*. New Zealand Energy Research and Development Committee, Wellington. 86 p.
JOHNSON, L. A. S. 1980: Notes on Casuarinaceae. *Telopea 2*: 83–84.
JOHNSON, L. A. S. 1982: Notes on Casuarinaceae II. *Journal of the Adelaide Botanical Garden 6*: 73–87.
JOHNSON, L. A. S. 1988: Notes on Casuarinaceae III. The new genus *Ceuthostoma*. *Telopea 3*: 133–137.
JOHNSON, L. A. S. 1991: Casuarinaceae – some clarifications. *Australian Systematic Botany Society Newsletter 67*: 25–26.
JOHNSON, L. A. S.; BRIGGS, B. G. 1975: On the Proteaceae – the evolution and classification of a southern family. *Botanical Journal of the Linnean Society 70*: 83–182.
JOHNSTON, M. R.; RAINE, J. I.; WATTERS, A. 1987: Drumduan group of east Nelson, New Zealand: plant-bearing Jurassic arc rocks metamorphosed during terrane interaction. *Journal of the Royal Society of New Zealand 17*: 275–301.
JORDAN, G. J.; BANNISTER, J. M.; MILDENHALL, D. C.; ZETTER, R.; LEE, D. E. 2010: Fossil Ericaceae from New Zealand – deconstructing the use of fossil evidence in historical biogeography. *American Journal of Botany 97*: 59–70.
KELCH, D. G. 1997: The phylogeny of the Podocarpaceae based on morphological evidence. *Systematic Botany 22*: 113–131.
KELCH, D. G. 1998: Phylogeny of Podocarpaceae: Comparison of evidence from morphology and 18S rDNA. *American Journal of Botany 85*: 975–985.
KENNEDY, E. M. 1993: Palaeo-environment of an Haumurian plant fossil locality within the Pakawau Group, North West Nelson, New Zealand. Unpublished MSc in Geology, University of Canterbury, Christchurch.
KENNEDY, E. M. 1998: Cretaceous and Tertiary megafloras from New Zealand and their climate signals. Unpublished PhD Department of Earth Sciences, The Open University, Milton Keynes.
KENNEDY, E. M.; ALLOWAY, B.V.; MILDENHALL, D. C.; COCHRAN, U.; PILLANS, B. 2008: An integrated terrestrial paleoenvironmental record from the Mid-Pleistocene transition, eastern North Island, New Zealand. *Quaternary International 178*: 146–166.
KIDSTON, R.; GWYNNE-VAUGHAN, D. T. 1907: On the fossil Osmundaceae. Part I. *Transactions of the Royal Society of Edinburgh 45*: 759–780.
KIDSTON, R.; GWYNNE-VAUGHAN, D. T. 1908: On the fossil Osmundaceae. Part II. *Transactions of the Royal Society of Edinburgh 46*: 213–232.
KIDSTON, R.; GWYNNE-VAUGHAN, D. T. 1909: On the fossil Osmundaceae. Part III. *Transactions of the Royal Society of Edinburgh 46*: 651–667.
KIDSTON, R.; GWYNNE-VAUGHAN, D. T. 1910: On the fossil Osmundaceae. Part IV. *Transactions of the Royal Society of Edinburgh 47*: 455–477.
KOVAR, J. B.; CAMPBELL, J. D.; HILL, R. S. 1987: *Nothofagus ninnisiana* (Unger) Oliver from Waikato Coal Measures Eocene–Oligocene) at Drury, Auckland, New Zealand. *New Zealand Journal of Botany 25*: 79–85.
LEE, D. E.; BANNISTER, J. M.; LINDQVIST, J. K. 2007: Late Oligocene–Early Miocene leaf macrofossils confirm a long history of *Agathis* in New Zealand. *New Zealand Journal of Botany 45*: 565–578.
LEE, D. E.; BANNISTER, J. M.; RAINE, J. I.; CONRAN, J. G. 2010: Euphorbiaceae: Acalyphoideae fossils from Early Miocene New Zealand: *Mallotus-Macaranga* leaves, fruits, and inflorescence with in situ *Nyssapollenites endobalteus* pollen. *Review of Palaeobotany and Palynology 163*: 127–138.
LINDQVIST, J. K. 1994: Lacustrine stromatolites and oncoids: Manuherikia Group (Miocene), New Zealand. Pp. 227–254 *in*: Bertrand-Safati, J.; Monty, C. (eds), *Phanerozoic Stromatolites II*. Kluwer Academic Publishers, Dordrecht. 471 p.
LINDQVIST, J. K.; ISAAC, M. J. 1991: Silicified conifer forests and potential mining problems in seam M2 of the Gore Lignite Measures (Miocene), Southland, New Zealand. *International Journal of Coal Geology 17*: 149–169.
LINTOTT, W. H.; BURROWS, C. J. 1973: A pollen diagram and macrofossils from Kettlehole Bog, Cass, South Island, New Zealand. *New Zealand Journal of Botany 11*: 269–82.
MARSHALL, P. 1926: A new species of *Osmundites* from Kawhia, New Zealand. *Transactions of the New Zealand Institute 56*: 210–203.
McGLONE, M. S.; MILDENHALL, D. C.; POLE, M. 1996: History and paleoecology of New Zealand *Nothofagus* forests. Pp. 83–130 *in*: Veblen, T. T.; Hill, R. S.; Read, J. (eds), *The Ecology and Biogeography of* Nothofagus *Forest*. Yale University Press, New Haven. 403 p.
McLOUGHLIN, S. M.; DRINNAN, A. N. 1996: Anatomically preserved Permian *Noeggerathiopsis* leaves from East Antarctica. *Review of Palaeobotany and Palynology 92*: 207–227.
McNEILL, J.; BARRIE, F. R.; BURDET, H. M.; DEMOULIN, V.; HAWKSWORTH, D. L.; MARHOLD, K.; NICOLSON, D. H.; PRADO, J.; SILVA, P. C.; SKOG, J. E.; WIERSEMA, J. H.; TURLAND, N. J. (Eds, Comps) 2007: *International Code of Botanical Nomenclature (Vienna Code) adopted by the Seventeenth International Botanical Congress Vienna, Austria, July 2005*. [*Regnum Vegetabile 146*.] Gantner, Ruggell, Liechtenstein. 568 p.
McQUEEN, D. R. 1953: A fossil flora from the Upper Pliocene of Rangitikei Valley. *New Zealand Journal of Science and Technology 35B*: 134–140.
McQUEEN, D. R. 1954a: Upper Palaeozoic plant fossils from South Island, New Zealand. *Transactions of the Royal Society of New Zealand 82*: 231–236.
McQUEEN, D. R. 1954b Palaeozoic plants from New Zealand. *Nature 173*: 88.
McQUEEN, D. R. 1954c: Fossil leaves, fruits and seeds from the Wanganui Series (Plio-Pleistocene) of New Zealand. *Transactions of the Royal Society of New Zealand 82*: 667–676.
McQUEEN, D. R. 1956: Leaves of Middle and Upper Cretaceous pteridophytes and cycads from New Zealand. *Transactions of the Royal Society of New Zealand 83*: 673–85.
MEYEN, S.V. 1987: *Fundamentals of Palaeobotany*. Chapman and Hall, London. 432 p.
MEYER-BERTHAUD, B.; TAYLOR, T. N. 1991: A probable conifer with podocarpacean affinities from the Triassic of Antarctica. *Review of Palaeobotany and Palynology 67*: 179–198.
MILDENHALL, D. C. 1968: A note on a fossil fruit found near Lyell. *Transactions of the Royal Society of New Zealand 6*: 131–132.
MILDENHALL, D. C. 1968: The fossil flora of the Pakawau Group, NW Nelson, New Zealand. Unpublished MSc thesis, Victoria University of Wellington, Wellington.
MILDENHALL, D. C. 1970a: Discovery of a New Zealand member of the Permian *Glossopteris* flora. *The Australian Journal of Science 32*: 474–475.
MILDENHALL, D. C. 1970b: Checklist of valid and invalid plant macrofossils from New Zealand. *Transactions of the Royal Society of New Zealand*

8: 77–89.
MILDENHALL, D. C. 1972: New name for the plant fossil *Haastia* Ettingshausen 1887 non *Haastia* Hook. f. 1864. *New Zealand Journal of Geology and Geophysics 15*: 181–182.
MILDENHALL, D. C. 1975a: *Glossopteris ampla* Dana from New Zealand Permian sediments. *New Zealand Journal of Geology and Geophysics 19*: 130–132.
MILDENHALL, D. C. 1975b[1976]: Early Cretaceous podocarp megastrobilus. *New Zealand Journal of Geology and Geophysics 19*: 389–391.
MILDENHALL, D. C.; JOHNSTON, M. R. 1971: A megastrobilus belonging to the genus *Araucarites* from the Upper Motuan (Upper Albian), Wairarapa, North Island, New Zealand. *New Zealand Journal of Botany 9*: 67–79.
MILLER, C. N. 1967: Evolution in the fern genus *Osmunda*. *Contributions Museum of Paleontology of the University of Michigan 21*: 139–203.
MILLER, C. N. Jr 1971: Evolution of the fern family Osmundaceae based on anatomical studies. *Contributions. Museum of Paleontology of the University of Michigan 23*: 105–169.
MILLER, C. N. Jr 1999: Implications of fossil conifers for the phylogenetic relationships of living families. *Botanical Review 65*: 239–277.
MOORE, P. R.; WALLACE, R. 2000: Petrified wood from the Miocene volcanic sequence of Coromandel Peninsula, northern New Zealand. *Journal of the Royal Society of New Zealand 30*: 115–130.
NEWNHAM, R.; LUSK, C. 1990: Comparison of plant micro- and macrofossils, Kariotahi, Awhitu Peninsula. *Tane 32*: 171–178.
OLIVER, W. R. B. 1928: The flora of the Waipaoa Series (Later Pliocene) of New Zealand. *Transactions of the New Zealand Institute 59*: 287–303.
OLIVER, W. R. B. 1936: The Tertiary flora of the Kaikorai Valley, Otago, New Zealand. *Transactions of the Royal Society of New Zealand 66*: 284–304.
OLIVER, W. R. B. 1950: The fossil flora of New Zealand. *Tuatara 3*: 1–11.
OLIVER, W. R. B. 1955: History of the flora of New Zealand. *Svensk Botanisk Tidskrift 49*: 11–18.
PARRIS, K. M.; DRINNAN, A. N.; CANTRILL, D. J. 1995: *Palissya* cones from the Mesozoic of Australia and New Zealand. *Alcheringa 19*: 87–111.
PARRISH, J. T.; DANIEL, I. L.; KENNEDY, E. M.; SPICER, R. A. 1998: Paleoclimatic significance of mid-Cretaceous floras from the Middle Clarence Valley, New Zealand. *Palaios 13*: 149–159.
PETERS, M. D.; CHRISTOPHEL, D. C. 1978: *Austrosequoia wintonensis*, a new taxodiaceous cone from Queensland, Australia. *Canadian Journal of Botany 56*: 3119–3128.
POLE, M. S. 1992a: Cretaceous macrofloras of eastern Otago, New Zealand: angiosperms. *Australian Journal of Botany 40*: 169–206.
POLE, M. S. 1992b: Early Miocene flora of the Manuherikia Group, New Zealand. 1. Ferns. *Journal of the Royal Society of New Zealand 22*: 279–286.
POLE, M. S. 1992c: Early Miocene flora of the Manuherikia Group, New Zealand. 2. Conifers. *Journal of the Royal Society of New Zealand 22*: 287–302.
POLE, M. S. 1992d: Early Miocene flora of the Manuherikia Group, New Zealand. 3. Possible cycad. *Journal of the Royal Society of New Zealand 22*: 303–306.
POLE, M. S. 1992e: Fossils of Leguminosae from the Miocene Manuherikia Group of New Zealand. Pp. 251–258 *in*: Herendeen, P. S.; Dilcher, D. L. (eds), *Advances in Legume Systematics: Part 4. The Fossil Record.* The Royal Botanic Gardens, Kew. 336 p.
POLE, M. S. 1993a: *Nothofagus* from the Dunedin volcanic group (Mid–Late Miocene), New Zealand. *Alcheringa 17*: 77–90.
POLE, M. S. 1993b: Miocene broad-leaved *Podocarpus* from Foulden Hills, New Zealand. *Alcheringa 17*: 173–177.
POLE, M. S. 1993c: Early Miocene flora of the Manuherikia Group, New Zealand. 4. Palm remains. *Journal of the Royal Society of New Zealand 23*: 283–288.
POLE, M. S. 1993d: Early Miocene flora of the Manuherikia Group, New Zealand. 5. Smilacaceae, Polygonaceae, Elaeocarpaceae. *Journal of the Royal Society of New Zealand 23*: 289–302.
POLE, M. S. 1993e: Early Miocene flora of the Manuherikia Group, New Zealand. 6. Lauraceae. *Journal of the Royal Society of New Zealand 23*: 303–312.
POLE, M. S. 1993f: Early Miocene floras of the Manuherikia Group, New Zealand. 7. Myrtaceae, including *Eucalyptus*. *Journal of the Royal Society of New Zealand 23*: 313–328.
POLE, M. S. 1993g: Early Miocene flora of the Manuherikia Group, New Zealand. 8. *Nothofagus*. *Journal of the Royal Society of New Zealand 23*: 329–344.
POLE, M. S. 1993h: Early Miocene flora of the Manuherikia Group, New Zealand. 9. Miscellaneous leaves and reproductive structures. *Journal of the Royal Society of New Zealand 23*: 345–391.
POLE, M. S. 1994a: An Eocene macroflora from the Taratu Formation at Livingstone, North Otago, New Zealand. *Australian Journal of Botany 42*: 341–367.
POLE, M. 1994b: The New Zealand flora – entirely long-distance dispersal? *Journal of Biogeography 21*: 625–635.
POLE, M. S. 1995: Late Cretaceous macrofloras of Eastern Otago, New Zealand: Gymnosperms. *Australian Systematic Botany 8*: 1067–1106.
POLE, M. S. 1996: Plant macrofossils from the Foulden Hills Diatomite (Miocene), Central Otago, New Zealand. *Journal of the Royal Society of New Zealand 26*: 1–39.
POLE, M. S. 1997a: Paleocene plant macrofossils from Kakahu, south Canterbury, New Zealand. *Journal of the Royal Society of New Zealand 27*: 371–400.
POLE, M. S. 1997b: Miocene conifers from the Manuherikia Group, New Zealand. *Journal of the Royal Society of New Zealand 27*: 355–370.
POLE, M. S. 1998a: The Proteaceae record in New Zealand. *Australian Systematic Botany 11*: 343–372.
POLE, M. S. 1998b: Paleocene gymnosperms from Mount Somers, New Zealand. *Journal of the Royal Society of New Zealand 28*: 375–403.
POLE, M. 2007a: Conifer and cycad distribution in the Miocene of southern New Zealand. *Australian Journal of Botany 55*: 143–164.
POLE, M. S. 2007b: Plant-macrofossil assemblages during Pliocene uplift, South Island, New Zealand. *Australian Journal of Botany 55*: 118–142.
POLE, M. 2007c: Lauraceae macrofossils and dispersed cuticle from the Miocene of Southern New Zealand. *Palaeontologia Electronica 10.1, 3A*: 1–38.
POLE, M. 2007d: Monocot macrofossils from the Miocene of Southern New Zealand. *Palaeontologia Electronica 10.3, 15A*:1–21.
POLE, M. 2008: The record of Araucariaceae macrofossils in New Zealand. *Alcheringa* 32: 409–430.
POLE, M. S. 2008: Dispersed leaf cuticle from the Early Miocene of southern New Zealand. *Palaeontologica Electronica 11.3, 15A*: 1–117.
Pole, M. 2009: Was New Zealand a primary source for the New Caledonian flora? *Alcheringa 34*: 64–74.
POLE, M. 2010. Ecology of Paleocene–Eocene Vegetation at Kakahu, South Canterbury, New Zealand. *Palaeontologia Electronica 13.2, 14A*: 1–29.
POLE, M. S.; CAMPBELL, J. D.; HOLDEN, A. M. 1989: Fossil legumes from the Manuherikia Group (Miocene), Central Otago, New Zealand. *Journal of the Royal Society of New Zealand 19*: 225–228.
POLE, M.; DAWSON, J.; DENTON, P; 2008. Fossil Myrtaceae from the Early Miocene of southern New Zealand. *Australian Journal of Botany* 56: 67–81.
POLE, M. S.; DOUGLAS, B. J. 1999: Plant macrofossils of the Upper Cretaceous Kaitangata Coalfield, New Zealand. *Australian Systematic Botany 12*: 331–364.
POLE, M. S.; HOLDEN, A. M.; CAMPBELL, J. D. 1989: Fossil legumes from Manuherikia Group sediments (Miocene), Central Otago, New Zealand. *Journal of the Royal Society of New Zealand 19*: 225–228.
POLE, M.; MOORE, P. R. 2011: A Late Miocene leaf assemblage from Coromandel Peninsula, New Zealand, and its climatic implications. *Alcheringa 35*: 103–121.
POLE, M. S.; RAINE, J. I. 1994: Triassic plant fossils from Pollock Road, Southland, New Zealand. *Alcheringa 18*: 147–159.
POLE, M. S., DOUGLAS, B. J., MASON, G. 2003: The terrestrial Miocene biota of southern New Zealand. *Journal of the Royal Society of New Zealand* 33: 415–426.
POLE, M.; VAJDA, V. 2009: A new terrestrial Cretaceous–Paleogene site in New Zealand – turnover in macroflora confirmed by palynology. *Cretaceous Research 30*: 917–938.
RAINE, J. I. 1982: Observations on Arber's (1917) collections of Triassic–Jurassic plant macrofossils from New Zealand. *New Zealand Geological Survey Report PAL 58*: 1–8.
RAINE, J. I.; POCKNALL, D. T. 1983: Plant fossils. Pp. 35–46 *in*: Brownsey, P. J.; Baker, A. N. (eds), *The New Zealand Biota – What do we know after 200 years? National Museum of New Zealand Miscellaneous Series 7*: 35–46
RAINE, J. I.; POLE, M. S. 1988: Middle Jurassic forest beds, Curio Bay. *New Zealand Geological Survey Record 33*: 47–57.
REES, P. M. 1993: Dipterid ferns from the Mesozoic of Antarctica and New Zealand and their stratigraphical significance. *Palaeontology 36*: 637–656.
RETALLACK, G. J. 1977: Reconstructing Triassic vegetation of eastern Australasia: a new approach for the biostratigraphy of Gondwanaland. *Alcheringa 1*: 247–277.
RETALLACK, G. J. 1979: Middle Triassic coastal outwash plain deposits in Tank Gully, Canterbury, New Zealand. *Journal of the Royal Society of New Zealand 9*: 397–414.
RETALLACK, G. J. 1980: Middle Triassic megafossil

plants and trace fossils from Tank Gully, Canterbury, New Zealand. *Journal of the Royal Society of New Zealand 10*: 31–63.
RETALLACK, G. J. 1981: Middle Triassic megafossil plants from Long Gully, near Otematata, north Otago, New Zealand. *Journal of the Royal Society of New Zealand 11*: 167–200.
RETALLACK, G. J. 1983a: Middle Triassic estuarine deposits near Benmore Dam, southern Canterbury and northern Otago, New Zealand. *Journal of the Royal Society of New Zealand 13*: 107–127.
RETALLACK, G. J. 1983b: Middle Triassic megafossil marine algae and land plants from near Benmore Dam, southern Canterbury, New Zealand. *Journal of the Royal Society of New Zealand 13*: 129–154.
RETALLACK, G. J. 1985: Triassic fossil plant fragments from shallow marine rocks of the Murihiku supergroup, New Zealand. *Journal of the Royal Society of New Zealand. 15*: 1–26.
RETALLACK, G. J.; RYBURN, R. J. 1982: Middle Triassic deltaic deposits in Long Gully, near Otematata, north Otago, New Zealand. *Journal of the Royal Society of New Zealand 12*: 207–227.
ROZEFELDS, A. C.; CHRISTOPHEL, D. C.; ALLEY, N. F. 1992: Tertiary occurrence of the fern *Lygodium* (Schizaeaceae) in Australia and New Zealand. *Memoirs of the Queensland Museum 32*: 203–222.
SEWARD, A. C. 1919: *Fossil Plants. Vol. 4. Ginkgoales, Coniferales, Gnetales*. Cambridge University Press, Cambridge. 543 p.
STOPES, M. C. 1914: A new *Araucarioxylon* from New Zealand. *Annals of Botany 28*: 341–350.
STOPES, M. C. 1916: An early type of the Abietineae(?) from the Cretaceous of New Zealand. *Annals of Botany 30*: 111–125.
SUTHERLAND, J. I. 1994: Some aspects of petrifaction of Miocene wood from Kaipara Harbour, New Zealand. *Geoscience Reports Shizuoka University 20*: 143–151.
SUTHERLAND, J. I. 2003: Miocene petrified wood and associated boring and termite faecal pellets from Hulatere Peninsula, North Auckland, New Zealand. *Journal of the Royal Society of New Zealand 33*: 395–414.
TIDWELL, W. D. 1994: *Ashicaulis*, a new genus for some species of *Millerocaulis* (Osmundaceae). *Sida 16*: 253–261.
TOWNROW, J. A. 1967: On *Rissikia* and *Mataia*, podocarpaceous conifers from the lower Mesozoic of southern lands. *Papers and Proceedings of the Royal Society of Tasmania 101*: 103–136.
UNGER, F. 1864: Fossile Pflanzenreste aus Neu–Seeland. *Paläontologie von Neu–Seeland, Novara–Expedition. Th. 1: Bd. 2*: 1–13.
VAN KONIJNENBURG-VAN CITTERT, J. H. A. 2008: A *Palissya* from the Jurassic flora of Yorkshire, with in situ pollen. [12th International Palynological Congress (IPC–XII). 8th International Organisation of Palaeobotany Conference (IOPC–VIII). Abstract Volume.] *Terra Nostra 2008(2)*: 292–293.
WAGSTAFF, S. J. 2004: Evolution and biogeography of the austral genus *Phyllocladus* (Podocarpaceae). *Journal of Biogeography 31:* 1569–1577.
WATSON, J.; SINCOCK, C. A. 1992: Bennettitales of the English Wealden. *Monograph of the Palaeontological Society, London 588* [*Part Vol. 145, 1991*]: 1–228.
WILSON, K. L.; JOHNSON, L. A. S. 1989: Casuarinaceae. Pp. 100–202 in: George, A. S. (ed.), *Flora of Australia, Volume 3, Hamamelidales to Casuarinales*. AGPS, Canberra. 219 p.

Checklist of New Zealand plant macrofossils

Family placement of conifer foliage is considered 'indet.' (indeterminable) in the absence of epidermal information, with some exceptions (e.g. *Phyllocladus*, which has a distinctive morphology). Family placement of angiosperms generally follows the original author, adapted to the Catalogue of Life and APG III systems. Those published generic records here considered to be so poorly identified with extant genera as to be worthless (and therefore possibly of dubious family placement), are listed in inverted commas. Where extinct genera have been described and placed in extant families with little comparative basis, the entire binomial is in inverted commas. The first symbols following the author citation refer to the age of the fossil. These are the standard abbreviations of the New Zealand Stage system (Edwards et al. 1988) or a more general International age. E signifies a potentially endemic species. This has not been applied to any taxon in inverted commas (which are all technically endemic but unreliable), to conifer foliage without epidermal detail, and to wood (which needs modern revision).

KINGDOM PLANTAE
SUBKINGDOM VIRIDIPLANTAE
INFRAKINGDOM CHLOROPHYTA
PHYLUM CHLOROPHYTA
Class ULVOPHYTA
Order SIPHONALES
CODIACEAE
Shonabellia verrucosa Retallack 1983 Gk E

INFRAKINGDOM STREPTOPHYTA
PHYLUM BRYOPHYTA
Class MARCHANTIOPSIDA
Order MARCHANTIALES
Marchantites sp. Bo

PHYLUM TRACHEOPHYTA
SUBPHYLUM LYCOPHYTINA
Class LYCOPODIOPSIDA
Order LYCOPODIALES
Lycopodites arberi Edwards 1934 Kt E
Lycopodium cf. *volubile* Forst. f. 1956 Cn-Ar

Order SELAGINELLALES
SELAGINELLACEAE
Selaginella sp. Ge

SUBPHYLUM EUPHYLLOPHYTINA
INFRAPHYLUM MONILOPHYTAE
Class EQUISETOPSIDA
Asterotheca hilariensis Menendez 1957 Ge
Equisetites sp. Permian
Equisetites hollowayi Edwards 1934 Kt E
Equisetites minuta (Arber) Townrow LTri-EJur E
Equisetites nicoli Arber 1917 Kt-Kh E
Equisetum sp. Cn-Ar
Neocalamites carrerei (Zeiller) Halle 1908 Ge-Gk
Neocalamostachys cf. *carrerei* (Zeiller) Halle 1908 Ge
Phyllotheca minuta Arber 1917 Bo EJur E

Class POLYPODIOPSIDA
ADIANTACEAE
'Pteris' pterioides (Ettingshausen) Oliver 1950 Mh-Dt
ASPLENIACEAE
'Aspidium' otagoicum Ettingshausen 1887 Mp
'Aspidium' tertiario-zeelandicum Ettingshausen 1887 Mp
'Asplenium' palaeopteris Unger 1864 Latest Jur-earliest Cre
BLECHNACEAE
Blechnum cf. *procerum* Po-Pl
'Blechnum' priscum Ettingshausen 1887 Mh-Dt
DAVALLIACEAE
Davallia walkeri Conran et al. 2010 Lw-Pl
DENNSTAEDTIACEAE
Pteridium esculentum (Forst.) Ckn. 1928 Wc
DICKSONIACEAE
'Dicksonia' pterioides Ettingshausen 1887 Mh-Dt
DIPTERIDACEAE
Dictyophyllum acutilobum (Braun) Schenk 1917 LTri-EJur
Dictyophyllum obtusilobum? (Braun) Schenk 1917 Kt-Kh
Dictyophyllum rugosum L. & H. 1950 Late Kt
Goeppertella cf. *woodii* P. McA. Rees 1993 LTri-EJur
GLEICHENIACEAE
'Gleichenia' obscura Ettingshausen 1887 Mh-Dt
'Sticherus' obscurus (Ettingshausen) Oliver 1950 Mh-Dt
OSMUNDACEAE
Ashicaulis gibbiana (Kidston & Gwynne-Vaughan) Tidwell 1907 Jur E
Cladophlebis sp. *'australis'* (Morris) Seward 1917 Jur–Cre
Millerocaulis dunlopi (Kidston & Gwynne-Vaughan) Tidwell 1907 Jur E
Todites maoricus Retallack 1981 Gk E
POLYPODIACEAE

Platycerium morgani Oliver 1928 Wc E
'Polypodium' hochstetteri Unger 1864 Latest Jur-earliest Cre
'Psaronius' huttonianus Crie 1889 Late Kt
SCHIZAEACEAE
Lygodium sp. Dt
THELYPTERIDACEAE
'Cyclosorus' cretaceo-zeelandicus Oliver 1950 Mh-Dt
Pneumatopteris pennigera (Forst. f.) Holttum Po-Pl
INCERTAE SEDIS
Chiropteris biloba Bell 1956 Gk E
Chiropteris lacerata Arber 1917 Bo E
Chiropteris waitakiensis Bell 1956 Gk E
Cladophlebis cf. *albertsi* (Dunker) Seward 1917 Latest Jur-earliest Cre
Cladophlebis cf. *takezakii* Oishi 1940 Gk
Cladophlebis antarctica Nathorst 1934 Kt-earliest Cre
Cladophlebis australis (Morris) Halle 1913 Gk
Cladophlebis carnei Holmes & Ash 1979 Gm-Ge
Cladophlebis denticulata (Brongniart) Seward 1917 K-O
Cladophlebis indica (Oldham & Morris) Sahni & Rao 1933 Ge-Gk
Cladophlebis obscura (Ettingshausen) McQueen 1956 Mh-Dt E
Cladophlebis prisca (Ettingshausen) McQueen 1956 Mh-Dt E
Cladophlebis reversa (Feistmantel) Seward & Holttum 1934 Late Kt to earliest Cre
Cladophlebis roylei Arber 1954 Perm
Cladophlebis wellmanii McQueen 1956 Mh-Dt E
Coniopteris hymenophylloides (Brongniart) Seward 1900 K-O
Coniopteris? lobata (Oldham & Morris) Edwards 1934 Latest Jur-Dt
Lobifolia dejerseyi Retallack 1977 Ge-Gk
Microphyllopteris pectinata Hector ex Arber 1917 Late Kt
Pachypteris sp.
Phyllopteroides cf. *dentata* Medwell 1954 C
Phyllopteroides cf. *laevis* Cantrill & Webb 1987 C
Phyllopteroides cf. *lanceolata* (Walkom) Medwell 1954 C
Phyllopteroides cf. *serrata* Cantrill & Webb 1987 C
Rufordia goepperti Seward 1950 Kt-Kh
Sphenopteris currani? (Tenison-Woods) Arber 1917 Kt-Kh
Sphenopteris fittoni Seward 1894 Latest Jur-earliest Cre
Sphenopteris gorensis Arber 1917 Kt-Kh E
Sphenopteris mackayi McQueen 1956 Cn-Mh E
Sphenopteris otagoensis Arber 1917 Late Kt E
Sphenopteris owakaensis Arber 1917 Late Kt E
Sphenopteris pterioides (Ettingshausen) McQueen 1956 Mp-Dt E
Sphenopteris cf. *lobifolia* Morris 1954 Perm
Taeniopteris batleyensis Edwards 1926 Mp E
Thinnfeldia feistmanteli? (Gothan) 1917 Kt
Thinnfeldia lancifolia (Morris) 1917 Bo-late Kt
Thinnfeldia odontopteroides (Morris) Arber 1917 Late Kt
Thinnfeldia cf. *argentinica* 1917 LTri-EJur
Yabiella sp. Bw

INFRAPHYLUM SPERMATOPHYTAE
Class CYCADOPSIDA
Order CYCADALES
ZAMIACEAE
Macrozamia sp. Mh
Pterostoma sp. Mh
Pterostoma douglasii Hill & Pole 1994 Po-Pl E

Class GINKGOOPSIDA
Order GINKGOALES
GINKGOACEAE
Ginkgo digitata (Brongn.) Heer 1956 Gk
Ginkgo sp. Mh
KARKENIACEAE
Karkenia fecunda Retallack 1981 Gk E
INCERTAE SEDIS
Baiera robusta Arber 1917 Bo
Sphenobaiera robusta (Arber) Florin 1936 Ge-Gk

Class PINOPSIDA
Order BENNETTITALES
Ptilophyllum acutifolium Morris 1917 Kt
Ptilophyllum seymouricum McQueen 1956 C E
Pterophyllum clarencianum McQueen 1956 C E
Pterophyllum matauriensis Hector ex Arber 1917 Late Kt E
Zamites cf. *takurarensis* Walkom 1919 Alb

Order CORDAITALES
Noeggerathiopsis hislopii (Bunbury) Walkom 1954 Perm

Order CORYSTOSPERMALES
CORYSTOSPERMACEAE
Dicroidium cf. *dubium* var. *dubium* Anderson & Anderson 1983 Bo
Dicroidium dubium var. *tasmaniense* (Anderson & Anderson) Retallack 1977 Gm-Gk
Dicroidium eskense (Walkom) Jacob & Jacob 1950 Ge-Gk
Dicroidium incisum (Frenguelli) Anderson & Anderson 1970 Ge-Gk
Dicroidium lancifolium var. *lancifolium* (Morris) Gotham 1912 Ge-Gk
Dicroidium odontopteroides var. *argentum* Retallack 1977 Ge-Gk
— var. *moltenense* Retallack 1977 Ge-Gk
— var. *odontopteroides* (Morris) Gotham 1912 Ge-Gk
— var. *remotum* (Szajnocha) Retallack 1977 Ge-Gk
Dicroidium prolungatum (Menendez) Retallack 1977 Ge-Gk
Dicroidium zuberi var. *feistmantelii* (Johnston) Retallack 1977 Ge
— var. *papillatum* (Townrow) Retallack 1977 Ge-Gk
— var. *sahni* (Seward) Retallack 1977 Ge-Gk
Dicroidium stelznerianum var. *stelznerianum* (Geinitz) Frenguelli 1941 Bw
Johnstonia dutoitii Retallack 1977 Ge-Gk
Johnstonia stelzneriana var. *stelzneriana* Ge-Gk
Pteruchus cf. *geminatus* Thomas 1933 Ge
?*Pteruchus barrealensis* var. *feistmantelii* Holmes et Ash 1979 Gm-Ge
Pteruchus dubius Thomas 1933 Gk
Pteruchus johnstonii (Feistmantel) Townrow 1962 Ge-Gk
Pilophorosperma sp. A Retallack 1980 Gk
Pilophorosperma cf. *costulatum* Ge-Gk
Umkomasia cf. *macleanii* Ge
?*Umkomasia* sp. Bo

Order GLOSSOPTERIDALES
Glossopteris ampla Dana 1849 Perm

Order PELTASPERMALES
PELTASPERMACEAE
Antevsia sp. Bo
Lepidopteris madagascariensis Carpentier 1935 Gk
Peltaspermum sp. indet. Gk
Peltaspermum cournanei Pole & Raine Bo E
Pachydermophyllum dubium (Burges) Retallack 1981 Gk
Pachydermophyllum praecordillerae (Frenguelli) Retallack 1981 Gm-Bo
Pachydermophyllum sp. indet. Gk
Townrovia petasata (Townrow) Retallack 1981 Gk

Order PENTOXYLALES
PENTOXYLACEAE
Carnoconites cranwelli T.M.Harris 1962 Latest Jur-earliest Cre E
Taeniopteris daintreei McCoy 1917 Jur

Order INCERTAE SEDIS
Carpolithus mackayi Arber 1917 Gk-Bw
Ginkgophytopsis sp. Bo
Ginkgophytopsis cuneata (Carruthers) Retallack 1980 Gk
Ginkgophytopsis lacerata (Arber) Retallack 1980 Gk E
Ginkgophytopsis tasmanica (Walkom) Retallack 1980 Gk
Linguifolium arctum Menendez 1951 Gk-Bo
Linguifolium lilleanum Arber 1913 Gk-Bo E
Linguifolium lillieanum Arber 1917 Bo-Kh E
Linguifolium steinmannii (Solms-Laubach) Frenguelli 1941 Gk-Bo
Linguifolium tenison-woodsii (Etheridge) Retallack 1980 Gk
Linguifolium waitakiense Bell 1956 Gk E
cf. *Linguifolium* Perm
Nageiopteris longifolia? Fontaine 1917 Latest Jur-earliest Cre
Nicola zelandica Unger 1864 Mid Cenozoic
Nilssonia compta? (Phillips) Schenk 1883 Kt-Kh
Nilssonia elegans Arber 1917 Late Kt E
Palissya bartrumi Edwards 1934 Mid Jur-earliest Cre E
Phylopteris expansa Walkom 1956 Cn-Dt
Phylopteris lanceolata Walkom 1956 Mh
Taeniopteris carruthersii Tennison-Woods 1883 Gm-Ge
Taeniopteris crassinervis (Feistmantel) 1917 Late Kt
Taeniopteris lentriculiformis (Etheridge) Walkom 1917 Ge
Taeniopteris stipulata Hector 1956 Cn-Mh E
Taeniopteris thomsoniana Arber 1917 emend. Blaschke & Grant-Mackie 1976 Bo E
Taeniopteris vittata Brongniart 1917 Kt
Taeniopteris cf. *lentriculiformis* (Etheridge) Walkom Gm-Ge
Taeniopteris sp. Gk

Order CONIFERALES
ARAUCARIACEAE
Agathis sp. Po
Araucaria sp. sect. *Eutacta* Po-Pl
Araucaria danai Ettingshausen 1887 Mp E
Araucaria desmondii Pole 1995 Cen E
Araucaria haastii Ettingshausen 1887 Mp E
Araucaria oweni Ettingshausen 1887 Mp E
Araucaria taieriensis Pole 1995 Mh E
Araucarioides falcata Pole 1995 Mp E
Araucarites cutchensis Feistmantel 1917 Kt-Kh
Araucarites grandis Walkom 1934 Kt
Araucarites marshalli Edwards 1926 Mp-Mh E
Dammara mantelli Ettingshausen 1887 Mh-Dt
Wairarapaia mildenhallii Cantrill et Raine Cen E
CUPRESSACEAE
Libocedrus cf. *bidwillii* Dt
?PODOCARPACEAE 'STEM-GROUP'
Mataia podocarpoides Townrow 1967 Jur
Telemachus lignosus Retallack 1981 Gk
PODOCARPACEAE
Acmopyle masonii Pole 1997 Po-Pl E
Dacrycarpus dacrydioides (A.Rich) de Laubenfels Po-Pl E
cf. *Dacrycarpus* Pole & Douglas 1999 Mh
Ginkgocladus novae-zeelandiae Ettingshausen 1887 LEoc-EOli

Kaia minuta Pole 1995 Mh E
Kakahuia sp. 'Kai Point' Pole & Douglas 1999 Mh E
Kakahuia drinnanii Pole 1998 Dt E
Kakahuia campbellii Pole 1997 Dt E
Katikia inordinata Pole 1995 C E
Lepidothamnus intermedius (T.Kirk) Quinn Po-Pl E
Mumu somerensis Pole 1998 Dt E
Phyllocladus sp. Po-Pl
Phyllocladus palmerii Pole & Moore Tt-Tk
Podocarpus alwyniae Pole 1992 Po-Pl E
Podocarpus travisiae Pole 1996 Po E
Podocarpus sp. 'Mata Creek' Pole 1997 Po-Pl
Prumnopitys limaniae Pole 1998 Dt E
Prumnopitys opihiensis Pole 1997 Dt E
Prumnopitys taxifolia (D.Don) de Laubenfels Po-Pl E
Prumnopitys sp. 'Mt Somers' Pole 1998 Dt
Retrophyllum vulcanense Pole 1992 Po-Pl E
PODOCARPACEAE or TAXACEAE
Podocarpaceae/Taxaceae gen. et sp. indet Pole 1997 Po–Pl
TAXACEAE or TAXODIACEAE
Paahake papillatus Pole 1998 Dt E
Paahake cf. *papillatus* Pole & Douglas 1999 Mh
Mataoraphyllum miocenicus Pole & Moore Tt-Tk E
TAXODIACEAE
Maikuku stephaniae Pole & Douglas 1999 Mh E
Otakauia lanceolata Pole 1995 Mh E
Sequoiadendron novae-zeelandiae (Ettingshausen) Pole 1995 Mp E
Waro riderensis Pole & Douglas 1999 Mh E
INCERTAE SEDIS
Araucarioxylon australe Crie 1889 Kt
Araucarioxylon novae-zeelandiae Stopes 1914 Mp
Dacrydinium cupressinum Ettingshausen 1887 Mh-Dt
'*Dacrydium*' *praecupressinum* Ettingshausen 1887 Mp
Dadoxylon australe (Crie) Seward 1934 Kt
Dadoxylon kaiparaensis Edwards 1926 Mp-Mh
Desmiopshyllum elongatum (Morris) Retallack 1980 Ge-Gk
Desmiopshyllum cf. *indicum* Sahni 1928 Ge-Bo
Desmiopshyllum sp. Ge-Gk
Elatocladus conferta (Oldham & Morris) 1917 Jurassic
Elatocladus plana (Feistmantel) 1934 Latest Jur-earliest Cre
Elatocladus tenuifolia (McCoy) Arber 1940 Latest Jur-earliest Cre
Heidiphyllum elongatum (Morris) Retallack 1981 Ge-Gk
Hoiki mcqueenii Pole 1998 Dt E
Podozamites gracilis Arber 1917 Kt
Planoxylon hectori Stopes 1916 Mp
Pagiophyllim peregrinum (Lindley & Hutton) 1917 Late Kt
Podocarpites tolleyi McQueen 1954 Late Wc
Podocarpium cupressinum Ettingsahausen 1887 Mh-EOli
Podocarpium dacrydioides Unger 1864
Podocarpium praedacrydioides Ettingshausen 1864 Mh-Dt
Podocarpium tenuifolium Ettingshausen 1887 Mh-Dt
Podocarpium ungeri Ettingshausen 1887 Mh-Dt
'*Podocarpus*' *hochstetteri* Ettingshausen 1887 Mp
'*Podocarpus*' *obtusifolius* Oliver 1936 Sl-Sw
'*Podocarpus*' *parkeri* Ettingshausen 1887 Mp
'*Podocarpus*' *praedacrydioides* (Ettingshausen) 1950 Mh-Dt
'*Podocarpus*' *trinerva* (Ettingshausen) 1950 LEoc-EOli
'*Taxodium*' *distichum eocenicum* Ettingshausen 1887 Mp

Class MAGNOLIOPSIDA
MONOCOTS
ARECACEAE
'*Cocos*' *zeylandica* Berry 1926 Sw-Tk
'*Flabellaria sublongirhachis*' Ettingshausen 1887 MEoc
Lepidocaryopsis zeylanicus Hartwich et al. Ar E
Phoenicites zeelandica (Ettingshausen) Pole 1993 Po-Pl
Rhopalostylis sapida (Forst.) Wendl. et Drude 1928 Wc E
LILIACEAE S.L.
'*Cinnamomum miocenicum*' Holden 1982
Ripogonum scandens J.R. et G.Forst. Po-Pl E
MUSACEAE
Pakawaua speciosa (Ettingshausen) Oliver ex Mildenhall 1972 Mh-Dt E
ORCHIDACEAE
Dendrobium winikaphyllum Conran, Bannister & Lee Po E
Earina fouldenensis Conran, Bannister & Lee Po E
POACEAE
'*Bambusites australis*' Ettingshausen 1887 Mh-Dt
'*Poacites nelsonicus*' Ettingshausen 1887 LEoc-EOli

EUDICOTS
ACERACEAE
'*Acer*' *subtrilobatum* Ettingshausen 1887 Mp
'*Acer*' *tasmani* (Ettingshausen) 1950 Mp
APOCYNACEAE
'*Apocynophyllum affine*' Ettingshausen 1887 Early Miocene
'*Apocynophyllum elegans*' Ettingshausen 1887 EMio
ARALIACEAE
'*Aralia*' *tasmani* Ettingshausen 1887 Mp
'*Nothopanax*' *distans* Oliver 1936 Sl-Sw
'*Nothopanax*' *edgerleyi* Harms 1954 E-mid Ple
Pseudopanax crassifolium (Sol. ex A. Cunn.) 1928 Wc E
ASTERACEAE
'*Senecio*' *pliocenicus* Oliver 1936 Sl-Sw
BETULACEAE
'*Alnus*' *novae-zeelandiae* Ettingshausen 1887 Mp
Betulites couperi McQueen 1954 Wo
?Betulaceae indet.Mh
CASUARINACEAE
Casuarina avenacea Campbell & Holden 1984 Po-Pl E
'*Casuarina*' *deleta* Ettingshausen 1887 Mp
Casuarina stellata Campbell & Holden 1984 Po-Pl E
'*Casuarinites cretaceus*' Ettingshausen 1887 Mh-MEoc
CELASTRACEAE
'*Celastrophyllum australe*' Ettingshausen 1887 MEoc
'*Elaeodendron*' *rigidum* Ettingshausen 1887 EMio
CORIARIACEAE
'*Coriaria*' *arborea* Lindsay 1928 Wc
'*Coriaria*' *latepetiolata* Oliver 1936 Sl-Sw
CUNONIACEAE
'*Ceratopetalum*' *kaikoraiensis* Oliver 1936 Sl-Sw
'*Ceratopetalum*' *pacificum* Oliver 1928 Wc
'*Ceratopetalum*' *rivulare* Ettingshausen 1887 MEoc
?*Weinmannia racemosa* Po-Pl
EBENACEAE
'*Diospyros*' *novae-zeelandiae* Ettingshausen 1887 Mp
ELAEOCARPACEAE
aff. *Elaeocarpus*/*Sloanea* indet. Po-Pl
Sloanea sp. Dw-Ab
ERICACEAE
Cyathodophyllum novae-zelandiae Jordan et al. 2010 Lw-Po E
Richeaphyllum waimumuensis Jordan et al. 2010 Lw-Po E
EUPHORBIACEAE
'*Euphorbia*' *lingua* (Unger) Oliver 1950 Eoc-Oli
Malloranga fouldenensis Lee et al. 2010 Po E
FABACEAE
'*Cassia*' *pluvalis* Penseler 1930 Ar
'*Cassia*' *pseudomemnonia* Ettingshausen 1887 Mp
'*Cassia*' *pseudophaseolites* Ettingshausen 1887 Mp
'*Carmichaelia*' *australis* R. Br. 1928 Wc
'*Dalbergia*' *australis* Ettingshausen 1887 Mp
'*Dalbergiophyllum nelsonicum*' Ettingshausen 1887 LEoc-EOli
'*Dalbergiophyllum rivulare*' Ettingshausen 1887 MEoc
'*Palaeocassia phaseolitoides*' Ettingshausen 1887 MEoc
Parvileguminophyllum sp. Po-Pl
FAGACEAE
'*Dryophyllum dubium*' Ettingshausen 1887 EMio
'*Dryophyllum nelsonicum*' Ettingshausen 1887 Mp
'*Fagus*' *nelsonica* Ettingshausen 1887 LEoc-EOli
'*Fagus*' *producta* Ettingshausen Mh-Dt
'*Quercus*' *calliprinoides* Ettingshausen 1887 MEoc-EOli
'*Quercus*' *celastrifolia* Ettingshausen 1887 Mp
'*Quercus*' *deleta* Ettingshausen 1887 Mp
'*Quercus*' *lonchitoides* Ettingshausen 1887 Mp
'*Quercus*' *nelsonica* Ettingshausen 1887 LEoc-EOli
'*Quercus*' *pachyphylla* Ettingshausen 1887 MEoc
'*Quercus*' *parkeri* Ettingshausen 1887 Mp
GROSSULARIACEAE
'*Ixerba*' *semidentata* Oliver 1936 Sl-Sw
LAURACEAE
Beilschmiedia ovata Oliver 1928 Wc E
'*Beilschmiedia*' *tarairoides* Penseler 1930 Ar
'*Cinnamomum*' *haastii* Ettingshausen 1807 Mh-Dt
'*Cinnamomum*' *intermedium* Ettingshausen 1807 Mp
'*Cinnamomum*' *waikatoensis* Penseler 1930 Ar
Cryptocarya sp. Pole 2007c EMio
Daphnophyllum australe Ettingshausen 1887 Mh
Endiandra sp. Pole 2007c EMio
Laurophyllum longfordiensis (Holden) Pole 1993 S
Laurophyllum tenuinerve Ettingshausen 1887 Mp
?*Litsea dawsoniana* Holden 1982 S
'*Litsea*' *calicaris* (A.Cunn.) Benth. et Hook. f. 1928 Wc
LOGANIACEAE
'*Geniostoma*' *apicullata* Penseler 1930 Ar
'*Geniostoma*' *oblonga* Oliver 1936 Sl-Sw
LORANTHACEAE
'*Loranthophyllum dubium*' Unger 1864 Eoc-Oli
'*Loranthophyllum griselinia*' Unger 1930?
'*Loranthus*' *otagoicus* Ettingshausen 1887 Mp
MALVACEAE
Plagianthus antiquus Oliver 1928 Wc E
Plagianthus betulinus A. Cunn. 1954 Wc E
ATHEROSPERMATACEAE
'*Laurelia*' *cuneata* Oliver 1936 Sl-Sw
MORACEAE
'*Ficus*' *similis* Ettingshausen 1887 LEoc
'*Ficus*' *sublanceolata* Ettingshausen 1887 Mp
MYRICACEAE
'*Myrica*' *proxima* Ettingshausen 1887 Mp-Mh
'*Myrica*' *subintegrifolia* Ettingshausen 1887 Mp
MYRSINACEAE
Myrsine waihiensis Pole & Moore Tt-Tk E
MYRTACEAE
'*Eucalyptus*' *dubia* Ettingshausen 1887 Mp
Eucalyptus sp. Po-Pl
aff. *Eucalyptus* sp. Pole 1994 Dw-Ab
Leptospermum ericoides A. Rich. 1954 Wc
Metrosideros sp. Po-Pl
'*Metrosideros*' *laeta* Oliver 1936 Sl-Sw
'*Metrosideros*' *pliocenica* Oliver 1936 Sl-Sw
aff. *Rhodomyrtus-Rhodamnia* sp. Pole 1994 Dw-Ab
NOTHOFAGACEAE
Nothofagus ninnisiana (Unger) Oliver 1864 Ar E

Nothofagus novae-zeelandiae (Oliver) Holden 1982 emend. Pole 1936 Po-Pl to Sl-Sw E
Nothofagus bidentatus Holden 1982 S E
Nothofagus oblonga Holden 1982 S E
Nothofagus oliveri Holden 1982 S E
Nothofagus ulmifolia (Ettingshausen) Pole 1992 Mp E
Nothofagus praequercifolia (Ettingshausen) Pole 1992 Mp E
Nothofagus cf. *cliffortioides* x *fusca* 1954 Wn E
Nothofagus lendenfeldi (Ettingshausen) Oliver 1950 Mp-Mh E
Nothofagus ulmifolia (Ettingshausen) 1950 Mp E
Nothofagus pinnata (Oliver) Pole 1936 Sl-Sw E
Nothofagus azureus Pole 1993 Po-Pl E
Nothofagus melanoides Pole 1992 Po-Pl E
NYCTAGINACEAE
'Calpida' zealandica Oliver 1936 Sl-Sw
'Pisonia' oliveri Penseler 1930 Ar
'Pisonia' purchasi (Ung.) Penseler 1930 Ar
PENNANTIACEAE
Pennantia corymbosa Forst. 1928 Wc E
PIPERACEAE
Piper erewhonensis McQueen 1954 Wo E
PITTOSPORACEAE
Pittosporum cf. *colensoi* Hook. f. 1954 Wc
POLYGONACEAE
Muehlenbeckia aff. *M. australis* Meissn. Po-Pl
Rumex pachyperianthus McQueen 1954 Late Wc E
PROTEACEAE
Banksia novae-zelandiae Carpenter et al. Lw E
'Dryandra' comptoniaefolia Ettingshausen 1887 Mh-Dt
Dryandroides pakawauica Ettingshausen 1807 Mh-Dt E
Embothrieae gen. et sp. indet. Po-Pl
Gevuininae–*Hicksbeachia* Po-Pl
Helicia sp. Po-Pl
'?Kermadecia merytifolia' Holden 1982 S
'Knightia' excelsa R. Br. 1936 Sl-Sw
'Knightia' oblonga Oliver 1936 Sl-Sw
'Knightiophyllum primaevum' Ettingshausen 1887 MEoc
Lomatia novae-zelandiae Pole 1997 Dt
Longfordia banksiaefolia Holden 1982 S
Macadamia sp. Po-Pl
Musgravea sp. Po-Pl
cf. *Orites excelsa* R.Br. Dw-Ab
Persoonieaephyllum villosum Carpenter et al Lw E
PLATANACEAE
Platanaceae gen. et sp. indet. Cen
Proteoides ohukaensis McQueen 1954 ?Wn E
RANUNCULACEAE
Clematis obovata Oliver 1928 Wc E
RUBIACEAE
'Coprosma' australis (A.Rich.) Robinson 1954 Wc
'Coprosma' pliocenica Oliver 1936 Sl-Sw
'Coprosma' praerepens Oliver 1936 Sl-Sw
'Coprosma' pseudoretusa Penseler 1930 Ar
RUTACEAE
Melicope ternata Forst. 1954 Wc E
ROSACEAE
Rubus australis Oliver 1928 Wc E
SANTALACEAE
'Santalum' subacheronticum Ettingshausen 1887 Mp
SAPINDACEAE
'Cupanites novae-zeelandiae' Ettingshausen 1887 MEoc
'Sapindophyllum coriaceum' Ettingshausen 1887 LEoc-EOli
'Sapindus' subfalcifolius Ettingshausen 1887 LEoc-EOli

TWENTY-SEVEN

Plant microfossils

POLLEN AND SPORES

DALLAS C. MILDENHALL, J. IAN RAINE

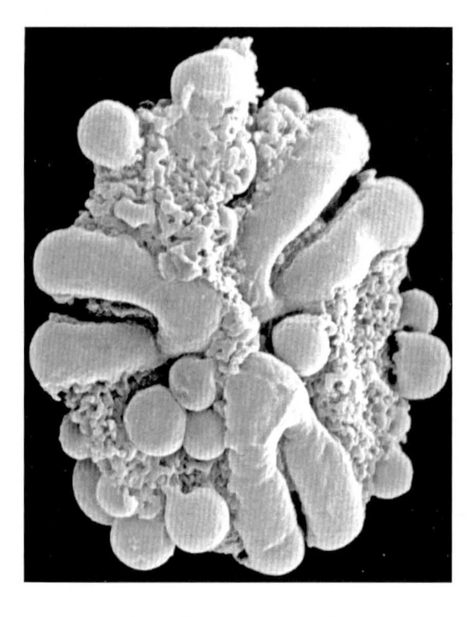

Oacolpopollenites sp., a tricolporate legume pollen (Angiospermae, Fabaceae, Caesalpinioideae?), 33 micrometres diameter. Late Oligocene, Southland.

From Pocknall 1982

Among the most resistant, and therefore fossilisable, parts of plants are the microscopic grains known as pollen and spores. Pollen is derived from flowering and coniferous plants, while ferns and lower plants produce spores. In fossil material, only the resistant cell wall (exine) of pollen is preserved, but this usually has the important structural and sculptural features that are characteristic of species or higher taxonomic levels. Most pollen grains and spores are between about 10–80 micrometres in size and can be studied satisfactorily with an optical microscope, although scanning-electron-microscope (SEM) observations are usually advisable for the description of new taxa and accurate determination of botanical affinity.

Spores are generally larger than pollen grains in size and are classified according to whether they have monolete or trilete scar marks or no scar marks at all. Monolete spores have a single scar-line on the spore indicating the vertical axis along which the mother spore was split into four; in trilete spores, the initial contact of the four spores to each other is such that, when they separate, each shows three lines radiating from a centre pole. Spores have a variety of shapes but most are pyramidal or boat-shaped.

Gymnosperm pollen grains are generally classified according to the presence or absence of sacci (bladders), and their number. Angiosperm pollen grains are classified according to the number and arrangement of their furrows (colpate) and pores (porate). Some grains lack apertures, for example *Beilschmiedia* pollen, or have only vague apertural areas, like the pollen of Cyperaceae. Most grains can therefore be described according to apertural classes, with grains having usually 1–4 or multiple furrows or pores or combinations of both with pores within the furrows. Grains are also recognised on the basis of details of exine structure and sculpture, for which there is a specialised terminology (e.g. Moar 1993).

This review describes what is known about the diversity, distribution, special features, and use of fossil pollen and spores in the New Zealand context. Approximately 750 species of fossil spores and pollen have been identified from the mainland and surrounding islands. Terrestrial palynomorphs (a palynomorph is any pollen-like object, regardless of origin) can be found in most sedimentary rocks, both marine and terrestrial, with research being concentrated on those spores and pollen obtained from rocks about 300 million years old and younger. The prime purposes of these studies are to provide information useful for mapping rocks and for providing information on biostratigraphy, climate change, vegetation history, evolution, depositional environments, paleoecology, and biogeography (see below).

New Zealand is in a relatively unusual position in having an almost complete sedimentary sequence of rocks covering the last 300 million years with just a few small gaps in the record. These rocks document notable events in the history of plant evolution including the appearance and rise of the flowering plants and the major biotic changes at the Permian–Triassic, Triassic–Jurassic, Cretaceous–Cenozoic, Paleogene–Neogene, and Pliocene–Pleistocene boundaries.

The current indigenous New Zealand spore- and pollen-producing vegetation comprises more than 3200 taxa – about 2180 vascular plants (Cockayne 1928; Wardle 1991; Breitwieser et al. this volume) plus 1064 species of bryophytes. In any one pollen diagram from a Late Holocene (last 5000 years) sequence (e.g. Newnham et al. 1998), however, fewer than 100 taxa are recognised. Although fossil sites represent a minimal range of ecological niches, this low figure highlights the difficulty of accurately determining diversity.

Statistical differences between fossil sequences can reflect diversity, although apparent differences may have many causes, including preservational, dispersal, and environmental. However, the New Zealand fossil record suggests that there have been major changes in diversity with time, resulting from causes as varied as climate change on the one hand, and varying land area and habitat availability on the other. For example, optimum growing conditions appear to have occurred in the Early to Middle Miocene about 25–15 million years ago (Ma), where individual pollen slides contain over 120 different identified taxa, plus a number of undescribed ones. Other peaks in diversity occurred in the Eocene and Late Cretaceous, whereas Paleocene and Oligocene floras seem to have been relatively less diverse.

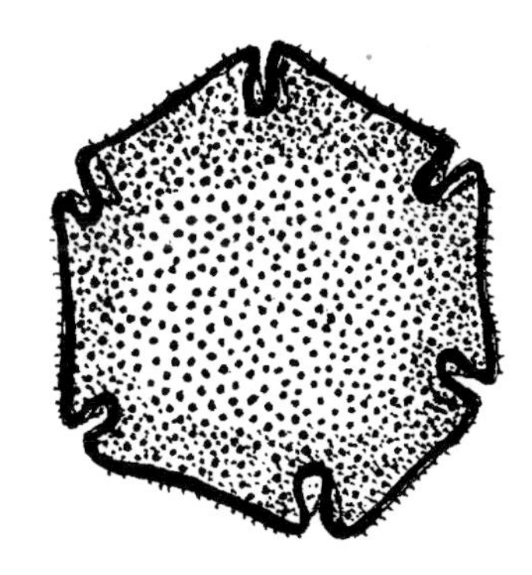

Nothofagidites cranwelliae (Angiospermae, Nothofagaceae, *Nothofagus* subg. *Brassospora*), Late Miocene.

From Couper 1953

Why are these microfossils worthy of study?

History of vegetation

Palynology provides information on the development and history of present-day New Zealand vegetation and the events that led to its composition, formation, and distribution. It is quite clear that the current vegetational patterns are an after-effect of the last major glaciation that concluded about 14,000 years ago, and sections of the vegetation are still adjusting. For example, the *Nothofagus* gap on the west coast of South Island (McGlone et al. 1996) has yet to be occupied by *Nothofagus* following the Last Glaciation.

Evolution

Throughout geological time, plants have appeared and disappeared. By studying these spores and pollen an idea of the source and evolution of current taxa can be derived and theories based on study of morphology, DNA, etc. of modern plants can be tested. For example, current theories about the proteaceous genus *Toronia* suggest that it is an ancient lineage in New Zealand (Johnson & Briggs 1975). However, palynology has shown that *Toronia* did not appear in New Zealand until very late in geological time, about 2–3 Ma (e.g. Nelson et al. 1988). The family Proteaceae, however, appeared well over 65 Ma, and New Zealand has an extensive record of extinct proteaceous taxa.

Paleoecology

The more we know about the history of New Zealand's flora, the more we begin to understand the complex interactions between the flora and fauna and physical variables related to climate. For example, moas are large birds and some appear to have been grassland dwellers (e.g. Stevens 1980), but all food found in stomach remains of fossil moas indicate that they were forest dwellers (Burrows et al. 1981). They appear to have been in New Zealand for the last 30 Ma or so based on molecular biology (Cooper & Cooper 1995) but palynological data show that New Zealand grasslands are a very recent phenomenon associated with the deteriorating climatic conditions that began about 10 Ma. So either the moas were originally forest-dwelling and have entered low-lying vegetation associations only relatively recently, or they were always forest-dwelling but periodically ventured into grassland/shrubland vegetation. Certainly during the last 2 Ma New Zealand had, for long periods of time, substantial grassland/

Coniopteris hymenophylloides (Polypodiopsida, Dicksoniaceae), Middle Jurassic.
From Edwards 1934

shrubland environments that moas must have survived in (e.g. McGlone & Webb 1981).

Biogeography

Great interest is always aroused when the source of New Zealand's vegetation is discussed. Did New Zealand's vegetation arrive on 'floating' islands now welded together to form the current landmass (e.g. Craw & Heads 1998)? Did it arrive via dispersal from distant sources, or did it simply evolve from the vegetation that was part of the original Cretaceous Gondwana landmass from which New Zealand separated about 80 Ma (Fleming 1962, 1975)? The answer is probably a combination of all three. The debate is over which of these three predominates. Botanical studies indicate different sources for New Zealand plants. As one would expect, some of the extinct taxa described are cosmopolitan, other taxa are austral (southern-hemispheric), others Asian, and some appear to be endemic. Quantifying these sources is not yet a useful exercise for extinct terrestrial taxa, as many of them have no known botanical affinity. What we do know is that the New Zealand vegetation suffered a crisis about 30–35 Ma when the landmass was reduced to a series of islands not much bigger than the province of Canterbury (Cooper & Cooper 1995). Speciation was rapid after this event with high diversity occurring about 15–25 Ma. Then New Zealand suffered from the effects of periodic glaciations, with increasing intensity, from about 10 Ma and many tropical and subtropical taxa became extinct. The current vegetational associations are the result of these glaciations and a function in particular of the Last Glaciation.

Paleoenvironments

Palynology is important in determining the various terrestrial environments that have occurred in the past, especially in exploration for economic minerals including peat, coal, lignite, gas, oil, and water (e.g. Beggs & Pocknall 1992).

Climate change

Palynology has an important contribution to make in determining the frequency and intensity of climate change. For example, the change between the end of the last glaciation and the arrival of full forest conditions in the Wellington region was sudden with no, or a very short, intervening shrub phase (e.g. Lewis & Mildenhall 1985). There is also evidence that changes of climate are sudden and dramatic (Bluemle et al. 1999), but that is not apparent from long sequences near Wellington (Mildenhall 1995), and research needs to involve paleopalynology to determine what causes these changes, and their intensity and frequency.

Stratigraphy

The determination of New Zealand's geological history is based to a large degree on the dating of rocks using the known sequence of fossils as they appear and disappear with time. Fossil spores and pollen play the same role, with the science of palynology contributing a great deal to the development of our understanding of the different periods of time in which, for example, coal has formed. The geological processes that lead to changes in coal quality can also be assessed, as can changes in the environment in which the coal formed. Since spores and pollen can be found in all kinds of sediments this leads to the possibility that biostratigraphic zonations using marine fossils can be correlated with biostratigraphic zonations developed in terrestrial sediments where marine fossils are never found *in situ*.

Mapping

One of the prime services of a national geological organisation is to produce maps at various scales. Maps require the information developed in the study of the topics above. This means that palynologists rarely work in isolation from researchers in other geological disciplines. The palynologist's prime service here is to provide accurate dates, controls on the timing of various geological events, determination of missing time, paleoenvironmental information, and many other data that can be summarised in map form.

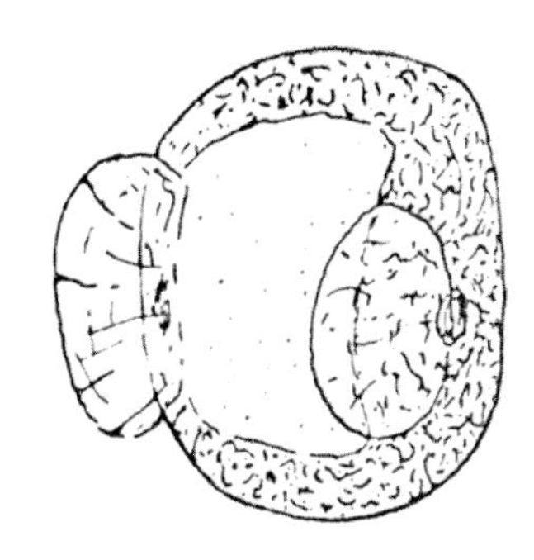

Phyllocladidites mawsonii (Gymnospermae, Podocarpaceae, aff. *Lagarostrobos franklinii*), Late Cretaceous.
From Te Punga 1949

History of study in New Zealand

Considering that serious studies of fossil pollen first commenced overseas in 1916, New Zealand has had a surprisingly long history of palynological exploration and local scientists have always been at the forefront of research into different aspects of the science. This history of study was comprehensively reviewed by Harris (1953, 1968), Raine and Pocknall (1983), and Moar (1993) and is briefly summarised below.

Fossil plants were first recognised in New Zealand well over 200 years ago during James Cook's first voyage, when kauri gum was found in the Northland peatlands (Raine & Pocknall 1983), but the existence of fossil spores and pollen in New Zealand sediments was not fully demonstrated until the early part of the 20th century when Edwards (1934) illustrated spores from fertile pinnae of the Jurassic fern *Coniopteris hymenophylloides* from Waikawa. This paper was closely followed by one dealing with six different Pleistocene localities around New Zealand (Cranwell & von Post 1936) in which the pollen diagrams produced were the first from the Southern Hemisphere. Erdtman (1924) had previously commented on pollen in peats from Central Otago and Chatham Island, but did not produce pollen diagrams or illustrations. Prior to these studies, some investigations of pollen from extant plants had centred around small projects on melissopalynology (grains found in honey) and aeropalynology (grains trapped from the air) (e.g. Waters 1915; Watt 1929).

Detailed studies of fossil spores and pollen did not really get underway until Martin Te Punga (1948, 1949) and Ashley Couper (e.g. 1951–54) produced their papers and bulletins, followed by papers from a succession of stratigraphic palynologists employed by New Zealand Geological Survey and its successor organisations. At the same time, William (Bill) Harris (Soil Bureau and New Zealand Geological Survey of DSIR) and Neville Moar (Botany Division, DSIR), aided by the pioneering descriptive work on modern New Zealand pollen by Lucy Cranwell (1940, 1942, 1953), concentrated on late Pleistocene palynology of New Zealand and its offshore islands. They produced a series of pollen diagrams representative of much of the country's Holocene and Last Glacial vegetation (e.g. Harris 1951, 1953, 1955; Moar 1958a,b, 1959).

The emphasis of the Pleistocene botanists was on climate and vegetational changes through time, leading to an understanding of the distribution and character of today's plant associations. The emphasis of the palynologists working on pre-Pleistocene sediments was more towards the use of spores and pollen as stratigraphic indicators of time. The fossils described tended to be those that were common, had obvious botanical affinities, were relatively large or unusual, and were apparently restricted in stratigraphic range.

Currently, New Zealand Pleistocene palynological studies are being undertaken in several universities and a number of Crown Research Institutes, as well as by private consultants and palynologists stationed overseas. However, pre-Pleistocene studies have been carried out mainly at GNS Science.

The review that follows concentrates on spores produced by tracheophytes and bryophytes. Fungi are also preserved, not only as spores but as small macrofossils in the form of other types of fruiting body. The only fungal reproductive body described from New Zealand is *Microthyriacites grandis* from Late Cretaceous coal measures at Ohai (Cookson 1947). It should be noted that the fossil record contains many other plant remains in fragmentary form, for example leaf hairs, cuticles, resin, wood fibres and vessels, and algal cysts, plus other sexual and asexual reproductive bodies, and, in the fungal kingdom (q.v.), hyphae and sclerotia.

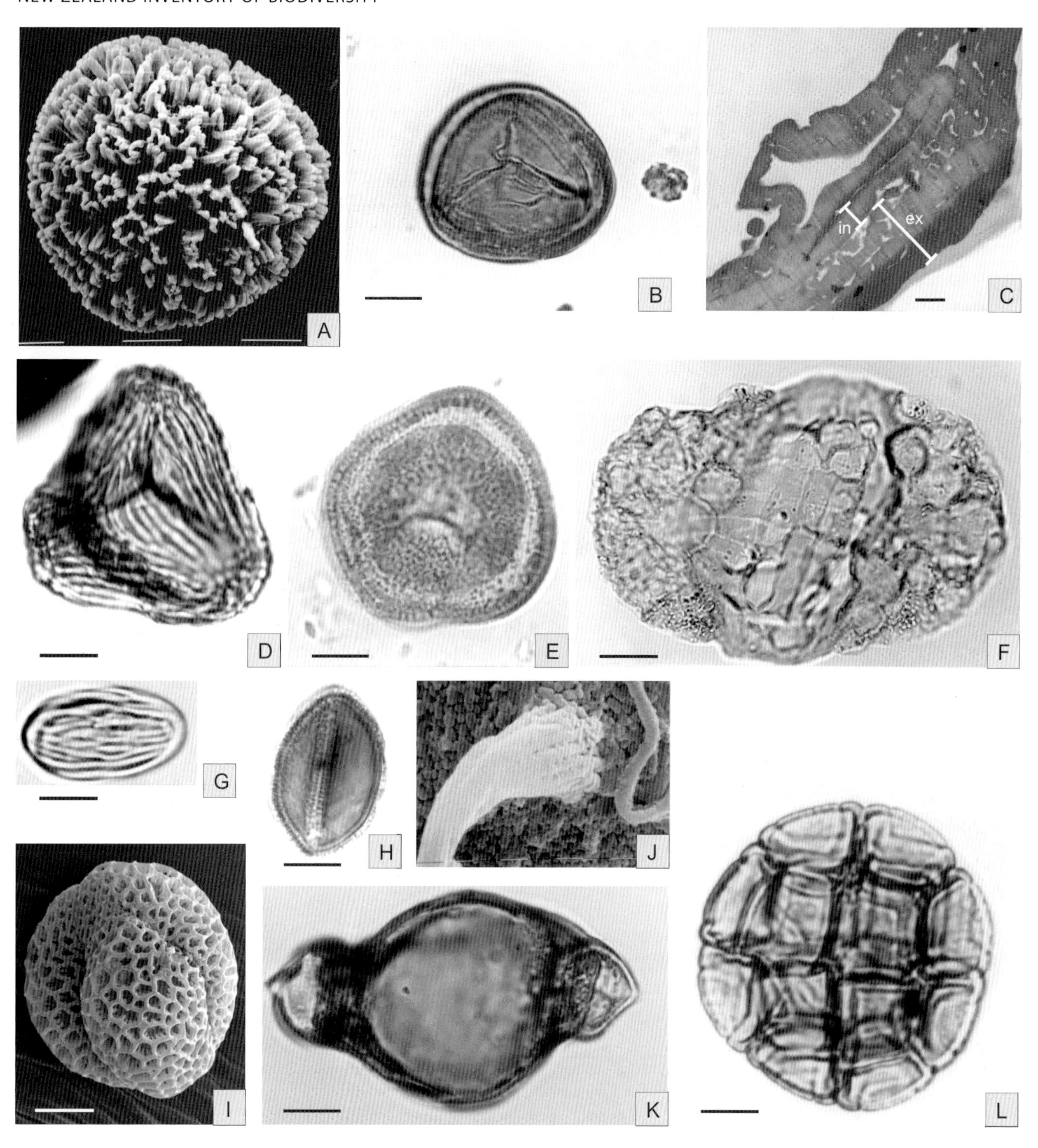

A selection of fossil spores and pollen from various plant groups and geological ages, reillustrated from published works; scale bars 10 µm unless noted. A: *Bryosporis anisopolaris* (Marchantiopsida, Fossombroniaceae?), Pleistocene, SEM, scalebar divisions 10 µm. From Bussell & Mildenhall 1990. B, C: *Densoisporites psilatus* (Lycopsida, Pleuromeiaceae), Late Triassic, LM and TEM, the latter showing the spongy outer wall (ex) and homogeneous inner wall layer (in); scale bar 1 µm. From Raine et al. 1988. D: *Ruffordiaspora australiensis* (Polypodiopsida, Schizaeaceae), Early Cretaceous, LM. From Raine et al. 1981. E: *Corollina* cf. *chateaunovi* (Gymnospermae, Cheirolepidaceae), Late Cretaceous, LM. From Mildenhall 1994. F: *Distriatites insolitus* (Gymnospermae, Glossopteridales?), Middle Triassic, LM. From Jeans et al. 2003. G: *Equisetosporites notensis* (Gymnospermae, Ephedraceae, *Ephedra*), Late Oligocene, LM. From Couper 1960. H: *Clavatipollenites hughesii* (Angiospermae, Chloranthaceae?), Early Cretaceous, LM. From Raine et al. 1981. I: *Fouldenia staminosa* (Angiospermae, Rutaceae?), Early Miocene, SEM. From Bannister et al. 2005. J, K: *Diporites aspis* (Angiospermae, Onagraceae, *Fuchsia*), Early Miocene, SEM detail of root of viscin thread and exine surface, scalebar divisions 1 µm, and LM of whole pollen grain. From Pocknall & Mildenhall 1984. L: *Acaciapollenites myriosporites* (Angiospermae, Fabaceae, *Acacia*), Pleistocene, LM of pollen tetrad. From Mildenhall 1972b. [LM, light microscope; SEM, scanning electron microscope; TEM, transmission electron microscope]

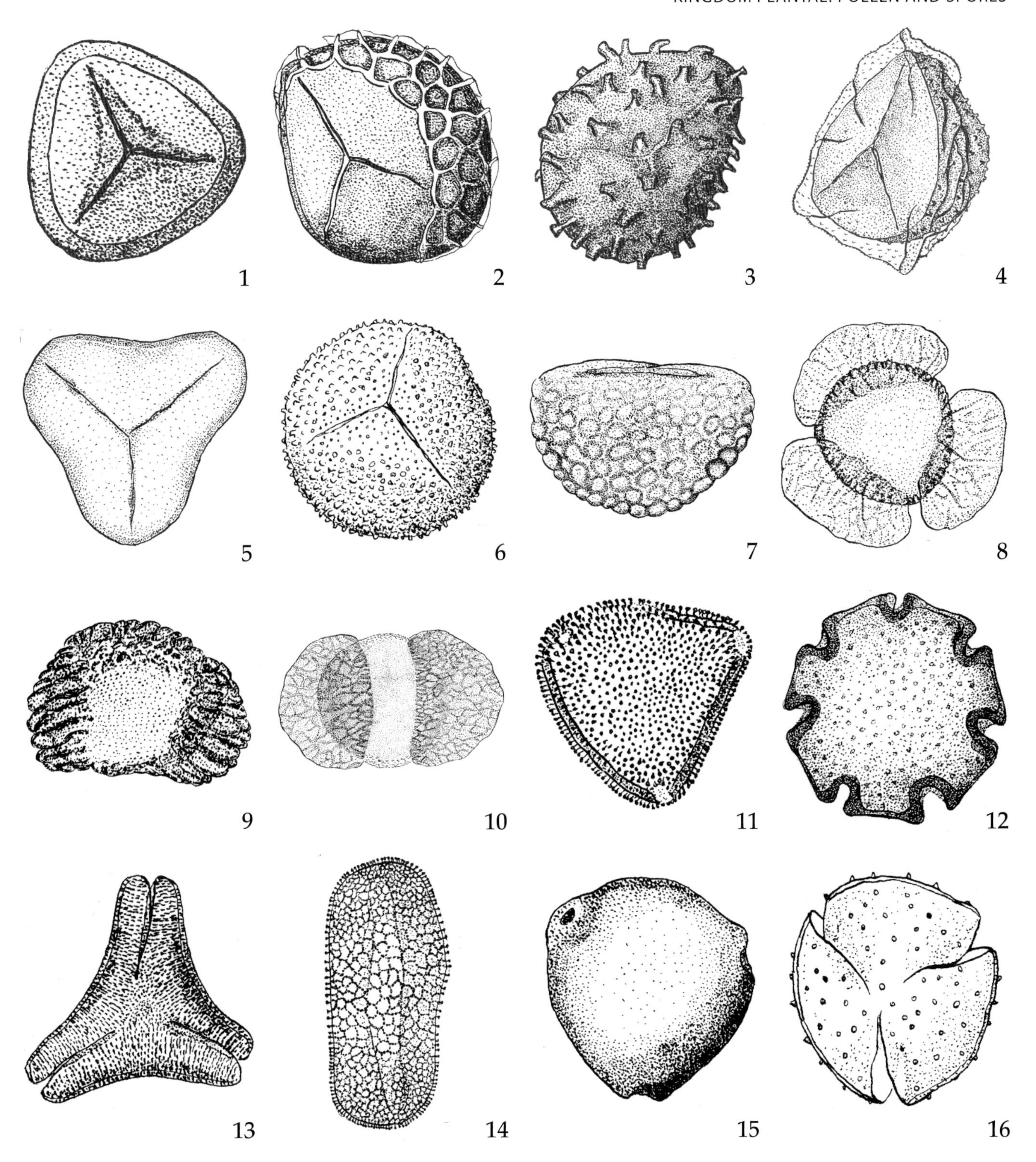

1. *Stereisporites antiquasporites* (Bryophytina, Sphagnaceae), Late Miocene; 2. *Lycopodiumsporites fastigioides* (Lycopsida, Lycopodiopsidaceae), Late Cretaceous; 3. *Neoraistrickia neozealandica* (Lycopsida, Selaginellaceae), Early Cretaceous; 4. *Perotrilites granulatus* (Lycopsida, Selaginellaceae), Early Cretaceous; 5. *Cyathidites australis* (Polypodiopsida, Cyatheaceae?), Early Cretaceous; 6. *Osmundacidites wellmanii* (Polypodiopsida, Osmundaceae), Early Cretaceous; 7. *Polypodiisporites inangahuensis* (Polypodiopsida, Polypodiaceae), Early Cretaceous; 8. *Dacrycarpites australiensis* (Gymnospermae, Podocarpaceae, *Dacrycarpus*), Pliocene; 9. *Dacrydiumites praecupressinoides* (Gymnospermae, Podocarpaceae, *Dacrydium*), Late Cretaceous; 10. *Podocarpidites marwickii* (Gymnospermae, Podocarpaceae), Late Cretaceous; 11. *Proteacidites palisadus* (Angiospermae, Proteaceae), Late Cretaceous; 12. *Nothofagidites flemingii* (Angiospermae, Nothofagaceae, *Nothofagus* subg. *Fuscospora*), Middle Eocene; 13. *Gothanipollis perplexus* (Angiospermae, Loranthaceae, *Elytranthe*), Late Miocene; 14. *Liliacidites kaitangataensis* (Angiospermae, Liliaceae), Late Cretaceous; 15. *Myricipites harrisii* (Angiospermae, Casuarinaceae), Late Oligocene/Early Miocene; 16. *Tricolpites lilliei* (Angiospermae, eudicot angiosperm of uncertain affinity), Late Cretaceous.

All images from Couper 1953

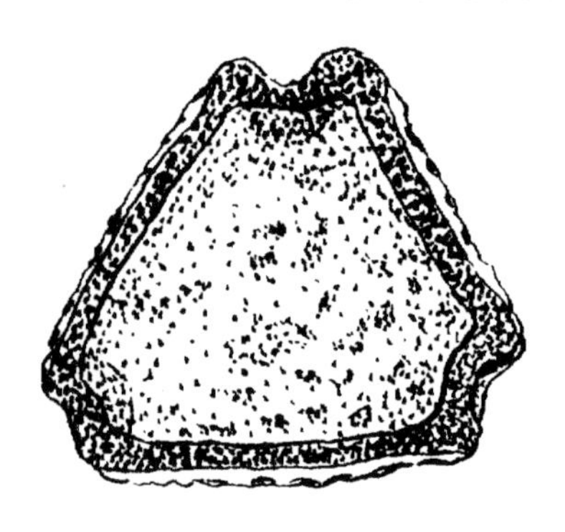

Bluffopollis scabratus (Angiospermae, Strasburgeriaceae), Pliocene.
From Couper 1954

Major repositories of New Zealand specimens

Most type specimens of form taxa of fossil spores and pollen described from New Zealand are stored as strew and single mounts in the Department of Paleontology, GNS Science. Different departments at Auckland, Waikato, Massey, Victoria, Canterbury, and Otago universities also hold collections of fossil spores and pollen, mainly from Pleistocene sequences, as does the National Institute for Water & Atmospheric Research (NIWA) and Landcare Research. None of these institutions holds important type specimens of described New Zealand spore/pollen taxa, although both Auckland University and Landcare Research hold some topotype specimens of fossil terrestrial palynomorphs.

Fossil pollen has been collected and described by overseas scientists or by New Zealand scientists stationed overseas, but as far as we are aware no type specimens originating from New Zealand are housed overseas, with the exception of the Late Triassic to Jurassic material described by Zhang and Grant-Mackie (1997), which is currently held in China.

Classification of spores and pollen

Spores and pollen are but a very small part of the parent plant. Many Pleistocene specimens can be directly related to source species or genera, but this is not true of most pre-Pleistocene spore and pollen types. It is even harder to relate palynomorphs to other fossil-plant remains unless they are found in attached flowers or other reproductive structures. As a result, a system of form genera and species was set up, principally led by Robert Potonié in West Germany, classifying spores and pollen on their morphological differences (e.g. Potonié 1960). This system was initially outside the International Code of Botanical Nomenclature (ICBN), but now palynomorph classification is based on rigorous application of the ICBN. This means that the classification adheres to nomenclatural priority and logic, type specimens, and other rules of typification.

Potonié's classification uses a hierarchical system of suprageneric categories, so that instead of Phylum (Division), Class, Order, and Family he used Anteturma, Turma, Subturma, Infraturma and Subinfraturma. This permits an unknown palynomorph to be easily placed in a group containing palynomorphs with similar morphological characters. A disadvantage of this strictly morphological approach is that it disregards phylogeny and phyletic relationships. However, because of the very conservative nature of many spore and pollen types, close phyletic relationships may never be ascertained, and also similar morphology occasionally indicates phylogeny. Thus palynomorphs are arranged in a morphological order and identified under fossil generic names. For example, the form-genus name *Nothofagidites* is used rather than the Linnean name *Nothofagus*, even though there is little doubt about the botanical affinity in this case. Another example is the fossil genus *Parsonsidites*. This genus was originally thought to represent fossil pollen from *Parsonsia* (Apocynaceae) but it probably contains representatives of the families Balanophoraceae and Malvaceae instead.

Another problem with the strictly morphological approach is that it restricts paleoecological interpretation. Can we justify a paleoecological interpretation of a fossil-pollen assemblage that has been classified under the fossil form-generic concept with its implied imprecision? The answer is yes, but we must keep in mind that we are dealing with dispersed organs and not the parent plants. Paleoecological interpretation must take into account that, for example, *Pseudowinterapollis* could represent *Drimys* or *Bubbia* as well as *Pseudowintera* (all genera of Winteraceae). Also many fossil spores and pollen do not have an obvious botanical affinity, and this is particularly true of many of the tricolpate and tricolporate pollen types.

The end-chapter checklist of spore and pollen taxa is organised using Potonié's system as revised by Burger (1994) but the broad phylogenetic affinity of each of the species listed is indicated after each binominal.

Diversity of New Zealand spores and pollen

Approximately 760 species of fossil spores and pollen have been identified from New Zealand and its surrounding islands. Terrestrial palynomorphs can be found in most sedimentary rocks, both marine and terrestrial. Most of the common spores and pollen found in Permian and younger rocks have been described, but there is a large number of more infrequent and rare spores and pollen that have yet to be described. Some groups of palynomorphs need revision using modern techniques.

The checklist of spores and pollen does not include taxonomic names of extant plants found as fossils. Most Pleistocene spores and pollen have a modern botanical affinity, as do some pre-Pleistocene taxa, but some fossil taxa with a clear modern botanical affinity have been given a 'fossil name' and described separately from that of their modern equivalents, and these have been included. Identification of the precise geological range of extant species or higher taxa would also have been useful, but fell outside the resources of the present review. Some information is available in Mildenhall (1980).

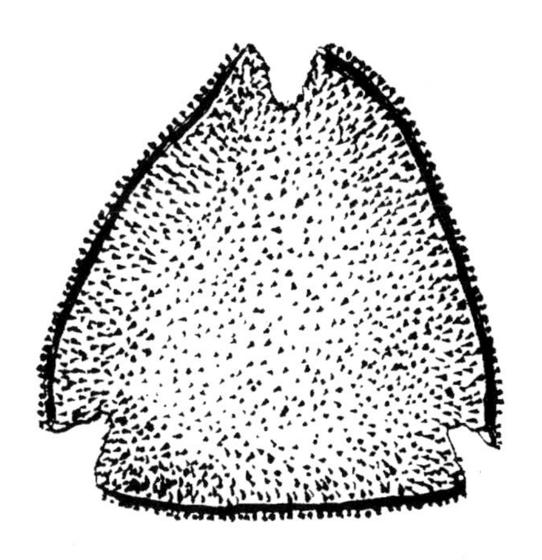

Beaupreaidites elegansiformis (Angiospermae, Proteaceae, *Beauprea*), Late Miocene.

From Couper 1954

Before commenting on the numbers of families, genera, and species it must be stressed again that palynologists deal with only a small part of the parent plant, and a morphologically conservative part at that. Because of the dispersed nature of spores and pollen they will always be described separately from most other fossil-plant organs. Also, many different species within the same genus, many genera, and even some families produce pollen or spores that are so alike they are very difficult or impossible to distinguish today, let alone in the fossil record.

On the other hand, fortunately unusual, some species produce pollen that is so variable that different specimens, when found as fossils may be given different binominals. The scope of fossil spore and pollen taxa is therefore somewhat different from modern plant taxa and the statistics produced cannot be directly compared with those of other fossil or modern groups of organisms.

A breakdown of the number of palynomorph taxa described from each of the listed major plant groups is given on page 489. Many of the taxa listed as ferns could prove to be from plants only distantly related to ferns, but which have similar spore morphology. A better idea of the modern botanical affinities of many of these taxa is usually given in the original description or in reviews (e.g. Mildenhall 1980, 1989; Pocknall 1989).

A summary of the number of genera and species of gymnosperms and angiosperms in each family recognised from the fossil record shows that some families have been centres of evolution in the New Zealand (or Australasian) region. The tabulation overleaf includes only those gymnosperm and angiosperm pollen that have a known Linnean family or order as their botanical affinity. Note, however, that some form genera are used for pollen from more than one family or order. The list excludes modern taxa found as fossils.

In the table, Podocarpaceae, Nothofagaceae, and Proteaceae clearly show a high degree of diversity in New Zealand, with their numbers of species disproportionate to those for other taxa. However, some taxa also show a high degree of diversity in the number of genera recognised within the family and this may indicate that these taxa formed centres of diversity in prehistoric New Zealand. This is especially true for Arecaceae, Loranthaceae, Rubiaceae, Sapindaceae, and Sterculiaceae. Yet others show a degree of species diversity but little obvious separation of genera appears warranted, e.g. Araucariaceae, Liliaceae, Aquifoliaceae, Asteraceae, Loranthaceae, Myrtaceae, Sapindaceae, and Winteraceae. The wholly extinct gymnosperm family Cheirolepidaceae contains a relatively large number of species, but it is highly likely that many of these names are synonyms.

Only a single genus and species have been recognised in 31 (44%) of the 70 angiosperm families listed, with a further seven families (10%) having two species in the genus and 10 families (14%) with two genera, each monospecific.

Diversity and taxonomic affinity of fossil spores and pollen in New Zealand

(Order/Family)	No. of genera	No. of species
GYMNOSPERMAE		
Araucariaceae	2	5
Caytoniales	9	19
Cheirolepidaceae	4	15
Coniferales (other)	5	14
Cordaitales	3	4
Cycadaceae	1	4
Ephedrales	4	8
Podocarpaceae	10	29
Voltziales	6	14
ANGIOSPERMAE ('MONOCOTS')		
Agavaceae	1	4
Amaryllidaceae	1	1
Arecaceae	5	10
Cyperaceae	1	1
Liliaceae	2	6
Pandanaceae	2	2
Poaceae	2	2
Restionaceae	1	2
Smilacaceae	1	1
Sparganiaceae	1	3
Typhaceae	1	1
ANGIOSPERMAE ('DICOTS')		
Apocynaceae	1	1
Aquifoliaceae	2	5
Araliaceae	2	2
Asteraceae	2	5
Balanophoraceae	1	1
Bignoniaceae	1	1
Bombacaceae	1	1
Buxaceae	2	2
Caesalpiniaceae	1	1
Caryophyllaceae	2	2
Casuarinaceae	2	2
Callitrichaceae	1	1
Chenopodiaceae	1	2
Chloranthaceae	2	4
Convolvulaceae	1	1
Corynocarpaceae	1	1
Cunoniaceae	1	1
Didymelaceae	1	1
Droseraceae	1	2
Epacridaceae	1	1
Ericaceae	1	2
Escalloniaceae	2	2
Euphorbiaceae	2	4
Gentianaceae	1	1
Geraniaceae	1	1
Goodeniaceae	1	1
Gunneraceae	1	1
Gyrostemonaceae	1	1
Haloragaceae	2	4
Lamiaceae	1	1
Loranthaceae	3	6
Malvaceae	2	2
Meliaceae	1	1
Mimosaceae	1	3
Moraceae	1	1
Myoporaceae	1	1
Myricaceae	1	1
Myrtaceae	1	7
Nothofagaceae	1	17
Nyctaginaceae	2	2
Olacaceae	1	3
Onagraceae	2	3
Plantaginaceae	1	1
Polygalaceae	2	2
Polygonaceae	1	1
Proteaceae	4	58
Rosaceae	1	2
Rubiaceae	4	4
Santalaceae	1	1
Sapindaceae	3	7
Sapotaceae	2	3
Scrophulariaceae	1	1
Sterculiaceae	4	5
Strasburgeriaceae	2	4
Symplocaceae	1	1
Thymelaeaceae	1	1
Trimeniaceae	1	2
Violaceae	1	2
Winteraceae	1	5

More detailed taxonomic work is required on many of these genera as some include quite a range of morphologies and in many cases affinities are not proven.

For this reason we have decided to list just broad botanical affinities for New Zealand's fossil spores and pollen. An example of this is the subdivision of the pollen types placed in the form-taxon *Beaupreaidites elegansiformis,* which has been shown to include a number of different morphological types (Milne 1998).

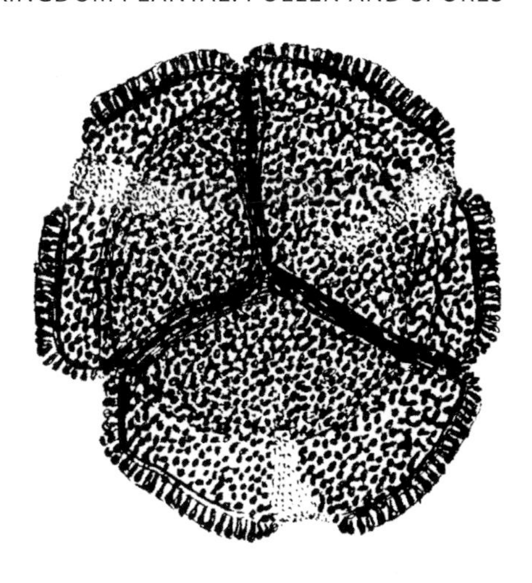

Dicotetradites clavatus (Angiospermae, eudicot angiosperm of uncertain affinity), Middle Eocene.

From Couper 1953

Endemism

For New Zealand's indigenous vascular plants, the present-day level of species-level endemism is about 85%. Most endemic species occur in families whose pollen is very difficult to differentiate or is rarely recorded as fossils – Asteraceae, Apiaceae, Cyperaceae, Epacridaceae, Orchidaceae, Poaceae, Rubiaceae, and Scrophulariaceae (Wardle 1991). This level of endemism and the conservative nature of the pollen grains within these families is probably a result of recent rapid evolution associated with the appearance of new ecological niches. These became available as a result of extreme Pleistocene climatic fluctuations and rapid Neogene tectonic uplift. The discontinuous nature of Pleistocene habitats has contributed towards genetic isolation and speciation. No extant families are endemic.

Diversity and taxonomic affinity of fossil spores and pollen in New Zealand

It is difficult to determine figures for endemism from other parts of the world for fossil taxa. Generally speaking, the percentage of taxa described from New Zealand and not yet identified outside the New Zealand landmass increases towards the present day. This could be partly a function of the amount of taxonomic work undertaken and published, as the more recent sediments have been studied in greater detail. However, it is mainly a function of increased isolation with time and the ongoing evolution of plants within geographically isolated habitats. The following figures, based solely on extinct taxa and including some non-vascular plants, comprise the percentage of endemic taxa currently recognised as occurring in New Zealand. In some cases these taxa have been reported outside New Zealand but those without adequate supporting evidence, such as descriptions and photographs, are not regarded as proven.

- Pleistocene (last 1.6 Ma) – c. 50% endemic, but many have close relatives in Australia and South America that may prove to be conspecific.
- Rest of Neogene (38–1.6 Ma) – c. 45%, same comment applies.
- Paleogene (65–38 Ma) – about 30%.
- Cretaceous (144–65 Ma) – about 12%.

Since spores and pollen are conservative in their morphology, the above values are probably best regarded as being on the generic level when compared to percentages for present-day New Zealand endemism. This can be illustrated by the modern New Zealand tree fern *Cyathea*. This genus produces smooth, pyramidal, trilete spores. The same kind of spores are found in sediments as old as Triassic and Jurassic and are given the same form-generic binominals, e.g. *Cyathidites australis* and *C. minor* for specimens obtained from both northern- and southern- hemisphere Jurassic–Cenozoic sediments. These spores certainly represent many different species and probably genera of ferns both in time and space, all producing very similar spores. The pre-Late Cretaceous flora is strongly Gondwanan in character, and the degree of endemism is apparently very low (less than 10%).

Significant features of distribution

Only a small range of potential fossil taxa is found fossilised in any numbers because a small range of environments is preserved in the geological record, even though some pollen comes from regional sources. Ancient spore- and pollen-

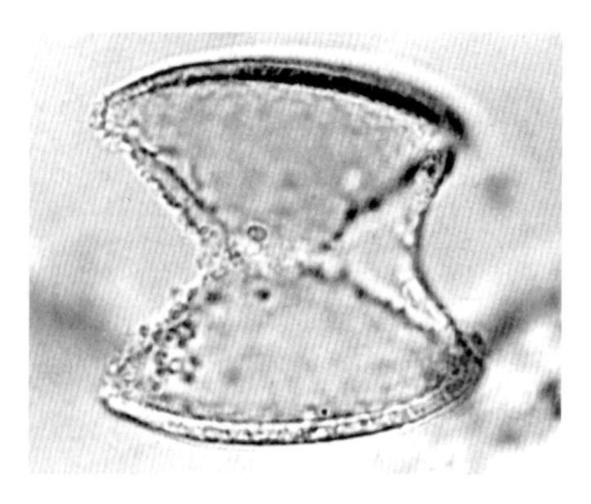

Pollen grain of the palm *Dicolpopollis* cf. *metroxylonoides* (Angiospermae, Arecaceae), Early to Middle Miocene.
From Mildenhall & Pocknall 1989

bearing sediments usually represent coastal, lowland, or at best lower-montane swamp habitats, especially in older rocks. A more varied set of environments is found preserved in the youngest rocks; even alpine environments can contain fossil material. There is no fossil-plant evidence in New Zealand of alpine or subalpine environments older than latest Miocene even though there is geological evidence of the existence of mountains in older rocks.

Many New Zealand flowering plants are insect-, bird-, or bat-pollinated. Pollen from these plants is rarely recorded in the fossil record, more because they are produced in small numbers rather than because they are not resistant to chemical or mechanical attack. Aquatic plants that produce thin-walled unornamented pollen grains are also rarely recorded as fossils. The most abundant palynomorphs found in fossil pollen samples are those that are dispersed by the wind, for example pollen of the various grasses, daisies, beeches, myrtles, and podocarps. Because of the small size of their pollen, some insect-pollinated plants also have their pollen distributed serendipitously by wind, and such was probably the case with many of the fossil proteas. Spores of vascular cryptogams (ferns and allies) are generally produced copiously and are well represented in sedimentary deposits. Bryophyte spores, on the other hand, are not so abundant, perhaps because of poor dispersal.

From the end-chapter checklist it can be seen that several genera have been unusually diverse in New Zealand in the past. This is particular true of the genera *Proteacidites* and *Nothofagidites*, representative of the families Proteaceae and Nothofagaceae respectively. The Proteaceae originated in the Late Cretaceous, became very diverse in the Paleogene, and gradually declined. There are now in New Zealand only two monospecific genera. Cretaceous strata contain c. 10 proteaceous taxa, Paleogene c. 35, Neogene c. 25, and Pleistocene five taxa.

The Nothofagaceae originated in the Late Cretaceous at about the same time as the Proteaceae. They also follow a pattern of increasing diversity followed by gradual decline from Middle to Late Neogene time onwards. This is demonstrated by McGlone et al. (1996) – three taxa occur in the Cretaceous, eight in the Palaeogene, 13 in the Neogene, and four in the Pleistocene. Different species of *Nothofagus* dominate pollen assemblages from different parts of the Cenozoic stratigraphic column so that it is relatively easy to get a good idea of the part of the geological column being dated from a sample of unknown age. Pollen from the five extant species of New Zealand *Nothofagus* would fall into only two fossil taxa as the pollen of the four species of the *Fuscospora* group are very difficult to separate. Pollen of extant *N. menziesii* may be impossible to separate from extinct pollen species of the *Lophozonia* group.

Casuarinaceae is another family that dominates many Cenozoic samples. The pollen of this family appears to be very conservative in nature, and only about four taxa are currently recognised. Since pollen of modern species of *Casuarina* and the closely related *Allocasuarina* are difficult to distinguish, then fossil taxa are likely to be even harder to distinguish. The four fossil-pollen species probably represent numerous different species or even genera of parent plants. Species in the fossil genera *Casuarinidites, Haloragacidites, Triorites,* and *Myricipites* are believed to represent this family.

The checklist includes a number of informal taxa whose species have yet to be described. As mentioned below, probably c. 25% of the total potentially useful terrestrial palynomorphs have been described. In the current science-funding climate, the formal documentation of unknown and undescribed palynomorphs has almost ceased.

Features of the New Zealand spore and pollen flora

Taphonomy

Generally speaking, spores and pollen can be found in all terrestrial sedimentary facies and in many marine sedimentary environments. Wind, water, and some other biological agents, usually insects, disperse both spores and pollen. However, the greatest concentrations are found in carbonaceous sediments – peats, lignites, and coals. In these, much of the pollen tends to be locally derived, whereas in other sediments the spores and pollen tend to be regionally derived. This is particularly true of spores and pollen in marine sediments, where climatic and environmental signals are muted by the incorporation of taxa from many different environments, by mechanical and chemical damage during transportation, and by water sorting of the grains into specific size or weight grades. For example, in the recent Ocean Drilling Programme it was found that Pliocene–Recent sediments (last five Ma) 1500 km offshore of New Zealand contained abundant terrestrial palynomorphs with 60 or more taxa in some samples (Mildenhall 2003).

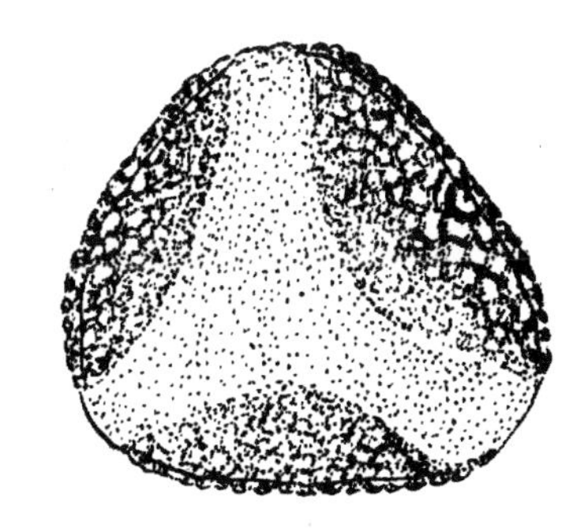

Phormium sp. (Angiospermae, Xanthorrhoeaceae), Pleistocene.

From Couper 1953

Ecological associations

Peats and lignites trap spores and pollen from both the immediate swamp area, providing clues about the development of the peat swamp, and the regional vegetation, which can in turn provide clues about changing climate. Spores and pollen from other sediments tend to represent regional lowland or lower-montane vegetation associations. Statistical analyses can help to decipher associations and environmental signals.

The advantage of paleopalynology is that it records the constant changes that vegetation undergoes as climate and local depositional conditions change. It can also detect rare catastrophic events such as volcanic disruption, storm events, tsunamis, and flooding. When good macrofossil assemblages are located, it is possible to get a very good picture of local vegetation at a specific point in time. However, these fossils are often collected from different stratigraphic horizons at any one outcrop in order to get a statistically useful assemblage.

Variation through geological time

New Zealand has benefited from its situation on an active plate boundary, in that sedimentation has been almost continuous for the last 150 Ma, not in any one spot, but always somewhere on the New Zealand continental land mass, 93% of which is currently under water. Relative to most other countries, New Zealand has few significant gaps in the record of continuous plant life. Also, from older strata we have good sequences that give an accurate picture of plant life up to c. 300 million years ago (Permian period).

Different spore and pollen groups dominate different parts of the geological column. Using the time scale of Gradstein & Ogg (1997), very little Permian (248–290 Ma) terrestrial material has been studied because most, if not all, rocks are marine. Spores from ferns and fern-like plants and pollen from conifers, pteridosperms, cycads, and cycad-like plants like *Glossopteris*, are dominant in the New Zealand Permian (Crosbie 1986).

Raine (1987), de Jersey and Raine (1990), and Zhang and Grant-Mackie (1997) have studied Jurassic (142–206 Ma) and Triassic (206–248 Ma) microfossil floras. Spores from lycopods, ferns, and fern-like plants and pollen from pteridosperms and conifers (*Alisporites*, *Araucariacites*, *Callialasporites*, *Corollina* or *Classopollis*, *Podocarpidites*, etc.) dominate these fossil assemblages.

The Cretaceous (65–142 Ma) has been extensively studied relative to the preceding periods of time, for a number of reasons (e.g. Raine et al. 1981; Raine 1984). These reasons include the widespread distribution and large thicknesses of Cretaceous sediment, the presence of economically important coal-bearing sediments, which may also be the source of petroleum in younger rocks, and the interest in the appearance, origins, and evolution of the angiosperms. At the beginning of the Cretaceous, spores of ferns and lycopods and pollen from conifers dominate samples (Couper 1953, 1960). Angiosperm pollen first appears in the middle Cretaceous at about 94 Ma and stays as a minor component of the flora until the late Cretaceous (75–82 Ma). Caryophyllaceae,

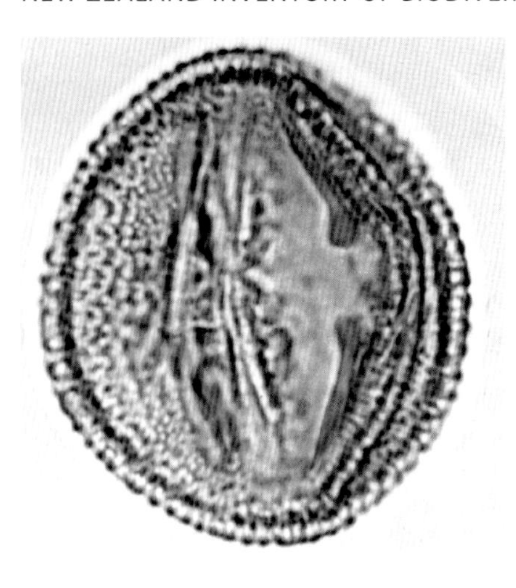

Rhoipites alveolatus (Angiospermae, Euphorbiaceae), Early Miocene.
From Pocknall & Crosby 1982

Chloranthaceae, Epacridaceae, Liliaceae, Loranthaceae, Nothofagaceae, Proteaceae, Ranunculaceae?, and Winteraceae (Mildenhall 1980) are among the modern angiosperm families to appear, but many other angiosperm pollen types have no known familial affinity.

During the last 65 Ma (Cenozoic), the pollen of angiosperms dominates in abundance and diversity (Couper 1953, 1960; Mildenhall 1980). Many different angiosperm families predominate at different times, but the most abundant pollen comes from species of *Nothofagus* and *Casuarina* (Mildenhall 1989; Pocknall 1989, 1990). Pollen from conifers is more abundant in samples from the beginning and end of the period, and spores of ferns rarely dominate except in the Pleistocene and Pliocene (last 5 Ma) (Mildenhall & Pocknall 1983).

Phylogeny

Although, as mentioned above, the (artificial) classification of spores and pollen is morphographic, there are nevertheless some parallels with phylogenetic classification, as might be expected. The fossil record of spores and pollen also reflects the 'punctuated equilibrium' mode of speciation seen in other groups of organisms, with discrete species morphologies rather than continuous intergradation being the general rule.

No work has been done in New Zealand on the gradual evolutionary sequences that probably occur in a number of taxa. Certainly the evolutionary sequence determined by de Jersey (1982) in the Triassic spore genus *Aratrisporites* also occurs in the New Zealand Triassic. There is also a gradual development from thin-walled and poorly ornamented to thick-walled and well-ornamented exine in the bryophyte species *Cingulatisporites bifurcatus* (Mildenhall pers. obs.), and other changes with time have been noted in some publications (e.g. for *Bluffopollis* – Pocknall & Mildenhall 1984) without detailed study being applied. However, as spores and pollen form but a small part of the original plant, such phylogenetic changes that would be apparent in the parent will rarely be detectable in the spores and pollen they produce.

Origins

Terrestrial palynomorphs have been studied primarily for their local age range and little work has been done in recent years on the origins of individual taxa. Some families, like the Asteraceae, appear suddenly in many parts of the world at about the same time, approximately 25 million years ago. Mildenhall (1980) has discussed some taxa. For example, *Dacrycarpus*-like fossils are known from the Lower Cretaceous and Jurassic (150 Ma) of India (e.g. Bera & Banerjee 1997) but fossil pollen of *Dacrycarpus* in New Zealand dates from the Eocene. Overseas authors have discussed other taxa, for example *Phyllocladidites mawsonii*, the fossil pollen of the Australian Huon pine (Playford & Dettmann 1978). This species first appeared in the Australian Turonian, at about 92 Ma, and quickly became widespread throughout southern Australia, Kerguelen Islands, Antarctica, New Zealand, and southern South America.

Gaps in knowledge for servicing economic and scientific needs

Determination of the total number of species in each taxon is not a useful exercise because we are dealing with morphological rather than Linnean taxonomic entities. Many fossil species, of course, will never be described. This is because they are found so rarely that description will not be warranted or because they lack a perceived utility, and also because of the difficulty of distinguishing slight morphological variations in entities that are only micrometres in size, even with a scanning electron microscope. The diversity of the various groups of spores and pollen taxa is summarised in the table at right.

There will always be a need to refine palynological knowledge, given that new taxa will continue to be discovered and sediments will need to be studied in more detail as time goes on. New Zealand paleopalynology is still in an exploratory phase analogous to that of neobotany in the early 19th century. Systematic work will probably still be required and needed for the foreseeable future, irrespective of the number of researchers working in the field. It is anticipated that the above figures represent less than 25% of the total number of fossil terrestrial palynomorphs that can and need to be described to provide the level of biostratigraphic control that is required at the present time and in the near future. A great deal of taxonomic work must be undertaken in order to compete with current geophysical techniques and to upgrade the quality of New Zealand's biostratigraphic servicing work.

There are a number of knowledge gaps that require filling if we are to obtain comprehensive first-order knowledge of the vegetation of the last 300 Ma. Fossil fungal and bryophyte spores (modern and fossil) can give us a good idea of environments but little work has been done on the fossil material. Any studies of fossil material, especially Pleistocene, require corresponding studies of modern material because precise botanical affinities are required if the environment is to be accurately determined. A morphological study of New Zealand bryophyte spores would complement the comprehensive morphological studies of spores of vascular cryptogams (Large & Braggins 1991) and pollen of monocotyledonous (Cranwell 1953) and dicotyledonous angiosperms (Moar 1993). Fungal spores in particular are important to study because they are on occasion the only microfossil plants to be found in some sediments, owing to their small size and thick, highly acid-resistant exines.

Studies of a number of New Zealand taxa and time periods are required to improve scientific service both to industry and other scientists. The New Zealand Paleozoic (older than 248 Ma) is almost totally unexplored palynologically as is the Late Jurassic (159–142 Ma). A large but undocumented number of taxa are yet to be described from Late Cretaceous, Eocene, and Late Miocene–Pliocene strata. As yet, no palynozonation of the Late Miocene–Pliocene has been developed and the zonation for the Eocene and Late Cretaceous could be greatly improved with further combined biostratigraphic-taxonomic work.

There are two types of collection needed for paleontological work. Slides of fossil assemblages from about 20,000 localities and slides of type and reference fossils prepared as single or strew mounts form the active working and reference collection for GNS Science. The other collection is of modern pollen types from New Zealand and around the world that need to be available with which to compare the fossil material. The GNS Science collection is just adequate for New Zealand pollen types, but is poor for pollen types from overseas. Published and electronic pollen catalogues can often assist in the determination of botanical affinity, but this comes a poor second to physical examination of a modern reference specimen.

Databasing, monographs, computer keys

There are a number of databases relevant to the study of fossil spores and pollen. The *New Zealand Fossil Record File* is a unique system run by the Geoscience Society of New Zealand. It catalogues all fossil localities and in some cases the fossils they contain. The lists of fossils found at these localities are the property of the institutions employing the individual researcher.

In New Zealand, monographs on fossil spores and pollen have been produced only by GNS Science and its predecessors. They are no longer being prepared in the numbers previously produced because little taxonomic work is being done, and the cost of production, or more accurately consumer resistance to paying for the high cost of these types of publication, means that production costs are difficult to recover.

Computer keys and identification systems are being set up by GNS Science, and Massey University, Palmerston North, is researching computer recognition of pollen grains under the guidance of Prof. John Flenley.

Spore/pollen taxon diversity

Group	Fossil (spore/pollen) taxa
Chlorophyta	1
Bryophyta	55
Pteridophyta	282
Lycopsida	88
Psilopsida	0
Sphenopsida	1
Filicopsida	193
Spermatophyta	425
Gymnospermae	102
Angiospermae	323
Totals	763

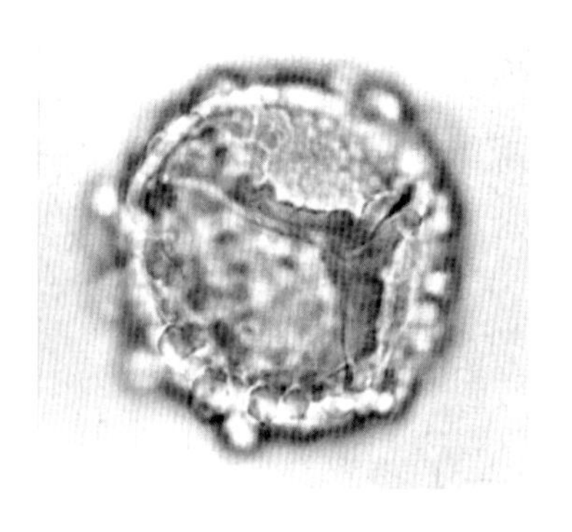

Gemmatriletes multiglobus (Polypodiopsida, Grammitidaceae), Early to Middle Miocene.
From Mildenhall & Pocknall 1989

Servicing scientific needs

The fossil collections held at GNS Science are adequate for the current servicing of scientific needs but much more taxonomic work needs to be done to improve the precision of this servicing. Commercial pressures to produce finer biostratigraphic zonations are always present, coupled with the need for more information about the depositional environments of potential economic mineral deposits, including coal, oil, gas, and water. But refined biostratigraphy can only result from refined taxonomy, for which research funding is inadequate. Paradoxically, such work would improve service to the range of clients that use the results of palynological research.

Note

This paper was written in 1998–99 and has not been substantially revised since that time. A number of important geological and botanical discoveries have occurred in the last 11 years that bear upon discussions presented above. An illustrated and annotated checklist of published records of pre-Pleistocene fossil spore and pollen species is now online at www://data.gns.cri.nz/sporepollen (Raine et al. 2008). This incorporates records published since the checklist attached to this chapter was compiled in 2000, and includes about 900 taxa compared to 760 in 2000.

Acknowledgement

The authors thank GNS Science for permission and funding to produce the checklist of spores and pollen.

Authors

Dr Dallas C. Mildenhall GNS Science, P.O. Box 30-368, Lower Hutt, New Zealand [d.mildenhall@gns.cri.nz]

Dr J. Ian Raine GNS Science, P.O. Box 30-368, Lower Hutt, New Zealand [i.raine@gns.cri.nz]

References

The reference list includes some papers (indicated by an asterisk) not cited in the text, including papers published between 2000 and 2011. These pertain to fossil spores and pollen, comprising taxonomic papers containing either original descriptions, comments on the identification and morphology of fossil spores and pollen found in New Zealand, or checklists (but not papers reporting subsequent taxonomic changes, mostly by authors outside of New Zealand, or papers on New Zealand taxa originally described from overseas localities).

BANNISTER, J. M.; LEE, D. E.; RAINE, J. I. 2005: Morphology and palaeoenvironmental context of *Fouldenia staminosa*, a fossil flower with associated pollen from the Early Miocene of New Zealand. *New Zealand Journal of Botany 43*: 515–525.*

BEGGS, J. M.; POCKNALL, D. T. 1992: Sequence stratigraphic controls on reservoir architecture, Lower Eocene Kapuni Group, Maui Field, Taranaki Basin, New Zealand. *New Zealand Geological Survey Report G162*: 1–28.

BERA, S.; BANERJEE, M. 1997: Palynostratigraphy of Mesozoic sediments from Western Bengal Basin, India. *Journal of Palynology 33*: 39–82.

BLUEMLE, J. P.; SABEL, J. M.; KARLEN, W. 1999: Rate and magnitude of past global climate changes. *Environmental Geosciences 6*: 63–75.

BROWNE, G. H.; KENNEDY, E. M.; CONSTABLE, R. M.; RAINE, J. I.; CROUCH, E. M.; SYKES, R. 2008: An outcrop-based study of the economically significant Late Cretaceous Rakopi Formation, northwest Nelson, Taranaki basin, New Zealand. *New Zealand Journal of Geology and Geophysics 51*: 295–315.*

BURGER, D. 1994: Guide to general file of fossil spores and pollen of Jansonius & Hills (1976). *Australian Geological Survey Organisation Publication 94/046A*: 1–36.

BURROWS, C. J.; McCULLOCH, B.; TROTTER, M. M. 1981: The diet of moas based on gizzard contents samples from Pyramid Valley, North Canterbury, and Scaifes Lagoon, Lake Wanaka, Otago. *Records of the Canterbury Museum 9*: 309–336.

BUSSELL, M. R.; MILDENHALL, D. C. 1990: Extinct palynomorphs from middle and late Pleistocene terrestrial sediments, South Wanganui Basin, New Zealand. *New Zealand Journal of Geology and Geophysics 33*: 439–447.*

CAMPBELL, H. A.; MORTIMER, N.; RAINE, J. I. 2001: Geology of the Permian Kuriwao Group, Murihiku Terrane, Southland, New Zealand. *New Zealand Journal of Geology and Geophysics 44*: 485–500.*

CARPENTER, R. J.; JORDAN, G. J.; MILDENHALL, D. C.; LEE, D. E. 2011: Leaf fossils of the ancient Tasmanian relict *Microcachrys* (Podocarpaceae) from New Zealand. *American journal of botany 98*: 1164–1172.

COCKAYNE, L. 1928: *The Vegetation of New Zealand*. 2nd edn. Verlag von Wilhelm Engelmann, Leipzig. 456 p.

CONRAN, J. G.; KAULFUSS, U.; BANNISTER, J. M.; MILDENHALL, D. C.; LEE, D. E. 2010: *Davallia* (Polypodiales: Davalliaceae) macrofossils from Early Miocene Otago

(New Zealand) with *in situ* spores. *Review of Palaeobotany and Palynology 162*: 84–94.*
COOKSON, I. C. 1947: Fossil fungi from Tertiary deposits in the Southern Hemisphere. Part 1. *Proceedings of the Linnean Society of New South Wales 72*: 207–214.*
COOPER, A.; COOPER, R. A. 1995: The Oligocene bottleneck and New Zealand biota: genetic record of a past environmental crisis. *Proceedings of the Royal Society of London, B 261*: 293–302.
COUPER, R. A. 1951: Microflora of a submarine lignite from Toetoes Bay, near Bluff, New Zealand. *New Zealand Journal of Science and Technology, B33*: 179–186.
COUPER, R. A. 1952: The spore and pollen flora of the *Cocos*-bearing beds, Mangonui, North Auckland. *Transactions of the Royal Society of New Zealand 79*: 340–348.
COUPER, R. A. 1953: Upper Mesozoic and Cainozoic spores and pollen grains from New Zealand. *New Zealand Geological Survey Paleontological Bulletin 22*: 1–77.*
COUPER, R. A. 1954: Plant microfossils from New Zealand no. 1. *Transactions of the Royal Society of New Zealand 81*: 479–483.*
COUPER, R. A. 1960: New Zealand Mesozoic and Cainozoic plant microfossils. *New Zealand Geological Survey Paleontological Bulletin 32*: 1–87.*
CRANWELL, L. M. 1940: Pollen grains of New Zealand conifers. *The New Zealand Journal of Science and Technology 22, 1B*: 1–17.
CRANWELL, L. M. 1942: New Zealand pollen studies. I. Key to the pollen grains of families and genera in the native flora. *Records of the Auckland Institute and Museum 2*: 280–308.
CRANWELL, L. M. 1953: New Zealand pollen studies. The Monocotyledons. *Bulletin of the Auckland Institute and Museum 3*: 1–91.
CRANWELL, L. M.; von POST, L. 1936: Pollen diagrams from the Southern Hemisphere, 1. New Zealand. *Geografiska Annaler 18*: 30–347.
CRAW, R.; HEADS, M. 1998: *Panbiogeography: Tracking the History of Life*. Oxford University Press, New York. 229 p.
CROSBIE, Y. M. 1986: Permian palynomorphs from the Kuriwao Group, Southland, New Zealand. *New Zealand Geological Survey Record 8*: 109–119.
CROSBIE, Y. M.; CLOWES, C. D. 1980: Revision of the fossil tetrad-pollen genus *Dicotetradites* Couper 1953. *New Zealand Journal of Botany 18*: 459–462.*
DAGHLIAN, C. P.; SKVARLA, J. J.; POCKNALL, D. T.; RAVEN, P. H. 1984: *Epilobium* pollen from Oligocene sediments in New Zealand. *New Zealand Journal of Botany 22*: 285–294.*
DAGHLIAN, C. P.; SKVARLA, J. J.; POCKNALL, D. T.; RAVEN, P. H., 1985: *Fuchsia* pollen from the Early Miocene of New Zealand. *American Journal of Botany 72*: 1039–1047.*
de JERSEY, N. J. 1982: An evolutionary sequence in *Aratrisporites* miospores from the Triassic of Queensland, Australia. *Palaeontology 25*: 665–672.
de JERSEY, N. J.; RAINE, J. I. 1990: Triassic and earliest Jurassic miospores from the Murihiku Supergroup, New Zealand. *New Zealand Geological Survey Paleontological Bulletin 62*: 1–164.*
DETTMANN, M. E.; POCKNALL, D. T.; ROMERO, E. J.; ZAMALOA, M.delC. 1990: *Nothofagidites* Erdtman ex Potonié, 1960; a catalogue of species with notes on the paleogeographic distribution of *Nothofagus* Bl. (southern beech). *New Zealand Geological Survey Paleontological Bulletin 60*: 1–79.*
EDWARDS, W. N. 1934: Jurassic plants from New Zealand. *Annals and Magazine of Natural History, ser. 10, 13*: 81–109.
ERDTMAN, O. G. E. (G.) 1924: Studies in micro-palaeontology. Part IV. Peat from the Chatham Islands and the Otago District, New Zealand. *Geologiska Foreningens i Stockholm Forhandlingar 46*: 679–681.
FERGUSON, D. K.; LEE, D. E.; BANNISTER, J. M.; ZETTER, R.; JORDAN, G. J.; VAVRA, N.; MILDENHALL, D. C. 2010: The taphonomy of a remarkable leaf bed assemblage from the Late Oligocene-Early Miocene Gore Lignite Measures, southern New Zealand. *International Journal of Coal Geology 83*: 173–181.*
FLEMING, C. A. 1962: New Zealand biogeogeography: a paleontologist's approach. *Tuatara 10*: 53–108.
FLEMING, C. A. 1975: The geological history of New Zealand and its biota. Pp. 1–86 *in*: Kuschel, G. (ed.), *Biogeography and Ecology in New Zealand*. Dr W. Junk, The Hague. xvi + 689 p., + folding charts.
GRADSTEIN, F. M.; OGG, J. G. 1996: A Phanerozoic time scale. *Episodes 19*: 3–5.
HARRIS, W. F. 1951: Unravelling forest history in New Zealand. New Zealand plants and their story: clues to the past. *New Zealand Science Review 9*: 5–9.
HARRIS, W. F. 1953: Palynology as key to the history of New Zealand vegetation. *Proceedings of the 7th Pacific Science Congress, 5 (Botany)*: 164–178.
HARRIS, W. F. 1955: Progress report on pollen statistics from Pyramid Valley Swamp. *Records of the Canterbury Museum 6*: 279–290.
HARRIS, W. F. 1968: Palaeobotany and palynology in New Zealand: a historical review. *Review of Palaeobotany and Palynology 6*: 137–145.
JARZEN, D. M.; POCKNALL, D. T. 1993: Tertiary *Bluffopollis scabratus* (Couper) Pocknall & Mildenhall, 1984 and modern *Strasburgeria* pollen: a botanical comparison. *New Zealand Journal of Botany 31*: 185–192.*
JEANS, C. V.; FISHER, M. J.; RAINE, J. I.; MERRIMAN, R. J.; CAMPBELL, H. J.; FALLICK, A. E.; CARR, A. D., KEMP, S. J. 2003: Triassic sediments of the Kaka Point Structural Belt, South Island, New Zealand, and their relationship to the Murihiku Terrane. *Journal of the Royal Society of New Zealand 33*: 57–84.*
JOHNSON, L. A. S.; BRIGGS, B. G. 1975: On the Proteaceae – the evolution and classification of a southern family. *Botanical Journal of the Linnean Society 70*: 83–182.
JOHNSTON, M. R.; RAINE, J. I.; WATTERS, A. 1987: Drumduan group of east Nelson, New Zealand: plant-bearing Jurassic arc rocks metamorphosed during terrane interaction. *Journal of the Royal Society of New Zealand 17*: 275–301.
JORDAN, G. J.; BANNISTER, J. M.; MILDENHALL, D. C.; ZETTER, R.; LEE, D. E. 2010: Fossil Ericaceae from New Zealand: deconstructing the use of fossil evidence in historical biogeography. *American Journal of Botany 97*: 59–70.*
KRUTZSCH, W. 1970: Information on fossil dispersed tetrad pollen. *Palaeontologische Abhandlungen B., Palaeobotanik 3*: 399–433.*
LANDIS, C. A.; CAMPBELL, H. J.; ASLUND, T.; CAWOOD, P. A.; DOUGLAS, A.; KIMBROUGH, D. L.; PILLAI, D. D. L.; RAINE, J. I.; WILLSMAN, A. 1999: Permian-Jurassic strata at Productus Creek, Southland, New Zealand: implications for terrane dynamics of the eastern Gondwanaland margin. *New Zealand Journal of Geology and Geophysics 42*: 255–278.*
LARGE, M. F.; BRAGGINS, J. E. 1991: *Spore Atlas of New Zealand Ferns and Fern Allies*. SIR Publishing, Wellington. 167 p.
LEE, D. E.; BANNISTER, J. M.; RAINE, J. I.; CONRAN, J. G. 2010: Euphorbiaceae: Acalyphoideae fossils from early Miocene New Zealand: *Mallotus-Macaranga* leaves, fruits, and inflorescence with *in situ Nyssapollenites endobalteus* pollen. *Review of Palaeobotany and Palynology 163*: 127–138.*
LEWIS, K. J. B.; MILDENHALL, D. C. 1985: The late Pleistocene seismic, sedimentary and palynological stratigraphy beneath Evans Bay, Wellington Harbour. *New Zealand Journal of Geology and Geophysics 28*: 129–152.
McGLONE, M. S.; MILDENHALL, D. C.; POLE, M. 1996: History and paleoecology of New Zealand *Nothofagus* forests. Pp. 83–130 *in*: Veblen, T. T.; Hill, R. S.; Read, J. (eds), *The Ecology and Biogeography of* Nothofagus *Forest*. Yale University Press, New Haven. 403 p.
McGLONE, M. S.; WEBB, C. J. 1981: Selective forces influencing the evolution of divaricating shrubs. *New Zealand Journal of Ecology 4*: 20–28.
McINTYRE, D. J. 1965: Some new pollen species from New Zealand Tertiary deposits. *New Zealand Journal of Botany 3*: 204–214.*
McINTYRE, D. J. 1968: Further new pollen species from New Zealand Tertiary and uppermost Cretaceous deposits. *New Zealand Journal of Botany 6*: 177–204.*
MILDENHALL, D. C. 1972b: Fossil pollen of *Acacia* type from New Zealand. *New Zealand Journal of Botany 10*: 485–494.*
MILDENHALL, D. C. 1975: New fossil spore from the Pakihikura Pumice (Okehuan, Pleistocene), Rangitikei Valley, New Zealand. *New Zealand Journal of Geology and Geophysics 18*: 667–674.*
MILDENHALL, D. C. 1978: *Cranwellia costata* and *Podosporites erugatus* n.sp. from middle Pliocene (?early Pleistocene) sediments, South Island, New Zealand. *Journal of the Royal Society of New Zealand 8*: 253–274.*
MILDENHALL, D. C. 1980: New Zealand Late Cretaceous and Cenozoic plant biogeography: a contribution. *Palaeogeography, Palaeoclimatology, Palaeoecology 31*: 197–233.
MILDENHALL, D. C. 1989: Summary of the age and paleoecology of the Miocene Manuherikia Group, Central Otago, New Zealand. *Journal of the Royal Society of New Zealand 19*: 19–29.
MILDENHALL, D. C. 1994: Palynological reconnaissance of early Cretaceous to Holocene sediments, Chatham Islands, New Zealand. *Institute of Geological and Nuclear Sciences Monograph 7*: 1–206.*
MILDENHALL, D. C. 1995: Pleistocene palynology of the Petone and Seaview drillholes, Petone, Lower Hutt Valley, North Island, New Zealand. *Journal of the Royal Society of New Zealand 25*: 207–262.
MILDENHALL, D. C. 2003: Deep-sea record of Pliocene and Pleistocene terrestrial palynomorphs from offshore eastern New Zealand (ODP Site 1123, leg 181). *New Zealand Journal of Geology and Geophysics 46*: 343–361.
MILDENHALL, D. C.; CROSBIE, Y. M. 1980: Some porate pollen from the upper Tertiary of New Zealand. *New Zealand Journal of Geology and Geophysics 22*: 499–508.*
MILDENHALL, D. C.; ALLOWAY, B.V. 2008: A widespread ca. 1.1 Ma TVZ silicic tephra

preserved near Wellington, New Zealand: implications for regional reconstruction of mid-Pleistocene vegetation. *Quaternary International 178*: 167–182.*

MILDENHALL, D. C.; BYRAMI, M. L. 2003: A redescription of *Podosporites parvus* (Couper) Mildenhall emend. Mildenhall & Byrami from the Early Pleistocene, and late extinction of plant taxa in northern New Zealand. *New Zealand Journal of Botany 41*: 147–160.*

MILDENHALL, D. C.; HARRIS, W. F. 1971: Status of *Haloragacidites* (al. *Triorites*) *harrisii* (Couper) Harris comb. nov. and *Haloragacidites trioratus* Couper, 1953. *New Zealand Journal of Botany 9*: 297–306.*

MILDENHALL, D. C.; POCKNALL, D. T. 1983: Paleobotanical evidence for changes in Miocene and Pliocene climates in New Zealand. Pp. 159–171 *in*: Vogel, J. C. (ed.), *Late Cainozoic Palaeoclimates of the Southern Hemisphere*. A. A. Balkema, Rotterdam. 536 p.

MILDENHALL, D. C.; POCKNALL, D. T. 1989: Miocene–Pleistocene spores and pollen from Central Otago, South Island, New Zealand. *New Zealand Geological Survey Paleontological Bulletin 59*: 1–128.*

MILNE, L. A. 1998: Tertiary palynology: *Beaupreaidites* and new Conospermae (Proteoideae) affiliates. *Australian Systematic Botany 11*: 55–603.

MOAR, N. T. 1958a: Contributions to the Pleistocene history of the New Zealand flora. 1. Auckland Island peat studies. *New Zealand Journal of Science 1*: 449–465.

MOAR, N. T. 1958b: Contributions to the Pleistocene history of the New Zealand flora. 2. Plant remains from a buried peat layer at Bowenvale, Christchurch. *New Zealand Journal of Science 1*: 480–486.

MOAR, N. T. 1959: Contributions to the Pleistocene history of the New Zealand flora. 3. Pollen analysis of a peat profile from Antipodes Island. *New Zealand Journal of Science 2*: 35–40.

MOAR, N. T. 1993: *Pollen Grains of New Zealand Dicotyledonous Plants*. Manaaki Whenua Press, Lincoln. 200 p.

NELSON, C. S.; MILDENHALL, D. C.; TODD, A. J.; POCKNALL, D. T. 1988: Subsurface stratigraphy, paleoenvironments, palynology, and depositional history of the Late Neogene Tauranga Group at Ohinewai, Lower Waikato Lowland, South Auckland, New Zealand. *New Zealand Journal of Geology and Geophysics 31*: 21–40.

NEWNHAM, R. M.; LOWE, D. J.; MATTHEWS, B. W. 1998: A late-Holocene and prehistoric record of environmental change from Lake Waikaremoana, New Zealand. *The Holocene 8*: 443–454.

NORRIS, G. 1968: Plant microfossils from the Hawks Crag Breccia, south-west Nelson, New Zealand. *New Zealand Journal of Geology and Geophysics 11*: 312–344.*

PLAYFORD, G.; DETTMANN, M. E. 1978: Pollen of *Dacrydium franklinii* Hook. F. and comparable early Tertiary microfossils. *Pollen et Spores 20*: 513–534.

POCKNALL, D. T. 1982: Palynology of late Oligocene Pomahaka Bed sediments, Waikoikoi, Southland, New Zealand. *New Zealand Journal of Botany 20*: 263–287.*

POCKNALL, D. T. 1982: Palynology of the Bluecliffs Siltstone (Early Miocene), Otaio River, South Canterbury, New Zealand. *New Zealand Geological Survey Report PAL 55*: 1–24.*

POCKNALL, D. T. 1985: Palynology of Waikato Coal Measures (Late Eocene–Late Oligocene) from the Raglan area, North Island, New Zealand. *New Zealand Journal of Geology and Geophysics 28*: 329–349.*

POCKNALL, D. T. 1989: Late Eocene to Early Miocene vegetation and climate history of New Zealand. *Journal of the Royal Society of New Zealand 19*: 1–18.

POCKNALL, D. T. 1990: Palynological evidence for the Early to Middle Eocene vegetation and climate history of New Zealand. *Review of Palaeobotany and Palynology 65*: 57–69.

POCKNALL, D. T.; CROSBIE, Y. M. 1982: Taxonomic revision of some Tertiary tricolporate and tricolpate pollen grains from New Zealand. *New Zealand Journal of Botany 20*: 7–15.*

POCKNALL, D. T.; CROSBIE, Y. M. 1988: Pollen morphology of *Beauprea* (Proteaceae): modern and fossil. *Review of Palaeobotany and Palynology 53*: 305–327.*

POCKNALL, D. T.; MILDENHALL, D. C. 1984: Late Oligocene–Early Miocene spores and pollen from Southland, New Zealand. *New Zealand Geological Survey Paleontological Bulletin 51*: 1–66.*

POLE, M. S.; VAJDA, V. M. 2009: A new Cretaceous-Paleogene site in New Zealand – turnover in macroflora confirmed by palynology. *Cretaceous Research 30*: 917–938.*

POTONIÉ, R. 1960: Synopsis der Gattungen der Sporae Dispersae. III. Teil: Nachträge Sporites, Fortsetzung Pollenites mit Generalregister zu Teil I-III. *Beihefte zum Geologischen Jahrbuch 39*: 1–189.

RAINE, J. I. 1984: Outline of a palynological zonation of Cretaceous to Paleogene terrestrial sediments in West Coast region, South Island, New Zealand. *New Zealand Geological Survey Report 109*: 1–82.*

RAINE, J. I. 1987: Jurassic plant microfossils and macrofossils from Murihiku Supergroup, Manganui Valley, Awakino District (North Island, New Zealand). *New Zealand Geological Survey Record 20*: 127–138.

RAINE, J. I. 2008: Zonate lycophyte spores from New Zealand Cretaceous to Palaeogene strata. *Alcheringa 32*: 99–127.*

RAINE, J. I.; de JERSEY, N. J.; RYAN, K. G. 1988: Ultrastructure and lycopod affinity of *Densoisporites psilatus* (de Jersey) comb. nov. from the Triassic of New Zealand and Queensland. *Memoir of the Association of Australasian Palaeontologists 5*: 79–88.*

RAINE, J. I.; MILDENHALL, D. C.; KENNEDY, E. M. 2008: New Zealand fossil spores and pollen: an illustrated catalogue. 3rd ed. *GNS Science Miscellaneous Series 4*. http://www.gns.cri.nz/what/earthhist/fossils/spore_pollen/catalog/index.htm.*

RAINE, J. I.; POCKNALL, D. T. 1983: Plant fossils. Pp. 35–46 *in*: Brownsey, P. J.; Baker, A. N. (eds), *The New Zealand Biota – What do we know after 200 years? National Museum of New Zealand Miscellaneous Series 7*: 35–46

RAINE, J. I.; POLE, M. S. 1988: Middle Jurassic forest beds, Curio Bay. *New Zealand Geological Survey Record 33*: 47–57.

RAINE, J. I.; SPEDEN, I. G.; STRONG, C. P. 1981: New Zealand. Pp. 221–267 *in*: Reyment, R. A.; Bengtson, P. (eds), *Aspects of Mid-Cretaceous Regional Geology*. Academic Press, London. ix + 327 p.

STEVENS, G. R. 1980: Southwest Pacific faunal paleobiogeography in Mesozoic and Cenozoic times: a review. *Palaeogeography, Palaeoclimatology, Palaeoecology 31*: 153–196.

STOVER, L. E.; PARTRIDGE, A. D. 1984: A new Late Cretaceous megaspore with grapnel-like appendage tips from Australia and New Zealand. *Palynology 8*: 139–144.

TE PUNGA, M. T. 1948: *Nothofagus* pollen from the Cretaceous Coal Measures at Kaitangata, Otago, New Zealand. *New Zealand Journal of Science and Technology 29B*: 32–35.

TE PUNGA, M. T. 1949: Fossil spores from New Zealand coals. *Transactions of the Royal Society of New Zealand 77*: 289–296.TURNBULL, I. M.; LINDQVIST, J. K.; MILDENHALL, D. C.; HORNIBROOK, N. deB.; BEU, A. G. 1985: Stratigraphy and paleontology of Pliocene–Pleistocene sediments on Five Fingers Peninsula, Dusky Sound, Fiordland. *New Zealand Journal of Geology and Geophysics 28*: 217–231.*

VAJDA, V.; RAINE, J. I. 2003: Pollen and spores in marine Cretaceous/Tertiary boundary sediments at mid-Waipara River, North Canterbury, New Zealand. *New Zealand Journal of Geology and Geophysics 46*: 255–273.*

VAJDA, V.; RAINE, J. I. 2010: A palynological investigation of plesiosaur-beariing rocks from the Upper Cretaceous Tahora Formation, Mangahouanga, New Zealand. *Alcheringa 34*: 359–374.*

WARDLE, P. 1991: *Vegetation of New Zealand*. Cambridge University Press, Cambridge. 672 p.

WATERS, R. 1915: Pollen grains as source-indicators of honey. *The Journal of Agriculture 11*: 384–388.

WATT, M. N. 1929: Summary of report on an atmospheric pollen survey, Dunedin, summer 1928–29. *New Zealand Department of Health Report 1928–29*: 69–71.

ZHANG, W.-P; GRANT-MACKIE, J. A. 1997: Late Triassic–Early Jurassic sporo-pollen assemblages of New Zealand and the synchronous sporo-pollen assemblages correlation between New Zealand and China. Pp. 1–80 *in*: Zhang, W.-P.; Grant-Mackie, J. A.; Yao, H.-Z. *Late Triassic–Early Jurassic Stratigraphy and Paleontology of the Circum-Pacific Region, China and New Zealand (2)*. Seismological Press, Beijing. 131 p.

ZHANG, W.-P.; GRANT-MACKIE, J. A. 2001: Late Triassic-Early Jurassic palynofloral assemblages from Murihiku strata of New Zealand, and comparisons with China. *Journal of the Royal Society of New Zealand 31*: 575–683.

Checklist of New Zealand fossil spores and pollen

The spores and pollen listed below are classified according to the system of Potonié (1960). This system uses a hierarchical system of suprageneric categories, so that instead of Phylum (Division), Class, Order, and Family there are Anteturma, Turma, Subturma, Infraturma and Subinfraturma. The plant group is indicated following each species (CH Chlorophyta, BR Bryophyta, LY Lycopsida, FI Filicopsida, SP Sphenopsida, GY Gymnospermae sensu lato, GN Gnetales, AN Angiospermae incertae sedis, MO monocotyledonous angiosperm, DI dicotyledonous angiosperm) and the stratigraphic range given (based on the New Zealand geological time scale (pages 14 and 15 of this volume). Inferred original ages of taxa known only from specimens redeposited in younger sediments are enclosed in parentheses.

ANTETURMA SPORITES
Turma MONOLETES
Suprasubturma ACAVATOMONOLETES
Subturma AZONOMONOLETES
Infraturma LAEVIGATOMONOLETI
Laevigatosporites major (Cookson 1947) Krutzsch 1959 FI K-Ng
Laevigatosporites ovatus Wilson & Webster 1946 FI K-Ng
Laevigatosporites vulgaris (Ibrahim *in* Potonié, Ibrahim & Loose 1932) Ibrahim 1933 FI Tr
Monolites alveolatus Couper 1960 FI Pg-Ng

Infraturma APICULATIMONOLETI
Echinosporis sp. Raine 1981 FI K
Hazaria sp. Raine 1982 FI Pg
"Hypolepis" spinysporis Martin 1973 FI Ng
Polypodiidites sp. A Raine 1982 FI Pg
Polypodiisporites favus (Potonié 1931) Potonié 1934 FI Pg-Ng
Polypodiisporites histiopteroides (Krutzsch 1962) Nagy 1973 FI Ng
Polypodiisporites inangahuensis (Couper 1953) Potonié 1956 FI Pg-Ng
Polypodiisporites minimus (Couper 1960) Khan & Martin 1971 FI K-Ng
Polypodiisporites perverrucatus (Couper 1953) Khan & Martin 1971 FI Pg-Ng
Polypodiisporites radiatus Pocknall & Mildenhall 1984 FI Pg-Ng
Polypodiisporites variscabratus Mildenhall & Pocknall 1989 FI Pg-Ng
Polypodiisporites sp. Pocknall 1982 FI Ng
Punctatosporites scabratus (Couper 1958) Norris 1965 FI J
Punctatosporites walkomii de Jersey 1962 LY? Tr
Thymospora ipsviciensis (de Jersey 1962) Jain 1965 FI P-Tr
Tuberculatosporites aberdarensis de Jersey 1962 FI Tr
Verrucatosporites sp. A Raine 1982 FI Pg

Infraturma MURORNATIMONOLETI
Microfoveolatosporis canaliculatus Dettmann 1963 FI K
Microfoveolatosporites fromensis (Cookson 1957) Harris 1965 FI K
Microfoveolatosporis sp. B Raine 1982 FI Pg
Reticuloidosporites arcus (Balme 1957) Dettmann 1963 FI K

Suprasubturma PERINOMONOLETES
Aratrisporites sp Zhang & Grant-Mackie 1997 LY Tr
Aratrisporites banksii Playford 1965 LY Tr
Aratrisporites fischeri (Klaus 1960) Playford & Dettmann 1965 LY Tr
Aratrisporites flexibilis Playford & Dettmann 1965 LY Tr
Aratrisporites granulatus (Klaus 1960) Playford & Dettmann 1965 LY Tr
Aratrisporites paenulatus Playford & Dettmann 1965 LY Tr
Aratrisporites paraspinosus Klaus 1960 LY Tr
Aratrisporites parvispinosus Leschik 1955 LY Tr
Aratrisporites rotundus Madler 1964 LY Tr
Aratrisporites rugulatus de Jersey 1970 LY Tr
Aratrisporites strigosus Playford 1965 LY (Tr)
Aratrisporites wollariensis Helby 1967 LY Tr
Perinomonoletes fsp. A Krutzsch 1967 FI Pg-Ng
Peromonolites bowenii Couper 1953 FI K-Pg
Peromonolites densus Harris 1965 FI K-Pg
Peromonolites vellosus Partridge *in* Stover & Partridge 1973 FI Pg-Ng

Turma TRILETES
Suprasubturma ACAVATITRILETES
Subturma AZONATI
Infraturma LAEVIGATI
Auritulinasporites scanicus Nilsson 1958 FI Tr/J
Auritulinasporites triclavis Nilsson 1958 FI Tr/J
Biretisporites modestus McKellar 1974 FI Tr/J
Biretisporites spectabilis Dettmann 1963 FI K
Biretisporites sp. de Jersey & Raine 1990 FI Tr
Biretisporites sp. Mildenhall 1994 FI K
Calamospora mesozoica Couper 1958 SP Tr
Cibotiumspora intrastriatus (Nilsson 1958) Zhang & Grant-Mackie 1997 FI Tr/J
Cibotiumspora juncta (Kara-Murza 1956) Zhang 1978 FI Tr/J
Cibotiumspora juriensis (Balme 1957) Filatoff 1975 FI Tr/J
Cibotiumspora sp. de Jersey & Raine 1990 FI Tr
Cibotiumspora sp. Zhang & Grant-Mackie 1997 FI Tr/J
Concavisporites bohemiensis Thiergart 1953 FI Tr/J
Concavisporites (Obtusisporis) sinuatus (Couper 1953) Krutzsch 1959 FI K
Cyathidites asper (Bolkhovitina 1953) Dettmann 1963 FI K
Cyathidites australis Couper 1953 FI J-K
Cyathidites concavus (Bolkhovitina 1953) Dettmann 1963 FI K
Cyathidites minor Couper 1953 FI Tr-Ng
Cyathidites punctatus (Delcourt & Sprumont 1955) Delcourt et al. 1963 FI K
Cyathidites splendens Harris 1965 FI Pg
Cyathidites subtilis Partridge 1973 FI Pg-Ng
Dictyophyllidites arcuatus Pocknall & Mildenhall 1984 FI Pg-Ng
Dictyophyllidites atraktos Stevens 1981 FI Tr-J
Dictyophyllidites concavus Harris 1965 FI Pg
Dictyophyllidites harrisii Couper 1958 FI J
Dictyophyllidites mortonii (De Jersey 1959) Playford & Dettmann 1965 FI Tr-J
Dictyophyllidites sp. Mildenhall 1994 FI K
Dictyophyllidites sp. Raine 1987 FI J
Leiotriletes blairatholensis Foster 1975 FI? (P)
Leiotriletes directus Balme & Hennelly 1956 FI P-J
Leiotriletes magna (de Jersey 1959) Norris 1963 FI Tr-J
Punctatisporites sp. Zhang & Grant-Mackie 1997 FI Tr/J
Trilites fragilis Couper 1953 FI? K
Trilites sp. Mildenhall 1994 FI K
Todisporites major Couper 1958 FI J-K
Todisporites minor Couper 1958 FI K
Todisporites rotundiformis (Maljavkina 1943) Pocock 1970 FI Tr/J

Infraturma APICULATI
Acanthotriletes bradiensis Playford 1965 LY? Tr/J
Acanthotriletes levidensis Balme 1957 LY? J
Acanthotriletes filiformis (Balme & Hennelly 1956) Tiwari 1965 LY? Tr
Acanthotriletes pallidus de Jersey 1960 LY? Tr/J
Acanthotriletes superbus Foster 1979 LY? (P)
Acanthotriletes tereteangulatus Balme & Hennelly 1956 LY? P?-Tr
Acanthotriletes sp. Mildenhall 1994 LY? (P-Tr?)
Anapiculatisporites cooksonae Playford 1965 BR? Tr/J
Anapiculatisporites pristidentatus Reiser & Williams 1969 BR? Tr-J
Apiculatisporis clematisi de Jersey 1968 FI? Tr
Apiculatisporis cornutus (Balme & Hennelly 1956) Hoegg & Bose 1960 FI? (P)
Apiculatisporis globosus (Leschik 1955) Playford & Dettmann 1965 FI? Tr
Apiculatisporis lentus Playford 1982 FI? Tr/J
Apiculatisporis otapiriensis de Jersey & Raine 1990 FI? Tr-J
Apiculatisporis ovalis (Nilsson 1958) Norris 1965 FI? Tr/J
Apiculatisporis sp. A. de Jersey & Raine 1990 FI? Tr
Apiculatisporis sp. Mildenhall 1994 FI? (P/Tr?)
Apiculatisporis sp. Waterhouse & Norris 1972 FI? J
Baculatisporites comaumensis (Cookson 1953) Potonié 1956 FI Tr-Ng
Baculatisporites disconformis Stover *in* Stover & Partridge 1973 FI Pg-Ng
Baculatisporites sp. Pocknall 1982 FI Ng
Baculatisporites sp. Raine 1987 FI J
Brevitriletes bulliensis (Helby 1973 ex de Jersey 1979) de Jersey & Raine 1990 FI? Tr
Brevitriletes levis (Balme & Hennelly 1956) Bharadwaj & Srivastava 1969 FI? (P/Tr?)
Brevitriletes sp. Mildenhall 1994 FI? (P/Tr?)
Cadargasporites baculatus (de Jersey & Paten 1964) Reiser & Williams 1969 FI? Tr/J
Cadargasporites granulatus (de Jersey & Paten 1964) Reiser & Williams 1969 FI? Tr/J
Ceratosporites equalis Cookson & Dettmann 1958 LY J-Ng
Ceratosporites masculus Norris 1968 LY K
Clavatisporites sp. A de Jersey & Raine 1990 FI Tr
Clavatisporites sp. B de Jersey & Raine 1990 FI Tr

Clavatisporites sp. Zhang & Grant-Mackie 1997 FI Tr/J
Clavatisporites conspicuus Playford 1982 FI Tr-J
Conbaculatisporites sp. A de Jersey & Raine 1990 FI Tr/J
Concavissimisporites grumulus Foster 1979 FI (P-Tr?)
Concavissimisporites penolaenis Dettmann 1963 FI K
Converrucosisporites sp. A de Jersey & Raine 1990 FI Tr-J
Converrucosisporites cameronii (de Jersey 1962) Playford & Dettmann 1965 FI-Tr
Converrucosisporites rewanensis de Jersey 1970 FI Tr
Craterisporites rotundus de Jersey 1970 FI Tr
Cyclogranisporites gondwanaensis Bharadwaj & Salujha 1964 FI? P
Didecitriletes ericianus (Balme & Hennelly 1956) Venkatachala & Kar 1965 FI? P
Didecitriletes uncinatus (Balme & Hennelly 1956) Venkatachala & Kar 1965 FI? P
Distaverrusporites visscheri (de Jersey 1968) de Jersey & Raine 1990 LY? Tr
Gemmatriletes multiglobus Mildenhall & Pocknall 1989 FI Ng
Granulatisporites micronodosus Balme & Hennelly 1956 FI? (P)
Granulatisporites minor de Jersey 1960 FI? Tr/J
Granulatisporites quadruplex Segroves 1970 FI? P
Granulatisporites trisinus Balme & Hennelly 1956 FI? P
Granulatisporites sp. A Raine 1987 FI? J
Granulatisporites sp. B Raine 1987 FI? J
Grapnelispora evansii Stover & Partridge 1984 FI? K
Herkosporites elliottii Stover 1973 LY Pg
Herkosporites proxistriatus Burger 1976 LY K
Horriditriletes sp. Mildenhall 1994 LY (P?)
Interradispora robusta (Foster 1975) Foster 1979 FI? P
Interradispora daedala Foster 1979 FI? P
Leptolepidites verrucatus Couper 1953 LY J-K
Lophotriletes bauhiniae de Jersey & Hamilton 1967 FI? P?-Tr
Lophotriletes novicus Singh 1964 FI? P-Tr?
Lophotriletes sp. Mildenhall 1994 FI? (P/Tr?)
Lycopodiacidites asperatus Dettmann 1963 FI K
Lycopodiacidites dettmannae Burger 1980 FI K
Microbaculispora tentula Tiwari 1965 FI? P
Neoraistrickia neozealandica (Couper 1953) Potonié 1956 LY K
Neoraistrickia ramosus (Balme & Hennelly 1956) Hart 1960 LY P-J
Neoraistrickia truncata (Cookson 1953) Potonié 1956 LY J-K
Osmundacidites fissus (Leschik 1955) Playford 1965 FI Tr-J
Osmundacidites parvus de Jersey 1962 FI Tr/J
Osmundacidites sparsituberculatus (Klimko 1961) Zhang & Grant-Mackie 1997 FI Tr/J
Osmundacidites wellmanii Couper 1953 FI Tr-K
Pilosisporites sp. Mildenhall 1994 FI K
Pustulatisporites blackstonensis de Jersey 1970 FI Tr
Rattiganispora sp. Mildenhall 1994 FI? (P-Tr?)
Rubinella major (Couper 1958) Norris 1968 FI J-K
Secarisporites bullatus (Balme & Hennelly 1956) Smith 1971 FI? (P-Tr?)
Stoverisporites microverrucatus Burger 1975 (1976) BR K
Stoverisporites verrucolabrus Kemp 1977 BR Pg-Ng
Stoverisporites sp. A Pocknall & Turnbull 1989 BR Pg
Toripustulatisporites hokonuiensis de Jersey 1990 FI J
Trilites hayii Couper 1953 FI Ng
Trilites morleyi Couper 1953 LY K
Trilites verrucatus Couper 1953 FI J-K
Uvaesporites argenteaeformis (Bolkhovitina 1953) Schulz 1967 LY Tr/J
Uvaesporites miniscultus Lu & Wang 1980 LY Tr/J
Uvaesporites projectus Zhang & Grant-Mackie 1997 LY Tr/J
Uvaesporites verrucosus (de Jersey 1964) Helby *in* de Jersey 1971 LY Tr-J
Uvaesporites viriosus Zhang & Grant-Mackie 1997 LY Tr/J
Uvaesporites sp. Zhang & Grant-Mackie 1997 LY Tr/J
Verrucosisporites cristatus Partridge 1973 FI Ng
Verrucosisporites kopukuensis (Couper 1960) Stover *in* Stover & Partridge FI Pg-Ng
Verrucosisporites sp. Crosbie 1986 FI P
Verrucosisporites sp. Zhang & Grant-Mackie 1997 FI Tr/J

Infraturma MURORNATI
Subinfraturma RETICULATI
Balmeisporites glenelgensis Cookson & Dettmann 1958 FI K
Balmeisporites holodictyus Cookson & Dettmann 1958 FI K
Balmeisporites tridictyus Cookson & Dettmann 1958 FI K
Dictyotosporites complex Cookson & Dettmann 1958 LY K
Dictyotosporites speciosus Cookson & Dettmann 1958 LY K
Dictyotosporites sp. Norris 1968 LY? K
Dictyotosporites sp. Pocknall & Lindqvist 1988 LY K
Foveosporites canalis Balme 1957 FI K
Foveosporites moretonensis de Jersey 1964 FI Tr-J
Foveotriletes balteus Partridge 1973 FI Pg-Ng
Foveotriletes crater Partridge 1973 FI Pg-Ng
Foveotriletes labrus Mildenhall & Pocknall 1989 FI Ng
Foveotriletes lacunosus Partridge 1973 LY Pg-Ng
Foveotriletes palaequetrus Partridge 1973 LY Pg-Ng
Foveotriletes parviretus (Balme 1957) Dettmann 1963 FI K
Foveotriletes verrucosus Pocknall & Mildenhall 1984 FI Pg-Ng
Foveotriletes sp. A Pocknall & Turnbull 1989 FI Pg
Interulobites intraverrucatus (Brenner 1963) Phillips 1972 BR? K
Ischyosporites gremius Stover 1973 FI Pg
Ischyosporites marburgensis (de Jersey 1963) de Jersey & Paten 1964 FI J
Ischyosporites volkheimeri Filatoff 1975 FI Tr-J
Kuylisporites lunaris Cookson & Dettmann 1958 FI K-Ng
Kuylisporites waterbolkii Potonié 1956 FI Pg-Ng
Lycopodiacidites bullerensis Couper 1953 LY K
Lycopodiumsporites saturnalis Norris 1968 LY K
Lycopodiumsporites sp. Pocknall & Lindqvist 1988 LY Pg-Ng
Microfoveolatispora explicita Foster 1979 FI? P
Retitriletes austroclavatidites (Cookson 1953) Döring et al. *in* Krutzsch 1963 LY J-Pg
Retitriletes circolumenus (Cookson & Dettmann 1958) Backhouse 1978 LY J-K
Retitriletes facetus (Dettmann 1963) Srivastava 1972 LY K
Retitriletes nodosus (Dettmann 1963) Srivastava 1972 LY K
Retitriletes reticulumsporites (Rouse 1959) Döring et al. *in* Krutzsch 1963 LY J-K
Retitriletes rosewoodensis (de Jersey 1959) McKellar 1974 LY Tr-J
Retitriletes tenuis (Balme 1957) Backhouse 1988 LY J-K
Rudolphisporis rudolphi (Krutzsch 1959) Krutzsch & Pacltova 1963 BR Ng

Subinfraturma RUGULATI
Lycopodiacidites cristatus Couper 1953 LY K
Retitriletes semimuris (Danzé-Corsin & Laveine 1963) McKellar 1974 LY J
Rugulatisporites nelsonensis de Jersey & Raine 1990 FI Tr
Rugulatisporites sp. A de Jersey & Raine 1990 FI Tr
Tripartina variabilis Maljavkina 1949 FI J

Subinfraturma STRIATITI
Cicatricosisporites australiensis (Cookson 1953) Potonié 1956 FI J-K
Cicatricosisporites cuneiformis Pocock 1965 FI K
Cicatricosisporites hughesii Dettmann 1963 FI K
Cicatricosisporites ludbrookiae Dettmann 1963 FI K
Cicatricosisporites venustus Deak 1963 FI K
Crassoretitriletes vanraadshoovenii Germeraad, Hopping & Muller 1968 FI Pg
Crassoretitriletes sp. Pocknall 1982 FI Pg-Ng
Cyclosporites hughesii (Cookson & Dettmann 1958) Cookson & Dettmann 1959 BR? K
Staplinisporites caminus (Balme 1957) Pocock 1962 LY? J
Triplexisporites playfordii (de Jersey & Hamilton 1967) Foster 1979 FI Tr

Subinfraturma ANNULATI
Taurocusporites sp A. Zhang & Grant-Mackie 1997 FI? Tr/J

Subturma AURICULATI
Infraturma LAEVIGATI
Hedlundisporites sp. de Jersey & Raine 1990 FI? Tr
Triancoraesporites sp. Raine & Wilson 1988 FI Pg
Triquitrites sp. Zhang & Grant-Mackie 1997 FI Tr/J

Infraturma MURORNATI
Appendicisporites distocarinatus Dettmann & Playford 1968 FI K
Cibotiidites tuberculiformis (Cookson 1947) Skarby 1974 FI J-Ng
Indospora clara Bharadwaj 1962 LY? P-Tr
Indospora reticulata de Jersey 1968 LY? (P-Tr)
Matonisporites ornamentalis (Cookson 1947) Partridge *in* Stover & Partridge 1973 FI Pg
Rugulatisporites cowrensis (Martin 1973) Mildenhall & Pocknall 1989 FI Ng
Rugulatisporites mallatus Stover *in* Stover & Partridge 1973 FI Pg-Ng
Rugulatisporites trisinus de Jersey & Hamilton 1967 FI Tr
Rugulatisporites trophus Partridge *in* Stover & Partridge 1973 FI Pg-Ng
Rugulatisporites sp. B de Jersey & Raine 1990 FI Tr

Subturma TRICRASSATI
Infraturma LAEVIGATI
Clavifera triplex (Bolkhovitina 1953) Bolkhovitina 1966 FI K-Pg
Gleicheniidites senonicus Ross 1949 FI Tr-Ng
Gleicheniidites feronensis (Delcourt & Sprumont 1955) Delcourt & Sprumont 1959 FI K
Plicifera sp. Raine 1984 FI Pg
Rotaspora laevigata (Schulz 1967) de Jersey & Raine 1990 LY? J

Infraturma APICULATI
Diatomozonotriletes sp. Mildenhall 1994 FI? (P)
Peregrinisporis sp. Raine 1984 FI K

Infraturma MURORNATI
Camarozonosporites australiensis Burger 1976 LY K-Pg
Camarozonosporites bullatus Harris 1965 LY Pg
Camarozonosporites ohaiensis (Couper 1953) Dettmann & Playford 1968 LY K
Camarozonosporites rudis (Leschik 1955) Klaus 1960 LY Tr

Camarozonosporites sp. a Raine & Wilson 1988 LY Pg
Camarozonosporites sp. d Raine & Wilson 1988 LY Pg
Camarozonosporites sp. B Raine 1982 LY Pg
Camarozonosporites sp. H Raine 1982 LY Pg
Clavifera rudis Bolkhovitina 1966 FI K-Pg
Foveogleicheniidites atavus Raine *in* de Jersey & Raine 1990 FI Tr
Foveogleicheniidites sp. de Jersey & Raine 1990 FI Tr
Foveogleicheniidtes sp. Raine 1982 FI Pg
Latrobosporites marginis Mildenhall & Pocknall 1989 LY Ng
Lycopodiacidites cerniidites (Ross 1949) Brenner 1963 LY K
Lycopodiacidites cernuoides Couper 1954 LY Pg
Toricingulatisporites sp. Raine 1982 FI Pg
Zebrasporites interscriptus (Thiergart 1949) Klaus 1960 LY? J

Subturma CINGULATI
Infraturma LAEVIGATI
Annulispora folliculosa (Rogalska 1954) de Jersey 1959 BR Tr-K
Annulispora microannulata de Jersey 1962 BR Tr-J
Annulispora minima Zhang 1990 BR Tr/J
Cingutriletes cestus Stevens 1981 BR J
Cingulatisporites bifurcatus (Couper 1960) Martin 1973 BR Ng
Cingulatisporites lachlanae (Couper 1953) Krutzsch 1959 BR Ng
Cingutriletes sp. Zhang & Grant-Mackie 1997 BR Tr/J
Lycospora pallida (de Jersey 1962) de Jersey 1971 LY Tr/J
Stereisporites sp. Norris 1968 BR K
Stereisporites antiquasporites (Wilson & Webster 1946) Dettmann 1963 BR Tr-Ng
Stereisporites compactus (Bolkhovitina 1956) Iljina 1985 BR Tr/J
Stereisporites minor Zhang 1990 BR Tr/J
Stereisporites psilatus (Ross 1943) Pflug *in* Thomson & Pflug 1953 BR Tr-J
Stereisporites cf. *stictus* (Wolff 1934) Krutzsch 1959 BR Pg
Undulatisporites fossulatus Singh 1971 FI K

Infraturma APICULATI
Antulsporites clavus (Balme 1957) Filatoff 1975 BR J-K
Antulsporites varigranulatus (Levet-Carette 1964) Reiser & Williams 1969 BR J
Antulsporites verrucatus Zhang & Grant-Mackie 1997 BR J
Antulsporites sp. Mildenhall 1994 BR (J?)
Antulsporites sp a McKellar 1974 BR Tr/J
Antulsporites sp. B Zhang & Grant-Mackie 1997 BR Tr/J
Antulsporites sp. Zhang & Grant-Mackie 1997 BR Tr/J
Foraminisporis asymmetricus (Cookson & Dettmann 1958) Dettmann 1963 BR K
Foraminisporis dailyi (Cookson & Dettmann 1958) Dettmann 1963 BR K
Foraminisporis wonthaggiensis (Cookson & Dettmann 1958) Dettmann 1963 BR K
Indotriradites splendens (Balme & Hennelly 1956) Foster 1979 FI? (P)
Kraeuselisporites papillatus Harris 1965 LY Pg
Limatulasporites limatulus (Playford 1965) Helby & Foster 1979 BR Tr
Nevesisporites vallatus de Jersey & Paten 1964 BR J-K
Perotrilites granulatus Couper 1953 LY K
Perotrilites linearis (Cookson & Dettmann 1958) Evans 1970 LY K
Stereisporites (*Distgranisporis*) sp. Crosbie 1986 BR P
Stereisporites regium (Drozhastichich 1961) Drugg 1967 BR J-Pg

Infraturma MURORNATI
Contignisporites glebulentus Dettmann 1963 FI J-K
Cyatheacidites annulatus Cookson 1947 FI K-Pg
Kyrtomisporis elsendoornii (Van Erve 1977) Zhang & Grant-Mackie 1997 FI J
Kyrtomisporis minor Zhang 1978 FI J
Perotrilites laceratus (Norris 1968) Dettmann 1987 LY K
Perotrilites majus (Cookson & Dettmann 1958) Evans 1970 LY K
Perotrilites 'senonicus' MS Raine 1984 LY K
Polycingulatisporites crenulatus Playford & Dettmann 1965 BR Tr-J
Polycingulatisporites dejerseyi Helby ex De Jersey 1979 BR Tr
Polycingulatisporites mooniensis de Jersey & Paten 1964 BR Tr-J
Polycingulatisporites radiatus Zhang & Grant-Mackie 1997 BR J
Polycingulatisporites sp. Pocknall & Lindqvist 1988 BR K
Polycingulatisporites sp. Mildenhall 1994 BR (Tr-J?)
Polycingulatisporites sp. A Zhang & Grant-Mackie 1997 BR Tr/J
Polypodiaceoisporites papuanus (Khan 1976) Pocknall 1985 FI Pg-Ng
Polypodiaceoisporites retirugatus Muller FI Pg
Polypodiaceoisporites tumulatus Partridge *in* Stover & Partridge 1973 FI Pg-Ng
Reticulatisporites castellatus Pocock 1962 FI? J
Reticulatisporites pudens Balme 1957 FI? J-K
Rogalskaisporites ambientis (Li & Shang 1980) Bai 1983 BR Tr
Rogalskaisporites bujargiensis (Bolkhovitina 1956) Zhang & Grant-Mackie 1997 BR Tr
Rogalskaisporites cicatricosus Rogalska 1954 ex Danzé-Corsin & Laveine 1963 BR Tr
Rogalskaisporites sp. A Zhang & Grant-Mackie 1997 BR Tr
Stereisporites (*Tripunctisporis*) sp. Stover & Evans 1973 BR K-Pg
Striatella seebergensis Madler 1964 FI Tr-J

Suprasubturma PERINOTRILETES
Subturma AZONOPEROTRILETES
Crybelosporites berberioides Burger 1976 FI? K
Crybelosporites striatus (Cookson & Dettmann 1958) Dettmann 1963 FI? K
Crybelosporites stylosus Dettmann 1963 FI? K
Grandispora sp. Crosbie 1986 LY? P
Playfordiaspora crenulata (Wilson 1962) Foster 1979 LY? P-Tr
Velosporites triquetrus (Lantz 1958) Dettmann 1963 FI? K

Subturma ZONOPEROTRILETES
Bryosporis anisopolaris Bussell & Mildenhall 1990 BR Ng
Bryosporis problematicus (Couper 1953) Mildenhall 1990 BR Ng
Densoisporites microrugulatus Brenner 1963 LY K
Densoisporites playfordii (Balme 1963) Dettmann 1963 LY Tr
Densoisporites psilatus (de Jersey 1964) Raine & de Jersey 1988 LY Tr-J
Indotriradites sp. Crosbie 1986 LY? P
Limbosporites antiquus (de Jersey 1964) de Jersey & Raine 1990 LY Tr
Limbosporites balmei Foster 1979 LY (P?)
Limbosporites denmeadii (de Jersey 1962) de Jersey & Raine 1990 LY Tr
Lundbladispora sp. Mildenhall 1994 LY (P-Tr?)

Turma HILATES
Aequitriradites spinulosus (Cookson & Dettmann 1958) Cookson & Dettmann 1961 BR K
Aequitriradites sp. Mildenhall 1994 BR K
Coptospora striata Dettmann 1963 BR K
Coptospora sp. Mildenhall 1994 BR K
Ricciaesporites kawaraensis Mildenhall & Pocknall 1989 BR Ng

ANTETURMA POLLENITES
Turma SACCITES
Subturma MONOSACCITES
Infraturma INAPERTURATI
Callialasporites dampieri (Balme 1957) Dev 1961 GY J-K
Callialasporites microvelatus Schulz 1967 GY J
Callialasporites segmentatus (Balme 1957) Srivastava 1963 GY J-K
Callialasporites trilobatus (Balme 1957) Dev 1961 GY J-K

Infraturma MONOLETI-TRILETI
Cannanoropollis janakii Potonié & Sah 1960 GY Tr
Cordaitina sp. de Jersey & Raine 1990 GY Tr
Plicatipollenites sp. Mildenhall 1994 GY (P)
Plicatipollenites densus Srivastava 1970 GY (P)

Infraturma SULCATI
Chasmatosporites canadensis Pocock 1970 GY J
Patinasporites densus Leschik 1955 GY Tr
Walchiites sp. Zhang & Grant-Mackie 1997 GY Tr/J

Subturma DISACCITES
Infraturma TRILETI
Limitisporites sp. Crosbie 1986 GY P
Walchiites sp. Zhang & Grant-Mackie 1997 GY Tr-J

Infraturma SULCATI
Alisporites australis de Jersey 1962 GY P-J
Alisporites grandis (Cookson 1953) Dettmann 1963 GY K
Alisporites lowoodensis de Jersey 1963 GY Tr-J
Alisporites parvus de Jersey 1962 GY Tr/J
Alisporites similis (Balme 1957) Dettmann 1963 GY Tr-K
Alisporites warepanus Raine *in* de Jersey & Raine 1990 GY Tr
Alisporites sp. A de Jersey & Raine 1990 GY Tr
Ashmoripollis woodywisei Mohr & Gee 1992 GY K-Pg
Cedripites priscus Balme 1970 GY P
Chordasporites australiensis de Jersey 1962 GY Tr-J
Dacrydiumites praecupressinoides (Couper 1953) Truswell 1983 GY K-Ng
Falcatisporites stabilis Balme 1970 GY P
Indusiisporites parvisaccatus (de Jersey 1959) de Jersey 1963 GY Tr-J
Klausipollenites sp. A de Jersey & Raine 1990 GY Tr
Klausipollenites sp. B de Jersey & Raine 1990 GY Tr
Microalatidites paleogenicus (Cookson & Pike 1954) Mildenhall & Pocknall 1989 GY Pg-Ng
Microalatidites varisaccatus Mildenhall & Pocknall 1989 GY Ng
Parvisaccites sp. Raine 1987 GY J
Parvisaccites catastus Partridge *in* Stover & Partridge 1973 GY Pg-Ng
Phyllocladidites mawsonii Cookson 1947 ex Couper 1953 GY K-Ng
Phyllocladidites verrucosus (Cookson 1957) Stover & Evans 1973 GY Pg
Platysaccus queenslandi de Jersey 1962 GY Tr
Podocarpidites ellipticus Cookson 1947 GY K-Ng

Podocarpidites exiguus Harris 1965 GY Pg
Podocarpidites major Couper 1953 GY J-K
Podocarpidites marwickii Couper 1953 GY K-Pg
Podocarpidites microreticuloidatus Cookson 1947 GY K-Pg
Podocarpidites multesimus (Bolkhovitina 1956) Pocock 1962 GY K
Podocarpidites otagoensis Couper 1953 GY K
Podocarpidites puteus Mildenhall & Pocknall 1989 GY Ng
Podocarpidites rugulatus Pocknall & Mildenhall 1984 GY Pg-Ng
Podocarpidites torquatus Mildenhall & Pocknall 1989 GY Ng
Podocarpidites verrucosus Volkheimer 1972 GY J
Quadraeculina sp. Zhang & Grant-Mackie 1997 GY Tr
Scheuringipollenites maximus (Hart 1960) Tiwari 1973 GY (P)
Scheuringipollenites ovatus (Balme & Hennelly 1955) Foster 1979 GY P
Sulcosaccispora alaticonformis (Malyavkina 1964) de Jersey 1968 GY Tr-J
Vitreisporites signatus Leschik 1955 GY P-K

Infraturma MURORNATI-COSTATI
Distriatites insolitus Bharadwaj & Salujha 1964 GY Tr
Lueckisporites virkkiae Potonié & Klaus 1954 GY (P)
Lunatisporites sp. Crosbie 1986 GY P
Lunatisporites noviaulensis (Leschik 1956) de Jersey 1979 GY (P-Tr)
Lunatisporites pellucidus (Goubin 1965) Helby 1972 GY P
Protohaploxypinus sp. Zhang & Grant-Mackie 1997 GY (P-Tr)
Protohaploxypinus amplus (Balme & Hennelly 1955) Hart 1964 GY (P)
Protohaploxypinus limpidus (Balme & Hennelly 1955) Balme & Playford 1967 GY (P-Tr)
Protohaploxypinus samoilovichii (Jansonius 1962) Hart 1964 GY (P?-)Tr
Striatoabieites multistriatus (Balme & Hennelly 1955) Hart 1964 GY (P)
Striatopodocarpites cancellatus (Balme & Hennelly 1955) Hart 1963 GY P
Striatopodocarpites fusus (Balme & Hennelly 1955) Potonié 1958 GY P

Subturma POLYSACCITES
Dacrycarpites australiensis Cookson & Pike 1953 GY Pg-Ng
Microcachryidites antarcticus Cookson 1947 GY J-Ng
Podosporites brevisaccatus (Couper 1960) Mildenhall 1978 GY Pg-Ng
Podosporites castellanosii (Menendez 1968) Filatoff 1975 GY J
Podosporites erugatus Mildenhall 1978 GY Ng
Podosporites ohikaensis (Couper 1953) Pocock 1962 GY K
Podosporites parvus (Couper 1960) Mildenhall 1978 GY Pg-Ng
Trichotomosulcites subgranulatus Couper 1953 GY K-Pg

Turma ALETES-INAPERTURATES
Suprasubturma ACAVATALETES
Subturma AZONALETES
Infraturma LAEVIGATI
Acaciapollenites miocenicus Mildenhall & Pocknall 1989 DI Pg?-Ng
Acaciapollenites myriosporites (Cookson 1954) Mildenhall 1972 DI Ng
Acaciapollenites octosporites (Cookson 1954) Mildenhall 1972 DI Ng
Inaperturopollenites dubius (Potonié & Venitz 1934) Thomson & Pflug 1953 GY J-K
Spheripollenites classopolloides (Nilsson 1958) Playford & Dettmann 1965 GY J
Spheripollenites psilatus Couper 1958 GY J-K
Spheripollenites subgranulatus Couper 1958 GY K
Taxodiaceaepollenites hiatus (Potonié 1931) Kremp 1949 GY K-Ng

Infraturma APICULATI
Araucariacites australis Cookson 1947 GY Tr-Ng
Araucariacites fissus Reisser & Williams 1969 GY Tr-J
Assamiapollenites incognitus Pocknall & Mildenhall 1984 AN Ng
Assamiapollenites inanis Pocknall & Mildenhall 1984 AN Pg-Ng
Conaletes sp. de Jersey & Raine 1990 GY Tr
Dilwynites granulatus Harris 1965 GY K-Ng

Infraturma MURORNATI
Maculatasporites delicatus Foster 1975 CH? P
Rugaletes awakinoensis Raine *in* de Jersey & Raine 1990 GY Tr-J
Rugaletes intestiniformis Zhang & Grant-Mackie 1997 GY Tr/J

Suprasubturma PERINOALETES
Subturma ZONOPERINALETES
Triporoletes radiatus (Dettmann 1963) Playford 1971 BR K
Triporoletes reticulatus (Pocock 1962) Playford 1971 BR K
Triporoletes simplex (Cookson & Dettmann 1958) Playford 1971 BR K

Turma POLYPLICATES
Equisetosporites sp. Mildenhall 1994 GN K
Equisetosporites sp. Zhang & Grant-Mackie 1997 GN Tr/J
Equisetosporites notensis (Cookson 1957) Romero 1977 GN K-Ng
Equisetosporites steevesii (Jansonius 1962) de Jersey 1968 GN Tr
Steevesipollenites claviger de Jersey & Raine 1990 GN Tr-J
Weylandites lucifer (Bharadwaj & Salujha 1964) Foster 1975 GN? P
Vittatina sp. Mildenhall 1994 GN? (P?)

Turma MONOSULCATES
Arecipites minutiscabratus (McIntyre 1968) Milne 1988 MO Pg
Arecipites otagoensis (Couper 1960) Mildenhall & Pocknall 1989 MO Pg-Ng
Arecipites subverrucatus (Pocknall) Midenhall & Pocknall 1989 MO Pg
Arecipites waitakiensis (McIntyre 1968) Mildenhall & Pocknall 1989 MO Pg-Ng
Clavatipollenites ascarinoides McIntyre 1968 DI K-Ng
Clavatipollenites hughesii Couper 1958 DI? K
Cycadopites follicularis Wilson & Webster 1946 GY Tr-K
Cycadopites granulatus (de Jersey 1962) de Jersey 1964 GY Tr
Cycadopites sp. Pocknall 1985 GY Pg
Cycadopites sp. A de Jersey & Raine 1990 GY Tr
Dryptopollenites semilunatus Stover *in* Stover & Partridge 1973 MO Pg
Liliacidites aviemorensis McIntyre 1968 MO Pg-Ng
Liliacidites intermedius Couper 1953 MO K-Ng
Liliacidites kaitangataensis Couper 1953 MO K
Liliacidites perforatus Pocknall 1982 MO Pg
Liliacidites variegatus Couper 1953 MO Ng
Marsupipollenites striatus (Balme & Hennelly 1956) Foster 1975 GY P
Marsupipollenites triradiatus Balme & Hennelly 1956 GY (P)
Monogemmites sp. Krutzsch 1970 MO Ng
Monogemmites gemmatus (Couper 1960) Krutzsch 1970 MO Pg-Ng
Monosulcites cf. *Rhipogonum scandens* Forst. MO Ng
Monosulcites cf. *Rhopalostylis sapida* Wendl. & Drude MO Pg-Ng
Nupharipollis mortonensis Pocknall & Mildenhall 1984 MO Pg-Ng
Palmidites maximus Couper 1953 MO Ng
Praecolpatites sinuosus (Balme & Hennelly 1956) Bharadwaj & Srivastava 1969 GY P
Retimonocolpites peroreticulatus (Brenner 1963) Doyle 1975 AN K
Retimonocolpites textus (Norris 1967) Singh 1983 AN K
Spinizonocolpites sp. A Raine 1982 MO Pg
Spinizonocolpites prominatus (McIntyre 1965) Stover & Evans 1973 MO Pg

Turma TRICHOTOMOSULCITES
Asteropollis asteroides Hedlund & Norris 1968 DI K
Luminidites phormoides (Stover & Partridge 1982) Pocknall & Mildenhall 1984 MO Ng
Luminidites reticulatus (Couper 1960) Pocknall & Mildenhall 1984 MO Pg-Ng

Turma DISULCITES-DICOLPATES
Dicolpopollis cf. *metroxylonoides* Khan 1976 MO Ng
Dicolpopollis sp. Mildenhall 1994 MO Ng

Turma ZONO-ANNULICOLPATES
Corollina cf. *chateaunovi* (Reyre 1970) Courtinat & Algouti 1985 GY Tr-K
Corollina simplex (Danzé-Corsin & Laveine 1963) Cornet & Traverse 1975 GY J-K
Corollina meyeriana (Klaus 1960) Venkatachala & Goczan 1964 GY J
Corollina sp. A de Jersey & Raine 1990 GY J

Turma TRICOLPATES
Beaupreaidites diversiformis Milne 1998 DI Pg-Ng
Beaupreaidites elegansiformis Cookson 1950 DI K-Ng
Beaupreaidites orbiculatus Dettmann & Jarzen 1988 DI K
Beaupreaidites verrucosus Cookson 1950 DI K-Ng
Beaupreaidites n. sp. Raine 1981 DI K
Cranwellia striata (Couper 1953) Srivastava 1966 DI Pg-Ng
Cupuliferoidaepollenites parvulus (Groot & Penny 1960) Dettmann 1973 DI K
Cupuliferoidaepollenites sp. Mildenhall 1994 DI K
Dactylopollis magnificus Muller 1968 DI K
Dicotetradites clavatus Couper 1960 DI Pg-Ng
Forcipites sabulosus (Dettmann & Playford 1968) Dettmann & Jarzen 1988 DI K
Guettardidites ivirensis Khan 1976 DI Ng
Gyropollis psilatus Mildenhall & Pocknall 1989 DI Ng
Ilexpollenites anguloclavatus McIntyre 1968 DI Pg-Ng
Ilexpollenites clifdenensis McIntyre 1968 DI K-Ng
Ilexpollenites megagemmatus McIntyre 1968 DI Pg-Ng
Ilexpollenites verrucosus Pocknall & Mildenhall 1984 DI Ng
Peninsulapollis askiniae Dettmann & Jarzen 1988 DI K
Peninsulapollis gillii (Cookson 1957) Dettmann & Jarzen 1988 DI K-Pg
Peninsulapollis truswelliae Dettmann & Jarzen 1988 DI K

Perfotricolpites digitatus Gonzalez Guzman 1967 DI Pg
Phimopollenites augathellaensis (Burger 1970) Dettmann 1973 DI K
Phimopollenites pannosus (Dettmann & Playford 1968) Dettmann 1973 DI K
Rousea georgensis (Brenner 1963) Dettmann 1973 DI K
Spinitricolpites latispinosus (McIntyre 1968) Mildenhall & Pocknall 1989 DI Ng
Striatopollis paraneus (Norris 1967) Singh 1971 DI K
Tricolpites asperamarginis McIntyre 1968 DI Pg
Tricolpites bathyreticulatus Stanley 1965 DI Ng
Tricolpites 'brunnerensis' Raine MS 1982 DI Pg
Tricolpites confessus Stover *in* Stover & Partridge 1973 DI K-Pg
Tricolpites delicatulus Couper 1960 DI Ng
Tricolpites densifoveatus McIntyre 1968 DI Ng
Tricolpites densipunctatus McIntyre 1968 DI Ng
Tricolpites discus Harris 1977 DI Pg-Ng
Tricolpites fissilis Couper 1960 DI K-Pg
Tricolpites geranioides Couper 1960 DI Ng
Tricolpites inargutus McIntyre 1968 DI Pg-Ng
Tricolpites inconspicuus Mildenhall & Pocknall 1989 DI Ng
Tricolpites minutus (Brenner 1963) Dettmann 1973 DI K
Tricolpites pachyexinus Couper 1953 DI K-Pg
Tricolpites patulus Truswell & Owen 1988 DI Ng
Tricolpites perimarginatus McIntyre 1968 DI Pg-Ng
Tricolpites perlongicolpus Pocknall & Mildenhall 1984 DI Pg-Ng
Tricolpites phillipsii Stover 1973 DI Pg
Tricolpites punctaticulus McIntyre 1968 DI Ng
Tricolpites reticulatus Cookson 1947 ex Couper 1953 DI K-Ng
Tricolpites variogranulosus McIntyre 1965 DI K-Pg
Tricolpites waitunaensis (Couper 1960) Pocknall & Crosbie 1982 DI Ng
Tricolpites trioblatus Mildenhall & Pocknall 1989 DI Ng
Tricolpites 'punctate' Raine 1981 DI Pg
Tricolpites sp. Pocknall 1985 DI Pg
Tricolpites sp. Raine et al. 1981 DI K
«*Tricolpites*» *lilliei* Couper 1953 DI K

Turma STEPHANOCOLPATES
Nothofagidites asperus (Cookson 1959) Romero 1973 DI Pg-Ng
Nothofagidites brachyspinulosus (Cookson 1959) Harris 1965 DI Pg-Ng
Nothofagidites cranwelliae (Couper 1953) Mildenhall & Pocknall 1989 DI Pg-Ng
Nothofagidites deminutus (Cookson 1959) Stover & Evans 1973 DI Pg
Nothofagidites emarcidus (Cookson 1959) Harris 1965 DI Ng
Nothofagidites falcatus (Cookson 1959) Hekel 1972 DI Ng
Nothofagidites flemingii (Couper 1953) Potonié 1960 DI Pg-Ng
Nothofagidites kaitangata (Te Punga 1948) Romero 1973 DI K-Pg
Nothofagidites lachlaniae (Couper 1953) Pocknall & Mildenhall 1984 DI Pg-Ng
Nothofagidites longispinosus (Couper 1960) Pocknall *in* Dettmann et al. 1990 DI Ng
Nothofagidites matauraensis (Couper 1953) Hekel 1972 DI Pg-Ng
Nothofagidites senectus Dettmann & Playford 1968 DI K
Nothofagidites spinosus (Couper 1960) Mildenhall & Pocknall 1989 DI Ng
Nothofagidites suggatei (Couper 1953) Hekel 1972 DI Ng
Nothofagidites vansteenisii (Cookson 1959) Stover & Evans 1973 DI Pg-Ng
Nothofagidites waipawaensis (Couper 1960) Fasola 1969 DI Pg
Nothofagidites sp. Pocknall 1985 DI Pg
Polycolpites clavatus Couper 1953 DI K
Polygalacidites sp. Turnbull et al. 1985 DI Ng
Ranunculacidites sp. Raine 1984 DI Pg
Reevesiapollis reticulatus (Couper 1960) Krutzsch 1970 DI Ng
Stephanocolpites oblatus Martin 1973 DI Ng
Stephanocolpites sphericus (Couper 1960) Mildenhall & Pocknall 1989 DI Pg-Ng
Tetracolpites sp. Mildenhall 1994 DI K

Turma POLYCOLPATES
Glencopollis ornatus Pocknall & Mildenhall 1984 DI Ng
Lymingtonia cenozoica Pocknall & Mildenhall 1984 DI Pg-Ng
'*Syndemicolpites*' sp. a Raine & Wilson 1988 DI Pg
'*Syndemicolpites*' sp. b Raine & Wilson 1988 DI K-Pg

Turma SYNCOLPORATES
Cupanieidites insularis Mildenhall & Pocknall 1989 DI Ng
Cupanieidites major Cookson & Pike 1954 DI Pg-Ng
Cupanieidites orthoteichus Cookson & Pike 1954 DI Pg-Ng
Cupanieidites reticularis Cookson & Pike 1954 DI Pg-Ng
Gothanipollis bassensis Stover *in* Partridge & Stover 1973 DI Pg-Ng
Gothanipollis gothani Krutzsch 1959 DI Ng
Gothanipollis perplexus Pocknall & Mildenhall 1984 DI Pg-Ng
Myrtaceidites eucalyptoides Cookson & Pike 1954 DI Ng
Myrtaceidites eugeniioides Cookson & Pike 1954 DI Pg-Ng
Myrtaceidites mesonesus Cookson & Pike 1954 DI Pg-Ng
Myrtaceidites parvus Cookson & Pike 1954 DI Pg-Ng
Myrtaceidites protrudiporens Martin 1973 DI Pg-Ng
Myrtaceidites verrucosus Partridge *in* Stover & Partridge 1973 DI Pg-Ng
Striasyncolpites laxus Mildenhall & Pocknall 1989 DI Ng

Turma TRICOLPORATES
Ailanthipites paenestriatus (Stover *in* Stover & Partridge 1973) Milne 1988 DI Ng
Anisotricolporites truncatus Pocknall & Mildenhall 1984 DI Ng
Anisotricolporites sp. Nelson, Mildenhall, Todd & Pocknall 1988 DI Ng
Biplanipollis whidbeyensis Mildenhall 1985 DI Ng
Bluffopollis dubius (Couper 1954) Pocknall & Mildenhall 1984 DI Pg
Bluffopollis maculatus Pocknall & Mildenhall 1984 DI Pg-Ng
Bluffopollis scabratus (Couper 1954) Pocknall & Mildenhall 1984 DI Pg-Ng
Bombacacidites bombaxoides Couper 1960 DI Pg-Ng
Bombacacidites isoreticulatus McIntyre 1965 DI Pg
Bombapollis hectorii Pocknall 1982 DI Pg-Ng
Bombapollis sp. Pocknall 1985 DI Pg
Dicrassipollis balteus Pocknall & Mildenhall 1984 DI Pg-Ng
Dicrassipollis rimulatus (Pocknall 1982) Mildenhall & Pocknall 1989 DI Pg-Ng
Dryadopollis retequetrus (Partridge *in* Stover & Partridge 1973) Pocknall & Mildenhall 1984 DI Pg-Ng
Ericipites crassiexinus Harris 1972 DI Pg
Ericipites longisulcatus Wodehouse 1933 DI Pg-Ng
Gemmapollis raglanensis Pocknall 1985 DI Pg-Ng
Integricorpus sp. Pocknall, Strong & Wilson 1989 DI Pg
Intratriporopollenites notabilis (Harris 1956) Stover *in* Stover & Partridge DI Pg-Ng
Keilmeyerapollenites sp. Raine 1984 DI K
Margocolporites cribellatus Srivastava 1972 DI Pg
Margocolporites scabratus Pocknall & Mildenhall 1984 DI Pg
Margocolporites spheripunctatus Mildenhall & Pocknall 1989 DI Ng
Margocolporites vanwijhei Germeraad, Hopping & Muller 1968 DI Pg
Nuxpollenites varicosus Pocknall & Mildenhall 1984 DI Ng
Nuxpollenites sp. Mildenhall 1989 DI Ng
Nyssapollenites chathamicus Mildenhall 1994 DI K
Nyssapollenites endobalteus (McIntyre 1965) Kemp & Harris 1977 DI Pg-Ng
Nyssapollenites squamosus Dettmann 1973 DI K
Oacolpopollenites sp. Pocknall 1982 DI Pg
Palaeocoprosmadites zelandiae Pocknall 1982 DI Pg-Ng
Paripollis sp. Raine 1984 DI K
Poluspissusites ramus Pocknall 1982 DI Pg-Ng
Rhoipites abnormis Mildenhall & Pocknall 1989 DI Ng
Rhoipites aequatorius Pocknall 1985 DI Pg
Rhoipites alveolatus (Couper 1953) Pocknall & Crosbie 1982 DI Pg-Ng
Rhoipites angurium (Partridge 1973) Pocknall & Mildenhall 1984 DI Pg-Ng
Rhoipites aralioides Pocknall & Mildenhall 1984 DI Pg-Ng
Rhoipites couperi Pocknall 1985 DI Pg-Ng
Rhoipites exiguus Pocknall 1982 DI Pg
Rhoipites fragilis Mildenhall & Pocknall 1989 DI Ng
Rhoipites hawkdunensis Mildenhall & Pocknall 1989 DI Ng
Rhoipites hekelii Mildenhall & Pocknall 1989 DI Ng
Rhoipites karamuensis Pocknall 1985 DI Pg
Rhoipites microluminus Kemp *in* Kemp & Harris 1977 DI Ng
Rhoipites retiformis Pocknall & Mildenhall 1984 DI Pg-Ng
Rhoipites rhomboidaliformis (McIntyre 1968) Mildenhall & Pocknall 1989 DI Ng
Rhoipites robustiexinus Mildenhall & Pocknall 1989 DI Ng
Rhoipites sphaerica (Cookson 1947) Pocknall & Crosbie 1982 DI Pg-Ng
Rhoipites titokioides Mildenhall & Pocknall 1989 DI Ng
Rhoipites waimumuensis (Couper 1953) Pocknall & Crosbie 1982 DI Pg-Ng
"*Rhoipites*" *triangulatus* Pocknall & Mildenhall 1984 DI Pg
Rubipollis oblatus (Pocknall & Mildenhall 1984) Mildenhall & Pocknall 1989 DI Pg-Ng
Santalumidites cainozoicus Cookson & Pike 1954 DI Pg-Ng
Siberiapollis sp. Raine 1984 DI K
Striatricolporites pseudostriatus (McIntyre 1968) Mildenhall & Pocknall 1989 DI Pg-Ng
Striatricolporites striatus (Couper 1954) Mildenhall & Pocknall 1989 DI K-Pg
Suprapollis variabilis Pocknall 1982 DI Pg
Symplocoipollenites austellus Partridge *in* Stover & Partridge 1973 DI Ng
Tricolpites brevicolpus Couper 1960 DI K-Pg
Tricolpites coprosmoides Couper 1960 DI Pg

Tricolpites membranus Couper 1960 DI Ng
Tricolpites secarius McIntyre 1965 DI Pg
Tricolpites variofoveatus McIntyre 1965 DI K-Pg
Tricolporites adelaidensis Harris 1970 ex Stover & Partridge 1982 DI Pg
Tricolporites scabratus Harris 1965 DI Pg-Ng
"*Tricolporites*" *leuros* Partridge *in* Stover & Partridge 1973 DI Ng
Tubulifloridites antipodica Cookson 1947 DI Pg-Ng
Tubulifloridites pleistocenicus Martin 1973 DI Ng
Tubulifloridites simplis Martin 1973 DI Ng

Turma TRICOLPIPOLYPORATES
Schizocolpus marlinensis Stover *in* Stover & Partridge 1973 DI Pg

Turma STEPHANOCOLPORATES
Polycolporopollenites esobalteus (McIntyre 1968) Pocknall & Mildenhall 1984 DI Pg-Ng
Psilastephanocolporites micus Partridge *in* Stover & Partridge 1973 DI Ng
Quintiniapollis psilatispora (Martin 1973) Mildenhall & Pocknall 1989 DI Pg-Ng
Sapotaceoidaepollenites latizonatus (McIntyre 1965) Pocknall & Mildenhall 1984 DI Pg-Ng
Sapotaceoidaepollenites rotundus Harris 1972 DI Ng
Tetracolporites ixerboides Couper 1960 DI Pg-Ng
Tetracolporites oamaruensis Couper 1953 DI Pg
Tetracolporites spectabilis Pocknall & Mildenhall 1984 DI Pg-Ng
Tetracolporopollenites costatus Pocknall & Mildenhall 1984 DI Ng

Turma MONOPORATES
Exesipollenites tumulus Balme 1957 GY J
Graminidites media Cookson 1947 MO Pg-Ng
Harrisipollenites annulatus Mildenhall & Crosbie 1980 DI Ng
Harrisipollenites kapukaensis Pocknall & Mildenhall 1984 DI Pg-Ng
Lateropora glabra Pocknall & Mildenhall 1984 MO Pg-Ng
Milfordia homeopunctata (McIntyre 1965) Partridge *in* Stover & Partridge 1973 MO Pg-Ng
Milfordia hypolaenoides Erdtman 1960 MO Ng
Monoporopollenites fossulatus McIntyre 1968 MO Pg-Ng
Perinopollenites elatoides Couper 1958 GY Tr-J
Pseudowinterapollis couperi Krutzsch 1970 DI Pg-Ng
Pseudowinterapollis cranwellae (Stover *in* Stover & Partridge) Mildenhall *in* Mildenhall& Crosbie 1980 DI Ng
Pseudowinterapollis wahooensis (Stover *in* Stover & Partridge) Mildenhall *in* Mildenhall& Crosbie 1980 DI K-Pg
Sparganiaceaepollenites barungensis Harris 1972 MO Pg-Ng
Sparganiaceaepollenites sparganioides (Meyer 1956) Krutzsch 1970 MO Ng
Sparganiaceaepollenites sphericus (Couper 1960) Mildenhall *in* Mildenhall & Crosbie 1979 MO Pg-Ng

Turma DIPORATES
Diporites aspis Pocknall & Mildenhall 1984 DI Pg-Ng
Banksieaeidites arcuatus Stover *in* Stover & Partridge 1973 DI Pg
Banksieaeidites elongatus Cookson 1950 ex Potonié 1960 DI Pg

Turma TRIPORATES
Bysmapollis emaciatus Partridge *in* Stover & Partridge 1973 DI Pg
Bysmapollis pergranulatus (Couper 1960) Crosbie & Clowes 1980 DI Pg
Bysmapollis sp. Raine 1984 DI Pg
Canthiumidites bellus (Partridge *in* Stover & Partridge1973) Mildenhall & Pocknall 1989 DI Ng
Corsinipollenites epilobioides Krutzsch 1968 DI Ng
Corsinipollenites oculusnoctis (Thiergart 1940) Nakoman 1965 DI Pg
Gambierina edwardsii (Cookson & Pike 1954) Harris 1972 DI Pg
Gambierina rudata Stover *in* Partridge & Stover 1973 DI K
Haloragacidites amolosus Partridge *in* Stover & Partridge 1973 DI Ng
Haloragacidites canacomyricoides (McIntyre 1968) Stover *in* Stover & Partridge 1973 DI Pg-Ng
Haloragacidites haloragoides Cookson & Pike 1954 DI Ng
Haloragacidites myriophylloides Cookson & Pike 1954 DI Ng
Myricipites harrisii (Couper 1953) Dutta & Sah 1970 DI Pg-Ng
Myrtaceoipollenites australis Harris 1965 DI Pg
Proteacidites adenanthoides Cookson 1950 DI Pg
Proteacidites alveolatus Stover *in* Stover & Partridge 1973 DI Ng
Proteacidites amolosexinus Dettmann & Playford 1968 DI K
Proteacidites annularis Cookson 1950 DI Pg-Ng
Proteacidites asperatus McIntyre 1968 DI Pg
Proteacidites beddoesii Stover *in* Stover & Partridge 1973 DI Pg
Proteacidites callosus Cookson 1950 DI Pg
Proteacidites crassus Cookson 1950 DI Pg
Proteacidites destructioris Mildenhall & Pocknall 1989 DI Ng
Proteacidites franktonensis Couper 1960 DI Ng
Proteacidites fromensis Harris 1972 DI Pg
Proteacidites grandis Cookson 1950 DI Pg
Proteacidites granoratus Couper 1960 DI K
Proteacidites hakeoides Couper 1960 DI Pg
Proteacidites incurvatus Cookson 1950 DI Pg
Proteacidites isopogiformis Couper 1960 DI Pg?-Ng
Proteacidites latrobensis Harris 1966 DI Ng
Proteacidites minimus Couper 1954 DI Pg-Ng
Proteacidites 'minutiporus' MS Pocknall & Turnbull 1989 DI Pg
Proteacidites nexinus Pocknall & Mildenhall 1984 DI Ng
Proteacidites obscurus Cookson 1950 DI Pg-Ng
Proteacidites palisadus Couper 1953 DI K
Proteacidites parvus Cookson 1950 DI K-Ng
Proteacidites polymorphus Couper 1960 DI Pg
Proteacidites pseudomoides Stover *in* Stover & Partridge 1973 DI Pg-Ng
Proteacidites rectomarginis Cookson 1950 DI Pg
Proteacidites rectus Pocknall & Mildenhall 1984 DI Pg-Ng
Proteacidites reflexus Partridge *in* Stover & Partridge 1973 DI Pg
Proteacidites reticuloscabratus Harris 1965 DI Pg
Proteacidites retiformis Couper 1960 DI K
Proteacidites scaboratus Couper 1960 DI K-Pg
Proteacidites similis Harris 1965 DI Pg
Proteacidites spiniferus McIntyre 1968 DI Pg-Ng
Proteacidites stratosus Pocknall & Mildenhall 1984 DI Pg-Ng
Proteacidites subpalisadus Couper 1953 DI K-Pg
Proteacidites subscabratus Couper 1960 DI Pg-Ng
Proteacidites symphyonemoides Cookson 1950 DI Pg-Ng
Proteacidites tenuiexinus Stover *in* Stover & Partridge 1973 DI Pg-Ng
Proteacidites tripartitus Harris 1972 DI Pg
Proteacidites truncatus Cookson 1950 DI Pg-Ng
Proteacidites tuberculatus Cookson 1950 DI Pg
Proteacidites sp. E Raine 1982 DI Pg
Proteacidites sp. M Raine 1982 DI Pg
Proteacidites sp. N Raine 1982 DI Pg
Proteacidites sp. indet. Turnbull et al. 1985 DI Ng
Triorites fragilis Couper 1953 DI K
Triorites minisculus McIntyre 1965 DI Pg-Ng
Triorites introlimbatus McIntyre 1968 DI Pg-Ng
Triorites minor Couper 1953 DI K-Pg
Triorites orbiculatus McIntyre 1965 DI Pg-Ng
Triorites spinosus Couper 1954 DI Pg
Triorites subalveolatus Couper 1960 DI K
Triorites subspinosus Couper 1960 DI Pg
Triporopollenites ambiguus Stover *in* Stover & Partridge 1973 DI Pg-Ng

Turma STEPHANOPORATES
Droseridites spinosa Cookson 1947 DI Ng
Droseridites sp. Mildenhall & Pocknall 1989 DI Ng
Malvacipollis diversus Harris 1965 DI Pg
Malvacipollis spinyspora (Martin 1973) Mildenhall & Pocknall 1989 DI Ng
Malvacipollis subtilis Stover *in* Stover & Partridge 1973 DI Pg-Ng

Turma POLYPORATES
Anacolosidites acutullus Cookson & Pike 1954 DI Pg
Anacolosidites luteoides Cookson & Pike 1954 DI Pg
Anacolosidites sectus Partridge *in* Stover & Partridge 1973 DI Pg
Chenopodipollis bipatterna (Martin 1973) Mildenhall & Pocknall 1989 DI Ng
Chenopodipollis chenopodiaceoides (Martin 1973) Truswell *in* Truswell et al. 1985 DI Ng
Chenopodipollis granulata (Martin 1973) Mildenhall & Pocknall 1989 DI Ng
Cyperacidites neogenicus Krutzsch 1970 MO Pg-Ng
Malvacearumpollis mannanensis Wood 1986 DI Pg-Ng
Parsonsidites multiporus Mildenhall & Crosbie 1980 DI Ng
Parsonsidites psilatus Couper 1960 DI Pg-Ng
Periporopollenites demarcatus Stover *in* Stover & Partridge 1973 DI Pg
Periporopollenites polyoratus (Couper 1960) Stover *in* Stover & Partridge 1973 DI K-Ng
Periporopollenites vesicus Partridge *in* Stover & Partridge 1973 DI Ng
Quadraplanus brossus Stover *in* Stover & Partridge 1973 DI K
Quadraplanus sp. Raine 1984 DI K
Roxburghpollis giganteus Mildenhall & Pocknall 1989 DI Ng
Sparsipollis acuminatus Pocknall 1985 DI Pg
Sparsipollis papillatus Mildenhall & Crosbie 1980 DI Ng

TWENTY-EIGHT

Kingdom
FUNGI
INTRODUCTION

PETER K. BUCHANAN, ROSS E. BEEVER, TRAVIS R. GLARE,
PETER R. JOHNSTON, ERIC H. C. McKENZIE,
BARBARA C. PAULUS, SHAUN R. PENNYCOOK,
GEOFF S. RIDLEY, J. M. B. (SANDY) SMITH

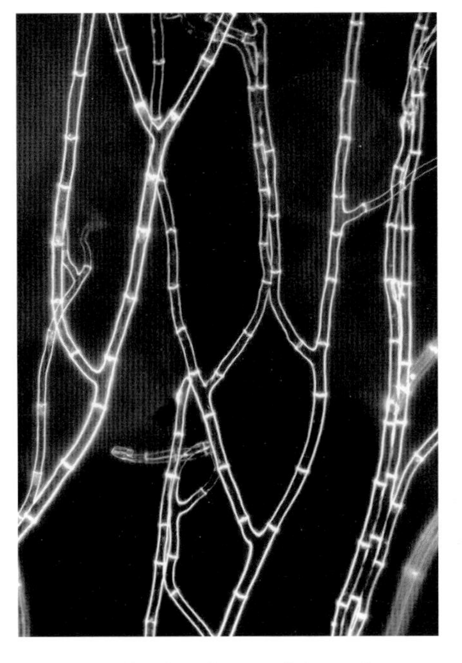

Branched hyphae, the growth form of most fungi, in *Sordaria fimicola* (Sordariomycetes: Sordariales).
George L. Barron, University of Guelph

Fungi are among the most diverse organisms on Earth and occur in virtually all ecosystems – on land, in fresh water, in the sea, and in the air. All fungi are heterotrophic and cannot photosynthesise. They feed and grow as either a spreading network of tube-like hyphae (mycelium) or as unicellular budding cells (yeasts). The cell wall is composed largely of the polysaccharide chitin, although glucans and mannoproteins are significant in some fungal cell walls. Macroscopic fruiting structures such as mushrooms consist of precisely aggregated hyphae, although such structures are produced by only a minority of fungi; most species are microscopic in both their vegetative and reproductive forms. Fungi are disseminated by sexual (meiospores) and asexual spores (mitospores), which may be produced in very large numbers. Fungal life-cycles vary from those with complex sexual cycles, with multiple mating types, to others that have dispensed with sex altogether and have evolved asexually.

Fungi are associated with almost all forms of life and contribute to ecosystems in a myriad of different roles. They live either saprobically on dead organic matter or symbiotically in association with living cells of other organisms. Symbiotic associations may be mutually beneficial (mutualistic, e.g. lichens, mycorrhizas), apparently not harmful to either partner (commensalistic, e.g. endophytes within living plant tissues), or harmful to the other partner (parasitic, e.g. animal and plant parasites, mycoparasites). Symbiotic associations with plant roots as mycorrhizas are one of the keys to the evolutionary success of the land plants, with fossil evidence indicating that migration of plants from sea to land around 400 million years ago was facilitated by glomeralean fungi (Redecker et al. 2000).

Almost 8400 species of fungi are known in New Zealand (Pennycook & Galloway 2004). Of the approximately 6500 non-lichenised species, about 4500 are indigenous (50% endemic), and 2000 exotic (Johnston 2006). New Zealand's currently recorded fungal biodiversity represents about 36% of an expected total of 23,300 species, not including Microsporidia. This estimate is conservatively derived using the ratio of vascular plants to fungi (from all substrates) of 1:6, first proposed by Hawksworth (1991) from collated knowledge of vascular plants and fungi of the British Isles. Globally, such a ratio predicts fungal species numbers in excess of 1.5 million taxa (Hawksworth 1997a), although by 2008 only 98,000 species (Kirk et al. 2008) or 6.5% had been described. Vast numbers of species await discovery in poorly explored habitats, for example tropical forests, and in association with other speciose groups such as arthropods. New Zealand records of 838 species of fungi on five indigenous taxa of *Nothofagus* (McKenzie et al. 2000) and 209 fungi on 10 indigenous taxa of *Metrosideros* (McKenzie et al. 1999) are indicative of high species diversity in this country.

Plectania campylosopora (Pezizomycetes: Pezizales), a cup fungus and member of the largest fungal phylum, Ascomycota.
Peter Johnston, Landcare Research

The precise definition of fungi, and of groups within the fungi, has been much debated although there is now wide acceptance that some organisms previously considered to be fungi belong in other kingdoms. In line with other New Zealand resources such as the book series *The Fungi of New Zealand* (McKenzie 2004) and the web-based NZFungi database (2001–2011), we have applied the widely accepted classification system of the *Dictionary of Fungi*, now in its tenth edition (Kirk et al. 2008) rather than the classification system proposed by Cavalier-Smith (2001). There are eight phyla within kingdom Fungi (Cavalier-Smith 1998; Kirk et al. 2008), each of which is treated in this volume as a separate chapter followed by a checklist of relevant species, except for the combined chapters on Chytridiomycota *sensu lato* and Zygomycota and Glomeromycota.

The true fungi comprise two major groups, the Ascomycota (e.g. cup fungi) with about 5300 species recorded in New Zealand, and the Basidiomycota (e.g. mushrooms, polypores) with almost 2800 species. Anamorphic (asexual) fungi (e.g. hyphomycetes, moulds) have no known teleomorph (sexual) stage; genetically, most are ascomycetes while others are basidiomycetes. In addition, there is a further chapter (34) on lichens, treating lichenised members of kingdom Fungi belonging to Ascomycota and (to a lesser extent) Basidiomycota. The Zygomycota and Glomeromycota (e.g. pin moulds), Chytridiomycota, and Microsporidia are smaller phyla. We accept Cavalier-Smith's (1998) proposal that the Microsporidia (minute intracellular parasites of animals traditionally treated as protozoans) are degenerate fungi that have secondarily lost their mitochondria, whereas Kirk et al. (2008) are equivocal about its kingdom alignment.

Fungoid organisms like slime moulds (phyla Amoebozoa and Percolozoa), Amoebidiales and Eccrinales (Choanozoa), plasmodiophoras (phylum Cercozoa), downy mildews and water moulds (Oomycota), and labryinthulids and thraustochytrids (Bigyra) are discussed under their relevant kingdoms. The obligate human pathogen *Pneumocystis jiroveci* (syn. *P. carinii*) and related species have been transferred from the Protozoa to Fungi mainly on the basis of cell membrane and DNA structure. *Helicosporidium*, an invertebrate pathogen previously thought to be a fungus, is a green alga (Chlorophyta).

Comparison of recorded global numbers of fungal genera and species (Kirk et al. 2008) with numbers recorded from New Zealand

	Global no. of genera	Global no. of species	No. of NZ genera	No. of NZ species	NZ genera as % of global no.	NZ species as % of global no.
KINGDOM FUNGI						
Phylum Chytridiomycota *s.l.*	123	914	40	157	33	17
Phylum Microsporidia	117	1,000	9	18	8	2
Phylum Zygomycota *s.l.*	181	1,090	42	128	23	12
Phylum Ascomycota	3,409	32,739	868	5,317	26	10
Phylum Basidiomycota	4,238	45,857	1,019	2,778	24	8
Agaricomycotina	1,037	20,391	412	2,372	40	8
Pucciniomycotina	195	8,057	30	297	15	3
Ustilaginomycotina	119	1,464	34	109	29	8
[Anamorphic fungi	2,887	15,945	543	1,855	19]	9
Totals	**12,306**	**127,457**	**1,977**	**8,398***	**24**	**9**

* Includes three species incertae sedis

Significance of fungi

Fungi play significant roles in human affairs, including uses as food, in industry and medicine, through impacts of mycotoxins, biodegradation, and disease, and fill vital roles in ecosystems.

Human interactions

Indigenous people's use of fungi as food, medicine, and tinder is well documented in many Asian and European countries but relatively little is known from the literature of Maori knowledge. Examples of fungi consumed by Maori, particularly in times of scarcity, include the wood ear fungus *Auricularia cornea* (hakeke), puffballs (pukurau) including *Langermannia gigantea*, the hydnoid icicle fungus *Hericium coralloides* (pekepeke-kiore), and the bootlace mushroom *Armillaria novaezelandiae* (harore) (Fuller et al. 2004). Some fungi have also been used for their medicinal properties and as environmental indicators. The use of ink produced from ground charcoal of *Cordyceps robertsii* (awheto, vegetable caterpillar) for tattooing (ta-moko) has been well documented, and *Laetiporus portentosus* (putawa) was used for lighting fires, as a source of live fire, and as a torch (Best 1942; Fuller et al. 2004). During the late 19th and early 20th centuries, the wood ear fungus represented an important source of income to both Maori and European families, particularly in the Taranaki region, as it was collected on fallen wood and exported in large quantities to China (McKenzie 2004).

The only commonly collected mushroom today is the field mushroom (*Agaricus campestris*), but the mushroom industry in New Zealand, once restricted to cultivation of button mushrooms (*Agaricus bisporus, A. bitorquis*), has diversified in recent years and now produces oyster mushrooms (*Pleurotus pulmonarius*), shiitake (*Lentinula edodes*), and enokitake (*Flammulina velutipes*). There is an emerging industry in the production of high-value edible ectomycorrhizal mushrooms, including the black truffle (*Tuber melanosporum*) which grows in association with oaks or hazels (Hall et al. 1998), and saffron milk cap *Lactarius deliciosus* under pines (Hall 2011). Fungi are also vital in food and drink manufacture, with yeasts integral to bread, beer and wine making. Late-harvest grapes naturally infected, in favourable years, with the hyphomycete *Botrytis cinerea* are prized for production of a sweet 'noble' or 'botrytised' wine. Species of *Penicillium* are used to ripen cheeses such as Camembert, Brie, and Roquefort.

Mushroom poisonings are rare in New Zealand. A number have involved the death-cap mushroom *Amanita phalloides*, localised in New Zealand in some parks under oak and other exotic ectomycorrhizal trees (e.g. Nicholls et al. 1995; Anon. 2000). Several fungi produce potent mycotoxins, such as aflatoxins and ergot alkaloids, some of which are involved in food spoilage. A range of other bioactive fungal secondary metabolites are medically important, including the antibiotics penicillin and cephalosporin, anti-fungal griseofulvin, and the immunosuppressant cyclosporine (Kendrick 1992). Wood-rots (e.g. 'dry rot' of housing timbers) and other fungi have been implicated in the biodeterioration of houses and in leaky building syndromes.

Upper: *Auricularia cornea* (hakeke, wood ear, 'Taranaki wool'), a native wood-decay fungus that was harvested and exported in quantity to China as an edible mushroom (Agaricomycetes: Auriculariales). Lower: *Armillaria novaezelandiae* (harore, bootlace mushroom), common on fallen wood in forests in early winter, and also a pathogen of some crop plants including pines and kiwifruit. Eaten by Maori, and harore adopted as a generic Maori word for all fungi (Agaricomycetes: Agaricales).

Peter Buchanan, Landcare Research

Ecologically defined groups of fungi

Attributes of fungi, such as their ecological roles and host relationships, frequently form the basis for research initiatives rather than taxonomy per se. Ecologically related taxa are likely to span phyla or even kingdoms. A number of the groups discussed below are the subjects of current research in New Zealand or are of particular importance for New Zealand ecosystems.

Plant parasitic fungi cause New Zealand growers, gardeners, foresters, and farmers millions of dollars annual loss through reduced yields, poorer quality, and increased production costs – costs that often flow on to the consumer. Parasitic fungi may attack any part of a plant, from pre-emergent seedlings through to

Teliospores of the gladiolus leaf rust *Uromyces transversalis* (Pucciniomycetes: Pucciniales).
Eric McKenzie, Landcare Research

mature individuals. Symptoms include pre-and post-emergence damping-off, root rots, butt rots, stem rots, wood rots, vascular wilts, leaf spots and blotches, flower rots, and fruit rots. Individual parasites range from facultative saprobes, to facultative parasites, to obligate parasites.

Plant parasites are found within all the major groups of fungi. Rust and smut fungi, in particular, have been recognised for centuries as important plant parasites (see chapter 33, Basidiomycota: Pucciniomycotina and Ustilaginomycotina). Published records of fungal and fungoid plant parasites in New Zealand (Pennycook 1989) have been updated and are incorporated in the NZFungi database (Anon. 2001–2011). This database contains over 1200 New Zealand disease references, more than 1500 accepted fungal/fungoid-parasite names, and approximately 4500 host x parasite combinations.

Most diseases of exotic crop, pasture, and forest plants in New Zealand are caused by introduced fungi, many of which were accidentally introduced with planting material, often before realisation of the importance of plant quarantine. These include, for example, leaf rust of wheat (*Puccinia recondita*), the die-back (*Wilsonomyces carpophilus*) and brown rot (*Monilinia fructicola*) of peach trees (Dingley 1969), and the covered smuts (*Tilletia* spp.) of cereals. Other diseases are assumed to have been introduced from Australia in wind currents, such as stripe rust of wheat (*Puccinia striiformis*), rust of castor oil plant (*Melampsora ricini*), and rust on *Gladiolus* spp. (*Uromyces transversalis*) (McKenzie 2000). In addition, insects may be vectors of fungal plant parasites, e.g. the destructive white rot disease of *Pinus radiata* caused by the corticioid fungus *Amylostereum areolatum* that has a symbiotic association with the woodwasp *Sirex noctilio* (Gilbertson 1984; Morgan 1989; Gadgil 2005). While there are many native plant-parasitic fungi, most are in a biological balance with their host plant. Examples of native parasites on native hosts include many rust and smut fungi, various leaf-spotting fungi, and the obvious (and spectacular) 'beech strawberries' (*Cyttaria* spp.) on *Nothofagus menziesii*. Few indigenous fungi parasitise exotic plants, although two native species of *Armillaria* (*A. limonea* and *A. novaezelandiae*) cause root rot and death of radiata pine seedlings when they are planted on land recently cleared from native forests (Hood 1989, 1992). Two indigenous non-gilled macrofungi, *Gloeocystidiellum sacratum* and *Rigidoporus vinctus*, both recorded as saprobes on many indigenous hosts, cause root and stem canker of exotic pines and a root rot disease in shelter belts and plantations, respectively (Dick 1983, 1987).

With international moves to reduce pesticide use, biological control using parasitic fungi is becoming an increasingly attractive alternative for pest control, especially for control of plant weeds and other parasitic fungi. Research in weed biocontrol in New Zealand includes evaluation of several naturalised fungi: *Fusarium tumidum* for control of gorse (Fröhlich et al. 2000), *Chondrostereum purpureum*, itself the cause of silver leaf disease of fruit trees, for biocontrol of broom and gorse (de Jong & Bourdot 1998; de Jong 2000), *Puccinia hieracii* var. *piloselloidarum* for control of hieracium (*Hieracium pilosella*) (Morin et al. 1997), and *Phragmidium violaceum* for control of blackberry (Hayes 1999, 2011). The smut *Entyloma ageratinae* was the first new weed biocontrol fungus to have been legally introduced to New Zealand. It has established and is spreading on its host, mistflower (*Ageratina riparia*), an urban and rural weed in the North Island (Fröhlich et al. 1999). Other fungi are of interest for their capability to induce plant defence mechanisms in order to reduce the effects of parasitic species (Hill et al. 2000).

The beech strawberry *Cyttaria nigra* (Leotiomycetes: Cyttariales), a native pathogen of silver beech *Nothofagus menziesii*. The fungus induces the tree branches to develop galls from which strawberry-like fruiting bodies form in spring and fall at maturity (as in photo). These fruiting bodies are apparently eaten by native birds and possums.
Peter Buchanan, Landcare Research

Fungi affecting human and animal health. More than 90 fungal species, covering a wide variety of taxa including ascomycetes, basidiomycetes, chytridiomycetes, zygomycetes, and microsporidia have been associated with human and/or animal disease in New Zealand. The list of fungi associated with disease in humans will undoubtedly increase as a result of increased mobility among the population and advances of modern medicine (including organ and bone marrow transplants) that enable or prolong the survival of severely immunocompromised patients. Almost any fungus capable of growth at around 37° C (i.e. the body temperature of mammals) is potentially capable of causing disease; these fungi are termed 'opportunists', and the diseases they cause, opportunistic mycoses.

Pulmonary opportunistic mycoses may be caused by a variety of fungi (Anon. 2011) including *Pneumocystis jiroveci* (syn. *P. carinii*) and aspergilli (e.g. *Aspergillus fumigatus, A. flavus, A. niger, A. nidulans*); the latter may also affect a variety of native and exotic avian species, and placentitis and abortion in cattle (Fairley 1998). Environmental cryptococci, such as *Cryptococcus neoformans*, are significant human (e.g. meningitis in AIDS patients) and animal (nasal and central nervous system disease in cat) parasites. This yeast is associated with old, dry avian droppings but it is possibly a plant parasite. Other pathogenic cryptococci (e.g. *Cryptococcus gattii*) are present in New Zealand and are often associated with decaying tree (e.g. eucalypt) hollows. The commensal yeast *Candida albicans* is associated with a variety of mucosal, cutaneous and systemic diseases in humans and animals, e.g. pigs. Possibly influenced by the widespread use of oral antifungal fluconazole, other *Candida* less susceptible to fluconazole, e.g. *C. glabrata* and *C. krusei*, are now becoming more prevalent in human disease. Other yeasts, for example *Pichia guilliermondii* (formerly classified in the genus *Candida*), have been isolated from a case of bovine mastitis (Fairley 1998).

Sporulating head of *Aspergillus niger* (Eurotiomycetes: Eurotiales).
Nancy Allin and George L. Barron, Guelph University

Rarely seen opportunistic infections include an increasing variety of environmental ascomycetous fungi, such as *Fusarium solani, Paecilomyces lilacinus*, various black moulds such as *Exophiala jeanselme*, and even basidiomycetes such as *Schizophyllum commune*, which was recently recorded causing maxillary sinusitis in a diabetic patient (Sigler et al. 1999). Various zygomycetes (order Mucorales) belonging to the genera *Absidia, Rhizopus*, and *Mortierella* have been also recovered from opportunistic systemic disease in humans and animals (Austwick 1976). Microsporidia are obligate parasites, which infect every major animal group, especially insects, fish and mammals (Wittner 1999), and in recent years, immunocompromised patients (Everts et al. 1997). In New Zealand, they have been reported from a range of pests such as grass grubs, porina, and gastropods (e.g. Malone 1990; Malone & McIvor 1996; Barker 2004).

Dermatophytes, specialised in colonising the cutaneous regions or nails of humans and animals, include genera such as *Trichophyton, Epidermophyton*, and *Microsporum*. Some dermatophytes can be transmitted from animals to humans e.g. ringworm including *M. canis* from cats and *T. mentagrophytes* var. *erinacei* from hedgehogs. The lipophilic yeast *Malassezia furfur* is a significant human cutaneous parasite (tinea versicolor) while *M. pachydermatis* is an important animal parasite, e.g. in ears of dogs. Apart from known dermatophytes, an increasing variety of environmental fungi have been recovered from human nail infections.

Mycotoxicoses are the poisoning of humans and animals by toxin-producing microfungi. The list of fungal secondary metabolites potentially toxic to humans is extensive and includes, for example, the liver carcinogen aflatoxin produced by *Aspergillus flavus*, a fungus that grows well on damp peanuts. Mycotoxins are involved in the development of facial eczema of sheep and sometimes cattle, following ingestion of grass contaminated by spores of the saprobic fungus *Pithomyces chartarum* (Thornton & Percival 1959). Other examples of mycotoxicosis in animals include ryegrass staggers (Fletcher & Harvey 1981) and fescue toxicosis (Siegel et al. 1987). In contrast to pathogenic effects, a beneficial role has been suggested for anaerobic saprobic chytridiomycetes, which have been discovered living in the rumen of herbivores, where they aid lignocellulose digestion by their hosts (Powell 1993).

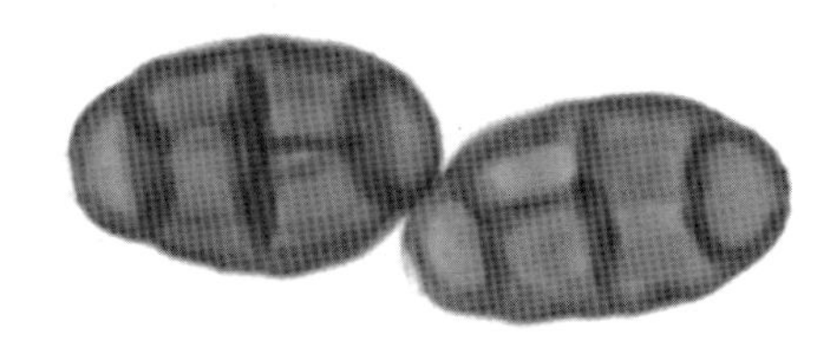

Grenade-shaped conidia of *Pithomyces chartarum*, the cause of facial eczema of sheep and cattle.
Eric McKenzie, Landcare Research

Entomogenous fungi are associated with insects and include representatives of all major groups of fungi. They range from lethal pathogens through ectoparasites, which do not seriously affect their hosts, to commensals. With the exception of a few species studied as potential biocontrol agents, entomogenous fungi are

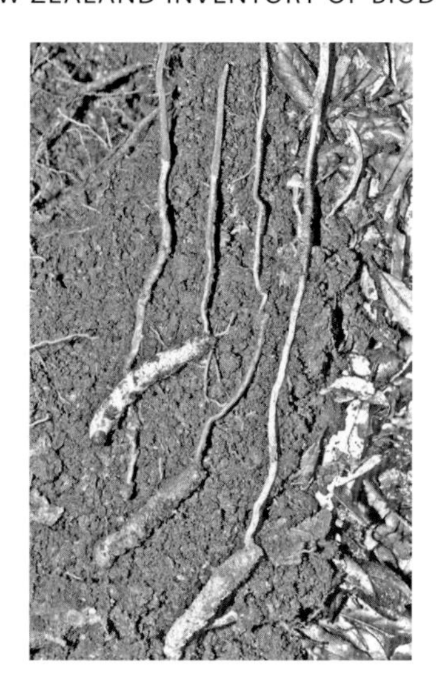

Awheto or vegetable caterpillar *Cordyceps robertsii* (Sordariomycetes: Hypocreales). This native fungus is parasitic on hepialid caterpillars, mummifying the larva in its burrow. The fungal fruiting body develops from the larval head, extending above ground to sporulate. Maori collected awheto as a food and as a source of black pigment, prepared from its powdered charcoal, for ta moko (tattooing).

Peter Buchanan, Landcare Research

known only from limited studies, which may explain why the known species diversity in New Zealand is relatively small by world standards.

The most conspicuous entomoparasitic fungi are the so-called vegetable caterpillars of the ascomycete genus *Cordyceps*. *Cordyceps robertsii,* which is parasitic on Hepialidae caterpillars ('porina'), was the first fungus to be described from New Zealand (Hooker 1836). At least fourteen species of *Cordyceps* have been reported from New Zealand to date. Some species, such as *C. novaezelandiae* and *C. hauturu,* are probably found only in New Zealand while others are distributed more widely. Affiliated anamorph genera including *Hymenostilbe, Beauveria, Metarhizium, Hirsutella, Paecilomyces,* and *Tolypocladium* (Stensrud et al. 2005) have been also reported from New Zealand in a checklist of entomoparasitic microbes and nematodes (Glare et al. 1993). This checklist recorded over 50 named and many undescribed species of insect parasitic fungi, in particular hyphomycetes. Among the zygomycetes, a number of the entomoparasitic Entomophthorales and allied members have been recorded in New Zealand, which mainly include aphid parasites such as *Entomophthora planchoniana* and *Zoophthora phalloides*. Parasitic chytridiomycetes attack invertebrate hosts including nematodes, rotifers, and insects; for example *Olpidium gregarium,* was recorded from New Zealand on rotifer eggs (Karling 1966), and *Coelomycidium* sp. and *Coelomomyces psorophorae* var. *tasmaniensis* from dipterans in an aquatic environment. The biology of *Coelomomyces,* parasitic on mosquito larvae, has been well studied overseas as a potential mycoinsecticide (Powell 1993).

Laboulbeniales is an order of ascomycetes whose species live as ectoparasites of arthropods. Associations are most frequent with beetles, but these fungi also occur on flies, millipedes, mites, true bugs, and mole crickets. Many are host specific and, as there is a high level of endemism in the insect fauna of New Zealand, the associated Laboulbeniales are also expected to show high endemism (Hughes et al. 2004a,b). Commensal entomogenous fungi include members of the Trichomycetes (Zygomycota), which live within the digestive tract of arthropods. Hosts include freshwater aquatic insect larvae and nymphs, freshwater crustacea, marine crabs, terrestrial millipedes, beetles, and crustacea (Williams & Lichtwardt 1990). Many species are host-specific, and all are critically dependent on their hosts for survival. Dispersal of propagules is poorly understood. Some, such as species of the Harpellales, form trichospores, asexual spores that have fine basal appendages thought to aid entrapment on substrates in streams following spore release from larvae (Williams & Lichtwardt 1990).

Arbuscular mycorrhizal fungi form thick-walled, very large (40–800 micrometres diameter) resting spores in the soil. These spores, containing hundreds to thousands of nuclei, germinate to form hyphae that grow towards a nearby root in response to root exudates. The 'infection' process is not damaging to the root.

George L. Barron, University of Guelph

Endophytic fungi exist symptomless within plants, and may include latent pathogens, which cause disease only under specific environmental conditions, or latent saprobes. The relationship between plant and fungus is not well understood for many groups of plants but may be mutual or commensal. Studies from other parts of the world indicate that endophytes occur in most plants studied to date and in all major plant organs such as leaves, stems and roots. In New Zealand, endophytes of grasses have received much attention owing to their major impact on the performance of pastures. Some grasses contain specific fungal endophytes, such as *Neotyphodium lolii* (e.g. Eerens et al. 1998). The association between the grasses and the fungus is mutualistic, with the host grass able to benefit from endophyte presence through reduced herbivore feeding, increased resistance to insects, improved plant growth, and possibly disease resistance. However, endophytes in grasses are also associated with a number of animal toxicoses including ryegrass staggers and fescue toxicosis.

In contrast to endophytic fungi in grasses, limited information is available on the endophytic mycobiota in native plants. Endophytes have been studied in *Leptospermum scoparium* (Johnston 1998), *Metrosideros excelsa* (McKenzie et al. 1999), and four native species of Podocarpaceae and *Kunzea ericoides* (Joshee et al. 2009). There is some evidence that communities of leaf endophytes in natural stands develop over time (Johnston 1998) and that their distribution within individual leaves is highly structured. Some fungi, for example *Coccomyces cupressinum*, *Hypoderma bidwillii*, and *Lophodermium kaikawakae*, which are consistently associated with fallen leaves of a single host (Johnston 1992), may have an endophytic phase for at least part of their life-cycle. More research is required to answer some intriguing questions about the ecology of endophytic fungi in New Zealand.

Cortinarius porphyroideus (formerly *Thaxterogaster porphyreus*) (Agaricomycetes: Agaricales), mycorrhizal with species of *Nothofagus* (southern beech).

Peter Buchanan, Landcare Research

Mycorrhizal fungi. The mycorrhizal ('fungus-root') habit, a feature of many species among the agarics, aphyllophorales, and zygomycetes, benefits the health and nutrition of most plant species and may also provide protection against drought or pathogenic attack. The fungal partner 'infects' the host plant roots and obtains carbohydrates from the plant. In return, the fungus assists nutrient and water uptake by the plant, by hyphal capture from the surrounding soil of immobile nutrients such as phosphorus. Different types of mycorrhizae are recognised, including ectomycorrhizae, vesicular-arbuscular mycorrhizae, and ericoid and orchid mycorrhizae.

In ectomycorrhizal associations (mainly involving basidiomycetes and a few ascomycetes such as truffles) the fungus encases the root with a thick mantle of hyphae that also penetrate between the outer cortical root cells to form the 'Hartig net'. This mycorrhizal type is limited in New Zealand to three genera of native trees, *Nothofagus*, *Leptospermum*, and *Kunzea*, genera of prime importance in native forests, but the diversity of fungal partners is high with 76 genera of known or probable ectomycorrhizal fungi reported from New Zealand (Orlovich & Cairney 2004). Dominant fungal genera include *Cortinarius* (and relatives), *Amanita*, and *Russula* among the agarics, and *Ramaria* and *Phellodon* among the aphyllophorales. It is likely that native members of other genera known to be ectomycorrhizal elsewhere (e.g. the corticioid *Tomentella*) will be found to form similar associations in New Zealand.

Some ectomycorrhizal fungi have been introduced to New Zealand, most likely along with seedlings of their host trees. At least two agaric species have become invasive in New Zealand's indigenous *Nothofagus* forests – the northern-hemisphere ectomycorrhizal species *Amanita muscaria* (fly agaric) and *Chalciporus piperatus*. These species appear to be displacing indigenous ectomycorrhizal fungi and may also be affecting health and nutrition of host trees. Other introduced ectomycorrhizal fungi, such as *Thelephora terrestris*, have so far been associated only with exotic trees in New Zealand.

Vesicular-arbuscular mycorrhizae are formed by 90% of all higher plants, including many tree species in New Zealand's podocarp-broadleaf forests, and hence are of great importance for plant nutrition. In this type of mycorrhiza, the fungus colonises the outer cortical root cells following germination of the thick-walled resting spore or after root to root contact (Hall 1977). It then forms finely branched 'arbuscules' within the cells although enveloped by the cell membrane; a loose weft of hyphae also surrounds the root. The fungal partners are members of the Glomerales (phylum Glomeromycota), which function as obligate vesicular-arbuscular mycorrhizal associates. Most species among the Glomerales have a broad host range, which extends beyond vascular plants to include New Zealand liverworts (Russell & Bulman 2005). Dispersal of glomeralean fungi is presumably by insects and ground-feeding birds in New Zealand, in the absence of native terrestrial mammals. Possums are known to feed on the large (up to 20 millimetres diameter) sporocarps of *Glomus macrocarpus* (Cowan 1989).

Ericoid mycorrhizae involve mainly discomycete fungi in association with roots of ericaceous plants, but sequencing studies indicate that other fungi are also participants (Monreal et al. 1999). In some cases, such as the mycorrhizae of Epacridoideae (Ericales), their occurrence in New Zealand is simply inferred based on research from Australia and other parts of the world. Orchids have a special mycorrhizal relationship with fungi such as *Rhizoctonia*, often depending on infection by the fungus for seed germination and for supply of carbohydrates

Amanita nothofagi (Agaricomycetes: Agaricales), mycorrhizal with species of *Nothofagus* and the teatrees *Leptospermum* and *Kunzea*.

Peter Buchanan, Landcare Research

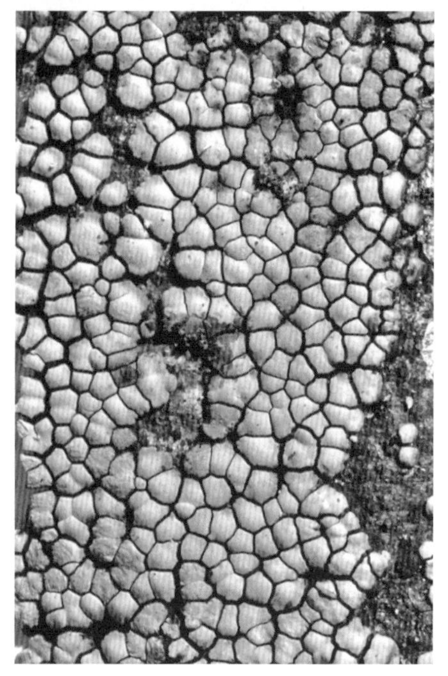

Upper: *Crucibulum laeve* (Agaricomycetes: Agaricales), a bird's-nest fungus, degrading wood mulch spread on garden beds.
Lower: *Aleurodiscus berggrenii* (Agaricomycetes: Russulales), a wood-decay corticioid fungus on dead wood.

Peter Buchanan, Landcare Research

during part or all of the orchid's growth. The fungus grows saprobically but is virtually parasitised by the orchid. Some orchids and other non-photosynthetic plants have mycorrhizal symbionts among the wood-rotting basidiomycetes; an example is *Gastrodia cunninghamii* with *Armillaria* sp. (Campbell 1962).

Wood- and litter-decaying fungi. Apart from forming associations with other organisms, fungi also contribute to ecosystems through their roles in nutrient cycling by degrading wood, litter, and other organic matter in both terrestrial and aquatic habitats. Saprobic decomposer fungi occur in most major groups of fungi. Along with wood-degrading ascomycetes and agarics, non-gilled basidiomycetes are the main agents of decomposition and nutrient cycling of wood in forests. Some species of New Zealand wood-decay fungi are cosmopolitan, but many species are endemic or indigenous, and some have probably been introduced recently; for example, the highly visible orange pore fungus *Favolaschia calocera*, probably introduced from Madagascar or Asia during the 1960s, has become a fungal weed and has invaded all forest types (Johnston et al. 1998).

Depending on the type of enzymatic degradation, wood-decomposing basidiomycete species cause either a white rot, in which the fungus degrades cellulose, hemicellulose, and lignin from the wood cell wall, or a brown rot, where the lignin remains largely intact. White-rot species are commoner, with brown-rot species mostly associated with gymnospermous hosts. Information about the ecology of New Zealand wood-decay fungi is still limited but a recent study has shown that the decomposition of *Nothofagus* logs involves complex changes in species composition as decay proceeds (Allen et al. 2000). Some wood-decay communities may be quite sensitive to disturbance, such as selective logging, and may take a considerable time to recover (unpublished data).

Litter decomposers include microfungi and larger fungi, which form mycelial mats within litter layers. Apart from some taxonomic treatments, for example for microfungi (Hughes 1971), knowledge of New Zealand litter fungi is limited. Litter fungi are highly diverse (Bridge & Spooner 2001) and include, in particular, anamorphic ascomycetes and zygomycetes, especially Mucorales (Thornton 1958). Chytridiomycetes in soils have specialised abilities to metabolise polymers such as chitin, keratin, and cellulose and to decay plant pollen (Karling 1966). Some saprobic chytrids may be particularly abundant; *Karlingia rosea* was reported to be the commonest chytridiomycete in New Zealand, found in almost 95% of 140 soils sampled from throughout the country (Karling 1968a).

Truffles and truffle-like fungi. Truffles in the strict sense are the underground fruiting bodies of certain ascomycetes, especially those belonging to the genus *Tuber*. Although there are no native species of *Tuber* in New Zealand, there is a nascent industry growing the highly regarded black truffle *Tuber melanosporum* (Hall et al. 1998). In the broader sense, truffles include all macroscopic fungi that fruit underground (hypogeous fungi) and include certain basidiomycetes (false truffles) and zygomycetes. The term 'truffle-like fungi' (also referred to as sequestrate fungi) is even broader, encompassing species that either fruit underground (truffles) or at the ground surface (epigeous). They differ from other macrofungi in that the spores remain indefinitely in the fruit bodies, which lack a dehiscence mechanism. Dispersal of spores is mediated by animals such as mice, squirrels, pigs, and deer in the Northern Hemisphere, and bettongs and potoroos in Australia. Little information is available in New Zealand but it has been suggested that many are dispersed by ground-feeding birds (Beever 1999), although invertebrates, including slugs and beetles, may also play a role.

New Zealand has perhaps 100–125 species of indigenous truffle-like fungi. Present knowledge indicates that many species are endemic, but some are shared with Australia and a few extend even further afield. A small group is introduced, including species of *Rhizopogon* that are associated with pines and Douglas fir. Truffle-like fungi belonging to the ascomycetes and basidiomycetes often form ectomycorrhizal associations with *Kunzea*, *Leptospermum*, and *Nothofagus*. Fruiting bodies of these fungi can reach the size of a tennis ball, and some are brightly coloured, such as the violet potato fungus *Gallacea scleroderma* (Taylor 1981; Castellano & Beever 1994). Truffle-like fungi belonging to the Glomeromycota (e.g. *Glomus* species) typically form endomycorrhizae and associate with hosts such as kauri, podocarps, and most broadleaf trees and shrubs. Generally, the fruiting bodies formed by these fungi are relatively small and inconspicuous. Other truffle-like ascomycetes and basidiomycetes are saprobic on dead plant material. Included among these is the white *Protubera parvispora* that looks rather like the unhatched 'eggs' of the basket fungus (Castellano & Beever 1994), the bright red *Leratiomyces erythrocephalus* and the bluish-green *Clavogaster novozelandicus* (formerly *Weraroa virescens*) (Taylor 1981).

Clavogaster novozelandicus (formerly *Weraroa virescens*) (Agaricomycetes: Agaricales), a sky blue truffle-like native fungus.

Peter Buchanan, Landcare Research

Much remains to be discovered about truffle-like fungi in New Zealand, including how many species of these somewhat reclusive fungi are present, defining their roles in supporting tree and shrub growth, and determining their interactions with birds, invertebrates, and other fungi.

Dung and ammonia fungi. The coprophilous ('dung-loving') fungi of mammalian dung are one of the best known fungal ecological groups in New Zealand. The group comprises members of the ascomycetes, basidiomycetes, and zygomycetes. When ingested along with plant material by herbivores, spores of coprophilous fungi are able to pass through the gut, withstand the action of gut enzymes, and germinate in the dung. Although spores of coprophilous species are present in fresh dung, fungal fruiting follows a successional pattern, which is controlled by complex interactions of factors including nutritional requirements of different fungi, competition and antagonism between species, and effects of other organisms inhabiting dung (Bell 1975). Generally, early-fruiting species are the zygomycetes (e.g. *Mucor*) and anamorphic fungi, followed later by ascomycetes and finally the basidiomycetes (Bell 1983). Although New Zealand dung fungi are well documented for a broad range of introduced herbivores (Bell 1983), it is not possible to distinguish indigenous from introduced species.

Ammonia fungi (Sagara 1975) are a largely cosmopolitan group of species that grow in ammonia-nitrogen-rich sites, such as near urine or dead-animal remains. Distinct successions of fungal fruiting bodies, beginning with ascomycetes and culminating in mycorrhizal agarics, have been observed in urea-treated plots in East Asia, Europe, Australia, and New Zealand (Suzuki 2000). Local studies have also demonstrated this succession *in vitro* (early stages only) and in forests near possum remains (Sagara et al. 1993; Suzuki & Tsuda 1993; Suzuki et al. 2002).

Sooty-mould fungi are a conspicuous feature of many *Nothofagus* forests and comprise assemblages of taxonomically diverse species that are supported by the sugar-rich honeydew excreted by scale insects (Hughes 1966, 1976). New Zealand has the greatest diversity of sooty moulds known in the world (Pirozynski & Weresub 1979). They form part of highly complex biological communities comprising sap-feeding insects, parasitic insects, fungal-feeding insects, saprobic fungi, and fungi parasitic to insects and other fungi. Unusual for terrestrial ecosystems, the physical structure of a sooty-mould community is provided not by vascular plants but by darkly pigmented fungal hyphae. Sooty moulds also include a pathogenic species, *Capnodium walteri*, which grows in association with the scale insect *Eriococcus orariensis* and causes manuka blight (Dingley 1969).

Pilobolus crystallinus on dung of the brown hare *Lepus europaeus occidentalis*.

Jerry A. Cooper, Landcare Research

Predacious fungi. A small and taxonomically diverse group of fungi has developed predacious habits, targeting small soil animals such as nematodes. Various mechanisms are used to achieve infection including ingested motile spores or conidia, detachable and constricting rings, toxins, and adhesive spores, hyphae, knobs, and nets (Kendrick 1992). In a New Zealand study of both modified and pristine habitats, no endemic taxa were discovered among the 15 predacious species recorded from surface soil and organic matter. However, predacious species appear to be widespread with about 56% of samples yielding such fungi (Fowler 1970). The prospects of using predacious fungi in the control of animal-parasitic nematodes have been investigated (Larsen 2000). Some species of *Pleurotus*, which includes many edible and commercially cultivated 'oyster mushrooms', produce hyphal droplets that are toxic to nematodes (Thorn & Tsuneda 1993). Such droplets were evident in cultures of four New Zealand species of *Pleurotus*, although nematode toxicity of these species was not tested (Segedin et al. 1995).

Aquatic and marine fungi. While the majority of fungi are terrestrial, fungi are also common in fresh water, and some are known from marine environments. Many aquatic fungi are ascomycetes and their anamorphs, most of which have spores adapted for dispersal in an aquatic habitat. Aimer and Segedin (1985) identified 49 species of aquatic hyphomycetes from New Zealand streams. Although no endemic taxa were identified in their studies, the undetermined species reported may represent new taxa. Aero-aquatic fungi grow underwater but produce their spores in the air. The diversity of New Zealand aero-aquatic fungi was highlighted by the report of six new species and 38 new records of mostly aero-aquatic hyphomycetes (Cooper 2005). Fresh water is also the main habitat of many Chytridiomycota (e.g. Karling 1968b).

Most genera isolated from the marine environment are thought to be facultatively marine, being originally terrestrial but able to grow in marine conditions (Kohlmeyer & Kohlmeyer 1979). Some marine fungi may have significant ecological roles. For example, as parasites of algae, chytridiomycetes are implicated in control of phytoplankton population levels. Examples of anamorphic fungi adapted to the marine environment include three anamorphic fungi from the epizoic microbial mat of a New Zealand bryozoan, two being common soil fungi while the third species, in the anamorphic black yeast genus *Exophiala*, appears to be a bryozoan associate (Sterflinger & Scholz 1997).

Global declines in amphibian populations have been attributed to chytridiomycosis. First described in 1998, this disease has been recorded in native frogs in New Zealand and appears to be caused by the chytrid fungus, *Batrachochytrium dendrobatidis* (Bishop et al. 2009).

Upper: *Flabellospora acuminata* (Pezizomycotina, incertae sedis), an Ingoldian (aquatic) fungus sporulating on submerged litter in woodlands. Ingoldian fungi often produce radiate asexual spores and masses of them create a foam that accumulates in woodland streams.
Lower: *Helicoon* sp. (Dothideomycetes: Pleosporales), a typical aero-aquatic fungus of submerged litter in stagnant pools. The asexual spores are formed above the water surface and trap air bubbles as a flotation device to assist dispersal.
Jerry Cooper, Landcare Research

Fossil fungi. Only one fossil fungus, *Microthyriacites grandis* from Tertiary coal deposits (Cookson 1947), has been described as new from New Zealand, mainly owing to a paucity of studies on the subject. Penseler (1930, fig. 7) reported an apparent bracket fungus from the Waikato coalfields. Fungal spores resembling those of *Glomus* (Glomeromycota) were illustrated by Lindqvist (1994) from Miocene deposits, and Sutherland (2003) reported fungal hyphae on termite faecal pellets and fungal wood rot also from the Miocene. Currently, several epiphyllous late Eocene (35 mya) fungi are being investigated from angiosperm leaf litter preserved as carbonaceous material, from Pikipiko, Southland (J. M. Bannister pers. comm.). Fungal spores of *Monoporisporites* spp. and aff. *Pluricellaesporites* spp. were recorded in a 'fungal spike', suggesting a brief abundance of saprobic fungi, at the Cretaceous–Tertiary (K-T) extinction event 65 million years ago (Vajda & McLoughlin 2004). The K-T boundary appears to have been characterised by a sequence of major dieback of photosynthetic plants, leading to an abundant substrate for fungi, and then rapid establishment of ferns and other pteridophytes. Fossil fungi are not included in the fungal checklist. [See chapters 26 and 27 in this volume for further reports of fossil fungi.]

Fungal conservation

Conservation in New Zealand, as elsewhere, has focused mainly on large, iconic, and attractive animals, less so on plants, and until recently has ignored most other groups of organisms. The move from species-based to ecosystem- and landscape-based conservation (Park 2000) has likely assisted conservation of 'non-target' organisms, although its efficacy for fungi has not been assessed. The challenge is to move towards conservation programmes that address not only fauna and flora but also fungi (Hawksworth 1997b).

Formal recognition of New Zealand's threatened fungi occurred first in 2002, with the inclusion of fungi in the *New Zealand Threat Classification System Lists* (Hitchmough 2002; Buchanan & May 2003; Buchanan et al. 2010). Of the 312 species judged most threatened (i.e. 'Nationally Critical', Hitchmough 2002), 50 species (16%) were fungi. In a 2011 reassessment for Department of Conservation, 62 fungal species were considered Nationally Critical, with 43 species listed under other threat categories. Some of the latter are associated with threatened host plants; for example, *Puccinia embergeriae* is considered threatened as its only host plant, the Chatham Islands giant sow thistle (*Embergeria grandifolia*), is vulnerable to extinction.

For many fungi, however, present knowledge is insufficient to determine threat status, and over 1100 fungal taxa are listed as 'Data Deficient' (fungi comprise 70% of this category). Included as 'Data Deficient' are over 200 species that were described from New Zealand type specimens but for which no material is held in New Zealand. In a survey of holdings of the New Zealand Fungal and Plant Disease Collection (PDD), over 1500 species are known from only a single New Zealand collection and a further 1000 species from fewer than four collections. For example, the hypogeous (truffle-like) species *Claustula fischeri*, the sole member of the family Claustulaceae, is known from only a few collections from Nelson, the Wairarapa, and Tasmania. Owing to inadequate collecting and knowledge of distribution of most microfungi, the conservation assessment of New Zealand fungi is biased towards the macrofungi.

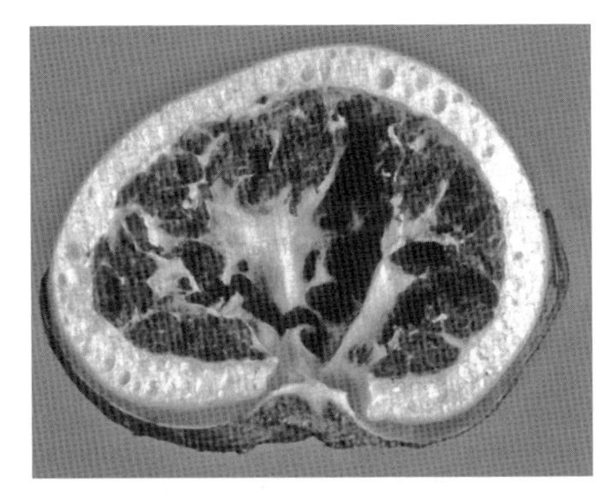

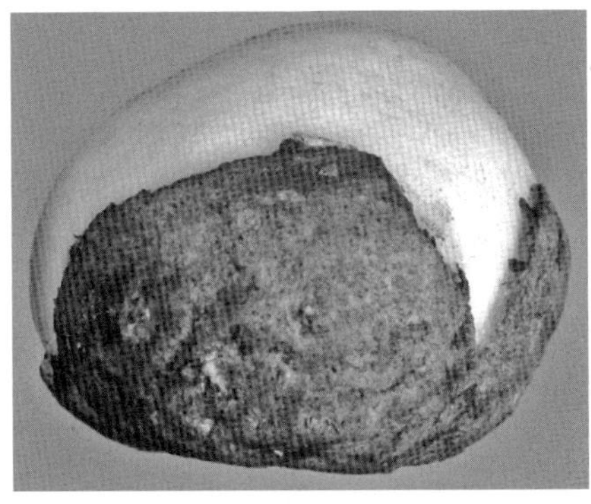

Claustula fischeri (Fischer's egg), a nationally critical fungus known from only two locations in New Zealand (Agaricomycetes: Phallales), shown sectioned and whole.

Ross Beever, ex Landcare Research

A constant challenge for fungal conservation is the huge biodiversity and large percentage of unrecognised taxa. However, those species recently designated as 'Nationally Critical' will stimulate increased awareness of threats to populations of all organisms and promote conservation of both these species and their ecosystems.

It is to be hoped that most of New Zealand's estimated 14,000 unrecognised fungal taxa will 'await' their discovery by science and that conservation of a range of habitats will adequately protect fungal diversity. This assumption can be questioned, however, in the absence of adequate knowledge about fungus-host relationships and fungal interactions, especially obligate, with other organisms. Since many fungi are endemic, and some species are known to be rare, it is possible that species extinction will occur without our knowledge.

Historical overview

New Zealand has a long history of mycological investigation. The Royal Botanic Gardens, Kew, was the centre for early study and description of New Zealand's fungi. The 1841 Antarctic Expedition of J.C. Ross was a stimulus for targeted collecting of fungi, promoted by the expedition's botanist and assistant surgeon J.D. Hooker. Another significant fungal collector was W. Colenso, resident in New Zealand from 1834 who, together with Lyall and other visiting or resident naturalists, sent large numbers of collections to Kew.

There, specimens were determined by Hooker, contributing to the 217 species of fungi documented in Hooker's (1867) *Handbook of the New Zealand Flora*, with later contributions by Hooker's successors, M. C. Cooke and G. Massee. Duplicates of Colenso's collections were retained in New Zealand and are presently held in the New Zealand Fungal and Plant Disease Collection (PDD), on loan from Museum of New Zealand. Other contributors to New Zealand mycology in the 19th Century included the visitors W. L. Lindsay and S. Berggren, and resident museum botanists T. Kirk and J. Buchanan. In the early

20th century, New Zealand collectors of wood-decay fungi sent collections to C. G. Lloyd in Ohio for determination; descriptions of New Zealand taxa are scattered through his mycological writings from 1909 to 1924.

The first professional resident mycologist was G. H. Cunningham, who began in 1920 to undertake a succession of floristic treatments of major groups of New Zealand and Australian fungi (Ramsbottom 1964). His most significant works were on the rust fungi (Cunningham 1931), gasteromycetes (Cunningham 1944), and non-gilled wood-decay fungi ('Thelephoraceae'– Cunningham 1963; 'Polyporaceae' – Cunningham 1965); the latter two volumes were published posthumously. Cunningham's early collections formed the basis of the current New Zealand Fungal and Plant Disease Collection (PDD) in Auckland. The herbarium acronym relates to the name of the Plant Diseases Division, DSIR, of which Cunningham was first director.

By the 1940s, J. M. Dingley was a co-worker with Cunningham, specialising in description and records of plant pathogenic species (Thompson 1998). Her list of plant diseases of New Zealand (Dingley 1969) covered 725 fungal pathogens. Subsequently, Pennycook (1989) documented plant diseases caused by 1102 fungal species, and the updated content of that list is now incorporated into the NZFungi database (http://nzfungi.LandcareResearch.co.nz) (Anon. 2001–2011).

Other resident mycologists active during the 1960s included G. Stevenson who published extensively on agarics (e.g. Stevenson 1962; Thompson 1999) and R. F. R. McNabb on agarics and tremelloid fungi (e.g. McNabb 1968, 1969; Thomson 1973). Research on the agarics was continued until late last century by J. M. Taylor (e.g. Taylor 1981; Buchanan 1999) and by B. P. Segedin (e.g. Segedin 1987). Over the past 40 years several visiting overseas mycologists have contributed significant treatments on New Zealand fungi, including S. J. Hughes (Canada) on hyphomycetes and sooty moulds (e.g. Hughes 1976; Hughes & Pirozynski 1994), E. Horak (Switzerland) on agarics (e.g. Horak 1973, 2008), R. H. Petersen (USA) on clavarioid fungi (e.g. Petersen 1988), M. A. Castellano and J. M Trappe (USA) on truffle-like fungi (e.g. Castellano & Trappe 1992), and M. Réblová on wood-inhabiting Sordariomycetes (e.g. Réblová & Seifert 2007). G. J. Samuels (USA) was employed in New Zealand during the 1970s and 1980s, studying Xylariales, Nectriaceae, and other ascomycetes (e.g. Rogers & Samuels 1987). Since 1970, overseas collaborating mycologists have described over 50% of all new species (Johnston 2006). Increasingly, studies have involved global-scale assessments of phylogeny and taxonomic relationships, facilitated by increased access to specimen data in collections, through electronic catalogues and web delivery, along with molecular systematics.

National resources in mycology

Fungal specimens, cultures, and associated literature are held primarily in two dried collections and three culture collections in New Zealand. In addition, significant numbers of early New Zealand fungal collections are held overseas at Kew (K).

New Zealand Fungal and Plant Disease Collection (PDD), Landcare Research, Auckland, comprises 90,000 dried collections, with New Zealand collections dating from the 1850s and including important collections by P. K. C. Austwick (polypores), W. Colenso (general), G. H. Cunningham (aphyllophorales, gasteromycetes, rusts), J. M. Dingley (ascomycetes), E. Horak (agarics), S. J. Hughes (hyphomycetes, sooty moulds), R. F. R. McNabb (agarics, boletes, Dacrymycetaceae, Strobilomycetaceae, Tremellaceae), R. H. Petersen (clavarioid fungi), G. J. Samuels (ascomycetes), B. P. Segedin (agarics), S. L. Stephenson (myxomycetes), G. Stevenson (agarics), G. M. Taylor (agarics), and K. Vánky (smut fungi). Collections from recent or current staff include those made by R. E. Beever (hypogeous fungi), P. K. Buchanan (wood-rotting basidiomycetes), P.

R. Johnston (ascomycetes, especially inoperculate discomycetes), and E. H. C. McKenzie (hyphomycetes, rusts, smuts). The collection also contains some of the fungal specimens collected by W. Colenso, on loan from Te Papa, Museum of New Zealand, and by K. Curtis, donated from the Cawthron Institute. Fungal specimens from the collection of the former Plant Health and Diagnostic Station (LEV) have been incorporated into PDD. The coprophilous fungi of A. Bell and D. P. Mahoney, located at Lower Hutt, are also part of PDD. All specimens are catalogued electronically and the data publicly available (Anon. 2001–2011). An extensive collection of mycological books and journals has been assembled since the 1920s in conjunction with Collection PDD.

New Zealand Forest Research Institute Mycological Herbarium (NZFRI-M), Scion, Rotorua, comprises more than 4500 dried and a small number of wet collections, with New Zealand collections dating from 1919. The collection comprises mainly pathogens of woody plants, wood-decay fungi, and ectomycorrhizal fungi of temperate New Zealand native and exotic trees. The collection contains important specimens from early forest pathologists including T. T. C. Birch, G. B. Rawlings, and J. W. Gilmour, as well as more recent material from M. A. Dick and K. Dobbie (pathogens), I. A. Hood (wood-decay fungi), and G. S. Ridley (agarics).

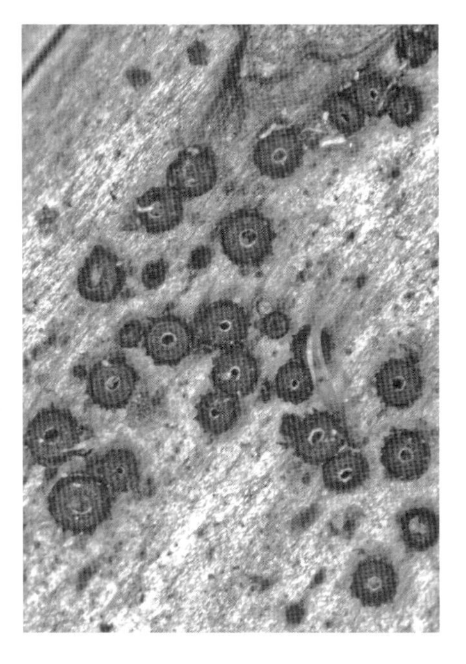

Asterocalyx mirabilis (Leiotiomycetes: Erysiphales) – its ascocarps (fruiting structures), shown here, have been found only on species of the tree-fern genera *Cyathea* and *Dicksonia*.

Peter Johnston

Royal Botanic Gardens, Kew (K), United Kingdom: The majority of pre-1920 collections from New Zealand are housed at Kew. Early New Zealand holdings of polypore, corticioid, and hydnoid fungi at K were listed by Cunningham (1950, 1953). Subsequently, some duplicate PDD collections were sent to K.

U.S. National Fungus Collections (BPI), Beltsville, U.S.A.: C. G. Lloyd's herbarium contains many early New Zealand collections. In addition, many duplicates of PDD collections are held in BPI, including a collection of rust fungi.

International Collection of Micro-organisms from Plants (ICMP), Landcare Research, Auckland, holds 18,000 live strains of fungi and plant-associated bacteria; about 50% are fungi and of these the majority are sourced from New Zealand. Most fungal strains are stored in liquid nitrogen. Data of all strains are available electronically (Anon. 2001–2011).

Mycology Culture Collection, New Zealand Mycology Reference Laboratory, LabPlus, Auckland City Hospital has about 1000 strains of fungi of medical and veterinary importance, established in Wellington by F. Rush-Munro and A. Woodgyer before transfer to Auckland Hospital.

Forest Research Culture Collection (NZFS), Scion, Rotorua, comprises about 2500 live strains of fungi of a similar range to those in Herbarium NZFRI-M.

Publications and databases. New Zealand has a well-developed infrastructure of literature, collection resources, and specialist expertise on which to expand studies in mycology to treat the many neglected taxonomic groups. The first volumes of the *Fungi of New Zealand* series include definitive treatments for the smuts and rusts (Vánky & McKenzie 2002), slime moulds (Stephenson 2003), pink-spored agarics (Horak 2008) and fungi on trees and shrubs (Gadgil 2005), as well as an introductory volume with emphasis on the agarics (McKenzie 2004). The next volumes are planned for Laboulbeniales, rust fungi, and Leotiaceae. The NZFungi database provides on-line and updated access to information about fungal collections and cultures, their names, illustrations, biostatus, and relevant literature as well as keys to many groups of fungi (Anon. 2001–2011).

Conclusions

Considerable progress has been made in recent years to collate knowledge of New Zealand's fungi, by way of taxonomic studies, popular books, reviews, databases, and checklists. The checklists in the succeeding chapters reflect current understanding of New Zealand's fungal diversity, with subsequent additions to be provided through the NZFUNGI website (Anon. 2001–2011) to the New Zealand Organisms Register, the latter to be established in 2012 and covering all biodiversity. Numerically, the majority of New Zealand's expected fungal diversity has yet to be recorded, and only a few taxonomic groups of New Zealand fungi are considered well known. Work to date has largely focussed on ecologically defined groups, such as lichens and plant parasites, and on some macrofungal groups. Among ascomycetes there is good knowledge of lichenised species (lichens) that are spread widely across the group. The non-lichenised ascomycetes are much more diverse and contain the largest numbers of unrecognised species. The study of their taxonomy, ecology, and relevance in applied sciences deserves priority when new resources are identified. Another group currently lacking New Zealand-funded research is the agarics, unquestionably important in nutrient recycling and mycorrhizal associations with ectomycorrhizal trees. Across all fungi, there is moderate knowledge of species in New Zealand that cause plant and animal diseases, and a limited knowledge of the vast number of saprobic species. About 25% of genera of fungi known globally have been recorded in New Zealand and about 9% of known species. These figures reflect a combination of the extent of research on fungi, globally and in New Zealand, plus New Zealand's share of total global biodiversity. Aside from taxonomy, knowledge of biology, distribution, and ecological interactions is known for few taxa.

Technological developments in pyrosequencing and environmental sampling from bulk DNA extracts, however, will enable a large number of fungal species to be recognised and recorded in the coming years, but providing names for each of those species remains a real problem (Hibbett et al. 2011). Names matter because they are the key to relevant literature and allow observations from specific experimental sites to be expanded to a landscape-scale understanding of fungal diversity. Current work using combined taxonomic and environmental molecular datasets hints at the potential of the more powerful pyrosequencing technologies to rapidly catalogue fungal diversity and distribution.

Authors

Dr Peter K. Buchanan Landcare Research, Private Bag 92170, Auckland, New Zealand [buchananp@landcareresearch.co.nz]

Dr Ross E. Beever Landcare Research, Private Bag 92170, Auckland, New Zealand. Deceased.

Dr Travis R. Glare AgResearch, P.O. Box 60, Lincoln, New Zealand [travis.glare@agresearch.co.nz]

Dr Peter R. Johnston Landcare Research, Private Bag 92170, Auckland, New Zealand [johnstonp@landcareresearch.co.nz]

Dr Eric H. C. McKenzie Landcare Research, Private Bag 92170, Auckland, New Zealand [mckenziee@landcareresearch.co.nz]

Dr Barbara C. Paulus RD 2, Masterton, New Zealand [bcpaulus@hotmail.com]

Dr Shaun R. Pennycook Landcare Research, Private Bag 92170, Auckland, New Zealand [pennycooks@landcareresearch.co.nz]

Dr Geoff S. Ridley Environmental Protection Authority, PO Box 131, Wellington [Geoff.Ridley@epa.govt.nz]

Dr J. M. B. (Sandy) Smith Department of Microbiology, University of Otago, PO Box 56, Dunedin, New Zealand. Deceased.

References

AIMER, R. D.; SEGEDIN, B. P. 1985: Some aquatic hyphomycetes from New Zealand streams. *New Zealand Journal of Botany 23*: 273–299.

ALLEN, R. B.; BUCHANAN, P. K.; CLINTON, P. W.; CONE, A. J. 2000: Composition and diversity of fungi on decaying logs in a New Zealand temperate beech (*Nothofagus*) forest. *Canadian Journal of Forest Research 30*: 1025–1033.

ANON. 1997: The State of New Zealand's Environment. Ministry for the Environment, Wellington.

ANON. 2000: Man's death may have been caused by magic mushrooms. *New Zealand Herald*, 24 Nov. 2000.

ANON. 2001–2011: NZ FUNGI Database of New Zealand Fungi. Landcare Research, New Zealand. http//nzfungi.landcare research.co.nz, October 2011, Version 2011.9.21.0

ANON. 2011: Skin manifestations of systemic mycoses. http://dermnetnz.org/fungal/systemic-mycoses.html. 29 June 2011.

AUSTWICK, P. K. C. 1976: Environmental aspects of *Mortierella wolfii* infection in cattle. *New Zealand Journal of Agricultural Research 19*: 25–33.

BARKER, G. M. (Ed.) 2004: *Natural Enemies of Terrestrial Molluscs*. CAB International, Wallingford, U.K. 644 p.

BEEVER, R. E. 1999: Dispersal of New Zealand sequestrate fungi. P. 90 *in*: 9th International Congress of Mycology, Sydney, 16–20 August 1999, Abstracts.

BELL, A. 1975: Fungal succession on dung of the brush-tailed opossum in New Zealand. *New Zealand Journal of Botany 13*: 437–462.

BELL, A. 1983: *Dung Fungi, an Illustrated Guide to Coprophilous Fungi in New Zealand*. Victoria University Press, Wellington. 88 p.

BEST, E. 1942 [1977 reprint]: *Forest Lore of the Maori: with methods of snaring, trapping and preserving birds and rats, uses of berries, roots, fern-roots and forest products, with mythological notes on origins, karakia used etc.* Government Printer, Wellington. 421 p.

BISHOP, P. J.; SPEARE, R.; POULTER, R.; BUTLER, M.; SPEARE, B. J.; HYATT, A.; OLSEN, V.; HAIGH, A. 2009: Elimination of the amphibian chytrid fungus *Batrachochytrium dendrobatidis* by Archey's frog *Leiopelma archeyi*. *Diseases of Aquatic Organisms 84*: 9–15.

BRIDGE, P.; SPOONER, B. 2001: Soil fungi: diversity and detection. *Plant and Soil 232*: 147–154.

BUCHANAN, P. K. 1999: Obituary Grace Marie Taylor, née Bulmer (1930-1999). *Australasian Mycologist 18*: 51–53.

BUCHANAN, P.; JOHNSTON, P.; DAVIS, J.; HITCHMOUGH, R.; MALONEY, R. 2010: New Zealand conservation strategies address fauna, flora, and fungi. *Mycologia Balcanica 7*: 49–51.

BUCHANAN, P. K.; MAY, T. W. 2003: Conservation of New Zealand and Australian fungi. *New Zealand Journal of Botany 41*: 407–421.

CAMPBELL, E. O. 1962: The mycorrhiza of *Gastrodia cunninghamii* Hook.f. *Transactions of the Royal Society of New Zealand, Botany 1*: 289–296.

CASTELLANO, M. A.; TRAPPE, J. M. 1992: Australasian truffle-like fungi. V. Nomenclatural bibliography of type descriptions of Ascomycotina and Zygomycotina. *Australian Systematic Botany* 5: 631–638.

CASTELLANO, M. A.; BEEVER, R. E. 1994: Truffle-like Basidiomycotina of New Zealand: *Gallacea*, *Hysterangium*, *Phallobata*, and *Protubera*. *New Zealand Journal of Botany 32*: 305–328.

CAVALIER-SMITH, T. 1998: A revised six-kingdom system of life. *Biological Reviews 73*: 203–266.

CAVALIER SMITH, T. 2001: What are fungi? Pp. 3–38 *in*: McLaughlin, D. J.; McLaughlin, E. G.; Lemke; P. A. (eds), *The Mycota VII. Systematics and Evolution Part A*. Springer-Verlag, Berlin. 366 p.

COOKSON, I. C.1947: Fossil fungi from the Tertiary deposits of the Southern Hemisphere. Part 1. *Proceedings of the Linnean Society of New South Wales. ser. 2, 72*: 207–214.

COOPER, J. 2005: New Zealand hyphomycete fungi: additional records, new species and notes on interesting collections. *New Zealand Journal of Botany 43*: 323–349.

COWAN, P. E. 1989: A vesicular-arbuscular fungus in the diet of brushtail possums, *Trichosurus vulpecula*. *New Zealand Journal of Botany 27*: 129–131.

CUNNINGHAM, G. H. 1931: *The Rust Fungi of New Zealand, together with the biology, cytology and therapeutics of the Uredinales*. John McIndoe, Dunedin. 261 p.

CUNNINGHAM, G. H. 1944: *The Gasteromycetes of Australia and New Zealand*. John McIndoe, Dunedin. xv + 236 p.

CUNNINGHAM, G. H. 1950: Australian Polyporaceae in herbaria of Royal Botanic Gardens, Kew, and British Museum of Natural History. *Proceedings of the Linnean Society of New South Wales 75*: 214–249.

CUNNINGHAM, G.H. 1953 [1952]: Revision of Australian and New Zealand species of Thelephoraceae and Hydnaceae in the Herbarium of the Royal Botanic Gardens, Kew. *Proceedings of the Linnean Society of New South Wales 77*: 275–299.

CUNNINGHAM, G. H.1963: The Thelephoraceae of Australia and New Zealand. *New Zealand Department of Scientific and Industrial Research Bulletin 145*: 1–359.

CUNNINGHAM, G. H. 1965. Polyporaceae of New Zealand. *New Zealand Department of Scientific and Industrial Research Bulletin 164*: 1–304.

de JONG, M. D. 2000: The BioChon story: deployment of *Chondrostereum purpureum* to suppress stump sprouting in hardwoods. *Mycologist 14*: 58–62.

de JONG, M. D.; BOURDOT, G. W. 1998: Biocontrol using pathogens with wide host-ranges like *Chondrostereum purpureum* and *Sclerotinia sclerotiorum*: special problems with registration. British Mycological Society International Symposium, The Future of Fungi in the Control of Pests, Weeds and Diseases, 5–9 April 1998, Southampton. Abstract.

DICK, M. 1983: *Peniophora* root and stem canker. *Forest Pathology in New Zealand 3*: 1–4.

DICK, M. 1987: *Junghuhnia* root disease. *Forest Pathology in New Zealand 18*: 1–4.

DINGLEY, J. M. 1969: Records of plant diseases in New Zealand. *New Zealand Department of Scientific and Industrial Research Bulletin 192*: 1–298.

EERENS, J. P. J.; LUCAS, R. J.; EASTON, H. S.; WHITE, J. G. H. 1998: Influence of the ryegrass endophyte (*Neotyphodium lolii*) in a cool moist environment. I. Pasture production. *New Zealand Journal of Agricultural Research 41*: 39–48.

EVERTS, R.; CHAMBERS, S. T.; PALTRIDGE, G.; NEWHOOK, C. 1997: Microsporidiosis in New Zealand. *New Zealand Medical Journal 110*: 83.

FAIRLEY, R. 1998: Invasive fungi in New Zealand livestock. *Surveillance* 25: 19.

FLETCHER, L. R.; HARVEY, I. C. 1981: An association of a *Lolium* endophyte with ryegrass staggers. *New Zealand Veterinary Journal 29*: 185–186.

FOWLER, M. 1970: New Zealand predacious fungi. *New Zealand Journal of Botany 8*: 283–302.

FRÖHLICH, J.; FOWLER, S.; GIANOTTI, A.; HILL, R.; KILLGORE, E.; MORIN, L.; SUGIYAMA, L.; WINKS, C. 1999: Biological control of mist flower (*Ageratina riparia*, Asteraceae) in New Zealand. *Proceedings of the Fifty Second New Zealand Plant Protection Conference 1999*: 6–11.

FRÖHLICH, J.; ZABKIEWICZ, J. A.; GIANOTTI, A. F.; RAY, J. W.; VANNER, A. L.; LIU, Z. Q.; GOUS, S. 2000: Field evaluation of *Fusarium tumidum* as a bioherbicide against gorse and broom. *New Zealand Plant Protection 53*: 59–65.

FULLER, J. M.; BUCHANAN, P. K. B.; ROBERTS, M. 2004: Māori knowledge of fungi/Mātauranga of Ngā Harore. Pp. 81–118 *in*: McKenzie, E. H. C. (ed.) *The Fungi of New Zealand/Ngā Harore o Aotearoa, Volume 1. Introduction to Fungi of New Zealand*. Fungal Diversity Press, Hong Kong. 500 p.

GADGIL, P. D. (with DICK, M. A.; HOOD, I. A.; PENNYCOOK, S. R.) 2005: *The Fungi of New Zealand/Ngā Harore o Aotearoa Volume 4. Fungi on Trees and Shrubs in New Zealand*. Fungal Diversity Press, Hong Kong. 437 p.

GILBERTSON, R. L. 1984: Relationships between insects and wood-rotting Basidiomycetes. Pp. 130–165 *in*: Wheeler, Q.; Blackwell, M. (eds), *Fungus-Insect Relationships: Perspectives in Ecology and Evolution*. Columbia University Press, New York. 514 p.

GLARE, T. R.; O'CALLAGHAN, M.; WIGLEY, P. J. 1993: Checklist of naturally-occurring entomopathogenic microbes and nematodes in New Zealand. *New Zealand Journal of Zoology 20*: 95–120.

HALL, I. R. 1977: Species and mycorrhizal infections of New Zealand Endogonaceae. *Transactions of the British Mycological Society 68*: 341–356.

HALL, I. R. 2011. Saffron milk cap (*Lactarius deliciosus*). http://www.trufflesandmushrooms.co.nz/Lactarius%20deliciosus%20web.pdf. 22 Sept., 2011.

HALL, I.; BUCHANAN, P. K.; WANG, Y.; COLE, A. L. J. 1998: *Edible and Poisonous Mushrooms: an Introduction*. New Zealand Institute for Crop & Food Research, Christchurch. 189 p.

HAWKSWORTH, D. L. 1991: The fungal dimension of biodiversity: magnitude, significance, and conservation. *Mycological Research 95*: 641–655.

HAWKSWORTH, D. L. 1997a: The fascination of fungi: exploring fungal diversity. *Mycologist 11*: 18–22.

HAWKSWORTH, D. L. 1997b: Orphans in 'botanical' diversity. *Muelleria 10*: 111–123.

HAYES, L. 1999: Blackberry rust *Phragmidium violaceum*. *In*: *The Biological Control of Weeds Book – A New Zealand Guide*. Landcare Research, Lincoln.

HAYES, L. 2011: Biocontrol agents for weeds in New Zealand – a quick reference guide. http://www.landcareresearch.co.nz/research/biocons/weeds/documents/biocontrol_agents_reference_2011.pdf. June 2011.

HIBBETT, D. S.; OHMAN, A.; GLOTZER, D.;

NUHN, M.; KIRK, P.; NILSSON, R. H. 2011: Progress in molecular and morphological taxon discovery in Fungi and options for formal classification of environmental sequences. *Fungal Biology Reviews 25*: 38-47.

HILL, R. A.; KAY, S. J.; VANNESTE, J. L. 2000: Biological control of sapstain fungi with natural products and biological control agents. P. 46 *in*: Asian Mycological Congress 2000, 9–13 July 2000, Hong Kong. Abstracts.

HITCHMOUGH, R. 2002: New Zealand Threat Classification System lists 2002. *Threatened Species Occasional Publication 23*: 1–16.

HOOD, I. A. 1989: *Armillaria* root disease in New Zealand forest. *New Zealand Journal of Forestry Science 19*: 180–197.

HOOD, I. A. 1992: *An Illustrated Guide to Fungi on Wood in New Zealand*. Auckland University Press, Auckland. 424 p.

HOOKER, W. J. 1836: *Sphaeria Robertsii*. Pl. 11 *in*: *Icones Plantarum; or figures with brief descriptive characters and remarks of new or rare plants selected from the author's herbarium*. Volume 1. Longman, Rees, Orme, Brown, Green, & Longman, London.

HOOKER, J. D. 1867: *Handbook of the New Zealand Flora*. Reeve & Co., London. Vol. 2, pp. 393–793. [Fungi, pp. 595–637]

HORAK, E. 1973: Fungi Agaricini Novazelandiae I-V. *Beihefte zur Nova Hedwigia 43*: 1–200.

HORAK, E. 2008: *The Fungi of New Zealand/ Ngā Harore o Aotearoa Volume 5. Agaricales of New Zealand. 1. Pluteaceae (Pluteus, Volvariella); Entolomataceae (Claudopus, Clitopilus, Entoloma, Pouzarella, Rhodocybe, Richoniella)*. Fungal Diversity Press, Hong Kong. 305 p.

HUGHES, M.; WEIR, A.; GILLEN, B.; ROSSI, W. 2004a: *Stigmatomyces* from New Zealand and New Caledonia: new records, new species and two new host families. *Mycologi 96*: 834–844.

HUGHES, M.; WEIR, A.; LESCHEN, R.; JUDD, C.; GILLEN, B. 2004b: New species and records of Laboulbeniales from the subantarctic islands of New Zealand. *Mycologia 96*: 1355–1369.

HUGHES, S.J. 1966: New Zealand fungi. 7. *Capnocybe* and *Capnophialophora*, new form genera of sooty moulds. *New Zealand Journal of Botany 4*: 333–353.

HUGHES, S. J. 1971: New Zealand fungi. 16. *Brachydesmiella, Ceratosporella*. *New Zealand Journal of Botany 9*: 351–354.

HUGHES, S. J. 1976: Sooty moulds. *Mycologia 68*: 693–820.

HUGHES, S. J.; PIROZYNSKI, K. A. 1994: New Zealand fungi. 34. *Endomeliola dingleyae*, a new genus and new species of Meliolaceae. *New Zealand Journal of Botany 32*: 53–59.

JOHNSTON, P. R. 1992: Rhytismataceae in New Zealand. 6. Checklist of species and hosts, with keys to species on each host genus. *New Zealand Journal of Botany 30*: 329–351.

JOHNSTON, P. R. 1998: Leaf endophytes of manuka (*Leptospermum scoparium*). *Mycological Research 102:* 1009–1018.

JOHNSTON, P. R. 2006: New Zealand's non-lichenised fungi – where they came from, who collected them, where they are now. *National Science Museum Monographs 34*: 37–49.

JOHNSTON, P. R.; BUCHANAN, P. K; LEATHWICK, J.; MORTIMER, S. 1998: Fungal invaders. *Australasian Mycological Newsletter 17*: 48–52.

JOSHEE, S.; PAULUS, B. C.; PARK, D.; JOHNSTON, P. R. 2009: Diversity and distribution of fungal foliar endophytes in New Zealand Podocarpaceae. *Mycological Research 113*: 1003–1015.

KARLING, J. S. 1966 [1965]: Some zoosporic fungi of New Zealand. I. *Sydowia 19*: 213–226.

KARLING, J. S. 1968a [1966]: Some zoosporic fungi of New Zealand. XII. Olpidiopsidaceae, Sirolpidiaceae and Lagenidiaceae. *Sydowia 20*: 190–199.

KARLING, J. S. 1968b [1966]: Some zoosporic fungi of New Zealand. III. *Phlyctidium, Rhizophydium, Septosperma*, and *Podochytrium*. *Sydowia 20*: 74–85.

KENDRICK, B. 1992: *The Fifth Kingdom*. 2nd Edn. Focus Publishing/R. Pullins, Newburyport. 414 p.

KIRK, P. M.; CANNON, P. F.; MINTER, D. W.; STALPERS, J. A. 2008: *Ainsworth & Bisby's Dictionary of the Fungi*. 10th Edn. Wallingford, CAB International. 771 p.

KOHLMEYER, J.; KOHLMEYER, E. 1979: *Marine Mycology – The Higher Fungi*. New York, Academic Press. xiv + 690 p.

LARSEN, M. 2000: Prospects for controlling animal parasitic nematodes by predacious micro fungi. *Parasitology 120*: S120–S131.

LINDQVIST, J. 1994: Lacustrine stromatolites and oncoids: Manuherikia Group (Miocene), New Zealand. Pp. 227–254 *in*: Bertrand-Sarfati, J.; Monty, C. (eds), *Phanerozoic Stromatolites II*. Dordrecht, Kluwer Academic Publishers. x + 249 p.

MALONE, L. A. 1990: Ultrastructure and tissue specificity of *Nosema costelytrae* and *Nosema takapauensis*, microsporidian pathogens of grass grubs, *Costelytra zealandica* (Coleoptera: Scarabaeidae) in New Zealand. *New Zealand Journal of Agricultural Research 33*: 459–465.

MALONE, L. A.; McIVOR, C. A. 1996: Use of nucleotide sequence data to identify a microsporidian pathogen of *Pieris rapae* (Lepidoptera, Pieridae). *Journal of Invertebrate Pathology 68*: 231–238.

McKENZIE, E. H. C. 2000: Plant disease record. *Uromyces transversalis*, rust fungus found infecting Iridaceae in New Zealand. *New Zealand Journal of Crop and Horticultural Science 28*: 289–291.

McKENZIE, E. H. C. (Ed.) 2004: *The Fungi of New Zealand/Ngā Harore o Aotearoa, Volume 1. Introduction to Fungi of New Zealand*. Fungal Diversity Press, Hong Kong. 500 p.

McKENZIE, E. H. C.; BUCHANAN, P. K.; JOHNSTON, P.R. 1999: Fungi on pohutukawa and other *Metrosideros* species in New Zealand. *New Zealand Journal of Botany 37*: 335–354.

McKENZIE, E. H. C.; BUCHANAN, P. K.; JOHNSTON, P.R. 2000: Checklist of fungi on *Nothofagus* species in New Zealand. *New Zealand Journal of Botany 38*: 635–720.

McNABB, R. F. R. 1968: The Boletaceae of New Zealand. *New Zealand Journal of Botany 6*: 137–176.

McNABB, R. F. R. 1969: New Zealand Tremellales – III. *New Zealand Journal of Botany 7*: 241–261.

MONREAL, M.; BERCH, S. M.; BERBEE, M. 1999: Molecular diversity of ericoid mycorrhizal fungi. *Canadian Journal of Botany 77*: 1580–1594.

MORGAN, F. D. 1989: Forty years of *Sirex noctilio* and *Ips grandicollis* in Australia. *New Zealand Journal of Forestry Science 19*: 198–209.

MORIN, L. M.; SMITH, L. A.; SYRETT, P. 1997: Field establishment of the rust *Puccinia hieracii* var. *piloselloidarum* on *Hieracium pilosella* in the South Island. *Landcare Research Contract Report LC9697/85*: 1–17.

NICHOLLS, D. W.; HYNE, B. E. B.; BUCHANAN, P. K. 1995: Death cap mushroom poisoning. Letter to the Editor. *New Zealand Medical Journal 108(1001)*: 234.

ORLOVICH, D. A.; CAIRNEY, J. W. G. 2004: Ectomycorrhizal fungi in New Zealand: current perspectives and future directions. *New Zealand Journal of Botany 42*: 721–738.

PARK, G. N. 2000: *New Zealand as Ecosystems: the Ecosystem Concept as a Tool for Environmental Management and Conservation*. Wellington, Department of Conservation. 96 p.

PENNYCOOK, S. R. 1989: *Plant Diseases Recorded in New Zealand*. Plant Diseases Division, DSIR, Auckland. Vol. 1, 276 p.; Vol. 2, 502 p.; Vol. 3, 180 p.

PENNYCOOK, S. R.; GALLOWAY, D. J. 2004: Checklist of New Zealand 'Fungi'. Pp. 401–488 *in*: McKenzie, E. H. C. (ed.), *The Fungi of New Zealand/Ngā Harore o Aotearoa, Volume 1. Introduction to Fungi of New Zealand*. Fungal Diversity Press, Hong Kong. 500 p.

PENSELER, W. H. A. 1930: Fossil leaves from the Waikato district. *Transactions of the New Zealand Institute 61*: 452–477.

PETERSEN, R. H. 1988: The clavarioid fungi of New Zealand. *New Zealand Department of Scientific and Industrial Research Bulletin 236*: 1–170.

PIROZYNSKI, K. A.; WERESUB, L. K. 1979: A biogeographic view of the history of ascomycetes and the development of their pleomorphism. Pp. 93–123 *in*: Kendrick, B. (ed.), *The Whole Fungus. Volume 1*. National Museums of Canada & The Kananaskis Foundation, Ottawa, Canada. 412 p.

POWELL, M. J. 1993: Looking at mycology with a Janus face: a glimpse at Chytridiomycetes active in the environment. *Mycologia 85*: 1–20.

RAMSBOTTOM, J. 1964: Gordon Herriot Cunningham, 1892–1962. *Bibliographical Memoirs of Fellows of the Royal Society 10*: 15–37.

RÉBLOVÁ, M.; SEIFERT, K. A. 2007: A new fungal genus, *Teracosphaeria*, with a phialophora-like anamorph (Sordariomycetes, Ascomycota). *Mycological Research 111*: 287–298.

REDECKER, D.; KODNER, R.; GRAHAM, L. E. 2000: Glomalean fungi from the Ordovician. *Science 289*: 1920–1921.

ROGERS, J. D.; SAMUELS, G. J. 1987 [1986]: Ascomycetes of New Zealand 8. *Xylaria*. *New Zealand Journal of Botany 24*: 615–650.

RUSSELL, J.; BULMAN, S. 2005: The liverwort *Marchantia foliacea* forms a specialized symbiosis with arbuscular mycorrhizal fungi in the genus *Glomus*. *New Phytologist 165*: 345–349.

SAGARA, N. 1975: Ammonia fungi: a chemoecological grouping of terrestrial fungi. *Contributions from the Biological Laboratory Kyoto University 24*: 205–276.

SAGARA, N.; BUCHANAN, P.; SUZUKI, A.; SEGEDIN, B.; TSUDA, M. 1993: Species composition of the ammonia fungi in New Zealand. P. 36 *in*: *Proceedings of the 37th Annual Meeting of the Mycological Society of Japan. Abstracts of submitted papers.*

SEGEDIN, B. P. 1987: An annotated checklist of agarics and boleti recorded from New Zealand. *New Zealand Journal of Botany 25*: 185–215.

SEGEDIN, B. P.; BUCHANAN, P. K.; WILKIE, J. P. 1995: Studies in the Agaricales of New Zealand: new species, new records and renamed species of *Pleurotus* (Pleurotaceae). *Australian Systematic Botany 8*: 453–482.

SIEGEL, M. R.; LATCH, G. C. M.; JOHNSON, M. C. 1987: Fungal endophytes of grasses. *Annual*

Review of Phytopathology 25: 293–315.
SIGLER, L.; BARTLEY, J. R.; PARR, D. H.; MORRIS, A. J. 1999: Maxillary sinusitis caused by medusoid form of *Schizophyllum commune*. *Journal of Clinical Microbiology 37*: 3395–3398.
STENSRUD, Ø.; HYWEL-JONES, N.; SCHUMACHER, T. 2005: Towards a phylogenetic classification of *Cordyceps:* ITS nrDNA sequence data confirm divergent lineages and paraphyly. *Mycological Research 109*: 41–56.
STEPHENSON, S. L. 2003: *The Fungi of New Zealand/Ngā Harore o Aotearoa, Volume 3. Myxomycetes of New Zealand*. Fungal Diversity Press, Hong Kong. 238 p.
STERFLINGER, K.; SCHOLZ, J. 1997: Fungal infection and bryozoan morphology. *Courier Forschungsinstitut Senckenberg 201*: 433–447.
STEVENSON, G. 1962 [1961]: The Agaricales of New Zealand: I. *Kew Bulletin 15*: 381–385.
SUTHERLAND, J. I. 2003: Miocene petrified wood and associated borings and termite faecal pellets from Hukatere Peninsula, Kaipara Harbour, North Auckland, New Zealand. *Journal of the Royal Society of New Zealand 33*: 395–414.
SUZUKI, A. 2000: Geographic distribution of ammonia fungi and their role in terrestrial ecosystems. P. 21 *in*: Asian Mycological Congress 2000, 9–13 July 2000, Hong Kong. Abstracts.
SUZUKI, A.; TANAKA, C.; BOUGHER, N. L.; TOMMERUP, I. C.; BUCHANAN, P. K.; FUKIHARU, T.; TSUCHIDA, S.; TSUDA, M.; ODA, T.; FUKADA, J.; SAGARA, N. 2002: ITS rDNA variation of the *Coprinopsis phlyctidospora* (syn. *Coprinus phlyctidosporus*) complex in the Northern and Southern Hemisphere. *Mycoscience 43*: 229–238.
SUZUKI, A.; TSUDA, M. 1993: [Fungi met in New Zealand. The second report]. *Biosphere 1*: 15–16. [In Japanese]
TAYLOR, G. M. 1981: *Mobil New Zealand Nature Series. Mushrooms and Toadstools*. A.H. & A.W. Reed, Wellington. 79 p.
THOMPSON, A .D. 1973: Robert Francis Ross McNabb (1934–72). *New Zealand Journal of Botany 11*: 799–802.
THOMPSON, A. D. 1998: Tribute to pioneer plant pathologist and mycologist, Dr Joan Dingley. *New Zealand Botanical Society Newsletter 54*: 13–17.
THOMPSON, A. D. 1999: Tribute to pioneer botanist, mycologist and mountaineer, Dr Greta Stevenson Cone. *New Zealand Botanical Society Newsletter 56*: 25–27.
THORN, G.; TSUNEDA, A. 1993: Interactions between *Pleurotus* species, nematodes, and bacteria on agar and in wood. *Transactions of the Mycological Society of Japan 34*: 449–464.
THORNTON, R. H. 1958: Biological studies of some tussock-grassland soils. 2. Fungi. *New Zealand Journal of Agricultural Research 1*: 922–938.
THORNTON, R. H.; PERCIVAL, J. C. 1959: An hepatotoxin from *Sporodesmium bakeri*, capable of producing facial eczema disease in sheep. *Nature 183*: 63.
VAJDA, V.; McLOUGHLIN, S. 2004: Fungal proliferation at the Cretaceous-Tertiary Boundary. *Science 303*: 1489.
VÁNKY, K.; McKENZIE, E. H. C. 2002: *The Fungi of New Zealand/Ngā Harore o Aotearoa, Volume 2. Smut Fungi of New Zealand*. Fungal Diversity Press, Hong Kong. 273 p.
WILLIAMS, M. C.; LICHTWARDT, R. W. 1990: Trichomycete gut fungi in New Zealand aquatic insect larvae. *Canadian Journal of Botany 68*: 1045–1056.
WITTNER, M. 1999. Historical perspective on the microsporidia: Expanding horizons. Pp. 1–6 *in*: Wittner, M.; Weiss, L. M. (eds), *The Microsporidia and Microsporidiosis*. ASM Press, Washington, D. C. 553 p.

TWENTY-NINE

Phylum CHYTRIDIOMYCOTA s. l. chytrids

ROSS E. BEEVER, PETER K. BUCHANAN, ERIC H. C. McKENZIE

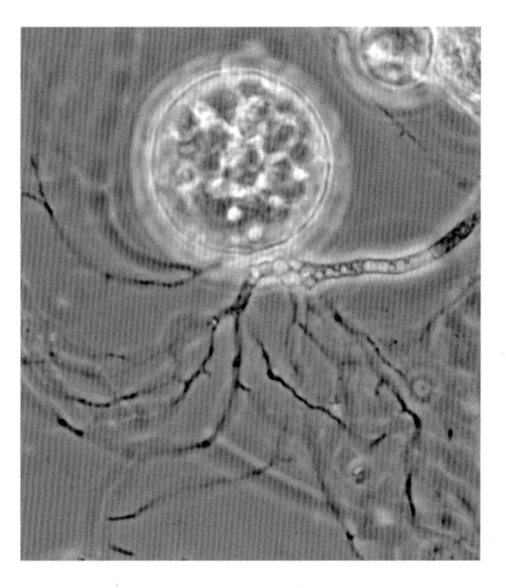

A non-New Zealand chytrid showing typical morphology of a thallus/sporangium with rhizoids.

David J. Patterson & Aimee Ladermann, per Micro*scope (MBL)

Chytridiomycota (also known as chytridiomycetes or chytrids) are microscopic fungi, globally numbering about 1000 described species, and occupying a basal position in the phylogenetic fungal tree (James et al. 2006). The smallest forms are reduced to a single cell but it may have rhizoids that penetrate the substrate as well as slender hyphae attached to sporangia. Species of Blastocladiales can produce a true mycelium. As with other true fungi, chytrids have chitin as their main cell-wall component. They are well characterised as a group of fungi by producing zoospores, each having one (rarely more) posteriorly directed whiplash cilium. Identification of species based on field collections or cultures remains problematic because of their small size, general lack of distinctive characters, and their morphological variability in different growing conditions (Barr 2001; Longcore 2005). In the absence of diagnostic morphological and biochemical characters, species and their relationships are increasingly being redefined by gene sequencing (Nagahama et al. 1995).

Zoospore ultrastructure and morphology help discriminate among the majority of chytrids that occupy 11 subclades within the 'core chytrid' group, but a further three separate phylogenetic lineages have been recently recognised – *Rozella* spp. (the earliest diverging lineage in the fungi), *Olpidium brassicae*, and the Blastocladiales (James et al. 2006). Sequence data of 18S ribosomal DNA indicates that Chytridiomycota might not be monophyletic (James et al. 2000). On this basis, Blastochytridiomycota has been proposed as a new phylum, distinguished by its life-cycle, ultrastructural features, and phylogeny (James et al. 2006). In New Zealand, early knowledge was limited to those species pathogenic on crops and other plants. Subsequently, a series of papers by Karling (e.g. Karling 1966, 1967a-e, 1968a-d, 1969, 1970) documented almost 200 species (150 chytridiomycetes, and other zoosporic fungi) from a diverse range of substrates and habitats. Caution is needed in interpreting Karling's taxonomic arrangement of records, however, owing to subsequent alteration of hierarchies of genera. For example, whereas the genera *Olpidium*, *Rozella*, and *Sphaerita* appear together as members of the Olpidiaceae in Karling (1966), the latter genus is now considered to belong in kingdom Chromista among the oomycete fungi (Kirk et al. 2008) while *Rozella* is now considered to be in a separate lineage in the Choanozoa (Protozoa).

Chytrids are a diverse group in form, habitat, and ecology, common in aquatic and moist habitats (Powell 1993) such as lakes, streams, and moist soil, but also found in aerial parts of plants and tree-canopy detritus. Longcore (2005) baited organic detritus from the tree canopy of a South Island lowland forest and recovered five species of chytrid; most canopy chytrids have also been recorded from terrestrial soils. As saprobes, chytrids are important in the decomposition of plant and animal remains, especially owing to their specialised ability to metabolise polymers such as chitin, keratin, and cellulose that are resistant to degradation by many micro-organisms. Chytrids and bacteria are among the few organisms able to decay plant pollen (Karling 1966). A small number are marine, such as the New Zealand species *Rhizophydium novozeylandiense* (= *Phlyctidium marinum* Karling 1967b) isolated from beach sand using dead pollen as a bait.

Some saprobic species may be particularly abundant. *Karlingia rosea*, reportedly the commonest chytridiomycete in New Zealand, was found in almost 95% of 140 soils sampled throughout the country (Karling 1968a). A range of anaerobic saprobic chytridiomycetes live in the rumen of herbivores (Powell 1993). Recent analyses of high-elevation soils devoid of plant life from the Rockies, Andes, Himalayas, and New Zealand mountains showed a surprisingly rich diversity of chytrid fungi (Freeman et al. 2009) including many taxa with novel DNA sequences. These chytrids are thought to be utilising pollen and microbial phototrophs in high-elevation soils as food. Golubeva and Stephenson (2003) provided the first records of chytrids from subantarctic Campbell Island.

As parasites, chytridiomycetes often show host specificity as zoospores respond to particular chemicals released from their hosts. Endoparasitic genera such as *Physoderma*, *Synchytrium*, and *Olpidium* are of economic significance, affecting higher plants, ferns, mosses, and algae. Zoospores encyst on a plant surface and inject their cell contents into the host cytoplasm. Karling (1967a) recorded ten *Synchytrium* species as plant parasites in New Zealand. Damage to plants by obligately pathogenic *Olpidium* species is mainly as a consequence of their role as virus vectors. *Olpidium brassicae*, a root pathogen of lettuce and introduced weed species in New Zealand (Dingley 1969), transmits lettuce big vein and other viruses via the zoospore (Beever & Fry 1970).

Other parasitic chytridiomycetes attack invertebrate hosts including nematodes, rotifers, and insects, e.g. *Olpidium gregarium* recorded from New Zealand on rotifer eggs (Karling 1966). *Coelomomyces*, parasitic on mosquito larvae, has been well studied overseas as a potential mycoinsecticide (Powell 1993). As parasites of algae, chytridiomycetes are implicated in control of phytoplankton population levels, and some are pathogenic on other fungi, particularly other chytridiomycetes, and members of the oomycetes (Chromista).

Of concern are recent records of the first chytridiomycete parasite of vertebrates. *Batrachochytrium dendrobatidis* has been associated with the skin disease chytridiomycosis, which caused death of the blue poison-dart frog in a Washington zoo (Longcore et al. 1999), and is implicated in declines of wild populations of amphibians noted in Central America and Australia (Berger et al. 1998) as well as in New Zealand. Chytridiomycosis caused by *B. dendrobatidis* has been recorded in New Zealand's endemic Archey's frog (*Leiopelma archeyi*) (Barnett 2000; Anon. 2001; Bishop et al. 2009), which has declined significantly in the Coromandel Ranges. A trial involving a small number of naturally infected frogs captured from the wild indicated a low level of susceptibility to the clinical effects of the disease, and infected individuals were effectively treated with topical chloromphenicol without negative side effects (Bishop et al. 2009). Moreno et al. (2011) surveyed two populations of Hochstetter's frog (*Leiopelma hochstetteri*) near Auckland and found no occurrence of *B. dendrobatidis*.

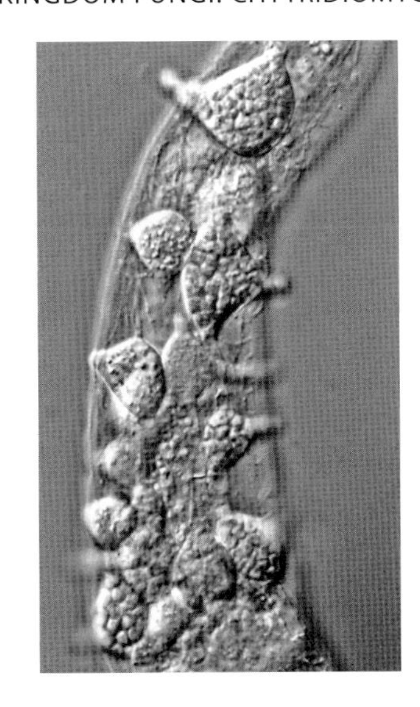

Zoosporangia of *Catenaria anguillulae* (Blastocladiomycetes: Blastocladiales) form inside an infected eel worm (Nematoda) and produce swimming zoospores that exit from the open-pored neck to locate other nematodes. Fine thread-like rhizoids absorb nutrients from the worm body.
George L. Barron, University of Guelph

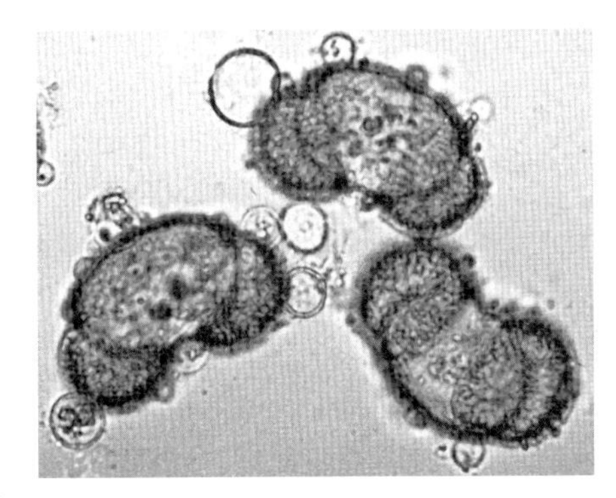

The chytrid *Rhizophydium pollonis* (Chytridiomycetes: Rhizophydiales) colonises pine pollen, producing swimming zoospores from blister-like zoosporangia (several shown empty) that form on the outside of the pollen grain.
George L. Barron, University of Guelph

Summary of New Zealand Chytridiomycota *sensu lato* diversity

Taxon	Described species	Known undescribed/ undetermined species	Estimated undiscovered species	Adventive species	Endemic species	Endemic genera
Blastocladiomycota	12+1	1	15	3	0	0
Chytridiomycota	133+1	0	150	9	7	0
Neocallimastigomycota	5	0	10	1	0	0
Totals	**150+2**	**1**	**175**	**13**	**7**	**0**

Summary of New Zealand Chytridiomycota *sensu lato* diversity by major environment

Taxon	Marine	Fresh-water	Terrestrial
Blastocladiomycota	0	3	10
Chytridiomycota	6	36	91
Neocallimastigomycota	0	0	5
Totals	6	39	106

Authors

Dr Ross E. Beever Landcare Research, Private Bag 92170, Auckland, New Zealand. Deceased.

Dr Peter K. Buchanan Landcare Research, Private Bag 92170, Auckland, New Zealand [buchananp@LandcareResearch.co.nz]

Dr Eric H.C. McKenzie Landcare Research, Private Bag 92170, Auckland, New Zealand [mckenziee@LandcareResearch.co.nz]

References

ANON. 2001: Killer fungus threatens New Zealand native frogs. *Biosecurity 31*: 7.

BARNETT, S. 2000: The trouble with frogs. *Forest & Bird 295*: 14–17.

BARR, D. J. S. 2001: Chytridiomycota. Pp. 93–112 *in*: Esser, K.; Lemke, P. A. (eds), *The Mycota VII. Systematics and Evolution Part A*. Springer, Berlin. 366 p.

BEEVER, J. E.; FRY, P. R. 1970: The effect of light on the transmission of tobacco necrosis virus by *Olpidium brassicae*. *Virology 40*: 357–362.

BERGER, L.; SPEARE, R.; DASZAK, P.; GREEN, E,; CUNNINGHAM, A. A.; GOGGIN, C. L.; SLOCOMBE, R.; RAGAN, M. A.; HYATT, A. D.; McDONALD, K. R.; HINES, H. B.; LIPS, K. R.; MARANTELLI, G.; PARKES, H. 1998: Chytridiomycosis causes amphibian mortality associated with population declines in the rain forests of Australia and Central America. *Proceedings of the National Academy of Sciences of the United States of America 95*: 9031–9036

BISHOP, P. J.; SPEARE, R.; POULTER, R.; BUTLER, M.; SPEARE, B. J.; HYATT, A.; OLSEN, V.; HAIGH, A. 2009: Elimination of the amphibian chytrid fungus *Batrachochytrium dendrobatidis* by Archey's frog *Leiopelma archeyi*. *Diseases of Aquatic Organisms 84*: 9–15.

DINGLEY, J. M. 1969: Records of plant diseases in New Zealand. *New Zealand Department of Scientific and Industrial Research Bulletin 192*: 1–298.

FREEMAN, K. R.; MARTIN, A. P.; KARKI, D.; LYNCH, R. C.; MITTER, M. S.; MEYER, A. F.; LONGCORE, J. E.; SIMMONS, D. R.; SCHMIDT, S. K. 2009: Evidence that chytrids dominate fungal communities in high-elevation soils. *Proceedings of the National Academy of Sciences of the United States of America 106*: 18315–18320.

GOLUBEVA, O. G.; STEPHENSON, S. L. 2003: Zoosporic fungi from subantarctic Campbell Island. *New Zealand Journal of Botany 41*: 319–324.

JAMES, T.Y.; LETCHER, P. M.; LONGCORE, J. E.; MOZLEY-STANDRIDGE, S. E.; PORTER, D.; POWELL, M. J.; GRIFFITH, G. W.; VILGALYS, R. 2006. A molecular phylogeny of the flagellated fungi (Chytridiomycota) and description of a new phylum (Blastocladiomycota). *Mycologia 98*: 860–871.

JAMES, T.Y.; PORTER, D.; LEANDER, C. A.; VILGALYS, R.; LONGCORE; J. E. 2000: Molecular phylogenetics of the Chytridiomycota supports the utility of ultrastructural data in chytrid systematics. *Canadian Journal of Botany 78*: 1–15.

KARLING, J. S. 1966 [1965]: Some zoosporic fungi of New Zealand. I. *Sydowia 19*: 213–226.

KARLING, J. S. 1967a [1966]: Some zoosporic fungi of New Zealand. II. Synchytriaceae. *Sydowia 20*: 51–66.

KARLING, J. S. 1967b [1966]: Some zoosporic fungi of New Zealand. III. *Phlyctidium, Rhizophydium, Septosperma*, and *Podochytrium*. *Sydowia 20*: 74–85.

KARLING, J. S. 1967c [1966]: Some zoosporic fungi of New Zealand. IV. *Polyphlyctis* gen. nov., *Phlyctochytrium* and *Rhizidium*. *Sydowia 20*: 86–95.

KARLING, J. S. 1967d [1966]: Some zoosporic fungi of New Zealand. V. Species of *Asterophlyctis, Obelidium, Rhizoclosmatium, Siphonaria* and *Rhizophlyctis*. *Sydowia 20*: 96–108.

KARLING, J. S. 1967e [1966]: Some zoosporic fungi of New Zealand. VI. *Entophlyctis, Diplophlyctis, Nephrochytrium* and *Endochytrium*. *Sydowia 20*: 109–118.

KARLING, J. S. 1968a [1966]: Some zoosporic fungi of New Zealand. VII. Additional monocentric operculate species. *Sydowia 20*: 119–128.

KARLING, J. S. 1968b [1966]: Some zoosporic fungi of New Zealand. VIII. Cladochytriaceae and Physodermataceae. *Sydowia 20*: 129–136.

KARLING, J. S. 1968c [1966]: Some zoosporic fungi of New Zealand. X. Blastocladiales. *Sydowia 20*: 144–150.

KARLING, J. S. 1968d [1966]: Some zoosporic fungi of New Zealand. XII. Olpidiopsidaceae, Sirolpidiaceae and Lagenidiaceae. *Sydowia 20*: 190–199.

KARLING, J. S. 1969: Zoosporic fungi of Oceania. VII. Fusions in *Rhizophlyctis*. *American Journal of Botany 56*: 211–221.

KARLING, J. S. 1970: Some zoosporic fungi of New Zealand. XIV. Additional species. *Archiv für Mikrobiologie 70*: 266–287.

KIRK, P. M.; CANNON, P. F.; MINTER, D. W.; STALPERS, J. A. 2008: *Ainsworth & Bisby's Dictionary of the Fungi*. 10th Edn. Wallingford, CAB International. 771 p.

LONGCORE, J. E. 2005: Zoosporic fungi from Australian and New Zealand tree-canopy detritus. *Australian Journal of Botany 53*: 259–272.

LONGCORE, J. E.; PESSIER, A. P.; NICHOLS, D. K. 1999: *Batrachochytrium dendrobatidis* gen. et sp. nov., a chytrid pathogenic to amphibians. *Mycologia 91*: 219–227.

MORENO, V.; AGUAYO, C. A.; BRUNTON, D. H. 2011: A survey for the amphibian chytrid fungus *Batrachochytrium dendrobatidis* in New Zealand's endemic Hochstetter's frog (*Leiopelma hochstetteri*). *New Zealand Journal of Zoology 38*: 181–184.

NAGAHAMA, T.; SATO, H.; SHIMAZU, M.; SUGIYAMA, J. 1995: Phylogenetic divergence of the entomophthoralean fungi: evidence from nuclear 18S ribosomal RNA gene sequences. *Mycologia 87*: 203–209.

POWELL, M.J. 1993: Looking at mycology with a Janus face: a glimpse at Chytridiomycetes active in the environment. *Mycologia 85*: 1–20.

Checklist of New Zealand Chytridiomycota

This checklist is arranged according to the taxonomic hierarchy in Kirk et al. (2008) where Chytridiomycota *sensu lato* is considered to comprise three phyla – Blastocladiomycota, Chytridiomycota *sensu stricto*, and Neocallimastigomycota. Species status: A, adventive (naturalised alien); E, endemic; U, status unknown (the majority); UO, occurrence uncertain in New Zealand; species not indicated by any notation are native (naturally indigenous but not endemic). Habitat codes: F, freshwater; M, marine; the remainder are terrestrial but require wet soil to complete their life-cycle.

KINGDOM FUNGI
PHYLUM BLASTOCLADIOMYCOTA
Class BLASTOCLADIOMYCETES
Order BLASTOCLADIALES
BLASTOCLADIACEAE
Blastocladiella britannica Horenst. & Cantino U
Blastocladiella microcystogena Whiffen U
Blastocladiella novae-zeylandiae Karling U
Blastocladiella simplex V.D. Matthews U F
CATENARIACEAE
Catenaria anguillulae Sorokīn U
Catenaria verrucosa Karling U
Catenomyces persicinus A.M. Hanson U
Catenophlyctis variabilis (Karling) Karling U F
COELOMOMYCETACEAE
Coelomomyces dodgei Couch (nom. inv.) OU F
Coelomomyces psorophorae Couch
Coelomomyces psorophorae var. *tasmaniensis* Couch

CATENARIACEAE
Catenaria anguillulae Sorokīn U
Catenaria verrucosa Karling U
Catenomyces persicinus A.M. Hanson U
Catenophlyctis variabilis (Karling) Karling U F
COELOMOMYCETACEAE
Coelomomyces dodgei Couch (nom. inv.) OU F
Coelomomyces psorophorae var. *tasmaniensis* Couch & Laird (nom. inv.)
PHYSODERMATACEAE
Physoderma alfalfae (Lagerh.) Karling A
Physoderma debeauxii Bubák A
Physoderma potteri (A.W. Bartlett) Karling A
INCERTAE SEDIS
Coelomycidium sp.

PHYLUM CHYTRIDIOMYCOTA
Class CHYTRIDIOMYCETES
OLPIDIACEAE
Olpidium appendiculatum Karling U
Olpidium brassicae (Woronin) P.A. Dang. A
Olpidium entophytum (A. Braun) Rabenh. U F
Olpidium gregarium (Nowak.) J. Schröt. U F
Olpidium luxurians (Tomaschek) A. Fisch. U
Olpidium pendulum Zopf U
Olpidium radicale Schwartz & Ivimey Cook U
Olpidium saccatum Sorokīn U F

Order CHYTRIDIALES
CHYTRIDIACEAE
Blyttiomyces verrucosus Dogma M
Chytridium oedogonii Couch F
Chytridium parasiticum Willoughby U
Chytridium proliferum Karling U
Chytriomyces appendiculatus Karling U
Chytriomyces aureus Karling U
Chytriomyces hyalinus Karling U F
— var. *granulatus* Karling E F
Chytriomyces parasiticus Karling U
Chytriomyces rotoruaensis Karling E
Cylindrochytridium johnstonii Karling U F
Karlingiomyces asterocystis (Karling) Sparrow U
Karlingiomyces curvispinosus (Karling) Sparrow U
Karlingiomyces dubius (Karling) Sparrow U
Karlingiomyces marylandicus (Karling) Sparrow U
Obelidium mucronatum Nowak. U F
Phlyctochytrium aestuarii Ulken U M
Phlyctochytrium bryopsidis Kobayasi & M. Ôkubo U M
Phlyctochytrium bullatum Sparrow U F
Phlyctochytrium chaetiferum Karling U F
Phlyctochytrium hirsutum Karling U
Phlyctochytrium indicum Karling U
Phlyctochytrium lagenaria (Schenk) Domjan U F
Phlyctochytrium mucronatum Canter U F
Phlyctochytrium planicorne G.F. Atk. U
Phlyctochytrium quadricorne (de Bary) J. Schröt. U F
Phlyctochytrium reinboldtiae Persiel U
Phlyctochytrium zygnematis (F. Rosen) J. Schröt. U F
Podochytrium chitinophilum Willoughby U F
Podochytrium emmanuelense (Sparrow) Sparrow & R.A. Paterson U F
Polyphlyctis unispina (R.A. Paterson) Karling U
Rhizidium chitinophilum Sparrow U
Rhizidium elongatum Karling U
Rhizidium reniforme Karling U
Rhizidium richmondense Willoughby U
Rhizidium varians Karling U
Rhizidium verrucosum Karling U
Rhizoclosmatium globosum H.E. Petersen U F
Rhizoclosmatium hyalinum Karling U F
Septosperma irregulare (Karling) Dogma
Septosperma rhizophydii Whiffen ex R.L. Seym. U
Siphonaria sparrowii Karling U F
Siphonaria variabilis H.E. Petersen U F
Sparrowia parasitica Willoughby F
CLADOCHYTRIACEAE
Cladochytrium aurantiacum M. Richards U
Cladochytrium caespitis Griffon & Maubl. U
Cladochytrium hyalinum Berdan U
Cladochytrium replicatum Karling U
Cladochytrium tenue Nowak. U F
Nowakowskiella atkinsii Sparrow U
Nowakowskiella crassa Karling U
Nowakowskiella elegans (Nowak.) J. Schröt. U
Nowakowskiella granulata Karling U
Nowakowskiella hemisphaerospora Shanor U
Nowakowskiella macrospora Karling U
Nowakowskiella multispora Karling U
Nowakowskiella profusa Karling U
Septochytrium variabile Berdan U
ENDOCHYTRIACEAE
Catenochytridium carolinianum Berdan U
Catenochytridium laterale A.M. Hanson U
Diplophlyctis chitinophila Willoughby U F
Diplophlyctis intestina (Schenk) J. Schröt. U F
Diplophlyctis nephrochytrioides Karling U F
Diplophlyctis sarcoptoides (H.E. Petersen) Dogma U F
Endochytrium operculatum (De Wild.) Karling U F
Entophlyctis confervae-glomeratae (Cienk.) Sparrow U F
Entophlyctis crenata Karling U F
Entophlyctis helioformis (P.A. Dang.) Ramsb. U F
Entophlyctis texana Karling U
Nephrochytrium appendiculatum Karling U F
SYNCHYTRIACEAE
Synchytrium aureum J. Schröt. A
Synchytrium australe Speg. A
Synchytrium cotulae du Plessis
Synchytrium endobioticum (Schilb.) Percival A
Synchytrium epilobii Karling
Synchytrium erieum Karling A
Synchytrium globosum J. Schröt.
Synchytrium hypochaeridis Karling A
Synchytrium leontodontis Karling A
Synchytrium limosellae Karling
Synchytrium macrosporum Karling U
Synchytrium melicopes Cooke & Massee E
Synchytrium papillatum Farl. A
Synchytrium taraxaci de Bary & Woronin A

Order RHIZOPHYDIALES
Batrachochytrium dendrobatidis Longcore, Pessier & D.K. Nichols U F
RHIZOPHYDIACEAE
Rhizophydium amoebae Karling U
Rhizophydium bullatum Sparrow U
Rhizophydium carpophilum (Zopf) A. Fisch. U
Rhizophydium chaetiferum Sparrow U F
Rhizophydium chitinophilum Antik. U
Rhizophydium chytriomycetis Karling U
Rhizophydium clavatum Karling U
Rhizophydium collapsum Karling U
Rhizophydium condylosum Karling U
Rhizophydium coronum A.M. Hanson U
Rhizophydium ellipsoideum Karling
Rhizophydium elyense Sparrow U
Rhizophydium gibbosum (Zopf) A. Fisch. U F
Rhizophydium globosum (A. Braun) Rabenh. U F
Rhizophydium keratinophilum Karling U
Rhizophydium macroporosum Karling E
Rhizophydium mycetophagum Karling U
Rhizophydium nodulosum Karling U
Rhizophydium novozeylandiense Karling E F
Rhizophydium obpyriforme Karling
Rhizophydium pollinis-pini (A. Braun) Zopf
Rhizophydium polystomum Karling E
Rhizophydium pythii De Wild. U
Rhizophydium racemosum A. Gaertn. U
Rhizophydium sphaerocarpum (Zopf) A. Fisch.
Rhizophydium sphaerotheca Zopf U
Rhizophydium stipitatum Sparrow U
Rhizophydium utriculare Uebelm. (nom. inv.) U

Order SPIZELLOMYCETALES
SPIZELLOMYCETACEAE
Karlingia chitinophila Karling U
Karlingia rosea (de Bary & Woronin) A.E. Johanson U F
Karlingia spinosa Karling U
Rhizophlyctis bonseyi Sparrow U
Rhizophlyctis fusca Karling U
Rhizophlyctis hirsuta Karling U
Rhizophlyctis ingoldii Sparrow U
Rhizophlyctis lovettii Karling U
Rhizophlyctis mastigotrichis (Nowak.) A. Fisch. U F
Rhizophlyctis oceanica Karling U M
Rhizophlyctis petersenii var. *appendiculata* Karling U
Rhizophlyctis variabilis Karling U

Class MONOBLEPHARIDOMYCETES
Order MONOBLEPHARIDALES
MONOBLEPHARIDACEAE
Monoblepharis thalassinosa M.K. Elias E M

PHYLUM NEOCALLIMASTIGOMYCOTA
Class NEOCALLIMASTIGOMYCETES
Order NEOCALLIMASTIGALES
NEOCALLIMASTIGACEAE
Caecomyces communis J.J. Gold, I.B. Heath & Bauchop U
Neocallimastix frontalis (R.A. Braune) I.B. Heath A
Piromyces communis J.J. Gold, I.B. Heath & Bauchop U
Piromyces dumbonicus J.L. Li U
Piromyces mae J.L. Li U

THIRTY

Phylum
MICROSPORIDIA
microsporidian fungi

LOUISE A. MALONE, TONY CHARLESTON

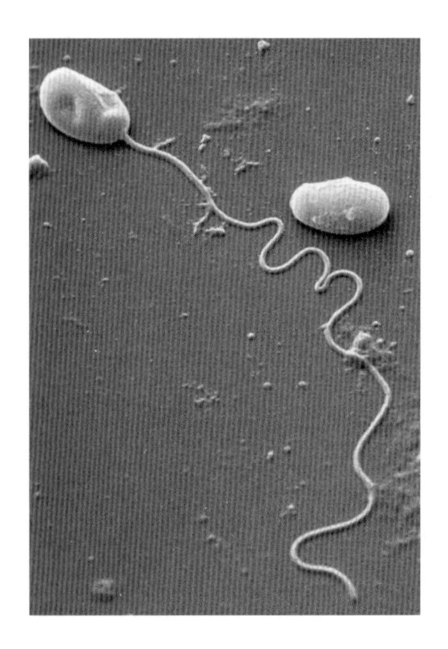

Spores of *Nosema tractabile*, a non-New Zealand species, one of them with an ejected polar tube (an infection tube) through which the cell content (sporoplasm) is discharged into a host.
Ronny Larsson, Lund University

Microsporidia are unicellular spore-forming parasites that used to be classified as protozoans but are now known to be highly reduced fungi (Cavalier-Smith (1998, 2001; Kirk et al. 2001). Some 1500 species have been discriminated worldwide but extrapolations of possible diversity based on potential hosts suggest there could be many hundreds of thousands of species. They are distinctive in lacking certain cell organelles (mitochondria and peroxisomes) typical of other eukaryotes and can be discriminated from all other fungi by the characteristic long, coiled, polar tube that forms within the spore and is everted on spore germination (Cavalier-Smith 1998). If the tube contacts a host cell and penetrates the membrane, the contents of the microsporidian cell can pass through into the host cell. Gene sequencing indicates that microsporidia evolved after the divergence of the chytrids, countering earlier interpretations of primitive eukaryote origins (Hirt et al. 1999; Keeling & McFadden 1998; Fast et al. 1999; Keeling et al. 2000, van de Peer 2000). Microsporidia have a very small genome, with DNA content similar to or smaller than that of the bacterium *Escherichia coli*.

Microsporidia are traditionally studied by zoologists because of their role as obligate, intracellular parasites of animals and, rarely, non-photosynthetic unicellular chromists (gregarines and ciliates). A wide range of invertebrate phyla host microsporidia (Cnidaria, Platyhelminthes, Dicyemida, Gnathifera, Mollusca, Bryozoa, Annelida, Arthropoda, Nematoda) and about ten per cent of microsporidian species infect vertebrates, including, opportunistically, immunocompromised humans (Keeling & McFadden 1998).

They are found in all environments and their spores are strongly adapted to the habitat of the host. For example, many species with aquatic hosts have spores with propeller-like wings, which cause the spore to spin into the filter-feeding mechanism of the host. In contrast, spores of those microsporidia with terrestrial hosts usually have a comparatively smooth exospore and a thick protective inner endospore (Barker 2004). Classification is challenging because many characters are cytological and difficult to study. The classification used here is conventional, based on that of Sprague (1977) using microscopic and life-history characters. Sprague et al. (1992) proposed a different classification system, based on whether a species is diplokaryotic (having twice the diploid number of chromosomes) at some point in its life cycle (Dihaplophasea) or is uninucleate throughout its life cycle (Haplophasea). Vossbrink et al. (2005) introduced yet another scheme based entirely on molecular analysis (specifically sequencing the small-subunit ribosomal DNA gene), which they found to coincide with habitat and host. Based on this finding they introduced three new classes – Aquasporidia, found mostly in freshwater hosts, Marinosporidia (mostly marine hosts), and Terresporidia (mostly terrestrial hosts). This scheme was critiqued by Larsson (2005), who noted that their analysis was based only on about ten per cent of known microsporidia, there were many exceptions to the putative primary habitat, and the new classes were not formally diagnosed, making them nomina nuda.

Ecological and economic importance

Arthropod hosts of microsporidia in New Zealand include insects such as grass grubs, porina, Argentine stem weevil, codling moth, and cabbage white butterfly (e.g. Malone 1990; Malone & McIvor 1996). *Nosema apis* and *N. bombi* infect honey bees and bumble bees, respectively (Malone et al. 1992; McIvor & Malone 1995), with *N. apis* causing substantial losses in honey production and pollination efficiency. Rose et al. (1999) listed two unidentified *Nosema* species in introduced vespid wasps in a paper on potential biological-control agents.

In lowland pastures of northern New Zealand, disease caused by *Microsporidium novacastriensis* (host-specific to the pestiferous introduced slug *Deroceras reticulatum*) was shown to be responsible for the density-dependent regulation in thise gastropod species (Barker 2004).

In the New Zealand marine environment, *Microsporidium rapuae* has been described from connective tissue surrounding the gut epithelium of the Bluff oyster *Ostrea chilensis* (Jones 1981). Its incidence in the native population is only about ten per cent and, except for encapsulation of the cysts by fibroblast cells of the oyster, no pathogenic (disease) effects were observed. *Octosporea* sp. has been found infecting the cranial nerves of the jack mackerel *Trachurus declivis* (Jones 1990; Hine et al. 2000).

New Zealand scampi, *Metanephrops challengeri*, has recently been discovered to host a new genus and species of microsporidian – *Myospora metanephrops*. Its likely impact on the fishery is not known but it causes lethargy in the scampi owing to infection of the musculature. Cytological and genetic analyses illustrate the difficulty of integrating these two sources of characters – whereas morphological features are consistent with members of the family Nosematidae, molecular data place the parasite closer to members of the family Thelohaniidae. The authors created a new family, Myosporidae, in the Vossbrinck et al. (2005) class Marinosporidia, for which they provided a brief diagnosis (Stentiford et al. 2010).

In the freshwater environment, a small population of the endemic crayfish *Paranephrops zealandicus* in the Leith Stream, Dunedin, was found to be infected by a microsporidian that caused muscle cells to atrophy, leading to premature death. Quilter (1976) attributed the microsporidian to the European species *Thelohania contejeani*, but after studying microsporidia in both *P. zealandicus* and *P. planifrons*, Jones (1979) concluded that a separate species of *Thelohania* parasitises each of these crayfish and that both differ from *T. contejeani*. Hine et al. (2000) reported an unidentified species attributed to *Microsporidium* sp. that affects gills of the redfin bully *Gobiomorphus huttoni*. *Microsporidium* sp. (unlikely to be the same species) was also noted in the same work to affect gills and fins of the yellowbelly flounder *Rhombosolea leporina*. Diggles (2003) reported xenomas (hypertrophic lesions) caused by a *Glugea*-like microsporidian in smelt (*Retropinna retropinna*) from the Waikato River.

Human microsporidiosis, a disease which is primarily seen in individuals infected with human immunodeficiency virus (HIV), manifests itself as an

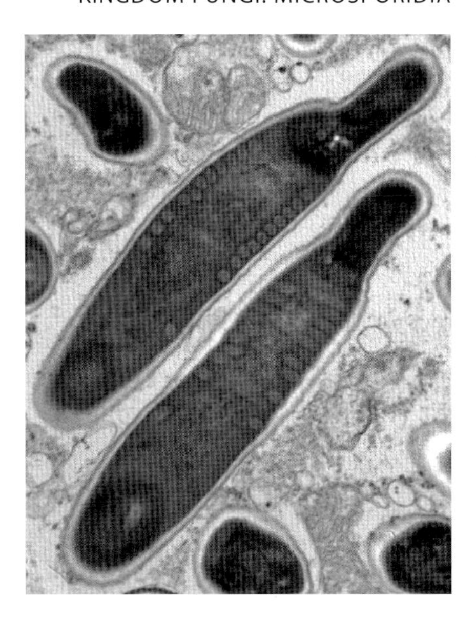

Transmission electron microscope image of mature spores of *Myospora metanephrops*. Sections of the coiled, undischarged polar tube can be seen adjacent to the wall of each cell.
Grant D. Stentiford, CEFAS, Weymouth Laboratory, UK

Summary of New Zealand microsporidian diversity

Taxon	Described species	Known undescribed/ undetermined species	Estimated undiscovered species	Adventive species	Endemic species	Endemic genera
Microsporidia	12	6	9,000	6	5	1

Summary of New Zealand Microsporidia by major environment of host

Taxon	Marine	Freshwater	Terrestrial
Microsporidia	3	3	12

infection of the intestine, lung, kidney, brain, sinuses, muscles, and eyes. Infections of people with normally functioning immune systems are generally asymptomatic or of minor significance and self-limiting. Didier et al. (2004) listed 14 named microsporidia recorded from humans. The most prevalent is *Enterocytozoon bieneusi*, which has a very wide host range in domestic and wild animals including cattle, sheep, pigs, dogs, cats, and chickens, and has been detected in natural water and a swimming pool. Other pathogens in humans include *Encephalitozoon cuniculi* and *E. intestinalis*. Hosts serve as reservoirs of further infection. Microsporidian spores are released from the faeces and urine of infected animals and can be consumed or inhaled by animals and humans. Once within a cell, the microsporidia develop and multiply, producing more spores. The infective spores are then released when the cell expands and bursts.

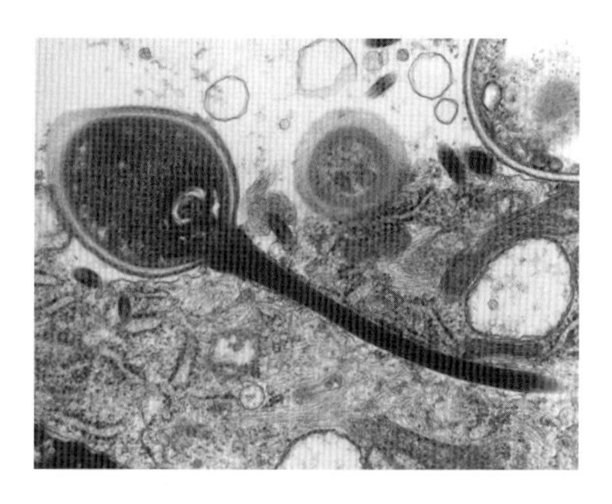

Transmission electron microscope image of a microsporidian spore with its extruded polar tube inserted into a host cell.

Massimo Scaglia, Institute of Infectious Diseases, University of Pavia-IRCC, Italy (per DPDx-DCD Image Library, Atlanta)

Very little is known about microsporidian infections of birds, mammals, or humans in New Zealand. *Enterocytozoon bieneusi* has been reported once in an HIV-positive patient (Everts et al. 1997) but there are no reports of its occurrence in domestic or wild animals. There is evidence that *Encephalitozoon cuniculi* is present in rabbits; this is based on observed histopathological changes and, in one instance, organisms staining with Goodpasture's stain, in the brains of rabbits that had shown neurological signs before death (Anon 1980; Smits 2001). There are, however, no published reports of the organisms being identified by other means or isolated, or of serological evidence of the infection in New Zealand rabbits.

Acknowledgements

The authors thank Shaun Pennycook and the late Ross Beever (Landcare Research, Auckland) for numerical data and editorial comment, respectively.

Authors

Dr Louise Malone New Zealand Insititute for Plant and Food Research, Private Bag 92169, Auckland, New Zealand [louise.malone@plantandfood.co.nz]

Dr W. A. G. (Tony) Charleston Institute of Veterinary, Animal & Biomedical Sciences, Massey University, Palmerston North (Present address: 488 College Street, Palmerston North, New Zealand) [charleston@inspire.net.nz]

References

ANON 1980: Ruakura Animal Health Laboratory Report. *Surveillance 7*(4): 9–16.

BARKER, G. M. (Ed.) 2004: *Natural Enemies of Terrestrial Molluscs*. CAB International, Wallingford, U. K. 644 p.

BULLA, L. A.; CHENG, T. C. 1977: *Comparative Pathobiology Volume 2. Systematics of the Microsporidia*. Plenum Press, New York & London. xi + 510 p.

CAVALIER-SMITH, T. 1998: A revised six-kingdom system of life. *Biological Reviews 73*: 203–266.

CAVALIER-SMITH, T. 2001: What are fungi? Pp. 3–38 *in*: Esser, K.; Lemke, P. A. (eds), *The Mycota VII. Systematics and Evolution Part A*. Springer-Verlag, Berlin. 366 p.

DIDIER, E. S.; STOVALL, M. E.; GREEN, L. C.; BRINDLEY, P. J.; SESTAK, K.; DIDIER, P. J. 2004: Epidemiology of microsporidiosis: sources and modes of transmission. *Veterinary Parasitology 126*: 145–166.

DIGGLES, B. K. 2003: Some pathological abnormalities of New Zealand fishes. *New Zealand Journal of Marine and Freshwater Research 37*: 705–713.

EVERTS, R.; CHAMBERS, S. T.; PALTRIDGE, G.; NEWHOOK, C. 1997: Microsporidiosis in New Zealand. *New Zealand Medical Journal 110*: 83.

FAST, N. M.; LOGSDON, J. M. Jr; DOOLITTLE, W. F. 1999: Phylogenetic analysis of the TATA box binding protein (TBP) gene from *Nosema locustae*: evidence for a microsporidia-fungi relationship and spliceosomal intron loss. *Molecular Biology and Evolution 16*: 1415–1419.

HINE, P. M.; JONES, J. B.; DIGGLES, B. K. 2000: A checklist of the parasites of New Zealand fishes, including previously unpublished records. *NIWA Technical Report 75*: 1–95.

HIRT, R. P.; LOGSDON, J. M.; HEALY, B.; DOREY, M. W.; DOOLITTLE, W. F.; EMBLEY, T. M. 1999: Microsporidia are related to fungi: Evidence from the largest subunit of RNA polymerase II and other proteins. *Proceedings of the National Academy of Science USA 96*: 580–585.

JONES, J. B. 1979: [Abstract] Trouble with microsporidia. *New Zealand Journal of Zoology 6*: 648.

JONES, J. B. 1981: A new *Microsporidium* from the oyster *Ostrea lutaria* in New Zealand. *Journal of Invertebrate Pathology 38*: 67–70.

JONES, J. B. 1990: Jack mackerels (*Trachurus* sp.) in New Zealand waters. *New Zealand Fisheries Technical Report 23*: 1–28.

KEELING, P. J.; MCFADDEN, G. I. 1998: Origins of microsporidia. *Trends in Microbiology 6*: 19–23.

KEELING, P. J., LUKER, M. A., PALMER, J. D. 2000: Evidence from beta-tubulin phylogeny that microsporidia evolved from within the fungi. *Molecular Biology and Evolution 17*: 23–31.

KIRK, P. M.; CANNON, P. F.; DAVID, J. C.; STALPERS, J. A. 2001: *Ainsworth & Bisby's Dictionary of the Fungi*. 9th Edn. Wallingford, CAB International. 655 p.

LARSSON, J. I. R. 2005: Molecular versus morphological approach to microsporidian classification. *Folia Parasitolica 52*: 143–144.

MALONE, L. A. 1990: Ultrastructure and tissue specificity of *Nosema costelytrae* and *Nosema takapauensis*, microsporidian pathogens of grass grubs, *Costelytra zealandica* (Coleoptera: Scarabaeidae) in New Zealand. *New Zealand Journal of Agricultural Research 33*: 459–465.

MALONE, L. A.; GIACON, H. A.; HUNAPO, R. J.; McIVOR, C .A. 1992: Response of New Zealand honey bee colonies to *Nosema apis*. *Journal of Agricultural Research 31*: 135–140.

MALONE, L. A.; McIVOR, C. A. 1996: Use of nucleotide sequence data to identify a microsporidian pathogen of *Pieris rapae* (Lepidoptera, Pieridae). *Journal of Invertebrate Pathology 68*: 231–238.

McIVOR, C. A.; MALONE, L. A. 1995: *Nosema bombi*, a microsporidian pathogen of the bumble bee *Bombus terrestris* (L.). *New Zealand Journal of Zoology 22*: 25–31.

QUILTER, C. G. 1976: Microsporidan parasite *Thelohania contejeani* Henneguy from New Zealand freshwater crayfish. *New Zealand Journal of Marine and Freshwater Research 10*: 225–231.

ROSE, E. A. F.; HARRIS, R. J.; GLARE, T. R. 1999: Possible pathogens of social wasps (Hymenoptera: Vespidae) and their potential as biological control agents. *New Zealand Journal of Zoology 26*: 179–190.

SMITS, B. 2001: Quarterly review of diagnostic cases – April to June 2001: Alpha Scientific Ltd. *Surveillance 28*(3): 21.

SPRAGUE, V. 1977: Classification and phylogeny of the microsporidia. Pp. 1–30 *in:* Bulla, L. A.; Cheng, T. C. (eds), *Comparative Pathobiology Volume 2. Systematics of the Microsporidia*. Plenum Press, New York & London. xi + 510 p.

SPRAGUE, V.; BECNEL, J. J.; HAZARD, E. I. 1992: Taxonomy of phylum Microspora. *Critical Review of Microbiology 18*: 285–395.

STENTIFORD, G. D.; BATEMAN, K. S.; SMALL, H. J.; MOSS, J.; SHIELDS, J. D.; REECE, K. S.; TUCK. I. 2010: *Myospora metanephrops* (n. g., n. sp.) from marine lobsters and a proposal for erection of a new order and family (Crustaceacida; Myosporidae) in the Class Marinosporidia (Phylum Microsporidia). *International Journal of Parasitology 40*: 1433–1446.

VAN DE PEER, Y.; BEN ALI, A.; MEYER, A. 2000: Microsporidia: accumulating molecular evidence that a group of amitochondriate and suspectedly primitive eukaryotes are just curious fungi. *Gene 246*: 1–8.

VOSSBRINCK, C. R.; DEBRUNNER-VOSSBRINCK, B. A. 2005. Molecular phylogeny of the Microsporidia: ecological, ultrastructural and taxonomic considerations. *Folia Parasitologica 52*: 131–142.

Checklist of New Zealand Microsporidia

Classification is based on that given by Sprague in Bulla and Cheng (1977). Major environments are denoted by: F, freshwater; M, marine; T, terrestrial; F/T denotes a species that lives in a moist or wet terrestrial setting. Host: B, bivalve; C, crustacean; G, gastropod; H, hexapod; V, vertebrate. Species status: A, adventive (naturalized alien); E, endemic; species not indicated by either notation are native (naturally indigenous but not endemic). Endemic genera are underlined.

PHYLUM MICROSPORIDIA
Class MICROSPOREA
Order MICROSPORIDA
Suborder PANSPOROBLASTINA
BURENELLIDAE
Vairimorpha mesnili (A.Paill.) Malone & C.A.McIvor H T
GLUGEIDAE
?*Glugea* sp. Diggles 2003 V F
Vavraia oncoperae (Milner & C.D.Beaton) Malone, Wigley & Dhana H T
THELOHANIIDAE
Thelohania sp. 1 Jones 1979 C F
Thelohania sp. 2 Jones 1979 C F

Suborder APANSPOROBLASTINA
PSEUDOPLEISTOPHORIDAE
Octosporea sp. V M
ENTEROCYTOZOONIDAE
Enterocytozoon bieneusi Desportes, Le Charpentier, Galian, Bernard, Cochand-Priollet, Lavergne, Ravisse & Modigliani V T A
NOSEMATIDAE
<u>*Myospora*</u> *metanephrops* Stentiford, Bateman, Small, Moss, Shields, Reece & Tuck C M E
Nosema apis E.Zander H T A
Nosema bombi Fantham & A.Porter H T A
Nosema carpocapsae A.Paill. H T
Nosema costelytrae I.M.Hall, E.H.A.Oliv. & B.B.Given H T E
Nosema takapauensis I.M.Hall, E.H.A.Oliv. & B.B.Given H T E
Nosema spp. indet. (2) P. Wigley & P. Scotti in Rose et al. 1999 H T 2A
INCERTAE SEDIS
Microsporidium itiiti Malone H T E
Microsporidium novacastriensis Jones & Selman G G T A
Microsporidium rapuae Jones B M E

THIRTY-ONE

Phyla
ZYGOMYCOTA and GLOMEROMYCOTA
pin moulds, arbuscular mycorrhizal fungi

PETER K. BUCHANAN

Pilobolus kleinii (Mucoromycotina: Mucorales), a common fungus on herbivore dung.
George L. Barron, University of Guelph

Zygomycota *sensu lato* is an ecologically diverse group of fungi (James & O'Donnell 2007), most familiar to people as the pin moulds that form on stale bread and rotting fruit and vegetables. 'Pin mould' alludes to the tall sporangiophores, each tipped with a pinhead-like sporangium, that form a mat of filaments above the host surface. Collectively, however, the group includes not only saprobes but also insect parasites and soil-inhabiting mycorrhizal forms, the latter now separated out as a distinct phylum, Glomeromycota (Schüßler et al. 2001; Redecker 2008).

Worldwide, some 1065 species of Zygomycota *sensu stricto* have been described (Kirk et al. 2008), including 97 species known from New Zealand. Considerable changes in classification have resulted from recent phylogenetic studies, with the monophyletic status of phylum Zygomycota now in question (James & O'Donnell 2007). Four subphyla are accepted by Kirk et al. (2008): Entomophthoromycotina, Kickxellomycotina, Mucoromycotina, and Zoopagomycotina. Mostly terrestrial, Zygomycota are characterised by a sexual cycle in which morphologically similar gametangia fuse (conjugate) to form a zygosporangium. In nature, however, the anamorphic (asexual) phase dominates.

Many members of the Entomophthoromycotina and Kickxellomycotina live in a commensal relationship within the digestive tract of arthropods. The only detailed published study of New Zealand arthropod-inhabiting species (earlier known as trichomycetes) documented 16 species in eight genera from aquatic insect larvae, including seven new species (Williams & Lichtwardt 1990). With the trichomycetes recognised as an artificial polyphyletic group, four of those New Zealand species are now placed in phylum Choanozoa (Protozoa) while the remainder are in Kickxellomycotina. About half of the New Zealand species were recognised to be geographically widely distributed (e.g. species of *Smittium*), living within the guts of endemic insects of several families. All of the newly described species are more restricted in distribution and are thought to be endemic. As yet undescribed are four additional new species of *Plecopteromyces* that were found in the South Island (Ferrington et al. 2005). Other members of the Entomophthoromycotina are insect pathogens (see section on Entomogenous fungi on page 503).

Commonly encountered mucoraceous fungi (Mucoromycotina) include fast-growing species, such as the bread mould *Rhizopus stolonifer*, that also live on fruit, soil, and dung. As decomposers in soil and dung, mucoraceous fungi have a significant role in the carbon cycle (James & O'Donnell 2007). The group has not been studied actively in New Zealand, although several species have been recorded from native soils (e.g. Thornton 1958; Domsch et al. 1980; Mahoney et al. 2004). *Mortierella* and *Mucor* species are common and widespread, although the latter occur mainly on organic matter such as dung. Bell (1975, 1983) reported several mucoraceous zygomycetes as being among the first fungi to fruit on dung of possums and other herbivores in New Zealand. Sporangiophores of the atypical *Pilobolus crystallinus* are large (two centimetres or more tall) and phototropic; the fungus is known as the 'cap thrower' because the sporangial head is explosively shot from the sporangiophore to a considerable distance following the rupture of the supporting vesicle. Some Mucoromycotina and Zoopagomycotina are mycoparasitic on similar other fungi. For example, *Piptocephalis microcephala* (Zoopagomycotina) parasitises *Mucor mucedo* on dung (Bell 1975). Mucoraceous species have also been isolated in association with animal and human diseases, while others cause plant diseases (see the sections on Plant-parasitic fungi and Fungi affecting human and animal health, respectively, on pages 501 and 502). Members of *Endogone* (Mucoromycotina) include species capable of forming ectomycorrhizal relationships with roots of gymnosperms, while others are saprobic (James & O'Donnell 2007).

The Glomeromycota, though largely unseen, are important globally to the nutrition of the majority of plant species, including most crop plants, because of their mutualistic symbioses with plant roots. Members of this phylum form arbuscular mycorrhizas (AM) with roots of most herbaceous plants and many tree species. In New Zealand, apart from *Nothofagus*, *Kunzea*, and *Leptospermum* that have ectomycorrhizal fungal associates (mostly Basidiomycota), most other native tree species are associated with Glomeromycota. They are thus common soil fungi, functioning as obligate mycorrhizal associates (see also Mycorrhizal fungi on page 505). The importance of AM fungi in New Zealand was championed by Prof. G. T. S. Baylis (Otago) (e.g. Baylis 1978). New Zealand members of Glomeromycota were surveyed by Hall (1977, as Endogonales) who described five new species. Johnson (1977) studied host relationships and soil distribution of Glomeromycota. Six genera in Glomeromycota are known from New Zealand – *Acaulospora*, *Entrophospora*, *Glomus*, *Gigaspora*, *Sclerocystis*, and *Scutellospora*, of which *Glomus* is the most diverse. Host specificity of AM fungi is typically low, and plants are often colonised by several AM species (Redecker 2008). Aquatic plants are also hosts to AM fungi; Clayton & Bagyaraj (1984) found AM fungi associated with several submerged water-plant species in many New Zealand lakes.

The arbuscule of AM fungi refers to the tree-like fungal structure that can largely fill a host root cell without breaking plant cell walls or membranes. The large surface area of the arbuscule facilitates movement of water, minerals, and carbohydrates between fungus and plant. Large asexual spores with layered walls and containing hundreds to thousands of nuclei are produced by these fungi but no species is known to reproduce sexually (Redecker 2008).

AM fungi may also provide underground pathways between plants. Kelly (1994) suspected that the unusual New Zealand fern, *Botrychium australe*, might be mycorrhizal, since it has only one frond above ground and the thick fleshy roots are hairless. Subsequent studies of North American species of *Botrychium* (Winther & Friedman 2007) confirmed that these ferns and neighbouring

Phycomyces nitens on dung.

Jerry A. Cooper, Landcare Research

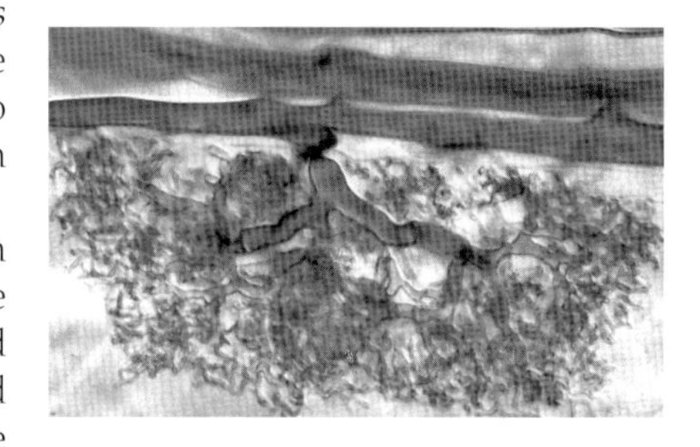

The large surface area of the tree-like 'arbuscule' of an arbuscular mycorrhizal fungus allows movement of minerals and water from the fungus to the root and of carbohydrates from the root to the fungus. The arbuscule rests against the intact invaginated cell membrane.

Mark Brundrett per George L. Barron, University of Guelph

Summary of diversity of New Zealand Zygomycota and Glomeromycota

Taxon	Described species	Known undescribed/ undetermined species	Estimated undiscovered species	Adventive species	Endemic species	Endemic genera
Zygomyota	89	0	170	19	10	0
Glomeromycota	38	1	5	2	0	0
Totals	**127**	**1**	**175**	**21**	**10**	**0**

Summary of diversity by major environment

Taxon	Marine	Fresh-water	Terrestrial
Zygomycota	0	17	72
Glomeromycota	0	0	39
Totals	0	17	111

photosynthetic plants share an association with AM *Glomus* species. The fungus can thus provide a pathway for carbohydrates from photosynthetic plants to the mycoheterotrophic *Botrychium*. This relationship, however, has yet to be confirmed in New Zealand.

Acknowledgements

I thank Shaun Pennycook and Jerry Cooper (Landcare Research) for preparation of the checklist.

Author

Dr Peter K. Buchanan Landcare Research, Private Bag 92170, Auckland, New Zealand [buchananp@LandcareResearch.co.nz]

References

BAYLIS, G. T. S. 1978: Endomycorrhizas in our native flora. *Journal of the Royal New Zealand Institute of Horticulture 6*: 63–69.

BELL, A. 1975: Fungal succession on dung of the brush-tailed opossum in New Zealand. *New Zealand Journal of Botany 13*: 437–462.

BELL, A. 1983: *Dung Fungi, an Illustrated Guide to Coprophilous Fungi in New Zealand*. Victoria University Press, Wellington. 88 p.

CLAYTON, J. S.; BAGYARAJ, D. J. 1984: Vesicular-arbuscular mycorrhizas in submerged aquatic plants of New Zealand. *Aquatic Botany 19*: 251–262.

DOMSCH, K. H.; GAMS, W.; ANDERSON, T.-H. 1980: *Compendium of Soil Fungi*. Academic Press, London. Vol. 1, viii + 860 p.; Vol. 2, vi + 406 p.

FERRINGTON, L. C.; LICHTWARDT, R. W.; HAYFORD, B.; WILLIAMS, M. C. 2005: Symbiotic Harpellales (Trichomycetes) in Tasmanian aquatic insects. *Mycologia 97*: 254–262.

HALL, I. R. 1977: Species and mycorrhizal infections of New Zealand Endogonaceae. *Transactions of the British Mycological Society 68*: 341–356.

JAMES, T. Y.; O'DONNELL, K. 2007. Zygomycota. Microscopic 'pin' or 'sugar' molds. Version 13 July 2007 (under construction). http://tolweb.org/Zygomycota/20518/2007.07.13 in The Tree of Life Web Project, http://tolweb.org/

JOHNSON, P. N. 1977: Mycorrhizal Endogonaceae in a New Zealand forest. *New Phytologist 78*: 161–170.

KELLY, D. 1994: Demography and conservation of *Botrychium australe*, a peculiar, sparse mycorrhizal fern. *New Zealand Journal of Botany 32*: 393–400.

KIRK, P. M.; CANNON, P. F.; MINTER, D. W.; STALPERS, J. A. 2008: *Ainsworth & Bisby's Dictionary of the Fungi*. 10th Edn. Wallingford, CAB International. 771 p.

MAHONEY, D.; GAMS, W.; MEYER, W.; STARINK-WILLEMSE, M. 2004: *Umbelopsis dimorpha* sp. nov., a link between *U. vinacea* and *U. versiformis*. *Mycological Research 108*: 107–111.

REDECKER, D. 2008: Glomeromycota. Arbuscular mycorrhizal fungi and their relative(s). Version 14 January 2008. http://tolweb.org/Glomeromycota/28715/2008.01.14 in The Tree of Life Web Project, http://tolweb.org/

SCHÜßLER, A.; SCHWARZOTT, D.; WALKER. C. 2001: A new fungal phylum, the Glomeromycota: phylogeny and evolution. *Mycological Research 105*: 1413–1421.

THORNTON, R. H. 1958: Biological studies of some tussock-grassland soils. 2. Fungi. *New Zealand Journal of Agricultural Research 1*: 922–938.

WILLIAMS, M. C.; LICHTWARDT, R. W. 1990: Trichomycete gut fungi in New Zealand aquatic insect larvae. *Canadian Journal of Botany 68*: 1045–1056.

WINTHER, J. L.; FRIEDMAN, W. E. 2007: Arbuscular mycorrhizal symbionts in *Botrychium* (Ophioglossaceae). *American Journal of Botany 94*: 1248–1255.

Checklist of New Zealand Zygomycota and Glomeromycota

This checklist is arranged according to the taxonomic hierarchy in Kirk et al. (2008). Species status: A, adventive (naturalised alien); E, endemic; F, freshwater (all other species are terrestrial); U, status unknown; UO, occurrence uncertain in New Zealand; species not indicated by any notation are native (naturally indigenous but not endemic).

KINGDOM FUNGI
PHYLUM ZYGOMYCOTA
SUBPHYLUM ENTOMOPHTHOROMYCOTINA
Order ENTOMOPHTHORALES
ANCYLISTACEAE
Conidiobolus apiculatus (Thaxt.) Remaud. & S. Keller U
Conidiobolus obscurus (I.M. Hall & P.H. Dunn) Remaud. & S. Keller A
ENTOMOPHTHORACEAE
Entomophthora aphidis Hoffm. U
Entomophthora conica Nowak. U F
Entomophthora muscae (Cohn) G. Winter U
Entomophthora planchoniana Cornu U
Entomophthora rhizospora (Thaxt.) Sacc. F
Entomophthora sphaerosperma Fresen. OU
Massospora cicadina Peck
Pandora neoaphidis (Remaud. & Hennebert) Humber A
Zoophthora phalloides A. Batko A
Zoophthora radicans (Bref.) A. Batko U
NEOZYGITACEAE
Neozygites fumosus (Speare) Remaud. & S. Keller (nom. inv.) U
Neozygites parvispora (D.M. MacLeod & K.P. Carl) Remaud. & S. Keller U

SUBPHYLUM KICKXELLOMYCOTINA
Order HARPELLALES
HARPELLACEAE
Harpella melusinae L. Léger & Duboscq F
Stachylina grandispora Lichtw. F
Stachylina minima M.C. Williams & Lichtw. E F
Stachylina nana Lichtw. F
LEGERIOMYCETACEAE
Austrosmittium kiwiorum M.C. Williams & Lichtw. (nom. inv.) E F
Austrosmittium norinsulare Lichtw. E F
Glotzia plecopterorum Lichtw. E F
Pennella asymmetrica M.C. Williams & Lichtw. E F
Smittium bullatum Lichtw. & M.C. Williams E F
Smittium culicis Tuzet & Manier ex Kobayasi F
Smittium culisetae Lichtw. F
Smittium elongatum Lichtw. F
Smittium rarum Lichtw. (nom. inv.) E F
Smittium simulii Lichtw. F

SUBPHYLUM MUCOROMYCOTINA
Order ENDOGONALES
ENDOGONACEAE
Endogone aggregata P.A. Tandy
Endogone flammicorona Trappe & Gerd. A
Endogone lactiflua Berk. & Broome A
Endogone verrucosa Gerd. & Trappe A

Order MORTIERELLALES
MORTIERELLACEAE
Aquamortierella elegans Embree & Indoh E F
Modicella malleola (Harkn.) Gerd. & Trappe U
Mortierella alpina Peyronel
Mortierella ambigua B.S. Mehrotra
Mortierella elongata Linnem.
Mortierella exigua Linnem.
Mortierella globulifera O. Rostr.
Mortierella horticola Linnem.
Mortierella humicola Oudem.
Mortierella humilis Linnem. ex W. Gams
Mortierella hyalina (Harz) W. Gams
Mortierella longicollis Dixon-Stew.
Mortierella minutissima Tiegh.
Mortierella mutabilis Linnem.
Mortierella parvispora Linnem.
Mortierella polycephala Coem.
Mortierella stylospora Dixon-Stew.
Mortierella traversiana Peyronel
Mortierella wolfii B.S. Mehrotra & Baijal
Mortierella zychae Linnem.

Order MUCORALES
CUNNINGHAMELLACEAE
Cunninghamella echinulata (Thaxt.) Thaxt. ex Blakeslee A
Cunninghamella elegans Lendn. U
Gongronella butleri (Lendn.) Peyronel & Dal Vesco A
MUCORACEAE
Absidia corymbifera (Cohn) Sacc. & Trotter U
Absidia glauca Hagem U
Absidia repens Tiegh. U
Absidia spinosa Lendn. U
Backusella lamprospora (Lendn.) Benny & R.K. Benj. U
Mucor circinelloides Tiegh.
Mucor hiemalis Wehmer
Mucor mucedo Fresen. A
Mucor piriformis A. Fisch. U
Mucor plumbeus Bonord. U
Mucor racemosus f. *sphaerosporus* (Hagem) Schipper (nom. inv.) U
Mucor strictus Hagem U
Pilaira anomala (Ces.) J. Schröt. U
Rhizomucor pusillus (Lindt) Schipper U
Rhizopus microsporus var. *rhizopodiformis* (Cohn) Schipper & Stalpers U
Rhizopus stolonifer (Ehrenb.) Vuill. A
Zygorhynchus exponens Burgeff
Zygorhynchus moelleri Vuill.
PHYCOMYCETACEAE
Phycomyces blakesleeanus Burgeff A
Phycomyces nitens (C. Agardh) Kunze A
PILOBOLACEAE
Pilobolus crystallinus (F.H. Wigg.) Tode A
Pilobolus kleinii Tiegh. A
Sacidium ixerbae Cooke E
SYNCEPHALASTRACEAE
Syncephalastrum racemosum Cohn ex J. Schröt.
Thamnostylum piriforme (Bainier) Arx & H.P. Upadhyay A
UMBELOPSIDACEAE
Umbelopsis dimorpha Mahoney & W. Gams E
Umbelopsis isabellina (Oudem.) W. Gams
Umbelopsis nana (Linnem.) Arx
Umbelopsis ramanniana (Möller) W. Gams

SUBPHYLUM ZOOPAGOMYCOTINA
Order ZOOPAGALES
PIPTOCEPHALIDACEAE
Piptocephalis lepidula (Marchal) R.K. Benj. A
Piptocephalis microcephala Tiegh. A
ZOOPAGACEAE
Acaulopage pectospora Drechsler A
Stylopage grandis Dudd. A
Zoophagus insidians Sommerst.

PHYLUM GLOMEROMYCOTA
Class GLOMEROMYCETES
Order ARCHAEOSPORALES
ARCHAEOSPORACEAE
Archaeospora gerdemannii (S.L. Rose, B.A. Daniels & Trappe) J.B. Morton & D. Redecker OU
Archaeospora leptoticha (N.C. Schenck & G.S. Sm.) J.B. Morton & D. Redecker OU
Archaeospora trappei (R.N. Ames & Linderman) J.B. Morton & D. Redecker OU

Order DIVERSISPORALES
ACAULOSPORACEAE
Acaulospora dilatata J.B. Morton U
Acaulospora excavata Ingleby & C. Walker U
Acaulospora lacunosa J.B. Morton U
Acaulospora laevis Gerd. & Trappe U
Entrophospora infrequens (I.R. Hall) R.N. Ames & R.W. Schneid. U
GIGASPORACEAE
Gigaspora gigantea (T.H. Nicolson & Gerd.) Gerd. & Trappe U
Gigaspora margarita W.N. Becker & I.R. Hall A
Scutellospora aurigloba (I.R. Hall) C. Walker & F.E. Sanders

Order GLOMERALES
GLOMERACEAE
Glomus aggregatum N.C. Schenck & G.S. Sm. U
Glomus australe (Berk.) S.M. Berch U
Glomus caledonium (T.H. Nicolson & Gerd.) Trappe & Gerd. U
Glomus claroideum N.C. Schenck & G.S. Sm. U
Glomus convolutum Gerd. & Trappe U
Glomus coremioides OU
Glomus diaphanum J.B. Morton & C. Walker A
Glomus fasciculatum (Thaxt.) Gerd. & Trappe U
Glomus fuegianum (Speg.) Trappe & Gerd. U
Glomus geosporum (T.H. Nicolson & Gerd.) C. Walker U
Glomus intraradices N.C. Schenck & G.S. Sm. OU
Glomus invermaium I.R. Hall U
Glomus macrocarpum Tul. & C. Tul. U
Glomus magnicaule I.R. Hall U
Glomus microcarpum Tul. & C. Tul. U
Glomus monosporum Gerd. & Trappe U
Glomus mosseae (T.H. Nicolson & Gerd.) Gerd. & Trappe U
Glomus pallidum I.R. Hall U
Glomus proliferum OU
Glomus pulvinatum (Henn.) Trappe & Gerd. U
Glomus rubiforme (Gerd. & Trappe) R.T. Almeida & N.C. Schenck
Glomus sinuosum (Gerd. & B.K. Bakshi) R.T. Almeida & N.C. Schenck U
Glomus sp. W3347 (nom. ined.) OU
Glomus tenue (Greenall) I.R. Hall U
Glomus vesiculiferum (Thaxt.) Gerd. & Trappe U
Sclerocystis coremioides Berk. & Broome
Sclerocystis dussii (Pat.) Höhn.

THIRTY-TWO

Phylum
ASCOMYCOTA

yeasts, sac fungi, truffles, and kin

ROSS E. BEEVER, MARGARET DI MENNA, PETER R. JOHNSTON, SHAUN R. PENNYCOOK, JERRY A. COOPER, AARON D. WILTON

Aleuria aurantia (in the broad sense) (Pezizomycetes: Pezizales), a bright orange cup fungus of disturbed ground.
Robyn Simcock, Landcare research

Ascomycota is the largest phylum in the fungal kingdom. Approximately half of the described species of fungi in the world are ascomycetes (64,056 species; Kirk et al. 2008) as are about two-thirds (5317 species) of New Zealand's fungi, including lichenised (almost 2000 species) and non-lichenised taxa. This known diversity represents perhaps one third of the species of ascomycetes estimated to occur in New Zealand.

Ascomycetes are characterised by their sexual spores being produced in a sac-like structure, the ascus. Each ascus typically contains eight spores, the result of one meiotic and one mitotic cell division. Many ascomycetes also form asexual spores at some stage of their lifecycle. The term teleomorph is used for the sexual spore state, and anamorph for the asexual state. Some fungi produce both sexual and asexual spores, the two spore states often having separate names, but many fungi (around 1465 species in New Zealand) are known from only the asexual state. For fungi where both anamorphic and teleomorphic states have been named, the checklist provides the teleomorph binominal for both states collectively. Classification of ascomycetes above the level of genus has traditionally been based on features of the teleomorph. In other recent compilations and checklists of New Zealand's fungi (e.g. McKenzie et al. 2000; Pennycook & Galloway 2004) anamorphic fungi have been treated separately in two informal form-taxa, the coelomycetes (for fungi with asexual spores forming within a defined fruiting body) and the hyphomycetes (for fungi with asexual spores forming directly on hyphae, with no fruiting body). In this checklist they are incorporated into the Ascomycota hierarchy following Kirk et al. (2008), the links between species forming only asexual spores and their sexual relatives often based on molecular studies. A recent change in the rules relating to the naming of fungi (International Code of Nomenclature for Algae, Fungi and Plants), will be the discontinuation of dual nomenclature for fungi with morphologically distinct asexual and sexual states (Hawksworth 2011). The incorporation of the names of asexual states into a phylogeny-based higher classification, as adopted in this list, is a step in this direction. Many of the generic names that have been applied to asexual fungi, especially those with a very simple morphology, are being shown with molecular studies to be polyphyletic. It is inevitable that there will be many changes to the names and the classification of fungi known only from asexual forms through increasing knowledge of their phylogenetic placement based on the availability of species-level DNA sequence data.

Exotic species make up about half of the non-lichenised ascomycete diversity of New Zealand. Many of these species are pathogens of cultivated crops, their economic importance meaning they have been studied more intensively than the indigenous ascomycetes. Included among the exotic ascomycetes are some choice edible species such as the truffles (*Tuber* spp.), introduced deliberately to establish a new industry (Hall et al. 1998), and the morels (*Morchella* spp.), self-introduced perhaps with contaminated soil early in New Zealand's colonial history. As New Zealand's indigenous fungal diversity becomes better catalogued (Johnston 2005) the proportion represented by exotic species is likely to decrease.

Some of New Zealand's ascomycete fungi have been quite intensively studied, such as the Hypocreales (e.g. Dingley 1951–1953), Capnodiaceae (e.g. Hughes 1966, 1970), *Rosellinia* (Petrini 2003), *Xylaria* (Rogers & Samuels 1987), Rhytismatales (Johnston 1992), Pezizales (Rifai 1968), and some inoperculate discomycete families (Spooner 1987). Most, however, have never been targeted for collecting in any systematic way. Many genera and families are represented in the New Zealand Fungal and Plant Disease Collection (PDD) by only one or two specimens, and many others almost certainly present in New Zealand are not represented at all. Of those taxa in the PDD collection, most have not been critically examined below the level of family or genus, and for those that have been more extensively studied, there are still major gaps in our basic knowledge. For example, the Sclerotiniaceae, a discomycete family included in the study of Spooner (1987), has collections in PDD representing at least ten undescribed species. Many of these discomycetes are comparatively large, conspicuous and common. One major gap has been in the study of New Zealand's bitunicate ascomycetes (Dothidiomycetes). Although many introduced bitunicate species have been reported as parasites of introduced crop plants (Pennycook 1989), there have been few reports of species in native ecosystems, one rare example being Ridley (1988). Despite this, they are common in native forests. For example, there are PDD specimens representing at least 10 undescribed *Mycosphaerella* species, all associated with distinctive foliar diseases of native plants.

The ascomycetes are extremely diverse, both genetically and ecologically. Most of the lichenised fungi and the yeasts are ascomycetes. The end-chapter checklist integrates non-lichenised and lichenised species of fungi in a uniform taxonomic arrangement but as they are unique life forms, lichens are treated

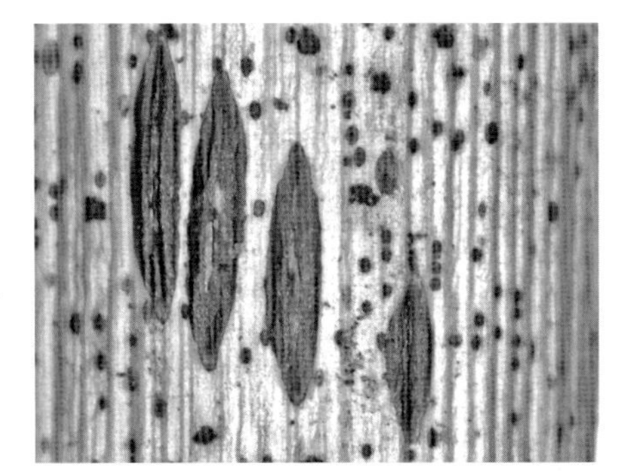

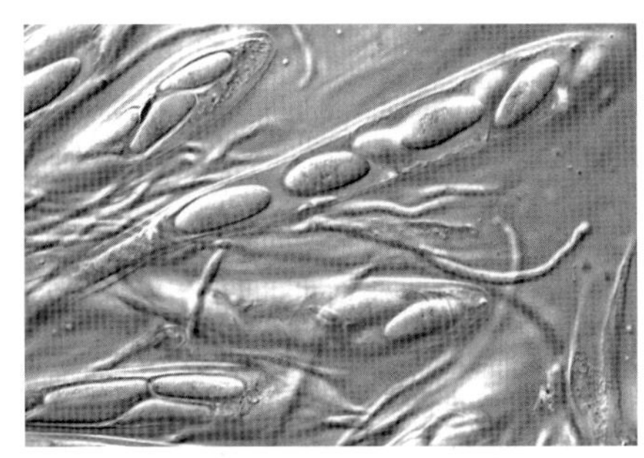

Hypoderma cordylines (Leiotiomycetes: Rhytismatales), showing (upper photo) elongate ascocarps (fruiting structures) that open by a narrow central slit and (lower photo) the sac-like asci, each of which has eight ascospores).

Peter Johnston, Landcare Research

Summary of New Zealand Ascomycota diversity

Taxon	Described species + additional subspecific taxa	Known undescribed/ undetermined species	Estimated undiscovered species	Adventive species + subspecies	Endemic species + subspecies	Endemic genera	Lichenised taxa
PEZIZOMYCOTINA	5,153+99	79	8,950	1,453+47	780+6	13	1,648
Incertae sedis	2,96+1	7	2,500	61	65	2	1
Arthoniomycetes	102	1	100	0	34	0	102
Dothideomycetes	1,155+11	10	2,500	660+7	157+2	1	49
Eurotiomycetes	287+6	2	500	51+2	16	0	106
Laboulbeniomycetes	33	4	200	0	25	1	0
Lecanoromycetes	1,406+28	3	100	2	162+2	1	1,379
Leotiomycetes	5,09+9	24	500	177+2	138+1	5	0
Lichinomycetes	11	0	10	0	0	0	11
Orbiliomycetes	22	0	20	8	1	0	0
Pezizomycetes	167	1	20	78	19	1	0
Sordariomycetes	1,165+44	27	2,500	416+36	163+1	2	0
SACCHAROMYCOTINA	77	0	100	26	0	0	0
Saccharomycetes	77	0	100	26	0	0	0
TAPHRINOMYCOTINA	8	0	15	6	0	0	0
Pneumocystidomycetes	1	0	5	1	0	0	0
Schizosaccharomycetes	1	0	5	0	0	0	0
Taphrinomycetes	6	0	5	5	0	0	0
Totals	5,238+99	79	9,065	1,485+37	780+6	13	1,648

Cyttaria gunnii (Leiotiomycetes: Cyttariales), a native pathogen of silver beech (*Nothofagus menziesii*). The fungus induces the tree branches to develop galls, from which yellow, strawberry-like fruiting bodies form in spring.
Peter Johnston, Landcare Research

separately, in chapter 34. The group also contains many plant parasites that attack leaves, wood, and roots, specialised grass parasites and endophytes (e.g. Clavicipitales), mycorrhizal symbionts (e.g. Epacridoideae (Ericaceae)-inhabiting Helotiales), leaf endophytes of trees, white rot and other nutrient-recycling fungi (e.g. Xylariaceae), sap-stain fungi (e.g. *Ophiostoma*), sooty moulds (e.g. Euantennariaceae and Metacapnodiaceae), and others. Those inhabiting litter and fallen wood in forests provide a food source for many invertebrates while sooty-mould fungi grow on sugary excrements of scale insects (e.g. Hughes 1976; Hughes & Pirozynski 1994). For a group of specialised insect ectoparasites, the Laboulbeniales, the combination of host-specific relationships and high levels of endemism among the entomological fauna of New Zealand suggest an estimated diversity of about 200 New Zealand species in this order (e.g. Rossi & Weir 1997; 1998; Weir 2001; Hughes et al. 2004a,b). Despite sketchy knowledge about the diversity of fungi in most of these ecological groups, the role of the ascomycetes in ecosystem processes is certain to be great.

The phylogenetic relationships of New Zealand's ascomycetes are just starting to be explored. Many indigenous morphospecies are shared with other parts of the world, especially Australia and tropical Asia. The genetic relationship between these respective populations is unknown, but species' distributional patterns suggest that it is likely to be close. Some of these geographically separated species may even form single interbreeding populations. Research on a group of plant parasites of horticultural importance, *Glomerella* and its asexual state *Colletotrichum*, suggests that genetically and biologically specialised populations of ascomycetes can evolve rapidly under New Zealand conditions (Lardner et al. 1999). Human modifications to New Zealand's environment, with the introduction of new hosts and new fungi, appear to have provided conditions ideal for rapid evolutionary change (Johnston 2010). Whether New Zealand has any truly 'ancient' ascomycetes remains debatable. For example, the beech strawberry *Cyttaria*, species of which are found on southern beech (*Nothofagus*) species across the Southern Hemisphere, were in the past thought to have been in New Zealand since before the break-up of Gondwana. Recent studies, however, have shown that both *Cyttaria* and the *Nothofagus* species that it attacks have arrived in New Zealand through transoceanic dispersal from Australia (Knapp et al. 2005; Peterson et al. 2010). Other possible ancient ascomycetes could include *Corynelia*, a parasite of *Podocarpus*, species of which are found throughout the range of its host, and some *Nothofagus*-associated *Torrendiella* spp. (Johnston 2010). Studies on the genetic relationships between the New Zealand species in these potentially ancient genera and those found elsewhere, and the closely related populations of apparently geographically widespread species, would provide a start to understanding the factors governing the geographic distribution of New Zealand's fungi and the evolution of a distinct New Zealand fungal biota.

New Zealand has 12 endemic genera of ascomycetes, five of these in the Helotiales. Among the Helotiales is the genetically isolated *Orbiliopsis callistea*, a spectacular leaf pathogen of *Veronica elliptica* and *V. subalpina*, common only in Stewart Island and the subantarctic islands, and *Colensoniella torulispora*, a large, dark saprobe of fallen wood.

Summary of New Zealand Ascomycota diversity by major environment

Taxon	Marine	Fresh-water	Terrestrial
Pezizomycotina	77	128	5,125
Saccharomycotina	0	0	77
Taphrinomycotina	0	0	8
Totals	77	128	5,210

Yeasts

Yeasts are a polyphyletic taxonomic unit, comprising members of Ascomycota and Basidiomycota, but most are ascomycetes. They can be defined as fungi having a predominantly unicellular growth form and multiplying by budding and/or fission. As yeast morphology is simple, much of their identification to species is based on physiological properties such as use of nitrate nitrogen, growth without added vitamins, and use of a range of carbon compounds. Some species can use little more than glucose but others assimilate various sugars, organic acids, alcohols, aromatic compounds, and short-chain hydrocarbons, although so far none has been reported to use cellulose or chitin. Yeasts are aerobic but can grow under micro-aerobic conditions, need significant amounts of organic carbon for growth, cannot fix atmospheric nitrogen, and have relatively low (ca. 20–40° C) maximum growth temperatures. Usually excluded from the yeasts are the black-pigmented yeast-like states of hyphomycete genera such as *Aureobasidium* and *Exophiala*, and the saprobic yeast-like phase of smut fungi.

The metabolic versatility of yeasts enables them to be harnessed for many industrial and medical purposes, far beyond their traditional role in wine, beer, and bread making. Yeasts are being investigated internationally for properties such as the production of collagen for surgical and medical use, pigments for food colourings and cosmetics, and enzymes to remove fat stains (Dixon 2000). The yeast *Saccharomyces cerevisiae*, well studied phenotypically and as a model organism, was the first eukaryote to have its complete genomic DNA sequence analysed. Knowledge at the genomic level will ensure further rapid expansion in applied uses of yeasts.

Some yeasts are parasitic on animals, both vertebrate (see the section on Fungi affecting human and animal health on page 502) and invertebrate, and a few are associated with plant diseases. Most, however, are saprobes. They are present in a wide variety of substrates, including those rich in carbon compounds such as fruit and silage or poor such as mineral soils and sea water.

Nearly 700 species belonging to about 90 genera are recognised in the most recent revision of yeast taxonomy. Of these, about 100 species belonging to 21 genera have been recorded from New Zealand sources. This low number is not indicative of a limited yeast flora, but rather the small number of habitats investigated in this country; these include the mammalian body (Parle 1957; Manktelow 1960; Clarke & di Menna 1961), soil (di Menna 1965, 1966), leaves (di Menna 1959; Hamamoto & Nakase 1995, 1996), silage (Barry et al. 1980), and wine (Wright & Parle 1974). In overseas studies, rotten wood, tree exudates, and insect frass, for instance, have been found to be abundant sources of yeast taxa and these three are only a fraction of the unexplored New Zealand habitats.

If there are yeast taxa indigenous to New Zealand, it is most likely that they will be obligate commensals of indigenous animals or plants. It is also likely that most described species are present in this country. When new species from New Zealand soils and plant leaves were described, reports of their occurrence in other countries soon followed. On a global scale, the number of undescribed yeast species is probably considerable, even though the number will in part depend on the vexed question of what is a species.

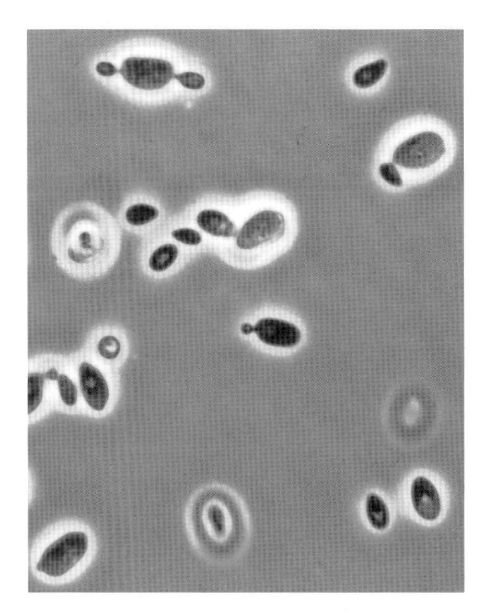

Brewer's yeast or baker's yeast, *Sacchomyces cerevisiae* (Saccharomycetes: Saccharomycetales). Yeast cells bud from one or both ends or from the sides. The yeast is fundamental to fermentation in the production of beer and wine and in raising bread (through production of carbon dioxide).

George L. Barron, University of Guelph

Authors

Dr Ross E. Beever Landcare Research, Private Bag 92170, Auckland, New Zealand. Deceased.

Dr Margaret di Menna New Zealand Pastoral Agriculture Research Institute, Private Bag 3123, Hamilton, New Zealand [margaret.dimenna@agresearch.co.nz]

Dr Peter R. Johnston Landcare Research, Private Bag 92170, Auckland, New Zealand [johnstonp@landcareresearch.co.nz]

Dr Shaun R. Pennycook Landcare Research, Private Bag 92170, Auckland, New Zealand [pennycooks@landcareresearch.co.nz]

Dr Jerry A. Cooper Landcare Research, PO Box 40, Lincoln 7640, New Zealand [cooperj@landcareresearch.co.nz]

Dr Aaron D. Wilton Landcare Research, PO Box 40, Lincoln 7640, New Zealand [wiltona@landcareresearch.co.nz]

Acknowledgements

The authors thank Barbara Paulus (Masterton) for numerical data and compilation of tables.

References

BARRY, T. N.; MENNA, M. E. di; WEBB, P. R.; PARLE, J. N. 1980: Some observations on aerobic deterioration in untreated silages and in silages made with formaldehyde-containing additives. *Journal of the Science of Food and Agriculture 32*: 1151–1156.

CLARKE, R. T. J.; MENNA, M. E. di 1961: Yeasts from the bovine rumen. *Journal of General Microbiology 25*: 113–117.

DINGLEY, J. M. 1951: The Hypocreales of New Zealand. II. The genus *Nectria*. *Transactions and Proceedings of the Royal Society of New Zealand 79*: 177–202.

DINGLEY, J. M. 1952: The Hypocreales of New Zealand. III. The genus *Hypocrea*. *Transactions and Proceedings of the Royal Society of New Zealand 79*: 323–337.

DINGLEY, J. M. 1953: The Hypocreales of New Zealand. V. The genera *Cordyceps* and *Torrubiella*. *Transactions and Proceedings of the Royal Society of New Zealand 81*: 329–343.

DIXON, B. 2000: Yeast as factory and factotum. *Biologist 47*: 15–18.

HALL, I.; BUCHANAN, P. K.; WANG, Y.; COLE, A. L. J. 1998: *Edible and Poisonous Mushrooms: an introduction*. New Zealand Institute for Crop & Food Research, Christchurch. 189 p.

HAMAMOTO, M.; NAKASE, T. 1995: Ballistosporous yeasts found on the surface of plant materials collected in New Zealand. 1. Six new species in the genus *Sporobolomyces*. *Antonie van Leeuwenhoek 67*: 151–171.

HAMAMOTO, M.; NAKASE, T. 1996: Ballistosporous yeasts found on the surface of plant materials collected in New Zealand. The genera *Bensingtonia* and *Bullera* with descriptions of five new species. *Antonie van Leeuwenhoek 69*: 279–291.

HAWKSWORTH, D. L. 2011: A new dawn for the naming of fungi: impacts of decisions made in Melbourne in July 2011 on the future publication and regulation of fungal names. *MycoKeys* 1: 7–20.

HUGHES, M. B.; WEIR, A.; GILLEN, B.; ROSI, W. 2004a: *Stigmatomyces* from New Zealand and New Caledonia: new records, new species and two new host families. *Mycologia 96*: 834–844.

HUGHES, M. B.; WEIR, A.; LESCHEN, R.; JUDD, C.; GILLEN, B. 2004b: New species and records of Laboulbeniales from the subantarctic islands of New Zealand. *Mycologia 96:* 1355–1369.

HUGHES, S. J. 1966: New Zealand fungi. 7. *Capnocybe* and *Capnophialophora*, new form genera of sooty moulds. *New Zealand Journal of Botany 4*: 333–353.

HUGHES, S. J. 1970: New Zealand fungi. 14. *Antennaria, Antennularia, Antennatula, Hyphosoma, Hormisciella,* and *Capnobotrys* gen. nov. *New Zealand Journal of Botany 8*: 153–209.

HUGHES, S. J. 1976: Sooty moulds. *Mycologia 68*: 693–820.

HUGHES, S. J.; PIROZYNSKI, K. A. 1994: New Zealand fungi. 34. *Endomeliola dingleyae,* a new genus and new species of Meliolaceae. *New Zealand Journal of Botany 32*: 53–59.

JOHNSTON, P. R. 1992: Rhytismataceae in New Zealand. 6. Checklist of species and hosts, with keys to species on each host genus. *New Zealand Journal of Botany 30*: 329–351.

JOHNSTON, P. R. 2005: New Zealand's non-lichenised fungi – where they came from, who collected them, where they are now. *National Science Museum Monographs 34*: 37–49.

JOHNSTON, P. R. 2010: Causes and consequences of changes to New Zealand's fungal biota. *New Zealand Journal of Ecology 34*: 175–184.

KNAPP, M.; STOCKLER, K.; HAVELL, D.; DELSUC, F.; SEBASTINI, F.; LOCKHART, P. J. 2005: Relaxed molecular clock provides evidence for long-distance dispersal of *Nothofagus* (Southern Beech). *PLOS Biology 3(1)*: e14.

KIRK, P. M.; CANNON, P. F.; DAVID, J. C.; STALPERS, J. A. 2001: *Ainsworth & Bisby's Dictionary of the Fungi*. 9th Edn. Wallingford, CAB International. 655 p.

LARDNER, R.; JOHNSTON, P. R.; PLUMMER, K. M.; PEARSON, M. 1999: Morphological and molecular analysis of *Colletotrichum acutatum sensu lato*. *Mycological Research 103*: 275–285.

McKENZIE, E. H. C.; BUCHANAN, P. K.; JOHNSTON, P.R. 2000: Checklist of fungi on *Nothofagus* species in New Zealand. *New Zealand Journal of Botany 38*: 635–720.

MANKTELOW, B. W. 1960: Yeasts of the genus *Pityrosporum* in the mammalian external auditory canal with special reference to the dog. *New Zealand Veterinary Journal 8*: 76–78.

MENNA, M. E. di 1959: Yeasts from the leaves of pasture plants. *New Zealand Journal of Agricultural Research 2*: 394–405.

MENNA, M. E. di 1965: Yeasts in New Zealand soils. *New Zealand Journal of Botany 3*: 194–203.

MENNA, M. E. di 1966: Yeasts in soils spray-irrigated with dairy factory wastes. *New Zealand Journal of Agricultural Research 9*: 576–589.

PARLE, J. N. 1957: Yeasts isolated from the mammalian alimentary tract. *Journal of General Microbiology 17*: 363–367.

PENNYCOOK, S. R. 1989: *Plant Diseases Recorded in New Zealand*. Plant Diseases Division, DSIR, Auckland. Vol. 1, 276 p.; Vol. 2, 502 p.; Vol. 3, 180 p.

PENNYCOOK, S. R.; GALLOWAY, D. J. 2004: Checklist of New Zealand 'Fungi'. Pp. 401–488 *in*: McKenzie, E. H. C. (ed.), *The Fungi of New Zealand/Nga Harore o Aotearoa, Volume 1. Introduction to Fungi of New Zealand*. Fungal Diversity Press, Hong Kong. 500 p.

PETERSON, K. R.; PFISTER, D. H.; BELL, C. D. 2010: Cophylogeny and biogeography of the fungal parasite *Cyttaria* and its host *Nothofagus*, southern beech. *Mycologia 102*: 1417-1425

PETRINI, L. E. 2003: *Rosellinia* and related genera in New Zealand. *New Zealand Journal of Botany 41:* 71–138.

RIDLEY, G. S. 1988. New records and species of loculoascomycetes from New Zealand. *New Zealand Journal of Botany 26*: 409–422.

RIFAI, M. A. 1968: The Australasian Pezizales in the Herbarium of the Royal Botanic Gardens Kew. *Verhandelingen der Koninklijke Nederlandse Akademie van Wetenschappen, Afd. Natuurkunde 57*(3): 1–295.

ROGERS, J. D.; SAMUELS, G. J. 1987 [1986]: Ascomycetes of New Zealand 8. *Xylaria*. *New Zealand Journal of Botany 24*: 615–650.

ROSSI, W.; WEIR, A. 1997: New and interesting Laboulbeniales from New Zealand. *Canadian Journal of Botany 75*: 791–798.

ROSSI, W.; WEIR, A. 1998: *Triainomyces*, a new genus of Laboulbeniales on the pill-millipede *Procyliosoma tuberculatum* from New Zealand. *Mycologia 90:* 282–289.

SPOONER, B. M. 1987: Helotiales of Australasia: Geoglossaceae, Orbiliaceae, Sclerotiniaceae, Hyaloscyphaceae. *Bibliotheca Mycologica 116*: 1–711.

WEIR, A. 2001: *Histeridomyces tishechkinii* sp. nov., a new species of Laboulbeniales (Ascomycetes) from New Zealand. *Mycotaxon 79*: 81–86.

WRIGHT, J. M.; PARLE, J. N. 1974: *Brettanomyces* in the New Zealand wine industry. *New Zealand Journal of Agricultural Research 17*: 273–278.

Checklist of New Zealand Ascomycota

Species status: A, adventive (naturalised alien); E, endemic; U, origin uncertain; species not indicated by either notation are native (naturally indigenous but not endemic); L, lichenised species; AN, known from asexual state only; OU, occurrence uncertain in New Zealand. Endemic genera are underlined. Habitat codes include F (freshwater) and M (marine). A number of freshwater fungi are also found in terrestrial environments and are not obligately aquatic. Lichens coded M are not submerged but are generally high-tidal or above.

KINGDOM FUNGI
INCERTAE SEDIS
Dictyonema moorei (Nyl.) Henssen
Dictyonema sericeum (Sw.) Berk.
Gowardia nigricans (Ach.) Halonen, Myllys, Velmala & Hyvärinen L

PHYLUM ASCOMYCOTA
SUBPHYLUM PEZIZOMYCOTINA
INCERTAE SEDIS
Abrothallus curreyi Linds.
Abrothallus microspermus Tul.
Abrothallus parmeliarum (Sommerf.) Nyl.
Abrothallus tulasnei M. Cole & D. Hawksw.
Abrothallus usneae Rabenh.
Acarocybellina arengae (Matsush.) Subram. AN
Acrodontium crateriforme (J.F.H. Beyma) de Hoog AN E
Acrodontium hydnicola (Peck) de Hoog AN

Acrophialophora fusispora (S.B. Saksena) Samson AN U
Acrospeira mirabilis Berk. & Broome AN U
Actinocladium rhodosporum Ehrenb. AN
Amblyosporium botrytis Fresen. AN OU
Annellospermosporella meliolinae P.R. Johnst. AN E
Ardhachandra cristaspora (Matsush.) Subram. & Sudha AN U
Arxiella terrestris Papendorf AN A
Asteromyces cruciatus Moreau & F. Moreau ex Hennebert AN M
Bactridium clavatum Berk. & Broome AN
Bactridium novae-zelandiae S. Hughes AN E
Bactrodesmiella novae-zelandiae S. Hughes AN E
Beltrania querna Harkn. A
Beltrania rhombica Penz. AN
Beverwykella pulmonaria (Beverw.) Tubaki AN F
Beverwykella sp. 'Erua Forest (PDD80884)' J.A. Cooper AN (nom. ined.) F
Biatoridium delitescens (Arnold) Hafellner 1994 L
Bispora betulina (Corda) S. Hughes AN U
Bispora novae-zelandiae Matsush. AN E
Botryosporium longibrachiatum (Oudem.) Maire AN A
Botryosporium pulchrum Corda U
Brachydesmiella biseptata G. Arnaud ex S. Hughes AN
Brachydesmiella orientalis (V. Rao & de Hoog) Goh AN
Camposporium antennatum Harkn. AN F
Camposporium cambrense S. Hughes AN F
Camposporium laundonii M.B. Ellis AN A
Camposporium pellucidum (Grove) S. Hughes AN F
Campylospora filicladia Nawawi AN F
Candelabrum brocchiatum Tubaki AN U F
Candelabrum clathrosphaeroides Voglmayr AN U F
Candelabrum japonense Tubaki AN F
Candelabrum microsporum R.F. Castañeda & W.B. Kendr. AN F
Candelabrum sp. 'Okuro (PDD75020)' J.A. Cooper AN (nom. ined.) F
Candelabrum spinulosum Beverw. AN F
Catenophoropsis eucalypticola Nag Raj & W.B. Kendr. AN A
Ceratosporella novae-zelandiae S. Hughes AN
Ceratosporella stipitata (Goid.) S. Hughes AN
Chaetopsis proboscioph ora DiCosmo, S.M. Berch & W.B. Kendr. AN E
Chaetospermum chaetosporum (Pat.) A.L. Sm. & Ramsb. AN A
Chalara acuaria Cooke & Ellis AN
Chalara affinis Sacc. & Berl. AN
Chalara africana (B. Sutton & Piroz.) P.M. Kirk AN
Chalara agathidis Nag Raj & W.B. Kendr. AN
Chalara angionacea Nag Raj & W.B. Kendr. AN E
Chalara aotearoa Nag Raj & S. Hughes AN E
Chalara aurea (Corda) S. Hughes AN
Chalara bicolor S. Hughes AN E
Chalara brevipes Nag Raj & W.B. Kendr. AN E
Chalara brunnipes Nag Raj & W.B. Kendr. AN E
Chalara constricta Nag Raj & W.B. Kendr. AN
Chalara curvata Nag Raj & W.B. Kendr. AN E
Chalara cylindrosperma (Corda) S. Hughes AN
Chalara dictyoseptata Nag Raj & S. Hughes AN E
Chalara distans McKenzie AN E
Chalara dracophylli McKenzie AN E
Chalara dualis Aramb. & Gamundí AN
Chalara fusidioides (Corda) Rabenh. AN
Chalara gracilis Nag Raj & W.B. Kendr. AN E
Chalara graminicola McKenzie AN E
Chalara hughesii Nag Raj & W.B. Kendr. AN
Chalara inaequalis Nag Raj & W.B. Kendr. AN E
Chalara myrsines Gadgil & M. Dick AN E
Chalara nothofagi Nag Raj & W.B. Kendr. AN E
Chalara novae-zelandiae Nag Raj & W.B. Kendr. AN E
Chalara parvispora Nag Raj & S. Hughes AN E
Chalara pteridina Syd. AN
Chalara pulchra Nag Raj & S. Hughes AN E
Chalara rhynchophialis Nag Raj & W.B. Kendr. AN
Chalara scabrida Nag Raj & W.B. Kendr. AN E
Chalara sessilis Nag Raj & W.B. Kendr. AN
Chalara sp. 1 sensu McKenzie, P.K. Buchanan & P.R. Johnst. (nom. ined.) E
Chalara stipitata Nag Raj & W.B. Kendr. AN E
Chalara tubifera Nag Raj & W.B. Kendr. AN E
Chalara unicolor S. Hughes & Nag Raj AN E
Chalara urceolata Nag Raj & W.B. Kendr. AN
Chalarodes bisetis McKenzie AN E
Cheilaria agrostidis Lib. AN A
Chionomyces meliolicola (Cif.) Deighton & Piroz. AN
Chromosporium pallescens Cooke & Massee AN E
Circinotrichum chathamiense McKenzie AN E
Circinotrichum falcatisporum Piroz. AN
Circinotrichum maculiforme Nees AN
Circinotrichum olivaceum (Speg.) Piroz. AN
Circinotrichum papakurae S. Hughes & Piroz. AN
Cirrosporium novae-zelandiae S. Hughes AN E
Coleophoma cylindrospora (Desm.) Höhn. AN A
Cordana abramovii E.O. Semen & Davydkina AN
Cordana ellipsoidea de Hoog AN F
Coremiella cubispora (Berk. & M.A. Curtis) M.B. Ellis AN A
Cornutispora ciliata Kalb AN
Cornutispora lichenicola D. Hawksw. & B. Sutton AN
Crucellisporiopsis gelatinosa Nag Raj AN E
Cryptocoryneum rilstonii M.B. Ellis AN
Cryptomycina pteridis (Rebent.) Höhn.
Cryptophiale insularis McKenzie AN E
Cryptophiale orthospora McKenzie AN
Cryptophiale pusilla McKenzie AN E
Cryptophiale udagawae Piroz. & Ichinoe AN F
Culicidospora aquatica R.H. Petersen AN F
Cytosporella anisotomes Aa, Vanev, Mel'nik & S.L. Stephenson AN E
Cytostagonospora photiniicola Bubák AN A
Dendrospora juncicola S.H. Iqbal AN F
Dendrosporium sp. 'Lake Stream Track (PDD76603)' J.A. Cooper AN (nom. ined.) F
Diplocladiella scalaroides G. Arnaud ex M.B. Ellis AN F
Diplorhynchus bilobus G. Arnaud AN (nom. inv.)
Discogloeum veronicae (Lib.) Petr. AN A
Endocalyx melanoxanthus (Berk. & Broome) Petch AN
Endophragmia verruculosa M.B. Ellis AN
Enthallopycnidium gouldiae F. Stevens AN
Eriomycopsis meliolinae Hansf. AN E
Exserticlava vasiformis (Matsush.) S. Hughes AN
Fairmaniella leprosa (Fairm.) Petr. & Syd. AN A
Feltgeniomyces physciae Etayo & Breuss AN
Flabellospora acuminata Descals AN F
Flabellospora verticillata Alas. AN F
Fontanospora eccentrica (R.H. Petersen) Dyko AN F
Fusariella sarniensis M.B. Ellis AN A
Fusichalara dimorphospora S. Hughes & Nag Raj AN E
Fusichalara novae-zelandiae S. Hughes & Nag Raj AN E
Fusidium aeruginosum Link AN A
Fusidium griseum Link AN
Gilmaniella humicola G.L. Barron AN A
Gloeodes pomigena (Schwein.) Colby AN A
Goniopila monticola (Dyko) Marvanová & Descals AN
Guedea novae-zelandiae S. Hughes AN E
Gyoerffyella biappendiculata (G.R.W. Arnold) Ingold AN F
Gyoerffyella craginiformis (R.H. Petersen) Marvanová AN F
Gyoerffyella entomobryoides (Boerema & Arx) Marvanová AN F
Gyoerffyella gemellipara Marvanová AN F
Gyoerffyella rotula (Höhn.) Marvanová AN F
Gyoerffyella speciosa (K. Miura) Dudka AN OU F
Gyrothrix circinata (Berk. & M.A. Curtis) S. Hughes AN
Gyrothrix citricola Piroz. AN
Gyrothrix grisea Piroz. AN
Gyrothrix microsperma (Höhn.) Piroz. AN
Gyrothrix pediculata J.L. Cunn. AN A
Gyrothrix podosperma (Corda) Rabenh. AN
Gyrothrix verticiclada (Goid.) S. Hughes & Piroz. AN
Hadrosporium dingleyae S. Hughes AN E
Hansfordia grisella (Sacc.) S. Hughes AN
Harpographium graminum Cooke & Massee AN U
Heliscina campanulata Marvanová AN F
Hemibeltrania mitrata P.M. Kirk AN F
Hormiactella fusca (Preuss) Sacc. AN A
Intralichen christiansenii (D. Hawksw.) D. Hawksw. & M.S. Cole AN
Janetia capnophila S. Hughes AN
Kendrickomyces indicus B. Sutton, V.G. Rao & Mhaskar AN
Kramasamuha sibika Subram. & Vittal AN A
Kumanasamuha novozelandica L.J. Hunter & W.B. Kendr. E
Lateriramulosa uniinflata Matsush. AN F
Lauriomyces bellulus Crous & M.J. Wingf. AN
Leptomelanconium australiense B. Sutton AN A
Leptothyrium coriariae (Berk.) Sacc. AN E
Leptothyrium panacis Cooke AN E
Leptothyrium pomi (Mont. & Fr.) Sacc. AN A
Lichenoconium cargillianum (Linds.) D. Hawksw. AN
Lichenoconium plectocarpoides S.Y. Kondr., D.J. Galloway & D. Hawksw. AN E
Lichenoconium usneae (Anzi) D. Hawksw. AN
Lichenodiplis lecanorae (Vouaux) Dyko & D. Hawksw. AN
Lichenodiplis pertusariicola (Nyl.) Diederich AN
Lichenodiplis poeltii S.Y. Kondr. & D. Hawksw. AN
Lunulospora curvula Ingold AN F
Lunulospora cymbiformis K. Miura AN
Lylea tetracoila (Corda) Hol.-Jech. AN
Mammariopsis variospora L.J. Hutchison & J. Reid AN E
Mastigosporium album Riess AN A
Mastigosporium muticum (Sacc.) Gunnerb. AN A
Mastigosporium rubricosum (Dearn. & Barthol.) Nannf. AN A
Melanocephala australiensis (G.W. Beaton & M.B. Ellis) S. Hughes AN
Melanocephala cupulifera S. Hughes AN E
Melanographium citri (Gonz. Frag. & Cif.) M.B. Ellis AN
Monilochaetes infuscans Ellis & Halst. ex Harter AN A
Monotosporella setosa (Berk. & M.A. Curtis) S. Hughes var. *setosa* AN
— var. *macrospora* S. Hughes AN E
Morrisographium fusisporium (A.L. Sm.) Illman & G.P. White AN
Mycofalcella calcarata Marvanová, Om-K. Khattab & J. Webster AN F
Myriellina cydoniae (Desm.) Höhn. AN A
Myxosporium necans Peck AN A
Nigropuncta rugulosa D. Hawksw. AN
Ochroconis constricta (E.V. Abbott) de Hoog &

Arx AN A
Ochroconis gallopava (W.B. Cooke) de Hoog AN U
Olpitrichum macrosporum (Farl.) Sumst. AN A
Oncopodiella trigonella (Sacc.) Rifai AN U
Parasympodiella laxa (Subram. & Vittal) Ponnappa AN A
Penzigomyces australiensis (M.B. Ellis) Subram. AN U
Penzigomyces cookei (S. Hughes) Subram. AN
Penzigomyces flagellatus (S. Hughes) Subram. AN
Penzigomyces parvus (S. Hughes) Subram. AN U
Pestalozziella subsessilis Sacc. & Ellis AN A
Phaeocandelabrum joseiturriagae R.F. Castañeda, Iturr., Heredia & M. Stadler AN F
Phaeocytostroma ambiguum (Mont.) Petr. AN A
Phaeoisaria clematidis (Fuckel) S. Hughes AN F
Phaeoisaria sparsa B. Sutton AN
Phaeosporobolus alpinus R. Sant., Alstrup & D. Hawksw. AN
Phaeosporobolus usneae D. Hawksw. & Hafellner AN
Phialea cyathoidea (Bull.) Gillet U
Phialea gnaphalii G. Cunn. (nom. ined.)
Phialomyces macrosporus P.C. Misra & P.H.B. Talbot AN U
Phragmocephala atra var. *stenophora* S. Hughes AN E
Phragmocephala prolifera (Sacc., M. Rousseau & E. Bommer) S. Hughes AN
Phragmocephala stemphylioides (Corda) S. Hughes AN A
Pirostoma viridisporum (Cooke) Grove
Pleiochaeta setosa (Kirchn.) S. Hughes AN A
Pleosphaeria otagensis (Linds.) Sacc. E
Pleuropedium tricladioides Marvanová & S.H. Iqbal AN F
Pleurophoma cava (Schulzer) Boerema, Loer. & Hamers AN A
Pleurophragmium bitunicatum Matsush. AN A
Pleurophragmium maculosum (Sacc.) S. Hughes AN
Pleurotheciopsis bramleyi B. Sutton AN A F
Polyschema yakuensis Matsush. AN
Polyscytalum ciliatum J.A. Cooper AN E
Polyscytalum gracilisporum (Matsush.) B. Sutton & Hodges AN
Polyscytalum pustulans (M.N. Owen & Wakef.) M.B. Ellis AN A
Polyscytalum truncatum B. Sutton & Hodges AN U
Pseudoacrodictys deightonii (M.B. Ellis) W.A. Baker & Morgan-Jones AN U
Pseudoclathrosphaerina spiralis J.A. Cooper AN F
Pseudohansfordia mycophila (Tubaki) de Hoog AN U
Pseudoseptoria bromigena (Sacc.) B. Sutton AN A
Pseudoseptoria donacis (Pass.) B. Sutton AN A
Pseudoseptoria stomaticola (Bäumler) B. Sutton AN U
Refractohilum galligenum D. Hawksw. AN
Remersonia thermophila (Fergus) Seifert & Samson AN A
Rhexoacrodictys fuliginosa (B. Sutton) W.A. Baker & Morgan-Jones
Rhinocladium dingleyae S. Hughes AN
Rhinocladium pulchrum S. Hughes & Hol.-Jech. AN
Rutola graminis (Desm.) J.L. Crane & Schokn. AN
Sarcopodium circinatum Ehrenb. AN A
Sarcopodium tortuosum (Wallr.) S. Hughes AN
Seifertia azaleae (Peck) Partridge & Morgan-Jones AN A
Septocyta ruborum (Lib.) Petr. A
Septosporium bulbotrichum Corda AN
Septotrullula bacilligera Höhn. A
Spegazzinia tessarthra (Berk. & M.A. Curtis) Sacc. AN A
Spermospora bromivora (Latch) Deighton AN A
Spermospora ciliata (R. Sprague) Deighton AN A
Spermospora lolii MacGarvie & O'Rourke AN A
Sphaeronaema solandri Cooke AN E
Spiropes capensis (Thüm.) M.B. Ellis AN
Spiropes dictyosporus Seifert & S. Hughes AN E
Spiropes helleri (F. Stevens) M.B. Ellis AN
Spiropes palmetto (W.R. Gerard) M.B. Ellis AN
Spondylocladiopsis cupulicola M.B. Ellis AN OU
Stachylidium bicolor Link AN
Staphylotrichum coccosporum J.A. Mey. & Nicot AN A
Stenocephalopsis subalutacea (Peck) Chamuris & C.J.K. Wang AN
Sterigmatobotrys macrocarpa (Corda) S. Hughes
Strasseria geniculata (Berk. & Broome) Höhn. AN A
Subramaniomyces fusisaprophyticus (Matsush.) P.M. Kirk A
Subramaniomyces simplex U. Braun & C.F. Hill AN U
Subulispora britannica B. Sutton F
Sympodiella fasciculata Aramb. & W.B. Kendr. AN, (nom. ined.) E
Sympodiella fragilis Aramb. & W.B. Kendr. AN, (nom. ined.) E
Sympodiella lobata Aramb. & W.B. Kendr. AN, (nom. ined.) E
Sympodiella nodosa Aramb. & W.B. Kendr. AN, (nom. ined.) E
Techonidula hippocrepida Réblová (nom. ined.) E
Tetrachaetum elegans Ingold AN F
Thamnogalla crombiei (Mudd) D. Hawksw.
Tompetchia webberi (H.S. Fawc.) Subram. AN
Torula herbarum (Pers.) Link AN U
Tracylla aristata (Cooke) Tassi AN
Tretospeira sp. 'Fantail Falls (PDD76585)' J.A. Cooper AN (nom. ined.)
Tricellula aurantiaca (Haskins) Arx AN F
Triscelophorus acuminatus Nawawi AN F
Triscelophorus sp. 1 sensu Aimer & Segedin (nom. ined.) AN F
Tritirachium oryzae (Vincens) de Hoog AN U
Troposporella monospora (W.B. Kendr.) M.B. Ellis AN
Ulocoryphus mastigophorus Michaelides, L.J. Hunter & W.B. Kendr. AN E
Vermisporium acutum H.J. Swart & M.An. Will. AN A
Vermisporium brevicentrum H.J. Swart & M.An. Will. AN A
Vermisporium cylindrosporum (H.J. Swart) Nag Raj AN A
Vermisporium eucalypti (McAlpine) Nag Raj AN A
Vermisporium falcatum (B. Sutton) Nag Raj AN A
Vermisporium obtusum H.J. Swart & M.An. Will. AN A
Vermisporium verrucisporum Nag Raj AN A
Veronaea botryosa Cif. & A.M. Corte AN U
Veronaea filicina Dingley AN E
Volucrispora sp. sensu Aimer & Segedin (nom. ined.) AN U
Vouauxiomyces santessonii D.Hawksw. 1981
Waydora typica (Rodway) B. Sutton AN
Wiesneriomyces laurinus (Tassi) P.M. Kirk AN
Xanthoriicola physciae (Kalchbr.) D. Hawksw. AN
Xylohypha curta (Corda) S. Hughes ex Deighton AN
Xylohypha ferruginosa (Corda) S. Hughes ex Deighton AN
Xylohypha nigrescens (Pers.) E.W. Mason ex Deighton AN A
Xylohypha novae-zelandiae S. Hughes & Sugiy. AN E
Xylohypha palmicola S. Hughes & Sugiy. AN E
Zanclospora novae-zelandiae S. Hughes & W.B. Kendr. AN
Zanclospora urewera J.A. Cooper AN E
Zebrospora bicolor McKenzie AN
Zelandiocoela ambigua (Nag Raj & W.B. Kendr.) Nag Raj AN E
Zetesimomyces setibicolor (R.F. Castañeda) Nag Raj AN
Zygosporium bioblitzi McKenzie, Thongkantha & Lumyong AN
Zygosporium gibbum (Sacc., M. Rousseau & E. Bommer) S. Hughes AN
Zygosporium minus S. Hughes AN
Zygosporium oscheoides Mont. AN

Order TRIBLIDIALES
TRIBLIDIACEAE
Blitridium nigrocinnabarinum (Schwein.) Sacc. A
Pseudographis ixerbae Henn. E

Class ARTHONIOMYCETES
Order ARTHONIALES
Arthothelium ampliatum (C.Knight & Mitt.) Müll. Arg. L
Arthothelium endoaurantiacum Makhija & Patw. 1995 E L
Arthothelium fusconigrum (Nyl.) Müll.Arg. L
Arthothelium interveniens (Nyl.) Zahlbr. L
Arthothelium obtusulum (Nyl.) Müll.Arg. E L
Arthothelium pellucidum (C.Knight) Müll.Arg. E L
Arthothelium spadiceum (C.Knight) Müll.Arg. E L
Arthothelium stirtianum Müll.Arg. E L
Arthothelium suffusum (C.Knight) Müll.Arg. E L
ARTHONIACEAE
Arthonia anjutiae S.Y. Kondr. & Alstrup L
Arthonia cinerascens Kremp. L
Arthonia cinereopruinosa Schaer. L
Arthonia cinnabarina (DC.) Wallr. L
Arthonia clemens (Tul.) Th. Fr. L
Arthonia conspicua Nyl. E L
Arthonia cyanea Müll.Arg. [
Arthonia diaphora Stirt. E L
Arthonia dispersa (Schrad.) Nyl. L
Arthonia epiodes Nyl. E L
Arthonia epiphyscia Nyl. L
Arthonia fuscopurpurea (Tul.) R. Sant. L
Arthonia galactinaria Leight. L
Arthonia glaucomaria (Nyl.) Nyl. L
Arthonia haematommatum Kalb & Hafellner L
Arthonia indistincta C.Knight & Mitt. E L
Arthonia lapidicola (Taylor) Branth & Rostr. L
Arthonia maculiformis Wedin & Hafellner E L
Arthonia molendoi (Heufl. ex Frauenf.) R. Sant. L
Arthonia nigrocincta C.Knight & Mitt. E L
Arthonia pelveti (Hepp) Almq. L
Arthonia peraffinis Nyl. L
Arthonia perparva (Zahlbr.) Matzer E L
Arthonia phymatodes C.Knight E L
Arthonia platygraphella Nyl. E L
Arthonia polymorpha Ach. L
Arthonia pseudocyphellariae Wedin L
Arthonia radiata (Pers.) Ach. L
Arthonia santessoniana Wedin & Hafellner L
Arthonia stictaria Nyl. E L
Arthonia subfuscicola (Linds.) Triebel L
Arthonia sytnikii S.Y. Kondr. L
Arthonia tasmanica Kantvilas & Vězda L
Arthonia vinosa Leight. L
Cryptothecia bartlettii G.Thor 1997 E L
CHRYSOTHRICACEAE
Chrysothrix candelaris (L.) J.R.Laundon L
ROCCELLACEAE
Bactrospora arthonioides Egea & Torrente L

Bactrospora metabola (Nyl.) Egea & Torrente L
Bactrospora pleistophragmoides (Nyl.) Egea & Torrente E L
Chiodecton colensoi (A.Massal.) Müll.Arg. L
Chiodecton montanum G.Thor L
Cresponea plurilocularis (Nyl.) Egea & Torrente L
Dictyographa cinerea (C.Knight & Mitt.) Müll. Arg. L
Dirina neozelandica (Redinger) Sparrius E L
Enterographa bartlettii Sérus. E L
Enterographa bella R.Sant. L
Enterographa crassa sensu Galloway (nom. inv.) L
Enterographa pallidella (Nyl.) Redinger L
Enterographa subgelatinosa (Stirt.) Redinger 1938 E L
Enterographa subserialis (Nyl.) Redinger L
Lecanactis abietina (Ach.) Körb. L
Lecanactis exigua Egea & Torrente E L
Lecanactis neozelandica Egea & Torrente L
Lecanactis subfarinosa (C.Knight) Hellb. E L
Lecanactis tibelliana Egea & Torrente E L
Lecanactis totarae Zahlbr. E L
Lecanographa abscondita (Th.Fr.) Egea & Torrente L
Mazosia melanophthalma (Müll.Arg.) R.Sant. L
Mazosia phyllosema (Nyl.) Zahlbr. L
Opegrapha agelaeoides Nyl. L
Opegrapha atra Pers. L
Opegrapha bonplandii Fée L
Opegrapha brevissima Kalb & Hafellner E L
Opegrapha concrucians Kremp. E L
Opegrapha devia (C.Knight & Mitt.) Nyl. E L
Opegrapha diaphoriza Nyl. E L
Opegrapha foreaui (Moreau) Hafellner & R. Sant. L
Opegrapha geographicola (Arnold) Hafellner L
Opegrapha intertexta C.Knight E L
Opegrapha maligna Triebel L
Opegrapha murina Kremp. E L
Opegrapha puiggarii Müll.Arg. L
Opegrapha rupestris Pers. L
Opegrapha spodopolia Nyl. E L
Opegrapha stellata C.Knight L
Opegrapha thelotrematis Coppins L
Opegrapha trassii S.Y. Kondr. & Coppins L
Perigrapha nitida Ertz, Diederich, Christnach & Wedin L
Perigrapha superveniens (Nyl.) Hafellner L
Plectocarpon bunodophori Wedin, Ertz & Diederich L
Plectocarpon concentricum Ertz, Diederich & Wedin L
Plectocarpon gallowayi (S.Y.Kondr.) Ertz & Diederich L
Plectocarpon lichenum (Sommerf.) D. Hawksw. L
Plectocarpon opegraphoideum Christnach, Ertz, Diederich & Wedin L
Plectocarpon pseudosticta (Fée) Fée U L
Plectocarpon sticticola Ertz, Wedin & Diederich L
Plectocarpon tibellii Ertz & Diederich L
Roccellina exspectata Tehler L
Sagenidium molle Stirt. L
Schismatomma atratum (Stirt.) Zahlbr. L
Schismatomma occultum (C.Knight & Mitt.) Zahlbr. L
MELASPILEACEAE
Encephalographa otagensis (Linds.) Müll. Arg. E L
Melaspilea gallowayi S.Y. Kondr. L
Melaspilea subeffigurans (Nyl.) Müll.Arg. E L

Class DOTHIDEOMYCETES
Acrogenotheca elegans (L.R. Fraser) Cif. & Bat.
Asteromella bellunensis Syd. AN U
Asteromella myriadea Cooke AN E
Bactrodesmium abruptum (Berk. & Broome) E.W. Mason & S. Hughes AN
Bactrodesmium atrum M.B. Ellis AN
Bactrodesmium betulicola M.B. Ellis AN A
Bactrodesmium biformatum (Höhn.) S. Hughes AN
Bactrodesmium globosum Hol.-Jech. AN
Bactrodesmium nothofagi J.A. Cooper AN E
Bryorella compressa Döbbeler
Cenococcum geophilum Fr.
Cercidospora trypetheliza (Nyl.) Hafellner & Obermayer
Cercidospora verrucosaria (Linds.) Arnold
Clypeostroma hemisphaericum (Berk.) Theiss. & Syd. E
Clypeostroma spilomeum (Berk.) Theiss. & Syd. E
Dawsophila pygmaea Döbbeler
Dendryphiopsis arbuscula (Berk. & M.A. Curtis) S. Hughes AN
Endococcus macrosporus (Arnold) Nyl.
Endococcus parietinarius (Linds.) Clauzade & Cl. Roux
Endococcus perpusillus Nyl.
Endococcus ramalinarius (Linds.) D. Hawksw. E
Endococcus rugulosus Nyl.
Gloeodiscus nigrorufus (Berk.) Dennis E
Heptameria obesa (Durieu & Mont.) Sacc. A
Kirschsteiniothelia incrustans (Ellis & Everh.) C.Y. Chen & W.H. Hsieh
Lidophia graminis (Sacc.) J. Walker & B. Sutton A
Monodictys lepraria (Berk.) M.B. Ellis AN
Monodictys levis (Wiltshire) S. Hughes AN U
Mycothyridium lividum (Pers.) Petr. U
Myxophora apotheciicola Nik. Hoffm. & Hafellner E
Punctillum hepaticarum (Cooke) Petr. & Syd.
Rosellinula lopadii (Vouaux) D.J. Galloway
Xylobotryum andinum Pat.

Order BOTRYOSPHAERIALES
Camarosporium propinquum Sacc. AN OU
Camarosporium pusillum Cooke AN E
Camarosporium solandri Cooke AN E
BOTRYOSPHAERIACEAE
Botryosphaeria acaciae (Hansf.) Dingley A
Botryosphaeria australis Slippers, Crous & M.J. Wingf. A
Botryosphaeria dothidea (Moug.) Ces. & De Not. A
Botryosphaeria lutea A.J.L. Phillips A
Botryosphaeria obtusa (Schwein.) Shoemaker A
Botryosphaeria parva Pennycook & Samuels A
Botryosphaeria rhodina (Berk. & M.A. Curtis) Arx A
Botryosphaeria ribis Grossenb. & Duggar A
Botryosphaeria sarmentorum A.J.L. Phillips, A. Alves & J. Luque A OU
Botryosphaeria stevensii Shoemaker A
Dichomera eucalypti (G. Winter) B. Sutton AN A
Diplodia crataegi Westend. AN A
Diplodia salicina Lév. AN A
Diplodia taxi (Sowerby) De Not. AN A
Discochora yuccae Bissett A
Dothiorella ellisii Arx AN E
Dothiorella pyrenophora Sacc. AN, (nom. illegit.) A
Fusicoccum bacillare Sacc. & Penz. AN A
Guignardia camelliae (Cooke) E.J. Butler A
Guignardia cytisi (Fuckel) Arx & E. Müll. A
Guignardia korthalsellae A. Sultan, P.R. Johnst., D.-C. Park & A.W. Robertson E
Guignardia mangiferae A.J. Roy A
Guignardia philoprina (Berk. & M.A. Curtis) Aa A
Guignardia rhodorae (Cooke) B.H. Davis A
Guignardia sawadae Tak. Kobay. A
Guignardia vaccinii Shear A
Macrophoma chenopodii Oudem. AN A
Macrophoma eugeniae Tassi AN A
Macrophoma salicaria (Sacc.) Berl. & Voglino AN A
Macrophomina phaseolina (Tassi) Goid. AN A
Neofusicoccum macroclavatum (T.I. Burgess, P.A. Barber & G.E. Hardy) T.I. Burgess, P.A. Barber & G.E. Hardy AN A
Neofusicoccum mangiferae (Syd. & P. Syd.) Crous, Slippers & A.J.L. Phillips AN A
Neoscytalidium dimidiatum (Penz.) Crous & Slippers AN A
Otthia spiraeae (Fuckel) Fuckel A
Phaeobotryosphaeria citrigena A.J.L. Phillips, P.R. Johnst. & Pennycook A
Phyllosticta abietis Bissett & M.E. Palm AN A
Phyllosticta acetosae Sacc. AN A
Phyllosticta aucubae Sacc. & Speg. AN A
Phyllosticta cordylinophila P.A. Young AN
Phyllosticta cyclaminis Brunaud AN A
Phyllosticta cynarae Westend. AN A
Phyllosticta grossulariae Sacc. AN A
Phyllosticta longispora McAlpine AN A
Phyllosticta polygonorum Sacc. AN A
Phyllosticta primulicola Desm. AN A
Phyllosticta ranunculorum Sacc. & Speg. AN A
Phyllosticta rhododendri Westend. AN A
Phyllosticta sp. 1 sensu McKenzie (nom. ined.) AN
Phyllosticta spinarum (Died.) Nag Raj & M. Morelet AN A
Phyllosticta tabaci Pass. AN A
Phyllosticta variabilis Peck AN
Phyllosticta violae Desm. AN A
Sphaeropsis cordylines G.F. Laundon AN, (nom. illegit.) E
Sphaeropsis sapinea (Fr.) Dyko & B. Sutton AN A
Wilsonomyces carpophilus (Lév.) Adas., J.M. Ogawa & E.E. Butler AN A

Order ACROSPERMALES
ACROSPERMACEAE
Acrospermum compressum Tode U
Gonatophragmium epilobii U. Braun & C.F. Hill AN A
Gonatophragmium obscurum U. Braun & C.F. Hill AN A
ASPIDOTHELIACEAE
Aspidothelium cinerascens Vain. L

Order CAPNODIALES
Cystocoleus ebeneus (Dillwyn) Thwaites L
Phyllachora bulbosa Parbery A
Phyllachora cladii-glomerati Hansf.
Phyllachora cyperi Rehm U
Phyllachora fuscescens (Speg.) Speg. A
Phyllachora hauturu P.R. Johnst. & P.F. Cannon subsp. *hauturu* E
— subsp. *lanceshawii* P.R. Johnst. & P.F. Cannon E
— subsp. *rekohu* P.R. Johnst. & P.F. Cannon E
Phyllachora manuka P.R. Johnst. & P.F. Cannon E
Phyllachora rostkoviae P.R. Johnst. & P.F. Cannon E
Phyllachora setariicola Speg. A
Phyllachora sylvatica Sacc. & Speg. A
Rachicladosporium luculiae Crous, U. Braun & C.F. Hill A
Racodium rupestre Pers. 1797 L
ANTENNULARIELLACEAE
Antennulariella concinna (L.R. Fraser) S. Hughes
Capnofrasera dendryphioides S. Hughes AN
ASTERINACEAE
Asterina effusa Cooke & Massee E
Asterina fragilissima Berk. E
Asterina geniostomatis Hansf.
Asterina sublibera Berk. E
Asterina torulosa Berk. E
Aulographina eucalypti (Cooke & Massee) Arx & E. Müll. A
Dimerosporium excelsum Cooke E
Dothidasteromella systema-solare (Massee) H.J. Swart A
Lembosia pandani (Rostr.) Theiss.
Placoasterella baileyi (Berk. & Broome) Arx A
Placosoma nothopanacis Syd. E

Thallochaete baileyi (Berk. & Broome) Hansf. A
Triposporium elegans Corda AN
Triposporium verruculosum R.F. Castañeda, Gené & Guarro AN U
CAPNODIACEAE
Capnodium annonae Pat. A
Capnodium australe Mont. U
Capnodium fibrosum Berk. E
Capnodium fuliginoides Rehm
Capnodium salicinum Mont. A
Capnodium uniseptatum (L.R. Fraser) S. Hughes
Capnodium walteri Sacc.
Ciferrioxyphium chaetomorphum (Speg.) S. Hughes AN
Scorias spongiosa (Schwein.) Fr.
Tripospermum juglandis (Thüm.) Speg. AN
Tripospermum myrti (Lind) S. Hughes AN
DAVIDIELLACEAE
Acroconidiella eschscholziae (Harkn.) M.B. Ellis AN A
Acroconidiella tropaeoli (T.E.T. Bond) J.C. Lindq. & Alippi AN A
Cladosporium allii (Ellis & G. Martin) P.M. Kirk & J.G. Crompton AN A
Cladosporium arthropodii K. Schub. & C.F. Hill E
Cladosporium avellaneum G.A. de Vries AN A
Cladosporium brunneoatrum McAlpine AN A
Cladosporium cladosporioides (Fresen.) G.A. de Vries AN U
Cladosporium colocasiae Sawada AN A
Cladosporium cucumerinum Ellis & Arthur AN A
Cladosporium herbarum var. *macrocarpum* (Preuss) M.H.M. Ho & Dugan AN A
Cladosporium hillianum Bensch, Crous & U. Braun E
Cladosporium jacarandicola K. Schub., U. Braun & C.F. Hill AN A
Cladosporium lacroixii Desm. AN A
Cladosporium oncobae K. Schub. & C.F. Hill A
Cladosporium oxysporum Berk. & M.A. Curtis AN A
Cladosporium perangustum Bensch, Crous & U. Braun U
Cladosporium phlei (C.T. Greg.) G.A. de Vries AN A
Cladosporium pseudiridis K. Schub., C.F. Hill, Crous & U. Braun AN A
Cladosporium pseudocladosporioides Bensch, Crous & U. Braun U
Cladosporium sinuosum K. Schub., C.F. Hill, Crous & U. Braun AN E
Cladosporium sphaeroideum Cooke AN U
Cladosporium sphaerospermum Penz. AN U
Cladosporium tenuissimum Cooke AN U
Cladosporium uredinicola Speg. AN U
Davidiella allicina (Fr.) Crous & Aptroot A
Davidiella dianthi (C.C. Burt) Crous & U. Braun A
Davidiella macrospora (Kleb.) Crous & U. Braun A
Davidiella tassiana (De Not.) Crous & U. Braun A
Graphiopsis chlorocephala (Fresen.) Trail AN A
EUANTENNARIACEAE
Antennatula dingleyae S. Hughes AN E
Antennatula fisherae S. Hughes AN E
Antennatula fraserae S. Hughes AN E
Antennatula triseptata S. Hughes AN E
Capnokyma corticola S. Hughes AN E
Euantennaria caulicola S. Hughes E
Euantennaria mucronata (Mont.) S. Hughes
Euantennaria novae-zelandiae S. Hughes E
Euantennaria pacifica S. Hughes E
Trichopeltheca asiatica Bat., C.A.A. Costa & Cif.
METACAPNODIACEAE
Capnobotrys atroolivacea S. Hughes AN E
Capnobotrys australis S. Hughes AN E
Capnobotrys laterivecta S. Hughes AN E
Capnobotrys pacifica S. Hughes AN
Capnobotrys paucispora S. Hughes AN E
Capnocybe novae-zelandiae S. Hughes AN E
Metacapnodium dingleyae S. Hughes
Metacapnodium fraserae (S. Hughes) S. Hughes
Metacapnodium moniliforme (L.R. Fraser) S. Hughes
Ophiocapnocoma batistae S. Hughes E
Ophiocapnocoma phloiophilia (E.E. Fisher) S. Hughes
MYCOSPHAERELLACEAE
Cercospora alchemillicola U. Braun & C.F. Hill AN A
Cercospora althaeina Sacc. AN A
Cercospora apii Fresen. AN A
Cercospora armoraciae Sacc. AN A
Cercospora beticola Sacc. AN A
Cercospora canescens Ellis & G. Martin AN A
Cercospora carotae (Pass.) Kazn. & Siemaszko AN A
Cercospora chrysanthemi Heald & F.A. Wolf AN A
Cercospora cyperigena U. Braun & Crous AN U
Cercospora depazeoides (Desm.) Sacc. AN A
Cercospora deutziae Ellis & Everh. AN A
Cercospora duddiae Welles AN A
Cercospora eragrostidis McKenzie & Latch AN A
Cercospora erysimi Davis A
Cercospora ipomoeae G. Winter AN A
Cercospora loti Hollós AN A
Cercospora microlaenae McKenzie & Latch AN E
Cercospora modiolae Tharp AN A
Cercospora moluccellae Bremer & Petr. AN A
Cercospora nasturtii Pass. AN A
Cercospora pantoleuca Sacc. AN A
Cercospora physalidis Ellis AN A
Cercospora resedae Fuckel AN A
Cercospora ribis Earle AN A
Cercospora setariae G.F. Atk. AN A
Cercospora sorghi Ellis & Everh. AN A OU
Cercospora statices Pesante AN, (nom. illegit.) A
Cercospora tetragoniae (Speg.) Siemaszko
Cercospora uredinophila (Sacc.) Deighton AN A
Cercospora violae Sacc. AN A
Cercospora zebrina Pass. AN A
Cercosporella primulae Allesch. AN A
Cercosporella rubi (G. Winter) Plakidas AN A
Deightoniella torulosa (Syd.) M.B. Ellis AN A
Diplochorella colensoi (Berk.) P.R. Johnst. & P.F. Cannon E
Discella lignicola Cooke AN E
Dissoconium aciculare de Hoog, Oorschot & Hijwegen AN A
Dissoconium dekkeri de Hoog & Hijwegen AN A
Distocercospora livistonae U. Braun & C.F. Hill A
Eriocercospora olivacea Piroz. AN
Mycophycias apophlaeae (Kohlm.) Kohlm. & Volkm.-Kohlm. M
Mycosphaerella 'nubilosa' sensu Crous et al. A
Mycosphaerella africana Crous & M.J. Wingf. A
Mycosphaerella aleuritis S.H. Ou A
Mycosphaerella anethi (Pers.) Petr. A
Mycosphaerella aristolochiae Cooke E
Mycosphaerella bolleana B.B. Higgins A
Mycosphaerella brassicicola (Duby) Lindau A
Mycosphaerella capsellae A.J. Inman & Sivan. A
Mycosphaerella carinthiaca Jaap A
Mycosphaerella cercidicola (Ellis & Kellerm.) F.A. Wolf A
Mycosphaerella coacervata Syd. E
Mycosphaerella communis Crous & Mansilla A
Mycosphaerella confusa F.A. Wolf A
Mycosphaerella cunninghamii Syd. E
Mycosphaerella davisii F.R. Jones U
Mycosphaerella effigurata (Schwein.) House A
Mycosphaerella enteleae (Dingley) Sivan. E
Mycosphaerella filipendulae-denudatae Kamilov A
Mycosphaerella fragariae (Tul. & C. Tul.) Johanson ex Oudem. A
Mycosphaerella graminicola (Fuckel) J. Schröt. (nom. illegit.) A
Mycosphaerella handelii Crous & U. Braun A
Mycosphaerella hieracii (Sacc. & Briard) Jaap A
Mycosphaerella hondai (I. Miyake) Tomilin A
Mycosphaerella intermedia M. Dick & K. Dobbie A
Mycosphaerella killianii Petr. A
Mycosphaerella linicola Naumov A
Mycosphaerella mariae (Sacc. & E. Bommer) Lindau A
Mycosphaerella marksii Carnegie & Keane A
Mycosphaerella metrosideri F. Stevens & P.A. Young
Mycosphaerella mori F.A. Wolf A
Mycosphaerella nebulosa (Pers.) Lindau OU
Mycosphaerella panacis (Cooke) Tomilin (nom. illegit.)
Mycosphaerella personata B.B. Higgins A
Mycosphaerella pini Rostr. ex Munk A
Mycosphaerella pittospori (Cooke) F.A. Weiss & M.J. O'Brien
Mycosphaerella pomi (Pass.) Lindau A
Mycosphaerella punctiformis (Pers.) J. Schröt. A
Mycosphaerella ribis (Fuckel) Kleb. A
Mycosphaerella rosicola B.H. Davis ex Deighton A
Mycosphaerella rubi Roark A
Mycosphaerella salicorniae (Auersw.) Lindau U M
Mycosphaerella spissa Syd. E
Mycosphaerella superflua (Fuckel) Petr. U
Mycosphaerella swartii R.F. Park & Keane A
Mycosphaerella telopeae M.E. Palm & Crous A
Mycosphaerella ulmi Kleb. A
Mycosphaerella walkeri R.F. Park & Keane A
Oligostroma zelandicum Syd. E
Ovularia malorum Cooke AN A
Passalora assamensis (S. Chowdhury) U. Braun & Crous AN A
Passalora brachycarpa (Syd.) U. Braun & Crous AN A
Passalora dubia (Riess) U. Braun AN A
Passalora fulva (Cooke) U. Braun & Crous AN A
Passalora fusimaculans (G.F. Atk.) U. Braun & Crous AN A
Passalora graminis (Fuckel) Höhn. AN A
Passalora omphacodes (Ellis & Holw.) Crous & U. Braun AN A
Passalora verbeniphila (Speg.) Crous & U. Braun AN A
Passalora vexans (C. Massal.) U. Braun & Crous AN A
Periconiella araliacearum G.F. Laundon AN E
Periconiella coprosmae G.F. Laundon AN E
Periconiella cordylines McKenzie AN E
Periconiella liberatas McKenzie AN E
Periconiella phormii M.B. Ellis AN
Periconiella pomaderris McKenzie AN E
Phaeoisariopsis griseola (Sacc.) Ferraris AN A
Phaeophleospora atkinsonii (Syd.) Pennycook & McKenzie AN
Phaeophleospora phormii (Naito) Crous, F.A. Ferreira & B. Sutton AN
Phloeospora crescentium (Barthol.) E.A. Riley AN A
Phloeospora robiniae (Lib.) Höhn. AN A
Pseudocercospora acerosa U. Braun & M. Dick AN A
Pseudocercospora ackamae U. Braun & C.F. Hill AN E
Pseudocercospora arecacearum U. Braun & C.F. Hill E
Pseudocercospora aristoteliae (Cooke) Deighton AN
Pseudocercospora atromarginalis (G.F. Atk.) Deighton AN A
Pseudocercospora beilschmiediae U. Braun & C.F. Hill AN E
Pseudocercospora camelliae (Deighton) U. Braun

AN A
Pseudocercospora camelliicola U. Braun & C.F. Hill AN A
Pseudocercospora ceanothi (Kellerm. & Swingle) Y.L. Guo & X.J. Liu AN A
Pseudocercospora cladosporioides (Sacc.) U. Braun AN A
Pseudocercospora coprosmae U. Braun & C.F. Hill AN E
Pseudocercospora crousii U. Braun & M. Dick AN A
Pseudocercospora cymbidiicola U. Braun & C.F. Hill AN A
Pseudocercospora dianellae U. Braun & C.F. Hill AN E
Pseudocercospora dingleyae U. Braun & C.F. Hill AN E
Pseudocercospora dodonaeae Boesew. AN
Pseudocercospora escalloniae (Marchion.) U. Braun & C.F. Hill AN A
Pseudocercospora eucalyptorum Crous, M.J. Wingf., Marasas & B. Sutton AN A
Pseudocercospora filipendulae-ulmariae U. Braun & C.F. Hill AN A
Pseudocercospora fuligena (Roldan) Deighton AN A
Pseudocercospora geicola U. Braun AN A
Pseudocercospora gunnerae U. Braun & C.F. Hill A
Pseudocercospora hebicola U. Braun & C.F. Hill AN E
Pseudocercospora karaka (G.F. Laundon) Deighton AN E
Pseudocercospora kurimensis (Fukui) U. Braun AN A
Pseudocercospora libertiae U. Braun & C.F. Hill AN E
Pseudocercospora lonicerae (Chupp) U. Braun & Crous AN A
Pseudocercospora lonicericola (W. Yamam.) Deighton AN A
Pseudocercospora lupini (Cooke) Deighton AN A
Pseudocercospora melicyti U. Braun & C.F. Hill AN E
Pseudocercospora metrosideri U. Braun AN
Pseudocercospora myrticola (Speg.) Deighton AN A
Pseudocercospora nandinae (Nagat.) X.J. Liu & Y.L. Guo AN A
Pseudocercospora nogalesii (Urries) U. Braun & M. Dick AN A
Pseudocercospora ocimicola (Petr. & Cif.) Deighton AN A
Pseudocercospora odontoglossi (Prill. & Delacr.) U. Braun AN A
Pseudocercospora oleariae U. Braun & M. Dick AN E
Pseudocercospora pandoreae U. Braun & C.F. Hill A
Pseudocercospora platensis (Speg.) U. Braun AN
Pseudocercospora pomaderridis U. Braun & C.F. Hill AN E
Pseudocercospora pseudoeucalyptorum Crous AN A
Pseudocercospora rhabdothamni U. Braun & C.F. Hill AN E
Pseudocercospora rumohrae W.H. Hsieh & Goh AN A
Pseudocercospora sawadae (W. Yamam.) Goh & W.H. Hsieh AN A
Pseudocercospora subulata Z.Q. Yuan, de Little & C. Mohammed AN A
Pseudocercospora tibouchinae (Viégas) Deighton A
Pseudocercospora varia (Peck) J.K. Bai & X.J. Cheng AN A
Pseudocercospora virgiliae U. Braun & C.F. Hill AN A
Pseudocercosporella bakeri (Syd. & P. Syd.) Deighton AN A
Pseudocercosporella helenii U. Braun & C.F. Hill AN A
Pseudocercosporella myopori U. Braun & C.F. Hill AN E
Pseudocercosporella pastinacae (P. Karst.) U. Braun AN A
Pseudocercosporella persicariicola U. Braun & C.F. Hill AN A
Pseudocercosporella spiraeigena U. Braun & C.F. Hill AN A
Ramichloridium anceps (Sacc. & Ellis) de Hoog AN
Ramichloridium carlinae (M.B. Ellis) de Hoog AN
Ramichloridium cerophilum (Tubaki) de Hoog AN
Ramichloridium subulatum de Hoog AN U
Ramularia acris Lindr. AN A
Ramularia ajugae (Niessl) Sacc. AN A
Ramularia armoraciae Fuckel AN A
Ramularia bellunensis Speg. AN A
Ramularia beticola Fautrey & Lambotte AN A
Ramularia coleosporii Sacc. AN A
Ramularia collo-cygni B. Sutton & J.M. Waller AN A
Ramularia coprosmae U. Braun & C.F. Hill AN E
Ramularia deusta var. *alba* U. Braun AN A
Ramularia didyma Unger var. *didyma* AN A
— var. *exigua* (U. Braun) U. Braun AN A
Ramularia gunnerae (Speg.) U. Braun AN A
Ramularia hellebori Fuckel AN A
Ramularia holci-lanati (Cavara) Deighton AN A
Ramularia hydrangeae-macrophyllae U. Braun & C.F. Hill AN A
Ramularia lactea (Desm.) Sacc. AN A
Ramularia lamii var. *minor* U. Braun AN OU
Ramularia lapsanae (Desm.) Sacc. AN A
Ramularia occidentalis var. *indica* (K.L. Kothari, M.K. Bhatn. & N.S. Bhatt) U. Braun AN A
Ramularia persicariicola U. Braun & C.F. Hill AN
Ramularia primulae Thüm. AN A
Ramularia pusilla Unger AN A
Ramularia rhabdospora (Berk. & Broome) Nannf. AN A
Ramularia rollandii Fautrey AN A
Ramularia rubella (Bonord.) Nannf. AN A
Ramularia rufomaculans Peck AN A
Ramularia schulzeri Bäumler AN A
Ramularia simplex Pass. AN A
Ramularia sphaeroidea Sacc. AN A
Ramularia spiraeae Peck A
Ramularia subtilis U. Braun & C.F. Hill A
Ramularia tenella U. Braun & C.F. Hill A
Ramularia valerianae var. *centranthi* (Brunaud) U. Braun AN A
Ramularia veronicae Fuckel AN A
Rhabdospora bambusae Viega AN U
Septoria aciculosa Ellis & Everh. AN A
Septoria alnifolia Ellis & Everh. AN A
Septoria alpicola Sacc. AN OU
Septoria antirrhini Desm. AN A
Septoria apiicola Speg. AN A
Septoria azaleae Voglino AN A
Septoria betae Westend. AN A
Septoria betulae Pass. AN A
Septoria bromi Sacc. AN A
Septoria caespitulosa Sacc. AN A
Septoria cerastii Desm. AN A
Septoria cercidis Fr. ex Lév. AN A
Septoria chrysanthemella Sacc. AN A
Septoria cirsii Niessl AN A
Septoria citri Pass. AN A
Septoria colensoi Cooke AN
Septoria convolvuli Desm. AN A
Septoria coprosmae Cooke AN E
Septoria coriariae Pass. AN
Septoria cucurbitacearum Sacc. AN A
Septoria cunninghamii Syd. AN E
Septoria cyclaminis Durieu & Mont. AN A
Septoria dianthi Desm. AN A
Septoria elymi Ellis & Everh. AN A
Septoria eucalypti G. Winter & Roum. A
Septoria exotica Speg. AN
Septoria ficariae Desm. AN A
Septoria fuchsiicola P. Syd. AN A
Septoria gerberae Syd. & P. Syd. AN A
Septoria gladioli Pass. AN A
Septoria hedericola (Fr.) Jørst. AN A
Septoria henriquesii Thüm. AN A
Septoria humuli Westend. AN A
Septoria lactucae Pass. AN A
Septoria lamii Pass. AN A
Septoria leucanthemi Sacc. & Speg. AN A
Septoria lycopersici Speg. AN A
Septoria macropoda Pass. AN A
Septoria magnoliae Cooke AN A
Septoria malvicola Ellis & G. Martin AN A
Septoria nesodes Kalchbr. AN A
Septoria oenotherae (Lasch) Westend. AN A
Septoria orchidearum Westend. AN A
Septoria paeoniae Westend. AN A
Septoria passerinii Sacc. AN A
Septoria passifloricola Punith. AN A
Septoria pepli D.E. Shaw AN A
Septoria perforans McAlpine AN A
Septoria persicariae O'Gara AN A
Septoria petroselini (Lib.) Desm. AN A
Septoria phlogis Sacc. & Speg. AN A
Septoria phyllodiorum Sacc. A
Septoria pisi Westend. AN A
Septoria plantaginis (Ces.) Sacc. AN A
Septoria polygonorum Desm. AN A
Septoria ranunculacearum Lév. A
Septoria rosae Desm. AN A
Septoria rubi Westend. AN A
Septoria scabiosicola (Desm.) Desm. AN A
Septoria selenophomoides E.K. Cash & A.J. Watson AN A
Septoria silenes Westend. AN A
Septoria silenes-nutantis C. Massal. ex Sacc. AN A
Septoria sisymbrii Ellis AN A
Septoria sisyrinchii Speg. AN A
Septoria slaptonensis D. Hawksw. & Punith. AN A
Septoria sonchi Sacc. AN A
Septoria stachydis Desm. AN A
Septoria stellariae Desm. AN A
Septoria triseti Speg. AN A
Septoria tritici var. *loliicola* R. Sprague & Aar.G. Johnson AN A
Septoria typica Gadgil & M. Dick AN A
Septoria unedonis Desm. AN A
Septoria urticae Roberge ex Desm. AN A
Septoria verbenae Desm. AN A
Sphaerella aristoteliae Cooke E
Sphaerella depressa (Berk.) Cooke E
Sphaerella junciginea Cooke E
Sphaerella weinmanniae Cooke E
Sphaerulina todeae (Cooke) Berl. & Voglino E
Stenella aucklandica U. Braun & C.F. Hill AN E
Stenella gahniae McKenzie AN E
Stenella novae-zelandiae Matsush. AN E
Stenella palmicola Matsush. AN E
Stenella pittospori U. Braun
Stenella sinuosogeniculata U. Braun & C.F. Hill AN E
Stenella tristaniae B. Huguenin AN A
Stigmidium congestum (Körb.) Triebel
Stigmidium frigidum (Sacc.) Alstrup & D. Hawksw.
Stigmidium peltideae (Vain.) R. Sant.
Stigmidium pumilum (Lettau) Matzer & Hafellner
Stigmidium schaereri (A. Massal.) Trevis.
Stigmidium xanthoparmeliarum Hafellner
Stigmina longispora (M.B. Ellis) S. Hughes AN F
Stigmina palmivora (Sacc.) S. Hughes AN A
Stigmina platani (Fuckel) Sacc. AN A

Stigmina thujina (Dearn.) B. Sutton AN A
SCHIZOTHYRIACEAE
Microthyriella hibisci F. Stevens A
Microthyriella phaeospora Hansf. E
Zygophiala jamaicensis E.W. Mason AN A
TERATOSPHAERIACEAE
Devriesia hilliana Crous & U. Braun AN A
Hortaea werneckii (Horta) Nishim. & Miyaji AN A
Readeriella mirabilis Syd. & P. Syd. AN A
Readeriella novae-zelandiae Crous AN A
Teratosphaeria concentrica (Racib.) E. Müll.

Order DOTHIDEALES
Colensoniella torulispora (W. Phillips) Hafellner E
Phaeocryptopus gaeumannii (T. Rohde) Petr. A
DOTHIDEACEAE
Dothidea filicina Mont. ex Berk. (nom. illegit.) OU
Dothidea vacciniorum Lév. E
Pachysacca pusilla H.J. Swart A
Vestergrenia leucopogonis Hansf.
DOTHIORACEAE
Aureobasidium caulivorum (Kirchn.) W.B. Cooke AN A
Aureobasidium pullulans (de Bary) G. Arnaud var. *pullulans* AN U
— var. *melanigenum* Herm.-Nijh. AN A
Aureobasidium zeae (Narita & Y. Hirats.) Dingley ex Herm.-Nijh. AN A
Discosphaerina fulvida (F.R. Sand.) Sivan. A
Dothichiza populea Sacc. & Briard AN A
Kabatiella borealis (Ellis & Everh.) Arx AN A
Kabatina thujae R. Schneid. & Arx AN A
Metasphaeria lindsayana (Curr.) Sacc. E
Metasphaeria orthospora Sacc. A
Plowrightia abietis (M.E. Barr) M.E. Barr A
Plowrightia ribesia (Pers.) Sacc. A
Sydowia polyspora (Bref. & Tavel) E. Müll. A
EPIGLOEACEAE
Epigloea soleiformis Döbbeler 1984
EREMOMYCETACEAE
Arthrographis kalrae (R.P. Tewari & Macph.) Sigler & J.W. Carmich. AN U

Order HYSTERIALES
HYSTERIACEAE
Acrogenospora gigantospora S. Hughes AN F
Acrogenospora novae-zelandiae S. Hughes AN E
Acrogenospora sphaerocephala (Berk. & Broome) M.B. Ellis AN
Gloniella pseudocomma Rehm E
Gloniopsis praelonga (Schwein.) Underw. & Earle A
Glonium hysterinum Rehm
Glonium sp. 'desmoschoenus' (nom. ined.) E
Hysterium angustatum Pers.
Hysterium barrianum E.W.A. Boehm, A.N. Mill., Mugambi, Huhndorf, C.L. Scoch
Hysterium insidens Schwein. A
Hysterium pulicare Pers. OU
Hysterium sinuosum Cooke E
Hysterobrevium constrictum (N. Amano) E.W.A. Boehm & C.L. Schoch
Hysterographium fraxini (Pers.) De Not. A

Order JAHNULALES
ALIQUANDOSTIPITACEAE
Xylomyces aquaticus (Dudka) K.D. Hyde & Goh AN F
LICHENOTHELIACEAE
Lichenostigma cosmopolites Haffelner & Calatayud
Lichenostigma elongatum Nav.-Ros. & Hafellner U
Lichenostigma rugosum G. Thor
MASTODIACEAE
Mastodia tessellata (Hook.f. & Harv.) Hook.f. & Harv. L M

Order MELIOLALES
MELIOLACEAE
Appendiculella acaenae Hansf.
Asteridiella coprosmae Hansf.
Asteridiella knightiae S. Hughes E
Endomeliola dingleyae S. Hughes & Piroz. E
Irenopsis hoheriae Hansf. E
Meliola argentina Speg.
Meliola beilschmiediae W. Yamam.
Meliola capensis (Kalchbr. & Cooke) Theiss. OU
Meliola cookeana Speg.
Meliola cyathodis Hansf.
Meliola formosensis W. Yamam.
Meliola leptospermi Hansf. OU
Meliola notelaeae Hansf.
Meliola panici Earle
Meliola peltata Doidge
Meliola pomaderridis Hansf. A
Meliola ripogoni Hansf.
MELIOLINACEAE
Meliolina leptospermi S. Hughes E
Meliolina metrosideri S. Hughes E
Meliolina novae-zealandiae Hansf. E
MICROPELTIDACEAE
Micropeltis applanata Mont. U

Order MICROTHYRIALES
AULOGRAPHACEAE
Aulographum bromi Berk. U
MICROTHYRIACEAE
Asterinella intensa (Cooke & Massee) Theiss.
Lichenopeltella epiphylla R. Sant.
Lichenopeltella maculans (Zopf) Höhn. U
Microthyriacites grandis Cookson E Fo
Microthyrium microscopicum Desm. A
Phaeothyriolum microthyrioides (G. Winter) H.J. Swart A
Trichopeltula hedycaryae Theiss. E
Trichothyrium asterophorum (Berk. & Broome) Höhn.
Trichothyrium reptans (Berk. & M.A. Curtis) S. Hughes

Order MYRIANGIALES
ELSINOACEAE
Elsinoe ampelina Shear A
Elsinoe dracophylli P.R. Johnst. & Beever E
Elsinoe eucalypti Hansf. A
Elsinoe fawcettii Bitanc. & Jenkins A
Elsinoe mattirolianum G. Arnaud & Bitanc. A
Elsinoe parthenocissi Jenkins & Bitanc. A
Elsinoe pyri (Woron.) Jenkins A
Elsinoe rosarum Jenkins & Bitanc. A
Elsinoe takoropuku G.S. Ridl. & Ramsfield E
Elsinoe tiliae Creelman A
Elsinoe veneta (Burkh.) Jenkins A
Melanophora sorbi (Rostr.) Arx AN A
Sphaceloma hederae Bitanc. & Jenkins AN A
Sphaceloma murrayae Grodz. & Jenkins AN A
Sphaceloma populi (Sacc.) Jenkins AN A
Sphaceloma psidii Bitanc. & Jenkins AN A
Sphaceloma sorbi (Rostr.) Jenkins AN A
Sphaceloma violae Jenkins AN A
MYRIANGIACEAE
Myriangium duriaei Mont. & Berk. U
MYXOTRICHACEAE
Malbranchea cinnamomea (Lib.) Oorschot & de Hoog AN U
Malbranchea sclerotica Guarro, Gené & De Vroey AN U
Myxotrichum deflexum Berk. U
Oidiodendron flavum Svilv. AN U
Oidiodendron griseum Robak AN U
Oidiodendron rhodogenum Robak AN A
Oidiodendron truncatum G.L. Barron AN U

PARMULARIACEAE
Clypeum peltatum Massee E
Hemigrapha asteriscus (Müll. Arg.) R. Sant. ex D. Hawksw.
Hemigrapha nephromatis Wedin & Diederich
Lauterbachiella dicksoniifolia Dingley E
Lauterbachiella filicina (Berk. & Broome) Dingley
Rhagadolobium bakerianum Sacc.
Rhagadolobium hemiteliae Henn. & Lindau
PARODIOPSIDACEAE
Dimeriella ammophilae Hansf.

Order PATELLARIALES
PATELLARIACEAE
Brunaudia phormiigena (Cooke) Kuntze E
Patellaria atrata (Hedw.) Fr. U
Rhizodiscina lignyota (Fr.) Hafellner U
Rhytidhysteron rufulum (Spreng.) Speg.

Order PLEOSPORALES
Amarenomyces ammophilae (Lasch) O.E. Erikss. U M
Anguillospora crassa Ingold AN F
Anguillospora filiformis Greath. AN F
Anguillospora longissima (Sacc. & P. Syd.) Ingold AN F
Anteaglonium abbreviatum (Schwein.) Mugambi & Huhndorf U
Anteaglonium parvulum (W.R. Gerard) Mugambi & Huhndorf
Ascochyta avenae (Petr.) R. Sprague & Aar.G. Johnson AN A
Ascochyta caricae Rabenh. AN A
Ascochyta caulicola Laubert AN A
Ascochyta chenopodii Rostr. AN A
Ascochyta corticola McAlpine AN A
Ascochyta desmazieresii Cavara AN A
Ascochyta fabae f.sp. *lentis* Gossen, Sheard, C.J. Beauch. & Morrall AN A
Ascochyta gerberae Maffei AN A
Ascochyta graminicola Sacc. AN U
Ascochyta leptospora (Trail) Hara AN A
Ascochyta manawaorae Verkley, Woudenberg & Gruyter AN E
Ascochyta nebulosa Sacc. & Berl. AN A
Ascochyta paspali (Syd.) Punith. AN A
Ascochyta sorghi Sacc. AN A
Ascochyta stilbocarpae Syd. AN E
Ascochyta subalpina R. Sprague & Aar.G. Johnson AN A
— f. *penniseti* Punith. AN A
Ascochyta ulicis (Grove) P.K. Buchanan AN A
Ascochyta viciae Lib. AN A
Ascochytella stagonosporoidea Gonz. Frag. AN A
Boeremia exigua (Desm.) Aveskamp, Gruyter & Verkley var. *exigua* AN A
— var. *heteromorpha* (Schulzer & Sacc.) Aveskamp, Gruyter & Verkley AN A
— var. *linicola* (Naumov & Vassiljevsky) Aveskamp, Gruyter & Verkley AN A
Boeremia foveata (Foister) Aveskamp, Gruyter & Verkley AN A
Boeremia hedericola (Durieu & Mont.) Aveskamp, Gruyter & Verkley AN A
Boeremia strasseri (Moesz) Aveskamp, Gruyter & Verkley AN A
Clavariopsis aquatica De Wild. AN F
Coronospora novae-zelandiae Matsush. AN E
Dictyosporium elegans Corda AN M
Dictyosporium foliicola P.M. Kirk AN OU
Dictyosporium freycinetiae McKenzie AN E
Dictyosporium heptasporum (Garov.) Damon AN
Dictyosporium hughesii McKenzie AN E
Dictyosporium pelagicum (Linder) G.C. Hughes ex E.B.G. Jones AN

Dictyosporium rhopalostylidis McKenzie AN E
Dictyosporium toruloides (Corda) Guég. AN A F
Didymella applanata (Niessl) Sacc. A
Didymella carduicola (Cooke) Sacc. U
Didymella fabae G.J. Jellis & Punith. A
Didymella graminicola Punith. A
Didymella hierochloes Petr. E
Didymella lycopersici Kleb. A
Didymella nigrella (Fr.) Sacc. A
Didymella phleina Punith. & Årsvoll A
Didymella pisi M.I. Chilvers, J.D. Rogers & T.L. Peever A
Didymella rabiei Arx (nom. inv.) A
Digitodesmium elegans P.M. Kirk AN U
Farlowiella carmichaeliana (Berk.) Sacc.
Herpotrichia herpotrichoides (Fuckel) P.F. Cannon U
Leptosphaerulina argentinensis (Speg.) J.H. Graham & Luttr. A
Leptosphaerulina australis McAlpine A
Leptosphaerulina chartarum Cec. Roux A
Leptosphaerulina trifolii (Rostr.) Petr. A
Mycocentrospora acerina (R. Hartig) Deighton AN A F
Mycocentrospora sp. sensu Aimer & Segedin (nom. ined.) AN U F
Mycocentrospora varians R.C. Sinclair & Morgan-Jones AN A F
Periconia atra Corda AN A
Periconia byssoides Pers. AN A
Periconia cambrensis E.W. Mason & M.B. Ellis AN A
Periconia echinochloae (Bat.) M.B. Ellis AN (nom. illegit.) A
Periconia hispidula (Pers.) E.W. Mason & M.B. Ellis AN U
Periconia lateralis Ellis & Everh. AN A
Periconia macrospinosa Lefebvre & Aar.G. Johnson AN A
Periconia minutissima Corda AN
Periconia saraswatipurensis Bilgrami AN A
Phoma apiicola Kleb. AN A
Phoma bismarckii Kidd & Beaumont AN A
Phoma chrysanthemicola Hollós AN A
Phoma clematidina (Thüm.) Boerema AN A
Phoma colensoi Cooke AN A
Phoma commelinicola (E. Young) Gruyter AN A
Phoma costaricensis Echandi AN A
Phoma cytospora (Vouaux) D.Hawksw. AN A
Phoma digitalis Boerema AN A
Phoma drobnjacensis Bubák AN A
Phoma dubia (Linds.) Sacc. & Trotter AN E
Phoma enteroleuca var. *influorescens* Boerema & Loer. AN A
Phoma eupyrena Sacc. AN A
Phoma exigua var. *inoxydabilis* Boerema & Vegh AN A
Phoma fallax Berk. AN E
Phoma fallens Sacc. AN A
Phoma fimeti Brunaud AN A
Phoma galegae Thüm. AN A
Phoma haematocycla (Berk.) Aa & Boerema AN A
Phoma herbarum Westend. AN A
Phoma huancayensis Turkenst. AN A
Phoma laundoniae Boerema & Gruyter AN A
Phoma leveillei Boerema & G.J. Bollen AN A
Phoma macrostoma Mont. var. *macrostoma* AN A
— var. *incolorata* (A.S. Horne) Boerema & Dorenb. AN A
Phoma medicaginis Malbr. & Roum. AN A
Phoma omnivora McAlpine AN A
Phoma paspali P.R. Johnst. AN A
Phoma physciicola Keissl. AN
Phoma plurivora P.R. Johnst. AN A
Phoma podocarpi Massee AN E
Phoma pratorum P.R. Johnst. & Boerema AN A
Phoma rhei (Ellis & Everh.) Aa & Boerema AN A
Phoma rhodorae Cooke AN A
Phoma rumicicola Boerema & Loer. AN A
Phoma senecionis P. Syd. AN A
Phoma violicola P. Syd. AN A
Phoma viridis Cooke AN A
Phoma wasabiae Yokogi A
Pyrenochaeta lycopersici R. Schneid. & Gerlach AN A
Pyrenochaeta terrestris (H.N. Hansen) Gorenz, J.C. Walker & Larson AN A
Sporidesmium fragilissimum (Berk. & M.A. Curtis) M.B. Ellis AN
Sporidesmium goidanichii (Rambelli) S. Hughes A
Sporidesmium hyalospermum (Corda) S. Hughes var. *hyalospermum* AN
Sporidesmium leptosporum (Sacc. & Roum.) S. Hughes AN
Sporidesmium omahutaense Matsush. AN E
Sporidesmium paludosum M.B. Ellis AN A F
Sporidesmium parvum (S. Hughes) M.B. Ellis AN A
Sporidesmium pedunculatum (Peck) M.B. Ellis AN
Sporidesmium verrucisporum M.B. Ellis AN
Stagonosporopsis aquilegiae (Rabenh.) Boerema, Gruyter & Noordel. AN A
Stagonosporopsis cucurbitacearum (Fr.) Aveskamp, Gruyter & Verkley A
Stagonosporopsis hortensis (Sacc. & Malbr.) Petr. AN A
Stagonosporopsis ligulicola (K.F. Baker, Dimock & L.H. Davis) Aveskamp, Gruyter & verkley A
Stagonosporopsis loticola (Died.) Aveskamp, Gruyter & Verkley AN A
AMNICULICOLACEAE
Neomassariosphaeria typhicola (P. Karst.) Yin. Zhang, J. Fourn. & K.D. Hyde M
ARTHOPYRENIACEAE
Arthopyrenia cinereopruinosa (Schaer.) A.Massal. L
Arthopyrenia gemellipara (C.Knight) Müll.Arg. L
Arthopyrenia leptiza (Stirt.) Müll.Arg. E L
Arthopyrenia minutella (C.Knight) Müll.Arg. E L
Arthopyrenia peltigerella Zahlbr. E L
Mycomicrothelia minutissima (C.Knight) D.Hawksw. E L
Mycomicrothelia striguloides Sérus. & Aptroot E L
CORYNESPORASCACEAE
Corynespora cassiicola (Berk. & M.A. Curtis) C.T. Wei AN A
Corynespora citricola M.B. Ellis AN A
Corynespora lanneicola Deighton & M.B. Ellis AN
Corynespora leptoderridicola Deighton & M.B. Ellis AN
Corynespora pruni (Berk. & M.A. Curtis) M.B. Ellis AN A
Corynespora ripogoni McKenzie AN E
Corynespora smithii (Berk. & Broome) M.B. Ellis AN
CUCURBITARIACEAE
Rhytidiella beloniza (Stirt.) M.B. Aguirre E L
Rhytidiella hebes P.R. Johnst. E
DACAMPIACEAE
Byssothecium circinans Fuckel A
Clypeococcum grossum (Körb.) D. Hawksw.
Polycoccum crespoae Váczi & D. Hawksw.
Polycoccum galligenum Vězda U
Polycoccum jamesii D. Hawksw.
Polycoccum pulvinatum (Eitner) R. Sant.
Polycoccum rugulosarium (Linds.) D. Hawksw.
Polycoccum squamarioides (Mudd) Arnold
Polycoccum stictaria (Linds.) D.J. Galloway
Polycoccum vermicularium (Linds.) D. Hawksw.
Pyrenidium actinellum Nyl.
Weddellomyces aspiciliicola Alstrup
DELITSCHIACEAE
Delitschia canina Mouton A
Delitschia didyma Auersw. A
Delitschia marchalii Berl. & Voglino A
Delitschia pachylospora Luck-Allen & Cain A
Delitschia patagonica Speg. A
Delitschia tomentosa Luck-Allen & Cain A
Delitschia winteri (W. Phillips & Plowr.) Sacc. A
DIADEMACEAE
Platyspora pentamera (P. Karst.) Wehm. A
DIDYMOSPHAERIACEAE
Didymosphaeria conoidella Sacc. & Berl.
Didymosphaeria futilis (Berk. & Broome) Rehm
DOTHIDOTTHIACEAE
Spencermartinsia viticola (A.J.L. Phillips & J. Luque) A.J.L. Phillips, A. Alves & Crous A
LEPTOSPHAERIACEAE
Coniothyrium celmisiae Syd. AN E
Coniothyrium celtidis Brunaud AN A
Coniothyrium concentricum (Desm.) Sacc. AN A
Coniothyrium dracaenae F. Stevens & Weedon AN A
Coniothyrium hellebori Cooke & Massee AN A
Coniothyrium ovatum H.J. Swart AN A
Coniothyrium rhododendri Henn. AN A
Coniothyrium sphaerospermum Fuckel AN A
Coniothyrium sporulosum (W. Gams & Domsch) Aa AN A
Leptosphaeria avenaria G.F. Weber A
Leptosphaeria avicenniae Kohlm. & E. Kohlm. M
Leptosphaeria caricicola Fautrey U
Leptosphaeria complanata (Tode) Ces. & De Not. U
Leptosphaeria coniothyrium (Fuckel) Sacc. A
Leptosphaeria doliolum (Pers.) Ces. & De Not. A
Leptosphaeria eustoma (Fr.) Sacc. A
Leptosphaeria maculans (Desm.) Ces. & De Not. A
Leptosphaeria nesodes (Berk. & Broome) Sacc.
Leptosphaeria orae-maris Linder OU M
Leptosphaeria pelagica E.B.G. Jones OU M
Leptosphaeria petkovicensis Bubák & Ranoj. OU
Leptosphaeria plagia (Cooke & Massee) L. Holm U M
Leptosphaeria praetermissa (P. Karst.) Sacc. OU
Leptosphaeria pratensis Sacc. & Briard A
Leptosphaeria reidiana Syd. E
Leptosphaeria typhae (Auersw.) Sacc.
Peyronellaea aurea (Gruyter, Noordel. & Boerema) Aveskamp, Gruyter & Verkley AN A
Peyronellaea australis Aveskamp, Gruyter & Verkley AN A
Peyronellaea curtisii (Berk.) Aveskamp, Gruyter & Verkley AN A
Peyronellaea glomerata (Corda) Goid. ex Togliani AN A
Peyronellaea musae P. Joly AN A
Peyronellaea pinodella (L.K. Jones) Aveskamp, Gruyter & Verkley AN A
Peyronellaea pinodes (Berk. & A. Bloxam) Aveskamp, Gruyter & Verkley A
Peyronellaea pomorum (Thüm.) Aveskamp, Gruyter & Verkley AN A
Plenodomus destruens Harter AN A
LOPHIOSTOMATACEAE
Ascocratera manglicola Kohlm. M
Byssolophis sphaerioides (P. Karst.) E. Müll.
Entodesmium niesslianum (Rabenh. ex Niessl) L. Holm A
Lophiostoma appendiculatum Fuckel U
Lophiostoma caudatum Fabre A
Lophiostoma caulium (Fr.) Ces. & De Not. A
Lophiostoma dacryosporum Fabre A
Lophiostoma corticola (Fuckel) E.C.Y. Liew, Aptroot & K.D. Hyde A
Lophiostoma semiliberum (Desm.) Ces. & De Not. A
Lophiostoma tetraploa (Scheuer) Aptroot & K.D. Hyde F
MASSARINACEAE

Helminthosporium dictyoseptatum S. Hughes AN E
Helminthosporium foveolatum Pat. AN A
Helminthosporium novae-zelandiae S. Hughes AN E
Helminthosporium palmigenum Matsush. AN
Helminthosporium solani Durieu & Mont. AN A
Helminthosporium velutinum Link AN U
Keissleriella culmifida (P. Karst.) S.K. Bose A
Keissleriella taminensis (H. Wegelin) S.K. Bose U
Massarina contraria (Syd.) Arx & E. Müll. E
Massarina waikanaensis (G.S. Ridl.) Shoemaker & C.E. Babc.
Massarina walkeri Shoemaker, C.E. Babc. & J.A.G. Irwin A
Pseudodictyosporium wauense Matsush. AN U
MELANOMMATACEAE
Astrosphaeriella aosimensis Hino & Katumoto
Astrosphaeriella bakeriana (Sacc.) K.D. Hyde & J. Fröhl. U
Astrosphaeriella samuelsii (Boise) K.D. Hyde & J. Fröhl.
Byssosphaeria rhodomphala (Berk.) Cooke U
Byssosphaeria schiedermayeriana (Fuckel) M.E. Barr
Melanomma cinereum (P. Karst.) Sacc. A
Melanomma martinianum (Linds.) Sacc. E
Melanomma pulvis-pyrius (Pers.) Fuckel
Melanomma sparsum Fuckel OU
Ohleria brasiliensis Starbäck
Sporidesmiella hyalosperma (Corda) P.M. Kirk var. *hyalosperma* AN
— var. *novae-zelandiae* (S. Hughes) P.M. Kirk AN
Sporidesmiella parva (M.B. Ellis) P.M. Kirk var. *parva* AN A
Xenolophium lanuginosum A.E. Bell & Mahoney E
Xenolophium pseudotrichioides A.E. Bell & Mahoney E
MONTAGNULACEAE
Bimuria novae-zelandiae D. Hawksw., Chea & Sheridan A
Microsphaeropsis conielloides B. Sutton AN A
Microsphaeropsis olivacea (Bonord.) Höhn. AN U
Microsphaeropsis onychiuri (Punith.) Morgan-Jones AN E
Microsphaeropsis pittospororum (Sacc.) G.F. Laundon AN
Montagnula rhodophaea (Bizz.) Leuchtm. U
Paraconiothyrium minitans (W.A. Campb.) Verkley AN A
Paraphaeosphaeria michotii (Westend.) O.E. Erikss. A
MYTILINIDIACEAE
Glyphium elatum (Grev.) H. Zogg OU
Lophium mytilinum (Pers.) Fr. A
Mytilinidion mytilinellum (Fr.) H. Zogg U
NAETROCYMBACEAE
Leptorhaphis haematommatum Hafellner & Halb L
Naetrocymbe punctiformis (Pers.) R.C.Harris L
PARODIELLACEAE
Parodiella maculata Massee E
PHAEOSPHAERIACEAE
Ampelomyces quisqualis Ces. AN A
Cicinobolus euonymi-japonici Arcang. AN A
Eudarluca australis Speg. OU
Eudarluca caricis (Fr.) O.E. Erikss.
Hendersonia grossulariae Oudem. AN A
Hendersonia microsticta Berk. E
Hendersonia sarmentorum Westend. AN OU
Nodulosphaeria erythrospora (Riess) L. Holm U
Ophiosphaerella herpotricha (Fr.) J. Walker A
Ophiosphaerella korrae (J. Walker & A.M. Sm.) Shoemaker & C.E. Babc. A
Ophiosphaerella narmari (J. Walker & A.M. Sm.) H.C. Wetzel, Hulbert & Tisserat A
Phaeoseptoria airae (Grove) R. Sprague AN A
Phaeosphaeria culmorum (Auersw.) Leuchtm. A
Phaeosphaeria eustoma (Fuckel) L. Holm A
Phaeosphaeria fuckelii (Niessl) L. Holm A
Phaeosphaeria halima (T.W. Johnson) Shoemaker & C.E. Babc. M
Phaeosphaeria kukutae G.S. Ridl. E
Phaeosphaeria lactuosa (Niessl) Y. Otani & Mikawa A
Phaeosphaeria microscopica (P. Karst.) O.E. Erikss. A
Phaeosphaeria nigrans (Desm.) L. Holm A M
Phaeosphaeria nodorum (E. Müll.) Hedjar. A
Phaeosphaeria sylvatica (Pass.) Hedjar. A
Phaeosphaeria typharum (Desm.) L. Holm A M
Phaeosphaeria vagans (Niessl) O.E. Erikss. A
Stagonospora arenaria (Sacc.) Sacc. AN A
Stagonospora atriplicis Lind AN A
Stagonospora calystegiae (Westend.) Grove AN A
Stagonospora hyalospora (Berk.) Sacc. AN E
Stagonospora innumerosa (Desm.) Sacc. AN U
Stagonospora maculata (Grove) R. Sprague AN A
Stagonospora paludosa (Sacc. & Speg.) Sacc. AN A
Stagonospora samararum (Desm.) Boerema AN A
Stagonospora triodiae Petr. AN E
PHAEOTRICHACEAE
Trichodelitschia bisporula (P. Crouan & H. Crouan) Munk A
Trichodelitschia munkii N. Lundq. A
PLEOMASSARIACEAE
Trematosphaeria crassiseptata Kaz. Tanaka, Y. Harada & M.E. Barr
PLEOSPORACEAE
Alternaria alternata (Fr.) Keissl. AN
Alternaria anagallidis A. Raabe AN A
Alternaria arborescens E.G. Simmons A
Alternaria argyranthemi E.G. Simmons & C.F. Hill AN A
Alternaria armoraciae E.G. Simmons & C.F. Hill AN A
Alternaria ascaloniae E.G. Simmons & C.F. Hill AN A
Alternaria axiaeriisporifera E.G. Simmons & C.F. Hill AN A
Alternaria beticola E.G. Simmons & C.F. Hill AN A
Alternaria brassicae (Berk.) Sacc. AN A
Alternaria brassicicola (Schwein.) Wiltshire AN A
Alternaria calendulae Ondřej AN, (nom. illegit.) A
Alternaria carotiincultae E.G. Simmons A
Alternaria cerealis E.G. Simmons & C.F. Hill AN A
Alternaria cheiranthi (Lib.) P.C. Bolle AN A
Alternaria cinerariae Hori & Enjoji AN A
Alternaria citri Ellis & N. Pierce AN A
Alternaria crassa (Sacc.) Rands AN A
Alternaria cucumericola E.G. Simmons & C.F. Hill AN A
Alternaria cucumerina (Ellis & Everh.) J.A. Elliott AN A
Alternaria cyphomandrae E.G. Simmons AN A
Alternaria dauci (J.G. Kühn) J.W. Groves & Skolko AN A
Alternaria dianthicola Neerg. AN A
Alternaria dichondrae Gambogi, Vannacci & Triolo AN A
Alternaria frumenti E.G. Simmons & C.F. Hill AN A
Alternaria gaurae E.G. Simmons & C.F. Hill A
Alternaria geniostomatis E.G. Simmons & C.F. Hill AN
Alternaria glyceriae E.G. Simmons & C.F. Hill AN A
Alternaria herbiculinae E.G. Simmons AN A
Alternaria hordeiseminis E.G. Simmons & G.F. Laundon AN A
Alternaria iridiaustralis E.G. Simmons, Alcorn & C.F. Hill AN A
Alternaria japonica Yoshii AN A
Alternaria linicola J.W. Groves & Skolko A
Alternaria longipes (Ellis & Everh.) E.W. Mason AN A
Alternaria merytae E.G. Simmons AN E
Alternaria nobilis (Vize) E.G. Simmons AN A
Alternaria novae-zelandiae E.G. Simmons AN A
Alternaria panax Whetzel AN
Alternaria passiflorae J.H. Simmonds AN A
Alternaria porri (Ellis) Cif. AN A
Alternaria radicina Meier, Drechsler & E.D. Eddy AN A
Alternaria rosae E.G. Simmons & C.F. Hill AN A
Alternaria rosifolii E.G. Simmons & C.F. Hill AN A
Alternaria selini E.G. Simmons AN A
Alternaria senecionicola E.G. Simmons & C.F. Hill AN A
Alternaria solani Sorauer AN A
Alternaria solani-nigri R. Dubey, S.K. Singh & Kamal AN A
Alternaria sonchi Davis AN A
Alternaria tenuissima (Kunze) Wiltshire AN A
Alternaria tomatophila E.G. Simmons AN A
Alternaria viciae-fabae E.G. Simmons & G.F. Laundon AN A
Alternaria zinniae M.B. Ellis AN A
Alternariaster helianthi (Hansf.) E.G. Simmons AN A
Bipolaris brizae (Nisik.) Shoemaker AN A
Bipolaris iridis (Oudem.) C.H. Dickinson AN A
Bipolaris portulacae (Rader) Alcorn AN A
Bipolaris sacchari (E.J. Butler) Shoemaker AN A
Cerebella andropogonis Ces. AN A
Chalastospora gossypii (Jacz.) U. Bruan & Crous A
Cochliobolus australiensis (Tsuda & Ueyama) Alcorn A
Cochliobolus bicolor A.R. Paul & Parbery A
Cochliobolus carbonum R.R. Nelson A
Cochliobolus cynodontis R.R. Nelson A
Cochliobolus geniculatus R.R. Nelson A
Cochliobolus hawaiiensis Alcorn A
Cochliobolus heterostrophus (Drechsler) Drechsler A
Cochliobolus intermedius R.R. Nelson A
Cochliobolus lunatus R.R. Nelson & F.A. Haasis A
Cochliobolus nodulosus Luttr. A
Cochliobolus ravenelii Alcorn A
Cochliobolus sativus (S. Ito & Kurib.) Drechsler ex Dastur A
Cochliobolus spicifer R.R. Nelson A
Cochliobolus verruculosus (Tsuda & Ueyama) Sivan. A
Crivellia papaveracea (De Not.) Shoemaker & Inderbitzin A
Curvularia brachyspora Boedijn AN A
Curvularia clavata B.L. Jain AN A
Curvularia gladioli Boerema & Hamers AN A
Curvularia inaequalis (Shear) Boedijn AN A
Curvularia protuberata R.R. Nelson & Hodges AN A
Curvularia trifolii (Kauffman) Boedijn AN A
Dendryphiella vinosa (Berk. & M.A. Curtis) Reisinger AN
Dendryphion comosum Wallr. AN
Dendryphion nanum (Nees) S. Hughes AN
Drechslera andersenii A. Lam AN A
Drechslera fugax (Wallr.) Shoemaker AN A
Drechslera nobleae McKenzie & D. Matthews AN A
Drechslera phlei (J.H. Graham) Shoemaker AN A
Drechslera poae (Baudyš) Shoemaker AN A
Embellisia abundans E.G. Simmons AN A
Embellisia allii (Campan.) E.G. Simmons AN A
Embellisia leptinellae E.G. Simmons & C.F. Hill AN E
Embellisia lolii E.G. Simmons & C.F. Hill AN A
Embellisia novae-zelandiae E.G. Simmons & C.F. Hill AN E
Epicoccum pallescens Berk. AN E
Epicoccum purpurascens Ehrenb. AN
Lewia infectoria (Fuckel) M.E. Barr & E.G. Simmons A

Macrospora scirpicola (DC.) Fuckel
Macrosporium obtusum Berk. AN E
Marielliottia biseptata (Sacc. & Roum.) Shoemaker AN A
Marielliottia dematioidea (Bubák & Wróbl.) Shoemaker AN A
Marielliottia triseptata (Drechsler) Shoemaker AN A
Nimbya alternantherae (Holcomb & Antonop.) E.G. Simmons & Alcorn AN A
Pithomyces chartarum (Berk. & M.A. Curtis) M.B. Ellis AN U
Pithomyces maydicus (Sacc.) M.B. Ellis AN A
Pithomyces valparadisiacus (Speg.) P.M. Kirk AN A
Pleospora allii (Rabenh.) Ces. & De Not. A
Pleospora bjoerlingii Byford A
Pleospora calvescens (Fr.) Tul. A
Pleospora chlamydospora Sacc. A
Pleospora eturmiuna E.G. Simmons A
Pleospora herbarum (Fr.) Rabenh. A
— f. *lactucum* Padhi & Snyder A
Pleospora penicillus Fuckel U
Pleospora scirpicola (DC.) P. Karst. U
Pleospora scrophulariae (Desm.) Höhn. OU
Pleospora sedicola E.G. Simmons A
Pleospora straminis Sacc. A
Pleospora tarda E.G. Simmons U
Pleospora valesiaca (Niessl) E. Müll. U
Pyrenophora chaetomioides Speg. A
Pyrenophora dactylidis Ammon A
Pyrenophora dictyoides A.R. Paul & Parbery A
Pyrenophora erythrospila A.R. Paul A
Pyrenophora graminea S. Ito & Kurib. A
Pyrenophora japonica S. Ito & Kurib. A
Pyrenophora lolii Dovaston A
Pyrenophora nuda Cooke E
Pyrenophora polytricha (Wallr.) Wehm. A
Pyrenophora semeniperda (Brittleb. & D.B. Adam) Shoemaker A
Pyrenophora teres Drechsler A
Pyrenophora tetrarrhenae A.R. Paul U
Pyrenophora tritici-repentis (Died.) Drechsler A
Setosphaeria rostrata K.J. Leonard A
Setosphaeria turcica (Luttr.) K.J. Leonard & Suggs A
Stemphylium lancipes (Ellis & Everh.) E.G. Simmons AN A
Stemphylium lycopersici (Enjoji) W. Yamam. AN A
Stemphylium sarciniforme (Cavara) Wiltshire AN A
Stemphylium symphyti E.G. Simmons AN A
Ulocladium atrum Preuss AN U
Ulocladium botrytis Preuss AN A
Ulocladium chartarum (Preuss) E.G. Simmons AN A
Ulocladium consortiale (Thüm.) E.G. Simmons AN A
Ulocladium cucurbitae (Letendre & Roum.) E.G. Simmons AN A
Ulocladium obovoideum E.G. Simmons A
Ulocladium oudemansii E.G. Simmons AN A
Ulocladium tuberculatum E.G. Simmons AN A
SPORORMIACEAE
Preussia aemulans (Rehm) Arx U
Preussia fleischhakii (Auersw.) Cain U
Preussia typharum (Sacc.) Cain U
Sporormiella australis (Speg.) S.I. Ahmed & Cain A
Sporormiella cylindrospora S.I. Ahmed & Cain A
Sporormiella heptamera (Auersw.) S.I. Ahmed & Cain A
Sporormiella intermedia (Auersw.) S.I. Ahmed & Cain ex Kobayasi A
Sporormiella minima (Auersw.) S.I. Ahmed & Cain A
Sporormiella minimoides S.I. Ahmed & Cain A
Sporormiella octomera (Auersw.) S.I. Ahmed & Cain ex Kobayasi A
Sporormiella pilosa (Cain) S.I. Ahmed & Cain A
Sporormiella teretispora S.I. Ahmed & Cain A
TESTUDINACEAE
Verruculina enalia (Kohlm.) Kohlm. & Volkm.-Kohlm. M
TUBEUFIACEAE
Acanthostigma affine Sacc. & Berl.
Acanthostigma minutum (Fuckel) Sacc.
Acanthostigma scopulum (Cooke & Peck) Peck
Boerlagiomyces laxus (Penz. & Sacc.) Butzin
Helicoma atroseptatum Linder AN
Helicoma curtisii Berk. AN
Helicoma monilipes Ellis & L.N. Johnson AN
Helicoma perelegans Thaxt. AN
Helicomyces ambiguus (Morgan) Linder AN F
Helicomyces colligatus R.T. Moore AN F
Helicomyces tenuis Speg. AN F
Helicomyces torquatus L.C. Lane & Shearer AN F
Helicoon ellipticum (Peck) Morgan AN F
Helicoon farinosum Linder AN U F
Helicoon fuscosporum Linder AN F
Helicoon maioricense Abdullah, Cano, Descals & Guarro AN OU F
Helicoon pluriseptatum Beverw. AN F
Helicoon richonis (Boud.) Linder AN F
Helicoon sp. 'Lake Stream Track (PDD76601)' J.A. Cooper AN (nom. ined.) F
Helicoon sp. 'Okuru Estuary (PDD80902)' J.A. Cooper AN (nom. ined.) F
Helicoon sp. 'Travis Wetland (PDD87028)' J.A. Cooper AN (nom. ined.) A F
Helicoon sp. 'Waihi Beach (PDD80026)' J.A. Cooper AN (nom. ined.) F
Helicosporium griseum Berk. & M.A. Curtis AN
Helicosporium lumbricopsis Linder AN
Helicosporium pilosum Ellis & Everh. AN
Malacaria meliolinae Hansf.
Pendulispora venezuelanica M.B. Ellis AN
Podonectria coccicola (Ellis & Everh.) Petch
Podonectria coccorum (Petch) Rossman
Podonectria gahnia Dingley E
Podonectria larvaespora (Cooke & Massee) Rossman
Podonectria novae-zealandiae Dingley E
Thaxteriella helicoma (W. Phillips & Plowr.) J.L. Crane, Shearer & M.E. Barr F
Thaxteriella ovata (Rossman) J.L. Crane, Shearer & M.E. Barr
Thaxteriella pezizula (Berk. & M.A. Curtis) Petr.
Tubeufia aurantiella (Penz. & Sacc.) Rossman
Tubeufia cerea (Berk. & M.A. Curtis) Höhn.
Tubeufia cylindrothecia (Seaver) Höhn.
Tubeufia paludosa (P. Crouan & H. Crouan) Rossman
Xenosporium berkeleyi (M.A. Curtis) Piroz. AN
Xenosporium boivinii S. Hughes AN
Xenosporium larvale (Morgan) Piroz. AN U
Xenosporium mirabile Penz. & Sacc. AN A F
Xenosporium thaxteri (Linder) Piroz. AN
VENTURIACEAE
Antennaria scoriadea Berk. E
Antennularia ericophila (Link) Höhn. OU
Anungitea fragilis B. Sutton AN A
Anungitea heterospora P.M. Kirk AN
Coleroa circinans (Fr.) G. Winter A
Coleroa potentillae (Wallr.) G. Winter A
Coleroa senniana (Sacc.) Arx A
Fusicladium convolvularum Ondřej AN A
Fusicladium eriobotryae (Cavara) Cavara AN OU
Fusicladium matsushimae (U. Braun & C.F. Hill) Crous, U. Braun & K. Schub. AN E
Fusicladium oleagineum (Castagne) Ritschel & U. Braun AN A
Fusicladium orchidis (E.A. Ellis & M.B. Ellis) K. Schub. & U. Braun AN A
Fusicladium scillae (Deighton) U. Braun & K. Schub. AN A
Gibbera selaginellae M.L. Farr & Horner OU
Piggotia nothofagi P.R. Johnst. AN E
Rhizosphaera kalkhoffii Bubák AN A
Rhizosphaera pini (Corda) Maubl. A
Rosenscheldiella brachyglottidis G.F. Laundon & Sivan. E
Rosenscheldiella korthalsellae A. Sultan, P.R. Johnst., D.-C. Park & A.W. Robertson E
Rosenscheldiella pullulans (Berk.) Hansf.
Rosenscheldiella styracis (Henn.) Theiss. & Syd. OU
Venturia asperata Samuels & Sivan. A
Venturia carpophila E.E. Fisher A
Venturia cerasi Aderh. A
Venturia chlorospora (Ces.) P. Karst. A
Venturia geranii (Fr.) G. Winter A
Venturia inaequalis (Cooke) G. Winter A
Venturia pyrina Aderh. A
Venturia rumicis (Desm.) G. Winter A
PSEUDEUROTIACEAE
Pseudeurotium zonatum J.F.H. Beyma U
Pseudogymnoascus roseus Raillo U
PSEUDOPERISPORIACEAE
Bryochiton monascus Döbbeler & Poelt
Bryochiton perpusillus Döbbeler
Bryomyces sp. Döbbeler
Epibryon dawsoniae Döbbeler
Epibryon elegantissimum Döbbeler
Epibryon interlamellare (Racov.) Döbbeler
Epibryon pogonati-urnigeri Döbbeler
Epibryon pulchellum Döbbeler E
Epibryon sp. 1 sensu Döbbeler
Epibryon sp. 3 sensu Döbbeler E
Episphaerella dodonaeae (F. Stevens) M.L. Farr, Schokn. & J.L. Crane
Eudimeriolum podocarpi (Syd.) Hansf.
Lizonia fragilis (Berk.) Sacc. E
Wentiomyces melioloides (Berk. & M.A. Curtis) E. Müll.
Wentiomyces tatjanae S.Y. Kondr. U
PYRENOTHRICACEAE
Pyrenothrix nigra Riddle
SACCARDIACEAE
Angatia thwaitesii (Petch) Arx
SEURATIACEAE
Atichia millardeti Racib. AN A
STRIGULACEAE
Phylloporis viridis Lücking L
Strigula affinis (A.Massal.) R.C.Harris L
Strigula albicascens (Nyl.) R.C.Harris L
Strigula australiensis P.M.McCarthy L
Strigula decipiens (Malme) P.M.McCarthy L
Strigula delicata Sérus. L
Strigula fossulicola P.M.McCarthy, Streimann & Elix L
Strigula fracticonidia R.C.Harris L
Strigula indutula (Nyl.) R.C.Harris L
Strigula johnsonii P.M.McCarthy E L
Strigula kaitokensis Sérus. & Polly L
Strigula melanobapha (Kremp.) R.Sant. L
Strigula minutula P.M.McCarthy L
Strigula nemathora Mont. L
Strigula novae-zelandiae (Nag Raj) Sérus. L
Strigula occulta P.M.McCarthy & Malcolm L
Strigula oceanica P.M.McCarthy, Streimann & Elix L
Strigula orbicularis Fr.:Fr. L
Strigula prasina Müll.Arg. L
Strigula schizospora R.Sant. L
Strigula smaragdula Fr.:Fr. L
Strigula subelegans Vain. L
Strigula subsimplicans (Nyl.) R.C.Harris L
Strigula subtilissima (Fée) Müll.Arg. L

THELENELLACEAE
Thelenella luridella (Nyl.) H.Mayrhofer L
THROMBIACEAE
Thrombium epigaeum (Pers.) Wallr. L

Order TRYPETHELIALES
TRYPETHELIACEAE
Aptrootia elatior (Stirt.) Aptroot L
Laurera cumingii (Mont.) Zahlbr. L
Laurera madreporiformis (Eschw.) Riddle L
Polymeridium catapastum (Nyl.) R.C.Harris L
Polymeridium quinqueseptatum (Nyl.) R.C. Harris U L
Trypethelium bicolor var. *pyrenuloides* Knight L
Trypethelium variolosum Ach. L
VIZELLACEAE
Vizella metrosideri P.R. Johnst. E
Vizella tunicata Gadgil E

Class EUROTIOMYCETES
AMORPHOTHECACEAE
Amorphotheca resinae Parbery A
MONASCACEAE
Monascus purpureus Went U
Monascus ruber Tiegh. U
Xeromyces bisporus L.R. Fraser A

Order ARACHNOMYCETALES
ARACHNOMYCETACEAE
Arachnomyces minimus Malloch & Cain A
Arachnomyces nodosetosus Sigler & S.P. Abbott A

Order ASCOSPHAERALES
ASCOSPHAERACEAE
Ascosphaera apis (Maasen ex Claussen) L.S. Olive & Spiltoir A
Ascosphaera scaccaria Pinnock, R.B. Coles & B. Donovan A
Bettsia alvei (Betts) Skou U

Order CHAETOTHYRIALES
Coniosporium culmigenum (Berk.) Sacc. AN U
Staninwardia breviuscula B. Sutton AN A
CHAETOTHYRIACEAE
Chaetothyrium guaraniticum Speg.
Chaetothyrium strigosum L.R. Fraser
Cyphellophora laciniata G.A. de Vries AN U
HERPOTRICHIELLACEAE
Capronia acutiseta Samuels E
Capronia coronata Samuels E
Capronia normandinae R. Sant. & D. Hawksw.
Capronia pilosella (P. Karst.) E. Müll., L.E. Petrini, P.J. Fisher, Samuels & Rossman
Capronia villosa Samuels E
Cladophialophora bantiana (Sacc.) de Hoog, Kwon-Chung & McGinnis AN U
Exophiala jeanselmei (Langeron) McGinnis & A.A. Padhye AN A
Exophiala pisciphila McGinnis & Ajello AN A
Fonsecaea pedrosoi (Brumpt) Negroni AN U
Phaeomoniella chlamydospora (W. Gams, Crous, M.J. Wingf. & L. Mugnai) Crous & W. Gams AN A
Phialophora alba J.F.H. Beyma AN OU
Phialophora asteris (Dowson) Burge & I. Isaac AN A
Phialophora cinerescens (Wollenw.) J.F.H. Beyma AN A
Phialophora cyclaminis J.F.H. Beyma AN A
Phialophora mustea Neerg. AN A
Phialophora radicicola Cain AN A
Phialophora richardsiae (Nannf.) Conant AN A
Phialophora sp. Latch, M.J. Chr. & Samuels (nom. ined.) AN A
Wangiella dermatitidis (Kano) McGinnis AN A

Order CORYNELIALES
CORYNELIACEAE
Corynelia tropica (Auersw. & Rabenh.) Starbäck

Order EUROTIALES
Thermomyces lanuginosus Tsikl. AN A
Thermomyces stellatus (Bunce) Apinis AN A
TRICHOCOMACEAE
Aspergillus alutaceus Berk. & M.A. Curtis AN U
Aspergillus caesiellus Saito AN U
Aspergillus candidus Link AN U
Aspergillus cervinus Massee AN U
Aspergillus clavatus Desm. AN U
Aspergillus flavus Link AN U
Aspergillus fumigatus Fresen. var. *fumigatus* AN A
— var. *ellipticus* Raper & Fennell AN A
Aspergillus microviridocitrinus Costantin & Lucet U
Aspergillus niger Tiegh. AN A
Aspergillus oryzae (Ahlb.) Cohn AN A
Aspergillus restrictus G. Sm. AN U
Aspergillus sulphureus (Fresen.) Wehmer AN U
Aspergillus sydowii (Bainier & Sartory) Thom & Church AN U M
Aspergillus terreus Thom AN U
Aspergillus ustus (Bainier) Thom & Church AN U
Aspergillus versicolor (Vuill.) Tirab. AN U
Aspergillus wentii Wehmer AN U
Byssochlamys nivea Westling U
Emericella nidulans (Eidam) Vuill.
Emericella unguis Malloch & Cain U
Eupenicillium crustaceum F. Ludw.
Eupenicillium senticosum D.B. Scott
Eurotium chevalieri L. Mangin
Eurotium halophilicum C.M. Chr., Papav. & C.R. Benj. A
Eurotium herbariorum Link
Eurotium rubrum Jos. König, Spieck. & W. Bremer
Fennellia flavipes B.J. Wiley & E.G. Simmons U
Fennellia nivea (B.J. Wiley & E.G. Simmons) Samson U
Hemicarpenteles paradoxus A.K. Sarbhoy & Elphick A
Neosartorya fischeri (Wehmer) Malloch & Cain
Paecilomyces breviramosus Bissett AN
Paecilomyces carneus (Duché & R. Heim) A.H.S. Br. & G. Sm. AN U
Paecilomyces lilacinus (Thom) Samson AN U
Paecilomyces marquandii (Massee) S. Hughes AN U
Paecilomyces pascuus Pitt & A.D. Hocking AN A
Paecilomyces variotii Bainier AN U
Penicillium aculeatum Raper & Fennell AN U
Penicillium aurantiogriseum Dierckx AN U
Penicillium bilaiae Chalab. AN U
Penicillium brevicompactum Dierckx AN U
Penicillium camemberti Thom AN A
Penicillium canescens Sopp AN U
Penicillium capsulatum Raper & Fennell AN U
Penicillium chrysogenum Thom AN U
Penicillium citreonigrum Dierckx AN U
Penicillium citrinum Thom AN U
Penicillium commune Thom AN OU
Penicillium coprophilum (Berk. & M.A. Curtis) Seifert & Samson AN A
Penicillium corylophilum Dierckx AN U
Penicillium crustosum Thom AN U
Penicillium decumbens Thom AN U
Penicillium digitatum (Pers.) Sacc. AN A
Penicillium diversum Raper & Fennell AN U
Penicillium expansum Link AN A
Penicillium fellutanum Biourge AN U
Penicillium funiculosum Thom AN U
Penicillium glabrum (Wehmer) Westling AN U
Penicillium griseofulvum Dierckx AN U
Penicillium herquei Bainier & Sartory AN U
Penicillium inflatum Stolk & Malla AN U
Penicillium italicum Wehmer AN A
Penicillium janczewskii K.M. Zalessky AN U
Penicillium janthinellum Biourge AN U
Penicillium jensenii K.M. Zalessky AN U
Penicillium lividum Westling AN U
Penicillium loliense Pitt AN A
Penicillium melinii Thom AN U
Penicillium miczynskii K.M. Zalessky AN U
Penicillium minioluteum Dierckx AN U
Penicillium montanense M. Chr. & Backus AN A
Penicillium nikau F.J. Morton AN (nom. ined.) U
Penicillium novae-zeelandiae J.F.H. Beyma AN A
Penicillium oxalicum Currie & Thom AN U
Penicillium palitans Westling AN U
Penicillium paxilli Bainier AN U
Penicillium purpurascens (Sopp) Biourge AN U
Penicillium purpurogenum Stoll AN U
Penicillium raistrickii G. Sm. AN U
Penicillium resedanum McLennan & Ducker AN U
Penicillium restrictum J.C. Gilman & E.V. Abbott AN U
Penicillium roqueforti Thom AN A
Penicillium roseopurpureum Dierckx AN U
Penicillium rugulosum Thom AN U
Penicillium sclerotiorum J.F.H. Beyma AN U
Penicillium simplicissimum (Oudem.) Thom AN U
Penicillium spinulosum Thom AN U
Penicillium sublateritium Biourge AN U
Penicillium thomii Maire AN U
Penicillium verruculosum Peyronel AN U
Penicillium viridicatum Westling AN U
Penicillium vulpinum (Cooke & Massee) Seifert & Samson AN U
Penicillium waksmanii K.M. Zalessky AN U
Sclerocleista thaxteri Subram. U
Talaromyces flavus (Klöcker) Stolk & Samson var. *flavus* A
Talaromyces luteus (Zukal) C.R. Benj.
Talaromyces thermophilus Stolk
Talaromyces wortmannii (Klöcker) C.R. Benj.
Thysanophora glaucoalbida (Desm.) M. Morelet AN U
Torulomyces lagena Delitsch AN
Trichocoma paradoxa Jungh.

Order MYCOCALICIALES
MYCOCALICIACEAE
Chaenothecopsis brevipes Tibell A L
Chaenothecopsis debilis (Turner & Borrer ex Sm.) Tibell L
Chaenothecopsis haematopus Tibell L
Chaenothecopsis lignicola (Nádv.) Alb. Schmidt L
Chaenothecopsis nana Tibell L
Chaenothecopsis nigra Tibell *L
Chaenothecopsis nigropedata Tibell L
Chaenothecopsis nivea (F. Wilson) Tibell L
Chaenothecopsis pusilla (Flörke) A. Schmidt L
Chaenothecopsis sagenidii Tibell A L
Chaenothecopsis sanguinea Tibell L
Chaenothecopsis savonica (Räsänen) Tibell L
Chaenothecopsis schefflerae (Samuels & D.E. Buchanan) Tibell E
Chaenothecopsis tasmanica Tibell A L
Chaenothecopsis viridireagens (Nádv.) Alb. Schmidt L
Mycocalicium albonigrum (Nyl.) Tibell L
Mycocalicium subtile (Pers.) Szatala L
Mycocalicium victoriae (C.Knight ex F.Wilson) Tibell L
Phaeocalicium asciiforme Tibell L
Stenocybe bartlettii Tibell E L
SPHINCTRINACEAE
Sphinctrina tubaeformis A. Massal. L

Order ONYGENALES
Myceliophthora vellerea (Sacc. & Speg.) Oorschot AN
AJELLOMYCETACEAE
Emmonsia parva (C.W. Emmons & Ashburn) Cif. & A.M. Corte var. parva AN U
— var. *crescens* (C.W. Emmons & Jellison) Oorschot AN U
Histoplasma capsulatum Darling AN U
ARTHRODERMATACEAE
Arthroderma benhamiae Ajello & S.L. Cheng U
Arthroderma cajetani (Ajello) Ajello, Weitzman, McGinnis & A.A. Padhye U
Arthroderma ciferrii Varsavsky & Ajello U
Arthroderma curreyi Berk. U
Arthroderma fulvum (Stockdale) Weitzman, McGinnis, A.A. Padhye & Ajello U
Arthroderma gypseum (Nann.) Weitzman, McGinnis, A.A. Padhye & Ajello U
Arthroderma incurvatum (Stockdale) Weitzman, McGinnis, A.A. Padhye & Ajello U
Arthroderma obtusum (C.O. Dawson & Gentles) Weitzman, McGinnis, A.A. Padhye & Ajello U
Arthroderma otae (A. Haseg. & Usui) McGinnis, WeitzmAN A.A. Padhye & Ajello U
Arthroderma persicolor (Stockdale) Weitzman, McGinnis, A.A. Padhye & Ajello U
Arthroderma quadrifidum C.O. Dawson & Gentles U
Arthroderma uncinatum C.O. Dawson & Gentles U
Arthroderma vanbreuseghemii Takashio U
Epidermophyton floccosum (Harz) Langeron & Miloch. AN A
Microsporum audouinii Gruby AN A
Microsporum canis var. *distortum* (di Menna & Marples) Tad. Matsumoto, A.A. Padhye & Ajello AN A
Microsporum equinum (Delacr. & E. Bodin) Guég. AN A
Microsporum nanum C.A. Fuentes AN A
Microsporum praecox Rivalier ex A.A. Padhye, Ajello & McGinnis AN A
Trichophyton concentricum R. Blanch. AN A
Trichophyton equinum Gedoelst var. *equinum* AN A
— var. *autotrophicum* J.M.B. Sm., Jolly, Georg & Connole AN A
Trichophyton krajdenii J. Kane, J.A. Scott & Summerb. AN A
Trichophyton mentagrophytes var. *erinacei* J.M.B. Sm. & Marples AN A
— var. *interdigitale* (Priestley) Moraes AN A
Trichophyton rubrum (Castell.) Sabour. var. *rubrum* AN A
— var. *oceanicum* Dompmartin, Drouhet & Moreau AN, (nom. inv.) A
Trichophyton soudanense Joyeux AN A
Trichophyton tonsurans Malmsten AN A
Trichophyton verrucosum E. Bodin AN A
Trichophyton violaceum Sabour. ex E.Bodin AN A
GYMNOASCACEAE
Gymnascella dankaliensis (Castell.) Currah U
ONYGENACEAE
Aphanoascus fulvescens (Cooke) Apinis A
Aphanoascus keratinophilus Punsola & Cano U
Aphanoascus reticulisporus (Routien) Hubálek A
Auxarthron californiense G.F. Orr & Kuehn A
Auxarthron umbrinum (Boud.) G.F. Orr & Plunkett A
Chrysosporium inops J.W. Carmich. AN A
Chrysosporium luteum (Costantin) Carmichael AN A
Chrysosporium pannicola (Corda) Oorschot & Stalpers AN A

Order PYRENULALES
MONOBLASTIACEAE
Acrocordia gemmata (Ach.) A.Massal. L
Anisomeridium biforme (Borrer) R.C.Harris L
Anisomeridium carinthiacum (J.Steiner) R.C.Harris L
Anisomeridium laevigatum (P.M.McCarthy) R.C.Harris E L
Anisomeridium magnosporum (C.Knight) D.Hawksw. E L
Anisomeridium subatomarium (C.Knight) R.C.Harris E L
Anisomeridium subbiforme (C.Knight) R.C.Harris E L
Anisomeridium subprostans (Nyl.) R.C.Harris L
Caprettia setifera (Malcolm & Vězda) Sérus. & Lücking L
PYRENULACEAE
Anthracothecium cellulosum (C.Knight) Müll.Arg. E L
Lithothelium australe Aptroot & H.Mayrhofer E L
XANTHOPYRENIACEAE
Collemopsidium sublitorale (Leight.) Grube & B.D.Ryan L M
Zwackhiomyces dispersus (J.Lahm ex Körb.) Triebel & Grube
Zwackhiomyces lecanorae (Stein) Nik. Hoffm. & Hafellner

Order VERRUCARIALES
Agonimia pacifica (H.Harada) Diederich L
Pocsia dispersa Vězda L
Staurothele fissa (Taylor) Zwackh L
ADELOCOCCACEAE
Sagediopsis campsteriana (Linds.) D. Hawksw. & R. Sant.
VERRUCARIACEAE
Bagliettoa baldensis (A.Massal) Vězda L
Catapyrenium cinereum (Pers.) Körb. L
Catapyrenium daedaleum (Kremp.) Stein L
Catapyrenium psoromoides (Borrer) R.Sant. L
Dermatocarpon luridum sensu Malcolm W.M. & Galloway, D.J. 1997 (nom. inv.) L
Dermatocarpon miniatum var. *complicatum* (Lightf.) Th.Fr. L
Endocarpon adscendens (Anzi) Müll.Arg. L
Endocarpon pusillum Hedw. L
Endocarpon simplicatum (Nyl.) Nyl. L
Heteroplacidium podolepis (Breuss) Breuss L
Lauderlindsaya borreri (Tul.) J.C. David & D. Hawksw.
Macentina stigonemoides Orange L
Muellerella lichenicola (Sommerf.) D. Hawksw.
Muellerella pygmaea (Körb.) D.Hawksw.
Normandina pulchella (Borrer) Nyl. L
Phaeospora perrugosaria (Linds.) R. Sant.
Placidium lacinulatum (Ach.) Breuss L
Placidium squamulosum (Ach.) Breuss L
Polyblastia cruenta (Körb.) P.James & Swinscow L
Polyblastia melaspora (Taylor) Zahlbr. L
Polyblastia trachyspora (C.Knight) Müll.Arg. L
Pyrenula concatervans (Nyl.) R.C.Harris L
Pyrenula crassescens (Stirt.) Müll.Arg. L
Pyrenula cyrtospora (Stirt.) Müll.Arg. L
Pyrenula dealbata (C.Knight) Müll.Arg. L
Pyrenula deliquescens (C.Knight) Müll.Arg. L
Pyrenula deprimens (C.Knight) D.J.Galloway L
Pyrenula homalisma (C.Knight) D.J.Galloway L
Pyrenula knightiana Müll.Arg. L
Pyrenula moniliformis (C.Knight) Müll.Arg. L
Pyrenula occulta (C.Knight) Müll.Arg. L
Pyrenula prostrata (Stirt.) D.J.Galloway L
Pyrenula pseudonitidella (C.Knight) D.J.Galloway L
Pyrenula pyrenastroides (C.Knight) D.J.Galloway L
Thelidium calcareum (C.Knight) Hellb. L
Thelidium maurospilum (Nyl.) Hellb. L
Thelidium neozelandicum Zahlbr. E L
Thelidium papulare (Fr.) Arnold L
Thelidium pluvium A.Orange L
Verrucaria adguttata Zahlbr. E L M
Verrucaria amnica P.M.McCarthy & P.N.Johnson L
Verrucaria aquatilis Mudd L
Verrucaria aucklandica Zahlbr. L M
Verrucaria austroschisticola P.M.McCarthy & P.N.Johnson L
Verrucaria bubalina P.M.McCarthy L M
Verrucaria calciseda DC. L
Verrucaria ceuthocarpa Wahlenb. L M
Verrucaria compacta (A.Massal.) Jatta L
Verrucaria cramba Stirt. L
Verrucaria dolosa Hepp L
Verrucaria dufourii DC. L
Verrucaria durietzii I.M.Lamb L M
Verrucaria fiordlandica P.M.McCarthy & P.N.Johnson L
Verrucaria fusconigrescens Nyl. L M
Verrucaria glaucina Ach. L
Verrucaria halizoa Leight. L M
Verrucaria hydrela Ach. L
Verrucaria inconstans P.M.McCarthy L
Verrucaria macrostoma Dufour ex DC. L
Verrucaria margacea (Wahlenb.) Wahlenb. L
Verrucaria maura Wahlenb. L M
Verrucaria microsporoides Nyl. L M
Verrucaria mucosa Wahlenb. L M
Verrucaria muralis Ach. L
Verrucaria nigrescens Pers. L
Verrucaria phaeoderma P.M.McCarthy L
Verrucaria praetermissa (Trevis.) Anzi L
Verrucaria prominula Nyl. L
Verrucaria rheitrophila Zschacke L
Verrucaria serpuloides I.M.Lamb L M
Verrucaria sessilis P.M.McCarthy E L M
Verrucaria striatula Wahlenb. subsp. *striatula* L M
— subsp. *australis* R.Sant. L
Verrucaria subdiscreta P.M.McCarthy L M
Verrucaria tessellatula Nyl. L M

Class LABOULBENIOMYCETES
Order LABOULBENIALES
LABOULBENIACEAE
Corethromyces bicolor Thaxt. E
Cucujomyces bilobatus Thaxt. E
Cucujomyces phycophilus A. Weir & W. Rossi E
Diaphoromyces kuschelii A. Weir & W. Rossi E
Diphymyces bidentatus (Thaxt.) I.I. Tav. E
Diphymyces curvatus (Thaxt.) I.I. Tav. E
Diphymyces depressus M.B. Hughes, A. Weir & C. Judd E
Diphymyces leschenii M.B. Hughes, A. Weir & C. Judd E
Diphymyces penicillifer A. Weir & W. Rossi E
Histeridomyces tishechkinii A. Weir E
Laboulbenia flagellata Peyr. U
Laboulbenia loxomeri M.B. Hughes & A. Weir E
Laboulbenia oopteri Thaxt. E
Laboulbenia sp. 1 M.B. Hughes, A. Weir, R. Leschen, C. Judd & B. Gillen (nom. ined.) E
Laboulbenia subantarctica M.B. Hughes, A. Weir & C. Judd E
Monoicomyces zealandicus Thaxt. E
Rhachomyces kenodactyli Balazuc & W. Rossi E
Rhachomyces sp. 1 sensu M.B. Hughes, A. Weir, R. Leschen, C. Judd & B. Gillen (non. ined.) E
Smeringomyces trinitatis Thaxt.
Stigmatomyces australis M.B. Hughes, A. Weir, Gillen & W. Rossi E
Stigmatomyces baeopteri M.B. Hughes, A. Weir & W. Rossi E
Stigmatomyces ceratophorus Whisler A

Stigmatomyces crassicollis Thaxt.
Stigmatomyces ephydrae L. Mercier & R. Poiss.
Stigmatomyces hydrelliae Thaxt.
Stigmatomyces ilytheae Thaxt.
Stigmatomyces limosinae Thaxt.
Stigmatomyces novozelandicus A. Weir & W. Rossi E
Stigmatomyces purpureus Thaxt.
Stigmatomyces rugosus Thaxt.
Stigmatomyces spiralis Thaxt.
Teratomyces insignis Thaxt. E
Teratomyces petiolatus Thaxt. E
Teratomyces sp. 1 sensu M.B. Hughes, A. Weir, R. Leschen, C. Judd & B. Gillen E
Teratomyces zealandicus Thaxt. E
Triainomyces hollowayanus W. Rossi & A. Weir E

Class LECANOROMYCETES
Podotara pilophoriformis Malcolm & Vězda L
Order ACAROSPORALES
ACAROSPORACEAE
Acarospora badiofusca (Nyl.) Th.Fr. L
Acarospora fuscata (Nyl.) Arnold L
Acarospora gallica H.Magn. L
Acarospora glaucocarpa (Ach.) Körb. L
Acarospora gyrodes H.Magn. E L
Acarospora murorum A.Massal. L
Acarospora nodulosa (Dufour) Hue L
Acarospora otagensis H.Magn. E L
Acarospora schleicheri (Ach.) A.Massal. L
Acarospora umbilicata Bagl. L
Acarospora veronensis A.Massal. L
Polysporina simplex (Davies) Vězda L
Sarcogyne regularis Körb. L

Order AGYRIALES
AGYRIACEAE
Trapeliopsis colensoi (C.Bab.) Gotth.Schneid. L
Trapeliopsis congregans (Zahlbr.) Brako L
Trapeliopsis flexuosa (Fr.) Coppins & P.James L
Trapeliopsis granulosa (Hoffm.) Lumbsch L
Trapeliopsis pseudogranulosa Coppins & P.James L
Xylographa parallela (Ach.) Fr. L
Xylographa perangusta (Stirt.) Müll.Arg. 1894 L
TRAPELIACEAE
Aspiciliopsis macrophthalma (Hook.f. & Taylor) de Lesd. L
Lithographa graphidioides (Cromb.) Imshaug ex Coppins & Fryday L
Lithographa olivacea Fryday L
Lithographa serpentina Coppins & Fryday E L
Placopsis ampliata (I.M.Lamb) D.J.Galloway E L
Placopsis bicolor (Tuck.) de Lesd. L
Placopsis brevilobata (Zahlbr.) I.M.Lamb L
Placopsis centrifuga D.J.Galloway E L
Placopsis clavifera (I.M.Lamb) D.J.Galloway L
Placopsis cribellans (Nyl.) Räsänen L
Placopsis dennanensis (Zahlbr.) I.M.Lamb ex D.J.Galloway E L
Placopsis durietziorum D.J.Galloway L
Placopsis dusenii I.M.Lamb L
Placopsis elixii D.J.Galloway E L
Placopsis fuscidula I.M.Lamb ex Räsänen L
Placopsis fusciduloides D.J.Galloway L
Placopsis gelida (L.) Linds. L
Placopsis gelidioides Du Rietz ex I.M.Lamb L
Placopsis hertelii D.J.Galloway E L
Placopsis illita (C.Knight) I.M.Lamb L
Placopsis lambii Hertel & V.Wirth L
Placopsis lateritioides I.M.Lamb L
Placopsis macrophthalma (Hook.f. & Taylor) Nyl. L
Placopsis macrospora D.J.Galloway E L
Placopsis microphylla (I.M.Lamb.) D.J.Galloway L
Placopsis murrayi D.J.Galloway E L
Placopsis perrugosa (Nyl.) Nyl. L
Placopsis polycarpa D.J.Galloway E L
Placopsis pruinosa D.J.Galloway E L
Placopsis rhodocarpa (Nyl.) Nyl. L
Placopsis rhodophthalma (Müll.Arg.) Räsänen L
Placopsis salazina I.M.Lamb L
Placopsis stenophylla (Hue) I.M.Lamb L
Placopsis subcribellans (I.M.Lamb) D.J.Galloway L
Placopsis subgelida (Nyl.) Nyl. L
Placopsis subparellina Nyl. L
Placopsis tararuana (Zahlbr.) D.J.Galloway E L
Placopsis trachyderma (Kremp.) P.James L
Placopsis venosa Imshaug ex D.J.Galloway L
Placynthiella oligotropha (J.R.Laundon) Coppins & P.James L
Placynthiella uliginosa (Schrad.) Coppins & P.James L
Trapelia coarctata (Turner ex Sm.) M.Choisy L
Trapelia corticola Coppins & P.James L
Trapelia herteliana Fryday L
Trapelia macrospora Fryday L
ARTHRORHAPHIDACEAE
Arthrorhaphis alpina (Schaer.) R.Sant. L
Arthrorhaphis citrinella (Ach.) Poelt var. *citrinella* L
— var. *catolechioides* Obermayer L
Arthrorhaphis grisea Th. Fr.

Order BAEOMYCETALES
BAEOMYCETACEAE
Baeomyces heteromorphus Nyl. ex C.Bab. & Mitt. L

Order CANDELARIALES
CANDELARIACEAE
Candelaria concolor (Dicks.) Arnold L
Candelariella aurella (Hoffm.) Zahlbr. L
Candelariella coralliza (Nyl.) H.Magn. L
Candelariella reflexa (Nyl.) Lettau L
Candelariella subdeflexa (Nyl.) Lettau L
Candelariella vitellina (Ehrh.) Müll.Arg. L
Candelariella xanthostigma (Pers. ex Ach.) Lettau L
CONIOCYBACEAE
Chaenotheca brunneola (Ach.) Müll. Arg. L
Chaenotheca chlorella (Ach.) Müll.Arg. L
Chaenotheca chrysocephala (Turner ex Ach.) Th.Fr. L
Chaenotheca citriocephala (F.Wilson) Tibell L
Chaenotheca confusa Tibell 1998 L
Chaenotheca degelii Tibell 1983 L
Chaenotheca deludens Tibell 1987 L
Chaenotheca ferruginea (Turner ex Sm.) Mig. L
Chaenotheca gracillima (Vain.) Tibell L
Chaenotheca hispidula (Ach.) Zahlbr. L
Chaenotheca stemonea (Ach.) Müll.Arg. L
Chaenotheca trichialis (Ach.) Th.Fr. L
Chaenotheca xyloxena Nádv. L
Coniocybe otagoensis Js. Murray E
Sclerophora amabilis (Tibell) Tibell L
Sclerophora sanguinea (Tibell) Tibell L

Order LECANORALES
Bartlettiella fragilis D.J.Galloway & P.M.Jørg. L
Bilimbia lobulata (Sommerf.) Hafellner & Coppins L
Bilimbia sabuletorum (Schreb.) Arnold L
Corticifraga fuckelii (Rehm) D. Hawksw. & R. Sant.
Lecania cyrtella (Ach.) Th.Fr. L
Lecania erysibe (Ach.) Mudd L
Lecania fructigena Zahlbr. L
Lecania inundata (Hepp ex Körb.) M.Mayrhofer L
Lecania naegelii (Hepp) Diederich & Van den Boom L
Lecania nylanderiana A.Massal L
Lecania rabenhorstii (Hepp) Arnold L
Lecania turicensis (Hepp) Müll.Arg. 1862 var. *turicensis* L
Lecania vallata (Stirt.) Müll.Arg. L
Lecidoma demissum (Rutstr.) Gotth. Schneid. & Hertel L
Leprocaulon arbuscula (Nyl.) Nyl. L
Mycobilimbia australis Kantvilas & Messuti L
Psilolechia clavulifera (Nyl.) Coppins L
Psilolechia lucida (Ach.) M.Choisy L
Scoliciosporum lividum Malcolm & Vězda L
Scoliciosporum umbrinum (Ach.) Arnold L
Solenopsora sordida (C.W.Dodge) D.J.Galloway L
Strangospora deplanata (Almq.) Clauzade & Cl.Roux L
Tremolecia atrata (Ach.) Hertel L
ARCTOMIACEAE
Wawea fruticulosa Henssen & Kantvilas L
BIATORELLACEAE
Biatorella desmaspora (C.Knight) Hellb. L
Biatorella epiphysa (Stirt.) Hellb. E
BRIGANTIAEACEAE
Argopsis megalospora Th.Fr. E L
Brigantiaea chrysosticta (Hook.f. & Taylor) Hafellner & Bellem. L
Brigantiaea fuscolutea (Dicks.) R.Sant. L
Brigantiaea lobulata F.J.Walker & Hafellner L
Brigantiaea phaeomma (Nyl.) Hafellner L
CALYCIDIACEAE
Calycidium cuneatum Stirt. L
Calycidium polycarpum (Colenso) Wedin L
CLADONIACEAE
Cladia aggregata (Sw.) Nyl. L
Cladia fuliginosa Filson L
Cladia inflata (F.Wilson) D.J.Galloway L
Cladia retipora (Labill.) Nyl. L
Cladia schizopora (Nyl.) Nyl. L
Cladia sullivanii (Müll.Arg.) W.Martin L
Cladonia archeri S.Stenroos L
Cladonia aspera Ahti & Kashiw. L
Cladonia aueri Räsänen L
Cladonia bimberiensis A.W.Archer L
Cladonia calycantha sensu GAlloway (nom. inv.) L
Cladonia capitellata (Hook.f. & Taylor) C.Bab. var. *capitellata* L
— var. *interhiascens* (Nyl.) Sandst. L
— var. *squamatica* A.W.Archer L
Cladonia carneola (Fr.) Fr. L
Cladonia cervicornis (Ach.) Flot. subsp. *cervicornis* L
— subsp. *verticillata* (Hoffm.) Ahti L
Cladonia chlorophaea (Flörke ex Sommerf.) Spreng. L
Cladonia coccifera (L.) Willd. L
Cladonia confusa R. Sant. L
Cladonia corniculata Ahti & Kashiw. L
Cladonia corymbescens Nyl. ex Leight. L
Cladonia crispata (Ach.) Flot. var. *crispata* L
— var. *cetrariiformis* (Delise) Vain. L
Cladonia cryptochlorophaea Asahina L
Cladonia cucullata S.Hammer L
Cladonia cyanopora S.Hammer L
Cladonia darwinii S.Hammer L
Cladonia deformis (L.) Hoffm. L
Cladonia ecmocyna Leight. L
Cladonia elixii Ahti & V.Wirth E L
Cladonia enantia Nyl. L
Cladonia fimbriata (L.) Fr. L
Cladonia floerkeana (Fr.) Flörke L
Cladonia fruticulosa Kremp. L
Cladonia furcata (Huds.) Schrad. L
Cladonia fuscofunda S.Hammer L
Cladonia gallowayi S.Hammer E L
Cladonia glebosa S.Hammer L
Cladonia gracilis (L.) Willd. subsp. *gracilis* L
— subsp. *turbinata* (Ach.) Ahti L
— subsp. *vulnerata* Ahti L
Cladonia humilis (With.) J.R.Laundon var. *humilis* L

— var. *bourgeanica* A.W.Archer L
Cladonia imbricata S.Hammer L
Cladonia incerta S.Hammer E L
Cladonia krempelhuberi (Vain.) Zahlbr. L
Cladonia macilenta Hoffm. L
Cladonia melanopoda Ahti L
Cladonia merochlorophaea Asahina L
Cladonia mitis Sandst. L
Cladonia murrayi W.Martin L
Cladonia neozelandica Vain. var. *neozelandica* L
— var. *lewis-smithii* Ahti, Elix & Øvstedal L
Cladonia nitidella S.Hammer E L
Cladonia novochlorophaea (Sipman) Brodo & Ahti L
Cladonia nudicaulis S.Hammer L
Cladonia ochrochlora Flörke L
Cladonia pertricosa Kremp. L
Cladonia pleurota (Flörke) Schaer. L
Cladonia pocillum (Ach.) O.J.Rich. L
Cladonia polycarpoides Nyl. L
Cladonia praetermissa A.W.Archer L
Cladonia pulchra S.Hammer E L
Cladonia pyxidata (L.) Hoffm. L
Cladonia rei Schaer. L
Cladonia rigida (Hook.f. & Taylor) Hampe L
Cladonia sarmentosa (Hook.f. & Taylor) C.W.Dodge L
Cladonia scabriuscula (Delise) Nyl. L
Cladonia southlandica W.Martin L
Cladonia squamosa sensu Galloway 1985 (nom. inv.) L
Cladonia strangulata S.Hammer E L
Cladonia subsubulata Nyl. L
Cladonia subulata (L.) F.H.Wigg. L
Cladonia sulcata A.W.Archer var. *sulcata* L
— var. *striata* A.W.Archer L
— var. *wilsonii* (A.W.Archer) A.W.Archer L
Cladonia sulphurina (Michx.) Fr. L
Cladonia tenerrima (Ahti) S.Hammer L
Cladonia tessellata Ahti & Kashiw. L
Cladonia uncialis (L.) F.H.Wigg. L
Cladonia ustulata (Hook.f. & Taylor) Leight. L
Cladonia weymouthii A.W.Archer L
Heterodea muelleri (Hampe) Nyl. L
Metus conglomeratus (F.Wilson) D.J.Galloway & P.James L
Notocladonia cochleata (Müll.Arg.) S.Hammer L
Notocladonia undulata S.Hammer L
Pycnothelia caliginosa D.J.Galloway & P.James L
Thysanothecium hookeri Mont. & Berk. L
Thysanothecium scutellatum (Fr.) D.J.Galloway L
DACTYLOSPORACEAE
Dactylospora acarosporae (H. Magn.) Hafellner
Dactylospora australis Triebel & Hertel
Dactylospora davidii Hafellner & H. Mayrhofer E
Dactylospora frigida Hafellner
Dactylospora lobariella (Nyl.) Hafellner
Dactylospora parasitica (Flörke) Zopf
ECTOLECHIACEAE
Badimiella pteridophila (Sacc.) Garn.-Jones & Malcolm L
Calopadia puiggarii (Müll.Arg.) Vězda L
Calopadia subcoerulescens (Zahlbr.) Vězda L
Lopadium coralloideum (Nyl.) Lynge
Lopadium monosporum (C.Knight) Hellb. E L
Pyrenotrichum splitgerberi Mont. AN U
Sporopodium phyllocharis (Mont.) A.Massal. L
Tapellaria phyllophila (Stirt.) R.Sant. L
HAEMATOMMATACEAE
Haematomma alpinum R.W.Rogers E L
Haematomma babingtonii A.Massal. E L
Haematomma fenzlianum A.Massal. L
Haematomma hilare Zahlbr. L
Haematomma nothofagi Kalb & Staiger L
Haematomma sorediatum R.W.Rogers L
HYMENELIACEAE
Ionaspis lacustris (With.) Lutzoni L
LECANORACEAE
Bryonora castanea (Hepp) Poelt L
Carbonea assentiens (Nyl.) Hertel L
Carbonea intrudens (H. Magn.) Hafellner
Carbonea phaeostoma (Nyl.) Hertel L
Carbonea vitellinaria (Nyl.) Hertel
Carbonea vorticosa (Flörke) Hertel L
Clauzadeana macula (Taylor) Coppins & Rambold L
Lecanora achroa Nyl. L
Lecanora aghardiana Ach. L
Lecanora albescens (Hoffm.) Branth & Rostr. L
Lecanora argentata (Ach.) Degel. L
Lecanora austrooceanica Hertel & Leuckert L
Lecanora bicincta Ramond L
Lecanora caesiorubella Ach. L
Lecanora capistrata (Darb.) Zahlbr. L
Lecanora carpinea (L.) Vain. L
Lecanora cavicola Creveld L
Lecanora cenisioides Lumbsch E L
Lecanora conizaeoides Nyl. ex Cromb. L
Lecanora crenulata Hook. L
Lecanora demersa (Kremp.) Hertel & Rambold L
Lecanora dispersa (Pers.) Sommerf. L
Lecanora elatinoides Räsänen L
Lecanora elixii Lumbsch L
Lecanora epibryon subsp. *broccha* (Nyl.) Lumbsch L
— subsp. *xanthophora* Lumbsch L
Lecanora expallens Ach. L
Lecanora farinacea Fée L
Lecanora fertilissima Zahlbr. E L
Lecanora flavidofusca Müll.Arg. L
Lecanora flavidomarginata B.de Lesd. L
Lecanora flavopallida Stirt. L
Lecanora flotoviana Spreng. L
Lecanora galactiniza Nyl. L
Lecanora helva Stizenb. L
Lecanora interjecta Müll.Arg. L
Lecanora intricata (Ach.) Ach. L
Lecanora intumescens (Rebent.) Rabenh. L
Lecanora lugubris (C.W.Dodge) D.J.Galloway & P.M.Jorg. L
Lecanora melacarpella Müll.Arg. L
Lecanora novaehollandiae Lumbsch L
Lecanora oreinoides (Körb.) Hertel & Rambold L
Lecanora physcielloides Fryday E L
Lecanora plumosa Müll.Arg. L
Lecanora polytropa (Hoffm.) Rabenh. L
Lecanora pruinosa Chaub. L
Lecanora pseudistera Nyl. L
Lecanora pyreniospora Nyl. L
Lecanora queenslandica C.Knight L
Lecanora rupicola (L.) Zahlbr. L
Lecanora subcoarctata (C.Knight) Hertel L
Lecanora subimmergens Vain. L
Lecanora subumbrina Müll.Arg. L
Lecanora swartzii (Ach.) Ach. L
Lecanora symmicta (Ach.) Ach. L
Lecanora umbrina (Ach.) A.Massal. L
Lecanora xylophila Hue L
Lecidella carpathica Körb. L
Lecidella commutata Knoph & Leuckert L
Lecidella effugiens (Nilson) Knoph & Hertel L
Lecidella elaeochroma (Ach.) Hazsl. L
Lecidella granulosila (Nyl.) Knoph & Leuckert L
Lecidella schistiseda (Zahlbr.) Hertel E L
Lecidella stigmatea (Ach.) Hertel & Leuckert L
Lecidella sublapicida (C.Knight) Hertel L
Lecidella wulfenii (Hepp) Körb. L
Miriquidica deusta (Stenh.) Hertel & Rambold L
Miriquidica nigroleprosa (Vain.) Hertel & Rambold L
Pyrrhospora laeta (Stirt.) Hafellner L
Pyrrhospora sanguinolenta (Kremp.) Rambold & Hafellner L
Ramboldia petraeoides (Nyl. ex C.Bab. & Mitt.) Kantvilas & Elix L
Ramboldia stuartii (Hampe) Kantvilas & Elix L
MEGALARIACEAE
Megalaria imshaugii Fryday E L
Megalaria grossa (Pers. ex Nyl.) Hafellner L
Megalaria macrospora Fryday E L
Megalaria maculosa (Stirt.) D.J.Galloway E L
Megalaria melanotropa (Nyl.) D.J.Galloway E L
Megalaria pulverea (Borrer) Hafellner & E.Schreiner L
Megalaria semipallida (C.Knight) D.J.Galloway E L
Megalaria spodophana (Nyl.) D.J.Galloway E L
Megalaria subcarnea (Müll.Arg.) D.J.Galloway E L
Megalaria sublivens (Nyl.) D.J.Galloway E *L
Megalaria variegata (Müll.Arg.) D.J.Galloway E L
MILTIDEACEAE
Miltidea ceroplasta (C.Bab.) D.J.Galloway & Hafellner L
MYCOBLASTACEAE
Mycoblastus campbellianus (Nyl.) Zahlbr. L
Mycoblastus hypomelinus (Stirt.) Müll.Arg. L
PARMELIACEAE
Alectoria nigricans (Ach.) Nyl.. L
Anzia entingiana Elix E L
Anzia gallowayi Elix E L
Anzia jamesii D.J.Galloway E L
Bryoria austromontana P.M.Jørg. & D.J.Galloway L
Bryoria indonesica (P.M.Jørg.) Brodo & D.Hawksw. L
Canoparmelia norpruinata Elix & J.Johnst. L
Canoparmelia pustulescens (Kurok.) Elix L
Canoparmelia subtiliacea (Nyl.) Elix & Hale L
Canoparmelia texana (Tuck.) Elix & Hale L
Cetraria aculeata (Schreb.) Fr. L
Cetraria islandica subsp. *antarctica* Kärnefelt L
Cetraria muricata (Ach.) Eckfeldt L
Cetrariella delisei (Bory ex Schaer.) Kärnefelt & A.Thell L
Cetrelia braunsiana (Müll.Arg.) W.L.Culb. & C.F.Culb. L
Everniastrum sorocheilum (Vain.) Hale ex Sipman L
Flavoparmelia haysomii (C.W.Dodge) Hale L
Flavoparmelia haywardiana Elix & J.Johnst. L
Flavoparmelia soredians (Nyl.) Hale L
Hypogymnia billardierei (Mont.) Filson L
Hypogymnia kosciuskoensis Elix L
Hypogymnia lugubris (Pers.) Krog var. *lugubris* L
— var. *compactior* (Zahlbr.) Elix L
— var. *sublugubris* (Müll.Arg.) Elix L
Hypogymnia pulchrilobata (Bitter) Elix L
Hypogymnia pulverata (Nyl. ex Cromb.) Elix L
Hypogymnia subphysodes (Kremp.) Filson var. *subphysodes* L
— var. *austeroides* Elix L
Hypogymnia turgidula (Bitter) Elix L
Hypotrachyna dactylifera (Vain.) Hale L
Hypotrachyna ensifolia (Kurok.) Hale L
Hypotrachyna exsecta (Taylor) Hale L
Hypotrachyna imbricatula (Zahlbr.) Hale L
Hypotrachyna immaculata (Kurok.) Hale L
Hypotrachyna laevigata (Sm.) Hale L
Hypotrachyna neodissecta (Hale) Hale L
Hypotrachyna osseoalba (Vain.) Y.S.Park & Hale L
Hypotrachyna producta Hale L
Hypotrachyna pseudosinuosa (Asahina) Hale L
Hypotrachyna revoluta (Flörke) Hale L
Hypotrachyna rockii (Zahlbr.) Hale L
Hypotrachyna sinuosa (Sm.) Hale L
Hypotrachyna thysanota (Kurok.) Hale L
Melanelia calva (Essl.) Essl. E L
Melanelia glabratuloides (Essl.) Essl. E L
Melanelia subglabra (Räsänen) Essl. L
Melanohalea inactiva (P.M.Jørg.) O.Blanco, A.Crespo, Divakar, Essl., D.Hawksw. & Lumbsch L

Melanohalea zopheroa (Essl.) O.Blanco, A.Crespo, Divakar, Essl., D.Hawksw. & Lumbsch L
Menegazzia aeneofusca (Müll.Arg.) R.Sant. L
Menegazzia aucklandica (Zahlbr.) P.James & D.J.Galloway E L
Menegazzia caliginosa P.James & D.J.Galloway L
Menegazzia castanea P.James & D.J.Galloway L
Menegazzia dielsii (Hillmann) R.Sant. E L
Menegazzia eperforata P.James & D.J.Galloway L
Menegazzia foraminulosa (Kremp.) Bitter E L
Menegazzia globulifera R.Sant. L
Menegazzia hypernota Bjerke E L
Menegazzia inactiva P.James & Kantvilas L
Menegazzia inflata (Hillmann) P.James & D.J.Galloway E L
Menegazzia kantvilasii P.James L
Menegazzia lucens P.James & D.J.Galloway E L
Menegazzia neozelandica (Zahlbr.) P.James L
Menegazzia nothofagi (Zahlbr.) P.James & D.J.Galloway L
Menegazzia pertransita (Stirt.) R.Sant. L
Menegazzia pulchra P.James & D.J.Galloway E L
Menegazzia stirtonii (Zahlbr.) Kantvilas & Louwhoff E L
Menegazzia subpertusa P.James & D.J.Galloway L
Menegazzia testacea P.James & D.J.Galloway L
Menegazzia ultralucens P.James & D.J.Galloway L
Nesolechia oxyspora (Tul.) A. Massal.
Pannoparmelia angustata (Pers.) Zahlbr. L
Pannoparmelia wilsonii (Räsänen) D.J.Galloway L
Parmelia congesta Kurok. & Filson L
Parmelia crambidiocarpa Zahlbr. L
Parmelia cunninghamii Cromb. L
Parmelia erumpens Kurok. L
Parmelia norcrambidiocarpa Hale L
Parmelia nortestacea E L
Parmelia novae-zelandiae Hale E L
Parmelia protosignifera Elix & J.Johnst. L
Parmelia protosulcata Hale L
Parmelia salcrambidiocarpa Hale L
Parmelia saxatilis (L.) Ach. L
Parmelia signifera Nyl. L
Parmelia subtestacea Hale L
Parmelia sulcata Taylor L
Parmelia tenuirima Hook.f. & Taylor L
Parmelia testacea Stirt. L
Parmelina conlabrosa (Hale) Elix & J.Johnst. L
Parmelina labrosa (Zahlbr.) Elix & J.Johnst. L
Parmelina pseudorelicina (Jatta) Kantvilas & Elix L
Parmelina quercina (Willd.) Hale L
Parmelinopsis afrorevoluta (Krog & Swinscow) Elix & Hale L
Parmelinopsis horrescens (Taylor) Elix & Hale L
Parmelinopsis jamesii (Hale) Elix & Hale L
Parmelinopsis minarum (Vain.) Elix & Hale L
Parmelinopsis spathulata (Kurok.) Elix & Hale L
Parmelinopsis spumosa (Asahina) Elix & Hale L
Parmelinopsis subfatiscens (Kurok.) Elix & Hale L
Parmelinopsis swinscowii (Hale) Elix & Hale L
Parmotrema arnoldii (Du Rietz) Hale L
Parmotrema austrocetratum Elix & J.Johnst. L
Parmotrema cetratum (Ach.) Hale L
Parmotrema crinitum (Ach.) M.Choisy L
Parmotrema cristiferum (Taylor) Hale L
Parmotrema dilatatum (Vain.) Hale L
Parmotrema grayanum (Hue) Hale L
Parmotrema lophogenum (Abbayes) Hale L
Parmotrema mellissii (C.W.Dodge) Hale L
Parmotrema perlatum (Huds.) M.Choisy L
Parmotrema reparatum (Stirt.) O.Blanco, A.Crespo, Divakar, Elix & Lumbsch L
Parmotrema reticulatum (Taylor) M.Choisy L
Parmotrema robustum (Degel.) Hale L
Parmotrema tinctorum (Despr. ex Nyl.) Hale L
Parmotrema zollingeri (Hepp) Hale L
Protoparmelia badia (Hoffm.) Hafellner L
Pseudephebe minuscula (Nyl. ex Arnold) Brodo & D.Hawksw. L
Pseudephebe pubescens (L.) M.Choisy L
Punctelia borreri (Sm.) Krog L
Punctelia novozelandica Elix & J.Johnst. L
Punctelia perreticulata (Räsänen) G.Wilh. & Ladd L
Punctelia subalbicans (Stirt.) D.J.Galloway & Elix L
Punctelia subflava (Taylor) Elix & J.Johnst. L
Punctelia subrudecta (Nyl.) Krog L
Punctelia transtasmanica Elix & Kantvilas L
Tuckermanopsis chlorophylla (Willd.) Hale L
Usnea acromelana Stirt. L
Usnea angulata Ach. L
Usnea articulata (L.) Hoffm. L
Usnea baileyi (Stirt.) Zahlbr. L
Usnea ciliata (Nyl.) Du Rietz L
Usnea ciliifera Motyka L
Usnea contexta Motyka L
Usnea cornuta Körb. L
Usnea dasypogoides Nyl. L
Usnea inermis Motyka L
Usnea maculata Stirt. L
Usnea molliuscula Stirt. L
Usnea nidifica Taylor L
Usnea oncodes Stirt. L
Usnea pseudocapillaris F.J.Walker L
Usnea pusilla (Räsänen) Räsänen L
Usnea rubicunda Stirt. L
Usnea rubrotincta Stirt. L
Usnea simplex Motyka L
Usnea subcapillaris (D.J.Galloway) F.J.Walker L
Usnea subeciliata (Motyka) Swinscow & Krog L
Usnea tenerior (Nyl.) Hue L
Usnea torulosa (Müll.Arg.) Zahlbr. L
Usnea trichodeoides Motyka L
Usnea undulata Stirt. L
Usnea wirthii P.Clerc L
Usnea xanthopoga Nyl. L
Xanthoparmelia adpicta (Zahlbr.) O.Blanco, A.Crespo, Elix, D.Hawksw. & Lumbsch L
Xanthoparmelia alexandrensis Elix & J.Johnst. L
Xanthoparmelia amplexula (Stirt.) Elix & J.Johnst. L
Xanthoparmelia arapilensis (Elix & P.M.Armstr.) Filson L
Xanthoparmelia atrobarbatica (Elix) O.Blanco, A.Crespo, Elix, D.Hawksw. & Lumbsch L
Xanthoparmelia atrocapnodes (Elix & J.Johnst.) Elix L
Xanthoparmelia australasica D.J.Galloway L
Xanthoparmelia barbellata (Kurok.) Hale L
Xanthoparmelia brattii (Essl.) O.Blanco, A.Crespo, Elix, D.Hawksw. & Lumbsch L
Xanthoparmelia bulfiniana (Elix) O.Blanco, A.Crespo, Elix, D.Hawksw. & Lumbsch L
Xanthoparmelia cheelii (Gyeln.) Hale L
Xanthoparmelia concomitans Elix & J.Johnst. L
Xanthoparmelia congesta (Kurok. & Filson) Elix & J.Johnst. L
Xanthoparmelia cordillerana (Gyeln.) Hale L
Xanthoparmelia depsidella (Elix) O.Blanco, A.Crespo, Elix, D.Hawksw. & Lumbsch L
Xanthoparmelia dichotoma (Müll.Arg.) Hale L
Xanthoparmelia digitiformis (Elix & P.M.Armstr.) Filson L
Xanthoparmelia elixii Filson L
Xanthoparmelia epheboides (Zahlbr.) O.Blanco, A.Crespo. Elix, D.Hawksw. & Lumbsch L
Xanthoparmelia filarszkyana (Gyeln.) Hale L
Xanthoparmelia flavescentireagens (Gyeln.) D.J.Galloway L
Xanthoparmelia flindersiana (Elix & P.M.Armstr.) Elix & J.Johnst. L
Xanthoparmelia furcata (Müll.Arg.) Hale L
Xanthoparmelia glabrans (Nyl.) O.Blanco, A.Crespo, Elix, D.Hawksw. & Lumbsch L
Xanthoparmelia glareosa (Kurok. & Filson) Elix & J.Johnst. L
Xanthoparmelia imitatrix (Taylor) O.Blanco, A.Crespo, Elix, D.Hawksw. & Lumbsch L
Xanthoparmelia incerta (Kurok. & Filson) Elix & J.Johnst. L
Xanthoparmelia isidiigera (Müll.Arg.) Elix & J.Johnst. L
Xanthoparmelia isidiotegeta Elix & Kantvilas L
Xanthoparmelia lineola (E.C.Berry) Hale L
Xanthoparmelia loxodella (Essl.) O.Blanco, A.Crespo, Elix, D.Hawksw. & Lumbsch L
Xanthoparmelia luteonotata (J.Steiner) O.Blanco, A.Crespo, Elix, D.Hawksw. & Lumbsch L
Xanthoparmelia malcolmii (Elix) O.Blanco, A.Crespo, Elix, D.Hawksw. & Lumbsch L
Xanthoparmelia martinii (Essl.) O.Blanco, A.Crespo, Elix, D.Hawksw. & Lumbsch L
Xanthoparmelia melanobarbatica (Essl.) O.Blanco, A.Crespo, Elix, D.Hawksw. & Lumbsch L
Xanthoparmelia metaclystoides (Kurok. & Filson) Elix & J.Johnst. L
Xanthoparmelia metamorphosa (Gyeln.) Hale L
Xanthoparmelia mexicana (Gyeln.) Hale L
Xanthoparmelia minutella (Essl.) O.Blanco, A.Crespo, Elix, D.Hawksw. & Lumbsch L
Xanthoparmelia molliuscula (Ach.) Hale L
Xanthoparmelia mougeotina (Nyl.) D.J.Galloway L
Xanthoparmelia murina (Kurok.) Elix L
Xanthoparmelia nebulosa (Kurok. & Filson) Elix & J.Johnst. L
Xanthoparmelia neotinctina (Elix) Elix & J.Johnst. L
Xanthoparmelia norcapnodes (Elix & J.Johnst.) Elix L
Xanthoparmelia notata (Kurok.) Hale L
Xanthoparmelia oleosa (Elix & P.M.Armstr.) Elix & T.H.Nash L
Xanthoparmelia olivetoricella O.Blanco, A.Crespo, Elix, D.Hawksw. & Lumbsch L
Xanthoparmelia peloloba (Essl.) O.Blanco, A.Crespo, Elix, D.Hawksw. & Lumbsch L
Xanthoparmelia petriseda (Zahlbr.) O.Blanco, A.Crespo, Elix, D.Hawksw. & Lumbsch L
Xanthoparmelia philippsiana (Filson) Elix & J.Johnst. L
Xanthoparmelia pictada (Essl.) O.Blanco, A.Crespo. Elix, D.Hawksw. & Lumbsch L
Xanthoparmelia plana (Essl.) O.Blanco, A.Crespo. Elix, D.Hawksw. & Lumbsch L
Xanthoparmelia pulla (Ach.) O.Blanco, A.Crespo. Elix, D.Hawksw. & Lumbsch L
Xanthoparmelia pustuliza (Elix) Elix & J.Johnst. L
Xanthoparmelia reptans (Kurok.) Elix & J.Johnst. L
Xanthoparmelia rubrireagens (Gyeln.) Hale L
Xanthoparmelia scabrosa (Taylor) Hale L
Xanthoparmelia scotophylla (Kurok.) Elix L
Xanthoparmelia semiviridis (F.Muell. ex Nyl.) O.Blanco, A.Crespo. Elix, D.Hawksw. & Lumbsch L
Xanthoparmelia sorediata (Elix & P.Child) O.Blanco, A.Crespo. Elix, D.Hawksw. & Lumbsch L
Xanthoparmelia squamans (Stizenb.) O.Blanco, A.Crespo. Elix, D.Hawksw. & Lumbsch L
Xanthoparmelia squamariatella (Elix) O.Blanco, A.Crespo. Elix, D.Hawksw. & Lumbsch L
Xanthoparmelia streimannii (Elix & P.M.Armstr.) Elix & J.Johnst. L
Xanthoparmelia stygiodes (Nyl. ex Cromb.) O.Blanco, A.Crespo. Elix, D.Hawksw. & Lumbsch L
Xanthoparmelia suberadicata (Abbayes) Hale L

Xanthoparmelia subhosseana (Essl.) O.Blanco, A.Crespo, Elix, D.Hawksw. & Lumbsch L
Xanthoparmelia subimitatrix (Essl.) O.Blanco, A.Crespo, Elix, D.Hawksw. & Lumbsch L
Xanthoparmelia subnuda (Kurok.) Hale L
Xanthoparmelia substrigosa (Hale) Hale L
Xanthoparmelia taractica (Kremp.) Hale L
Xanthoparmelia tasmanica (Hook.f. & Taylor) Hale L
Xanthoparmelia tegeta Elix & J.Johnst. L
Xanthoparmelia thamnoides (Kurok.) Hale L
Xanthoparmelia ustulata (Kurok. & Filson) Elix & J.Johnst. L
Xanthoparmelia verdonii Elix & J.Johnst. L
Xanthoparmelia verisidiosa (Essl.) O.Blanco, A.Crespo, Elix, D.Hawksw. & Lumbsch L
Xanthoparmelia verrucella (Essl.) O.Blanco, A.Crespo, Elix, D.Hawksw. & Lumbsch L
Xanthoparmelia waiporiensis (Hillmann) O.Blanco, A.Crespo, Elix, D.Hawksw. & Lumbsch L
Xanthoparmelia xanthomelaena (Müll.Arg.) Hale L
PILOCARPACEAE
Bapalmuia buchananii (Stirt.) Kalb & Lücking L
Byssoloma adspersum Malcolm & Vězda E L
Byssoloma leucoblepharum (Nyl.) Vain. L
Byssoloma octomerum Malcolm & Vězda E L
Byssoloma subdiscordans (Nyl.) P.James L
Byssoloma subundulatum (Stirt.) Vězda L
Fellhanera bouteillei (Desm.) Vězda L
Fellhanera semecarpi (Vain.) Vězda L
Micarea erratica (Körb.) Hertel, Rambold & Pietschm. L
Micarea flagellispora Coppins & Kantvilas L
Micarea isabellina Coppins & Kantvilas L
Micarea magellanica (Müll.Arg.) Fryday L
Micarea nitschkeana (J.Lahm ex Rabenh.) Harm. L
Micarea pannarica Fryday E L
Micarea peliocarpa (Anzi) Coppins & R.Sant. L
Micarea prasina Fr. E L
Roccellinastrum flavescens Kantvilas L
Roccellinastrum neglectum Henssen & Vobis L
PSORACEAE
Protoblastenia rupestris (Scop.) J.Steiner L
Psora crenata (Taylor) Reinke L
Psora crystallifera (Taylor) Müll.Arg. L
Psora decipiens (Hedw.) Hoffm. L
RAMALINACEAE
Bacidia albicerata (Kremp.) Zahlbr. E L
Bacidia albidoprasina C.Knight E L
Bacidia allotropa (Nyl.) Zahlbr. E L
Bacidia bagliettoana (A.Massal. & De Not.) Jatta L
Bacidia curvispora Coppins & Fryday E L
Bacidia gallowayi Coppins & Fryday E L
Bacidia glomerulosa C.Knight E L
Bacidia killiasii (Hepp) D. Hawksw. L
Bacidia laurocerasi (Delise ex Duby) Vain. L
Bacidia leucocarpa C.Knight L
Bacidia leucothalamia (Nyl.) Hellb. L
Bacidia macrospora (C.Knight) Zahlbr. E L
Bacidia minutissima C.Knight E L
Bacidia placodioides Coppins & Fryday E L
Bacidia plesia (C.Knight) Zahlbr. E L
Bacidia subcerina Zahlbr. L
Bacidia superula (Nyl.) Hellb. L
Bacidia tholera Zahlbr. E L
Bacidia wellingtonii (Stirt.) D.J.Galloway E L
Bacidina apiahica (Müll.Arg.) Vězda L
Bacidina phacodes (Körb.) Vězda L
Biatora albipraetexta (C.Knight) Hellb. E L
Cliostomum griffithii (Sm.) Coppins L
Frutidella caesioatra (Schaer.) Kalb L
Herteliana australis Fryday E L
Malcolmiella cinereovirens Vězda E L
Phyllopsora buettneri var. *glauca* (de Lesd.) Brako L
Phyllopsora corallina (Eschw.) Müll.Arg. L
Phyllopsora furfuracea (Pers.) Zahlbr. L
Phyllopsora malcolmii Vězda & Kalb L
Phyllopsora microdactyla (C.Knight) D.J.Galloway L
Ramalina australiensis Nyl. L
Ramalina canariensis J.Steiner L
Ramalina celastri (Spreng.) Krog & Swinscow L
Ramalina erumpens Blanchon, Braggins & Alison Stewart L
Ramalina exiguella Stirt. L
Ramalina fimbriata Krog & Swinscow L
Ramalina geniculata Hook.f. & Taylor L
Ramalina glaucescens Kremp. L
Ramalina inflata (Hook.f. & Taylor) Hook.f. & Taylor L
Ramalina inflexa Blanchon, Braggins & Alison Stewart L
Ramalina luciae Molho, Brodo, W.L.Culb. & C.F.Culb. L
Ramalina meridionalis D.Blanchon & Bannister L
Ramalina pacifica Asahina L
Ramalina peruviana Ach. L
Ramalina pollinaria (Westr.) Ach. L
Ramalina riparia Blanchon, Braggins & Alison Stewart L
Ramalina unilateralis F.Wilson L
Scutula miliaris (Wallr.) Trevis. U
Scutula tuberculosa (Th. Fr.) Rehm
Stirtoniella kelica (Stirt.) D.J.Galloway, Hafellner & Elix L
Tylothallia pahiensis (Zahlbr.) Hertel and H.Kilias L
SARRAMEANACEAE
Loxospora cyamidia (Stirt.) Kantvilas L
Loxospora septata (Sipman & Aptroot) Kantvilas L
Loxospora solenospora (Müll.Arg.) Kantvilas L
SPHAEROPHORACEAE
Austropeltum glareosum Henssen, H.Döring & Kantvilas L
Bunodophoron agnetae Wedin L
Bunodophoron australe (Laurer) A.Massal. L
Bunodophoron flaccidum (Kantvilas & Wedin) Wedin L
Bunodophoron imshaugii (Ohlsson) Wedin L
Bunodophoron insigne (Laurer) Wedin L
Bunodophoron macrocarpum (Ohlsson) Wedin L
Bunodophoron microsporum (Ohlsson) Wedin E L
Bunodophoron murrayi (Ohlsson) Wedin L
Bunodophoron notatum (Tibell) Wedin L
Bunodophoron ohlssonii (Wedin) Wedin E L
Bunodophoron palmatum (Js.Murray) Wedin E L
Bunodophoron patagonicum (C.W.Dodge) Wedin L
Bunodophoron ramuliferum (I.M.Lamb) Wedin L
Bunodophoron scrobiculatum (C.Bab.) Wedin L
Bunodophoron tibellii (Wedin) Wedin L
Bunodophoron whakapapaense (Wedin) Wedin L
Leifidium tenerum (Laurer) Wedin L
Sphaerophorus stereocauloides Nyl. L
STEREOCAULACEAE
Lepraria eburnea J.R.Laundon L
Lepraria incana (L.) Ach. L
Lepraria lobificans Nyl. L
Lepraria membranacea (Dicks.) Vain. L
Lepraria neglecta (Nyl.) Lettau L
Lepraria vouauxii (Hue) R.C.Harris L
Stereocaulon argus Hook.f. & Taylor L
Stereocaulon caespitosum Redinger L
Stereocaulon colensoi C.Bab. L
Stereocaulon corticatulum Nyl. L
Stereocaulon delisei Bory ex Duby L
Stereocaulon fronduliferum I.M.Lamb L
Stereocaulon gregarium Redinger L
Stereocaulon loricatum I.M.Lamb L
Stereocaulon ramulosum Räuschel L
Stereocaulon trachyphloeum I.M.Lamb L
Stereocaulon vesuvianum Pers. L
Stereocaulon wadei I.M.Lamb L
TEPHROMELATACEAE
Tephromela atra (Huds.) Hafellner L

Order LECIDEALES
LECIDEACEAE
Lecidea atromorio C.Knight L
Lecidea aucklandica Zahlbr. E L
Lecidea canorufescens Kremp. L
Lecidea capensis Zahlbr. L
Lecidea cerinocarpa C.Knight L
Lecidea coccodes C.Knight (nom. illegit.) E L
Lecidea conisalea C.Knight E L
Lecidea dacrydii Müll.Arg. E L
Lecidea diducens Nyl. L
Lecidea dracophylli Zahlbr. E L
Lecidea endochlora (Hook.f. & Taylor) Tuck. L
Lecidea fuscoatra (L.) Ach. L
Lecidea fuscoatrula Nyl. L
Lecidea fuscocincta Stirt. E L
Lecidea lapicida (Ach.) Ach. var. *lapicida* L
— var. *maungahukae* Hertel E L
— var. *pantherina* Ach. L
Lecidea lygomma Nyl. var *lygomma* L
— var. *crassilabra* (Müll.Arg.) Hertel & Rambold L
Lecidea miscescens Nyl. L
Lecidea nigratula Müll.Arg. E L
Lecidea ochroleuca Pers. L
Lecidea plana (J.Lahm) Nyl. L
Lecidea sarcogynoides Körb. L
Lecidea senescens Zahlbr. E L
Lecidea spheniscidarum Hertel L
Lecidea subsericea Zahlbr. E L
Lecidea swartzioidea Nyl. L
Lecidea taitensis (Mont.) Nyl. L
Lecidea thomsonii Zahlbr. E L
Lecidea verruca Poelt
Rhizolecia hybrida (Zahlbr.) Hertel L
PORPIDIACEAE
Bellemerea alpina (Sommerf.) Clauzade & Cl.Roux L
Bellemerea subsorediza (Lynge) R.Sant. L
Clauzadea monticola (Ach.) Hafellner & Bellem. L
Immersaria athroocarpa (Ach.) Rambold & Pietschm. L
Labyrintha implexa Malcolm, Elix & Owe-Larss. E L
Paraporpidia glauca (Taylor) Rambold L
Paraporpidia leptocarpa (Nyl.) Rambold & Hertel L
Poeltiaria coromandelica (Zahlbr.) Rambold & Hertel L
Poeltiaria corralensis (Räsänen) Hertel L
Poeltiaria turgescens (Körb.) Hertel L
Poeltidea perusta (Nyl.) Hertel & Hafellner L
Porpidia albocaerulescens (Wulfen) Hertel & Knoph L
Porpidia crustulata (Ach.) Hertel & Knoph L
Porpidia macrocarpa (DC.) Hertel & A.J.Schwab L
Porpidia platycarpoides (Bagl.) Hertel L
Porpidia skottsbergiana Hertel L
Porpidia speirea (Ach.) Kremp. L
Porpidia superba (Körb.) Hertel & Knoph L

Order OSTROPALES
ASTEROTHYRIACEAE
Gyalidea cerina Malcolm & Vězda E L
Gyalidea hensseniae Hafellner, Poelt & Vězda L
Gyalidea hyalinescens (Nyl.) Vězda L
Gyalidea lecanorina (C.Knight) P.James L
COENOGONIACEAE
Coenogonium fallaciosum (Müll.Arg.) Kalb & Lücking L
Coenogonium flavum (Malcolm & Vězda) Malcolm L
Coenogonium fuscescens (Vězda & Malcolm)

Malcolm L
Coenogonium implexum Nyl. L
Coenogonium lutescens (Vězda & Malcolm) Malcolm E L
Coenogonium luteum (Dicks.) Kalb & Lücking L
Coenogonium queenslandicum (Kalb & Vězda) Lücking L
Coenogonium rubrifuscum (Malcolm & Vězda) Malcolm E L
Coenogonium zonatum (Müll.Arg.) Kalb & Lücking L
GOMPHILLACEAE
Aderkomyces albostrigosus (R.Sant.) Lücking, Sérus. & Vězda L
Aulaxina quadrangula (Stirt.) R.Sant. L
Calenia microcarpa Vězda L
Jamesiella anastomosans (P.James & Vězda) Lücking, Sérus. & Vězda L
Lithogyalideopsis zeylandica (Vězda & Malcolm) Lücking, Sérus. & Vězda E L
GRAPHIDACEAE
Fissurina confraga Kremp. E L
Fissurina incrustans Fée L
Fissurina inquinata C.Knight & Mitt. L
Fissurina insidiosa C.Knight & Mitt. L
Fissurina monospora C.Knight E L
Fissurina novae-zelandiae C.Knight E L
Fissurina subcontexta (Nyl.) Nyl. L
Fissurina triticea (Nyl.) Staiger L
Glyphis cicatricosa Ach. L
Graphis anfractuosa (Eschw.) Eschw. L
Graphis elegans (Sm.) Ach. L
Graphis librata C.Knight L
Graphis tenella Ach. L
Leiorreuma exaltatum (Mont. & Bosch) Staiger L
Phaeographina arechavaletae Müll.Arg. L
Phaeographis intricans (Nyl.) Staiger L
Phaeographis inusta (Ach.) Müll.Arg. L
Phaeographis mucronata (Stirt.) Zahlbr. L
Sarcographa labyrinthica (Ach.) Müll.Arg. L
Thalloloma subvelata (Stirt.) D.J.Galloway L
GYALECTACEAE
Belonia pellucida Coppins & Malcolm E L
Belonia vězdana Malcolm & Coppins E L
Cryptolechia myriadella (Nyl.) D.Hawksw. & Dibben E L
Gyalecta truncigena (Ach.) Hepp L
Pachyphiale carneola (Ach.) Arnold L
ODONTOTREMATACEAE
Potriphila epiphylla Döbbeler
Potriphila navicularis Döbbeler
Skyttea mayerhoferi Diederich & Etayo L
PHLYCTIDACEAE
Phlyctis longifera (Nyl.) D.J.Galloway & Guzmán L
Phlyctis megalospora (P.James) D.J.Galloway & Guzmán L
Phlyctis oleosa Stirt. L
Phlyctis sordida C.Knight L
Phlyctis subuncinata Stirt. L
Phlyctis uncinata Stirt. L
PORINACEAE
Porina ahlesiana (Körb.) Zahlbr. L
Porina aptrootii P.M.McCarthy L
Porina atrocoerulea Müll.Arg. L
Porina cerina (Zahlbr.) R.Sant. L
Porina chlorotica (Ach.) Müll.Arg. L
Porina chrysophora (Stirt.) R.Sant. L
Porina cinereonigrescens (Stirt.) Müll.Arg. L
Porina constrictispora P.M.McCarthy & Kantvilas L
Porina corrugata Müll.Arg. L
Porina decrescens P.M.McCarthy & Kantvilas L
Porina diffluens Malcom & Vězda L
Porina elegantula Müll.Arg. L
Porina emiscens (Nyl.) Müll.Arg. L
Porina epiphylla (Fée) Fée L
Porina exacta Malcolm, P.M.McCarthy & Kantvilas L
Porina exocha (Nyl.) P.M.McCarthy L
Porina fluminea P.M.McCarthy & P.N.Johnson L
Porina guentheri (Flot.) Zahlbr. L
Porina kantvilasii P.M.McCarthy L
Porina lamprocarpa (Stirt.) Müll.Arg. L
Porina leptalea (Durieu & Mont.) A.L.Sm. L
Porina leptaleina (Nyl.) Müll.Arg. L
Porina leptosperma Müll.Arg. L
Porina leptostegia (C.Knight) Müll.Arg. L
Porina mastoidea (Ach.) Müll.Arg. L
Porina nucula Ach. L
Porina otagensis P.M.McCarthy L
Porina palmicola Malcolm & Vězda L
Porina partita P.M.McCarthy L
Porina psilocarpa P.M.McCarthy L
Porina rhaphidiophora (Nyl.) Müll.Arg. L
Porina rubella (Malcolm & Vězda) Lücking L
Porina rubrofusca (Malcolm & Vězda) Lücking L
Porina rufula (Kremp.) Vain. L
Porina semecarpi Vain. L
Porina speciosa P.M.McCarthy & Malcolm L
Porina subapplanata Malcolm, Vězda, P.M.McCarthy & Kantvilas L
Porina sylvatica P.M.McCarthy & Kantvilas L
Porina tetramera (Malme) R.Sant. L
Trichothelium alboatrum Vain. L
Trichothelium assurgens (Cooke) Aptroot & Lücking L
Trichothelium javanicum (F.Schill.) Vězda L
STICTIDACEAE
Acarosporina hyalina P.R. Johnst. E
Conotremopsis weberiana Vězda L
Cryptodiscus pallidus (Pers.) Corda
Delpontia pulchella Penz. & Sacc.
Schizoxylon lividum McAlpine A
Stictis arundinacea Pers. OU
Stictis asteliae P.R. Johnst. E
Stictis brachyspora (Sacc. & Berl.) Sherwood
Stictis carnea Seaver & Waterston
Stictis clavata P.R. Johnst. E
Stictis collospermi P.R. Johnst. E
Stictis cordylines P.R. Johnst. E
Stictis dealbata P.R. Johnst. E
Stictis dicksoniae Sherwood E
Stictis dumontii Sherwood
Stictis filicicola Seaver & Waterston OU
Stictis fuscella Sherwood
Stictis gigantea Sherwood OU
Stictis hawaiiensis E.K. Cash
Stictis inconstans P.R. Johnst. E
Stictis laciniata P.R. Johnst. E
Stictis lata P.R. Johnst. E
Stictis lupini W. Phillips & Harkn. OU
Stictis paucula P.R. Johnst. E
Stictis prominens Sherwood
Stictis pusilla Speg. OU
Stictis pustulata Ellis OU
Stictis radiata Pers.
Stictis ramuligera Starbäck var. *ramuligera*
— var. *minor* P.R. Johnst. E
Stictis serpentaria Ellis & Everh.
Stictis stellata Wallr. A
Stictis subiculata P.R. Johnst.
Stictis tortilis P.R. Johnst. E
Stictis trinervia P.R. Johnst. E
Stictis virginea Cooke & W. Phillips E
Topelia rosea (Servít) P.M.Jørg. & Vězda L
THELOTREMATACEAE
Chapsa asteliae (Kantvilas & Vězda) Mangold L
Chapsa lamellifera (Kantvilas & Vězda) Mangold L
Chapsa megalophthalma (Müll.Arg.) Mangold L
Chroodiscus macrocarpus (C.W.Dodge) D.J.Galloway L
Diploschistes actinostomus (Pers.) Zahlbr. L
Diploschistes euganeus (A.Massal.) J.Steiner L
Diploschistes gypsaceus (Ach.) Zahlbr. L
Diploschistes hensseniae Lumbsch & Elix L
Diploschistes muscorum (Scop.) R.Sant. subsp. *muscorum* L
— subsp. *bartlettii* Lumbsch L
Diploschistes ocellatus (Vill.) Norman L
Diploschistes scruposus (Schreb.) Norman L
Diploschistes sticticus (Körb.) Müll.Arg. L
Ingvariella bispora (Bagl.) Guderley & Lumbsch L
Ocellularia allosporoides (Nyl.) Patw. & C.R.Kulk. L
Ocellularia concentrica (Stirt.) Sherwood E L
Ocellularia hians (Stirt.) Müll.Arg. E L
Ocellularia monosporoides (Nyl.) Hale L
Thelotrema circumscriptum C.Knight L
Thelotrema farinaceum C.Knight L
Thelotrema lepadinum (Ach.) Ach. L
Thelotrema monosporum Nyl. L
Thelotrema novae-zelandiae Szatala L
Thelotrema porinoides Mont. & Bosch L
Thelotrema saxatile C.Knight L
Thelotrema subtile Tuck. L
Thelotrema weberi Hale L
Topeliopsis decorticans (Müll.Arg.) Frisch & Kalb L
Topeliopsis muscigena (Stizenb.) Kalb L
Topeliopsis subdenticulata (Zahlbr.) A.Frisch & Kalb L
Tremotylium occultum Stirt. L
Tremotylium subocculltum Stirt. L

Order PELTIGERALES
COCCOCARPIACEAE
Coccocarpia erythroxyli (Spreng.) Swinscow & Krog L
Coccocarpia palmicola (Spreng.) Arv. & D.J.Galloway L
Coccocarpia pellita (Ach.) Müll.Arg. L
Peltularia crassa P.M.Jørg. & D.J.Galloway L
Spilonema dendroides Henssen L
Steinera neozelandica C.W.Dodge L
Steinera polymorpha P.James & Henssen E L
Steinera radiata P.James & Henssen subsp. *radiata* E L
— subsp. *aucklandica* P.James & Henssen E L
Steinera sorediata P.James & Henssen L
COLLEMATACEAE
Collema coccophorum Tuck. L
Collema crispum (Huds.) Weber ex F.H.Wigg. L
Collema durietzii Degel. L
Collema fasciculare (L.) Weber ex F.H.Wigg. var. *fasciculare* L
— var. *colensoi* C.Bab. L
— var. *microcarpum* (Müll.Arg.) Degel. L
Collema fragrans var. *contiguum* (C.Knight & Mitt.) Degel. E L
Collema glaucophthalmum Nyl. L
Collema japonicum (Müll.Arg.) Hue L
Collema kauaiense H.Magn. L
Collema laeve Hook.f. & Taylor L
Collema leptaleum Tuck. L
Collema leucocarpum Hook.f. & Taylor L
Collema novozelandicum Degel. L
Collema quadriloculare var. *tasmaniae* F.Wilson L
Collema subconveniens Nyl. L
Collema subflaccidum Degel. L
Collema subfragrans Degel. L
Collema subundulatum Degel. E L
Leptogium aucklandicum Zahlbr. E L
Leptogium australe (Hook.f. & Taylor) Müll.Arg. L
Leptogium austroamericanum (Malme) C.W.Dodge L
Leptogium biloculare F.Wilson L

Leptogium burgessii (L.) Mont. L
Leptogium coralloideum (Meyen & Flot.) Vain. L
Leptogium crispatellum Nyl. L
Leptogium cyanescens (Rabenh.) Körb. L
Leptogium cyanizum (Nyl.) Nyl. L
Leptogium denticulatum Nyl. L
Leptogium laceroides B.de Lesd. L
Leptogium limbatum F.Wilson L
Leptogium malmei P.M.Jørg. L
Leptogium menziesii (Ach.) Mont. L
Leptogium pecten F.Wilson L
Leptogium philorheuma F.Wilson L
Leptogium phyllocarpum (Pers.) Mont. L
Leptogium plicatile (Ach.) Leight. L
Leptogium propaguliferum Vain. L
Leptogium victorianum F.Wilson L
Physma byrsaeum (Ach.) Tuck. L
Physma chilense Hue L
Ramalodium dumosum Henssen L
Ramalodium fecundissimum Henssen L
LOBARIACEAE
Dendriscocaulon dendriothamnodes Dughi ex D.J.Galloway (nom. inv.) L
Dendriscocaulon dendroides (Nyl.) R.Sant. ex H.Magn. 1950 (nom. inv.) E L
Lobaria adscripta (Nyl.) Hue E L
Lobaria asperula (Stirt.) Yoshim. A L
Lobaria dictyophora (Müll.Arg.) D.J.Galloway E L
Lobaria retigera (Bory) Trevis. L
Lobarina scrobiculata (Scop.) Nyl. L
Pseudocyphellaria ardesiaca D.J.Galloway L
Pseudocyphellaria argyracea (Delise) Vain. L
Pseudocyphellaria aurata (Ach.) Vain. L
Pseudocyphellaria bartlettii D.J.Galloway L
Pseudocyphellaria billardierei Delise Räsänen L
Pseudocyphellaria carpoloma (Delise) Vain. L
Pseudocyphellaria chloroleuca (Hook.f. & Taylor) Du Rietz L
Pseudocyphellaria cinnamomea (A.Rich.) Vain. E L
Pseudocyphellaria colensoi (C. Bab.) Vain. L
Pseudocyphellaria corbettii D.J.Galloway L
Pseudocyphellaria coriacea (Hook.f. & Taylor) D.J.Galloway & P.James L
Pseudocyphellaria coronata (Müll.Arg.) Malme L
Pseudocyphellaria crassa D.J.Galloway L
Pseudocyphellaria crocata (L.) Vain. L
Pseudocyphellaria crocatoides D.J.Galloway L
Pseudocyphellaria degelii D.J.Galloway & P.James L
Pseudocyphellaria dissimilis (Nyl.) D.J.Galloway & P.James L
Pseudocyphellaria durietzii D.J.Galloway L
Pseudocyphellaria episticta (Nyl.) Vain. L
Pseudocyphellaria faveolata (Delise) Malme L
Pseudocyphellaria fimbriata D.J.Galloway & P.James L
Pseudocyphellaria fimbriatoides D.J.Galloway & P.James L
Pseudocyphellaria glabra (Hook.f. & Taylor) C.W.Dodge L
Pseudocyphellaria granulata (C.Bab.) Malme L
Pseudocyphellaria gretae D.J.Galloway L
Pseudocyphellaria halei D.J.Galloway L
Pseudocyphellaria haywardiorum D.J.Galloway L
Pseudocyphellaria homoeophylla (Nyl.) C.W.Dodge L
Pseudocyphellaria hookeri (C.Bab.) D.J.Galloway & P.James L
Pseudocyphellaria intricata (Delise) Vain. L
Pseudocyphellaria jamesii D.J.Galloway L
Pseudocyphellaria lindsayi D.J.Galloway L
Pseudocyphellaria lividofusca (Kremp.) D.J.Galloway & P.James L
Pseudocyphellaria maculata D.J.Galloway L
Pseudocyphellaria mallota (Tuck.) H.Magn. L
Pseudocyphellaria margaretiae D.J.Galloway L
Pseudocyphellaria montagnei (C.Bab.) D.J.Galloway & P.James L
Pseudocyphellaria multifida (Nyl.) D.J.Galloway & P.James L
Pseudocyphellaria neglecta (Müll.Arg.) H.Magn. L
Pseudocyphellaria nermula D.J.Galloway L
Pseudocyphellaria physciospora (Nyl.) Malme L
Pseudocyphellaria pickeringii (Tuck.) D.J.Galloway L
Pseudocyphellaria poculifera (Müll.Arg.) D.J.Galloway & P.James L
Pseudocyphellaria pubescens (Müll.Arg.) D.J.Galloway & P.James L
Pseudocyphellaria rubella (Hook.f. & Taylor) D.J.Galloway & P.James L
Pseudocyphellaria rufovirescens (C.Bab.) D.J.Galloway E L
Pseudocyphellaria sericeofulva D.J.Galloway L
Pseudocyphellaria wilkinsii D.J.Galloway L
Sticta babingtonii D.J.Galloway L
Sticta caliginosa D.J.Galloway L
Sticta cinereoglauca Hook.f. & Taylor L
Sticta colinii D.J.Galloway L
Sticta filix (Sw.) Nyl. L
Sticta fuliginosa (Hoffm.) Ach. L
Sticta lacera (Hook.f. & Taylor) Müll.Arg. L
Sticta latifrons A.Rich. L
Sticta limbata (Sm.) Ach. L
Sticta livida Kremp. L
Sticta martinii D.J.Galloway L
Sticta squamata D.J.Galloway L
Sticta subcaperata (Nyl.) Nyl. L
Sticta sublimbata (J.Steiner) Swinscow & Krog L
MASSALONGIACEAE
Massalongia carnosa (Dicks.) Körb. L
Polychidium contortum Henssen L
NEPHROMATACEAE
Nephroma australe A.Rich. L
Nephroma cellulosum (Ach.) Ach. var. *cellulosum* L
— var. *isidioferum* Js.Murray L
Nephroma helveticum Ach. L
Nephroma plumbeum (Mont.) Mont. var. *plumbeum* L
— var. *isidiatum* (Js.Murray) F.J.White & P.James L
Nephroma rufum (C.Bab.) P.James L
Nephromium helveticum Ach. L
PANNARIACEAE
Austrella brunnea (P.M.Jørg.) P.M.Jørg. E L
Degelia crustacea P.M.Jørg. & D.J.Galloway L
Degelia duplomarginata (P.James & Henssen) Arv. & D.J.Galloway L
Degelia durietzii Arv. & D.J.Galloway L
Degelia gayana (Mont.) Arv. & D.J.Galloway L
Degelia periptera (C.Knight) P.M.Jørg. & P.James L
Degeliella rosulata (P.M.Jørg. & D.J.Galloway) P.M.Jørg. L
Degeliella versicolor (Hook.f. & Taylor) P.M.Jørg. L
Erioderma leylandii (Taylor) Müll.Arg. subsp. *leylandii* L
Erioderma sorediatum D.J.Galloway & P.M.Jørg. L
Fuscoderma amphibolum (C.Knight) P.M.Jørg. & D.J.Galloway L
Fuscoderma applanatum (D.J.Galloway & P.M.Jørg.) P.M.Jørg. & D.J.Galloway E L
Fuscoderma limbatum P.M.Jørg. & D.J.Galloway L
Fuscoderma pyxinoides P.M.Jørg. L
Fuscopannaria crustata (Stirt.) P.M.Jørg. L
Fuscopannaria granulans P.M.Jørg. L
Fuscopannaria minor (Darb.) P.M.Jørg. L
Fuscopannaria subimmixta (C.Knight) P.M.Jørg. L
Leioderma duplicatum (Müll.Arg.) D.J.Galloway & P.M.Jørg. L
Leioderma erythrocarpum (Delise ex Nyl.) D.J.Galloway & P.M.Jørg. L
Leioderma pycnophorum Nyl. L
Leioderma sorediatum D.J.Galloway & P.M.Jørg. L
Pannaria allorhiza (Nyl.) Elvebakk & D.J.Galloway L
Pannaria araneosa (C.Bab.) Hue E L
Pannaria athroophylla (Stirt.) Elvebakk & D.J.Galloway E L
Pannaria centrifuga P.M.Jørg. L
Pannaria crenulata P.M.Jørg. L
Pannaria delicata P.M.Jørg. & D.J.Galloway E L
Pannaria dichroa (Hook.f. & Taylor) Cromb. L
Pannaria durietzii (P.James & Henssen) Elvebakk & D.J.Galloway L
Pannaria elixii P.M.Jørg. & D.J.Galloway L
Pannaria euphylla (Nyl.) Elvebakk & D.J.Galloway L
Pannaria farinosa Elvebakk & Fritt-Rasm. L
Pannaria fulvescens (Mont.) Nyl. L
Pannaria globuligera Hue L
Pannaria hookeri (Borrer ex Sm.) Nyl. L
Pannaria immixta Nyl. L
Pannaria isidiosa Elvebakk & Elix L
Pannaria leproloma (Nyl.) P.M.Jørg. L
Pannaria microphyllizans (Nyl.) P.M.Jørg. L
Pannaria pallida (Nyl.) Hue L
Pannaria patagonica (Malme) Elvebakk & J.J.Galloway L
Pannaria sphinctrina (Mont.) Hue L
Pannaria subcrustacea (Räsänen) P.M.Jørg. L
Pannaria xanthomelana (Nyl.) Hue E L
Parmeliella aggregata P.M.Jørg. & D.J.Galloway E L
Parmeliella concinna I.M.Lamb L
Parmeliella crassa P.M.Jørg. & D.J.Galloway E L
Parmeliella granulata I.M.Lamb L
Parmeliella gymnocheila (Nyl.) Müll.Arg. L
Parmeliella ligulata P.M.Jørg. & D.J.Galloway L
Parmeliella nigrata (Müll.Arg.) P.M.Jørg. & D.J.Galloway L
Parmeliella nigrocincta (Mont.) Müll.Arg. L
Parmeliella parvula P.M.Jørg. L
Parmeliella rakiurae P.M.Jørg. & D.J.Galloway L
Parmeliella subgranulata D.J.Galloway & P.M.Jørg. L
Parmeliella subtilis P.M.Jørg. & P.James E L
Parmeliella thysanota (Stirt.) Zahlbr. L
Parmeliella triptophylla (Ach.) Müll.Arg. L
Parmeliella variegata (Stirt.) Müll.Arg. L
Psoroma angustisectum Zahlbr. L
Psoroma asperellum Nyl. L
Psoroma buchananii (C.Knight) Nyl. L
Psoroma caliginosum Stirt. L
Psoroma coralloideum Nyl. L
Psoroma cyanosorediatum P.M. Jørg. L
Psoroma fruticulosum P.James & Henssen L
Psoroma geminatum P.M.Jørg. L
Psoroma hypnorum (Vahl) Gray L
Psoroma implexum Stirt. L
Psoroma melanizum Zahlbr. L
Psoroma paleaceum (Fr.) Timdal & Tønsberg L
Psoroma patagonicum Malme L
Psoroma pholidotoides (Nyl.) Trevis. L
Psoroma rubromarginatum P.James & J.S.Murray L
Psoromidium aleuroides (Stirt.) D.J.Galloway
Santessoniella pulchella P.M.Jørg. L
Siphulastrum mamillatum (Hook.f. & Taylor) D.J.Galloway L
Siphulastrum triste Müll.Arg. L
Xanthopsoroma contextum (Stirt.) Elvebakk L
Xanthopsoroma soccatum (R.Br. ex Cromb.) Elvebakk L
PELTIGERACEAE
Peltigera canina (L.) Willd. L
Peltigera didactyla (With.) J.R.Laundon L
Peltigera dilacerata (Gyeln.) Gyeln. L

Peltigera dolichorhiza (Nyl.) Nyl. L
Peltigera elisabethae Gyeln. L
Peltigera hymenina (Ach.) Delise L
Peltigera lepidophora (Vain.) Bitter L
Peltigera malacea (Ach.) Funck L
Peltigera membranacea (Ach.) Nyl. L
Peltigera nana Vain. L
Peltigera neckeri Hepp ex Müll.Arg. L
Peltigera neopolydactyla (Gyeln.) Gyeln. L
Peltigera polydactylon (Neck.) Hoffm. L
Peltigera praetextata (Flörke ex Sommerf.) Zopf L
Peltigera rufescens (Weiss) Humb. L
Peltigera tereziana Gyeln. L
Peltigera ulcerata Müll.Arg. L
Solorina crocea (L.) Ach. L
Solorina spongiosa (Sm.) Anzi L
PLACYNTHIACEAE
Hertella neozelandica Henssen E L
Placynthium nigrum (Huds.) Gray L
Placynthium rosulans (Th.Fr.) Zahlbr. L

Order PERTUSARIALES
COCCOTREMATACEAE
Coccotrema cucurbitula (Mont.) Müll.Arg. L
Coccotrema porinopsis (Nyl.) Imshaug ex Yoshim. L
ICMADOPHILACEAE
Dibaeis absoluta (Tuck.) Kalb & Gierl L
Dibaeis arcuata (Stirt.) Kalb & Gierl L
Icmadophila ericetorum (L.) Zahlbr. L
Icmadophila splachnirima (Hook.f. & Taylor) D.J.Galloway L
Thamnolia vermicularis (Sw.) Ach. ex Schaer. L
MEGASPORACEAE
Aspicilia aquatica Körb. L
Aspicilia caesiocinerea (Nyl.) Arnold L
Aspicilia calcarea (L.) Mudd. L
Aspicilia cinerea (L.) Körb. L
Aspicilia contorta (Hoffm.) Kremp. L
Circinaria calcarea (L.) A.Nordin, S.Savič & Tibell L
Lobothallia radiosa (Hoffm.) Hafellner L
Megaspora verrucosa (Ach.) Hafellner & V.Wirth L
OCHROLECHIACEAE
Ochrolechia frigida (Sw.) Lynge L
Ochrolechia pallescens (L.) A.Massal. L
Ochrolechia parella (L.) A.Massal. L
Ochrolechia tartarea (L.) A.Massal. L
Ochrolechia thelotremoides (Nyl.) Zahlbr. E L
Ochrolechia xanthostoma (Sommerf.) K.Schmitz & Lumbsch L
PERTUSARIACEAE
Pertusaria albissima Müll.Arg. L
Pertusaria alboatra Zahlbr. L
Pertusaria allanii Zahlbr. L
Pertusaria barbatica A.W.Archer & Elix L
Pertusaria bartlettii A.W.Archer & Elix L
Pertusaria celata A.W.Archer & Elix L
Pertusaria circumcincta Stirt. L
Pertusaria dactylina (Ach.) Nyl. L
Pertusaria dennistonensis Elix & A.W.Archer E L
Pertusaria duppensis A.W.Archer & Malcolm L
Pertusaria erubescens (Taylor) Nyl. L
Pertusaria erumpescens Nyl. L
Pertusaria flavovelata Elix & Malcolm L
Pertusaria graphica C.Knight L
Pertusaria gymnospora Kantvilas L
Pertusaria hadrospora A.W.Archer & Elix L
Pertusaria hypoxantha Malme L
Pertusaria jamesii Kantvilas L
Pertusaria knightiana Müll.Arg. L
Pertusaria laevis C.Knight L
Pertusaria lavata Müll.Arg. L
Pertusaria leucodes C.Knight L
Pertusaria leucoplaca Müll.Arg.
Pertusaria lophocarpa Körb. L
Pertusaria macloviana Müll.Arg. L
Pertusaria melaleucoides Müll.Arg. L
Pertusaria melanospora Nyl. L
Pertusaria micropora Kremp. L
Pertusaria monticola Messuti L
Pertusaria muricata J.C.David L
Pertusaria murrayi Elix & A.W.Archer L
Pertusaria novaezelandiae Szatala L
Pertusaria otagoana D.J.Galloway L
Pertusaria paratropa Müll.Arg. L
Pertusaria parvula A.W.Archer & Elix L
Pertusaria perrimosa Nyl. L
Pertusaria petrophyes C.Knight L
Pertusaria psoromica A.W.Archer & Elix L
Pertusaria scottii Elix & A.W.Archer L
Pertusaria scutellifera A.W.Archer & Elix L
Pertusaria sorodes Stirt. L
Pertusaria spilota A.W.Archer & Malcolm L
Pertusaria sporellula A.W.Archer & Elix L
Pertusaria subisidiosa A.W.Archer L
Pertusaria subplanaica A.W.Archer & Elix L
Pertusaria subventosa Malme L
Pertusaria subverrucosa Nyl. L
Pertusaria thamnolica A.W.Archer L
Pertusaria theochroa Kremp. L
Pertusaria thiospoda C.Knight L
Pertusaria truncata Kremp. L
Pertusaria tyloplaca Nyl. L
Pertusaria vallicola Elix & Malcolm L
Pertusaria velata (Turner) Nyl. L
Pertusaria xanthoplaca Müll.Arg. L

Order RHIZOCARPALES
CATILLARIACEAE
Catillaria chalybeia (Borrer) A.Massal L
Catillaria contristans (Nyl.) Zahlbr. L
Catillaria glaucogrisea Fryday E L
Halecania australis Lumbsch L
Halecania ralfsii (Salwey) M.Mayrhofer L
Sporastatia testudinea (Ach.) A.Massal. L
Toninia aromatica (Sm.) A.Massal. L
Toninia australis Timdal L
Toninia bullata (Meyen & Flot.) Zahlbr. L
Toninia glaucocarpa Timdal L
Toninia sedifolia (Scop.) Timdal L
RHIZOCARPACEAE
Rhizocarpon copelandii (Körb.) Th.Fr. L
Rhizocarpon disporum (Nägeli ex Hepp) Müll. Arg. L
Rhizocarpon distinctum Th.Fr. L
Rhizocarpon eupetraeum (Nyl.) Arnold L
Rhizocarpon geminatum Körb. L
Rhizocarpon geographicum (L.) DC. subsp. *geographicum* L
— subsp. *arcticum* (Runemark) Hertel L
Rhizocarpon grande (Flörke) Arnold L
Rhizocarpon hochstetteri (Körb.) Vain. L
Rhizocarpon lavatum (Fr.) Hazsl. L
Rhizocarpon lecanorinum Anders L
Rhizocarpon oxydatum Fryday L
Rhizocarpon petraeum (Wulfen) A.Massal. L
Rhizocarpon polycarpum (Hepp) Th.Fr. L
Rhizocarpon postumum (Nyl.) Arnold L
Rhizocarpon purpurescens Fryday L
Rhizocarpon pusillum Runemark L
Rhizocarpon reductum Th.Fr. L
Rhizocarpon submodestum (Vain.) Vain. L
Rhizocarpon subpostumum (Nyl.) Arnold L
Rhizocarpon superficiale (Schaer.) Malme L
Rhizocarpon viridiatrum (Wulfen) Körb.

Order TELOSCHISTALES
CALICIACEAE
Amandinea adjuncta (Th.Fr.) Hafellner
Amandinea decedens (Nyl.) Blaha & H.Mayrhofer L
Amandinea diorista var. *hypopelidna* (Stirt.) Marbach & Kalb E L
Amandinea insperata (Nyl.) H.Mayrhofer & Ropin L
Amandinea lecideina (H.Mayrhofer & Poelt) Scheid. & H.Mayrhofer L
Amandinea otagensis (Zahlbr.) Blaha & H.Mayrhofer E L
Amandinea punctata (Hoffm.) Coppins & Scheid. L
Buellia aethalea (Ach.) Th.Fr. L
Buellia albula (Nyl.) Müll.Arg. L
Buellia alutacea Zahlbr. L
Buellia cranwelliae Zahlbr. E L
Buellia demutans (Stirt.) Zahlbr. L
Buellia disciformis (Fr.) Mudd L
Buellia dunedina Zahlbr. E L
Buellia ferax Müll.Arg. E L
Buellia fuscoatratula Zahlbr. L
Buellia griseovirens (Turner & Borrer ex Sm.) Almb. L
Buellia macularis Zahlbr. E L
Buellia porulosa Müll.Arg. E L
Buellia spuria (Schaer.) Anzi L
Buellia subbadioatra (C.Knight) Müll.Arg. E L
Buellia tetrapla (Nyl.) Müll.Arg. L
Calicium abietinum Pers. L
Calicium adspersum subsp. *australe* Tibell L
Calicium chlorosporum F.Wilson L
Calicium glaucellum Ach. L
Calicium hyperelloides Nyl. L
Calicium lenticulare Ach. L
Calicium robustellum Nyl. L
Calicium salicinum Pers. L
Calicium trabinellum (Ach.) Ach. L
Calicium tricolor F.Wilson L
Calicium victorianum (F.Wilson) Tibell L
Cyphelium inquinans (Sm.) Trevis. L
Diplotomma alboatrum (Hoffm.) Flot. L
Diplotomma canescens (Dicks.) Flot. 1849 subsp. *canescens* L
— subsp. *australasica* (Elix & Lumbsch) D.J.Galloway L
Diplotomma chlorophaeum (Hepp ex Leight.) Szatala L
Dirinaria aegialita (Afzel. ex Ach.) B.J.Moore L
Dirinaria applanata (Fée) D.D.Awasthi L
Dirinaria picta (Sw.) Clem. & Shear L
Pyxine cocoës (Sw.) Nyl. L
Pyxine subcinerea Stirt. L
Tetramelas confusus Nordin L
Thelomma ocellatum (Körb.) Tibell L
MEGALOSPORACEAE
Austroblastenia pauciseptata (Shirley) Sipman L
Austroblastenia pupa Sipman L
Megaloblastenia flavidoatra (Nyl.) Sipman L
Megaloblastenia marginiflexa (Hook.f. & Taylor) Sipman L
Megalospora atrorubicans (Nyl.) Zahlbr. L
Megalospora bartlettii Sipman L
Megalospora campylospora (Stirt.) Sipman L
Megalospora disjuncta Sipman L
Megalospora gompholoma (Müll.Arg.) C.W.Dodge subsp. *gompholoma* E L
— subsp. *fuscolineata* Sipman L
Megalospora knightii Sipman E L
Megalospora lopadioides Sipman L
Megalospora subtuberculosa (C.Knight) Sipman L
MICROCALICIACEAE
Microcalicium arenarium (Hampe ex Massal.) Tibell
Microcalicium conversum Tibell
PHYSCIACEAE
Heterodermia appendiculata (Kurok.) Swinscow &

Krog L
Heterodermia casarettiana (A.Massal.) Trevis. L
Heterodermia chilensis (Kurok.) Swinscow & Krog L
Heterodermia isidiophora (Nyl.) D.D.Awasthi L
Heterodermia japonica (M.Sâto) Swinscow & Krog L
Heterodermia leucomela (L.) Poelt L
Heterodermia lutescens (Kurok.) Follmann L
Heterodermia microphylla (Kurok.) Swinscow & Krog L
Heterodermia obscurata (Nyl.) Trevis. L
Heterodermia podocarpa (Bél.) D.D.Awasthi L
Heterodermia spathulifera Moberg & Purvis L
Heterodermia speciosa (Wulfen) Trevis. L
Hyperphyscia adglutinata (Flörke) H.Mayrhofer & Poelt L
Hyperphyscia plinthiza (Nyl.) Müll.Arg. E L
Monerolechia badia (Fr.) Kalb
Phaeophyscia adiastola (Essl.) Essl. L
Phaeophyscia endococcina var. *endococcinodes* (Poelt) Moberg L
Phaeophyscia hispidula (Ach.) Essl. L
Phaeophyscia orbicularis (Neck.) Moberg L
Phaeophyscia sciastra (Ach.) Moberg L
Physcia adscendens (Fr.) H.Olivier L
Physcia albata (F.Wilson) Hale L
Physcia atrostriata Moberg L
Physcia caesia (Hoffm.) Fürnr. L
Physcia crispa Nyl. L
Physcia dubia (Hoffm.) Lettau L
Physcia erumpens Moberg L
Physcia integrata Nyl. L
Physcia jackii Moberg L
Physcia nubila Moberg L
Physcia poncinsii Hue L
Physcia tribacia (Ach.) Nyl. L
Physcia tribacoides Nyl. L
Physcia undulata Moberg L
Rinodina bischoffii (Hepp) A.Massal. L
Rinodina blastidiata Matzer & H.Mayrhofer L
Rinodina boleana Giralt & H.Mayrhofer L
Rinodina cacaotina Zahlbr. L
Rinodina capensis Hampe L
Rinodina confragosula (Nyl.) Müll.Arg. L
Rinodina conradii Körb. L
Rinodina exigua (Ach.) Gray L
Rinodina gallowayii H.Mayrhofer L
Rinodina herteliana Kaschik E L
Rinodina immersa (Körb.) Arnold L
Rinodina inflata Kalb L
Rinodina insularis (Arnold) Hafellner L
Rinodina jamesii H.Mayrhofer L
Rinodina luridata (Körb.) H.Mayrhofer L
Rinodina murrayii H.Mayrhofer L
Rinodina nigricans H.Mayrhofer L
Rinodina olivaceobrunnea C.W. Dodge & G.E. Baker L
Rinodina peloleuca (Nyl.) Müll.Arg. L
Rinodina pyrina (Ach.) Arnold L
Rinodina reagens Matzer & H.Mayrhofer L
Rinodina septentrionalis Malme L
Rinodina subtubulata (C.Knight) Zahlbr. L
Rinodina thiomela (Nyl.) Müll.Arg. L
TELOSCHISTACEAE
Caloplaca acheila Zahlbr. E L
Caloplaca allanii Zahlbr. E L
Caloplaca ammiospila (Wahlenb.) H.Oliver L
Caloplaca biatorina (A.Massal.) J.Steiner L
Caloplaca caesiorufella (Nyl.) Zahlbr. L
Caloplaca cerina (Ehrh. ex Hedw.) Th.Fr. L
Caloplaca cerinella (Nyl.) Flagey L
Caloplaca cerinelloides (Erichsen) Poelt L
Caloplaca chrysodeta (Vain. ex Räsänen) Dombr. L
Caloplaca chrysophthalma Degel. L
Caloplaca cinnabarina (Ach.) Zahlbr. L
Caloplaca circumlutosa Zahlbr. E L
Caloplaca cirrochrooides (Vain.) Zahlbr. L
Caloplaca citrina (Hoffm.) Th.Fr. L
Caloplaca concilians (Nyl.) H.Olivier L
Caloplaca crenulatella (Nyl.) H.Olivier L
Caloplaca cribrosa (Hue) Zahlbr. L
Caloplaca decipiens (Arnold) Blomb. & Forssell L
Caloplaca erecta Arup & H.Mayrhofer E L
Caloplaca ferruginea (Huds.) Th.Fr. L
Caloplaca flavorubescens (Huds.) J.R.Laundon L
Caloplaca flavovirescens (Wulfen) Dalla Torre & Sarnth. L
Caloplaca holocarpa (Hoffm.) A.E.Wade L
Caloplaca homologa (Nyl.) Hellb. L
Caloplaca irrubescens (Nyl.) Zahlbr. L
Caloplaca lactea (A.Massal.) Zahlbr. L
Caloplaca maculata D.J.Galloway E L
Caloplaca mooreae D.J.Galloway L
Caloplaca murrayi D.J.Galloway E L
Caloplaca ochracea (Schaer.) Flagey L
Caloplaca papanui D.J.Galloway E L
Caloplaca perileuca Zahlbr. E L
Caloplaca rosei Hasse L
Caloplaca rubentior (Zahlbr.) D.J.Galloway E L
Caloplaca saxicola (Hoffm.) Nordin L
Caloplaca schisticola D.J.Galloway E L
Caloplaca sublobulata (Nyl.) Zahlbr. L
Caloplaca subpyracea (Nyl.) Zahlbr. E L
Caloplaca tornoënsis H.Magn. L
Caloplaca vitellinula sensu Galloway L
Caloplaca xantholyta (Nyl.) Jatta L
Fulgensia bracteata (Hoffm.) Räsänen L
Fulgensia fulgens (Sw.) Elenkin L
Siphula complanata (Hook.f. & Taylor) R.Sant. L
Siphula coriacea Nyl. L
Siphula decumbens Nyl. L
Siphula dissoluta Nyl. L
Siphula elixii Kantvilas L
Siphula fastigiata (Nyl.) Nyl. L
Siphula foliacea D.J.Galloway L
Siphula fragilis (Hook.f. & Taylor) J.S.Murray L
Siphula georginae Kantvilas L
Siphula gracilis Kantvilas L
Siphula jamesii Kantvilas L
Siphula pickeringii Tuck. L
Teloschistes chrysophthalmus (L.) Th.Fr. L
Teloschistes fasciculatus Hillmann L
Teloschistes flavicans (Sw.) Norman L
Teloschistes sieberianus (Laurer) Hillmann L
Teloschistes spinosus (Hook.f. & Taylor) Js.Murray L
Teloschistes velifer F.Wilson L
Teloschistes xanthorioides Js.Murray L
Xanthomendoza novozelandica (Hillmann) Søchting, Kärnefelt & S.Y.Kondr. L
Xanthoria candelaria (L.) Th.Fr. L
Xanthoria elegans (Link) Th.Fr. L
Xanthoria incavata (Stirt.) Zahlbr. L
Xanthoria ligulata (Körb.) P.James L
Xanthoria parietina (L.) Th.Fr. L
Xanthoria polycarpa (Hoffm.) Th.Fr. ex Rieber L

Order UMBILICARIALES
FUSCIDEACEAE
Fuscidea asbolodes (Nyl.) Hertel & V.Wirth l
Fuscidea cf. *cyathoides* (Ach.) V.Wirth & Vězda l
Fuscidea impolita (Müll.Arg.) Hertel l
Fuscidea subasbolodes Kantvilas l
Maronea constans (Nyl.) Hepp L
Sarrameana albidoplumbea (Hook.f. & Taylor) Farkas L
OPHIOPARMACEAE
Hypocenomyce australis Timdal L
Hypocenomyce scalaris (Ach. ex Lilj.) M.Choisy L
UMBILICARIACEAE
Umbilicaria cylindrica (L.) Delise ex Duby L
Umbilicaria decussata (Vill.) Zahlbr. L
Umbilicaria deusta (L.) Baumg. L
Umbilicaria durietzii Frey L
Umbilicaria hyperborea (Ach.) Hoffm. L
Umbilicaria kraschenínnikovii (Savicz) Zahlbr. L
Umbilicaria murihikuana D.J.Galloway & L.G.Sancho E L
Umbilicaria nylanderiana (Zahlbr.) H.Magn. L
Umbilicaria polyphylla (L.) Baumg. L
Umbilicaria robusta (Llano) D.J.Galloway & Sancho E L
Umbilicaria subaprina Frey L
Umbilicaria subglabra (Nyl.) Harm. L
Umbilicaria vellea (L.) Ach. L
Umbilicaria zahlbruckneri Frey L

Class LEOTIOMYCETES
Chaetomella raphigera Swift
Cyclaneusma minus (Butin) DiCosmo, Peredo & Minter A
Cyclaneusma niveum (Pers.) DiCosmo, Peredo & Minter A
Discohainesia oenotherae (Cooke & Ellis) Nannf. A
Geomyces pannorum (Link) Sigler & J.W. Carmich. AN A
Sarea resinae (Fr.) Kuntze A

Order CYTTARIALES
CYTTARIACEAE
Cyttaria gunnii sensu auct. NZ E
Cyttaria nigra Rawlings E
Cyttaria pallida Rawlings E
Cyttaria purdiei Buchanan E

Order ERYSIPHALES
ERYSIPHACEAE
Arthrocladiella mougeotii (Lév.) Vassilkov A
Blumeria graminis (DC.) Speer A
Erysiphe alphitoides (Griffon & Maubl.) U. Braun & S. Takam. A
Erysiphe aquilegiae DC. var. *aquilegiae* A
— var. *ranunculi* (Grev.) R.Y. Zheng & G.Q. Chen A
Erysiphe betae (Va ha) Weltzien A
Erysiphe carpophila Syd. E
Erysiphe cruciferarum Opiz ex L. Junell A
Erysiphe densa Berk. E
Erysiphe helichrysi U. Braun (nom. inv.) A
Erysiphe heraclei DC. A
Erysiphe howeana U. Braun A
Erysiphe pisi DC. A
Erysiphe platani (Howe) U. Braun & S. Takam. A
Erysiphe rubicola (B.J. Murray) Boesew. E
Golovinomyces biocellatus (Ehrenb.) V.P. Gelyuta A
Golovinomyces cichoracearum (DC.) V.P. Gelyuta var. *cichoracearum* A
Golovinomyces galeopsidis (DC.) V.P. Gelyuta A
Golovinomyces magnicellulatus (U. Braun) V.P. Gelyuta var. *magnicellulatus* A
Golovinomyces orontii (Castagne) V.P. Gelyuta A
Golovinomyces sordidus (L. Junell) V.P. Gelyuta A
Golovinomyces verbasci (Jacz.) V.P. Gelyuta A
Leveillula taurica (Lév.) G. Arnaud A
Microsphaera alni (DC. ex Wallr.) G. Winter A
Microsphaera alphitoides Griffon & Maubl. var. *alphitoides* A
Microsphaera begoniae Sivan. A
Microsphaera euonymi-japonici Vienn.-Bourg. A
Microsphaera grossulariae (Wallr.) Lév. A
Microsphaera hypophylla Nevod. A
Microsphaera penicillata (Wallr.) Lév. A
Microsphaera russellii Clinton A
Microsphaera sparsa Howe A
Microsphaera trifolii (Grev.) U. Braun var. *trifolii* A

Neoerysiphe geranii A
Oidium chrysanthemi Rabenh. AN A
Oidium hardenbergiae Boesew. AN A
Oidium helichrysi Boesew. AN A
Oidium hortensiae Jørst. AN A
Oidium oxalidis McAlpine AN A
Podosphaera clandestina (Wallr.) Lév. A
Podosphaera euphorbiae (Castagne) U. Braun & S. Takam. A
Podosphaera leucotricha (Ellis & Everh.) E.S. Salmon A
Podosphaera tridactyla (Wallr.) de Bary A
Sawadaea bicornis (Wallr.) Miyabe U
Sphaerotheca aphanis (Wallr.) U. Braun var. *aphanis* A
Sphaerotheca fugax Penz. & Sacc. A
Sphaerotheca fuliginea (Schltdl.) Pollacci A
Sphaerotheca mors-uvae (Schwein.) Berk. & M.A. Curtis A
Sphaerotheca pannosa (Wallr.) Lév. var. *pannosa* A
— var. *persicae* Woron. A
Sphaerotheca verbenae Săvul. & Negru A
Uncinula necator (Schwein.) Burrill var. *necator* A
Uncinuliella australiana (McAlpine) R.Y. Zheng & G.Q. Chen A

Order HELOTIALES
Ascocoryne cylichnium (Tul.) Korf
Ascocoryne sarcoides (Jacq.) J.W. Groves & D.E. Wilson
Bisporella citrina (Hedw.) Korf & S.E. Carp.
Bisporella claroflava (Grev.) Lizoň & Korf
Bisporella sulfurina (Quél.) S.E. Carp. U OU
Cadophora fastigiata Lagerb. & Melin AN A
Cadophora luteoolivacea (J.F.H. Beyma) T.C. Harr. & McNew A
Cadophora malorum (Kidd & Beaumont) W. Gams A
Cadophora sp. sensu Blanchette et al.
Chlorociboria aeruginascens subsp. *australis* P.R. Johnst. E
Chlorociboria albohymenia P.R. Johnst. E
Chlorociboria argentinensis J.R. Dixon OU
Chlorociboria awakinoana P.R. Johnst. E
Chlorociboria campbellensis P.R. Johnst. E
Chlorociboria clavula P.R. Johnst. E
Chlorociboria colubrosa P.R. Johnst. E
Chlorociboria duriligna P.R. Johnst. E
Chlorociboria halonata P.R. Johnst. E
Chlorociboria macrospora P.R. Johnst. E
Chlorociboria pardalota P.R. Johnst. E
Chlorociboria poutoensis P.R. Johnst. E
Chlorociboria procera P.R. Johnst. E
Chlorociboria spathulata P.R. Johnst. E
Chlorociboria spiralis P.R. Johnst. E
Clathrosporium intricatum Nawawi & Kuthub. AN F
Clathrosporium olivatrum (Sacc.) Hennebert AN
Cordierites acanthophora Samuels & L.M. Kohn E
Cylindrosporium betulae Davis AN A
Cylindrosporium nanum Cooke AN U
Dactylaria acerosa Matsush. AN A
Dactylaria appendiculata Cazau, Aramb. & Cabella AN A F
Dactylaria candidula (Höhn.) G.C. Bhatt & W.B. Kendr. AN
Dactylaria dimorpha Matsush. A
Dactylaria fusiformis Shearer & J.L. Crane AN
Dactylaria higginsii (Luttr.) M.B. Ellis AN
Dactylaria junci M.B. Ellis AN
Dactylaria leptospermi J.A. Cooper AN E
Dactylaria obtriangularia Matsush. AN U
Dactylaria parvispora (Preuss) de Hoog & Arx AN A
Dawsicola neglecta Döbbeler
Endoscypha perforans Syd. E
Gloeotinia temulenta (Prill. & Delacr.) M. Wilson, Noble & E.G. Gray A
Helicodendron conglomeratum Glen-Bott AN F
Helicodendron hyalinum Linder AN F
Helicodendron luteoalbum Glen-Bott AN F
Helicodendron multiseptatum Abdullah AN F
Helicodendron paradoxum Peyronel AN U F
Helicodendron triglitziense (Jaap) Linder AN F
Helicodendron websteri Voglmayr & P.J. Fisher AN F
Helicodendron westerdijkae Beverw. AN F
Hysteropezizella diminuens (P. Karst.) Nannf.
Hysteropezizella phragmitina (P. Karst. & Starbäck) Nannf. F
Lemonniera aquatica De Wild. AN F
Lemonniera centrosphaera Marvanová AN OU F
Leptodontidium elatius (F. Mangenot) de Hoog var. *elatius* U
Llimoniella ramalinae (Müll. Arg.) Etayo & Diederich
Mirandina corticola G. Arnaud ex Matsush. AN U
Mollisiella ilicincola (Berk. & Broome) Seaver OU
Monostichella nothofagi P.R. Johnst. AN E
Monostichella robergei (Desm.) Höhn. AN A
Orbiliopsis callistea Syd. E
Phaeopyxis punctum (A.Massal) Rambold, Triebel & Coppins 1990
Pilidium acerinum (Alb. & Schwein.) Kunze AN U
Pirottaea horoeka P.R. Johnst. E
Pirottaea mahinapua P.R. Johnst. E
Pirottaea nidulans (Syd.) P.R. Johnst. E
Pirottaea palmicola P.R. Johnst. E
Pyrenopeziza atrata (Pers.) Fuckel A
Pyrenopeziza brassicae B. Sutton & Rawl. A
Rhexocercosporidium panacis Reeleder AN A
Rhodesia subtecta (Roberge) Grove OU
Rhynchosporium orthosporum Caldwell AN A
Rhynchosporium secalis (Oudem.) Davis AN A
Scytalidium thermophilum (Cooney & R. Emers.) Austwick AN, (nom. inv.) U
Sphaerographium tenuirostrum Verkley AN A
Spirosphaera beverwijkiana Hennebert AN U F
Spirosphaera caricis-graminis Voglmayr U F
Spirosphaera floriformis Beverw. AN F
Spirosphaera minuta Hennebert AN F
Tapesia yallundae Wallwork & Spooner A
Tetracladium marchalianum De Wild. AN U F
Tetracladium maxilliforme (Rostr.) Ingold AN F
Tetracladium setigerum (Grove) Ingold AN U F
Tetracladium sp. 1 sensu Aimer & Segedin (nom. ined.) AN U F
Tetracladium sp. 2 sensu Aimer & Segedin (nom. ined.) AN U F
Tiarosporella madreeya (Subram. & K. Ramakr.) Nag Raj AN
Tiarosporella paludosa (Sacc. & Fiori) Höhn. AN
Trimmatostroma betulinum (Corda) S. Hughes A
Trimmatostroma bifarium Gadgil & M. Dick AN A
Trimmatostroma scutellare (Berk. & Broome) M.B. Ellis AN A
Trochila ilicina sensu (Nees ex Fr.) Courtec. A
DERMATEACEAE
Ankistrocladium fuscum Perrott AN
Casaresia sphagnorum Gonz. Frag. AN F
Cashiella montiicola (Berk.) Dennis E
Cashiella sticheri Gadgil & M. Dick E
Chlorosplenium flavovirens Massee (nom. ined.) E
Cryptohymenium pycnidiophorum Samuels & L.M. Kohn E
Cryptosporiopsis actinidiae P.R. Johnst., M.A. Manning & X. Meier AN A
Cryptosporiopsis citri P.R. Johnst. & Full. AN A
Cryptosporiopsis edgertonii Gadgil & M. Dick AN U
Diplocarpon earlianum (Ellis & Everh.) F.A. Wolf A
Diplocarpon mespili (Sorauer) B. Sutton A
Diplocarpon rosae F.A. Wolf A
Drepanopeziza populi-albae (Kleb.) Nannf. A
Drepanopeziza ribis (Kleb.) Höhn. A
Drepanopeziza sphaerioides (Pers.) Höhn. A
Drepanopeziza tremulae Rimpau A
Ephelina gregaria (Berk.) Sacc. E
Leptotrochila cerastiorum (Wallr.) Schüepp A
Leptotrochila medicaginis (Fuckel) Schüepp A
Leptotrochila porri Arx & Boerema A
Leptotrochila verrucosa (Wallr.) Schüepp OU
Marssonina betulae (Lib.) Magnus AN A
Marssonina daphnes (Desm.) Magnus AN A
Mollisia cinerea (Batsch) P. Karst.
Mollisia coprosmae Dennis E
Mollisia dextrinospora Korf U
Mollisia discolor (Mont. & Fr.) W. Phillips var. *discolor* OU
— var. *longispora* Le Gal U OU
Mollisia ventosa (P. Karst.) P. Karst. U
Neofabraea alba (E.J. Guthrie) Verkley A
Neofabraea malicorticis H.S. Jacks. A OU
Neofabraea malicorticis sensu auct. NZ A OU
Neofabraea perennans Kienholz A OU
Pezicula cinnamomea (DC.) Sacc.
Pezicula corticola (C.A. Jorg.) Nannf. OU
Pezicula livida (Berk. & Broome) Rehm OU
Pezicula sessilis (Rodway) Dennis OU
Pezicula sp. 1 Abeln, Patger & Verkley (nom. ined.) A
Phlyctema asparagi Fautrey & Roum. AN A
Phlyctema caulium (Lib.) Petr. AN A
Phlyctema dissepta Berk. AN E
Pseudopeziza colensoi (Berk.) Massee E
Pseudopeziza geranii Rodway A
Pseudopeziza medicaginis (Lib.) Sacc. A
Pseudopeziza trifolii (Biv.) Fuckel A
GEOGLOSSACEAE
Corynetes atropurpureus (Pers.) E.J. Durand OU
Geoglossum alveolatum (E.J. Durand ex Rehm) E.J. Durand OU
Geoglossum australe Cooke
Geoglossum cookeanum Nannf. (nom. inv.)
Geoglossum glabrum Pers. OU
Geoglossum glutinosum Pers.:Fr. OU
Geoglossum muelleri Berk.
Geoglossum nigritum (Fr.) Cooke
Geoglossum umbratile Sacc.
Thuemenidium berteroi (Mont.) Gamundí
Trichoglossum confusum E.J. Durand OU
Trichoglossum farlowii (Cooke) E.J. Durand
Trichoglossum hirsutum (Pers.) Boud. var. *hirsutum*
— var. *longisporum* (Tai) Mains OU
Trichoglossum octopartitum Mains OU
Trichoglossum variabile (E.J. Durand) Nannf.
Trichoglossum walteri (Berk.) E.J. Durand
HELOTIACEAE
Articulospora moniliformis Ranzoni AN F
Articulospora tetracladia Ingold AN F
Ascotremella faginea (Peck) Seaver OU
Bloxamia truncata Berk. & Broome AN
Bulgariella pulla (Fr.) P. Karst.
Calycella brevispora (Cooke & W. Phillips) Dennis E
Calycella ochracea Boud. OU
Claussenomyces dacrymycetoideus Ouell. & Korf OU
Coryne rugipes Cooke AN E
Coryne tasmanica (Rodway) Dennis AN OU
Coryne urnalis (Nyl.) Sacc. AN
Crocicreas coronatum (Bull.) S.E. Carp. OU
Crocicreas cyathoideum (Bull.) S.E. Carp. var. *cyathoideum* OU
— var. *cacaliae* (Pers.) S.E. Carp. OU

Crocicreas dolosellum (P. Karst.) S.E. Carp. OU
Crocicreas epitephrum (Berk.) S.E. Carp.
Crocicreas hieraciifolium P.R. Johnst. E
Crocicreas ilicifolium P.R. Johnst. E
Crocicreas macrostipitatum P.R. Johnst. E
Crocicreas multicuspidatum (Rodway) S.E. Carp.
Cudoniella acicularis (Bull.) J. Schröt. OU
Dendrostilbella mycophila (Pers.) Seifert AN
Dimorphospora foliicola Tubaki AN F
Helotium boudieri Sacc. OU
Helotium citrinum var. pallidum Cooke (nom. inv.) E
Helotium elaeocarpi Dennis
Helotium pateriforme (Berk.) Cooke OU
Helotium phormium Cooke E
Helotium scutula P. Karst. A
Helotium subsordidum Dennis E
Heteropatella antirrhini Buddin & Wakef. AN A
Heteropatella eriophila Syd. AN E
Heteropatella valtellinensis (Traverso) Wollenw. AN A
Hymenoscyphus caudatus (P. Karst.) Dennis OU
Hymenoscyphus epiphyllus (Pers.) Rehm OU
Hymenoscyphus erythropus Döbbeler
Hymenoscyphus gregarius (Boud.) Gamundí & Giaiotti
Hymenoscyphus lacteus (Cooke) Kuntze E
Hymenoscyphus leucopus (Mont.) Dennis
Hymenoscyphus pezizoideus (Cooke & W. Phillips) Gamundí
Hymenoscyphus pseudociliatus (W. Phillips) S.E. Carp. E
Hymenoscyphus quintiniae (Dennis) Dennis E
Hymenoscyphus scutula (Pers.) Gillet A
Idriella australiensis B. Sutton, Piroz. & Deighton AN E
Idriella vandalurensis Vittal AN
Neobulgaria alba P.R. Johnst., D.C. Park & M.A. Manning E
Neobulgaria pura (Fr.) Petr. OU
Ombrophila microspora (Ellis & Everh.) Sacc. & P. Syd. OU
Ombrophila violacea (Hedw.) Fr. OU
Phaeohelotium luteum (Rick) Dumont
Pseudohelotium haematoideum (Cooke & W. Phillips) Dennis E
Pseudospiropes novae-zealandiae S. Hughes AN E
Pseudospiropes rousselianus (Mont.) M.B. Ellis AN
Pseudospiropes variabilis S. Hughes AN E
Rhizoscyphus ericae (D.J. Read) W.Y. Zhuang & Korf A
Strossmayeria atriseda (Saut.) Iturr.
Strossmayeria basitricha (Sacc.) Dennis
Strossmayeria ochrospora Iturr.
Tricladium attenuatum S.H. Iqbal AN F
Tricladium castaneicola B. Sutton AN
Tricladium chaetocladium Ingold AN F
Tricladium curvisporum Descals AN F
Tricladium sp. 1 sensu Aimer & Segedin (nom. ined.) AN F
Tricladium sp. 2 sensu Aimer & Segedin (nom. ined.) AN F
Tricladium sp. 3 sensu Aimer & Segedin (nom. ined.) AN F
Tricladium splendens Ingold AN U F
Unguiculariopsis triregia S.Y. Kondr. & D.J. Galloway
Varicosporium elodeae W. Kegel AN F
Varicosporium giganteum J.L. Crane AN F
Velutarina rufoolivacea (Alb. & Schwein.) Korf
HEMIPHACIDIACEAE
Chlorencoelia torta (Schwein.) J.R. Dixon
Chlorencoelia versiformis (Pers.) J.R. Dixon
Didymascella thujina (E.J. Durand) Maire A
Meria laricis Vuill. AN A
HYALOSCYPHACEAE
Arachnopeziza araneosa (Sacc.) Korf
Arachnopeziza aurelia (Pers.) Fuckel
Arenaea javanica sensu Spooner
Asperopilum juncicola (Dennis) Spooner
Austropezia samuelsii (Korf) Spooner E
Brunnipila clandestina (Bull.) Baral A
Calycellina carolinensis Nag Raj & W.B. Kendr.
Cheiromycella microscopica (P. Karst.) S. Hughes AN A
Dasyscypha carneola (Sacc.) Sacc. OU
Dasyscypha emerici (Berk. & W. Phillips) Sacc. OU
Dasyscypha fimbriifera (Berk. & M.A. Curtis) Sacc. var. fimbriifera OU
— var. singerianus (Dennis) J.H. Haines OU
Dasyscypha glabrescens (Cooke & W. Phillips) Sacc.
Dasyscyphella nivea (R. Hedw.) Raitv.
Dasyscyphus dumorum (Roberge ex Desm.) Massee A
Dasyscyphus pudibundus (Quél.) Sacc. A
Dasyscyphus sulphureus Rick OU
Dematioscypha dematiicola (Berk. & Broome) Svrček
Dimorphotricha australis Spooner
Hamatocanthoscypha laricionis (Velen.) Svrček A
Hamatocanthoscypha ocellata Huhtinen A
Haplographium bicolor Grove AN A
Hispidula dicksoniae (G.W. Beaton & Weste) P.R. Johnst.
Hispidula pounamu P.R. Johnst. E
Hispidula rubra P.R. Johnst. E
Hispidula tokerau P.R. Johnst. E
Hyalopeziza sp. 'urceolate' P.R. Johnst. (nom. ined.)
Hyaloscypha albohyalina var. spiralis (Velen.) Huhtinen F
Hyaloscypha aureliella (Nyl.) Huhtinen A
Hyaloscypha carnosa (Rodway) Spooner
Hyaloscypha zalewskii Descals & J. Webster F
Lachnellula calyciformis (Batsch) Dharne AN A
Lachnellula calycina (Schumach.) Sacc. A
Lachnellula hahniana (Seaver) Dennis A
Lachnellula occidentalis (G.G. Hahn & Ayers) Dharne A
Lachnellula pseudotsugae (G.G. Hahn) Dennis A
Lachnellula rhopalostylidis (Dennis) Korf E
Lachnellula subtilissima (Cooke) Dennis A
Lachnellula suecica (de Bary ex Fuckel) Nannf. U
Lachnum abnorme (Mont.) J.H. Haines & Dumont
Lachnum apalum (Berk. & Broome) Nannf. var. apalum
— var. beatonii Spooner
Lachnum berggrenii Spooner E
Lachnum brevipilosum Baral
Lachnum callimorphum (P. Karst.) P. Karst. OU
Lachnum confertum Spooner OU
Lachnum controversum (Cooke) Rehm
Lachnum correae Spooner
Lachnum ellipsosporum Spooner OU
Lachnum enzenspergerianum Henn.
Lachnum filiceum (Cooke & W. Phillips) Spooner E
Lachnum gahniae Spooner OU
Lachnum hyalopus (Cooke & Massee) Spooner E
Lachnum juncinum Spooner
Lachnum melanophthalmum (Dennis) Spooner E
Lachnum nothofagi (Dennis) Spooner
Lachnum palmae (Kanouse) Spooner
Lachnum pinnicola Spooner OU
Lachnum pritzelianum (Henn.) Spooner
Lachnum pteridicola (Dennis) Spooner E
Lachnum pteridophyllum (Rodway) Spooner
Lachnum sp. 'blechnum' P.R. Johnst. (nom. ined.) E
Lachnum varians (Rehm) M.P. Sharma
Lachnum virgineum (Batsch) P. Karst.
Lachnum willisii (G.W. Beaton) Spooner
Lasiobelonium subflavidum Ellis & Everh.
Lasiobelonium variegatum (Fuckel) Raitv. OU
Neodasyscypha cerina (Pers.) Spooner
Perrotia alba Dennis E
Perrotia apiculata (Dennis) Spooner E
Perrotia gallica var. phyllocladi (Dennis) Spooner E
Perrotia lutea (W. Phillips) Dennis
Perrotia microspora Verkley E
Pezizella alniella (Nyl.) Dennis A
Pezizella australis Dennis E
Pezizella crispa (Cooke & W. Phillips) Sacc. E
Pezizella eburnea (Roberge) Dennis A
Phaeoscypha cladii (Nag Raj & W.B. Kendr.) Spooner
Polydesmia fructicola Korf OU
Proliferodiscus dingleyae Spooner
Pseudaegerita conifera Abdullah, Gené & Guarro AN A F
Pseudaegerita foliicola Abdullah ex J.A. Cooper AN F
Pseudaegerita viridis (Bayl. Ell.) Abdullah & J. Webster AN F
Trichoscyphella calycina (Fr.) Nannf. A
PHACIDIACEAE
Allantophomopsis lycopodina (Höhn.) Carris AN A
Ascocoma eucalypti (Hansf.) H.J. Swart A
Ceuthospora foliicola (Lib.) Cooke AN
Ceuthospora laurocerasi Grove AN A
Ceuthospora molleriana (Thüm.) Petr. AN A
Phacidium betulinum Mouton A
Phacidium coniferarum (G.G. Hahn) DiCosmo, Nag Raj & W.B. Kendr. A
Phacidium sp. sensu McKenzie, P.K. Buchanan & P.R. Johnst. (nom. ined.)
RUTSTROEMIACEAE
Dicephalospora chrysotricha (Berk.) Verkley E
Dicephalospora rufocornea (Berk. & Broome) Spooner
Lambertella corni-maris Höhn.
Lambertella tubulosa Abdullah & J. Webster F
Lanzia allantospora (Dennis) Spooner E
Lanzia berggrenii (Cooke & W. Phillips) Spooner var. berggrenii
— var. metrosideri (Dennis) Spooner E
Lanzia griseliniae (Dennis) Dumont E
Lanzia lanaripes (Dennis) Spooner
Lanzia novae-zelandiae (Dennis) J.A. Simpson & Grgur.
Lanzia ovispora Spooner E
Lanzia sp. 'aureus chathamica' P.R. Johnst. (nom. ined.) E
Lanzia sp. 'aureus pohutukawa' P.R. Johnst. (nom. ined.) E
Lanzia sp. 'aureus' P.R. Johnst. (nom. ined.) E
Lanzia sp. 'big red' P.R. Johnst. (nom. ined.) E
Lanzia sp. 'quadrispora' P.R. Johnst. (nom. ined.) E
Moellerodiscus coprosmae P.R. Johnst. E
Moellerodiscus griseliniae P.R. Johnst. E
Moellerodiscus microcoprosmae P.R. Johnst. E
Poculum fuegianum (Speg.) S.E. Carp. OU
Poculum sp. 'hinau' P.R. Johnst. (nom. ined.) E
Poculum sp. 'rata' P.R. Johnst. (nom. ined.)
Poculum subcinnabarinum (Dennis) Dumont E
Rutstroemia macrospora (Peck) Kanouse
— f. gigaspora Korf
SCLEROTINIACEAE
Asterocalyx mirabilis Höhn.
Botryotinia draytonii (Buddin & Wakef.) Seaver A
Botryotinia fabae J.Y. Lu & T.H. Wu A
Botryotinia fuckeliana (de Bary) Whetzel A
Botryotinia narcissicola (P.H. Greg.) N.F. Buchw. A
Botryotinia porri (J.F.H. Beyma) Whetzel A
Botryotinia sphaerosperma (P.H. Greg.) N.F.

Buchw. A
Botryotinia squamosa Vienn.-Bourg. A
Botrytis aclada Fresen. AN A
Botrytis anthophila Bondartsev AN A
Botrytis byssoidea J.C. Walker AN A
Botrytis elliptica (Berk.) Cooke AN A
Botrytis galanthina (Berk. & Broome) Sacc. AN A
Botrytis paeoniae Oudem. AN A
Botrytis tulipae Lind AN A
Ciborinia allii (Sawada) L.M. Kohn OU
Ciborinia camelliae L.M. Kohn A
Mitrulinia ushuaiae (Rehm) Spooner
Monilia aurea Pers. AN A
Monilia carbonaria Cooke AN E
Monilinia fructicola (G. Winter) Honey A
Monilinia laxa (Aderh. & Ruhland) Honey A
Ovulinia azaleae F.A. Weiss A
Penicillium brocae S.W.Peterson, J.Pérez, Vega & Infante AN U
Sclerocrana atra Samuels & L.M. Kohn E
Sclerotinia homoeocarpa F.T. Benn. A
Sclerotinia minor Jagger A
Sclerotinia sclerotiorum (Lib.) de Bary A
Sclerotinia spermophila Noble A
Sclerotinia trifoliorum Erikss. A
Stromatinia gladioli (Drayton) Whetzel A
Torrendiella brevisetosa P.R. Johnst. & Gamundí E
Torrendiella cannibalensis P.R. Johnst. & Gamundí E
Torrendiella dingleyae P.R. Johnst. & Gamundí E
Torrendiella eucalypti (Berk.) Spooner A
Torrendiella madsenii (G.W. Beaton & Weste) Spooner
Torrendiella sp. 'coriaria' P.R. Johnst. (nom. ined.) E
VIBRISSEACEAE
Chlorovibrissea bicolor (G.W. Beaton & Weste) L.M. Kohn OU
Chlorovibrissea melanochlora (G.W. Beaton & Weste) L.M. Kohn OU
Chlorovibrissea phialophora Samuels & L.M. Kohn E
Chlorovibrissea tasmanica (Rodway) L.M. Kohn
Phialocephala humicola S.C. Jong & E.E. Davis AN
Vibrissea albofusca G.W. Beaton E
Vibrissea filisporia (Bonord.) A. Sánchez & Korf f. *filisporia* A F

Order LEOTIALES
BULGARIACEAE
Bulgaria inquinans (Pers.) Fr.
Phacidiopycnis tuberivora (Güssow & W.R. Foster) B. Sutton AN A
LEOTIACEAE
Alatospora acuminata Ingold AN F
Leotia lubrica (Scop.) Pers.
Microglossum olivaceum (Pers.) Gillet OU
Microglossum rufum (Schwein.) Underw.
Microglossum viride (Schrad.) Gillet OU
Potridiscus polymorphus Döbbeler & Triebel

Order RHYTISMATALES
RHYTISMATACEAE
Bivallum zelandicum P.R. Johnst. E
Coccomyces clavatus P.R. Johnst.
Coccomyces crystalligerus Sherwood
Coccomyces cupressini P.R. Johnst. E
Coccomyces globosus P.R. Johnst.
Coccomyces lauraceus P.R. Johnst.
Coccomyces libocedri P.R. Johnst. E
Coccomyces limitatus (Berk. & M.A. Curtis) Sacc.
Coccomyces longwoodicus P.R. Johnst. E
Coccomyces phyllocladi P.R. Johnst.
Coccomyces radiatus Sherwood
Colpoma agathidis P.R. Johnst. E
Colpoma nothofagi P.R. Johnst. E
Colpoma quercinum (Pers.) Wallr. A
Davisomycella ampla (Davis) Darker A
Hypoderma alborubrum P.R. Johnst. E
Hypoderma bidwillii P.R. Johnst. E
Hypoderma bihospitum P.R. Johnst. E
Hypoderma campanulatum P.R. Johnst. E
Hypoderma carinatum P.R. Johnst. E
Hypoderma cookianum P.R. Johnst. E
Hypoderma cordylines P.R. Johnst. E
Hypoderma dundasicum P.R. Johnst. E
Hypoderma liliense P.R. Johnst. E
Hypoderma obtectum P.R. Johnst. E
Hypoderma rubi (Pers.) DC. ex Chevall.
Hypoderma sigmoideum P.R. Johnst. E
Hypoderma sp. 'flax' P.R. Johnst. (nom. ined.) E
Hypoderma sp. 'nothofagus' P.R. Johnst. (nom. ined.) E
Hypoderma sticheri P.R. Johnst. E
Hypohelion parvum P.R. Johnst. E
Leptostroma litigiosum var. *exasperatum* Berk. AN E
Leptostroma pinastri Desm. AN A
Lophodermium actinothyrium Fuckel OU
Lophodermium agathidis Minter & Hettige
Lophodermium alpinum (Rehm) Weese
Lophodermium atrum P.R. Johnst. E
Lophodermium brunneolum P.R. Johnst. E
Lophodermium conigenum (Brunaud) Hilitzer A
Lophodermium croesicum P.R. Johnst. E
Lophodermium culmigenum (Fr.) De Not.
Lophodermium eucalypti (Rodway) P.R. Johnst.
Lophodermium hauturuanum P.R. Johnst. E
Lophodermium inclusum P.R. Johnst. E
Lophodermium irregulare P.R. Johnst. E
Lophodermium kaikawakae P.R. Johnst. E
Lophodermium mangatepopense P.R. Johnst. E
Lophodermium medium P.R. Johnst. E
Lophodermium molitoris Minter A
Lophodermium nigrofactum P.R. Johnst. E
Lophodermium pinastri (Schrad.) Chevall. A
Lophodermium rectangulare P.R. Johnst.
Lophodermium richeae Petr.
Lophodermium rubrum P.R. Johnst. E
Lophodermium sieglingiae Hilitzer
Lophodermium tindalii P.R. Johnst. E
Lophodermium unciniae P.R. Johnst. E
Marthamyces dendrobii (P.R. Johnst.) Minter E
Marthamyces desmoschoeni (P.R. Johnst.) Minter E
Marthamyces dracophylli (P.R. Johnst.) Minter E
Marthamyces emarginatus (Cooke & Massee) Minter
Marthamyces quadrifidus (Lév.) Minter
Meloderma desmazieresii (Duby) Darker A
Meloderma dracophylli P.R. Johnst. E
Propolis farinosa (Pers.) Fr.
Pureke zelandicum P.R. Johnst.
Terriera asteliae (P.R. Johnst.) P.R. Johnst. E
Terriera brevis (Berk.) P.R. Johnst. E
Terriera minor (Tehon) P.R. Johnst.
Terriera nematoidea (P.R. Johnst.) P.R. Johnst. E

Order THELEBOLALES
THELEBOLACEAE
Ascozonus woolhopensis (Renny) Boud. A
Coprotus aurora (P. Crouan & H. Crouan) K.S. Thind & Waraitch A
Coprotus disculus Kimbr., Luck-Allen & Cain A
Coprotus glaucellus (Rehm) Kimbr. A
Coprotus granuliformis (P. Crouan & H. Crouan) Kimbr. A
Coprotus lacteus (Cooke & W. Phillips) Kimbr., Luck-Allen & Cain A
Coprotus leucopocillum Kimbr., Luck-Allen & Cain A
Coprotus luteus Kimbr., Luck-Allen & Cain A
Coprotus sexdecimsporus (P. Crouan & H. Crouan) Kimbr. & Korf A
Coprotus trichosuri A.E. Bell & Kimbr. A
Thelebolus crustaceus (Fuckel) Kimbr. A
Thelebolus microsporus (Berk. & Broome) Kimbr. A
Thelebolus stercoreus Tode A

Class LICHINOMYCETES
Order LICHINALES
LICHINACEAE
Digitothyrea rotundata (Büdel, Hensson & Wessels) P.P.Moreno & Egea L
Ephebe fruticosa Henssen L
Ephebe ocellata Henssen L
Lempholemma cladodes (Tuck.) Zahlbr. L
Lichina pygmaea (Lightfoot) C.Agardh L M
Lichina minutissima Henssen L
Phyllisciella aotearoa Henssen & J.K.Bartlett L
Pyrenopsis tasmanica Nyl. L
Zahlbrucknerella calcarea (Herre) Herre L
Zahlbrucknerella compacta Henssen L
PELTULACEAE
Peltula euploca (Ach.) Ozenda & Clauzade L

Class ORBILIOMYCETES
Order ORBILIALES
ORBILIACEAE
Arthrobotrys cladodes Drechsler AN U
Arthrobotrys conoides Drechsler AN A
Arthrobotrys gephyropaga (Drechsler) Y. Li AN A
Arthrobotrys robusta Dudd. AN A
Dactylellina candida (Nees) Y. Li AN U
Dactylellina parvicolle (Drechsler) Y. Li AN A
Drechslerella brochopoda (Drechsler) M. Scholler, Hagedorn & A. Rubner AN A
Drechslerella dactyloides (Drechsler) M. Scholler, Hagedorn & A. Rubner AN A
Duddingtonia flagrans (Dudd.) R.C. Cooke AN
Dwayaangam heterospora G.L. Barron AN E F
Hyalorbilia inflatula (P. Karst.) Baral & G. Marson OU
Monacrosporium bembicodes (Drechsler) Subram. AN U
Monacrosporium copepodorum (G.L. Barron) A. Rubner AN U
Monacrosporium ellipsosporum (Preuss) R.C. Cooke & C.H. Dickinson AN U
Monacrosporium eudermatum (Drechsler) Subram. AN A M
Monacrosporium leptosporum (Drechsler) A. Rubner AN U
Orbilia auricolor (A. Bloxam) Sacc.
Orbilia cunninghamii Syd.
Orbilia delicatula (P. Karst.) P. Karst. (nom. inv.)
Orbilia juruensis Henn. OU
Orbilia luteorubella (Nyl.) P. Karst. A F
Orbilia vinosa (Alb. & Schwein.) P. Karst. OU

Class PEZIZOMYCETES
Order PEZIZALES
Berggrenia aurantiaca Cooke E
Berggrenia cyclospora (Cooke) Sacc. E
Cephaliophora longispora G.L. Barron, C. Morik. & Saikawa AN
Cephaliophora muscicola G.L. Barron, C. Morik. & Saikawa AN
Diehliomyces microsporus (Diehl & E.B. Lamb.) Gilkey A
Discinella confusa Dennis E
Discinella terrestris (Berk. & Broome) Dennis
Oedocephalum fimetarium (Riess) Sacc. AN A
Oedocephalum glomerulosum (Bull.) Sacc.
Pulvinula globifera (Berk. & M.A. Curtis) Le Gal U
Pulvinula miltina (Berk.) Rifai
Trichobolus sphaerosporus Kimbr. A
ASCOBOLACEAE
Ascobolus albidus P. Crouan & H. Crouan A

Ascobolus bistisii Gamundí & Ranalli A
Ascobolus crenulatus P. Karst. A
Ascobolus degluptus Brumm. A
Ascobolus denudatus Fr. A
Ascobolus epimyces (Cooke) Seaver A
Ascobolus foliicola Berk. & Broome A
Ascobolus furfuraceus Pers. A
Ascobolus hansenii M.D. Paulsen & Dissing A
Ascobolus hawaiiensis Brumm. A
Ascobolus immersus Pers. A
Ascobolus lineolatus Brumm. A
Ascobolus mancus (Rehm) Brumm. A
Ascobolus minutus Boud. A
Ascobolus perplexans Massee & E.S. Salmon A
Ascobolus stictoideus Speg. A
Saccobolus citrinus Boud. & Torrend A
Saccobolus depauperatus (Berk. & Broome) E.C. Hansen A
Saccobolus glaber (Pers.) Lambotte A
Saccobolus minimus Velen. A
Saccobolus saccoboloides (Seaver) Brumm. A
Saccobolus thaxteri Brumm. A
Saccobolus verrucisporus Brumm. A
Saccobolus versicolor (P. Karst.) P. Karst. A
Thecotheus agranulosus Kimbr. A
Thecotheus cinereus (P. Crouan & H. Crouan) Chenant. A
Thecotheus crustaceus (Starbäck) Aas & N. Lundq. A
Thecotheus holmskjoldii (E.C. Hansen) Eckblad A
ASCODESMIDACEAE
Lasiobolus ciliatus (J.C. Schmidt) Boud. A
Lasiobolus cuniculi Velen. A
Lasiobolus diversisporus (Fuckel) Sacc. A
Lasiobolus intermedius J.L. Bezerra & Kimbr. A
Lasiobolus macrotrichus Rea A
CHORIOACTIDACEAE
Wolfina aurantiopsis (Ellis) Seaver OU
DISCINACEAE
Gyromitra esculenta (Pers.) Fr. A OU
Gyromitra infula (Schaeff.) Quél. A
Gyromitra tasmanica Berk. & Cooke
HELVELLACEAE
Helvella crispa (Scop.) Fr. A
Helvella engleriana Henn. U
MORCHELLACEAE
Costantinella micheneri (Berk. & M.A. Curtis) S. Hughes AN
Morchella conica Pers. A
Morchella crassipes (Vent.) Pers. A
Morchella elata Fr. A
Morchella esculenta (L.) Pers. A
PEZIZACEAE
Aleuria aurantia (Pers.) Fuckel OU
Chromelosporium carneum (Pers.) Hennebert AN
Iodophanus carneus (Pers.) Korf A
Iodophanus testaceus (Moug.) Korf A
Marcelleina atroviolacea (Delile ex De Seynes) Brumm.
Pachyella babingtonii (Berk. & Broome) Boud. OU F
Peziza ammophila Durieu & Lév.
Peziza arvernensis Boud. A
Peziza austrogeaster (Rodway) Rifai
Peziza badioconfusa Korf U
Peziza bovina W. Phillips A
Peziza cerea Sowerby A
Peziza domiciliana Cooke OU
Peziza fimeti (Fuckel) Seaver OU
Peziza moravecii (Svrček) Donadini
Peziza natrophila N. Khan
Peziza nivalis (R. Heim & L. Remy) M.M. Moser
Peziza ostracoderma Korf U
Peziza petersii Berk. U
Peziza repanda Wahlenb. U
Peziza rifaii J. Moravec & Spooner
Peziza saniosa Schrad.
Peziza sp. 'Rotokuru Lakes (PDD80842)' J.A. Cooper (nom. ined.)
Peziza vesiculosa Bull. A
Peziza violacea Pers. OU
Plicaria badia (Pers.) Fuckel U
Plicaria endocarpoides (Berk.) Rifai U
PYRONEMATACEAE
Aleuria rhenana Fuckel
Aleurina calospora (Rifai) Korf & W.Y. Zhuang
Aleurina ferruginea (W. Phillips ex Cooke) W.Y. Zhuang & Korf
Aleurina magnicellula Dissing, Korf & W.Y. Zhuang E
Anthracobia macrocystis (Cooke) Boud. A
Anthracobia melaloma (Alb. & Schwein.) Arnould A
Anthracobia muelleri (Berk.) Rifai
Byssonectria chrysocoma Cooke & Harkn.
Cheilymenia catenipila J. Moravec E
Cheilymenia ciliata (Bull.) Maas Geest. A
Cheilymenia coprinaria (Cooke) Boud. A
Cheilymenia fimicola (De Not. & Bagl.) Dennis A
Cheilymenia magnipila J. Moravec OU
Cheilymenia pallida A.E. Bell & Dennis A
Cheilymenia raripila (W. Phillips) Dennis A
Cheilymenia stercorea (Pers.) Boud. A
Cheilymenia theleboloides (Alb. & Schwein.) Boud. A
Cheilymenia vitellina (Pers.) Dennis A
Complexipes moniliformis C. Walker AN A
Coprobia granulata (Bull.) Boud. A
Geneosperma laevisporum Korf & W.Y. Zhuang E
Humaria novozeelandica Henn. E
Humaria stromella Cooke & W. Phillips E
Humaria velenovskyi (Vacek ex Svrček) Korf & Sagara
Humarina leucoloma (Hedw.) Seaver OU
Humarina pachyderma Cooke (nom. ined.) U
Inermisia fusispora (Berk.) Rifai U
Lachnea dalmeniensis (Cooke) Sacc.
Lachnea scutellata (L.) Gillet U
Leucoscypha virginea Rifai E
Miladina lecithina (Cooke) Svrček A
Orbicula parietina (Schrad.) S. Hughes A
Otidea alutacea (Pers.) Massee A
Paurocotylis pila Berk. E
Pseudombrophila hepatica (Batsch) Brumm. A
Pseudombrophila theioleuca Rolland A
Pyronema domesticum (Sowerby) Sacc. U
Pyronema omphalodes (Bull.) Fuckel
Scutellinia badioberbis (Cooke) Kuntze
Scutellinia cejpii (Velen.) Svrček
Scutellinia colensoi Massee ex Le Gal
Scutellinia crinita (Bull.) Lambotte
Scutellinia crucipila (Cooke & W. Phillips) J. Moravec
Scutellinia kerguelensis (Berk.) Kuntze
Scutellinia lusatiae (Cooke) Kuntze
Scutellinia marginata Gamundí
Scutellinia nigrohirtula (Svrček) Le Gal
Scutellinia olivascens (Cooke) Kuntze
Scutellinia patagonica (Rehm) Gamundí
Scutellinia pseudomagaritacea Le Gal OU
Scutellinia scutellata (L.) Lambotte
Scutellinia subbadioberbis Le Gal E
Scutellinia subhirtella Svrček
Scutellinia totaranuiensis J. Moravec E
Scutellinia trechispora (Berk. & Broome) Lambotte OU
Tarzetta cupularis (L.) Svrček OU
Tarzetta jafneospora W.Y. Zhuang & Korf E
Wilcoxina mikolae (Chin S. Yang & H.E. Wilcox) Chin S. Yang & Korf A
SARCOSCYPHACEAE
Cookeina colensoi (Berk.) Seaver
Sarcoscypha excelsa Syd. E
SARCOSOMATACEAE
Conoplea fusca Pers. AN A
Conoplea novae-zelandiae S. Hughes AN E
Conoplea tortuosa S. Hughes AN E
Plectania campylospora (Berk.) Nannf.
Plectania melastoma (Sowerby) Fuckel
Plectania platensis (Speg.) Rifai
Plectania rhytidia (Berk.) Nannf. & Korf
Pseudoplectania nigrella (Pers.) Fuckel U
Sarcosoma orientale Pat.
Sarcosoma zelandicum Lloyd (nom. inv.) E
Strumella hysterioidea Cooke & Massee AN, (nom. inv.)
TUBERACEAE
Dingleya phymatodea (B.C. Zhang & Minter) Trappe, Castellano & Malajczuk
Dingleya turbinata (B.C. Zhang & Minter) Trappe, Castellano & Malajczuk E
Dingleya verrucosa Trappe E
Labyrinthomyces varius (Rodway) Trappe
Tuber aestivum Vittad. A
Tuber borchii Vittad. A
Tuber californicum Harkn. A
Tuber dryophilum Tul. & C. Tul. A
Tuber foetidum Vittad. A
Tuber levissimum Gilkey A
Tuber maculatum Vittad. A
Tuber melanosporum Vittad. A
Tuber rufum Picco A
Tuber separans Gilkey A

Class SORDARIOMYCETES
Adomia avicenniae S. Schatz M
Biconiosporella corniculata Schaumann AN M
Ellisembia adscendens (Berk.) Subram. AN
Ellisembia brachypus (Ellis & Everh.) Subram. AN
Ellisembia folliculata (Corda) Subram. AN
Ellisembia sp. 'Arthur's Pass (PDD80069)' J.A. Cooper AN (nom. ined.)
Ellisembia vaga (Nees & T. Nees) Subram. AN OU
Globosphaeria jamesii D. Hawksw. U
Leptosporella gregaria Penz. & Sacc.
Linocarpon pandani (Syd. & P. Syd.) Syd. & P. Syd.
Natantiella ligneola (Berk. & Broome) Réblová
Rimaconus coronatus Huhndorf & A.N. Mill. E
Savoryella lignicola E.B.G. Jones & R.A. Eaton M
Selenosporella curvispora G. Arnaud ex MacGarvie
Selenosporella verticillata B. Sutton & Hodges AN U F
Sporoschismopsis dingleyae S. Hughes & Hennebert AN
Teracosphaeria petroica Réblová & Seifert E
Torpedospora radiata Meyers M
Xylomelasma novae-zelandiae Réblová E
Xylomelasma sordida Réblová E
ANNULATASCACEAE
Annulusmagnus triseptatus (S.W. Wong, K.D. Hyde & E.B.G. Jones) J. Campb. & Shearer
Ceratostomella cuspidata (Fr.) Réblová
APIOSPORACEAE
Apiospora bambusae (Turconi) Sivan. A
Apiospora camptospora Penz. & Sacc. A
Apiospora montagnei Sacc. A
Apiospora setosa Samuels, McKenzie & D.E. Buchanan A
Arthrinium phaeospermum (Corda) M.B. Ellis AN A
Arthrinium sp. (*Cordella johnstonii*) AN A
Arthrinium sporophleum Kunze AN A
THYRIDIACEAE
Thyridium vestitum (Fr.) Fuckel A

Order BOLINIALES

BOLINIACEAE
Camarops flava Samuels & J.D. Rogers E
Camarops polysperma (Mont.) J.H. Mill.
Lentomitella cirrhosa (Pers.) Réblová
Lentomitella crinigera (Cooke) Réblová
Lentomitella tomentosa Réblová & J. Fourn. OU
CATABOTRYDACEAE
Catabotrys decidua (Berk. & Broome) Seaver & Waterston

Order CALOSPHAERIALES
CALOSPHAERIACEAE
Tectonidula hippocrepica Réblová E
Togniniella acerosa Réblová, L. Mostert, W. Gams & Crous
Wegelina saccardiana Berl.

Order CHAETOSPHAERIALES
CHAETOSPHAERIACEAE
Catenularia longispora S. Hughes AN E
Catenularia macrospora S. Hughes AN
Chaetosphaeria acutata Réblová & W. Gams OU
Chaetosphaeria albida T.J. Atk., A.N. Mill. & Huhndorf E
Chaetosphaeria bombycina T.J. Atk., A.N. Mill. & Huhndorf E
Chaetosphaeria brevispora Shoemaker E
Chaetosphaeria callimorpha (Mont.) Sacc.
Chaetosphaeria capitata Sivan. & H.S. Chang OU
Chaetosphaeria chaetosa Kohlm. M
Chaetosphaeria ciliata Réblová & Seifert
Chaetosphaeria coelestina Höhn.
Chaetosphaeria cupulifera (Berk. & Broome) Sacc. OU
Chaetosphaeria curvispora Réblová E
Chaetosphaeria dingleyae S. Hughes, W.B. Kendr. & Shoemaker
Chaetosphaeria ellisii (M.E. Barr) Huhndorf & F.A. Fernández U
Chaetosphaeria fuegiana Réblová
Chaetosphaeria fusichalaroides Réblová
Chaetosphaeria gallica (Sacc. & Flageolet) Réblová
Chaetosphaeria hebetiseta Réblová & W. Gams E
Chaetosphaeria metallicans T.J. Atk., A.N. Mill. & Huhndorf E
Chaetosphaeria myriocarpa (Fr.) C. Booth U
Chaetosphaeria novae-zelandiae S. Hughes & Shoemaker
Chaetosphaeria ovoidea (Fr.) Constant., K. Holm & L. Holm
Chaetosphaeria phaeostalacta Réblová E
Chaetosphaeria phaeostromoides (Peck) Sacc.
Chaetosphaeria preussii W. Gams & Hol.-Jech.
Chaetosphaeria pulchriseta S. Hughes, W.B. Kendr. & Shoemaker
Chaetosphaeria raciborskii (Penz. & Sacc.) A.N. Mill. & Huhndorf
Chaetosphaeria rubicunda Huhndorf & F.A. Fernández A
Chaetosphaeria talbotii S. Hughes, W.B. Kendr. & Shoemaker
Chloridium virescens (Pers.) W. Gams & Hol.-Jech. var. *virescens* AN
— var. *chlamydosporum* (J.F.H. Beyma) W. Gams & Hol.-Jech. AN U
Cylindrotrichum clavatum W. Gams AN U
Cylindrotrichum oligospermum (Corda) Bonord. AN
Dictyochaeta australiensis (B. Sutton) Whitton, McKenzie & K.D. Hyde AN
Dictyochaeta botulispora (S. Hughes & W.B. Kendr.) Whitton, McKenzie & K.D. Hyde AN E
Dictyochaeta brevisetula (S. Hughes & W.B. Kendr.) Whitton, McKenzie & K.D. Hyde AN E
Dictyochaeta britannica (M.B. Ellis) Whitton, McKenzie & K.D. Hyde AN
Dictyochaeta fertilis (S. Hughes & W.B. Kendr.) Hol.-Jech. AN
Dictyochaeta longispora (S. Hughes & W.B. Kendr.) Whitton, McKenzie & K.D. Hyde AN E
Dictyochaeta obesispora (S. Hughes & W.B. Kendr.) Whitton, McKenzie & K.D. Hyde AN E
Dictyochaeta parva (S. Hughes & W.B. Kendr.) Hol.-Jech. AN
Dictyochaeta setosa (S. Hughes & W.B. Kendr.) Whitton, McKenzie & K.D. Hyde AN E
Dictyochaeta simplex (S. Hughes & W.B. Kendr.) Hol.-Jech. AN
Dictyochaeta vulgaris (S. Hughes & W.B. Kendr.) Whitton, McKenzie & K.D. Hyde AN
Dischloridium laeense (Matsush.) B. Sutton AN
Gonytrichum macrocladum (Sacc.) S. Hughes AN
Kylindria ellisii (Morgan-Jones) DiCosmo, S.M. Berch & W.B. Kendr. AN U
Melanochaeta aotearoae (S. Hughes) E. Müll., Harr & Sulmont F
Melanopsammella inaequalis (Grove) Höhn.
Melanopsammella vermicularioides (Sacc. & Roum.) Réblová, M.E. Barr & Samuels
Menispora uncinata S. Hughes & W.B. Kendr. AN
Menisporopsis novae-zelandiae S. Hughes & W.B. Kendr. AN
Miyoshiella larvata Réblová U
Phaeostalagmus cyclosporus (Grove) W. Gams AN
Phaeostalagmus novae-zelandiae S. Hughes AN E
Sporoschisma nigroseptatum D. Rao & P. Rag. Rao AN
Stanjehughesia caespitulosa (Ellis & Everh.) Subram.
Thozetella havanensis R.F. Castañeda AN A
Thozetella tocklaiensis (Agnihothr.) Piroz. & Hodges AN U
Umbrinosphaeria caesariata (Clinton & Peck) Réblová
Zignoella macrospora Sacc.

Order CONIOCHAETALES
Porosphaerella cordanophora E. Müll. & Samuels U
Porosphaerella setosa A.I. Romero & Samuels
Pseudobotrytis terrestris (Timonin) Subram. AN F
Wallrothiella congregata (Wallr.) Sacc.
Wallrothiella subiculosa Höhn.
CONIOCHAETACEAE
Coniochaeta discospora (Auersw.) Cain A
Coniochaeta hansenii (Oudem.) Cain A
Coniochaeta leucoplaca (Berk. & Ravenel) Cain A
Coniochaeta philocoproides (Griffiths) Cain A
Coniochaeta pulveracea (Ehrh.) Munk
Coniochaeta velutina (Fuckel) Cooke
Lecythophora mutabilis (J.F.H. Beyma) W. Gams & McGinnis AN U

Order CORONOPHORALES
BERTIACEAE
Bertia moriformis (Tode) De Not.
CHAETOSPHAERELLACEAE
Crassochaeta fusispora (Sivan.) Réblová
NITSCHKIACEAE
Fracchiaea rasa (Berk.) Sacc. E
Nitschkia acanthostroma (Mont.) Nannf.
Nitschkia collapsa (Romell) Chenant.
Nitschkia macrospora Teng U

Order DIAPORTHALES
Botryodiplodia ramulicola (Desm.) Petr. AN A
Harknessia eucrypta (Cooke & Massee) Nag Raj & DiCosmo AN A
Harknessia globosa B. Sutton AN A
Macrohilum eucalypti H.J. Swart AN A
Pseudoplagiostoma eucalypti Cheewangkoon, M.J. Wingf. & Crous A
Stenocarpella maydis (Berk.) B. Sutton AN A
Valsaria rubricosa (Fr.) Sacc.
Wuestneia epispora Z.Q. Yuan A
Wuestneia karwarrae (B. Sutton & Pascoe) Z.Q. Yuan A
CRYPHONECTRIACEAE
Amphilogia gyrosa (Berk. & Broome) Gryzenh., Glen & M.J. Wingf.
Amphilogia major Gryzenh., Glen & M.J. Wingf. E
Cryphonectria radicalis (Schwein.) M.E. Barr
Endothia fluens (Sowerby) Shear & N.E. Stevens
Endothia gyrosa (Schwein.) Fr.
Rostraureum longirostre (Earle) Gryzenh. & M.J. Wingf.
DIAPORTHACEAE
Diaporthe abdita Sacc. & Speg. A
Diaporthe actinidiae N.F. Sommer & Beraha A
Diaporthe adunca (Desm.) Niessl A
Diaporthe arctii (Lasch) Nitschke A
Diaporthe citri F.A. Wolf A
Diaporthe eres Nitschke A
Diaporthe helianthi Munt.-Cvetk., Mihaljč. & M. Petrov A OU
Diaporthe juglandina (Fuckel) Nitschke A
Diaporthe leiphaemia (Fr.) Sacc. A
Diaporthe lupini Harkn. A
Diaporthe medusaea Nitschke A
Diaporthe oncostoma (Duby) Fuckel A
Diaporthe perniciosa Marchal & É.J. Marchal A
Diaporthe phaseolorum (Cooke & Ellis) Sacc. var. *phaseolorum* A
— var. *batatas* (Harter & E.C. Field) Wehmeyer A
— var. *phaseolorum* A
— var. *sojae* (Lehman) Wehm. A
Diaporthe sophorae Sacc. A
Diaporthe sp. 1 (nom. ined.)
Diaporthe sp. 2 (nom. ined.) E
Diaporthe viticola Nitschke A
Diaporthe woodii Punith. A
Phomopsis abdita (Sacc.) Traverso AN A
Phomopsis acmella (Berk.) Petr. & Syd. AN E
Phomopsis amygdali (Delacr.) Tuset & Portilla AN A
Phomopsis castanea (Sacc.) Petr. AN A
Phomopsis cunninghamii Syd. AN E
Phomopsis eucommiicola C.Q. Chang, Z.D. Jiang & P.K. Chi AN OU
Phomopsis juniperivora G.G. Hahn AN A
Phomopsis lathyrina Grove AN A
Phomopsis limonii I.C. Harv., E.R. Morgan & G.K. Burge AN A
Phomopsis nux H.C. Sm. (nom. ined.) U
Phomopsis obscurans (Ellis & Everh.) B. Sutton AN A
Phomopsis sclerotioides Kesteren AN A
Phomopsis sophorae (Sacc.) Traverso AN
Phomopsis sp. 'corokia' (nom. ined.) E
Phomopsis subordinaria (Desm.) Traverso AN A
Phomopsis viticola (Sacc.) Sacc. AN A
GNOMONIACEAE
Apiognomonia errabunda (Roberge ex Desm.) Höhn. A
Apiognomonia veneta (Sacc. & Speg.) Höhn. A
Asteroma dilatatum Berk. AN E
Cryptodiaporthe salicina Wehm. A
Cryptosporella hypodermia (Fr.) Sacc.
Diplodina eurhododendri W. Voss AN A
Discula brenckleana (Sacc., Syd. & P. Syd.) Petr. AN A
Discula microsperma (Berk. & R. Br.) Sacc. AN OU
Gloeosporidium coprosmae Syd. AN E
Gnomonia intermedia Rehm A
Gnomonia pascoei H.C. Sm. (nom. ined.) A
Greeneria uvicola (Berk. & M.A. Curtis) Punith.

AN OU
Ophiovalsa betulae (Tul. & C. Tul.) Petr. A
MELANCONIDACEAE
Dicarpella dryina Belisario & M.E. Barr A
Melanconis alni Tul. A
Melanconis carthusiana Tul. A
Melanconis juglandis (Ellis & Everh.) A.H. Graves A
Melanconis stilbostoma (Fr.) Tul. A
Melanconium atrum Link AN A
PSEUDOVALSACEAE
Coryneum betulinum Schulzer AN A
Coryneum modonium (Sacc.) Griffon & Maubl. AN A
Coryneum salicis Tognini AN A
Pseudovalsa lanciformis (Fr.) Ces. & De Not A
Pseudovalsa longipes (Tul.) Sacc. A
SCHIZOPARMACEAE
Schizoparme straminea Shear U
TOGNINIACEAE
Phaeoacremonium armeniacum A.B. Graham, P.R. Johnst. & B. Weir AN A
Phaeoacremonium globosum A.B. Graham, P.R. Johnst. & B. Weir AN A
Phaeoacremonium mortoniae Crous & W. Gams AN A
Phaeoacremonium occidentale A.B. Graham, P.R. Johnst. & B. Weir AN A
Phaeoacremonium rubrigenum W. Gams, Crous & M.J. Wingf. AN A
Togninia minima (Tul. & C. Tul.) Berl. A
Togninia novae-zealandiae Hausner, Eyjólfsdóttir & J. Reid A
VALSACEAE
Apioporthe vepris (Delacr.) Wehm. A
Cytospora metrosideri Rabenh. AN E
Leucostoma niveum (Hoffm.) Höhn. A
Leucostoma persoonii Höhn.
Leucostoma sp. 1 (nom. ined.) E
Leucostoma sp. 2 (nom. ined.) E
Valsa ambiens (Pers.) Fr. subsp. *ambiens* A
Valsa ceratophora Tul. & C. Tul. A
Valsa ceratosperma (Tode) Maire A
Valsa cincta Fr. A
Valsa fabianae G.C. Adams, M.J. Wingf. & Jol. Roux A
Valsa sordida Nitschke A
GLOMERELLACEAE
Colletotrichum acutatum f.sp. *pineum* Dingley & J.W. Gilmour AN A
Colletotrichum boninense Moriwaki, Toy. Sato & Tsukib. AN A
Colletotrichum circinans (Berk.) Voglino AN A
Colletotrichum coccodes (Wallr.) S. Hughes AN A
Colletotrichum crassipes (Speg.) Arx AN A
Colletotrichum dematium (Pers.) Grove AN A
Colletotrichum fioriniae (J.A.P. Marcelino & S. Gouli) R.G. Shivas & Y.P. Tan U
Colletotrichum fuscum Laubert AN A
Colletotrichum gloeosporioides var. *hederae* Pass. AN A
Colletotrichum godetiae Neerg. AN A
Colletotrichum higginsianum Sacc. AN A
Colletotrichum horii B. Weir & P.R. Johnst. AN A
Colletotrichum lindemuthianum (Sacc. & Magnus) Briosi & Cavara AN A
Colletotrichum linicola Pethybr. & Laff. AN A
Colletotrichum lupini (Bondar) Nirenberg, Feiler & Hagedorn AN U
Colletotrichum malvarum (A. Braun & Casp.) Southw. AN A
Colletotrichum musae (Berk. & M.A. Curtis) Arx AN A
Colletotrichum orbiculare (Berk. & Mont.) Arx AN A
Colletotrichum spaethianum (Allesch.) Damm, P.F. Cannon & Crous A
Colletotrichum spinaciae Ellis & Halst. A
Colletotrichum trichellum (Fr.) Duke AN A
Colletotrichum trifolii Bain AN A
Colletotrichum truncatum (Schwein.) Andrus & W.D. Moore AN A
Glomerella acutata J.C. Guerber & J.C. Correll A
Glomerella cingulata Stoneman U
— f.sp. *camelliae* Dickens & R.T.A. Cook A
Glomerella glycines Lehman & F.A. Wolf A
Glomerella graminicola D.J. Politis A
Glomerella miyabeana (Fukushi) Arx A
Glomerella phormii (J. Schröt.) D.F. Farr & Rossman
Glomerella tucumanensis (Speg.) Arx & E. Müll. A
Gnomoniopsis comari (P. Karst.) Sogonov A

Order HYPOCREALES
Acremonium alternatum Link AN A
Acremonium curvulum W. Gams AN U
Acremonium falciforme (Carrión) W. Gams AN U
Acremonium kiliense Grütz AN U
Acremonium luzulae (Fuckel) W. Gams AN U
Acremonium murorum (Corda) W. Gams AN
Acremonium novae-zelandiae (S. Hughes & C.H. Dickinson) W. Gams AN E
Acremonium persicinum (Nicot) W. Gams AN U
Acremonium strictum W. Gams AN U
Acremonium tubakii W. Gams AN U M
Calcarisporium arbuscula Preuss AN
Cancellidium applanatum Tubaki AN U F
Geosmithia argillacea (Stolk, H.C. Evans & T. Nilsson) Pitt AN A
Geosmithia putterillii (Thom) Pitt AN
Haptospora appendiculata G.L. Barron AN E
Illosporium carneum Fr. AN
Melanopsamma pomiformis (Pers.) Sacc.
Memnoniella echinata (Rivolta) L.D. Galloway AN A
Memnoniella subsimplex (Cooke) Deighton AN A
Myrothecium carmichaelii Grev. AN U
Myrothecium cinctum (Corda) Sacc. AN A
Myrothecium gramineum Lib. AN A
Myrothecium leucotrichum (Peck) M.C. Tulloch AN
Myrothecium roridum Tode AN A
Myrothecium verrucaria (Alb. & Schwein.) Ditmar AN A
Stachybotrys breviuscula McKenzie AN
Stachybotrys chartarum (Ehrenb.) S. Hughes AN U
Stachybotrys cordylines McKenzie AN, (nom. ined.) E
Stachybotrys dichroa Grove AN
Stachybotrys freycinetiae McKenzie AN E
Stachybotrys kampalensis Hansf. AN
Stachybotrys kapiti Whitton, McKenzie & K.D. Hyde AN
Stachybotrys microspora (B.L. Mathur & Sankhla) S.C. Jong & E.E. Davis AN
Stachybotrys nephrodes McKenzie AN U
Stachybotrys nephrospora Hansf. AN A
Stachybotrys nilagirica Subram. AN
Stachybotrys parvispora S. Hughes AN
Stachybotrys ruwenzoriensis Matsush. AN U
Stachybotrys waitakere Whitton, McKenzie & K.D. Hyde AN E
Stilbella albocitrina (Ellis & Everh.) Seifert AN
Stilbella fimetaria (Pers.) Lindau AN A
Trichothecium roseum (Pers.) Link AN A
BIONECTRIACEAE
Bionectria aurantia (Penz. & Sacc.) Rossman, Samuels & Lowen
Bionectria aureofulvella Schroers & Samuels
Bionectria byssicola (Berk. & Broome) Schroers & Samuels
Bionectria grammicosporopsis (Samuels) Schroers & Samuels
Bionectria kowhai (Dingley) Schroers E
Bionectria ochroleuca (Schwein.) Schroers & Samuels U
Bionectria ralfsii (Berk. & Broome) Schroers & Samuels
Bionectria subquaternata (Berk. & Broome) Schroers & Samuels
Bionectria verrucispora Schroers & Samuels E
Bionectria zelandiae-novae Schroers E
Bryonectria disciformis Döbbeler E
Clonostachys candelabrum (Bonord.) Schroers AN A
Clonostachys solani f. *nigrovirens* (J.F.H. Beyma) Schroers AN A
Dendrodochium ellipticum Cooke & Massee AN
Hydropisphaera arenula (Berk. & Broome) Rossman & Samuels
Hydropisphaera arenuloides (Samuels) Rossman & Samuels E
Hydropisphaera boothii (D. Hawksw.) Rossman & Samuels
Hydropisphaera cyatheae (Dingley) Rossman & Samuels E
Hydropisphaera erubescens (Desm.) Rossman & Samuels
Hydropisphaera macrarenula (Samuels) Rossman & Samuels
Hydropisphaera multiloculata (Samuels) Rossman & Samuels E
Hydropisphaera multiseptata (Samuels) Rossman & Samuels E
Hydropisphaera peziza (Tode) Dumort.
Hydropisphaera suffulta (Berk. & M.A. Curtis) Rossman & Samuels
Ijuhya corynospora (Samuels) Rossman & Samuels E
Ijuhya dentifera (Samuels) Rossman & Samuels E
Ijuhya peristomialis (Berk. & Broome) Rossman & Samuels
Lasionectria sylvana (Mouton) Rossman & Samuels
Lasionectria vulpina (Cooke) Rossman & Samuels U
Nectriella bloxamii (Berk. & Broome) Fuckel
Nectriopsis candicans (Plowr.) Maire
Nectriopsis exigua (Pat.) W. Gams
Nectriopsis hirsuta (Samuels) Samuels
Nectriopsis hyperbiota Samuels E
Nectriopsis lasiodermopsis Samuels
Nectriopsis oropensoides (Rehm) Samuels
Nectriopsis perpusilla (Mont.) Samuels
Nectriopsis puiggarii (Speg.) Samuels
Nectriopsis septofusidiae Samuels
Nectriopsis sibicola Samuels E
Nectriopsis sororicola Samuels E
Nectriopsis squamulosa (Ellis) Samuels
Paranectria alstrupii Zhurb.
Pronectria subimperspicua (Speg.) Lowen
Protocreopsis freycinetiae (Samuels) Samuels & Rossman E
Protocreopsis pertusa (Pat.) Samuels & Rossman
Protocreopsis pertusoides (Samuels) Samuels & Rossman E
Protocreopsis phormiicola (Samuels) Samuels & Rossman E
Roumegueriella rufula (Berk. & Broome) Malloch & Cain U
Selinia pulchra (G. Winter) Sacc. A
Stephanonectria keithii (Berk. & Broome) Schroers & Samuels
Stilbocrea gracilipes (Tul. & C. Tul.) Samuels & Seifert

Stilbocrea macrostoma (Berk. & M.A. Curtis) Höhn.
Trichonectria horrida Samuels E
CLAVICIPITACEAE
Aschersonia australiensis Henn. AN
Berkelella stilbigera (Berk. & Broome) Sacc.
Claviceps nigricans Tul. U
Claviceps paspali F. Stevens & J.G. Hall A
Claviceps purpurea (Fr.) Tul. A
Conoideocrella luteorostrata (Zimm.) D. Johnson, G.H. Sung, Hywel-Jones & Spatafora
Drechmeria coniospora (Drechsler) W. Gams & H.-B. Jansson AN U
Epichloe festucae Leuchtm., Schardl & M.R. Siegel A
Harposporium anguillulae Lohde AN U
Harposporium bysmatosporum Drechsler AN
Harposporium helicoides Drechsler AN U
Harposporium leptospira Drechsler AN U
Harposporium lilliputanum S.M. Dixon AN A
Harposporium trigonosporum G.L. Barron & Szijarto AN
Hypocrella duplex Petch
Metacordyceps chlamydosporia (H.C. Evans) G.H. Sung, J.M. Sung, Hywel-Jones & Spatafora A
Metarhizium anisopliae (Metschn.) Sorok n var. *anisopliae* AN
Metarhizium flavoviride W. Gams & Rozsypal var. *flavoviride* AN U
— var. *novozealandicum* Driver & Milner AN
Neobarya agaricola (Berk.) Sameuls & Lowen
Neobarya parasitica (Fuckel) Lowen
Neotyphodium aotearoae C.D. Moon, C.O. Miles & Schardl AN
Neotyphodium coenophialum (Morgan-Jones & W. Gams) Glenn, C.W. Bacon & Hanlin AN A
Neotyphodium lolii (Latch, M.J. Chr. & Samuels) Glenn, C.W. Bacon & Hanlin AN A
Neotyphodium occultans C.D. Moon, B. Scott & M.J. Chr. AN A
Neotyphodium sp. FaTG-3 sensu M.J. Chr., Leuchtm., D.D. Rowan & Tapper AN U
Neotyphodium sp. LpTG-2 sensu M.J. Chr., Leuchtm., D.D. Rowan & Tapper AN U
Neotyphodium uncinatum (W. Gams, Petrini & D. Schmidt) Glenn, C.W. Bacon & Hanlin AN A
Nomuraea atypicola (Yasuda) Samson AN U
Nomuraea rileyi (Farl.) Samson AN U
Pochonia bulbillosa (W. Gams & Malla) Zare & W. Gams AN U
Podocrella harposporifera (Samuels) P. Chaverri & Samuels E
CORDYCIPITACEAE
Akanthomyces aranearum (Petch) Mains AN
Beauveria brongniartii (Sacc.) Petch AN U
Beauveria cf. *bassiana* S.A. Rehner & E. Buckley (nom. ined.) U
Cordyceps bassiana Z.Z. Li, C.R. Li, B. Huang & M.Z. Fan
Cordyceps gunnii (Berk.) Berk.
Cordyceps hauturu Dingley E
Cordyceps kirkii G. Cunn. E
Cordyceps lateritia Dingley E
Cordyceps martialis Speg. OU
Cordyceps militaris (L.) Fr. OU
Cordyceps novae-zealandiae Dingley E
Cordyceps scarabaeicola Kobayasi U
Cordyceps tuberculata (Lebert) Maire
Engyodontium aranearum (Cavara) W. Gams, de Hoog & Samson AN A
Gibellula pulchra (Sacc.) Cavara AN
Isaria aggregata Cooke & Massee AN E
Isaria farinosa (Holmsk.) Fr. AN A
Isaria felina (DC.) Fr. AN A
Isaria fumosorosea Wize AN U
Isaria javanica (Frieder. & Bally) Samson & Hywel-Jones AN U
Isaria sinclairii (Berk.) Lloyd AN
Isaria sulphurea Fiedl. AN A
Isaria tenuipes Peck AN U
Lecanicillium psalliotae (Treschew) Zare & W. Gams AN A
Rotiferophthora angustispora (G.L. Barron) G.L. Barron AN OU
Rotiferophthora asymmetrica (G.L. Barron) G.L. Barron AN OU
Rotiferophthora biconica G.L. Barron AN E
Rotiferophthora brevipes G.L. Barron AN E
Rotiferophthora cylindrospora (G.L. Barron) G.L. Barron AN OU
Rotiferophthora denticulispora G.L. Barron AN E
Rotiferophthora globispora G.L. Barron AN E
Rotiferophthora guttulaspora (G.L. Barron) G.L. Barron AN OU
Rotiferophthora humicola (G.L. Barron) G.L. Barron AN OU
Rotiferophthora intermedia (G.L. Barron) G.L. Barron AN OU
Rotiferophthora microspora (G.L. Barron) G.L. Barron AN OU
Rotiferophthora ovispora (G.L. Barron) G.L. Barron AN OU
Rotiferophthora reniformis (G.L. Barron) G.L. Barron AN OU
Rotiferophthora rotiferorum (G.L. Barron) G.L. Barron AN OU
Rotiferophthora tagenophora (Drechsler) G.L. Barron AN OU
Rotiferophthora turbinaspora (G.L. Barron) G.L. Barron AN OU
Rotiferophthora zeaspora (G.L. Barron) G.L. Barron AN OU
Torrubiella arachnophila (J.R. Johnst.) Mains U
Torrubiella confragosa Mains
Torrubiella gibellulae Petch
Torrubiella lloydii (Mains) Rossman
Torrubiella sublintea Petch
Torrubiella tomentosa Pat.
HYPOCREACEAE
Acrostalagmus obovatus Drechsler AN A
Arachnocrea scabrida Yoshim. Doi U
Cladobotryum apiculatum (Tubaki) W. Gams & Hooz. AN A
Cladobotryum multiseptatum de Hoog AN U
Cladobotryum verticillatum (Link) S. Hughes AN A
Gliocladium sp. sensu Latch, M.J. Chr. & Samuels AN A
Hypocrea ascoboloides Rehm
Hypocrea atrogelatinosa Dingley E
Hypocrea atroviridis Dodd, Lieckf. & Samuels
Hypocrea aureoviridis Plowr. & Cooke U
Hypocrea austrokoningii Samuels & Druzhinina
Hypocrea carnea Kalchbr. & Cooke
Hypocrea cerebriformis Berk.
Hypocrea citrina (Pers.) Fr.
Hypocrea colensoi Lloyd
Hypocrea coprosma Dingley E
Hypocrea crassa P. Chaverri & Samuels U
Hypocrea cremea P. Chaverri & Samuels AN
Hypocrea dingleyae Samuels & Dodd E
Hypocrea dorotheae Samuels & Dodd
Hypocrea farinosa Berk. & Broome
Hypocrea gelatinosa (Tode) Fr.
Hypocrea hunua Dingley E
Hypocrea koningii Lieckf., Samuels & W. Gams
Hypocrea lactea (Fr.) Fr. OU
Hypocrea lacuwombatensis B.S. Lu, Druzhinina & Samuels E
Hypocrea lixii Pat.
Hypocrea lutea (Tode) Petch A
Hypocrea macrospora Dingley E
Hypocrea manuka Dingley E
Hypocrea megalosulphurea Yoshim. Doi OU
Hypocrea nebulosa Massee
Hypocrea novae-zelandiae Samuels & Petrini E
Hypocrea orientalis Samuels & Petrini
Hypocrea pachybasioides Yoshim. Doi
Hypocrea patella Cooke & Peck
Hypocrea pseudokoningii Samuels & Petrini
Hypocrea rufa (Pers.) Fr. A
Hypocrea semiorbis (Berk.) Berk.
Hypocrea sp. Vd 2 Jaklitsch et al. (nom. ined.) U
Hypocrea stellata B.S. Lu, Druzhinina & Samuels E
Hypocrea sulfurella Kalchbr. & Cooke
Hypocrea sulphurea (Schwein.) Sacc. OU
Hypocrea tawa Dingley
Hypocrea toro Dingley E
Hypocrea vinosa Cooke
Hypocrea virens P. Chaverri, Samuels & E.L. Stewart
Hypocrea viridescens Jaklitsch & Samuels
Hypocreopsis amplectens T.W. May & P.R. Johnst.
Hypomyces armeniacus Tul. & C. Tul.
Hypomyces aurantius (Pers.) Tul. & C. Tul.
Hypomyces badius Rogerson & Samuels
Hypomyces boletiphagus Rogerson & Samuels
Hypomyces chlorinigenus Rogerson & Samuels
Hypomyces chrysospermus Tul. & C. Tul.
Hypomyces chrysostomus Berk. & Broome
Hypomyces dactylarioides G.R.W. Arnold E
Hypomyces fulgens (Fr.) P. Karst.
Hypomyces geoglossi Ellis & Everh. U
Hypomyces lateritius (Fr.) Tul. & C. Tul.
Hypomyces leotiicola Rogerson & Samuels
Hypomyces microspermus Rogerson & Samuels
Hypomyces novae-zealandiae Dingley E
Hypomyces pallidus Petch
Hypomyces papulasporae Rogerson & Samuels var. *papulasporae* E
Hypomyces petchii G.R.W. Arnold E
Hypomyces polyporinus Peck
Hypomyces rosellus (Alb. & Schwein.) Tul. & C. Tul.
Hypomyces semitranslucens G.R.W. Arnold
Hypomyces sibirinae Rogerson & Samuels
Hypomyces subiculosus (Berk. & M.A. Curtis) Höhn.
Hypomyces tremellicola (Ellis & Everh.) Rogerson
Hypomyces violaceus (J.C. Schmidt) Tul. & C. Tul.
Mycogone novae-zelandiae Matsush. AN E
Mycogone perniciosa (Magnus) ?Costantin & L.M. Dufour 1892 OR ?Delacr. AN U
Mycogone rosea Link AN A
Pseudohypocrea citrinella (Ellis) Yoshim. Doi U
Sepedonium ampullosporum Damon AN
Sepedonium chalcipori Helfer AN U
Sepedonium laevigatum Sahr & Ammer AN U
Sphaerostilbella aureonitens (Tul. & C. Tul.) Seifert, Samuels & W. Gams
Sphaerostilbella berkeleyana (Plowr. & Cooke) Samuels & Cand.
Sphaerostilbella lutea (Henn.) Sacc. & D. Sacc.
Sphaerostilbella novae-zelandiae Seifert, Samuels & W. Gams E
Stephanoma tetracoccum Zind.-Bakker AN
Trichoderma asperellum Samuels, Lieckf. & Nirenberg. AN
Trichoderma fertile Bissett AN
Trichoderma hamatum (Bonord.) Bainier AN
Trichoderma sp. DAOM 175924 Hatvani et al. (nom. ined.) U
Trichoderma sp. Ve Jaklitsch et al. (nom. ined.) U
NECTRIACEAE
Albonectria albosuccinea (Pat.) Rossman & Samuels OU

Albonectria rigidiuscula (Berk. & Broome) Rossman & Samuels
Aphanocladium album (Preuss) W. Gams AN
Aphanocladium aranearum (Petch) W. Gams AN A
Calonectria diploa (Berk. & M.A. Curtis) Wollenw. A
Calonectria kyotensis Terash. A
Calonectria morganii Crous, Alfenas & M.J. Wingf. A
Calonectria ochraceopallida (Berk. & Broome) Sacc. A
Calonectria pauciramosa C.L. Schoch & Crous A
Calonectria spathiphylli El-Gholl, J.Y. Uchida, Alfenas, T.S. Schub., Alfieri & A.R. Chase A
Corallomycetella repens (Berk. & Broome) Rossman & Samuels
Cosmospora aurantiicola (Berk. & Broome) Rossman & Samuels A
Cosmospora chaetopsinae (Samuels) Rossman & Samuels
Cosmospora chaetopsinae-penicillatae (Samuels) Rossman & Samuels
Cosmospora consors (Ellis & Everh.) Rossman & Samuels
Cosmospora dingleyae Lowen E
Cosmospora diploa (Berk. & M.A. Curtis) Rossman & Samuels U
Cosmospora episphaeria (Tode) Rossman & Samuels
Cosmospora flammea (Tul. & C. Tul.) Rossman & Samuels
Cosmospora geastroides (Samuels) Rossman & Samuels OU
Cosmospora joca (Samuels) Rossman & Samuels OU
Cosmospora nothepisphaeria (Samuels) Rossman & Samuels E
Cosmospora obscura Lowen
Cosmospora pseudoflavoviridis (Lowen & Samuels) Rossman & Samuels E
Cosmospora purtonii (Grev.) Rossman & Samuels A OU
Cosmospora rickii (Rehm) Rossman & Samuels OU
Cosmospora vilior (Starbäck) Rossman & Samuels
Cosmospora wegeliniana (Rehm) Rossman & Samuels
Cylindrocarpon destructans (Zinssm.) Scholten var. *destructans* AN
Cylindrocarpon didymum (Harting) Wollenw. AN A
Cylindrocarpon ianthothele Wollenw. AN U
Cylindrocarpon macrosporum (Speg.) Deighton & Piroz. AN A
Cylindrocarpon obtusisporum (Cooke & Harkn.) Wollenw. A
Cylindrocarpon theobromicola C. Booth AN A
Cylindrocladiella camelliae (Venkataram. & C.S.V. Ram) Boesew. AN A
Cylindrocladiella novae-zelandiae (Boesew.) Boesew. AN A
Cylindrocladiella parva (P.J. Anderson) Boesew. AN A
Cylindrocladium pacificum J.C. Kang, Crous & C.L. Schoch AN A
Cylindrocladium pseudonaviculatum Crous, J.Z. Groenew. & C.F. Hill AN A
Flagellospora penicillioides Ingold AN F
Fusarium anthophilum (A. Braun) Wollenw. AN A
Fusarium aquaeductuum (Radlk. & Rabenh.) Lagerh. AN A F
Fusarium arthrosporioides Sherb. AN A
Fusarium bulbicola Nirenberg & O'Donnell AN A
Fusarium camptoceras Wollenw. & Reinking AN A
Fusarium cerealis (Cooke) Sacc. A
Fusarium chlamydosporum Wollenw. & Reinking var. *chlamydosporum* AN A
— var. *fuscum* Gerlach AN A
Fusarium compactum (Wollenw.) W.L. Gordon AN A
Fusarium concolor Reinking AN A
Fusarium cortaderiae O'Donnell, T. Aoki, Kistler & Geiser AN A
Fusarium culmorum (W.G. Sm.) Sacc. AN A
Fusarium elongatum Cooke AN
Fusarium flocciferum Corda AN A
Fusarium graminum Corda AN A
Fusarium heterosporum Nees & T. Nees var. *heterosporum* AN A
— var. *congoense* (Wollenw.) Wollenw. AN A
Fusarium lateritium var. *longum* Wollenw. AN A
Fusarium limonis Briosi AN A
Fusarium longisporum Cooke & Massee AN A
Fusarium merismoides Corda AN A
Fusarium oxysporum Schltdl. var. *oxysporum* AN A
— var. *aurantiacum* (Link) Wollenw. AN A
— f.sp. *asparagi* sensu S.I. Cohen AN A
— f.sp. *basilici* Tamietti & Matta A
— f.sp. *batatas* (Wollenw.) W.C. Snyder & H.N. Hansen AN A OU
— f.sp. *callistephi* (Beach) W.C. Snyder & H.N. Hansen AN A
— f.sp. *cattleyae* Foster A OU
— f.sp. *cepae* (Hanzawa) W.C. Synder & H.N. Hansen AN A
— f.sp. *conglutinans* (Wollenw.) W.C. Snyder & H.N. Hansen AN A
— f.sp. *cucumerinum* J.H. Owen AN A
— f.sp. *cyclaminis* Gerlach AN A
— f.sp. *dianthi* (Prill. & Delacr.) W.C. Snyd. & Hansen AN A
— f.sp. *gladioli* (Massey) W.C. Synder & H.N. Hansen AN A
— f.sp. *hebes* R.D. Raabe (nom. inv.) OU
— f.sp. *lilii* Imle A
— f.sp. *lini* (Bolley) W.C. Snyder & H.N. Hansen AN A
— f.sp. *lycopersici* (Sacc.) W.C. Snyder & H.N. Hansen AN A
— f.sp. *matthiolae* K.F. Baker AN A
— f.sp. *melonis* W.C. Snyder & H.N. Hansen AN A
— f.sp. *narcissi* W.C. Snyder & H.N. Hansen AN A
— f.sp. *niveum* (E.F. Sm.) W.C. Snyder & H.N. Hansen AN A
— f.sp. *perniciosum* (Hepting) Toole AN A
—f.sp. *phaseoli* J.B. Kendr. & W.C. Snyder OU
— f.sp. *pisi* (C.J.J. Hall) W.C. Snyder & H.N. Hansen AN A
Fusarium pallidoroseum (Cooke) Sacc. AN A
Fusarium phyllophilum Nirenberg & O'Donnell AN A
Fusarium poae (Peck) Wollenw. AN A
Fusarium proliferatum (Matsush.) Nirenberg ex Gerlach & Nirenberg AN A
Fusarium pseudograminearum O'Donnell & T. Aoki AN A
Fusarium redolens Wollenw. AN A
Fusarium sacchari (E.J. Butler) W. Gams var. *sacchari* AN A
Fusarium semitectum Berk. & Ravenel var. *semitectum* AN A
— var. *majus* Wollenw. AN A
Fusarium solani var. *coeruleum* (Lib. ex Sacc.) C. Booth AN A
— f.sp. *phaseoli* (Burkh.) W.C. Snyder & H.N. Hansen AN A
— f.sp. *radicicola* (Wollenw.) W.C. Snyder & H.N. Hansen AN A
Fusarium sporotrichioides Sherb. AN A
Fusarium sterilihyphosum Britz, Marasas & M.J. Wingf. AN A
Fusarium trichothecioides Wollenw. AN A
Fusarium urticearum (Corda) Sacc. AN A
Fusarium vasinfectum G.F. Atk. AN A
Fusarium venenatum Nirenberg AN A
Gibberella acuminata Wollenw. A
Gibberella avenacea R.J. Cook A
Gibberella baccata (Wallr.) Sacc.
Gibberella cyanea (Sollm.) Wollenw. U
Gibberella cyanogena (Desm.) Sacc. A
Gibberella flacca (Wallr.) Sacc. A
Gibberella fujikuroi (Sawada) Wollenw. var. *fujikuroi* A
— var. *subglutinans* E.T. Edwards A
Gibberella intricans Wollenw. A
Gibberella macrolopha (Syd.) Dingley E
Gibberella pulicaris (Fr.) Sacc. var. *pulicaris* A
— var. *minor* Wollenw. A
Gibberella stilboides W.L. Gordon ex C. Booth A
Gibberella tricincta El-Gholl, McRitchie, Schoult. & Ridings A
Gibberella tumida P.G. Broadh. & P.R. Johnst. A
Gibberella zeae (Schwein.) Petch A
Gliocephalotrichum bulbilium J.J. Ellis & Hesselt. AN U
Glionectria tenuis Crous & C.L. Schoch U
Haematonectria haematococca (Berk. & Broome) Samuels & Nirenberg var. *haematococca* A
Haematonectria illudens (Berk.) Samuels & Nirenberg A
Haematonectria ipomoeae (Halst.) Samuels & Nirenberg A
Hyaloflorea ramosa Bat. & H. Maia AN
Lanatonectria flocculenta (Henn. & E. Nyman) Samuels & Rossman
Lisea nemorosa Sacc. OU
Mariannaea elegans (Corda) Samson var. *elegans* AN A
Moeszia cylindroides Bubák AN
Nalanthamala vermoesenii (Biourge) Schroers AN A
Nectria aurantiaca (Tul.) Jacz. A
Nectria austroradicicola Samuels & Brayford E
Nectria byssophila Rossman
Nectria calami (Henn. & E. Nyman) Rossman A
Nectria chaetostroma Ellis & T. Macbr. OU
Nectria cinnabarina (Tode) Fr. A
Nectria cucurbitula (Tode) Fr. A
Nectria egmontensis Dingley E
Nectria fibropapillata Samuels
Nectria flammeola Weese
Nectria flavolanata Berk. & Broome U
Nectria flavoviridis (Fuckel) Wollenw.
Nectria foliicola Berk. & M.A. Curtis
Nectria fragilis Dingley E
Nectria haematococca var. *brevicona* (Wollenw.) Gerlach A
— f.sp. *cucurbitae* (W.C. Snyder & H.N. Hansen) Dingley A
Nectria hoheriae Dingley E
Nectria lichenicola (Ces.) Sacc.
Nectria lugdunensis J. Webster U F
Nectria magnusiana Rehm ex Sacc.
Nectria manuka Dingley E
Nectria nothofagi Dingley (nom. ined.) E
Nectria novae-zealandiae (Dingley) Rossman E
Nectria otagensis Linds. E
Nectria parilis Syd. OU
Nectria perpusilla Mouton
Nectria plagianthi Dingley E
Nectria pseudopeziza (Desm.) Rossman A
Nectria pseudotrichia Berk. & M.A. Curtis
Nectria pyrrhochlora Auersw. U
Nectria quisquiliaris Cooke
Nectria rishbethii C. Booth
Nectria ruapehu Dingley E
Nectria sanguinea (Bolt.) Fr.

Nectria sinopica (Fr.) Fr. A
Nectria stenospora Berk. & Broome
Nectria sulcispora Petch
Nectria vulpina (Cooke) Ellis
Nectricladiella infestans Crous & C.L. Schoch A
Neonectria coccinea (Pers.) Rossman & Samuels A
Neonectria discophora (Mont.) Mantiri & Samuels var. *discophora*
— var. *rubi* (Osterw.) Brayford & Samuels A
Neonectria ditissma (Tul. & C. Tul.) Samuels & Rossman A
Neonectria fuckeliana (C. Booth) Castl. & Rossman A
Neonectria punicea (J.C. Schmidt) Castl. & Rossman
Pleurocolla compressa (Ellis & Everh.) Diehl AN
Pseudonectria rousseliana (Mont.) Wollenw. A
Septofusidium herbarum (A.H.S. Br. & G. Sm.) Samson AN
Volutella ciliata (Alb. & Schwein.) Fr. AN
NIESSLIACEAE
Niesslia exosporioides (Desm.) G. Winter
Niesslia fuegiana (Speg.) W. Gams
Niesslia subiculosa Syd.
Trichosphaerella decipiens E. Bommer, M. Rousseau & Sacc. A
Trichosphaerella tuberculata Samuels E
Valetoniella crucipila Höhn.
OPHIOCORDYCIPITACEAE
Haptocillium balanoides (Drechsler) Zare & W. Gams U
Hirsutella aphidis Petch AN A
Hirsutella citriformis Speare AN
Hirsutella eleutheratorum (Nees) Petch AN A
Hirsutella rhossiliensis Minter & B.L. Brady AN U
Hirsutella saussurei (Cooke) Speare AN A
Hirsutella subulata Petch AN A
Ophiocordyceps aphodii (Mathieson) G.H. Sung, J.M. Sung, Hywel-Jones & Spatafora A
Ophiocordyceps clavulata (Schwein.) Petch U
Ophiocordyceps dipterigena (Berk. & Broome) G.H. Sung, J.M. Sung, Hywel-Jones & Spatafora OU
Ophiocordyceps dovei (Rodway) G.H. Sung, J.M. Sung, Hywel-Jones & Spatafora
Ophiocordyceps gracilis (Grev.) G.H. Sung, J.M. Sung, Hywel-Jones & Spatafora
Ophiocordyceps michiganensis (Mains) G.H. Sung, J.M. Sung, Hywel-Jones & Spatafora
Ophiocordyceps robertsii (Hook.) G.H. Sung, J.M. Sung, Hywel-Jones & Spatafora
Ophiocordyceps stylophora (Berk. & Broome) G.H. Sung, J.M. Sung, Hywel-Jones & Spatafora U
Polycephalomyces formosus Kobayasi AN U
Tolypocladium cylindrosporum W. Gams AN U
Tolypocladium extinguens Samson & Soares AN
Tolypocladium inflatum W. Gams AN

Order LULWORTHIALES
Haloguignardia tumefaciens (Cribb & J.W. Herb.) Cribb & J.W. Cribb M
LULWORTHIACEAE
Lulworthia floridana Meyers M
Lulworthia fucicola G.K. Sutherl. M
Lulworthia purpurea (I.M. Wilson) T.W. Johnson M
Lulworthia uniseptata Nakagiri M
Zalerion arboricola Buczacki AN U
SPATHULOSPORACEAE
Spathulospora calva Kohlm. M
Spathulospora lanata Kohlm. M
Spathulospora phycophila A.R. Caval. & T.W. Johnson M
MAGNAPORTHACEAE
Buergenerula zelandica McKenzie E
Clasterosporium flagellatum (Syd. & P. Syd.) M.B. Ellis
Gaeumannomyces cylindrosporus D. Hornby, Slope, Gutter. & Sivan. A
Gaeumannomyces graminis (Sacc.) Arx & D.L. Olivier var. *graminis* A
— var. *avenae* (E.M. Turner) Dennis A
— var. *tritici* J. Walker A
Magnaporthe grisea (T.T. Hebert) M.E. Barr A
Mycoleptodiscus indicus (V.P. Sahni) B. Sutton AN U
Mycoleptodiscus terrestris (Gerd.) Ostaz. AN U
Pyricularia caricis G. Arnaud (nom. inv.)
Pyricularia cortaderiae McKenzie AN E
Pyricularia zizaniicola Hashioka OU

Order MELANOSPORALES
CERATOSTOMATACEAE
Gonatobotrys simplex Corda AN A
Harzia acremonioides (Harz) Costantin AN
Melanospora brevirostris (Fuckel) Höhn. A
Melanospora caprina (Fr.) Sacc.
Melanospora lagenaria (Pers.) Fuckel
Melanospora parasitica (Tul.) Tul. & C. Tul.
Melanospora zamiae Corda
Microthecium geoporae (W. Oberm.) Höhn.
Persiciospora moreaui P.F. Cannon & D. Hawksw.
Syspastospora parasitica (Tul.) P.F. Cannon & D. Hawksw.

Order MICROASCALES
Cornuvesica falcata (E.F. Wright & Cain) Viljoen, M.J. Wingf. & K. Jacobs A
Sphaeronaemella filicina Cooke & Massee E
Sphaeronaemella fimicola Marchal A
CERATOCYSTIDACEAE
Ceratocystis autographa B.K. Bakshi A
Ceratocystis coerulescens (Münch) B.K. Bakshi A
Ceratocystis fimbriata Ellis & Halst. A
Ceratocystis musarum Riedl A
Ceratocystis paradoxa (Dade) C. Moreau A
Sporendocladia truncata (B. Sutton) M.J. Wingf. AN U
Thielaviopsis basicola (Berk. & Broome) Ferraris AN A
HALOSPHAERIACEAE
Appendichordella amicta (Kohlm.) R.G. Johnson, E.B.G. Jones & S.T. Moss M
Carbosphaerella leptosphaerioides I. Schmidt M
Ceriosporopsis halima Linder M
Ceriosporopsis tubulifera (Kohlm.) P.W. Kirk ex Kohlm. M
Chadefaudia corallinarum (P. Crouan & H. Crouan) E. Müll. & Arx M
Chadefaudia gymnogongri (Feldmann) Kohlm. M
Cirrenalia macrocephala (Kohlm.) Meyers & R.T. Moore AN M
Clavatospora longibrachiata (Ingold) Sv. Nilsson ex Marvanová & Sv. Nilsson AN F
Corollospora lacera (Linder) Kohlm. M
Corollospora maritima Werderm. M
Corollospora trifurcata (Höhnk) Kohlm. M
Haligena elaterophora Kohlm. M
Halosarpheia fibrosa Kohlm. & E. Kohlm. M
Halosarpheia trullifera (Kohlm.) E.B.G. Jones, S.T. Moss & Cuomo M
Halosphaeria appendiculata Linder M
Halosphaeria hamata (Höhnk) Kohlm. M
Halosphaeriopsis mediosetigera (Cribb & J.W. Cribb) T.W. Johnson M
Lignincola laevis Höhnk M
Marinospora calyptrata (Kohlm.) A.R. Caval. M
Nais inornata Kohlm. M
Nereiospora comata (Kohlm.) E.B.G. Jones, R.G. Johnson & S.T. Moss M
Nereiospora cristata (Kohlm.) E.B.G. Jones, R.G. Johnson & S.T. Moss M
Ondiniella torquata (Kohlm.) E.B.G. Jones, R.G. Johnson & S.T. Moss M
Remispora maritima Linder M
Remispora quadriremis (Höhnk) Kohlm. M
Remispora stellata Kohlm. M
MICROASCACEAE
Cephalotrichum microsporum (Sacc.) P.M. Kirk AN U
Cephalotrichum nanum (Ehrenb.) S. Hughes AN U
Cephalotrichum stemonitis (Pers.) Link AN U
Graphium calicioides (Fr.) Cooke & Massee AN
Graphium penicillioides Corda AN U
Graphium putredinis (Corda) S. Hughes AN U
Graphium tectonae C. Booth AN OU
Microascus brevicaulis S.P. Abbott
Microascus cirrosus Curzi A
Microascus manginii (Loubière) Curzi A
Microascus trigonosporus C.W. Emmons & B.O. Dodge A
Pithoascus schumacheri (E.C. Hansen) Arx A
Pseudallescheria boydii (Shear) McGinnis, A.A. Padhye & Ajello U
Scedosporium prolificans (Hennebert & B.G. Desai) E. Guého & de Hoog AN U
Scopulariopsis acremonium (Delacr.) Vuill. AN U
Scopulariopsis brumptii Salv.-Duval AN U
Scopulariopsis fusca Zach AN U
Trichurus gorgonifer Bainier AN U
Trichurus spiralis Hasselbr. AN U
Trichurus terrophilus Swift & Povah U

Order OPHIOSTOMATALES
OPHIOSTOMATACEAE
Grosmannia galeiformis (B.K. Bakshi) Zipfel, Z.W. Beer & M.J. Wingf. A
Grosmannia huntii (Rob.-Jeffr.) Zipfel, Z.W. Beer & M.J. Wingf. A
Grosmannia radiaticola (J.J. Kim, Seifert & G.H. Kim) Z.W. Beer & M.J. Wingf. A
Leptographium alethinum K. Jacobs, M.J. Wingf. & Uzunovic AN A
Leptographium bistatum J.J. Kim & G.H. Kim AN OU
Leptographium euphyes K. Jacobs & M.J. Wingf. AN A
Leptographium procerum (W.B. Kendr.) M.J. Wingf. AN A
Leptographium truncatum (M.J. Wingf. & Marasas) M.J. Wingf. AN A
Ophiostoma brevicolle (R.W. Davidson) de Hoog & R.J. Scheff. A
Ophiostoma coronatum (Olchow. & J. Reid) M. Villarreal U
Ophiostoma floccosum Math.-Käärik A
Ophiostoma ips (Rumbold) Nannf. A
Ophiostoma narcissi Limber A
Ophiostoma nigrocarpum (R.W. Davidson) de Hoog A
Ophiostoma novo-ulmi Brasier A
Ophiostoma piceae (Münch) Syd. & P. Syd. OU
Ophiostoma piliferum (Fr.) Syd. & P. Syd.
Ophiostoma pluriannulatum (Hedgc.) Syd. & P. Syd. A
Ophiostoma quercus (Georgev.) Nannf. A
Ophiostoma rostrocoronatum (R.W. Davidson & Eslyn) de Hoog & R.J.Scheff. A
Ophiostoma setosum Uzunovic, Seifert, S.H. Kim & C. Breuil A
Ophiostoma sp. A sensu G.H. Kim, J.J. Kim, Y.W. Lim & C. Breuil OU
Ophiostoma sp. E Uzunovic et al. (nom. ined.) A
Ophiostoma stenoceras (Robak) Nannf. A
Ophiostoma ulmi (Buisman) Nannf. A
Pesotum fragrans (Math.-Käärik) G. Okada &

Seifert AN A
Sporothrix schenckii Hektoen & C.F. Perkins AN U
Sporothrix sp. A sensu Schirp, R. Farrell & Kreber AN U
Sporothrix sp. B sensu Schirp, R. Farrell & Kreber AN U
Sporothrix sp. C sensu Schirp, R. Farrell & Kreber AN U
Verticicladiella serpens (Goid.) W.B. Kendr. AN U

Order PHYLLACHORALES
Lichenochora xanthoriae Triebel & Rambold
PHYLLACHORACEAE
Linochora aberrans Syd. AN E
Polystigma apophlaeae Kohlm. M
Schizochora calocarpa Syd. E
Sphaerodothella danthoniae (McAlpine) C.A. Pearce & K.D. Hyde
Telimena graminis (Höhn.) Theiss. & Syd. A
Trabutia nothofagi Syd. E
Trabutia sp. sensu P.R. Johnst.
PLECTOSPHAERELLACEAE
Plectosphaerella cucumerina (Lindf.) W. Gams A
Plectosporium alismatis (Oudem.) W.M. Pitt, W. Gams & U. Braun A
Verticillium alboatrum Reinke & Berthold AN A
Verticillium dahliae Kleb. AN A
Verticillium fungicola (Preuss) Hassebr. AN A
Verticillium insectorum (Petch) W. Gams AN U
Verticillium luteoalbum (Link) Subram. U
Verticillium nigrescens Pethybr. AN A
Verticillium sphaerosporum Goodey AN U
Verticillium theobromae (Turconi) E.W. Mason & S. Hughes AN A
Verticillium tricorpus I. Isaac AN A
Verticillium tubercularioides Speg. AN A

Order SORDARIALES
Brachysporiella gayana Bat. AN
Carpoligna pleurothecii F.A. Fernández & Huhndorf F
Edmundmasonia pulchra Subram. AN
Pleurothecium leptospermi J.A. Cooper AN E
Roselliniella coccocarpiae (Pat.) Matzer & R. Sant.
CEPHALOTHECACEAE
Phialemonium dimorphosporum W. Gams & W.B. Cooke AN A
CHAETOMIACEAE
Botryotrichum piluliferum Sacc. & Marchal AN U
Chaetomidium fimeti (Fuckel) Sacc. A
Chaetomium ampullare Chivers A
Chaetomium anguipilium L.M. Ames U
Chaetomium bostrychodes Zopf A
Chaetomium cochliodes Palliser U
Chaetomium crispatum Fuckel A
Chaetomium elatum Kunze
Chaetomium funicola Cooke U
Chaetomium globosum Kunze A
Chaetomium indicum Corda U
Chaetomium pannosum Wallr. A
Chaetomium pulchellum L.M. Ames A
Chaetomium spinosum Chivers A
Chaetomium spirale Zopf A
Chaetomium subaffine Sergeeva
Chaetomium trigonosporum (Marchal) Chivers A
Farrowia seminuda (L.M. Ames) D. Hawksw. U
Humicola grisea Traaen AN
Thielavia terrestris (Apinis) Malloch & Cain A
Trichocladium alopallonellum (Meyers & R.T. Moore) Kohlm. & Volkm.-Kohlm. M
Trichocladium asperum Harz AN A
Trichocladium macrosporum P.M. Kirk AN
Trichocladium novae-zelandiae S. Hughes AN E
Trichocladium opacum (Corda) S. Hughes AN U
Trichocladium uniseptatum (Berk. & Broome) S. Hughes & Piroz. AN
HELMINTHOSPHAERIACEAE
Ceratosporium fuscescens Schwein. AN
Ceratosporium rilstonii S. Hughes AN
Echinosphaeria cincinnata A.E. Bell E
Echinosphaeria medusa A.E. Bell & Mahoney E
Endophragmiella biseptata (Peck) S. Hughes AN
Endophragmiella boewei (J.L. Crane) S. Hughes AN F
Endophragmiella boothii (M.B. Ellis) S. Hughes AN
Endophragmiella dingleyae S. Hughes AN E
Endophragmiella ellisii S. Hughes AN
Endophragmiella novae-zelandiae S. Hughes AN E
Endophragmiella pinicola (M.B. Ellis) S. Hughes AN A
Endophragmiella tripartita S. Hughes AN E
Endophragmiella valdiviana (Speg.) S. Hughes AN
Spadicoides atra (Corda) S. Hughes AN
Spadicoides obovata (Cooke & Ellis) S. Hughes AN
Spadicoides sphaerosperma McKenzie AN E
LASIOSPHAERIACEAE
Apiosordaria verruculosa (C.N. Jensen) Arx & W. Gams A
Arnium cirriferum (Speg.) J.C. Krug & Cain A
Arnium hirtum (E.C. Hansen) N. Lundq. & J.C. Krug A
Arnium leporinum (Cain) N. Lundq. & J.C. Krug A
Arnium mendax N. Lundq. A
Arnium sudermanniae N. Lundq. A
Bombardia bombarda (Batsch) J. Schröt.
Bombardioidea stercoris (DC.) N. Lundq.
Cercophora albicollis N. Lundq.
Cercophora ambigua (Sacc.) R. Hilber
Cercophora californica (Plowr.) N. Lundq.
Cercophora coprophila (Fr.) N. Lundq. A
Cercophora mirabilis Fuckel
Cercophora samala Udagawa & T. Muroi
Cercophora silvatica N. Lundq.
Cercophora solaris (Cooke & Ellis) R. Hilber & O. Hilber U
Cercophora sulphurella (Sacc.) R. Hilber U
Immersiella caudata (Curr.) A.N. Mill. & Huhndorf U
Lasiosphaeria depilata Fuckel
Lasiosphaeria glabrata (Fr.) Munk (nom. inv.)
Lasiosphaeria moseri O. Hilber OU
Lasiosphaeria mucida (nom. ined.) U
Lasiosphaeria ovina (Pers.) Ces. & De Not
Lasiosphaeria sorbina (Nyl.) P. Karst. U
Lasiosphaeria vestita Cand., J. Fourn. & Magni A
Lasiosphaeris hirsuta (Fr.) A.N. Mill. & Huhndorf
Lasiosphaeris hispida (Tode) Clem.
Podospora aloides (Fuckel) J.H. Mirza & Cain A
Podospora anserina (Rabenh.) Niessl A
Podospora appendiculata (Auersw. ex Niessl) Niessl A
Podospora austrohemisphaerica N. Lundq. A
Podospora bifida N. Lundq. A
Podospora communis (Speg.) Niessl A
Podospora conica (Fuckel) A.E. Bell & Mahoney A
Podospora curvicolla (G. Winter) Niessl A
Podospora curvula (de Bary) Niessl A
Podospora curvuloides Cain A
Podospora dactylina N. Lundq. A
Podospora dakotensis (Griffiths) J.H. Mirza & Cain A
Podospora decipiens (G. Winter) Niessl A
Podospora dolichopodalis J.H. Mirza & Cain A
Podospora ellisiana (Griffiths) J.H. Mirza & Cain A
Podospora excentrica N. Lundq. A
Podospora fimiseda (Ces. & De Not.) Niessl A
Podospora gigantea J.H. Mirza & Cain A
Podospora globosa (Massee & E.S. Salmon) Cain A
Podospora glutinans (Cain) Cain A
Podospora hyalopilosa (R. Stratton) Cain A
Podospora intestinacea N. Lundq. A
Podospora miniglutinans J.H. Mirza & Cain A
Podospora minuta (Fuckel) Niessl A
Podospora myriaspora (P. Crouan & H. Crouan) Niessl A
Podospora nannopodalis Cain A
Podospora pauciseta (Ces.) Traverso A
Podospora pectinata N. Lundq. A
Podospora perplexens (Cain) Cain A
Podospora pleiospora (G. Winter) Niessl A
Podospora pyriformis (A. Bayer) Cain A
Podospora setosa (G. Winter) Niessl A
Podospora similis (E.C. Hansen) Niessl A
Podospora tetraspora (G. Winter) Cain A
Podospora vesticola (Berk. & Broome) J.H. Mirza & Cain ex Kobayasi A
Ruzenia spermoides (Hoffm.) O. Hilber ex A.N. Mill. & Huhndorf
Zygopleurage zygospora (Speg.) Boedijn A
Zygospermella insignis (Mouton) Cain
SORDARIACEAE
Asordaria arctica (Cain) Arx & Guarro A
Asordaria humana (Fuckel) Arx & Guarro A
Gelasinospora adjuncta Cain U
Gelasinospora tetrasperma Dowding U
Neurospora discreta D.D. Perkins & N.B. Raju
Neurospora intermedia F.L. Tai
Neurospora sitophila Shear & B.O. Dodge
Neurospora tetrasperma Shear & B.O. Dodge
Sordaria fimicola (Desm.) Ces. & De Not. A
Sordaria lappae Potebnia A
Sordaria macrospora Auersw. A
Sordaria superba De Not. A

Order TRICHOSPHAERIALES
Khuskia oryzae H.J. Huds. A
Nigrospora sacchari (Speg.) E.W. Mason AN A
Nigrospora sphaerica (Sacc.) E.W. Mason AN A
TRICHOSPHAERIACEAE
Brachysporium dingleyae S. Hughes AN A
Brachysporium novae-zelandiae S. Hughes AN E
Cryptadelphia polyseptata Réblová & Seifert OU
Malacosphaeria scabrosa Syd. E
Trichosphaeria pilosa (Pers.) Fuckel

Order XYLARIALES
Dinemasporium graminum var. *strigosulum* P. Karst. AN OU
Fasciatispora nypae K.D. Hyde M
Leiosphaerella cocoes (Petch) Samuels & Rossman A
Microdochium bolleyi (R. Sprague) de Hoog & Herm.-Nijh. AN A
Microdochium dimerum (Penz.) Arx AN A
Microdochium panattonianum (Berl.) B. Sutton, Galea & T.V. Price AN A
Microdochium queenslandicum Matsush. AN A
Monographella nivalis (Schaffnit) E. Müll. A
Monographella passiflorae Samuels, E. Müll. & Petrini A
Oxydothis opaca (Berk.) K.D. Hyde
Oxydothis selenosporellae Samuels & Rossman E
Phaeotrichosphaeria britannica Sivan. U
Phaeotrichosphaeria hymenochaeticola Sivan. E
Phaeotrichosphaeria minor A.I. Romero & Carmarán
Phomatospora dinemasporium J. Webster A
AMPHISPHAERIACEAE
Amphisphaeria multipunctata (Fuckel) Petr. A
Diploceras hypericinum (Ces.) Died. AN A
Discosia novae-zelandiae Nag Raj AN E
Discostroma callistemonis (H.J. Swart) Sivan. A
Discostroma corticola (Fuckel) Brockmann A
Discostroma leptospermi (H.J. Swart) Sivan.
Discostroma stoneae (H.J. Swart) Sivan. A
Lepteutypa cupressi (Nattras, C. Booth & B.

Sutton) H.J. Swart A
Lepteutypa podocarpi (Butin) Aa
Monochaetia concentrica (Berk. & Broome) Sacc. & D. Sacc. AN A
Monochaetia monochaeta (Desm.) Allesch. AN U
Monochaetia saccardoi (Speg.) Allesch. AN A OU
Pestalosphaeria hansenii Shoemaker & J.A. Simpson A
Pestalosphaeria leucospermi Samuels, E. Müll. & Petrini A
Pestalotia brassicae Guba AN A
Pestalotia photiniae Thüm. AN A
Pestalotia vaccinii (Shear) Guba AN A
Pestalotia watsoniae Verwoerd & Dippen. AN A
Pestalotiopsis adusta (Ellis & Everh.) Steyaert AN A
Pestalotiopsis antenniformis (B.J. Murray) Y.X. Chen AN
Pestalotiopsis clavispora (G.F. Atk.) Steyaert AN A
Pestalotiopsis disseminata (Thüm.) Steyaert AN A
Pestalotiopsis funerea (Desm.) Steyaert AN A
Pestalotiopsis karstenii (Sacc. & P. Syd.) Steyaert AN A
Pestalotiopsis leucopogonis Nag Raj AN
Pestalotiopsis maculans (Corda) Nag Raj AN A
Pestalotiopsis palmarum (Cooke) Steyaert AN A
Pestalotiopsis sp. 'desmoschoenus' (nom. ined.) E
Pestalotiopsis stevensonii (Peck) Nag Raj AN A
Pestalotiopsis sydowiana (Bres.) B. Sutton AN A
Pestalotiopsis uvicola (Speg.) Bissett AN U
Pestalotiopsis versicolor (Speg.) Steyaert AN A
Sarcostroma arbuti (Bonar) M. Morelet A
Sarcostroma brevilatum (H.J. Swart & D.A. Griffiths) Nag Raj AN A
Sarcostroma grevilleae (Loos) M. Morelet AN A
Sarcostroma hakeae (B. Sutton) M. Morelet AN A
Sarcostroma kennedyae (McAlpine) M. Morelet AN A
Sarcostroma mahinapuense Gadgil & M. Dick AN A
Seimatosporium salicinum (Corda) Nag Raj AN U
Seimatosporium sp. 'desmoschoenus' (nom. ined.) E
Seiridium cardinale (W.W. Wagener) B. Sutton & I.A.S. Gibson AN A
Seiridium eucalypti Nag Raj AN A
Sporocadus pestalozzioides (Sacc.) M. Morelet AN A
Sporocadus rhododendri (Schwein.) M. Morelet AN A
Sporocadus vaccinii (Fuckel) M. Morelet AN A
Truncatella angustata (Pers.) S. Hughes AN A
Truncatella conorum-piceae (Tubeuf) Steyaert A
Truncatella laurocerasi (Westend.) Steyaert AN A
Truncatella sp. 'desmoschoenus' (nom. ined.) E
Zetiasplozna thuemenii (Speg.) Nag Raj AN U
CLYPEOSPHAERIACEAE
Clypeosphaeria stevensii Syd.
DIATRYPACEAE
Diatrype aurantii (De Not.) Sacc. A
Diatrype bullata (Hoffm.) Fr. A
Diatrype disciformis (Hoffm.) Fr. OU
Diatrype elliptica Cooke & Massee E
Diatrype flavovirens (Pers.) Fr. A
Diatrype glomeraria Berk. E
Diatrype microstega Ellis & Everh.
Diatrype princeps Penz. & Sacc.
Diatrype stigma (Hoffm.) Fr.
Eutypa lata (Pers.) Tul. & C. Tul. A
Eutypella dissepta (Fr.) Rappaz A
Eutypella juglandina (Cooke & Ellis) Sacc. A
Eutypella leprosa (Pers. ex Fr.) Berl. OU
Eutypella prunastri (Pers.) Sacc. OU
Eutypella quaternata (Pers.) Rappaz OU
Eutypella scoparia (Schwein.) Ellis & Everh.
Eutypella stellulata (Fr.) Sacc. A
Eutypella vitis (Schwein.) Ellis & Everh. A
Peroneutypa heteracantha (Sacc.) Berl. U
HYPONECTRIACEAE
Beltraniella portoricensis (F. Stevens) Piroz. & S.D. Patil AN
Physalospora cupressi (Berk. & M.A. Curtis) Sacc. A
Physalospora vaccinii (Shear) Arx & E. Mull. A
Pseudomassaria islandica (Johanson) M.E. Barr OU
Pseudomassaria sepincoliformis (De Not.) Arx A
IODOSPHAERIACEAE
Iodosphaeria phyllophila (Mouton) Samuels, E. Müll. & Petrini
Iodosphaeria ripogoni Samuels, E. Müll. & Petrini E
XYLARIACEAE
Annulohypoxylon annulatum (Schwein.) Y.M. Ju, J.D. Rogers & H.M. Hsieh OU
Annulohypoxylon archeri (Berk.) Y.M. Ju, J.D. Rogers & H.M. Hsieh
Annulohypoxylon bovei (Speg.) Y.M. Ju, J.D. Rogers & H.M. Hsieh var. *bovei*
— var. *microsporum* (J.H. Mill.) Y.M. Ju, J.D. Rogers & H.M. Hsieh
Annulohypoxylon moriforme (Henn.) Y.M. Ju, J.D. Rogers & H.M. Hsieh
Annulohypoxylon nothofagi (Y.M. Ju & J.D. Rogers) Y.M. Ju, J.D. Rogers & H.M. Hsieh E
Annulohypoxylon truncatum (Schwein.) Y.M. Ju, J.D. Rogers & H.M. hsieh OU
Anthostomella kapiti Whitton, K.D. Hyde & McKenzie E
Anthostomella lucens Sacc.
Anthostomella ludoviciana Ellis & Langl.
Anthostomella manawatu Whitton, K.D. Hyde & McKenzie E
Anthostomella okatina Whitton, K.D. Hyde & McKenzie E
Anthostomella phaeosticta (Berk.) Sacc.
Anthostomella phormiicola (Cooke) Berl. & Voglino E
Anthostomella puiggarii Speg.
Anthostomella rehmii (Thüm.) Rehm
Anthostomella tenacis (Cooke) Sacc.
Anthostomella umbrinella (de Not.) Sacc. U
Astrocystis dealbata L.E. Petrini E
Astrocystis rachidis (Pat.) K.D. Hyde & J. Fröhl. OU
Biscogniauxia capnodes (Berk.) Y.M. Ju & J.D. Rogers var. *capnodes*
— var. *rumpens* (Cooke) Y.M. Ju & J.D. Rogers
Biscogniauxia zelandica Y.M. Ju & J.D. Rogers E
Camillea sagraeana (Mont.) Berk. & M.A. Curtis OU
Creosphaeria sassafras (Schwein.) Y.M. Ju, F. San Martín & J.D. Rogers
Daldinia bakeri Lloyd
Daldinia childiae J.D. Rogers & Y.M. Ju
Daldinia dennisii Wollw., J.A. Simpson & M. Stadler var. *dennisii*
— var. *microspora* Wollw. & M. Stadler E
Daldinia eschscholzii (Ehrenb.) Rehm OU
Daldinia fissa Lloyd OU
Daldinia grandis Child OU
Daldinia novae-zelandiae Wollw. & M. Stadler E
Daldinia sp. 1 sensu P.R. Johnst. et al.
Dicyma ovalispora (S. Hughes) Arx AN A
Dicyma pulvinata (Berk. & M.A. Curtis) Arx AN
Helicogermslita aucklandica (Rabenh.) L.E. Petrini E
Helicogermslita gisbornia L.E. Petrini E
Helicogermslita johnstonii L.E. Petrini
Helicogermslita mckenziei L.E. Petrini E
Hypocopra cataphracta J.C. Krug & Cain OU
Hypocopra equorum (Fuckel) G. Winter OU
Hypocopra keniensis J.C. Krug & Cain A
Hypocopra merdaria (Fr.) J. Kickx f. A
Hypocopra ornithophila Speg. A
Hypocopra stercoraria (Sowerby) Fuckel A
Hypoxylon allantoideum Cooke (nom. illegit.) E
Hypoxylon anthochroum Berk. & Broome
Hypoxylon aucklandiae Y.M. Ju & J.D. Rogers E
Hypoxylon carneum Petch
Hypoxylon chathamense Y.M. Ju & J.D. Rogers E
Hypoxylon cinnabarinum (Henn.) Y.M. Ju & J.D. Rogers
Hypoxylon conostomum (Mont.) Mont. OU
Hypoxylon crocopeplum Berk. & M.A. Curtis
Hypoxylon diatrypeoides Rehm OU
Hypoxylon dingleyae Y.M. Ju & J.D. Rogers
Hypoxylon fendleri Berk. ex Cooke
Hypoxylon fuscum (Pers.) Fr. OU
Hypoxylon glomeratum Cooke OU
Hypoxylon howeanum Peck
Hypoxylon hughesii Y.M. Ju & J.D. Rogers E
Hypoxylon hypomiltum Mont. OU
Hypoxylon hypophloeum (Berk. & Ravenel) J.H. Mill. OU
Hypoxylon investiens (Schwein.) M.A. Curtis OU
Hypoxylon nothofagicola Y.M. Ju & J.D. Rogers E
Hypoxylon novemexicanum J.H. Mill. OU
Hypoxylon perforatum (Schwein.) Fr.
Hypoxylon placentiforme Berk. & M.A. Curtis
Hypoxylon punctulatum (Berk. & Ravenel) Cooke OU
Hypoxylon rubiginosum (Pers.) Fr. OU
Hypoxylon samuelsii Y.M. Ju & J.D. Rogers
Hypoxylon semiimmersum Nitschke
Hypoxylon subcorticeum Y.M. Ju & J.D. Rogers E
Hypoxylon subcrocopeplum Y.M. Ju & J.D. Rogers E
Hypoxylon subrutiloides Y.M. Ju & J.D. Rogers E
Hypoxylon torrendii Bres.
Induratia apiospora Samuels, E. Müll. & Petrini E
Kretzschmaria clavus (Fr.) Sacc. OU
Kretzschmaria colensoi (Berk.) Sacc. E
Kretzschmaria deusta (Hoffm.) P.M.D. Martin OU
Kretzschmaria pavimentosa (Ces.) P.M.D. Martin
Kretzschmaria zelandica J.D. Rogers & Y.M. Ju E
Kretzschmariella culmorum (Cooke) Y.M.Ju & J.D.Rogers
Nemania bipapillata (Berk. & M.A. Curtis) Pouzar
Nemania caries (Schwein.) Y.M.Ju & J.D.Rogers
Nemania diffusa (Sowerby) Gray
Nemania maritima Y.M. Ju & J.D. Rogers M
Nemania serpens (Pers.) Gray var. *serpens*
— var. *colliculosa* (Schwein.:Fr.) Y.M. Ju & J.D. Rogers
Nodulisporium corticioides (Ferraris & Sacc.) S. Hughes AN
Penzigia arntzenii Theiss.
Penzigia discolor (Berk. & Broome) J.H. Mill.
Penzigia placenta Petch (nom. inv.)
Poronia oedipus (Mont.) Mont. A
Rosellinia aquila (Fr.) De Not. OU
Rosellinia arcuata Petch
Rosellinia bunodes (Berk. & Broome) Sacc. OU
Rosellinia callosa G. Winter OU
Rosellinia caudata Petch OU
Rosellinia chusqueae Pat.
Rosellinia colensoi Cooke E
Rosellinia communis L.E. Petrini E
Rosellinia corticium (Schwein.) Sacc. OU
Rosellinia dingleyae L.E. Petrini E
Rosellinia freycinetiae L.E. Petrini E
Rosellinia gisbornia L.E. Petrini E
Rosellinia hughesii L.E. Petrini E
Rosellinia johnstonii L.E. Petrini E
Rosellinia longispora Rick
Rosellinia mammiformis sensu (Pers.) Ces. & De Not. OU
Rosellinia mammoidea (Cooke) Sacc. E
Rosellinia necatrix Berl. ex Prill. A
Rosellinia nothofagi L.E. Petrini E
Rosellinia novae-zelandiae L.E. Petrini A
Rosellinia palmae L.E. Petrini E

Rosellinia pepo Pat. OU
Rosellinia pulveracea (Ehrenb.) Fuckel OU
Rosellinia radiciperda Massee A
Rosellinia rhopalostylidicola L.E. Petrini E
Rosellinia rosarum Niessl OU
Rosellinia samuelsii L.E. Petrini E
Rosellinia stenasca Rick
Rosellinia subiculata (Schwein.) Sacc. OU
Rosellinia victoriae Syd. & P. Syd.
Stilbohypoxylon novae-zelandiae L.E. Petrini
Virgaria nigra (Link) Nees AN
Xylaria anisopleura (Mont.) Fr.
Xylaria apiculata Cooke E
Xylaria arbuscula Sacc.
Xylaria avellana (Ces.) P.M.D. Martin OU
Xylaria berkeleyi Mont. OU
Xylaria castorea Berk. E
Xylaria corniformis (Fr.) Fr. OU
Xylaria cubensis (Mont.) Fr.
Xylaria feejeensis (Berk.) Fr. OU
Xylaria furcata Fr. OU
Xylaria hypoxylon (L.) Grev. F
Xylaria luteostromata var. *macrospora* J.D. Rogers & Samuels E
Xylaria multiplex (Kunze) Fr.
Xylaria myosurus Mont. OU
Xylaria palmicola G. Winter
Xylaria polytricha Colenso E
Xylaria psamathos Boise U
Xylaria rimulata Lloyd
Xylaria schreuderiana Van der Byl
Xylaria sp. 1 sensu Rogers, J.D.; Samuels, G.J.
Xylaria sp. 2 sensu Rogers, J.D.; Samuels, G.J.
Xylaria theissenii var. *macrospora* J.D. Rogers & Samuels E
Xylaria tuberiformis Berk. E
Xylaria wellingtonensis J.D. Rogers & Samuels E
Xylaria zealandica Cooke E

SUBPHYLUM SACCHAROMYCOTINA
Class SACCHAROMYCETES
Order SACCHAROMYCETALES
Ambrosiozyma platypodis (J.M. Baker & Kreger) Van der Walt
Candida albicans (C.P. Robin) Berkhout AN A
Candida anatomiae (Zwillenb.) S.A. Mey. & Yarrow AN A
Candida berthetii Boidin, Pignal, Mermiér & Arpin AN A
Candida boidinii C. Ramírez AN A
Candida catenulata Diddens & Lodder AN A
Candida diversa Uden & H.R. Buckley AN A
Candida dublinionensis D.J. Sullivan, Westerneng, K.A. Haynes, Dés.E. Benn. & D.C. Coleman AN A
Candida glabrata (H.W. Anderson) S.A. Mey. & Yarrow AN A
Candida haemulonis (Uden & Kolip.) S.A. Mey. & Yarrow AN A
Candida inconspicua (Lodder & Kreger) S.A. Mey. & Yarrow AN A
Candida intermedia (Cif. & Ashford) Langeron & Guerra AN A
Candida membranifaciens (Lodder & Kreger) Wick. & K.A. Burton AN U
Candida oleophila Montrocher AN OU
Candida parapsilosis (Ashford) Langeron & Talice AN A
Candida pintolopesii var. *slooffiae* (Uden & Carmo Souza) Mend.-Hagler & Phaff AN A
Candida pinus (Lodder & Kreger) S.A. Mey. & Yarrow A OU
Candida railenensis C. Ramírez & A.E. González AN
Candida robusta Diddens & Lodder AN A
Candida rugosa (H.W. Anderson) Diddens & Lodder AN A
Candida sake (Saito & M. Ota) Uden & H.R. Buckley ex S.A. Mey. & Ahearn AN U
Candida stellata (Kroemer & Krumbholz) S.A. Mey. & Yarrow U
Candida tenuis Diddens & Lodder AN A
Candida tropicalis (Castell.) Berkhout AN A
Candida vini (J.N. Vallot ex Desm.) Uden & H.R. Buckley ex S.A. Mey. & Ahearn AN A
Candida zemplinina Sipiczki U
Candida zeylanoides (Castell.) Langeron & Guerra AN A
Cyniclomyces guttulatus (C.P. Robin) Van der Walt & D.B. Scott U
Debaryomyces carsonii (Phaff & E.P. Knapp) Y. Yamada, K. Maeda, I. Banno & Van der Walt
Debaryomyces hansenii (Zopf) Lodder & Kreger
Debaryomyces marama di Menna
Debaryomyces polymorphus (Klöcker) C.W. Price & Phaff
Debaryomyces subglobosus (Zach) Lodder & Kreger
Schizoblastosporion starkeyi-henricii Cif. AN
Yarrowia lipolytica (Wick., Kurtzman & E.A. Herrm.) Van der Walt & Arx
DIPODASCACEAE
Dipodascus ovetensis (Peláez & C. Ramírez) Arx U
Galactomyces geotrichum (E.E. Butler & L.J. Petersen) Redhead & Malloch A
ENDOMYCETACEAE
Endomyces fibuliger Lindner U
METSCHNIKOWIACEAE
Clavispora lusitaniae Rodr. Mir. A
Metschnikowia bicuspidata (Metschn.) T. Kamieski var. *bicuspidata*
— var. *chathamia* Fell & Pitt
Metschnikowia pulcherrima Pitt & M.W. Mill.
PICHIACEAE
Dekkera bruxellensis Van der Walt U
Issatchenkia orientalis Kudrjavzev
Issatchenkia terricola (Van der Walt) Kurtzman, M.J. Smiley & C.J. Johnson (nom. inv.)
Pichia anomala (E.C. Hansen) Kurtzman U
Pichia bimundalis (Wick. & Santa María) Kurtzman U
Pichia burtonii Boidin, Pignal, Lehodey, Vey & Abadie U
Pichia canadensis (Wick.) Kurtzman U
Pichia fermentans Lodder U
Pichia guilliermondii Wick. U
Pichia holstii (Wick.) Kurtzman U
Pichia jadinii (Sartory, R. Sartory, J. Weill & J. Mey.) Kurtzman U
Pichia kluyveri Bedford (nom. inv.) U
Pichia membranifaciens (E.C. Hansen) E.C. Hansen U
Pichia norvegensis Leask & Yarrow U
Pichia ohmeri (Etchells & T.A. Bell) Kreger U
Pichia pastoris (Guilliermond.) Phaff U
Pichia trehalophila sensu Phaff, M.W. Mill. & J.F.T. Spencer U
SACCHAROMYCETACEAE
Arxiozyma telluris (Van der Walt) Van der Walt & Yarrow U
Kluyveromyces lactis (Boidin, Abadie, J.L. Jacob & Pignal) Van der Walt
Kluyveromyces marxianus (E.C. Hansen) Van der Walt
Kluyveromyces thermotolerans (Filippov) Yarrow
Saccharomyces bayanus Sacc. U
Saccharomyces cerevisiae Meyen ex E.C. Hansen
Saccharomyces exiguus Reess ex E.C. Hansen
Torulaspora delbrueckii (Lindner) Lindner A
Williopsis californica (Lodder) Arx
Williopsis saturnus (Klöcker) Zender var. *saturnus*
— var. *mrakii* (Wick.) Kurtzman
Zygosaccharomyces bailii (Lindner) Guilliermond. U
Zygosaccharomyces cidri (Legakis) Yarrow
Zygosaccharomyces rouxii (Boutroux) Yarrow A
SACCHAROMYCODACEAE
Hanseniaspora osmophila (Niehaus) Phaff, M.W. Mill. & Shifrine ex M.T. Sm.
Hanseniaspora uvarum (Niehaus) Shehata, Mrak & Phaff ex M.T. Sm. A
Hanseniaspora vineae Van der Walt & Tscheuschner A
Saccharomycodes ludwigii (E.C. Hansen) E.C. Hansen U

SUBPHYLUM TAPHRINOMYCOTINA
Class PNEUMOCYSTIDOMYCETES
Order PNEUMOCYSTIDALES
PNEUMOCYSTIDACEAE
Pneumocystis jirovecii Frenkel A

Class SCHIZOSACCHAROMYCETES
Order SCHIZOSACCHAROMYCETALES
SCHIZOSACCHAROMYCETACEAE
Schizosaccharomyces pombe Lindner U

Class TAPHRINOMYCETES
Order TAPHRINALES
TAPHRINACEAE
Taphrina betulina Rostr. A
Taphrina cornu-cervae Giesenh.
Taphrina deformans (Berk.) Tul. A
Taphrina populina (Fr.) Fr. A
Taphrina pruni (Fuckel) Tul. A
Taphrina wiesneri (Ráthay) Mix A

THIRTY-THREE

Phylum

BASIDIOMYCOTA

mushrooms, rusts, smuts, and kin

PETER K. BUCHANAN, ROSS E. BEEVER, ERIC. H. C. McKENZIE, BARBARA C. PAULUS, SHAUN R. PENNYCOOK, GEOFF S RIDLEY, MAHAJABEEN PADAMSEE, JERRY A. COOPER

Pholiota aurivella (Agaricomycetes: Agaricales), a native mushroom usually fruiting from decaying parts of living trees.
Peter K. Buchanan

Basidiomycota (basidiomycetes in a general sense) are characterised by structures called basidia, from which basidiospores develop externally following meiosis. For the most conspicuous basidiomycete groups, which form mushrooms and other kinds of large fruiting bodies (agarics, boletes, non-gilled macrofungi), the typical basidium is a single cell producing four haploid, unicellular, wind-dispersed basidiospores each at the tip of a sterigma. The growth stage is often mycelial and dikaryotic, mostly with characteristic crosswalls that have a barrel-shaped swelling around a central pore as seen under an electron microscope. These 'dolipore septa' usually prevent the passage of nuclei and mitochondria, and may either be simple in form or have a clamp connection. Clamp connections are found only in the Basidiomycota and have developed to carefully sort the two pairs of daughter nuclei as a hyphal tip cell is partitioned. The 'clamp' comprises a small hooked hyphal outgrowth that rejoins the hypha beyond the crosswall. One daughter nucleus moves into the hooked outgrowth and transfers via the hook to the proximal newly walled-off cell, while the other pair of nuclei is separated as the original crosswall forms.

A few described basidiomycetes are single-celled yeasts or have yeast-like states and some cause severe diseases in humans (e.g. *Cryptococcus*, yeast-forming species within the Tremellomycetes). Compared to members of the Ascomycota, fewer taxa have a known anamorph (asexual stage), and there has been a long tradition of not giving a separate name to anamorphic stages in the Basidiomycota.

The Basidiomycota is the second-largest fungal phylum (31,515 species globally; Kirk et al. 2008). About one-third of the species so far described for New Zealand (2837 species) are Basidiomycota.

The classification of basidiomycetes has undergone a period of major revision, with giant leaps in our understanding of evolutionary relationships (e.g. Bauer et al. 2006; Hibbett et al. 2007). Here, the more recent hierarchy of the *Dictionary of the Fungi* (Kirk et al. 2008) is followed instead of the system proposed by Cavalier-Smith (1998, 2001). In Basidiomycota, the clades formerly known as Basidiomycetes, Urediniomycetes, and Ustilaginomycetes (Kirk et al. 2001) are now respectively referred to as subphyla Agaricomycotina (e.g. agarics, boletes, non-gilled macrofungi), Pucciniomycotina (e.g. rusts), and Ustilaginomycotina (smuts) (Hibbett et al. 2007). The Entorrhizomycetes have also been included in Basidiomycota as a class *incertae sedis* (Hibbett et al. 2007). The three subphyla are separated on the basis of cell-wall composition and secondary structure of 5S RNA (Bauer et al. 2006). In the discussion below, the Pucciniomycotina and Ustilaginomycotina are treated together.

Subphylum Agaricomycotina

Class Agaricomycetes

The Agaricomycetes comprises two subclasses (Agaricomycetidae and Phallomycetidae) and 10 orders that are *incertae sedis* within the class (Hibbett et al. 2007). There are 2289 species in the Agaricomycetes in the end-chapter checklist. The class is treated here as two broadly based and easily recognised groups – agarics and boletes on the one hand and the non-gilled fungi on the other.

The agarics and boletes comprise an artificial collection of basidiomycetes that share certain conspicuous morphological characters. These include a generally soft and fleshy fruit body divided into a pileus and stipe, and an exposed hymenium either spread over the surface of radiating lamellae or gills in agarics or lining the inner surface of tubes in boletes (Ramsbottom 1923; Lancaster 1955). Previous studies of the 'gasteromycetes' – basidiomycetes with enclosed hymenia – have shown that they are a polyphyletic group, having evolved from a range of agaric and bolete ancestors (Thiers 1984; Moncalvo et al. 2002). For instance, species of *Thaxterogaster*, put in the gasteromycete genus *Secotium* by Cunningham (1944), are now firmly placed in the genus *Cortinarius* (Peintner et al. 2002).

The first full list of New Zealand agarics was compiled by Massee (1899), when he recorded 129 species. Horak's (1971) annotated checklist, which marked the start of modern agaricology in New Zealand, comprised 191 species and covered the publications of Stevenson (1962) and McNabb (1967, 1968, 1969b). Segedin's (1987) checklist recorded 612 species, and Segedin and Pennycook (2001) listed more than 800 species in 178 genera, including many gasteromycetes excluded from the earlier lists. The latter list formed the basis for the associated checklist. There has been a significant increase in the number of species since Massee's 1899 list. Most of this increase in known species numbers has been subsequent to the pioneering paper of Stevenson (1962) and provides an indication of the extent of alpha taxonomy still required. In addition, many well-represented genera that have not been critically studied or have accumulated names from a number of sources, are in need of taxonomic re-evaluation. The most recent key to agarics, boletes, and related fungi was by Horak (2001).

Mycena sp. (Agaricomycetes: Agaricales), a tiny bioluminescent species.

Barbara Paulus

Summary of New Zealand Basidiomycota diversity

Taxon	Described species + sub/infra specific taxa	Known undescribed/ undeterm. species	Estimated undiscovered species	Adventive species + sub/infra specific taxa	Endemic species+ sub/infra specific taxa	Endemic genera
AGARICOMYCOTINA	2,113+37	259	5,045	292+6	753+18	6
Agaricomycetes	2,030+35	259	5,000	282+6	740+17	6
Dacrymycetes	27+1	0	20	0	4+1	0
Tremellomycetes	49+1	0	20	9	7	0
Entorhizomycetes	7	0	5	1	2	0
PUCCINIOMYCOTINA	295+13	2	50	127+11	92+1	0
Agaricostilbomycetes	6	0	5	0	1	0
Atractiellomycetes	5	0	5	0	0	0
Classiculomycetes	1	0	5	0	0	0
Microbotryomyctes	30	0	20	6	0	0
Puccinomycetes	253+13	2	15	121+11	91+1	0
USTILAGINOMYCOTINA	107	2	15	62	16	0
Exobasidiomycetes	51	0	5	37	5	0
Ustilaginomycetes	56	2	10	25	11	0
Totals	**2,515+50**	**263**	**5,110**	**481+17**	**861+19**	**6**

Summary of New Zealand Basidiomycota diversity by major environment

Taxon	Marine	Fresh-water	Terrestrial
AGARICOMYCOTINA	3	3	2,366
PUCCINIOMYCOTINA	0	1	296
USTILAGINOMYCOTINA	0	0	109
Totals	3	4	2,771

The earth star *Geastrum triplex* (Agaricomycetes: Geastrales).
Ross E. Beever

The agarics and boletes include the majority of species that form ectomycorrhizal associations with *Kunzea*, *Leptospermum*, and *Nothofagus* (e.g. species of *Amanita*, *Cortinarius*, and *Russula*; see section on Mycorrhizal fungi, page 505). Levels of endemism are highest among these host-specialised taxa. Of 205 ectomycorrhizal agarics recorded on *Nothofagus*, about 90% are indigenous and most of these are considered to be endemic (McKenzie et al. 2000). In total, an estimated 30% of New Zealand agarics are introduced taxa. Cunningham (1944) estimated that 22% of gasteromycetes were endemic; this group also includes several ectomycorrhizal species. Other agarics and boletes grow saprobically on soil, litter, and wood (see section on Wood and litter-decaying fungi, page 506), contributing to decomposition of organic matter and nutrient cycling, while a few such as *Armillaria* spp. are pathogenic on plants and some, such as *Squamanita squarrulosa*, attack other agarics (Ridley 1988).

The non-gilled macrofungi (historically referred to as aphyllophorales) are circumscribed by the absence of gills. Fruiting bodies take various forms, such as a thin layer attached to the plant substrate with or without a reflexed margin, pileate (shelf-like) with direct lateral attachment or with a lateral to central stalk, club-shaped, cup-shaped, tubular, or coral-like. For convenience, common names for broad groups of these fungi are based on the surfaces from which basidia and spores develop. These may be polypore fungi (with an inner surface of vertical tubes), corticioid or crust fungi (a smooth or roughened surface attached to wood), clavarioid fungi (a simple or branched club), hydnoid fungi (teeth), or cantharelloid (folded) or cupuloid (cup-shaped) surface. However, these are entirely artificial groupings. For example, molecular studies, which focused on representatives of corticioid fungi, indicated that many group in clades that include agarics and boletes. Additional clades have also become apparent that currently comprise only resupinate forms; these include for example the atheloid, trechisporoid, and phleboid clades (Larsson et al. 2004; Binder et al. 2005). Results of some molecular studies are yet to be integrated in classification systems.

Members of this group are abundant in forests; over 220 species in 85 genera have been recorded on *Nothofagus* alone in New Zealand (McKenzie et al. 2000). Most species are saprobic on wood or litter. Several genera of non-gilled fungi form ectomycorrhizal associations with three host genera, *Kunzea*, *Leptospermum*, and *Nothofagus*, for example, the clavarioid genus *Ramaria* and the hydnoid genus *Phellodon*. It is likely that native members of other genera known to be ectomycorrhizal from other regions (e.g. the corticioid *Tomentella*) will be found to form similar associations in New Zealand.

Floristic treatments have been published for the polypores (Cunningham 1965), corticioids (Cunningham 1963), and clavarioid fungi (Petersen 1988). Hydnoid species were treated by Cunningham (1958, 1959), McNabb (1971), and Maas Geesteranus (1971). At all hierarchic levels, the taxonomy of non-gilled fungi has changed dramatically in the last decades. For example, 71% of Cunningham's corticioid taxa have been recombined in other genera or have been synonymised, the taxonomic position of 11 % of taxa are uncertain, and only 18% are currently retained as 'good' taxa. Given the changed species and generic concepts, additional New Zealand species with northern-hemisphere names may represent unrecognised endemic taxa. While generic concepts and species names have altered considerably since Cunningham's publications (Stalpers 1985; Stalpers & Buchanan 1991; Buchanan & Ryvarden 2000), these publications nevertheless remain invaluable for their diagnostic keys and illustrated descriptions of most recorded species.

Artomyces turgidus (Agaricomycetes: Russulales), a wood-decaying coral fungus.
Peter K. Buchanan

With the patchy state of knowledge of non-gilled fungi across the Southern Hemisphere, only tentative estimates of endemism are possible. On the basis of total numbers of indigenous and introduced species, Cunningham (1963, 1965) estimated that 15% of polypore fungi and 41% of corticioid fungi were endemic. It is likely that the level of endemism is an overestimate and that the actual level will prove to be lower when our knowledge of these groups has increased for both New Zealand and neighbouring Australia. For example, endemism among corticioid fungi from Patagonia, where increased research effort was placed in recent years, was estimated as 20% (Greslebin & Rajchenberg 2003). The monotypic clavarioid genus *Setigeroclavula* (Petersen 1988) may be the only currently recognised endemic genus of non-gilled macrofungi.

Class Tremellomycetes and Dacrymycetes: jelly fungi

The jelly fungi, traditionally known as heterobasidiomycetes or phragmobasidiomycetes, are a small group characterised by mostly gelatinous to rubbery fruiting bodies and septate basidia, in addition to unique features of the septal pore apparatus and spindle pole body (Wells 1994). However, the fruiting body form is variable, sometimes mimicking members of the non-gilled macrofungi, as in poroid (e.g. *Protomerulius*), hydnoid (e.g. *Pseudohydnum*), corticioid (e.g. *Exidiopsis*) or clavarioid (e.g. *Tremellodendropsis*) forms; some are also yeast-like (e.g. *Cryptococcus*).

Most occur on dead wood and are saprobic such as New Zealand's most conspicuous member of the jelly fungi, *Auricularia cornea* (wood ear; see section on Human interactions, page 501). Others may fruit on wood but are actually mycoparasitic on wood-inhabiting ascomycete and basidiomycete fungi (Zugmaier et al. 1994), for example, the edible *Tremella fuciformis* (white jelly fungus). Thirty-four New Zealand species of Tremellales were described by McNabb (1964, 1966, 1969a). In addition, the Dacrymycetales are a particularly well-studied order with a series of monographic papers of genera prepared by McNabb (e.g. McNabb 1973).

Subphyla Pucciniomycotina and Ustilaginomycotina: rust and smut fungi

Rust and smut fungi cause many serious diseases of cultivated plants in farms and gardens, early leaf fall and debilitation of valued erosion-control trees, and reduced value of timber (see section on Plant-parasitic fungi, page 501). Though belonging in separate subphyla, the two groups are often treated together because of their shared importance in plant pathology. In addition, some smut-like fungi in the Microbotryales are members of the Pucciniomycetes, differing from the Ustilaginomycetes in 5S rDNA secondary structure, cell wall carbohydrates, and morphology (Bauer et al. 2000).

Rust fungi can have complex life-cycles with up to five spore stages. The life-cycle may be completed on one host or on two unrelated hosts. They are near-obligately biotrophic on vascular plants infecting angiosperms, gymnosperms, and ferns. Typically they produce 'rust-coloured' (often orange or yellow) pustules on above-ground organs. Pustules are usually circular or elongate and up to a few millimetres in diameter. Some rusts are responsible for galls, blisters, and witches' brooms. Although a few species have a wide host range, many are restricted to parasitising a single host family, genus, or species.

Smut fungi are facultative biotrophs, producing a yeast-like phase in culture and in nature. They infect mainly angiosperms, apart from two species on *Selaginella*, one species on *Osmunda*, and two species on *Araucaria*, and are particularly common on grasses and sedges. Often they infect seeds or flower parts, replacing them with masses of black or dark brown spores. Other species produce dark pustules on leaves or stems.

Worldwide there are approximately 5000 species of rust fungi in about 140 genera (Cummins & Hiratsuka 2003); 3000 of the species are in the genus *Puccinia*. There are about 1450 smut species in 77 genera (Vánky 2002); 230 species are in the genus *Ustilago*. In New Zealand there are approximately 250 species in 22 genera of rust fungi and 100 smut fungi in 34 genera. Of these, about 54% of the rusts and 40% of the smuts are native (McKenzie 1998; Vánky & McKenzie 2002). The native species usually parasitise native plants and co-exist in a biologically balanced relationship. A native smut fungus (*Mundkurella schefflerae*) described

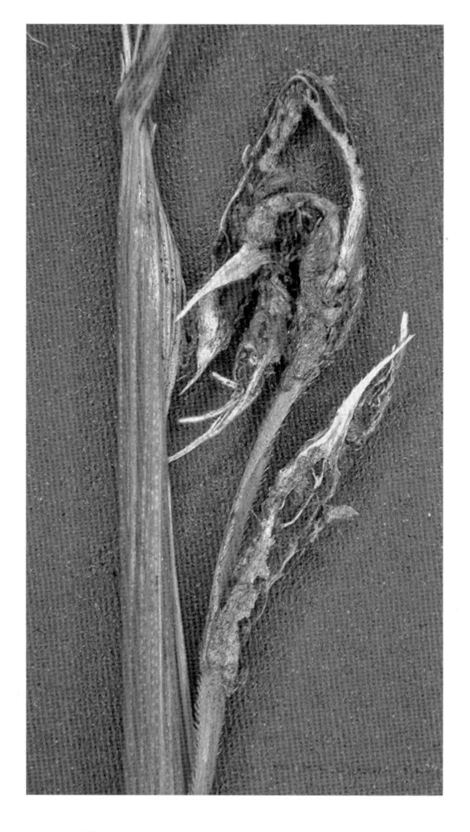

Ustilago agropyri (Ustilaginomycetes: Ustilaginales), a smut parasite of certain grasses.
Jerry A. Cooper

Puccinia caricina (Pucciniomycetes: Pucciniales), a white rust.
Jerry A. Cooper

from pate (*Schefflera digitata*) is one of the few smut species known to infect woody plants anywhere in the world (Vánky & McKenzie 2002).

Extensive collecting in New Zealand suggests that most of the indigenous species of rust fungi are known. However, some new species have been found on endemic plants in recent years (e.g. *Puccinia chathamica* on *Carex chathamica*, *P. embergeriae* on *Embergeria grandifolia*); re-examination of previously described species has sometimes revealed additional spore stages that have necessitated reclassification of the rust; and additional features have been occasionally found that have allowed splitting of a single species into two entities. Smut fungi are often cryptic and it is probable that additional native and introduced species will be found. Indigenous rusts and smuts are shared mainly with Australia, and closer comparison of New Zealand and Australian species may reveal synonymies and further species in common.

G. H. Cunningham produced a monograph of the New Zealand rust fungi (Cunningham 1931), and in the 1920s and 1930s a series of papers describing smut fungi (e.g. Cunningham 1930). A checklist of the smut fungi (McKenzie & Vánky 2001) was followed by an illustrated monograph of all 93 recorded smut species (Vánky & McKenzie 2002) as the first published volume of the series *The Fungi of New Zealand*. A volume in this series on rust fungi is now in preparation. It should soon, therefore, be possible to examine and readily name both groups of these fungi in New Zealand.

Authors

Dr Peter K. Buchanan Landcare Research, Private Bag 92170, Auckland, New Zealand [buchananp@landcareresearch.co.nz]

Dr Ross E. Beever Landcare Research, Private Bag 92170, Auckland, New Zealand. Deceased.

Dr Eric H. C. McKenzie Landcare Research, Private Bag 92170, Auckland, New Zealand [mckenziee@landcareresearch.co.nz]

Dr Barbara C. Paulus RD2 Masterton [bcpaulus@hotmail.com]

Dr Shaun R. Pennycook Landcare Research, Private Bag 92170, Auckland, New Zealand [pennycooks@landcareresearch.co.nz]

Dr Geoff S. Ridley Environmental Protection Authority, PO Box 131, Wellington [geoff.ridley@epa.govt.nz]

Dr Mahajabeen Padamsee Landcare Research, Private Bag 92170, Auckland, New Zealand [padamseem@landcareresearch.co.nz]

Dr Jerry A. Cooper Landcare Research, PO Box 40, Lincoln 7640, New Zealand [cooperj@landcareresearch.co.nz]

References

BAUER, R.; BEGEROW, D.; OBERWINKLER, F.; PIEPENBRING, M; BERBEE, M. L. 2000: Ustilaginomycetes. Pp. 57–83 *in*: McLaughlin, D. J.; McLaughlin, E. G.; Lemke, P. A. (eds), *The Mycota VII: Part B.Systematics and evolution*. Springer-Verlag, Berlin. 259 p.

BAUER, R.; BEGEROW, D.; SAMPAIO, J. P.; WEIß, M.; OBERWINKLER, F. 2006: The simple-septate basidiomycetes: a synopsis. *Mycological Progress 5*: 41-46

BINDER, M.; HIBBETT, D. S.; LARSSON, K.-H.; LARSSON, E.; LANGER, E.; LANGER, G. 2005: The phylogenetic distribution of resupinate forms across the major clades of mushroom-forming fungi (Homobasidiomycetes). *Systematics and Biodiversity 3*: 113–157.

BUCHANAN, P. K.; RYVARDEN, L. 2000: An annotated checklist of polypore and polypore-like fungi recorded from New Zealand. *New Zealand Journal of Botany 38*: 265–323.

CAVALIER-SMITH, T. 1998: A revised six-kingdom system of life. *Biological Reviews 73*: 203-266.

CAVALIER-SMITH, T. 2001. What are fungi? Pp. 3–38 *in*: Esser, K.; Lemke, P. A. (eds), *The Mycota VII: Part A. Systematics and evolution*. Springer-Verlag, Berlin. 365 p.

CUMMINS, G. B; HIRATSUKA, Y. 2003: *Illustrated Genera of Rust Fungi*. 3rd edn. APS Press, St Paul. 240 p.

CUNNINGHAM, G. H. 1930: Seventh supplement to the Uredinales and Ustilaginales of New Zealand. *Transactions and Proceedings of the New Zealand Institute 61*: 402–418.

CUNNINGHAM, G. H. 1931: *The Rust Fungi of New Zealand, together with the biology, cytology and therapeutics of the Uredinales*. John McIndoe, Dunedin. xx + 261 p.

CUNNINGHAM, G. H. 1944: *The Gasteromycetes of Australia and New Zealand*. John McIndoe, Dunedin. xv + 236 p.

CUNNINGHAM, G. H. 1958: Hydnaceae of New Zealand. Part I. The pileate genera *Beenakia, Dentinum, Hericium, Hydnum, Phellodon* and *Steccherinum*. *Transactions of the Royal Society of New Zealand 85*: 585–601.

CUNNINGHAM, G. H. 1959: Hydnaceae of New Zealand. Part II. The genus *Odontia*. *Transactions of the Royal Society of New Zealand 86*: 65–103.
CUNNINGHAM, G. H. 1963: The Thelephoraceae of Australia and New Zealand. *New Zealand Department of Scientific and Industrial Research, Bulletin 145*: 1–359.
CUNNINGHAM, G. H. 1965: Polyporaceae of New Zealand. *New Zealand Department of Scientific and Industrial Research, Bulletin 164*: 1–304.
GRESLEBIN, A. G.; RAJCHENBERG, M. 2003: Diversity of Corticiaceae sens. lat. in Patagonia, Southern Argentina. *New Zealand Journal of Botany 41*: 437–446.
HIBBETT, D. S.; THORN, R. G. 2001. Basidiomycota: Homobasidiomycetes. Pp. 121–170 *in*: McLaughlin, D. J.; McLaughlin, E. G.; Lemke; P. A. (eds) *The Mycota VII: Part B.Systematics and evolution*. Springer-Verlag, Berlin. 259 p.
HIBBETT, D. S.; BINDER, M.; BISCHOFF, J. F.; BLACKWELL, M.; CANNON, P. F. et al. 2007: A higher-level phylogenetic classification of the Fungi. *Mycological Research 111*: 509–547.
HORAK, E. 1971: A contribution towards the revision of the Agaricales (Fungi) from New Zealand. *New Zealand Journal of Botany 9*: 403–462.
HORAK, E. 2001: Keys to genera of agarics, boletes, and related fungi in New Zealand. Pp 119–164 *in*: McKenzie, E. H. C. (ed.), *The Fungi of New Zealand/Nga Harore o Aotearoa, Volume 1. Introduction to Fungi of New Zealand*. Fungal Diversity Press, Hong Kong. 500 p.
KIRK, P. M.; CANNON, P. F.; DAVID, J. C.; STALPERS, J. A. 2001: Ainsworth & Bisby's Dictionary of the Fungi. Ninth Edition. CAB International, Wallingford. 624 p.
KIRK, P. M.; CANNON, P. F.; MINTER, D. W.; STALPERS, J. A. 2008: *Ainsworth & Bisby's Dictionary of the Fungi*. 10th edn. CAB International, Wallingford. 771 p.
LANCASTER, M. E. 1955: *Forest Fungi*. Government Printer, Wellington. 96 p.
LARSSON, K.-H., LARSSON, E.; KÕLJALG, U. 2004: High phylogenetic diversity among corticioid homobasidiomycetes. *Mycological Research 108*: 983–1002.
MAAS GEESTERANUS, R. A. 1971: Hydnaceous fungi of the eastern Old World. *Verhandelingen der Koninklijke Nederlandsche Akademie van Wetenschappen. Afd. Natuurkunde, ser. 2, 60(3)*: 1–176.
MASSEE, G. [1898] 1899: The fungus flora of New Zealand. *Transactions and Proceedings of the New Zealand Institute 31*: 282–349.
McKENZIE, E. H. C. 1998: Rust fungi of New Zealand – An introduction, and list of recorded species. *New Zealand Journal of Botany 36*: 233–271.
McKENZIE, E. H. C.; BUCHANAN, P. K.; JOHNSTON, P. R. 2000: Checklist of fungi on *Nothofagus* species in New Zealand. *New Zealand Journal of Botany 38*: 635–720.
McKENZIE, E. H. C.; VÁNKY, K. 2001: Smut fungi of New Zealand: an introduction, and list of recorded species. *New Zealand Journal of Botany 39*: 501–515.
McNABB, R. F. R. 1964: New Zealand Tremellales – I. *New Zealand Journal of Botany 2*: 403–414.
McNABB, R. F. R. 1966: New Zealand Tremellales – II. *New Zealand Journal of Botany 4*: 533–545.
McNABB, R. F. R. 1967: The Strobilomycetaceae of New Zealand. *New Zealand Journal of Botany 5*: 532–547.
McNABB, R. F. R. 1968: The Boletaceae of New Zealand. *New Zealand Journal of Botany 6*: 137–176.
MCNABB, R. F. R. 1969a: New Zealand Tremellales – III. *New Zealand Journal of Botany 7*: 241–261.
McNABB, R. F. R. 1969b: The Paxillaceae of New Zealand. *New Zealand Journal of Botany 7*: 349–362.
McNABB, R. F. R. 1971: Some new and revised taxa of New Zealand basidiomycetes (fungi). *New Zealand Journal of Botany 9*: 355–370.
McNABB, R. F. R. 1973: Taxonomic studies in the Dacrymycetaceae VIII. *Dacrymyces* Nees ex Fries. *New Zealand Journal of Botany 11*: 461–524.
MONCALVO, J.-M.; VILGALYS, R.; REDHEAD, S. A.; JOHNSON, J. E.; JAMES, T. Y.; AIME, M. C.; HOFSTETTER, V.; VERDUIN, S. J. W.; LARSSON, E.; BARONI, T. J.; THORN, R. G.; JACOBSSON, S.; CLEMENCON, H.; MILLER O. K. Jr 2002: One hundred and seventeen clades of euagarics. *Molecular Phylogenetics and Evolution 23*: 357–400.
PEINTNER, U.; MOSER, M.; VILGALYS, R. 2002: *Thaxterogaster* is a taxonomic synonym of *Cortinarius*: new combinations and new names. *Mycotaxon 81*: 177–184.
PETERSEN, R. H. 1988: The clavarioid fungi of New Zealand. *New Zealand Department of Scientific and Industrial Research Bulletin 236*: 1–170.
RAMSBOTTOM, J. 1923: *A Handbook of the Larger British Fungi*. Trustees of the British Museum (Natural History), London. 222 p.
RIDLEY, G. S. 1988: *Squamanita squarrulosa*, a new species from New Zealand. *Persoonia 13*: 459–462.
SEGEDIN, B. P. 1987: An annotated checklist of agarics and boleti recorded from New Zealand. *New Zealand Journal of Botany 25*: 185–215.
SEGEDIN, B. P.; PENNYCOOK, S. R. 2001: A nomenclatural checklist of agarics, boletes, and related secotioid and gasteromycetous fungi recorded from New Zealand. *New Zealand Journal of Botany 39*: 285–348.
STALPERS, J. A. 1985: Type studies of the species of *Corticium* described by G.H. Cunningham. *New Zealand Journal of Botany 23*: 301–310.
STALPERS, J. A.; BUCHANAN, P. K. 1991: Type studies of the species of *Pellicularia* and *Peniophora* described by G.H. Cunningham. *New Zealand Journal of Botany 29*: 331–340.
STEVENSON, G. [1961] 1962: The Agaricales of New Zealand: I. *Kew Bulletin 15*: 381–385.
THIERS, H. D. 1984: The secotioid syndrome. *Mycologia 76*: 1–8.
VÁNKY, K. 2002: *Illustrated Genera of Smut Fungi*. Second Edition. APS Press, St Paul, Minnesota.
VÁNKY, K.; McKENZIE, E. H. C. 2002: *The Fungi of New Zealand/Nga Harore o Aotearoa, Volume 2. Smut Fungi of New Zealand*. Fungal Diversity Press, Hong Kong. 273 p.
WELLS, K. 1994: Jelly fungi, then and now! *Mycologia 86*: 18–48.
ZUGMAIER, W.; BAUER, R.; OBERWINKLER, F. 1994: Mycoparasitism of some *Tremella* species. *Mycologia 86*: 49–56.

Checklist of New Zealand Basidiomycota

This checklist is arranged according to the taxonomic hierarchy in Kirk et al. (2008). Species status: A, adventive (naturalised alien); AN, anamorphic; E, endemic; L, lichenised; U, status unknown; UO, occurrence uncertain in New Zealand; species not indicated by any notation are native (naturally indigenous but not endemic). Endemic genera are underlined (first entry only). Major habitats are indicated by M (marine) and F (freshwater); all other species are terrestrial.

KINGDOM FUNGI
PHYLUM BASIDIOMYCOTA
SUBPHYLUM AGARICOMYCOTINA
Class AGARICOMYCETES
Cyphellostereum laeve (Fr.) D.A. Reid
Dictyonema moorei (Nyl.) Henssen L
Dictyonema sericeum (Sw.) Berk. L
Grandinia farinacea (Pers.) Bourdot & Galzin U
Intextomyces contiguus (P. Karst.) J. Erikss. & Ryvarden
Oxyporus spiculifer (G. Cunn.) P.K. Buchanan & Ryvarden E
Peniophorella flagellata (G. Cunn.) K.H. Larss. E
Peniophorella praetermissa (P. Karst.) K.H. Larss.
Repetobasidium canadense J. Erikss. & Hjortstam U
Repetobasidium glaucocanum (G. Cunn.) Stalpers E
Resinicium bicolor (Alb. & Schwein.) Parmasto U
Resinicium friabile Hjortstam & Melo A
Resinicium pinicola (J. Erikss.) J. Erikss. & Hjortstam OU
Rickenella fibula (Bull.) Raithelh. U
Rickenella swartzii (Fr.) Kuyper A
Timgrovea reticulata (G. Cunn.) Bougher & Castellano

Order AGARICALES
Anellaria fimiputris sensu Massee U
Anellaria semiovata A
Anellaria separata sensu Massee U
Copelandia cyanescens (Berk. & Broome) Singer A
Panaeolina foenisecii (Pers.) Maire A
Panaeolus ater (J.E. Lange) Kühner & Romagn. OU
Panaeolus campanulatus (L.) Quél. A
Panaeolus cinctulus (Bolton) Britzelm. A
Panaeolus olivaceus F.H. Møller A
Panaeolus papilionaceus sensu Massee A
Panaeolus rickenii Hora OU
Panaeolus semiovatus (With.) S. Lundell A
Panaeolus speciosus P.D. Orton A
Panaeolus sphinctrinus (Fr.) Quél. A
Plicaturopsis scarlatina P.K. Buchanan & Hood E
AGARICACEAE

Abstoma purpureum (Lloyd) G. Cunn. E
Abstoma stuckertii (Speg.) J.E. Wright & V.L. Suarez
Agaricus arvensis Schaeff. A
Agaricus augustus Fr. A
Agaricus auratus sensu J.B. Armstrong U
Agaricus bambusae Beeli A var. *bambusae*
— var. *australis* Heinem. E
Agaricus bernardii (Quél.) Sacc. A
Agaricus bisporus (J.E. Lange) Imbach var. *bisporus* A
Agaricus bitorquis (Quél.) Sacc. A
Agaricus campbellensis Geml, Laursen & D. Lee Taylor E
Agaricus campestris L. var. *campestris* A
— var. *floccipes* (F.H. Møller) Pilát OU
Agaricus campigenus Berk.
Agaricus cupreobrunneus (F.H. Møller) Pilát A
Agaricus dulcidulus Schulzer OU
Agaricus gennadii (Chatin & Boud.) P.D. Orton OU
Agaricus horakii Heinem. E
Agaricus hypnorum var. *sphagnicola* sensu Linds. (nom. inv.) OU
Agaricus impudicus (Rea) Pilát A
Agaricus kroneanus Rabenh.
Agaricus lanatoniger Heinem. E
Agaricus lanipes (F.H. Møller & Jul. Schäff.) Hlaváček A
Agaricus moelleri Wasser OU
Agaricus oligocystis Heinem. E
Agaricus placomyces Peck OU
Agaricus porphyrocephalus F.H. Møller A
Agaricus praeclaresquamosus A.E. Freeman OU
Agaricus purpureoniger Heinem. E
Agaricus semotus Fr. A
Agaricus silvicola (Vittad.) Peck OU
Agaricus sp. 'Prices Valley (PDD87152)' J.A. Cooper (nom. ined.)
Agaricus sp. 'Rotokuru Lakes (PDD80826)' J.A. Cooper (nom. ined.)
Agaricus subantarcticus Geml, Laursen & D. Lee Taylor E
Agaricus subperonatus (J.E. Lange) Singer A
Agaricus subrutilescens (Kauffman) Hotson & D.E. Stuntz OU
Agaricus sylvaticus Schaeff. OU
Agaricus urinascens (Jul. Schäff. & F.H. Møller) Singer var. *urinascens* A
— var. *excellens* (F.H. Møller) Nauta A
Agaricus viridopurpurascens Heinem. E
Agaricus xanthoderma Genev. A
Agaricus xantholepis (F.H. Møller) F.H. Møller OU
Bovista aestivalis (Bonord.) Demoulin
Bovista brunnea Berk.
Bovista gunnii (Berk.) Kreisel
Bovista plumbea Pers.
Calvatia caelata sensu G. Cunn.
Calvatia candida (Rostk.) Hollós U var. *candida*
— var. *fusca* G. Cunn. U
— var. *rubroflava* (Cragin) G. Cunn. U
Calvatia cretacea sensu G. Cunn.
Calvatia cyathiformis (Bosc) Morgan
Calvatia excipuliformis (Scop.) Perdeck OU
Calvatia gigantea (Batsch) Lloyd A
Calvatia lilacina sensu G. Cunn.
Calvatia utriformis (Bull.) Jaap
Chlorophyllum rachodes (Vittad.) Vellinga A
<u>*Clavogaster*</u> *novozelandicus* Henn. E
Coprinus colensoi Berk. E
Coprinus comatus (O.F. Müll.) Pers. A
Coprinus ephemerus sensu Colenso U
Coprinus impatiens (Fr.) Quél. OU
Coprinus patouillardii Quél. A
Coprinus poliomallus Romagn. OU
Coprinus velox Godey A
Crucibulum laeve (Huds.) Kambly
Crucibulum vulgare var. *lanosum* Cooke E
Crucispora naucorioides E. Horak
Cyathus colensoi Berk.
Cyathus hookeri Berk.
Cyathus novae-zeelandiae Tul. & C. Tul. E
Cyathus olla (Batsch) Pers.
Cyathus stercoreus (Schwein.) De Toni A
Cyathus striatus (Huds.) Willd. A
Cystoderma amianthinum (Scop.) Konrad & Maubl. OU
Cystoderma cinnabarinum (Alb. & Schwein.) Fayod (nom. inv.) OU
Cystoderma clastotrichum (G. Stev.) E. Horak E
Cystoderma muscicola (Cleland) Grgur. U
Cystoderma terreyi (Berk. & Broome) Harmaja OU
Cystolepiota hetieri (Boud.) Singer A
Cystolepiota seminuda (Lasch) Bon OU
Cystolepiota sp. 'Nile River (PDD87126)' J.A. Cooper (nom. ined.)
Disciseda candida (Schwein.) Lloyd
Disciseda cervina sensu G. Cunn.
Disciseda verrucosa G. Cunn.
Lepiota adusta (E. Horak) E. Horak E
Lepiota agricola (Murrill) Sacc. & Trotter A
Lepiota alba sensu auct. NZ
Lepiota alopochroa (Berk. & Broome) Sacc. A
Lepiota aspera (Pers.) Quél. A
Lepiota brunneum (Farl. & Burt) Vellinga A
Lepiota calcarata (E. Horak) E. Horak E
Lepiota castanea Quél. A
Lepiota clypeolaria sensu Massee U
Lepiota cristata (Bolton) P. Kumm. A
Lepiota exstructa (Berk.) Sacc. A
Lepiota grangei (Eyre) J.E. Lange A
Lepiota mesomorpha sensu Massee U
Lepiota purpurata (G. Stev.) E. Horak E
Lepiota rufipes Morgan U
Lepiota subincarnata J.E. Lange A
Leucoagaricus badhamii (Berk. & Broome) Singer A
Leucoagaricus bresadolae var. *biornatus* (Berk. & Broome) Bon OU
Leucoagaricus croceovelutinus Bon
Leucoagaricus gauguei Bon & Boiffard OU
Leucoagaricus leucothites (Vittad.) Wasser var. *leucothites* A
Leucoagaricus menieri (Sacc.) Singer A
Leucoagaricus purpureorimosus Bon & Boiffard A
Leucoagaricus serenus (Fr.) Bon & Boiffard OU
Leucoagaricus sp. 'Erua Forest (PDD80769)' J.A. Cooper (nom. ined.)
Leucoagaricus sp. 'Gypsy Glen (PDD87679)' J.A. Cooper (nom. ined.)
Leucoagaricus sp. 'Hay Reserve (PDD87677)' J.A. Cooper (nom. ined.)
Leucoagaricus sp. 'Kaituna Valley (PDD86991)' J.A. Cooper (nom. ined.)
Leucoagaricus sp. 'Mt Bruce (PDD87444)' J.A. Cooper (nom. ined.)
Leucoagaricus sp. 'Okuti Valley (PDD87672)' J.A. Cooper (nom. ined.)
Leucoagaricus sp. 'Rotokuru Lakes(PDD80831)' J.A. Cooper (nom. ined.)
Leucoagaricus sp. 'Waiohine Gorge (PDD87425)' J.A. Cooper (nom. ined.)
Leucocoprinus birnbaumii (Corda) Singer A
Leucocoprinus brebissonii (Godey) Locq. U
Leucocoprinus cepistipes (Sowerby) Pat. U
Leucocoprinus fragilissimus (Ravenel) Pat. U
Leucocoprinus sp. 'Opawa (PDD87349)' J.A. Cooper (nom. ined.) A
Leucocoprinus straminellus (Bagl.) Narducci & Caroti A
Lycoperdon glabrescens Berk.
Lycoperdon glabrescens sensu G. Cunn.
Lycoperdon microspermum Berk.
Lycoperdon nigrescens Pers. OU
Lycoperdon nitidum Lloyd
Lycoperdon perlatum sensu G.Cunn.
Lycoperdon polymorphum Vittad. (nom. illegit.)
Lycoperdon pyriforme sensu G. Cunn.
Lycoperdon scabrum (Lloyd) G. Cunn.
Lycoperdon sp. G.M. Taylor (nom. ined.) U
Lycoperdon spadiceum Schaeff.
Lycoperdon spadiceum sensu G. Cunn.
Macrolepiota clelandii Grgur.
Macrolepiota gracilenta (Krombh.) Wasser U
Macrolepiota procera (Scop.) Singer U
Macrolepiota rachodes var. *hortensis* (Pilát) Wasser (nom. inv.) OU
Melanophyllum echinatum (Roth) Singer U
Morganella compacta (G. Cunn.) Kreisel & Dring E
Morganella pyriformis (Schaeff.) Kreisel & D. Krüger
Mycenastrum corium (Guers. ex DC.) Desv.
Nidula candida (Peck) V.S. White
Nidula emodensis (Berk.) Lloyd U
Nidula niveotomentosa (Henn.) Lloyd
Nidularia deformis (Willd.) Fr.
<u>*Notholepiota*</u> *areolata* (G. Cunn.) E. Horak E
Secotium ochraceum Rodway U
Tulostoma adhaerens Lloyd U
Tulostoma album Massee
Tulostoma leprosum (Kalchbr.) De Toni
Tulostoma pulchellum Sacc.
Tulostoma simulans Lloyd
Tulostoma striatum G. Cunn.
Vascellum pratense (Pers.) Kreisel A
AMANITACEAE
Amanita australis G. Stev. E
Amanita excelsa (Fr.) Bertill. OU var. *excelsa*
— var. *spissa* (Fr.) Neville & Poumarat A
Amanita inopinata D.A. Reid & Bas A
Amanita karea G.S. Ridl. E
Amanita mumura G.S. Ridl. E
Amanita muscaria (L.) Lam. A
Amanita nauseosa (Wakef.) D.A. Reid A
Amanita nehuta G.S. Ridl. E
Amanita nigrescens G. Stev. E
Amanita nothofagi G. Stev. E
Amanita pantherina (DC.) Krombh. A OU
Amanita pareparina G.S. Ridl. E
Amanita pekeoides G.S. Ridl. E
Amanita phalloides (Fr.) Link A
Amanita pumatona G.S. Ridl. E
Amanita rubescens Pers. A OU
Amanita solitaria (Bull.) Fr. A
Amanita sp. [fig. 55–56] sensu Hall, I.R.; Buchanan, P.K.; Wang, Y.; Cole, A.L.J. U
Amanita sp. 2 G.S. Ridl. (nom. ined.) A
Amanita taiepa G.S. Ridl. E
Limacella whereoparaonea G.S. Ridl. E
BOLBITIACEAE
Bolbitius muscicola (G. Stev.) Watling E
Bolbitius reticulatus (Pers.) Ricken OU
Bolbitius sp. 1 sensu Watling & G.M. Taylor
Bolbitius sp. 2 sensu Watling & G.M. Taylor
Bolbitius titubans (Bull.) Fr. A
Bolbitius vitellinus (Pers.) Fr. A
Conocybe coprophila (Kühner) Kühner OU
Conocybe filaris (Fr.) Kühner U
Conocybe gracilenta Watling & G.M. Taylor E
Conocybe horakii Watling & G.M. Taylor E
Conocybe huijsmanii Watling U
Conocybe lactea (J.E. Lange) Métrod A
Conocybe mesospora Kühner ex Kühner & Watling U
Conocybe novae-zelandiae Watling & G.M. Taylor E
Conocybe piloselloides Watling OU
Conocybe pubescens (Gillet) Kühner A
Conocybe pubescens sensu A.E. Bell A
Conocybe rickeniana P.D. Orton OU
Conocybe rickenii sensu A.E. Bell A

Conocybe rugosa (Peck) Watling U
Conocybe sp. 1 sensu Watling & G.M. Taylor U
Conocybe sp. 2 sensu Watling & G.M. Taylor U
Conocybe sp. 3 sensu Watling & G.M. Taylor U
Conocybe sp. 4 sensu Watling & G.M. Taylor U
Conocybe vexans P.D. Orton OU
Pholiotina utricystidiata Enderle & H.-J. Hübner A
Tympanella galanthina (Cooke & Massee) E. Horak E
CLAVARIACEAE
Clavaria acuta Sowerby OU
Clavaria alboglobospora R.H. Petersen E
Clavaria alliacea Corner OU
Clavaria amoena Zoll. & Moritzi
Clavaria arborescens Berk. E
Clavaria archeri Berk.
Clavaria ardosiaca R.H. Petersen E
Clavaria argillacea Fr. OU
Clavaria cinerea (Bull.) Fr. OU
Clavaria corallinorosacea Cleland
Clavaria crispula sensu Berk.
Clavaria cupreicolor R.H. Petersen E
Clavaria echinobrevispora R.H. Petersen E
Clavaria echinonivosa R.H. Petersen E
Clavaria echinoolivacea R.H. Petersen E
Clavaria flavopurpurea R.H. Petersen E
Clavaria fusiformis Fr. OU
Clavaria gibbsiae Ramsb.
Clavaria inaequalis sensu Berk.
Clavaria lutea sensu Berk.
Clavaria luteostirpata S.G.M. Fawc. OU
Clavaria megaspinosa R.H. Petersen E
Clavaria mima R.H. Petersen E
Clavaria muscula R.H. Petersen
Clavaria musculospinosa R.H. Petersen E
Clavaria novozealandica R.H. Petersen E
Clavaria phoenicea Zoll. & Mor. OU
— var. *persicina* R.H. Petersen
Clavaria plumbeoargillacea R.H. Petersen E
Clavaria redoleoalii R.H. Petersen E
Clavaria rosea Fr. OU
Clavaria roseoviolacea R.H. Petersen E
Clavaria rubicundula Leathers OU
Clavaria sp. 1 sensu R.H. Petersen
Clavaria sp. 2 sensu R.H. Petersen
Clavaria sp. 3 sensu R.H. Petersen
Clavaria sp. 4 sensu R.H. Petersen
Clavaria sp. 5 sensu R.H. Petersen
Clavaria subacuta S. Ito & S. Imai OU
Clavaria subsordida R.H. Petersen E
Clavaria subviolacea R.H. Petersen E
Clavaria sulcata (Overeem) R.H. Petersen
Clavaria tuberculospora R.H. Petersen E
Clavaria ypsilonidia R.H. Petersen E
Clavaria zollingeri Lév.
Clavulinopsis fusiformis (Sowerby) Corner OU
Clavulinopsis miniata (Berk.) Corner OU
— var. *rosacea* Corner OU
Clavulinopsis spiralis (Jungh.) Corner U OU
Mucronella calva (Alb. & Schwein.) Fr. U
Mucronella ulmi Peck U
Ramaria pusilla Corner OU
Ramariopsis agglutinata R.H. Petersen E
Ramariopsis alutacea R.H. Petersen E
Ramariopsis antillarum (Pat.) R.H. Petersen
Ramariopsis aurantioolivacea R.H. Petersen E
Ramariopsis avellanea R.H. Petersen E
Ramariopsis avellaneoinversa R.H. Petersen E
Ramariopsis bicolor R.H. Petersen E
Ramariopsis cinnamomea (S.G.M. Fawc.) R.H. Petersen OU
Ramariopsis cremicolor R.H. Petersen E
Ramariopsis crocea (Pers.) Corner
Ramariopsis depokensis (Overeem) R.H. Petersen
Ramariopsis depokensis f. *persicina* R.H. Petersen E
Ramariopsis junquillea R.H. Petersen E
Ramariopsis laeticolor (Berk. & M.A. Curtis) R.H. Petersen
Ramariopsis longipes R.H. Petersen E
Ramariopsis luteotenerrima (Overeem) R.H. Petersen
Ramariopsis minutula (Bourdot & Galzin) R.H. Petersen OU
Ramariopsis novohibernica Corner
Ramariopsis ovispora R.H. Petersen E
Ramariopsis pulchella (Boud.) Corner
Ramariopsis ramarioides R.H. Petersen E
Ramariopsis simplex R.H. Petersen (nom. inv.)
Ramariopsis sp. 1 sensu R.H. Petersen
Ramariopsis sp. 2 sensu R.H. Petersen
Ramariopsis sp. 3 sensu R.H. Petersen
Ramariopsis sp. 4 sensu R.H. Petersen
Ramariopsis sp. 5 sensu R.H. Petersen
Ramariopsis tortuosa R.H. Petersen E
Scytinopogon angulisporus (Pat.) Corner U OU
— var. *gracilis* Corner U
Scytinopogon pallescens (Bres.) Singer
Setigeroclavula ascendens R.H. Petersen E
CORTINARIACEAE
Cortinarius achrous E. Horak, Peintner, M.M. Moser & Vilgalys E
Cortinarius acutus (Pers.) Fr. OU
Cortinarius aegrotus E. Horak E
Cortinarius aerugineoconicus E. Horak E
Cortinarius alboaggregatus Soop E
Cortinarius alboroseus (R. Heim) Peintner, E. Horak, M.M. Moser & Vilgalys
Cortinarius anauensis Soop E
Cortinarius anisodorus (E. Horak) Peintner & M.M. Moser E
Cortinarius anomalus (Fr.) Fr. OU
Cortinarius atrolazulinus M.M. Moser E
Cortinarius atroviolaceus M.M. Moser
Cortinarius aurantioferreus Soop E
Cortinarius australiensis (Cleland & Cheel) E. Horak
Cortinarius austroalbidus Cleland & J.R. Harris
Cortinarius austrocyanites Soop
Cortinarius basifibrillosus E. Horak (nom. ined.) E
Cortinarius bellus E. Horak E
Cortinarius cagei Melot OU
Cortinarius carbonellus Soop E
Cortinarius cartilagineus (G. Cunn.) Peintner & M.M. Moser
Cortinarius caryotis Soop E
Cortinarius castaneiceps E. Horak E
Cortinarius castoreus Soop E
Cortinarius chalybeus Soop E
Cortinarius chrysma Soop E
Cortinarius collybianus Soop E
Cortinarius coneae (R. Heim) Peintner & M.M. Moser E
Cortinarius crassus Fr. OU
Cortinarius cremeolinus Soop E var. *cremeolinus*
— var. *subpicoides* Soop (nom. ined.) E
Cortinarius cretax Soop E
Cortinarius cucumeris E. Horak E
Cortinarius cupreonatus Soop E
Cortinarius cycneus E. Horak E
Cortinarius decipiens (Pers.) Fr. A
Cortinarius decumbens (Pers.) Fr. OU
Cortinarius dulciolens E. Horak, M.M. Moser, Peintner & Vilgalys E
Cortinarius dulciorum Soop E
Cortinarius dysodes Soop E
Cortinarius elacatipus E. Horak, Peintner, M.M. Moser & Vilgalys
Cortinarius elaiochrous E. Horak, M.M. Moser, Peintner & Vilgalys E var. *elaiochrous*
— var. *leontis* Soop E
Cortinarius elaiops Soop E
Cortinarius epiphaeus (E. Horak) Peintner & M.M. Moser E
Cortinarius erugatus (Weinm.) Fr. OU
Cortinarius eutactus Soop E
Cortinarius evernius Fr. OU
Cortinarius exlugubris Soop E
Cortinarius flammuloides E. Horak & M.M. Moser OU
Cortinarius flavidulus Peintner & M.M. Moser E
Cortinarius gamundiae (E. Horak) E. Horak, Peintner, M.M. Moser & Vilgalys OU
Cortinarius gemmeus E. Horak E
Cortinarius gymnocephalus Soop E
Cortinarius gymnopiloides E. Horak (nom. inv.)
Cortinarius hebelomaticus E. Horak & Soop (nom. ined.) E
Cortinarius ignellus Soop E
Cortinarius ignotus E. Horak E
Cortinarius incensus Soop E
Cortinarius indolicus E. Horak E
Cortinarius infractus (Pers.) Fr. OU
Cortinarius ionomataius Soop E
Cortinarius iringa Soop E
Cortinarius kaimanawa Soop E
Cortinarius lamproxanthus Soop E
Cortinarius laquellus Soop E
Cortinarius largus Fr. OU
Cortinarius lavendulensis Cleland OU
Cortinarius leucocephalus (Massee) Peintner & M.M. Moser
Cortinarius lubricanescens Soop E
Cortinarius luteinus Soop E
Cortinarius luteobrunneus Peintner & M.M. Moser E
Cortinarius magellanicus sensu auct. NZ
Cortinarius malosinae Soop E
Cortinarius mariae (E. Horak) E. Horak, Peintner, M.M. Moser & Vilgalys
Cortinarius marmoratus E. Horak E
Cortinarius meleagris (E. Horak & G.M. Taylor) E. Horak, Peintner, M.M. Moser & Vilgalys
Cortinarius melimyxa E. Horak E
Cortinarius melleomitis M.M. Moser & E. Horak
Cortinarius memoria-annae Gasparini
Cortinarius minoscaurus Soop E
Cortinarius myxenosma Soop E
Cortinarius naphthalinus Soop E
Cortinarius napivelatus (E. Horak) Peintner & M.M. Moser E
Cortinarius neccesarius E. Horak OU
Cortinarius nivalis (E. Horak) Peintner & M.M. Moser E
Cortinarius ohauensis (Soop) Peintner & M.M. Moser E
Cortinarius obtusus (Fr.) Fr. OU
Cortinarius olorinatus E. Horak E
Cortinarius opaca Soop (nom. ined.) E
Cortinarius ophryx Soop (nom. ined.) E
Cortinarius orixanthus Soop E
Cortinarius papaver Soop
Cortinarius paraonui Soop E
Cortinarius paraxanthus Soop E
Cortinarius pectochelis Soop E
Cortinarius peraurantiacus Peintner & M.M. Moser E
Cortinarius peraureus Soop E
Cortinarius perelegans Soop E
Cortinarius periclymenus Soop E
Cortinarius persicanus Soop
Cortinarius persplendidus Gasparini
Cortinarius phaeochlorus E. Horak E
Cortinarius phaeomyxa (E. Horak) Peintner, E. Horak, M.M. Moser & Vilgalys E
Cortinarius pholideus Fr. OU
Cortinarius pholiotellus Soop E
Cortinarius picoides Soop E
Cortinarius pisciodorus (E. Horak) Peintner & M.M.

Moser E
Cortinarius porphyroideus Peintner & M.M. Moser E
Cortinarius porphyrophaeus E. Horak E
Cortinarius pselioticton Soop E
Cortinarius rattinoides Soop E
Cortinarius rattinus Soop E
Cortinarius retipes E. Horak & Soop (nom. ined.) E
Cortinarius rhipiduranus Soop
Cortinarius rotundisporus Cleland & Cheel
Cortinarius rubrocastaneus (Soop) A.-M.B. Oliv. & Orlovich E
Cortinarius rugosiceps (E. Horak & G.M. Taylor) Peintner, E. Horak, M.M. Moser & Vilgalys E
Cortinarius salor Fr. A
Cortinarius sarcinochrous Peintner & M.M. Moser E
Cortinarius saturniorum Soop E
Cortinarius scaurus (Fr.) Fr. OU
Cortinarius semisanguineus (Brig.) Maire OU
Cortinarius sinapicolor Cleland
Cortinarius singularis Soop E
Cortinarius sp. [PDD 77486] sensu Garnica et al. (nom. ined.) E
Cortinarius sp. [ZT NZ8682] Peintner et al. (nom. ined.) E
Cortinarius sterilis Kauffman OU
Cortinarius subarcheri Cleland OU
Cortinarius subargentatus P.D. Orton (nom. illegit.) OU
Cortinarius subcalyptrosporus M.M. Moser
Cortinarius subcastanellus E. Horak, Peintner, M.M. Moser & Vilgalys E
Cortinarius subgemmeus Soop E
Cortinarius suecicolor Soop E
Cortinarius taylorianus E. Horak E
Cortinarius tessiae Soop E
Cortinarius thaumastus Soop E
Cortinarius tigrellus Soop E
Cortinarius trichocarpus Soop (nom. ined.)
Cortinarius ursus Soop E
Cortinarius varius (Schaeff.) Fr. OU
Cortinarius vernicifer Soop E
Cortinarius vernus H. Lindstr. & Melot A
Cortinarius veronicae Soop var. *veronicae*
— var. *dilutus* Soop E
Cortinarius violaceovolvatus (E. Horak) Peintner & M.M. Moser E
Cortinarius violaceus (L.) Gray OU
Cortinarius viscilaetus Soop E
Cortinarius viscoviridis E. Horak E
Cortinarius viscostriatus E. Horak E
Cortinarius vitreopileatus E. Horak E
Cortinarius xenosma Soop E
Cortinarius xiphidipus M.M. Moser & E. Horak OU
Dermocybe alienata E. Horak E
Dermocybe aurantiella E. Horak E
Dermocybe canaria E. Horak
Dermocybe cardinalis E. Horak E
Dermocybe castaneodisca E. Horak E
Dermocybe cinnabarina (Fr.) Wünsche
Dermocybe cinnamomea (L.) Wünsche OU
Dermocybe cramesina E. Horak
Dermocybe crocea (Schaeff.) M.M. Moser OU
Dermocybe egmontiana E. Horak E
Dermocybe icterinoides E. Horak E
Dermocybe indotata E. Horak E
Dermocybe largofulgens E. Horak E
Dermocybe leptospermorum E. Horak E
Dermocybe olivaceonigra E. Horak E
Dermocybe parietalis Høil. (nom. inv.)
Dermocybe purpurata E. Horak & Gerw. Keller E
Dermocybe sanguinea (Wulfen) Wünsche OU
Dermocybe vinicolor E. Horak E
Descolea antarctica OU
Descolea gunnii (Massee) E. Horak
Descolea majestatica E. Horak E
Descolea phlebophora E. Horak
Descomyces albellus (Massee & Rodway) Bougher & Castellano
Descomyces albus (Klotzsch) Bougher & Castellano A
Descomyces giachinii Trappe, V.L. Oliveira, Castellano & Claridge
Descomyces sp. sensu Soop
Phaeocollybia longipes E. Horak E
Phaeocollybia minuta E. Horak E
Phaeocollybia ratticauda E. Horak
Protoglossum luteum Massee U
Protoglossum sp. 1 sensu Peintner et al. (nom. ined.)
Protoglossum violaceum (Massee & Rodway) T.W. May OU
Pyrrhoglossum viriditinctum E. Horak E
Rozites cyanoxanthus Soop (nom. ined.) E
Rozites sp. sensu Soop (nom. ined.)
Thaxterogaster viola Soop E
CYPHELLACEAE
Cheimonophyllum candidissimum (Berk. & M.A. Curtis) Singer
Cheimonophyllum roseum Segedin E
Cheimonophyllum sp.'Kennedy's Bush (PDD87367)' J.A. Cooper (nom. ined.)
Chondrostereum purpureum (Pers.) Pouzar A
Chondrostereum vesiculosum (G. Cunn.) Stalpers & P.K. Buchanan E
Cyphella filicicola sensu Massee
Incrustocalyptella pseudopanacis (Agerer) Agerer E
Sphaerobasidioscypha citrispora Agerer E
ENTOLOMATACEAE
Claudopus byssisedus (Pers.) Gillet
Claudopus depluens sensu Massee U
Clitopilus albovelutinus (G. Stev.) Noordel. & Co-David E
Clitopilus argentinus Singer OU
Clitopilus conchatus (E. Horak) Noordel. & Co-David E
Clitopilus dingleyae (E. Horak) Noordel. & Co-David E
Clitopilus fuligineus (E. Horak) Noordel. & Co-David E
Clitopilus geminus (Paulet) Noordel. & Co-David OU
Clitopilus hobsonii (Berk.) P.D. Orton
Clitopilus iti (E. Horak) Noordel. & Co-David E
Clitopilus maleolens (E. Horak) Noordel. & Co-David E
Clitopilus multilamellatus (E. Horak) Noordel. & Co-David E
Clitopilus muritai (G. Stev.) Noordel. & Co-David E
Clitopilus piperitus (G. Stev.) Noordel. & Co-David E
Clitopilus sp. [PDD 88168] sensu E. Horak A
Clitopilus sp. [PDD 88169] sensu E. Horak A
Eccilia haeusleriana Henn. E
Entoloma aberrans E. Horak E
Entoloma acuminatum E. Horak E
Entoloma aethiops sensu G. Stev. (nom. inv.)
Entoloma aromaticellum E. Horak E
Entoloma aromaticum E. Horak
Entoloma asprelloides G. Stev.
Entoloma asprellum sensu G. Stev. (nom. inv.) U
Entoloma atrellum E. Horak E
Entoloma blandiodorum E. Horak E
Entoloma brunneolilacinum E. Horak E
Entoloma canoconicum E. Horak E
Entoloma captiosum E. Horak E
Entoloma cavipes E. Horak E
Entoloma cerifactum E. Horak E
Entoloma cerinum E. Horak E
Entoloma chloroxanthum G. Stev. E
Entoloma clypeatum (L.) P. Kumm. A
Entoloma colensoi G. Stev. E
Entoloma confusum E. Horak E
Entoloma congregatum G. Stev. A
Entoloma consanguineum E. Horak E
Entoloma convexum G. Stev. E
Entoloma corneum E. Horak E
Entoloma crinitum E. Horak E
Entoloma croceum E. Horak E
Entoloma cucurbita E. Horak
Entoloma deceptivum E. Horak E
Entoloma deprensum E. Horak E
Entoloma distinctum E. Horak E
Entoloma dupocoloratum E. Horak E
Entoloma elegantissimum E. Horak E
Entoloma fabulosum E. Horak E
Entoloma farinolens E. Horak E
Entoloma gasteromycetoides Co-David & Noordel.
Entoloma gelatinosum E. Horak E
Entoloma glaucoroseum E. Horak E
Entoloma gracile G. Stev. E
Entoloma haastii G. Stev.
Entoloma hochstetteri (Reichardt) G. Stev.
Entoloma horakii Courtec.
Entoloma imbecille (E. Horak) E. Horak ex Segedin & Pennycook
Entoloma improvisum E. Horak E
Entoloma inops E. Horak E
Entoloma inventum E. Horak E
Entoloma lampropus sensu G. Stev. (nom. inv.) U
Entoloma latericolor E. Horak E
Entoloma mancum E. Horak E
Entoloma mariae G. Stev.
Entoloma mcnabbianum E. Horak E
Entoloma melanocephalum G. Stev.
Entoloma melleum E. Horak E
Entoloma minutoalbum E. Horak
Entoloma neosericellum E. Horak E
Entoloma nothofagi G. Stev. E
Entoloma nubigenum (Singer) Garrido OU
Entoloma obrusseum E. Horak E
Entoloma orichalceum E. Horak E
Entoloma parasericeum E. Horak E
Entoloma parsonsiae G. Stev.
Entoloma pascuum (Pers.) Donk OU
Entoloma peraffine E. Horak E
Entoloma peralbidum E. Horak E
Entoloma perconfusum E. Horak E
Entoloma perplexum E. Horak E
Entoloma persimile E. Horak E
Entoloma phaeocyathus Noordel. A
Entoloma phaeomarginatum E. Horak
Entoloma placidum (Fr.) Noordel. OU
Entoloma pluteimorphum E. Horak E
Entoloma porphyrescens E. Horak
Entoloma procerum G. Stev. E
Entoloma psittacinum (Romagn.) E. Horak
Entoloma pumilum E. Horak E
Entoloma rancidulum E. Horak E
Entoloma readiae G. Stev. A
Entoloma rusticoides (Gillet) Noordel. A
Entoloma scabripes E. Horak E
Entoloma sericellum (Fr.) P. Kumm. A
Entoloma sericeum Quél. OU
Entoloma serratomarginatum Horak
Entoloma squamiferum E. Horak E
Entoloma stramineum E. Horak E
Entoloma strictum G. Stev. E
Entoloma sulphureum E. Horak E
Entoloma tectum E. Horak E
Entoloma translucidum E. Horak
Entoloma uliginicola E. Horak E
Entoloma virescens (Berk. & M.A. Curtis) E. Horak ex Courtec.
Entoloma viridomarginatum (Cleland) E. Horak E
Entoloma vulsum E. Horak E
Entoloma waikaremoana E. Horak E

Rhodocybe antipoda (G. Stev.) E. Horak E
Richoniella pumila f. *bispora* E. Horak E
FISTULINACEAE
Porodisculus pendulus (Schwein.) Murrill
GIGASPERMACEAE
Gigasperma cryptica E. Horak E
HYDNANGIACEAE
Hydnangium carneum Wallr.
Hydnangium sp.'Hinewai (PDD86860)' J.A. Cooper (nom. ined.) E
Laccaria amethystina Cooke
Laccaria canaliculata (Sacc.) Massee
Laccaria echinospora (Speg.) Singer A
Laccaria fibrillosa McNabb E
Laccaria fraterna (Sacc.) Pegler
Laccaria glabripes McNabb E
Laccaria impolita Vellinga & G.M. Muell. OU
Laccaria laccata (Scop.) Cooke OU var. *laccata*
— var. *pallidifolia* (Peck) Peck A
Laccaria lilacina G. Stev. E
Laccaria masoniae G. Stev. var. *masoniae*
— var. *brevispinosa* McNabb E
Laccaria ohiensis (Mont.) Singer A var. *ohiensis*
— var. *paraphysata* McNabb E
Laccaria proxima (Boud.) Pat. A
Laccaria pumila Fayod A
Laccaria sp.'Lincoln (PDD80630)' J.A. Cooper (nom. ined.)
Laccaria violaceonigra G. Stev. E
Podohydnangium australe G.W. Beaton, Pegler & T.W.K. Young U
HYGROPHORACEAE
Bertrandia astatogala R. Heim
Camarophyllus apricosus (E. Horak) E. Horak E
Camarophyllus aurantiopallens E. Horak
Camarophyllus canus E. Horak E
Camarophyllus delicatus E. Horak E
Camarophyllus griseorufescens E. Horak E
Camarophyllus impurus E. Horak E
Camarophyllus muritaiensis (G. Stev.) E. Horak E
Camarophyllus patinicolor E. Horak E
Camarophyllus pratensis (Pers.) P. Kumm. var. *pratensis*
— var. *gracilis* E. Horak E
Chrysomphalina chrysophylla (Fr.) Clémençon OU
Gliophorus chromolimoneus (G. Stev.) E. Horak
Gliophorus fumosogriseus E. Horak E
Gliophorus graminicolor E. Horak
Gliophorus lilacinoides E. Horak E
Gliophorus lilacipes E. Horak E
Gliophorus luteoglutinosus E. Horak E
Gliophorus ostrinus E. Horak E
Gliophorus subheteromorphus (Singer) E. Horak
Gliophorus sulfureus (G. Stev.) E. Horak E
Gliophorus versicolor E. Horak E
Gliophorus viridis (G. Stev.) E. Horak
Gliophorus viscaurantius E. Horak E
Humidicutis conspicua (E. Horak) E. Horak E
Humidicutis lewelliniae (Kalchbr.) A.M. Young OU
Humidicutis luteovirens (E. Horak) E. Horak E
Humidicutis mavis (G. Stev.) A.M. Young
Humidicutis multicolor (Berk. & Broome) E. Horak
Humidicutis rosella (E. Horak) E. Horak E
Hygrocybe blanda E. Horak E
Hygrocybe brunnea (Cleland) Grgur.
Hygrocybe canescens (A.H. Sm. & Hesler) P.D. Orton U
Hygrocybe cantharellus (Schwein.) Murrill
Hygrocybe cavipes E. Horak E
Hygrocybe ceracea (Wulfen) P. Kumm.
Hygrocybe cerinolutea E. Horak E
Hygrocybe coccinea (Schaeff.) P. Kumm. OU
Hygrocybe conica (Scop.) P. Kumm. A
Hygrocybe conicoides (P.D. Orton) P.D. Orton & Watling OU
Hygrocybe elegans E. Horak E
Hygrocybe firma (Berk. & Broome) Singer
Hygrocybe fuliginata E. Horak E
Hygrocybe fuscoaurantiaca (G. Stev.) E. Horak E
Hygrocybe julietae (G. Stev.) E. Horak E
Hygrocybe keithgeorgei (G. Stev.) E. Horak E
Hygrocybe laeta (Pers.) P. Kumm. U
Hygrocybe lilaceolamellata (G. Stev.) E. Horak E
Hygrocybe miniata (Fr.) P. Kumm.
Hygrocybe miniatoaurantiaca Hongo E
Hygrocybe miniceps (G. Stev.) E. Horak E
Hygrocybe pantoleuca (Hongo) Hongo OU
Hygrocybe procera (G. Stev.) E. Horak E
Hygrocybe radiata Arnolds U
Hygrocybe rubrocarnosa (G. Stev.) E. Horak E
Hygrocybe russocoriacea (Berk. & T.K. Mill.) P.D. Orton & Watling A
Hygrocybe striatolutea E. Horak E
Hygrophorus brunneus sensu G. Stev.
Hygrophorus carcharias E. Horak E
Hygrophorus coccineus sensu Massee
Hygrophorus gloriae G. Stev. E
Hygrophorus involutus G. Stev. E
Hygrophorus laetus sensu E. Shaw
Hygrophorus nigricans Berk.
Hygrophorus niveus sensu Colenso
Hygrophorus pudorinus sensu J.E. Lange OU
Hygrophorus salmonipes G. Stev. E
Hygrophorus segregatus E. Horak E
Hygrophorus turundus Fr. OU
Hygrophorus waikanaensis G. Stev. E
Hygrotrama roseolum (G. Stev.) E. Horak E
Lichenomphalia alpina (Britzelm.) Redhead, Lutzoni, Moncalvo & Vilgalys L
Lichenomphalia umbellifera (L.) Redhead, Lutzoni, Moncalvo & Vilgalys L
Neohygrocybe innata E. Horak E
Neohygrocybe squarrosa E. Horak E
INOCYBACEAE
Astrosporina asterospora (Quél.) Rea
Astrosporina avellana E. Horak
Chromocyphella galeata sensu W.B. Cooke U
Chromocyphella muscicola (Fr.) Donk
Crepidotus applanatus (Pers.) P. Karst. A OU
Crepidotus fulvifibrillosus Murrill OU var. *fulvifibrillosus*
— var. *meristocystis* (Singer) E. Horak
Crepidotus improvisus (E. Horak) T.W. May & A.E. Wood
Crepidotus mollis sensu Massee U
Crepidotus nanicus E. Horak E
Crepidotus novae-zealandiae Pilát E
Crepidotus parietalis E. Horak
Crepidotus pezizoides sensu E. Horak U
Crepidotus phillipsii Berk. & Broome OU
Crepidotus sp.'Nile River (PDD87119)' J.A. Cooper (nom. ined.)
Crepidotus subhaustellaris Cleland A OU
Crepidotus uber (Berk. & M.A.Curtis) Sacc. U
Crepidotus variabilis sensu E. Horak U
Crepidotus variegatus E. Horak (nom. ined.) E
Flammulaster disseminatus (E. Horak) E. Horak
Flammulaster foliicola E. Horak E
Flammulaster pulveraceus E. Horak E
Inocybe aequalis (E. Horak) Garrido E
Inocybe albovestita E. Horak E
Inocybe amygdalina (E. Horak) Garrido E
Inocybe calamistratoides E. Horak E
Inocybe cerea E. Horak E
Inocybe ciliata E. Horak
Inocybe corydalina Quél. OU
Inocybe curvipes P. Karst. U
Inocybe destruens E. Horak E
Inocybe geophila (Bull.) P. Kumm. A
Inocybe graveolens (E. Horak) Garrido E
Inocybe horakomyces Garrido
Inocybe irregularis E. Horak (nom. ined.) E
Inocybe jurana (Pat.) Sacc. OU
Inocybe kuehneri Stangl & J. Veselský OU
Inocybe lacera (Fr.) P. Kumm. A var. *lacera*
— var. *helobia* Kuyper A
Inocybe lanuginosa (Bull.) Quél. OU
Inocybe latericia E. Horak
Inocybe leptospermi (E. Horak) Garrido E
Inocybe luteobulbosa E. Horak var. *luteobulbosa* E
— var. *volvata* E. Horak E
Inocybe maculata sensu auct. NZ A
Inocybe magnibulbosa E. Horak (nom. ined.) E
Inocybe manukanea (E. Horak) Garrido E
Inocybe mendica E. Horak E
Inocybe paracerasphora (E. Horak) Garrido E
Inocybe phaeosquarrosa E. Horak E
Inocybe poculata E. Horak (nom. ined.) E
Inocybe posterula (Britzelm.) Sacc. A
Inocybe renispora E. Horak E
Inocybe rimosa (Bull.) P. Kumm. A
Inocybe sambucina (Fr.) Quél. OU
Inocybe scabriuscula E. Horak E
Inocybe scissa (E. Horak) Garrido E
Inocybe sindonia (Fr.) P. Karst. A
Inocybe straminea (E. Horak) Garrido E
Inocybe strobilomyces E. Horak E
Inocybe subclavata (E. Horak) Garrido E
Inocybe umbrosa E. Horak E
Inocybe viscata (E. Horak) Garrido E
Phaeomarasmius aureosimilis E. Horak E
Phaeosolenia densa (Berk.) W.B. Cooke
Pleuroflammula praestans E. Horak E
Simocybe austrorubi E. Horak
Simocybe luteomellea E. Horak E
Simocybe phlebophora E. Horak E
Simocybe pruinata E. Horak E
Simocybe tabacina E. Horak E
Simocybe unica E. Horak E
Tubaria crobula sensu Massee
Tubaria furfuracea (Pers.) Gillet A
Tubaria hiemalis Romagn. ex Bon OU
Tubaria inquilina sensu Massee
Tubaria rufofulva (Cleland) D.A. Reid & E. Horak
LYOPHYLLACEAE
Calocybe onychina (Fr.) Donk
Calocybe readiae (G. Stev.) E. Horak E
Lyophyllum connatum (Schumach.) Singer OU
Tephrocybe atrata (Fr.) Donk U
Tephrocybe confusa (P.D. Orton) P.D. Orton OU
Tephrocybe tesquorum (Fr.) M.M. Moser
MARASMIACEAE
Anastrophella macrospora E. Horak & Desjardin E
Anthracophyllum archeri (Berk.) Pegler
Anthracophyllum glaucophyllum (Cooke & Massee) Segedin E
Anthracophyllum pallidum Segedin E
Calyptella hebe (G. Cunn.) W.B. Cooke E
Calyptella totara (G. Cunn.) W.B. Cooke E
Campanella fimbriata Segedin E
Campanella olivaceonigra (E. Horak) T.W. May & A.E. Wood
Campanella tristis (G. Stev.) Segedin
Campanella vinosolivida Segedin E
Chaetocalathus cocciformis (Berk.) E. Horak E
Clitocybula grisella (G. Stev. & G.M. Taylor) E. Horak E
Collybiopsis rimutaka (G. Stev.) E. Horak E
Crinipellis filiformis G. Stev. E
Crinipellis procera G. Stev. E
Crinipellis roseola G. Stev. E
Crinipellis scabella (Alb. & Schwein.) Murrill A
Crinipellis substipitaria G. Stev. E
Gerronema fibula sensu E. Horak
Gerronema sp.'Ohakune (PDD80757)' J.A. Cooper

(nom. ined.)
Gymnopus benoistii (Boud.) Antonín & Noordel. OU
Gymnopus biformis (Peck) Halling
Gymnopus foetidus (Sowerby) J.L. Mata & R.H. Petersen U OU
Gymnopus readiae (G. Stev.) J.L. Mata E
Gymnopus sp.'Lincoln (PDD81072)' J.A. Cooper (nom. ined.)
Gymnopus sp.'Pororari River (PDD87069)' J.A. Cooper (nom. ined.)
Gymnopus villosipes (Cleland) Desjardin, Halling & B.A. Perry A
Henningsomyces candidus (Pers.) Kuntze
Hydropus ardesiacus (G. Stev. & G.M. Taylor) Singer E
Hydropus funebris (Speg.) Singer
Hydropus sp.'Kennedy's Bush (PDD86896)' J.A. Cooper (nom. ined.)
Lentinula novae-zelandiae (G. Stev.) Pegler E
Macrocystidia reducta E. Horak & Capellano E
Marasmiellus affixus (Berk.) Singer OU
Marasmiellus bonii Segedin E
Marasmiellus omphaloides G. Stev. E
Marasmiellus pseudoparaphysatus U
Marasmiellus sp.'Ahuriri Reserve (PDD80989)' J.A. Cooper (nom. ined.)
Marasmiellus sp.'Hinewai (PDD86816)' J.A. Cooper (nom. ined.)
Marasmiellus sp.'McCleans Island (PDD87681)' J.A. Cooper (nom. ined.)
Marasmiellus sp.'Okuti Valley (PDD87290)' J.A. Cooper (nom. ined.)
Marasmiellus vaillantii (Pers.) Singer OU
Marasmiellus violaceogriseus (G. Stev.) E. Horak E
Marasmius aciculiformis var. *albus* Dennis OU
Marasmius alliaceus (Jacq.) Fr. U
Marasmius androsaceus sensu Colenso
Marasmius atrocastaneus G. Stev. E
Marasmius aucklandicus Henn. E
Marasmius aurantiobasalis Desjardin & E. Horak var. *aurantiobasalis*
Marasmius bellus Berk. U
Marasmius caperatus sensu Berk. U
Marasmius croceus G. Stev.
Marasmius curraniae G. Stev. E
Marasmius delicatus (G. Stev.) E. Horak E
Marasmius elegans (Cleland) Grgur.
Marasmius erythropus sensu Massee
Marasmius exocarpi sensu Colenso
Marasmius exustoides Desjardin & E. Horak E
Marasmius fishii G. Stev. & G.M. Taylor E
Marasmius foetidus sensu Colenso
Marasmius gelatinosipes Desjardin & E. Horak E
Marasmius haematocephalus (Mont.) Fr. OU
Marasmius haematocephalus sensu Colenso
Marasmius impudicus sensu Massee U
Marasmius insititius sensu Colenso
Marasmius kanukaneus G. Stev.
Marasmius masoniae G. Stev. E
Marasmius meridionalis E. Horak & Desjardin E
Marasmius micraster Petch
Marasmius oreades (Bolton) Fr. A
Marasmius otagensis G. Stev. E
Marasmius pallenticeps Singer
Marasmius perpusillus Desjardin & E. Horak E
Marasmius podocarpicola Pennycook E
Marasmius pusillissimus Desjardin & R.H. Petersen E
Marasmius pusio Berk. & M.A. Curtis
Marasmius ramealis sensu Colenso U
Marasmius rhombisporus Desjardin & E. Horak E
Marasmius rhopalostylidis Desjardin & E. Horak E
Marasmius rimuphilus Desjardin & E. Horak E
Marasmius rosulatus Desjardin & R.H. Petersen E
Marasmius rotula (Scop.) Fr. OU
Marasmius sp.'Rotokuru Lakes (PDD80829)' J.A. Cooper (nom. ined.)
Marasmius spaniophyllus sensu Colenso
Marasmius sphaerodermus Speg.
Marasmius subsupinus Berk.
Marasmius subsupinus sensu Massee U
Marasmius tinctorius Massee
Marasmius unilamellatus Desjardin & E. Horak E
Marasmius vaillantii sensu Colenso U
Micromphale sp.'Erua Forest (PDD80771)' J.A. Cooper (nom. ined.)
Micromphale sp.'Kennedy's Bush (PDD81086)' J.A. Cooper (nom. ined.)
Micromphale sp.'Lewis Pass (PDD80154)' J.A. Cooper (nom. ined.)
Micromphale sp.'Lyell Walkway (PDD80157)' J.A. Cooper (nom. ined.)
Micromphale sp.'Mt Holdsworth (PDD87382)' J.A. Cooper (nom. ined.)
Rectipilus fasciculatus (Pers.) Agerer
Rectipilus sulphureus (Sacc. & Ellis) W.B. Cooke
Rhodocollybia butyracea (Bull.) Lennox
Rhodocollybia sp.'Waiora (PDD87541)' J.A. Cooper (nom. ined.)
Tetrapyrgos subdendrophora (Redhead) E. Horak
MYCENACEAE
Collopus epipterygius (Scop.) Earle
Collopus subviscosus (G. Stev.) E. Horak E
Favolaschia austrocyatheae P.R. Johnst. E
Favolaschia calocera R. Heim A
Favolaschia cyatheae P.R. Johnst. E
Favolaschia pustulosa (Jungh.) Kuntze
Heimiomyces atrofulvus (G. Stev.) E. Horak E
Heimiomyces neovelutipes (Hongo) E. Horak
Hemimycena cephalotricha (Joss. ex Redhead) Singer OU
Hemimycena hirsuta (Tode) Singer
Hemimycena reducta E. Horak & Desjardin E
Hemimycena tortuosa (P.D. Orton) Redhead
Insiticia flavovirens (Sacc.) E. Horak ex Segedin E
Insiticia roseoflava (G. Stev.) E. Horak A
Mycena abramsii (Murrill) Murrill OU
Mycena acicula (Schaeff.) P. Kumm. OU
Mycena adscendens (Lasch) Maas Geest. OU
Mycena aetites (Fr.) Quél. OU
Mycena alba (Bres.) Kühner OU
Mycena amicta (Fr.) Quél. A
Mycena ammoniaca Fr. U
Mycena atrocyanea sensu E. Horak U
Mycena atroincrustata Singer
Mycena austroavenacea Singer
Mycena austrofilopes Grgur.
Mycena austrororida Singer
Mycena avellanea Murrill OU
Mycena avenacea sensu G. Stev. U
Mycena bulbosa (Cejp) Kühner OU
Mycena capillaripes Peck A
Mycena capillaris (Schumach.) P. Kumm. A OU
Mycena carmeliana Grgur.
Mycena citrinomarginata Gillet OU
Mycena conicola G. Stev. U
Mycena cystidiosa (G. Stev.) E. Horak
Mycena detrusa Maas Geest. & E. Horak U OU
Mycena dorotheae G.M. Taylor (nom. ined.) U
Mycena filopes sensu Massee U
Mycena galericulata sensu Massee
Mycena galopus (Pers.) P. Kumm.
Mycena globuliformis Segedin E
Mycena helminthobasis var. *novae-zelandiae* E. Horak E
Mycena hiemalis sensu Massee U
Mycena hygrophora G. Stev. E
Mycena inclinata sensu G. Stev. U
Mycena interrupta (Berk.) Sacc.
Mycena kurramulla Grgur. OU
Mycena lactea sensu E. Horak U
Mycena leaiana var. *australis* Dennis
Mycena leptocephala (Pers.) Gillet OU
Mycena lividorubra Segedin E
Mycena mamaku Segedin E
Mycena mariae G. Stev. E
Mycena metata (Fr.) P. Kumm. U
Mycena minirubra G. Stev. & G.M. Taylor E
Mycena miriamae G. Stev. E
Mycena morris-jonesii G. Stev. E
Mycena mucor sensu G. Stev. U
Mycena munyozii Singer
Mycena ochracea (G. Stev.) E. Horak E
Mycena olivaceomarginata (Massee) Massee A
Mycena oratiensis Segedin E
Mycena papillata sensu G. Stev. U
Mycena parabolica sensu G. Stev. U
Mycena parsonsii G. Stev. E
Mycena pelianthina (Fr.) Quél. OU
Mycena pinicola G. Stev. U
Mycena podocarpi Segedin E
Mycena polygramma sensu E. Horak U
Mycena primulina G. Stev. E
Mycena pura (Pers.) P. Kumm. U OU
Mycena rubroglobulosa Segedin E
Mycena sanguinolenta (Alb. & Schwein.) P. Kumm. U
Mycena simia Kühner OU
Mycena sp.'Ahuriri Reserve (PDD80918)' J.A. Cooper (nom. ined.)
Mycena sp.'Crystal Falls (PDD87606)' J.A. Cooper (nom. ined.)
Mycena sp.'Kennedy's Bush (PDD80686)' J.A. Cooper (nom. ined.)
Mycena sp.'Mt Grey (PDD87373)' J.A. Cooper (nom. ined.)
Mycena sp.'Mt Holdsworth (PDD87426)' J.A. Cooper (nom. ined.) E
Mycena sp.'Oparara Arches (PDD87085)' J.A. Cooper (nom. ined.)
Mycena sp.'Perseverence Road (PDD87228)' J.A. Cooper (nom. ined.)
Mycena sp.'Pororari (PDD87077)' J.A. Cooper (nom. ined.)
Mycena sp.'Rotokuru Lakes (PDD80841)' J.A. Cooper (nom. ined.)
Mycena sp.'Waiohine Gorge (PDD87377)' J.A. Cooper (nom. ined.)
Mycena sp.'Whakapapa (PDD80862)' J.A. Cooper (nom. ined.)
Mycena speirea (Fr.) Gillet A
Mycena stylobates (Pers.) P. Kumm. OU
Mycena subdebilis G. Stev. E
Mycena subfragillima G. Stev. E
Mycena ura Segedin E
Mycena vinacea Cleland OU
Mycena vinaceipora Segedin E
Mycenula fuscovinacea (G. Stev.) E. Horak E
Panellus crawfordiae (G. Stev.) Segedin, P.K. Buchanan & J.P. Wilkie E
Panellus ligulatus E. Horak
Panellus minimus (Jungh.) P.R. Johnst. & Moncalvo
Panellus niger G. Stev. E
Panellus pusillus (Pers. ex Lév.) Burds. & O.K. Mill.
Panellus sp.'Montgomery Park (PDD87050)' J.A. Cooper (nom. ined.)
Panellus sp.'Waiohine Gorge (PDD87405)' J.A. Cooper (nom. ined.)
Panellus stypticus (Bull.) P. Karst.
Xeromphalina leonina (Massee) E. Horak
Xeromphalina podocarpi E. Horak E
Xeromphalina testacea E. Horak E
NIACEAE
Flagelloscypha aotearoa (G. Cunn.) Agerer E
Flagelloscypha pseudopanax (G. Cunn.) Agerer E

Flagelloscypha tongariro (G. Cunn.) Agerer E
Lachnella alboviolascens (Alb. & Schwein.) Fr.
Lachnella coprosmae G. Cunn. E
Lachnella nikau G. Cunn. E
Lachnella pyriformis (G. Cunn.) W.B. Cooke E
Lachnella snaresensis W.B. Cooke E
Lachnella turbinata (G. Cunn.) W.B. Cooke E
Lachnella villosa (Pers.) Gillet A
Merismodes anomala (Pers.) Singer
Merismodes bresadolae (Grélet) Singer
Nia vibrissa R.T. Moore & Meyers M
Peyronelina glomerulata G. Arnaud ex P.J. Fisher, J. Webster & D.F. Kane AN, U F
PHYSALACRIACEAE
Armillaria hinnulea Kile & Watling A
Armillaria limonea (G. Stev.) Boesew. E
Armillaria novae-zelandiae (G. Stev.) Herink (nom. illegit.)
Armillaria sp. sensu Hood U
Armillaria sp. ('mellea') sensu auct. NZ U
Armillaria sp.'Orokuni (PDD87529)' J.A. Cooper (nom. ined.)
Cylindrobasidium coprosmae (G. Cunn.) Hjortstam E
Cylindrobasidium evolvens (Fr.) Jülich OU
Cyptotrama asprata (Berk.) Redhead & Ginns
Flammulina stratosa Redhead, R.H. Petersen & Methven E
Flammulina velutipes (Curtis) P.Karst. ex Singer A
Gloiocephala gracilis Desjardin & E. Horak E
Gloiocephala nothofagi Desjardin & E. Horak E
Gloiocephala phormiorum E. Horak & Desjardin E
Gloiocephala rubescens (Segedin) Desjardin & E. Horak
Gloiocephala tibiicystis E. Horak & Desjardin E
Gloiocephala xanthocephala (G. Stev.) Desjardin & E. Horak E
Oudemansiella longipes (Bull.) M.M. Moser U
Oudemansiella radicata (Relhan) Singer OU
Oudemansiella sp. sensu Hood (nom. ined.) U
Physalacria australiensis Corner OU
Physalacria concinna Syd. OU
Physalacria pseudotropica Berthier E
Physalacria stilboidea (Cooke) Sacc.
Xerulina asprata (Berk.) Pegler
PLEUROTACEAE
Hohenbuehelia brunnea G. Stev. E
Hohenbuehelia culmicola Bon A
Hohenbuehelia cyphelliformis (Berk.) O.K. Mill. A
Hohenbuehelia geogenia (DC.) Singer
Hohenbuehelia luteohinnulea (G. Stev.) E. Horak E
Hohenbuehelia luteola G. Stev. E
Hohenbuehelia metuloidea (G. Stev.) E. Horak E
Hohenbuehelia nothofaginea G. Stev. E
Hohenbuehelia petalodes (Bull.) Schulzer A
Hohenbuehelia podocarpinea G. Stev. E
Nematoctonus concurrens Drechsler AN A
Nematoctonus leiosporus Drechsler AN A
Nematoctonus robustus F.R. Jones AN A
Pleurotus affixus sensu Massee U
Pleurotus applicatus sensu Massee U
Pleurotus australis (Cooke & Massee) Sacc.
Pleurotus chioneus sensu E. Horak U
Pleurotus diversipes sensu Massee U
Pleurotus djamor (Rumph. ex Fr.) Boedijn
Pleurotus floridanus Singer U
Pleurotus mitis sensu Massee U
Pleurotus novae-zelandiae (Berk.) Sacc.
Pleurotus ostreatus sensu Massee U
Pleurotus parsonsiae G. Stev. E
Pleurotus porrigens sensu Massee U
Pleurotus pulmonarius (Fr.) Quél. A
Pleurotus purpureoolivaceus (G. Stev.) Segedin, P.K. Buchanan & J.P. Wilkie E
Pleurotus sajor-caju (Fr.) Singer A
Pleurotus scabriusculus sensu Massee U
Pleurotus serotinus (Pers.) P. Kumm. OU
Pleurotus velatus Segedin, P.K. Buchanan & J.P. Wilkie E
Pleurotus viscidulus (Berk. & Broome) Cleland U
PLUTEACEAE
Pluteus atricapillus (Batsch) Fayod OU
Pluteus atromarginatus (Konrad) Kühner A
Pluteus cervinus (Schaeff.) P. Kumm. A
Pluteus cervinus sensu Massee U
Pluteus concentricus E. Horak E
Pluteus decoloratus E. Horak E
Pluteus flammipes var. *depauperatus* E. Horak
Pluteus hispidilacteus E. Horak E
Pluteus inconspicuus E. Horak U
Pluteus microspermus E. Horak E
Pluteus minor G. Stev. E
Pluteus nanus (Pers.) P. Kumm. A
Pluteus paradoxus E. Horak E
Pluteus pauperculus E. Horak
Pluteus perroseus E. Horak
Pluteus petasatus (Fr.) Gillet U
Pluteus phlebophoroides Henn. E
Pluteus readiarum G. Stev. E
Pluteus similis E. Horak U
Pluteus spegazzinianus Singer U
Pluteus subantarcticus E. Horak E
Pluteus terricola E. Horak E
Pluteus umbrosus (Pers.) P.Kumm. OU
Pluteus velutinornatus G. Stev. E
Volvaria primulina (Cooke & Massee) Sacc.
Volvariella gloiocephala (DC.) Boekhout & Enderle A
Volvariella hypopithys (Fr.) M.M. Moser A
Volvariella surrecta (Knapp) Singer A
Volvariella taylorii (Berk. & Broome) Singer A OU
PSATHYRELLACEAE
Coprinellus deliquescens (Bull.) P. Karst. OU
Coprinellus disseminatus (Pers.) J.E. Lange OU
Coprinellus heptemerus (M. Lange & A.H. Sm.) Vilgalys, Hopple & Jacq. Johnson A
Coprinellus micaceus (Bull.) Vilgalys, Hopple & Jacq. Johnson A
Coprinellus radians (Desm.) Vilgalys, Hopple & Jacq. Johnson A OU
Coprinopsis ammophilae (Courtec.) Redhead, Vilgalys & Moncalvo OU
Coprinopsis atramentaria (Bull.) Redhead, Vilgalys & Moncalvo A
Coprinopsis austrophlyctidospora Fukiharo
Coprinopsis cinerea (Schaeff.) Redhead, Vilgalys & Moncalvo A
Coprinopsis echinospora (Buller) Redhead, Vilgalys & Moncalvo U
Coprinopsis filamentifera (Kühner) Redhead, Vilgalys & Moncalvo OU
Coprinopsis friesii (Quél.) P. Karst. A
Coprinopsis laanii (Kits van Wav.) Redhead, Vilgalys & Moncalvo OU
Coprinopsis lagopus (Fr.) Redhead, Vilgalys & Moncalvo U
Coprinopsis macrocephala (Berk.) Redhead, Vilgalys & Moncalvo A
Coprinopsis nivea (Pers.) Redhead, Vilgalys & Moncalvo A
Coprinopsis phlyctidospora (Romagn.) Redhead, Vilgalys & Moncalvo
Coprinopsis radiata (Bolton) Redhead, Vilgalys & Moncalvo A
Coprinopsis semitalis (P.D. Orton) Redhead, Vilgalys & Moncalvo U
Coprinopsis stercorea (Fr.) Redhead, Vilgalys & Moncalvo A
Cystoagaricus strobilomyces (Murrill) Singer
Lacrymaria lacrymabunda (Bull.) Pat. A
Parasola auricoma (Pat.) Redhead, Vilgalys & Hopple A
Parasola hemirobia (Fr.) Redhead, Vilgalys & Hopple A
Parasola misera (P. Karst.) Redhead, Vilgalys & Hopple A
Parasola plicatilis (Curtis) Redhead, Vilgalys & Hopple U
Psathyrella ammophila (Lév. & Durieu) P.D. Orton U
Psathyrella asperospora (Cleland) Guzmán, Band.-Muñoz & Montoya OU
Psathyrella bipellis aff. (Quél.) A.H. Sm. U OU
Psathyrella candolleana (Fr.) Maire A
Psathyrella conopila (Fr.) Konrad & Maubl. U
Psathyrella echinata (Cleland) Grgur.
Psathyrella gracilis (Fr.) Quél. A
Psathyrella macquariensis Singer
Psathyrella microrhiza (Lasch) Konrad & Maubl. OU
Psathyrella multipedata (Peck) A.H. Sm. U
Psathyrella prona (Fr.) Gillet U
Psilocybe brunneoalbescens Y.S. Chang ex Ratkowsky & G.M. Gates
Psilocybe subviscida (Peck) Kauffman U
PTERULACEAE
Deflexula fascicularis (Bres. & Pat.) Corner
Pterula epiphylloides Corner OU
Pterula stipata Corner
Pterula tenuissima (M.A. Curtis) Corner
Pterula verticillata Corner
Radulomyces confluens (Fr.) M.P. Christ.
Radulomyces rickii (Bres.) M.P. Christ.
SCHIZOPHYLLACEAE
Auriculariopsis ampla (Lév.) Maire
Schizophyllum commune Fr. U
STROPHARIACEAE
Agrocybe acericola (Peck) Singer A
Agrocybe broadwayi (Murrill) Dennis A
Agrocybe erebia (Fr.) Kühner U
Agrocybe howeana (Peck) Singer U
Agrocybe molesta (Lasch) Singer U
Agrocybe olivacea Watling & G.M. Taylor E
Agrocybe parasitica G. Stev.
Agrocybe pediades (Fr.) Fayod A
Agrocybe praecox (Pers.) Fayod (nom. inv.) A
Agrocybe puiggarii (Speg.) Singer OU
Agrocybe putaminum (Maire) Singer A
Agrocybe semiorbicularis (Bull.) Fayod A
Agrocybe sp. 2 sensu Watling & G.M. Taylor U
Flammula chrysotricha sensu Segedin
Flammula croesus (Berk. & M.A. Curtis) Sacc. E
Flammula hyperion sensu Massee
Flammula inopus sensu Massee
Flammula purpureonitens sensu Massee
Flammula sapinea sensu Massee
Flammula schinziana Henn. E
Flammula spumosa sensu Massee U
Flammula tilopus sensu Massee
Flammula vinosa sensu Massee
Galerina atkinsoniana A.H. Sm. OU
Galerina austrocalyptrata A.H. Sm. & Singer OU
Galerina calyptrata P.D. Orton OU
Galerina decipiens A.H. Sm. & Singer OU
Galerina excentrica E. Horak E
Galerina hypnorum (Schrank) Kühner OU
Galerina nana (Petri) Kühner
Galerina nasuta (Kalchbr.) Pegler OU
Galerina nothofaginea E. Horak E
Galerina paludosa (Fr.) Kühner OU
Galerina patagonica Singer
Galerina sp. 1 sensu Watling & G.M. Taylor
Galerina sp. 2 sensu Watling & G.M. Taylor
Galerina sp. 3 sensu Watling & G.M. Taylor
Galerina sp.'Blyth Track (PDD80792) J.A. Cooper (nom. ined.)
Galerina unicolor (Vahl) Singer OU
Galerina vittiformis (Fr.) Earle OU

Gymnopilus chrysopellus (Berk. & M.A. Curtis) Murrill OU
Gymnopilus crociphyllus (Sacc.) Pegler
Gymnopilus junonius (Fr.) P.D. Orton U
Gymnopilus mesosporus E. Horak E
Gymnopilus penetrans (Fr.) Murrill OU
Gymnopilus purpuratus (Cooke & Massee) Singer A
Gymnopilus pyrrhum (Berk. & M.A. Curtis) B.J. Rees
Gymnopilus sapineus (Fr.) Maire OU
Hebeloma aminophilum R.N. Hilton & O.K. Mill.
Hebeloma cavipes Huijsman OU
Hebeloma crustuliniforme (Bull.) Quél. A
Hebeloma hiemale Bres. U
Hebeloma leucosarx P.D. Orton OU
Hebeloma lutense Romagn. OU
Hebeloma mediorufum Soop E
Hebeloma mesophaeum (Pers.) Quél. A
Hebeloma populinum Romagn. A
Hebeloma sacchariolens Quél. A
Hebeloma sinapizans (Paulet) Gillet OU
Hebeloma stenocystis J. Favre OU
Hebeloma vejlense Vesterh. OU
Hebeloma velutipes Bruchet OU
Hebeloma victoriense A.A. Holland & Pegler
Hebeloma vinosophyllum Hongo OU
Hymenogaster viscidus (Massee & Rodway) C.W. Dodge & Zeller
Hypholoma acutum (Cooke) E. Horak E
Hypholoma appendiculatum sensu Massee
Hypholoma brunneum (Massee) D.A. Reid
Hypholoma capnoides (Fr.) P. Kumm. OU
Hypholoma elongatum (Pers.) Ricken OU
Hypholoma fasciculare (Huds.) P. Kumm.
Hypholoma frowardii (Speg.) Garrido OU
Hypholoma marginatum (Fr.) J. Schröt. OU
Hypholoma stuppeum (Berk.) Sacc. E
Hypholoma sublateritium (Fr.) Quél. A
Hypholoma tuberosum Redhead & Kroeger A
Kuehneromyces mutabilis (Schaeff.) Singer & A.H. Sm. U
Leratiomyces ceres (Cooke & Massee) Bridge & Spooner A
Leratiomyces erythrocephalus (Tul. & C. Tul.) Beever & D.-C. Park
Melanotus citrisporus E. Horak E
Melanotus hepatochrous (Berk.) Singer
Melanotus phillipsii (Berk. & Broome) Singer A
Melanotus vorax E. Horak E
Naucoria aurora Sacc.
Naucoria cerodes sensu E. Horak U
Naucoria erinacea sensu E. Horak U
Naucoria escharoides (Fr.) P. Kummer A
Naucoria fraterna sensu Massee U
Naucoria melinoides sensu Massee U
Naucoria salicis P.D. Orton U
Naucoria sideroides sensu Massee U
Naucoria siparia sensu Massee U
Naucoria temulenta sensu Massee U
Nivatogastrium baylisianum E. Horak E
Nivatogastrium lignicola E. Horak E
Nivatogastrium sp.'Pororari (PDD87061)' J.A. Cooper (nom. ined.)
Nivatogastrium sulcatum E. Horak E
Pholiota aegerita (V. Brig.) Quél. OU
Pholiota alnicola (Fr.) Singer U
Pholiota aurivella (Batsch) P. Kumm. A
Pholiota carbonaria A.H. Sm. A
Pholiota cerea E. Horak (nom. ined.)
Pholiota chrysmoides Soop E
Pholiota conissans (Fr.) M.M. Moser U
Pholiota glutinosa (Massee) E. Horak E
Pholiota gummosa (Lasch) Singer OU
Pholiota heteroclita sensu E. Horak U
Pholiota lenta (Pers.) Singer U
Pholiota malicola (Kauffman) A.H. Sm. OU
Pholiota multicingulata E. Horak var. *multicingulata*
— var. *hanmerensis* Soop E
Pholiota mutabilis sensu Massee U
Pholiota psathyrelloides Singer
Pholiota pumila sensu Massee U
Pholiota scamba (Fr.) M.M. Moser OU
Pholiota sp. sensu Soop
Pholiota sp.'Hinewai (PDD80269)' J.A. Cooper (nom. ined.)
Pholiota spumosa (Fr.) Singer OU
Pholiota squarrosa (Vahl) P. Kumm. U
Pholiota squarrosa sensu Massee U
Pholiota squarrosipes Cleland U
Pholiota squarrosoides (Peck) Sacc. U
Pholiota subflammans (Speg.) Sacc.
Psilocybe argentina (Speg.) Singer A
Psilocybe aucklandiae Guzmán, C.C. King & Bandala U
Psilocybe coprophila (Bull.) P. Kumm. A
Psilocybe makarorae P.R. Johnst. & P.K. Buchanan E
Psilocybe merdaria (Fr.) Ricken A
Psilocybe novae-zelandiae Guzmán & E. Horak E
Psilocybe physaloides (Bull.) Quél. A OU
Psilocybe sabulosa Peck OU
Psilocybe semilanceata (Fr.) P. Kumm. U
Psilocybe sp. 1 sensu Watling & G.M. Taylor
Psilocybe sp. 2 sensu Watling & G.M. Taylor
Psilocybe subaeruginosa Cleland A
Psilocybe subcoprophila (Britzelm.) Sacc. A
Stropharia aeruginosa (Curtis) Quél. A
Stropharia caerulea Kreisel (nom. illegit.) A OU
Stropharia coronilla (Bull. ex DC.) Quél. A
Stropharia lepiotiformis (Cooke & Massee) Sacc. U
Stropharia pseudocyanea (Desm.) Morgan OU
Stropharia rugosoannulata Farl. ex Murrill A
Stropharia semiglobata (Batsch) Quél. A OU
Stropharia semiglobata sensu Massee U
Stropharia sp.'Kennedy's Bush (PDD79791)' J.A. Cooper (nom. ined.)
TRICHOLOMATACEAE
Aeruginospora furfuracea E. Horak E
Cantharellula alpina G. Stev. E
Cantharellula waiporiensis (G. Stev.) E. Horak E
Cellypha goldbachii (Weinm.) Donk A
Clitocybe albida (G. Stev.) E. Horak E
Clitocybe clitocyboides (Cooke & Massee) Pegler
Clitocybe dealbata (Sowerby) P. Kumm. A
Clitocybe ericetorum Quél. OU
Clitocybe fragrans (With.) P. Kumm. A
Clitocybe infundibuliformis sensu Massee U
Clitocybe odora (Bull.) P. Kumm. OU
Clitocybe paraditopa Cleland & Cheel
Clitocybe phaeophthalma (Pers.) Kuyper OU
Clitocybe vibecina (Fr.) Quél. OU
Clitocybe wellingtonensis G.M. Taylor & G. Stev. E
Collybia acervata sensu Massee U
Collybia cockaynei (G. Stev.) Desjardin & E. Horak E
Collybia distorta sensu Massee U
Collybia druceae (G. Stev.) E. Horak E
Collybia dryophila sensu Massee U
Collybia incarnata G. Stev. E
Collybia inolens sensu G. Stev. U
Collybia kidsoniae (G. Stev.) E. Horak E
Collybia laccatina sensu Massee U
Collybia nummularia sensu Massee U
Collybia vinacea (G. Stev.) E. Horak E
Collybia xanthopus sensu Massee U
Conchomyces bursiformis (Berk.) E. Horak
Dermoloma hemisphaericum (G. Stev.) E. Horak E
Dermoloma murinum (G.M. Taylor & G. Stev.) E. Horak E
Fayodia granulospora G. Stev. A
Fayodia pseudoclusilis (Joss. & Konrad) Singer OU
Lepista caespitosa (Bres.) Singer A
Lepista fibrosissima Singer A
Lepista irina (Fr.) H.E. Bigelow A
Lepista luscina (Fr.) Singer OU
Lepista nebularis (Batsch) Harmaja A
Lepista nuda (Bull.) Cooke A
Lepista panaeolus (Fr.) P. Karst. U
Lepista sp.'Scarboro Cliffs (PDD87164)' J.A. Cooper (nom. ined.)
Leucopaxillus cerealis var. *piceinus* (Peck) H.E. Bigelow A
Leucopaxillus giganteus (Sibth.) Singer A
Leucopaxillus lilacinus Bougher
Melanoleuca arcuata (Bull.) Singer A
Melanoleuca brevipes (Bull.) Pat. OU
Melanoleuca exscissa (Fr.) Singer var. *exscissa* OU
Melanoleuca vinosa (G. Stev.) E. Horak E
Melanoleuca vulgaris (Pat.) Pat. A OU
Mniopetalum bisporum Singer
Mniopetalum bryophilum (Pers.) Donk
Mniopetalum megalosporum Singer
Mycenella margaritispora (J.E. Lange) Singer
Mycenella minima Singer
Omphalia colensoi (Berk.) Sacc.
Omphalia umbellifera sensu Massee U
Omphalina foetida (G. Stev.) E. Horak E
Omphalina nothofaginea (G. Stev.) E. Horak E
Omphalina pyxidata sensu E. Horak U
Omphalina wellingtonensis G. Stev. E
Phaeomycena fusca G. Stev. & G.M. Taylor E
Pleurocollybia cremea (G. Stev.) E. Horak E
Pleurotopsis longinqua (Berk.) E. Horak
Pleurotopsis roseola (G. Stev.) E. Horak E
Pleurotopsis subgrisea (G. Stev.) E. Horak E
Porpoloma amyloideum (G. Stev.) E. Horak E
Pseudoarmillariella fistulosa (G. Stev.) E. Horak E
Resupinatus applicatus (Batsch) Gray OU
Resupinatus huia (G. Cunn.) Thorn, Moncalvo & Redhead E
Resupinatus merulioides Redhead & Nagas. U
Resupinatus poriaeformis (Pers.) Thorn, Moncalvo & Redhead
Resupinatus trichotis (Pers.) Singer
Rhodocyphella cupuliformis (Berk. & Ravenel) W.B. Cooke
Ripartites helomorphus (Fr.) P. Karst. U
Squamanita squarrulosa G.S. Ridl. E
Tricholoma albobrunneum (Pers.) P. Kumm. A
Tricholoma atrosquamosum (Chevall.) Sacc. A
Tricholoma brevipes sensu Massee U
Tricholoma bubalinum (G. Stev.) E. Horak E
Tricholoma carneum sensu Massee U
Tricholoma cartilagineum sensu Massee U
Tricholoma elegans G. Stev. E
Tricholoma flavobrunneum (Fr.) P. Kumm.
Tricholoma fracticum (Britzelm.) Kreisel A
Tricholoma intermedium Peck U OU
Tricholoma matsutake (S. Ito & S. Imai) Singer A
Tricholoma persicinum (Fr.) Quél. U OU
Tricholoma pessundatum (Fr.) Quél. A
Tricholoma psammopus (Kalchbr.) Quél. A
Tricholoma saponaceum (Fr.) P. Kumm. A var. *saponaceum*
— var. *squamosum* (Cooke) Rea U
Tricholoma stans (Fr.) Sacc. A
Tricholoma sulphureum (Bull.) P. Kumm. A
Tricholoma terreum (Schaeff.) P. Kumm. A
Tricholoma terreum sensu Massee U
Tricholoma testaceum G. Stev. A
Tricholoma ustale (Fr.) P. Kumm. A OU
Tricholoma viridiolivaceum G. Stev. E
Tricholomopsis ornaticeps (G. Stev.) E. Horak E
Tricholomopsis rutilans (Schaeff.) Singer U
TYPHULACEAE
Macrotyphula defibulata R.H. Petersen E
— f. *pallida* R.H. Petersen E

Macrotyphula juncea (Fr.) Berthier OU
Macrotyphula rhizomorpha R.H. Petersen E
Pistillaria ovata (Pers.) Fr. U
Typhula setipes (Grev.) Berthier OU

Order ATHELIALES
ATHELIACEAE
Amphinema byssoides (Pers.) J. Erikss.
Athelia acrospora Jülich OU
Athelia arachnoidea (Berk.) Jülich U
Athelia decipiens (Höhn. & Litsch.) J. Erikss. OU
Athelia epiphylla Pers.
Athelia fibulata M.P. Christ. OU
Athelia pellicularis (P. Karst.) Donk A
Athelia rolfsii (Curzi) C.C. Tu & Kimbr. A
Athelopsis bananispora (Boidin & Gilles) Hjortstam
Athelopsis glaucina (Bourdot & Galzin) Parmasto OU
Athelopsis lembospora (Bourdot) Oberw.
Digitatispora marina Doguet M
Hypochniciellum oblongisporum (G. Cunn.) Gresl. & Rajchenb.
Leptosporomyces galzinii (Bourdot) Jülich
Leptosporomyces mutabilis (Bres.) Krieglst.
Melzericium udicola (Bourdot) Hauerslev
Piloderma byssinum (P. Karst.) Jülich U
Taeniospora gracilis Marvanová AN

Order AURICULARIALES
Basidiodendron caesiocinereum (Höhn. & Litsch.) Luck-Allen
Basidiodendron cinereum (Bourdot & Galzin) Luck-Allen
Basidiodendron cremeum (McNabb) K. Wells & Raitv.
Basidiodendron minutisporum (McNabb) Wojewoda E
Basidiodendron nikau (McNabb) Wojewoda E
Basidiodendron pini (H.S. Jacks. & G.W. Martin) Luck-Allen A
Bourdotia galzinii (Bres.) Bres. & Torrend
Ductifera sucina (Möller) K. Wells U
Heterochaetella brachyspora Luck-Allen OU
Protomerulius caryae (Schwein.) Ryvarden
Pseudohydnum gelatinosum (Scop.) P. Karst.
Stypella dubia (Bourdot & Galzin) P. Roberts
Stypella grilletii (Boud.) P. Roberts
Stypella subgelatinosa (P. Karst.) P. Roberts U
Stypella vermiformis (Berk. & Broome) D.A. Reid OU
Tremellodendropsis flagelliformis (Berk.) D.A. Crawford
Tremellodendropsis inflata (D.A. Crawford) R.H. Petersen
Tremellodendropsis pusio (Berk.) D.A. Crawford
Tremellodendropsis sp. 1 sensu R.H. Petersen E
Tremellodendropsis sp. 2 sensu R.H. Petersen
Tremellodendropsis tuberosa (Grev.) D.A. Crawford OU
AURICULARIACEAE
Auricularia auricula (L.) Underw.
Auricularia cornea Ehrenb.
Auricularia delicata (Fr.) Henn. OU
Auricularia fuscosuccinea (Mont.) Henn. A
Eichleriella hoheriae McNabb E
Eichleriella spinulosa (Berk. & M.A. Curtis) Burt
Eichleriella subleucophaea McNabb E
Exidia albida (Huds.) Bref. OU
Exidia glandulosa (Bull.) Fr.
Exidia nigricans (With.) P. Roberts
Exidia novozealandica Lloyd E
Exidia nucleata (Schwein.) Burt
Exidiopsis grisea (Pers.) Bourdot & L. Maire
Exidiopsis mucedinea (Pat.) K. Wells
Exidiopsis novae-zelandiae (McNabb) Wojewoda E
Exidiopsis tawa (McNabb) Wojewoda E
Heterochaete delicata Bres.
Pseudostypella nothofagi McNabb E
Pseudostypella translucens (H.D. Gordon) D.A. Reid & Minter

Order BOLETALES
AMYLOCORTICIACEAE
Amyloathelia amylacea (Bourdot & Galzin) Hjortstam & Ryvarden
Amyloathelia crassiuscula Hjortstam & Ryvarden OU
Amylocorticium cebennense (Bourdot) Pouzar
Ceraceomyces cerebrosus (G. Cunn.) Stalpers & P.K. Buchanan
Ceraceomyces tessulatus (Cooke) Jülich OU
Ceraceomyces variicolor (G. Cunn.) Stalpers E
Podoserpula pusio (Berk.) D.A. Reid var. *pusio*
— var. *tristis* D.A. Reid E
BOLETACEAE
Austroboletus lacunosus (Kuntze) T.W. May & A.E. Wood
Austroboletus niveus (G. Stev.) Wolfe E
Boletellus ananas (M.A. Curtis) Murrill A
Boletus brunneus Cooke & Massee OU
Boletus edulis Bull. A
Boletus leptospermi McNabb E
Boletus novae-zelandiae McNabb E
Boletus paradisiacus R. Heim (nom. inv.) E
Boletus rawlingsii McNabb E
Boletus scaber aff. Rawlings
Chalciporus aurantiacus (McNabb) Pegler & T.W.K. Young E
Chalciporus piperatus (Bull.) Bataille A
Fistulinella lutea Redeuilh & Soop E
Leccinum scabrum (Bull.) Gray A
Mucilopilus violaceiporus (G. Stev.) Wolfe
Octaviania hinsbyi (Rodway) G. Cunn. U
Octaviania tasmanica (Kalchbr. ex Massee) Lloyd
Phylloporus novae-zelandiae McNabb E
Tylopilus brunneus (McNabb) Wolfe E
Tylopilus formosus G. Stev.
Xerocomus badius (Fr.) Kühner ex J.-E. Gilbert A OU
Xerocomus chrysenteron (Bull.) Quél. A
Xerocomus cisalpinus Simonini, H. Ladurner & Peintner A
Xerocomus griseoolivaceus McNabb E
Xerocomus lentistipitatus (G. Stev.) McNabb E
Xerocomus macrobbii McNabb E
Xerocomus nothofagi McNabb E
Xerocomus porosporus Imler (nom. inv.) A
Xerocomus ripariellus Redeuilh A OU
Xerocomus rubellus (Krombh.) Moser A
Xerocomus rufostipitatus McNabb E
Xerocomus scabripes McNabb E
Xerocomus squamulosus McNabb E
BOLETINELLACEAE
Phaeogyroporus portentosus (Berk. & Broome) McNabb U
CALOSTOMATACEAE
Calostoma fuscum (Berk.) Massee
Calostoma rodwayi Lloyd
CONIOPHORACEAE
Coniophora arida (Fr.) P. Karst. var. *arida*
— var. *suffocata* (Peck) Ginns A
Coniophora dimitica G. Cunn.
Coniophora minor G. Cunn. E
Coniophora olivacea (Fr.) P. Karst. OU
Coniophora puteana (Schumach.) P. Karst. var. *puteana*
GOMPHIDIACEAE A
Gomphidius maculatus (Scop.) Fr. A
GYROPORACEAE
Gyroporus castaneus (Bull.) Quél.
HYGROPHOROPSIDACEAE
Hygrophoropsis aurantiaca (Wulfen) Maire
Hygrophoropsis umbriceps (Cooke) McNabb E
Leucogyrophana mollusca (Fr.) Pouzar A
Leucogyrophana pinastri (Fr.) Ginns & Weresub A
Leucogyrophana romellii Ginns A OU
Leucogyrophana sororia (Burt) Ginns
PAXILLACEAE
Austrogaster novae-zelandiae D.A. Reid E
Melanogaster ambiguus (Vittad.) Tul. & C. Tul. A
Paxillus involutus (Batsch) Fr. A
Paxillus panuoides (Fr.) Fr. A
RHIZOPOGONACEAE
Rhizopogon hawkerae A.H. Sm. A
Rhizopogon luteolus Fr. A
Rhizopogon parksii A.H. Sm. A
Rhizopogon pseudoroseolus A.H. Sm. A
Rhizopogon roseolus (Corda) Th. Fr. A
Rhizopogon rubescens (Tul. & C. Tul.) Tul. & C. Tul. var. *rubescens* A
Rhizopogon villosulus Zeller A
Rhizopogon vinicolor A.H. Sm. A
Rhizopogon vulgaris (Vittad.) M. Lange U
SCLERODERMATACEAE
Pisolithus albus (Cooke & Massee) Priest ex Bougher & K. Syme (nom. inv.)
Pisolithus marmoratus (Berk.) E. Fisch.
Pisolithus sp. 10 sensu F. Martin, J. Díez, B. Dell & Delaruelle
Pisolithus tinctorius sensu G. Cunn.
Scleroderma albidum Pat. & Trab.
Scleroderma areolatum Ehrenb.
Scleroderma aurantium (L.) Pers. A
Scleroderma bovista Fr. A
Scleroderma cepa Pers. U
Scleroderma verrucosum (Bull.) Pers.
SERPULACEAE
Austropaxillus macnabbii (Singer, J. García & L.D. Gómez) Jarosch E
Austropaxillus nothofagi (McNabb) Bresinsky & Jarosch E
Austropaxillus squarrosus (McNabb) Bresinsky & Jarosch E
Serpula himantioides (Fr.) P. Karst. A
Serpula lacrymans (Wulfen) J. Schröt. U
SUILLACEAE
Suillus brevipes (Peck) Kuntze A
Suillus cavipes (Opat.) A.H. Sm. & Thiers A OU
Suillus granulatus (L.) Roussel A
Suillus grevillei (Klotzsch) Singer A
Suillus lakei (Murrill) A.H. Sm. & Thiers A
Suillus luteus (L.) Roussel A
Suillus pungens Thiers & A.H. Sm. A OU
Suillus subacerbus McNabb A
Suillus variegatus (Sw.) Kuntze A OU

Order CANTHARELLALES
APHELARIACEAE
Aphelaria dendroides (Jungh.) Corner
Tumidapexus ravus D.A. Crawford E
BOTRYOBASIDIACEAE
Botryobasidium aureum Parmasto
Botryobasidium chilense Hol.-Jech.
Botryobasidium conspersum J. Erikss. U
Botryobasidium obtusisporum J. Erikss. OU
Botryobasidium pruinatum (Bres.) J. Erikss.
Botryobasidium subcoronatum (Höhn. & Litsch.) Donk
Botryobasidium vagum (Berk. & M.A. Curtis) D.P. Rogers OU
Botryohypochnus isabellinus (Fr.) J. Erikss.
CANTHARELLACEAE
Cantharellus attenuatus Cleland OU
Cantharellus elsae (G. Stev.) E. Horak E
Cantharellus infundibuliformis Fr. OU
Cantharellus insignis (Cooke) Corner

Cantharellus rugosus Cleland OU
Cantharellus variabilis Schulzer OU
Cantharellus wellingtonensis McNabb E
CERATOBASIDIACEAE
Ceratobasidium cornigerum (Bourdot) D.P. Rogers A
Scotomyces subviolaceus (Peck) Jülich
Thanatephorus cucumeris (A.B. Frank) Donk A
Thanatephorus fusisporus (J. Schröt.) Hauerslev & P. Roberts U
Uthatobasidium ochraceum (Massee) Donk
CLAVULINACEAE
Clavulicium delectabile (H.S. Jacks.) Hjortstam OU
Clavulina alutaceosiccescens R.H. Petersen E
Clavulina amethystina (Bull.) Donk OU
Clavulina brunneocinerea R.H. Petersen E
Clavulina cavipes Corner (nom. ined.) OU
Clavulina cinerea (Bull.) J. Schröt. OU
Clavulina copiosocystidiata R.H. Petersen E
Clavulina cristata var. *zealandica* R.H. Petersen E
Clavulina floridana (Singer) Corner
Clavulina geoglossoides Corner
Clavulina hispidulosa Corner, K.S. Thind & Anand
Clavulina humilis (Cooke) Corner
Clavulina leveillei (Sacc.) Overeem var. *leveillei*
— var. *atricha* Corner
Clavulina mussooriensis Corner, K.S. Thind & Dev OU
Clavulina purpurea R.H. Petersen E
Clavulina rugosa (Bull.) J. Schröt. OU
Clavulina samuelsii R.H. Petersen E
Clavulina septocystidiata R.H. Petersen E
Clavulina subrugosa (Cleland) Corner var. *subrugosa*
— var. *tenuis* R.H. Petersen E
Clavulina urnigerobasidiata R.H. Petersen E
Clavulina vinaceocervina var. *avellanea* R.H. Petersen E
Multiclavula coronilla (G.W. Martin) R.H. Petersen U L
Multiclavula corynoides (Peck) R.H. Petersen L
Multiclavula mucida (Pers.) R.H. Petersen U L
Multiclavula samuelsii R.H. Petersen E L
HYDNACEAE
Hydnum crocidens Cooke
— var. *badium* McNabb E
— var. *crocidens*
— var. *wellingtonii* (Lloyd) McNabb E
Hydnum molluscum Fr. OU
Hydnum repandum L. OU var. *repandum*
— var. *albidum* (Quél.) Rea OU
Hydnum sordidum W. Phillips (nom. illegit.)
Hydnum umbilicatum Peck
Sistotrema brinkmannii (Bres.) J. Erikss.
Sistotrema muscicola (Pers.) S. Lundell A
Sistotrema otagense (G. Cunn.) Stalpers & P.K. Buchanan E
Sistotrema resinicystidium Hallenb. OU
Sistotrema sernanderi (Litsch.) Donk OU

Order CORTICIALES
CORTICIACEAE
Corticium caeruleum (Schrad.) Fr. OU
Corticium filicinum sensu G. Cunn. U
Corticium kauri G. Cunn. E
Corticium lividum sensu G. Cunn.
Corticium lividum sensu G. Cunn.
Corticium otagense G. Cunn.
Corticium perenne G. Cunn. E
Corticium permodicum sensu G. Cunn.
Corticium protrusum Burt OU
Corticium protrusum sensu G. Cunn.
Corticium pteridophilum G. Cunn.
Corticium roseum Pers. OU
Corticium scutellare sensu G. Cunn.
Corticium utriculicum G. Cunn.
Corticium vescum Burt
Dendrothele ampullospora (G. Cunn.) Nakasone & Burds. E
Dendrothele arachispora Nakasone & Burds. E
Dendrothele aucklandica Nakasone & Burds. E
Dendrothele australis Nakasone & Burds. E
Dendrothele biapiculata (G. Cunn.) P.A. Lemke
Dendrothele commixta (Höhn. & Litsch.) J. Erikss. & Ryvarden
Dendrothele corniculata (G. Cunn.) Stalpers E
Dendrothele cymbiformis Nakasone & Burds. E
Dendrothele incrustans (P.A. Lemke) P.A. Lemke
Dendrothele leptostachys Nakasone & Burds. E
Dendrothele magninavicularis Nakasone & Burds. E
Dendrothele navicularis Nakasone & Burds. E
Dendrothele nivosa (Berk. ex Höhn. & Litsch.) P.A. Lemke
Dendrothele nivosa sensu G. Cunn. E
Dendrothele novae-zelandiae Nakasone & Burds. E
Dendrothele pulvinata (G. Cunn.) P.A. Lemke E
Dendrothele subellipsoidea Nakasone & Burds. E
Erythricium salmonicolor (Berk. & Broome) Burds.
Galzinia pedicellata Bourdot
Laetisaria arvalis Burds. U
Laetisaria fuciformis (McAlpine) Burds. A
Punctularia strigosozonata (Schwein.) P.H.B. Talbot
Waitea circinata Warcup & P.H.B. Talbot A

Order GEASTRALES
GEASTRACEAE
Geastrum australe Berk.
Geastrum coronatum Pers. A
Geastrum fimbriatum Fr.
Geastrum floriforme Vittad. A
Geastrum minimum Schwein.
Geastrum morganii Lloyd
Geastrum pectinatum Pers.
Geastrum rufescens Pers. A
Geastrum saccatum Fr.
Geastrum saccatum sensu G. Cunn.
Geastrum smardae V.J. Stan k
Geastrum striatum DC. OU
Geastrum triplex Jungh.
Geastrum velutinum Morgan
Sphaerobolus stellatus Tode

Order GLOEOPHYLLALES
GLOEOPHYLLACEAE
Gloeophyllum abietinum (Bull.) P. Karst. A
Gloeophyllum sepiarium (Wulfen) P. Karst. A
Gloeophyllum trabeum (Pers.) Murrill U
Mycothele disciformis (G. Cunn.) Jülich E
Veluticeps fusispora (G. Cunn.) Hjortstam & Ryvarden E

Order GOMPHALES
CLAVARIADELPHACEAE
Beenakia dacostae D.A. Reid
GOMPHACEAE
Gautieria costata G. Cunn. OU
Gloeocantharellus dingleyae (Segedin) Giachini E
Gloeocantharellus novae-zelandiae (Segedin) Giachini E
Phaeoclavulina flaccida (Fr.) Giachini (nom. ined.)
Phaeoclavulina gigantea (Pat.) Giachini
Phaeoclavulina ochracea (Bres.) Giachini
Phaeoclavulina zealandica (R.H. Petersen) Giachini (nom. ined.) E
Ramaria ambigua R.H. Petersen E
Ramaria anziana R.H. Petersen E
Ramaria aureorhiza R.H. Petersen E
Ramaria australiana (Cleland) R.H. Petersen OU
Ramaria avellaneovertex R.H. Petersen E
Ramaria basirobusta R.H. Petersen E
Ramaria filicicola (S.G.M. Fawc.) Corner
Ramaria formosa (Pers.) Quél. OU
Ramaria fragillima (Sacc. & P. Syd.) Corner
Ramaria gracilis (Fr.) Quél. OU
Ramaria junquilleovertex R.H. Petersen E
Ramaria lorithamnus (Berk.) R.H. Petersen
Ramaria piedmontiana R.H. Petersen
Ramaria polypus Corner
Ramaria purpureopallida R.H. Petersen E
Ramaria rotundispora R.H. Petersen E
Ramaria rubripermanens Marr & D.E. Stuntz A
Ramaria samuelsii R.H. Petersen E
Ramaria sclerocarnosa R.H. Petersen E
Ramaria zippelii var. *gracilis* Corner OU
Ramaricium polyporoideum (Berk. & M.A. Curtis) Ginns
LENTARIACEAE
Kavinia himantia (Schwein.) J. Erikss.
Lentaria boletosporioides R.H. Petersen E
Lentaria glaucosiccescens R.H. Petersen E
Lentaria surculus (Berk.) Corner

Order HYMENOCHAETALES
Sidera lenis (P. Karst.) Miettinen
Sidera lowei (Rajchenb.) Miettinen
Sidera vulgaris (Fr.) Miettinen U
HYMENOCHAETACEAE
Coltricia cinnamomea (Jacq.) Murrill
Coltricia perennis (L.) Murrill
Coltricia salpincta (Cooke) G. Cunn.
Coltricia strigosa G. Cunn. E
Coltriciella dependens (Berk. & M.A. Curtis) Murrill
Cyclomyces tabacinus (Mont.) Pat.
Fuscoporia tawa in Herb. K (nom. ined.)
Hymenochaete bispora G. Cunn. E
Hymenochaete cervina Berk. & M.A. Curtis
Hymenochaete cinnamomea (Pers.) Bres.
Hymenochaete contiformis G. Cunn.
Hymenochaete corrugata (Fr.) Lév.
Hymenochaete cruenta (Pers.) Donk
Hymenochaete dictator G. Cunn. E
Hymenochaete dissimilis G. Cunn.
Hymenochaete dura Berk. & M.A. Curtis
Hymenochaete fuliginosa (Pers.) Lév.
Hymenochaete gladiola G. Cunn. E
Hymenochaete innexa G. Cunn.
Hymenochaete lictor Petch
Hymenochaete magnahypha G. Cunn.
Hymenochaete minuscula G. Cunn.
Hymenochaete nothofagicola Parmasto E
Hymenochaete patelliformis G. Cunn. E
Hymenochaete plurimaesetae G. Cunn. E
Hymenochaete rhabarbarina (Berk.) Cooke
Hymenochaete rheicolor (Mont.) Lév.
Hymenochaete rubiginosa (Dicks.) Lév.
Hymenochaete semistupposa Petch
Hymenochaete separata G. Cunn.
Hymenochaete stratura G. Cunn. E
Hymenochaete tabacina (Sowerby) Lév.
Hymenochaete tasmanica Massee
Hymenochaete tenuissima (Berk.) Berk.
Hymenochaete unicolor Berk. & M.A. Curtis
Hymenochaete vaginata G. Cunn. E
Hymenochaete vallata G. Cunn.
Hymenochaete villosa (Lév.) Bres.
Inonotus chondromyelus Pegler
Inonotus glomeratus sensu G. Cunn.
Inonotus lloydii (Cleland) P.K. Buchanan & Ryvarden
Inonotus nothofagi G. Cunn. E
Inonotus rodwayi D.A. Reid OU
Phellinus conchatus sensu G. Cunn. E
Phellinus contiguus (Pers.) Pat.
Phellinus dingleyae P.K. Buchanan & Ryvarden E
Phellinus ferreus (Pers.) Bourdot & Galzin
Phellinus gilvus (Schwein.) Pat.

Phellinus kamahi (G. Cunn.) P.K. Buchanan & Ryvarden E
Phellinus nothofagi (G. Cunn.) Ryvarden E
Phellinus punctatus (P. Karst.) Pilát
Phellinus robustus (P. Karst.) Bourdot & Galzin
Phellinus senex (Nees & Mont.) Imazeki
Phellinus sublaevigatus (Cleland & Rodway) P.K. Buchanan & Ryvarden
Phellinus tawhai (G. Cunn.) G. Cunn. E
Phellinus wahlbergii (Fr.) D.A. Reid
Phylloporia pectinata (Klotzsch) Ryvarden
Polystictus imbricatus Lloyd U
Polystictus pergamenus sensu Colenso U
SCHIZOPORACEAE
Alutaceodontia alutacea (Fr.) Hjortstam & Ryvarden U
Hyphodontia alienata (S. Lundell) J. Erikss. OU
Hyphodontia alutaria (Burt) J. Erikss. U
Hyphodontia arguta (Fr.) J. Erikss.
Hyphodontia australis (Berk.) Hjortstam
Hyphodontia barba-jobi (Bull.) J. Erikss.
Hyphodontia brevidens (Pat.) Ryvarden OU
Hyphodontia crustosa (Pers.) J. Erikss.
Hyphodontia cunninghamii Gresl. & Rajchenb. E
Hyphodontia fimbriata Sheng H. Wu OU
Hyphodontia lanata Burds. & Nakasone
Hyphodontia nespori (Bres.) J. Erikss. & Hjortstam OU
Hyphodontia pallidula (Bres.) J. Erikss.
Hyphodontia papillosa (Fr.) J. Erikss.
Hyphodontia sambuci (Pers.) J. Erikss.
Hyphodontia serpentiformis Langer OU
Hyphodontia subalutacea (P. Karst.) J. Erikss.
Hyphodontia subscopinella (G. Cunn.) Gresl. & Rajchenb. E
Palifer verecundus (G. Cunn.) Stalpers & P.K. Buchanan E
Rogersella griseliniae (G. Cunn.) Stalpers
Schizopora nothofagi (G. Cunn.) P.K. Buchanan & Ryvarden E
Schizopora radula (Pers.) Hallenb. A
Schizopora sp. sensu B.C. Paulus U
Xylodon scopinellus (Berk.) Hjortstam & Ryvarden E

Order HYSTERANGIALES
GALLACEACEAE
Austrogautieria clelandii G. Cunn. ex E.L. Stewart & Trappe
Austrogautieria costata G. Cunn. ex E.L. Stewart & Trappe OU
Gallacea dingleyae Castellano & Beever E
Gallacea eburnea Castellano & Beever E
Gallacea scleroderma (Cooke) Lloyd E
HYSTERANGIACEAE
Hysterangium gelatinosporum Cribb
Hysterangium inflatum Rodway A
Hysterangium neotunicatum Castellano & Beever
Hysterangium rugisporum Castellano & Beever E
Hysterangium rupticutis Castellano & Beever A
Hysterangium sp. 1 sensu Chu-Chou & Grace U
Hysterangium sp. 6 sensu Chu-Chou & Grace U
Hysterangium youngii Castellano & Beever E
MESOPHELLIACEAE
Chondrogaster pachysporus Maire A
Malajczukia novae-zelandiae (G. Cunn.) Trappe & Castellano E
Mesophellia arenaria Berk. A
Mesophellia glauca (Cooke & Massee) D.A. Reid A
Mesophellia oleifera Trappe, Castellano & Malajczuk U
PHALLOGASTRACEAE
Protubera hautuensis Castellano & Beever E
Protubera nothofagi Castellano & Beever E
Protubera parvispora Castellano & Beever E
TRAPPEACEAE
Phallobata alba G. Cunn. E

Order PHALLALES
CLAUSTULACEAE
Claustula fischeri K.M. Curtis
PHALLACEAE
Aseroe rubra Labill.
Clathrus archeri (Berk.) Dring
Clathrus chrysomycelinus Möller
Clathrus columnatus Bosc
Ileodictyon cibarium Tul. & C. Tul. E
Ileodictyon gracile Berk.
Lysurus cruciatus (Lepr. & Mont.) Henn. A
Mutinus bambusinus (Zoll.) E. Fisch.
Mutinus borneensis Ces.
Mutinus caninus (Huds.) Fr.
Mutinus curtus (Berk.) E. Fisch.
Mutinus ravenelii (Berk. & M.A. Curtis) E. Fisch. A
Phallus hadriani Vent. U
Phallus impudicus L. var. *impudicus*
— var. *togatus* (Kalchbr.) Costantin & L.M. Dufour
Pseudocolus fusiformis (E. Fisch.) Lloyd

Order POLYPORALES
Phlebiella allantospora (Oberw.) K.H. Larss. & Hjortstam
Phlebiella filicina (Bourdot) K.H. Larss. & Hjortstam
Phlebiella sulphurea (Pers.) Ginns & M.N.L. Lefebvre
Phlebiella tulasnelloidea (Höhn. & Litsch.) Ginns & M.N.L. Lefebvre
CYSTOSTEREACEAE
Cystidiodontia artocreas (Berk. & M.A. Curtis ex Cooke) Hjortstam
Cystostereum murrayi (Berk. & M.A. Curtis) Pouzar
Parvobasidium lianicola (G. Cunn.) Stalpers E
FOMITOPSIDACEAE
Anomoporia albolutescens (Romell) Pouzar OU
Anomoporia myceliosa (Peck) Pouzar
Antrodia albida (Fr.) Donk
Antrodia albida sensu sensu auct. NZ
Antrodia malicola (Berk. & M.A. Curtis) Donk
Antrodia novae-zelandiae P.K. Buchanan & Ryvarden E
Antrodia serialis (Fr.) Donk A
Antrodia vaillantii (DC.) Ryvarden A
Antrodia xantha (Fr.) Ryvarden A
Dacryobolus sudans (Alb. & Schwein.) Fr.
Daedalea confragosa sensu Berk.
Fomitopsis lilacinogilva (Berk.) J.E. Wright & J.R. Deschamps A
Fomitopsis maire (G. Cunn.) P.K. Buchanan & Ryvarden E
Fomitopsis nivosa (Berk.) Gilb. & Ryvarden A
Gloeocystidium argillaceum (Bres.) Höhn. & Litsch.
Ischnoderma rosulatum (G. Cunn.) P.K. Buchanan & Ryvarden U
Laetiporus portentosus (Berk.) Rajchenb.
Phaeolus schweinitzii (Fr.) Pat. A
Phaeolus schweinitzii sensu Hood
Postia atrostrigosa (Cooke) Rajchenb. E
Postia brunnea Rajchenb. & P.K. Buchanan
Postia caesia (Schrad.) P. Karst.
Postia dissecta (Cooke) Rajchenb.
Postia fragilis (Fr.) Jülich A
Postia globicystidia P.K. Buchanan & Ryvarden E
Postia manuka (G. Cunn.) P.K. Buchanan & Ryvarden E
Postia pelliculosa (Berk.) Rajchenb.
Postia subcaesia (A. David) Jülich
Postia tephroleuca (Fr.) Jülich U
Postia venata (Rajchenb. & J.E. Wright) Rajchenb.
Sporotrichum dimorphosporum Arx AN, U
GANODERMATACEAE
Amauroderma calcigenum (Berk.) Torrend
Ganoderma applanatum sensu Wakef. E
Ganoderma australe (Fr.) Pat.
Ganoderma chilense (Fr.) Pat. U
Ganoderma sp.'Awaroa' (nom. ined.)
Ganoderma sp. (lucidum)
MERIPILACEAE
Grifola colensoi (Berk.) G. Cunn. E
Grifola frondosa (Dicks.) Gray OU
Grifola gargal Singer OU
Grifola sordulenta (Mont.) Singer
Grifola sp. [fig. 85] sensu Hood
Meripilus giganteus (Pers.) P. Karst. A
Physisporinus rivulosus (Berk. & M.A. Curtis) Ryvarden U
Porotheleum fimbriatum (Pers.) Fr. U
Rigidoporus albostygius (Berk. & M.A. Curtis) Rajchenb.
Rigidoporus aureofulvus (Lloyd) P.K. Buchanan & Ryvarden E
Rigidoporus cartilagineus (Berk. & Broome) Ginns
Rigidoporus concrescens (Mont.) Rajchenb.
Rigidoporus laetus (Cooke) P.K. Buchanan & Ryvarden
Rigidoporus longicystidius P.K. Buchanan & Ryvarden E
Rigidoporus microporus (Sw.) Overeem OU
Rigidoporus vinctus (Berk.) Ryvarden
MERULIACEAE
Abortiporus biennis (Bull.) Singer A
Bjerkandera adusta (Willd.) P. Karst.
Bjerkandera fumosa (Fr.) P. Karst. OU
Bulbillomyces farinosus (Bres.) Jülich F
Gloeoporus dichrous (Fr.) Bres. U
Gloeoporus phlebophorus (Berk.) G. Cunn.
Gloeoporus taxicola (Pers.) Gilb. & Ryvarden
Hyphoderma assimile (H.S. Jacks. & Dearden) Donk
Hyphoderma cremeoalbum (Höhn. & Litsch.) Jülich
Hyphoderma hjortstamii Sheng H. Wu U
Hyphoderma incrustatum K.H. Larss. OU
Hyphoderma litschaueri (Burt) J. Erikss. & Å. Strid
Hyphoderma medioburiense (Burt) Donk
Hyphoderma nudicephalum Gilb. & M. Blackw. OU
Hyphoderma obtusiforme J. Erikss. & Å. Strid
Hyphoderma obtusum J. Erikss. U
Hyphoderma orphanellum (Bourdot & Galzin) Donk OU
Hyphoderma puberum (Fr.) Wallr. OU
Hyphoderma setigerum (Fr.) Donk
Hyphoderma subdefinitum J. Erikss. & Å. Strid OU
Hyphoderma utriculosum (G. Cunn.) Stalpers & P.K. Buchanan E
Hypochnicium aotearoae B.C. Paulus, Nilsson & Hallenb. E
Hypochnicium lyndoniae (D.A. Reid) Hjortstam
Hypochnicium polonense (Bres.) Å. Strid U
Hypochnicium zealandicum (G. Cunn.) Hjortstam E
Irpex archeri sensu G. Cunn.
Irpex flavus sensu Colenso
Irpex ochrosimilis Lloyd E
Irpex sp. sensu Hood, Sandberg & Kimberley U
Junghuhnia brownii (Humb.) Niemelä
Junghuhnia meridionalis (Rajchenb.) Rajchenb.
Junghuhnia nitida (Pers.) Ryvarden
Junghuhnia rhinocephalus (Berk.) Ryvarden
Junghuhnia separabilima (Pouzar) Ryvarden U
Kneiffia subtilis Berk. ex Cooke (nom. inv., nom. illegit.)
Laschia thwaitesii sensu Massee U
Merulius debriscola Lloyd
Mycoacia columellifera (G. Cunn.) Hjortstam E
Mycoacia lutea (G. Cunn.) Hjortstam E
Odontia flexibilis G. Cunn. E
Odontia fragilis G. Cunn. E
Odontia novae-zealandiae G. Cunn. E

Odontia oleifera G. Cunn. E
Odontia stratosa G. Cunn. E
Odontia tessellata G. Cunn. E
Phlebia acanthocystis Gilb. & Nakasone
Phlebia celtidis W.B. Cooke OU
Phlebia chrysocreas (Berk. & M.A. Curtis) Burds.
Phlebia femsioeensis (Litsch. & S. Lundell) J. Erikss. & Hjortstam U
Phlebia leptospermi (G. Cunn.) Stalpers E
Phlebia lilascens (Bourdot) J. Erikss. & Hjortstam
Phlebia livida (Pers.) Bres.
Phlebia longicystidia (Litsch.) Hjortstam & Ryvarden
Phlebia nothofagi (G. Cunn.) Nakasone
Phlebia queletii (Bourdot & Galzin) M.P. Christ.
Phlebia rufa (Pers.) M.P. Christ.
Phlebia sp. [1] sensu Stalpers (nom. ined.)
Phlebia sp. [2] sensu Stalpers
Phlebia subceracea (Wakef.) Nakasone
Phlebia subfascicularis (Wakef.) Nakasone & Gilb.
Phlebia subochracea (Alb. & Schwein.) J. Erikss. & Ryvarden
Phlebia subserialis (Bourdot & Galzin) Donk U
Phlebia totara (G. Cunn.) Stalpers & P.K. Buchanan E
Phlebia tuberculata (Hallenb. & E. Larss.) Ghobad-Nejhad (nom. ined.)
Podoscypha involuta (Klotzsch ex Fr.) Imazeki U
Podoscypha nitidula (Berk.) Pat. U
Podoscypha pergamena (Berk. & M.A. Curtis) Pat.
Podoscypha petalodes (Berk.) Pat.
— subsp. *floriformis* (Bres.) D.A. Reid
Podoscypha venustula (Speg.) D.A. Reid U
Podoscypha venustula (Speg.) D.A. Reid subsp. *venustula* OU
— subsp. *cuneata* D.A. Reid
Scopuloides hydnoides (Cooke & Massee) Hjortstam & Ryvarden
Steccherinum fimbriatum (Pers.) J. Erikss.
Steccherinum ochraceum (Pers.) Gray OU
Steccherinum rawakense (Pers.) Banker
Steccherinum resupinatum G. Cunn. E
Stereopsis hiscens (Berk. & Ravenel) D.A. Reid
PHANEROCHAETACEAE
Antrodiella citrea (Berk.) Ryvarden
Antrodiella hunua (G. Cunn.) Ryvarden E
Antrodiella hydrophila (Berk. & M.A. Curtis) Ryvarden
Antrodiella rata (G. Cunn.) P.K. Buchanan & Ryvarden E
Antrodiella sp. sensu McKenzie
Antrodiella sp. [*Poria undata* sensu G. Cunn.]
Antrodiella zonata (Berk.) Ryvarden
Australicium singulare (G. Cunn.) Hjortstam & Ryvarden
Australohydnum dregeanum (Berk.) Hjortstam & Ryvarden
Byssomerulius corium (Pers.) Parmasto
Byssomerulius miniatus (Wakef.) Hjortstam
Byssomerulius psittacinus P.K. Buchanan, Ryvarden & M. Izawa E
Candelabrochaete eruciformis (G. Cunn.) Stalpers & P.K. Buchanan E
Ceriporia otakou (G. Cunn.) P.K. Buchanan & Ryvarden E
Ceriporia sp. sensu P.K. Buchanan & Ryvarden
Ceriporia spissa (Schwein.) Rajchenb. OU
Ceriporia tarda (Berk.) Ginns
Ceriporia totara (G. Cunn.) P.K. Buchanan & Ryvarden E
Ceriporiopsis merulinus (Berk.) Rajchenb.
Ceriporiopsis rivulosa var. *valdiviana* Rajchenb.
Ceriporiopsis sp. sensu P.K. Buchanan & Ryvarden U
Hjortstamia crassa (Lév.) Boidin & Gilles
Hyphodermella corrugata (Fr.) J. Erikss. & Ryvarden U
Phanerochaete areolata (G. Cunn.) Hjortstam & Ryvarden E
Phanerochaete avellanea (Bres. ex Bourdot & Galzin) J. Erikss. & Hjortstam
Phanerochaete citrina Burds. (nom. ined.) E
Phanerochaete cordylines (G. Cunn.) Burds. E
Phanerochaete corymbata (G. Cunn.) Burds.
Phanerochaete luteoaurantiaca (Wakef.) Burds. E
Phanerochaete monomitica (G. Cunn.) Sheng H. Wu & Popoff
Phanerochaete rosea (Henn.) P.K. Buchanan & Hood A
Phanerochaete sanguinea (Fr.) Pouzar OU
Phanerochaete sordida (P. Karst.) J. Erikss. & Ryvarden
Phanerochaete tuberculata (P. Karst.) Parmasto OU
Phanerochaete viticola (Schwein.) Parmasto OU
Phlebiopsis gigantea (Fr.) Jülich A
Porostereum fulvum (Lév.) Boidin & Gilles
Pseudolagarobasidium calcareum (Cooke & Massee) Sheng H. Wu
Rhizochaete filamentosa (Berk. & M.A. Curtis) Gresl., Nakasone & Rajchenb.
Rhizochaete radicata (Henn.) Gresl., Nakasone & Rajchenb.
POLYPORACEAE
Aurantiporus pulcherrimus (Rodway) P.K. Buchanan & Hood
Australoporus tasmanicus (Berk.) P.K. Buchanan & Ryvarden
Coriolopsis floccosa (Jungh.) Ryvarden
Coriolopsis salebrosa OU
Coriolopsis sanguinaria (Klotzsch) Teng OU
Coriolopsis strumosa (Fr.) Ryvarden A
Datronia scutellata (Schwein.) Gilb. & Ryvarden
Dichomitus leucoplacus (Berk.) Ryvarden E
Dichomitus newhookii P.K. Buchanan & Ryvarden E
Diplomitoporus cunninghamii P.K. Buchanan & Ryvarden E
Echinochaete russiceps (Berk. & Broome) D.A. Reid
Epithele fasciculata (G. Cunn.) Boidin & Gilles E
Epithele nikau G. Cunn.
Epithelopsis fulva (G. Cunn.) Jülich E
Fomes fomentarius sensu Colenso U
Fomes haeuslerianus Henn.
Fomes hemitephrus (Berk.) Cooke
Hapalopilus albocitrinus (Petch) Ryvarden
Hapalopilus nidulans (Fr.) P. Karst. A
Laccocephalum mylittae (Cooke & Massee) Núñez & Ryvarden
Lentinus lepideus sensu Colenso U
Lentinus punctaticeps Berk. & Broome OU
Lentinus strigosus (Schwein.) Fr. OU
Lentinus suffrutescens (Brot.) Fr. A
Lentinus zelandicus Sacc. & Cub. U
Lenzites acutus Berk.
Lenzites betulinus (L.) Fr.
Lenzites repandus sensu Berk.
Lenzites vespaceus (Pers.) Ryvarden U
Lopharia cinerascens (Schwein.) G. Cunn. U
Lopharia involuta (Fr.) G. Cunn. U
Loweporus roseoalbus (Jungh.) Ryvarden
Macrohyporia dictyopora (Cooke) I. Johans. & Ryvarden
Microporus microloma (Lév.) G. Cunn. U
Nigrofomes melanoporus (Mont.) Murrill
Oligoporus hibernicus (Berk. & Broome) Gilb. & Ryvarden U
Pachykytospora papyracea (Cooke) Ryvarden
Panus incandescens sensu Colenso U
Panus maculatus Berk. E
Panus purpuratus G. Stev. E
Panus stypticus sensu Berk. U
Panus tahitensis sensu Colenso U
Panus viscidulus sensu Colenso U
Perenniporia clelandii (Lloyd) Ryvarden U
Perenniporia cunninghamii Decock, P.K. Buchanan & Ryvarden E
Perenniporia ochroleuca (Berk.) Ryvarden
Perenniporia oviformis G. Cunn. ex P.K. Buchanan & Ryvarden
Perenniporia podocarpi P.K. Buchanan & Hood E
Polyporus albatrellus OU
Polyporus arcularius (Batsch) Fr.
Polyporus benzoinus (Wahlenb.) Fr. A
Polyporus borealis sensu Hook. f. U
Polyporus dictyopus Mont.
Polyporus hypomelanus Berk. ex Cooke
Polyporus infernalis Berk.
Polyporus melanopus (Pers.) Fr. A
Polyporus nigrocristatus E. Horak & Ryvarden E
Polyporus occidentalis sensu Colenso U
Polyporus semipileatus sensu J.A. Butcher U
Polyporus septosporus P.K. Buchanan & Ryvarden E
Polyporus sp. (unnamed) sensu P.K. Buchanan & Ryvarden
Polyporus sp.'Kennedy's Bush (PDD86897)' J.A. Cooper (nom. ined.)
Polyporus sulphureus sensu Lloyd U
Polyporus varius (Pers.) Fr. A
Polyporus xerophyllus Berk. E
Poria byssina sensu G. Cunn. U
Poria medullaris Gray A
Pycnoporus coccineus (Fr.) Bondartsev & Singer
Ryvardenia campyla (Berk.) Rajchenb.
Skeletocutis alutacea (J. Lowe) Jean Keller
Skeletocutis amorpha (Fr.) Kotl. & Pouzar A
Skeletocutis nivea (Jungh.) Jean Keller
Skeletocutis novae-zelandiae (G. Cunn.) P.K. Buchanan & Ryvarden E
Skeletocutis stramentica (G. Cunn.) Rajchenb.
Trametes conchifera (Schwein.) Pilát
Trametes hirsuta (Wulfen) Pilát
Trametes menziesii (Berk.) Ryvarden OU
Trametes scabrosa (Pers.) G. Cunn. A
Trametes velutina (Pers.) G. Cunn. A
Trametes versicolor (L.) Lloyd
Trametes zonata sensu G. Cunn. E
Trichaptum byssogenum (Jungh.) Ryvarden
Tyromyces albidus sensu G. Cunn. A
Tyromyces fissilis (Berk. & M.A. Curtis) Donk A
Tyromyces floriformis (Quél.) Bondartsev & Singer U
Tyromyces fuscolineatus sensu G. Cunn. E
Tyromyces guttulatus sensu G. Cunn.
Tyromyces hypolateritius (Cooke) Ryvarden
Tyromyces setiger (Cooke) Teng
Tyromyces spathulatus sensu G. Cunn. E
Tyromyces spiculifer (Cooke) G. Cunn. OU
Tyromyces tephroleucus sensu G. Cunn.
Tyromyces toatoa G. Cunn. E
TUBULICRINACEAE
Tubulicrinis callosus G. Cunn. E
Tubulicrinis calothrix (Pat.) Donk U
Tubulicrinis cinctus G. Cunn. E
Tubulicrinis gracillimus (D.P. Rogers & H.S. Jacks.) G. Cunn.
Tubulicrinis thermometrus (G. Cunn.) M.P. Christ. E
Tubulicrinis umbraculus (G. Cunn.) G. Cunn. E
XENASMATACEAE
Xenasma praeteritum (H.S. Jacks.) Donk
Xenasma pulverulentum (Litsch.) Donk
Xenasma rimicola (P. Karst.) Donk
Xenasma subflavidogriseum (Litsch.) Parmasto OU
Xenasma umbonatum (G. Cunn.) Hjortstam

Order RUSSULALES
Gloeohypochnicium analogum (Bourdot & Galzin) Hjortstam
Gloeopeniophorella sacrata (G. Cunn.) Hjortstam &

Ryvarden
Scytinostromella heterogenea (Bourdot & Galzin) Parmasto U
AMYLOSTEREACEAE A
Amylostereum areolatum (Chaillet ex Fr.) Boidin A
Amylostereum chailletii (Pers.) Boidin A
AURISCALPIACEAE
Artomyces austropiperatus Lickey
Artomyces candelabrum (Massee) Jülich
Artomyces colensoi (Berk.) Jülich
Artomyces novae-zelandiae Lickey E
Artomyces turgidus (Lév.) Jülich
Auriscalpium umbella Maas Geest. E
Clavicorona piperata (Kauffman) Leathers & A.H. Sm. OU
Clavicorona pyxidata (Pers.) Doty U OU
Lentinellus castoreus (Fr.) Kühner & Maire
Lentinellus crawfordiae G. Stev. E
Lentinellus novae-zelandiae (Berk.) R.H. Petersen E
Lentinellus pulvinulus (Berk.) Pegler
BONDARZEWIACEAE
Bondarzewia berkeleyi (Fr.) Bondartsev & Singer
Heterobasidion araucariae P.K. Buchanan
Stecchericium seriatum (Lloyd) Maas Geest.
Wrightoporia micropora P.K. Buchanan & Ryvarden E
Wrightoporia novae-zelandiae Rajchenb. & A. David E
Wrightoporia subrutilans sensu P.K. Buchanan & Ryvarden
HERICIACEAE
Dentipellis leptodon (Mont.) Maas Geest. OU
Hericium coralloides (Scop.) Pers.
LACHNOCLADIACEAE
Asterostroma andinum Pat.
Asterostroma cervicolor (Berk. & M.A. Curtis) Massee
Asterostroma muscicola (Berk. & M.A. Curtis) Massee
Asterostroma persimile Wakef.
Dichostereum granulosum (Pers.) Boidin & Lanq. OU
Dichostereum rhodosporum (Wakef.) Boidin & Lanq.
Scytinostroma duriusculum (Berk. & Broome) Donk U
Scytinostroma portentosum (Berk. & M.A. Curtis) Donk
Scytinostroma praestans (H.S. Jacks.) Donk U
Scytinostroma sp. ['PDD 14.170'] sensu Boidin & Lanq.
Vararia cunninghamii Boidin & Lanq. E
Vararia ellipsospora G. Cunn. E
Vararia fusispora G. Cunn. E
Vararia incrustata Gresl. & Rajchenb. OU
Vararia investiens (Schwein.) P. Karst.
Vararia ochroleuca sensu G. Cunn. E
Vararia protrusa G. Cunn. E
PENIOPHORACEAE
Amylofungus corrosus (G. Cunn.) Sheng H. Wu
Duportella sphaerospora G. Cunn. E
Metulodontia nivea (P. Karst.) Parmasto OU
Peniophora cinerea (Pers.) Cooke
Peniophora coprosmae G. Cunn.
Peniophora crustosa Cooke
Peniophora incarnata (Pers.) P. Karst.
Peniophora lycii (Pers.) Höhn. & Litsch.
Peniophora nuda (Fr.) Bres.
Peniophora scintillans G. Cunn.
RUSSULACEAE
Boidinia crystallitecta (G. Cunn.) Sheng H. Wu & P.K. Buchanan E
Boidinia lacticolor (Bres.) Hjortstam & Ryvarden OU
Cystangium seminudum (Massee & Rodway) T. Lebel & Castellano
Gymnomyces cristatus T. Lebel E
Gymnomyces fuscus T. Lebel E
Gymnomyces leucocarpus T. Lebel E
Gymnomyces pallidus Massee & Rodway
Gymnomyces parvisaxoides T. Lebel E
Gymnomyces redolens (G. Cunn.) Pfister E
Lactarius blennius (Fr.) Fr. OU
Lactarius clarkeae Cleland var. *clarkeae*
— var. *aurantioruber* McNabb E
Lactarius deliciosus (L.) Gray A
Lactarius fuliginosus (Fr.) Fr. OU
Lactarius glyciosmus (Fr.) Fr. A
Lactarius maruiaensis McNabb E
Lactarius nothofagi R. Heim (nom. inv.) E
Lactarius novae-zelandiae McNabb E
Lactarius hepaticus Plowr. OU
Lactarius pubescens Fr. A
Lactarius rufus (Scop.) Fr. A
Lactarius sepiaceus McNabb E
Lactarius tawai McNabb E
Lactarius turpis (Weinm.) Fr. A
Lactarius umerensis McNabb E
Lactarius zonarius (Bull.) Fr. OU
Russula acrolamellata McNabb E
Russula aeruginea Lindblad ex Fr. OU
Russula albolutescens McNabb E
Russula allochroa McNabb E
Russula amoenolens Romagn. A
Russula atropurpurea (Krombh.) Britzelm. (nom. illegit.) OU
Russula atroviridis Buyck E
Russula aucklandica McNabb E
Russula australis McNabb E
Russula cremeoochracea McNabb E
Russula cyanoxantha (Schaeff.) Fr. OU
Russula drimeia sensu Dalrymple
Russula erumpens Cleland & Cheel OU
Russula foetens Pers. OU
Russula griseobrunnea McNabb E
Russula griseostipitata McNabb E
Russula griseoviolacea McNabb E
Russula griseoviridis McNabb E
Russula inquinata McNabb E
Russula kermesina T. Lebel E
Russula littorea Pennycook E
Russula macrocystidiata McNabb E
Russula mariae Peck OU
Russula miniata McNabb E
Russula multicystidiata McNabb E
Russula nitida (Pers.) Fr. A
Russula novae-zelandiae McNabb E
Russula papakaiensis McNabb E
Russula pectinata sensu G. Stev. OU
Russula pectinatoides sensu G. Stev. OU
Russula pilocystidiata McNabb E
Russula pleurogena Buyck & E. Horak E
Russula pseudoareolata McNabb E
Russula pudorina McNabb E
Russula purpureotincta McNabb E var. *purpureotincta*
— var. *alba* P. Leonard (nom. ined.) E
Russula rimulosa Pennycook E
Russula roseopileata McNabb E
Russula roseostipitata McNabb E
Russula rubrolutea (T. Lebel) T. Lebel E
Russula solitaria McNabb E
Russula sororia (Fr.) Romell OU
Russula subvinosa McNabb E
Russula tapawera (T. Lebel) T. Lebel E
Russula tawai McNabb E
Russula tricholomopsis McNabb E
Russula umerensis McNabb E
Russula vinaceocuticulata McNabb E
Russula viridis Cleland (nom. illegit.) OU
Russula vivida McNabb E
Zelleromyces australiensis (Berk. & Broome) Pegler & T.W.K. Young
STEPHANOSPORACEAE
Stephanospora flava (Rodway) G.W. Beaton, Pegler & T.W.K. Young
STEREACEAE
Aleurobotrys botryosus (Burt) Boidin, Lanq. & Gilles U
Aleurocystis habgallae (Berk. & Broome) G. Cunn.
Aleurodiscus aberrans G. Cunn. E
Aleurodiscus apricans Bourdot A
Aleurodiscus aurantius (Pers.) J. Schröt.
Aleurodiscus berggrenii (Cooke) G. Cunn. E
Aleurodiscus coralloides G. Cunn. E
Aleurodiscus coronatus G. Cunn. E
Aleurodiscus limonisporus D.A. Reid
Aleurodiscus mirabilis (Berk. & M.A. Curtis) Höhn.
Aleurodiscus ochraceoflavus Lloyd E
Aleurodiscus parmuliformis G. Cunn. E
Aleurodiscus patelliformis G. Cunn. E
Aleurodiscus sparsus (Berk.) Höhn. & Litsch.
Aleurodiscus zealandicus (Cooke & W. Phillips) G. Cunn.
Gloeocystidiellum fistulatum (G. Cunn.) Boidin E
Gloeocystidiellum inconstans (G. Cunn.) Stalpers & P.K. Buchanan E
Gloeocystidiellum porosum (Berk. & M.A. Curtis) Donk A
Gloeocystidiellum sp. sensu Sheng H. Wu & P.K. Buchanan E
Megalocystidium afibulatum (G. Cunn.) Boidin, Lanq. & Gilles
Scotoderma viride (Sacc.) Jülich E
Stereum aotearoa G. Cunn.
Stereum hirsutum (Willd.) Pers.
Stereum illudens Berk.
Stereum molle Sacc.
Stereum obliquum sensu auct. NZ
Stereum ochraceoflavum (Schwein.) Sacc.
Stereum ostrea (Blume & T. Nees) Fr.
Stereum rugosum Pers. OU
Stereum sanguinolentum (Alb. & Schwein.) Fr. A
Stereum scutellatum G. Cunn. E
Stereum vellereum Berk.

Order SEBACINALES
SEBACINACEAE
Sebacina calcea (Pers.) Bres. U
Sebacina epigaea (Berk. & Broome) Rea
Sebacina filicola McNabb E
Sebacina pteridicola McNabb E

Order THELEPHORALES
BANKERACEAE
Hydnellum scrobiculatum (Fr.) P. Karst.
Phellodon maliensis (Lloyd) Maas Geest.
Phellodon nothofagi McNabb E
Phellodon sinclairii (Berk.) G. Cunn. E
Sarcodon ionides (Pass.) Bataille
Sarcodon sp. 1 sensu Maas Geest. U
Sarcodon thwaitesii (Berk. & Broome) Maas Geest.
THELEPHORACEAE
Amaurodon viridis (Alb. & Schwein.) J. Schröt.
Pseudotomentella tristis (P. Karst.) M.J. Larsen A
Thelephora griseozonata Cooke
Thelephora pedicellata sensu auct. NZ U
Thelephora terrestris Ehrh. A
Tomentella crinalis (Fr.) M.J. Larsen OU
Tomentella ellisii (Sacc.) Jülich & Stalpers A
Tomentella ferruginea (Pers.) Pat.
Tomentella lateritia Pat.
Tomentella pilosa (Burt) Bourdot & Galzin U
Tomentella scobinella G. Cunn. E
Tomentella sublilacina (Ellis & Holw.) Wakef. A

Order TRECHISPORALES
HYDNODONTACEAE
Litschauerella abietis (Bourdot & Galzin) Oberw. ex Jülich A
Litschauerella gladiola (G. Cunn.) Stalpers & P.K. Buchanan E
Litschauerella hastata (G. Cunn.) Stalpers & P.K. Buchanan E
Porpomyces mucidus (Pers.) Jülich
Sistotremella perpusilla Hjortstam
Subulicystidium longisporum (Pat.) Parmasto F
Subulicystidium nikau (G. Cunn.) Jülich E
Trechispora coronifera sensu G. Cunn.
Trechispora farinacea (Pers.) Liberta
Trechispora mollusca (Pers.) Liberta OU
Trechispora regularis (Murrill) Liberta
Trechispora stevensonii (Berk. & Broome) K.H. Larss.
Trechispora verruculosa (G. Cunn.) K.H. Larss.
Tubulicium dussii (Pat.) Oberw. ex Jülich
Tubulicium filicicola (G. Cunn.) Oberw. E
Tubulicium vermiferum (Bourdot) Oberw. ex Jülich

Class DACRYMYCETES
Order DACRYMYCETALES
DACRYMYCETACEAE
Calocera australis McNabb
Calocera cornea (Batsch) Fr. OU
Calocera furcata (Fr.) Fr. OU
Calocera fusca Lloyd
Calocera glossoides (Pers.) Fr. E
— var. *spathulata* Cooke (nom. inv.) E
Calocera guepinioides Berk.
Calocera lutea (Massee) McNabb
Calocera sinensis McNabb U
Calocera viscosa (Pers.) Fr. OU
Cerinomyces lagerheimii (Pat.) McNabb
Cerinomyces pallidus G.W. Martin
Dacrymyces capitatus Schwein.
Dacrymyces cupularis Lloyd
Dacrymyces flabelliformis Burds. & Laursen E
Dacrymyces lacrymalis (Pers.) Sommerf.
Dacrymyces minor Peck
Dacrymyces novae-zelandiae McNabb E
Dacrymyces paraphysatus L.S. Olive
Dacrymyces punctiformis Neuhoff
Dacrymyces stillatus Nees
Dacrymyces subantarcticensis Burds. & Laursen E
Dacrymyces tortus (Willd.) Fr.
Dacryopinax spathularius (Schwein.) G.W. Martin
Guepiniopsis buccina (Pers.) L.L. Kenn.
Heterotextus luteus (Bres.) McNabb
Heterotextus miltinus (Berk.) McNabb
Heterotextus pezizaeformis (Berk.) Lloyd

Class TREMELLOMYCETES
Order CYSTOFILOBASIDIALES
CYSTOFILOBASIDIACEAE
Cystofilobasidium capitatum (Fell, I.L. Hunter & Tallman) Oberw. & Bandoni U
Cystofilobasidium infirmominiatum (Fell, I.L. Hunter & Tallman) Hamam., Sugiy. & Komag. U
Itersonilia pastinacae Channon AN A
Itersonilia perplexans Derx AN A
Udeniomyces pyricola (F. Stadelmann) Nakase & Takem. A

Order TREMELLALES
SIROBASIDIACEAE
Sirobasidium rubrofuscum (Berk.) P. Roberts
TREMELLACEAE
Biatoropsis usnearum Räsänen OU
Bullera coprosmae Hamam. & Nakase AN E
Bullera crocea Buhagiar AN
Bullera hannae Hamam. & Nakase AN E
Bullera huiaensis Hamam. & Nakase AN E
Bullera mrakii Hamam. & Nakase AN E
Bullera unica Hamam. & Nakase AN E
Bullera variabilis Nakase & M. Suzuki AN
Bulleromyces albus Boekhout & A. Fonseca
Cryptococcus aerius (Saito) Nann. AN
Cryptococcus albidus (Saito) C.E. Skinner AN
Cryptococcus curvatus (Diddens & Lodder) Golubev AN (nom. inv.)
Cryptococcus diffluens (Zach) Lodder & Kreger AN
Cryptococcus dimennae Fell & Phaff AN
Cryptococcus festucosus Golubev & J.P. Samp. U
Cryptococcus flavus (Saito) Phaff & Fell AN
Cryptococcus gastricus Reiersöl & di Menna AN
Cryptococcus humicola (Dasz.) Golubev AN
Cryptococcus hungaricus (Zsolt) Phaff & Fell AN U
Cryptococcus laurentii (Kuff.) C.E. Skinner AN
Cryptococcus luteolus (Saito) C.E. Skinner AN
Cryptococcus macerans (P.S. Fred.) Phaff & Fell AN
Cryptococcus oeirensis Á. Fonseca, Scorzetti & Fell AN U
Cryptococcus podzolicus (Babeva & Reshetova) Golubev AN U
Cryptococcus terreus di Menna AN
Cryptococcus victoriae M.J. Montes, Belloch, Galiana, M.D. García, C. Andrés, S. Ferrer, Torr.-Rodr. & J. Guinea AN U
Filobasidiella neoformans Kwon-Chung var. *neoformans* U
— var. *bacillispora* (Kwon-Chung) Kwon-Chung U
Tremella foliacea Pers. OU
Tremella fuciformis Berk.
Tremella lobariacearum Diederich & M.S. Christ.
Tremella lutescens Pers.
Tremella mesenterica Retz. OU
Tremella muelleri Berk. OU
Tremella obscura (L.S. Olive) M.P. Christ. U
Tremella ramalinae Diederich 1996 U
Tremella tawa McNabb E
Tremella vesiculosa McNabb E
TRICHOSPORONACEAE A
Trichosporon asahii Akagi ex Sugita, Nishikawa & Shinoda AN A
Trichosporon capitatum Diddens & Lodder AN A
Trichosporon cutaneum (Beurm., Gougerot & Vaucher) M. Ota AN A
Trichosporon inkin (Oho) Carmo Souza & Uden AN A
Trichosporon niger G.H. Green AN A
Trichosporon pullulans (Lindner) Diddens & Lodder AN A

Class ENTORRHIZOMYCETES
Order ENTORRHIZALES
ENTORRHIZACEAE
Entorrhiza aschersoniana (Magnus) Lagerh. U
Entorrhiza caricicola Ferd. & Winge U
Entorrhiza casparyana (Magnus) Lagerh. A
Entorrhiza casparyanella Vánky E
Entorrhiza citriformis Vánky & McKenzie E
Entorrhiza fineraniae Vánky
Entorrhiza scirpicola (Correns) Sacc. & P. Syd.

SUBPHYLUM PUCCINIOMYCOTINA
Class AGARICOSTILBOMYCETES
Order AGARICOSTILBALES
Mycogloea macrospora (Berk. & Broome) McNabb
AGARICOSTILBACEAE
Agaricostilbum novozelandicum W.B. Kendr. & X.D. Gong (nom. inv.) E
Agaricostilbum pulcherrimum (Berk. & Broome) B.L. Brady, B. Sutton & Samson OU
Bensingtonia ingoldii Nakase & Itoh AN
Bensingtonia naganoensis (Nakase & M. Suzuki) Nakase & Boekhout AN

CHIONOSPHAERACEAE
Stilbum rigidum Pers. U

Class ATRACTIELLOMYCETES
Order ATRACTIELLALES
PHLEOGENACEAE
Helicogloea alba (Burt) Couch
Helicogloea farinacea (Höhn.) D.P. Rogers
Helicogloea lagerheimii Pat.
Phleogena faginea (Fr.) Link
SACCOBLASTIACEAE
Infundibura adhaerens Nag Raj & W.B. Kendr. AN F

Class CLASSICULOMYCETES
Order CLASSICULALES
CLASSICULACEAE
Jaculispora submersa H.J. Huds. & Ingold AN U

Class MICROBOTRYOMYCETES
Order LEUCOSPORIDIALES
LEUCOSPORIDIACEAE
Leucosporidiella fragaria (J.A. Barnett & Buhagiar) J.P. Samp. AN U
Leucosporidiella muscorum (di Menna) J.P. Samp. AN
Leucosporidium scottii Fell, Statzell, I.L. Hunter & Phaff

Order MICROBOTRYALES
MICROBOTRYACEAE
Bauerago abstrusa (Malençon) Vánky
Bauerago gardneri (McKenzie & Vánky) Vánky
Microbotryum cordae (Liro) G. Deml & Prillinger A
Microbotryum dianthorum (Liro) H. Scholz & I. Scholz A
Microbotryum nivale (Liro) Vánky A
Microbotryum tenuisporum (Cif.) Vánky A
Sphacelotheca hydropiperis (Schumach.) de Bary A
Sphacelotheca polygoni-serrulati Maire A

Order SPORIDIOBOLALES
Rhodotorula aurantiaca (Saito) Lodder AN
Rhodotorula glutinis (Fresen.) F.C. Harrison AN
Rhodotorula graminis di Menna AN
Rhodotorula ingeniosa (di Menna) Arx & Weijman AN
Rhodotorula marina Phaff, Mrak & O.B. Williams AN M
Rhodotorula minuta (Saito) F.C. Harrison AN
Rhodotorula mucilaginosa (A. Jörg.) F.C. Harrison AN U
Sporobolomyces coprosmae Hamam. & Nakase AN
Sporobolomyces coprosmicola Hamam. & Nakase AN
Sporobolomyces dimennae Hamam. & Nakase AN U
Sporobolomyces dracophylli Hamam. & Nakase AN
Sporobolomyces gracilis Derx AN
Sporobolomyces inositophilus Nakase & M. Suzuki AN U
Sporobolomyces novozealandicus Hamam. & Nakase AN
Sporobolomyces roseus Kluyver & C.B. Niel AN U
Sporobolomyces sasicola Nakase & M. Suzuki AN
Sporobolomyces taupoensis Hamam. & Nakase AN U
SPORIDIOBOLACEAE
Sporidiobolus pararoseus Fell & Tallman
Sporidiobolus salmonicolor Fell & Tallman

Class PUCCINIOMYCETES
Order HELICOBASIDIALES A
HELICOBASIDIACEAE A
Helicobasidium purpureum Pat. A
Tuberculina persicina (Ditmar) Sacc. AN A

Order PLATYGLOEALES
PLATYGLOEACEAE
Insolibasidium deformans (C.J. Gould) Oberw. & Bandoni A
Platygloea australis McNabb E

Order PUCCINIALES
Aecidium celmisiae Rodway
Aecidium celmisiae-discoloris G. Cunn. E
Aecidium disciforme McAlpine
Aecidium hebe G. Cunn.
Aecidium hupiro G. Cunn. E
Aecidium microstomum Berk.
Aecidium milleri G. Cunn. E
Aecidium monocystis Berk.
Aecidium myopori G. Cunn. E
Aecidium myrsines McNabb E
Aecidium otagense Linds. E
Aecidium otira G. Cunn. E
Aecidium plantaginis-variae McAlpine A
Aecidium ranunculi-depressi G. Cunn. A
Aecidium ranunculi-insignis G. Cunn. E
Aecidium ranunculi-lyallii G. Cunn. E
Aecidium ranunculi-monroi G. Cunn. E
Aecidium traversiae G. Cunn. E
Aecidium westlandicum G. Cunn. E
Caeoma kaiku G. Cunn. E
Hypodermium orchidearum Cooke & Massee AN A
Uredo brownii Syd. & P. Syd. E
Uredo cheesemanii G. Cunn. E
Uredo dianellae Dietel
Uredo forsterae G. Cunn. E
Uredo histiopteridis (G. Cunn.) Hirats. f.
Uredo horopito G. Cunn. E
Uredo karetu G. Cunn. E
Uredo lindsaeae Henn.
Uredo novae-zelandiae G.F. Laundon E
Uredo oleariae Cooke E
Uredo phormii G. Cunn. E
Uredo polygonorum DC. A
Uredo puawhananga G.T.S. Baylis E
Uredo rhagodiae Cooke & Massee
Uredo salicorniae G. Cunn. E
Uredo scirpi-nodosi McAlpine
Uredo spyridii Cooke & Massee
Uredo toetoe G. Cunn. E
Uredo tupare G. Cunn. E
Uredo wharanui G. Cunn. E
COLEOSPORIACEAE A
Chrysomyxa ledi de Bary A var. *ledi*
— var. *rhododendri* (de Bary) Savile A
Chrysomyxa reticulata P.E. Crane A
Coleosporium campanulae (F. Strauss) Tul. A
Coleosporium senecionis J. Kickx f. A
MELAMPSORACEAE
Melampsora ×medusae-populina Spiers A
Melampsora coleosporioides Dietel A
Melampsora epitea Thüm. var. *epitea* A
Melampsora euphorbiae (C. Schub.) Castagne A
Melampsora hypericorum G. Winter A
Melampsora kusanoi Dietel
Melampsora larici-populina Kleb. A
Melampsora lini (Ehrenb.) Desm. A
Melampsora medusae Thüm. A
Melampsora novae-zelandiae G. Cunn. E
Melampsora ricini E.A. Noronha A
Melampsora sp. [1] sensu Spiers & Hopcroft A
Melampsora sp. [2] sensu Spiers & Hopcroft A
MIKRONEGERIACEAE
Mikronegeria fuchsiae P.E. Crane & R.S. Peterson E
Petersonia dracophylli McKenzie E
PHAKOPSORACEAE A
Cerotelium fici (E.J. Butler) Arthur A
Phakopsora apoda (Har. & Pat.) Mains A
PHRAGMIDIACEAE
Frommea obtusa (F. Strauss) Arthur A
Frommeella mexicana (Mains) J.W. McCain & J.F. Hennen A var. *mexicana*
— var. *indicae* J.W. McCain & J.F. Hennen A
Frommeella tormentillae (Fuckel) Cummins & Y. Hirats. A
Hamaspora australis G. Cunn.
Kuehneola uredinis (Link) Arthur A
Phragmidium acaenae G. Cunn. E
Phragmidium constrictosporum G.F. Laundon E
Phragmidium mucronatum (Pers.) Schltdl. A
Phragmidium novae-zelandiae G. Cunn. E
Phragmidium rubi-idaei (DC.) P. Karst. A
Phragmidium subsimile G. Cunn. E
Phragmidium tuberculatum J.H.H. Müll. A
Phragmidium violaceum (Schultz) G. Winter A
PILEOLARIACEAE A
Uromycladium acaciae (Cooke) P. Syd. & Syd. A
Uromycladium alpinum McAlpine A
Uromycladium bisporum McAlpine A
Uromycladium maritimum McAlpine A
Uromycladium notabile McAlpine A
Uromycladium robinsonii McAlpine A
Uromycladium simplex McAlpine A
Uromycladium tepperianum (Sacc.) McAlpine A
PUCCINIACEAE
Cumminsiella mirabilissima (Peck) Nannf. A
Miyagia pseudosphaeria (Mont.) Jørst. A
Puccinia acetosae (Schumach.) Korn. A
Puccinia akiraho G. Cunn. E
Puccinia alboclava G.T.S. Baylis E
Puccinia allii F. Rudolphi A
Puccinia anisotomes G. Cunn. E
Puccinia antirrhini Dietel & Holw. A
Puccinia arenariae (Schumach.) G. Winter A
Puccinia arnaudensis G. Cunn. E
Puccinia atkinsonii G. Cunn. E
Puccinia aucta Berk. & F. Muell.
Puccinia austrina McKenzie E
Puccinia brachypodii G.H. Otth A var. *brachypodii*
— var. *poae-nemoralis* (G.H. Otth) Cummins & H.C. Greene A
Puccinia calcitrapae DC. var. *calcitrapae* A
— var. *centaureae* (DC.) Cummins A
Puccinia caricina DC.
Puccinia celmisiae G. Cunn. E
Puccinia chathamica McKenzie E
Puccinia chrysanthemi Roze A
Puccinia chrysanthemicola Sousa da Câmara, Oliveira & Luz A
Puccinia clavata P. Syd. & Syd. E
Puccinia cnici H. Mart. A
Puccinia cockaynei G. Cunn. E
Puccinia conii Lagerh. A
Puccinia contegens G. Cunn. E
Puccinia coprosmae Cooke
Puccinia coronata Corda A
— f.sp. *holci* (?Erikss.) ? OU
Puccinia crepidicola Syd. & P. Syd. A
Puccinia crinitae McNabb E
Puccinia cruciferarum subsp. *inornata* (G. Cunn.) J. Walker E
Puccinia cuniculi G. Cunn. E
Puccinia cyani Pass. A
Puccinia cynodontis Lacroix A
Puccinia dichondrae Mont.
Puccinia egmontensis G. Cunn. E
Puccinia embergeriae McKenzie & P.R. Johnst. E
Puccinia euphrasiana G. Cunn.
Puccinia festucae Plowr. A
Puccinia flaccida Berk. & Broome A
Puccinia flavescens McAlpine A
Puccinia fodiens G. Cunn. E
Puccinia foyana G. Cunn. E
Puccinia freycinetiae McKenzie E
Puccinia gahniae Dingley E
Puccinia gei McAlpine
Puccinia gei-parviflori McNabb E
Puccinia geranii-pilosi McAlpine
Puccinia gnaphaliicola Henn. A
Puccinia grahamii G. Cunn. E
Puccinia graminis Pers. A
— subsp. *graminicola* Z. Urb.
Puccinia hectorensis G. Cunn. E
Puccinia hederaceae McAlpine
Puccinia helianthi Schwein. A
Puccinia hieracii (Röhl.) H. Mart. var. *hieracii* A
— var. *hypochaeridis* (Oudem.) Jørst. A
— var. *piloselloidarum* (Probst) Jørst. A
Puccinia hordei G.H. Otth A
Puccinia horiana Henn. A
Puccinia hydrocotyles Cooke
Puccinia iridis Wallr. A
Puccinia junciphila Cooke & Massee
Puccinia keae G. Cunn. E
Puccinia kenmorensis Cummins A
Puccinia kinseyi G. Cunn. E
Puccinia kirkii G. Cunn. E
Puccinia koherika G. Cunn. E
Puccinia kopoti G. Cunn. E
Puccinia lagenophorae Cooke
Puccinia levis (Sacc. & Bizz.) Magnus A
— var. *panici-sanguinalis* (Rangel) Ramachar & Cummins A
Puccinia liberta F. Kern E
Puccinia ludwigii Tepper A
Puccinia malvacearum Bertero ex Mont. A
Puccinia mania G. Cunn. E
Puccinia maurea G. Cunn. E
Puccinia menthae sensu G. Cunn. var. *menthae* A
— var. *pseudomenthae* (G. Cunn.) J.W. Baxter E
Puccinia microspora Dietel A
Puccinia morrisonii McAlpine
Puccinia moschata G. Cunn. E
Puccinia muehlenbeckiae (Cooke) P. Syd. & Syd. A
Puccinia myrsiphylli G. Winter A
Puccinia nakanishikii Dietel A
Puccinia namua G. Cunn. E
Puccinia novozelandica Bubák E
Puccinia oahuensis Ellis & Everh. A
Puccinia obtectella Cummins A
Puccinia oreoboli Cummins A
Puccinia oxalidis Dietel & Ellis A
Puccinia paspalina Cummins A
Puccinia pedatissima G. Cunn. E
Puccinia pelargonii-zonalis Doidge A
Puccinia perlaevis G. Cunn. E
Puccinia plagianthi McAlpine
Puccinia polygoni-amphibii Pers. A
Puccinia polypogonobia McKenzie E
Puccinia pounamu G. Cunn. E
Puccinia pulverulenta Grev.
Puccinia punctata Link
Puccinia punctiformis (F. Strauss) Röhl. A
Puccinia pygmaea Erikss. A
Puccinia rara McKenzie E
Puccinia rautahi G. Cunn. E
Puccinia recondita Desm. A
Puccinia reidii G. Cunn. E
Puccinia rhei-undulati Hirats. f. A
Puccinia ruizensis Mayor
Puccinia schoenus G. Cunn. E
Puccinia scirpi DC. A
Puccinia sessilis W.G. Schneid. A
Puccinia sorghi Schwein. A
Puccinia stenotaphricola J. Walker A
Puccinia striiformis Westend. var. *striiformis* A
— f. *tritici* Erikss. A
Puccinia striiformoides M. Abbasi, Hedjar. & M. Scholler A

Puccinia tararua G. Cunn. E
Puccinia tekapo McNabb E
Puccinia tenuispora McAlpine
Puccinia tetragoniae var. *novae-zelandiae* McKenzie E
Puccinia thuemenii McAlpine
Puccinia tiritea G. Cunn.
Puccinia toa G. Cunn. E
Puccinia triticina Erikss. A
Puccinia unciniarum Dietel & Neger
Puccinia variabilis Grev. A var. *variabilis*
— var. *lapsanae* (Fuckel) Cummins A
Puccinia violae DC. A
Puccinia wahlenbergiae G. Cunn. E
Puccinia whakatipu G. Cunn. E
Puccinia zoysiae Dietel A
Uromyces anthyllidis J. Schröt. A
Uromyces appendiculatus (Pers.) Unger A
Uromyces armeriae J. Kickx f. A
Uromyces azorellae Cooke E
Uromyces betae J. Kickx f. A
Uromyces bidenticola Arthur A
Uromyces dactylidis G.H. Otth A
Uromyces danthoniae McAlpine
Uromyces dianthi (Pers.) Niessl A
Uromyces discariae G. Cunn. E
Uromyces edwardsiae G. Cunn. E
Uromyces ehrhartae McAlpine
Uromyces inflatus (Cooke) McKenzie E
Uromyces junci (Desm.) Tul.
Uromyces macnabbii Cummins E
Uromyces magnusii Kleb. A
Uromyces microtidis Cooke
Uromyces minor J. Schröt. A
Uromyces muscari (Duby) Lév. A
Uromyces otakou G. Cunn. E
Uromyces pisi (DC.) G.H. Otth OU
Uromyces polygoni-avicularis (Pers.) P. Karst. A
Uromyces rumicis (Schumach.) G. Winter A
Uromyces scaevolae G. Cunn. E
Uromyces sellierae G. Cunn. E
Uromyces striatus J. Schröt. var. *striatus* A
— var. *loti* (A. Blytt) Arthur A
Uromyces tenuicutis McAlpine A
Uromyces thelymitrae McAlpine
Uromyces transversalis (Thüm.) G. Winter A
Uromyces trifolii (R. Hedw.) Fuckel A
Uromyces trifolii-repentis Liro var. *trifolii-repentis* A
— var. *fallens* (Arthur) Cummins A
Uromyces viciae-fabae (Pers.) J. Schröt. A
Uromyces waipoua McNabb E
PUCCINIASTRACEAE
Hyalopsora polypodii (Dietel) Magnus A
Melampsoridium betulinum Kleb. A
Milesia polystichi-vestiti McKenzie E
Naohidemyces vacciniorum (J. Schröt.) Spooner A
Pucciniastrum pustulatum (Pers.) Dietel A
RAVENELIACEAE
Endoraecium digitatum (G. Winter) M. Scholler & Aime A
UROPYXIDACEAE
Tranzschelia discolor (Fuckel) Tranzschel & M.A. Litv. A
Tranzschelia pruni-spinosae (Pers.) Dietel A

Order SEPTOBASIDIALES
SEPTOBASIDIACEAE
Septobasidium bogoriense Pat. OU
Septobasidium pedicellatum Pat. OU
Septobasidium simmondsii Couch ex L.D. Gómez & Henk

SUBPHYLUM USTILAGINOMYCOTINA
Class EXOBASIDIOMYCETES
Tilletiopsis lilacina Tubaki AN U
Tilletiopsis minor Nyland AN

Order DOASSANSIALES A
RHAMPHOSPORACEAE A
Rhamphospora nymphaeae D.D. Cunn. A

Order ENTYLOMATALES
ENTYLOMATACEAE
Entyloma ageratinae R.W. Barreto & H.C. Evans A
Entyloma australe Speg. A
Entyloma brizae Unamuno & Cif. A
Entyloma calendulae (Oudem.) de Bary A
Entyloma compositarum Farl. A
Entyloma dahliae Syd. & P. Syd. A
Entyloma echinaceae Vánky & McKenzie A
Entyloma eschscholziae Harkn. A
Entyloma fergussonii (Berk. & Broome) Plowr. A
Entyloma ficariae A.A. Fisch. Waldh. A
Entyloma fuscum J. Schröt. A
Entyloma gaillardianum Vánky A
Entyloma microsporum (Unger) J. Schröt. A
Entyloma novae-zelandiae McKenzie & Vánky E
Entyloma parietariae Rayss A
Entyloma picridis Rostr. A
Entyloma ranunculi (Bon) J. Schröt. OU
Entyloma ranunculi-repentis Sternon A
Entyloma saccardianum Scalia ex Cif.
Entyloma schinzianum (Magnus) Bubák A
Entyloma serotinum J. Schröt. A
Entylomella geranii U. Braun & C.F. Hill AN, A

Order EXOBASIDIALES
EXOBASIDIACEAE
Exobasidium camelliae Shirai A
Exobasidium dracophylli McNabb E
Exobasidium fraseri McNabb E
Exobasidium gaultheriae Sawada
Exobasidium gracile (Shirai) Syd. & P. Syd. A
Exobasidium novae-zealandiae McNabb E
Exobasidium pentachondrae McNabb E
Exobasidium vaccinii (Fuckel) Woronin A
GRAPHIOLACEAE A
Graphiola phoenicis (Moug.) Poit. A

Order GEORGEFISCHERIALES A
GEORGEFISCHERIACEAE A
Jamesdicksonia dactylidis (Pass.) R. Bauer, Begerow, A. Nagler & Oberw. A
Jamesdicksonia irregularis (Johanson) R. Bauer, Begerow, A. Nagler & Oberw. A

Order MICROSTROMATALES A
MICROSTROMATACEAE A
Microstroma album (Desm.) Sacc. A
Microstroma juglandis (Berenger) Sacc. A

Order TILLETIALES
TILLETIACEAE
Tilletia anthoxanthi A. Blytt A
Tilletia bromi (Brockm.) Brockm. A
Tilletia caries (DC.) Tul. & C. Tul. A
Tilletia cathcartae Durán & G.W. Fisch.
Tilletia holci (Westend.) J. Schröt. A
Tilletia inolens McAlpine
Tilletia laevis J.G. Kühn A
Tilletia lolii Auersw. ex G.Winter A
Tilletia sphaerococca (Wallr.) A.A. Fisch. Waldh. A
Tilletia walkeri Castlebury & Carris A

Order MALASSEZIALES
Malassezia furfur (C.P. Robin) Baill. AN, U
Malassezia pachydermatis (Weidman) C.W. Dodge AN U
Pityrosporum ovale (Bizz.) Castell. & Chalm. A

Class USTILAGINOMYCETES
Order UROCYSTIDIALES
DOASSANSIOPSIDACEAE A
Doassansiopsis hydrophila (A. Dietr.) Lavrov A
FLOROMYCETACEAE A
Antherospora vaillantii (Tul. & C. Tul.) R. Bauer, M. Lutz, Begerow, Piatek & Vánky A
UROCYSTIDACEAE
Mundkurella schefflerae Vánky, C. Vánky & McKenzie E
Urocystis agropyri (Preuss) A.A. Fisch. Waldh. A
Urocystis agrostidis (Lavrov) Zundel A
Urocystis alopecuri A.B. Frank A
Urocystis bolivarii Bubák & Gonz. Frag. A
Urocystis junci Lagerh. A
Urocystis magica Pass. A
Urocystis novae-zelandiae (G. Cunn.) G. Cunn. E
Urocystis ranunculi (Lib.) Moesz
Urocystis roivainenii (Liro) Zundel A
Urocystis tothii Vánky A
Urocystis ulei Magnus A

Order USTILAGINALES
ANTHRACOIDEACEAE
Anthracoidea carphae (Speg.) Vánky
Anthracoidea heterospora (B. Lindeb.) Kukkonen
Anthracoidea schoenus (G. Cunn.) Vánky
Anthracoidea sclerotiformis (Cooke & Massee) Kukkonen
Anthracoidea wakatipu Vánky E
Cintractia oreoboli Vánky & McKenzie
Cintractia solida (Berk.) M. Piepenbr.
Farysia catenata (F. Ludw.) Syd.
Farysia longispora Vánky & McKenzie E
Farysia microspora Vánky & McKenzie E
Farysia nigra (G. Cunn.) G. Cunn. E
Farysia sp. 1 sensu Vánky & McKenzie E
Farysia sp. 2 sensu Vánky & McKenzie E
Farysia thuemenii (A.A. Fisch. Waldh.) Nannf.
Farysia zeylanica Liro E
Farysporium endotrichum (Berk.) Vánky
Heterotolyposporium piluliforme (Berk.) Vánky
Moreaua kochiana (Gäum.) Vánky
Moreaua littoralis (G. Cunn.) Vánky
Moreaua rodwayi (McAlpine) Vánky
Moreaua schoeni (Vánky & McKenzie) Vánky
Tolyposporium neillii (G. Cunn.) Vánky & McKenzie E
USTILAGINACEAE
Moesziomyces bullatus (J. Schröt.) Vánky A
Sporisorium destruens (Schltdl.) Vánky A
Sporisorium reilianum (J.G. Kühn) Langdon & Full. A
Tranzscheliella comburens (F. Ludw.) Vánky & McKenzie
Tranzscheliella hypodytes (Schltdl.) Vánky & McKenzie
Tranzscheliella williamsii (Griffiths) Dingley & Versluys
Ustilago agropyri McAlpine
Ustilago avenae (Pers.) Rostr. A
Ustilago bromivora (Tul. & C. Tul.) A.A. Fisch. Waldh. A
Ustilago bullata Berk.
Ustilago caricis (Pers.) Fuckel
Ustilago cynodontis (Henn.) Henn. A
Ustilago filiformis (Schrank) Rostr. A
Ustilago hordei (Pers.) Lagerh. A
Ustilago maydis (DC.) Corda A
Ustilago nuda (J.L. Jensen) Kellerm. & Swingle A
Ustilago serpens (P. Karst.) B. Lindeb. A
Ustilago spinificis F. Ludw.
Ustilago striiformis (Westend.) Niessl A
Ustilago trichophora (Link) Körn. A
Ustilago tritici (Pers.) Rostr. A
WEBSDANEACEAE
Restiosporium dissimile Vánky & McKenzie E

THIRTY-FOUR

LICHENS

DAVID J. GALLOWAY

Lichens are stable, self-supporting associations of a usually dominant fungus (mycobiont) and an alga or cyanobacterium (photobiont). More precisely, a lichen is an ecologically obligate, stable mutualism between an exhabitant fungal partner and an inhabitant population of extracellularly located unicellular or filamentous algal or cyanobacterial cells (Hawksworth 1988; Hawksworth & Honegger 1994; Nash 2008a; Friedl & Büdel 2008; Honegger 2008a). Lichens are a biological group, not a taxonomic one, and are unique in that in many cases the resulting life forms and behaviour differ markedly from those of the isolated components.

This growing together of dissimilar organisms, a photosynthetic producer as minor partner, and a heterotrophic consumer, is known as symbiosis. Symbiosis is a major source of evolutionary innovation in the living world and the lichen symbiosis is one of the most successful and diverse (Ahmadjian & Paracer 1987; Smith & Douglas 1987; Ahmadjian 1993; Margulis 1993; Richardson 1999; Sanders 2001, 2006, 2010; Schwartzman 2010). Lichenisation is a major nutritional strategy of fungi, adopted over a long evolutionary history by many diverse and quite unrelated fungal groups in the Ascomycota and occasionally also in the Basidiomycota. It is estimated that one out of five fungi is lichenised (Honegger 1991–93, 1998; Kirk et al. 2008).

Steinera sorediata (Lecanoromycetes: Peltigerales).

Janet Ledingham, University of Otago

In the process of lichenisation, the photosynthetic products of the photobiont are used by the mycobiont for growth, and, in return for a continuing food supply, the mycobiont that surrounds the photobiont provides it with a sheltered environment, secures adequate illumination, and facilitates gas exchange (Honegger 1996, 1998, 2001, 2008b). Lichens are the 'farmers' of the fungal kingdom. While their non-symbiotic relatives continue as 'hunter-gatherers' of transient carbon sources, the lichen fungi have become indoor gardeners, cultivating and perpetuating their internalised source of food. This agrarian control over food resources confers both stability and the potential to occupy entirely new ecological niches. In human development, agriculture permitted the rise of populous, sedentary, highly complex civilisations by providing a resource base far larger and more reliable than that available from the unmanipulated environment (Sanders 2001; Lücking et al. 2009). For the fungi, 'algaculture' has led to the development of the structurally elaborate, self-sufficient, long-lived thalli that we call lichens (Honegger 1998, 2008b; Sanders 2001; Goward 2008-2011).

The polyphyletic origin of lichens has been generally accepted by ascomycete systematists since the mid-1970s, and is now indisputably confirmed by molecular data (Gargas et al. 1995; Grube & Kroken 2001). Recent molecular studies have shown that the lichen-forming fungi are much older than was previously thought, with many lines of free-living fungi being derived from lichen-like ancestors (Lutzoni et al. 2001).

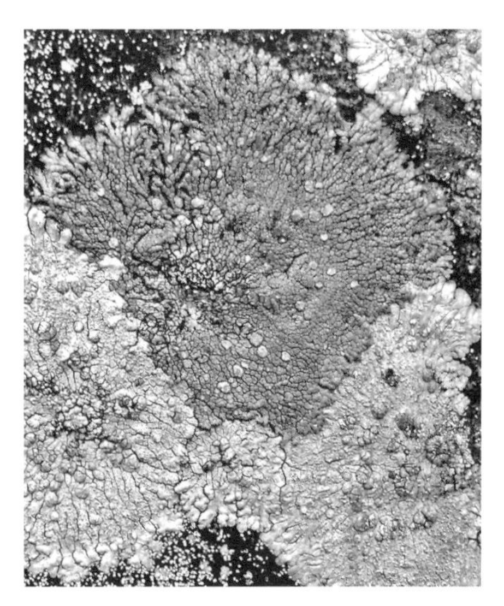

Placopsis perrugosa and, at bottom, *P. cribellans* (Lecanoromycetes: Argyriales).
Janet Ledingham

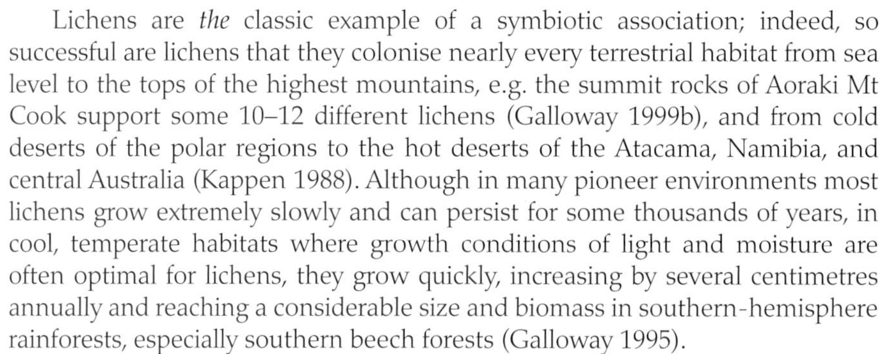

Lichens are *the* classic example of a symbiotic association; indeed, so successful are lichens that they colonise nearly every terrestrial habitat from sea level to the tops of the highest mountains, e.g. the summit rocks of Aoraki Mt Cook support some 10–12 different lichens (Galloway 1999b), and from cold deserts of the polar regions to the hot deserts of the Atacama, Namibia, and central Australia (Kappen 1988). Although in many pioneer environments most lichens grow extremely slowly and can persist for some thousands of years, in cool, temperate habitats where growth conditions of light and moisture are often optimal for lichens, they grow quickly, increasing by several centimetres annually and reaching a considerable size and biomass in southern-hemisphere rainforests, especially southern beech forests (Galloway 1995).

Lichens are classified as fungi (Kirk et al. 2008; Tehler & Wedin 2008; Lumbsch & Huhndorf 2010) and are divided into genera, species, and subspecies. Thus the name of the lichen is attached to the fungal partner, and is based on morphological, reproductive, chemical, and molecular characters. [For a stimulating series of essays on the lichenised condition see Goward (2008–2011). Worldwide, the number of lichen species is perhaps 15,000, with reports fluctuating between 13,500 and 24,000 (Galloway 1992b; Sipman & Aptroot 2001; Lücking 2008), distributed among 500–600 genera. Presently, it is estimated that there are some 10,000 undescribed species of lichenised fungi, which represents a massive task in rapid identification for the ever-dwindling number of competent lichen taxonomists (Lumbsch et al. 2011).

Progress in lichen systematics at all taxonomic levels is currently dynamic and rapid. The refined understanding of natural relationships provided by new character complexes derived directly from the genome is likely to lead to future dramatic changes in our classifications and concepts of taxa. Many traditional lichen systematists are now utilising molecular techniques and phylogenetic analysis is now widely accepted as the most useful method for analysing natural relationships in lichens (see for example Thomas et al. 2002; Miadlikowska & Lutzoni 2004; Miadlikowska et al. 2006; Högnabba et al. 2009; Thell et al. 2009; Otálora et al. 2010; Divakar et al. 2010; Cornejo & Scheidegger 2010; Werth 2010). Lichen-forming fungi are being used in the AFTOL (Assembling the Fungal Tree of Life) Project (see Lutzoni et al. 2004; Spatafora 2005; Hibbett et al. 2007; Schoch et al. 2009; McLaughlin et al. 2009; Lumbsch et al. 2011).

Diversity of New Zealand lichens

A total of 1767 lichen taxa in 322 genera is currently accepted as occurring in New Zealand (from the Kermadec Islands to the Auckland Islands and Campbell Island and including the Chatham Islands, though lichen mycobiotas of the last three named are still very imperfectly documented). This number compares with 2258 taxa in 448 genera [including 298 lichen-forming and 131 lichenicolous genera] known from the United Kingdom and Ireland (Coppins et al. 2005; Smith et al. 2009); 2968 taxa in 449 genera known from Sweden and Norway (Santesson et al. 2004); 2345 taxa in 305 genera known from Italy (Nimis 1993; Nimis & Martellos 2003); 5246 taxa from 646 genera known from the USA and Canada (Esslinger 2011); 3616 taxa in 484 genera known from Australia (McCarthy 2011); 1383 taxa in 281 genera known from Chile (Galloway & Quilhot 1999); and 420 taxa in 124 genera known from Antarctica and South Georgia (Øvstedal and Lewis Smith 2001).

Lichenomphalia alpina, a basiomycetous lichen (Agaricomycetes: Agaricales).
Janet Ledingham

These 1767 taxa (the number continues to increase annually) are accommodated in nine orders and 78 families (Lumbsch & Huhndorf 2010). A growing number of lichenicolous fungi (marked with an asterisk in the Ascomycota checklist (chapter 32), are also recognised as parasites or commensals on a variety of New Zealand lichen hosts, these lichenicolous taxa comprising 140 species in 51 genera. The order Lecanorales in New Zealand comprises 45 families and contains the bulk of the lichen species, its major families being

Parmeliaceae (240 species), Physciaceae (99 species), Pannariaceae (89 species), Lobariaceae (79 species), and Cladoniaceae (102 taxa). Two other large families are the Verrucariaceae (57 species) (order Verrucariales) and Pertusariaceae (66 species) (order Pertusariales). New Zealand's most speciose genera are *Cladonia* (88 taxa), *Xanthoparmelia* (81 taxa), *Pertusaria* (60 taxa), *Pseudocyphellaria* (54 taxa), and *Placopsis* (39 taxa). The most diverse of the orders after Lecanorales, in terms of numbers of included families, are the Arthoniales (five families), Ostropales (four families), Pyrenulales (three families), Pertusariales (two families), and the Gyalectales, Patellariales, Trichotheliales, and Verrucariales with one family apiece.

Olaf Swartz's illustration of *Sticta filix* (Lecanoromycetes: Peltigerales) from Dusky Sound.

Historical synopsis of New Zealand lichen taxonomy

New Zealand has a venerable lichenological tradition dating from the 18th century (Galloway 1985a,b, 1998a, 2008b). Banks and Solander, botanists with Cook on the *Endeavour* voyage, collected several lichens that are preserved in the lichen herbarium of the Natural History Museum in London. From Cook's second voyage, a lichen collected by Anders Sparrmann and J. R. & J. G. A. Forster from Dusky Sound was examined by the Swedish botanist Olof Swartz, and described as *Lichen filix* [= *Sticta filix* (Sw.) Ach.] in his doctoral thesis (Swartz 1781; Galloway 2012). The major 18th-century New Zealand collection was that made Archibald Menzies, naturalist to Vancouver's *Discovery* expedition that visited Dusky Sound in 1791 (Galloway & Groves 1987) when he collected 14 taxa in seven genera (Galloway 1995a, 1999a).

In the early part of the 19th century, French botanists made landfalls in both North and South Islands and in the subantarctic islands collecting a number of lichens (Galloway 2000), with Achille Richard recording 27 species, five of which were newly described and a number illustrated with hand-coloured engravings (Richard 1832).

Lichenology in New Zealand received considerable impetus from the visit of Joseph Hooker (1817–1911) to the Bay of Islands in 1841, during the visit there of the Antarctic expedition ships *Erebus* and *Terror* under the command of Captain James Clark Ross. Hooker was assistant surgeon on the *Erebus*, and in company with William Colenso, David Lyall (assistant surgeon on the *Terror*), and Andrew Sinclair (later Colonial Secretary), collected many lichens from the Bay of Islands. Hooker also collected on the Auckland Islands and on Campbell Island. On his return to England in 1843, Hooker enlisted the help of the Irish cryptogamist Thomas Taylor in preparing the lichenological results of the Antarctic voyage for publication. Together they published a preliminary account of Antarctic lichens, which included many taxa from New Zealand (Hooker and Taylor 1844), followed by a more-detailed treatment of Auckland and Campbell Island lichens in which four pages of illustrations were published (Taylor & Hooker 1845). Hooker encouraged several local botanists (among them Colenso, Haast, Hector, Knight, Lyall, Monro, Sinclair, and Travers) to collect New Zealand lichens and send them to Kew. These he then passed on to Churchill Babington for identification. Babington (1821–1889), a Fellow of St John's College, was Disney Professor of Archaeology and Lecturer in Theology in the University of Cambridge and was the finest lichenologist working in Britain in the 1850s (Galloway 1991b). His account of New Zealand lichens appeared in volume two of Hooker's *Flora Novae Zelandiae* (Babington 1855) and discussed 150 taxa in 33 genera, magnificently illustrated with 20 plates drawn by Walter Hood Fitch from Joseph Hooker's pencil drawings made during the expedition. Hooker made a later account of New Zealand lichens in his *Handbook* (Hooker 1867; Galloway 1998c), this neglected work being the first true lichen flora of New Zealand, discussing 213 taxa in 44 genera and appending brief habitat notes and known distributions.

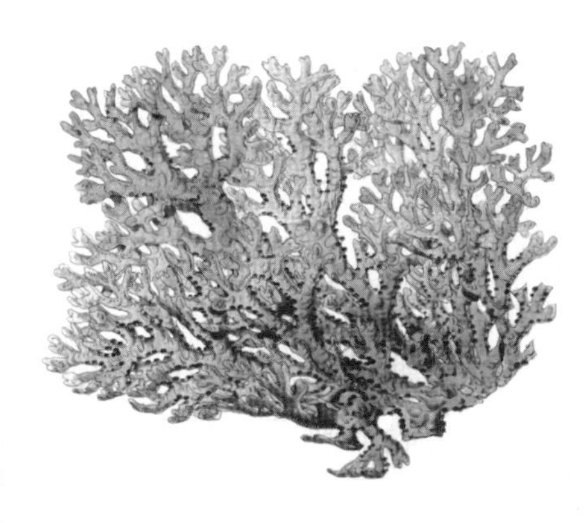

Fitch's illustration of *Pseudocyphellaria carpoloma* (Lecanoromycetes: Peltigerales) in *Flora Novae Zealandiae*.

From Babington 1855

The missionary William Colenso (1811–1899), who was encouraged to collect lichens by Joseph Hooker, sent many specimens to Kew for identification. These were published by Babington (1855), Leighton (1869), and Müller Argoviensis (1896). Colenso's lichens are held at BM and his own herbarium at WELT (St George 2009).

The Scottish physician William Lauder Lindsay (1829–1880) visited Otago in 1861–62, staying for several months with William Martin at his farm/nursery, *Fairfield*, south of Dunedin. Linsday collected many lichens from the environs of Dunedin, inland from the goldfields of the Tuapeka, and as far south as the Nuggets. On his return to Scotland, Lindsay sent his Otago lichens to Paris for identification by William Nylander. Duplicates of Lindsay's Otago collections are in Nylander's herbarium, now in Helsinki (H-NYL), in the Hookerian herbarium (BM), and in his own herbarium at Edinburgh (E). The main systematic accounts of the Otago lichens were those of Nylander (1865, 1866) and Lindsay (1866a,b, 1867, 1868, 1869) and Joseph Hooker made use of these in his account of New Zealand lichens in the *Handbook* (Hooker 1867; Galloway 1998c).

John Buchanan (1819–1898), another Scot, but resident in New Zealand from February 1852 (Adams 2002), collected lichens, mainly in Wellington in the 1870s, sending his material (some 180 numbers) from the Colonial Museum in Wellington, to Glasgow to be identified by James Stirton (Stirton 1873–75,1877a,b). Buchanan was a fine collector, Stirton commenting '... to whose assiduity and enthusiasm I can bear ample witness' (Stirton 1873, p. 15), and from his collections (held in BM, CHR, GLAM, OTA and WELT) in excess of 100 new lichens were named by Stirton. Stirton's name is honoured in the Australasian corticolous, monospecific genus *Stirtoniella* (Galloway et al. 2005).

Auditor-General Charles Knight (1808–1891) began collecting lichens in the 1850s when still living in Auckland (Galloway 1998c) and published two papers on pyrenocarpous lichens found there (Knight 1860; Knight & Mitten 1860). On moving to Wellington he established links with many of the leading European lichenologists of the day, elaborated a substantial lichen library, and bought most of the available 19th-century lichen exsiccati (Galloway 1990). In his capacity as the senior public servant he travelled to many parts of the country and no doubt took opportunities to collect lichens at these times, though his extensive herbarium (WELT) rarely documents localities, the majority of his specimens being simply labelled Nova Zelandia. He was an acute observer and his specimens frequently have his attractive microscopic dissections attached. He published many new taxa, most of which are still valid (Knight 1875–77, 1880, 1883, 1884) and his collections were also named by Nylander (1988) and Müller Argoviensis 1892). He was without doubt the major 19th-century resident lichenologist.

At the end of the century Christchurch bryologist T. W. N. Beckett (1838–1906) sent some lichen material to James Stirton in Glasgow (Galloway 1998b), which eventually provided the basis for two papers on new lichens (Stirton 1898, 1900). Beckett's lichens, formerly in the Canterbury Museum herbarium, are held at Landcare Research (CHR). Additional Beckett lichens are found in Stirton's herbarium in Glasgow (GLAM) and at the Natural History Museum in London (BM), and a set of Canterbury lichens that Beckett sent to the Italian lichenologist Antonio Jatta is in the herbarium of the University of Florence (FI).

In the early 20th century lichenology in New Zealand lapsed until the arrival from Uppsala in November 1926 of the Swedish lichenologist G. Einar Du Rietz (1895–1967) and his wife, Greta Sernander Du Rietz (1897–1981), for the express purpose of collecting lichens. During their visit of several months, they collected from North Auckland to Fiordland and also visited the subantarctic islands, amassing a collection of some 4000 numbers (held in UPS), probably the most extensive collection of New Zealand lichens ever made at one time (Galloway 2004e). Although Du Rietz published relatively little on his New Zealand collections (Du Rietz 1929), he curated this large collection to a very high standard and had hopes of working it up in his retirement, an eventuality that was frustrated by his death in 1967. He was, however, an important influence in the rejuvenation of interest in New Zealand lichens and he actively encouraged

Xanthoria elegans (Lecanoromycetes: Teloschistales).
Barrie Wills

H. H. Allan, J. S. Thomson, and others to collect lichens and to send them to Europe for identification (Allan 1927; Galloway 1976a, 2004e).

Encouraged by Du Rietz, the first Director of Botany Division, DSIR, Dr H. H. Allan (1882–1957), took up the study of New Zealand lichens, collecting from mainly North Island localities and encouraging collectors such as Lucy Cranwell (Galloway 2004c), Lucy Moore, Ted Chamberlain, and others to bring back lichens from botanical trips to far-flung places in North Island, and Jack Scott Thomson from the mountains of South Island. These he then sent to the lichenologists Zahlbruckner in Vienna, Hillmann in Berlin and Gyelnik in Budapest for identification, the results of this burst of 1930s lichen collecting being Zahlbruckner's posthumous *Lichenes Novae Zealandiae* (Zahlbruckner 1941), which contains many new names and is the major taxonomic compendium of the first half of the 20th century. Allan's interest in lichens declined thereafter but he also published several introductory papers on various groups (Allan 1948, 1949a,b, 1951).

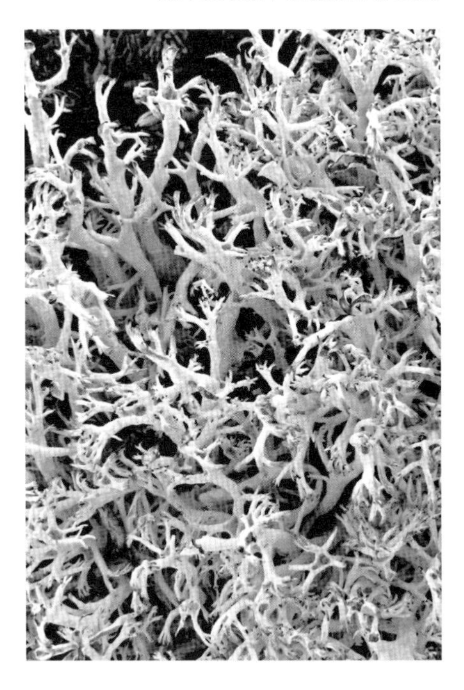

Cladonia mitis (Lecanoromycetes: Lecanorales).

Janet Ledingham

The Dunedin businessman, botanist, photographer, and mountaineer Jack Scott Thomson (1883–1943) met Einar and Greta Du Rietz in Dunedin in 1927 and, during several joint field trips together, they became fast friends, with Du Rietz encouraging Thomson to collect lichens seriously. Over the space of 10 years, Thomson collected from a wide range of localities from Nelson to Stewart Island, and from sea level to the summit of the Inland Kaikouras, the collection amounting to just over 3000 numbers (Bannister 2000). Many of his collections were named by Zahlbruckner (1941). Duplicates of his collections were sent to Allan who then posted them off to Vienna to Zahlbruckner. The major repositories of Thomson's lichens are Vienna (W), Lincoln (CHR), and Thomson's personal herbarium (which languished for years in several thousand tobacco tins) in Dunedin (OTA). A small collection was also sent to the Italian poet and lichenologist Camillo Sbarbaro and is held in the herbarium of the Museo Civico di Storia Naturale Giacomo Doria in Genoa (Galloway 1985a).

After the Second World War, lichenology was revived in Dunedin by the organic chemist James Murray (1923–1961) and the retired schoolteacher William Martin (1886–1975). They were together at Port Pegasus in southern Stewart Island in January 1949, Martin gathering bryophytes and Murray collecting yellow-medulla species of *Pseudocyphellaria* (*P. colensoi* and *P. coronata*) that were part of his experimental work on lichens (Murray 1952). Murray returned from Cambridge (where he did a PhD in organic chemistry in 1950–1952), eager to come to grips in the field with the various taxonomic problems that his work on lichen chemistry had raised. All of the New Zealand genera were in desperate need of revision and the first necessity was a systematic collection of specimens. Martin was keen to collaborate and also had time for South Island-wide collecting. It was arranged that Martin would study the genus *Cladonia*, and that Murray would undertake revisions of everything else, each making his collections available to the other. The system worked well and soon Martin had amassed a very large lichen herbarium. His work on *Cladonia* was careful, detailed, and thorough. In it he enjoyed the constant encouragement of the leading North American student of *Cladonia*, A. W. Evans of Yale University. Martin's work culminated in the publication of 'The *Cladoniae* of New Zealand' (Martin 1958), to be followed by subsequent papers on the genus and on *Cladia* (Martin 1960, 1962, 1965).

Coralline *Cladia sullivanii* (Lecanoromycetes: Lecanorales).

Janet Ledingham

Murray meanwhile published the first of a projected series of revisions of New Zealand lichen genera (Murray 1960a-c), and began work on lichens collected for him from the New Zealand sector of Antarctica (Murray 1963a) and from Campbell Island. He was acutely aware of the difficulties of attempting to do taxonomic lichenology in New Zealand, remote from original material and sources of literature, so in 1960 he visited most of the important European herbaria while he was based in London at Imperial College. His searching of New Zealand lichen types was exceptionally thorough, as a perusal of his notebooks

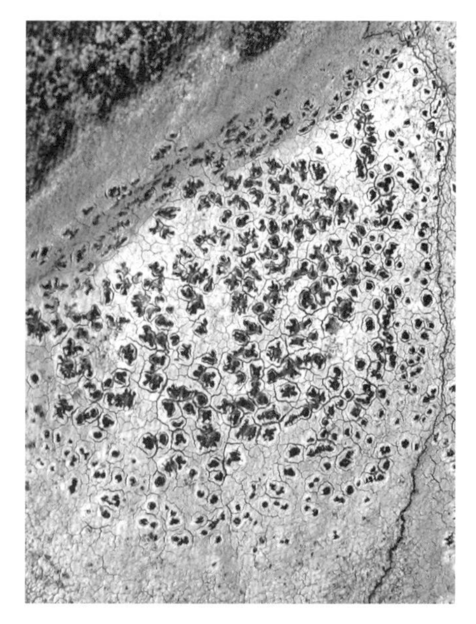

Pertusaria graphica, a maritime lichen that can be found just above high-tidal rocks (Lecanoromycetes: Pertusariales).
Janet Ledingham

shows, and he returned to New Zealand eager to reorganise his herbarium. While in London, he worked closely on *Psoroma* (and other genera in the Pannariaceae), *Pseudocyphellaria*, *Sticta*, and *Menegazzia*, mainly in collaboration with Peter James at the Natural History Museum. His death in a car accident in 1961 robbed New Zealand of its most able local lichenologist who, had he lived, would have made major and wide-ranging contributions to New Zealand lichenology. Several of his incomplete papers were prepared for publication by the late Dr G. A. M. Scott and Peter James (Murray 1962, 1963b–d).

After James Murray's death, William Martin carried on alone with the collection and study of New Zealand lichens. His careful sifting of such lichen literature as was available to him resulted in the publication of his 'Census Catalogue' (Martin 1966) and its supplement two years later (Martin 1968), the first systematic recording of all published New Zealand lichen taxa since Müller Argoviensis's 'Conspectus' of 72 years before (Müller Argoviensis 1894). He published an account of the lichens of the Dunedin district (Martin 1970) and enjoyed a fruitful collaboration with John Child (1923–1986), a keen collector and photographer, resulting in publication of the highly successful introductory work 'Lichens of New Zealand' (Martin & Child 1972), which made the New Zealand lichen mycobiota accessible to a public far beyond the local audience for whom the book was originally planned (Galloway 2010).

The Canterbury botanist and mountaineer Brian Fineran collected lichens in the 1960s from several southern offshore-island habitats, including Bird Island in Foveaux Strait, Codfish Island, The Snares, and the Auckland Islands (Fineran 1966a,b, 1969, 1971; Dodge 1971), and from high-alpine habitats in Canterbury from the Mt Cook region to Arthur's Pass (Fineran & Dodge 1970, 1973; Dodge 1971). Fineran's lichens were identified by Carroll Dodge and included several new taxa (Fineran 1969, Dodge 1971; Galloway 2004c).

The American lichenologist Henry Imshaug (1925–2010), from Michigan State University, visited South Island and Campbell Island in 1969–1970 (Imshaug 1970) and the Auckland Islands in 1972–73 (Imshaug 1973), making extensive lichen collections from both subantarctic island areas. In subsequent years he meticulously curated these collections and recent taxonomic work on this major lichenological resource has begun in earnest (Fryday & Prather 2001), with many novelties being uncovered (Fryday 2004, 2007a, 2007b, 2008, 2011; Coppins & Fryday 2006, 2007; Øvstedal & Fryday 2011).

Dibaeis arcuata (Lecanoromycetes: Pertusariales).
Janet Ledingham

David Galloway began collecting lichens as a schoolboy in Invercargill in 1957. He met James Murray at Otago University in 1961 and they worked together for several months before Murray's death in June of that year. He was employed as a research assistant to Peter James from London's Natural History Museum, who was brought out to Dunedin (November 1962–March 1963) to curate Murray's extensive but largely unidentified lichen herbarium (now in OTA). Encouragement came from Peter James (London) and from visiting lichenologists Masami Sato (1910–1985) (from Japan in 1964), Gunnar Degelius (1903–1993) (from Sweden in 1970), and Fritz Mattick (1901–1984) (from Germany in 1971), and over 10 years he collected throughout New Zealand from the Three Kings Islands to Stewart Island. In 1972 he transferred to Botany Division, DSIR, and was seconded to the Natural History Museum in 1973 to undertake compilation of a New Zealand lichen flora, which was eventually completed in 1983 and published in 1985 (Galloway 1985a). He worked on various southern-hemisphere lichen projects at the Natural History Museum in London from 1985 until 1994, returning permanently to New Zealand in November of that year. Since then, he has continued collecting and publishing on New Zealand lichens and encouraging others to do likewise. A major project was the preparation of a revised, second edition of *Flora of New Zealand: Lichens* (Galloway 2007), incorporating all the changes and additions to New Zealand's lichen mycobiota that occurred since 1985. It is part of Landcare Research's *Flora* series (Garnock-Jones & Breitwieser 1998). A review of austral lichenology

(Galloway 2008b) sets research on the New Zealand lichen mycobiota in its southern-hemisphere regional perspective.

In the 1970s Bruce and Glenys Hayward became interested in the lichens of northern offshore islands and published many papers on these island lichen mycobiotas (e.g. Hayward & Hayward 1986; Hayward et al. 1986, Hayward & Wright 1991). Glenys Hayward published a study on New Zealand taxa in the families Graphidaceae and Opegraphaceae (Hayward 1977) under the guidance of the late Mason Hale (1928–1990) at the Smithsonian Institution in Washington, D. C. Bruce Hayward also published on the lichens of southern Stewart Island (Hayward & Lumbsch 1992).

Teloschistes sieberianus (Lecanoromycetes: Teloschistales).
Rick Kooperberg

Auckland schoolteacher John Bartlett (1945–1986) began collecting lichens from northern New Zealand habitats in 1977 at the suggestion of David Galloway, and within a few years he emerged as one of the finest collectors the country has yet had. He collected widely and in depth from North Cape to Fiordland and added many taxa to the New Zealand mycobiota (see Galloway 1985a). His tragically early death in 1986 robbed the country of its finest collector since Colenso (Galloway 1987). John Bartlett's large lichen collection, now beautifully curated, is housed in the herbarium of the Auckland Museum (AK). His preliminary account of the lichens of the Waitakere Ranges, west Auckland, was published posthumously (Bartlett 1988). His name is commemorated in the lichen genus *Bartlettiella* (Galloway & Jørgensen 1990).

The 1990s saw a great expansion of local interest in New Zealand lichens. Several lichen workshops were held (Cass; Lake Rotoiti, Nelson; Wellington; Dunedin) and this custom is now well established and has recently become part of the expanded John Child Bryological and Lichenological Workshop series held annually in different parts of the country. Strong links are maintained with lichen taxonomists and collectors in Australia, many of whom attend the New Zealand workshop meetings, and the journal *Australasian Lichenology* is now edited and produced in Nelson by Dr Bill Malcolm, who has also emerged in recent years as a champion of little-studied crustose groups and foliicolous lichens (see for example Malcolm et al. 1995; Malcolm & Vězda 1995a,b, 1996). His lichen photographs and illustrations are particularly noteworthy, and his illustrated checklist (Malcolm & Galloway 1997) is one of the finest illustrated lichen books yet produced. He and his wife Nancy have produced two beautifully illustrated books from their Micro-Optics Press in Nelson – *New Zealand Lichens* (Malcolm & Malcolm 2000) and *New Zealand's Leaf-dwelling Lichens* (Malcolm & Malcolm (2001). Bill Malcolm's contribution to lichenology is recognised in the genus name *Malcolmiella* (Vězda 1997). Lichen dating of earthquake-generated regional rockfall events in the Southern Alps of South Island was discussed in a comprehensive paper by Bull and Brandon (1998), extending the earlier lichenometric work of Orwin (1970, 1971). Blanchon et al. (2007) recently produced a nicely illustrated account of the lichens of Rangitoto Island. Recent work on lichenology in New Zealand is summarised in Galloway (2008b). A listing of New Zealand's lichen and lichenicolous fungus mycobiota, using the threat classification system of Townsend et al. (2008) is in progress under the vigorous chairmanship of Dr Peter de Lange (DOC) and will be published shortly.

Nephroma australe (Lecanoromycetes: Peltigerales), reproduced from the painting on the cover of *Australasian Lichenology*, number 41, July 1997.
Bill Malcolm

Herbaria

New Zealand lichens are now well catered for in each of the main regional herbaria AK, WELT, CHR and OTA.

AK: This is a very well-curated lichen herbarium of mainly northern-New Zealand lichens (including lichens formerly in AKU). The major collection held here is the very extensive New Zealand-wide herbarium of the late John Bartlett (1945–1986), which was bequeathed to the Museum after his death in 1986 (Galloway 1987). It is actively being added to and is accessible online.

WELT: The major 19th-century collections of William Colenso, John Buchanan

Placopsis salazina (Lecanoromycetes: Argyriales).
Janet Ledingham

(transferred on permanent loan from the Otago Museum via CHR), and Charles Knight are held here. The Knight collection (complete with its wide range of some 30 European exsiccati) is particularly extensive and rich in type specimens. It was originally given to the Biology Department of Victoria University but was subsequently transferred to the Botany Department of the Dominion Museum (now Te Papa) in 1939. These historical collections have high taxonomic value and still require proper cataloguing by a competent lichenologist. Besides the historical collections, WELT has a growing regional lichen collection that is being continually added to, and additionally holds an important collection of lichens from Antarctica. The collections are databased and available on-line.

CHR: This is the principal lichen collection in the country, being formed by Dr H. H. Allan, first Director of Botany Division, DSIR, and maintained as a working collection from the late 1920s onwards. It is particularly rich in type material of taxa described by Zahlbruckner (1941) and was the foundation collection for the 1985 lichen Flora. Important collections are those of H. H. Allan, J. S. Thomson, W. Martin, J. Child, D. J. Galloway, and a small historical collection of T. W. N. Beckett (Galloway 1998b). The collections are continually being added to by both local and overseas lichenologists. It has a wide geographical coverage and a considerable amount of foreign material as well. This important lichen herbarium is now being databased, and curation and identification of the large, unidentified backlog is being actively addressed.

OTA has the lichen collections of Jack Scott Thomson (Bannister 2000) and James Murray, the latter collection forming the major part of the herbarium. In addition, it has important Otago-Southland collections of W. Martin, A. F. Mark, J. & P. Child, and D. J. Galloway. It is being actively maintained and added to by J. Bannister, A. Knight, D. J. Galloway, and others. It holds important type material from both Murray and Thomson collections.

Several smaller lichen collections are held in Rotorua (FRI) in the Canterbury University Botany Department (CANU), and some private individuals also maintain private herbaria. The CANU herbarium houses the lichen collection of Brian Fineran and many specimens from this collection, from both high-alpine areas of Canterbury and from the subantarctic islands, were examined by the American lichenologist Carrol W. Dodge (1895–1988), who named many new taxa from Fineran's collections (see Fineran 1969; Dodge 1971; Fineran & Dodge 1970, 1973; Galloway 2004c). The CANU lichen herbarium consequently holds many type specimens of New Zealand material named by Dodge, with Dodge's duplicates of this material now deposited in the Farlow Herbarium (FH) at Harvard.

New Zealand lichens are widely scattered throughout northern-hemisphere herbaria (Galloway 1985a, pp. xxiv-xxvii, xv) with major historical collections held at BM, E, FH, G, GLAM, H, H-NYL, M, PC, S, UPS and W.

Hypogymnia turgidula, an Australasian species (Lecanoromycetes: Lecanorales).
Janet Ledingham

Biogeographical relationships of New Zealand lichens

Biogeographical elements (Galloway 1985b, 1991a, 1996, 2008a,b) recognised in the New Zealand lichen biota are briefly defined as follows.

Endemic. These are taxa that are confined to New Zealand and its subantarctic islands. At present only five genera are endemic to New Zealand (2% of the total), and about 380 endemic species are known (~ 30% of the total) though this latter figure may be expected to change once New Zealand and southern-hemisphere microlichens are better known.

Australasian. These are taxa occurring in New Zealand and Australia. Three subgroups are discerned here: 1) New Zealand, Tasmania and the subantarctic islands; 2) New Zealand, Australia and the subantarctic islands; 3) the subantarctic islands and Tasmania. At the generic level, 173 genera are common to both New Zealand and Tasmania while at the species level about 450 taxa are common to both New Zealand and Tasmania.

Austral: These are taxa occurring in New Zealand, the subantarctic islands, and southern South America. At the generic level, 70% of lichen genera found in New Zealand also occur in Chile, but at the species level only about 390 taxa are common to both Chile and New Zealand. These taxa include both recently transported (neoaustral) elements as well as Gondwanan (palaeoaustral) elements (Galloway 2008b).

Cosmopolitan. These are taxa occurring on all known landmasses. In the New Zealand lichen mycobiota about 370 taxa (22% of the total) are cosmopolitan.

Bipolar. These taxa occur in New Zealand (usually on old alpine surfaces such as the Central Otago mountains) and in boreal localities in the Northern Hemisphere (Galloway & Bartlett 1986; Galloway & Aptroot 1995; Galloway 2003, 2008b). Notable examples are *Caloplaca tornoensis, Frutidella caesioatra, Lecanora cavicola, Pannaria hookeri, Placynthium rosulans*, and *Solorina crocea*, the number increasing as high-alpine localities are scrutinised more carefully. At the species level, about 180 taxa of bipolar lichens and lichenicolous fungi are recorded from New Zealand, comprising about 10% of the lichen mycobiota (Galloway 2003, 2008b).

Western Pacific (*Indo-Malaysian/Indo-Pacific*). These are taxa occurring in mainland coastal Asia, Japan, Philippines, Indonesia, New Guinea, New Caledonia, the east coast of Australia, and the west coast of New Zealand. A small number of species such as *Thysanothecium scutellatum* belong here.

The widely distributed beard lichen *Usnea ciliifera* (Lecanoromycetes: Lecanorales).
Janet Ledingham

Pantropical. Taxa occurring in all major tropical regions of the world.

Paleotropical. Taxa east of and including Easter Island, to St Helena; most of the Pacific islands are included here.

Circum-Pacific. Taxa occurring on the seaboards of North and South America and of Asia and including Australia and New Zealand. Several lichens show this wide Pacific distribution, including *Leioderma sorediatum, Placopsis cribellans*, and *Turgidosculum complicatulum*.

Southern xeric. Taxa occurring in South Africa, Western Australia, South Australia, and New Zealand (in dry intermontane basins). A few species such as *Xanthoparmelia molliuscula* (the type locality is Table Mountain) occur here.

Heads (1997) used distribution patterns in *Pseudocyphellaria* to illustrate regional patterns of diversity and noted disjunct lichen distributions along the Alpine Fault (Heads 1998). The disjunction of the calcicole *Solorina spongiosa* in Nelson and Fiordland was most probably caused by the 440 kilometres of lateral displacement on the fault.

Ecology and distribution in New Zealand

Although *Flora of New Zealand: Lichens* (Galloway 1985a, 2007) gave notes on the distribution of most of the taxa recorded therein, for the vast majority of New Zealand lichens detailed distributions and defined ecological requirements are still lacking. Regional checklists are currently being compiled for Wellington and Dunedin but there are still no modern lichen checklists for any national park or similar regional area. Work is in progress on documenting the high-alpine lichen mycobiota of the Central Otago mountains and on lichen communities developed on limestone exposures, but what is needed is a series of regional-survey initiatives analogous to the British Lichen Society mapping scheme. A recent paper giving detailed distribution maps for the 14 species of *Ramalina* occurring in New Zealand (Bannister et al. 2004) is an excellent model for future, in-depth studies to emulate. Synecological analysis of corticolous lichen communities in New Zealand rainforests, analogous to the study made by Kantvilas (1988) in Tasmanian rainforest, have not yet been widely attempted here, though a PhD study in the Craigieburn Forest catchment (Tara Schoenwetter pers. comm.) was recently completed and will soon be published. The ecology of aquatic lichens

Ramalina glaucescens (Lecanoromycetes: Lecanorales).
Janet Ledingham

Icmadophila splachnirima (Lecanoromycetes: Pertusariales).
Janet Ledingham

of alpine streams has been published (Johnson & McCarthy 1997; McCarthy & Johnson 1998). Some 41 genera (about 14% of the total) and 235 species (about 16% of the total) in the New Zealand lichen mycobiota have cyanobacteria as their primary photobiont. Of these potential diazotrophs, the majority are found in either (or both) forest or grassland biomes. Since both biomes are commonly nitrogen-limited, it seems likely that diazotrophic lichens contribute substantial amounts of fixed nitrogen to these ecosystems. For discussion of the role of lichens in ecosystem processes see Seaward (1996, 2008) and Hawksworth (1991, 1994, 1997). Galloway (2008b) reviewed local and regional lichen-based studies relating to climate change, nutrient cycling, soil consolidation, and lichenometry. Lars Ludwig has recently (March 2011) begun a doctoral study of the reproductive ecology of the lichen *Icmadophila splachnirima* (see Ludwig 2011), the first such detailed research agenda using New Zealand lichen communities, which already appears to be a very promising and productive field of enquiry.

Lichens as bioindicators

Despite the apparent hardiness of lichens and their longevity in harsh environments, the extremely close anatomical and physiological interactions between the photobiont and mycobiont partners in lichens make the lichen symbiosis extremely sensitive to various kinds of environmental perturbations. Lichens have no protective cuticle and absorb water and nutrients largely from the air and from the surfaces on which they grow. This, and their ability to accumulate but not excrete substances that might be toxic (e.g. heavy metals, radionuclides), contribute to their sensitivity, with particular impacts on photosynthesis, respiration, and nitrogen fixation. Lichens have wide application as biomonitors, especially of atmospheric pollution (Hawksworth 1992, 1994; Seaward 1996, 1997, 2004; Seaward & Coppins 2004; Nash 2008b). Methodologies for using lichens as indicators of air pollution are well established in the northern hemisphere (see Richardson 1992) but very few such studies have yet been carried out in New Zealand, with the notable exception of Daly (1970), who showed that in Christchurch City both lichen abundance and diversity decreased along a gradient of increasing winter sulphur dioxide levels towards the city centre. The reasons why lichens have scarcely been used to date as bioindicators of air pollution in New Zealand include the paucity, until recently, of many trained lichenologists and the lack of knowledge of lichens from urban and industrial habitats and their sensitivity to environmental contaminants. A study of lichens in urban and industrial sites from Whangarei to Bluff was undertaken in the 1990s under the auspices of the Ministry for the Environment and the preliminary results reported on (Johnson et al. 1998), demonstrating that, in a New Zealand setting, lichens have considerably utility in pollution monitoring. An identification guide to lichens occurring on urban and rural trees and used in pollution monitoring studies was also developed (Johnson & Galloway 1999). The rapid growth and colonisation of lichens in urban environments was noted recently (Galloway 2009) and is a trend worth monitoring as it may well be a consequence of, or response to, global warming and regional climate change.

Usnea (Lecanoromycetes: Lecanorales) on red beech (*Nothofagus fusca*) in Eglinton Valley, Fiordland.
Janet Ledingham

Lichen chemistry

Lichens produce some 700 unique secondary compounds (Huneck & Yoshimura 1996; Elix & Stocker-Wörgötter 2008), many of which are routinely used in lichen identification, especially at the species level (Lumbsch 1988a,b). Many lichen compounds also have well-documented biological activity as antibiotic, antiherbivore, and antimicrobial compounds (Lawrey 1986), with traditional use of lichens in dyeing and in folk medicine having a long history (Richardson 1975, 1991). New Zealand Maori traditionally used pendulous species of *Usnea*, especially *U. angulata*, for nappies and sanitary pads (Galloway 1985; Riley 1994;

Fuller et al. 2004), which took advantage of the antibiotic properties of usnic acid that is abundant in these lichens. The genus *Pseudocyphellaria* has provided a rich source of novel lichen compounds, with the studies of the late Dr James Murray and Professor R. E. Corbett and his students at Otago University from 1952 until the late 1980s (for a review see Galloway 1988, pp. 26–35) on terpenoid chemistry being especially noteworthy. A list of the compounds reported from New Zealand lichens was given by Walker and Lintott (1997) and a few were used in screening for bioactive compounds. Antileukaemic activity of polyporic acid extracted from *Pseudocyphellaria coronata* was reported by Burton and Cain (1959), and more recently Calder et al. (1986) included 13 New Zealand lichens in their search for antibiotic activity. A comprehensive screening of 69 New Zealand lichens for antimicrobial, antiviral, and cytotoxic activity was reported by Perry et al. (1999). The importance of lichen secondary compounds as high-light screens and UV filters has attracted great interest as a biological response to increased levels of UVB as a consequence of ozone thinning (Galloway 1993; Quilhot et al. 1994, 1998; Rikkinen 1995; Solhaug et al. 2003; Bjerke et al. 2005), the amounts of the secondary compounds having the potential also to be used as biomonitors of UV levels

Umbilicaria hyperborea (Lecanoromycetes: Umbilicariales).

Janet Ledingham

The future

I believe that New Zealand lichenology has an assured and exciting future. It has at the moment a sound taxonomic base with a published flora and a modern revision, both accessible online (Galloway 1985a, 2007), updated checklists (Galloway 1992a; Malcolm & Galloway 1997; Pennycook & Galloway 2004), and a continuing bibliography (Galloway 1974, 1985b, 1994), supported by well-curated regional lichen herbaria and for the foreseeable future a commitment to employment of one or more professional lichenologists. Internationally the workforce of full-time lichenologists is a diminishing resource, and this is a cause of much concern both locally and worldwide (Hawksworth 1999; Lumbsch et al. 2011).

There are at present no lichenology courses offered at university level in New Zealand to train replacement professional lichenologists, and the discipline is vulnerable to loss of local expertise when present lichenologists retire. The Australasian Lichen Society, however, addresses this need in now well-established biennial lichen workshops, and both New Zealand and Australian lichenologists are well catered for with a twice-yearly journal, *Australasian Lichenology*, currently edited and produced to a very high standard by Dr Bill Malcolm in Nelson. But lichenology always has been and will continue to be (both in New Zealand and worldwide) a field of profound interest to a variety of keen amateurs and societies (The British Lichen Society is a good example), and workshops, newsletters, field trips, herbarium assistance, and many other activities find ready help from amateur lichenologists who continue to provide the backbone of interest and support for the group. Their interests are catered for in currently available, accessibly written, and attractively presented introductory books (see for instance Malcolm & Galloway 1997; Kantvilas & Jarman 1999; Purvis 2000; Gilbert 2000; Malcolm & Malcolm 2000, 2001; Brodo et al. 2001; Lumbsch et al. 2001; McCarthy & Malcolm 2004; Dobson 2005; Schumm & Aptroot 2010; Schumm 2011). Lichen websites are now well established and easily accessible, with news and up-to-date information.

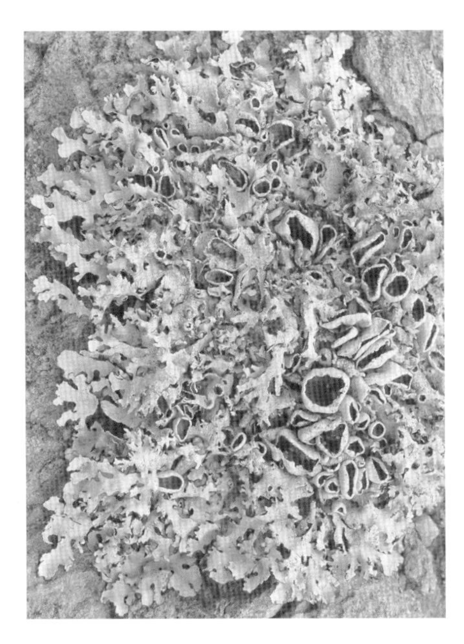

Xanthoparmelia tasmanica (Lecanoromycetes: Lecanorales).

Janet Ledingham

Long term, I expect applied aspects of lichenology to dominate over systematic work, with the emergence of exciting fields of bioindication, biomonitoring, and habitat and ecological restoration, in which systematic lichenology has application in a variety of ongoing applied fields (Galloway 2004b). However, much still remains to be done on the systematics of microlichens, of lichenicolous fungi (fungal parasites coevolved with lichens), and of fast-growing muscicolous and foliicolous lichens, so that training and recruitment of lichen taxonomists is a necessary condition of the future development of New Zealand lichenology. Also, the now-commonplace utilisation of molecular methods of analysis is marking the emergence of a new and reformed lichen taxonomy in nearly all systematic groups. Genera well represented in New Zealand are expected to produce many fascinating groupings as local populations are investigated. The biological role of lichens in New Zealand biomes where they are well developed should provide ample scope for further joint ventures between lichenology and many other disciplines.

Acknowledgements

I acknowledge the support of the Foundation for Research, Science and Technology (Contract 09618) and the Mardsen Fund (Contract UOO805).

Author

Dr David J. Galloway Landcare Research NZ Ltd, Private Bag 1930, Dunedin, New Zealand [gallowayd@landcareresearch.co.nz]

References

ADAMS, N. M. 2002: John Buchanan F.L.S. botanist and artist (1819–1898). *Tuhinga 13*: 71–115.

AHMADJIAN, V. 1993: *The Lichen Symbiosis*. Revised edn. John Wiley & Sons, New York. 250 p.

AHMADJIAN, V.; PARACER, S. 1987 [1986]: *Symbiosis. An introduction to biological associations*. University Press of New England, Hanover. xii + 212 p.

ALLAN, H. H. 1927: Lichens, mosses and ferns of Canterbury. Pp. 160–166 *in*: Speight, R.; Wall, A.; Laing, R. M. (eds), *Natural History of Canterbury*. Simpson and Williams, Christchurch. x + 299 p.

ALLAN, H. H. 1948: A note on lichens with a key to the commoner New Zealand genera. *Tuatara 1*: 20–35.

ALLAN, H. H. 1949: A note on the crustaceous lichens of New Zealand. *Tuatara 2*: 15–21.

ALLAN, H. H. 1949b: A key to the Stictaceae of New Zealand. *Tuatara 2*: 97–101

ALLAN, H. H. 1951: New Zealand lichens – a key to the family Umbilicariaceae. *Tuatara 4*: 59–62.

BABINGTON, C. 1855. Lichenes. Pp. 266–311 *in*: Hooker, J. D. (ed.), *Botany of the Antarctic Voyage of H. M. Discovery Ships Erebus and Terror, in the years 1839–1843. II. Flowerless Plants*. Lovell Reeve, London. 378 p.

BANNISTER, J. 2000: Profile of a botanist: John Scott Thomson and Thomson's lichen collection in OTA. *Botanical Society of Otago Newsletter 16*: 9–12.

BANNISTER, P.; BANNISTER, J. M.; BLANCHON, D. J. 2004: Distribution, habitat, and relation to climatic factors of the lichen genus *Ramalina* in New Zealand. *New Zealand Journal of Botany 42*: 121–138.

BARTLETT, J. K. 1988: Lichens of the Waitakere Range. *Auckland Botanical Society Bulletin 17*: 1–29.

BJERKE, J. W.; ELVEBAKK, A.; DOMINGUEZ, E.; DAHLBACK, A. 2005: Seasonal trends in usnic acid concentrations of Arctic, alpine and Patagonian populations of the lichen *Flavocetraria nivalis*. *Phytochemistry 66*: 337–344.

BJERKE, J. W.; GWYNN-JONES, D.; CALLAGHAN, T. V. 2005: Effects of enhanced UV-B radiation in the field on the concentration of phenolics and chlorophyll fluorescence in two boreal and arctic-alpine lichens. *Environmental and Experimental Botany 53*: 139–149.

BLANCHON, D. J.; KOOPERBERG, W. F.; LOCKETT, C. E. A. 2007: Lichens. Pp. 135–144 *in*: Wilcox, M. J. (ed.), *Natural History of Rangitoto Island, Hauraki Gulf, Auckland, New Zealand*. Auckland Botanical Society Inc., Auckland. 192 p.

BULL, W. B.; BRANDON, M. T. 1998: Lichen dating of earthquake-generated regional rock-fall events, Southern Alps, New Zealand. *Geological Society of America Bulletin 110*: 60–84.

BURTON, J. F.; CAIN, B. F. 1959: Antileukaemic activity of polyporic acid. *Nature 184*: 1326–1327.

CALDER, V. L.; COLE, A. L. J.; WALKER, J. R. L. 1986: Antibiotic compounds from New Zealand plants. III: a survey of some New Zealand plants for antibiotic substances. *Journal of the Royal Society of New Zealand 16*: 169–181.

COPPINS, B. J.; FRYDAY, A. M. 2006: New or previously misunderstood species of *Lithographa* and *Rimularia* (*Agyriaceae*) from the southern subpolar region and western Canada. *Lichenologist 38*: 93–107.

COPPINS, B. J.; FRYDAY, A. M. 2007: Three new species of *Bacidia* s.lat. (Ramalinaceae) from Campbell Island (New Zealand). *Bibliotheca Lichenologica 95*: 155–164.

COPPINS, B. J.; SEAWARD, M. R. D.; SIMKIN, J. 2005: British Isles list of lichens and lichenicolous fungi (with B.L.S. recording code numbers). http://www.thebls.org.uk/ checklist.html

CORNEJO, C.; SCHEIDEGGER, C. 2010: *Lobaria macaronesica* sp. nov., and the phylogeny of *Lobaria* sect. *Lobaria* (Lobariaceae) in Macaronesia. *Bryologist 113*: 590–604.

DALY, G. T. 1970: Bryophyte and lichen indicators of air pollution in Christchurch. New Zealand. *Proceedings of the New Zealand Ecological Society 17*: 70–79.

DIVAKAR, P. K.; LUMBSCH, H. T.; FERENCOVA, Z.; DEL PRADO, R.; CRESPO, A. 2010: *Remototrachyna*, a newly recognized tropical lineage of lichens in the *Hypotrachyna* clade (Parmeliaceae, Ascomycota), originated in the Indian subcontinent. *American Journal of Botany 97*: 579–590.

DOBSON, F. S. 2005: *Lichens. An Illustrated Guide to the British and Irish Species*. 5th edn. The Richmond Publishing Co. Ltd, Slough. 480 p.

DODGE, C. W. 1971 [1970]: Lichenological notes on the flora of the Antarctic continent and the Subantarctic Islands. *Nova Hedwigia 19*: 439–502.

DU RIETZ, G. E. 1929: The discovery of an arctic element in the lichen flora of New Zealand, and its plant-geographical consequences. *Report of the Australasian Association for the Advancement of Science 19*: 628–635.

ELIX, J. A.; STOCKER-WÖRGÖTTER, E. 2008: Biochemistry and secondary metabolites. Pp. 104–133 *in*: Nash, T. H. III (ed.), *Lichen Biology*. 2nd edn. Cambridge University Press, Cambridge. 486 p.

ESSLINGER, T. L. 2011 *A cumulative checklist for the lichen-forming, lichenicolous, and allied fungi of the continental United States and Canada*. http://www. ndsu.nodak.edu/instruct/ esslinge/ chcklst7.htm (First posted 1 December 1997, Most recent Version (#17) 16 May 2001), Fargo, North Dakota.

FINERAN, B. A. 1966: The vegetation and flora of Bird Island, Foveaux Strait. *New Zealand Journal of Botany 4*: 133–146.

FINERAN, B. A. 1966b: Contributions to the botany of Codfish Island, Stewart Island. *Transactions of the Royal Society of New Zealand (Botany) 3*: 111–122.

FINERAN, B. A. 1969: The flora of the Snares Islands, New Zealand. *Transactions of the Royal Society of New Zealand (Botany) 3*: 237–270.

FINERAN, B. A. 1971: A catalogue of the bryophytes, lichens, and fungi collected on the Auckland Islands. *Journal of the Royal Society of New Zealand 1*: 215–229.

FINERAN, B. A.; DODGE, C. W. 1970: Lichens from the Southern Alps, New Zealand: records from Phipps Peak, the Two Thumbs, and the Tasman Valley. *Pacific Science 24*: 401–408.

FINERAN, B. A.; DODGE, C. W. 1973: Lichens from the Southern Alps, New Zealand II. Records from the Mt Cook district. *Pacific Science 27*: 274–280.

FRIEDL, T.; BÜDEL, B. 2008: Photobionts. Pp. 9–26 *in*: Nash, T.H.III (ed.), *Lichen Biology*. 2nd edn. Cambridge University Press, Cambridge. 486 p.

FRYDAY, A. M. 2004: New species and records of lichenized fungi from Campbell Island and the Auckland Islands, New Zealand. *Bibliotheca Lichenologica 88*: 127–146.

FRYDAY, A. M. 2007a: Additional lichen records from New Zealand 46. *Degelia symptychia* (Tuck.) P.M.Jørg. *Australasian Lichenology 60*: 4–5.

FRYDAY, A. M. 2007b: Additional lichen records from New Zealand 47. *Coccotrema corallinum* Messuti and *C. pocillarum* (C.E.Cumm.) Brodo. *Australasian Lichenology 61*: 3–5.

FRYDAY, A. M. 2008: Three new species of lichenized fungi with cephalodia from the southern New Zealand shelf islands (Campbell Plateau). *Lichenologist 40*: 283–294.

FRYDAY, A. M. 2011: New species and combinations in *Calvitimela* and *Tephromela* from the southern subpolar region. *Lichenologist 43*: 225–239.

FRYDAY, A. M.; PRATHER, L. A. 2001: The lichen collections of Henry Imshaug at the Michigan State University Herbarium (MSC). *Bryologist 104*: 464–467.

FULLER, R. J. M.; BUCHANAN, P. K.; ROBERTS, M. 2004: Maori knowledge of fungi. Matauranga o Nga Harore. Pp. 81–118 *in*: McKenzie, E. H. C. (ed.), *The Fungi of New Zealand/Ng Harore o Aotearoa, Volume 1. Introduction to Fungi of New Zealand*. Fungal Diversity Press, Hong Kong. 500 p.

GALLOWAY, D. J. 1974: A bibliography of New Zealand lichenology. *New Zealand Journal of Botany 12*: 397–422.

GALLOWAY, D. J. 1976a: H. H. Allan's early collections of New Zealand lichens. *New Zealand Journal of Botany 14*: 225–230.

GALLOWAY, D. J. 1976b: Obituary. William Martin, B.Sc. 1886–1975. *New Zealand Journal of Botany 14*: 367–374.

GALLOWAY, D. J. 1981: Erik Acharius, Olof Swartz and the evolution of generic concepts in lichenology. *In*: Wheeler, A.; Price, J. H. (eds), *History in the Service of Systematics. Society for the Bibliography of Natural History, Special Publication 1*: 119–127.

GALLOWAY, D. J. 1985a: *Flora of New Zealand: Lichens*. Government Printer, Wellington. 662 p.

GALLOWAY, D. J. 1985b: Lichenology in the South Pacific, 1790-1840. *In*: Wheeler, A.; Price, J. H. (eds), *From Linnaeus to Darwin: commentaries on the history of biology and geology. Society for the Bibliography of Natural History, Special Publication 2*: 205–214.

GALLOWAY, D. J. 1985c: A bibliography of New Zealand lichenology. 2 Additions and alterations, 1974–1984. *New Zealand Journal of Botany 23*: 351–359.

GALLOWAY, D. J. 1987: Obituary. John Kenneth Bartlett, M.Sc. (Sydney) 1945–1986: An appreciation. *New Zealand Journal of Botany 25*: 173–176.

GALLOWAY, D. J. 1988: Studies in *Pseudocyphellaria* (lichens) I. The New Zealand species. *Bulletin of the British Museum (Natural History), Botany 17*: 1–267.

GALLOWAY, D. J. 1990: Knight, Charles 1808?–1891. Doctor, public servant, botanist. *The Dictionary of New Zealand Biography 1*: 229.

GALLOWAY, D. J. 1991a: Phytogeography of Southern Hemisphere lichens. Pp. 233–262 *in*: Nimis, P. L.; Crovello, T. J. (eds), *Quantitative Approaches to Phytogeography*. Kluwer Academic Publishers, Dordrecht. 304 p.

GALLOWAY, D. J. 1991b: Churchill Babington MA, DD, FLS (1821–1889), theologian, archaeologist, rector: a forgotten Victorian lichenologist. *British Lichen Society Bulletin 69*: 1–7.

GALLOWAY, D. J. 1992a: Checklist of New Zealand lichens. *DSIR Land Resources Scientific Report 26*: 1–58.

GALLOWAY, D. J. 1992b: Biodiversity: a lichenological perspective. *Biodiversity and Conservation 1*: 312–323.

GALLOWAY, D. J. 1993: Global environmental change: lichens and chemistry. *Bibliotheca Lichenologica 53*: 87–95.

GALLOWAY, D. J. 1994: A bibliography of New Zealand lichenology 3. Additions, 1985–92. *New Zealand Journal of Botany 32*: 1–10.

GALLOWAY, D. J 1995a: Lichens in Southern Hemisphere temperate rainforests and their role in the maintenance of biodiversity. Pp. 125–135 *in*: Alsopp, D.; Hawksworth, D. L.; Colwell, R. R. (eds), *Microbial Diversity and Ecosystem Function*. CAB International, Wallingford. 482 p.

GALLOWAY, D. J. 1995b: The extra-European lichen collections of Archibald Menzies MD, FLS (1754–1842). *Edinburgh Journal of Botany 52*: 95–139.

GALLOWAY, D. J. 1996: Lichen biogeography. Pp. 199–216 *in*: Nash, T. H. III (ed.), *Lichen Biology*. Cambridge University Press, Cambridge. 303 p.

GALLOWAY, D. J. 1997: Studies on the lichen genus *Sticta* (Schreber) Ach. IV. New Zealand species. *Lichenologist 29*: 105–168.

GALLOWAY, D. J. 1998a: Contributions to a history of New Zealand lichenology 1. Cook's botanists. *Australasian Lichenology 43*: 20–26.

GALLOWAY, D. J. 1998b: Banks Peninsula lichens: Raoul's legacy.pp. 59–64 *in*: Burrows, C. J. (ed.), *Etienne Raoul and Canterbury Botany 1840–1996*. Canterbury Botanical Society/Manuka Press, Christchurch. 130 p.

GALLOWAY, D. J. 1998c: Joseph Hooker, Charles Knight, and the commissioning of New Zealand's first popular flora: Hooker's Handbook of the New Zealand Flora (1864–1867). *Tuhinga 10*: 31–62.

GALLOWAY, D. J. 1999a: Contributions to a history of New Zealand lichenology 2."A diligent and persevering zeal ...", Archibald Menzies (1754–1842). *Australasian Lichenology 45*: 28–35.

GALLOWAY, D. J. 1999b: Alpine lichens of New Zealand. *New Zealand Alpine Journal 51*: 109–111.

GALLOWAY, D. J. 2000: Contributions to a history of New Zealand lichenology 3. The French. *Australasian Lichenology 46*: 7–17.

GALLOWAY, D. J. 2004a:. New lichen taxa and names in the New Zealand mycobiota. *New Zealand Journal of Botany 42*: 105–120.

GALLOWAY, D. J. 2004b: Lichens in our landscape. *Te Taiao 3*: 8–9.

GALLOWAY, D. J. 2004c: Notes on some lichen names recorded from the Snares Islands, southern New Zealand. Australasian *Lichenology 55*: 21–25.

GALLOWAY, D. J. 2004d: Lucy Cranwell Lecture 2003 – The Kew connection: the Hookers and New Zealand Botany. *Auckland Botanical Society Bulletin 59*: 2–9.

GALLOWAY, D. J. 2004e: The Swedish connection in New Zealand lichenology, 1769–2004. *Symbolae Botanicae Upsalienses 34*: 63–85.

GALLOWAY, D. J. 2007: *Flora of New Zealand Lichens, Revised second edition: Including lichen-forming and lichenicolous fungi*. Manaaki Whenua Press, Lincoln. cxxx + 2261 p.

GALLOWAY, D. J. 2008a: Lichen biogeography. Pp. 315–335 *in*: Nash, T.H.III (ed.), *Lichen Biology*. 2nd edn. Cambridge University Press, Cambridge. 486 p.

GALLOWAY, D.J. 2008b: Godley Review. Austral lichenology: 1690–2008. *New Zealand Journal of Botany 46*: 433–521.

GALLOWAY, D. J. 2009: Lichen notes 2: Lichens on the move – are they telling us something? *New Zealand Botanical Society Newsletter 98*: 15–19.

GALLOWAY, D. J. 2010: Lichenological memories of John Child (1922–1984) and notes on the publication of Martin & Child (1972). *New Zealand Botanical Society Newsletter 102*: 14–19.

GALLOWAY, D. J. 2012: Olof Swartz's contributions to lichenology, 1781–1811. *Archives of Natural History* (in press).

GALLOWAY, D. J.; APTROOT, A. 1995: bipolar lichens: a review. *Cryptogamic Botany 5*: 184–191.

GALLOWAY, D. J.; BARTLETT, J. K. 1986: *Arthrorhaphis* Th.Fr. (lichenised Ascomycotina) in New Zealand. *New Zealand Journal of Botany 24*: 393–402.

GALLOWAY, D. J.; HAFELLNER, J.; ELIX, J. A. 2005: *Stirtoniella*, a new genus for *Catillaria kelica* (*Lecanorales*: *Ramalinaceae*). *Lichenologist 37*: 261–271.

GALLOWAY, D. J.; GROVES, E. W. 1987: Archibald Menzies MD, FLS (1754–1842), aspects of his life, travels and collections. *Archives of Natural History 14*: 3–43.

GALLOWAY, D. J.; JØRGENSEN, P. M. 1990: *Bartlettiella*, a new lichen genus from New Zealand, with notes on a new species of *Melanelia*, and a new chemodeme of *Bryoria indonesica* in New Zealand. *New Zealand Journal of Botany 28*: 5–12.

GALLOWAY, D. J.; QUILHOT, W. 1999 [1998]: Checklist of Chilean lichen-forming and lichenicolous fungi. *Gayana (Botanica) 55*(2): 11–185.

GARGAS, A.; DE PRIEST, P. T.; GRUBE, M.; TEHLER, A. 1995: Multiple origins of lichen symbioses in fungi suggested by SSU rDNA phylogeny. *Science 268*: 1492–1495.

GARNOCK-JONES, P. J.; BREITWIESER, I. 1998: New Zealand floras and systematic botany: progress and prospects. *Australian Systematic Botany 11*: 175–184.

GILBERT, O. [L.] 2000: Lichens. *New Naturalist 86*: 1–288.

GOWARD, T. 2008: Twelve readings on the lichen thallus II"Nameless little things". *Evansia 25* (3): 54-56.

GOWARD, T. 2009a: Twelve readings on the lichen thallus IV– Re-emergence. *Evansia 26*: 1–6.

GOWARD, T. 2009b: Twelve readings on the lichen thallus V. Conversational. *Evansia 26*: 31–37.

GOWARD, T. 2009c: Twelve readings on the lichen thallus VI. Reassembly. *Evansia 26*: 91–97.

GOWARD, T. 2009d: Twelve readings on the lichen thallus VII. Species. *Evansia 26*: 153–162.

GOWARD, T. 2010a: Twelve readings on the lichen thallus VIII. Theoretical. *Evansia 27*: 2–10.

GOWARD, T. 2010b: Twelve readings on the lichen thallus IX. Paralichens. *Evansia 27*:

40–46.
GOWARD, T. 2010c: Twelve readings on the lichen thallus X. Homeostasis. *Evansia 27*: 71–81.
GOWARD, T. 2011: Twelve readings on the lichen thallus XI. Preassembly. *Evansia 28*: 1–17.
HAWKSWORTH, D. L. 1988: The variety of fungal-algal symbioses, their evolutionary significance, and the nature of lichens. *Botanical Journal of the Linnean Society 96*: 3–20.
HAWKSWORTH, D. L. 1991: The fungal dimension of biodiversity: magnitude, significance, and conservation. *Mycological Research 95*: 641–655.
HAWKSWORTH, D. L. 1992: Litmus test for ecosystem health: the potential of bioindicators in the monitoring of biodiversity. Pp. 184–204 *in*: Swaminathan, M. S.; Jana, S. (eds), *Biodiversity: Implications for Global Food Security*. Macmillan India, Madras. 330 p.
HAWKWORSTH, D. L. 1994: The recent evolution of lichenology: a science for our times. *Cryptogamic Botany 4*: 117–129.
HAWKSWORTH, D. L. 1997: Orphans in 'botanical' diversity. *Muelleria 10*: 111–123.
HAWKSWORTH, D. L. 1999: Visions of systematic and organismal lichenology in the next century. *British Lichen Society Bulletin 84*: 1–4.
HAWKSWORTH, D. L. 2003: The lichenicolous fungi of Great Britain and Ireland: an overview and annotated checklist. *Lichenologist 35*: 191–232.
HAWKSWORTH, D. L. 2005: Life-style choices in lichen-forming and lichen-dwelling fungi. *Mycological Research 109*: 135–136.
HAWKSWORTH, D. L.; HONEGGER, R. 1994: The lichen thallus: a symbiotic phenotype of nutritionally specialized fungi and its response to gall producers. Pp. 77–98 *in*: Williams, M. A. J. (ed.), *Plant Galls: organisms, interactions, populations*. [Systematics association Special Volume 49.] Clarendon Press, Oxford. xiv + 488 p.
HAYWARD, B. W.; HAYWARD, G. C. 1984: Lichens of the Chickens Islands, northern New Zealand. *Tane (Journal of the Auckland University Field Club) 30*: 43–51.
HAYWARD, B. W.; HAYWARD, G. C. 1986: Lichen flora of the offshore islands of northern New Zealand. *In*: Wright, A. E.; Beever, R. E. (eds), *The Offshore Islands of New Zealand. New Zealand Department of Lands and Survey Information Series 16*: 153–160.
HAYWARD, B. W.; HAYWARD, G. C.; GALLOWAY, D. J. 1986: Lichens of Great Barrier and adjacent islands, northern New Zealand. *Journal of the Royal Society of New Zealand 16*: 121–137.
HAYWARD, B. W.; LUMBSCH, H. T. 1992. Lichens of south-east Stewart Island. *New Zealand Natural Sciences 19*: 69–78.
HAYWARD, B. W.; WRIGHT, A. E. 1991: Lichens from the Poor Knights Islands, northern New Zealand – additions and an updated species list. *Tane (Journal of the Auckland University Field Club) 33*: 39–48.
HAYWARD, G. C. 1977: Taxonomy of the lichen families Graphidaceae and Opegraphaceae in New Zealand. *New Zealand Journal of Botany 15*: 565–584.
HEADS, M. J. 1997: Regional patterns of biodiversity in New Zealand: one degree grid analysis of plant and animal distributions. *Journal of the Royal Society of New Zealand 27*: 337–354.
HEADS, M. 1998: Biogeographic disjunction along the Alpine fault, New Zealand. *Biological Journal of the Linnean Society 63*: 161–176.
HELLBOM, P. J. 1896: Lichenaea neo-zeelandica seu lichenes Novae-Zeelandiae a Sv, Berggren annis 1874–75 collecti. *Bihang till Konglica Svenska Vetenskaps-Akademiens Handlingar 21(3/13)*: 1–150.
HIBBETT, D. S.; BINDER, M.; BISCHOFF, J. F.; BLACKWELL, M.; CANNON, P. F. et al [with 61 other authors] 2007: A higher level phylogenetic classification of the Fungi. *Mycological Research 111*: 509–547.
HÖGNABBA, F.; STENROOS, S.; THELL, A. 2009: Phylogenetic relationships and evolution of photobiont associations in the Lobariaceae (Peltigerales, Lecanoromycetes, Ascomycota). *Bibliotheca Lichenologica 100*: 157–187.
HONEGGER, R. 1991: Functional aspects of the lichen symbiosis. *Annual Review of Plant Physiology and Plant Molecular Biology 42*: 553–578.
HONEGGER, R. 1992: Lichens: mycobiont-photobiont relationships. Pp. 255–275 *in*: Reisser, W. (ed.), *Algae and Symbioses: Plants, animals, fungi*. Biopress, Bristol. 746 p.
HONEGGER, R. 1993: Developmental biology of lichens. *New Phytologist 125*: 659–677.
HONEGGER, R. 1996: Mycobionts. Pp. 24–36 *in*: Nash, T. H. III (ed.), *Lichen Biology*. Cambridge University Press, Cambridge. 303 p.
HONEGGER, R. 1998: The lichen symbiosis – what is so spectacular about it? *Lichenologist 30*: 193–212.
HONEGGER, R. 2008a: Mycobionts. Pp. 27–39 *in*: Nash, T. H. III (ed.), *Lichen Biology*. 2nd edn. Cambridge University Press, Cambridge. 486 p.
HONEGGER, R. 2008b: Morphogenesis. Pp. 69–93 *in*: Nash, T. H. III (ed.), *Lichen Biology*. 2nd edn. Cambridge University Press, Cambridge. 486 p.
HOOKER, J. D. 1867: *Handbook of the New Zealand Flora*. Reeve & Co., London. Vol. 2, pp. 393–793.
HOOKER, J. D.; TAYLOR, T. 1844: *Lichenes Antarctici*; being characters and brief descriptions of the new lichens discovered in the southern circumpolar regions, Van Diemen's Land and New Zealand during the voyage of H. M. Discovery Ships *Erebus* and *Terror*. *London Journal of Botany 3*: 634–658.
HUNECK, S.; YOSHIMURA, I. 1996: *Identification of Lichen Substances*. Springer, Berlin. xi + 493 p.
IMSHAUG, H. A. 1970: Campbell Island Expedition 1969–1970. *Antarctic Journal of the United States 5*: 117–118.
IMSHAUG, H. A. 1973: Auckland Islands Expedition, 1972–1973. *Antarctic Journal of the United States 8*: 187–188.
JOHNSON, P. N.; GALLOWAY, D. J. 1999: Lichens on trees: identification guide to common lichens and plants on urban and rural trees in New Zealand. *Landcare Research Contract Report LC9899/071*: 1–33.
JOHNSON, P. N.; BURROWS, L. E.; GALLOWAY, D. J. 1998: Lichens as air pollution indicators. *Landcare Research Contract Report LC9899.004*: 1–42.
JOHNSON, P. N.; GALLOWAY, D. J. 2002: Lichens and their conservation needs in New Zealand. *Landcare Research Contract Report LC0102/132*: 1-69.
KANTVILAS, G. 1988: Tasmanian rainforest lichen communities: a preliminary classification. *Phytocoenologia 16*: 391–428.
KANTVILAS, G.; JARMAN, S. J. 1999: Lichens of rainforest in Tasmania and south-eastern Australia. *Flora of Australia, Supplementary Series 9*: i-xi + 1–212.
KIRK, P. M.; CANNON, P. F.; STALPERS, J. A.; MINTER, D. W. 2008: *Ainsworth & Bisby's Dictionary of the Fungi*. 10th Edn. Wallingford, CAB International. 784 p.
KNIGHT, C. 1860: On some New Zealand *Verrucariae*. *Transactions of the Linnean Society 23*: 99–100.
KNIGHT, C. 1875: Description of some New Zealand lichens. *Transactions of the New Zealand Institute 7*: 356–367.
KNIGHT, C. 1876: Further contributions to the lichen flora of New Zealand. *Transactions of the New Zealand Institute 8*: 313–328.
KNIGHT, C. 1877: Contribution to the lichenographia of New Zealand. *Transactions of the Linnean Society, ser. 2, Botany 1*: 275–283.
KNIGHT, C. 1880: Contribution to the lichenographia of New Zealand. *Transactions of the New Zealand Institute 12*: 367–379.
KNIGHT, C. 1883: On the lichenographia of New Zealand. *Transactions of the New Zealand Institute 15*: 346–358.
KNIGHT, C. 1884: On the lichenographia of New Zealand. *Transactions of the New Zealand Institute 16*: 400–408.
KNIGHT, C.; MITTEN, W. 1860: Contributions to the lichenographia of New Zealand; being an account, with figures, of some new species of Graphideae and allied lichens. *Transactions of the Linnean Society 23*: 101–106.
KUROKAWA, S. (Ed.) 2000: *Checklist of Japanese Lichens*. National Science Museum, Tokyo. 128 p.
LAWREY, J. D. 1986: Biological role of lichen substances. *Bryologist 89*: 111–122.
LAWREY, J. D.; DIEDERICH, P. 2003: Lichenicolous fungi: interactions, evolution and biodiversity. *Bryologist 106*: 80-120.
LEIGHTON, W. A. 1869: Additions to the lichens of New Zealand. *Botanical Journal of the Linnean Society 10*: 30–33.
LINDSAY, W. L. 1866a: List of lichens collected in Otago, New Zealand. *Transactions of the Botanical Society of Edinburgh 8*: 349–358.
LINDSAY, W. L. 1866b: Observations on new lichens and fungi from Otago, New Zealand. *Proceedings of the Royal Society of Edinburgh 5*: 244–259.
LINDSAY, W. L. 1867: Observations on new lichens and fungi collected in Otago, New Zealand. *Transactions of the Royal Society of Edinburgh 24*: 407–456.
LINDSAY, W. L. 1868: *Contributions to New Zealand Botany*. Williams and Norgate, Edinburgh & London. 102 p.
LINDSAY, W. L. 1869: Observations on New Zealand lichens. *Transactions of the Linnean Society 25*: 493–560.
LÜCKING, R.; LAWREY, J. D.; SIKAROODI, M.; GILLEVET, P. M.; CHAVES, J. L.; SIPMAN, H. J. M.; BUNGARTZ, F. 2009: Do lichens domesticate photobionts like farmers domesticate crops? Evidence from a previously unrecognized lineage of filamentous cyanobacteria. *American Journal of Botany 96*: 1409–1418.
LUDWIG, L. 2011: Marginal soralia and conidiomata in *Icmadophila splachnirima* (Icmadophilaceae) from southern New Zealand. *Australasian Lichenology 68*: 4–11.
LUMBSCH, H. T. 1998a: The taxonomic use of metabolic data in lichen-forming fungi. Pp. 345–387 in: Frisvad, J. C.; Bridge, P. D.; Akora,

D. K. (eds), *Chemical Fungal Taxonomy*. Marcel Dekker, New York. viii + 398 p.

LUMBSCH, H. T. 1998b: The use of metabolic data in lichenology at the species and subspecific levels. *Lichenologist 30*: 357–367.

LUMBSCH, H. T. 2002: How objective are genera in euascomycetes? *Perspectives in Plant Ecology, Evolution and Systematics 5*: 91–101.

LUMBSCH, H. T.; AHTI, T.; ALTERMANN, S.; AMO DE PAZ, G.; APTROOT, A. et al. [with 96 other authors] 2011: One hundred new species of lichenized fungi: a signature of undiscovered global diversity. *Phytotaxa 18*: 1–127.

LUMBSCH, H. T.; HUHNDORF, S. M. (Eds) 2010: Myconet Volume 14, Part One. Outline of Ascomycota-2009. *Fieldiana: Life and Earth Sciences, N.S. 1*: 1–42.

LUMBSCH, H. T.; McCARTHY, P. M.; MALCOLM, W. M. 2001: Key to the genera of Australian lichens. Apothecial crusts. *Flora of Australia, Supplementary Series 11*: 1–64.

LUTZONI, F.; KAUFF, F.; COX, C. J.; McLAUGHLIN, D.; CELIO, G. et al. [with 39 other authors] 2004: Assembling the fungal tree of life: progress, classification, and evolution of subcellular traits. *American Journal of Botany 91*: 1446–1480.

LUTZONI, F.; PAGEL, M.; REEB, V. 2001: Major fungal lineages are derived from lichen symbiotic ancestors. *Nature 411*: 937–940.

MALCOLM, W. M.; ELIX, J. A.; OWE-LARSSON, B. 1995: *Labyrintha implexa* (Porpidiaceae), a new genus and species from New Zealand. *Lichenologist 27*: 241–248.

MALCOLM, W. M.; GALLOWAY, D. J. 1997: New Zealand Lichens. Checklist, Key, and Glossary. Museum of New Zealand Te Papa Tongarewa, Wellington. 192 p.

MALCOLM, B. [W. M.]; MALCOLM, N. 2000: *New Zealand Lichens*. Micro-Optics Press, Nelson. 134 p.

MALCOLM, B. [W. M.]; MALCOLM, N. 2001: *New Zealand's Leaf-Dwelling Lichens*. Micro-Optics Press, Nelson. vi + 73 p.

MALCOLM, W. M.; VĚZDA, A. 1995a: New foliicolous lichens from New Zealand 1. *Folia Geobotanica et Phytotaxonomica 30*: 91–96.

MALCOLM, W. M.; VĚZDA, A. 1995b: New foliicolous lichens from New Zealand 2. *Folia Geobotanica et Phytotaxonomica 30*: 315–318.

MALCOLM, W. M.; VĚZDA, A. 1996: New foliicolous lichens from New Zealand 3. *Folia Geobotanica et Phytotaxonomica 31*: 263–268.

MARGULIS, L. 1993: *Symbiosis in Cell Evolution. Microbial communities in the Archean and Proterozoic eons*. 2nd edn. W. H. Freeman & Co., New York. xxvii + 452 p.

MARTIN, W. 1958: The *Cladoniae* of New Zealand. *Transactions of the Royal Society of New Zealand 85*: 603–632.

MARTIN, W. 1960: The lichen genus *Cladonia*, subsection *Cladina*, in New Zealand. *Transactions of the Royal society of New Zealand 88*: 169–175.

MARTIN, W. 1962: Notes on some New Zealand species of *Cladonia* with descriptions of two new species and one new form. *Transactions of the Royal Society of New Zealand* (*Botany*) 2: 39–44.

MARTIN, W. 1965: The lichen genus *Cladia*. *Transactions of the Royal Society of New Zealand* (*Botany*) 3: 7–12.

MARTIN, W. 1966: Census catalogue of the lichen flora of New Zealand. *Transactions of the Royal Society of New Zealand* (*Botany*) 3: 139–159.

MARTIN, W. 1968: Supplement to census catalogue of New Zealand lichens. *Transactions of the Royal Society of New Zealand* (*Botany*) 3: 203–208.

MARTIN, W. 1970: The lichen flora of the Dunedin Botanical Subdistrict. *Transactions of the Royal Society of New Zealand* (*Biological Sciences*) *11*: 243–255.

MARTIN, W.; CHILD, J. 1972: *Lichens of New Zealand*. A. H. & A. W. Reed, Wellington. ix + 193 p.

McCARTHY, P. M. 2011: *Checklist of the Lichens of Australia and its Island Territories*. Australian Biological Resources Study, Canberra. Version 23 May 2011 http://www.anbg.gov.au/abrs/lichenlist/introduction.html.

McCARTHY, P. M.; MALCOLM, W. M. 2004: Key to the genera of Australian macrolichens. *Flora of Australia, Supplementarty Series 23*: 1–63.

McLAUGHLIN, D. J.; HIBBETT, D. S.; LUTZONI, F.; SPATAFORA, J. W.; VILGALYS, R. 2009: The search for the fungal tree of life. *Trends in Microbiology 17*: 488–497.

MIADŁIKOWSKA, J.; LUTZONI, F. 2004: Phylogenetic classification of Peltigeralean fungi (Peltigerales, Ascomycota) based on ribosomal RNA small and lage subunits. *American Journal of Botany 91*: 449–464.

MIADŁIKOWSKA, J.; KAUFF, F.; HOFSTETTER, V.; FRAKER, E.; GRUBE, M. et al. [with 24 other authors] 2006: New insights into classification and evolution of the Lecanoromycetes (Pezizomycotina, Ascomycota) from phylogenetic analyses of three ribosomal RNA- and two protein-coding genes. *Mycologia 98*: 1088–1103.

MÜLLER ARGOVIENSIS, J. 1894: Conspectus sytematicus lichenum Novae Zealandiae. *Bulletin de l'Herbier Boissier 2, Appendix 1*: 1–114.

MÜLLER ARGOVIENSIS, J. [1895] 1896: Lichenes Colensoani a Reverendiss. Colenso in Novâ Zelandiâ septentrionali prope Napier lecti, et nuperius missi in Herbario Reg. Kewensi servati. *Botanical Journal of the Linnean Society 32*: 197–208.

MURRAY, J. 1952: Lichens and fungi. Part I. Polyporic acid in Stictae. *Journal of the Chemical Society 1952*: 1345–1350.

MURRAY, J. 1960a: Studies of New Zealand lichens. I – The Coniocarpineae. *Transactions of the Royal Society of New Zealand 88*: 169–175.

MURRAY, J. 1960b: Studies on New Zealand lichens. II – The Teloschistaceae. *Transactions of the Royal Society of New Zealand 88*: 197–210.

MURRAY, J. 1960c: Studies on New Zealand lichens. Part III – The family Peltigeraceae. *Transactions of the Royal Society of New Zealand 88*: 381–399.

MURRAY, J. 1962: Keys to New Zealand lichens. Part 1. *Tuatara 10*: 120–128.

MURRAY, J. 1963: Lichens from Cape Hallett area, Antarctica. *Transactions of the Royal Society of New Zealand, Botany* 2: 59–72.

MURRAY, J. 1963b: Keys to New Zealand lichens. Part 2. *Tuatara 11*: 46–56.

MURRAY, J. 1963c: Keys to New Zealand lichens. Part 3. *Tuatara 11*: 98–109.

MURRAY, J. 1963d: Vegetation studies on Secretary Island, Fiordland. Part 7. Bryophytes and lichens. *New Zealand Journal of Botany 1*: 221–235.

NASH, T. H. III 2008a. Introduction. Pp. 1–8 *in*: Nash, T. H. III (ed.), *Lichen Biology*. 2nd edn. Cambridge University Press, Cambridge. 486 p.

NASH, T. H. III 2008b: Lichen sensitivity to air pollution. Pp. 299–314 *in*: Nash, T. H. III (ed.), *Lichen Biology*. 2nd edn. Cambridge University Press, Cambridge. 486 p.

NIMIS, P. L. 1993: The lichens of Italy. An annotated catalogue. *Museo Regionale di Scienze Naturali, Torino, Monographie 12*: 1–897.

NIMIS, P. L.; MARTELLOS, S. 2003: A second checklist of the lichens of Italy with a thesaurus of synonyms. *Museo Regionale di Scienze Naturali Saint-Pierre, Valle d-Aosta, Monografie 4*: 1–191.

NYLANDER, W. 1866: Lichenes Novae Zelandiae, quos ibi legit anno 1861 Dr Lauder Lindsay. *Botanical Journal of the Linnean Society 9*: 244–259.

NYLANDER, W. 1888: *Lichenes Novae Zelandiae*. P. Schmidt, Paris. 156 p.

ORWIN, J. 1970: Lichen succession on recently deposited rock surfaces. *New Zealand Journal of Botany 8*: 452–477.

ORWIN, J. 1971: The effect of environment on assemblages of lichens growing on rock surfaces. *New Zealand Journal of Botany 10*: 37–47.

OTÁLORA, M. A. G.; MARTÍNEZ, I.; O'BRIEN, H.; MOLINA, M. A.; ARAGÓN, G.; LUTZONI, F. 2010: Multiple origins of high reciprocal symbiotic specificity at an intercontinental spatial scale among gelatinous lichens (Collemataceae, Lecanoromycetes). *Molecular Phylogenetics and Evolution 56*: 1089–1095.

ØVSTEDAL, D. O.; FRYDAY, A. M. 2011: A new species of *Protopannaria* (Pannariaceae, Ascomycota) from the southern New Zealand shelf islands, and additional records from South America. *Australasian Lichenology 68*: 12–15.

ØVSTEDAL, D. O.; LEWIS SMITH, R. I. 2001: *Lichens of Antarctica and South Georgia. A guide to their identification and ecology*. Cambridge University Press, Cambridge. 411 p.

PENNYCOOK, S. R.; GALLOWAY, D. J. 2004: Checklist of New Zealand 'Fungi'. Pp. 401–488 *in*: McKenzie, E. H. C. (ed.), *The Fungi of New Zealand/Nga Harore o Aotearoa, Volume 1. Introduction to Fungi of New Zealand*. Fungal Diversity Press, Hong Kong. 500 p.

PERRY, N. B.; BENN, M. H.; BRENNAN, N. J.; BURGESS, E. J.; ELLIS, G.; GALLOWAY, D. J.; LORIMER, S. D.; TANGNEY, R. S. 1999: Antimicrobial, antiviral and cytotoxic activity of New Zealand lichens. *Lichenologist 31*: 627–636.

PURVIS, [O.] W. 2000: *Lichens*. The Natural History Museum, London. 112 p.

PURVIS, O. W.; COPPINS, B. J.; HAWKSWORTH, D. L.; JAMES, P. W.; MOORE, D. M. (Eds) 1992: *The Lichen Flora of Great Britain and Ireland*. The Natural History Museum, London. 710 p.

QUILHOT, W.; FERNÁNDEZ, E.; HIDALGO, M. E. 1994: Photoprotection mechanisms in lichens against UV radiation. *British Lichen Society Bulletin 75*: 1–5.

QUILHOT, W.; FERNÁNDEZ, E.; RUBIO, C.; GODARD, M.; HIDALGO, M. E. 1998: Lichen secondary products and their importance in environmental studies. Pp. 171–179 *in*: Marcelli, M. P.; Seaward, M. R. D. (eds), *Lichenology in Latin America: History, current knowledge and applications*. CETESB, São Paulo. viii + 179 p.

RAMBOLD, G. (Ed.) 2004: *LIAS – a global information system for lichenized and non-lichenized Ascomycetes. A multi-authored distributed internet project containing information about phylogeny and biodiversity*. http://www.checklists.lias.net

RICHARD, A. 1832: Essai d'une flore de la Nouvelle-Zélande. *In*: Lesson, A.; Richard, A.

eds), *Voyage de découvertes de l'Astrolabe execute ... pendant les années 1826–1827–1828–1829, sous le commandement de M. J. Dumont d'Urville. Botanique, par MM. A. Lesson et A. Richard.* J. Tastu, Paris. xvi + 376 p.

RIKKINEN, J. 1995: What's behind the pretty colours? A study on the photobiology of lichens. *Bryobrothera 4*: 1–239.

RILEY, M. 1994: *Maori Healing and Herbal.* Viking Seven Seas N.Z. Ltd, Paraparaumu. 529 p.

RICHARDSON, D. H. S. 1975: *The vanishing Lichens: Their history, biology and importance.* David and Charles, Abbott Newton. 231 p.

RICHARDSON, D. H. S. 1991: Lichens and man. Pp. 187–210 *in*: Hawksworth, D. L. (ed.), *Frontiers in Mycology*. CAB International, Wallingford. 260 p.

RICHARDSON, D. H. S. 1999: War in the world of lichens: parasitism and symbiosis as exemplified by lichens and lichenicolous fungi. *Mycological Research 103*: 641–650.

SANDERS, W. B. 2001: Lichens: the interface between mycology and plant morphology. *Bioscience 51*: 1025–1035.

SANDERS, W. B. 2006: A feeling for the superorganism: expression of plant form in the lichen thallus. *Botanical Journal of the Linnean Society 15*: 89–99.

SANDERS, W. B. 2010: Together and separate: Reconstructing life histories of lichen symbionts. *Bibliotheca Lichenologica 105*: 1–16.

SANDERS, W. B.; LÜCKING, R. 2002: Reproductive strategies, relichenization and thallus development observed *in situ* in leaf-dwelling lichen communities. *New Phytologist 155*: 425–435.

SANTESSON, R.; MOBERG, R.; NORDIN, A.; TØNSBERG, T.; VITIKAINEN, O. 2004: *Lichen-forming and lichenicolous fungi of Fennoscandia.* Museum of Evolution, Uppsala. 358 p.

SCHOCH, C. L.; SUNG, G.-H.; LÓPEZ-GIRÁLDEZ, F.; TOWNSEND, J. P.; MIADŁIKOWSKA, J. et al. [with 59 other authors] 2009: The Ascomycota Tree of Life: A phylum-wide phylogeny clarifies the origin and evolution of fundamental reproductive and ecological traits. *Systematic Biology 58*: 224–239.

SCHUMM, F. 2011: *Kalkflechten der Schwäbischen Alb ein mikroskopisch anatomischer Atlas.* Herstellung und Verlag: Books on Demand GmbH, Norderstedt. 410 p.

SCHUMM, F.; APTROOT, A. 2010: *Seychelles Lichen Guide*. Beck, OHG 73079 Süssen. 404 p.

SCHWARTZMAN, D. W. 2010: Was the origin of the lichen symbiosis triggered by declining atmospheric carbon dioxide levels? *Bibliotheca Lichenologica 105*: 191–196.

SEAWARD, M. R. D. 1996: Lichens and the environment. Pp. 293–320 *in*: Sutton, B. C. (ed.), *A Century of Mycology*. Cambridge University Press, Cambridge. 308 p.

SEAWARD, M. R. D. 1997: Urban deserts bloom: a lichen renaissance. *Bibliotheca Lichenologica 67*: 297–309.

SEAWARD, M. R. D. 2004: The use of lichen for environmental impact assessment. *Symbiosis 37*: 293–305.

SEAWARD, M.R.D. 2008: Environmental role of lichens. Pp. 274–298 *in*: Nash, T. H. III (ed.), *Lichen Biology*. 2nd edn. Cambridge University Press, Cambridge. 486 p.

SEAWARD, M. R. D.; COPPINS, B. J. 2004: Lichens and hypertrophication. *Bibliotheca Lichenologica 88*: 561–572.

SIPMAN, H. J. M.; APTROOT, A. 2001: Where are the missing lichens? *Mycological Research 105*: 1433–1439.

SMITH, C. W.; APTROOT, A.; COPPINS, B. J.; FLETCHER, A.; GILBERT, O. L.; JAMES, P. W.; WOLSELEY, P. A. 2009: *The Lichens of Great Britain and Ireland.* The British Lichen Society, London. 1046 p.

SMITH, D. C.; DOUGLAS, A. E. 1987: *The Biology of Symbiosis*. Edward Arnold, London. 320 p.

SOLHAUG, K. A.; GAUSLAA, Y.; NYBAKKEN, L.; BILGER, W. 2003: UV-induction of sun-screening pigments in lichens. *New Phytologist 158*: 91–100.

SPATAFORA, J. 2005: Assembling the Fungal Tree of Life (AFTOL). *Mycological Research 109*: 755–756.

ST GEORGE, I. 2009: *Colenso's Collections, including the unpublished work of the late Bruce Hamlin on William Colenso's New Zealand plants held at Te Papa.* The New Zealand Native Orchid Group, Wellington. 412 p.

STIRTON, J. 1873: Additions to the lichen flora of New Zealand. *Reports and Transactions of the Glasgow Society of Field Naturalists 1*: 15–23.

STIRTON, J. 1874: Descriptions of some New Zealand lichens, collected by John Buchanan in the province of Wellington. *Transactions of the New Zealand Institute 6*: 235–241.

STIRTON, J. 1875: Additions to the lichen flora of New Zealand. *Botanical Journal of the Linnean Society 14*: 458–474.

STIRTON, J. 1877: On new genera and species of lichens from New Zealand. *Proceedings of the Philosophical Society of Glasgow* 10: 285–306.

STIRTON, J. 1877b: Additions to the lichen flora of New Zealand and the Chatham Islands. *Grevillea 5*: 147–148.

STIRTON, J. 1898: On new Australian and New Zealand lichens. *Transactions of the New Zealand Institute 30*: 382–393.

STIRTON, J. 1900: On new lichens from Australia and New Zealand. *Transactions of the New Zealand Institute 32*: 70–82.

SWARTZ, O. P. 1781: *Methodus Muscoroum Illustrata*. J. Edman, Uppsaliae. 38 p., 2 pls.

TAYLOR, T.; HOOKER, J. D. 1845: Lichenes. Pp. 194–200 *in*: Hooker, J. D. (ed.), *Botany of the Antarctic Voyage of H. M. Discovery Ships Erebus and Terror in the years 1839–1843. I. Flora Antarctica. Part I. Botany of Lord Auckland's Group and Campbell's Island*. Lovell Reeve, London. 574 p.

TEHLER, A.; WEDIN, M. 2008: Systematics of lichenized fungi. Pp. 336–352 *in*: Nash, T. H. III (ed.), *Lichen Biology*. 2nd edn. Cambridge University Press, Cambridge. 486 p.

THELL, A.; HÖGNABBA, F.; ELIX, J. A.; FEUERER, T.; KÄRNEFELT, I.; MYLLYS, L.; RANDLANE, T.; SAAG, A.; STENROOS, S.; AHTI, T.; SEAWARD, M. R. D. 2009: Phylogeny of the cetrarioid core (Parmeliaceae) based on five genetic markers. *Lichenologist 41*: 489–511.

THOMAS, M. A.; RYAN, D. J.; FARNDEN, K. J.; GALLOWAY, D. J. 2002: Observations on phylogenetic relationships within Lobariaceae Chevall. (Lecanorales, Ascomycota) in New Zealand, based on ITS-5.8S molecular sequence data. *Bibliotheca Lichenologica 82*: 123–138.

TIBELL, L. 1987: Australasian Caliciales. *Symbolae Botanicae Upsalienses 27*: 1–279.

TOWNSEND, A. J.; de LANGE, P. J.; DUFFY, C. A. J.; MISKELLY, C. M.; MOLLOY, J.; NORTON, D. 2008: *New Zealand Threat Classification System, Manual*. Department of Conservation, Wellington. 35p.

VĚZDA, A. 1997: [*Scheda*] *Lichenes Rariores Exsiccati*. Fasciculus septimus vicesimus (numeris 261–270). Brno. 7 p.

WALKER, J. R. L.; LINTOTT, E. A. 1997: A phytochemical register of New Zealand lichens. *New Zealand Journal of Botany 35*: 369–384.

WEDIN, M. 1995: The lichen family Sphaerophoraceae (Caliciales, Ascomycotina) in temperate areas of the Southern Hemisphere. *Symbolae Botanicae Upsalienses 31*: 1–102.

WERTH, S. 2010: Population genetics of lichen-forming fungi – a review. *Lichenologist 42*: 499–519.

ZAHLBRUCKNER, A. 1941: Lichenes Novae Zelandiae a cl. H. H. Allan eiusque collaboratoribus lecti. *Denkschriften der Akademie der Wissenschaften Wien mathematisch-naturwissenschafltliche Klasse 104*: 249–380.

CREDITS AND ACKNOWLEDGEMENTS

Captions and credits for thumbnail images on page 16

Top to bottom, left to right:

BACTERIA, Cyanobacteria: *Anabaena inaequalis* (Faradina Merican, University of Canterbury, Christchurch)
PROTOZOA: Euglenozoa: *Phacus gigas* (Richard E. Triemer, Michigan State University, East Lansing, MI)
PROTOZOA: Percolozoa: *Naegleria fowleri* (Francine Cabral, Virginia Commonweath University, Richmond, VA)
PROTOZOA: Loukozoa: *Reclinomonas americana* (David J. Patterson, Marine Biological Laboratory, Woods Hole, MA)
PROTOZOA: Metamonada: *Giardia duodenalis* (W. Zachaeus Cande, University of California, Berkeley, CA)
PROTOZOA: Amoebozoa: *Ceratiomyxa fruticulosa* (Clive Shirley, Mt Wellington, Auckland)
PROTOZOA: Apusozoa: *Collodictyon triciliatum* (Protist Information Server, Hosei University, Tokyo)
PROTOZOA: Choanozoa: *Codosiga botrytis* (Protist Information Server, Hosei University, Tokyo)
CHROMISTA: Ochrophyta: *Didymosphenia geminata* (Cathy Kilroy, NIWA, Christchurch)
CHROMISTA: Oomycota: *Phytophthora fallax* (Margaret A. Dick, Scion, Rotorua)
CHROMISTA: Bigyra: *Actinophrys sol* (David J. Patterson, Marine Biological Laboratory, Woods Hole, MA)
CHROMISTA: Myzozoa: *Akashiwo sanguinea* (Hwan Su Yoon, Bigelow Laboratory for Ocean Sciences, Portland, ME)
CHROMISTA: Ciliophora: *Loxodes rostrum* (Martin Kreutz, Private Laboratory, Konstanz)
CHROMISTA: Cercozoa: *Biomyxa vagans* (David J. Patterson, Marine Biological Laboratory, Woods Hole, MA)
CHROMISTA: Foraminifera: *Globigerinella* sp. (O. Roger Anderson, Columbia University, New York, NY)
CHROMISTA: Radiozoa: *Euchitonia elegans* (Rie S. Hori, Ehime University, Matsuyama & Akihiro Tuji, NMNS, Tsukuba)
CHROMISTA: Cryptophyta: *Cryptomonas paramoecium* (David J. Patterson, Marine Biological Laboratory, Woods Hole, MA)
CHROMISTA: Haptophyta: *Emiliania huxleyi* (F. Hoe Chang, NIWA, Wellington)
CHROMISTA: Heliozoa: *Raphidiophrys elegans* (Martin Kreutz, Private Laboratory, Konstanz)
PLANTAE: Glaucophyta: *Glaucocystis nostochinearum* (Protist Information Server, Hosei University, Tokyo)
PLANTAE: Rhodophyta: *Corallina* sp. (Kate Neill, NIWA, Wellington)
PLANTAE: Chlorophyta: *Chaetomorpha coliformis* (Wendy A. Nelson, NIWA, Wellington)
PLANTAE: Bryophyta: *Lepidolaena clavigera* (Bill Malcolm, Micro-Optics Ltd, Nelson)
PLANTAE: Tracheophyta: *Alseuosmia quercifolia* (Dennis P. Gordon, NIWA, Wellington)
PLANTAE: Fossils: *Araucaria* sp. (Michael S. Pole, Brisbane Botanic Gardens, QLD)
FUNGI: Introduction: *Sordaria fimicola* (George L. Barron, University of Guelph, ON)
FUNGI: Chytridiomycota: Unidentified chytrid (David J. Patterson & Aimee Ladermann, Marine Biological Laboratory, Woods Hole, MA)
FUNGI: Microsporidia: Undescribed microsporidian (Miroslav Hylis, Charles University, Prague)
FUNGI: Zygomycota: *Pilobolus kleinii* (George L. Barron, University of Guelph, ON)
FUNGI: Ascomycota: *Cyttaria gunnii* (Peter R. Johnston, Landcare Research, Auckland)
FUNGI: Basidiomycota: *Clathrus archeri* (Dennis P. Gordon, NIWA, Wellington)
FUNGI: Lichens: *Xanthoria elegans* (Janet M. Ledingham, University of Otago, Dunedin)

Identifications of the foraminifera illustrated in colour plate 8a

The foraminifera illustrated in the plate are a mixed collection of shallow and deep-sea species from the Southwest Pacific, taken during the 1950–52 round-the-world Danish Deep-Sea Expedition. All but 9, 10, and 12 are found living in the New Zealand EEZ. The photo was taken by John Whalan of the former New Zealand Department of Scientific and Industrial Research (DSIR).

1. *Cyclammina cancellata* Brady, 1879 – Lituolida: Cyclamminidae
2. *Dentalina cuvieri* (d'Orbigny, 1826) – Lagenida: Nodosariidae
3. *Frondicularia* sp. – Lagenida: Nodosariidae
4. *Cornuloculina inconstans* (Brady, 1879) – Miliolida: Ophthalmidiidae
5. *Saccammina sphaerica* Brady, 1871 – Astrorhizida: Saccamminidae
6. *Hormosina globulifera* Bardy, 1879 – Lituolida: Hormosinidae
7. *Hoeglundina elegans* (d'Orbigny, 1826) – Robertinida: Epistominidae
8. *Cornuspira foliacea* (Philippi, 1844) – Miliolida: Cornuspiridae
9. *Cornuspira involvens* (Reuss, 1850) – Miliolida: Cornuspiridae
10. *Rhabdammina major* de Folin, 1887 – Astrorhizida: Rhabdamminidae
11. *Miliolinella vigilax* Vella, 1957 – Miliolida: Hauerinidae
12. *Baculogypsina sphaerulata* (Parker & Jones, 1860) – Rotaliida: Calcarinidae
13. *Lenticulina cultrata* Montfort, 1808 – Lagenida: Vaginulinidae
14. *Ammodiscus exsertus* Cushman, 1910 – Lituolida: Ammodiscidae

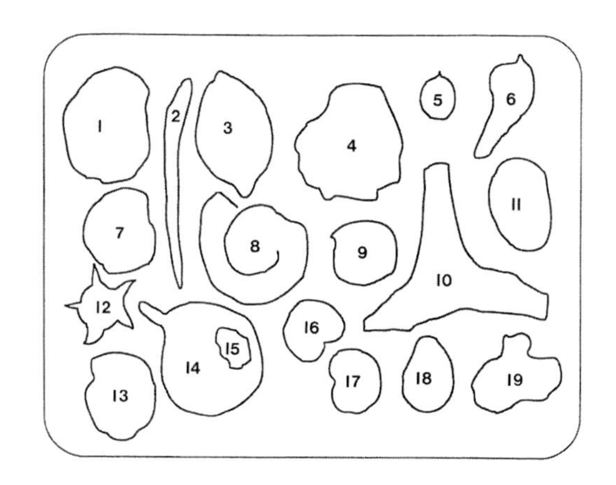

15. *Hemisphaerammina depressa* Heron-Allen & Earland, 1932 – Astrorhizida: Hemisphaeramminidae
16. *Laticarinina pauperata* (Parker & Jones, 1865) – Rotaliida: Discorbinellidae
17. *Cribrostomoides subglobosus* (Cushman, 1910) – Lituolida: Haplophragmoididae
18. *Pyrgo inornata* (d'Orbigny, 1846) – Miliolida: Hauerinidae
19. *Miniacina miniacea* (Pallas, 1766) – Rotaliida: Homotrematidae

Credits for other figures

All illustrations are acknowledged as to source. In many cases, authors or authors' colleagues freely supplied photographs and drawings. Illustrations requiring permission to be reproduced or adapted from published literature were obtained by authors. Landcare Research, in particular, is thanked for permission to reproduce illustrations of desmids from volumes I (1986), II (1988), and III (1994) of *Flora of New Zealand: Desmids* (see pages 354–356 of this volume) and of vascular plants from *Poole and Adams' Trees and Shrubs of New Zealand* (Government Printer, 1964).

Images were also sourced, with permission, from the following web sites, which readers are encouraged to visit:

Japan

Protist Information Server – comprising databases of many protist images collected from field and digital specimen archives: http://protist.i.hosei.ac.jp

New Zealand

The Hidden Forest – Clive Shirley's website of images of slime moulds, bryophytes, fungi (including lichens), and lower tracheophytes: http://hiddenforest.co.nz/index.htm

USA

Centers for Disease Control & Prevention, Laboratory Identification of Parasites of Public Health Concern: http://dpd.cdc.gov/DPDx/default.htm

Micro*scope – a communal website that provides descriptive information about all kinds of microbes. It combines locally assembled content with links to other expert sites on the internet. Information is assembled in collections provided by various contributors: http://starcentral.mbl.edu/microscope/portal.php

Captions and credits for images on the back covers of all three volumes

Top left to bottom right:

Volume 1

Marine bristleworm *Eunice laticeps* (Annelida): Geoffrey B. Read, NIWA, Wellington
Southern royal albatross *Diomedea epomophora* (Chordata): David R. Thompson, NIWA, Wellington
Unidentified polyclad flatworm (Platyhelminthes): Geoffrey B. Read, NIWA, Wellington
Acorn worm *Saccoglossus otagoensis* (Hemichordata): Geoffrey B. Read, NIWA, Wellington
Sea gooseberry *Pleurobrachia pileus* (Ctenophora): Len Doel, Papatoetoe East Primary School, Auckland
Unidentified ribbon worm (Nemertea): Geoffrey B. Read, NIWA, Wellington
Sea urchin *Tripneustes gratilla* (Echinodermata): Roger V. Grace, Environmental Consultant, Leigh

Volume 2

Water bear *Echiniscus elaeinae* (Tardigrada): Diane Nelson, East Tennessee State University, Johnson City
Marine round worm *Pselionema* sp. (Nematoda): Daniel Leduc, NIWA, Wellington
Water spider *Dolomedes* sp. (Arthropoda, Arachnida): Steven Moore, Landcare Research, Auckland
Pill millipede *Procyliosoma striolatum* (Arthropoda, Diplopoda): Alastair Robertson & Maria Minor, Massey University, Palmerston North
Damselfly *Ischnura aurora* (Arthropoda, Insecta): Steven Moore, Landcare Research, Auckland
Weta *Hemideina maori* (Arthropoda, Insecta): Alastair Robertson & Maria Minor, Massey University, Palmerston North
Unidentified seed shrimp (Arthropoda, Ostracoda): Steven Moore, Landcare Research, Auckland

Volume 3

Freshwater protist *Actinophyrs sol* (Bigyra): David A. Patterson, Marine Biological Laboratory, Woods Hole, MA
Octopus stinkhorn *Clathrus archeri* (Basidiomycota): Dennis P. Gordon, NIWA, Wellington
Green seaweed *Chaetomorpha coliformis* (Chlorophyta): Wendy A. Nelson, NIWA, Wellington
Saprophyte *Thismia rodwayi* (Tracheophyta): Kenneth R. Grange, NIWA, Nelson
Rock snot diatom *Didymosphenia geminata* (Ochrophyta): Cathy Kilroy, NIWA, Christchurch

Kidney fern *Cardiomanes reniforme* (Tracheophyta): Leon R. Perrie, Te Papa Tongarewa Museum of New Zealand, Wellington
Bluegreen alga *Godleya alpina* (Cyanobacteria): Philip M. Novis, Landcare Research, Lincoln

Box for boxed set
Back of box
Colonial sea squirt *Hypsistozoa fasmeriana* (Tunicata): Darryl Torckler Photography, Warkworth
Fossil trilobite *Koptura* sp. (Arthropoda, Trilobita): Marianna Terezow, GNS Science, Lower Hutt
Cinnabar looper moth *Asaphodes cinnabari* (Arthropoda, Insecta): Brian Patrick, Central Stories Museum & Art Gallery, Alexandra
Arrow worm *Pterosagitta draco* (Chaetognatha): Cheryl Clarke, University of Alaska, Fairbanks
Green seaweed *Chaetomorpha coliformis* (Chlorophyta): Wendy A. Nelson, NIWA, Wellington
Intestinal parasite *Giardia duodenalis* (Metamonada): W. Zachaeus Cande, University of California, Berkeley, CA
Velvet worm *Peripatoides novaezelandiae* (Onychophora): Hilke Ruhberg, Universität Hamburg, Hamburg
Southern royal albatross *Diomedea epomophora* (Chordata): David R. Thompson, NIWA, Wellington

Spine
Dipluran *Heterojapyx novaezealandiae* (Arthropoda, Diplura): Alastair Robertson & Maria Minor, Massey University, Palmerston North

Funding

Dennis Gordon acknowledges support for the inventory project from the former New Zealand Foundation for Research, Science & Technology and from the Ministry for Science and Innovation (Contracts C01421, C01X0219, C01X0026, C01X0502).

SPECIES INDEX

To keep this index to a manageable size, it has been applied to the main text and figures only, and not to the checklists or summary tables. Full binominals are given where they appear in the text, e.g. *Alseuosmia quercifolia*; otherwise, the generic name only is given without the qualifier 'sp.', e.g. *Giardia*. Names of higher taxa like phyla, classes, orders, and families are not indexed, e.g. Rhodophyta, Bacillariophyceae, Trypanosomatida, Isoetaceae, but the common names for them are, e.g. red algae, diatoms, trypanosomes, quillworts. Where such groups appear in the index, the page reference is to the first that deals substantively with that group, e.g. diatoms 127 signifies that the main description of diatoms begins on page 127. Common names for individual species have not normally been indexed.